Developmental Mathematics

Developmental Mathematics

Third Edition

Elayn Martin-Gay

University of New Orleans

PEARSON

Boston   Columbus   Hoboken   Indianapolis   New York   San Francisco
Amsterdam   Cape Town   Dubai   London   Madrid   Milan   Munich   Paris   Montréal   Toronto
Delhi   Mexico City   São Paulo   Sydney   Hong Kong   Seoul   Singapore   Taipei   Tokyo

Editorial Director, Mathematics: *Christine Hoag*
Editor-in-Chief: *Michael Hirsch*
Acquisitions Editor: *Mary Beckwith*
Project Team Lead: *Christina Lepre*
Project Manager: *Lauren Morse*
Assistant Editor: *Matthew Summers*
Editorial Assistant: *Megan Tripp*
Development Editor: *Dawn Nuttall*
Program Team Lead: *Karen Wernholm*
Program Manager: *Patty Bergin*
Cover and Illustration Design: *Tamara Newnam*
Program Design Lead: *Heather Scott*
Director, Course Production: *Ruth Berry*
Executive Content Manager, MathXL: *Rebecca Williams*
Senior Content Developer, TestGen: *John Flanagan*
Media Producer: *Audra Walsh*
Director of Marketing, Mathematics: *Roxanne McCarley*
Senior Marketing Manager: *Rachel Ross*
Marketing Assistant: *Kelly Cross*
Senior Author Support/Technology Specialist: *Joe Vetere*
Procurement Specialist: *Carol Melville*
Interior Design, Production Management, Answer Art,
   and Composition: *Integra Software Services Pvt. Ltd.*
Text Art: *Scientific Illustrators*

Acknowledgments of third party content appear on page P1, which constitutes an extension of this copyright page.

PEARSON, ALWAYS LEARNING, and MYMATHLAB are exclusive trademarks in the U.S. and/or other countries owned by Pearson Education, Inc. or its affiliates.

Library of Congress Cataloging-in-Publication Data
Martin-Gay, K. Elayn
   Developmental mathematics / Elayn Martin-Gay, University of New Orleans. – 3rd edition.
     pages cm
   ISBN-13: 978-0-321-93687-5 (alk. paper)
   ISBN-10: 0-321-93687-6 (alk. paper)
1. Mathematics–Textbooks.   2. Arithmetic–Textbooks.   3. Algebra–Textbooks.   4. Problem solving–Textbooks.   I. Title.
   QA107.2.M368 2015
   510–dc23   2013043429

1 2 3 4 5 6 7 8 9 10—CRK—19 18 17 16 15

www.pearsonhighered.com

ISBN-10: 0-321-93687-6 (Student Edition)
ISBN-13: 978-0-321-93687-5

This book is dedicated to students everywhere—and we should all be students. After all, is there anyone among us who really knows too much? Take that hint and continue to learn something new every day of your life.

Best of wishes from a fellow student:
Elayn Martin-Gay

Contents

Preface

Developmental Mathematics, **Third Edition** was written to provide a solid foundation in arithmetic and algebra as well as to develop problem-solving skills. It is intended for basic math and introductory algebra courses; however, all of the necessary intermediate topics are included in the appendices for those wishing to extend the course to intermediate algebra. Specific care was taken to make sure students have the most up-to-date relevant text preparation for their next mathematics course or for nonmathematical courses that require an understanding of algebraic fundamentals. I have tried to achieve this by writing a user-friendly text that is keyed to objectives and contains many worked-out examples. As suggested by AMATYC and the NCTM Standards (plus Addenda), real-life and real-data applications, data interpretation, conceptual understanding, problem solving, writing, cooperative learning, appropriate use of technology, number sense, estimation, critical thinking, and geometric concepts are emphasized and integrated throughout the book.

The many factors that contributed to the success of the previous edition have been retained. In preparing the Third Edition, I considered comments and suggestions of colleagues, students, and many users of the prior edition throughout the country.

What's New in the Third Edition?

- **The Martin-Gay Program** has been revised and enhanced with a new design in the text and MyMathLab® to actively encourage students to use the text, video program, Video Organizer, and Student Organizer as an integrated learning system.

- **The new Video Organizer** is designed to help students take notes and work practice exercises while watching the Interactive Lecture Series videos (available in MyMathLab and on DVD). All content in the Video Organizer is presented in the same order as it is presented in the videos, making it easy for students to create a course notebook and build good study habits.

  — Covers all of the video examples in order.

  — Provides ample space for students to write down key definitions and properties.

  — Includes "Play" and "Pause" button icons to prompt students to follow along with the author for some exercises while they try others on their own.

  The Video Organizer is available in a loose-leaf, notebook-ready format. It is also available for download in MyMathLab.

- **Vocabulary, Readiness & Video Check** questions have been added prior to every section exercise set. These exercises quickly check a student's understanding of new vocabulary words. The **readiness** exercises center on a student's understanding of a concept that is necessary in order to continue to the exercise set. **New Video Check questions for the Martin-Gay Interactive Lecture videos** are now included in every section for each learning objective. **These exercises are all available for assignment in MyMathLab** and are a great way to assess whether students have viewed and understood the key concepts presented in the videos.

- **New Student Success Tips Videos** are 3- to 5-minute video segments designed to be daily reminders to students to continue practicing and maintaining good organizational and study habits. They are organized in three categories and are available in MyMathLab and the Interactive Lecture Series. The categories are:

1. Success Tips that apply to any course in college in general, such as Time Management.

2. Success Tips that apply to any mathematics course. One example is based on understanding that mathematics is a course that requires homework to be completed in a timely fashion.

3. Section- or Content-specific Success Tips to help students avoid common mistakes or to better understand concepts that often prove challenging. One example of this type of tip is how to apply the order of operations to simplify an expression.

- **Interactive DVD Lecture Series**, featuring your text author (Elayn Martin-Gay), provides students with active learning at their own pace. The videos offer the following resources and more:

   A complete lecture for each section of the text highlights key examples and exercises from the text. "Pop-ups" reinforce key terms, definitions, and concepts.

   An interface with menu navigation features allows students to quickly find and focus on the examples and exercises they need to review.

   Interactive Concept Check exercises measure students' understanding of key concepts and common trouble spots.

   New Student Success Tips Videos

- **The Interactive DVD Lecture Series** also includes the following resources for test prep:

   The Chapter Test Prep Videos help students during their most teachable moment—when they are preparing for a test. This innovation provides step-by-step solutions for the exercises found in each Chapter Test. For the Third Edition, the chapter test prep videos are also available on YouTube™. The videos are captioned in English and Spanish.

   The Practice Final Exam Videos help students prepare for an end-of-course final. Students can watch full video solutions to each exercise in the Practice Final Exam at the end of this text.

- **The Martin-Gay MyMathLab** course has been updated and revised to provide more exercise coverage, including assignable video check questions and an expanded video program. There are section lecture videos for every section, which students can also access at the specific objective level; Student Success Tips videos; and an increased number of watch clips at the exercise level to help students while doing homework in MathXL. Suggested homework assignments have been premade for assignment at the instructor's discretion.

- **New MyMathLab Ready to Go courses** (access code required) provide students with all the same great MyMathLab features that you're used to, but make it easier for instructors to get started. Each course includes preassigned homework and quizzes to make creating your course even simpler. Ask your Pearson representative about the details for this particular course or to see a copy of this course.

Key Pedagogical Features

The following key features have been retained and/or updated for the Third Edition of the text:

Problem-Solving Process   This is formally introduced in Chapter 1 with a four-step process that is integrated throughout the text. The four steps are **Understand, Translate,**

Solve, and **Interpret.** The repeated use of these steps in a variety of examples shows their wide applicability. Reinforcing the steps can increase students' comfort level and confidence in tackling problems.

Exercise Sets Revised and Updated   The exercise sets have been carefully examined and extensively revised. Special focus was placed on making sure that even- and odd-numbered exercises are paired and that real-life applications were updated.

Examples   Detailed, step-by-step examples were added, deleted, replaced, or updated as needed. Many examples reflect real life. Additional instructional support is provided in the annotated examples.

Practice Exercises   Throughout the text, each worked-out example has a parallel Practice exercise. These invite students to be actively involved in the learning process. Students should try each Practice Exercise after finishing the corresponding example. Learning by doing will help students grasp ideas before moving on to other concepts. Answers to the Practice Exercises are provided at the bottom of each page.

Helpful Hints   Helpful Hints contain practical advice on applying mathematical concepts. Strategically placed where students are most likely to need immediate reinforcement, Helpful Hints help students avoid common trouble areas and mistakes.

Concept Checks   This feature allows students to gauge their grasp of an idea as it is being presented in the text. Concept Checks stress conceptual understanding at the point-of-use and help suppress misconceived notions before they start. Answers appear at the bottom of the page. Exercises related to Concept Checks are included in the exercise sets.

Mixed Practice Exercises   In the section exercise sets, these exercises require students to determine the problem type and strategy needed to solve it just as they would need to do on a test.

Integrated Reviews   This unique mid-chapter exercise set helps students assimilate new skills and concepts that they have learned separately over several sections. These reviews provide yet another opportunity for students to work with "mixed" exercises as they master the topics.

Vocabulary Check   This feature provides an opportunity for students to become more familiar with the use of mathematical terms as they strengthen their verbal skills. These appear at the end of each chapter before the Chapter Highlights. Vocabulary, Readiness & Video exercises provide practice at the section level.

Chapter Highlights   Found at the end of every chapter, these contain key definitions and concepts with examples to help students understand and retain what they have learned and help them organize their notes and study for tests.

Chapter Review   The end of every chapter contains a comprehensive review of topics introduced in the chapter. The Chapter Review offers exercises keyed to every section in the chapter, as well as Mixed Review exercises that are not keyed to sections.

Chapter Test and Chapter Test Prep Videos   The Chapter Test is structured to include those problems that involve common student errors. The **Chapter Test Prep Videos** gives students instant access to a step-by-step video solution of each exercise in the Chapter Test.

Cumulative Review   This review follows every chapter in the text (except Chapter 1). Each odd-numbered exercise contained in the Cumulative Review is an earlier worked example in the text that is referenced in the back of the book along with the answer.

Writing Exercises ＼ These exercises occur in almost every exercise set and require students to provide a written response to explain concepts or justify their thinking.

Applications Real-world and real-data applications have been thoroughly updated, and many new applications are included. These exercises occur in almost every exercise set and show the relevance of mathematics and help students gradually and continuously develop their problem-solving skills.

Review Exercises These exercises occur in each exercise set (except in Chapter 1) and are keyed to earlier sections. They review concepts learned earlier in the text that will be needed in the next section or chapter.

Exercise Set Resource Icons Located at the opening of each exercise set, these icons remind students of the resources available for extra practice and support:

MyMathLab®

See Student Resources descriptions on page xvii for details on the individual resources available.

Exercise Icons These icons facilitate the assignment of specialized exercises and let students know what resources can support them.

- ▶ DVD Video icon: exercise worked on the Interactive DVD Lecture Series.
- △ Triangle icon: identifies exercises involving geometric concepts.
- ＼ Pencil icon: indicates a written response is needed.
- ▦ Calculator icon: optional exercises intended to be solved using a scientific or graphing calculator.

Group Activities Found at the end of each chapter, these activities are for individual or group completion, and are usually hands-on or data-based activities that extend the concepts found in the chapter, allowing students to make decisions and interpretations and to think and write about algebra.

Optional: Calculator Exploration Boxes and Calculator Exercises The optional Calculator Explorations provide keystrokes and exercises at appropriate points to give students an opportunity to become familiar with these tools. Section exercises that are best completed by using a calculator are identified by ▦ for ease of assignment.

Student and Instructor Resources

STUDENT RESOURCES

Student Organizer Guides students through the 3 main components of studying effectively—notetaking, practice, and homework. The Organizer includes before-class preparation exercises, notetaking pages in a 2-column format for use in class, and examples paired with exercises for practice for each section. Includes an outline and questions for use with the Student Success Tip Videos. It is 3-hole-punched. Available in loose-leaf, notebook-ready format and in MyMathLab.	**Student Solutions Manual** Provides completely worked-out solutions to the odd-numbered section exercises; all exercises in the Integrated Reviews, Chapter Reviews, Chapter Tests, and Cumulative Reviews
Interactive DVD Lecture Series Videos Provides students with active learning at their pace. The videos offer: <ul><li>A complete lecture for each text section. The interface allows easy navigation to examples and exercises students need to review.</li><li>Interactive Concept Check exercises</li><li>Student Success Tips Videos</li><li>Practice Final Exam</li><li>Chapter Test Prep Videos</li></ul>	**Video Organizer** Designed to help students take notes and work practice exercises while watching the Interactive Lecture Series videos. The Video Organizer: <ul><li>Covers all of the video examples in order</li><li>Provides ample space for students to write down key definitions and rules</li><li>Includes "Play" and "Pause" button icons to prompt students to follow along with the author for some exercises while they try others on their own</li><li>Includes Student Success Tips Outline and Questions</li></ul>Available in loose-leaf, notebook-ready format and in MyMathLab. Answers to exercises available to instructors in MyMathLab.

INSTRUCTOR RESOURCES

Annotated Instructor's Edition Contains all the content found in the student edition, plus the following: <ul><li>Answers to exercises on the same text page</li><li>Teaching Tips throughout the text placed at key points</li></ul>	**Instructor's Resource Manual with Tests and Mini-Lectures** <ul><li>Mini-lectures for each text section</li><li>Additional practice worksheets for each section</li><li>Several forms of tests per chapter—free response and multiple choice</li><li>Answers to all items</li></ul>**Instructor's Solutions Manual** **TestGen**® (Available for download from the IRC)
Instructor-to-Instructor Videos—available in the Instructor Resources section of the MyMathLab course.	**Online Resources** **MyMathLab**® (access code required) **MathXL**® (access code required)

Acknowledgments

There are many people who helped me develop this text, and I will attempt to thank some of them here. Emily Keaton, Courtney Slade, and Cindy Trimble were *invaluable* for contributing to the overall accuracy of the text. Dawn Nuttall was *invaluable* for her many suggestions and contributions during the development and writing of this Third Edition. Allison Campbell and Lauren Morse provided guidance throughout the production process.

A very special thank you goes to my editor, Mary Beckwith, for being there 24/7/365, as my students say. And, my thanks to the staff at Pearson for all their support: Heather Scott, Patty Bergin, Matt Summers, Michelle Renda, Rachel Ross, Michael Hirsch, Chris Hoag, and Greg Tobin.

I would like to thank the following reviewers for their input and suggestions:

Anita Aikman, *Collin County Community College*
Sheila Anderson, *Housatonic Community College*
Adrianne Arata, *College of the Siskyous*
Cedric Atkins, *Mott Community College*
Laurel Berry, *Bryant & Stratton College*
Connie Buller, *Metropolitan Community College*
Lisa Feintech, *Cabrillo College*
Chris Ford, *Shasta College*
Cindy Fowler, *Central Piedmont Community College*
Pam Gerszewski, *College of the Albemarle*
Doug Harley, *Del Mar College*

Sonya Johnson, *Central Piedmont Community College*
Deborah Jones, *High Tech College*
Nancy Lange, *Inver Hills Community College*
Jean McArthur, *Joliet Junior College*
Carole Shapero, *Oakton Community College*
Jennifer Strehler, *Oakton Community College*
Tanomo Taguchi, *Fullerton College*
Leigh Ann Wheeler, *Greenville Technical Community College*
Valerie Wright, *Central Piedmont Community College*

I would also like to thank the following dedicated group of instructors who participated in our focus groups, Martin-Gay Summits, and our design review for the series. Their feedback and insights have helped to strengthen this edition of the text. These instructors include:

Billie Anderson, *Tyler Junior College*
Cedric Atkins, *Mott Community College*
Lois Beardon, *Schoolcraft College*
Laurel Berry, *Bryant & Stratton College*
John Beyers, *University of Maryland*
Bob Brown, *Community College of Baltimore County–Essex*
Lisa Brown, *Community College of Baltimore County–Essex*
NeKeith Brown, *Richland College*
Gail Burkett, *State College of Florida*
Cheryl Cantwell, *Seminole Community College*
Ivette Chuca, *El Paso Community College*
Jackie Cohen, *Augusta State College*
Julie Dewan, *Mohawk Valley Community College*
Monette Elizalde, *Palo Alto College*
Kiel Ellis, *Delgado Community College*
Janice Ervin, *Central Piedmont Community College*
Richard Fielding, *Southwestern College*

Dena Frickey, *Delgado Community College*
Cindy Gaddis, *Tyler Junior College*
Gary Garland, *Tarrant County Community College*
Kim Ghiselin, *State College of Florida*
Nita Graham, *St. Louis Community College*
Kim Granger, *St. Louis Community College*
Pauline Hall, *Iowa State College*
Pat Hussey, *Triton College*
Dorothy Johnson, *Lorain County Community College*
Sonya Johnson, *Central Piedmont Community College*
Ann Jones, *Spartanburg Community College*
Irene Jones, *Fullerton College*
Paul Jones, *University of Cincinnati*
Mike Kirby, *Tidewater Community College*
Kathy Kopelousous, *Lewis and Clark Community College*

Tara LaFrance, *Delgado Community College*

John LaMaster, *Indiana Purdue University Fort Wayne*

Nancy Lange, *Inver Hills Community College*

Judy Langer, *Westchester Community College*

Lisa Lindloff, *McLennan Community College*

Sandy Lofstock, *St. Petersburg College*

Kathy Lovelle, *Westchester Community College*

Nicole Mabine, *North Lake College*

Jean McArthur, *Joliet Junior College*

Kevin McCandless, *Evergreen Valley College*

Ena Michael, *State College of Florida*

Daniel Miller, *Niagara County Community College*

Marica Molle, *Metropolitan Community College*

Carol Murphy, *San Diego Miramar College*

Charlotte Newsom, *Tidewater Community College*

Cao Nguyen, *Central Piedmont Community College*

Greg Nguyen, *Fullerton College*

Eric Oilila, *Jackson Community College*

Linda Padilla, *Joliet Junior College*

Armando Perez, *Laredo Community College*

Davidson Pierre, *State College of Florida*

Marilyn Platt, *Gaston College*

Chris Riola, *Moraine Valley College*

Carole Shapero, *Oakton Community College*

Janet Sibol, *Hillsborough Community College*

Anne Smallen, *Mohawk Valley Community College*

Barbara Stoner, *Reading Area Community College*

Jennifer Strehler, *Oakton Community College*

Ellen Stutes, *Louisiana State University Eunice*

Tanomo Taguchi, *Fullerton College*

Robyn Toman, *Anne Arundel Community College*

MaryAnn Tuerk, *Elsin Community College*

Walter Wang, *Baruch College*

Leigh Ann Wheeler, *Greenville Technical Community College*

Darlene Williams, *Delgado Community College*

Valerie Wright, *Central Piedmont Community College*

A special thank you to those students who participated in our design review: Katherine Browne, Mike Bulfin, Nancy Canipe, Ashley Carpenter, Jeff Chojnachi, Roxanne Davis, Mike Dieter, Amy Dombrowski, Kay Herring, Todd Jaycox, Kaleena Levan, Matt Montgomery, Tony Plese, Abigail Polkinghorn, Harley Price, Eli Robinson, Avery Rosen, Robyn Schott, Cynthia Thomas, and Sherry Ward.

Elayn Martin-Gay

About the Author

Elayn Martin-Gay has taught mathematics at the University of New Orleans for more than 25 years. Her numerous teaching awards include the local University Alumni Association's Award for Excellence in Teaching, and Outstanding Developmental Educator at University of New Orleans, presented by the Louisiana Association of Developmental Educators.

Prior to writing textbooks, Elayn Martin-Gay developed an acclaimed series of lecture videos to support developmental mathematics students in their quest for success. These highly successful videos originally served as the foundation material for her texts. Today, the videos are specific to each book in the Martin-Gay series. The author has also created Chapter Test Prep Videos to help students during their most "teachable moment"—as they prepare for a test—along with Instructor-to-Instructor videos that provide teaching tips, hints, and suggestions for each developmental mathematics course, including basic mathematics, prealgebra, beginning algebra, and intermediate algebra.

Elayn is the author of 12 published textbooks as well as multimedia, interactive mathematics, all specializing in developmental mathematics courses. She has also published series in Algebra 1, Algebra 2, and Geometry. She has participated as an author across the broadest range of educational materials: textbooks, videos, tutorial software, and courseware. This provides an opportunity of various combinations for an integrated teaching and learning package offering great consistency for the student.

Applications Index

1

The Whole Numbers

A Selection of Resources for Success in This Mathematics Course

Developmental Mathematics
Third Edition
Elayn Martin-Gay

Textbook

Instructor

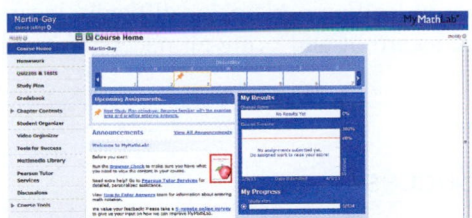

MyMathLab and MathXL

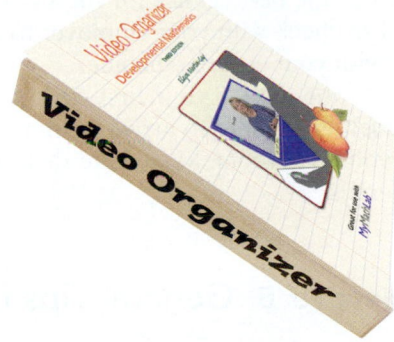

Video Organizer

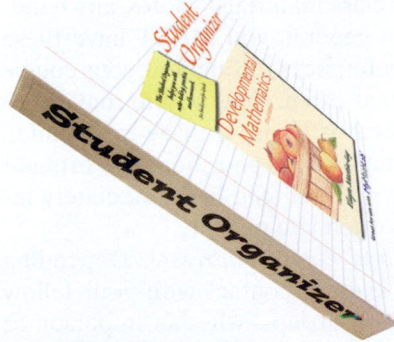

Student Organizer

Interactive Lecture Series

For more information about the resources illustrated above, read Section 1.1.

Whole numbers are the basic building blocks of mathematics. The whole numbers answer the question "How many?"

This chapter covers basic operations on whole numbers. Knowledge of these operations provides a good foundation on which to build further mathematical skills.

1.1 Study Skill Tips for Success in Mathematics

Objectives

A Get Ready for This Course. ▶

B Understand Some General Tips for Success. ▶

C Know How to Use This Text. ▶

D Know How to Use Video and Notebook Organizer Resources. ▶

E Get Help as Soon as You Need It. ▶

F Learn How to Prepare for and Take an Exam. ▶

G Develop Good Time Management. ▶

Before reading Section 1.1, you might want to ask yourself a few questions.

1. When you took your last math course, were you organized? Were your notes and materials from that course easy to find, or were they disorganized and hard to find—if you saved them at all?

2. Were you satisfied—really satisfied—with your performance in that course? In other words, do you feel that your outcome represented your best effort?

If the answer is "no" to these questions, then it is time to make a change. Changing to or resuming good study skill habits is not a process you can start and stop as you please. It is something that you must remember and practice each and every day. To begin, continue reading this section.

Objective A Getting Ready for This Course

Now that you have decided to take this course, remember that a *positive attitude* will make all the difference in the world. Your belief that you can succeed is just as important as your commitment to this course. Make sure you are ready for this course by having the time and positive attitude that it takes to succeed.

Make sure that you are familiar with the way that this course is being taught. Is it a traditional course, in which you have a printed textbook and meet with an instructor? Is it taught totally online, and your textbook is electronic and you e-mail your instructor? Or is your course structured somewhere in between these two methods? (Not all of the tips that follow will apply to all forms of instruction.)

Also make sure that you have scheduled your math course for a time that will give you the best chance for success. For example, if you are also working, you may want to check with your employer to make sure that your work hours will not conflict with your course schedule.

On the day of your first class period, double-check your schedule and allow yourself extra time to arrive on time in case of traffic problems or difficulty locating your classroom. Make sure that you are aware of and bring all necessary class materials.

Objective B General Tips for Success

Below are some general tips that will increase your chance for success in a mathematics class. Many of these tips will also help you in other courses you may be taking.

Most important! Organize your class materials. In the next couple pages, many ideas will be presented to help you organize your class materials—notes, any handouts, completed homework, previous tests, etc. In general, you MUST have these materials organized. All of them will be valuable references throughout your course and when studying for upcoming tests and the final exam. One way to make sure you can locate these materials when you need them is to use a three-ring binder. This binder should be used solely for your mathematics class and should be brought to each and every class and/or lab. This way, any material can be immediately inserted in a section of this binder and will be there when you need it.

Form study groups and/or exchange names and e-mail addresses. Depending on how your course is taught, you may want to keep in contact with your fellow students. Some ways of doing this are to form a study group—whether in person or through the Internet. Also, you may want to ask if anyone is interested in exchanging e-mail addresses or any other form of contact.

Helpful Hint

MyMathLab® and MathXL® When assignments are turned in online, keep a hard copy of your complete written work. You will need to refer to your written work to be able to ask questions and to study for tests later.

Choose to attend all class periods. If possible, sit near the front of the classroom. This way, you will see and hear the presentation better. It may also be easier for you to participate in classroom activities.

Do your homework. You've probably heard the phrase "practice makes perfect" in relation to music and sports. It also applies to mathematics. You will find that the more time you spend solving mathematics exercises, the easier the process becomes. Be sure to schedule enough time to complete your assignments before the due date assigned by your instructor.

Check your work. Review the steps you took while working a problem. Learn to check your answers in the original exercises. You may also compare your answers with the "Answers to Selected Exercises" section in the back of the book. If you have made a mistake, try to figure out what went wrong. Then correct your mistake. If you can't find what went wrong, **don't** erase your work or throw it away. Show your work to your instructor, a tutor in a math lab, or a classmate. It is easier for someone to find where you had trouble if he or she looks at your original work.

Learn from your mistakes and be patient with yourself. Everyone, even your instructor, makes mistakes. (That definitely includes me—Elayn Martin-Gay.) Use your errors to learn and to become a better math student. The key is finding and understanding your errors.

Was your mistake a careless one, or did you make it because you can't read your own math writing? If so, try to work more slowly or write more neatly and make a conscious effort to carefully check your work.

Did you make a mistake because you don't understand a concept? Take the time to review the concept or ask questions to better understand it.

Did you skip too many steps? Skipping steps or trying to do too many steps mentally may lead to preventable mistakes.

Know how to get help if you need it. It's all right to ask for help. In fact, it's a good idea to ask for help whenever there is something that you don't understand. Make sure you know when your instructor has office hours and how to find his or her office. Find out whether math tutoring services are available on your campus. Check on the hours, location, and requirements of the tutoring service.

Don't be afraid to ask questions. You are not the only person in class with questions. Other students are normally grateful that someone has spoken up.

Turn in assignments on time. This way, you can be sure that you will not lose points for being late. Show every step of a problem and be neat and organized. Also be sure that you understand which problems are assigned for homework. If allowed, you can always double-check the assignment with another student in your class.

Objective C Knowing and Using Your Text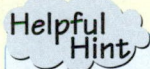

Flip through the pages of this text or view the e-text pages on a computer screen. Start noticing examples, exercise sets, end-of-chapter material, and so on. Every text is organized in some manner. Learn the way this text is organized by reading about and then finding an example in your text of each type of resource listed below. Finding and using these resources throughout your course will increase your chance of success.

- *Practice Exercises.* Each example in every section has a parallel Practice exercise. As you read a section, try each Practice exercise after you've finished the corresponding example. Answers are at the bottom of the page. This "learn-by-doing" approach will help you grasp ideas before you move on to other concepts.

- *Symbols at the Beginning of an Exercise Set.* If you need help with a particular section, the symbols listed at the beginning of each exercise set will remind you of the resources available.

Helpful Hint

MyMathLab® and MathXL® If you are doing your homework online, you can work and re-work those exercises that you struggle with until you master them. Try working through all the assigned exercises twice before the due date.

Helpful Hint

MyMathLab® and MathXL® If you are completing your homework online, it's important to work each exercise on paper before submitting the answer. That way, you can check your work and follow your steps to find and correct any mistakes.

Helpful Hint

MyMathLab® and MathXL® Be aware of assignments and due dates set by your instructor. Don't wait until the last minute to submit work online.

- *Objectives.* The main section of exercises in each exercise set is referenced by an objective, such as **A** or **B**, and also an example(s). There is also often a section of exercises entitled "Mixed Practice," which is referenced by two or more objectives or sections. These are mixed exercises written to prepare you for your next exam. Use all of this referencing if you have trouble completing an assignment from the exercise set.

- *Icons (Symbols).* Make sure that you understand the meaning of the icons that are beside many exercises. ▶ tells you that the corresponding exercise may be viewed on the video Lecture Series that corresponds to that section. ✎ tells you that this exercise is a writing exercise in which you should answer in complete sentences. △ tells you that the exercise involves geometry.

- *Integrated Reviews.* Found in the middle of each chapter, these reviews offer you a chance to practice—in one place—the many concepts that you have learned separately over several sections.

- *End-of-Chapter Opportunities.* There are many opportunities at the end of each chapter to help you understand the concepts of the chapter.

  Vocabulary Checks contain key vocabulary terms introduced in the chapter.

  Chapter Highlights contain chapter summaries and examples.

  Chapter Reviews contain review problems. The first part is organized section by section and the second part contains a set of mixed exercises.

  Chapter Tests are sample tests to help you prepare for an exam. The Chapter Test Prep Videos found in the Interactive Lecture Series, MyMathLab, and YouTube provide the video solution to each question on each Chapter Test.

  Cumulative Reviews start at Chapter 2 and are reviews consisting of material from the beginning of the book to the end of that particular chapter.

- *Student Resources in Your Textbook.* You will find a **Student Resources** section at the back of this textbook. It contains the following to help you study and prepare for tests:

  Study Skill Builders contain study skills advice. To increase your chance for success in the course, read these study tips and answer the questions.

  Bigger Picture—Study Guide Outline provides you with a study guide outline of the course, with examples.

  Practice Final provides you with a Practice Final Exam to help you prepare for a final.

- *Resources to Check Your Work.* The **Answers to Selected Exercises** section provides answers to all odd-numbered section exercises and to all integrated review, chapter review, chapter test, and cumulative review exercises. Use the **Solutions to Selected Exercises** to see the worked-out solution to every other odd-numbered exercise.

Helpful Hint

MyMathLab®

In MyMathLab, you have access to the following video resources:

- Lecture Videos for each section
- Chapter Test Prep Videos

Use these videos provided by the author to prepare for class, review, and study for tests.

Objective **D** Knowing and Using Video and Notebook Organizer Resources ▶

Video Resources

Below is a list of video resources that are all made by me—the author of your text, Elayn Martin-Gay. By making these videos, I can be sure that the methods presented are consistent with those in the text.

- *Interactive DVD Lecture Series.* Exercises marked with a ▶ are fully worked out by the author on the DVDs and within MyMathLab. The lecture series provides approximately 20 minutes of instruction per section and is organized by Objective.

- *Chapter Test Prep Videos.* These videos provide solutions to all of the Chapter Test exercises worked out by the author. They can be found in MyMathLab, the Interactive Lecture series, and You Tube. This supplement is very helpful before a test or exam.
- *Student Success Tips.* These video segments are about 3 minutes long and are daily reminders to help you continue practicing and maintaining good organizational and study habits.
- *Final Exam Videos.* These video segments provide solutions to each question. These videos can be found within MyMathLab and the Interactive Lecture Series.

Notebook Organizer Resources

The resources below are in three-ring notebook ready form. They are to be inserted in a three-ring binder and completed. Both resources are numbered according to the sections in your text to which they refer.

- *Video Organizer.* This organizer is closely tied to the Interactive Lecture (Video) Series. Each section should be completed while watching the lecture video on the same section. Once completed, you will have a set of notes to accompany the Lecture (Video) Series section by section.
- *Student Organizer.* This organizer helps you study effectively through note-taking hints, practice, and homework while referencing examples in the text and examples in the Lecture Series.

Objective E Getting Help

If you have trouble completing assignments or understanding the mathematics, get help as soon as you need it! This tip is presented as an objective on its own because it is so important. In mathematics, usually the material presented in one section builds on your understanding of the previous section. This means that if you don't understand the concepts covered during a class period, there is a good chance that you will not understand the concepts covered during the next class period. If this happens to you, get help as soon as you can.

Where can you get help? Many suggestions have been made in this section on where to get help, and now it is up to you to get it. Try your instructor, a tutoring center, or a math lab, or you may want to form a study group with fellow classmates. If you do decide to see your instructor or go to a tutoring center, make sure that you have a neat notebook and are ready with your questions.

Objective F Preparing for and Taking an Exam

Make sure that you allow yourself plenty of time to prepare for a test. If you think that you are a little "math anxious," it may be that you are not preparing for a test in a way that will ensure success. The way that you prepare for a test in mathematics is important. To prepare for a test:

1. Review your previous homework assignments.
2. Review any notes from class and section-level quizzes you have taken. (If this is a final exam, also review chapter tests you have taken.)
3. Review concepts and definitions by reading the Chapter Highlights at the end of each chapter.
4. Practice working out exercises by completing the Chapter Review found at the end of each chapter. (If this is a final exam, go through a Cumulative Review. There is one found at the end of each chapter except Chapter 1. Choose the review found at the end of the latest chapter that you have covered in your course.) *Don't stop here!*

Helpful Hint

MyMathLab® and MathXL®

- Use the **Help Me Solve This** button to get step-by-step help for the exercise you are working. You will need to work an additional exercise of the same type before you can get credit for having worked it correctly.
- Use the **Video** button to view a video clip of the author working a similar exercise.

Helpful Hint

MyMathLab® and MathXL®

Review your written work for previous assignments. Then, go back and re-work previous assignments. Open a previous assignment, and click **Similar Exercise** to generate new exercises. Re-work the exercises until you fully understand them and can work them without help features.

5. It is important that you place yourself in conditions similar to test conditions to find out how you will perform. In other words, as soon as you feel that you know the material, get a few blank sheets of paper and take a sample test. There is a Chapter Test available at the end of each chapter, or you can work selected problems from the Chapter Review. Your instructor may also provide you with a review sheet. During this sample test, do not use your notes or your textbook. Then check your sample test. If your sample test is the Chapter Test in the text, don't forget that the video solutions are in MyMathLab, the Interactive Lecture Series, and YouTube. If you are not satisfied with the results, study the areas that you are weak in and try again.

6. On the day of the test, allow yourself plenty of time to arrive at where you will be taking your exam.

When taking your test:

1. Read the directions on the test carefully.
2. Read each problem carefully as you take the test. Make sure that you answer the question asked.
3. Watch your time and pace yourself so that you can attempt each problem on your test.
4. If you have time, check your work and answers.
5. Do not turn your test in early. If you have extra time, spend it double-checking your work.

Objective **G** Managing Your Time

As a college student, you know the demands that classes, homework, work, and family place on your time. Some days you probably wonder how you'll ever get everything done. One key to managing your time is developing a schedule. Here are some hints for making a schedule:

1. Make a list of all of your weekly commitments for the term. Include classes, work, regular meetings, extracurricular activities, etc. You may also find it helpful to list such things as laundry, regular workouts, grocery shopping, etc.
2. Next, estimate the time needed for each item on the list. Also make a note of how often you will need to do each item. Don't forget to include time estimates for the reading, studying, and homework you do outside of your classes. You may want to ask your instructor for help estimating the time needed.
3. In the exercise set that follows, you are asked to block out a typical week on the schedule grid given. Start with items with fixed time slots like classes and work.
4. Next, include the items on your list with flexible time slots. Think carefully about how best to schedule items such as study time.
5. Don't fill up every time slot on the schedule. Remember that you need to allow time for eating, sleeping, and relaxing! You should also allow a little extra time in case some items take longer than planned.
6. If you find that your weekly schedule is too full for you to handle, you may need to make some changes in your workload, classload, or other areas of your life. You may want to talk to your advisor, manager or supervisor at work, or someone in your college's academic counseling center for help with such decisions.

1.1 Exercise Set MyMathLab®

1. What is your instructor's name?

2. What are your instructor's office location and office hours?

3. What is the best way to contact your instructor?

4. Do you have the name and contact information of at least one other student in class?

5. Will your instructor allow you to use a calculator in this class?

6. Why is it important that you write step-by-step solutions to homework exercises and keep a hard copy of all work submitted online?

7. Is there a tutoring service available on campus? If so, what are its hours? What services are available?

8. Have you attempted this course before? If so, write down ways that you might improve your chances of success during this attempt.

9. List some steps that you can take if you begin having trouble understanding the material or completing an assignment. If you are completing your homework in MyMathLab® and MathXL®, list the resources you can use for help.

10. How many hours of studying does your instructor advise for each hour of instruction?

11. What does the ✎ icon in this text mean?

12. What does the △ icon in this text mean?

13. What does the ▶ icon in this text mean?

14. Search the minor columns in your text. What are Practice exercises?

15. When might be the best time to work a Practice exercise?

16. Where are the answers to Practice exercises?

17. What answers are contained in this text and where are they?

18. What are Study Skill Tips of the Day and where are they?

19. What and where are Integrated Reviews?

20. How many times is it suggested that you work through the homework exercises in MathXL® before the submission deadline?

21. How far in advance of the assigned due date is it suggested that homework be submitted online? Why?

22. Chapter Highlights are found at the end of each chapter. Find the Chapter 1 Highlights and explain how you might use it and how it might be helpful.

23. Chapter Reviews are found at the end of each chapter. Find the Chapter 1 Review and explain how you might use it and how it might be helpful.

24. Chapter Tests are found at the end of each chapter. Find the Chapter 1 Test and explain how you might use it and how it might be helpful when preparing for an exam on Chapter 1. Include how the Chapter Test Prep Videos may help. If you are working in MyMathLab® and MathXL®, how can you use previous homework assignments to study?

25. What is the Video Organizer? Explain the contents and how it might be used.

26. What is the Student Organizer? Explain the contents and how it might be used.

27. Read or reread objective G and fill out the schedule grid on the next page.

	Monday	Tuesday	Wednesday	Thursday	Friday	Saturday	Sunday
4:00 a.m.							
5:00 a.m.							
6:00 a.m.							
7:00 a.m.							
8:00 a.m.							
9:00 a.m.							
10:00 a.m.							
11:00 a.m.							
12:00 p.m.							
1:00 p.m.							
2:00 p.m.							
3:00 p.m.							
4:00 p.m.							
5:00 p.m.							
6:00 p.m.							
7:00 p.m.							
8:00 p.m.							
9:00 p.m.							
10:00 p.m.							
11:00 p.m.							
Midnight							
1:00 a.m.							
2:00 a.m.							
3:00 a.m.							

1.2 Place Value, Names for Numbers, and Reading Tables ▶

Objectives

A Find the Place Value of a Digit in a Whole Number. ▶

B Write a Whole Number in Words and in Standard Form. ▶

C Write a Whole Number in Expanded Form. ▶

D Read Tables. ▶

The **digits** 0, 1, 2, 3, 4, 5, 6, 7, 8, and 9 can be used to write numbers. For example, the **whole numbers** are

0, 1, 2, 3, 4, 5, 6, 7, 8, 9, 10, 11, . . .

and the **natural numbers** are 1, 2, 3, 4, 5, 6, 7, 8, 9, 10, 11, . . .

The three dots (. . .) after each 11 means that these lists continue indefinitely. That is, there is no largest whole number. The smallest whole number is 0. Also, there is no largest natural number. The smallest natural number is 1.

Objective **A** Finding the Place Value of a Digit in a Whole Number ▶

The position of each digit in a number determines its **place value.** For example, the distance (in miles) between the planet Mercury and the planet Earth can be represented by the whole number 48,337,000. Next is a place-value chart for this whole number.

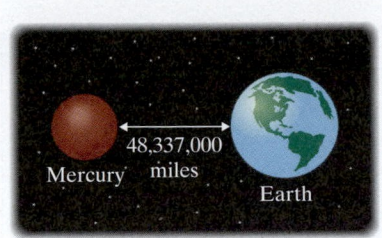

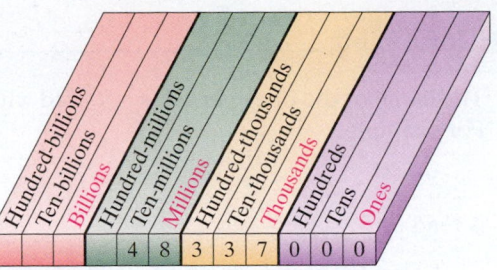

The two 3s in 48,337,000 represent different amounts because of their different placements. The place value of the 3 on the left is hundred-thousands. The place value of the 3 on the right is ten-thousands.

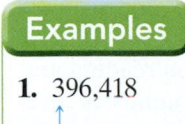 **Examples** Find the place value of the digit 3 in each whole number.

1. 396,418

hundred-thousands

2. 93,192

thousands

3. 534,275,866

ten-millions

▶ **Work Practice 1–3**

Objective **B** Writing a Whole Number in Words and in Standard Form ▶

A whole number such as 1,083,664,500 is written in **standard form.** Notice that commas separate the digits into groups of three, starting from the right. Each group of three digits is called a **period.** The names of the first four periods are shown in red.

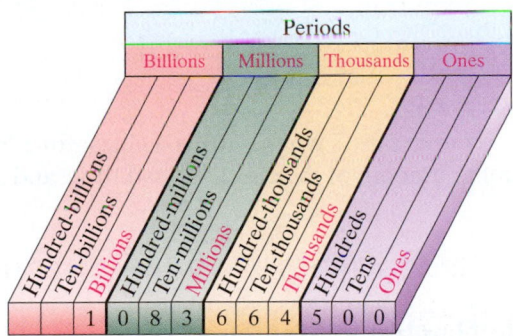

Writing a Whole Number in Words

To write a whole number in words, write the number in each period followed by the name of the period. (The ones period is usually not written.) This same procedure can be used to read a whole number.

For example, we write 1,083,664,500 as

one **billion,**
eighty-three **million,**
six hundred sixty-four **thousand,**
five hundred

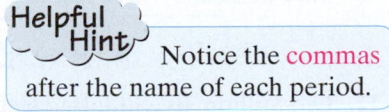 Notice the **commas** after the name of each period.

Practice 1–3

Find the place value of the digit 8 in each whole number.
1. 38,760,005
2. 67,890
3. 481,922

Answers
1. millions **2.** hundreds
3. ten-thousands

Helpful Hint

The name of the ones period is not used when reading and writing whole numbers. For example,

9,265

is read as

"nine **thousand,** two hundred sixty-five."

Practice 4–6

Write each number in words.

4. 67
5. 395
6. 12,804

Examples Write each number in words.

4.	85	eighty-five
5.	126	one hundred twenty-six
6.	27,034	twenty-seven thousand, thirty-four

■ Work Practice 4–6

Helpful Hint The word "and" is *not* used when reading and writing whole numbers. It is used when reading and writing mixed numbers and some decimal values, as shown later in this text.

Practice 7

Write 321,670,200 in words.

Example 7 Write 106,052,447 in words.

Solution: 106,052,447 is written as

one hundred six **million,** fifty-two **thousand,** four hundred forty-seven

■ Work Practice 7

✓**Concept Check** True or false? When writing a check for $2600, the word name we write for the dollar amount of the check is "two thousand sixty." Explain your answer.

Writing a Whole Number in Standard Form

To write a whole number in standard form, write the number in each period, followed by a comma.

Practice 8–11

Write each number in standard form.

8. twenty-nine
9. seven hundred ten
10. twenty-six thousand, seventy-one
11. six million, five hundred seven

Examples Write each number in standard form.

8. sixty-one   61   9. eight hundred five   805

10. nine thousand, three hundred eighty-six

9,386   or   9386

11. two million, five hundred sixty-four thousand, three hundred fifty

2,564,350

■ Work Practice 8–11

Answers

4. sixty-seven   5. three hundred ninety-five   6. twelve thousand, eight hundred four   7. three hundred twenty-one million, six hundred seventy thousand, two hundred   8. 29   9. 710   10. 26,071   11. 6,000,507

✓**Concept Check Answer**

false

Helpful Hint

A comma may or may not be inserted in a four-digit number. For example, both

   9,386   and   9386

are acceptable ways of writing nine thousand, three hundred eighty-six.

Objective C Writing a Whole Number in Expanded Form

The place value of a digit can be used to write a number in expanded form. The **expanded form** of a number shows each digit of the number with its place value. For example, 5672 is written in expanded form as

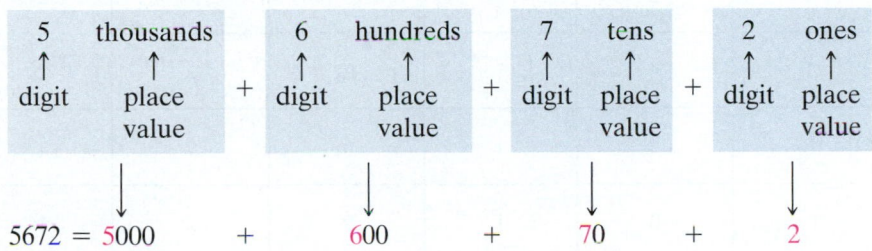

5	thousands		6	hundreds		7	tens		2	ones
↑	↑	+	↑	↑	+	↑	↑	+	↑	↑
digit	place value		digit	place value		digit	place value		digit	place value

$$5672 = 5000 + 600 + 70 + 2$$

> **Example 12**   Write 2,706,449 in expanded form.
>
> **Solution:**   2,000,000 + 700,000 + 6000 + 400 + 40 + 9
>
> ■ **Work Practice 12**

Practice 12

Write 1,047,608 in expanded form.

We can visualize whole numbers by points on a line. The line below is called a **number line.** This number line has equally spaced marks for each whole number. The arrow to the right simply means that the whole numbers continue indefinitely. In other words, there is no largest whole number.

Number Line

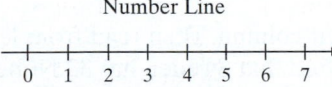

0   1   2   3   4   5   6   7

We will study number lines further in Section 1.5.

Objective D Reading Tables

Now that we know about place value and names for whole numbers, we introduce one way that whole numbers may be presented. **Tables** are often used to organize and display facts that involve numbers. The table on the next page shows the ten countries with the most Nobel Prize winners since the inception of the Nobel Prize in 1901, and the categories of the prizes. The numbers for the Economics prize reflect the winners since 1969, when this category was established. (The numbers may seem large for two reasons: first, the annual Nobel Prize is often awarded to more than one individual, and second, several award winners hold dual citizenship, so they are counted in two countries.)

Answer

12. 1,000,000 + 40,000 + 7000 + 600 + 8

Countries with Most Nobel Prize Winners, 1901–2013

Country	Chemistry	Economics	Literature	Peace	Physics	Physiology & Medicine	Total
United States	68	54	11	22	90	97	342
United Kingdom	27	8	8	9	23	31	106
Germany	27	1	8	4	22	17	79
France	7	1	15	8	13	11	55
Sweden	4	2	9	5	4	8	32
Switzerland	6	0	2	3	4	6	21
Russia (USSR)	1	1	4	2	10	2	20
Japan	6	0	2	1	6	2	17
Netherlands	3	1	0	1	9	2	16
Italy	1	0	6	1	4	3	15

Source: Based on data from Encyclopaedia Britannica, Inc.

For example, by reading from left to right along the row marked "United States," we find that the United States has 68 Chemistry, 54 Economics, 11 Literature, 22 Peace, 90 Physics, and 97 Physiology and Medicine Nobel Prize winners.

Example 13   Use the Nobel Prize Winner table to answer each question.

a. How many total Nobel Prize winners are from Sweden?

b. Which countries shown have fewer Nobel Prize winners than Russia?

Solution:

a. Find "Sweden" in the left column. Then read from left to right until the "Total" column is reached. We find that Sweden has 32 Nobel Prize winners.

b. Russia has 20 Nobel Prize winners. Japan has 17, Netherlands has 16, and Italy has 15, so they have fewer Nobel Prize winners than Russia.

■ **Work Practice 13**

Practice 13

Use the Nobel Prize Winner table to answer the following questions:

a. How many Nobel Prize winners in Literature come from France?

b. Which countries shown have more than 60 Nobel Prize winners?

Answers

13. a. 15   **b.** United States, United Kingdom, and Germany

Vocabulary, Readiness & Video Check

Use the choices below to fill in each blank.

standard form	period	whole
expanded form	place value	words

1. The numbers 0, 1, 2, 3, 4, 5, 6, 7, 8, 9, 10, 11, 12, … are called _____ numbers.

2. The number 1,286 is written in _____.

3. The number "twenty-one" is written in _____.

4. The number $900 + 60 + 5$ is written in _____.

5. In a whole number, each group of three digits is called a(n) _____.

6. The _____ of the digit 4 in the whole number 264 is ones.

Martin-Gay Interactive Videos   *Watch the section lecture video and answer the following questions.*

Objective A 7. In ▣ Example 1, what is the place value of the digit 6? ▶

Objective B 8. Complete this statement based on ▣ Example 3: To read (or write) a number, read from _____ to _____. ▶

Objective C 9. In ▣ Example 5, what is the expanded-form value of the digit 8? ▶

Objective D 10. Use the table given in ▣ Example 6 to determine which mountain in the table is the shortest. ▶

See Video 1.2 🍎

1.2 Exercise Set MyMathLab® ▶

Objective A *Determine the place value of the digit 5 in each whole number. See Examples 1 through 3.*

▶ **1.** 657

2. 905

▶ **3.** 5423

4. 6527

5. 43,526,000

6. 79,050,000

7. 5,408,092

8. 51,682,700

Objective B *Write each whole number in words. See Examples 4 through 7.*

9. 354

10. 316

11. 8279

12. 5445

▶ **13.** 26,990

14. 42,009

15. 2,388,000

16. 3,204,000

17. 24,350,185

18. 47,033,107

Write each number in the sentence in words. See Examples 4 through 7.

19. As of July 2013, the population of Iceland was 315,281. (*Source:* CIA World Factbook)

20. The land area of Belize is 22,806 square kilometers. (*Source:* CIA World Factbook)

21. The Burj Khalifa, in Dubai, United Arab Emirates, a hotel and office building, is the world's tallest building at a height of 2717 feet. (*Source:* Council on Tall Buildings and Urban Habitat)

22. As of October 2013, there were 119,948 patients in the United States waiting for an organ transplant. (*Source:* Organ Procurement and Transplantation Network)

23. In 2012, UPS received an average of 32,100,000 online tracking requests each day. (*Source:* UPS)

24. Each year, an estimated 500,000,000 Americans visit carnivals, fairs, and festivals. (*Source:* Outdoor Amusement Business Association)

25. The highest point in Colorado is Mount Elbert, at an elevation of 14,433 feet. (*Source:* U.S. Geological Survey)

26. The highest point in Oregon is Mount Hood, at an elevation of 11,239 feet. (*Source:* U.S. Geological Survey)

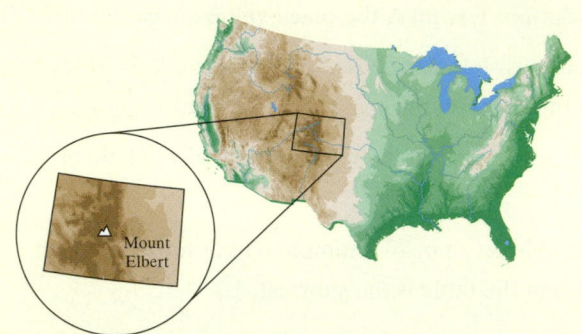

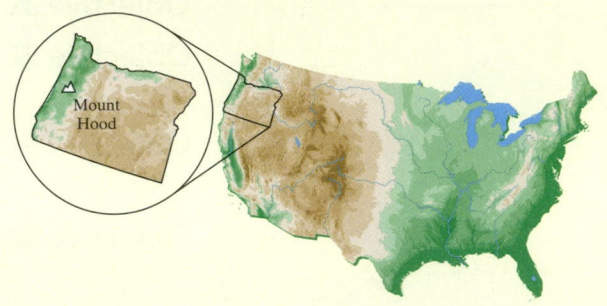

27. In 2013, the Great Internet Mersenne Prime Search, a cooperative computing project, helped find a prime number that has over 17,000,000 digits. (*Source:* Mersenne Research, Inc.)

28. The Goodyear blimp *Eagle* holds 202,700 cubic feet of helium. (*Source:* The Goodyear Tire & Rubber Company)

Write each whole number in standard form. See Examples 8 through 11.

29. Six thousand, five hundred eighty-seven

30. Four thousand, four hundred sixty-eight

31. Fifty-nine thousand, eight hundred

32. Seventy-three thousand, two

33. Thirteen million, six hundred one thousand, eleven

34. Sixteen million, four hundred five thousand, sixteen

35. Seven million, seventeen

36. Two million, twelve

37. Two hundred sixty thousand, nine hundred ninety-seven

38. Six hundred forty thousand, eight hundred eighty-one

Write the whole number in each sentence in standard form. See Examples 8 through 11.

39. After an orbit correction in October 2013, the International Space Station orbited Earth at an average altitude of about four hundred eighteen kilometers. (*Source:* Heavens Above)

40. The average distance between the surfaces of Earth and the Moon is about two hundred thirty-four thousand miles.

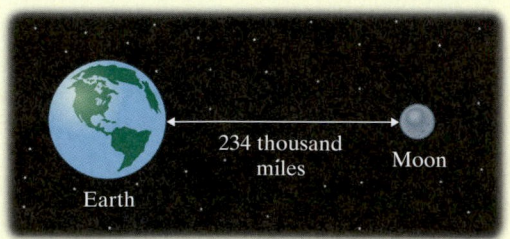

41. La Rinconada, Peru, is the highest town in the world. It is located sixteen thousand, seven hundred thirty-two feet above sea level. (*Source:* Russell Ash: *Top 10 of Everything*)

42. The world's tallest freestanding tower is the Tokyo Sky Tree in Japan. Its height is two thousand eighty feet tall. (*Source:* Council on Tall Buildings and Urban Habitat)

43. The Warner Bros. film *Harry Potter and the Deathly Hallows Part 2* holds the record for U.S./Canada opening day box office gross when it took in approximately ninety-one million, seventy-one thousand dollars on its opening day in 2011. (*Source:* Box Office Mojo)

44. The Buena Vista film *Marvel's The Avengers* set the U.S./Canada record for second-highest opening day box office gross when it took in approximately eighty million, eight hundred fourteen thousand dollars on its opening day in 2012. (*Source:* Box Office Mojo)

45. In 2012, the UPS delivery fleet consisted of one hundred one thousand vehicles. (*Source:* UPS)

46. Morten Andersen, who played football for New Orleans, Atlanta, N.Y. Giants, Kansas City, and Minnesota between 1982 and 2007, holds the record for the most points scored in a career. Over his 25-year career he scored two thousand, five hundred forty-four points. (*Source:* NFL.com)

Objective C *Write each whole number in expanded form. See Example 12.*

47. 406

48. 789

49. 3470

50. 6040

51. 80,774

52. 20,215

53. 66,049

54. 99,032

55. 39,680,000

56. 47,703,029

Objectives B C D Mixed Practice *The table shows the six tallest mountains in New England and their elevations. Use this table to answer Exercises 57 through 62. See Example 13.*

Mountain (State)	Elevation (in feet)
Boott Spur (NH)	5492
Mt. Adams (NH)	5774
Mt. Clay (NH)	5532
Mt. Jefferson (NH)	5712
Mt. Sam Adams (NH)	5584
Mt. Washington (NH)	6288
Source: U.S. Geological Survey	

Elevation in feet

57. Write the elevation of Mt. Clay in standard form and then in words.

58. Write the elevation of Mt. Washington in standard form and then in words.

59. Write the height of Boott Spur in expanded form.

60. Write the height of Mt. Jefferson in expanded form.

61. Which mountain is the tallest in New England?

62. Which mountain is the second tallest in New England?

The table shows the top ten museums in the world in 2012. Use this table to answer Exercises 63 through 68. See Example 13.

Top 10 Museums Worldwide in 2012

Museum	Location	Visitors
Louvre	Paris, France	9,720,000
National Museum of Natural History	Washington, DC, United States	7,600,000
National Air and Space Museum	Washington, DC, United States	6,800,000
The Metropolitan Museum of Art	New York, NY, United States	6,116,000
British Museum	London, United Kingdom	5,576,000
Tate Modern	London, United Kingdom	5,319,000
National Gallery	London, United Kingdom	5,164,000
Vatican Museums	Vatican City	5,065,000
American Museum of Natural History	New York, NY, United States	5,000,000
Natural History Museum	London, United Kingdom	4,936,000

(*Source:* Themed Entertainment Association)

63. Which museum had fewer visitors, the National Gallery in London or the National Air and Space Museum in Washington, DC?

64. Which museum had more visitors, the British Museum in London or the Louvre in Paris?

65. How many people visited the Vatican Museums? Write the number of visitors in words.

66. How many people visited The Metropolitan Museum of Art? Write the number of visitors in words.

67. How many of 2012's top ten museums in the world were located in the United States?

68. How many of 2012's top ten museums in the world were visited by fewer than 6,000,000 people?

Concept Extensions

69. Write the largest four-digit number that can be made from the digits 1, 9, 8, and 6 if each digit must be used once. ____ ____ ____ ____

70. Write the largest five-digit number that can be made using the digits 5, 3, and 7 if each digit must be used at least once. ____ ____ ____ ____ ____

Check to see whether each number written in standard form matches the number written in words. If not, correct the number in words. See the Concept Check in this section.

71.

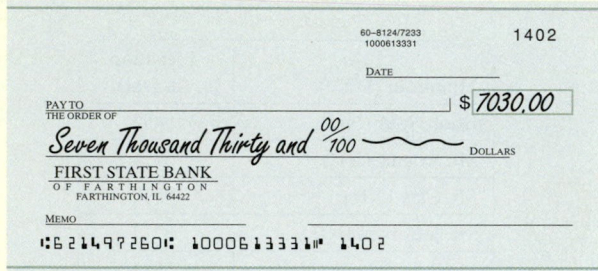

72.

73. If a number is given in words, describe the process used to write this number in standard form.

74. If a number is written in standard form, describe the process used to write this number in expanded form.

75. In June 2013, the MilkyWay-2, a high-speed computer built by China's National University of Defense Technology, was ranked as the world's fastest computer. Its speed was clocked at nearly 34 petaflops, or more than 34 quadrillion arithmetic operations per second. Look up "quadrillion" (in the American system) and use the definition to write this number in standard form. (*Source:* top500.org)

76. As of December 2012, the national debt of France was approximately $5 trillion. Look up "trillion" (in the American system) and use the definition to write 5 trillion in standard form. (*Source:* CIA World Factbook)

77. The Pro Football Hall of Fame was established on September 7, 1963, in this town. Use the information and the diagram to the right to find the name of the town.
- Alliance is east of Massillon.
- Dover is between Canton and New Philadelphia.
- Massillon is not next to Alliance.
- Canton is north of Dover.

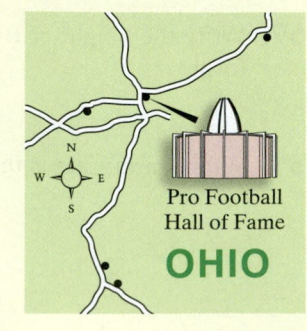

Objective A Adding Whole Numbers

According to Gizmodo, the iPod nano (currently in its seventh generation) is still the best overall MP3 player.

Suppose that an electronics store received a shipment of two boxes of iPod nanos one day and an additional four boxes of iPod nanos the next day. The **total** shipment in the two days can be found by adding 2 and 4.

2 boxes of iPod nanos + 4 boxes of iPod nanos = 6 boxes of iPod nanos

The **sum** (or total) is 6 boxes of iPod nanos. Each of the numbers 2 and 4 is called an **addend,** and the process of finding the sum is called **addition.**

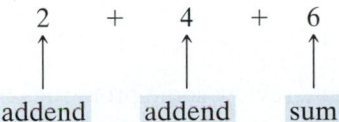

To add whole numbers, we add the digits in the ones place, then the tens place, then the hundreds place, and so on. For example, let's add $2236 + 160$.

$$\begin{array}{r} 2236 \\ + \ 160 \\ \hline 2396 \end{array}$$

Line up numbers vertically so that the place values correspond. Then add digits in corresponding place values, starting with the ones place.

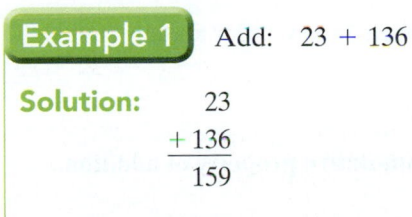

sum of ones
sum of tens
sum of hundreds
sum of thousands

Example 1 Add: $23 + 136$

Solution:
$$\begin{array}{r} 23 \\ + 136 \\ \hline 159 \end{array}$$

■ Work Practice 1

When the sum of digits in corresponding place values is more than 9, **carrying** is necessary. For example, to add $365 + 89$, add the ones-place digits first.

Carrying
$$\begin{array}{r} \overset{1}{3}65 \\ + \ 89 \\ \hline 4 \end{array}$$

5 ones + 9 ones = **14 ones** or **1 ten** + **4 ones**
Write the 4 ones in the ones place and carry the 1 ten to the tens place.

Next, add the tens-place digits.

$$\begin{array}{r} \overset{1\ 1}{3}65 \\ + \ 89 \\ \hline 54 \end{array}$$

1 ten + 6 tens + 8 tens = **15 tens** or **1 hundred** + **5 tens**
Write the 5 tens in the tens place and carry the 1 hundred to the hundreds place.

Next, add the hundreds-place digits.

$$\begin{array}{r} \overset{1\ 1}{3}65 \\ + \ 89 \\ \hline 454 \end{array}$$

1 hundred + 3 hundreds = **4 hundreds**
Write the 4 hundreds in the hundreds place.

Objectives

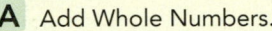

A Add Whole Numbers.

B Find the Perimeter of a Polygon. ▶

C Solve Problems by Adding Whole Numbers. ▶

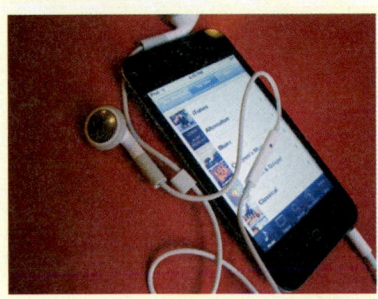

Practice 1

Add: $7235 + 542$

Answer
1. 7777

Practice 2

Add: $27{,}364 + 92{,}977$

Copyright 2015 Pearson Education, Inc.

Example 2 Add: $34{,}285 + 149{,}761$

Solution:

$$
\begin{array}{r}
\overset{11\ 1}{34{,}285} \\
+\ 149{,}761 \\
\hline
184{,}046
\end{array}
$$

■ **Work Practice 2**

✓**Concept Check** What is wrong with the following computation?

$$
\begin{array}{r}
394 \\
+\ 283 \\
\hline
577
\end{array}
$$

Before we continue adding whole numbers, let's review some properties of addition that you may have already discovered. The first property that we will review is the **addition property of 0.** This property reminds us that the sum of 0 and any number is that same number.

> **Addition Property of 0**
>
> The sum of 0 and any number is that number. For example,
>
> $7 + 0 = 7$
>
> $0 + 7 = 7$

Next, notice that we can add any two whole numbers in any order and the sum is the same. For example,

$4 + 5 = 9$ and $5 + 4 = 9$

We call this special property of addition the **commutative property of addition.**

> **Commutative Property of Addition**
>
> Changing the **order** of two addends does not change their sum. For example,
>
> $2 + 3 = 5$ and $3 + 2 = 5$

Another property that can help us when adding numbers is the **associative property of addition.** This property states that when adding numbers, the grouping of the numbers can be changed without changing the sum. We use parentheses to group numbers. They indicate which numbers to add first. For example, let's use two different groupings to find the sum of $2 + 1 + 5$.

$$(2 + 1) + 5 = 3 + 5 = 8$$

Also,

$$2 + (1 + 5) = 2 + 6 = 8$$

Both groupings give a sum of 8.

Answer

2. 120,341

✓**Concept Check Answer**

forgot to carry 1 hundred to the hundreds place

Associative Property of Addition

Changing the **grouping** of addends does not change their sum. For example,

$$3 + (5 + 7) = 3 + 12 = 15 \quad \text{and} \quad (3 + 5) + 7 = 8 + 7 = 15$$

The commutative and associative properties tell us that we can add whole numbers using any order and grouping that we want.

When adding several numbers, it is often helpful to look for two or three numbers whose sum is 10, 20, and so on. Why? Adding multiples of 10 such as 10 and 20 is easier.

Example 3   Add: $13 + 2 + 7 + 8 + 9$

Solution:

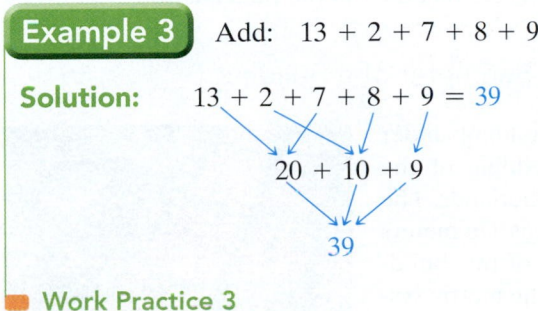

■ **Work Practice 3**

Practice 3

Add: $11 + 7 + 8 + 9 + 13$

Feel free to use the process of Example 3 anytime when adding.

Example 4   Add: $1647 + 246 + 32 + 85$

Solution:

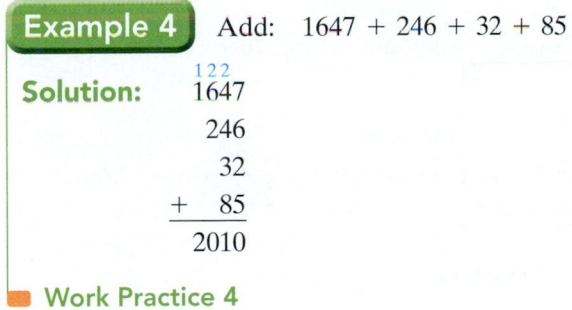

■ **Work Practice 4**

Practice 4

Add: $19 + 5042 + 638 + 526$

Objective B Finding the Perimeter of a Polygon ▶

In geometry addition is used to find the perimeter of a polygon. A **polygon** can be described as a flat figure formed by line segments connected at their ends. (For more review, see Appendix A.3.) Geometric figures such as triangles, squares, and rectangles are called polygons.

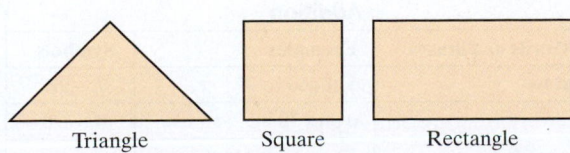

Triangle   Square   Rectangle

The **perimeter** of a polygon is the *distance around* the polygon. This means that the perimeter of a polygon is the sum of the lengths of its sides.

Practice 5

Find the perimeter of the polygon shown. (A centimeter is a unit of length in the metric system.)

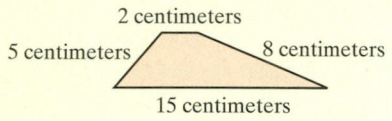

5 centimeters / 2 centimeters / 8 centimeters
15 centimeters

Example 5 Find the perimeter of the polygon shown.

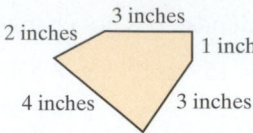

3 inches
2 inches   1 inch
4 inches   3 inches

Solution: To find the perimeter (distance around), we add the lengths of the sides.

2 in. + 3 in. + 1 in. + 3 in. + 4 in. = 13 in.

The perimeter is 13 inches.

■ **Work Practice 5**

To make the addition appear simpler, we will often not include units with the addends. If you do this, make sure units are included in the final answer.

Practice 6

A park is in the shape of a triangle. Each of the park's three sides is 647 feet. Find the perimeter of the park.

Example 6 Calculating the Perimeter of a Building

The world's largest commercial building under one roof is the flower auction building of the cooperative VBA in Aalsmeer, Netherlands. The floor plan is a rectangle that measures 776 meters by 639 meters. Find the perimeter of this building. (A meter is a unit of length in the metric system.) (*Source: The Handy Science Answer Book,* Visible Ink Press)

Solution: Recall that opposite sides of a rectangle have the same length. To find the perimeter of this building, we add the lengths of the sides. The sum of the lengths of its sides is

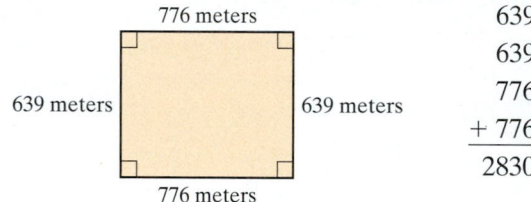

776 meters

639 meters   639 meters

776 meters

$$\begin{array}{r} 639 \\ 639 \\ 776 \\ + 776 \\ \hline 2830 \end{array}$$

The perimeter of the building is 2830 meters.

■ **Work Practice 6**

Objective C Solving Problems by Adding

Often, real-life problems occur that can be solved by adding. The first step in solving any word problem is to *understand* the problem by reading it carefully.

Descriptions of problems solved through addition *may* include any of these key words or phrases:

Addition		
Key Words or Phrases	**Examples**	**Symbols**
added to	5 added to 7	7 + 5
plus	0 plus 78	0 + 78
increased by	12 increased by 6	12 + 6
more than	11 more than 25	25 + 11
total	the total of 8 and 1	8 + 1
sum	the sum of 4 and 133	4 + 133

Answers
5. 30 cm **6.** 1941 ft

To solve a word problem that involves addition, we first use the facts given to write an addition statement. Then we write the corresponding solution of the real-life problem. It is sometimes helpful to write the statement in words (brief phrases) and then translate to numbers.

Example 7 Finding the Number of Vehicles Sold in the United States

In 2011, a total of 12,734,424 passenger vehicles were sold in the United States. In 2012, total passenger vehicle sales in the United States had increased by 1,705,636 vehicles. Find the total number of passenger vehicles sold in the United States in 2012. (*Source:* Alliance of Automobile Manufacturers)

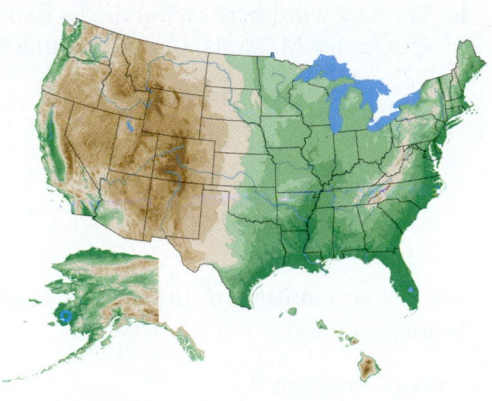

Solution: The key phrase here is "had increased by," which suggests that we add. To find the number of vehicles sold in 2012, we add the increase, 1,705,636, to the number of vehicles sold in 2011.

In Words		Translate to Numbers
Number sold in 2011	⟶	12,734,424
+ increase	⟶	+ 1,705,636
Number sold in 2012	⟶	14,440,060

The number of passenger vehicles sold in the United States in 2012 was 14,440,060.

■ **Work Practice 7**

Graphs can be used to visualize data. The graph shown next is called a **bar graph.** For this bar graph, the height of each bar is labeled above the bar. To check this height, follow the top of each bar to the vertical line to the left. For example, the first bar is labeled 185. Follow the top of that bar to the left until the vertical line is reached, between 180 and 200, but closer to 180, or 185.

Example 8 Reading a Bar Graph

In the following graph, each bar represents a country and the height of each bar represents the number of threatened mammal species identified in that country.

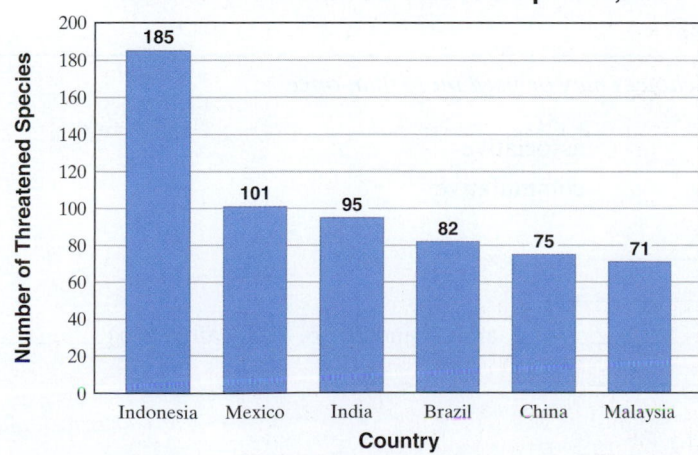

Number of Threatened Mammal Species, 2013

Source: International Union for Conservation of Nature

(Continued on next page)

Practice 7

Georgia produces 70 million pounds of freestone peaches per year. The second largest U.S. producer of peaches, South Carolina, produces 50 million more freestone peaches than Georgia. How much does South Carolina produce? (*Source:* farms.com)

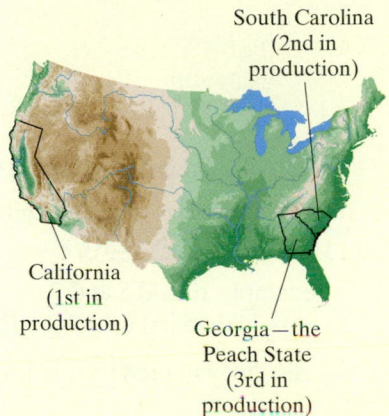

South Carolina
(2nd in
production)

California
(1st in
production)

Georgia—the
Peach State
(3rd in
production)

Practice 8

Use the graph in Example 8 to answer the following:

a. Which country shown has the fewest threatened mammal species?

b. Find the total number of threatened mammal species for Brazil, India, and Mexico.

Answers

7. 120 million lb

8. a. Malaysia   **b.** 278

a. Which country shown has the greatest number of threatened mammal species?

b. Find the total number of threatened mammal species for Malaysia, China, and Indonesia.

Solution:

a. The country with the greatest number of threatened mammal species corresponds to the tallest bar, which is Indonesia.

b. The key word here is "total." To find the total number of threatened mammal species for Malaysia, China, and Indonesia, we add.

In Words		Translate to Numbers
Malaysia	$\longrightarrow$	71
China	$\longrightarrow$	75
Indonesia	$\longrightarrow$	+ 185
	Total	331

The total number of threatened mammal species for Malaysia, China, and Indonesia is 331.

■ **Work Practice 8**

Calculator Explorations Adding Numbers

To add numbers on a calculator, find the keys marked ⊞ and ⊟ or ⎡ENTER⎤.

For example, to add 5 and 7 on a calculator, press the keys ⎡5⎤ ⊞ ⎡7⎤ then ⊟ or ⎡ENTER⎤.

The display will read ⎡   12   ⎤.

Thus, 5 + 7 = 12.

To add 687, 981, and 49 on a calculator, press the keys ⎡687⎤ ⊞ ⎡981⎤ ⊞ ⎡49⎤ then ⊟ or ⎡ENTER⎤.

The display will read ⎡   1717   ⎤.

Thus, 687 + 981 + 49 = 1717. (Although entering 687, for example, requires pressing more than one key, here numbers are grouped together for easier reading.)

Use a calculator to add.

1. 89 + 45

2. 76 + 97

3. 285 + 55

4. 8773 + 652

5.
```
    985
   1210
    562
 +   77
```

6.
```
    465
   9888
    620
 + 1550
```

Vocabulary, Readiness & Video Check

Use the choices below to fill in each blank. Some choices may be used more than once.

sum	order	addend	associative
perimeter	number	grouping	commutative

1. The sum of 0 and any number is the same _____.

2. The sum of any number and 0 is the same _____.

3. In 35 + 20 = 55, the number 55 is called the _____ and 35 and 20 are each called a(n) _____.

4. The distance around a polygon is called its _____.

5. Since $(3 + 1) + 20 = 3 + (1 + 20)$, we say that changing the _____ in addition does not change the sum. This property is called the _____ property of addition.

6. Since $7 + 10 = 10 + 7$, we say that changing the _____ in addition does not change the sum. This property is called the _____ property of addition.

Martin-Gay Interactive Videos   *Watch the section lecture video and answer the following questions.*

See Video 1.3

Objective A 7. Complete this statement based on the lecture before Example 1: To add whole numbers, we line up _____ values and add from _____ to _____.

Objective B 8. In Example 4, the perimeter of what type of polygon is found? How many addends are in the resulting addition problem?

Objective C 9. In Example 6, what key word or phrase indicates addition?

1.3 Exercise Set MyMathLab®

Objective A *Add. See Examples 1 through 4.*

1. 14
    + 22

2. 27
    + 31

3.   62
    + 230

4.   37
    + 542

5.   12
     13
    + 24

6. 23
    45
    + 30

7. 5267
    + 132

8.   236
    + 6243

9. 53 + 64

10. 41 + 74

11. 22 + 490

12. 35 + 470

13. 22,781 + 186,297

14. 17,427 + 821,059

15. 8
     9
     2
     5
    + 1

16. 3
     5
     8
     5
    + 7

17.   6
     21
     14
      9
    + 12

18. 12
      4
      8
     26
    + 10

19. 81
     17
     23
     79
    + 12

20. 64
     28
     56
     25
    + 32

21. 62 + 18 + 14

22. 23 + 49 + 18

23. 40 + 800 + 70

24. 30 + 900 + 20

25. 7542 + 49 + 682

26. 1624 + 32 + 976

27. 24 + 9006 + 489 + 2407   **28.** 16 + 1056 + 748 + 7770

29.
```
  627
  628
+ 629
```

30.
```
  427
  383
+ 229
```

31.
```
 6820
 4271
+5626
```

32.
```
 6789
 4321
+5555
```

33.
```
 507
 593
+ 10
```

34.
```
 864
  33
+356
```

35.
```
 4200
 2107
+2692
```

36.
```
 5000
 1400
+3021
```

37.
```
     49
    628
   5 762
+29,462
```

38.
```
     26
    582
   4 763
+62,511
```

39.
```
 121,742
  57,279
  26,586
+426,782
```

40.
```
 504,218
 321,920
  38,507
+594,687
```

Objective B   Find the perimeter of each figure. See Examples 5 and 6.

△ **41.**

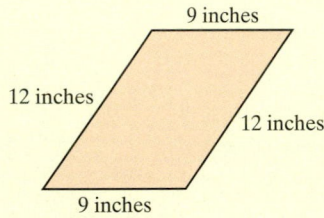

△ **42.**

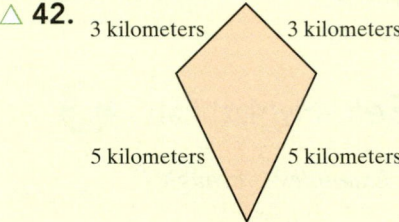

△ **43.**

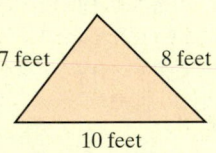

△ **44.**

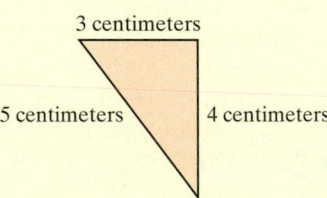

△ **45.**

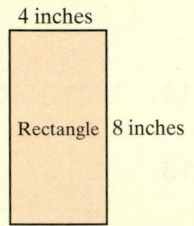

△ **46.**

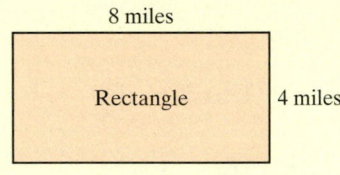

△ **47.**

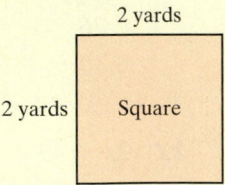

△ **48.**

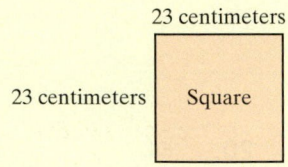

△ **49.**

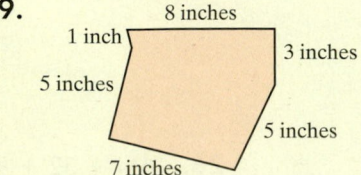

△ **50.**

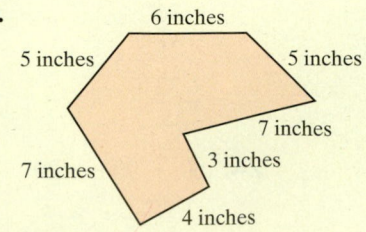

△ **51.**

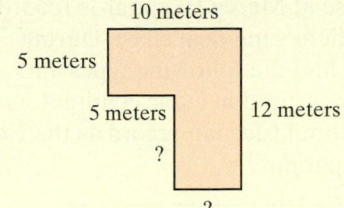

10 meters
5 meters
5 meters   12 meters
?
?

△ **52.**

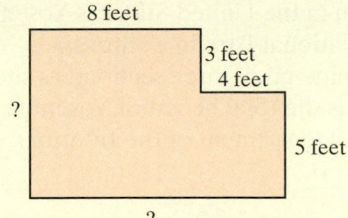

8 feet
3 feet
4 feet
?
5 feet
?

Objectives A B C Mixed Practice–Translating *Solve. See Examples 1 through 8.*

53. Find the sum of 297 and 1796.

54. Find the sum of 802 and 6487.

55. Find the total of 76, 39, 8, 17, and 126.

56. Find the total of 89, 45, 2, 19, and 341.

57. What is 452 increased by 92?

58. What is 712 increased by 38?

59. What is 2686 plus 686 plus 80?

60. What is 3565 plus 565 plus 70?

61. The estimated population of Florida was 19,318 thousand in 2012. If is projected to increase by 1823 thousand by 2020. What is Florida's projected population in 2020? (*Source:* U.S. Census Bureau, Florida Office of Economic & Demographic Research)

62. The estimated population of California was 38,041 thousand in 2012. It is projected to increase by 2603 thousand by 2020. What is California's projected population in 2020? (*Source:* U.S. Census Bureau, California Department of Finance)

63. The highest point in South Carolina is Sassafras Mountain at 3560 feet above sea level. The highest point in North Carolina is Mt. Mitchell, whose peak is 3124 feet increased by the height of Sassafras Mountain. Find the height of Mt. Mitchell. (*Source:* U.S. Geological Survey)

64. The distance from Kansas City, Kansas, to Hays, Kansas, is 285 miles. Colby, Kansas, is 98 miles farther from Kansas City than Hays. Find the total distance from Kansas City to Colby.

△ **65.** Leo Callier is installing an invisible fence in his backyard. How many feet of wiring are needed to enclose the yard below?

70 feet   78 feet
90 feet
102 feet

△ **66.** A homeowner is considering installing gutters around her home. Find the perimeter of her rectangular home.

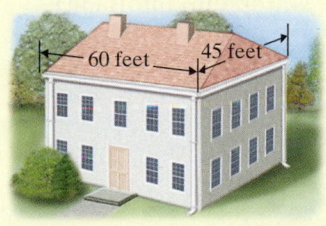

60 feet   45 feet

67. The tallest waterfall in the United States is Yosemite Falls in Yosemite National Park in California. Yosemite Falls is made up of three sections, as shown in the graph. What is the total height of Yosemite Falls? (*Source:* U.S. Department of the Interior)

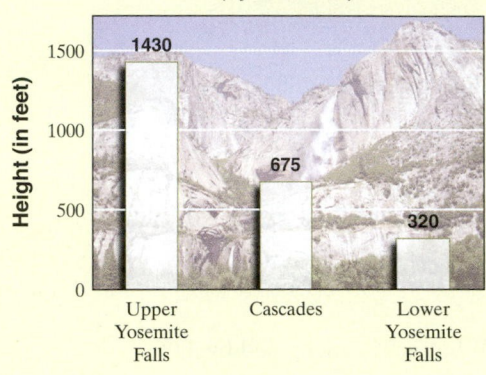

Tallest U.S. Waterfall
(by Sections)

68. Jordan White, a nurse at Mercy Hospital, is recording fluid intake on a patient's medical chart. During his shift, the patient had the following types and amounts of intake measured in cubic centimeters (cc). What amount should Jordan record as the total fluid intake for this patient?

Oral	Intravenous	Blood
240	500	500
100	200	
355		

69. In 2012, Harley-Davidson sold 172,251 of its motorcycles domestically. In addition, 77,598 Harley-Davidson motorcycles were sold internationally. What was the total number of Harley-Davidson motorcycles sold in 2012? (*Source:* Harley-Davidson, Inc.)

70. Hank Aaron holds Major League Baseball's record for the most runs batted in over his career. He batted in 1305 runs from 1954 to 1965. He batted in another 992 runs from 1966 until he retired in 1976. How many total runs did Hank Aaron bat in during his career in professional baseball?

71. During August 2013, a total of 999,040 vehicles were produced in the United States. During the same period, a total of 475,671 vehicles were produced in Canada and Mexico. What was the total number of vehicles produced in North America in August 2013? (*Source:* WardsAuto InfoBank)

72. In 2012, the country of New Zealand had 26,767,000 more sheep than people. If the human population of New Zealand in 2012 was 4,433,000, what was the sheep population? (*Source:* Statistics New Zealand)

73. The largest permanent Monopoly board is made of granite and located in San Jose, California. Find the perimeter of the square playing board.

31 ft

31 ft

74. The smallest commercially available jigsaw puzzle (with a minimum of 1000 pieces) is manufactured in Hong Kong, China. (*Source: Guinness World Records*) Find the exact perimeter of this rectangular-shaped puzzle in millimeters.

182 millimeters (about 7 in.)

257 millimeters (about 10 in.)

75. In 2013, there were 2657 Gap Inc. (Gap, Banana Republic, Old Navy) stores located in the United States and 438 located outside the United States. How many Gap Inc. stores were located worldwide? (*Source:* Gap Inc.)

76. Wilma Rudolph, who won three gold medals in track and field events in the 1960 Summer Olympics, was born in 1940. Allyson Felix, who also won three gold medals in track and field events but in the 2012 Summer Olympics, was born 45 years later. In what year was Allyson Felix born?

The table shows the number of Target stores in ten states. Use this table to answer Exercises 77 through 82.

| The Top States for Target Stores in 2012 ||
State	Number of Stores
California	257
Florida	123
Illinois	89
Michigan	59
Minnesota	75
New York	67
Ohio	64
Pennsylvania	63
Texas	149
Virginia	57

(*Source:* Target Corporation)

77. Which state has the most Target stores?

78. Which of the states listed in the table has the fewest Target stores?

79. What is the total number of Target stores located in the three states with the most Target stores?

80. How many Target stores are located in the ten states listed in the table?

81. Which pair of neighboring states has more Target stores combined, New York and Pennsylvania or Michigan and Ohio?

82. Target operates stores in 49 states. There are 775 Target stores located in the states not listed in the table. How many Target stores are in the United States?

83. The state of Delaware has 2997 miles of urban high-ways and 3361 miles of rural highways. Find the total highway mileage in Delaware. (*Source:* U.S. Federal Highway Administration)

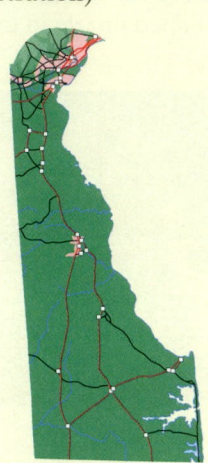

84. The state of Rhode Island has 5260 miles of urban highways and 1225 miles of rural highways. Find the total highway mileage in Rhode Island. (*Source:* U.S. Federal Highway Administration)

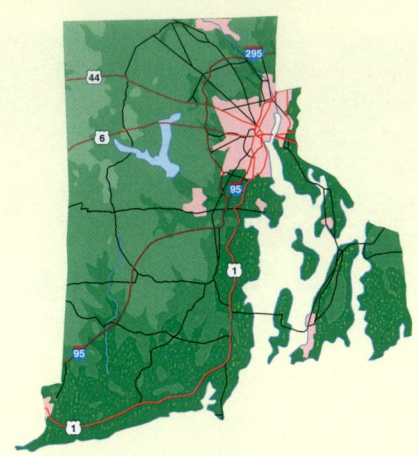

Concept Extensions

85. In your own words, explain the commutative property of addition.

86. In your own words, explain the associative property of addition.

87. Give any three whole numbers whose sum is 100.

88. Give any four whole numbers whose sum is 25.

89. Add: 56,468,980 + 1,236,785 + 986,768,000

90. Add: 78,962 + 129,968,350 + 36,462,880

Check each addition below. If it is incorrect, find the correct answer. See the Concept Check in this section.

91.
```
   566
   932
 +871
 2369
```

92.
```
   773
   659
 +481
 1913
```

93.
```
   14
  173
   86
+257
  520
```

94.
```
   19
  214
   49
+651
  923
```

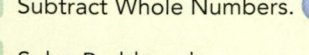

1.4 Subtracting Whole Numbers

Objectives

A Subtract Whole Numbers.

B Solve Problems by Subtracting Whole Numbers.

Objective A Subtracting Whole Numbers

If you have $5 and someone gives you $3, you have a total of $8, since 5 + 3 = 8. Similarly, if you have $8 and then someone borrows $3, you have $5 left. **Subtraction** is finding the **difference** of two numbers.

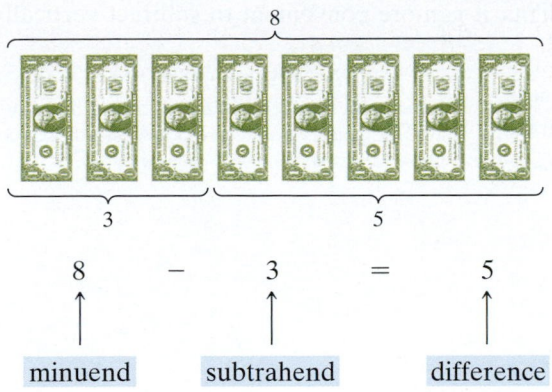

In this example, 8 is the **minuend,** and 3 is the **subtrahend.** The **difference** between these two numbers, 8 and 3, is 5.

Notice that addition and subtraction are very closely related. In fact, subtraction is defined in terms of addition.

$$8 - 3 = 5 \text{ because } 5 + 3 = 8$$

This means that subtraction can be *checked* by addition, and we say that addition and subtraction are reverse operations.

Example 1 Subtract. Check each answer by adding.

a. $12 - 9$ **b.** $22 - 7$ **c.** $35 - 35$ **d.** $70 - 0$

Solution:

a. $12 - 9 = 3$ because $3 + 9 = 12$
b. $22 - 7 = 15$ because $15 + 7 = 22$
c. $35 - 35 = 0$ because $0 + 35 = 35$
d. $70 - 0 = 70$ because $70 + 0 = 70$

■ **Work Practice 1**

Look again at Examples 1(c) and 1(d).

1(c) $35 - 35 = 0$

same number   difference is 0

1(d) $70 - 0 = 70$

a number minus 0   difference is the same number

These two examples illustrate the subtraction properties of 0.

Subtraction Properties of 0

The difference of any number and that same number is 0. For example,

$$11 - 11 = 0$$

The difference of any number and 0 is that same number. For example,

$$45 - 0 = 45$$

To subtract whole numbers we subtract the digits in the ones place, then the tens place, then the hundreds place, and so on. When subtraction involves numbers

Practice 1

Subtract. Check each answer by adding.

a. $14 - 6$
b. $20 - 8$
c. $93 - 93$
d. $42 - 0$

Answers
1. a. 8 **b.** 12 **c.** 0 **d.** 42

of two or more digits, it is more convenient to subtract vertically. For example, to subtract $893 - 52$,

$$893 \leftarrow \text{minuend}$$
$$-52 \leftarrow \text{subtrahend}$$
$$841 \leftarrow \text{difference}$$

Line up the numbers vertically so that the minuend is on top and the place values correspond. Subtract in corresponding place values, starting with the ones place.

$3 - 2$
$9 - 5$
$8 - 0$

To check, add.

$$\begin{array}{r} \text{difference} \\ + \text{ subtrahend} \\ \hline \text{minuend} \end{array} \quad \text{or} \quad \begin{array}{r} 841 \\ +\ 52 \\ \hline 893 \end{array}$$

$\leftarrow$ Since this is the original minuend, the problem checks.

Practice 2

Practice 2

Subtract. Check by adding.
a. $9143 - 122$
b. $978 - 851$

Example 2 Subtract: $7826 - 505$. Check by adding.

Solution:
$$\begin{array}{r} 7826 \\ -\ 505 \\ \hline 7321 \end{array}$$

Check:
$$\begin{array}{r} 7321 \\ +\ 505 \\ \hline 7826 \end{array}$$

■ **Work Practice 2**

Subtracting by Borrowing

When subtracting vertically, if a digit in the second number (subtrahend) is larger than the corresponding digit in the first number (minuend), **borrowing** is necessary. For example, consider

$$\begin{array}{r} 8\,|1| \\ -6\,|3| \end{array}$$

Since the 3 in the ones place of 63 is larger than the 1 in the ones place of 81, borrowing is necessary. We borrow 1 ten from the tens place and add it to the ones place.

Borrowing

$$8 - 1 = 7 \rightarrow \overset{7\ 11}{\cancel{8}\cancel{1}} \leftarrow 1 \text{ ten} + 1 \text{ one} = 11 \text{ ones}$$
$$\text{tens}\quad\text{ten}\quad\text{tens}$$
$$-6\ 3$$

Now we subtract the ones-place digits and then the tens-place digits.

$$\overset{7\ 11}{\cancel{8}\cancel{1}}$$
$$-6\ 3$$
$$\hline 1\ 8 \leftarrow 11 - 3 = 8$$
$$\quad\ \ 7 - 6 = 1$$

Check:
$$\begin{array}{r} 18 \\ +63 \\ \hline 81 \end{array}$$ The original minuend.

Practice 3

Subtract. Check by adding.
a. $\begin{array}{r} 697 \\ -\ 49 \end{array}$
b. $\begin{array}{r} 326 \\ -245 \end{array}$
c. $\begin{array}{r} 1234 \\ -\ 822 \end{array}$

Example 3 Subtract: $543 - 29$. Check by adding.

Solution:
$$\begin{array}{r} \overset{3\ 13}{5\cancel{4}\cancel{3}} \\ -\ 29 \\ \hline 5\ 1\ 4 \end{array}$$

Check:
$$\begin{array}{r} 514 \\ +\ 29 \\ \hline 543 \end{array}$$

■ **Work Practice 3**

Answers
2. a. 9021 **b.** 127
3. a. 648 **b.** 81 **c.** 412

Sometimes we may have to borrow from more than one place. For example, to subtract $7631 - 152$, we first borrow from the tens place.

$$\begin{array}{r} 2\ 11 \\ 76\cancel{3}\cancel{1} \\ -\ \ 1\ 5\ 2 \\ \hline 9 \end{array} \leftarrow 11 - 2 = 9$$

In the tens place, 5 is greater than 2, so we borrow again. This time we borrow from the hundreds place.

6 hundreds − **1 hundred** = 5 hundreds

1 hundred + 2 tens

or

10 tens + 2 tens = 12 tens

$$\begin{array}{r} 5\ \overset{12}{2}\ 11 \\ 7\cancel{6}\cancel{3}\cancel{1} \\ -\ \ 1\ 5\ 2 \\ \hline 7\ 4\ 7\ 9 \end{array}$$

Check:
$$\begin{array}{r} 7479 \\ +\ 152 \\ \hline 7631 \end{array}$$ The original minuend.

Example 4 Subtract: $900 - 174$. Check by adding.

Solution: In the ones place, 4 is larger than 0, so we borrow from the tens place. But the tens place of 900 is 0, so to borrow from the tens place we must first borrow from the hundreds place.

$$\begin{array}{r} 8\ 10 \\ \cancel{9}\ \cancel{0}\ 0 \\ -\ 1\ 7\ 4 \end{array}$$

Now borrow from the tens place.

$$\begin{array}{r} 8\ \overset{9}{\cancel{10}}\ 10 \\ \cancel{9}\ \cancel{0}\ \cancel{0} \\ -\ 1\ 7\ 4 \\ \hline 7\ 2\ 6 \end{array}$$

Check:
$$\begin{array}{r} 726 \\ +174 \\ \hline 900 \end{array}$$

■ **Work Practice 4**

Objective B Solving Problems by Subtracting ▶

Often, real-life problems occur that can be solved by subtracting. The first step in solving any word problem is to *understand* the problem by reading it carefully.

Descriptions of problems solved through subtraction *may* include any of these key words or phrases:

Subtraction		
Key Words or Phrases	**Examples**	**Symbols**
subtract	subtract 5 from 8	$8 - 5$
difference	the difference of 10 and 2	$10 - 2$
less	17 less 3	$17 - 3$
less than	2 less than 20	$20 - 2$
take away	14 take away 9	$14 - 9$
decreased by	7 decreased by 5	$7 - 5$
subtracted from	9 subtracted from 12	$12 - 9$

Practice 4

Subtract. Check by adding.

a. $\begin{array}{r} 400 \\ -164 \end{array}$

b. $\begin{array}{r} 1000 \\ -\ 762 \end{array}$

Helpful Hint Be careful when solving applications that suggest subtraction. Although order *does not* matter when adding, order *does* matter when subtracting. For example, $20 - 15$ and $15 - 20$ do not simplify to the same number.

Answers
4. a. 236 **b.** 238

✓**Concept Check** In each of the following problems, identify which number is the minuend and which number is the subtrahend.

a. What is the result when 6 is subtracted from 40?

b. What is the difference of 15 and 8?

c. Find a number that is 15 fewer than 23.

To solve a word problem that involves subtraction, we first use the facts given to write a subtraction statement. Then we write the corresponding solution of the real-life problem. It is sometimes helpful to write the statement in words (brief phrases) and then translate to numbers.

Practice 5

The radius of Uranus is 15,759 miles. The radius of Neptune is 458 miles less than the radius of Uranus. What is the radius of Neptune? (*Source:* National Space Science Data Center)

Example 5   Finding the Radius of a Planet

The radius of Jupiter is 43,441 miles. The radius of Saturn is 7257 miles less than the radius of Jupiter. Find the radius of Saturn. (*Source:* National Space Science Data Center)

43,441 miles   ?

Jupiter   Saturn

Solution:

In Words		Translate to Numbers
radius of Jupiter	⟶	43,441
− 7257	⟶	− 7 257
radius of Saturn	⟶	36,184

The radius of Saturn is 36,184 miles.

■ **Work Practice 5**

> **Helpful Hint**   Since subtraction and addition are reverse operations, don't forget that a subtraction problem can be checked by adding.

Practice 6

During a sale, the price of a new suit is decreased by $47. If the original price was $92, find the sale price of the suit.

Example 6   Calculating Miles per Gallon

A subcompact car gets 42 miles per gallon of gas. A full-size car gets 17 miles per gallon of gas. Find the difference between the subcompact car miles per gallon and the full-size car miles per gallon.

Solution:

In Words		Translate to Numbers
subcompact miles per gallon	⟶	42
− full-size miles per gallon	⟶	−1 7
difference in miles per gallon		2 5

The difference in the subcompact car miles per gallon and the full-size car miles per gallon is 25 miles per gallon.

■ **Work Practice 6**

Answers

5. 15,301 miles   **6.** $45

✓**Concept Check Answers**

a. minuend: 40; subtrahend: 6

b. minuend: 15; subtrahend: 8

c. minuend: 23; subtrahend: 15

Helpful Hint

Once again, because subtraction and addition are reverse operations, don't forget that a subtraction problem can be checked by adding.

Calculator Explorations Subtracting Numbers

To subtract numbers on a calculator, find the keys marked $-$ and $=$ or ENTER.

For example, to find $83 - 49$ on a calculator, press the keys 83 $-$ 49 then $=$ or ENTER.

The display will read [34].

Thus, $83 - 49 = 34$.

Use a calculator to subtract.

1. $865 - 95$ **2.** $76 - 27$

3. $147 - 38$ **4.** $366 - 87$

5. $9625 - 647$ **6.** $10,711 - 8925$

Vocabulary, Readiness & Video Check

Use the choices below to fill in each blank.

0   minuend   difference

number   subtrahend

1. The difference of any number and that same number is _____.

2. The difference of any number and 0 is the same _____.

3. In $37 - 19 = 18$, the number 37 is the _____, and the number 19 is the _____.

4. In $37 - 19 = 18$, the number 18 is called the _____.

Find each difference.

5. $6 - 6$ **6.** $93 - 93$ **7.** $600 - 0$ **8.** $5 - 0$

Martin-Gay Interactive Videos

See Video 1.4

Watch the section lecture video and answer the following questions.

Objective A **9.** In ▤ Example 2, explain how we end up subtracting 7 from 12 in the ones place. ▶

Objective B **10.** Complete this statement based on ▤ Example 4: Order does not matter when _____, but order does matter when _____. ▶

1.4 Exercise Set MyMathLab®

Objective A *Subtract. Check by adding. See Examples 1 and 2.*

1. 67
     − 23

2. 72
     − 41

3. 389
     − 124

4. 572
     − 321

5. 167
     − 32

6. 286
     − 45

7. 2677 − 423

8. 5766 − 324

9. 6998 − 1453

10. 4912 − 2610

11. 749
      − 149

12. 257
      − 257

Subtract. Check by adding. See Examples 1 through 4.

13. 62
      − 37

14. 55
      − 29

15. 70
      − 25

16. 80
      − 37

17. 938
      − 792

18. 436
      − 275

19. 922
      − 634

20. 674
      − 299

21. 600
      − 432

22. 300
      − 149

23. 142
      − 36

24. 773
      − 29

25. 923
      − 476

26. 813
      − 227

27. 6283
      − 560

28. 5349
      − 720

29. 533
      − 29

30. 724
      − 16

31. 200
      − 111

32. 300
      − 211

33. 1983
      − 1904

34. 1983
      − 1914

35. 56,422
      − 16,508

36. 76,652
      − 29,498

37. 50,000 − 17,289

38. 40,000 − 23,582

39. 7020 − 1979

40. 6050 − 1878

41. 51,111 − 19,898

42. 62,222 − 39,898

Objective B *Solve. See Examples 5 and 6.*

43. Subtract 5 from 9.

44. Subtract 9 from 21.

45. Find the difference of 41 and 21.

46. Find the difference of 16 and 5.

47. Subtract 56 from 63.

48. Subtract 41 from 59.

49. Find 108 less 36.

50. Find 25 less 12.

51. Find 12 subtracted from 100.

52. Find 86 subtracted from 90.

53. Professor Graham is reading a 503-page book. If she has just finished reading page 239, how many more pages must she read to finish the book?

54. When a couple began a trip, the odometer read 55,492. When the trip was over, the odometer read 59,320. How many miles did they drive on their trip?

55. In 2008, the hole in the Earth's ozone layer over Antarctica was about 25 million square kilometers in size. By 2012, the hole had shrunk to about 18 million square kilometers. By how much did the hole shrink from 2008 to 2012? (*Source:* NASA Ozone Watch)

56. Bamboo can grow to 98 feet while Pacific giant kelp (a type of seaweed) can grow to 197 feet. How much taller is the kelp than the bamboo?

Bamboo

Kelp

A river basin is the geographic area drained by a river and its tributaries. The Mississippi River Basin is the third largest in the world and is divided into six sub-basins, whose areas are shown in the following bar graph. Use this graph for Exercises 57 through 60.

57. Find the total U.S. land area drained by the Upper Mississippi and Lower Mississippi sub-basins.

58. Find the total U.S. land area drained by the Ohio and Tennessee sub-basins.

59. How much more land is drained by the Missouri sub-basin than the Arkansas Red-White sub-basin?

60. How much more land is drained by the Upper Mississippi sub-basin than the Lower Mississippi sub-basin?

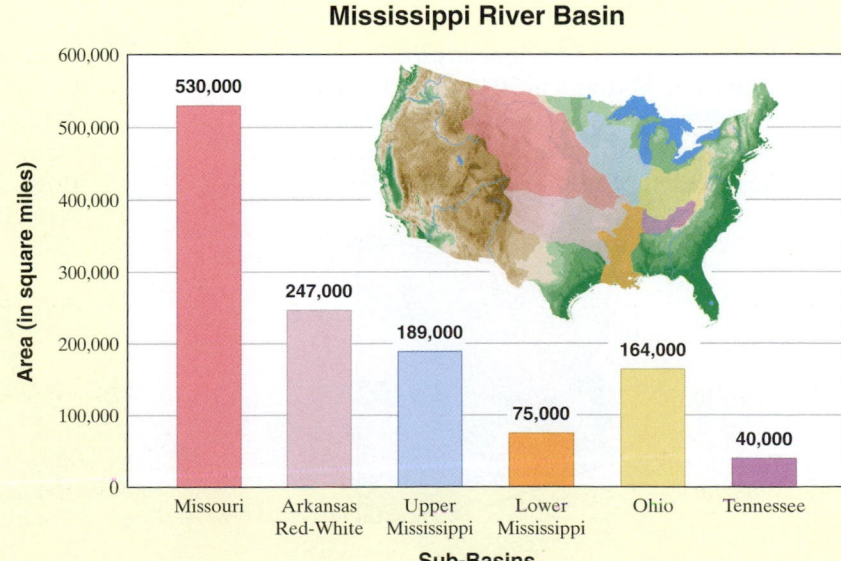

Mississippi River Basin

61. The peak of Mt. McKinley in Alaska is 20,320 feet above sea level. The peak of Long's Peak in Colorado is 14,255 feet above sea level. How much higher is the peak of Mt. McKinley than Long's Peak? (*Source:* U.S. Geological Survey)

Mt. McKinley, Alaska   Long's Peak, Colorado

62. On January 12, 1916, the city of Indianapolis, Indiana, had the greatest temperature change in a day. It dropped 58 degrees. If the high temperature was 68° Fahrenheit, what was the low temperature?

63. The Oroville Dam, on the Feather River, is the tallest dam in the United States at 754 feet. The Hoover Dam, on the Colorado River, is 726 feet high. How much taller is the Oroville Dam than the Hoover Dam? (*Source:* U.S. Bureau of Reclamation)

64. A new iPhone with 32 GB costs $299. Jocelyn Robinson has $713 in her savings account. How much will she have left in her savings account after she buys the iPhone? (*Source:* Apple, Inc.)

65. The distance from Kansas City to Denver is 645 miles. Hays, Kansas, lies on the road between the two and is 287 miles from Kansas City. What is the distance between Hays and Denver?

66. Pat Salanki's blood cholesterol level is 243. The doctor tells him it should be decreased to 185. How much of a decrease is this?

67. A new DVD player with remote control costs $295. A college student has $914 in her savings account. How much will she have left in her savings account after she buys the DVD player?

68. A stereo that regularly sells for $547 is discounted by $99 in a sale. What is the sale price?

69. The population of Arizona is projected to grow from 6499 thousand in 2012 to 9129 thousand in 2032. What is Arizona's projected population increase over this time? (*Source:* Arizona Department of Administration)

70. In 1996, the centennial of the Boston Marathon, the official number of participants was 38,708. In 2013, there were 11,869 fewer participants. How many official participants were there for the 2013 Boston Marathon? (*Source:* Boston Athletic Association)

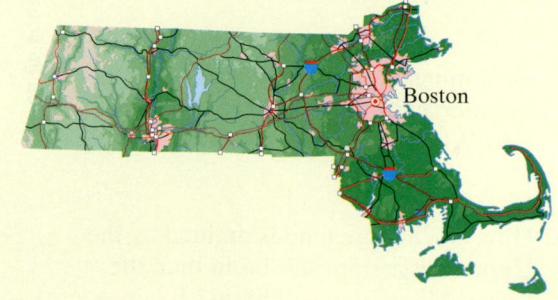

The decibel (dB) is a unit of measurement for sound. Every increase of 10 dB is a tenfold increase in sound intensity. The bar graph below shows the decibel levels for some common sounds. Use this graph for Exercises 71 through 74.

71. What is the dB rating for live rock music?

72. Which is the quietest of all the sounds shown in the graph?

73. How much louder is the sound of snoring than normal conversation?

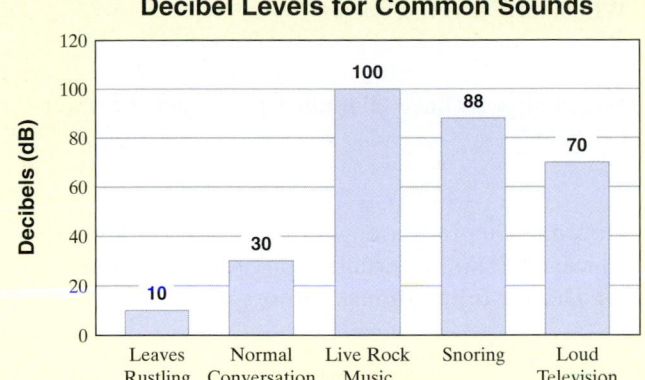

Decibel Levels for Common Sounds

74. What is the difference in sound intensity between live rock music and loud television?

75. The 113th Congress has 535 senators and representatives. Of these, 189 were registered Boy Scouts at some time in their lives. How many members of the 113th Congress were never Boy Scouts? (*Source: Boy Scouts of America*)

76. In the United States, there were 28,799 tornadoes from 1990 through 2012. In all, 13,205 of these tornadoes occurred from 1990 through 2000. How many tornadoes occurred during the period after 2000? (*Source: Storm Prediction Center, National Weather Service*)

77. Until recently, the world's largest permanent maze was located in Ruurlo, Netherlands. This maze of beech hedges covers 94,080 square feet. A new hedge maze using hibiscus bushes at the Dole Plantation in Wahiawa, Hawaii, covers 100,000 square feet. How much larger is the Dole Plantation maze than the Ruurlo maze? (*Source: The Guinness Book of Records*)

78. There were only 27 California condors in the entire world in 1987. To date, the number has increased to an estimated 223 living in the wild. How much of an increase is this? (*Source: California Department of Fish and Wildlife*)

The bar graph shows the top five U.S. airports according to number of passengers arriving and departing in 2013. Use this graph to answer Exercises 79 through 82.

79. Which airport is the busiest?

80. Which airports have 60 million passengers or fewer per year?

81. How many more passengers per year does the Chicago O'Hare International Airport have than the Denver International Airport?

82. How many more passengers per year does the Hartsfield-Jackson Atlanta International Airport have than the Dallas/Ft. Worth International Airport?

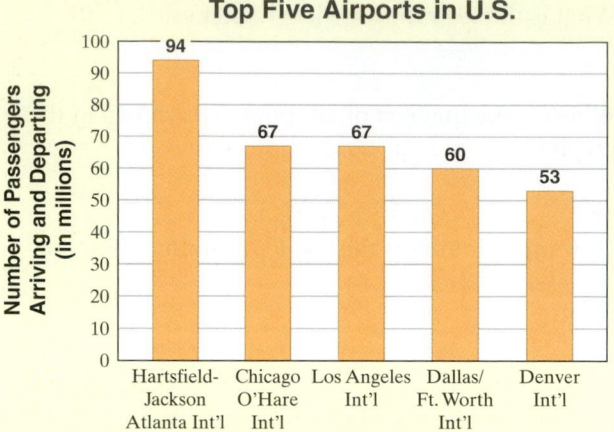

Top Five Airports in U.S.

Source: Airports Council International

Solve.

83. Two seniors, Jo Keen and Trudy Waterbury, were candidates for student government president. Who won the election if the votes were cast as follows? By how many votes did the winner win?

	Candidate	
Class	**Jo**	**Trudy**
Freshman	276	295
Sophomore	362	122
Junior	201	312
Senior	179	18

84. Two students submitted advertising budgets for a student government fund-raiser.

	Student A	**Student B**
Radio ads	$600	$300
Newspaper ads	$200	$400
Posters	$150	$240
Handbills	$120	$170

If $1200 is available for advertising, how much excess would each budget have?

Mixed Practice (Sections 1.3 and 1.4) *Add or subtract as indicated.*

85.
$$\begin{array}{r} 986 \\ + \ 48 \\ \hline \end{array}$$

86.
$$\begin{array}{r} 986 \\ - \ 48 \\ \hline \end{array}$$

87. $76 - 67$

88. $80 + 93 + 17 + 9 + 2$

89.
$$\begin{array}{r} 9000 \\ - \ 482 \\ \hline \end{array}$$

90.
$$\begin{array}{r} 10{,}000 \\ - \ 1786 \\ \hline \end{array}$$

91.
$$\begin{array}{r} 10{,}962 \\ 4851 \\ + \ 7063 \\ \hline \end{array}$$

92.
$$\begin{array}{r} 12{,}468 \\ 3211 \\ + \ 1988 \\ \hline \end{array}$$

Concept Extensions

For each exercise, identify which number is the minuend and which number is the subtrahend. See the Concept Check in this section.

93.
$$\begin{array}{r} 48 \\ - \ 1 \\ \hline \end{array}$$

94.
$$\begin{array}{r} 2863 \\ -1904 \\ \hline \end{array}$$

95. Subtract 7 from 70.

96. Find 86 decreased by 25.

Identify each answer as correct or incorrect. Use addition to check. If the answer is incorrect, then write the correct answer.

97.   741
        − 56
    ‾‾‾‾‾‾
        675

98.   478
        − 89
    ‾‾‾‾‾‾
        389

99.   1029
        − 888
    ‾‾‾‾‾‾‾
        141

100.   7615
         − 547
    ‾‾‾‾‾‾‾
         7168

Fill in the missing digits in each problem.

101.   526_
        − 2_85
    ‾‾‾‾‾‾‾‾
        28_4

102.   10,_4_
        − 8_5_4
    ‾‾‾‾‾‾‾‾
        _710

103. Is there a commutative property of subtraction? In other words, does order matter when subtracting? Why or why not?

104. Explain why the phrase "Subtract 7 from 10" translates to "10 − 7."

105. The local college library is having a Million Pages of Reading promotion. The freshmen have read a total of 289,462 pages; the sophomores have read a total of 369,477 pages; the juniors have read a total of 218,287 pages; and the seniors have read a total of 121,685 pages. Have they reached a goal of one million pages? If not, how many more pages need to be read?

1.5 Rounding and Estimating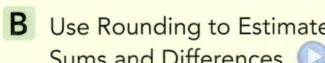

Objective A Rounding Whole Numbers

Rounding a whole number means approximating it. A rounded whole number is often easier to use, understand, and remember than the precise whole number. For example, instead of trying to remember the Minnesota state population as 5,197,621, it is much easier to remember it rounded to the nearest million: 5,000,000, or 5 million people. (*Source: World Almanac*)

Recall from Section 1.2 that the line below is called a number line. To **graph** a whole number on this number line, we darken the point representing the location of the whole number. For example, the number 4 is graphed below.

Objectives

A Round Whole Numbers.

B Use Rounding to Estimate Sums and Differences.

C Solve Problems by Estimating.

Minnesota
Population:
5,197,621
or
about
5 million

On a number line, the whole number 36 is closer to 40 than 30, so 36 rounded to the nearest ten is 40.

The whole number 52 is closer to 50 than 60, so 52 rounded to the nearest ten is 50.

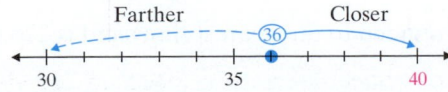

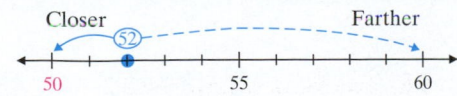

In trying to round 25 to the nearest ten, we see that 25 is halfway between 20 and 30. It is not closer to either number. In such a case, we round to the larger ten, that is, to 30.

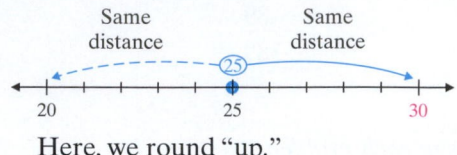

Here, we round "up."

To round a whole number without using a number line, follow these steps:

Rounding Whole Numbers to a Given Place Value

Step 1: Locate the digit to the right of the given place value.

Step 2: If this digit is 5 or greater, add 1 to the digit in the given place value and replace each digit to its right by 0.

Step 3: If this digit is less than 5, replace it and each digit to its right by 0.

Practice 1

Round to the nearest ten.

a. 57

b. 641

c. 325

Example 1 Round 568 to the nearest ten.

Solution: 5 6 ⑧   The digit to the right of the tens place is the ones place, which is circled.
↑
tens place

5 6 ⑧   Since the circled digit is 5 or greater, add 1 to the 6 in the tens place and replace the digit to the right by 0.
↑   ↖
Add 1.   Replace with 0.

We find that 568 rounded to the nearest ten is 570.

■ **Work Practice 1**

Practice 2

Round to the nearest thousand.

a. 72,304

b. 9222

c. 671,800

Example 2 Round 278,362 to the nearest thousand.

Solution: Thousands place
↓   ┌─ 3 is less than 5.
278,③62
↑   ↑
Do not add 1.   Replace with zeros.

The number 278,362 rounded to the nearest thousand is 278,000.

■ **Work Practice 2**

Example 3 Round 248,982 to the nearest hundred.

Solution:

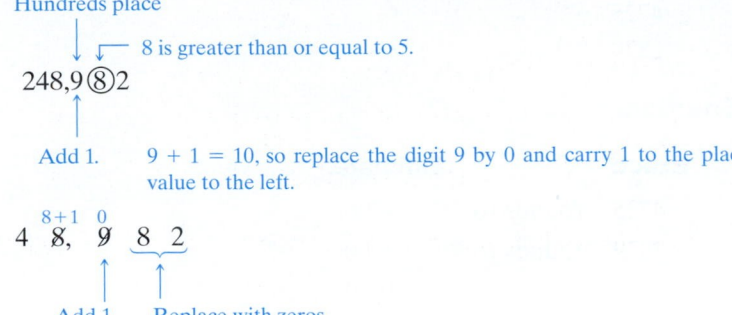

Hundreds place

8 is greater than or equal to 5.

248,9⑧2

Add 1.   9 + 1 = 10, so replace the digit 9 by 0 and carry 1 to the place value to the left.

2 4 $\overset{8+1}{8},\ \overset{0}{9}\ 8\ 2$

Add 1.   Replace with zeros.

The number 248,982 rounded to the nearest hundred is 249,000.

■ **Work Practice 3**

✓**Concept Check** Round each of the following numbers to the nearest *hundred*. Explain your reasoning.

a. 59   **b.** 29

Objective B Estimating Sums and Differences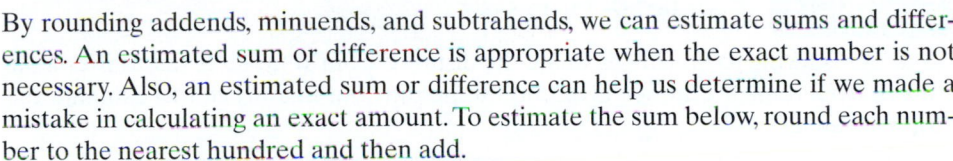

By rounding addends, minuends, and subtrahends, we can estimate sums and differences. An estimated sum or difference is appropriate when the exact number is not necessary. Also, an estimated sum or difference can help us determine if we made a mistake in calculating an exact amount. To estimate the sum below, round each number to the nearest hundred and then add.

768	rounds to	800
1952	rounds to	2000
225	rounds to	200
+ 149	rounds to	+ 100
		3100

The estimated sum is 3100, which is close to the **exact** sum of 3094.

Example 4 Round each number to the nearest hundred to find an estimated sum.

294
625
1071
+ 349

Solution:

Exact:		**Estimate:**
294	rounds to	300
625	rounds to	600
1071	rounds to	1100
+ 349	rounds to	+ 300
		2300

The estimated sum is 2300. (The exact sum is 2339.)

■ **Work Practice 4**

Practice 3

Round to the nearest hundred.

a. 3474

b. 76,243

c. 978,965

Practice 4

Round each number to the nearest ten to find an estimated sum.

49
25
32
51
+ 98

Answers

3. a. 3500 **b.** 76,200 **c.** 979,000

4. 260

✓**Concept Check Answers**
a. 100 **b.** 0

Practice 5

Round each number to the nearest thousand to find an estimated difference.

$$3785$$
$$-\,2479$$

Example 5   Round each number to the nearest hundred to find an estimated difference.

$$4725$$
$$-\,2879$$

Solution:

Exact:		Estimate:
4725	rounds to	4700
$-\,2879$	rounds to	$-\,2900$
		1800

The estimated difference is 1800. (The exact difference is 1846.)

■ Work Practice 5

Objective C Solving Problems by Estimating

Making estimates is often the quickest way to solve real-life problems when solutions do not need to be exact.

Practice 6

Tasha Kilbey is trying to estimate how far it is from Gove, Kansas, to Hays, Kansas. Round each given distance on the map to the nearest ten to estimate the total distance.

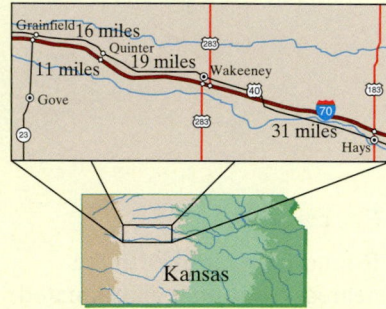

Example 6   Estimating Distances

A driver is trying to quickly estimate the distance from Temple, Texas, to Brenham, Texas. Round each distance given on the map to the nearest ten to estimate the total distance.

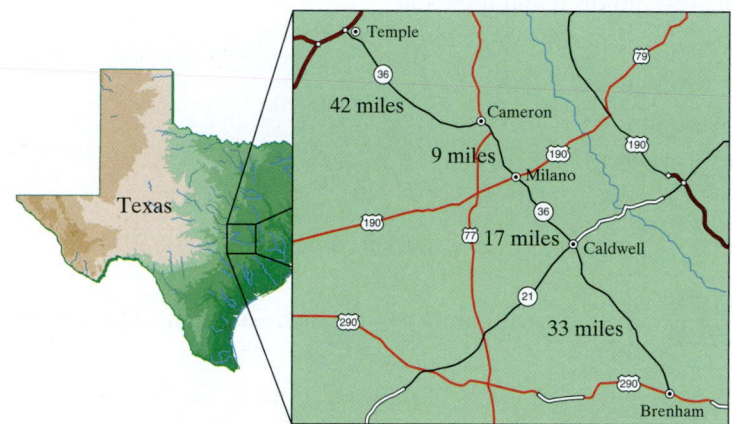

Solution:

Exact Distance:		Estimate:
42	rounds to	40
9	rounds to	10
17	rounds to	20
$+\,33$	rounds to	$+\,30$
		100

It is approximately 100 miles from Temple to Brenham. (The exact distance is 101 miles.)

■ Work Practice 6

Answers

5. 2000   **6.** 80 mi

Example 7 Estimating Data

In three recent months, the numbers of tons of mail that went through Hartsfield-Jackson Atlanta International Airport were 635, 687, and 567. Round each number to the nearest hundred to estimate the tons of mail that passed through this airport.

Solution:

Exact Tons of Mail:		Estimate:
635	rounds to	600
687	rounds to	700
+567	rounds to	+600
		1900

The approximate tonnage of mail that moved through Atlanta's airport over this period was 1900 tons. (The exact tonnage was 1889 tons.)

■ **Work Practice 7**

Practice 7

In 2012, Ecuador topped the International Union for Conservation of Nature's Red List of Threatened Species. At that time, Ecuador was home to 139 threatened bird and mammal species, 316 threatened other animal species, and 1842 threatened plant species. Round each number to the nearest hundred to estimate the total number of threatened species in Ecuador. (*Source: International Union for Conservation of Nature*)

Answer

7. 2200 total threatened species

Vocabulary, Readiness & Video Check

Use the choices below to fill in each blank.

60	rounding	exact
70	estimate	graph

1. To _____ a number on a number line, darken the point representing the location of the number.

2. Another word for approximating a whole number is _____.

3. The number 65 rounded to the nearest ten is _____, but the number 61 rounded to the nearest ten is _____.

4. A(n) _____ number of products is 1265, but a(n) _____ is 1000.

Martin-Gay Interactive Videos

See Video 1.5 🍎

Watch the section lecture video and answer the following questions.

Objective A 5. In ⊞ Example 1, when rounding the number to the nearest ten, why do we replace the digit 3 with a 4? ▶

Objective B 6. As discussed in ⊞ Example 3, explain how a number line can help us understand how to round 22 to the nearest ten. ▶

Objective C 7. What is the significance of the circled digit in each height value in ⊞ Example 5? ▶

1.5 Exercise Set MyMathLab®

Objective A *Round each whole number to the given place. See Examples 1 through 3.*

1. 423 to the nearest ten

2. 273 to the nearest ten

▶ 3. 635 to the nearest ten

4. 846 to the nearest ten

5. 2791 to the nearest hundred

6. 8494 to the nearest hundred

7. 495 to the nearest ten

8. 898 to the nearest ten

9. 21,094 to the nearest thousand

10. 82,198 to the nearest thousand

11. 33,762 to the nearest thousand

12. 42,682 to the nearest ten-thousand

13. 328,495 to the nearest hundred

14. 179,406 to the nearest hundred

▶ 15. 36,499 to the nearest thousand

16. 96,501 to the nearest thousand

17. 39,994 to the nearest ten

18. 99,995 to the nearest ten

19. 29,834,235 to the nearest ten-million

20. 39,523,698 to the nearest million

Complete the table by estimating the given number to the given place value.

		Tens	Hundreds	Thousands
21.	5281			
22.	7619			
23.	9444			
24.	7777			
25.	14,876			
26.	85,049			

Round each number to the indicated place.

27. The University of California, Los Angeles, had a total undergraduate enrollment of 27,941 students in fall 2012. Round this number to the nearest thousand. (*Source:* UCLA)

28. In 2012, there were 12,997 Burger King restaurants worldwide. Round this number to the nearest thousand. (*Source:* Burger King Worldwide, Inc.)

29. Kareem Abdul-Jabbar holds the NBA record for points scored, a total of 38,387 over his NBA career. Round this number to the nearest thousand. (*Source:* National Basketball Association)

30. It takes 60,149 days for Neptune to make a complete orbit around the Sun. Round this number to the nearest hundred. (*Source:* National Space Science Data Center)

31. In 2013, the most valuable brand in the world was Apple, having just overtaken the longtime leader, Coca-Cola. The estimated brand value of Apple was $98,316,000,000. Round this to the nearest billion. (*Source:* Interbrand)

32. According to the 2013 Population Clock, the population of the United States was 316,797,189 in October 2013. Round this population figure to the nearest million. (*Source:* U.S. Census population clock)

33. The average salary for a Boston Red Sox baseball player during the 2013 season was $5,021,850. Round this average salary to the nearest thousand. (*Source:* CBSSports.com)

34. In FY 2013, the Procter & Gamble Company had $84,167,000,000 in sales. Round this sales figure to the nearest billion. (*Source:* Procter & Gamble)

35. In the United States in 2012, the travel industry generated $128,800,000,000 in tax revenue for local, state, and federal governments. Round this travel-related tax revenue figure to the nearest billion. (*Source:* U.S. Travel Association)

36. U.S. farms produced 3,149,166,000 bushels of soybeans in 2013. Round the soybean production figure to the nearest ten-million. (*Source:* U.S. Department of Agriculture)

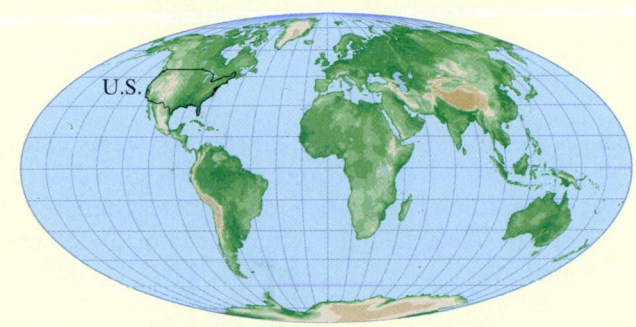

Objective B Estimate the sum or difference by rounding each number to the nearest ten. See Examples 4 and 5.

37.
```
  39
  45
  22
+ 17
```

38.
```
  52
  33
  15
+ 29
```

39.
```
  449
- 373
```

40.
```
  555
- 235
```

Estimate the sum or difference by rounding each number to the nearest hundred. See Examples 4 and 5.

41.
```
  1913
  1886
+ 1925
```

42.
```
  4050
  3133
+ 1220
```

43.
```
  1774
- 1492
```

44.
```
  1989
- 1870
```

45.
```
  3995
  2549
+ 4944
```

46.
```
  799
  1655
+ 271
```

Three of the given calculator answers below are incorrect. Find them by estimating each sum.

47. 463 + 219   602

48. 522 + 785   1307

49. 229 + 443 + 606   1278

50. 542 + 789 + 198   2139

51. 7806 + 5150   12,956

52. 5233 + 4988   9011

> **Helpful Hint**
>
> Estimation is useful to check for incorrect answers when using a calculator. For example, pressing a key too hard may result in a double digit, while pressing a key too softly may result in the digit not appearing in the display.

Objective C   *Solve each problem by estimating. See Examples 6 and 7.*

53. An appliance store advertises three refrigerators on sale at $899, $1499, and $999. Round each cost to the nearest hundred to estimate the total cost.

54. Suppose you scored 89, 97, 100, 79, 75, and 82 on your biology tests. Round each score to the nearest ten to estimate your total score.

55. The distance from Kansas City to Boston is 1429 miles and from Kansas City to Chicago is 530 miles. Round each distance to the nearest hundred to estimate how much farther Boston is from Kansas City than Chicago is.

56. The Gonzales family took a trip and traveled 588, 689, 277, 143, 59, and 802 miles on six consecutive days. Round each distance to the nearest hundred to estimate the distance they traveled.

▶ 57. The peak of Mt. McKinley, in Alaska, is 20,320 feet above sea level. The top of Mt. Rainier, in Washington, is 14,410 feet above sea level. Round each height to the nearest thousand to estimate the difference in elevation of these two peaks.(*Source:* U.S. Geological Survey)

58. A student is pricing new car stereo systems. One system sells for $1895 and another system sells for $1524. Round each price to the nearest hundred dollars to estimate the difference in price of these systems.

59. In 2012, the United States Postal Service delivered 159,859,000,000 pieces of mail. In 2011, it delivered 168,297,000,000 pieces of mail. Round each number to the nearest billion to estimate how much the mail volume decreased from 2011 to 2012. (*Source:* United States Postal Service)

60. Round each distance given on the map to the nearest ten to estimate the total distance from North Platte, Nebraska, to Lincoln, Nebraska.

61. Head Start is a national program that provides developmental and social services for America's low-income preschool children ages three to five. Enrollment figures in Head Start programs showed an increase from 1,125,209 in 2011 to 1,128,030 in 2012. Round each number of children to the nearest thousand to estimate this increase. (*Source:* U.S. Department of Health and Human Services)

62. Enrollment figures at a local community college showed an increase from 49,713 credit hours in 2012 to 51,746 credit hours in 2013. Round each number to the nearest thousand to estimate the increase.

Mixed Practice (Sections 1.2 and 1.5) *The following table shows the top five leading U.S. advertisers in 2012 and the amount of money spent that year on advertising. Complete this table. The first line is completed for you. (Source: Ad Age DataCenter)*

Advertiser	Amount Spent on Advertising in 2012 (in millions of dollars)	Amount Written in Standard Form	Standard Form Rounded to Nearest Ten-Million	Standard Form Rounded to Nearest Hundred-Million
Procter & Gamble Co.	$4829	$4,829,000,000	$4,830,000,000	$4,800,000,000
63. General Motors Co.	$3067			
64. Comcast Corp.	$2989			
65. AT&T	$2910			
66. Verizon Communications	$2381			

Concept Extensions

67. Find one number that when rounded to the nearest hundred is 5700.

68. Find one number that when rounded to the nearest ten is 5700.

69. A number rounded to the nearest hundred is 8600.
   a. Determine the smallest possible number.
   b. Determine the largest possible number.

70. On August 23, 1989, it was estimated that 1,500,000 people joined hands in a human chain stretching 370 miles to protest the fiftieth anniversary of the pact that allowed what was then the Soviet Union to annex the Baltic nations in 1939. If the estimate of the number of people is to the nearest hundred-thousand, determine the largest possible number of people in the chain.

71. In your own words, explain how to round a number to the nearest thousand.

72. In your own words, explain how to round 9660 to the nearest thousand.

73. Estimate the perimeter of the rectangle by first rounding the length of each side to the nearest ten.

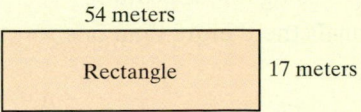

54 meters
Rectangle   17 meters

74. Estimate the perimeter of the triangle by first rounding the length of each side to the nearest hundred.

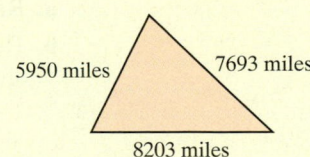

5950 miles   7693 miles
8203 miles

Objectives

A Use the Properties of Multiplication. ▶

B Multiply Whole Numbers. ▶

C Multiply by Whole Numbers Ending in Zero(s). ▶

D Find the Area of a Rectangle. ▶

E Solve Problems by Multiplying Whole Numbers. ▶

Multiplication Shown as Repeated Addition Suppose that we wish to count the number of laptops provided in a computer class. The laptops are arranged in 5 rows, and each row has 6 laptops.

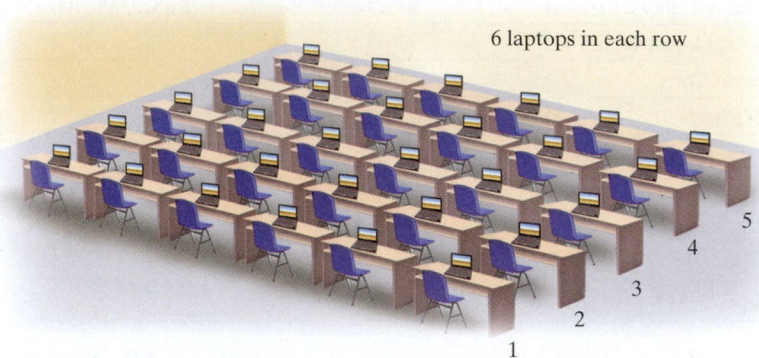

6 laptops in each row

Adding 5 sixes gives the total number of laptops. We can write this as $6 + 6 + 6 + 6 + 6 = 30$ laptops. When each addend is the same, we refer to this as **repeated addition.**

Multiplication is repeated addition but with different notation.

$$6 + 6 + 6 + 6 + 6 \quad = \quad 5 \quad \times \quad 6 \quad = \quad 30$$

| 5 addends; each addend is 6 | (number of addends) factor | (each addend) factor | product |

The × is called a **multiplication sign.** The numbers 5 and 6 are called **factors.** The number 30 is called the **product.** The notation 5×6 is read as "five times six." The symbols · and ( ) can also be used to indicate multiplication.

$$5 \times 6 = 30, \quad 5 \cdot 6 = 30, \quad (5)(6) = 30, \quad \text{and} \quad 5(6) = 30$$

✓Concept Check

a. Rewrite $5 + 5 + 5 + 5 + 5 + 5 + 5$ using multiplication.

b. Rewrite 3×16 as repeated addition. Is there more than one way to do this? If so, show all ways.

Objective **A** Using the Properties of Multiplication

As with addition, we memorize products of one-digit whole numbers and then use certain properties of multiplication to multiply larger numbers. (If necessary, review the multiplication of one-digit numbers in Appendix A.2)

Notice that when any number is multiplied by 0, the result is always 0. This is called the **multiplication property of 0.**

✓Concept Check Answers
a. $7 \times 5 = 35$
b. $16 + 16 + 16 = 48$; yes,
$3 + 3 + 3 + 3 + 3 + 3 + 3 + 3 +$
$3 + 3 + 3 + 3 + 3 + 3 + 3 + 3 = 48$

Multiplication Property of 0

The product of 0 and any number is 0. For example,

$5 \cdot 0 = 0$ and $0 \cdot 8 = 0$

Also notice in Appendix A.2 that when any number is multiplied by 1, the result is always the original number. We call this result the **multiplication property of 1.**

Multiplication Property of 1

The product of 1 and any number is that same number. For example,

$1 \cdot 9 = 9$ and $6 \cdot 1 = 6$

Example 1   Multiply.

a. 6×1 **b.** $0(18)$ **c.** $1 \cdot 45$ **d.** $(75)(0)$

Solution:

a. $6 \times 1 = 6$ **b.** $0(18) = 0$

c. $1 \cdot 45 = 45$ **d.** $(75)(0) = 0$

■ **Work Practice 1**

Practice 1

Multiply.
a. 3×0
b. $4(1)$
c. $(0)(34)$
d. $1 \cdot 76$

Like addition, multiplication is commutative and associative. Notice that when multiplying two numbers, the order of these numbers can be changed without changing the product. For example,

$3 \cdot 5 = 15$ and $5 \cdot 3 = 15$

This property is the **commutative property of multiplication.**

Commutative Property of Multiplication

Changing the **order** of two factors does not change their product. For example,

$9 \cdot 2 = 18$ and $2 \cdot 9 = 18$

Another property that can help us when multiplying is the **associative property of multiplication.** This property states that when multiplying numbers, the grouping of the numbers can be changed without changing the product. For example,

$(2 \cdot 3) \cdot 4 = 6 \cdot 4 = 24$

Also,

$2 \cdot (3 \cdot 4) = 2 \cdot 12 = 24$

Both groupings give a product of 24.

Associative Property of Multiplication

Changing the **grouping** of factors does not change their product. From the previous page, we know that for example,

$$(2 \cdot 3) \cdot 4 = 2 \cdot (3 \cdot 4)$$

With these properties, along with the **distributive property,** we can find the product of any whole numbers. The distributive property says that multiplication **distributes** over addition. For example, notice that $3(2 + 5)$ simplifies to the same number as $3 \cdot 2 + 3 \cdot 5$.

$$3(2 + 5) = 3(7) = 21$$

$$3 \cdot 2 + 3 \cdot 5 = 6 + 15 = 21$$

Since $3(2 + 5)$ and $3 \cdot 2 + 3 \cdot 5$ both simplify to 21, then

$$3(2 + 5) = 3 \cdot 2 + 3 \cdot 5$$

Notice in $3(2 + 5) = 3 \cdot 2 + 3 \cdot 5$ that each number inside the parentheses is multiplied by 3.

Distributive Property

Multiplication distributes over addition. For example,

$$2(3 + 4) = 2 \cdot 3 + 2 \cdot 4$$

Practice 2

Rewrite each using the distributive property.

a. $5(2 + 3)$

b. $9(8 + 7)$

c. $3(6 + 1)$

Example 2   Rewrite each using the distributive property.

a. $3(4 + 5)$ **b.** $10(6 + 8)$ **c.** $2(7 + 3)$

Solution: Using the distributive property, we have

a. $3(4 + 5) = 3 \cdot 4 + 3 \cdot 5$

b. $10(6 + 8) = 10 \cdot 6 + 10 \cdot 8$

c. $2(7 + 3) = 2 \cdot 7 + 2 \cdot 3$

■ **Work Practice 2**

Objective B   Multiplying Whole Numbers

Let's use the distributive property to multiply $7(48)$. To do so, we begin by writing the expanded form of 48 (see Section 1.2) and then applying the distributive property.

$$
\begin{aligned}
7(48) &= 7(40 + 8) &&\text{Write 48 in expanded form.}\\
&= 7 \cdot 40 + 7 \cdot 8 &&\text{Apply the distributive property.}\\
&= 280 + 56 &&\text{Multiply.}\\
&= 336 &&\text{Add.}
\end{aligned}
$$

Answers

2. a. $5(2 + 3) = 5 \cdot 2 + 5 \cdot 3$

    b. $9(8 + 7) = 9 \cdot 8 + 9 \cdot 7$

    c. $3(6 + 1) = 3 \cdot 6 + 3 \cdot 1$

This is how we multiply whole numbers. When multiplying whole numbers, we will use the following notation.

First:

$$\begin{array}{r} \overset{5}{4}8 \\ \times\ 7 \\ \hline 336 \end{array}$$

← 7·8 = 56 Write 6 in the ones place and carry 5 to the tens place.

Next:

$$\begin{array}{r} \overset{5}{4}8 \\ \times\ 7 \\ \hline 336 \end{array}$$

7·4 + 5 = 28 + 5 = 33

The product of 48 and 7 is 336.

Example 3 Multiply:

a. 25
   × 8

b. 246
   × 5

Solution:

a. $\overset{4}{2}5$
   × 8
   ——
   200

b. $\overset{2\,3}{2}46$
   × 5
   ——
   1230

■ **Work Practice 3**

To multiply larger whole numbers, use the following similar notation. Multiply 89 × 52.

Step 1

$$\begin{array}{r} \overset{1}{8}9 \\ \times\ 52 \\ \hline 178 \end{array}$$ ← Multiply 89 × 2.

Step 2

$$\begin{array}{r} \overset{4}{8}9 \\ \times\ 52 \\ \hline 178 \\ 4450 \end{array}$$ ← Multiply 89 × 50.

Step 3

$$\begin{array}{r} 89 \\ \times\ 52 \\ \hline 178 \\ 4450 \\ \hline 4628 \end{array}$$ ← Add.

The numbers 178 and 4450 are called **partial products.** The sum of the partial products, 4628, is the product of 89 and 52.

Example 4 Multiply: 236 × 86

Solution:

$$\begin{array}{r} 236 \\ \times\ 86 \\ \hline 1416 \\ 18880 \\ \hline 20{,}296 \end{array}$$

← 6(236)
← 80(236)
Add.

■ **Work Practice 4**

Example 5 Multiply: 631 × 125

Solution:

$$\begin{array}{r} 631 \\ \times\ 125 \\ \hline 3155 \\ 12620 \\ 63100 \\ \hline 78{,}875 \end{array}$$

← 5(631)
← 20(631)
← 100(631)
Add.

■ **Work Practice 5**

Practice 3

Multiply.

a. 36
   × 4

b. 132
   × 9

Practice 4

Multiply.

a. 594
   × 72

b. 306
   × 81

Practice 5

Multiply.

a. 726
   × 142

b. 288
   × 4

Answers

3. a. 144   **b.** 1188
4. a. 42,768   **b.** 24,786
5. a. 103,092   **b.** 1152

✓**Concept Check** Find and explain the error in the following multiplication problem.

$$
\begin{array}{r}
102 \\
\times\ \ 33 \\
\hline
306 \\
306 \\
\hline
612
\end{array}
$$

Objective C Multiplying by Whole Numbers Ending in Zero(s) ▶

Interesting patterns occur when we multiply by a number that ends in zeros. To see these patterns, let's multiply a number, say 34, by 10, then 100, then 1000.

$$
\begin{array}{l}
\overset{\text{1 zero}}{\downarrow} \\
34 \cdot 10 = 340 \qquad \text{1 zero attached to 34.} \\[4pt]
\overset{\text{2 zeros}}{34 \cdot 100} = 3400 \qquad \text{2 zeros attached to 34.} \\[4pt]
\overset{\text{3 zeros}}{34 \cdot 1000} = 34{,}000 \qquad \text{3 zeros attached to 34.}
\end{array}
$$

These patterns help us develop a shortcut for multiplying by whole numbers ending in zeros.

> To multiply by 10, 100, 1000, and so on,
> Form the product by attaching the number of zeros in that number to the other factor.
> For example, $41 \cdot 100 = 4100$.
>   2 zeros ──┘

Practice 6–7

Practice 6–7

Multiply.

6. $75 \cdot 100$

7. $808 \cdot 1000$

Examples   Multiply.

6. $176 \cdot 1000 = 176{,}000$ Attach 3 zeros.

7. $2041 \cdot 100 = 204{,}100$ Attach 2 zeros.

■ Work Practice 6–7

We can use a similar format to multiply by any whole number ending in zeros. For example, since

$$15 \cdot 500 = 15 \cdot 5 \cdot 100,$$

we find the product by multiplying 15 and 5, then attaching two zeros to the product.

$$
\begin{array}{r}
\overset{2}{1}5 \\
\times\ 5 \\
\hline
75
\end{array}
\qquad 15 \cdot 500 = 7500
$$

Answers

6. 7500 **7.** 808,000

✓**Concept Check Answer**

$$
\begin{array}{r}
102 \\
\times\ \ 33 \\
\hline
306 \\
3060 \\
\hline
3366
\end{array}
$$

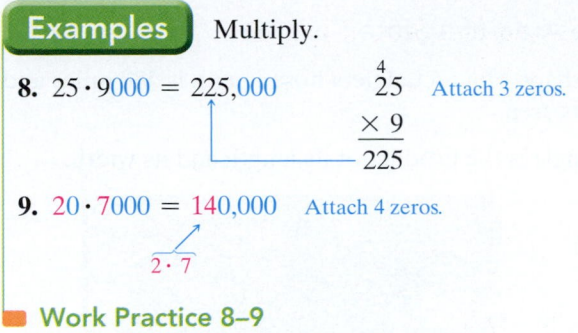

Examples Multiply.

8. $25 \cdot 9000 = 225{,}000$ $\overset{4}{25}$ Attach 3 zeros.
                                        $\underline{\times\ 9}$
                                           225

9. $20 \cdot 7000 = 140{,}000$ Attach 4 zeros.
              $2 \cdot 7$

■ **Work Practice 8–9**

Practice 8–9
Multiply.
8. $35 \cdot 3000$
9. $600 \cdot 600$

Objective D Finding the Area of a Rectangle ▶

A special application of multiplication is finding the **area** of a region. Area measures the amount of surface of a region. For example, we measure a plot of land or the living space of a home by its area. The figures below show two examples of units of area measure. (A centimeter is a unit of length in the metric system.)

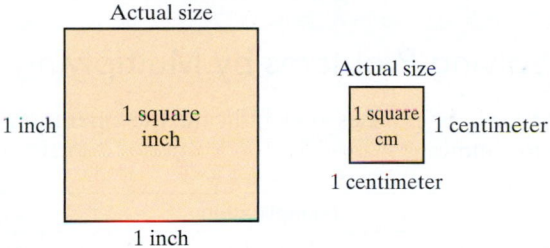

For example, to measure the area of a geometric figure such as the rectangle below, count the number of square units that cover the region.

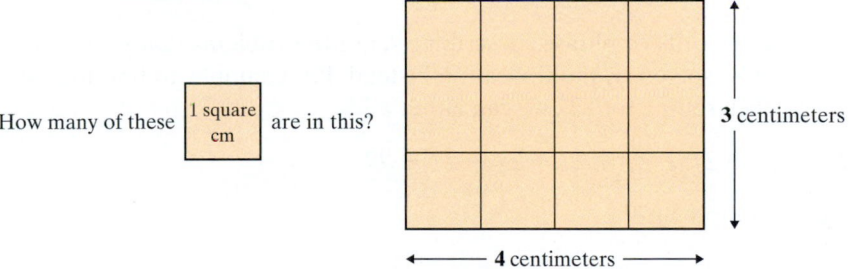

This rectangular region contains 12 square units, each 1 square centimeter. Thus, the area is 12 square centimeters. This total number of squares can be found by counting or by multiplying **4 · 3** (length · width).

> Area of a rectangle = length · width
> = (4 centimeters)(3 centimeters)
> = 12 square centimeters

In this section, we find the areas of rectangles only. In later sections, we will find the areas of other geometric regions.

Helpful Hint

Notice that area is measured in **square** units while perimeter is measured in units.

Answers
8. 105,000 **9.** 360,000

Practice 10

The state of Wyoming is in the shape of a rectangle whose length is 360 miles and whose width is 280 miles. Find its area.

Example 10   Finding the Area of a State

The state of Colorado is in the shape of a rectangle whose length is 380 miles and whose width is 280 miles. Find its area.

Solution: The area of a rectangle is the product of its length and its width.

Area = length · width
= (380 miles)(280 miles)
= 106,400 square miles

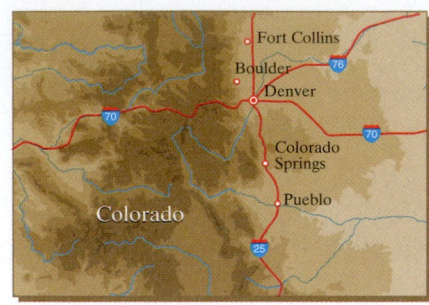

The area of Colorado is 106,400 square miles.

■ **Work Practice 10**

Objective E Solving Problems by Multiplying

There are several words or phrases that indicate the operation of multiplication. Some of these are as follows:

Multiplication		
Key Words or Phrases	**Examples**	**Symbols**
multiply	multiply 5 by 7	5 · 7
product	the product of 3 and 2	3 · 2
times	10 times 13	10 · 13

Many key words or phrases describing real-life problems that suggest addition might be better solved by multiplication instead. For example, to find the **total** cost of 8 shirts, each selling for $27, we can either add

27 + 27 + 27 + 27 + 27 + 27 + 27 + 27

or we can multiply 8(27).

Practice 11

A particular computer printer can print 16 pages per minute in color. How many pages can it print in 45 minutes?

Example 11   Finding DVD Space

A digital video disc (DVD) can hold about 4800 megabytes (MB) of information. How many megabytes can 12 DVDs hold?

Solution: Twelve DVDs will hold 12 × 4800 megabytes.

In Words	Translate to Numbers
megabytes per disc →	4800
× DVDs →	× 12
	9600
	48000
total megabytes	57,600

Twelve DVDs will hold 57,600 megabytes.

■ **Work Practice 11**

Answers
10. 100,800 sq mi   **11.** 720 pages

Example 12 Budgeting Money

Suzanne Scarpulla and a friend plan to take their children to the Georgia Aquarium in Atlanta, the world's largest aquarium. The peak hour ticket price for each child is $29 and for each adult, $35. If five children and two adults plan to go, how much money is needed for admission? (*Source:* GeorgiaAquarium.org)

Solution: If the price of one child's ticket is $29, the cost for 5 children is $5 \times 29 = \$145$. The price of one adult ticket is $35, so the cost for two adults is $2 \times 35 = \$70$. The total cost is:

In Words		**Translate to Numbers**
cost for 5 children	→	145
+ cost for 2 adults	→	+ 70
total cost		215

The total cost is $215.

■ **Work Practice 12**

Practice 12

Ken Shimura purchased DVDs and CDs through a club. Each DVD was priced at $11, and each CD cost $9. Ken bought eight DVDs and five CDs. Find the total cost of the order.

Example 13 Estimating Word Count

The average page of a book contains 259 words. Estimate, rounding each number to the nearest hundred, the total number of words contained on 212 pages.

Solution: The exact number of words is 259×212. Estimate this product by rounding each factor to the nearest hundred.

$$\begin{array}{ll} 259 & \text{rounds to} & 300 \\ \times\, 212 & \text{rounds to} & \times\, 200, \end{array}$$

$$300 \times 200 = 60,000$$
$$3 \cdot 2 = 6$$

There are approximately 60,000 words contained on 212 pages.

■ **Work Practice 13**

Practice 13

If an average page in a book contains 163 words, estimate, rounding each number to the nearest hundred, the total number of words contained on 391 pages.

Calculator Explorations Multiplying Numbers

To multiply numbers on a calculator, find the keys marked $\boxed{\times}$ and $\boxed{=}$ or $\boxed{\text{ENTER}}$. For example, to find $31 \cdot 66$ on a calculator, press the keys $\boxed{31}$ $\boxed{\times}$ $\boxed{66}$ then $\boxed{=}$ or $\boxed{\text{ENTER}}$. The display will read $\boxed{\quad 2046}$. Thus, $31 \cdot 66 = 2046$.

Use a calculator to multiply.

1. 72×48 **2.** 81×92

3. $163 \cdot 94$ **4.** $285 \cdot 144$

5. $983(277)$ **6.** $1562(843)$

Answers
12. $133   **13.** 80,000 words

Vocabulary, Readiness & Video Check

Use the choices below to fill in each blank.

area	grouping	commutative	1	product	length
factor	order	associative	0	distributive	number

1. The product of 0 and any number is _____.
2. The product of 1 and any number is the _____.
3. In $8 \cdot 12 = 96$, the 96 is called the _____ and 8 and 12 are each called a(n) _____.
4. Since $9 \cdot 10 = 10 \cdot 9$, we say that changing the _____ in multiplication does not change the product. This property is called the _____ property of multiplication.
5. Since $(3 \cdot 4) \cdot 6 = 3 \cdot (4 \cdot 6)$, we say that changing the _____ in multiplication does not change the product. This property is called the _____ property of multiplication.
6. _____ measures the amount of surface of a region.
7. Area of a rectangle = _____ $\cdot$ width.
8. We know $9(10 + 8) = 9 \cdot 10 + 9 \cdot 8$ by the _____ property.

Martin-Gay Interactive Videos   *Watch the section lecture video and answer the following questions.*

See Video 1.6

Objective A 9. The expression in ⊞ Example 3 is rewritten using what property? ▶

Objective B 10. During the multiplication process for ⊞ Example 5, why is a single zero placed at the end of the second partial product? ▶

Objective C 11. Explain two different approaches to solving the multiplication problem $50 \cdot 900$ in ⊞ Example 7. ▶

Objective D 12. Why are the units to the answer to ⊞ Example 8 not just meters? What are the correct units? ▶

Objective E 13. In ⊞ Example 9, why can "total" imply multiplication as well as addition? ▶

1.6 Exercise Set MyMathLab®

Objective A *Multiply. See Example 1.*

▶ **1.** $1 \cdot 24$

2. $55 \cdot 1$

▶ **3.** $0 \cdot 19$

4. $27 \cdot 0$

5. $8 \cdot 0 \cdot 9$

6. $7 \cdot 6 \cdot 0$

7. $87 \cdot 1$

8. $1 \cdot 41$

Use the distributive property to rewrite each expression. See Example 2.

9. $6(3 + 8)$

10. $5(8 + 2)$

11. $4(3 + 9)$

12. $6(1 + 4)$

▶ **13.** $20(14 + 6)$

14. $12(12 + 3)$

Objective **B** *Multiply. See Example 3.*

15. 64
× 8

16. 79
× 3

17. 613
× 6

18. 638
× 5

▶ 19. 277 × 6

20. 882 × 2

21. 1074 × 6

22. 9021 × 3

Objectives **A** **B** **Mixed Practice** *Multiply. See Examples 1 through 5.*

23. 89
× 13

24. 91
× 72

25. 421
× 58

26. 526
× 23

27. 306
× 81

28. 708
× 21

29. (780)(20)

30. (720)(80)

31. (495)(13)(0)

32. (593)(47)(0)

33. (640)(1)(10)

34. (240)(1)(20)

35. 1234 × 39

36. 1357 × 79

37. 609 × 234

38. 807 × 127

▶ 39. 8649
× 274

40. 1234
× 567

41. 589
× 110

42. 426
× 110

43. 1941
× 2035

44. 1876
× 1407

Objective **C** *Multiply. See Examples 6 through 9.*

▶ 45. 8 × 100

46. 6 × 100

47. 11 × 1000

48. 26 × 1000

49. 7406 · 10

50. 9054 · 10

51. 6 · 4000

52. 3 · 9000

▶ 53. 50 · 900

54. 70 · 300

55. 41 · 80,000

56. 27 · 50,000

Objective **D** **Mixed Practice (Section 1.3)** *Find the area and the perimeter of each rectangle. See Example 10.*

▶ △ 57.

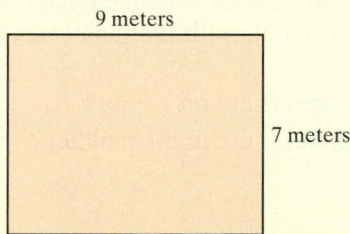

9 meters

7 meters

△ 58.

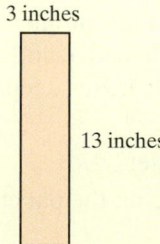

3 inches

13 inches

△ 59.

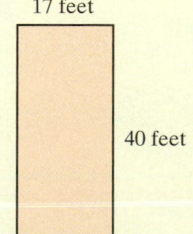

17 feet

40 feet

△ 60.

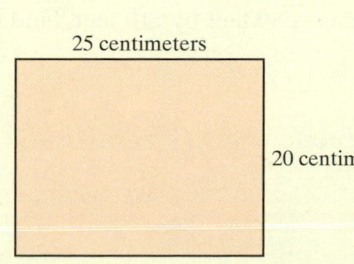

25 centimeters

20 centimeters

Objective E Mixed Practice (Section 1.5) *Estimate the products by rounding each factor to the nearest hundred. See Example 13.*

61. 576 × 354   **62.** 982 × 650   **63.** 604 × 451   **64.** 111 × 999

Without actually calculating, mentally round, multiply, and choose the best estimate.

65. 38 × 42 =
   a. 16
   b. 160
   c. 1600
   d. 16,000

66. 2872 × 12 =
   a. 2872
   b. 28,720
   c. 287,200
   d. 2,872,000

67. 612 × 29 =
   a. 180
   b. 1800
   c. 18,000
   d. 180,000

68. 706 × 409 =
   a. 280
   b. 2800
   c. 28,000
   d. 280,000

Objectives D E Mixed Practice–Translating *Solve. See Examples 10 through 13.*

69. Multiply 80 by 11.

70. Multiply 70 by 12.

71. Find the product of 6 and 700.

72. Find the product of 9 and 900.

73. Find 2 times 2240.

74. Find 3 times 3310.

75. One tablespoon of olive oil contains 125 calories. How many calories are in 3 tablespoons of olive oil? (*Source: Home and Garden Bulletin No. 72,* U.S. Department of Agriculture)

76. One ounce of hulled sunflower seeds contains 14 grams of fat. How many grams of fat are in 8 ounces of hulled sunflower seeds? (*Source: Home and Garden Bulletin No. 72,* U.S. Department of Agriculture)

77. The textbook for a course in biology costs $94. There are 35 students in the class. Find the total cost of the biology books for the class.

78. The seats in a lecture hall are arranged in 14 rows with 34 seats in each row. Find how many seats are in this room.

79. Cabot Creamery is packing a pallet of 20-lb boxes of cheddar cheese to send to a local restaurant. There are five layers of boxes on the pallet, and each layer is four boxes wide by five boxes deep.
   a. How many boxes are in one layer?
   b. How many boxes are on the pallet?
   c. What is the weight of the cheese on the pallet?

80. An apartment building has *three floors*. Each floor has five rows of apartments with four apartments in each row.
   a. How many apartments are on 1 floor?
   b. How many apartments are in the building?

△ **81.** A plot of land measures 80 feet by 110 feet. Find its area.

△ **82.** A house measures 45 feet by 60 feet. Find the floor area of the house.

△ **83.** The largest hotel lobby can be found at the Hyatt Regency in San Francisco, CA. It is in the shape of a rectangle that measures 350 feet by 160 feet. Find its area.

△ **84.** Recall from an earlier section that the world's largest commercial building under one roof is the flower auction building of the cooperative VBA in Aalsmeer, Netherlands. The floor plan is a rectangle that measures 776 meters by 639 meters. Find the area of this building. (*Source: The Handy Science Answer Book,* Visible Ink Press)

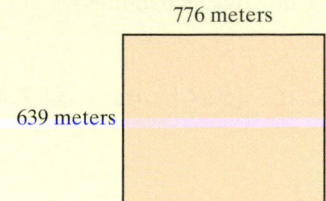

776 meters

639 meters

85. A pixel is a rectangular dot on a graphing calculator screen. If a graphing calculator screen contains 62 pixels in a row and 94 pixels in a column, find the total number of pixels on a screen.

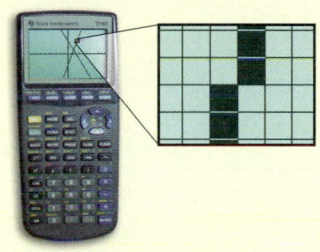

86. A certain compact disc (CD) can hold 700 megabytes (MB) of information. How many MB can 17 discs hold?

87. A line of print on a computer contains 60 characters (letters, spaces, punctuation marks). Find how many characters there are in 35 lines.

88. An average cow eats 3 pounds of grain per day. Find how much grain a cow eats in a year. (Assume 365 days in 1 year.)

89. One ounce of Planters® Dry Roasted Peanuts has 170 calories. How many calories are in 8 ounces? (*Source:* Kraft Foods)

90. One ounce of Planters® Dry Roasted Peanuts has 14 grams of fat. How many grams of fat are in 16 ounces? (*Source:* Kraft Foods)

91. The Thespian club at a local community college is ordering T-shirts. T-shirts size S, M, or L cost $10 each and T-shirts size XL or XXL cost $12 each. Use the table below to find the total cost. (The first row is filled in for you.)

T-Shirt Size	Number of Shirts Ordered	Cost per Shirt	Cost per Size Ordered
S	4	$10	$40
M	6		
L	20		
XL	3		
XXL	3		

92. The student activities group at North Shore Community College is planning a trip to see the local minor league baseball team. Tickets cost $5 for students, $7 for nonstudents, and $2 for children under 12. Use the following table to find the total cost.

Person	Number of Persons	Cost per Person	Cost per Category
Student	24	$5	$120
Nonstudent	4		
Children under 12	5		

93. Celestial Seasonings of Boulder, Colorado, is a tea company that specializes in herbal teas, accounting for over $100,000,000 in herbal tea blend sales in the United States annually. Their plant in Boulder has bagging machines capable of bagging over 1000 bags of tea per minute. If the plant runs 24 hours a day, how many tea bags are produced in one day? (*Source:* Celestial Seasonings)

94. The number of "older" Americans (ages 65 and older) has increased fourteenfold since 1900. If there were 3 million "older" Americans in 1900, how many were there in 2012? (*Source:* U.S. Census Bureau)

Mixed Practice (*Sections 1.3, 1.4, 1.6*) *Perform each indicated operation.*

95.
$$\begin{array}{r} 128 \\ + 7 \\ \hline \end{array}$$

96.
$$\begin{array}{r} 126 \\ - 8 \\ \hline \end{array}$$

97.
$$\begin{array}{r} 134 \\ \times 16 \\ \hline \end{array}$$

98. $47 + 26 + 10 + 231 + 50$

99. Find the sum of 19 and 4.

100. Find the product of 19 and 4.

101. Find the difference of 19 and 4.

102. Find the total of 19 and 4.

Concept Extensions

Solve. See the first Concept Check in this section.

103. Rewrite $7 + 7 + 7 + 7$ using multiplication.

104. Rewrite $11 + 11 + 11 + 11 + 11 + 11$ using multiplication.

105. a. Rewrite $3 \cdot 5$ as repeated addition.
b. Explain why there is more than one way to do this.

106. a. Rewrite $4 \cdot 5$ as repeated addition.
b. Explain why there is more than one way to do this.

Find and explain the error in each multiplication problem. See the second Concept Check in this section.

107.
$$\begin{array}{r} 203 \\ \times 14 \\ \hline 812 \\ 203 \\ \hline 1015 \end{array}$$

108.
$$\begin{array}{r} 31 \\ \times 50 \\ \hline 155 \end{array}$$

Fill in the missing digits in each problem.

109.
$$\begin{array}{r} 4\_ \\ \times \_3 \\ \hline 126 \\ 3780 \\ \hline 3906 \end{array}$$

110.
$$\begin{array}{r} \_7 \\ \times 6\_ \\ \hline 171 \\ 3420 \\ \hline 3591 \end{array}$$

111. Explain how to multiply two 2-digit numbers using partial products.

112. In your own words, explain the meaning of the area of a rectangle and how this area is measured.

113. A window washer in New York City is bidding for a contract to wash the windows of a 23-story building. To write a bid, the number of windows in the building is needed. If there are 7 windows in each row of windows on 2 sides of the building and 4 windows per row on the other 2 sides of the building, find the total number of windows.

114. During the 2012–2013 regular season, Kevin Durant of the Oklahoma City Thunder led the NBA in total points scored. He scored 139 three-point field goals, 592 two-point field goals, and 679 free throws (worth one point each). How many points did Kevin Durant score during the 2012–2013 regular season? (*Source:* NBA)

1.7 Dividing Whole Numbers

Suppose three people pooled their money and bought a raffle ticket at a local fund-raiser. Their ticket was the winner, and they won a $75 cash prize. They then divided the prize into three equal parts so that each person received $25.

Objectives

A Divide Whole Numbers.

B Perform Long Division.

C Solve Problems That Require Dividing by Whole Numbers.

D Find the Average of a List of Numbers.

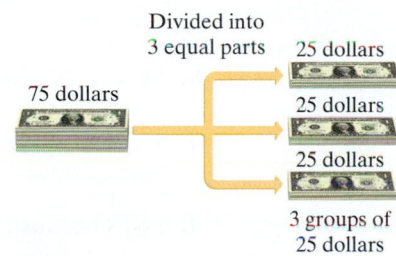

Divided into 3 equal parts — 25 dollars, 25 dollars, 25 dollars

75 dollars

3 groups of 25 dollars

Objective A Dividing Whole Numbers

The process of separating a quantity into equal parts is called **division.** The division above can be symbolized by several notations.

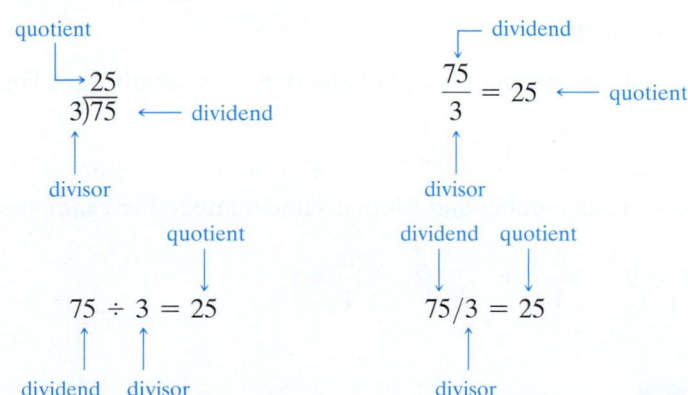

$$\begin{array}{c} \text{quotient} \\ \underset{3\overline{)75}}{\overset{25}{}} \leftarrow \text{dividend} \\ \uparrow \\ \text{divisor} \end{array}$$

$$\frac{75}{3} = 25 \leftarrow \text{quotient}$$

dividend, divisor

$$75 \div 3 = 25$$
dividend divisor quotient

$$75/3 = 25$$
dividend quotient divisor

(In the notation $\frac{75}{3}$, the bar separating 75 and 3 is called a **fraction bar.**) Just as subtraction is the reverse of addition, division is the reverse of multiplication. This means that division can be checked by multiplication.

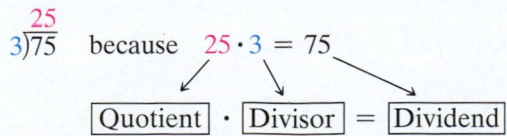

$$3\overline{)75} \quad \text{because} \quad 25 \cdot 3 = 75$$

$$\boxed{\text{Quotient}} \cdot \boxed{\text{Divisor}} = \boxed{\text{Dividend}}$$

Since multiplication and division are related in this way, you can use your knowledge of multiplication facts (or study Appendix A.2) to review quotients of one-digit divisors if necessary.

Practice 1

Find each quotient. Check by multiplying.

a. $9\overline{)72}$

b. $40 \div 5$

c. $\dfrac{24}{6}$

Example 1 Find each quotient. Check by multiplying.

a. $42 \div 7$ **b.** $\dfrac{64}{8}$ **c.** $3\overline{)21}$

Solution:

a. $42 \div 7 = 6$ because $6 \cdot 7 = 42$

b. $\dfrac{64}{8} = 8$ because $8 \cdot 8 = 64$

c. $3\overline{)21}^{\,7}$ because $7 \cdot 3 = 21$

■ **Work Practice 1**

Practice 2

Find each quotient. Check by multiplying.

a. $\dfrac{7}{7}$ **b.** $5 \div 1$

c. $1\overline{)11}$ **d.** $4 \div 1$

e. $\dfrac{10}{1}$ **f.** $21 \div 21$

Example 2 Find each quotient. Check by multiplying.

a. $1\overline{)7}$ **b.** $12 \div 1$ **c.** $\dfrac{6}{6}$ **d.** $9 \div 9$ **e.** $\dfrac{20}{1}$ **f.** $18\overline{)18}$

Solution:

a. $1\overline{)7}^{\,7}$ because $7 \cdot 1 = 7$ **b.** $12 \div 1 = 12$ because $12 \cdot 1 = 12$

c. $\dfrac{6}{6} = 1$ because $1 \cdot 6 = 6$ **d.** $9 \div 9 = 1$ because $1 \cdot 9 = 9$

e. $\dfrac{20}{1} = 20$ because $20 \cdot 1 = 20$ **f.** $18\overline{)18}^{\,1}$ because $1 \cdot 18 = 18$

■ **Work Practice 2**

Example 2 illustrates the important properties of division described next:

Division Properties of 1

The quotient of any number (except 0) and that same number is 1. For example,

$$8 \div 8 = 1 \qquad \dfrac{5}{5} = 1 \qquad 4\overline{)4}^{\,1}$$

The quotient of any number and 1 is that same number. For example,

$$9 \div 1 = 9 \qquad \dfrac{6}{1} = 6 \qquad 1\overline{)3}^{\,3} \qquad \dfrac{0}{1} = 0$$

Practice 3

Find each quotient. Check by multiplying.

a. $\dfrac{0}{7}$ **b.** $8\overline{)0}$

c. $7 \div 0$ **d.** $0 \div 14$

Example 3 Find each quotient. Check by multiplying.

a. $9\overline{)0}$ **b.** $0 \div 12$ **c.** $\dfrac{0}{5}$ **d.** $\dfrac{3}{0}$

Solution:

a. $9\overline{)0}^{\,0}$ because $0 \cdot 9 = 0$ **b.** $0 \div 12 = 0$ because $0 \cdot 12 = 0$

c. $\dfrac{0}{5} = 0$ because $0 \cdot 5 = 0$

Answers

1. a. 8 **b.** 8 **c.** 4 **2. a.** 1 **b.** 5
c. 11 **d.** 4 **e.** 10 **f.** 1 **3. a.** 0
b. 0 **c.** undefined **d.** 0

d. If $\dfrac{3}{0}$ = a *number,* then the *number* times 0 = 3. Recall from Section 1.6 that any number multiplied by 0 is 0 and not 3. We say, then, that $\dfrac{3}{0}$ is **undefined.**

■ **Work Practice 3**

Example 3 illustrates important division properties of 0.

Division Properties of 0

The quotient of 0 and any number (except 0) is 0. For example,

$$0 \div 9 = 0 \qquad \frac{0}{5} = 0 \qquad 14\overline{)0}$$

The quotient of any number and 0 is not a number. We say that

$$\frac{3}{0}, \quad 0\overline{)3}, \quad \text{and} \quad 3 \div 0$$

are **undefined.**

Objective B Performing Long Division

When dividends are larger, the quotient can be found by a process called **long division.** For example, let's divide 2541 by 3.

$$\text{divisor} \longrightarrow 3\overline{)2541}$$
$$\uparrow$$
$$\text{dividend}$$

We can't divide 3 into 2, so we try dividing 3 into the first two digits.

$$\begin{array}{r} 8 \\ 3\overline{)2541} \end{array}\qquad \begin{array}{l} 25 \div 3 = 8 \text{ with 1 left, so our best estimate is 8. We place 8 over} \\ \text{the 5 in 25.} \end{array}$$

Next, multiply 8 and 3 and subtract this product from 25. Make sure that this difference is less than the divisor.

$$\begin{array}{r} 8 \\ 3\overline{)2541} \\ -24 \\ \hline 1 \end{array}\qquad \begin{array}{l} 8(3) = 24 \\ 25 - 24 = 1, \text{ and 1 is less than the divisor 3.} \end{array}$$

Bring down the next digit and go through the process again.

$$\begin{array}{r} 84 \\ 3\overline{)2541} \\ -24\downarrow \\ \hline 14 \\ -12 \\ \hline 2 \end{array}\qquad \begin{array}{l} 14 \div 3 = 4 \text{ with 2 left} \\ \\ \\ 4(3) = 12 \\ 14 - 12 = 2 \end{array}$$

Once more, bring down the next digit and go through the process.

$$\begin{array}{r} 847 \\ 3\overline{)2541} \\ -24 \\ \hline 14 \\ -12\downarrow \\ \hline 21 \\ -21 \\ \hline 0 \end{array}\qquad \begin{array}{l} 21 \div 3 = 7 \\ \\ \\ \\ \\ 7(3) = 21 \\ 21 - 21 = 0 \end{array}$$

The quotient is 847. To check, see that $847 \times 3 = 2541$.

Practice 4

Divide. Check by multiplying.

a. $4908 \div 6$

b. $2212 \div 4$

c. $753 \div 3$

Example 4   Divide: $3705 \div 5$. Check by multiplying.

Solution:

$$\begin{array}{r} 7 \\ 5\overline{)3705} \\ -35\downarrow \\ \hline 20 \end{array}$$

$37 \div 5 = 7$ with 2 left. Place this estimate, 7, over the 7 in 37.

$7(5) = 35$

$37 - 35 = 2$, and 2 is less than the divisor 5.
Bring down the 0.

$$\begin{array}{r} 74 \\ 5\overline{)3705} \\ -35 \\ \hline 20 \\ -20\downarrow \\ \hline 05 \end{array}$$

$20 \div 5 = 4$

$4(5) = 20$

$20 - 20 = 0$. and 0 is less than the divisor 5.
Bring down the 5.

$$\begin{array}{r} 741 \\ 5\overline{)3705} \\ -35 \\ \hline 20 \\ -20\downarrow \\ \hline 5 \\ -5 \\ \hline 0 \end{array}$$

$5 \div 5 = 1$

$1(5) = 5$

$5 - 5 = 0$

Check:

$$\begin{array}{r} 741 \\ \times\ \ \ 5 \\ \hline 3705 \end{array}$$

Work Practice 4

Helpful Hint   Since division and multiplication are reverse operations, don't forget that a division problem can be checked by multiplying.

Practice 5

Divide and check by multiplying.

a. $7\overline{)2128}$

b. $9\overline{)45,900}$

Example 5   Divide and check: $1872 \div 9$

Solution:

$$\begin{array}{r} 208 \\ 9\overline{)1872} \\ -18\downarrow\downarrow \\ \hline 07 \\ -0\downarrow \\ \hline 72 \\ -72 \\ \hline 0 \end{array}$$

$2(9) = 18$

$18 - 18 = 0$; bring down the 7.

$0(9) = 0$

$7 - 0 = 7$; bring down the 2.

$8(9) = 72$

$72 - 72 = 0$

Check: $208 \cdot 9 = 1872$

Work Practice 5

Answers

4. a. 818   b. 553   c. 251

5. a. 304   b. 5100

Naturally, quotients don't always "come out even." Making 4 rows out of 26 chairs, for example, isn't possible if each row is supposed to have exactly the same number of chairs. Each of 4 rows can have 6 chairs, but 2 chairs are still left over.

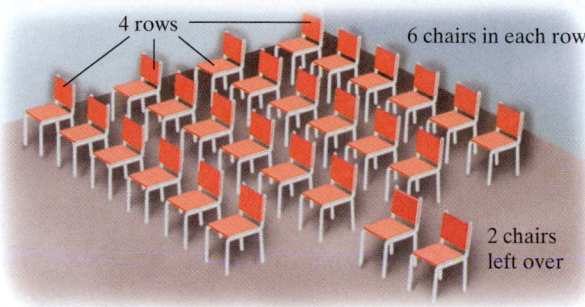

4 rows — 6 chairs in each row

2 chairs left over

We signify "leftovers" or **remainders** in this way:

$$4\overline{)26} \quad 6 \text{ R } 2$$

The **whole number part of the quotient** is 6; the **remainder part of the quotient** is 2. Checking by multiplying,

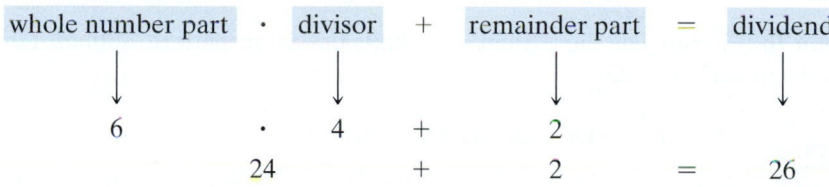

whole number part	·	divisor	+	remainder part	=	dividend
6	·	4	+	2		
	24		+	2	=	26

Example 6   Divide and check: $2557 \div 7$

Solution:

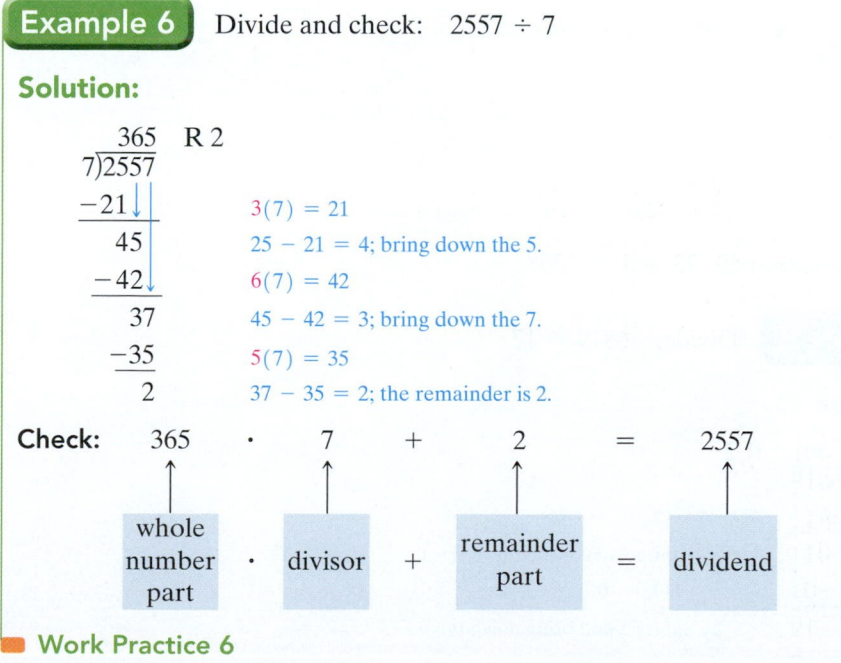

$$7\overline{)2557} \quad 365 \text{ R } 2$$

$$
\begin{array}{r}
-21 \\
\hline
45 \\
-42 \\
\hline
37 \\
-35 \\
\hline
2
\end{array}
$$

3(7) = 21
25 − 21 = 4; bring down the 5.
6(7) = 42
45 − 42 = 3; bring down the 7.
5(7) = 35
37 − 35 = 2; the remainder is 2.

Check: 365 · 7 + 2 = 2557

whole number part	·	divisor	+	remainder part	=	dividend

▪ **Work Practice 6**

Practice 6
Divide and check.
a. $4\overline{)939}$
b. $5\overline{)3287}$

Practice 7

Divide and check.

a. $9\overline{)81{,}605}$

b. $4\overline{)23{,}310}$

Example 7 Divide and check: $56{,}717 \div 8$

Solution:

$$
\begin{array}{r}
7089 \ \text{R } 5 \\
8\overline{)56717} \\
\end{array}
$$

$-56\downarrow\downarrow\downarrow$	$7(8) = 56$
07	Subtract and bring down the 7.
$-0\downarrow$	$0(8) = 0$
71	Subtract and bring down the 1.
$-64\downarrow$	$8(8) = 64$
77	Subtract and bring down the 7.
-72	$9(8) = 72$
5	Subtract. The remainder is 5.

Check: $7089 \quad \cdot \quad 8 \quad + \quad 5 \quad = \quad 56{,}717$

whole number part	$\cdot$	divisor	$+$	remainder part	$=$	dividend

■ **Work Practice 7**

When the divisor has more than one digit, the same pattern applies. For example, let's find $1358 \div 23$.

$$
\begin{array}{r}
5 \\
23\overline{)1358} \\
-115\downarrow \\
208 \\
\end{array}
$$

$135 \div 23 = 5$ with 20 left over. Our estimate is 5.

$5(23) = 115$

$135 - 115 = 20$. Bring down the 8.

Now we continue estimating.

$$
\begin{array}{r}
59 \ \text{R } 1 \\
23\overline{)1358} \\
-115 \\
208 \\
-207 \\
1 \\
\end{array}
$$

$208 \div 23 = 9$ with 1 left over.

$9(23) = 207$

$208 - 207 = 1$. The remainder is 1.

To check, see that $59 \cdot 23 + 1 = 1358$.

Practice 8

Divide: $8920 \div 17$

Example 8 Divide: $6819 \div 17$

Solution:

$$
\begin{array}{r}
401 \ \text{R } 2 \\
17\overline{)6819} \\
\end{array}
$$

$-68\downarrow$	$4(17) = 68$
01	Subtract and bring down the 1.
$-0\downarrow$	$0(17) = 0$
19	Subtract and bring down the 9.
-17	$1(17) = 17$
2	Subtract. The remainder is 2.

To check, see that $401 \cdot 17 + 2 = 6819$.

■ **Work Practice 8**

Answers

7. a. 9067 R 2   **b.** 5827 R 2

8. 524 R 12

Example 9   Divide: 51,600 ÷ 403

Solution:

$$
\begin{array}{r}
128 \text{ R } 16 \\
403 \overline{)51600} \\
-403 \downarrow \\
\hline
1130 \\
-806 \downarrow \\
\hline
3240 \\
-3224 \\
\hline
16
\end{array}
$$

1(403) = 403
Subtract and bring down the 0.
2(403) = 806
Subtract and bring down the 0.
8(403) = 3224
Subtract. The remainder is 16.

To check, see that $128 \cdot 403 + 16 = 51,600$.

■ **Work Practice 9**

Division Shown as Repeated Subtraction To further understand division, re-call from Section 1.6 that addition and multiplication are related in the following manner:

$$\underbrace{3 + 3 + 3 + 3}_{\text{4 addends; each addend is 3}} = 4 \times 3 = 12$$

In other words, multiplication is repeated addition. Likewise, division is repeated subtraction.

For example, let's find

$$35 \div 8$$

by repeated subtraction. Keep track of the number of times 8 is subtracted from 35. We are through when we can subtract no more because the difference is less than 8.

35 ÷ 8:   Repeated Subtraction

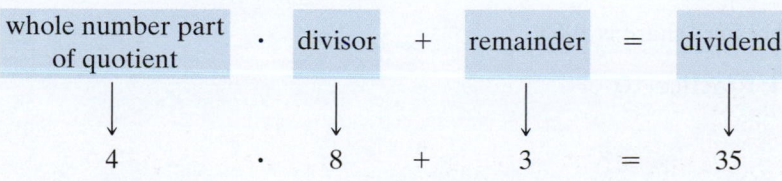

$$
\begin{array}{ll}
\left.\begin{array}{r} 35 \\ -8 \end{array}\right\} & \text{1 time} \\
\left.\begin{array}{r} 27 \\ -8 \end{array}\right\} & \text{2 times} \\
\left.\begin{array}{r} 19 \\ -8 \end{array}\right\} & \text{3 times} \\
\left.\begin{array}{r} 11 \\ -8 \end{array}\right\} & \text{4 times} \\
\hline
3 & \longleftarrow \text{ Remainder} \\
& \text{(We cannot subtract 8 again.)}
\end{array}
$$

8 dollars   1 time
8 dollars   2 times
35 dollars
8 dollars   3 times
8 dollars   4 times
3 dollars
left over

Thus, $35 \div 8 = 4 \text{ R } 3$.

To check, perform the same multiplication as usual and finish by adding in the remainder.

whole number part of quotient	·	divisor	+	remainder	=	dividend
4	·	8	+	3	=	35

Practice 9

Divide: 33,282 ÷ 678

Answer

9. 49 R 60

Objective C Solving Problems by Dividing ▶

Below are some key words and phrases that may indicate the operation of division:

Division		
Key Words or Phrases	**Examples**	**Symbols**
divide	divide 10 by 5	$10 \div 5$ or $\dfrac{10}{5}$
quotient	the quotient of 64 and 4	$64 \div 4$ or $\dfrac{64}{4}$
divided by	9 divided by 3	$9 \div 3$ or $\dfrac{9}{3}$
divided or shared equally among	$100 divided equally among five people	$100 \div 5$ or $\dfrac{100}{5}$
per	100 miles per 2 hours	$\dfrac{100 \text{ miles}}{2 \text{ hours}}$

✓**Concept Check** Determine whether each of the following is the correct way to represent "the quotient of 60 and 12." Explain your answer.

a. $12 \div 60$

b. $60 \div 12$

Practice 10

Three students bought 171 blank CDs to share equally. How many CDs did each person get?

Example 10   Finding Shared Earnings

Three college students share a paper route to earn money for expenses. The total in their fund after expenses was $2895. How much is each person's equal share?

Solution:

In words: Each person's share = total money ÷ number of persons

Translate: Each person's share = 2895 ÷ 3

Then

$$
\begin{array}{r}
965 \\
3\overline{)2895} \\
-27 \\
\hline
19 \\
-18 \\
\hline
15 \\
-15 \\
\hline
0
\end{array}
$$

Each person's share is $965.

Work Practice 10

Answer

10. 57 CDs

✓**Concept Check Answers**

a. incorrect   **b.** correct

Example 11 Dividing Number of Downloads

As part of a promotion, an executive receives 238 cards, each good for one free song download. If she wants to share them evenly with 19 friends, how many download cards will each friend receive? How many will be left over?

Solution:

In words:

| Number of cards for each person | = | number of cards | ÷ | number of friends |

Translate:

$$\text{Number of cards for each person} = 238 \div 19$$

$$\begin{array}{r} 12 \text{ R } 10 \\ 19\overline{)238} \\ -19 \\ \hline 48 \\ -38 \\ \hline 10 \end{array}$$

Each friend will receive 12 download cards. The cards cannot be divided equally among her friends since there is a nonzero remainder. There will be 10 download cards left over.

■ **Work Practice 11**

Objective D Finding Averages

A special application of division (and addition) is finding the average of a list of numbers. The **average** of a list of numbers is the sum of the numbers divided by the *number* of numbers.

$$\text{average} = \frac{\text{sum of numbers}}{\textit{number of numbers}}$$

Example 12 Averaging Scores

A mathematics instructor is checking a simple program she wrote for averaging the scores of her students. To do so, she averages a student's scores of 75, 96, 81, and 88 by hand. Find this average score.

Solution: To find the average score, we find the sum of the student's scores and divide by 4, the number of scores.

$$\begin{array}{r} 75 \\ 96 \\ 81 \\ +88 \\ \hline 340 \quad \text{sum} \end{array}$$

$$\text{average} = \frac{340}{4} = 85$$

$$\begin{array}{r} 85 \\ 4\overline{)340} \\ -32 \\ \hline 20 \\ -20 \\ \hline 0 \end{array}$$

The average score is 85.

■ **Work Practice 12**

Calculator Explorations **Dividing Numbers**

To divide numbers on a calculator, find the keys marked
$\boxed{\div}$ and $\boxed{=}$ or $\boxed{\text{ENTER}}$. For example, to find $435 \div 5$
on a calculator, press the keys $\boxed{435}$ $\boxed{\div}$ $\boxed{5}$ then $\boxed{=}$ or
$\boxed{\text{ENTER}}$. The display will read $\boxed{87}$. Thus, $435 \div 5 = 87$.

Use a calculator to divide.

1. $848 \div 16$ **2.** $564 \div 12$

3. $95\overline{)5890}$ **4.** $27\overline{)1053}$

5. $\dfrac{32{,}886}{126}$ **6.** $\dfrac{143{,}088}{264}$

7. $0 \div 315$ **8.** $315 \div 0$

Vocabulary, Readiness & Video Check

Use the choices below to fill in each blank. Some choices may be used more than once.

1	number	divisor	dividend
0	undefined	average	quotient

1. In $90 \div 2 = 45$, the answer 45 is called the _____ , 90 is called the _____ , and 2 is called the

_____ .

2. The quotient of any number and 1 is the same _____ .

3. The quotient of any number (except 0) and the same number is _____ .

4. The quotient of 0 and any number (except 0) is _____ .

5. The quotient of any number and 0 is _____ .

6. The _____ of a list of numbers is the sum of the numbers divided by the _____ of numbers.

Martin-Gay Interactive Videos

See Video 1.7 🍎

Watch the section lecture video and answer the following questions.

Objective  **7.** Look at ⊟ Examples 6–8. What number can never be the divisor in division? ▶

Objective  **8.** In ⊟ Example 10, how many 102s are in 21? How does this result affect the quotient? ▶

         9. What calculation would you use to check the answer in ⊟ Example 10? ▶

Objective  **10.** In ⊟ Example 11, what is the importance of knowing that the distance to each hole is the same? ▶

Objective D **11.** As shown in ⊟ Example 12, what two operations are used when finding an average? ▶

1.7 Exercise Set MyMathLab®

Objective A *Find each quotient. See Examples 1 through 3.*

1. $54 \div 9$ **2.** $72 \div 9$ **3.** $36 \div 3$ **4.** $24 \div 3$ **5.** $0 \div 8$

6. $0 \div 4$ **7.** $31 \div 1$ **8.** $38 \div 1$ **9.** $\dfrac{18}{18}$ **10.** $\dfrac{49}{49}$

11. $\dfrac{24}{3}$ **12.** $\dfrac{45}{9}$ **13.** $26 \div 0$ **14.** $\dfrac{12}{0}$ **15.** $26 \div 26$

16. $6 \div 6$ **17.** $0 \div 14$ **18.** $7 \div 0$ **19.** $18 \div 2$ **20.** $18 \div 3$

Objectives A B Mixed Practice *Divide and then check by multiplying. See Examples 1 through 5.*

21. $3\overline{)87}$ **22.** $5\overline{)85}$ **23.** $3\overline{)222}$ **24.** $8\overline{)640}$ **25.** $3\overline{)1014}$ **26.** $4\overline{)2104}$

27. $\dfrac{30}{0}$ **28.** $\dfrac{0}{30}$ **29.** $63 \div 7$ **30.** $56 \div 8$ **31.** $150 \div 6$ **32.** $121 \div 11$

Divide and then check by multiplying. See Examples 6 and 7.

33. $7\overline{)479}$ **34.** $7\overline{)426}$ **35.** $6\overline{)1421}$ **36.** $3\overline{)1240}$

37. $305 \div 8$ **38.** $167 \div 3$ **39.** $2286 \div 7$ **40.** $3333 \div 4$

Divide and then check by multiplying. See Examples 8 and 9.

41. $55\overline{)715}$ **42.** $23\overline{)736}$ **43.** $23\overline{)1127}$ **44.** $42\overline{)2016}$ **45.** $97\overline{)9417}$

46. $44\overline{)1938}$ **47.** $3146 \div 15$ **48.** $7354 \div 12$ **49.** $6578 \div 13$ **50.** $5670 \div 14$

51. $9299 \div 46$ **52.** $2505 \div 64$ **53.** $\dfrac{12{,}744}{236}$ **54.** $\dfrac{5781}{123}$ **55.** $\dfrac{10{,}297}{103}$

56. $\dfrac{23{,}092}{240}$ **57.** $20{,}619 \div 102$ **58.** $40{,}853 \div 203$ **59.** $244{,}989 \div 423$ **60.** $164{,}592 \div 543$

Divide. See Examples 1 through 9.

61. $7\overline{)119}$ **62.** $8\overline{)104}$ **63.** $7\overline{)3580}$ **64.** $5\overline{)3017}$

65. $40\overline{)85,312}$ **66.** $50\overline{)85,747}$ **67.** $142\overline{)863,360}$ **68.** $214\overline{)650,560}$

Objective C Translating *Solve. See Examples 10 and 11.*

69. Find the quotient of 117 and 5.

70. Find the quotient of 94 and 7.

71. Find 200 divided by 35.

72. Find 116 divided by 32.

73. Find the quotient of 62 and 3.

74. Find the quotient of 78 and 5.

75. Martin Thieme teaches American Sign Language classes for $65 per student for a 7-week session. He collects $2145 from the group of students. Find how many students are in the group.

76. Kathy Gomez teaches Spanish lessons for $85 per student for a 5-week session. From one group of students, she collects $4930. Find how many students are in the group.

77. The gravity of Jupiter is 318 times as strong as the gravity of Earth, so objects on Jupiter weigh 318 times as much as they weigh on Earth. If a person would weigh 52,470 pounds on Jupiter, find how much the person weighs on Earth.

78. Twenty-one people pooled their money and bought lottery tickets. One ticket won a prize of $5,292,000. Find how many dollars each person received.

79. An 18-hole golf course is 5580 yards long. If the distance to each hole is the same, find the distance between holes.

80. A truck hauls wheat to a storage granary. It carries a total of 5768 bushels of wheat in 14 trips. How much does the truck haul each trip if each trip it hauls the same amount?

81. There is a bridge over highway I-35 every three miles. The first bridge is at the beginning of a 265-mile stretch of highway. Find how many bridges there are over 265 miles of I-35.

82. The white stripes dividing the lanes on a highway are 25 feet long, and the spaces between them are 25 feet long. Let's call a "lane divider" a stripe followed by a space. Find how many whole "lane dividers" there are in 1 mile of highway. (A mile is 5280 feet.)

83. Ari Trainor is in the requisitions department of Central Electric Lighting Company. Light poles along a highway are placed 492 feet apart. The first light pole is at the beginning of a 1-mile strip. Find how many poles he should order for the 1-mile strip of highway. (A mile is 5280 feet.)

84. Professor Lopez has a piece of rope 185 feet long that she wants to cut into pieces for an experiment in her physics class. Each piece of rope is to be 8 feet long. Determine whether she has enough rope for her 22-student class. Determine the amount extra or the amount short.

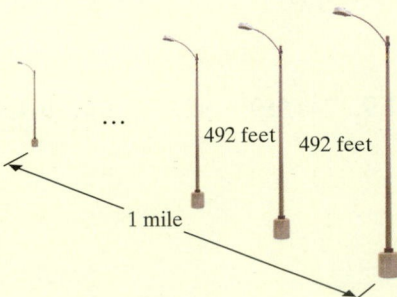

85. Broad Peak in Pakistan is the twelfth-tallest mountain in the world. Its elevation is 26,400 feet. A mile is 5280 feet. How many miles tall is Broad Peak? (*Source:* National Geographic Society)

86. Arian Foster of the Houston Texans led the NFL in touchdowns during the 2012 regular football season, scoring a total of 102 points from touchdowns. If a touchdown is worth 6 points, how many touchdowns did Foster make during 2012? (*Source:* National Football League)

87. Find how many yards are in 1 mile. (A mile is 5280 feet; a yard is 3 feet.)

88. Find how many whole feet are in 1 rod. (A mile is 5280 feet; 1 mile is 320 rods.)

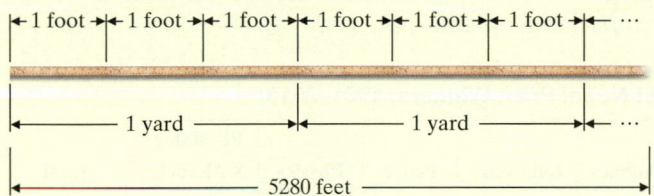

Objective D *Find the average of each list of numbers. See Example 12.*

89. 10, 24, 35, 22, 17, 12

90. 37, 26, 15, 29, 51, 22

91. 205, 972, 210, 161

92. 121, 200, 185, 176, 163

▶ **93.** 86, 79, 81, 69, 80

94. 92, 96, 90, 85, 92, 79

The normal monthly temperatures in degrees Fahrenheit for Salt Lake City, Utah, is given in the graph. Use this graph to answer Exercises 95 and 96. (Source: National Climatic Data Center)

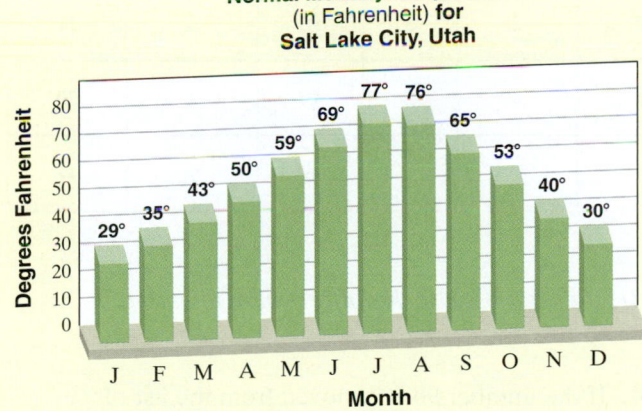

Normal Monthly Temperatures (in Fahrenheit) for Salt Lake City, Utah

95. Find the average temperature for June, July, and August.

96. Find the average temperature for October, November, and December.

Mixed Practice (*Sections 1.3, 1.4, 1.6, 1.7*) *Perform each indicated operation. Watch the operation symbol.*

97. 82 + 463 + 29 + 8704

98. 23 + 407 + 92 + 7011

99.
$$\begin{array}{r} 546 \\ \times\ 28 \end{array}$$

100.
$$\begin{array}{r} 712 \\ \times\ 54 \end{array}$$

101.
$$\begin{array}{r} 722 \\ -\ 43 \end{array}$$

102.
$$\begin{array}{r} 712 \\ -\ 54 \end{array}$$

103. $\dfrac{45}{0}$

104. $\dfrac{0}{23}$

105. 228 ÷ 24

106. 304 ÷ 31

Concept Extensions

Match each word phrase to the correct translation. (Not all letter choices will be used.) See the Concept Check in this section.

107. The quotient of 40 and 8   **108.** The quotient of 200 and 20   **a.** $20 \div 200$ **b.** $200 \div 20$

109. 200 divided by 20   **110.** 40 divided by 8   **c.** $40 \div 8$ **d.** $8 \div 40$

The following table shows the top eight countries with the most Nobel Prize winners through 2013. Use this table to answer Exercises 111 and 112. (Source: Based on data from Encyclopaedia Britannica, Inc.)

111. Find the average number of Nobel Prize winners for the countries shown.

112. Find the average number of Nobel Prize winners per category for the United States.

Countries with Most Nobel Prize Winners, 1901–2013							
	Country	Chemistry	Economics	Literature	Peace	Physics	Physiology & Medicine	Total
	United States	68	54	11	22	90	97	342
	United Kingdom	27	8	8	9	23	31	106
	Germany	27	1	8	4	22	17	79
	France	7	1	15	8	13	11	55
	Sweden	4	2	9	5	4	8	32
	Switzerland	6	0	2	3	4	6	21
	Russia (USSR)	1	1	4	2	10	2	20
	Japan	6	0	2	1	6	2	17

In Example 12 in this section, we found that the average of 75, 96, 81, and 88 is 85. Use this information to answer Exercises 113 and 114.

113. If the number 75 is removed from the list of numbers, does the average increase or decrease? Explain why.

114. If the number 96 is removed from the list of numbers, does the average increase or decrease? Explain why.

115. Without computing it, tell whether the average of 126, 135, 198, 113 is 86. Explain why it is possible or why it is not.

116. Without computing it, tell whether the average of 38, 27, 58, and 43 is 17. Explain why it is possible or why it is not.

117. If the area of a rectangle is 60 square feet and its width is 5 feet, what is its length?

118. If the area of a rectangle is 84 square inches and its length is 21 inches, what is its width?

119. Write down any two numbers whose quotient is 25.

120. Write down any two numbers whose quotient is 1.

121. Find $26 \div 5$ using the process of repeated subtraction.

122. Find $86 \div 10$ using the process of repeated subtraction.

Integrated Review

Operations on Whole Numbers

1.
$$\begin{array}{r} 23 \\ 46 \\ +79 \\ \hline \end{array}$$

2.
$$\begin{array}{r} 7006 \\ -\ 451 \\ \hline \end{array}$$

3.
$$\begin{array}{r} 36 \\ \times 45 \\ \hline \end{array}$$

4. $8\overline{)4496}$

5. $1 \cdot 79$

6. $\dfrac{36}{0}$

7. $9 \div 1$

8. $9 \div 9$

9. $0 \cdot 13$

10. $7 \cdot 0 \cdot 8$

11. $0 \div 2$

12. $12 \div 4$

13. $4219 - 1786$

14. $1861 + 7965$

15. $5\overline{)1068}$

16.
$$\begin{array}{r} 1259 \\ \times\ \ 63 \\ \hline \end{array}$$

17. $3 \cdot 9$

18. $45 \div 5$

19.
$$\begin{array}{r} 207 \\ -\ 69 \\ \hline \end{array}$$

20.
$$\begin{array}{r} 207 \\ +\ 69 \\ \hline \end{array}$$

21. $7\overline{)7695}$

22. $9\overline{)1000}$

23. $32\overline{)21{,}222}$

24. $65\overline{)70{,}000}$

25. $4000 - 2976$

26. $10{,}000 - 101$

27.
$$\begin{array}{r} 303 \\ \times 101 \\ \hline \end{array}$$

28. $(475)(100)$

29. Find the total of 57 and 8.

30. Find the product of 57 and 8.

Answers

1. _____
2. _____
3. _____
4. _____
5. _____
6. _____
7. _____
8. _____
9. _____
10. _____
11. _____
12. _____
13. _____
14. _____
15. _____
16. _____
17. _____
18. _____
19. _____
20. _____
21. _____
22. _____
23. _____
24. _____
25. _____
26. _____
27. _____
28. _____
29. _____
30. _____

31. _____

32. _____

33. _____

34. _____

35. _____

36. _____

37. _____

38. _____

39. _____

40. _____

41. _____

42. _____

43. _____

44. _____

45. _____

46. _____

31. Find the quotient of 62 and 9.

32. Find the difference of 62 and 9.

33. Subtract 17 from 200.

34. Find the difference of 432 and 201.

Complete the table by rounding the given number to the given place value.

	Tens	Hundreds	Thousands
35. 9735			
36. 1429			
37. 20,801			
38. 432,198			

Find the perimeter and area of each figure.

△ **39.**

Square   6 feet

△ **40.**

14 inches

Rectangle   7 inches

Find the perimeter of each figure.

△ **41.**

13 miles   9 miles

6 miles

△ **42.**

3 meters

4 meters

3 meters

3 meters

Find the average of each list of numbers.

43. 19, 15, 25, 37, 24

44. 108, 131, 98, 159

45. The Mackinac Bridge is a suspension bridge that connects the lower and upper peninsulas of Michigan across the Straits of Mackinac. Its total length is 26,372 feet. The Lake Pontchartrain Bridge is a twin concrete trestle bridge in Slidell, Louisiana. Its total length is 28,547 feet. Which bridge is longer and by how much? (_Sources:_ Mackinac Bridge Authority and Federal Highway Administration, Bridge Division)

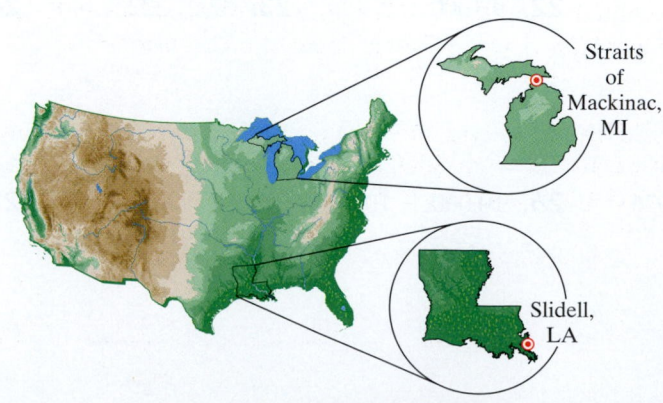

Straits of Mackinac, MI

Slidell, LA

46. The average teenage male American consumes 2 quarts of carbonated soft drinks per day. On average, how many quarts of carbonated soft drinks would be consumed in a year? (Use 365 for the number of days.) (_Source:_ American Beverage Association)

An Introduction to Problem Solving

Objective A Solving Problems Involving Addition, Subtraction, Multiplication, or Division

In this section, we decide which operation to perform in order to solve a problem. Don't forget the key words and phrases that help indicate which operation to use. Some of these are listed below and were introduced earlier in the chapter. Also included are several words and phrases that translate to the symbol " = ":

Addition (+)	Subtraction (−)	Multiplication (·)	Division (÷)	Equality (=)
sum	difference	product	quotient	equals
plus	minus	times	divide	is equal to
added to	subtract	multiply	shared equally	is/was
more than	less than	multiply by	among	yields
increased by	decreased by	of	divided by	
total	less	double/triple	divided into	

The following problem-solving steps may be helpful to you:

Problem-Solving Steps

1. **UNDERSTAND** the problem. Some ways of doing this are to read and reread the problem, construct a drawing, and look for key words to identify an operation.
2. **TRANSLATE** the problem. That is, write the problem in short form using words, and then translate to numbers and symbols.
3. **SOLVE** the problem. It is helpful to estimate the solution by rounding. Then carry out the indicated operation from step 2.
4. **INTERPRET** the results. *Check* the proposed solution in the stated problem and *state* your conclusions. Write your results with the correct units attached.

Example 1 Calculating the Length of a River

The Hudson River in New York State is 306 miles long. The Snake River in the northwestern United States is 732 miles longer than the Hudson River. How long is the Snake River? (*Source:* U.S. Department of the Interior)

Solution:

1. **UNDERSTAND.** Read and reread the problem, and then draw a picture. Notice that we are told that Snake River is 732 miles longer than the Hudson River. The phrase "longer than" means that we add.

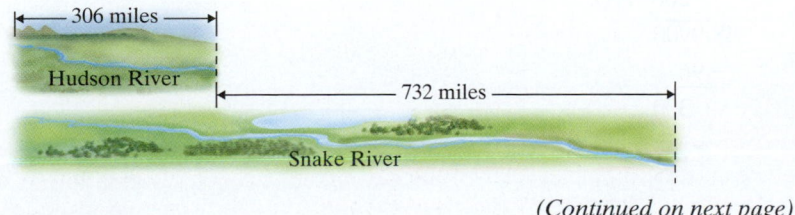

(Continued on next page)

Objectives

A Solve Problems by Adding, Subtracting, Multiplying, or Dividing Whole Numbers. ▶

B Solve Problems That Require More Than One Operation. ▶

Practice 1

The building called 555 California Street is the second-tallest building in San Francisco, California, at 779 feet. The tallest building in San Francisco is the Transamerica Pyramid, which is 74 feet taller than 555 California Street. How tall is the Transamerica Pyramid? (*Source: The World Almanac*)

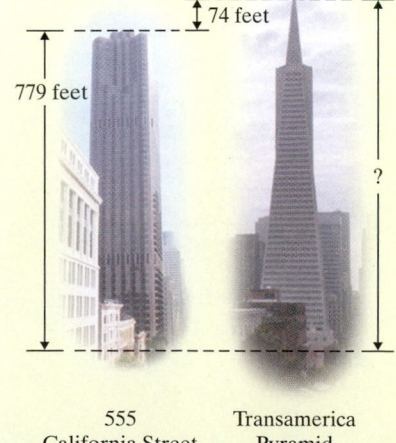

555 California Street   Transamerica Pyramid

Answer
1. 853 ft

2. TRANSLATE.

In words: Snake River   is   732 miles   longer than   the Hudson River
↓   ↓   ↓   ↓   ↓
Translate: Snake River   =   732   +   306

3. SOLVE: Let's see if our answer is reasonable by also estimating. We will estimate each addend to the nearest hundred.

732	rounds to	700
+306	rounds to	+300
1038	exact	1000 estimate

4. INTERPRET. *Check* your work. The answer is reasonable since 1038 is close to our estimated answer of 1000. *State* your conclusion: The Snake River is 1038 miles long.

■ **Work Practice 1**

Practice 2

Four friends bought a lottery ticket and won $65,000. If each person is to receive the same amount of money, how much does each person receive?

> **Example 2**   Filling a Shipping Order

How many cases can be filled with 9900 cans of jalapeños if each case holds 48 cans? How many cans will be left over? Will there be enough cases to fill an order for 200 cases?

Solution:

1. UNDERSTAND. Read and reread the problem. Draw a picture to help visualize the situation.

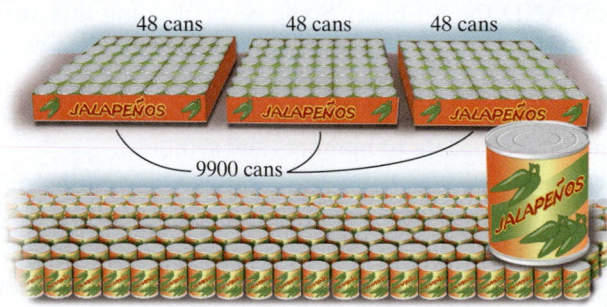

Since each case holds 48 cans, we want to know how many 48s there are in 9900. We find this by dividing.

2. TRANSLATE.

In words: Number of cases   is   9900   divided by   48
↓   ↓   ↓   ↓   ↓
Translate: Number of cases   =   9900   ÷   48

3. SOLVE: Let's estimate a reasonable solution before we actually divide. Since 9900 rounded to the nearest thousand is 10,000 and 48 rounded to the nearest ten is 50, 10,000 ÷ 50 = 200. Now find the exact quotient.

```
      206  R12
48)9900
   −96
    300
   −288
     12
```

Answer

2. $16,250

Copyright 2015 Pearson Education, Inc.

4. INTERPRET. *Check* your work. The answer is reasonable since 206 R 12 is close to our estimate of 200. *State* your conclusion: 206 cases will be filled, with 12 cans left over. There will be enough cases to fill an order for 200 cases.

■ **Work Practice 2**

Example 3 Calculating Budget Costs

The director of a learning lab at a local community college is working on next year's budget. Thirty-three new DVD players are needed at a cost of $187 each. What is the total cost of these DVD players?

Solution:

1. UNDERSTAND. Read and reread the problem, and then draw a diagram.

33 DVD Players

$ 187 $ 187   ... $ 187

From the phrase "total cost," we might decide to solve this problem by adding. This would work, but repeated addition, or multiplication, would save time.

2. TRANSLATE.

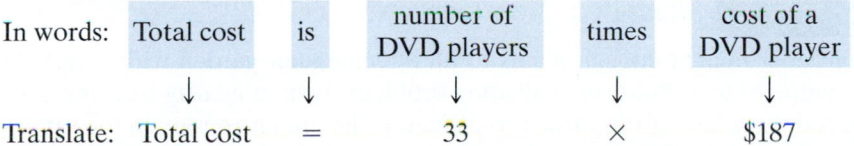

In words:	Total cost	is	number of DVD players	times	cost of a DVD player
	↓	↓	↓	↓	↓
Translate:	Total cost	=	33	×	$187

3. SOLVE: Once again, let's estimate a reasonable solution.

$$
\begin{array}{r}
187 \\
\times\ 33 \\
\hline
561 \\
5610 \\
\hline
6171
\end{array}
\quad
\begin{array}{l}
\text{rounds to} \\
\text{rounds to}
\end{array}
\quad
\begin{array}{r}
200 \\
\times\ 30 \\
\hline
6000
\end{array}
$$

187 rounds to 200
× 33 rounds to × 30
561   6000 estimate
5610
6171 exact

4. INTERPRET. *Check* your work. *State* your conclusion: The total cost of the DVD players is $6171.

■ **Work Practice 3**

Example 4 Calculating a Public School Teacher's Salary

In 2012, the average salary for a public school teacher in California was $69,324. For the same year, the average salary for a public school teacher in Iowa was $17,796 less than this. What was the average public school teacher's salary in Iowa? (*Source:* National Education Association)

Solution:

1. UNDERSTAND. Read and reread the problem. Notice that we are told that the Iowa salary is $17,796 less than the California salary. The phrase "less than" indicates subtraction.

(Continued on next page)

Practice 3

The director of the learning lab also needs to include in the budget a line for 425 blank CDs at a cost of $4 each. What is this total cost for the blank CDs?

Practice 4

In 2012, the average salary for a public school teacher in Alaska was $65,468. For the same year, the average salary for a public school teacher in Hawaii was $11,168 less than this. What was the average public school teacher's salary in Hawaii? (*Source:* National Education Association)

Answers
3. $1700   **4.** $54,300

2. TRANSLATE. Remember that order matters when subtracting, so be careful when translating.

In words: | Iowa salary | is | California salary | minus | $17,796 |

Translate: Iowa salary   =   69,324   −   17,796

3. SOLVE. This time, instead of estimating, let's check by adding.

$$\begin{array}{r} 69,324 \\ -17,796 \\ \hline 51,528 \end{array}$$ **Check:** $$\begin{array}{r} 51,528 \\ +17,796 \\ \hline 69,324 \end{array}$$

4. INTERPRET. *Check* your work. The check is above. *State* your conclusion: The average Iowa teacher's salary in 2012 was $51,528.

■ Work Practice 4

Objective B Solving Problems That Require More Than One Operation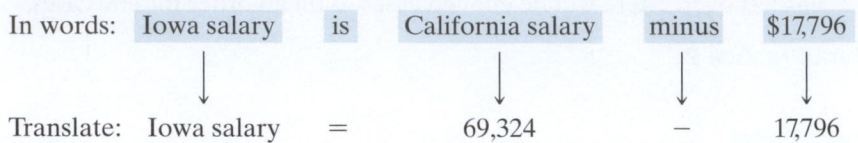

We must sometimes use more than one operation to solve a problem.

Practice 5

A gardener is trying to decide how much fertilizer to buy for his yard. He knows that his lot is in the shape of a rectangle that measures 90 feet by 120 feet. He also knows that the floor of his house is in the shape of a rectangle that measures 45 feet by 65 feet. How much area of the lot is not covered by the house?

Answer

5. 7875 sq ft

Example 5   Planting a New Garden

A gardener bought enough plants to fill a rectangular garden with length 30 feet and width 20 feet. Because of shading problems from a nearby tree, the gardener changed the width of the garden to 15 feet. If the area is to remain the same, what is the new length of the garden?

Solution:

1. UNDERSTAND. Read and reread the problem. Then draw a picture to help visualize the problem.

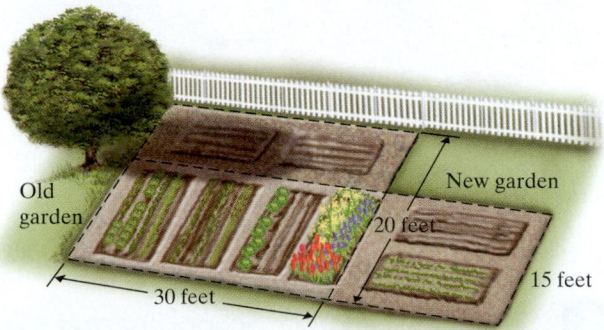

2. TRANSLATE. Since the area of the new garden is to be the same as the area of the old garden, let's find the area of the old garden. Recall that

Area = length × width = 30 feet × 20 feet = 600 square feet

Since the area of the new garden is to be 600 square feet also, we need to see how many 15s there are in 600. This means division. In other words,

In words: New length   =   Area of garden   ÷   New width

Translate: New length   =   600   ÷   15

3. SOLVE.

$$\begin{array}{r} 40 \\ 15\overline{)600} \\ -60 \\ \hline 00 \end{array}$$

4. INTERPRET. *Check* your work. *State* your conclusion: The length of the new garden is 40 feet.

■ **Work Practice 5**

Vocabulary, Readiness & Video Check

Martin-Gay Interactive Videos   Watch the section lecture video and answer the following questions.

Objective A 1. The answer to the calculations in ▥ Example 3 is 3500. What is the final interpreted solution, written as a sentence? ⟳

Objective B 2. What two operations are used to solve ▥ Example 4? ⟳

See Video 1.8 🍎

1.8 Exercise Set   MyMathLab® ⟳

Objective A *Solve. Exercises 1, 2, 11, and 12 have been started for you. See Examples 1 through 4.*

1. 41 increased by 8 is what number?

Start the solution:

1. UNDERSTAND the problem. Reread it as many times as needed.
2. TRANSLATE into an equation. (Fill in the blanks below.)

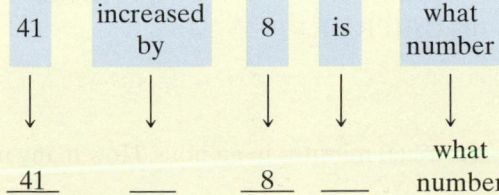

Finish with:

3. SOLVE

4. INTERPRET

2. What is 12 multiplied by 9?

Start the solution:

1. UNDERSTAND the problem. Reread it as many times as needed.
2. TRANSLATE into an equation. (Fill in the blanks below.)

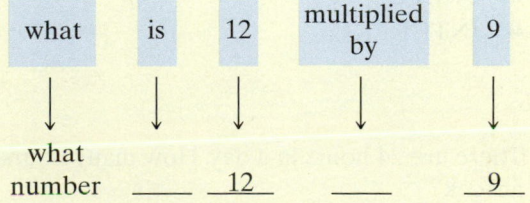

Finish with:

3. SOLVE

4. INTERPRET

▶ **3.** What is the quotient of 1185 and 5?

4. 78 decreased by 12 is what number?

▶ **5.** What is the total of 35 and 7?

6. What is the difference of 48 and 8?

7. 60 times 10 is what number?

8. 60 divided by 10 is what number?

△ **9.** A vacant lot in the shape of a rectangle measures 120 feet by 80 feet.

    a. What is the perimeter of the lot?

    b. What is the area of the lot?

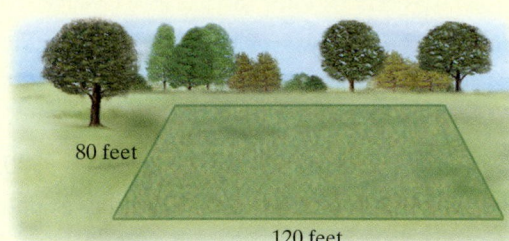

△ **10.** A parking lot in the shape of a rectangle measures 100 feet by 150 feet.

    a. What is the perimeter of the lot?

    b. What is the area of the parking lot?

11. A family bought a house for $185,700 and later sold the house for $201,200. How much money did they make by selling the house?

Start the solution:

1. UNDERSTAND the problem. Reread it as many times as needed.
2. TRANSLATE into an equation. (Fill in the blanks below.)

money made	is	selling price	minus	purchase price
↓	↓	↓	↓	↓
money made	= ____	____	−	____

Finish with:

3. SOLVE
4. INTERPRET

12. Three people dream of equally sharing a $147 million lottery. How much would each person receive if they have the winning ticket?

Start the solution:

1. UNDERSTAND the problem. Reread it as many times as needed.
2. TRANSLATE into an equation. (Fill in the blanks below.)

each person's share	is	lottery amount	divided by	number by persons
↓	↓	↓	↓	↓
each person's share	= ____	____	÷	____

Finish with:

3. SOLVE
4. INTERPRET

13. There are 24 hours in a day. How many hours are in a week?

14. There are 60 minutes in an hour. How many minutes are in a day?

15. The Verrazano Narrows Bridge is the longest bridge in New York, measuring 4260 feet. The George Washington Bridge, also in New York, is 760 feet shorter than the Verrazano Narrows Bridge. Find the length of the George Washington Bridge.

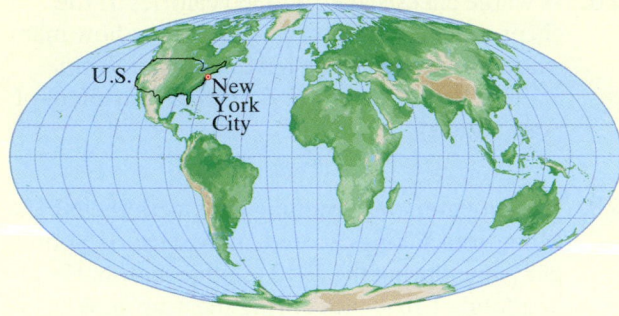

16. In 2013, the Goodyear Tire & Rubber Company began replacing its fleet of nonrigid GZ-20 blimps with new Goodyear NT semi-rigid airships. The new Goodyear NT airship can hold 297,527 cubic feet of helium. Its GZ-20 predecessor held 94,827 fewer cubic feet of helium. How much helium did a GZ-20 blimp hold? (*Source:* Goodyear Tire & Rubber Company)

17. Yellowstone National Park in Wyoming was the first national park in the United States. It was created in 1872. One of the more recent additions to the National Park System is First State National Monument. It was established in 2013. How much older is Yellowstone than First State? (*Source:* National Park Service)

18. Razor scooters were introduced in 2000. Radio Flyer Wagons were first introduced 83 years earlier. In what year were Radio Flyer Wagons introduced? (*Source:* Toy Industry Association, Inc.)

19. Since their introduction, the number of LEGO building bricks that have been sold is equivalent to the world's current population of approximately 6 billion people owning 62 LEGO bricks each. About how many LEGO bricks have been sold since their introduction? (*Source:* LEGO Company)

20. In 2012, the average weekly pay for a home health aide in the United States was about $420. At this rate, how much will a home health aide earn working a 52-week year? (*Source:* Bureau of Labor Statistics)

21. The three most common city names in the United States are Fairview, Midway, and Riverside. There are 287 towns named Fairview, 252 named Midway, and 180 named Riverside. Find the total number of towns named Fairview, Midway, or Riverside.

22. In the game of Monopoly, a player must own all properties in a color group before building houses. The yellow color-group properties are Atlantic Avenue, Ventnor Avenue, and Marvin Gardens. These cost $260, $260, and $280, respectively, when purchased from the bank. What total amount must a player pay to the bank before houses can be built on the yellow properties? (*Source:* Hasbro, Inc.)

23. In 2012, the average weekly pay for a correctional officer in the United States was $840. If such an officer works 40 hours in one week, what is his or her hourly pay? (*Source:* Bureau of Labor Statistics)

24. In 2012, the average weekly pay for a loan officer was $1360. If a loan officer works 40 hours in one week, what is his or her hourly pay? (*Source:* Bureau of Labor Statistics)

25. Three ounces of canned tuna in oil has 165 calories. How many calories does 1 ounce have? (*Source: Home and Garden Bulletin No. 72,* U.S. Department of Agriculture)

26. A whole cheesecake has 3360 calories. If the cheesecake is cut into 12 equal pieces, how many calories will each piece have? (*Source: Home and Garden Bulletin No. 72,* U.S. Department of Agriculture)

27. The average estimated 2012 U.S. population was 313,900,000. Between Memorial Day and Labor Day, 7 billion hot dogs are consumed. Approximately how many hot dogs were consumed per person between Memorial Day and Labor Day in 2012? Divide, but do not give the remainder part of the quotient. (*Source:* U.S. Census Bureau, National Hot Dog and Sausage Council)

28. Adrian Peterson, a running back with the NFL's Minnesota Vikings, scored an average of 5 points per game during the 2012 regular season. He played in a total of 16 games during the season. What was the total number of points he scored during the 2012 football season? (*Source:* National Football League)

29. Macy's, formerly the Federated Department Stores Company, operates a total of 843 Macy's and Bloomingdale's department stores around the country. In 2012, Macy's had sales of approximately $27,685,999,890. What is the average amount of sales made by each of the 843 stores? Divide, but do not give the remainder part of the quotient. (*Source:* Macy's)

30. In 2012, PetSmart employed approximately 52,000 associates and operated roughly 1300 stores. What is the average number of associates employed at each of its stores? (*Source:* PetSmart, Inc.)

31. In 2012, the Museum of Modern Art in New York welcomed 2,805,659 visitors. The J. Paul Getty Museum in Los Angeles received 1,590,608 visitors. How many more people visited the Museum of Modern Art than the Getty Museum? (*Source: The Art Newspaper*)

32. In 2012, Target Corporation operated 1778 stores in the United States. Of these, 149 were in Texas. How many Target Stores were located in states other than Texas? (*Source:* Target Corporation)

33. The length of the southern boundary of the conterminous United States is 1933 miles. The length of the northern boundary of the conterminous United States is 2054 miles longer than this. What is the length of the northern boundary? (*Source:* U.S. Geological Survey)

34. In humans, 14 muscles are required to smile. It takes 29 more muscles to frown. How many muscles does it take to frown?

2054 miles longer

1933 miles

35. An instructor at the University of New Orleans receives a paycheck every four weeks. Find how many paychecks he receives in a year. (A year has 52 weeks.)

36. A loan of $6240 is to be paid in 48 equal payments. How much is each payment?

Objective **B** *Solve. See Example 5.*

▶ **37.** Find the total cost of 3 sweaters at $38 each and 5 shirts at $25 each.

38. Find the total cost of 10 computers at $2100 each and 7 boxes of diskettes at $12 each.

39. A college student has $950 in an account. She spends $205 from the account on books and then deposits $300 in the account. How much money is now in the account?

40. The temperature outside was 57°F (degrees Fahrenheit). During the next few hours, it decreased by 18 degrees and then increased by 23 degrees. Find the new temperature.

The table shows the menu from a concession stand at the county fair. Use this menu to answer Exercises 41 and 42.

41. A hungry college student is debating between the following two orders:
   a. a hamburger, an order of onion rings, a candy bar, and a soda.
   b. a hot dog, an apple, an order of french fries, and a soda.
   Which order will be cheaper? By how much?

42. A family of four is debating between the following two orders:
   a. 6 hot dogs, 4 orders of onion rings, and 4 sodas.
   b. 4 hamburgers, 4 orders of french fries, 2 apples, and 4 sodas.
   Will the family save any money by ordering (b) instead of (a)? If so, how much?

Corky's Concession Stand Menu	
Item	**Price**
Hot dog	$3
Hamburger	$4
Soda	$1
Onion rings	$3
French fries	$2
Apple	$1
Candy bar	$2

Objectives **A** **B** **Mixed Practice** *Use the bar graph to answer Exercises 43 through 50. (Source: Internet World Stats)*

43. Which region of the world listed had the greatest number of Internet users in 2012?

44. Which region of the world listed had the least number of Internet users in 2012?

45. How many more Internet users (in millions) did the world region with the most Internet users have than the world region with the fewest Internet users?

46. How many more Internet users did Africa have than the Middle East in 2012?

47. How many more Internet users did North America have than Latin America/Caribbean?

48. Which region of the world had more Internet users, Europe or North America? How many more Internet users did it have?

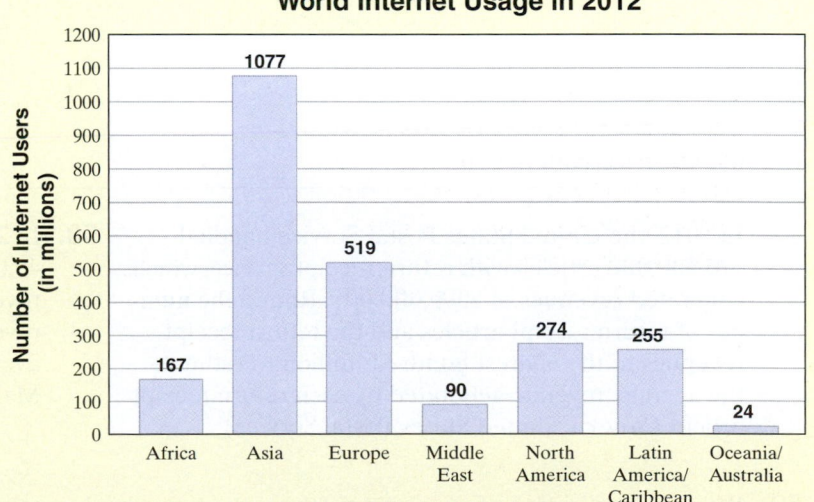

World Internet Usage in 2012

Find the average number of Internet users for the world regions listed in the graph.

49. The two world regions with the greatest number of Internet users.

50. The four world regions with the least number of Internet users.

Solve.

51. The learning lab at a local university is receiving new equipment. Twenty-two computers are purchased for $615 each and three printers for $408 each. Find the total cost for this equipment.

52. The washateria near the local community college is receiving new equipment. Thirty-six washers are purchased for $585 each and ten dryers are purchased for $388 each. Find the total cost for this equipment.

53. The American Heart Association recommends consuming no more than 2400 milligrams of salt per day. (This is about the amount in 1 teaspoon of salt.) How many milligrams of sodium is this in a week?

54. This semester a particular student pays $1750 for room and board, $709 for a meal ticket plan, and $2168 for tuition. What is her total bill?

△ **55.** The Meishs' yard is in the shape of a rectangle and measures 50 feet by 75 feet. In their yard, they have a rectangular swimming pool that measures 15 feet by 25 feet.
   a. Find the area of the entire yard.
   b. Find the area of the swimming pool.
   c. Find the area of the yard that is not part of the swimming pool.

56. The community is planning to construct a rectangular-shaped playground within the local park. The park is in the shape of a square and measures 100 yards on each side. The playground is to measure 15 yards by 25 yards.
   a. Find the area of the entire park.
   b. Find the area of the playground.
   c. Find the area of the park that is not part of the playground.

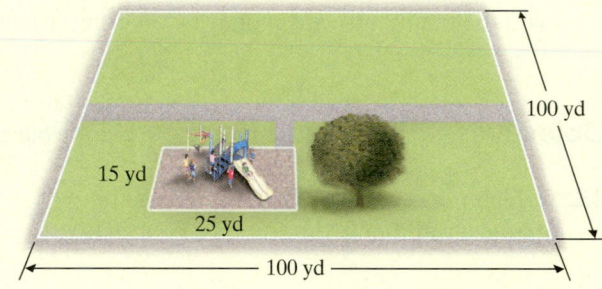

Concept Extensions

57. In 2012, the United States Postal Service handled 170,000,000 articles with return receipt service, which generated revenues of $399,000,000. Round the number of return receipt articles and the return receipt revenues to the nearest hundred-million to estimate the average revenue generated by each return receipt article. (*Source:* United States Postal Service)

58. In 2012, the United States Postal Service handled 40,000,000 pieces of Express Mail, which generated revenues of $802,000,000. Round the Express Mail revenues to the nearest ten-million to estimate the average revenue generated by each piece of Express Mail. (*Source:* United States Postal Service)

59. Write an application of your own that uses the term "bank account" and the numbers 1036 and 524.

1.9 Exponents, Square Roots, and Order of Operations

Objective A Using Exponential Notation

In the product $3 \cdot 3 \cdot 3 \cdot 3 \cdot 3$, notice that 3 is a factor several times. When this happens, we can use a shorthand notation, called an **exponent,** to write the repeated multiplication.

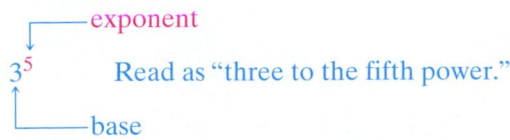

$\underbrace{3 \cdot 3 \cdot 3 \cdot 3 \cdot 3}$ can be written as

3 is a factor 5 times

3^5 Read as "three to the fifth power."

This is called **exponential notation.** The **exponent,** 5, indicates how many times the **base,** 3, is a factor.

The table below shows examples of reading exponential notation in words.

Expression	In Words
5^2	"five to the second power" or "five squared"
5^3	"five to the third power" or "five cubed"
5^4	"five to the fourth power"

Usually, an exponent of 1 is not written, so when no exponent appears, we assume that the exponent is 1. For example, $2 = 2^1$ and $7 = 7^1$.

Examples Write using exponential notation.

1. $7 \cdot 7 \cdot 7 = 7^3$
2. $3 \cdot 3 = 3^2$
3. $6 \cdot 6 \cdot 6 \cdot 6 \cdot 6 = 6^5$
4. $3 \cdot 3 \cdot 3 \cdot 3 \cdot 17 \cdot 17 \cdot 17 = 3^4 \cdot 17^3$

■ Work Practice 1–4

Objective B Evaluating Exponential Expressions

To **evaluate** an exponential expression, we write the expression as a product and then find the value of the product.

Examples Evaluate.

5. $9^2 = 9 \cdot 9 = 81$
6. $6^1 = 6$
7. $3^4 = 3 \cdot 3 \cdot 3 \cdot 3 = 81$
8. $5 \cdot 6^2 = 5 \cdot 6 \cdot 6 = 180$

■ Work Practice 5–8

Objectives

A Write Repeated Factors Using Exponential Notation.

B Evaluate Expressions Containing Exponents.

C Evaluate the Square Root of a Perfect Square.

D Use the Order of Operations.

E Find the Area of a Square.

Practice 1–4

Write using exponential notation.

1. $8 \cdot 8 \cdot 8 \cdot 8$
2. $3 \cdot 3 \cdot 3$
3. $10 \cdot 10 \cdot 10 \cdot 10 \cdot 10$
4. $5 \cdot 5 \cdot 4 \cdot 4 \cdot 4 \cdot 4 \cdot 4 \cdot 4$

Practice 5–8

Evaluate.

5. 4^2 6. 7^3
7. 11^1 8. $2 \cdot 3^2$

Answers

1. 8^4 2. 3^3 3. 10^5 4. $5^2 \cdot 4^6$
5. 16 6. 343 7. 11 8. 18

87

Example 8 illustrates an important property: An exponent applies only to its base. The exponent 2, in $5 \cdot 6^2$, applies only to its base, 6.

> **Helpful Hint**
>
> An exponent applies only to its base. For example, $4 \cdot 2^3$ means $4 \cdot 2 \cdot 2 \cdot 2$.

> **Helpful Hint**
>
> Don't forget that 2^4, for example, is *not* $2 \cdot 4$. The expression 2^4 means repeated multiplication of the same factor.
>
> $$2^4 = 2 \cdot 2 \cdot 2 \cdot 2 = 16, \quad \text{whereas } 2 \cdot 4 = 8$$

✓**Concept Check** Which of the following statements is correct?

a. 3^6 is the same as $6 \cdot 6 \cdot 6$.
b. "Eight to the fourth power" is the same as 8^4.
c. "Ten squared" is the same as 10^3.
d. 11^2 is the same as $11 \cdot 2$.

Objective C Evaluating Square Roots ▶

A **square root** of a number is one of two identical factors of the number. For example,

$$7 \cdot 7 = 49, \text{ so a square root of 49 is 7.}$$

We use this symbol $\sqrt{}$ (called a radical sign) for finding square roots. Since

$$7 \cdot 7 = 49, \text{ then } \sqrt{49} = 7.$$

Practice 9–11

Find each square root.
9. $\sqrt{100}$
10. $\sqrt{4}$
11. $\sqrt{1}$

> **Examples** Find each square root.
>
> 9. $\sqrt{25} = 5$ because $5 \cdot 5 = 25$
> 10. $\sqrt{81} = 9$ because $9 \cdot 9 = 81$
> 11. $\sqrt{0} = 0$ because $0 \cdot 0 = 0$

■ **Work Practice 9–11**

> **Helpful Hint**
>
> Make sure you understand the difference between squaring a number and finding the square root of a number.
>
> $$9^2 = 9 \cdot 9 = 81 \quad \sqrt{9} = 3 \text{ because } 3 \cdot 3 = 9$$

Answers
9. 10 **10.** 2 **11.** 1

✓**Concept Check Answer**
b

Not every square root simplifies to a whole number. We will study this more in a later chapter. In this section, we will find square roots of perfect squares only.

Objective D Using the Order of Operations

Suppose that you are in charge of taking inventory at a local cell phone store. An employee has given you the number of a certain cell phone in stock as the expression

$6 + 2 \cdot 30$

To calculate the value of this expression, do you add first or multiply first? If you add first, the answer is 240. If you multiply first, the answer is 66.

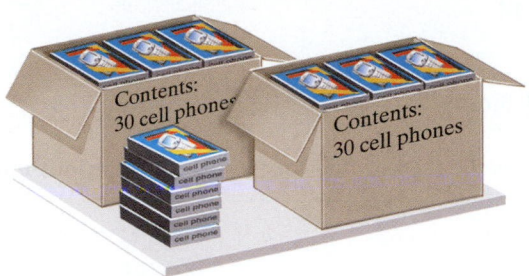

Contents: 30 cell phones

Contents: 30 cell phones

Mathematical symbols wouldn't be very useful if two values were possible for one expression. Thus, mathematicians have agreed that, given a choice, we multiply first.

$$6 + 2 \cdot 30 = 6 + 60 \quad \text{Multiply.}$$
$$= 66 \quad \text{Add.}$$

This agreement is one of several **order of operations** agreements.

Order of Operations

1. Perform all operations within parentheses ( ), brackets [ ], or other grouping symbols such as fraction bars or square roots, starting with the innermost set.
2. Evaluate any expressions with exponents.
3. Multiply or divide in order from left to right.
4. Add or subtract in order from left to right.

Below we practice using order of operations to simplify expressions.

Example 12 Simplify: $2 \cdot 4 - 3 \div 3$

Solution: There are no parentheses and no exponents, so we start by multiplying and dividing, from left to right.

$$2 \cdot 4 - 3 \div 3 = 8 - 3 \div 3 \quad \text{Multiply.}$$
$$= 8 - 1 \quad \text{Divide.}$$
$$= 7 \quad \text{Subtract.}$$

■ Work Practice 12

Practice 12
Simplify: $9 \cdot 3 - 8 \div 4$

Example 13 Simplify: $4^2 \div 2 \cdot 4$

Solution: We start by evaluating 4^2.

$$4^2 \div 2 \cdot 4 = 16 \div 2 \cdot 4 \quad \text{Write } 4^2 \text{ as 16.}$$

Next we multiply or divide *in order* from left to right. Since division appears before multiplication from left to right, we divide first, then multiply.

$$16 \div 2 \cdot 4 = 8 \cdot 4 \quad \text{Divide.}$$
$$= 32 \quad \text{Multiply.}$$

■ Work Practice 13

Practice 13
Simplify: $48 \div 3 \cdot 2^2$

Answers
12. 25 13. 64

Practice 14

Simplify: $(10 - 7)^4 + 2 \cdot 3^2$

Example 14 Simplify: $(8 - 6)^2 + 2^3 \cdot 3$

Solution: $(8 - 6)^2 + 2^3 \cdot 3 = 2^2 + 2^3 \cdot 3$ Simplify inside parentheses.

$$= 4 + 8 \cdot 3 \qquad \text{Write } 2^2 \text{ as 4 and } 2^3 \text{ as 8.}$$
$$= 4 + 24 \qquad \text{Multiply.}$$
$$= 28 \qquad \text{Add.}$$

■ Work Practice 14

Practice 15

Simplify:

$36 \div [20 - (4 \cdot 2)] + 4^3 - 6$

Example 15 Simplify: $4^3 + [3^2 - (10 \div 2)] - 7 \cdot 3$

Solution: Here we begin with the innermost set of parentheses.

$$4^3 + [3^2 - (10 \div 2)] - 7 \cdot 3 = 4^3 + [3^2 - 5] - 7 \cdot 3 \quad \text{Simplify inside parentheses.}$$

$$= 4^3 + [9 - 5] - 7 \cdot 3 \quad \text{Write } 3^2 \text{ as 9.}$$

$$= 4^3 + 4 - 7 \cdot 3 \quad \text{Simplify inside brackets.}$$

$$= 64 + 4 - 7 \cdot 3 \quad \text{Write } 4^3 \text{ as 64.}$$

$$= 64 + 4 - 21 \quad \text{Multiply.}$$

$$= 47 \quad \text{Add and subtract from left to right.}$$

■ Work Practice 15

Practice 16

Simplify: $\dfrac{25 + 8 \cdot 2 - 3^3}{2(3 - 2)}$

Example 16 Simplify: $\dfrac{7 - 2 \cdot 3 + 3^2}{5(2 - 1)}$

Solution: Here, the fraction bar is like a grouping symbol. We simplify above and below the fraction bar separately.

$$\frac{7 - 2 \cdot 3 + 3^2}{5(2 - 1)} = \frac{7 - 2 \cdot 3 + 9}{5(1)} \quad \text{Evaluate } 3^2 \text{ and } (2 - 1).$$

$$= \frac{7 - 6 + 9}{5} \quad \text{Multiply } 2 \cdot 3 \text{ in the numerator and multiply 5 and 1 in the denominator.}$$

$$= \frac{10}{5} \quad \text{Add and subtract from left to right.}$$

$$= 2 \quad \text{Divide.}$$

■ Work Practice 16

Practice 17

Simplify: $81 \div \sqrt{81} \cdot 5 + 7$

Example 17 Simplify: $64 \div \sqrt{64} \cdot 2 + 4$

Solution: $64 \div \sqrt{64} \cdot 2 + 4 = 64 \div 8 \cdot 2 + 4$ Find the square root.

$$= 8 \cdot 2 + 4 \quad \text{Divide.}$$
$$= 16 + 4 \quad \text{Multiply.}$$
$$= 20 \quad \text{Add.}$$

■ Work Practice 17

Answers

14. 99 **15.** 61 **16.** 7 **17.** 52

Objective E Finding the Area of a Square

Since a square is a special rectangle, we can find its area by finding the product of its length and its width.

Area of a rectangle = length · width

By recalling that each side of a square has the same measurement, we can use the following procedure to find its area:

Area of a square = length · width
= side · side
= (side)^2

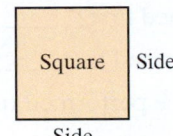

Helpful Hint

Recall from Section 1.6 that area is measured in **square** units while perimeter is measured in units.

Example 18 Find the area of a square whose side measures 4 inches.

Solution: Area of a square = (side)^2
= (4 inches)^2
= 16 square inches

4 inches

The area of the square is 16 square inches.

■ Work Practice 18

Practice 18

Find the area of a square whose side measures 12 centimeters.

Calculator Explorations Exponents

To evaluate an exponent such as 4^7 on a calculator, find the keys marked $\boxed{y^x}$ or $\boxed{\wedge}$ and $\boxed{=}$ or $\boxed{\text{ENTER}}$. To evaluate 4^7, press the keys $\boxed{4}$ $\boxed{y^x}$ (or $\boxed{\wedge}$) $\boxed{7}$ then $\boxed{=}$ or $\boxed{\text{ENTER}}$. The display will read $\boxed{16384}$. Thus, $4^7 = 16,384$.

Use a calculator to evaluate.

1. 4^6 **2.** 5^6 **3.** 5^5

4. 7^6 **5.** 2^{11} **6.** 6^8

Order of Operations

To see whether your calculator has the order of operations built in, evaluate $5 + 2 \cdot 3$ by pressing the keys $\boxed{5}$ $\boxed{+}$ $\boxed{2}$ $\boxed{\times}$ $\boxed{3}$ then $\boxed{=}$ or $\boxed{\text{ENTER}}$. If the display reads $\boxed{11}$, your calculator does have the order of operations built in. This means that most of the time,

you can key in a problem exactly as it is written and the calculator will perform operations in the proper order. When evaluating an expression containing parentheses, key in the parentheses. (If an expression contains brackets, key in parentheses.) For example, to evaluate $2[25 - (8 + 4)] - 11$, press the keys $\boxed{2}$ $\boxed{\times}$ $\boxed{(}$ $\boxed{25}$ $\boxed{-}$ $\boxed{(}$ $\boxed{8}$ $\boxed{+}$ $\boxed{4}$ $\boxed{)}$ $\boxed{)}$ $\boxed{-}$ $\boxed{11}$ then $\boxed{=}$ or $\boxed{\text{ENTER}}$.

The display will read $\boxed{15}$.

Use a calculator to evaluate.

7. $7^4 + 5^3$

8. $12^4 - 8^4$

9. $63 \cdot 75 - 43 \cdot 10$

10. $8 \cdot 22 + 7 \cdot 16$

11. $4(15 \div 3 + 2) - 10 \cdot 2$

12. $155 - 2(17 + 3) + 185$

Answer

18. 144 sq cm

Vocabulary, Readiness & Video Check

Use the choices below to fill in each blank.

addition	multiplication	exponent	base
subtraction	division	square root	

1. In $2^5 = 32$, the 2 is called the _____ and the 5 is called the _____.
2. To simplify $8 + 2 \cdot 6$, which operation should be performed first? _____
3. To simplify $(8 + 2) \cdot 6$, which operation should be performed first? _____
4. To simplify $9(3 - 2) \div 3 + 6$, which operation should be performed first? _____
5. To simplify $8 \div 2 \cdot 6$, which operation should be performed first? _____
6. The _____ of a whole number is one of two identical factors of the number.

Martin-Gay Interactive Videos

See Video 1.9 🍎

Watch the section lecture video and answer the following questions.

Objective A 7. In the ▥ Example 1 expression, what is the 3 called and what is the 12 called? ▶

Objective B 8. As mentioned in ▥ Example 4, what "understood exponent" does any number we've worked with before have? ▶

Objective C 9. From ▥ Example 7, how do we know that $\sqrt{64} = 8$? ▶

Objective D 10. List the three operations needed to evaluate ▥ Example 9 in the order they should be performed. ▶

Objective E 11. As explained in the lecture before ▥ Example 12, why does the area of a square involve an exponent whereas the area of a rectangle usually does not? ▶

1.9 Exercise Set MyMathLab®

Objective A *Write using exponential notation. See Examples 1 through 4.*

1. $4 \cdot 4 \cdot 4$ 2. $5 \cdot 5 \cdot 5 \cdot 5$ 3. $7 \cdot 7 \cdot 7 \cdot 7 \cdot 7 \cdot 7$ 4. $6 \cdot 6 \cdot 6 \cdot 6 \cdot 6 \cdot 6 \cdot 6$

▶ 5. $12 \cdot 12 \cdot 12$ 6. $10 \cdot 10 \cdot 10$ ▶ 7. $6 \cdot 6 \cdot 5 \cdot 5 \cdot 5$ 8. $4 \cdot 4 \cdot 3 \cdot 3 \cdot 3$

9. $9 \cdot 8 \cdot 8$ 10. $7 \cdot 4 \cdot 4 \cdot 4$ 11. $3 \cdot 2 \cdot 2 \cdot 2 \cdot 2$ 12. $4 \cdot 6 \cdot 6 \cdot 6 \cdot 6$

13. $3 \cdot 2 \cdot 2 \cdot 2 \cdot 2 \cdot 5 \cdot 5 \cdot 5 \cdot 5 \cdot 5$ 14. $6 \cdot 6 \cdot 2 \cdot 9 \cdot 9 \cdot 9 \cdot 9$

Objective B *Evaluate. See Examples 5 through 8.*

15. 8^2 **16.** 6^2 ▶ **17.** 5^3 **18.** 6^3 **19.** 2^5 **20.** 3^5

21. 1^{10} **22.** 1^{12} ▶ **23.** 7^1 **24.** 8^1 **25.** 2^7 **26.** 5^4

27. 2^8 **28.** 3^3 **29.** 4^4 **30.** 4^3 **31.** 9^3 **32.** 8^3

33. 12^2 **34.** 11^2 ▶ **35.** 10^2 **36.** 10^3 **37.** 20^1 **38.** 14^1

39. 3^6 **40.** 4^5 **41.** $3 \cdot 2^6$ **42.** $5 \cdot 3^2$ **43.** $2 \cdot 3^4$ **44.** $2 \cdot 7^2$

Objective C *Find each square root. See Examples 9 through 11.*

▶ **45.** $\sqrt{9}$ **46.** $\sqrt{36}$ ▶ **47.** $\sqrt{64}$ **48.** $\sqrt{121}$

49. $\sqrt{144}$ **50.** $\sqrt{0}$ **51.** $\sqrt{16}$ **52.** $\sqrt{169}$

Objective D *Simplify. See Examples 12 through 16. (This section does not contain square roots.)*

▶ **53.** $15 + 3 \cdot 2$ **54.** $24 + 6 \cdot 3$ ▶ **55.** $14 \div 7 \cdot 2 + 3$ **56.** $100 \div 10 \cdot 5 + 4$

57. $32 \div 4 - 3$ **58.** $42 \div 7 - 6$ **59.** $13 + \dfrac{24}{8}$ **60.** $32 + \dfrac{8}{2}$

61. $6 \cdot 5 + 8 \cdot 2$ **62.** $3 \cdot 4 + 9 \cdot 1$ **63.** $\dfrac{5 + 12 \div 4}{1^7}$ **64.** $\dfrac{6 + 9 \div 3}{3^2}$

65. $(7 + 5^2) \div 4 \cdot 2^3$ **66.** $6^2 \cdot (10 - 8)$ **67.** $5^2 \cdot (10 - 8) + 2^3 + 5^2$

68. $5^3 \div (10 + 15) + 9^2 + 3^3$ **69.** $\dfrac{18 + 6}{2^4 - 2^2}$ **70.** $\dfrac{40 + 8}{5^2 - 3^2}$

71. $(3 + 5) \cdot (9 - 3)$ **72.** $(9 - 7) \cdot (12 + 18)$ ▶ **73.** $\dfrac{7(9 - 6) + 3}{3^2 - 3}$

74. $\dfrac{5(12-7)-4}{5^2-18}$

75. $8 \div 0 + 37$

76. $18 - 7 \div 0$

77. $2^4 \cdot 4 - (25 \div 5)$

78. $2^3 \cdot 3 - (100 \div 10)$

79. $3^4 - [35 - (12 - 6)]$

80. $[40 - (8 - 2)] - 2^5$

⊙ **81.** $(7 \cdot 5) + [9 \div (3 \div 3)]$

82. $(18 \div 6) + [(3 + 5) \cdot 2]$

83. $8 \cdot \left[2^2 + (6 - 1) \cdot 2\right] - 50 \cdot 2$

84. $35 \div \left[3^2 + (9 - 7) - 2^2\right] + 10 \cdot 3$

85. $\dfrac{9^2 + 2^2 - 1^2}{8 \div 2 \cdot 3 \cdot 1 \div 3}$

86. $\dfrac{5^2 - 2^3 + 1^4}{10 \div 5 \cdot 4 \cdot 1 \div 4}$

Simplify. See Examples 12 through 17. (This section does contain square roots.)

87. $6 \cdot \sqrt{9} + 3 \cdot \sqrt{4}$

88. $3 \cdot \sqrt{25} + 2 \cdot \sqrt{81}$

89. $4 \cdot \sqrt{49} - 0 \div \sqrt{100}$

90. $7 \cdot \sqrt{36} - 0 \div \sqrt{64}$

91. $\dfrac{\sqrt{4} + 4^2}{5(20 - 16) - 3^2 - 5}$

92. $\dfrac{\sqrt{9} + 9^2}{3(10 - 6) - 2^2 - 1}$

93. $\sqrt{81} \div \sqrt{9} + 4^2 \cdot 2 - 10$

94. $\sqrt{100} \div \sqrt{4} + 3^3 \cdot 2 - 20$

95. $\left[\sqrt{225} \div (11 - 6) + 2^2\right] + \left(\sqrt{25} - \sqrt{1}\right)^2$

96. $\left[\sqrt{169} \div (20 - 7) + 2^5\right] - \left(\sqrt{4} + \sqrt{9}\right)^2$

97. $7^2 - \left\{18 - \left[40 \div (4 \cdot 2) + \sqrt{4}\right] + 5^2\right\}$

98. $29 - \left\{5 + 3\left[8 \cdot (10 - \sqrt{64})\right] - 50\right\}$

Objective **E** **Mixed Practice** (*Sections 1.3, 1.6*) *Find the area and perimeter of each square. See Example 18.*

⊙ △**99.**

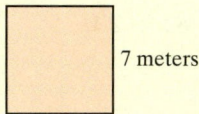

7 meters

△**100.**

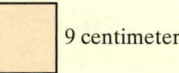

9 centimeters

△**101.**

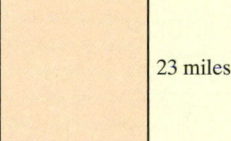

23 miles

△**102.**

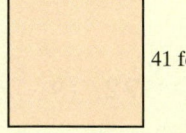

41 feet

Concept Extensions

Answer the following true or false. See the Concept Check in this section.

103. "Six to the fifth power" is the same as 6^5.

104. "Seven squared" is the same as 7^2.

105. 2^5 is the same as $5 \cdot 5$.

106. 4^9 is the same as $4 \cdot 9$.

Insert grouping symbols (parentheses) so that each given expression evaluates to the given number.

107. $2 + 3 \cdot 6 - 2$; evaluates to 28

108. $2 + 3 \cdot 6 - 2$; evaluates to 20

109. $24 \div 3 \cdot 2 + 2 \cdot 5$; evaluates to 14

110. $24 \div 3 \cdot 2 + 2 \cdot 5$; evaluates to 15

△ **111.** A building contractor is bidding on a contract to install gutters on seven homes in a retirement community, all in the shape shown. To estimate the cost of materials, she needs to know the total perimeter of all seven homes. Find the total perimeter.

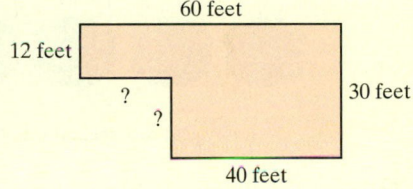

112. The building contractor from Exercise 111 plans to charge $4 per foot for installing vinyl gutters. Find the total charge for the seven homes given the total perimeter answer to Exercise 111.

Simplify.

113. $(7 + 2^4)^5 - (3^5 - 2^4)^2$

114. $25^3 \cdot (45 - 7 \cdot 5) \cdot 5$

115. Write an expression that simplifies to 5. Use multiplication, division, addition, subtraction, and at least one set of parentheses. Explain the process you would use to simplify the expression.

116. Explain why $2 \cdot 3^2$ is not the same as $(2 \cdot 3)^2$.

Chapter 1 Group Activity

Modeling Subtraction of Whole Numbers

A mathematical concept can be represented or modeled in many different ways. For instance, subtraction can be represented by the following symbolic model:

$$11 - 4$$

The following verbal models can also represent subtraction of these same quantities:

"Four subtracted from eleven" or
"Eleven take away four"

Physical models can also represent mathematical concepts. In these models, a number is represented by that many objects. For example, the number 5 can be represented by five pennies, squares, paper clips, tiles, or bottle caps.

A physical representation of the number 5

Take-Away Model for Subtraction: 11 − 4

- Start with 11 objects.
- Take 4 objects away.
- How many objects remain?

Start:

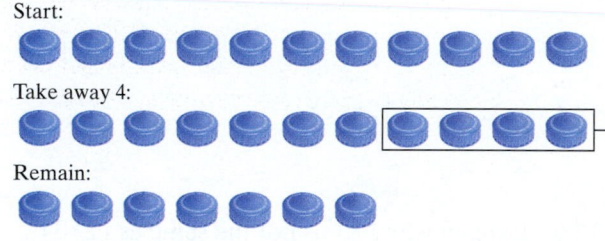

Take away 4:

Remain:

Comparison Model for Subtraction: 11 − 4

- Start with a set of 11 of one type of object and a set of 4 of another type of object.

- Make as many pairs that include one object of each type as possible.

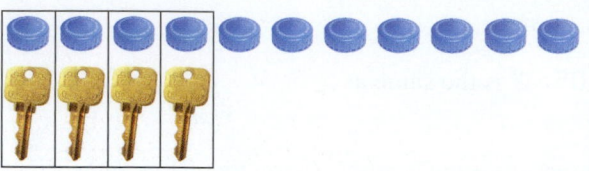

- How many more objects left are in the larger set?

Missing Addend Model for Subtraction: 11 − 4

- Start with 4 objects.
- Continue adding objects until a total of 11 is reached.
- How many more objects were needed to give a total of 11?

Start:

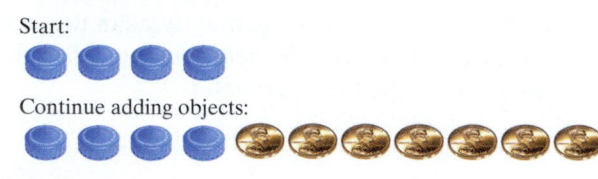

Continue adding objects:

Group Activity

Use an appropriate physical model for subtraction to solve each of the following problems. Explain your reasoning for choosing each model.

1. Sneha has assembled 12 computer components so far this shift. If her quota is 20 components, how many more components must she assemble to reach her quota?

2. Yuko has 14 daffodil bulbs to plant in her yard. She planted 5 bulbs in the front yard. How many bulbs does she have left for planting in the backyard?

3. Todd is 19 years old and his sister Tanya is 13 years old. How much older is Todd than Tanya?

Chapter 1 Vocabulary Check

Fill in each blank with one of the words or phrases listed below.

difference	area	square root
place value	factor	quotient
sum	whole numbers	perimeter

addend	divisor	minuend
subtrahend	exponent	digits
dividend	average	product

1. The _____ are 0, 1, 2, 3, ...

2. The _____ of a polygon is its distance around or the sum of the lengths of its sides.

3. The position of each digit in a number determines its _____.

4. A(n) _____ is a shorthand notation for repeated multiplication of the same factor.

5. To find the _____ of a rectangle, multiply length times width.

6. A(n) _____ of a number is one of two identical factors of the number.

7. The _____ used to write numbers are 0, 1, 2, 3, 4, 5, 6, 7, 8, and 9.

8. The _____ of a list of numbers is their sum divided by the number of numbers.

Use the facts below for Exercises 9 through 18.

$$2 \cdot 3 = 6 \quad 4 + 17 = 21 \quad 20 - 9 = 11 \quad 5\overline{)35}^{\,7}$$

9. The 5 above is called the _____.

10. The 35 above is called the _____.

11. The 7 above is called the _____.

12. The 3 above is called a(n) _____.

13. The 6 above is called the _____.

14. The 20 above is called the _____.

15. The 9 above is called the _____.

16. The 11 above is called the _____.

17. The 4 above is called a(n) _____.

18. The 21 above is called the _____.

Helpful Hint

▶ Are you preparing for your test? Don't forget to take the Chapter 1 Test on page 108. Then check your answers at the back of the text and use the Chapter Test Prep Videos to see the fully worked-out solutions to any of the exercises you want to review.

1 Chapter Highlights

Definitions and Concepts	Examples
Section 1.2 Place Value, Names for Numbers, and Reading Tables	
The **whole numbers** are 0, 1, 2, 3, 4, 5, ...	0, 14, 968, 5,268,619
The position of each digit in a number determines its **place value**. A place-value chart is shown next with the names of the periods given. 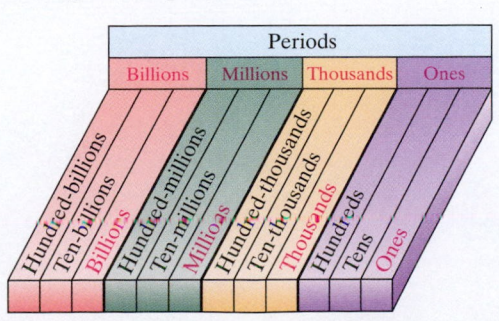	

(continued)

Definitions and Concepts	Examples

Section 1.2   Place Value, Names for Numbers, and Reading Tables (*continued*)

To write a whole number in words, write the number in each period followed by the name of the period. (The name of the ones period is not included.)

9,078,651,002 is written as nine billion, seventy-eight million, six hundred fifty-one thousand, two.

To write a whole number in standard form, write the number in each period, followed by a comma.

Four million, seven hundred six thousand, twenty-eight is written as 4,706,028.

Section 1.3   Adding Whole Numbers and Perimeter

To add whole numbers, add the digits in the ones place, then the tens place, then the hundreds place, and so on, carrying when necessary.

Find the sum:

$$
\begin{array}{r}
\overset{2\,1\,1}{2689} \leftarrow \text{addend} \\
1735 \leftarrow \text{addend} \\
+\ 662 \leftarrow \text{addend} \\
\hline
5086 \leftarrow \text{sum}
\end{array}
$$

The **perimeter** of a polygon is its distance around or the sum of the lengths of its sides.

△ Find the perimeter of the polygon shown.

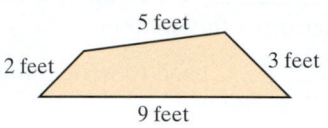

The perimeter is
5 feet + 3 feet + 9 feet + 2 feet = 19 feet.

Section 1.4   Subtracting Whole Numbers

To subtract whole numbers, subtract the digits in the ones place, then the tens place, then the hundreds place, and so on, borrowing when necessary.

Subtract:

$$
\begin{array}{r}
\overset{8\,15}{79\cancel{9}4} \leftarrow \text{minuend} \\
-5673 \leftarrow \text{subtrahend} \\
\hline
2281 \leftarrow \text{difference}
\end{array}
$$

Section 1.5   Rounding and Estimating

Rounding Whole Numbers to a Given Place Value

Step 1: Locate the digit to the right of the given place value.

Step 2: If this digit is 5 or greater, add 1 to the digit in the given place value and replace each digit to its right with 0.

Step 3: If this digit is less than 5, replace it and each digit to its right with 0.

Round 15,721 to the nearest thousand.

15,⑦21

Add 1 ⎯⎯ Replace with zeros.

Since the circled digit is 5 or greater, add 1 to the given place value and replace digits to its right with zeros.

15,721 rounded to the nearest thousand is 16,000.

Definitions and Concepts	**Examples**

Section 1.6   Multiplying Whole Numbers and Area

To multiply 73 and 58, for example, multiply 73 and 8, then 73 and 50. The sum of these partial products is the product of 73 and 58. Use the notation to the right.	$\begin{array}{r} 73 \leftarrow \text{factor} \\ \times\ 58 \leftarrow \text{factor} \\ \hline 584 \leftarrow 73 \times 8 \\ 3650 \leftarrow 73 \times 50 \\ \hline 4234 \leftarrow \text{product} \end{array}$

To find the **area** of a rectangle, multiply length times width.

△ Find the area of the rectangle shown.

11 meters

7 meters

$$\text{area of rectangle} = \text{length} \cdot \text{width}$$
$$= (11 \text{ meters})(7 \text{ meters})$$
$$= 77 \text{ square meters}$$

Section 1.7   Dividing Whole Numbers

Division Properties of 0

The quotient of 0 and any number (except 0) is 0.

The quotient of any number and 0 is not a number. We say that this quotient is undefined.

$$\frac{0}{5} = 0$$

$$\frac{7}{0} \text{ is undefined}$$

To divide larger whole numbers, use the process called **long division** as shown to the right.

$$\begin{array}{r} 507 \quad \text{R } 2 \leftarrow \text{quotient and remainder} \\ \text{divisor} \rightarrow 14\overline{)7100} \longleftarrow \text{dividend} \\ -70\downarrow \qquad\qquad 5(14) = 70 \\ \hline 10 \qquad\qquad \text{Subtract and bring down the 0.} \\ -0\downarrow \qquad\qquad 0(14) = 0 \\ \hline 100 \qquad\qquad \text{Subtract and bring down the 0.} \\ -98 \qquad\qquad 7(14) = 98 \\ \hline 2 \qquad\qquad \text{Subtract. The remainder is 2.} \end{array}$$

To check, see that $507 \cdot 14 + 2 = 7100$.

The **average** of a list of numbers is

$$\text{average} = \frac{\text{sum of numbers}}{\textit{number} \text{ of numbers}}$$

Find the average of 23, 35, and 38.

$$\text{average} = \frac{23 + 35 + 38}{3} = \frac{96}{3} = 32$$

Definitions and Concepts	Examples

Section 1.8 An Introduction to Problem Solving

Problem-Solving Steps

1. UNDERSTAND the problem.

2. TRANSLATE the problem.

3. SOLVE the problem.

4. INTERPRET the results.

Suppose that 225 tickets are sold for each performance of a play. How many tickets are sold for 5 performances?

1. UNDERSTAND. Read and reread the problem. Since we want the number of tickets for 5 performances, we multiply.

2. TRANSLATE.

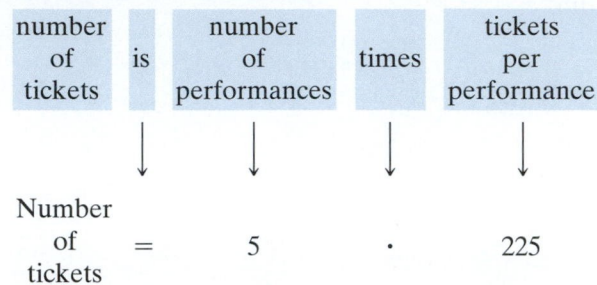

$$\text{Number of tickets} = 5 \cdot 225$$

3. SOLVE: See if the answer is reasonable by also estimating.

$$\begin{array}{r} \overset{12}{225} \\ \times\;\;5 \\ \hline 1125 \end{array} \text{ exact} \qquad \begin{array}{r} 200 \\ \times\;\;5 \\ \hline 1000 \end{array} \text{ estimate}$$

225 rounds to 200

4. INTERPRET. **Check** your work. The product is reasonable since 1125 is close to our estimated answer of 1000, and **state** your conclusion: There are 1125 tickets sold for 5 performances.

Section 1.9 Exponents, Square Roots, and Order of Operations

An **exponent** is a shorthand notation for repeated multiplication of the same factor.

A **square root** of a number is one of two identical factors of the number.

Order of Operations

1. Perform all operations within parentheses ( ), brackets [ ], or other grouping symbols such as square roots or fraction bars, starting with the innermost set.

2. Evaluate any expressions with exponents.

3. Multiply or divide in order from left to right.

4. Add or subtract in order from left to right.

The **area of a square** is $(\text{side})^2$.

$$3^4 = \underbrace{3 \cdot 3 \cdot 3 \cdot 3}_{\text{4 factors of 3}} = 81$$

base   exponent

$$\sqrt{36} = 6 \quad \text{because} \quad 6 \cdot 6 = 36$$
$$\sqrt{121} = 11 \quad \text{because} \quad 11 \cdot 11 = 121$$
$$\sqrt{0} = 0 \quad \text{because} \quad 0 \cdot 0 = 0$$

Simplify: $\dfrac{5 + 3^2}{2(7 - 6)}$

Simplify above and below the fraction bar separately.

$$\frac{5 + 3^2}{2(7 - 6)} = \frac{5 + 9}{2(1)} \quad \text{Evaluate } 3^2 \text{ above the fraction bar.}$$
$$\text{Subtract: } 7 - 6 \text{ below the fraction bar.}$$
$$= \frac{14}{2} \quad \text{Add.}$$
$$\text{Multiply.}$$
$$= 7 \quad \text{Divide.}$$

Find the area of a square with side length 9 inches.

$$\text{Area of the square} = (\text{side})^2$$
$$= (9 \text{ inches})^2$$
$$= 81 \text{ square inches}$$

Chapter 1   Review

(1.2) *Determine the place value of the digit 4 in each whole number.*

1. 7640

2. 46,200,120

Write each whole number in words.

3. 7640

4. 46,200,120

Write each whole number in expanded form.

5. 3158

6. 403,225,000

Write each whole number in standard form.

7. Eighty-one thousand, nine hundred

8. Six billion, three hundred four million

The following table shows the Internet and Facebook use of world regions as of June 2012. Use this table to answer Exercises 9 through 12.

World Region	Internet Users	Facebook Users
Africa	167,335,676	51,612,460
Asia	1,076,681,059	254,336,520
Europe	518,512,109	250,934,000
Middle East	90,000,455	23,811,620
North America	273,785,413	184,177,220
Latin America/ Caribbean	254,915,745	188,339,620
Oceania/Australia	24,287,919	14,614,780
(Source: Internet World Stats)		

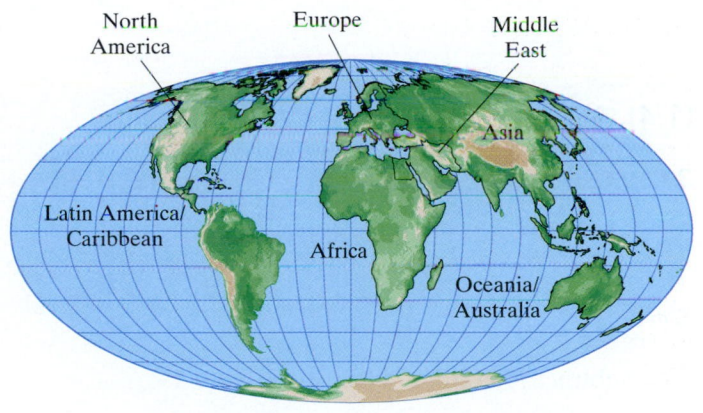

9. Find the number of Internet users in Europe.

10. Find the number of Facebook users in North America.

11. Which world region had the largest number of Facebook users?

12. Which world region had the smallest number of Internet users?

(1.3) *Add.*

13. 17 + 46

14. 28 + 39

15. 25 + 8 + 15

16. 27 + 9 + 41

17. 932 + 24

18. 819 + 21

19. 567 + 7383

20. 463 + 6787

21. 91 + 3623 + 497

22. 82 + 1647 + 238

Solve.

23. Find the sum of 86, 331, and 909.

24. Find the sum of 49, 529, and 308.

25. What is 26,481 increased by 865?

26. What is 38,556 increased by 744?

27. The distance from Chicago to New York City is 714 miles. The distance from New York City to New Delhi, India, is 7318 miles. Find the total distance from Chicago to New Delhi if traveling by air through New York City.

28. Susan Summerline earned salaries of $62,589, $65,340, and $69,770 during the years 2002, 2003, and 2004, respectively. Find her total earnings during those three years.

Find the perimeter of each figure.

 29.

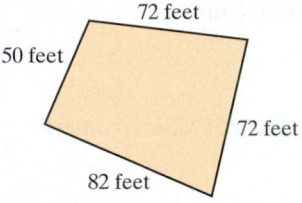

72 feet
50 feet
72 feet
82 feet

△ **30.**   11 kilometers   20 kilometers

35 kilometers

(1.4) *Subtract and then check.*

31. 93 − 79   **32.** 61 − 27   **33.** 462 − 397   **34.** 583 − 279   **35.** 4000 − 86   **36.** 8000 − 92

Solve.

37. Subtract 7965 from 25,862.

38. Subtract 4349 from 39,007.

39. Find the increase in population for San Antonio, Texas, from 2000 (population: 1,144,646) to 2012 (population: 1,382,951). (*Source:* U.S. Census Bureau)

40. Find the decrease in population for Detroit, Michigan, from 2000 (population: 951,270) to 2012 (population: 701,475). (*Source:* U.S. Census Bureau)

41. Bob Roma is proofreading the Yellow Pages for his county. If he has finished 315 pages of the total 712 pages, how many pages does he have left to proofread?

42. Shelly Winters bought a new car listed at $28,425. She received a discount of $1599 and a factory rebate of $1200. Find how much she paid for the car.

The following bar graph shows the monthly savings account balance for a freshman attending a local community college. Use this graph to answer Exercises 43 through 46.

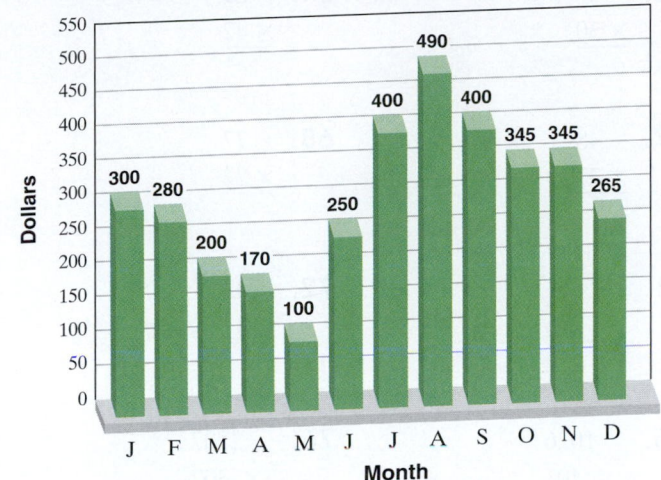

43. During what month was the balance the least?

44. During what month was the balance the greatest?

45. By how much did his balance decrease from February to April?

46. By how much did his balance increase from June to August?

(1.5) *Round to the given place.*

47. 93 to the nearest ten

48. 45 to the nearest ten

49. 467 to the nearest ten

50. 493 to the nearest hundred

51. 4832 to the nearest hundred

52. 57,534 to the nearest thousand

53. 49,683,712 to the nearest million

54. 768,542 to the nearest hundred-thousand

55. In 2012, 126,226,713 Americans cast a ballot in the presidential election. Round this number to the nearest million. (*Source:* CNN)

56. In 2011, there were 98,817 public elementary and secondary schools in the United States. Round this number to the nearest thousand. (*Source:* National Center for Education Statistics)

Estimate the sum or difference by rounding each number to the nearest hundred.

57. 4892 + 647 + 1876

58. 5925 − 1787

59. A group of students took a week-long driving trip and traveled 628, 290, 172, 58, 508, 445, and 383 miles on seven consecutive days. Round each distance to the nearest hundred to estimate the distance they traveled.

60. The estimated 2012 population of Houston, Texas, was 2,160,821, and for San Diego, California, it was 1,338,348. Round each number to the nearest hundred-thousand and estimate how much larger Houston is than San Diego. (*Source:* U.S. Census Bureau)

(1.6) *Multiply.*

61. $\begin{array}{r} 273 \\ \times\ \ 7 \\ \hline \end{array}$

62. $\begin{array}{r} 349 \\ \times\ \ 4 \\ \hline \end{array}$

63. $\begin{array}{r} 47 \\ \times\ 30 \\ \hline \end{array}$

64. $\begin{array}{r} 69 \\ \times\ 42 \\ \hline \end{array}$

65. 20(8)(5)

66. 25(9)(4)

67. $\begin{array}{r} 48 \\ \times\ 77 \\ \hline \end{array}$

68. $\begin{array}{r} 77 \\ \times\ 22 \\ \hline \end{array}$

69. $49 \cdot 49 \cdot 0$

70. $62 \cdot 88 \cdot 0$

71. $\begin{array}{r} 586 \\ \times\ 29 \\ \hline \end{array}$

72. $\begin{array}{r} 242 \\ \times\ 37 \\ \hline \end{array}$

73. $\begin{array}{r} 642 \\ \times\ 177 \\ \hline \end{array}$

74. $\begin{array}{r} 347 \\ \times\ 129 \\ \hline \end{array}$

75. $\begin{array}{r} 1026 \\ \times\ 401 \\ \hline \end{array}$

76. $\begin{array}{r} 2107 \\ \times\ 302 \\ \hline \end{array}$

77. $375 \cdot 1000$

78. $108 \cdot 1000$

79. $30 \cdot 400$

80. $50 \cdot 700$

81. $1700 \cdot 3000$

82. $1900 \cdot 4000$

Solve.

83. Find the product of 5 and 230.

84. Find the product of 6 and 820.

85. Multiply 9 and 12.

86. Multiply 8 and 14.

87. One ounce of Swiss cheese contains 8 grams of fat. How many grams of fat are in 3 ounces of Swiss cheese? (*Source: Home and Garden Bulletin No. 72,* U.S. Department of Agriculture)

88. The cost for a South Dakota resident to attend Black Hills State University full-time is $7617 per semester. Determine the cost for 20 students to attend full-time. (*Source:* Black Hills State University)

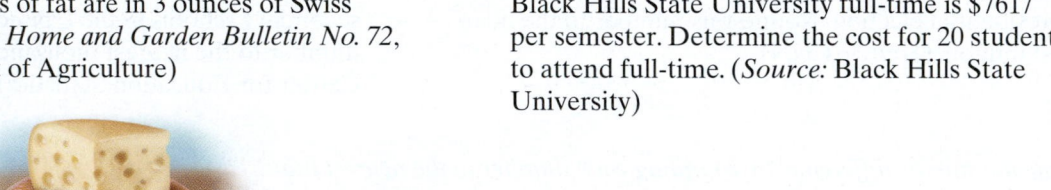

Find the area of each rectangle.

△**89.**

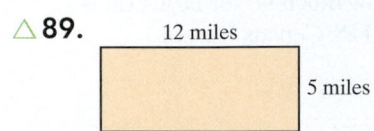

12 miles

5 miles

△**90.** 20 centimeters

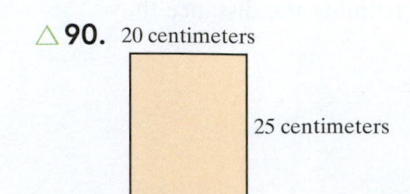

25 centimeters

(1.7) *Divide and then check.*

91. $\dfrac{18}{6}$

92. $\dfrac{36}{9}$

93. $42 \div 7$

94. $35 \div 5$

95. $27 \div 5$

96. $18 \div 4$

97. $16 \div 0$

98. $0 \div 8$

99. $9 \div 9$

100. $10 \div 1$

101. $0 \div 668$

102. $918 \div 0$

103. $5\overline{)167}$

104. $8\overline{)159}$

105. $26\overline{)626}$

106. $19\overline{)680}$

107. $47\overline{)23,792}$

108. $53\overline{)48,111}$

109. $207\overline{)578,291}$

110. $306\overline{)615,732}$

Solve.

111. Find the quotient of 92 and 5.

112. Find the quotient of 86 and 4.

113. One foot is 12 inches. Find how many feet there are in 5496 inches.

114. One mile is 1760 yards. Find how many miles there are in 22,880 yards.

115. Find the average of the numbers 76, 49, 32, and 47.

116. Find the average of the numbers 23, 85, 62, and 66.

(1.8) *Solve.*

117. A box can hold 24 cans of corn. How many boxes can be filled with 648 cans of corn?

118. If a ticket to a movie costs $6, how much do 32 tickets cost?

119. In 2012, U.S. companies spent $74 billion on television advertising. By comparison, U.S. companies spent only $69 billion on television advertising in 2011. How much more did U.S. companies spend on television advertising in 2012? (*Source:* Kantar Media)

120. The cost to banks when a person uses an ATM (Automatic Teller Machine) is 27¢. The cost to banks when a person deposits a check with a teller is 48¢ more. How much is this cost?

121. A golf pro orders shirts for the company sponsoring a local charity golfing event. Shirts size large cost $32 while shirts size extra-large cost $38. If 15 large shirts and 11 extra-large shirts are ordered, find the cost.

122. Two rectangular pieces of land are purchased: one that measures 65 feet by 110 feet and one that measures 80 feet by 200 feet. Find the total area of land purchased. (*Hint:* Find the area of each rectangle, then add.)

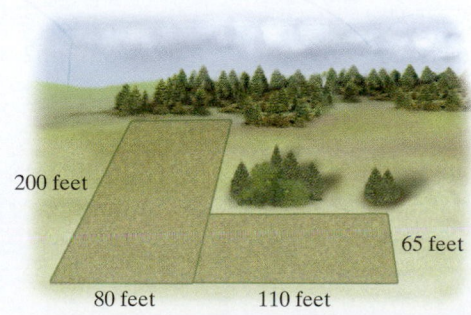

200 feet

65 feet

80 feet   110 feet

(1.9) *Simplify.*

123. 7^2 **124.** 5^3 **125.** $5 \cdot 3^2$ **126.** $4 \cdot 10^2$

127. $18 \div 3 + 7$ **128.** $12 - 8 \div 4$ **129.** $\dfrac{5(6^2 - 3)}{3^2 + 2}$ **130.** $\dfrac{7(16 - 8)}{2^3}$

131. $48 \div 8 \cdot 2$ **132.** $27 \div 9 \cdot 3$

133. $2 + 3[1^5 + (20 - 17) \cdot 3] + 5 \cdot 2$ **134.** $21 - [2^4 - (7 - 5) - 10] + 8 \cdot 2$

Simplify. (These exercises contain square roots.)

135. $\sqrt{81}$ **136.** $\sqrt{4}$ **137.** $\sqrt{1}$ **138.** $\sqrt{0}$

139. $4 \cdot \sqrt{25} - 2 \cdot 7$ **140.** $8 \cdot \sqrt{49} - 3 \cdot 9$

141. $(\sqrt{36} - \sqrt{16})^3 \cdot [10^2 \div (3 + 17)]$ **142.** $(\sqrt{49} - \sqrt{25})^3 \cdot [9^2 \div (2 + 7)]$

143. $\dfrac{5 \cdot 7 - 3 \cdot \sqrt{25}}{2(\sqrt{121} - 3^2)}$ **144.** $\dfrac{4 \cdot 8 - 1 \cdot \sqrt{121}}{3(\sqrt{81} - 2^3)}$

Find the area of each square.

△**145.** A square with side length of 7 meters.   △**146.**

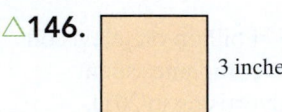

3 inches

Mixed Review

Perform the indicated operations.

147. $375 - 68$ **148.** $729 - 47$ **149.** 723×3 **150.** 629×4

151. $264 + 39 + 598$ **152.** $593 + 52 + 766$ **153.** $13\overline{)5962}$ **154.** $18\overline{)4267}$

155. 1968×36 **156.** 5324×18 **157.** $2000 - 356$ **158.** $9000 - 519$

Round to the given place.

159. 736 to the nearest ten

160. 258,371 to the nearest thousand

161. 1999 to the nearest hundred

162. 44,499 to the nearest ten thousand

Write each whole number in words.

163. 36,911

164. 154,863

Write each whole number in standard form.

165. Seventy thousand, nine hundred forty-three

166. Forty-three thousand, four hundred one

Simplify.

167. 4^3

168. 5^3

169. $\sqrt{144}$

170. $\sqrt{100}$

171. $24 \div 4 \cdot 2$

172. $\sqrt{256} - 3 \cdot 5$

173. $\dfrac{8(7-4)-10}{4^2-3^2}$

174. $\dfrac{(15+\sqrt{9}) \cdot (8-5)}{2^3+1}$

Solve.

175. 36 divided by 9 is what number?

176. What is the product of 2 and 12?

177. 16 increased by 8 is what number?

178. 7 subtracted from 21 is what number?

The following table shows the 2012 and 2013 average Major League Baseball salaries (rounded to the nearest thousand) for the five teams with the largest payrolls for 2013. Use this table to answer Exercises 179 and 180. (Source: CBSSports.com, Associated Press)

Team	2013 Average Salary	2012 Average Salary
Los Angeles Dodgers	$7,469,000	$3,264,000
New York Yankees	$7,151,000	$6,256,000
Philadelphia Phillies	$6,125,000	$5,798,000
Detroit Tigers	$5,708,000	$4,561,000
Boston Red Sox	$5,022,000	$5,094,000

179. How much more was the average salary for a Los Angeles Dodgers player in 2013 than in 2012?

180. How much less was the average Boston Red Sox salary than the average New York Yankee salary in 2013?

181. A manufacturer of drinking glasses ships his delicate stock in special boxes that can hold 32 glasses. If 1714 glasses are manufactured, how many full boxes are filled? Are there any glasses left over?

182. A teacher orders 2 small white boards for $27 each and 8 boxes of dry erase pens for $4 each. What is her total bill before taxes?

Chapter 1 — Test

Step-by-step test solutions are found on the Chapter Test Prep Videos. Where available: **MyMathLab®** or **You Tube**

Answers

Simplify.

1. Write 82,426 in words.

2. Write "four hundred two thousand, five hundred fifty" in standard form.

1. _____

2. _____

3. $59 + 82$

4. $600 - 487$

5. $\begin{array}{r} 496 \\ \times \ 30 \\ \hline \end{array}$

3. _____

4. _____

5. _____

6. $52{,}896 \div 69$

7. $2^3 \cdot 5^2$

8. $\sqrt{4} \cdot \sqrt{25}$

6. _____

7. _____

8. _____

9. $0 \div 49$

10. $62 \div 0$

11. $\left(2^4 - 5\right) \cdot 3$

9. _____

10. _____

11. _____

12. $16 + 9 \div 3 \cdot 4 - 7$

13. $\dfrac{64 \div 8 \cdot 2}{(\sqrt{9} - \sqrt{4})^2 + 1}$

12. _____

13. _____

14. $2\left[(6 - 4)^2 + (22 - 19)^2\right] + 10$

15. $5698 \cdot 1000$

14. _____

15. _____

16. $8000 \cdot 1400$

17. Round 52,369 to the nearest thousand.

16. _____

17. _____

18. _____

Estimate each sum or difference by rounding each number to the nearest hundred.

19. _____

18. $6289 + 5403 + 1957$

19. $4267 - 2738$

Solve.

20. Subtract 15 from 107.

21. Find the sum of 15 and 107.

22. Find the product of 15 and 107.

23. Find the quotient of 107 and 15.

24. Twenty-nine cans of Sherwin-Williams paint cost $493. How much was each can?

25. Jo McElory is looking at two new refrigerators for her apartment. One costs $599 and the other costs $725. How much more expensive is the higher-priced one?

26. One tablespoon of white granulated sugar contains 45 calories. How many calories are in 8 tablespoons of white granulated sugar? (*Source: Home and Garden Bulletin No. 72, U.S. Department of Agriculture*)

27. A small business owner recently ordered 16 digital cameras that cost $430 each and 5 printers that cost $205 each. Find the total cost for these items.

Find the perimeter and the area of each figure.

△**28.**

| Square | 5 centimeters |

△**29.**

20 yards

| Rectangle | 10 yards |

20. _____

21. _____

22. _____

23. _____

24. _____

25. _____

26. _____

27. _____

28. _____

29. _____

Multiplying and Dividing Fractions

Plateau

Fin: A thin wall of rock

Window: A natural hole

Hoodoo: A tall, thin tower of rock

Fractions are numbers, and like whole numbers, they can be added, subtracted, multiplied, and divided. Fractions are very useful and appear frequently in everyday language, in common phrases like "half an hour," "quarter of a pound," and "third of a cup." This chapter introduces the concept of fractions, presents some basic vocabulary, and demonstrates how to multiply and divide fractions.

What Are Hoodoos, and How Are They Formed?

A hoodoo is a tall thin tower of sedimentary rock that is topped by a piece of harder stone. The diagrams above show how a hoodoo is formed, with the two main eroding processes being rain and also freeze/thaw cycles. In Bryce Canyon, there are over 200 freeze/thaw cycles each year that slowly pry open cracks and create many hoodoos.

Hoodoos can be anywhere from 5 to 150 feet tall. Unfortunately, hoodoos don't last long because the erosion that formed them continues. The average rate of erosion is 2 to 4 feet every 100 years.

The National Park Service (NPS) is charged with the enormous task of managing and protecting our national resources such as Bryce Canyon. In Section 2.3 and throughout Chapter 3, we will explore fractions relating to the various types of national parks and monuments that are protected by the NPS.

Overnight Stays at National Parks

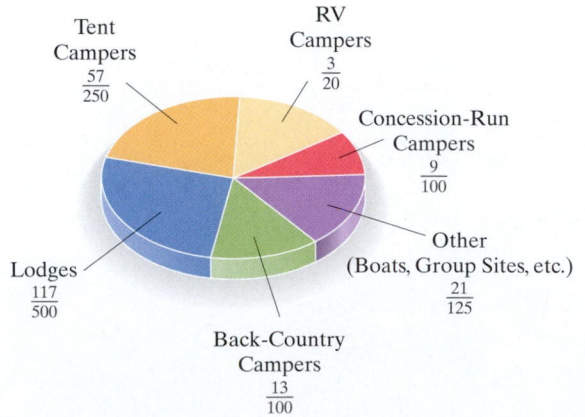

Tent Campers $\frac{57}{250}$

RV Campers $\frac{3}{20}$

Concession-Run Campers $\frac{9}{100}$

Other (Boats, Group Sites, etc.) $\frac{21}{125}$

Lodges $\frac{117}{500}$

Back-Country Campers $\frac{13}{100}$

Source: National Park Service

Introduction to Fractions and Mixed Numbers

Objective A Identifying Numerators and Denominators and Reviewing Division Properties of 0 and 1

Whole numbers are used to count whole things or units, such as cars, horses, dollars, and people. To refer to a part of a whole, fractions can be used. Here are some examples of **fractions.** Study these examples for a moment.

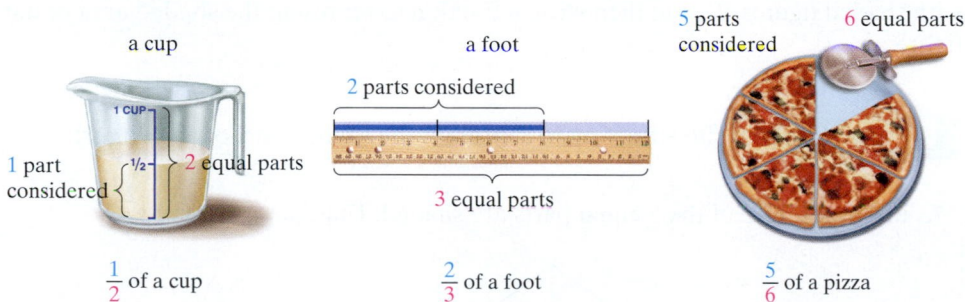

a cup

1 part considered ½ 2 equal parts

$\frac{1}{2}$ of a cup

a foot

2 parts considered

3 equal parts

$\frac{2}{3}$ of a foot

5 parts considered   6 equal parts

$\frac{5}{6}$ of a pizza

In a fraction, the top number is called the **numerator** and the bottom number is called the **denominator.** The bar between the numbers is called the **fraction bar.**

Names	Fraction	Meaning
numerator $\longrightarrow$	5	$\longleftarrow$ number of parts being considered
denominator $\longrightarrow$	6	$\longleftarrow$ number of equal parts in the whole

Examples Identify the numerator and the denominator of each fraction.

1. $\frac{3}{7}$ ← numerator
← denominator

2. $\frac{13}{5}$ ← numerator
← denominator

Work Practice 1–2

Helpful Hint

Notice the fraction $\frac{11}{1} = 11$, or also $11 = \frac{11}{1}$.

Before we continue further, don't forget from Section 1.7 that the fraction bar indicates division. Let's review some division properties of 1 and 0.

$\frac{9}{9} = 1$ because $1 \cdot 9 = 9$ $\frac{11}{1} = 11$ because $11 \cdot 1 = 11$

$\frac{0}{6} = 0$ because $0 \cdot 6 = 0$ $\frac{6}{0}$ *is undefined* because there is no number that when multiplied by 0 gives 6.

In general, we can say the following.

Let n be any whole number except 0.

$\frac{n}{n} = 1$ $\frac{0}{n} = 0$

$\frac{n}{1} = n$ $\frac{n}{0}$ is undefined.

Practice 1–2

Identify the numerator and the denominator of each fraction.

1. $\frac{9}{2}$ **2.** $\frac{10}{17}$

Practice 3–6

Simplify.

3. $\dfrac{0}{2}$ **4.** $\dfrac{8}{8}$

5. $\dfrac{4}{0}$ **6.** $\dfrac{20}{1}$

Examples Simplify.

3. $\dfrac{5}{5} = 1$ **4.** $\dfrac{0}{7} = 0$ **5.** $\dfrac{10}{1} = 10$ **6.** $\dfrac{3}{0}$ is undefined

■ **Work Practice 3–6**

Objective B Writing Fractions to Represent Parts of Figures or Real-Life Data ▶

One way to become familiar with the concept of fractions is to visualize fractions with shaded figures. We can then write a fraction to represent the shaded area of the figure.

Practice 7–8

Write a fraction to represent the shaded part of each figure.

7.

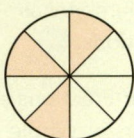

8.

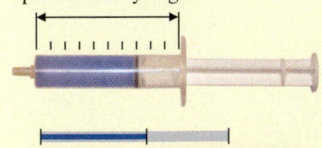

Examples Write a fraction to represent the shaded part of each figure.

7. In this figure, 2 of the 5 equal parts are shaded. Thus, the fraction is $\dfrac{2}{5}$.

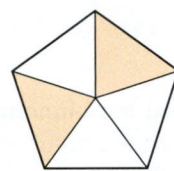

$\dfrac{2}{5}$ ← number of parts shaded
      ← number of equal parts

8. In this figure, 3 of the 10 rectangles are shaded. Thus, the fraction is $\dfrac{3}{10}$.

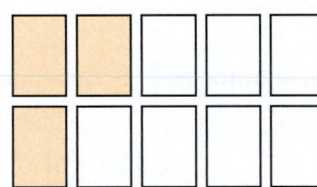

$\dfrac{3}{10}$ ← number of parts shaded
      ← number of equal parts

■ **Work Practice 7–8**

Practice 9–10

Write a fraction to represent the part of the whole shown.

9. Just consider this part of the syringe

10.

Examples Write a fraction to represent the shaded part of the diagram.

9.

The fraction is $\dfrac{3}{10}$.

10.

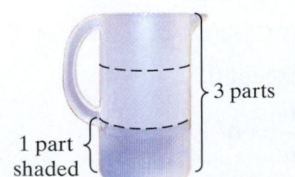

The fraction is $\dfrac{1}{3}$.

■ **Work Practice 9–10**

Answers

3. 0 4. 1 5. undefined 6. 20

7. $\dfrac{3}{8}$ 8. $\dfrac{1}{6}$ 9. $\dfrac{7}{10}$ 10. $\dfrac{9}{16}$

Examples Draw a figure and then shade a part of it to represent each fraction.

11. $\frac{5}{6}$ of a figure

We will use a geometric figure such as a rectangle. Since the denominator is 6, we divide it into 6 equal parts. Then we shade 5 of the equal parts.

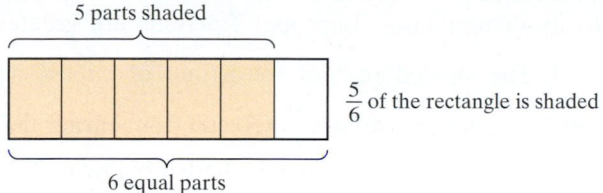

5 parts shaded

$\frac{5}{6}$ of the rectangle is shaded

6 equal parts

12. $\frac{3}{8}$ of a figure

If you'd like, our figure can consist of 8 triangles of the same size. We will shade 3 of the triangles.

3 triangles shaded

$\frac{3}{8}$ of the figure or diagram is shaded

8 triangles

■ **Work Practice 11–12**

✓**Concept Check** If  represents $\frac{6}{7}$ of a whole diagram, sketch the whole diagram.

Example 13 Writing Fractions from Real-Life Data

Of the eight planets in our solar system (Pluto is now a dwarf planet), three are closer to the Sun than Mars. What fraction of the planets are closer to the Sun than Mars?

Solution: The fraction of planets closer to the Sun than Mars is:

$\frac{3}{8}$ ← number of planets closer
     ← number of planets in our solar system

Thus, $\frac{3}{8}$ of the planets in our solar system are closer to the Sun than Mars.

■ **Work Practice 13**

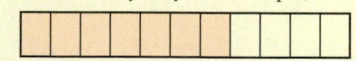

Objective C Identifying Proper Fractions, Improper Fractions, and Mixed Numbers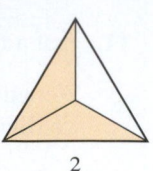

A **proper fraction** is a fraction whose numerator is less than its denominator. Proper fractions are less than 1. For example, the shaded portion of the triangle's area is represented by $\frac{2}{3}$.

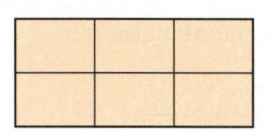

$\frac{2}{3}$

An **improper fraction** is a fraction whose numerator is greater than or equal to its denominator. Improper fractions are greater than or equal to 1. The shaded part of the group of circles' area below is $\frac{9}{4}$. The shaded part of the rectangle's area is $\frac{6}{6}$. (Recall from earlier that $\frac{6}{6}$ simplifies to 1 and notice that 1 whole figure or rectangle is shaded below.)

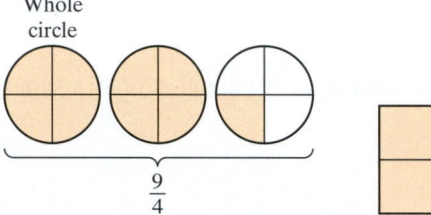

Whole circle

$\frac{9}{4}$

$\frac{6}{6}$

A **mixed number** contains a whole number and a fraction. Mixed numbers are greater than 1. Above, we wrote the shaded part of the group of circles below as the improper fraction $\frac{9}{4}$. Now let's write the shaded part as a mixed number. The shaded part of the group of circles' area is $2\frac{1}{4}$. (Read "two and one-fourth.")

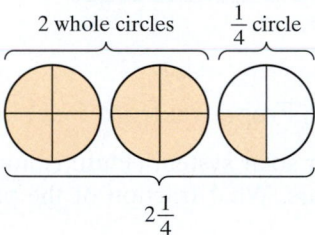

2 whole circles $\frac{1}{4}$ circle

$2\frac{1}{4}$

> **Helpful Hint**
>
> The mixed number $2\frac{1}{4}$ represents $2 + \frac{1}{4}$.

Practice 14

Identify each number as a proper fraction, improper fraction, or mixed number.

a. $\frac{5}{8}$ **b.** $\frac{7}{7}$

c. $\frac{14}{13}$ **d.** $\frac{13}{14}$

e. $5\frac{1}{4}$ **f.** $\frac{100}{49}$

Answers

14. a. proper fraction   **b.** improper fraction   **c.** improper fraction   **d.** proper fraction   **e.** mixed number   **f.** improper fraction

Example 14 Identify each number as a proper fraction, improper fraction, or mixed number.

a. $\frac{6}{7}$ is a proper fraction   **b.** $\frac{13}{12}$ is an improper fraction

c. $\frac{2}{2}$ is an improper fraction   **d.** $\frac{99}{101}$ is a proper fraction

e. $1\frac{7}{8}$ is a mixed number   **f.** $\frac{93}{74}$ is an improper fraction

■ **Work Practice 14**

Examples Represent the shaded part of each figure group's area as both an improper fraction and a mixed number.

15.

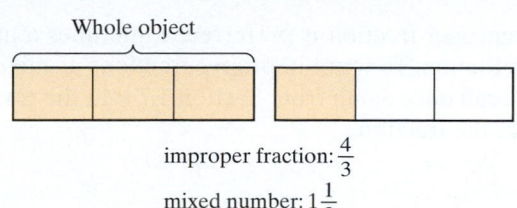

improper fraction: $\frac{4}{3}$

mixed number: $1\frac{1}{3}$

16.

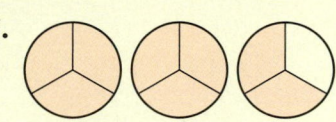

improper fraction: $\frac{5}{2}$

mixed number: $2\frac{1}{2}$

■ **Work Practice 15–16**

✓**Concept Check** If you were to estimate $2\frac{1}{8}$ by a whole number, would you choose 2 or 3? Why?

Objective D Writing Mixed Numbers as Improper Fractions ▶

Notice from Examples 15 and 16 that mixed numbers and improper fractions were both used to represent the shaded area of the figure groups. For example,

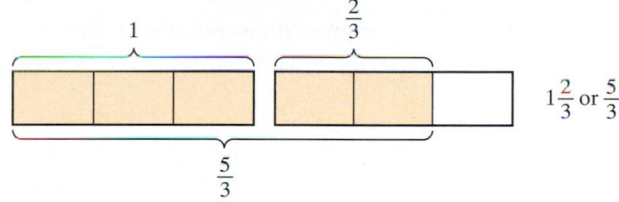

$1\frac{2}{3}$ or $\frac{5}{3}$

The following steps may be used to write a mixed number as an improper fraction:

Writing a Mixed Number as an Improper Fraction

To write a mixed number as an improper fraction:

Step 1: Multiply the denominator of the fraction by the whole number.

Step 2: Add the numerator of the fraction to the product from Step 1.

Step 3: Write the sum from Step 2 as the numerator of the improper fraction over the original denominator.

For example,

$$1\frac{2}{3} = \frac{\overset{\text{Step 1}}{3 \cdot 1} + \overset{\text{Step 2}}{2}}{3} = \frac{3 + 2}{3} = \frac{5}{3}$$

Step 3

Example 17 Write each as an improper fraction.

a. $4\frac{2}{9} = \frac{9 \cdot 4 + 2}{9} = \frac{36 + 2}{9} = \frac{38}{9}$

b. $1\frac{8}{11} = \frac{11 \cdot 1 + 8}{11} = \frac{11 + 8}{11} = \frac{19}{11}$

■ **Work Practice 17**

Objective E Writing Improper Fractions as Mixed Numbers or Whole Numbers ▶

Just as there are times when an improper fraction is preferred, sometimes a mixed or a whole number better suits a situation. To write improper fractions as mixed or whole numbers, we use division. Recall once again from Section 1.7 that the fraction bar means division. This means that the fraction

$$\dfrac{5}{3} \quad \substack{\text{numerator} \\ \text{denominator}} \qquad \text{means} \qquad 3\overline{)5} \; \substack{\uparrow\quad\uparrow \\ \text{numerator} \\ \text{denominator}}$$

Writing an Improper Fraction as a Mixed Number or a Whole Number

To write an improper fraction as a mixed number or a whole number:

Step 1: Divide the denominator into the numerator.

Step 2: The whole number part of the mixed number is the quotient. The fraction part of the mixed number is the remainder over the original denominator.

$$\text{quotient} \dfrac{\text{remainder}}{\text{original denominator}}$$

For example,

$$\underset{\text{Step 1}}{\dfrac{5}{3}} : \begin{array}{r} 1 \\ 3\overline{)5} \\ -3 \\ \hline 2 \end{array} \qquad \underset{\text{Step 2}}{\dfrac{5}{3}} = 1\dfrac{2}{3} \begin{array}{l} \leftarrow \text{remainder} \\ \leftarrow \text{original denominator} \end{array}$$

quotient

Practice 18

Write each as a mixed number or a whole number.

a. $\dfrac{9}{5}$ **b.** $\dfrac{23}{9}$ **c.** $\dfrac{48}{4}$

d. $\dfrac{62}{13}$ **e.** $\dfrac{51}{7}$ **f.** $\dfrac{21}{20}$

Example 18 Write each as a mixed number or a whole number.

a. $\dfrac{30}{7}$ **b.** $\dfrac{16}{15}$ **c.** $\dfrac{84}{6}$

Solution:

a. $\dfrac{30}{7}$: $\begin{array}{r} 4 \\ 7\overline{)30} \\ -28 \\ \hline 2 \end{array}$ $\dfrac{30}{7} = 4\dfrac{2}{7}$

b. $\dfrac{16}{15}$: $\begin{array}{r} 1 \\ 15\overline{)16} \\ -15 \\ \hline 1 \end{array}$ $\dfrac{16}{15} = 1\dfrac{1}{15}$

c. $\dfrac{84}{6}$: $\begin{array}{r} 14 \\ 6\overline{)84} \\ -6 \\ \hline 24 \\ -24 \\ \hline 0 \end{array}$ $\dfrac{84}{6} = 14$ Since the remainder is 0, the result is the whole number 14.

> **Helpful Hint** When the remainder is 0, the improper fraction is a whole number. For example, $\dfrac{92}{4} = 23$.
>
> $\begin{array}{r} 23 \\ 4\overline{)92} \\ -8 \\ \hline 12 \\ -12 \\ \hline 0 \end{array}$

■ **Work Practice 18**

Answers

18. a. $1\dfrac{4}{5}$ **b.** $2\dfrac{5}{9}$ **c.** 12   **d.** $4\dfrac{10}{13}$

e. $7\dfrac{2}{7}$ **f.** $1\dfrac{1}{20}$

Vocabulary, Readiness & Video Check

Use the choices below to fill in each blank.

improper	fraction	proper	is undefined	mixed number	= 0
greater than or equal to 1	denominator	= 1	less than 1	numerator	

1. The number $\frac{17}{31}$ is called a(n) _____. The number 31 is called its _____ and 17 is called its _____.

2. If we simplify each fraction, $\frac{9}{9}$ _____, $\frac{0}{4}$ _____, and we say $\frac{4}{0}$ _____.

3. The fraction $\frac{8}{3}$ is called a(n) _____ fraction, the fraction $\frac{3}{8}$ is called a(n) _____ fraction, and $10\frac{3}{8}$ is called a(n) _____ .

4. The value of an improper fraction is always _____, and the value of a proper fraction is always _____.

Martin-Gay Interactive Videos

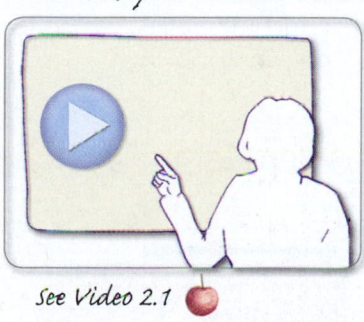

See Video 2.1

Watch the section lecture video and answer the following questions.

Objective A 5. From ▢ Example 3, what can you conclude about any fraction where the numerator and denominator are the same nonzero number? ▷

Objective B 6. In ▢ Example 8, what does the denominator 50 represent? ▷

Objective C 7. In ▢ Example 11, there are two shapes in the diagram, so why do the representative fractions have a denominator 3? ▷

Objective D 8. Complete this statement based on the lecture before ▢ Example 12: The operation of _____ is understood in a mixed number notation; for example, $1\frac{1}{3}$ means 1 _____ $\frac{1}{3}$.

Objective E 9. From the lecture before ▢ Example 15, what operation is used to write an improper fraction as a mixed number? ▷

2.1 Exercise Set MyMathLab ▷

Objectives A C Mixed Practice *Identify the numerator and the denominator of each fraction and identify each fraction as proper or improper. See Examples 1, 2, and 14.*

▷ 1. $\frac{1}{2}$

2. $\frac{1}{4}$

▷ 3. $\frac{10}{3}$

4. $\frac{53}{21}$

5. $\frac{15}{15}$

6. $\frac{26}{26}$

Objective A *Simplify. See Examples 3 through 6.*

7. $\frac{21}{21}$ 8. $\frac{14}{14}$ 9. $\frac{5}{0}$ 10. $\frac{1}{0}$ 11. $\frac{13}{1}$ 12. $\frac{14}{1}$

13. $\frac{0}{20}$ 14. $\frac{0}{17}$ 15. $\frac{10}{0}$ 16. $\frac{0}{18}$ 17. $\frac{16}{1}$ 18. $\frac{18}{18}$

Objective B *Write a fraction to represent the shaded part of each. See Examples 7 through 10.*

19. 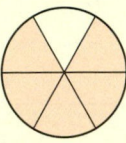   20. 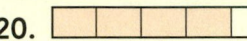   21.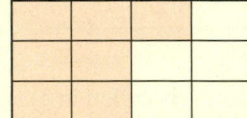

22. 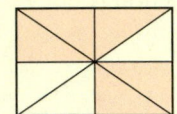   23. 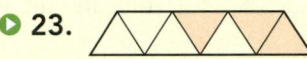   24.

25. 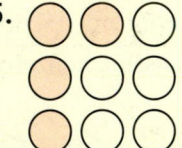   26. 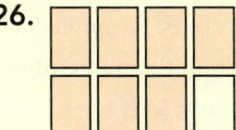   27.

28.    29.    30.

Draw and shade a part of a figure to represent each fraction. See Examples 11 and 12.

31. $\frac{1}{5}$ of a figure   32. $\frac{1}{16}$ of a figure   33. $\frac{7}{8}$ of a figure   34. $\frac{3}{5}$ of a figure

35. $\frac{6}{7}$ of a figure   36. $\frac{7}{9}$ of a figure   37. $\frac{4}{4}$ of a figure   38. $\frac{6}{6}$ of a figure

Write each fraction. See Example 13.

39. Of the 131 students at a small private school, 42 are freshmen. What fraction of the students are freshmen?

40. Of the 63 employees at a new biomedical engineering firm, 22 are men. What fraction of the employees are men?

41. Use Exercise 39 to answer a and b.

   a. How many students are *not* freshmen?

   b. What fraction of the students are *not* freshmen?

42. Use Exercise 40 to answer a and b.

   a. How many of the employees are women?

   b. What fraction of the employees are women?

43. As of the beginning of 2014, the United States has had 44 different presidents. A total of seven U.S. presidents were born in the state of Ohio, second only to the state of Virginia in producing U.S. presidents. What fraction of U.S. presidents were born in Ohio? (*Source: World Almanac and Book of Facts*)

44. Of the eight planets in our solar system, four have days that are longer than the 24-hour Earth day. What fraction of the planets have longer days than Earth has? (*Source:* National Space Science Data Center)

45. Hurricane Sandy, which struck the East Coast in October 2012, was the largest Atlantic hurricane ever documented. Sandy was just one of 19 named tropical storms that formed during the 2012 Atlantic hurricane season. A total of 10 of these tropical storms turned into hurricanes. What fraction of the 2012 Atlantic tropical storms escalated into hurricanes? (*Source:* National Oceanic and Atmospheric Administration)

46. There are 12 inches in a foot. What fractional part of a foot does 5 inches represent?

47. There are 31 days in the month of March. What fraction of the month does 11 days represent?

Mon.	Tue.	Wed.	Thu.	Fri.	Sat.	Sun.
					1	2
3	4	5	6	7	8	9
10	11	12	13	14	15	16
17	18	19	20	21	22	23
24	25	26	27	28	29	30
31						

48. There are 60 minutes in an hour. What fraction of an hour does 37 minutes represent?

49. In a basic college mathematics class containing 31 students, there are 18 freshmen, 10 sophomores, and 3 juniors. What fraction of the class is sophomores?

50. In a sports team with 20 children, there are 9 boys and 11 girls. What fraction of the team is boys?

51. Thirty-three out of the fifty total states in the United States contain federal Indian reservations.
   a. What fraction of the states contain federal Indian reservations?
   b. How many states do not contain federal Indian reservations?
   c. What fraction of the states do not contain federal Indian reservations? (*Source:* Tiller Research, Inc., Albuquerque, NM)

52. Consumer fireworks are legal in 46 out of the 50 total states in the United States. (*Source:* USA.gov)
   a. In what fraction of the states are consumer fireworks legal?
   b. In how many states are consumer fireworks illegal?
   c. In what fraction of the states are consumer fireworks illegal? (*Source:* United States Fireworks Safety Council)

53. A bag contains 50 red or blue marbles. If 21 marbles are blue,
   a. What *fraction* of the marbles are blue?
   b. How many marbles are red?
   c. What *fraction* of the marbles are red?

54. An art dealer is taking inventory. His shop contains a total of 37 pieces, which are all sculptures, watercolor paintings, or oil paintings. If there are 15 watercolor paintings and 17 oil paintings, answer each question.
   a. What fraction of the inventory is watercolor paintings?
   b. What fraction of the inventory is oil paintings?
   c. How many sculptures are there?
   d. What fraction of the inventory is sculptures?

Objective **C**   *Write the shaded area in each figure group as (a) an improper fraction and (b) a mixed number. See Examples 15 and 16.*

55.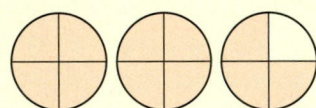

56.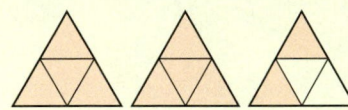

57.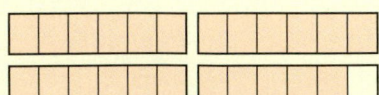

58.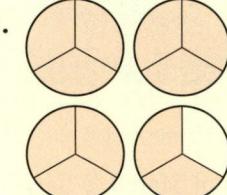

59.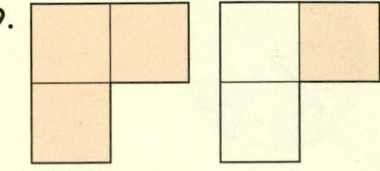

60.

61.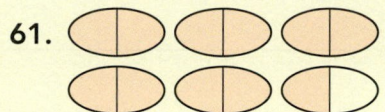

62.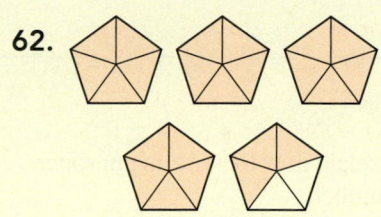

Objective D *Write each mixed number as an improper fraction. See Example 17.*

63. $2\frac{1}{3}$

64. $6\frac{3}{4}$

65. $3\frac{3}{5}$

66. $2\frac{5}{9}$

67. $6\frac{5}{8}$

68. $7\frac{3}{8}$

69. $2\frac{11}{15}$

70. $1\frac{13}{17}$

71. $11\frac{6}{7}$

72. $12\frac{2}{5}$

73. $6\frac{6}{13}$

74. $8\frac{9}{10}$

75. $4\frac{13}{24}$

76. $5\frac{17}{25}$

77. $17\frac{7}{12}$

78. $12\frac{7}{15}$

79. $9\frac{7}{20}$

80. $10\frac{14}{27}$

81. $2\frac{51}{107}$

82. $3\frac{27}{125}$

83. $166\frac{2}{3}$

84. $114\frac{2}{7}$

Objective E *Write each improper fraction as a mixed number or a whole number. See Example 18.*

85. $\frac{17}{5}$

86. $\frac{13}{7}$

87. $\frac{37}{8}$

88. $\frac{64}{9}$

89. $\frac{47}{15}$

90. $\frac{65}{12}$

91. $\frac{46}{21}$

92. $\frac{67}{17}$

93. $\frac{198}{6}$

94. $\frac{112}{7}$

95. $\frac{225}{15}$

96. $\frac{196}{14}$

97. $\frac{200}{3}$

98. $\frac{300}{7}$

99. $\frac{247}{23}$

100. $\frac{437}{53}$

101. $\frac{319}{18}$

102. $\frac{404}{21}$

103. $\frac{182}{175}$

104. $\frac{149}{143}$

105. $\frac{737}{112}$

106. $\frac{901}{123}$

Review

Simplify. See Section 1.9.

107. 3^2

108. 4^3

109. 5^3

110. 3^4

Write each using exponents.

111. $7 \cdot 7 \cdot 7 \cdot 7 \cdot 7$

112. $5 \cdot 5 \cdot 5 \cdot 5$

113. $2 \cdot 2 \cdot 2 \cdot 3$

114. $4 \cdot 4 \cdot 10 \cdot 10 \cdot 10$

Concept Extensions

Write each fraction.

115. In your own words, explain how to write an improper fraction as a mixed number.

116. In your own words, explain how to write a mixed number as an improper fraction.

Identify the larger fraction for each pair.

117. $\frac{1}{2}$ or $\frac{2}{3}$ (*Hint:* Represent each fraction by the shaded part of equivalent figures. Then compare the shaded areas.)

118. $\frac{7}{4}$ or $\frac{3}{5}$ (*Hint:* Identify each as a proper fraction or an improper fraction.)

Solve. See the first Concept Check in this section.

119. If ◯◯◯◯ represents $\frac{4}{9}$ of a whole diagram, sketch the whole diagram.

120. If △△ represents $\frac{1}{3}$ of a whole diagram, sketch the whole diagram.

121. IKEA Group employs workers in four different regions worldwide, as shown on the bar graph. What fraction of IKEA employees work in the North American region? (*Source:* IKEA Group)

122. The Public Broadcasting Service (PBS) provides programming to the noncommercial public TV stations of the United States. The bar graph shows a breakdown of the public television licensees by type. Each licensee operates one or more PBS member TV stations. What fraction of the public television licensees are universities or colleges? (*Source:* The Public Broadcasting Service)

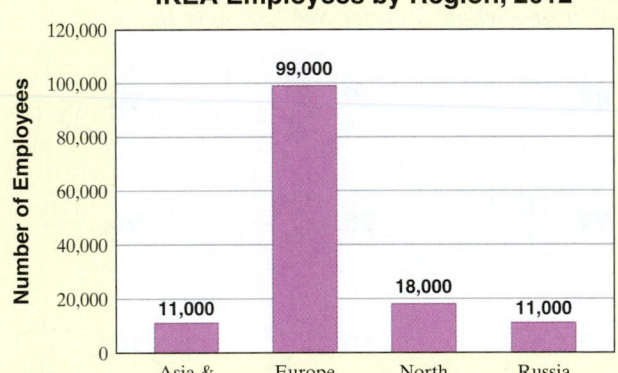

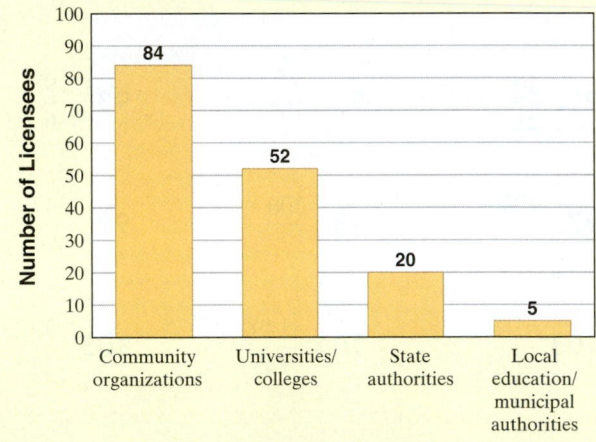

123. Heifer International is a nonprofit world hunger relief organization that focuses on sustainable agriculture programs. Currently Heifer International is working in 6 North American countries, 5 South American countries, 4 Central/Eastern European countries, 14 African countries, and 7 Asian/South Pacific countries. What fraction of the total countries in which Heifer International works is located in North America? (*Hint:* First find the total number of countries) (*Source:* Heifer International)

124. The United States Mint operates six facilities. One facility is headquarters, one facility is a depository, and four facilities mint coins. What fraction of the United States Mint facilities produces coins? (*Hint:* First find the total number of facilities.) (*Source:* United States Mint)

2.2 Factors and Prime Factorization

To perform many operations with fractions, it is necessary to be able to factor a number. In this section, only the **natural numbers**—1, 2, 3, 4, 5, and so on—will be considered.

✔**Concept Check** How are the natural numbers and the whole numbers alike? How are they different?

Objective A Finding Factors of Numbers

Recall that when numbers are multiplied to form a product, each number is called a factor. Since $5 \cdot 9 = 45$, both 5 and 9 are **factors** of 45, and $5 \cdot 9$ is called a **factorization** of 45.

The two-number factorizations of 45 are

$1 \cdot 45$ $3 \cdot 15$ $5 \cdot 9$

Thus, we say that the factors of 45 are $1, 3, 5, 9, 15,$ and 45.

> **Helpful Hint**
>
> From our definition of factor above, notice that a **factor** of a number divides the number evenly (with a remainder of 0). For example,
>
> $$\begin{array}{cccccc} 45 & 15 & 9 & 5 & 3 & 1 \\ 1\overline{)45} & 3\overline{)45} & 5\overline{)45} & 9\overline{)45} & 15\overline{)45} & 45\overline{)45} \end{array}$$

Example 1 Find all the factors of 20.

Solution: First we write all the two-number factorizations of 20.

$1 \cdot 20 = 20$

$2 \cdot 10 = 20$

$4 \cdot \ 5 = 20$

The factors of 20 are 1, 2, 4, 5, 10, and 20.

■ **Work Practice 1**

Objective B Identifying Prime and Composite Numbers

Of all the ways to factor a number, one special way is called the **prime factorization.** To help us write prime factorizations, we first review prime and composite numbers.

> **Prime Numbers**
>
> A **prime number** is a natural number that has exactly two different factors, 1 and itself.

The first several prime numbers are

2, 3, 5, 7, 11, 13, 17

It would be helpful to memorize these.

If a natural number other than 1 is not a prime number, it is called a **composite number.**

Objectives

A Find the Factors of a Number.

B Identify Prime and Composite Numbers.

C Find the Prime Factorization of a Number.

Practice 1

Find all the factors of each number.

a. 15 **b.** 7 **c.** 24

Answers

1. a. 1, 3, 5, 15 **b.** 1, 7
c. 1, 2, 3, 4, 6, 8, 12, 24

✔**Concept Check Answer**

answers may vary

123

Composite Numbers

A **composite number** is any natural number, other than 1, that is not prime.

Helpful Hint

The natural number 1 is neither prime nor composite.

Practice 2

Determine whether each number is prime or composite. Explain your answers.

21, 13, 18, 29, 39

Example 2 Determine whether each number is prime or composite. Explain your answers.

3, 9, 11, 17, 26

Solution: The number 3 is prime. Its only factors are 1 and 3 (itself).

The number 9 is composite. It has more than two factors: 1, 3, and 9.

The number 11 is prime. Its only factors are 1 and 11.

The number 17 is prime. Its only factors are 1 and 17.

The number 26 is composite. Its factors are 1, 2, 13, and 26.

■ Work Practice 2

Objective C Finding Prime Factorizations ▶

Now we are ready to find **prime factorizations** of numbers.

Prime Factorization

The **prime factorization** of a number is the factorization in which all the factors are prime numbers.

For example, the prime factorization of 12 is $2 \cdot 2 \cdot 3$ because

$$12 = \underbrace{2 \cdot 2 \cdot 3}$$

This product is 12 and each number is a prime number.

Every whole number greater than 1 has exactly one prime factorization.

Helpful Hint

Don't forget that multiplication is commutative, so $2 \cdot 2 \cdot 3$ can also be written as $2 \cdot 3 \cdot 2$ or $3 \cdot 2 \cdot 2$ or $2^2 \cdot 3$. Any one of these can be called *the prime factorization of* 12.

Practice 3

Find the prime factorization of 28.

Example 3 Find the prime factorization of 45.

Solution: The first prime number, 2, does not divide 45 evenly (with a remainder of 0). The second prime number, 3, does, so we divide 45 by 3.

$$\begin{array}{r} 15 \\ 3\overline{)45} \end{array}$$

Because 15 is not prime and 3 also divides 15 evenly, we divide by 3 again.

$$\begin{array}{r} 5 \\ 3\overline{)15} \\ 3\overline{)45} \end{array}$$

Answers

2. 13 and 29 are prime. 21, 18, and 39 are composite.   **3.** $2 \cdot 2 \cdot 7$ or $2^2 \cdot 7$

The quotient, 5, is a prime number, so we are finished. The prime factorization of 45 is

$$45 = 3 \cdot 3 \cdot 5 \quad \text{or} \quad 45 = 3^2 \cdot 5,$$

using exponents.

■ **Work Practice 3**

There are a few quick **divisibility tests** to determine whether a number is divisible by the primes 2, 3, or 5. (A number is divisible by 2, for example, if 2 divides it evenly.)

Divisibility Tests

A whole number is divisible by:
- 2 if the last digit is 0, 2, 4, 6, or 8.
    ↓

    13**2** is divisible by 2 since the last digit is a 2.
- 3 if the sum of the digits is divisible by 3.

    144 is divisible by 3 since $1 + 4 + 4 = 9$ is divisible by 3.

- 5 if the last digit is 0 or 5.
    ↓

    111**5** is divisible by 5 since the last digit is a 5.

Helpful Hint

Here are a few other divisibility tests you may find interesting. A whole number is divisible by:
- 4 if its last two digits are divisible by 4.

    17**12** is divisible by 4.
- 6 if it's divisible by 2 and 3.

    9858 is divisible by 6.
- 9 if the sum of its digits is divisible by 9.

    5238 is divisible by 9 since $5 + 2 + 3 + 8 = 18$ is divisible by 9.

We will usually begin the division process with the smallest prime number factor of the given number. Since multiplication is commutative, this is not necessary. As long as the divisor is any prime number factor, this process works.

Example 4 Find the prime factorization of 180.

Solution: We divide 180 by 2 and continue dividing until the quotient is no longer divisible by 2. We then divide by the next largest prime number, 3, until the quotient is no longer divisible by 3. We continue this process until the quotient is a prime number.

$$
\begin{array}{r}
5 \\
3\overline{)15} \\
3\overline{)45} \\
2\overline{)90} \\
2\overline{)180}
\end{array}
$$

(Continued on next page)

Practice 4

Find the prime factorization of 120.

Answer

4. $2 \cdot 2 \cdot 2 \cdot 3 \cdot 5$ or $2^3 \cdot 3 \cdot 5$

Thus, the prime factorization of 180 is

$$180 = 2 \cdot 2 \cdot 3 \cdot 3 \cdot 5 \quad \text{or} \quad 180 = 2^2 \cdot 3^2 \cdot 5,$$

using exponents.

■ **Work Practice 4**

Practice 5

Find the prime factorization of 756.

Example 5   Find the prime factorization of 945.

Solution: This number is not divisible by 2 but is divisible by 3. We will begin by dividing 945 by 3.

$$
\begin{array}{r}
7 \\
5)\overline{35} \\
3)\overline{105} \\
3)\overline{315} \\
3)\overline{945}
\end{array}
$$

Thus, the prime factorization of 945 is

$$945 = 3 \cdot 3 \cdot 3 \cdot 5 \cdot 7 \quad \text{or} \quad 945 = 3^3 \cdot 5 \cdot 7$$

■ **Work Practice 5**

Another way to find the prime factorization is to use a factor tree, as shown in the next example.

Practice 6

Use a factor tree to find the prime factorization of 45.

Example 6   Use a factor tree to find the prime factorization of 18.

Solution: We begin by writing 18 as a product of two natural numbers greater than 1, say $2 \cdot 9$.

$$
\begin{array}{c}
18 \\
\swarrow \; \searrow \\
2 \cdot 9.
\end{array}
$$

The number 2 is prime, but 9 is not. So we write 9 as $3 \cdot 3$.

$$
\begin{array}{c}
18 \\
\swarrow \; \searrow \\
2 \cdot 9 \\
\downarrow \;\; \downarrow\searrow \\
2 \cdot 3 \cdot 3
\end{array}
$$

Each factor is now prime, so the prime factorization is

$$18 = 2 \cdot 3 \cdot 3 \quad \text{or} \quad 18 = 2 \cdot 3^2,$$

using exponents.

■ **Work Practice 6**

In this text, we will write the factorization of a number from the smallest factor to the largest factor.

Answers

5. $2 \cdot 2 \cdot 3 \cdot 3 \cdot 3 \cdot 7$ or $2^2 \cdot 3^3 \cdot 7$

6. $3 \cdot 3 \cdot 5$ or $3^2 \cdot 5$

Example 7 Use a factor tree to find the prime factorization of 80.

Solution: Write 80 as a product of two numbers. Continue this process until all factors are prime.

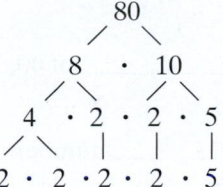

All factors are now prime, so the prime factorization of 80 is

$2 \cdot 2 \cdot 2 \cdot 2 \cdot 5$ or $2^4 \cdot 5$.

■ **Work Practice 7**

Helpful Hint

It makes no difference which factors you start with. The prime factorization of a number will be the same.

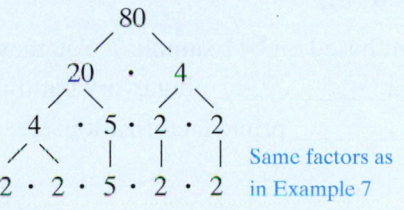

Same factors as in Example 7

✓**Concept Check** True or false? Two different numbers can have exactly the same prime factorization. Explain your answer.

Example 8 Use a factor tree to find the prime factorization of 175.

Solution: We begin by writing 175 as a product of two numbers greater than 1, say $7 \cdot 25$.

175
7 · 25
7 · 5 · 5

The prime factorization of 175 is

$175 = 5 \cdot 5 \cdot 7$ or $175 = 5^2 \cdot 7$

■ **Work Practice 8**

Practice 7
Use a factor tree to find the prime factorization of each number.
a. 30   **b.** 56   **c.** 72

Practice 8
Use a factor tree to find the prime factorization of 117.

Answers
7. a. $2 \cdot 3 \cdot 5$ **b.** $2 \cdot 2 \cdot 2 \cdot 7$ or $2^3 \cdot 7$
c. $2 \cdot 2 \cdot 2 \cdot 3 \cdot 3$ or $2^3 \cdot 3^2$
8. $3 \cdot 3 \cdot 13$ or $3^2 \cdot 13$

✓**Concept Check Answer**
false; answers may vary

Vocabulary, Readiness & Video Check

Use the choices below to fill in each blank.

factor(s)   prime factorization   prime
natural   composite

1. The number 40 equals $2 \cdot 2 \cdot 2 \cdot 5$. Since each factor is prime, we call $2 \cdot 2 \cdot 2 \cdot 5$ the _____ of 40.

2. A natural number, other than 1, that is not prime is called a(n) _____ number.

3. A natural number that has exactly two different factors, 1 and itself, is called a(n) _____ number.

4. The numbers $1, 2, 3, 4, 5, \ldots$ are called the _____ numbers.

5. Since $30 = 5 \cdot 6$, the numbers 5 and 6 are _____ of 30.

6. Answer true or false: $5 \cdot 6$ is the prime factorization of 30. _____

Martin-Gay Interactive Videos   Watch the section lecture video and answer the following questions.

See Video 2.2

Objective A 7. From Example 2, what aren't $3 \cdot 4$ and $4 \cdot 3$ considered different two-number factorizations of 12?

Objective B 8. From the lecture before Example 3, are all natural numbers either prime or composite?

Objective C 9. Complete this statement based on Example 7: You may write factors in different _____, but every natural number has only _____ prime factorization.

2.2 Exercise Set MyMathLab®

Objective A *List all the factors of each number. See Example 1.*

1. 8   **2.** 6   **3.** 25   **4.** 30   **5.** 4   **6.** 9

7. 18   **8.** 48   **9.** 29   **10.** 37   **11.** 80   **12.** 100

13. 12   **14.** 28   **15.** 34   **16.** 26

Objective B *Identify each number as prime or composite. See Example 2.*

17. 7   **18.** 5   **19.** 4   **20.** 10   **21.** 23   **22.** 13

23. 49   **24.** 45   **25.** 67   **26.** 89   **27.** 39   **28.** 21

29. 31   **30.** 27   **31.** 63   **32.** 51   **33.** 119   **34.** 147

Objective C *Find the prime factorization of each number. Write any repeated factors using exponents. See Examples 3 through 8.*

35. 32   **36.** 64   ▶ **37.** 15   **38.** 21   **39.** 40   **40.** 63

▶ **41.** 36   **42.** 80   **43.** 39   **44.** 56   **45.** 60   **46.** 84

47. 110   **48.** 130   **49.** 85   **50.** 93   **51.** 128   **52.** 81

53. 154   **54.** 198   **55.** 300   **56.** 360   ▶ **57.** 240   **58.** 836

59. 828   **60.** 504   **61.** 882   **62.** 405   **63.** 637   **64.** 539

Objectives B C Mixed Practice *Find the prime factorization of each composite number. Write any repeated factors using exponents. Write prime if the number is prime.*

65. 33   **66.** 48   **67.** 98   **68.** 54   **69.** 67   **70.** 59

71. 459   **72.** 208   **73.** 97   **74.** 103   **75.** 700   **76.** 1000

Review

Round each whole number to the indicated place value. See Section 1.5.

77. 4267 hundreds   **78.** 32,465 thousands   **79.** 7,658,240 ten-thousands

80. 4,286,340 tens   **81.** 19,764 thousands   **82.** 10,292,876 millions

The bar graph below shows the number of patents that Apple Inc. has been granted over a three-year period. Use this bar graph to answer the questions below. See Section 2.1. (Source: IFI CLAIMS Patent Services)

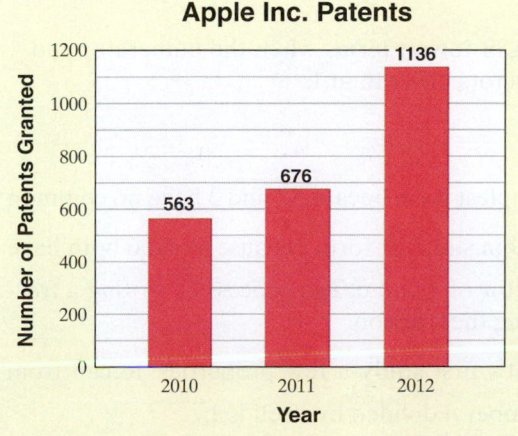

Apple Inc. Patents

83. Find the total number of patents received by Apple for the years shown.

84. How many fewer patents were granted in 2010 than in 2012?

85. What fraction of the patents were granted in 2012?

86. What fraction of the patents were granted in 2011?

Concept Extensions

Find the prime factorization of each number.

87. 34,020

88. 131,625

89. In your own words, define a prime number.

90. The number 2 is a prime number. All other even natural numbers are composite numbers. Explain why.

91. Why are we interested in the prime factorizations of nonzero whole numbers only?

92. Two students have different prime factorizations for the same number. Is this possible? Explain.

Objectives

A Write a Fraction in Simplest Form or Lowest Terms.

B Determine Whether Two Fractions Are Equivalent.

C Solve Problems by Writing Fractions in Simplest Form.

2.3 Simplest Form of a Fraction

Objective A Writing Fractions in Simplest Form

Fractions that represent the same portion of a whole are called **equivalent fractions.**

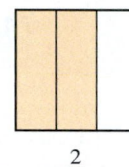   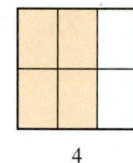   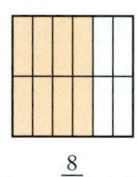

$$\frac{2}{3} \qquad \frac{4}{6} \qquad \frac{8}{12}$$

For example, $\frac{2}{3}, \frac{4}{6}$, and $\frac{8}{12}$ all represent the same shaded portion of the rectangle's area, so they are equivalent fractions.

$$\frac{2}{3} = \frac{4}{6} = \frac{8}{12}$$

There are many equivalent forms of a fraction. A special form of a fraction is called **simplest form.**

> **Simplest Form of a Fraction**
>
> A fraction is written in **simplest form** or **lowest terms** when the numerator and the denominator have no common factors other than 1.

For example, the fraction $\frac{2}{3}$ *is* in simplest form because 2 and 3 have no common factor other than 1. The fraction $\frac{4}{6}$ *is not* in simplest form because 4 and 6 both have a factor of 2. That is, 2 is a common factor of 4 and 6. The process of writing a fraction in simplest form is called **simplifying** the fraction.

To simplify $\frac{4}{6}$ and write it as $\frac{2}{3}$, let's first study a few properties. Recall from Section 2.1 that any nonzero whole number n divided by itself is 1.

Any nonzero number n divided by itself is 1.

$$\frac{5}{5} = 1, \ \frac{17}{17} = 1, \ \frac{24}{24} = 1, \text{ or, in general, } \frac{n}{n} = 1$$

Also, in general, if $\dfrac{a}{b}$ and $\dfrac{c}{d}$ are fractions (with b and d not 0), the following is true.

$$\frac{a \cdot c}{b \cdot d} = \frac{a}{b} \cdot \frac{c}{d}^{*}$$

These properties allow us to do the following:

$$\frac{4}{6} = \frac{2 \cdot 2}{2 \cdot 3} = \frac{2}{2} \cdot \frac{2}{3} = 1 \cdot \frac{2}{3} = \frac{2}{3} \qquad \text{When 1 is multiplied by a number, the result is the same number.}$$
$$\underset{\text{This is 1}}{\big\downarrow}$$

Note: We will study this concept further in the next section.

Example 1   Write in simplest form: $\dfrac{12}{20}$

Solution: Notice that 12 and 20 have a common factor of 4.

$$\frac{12}{20} = \frac{4 \cdot 3}{4 \cdot 5} = \frac{4}{4} \cdot \frac{3}{5} = 1 \cdot \frac{3}{5} = \frac{3}{5}$$

Since 3 and 5 have no common factors (other than 1), $\dfrac{3}{5}$ is in simplest form.

■ **Work Practice 1**

If you have trouble finding common factors, write the prime factorizations of the numerator and the denominator.

Example 2   Write in simplest form: $\dfrac{42}{66}$

Solution: Let's write the prime factorizations of 42 and 66.

$$\frac{42}{66} = \frac{2 \cdot 3 \cdot 7}{2 \cdot 3 \cdot 11} = \frac{2}{2} \cdot \frac{3}{3} \cdot \frac{7}{11} = 1 \cdot 1 \cdot \frac{7}{11} = \frac{7}{11}$$

■ **Work Practice 2**

In the example above, you may have saved time by noticing that 42 and 66 have a common factor of 6.

$$\frac{42}{66} = \frac{6 \cdot 7}{6 \cdot 11} = \frac{6}{6} \cdot \frac{7}{11} = 1 \cdot \frac{7}{11} = \frac{7}{11}$$

Helpful Hint

Writing the prime factorizations of the numerator and the denominator is helpful in finding any common factors.

Practice 1
Write in simplest form: $\dfrac{30}{45}$

Practice 2
Write in simplest form: $\dfrac{39}{51}$

Answers

1. $\dfrac{2}{3}$ **2.** $\dfrac{13}{17}$

Practice 3

Write in simplest form: $\dfrac{9}{50}$

Example 3 Write in simplest form: $\dfrac{10}{27}$

Solution:

$$\frac{10}{27} = \frac{2 \cdot 5}{3 \cdot 3 \cdot 3}$$ Prime factorizations of 10 and 27.

Since 10 and 27 have no common factors, $\dfrac{10}{27}$ is already in simplest form.

■ **Work Practice 3**

Practice 4

Write in simplest form: $\dfrac{49}{112}$

Example 4 Write in simplest form: $\dfrac{30}{108}$

Solution:

$$\frac{30}{108} = \frac{2 \cdot 3 \cdot 5}{2 \cdot 2 \cdot 3 \cdot 3 \cdot 3} = \frac{2}{2} \cdot \frac{3}{3} \cdot \frac{5}{2 \cdot 3 \cdot 3} = 1 \cdot 1 \cdot \frac{5}{18} = \frac{5}{18}$$

■ **Work Practice 4**

We can use a shortcut procedure with common factors when simplifying.

$$\frac{4}{6} = \frac{\overset{1}{\cancel{2}} \cdot 2}{\underset{1}{\cancel{2}} \cdot 3} = \frac{1 \cdot 2}{1 \cdot 3} = \frac{2}{3}$$ Divide out the common factor of 2 in the numerator and denominator.

This procedure is possible because dividing out a common factor in the numerator and denominator is the same as removing a factor of 1 in the product.

> ### Writing a Fraction in Simplest Form
>
> To write a fraction in simplest form, write the prime factorizations of the numerator and the denominator and then divide both by all common factors.

Practice 5

Write in simplest form: $\dfrac{64}{20}$

Example 5 Write in simplest form: $\dfrac{72}{26}$

Solution:

$$\frac{72}{26} = \frac{\overset{1}{\cancel{2}} \cdot 2 \cdot 2 \cdot 3 \cdot 3}{\underset{1}{\cancel{2}} \cdot 13} = \frac{1 \cdot 2 \cdot 2 \cdot 3 \cdot 3}{1 \cdot 13} = \frac{36}{13},$$

which can also be written as

$$2\frac{10}{13}$$

■ **Work Practice 5**

✓**Concept Check** Which is the correct way to simplify the fraction $\dfrac{15}{25}$? Or are both correct? Explain.

a. $\dfrac{15}{25} = \dfrac{3 \cdot \overset{1}{\cancel{5}}}{5 \cdot \underset{1}{\cancel{5}}} = \dfrac{3}{5}$ **b.** $\dfrac{\overset{1}{\cancel{15}}}{\underset{1}{\cancel{25}}} = \dfrac{11}{21}$

Answers

3. $\dfrac{9}{50}$ **4.** $\dfrac{7}{16}$ **5.** $\dfrac{16}{5}$ or $3\dfrac{1}{5}$

✓**Concept Check Answers**

a. correct **b.** incorrect

Example 6   Write in simplest form: $\dfrac{6}{60}$

Solution:

$$\dfrac{6}{60} = \dfrac{\overset{1}{\cancel{2}} \cdot \overset{1}{\cancel{3}}}{\underset{1}{\cancel{2}} \cdot 2 \cdot \underset{1}{\cancel{3}} \cdot 5} = \dfrac{1 \cdot 1}{1 \cdot 2 \cdot 1 \cdot 5} = \dfrac{1}{10}$$

■ **Work Practice 6**

Helpful Hint

Be careful when all factors of the numerator or denominator are divided out. In Example 6, the numerator was $1 \cdot 1 = 1$, so the final result was $\dfrac{1}{10}$.

In the fraction of Example 6, $\dfrac{6}{60}$, you may have immediately noticed that the largest common factor of 6 and 60 is 6. If so, you may simply divide out that largest common factor.

$$\dfrac{6}{60} = \dfrac{\overset{1}{\cancel{6}}}{\underset{1}{\cancel{6}} \cdot 10} = \dfrac{1}{1 \cdot 10} = \dfrac{1}{10} \qquad \text{\color{blue}Divide out the common factor of 6.}$$

Notice that the result, $\dfrac{1}{10}$, is in simplest form. If it were not, we would repeat the same procedure until the result was in simplest form.

Example 7   Write in simplest form: $\dfrac{45}{75}$

Solution: You may write the prime factorizations of 45 and 75 or you may notice that these two numbers have a common factor of 15.

$$\dfrac{45}{75} = \dfrac{3 \cdot \overset{1}{\cancel{15}}}{5 \cdot \underset{1}{\cancel{15}}} = \dfrac{3 \cdot 1}{5 \cdot 1} = \dfrac{3}{5}$$

The numerator and denominator of $\dfrac{3}{5}$ have no common factors other than 1, so $\dfrac{3}{5}$ is in simplest form.

■ **Work Practice 7**

Objective B Determining Whether Two Fractions Are Equivalent ▶

Recall that two fractions are equivalent if they represent the same part of a whole. One way to determine whether two fractions are equivalent is to see whether they simplify to the same fraction.

Practice 6

Write in simplest form: $\dfrac{8}{56}$

Practice 7

Write in simplest form: $\dfrac{42}{48}$

Answers

6. $\dfrac{1}{7}$ **7.** $\dfrac{7}{8}$

Practice 8

Determine whether $\dfrac{7}{9}$ and $\dfrac{21}{27}$ are equivalent.

Example 8 Determine whether $\dfrac{16}{40}$ and $\dfrac{10}{25}$ are equivalent.

Solution: Simplify each fraction.

$$\dfrac{16}{40} = \dfrac{\overset{1}{\cancel{8}} \cdot 2}{\underset{1}{\cancel{8}} \cdot 5} = \dfrac{1 \cdot 2}{1 \cdot 5} = \dfrac{2}{5}$$

$$\dfrac{10}{25} = \dfrac{2 \cdot \overset{1}{\cancel{5}}}{5 \cdot \underset{1}{\cancel{5}}} = \dfrac{2 \cdot 1}{5 \cdot 1} = \dfrac{2}{5}$$

Since these fractions are the same, $\dfrac{16}{40} = \dfrac{10}{25}$.

■ Work Practice 8

There is a shortcut method you may use to check or test whether two fractions are equivalent. In the example above, we learned that the fractions are equivalent, or

$$\dfrac{16}{40} = \dfrac{10}{25}$$

In this example above, we call $25 \cdot 16$ and $40 \cdot 10$ **cross products** because they are the products one obtains by multiplying across.

$$\overbrace{}^{\text{Cross Products}}$$

$$25 \cdot 16 \qquad\qquad 40 \cdot 10$$

$$\dfrac{16}{40} = \dfrac{10}{25}$$

Notice that these cross products are equal

$$25 \cdot 16 = 400, \quad 40 \cdot 10 = 400$$

In general, this is true for equivalent fractions.

Equivalent Fractions

$$8 \cdot 6 \qquad\qquad 24 \cdot 2$$

$$\dfrac{6}{24} \overset{?}{=} \dfrac{2}{8}$$

Since the cross products ($8 \cdot 6 = 48$ and $24 \cdot 2 = 48$) are equal, the fractions are equivalent.
  Note: If the cross products are not equal, the fractions are not equivalent.

Practice 9

Determine whether $\dfrac{4}{13}$ and $\dfrac{5}{18}$ are equivalent.

Helpful Hint "Not equal to" symbol.

Example 9 Determine whether $\dfrac{8}{11}$ and $\dfrac{19}{26}$ are equivalent.

Solution: Let's check cross products.

$$26 \cdot 8 \qquad\qquad 11 \cdot 19$$
$$= 208 \qquad \dfrac{8}{11} \overset{?}{=} \dfrac{19}{26} \qquad = 209$$

Since $208 \neq 209$, then $\dfrac{8}{11} \neq \dfrac{19}{26}$.

■ Work Practice 9

Answer

8. equivalent   **9.** not equivalent

Objective C Solving Problems by Writing Fractions in Simplest Form

Many real-life problems can be solved by writing fractions. To make the answers clearer, these fractions should be written in simplest form.

Example 10   Calculating Fraction of Parks in Pennsylvania

There are currently 46 national historical parks in the United States. Two of these historical parks are located in the state of Pennsylvania. What fraction of the United States' national historical parks can be found in Pennsylvania? Write the fraction in simplest form. (*Source:* National Park Service)

Solution: First we determine the fraction of parks found in Pennsylvania.

$$\dfrac{2}{46} \quad \begin{array}{l} \leftarrow \text{national historical parks in Pennsylvania} \\ \leftarrow \text{total national historical parks} \end{array}$$

Next we simplify the fraction.

$$\frac{2}{46} = \frac{\overset{1}{\cancel{2}}}{\underset{1}{\cancel{2}} \cdot 23} = \frac{1}{1 \cdot 23} = \frac{1}{23}$$

Thus, $\dfrac{1}{23}$ of the United States' national parks are in the state of Pennsylvania.

■ **Work Practice 10**

Practice 10

There are four national historical parks in the state of Virginia. See Example 10 and determine what fraction of the United States' national historical parks can be found in Virginia. Write the fraction in simplest form.

Answer

10. $\dfrac{2}{23}$

📟 Calculator Explorations Simplifying Fractions

Scientific Calculator

Many calculators have a fraction key, such as $\boxed{a\ _{b/c}}$, that allows you to simplify a fraction on the calculator. For example, to simplify $\dfrac{324}{612}$, enter

$$\boxed{3}\ \boxed{2}\ \boxed{4}\ \boxed{a\ _{b/c}}\ \boxed{6}\ \boxed{1}\ \boxed{2}\ \boxed{=}$$

The display will read

$$\boxed{\quad 9\ |\ 17\quad}$$

which represents $\dfrac{9}{17}$, the original fraction simplified.

> **Helpful Hint**   The Calculator Explorations boxes in this chapter provide only an introduction to fraction keys on calculators. Any time you use a calculator, there are both advantages and limitations to its use. Never rely solely on your calculator. It is very important that you understand how to perform all operations on fractions by hand in order to progress through later topics. For further information, talk to your instructor.

Use your calculator to simplify each fraction.

1. $\dfrac{128}{224}$ **2.** $\dfrac{231}{396}$ **3.** $\dfrac{340}{459}$ **4.** $\dfrac{999}{1350}$

5. $\dfrac{810}{432}$ **6.** $\dfrac{315}{225}$ **7.** $\dfrac{243}{54}$ **8.** $\dfrac{689}{455}$

Vocabulary, Readiness & Video Check

Use the choices below to fill in each blank.

    0   cross products   equivalent

    1   simplest form   *n*

1. In $\frac{11}{48}$, since 11 and 48 have no common factors other than 1, $\frac{11}{48}$ is in _____.

2. Fractions that represent the same portion of a whole are called _____ fractions.

3. In the statement $\frac{5}{12} = \frac{15}{36}$, $5 \cdot 36$ and $12 \cdot 15$ are called _____.

4. The fraction $\frac{7}{7}$ simplifies to _____.

5. The fraction $\frac{0}{7}$ simplifies to _____.

6. The fraction $\frac{n}{1}$ simplifies to _____.

Martin-Gay Interactive Videos

See Video 2.3

Watch the section lecture video and answer the following questions.

Objective A 7. Complete this statement based on the lecture before
Example 1: A special form of a(n) _____
form of a fraction is called simplest form. ▶

Objective B 8. The fractions in Example 5 are shown to be equivalent by
using cross products. Describe another way to tell whether
the two fractions are equivalent. ▶

Objective C 9. Why isn't $\frac{10}{24}$ the final answer to Example 7? What is the
final answer? ▶

2.3 Exercise Set MyMathLab®

Objective A *Write each fraction in simplest form. See Examples 1 through 7.*

1. $\frac{3}{12}$ **2.** $\frac{5}{30}$ **3.** $\frac{4}{42}$ **4.** $\frac{9}{48}$

▶ **5.** $\frac{14}{16}$ **6.** $\frac{22}{34}$ **7.** $\frac{20}{30}$ **8.** $\frac{70}{80}$

9. $\frac{35}{50}$ **10.** $\frac{25}{55}$ **11.** $\frac{63}{81}$ **12.** $\frac{21}{49}$

▶ **13.** $\frac{24}{40}$ **14.** $\frac{36}{54}$ **15.** $\frac{27}{64}$ **16.** $\frac{32}{63}$

17. $\frac{25}{40}$ **18.** $\frac{36}{42}$ **19.** $\frac{40}{64}$ **20.** $\frac{28}{60}$

21. $\frac{56}{68}$ **22.** $\frac{39}{42}$ **23.** $\frac{36}{24}$ **24.** $\frac{60}{36}$

25. $\frac{90}{120}$ **26.** $\frac{60}{150}$ ▶ **27.** $\frac{70}{196}$ **28.** $\frac{98}{126}$

29. $\dfrac{66}{308}$ **30.** $\dfrac{65}{234}$ **31.** $\dfrac{55}{85}$ **32.** $\dfrac{78}{90}$

33. $\dfrac{75}{350}$ **34.** $\dfrac{72}{420}$ **35.** $\dfrac{189}{216}$ **36.** $\dfrac{144}{162}$

37. $\dfrac{288}{480}$ **38.** $\dfrac{135}{585}$ **39.** $\dfrac{224}{16}$ **40.** $\dfrac{270}{15}$

Objective B *Determine whether each pair of fractions is equivalent. See Examples 8 and 9.*

41. $\dfrac{3}{6}$ and $\dfrac{4}{8}$ **42.** $\dfrac{3}{9}$ and $\dfrac{2}{6}$ ▶ **43.** $\dfrac{7}{11}$ and $\dfrac{5}{8}$

44. $\dfrac{2}{5}$ and $\dfrac{4}{11}$ **45.** $\dfrac{10}{15}$ and $\dfrac{6}{9}$ **46.** $\dfrac{4}{10}$ and $\dfrac{6}{15}$

▶ **47.** $\dfrac{3}{9}$ and $\dfrac{6}{18}$ **48.** $\dfrac{2}{8}$ and $\dfrac{7}{28}$ **49.** $\dfrac{10}{13}$ and $\dfrac{12}{15}$

50. $\dfrac{16}{20}$ and $\dfrac{9}{12}$ **51.** $\dfrac{8}{18}$ and $\dfrac{12}{24}$ **52.** $\dfrac{6}{21}$ and $\dfrac{14}{35}$

Objective C *Solve. Write each fraction in simplest form. See Example 10.*

53. A work shift for an employee at McDonald's consists of 8 hours. What fraction of the employee's work shift is represented by 2 hours?

54. Two thousand baseball caps were sold one year at the U.S. Open Golf Tournament. What fractional part of this total does 200 caps represent?

▶ **55.** There are 5280 feet in a mile. What fraction of a mile is represented by 2640 feet?

56. There are 100 centimeters in 1 meter. What fraction of a meter is 20 centimeters?

57. Sixteen states in the United States have Ritz-Carlton hotels. (*Source:* Ritz-Carlton Hotel Company, LLC)

   a. What fraction of states can claim at least one Ritz-Carlton hotel?

   b. How many states do not have a Ritz-Carlton hotel?

   c. Write the fraction of states without a Ritz-Carlton hotel.

58. There are 78 national monuments in the United States. Ten of these monuments are located in New Mexico. (*Source:* National Park Service)

   a. What fraction of the national monuments in the United States can be found in New Mexico?

   b. How many of the national monuments in the United States are found outside New Mexico?

   c. Write the fraction of national monuments found in states other than New Mexico.

▶ **59.** The outer wall of the Pentagon is 24 inches wide. Ten inches is concrete, 8 inches is brick, and 6 inches is limestone. What fraction of the wall is concrete? (*Source: USA Today*)

60. There are 35 students in a biology class. If 10 students made an A on the first test, what fraction of the students made an A?

Limestone (6 in.)

Brick (8 in.)

Concrete (10 in.)

61. Safeway Inc. operates grocery stores under multiple banners in 20 states in the United States. (*Source:* Safeway Inc.)

    a. How many states do not have a Safeway-owned store?

    b. What fraction of states do not have a Safeway-owned store?

62. Katy Biagini just bought a brand-new 2013 Toyota Camry hybrid for $28,000. Her old car was traded in for $12,000.

    a. How much of her purchase price was not covered by her trade-in?

    b. What fraction of the purchase price was not covered by the trade-in?

63. Of the 20 top-selling video games of 2011, eight had ESRB ratings of "Everyone." What fraction of 2011's top-selling video games had "Everyone" ratings? (*Source:* The NPD Group/Retail Tracking Service)

64. Worldwide, Hallmark employs about 12,000 full-time employees. About 3200 employees work at the Hallmark headquarters in Kansas City, Missouri. What fraction of Hallmark employees work in Kansas City? (*Source:* Hallmark Cards, Inc.)

Review

Multiply. See Section 1.6.

65.
$$\begin{array}{r} 91 \\ \times\ 4 \\ \hline \end{array}$$

66.
$$\begin{array}{r} 73 \\ \times\ 8 \\ \hline \end{array}$$

67.
$$\begin{array}{r} 387 \\ \times\ 6 \\ \hline \end{array}$$

68.
$$\begin{array}{r} 562 \\ \times\ 9 \\ \hline \end{array}$$

69.
$$\begin{array}{r} 72 \\ \times\ 35 \\ \hline \end{array}$$

70.
$$\begin{array}{r} 238 \\ \times\ 26 \\ \hline \end{array}$$

Concept Extensions

71. In your own words, define equivalent fractions.

72. Given a fraction, say $\frac{3}{8}$, how many fractions are there that are equivalent to it? Explain your answer.

Write each fraction in simplest form.

73. $\dfrac{3975}{6625}$

74. $\dfrac{9506}{12,222}$

There are generally considered to be eight basic blood types. The table shows the number of people with the various blood types in a typical group of 100 blood donors. Use the table to answer Exercises 75 through 78. Write each answer in simplest form.

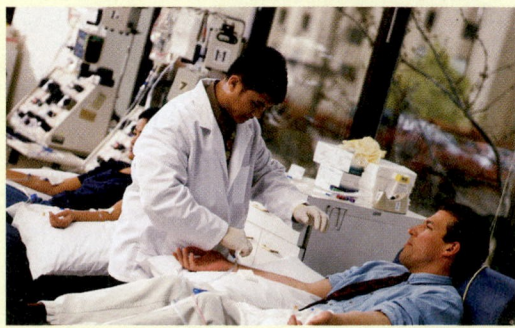

Distribution of Blood Types in Blood Donors	
Blood Type	**Number of People**
O Rh-positive	37
O Rh-negative	7
A Rh-positive	36
A Rh-negative	6
B Rh-positive	9
B Rh-negative	1
AB Rh-positive	3
AB Rh-negative	1
(*Source:* American Red Cross Biomedical Services)	

75. What fraction of blood donors have blood type A Rh-positive?

76. What fraction of blood donors have an O blood type?

77. What fraction of blood donors have an AB blood type?

78. What fraction of blood donors have a B blood type?

*The following graph is called a **circle graph** or **pie chart**. Each sector (shaped like a piece of pie) shows the fraction of entering college freshmen who expect to major in each discipline shown. The whole circle represents the entire class of college freshmen. Use this graph to answer Exercises 79 through 82.*

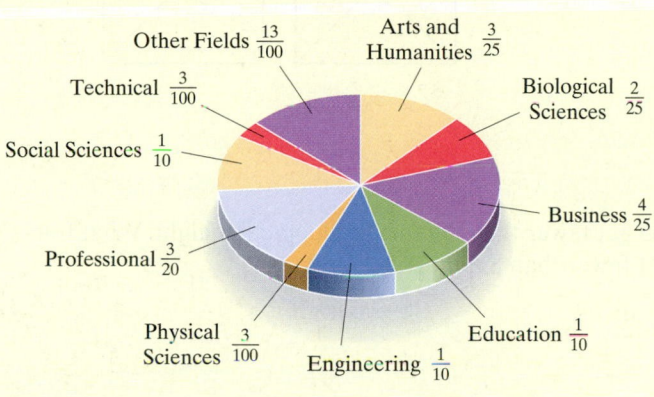

Other Fields $\frac{13}{100}$

Arts and Humanities $\frac{3}{25}$

Technical $\frac{3}{100}$

Biological Sciences $\frac{2}{25}$

Social Sciences $\frac{1}{10}$

Business $\frac{4}{25}$

Professional $\frac{3}{20}$

Physical Sciences $\frac{3}{100}$

Engineering $\frac{1}{10}$

Education $\frac{1}{10}$

Source: The Higher Education Research Institute

79. What fraction of entering college freshmen plan to major in education?

80. What fraction of entering college freshmen plan to major in biological sciences?

81. Why is the Social Sciences sector the same size as the Engineering sector?

82. Why is the Physical Sciences sector smaller than the Business sector?

Use this circle graph to answer Exercises 83 through 86.

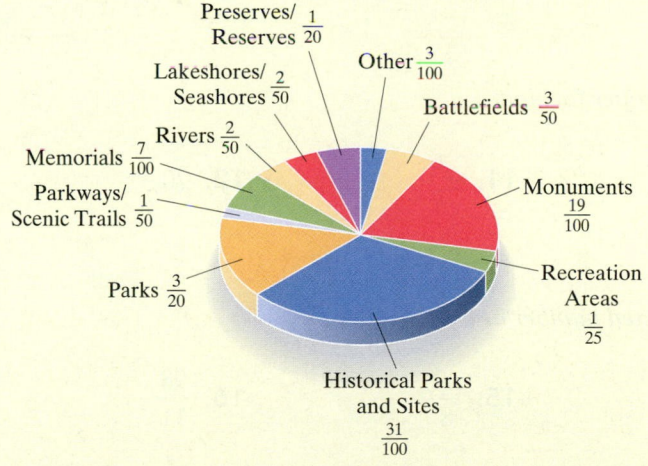

Areas Maintained by the National Park Service

Preserves/Reserves $\frac{1}{20}$

Lakeshores/Seashores $\frac{2}{50}$

Other $\frac{3}{100}$

Rivers $\frac{2}{50}$

Battlefields $\frac{3}{50}$

Memorials $\frac{7}{100}$

Parkways/Scenic Trails $\frac{1}{50}$

Monuments $\frac{19}{100}$

Parks $\frac{3}{20}$

Recreation Areas $\frac{1}{25}$

Historical Parks and Sites $\frac{31}{100}$

83. What fraction of National Park Service areas are National Memorials?

84. What fraction of National Park Service areas are National Parks?

85. Why is the National Battlefields sector smaller than the National Monuments sector?

86. Why is the National Lakeshores/National Seashores sector the same size as the National Rivers sector?

Use the following numbers for Exercises 87 through 90.

8691   786   1235   2235   85   105   22   222   900   1470

87. List the numbers divisible by both 2 and 3.

88. List the numbers that are divisible by both 3 and 5.

89. The answers to Exercise 87 are also divisible by what number? Tell why.

90. The answers to Exercise 88 are also divisible by what number? Tell why.

Integrated Review   Sections 2.1–2.3

Summary on Fractions, Mixed Numbers, and Factors

Answers

1. _____

2. _____

3. _____

4. _____

5. _____

6. _____

7. _____

8. _____

9. _____

10. _____

11. _____

12. _____

13. _____

14. _____

15. _____

16. _____

17. _____

18. _____

19. _____

20. _____

Use a fraction to represent the shaded area of each figure. If the fraction is improper, also write the fraction as a mixed number.

1.

2.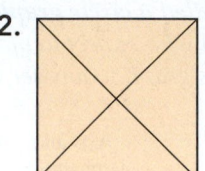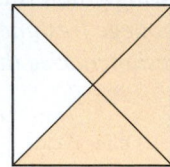

Solve.

3. In a survey, 73 people out of 85 get fewer than 8 hours of sleep each night. What fraction of people in the survey get fewer than 8 hours of sleep?

4. Sketch a diagram to represent $\dfrac{9}{13}$.

Simplify.

5. $\dfrac{11}{11}$ **6.** $\dfrac{17}{1}$ **7.** $\dfrac{0}{3}$ **8.** $\dfrac{7}{0}$

Write each mixed number as an improper fraction.

9. $3\dfrac{1}{8}$ **10.** $5\dfrac{3}{5}$ **11.** $9\dfrac{6}{7}$ **12.** $20\dfrac{1}{7}$

Write each improper fraction as a mixed number or a whole number.

13. $\dfrac{20}{7}$ **14.** $\dfrac{55}{11}$ **15.** $\dfrac{39}{8}$ **16.** $\dfrac{98}{11}$

List the factors of each number.

17. 35   **18.** 40

Determine whether each number is prime or composite.

19. 72   **20.** 13

140

Write the prime factorization of each composite number. Write prime if the number is prime.
Write any repeated factors using exponents.

21. 65   **22.** 70   **23.** 96   **24.** 132

25. 252   **26.** 31   **27.** 315   **28.** 441

29. 286   **30.** 41

Write each fraction in simplest form.

31. $\dfrac{2}{14}$ **32.** $\dfrac{24}{20}$ **33.** $\dfrac{18}{38}$ **34.** $\dfrac{42}{110}$

35. $\dfrac{56}{60}$ **36.** $\dfrac{72}{80}$ **37.** $\dfrac{54}{135}$ **38.** $\dfrac{90}{240}$

39. $\dfrac{165}{210}$ **40.** $\dfrac{245}{385}$

Determine whether each pair of fractions is equivalent.

41. $\dfrac{7}{8}$ and $\dfrac{9}{10}$ **42.** $\dfrac{10}{12}$ and $\dfrac{15}{18}$

Solve. Write fraction answers in simplest form.

43. Of the 50 states, 2 states are not adjacent to any other states.

    a. What fraction of the states are not adjacent to other states?

    b. How many states are adjacent to other states?

    c. What fraction of the states are adjacent to other states?

44. In a recent year, 540 films were released and rated. Of these, 145 were rated PG-13.
(*Source:* Nash Information, LLC.)

    a. What fraction were rated PG-13?

    b. How many films were rated other than PG-13?

    c. What fraction of films were rated other than PG-13?

21. _____
22. _____
23. _____
24. _____
25. _____
26. _____
27. _____
28. _____
29. _____
30. _____
31. _____
32. _____
33. _____
34. _____
35. _____
36. _____
37. _____
38. _____
39. _____
40. _____
41. _____
42. _____
43. _____
44. _____

2.4 Multiplying Fractions and Mixed Numbers

Objectives

A Multiply Fractions. ▶

B Multiply Fractions and Mixed Numbers or Whole Numbers. ▶

C Solve Problems by Multiplying Fractions. ▶

Objective A Multiplying Fractions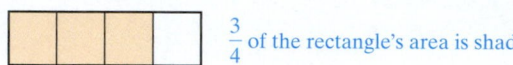

Let's use a diagram to discover how fractions are multiplied. For example, to multiply $\frac{1}{2}$ and $\frac{3}{4}$, we find $\frac{1}{2}$ of $\frac{3}{4}$. To do this, we begin with a diagram showing $\frac{3}{4}$ of a rectangle's area shaded.

$\frac{3}{4}$ of the rectangle's area is shaded.

To find $\frac{1}{2}$ of $\frac{3}{4}$, we heavily shade $\frac{1}{2}$ of the part that is already shaded.

By counting smaller rectangles, we see that $\frac{3}{8}$ of the larger rectangle is now heavily shaded, so that

$$\frac{1}{2} \text{ of } \frac{3}{4} \text{ is } \frac{3}{8}, \text{ or } \frac{1}{2} \cdot \frac{3}{4} = \frac{3}{8}$$ Notice that $\frac{1}{2} \cdot \frac{3}{4} = \frac{1 \cdot 3}{2 \cdot 4} = \frac{3}{8}.$

> ### Multiplying Fractions
>
> To multiply two fractions, multiply the numerators and multiply the denominators. If $a, b, c,$ and d represent positive whole numbers, we have
>
> $$\frac{a}{b} \cdot \frac{c}{d} = \frac{a \cdot c}{b \cdot d}$$

Practice 1–2

Multiply.

1. $\frac{3}{8} \cdot \frac{5}{7}$ **2.** $\frac{1}{3} \cdot \frac{1}{6}$

Examples   Multiply.

1. $\frac{2}{3} \cdot \frac{5}{11} = \frac{2 \cdot 5}{3 \cdot 11} = \frac{10}{33}$ Multiply numerators.
Multiply denominators.

This fraction is in simplest form since 10 and 33 have no common factors other than 1.

2. $\frac{1}{4} \cdot \frac{1}{2} = \frac{1 \cdot 1}{4 \cdot 2} = \frac{1}{8}$ This fraction is in simplest form.

■ Work Practice 1–2

Practice 3

Multiply and simplify: $\frac{6}{55} \cdot \frac{5}{8}$

Example 3   Multiply and simplify: $\frac{6}{7} \cdot \frac{14}{27}$

Solution:

$$\frac{6}{7} \cdot \frac{14}{27} = \frac{6 \cdot 14}{7 \cdot 27}$$

Answers

1. $\frac{15}{56}$ **2.** $\frac{1}{18}$ **3.** $\frac{3}{44}$

We can simplify by finding the prime factorizations and using our shortcut procedure of dividing out common factors in the numerator and denominator.

$$\frac{6 \cdot 14}{7 \cdot 27} = \frac{2 \cdot \overset{1}{\cancel{3}} \cdot 2 \cdot \overset{1}{\cancel{7}}}{\underset{1}{\cancel{7}} \cdot \underset{1}{\cancel{3}} \cdot 3 \cdot 3} = \frac{2 \cdot 2}{3 \cdot 3} = \frac{4}{9}$$

■ **Work Practice 3**

Helpful Hint

Remember that the shortcut procedure above is the same as removing factors of 1 in the product.

$$\frac{6 \cdot 14}{7 \cdot 27} = \frac{2 \cdot 3 \cdot 2 \cdot 7}{7 \cdot 3 \cdot 3 \cdot 3} = \frac{7}{7} \cdot \frac{3}{3} \cdot \frac{2 \cdot 2}{3 \cdot 3} = 1 \cdot 1 \cdot \frac{4}{9} = \frac{4}{9}$$

Helpful Hint

In simplifying a product, don't forget that it may be possible to identify common factors without actually writing the prime factorizations. For example,

$$\frac{10}{11} \cdot \frac{1}{20} = \frac{10 \cdot 1}{11 \cdot 20} = \frac{\overset{1}{\cancel{10}} \cdot 1}{11 \cdot \underset{1}{\cancel{10}} \cdot 2} = \frac{1}{11 \cdot 2} = \frac{1}{22}$$

Example 4 Multiply and simplify: $\dfrac{23}{32} \cdot \dfrac{4}{7}$

Solution: Notice that 4 and 32 have a common factor of 4.

$$\frac{23}{32} \cdot \frac{4}{7} = \frac{23 \cdot 4}{32 \cdot 7} = \frac{23 \cdot \overset{1}{\cancel{4}}}{\underset{1}{\cancel{4}} \cdot 8 \cdot 7} = \frac{23}{8 \cdot 7} = \frac{23}{56}$$

■ **Work Practice 4**

After multiplying two fractions, always check to see whether the product can be simplified.

Examples Multiply.

5. $\dfrac{3}{4} \cdot \dfrac{8}{5} = \dfrac{3 \cdot 8}{4 \cdot 5} = \dfrac{3 \cdot \overset{1}{\cancel{4}} \cdot 2}{\underset{1}{\cancel{4}} \cdot 5} = \dfrac{6}{5}$

6. $\dfrac{6}{13} \cdot \dfrac{26}{30} = \dfrac{6 \cdot 26}{13 \cdot 30} = \dfrac{\overset{1}{\cancel{6}} \cdot \overset{1}{\cancel{13}} \cdot 2}{\underset{1}{\cancel{13}} \cdot \underset{1}{\cancel{6}} \cdot 5} = \dfrac{2}{5}$

7. $\dfrac{1}{3} \cdot \dfrac{2}{5} \cdot \dfrac{9}{16} = \dfrac{1 \cdot 2 \cdot 9}{3 \cdot 5 \cdot 16} = \dfrac{1 \cdot \overset{1}{\cancel{2}} \cdot \overset{1}{\cancel{3}} \cdot 3}{\underset{1}{\cancel{3}} \cdot 5 \cdot \underset{1}{\cancel{2}} \cdot 8} = \dfrac{3}{40}$

■ **Work Practice 5–7**

Practice 4

Multiply and simplify: $\dfrac{4}{15} \cdot \dfrac{3}{8}$

Practice 5–7

Multiply.

5. $\dfrac{2}{5} \cdot \dfrac{20}{7}$

6. $\dfrac{4}{11} \cdot \dfrac{33}{16}$

7. $\dfrac{1}{6} \cdot \dfrac{3}{10} \cdot \dfrac{25}{16}$

Answers

4. $\dfrac{1}{10}$ 5. $\dfrac{8}{7}$ 6. $\dfrac{3}{4}$ 7. $\dfrac{5}{64}$

Objective B Multiplying Fractions and Mixed Numbers or Whole Numbers ▶

When multiplying a fraction and a mixed or a whole number, remember that mixed and whole numbers can be written as fractions.

> **Multiplying Fractions and Mixed Numbers or Whole Numbers**
>
> To multiply with mixed numbers or whole numbers, first write any mixed or whole numbers as fractions and then multiply as usual.

Practice 8

Multiply and simplify: $2\frac{1}{2} \cdot \frac{8}{15}$

Example 8 Multiply: $3\frac{1}{3} \cdot \frac{7}{8}$

Solution: The mixed number $3\frac{1}{3}$ can be written as the fraction $\frac{10}{3}$. Then,

$$3\frac{1}{3} \cdot \frac{7}{8} = \frac{10}{3} \cdot \frac{7}{8} = \frac{\overset{1}{2} \cdot 5 \cdot 7}{3 \cdot \underset{1}{2} \cdot 4} = \frac{35}{12} \quad \text{or} \quad 2\frac{11}{12}$$

■ **Work Practice 8**

Don't forget that a whole number can be written as a fraction by writing the whole number over 1. For example,

$$20 = \frac{20}{1} \quad \text{and} \quad 7 = \frac{7}{1}$$

Practice 9

Multiply.

$\frac{2}{3} \cdot 18$

Example 9 Multiply.

$$\frac{3}{4} \cdot 20 = \frac{3}{4} \cdot \frac{20}{1} = \frac{3 \cdot 20}{4 \cdot 1} = \frac{3 \cdot \overset{1}{4} \cdot 5}{\underset{1}{4} \cdot 1} = \frac{15}{1} \quad \text{or} \quad 15$$

■ **Work Practice 9**

When both numbers to be multiplied are mixed or whole numbers, it is a good idea to estimate the product to see if your answer is reasonable. To do this, we first practice rounding mixed numbers to the nearest whole. If the fraction part of the mixed number is $\frac{1}{2}$ or greater, we round the whole number part up. If the fraction part of the mixed number is less than $\frac{1}{2}$, then we do not round the whole number part up. Study the table below for examples.

Mixed Number	Rounding
$5\frac{1}{4}$ $\frac{1}{4}$ is less than $\frac{1}{2}$ $\frac{1}{4}$ $\frac{1}{2}$	Thus, $5\frac{1}{4}$ rounds to 5.
$3\frac{9}{16}$ ← 9 is greater than 8. → Half of 16 is 8.	Thus, $3\frac{9}{16}$ rounds to 4.
$1\frac{3}{7}$ ← 3 is less than $3\frac{1}{2}$. → Half of 7 is $3\frac{1}{2}$.	Thus, $1\frac{3}{7}$ rounds to 1.

Answers

8. $\frac{4}{3}$ or $1\frac{1}{3}$ **9.** 12

Examples Multiply. Check by estimating.

10. $1\frac{2}{3} \cdot 2\frac{1}{4} = \frac{5}{3} \cdot \frac{9}{4} = \frac{5 \cdot 9}{3 \cdot 4} = \frac{5 \cdot \overset{1}{\cancel{3}} \cdot 3}{\cancel{3} \cdot 4} = \frac{15}{4}$ or $3\frac{3}{4}$ Exact

Let's check by estimating.

$1\frac{2}{3}$ rounds to 2, $2\frac{1}{4}$ rounds to 2, and $2 \cdot 2 = 4$ Estimate

The estimate is close to the exact value, so our answer is reasonable.

11. $7 \cdot 2\frac{11}{14} = \frac{7}{1} \cdot \frac{39}{14} = \frac{7 \cdot 39}{1 \cdot 14} = \frac{\overset{1}{\cancel{7}} \cdot 39}{1 \cdot 2 \cdot \cancel{7}} = \frac{39}{2}$ or $19\frac{1}{2}$ Exact

To estimate,

$2\frac{11}{14}$ rounds to 3 and $7 \cdot 3 = 21$.   Estimate

The estimate is close to the exact value, so our answer is reasonable

■ **Work Practice 10–11**

Recall from Section 1.6 that 0 multiplied by any number is 0. This is true of fractions and mixed numbers also.

Examples Multiply.

12. $0 \cdot \frac{3}{5} = 0$

13. $2\frac{3}{8} \cdot 0 = 0$

■ **Work Practice 12–13**

✓**Concept Check** Find the error.

$2\frac{1}{4} \cdot \frac{1}{2} \,\cancel{=}\, 2\frac{1 \cdot 1}{4 \cdot 2} = 2\frac{1}{8}$

Objective C Solving Problems by Multiplying Fractions ▶

To solve real-life problems that involve multiplying fractions, we use our four problem-solving steps from Chapter 1. In Example 14, a key word that implies multiplication is used. That key word is "**of.**"

Helpful Hint

"of" usually translates to multiplication.

Practice 10–11

Multiply.

10. $3\frac{1}{5} \cdot 2\frac{3}{4}$ **11.** $5 \cdot 3\frac{11}{15}$

Practice 12–13

Multiply.

12. $\frac{9}{11} \cdot 0$

13. $0 \cdot 4\frac{1}{8}$

Answers

10. $\frac{44}{5}$ or $8\frac{4}{5}$ **11.** $\frac{56}{3}$ or $18\frac{2}{3}$

12. 0   **13.** 0

✓**Concept Check Answer**

forgot to change mixed number to fraction

Practice 14

Kings Dominion is an amusement park in Doswell, Virginia. Of its 48 rides, $\frac{5}{16}$ of them are roller coasters. How many roller coasters are in Kings Dominion? (*Source:* Cedar Fair Parks)

Example 14 Finding the Number of Roller Coasters in an Amusement Park

Cedar Point is an amusement park located in Sandusky, Ohio. Its collection of 72 rides is the largest in the world. Of the rides, $\frac{2}{9}$ are roller coasters. How many roller coasters are in Cedar Point's collection of rides? (*Source:* Cedar Fair Parks)

Solution:

1. UNDERSTAND the problem. To do so, read and reread the problem. We are told that $\frac{2}{9}$ of Cedar Point's rides are roller coasters. The word "of" here means multiplication.

2. TRANSLATE.

In words:	number of roller coasters	is	$\frac{2}{9}$	of	total rides at Cedar Point
	↓	↓	↓	↓	↓
Translate:	number of roller coasters	=	$\frac{2}{9}$	·	72

3. SOLVE: Before we solve, let's estimate a reasonable answer. The fraction $\frac{2}{9}$ is less than $\frac{1}{3}$ (draw a diagram, if needed), and $\frac{1}{3}$ of 72 rides is 24 rides, so the number of roller coasters should be less than 24.

$$\frac{2}{9} \cdot 72 = \frac{2}{9} \cdot \frac{72}{1} = \frac{2 \cdot 72}{9 \cdot 1} = \frac{2 \cdot \overset{1}{\cancel{9}} \cdot 8}{\underset{1}{\cancel{9}} \cdot 1} = \frac{16}{1} \quad \text{or} \quad 16$$

4. INTERPRET. *Check* your work. From our estimate, our answer is reasonable. *State* your conclusion: The number of roller coasters at Cedar Point is 16.

■ **Work Practice 14**

Helpful Hint

To help visualize a fractional part of a whole number, look at the diagram below.

$\frac{1}{5}$ of 60 = ?

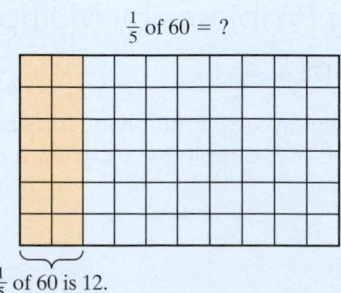

$\frac{1}{5}$ of 60 is 12.

Answer
14. 15 roller coasters

Vocabulary, Readiness & Video Check

Use the choices below to fill in each blank. Not all choices will be used.

multiplication $\dfrac{a \cdot d}{b \cdot c}$ $\dfrac{a \cdot c}{b \cdot d}$ $\dfrac{2 \cdot 2 \cdot 2}{7}$ $\dfrac{2}{7} \cdot \dfrac{2}{7} \cdot \dfrac{2}{7}$

division 0

1. To multiply two fractions, we write $\dfrac{a}{b} \cdot \dfrac{c}{d} = $ _____ .

2. Using the definition of an exponent, the expression $\dfrac{2^3}{7} = $ _____ while $\left(\dfrac{2}{7}\right)^3 = $ _____ .

3. The word "of" indicates _____ .

4. $\dfrac{1}{5} \cdot 0 = $ _____ .

Martin-Gay Interactive Videos   *Watch the section lecture video and answer the following questions.*

Objective A 5. In ⊞ Example 2, why isn't the multiplication of fractions done immediately?

Objective B 6. Why do we need to know how to multiply fractions to perform the operation in ⊞ Example 5? ▶

Objective C 7. What relationship between radius and diameter is used to solve ⊞ Example 7? ▶

See Video 2.4

2.4 Exercise Set MyMathLab® ▶

Objective A *Multiply. Write each answer in simplest form. See Examples 1 through 7 and 12.*

1. $\dfrac{1}{3} \cdot \dfrac{2}{5}$

2. $\dfrac{2}{3} \cdot \dfrac{4}{7}$

▶ 3. $\dfrac{6}{5} \cdot \dfrac{1}{7}$

4. $\dfrac{7}{3} \cdot \dfrac{1}{4}$

5. $\dfrac{3}{10} \cdot \dfrac{3}{8}$

6. $\dfrac{2}{5} \cdot \dfrac{7}{11}$

▶ 7. $\dfrac{2}{7} \cdot \dfrac{5}{8}$

8. $\dfrac{7}{8} \cdot \dfrac{2}{3}$

9. $\dfrac{16}{5} \cdot \dfrac{3}{4}$

10. $\dfrac{8}{3} \cdot \dfrac{5}{12}$

11. $\dfrac{5}{28} \cdot \dfrac{2}{25}$

12. $\dfrac{4}{35} \cdot \dfrac{5}{24}$

13. $0 \cdot \dfrac{8}{9}$

14. $\dfrac{11}{12} \cdot 0$

15. $\dfrac{1}{10} \cdot \dfrac{1}{11}$

16. $\dfrac{1}{9} \cdot \dfrac{1}{13}$

17. $\dfrac{18}{20} \cdot \dfrac{36}{99}$

18. $\dfrac{5}{32} \cdot \dfrac{64}{100}$

19. $\dfrac{3}{8} \cdot \dfrac{9}{10}$

20. $\dfrac{4}{5} \cdot \dfrac{8}{25}$

▶ 21. $\dfrac{11}{20} \cdot \dfrac{1}{7} \cdot \dfrac{5}{22}$

22. $\dfrac{27}{32} \cdot \dfrac{10}{13} \cdot \dfrac{16}{30}$

23. $\dfrac{1}{3} \cdot \dfrac{2}{7} \cdot \dfrac{1}{5}$

24. $\dfrac{3}{5} \cdot \dfrac{1}{2} \cdot \dfrac{3}{7}$

25. $\dfrac{9}{20} \cdot 0 \cdot \dfrac{4}{19}$

26. $\dfrac{8}{11} \cdot \dfrac{4}{7} \cdot 0$

27. $\dfrac{3}{14} \cdot \dfrac{6}{25} \cdot \dfrac{5}{27} \cdot \dfrac{7}{6}$

28. $\dfrac{7}{8} \cdot \dfrac{9}{20} \cdot \dfrac{12}{22} \cdot \dfrac{11}{14}$

Objective B *Round each mixed number to the nearest whole number. See the table at the bottom of page 144.*

29. $7\frac{7}{8}$ **30.** $11\frac{3}{4}$ **31.** $6\frac{1}{5}$ **32.** $4\frac{1}{9}$ **33.** $19\frac{11}{20}$ **34.** $18\frac{12}{22}$

Multiply. Write each answer in simplest form. For those exercises marked, find both an exact product and an estimated product. See Examples 8 through 13.

35. $12 \cdot \frac{1}{4}$ **36.** $\frac{2}{3} \cdot 6$ ▶ **37.** $\frac{5}{8} \cdot 4$ **38.** $10 \cdot \frac{7}{8}$ **39.** $1\frac{1}{4} \cdot \frac{4}{25}$

40. $\frac{3}{22} \cdot 3\frac{2}{3}$ **41.** $\frac{2}{5} \cdot 4\frac{1}{6}$ **42.** $2\frac{1}{9} \cdot \frac{6}{7}$ **43.** $\frac{2}{3} \cdot 1$ **44.** $1 \cdot \frac{5}{9}$

▶ **45.** $2\frac{1}{5} \cdot 3\frac{1}{2}$ **46.** $2\frac{1}{4} \cdot 7\frac{1}{8}$ **47.** $3\frac{4}{5} \cdot 6\frac{2}{7}$ **48.** $5\frac{5}{6} \cdot 7\frac{3}{5}$ **49.** $5 \cdot 2\frac{1}{2}$

Exact: Exact: Exact: Exact:

Estimate: Estimate: Estimate: Estimate:

50. $6 \cdot 3\frac{1}{3}$ **51.** $1\frac{1}{5} \cdot 12\frac{1}{2}$ **52.** $1\frac{1}{6} \cdot 7\frac{1}{5}$ **53.** $\frac{3}{4} \cdot 16 \cdot \frac{1}{2}$ **54.** $\frac{7}{8} \cdot 24 \cdot \frac{1}{3}$

55. $\frac{3}{10} \cdot 15 \cdot 2\frac{1}{2}$ **56.** $\frac{11}{20} \cdot 12 \cdot 3\frac{1}{3}$ **57.** $3\frac{1}{2} \cdot 1\frac{3}{4} \cdot 2\frac{2}{3}$ **58.** $4\frac{1}{2} \cdot 2\frac{1}{9} \cdot 1\frac{1}{5}$

Objectives A B **Mixed Practice** *Multiply and simplify. See Examples 1 through 13.*

59. $\frac{1}{4} \cdot \frac{2}{15}$ **60.** $\frac{3}{8} \cdot \frac{5}{12}$ **61.** $\frac{19}{37} \cdot 0$ **62.** $0 \cdot \frac{3}{31}$ **63.** $2\frac{4}{5} \cdot 1\frac{1}{7}$

64. $3\frac{1}{5} \cdot 2\frac{11}{32}$ **65.** $\frac{3}{2} \cdot \frac{7}{3}$ **66.** $\frac{15}{2} \cdot \frac{3}{5}$ **67.** $\frac{6}{15} \cdot \frac{5}{16}$ **68.** $\frac{9}{20} \cdot \frac{10}{90}$

69. $\frac{7}{72} \cdot \frac{9}{49}$ **70.** $\frac{3}{80} \cdot \frac{2}{27}$ **71.** $20 \cdot \frac{11}{12}$ **72.** $30 \cdot \frac{8}{9}$ **73.** $9\frac{5}{7} \cdot 8\frac{1}{5} \cdot 0$

74. $4\frac{11}{13} \cdot 0 \cdot 12\frac{1}{13}$ **75.** $12\frac{4}{5} \cdot 6\frac{7}{8} \cdot \frac{26}{77}$ **76.** $14\frac{2}{5} \cdot 8\frac{1}{3} \cdot \frac{11}{16}$

Objective C *Solve. Write each answer in simplest form. For Exercises 77 through 80, recall that "of" translates to multiplication. See Example 14.*

77. Find $\frac{1}{4}$ of 200.

78. Find $\frac{1}{5}$ of 200.

79. Find $\frac{5}{6}$ of 24.

80. Find $\frac{5}{8}$ of 24.

Solve. For Exercises 81 and 82, the solutions have been started for you. See Example 14.

81. In the United States, $\frac{4}{25}$ of college freshmen major in business. A community college in Pennsylvania has a freshman enrollment of approximately 800 students. How many of these freshmen might we expect to major in business?

Start the solution:

1. UNDERSTAND the problem. Reread it as many times as needed.
2. TRANSLATE into an equation. (Fill in the blank below.)

freshmen majoring in business	is	$\frac{4}{25}$	of	community college freshmen enrollment
↓	↓	↓	↓	↓

$$\text{freshmen majoring in business} = \frac{4}{25} \cdot \underline{\hspace{2cm}}$$

Finish with:

3. SOLVE
4. INTERPRET

82. A patient was told that, at most, $\frac{1}{5}$ of his calories should come from fat. If his diet consists of 3000 calories a day, find the maximum number of calories that can come from fat.

Start the solution:

1. UNDERSTAND the problem. Reread it as many times as needed.
2. TRANSLATE into an equation. (Fill in the blank below.)

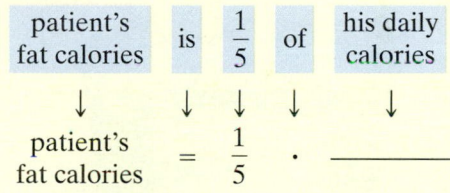

patient's fat calories	is	$\frac{1}{5}$	of	his daily calories
↓	↓	↓	↓	↓

$$\text{patient's fat calories} = \frac{1}{5} \cdot \underline{\hspace{2cm}}$$

Finish with:

3. SOLVE
4. INTERPRET

83. In 2012, there were approximately 225 million moviegoers in the United States and Canada. Of these, about $\frac{13}{25}$ were female. Find the approximate number of females who attended the movies in that year. (*Source:* Motion Picture Association of America)

84. In 2012, cinemas in the United States and Canada sold about 1400 million movie tickets. About $\frac{57}{100}$ of these tickets were purchased by frequent moviegoers who go to the cinema once or more per month. Find the number of tickets purchased by frequent moviegoers in 2012. (*Source:* Motion Picture Association of America)

85. The Oregon National Historic Trail is 2170 miles long. It begins in Independence, Missouri, and ends in Oregon City, Oregon. Manfred Coulon has hiked $\frac{2}{5}$ of the trail before. How many miles has he hiked?

(*Source:* National Park Service)

86. Each turn of a screw sinks it $\frac{3}{16}$ of an inch deeper into a piece of wood. Find how deep the screw is after 8 turns.

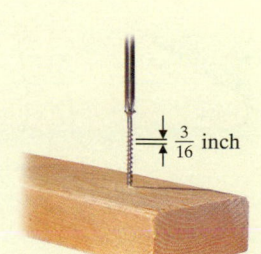

87. The radius of a circle is one-half of its diameter, as shown. If the diameter of a circle is $\frac{3}{8}$ of an inch, what is its radius?

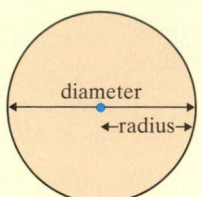

88. The diameter of a circle is twice its radius, as shown in the Exercise 87 illustration. If the radius of a circle is $\frac{7}{20}$ of a foot, what is its diameter?

89. A veterinarian's dipping vat holds 36 gallons of liquid. She normally fills it $\frac{5}{6}$ full of a medicated flea dip solution. Find how many gallons of solution are normally in the vat.

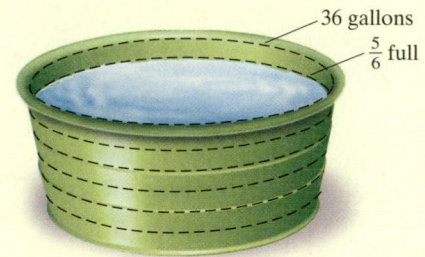

90. The plans for a deck call for $\frac{2}{5}$ of a 4-foot post to be underground. Find the length of the post that is to be buried.

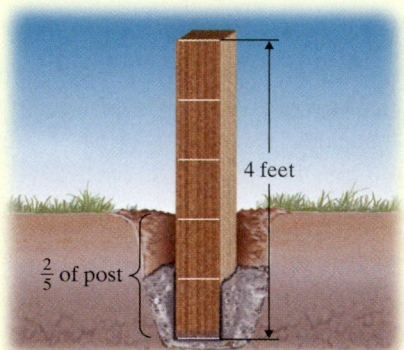

91. An estimate for the measure of an adult's wrist is $\frac{1}{4}$ of the waist size. If Jorge has a 34-inch waist, estimate the size of his wrist.

92. An estimate for an adult's waist measurement is found by multiplying the neck size (in inches) by 2. Jock's neck measures $17\frac{1}{2}$ inches. Estimate his waist measurement.

93. A sidewalk is built 6 bricks wide by laying each brick side by side. How many inches wide is the sidewalk if each brick measures $3\frac{1}{4}$ inches wide?

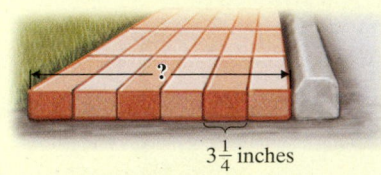

94. A recipe calls for $\frac{1}{3}$ of a cup of flour. How much flour should be used if only $\frac{1}{2}$ of the recipe is being made?

95. A Japanese company called Che-ez! manufactures a small digital camera, the SPYZ camera. The face of the camera measures $2\frac{9}{25}$ inches by $1\frac{13}{25}$ inches and is slightly bigger than a Zippo lighter. Find the area of the face of this camera. (Area = length · width)

$1\frac{13}{25}$ in.

$2\frac{9}{25}$ in.

96. As part of his research, famous tornado expert Dr. T. Fujita studied approximately 31,050 tornadoes that occurred in the United States between 1916 and 1985. He found that roughly $\frac{7}{10}$ of these tornadoes occurred during April, May, June, and July. How many of these tornadoes occurred during these four months? (*Source: U.S. Tornadoes Part 1*, T. Fujita, University of Chicago)

Find the area of each rectangle. Recall that area = length · width.

△ **97.**

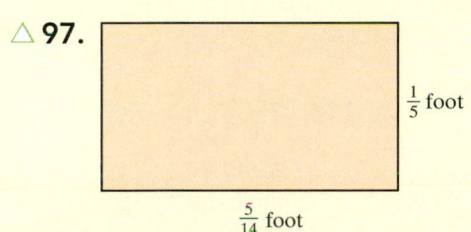

$\frac{1}{5}$ foot

$\frac{5}{14}$ foot

△ **98.**

$\frac{1}{2}$ mile

$\frac{3}{8}$ mile

△ **99.**

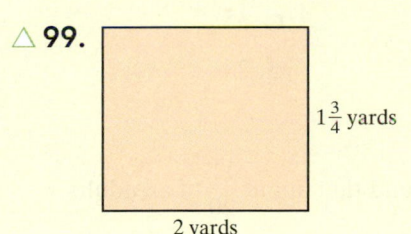

$1\frac{3}{4}$ yards

2 yards

△ **100.**

5 inches

$3\frac{1}{2}$ inches

*Recall that the following graph is called a **circle graph** or **pie chart**. Each sector (shaped like a piece of pie) shows the fractional part of a car's total mileage that falls into a particular category. The whole circle represents a car's total mileage.*

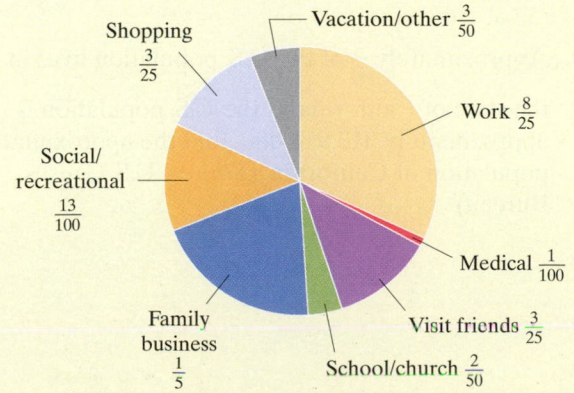

Shopping $\frac{3}{25}$

Vacation/other $\frac{3}{50}$

Work $\frac{8}{25}$

Social/recreational $\frac{13}{100}$

Medical $\frac{1}{100}$

Family business $\frac{1}{5}$

Visit friends $\frac{3}{25}$

School/church $\frac{2}{50}$

Source: The American Automobile Manufacturers Association and The National Automobile Dealers Association

In one year, a family drove 12,000 miles in the family car. Use the circle graph to determine how many of these miles might be expected to fall in the categories shown in Exercises 101 through 104.

101. Work

102. Shopping

103. Family business

104. Medical

Review

Divide. See Section 1.7.

105. $8\overline{)1648}$ **106.** $7\overline{)3920}$ **107.** $23\overline{)1300}$ **108.** $31\overline{)2500}$

Concept Extensions

109. In your own words, explain how to multiply
  a. fractions
  b. mixed numbers

110. In your own words, explain how to round a mixed number to the nearest whole number.

Find the error in each calculation. See the Concept Check in this section.

111. $3\frac{2}{3} \cdot 1\frac{1}{7} = 3\frac{2}{21}$

112. $5 \cdot 2\frac{1}{4} = 10\frac{1}{4}$

Choose the best estimate for each product.

113. $3\frac{1}{5} \cdot 4\frac{5}{8}$
  a. 7
  b. 15
  c. 8
  d. $12\frac{1}{8}$

114. $\frac{11}{12} \cdot 4\frac{1}{16}$
  a. 16
  b. 1
  c. 4
  d. 8

115. $9 \cdot \frac{10}{11}$
  a. 9
  b. 90
  c. 99
  d. 0

116. $7\frac{1}{4} \cdot 4\frac{1}{5}$
  a. 40
  b. $\frac{7}{5}$
  c. 35
  d. 28

117. If $\frac{3}{4}$ of 36 students on a first bus are girls and $\frac{2}{3}$ of the 30 students on a second bus are *boys,* how many students on the two buses are girls?

118. In 2013, a survey found that about $\frac{11}{20}$ of all adults in the United States owned a smartphone. There were roughly 240 million U.S. adults at that time. How many U.S. adults owned a smartphone in 2013? (*Source:* Pew Research Center, U.S. Census Bureau)

119. The estimated population of New Zealand was 4,433,000 in 2012. About $\frac{3}{20}$ of New Zealand's population is of Māori descent. How many Māori lived in New Zealand in 2012? (*Source:* Statistics New Zealand)

120. Approximately $\frac{1}{8}$ of the U.S. population lives in the state of California. If the U.S. population is approximately 313,914,000, find the approximate population of California. (*Source:* U.S. Census Bureau)

Objective A Finding Reciprocals of Fractions

Before we can divide fractions, we need to know how to find the **reciprocal** of a fraction or whole number.

Objectives

A Find the Reciprocal of a Fraction.

B Divide Fractions.

C Divide Fractions and Mixed Numbers or Whole Numbers.

D Solve Problems by Dividing Fractions.

Reciprocal of a Fraction

Two numbers are **reciprocals** of each other if their product is 1. The reciprocal of the fraction $\frac{a}{b}$ is $\frac{b}{a}$ because $\frac{a}{b} \cdot \frac{b}{a} = \frac{a \cdot b}{b \cdot a} = 1$.

Finding the Reciprocal of a Fraction

To find the reciprocal of a fraction, interchange its numerator and denominator.

For example,

The reciprocal of $\frac{2}{5}$ is $\frac{5}{2}$ because $\frac{2}{5} \cdot \frac{5}{2} = \frac{10}{10} = 1$.

The reciprocal of 7, or $\frac{7}{1}$, is $\frac{1}{7}$ because $7 \cdot \frac{1}{7} = \frac{7}{1} \cdot \frac{1}{7} = \frac{7}{7} = 1$.

Examples Find the reciprocal of each number.

1. The reciprocal of $\frac{5}{6}$ is $\frac{6}{5}$.
$$\frac{5}{6} \cdot \frac{6}{5} = \frac{5 \cdot 6}{6 \cdot 5} = \frac{30}{30} = 1$$

2. The reciprocal of $\frac{11}{8}$ is $\frac{8}{11}$.
$$\frac{11}{8} \cdot \frac{8}{11} = \frac{11 \cdot 8}{8 \cdot 11} = \frac{88}{88} = 1$$

3. The reciprocal of $\frac{1}{3}$ is $\frac{3}{1}$ or 3.
$$\frac{1}{3} \cdot \frac{3}{1} = \frac{1 \cdot 3}{3 \cdot 1} = \frac{3}{3} = 1$$

4. The reciprocal of 5, or $\frac{5}{1}$, is $\frac{1}{5}$.
$$\frac{5}{1} \cdot \frac{1}{5} = \frac{5 \cdot 1}{1 \cdot 5} = \frac{5}{5} = 1$$

■ **Work Practice 1–4**

Practice 1–4

Find the reciprocal of each number.

1. $\frac{4}{9}$ **2.** $\frac{15}{7}$

3. 9   **4.** $\frac{1}{8}$

Helpful Hint

Every number except 0 has a reciprocal. The number 0 has no reciprocal because there is no number that when multiplied by 0 gives a result of 1.

Objective B Dividing Fractions

Division of fractions has the same meaning as division of whole numbers. For example,

$10 \div 5$ means: How many 5s are there in 10?

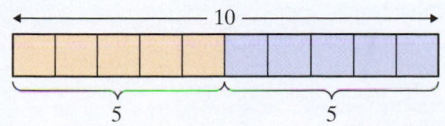

There are two 5s in 10, so $10 \div 5 = 2$.

Answers

1. $\frac{9}{4}$ **2.** $\frac{7}{15}$ **3.** $\frac{1}{9}$ **4.** 8

$\dfrac{3}{4} \div \dfrac{1}{8}$ means: How many $\dfrac{1}{8}$s are there in $\dfrac{3}{4}$?

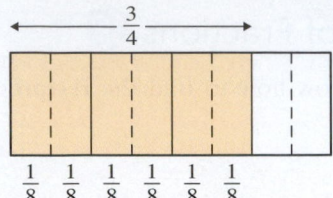

There are six $\dfrac{1}{8}$ s in $\dfrac{3}{4}$, so $\dfrac{3}{4} \div \dfrac{1}{8} = 6$.

We use reciprocals to divide fractions.

Dividing Fractions

To divide two fractions, multiply the first fraction by the reciprocal of the second fraction.

If a, b, c, and d represent numbers, and b, c, and d are not 0, then

$$\dfrac{a}{b} \div \dfrac{c}{d} = \dfrac{a}{b} \cdot \underbrace{\dfrac{d}{c}}_{\text{reciprocal}} = \dfrac{a \cdot d}{b \cdot c}$$

For example,

$$\dfrac{3}{4} \overset{\text{multiply by reciprocal}}{\div} \dfrac{1}{8} = \dfrac{3}{4} \cdot \dfrac{8}{1} = \dfrac{3 \cdot 8}{4 \cdot 1} = \dfrac{3 \cdot 2 \cdot \overset{1}{\cancel{4}}}{\cancel{4} \cdot 1} = \dfrac{6}{1} \text{ or } 6$$

Just as when you are multiplying fractions, always check to see whether your answer can be simplified when you divide fractions.

Practice 5–7

Divide and simplify.

5. $\dfrac{3}{2} \div \dfrac{14}{5}$ **6.** $\dfrac{8}{7} \div \dfrac{2}{9}$

7. $\dfrac{4}{9} \div \dfrac{1}{2}$

Examples   Divide and simplify.

5. $\dfrac{7}{8} \div \dfrac{2}{9} = \dfrac{7}{8} \cdot \dfrac{9}{2} = \dfrac{7 \cdot 9}{8 \cdot 2} = \dfrac{63}{16}$

6. $\dfrac{5}{16} \div \dfrac{3}{4} = \dfrac{5}{16} \cdot \dfrac{4}{3} = \dfrac{5 \cdot 4}{16 \cdot 3} = \dfrac{5 \cdot \overset{1}{\cancel{4}}}{\cancel{4} \cdot 4 \cdot 3} = \dfrac{5}{12}$

7. $\dfrac{2}{5} \div \dfrac{1}{2} = \dfrac{2}{5} \cdot \dfrac{2}{1} = \dfrac{2 \cdot 2}{5 \cdot 1} = \dfrac{4}{5}$

■ Work Practice 5–7

Helpful Hint

When dividing fractions, do *not* look for common factors to divide out until you rewrite the division as multiplication.

Do not try to divide out these two 2s.

$$\dfrac{1}{2} \div \dfrac{2}{3} = \dfrac{1}{2} \cdot \dfrac{3}{2} = \dfrac{3}{4}$$

Answers

5. $\dfrac{15}{28}$ **6.** $\dfrac{36}{7}$ **7.** $\dfrac{8}{9}$

Objective C Dividing Fractions and Mixed Numbers or Whole Numbers ▶

Just as with multiplying, mixed or whole numbers should be written as fractions before you divide them.

Dividing Fractions and Mixed Numbers or Whole Numbers

To divide with a mixed number or a whole number, first write the mixed or whole number as a fraction and then divide as usual.

Examples Divide.

8. $\dfrac{3}{4} \div 5 = \dfrac{3}{4} \div \dfrac{5}{1} = \dfrac{3}{4} \cdot \dfrac{1}{5} = \dfrac{3 \cdot 1}{4 \cdot 5} = \dfrac{3}{20}$

9. $\dfrac{11}{18} \div 2\dfrac{5}{6} = \dfrac{11}{18} \div \dfrac{17}{6} = \dfrac{11}{18} \cdot \dfrac{6}{17} = \dfrac{11 \cdot 6}{18 \cdot 17} = \dfrac{11 \cdot \overset{1}{\cancel{6}}}{\underset{1}{\cancel{6}} \cdot 3 \cdot 17} = \dfrac{11}{51}$

10. $5\dfrac{2}{3} \div 2\dfrac{5}{9} = \dfrac{17}{3} \div \dfrac{23}{9} = \dfrac{17}{3} \cdot \dfrac{9}{23} = \dfrac{17 \cdot 9}{3 \cdot 23} = \dfrac{17 \cdot \overset{3}{\cancel{9}} \cdot 3}{\underset{1}{\cancel{3}} \cdot 23} = \dfrac{51}{23} \text{ or } 2\dfrac{5}{23}$

■ **Work Practice 8–10**

Practice 8–10

Divide.

8. $\dfrac{4}{9} \div 7$ **9.** $\dfrac{8}{15} \div 3\dfrac{4}{5}$

10. $3\dfrac{2}{7} \div 2\dfrac{3}{14}$

Recall from Section 1.7 that the quotient of 0 and any number (except 0) is 0. This is true of fractions and mixed numbers also. For example,

$$0 \div \dfrac{7}{8} = 0 \cdot \dfrac{8}{7} = 0 \quad \text{\color{blue}Recall that 0 multiplied by any number is 0.}$$

Also recall from Section 1.7 that the quotient of any number and 0 is undefined. This is also true of fractions and mixed numbers. For example, to find $\dfrac{7}{8} \div 0$, or $\dfrac{7}{8} \div \dfrac{0}{1}$, we would need to find the reciprocal of 0 $\left(\text{or } \dfrac{0}{1}\right)$. As we mentioned in the helpful hint at the beginning of this section, 0 has no reciprocal because there is no number that when multiplied by 0 gives a result of 1. Thus,

$$\dfrac{7}{8} \div 0 \text{ is undefined.}$$

Examples Divide.

11. $0 \div \dfrac{2}{21} = 0 \cdot \dfrac{21}{2} = 0$ **12.** $1\dfrac{3}{4} \div 0 \text{ is undefined.}$

■ **Work Practice 11–12**

Practice 11–12

Divide.

11. $\dfrac{14}{17} \div 0$ **12.** $0 \div 2\dfrac{1}{8}$

Answers

8. $\dfrac{4}{63}$ **9.** $\dfrac{8}{57}$ **10.** $\dfrac{46}{31}$ or $1\dfrac{15}{31}$

11. undefined   **12.** 0

✔**Concept Check** Which of the following is the correct way to divide $\dfrac{2}{5}$ by $\dfrac{3}{4}$? Or are both correct? Explain.

a. $\dfrac{5}{2} \cdot \dfrac{3}{4}$

b. $\dfrac{2}{5} \cdot \dfrac{4}{3}$

✔**Concept Check Answers**

a. incorrect   **b.** correct

Objective D Solving Problems by Dividing Fractions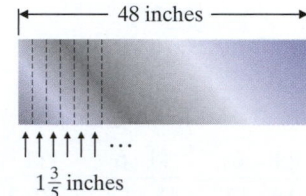

To solve real-life problems that involve dividing fractions, we continue to use our four problem-solving steps.

Example 13 Calculating Manufacturing Materials Needed

In a manufacturing process, a metal-cutting machine cuts strips $1\frac{3}{5}$ inches wide from a piece of metal stock. How many such strips can be cut from a 48-inch piece of stock?

Solution:

1. UNDERSTAND the problem. To do so, read and reread the problem. Then draw a diagram:

| ←——— 48 inches ———→ |

↑↑↑↑↑↑ ···
$1\frac{3}{5}$ inches

We want to know how many $1\frac{3}{5}$s there are in 48.

2. TRANSLATE.

In words:

Number of strips	is	48	divided by	$1\frac{3}{5}$
↓	↓	↓	↓	↓

Translate:

Number of strips	=	48	÷	$1\frac{3}{5}$

3. SOLVE: Let's estimate a reasonable answer. The mixed number $1\frac{3}{5}$ rounds to 2 and $48 \div 2 = 24$.

$$48 \div 1\frac{3}{5} = 48 \div \frac{8}{5} = \frac{48}{1} \cdot \frac{5}{8} = \frac{48 \cdot 5}{1 \cdot 8} = \frac{\overset{1}{\cancel{8}} \cdot 6 \cdot 5}{1 \cdot \underset{1}{\cancel{8}}} = \frac{30}{1} \text{ or } 30$$

4. INTERPRET. *Check* your work. Since the exact answer of 30 is close to our estimate of 24, our answer is reasonable. *State* your conclusion: Thirty strips can be cut from the 48-inch piece of stock.

■ Work Practice 13

Vocabulary, Readiness & Video Check

Use the choices below to fill in each blank. Not all choices will be used.

multiplication	$\dfrac{a \cdot d}{b \cdot c}$	$\dfrac{a \cdot c}{b \cdot d}$
division	0	reciprocals

1. Two numbers are _____ of each other if their product is 1.

2. Every number has a reciprocal except _____.

3. To divide two fractions, we write $\dfrac{a}{b} \div \dfrac{c}{d} =$ _____.

4. The word "per" usually indicates _____.

Martin-Gay Interactive Videos   Watch the section lecture video and answer the following questions.

Objective A 5. From Example 2, what can we conclude is the reciprocal of any nonzero number *n*? ▶

Objective B 6. From ⊞ Example 6, what number has no reciprocal? ▶

Objective C 7. In ⊞ Example 8, why can't we divide out common factors once we've written the mixed numbers as fractions? ▶

Objective C 8. In ⊞ Example 9, what phrase tells us that we have a division problem? ▶

See Video 2.5

2.5 Exercise Set MyMathLab® ▶

Objective A *Find the reciprocal of each number. See Examples 1 through 4.*

▶ **1.** $\dfrac{4}{7}$

2. $\dfrac{9}{10}$

3. $\dfrac{1}{11}$

4. $\dfrac{1}{20}$

▶ **5.** 15

6. 13

7. $\dfrac{12}{7}$

8. $\dfrac{10}{3}$

Objective B *Divide. Write each answer in simplest form. See Examples 5 through 7 and 11 and 12.*

▶ **9.** $\dfrac{2}{3} \div \dfrac{5}{6}$

10. $\dfrac{5}{8} \div \dfrac{2}{3}$

11. $\dfrac{8}{9} \div \dfrac{1}{2}$

12. $\dfrac{10}{11} \div \dfrac{4}{5}$

13. $\dfrac{3}{7} \div \dfrac{5}{6}$

14. $\dfrac{16}{27} \div \dfrac{8}{15}$

15. $\dfrac{3}{5} \div \dfrac{4}{5}$

16. $\dfrac{11}{16} \div \dfrac{13}{16}$

▶ **17.** $\dfrac{1}{10} \div \dfrac{10}{1}$

18. $\dfrac{3}{13} \div \dfrac{13}{3}$

19. $\dfrac{7}{9} \div \dfrac{7}{3}$

20. $\dfrac{6}{11} \div \dfrac{6}{5}$

21. $\dfrac{5}{8} \div \dfrac{3}{8}$

22. $\dfrac{7}{8} \div \dfrac{5}{6}$

23. $\dfrac{7}{45} \div \dfrac{4}{25}$

24. $\dfrac{14}{52} \div \dfrac{1}{13}$

25. $\dfrac{2}{37} \div \dfrac{1}{7}$

26. $\dfrac{100}{158} \div \dfrac{10}{79}$

▶ **27.** $\dfrac{3}{25} \div \dfrac{27}{40}$

28. $\dfrac{6}{15} \div \dfrac{7}{10}$

29. $\dfrac{11}{12} \div \dfrac{11}{12}$

30. $\dfrac{7}{13} \div \dfrac{7}{13}$

▶ **31.** $\dfrac{8}{13} \div 0$

32. $0 \div \dfrac{4}{11}$

33. $0 \div \dfrac{7}{8}$

34. $\dfrac{2}{3} \div 0$

35. $\dfrac{25}{126} \div \dfrac{125}{441}$

36. $\dfrac{65}{495} \div \dfrac{26}{231}$

Objective **C** *Divide. Write each answer in simplest form. See Examples 8 through 12.*

37. $\dfrac{2}{3} \div 4$

38. $\dfrac{5}{6} \div 10$

39. $8 \div \dfrac{3}{5}$

40. $7 \div \dfrac{2}{11}$

41. $2\dfrac{1}{2} \div \dfrac{1}{2}$

42. $4\dfrac{2}{3} \div \dfrac{2}{5}$

43. $\dfrac{5}{12} \div 2\dfrac{1}{3}$

44. $\dfrac{4}{15} \div 2\dfrac{1}{2}$

45. $3\dfrac{3}{7} \div 3\dfrac{1}{3}$

46. $2\dfrac{5}{6} \div 4\dfrac{6}{7}$

47. $1\dfrac{4}{9} \div 2\dfrac{5}{6}$

48. $3\dfrac{1}{10} \div 2\dfrac{1}{5}$

49. $0 \div 15\dfrac{4}{7}$

50. $\dfrac{33}{50} \div 1$

51. $1 \div \dfrac{13}{17}$

52. $0 \div 7\dfrac{9}{10}$

53. $1 \div \dfrac{18}{35}$

54. $\dfrac{17}{75} \div 1$

55. $10\dfrac{5}{9} \div 16\dfrac{2}{3}$

56. $20\dfrac{5}{6} \div 137\dfrac{1}{2}$

Objectives **B** **C** **Mixed Practice** *Divide. Write each answer in simplest form. See Examples 5 through 12.*

57. $\dfrac{6}{15} \div \dfrac{12}{5}$

58. $\dfrac{4}{15} \div \dfrac{8}{3}$

59. $\dfrac{11}{20} \div \dfrac{3}{11}$

60. $\dfrac{9}{20} \div \dfrac{2}{9}$

61. $12 \div \dfrac{1}{8}$

62. $9 \div \dfrac{1}{6}$

63. $\dfrac{3}{7} \div \dfrac{4}{7}$

64. $\dfrac{3}{8} \div \dfrac{5}{8}$

65. $2\dfrac{3}{8} \div 0$

66. $20\dfrac{1}{5} \div 0$

67. $\dfrac{11}{85} \div \dfrac{7}{5}$

68. $\dfrac{13}{84} \div \dfrac{3}{16}$

69. $4\dfrac{5}{11} \div 1\dfrac{2}{5}$

70. $8\dfrac{2}{7} \div 3\dfrac{1}{7}$

71. $\dfrac{27}{100} \div \dfrac{3}{20}$

72. $\dfrac{25}{128} \div \dfrac{5}{32}$

Objective D *Solve. For Exercises 73 and 74, the solutions have been started for you. Write each answer in simplest form. See Example 13.*

73. A heart attack patient in rehabilitation walked on a treadmill $12\frac{3}{4}$ miles over 4 days. How many miles is this per day?

Start the solution:

1. UNDERSTAND the problem. Reread it as many times as needed.
2. TRANSLATE into an equation. (Fill in the blanks.)

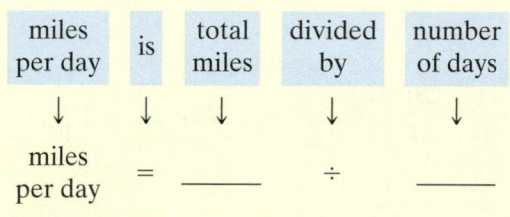

miles per day	is	total miles	divided by	number of days
↓	↓	↓	↓	↓

miles per day $=$ _____ $\div$ _____

Finish with:

3. SOLVE and
4. INTERPRET

74. A local restaurant is selling hamburgers from a booth on Memorial Day. A total of $27\frac{3}{4}$ pounds of hamburger have been ordered. How many quarter-pound hamburgers can this make?

Start the solution:

1. UNDERSTAND the problem. Reread it as many times as needed.
2. TRANSLATE into an equation. (Fill in the blanks.)

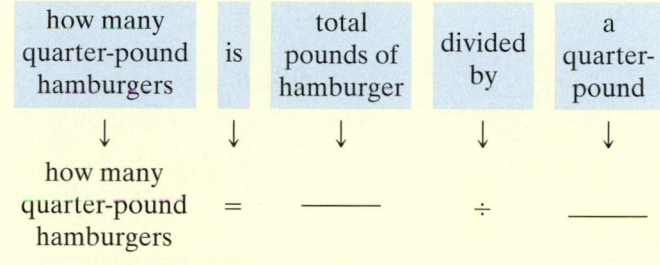

how many quarter-pound hamburgers	is	total pounds of hamburger	divided by	a quarter-pound
↓	↓	↓	↓	↓

how many quarter-pound hamburgers $=$ _____ $\div$ _____

Finish with:

3. SOLVE and
4. INTERPRET

75. A patient is to take $3\frac{1}{3}$ tablespoons of medicine per day in 4 equally divided doses. How much medicine is to be taken in each dose?

76. If there are $13\frac{1}{3}$ grams of fat in 4 ounces of lean hamburger meat, how many grams of fat are in an ounce?

77. The record for rainfall during a 24-hour period in Alaska is $15\frac{1}{5}$ inches. This record was set in Angoon, Alaska, in October 1982. How much rain fell per hour on average? (*Source:* National Climatic Data Center)

78. An order for 125 custom-made candle stands was placed with Mr. Levi, the manager of Just For You, Inc. The worker assigned to the job can produce $2\frac{3}{5}$ candle stands per hour. Using this worker, how many work hours will be required to complete the order?

79. In September 2013, the average price of aluminum was $82\frac{1}{10}$ ¢ per pound. During that time, a family received 1642¢ for aluminum cans that they sold for recycling at a scrap metal center. Assuming that they received the average price, how many pounds of aluminum cans did they recycle? (*Source:* London Metal Exchange)

80. Yoko's Fine Jewelry paid $450 for a $\frac{3}{4}$-carat gem. At this price, what is the cost of one carat?

△ 81. The area of the rectangle below is 12 square meters. If its width is $2\frac{4}{7}$ meters, find its length.

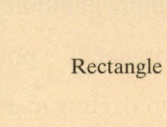

| Rectangle | $2\frac{4}{7}$ meters |

△ 82. The perimeter of the square below is $23\frac{1}{2}$ feet. Find the length of each side.

| Square |

Mixed Practice (Sections 2.4, 2.5)   *Perform the indicated operation.*

83. $\frac{2}{5} \cdot \frac{4}{7}$

84. $\frac{2}{5} \div \frac{4}{7}$

85. $2\frac{2}{3} \div 1\frac{1}{16}$

86. $2\frac{2}{3} \cdot 1\frac{1}{16}$

87. $5\frac{1}{7} \cdot \frac{2}{9} \cdot \frac{14}{15}$

88. $8\frac{1}{6} \cdot \frac{3}{7} \cdot \frac{18}{25}$

89. $\frac{11}{20} \div \frac{20}{11}$

90. $2\frac{1}{5} \div 1\frac{7}{10}$

Review

Perform each indicated operation. See Sections 1.3 and 1.4.

91.
$\begin{array}{r} 27 \\ 76 \\ + 98 \\ \hline \end{array}$

92.
$\begin{array}{r} 811 \\ 42 \\ + 69 \\ \hline \end{array}$

93.
$\begin{array}{r} 968 \\ -772 \\ \hline \end{array}$

94.
$\begin{array}{r} 882 \\ -773 \\ \hline \end{array}$

95.
$\begin{array}{r} 2000 \\ - 431 \\ \hline \end{array}$

96.
$\begin{array}{r} 500 \\ - 92 \\ \hline \end{array}$

Concept Extensions

A student asked you to find the error in the work below. Find the error and correct it. See the Concept Check in this section.

97. $20\frac{2}{3} \div 10\frac{1}{2} = 2\frac{1}{3}$

98. $6\frac{1}{4} \div \frac{1}{2} = 3\frac{1}{8}$

Choose the best estimate for each quotient.

99. $20\frac{1}{4} \div \frac{5}{6}$

    a. 5   **b.** $5\frac{1}{8}$ **c.** 20   **d.** 10

100. $\frac{11}{12} \div 16\frac{1}{5}$

    a. $\frac{1}{16}$ **b.** 4   **c.** 8   **d.** 16

101. $12\frac{2}{13} \div 3\frac{7}{8}$

    a. 4   **b.** 9   **c.** 36   **d.** 3

102. $10\frac{1}{4} \div 2\frac{1}{16}$

    a. 8   **b.** 5   **c.** 20   **d.** 12

Simplify.

103. $\dfrac{42}{25} \cdot \dfrac{125}{36} \div \dfrac{7}{6}$

104. $\left(\dfrac{8}{13} \cdot \dfrac{39}{16} \cdot \dfrac{8}{9}\right)^2 \div \dfrac{1}{2}$

105. The FedEx Express air fleet includes 64 Boeing MD11s. These Boeing MD11s make up $\dfrac{32}{317}$ of the FedEx Express fleet. How many aircraft make up the entire FedEx Express air fleet? (*Source:* FedEx Corporation)

106. One-third of all native flowering plant species in the United States are at risk of becoming extinct. That translates into 5144 at-risk flowering plant species. Based on this data, how many flowering plant species are native to the United States overall? (*Source:* The Nature Conservancy)

    (*Hint:* How many $\dfrac{1}{3}$s are in 5144?)

107. In your own words, describe how to find the reciprocal of a number.

108. In your own words, describe how to divide fractions.

Chapter 2 Group Activity

Blood and Blood Donation (Sections 2.1, 2.2, 2.3)

Blood is the workhorse of the body. It carries to the body's tissues everything they need, from nutrients to antibodies to heat. Blood also carries away waste products like carbon dioxide. Blood contains three types of cells—red blood cells, white blood cells, and platelets—suspended in clear, watery fluid called plasma. Blood is $\dfrac{11}{20}$ plasma, and plasma itself is $\dfrac{9}{10}$ water. In the average healthy adult human, blood accounts for $\dfrac{1}{11}$ of a person's body weight.

Roughly every 2 seconds someone in the United States needs blood. Although only $\dfrac{1}{20}$ of eligible donors donate blood, the American Red Cross is still able to collect nearly 6 million volunteer donations of blood each year. This volume makes Red Cross Biomedical Services the largest blood supplier for blood transfusions in the United States.

Group Activity

Contact your local Red Cross Blood Service office. Find out how many people donated blood in your area in the past two months. Ask whether it is possible to get a breakdown of the blood donations by blood type. (For more on blood types, see Exercises 75 through 78 in Section 2.3.)

1. Research the population of the area served by your local Red Cross Blood Service office. Write the fraction of the local population who gave blood in the past two months.

2. Use the breakdown by blood type to write the fraction of donors giving each type of blood.

Chapter 2 Vocabulary Check

Fill in each blank with one of the words or phrases listed below.

mixed number	equivalent	0	undefined
composite number	improper fraction	simplest form	prime factorization
prime number	proper fraction	numerator	denominator
reciprocals	cross products		

1. Two numbers are _____ of each other if their product is 1.

2. A(n) _____ is a natural number greater than 1 that is not prime.

3. Fractions that represent the same portion of a whole are called _____ fractions.

4. A(n) _____ is a fraction whose numerator is greater than or equal to its denominator.

5. A(n) _____ is a natural number greater than 1 whose only factors are 1 and itself.

6. A fraction is in _____ when the numerator and the denominator have no factors in common other than 1.

7. A(n) _____ is one whose numerator is less than its denominator.

8. A(n) _____ contains a whole number part and a fraction part.

9. In the fraction $\frac{7}{9}$, the 7 is called the _____ and the 9 is called the _____.

10. The _____ of a number is the factorization in which all the factors are prime numbers.

11. The fraction $\frac{3}{0}$ is _____.

12. The fraction $\frac{0}{5} = $ _____.

13. In $\frac{a}{b} = \frac{c}{d}$, $a \cdot d$ and $b \cdot c$ are called _____.

Helpful Hint ▸ Are you preparing for your test? Don't forget to take the Chapter 2 Test on page 168. Then check your answers at the back of the text and use the Chapter Test Prep Videos to see the fully worked-out solutions to any of the exercises you want to review.

2 Chapter Highlights

Definitions and Concepts	Examples
Section 2.1 Introduction to Fractions and Mixed Numbers	

A **fraction** is of the form: $\frac{\text{numerator}}{\text{denominator}}$ ← number of parts being considered ← number of equal parts in the whole	Write a fraction to represent the shaded part of the figure. 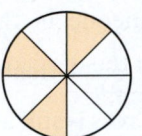 $\dfrac{3}{8}$ ← number of parts shaded ← number of equal parts
A fraction is called a **proper fraction** if its numerator is less than its denominator.	$\dfrac{1}{3}, \dfrac{2}{5}, \dfrac{7}{8}, \dfrac{100}{101}$
A fraction is called an **improper fraction** if its numerator is greater than or equal to its denominator.	$\dfrac{5}{4}, \dfrac{2}{2}, \dfrac{9}{7}, \dfrac{101}{100}$
A **mixed number** contains a whole number and a fraction.	$1\dfrac{1}{2}, 5\dfrac{7}{8}, 25\dfrac{9}{10}$

Definitions and Concepts	Examples

Section 2.1   Introduction to Fractions and Mixed Numbers (*continued*)

To Write a Mixed Number as an Improper Fraction

1. Multiply the denominator of the fraction by the whole number.

2. Add the numerator of the fraction to the product from step 1.

3. Write the sum from step 2 as the numerator of the improper fraction over the original denominator.

$$5\frac{2}{7} = \frac{7 \cdot 5 + 2}{7} = \frac{35 + 2}{7} = \frac{37}{7}$$

To Write an Improper Fraction as a Mixed Number or a Whole Number

1. Divide the denominator into the numerator.

2. The whole number part of the mixed number is the quotient. The fraction is the remainder over the original denominator.

$$\text{quotient} \frac{\text{remainder}}{\text{original denominator}}$$

$$\frac{17}{3} = 5\frac{2}{3}$$

$$\begin{array}{r} 5 \\ 3{\overline{)17}} \\ -15 \\ \hline 2 \end{array}$$

Section 2.2   Factors and Prime Factorization

A **prime number** is a natural number that has exactly two different factors, 1 and itself.

$2, 3, 5, 7, 11, 13, 17, \ldots$

A **composite number** is any natural number other than 1 that is not prime.

$4, 6, 8, 9, 10, 12, 14, 15, 16, \ldots$

The prime factorization of a number is the factorization in which all the factors are prime numbers.

Write the prime factorization of 60.
$$60 = 6 \cdot 10$$
$$= 2 \cdot 3 \cdot 2 \cdot 5 \quad \text{or} \quad 2^2 \cdot 3 \cdot 5$$

Section 2.3   Simplest Form of a Fraction

Fractions that represent the same portion of a whole are called **equivalent fractions.**

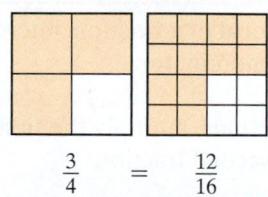

$$\frac{3}{4} = \frac{12}{16}$$

A fraction is in **simplest form** or **lowest terms** when the numerator and the denominator have no common factors other than 1.

The fraction $\frac{2}{3}$ is in simplest form.

To write a fraction in simplest form, write the prime factorizations of the numerator and the denominator and then divide both by all common factors.

Write in simplest form: $\dfrac{30}{36}$

$$\frac{30}{36} = \frac{2 \cdot 3 \cdot 5}{2 \cdot 2 \cdot 3 \cdot 3} = \frac{2}{2} \cdot \frac{3}{3} \cdot \frac{5}{2 \cdot 3} = 1 \cdot 1 \cdot \frac{5}{6} = \frac{5}{6}$$

$$\text{or} \quad \frac{30}{36} = \frac{2 \cdot \overset{1}{3} \cdot 5}{2 \cdot 2 \cdot \underset{1}{3} \cdot 3} = \frac{5}{6}$$

(continued)

Definitions and Concepts	Examples
Section 2.3 Simplest Form of a Fraction (continued)	

Two fractions are equivalent if

Determine whether $\dfrac{7}{8}$ and $\dfrac{21}{24}$ are equivalent.

Method 1.   They simplify to the same fraction.

Method 1. $\dfrac{7}{8}$ is in simplest form; $\dfrac{21}{24} = \dfrac{\overset{1}{\cancel{3}} \cdot 7}{\cancel{3} \cdot 8} = \dfrac{7}{8}$

Since both simplify to $\dfrac{7}{8}$, then $\dfrac{7}{8} = \dfrac{21}{24}$.

Method 2.   Their cross products are equal.

Method 2.

$24 \cdot 7$
$= 168$ $\dfrac{7}{8} = \dfrac{21}{24}$ $8 \cdot 21$
$= 168$

Since $168 = 168$, $\dfrac{7}{8} = \dfrac{21}{24}$

| **Section 2.4   Multiplying Fractions and Mixed Numbers** | |

To multiply two fractions, multiply the numerators and multiply the denominators.

Multiply.

$$\frac{7}{8} \cdot \frac{3}{5} = \frac{7 \cdot 3}{8 \cdot 5} = \frac{21}{40}$$

$$\frac{3}{4} \cdot \frac{1}{6} = \frac{3 \cdot 1}{4 \cdot 6} = \frac{\overset{1}{\cancel{3}} \cdot 1}{4 \cdot \underset{1}{\cancel{3}} \cdot 2} = \frac{1}{8}$$

To multiply with mixed numbers or whole numbers, first write any mixed or whole numbers as fractions and then multiply as usual.

$$2\frac{1}{3} \cdot \frac{1}{9} = \frac{7}{3} \cdot \frac{1}{9} = \frac{7 \cdot 1}{3 \cdot 9} = \frac{7}{27}$$

| **Section 2.5   Dividing Fractions and Mixed Numbers** | |

To find the **reciprocal** of a fraction, interchange its numerator and denominator.

The reciprocal of $\dfrac{3}{5}$ is $\dfrac{5}{3}$.

To divide two fractions, multiply the first fraction by the reciprocal of the second fraction.

Divide.

$$\frac{3}{10} \div \frac{7}{9} = \frac{3}{10} \cdot \frac{9}{7} = \frac{3 \cdot 9}{10 \cdot 7} = \frac{27}{70}$$

To divide with mixed numbers or whole numbers, first write any mixed or whole numbers as fractions and then divide as usual.

$$2\frac{5}{8} \div 3\frac{7}{16} = \frac{21}{8} \div \frac{55}{16} = \frac{21}{8} \cdot \frac{16}{55} = \frac{21 \cdot 16}{8 \cdot 55}$$

$$= \frac{21 \cdot 2 \cdot \overset{1}{\cancel{8}}}{\underset{1}{\cancel{8}} \cdot 55} = \frac{42}{55}$$

(2.1) *Determine whether each number is an improper fraction, a proper fraction, or a mixed number.*

1. $\frac{11}{23}$

2. $\frac{9}{8}$

3. $\frac{1}{2}$

4. $2\frac{1}{4}$

Write a fraction to represent the shaded area.

5.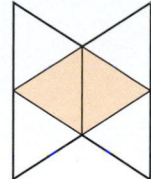

6.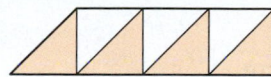

7.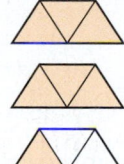

8.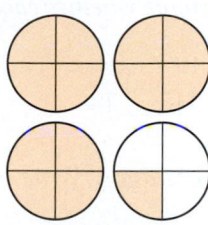

9. A basketball player made 11 free throws out of 12 during a game. What fraction of free throws did the player make?

10. A new car lot contained 23 blue cars out of a total of 131 cars.
   a. How many cars on the lot are not blue?
   b. What fraction of cars on the lot are not blue?

Write each improper fraction as a mixed number or a whole number.

11. $\frac{15}{4}$

12. $\frac{275}{6}$

13. $\frac{39}{13}$

14. $\frac{60}{12}$

Write each mixed number as an improper fraction.

15. $1\frac{1}{5}$

16. $1\frac{1}{21}$

17. $2\frac{8}{9}$

18. $3\frac{11}{12}$

(2.2) *Identify each number as prime or composite.*

19. 51

20. 17

List all factors of each number.

21. 42

22. 20

Find the prime factorization of each number.

23. 68

24. 90

25. 785

26. 255

(2.3) *Write each fraction in simplest form.*

27. $\frac{12}{28}$

28. $\frac{15}{27}$

29. $\frac{25}{75}$

30. $\frac{36}{72}$

31. $\frac{29}{32}$

32. $\frac{18}{23}$

33. $\frac{48}{6}$

34. $\frac{54}{9}$

Solve.

35. There are 12 inches in a foot. What fractional part of a foot does 8 inches represent?

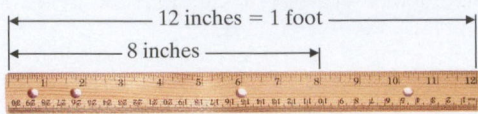

36. Six out of 15 cars are white. What fraction of the cars are *not* white?

Determine whether each two fractions are equivalent.

37. $\dfrac{10}{34}$ and $\dfrac{4}{14}$

38. $\dfrac{30}{50}$ and $\dfrac{9}{15}$

(2.4) *Multiply. Write each answer in simplest form. Estimate where noted.*

39. $\dfrac{3}{5} \cdot \dfrac{1}{2}$

40. $\dfrac{6}{7} \cdot \dfrac{5}{12}$

41. $\dfrac{24}{5} \cdot \dfrac{15}{8}$

42. $\dfrac{27}{21} \cdot \dfrac{7}{18}$

43. $5 \cdot \dfrac{7}{8}$

44. $6 \cdot \dfrac{5}{12}$

45. $\dfrac{39}{3} \cdot \dfrac{7}{13} \cdot \dfrac{5}{21}$

46. $\dfrac{42}{5} \cdot \dfrac{15}{6} \cdot \dfrac{7}{9}$

47. $1\dfrac{5}{8} \cdot 3\dfrac{1}{5}$

Exact:

Estimate:

48. $3\dfrac{6}{11} \cdot 1\dfrac{7}{13}$

Exact:

Estimate:

49. $\dfrac{3}{4} \cdot 8 \cdot 4\dfrac{1}{8}$

50. $2\dfrac{1}{9} \cdot 3 \cdot \dfrac{1}{38}$

51. There are $7\dfrac{1}{3}$ grams of fat in each ounce of hamburger. How many grams of fat are in a 5-ounce hamburger patty?

52. An art teacher needs 45 pieces of PVC piping for an art project. If each piece needs to be $\dfrac{3}{4}$ inch long, find the total length of piping she needs.

△ **53.** Find the area of each rectangle.

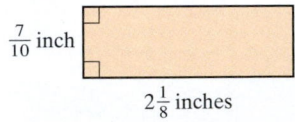

$\dfrac{7}{10}$ inch

$2\dfrac{1}{8}$ inches

△ **54.**

5 meters

$6\dfrac{7}{8}$ meters

(2.5) *Find the reciprocal of each number.*

55. 7

56. $\dfrac{1}{8}$

57. $\dfrac{14}{23}$

58. $\dfrac{17}{5}$

Divide. Write each answer in simplest form.

59. $\dfrac{3}{4} \div \dfrac{3}{8}$

60. $\dfrac{21}{4} \div \dfrac{7}{5}$

61. $\dfrac{5}{3} \div 2$

62. $5 \div \dfrac{15}{8}$

63. $6\dfrac{3}{4} \div 1\dfrac{2}{7}$

64. $5\dfrac{1}{2} \div 2\dfrac{1}{11}$

65. A truck traveled 341 miles on $15\frac{1}{2}$ gallons of gas. How many miles might we expect the truck to travel on 1 gallon of gas?

66. Herman Heltznutt walks 5 days a week for a total distance of $5\frac{1}{4}$ miles per week. If he walks the same distance each day, find the distance he walks each day.

Mixed Review

Determine whether each number is an improper fraction, a proper fraction, or a mixed number.

67. $\frac{0}{3}$

68. $\frac{12}{12}$

69. $5\frac{6}{7}$

70. $\frac{13}{9}$

Write each improper fraction as a mixed number or a whole number. Write each mixed number as an improper fraction.

71. $\frac{125}{4}$

72. $\frac{54}{9}$

73. $5\frac{10}{17}$

74. $7\frac{5}{6}$

Identify each number as prime or composite.

75. 27

76. 23

Find the prime factorization of each number.

77. 180

78. 98

Write each fraction in simplest form.

79. $\frac{45}{50}$

80. $\frac{30}{42}$

81. $\frac{140}{150}$

82. $\frac{84}{140}$

Multiply or divide as indicated. Write each answer in simplest form. Estimate where noted.

83. $\frac{7}{8} \cdot \frac{2}{3}$

84. $\frac{6}{15} \cdot \frac{5}{8}$

85. $\frac{18}{5} \div \frac{2}{5}$

86. $\frac{9}{2} \div \frac{1}{3}$

87. $4\frac{1}{6} \cdot 2\frac{2}{5}$

Exact:

Estimate:

88. $5\frac{2}{3} \cdot 2\frac{1}{4}$

Exact:

Estimate:

89. $\frac{7}{2} \div 1\frac{1}{2}$

90. $1\frac{3}{5} \div \frac{1}{4}$

△**91.** A slab of natural granite is purchased and a rectangle with length $7\frac{4}{11}$ feet and width $5\frac{1}{2}$ feet is cut from it. Find the area of the rectangle.

92. An area of Mississippi received $23\frac{1}{2}$ inches of rain in $30\frac{1}{2}$ hours. How many inches per 1 hour is this?

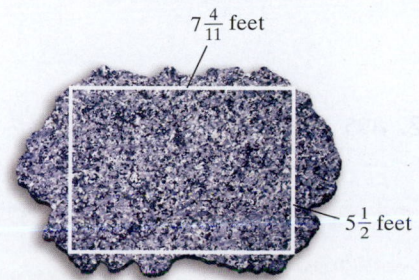

$7\frac{4}{11}$ feet

$5\frac{1}{2}$ feet

Answers

Write a fraction to represent the shaded area.

1.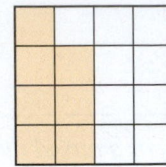

2.

1. _____

Write each mixed number as an improper fraction.

3. $7\frac{2}{3}$

4. $3\frac{6}{11}$

2. _____

3. _____

Write each improper fraction as a mixed number or a whole number.

5. $\frac{23}{5}$

6. $\frac{75}{4}$

4. _____

5. _____

6. _____

Write each fraction in simplest form.

7. $\frac{24}{210}$

8. $\frac{42}{70}$

7. _____

8. _____

9. _____

Determine whether these fractions are equivalent.

9. $\frac{5}{7}$ and $\frac{8}{11}$

10. $\frac{6}{27}$ and $\frac{14}{63}$

10. _____

11. _____

Find the prime factorization of each number.

11. 84

12. 495

12. _____

Perform each indicated operation. Write each answer in simplest form.

13. $\dfrac{4}{4} \div \dfrac{3}{4}$

14. $\dfrac{4}{3} \cdot \dfrac{4}{4}$

15. $2 \cdot \dfrac{1}{8}$

16. $\dfrac{2}{3} \cdot \dfrac{8}{15}$

17. $8 \div \dfrac{1}{2}$

18. $13\dfrac{1}{2} \div 3$

19. $\dfrac{3}{8} \cdot \dfrac{16}{6} \cdot \dfrac{4}{11}$

20. $5\dfrac{1}{4} \div \dfrac{7}{12}$

21. $\dfrac{16}{3} \div \dfrac{3}{12}$

22. $3\dfrac{1}{3} \cdot 6\dfrac{3}{4}$

23. $12 \div 3\dfrac{1}{3}$

24. $\dfrac{14}{5} \cdot \dfrac{25}{21} \cdot 2$

△**25.** Find the area of the figure.

$\frac{2}{3}$ mile ▭

$1\dfrac{8}{9}$ miles

26. During a 258-mile trip, a car used $10\dfrac{3}{4}$ gallons of gas. How many miles would we expect the car to travel on 1 gallon of gas?

27. How many square yards of artificial turf are necessary to cover a football field, *not* including the end zones and the sidelines? (*Hint:* A football field measures $100 \times 53\dfrac{1}{3}$ yards.)

$53\frac{1}{3}$ yards

100 yards

28. Prior to an oil spill, the stock in an oil company sold for $120 per share. As a result of the liability that the company incurred from the spill, the price per share fell to $\dfrac{3}{4}$ of the price before the spill. What did the stock sell for after the spill?

13. _____

14. _____

15. _____

16. _____

17. _____

18. _____

19. _____

20. _____

21. _____

22. _____

23. _____

24. _____

25. _____

26. _____

27. _____

28. _____

Answers

1. Find the place value of the digit 3 in the whole number 396,418.

2. Write 2036 in words.

1. _____

3. Write the number, eight hundred five, in standard form.

4. Add: $7 + 6 + 10 + 3 + 5$

2. _____

5. Add: $34,285 + 149,761$

6. Find the average of 56, 18, and 43.

3. _____

△ 7. Find the perimeter of the polygon shown.

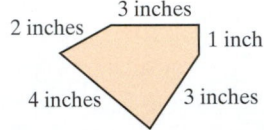

2 inches 3 inches 1 inch 4 inches 3 inches

8. Subtract 8 from 25.

4. _____

5. _____

9. In 2011, a total of 12,734,424 passenger vehicles were sold in the United States. In 2012, total passenger vehicle sales in the United States had increased by 1,705,636 vehicles. Find the total number of passenger vehicles sold in the United States in 2012. (*Source:* Alliance of Automobile Manufacturers)

6. _____

7. _____

10. Find $\sqrt{25}$.

11. Subtract: $7826 - 505$
Check by adding.

12. Find 8^2.

8. _____

13. In the following graph, each bar represents a country and the height of each bar represents the number of threatened mammal species identified in that country.

9. _____

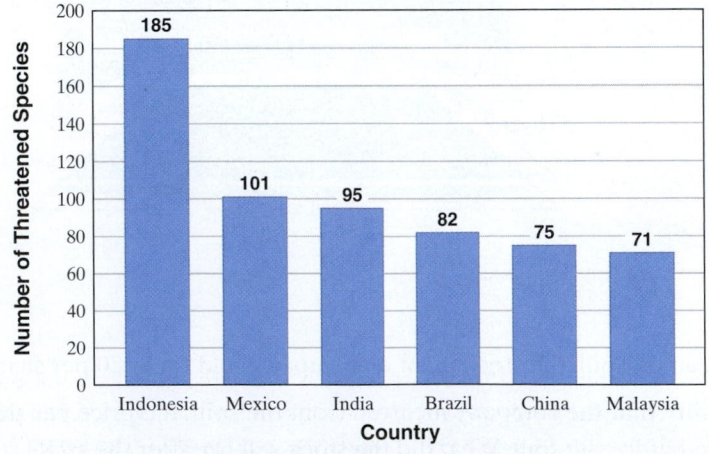

Number of Threatened Mammal Species, 2013

Source: International Union for Conservation of Nature

10. _____

11. _____

12. _____

13. a. _____

b. _____

a. Which country shown has the greatest number of threatened mammal species?
b. Find the total number of threatened mammal species for Malaysia, China, and Indonesia.

14. Find $205 \div 8$.

15. Round 568 to the nearest ten.

16. Round 2366 to the nearest hundred.

17. Round each number to the nearest hundred to find an estimated difference.

$$4725$$
$$-2879$$

18. Round each number to the nearest ten to find an estimated sum.
$38 + 43 + 126 + 92$

19. Multiply.
- **a.** 6×1
- **b.** $0(18)$
- **c.** $1 \cdot 45$
- **d.** $(75)(0)$

20. Simplify: $30 \div 3 \cdot 2$

21. Rewrite each using the distributive property.
- **a.** $3(4 + 5)$
- **b.** $10(6 + 8)$
- **c.** $2(7 + 3)$

22. Multiply: 12×15

23. Find each quotient. Check by multiplying.
- **a.** $9\overline{)0}$
- **b.** $0 \div 12$
- **c.** $\dfrac{0}{5}$
- **d.** $\dfrac{3}{0}$

24. Find the area.

7 miles | Rectangle

22 miles

25. Divide and check: $1872 \div 9$

26. Subtract: $5000 - 986$

27. As part of a promotion, an executive receives 238 cards, each good for one free song download. If she wants to share them evenly with 19 friends, how many download cards will each friend receive? How many will be left over?

28. Find the product of 9 and 7.

29. A gardener bought enough plants to fill a rectangular garden with length 30 feet and width 20 feet. Because of shading problems from a nearby tree, the gardener changed the width of the garden to 15 feet. If the area is to remain the same, what is the new length of the garden?

30. Find the sum of 9 and 7.

14. _____

15. _____

16. _____

17. _____

18. _____

19. a. _____

  b. _____

  c. _____

  d. _____

20. _____

21. a. _____

  b. _____

  c. _____

22. _____

23. a. _____

  b. _____

  c. _____

  d. _____

24. _____

25. _____

26. _____

27. _____

28. _____

29. _____

30. _____

31. _____

32. _____

33. _____

34. _____

35. _____

36. _____

37. _____

38. _____

39. a. _____

   b. _____

40. _____

41. _____

42. _____

43. _____

44. _____

45. _____

46. _____

47. _____

48. _____

49. _____

50. _____

Write using exponential notation.

31. $7 \cdot 7 \cdot 7$

32. $7 \cdot 7 \cdot 7 \cdot 7$

33. $3 \cdot 3 \cdot 3 \cdot 3 \cdot 17 \cdot 17 \cdot 17$

34. $2 \cdot 2 \cdot 3 \cdot 3 \cdot 3 \cdot 3$

35. Simplify: $2 \cdot 4 - 3 \div 3$

36. Simplify: $8 \cdot \sqrt{100} - 4^2 \cdot 5$

37. Write a fraction to represent the shaded part of the figure.

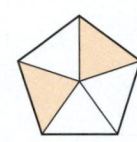

38. Write the prime factorization of 156.

39. Write each as an improper fraction.

   a. $4\frac{2}{9}$ **b.** $1\frac{8}{11}$

40. Write $7\frac{4}{5}$ as an improper fraction.

41. Find all the factors of 20.

42. Determine whether $\frac{8}{20}$ and $\frac{14}{35}$ are equivalent.

43. Write in simplest form: $\frac{42}{66}$

44. Write in simplest form: $\frac{70}{105}$

45. Multiply: $3\frac{1}{3} \cdot \frac{7}{8}$

46. Multiply: $\frac{2}{3} \cdot 4$

47. Find the reciprocal of $\frac{1}{3}$.

48. Find the reciprocal of 9.

49. Divide and simplify: $\frac{5}{16} \div \frac{3}{4}$

50. Divide: $1\frac{1}{10} \div 5\frac{3}{5}$

Adding and Subtracting Fractions

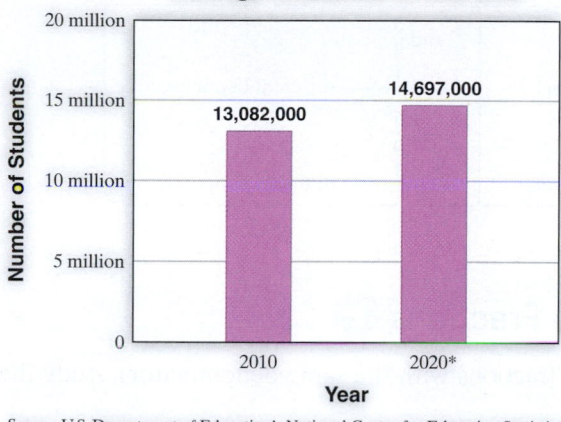

Increase in Full-Time College Students in the U.S.

Source: U.S. Department of Education's National Center for Education Statistics

* prediction

About $\frac{3}{4}$ of College Students Have a Job. What Else Do They Do?

The circle graph below shows how an average full-time college student spends his or her weekdays and the portion of each weekday that is devoted to each type of activity. For example, on average, a full-time U.S. college student spends $\frac{36}{100}$ of each weekday sleeping, and so on. In Section 3.1, Exercises 49–52, we use fractions to help us understand the time use illustrated in this circle graph.

Having learned what fractions are and how to multiply and divide them in Chapter 2, we are ready to continue our study of fractions. In this chapter, we learn how to add and subtract fractions and mixed numbers. We then conclude this chapter with solving problems using fractions.

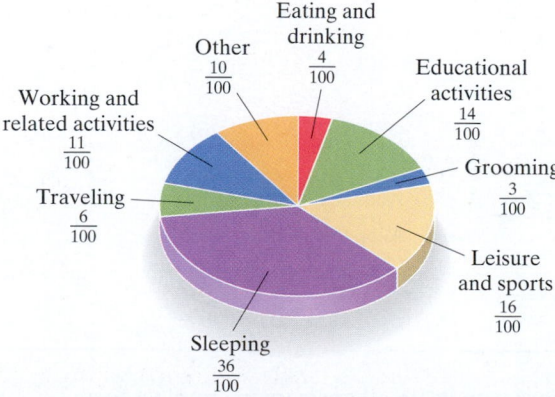

Full-Time College Students' Average Weekday Time Use

Source: Bureau of Labor Statistics

3.1 Adding and Subtracting Like Fractions

Objectives

A Add Like Fractions.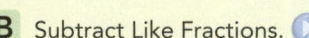

B Subtract Like Fractions.

C Solve Problems by Adding or Subtracting Like Fractions.

Fractions with the same denominator are called **like fractions.** Fractions that have different denominators are called **unlike fractions.**

Like Fractions	Unlike Fractions
$\frac{2}{5}$ and $\frac{3}{5}$ ↑↑ —— same denominator	$\frac{2}{5}$ and $\frac{3}{4}$ ↑↑ —— different denominators
$\frac{5}{21}, \frac{16}{21},$ and $\frac{7}{21}$ ↑↑↑ —— same denominator	$\frac{5}{7}$ and $\frac{5}{9}$ ↑↑ —— different denominators

Objective A Adding Like Fractions

To see how we add like fractions (fractions with the same denominator), study the figures below:

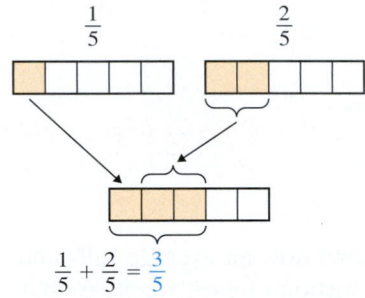

$$\frac{1}{5} + \frac{2}{5} = \frac{3}{5}$$

Adding Like Fractions (Fractions with the Same Denominator)

To add like fractions, add the numerators and write the sum over the common denominator.

If $a, b,$ and c represent nonzero whole numbers, we have

$$\frac{a}{c} + \frac{b}{c} = \frac{a+b}{c}$$

For example,

$$\frac{1}{4} + \frac{2}{4} = \frac{1+2}{4} = \frac{3}{4}$$ ← Add the numerators.
← Keep the denominator.

Helpful Hint

As usual, don't forget to write all answers in simplest form.

Examples Add and simplify.

1. $\dfrac{2}{7} + \dfrac{3}{7} = \dfrac{2+3}{7} = \dfrac{5}{7}$ ⟵ Add the numerators.
   ⟵ Keep the common denominator.

2. $\dfrac{3}{16} + \dfrac{7}{16} = \dfrac{3+7}{16} = \dfrac{10}{16} = \dfrac{\overset{1}{\cancel{2}}\cdot 5}{\underset{1}{\cancel{2}}\cdot 8} = \dfrac{5}{8}$

3. $\dfrac{7}{13} + \dfrac{6}{13} + \dfrac{3}{13} = \dfrac{7+6+3}{13} = \dfrac{16}{13}$ or $1\dfrac{3}{13}$

■ **Work Practice 1–3**

✓**Concept Check** Find and correct the error in the following:

$\dfrac{1}{5} + \dfrac{1}{5} \ \cancel{=} \ \dfrac{2}{10}$

Objective B Subtracting Like Fractions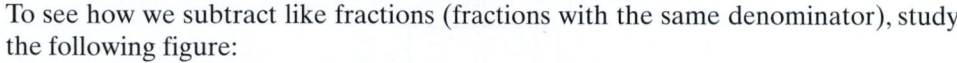

To see how we subtract like fractions (fractions with the same denominator), study the following figure:

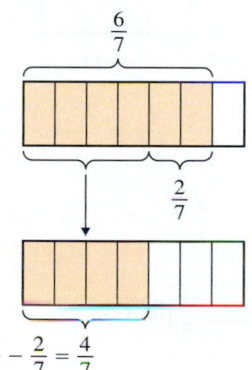

$\dfrac{6}{7}$

$\dfrac{2}{7}$

$\dfrac{6}{7} - \dfrac{2}{7} = \dfrac{4}{7}$

Subtracting Like Fractions (Fractions with the Same Denominator)

To subtract like fractions, subtract the numerators and write the difference over the common denominator.
   If a, b, and c represent nonzero whole numbers, then

$$\dfrac{a}{c} - \dfrac{b}{c} = \dfrac{a-b}{c}$$

For example,

$\dfrac{4}{5} - \dfrac{2}{5} = \dfrac{4-2}{5} = \dfrac{2}{5}$ ⟵ Subtract the numerators.
   ⟵ Keep the denominator.

Examples Subtract and simplify.

4. $\dfrac{8}{9} - \dfrac{1}{9} = \dfrac{8-1}{9} = \dfrac{7}{9}$ ⟵ Subtract the numerators.
   ⟵ Keep the common denominator.

5. $\dfrac{7}{8} - \dfrac{5}{8} = \dfrac{7-5}{8} = \dfrac{2}{8} = \dfrac{\overset{1}{\cancel{2}}}{\underset{1}{\cancel{2}}\cdot 4} = \dfrac{1}{4}$

■ **Work Practice 4–5**

Objective C Solving Problems by Adding or Subtracting Like Fractions ▶

Many real-life problems involve finding the perimeters of square or rectangular areas such as pastures, swimming pools, and so on. We can use our knowledge of adding fractions to find perimeters.

Practice 6

Find the perimeter of the square.

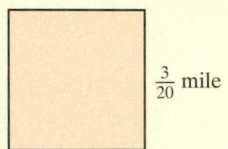

$\frac{3}{20}$ mile

Example 6 Find the perimeter of the rectangle.

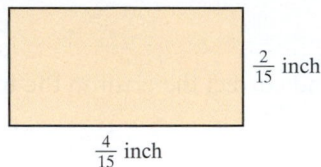

$\frac{2}{15}$ inch

$\frac{4}{15}$ inch

Solution: Recall that perimeter means distance around and that opposite sides of a rectangle are the same length.

$\frac{4}{15}$ inch

$\frac{2}{15}$ inch $\frac{2}{15}$ inch

$\frac{4}{15}$ inch

$$\text{Perimeter} = \frac{2}{15} + \frac{4}{15} + \frac{2}{15} + \frac{4}{15} = \frac{2+4+2+4}{15}$$

$$= \frac{12}{15} = \frac{\cancel{3} \cdot 4}{\cancel{3} \cdot 5} = \frac{4}{5}$$

The perimeter of the rectangle is $\frac{4}{5}$ inch.

■ **Work Practice 6**

We can combine our skills in adding and subtracting fractions with our four problem-solving steps from Chapter 1 to solve many kinds of real-life problems.

Practice 7

If a piano student practices the piano $\frac{3}{8}$ of an hour in the morning and $\frac{1}{8}$ of an hour in the evening, how long did she practice that day?

Example 7 Total Amount of an Ingredient in a Recipe

A recipe calls for $\frac{1}{3}$ of a cup of honey at the beginning and $\frac{2}{3}$ of a cup of honey later. How much total honey is needed to make the recipe?

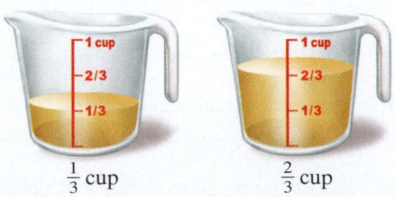

$\frac{1}{3}$ cup $\frac{2}{3}$ cup

Solution:

1. **UNDERSTAND** the problem. To do so, read and reread the problem. Since we are finding total honey, we add.

Answers

6. $\frac{3}{5}$ mi   **7.** $\frac{1}{2}$ hr

2. TRANSLATE.

In words:	total honey	is	honey at the beginning	added to	honey later
	↓	↓	↓	↓	↓
Translate:	total honey	=	$\frac{1}{3}$	+	$\frac{2}{3}$

3. SOLVE: $\dfrac{1}{3} + \dfrac{2}{3} = \dfrac{1+2}{3} = \dfrac{\overset{1}{\cancel{3}}}{\cancel{3}} = 1$

4. INTERPRET. *Check* your work. *State* your conclusion: The total honey needed for the recipe is 1 cup.

■ **Work Practice 7**

Example 8 Calculating Distance

The distance from home to the World Gym is $\dfrac{7}{8}$ of a mile and from home to the post office is $\dfrac{3}{8}$ of a mile. How much farther is it from home to the World Gym than from home to the post office?

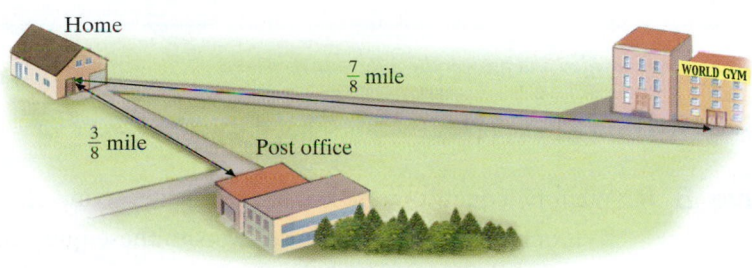

Solution:

1. UNDERSTAND. Read and reread the problem. The phrase "How much farther" tells us to subtract distances.

2. TRANSLATE.

In words:	distance farther	is	home to World Gym distance	minus	home to post office distance
	↓	↓	↓	↓	↓
Translate:	distance farther	=	$\frac{7}{8}$	−	$\frac{3}{8}$

3. SOLVE: $\dfrac{7}{8} - \dfrac{3}{8} = \dfrac{7-3}{8} = \dfrac{4}{8} = \dfrac{\overset{1}{\cancel{4}}}{2 \cdot \underset{1}{\cancel{4}}} = \dfrac{1}{2}$

4. INTERPRET. *Check* your work. *State* your conclusion: The distance from home to the World Gym is $\dfrac{1}{2}$ mile farther than from home to the post office.

■ **Work Practice 8**

Practice 8

A jogger ran $\dfrac{13}{4}$ miles on Monday and $\dfrac{7}{4}$ miles on Wednesday. How much farther did he run on Monday than on Wednesday?

Answer

8. $\dfrac{3}{2}$ or $1\dfrac{1}{2}$ mi

Vocabulary, Readiness & Video Check

Use the choices below to fill in each blank. Not all choices will be used.

perimeter   like $\dfrac{a-c}{b}$ $\dfrac{a+c}{b}$

equivalent   unlike

1. The fractions $\dfrac{9}{11}$ and $\dfrac{13}{11}$ are called _____ fractions while $\dfrac{3}{4}$ and $\dfrac{1}{3}$ are called _____ fractions.

2. $\dfrac{a}{b} + \dfrac{c}{b} = $ _____ .

3. $\dfrac{a}{b} - \dfrac{c}{b} = $ _____ .

4. The distance around a figure is called its _____ .

State whether the fractions in each list are like or unlike fractions.

5. $\dfrac{7}{8}, \dfrac{7}{10}$

6. $\dfrac{2}{3}, \dfrac{4}{9}$

7. $\dfrac{9}{10}, \dfrac{1}{10}$

8. $\dfrac{8}{11}, \dfrac{2}{11}$

9. $\dfrac{2}{31}, \dfrac{30}{31}, \dfrac{19}{31}$

10. $\dfrac{3}{10}, \dfrac{3}{11}, \dfrac{3}{13}$

11. $\dfrac{5}{12}, \dfrac{7}{12}, \dfrac{12}{11}$

12. $\dfrac{1}{5}, \dfrac{2}{5}, \dfrac{4}{5}$

Martin-Gay Interactive Videos

See Video 3.1 🍎

Watch the section lecture video and answer the following questions.

Objective A 13. In ▣ Example 2, why is $\dfrac{6}{9}$ not the final answer? ▶

Objective B 14. What two questions are asked during the solving of ▣ Example 5? What are the answers to these questions? ▶

Objective C 15. What is the perimeter equation used to solve ▣ Example 6? What is the final answer? ▶

3.1 Exercise Set MyMathLab®

Objective A *Add and simplify. See Examples 1 through 3.*

▶ 1. $\dfrac{1}{7} + \dfrac{2}{7}$

2. $\dfrac{9}{17} + \dfrac{2}{17}$

3. $\dfrac{1}{10} + \dfrac{1}{10}$

4. $\dfrac{1}{4} + \dfrac{1}{4}$

▶ 5. $\dfrac{2}{9} + \dfrac{4}{9}$

6. $\dfrac{3}{10} + \dfrac{2}{10}$

7. $\dfrac{6}{20} + \dfrac{1}{20}$

8. $\dfrac{2}{8} + \dfrac{3}{8}$

9. $\dfrac{3}{14} + \dfrac{4}{14}$

10. $\dfrac{5}{24} + \dfrac{7}{24}$

11. $\dfrac{10}{11} + \dfrac{3}{11}$

12. $\dfrac{13}{17} + \dfrac{9}{17}$

13. $\frac{4}{13} + \frac{2}{13} + \frac{1}{13}$

14. $\frac{5}{11} + \frac{1}{11} + \frac{2}{11}$

15. $\frac{7}{18} + \frac{3}{18} + \frac{2}{18}$

16. $\frac{7}{15} + \frac{4}{15} + \frac{1}{15}$

Objective B *Subtract and simplify. See Examples 4 and 5.*

17. $\frac{10}{11} - \frac{4}{11}$

18. $\frac{9}{13} - \frac{5}{13}$

19. $\frac{4}{5} - \frac{1}{5}$

20. $\frac{7}{8} - \frac{4}{8}$

21. $\frac{7}{4} - \frac{3}{4}$

22. $\frac{18}{5} - \frac{3}{5}$

23. $\frac{7}{8} - \frac{1}{8}$

24. $\frac{5}{6} - \frac{1}{6}$

25. $\frac{25}{12} - \frac{15}{12}$

26. $\frac{30}{20} - \frac{15}{20}$

27. $\frac{11}{10} - \frac{3}{10}$

28. $\frac{14}{15} - \frac{4}{15}$

29. $\frac{86}{90} - \frac{85}{90}$

30. $\frac{74}{80} - \frac{73}{80}$

31. $\frac{27}{33} - \frac{8}{33}$

32. $\frac{37}{45} - \frac{18}{45}$

Objectives A B Mixed Practice *Perform the indicated operation. See Examples 1 through 5.*

33. $\frac{8}{21} + \frac{5}{21}$

34. $\frac{7}{37} + \frac{9}{37}$

35. $\frac{99}{100} - \frac{9}{100}$

36. $\frac{85}{200} - \frac{15}{200}$

37. $\frac{13}{28} - \frac{13}{28}$

38. $\frac{15}{26} - \frac{15}{26}$

39. $\frac{3}{16} + \frac{7}{16} + \frac{2}{16}$

40. $\frac{5}{18} + \frac{1}{18} + \frac{6}{18}$

Objective C *Find the perimeter of each figure. (Hint: Recall that perimeter means distance around.) See Example 6.*

41.
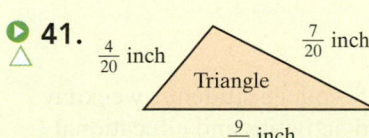
$\frac{4}{20}$ inch $\frac{7}{20}$ inch   Triangle $\frac{9}{20}$ inch

42.
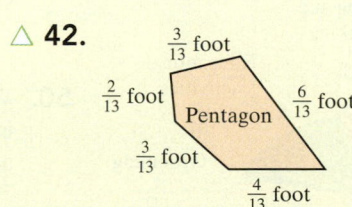
$\frac{3}{13}$ foot $\frac{2}{13}$ foot   Pentagon $\frac{6}{13}$ foot $\frac{3}{13}$ foot $\frac{4}{13}$ foot

43.
$\frac{5}{12}$ meter   Rectangle $\frac{7}{12}$ meter

44.
Square $\frac{1}{6}$ centimeter

Solve. For Exercises 45 and 46, the solutions have been started for you. Write each answer in simplest form. See Examples 7 and 8.

45. A railroad inspector must inspect $\frac{19}{20}$ of a mile of railroad track. If she has already inspected $\frac{5}{20}$ of a mile, how much more does she need to inspect?

Start the solution:

1. UNDERSTAND the problem. Reread it as many times as needed.

2. TRANSLATE into an equation. (Fill in the blanks.)

distance left to inspect	is	distance needed to inspect	minus	distance already inspected
↓	↓	↓	↓	↓

$$\text{distance left to inspect} = \underline{\qquad} - \underline{\qquad}$$

Finish with:

3. SOLVE. and
4. INTERPRET.

46. Scott Davis has run $\frac{11}{8}$ miles already and plans to complete $\frac{16}{8}$ miles. To do this, how much farther must he run?

Start the solution:

1. UNDERSTAND the problem. Reread it as many times as needed.

2. TRANSLATE into an equation. (Fill in the blanks.)

distance left to run	is	distance planned to run	minus	distance already run
↓	↓	↓	↓	↓

$$\text{distance left to run} = \underline{\qquad} - \underline{\qquad}$$

Finish with:

3. SOLVE. and
4. INTERPRET.

47. Emil Vasquez, a bodybuilder, worked out $\frac{7}{8}$ of an hour one morning before school and $\frac{5}{8}$ of an hour that evening. How long did he work out that day?

48. A recipe for Heavenly Hash cake calls for $\frac{3}{4}$ cup of sugar and later $\frac{1}{4}$ cup of sugar. How much sugar is needed to make the recipe?

The circle graph below shows full-time U.S. college students' time use on an average weekday. Use this graph for Exercises 49–52. Write your answers in simplest form.

Full-Time College Students' Average Weekday Time Use

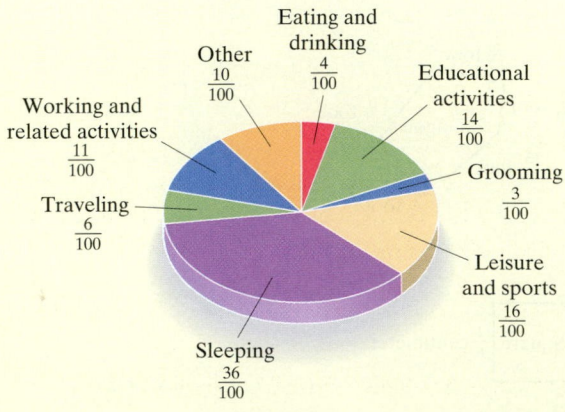

Working and related activities $\frac{11}{100}$

Other $\frac{10}{100}$

Eating and drinking $\frac{4}{100}$

Educational activities $\frac{14}{100}$

Grooming $\frac{3}{100}$

Leisure and sports $\frac{16}{100}$

Traveling $\frac{6}{100}$

Sleeping $\frac{36}{100}$

Source: Bureau of Labor Statistics

49. What fraction of a full-time U.S. college student's weekday is spent on eating and drinking, grooming, and sleeping?

50. What fraction of a full-time U.S. college student's weekday is spent on working and related activities and educational activities?

51. How much greater is the fractional part of a college student's weekday that is spent on leisure and sports than on traveling?

52. How much greater is the fractional part of a college student's weekday that is spent on sleeping than on educational activities?

Solve. Write your answers in simplest form.

53. In 2012, $\frac{5}{20}$ of Target's total retail sales were in the health, beauty, and household essentials category, and $\frac{4}{20}$ of Target's total retail sales were in the food and pet supplies category. What fraction of Target's total retail sales were made in these two categories combined? (*Source:* Target Corporation)

54. Road congestion can be caused by a variety of problems. Approximately $\frac{4}{10}$ of all road congestion in the United States is caused by bottlenecks, while $\frac{1}{10}$ of all road congestion in the United States is caused by road construction zones. What fraction of U.S. road congestion is caused by bottlenecks or construction zones? (*Source:* Nationwide Insurance)

55. Effective January 2014, the fraction of states in the United States with maximum interstate highway speed limits up to and including 70 mph was $\frac{34}{50}$. The fraction of states with 70 mph speed limits was $\frac{21}{50}$. What fraction of states had speed limits that were less than 70 mph? (*Source:* Insurance Institute for Highway Safety)

56. When people take aspirin, $\frac{31}{50}$ of the time it is used to treat some type of pain. Approximately $\frac{7}{50}$ of all aspirin use is for treating headaches. What fraction of aspirin use is for treating pain other than headaches? (*Source:* Bayer Market Research)

The map of the world below shows the fraction of the world's surface land area taken up by each continent. In other words, the continent of Africa makes up $\frac{20}{100}$ of the land in the world. Use this map for Exercises 57 through 60. Write your answers in simplest form.

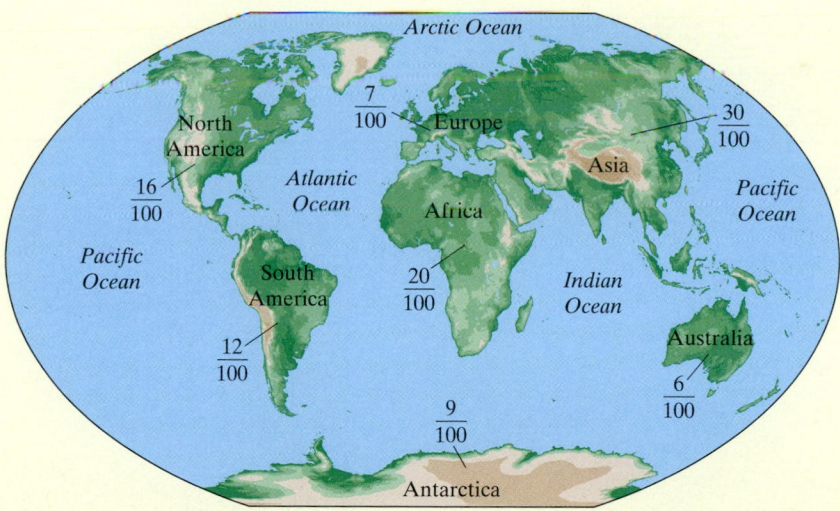

57. Find the fractional part of the world's land area within the continents of North America and South America.

58. Find the fractional part of the world's land area within the continents of Asia and Africa.

59. How much greater is the fractional part of the continent of Antarctica than the fractional part of the continent of Europe?

60. How much greater is the fractional part of the continent of Asia than the continent of Australia?

The theater industry is shifting away from analog movie screens toward digital and digital 3-D movie screens. Use the circle graph to answer Exercises 61 and 62. Write your answers in simplest form.

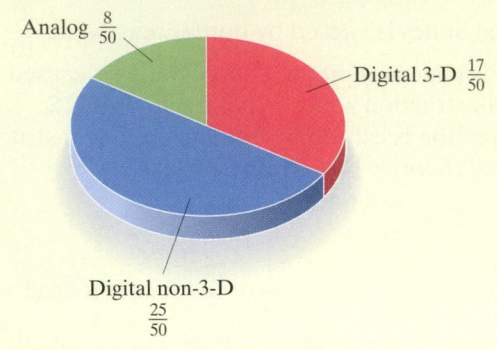

U.S. Theater Screens by Type, 2012

Analog $\frac{8}{50}$

Digital 3-D $\frac{17}{50}$

Digital non-3-D
$\frac{25}{50}$

Source: IHS Screen Digest

61. What fraction of U.S. theater screens are digital, either 3-D or non-3-D?

62. Home much greater is the fraction of screens that are digital 3-D than the fraction of screens that are analog?

Review

Write the prime factorization of each number. See Section 2.2.

63. 10

64. 12

65. 8

66. 20

67. 55

68. 28

Concept Extensions

Perform each indicated operation.

69. $\dfrac{3}{8} + \dfrac{7}{8} - \dfrac{5}{8}$

70. $\dfrac{12}{20} - \dfrac{1}{20} - \dfrac{3}{20}$

71. $\dfrac{4}{11} + \dfrac{5}{11} - \dfrac{3}{11} + \dfrac{2}{11}$

72. $\dfrac{9}{12} + \dfrac{1}{12} - \dfrac{3}{12} - \dfrac{5}{12}$

Find and correct the error. See the Concept Check in this section.

73. $\dfrac{2}{7} + \dfrac{9}{7} = \dfrac{11}{14}$

74. $\dfrac{3}{4} - \dfrac{1}{4} = \dfrac{2}{8} = \dfrac{1}{4}$

Solve. For Exercises 77 through 80, write each answer in simplest form.

75. In your own words, explain how to add like fractions.

76. In your own words, explain how to subtract like fractions.

77. Use the circle graph for Exercises 49 through 52 and find the sum of all the daily time-use fractions. Explain your answer.

78. Use the map of the world for Exercises 57 through 60 and find the sum of all the continents' fractions. Explain your answer.

79. Mike Cannon jogged $\frac{3}{8}$ of a mile from home and then rested. Then he continued jogging farther from home for another $\frac{3}{8}$ of a mile until he discovered his watch had fallen off. He walked back along the same path for $\frac{4}{8}$ of a mile until he found his watch. Find how far he was from his home.

80. A trim carpenter needs the following lengths of boards: $\frac{5}{4}$ feet, $\frac{15}{4}$ feet, $\frac{9}{4}$ feet, and $\frac{13}{4}$ feet. Is a 10-foot board long enough for the carpenter to cut these lengths? If not, how much more length is needed?

3.2 Least Common Multiple

Objective A Finding the Least Common Multiple Using Multiples

A multiple of a number is the product of that number and a natural number. For example, multiples of 5 are

$$\underbrace{5\cdot1}\quad\underbrace{5\cdot2}\quad\underbrace{5\cdot3}\quad\underbrace{5\cdot4}\quad\underbrace{5\cdot5}\quad\underbrace{5\cdot6}\quad\underbrace{5\cdot7}\quad\underbrace{5\cdot8}$$
$$\downarrow\quad\downarrow\quad\downarrow\quad\downarrow\quad\downarrow\quad\downarrow\quad\downarrow\quad\downarrow$$
$$5,\quad 10,\quad 15,\quad 20,\quad 25,\quad 30,\quad 35,\quad 40,\ldots$$

Multiples of 4 are

4, 8, 12, 16, 20, 24, 28, 32, 36, 40, 44, …

Common multiples of both 4 and 5 are numbers that are found in both lists above. If we study the lists of multiples and extend them we have

Common multiples of 4 and 5: 20, 40, 60, 80, …

We call the smallest number in the list of common multiples the **least common multiple (LCM).** From the list of common multiples of 4 and 5, we see that the LCM of 4 and 5 is 20.

Objectives

A Find the Least Common Multiple (LCM) Using Multiples.

B Find the LCM Using Prime Factorization.

C Write Equivalent Fractions.

Example 1 Find the LCM of 6 and 8.

Solution: Multiples of 6: 6, 12, 18, 24, 30, 36, 42, 48, …

Multiples of 8: 8, 16, 24, 32, 40, 48, 56, …

The common multiples are 24, 48, …. The least common multiple (LCM) is 24.

■ Work Practice 1

Practice 1

Find the LCM of 15 and 50.

Answer

1. 150

Listing all the multiples of every number in a list can be cumbersome and tedious. We can condense the procedure shown in Example 1 with the following steps:

> **Method 1: Finding the LCM of a List of Numbers Using Multiples of the Largest Number**
>
> **Step 1:** Write the multiples of the largest number (starting with the number itself) until a multiple common to all numbers in the list is found.
>
> **Step 2:** The multiple found in Step 1 is the LCM.

Example 2   Find the LCM of 9 and 12.

Solution: We write the multiples of 12 until we find a number that is also a multiple of 9.

$12 \cdot 1 = 12$ Not a multiple of 9.
$12 \cdot 2 = 24$ Not a multiple of 9.
$12 \cdot 3 = 36$ A multiple of 9.

The LCM of 9 and 12 is 36.

■ **Work Practice 2**

Practice 2
Find the LCM of 8 and 10.

Example 3   Find the LCM of 7 and 14.

Solution: We write the multiples of 14 until we find one that is also a multiple of 7.

$14 \cdot 1 = 14$ A multiple of 7

The LCM of 7 and 14 is 14.

■ **Work Practice 3**

Practice 3
Find the LCM of 8 and 16.

Example 4   Find the LCM of 12 and 20.

Solution: We write the multiples of 20 until we find one that is also a multiple of 12.

$20 \cdot 1 = 20$ Not a multiple of 12
$20 \cdot 2 = 40$ Not a multiple of 12
$20 \cdot 3 = 60$ A multiple of 12

The LCM of 12 and 20 is 60.

■ **Work Practice 4**

Practice 4
Find the LCM of 25 and 30.

Objective B Finding the LCM Using Prime Factorization

Method 1 for finding multiples works fine for smaller numbers, but may get tedious for larger numbers. A second method that uses prime factorization may be easier to use for larger numbers.

For example, to find the LCM of 270 and 84, let's look at the prime factorization of each.

$270 = 2 \cdot 3 \cdot 3 \cdot 3 \cdot 5$
$84 = 2 \cdot 2 \cdot 3 \cdot 7$

Answers
2. 40   **3.** 16   **4.** 150

Recall that the LCM must be a multiple of both 270 and 84. Thus, to build the LCM, we will circle the greatest number of factors for each different prime number. The LCM is the product of the circled factors.

Prime Number Factors

$270 = 2 \cdot \boxed{3 \cdot 3 \cdot 3} \cdot \boxed{5}$

$84 = \boxed{2 \cdot 2} \cdot 3 \cdot \boxed{7}$

Circle the greatest number of factors for each different prime number.

$\text{LCM} = 2 \cdot 2 \cdot 3 \cdot 3 \cdot 3 \cdot 5 \cdot 7 = 3780$

The number 3780 is the smallest number that both 270 and 84 divide into evenly.
This Method 2 is summarized below:

Method 2: Finding the LCM of a List of Numbers Using Prime Factorization

Step 1: Write the prime factorization of each number.

Step 2: For each different prime factor in step 1, circle the greatest number of times that factor occurs in any one factorization.

Step 3: The LCM is the product of the circled factors.

Example 5 Find the LCM of 72 and 60.

Solution: First we write the prime factorization of each number.

$72 = 2 \cdot 2 \cdot 2 \cdot 3 \cdot 3$
$60 = 2 \cdot 2 \cdot 3 \cdot 5$

For the prime factors shown, we circle the greatest number of prime factors found in either factorization.

$72 = \boxed{2 \cdot 2 \cdot 2} \cdot \boxed{3 \cdot 3}$
$60 = 2 \cdot 2 \cdot 3 \cdot \boxed{5}$

The LCM is the product of the circled factors.

$\text{LCM} = 2 \cdot 2 \cdot 2 \cdot 3 \cdot 3 \cdot 5 = 360$

The LCM is 360.

■ **Work Practice 5**

Practice 5
Find the LCM of 40 and 108.

Helpful Hint

If you prefer working with exponents, circle the factor with the greatest exponent.
Example 5:

$72 = \boxed{2^3} \cdot \boxed{3^2}$
$60 = 2^2 \cdot 3 \cdot \boxed{5}$

$\text{LCD} = 2^3 \cdot 3^2 \cdot 5 = 360$

Answer

5. 1080

Helpful Hint

If the number of factors of a prime number are equal, circle either one, but not both. For example,

$$12 = \boxed{2 \cdot 2} \cdot \boxed{3}$$

$$15 = 3 \cdot \boxed{5} \qquad \text{Circle either 3 but not both.}$$

The LCM is $2 \cdot 2 \cdot 3 \cdot 5 = 60$.

Practice 6

Find the LCM of 20, 24, and 45.

Example 6   Find the LCM of 15, 18, and 54.

Solution: $15 = 3 \cdot \boxed{5}$

$18 = \boxed{2} \cdot 3 \cdot 3$

$54 = 2 \cdot \boxed{3 \cdot 3 \cdot 3}$

The LCM is $2 \cdot 3 \cdot 3 \cdot 3 \cdot 5$ or 270.

■ **Work Practice 6**

Practice 7

Find the LCM of 7 and 21.

Example 7   Find the LCM of 11 and 33.

Solution: $11 = \boxed{11}$ ← It makes no difference

$33 = 3 \cdot \boxed{11}$ which 11 is circled.

The LCM is $3 \cdot 11$ or 33.

■ **Work Practice 7**

Objective C Writing Equivalent Fractions

To add or subtract unlike fractions in the next section, we first write equivalent fractions with the LCM as the denominator. Recall from Section 2.3 that fractions that represent the same portion of a whole are called "equivalent fractions."

$$\frac{1}{3} \quad = \quad \frac{2}{6} \quad = \quad \frac{4}{12}$$

To write $\frac{1}{3}$ as an equivalent fraction with a denominator of 12, we multiply by 1 in the form of $\frac{4}{4}$.

$$\frac{1}{3} = \frac{1}{3} \cdot 1 = \frac{1}{3} \cdot \frac{4}{4} = \frac{1 \cdot 4}{3 \cdot 4} = \frac{4}{12}$$

$$\frac{4}{4} = 1$$

So $\frac{1}{3} = \frac{4}{12}$.

Answers

6. 360   **7.** 21

To write an equivalent fraction,

$$\frac{a}{b} = \frac{a}{b} \cdot \frac{c}{c} = \frac{a \cdot c}{b \cdot c}$$

where a, b, and c are nonzero numbers.

✔**Concept Check** Which of the following is not equivalent to $\frac{3}{4}$?

a. $\frac{6}{8}$ **b.** $\frac{18}{24}$ **c.** $\frac{9}{14}$ **d.** $\frac{30}{40}$

Example 8   Write an equivalent fraction with the indicated denominator.

$$\frac{3}{4} = \frac{}{20}$$

Solution: In the denominators, since $4 \cdot 5 = 20$, we will multiply by 1 in the form of $\frac{5}{5}$.

$$\frac{3}{4} = \frac{3}{4} \cdot \frac{5}{5} = \frac{3 \cdot 5}{4 \cdot 5} = \frac{15}{20}$$

Thus, $\frac{3}{4} = \frac{15}{20}$.

■ **Work Practice 8**

Practice 8

Write an equivalent fraction with the indicated denominator:

$$\frac{7}{8} = \frac{}{56}$$

Helpful Hint

To check Example 8, write $\frac{15}{20}$ in simplest form.

$$\frac{15}{20} = \frac{3 \cdot \overset{1}{\cancel{5}}}{4 \cdot \underset{1}{\cancel{5}}} = \frac{3}{4}, \text{ the original fraction.}$$

If the original fraction is in lowest terms, we can check our work by writing the new equivalent fraction in simplest form. This form should be the original fraction.

✔**Concept Check** True or false? When the fraction $\frac{2}{9}$ is rewritten as an equivalent fraction with 27 as the denominator, the result is $\frac{2}{27}$.

Example 9   Write an equivalent fraction with the indicated denominator.

$$\frac{1}{2} = \frac{}{24}$$

Solution: Since $2 \cdot 12 = 24$, we multiply by 1 in the form of $\frac{12}{12}$.

$$\frac{1}{2} = \frac{1}{2} \cdot \frac{12}{12} = \frac{1 \cdot 12}{2 \cdot 12} = \frac{12}{24}$$

Thus, $\frac{1}{2} = \frac{12}{24}$.

■ **Work Practice 9**

Practice 9

Write an equivalent fraction with the indicated denominator.

$$\frac{3}{5} = \frac{}{15}$$

Answers

8. $\frac{49}{56}$ **9.** $\frac{9}{15}$

✔**Concept Check Answers**

c

false; the correct result would be $\frac{6}{27}$

Practice 10

Write an equivalent fraction with the given denominator.

$$4 = \frac{}{6}$$

Answer

10. $\dfrac{24}{6}$

Example 10   Write an equivalent fraction with the given denominator.

$$3 = \frac{}{7}$$

Solution: Recall that $3 = \dfrac{3}{1}$. Since $1 \cdot 7 = 7$, multiply by 1 in the form of $\dfrac{7}{7}$.

$$\frac{3}{1} = \frac{3}{1} \cdot \frac{7}{7} = \frac{3 \cdot 7}{1 \cdot 7} = \frac{21}{7}$$

■ **Work Practice 10**

Vocabulary, Readiness & Video Check

Use the choices below to fill in each blank.

least common multiple (LCM)   multiple   equivalent

1. Fractions that represent the same portion of a whole are called _____ fractions.
2. The smallest positive number that is a multiple of all numbers in a list is called the _____.
3. A(n) _____ of a number is the product of that number and a natural number.

Martin-Gay Interactive Videos   Watch the section lecture video and answer the following questions.

Objective **A** **4.** From the lecture before ⊞ Example 1, why do the multiples of a number continue on indefinitely? ▶

Objective **B** **5.** In ⊞ Example 2, why does it make sense that the LCM of 8 and 24 is 24? ▶

Objective **C** **6.** Why isn't the answer to ⊞ Example 5, $\dfrac{20}{35}$, simplified? ▶

See Video 3.2

3.2   **Exercise Set** MyMathLab® ▶

Objective **A** **B** **Mixed Practice** *Find the LCM of each list of numbers. See Examples 1 through 7.*

1. 3, 4	**2.** 4, 6	▶ **3.** 9, 15	**4.** 15, 20	**5.** 12, 18	**6.** 10, 15
7. 24, 36	**8.** 42, 70	**9.** 18, 21	**10.** 24, 45	**11.** 15, 25	**12.** 21, 14
▶ **13.** 8, 24	**14.** 15, 90	**15.** 6, 7	**16.** 13, 8	**17.** 8, 6, 27	**18.** 6, 25, 10
▶ **19.** 25, 15, 6	**20.** 4, 14, 20	**21.** 34, 68	**22.** 25, 175	**23.** 84, 294	**24.** 48, 54

25. 30, 36, 50   **26.** 21, 28, 42   **27.** 50, 72, 120   **28.** 70, 98, 100   **29.** 11, 33, 121   **30.** 10, 15, 100

31. 4, 6, 10, 15   **32.** 25, 3, 15, 10

Objective C *Write each fraction or whole number as an equivalent fraction with the given denominator. See Examples 8 through 10.*

33. $\dfrac{4}{7} = \dfrac{}{35}$ **34.** $\dfrac{3}{5} = \dfrac{}{20}$ **35.** $\dfrac{2}{3} = \dfrac{}{21}$ **36.** $6 = \dfrac{}{10}$ **37.** $5 = \dfrac{}{3}$

38. $\dfrac{9}{10} = \dfrac{}{70}$ **39.** $\dfrac{1}{2} = \dfrac{}{30}$ **40.** $\dfrac{1}{3} = \dfrac{}{30}$ **41.** $\dfrac{10}{7} = \dfrac{}{21}$ **42.** $\dfrac{5}{3} = \dfrac{}{21}$

43. $\dfrac{3}{4} = \dfrac{}{28}$ **44.** $\dfrac{4}{5} = \dfrac{}{45}$ **45.** $\dfrac{2}{3} = \dfrac{}{45}$ **46.** $\dfrac{2}{3} = \dfrac{}{75}$ **47.** $\dfrac{4}{9} = \dfrac{}{81}$

48. $\dfrac{5}{11} = \dfrac{}{88}$ **49.** $\dfrac{15}{13} = \dfrac{}{78}$ **50.** $\dfrac{9}{7} = \dfrac{}{84}$ **51.** $\dfrac{14}{17} = \dfrac{}{68}$ **52.** $\dfrac{19}{21} = \dfrac{}{126}$

A non-store retailer is a mail-order business that sells goods via catalogs, toll-free telephone numbers, or online media. The table shows the fraction of non-store retailers' goods that were sold online in 2011 by type of goods. Use this table to answer Exercises 53 through 56.

Type of Goods Sold by Non-Store Retailers	Fraction of Goods That Were Sold Online	Equivalent Fraction with a Denominator of 100
Books and magazines	$\dfrac{43}{50}$	
Clothing and accessories	$\dfrac{4}{5}$	
Computer hardware	$\dfrac{29}{50}$	
Computer software	$\dfrac{17}{25}$	
Drugs, health and beauty aids	$\dfrac{3}{25}$	
Electronics and appliances	$\dfrac{21}{25}$	
Food, beer, and wine	$\dfrac{17}{25}$	
Home furnishings	$\dfrac{81}{100}$	
Music and videos	$\dfrac{9}{10}$	
Office equipment and supplies	$\dfrac{79}{100}$	
Sporting goods	$\dfrac{37}{50}$	
Toys, hobbies, and games	$\dfrac{39}{50}$	

(*Source:* U.S. Census Bureau)

53. Complete the table by writing each fraction as an equivalent fraction with a denominator of 100.

54. Which of these types of goods has the largest fraction sold online?

55. Which of these types of goods has the smallest fraction sold online?

56. Which of the types of goods has **more than** $\dfrac{4}{5}$ of the goods sold online? (*Hint:* Write $\dfrac{4}{5}$ as an equivalent fraction with a denominator of 100.)

Review

Add or subtract as indicated. See Section 3.1.

57. $\dfrac{7}{10} - \dfrac{2}{10}$

58. $\dfrac{8}{13} - \dfrac{3}{13}$

59. $\dfrac{1}{5} + \dfrac{1}{5}$

60. $\dfrac{1}{8} + \dfrac{3}{8}$

61. $\dfrac{23}{18} - \dfrac{15}{18}$

62. $\dfrac{36}{30} - \dfrac{12}{30}$

63. $\dfrac{2}{9} + \dfrac{1}{9} + \dfrac{6}{9}$

64. $\dfrac{2}{12} + \dfrac{7}{12} + \dfrac{3}{12}$

Concept Extensions

Write each fraction as an equivalent fraction with the indicated denominator.

65. $\dfrac{37}{165} = \dfrac{}{3630}$

66. $\dfrac{108}{215} = \dfrac{}{4085}$

67. In your own words, explain how to find the LCM of two numbers.

68. In your own words, explain how to write a fraction as an equivalent fraction with a given denominator.

Solve. See the Concept Checks in this section.

69. Which of the following are equivalent to $\dfrac{2}{3}$?

    a. $\dfrac{10}{15}$ **b.** $\dfrac{40}{60}$

    c. $\dfrac{16}{20}$ **d.** $\dfrac{200}{300}$

70. True or False? When the fraction $\dfrac{7}{12}$ is rewritten with a denominator of 48, the result is $\dfrac{11}{48}$. If false, give the correct fraction.

3.3 Adding and Subtracting Unlike Fractions

Objectives

A Add Unlike Fractions.

B Subtract Unlike Fractions.

C Solve Problems by Adding or Subtracting Unlike Fractions.

Objective A Adding Unlike Fractions

In this section we add and subtract fractions with unlike denominators. To add or subtract these unlike fractions, we first write the fractions as equivalent fractions with a common denominator and then add or subtract the like fractions. The common denominator that we use is the least common multiple (LCM) of the denominators. This denominator is called the **least common denominator (LCD)**.

To begin, let's add the unlike fractions $\dfrac{3}{4} + \dfrac{1}{6}$. The LCM of denominators 4 and 6 is 12. This means that the number 12 is also the LCD. So we write each fraction as an equivalent fraction with a denominator of 12, then add as usual. This addition process is shown next and also illustrated by figures.

Add: $\dfrac{3}{4} + \dfrac{1}{6}$	The LCD is **12**.
Figures	**Algebra**
$\dfrac{3}{4}$ + $\dfrac{1}{6}$	$\dfrac{3}{4} = \dfrac{3}{4} \cdot \dfrac{3}{3} = \dfrac{9}{12}$ and $\dfrac{1}{6} = \dfrac{1}{6} \cdot \dfrac{2}{2} = \dfrac{2}{12}$
$\dfrac{9}{12}$ + $\dfrac{2}{12}$	Remember $\dfrac{3}{3} = 1$ and $\dfrac{2}{2} = 1$.
	Now we can add just as we did in Section 3.1.
$\dfrac{9}{12} + \dfrac{2}{12} = \dfrac{11}{12}$	$\dfrac{3}{4} + \dfrac{1}{6} = \dfrac{9}{12} + \dfrac{2}{12} = \dfrac{11}{12}$

Thus, the sum is $\dfrac{11}{12}$.

Adding or Subtracting Unlike Fractions

Step 1: Find the LCM of the denominators of the fractions. This number is the least common denominator (LCD).

Step 2: Write each fraction as an equivalent fraction whose denominator is the LCD.

Step 3: Add or subtract the like fractions.

Step 4: Write the sum or difference in simplest form.

Example 1   Add: $\dfrac{2}{5} + \dfrac{4}{15}$

Solution:

Step 1: The LCM of the denominators 5 and 15 is 15. Thus, the LCD is 15. In later examples, we shall simply say, for example, that the LCD of 5 and 15 is 15.

Step 2: $\dfrac{2}{5} = \dfrac{2}{5} \cdot \dfrac{3}{3} = \dfrac{6}{15}, \dfrac{4}{15} = \dfrac{4}{15}$ $\longleftarrow$ This fraction already has a denominator of 15.

         $\uparrow$ Multiply by 1 in the form $\dfrac{3}{3}$

Step 3: $\dfrac{2}{5} + \dfrac{4}{15} = \dfrac{6}{15} + \dfrac{4}{15} = \dfrac{10}{15}$

Step 4: Write in simplest form.

$$\dfrac{10}{15} = \dfrac{2 \cdot \overset{1}{\cancel{5}}}{3 \cdot \underset{1}{\cancel{5}}} = \dfrac{2}{3}$$

■ **Work Practice 1**

Practice 1

Add: $\dfrac{1}{6} + \dfrac{5}{18}$

Answer

1. $\dfrac{4}{9}$

Practice 2

Add: $\dfrac{5}{6} + \dfrac{2}{9}$

Example 2 Add: $\dfrac{11}{15} + \dfrac{3}{10}$

Solution:

Step 1: The LCD of 15 and 10 is 30.

Step 2: $\dfrac{11}{15} = \dfrac{11}{15} \cdot \dfrac{2}{2} = \dfrac{22}{30}$ $\dfrac{3}{10} = \dfrac{3}{10} \cdot \dfrac{3}{3} = \dfrac{9}{30}$

Step 3: $\dfrac{11}{15} + \dfrac{3}{10} = \dfrac{22}{30} + \dfrac{9}{30} = \dfrac{31}{30}$

Step 4: $\dfrac{31}{30}$ is in simplest form. We can write the sum as $\dfrac{31}{30}$ or $1\dfrac{1}{30}$.

■ Work Practice 2

Practice 3

Add: $\dfrac{2}{5} + \dfrac{4}{9}$

Example 3 Add: $\dfrac{2}{3} + \dfrac{1}{7}$

Solution: The LCD of 3 and 7 is 21.

$$\dfrac{2}{3} + \dfrac{1}{7} = \dfrac{2}{3} \cdot \dfrac{7}{7} + \dfrac{1}{7} \cdot \dfrac{3}{3}$$

$$= \dfrac{14}{21} + \dfrac{3}{21}$$

$$= \dfrac{17}{21} \qquad \text{Simplest form.}$$

■ Work Practice 3

Practice 4

Add: $\dfrac{1}{2} + \dfrac{4}{5} + \dfrac{7}{10}$

Example 4 Add: $\dfrac{1}{2} + \dfrac{2}{3} + \dfrac{5}{6}$

Solution: The LCD of 2, 3, and 6 is 6.

$$\dfrac{1}{2} + \dfrac{2}{3} + \dfrac{5}{6} = \dfrac{1}{2} \cdot \dfrac{3}{3} + \dfrac{2}{3} \cdot \dfrac{2}{2} + \dfrac{5}{6}$$

$$= \dfrac{3}{6} + \dfrac{4}{6} + \dfrac{5}{6}$$

$$= \dfrac{12}{6} = 2$$

■ Work Practice 4

Answers

2. $\dfrac{19}{18}$ or $1\dfrac{1}{18}$ **3.** $\dfrac{38}{45}$ **4.** 2

✔ **Concept Check Answer**

When adding unlike fractions, we don't add the denominators.
Correct solution:

$\dfrac{2}{9} + \dfrac{4}{11} = \dfrac{22}{99} + \dfrac{36}{99} = \dfrac{58}{99}$

✔ **Concept Check** Find and correct the error in the following:

$$\dfrac{2}{9} + \dfrac{4}{11} \;\;\cancel{=}\;\; \dfrac{6}{20} \;\;\cancel{=}\;\; \dfrac{3}{10}$$

Objective B Subtracting Unlike Fractions

As indicated in the box on page 191, we follow the same steps when subtracting unlike fractions as when adding them.

Example 5 Subtract: $\dfrac{2}{5} - \dfrac{3}{20}$

Solution:

Step 1: The LCD of 5 and 20 is 20.

Step 2: $\dfrac{2}{5} = \dfrac{2}{5} \cdot \dfrac{4}{4} = \dfrac{8}{20}$ $\dfrac{3}{20} = \dfrac{3}{20}$ ← The fraction already has a denominator of 20.

Step 3: $\dfrac{2}{5} - \dfrac{3}{20} = \dfrac{8}{20} - \dfrac{3}{20} = \dfrac{5}{20}$

Step 4: Write in simplest form.

$$\dfrac{5}{20} = \dfrac{\overset{1}{\cancel{5}}}{\underset{1}{\cancel{5}} \cdot 4} = \dfrac{1}{4}$$

■ **Work Practice 5**

Practice 5

Subtract: $\dfrac{7}{12} - \dfrac{5}{24}$

Example 6 Subtract: $\dfrac{10}{11} - \dfrac{2}{3}$

Solution:

Step 1: The LCD of 11 and 3 is 33.

Step 2: $\dfrac{10}{11} = \dfrac{10}{11} \cdot \dfrac{3}{3} = \dfrac{30}{33}$ $\dfrac{2}{3} = \dfrac{2}{3} \cdot \dfrac{11}{11} = \dfrac{22}{33}$

Step 3: $\dfrac{10}{11} - \dfrac{2}{3} = \dfrac{30}{33} - \dfrac{22}{33} = \dfrac{8}{33}$

Step 4: $\dfrac{8}{33}$ is in simplest form.

■ **Work Practice 6**

Practice 6

Subtract: $\dfrac{9}{10} - \dfrac{3}{7}$

Example 7 Subtract: $\dfrac{11}{12} - \dfrac{2}{9}$

Solution: The LCD of 12 and 9 is 36.

$$\dfrac{11}{12} - \dfrac{2}{9} = \dfrac{11}{12} \cdot \dfrac{3}{3} - \dfrac{2}{9} \cdot \dfrac{4}{4}$$

$$= \dfrac{33}{36} - \dfrac{8}{36}$$

$$= \dfrac{25}{36}$$

■ **Work Practice 7**

Practice 7

Subtract: $\dfrac{7}{8} - \dfrac{5}{6}$

Answers

5. $\dfrac{3}{8}$ 6. $\dfrac{33}{70}$ 7. $\dfrac{1}{24}$

✓**Concept Check** Find and correct the error in the following:

$$\dfrac{11}{12} - \dfrac{3}{4} = \dfrac{8}{8} = 1$$

✓**Concept Check Answer**

When subtracting unlike fractions, we don't subtract the denominators. Correct solution:

$$\dfrac{11}{12} - \dfrac{3}{4} = \dfrac{11}{12} - \dfrac{9}{12} = \dfrac{2}{12} = \dfrac{1}{6}$$

Objective C Solving Problems by Adding or Subtracting Unlike Fractions

Very often, real-world problems involve adding or subtracting unlike fractions.

Example 8 Finding Total Weight

A freight truck has $\frac{1}{4}$ ton of computers, $\frac{1}{3}$ ton of televisions, and $\frac{3}{8}$ ton of small appliances. Find the total weight of its load.

> $\frac{1}{4}$ ton of computers $\frac{1}{3}$ ton of televisions $\frac{3}{8}$ ton of appliances

Solution:

1. UNDERSTAND. Read and reread the problem. The phrase "total weight" tells us to add.

2. TRANSLATE.

In words:	total weight	is	weight of computers	plus	weight of televisions	plus	weight of appliances
	↓	↓	↓	↓	↓	↓	↓
Translate:	total weight	=	$\frac{1}{4}$	+	$\frac{1}{3}$	+	$\frac{3}{8}$

3. SOLVE: The LCD is 24.

$$\frac{1}{4} + \frac{1}{3} + \frac{3}{8} = \frac{1}{4} \cdot \frac{6}{6} + \frac{1}{3} \cdot \frac{8}{8} + \frac{3}{8} \cdot \frac{3}{3}$$

$$= \frac{6}{24} + \frac{8}{24} + \frac{9}{24}$$

$$= \frac{23}{24}$$

4. INTERPRET. *Check* the solution. *State* your conclusion: The total weight of the truck's load is $\frac{23}{24}$ ton.

■ Work Practice 8

Practice 8

To repair her sidewalk, a homeowner must pour small amounts of cement in three different locations. She needs $\frac{3}{5}$ of a cubic yard, $\frac{2}{10}$ of a cubic yard, and $\frac{2}{15}$ of a cubic yard for these locations. Find the total amount of cement the homeowner needs.

Answer

8. $\frac{14}{15}$ cu yd

Example 9 Calculating Flight Time

A flight from Tucson to Phoenix, Arizona, requires $\frac{5}{12}$ of an hour. If the plane has been flying $\frac{1}{4}$ of an hour, find how much time remains before landing.

Solution:

1. **UNDERSTAND.** Read and reread the problem. The phrase "how much time remains" tells us to subtract.

2. **TRANSLATE.**

In words:	time remaining	is	flight time from Tucson of Phoenix	minus	flight time already passed
	↓	↓	↓	↓	↓
Translate:	time remaining	=	$\frac{5}{12}$	−	$\frac{1}{4}$

3. **SOLVE:** The LCD is 12.

$$\frac{5}{12} - \frac{1}{4} = \frac{5}{12} - \frac{1}{4} \cdot \frac{3}{3}$$

$$= \frac{5}{12} - \frac{3}{12}$$

$$= \frac{2}{12}$$

$$= \frac{\overset{1}{\cancel{2}}}{\underset{1}{\cancel{2}} \cdot 6}$$

$$= \frac{1}{6}$$

4. **INTERPRET.** *Check* the solution. *State* your conclusion: The flight time remaining is $\frac{1}{6}$ of an hour.

■ **Work Practice 9**

Practice 9

Find the difference in length of two boards if one board is $\frac{4}{5}$ of a foot long and the other is $\frac{2}{3}$ of a foot long.

Answer

9. $\frac{2}{15}$ ft

Calculator Explorations Performing Operations on Fractions

Scientific Calculator

Many calculators have a fraction key, such as $\boxed{a_{b/c}}$, that allows you to enter fractions and perform operations on them, and then it gives the result as a fraction. If your calculator has a fraction key, use it to calculate

$$\frac{3}{5} + \frac{4}{7}$$

Enter the keystrokes

$\boxed{3}\,\boxed{a_{b/c}}\,\boxed{5}\,\boxed{+}\,\boxed{4}\,\boxed{a_{b/c}}\,\boxed{7}\,\boxed{=}$

The display should read $\boxed{1\_6\,|\,35}$, which represents the mixed number $1\frac{6}{35}$. Let's write the result as a fraction. To convert from mixed number notation to fractional notation, press

$\boxed{2^{\text{nd}}}\,\boxed{d/c}$

The display now reads $\boxed{41\,|\,35}$, which represents $\frac{41}{35}$, the sum in fractional notation.

Graphing Calculator

Graphing calculators also allow you to perform operations on fractions and will give exact fractional results. The fraction option on a graphing calculator may be found under the $\boxed{\text{MATH}}$ menu. To perform the addition to the left, try the keystrokes.

$\boxed{3}\,\boxed{\div}\,\boxed{5}\,\boxed{+}\,\boxed{4}\,\boxed{\div}\,\boxed{7}\,\boxed{\text{MATH}}\,\boxed{\text{ENTER}}$
$\boxed{\text{ENTER}}$

The display should read

$\boxed{3/5 + 4/7 \blacktriangleright \text{Frac } 41/35}$

Use a calculator to add the following fractions. Give each sum as a fraction.

1. $\dfrac{1}{16} + \dfrac{2}{5}$ 2. $\dfrac{3}{20} + \dfrac{2}{25}$ 3. $\dfrac{4}{9} + \dfrac{7}{8}$

4. $\dfrac{9}{11} + \dfrac{5}{12}$ 5. $\dfrac{10}{17} + \dfrac{12}{19}$ 6. $\dfrac{14}{31} + \dfrac{15}{21}$

Vocabulary, Readiness & Video Check

Use the choices below to fill in each blank. Any numerical answers are not listed.

least common denominator   equivalent

1. To add or subtract unlike fractions, we first write the fractions as _____ fractions with a common denominator. The common denominator we use is called the _____.

2. The LCD for $\dfrac{5}{8}$ and $\dfrac{1}{6}$ is _____.

3. $\dfrac{5}{8} + \dfrac{1}{6} = \dfrac{5}{8}\cdot\dfrac{3}{3} + \dfrac{1}{6}\cdot\dfrac{4}{4} = \dfrac{\ }{\ } + \dfrac{\ }{\ } = \dfrac{\ }{\ }$.

4. $\dfrac{5}{8} - \dfrac{1}{6} = \dfrac{5}{8}\cdot\dfrac{3}{3} - \dfrac{1}{6}\cdot\dfrac{4}{4} = \dfrac{\ }{\ } - \dfrac{\ }{\ } = \dfrac{\ }{\ }$.

Martin-Gay Interactive Videos   *Watch the section lecture video and answer the following questions.*

Objective A 5. In  Example 1, why does multiplying $\dfrac{2}{11}$ by $\dfrac{3}{3}$ not change the value of the fraction? ▶

Objective B 6. In  Example 3, how did we know we needed to multiply the two denominators to get the LCD? ▶

Objective C 7. What are the two forms of the answer to Example 5? ▶

See Video 3.3 🍎

3.3 Exercise Set MyMathLab®

Objective A *Add and simplify. See Examples 1 through 4.*

1. $\frac{2}{3} + \frac{1}{6}$

2. $\frac{5}{6} + \frac{1}{12}$

3. $\frac{1}{2} + \frac{1}{3}$

4. $\frac{2}{3} + \frac{1}{4}$

5. $\frac{2}{11} + \frac{2}{33}$

6. $\frac{5}{9} + \frac{1}{3}$

7. $\frac{3}{14} + \frac{3}{7}$

8. $\frac{2}{5} + \frac{2}{15}$

9. $\frac{11}{35} + \frac{2}{7}$

10. $\frac{4}{5} + \frac{3}{40}$

11. $\frac{8}{25} + \frac{7}{35}$

12. $\frac{5}{14} + \frac{10}{21}$

13. $\frac{7}{15} + \frac{5}{12}$

14. $\frac{5}{8} + \frac{3}{20}$

15. $\frac{2}{28} + \frac{2}{21}$

16. $\frac{6}{25} + \frac{7}{35}$

17. $\frac{9}{44} + \frac{17}{36}$

18. $\frac{2}{33} + \frac{2}{21}$

19. $\frac{5}{11} + \frac{3}{13}$

20. $\frac{3}{7} + \frac{9}{17}$

21. $\frac{1}{3} + \frac{1}{9} + \frac{1}{27}$

22. $\frac{1}{4} + \frac{1}{16} + \frac{1}{64}$

23. $\frac{5}{7} + \frac{1}{8} + \frac{1}{2}$

24. $\frac{10}{13} + \frac{7}{10} + \frac{1}{5}$

25. $\frac{5}{36} + \frac{3}{4} + \frac{1}{6}$

26. $\frac{7}{18} + \frac{2}{9} + \frac{5}{6}$

27. $\frac{13}{20} + \frac{3}{5} + \frac{1}{3}$

28. $\frac{2}{7} + \frac{13}{28} + \frac{2}{5}$

Objective B *Subtract and simplify. See Examples 5 through 7.*

29. $\frac{7}{8} - \frac{3}{16}$

30. $\frac{5}{13} - \frac{3}{26}$

31. $\frac{5}{6} - \frac{3}{7}$

32. $\frac{3}{4} - \frac{1}{7}$

33. $\frac{5}{7} - \frac{1}{8}$

34. $\frac{10}{13} - \frac{7}{10}$

35. $\frac{9}{11} - \frac{4}{9}$

36. $\frac{7}{18} - \frac{2}{9}$

37. $\frac{11}{35} - \frac{2}{7}$

38. $\frac{2}{5} - \frac{3}{25}$

39. $\frac{5}{12} - \frac{1}{9}$

40. $\frac{7}{12} - \frac{5}{18}$

41. $\frac{7}{15} - \frac{5}{12}$

42. $\frac{5}{8} - \frac{3}{20}$

43. $\frac{3}{28} - \frac{2}{21}$

44. $\frac{6}{25} - \frac{7}{35}$

45. $\frac{1}{100} - \frac{1}{1000}$

46. $\frac{1}{50} - \frac{1}{500}$

47. $\frac{21}{44} - \frac{11}{36}$

48. $\frac{7}{18} - \frac{2}{45}$

Objectives A B Mixed Practice *Perform the indicated operation. See Examples 1 through 7.*

49. $\dfrac{5}{12} + \dfrac{1}{9}$

50. $\dfrac{7}{12} + \dfrac{5}{18}$

51. $\dfrac{17}{35} - \dfrac{2}{7}$

52. $\dfrac{13}{24} - \dfrac{1}{6}$

53. $\dfrac{9}{28} - \dfrac{3}{40}$

54. $\dfrac{10}{26} - \dfrac{3}{8}$

55. $\dfrac{2}{3} + \dfrac{4}{45} + \dfrac{4}{5}$

56. $\dfrac{3}{16} + \dfrac{1}{4} + \dfrac{1}{16}$

Objective C *Find the perimeter of each geometric figure. (Hint: Recall that perimeter means distance around.)*

 57.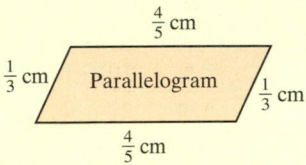

△58.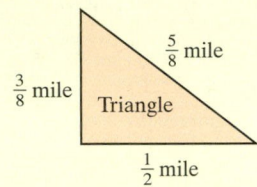

△59.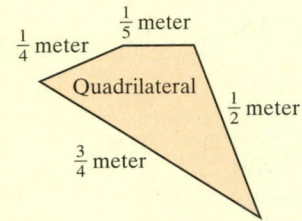

△60.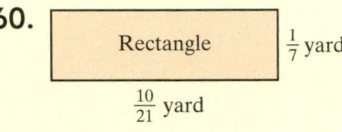

Solve. For Exercises 61 and 62, the solutions have been started for you. See Examples 8 and 9.

61. The slowest mammal is the three-toed sloth from South America. The sloth has an average ground speed of $\dfrac{1}{10}$ mph. In the trees, it can accelerate to $\dfrac{17}{100}$ mph. How much faster can a sloth travel in the trees? (*Source: Guinness World Records*)

Start the solution:

1. UNDERSTAND the problem. Reread it as many times as needed.
2. TRANSLATE into an equation. (Fill in the blanks.)

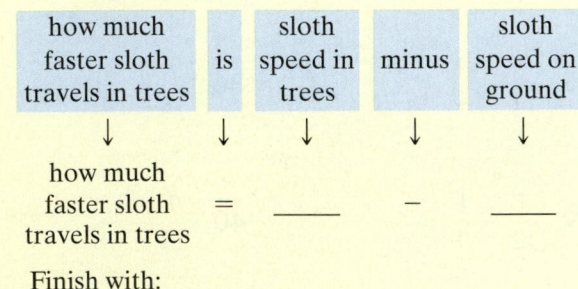

how much faster sloth travels in trees	is	sloth speed in trees	minus	sloth speed on ground
↓	↓	↓	↓	↓

how much faster sloth travels in trees $=$ ____ $-$ ____

Finish with:

3. SOLVE. and
4. INTERPRET.

62. Killer bees have been known to chase people for up to $\dfrac{1}{4}$ of a mile, while domestic European honey-bees will normally chase a person for no more than 100 feet, or $\dfrac{5}{264}$ of a mile. How much farther will a killer bee chase a person than a domestic honey-bee? (*Source:* Coachella Valley Mosquito & Vector Control District)

Start the solution:

1. UNDERSTAND the problem. Reread it as many times as needed.
2. TRANSLATE into an equation. (Fill in the blanks.)

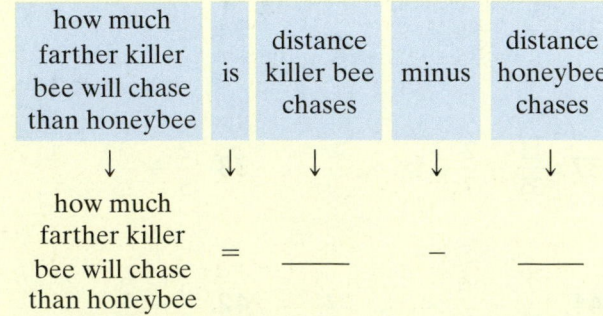

how much farther killer bee will chase than honeybee	is	distance killer bee chases	minus	distance honeybee chases
↓	↓	↓	↓	↓

how much farther killer bee will chase than honeybee $=$ ____ $-$ ____

Finish with:

3. SOLVE. and
4. INTERPRET.

63. Find the inner diameter of the washer. (*Hint:* Use the outer diameter and subtract the washer widths.)

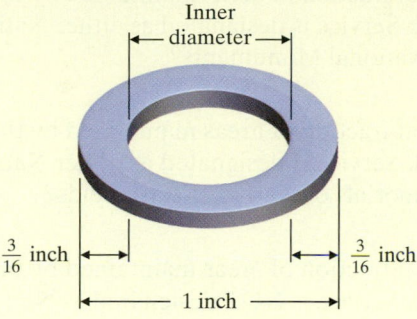

Inner diameter

$\frac{3}{16}$ inch $\frac{3}{16}$ inch

1 inch

64. Find the inner diameter of the tubing. (See the hint for Exercise 63.)

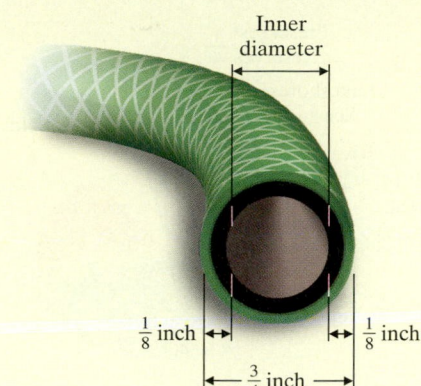

Inner diameter

$\frac{1}{8}$ inch $\frac{1}{8}$ inch

$\frac{3}{4}$ inch

65. Given the following diagram, find its total length. (*Hint:* Find the sum of the partial lengths.)

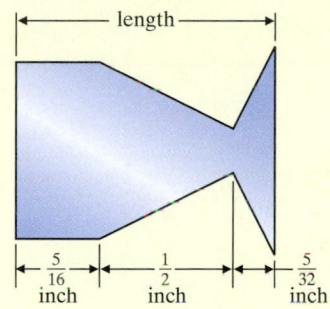

length

$\frac{5}{16}$ inch $\frac{1}{2}$ inch $\frac{5}{32}$ inch

66. Given the following diagram, find its total width. (*Hint:* Find the sum of the partial widths.)

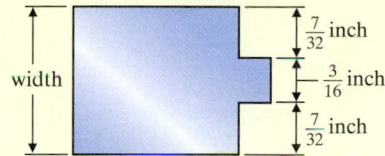

width

$\frac{7}{32}$ inch

$\frac{3}{16}$ inch

$\frac{7}{32}$ inch

67. Together, Thin Mints and Samoas account for $\frac{11}{25}$ of the Girl Scout cookies sold each year. Thin Mints alone account for $\frac{1}{4}$ of all Girl Scout cookie sales. What fraction of Girl Scout cookies sold are Samoas? (*Source:* Girl Scouts of the United States of America)

68. About $\frac{13}{20}$ of American students ages 10 to 17 name math, science, or art as their favorite subject in school. Art is the favorite subject for about $\frac{4}{25}$ of the American students ages 10 to 17. For what fraction of students this age is math or science their favorite subject? (*Source:* Peter D. Hart Research Associates for the National Science Foundation)

The table below shows the fraction of the Earth's water area taken up by each ocean. Use this table for Exercises 69 and 70.

Fraction of Earth's Water Area per Ocean	
Ocean	**Fraction**
Arctic	$\frac{1}{25}$
Atlantic	$\frac{13}{50}$
Pacific	$\frac{1}{2}$
Indian	$\frac{1}{5}$

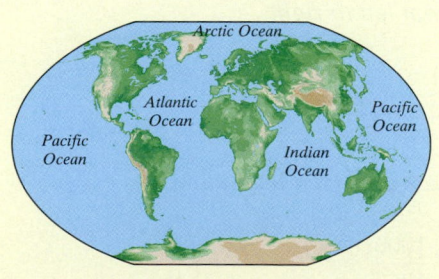

69. What fraction of the world's water surface area is accounted for by the Pacific and Atlantic Oceans?

70. What fraction of the world's water surface area is accounted for by the Arctic and Indian Oceans?

We first viewed this circle graph in Section 2.3. In this section we study it further. Use it to answer Exercises 71 through 74.

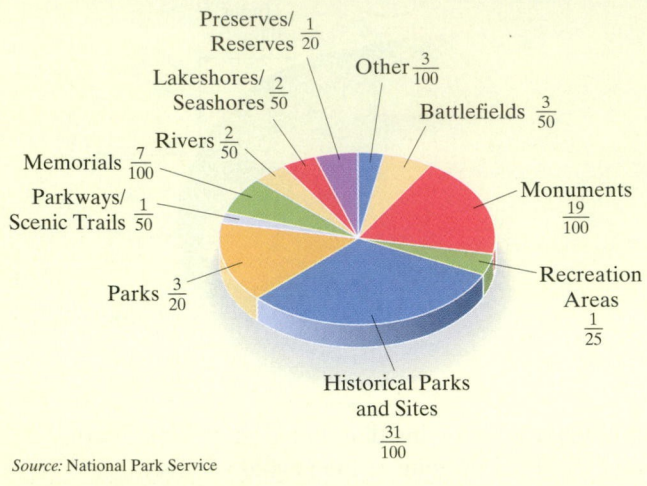

Areas Maintained by the National Park Service

Preserves/ Reserves $\frac{1}{20}$

Other $\frac{3}{100}$

Lakeshores/ Seashores $\frac{2}{50}$

Battlefields $\frac{3}{50}$

Rivers $\frac{2}{50}$

Memorials $\frac{7}{100}$

Parkways/ Scenic Trails $\frac{1}{50}$

Monuments $\frac{19}{100}$

Parks $\frac{3}{20}$

Recreation Areas $\frac{1}{25}$

Historical Parks and Sites $\frac{31}{100}$

Source: National Park Service

71. What fraction of areas maintained by the National Park Service is designated as either National Parks or National Monuments?

72. What fraction of areas maintained by the National Park Service is designated as either National Memorials or National Battlefields?

73. What fraction of areas maintained by the National Park Service is NOT designated as National Monuments?

74. What fraction of areas maintained by the National Park Service is NOT designated as National Preserves or National Reserves?

Review

Multiply or divide as indicated. See Sections 2.4 and 2.5.

75. $1\frac{1}{2} \cdot 3\frac{1}{3}$ **76.** $2\frac{5}{6} \div 5$ **77.** $4 \div 7\frac{1}{4}$ **78.** $4\frac{3}{4} \cdot 5\frac{1}{5}$ **79.** $3 \cdot 2\frac{1}{9}$ **80.** $6\frac{2}{7} \cdot 14$

Concept Extensions

For Exercises 81 and 82 below, do the following:

a. *Draw three rectangles of the same size and represent each fraction in the sum or difference, one fraction per rectangle, by shading.*

b. *Using these rectangles as estimates, determine whether there is an error in the sum or difference.*

c. *If there is an error, correctly calculate the sum or difference.*

See the Concept Checks in this section.

81. $\frac{3}{5} + \frac{4}{5} \stackrel{?}{=} \frac{7}{10}$

82. $\frac{3}{4} - \frac{5}{8} \stackrel{?}{=} \frac{2}{4}$

Subtract from left to right.

83. $\frac{2}{3} - \frac{1}{4} - \frac{2}{540}$

84. $\frac{9}{10} - \frac{7}{200} - \frac{1}{3}$

Perform each indicated operation.

85. $\frac{30}{55} + \frac{1000}{1760}$

86. $\frac{19}{26} - \frac{968}{1352}$

87. In your own words, describe how to add or subtract two fractions with different denominators.

88. Find the sum of the fractions in the circle graph above. Did the sum surprise you? Why or why not?

Operations on Fractions and Mixed Numbers

Answers

Find the LCM of each list of numbers.

1. 5, 6

2. 3, 7

3. 2, 14

4. 5, 25

5. 4, 20, 25

6. 6, 18, 30

Write each fraction as an equivalent fraction with the indicated denominator.

7. $\dfrac{3}{8} = \dfrac{}{24}$

8. $\dfrac{7}{9} = \dfrac{}{36}$

9. $\dfrac{1}{4} = \dfrac{}{40}$

10. $\dfrac{2}{5} = \dfrac{}{30}$

11. $\dfrac{11}{15} = \dfrac{}{75}$

12. $\dfrac{5}{6} = \dfrac{}{48}$

Add or subtract as indicated. Simplify if necessary.

13. $\dfrac{3}{8} + \dfrac{1}{8}$

14. $\dfrac{7}{10} - \dfrac{3}{10}$

15. $\dfrac{17}{24} - \dfrac{3}{24}$

16. $\dfrac{4}{15} + \dfrac{9}{15}$

17. $\dfrac{1}{4} + \dfrac{1}{2}$

18. $\dfrac{1}{3} - \dfrac{1}{5}$

19. $\dfrac{7}{9} - \dfrac{2}{5}$

20. $\dfrac{3}{10} + \dfrac{2}{25}$

21. $\dfrac{7}{8} + \dfrac{1}{20}$

1. _____

2. _____

3. _____

4. _____

5. _____

6. _____

7. _____

8. _____

9. _____

10. _____

11. _____

12. _____

13. _____

14. _____

15. _____

16. _____

17. _____

18. _____

19. _____

20. _____

21. _____

22. _____

23. _____

24. _____

25. _____

26. _____

27. _____

28. _____

29. _____

30. _____

31. _____

32. _____

33. _____

34. _____

35. _____

36. _____

37. _____

38. _____

39. _____

40. _____

41. _____

42. _____

43. _____

44. _____

45. _____

46. _____

22. $\dfrac{5}{12} - \dfrac{2}{18}$

23. $\dfrac{1}{11} - \dfrac{1}{11}$

24. $\dfrac{3}{17} - \dfrac{2}{17}$

25. $\dfrac{9}{11} - \dfrac{2}{3}$

26. $\dfrac{1}{6} - \dfrac{1}{7}$

27. $\dfrac{2}{9} + \dfrac{1}{18}$

28. $\dfrac{4}{13} + \dfrac{2}{26}$

29. $\dfrac{2}{9} + \dfrac{1}{18} + \dfrac{1}{3}$

30. $\dfrac{3}{10} + \dfrac{1}{5} + \dfrac{6}{25}$

Mixed Practice (*Sections 2.4, 2.5, 3.1, 3.2, 3.3*) *Perform the indicated operation.*

31. $\dfrac{9}{10} + \dfrac{2}{3}$

32. $\dfrac{9}{10} - \dfrac{2}{3}$

33. $\dfrac{9}{10} \cdot \dfrac{2}{3}$

34. $\dfrac{9}{10} \div \dfrac{2}{3}$

35. $\dfrac{21}{25} - \dfrac{3}{70}$

36. $\dfrac{21}{25} + \dfrac{3}{70}$

37. $\dfrac{21}{25} \div \dfrac{3}{70}$

38. $\dfrac{21}{25} \cdot \dfrac{3}{70}$

39. $3\dfrac{7}{8} \cdot 2\dfrac{2}{3}$

40. $3\dfrac{7}{8} \div 2\dfrac{2}{3}$

41. $\dfrac{2}{9} + \dfrac{5}{27} + \dfrac{1}{2}$

42. $\dfrac{3}{8} + \dfrac{11}{16} + \dfrac{2}{3}$

43. $11\dfrac{7}{10} \div 3\dfrac{3}{100}$

44. $7\dfrac{1}{4} \cdot 3\dfrac{1}{5}$

45. $\dfrac{14}{15} - \dfrac{4}{27}$

46. $\dfrac{9}{14} - \dfrac{11}{32}$

3.4 Adding and Subtracting Mixed Numbers

Objective A Adding Mixed Numbers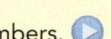

Recall that a mixed number has a whole number part and a fraction part.

$$\overset{\text{whole number}}{2\tfrac{3}{8}} \text{ means } 2 + \underset{\text{fraction}}{\tfrac{3}{8}}$$

✔**Concept Check** Which of the following are equivalent to 7?

a. $6\dfrac{5}{5}$ **b.** $6\dfrac{7}{7}$ **c.** $5\dfrac{8}{4}$

d. $6\dfrac{17}{17}$ **e.** all of these

Adding or Subtracting Mixed Numbers

To add or subtract mixed numbers, add or subtract the fraction parts and then add or subtract the whole number parts.

For example,

$$\begin{aligned}
&2\tfrac{2}{7}\\
+&6\tfrac{3}{7}\\
\hline
&8\tfrac{5}{7}
\end{aligned}$$
← Add the fractions:

└── then add the whole numbers

Example 1 Add: $2\dfrac{1}{3} + 5\dfrac{3}{8}$. Check by estimating.

Solution: The LCD of 3 and 8 is 24.

$$\begin{aligned}
2\dfrac{1\cdot 8}{3\cdot 8} &= 2\dfrac{8}{24}\\
+5\dfrac{3\cdot 3}{8\cdot 3} &= +5\dfrac{9}{24}\\
\hline
& \quad 7\dfrac{17}{24}
\end{aligned}$$
← Add the fractions

└── Add the whole numbers

To check by estimating, we round as usual. The fraction $2\dfrac{1}{3}$ rounds to 2, $5\dfrac{3}{8}$ rounds to 5, and $2 + 5 = 7$, our estimate.

Our exact answer is close to 7, so our answer is reasonable.

■ Work Practice 1

Objectives

A Add Mixed Numbers.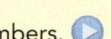

B Subtract Mixed Numbers.

C Solve Problems by Adding or Subtracting Mixed Numbers.

Practice 1

Add: $4\dfrac{2}{5} + 5\dfrac{1}{6}$

Answer

1. $9\dfrac{17}{30}$

✔**Concept Check Answer**

e

203

When adding or subtracting mixed numbers and whole numbers, it is a good idea to estimate to see if your answer is reasonable.

For the rest of this section, we leave most of the checking by estimating to you.

Practice 2

Add: $2\frac{5}{14} + 5\frac{6}{7}$

Example 2   Add: $3\frac{4}{5} + 1\frac{4}{15}$

Solution: The LCD of 5 and 15 is 15.

$$
\begin{array}{rcl}
3\dfrac{4}{5} & = & 3\dfrac{12}{15} \\[2mm]
+1\dfrac{4}{15} & = & +1\dfrac{4}{15} \\[2mm]
\hline
 & & 4\dfrac{16}{15}
\end{array}
$$

Add the fractions; then add the whole numbers.

Notice that the fraction part is improper.

Since $\frac{16}{15}$ is $1\frac{1}{15}$ we can write the sum as

$$4\frac{16}{15} = 4 + 1\frac{1}{15} = 5\frac{1}{15}$$

■ **Work Practice 2**

✔**Concept Check**   Explain how you could estimate the following sum:

$$5\frac{1}{9} + 14\frac{10}{11}.$$

Practice 3

Add: $10 + 2\frac{6}{7} + 3\frac{1}{5}$

Example 3   Add: $1\frac{4}{5} + 4 + 2\frac{1}{2}$

Solution: The LCD of 5 and 2 is 10.

$$
\begin{array}{rcl}
1\dfrac{4}{5} & = & 1\dfrac{8}{10} \\[2mm]
4 & = & 4 \\[2mm]
+2\dfrac{1}{2} & = & +2\dfrac{5}{10} \\[2mm]
\hline
 & & 7\dfrac{13}{10}
\end{array}
$$

$$7\frac{13}{10} = 7 + 1\frac{3}{10} = 8\frac{3}{10}$$

■ **Work Practice 3**

Answers

2. $8\frac{3}{14}$ 3. $16\frac{2}{35}$

✔**Concept Check Answer**

Round each mixed number to the nearest whole number and add. $5\frac{1}{9}$ rounds to 5 and $14\frac{10}{11}$ rounds to 15, and the estimated sum is $5 + 15 = 20$.

Objective B Subtracting Mixed Numbers

Example 4 Subtract: $9\frac{3}{7} - 5\frac{2}{21}$. Check by estimating.

Solution: The LCD of 7 and 21 is 21.

$$
\begin{array}{r}
9\frac{3}{7} = 9\frac{9}{21} \quad \leftarrow \text{The LCD of 7 and 21 is 21.}\\
-5\frac{2}{21} = -5\frac{2}{21}\\
\hline
4\frac{7}{21} \quad \leftarrow \text{Subtract the fractions.}\\
\uparrow
\end{array}
$$

Subtract the whole numbers.

Then $4\frac{7}{21}$ simplifies to $4\frac{1}{3}$. The difference is $4\frac{1}{3}$.

To check, $9\frac{3}{7}$ rounds to 9, $5\frac{2}{21}$ rounds to 5, and $9 - 5 = 4$, our estimate.

Our exact answer is close to 4, so our answer is reasonable.

■ **Work Practice 4**

When subtracting mixed numbers, borrowing may be needed, as shown in the next example.

Example 5 Subtract: $7\frac{3}{14} - 3\frac{6}{7}$

Solution: The LCD of 7 and 14 is 14.

$$
\begin{array}{r}
7\frac{3}{14} = 7\frac{3}{14}\\
-3\frac{6}{7} = -3\frac{12}{14}\\
\hline
\end{array}
$$

Notice that we cannot subtract $\frac{12}{14}$ from $\frac{3}{14}$, so we borrow from the whole number 7.

borrow 1 from 7

$$7\frac{3}{14} = 6 + 1\frac{3}{14} = 6 + \frac{17}{14} \text{ or } 6\frac{17}{14}$$

Now subtract.

$$
\begin{array}{r}
7\frac{3}{14} = 7\frac{3}{14} = 6\frac{17}{14}\\
-3\frac{6}{7} = -3\frac{12}{14} = -3\frac{12}{14}\\
\hline
3\frac{5}{14} \quad \leftarrow \text{Subtract the fractions.}\\
\uparrow
\end{array}
$$

Subtract the whole numbers.

■ **Work Practice 5**

✔**Concept Check** In the subtraction problem $5\frac{1}{4} - 3\frac{3}{4}$, $5\frac{1}{4}$ must be rewritten because $\frac{3}{4}$ cannot be subtracted from $\frac{1}{4}$. Why is it incorrect to rewrite $5\frac{1}{4}$ as $5\frac{5}{4}$?

Practice 4

Subtract: $29\frac{7}{9} - 13\frac{5}{18}$

Practice 5

Subtract: $9\frac{7}{15} - 5\frac{3}{5}$

Answers

4. $16\frac{1}{2}$ 5. $3\frac{13}{15}$

✔**Concept Check Answer**

Rewrite $5\frac{1}{4}$ as $4\frac{5}{4}$ by borrowing from the 5.

Practice 6

Subtract: $25 - 10\frac{2}{9}$

Example 6 Subtract: $12 - 8\frac{3}{7}$

Solution:

$$12 = 11\frac{7}{7} \quad \text{Borrow 1 from 12 and write it as } \frac{7}{7}.$$

$$-8\frac{3}{7} = -8\frac{3}{7}$$

$$3\frac{4}{7} \quad \leftarrow \text{Subtract the fractions.}$$

Subtract the whole numbers.

■ Work Practice 6

Objective C Solving Problems by Adding or Subtracting Mixed Numbers ▶

Now that we know how to add and subtract mixed numbers, we can solve real-life problems.

Practice 7

Two rainbow trout weigh $2\frac{1}{2}$ pounds and $3\frac{2}{3}$ pounds. What is the total weight of the two trout?

Example 7 Calculating Total Weight

Two packages of ground round are purchased. One package weighs $2\frac{3}{8}$ pounds and the other $1\frac{4}{5}$ pounds. What is the combined weight of the ground round?

Solution:

1. UNDERSTAND. Read and reread the problem. The phrase "combined weight" tells us to add.

2. TRANSLATE.

In words:	combined weight	is	weight of one package	plus	weight of second package
	↓	↓	↓	↓	↓
Translate:	combined weight	=	$2\frac{3}{8}$	+	$1\frac{4}{5}$

3. SOLVE: Before we solve, let's estimate. The fraction $2\frac{3}{8}$ rounds to 2, $1\frac{4}{5}$ rounds to 2, and $2 + 2 = 4$. The combined weight should be close to 4.

$$2\frac{3}{8} = 2\frac{15}{40}$$

$$+1\frac{4}{5} = +1\frac{32}{40}$$

$$3\frac{47}{40} = 4\frac{7}{40}$$

4. INTERPRET. *Check* your work. Our estimate of 4 tells us that the exact answer of $4\frac{7}{40}$ is reasonable. *State* your conclusion: The combined weight of the ground round is $4\frac{7}{40}$ pounds.

■ Work Practice 7

Answers

6. $14\frac{7}{9}$ 7. $6\frac{1}{6}$ lb

Example 8 Finding Legal Lobster Size

Lobster fishermen must measure the upper body shells of the lobsters they catch. Lobsters that are too small are thrown back into the ocean. Each state has its own size standard for lobsters to help control the breeding stock. Massachusetts divided its waters into four Lobster Conservation Management Areas, with a different minimum lobster size permitted in each area. In the off-shore area, the legal lobster size increased from $3\frac{13}{32}$ inches in 2006 to $3\frac{1}{2}$ inches in 2008. How much of an increase was this? (*Source:* Massachusetts Division of Marine Fisheries)

Solution:

1. **UNDERSTAND.** Read and reread the problem carefully. The word "increase" found in the problem might make you think that we add to solve the problem. But the phrase "how much of an increase" tells us to subtract to find the increase.

2. **TRANSLATE.**

In words:	increase	is	new lobster size	minus	old lobster size
	↓	↓	↓	↓	↓
Translate:	increase	=	$3\frac{1}{2}$	−	$3\frac{13}{32}$

3. **SOLVE.** Before we solve, let's estimate. The fraction $3\frac{1}{2}$ can be rounded to 4, $3\frac{13}{32}$ can be rounded to 3, and $4 - 3 = 1$. The increase is not 1, but will be smaller since we rounded $3\frac{1}{2}$ up more than we rounded $3\frac{13}{32}$ down.

$$3\frac{1}{2} = 3\frac{16}{32}$$
$$-3\frac{13}{32} = -3\frac{13}{32}$$
$$\overline{\frac{3}{32}}$$

4. **INTERPRET.** *Check* your work. Our estimate tells us that the exact increase of $\frac{3}{32}$ is reasonable. *State* your conclusion: The increase in lobster size is $\frac{3}{32}$ of an inch.

■ **Work Practice 8**

Vocabulary, Readiness & Video Check

Use the choices below to fill in each blank.

round   fraction   whole number

improper   mixed number

1. The number $5\frac{3}{4}$ is called a(n) _____.

2. For $5\frac{3}{4}$, the 5 is called the _____ part and $\frac{3}{4}$ is called the _____ part.

3. To estimate operations on mixed numbers, we _____ mixed numbers to the nearest whole number.

4. The mixed number $2\frac{5}{8}$ written as a(n) _____ fraction is $\frac{21}{8}$.

Choose the best estimate for each sum or difference.

5. $3\frac{7}{8} + 2\frac{1}{5}$

   a. 6   **b.** 5   **c.** 1   **d.** 2

6. $3\frac{7}{8} - 2\frac{1}{5}$

   a. 6   **b.** 5   **c.** 1   **d.** 2

7. $8\frac{1}{3} - 1\frac{1}{2}$

   a. 4   **b.** 10   **c.** 6   **d.** 16

8. $8\frac{1}{3} + 1\frac{1}{2}$

   a. 4   **b.** 10   **c.** 6   **d.** 16

Martin-Gay Interactive Videos

See Video 3.4 🍎

Watch the section lecture video and answer the following questions.

Objective A   9. In ▣ Example 2, why is the first form of the answer not in a good format? ▶

Objective B   10. In ▣ Example 3, how is 6 rewritten in the subtraction problem? ▶

Objective C   11. In ▣ Example 4, why can't we subtract immediately once we rewrite the fraction parts of the mixed numbers with the LCD? ▶

3.4   Exercise Set   MyMathLab®

Objective A   *Add. For those exercises marked, find an exact sum and an estimated sum. See Examples 1 through 3.*

1. $4\frac{7}{10}$
   $+2\frac{1}{10}$

   Exact:

   Estimate:

2. $7\frac{4}{9}$
   $+3\frac{2}{9}$

   Exact:

   Estimate:

▶ 3. $10\frac{3}{14}$
   $+3\frac{4}{7}$

   Exact:

   Estimate:

4. $12\frac{5}{12}$
   $+4\frac{1}{6}$

   Exact:

   Estimate:

5. $9\frac{1}{5}$
   $+8\frac{2}{25}$

6. $6\frac{2}{13}$

$+8\frac{7}{26}$

7. $3\frac{1}{2}$

$+4\frac{1}{8}$

8. $9\frac{3}{4}$

$+2\frac{1}{8}$

9. $1\frac{5}{6}$

$+5\frac{3}{8}$

10. $2\frac{5}{12}$

$+1\frac{5}{8}$

11. $8\frac{2}{5}$

$+11\frac{2}{3}$

12. $7\frac{3}{7}$

$+3\frac{3}{5}$

13. $11\frac{3}{5}$

$+7\frac{2}{5}$

14. $19\frac{7}{9}$

$+8\frac{2}{9}$

15. $40\frac{9}{10}$

$+15\frac{8}{27}$

16. $102\frac{5}{8}$

$+96\frac{21}{25}$

17. $3\frac{5}{8}$

$2\frac{1}{6}$

$+7\frac{3}{4}$

18. $4\frac{1}{3}$

$9\frac{2}{5}$

$+3\frac{1}{6}$

19. $12\frac{3}{14}$

10

$+25\frac{5}{12}$

20. $8\frac{2}{9}$

32

$+9\frac{10}{21}$

Objectives **B** *Subtract. For those exercises marked, find an exact difference and an estimated difference. See Examples 4 through 6.*

21. $4\frac{7}{10}$

$-2\frac{1}{10}$

Exact:

Estimate:

22. $7\frac{4}{9}$

$-3\frac{2}{9}$

Exact:

Estimate:

23. $10\frac{13}{14}$

$-3\frac{4}{7}$

Exact:

Estimate:

24. $12\frac{5}{12}$

$-4\frac{1}{6}$

Exact:

Estimate:

25. $9\frac{1}{5}$

$-8\frac{6}{25}$

26. $5\frac{2}{13}$

$-4\frac{7}{26}$

27. $5\frac{2}{3} - 3\frac{1}{5}$

28. $23\frac{3}{5}$

$-8\frac{8}{15}$

29. $15\frac{4}{7}$

$-9\frac{11}{14}$

30. $5\frac{3}{8} - 2\frac{13}{20}$

31. $47\frac{4}{18} - 23\frac{19}{24}$

32. $6\frac{1}{6} - 5\frac{11}{14}$

33. 10

$-8\frac{1}{5}$

34. 23

$-17\frac{3}{4}$

35. $11\frac{3}{5}$

$-9\frac{11}{15}$

36. $9\dfrac{1}{10}$

$-7\dfrac{2}{5}$

37. 6

$-2\dfrac{4}{9}$

38. 8

$-1\dfrac{7}{10}$

39. $63\dfrac{1}{6}$

$-47\dfrac{5}{12}$

40. $86\dfrac{2}{15}$

$-27\dfrac{3}{10}$

Objectives A B **Mixed Practice** *Perform the indicated operation. See Examples 1 through 6.*

41. $15\dfrac{1}{6}$

$+13\dfrac{5}{12}$

42. $21\dfrac{3}{10}$

$+11\dfrac{3}{5}$

43. $22\dfrac{7}{8}$

$-\ 7$

44. $27\dfrac{3}{21}$

$-\ 9$

45. $5\dfrac{8}{9} + 2\dfrac{1}{9}$

46. $12\dfrac{13}{16} + 7\dfrac{3}{16}$

47. $33\dfrac{11}{20} - 15\dfrac{19}{30}$

48. $54\dfrac{7}{30} - 38\dfrac{29}{50}$

Objective C *Solve. For Exercises 49 and 50, the solutions have been started for you. Write each answer in simplest form. See Examples 7 and 8.*

49. To prevent intruding birds, birdhouses built for Eastern Bluebirds should have an entrance hole measuring $1\dfrac{1}{2}$ inches in diameter. Entrance holes in birdhouses for Mountain Bluebirds should measure $1\dfrac{9}{16}$ inches in diameter. How much wider should entrance holes for Mountain Bluebirds be than for Eastern Bluebirds? (*Source:* North American Bluebird Society)

Start the solution:

1. UNDERSTAND the problem. Reread it as many times as needed.
2. TRANSLATE into an equation. (Fill in the blanks.)

how much wider	is	larger entrance hole	minus	smaller entrance hole

$$\text{how much wider} = \underline{\quad\quad} - \underline{\quad\quad}$$

Finish with:

3. SOLVE and
4. INTERPRET

50. If the total weight allowable without overweight charges is 50 pounds and the traveler's luggage weighs $60\dfrac{5}{8}$ pounds, on how many pounds will the traveler's overweight charges be based?

Start the solution:

1. UNDERSTAND the problem. Reread it as many times as needed.
2. TRANSLATE into an equation. (Fill in the blanks.)

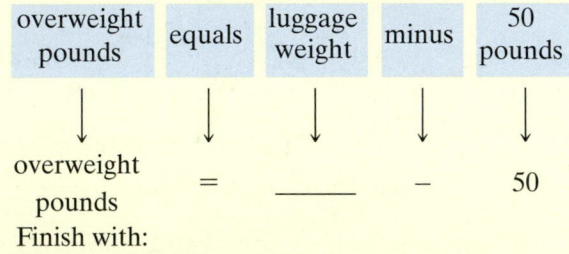

overweight pounds	equals	luggage weight	minus	50 pounds

$$\text{overweight pounds} = \underline{\quad\quad} - 50$$

Finish with:

3. SOLVE and
4. INTERPRET

51. Charlotte Dowlin has $15\frac{2}{3}$ feet of plastic pipe. She cuts off a $2\frac{1}{2}$-foot length and then a $3\frac{1}{4}$-foot length. If she now needs a 10-foot piece of pipe, will the remaining piece do? If not, by how much will the piece be short?

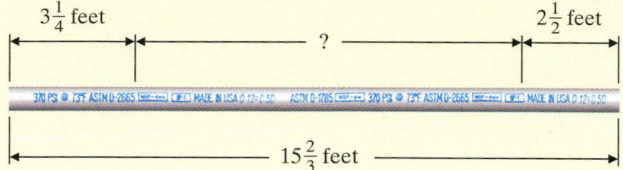

$3\frac{1}{4}$ feet　　?　　$2\frac{1}{2}$ feet

$15\frac{2}{3}$ feet

52. A trim carpenter cuts a board $3\frac{3}{8}$ feet long from one 6 feet long. How long is the remaining piece?

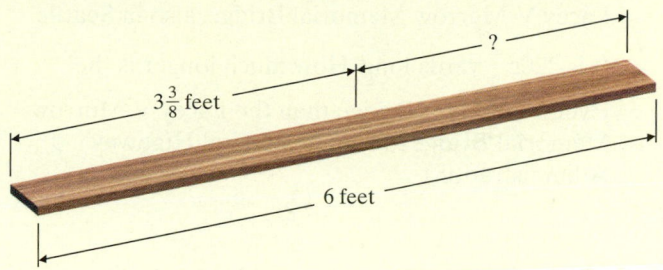

?

$3\frac{3}{8}$ feet

6 feet

53. If Tucson's average annual rainfall is $11\frac{1}{4}$ inches and Yuma's is $3\frac{3}{5}$ inches, how much more rain, on average, does Tucson get than Yuma?

54. A pair of crutches needs adjustment. One crutch is 43 inches and the other is $41\frac{5}{8}$ inches. Find how much the short crutch should be lengthened to make both crutches the same length.

55. On four consecutive days, a concert pianist practiced for $2\frac{1}{2}$ hours, $1\frac{2}{3}$ hours, $2\frac{1}{4}$ hours, and $3\frac{5}{6}$ hours. Find his total practice time.

56. A tennis coach was preparing her team for a tennis tournament and enforced this practice schedule: Monday, $2\frac{1}{2}$ hours; Tuesday, $2\frac{2}{3}$ hours; Wednesday, $1\frac{3}{4}$ hours; and Thursday, $1\frac{9}{16}$ hours. How long did the team practice that week before Friday's tournament?

57. Jerald Divis, a tax consultant, takes $3\frac{1}{2}$ hours to prepare a personal tax return and $5\frac{7}{8}$ hours to prepare a small business return. How much longer does it take him to prepare the small business return?

58. Jessica Callac takes $2\frac{3}{4}$ hours to clean her room. Her brother Matthew takes $1\frac{1}{3}$ hours to clean his room. If they start at the same time, how long does Matthew have to wait for Jessica to finish?

59. Located on an island in New York City's harbor, the Statue of Liberty is one of the largest statues in the world. The copper figure is $46\frac{1}{20}$ meters tall from feet to tip of torch. The figure stands on a pedestal that is $46\frac{47}{50}$ meters tall. What is the overall height of the Statue of Liberty from the base of the pedestal to the tip of the torch? (*Source:* National Park Service)

60. The record for largest rainbow trout ever caught is 48 pounds and was set in Saskatchewan in 2009. The record for largest brown trout ever caught is $42\frac{1}{16}$ pounds and was set in New Zealand in 2013. How much more did the record-setting rainbow trout weigh than the record-setting brown trout? (*Source:* International Game Fish Association)

61. The longest floating pontoon bridge in the United States is the Evergreen Point Bridge in Seattle, Washington. It is 2526 yards long. The second-longest pontoon bridge in the United States is the Lacey V. Murrow Memorial Bridge, also in Seattle. It is $2206\frac{2}{3}$ yards long. How much longer is the Evergreen Point Bridge than the Lacey V. Murrow Memorial Bridge? (*Source:* Federal Highway Administration)

62. What is the difference between interest rates of $11\frac{1}{2}$ percent and $9\frac{3}{4}$ percent?

The following table lists some upcoming total eclipses of the Sun that will be visible in North America. The duration of each eclipse is listed in the table. Use the table to answer Exercises 63 through 66.

Total Solar Eclipses Visible from North America	
Date of Eclipse	**Duration (in Minutes)**
August 21, 2017	$2\frac{2}{3}$
April 8, 2024	$4\frac{7}{15}$
August 12, 2026	$2\frac{3}{10}$

(*Source:* NASA/Goddard Space Flight Center)

63. What is the total duration for the three eclipses?

64. What is the total duration for the two eclipses occurring in even-numbered years?

65. How much longer will the April 8, 2024, eclipse be than the August 21, 2017, eclipse?

66. How much longer will the August 21, 2017, eclipse be than the August 12, 2026, eclipse?

Find the perimeter of each figure.

△ **67.**

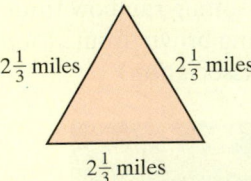

$2\frac{1}{3}$ miles $2\frac{1}{3}$ miles $2\frac{1}{3}$ miles

△ **68.**

7 inches $11\frac{1}{5}$ inches $12\frac{1}{3}$ inches

△ **69.**

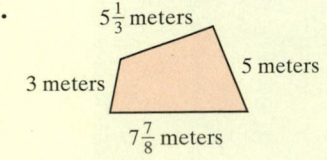

$5\frac{1}{3}$ meters   3 meters   5 meters $7\frac{7}{8}$ meters

△ **70.**

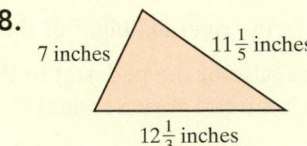

$3\frac{1}{4}$ yards $3\frac{1}{4}$ yards $3\frac{1}{4}$ yards $3\frac{1}{4}$ yards $3\frac{1}{4}$ yards

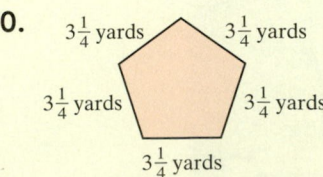

Review

Evaluate each expression. See Section 1.9.

71. 2^3

72. 3^2

73. 5^2

74. 2^5

75. $20 \div 10 \cdot 2$

76. $36 - 5 \cdot 6 + 10$

77. $2 + 3(8 \cdot 7 - 1)$

78. $2(10 - 2 \cdot 5) + 13$

Simplify. Write any mixed number whose fraction part is not a proper fraction in simplest form. See Sections 2.1 and 2.3.

79. $3\frac{5}{5}$

80. $10\frac{8}{7}$

81. $9\frac{10}{16}$

82. $6\frac{7}{14}$

Concept Extensions

Solve. See the Concept Checks in this section.

83. Which of the following are equivalent to 10?

    a. $9\frac{5}{5}$ **b.** $9\frac{100}{100}$ **c.** $6\frac{44}{11}$ **d.** $8\frac{13}{13}$

84. Which of the following are equivalent to $7\frac{3}{4}$?

    a. $6\frac{7}{4}$ **b.** $5\frac{11}{4}$ **c.** $7\frac{12}{16}$ **d.** all of them

85. Explain in your own words why $9\frac{13}{9}$ is equal to $10\frac{4}{9}$.

86. In your own words, explain

    a. when to borrow when subtracting mixed numbers, and

    b. how to borrow when subtracting mixed numbers.

Solve.

87. Carmen's Candy Clutch is famous for its "Nutstuff," a special blend of nuts and candy. A Supreme box of Nutstuff has $2\frac{1}{4}$ pounds of nuts and $3\frac{1}{2}$ pounds of candy. A Deluxe box has $1\frac{3}{8}$ pounds of nuts and $4\frac{1}{4}$ pounds of candy. Which box is heavier and by how much?

88. Willie Cassidie purchased three Supreme boxes and two Deluxe boxes of Nutstuff from Carmen's Candy Clutch. (See Exercise 87.) What is the total weight of his purchase?

3.5 Order, Exponents, and the Order of Operations

Objectives

A Compare Fractions.

B Evaluate Fractions Raised to Powers.

C Review Operations on Fractions.

D Use the Order of Operations.

Objective A Comparing Fractions

Recall that whole numbers can be shown on a number line using equally spaced distances.

$$\begin{array}{cccccccc} | & | & | & \bullet & | & \bullet & | & | \\ 0 & 1 & 2 & 3 & 4 & 5 & 6 & 7 \end{array}$$

From the number line, we can see the order of numbers. For example, we can see that 3 is less than 5 because 3 is to the left of 5.

For any two numbers on a number line, the number to the **left** is always the **smaller** number, and the number to the **right** is always the **larger** number.

We use the **inequality symbols** $<$ or $>$ to write the order of numbers.

> **Inequality Symbols**
>
> $<$ means *is less than.*
>
> $>$ means *is greater than.*

For example,

3 is less than 5   or   5 is greater than 3

$$3 < 5 \qquad\qquad 5 > 3$$

We can compare fractions the same way. To see fractions on a number line, divide the spaces between whole numbers into equal parts.

For example, let's compare $\dfrac{2}{5}$ and $\dfrac{4}{5}$.

$$\begin{array}{ccccccccc} | & | & \bullet & | & \bullet & | & | & | & | & | \\ 0 & \frac{1}{5} & \frac{2}{5} & \frac{3}{5} & \frac{4}{5} & 1 & & & 2 \end{array}$$

$$\frac{5}{5} = 1$$

Helpful Hint   Notice that to compare like fractions, we compare the numerators. The order of the like fractions is the same as the order of the numerators.

Since $\dfrac{4}{5}$ is to the right of $\dfrac{2}{5}$,

$$\frac{2}{5} < \frac{4}{5} \qquad \text{Notice that } 2 < 4 \text{ also.}$$

> **Comparing Fractions**
>
> To determine which of two fractions is greater,
>
> **Step 1:** Write the fractions as like fractions.
>
> **Step 2:** The fraction with the greater numerator is the greater fraction.

Practice 1

Insert $<$ or $>$ to form a true statement.

$$\frac{8}{9} \qquad \frac{10}{11}$$

Answer

1. $<$

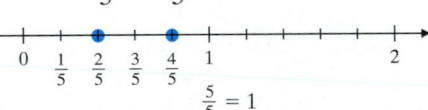 Insert $<$ or $>$ to form a true statement.

$$\frac{3}{10} \qquad \frac{2}{7}$$

Solution:

Step 1: The LCD of 10 and 7 is 70.

$$\frac{3}{10} = \frac{3}{10} \cdot \frac{7}{7} = \frac{21}{70}; \qquad \frac{2}{7} = \frac{2}{7} \cdot \frac{10}{10} = \frac{20}{70}$$

Step 2: Since $21 > 20$, then $\dfrac{21}{70} > \dfrac{20}{70}$ or

$$\dfrac{3}{10} > \dfrac{2}{7}$$

■ **Work Practice 1**

Example 2 Insert $<$ or $>$ to form a true statement.

$$\dfrac{9}{10} \qquad \dfrac{11}{12}$$

Solution:

Step 1: The LCD of 10 and 12 is 60.

$$\dfrac{9}{10} = \dfrac{9}{10} \cdot \dfrac{6}{6} = \dfrac{54}{60} \qquad \dfrac{11}{12} = \dfrac{11}{12} \cdot \dfrac{5}{5} = \dfrac{55}{60}$$

Step 2: Since $54 < 55$, then $\dfrac{54}{60} < \dfrac{55}{60}$ or

$$\dfrac{9}{10} < \dfrac{11}{12}$$

■ **Work Practice 2**

Helpful Hint

If we think of $<$ and $>$ as arrowheads, a true statement is always formed when the arrow points to the smaller number.

$$\dfrac{2}{3} > \dfrac{1}{3} \qquad\qquad \dfrac{5}{6} < \dfrac{7}{6}$$

points to smaller number   points to smaller number

Objective B Evaluating Fractions Raised to Powers

Recall from Section 1.9 that exponents indicate repeated multiplication.

exponent

$$5^3 = \underbrace{5 \cdot 5 \cdot 5}_{} = 125$$

base   3 factors of 5

Exponents mean the same when the base is a fraction. For example,

$$\left(\dfrac{1}{3}\right)^4 = \underbrace{\dfrac{1}{3} \cdot \dfrac{1}{3} \cdot \dfrac{1}{3} \cdot \dfrac{1}{3}}_{} = \dfrac{1}{81}$$

base   4 factors of $\dfrac{1}{3}$

Examples Evaluate each expression.

3. $\left(\dfrac{1}{4}\right)^2 = \dfrac{1}{4} \cdot \dfrac{1}{4} = \dfrac{1}{16}$

4. $\left(\dfrac{3}{5}\right)^3 = \dfrac{3}{5} \cdot \dfrac{3}{5} \cdot \dfrac{3}{5} = \dfrac{27}{125}$

5. $\left(\dfrac{1}{6}\right)^2 \cdot \left(\dfrac{3}{4}\right)^3 = \left(\dfrac{1}{6} \cdot \dfrac{1}{6}\right) \cdot \left(\dfrac{3}{4} \cdot \dfrac{3}{4} \cdot \dfrac{3}{4}\right) = \dfrac{1 \cdot 1 \cdot \overset{1}{\cancel{3}} \cdot \overset{1}{\cancel{3}} \cdot 3}{2 \cdot \underset{1}{\cancel{3}} \cdot 2 \cdot \underset{1}{\cancel{3}} \cdot 4 \cdot 4 \cdot 4} = \dfrac{3}{256}$

■ **Work Practice 3–5**

Practice 2

Insert $<$ or $>$ to form a true statement.

$$\dfrac{3}{5} \qquad \dfrac{2}{9}$$

Practice 3–5

Evaluate each expression.

3. $\left(\dfrac{1}{5}\right)^2$ **4.** $\left(\dfrac{2}{3}\right)^3$

5. $\left(\dfrac{1}{4}\right)^2\left(\dfrac{2}{3}\right)^3$

Answers

2. $>$ **3.** $\dfrac{1}{25}$ **4.** $\dfrac{8}{27}$ **5.** $\dfrac{1}{54}$

Objective C Reviewing Operations on Fractions

To get ready to use the order of operations with fractions, let's first review the operations on fractions that we have learned.

Review of Operations on Fractions		
Operation	**Procedure**	**Example**
Multiply	Multiply the numerators and multiply the denominators.	$\dfrac{5}{9} \cdot \dfrac{1}{2} = \dfrac{5 \cdot 1}{9 \cdot 2} = \dfrac{5}{18}$
Divide	Multiply the first fraction by the reciprocal of the second fraction.	$\dfrac{2}{3} \div \dfrac{11}{13} = \dfrac{2}{3} \cdot \dfrac{13}{11} = \dfrac{2 \cdot 13}{3 \cdot 11} = \dfrac{26}{33}$
Add or Subtract	1. Write each fraction as an equivalent fraction whose denominator is the LCD 2. Add or subtract numerators and write the result over the common denominator.	$\dfrac{3}{4} + \dfrac{1}{8} = \dfrac{3}{4} \cdot \dfrac{2}{2} + \dfrac{1}{8} = \dfrac{6}{8} + \dfrac{1}{8} = \dfrac{7}{8}$

Practice 6–9

Perform each indicated operation.

6. $\dfrac{3}{7} \div \dfrac{10}{11}$ **7.** $\dfrac{4}{15} + \dfrac{2}{5}$

8. $\dfrac{2}{3} \cdot \dfrac{9}{10}$ **9.** $\dfrac{11}{12} - \dfrac{2}{5}$

Examples   Perform each indicated operation.

6. $\dfrac{1}{2} \div \dfrac{8}{7} = \dfrac{1}{2} \cdot \dfrac{7}{8} = \dfrac{1 \cdot 7}{2 \cdot 8} = \dfrac{7}{16}$ To divide: multiply by the reciprocal.

7. $\dfrac{6}{35} + \dfrac{3}{7} = \dfrac{6}{35} + \dfrac{3}{7} \cdot \dfrac{5}{5} = \dfrac{6}{35} + \dfrac{15}{35} = \dfrac{21}{35}$ To add: need the LCD. The LCD is 35.

$= \dfrac{\overset{1}{\cancel{7}} \cdot 3}{\underset{1}{\cancel{7}} \cdot 5} = \dfrac{3}{5}$

8. $\dfrac{2}{9} \cdot \dfrac{3}{11} = \dfrac{2 \cdot 3}{9 \cdot 11} = \dfrac{2 \cdot \overset{1}{\cancel{3}}}{\underset{1}{\cancel{3}} \cdot 3 \cdot 11} = \dfrac{2}{33}$ To multiply: multiply numerators and multiply denominators.

9. $\dfrac{6}{7} - \dfrac{1}{3} = \dfrac{6}{7} \cdot \dfrac{3}{3} - \dfrac{1}{3} \cdot \dfrac{7}{7} = \dfrac{18}{21} - \dfrac{7}{21} = \dfrac{11}{21}$ To subtract: need the LCD. The LCD is 21.

■ **Work Practice 6–9**

Objective D Using the Order of Operations

The order of operations that we use on whole numbers applies to expressions containing fractions and mixed numbers also.

Order of Operations

1. Perform all operations within parentheses ( ), brackets [ ], or other grouping symbols such as square roots or fraction bars, starting with the innermost set.
2. Evaluate any expressions with exponents.
3. Multiply or divide in order from left to right.
4. Add or subtract in order from left to right.

Answers

6. $\dfrac{33}{70}$ **7.** $\dfrac{2}{3}$ **8.** $\dfrac{3}{5}$ **9.** $\dfrac{31}{60}$

Example 10 Simplify: $\frac{1}{5} \div \frac{2}{3} \cdot \frac{4}{5}$

Solution: Multiply or divide *in order* from left to right. We divide first.

$$\frac{1}{5} \div \frac{2}{3} \cdot \frac{4}{5} = \frac{1}{5} \cdot \frac{3}{2} \cdot \frac{4}{5}$$

To divide, multiply by the reciprocal.

$$= \frac{3}{10} \cdot \frac{4}{5}$$

$$= \frac{3 \cdot 4}{10 \cdot 5} \qquad \text{Multiply.}$$

$$= \frac{3 \cdot 2 \cdot \overset{1}{\cancel{2}}}{\underset{1}{\cancel{2}} \cdot 5 \cdot 5} \qquad \text{Simplify.}$$

$$= \frac{6}{25} \qquad \text{Simplify.}$$

■ **Work Practice 10**

Example 11 Simplify: $\left(\frac{2}{3}\right)^2 \div \left(\frac{8}{27} + \frac{2}{3}\right)$

Solution: Start within the right set of parentheses. We add.

$$\left(\frac{2}{3}\right)^2 \div \left(\frac{8}{27} + \frac{2}{3}\right) = \left(\frac{2}{3}\right)^2 \div \left(\frac{8}{27} + \frac{18}{27}\right) \qquad \text{The LCD is 27. Write } \frac{2}{3} \text{ as } \frac{18}{27}.$$

$$= \left(\frac{2}{3}\right)^2 \div \frac{26}{27} \qquad \text{Simplify inside the parentheses.}$$

$$= \frac{4}{9} \div \frac{26}{27} \qquad \text{Write } \left(\frac{2}{3}\right)^2 \text{ as } \frac{4}{9}.$$

$$= \frac{4}{9} \cdot \frac{27}{26}$$

$$= \frac{\overset{1}{\cancel{2}} \cdot 2 \cdot 3 \cdot \overset{1}{\cancel{9}}}{\underset{1}{\cancel{9}} \cdot \underset{1}{\cancel{2}} \cdot 13}$$

$$= \frac{6}{13}$$

■ **Work Practice 11**

✓ **Concept Check** What should be done first to simplify $3\left[\left(\frac{1}{4}\right)^2 + \frac{3}{2}\left(\frac{6}{7} - \frac{1}{3}\right)\right]$?

Recall from Section 1.7 that the average of a list of numbers is their sum divided by the number of numbers in the list.

Example 12 Find the average of $\frac{1}{3}, \frac{2}{5},$ and $\frac{2}{9}$.

Solution: The average is their sum, divided by 3.

$$\left(\frac{1}{3} + \frac{2}{5} + \frac{2}{9}\right) \div 3 = \left(\frac{15}{45} + \frac{18}{45} + \frac{10}{45}\right) \div 3 \qquad \text{The LCD is 45.}$$

$$= \frac{43}{45} \div 3 \qquad \text{Add.}$$

$$= \frac{43}{45} \cdot \frac{1}{3}$$

$$= \frac{43}{135} \qquad \text{Multiply.}$$

■ **Work Practice 12**

Practice 10

Simplify: $\frac{2}{9} \div \frac{4}{7} \cdot \frac{3}{10}$

Practice 11

Simplify: $\left(\frac{2}{5}\right)^2 \div \left(\frac{3}{5} - \frac{11}{25}\right)$

Practice 12

Find the average of $\frac{1}{2}, \frac{3}{8},$ and $\frac{7}{24}$.

Answers

10. $\frac{7}{60}$ **11.** 1 **12.** $\frac{7}{18}$

✓ **Concept Check Answer**

$\frac{6}{7} - \frac{1}{3}$

Vocabulary, Readiness & Video Check

Use the choices below to fill in each blank. Not all choices will be used.

addition   multiplication   evaluate the exponential expression

subtraction   division

1. To simplify $\dfrac{1}{2} + \dfrac{2}{3} \cdot \dfrac{7}{8}$, which operation do we perform first? _____

2. To simplify $\dfrac{1}{2} \div \dfrac{2}{3} \cdot \dfrac{7}{8}$, which operation do we perform first? _____

3. To simplify $\dfrac{7}{8} \cdot \left(\dfrac{1}{2} - \dfrac{2}{3} \right)$, which operation do we perform first? _____

4. To simplify $9 - \left(\dfrac{3}{4} \right)^2$, which operation do we perform first? _____

Martin-Gay Interactive Videos

See Video 3.5 🍎

Watch the section lecture video and answer the following questions.

Objective A   **5.** Complete this statement based on ⊟ Example 1: When comparing fractions, as long as the _____ are the same, we can just compare _____.

Objective B   **6.** Complete this statement based on ⊟ Example 3: The meaning of an exponent is the same whether the base is a _____ or the base is a _____. ▶

Objective C   **7.** Fraction operations are reviewed in the lecture before ⊟ Example 4. How are denominators treated differently when adding and subtracting fractions than when multiplying and dividing? ▶

Objective D   **8.** In ⊟ Example 6, why did we subtract before applying the exponent? ▶

3.5   Exercise Set   MyMathLab® ▶

Objective A   *Insert < or > to form a true statement. See Examples 1 and 2.*

1. $\dfrac{7}{9}$ $\dfrac{6}{9}$ **2.** $\dfrac{12}{17}$ $\dfrac{13}{17}$ ▶ **3.** $\dfrac{3}{3}$ $\dfrac{5}{3}$ **4.** $\dfrac{3}{23}$ $\dfrac{4}{23}$

5. $\dfrac{9}{42}$ $\dfrac{5}{21}$ **6.** $\dfrac{17}{32}$ $\dfrac{5}{16}$ **7.** $\dfrac{9}{8}$ $\dfrac{17}{16}$ **8.** $\dfrac{3}{8}$ $\dfrac{14}{40}$

9. $\dfrac{3}{4}$ $\dfrac{2}{3}$ **10.** $\dfrac{2}{5}$ $\dfrac{1}{3}$ ▶ **11.** $\dfrac{3}{5}$ $\dfrac{9}{14}$ **12.** $\dfrac{3}{10}$ $\dfrac{7}{25}$

13. $\dfrac{1}{10}$ $\dfrac{1}{11}$ **14.** $\dfrac{1}{13}$ $\dfrac{1}{14}$ **15.** $\dfrac{27}{100}$ $\dfrac{7}{25}$ **16.** $\dfrac{37}{120}$ $\dfrac{9}{30}$

Objective B *Evaluate each expression. See Examples 3 through 5.*

17. $\left(\dfrac{1}{2}\right)^4$

18. $\left(\dfrac{1}{7}\right)^2$

▶ 19. $\left(\dfrac{2}{5}\right)^3$

20. $\left(\dfrac{3}{4}\right)^3$

21. $\left(\dfrac{4}{7}\right)^3$

22. $\left(\dfrac{2}{3}\right)^4$

23. $\left(\dfrac{2}{9}\right)^2$

24. $\left(\dfrac{7}{11}\right)^2$

25. $\left(\dfrac{3}{4}\right)^2 \cdot \left(\dfrac{2}{3}\right)^3$

26. $\left(\dfrac{1}{6}\right)^2 \cdot \left(\dfrac{9}{10}\right)^2$

27. $\dfrac{9}{10}\left(\dfrac{2}{5}\right)^2$

28. $\dfrac{7}{11}\left(\dfrac{3}{10}\right)^2$

Objective C *Perform each indicated operation. See Examples 6 through 9.*

29. $\dfrac{2}{15} + \dfrac{3}{5}$

30. $\dfrac{5}{12} + \dfrac{5}{6}$

31. $\dfrac{3}{7} \cdot \dfrac{1}{5}$

32. $\dfrac{9}{10} \div \dfrac{2}{3}$

33. $1 - \dfrac{4}{9}$

34. $5 - \dfrac{2}{3}$

▶ 35. $4\dfrac{2}{9} + 5\dfrac{9}{11}$

36. $7\dfrac{3}{7} + 6\dfrac{3}{5}$

37. $\dfrac{5}{6} - \dfrac{3}{4}$

38. $\dfrac{7}{10} - \dfrac{3}{25}$

39. $\dfrac{6}{11} \div \dfrac{2}{3}$

40. $\dfrac{3}{8} \cdot \dfrac{1}{11}$

41. $0 \cdot \dfrac{9}{10}$

42. $\dfrac{5}{6} \cdot 0$

43. $0 \div \dfrac{9}{10}$

44. $\dfrac{5}{6} \div 0$

45. $\dfrac{20}{35} \cdot \dfrac{7}{10}$

46. $\dfrac{18}{25} \div \dfrac{3}{5}$

47. $\dfrac{4}{7} - \dfrac{6}{11}$

48. $\dfrac{11}{20} + \dfrac{7}{15}$

Objective D *Use the order of operations to simplify each expression. See Examples 10 and 11.*

▶ 49. $\dfrac{1}{5} + \dfrac{1}{3} \cdot \dfrac{1}{4}$

50. $\dfrac{1}{2} + \dfrac{1}{6} \cdot \dfrac{1}{3}$

51. $\dfrac{5}{6} \div \dfrac{1}{3} \cdot \dfrac{1}{4}$

52. $\dfrac{7}{8} \div \dfrac{1}{4} \cdot \dfrac{1}{7}$

53. $\dfrac{1}{5} \cdot \left(2\dfrac{5}{6} - \dfrac{1}{3}\right)$

54. $\dfrac{4}{7} \cdot \left(6 - 2\dfrac{1}{2}\right)$

55. $2 \cdot \left(\dfrac{1}{4} + \dfrac{1}{5}\right) + 2$

56. $\dfrac{2}{5} \cdot \left(5 - \dfrac{1}{2}\right) - 1$

57. $\left(\dfrac{3}{4}\right)^2 \div \left(\dfrac{3}{4} - \dfrac{1}{12}\right)$

58. $\left(\dfrac{8}{9}\right)^2 \div \left(2 - \dfrac{2}{3}\right)$

▶ 59. $\left(\dfrac{2}{3} - \dfrac{5}{9}\right)^2$

60. $\left(1 - \dfrac{2}{5}\right)^3$

61. $\dfrac{5}{9} \cdot \dfrac{1}{2} + \dfrac{2}{3} \cdot \dfrac{5}{6}$

62. $\dfrac{7}{10} \cdot \dfrac{1}{2} + \dfrac{3}{4} \cdot \dfrac{3}{5}$

63. $\dfrac{27}{16} \cdot \left(\dfrac{2}{3}\right)^2 - \dfrac{3}{20}$

64. $\dfrac{64}{27} \cdot \left(\dfrac{3}{4}\right)^2 - \dfrac{7}{10}$

65. $\dfrac{3}{13} \div \dfrac{9}{26} - \dfrac{7}{24} \cdot \dfrac{8}{14}$ **66.** $\dfrac{5}{11} \div \dfrac{15}{77} - \dfrac{7}{10} \cdot \dfrac{5}{14}$ **67.** $\dfrac{3}{14} + \dfrac{10}{21} \div \left(\dfrac{3}{7}\right)\left(\dfrac{9}{4}\right)$ **68.** $\dfrac{11}{15} + \dfrac{7}{9} \div \left(\dfrac{14}{3}\right)\left(\dfrac{2}{3}\right)$

69. $\left(\dfrac{3}{4} + \dfrac{1}{8}\right)^2 - \left(\dfrac{1}{2} + \dfrac{1}{8}\right)$ **70.** $\left(\dfrac{1}{6} + \dfrac{1}{3}\right)^3 + \left(\dfrac{2}{5} \cdot \dfrac{3}{4}\right)^2$

Find the average of each list of numbers. See Example 12.

71. $\dfrac{5}{6}$ and $\dfrac{2}{3}$

72. $\dfrac{1}{2}$ and $\dfrac{4}{7}$

73. $\dfrac{1}{5}, \dfrac{3}{10},$ and $\dfrac{3}{20}$

74. $\dfrac{1}{3}, \dfrac{1}{4},$ and $\dfrac{1}{6}$

Objective C D Mixed Practice

75. The average fraction of online sales of computer hardware is $\dfrac{23}{50}$, of computer software is $\dfrac{1}{2}$, and of movies and movies is $\dfrac{3}{5}$. Find the average of these fractions.

76. The average fraction of online sales of sporting goods is $\dfrac{12}{25}$, of toys and hobbies and games is $\dfrac{1}{2}$, and of computer hardware is $\dfrac{23}{50}$. Find the average of these fractions.

Review

Identify each key word with the operation it most likely translates to. After each word, write A for addition, S for subtraction, M for multiplication, and D for division. See Sections 1.3, 1.4, 1.6, and 1.7.

77. increased by

78. sum

79. triple

80. product

81. subtracted from

82. decreased by

83. quotient

84. divided by

85. times

86. difference

87. total

88. more than

Concept Extensions

Solve.

89. Calculate $\dfrac{2^3}{3}$ and $\left(\dfrac{2}{3}\right)^3$. Do both of these expressions simplify to the same number? Explain why or why not.

90. Calculate $\left(\dfrac{1}{2}\right)^2 \cdot \left(\dfrac{3}{4}\right)^2$ and $\left(\dfrac{1}{2} \cdot \dfrac{3}{4}\right)^2$. Do both of these expressions simplify to the same number? Explain why or why not.

Each expression contains one addition, one subtraction, one multiplication, and one division. Write the operations in the order that they should be performed. Do not actually simplify. See the Concept Check in this section.

91. $[9 + 3(4 - 2)] \div \dfrac{10}{21}$

92. $[30 - 4(3 + 2)] \div \dfrac{5}{2}$

93. $\dfrac{1}{3} \div \left(\dfrac{2}{3}\right)\left(\dfrac{4}{5}\right) - \dfrac{1}{4} + \dfrac{1}{2}$

94. $\left(\dfrac{5}{6} - \dfrac{1}{3}\right) \cdot \dfrac{1}{3} + \dfrac{1}{2} \div \dfrac{9}{8}$

Solve.

95. In 2012, about $\dfrac{4}{25}$ of the total weight of mail delivered by the United States Postal Service was first-class mail. That same year, about $\dfrac{81}{200}$ of the total weight of mail delivered by the United States Postal Service was standard mail. Which of these two categories accounts for a greater portion of the mail handled by weight? (*Source:* U.S. Postal Service)

96. The National Park System (NPS) in the United States includes a wide variety of park types. National military parks account for $\dfrac{11}{500}$ of all NPS parks, and $\dfrac{9}{200}$ of NPS parks are classified as national preserves. Which category, national military park or national preserve, is bigger? (*Source:* National Park Service)

97. A recent survey reported that $\dfrac{2}{5}$ of the average college student's spending is for discretionary purchases, such as clothing, entertainment, and technology. About $\dfrac{13}{50}$ of the average college student's spending is for room and board. In which category, discretionary purchases or room and board, does the average college student spend more? (*Source:* Nationwide Bank)

98. On a normal work day in 2012, the average working parent in the U.S. spent $\dfrac{1}{20}$ of his or her day caring for others and $\dfrac{77}{240}$ of his or her day sleeping. Which did the average working parent spend more time doing, caring for others or sleeping? (*Source:* U.S. Bureau of Labor Statistics)

3.6 Fractions and Problem Solving

Objective A Solving Problems Containing Fractions or Mixed Numbers

Now that we know how to add, subtract, multiply, and divide fractions and mixed numbers, we can solve problems containing these numbers.

Don't forget the key words and phrases listed below that help indicate which operation to use. Also included are several words and phrases that translate to the symbol "=".

Objective

A Solve Problems by Performing Operations on Fractions or Mixed Numbers.

Addition (+)	Subtraction (−)	Multiplication (·)	Division (÷)	Equality (=)
sum	difference	product	quotient	equals
plus	minus	times	divide	is equal to
added to	subtract	multiply	shared equally	is/was
more than	less than	multiply by	among	yields
increased by	decreased by	of	divided by	
total	less	double/triple	divided into	

Recall the following problem-solving steps introduced in Section 1.8. They may be helpful to you:

> ## Problem-Solving Steps
>
> 1. **UNDERSTAND** the problem. Some ways of doing this are to read and reread the problem, construct a drawing, and look for key words to identify an operation.
> 2. **TRANSLATE** the problem. That is, write the problem in short form using words, and then translate to numbers and symbols.
> 3. **SOLVE** the problem. It is helpful to estimate the solution by rounding. Then carry out the indicated operation from step 2.
> 4. **INTERPRET** the results. *Check* the proposed solution in the stated problem and *state* your conclusions. Write your results with the correct units attached.

In the first example, we find the volume of a box. Volume measures the space enclosed by a region and is measured in cubic units. We study volume further in a later chapter.

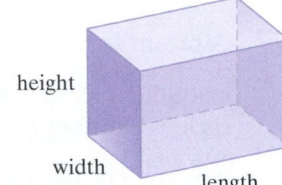

$$\text{Volume of a box} = \text{length} \cdot \text{width} \cdot \text{height}$$

Helpful Hint

Remember:

Perimeter measures the distance around a figure. It is measured in **units.**

☐ Perimeter

Area measures the amount of surface of a figure. It is measured in **square units.**

▭ Area

Volume measures the amount of space enclosed by a region. It is measured in **cubic units.**

 Volume

Practice 1

Find the volume of a box that measures $4\frac{1}{3}$ feet by $1\frac{1}{2}$ feet by $3\frac{1}{3}$ feet.

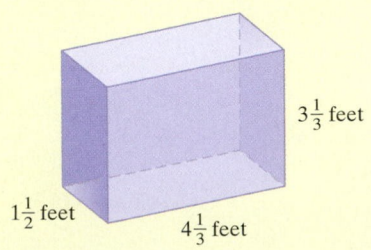

$3\frac{1}{3}$ feet

$1\frac{1}{2}$ feet $4\frac{1}{3}$ feet

Answer

1. $21\frac{2}{3}$ cu ft

> ## Example 1   Finding Volume of a Camcorder Box
>
> Toshiba recently produced a small camcorder. It measures approximately $1\frac{1}{2}$ inches by $1\frac{3}{5}$ inches by $1\frac{7}{25}$ inches. Find the volume of a box with these dimensions. (*Note:* The camcorder weighs only 65 grams—about the weight of 65 standard paper clips.) (*Source: Guinness World Records*)
>
> **Solution:**
>
> 1. **UNDERSTAND.** Read and reread the problem. The phrase "volume of a box" tells us what to do. The volume of a box is the product of its length, width, and height. Since we are multiplying, it makes no difference which measurement we call length, width, or height.

2. TRANSLATE.

In words:　　| volume of a box | is | length | · | width | · | height |

$$\downarrow \qquad \downarrow \quad \downarrow \qquad\quad \downarrow \qquad\quad \downarrow$$

Translate:　volume of a box $= 1\dfrac{1}{2}$ in. $\cdot\ 1\dfrac{3}{5}$ in. $\cdot\ 1\dfrac{7}{25}$ in.

3. SOLVE: Before we multiply, let's estimate by rounding each dimension to a whole number. The number $1\dfrac{1}{2}$ rounds to 2, $1\dfrac{3}{5}$ rounds to 2, and $1\dfrac{7}{25}$ rounds to 1, so our estimate is $2 \cdot 2 \cdot 1$ or 4 cubic inches.

$$1\dfrac{1}{2}\text{ in.}\cdot 1\dfrac{3}{5}\text{ in.}\cdot 1\dfrac{7}{25}\text{ in.} = \dfrac{3}{2}\cdot\dfrac{8}{5}\cdot\dfrac{32}{25} \quad \text{cubic inches}$$

$$= \dfrac{3\cdot \overset{4}{8}\cdot 32}{\underset{1}{2}\cdot 5\cdot 25} \quad \text{cubic inches}$$

$$= \dfrac{384}{125}\ \text{or}\ 3\dfrac{9}{125} \quad \text{cubic inches}$$

> **Helpful Hint**　　Notice a short-cut taken when simplifying the fraction. Here, a common factor of 2 is recognized in 8 and 2. Then $8 \div 2$ is 4 in the numerator and $2 \div 2$ is 1 in the denominator.

4. INTERPRET. *Check* your work. The exact answer is close to our estimate, so it is reasonable. *State* your conclusion: The volume of a box that measures $1\dfrac{1}{2}$ inches by $1\dfrac{3}{5}$ inches by $1\dfrac{7}{25}$ inches is $3\dfrac{9}{125}$ cubic inches.

■ **Work Practice 1**

Example 2　Finding Unknown Length

Given the following diagram, find its total length.

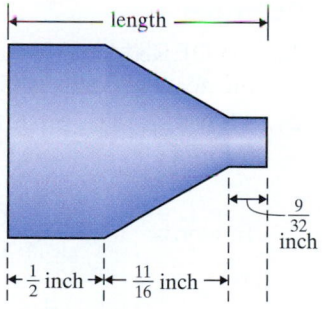

Practice 2

Given the following diagram, find its total width.

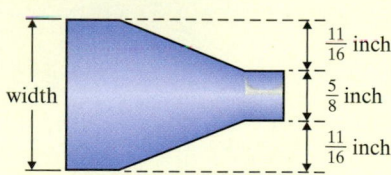

Solution:

1. UNDERSTAND. Read and reread the problem. Then study the diagram. The phrase "total length" tells us to add.

2. TRANSLATE. It makes no difference which length we call first, second, or third length.

In words:　| total length | is | first length | + | second length | + | third length |

$$\downarrow \qquad \downarrow \quad\ \downarrow \qquad\qquad \downarrow \qquad\qquad \downarrow$$

Translate:　total length $= \dfrac{1}{2}$ in. $+ \dfrac{11}{16}$ in. $+ \dfrac{9}{32}$ in.

(Continued on next page)

3. SOLVE:

$$\frac{1}{2} + \frac{11}{16} + \frac{9}{32} = \frac{1 \cdot 16}{2 \cdot 16} + \frac{11 \cdot 2}{16 \cdot 2} + \frac{9}{32}$$

$$= \frac{16}{32} + \frac{22}{32} + \frac{9}{32}$$

$$= \frac{47}{32} \text{ or } 1\frac{15}{32}$$

4. INTERPRET. *Check* your work. *State* your conclusion: The total length is $1\frac{15}{32}$ inches.

■ **Work Practice 2**

Many problems require more than one operation to solve, as shown in the next application.

Example 3 Acreage for Single-Family Home Lots

A contractor is considering buying land to develop a subdivision for single-family homes. Suppose she buys 44 acres and calculates that $4\frac{1}{4}$ acres of this land will be used for roads and a retention pond. How many $\frac{3}{4}$-acre lots can she sell using the rest of the acreage?

Solution:

1a. UNDERSTAND. Read and reread the problem. The phrase "using the rest of the acreage" tells us that initially we are to subtract.

2a. TRANSLATE. First, let's calculate the amount of acreage that can be used for lots.

In words:	acreage for lots	is	total acreage	minus	acreage for roads and a pond
	↓	↓	↓	↓	↓
Translate:	acreage for lots	=	44	−	$4\frac{1}{4}$

3a. SOLVE:

$$\begin{array}{rcl} 44 & = & 43\frac{4}{4} \\ -\ 4\frac{1}{4} & = & -\ 4\frac{1}{4} \\ \hline & & 39\frac{3}{4} \end{array}$$

1b. UNDERSTAND. Now that we know $39\frac{3}{4}$ acres can be used for lots, we calculate how many $\frac{3}{4}$ acres are in $39\frac{3}{4}$. This means that we divide.

2b. TRANSLATE.

In words:	number of $\frac{3}{4}$-acre lots	is	acreage for lots	divided by	size of each lot
	↓	↓	↓	↓	↓
Translate:	number of $\frac{3}{4}$-acre lots	=	$39\frac{3}{4}$	÷	$\frac{3}{4}$

3b. SOLVE:

$$39\frac{3}{4} \div \frac{3}{4} = \frac{159}{4} \cdot \frac{4}{3} = \frac{\overset{53}{\cancel{159}} \cdot \overset{1}{\cancel{4}}}{\underset{1}{\cancel{4}} \cdot \underset{1}{\cancel{3}}} = \frac{53}{1} \text{ or } 53$$

Helpful Hint See the Helpful Hint on page 223 for an explanation of a shortcut used to simplify this fraction.

4. INTERPRET. *Check* your work. *State* your conclusion: The contractor can sell 53 $\frac{3}{4}$-acre lots.

■ **Work Practice 3**

Vocabulary, Readiness & Video Check

Martin-Gay Interactive Videos Watch the section lecture video and answer the following question.

See Video 3.6

Objective A 1. From the lecture before ▣ Example 1, what's the purpose of interpreting the results when problem solving? ▷

3.6 Exercise Set MyMathLab® ▷

To prepare for problem solving, translate each phrase to an expression. Do not simplify the expression.

1. The sum of $\frac{1}{2}$ and $\frac{1}{3}$.

2. The product of $\frac{1}{2}$ and $\frac{1}{3}$.

3. The quotient of 20 and $6\frac{2}{5}$.

4. The difference of 20 and $6\frac{2}{5}$.

5. Subtract $\frac{5}{8}$ from $\frac{15}{16}$.

6. The total of $\frac{15}{36}$ and $\frac{18}{30}$.

7. $\frac{21}{68}$ increased by $\frac{7}{34}$.

8. $\frac{21}{68}$ decreased by $\frac{7}{34}$.

9. The product of $8\frac{1}{3}$ and $\frac{7}{9}$.

10. $37\frac{1}{2}$ divided by $9\frac{1}{2}$.

Objective **A** *Solve. Write any improper-fraction answers as mixed numbers. For Exercises 11 and 12, the solutions have been started for you. Write each answer in simplest form. See Examples 1 through 3.*

11. A recipe for brownies calls for $1\frac{2}{3}$ cups of sugar. If you are doubling the recipe, how much sugar do you need?

Start the solution:

1. UNDERSTAND the problem. Reread it as many times as needed.
2. TRANSLATE into an equation. (Fill in the blanks below.)

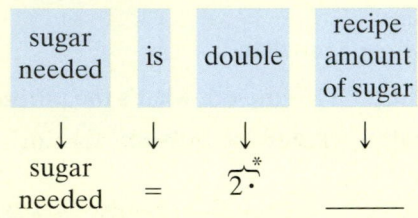

sugar needed	is	double	recipe amount of sugar
↓	↓	↓	↓

$$\text{sugar needed} = 2 \cdot\!{}^{*} \underline{\hspace{1cm}}$$

Finish with:

3. SOLVE
4. INTERPRET

*Note: Another way to double a number is to add the number to the same number.

12. A nacho recipe calls for $\frac{1}{3}$ cup cheddar cheese and $\frac{1}{2}$ cup jalapeño cheese. Find the total amount of cheese in the recipe.

Start the solution:

1. UNDERSTAND the problem. Reread it as many times as needed.
2. TRANSLATE into an equation. (Fill in the blanks below.)

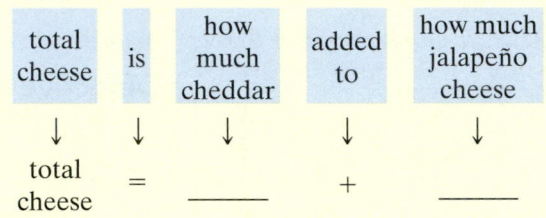

total cheese	is	how much cheddar	added to	how much jalapeño cheese
↓	↓	↓	↓	↓

$$\text{total cheese} = \underline{\hspace{1cm}} + \underline{\hspace{1cm}}$$

Finish with:

3. SOLVE
4. INTERPRET

13. A decorative wall in a garden is to be built using bricks that are $2\frac{3}{4}$ inches wide and mortar joints that are $\frac{1}{2}$ inch wide. Use the diagram to find the height of the wall.

height   Mortar joint

14. Suppose that the contractor building the wall in Exercise 13 decides that he wants one more layer of bricks with a mortar joint below and above that layer. Find the new height of the wall.

15. Doug and Claudia Scaggs recently drove $290\frac{1}{4}$ miles on $13\frac{1}{2}$ gallons of gas. Calculate how many miles per gallon they get in their vehicle.

16. A contractor is using 18 acres of his land to sell $\frac{3}{4}$-acre lots. How many lots can he sell?

○ 17. The life expectancy of a circulating coin is 30 years. The life expectancy of a circulating dollar bill is only $\frac{1}{20}$ as long. Find the life expectancy of circulating paper money. (*Source:* The U.S. Mint)

18. The Indian head one-cent coin of 1859–1864 was made of copper and nickel only. If $\frac{3}{25}$ of the coin was nickel, what part of the whole coin was copper? (*Source:* The U.S. Mint)

19. The Gauge Act of 1846 set the standard gauge for U.S. railroads at $56\frac{1}{2}$ inches. (See figure.) If the standard gauge in Spain is $65\frac{9}{10}$ inches, how much wider is Spain's standard gauge than the U.S. standard gauge? (*Source:* San Diego Railroad Museum)

20. The standard railroad track gauge (see figure) in Spain is $65\frac{9}{10}$ inches, while in neighboring Portugal it is $65\frac{11}{20}$ inches. Which gauge is wider and by how much? (*Source:* San Diego Railroad Museum)

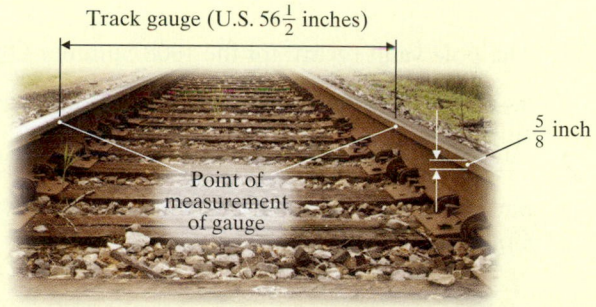

Track gauge (U.S. $56\frac{1}{2}$ inches)

$\frac{5}{8}$ inch

Point of measurement of gauge

21. Mark Nguyen is a tailor making costumes for a play. He needs enough material for 1 large shirt that requires $1\frac{1}{2}$ yards of material and 5 small shirts that each require $\frac{3}{4}$ yard of material. He finds a 5-yard remnant of material on sale. Is 5 yards of material enough to make all 6 shirts? If not, how much more material does he need?

22. A beanbag manufacturer makes a large beanbag requiring $4\frac{1}{3}$ yards of vinyl fabric and a smaller size requiring $3\frac{1}{4}$ yards. A 100-yard roll of fabric is to be used to make 12 large beanbags. How many smaller beanbags can be made from the remaining piece?

23. A plumber has a 10-foot piece of PVC pipe. How many $\frac{9}{5}$-foot pieces can be cut from the 10-foot piece?

24. A carpenter has a 12-foot board to be used to make windowsills. If each sill requires $2\frac{5}{16}$ feet, how many sills can be made from the 12-foot board?

○ 25. Suppose that the cross section of a piece of pipe looks like the diagram shown. Find the total outer diameter.

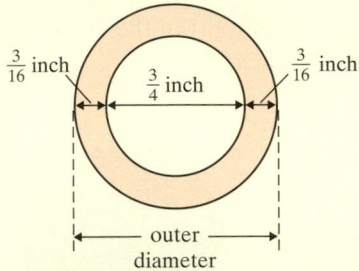

$\frac{3}{16}$ inch $\frac{3}{4}$ inch $\frac{3}{16}$ inch

outer diameter

26. Suppose that the cross section of a piece of pipe looks like the diagram shown. Find the inner diameter.

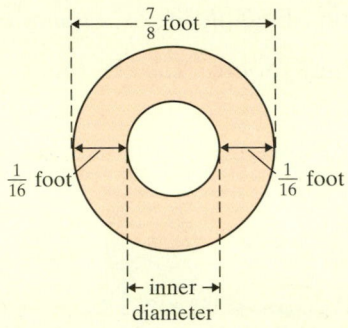

$\frac{7}{8}$ foot

$\frac{1}{16}$ foot $\frac{1}{16}$ foot

inner diameter

27. A recipe for chocolate chip cookies calls for $2\frac{1}{2}$ cups of flour. If you are making $1\frac{1}{2}$ recipes, how many cups of flour are needed?

28. A recipe for a homemade cleaning solution calls for $1\frac{3}{4}$ cups of vinegar. If you are tripling the recipe, how much vinegar is needed?

△ 29. The Polaroid Pop Shot, the world's first disposable instant camera, can take color photographs measuring $4\frac{1}{2}$ inches by $2\frac{1}{2}$ inches. Find the area of a photograph. (*Source: Guinness World Records*)

△ 30. A model for a proposed computer chip measures $\frac{3}{4}$ inch by $1\frac{1}{4}$ inches. Find its area.

31. A total solar eclipse on March 9, 2016, will last $4\frac{3}{20}$ minutes and can be viewed from Sumatra, Borneo, and Sulawesi. The next total solar eclipse, on August 21, 2017, will last $2\frac{2}{3}$ minutes and can be viewed in the northern Pacific Ocean, United States, and southern Atlantic Ocean. How much longer is the 2016 eclipse? (*Source:* NASA/Goddard Space Flight Center)

32. The pole vault record for the 1908 Summer Olympics was $12\frac{1}{6}$ feet. The record for the 2012 Summer Olympics was a little over $19\frac{9}{16}$ feet. Find the difference in the heights. (*Source:* International Olympic Committee)

△ 33. The world's lightest handheld mobile phone measures approximately $2\frac{4}{5}$ inches by $1\frac{1}{2}$ inches by $\frac{3}{10}$ inch. Find the volume of a box with those dimensions. (*Source: Guinness World Records*)

△ 34. Early cell phones were large and heavy. One early model measured approximately 8 inches by $2\frac{1}{2}$ inches by $2\frac{1}{2}$ inches. Find the volume of a box with those dimensions.

35. A stack of $\frac{5}{8}$-inch-wide sheetrock has a height of $41\frac{7}{8}$ inches. How many sheets of sheetrock are in the stack?

36. A stack of $\frac{5}{4}$-inch-thick books has a height of $28\frac{3}{4}$ inches. How many books are in the stack?

37. William Arcencio is remodeling his home. In order to save money, he is upgrading the plumbing himself. He needs 12 pieces of copper tubing, each $\frac{3}{4}$ of a foot long.

   a. If he has a 10-foot piece of tubing, will that be enough?

   b. How much more does he need or how much tubing will he have left over?

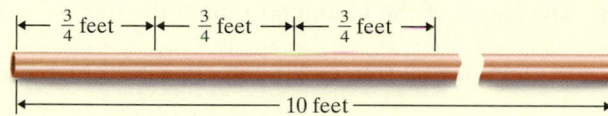

38. Trishelle Dallam is building a bookcase. Each shelf will be $2\frac{3}{8}$ feet long, and she needs wood for 7 shelves.

   a. How many shelves can she cut from an 8-foot board?

   b. Based on your answer for part a, how many 8-foot boards will she need?

Recall that the average of a list of numbers is their sum divided by the number of numbers in the list. Use this procedure for Exercises 39 and 40.

39. A female lion had 4 cubs. They weighed $2\frac{1}{8}$, $2\frac{7}{8}$, $3\frac{1}{4}$, and $3\frac{1}{2}$ pounds. What is the average cub weight?

40. Three brook trout were caught, tagged, and then released. They weighed $1\frac{1}{2}$, $1\frac{3}{8}$, and $1\frac{7}{8}$ pounds. Find their average weight.

Find the area and perimeter of each figure.

△**41.**
Rectangle $\frac{3}{16}$ inch
$\frac{3}{8}$ inch

△**42.**
Square $1\frac{7}{10}$ mile

△**43.**
Square $\frac{5}{9}$ meter

△**44.**
Rectangle 5 inches
$3\frac{1}{2}$ inches

For Exercises 45 through 48, see the diagram. (Source: www.usflag.org)

45. The length of the U.S. flag is $1\frac{9}{10}$ its width. If a flag is being designed with a width of $2\frac{1}{2}$ feet, find its length.

46. The width of the Union portion of the U.S. flag is $\frac{7}{13}$ of the width of the flag. If a flag is being designed with a width of $2\frac{1}{2}$ feet, find the width of the Union portion.

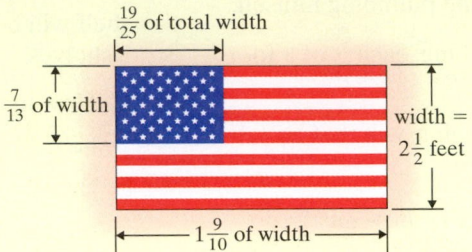

$\frac{19}{25}$ of total width

$\frac{7}{13}$ of width

width = $2\frac{1}{2}$ feet

$1\frac{9}{10}$ of width

47. There are 13 stripes of equal width in the flag. If the width of a flag is $2\frac{1}{2}$ feet, find the width of each stripe.

48. The length of the Union portion of the flag is $\frac{19}{25}$ of the total width. If the width of a flag is $2\frac{1}{2}$ feet, find the length of the Union portion.

Review

Simplify. See Section 1.9.

49. $\sqrt{9}$

50. $\sqrt{4}$

51. 9^2

52. 4^2

53. $8 \div 4 \cdot 2$

54. $20 \div 5 \cdot 2$

55. $3^2 - 2^2 + 5^2$

56. $8^2 - 6^2 + 7^2$

57. $5 + 3[14 - (12 \div 3)]$

58. $7 + 2[20 - (35 \div 5)]$

Concept Extensions

59. Suppose you are finding the average of $7\frac{1}{9}$ and $12\frac{19}{20}$. Can the average be $1\frac{1}{2}$? Can the average be $15\frac{1}{2}$? Why or why not?

60. Suppose that you are finding the average of $1\frac{3}{4}$, $1\frac{1}{8}$, and $1\frac{9}{10}$. Can the average be $2\frac{1}{4}$? Can the average be $\frac{15}{16}$? Why or why not?

The figure shown is for Exercises 61 and 62.

△**61.** Find the area of the figure. (*Hint:* The area of the figure can be found by finding the sum of the areas of the rectangles shown in the figure.)

△**62.** Find the perimeter of the figure.

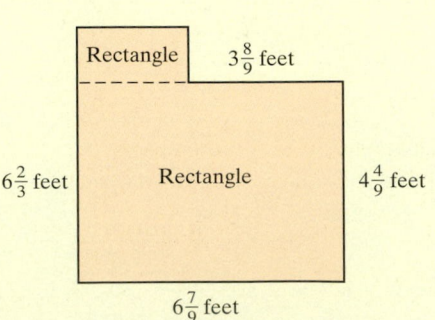

Rectangle $3\frac{8}{9}$ feet

$6\frac{2}{3}$ feet   Rectangle $4\frac{4}{9}$ feet

$6\frac{7}{9}$ feet

63. On a particular day, 240 customers ate lunch at a local restaurant. If $\frac{3}{10}$ of them ordered a $7 lunch, $\frac{5}{12}$ of them ordered a $5 lunch, and the remaining customers ordered a $9 lunch, how many customers ordered a $9 lunch?

64. Scott purchased a case of 24 apples. He used $\frac{1}{3}$ of them to make an apple pie, $\frac{1}{4}$ of them to make apple crisp, and kept the rest for after-school snacks for his children. How many apples did Scott keep for snacks?

65. Coins were practically made by hand in the late 1700s. Back then, it took 3 years to produce our nation's first million coins. Today, it takes only $\frac{11}{13,140}$ as long to produce the same amount. Calculate how long it takes today in hours to produce one million coins. (*Hint:* First convert 3 years to equivalent hours. Use 365 days for each of the 3 years.) (*Source:* The U.S. Mint)

66. The largest suitcase measures $13\frac{1}{3}$ feet by $8\frac{3}{4}$ feet by $4\frac{4}{25}$ feet. Find its volume. (*Source: Guinness World Records*)

Chapter 3 Group Activity

Sections 3.1–3.6

This activity may be completed by working in groups or individually.

Lobsters are normally classified by weight. Use the weight classification table to answer the questions in this activity.

Classification of Lobsters

Class	Weight (in Pounds)
Chicken	1 to $1\frac{1}{8}$
Quarter	$1\frac{1}{4}$
Half	$1\frac{1}{2}$ to $1\frac{3}{4}$
Select	$1\frac{3}{4}$ to $2\frac{1}{2}$
Large select	$2\frac{1}{2}$ to $3\frac{1}{2}$
Jumbo	Over $3\frac{1}{2}$

(*Source:* The Maine Lobster Promotion Council)

1. A lobster fisher has kept four lobsters from a lobster trap. Classify each lobster if they have the following weights:

  a. $1\frac{7}{8}$ pounds   **b.** $1\frac{9}{16}$ pounds

  c. $2\frac{3}{4}$ pounds   **d.** $2\frac{3}{8}$ pounds

2. A recipe requires 5 pounds of lobster. Using the minimum weight for each class, decide whether a chicken, half, and select lobster will be enough for the recipe, and explain your reasoning. If not, suggest a better choice of lobsters to meet the recipe requirements.

3. A lobster market customer has selected two chickens, a select, and a large select. What is the most that these four lobsters could weigh? What is the least that these four lobsters could weigh?

4. A lobster market customer wishes to buy four quarters. If lobsters sell for $7 per pound, how much will the customer owe for her purchase?

5. Why do you think there is no classification for lobsters weighing under 1 pound?

Chapter 3 Vocabulary Check

Fill in each blank with one of the words or phrases listed below.

equivalent	least common multiple	exponent	unlike
mixed number	<	>	least common denominator	like

1. Fractions that have the same denominator are called _____ fractions.

2. The _____ is the smallest number that is a multiple of all numbers in a list of numbers.

3. _____ fractions represent the same portion of a whole.

4. A(n)_____ has a whole number part and a fraction part.

5. The symbol _____ means is greater than.

6. The symbol _____ means is less than.

7. The LCM of the denominators in a list of fractions is called the _____.

8. Fractions that have different denominators are called _____ fractions.

9. A shorthand notation for repeated multiplication of the same factor is a(n) _____.

Helpful Hint

▶ Are you preparing for your test? Don't forget to take the Chapter 3 Test on page 240. Then check your answers at the back of the text and use the Chapter Test Prep Videos to see the fully worked-out solutions to any of the exercises you want to review.

3 Chapter Highlights

Definitions and Concepts	Examples
Section 3.1 Adding and Subtracting Like Fractions	
Fractions that have the same denominator are called **like fractions.**	$\frac{1}{3}$ and $\frac{2}{3}$; $\frac{5}{7}$ and $\frac{6}{7}$
To add or subtract like fractions, combine the numerators and place the sum or difference over the common denominator.	$\frac{2}{7} + \frac{3}{7} = \frac{5}{7}$ ← Add the numerators. ← Keep the common denominator.
	$\frac{7}{8} - \frac{4}{8} = \frac{3}{8}$ ← Subtract the numerators. ← Keep the common denominator.
Section 3.2 Least Common Multiple	
The **least common multiple (LCM)** is the smallest number that is a multiple of all numbers in a list of numbers.	The LCM of 2 and 6 is 6 because 6 is the smallest number that is a multiple of both 2 and 6.
Method 1 for Finding the LCM of a List of Numbers Using Multiples	Find the LCM of 4 and 6 using Method 1.
Step 1: Write the multiples of the largest number (starting with the number itself) until a multiple common to all numbers in the list is found.	$6 \cdot 1 = 6$ Not a multiple of 4 $6 \cdot 2 = 12$ A multiple of 4 The LCM is 12.
Step 2: The multiple found in step 1 is the LCM.	

Definitions and Concepts	Examples

Section 3.2   Least Common Multiple (*continued*)

Method 2 for Finding the LCM of a List of Numbers Using Prime Factorization

Step 1:   Write the prime factorization of each number.

Step 2:   For each different prime factor in step 1, circle the greatest number of times that factor occurs in any one factorization.

Step 3:   The LCM is the product of the circled factors.

Equivalent fractions represent the same portion of a whole.

Find the LCM of 6 and 20 using Method 2.

$$6 = 2 \cdot ③$$
$$20 = ②\cdot② \cdot ⑤$$

The LCM is

$$2 \cdot 2 \cdot 3 \cdot 5 = 60$$

Write an equivalent fraction with the indicated denominator.

$$\frac{2}{8} = \frac{}{16}$$

$$\frac{2 \cdot 2}{8 \cdot 2} = \frac{4}{16}$$

Section 3.3   Adding and Subtracting Unlike Fractions

To Add or Subtract Fractions with Unlike Denominators

Step 1:   Find the LCD.

Step 2:   Write each fraction as an equivalent fraction whose denominator is the LCD.

Step 3:   Add or subtract the like fractions.

Step 4:   Write the sum or difference in simplest form.

Add: $\dfrac{3}{20} + \dfrac{2}{5}$

Step 1:   The LCD of 20 and 5 is 20.

Step 2: $\dfrac{3}{20} = \dfrac{3}{20}, \dfrac{2}{5} = \dfrac{2}{5} \cdot \dfrac{4}{4} = \dfrac{8}{20}$

Step 3: $\dfrac{3}{20} + \dfrac{2}{5} = \dfrac{3}{20} + \dfrac{8}{20} = \dfrac{11}{20}$

Step 4: $\dfrac{11}{20}$ is in simplest form.

Section 3.4   Adding and Subtracting Mixed Numbers

To add or subtract with mixed numbers, add or subtract the fractions and then add or subtract the whole numbers.

Add: $2\dfrac{1}{2} + 5\dfrac{7}{8}$

$$2\dfrac{1}{2} = 2\dfrac{4}{8}$$

$$+5\dfrac{7}{8} = +5\dfrac{7}{8}$$

$$7\dfrac{11}{8} = 7 + 1\dfrac{3}{8} = 8\dfrac{3}{8}$$

Section 3.5   Order, Exponents, and the Order of Operations

To compare like fractions, compare the numerators. The order of the fractions is the same as the order of the numerators.

Compare $\dfrac{3}{10}$ and $\dfrac{4}{10}$.

$$\dfrac{3}{10} < \dfrac{4}{10} \text{ since } 3 < 4$$

(continued)

Definitions and Concepts	Examples

Section 3.5   Order, Exponents, and the Order of Operations (*continued*)

To compare unlike fractions, first write the fractions as like fractions. Then the fraction with the greater numerator is the greater fraction.

Compare $\dfrac{2}{5}$ and $\dfrac{3}{7}$.

$$\frac{2}{5} = \frac{2}{5} \cdot \frac{7}{7} = \frac{14}{35} \qquad \frac{3}{7} = \frac{3}{7} \cdot \frac{5}{5} = \frac{15}{35}$$

Since $14 < 15$, then

$$\frac{14}{35} < \frac{15}{35} \quad \text{or} \quad \frac{2}{5} < \frac{3}{7}$$

Exponents mean repeated multiplication whether the base is a whole number or a fraction.

$$\left(\frac{1}{2}\right)^3 = \frac{1}{2} \cdot \frac{1}{2} \cdot \frac{1}{2} = \frac{1}{8}$$

Order of Operations

1. Perform all operations within parentheses ( ), brackets [ ], or other grouping symbols such as square roots or fraction bars.

2. Evaluate any expressions with exponents.

3. Multiply or divide in order from left to right.

4. Add or subtract in order from left to right.

Perform each indicated operation.

$$\frac{1}{2} + \frac{2}{3} \cdot \frac{1}{5} = \frac{1}{2} + \frac{2}{15} \qquad \text{Multiply.}$$

$$= \frac{1}{2} \cdot \frac{15}{15} + \frac{2}{15} \cdot \frac{2}{2} \qquad \text{The LCD is 30.}$$

$$= \frac{15}{30} + \frac{4}{30}$$

$$= \frac{19}{30} \qquad \text{Add.}$$

Section 3.6   Fractions and Problem Solving

Problem-Solving Steps

A stack of $\dfrac{3}{4}$-inch plywood has a height of $50\dfrac{1}{4}$ inches. How many sheets of plywood are in the stack?

1. UNDERSTAND the problem.

1. UNDERSTAND. Read and reread the problem. We want to know how many $\dfrac{3}{4}$'s are in $50\dfrac{1}{4}$, so we divide.

2. TRANSLATE the problem.

2. TRANSLATE.

number of sheets in stack	is	height of stack	÷	height of a sheet

$$\text{number of sheets in stack} = 50\frac{1}{4} \div \frac{3}{4}$$

3. SOLVE the problem.

3. SOLVE. $50\dfrac{1}{4} \div \dfrac{3}{4} = \dfrac{201}{4} \cdot \dfrac{4}{3}$

$$= \frac{\overset{67}{\cancel{201}} \cdot \overset{1}{\cancel{4}}}{\underset{1}{\cancel{4}} \cdot \underset{1}{\cancel{3}}}$$

$$= 67$$

4. INTERPRET the results.

4. INTERPRET. *Check* your work and *state* your conclusion: There are 67 sheets of plywood in the stack.

(3.1) *Add or subtract as indicated. Simplify your answers.*

1. $\dfrac{7}{11} + \dfrac{3}{11}$

2. $\dfrac{4}{50} + \dfrac{2}{50}$

3. $\dfrac{11}{15} - \dfrac{1}{15}$

4. $\dfrac{4}{21} - \dfrac{1}{21}$

5. $\dfrac{4}{15} + \dfrac{3}{15} + \dfrac{2}{15}$

6. $\dfrac{3}{20} + \dfrac{7}{20} + \dfrac{2}{20}$

7. $\dfrac{1}{12} + \dfrac{11}{12}$

8. $\dfrac{3}{4} + \dfrac{1}{4}$

9. $\dfrac{11}{25} + \dfrac{6}{25} + \dfrac{2}{25}$

10. $\dfrac{4}{21} + \dfrac{1}{21} + \dfrac{11}{21}$

Solve.

11. One evening Mark Alorenzo did $\dfrac{3}{8}$ of his homework before supper, another $\dfrac{2}{8}$ of it while his children did their homework, and $\dfrac{1}{8}$ after his children went to bed. What part of his homework did he do that evening?

△ 12. The Simpsons will be fencing in their land, which is in the shape of a rectangle. In order to do this, they need to find its perimeter. Find the perimeter of their land.

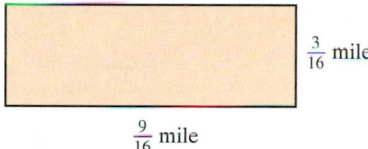

$\frac{3}{16}$ mile

$\frac{9}{16}$ mile

(3.2) *Find the LCM of each list of numbers.*

13. $5, 11$

14. $20, 30$

15. $20, 24$

16. $16, 5$

17. $12, 21, 63$

18. $6, 8, 18$

Write each fraction as an equivalent fraction with the given denominator.

19. $\dfrac{7}{8} = \dfrac{}{64}$

20. $\dfrac{2}{3} = \dfrac{}{30}$

21. $\dfrac{7}{11} = \dfrac{}{33}$

22. $\dfrac{10}{13} = \dfrac{}{26}$

23. $\dfrac{4}{15} = \dfrac{}{60}$

24. $\dfrac{5}{12} = \dfrac{}{60}$

(3.3) *Add or subtract as indicated. Simplify your answers.*

25. $\dfrac{7}{18} + \dfrac{2}{9}$

26. $\dfrac{4}{15} + \dfrac{1}{5}$

27. $\dfrac{4}{13} - \dfrac{1}{26}$

28. $\dfrac{7}{12} - \dfrac{1}{9}$

29. $\dfrac{1}{3} + \dfrac{9}{14}$

30. $\dfrac{7}{18} + \dfrac{5}{24}$

31. $\dfrac{11}{15} - \dfrac{4}{9}$

32. $\dfrac{9}{14} - \dfrac{3}{35}$

Find the perimeter of each figure.

△ **33.**

$\frac{2}{9}$ meter | Rectangle |

$\frac{5}{6}$ meter

△ **34.** $\frac{1}{5}$ foot $\frac{3}{5}$ foot

$\frac{7}{10}$ foot

35. Find the difference in length of two scarves if one scarf is $\frac{5}{12}$ of a yard long and the other is $\frac{2}{3}$ of a yard long.

36. Truman Kalzote cleaned $\frac{3}{5}$ of his house yesterday and $\frac{1}{10}$ of it today. How much of the house has been cleaned?

(3.4) *Add or subtract as indicated. Simplify your answers.*

37. $31\frac{2}{7} + 14\frac{10}{21}$

38. $24\frac{4}{5} + 35\frac{1}{5}$

39. $69\frac{5}{22} - 36\frac{7}{11}$

40. $36\frac{3}{20} - 32\frac{5}{6}$

41. $\begin{aligned}29\frac{2}{9}\\27\frac{7}{18}\\+54\frac{2}{3}\\\hline\end{aligned}$

42. $\begin{aligned}7\frac{3}{8}\\9\frac{5}{6}\\+3\frac{1}{12}\\\hline\end{aligned}$

43. $\begin{aligned}9\frac{3}{5}\\-4\frac{1}{7}\\\hline\end{aligned}$

44. $\begin{aligned}8\frac{3}{11}\\-5\frac{1}{5}\\\hline\end{aligned}$

Solve.

45. The average annual snowfall at a certain ski resort is $62\frac{3}{10}$ inches. Last year it had $54\frac{1}{2}$ inches. How many inches below average was last year's snowfall?

46. Dinah's homemade canned peaches contain $15\frac{3}{5}$ ounces per can. A can of Amy's brand contains $15\frac{5}{8}$ ounces per can. Amy's brand weighs how much more than Dinah's?

△ **47.** Find the perimeter of a sheet of shelf paper needed to fit exactly a square drawer $1\frac{1}{4}$ feet long on each side.

$1\frac{1}{4}$ feet

△ **48.** Find the perimeter of a rectangular sheet of gift wrap that is $2\frac{1}{4}$ feet by $3\frac{1}{3}$ feet.

$2\frac{1}{4}$ feet

$3\frac{1}{3}$ feet

(3.5) *Insert $<$ or $>$ to form a true statement.*

49. $\dfrac{5}{11}$ $\dfrac{6}{11}$

50. $\dfrac{4}{35}$ $\dfrac{3}{35}$

51. $\dfrac{5}{14}$ $\dfrac{16}{42}$

52. $\dfrac{6}{35}$ $\dfrac{17}{105}$

53. $\dfrac{7}{8}$ $\dfrac{6}{7}$

54. $\dfrac{7}{10}$ $\dfrac{2}{3}$

Evaluate each expression. Use the order of operations to simplify.

55. $\left(\dfrac{3}{7}\right)^2$

56. $\left(\dfrac{4}{5}\right)^3$

57. $\left(\dfrac{1}{2}\right)^4 \cdot \left(\dfrac{3}{5}\right)^2$

58. $\left(\dfrac{1}{3}\right)^2 \cdot \left(\dfrac{9}{10}\right)^2$

59. $\dfrac{5}{13} \div \dfrac{1}{2} \cdot \dfrac{4}{5}$

60. $\dfrac{8}{11} \div \dfrac{1}{3} \cdot \dfrac{11}{12}$

61. $\left(\dfrac{6}{7} - \dfrac{3}{14}\right)^2$

62. $\left(\dfrac{1}{3}\right)^2 - \dfrac{2}{27}$

63. $\dfrac{8}{9} - \dfrac{1}{8} \div \dfrac{3}{4}$

64. $\dfrac{9}{10} - \dfrac{1}{9} \div \dfrac{2}{3}$

65. $\dfrac{2}{7} \cdot \left(\dfrac{1}{5} + \dfrac{3}{10}\right)$

66. $\dfrac{9}{10} \div \left(\dfrac{1}{5} + \dfrac{1}{20}\right)$

67. $\left(\dfrac{3}{4} + \dfrac{1}{2}\right) \div \left(\dfrac{4}{9} + \dfrac{1}{3}\right)$

68. $\left(\dfrac{3}{8} - \dfrac{1}{16}\right) \div \left(\dfrac{1}{2} - \dfrac{1}{8}\right)$

69. $\dfrac{6}{7} \cdot \dfrac{5}{2} - \dfrac{3}{4} \cdot \dfrac{1}{2}$

70. $\dfrac{9}{10} \cdot \dfrac{1}{3} - \dfrac{2}{5} \cdot \dfrac{1}{11}$

Find the average of each list of fractions.

71. $\dfrac{2}{3}, \dfrac{5}{6}, \dfrac{1}{9}$

72. $\dfrac{4}{5}, \dfrac{9}{10}, \dfrac{3}{20}$

(3.6)

73. Our solar system has 146 known and officially confirmed moons. The planet Jupiter can claim $\frac{25}{73}$ of these moons. How many moons does Jupiter have? (*Source:* NASA)

74. James Hardaway just bought $5\frac{7}{8}$ acres of land adjacent to the $9\frac{3}{4}$ acres he already owned. How much land does he now own?

Find the unknown measurements.

△ **75.**

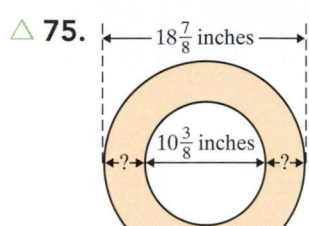

△ **76.**

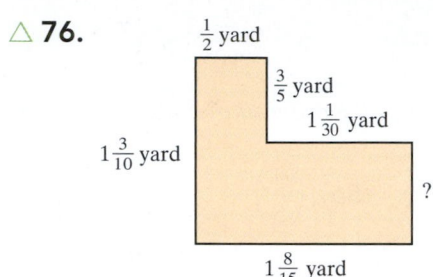

Find the perimeter and area of each rectangle. Attach the proper units to each. Remember that perimeter is measured in units and area is measured in square units.

△ **77.**

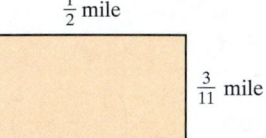

△ **78.**

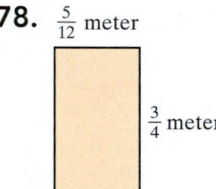

Mixed Review

Find the LCM of each list of numbers.

79. $15, 30, 45$

80. $6, 15, 20$

Write each fraction as an equivalent fraction with the given denominator.

81. $\frac{5}{6} = \frac{}{48}$

82. $\frac{7}{8} = \frac{}{72}$

Add or subtract as indicated. Simplify your answers.

83. $\frac{5}{12} - \frac{3}{12}$

84. $\frac{3}{10} - \frac{1}{10}$

85. $\frac{2}{3} + \frac{1}{4}$

86. $\frac{5}{11} + \frac{2}{55}$

87. $7\dfrac{3}{4}$
$+5\dfrac{2}{3}$

88. $2\dfrac{7}{8}$
$+9\dfrac{1}{2}$

89. $12\dfrac{3}{5}$
$-9\dfrac{1}{7}$

90. $32\dfrac{10}{21}$
$-24\dfrac{3}{7}$

Evaluate each expression. Use the order of operations to simplify.

91. $\dfrac{2}{5} + \left(\dfrac{2}{5}\right)^2 - \dfrac{3}{25}$

92. $\dfrac{1}{4} + \left(\dfrac{1}{2}\right)^2 - \dfrac{3}{8}$

93. $\left(\dfrac{5}{6} - \dfrac{3}{4}\right)^2$

94. $\left(2 - \dfrac{2}{3}\right)^3$

95. $\dfrac{2}{3} \div \left(\dfrac{3}{5} + \dfrac{5}{3}\right)$

96. $\dfrac{3}{8} \cdot \left(\dfrac{2}{3} - \dfrac{4}{9}\right)$

Insert $<$ or $>$ to form a true statement.

97. $\dfrac{3}{14}$ $\dfrac{2}{3}$

98. $\dfrac{7}{23}$ $\dfrac{3}{16}$

Solve.

99. Gregor Krowsky studied math for $\dfrac{3}{8}$ of an hour and geography for $\dfrac{1}{8}$ of an hour. How long did he study?

100. Two packages to be mailed weigh $3\dfrac{3}{4}$ pounds and $2\dfrac{3}{5}$ pounds. Find their combined weight.

101. A ribbon $5\dfrac{1}{2}$ yards long is cut from a reel of ribbon with 50 yards on it. Find the length of the piece remaining on the reel.

102. Linda Taneff has a board that is $10\dfrac{2}{3}$ feet in length. She plans to cut it into 5 equal lengths to use for a bookshelf. Find the length of each piece.

103. A recipe for pico de gallo calls for $1\dfrac{1}{2}$ tablespoons of cilantro. Five recipes will be made for a charity event. How much cilantro is needed?

104. Beryl Goldstein mixed $\dfrac{5}{8}$ of a gallon of water with $\dfrac{1}{8}$ of a gallon of punch concentrate. Then she and her friends drank $\dfrac{3}{8}$ of a gallon of the punch. How much of the punch was left?

Chapter 3 Test

Step-by-step test solutions are found on the Chapter Test Prep Videos. Where available: **MyMathLab®** or **You Tube**

Answers

1. Find the LCM of 4 and 15.

2. Find the LCM of 8, 9, and 12.

Insert < or > to form a true statement.

3. $\dfrac{5}{6}$ $\dfrac{26}{30}$

4. $\dfrac{7}{8}$ $\dfrac{8}{9}$

1. _____

2. _____

3. _____

Perform each indicated operation. Simplify your answers.

4. _____

5. $\dfrac{7}{9} + \dfrac{1}{9}$

6. $\dfrac{8}{15} - \dfrac{2}{15}$

7. $\dfrac{9}{10} + \dfrac{2}{5}$

5. _____

6. _____

7. _____

8. $\dfrac{1}{6} + \dfrac{3}{14}$

9. $\dfrac{7}{8} - \dfrac{1}{3}$

10. $\dfrac{17}{21} - \dfrac{1}{7}$

8. _____

9. _____

10. _____

11. $\dfrac{9}{20} + \dfrac{2}{3}$

12. $\dfrac{16}{25} - \dfrac{1}{2}$

13. $\dfrac{11}{12} + \dfrac{3}{8} + \dfrac{5}{24}$

11. _____

12. _____

13. _____

14. $\begin{array}{r} 3\frac{7}{8} \\[4pt] 7\frac{2}{5} \\[4pt] +2\frac{3}{4} \\ \hline \end{array}$

15. $\begin{array}{r} 8\frac{2}{9} \\[4pt] 12 \\[4pt] +10\frac{1}{15} \\ \hline \end{array}$

16. $\begin{array}{r} 5\frac{1}{6} \\[4pt] -3\frac{7}{8} \\ \hline \end{array}$

14. _____

15. _____

16. _____

17. _____

18. _____

17. $\begin{array}{r} 19 \\[4pt] -2\frac{3}{11} \\ \hline \end{array}$

18. $\dfrac{2}{7} \cdot \left(6 - \dfrac{1}{6}\right)$

19. $\left(\dfrac{2}{3}\right)^4$

19. _____

240

20. $\frac{1}{2} \div \frac{2}{3} \cdot \frac{3}{4}$

21. $\left(\frac{4}{5}\right)^2 + \left(\frac{1}{2}\right)^3$

22. $\left(\frac{3}{4}\right)^2 \div \left(\frac{2}{3} + \frac{5}{6}\right)$

23. Find the average of $\frac{5}{6}, \frac{4}{3}$, and $\frac{7}{12}$.

Solve.

24. A carpenter cuts a piece $2\frac{3}{4}$ feet long from a cedar plank that is $6\frac{1}{2}$ feet long. How long is the remaining piece?

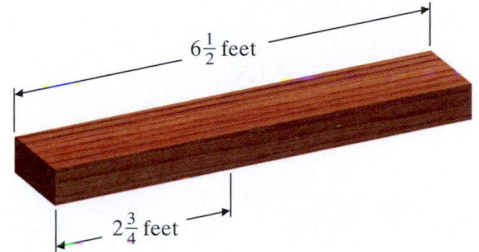

$6\frac{1}{2}$ feet

$2\frac{3}{4}$ feet

25. A small airplane used $58\frac{3}{4}$ gallons of fuel on a $7\frac{1}{2}$-hour trip. How many gallons of fuel were used for each hour?

The circle graph below shows us how the average consumer spends money. For example, $\frac{7}{50}$ of your spending goes for food. Use this information for Exercises 26 through 28.

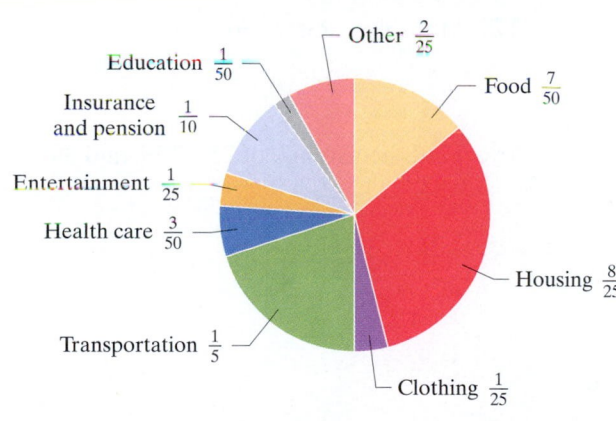

Consumer Spending

Other $\frac{2}{25}$
Education $\frac{1}{50}$
Insurance and pension $\frac{1}{10}$
Entertainment $\frac{1}{25}$
Health care $\frac{3}{50}$
Transportation $\frac{1}{5}$
Clothing $\frac{1}{25}$
Housing $\frac{8}{25}$
Food $\frac{7}{50}$

26. What fraction of spending goes for housing and food combined?

27. What fraction of spending goes for education, transportation, and clothing?

28. Suppose your family spent $47,000 on the items in the graph. How much might we expect was spent on health care?

Find the perimeter of each figure. For Exercise 29, find the area also.

△**29.**

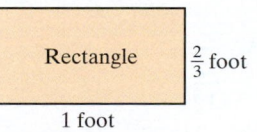

Rectangle $\frac{2}{3}$ foot

1 foot

△**30.**

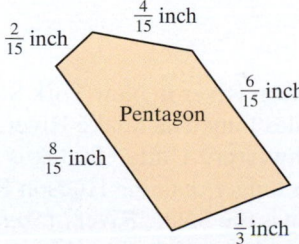

$\frac{4}{15}$ inch

$\frac{2}{15}$ inch

$\frac{6}{15}$ inch

Pentagon

$\frac{8}{15}$ inch

$\frac{1}{3}$ inch

20. _____

21. _____

22. _____

23. _____

24. _____

25. _____

26. _____

27. _____

28. _____

29. _____

30. _____

Answers

Write each number in words.

1. 85

2. 107

3. 126

4. 5026

5. Add: $23 + 136$

6. Find the perimeter.

3 in.   7 in.

9 in.

7. Subtract: $543 - 29$. Then check by adding.

8. Divide: $3268 \div 27$

9. Round 278,362 to the nearest thousand.

10. Find all the factors of 30.

11. Multiply: 236×86

12. Multiply: $236 \times 86 \times 0$

13. Find each quotient and then check the answer by multiplying.

   a. $1\overline{)7}$

   b. $12 \div 1$

   c. $\dfrac{6}{6}$

   d. $9 \div 9$

   e. $\dfrac{20}{1}$

   f. $18\overline{)18}$

14. Find the average of 25, 17, 19, and 39.

15. The Hudson River in New York State is 306 miles long. The Snake River, in the northwestern United States, is 732 miles longer than the Hudson River. How long is the Snake River? (*Source:* U.S. Department of the Interior)

16. Evaluate: $\sqrt{121}$

1. _____

2. _____

3. _____

4. _____

5. _____

6. _____

7. _____

8. _____

9. _____

10. _____

11. _____

12. _____

13. a. _____

   b. _____

   c. _____

   d. _____

   e. _____

   f. _____

14. _____

15. _____

16. _____

Evaluate.

17. 9^2 **18.** 5^3 **19.** 3^4 **20.** 10^3

Write the shaded part of each diagram as an improper fraction and a mixed number.

21.  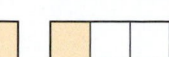   **22.**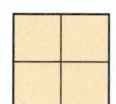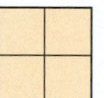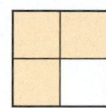

23. 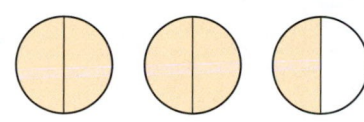   **24.**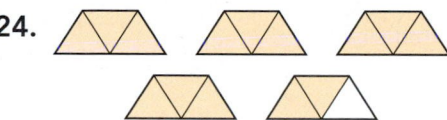

25. Of the numbers 3, 9, 11, 17, 26, which are prime and which are composite?

26. Simplify: $\dfrac{6^2 + 4 \cdot 4 + 2^3}{37 - 5^2}$

27. Find the prime factorization of 180.

28. Find the difference of 87 and 25.

29. Write $\dfrac{72}{26}$ in simplest form.

30. Write $9\dfrac{7}{8}$ as an improper fraction.

31. Determine whether $\dfrac{16}{40}$ and $\dfrac{10}{25}$ are equivalent.

32. Insert < or > to form a true statement. $\dfrac{4}{7} \quad \dfrac{5}{9}$

Multiply.

33. $\dfrac{2}{3} \cdot \dfrac{5}{11}$ **34.** $2\dfrac{5}{8} \cdot \dfrac{4}{7}$

35. $\dfrac{1}{4} \cdot \dfrac{1}{2}$ **36.** $7 \cdot 5\dfrac{2}{7}$

17. _____

18. _____

19. _____

20. _____

21. _____

22. _____

23. _____

24. _____

25. _____

26. _____

27. _____

28. _____

29. _____

30. _____

31. _____

32. _____

33. _____

34. _____

35. _____

36. _____

Divide.

37. $\dfrac{11}{18} \div 2\dfrac{5}{6}$

38. $\dfrac{15}{19} \div \dfrac{3}{5}$

39. $5\dfrac{2}{3} \div 2\dfrac{5}{9}$

40. $\dfrac{8}{11} \div \dfrac{1}{22}$

41. Add and simplify: $\dfrac{3}{16} + \dfrac{7}{16}$

42. Subtract and simplify: $\dfrac{11}{20} - \dfrac{7}{20}$

43. Find the LCM of 6 and 8.

44. Find the LCM of 7 and 5.

45. Add: $\dfrac{1}{2} + \dfrac{2}{3} + \dfrac{5}{6}$

46. Evaluate: $\left(\dfrac{5}{9}\right)^2$

47. Subtract: $9\dfrac{3}{7} - 5\dfrac{2}{21}$

48. Subtract: $\dfrac{31}{100} - \dfrac{5}{25}$

49. Simplify: $\left(\dfrac{2}{3}\right)^2 \div \left(\dfrac{8}{27} + \dfrac{2}{3}\right)$

50. $\dfrac{1}{10} \div \dfrac{7}{8} \cdot \dfrac{2}{5}$

37. _____

38. _____

39. _____

40. _____

41. _____

42. _____

43. _____

44. _____

45. _____

46. _____

47. _____

48. _____

49. _____

50. _____

Decimals

United Kingdom
London
English Channel
Netherlands
Brussels
Germany
Belgium
Paris
Le Mans
France
Switzerland

Layout of the 2014
Circuit de la Sarthe

4

245

What Sports Car Endurance Race Is the Oldest?

The 24 Hours of Le Mans is the oldest active sports car endurance race in the world. First held in 1923, the race is run each year at Circuit de la Sarthe near the village of Le Mans, France. On race day, each qualifying car is driven in roughly two-hour shifts by a team of three drivers. The team that completes the most laps of the circuit after 24 hours is declared the winner. Some of the world's most-recognized car manufacturers, such as Porsche and Audi, have taken top honors at Le Mans.

In Section 4.2, Exercises 69–72, we will explore the average race speeds of recent 24 Hours of Le Mans winners. We will revisit Le Mans in Exercise 69 of Section 4.5 and Exercise 36 of Section 4.6.

Decimal numbers represent parts of a whole, just like fractions. In this chapter, we learn to perform arithmetic operations using decimals and to analyze the relationship between fractions and decimals. We also learn how decimals are used in the real world.

Average Speed of 24 Hours of Le Mans Winners

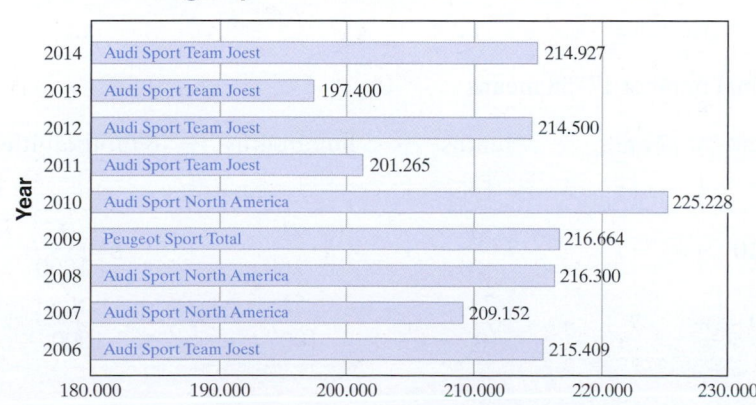

Year		Average Speed (in kilometers per hour)
2014	Audi Sport Team Joest	214.927
2013	Audi Sport Team Joest	197.400
2012	Audi Sport Team Joest	214.500
2011	Audi Sport Team Joest	201.265
2010	Audi Sport North America	225.228
2009	Peugeot Sport Total	216.664
2008	Audi Sport North America	216.300
2007	Audi Sport North America	209.152
2006	Audi Sport Team Joest	215.409

180.000   190.000   200.000   210.000   220.000   230.000

Average Speed (in kilometers per hour)

Source: lemans-history.com

Objectives

A Know the Meaning of Place Value for a Decimal Number, and Write Decimals in Words.

B Write Decimals in Standard Form.

C Write Decimals as Fractions.

D Write Fractions as Decimals.

Objective **A** Decimal Notation and Writing Decimals in Words

Like fractional notation, decimal notation is used to denote a part of a whole. Numbers written in decimal notation are called **decimal numbers,** or simply **decimals.** The decimal 17.758 has three parts.

$$1\ 7\ .\ 7\ 5\ 8$$

Whole number part ↑ Decimal part

Decimal point

In Section 1.2, we introduced place value for whole numbers. Place names and place values for the whole number part of a decimal number are exactly the same, as shown next. Place names and place values for the decimal part are also shown.

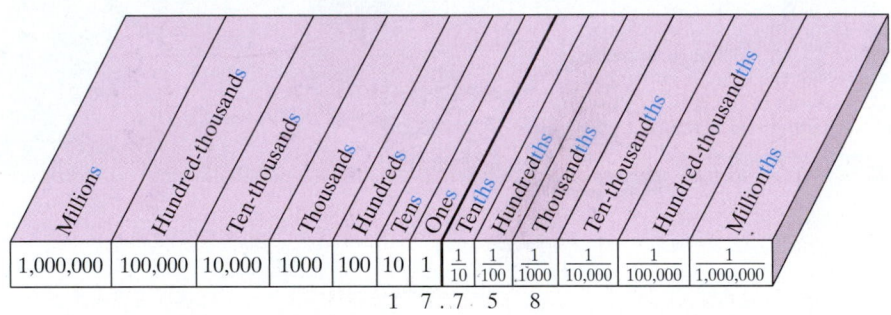

Helpful Hint Notice that place values to the left of the decimal point end in "s." Place values to the right of the decimal point end in "ths."

Notice that the value of each place is $\frac{1}{10}$ of the value of the place to its left. For example,

$$1 \cdot \frac{1}{10} = \frac{1}{10}$$

↑ ones   ↑ tenths

$$\frac{1}{10} \cdot \frac{1}{10} = \frac{1}{100}$$

↑ tenths   ↑ hundredths

The decimal number 17.758 means

1 ten	+	7 ones	+	7 tenths	+	5 hundredths	+	8 thousandths
↓ ↓		↓ ↓		↓ ↓		↓ ↓		↓

or $\quad 1 \cdot 10 \quad + \quad 7 \cdot 1 \quad + \quad 7 \cdot \dfrac{1}{10} \quad + \quad 5 \cdot \dfrac{1}{100} \quad + \quad 8 \cdot \dfrac{1}{1000}$

or $\quad 10 \quad + \quad 7 \quad + \quad \dfrac{7}{10} \quad + \quad \dfrac{5}{100} \quad + \quad \dfrac{8}{1000}$

Writing (or Reading) a Decimal in Words

Step 1: Write the whole number part in words.

Step 2: Write "and" for the decimal point.

Step 3: Write the decimal part in words as though it were a whole number, followed by the place value of the last digit.

Example 1 Write the decimal 1.3 in words.

Solution: one and three tenths

■ Work Practice 1

Practice 1

Write the decimal 8.7 in words.

Example 2

Write the decimal in the following sentence in words: The Golden Jubilee Diamond is a 545.67-carat cut diamond. (*Source: Guinness World Records*)

Solution: five hundred forty-five and sixty-seven hundredths

■ Work Practice 2

Practice 2

Write the decimal 97.28 in words.

Example 3 Write the decimal 19.5023 in words.

Solution: nineteen and five thousand twenty-three ten-thousandths

■ Work Practice 3

Practice 3

Write the decimal 302.105 in words.

Example 4

Write the decimal in the following sentence in words: The oldest known fragments of the Earth's crust are zircon crystals; they were discovered in Australia and are thought to be 4.276 billion years old. (*Source: Guinness World Records*)

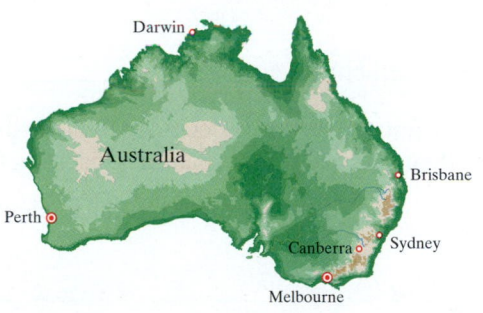

Solution: four and two hundred seventy-six thousandths

■ Work Practice 4

Practice 4

Write the decimal 72.1085 in words.

Suppose that you are paying $368.42 for an automotive repair job at Jake's Body Shop by writing a check. Checks are usually written using the following format.

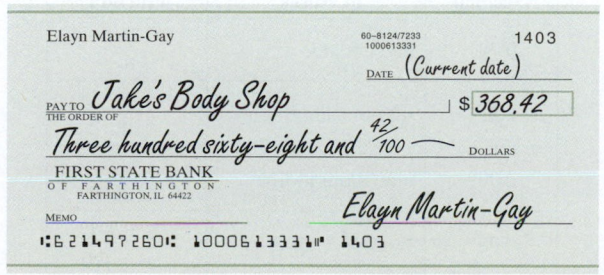

Answers

1. eight and seven tenths
2. ninety-seven and twenty-eight hundredths   **3.** three hundred two and one hundred five thousandths
4. seventy-two and one thousand eighty-five ten-thousandths

Practice 5

Fill in the check to CLECO (Central Louisiana Electric Company) to pay for your monthly electric bill of $207.40.

Example 5 Fill in the check to Camelot Music to pay for your purchase of $92.98.

Solution:

■ Work Practice 5

Objective B Writing Decimals in Standard Form

A decimal written in words can be written in standard form by reversing the preceding procedure.

Practice 6–7

Write each decimal in standard form.

6. Three hundred and ninety-six hundredths

7. Thirty-nine and forty-two thousandths

Examples Write each decimal in standard form.

6. Forty-eight and twenty-six hundredths is

48.26   hundredths place

7. Six and ninety-five thousandths is

6.095   thousandths place

■ Work Practice 6–7

Helpful Hint

When converting a decimal from words to decimal notation, make sure the last digit is in the correct place by inserting 0s if necessary. For example,

Two and thirty-eight thousandths is 2.038

thousandths place

Objective C Writing Decimals as Fractions

Once you master reading and writing decimals, writing a decimal as a fraction follows naturally.

Decimal	In Words	Fraction
0.7	seven tenths	$\frac{7}{10}$
0.51	fifty-one hundredths	$\frac{51}{100}$
0.009	nine thousandths	$\frac{9}{1000}$
0.05	five hundredths	$\frac{5}{100} = \frac{1}{20}$

Answers

5. CLECO; 207.40; Two hundred seven and $\frac{40}{100}$ **6.** 300.96   **7.** 39.042

Notice that the number of decimal places in a decimal number is the same as the number of zeros in the denominator of the equivalent fraction. We can use this fact to write decimals as fractions.

$$0.\underset{\substack{\uparrow \\ \text{2 decimal} \\ \text{places}}}{51} = \frac{51}{\underset{\substack{\uparrow \\ \text{2 zeros}}}{100}} \qquad 0.\underset{\substack{\uparrow \\ \text{3 decimal} \\ \text{places}}}{009} = \frac{9}{\underset{\substack{\uparrow \\ \text{3 zeros}}}{1000}}$$

Example 8 Write 0.43 as a fraction.

Solution: $0.43 = \frac{43}{100}$

$\underset{\substack{\uparrow \\ 2 \\ \text{decimal places}}}{} \quad \underset{\substack{\uparrow \\ 2 \\ \text{zeros}}}{}$

■ Work Practice 8

Practice 8

Write 0.037 as a fraction.

Example 9 Write 5.7 as a mixed number.

Solution: $5.7 = 5\frac{7}{10}$

$\underset{\substack{\uparrow \\ 1 \\ \text{decimal place}}}{} \quad \underset{\substack{\uparrow \\ 1 \\ \text{zero}}}{}$

■ Work Practice 9

Practice 9

Write 14.97 as a mixed number.

Examples Write each decimal as a fraction or a mixed number. Write your answer in simplest form.

10. $0.125 = \frac{125}{1000} = \frac{\overset{1}{\cancel{125}}}{8 \cdot \cancel{125}} = \frac{1}{8}$

11. $23.5 = 23\frac{5}{10} = 23\frac{\overset{1}{\cancel{5}}}{2 \cdot \cancel{5}} = 23\frac{1}{2 \cdot 1} = 23\frac{1}{2}$

12. $105.083 = 105\frac{83}{1000}$

■ Work Practice 10–12

Practice 10–12

Write each decimal as a fraction or mixed number. Write your answer in simplest form.
10. 0.12
11. 57.8
12. 209.986

Objective D Writing Fractions as Decimals ▶

If the denominator of a fraction is a power of 10, we can write it as a decimal by reversing the procedure above.

Examples Write each fraction as a decimal.

13. $\frac{8}{\underset{\substack{\uparrow \\ \text{1 zero}}}{10}} = 0.\underset{\substack{\uparrow \\ \text{1 decimal place}}}{8}$

14. $\frac{87}{\underset{\substack{\uparrow \\ \text{1 zero}}}{10}} = 8.\underset{\substack{\uparrow \\ \text{1 decimal place}}}{7}$

15. $\frac{18}{\underset{\substack{\uparrow \\ \text{3 zeros}}}{1000}} = 0.\underset{\substack{\uparrow \\ \text{3 decimal places}}}{018}$

16. $\frac{507}{\underset{\substack{\uparrow \\ \text{2 zeros}}}{100}} = 5.\underset{\substack{\uparrow \\ \text{2 decimal places}}}{07}$

■ Work Practice 13–16

Practice 13–16

Write each fraction as a decimal.

13. $\frac{58}{100}$ **14.** $\frac{59}{100}$

15. $\frac{6}{1000}$ **16.** $\frac{172}{10}$

Answers
8. $\frac{37}{1000}$ **9.** $14\frac{97}{100}$ **10.** $\frac{3}{25}$
11. $57\frac{4}{5}$ **12.** $209\frac{493}{500}$ **13.** 0.58
14. 0.59 **15.** 0.006 **16.** 17.2

Vocabulary, Readiness & Video Check

Use the choices below to fill in each blank.

words	decimals	and
tens	tenths	standard form

1. The number "twenty and eight hundredths" is written in _____ and "20.08" is written in _____.
2. Like fractions, _____ are used to denote parts of a whole.
3. When writing a decimal number in words, the decimal point is written as _____.
4. The place value _____ is to the right of the decimal point while _____ is to the left of the decimal point.

Determine the place value for the digit 7 in each number.

5. 70   6. 700   7. 0.7   8. 0.07

Martin-Gay Interactive Videos   Watch the section lecture video and answer the following questions.

See Video 4.1

Objective A 9. In ▣ Example 1, how is the decimal point written in words?

Objective B 10. Why is 9.8 not the correct answer to ▣ Example 3? What is the correct answer? ▶

Objective C 11. From ▣ Example 5, why does reading a decimal number correctly help us write it as an equivalent fraction? ▶

Objective D 12. In ▣ Examples 7–9, when writing a fraction as a decimal, and the denominator of the fraction is a power of 10, what similarity is there between the number of zeros in the denominator and the number of decimal places in the decimal? ▶

4.1 Exercise Set MyMathLab®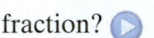

Objective A *Write each decimal number in words. See Examples 1 through 4.*

1. 6.52   2. 7.59   ▶ 3. 16.23   4. 47.65

5. 0.205   6. 0.495   ▶ 7. 167.009   8. 233.056

9. 200.005   10. 5000.02   11. 105.6   12. 410.30

13. The Akashi Kaikyo Bridge, between Kobe and Awaji-Shima, Japan, is approximately 2.43 miles long.

14. The English Channel Tunnel is 31.04 miles long. (*Source: Railway Directory & Year Book*)

15. Mercury makes a complete orbit of the Sun every 87.97 days. (*Source:* National Space Science Data Center)

16. Saturn makes a complete orbit of the Sun every 29.48 years. (*Source:* National Space Science Data Center)

17. The total number of television households within the United States for the 2013–2014 season was estimated at 115.6 million. (*Source:* Nielsen Media Research)

18. In 2012, the average retail customer's wait time in line at a U.S. post office was 2.6 minutes. (*Source:* USPS)

Fill in each check for the described purchase. See Example 5.

19. Your monthly car loan of $321.42 to R. W. Financial.

Your Preprinted Name Your Preprinted Address	60-8124/7233 1000613331	1407
	DATE	
PAY TO THE ORDER OF		$
		DOLLARS
FIRST STATE BANK OF FARTHINGTON FARTHINGTON, IL 64422		
MEMO		
⑆621497260⑆ 1000613331⑈ 1407		

20. Your part of the monthly apartment rent, which is $213.70. You pay this to Amanda Dupre.

Your Preprinted Name Your Preprinted Address	60-8124/7233 1000613331	1408
	DATE	
PAY TO THE ORDER OF		$
		DOLLARS
FIRST STATE BANK OF FARTHINGTON FARTHINGTON, IL 64422		
MEMO		
⑆621497260⑆ 1000613331⑈ 1408		

21. Your cell phone bill of $59.68 to Bell South.

Your Preprinted Name Your Preprinted Address	60-8124/7233 1000613331	1409
	DATE	
PAY TO THE ORDER OF		$
		DOLLARS
FIRST STATE BANK OF FARTHINGTON FARTHINGTON, IL 64422		
MEMO		
⑆621497260⑆ 1000613331⑈ 1409		

22. Your grocery bill of $87.49 to Albertsons.

Your Preprinted Name Your Preprinted Address	60-8124/7233 1000613331	1410
	DATE	
PAY TO THE ORDER OF		$
		DOLLARS
FIRST STATE BANK OF FARTHINGTON FARTHINGTON, IL 64422		
MEMO		
⑆621497260⑆ 1000613331⑈ 1410		

Objective B *Write each decimal number in standard form. See Examples 6 and 7.*

23. Six and five tenths

24. Three and nine tenths

25. Nine and eight hundredths

26. Twelve and six hundredths

27. Seven hundred five and six hundred twenty-five thousandths

28. Eight hundred four and three hundred ninety-nine thousandths

29. Forty-six ten-thousandths

30. Thirty-eight ten-thousandths

31. The record rainfall amount for a 24-hour period in Alabama is thirty-two and fifty-two hundredths inches. This record was set at Dauphin Island Sea Lab in 1997. (*Source:* National Climatic Data Center)

32. For model year 2013, the average fuel economy of all imported passenger cars sold in the United States was thirty-seven and one tenth miles per gallon. (*Source:* U.S. Department of Transporation)

33. The average IndyCar burns one and three-tenths gallons of fuel per lap at the Indianapolis Motor Speedway. (*Source:* Indianapolis Motor Speedway)

34. The IZOD IndyCar series races at the Mid-Ohio Sports Car Course each season. The track length there is two and two hundred fifty-eight thousandths miles. (*Source:* IndyCar.com)

Objective C *Write each decimal as a fraction or a mixed number. Write your answer in simplest form. See Examples 8 through 12.*

35. 0.3

36. 0.9

37. 0.27

38. 0.39

39. 0.8

40. 0.4

41. 0.15

42. 0.64

43. 5.47

44. 6.3

45. 0.048

46. 0.082

47. 7.008

48. 9.005

49. 15.802

50. 11.406

51. 0.3005

52. 0.2006

53. 487.32

54. 298.62

Objective D *Write each fraction as a decimal. See Examples 13 through 16.*

55. $\dfrac{6}{10}$

56. $\dfrac{3}{10}$

57. $\dfrac{45}{100}$

58. $\dfrac{75}{100}$

59. $\dfrac{37}{10}$

60. $\dfrac{28}{10}$

61. $\dfrac{268}{1000}$

62. $\dfrac{709}{1000}$

63. $\dfrac{9}{100}$

64. $\dfrac{7}{100}$

65. $\dfrac{4026}{1000}$

66. $\dfrac{3601}{1000}$

67. $\dfrac{28}{1000}$

68. $\dfrac{63}{1000}$

69. $\dfrac{563}{10}$

70. $\dfrac{206}{10}$

Objectives A B C D Mixed Practice *Fill in the chart. The first row is completed for you. See Examples 1 through 16.*

Decimal Number in Standard Form	In Words	Fraction
0.37	thirty-seven hundredths	$\frac{37}{100}$
71.		$\frac{43}{100}$
72.		$\frac{89}{100}$
73.	eight tenths	
74.	five tenths	
75. 0.077		
76. 0.019		

Review

Round 47,261 to the indicated place value. See Section 1.5.

77. tens

78. hundreds

79. thousands

80. ten-thousands

Concept Extensions

81. In your own words, describe how to write a decimal as a fraction or a mixed number.

82. In your own words, describe how to write a fraction as a decimal.

83. Write 0.00026849576 in words.

84. Write 0.00026849576 as a fraction. Do not simplify the resulting fraction.

85. Write $17\frac{268}{1000}$ as a decimal.

86. Write $7\frac{12}{100}$ as a decimal.

4.2 Order and Rounding

Objective A Comparing Decimals

One way to compare decimals is to compare their graphs on a number line. Recall from Section 3.5 that for any two numbers on a number line, the number to the left is smaller and the number to the right is larger. The decimals 0.5 and 0.8 are graphed as follows:

Objectives

A Compare Decimals.

B Round a Decimal Number to a Given Place Value.

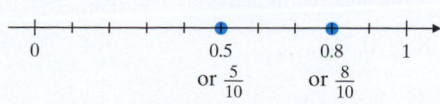

Comparing decimals by comparing their graphs on a number line can be time consuming. Another way to compare the size of decimals is to compare digits in corresponding places.

> ### Comparing Two Decimals
>
> Compare digits in the same places from left to right. When two digits are not equal, the number with the larger digit is the larger decimal. If necessary, insert 0s after the last digit to the right of the decimal point to continue comparing.
>
> Compare hundredths-place digits
>
> $$28.253 \qquad 28.263$$
> $$\uparrow \qquad\qquad \uparrow$$
> $$5 \quad < \quad 6$$
> $$\text{so } 28.253 \quad < \quad 28.263$$

Before we continue, let's take a moment and convince ourselves that inserting a zero after the last digit to the right of a decimal point does not change the value of the number.

For example, let's show that

$$0.7 = 0.70$$

If we write 0.7 as a fraction, we have

$$0.7 = \frac{7}{10}$$

Let's now multiply by 1. Recall that multiplying a number by 1 does not change the value of the number.

$$0.7 = \frac{7}{10} = \frac{7}{10} \cdot 1 = \frac{7}{10} \cdot \frac{10}{10} = \frac{7 \cdot 10}{10 \cdot 10} = \frac{70}{100} = 0.70$$

Thus $0.7 = 0.70$ and so on.

Helpful Hint

For any decimal, inserting 0s after the last digit to the right of the decimal point does not change the value of the number.

$$7.6 = 7.60 = 7.600, \text{ and so on}$$

When a whole number is written as a decimal, the decimal point is placed to the right of the ones digit.

$$25 = 25.0 = 25.00, \text{ and so on}$$

Practice 1

Insert $<$, $>$, or $=$ to form a true statement.

　　13.208　　13.281

Example 1 Insert $<$, $>$, or $=$ to form a true statement.

　　0.378　　0.368

Solution:

$$0.3\,\boxed{7}\,8 \qquad 0.3\,\boxed{6}\,8 \qquad \text{The tenths places are the same.}$$

$$0.3\,\boxed{7}\,8 \qquad 0.3\,\boxed{6}\,8 \qquad \text{The hundredths places are different.}$$

Since $7 > 6$, then $0.378 > 0.368$.

■ **Work Practice 1**

Answer

1. $<$

Example 2   Insert <, >, or = to form a true statement.

   0.052   0.236

Solution:

   0. **0** 52 < 0. **2** 36   0 is smaller than 2 in the tenths place.
      ↑   ↑

■ **Work Practice 2**

Example 3   Insert <, >, or = to form a true statement.

   0.52   0.063

Solution:

   0. **5** 2 > 0. **0** 63   5 is larger than 0 in the tenths place.
      ↑   ↑

■ **Work Practice 3**

Example 4   Write the decimals in order from smallest to largest.

   7.035, 8.12, 7.03, 7.1

Solution: By comparing the ones digits, the decimal 8.12 is the largest number. To write the rest of the decimals in order, we compare digits to the right of the decimal point. We will insert zeros to help us compare.

   7.035   7.030   7.100

> **Helpful Hint**
>
> You may also immediately notice that 7.1 is larger than both 7.035 and 7.03.

By comparing digits to the right of the decimal point, we can now arrange the decimals from smallest to largest.

   7.030,   7.035,   7.100,   8.12   or

   7.03,   7.035,   7.1,   8.12

■ **Work Practice 4**

Objective B Rounding Decimals

We **round the decimal part** of a decimal number in nearly the same way as we round whole numbers. The only difference is that we delete digits to the right of the rounding place, instead of replacing these digits by 0s. For example,

   24.954   rounded to the nearest hundredth is   24.95
      ↑
   hundredths place

Rounding Decimals to a Place Value to the Right of the Decimal Point

Step 1: Locate the digit to the right of the given place value.

Step 2: If this digit is 5 or greater, add 1 to the digit in the given place value and delete all digits to its right. If this digit is less than 5, delete all digits to the right of the given place value.

Practice 2

Insert <, >, or = to form a true statement.

   0.124   0.086

Practice 3

Insert <, >, or = to form a true statement.

   0.61   0.076

Practice 4

Write the decimals in order from smallest to largest.

   14.605, 14.65, 13.9, 14.006

Answers
2. >   **3.** >
4. 13.9, 14.006, 14.605, 14.65

Practice 5

Round 123.7814 to the nearest thousandth.

Example 5   Round 736.2359 to the nearest tenth.

Solution:

Step 1: We locate the digit to the right of the tenths place.

```
              ┌── tenths place
          736.2③59
                 └──→ digit to the right
```

Step 2: Since the digit to the right is less than 5, we delete it and all digits to its right.

Thus, 736.2359 rounded to the nearest tenth is 736.2.

■ **Work Practice 5**

Practice 6

Round 123.7817 to the nearest tenth.

Example 6   Round 736.2359 to the nearest hundredth.

Solution:

Step 1: We locate the digit to the right of the hundredths place.

```
                ┌── hundredths place
          736.23⑤9
                  └──→ digit to the right
```

Step 2: Since the digit to the right is 5, we add 1 to the digit in the hundredths place and delete all digits to the right of the hundredths place.

```
          736.23⑤9
               ↑ ↑└── Delete these digits.
              Add 1.
```

Thus, 736.2359 rounded to the nearest hundredth is 736.24.

■ **Work Practice 6**

Rounding often occurs with money amounts. Since there are 100 cents in a dollar, each cent is $\frac{1}{100}$ of a dollar. This means that if we want to round to the nearest cent, we round to the nearest hundredth of a dollar.

Practice 7

In Sandersville, the price of a gallon of premium gasoline is $3.1589. Round this to the nearest cent.

Example 7   The price of a gallon of premium gasoline in Cross City is currently $3.1779. Round this to the nearest cent.

Solution:

```
  hundredths place ──────┐ ┌── 7 is greater than 5
                  $3.17⑦9
                        ↑ └── Delete these digits.
                      Add 1.
```

Since the digit to the right is greater than 5, we add 1 to the hundredths digit and delete all digits to the right of the hundredths digit.

Thus, $3.1779 rounded to the nearest cent is $3.18.

■ **Work Practice 7**

Answers

5. 123.781   **6.** 123.8   **7.** $3.16

Example 8 Round $0.098 to the nearest cent.

Solution:

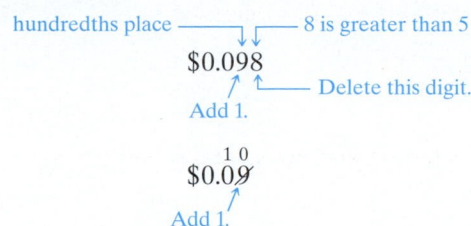

hundredths place —— 8 is greater than 5

$0.098

Add 1. Delete this digit.

$0.0\overset{1\ 0}{9}

Add 1.

$9 + 1 = 10$, so replace the digit 9 by 0 and carry the 1 to the place value to the left. Thus, $0.098 rounded to the nearest cent is $0.10.

■ **Work Practice 8**

✓**Concept Check** 1756.0894 rounded to the nearest ten is

a. 1756.1   **b.** 1760.0894   **c.** 1760   **d.** 1750

Example 9 Determining State Taxable Income

A high school teacher's taxable income is $41,567.72. The tax tables in the teacher's state use amounts to the nearest dollar. Round the teacher's income to the nearest dollar.

Solution: Rounding to the nearest dollar means rounding to the ones place.

ones place —— 7 is greater than 5

$41,567.72

Add 1. Delete these digits.

Thus, the teacher's income rounded to the nearest dollar is $41,568.

■ **Work Practice 9**

   In Section 4.4, we will introduce a formula for the distance around a circle. The distance around a circle is given the special name **circumference.**
   The symbol π is the Greek letter pi, pronounced "pie." We use π to denote the following constant:

$$\pi = \frac{\text{circumference of a circle}}{\text{diameter of a circle}}$$

   The value π is an **irrational number.** This means if we try to write it as a decimal, it neither ends nor repeats in a pattern.

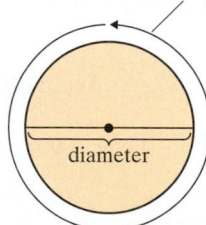

circumference

diameter

Example 10 $\pi \approx 3.14159265$. Round π to the nearest hundredth.

Solution:

hundredths place —— 1 is less than 5.

3.14159265

Delete these digits.

Thus, 3.14159265 rounded to the nearest hundredth is 3.14. In other words, $\pi \approx 3.14$.

■ **Work Practice 10**

Practice 8

Round $1.095 to the nearest cent.

Practice 9

Water bills in Gotham City are always rounded to the nearest dollar. Round a water bill of $24.62 to the nearest dollar.

Practice 10

$\pi \approx 3.14159265$. Round π to the nearest ten-thousandth.

Answers

8. $1.10   **9.** $25   **10.** $\pi \approx 3.1416$

✓**Concept Check Answer**

c

Vocabulary, Readiness & Video Check

Use the choices below to fill in each blank. Some choices may be used more than once or not at all.

before	7.0	diameter
after	0.7	circumference

1. Another name for the distance around a circle is its _____.

2. $\pi = \dfrac{\text{\underline{\hspace{2cm}} of a circle}}{\text{\underline{\hspace{2cm}} of a circle}}$

3. The decimal point in a whole number is _____ the last digit.

4. The whole number $7 = $ _____.

Martin-Gay Interactive Videos   *Watch the section lecture video and answer the following questions.*

Objective A   5. In Example 2, we compare place value by place value in which direction? ▶

Objective B   6. The decimal number in Example 5 is being rounded to the nearest thousandth. Why is the digit 5, which is not in the thousandth place, looked at? ▶

See Video 4.2 🍒

4.2   Exercise Set   MyMathLab® ▶

Objective A   *Insert $<$, $>$, or $=$ to form a true statement. See Examples 1 through 3.*

1. 0.15   0.16   **2.** 0.12   0.15   ▶**3.** 0.57   0.54   **4.** 0.59   0.52

5. 0.098   0.1   **6.** 0.0756   0.2   **7.** 0.54900   0.549   **8.** 0.98400   0.984

▶**9.** 167.908   167.980   **10.** 519.3405   519.3054   **11.** 420,000   0.000042   **12.** 0.000987   987,000

Write the decimals in order from smallest to largest. See Example 4.

13. 0.006, 0.06, 0.0061   **14.** 0.082, 0.008, 0.080   **15.** 0.042, 0.36, 0.03

16. 0.21, 0.056, 0.065   **17.** 1.1, 1.16, 1.01, 1.09   **18.** 3.6, 3.069, 3.09, 3.06

19. 21.001, 20.905, 21.03, 21.12   **20.** 36.050, 35.72, 35.702, 35.072

Objective B   *Round each decimal to the given place value. See Examples 5 through 10.*

21. 0.57, to the nearest tenth   **22.** 0.54, to the nearest tenth   ▶**23.** 0.234, to the nearest hundredth

24. 0.452, to the nearest hundredth   **25.** 0.5942, to the nearest thousandth   **26.** 63.4523, to the nearest thousandth

▶ **27.** 98,207.23, to the nearest ten   **28.** 68,934.543, to the nearest ten   **29.** 12.342, to the nearest tenth

30. 42.9878, to the nearest thousandth   **31.** 17.667, to the nearest hundredth   **32.** 0.766, to the nearest hundredth

33. 0.501, to the nearest tenth   **34.** 0.602, to the nearest tenth   ▶ **35.** 0.1295, to the nearest thousandth

36. 0.8295, to the nearest thousandth   **37.** 3829.34, to the nearest ten   **38.** 4520.876, to the nearest hundred

Round each monetary amount to the nearest cent or dollar as indicated. See Examples 7 through 9.

39. $0.067, to the nearest cent   **40.** $0.025, to the nearest cent   **41.** $42,650.14, to the nearest dollar

42. $768.45, to the nearest dollar   ▶ **43.** $26.95, to the nearest dollar   **44.** $14,769.52, to the nearest dollar

45. $0.1992, to the nearest cent   **46.** $0.7633, to the nearest cent

Round each number to the given place value. See Examples 5 through 10.

47. The latest generation Apple MacBook Air, at its thinnest point, measures 0.2794 cm. Round this number to the nearest tenth. (*Source:* Apple, Inc.)

48. A large tropical cockroach of the family Dictyoptera is the fastest-moving insect. This insect was clocked at a speed of 3.36 miles per hour. Round this number to the nearest tenth. (*Source:* University of California, Berkeley)

49. During the 2013 Boston Marathon, Hiroyuki Yamamoto of Japan was the first wheelchair competitor to cross the finish line. His time was 1.4256 hours. Round this time to the nearest hundredth. (*Source:* Boston Athletic Association)

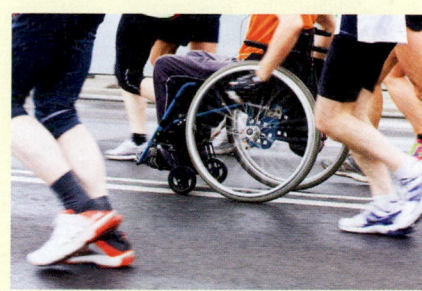

50. Lindsey Vonn of the U.S. ski team took first place in the women's giant slalom in the 2013 International Ski Federation World Cup. Her winning time was 2.372 minutes. Round this time to the nearest hundredth of a minute. (*Source:* International Ski Federation)

51. A used biology textbook is priced at $47.89. Round this price to the nearest dollar.

52. A used office desk is advertised at $49.95 by Drawley's Office Furniture. Round this price to the nearest dollar.

53. The 2012 estimated population density of the state of Louisiana is 106.5157 people per square mile. Round this population density to the nearest tenth. (*Source:* U.S. Census Bureau)

54. The 2012 estimated population density of the state of Arkansas is 56.675 people per square mile. Round this population density to the nearest tenth. (*Source:* U.S. Census Bureau)

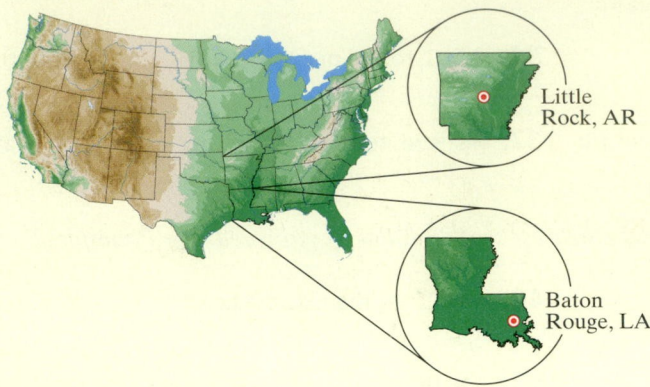

○ 55. The length of a day on Mars is 24.6229 hours. Round this figure to the nearest thousandth. (*Source:* National Space Science Data Center)

56. Venus makes a complete orbit around the Sun every 224.695 days. Round this figure to the nearest whole day. (*Source:* National Space Science Data Center)

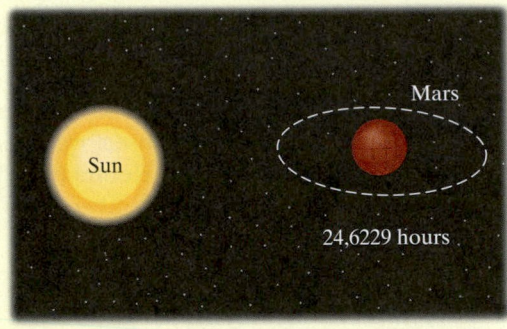

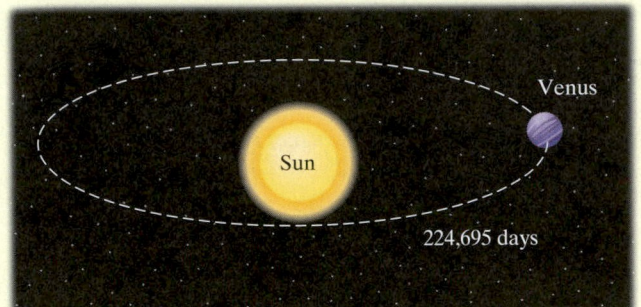

57. Kingda Ka is a hydraulic-launch roller coaster at Six Flags Great Adventure, an amusement park in Jackson, New Jersey. Currently it is the world's tallest roller coaster. A ride on the Kingda Ka lasts about 0.4667 minute. Round this figure to the nearest tenth. (*Source:* Roller Coaster DataBase)

58. During the 2012 NFL season, the average length of an Indianapolis Colts' punt was 47.6 yards. Round this figure to the nearest whole yard. (*Source:* National Football League)

Review

Perform each indicated operation. See Sections 1.3 and 1.4.

59. $3452 + 2314$

60. $8945 + 4536$

61. $94 - 23$

62. $82 - 47$

63. $482 - 239$

64. $4002 - 3897$

Concept Extensions

Solve. See the Concept Check in this section.

65. 2849.1738 rounded to the nearest hundred is
  a. 2849.17   **b.** 2800   **c.** 2850   **d.** 2849.174

66. 146.059 rounded to the nearest ten is
  a. 146.0   **b.** 146.1   **c.** 140   **d.** 150

67. 2849.1738 rounded to the nearest hundredth is

    a. 2849.17   **b.** 2800   **c.** 2850   **d.** 2849.174

68. 146.059 rounded to the nearest tenth is

    a. 146.0   **b.** 146.1   **c.** 140   **d.** 150

Mixed Practice (*Sections 4.1, 4.2*) *The table gives the average speed, in kilometers per hour, for the winners of the 24 Hours of Le Mans for each of the years listed. Use the table to answer Exercises 69 through 72. (Source: lemans-history.com)*

Year	Team	Average Speed (in kph)
2006	Audi Sport Team Joest	215.409
2007	Audi Sport North America	209.152
2008	Audi Sport North America	216.300
2009	Peugeot Sport Total	216.664
2010	Audi Sport North America	225.228
2011	Audi Sport Team Joest	201.265
2012	Audi Sport Team Joest	214.500
2013	Audi Sport Team Joest	197.400
2014	Audi Sport Team Joest	214.927

69. What is the fastest average speed on the list? Write this speed as a mixed number. Which team achieved this average speed?

70. What is the slowest average speed on the list? Write this speed as a mixed number. Which team achieved this average speed?

71. Make a list of the average winning speeds in order from fastest to slowest for the years 2010 through 2013.

72. Make a list of the average winning speeds in order from fastest to slowest for the years 2006 through 2009.

73. Write a 5-digit number that rounds to 1.7.

74. Write a 4-digit number that rounds to 26.3.

75. Write a decimal number that is greater than 8 but less than 9.

76. Write a decimal number that is greater than 48.1, but less than 48.2.

77. Which number(s) rounds to 0.26?

    0.26559   0.26499   0.25786   0.25186

78. Which number(s) rounds to 0.06?

    0.0612   0.066   0.0586   0.0506

Write these numbers from smallest to largest.

79. 0.9

    0.1038

    0.10299

    0.1037

80. 0.01

    0.0839

    0.09

    0.1

81. The all-time top six movies* (those that earned the most money in the United States) along with the approximate amount of money they have earned are listed in the table. Estimate the total amount of money that these movies have earned by first rounding each earning to the nearest hundred million. (*Source:* The Internet Movie Database)

Top All-Time American Movies	
Movie	**Gross Domestic Earnings**
Avatar (2009)	$760.5 million
Titanic (1997)	$658.7 million
Marvel's The Avengers (2012)	$623.4 million
The Dark Knight (2008)	$534.9 million
Star Wars: Episode I—The Phantom Menace (1999)	$474.5 million
Star Wars (1977)	$461.0 million
*Note: Many of these movies are still earning substantial amounts of money.	

82. In 2012, American manufacturers shipped approximately 6.2 million music videos to retailers. The value of these shipments was approximately $118.2 million. Estimate the value of an individual music video by rounding 118.2 and 6.2 to the nearest ten, then dividing. (*Source:* Recording Industry Association of America)

4.3 Adding and Subtracting Decimals

Objective A Adding Decimals

Adding decimals is similar to adding whole numbers. We add digits in corresponding place values from right to left, carrying if necessary. To make sure that digits in corresponding place values are added, we line up the decimal points vertically.

Adding or Subtracting Decimals

Step 1: Write the decimals so that the decimal points line up vertically.

Step 2: Add or subtract as with whole numbers.

Step 3: Place the decimal point in the sum or difference so that it lines up vertically with the decimal points in the problem.

In this section, we will insert zeros in decimal numbers so that place-value digits line up neatly. For instance, see Example 1.

Practice 1

Add.

a. $15.52 + 2.371$

b. $20.06 + 17.612$

c. $0.125 + 122.8$

Example 1 Add: $23.85 + 1.604$

Solution: First we line up the decimal points vertically.

$$
\begin{array}{r}
23.85\underline{0} \quad \text{Insert one 0 so that digits line up neatly.}\\
+\ 1.604 \\
\hline
\end{array}
$$

↑
Line up decimal points.

Then we add the digits from right to left as for whole numbers.

$$
\begin{array}{r}
\overset{1}{2}3.850 \\
+\ 1.604 \\
\hline
25.454
\end{array}
$$

└── Place the decimal point in the sum so that all decimal points line up.

■ **Work Practice 1**

Helpful Hint

Recall that 0's may be placed after the last digit to the right of the decimal point without changing the value of the decimal. This may be used to help line up place values when adding decimals.

$$
\begin{array}{r}
3.2 \\
15.567 \\
+\ 0.11 \\
\hline
\end{array}
\quad \text{becomes} \quad
\begin{array}{r}
3.2\underline{00} \quad \text{Insert two 0s.}\\
15.567 \\
+\ 0.11\underline{0} \quad \text{Insert one 0.}\\
\hline
18.877 \quad \text{Add.}
\end{array}
$$

Answers

1. a. 17.891 **b.** 37.672 **c.** 122.925

Example 2   Add: 763.7651 + 22.001 + 43.89

Solution: First we line up the decimal points.

$$
\begin{array}{r}
\overset{1\ 1\ 1}{763.7651} \\
22.0010 \quad \text{Insert one 0.} \\
+\ \ 43.8900 \quad \text{Insert two 0s.} \\
\hline
829.6561 \quad \text{Add.}
\end{array}
$$

■ **Work Practice 2**

Helpful Hint

Don't forget that the decimal point in a whole number is after the last digit.

Example 3   Add: 45 + 2.06

Solution:
$$
\begin{array}{r}
45.00 \quad \text{Insert a decimal point and two 0s.} \\
+\ 2.06 \quad \text{Line up decimal points.} \\
\hline
47.06 \quad \text{Add.}
\end{array}
$$

■ **Work Practice 3**

✓**Concept Check** What is wrong with the following calculation of the sum of 7.03, 2.008, 19.16, and 3.1415?

$$
\begin{array}{r}
7.03 \\
2.008 \\
19.16 \\
+\ 3.1415 \\
\hline
3.6042
\end{array}
$$

Objective B Subtracting Decimals

Subtracting decimals is similar to subtracting whole numbers. We line up digits and subtract from right to left, borrowing when needed.

Example 4   Subtract: 35.218 − 23.65. Check your answer.

Solution: First we line up the decimal points.

$$
\begin{array}{r}
\overset{\ \ \ 11}{\underset{}{4\ \ 1\ 11}} \\
3\cancel{5}.2\cancel{1}8 \\
-23.650 \quad \text{Insert one 0.} \\
\hline
11.568 \quad \text{Subtract.}
\end{array}
$$

Recall that we can check a subtraction problem by adding.

$$
\begin{array}{r}
\overset{1\ \ 1}{11.568} \quad \text{Difference} \\
+23.650 \quad \text{Subtrahend} \\
\hline
35.218 \quad \text{Minuend}
\end{array}
$$

■ **Work Practice 4**

Practice 5

Subtract. Check your answers.
a. $5.8 - 3.92$
b. $9.72 - 4.068$

Example 5   Subtract: $3.5 - 0.068$. Check your answer.

Solution:

$$
\begin{array}{r}
\overset{4\ \overset{9}{\cancel{10}}\ 10}{3.\cancel{5}\,0\,0} \\
-0.0\,6\,8 \\
\hline
3.4\,3\,2
\end{array}
$$

Insert two 0s.
Line up decimal points.
Subtract.

Check:

$$
\begin{array}{r}
\overset{1\ 1}{3.432} \\
+0.068 \\
\hline
3.500
\end{array}
$$

Difference
Subtrahend
Minuend

■ **Work Practice 5**

Practice 6

Subtract. Check your answers.
a. $53 - 29.31$
b. $120 - 68.22$

Example 6   Subtract: $85 - 17.31$. Check your answer.

Solution:

$$
\begin{array}{r}
\overset{7\ 14\ \overset{9}{\cancel{10}}\ 10}{\cancel{8}\,\cancel{5}.0\,0} \\
-1\,7.3\,1 \\
\hline
6\,7.6\,9
\end{array}
$$

Check:

$$
\begin{array}{r}
\overset{1\ 1\ 1}{67.69} \\
+17.31 \\
\hline
85.00
\end{array}
$$

Difference
Subtrahend
Minuend

■ **Work Practice 6**

Objective C Estimating When Adding or Subtracting Decimals ▶

To help avoid errors, we can also estimate to see if our answer is reasonable when adding or subtracting decimals. Although only one estimate is needed per operation, we show two to show variety.

Practice 7

Add or subtract as indicated. Then estimate to see if the answer is reasonable by rounding the given numbers and adding or subtracting the rounded numbers.

a. $48.1 + 326.97$
b. $18.09 - 0.746$

Example 7   Add or subtract as indicated. Then estimate to see if the answer is reasonable by rounding the given numbers and adding or subtracting the rounded numbers.

a. $27.6 + 519.25$

Solution:

Exact		Estimate 1		Estimate 2
$\overset{1}{27.60}$	rounds to	30		30
$+\,519.25$	rounds to	$+500$	or	$+520$
546.85		530		550

Since the exact answer is close to either estimate, it is reasonable. (In the first estimate, each number is rounded to the place value of the leftmost digit. In the second estimate, each number is rounded to the nearest ten.)

b. $11.01 - 0.862$

Solution:

Exact		Estimate 1		Estimate 2
$\overset{0\ \ 9\ 1010}{1\cancel{1}.\cancel{0}\cancel{1}\cancel{0}}$	rounds to	10		11
$-0.8\,6\,2$	rounds to	-1	or	-1
$1\,0.1\,4\,8$		9		10

In the first estimate, we rounded the first number to the nearest ten and the second number to the nearest one. In the second estimate, we rounded both numbers to the nearest one. Both estimates show us that our answer is reasonable.

■ **Work Practice 7**

> **Helpful Hint**   Remember that estimates are for our convenience to quickly check the reasonableness of an answer.

Answers

5. a. 1.88   **b.** 5.652
6. a. 23.69   **b.** 51.78
7. a. 375.07   **b.** 17.344

✓**Concept Check** Why shouldn't the sum $21.98 + 42.36$ be estimated as $30 + 50 = 80$?

Objective D Solving Problems by Adding or Subtracting Decimals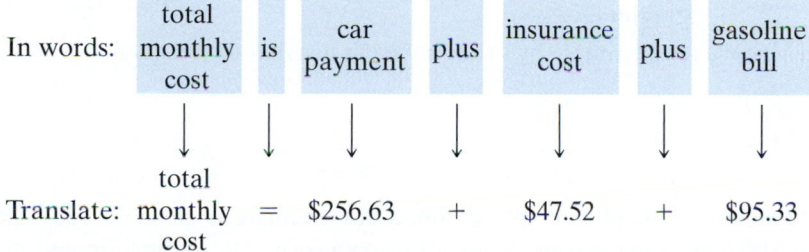

Decimals are very common in real-life problems.

Example 8 | Calculating the Cost of Owning an Automobile

Find the total monthly cost of owning and operating a certain automobile given the expenses shown.

Monthly car payment:	$256.63
Monthly insurance cost:	$47.52
Average gasoline bill per month:	$95.33

Solution:

1. **UNDERSTAND.** Read and reread the problem. The phrase "total monthly cost" tells us to add.
2. **TRANSLATE.**

In words:

total monthly cost	is	car payment	plus	insurance cost	plus	gasoline bill
↓	↓	↓	↓	↓	↓	↓

Translate:

$$\text{total monthly cost} = \$256.63 + \$47.52 + \$95.33$$

3. **SOLVE:** Let's also estimate by rounding each number to the nearest ten.

$$
\begin{array}{ll}
\overset{1\,1\,1}{256.63} & \text{rounds to} \quad 260 \\
47.52 & \text{rounds to} \quad 50 \\
\underline{+\ \ 95.33} & \text{rounds to} \quad \underline{100} \\
399.48 \ \ \text{Exact} & \phantom{\text{rounds to} \quad} 410 \quad \text{Estimate}
\end{array}
$$

4. **INTERPRET.** *Check* your work. Since our estimate is close to our exact answer, our answer is reasonable. *State* your conclusion: The total monthly cost is $399.48.

■ **Work Practice 8**

The bar graph in Example 9 has horizontal bars. To visualize the value represented by a bar, see how far it extends to the right. The value of each bar is labeled and we will study bar graphs further in a later chapter.

Example 9 | Comparing Average Heights

The bar graph on the next page shows the current average heights for male adults in various countries. How much greater is the average male height in Denmark than the average male height in the United States?

(*Continued on next page*)

Practice 8

Find the total monthly cost of owning and operating a certain automobile given the expenses shown.

Monthly car payment:	$536.52
Monthly insurance cost:	$52.68
Average gasoline bill per month:	$87.50

Answer
8. $676.70

✓**Concept Check Answer**
Each number is rounded incorrectly. The estimate is too high.

Practice 9

Use the bar graph in Example 9. How much greater is the average height in the Netherlands than the average height in Israel?

Average Male Adult Height

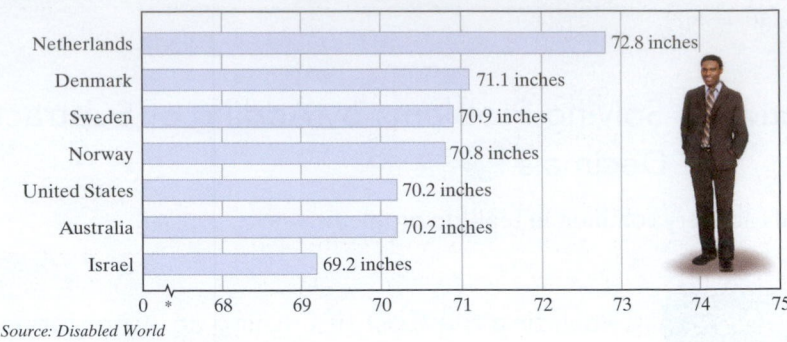

Netherlands — 72.8 inches
Denmark — 71.1 inches
Sweden — 70.9 inches
Norway — 70.8 inches
United States — 70.2 inches
Australia — 70.2 inches
Israel — 69.2 inches

0 * 68 69 70 71 72 73 74 75

Source: Disabled World

* The 〜 means that some numbers are purposefully missing on the axis.

Solution:

1. UNDERSTAND. Read and reread the problem. Since we want to know "how much greater," we subtract.

2. TRANSLATE.

In words:	How much greater	is	Denmark's average male height	minus	United States's average male height
	↓	↓	↓	↓	↓
Translate:	How much greater	=	71.1	−	70.2

3. SOLVE: We also estimate by rounding each number to the nearest whole.

$$
\begin{array}{r}
\overset{0\;11}{7\cancel{1}.\cancel{1}} \\
-\;70.2 \\
\hline
0.9
\end{array}
$$

7 1̸ . 1̸   rounds to   71
− 70 . 2   rounds to   − 70
  0 . 9   Exact   1   Estimate

4. INTERPRET. *Check* your work. Since our estimate is close to our exact answer, 0.9 inch is reasonable. *State* your conclusion: The average male height in Denmark is 0.9 inch greater than the U.S. average male height.

■ **Work Practice 9**

🖩 Calculator Explorations

Entering Decimal Numbers

To enter a decimal number, find the key marked ⸳ .
To enter the number 2.56, for example, press the keys
2 ⸳ 5 6 .

The display will read ⌐ 2.56 .

Operations on Decimal Numbers

Operations on decimal numbers are performed in the same way as operations on whole or signed numbers. For example, to find 8.625 − 4.29, press the keys
8.625 − 4.29 = or ENTER .

The display will read ⌐ 4.335 . (Although entering 8.625, for example, requires pressing more than one key, we group numbers together here for easier reading.)

Use a calculator to perform each indicated operation.

1. 315.782 + 12.96
2. 29.68 + 85.902
3. 6.249 − 1.0076
4. 5.238 − 0.682
5.
```
    12.555
   224.987
     5.2
 + 622.65
```
6.
```
    47.006
     0.17
   313.259
 + 139.088
```

Answer

9. 3.6 in.

Vocabulary, Readiness & Video Check

Use the choices below to fill in each blank. Not all choices will be used.

minuend	vertically	first	true	37.0	horizontally
difference	subtrahend	last	false	0.37	

1. The number 37 equals _____.
2. The decimal point in a whole number is positioned after the _____ digit.
3. In $89.2 - 14.9 = 74.3$, the number 74.3 is called the _____, 89.2 is the _____, and 14.9 is the _____.
4. To add or subtract decimals, we line up the decimal points _____.
5. True or false: The number 5.6 is closer to 5 than 6 on a number line. _____.
6. True or false: The number 10.48 is closer to 10 than 11 on a number line. _____.

Martin-Gay Interactive Videos   *Watch the section lecture video and answer the following questions.*

See Video 4.3

Objective A 7. From Example 2, what does lining up the decimal points also line up? Why is this important?

Objective B 8. From Example 3, where is the decimal point in a whole number?

Objective C 9. In Example 5, estimating is used to check whether the answer to the subtraction problem is reasonable, but what is the best way to fully check?

Objective D 10. Complete this statement based on Example 6: To calculate the amount of border material needed, we are actually calculating the _____ of the triangle.

4.3 Exercise Set MyMathLab®

Objectives A C Mixed Practice *Add. See Examples 1 through 3 and 7. For those exercises marked, also estimate to see if the answer is reasonable.*

1. $1.3 + 2.2$ **2.** $2.5 + 4.1$ **3.** $5.7 + 1.13$ **4.** $2.31 + 6.4$ **5.** $0.003 + 0.091$

6. $0.004 + 0.085$ **7.** $19.23 + 602.782$ **8.** $47.14 + 409.567$ **9.** $490 + 93.09$ **10.** $600 + 83.0062$

11.
$$234.89$$
$$+\ 230.67$$
Exact: Estimate:

12.
$$734.89$$
$$+\ 640.56$$
Exact: Estimate:

13.
$$100.009$$
$$6.08$$
$$+\quad 9.034$$
Exact: Estimate:

14.
$$200.89$$
$$7.49$$
$$+\ \ 62.83$$
Exact: Estimate:

15. $24.6 + 2.39 + 0.0678$

16. $32.4 + 1.58 + 0.0934$

17. Find the sum of 45.023, 3.006, and 8.403

18. Find the sum of 65.0028, 5.0903, and 6.9003

Objectives B C Mixed Practice *Subtract and check. See Examples 4 through 7. For those exercises marked, also estimate to see if the answer is reasonable.*

19. 8.8 − 2.3

20. 7.6 − 2.1

21. 18 − 2.7

22. 28 − 3.3

23.   654.9
         − 56.67

24.   863.23
         − 39.453

25. 5.9 − 4.07
Exact:
Estimate:

26. 6.4 − 3.04
Exact:
Estimate:

27. 923.5 − 61.9

28. 845.93 − 45.8

29. 500.34 − 123.45

30. 600.74 − 463.98

31.   1000
         − 123.4
Exact:

Estimate:

32.   2000
         − 327.47
Exact:

Estimate:

33. 200 − 5.6

34. 800 − 8.9

35. 3 − 0.0012

36. 7 − 0.097

37. Subtract 6.7 from 23.

38. Subtract 9.2 from 45.

Objectives A B Mixed Practice *Perform the indicated operation. See Examples 1 through 6.*

39. 86.05 + 1.978

40. 95.07 + 4.216

41. 86.05 − 1.978

42. 95.07 − 4.216

43. Add 150 and 93.17.

44. Add 250 and 86.07.

45. 150 − 93.17

46. 250 − 86.07

47. Subtract 8.94 from 12.1.

48. Subtract 6.73 from 20.2.

Objective D *Solve. For Exercises 49 and 50, the solutions have been started for you. See Examples 8 and 9.*

49. Ann-Margaret Tober bought a book for $32.48. If she paid with two $20 bills, what was her change?

Start the solution:

1. UNDERSTAND the problem. Reread it as many times as needed.
2. TRANSLATE into an equation. (Fill in the blank.)

change	is	two $20 bills	minus	cost of book
↓	↓	↓	↓	↓

change = 40 − _____

Finish with
3. SOLVE and 4. INTERPRET

50. Phillip Guillot bought a car part for $18.26. If he paid with two $10 bills, what was his change?

Start the solution:

1. UNDERSTAND the problem. Reread it as many times as needed.
2. TRANSLATE into an equation. (Fill in the blank.)

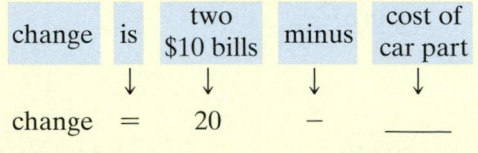

change	is	two $10 bills	minus	cost of car part
↓	↓	↓	↓	↓

change = 20 − _____

Finish with
3. SOLVE and 4. INTERPRET

51. Find the total monthly cost of owning and maintaining a car given the information shown.

Monthly car payment:	$275.36
Monthly insurance cost:	$ 83.00
Average cost of gasoline per month:	$ 81.60
Average maintenance cost per month:	$ 14.75

52. Find the total monthly cost of owning and maintaining a car given the information shown.

Monthly car payment:	$306.42
Monthly insurance cost:	$ 53.50
Average cost of gasoline per month:	$123.00
Average maintenance cost per month:	$ 23.50

53. Gasoline was $2.839 per gallon one week and $2.979 per gallon the next. By how much did the price change?

54. A pair of eyeglasses costs a total of $347.89. The frames of the glasses are $97.23. How much do the lenses of the eyeglasses cost?

55. Find the perimeter.

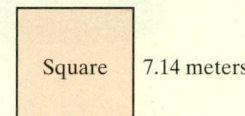

Square   7.14 meters

56. Find the perimeter.

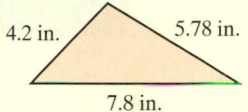

4.2 in.   5.78 in.   7.8 in.

57. The top face of the latest iPod nano measures 3.01 inches by 1.56 inches. Find the perimeter of the rectangular face. (*Source:* Apple Inc.)

58. The top face of the latest iPod touch measures 4.86 inches by 2.31 inches. Find the perimeter of this rectangular face. (*Source:* Apple Inc.)

59. The normal monthly average wind speed for April at the weather station on Mt. Washington in New Hampshire is 34.7 miles per hour. The highest speed ever recorded at the station, in April 1934, was 231.0 miles per hour. How much faster was the highest speed than the average April wind speed? (*Source:* Mount Washington Observatory)

60. In 2012, the average adult American spent 9.49 hours per day on personal care activities (including sleeping) and 5.37 hours per day on leisure and sports. How much more time on average do Americans spend on personal care activities than on leisure and sports? (*Source:* U.S. Bureau of Labor Statistics)

61. The average temperature for the contiguous United States during September 2013 was 67.3 degrees Fahrenheit. This is 2.5 degrees Fahrenheit above the 20th-century average temperature for September. What is the United States' 20th-century average temperature for September? (*Source:* National Climatic Data Center)

62. Historically, the average rainfall for the month of May in Omaha, Nebraska, is 4.79 inches. In May 2013, Omaha received 6.30 inches of rain. By how much was Omaha's May rainfall above average? (*Source:* National Oceanic and Atmospheric Administration)

63. Andy Green still holds the record for one-mile land speed. This record was 129.567 miles per hour faster than a previous record of 633.468 set in 1983. What was Green's record-setting speed? (*Source:* United States Auto Club; this record was made in October 1997)

64. It costs $5.80 to send a 2-pound package locally via Priority Mail at a U.S. Post Office. To send the same package as Express Mail, it costs $18.10. How much more does it cost to send a package as Express Mail? (*Source:* USPS)

65. The face of the Apple iPhone 5c is a rectangle measuring 4.90 inches by 2.33 inches. Find the perimeter of this rectangular phone. (*Source:* Apple Inc.)

66. The Samsung Galaxy S4, an Android-based smart-phone and a leading iPhone competitor, measures 5.38 inches by 2.75 inches. Find the perimeter of this rectangular phone. (*Source:* based on data from Samsung Electronics Co.)

67. The average U.S. movie theater ticket price in 2012 was $7.96. In 2011, it was $7.93. Find the increase in average movie theater ticket price from 2011 to 2012. (*Source:* MPAA)

68. The average U.S. movie theater ticket price in 2003 was $6.03. For 2013, it was estimated to be $8.05. Find the increase in average movie theater ticket price for this 10-year period. (*Source:* MPAA and Box Office Mojo)

This bar graphs shows the average number of text messages sent each month in the United States for the years shown. Use this graph to answer Exercises 69 and 70. (Source: CTIA — The Wireless Association)

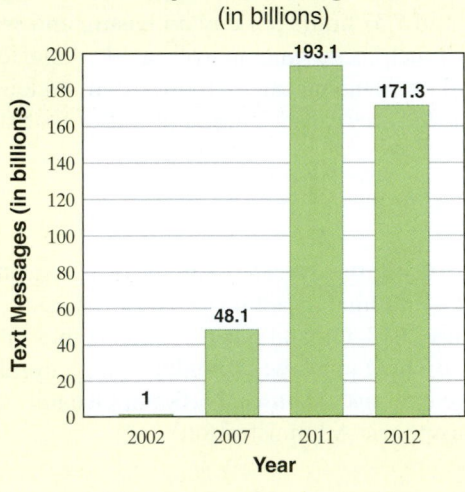

Monthly Text Messages in U.S.
(in billions)

69. Find the decrease in monthly text messages from 2011 to 2012.

70. Find the increase in monthly text messages from 2002 to 2007.

71. The snowiest city in the United States is Valdez, Alaska, which receives an average of 85.7 more inches of snow than the second snowiest city. The second snowiest city in the United States is Blue Canyon, California. Blue Canyon receives an average of 240.3 inches of snow annually. How much snow does Valdez receive on average each year? (*Source:* National Climatic Data Center)

72. The driest city in the world is Aswan, Egypt, which receives an average of only 0.02 inch of rain per year. Yuma, Arizona, is the driest city in the United States. Yuma receives an average of 2.63 more inches of rain each year than Aswan. What is the average annual rainfall in Yuma? (*Source:* National Climatic Data Center)

73. A landscape architect is planning a border for a flower garden shaped like a triangle. The sides of the garden measure 12.4 feet, 29.34 feet, and 25.7 feet. Find the amount of border material needed.

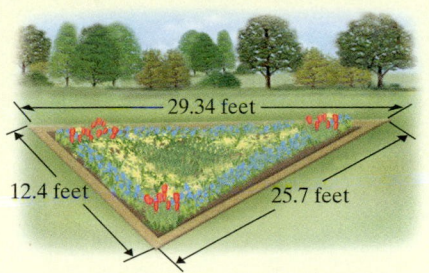

29.34 feet

12.4 feet   25.7 feet

74. A contractor purchased enough railing to completely enclose the newly built deck shown below. Find the amount of railing purchased.

15.7 feet

10.6 feet

The table shows the average retail price of a gallon of gasoline (all grades and formulations) in the United States in each of the years shown. Use this table to answer Exercises 75 and 76. (Source: Energy Information Administration)

Year	Gasoline Price (dollars per gallon)
2008	3.317
2009	2.401
2010	2.836
2011	3.577
2012	3.695

75. How much more was the average cost of a gallon of gasoline in 2012 than in 2009?

76. How much less was the average cost of a gallon of gasoline in 2010 than in 2008?

The following table shows spaceflight information for astronaut James A. Lovell. Use this table to answer Exercises 77 and 78.

Spaceflights of James A. Lovell		
Year	Mission	Duration (in hours)
1965	Gemini 6	330.583
1966	Gemini 12	94.567
1968	Apollo 8	147.0
1970	Apollo 13	142.9

(*Source:* NASA)

77. Find the total time spent in spaceflight by astronaut James A. Lovell.

78. Find the total time James A. Lovell spent in spaceflight on all Apollo missions.

The bar graph shows the top five chocolate-consuming nations in the world in 2012. Use this table to answer Exercises 79 through 84.

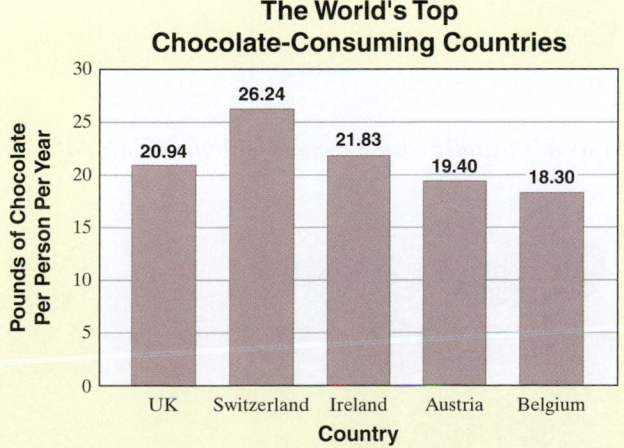

The World's Top Chocolate-Consuming Countries

Pounds of Chocolate Per Person Per Year

UK 20.94, Switzerland 26.24, Ireland 21.83, Austria 19.40, Belgium 18.30

Country

Source: Based on data from confectionarynews.com

79. Which country in the bar graph has the greatest chocolate consumption per person?

80. Which country in the bar graph has the least chocolate consumption per person?

81. How much more is the greatest chocolate consumption than the least chocolate consumption shown in the bar graph?

82. How much more chocolate does the average Irish citizen consume per year than the average Austrian?

83. Make a table listing the countries and their corresponding chocolate consumptions in order from greatest to least.

84. Find the sum of the five bar heights shown in the graph. What type of company might be interested in this sum?

Review

Multiply. See Sections 1.6 and 3.5.

85. $23 \cdot 2$

86. $46 \cdot 3$

87. $43 \cdot 90$

88. $30 \cdot 32$

89. $\left(\dfrac{2}{3}\right)^2$

90. $\left(\dfrac{1}{5}\right)^3$

Concept Extensions

A friend asks you to check his calculations for Exercises 91 and 92. Are they correct? If not, explain your friend's errors and correct the calculations. See the first Concept Check in this section.

91.
$$\begin{array}{r} \overset{1}{9}.2 \\ \overset{1}{8}.63 \\ +\ 4.005 \\ \hline 4.960 \end{array}$$

92.
$$\begin{array}{r} \overset{8\ 9\ 9\ 9}{9\,0\,0.0} \\ -\ 96.4 \\ \hline 8\,03.5 \end{array}$$

Find the unknown length in each figure.

△ **93.**

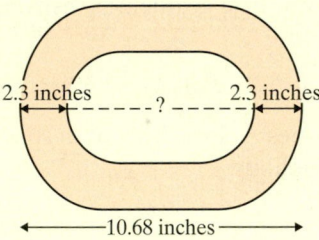

2.3 inches   ?   2.3 inches

10.68 inches

△ **94.**

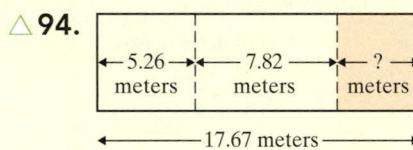

←5.26→ ←7.82→ ←?→
meters   meters   meters

←——— 17.67 meters ———→

Let's review the values of these common U.S. coins in order to answer the following exercises.

Penny Nickel Dime Quarter

$0.01 $0.05 $0.10 $0.25

For Exercises 95 and 96, write the value of each group of coins. To do so, it is usually easiest to start with the coin(s) of greatest value and end with the coin(s) of least value.

95.

96.

97. Name the different ways that coins can have a value of $0.17 given that you may use no more than 10 coins.

98. Name the different ways that coin(s) can have a value of $0.25 given that there are no pennies.

99. Why shouldn't the sum
$$82.95 + 51.26$$
be estimated as $90 + 60 = 150$?
See the second Concept Check in this section.

100. Laser beams can be used to measure the distance to the moon. One measurement showed the distance to the moon to be 256,435.235 miles. A later measurement showed that the distance is 256,436.012 miles. Find how much farther away the moon is in the second measurement as compared to the first.

101. Explain how adding or subtracting decimals is similar to adding or subtracting whole numbers.

102. Explain how adding or subtracting decimals is different from adding or subtracting whole numbers.

4.4 Multiplying Decimals and Circumference of a Circle ▶

Objective A Multiplying Decimals ▶

Multiplying decimals is similar to multiplying whole numbers. The only difference is that we place a decimal point in the product. To discover where a decimal point is placed in the product, let's multiply 0.6×0.03. We first write each decimal as an equivalent fraction and then multiply.

$$\underset{\substack{\uparrow \\ \text{1 decimal} \\ \text{place}}}{0.6} \quad \times \quad \underset{\substack{\uparrow \\ \text{2 decimal} \\ \text{places}}}{0.03} \quad = \frac{6}{10} \times \frac{3}{100} = \frac{18}{1000} = \underset{\substack{\uparrow \\ \text{3 decimal} \\ \text{places}}}{0.018}$$

Notice that $1 + 2 = 3$, the number of decimal places in the product. Now let's multiply 0.03×0.002.

$$\underset{\substack{\uparrow \\ \text{2 decimal} \\ \text{places}}}{0.03} \quad \times \quad \underset{\substack{\uparrow \\ \text{3 decimal} \\ \text{places}}}{0.002} \quad = \frac{3}{100} \times \frac{2}{1000} = \frac{6}{100,000} = \underset{\substack{\uparrow \\ \text{5 decimal} \\ \text{places}}}{0.00006}$$

Again, we see that $2 + 3 = 5$, the number of decimal places in the product.

Instead of writing decimals as fractions each time we want to multiply, we notice a pattern from these examples and state a rule that we can use:

Objectives

A Multiply Decimals. ▶

B Estimate When Multiplying Decimals. ▶

C Multiply by Powers of 10. ▶

D Find the Circumference of a Circle. ▶

E Solve Problems by Multiplying Decimals. ▶

Multiplying Decimals

Step 1: Multiply the decimals as though they are whole numbers.

Step 2: The decimal point in the product is placed so that the number of decimal places in the product is equal to the *sum* of the number of decimal places in the factors.

Practice 1

Multiply: 45.9×0.42

Example 1   Multiply: 23.6×0.78

Solution:

$$
\begin{array}{r}
23.6 \\
\times\ 0.78 \\
\hline
1888 \\
16520 \\
\hline
18.408
\end{array}
$$

23.6   1 decimal place
× 0.78   2 decimal places

Since $1 + 2 = 3$, insert the decimal point in the product so that there are 3 decimal places.

■ Work Practice 1

Practice 2

Multiply: 0.112×0.6

Example 2   Multiply: 0.283×0.3

Solution:

$$
\begin{array}{r}
0.283 \\
\times\ \ \ 0.3 \\
\hline
0.0849
\end{array}
$$

0.283   3 decimal places
×   0.3   1 decimal place

Since $3 + 1 = 4$, insert the decimal point in the product so that there are 4 decimal places.

Insert one 0 since the product must have 4 decimal places.

■ Work Practice 2

Practice 3

Multiply: 0.0721×48

Example 3   Multiply: 0.0531×16

Solution:

$$
\begin{array}{r}
0.0531 \\
\times\ \ \ \ 16 \\
\hline
3186 \\
5310 \\
\hline
0.8496
\end{array}
$$

0.0531   4 decimal places
×   16   0 decimal places

4 decimal places $(4 + 0 = 4)$

■ Work Practice 3

✓**Concept Check** True or false? The number of decimal places in the product of 0.261 and 0.78 is 6. Explain.

Objective B Estimating When Multiplying Decimals

Just as for addition and subtraction, we can estimate when multiplying decimals to check the reasonableness of our answer.

Practice 4

Multiply: 30.26×2.98. Then estimate to see whether the answer is reasonable.

Example 4   Multiply: 28.06×1.95. Then estimate to see whether the answer is reasonable by rounding each factor, then multiplying the rounded numbers.

Solution:

Exact:	Estimate 1	Estimate 2
28.06	28 Rounded to ones	30 Rounded to tens
× 1.95	× 2	× 2
14030	56	60
252540		
280600		
54.7170		

The answer 54.7170 (or 54.717) is reasonable.

■ Work Practice 4

As shown in Example 4, estimated results will vary depending on what estimates are used. Notice that estimating results is a good way to see whether the decimal point has been correctly placed.

Objective C Multiplying by Powers of 10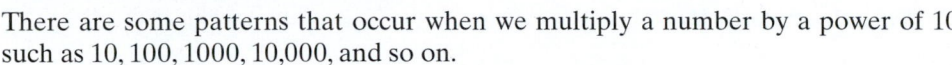

There are some patterns that occur when we multiply a number by a power of 10 such as 10, 100, 1000, 10,000, and so on.

$23.6951 \times 10 = 236.951$ Move the decimal point *1 place* to the *right*.

     1 zero

$23.6951 \times 100 = 2369.51$ Move the decimal point *2 places* to the *right*.

     2 zeros

$23.6951 \times 100,000 = 2,369,510.$ Move the decimal point *5 places* to the *right* (insert a 0).

     5 zeros

Notice that we move the decimal point the same number of places as there are zeros in the power of 10.

> **Multiplying Decimals by Powers of 10 such as 10, 100, 1000, 10,000...**
>
> Move the decimal point to the *right* the same number of places as there are *zeros* in the power of 10.

Examples   Multiply.

5. $7.68 \times 10 = 76.8$ 7.68

6. $23.702 \times 100 = 2370.2$ 23.702

7. $76.3 \times 1000 = 76,300$ 76.300

■ **Work Practice 5–7**

There are also powers of 10 that are less than 1. The decimals 0.1, 0.01, 0.001, 0.0001, and so on are examples of powers of 10 less than 1. Notice the pattern when we multiply by these powers of 10:

$569.2 \times 0.1 = 56.92$ Move the decimal point *1 place* to the *left*.

     1 decimal place

$569.2 \times 0.01 = 5.692$ Move the decimal point *2 places* to the *left*.

     2 decimal places

$569.2 \times 0.0001 = 0.05692$ Move the decimal point *4 places* to the *left* (insert one 0).

     4 decimal places

> **Multiplying Decimals by Powers of 10 such as 0.1, 0.01, 0.001, 0.0001...**
>
> Move the decimal point to the *left* the same number of places as there are *decimal places* in the power of 10.

Practice 5–7

Multiply.

5. 23.7×10

6. 203.004×100

7. 1.15×1000

Answers

5. 237   **6.** 20,300.4   **7.** 1150

Practice 8–10

Multiply.

8. 7.62×0.1
9. 1.9×0.01
10. 7682×0.001

Examples Multiply.

8. $42.1 \times 0.1 = 4.21$ 42.1
9. $76,805 \times 0.01 = 768.05$ 76,805.
10. $9.2 \times 0.001 = 0.0092$ 0009.2

■ Work Practice 8–10

Many times we see large numbers written, for example, in the form 451.8 million rather than in the longer standard notation. The next example shows us how to interpret these numbers.

Practice 11

In 2020, the population of the United States is projected to be 333.9 million. Write this number in standard notation. (*Source:* U.S. Census Bureau)

Example 11 In 2050, the population of the United States is projected to be 399.8 million. Write this number in standard notation. (*Source:* U.S. Census Bureau)

399.8 million in 2050

Solution: 399.8 million $= 399.8 \times 1$ million
$= 399.8 \times 1,000,000 = 399,800,000$

■ Work Practice 11

Objective D Finding the Circumference of a Circle

Recall from Section 1.3 that the distance around a polygon is called its perimeter. The distance around a circle is given the special name **circumference,** and this distance depends on the radius or the diameter of the circle.

Circumference of a Circle

Radius / Diameter

r / d

Circumference $= 2 \cdot \pi \cdot$ **radius**   or   **Circumference** $= \pi \cdot$ **diameter**

In Section 4.2, we learned about the symbol π as the Greek letter pi, pronounced "pie." It is a constant between 3 and 4.

Approximations for π

Two common approximations for π are:

$\pi \approx 3.14$ or $\pi \approx \dfrac{22}{7}$

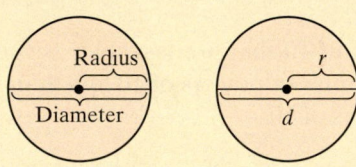
a decimal approximation   a fraction approximation

Answers

8. 0.762 **9.** 0.019 **10.** 7.682
11. 333,900,000

Example 12 Circumference of a Circle

Find the circumference of a circle whose radius is 5 inches. Then use the approximation 3.14 for π to approximate the circumference.

Solution: Circumference $= 2 \cdot \pi \cdot \text{radius}$

$$= 2 \cdot \pi \cdot 5 \text{ inches}$$
$$= 10\pi \text{ inches}$$

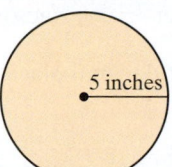

5 inches

Next, we replace π with the approximation 3.14.

$$\text{Circumference} = 10\pi \text{ inches}$$
$(\text{"is approximately"}) \rightarrow \approx 10(3.14) \text{ inches}$
$$= 31.4 \text{ inches}$$

The *exact* circumference or distance around the circle is 10π inches, which is *approximately* 31.4 inches.

■ Work Practice 12

Practice 12

Find the circumference of a circle whose radius is 11 meters. Then use the approximation 3.14 for π to approximate this circumference.

Objective E Solving Problems by Multiplying Decimals

The solutions to many real-life problems are found by multiplying decimals. We continue using our four problem-solving steps to solve such problems.

Example 13 Finding the Total Cost of Materials for a Job

A college student is hired to paint a billboard with paint costing $2.49 per quart. If the job requires 3 quarts of paint, what is the total cost of the paint?

Solution:

1. UNDERSTAND. Read and reread the problem. The phrase "total cost" might make us think addition, but since this problem requires repeated addition, let's multiply.
2. TRANSLATE.

In words:	Total cost	is	cost per quart of paint	times	number of quarts
	↓	↓	↓	↓	↓
Translate:	Total cost	=	2.49	×	3

3. SOLVE. We can estimate to check our calculations. The number 2.49 rounds to 2 and 2 × 3 = 6.

$$\begin{array}{r} \overset{1\ 2}{2.49} \\ \times \quad 3 \\ \hline 7.47 \end{array}$$

4. INTERPRET. *Check* your work. Since 7.47 is close to our estimate of 6, our answer is reasonable. *State* your conclusion: The total cost of the paint is $7.47.

■ Work Practice 13

Practice 13

A biology major is fertilizing her garden. She uses 5.6 ounces of fertilizer per square yard. The garden measures 60.5 square yards. How much fertilizer does she need?

Vocabulary, Readiness & Video Check

Use the choices below to fill in each blank.

circumference	left	sum	zeros
decimal places	right	product	factor

1. When multiplying decimals, the number of decimal places in the product is equal to the _____ of the number of decimal places in the factors.

2. In $8.6 \times 5 = 43$, the number 43 is called the _____, while 8.6 and 5 are each called a(n) _____.

3. When multiplying a decimal number by powers of 10, such as 10, 100, 1000, and so on, we move the decimal point in the number to the _____ the same number of places as there are _____ in the power of 10.

4. When multiplying a decimal number by powers of 10, such as 0.1, 0.01, and so on, we move the decimal point in the number to the _____ the same number of places as there are _____ in the power of 10.

5. The distance around a circle is called its _____.

Do not multiply. Just give the number of decimal places in the product. See the Concept Check in this section.

6.	0.46	7.	57.9	8.	0.428	9.	0.0073	10.	0.028	11.	5.1296
	$\times\, 0.81$		$\times\, 0.36$		$\times\;\; 0.2$		$\times\quad 21$		$\times\, 1.36$		$\times\, 7.3987$

Martin-Gay Interactive Videos

See Video 4.4

Watch the section lecture video and answer the following questions.

Objective A 12. From ▱ Example 1, explain the difference between multiplying whole numbers and multiplying decimal numbers. ▷

Objective B 13. From ▱ Example 2, what does estimating especially help us with? ▷

Objective C 14. In ▱ Example 3, why don't we multiply as we did in ▱ Example 2? ▷

Objective D 15. Why is 25.12 meters not the exact answer to ▱ Example 5? ▷

Objective E 16. In ▱ Example 6, why is 24.8 not the complete answer? What is the complete answer? ▷

4.4 Exercise Set MyMathLab®

Objectives A B Mixed Practice *Multiply. See Examples 1 through 4. For those exercises marked, also estimate to see if the answer is reasonable.*

1.	0.2	2.	0.7	▷3.	1.2	4.	6.8
	$\times 0.6$		$\times 0.9$		$\times 0.5$		$\times 0.3$

5. 0.26×5 6. 0.19×6

7. 5.3×4.2
   Exact:
   Estimate:

8. 6.2×3.8
   Exact:
   Estimate:

9.	0.576	10.	0.971	▷11.	1.0047	12.	2.0005
	$\times\;\; 0.7$		$\times\;\; 0.5$		$\times\quad 8.2$		$\times\quad 5.5$

11. Exact: Estimate:

12. Exact: Estimate:

13. 490.2
       $\times 0.023$

14. 300.9
       $\times 0.032$

15. Multiply 16.003 and 5.31

16. Multiply 31.006 and 3.71

Objective C *Multiply. See Examples 5 through 10.*

17. 6.5×10

18. 7.2×100

19. 6.5×0.1

20. 4.7×0.1

21. 7.2×0.01

22. 0.06×0.01

▶**23.** 7.093×100

24. 0.5×100

25. 6.046×1000

26. 9.1×1000

▶**27.** 37.62×0.001

28. 14.3×0.001

Objectives A B C Mixed Practice *Multiply. See Examples 1 through 10.*

29. 0.123×0.4

30. 0.216×0.3

31. 0.123×100

32. 0.216×100

33. 8.6×0.15

34. 0.42×5.7

35. 9.6×0.01

36. 5.7×0.01

37. 562.3×0.001

38. 993.5×0.001

39.   5.62
        $\times 7.7$

40.   8.03
        $\times 5.5$

Write each number in standard notation. See Example 11.

41. The storage silos at the main Hershey chocolate factory in Hershey, Pennsylvania, can hold enough cocoa beans to make 5.5 billion Hershey's milk chocolate bars. (*Source:* Hershey Foods Corporation)

42. The total domestic revenue collected by Netflix in 2012 was $3.322 billion. (*Source:* Netflix, Inc.)

43. The Racer is the most-ridden roller coaster at King's Island, an amusement park near Cincinnati, Ohio. Since 1972, it has given more than 97.8 million rides. (*Source:* Cedar Fair, L.P.)

44. About 45.6 million American households own at least one dog. (*Source:* American Pet Products Association)

45. The most-visited national park in the United States in 2012 was the Blue Ridge Parkway in Virginia and North Carolina. An estimated 292 thousand people visited the park each week that year. (*Source:* National Park Service)

46. In 2012, approximately 17.2 thousand vessels passed through the Suez Canal. (*Source:* suezcanal. gov)

Objective D *Find the circumference of each circle. Then use the approximation 3.14 for π and approximate each circumference. See Example 12.*

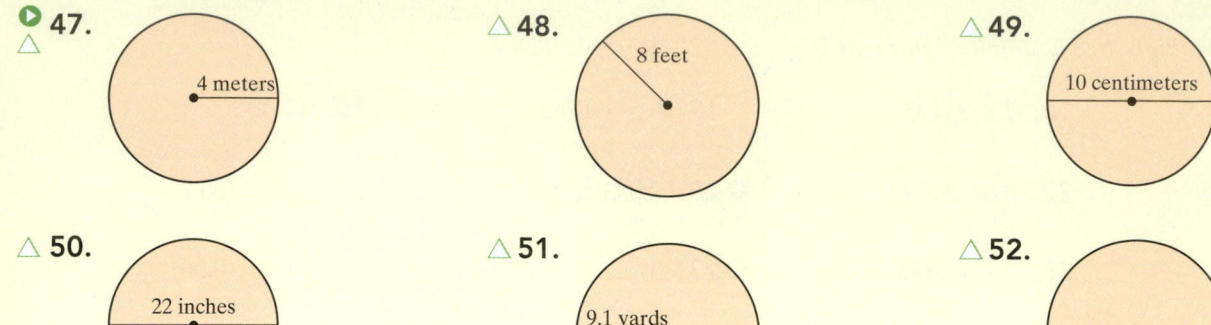

47. 4 meters

△ 48. 8 feet

△ 49. 10 centimeters

△ 50. 22 inches

△ 51. 9.1 yards

△ 52. 5.9 kilometers

Objectives D E **Mixed Practice** *Solve. For Exercises 53 and 54, the solutions have been started for you. See Examples 12 and 13. For circumference applications, find the exact circumference and then use 3.14 for π to approximate the circumference.*

53. An electrician for Central Power and Light worked 40 hours last week. Calculate his pay before taxes for last week if his hourly wage is $17.88.

Start the solution:

1. UNDERSTAND the problem. Reread it as many times as needed.
2. TRANSLATE into an equation. (Fill in the blanks.)

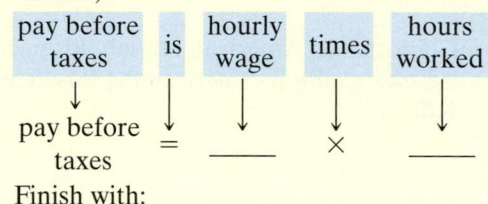

| pay before taxes | is | hourly wage | times | hours worked |

pay before taxes = _____ × _____

Finish with:
3. SOLVE and 4. INTERPRET.

54. An assembly line worker worked 20 hours last week. Her hourly rate is $19.52 per hour. Calculate her pay before taxes.

Start the solution:

1. UNDERSTAND the problem. Reread it as many times as needed.
2. TRANSLATE into an equation. (Fill in the blanks.)

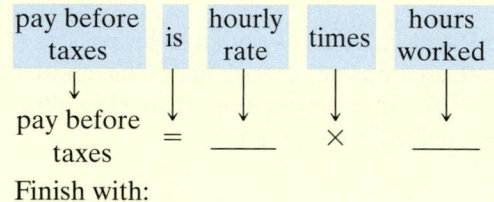

| pay before taxes | is | hourly rate | times | hours worked |

pay before taxes = _____ × _____

Finish with:
3. SOLVE and 4. INTERPRET.

55. Under certain conditions, the average cost of driving a medium sedan in 2012 was $0.61 per mile. How much would it have cost to drive such a car 15,000 miles in 2012? (*Source:* American Automobile Association)

56. In September 2013, a U.S. airline passenger paid an average of $0.1643, disregarding taxes and fees, to fly 1 mile. Use this number to calculate the cost before taxes and fees to fly from Atlanta, Georgia, to Minneapolis, Minnesota, a distance of 905 miles. Round to the nearest cent. (*Source:* Airlines for America)

57. A 1-ounce serving of cream cheese contains 6.2 grams of saturated fat. How much saturated fat is in 4 ounces of cream cheese? (*Source: Home and Garden Bulletin No. 72;* U.S. Department of Agriculture)

58. A 3.5-ounce serving of lobster meat contains 0.1 gram of saturated fat. How much saturated fat do 3 servings of lobster meat contain? (*Source:* The National Institutes of Health)

△ **59.** Recall that the face of the Apple iPhone 5c (see Section 4.3) is a rectangle measuring 4.90 inches by 2.33 inches. Find the area of the face of the Apple iPhone 5c. (*Source:* Apple Inc.)

△ **60.** Recall that the rectangular face of the Samsung Galaxy S4 smartphone (see Section 4.3) measures 5.38 inches by 2.75 inches. Find the area of the face of the Samsung Galaxy S4. (*Source:* Based on data from Samsung Electronics Co.)

△ **61.** In 1893, the first ride called a Ferris wheel was constructed by Washington Gale Ferris. Its diameter was 250 feet. Find its circumference. Give an exact answer and an approximation using 3.14 for π. (*Source: The Handy Science Answer Book,* Visible Ink Press, 1994)

△ **62.** The radius of Earth is approximately 3950 miles. Find the distance around Earth at the equator. Give an exact answer and an approximation using 3.14 for π. (*Hint:* Find the circumference of a circle with radius 3950 miles.)

△ **63.** The London Eye, built for the Millennium celebration in London, resembles a gigantic Ferris wheel with a diameter of 135 meters. If Adam Hawn rides the Eye for one revolution, find how far he travels. Give an exact answer and an approximation using 3.14 for π. (*Source:* Londoneye.com)

△ **64.** The world's longest suspension bridge is the Akashi Kaikyo Bridge in Japan. This bridge has two circular caissons, which are underwater foundations. If the diameter of a caisson is 80 meters, find its circumference. Give an exact answer and an approximation using 3.14 for π. (*Source: Scientific American;* How Things Work Today)

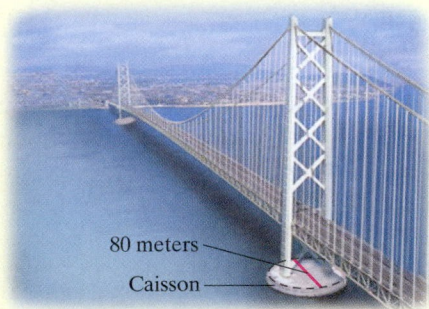

80 meters

Caisson

65. A meter is a unit of length in the metric system that is approximately equal to 39.37 inches. Sophia Wagner is 1.65 meters tall. Find her approximate height in inches.

66. The doorway to a room is 2.15 meters tall. Approximate this height in inches. (*Hint:* See Exercise 65.)

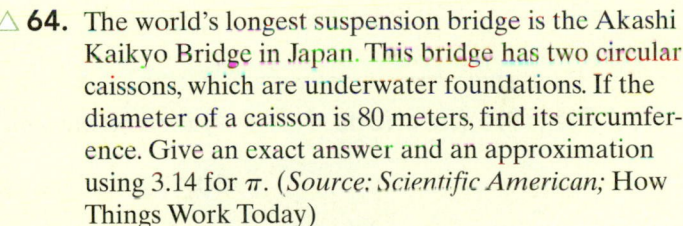

△ **67. a.** Approximate the circumference of each circle.

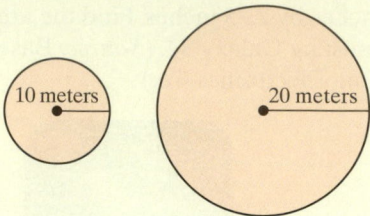

10 meters   20 meters

b. If the radius of a circle is doubled, is its corresponding circumference doubled?

△ **68. a.** Approximate the circumference of each circle.

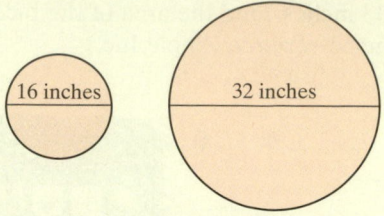

16 inches   32 inches

b. If the diameter of a circle is doubled, is its corresponding circumference doubled?

69. Recall that the top face of the latest iPod nano (see Section 4.3) measures 3.01 inches by 1.56 inches. Find the area of the face of the iPod nano. Round to the nearest hundredth. (*Source:* Apple Inc.)

70. Recall that the top face of the latest iPod touch (see Section 4.3) measures 4.86 inches by 2.31 inches. Find the area of the face of the iPod touch. Round to the nearest hundredth. (*Source:* Apple Inc.)

Review

Divide. See Sections 1.7 and 2.5.

71. $130 \div 5$

72. $486 \div 27$

73. $2016 \div 56$

74. $1863 \div 69$

75. $2920 \div 365$

76. $2916 \div 6$

77. $\dfrac{24}{7} \div \dfrac{8}{21}$

78. $\dfrac{162}{25} \div \dfrac{9}{75}$

Concept Extensions

Mixed Practice (Sections 4.3, 4.4) *Perform the indicated operations.*

79. $3.6 + 0.04$

80. $7.2 + 0.14 + 98.6$

81. $3.6 - 0.04$

82. $100 - 48.6$

83. 0.221×0.5

84. 3.6×0.04

85. Find how far radio waves travel in 20.6 seconds. (Radio waves travel at a speed of $1.86 \times 100{,}000$ miles per second.)

86. If it takes radio waves approximately 8.3 minutes to travel from the Sun to the Earth, find approximately how far it is from the Sun to the Earth. (*Hint:* See Exercise 85.)

87. In your own words, explain how to find the number of decimal places in a product of decimal numbers.

88. In your own words, explain how to multiply by a power of 10.

89. Write down two decimal numbers whose product will contain 5 decimal places. Without multiplying, explain how you know your answer is correct.

90. Explain the process for multiplying a decimal number by a power of 10.

Integrated Review

Operations on Decimals

Perform the indicated operations.

1. $1.6 + 0.97$

2. $3.2 + 0.85$

3. $9.8 - 0.9$

4. $10.2 - 6.7$

5. $\begin{array}{r} 0.8 \\ \times\, 0.2 \\ \hline \end{array}$

6. $\begin{array}{r} 0.6 \\ \times\, 0.4 \\ \hline \end{array}$

7. $8 + 2.16 + 0.9$

8. $6 + 3.12 + 0.6$

9. $\begin{array}{r} 9.6 \\ \times\, 0.5 \\ \hline \end{array}$

10. $\begin{array}{r} 8.7 \\ \times\, 0.7 \\ \hline \end{array}$

11. $\begin{array}{r} 123.6 \\ -\ 48.04 \\ \hline \end{array}$

12. $\begin{array}{r} 325.2 \\ -\ 36.08 \\ \hline \end{array}$

13. $25 + 0.026$

14. $0.125 + 44$

15. $100 - 17.3$

16. $300 - 26.1$

17. 2.8×100

18. 1.6×1000

19. $\begin{array}{r} 96.21 \\ 7.028 \\ +\ 121.7 \\ \hline \end{array}$

20. $\begin{array}{r} 0.268 \\ 1.93 \\ +\ 142.881 \\ \hline \end{array}$

21. Find the product of 1.2 and 5.

22. Find the sum of 1.2 and 5.

23. $\begin{array}{r} 12.004 \\ \times\ \ \ \ 2.3 \\ \hline \end{array}$

24. $\begin{array}{r} 28.006 \\ \times\ \ \ \ 5.2 \\ \hline \end{array}$

25. Subtract 4.6 from 10.

26. Subtract 0.26 from 18.

Answers

1. _____
2. _____
3. _____
4. _____
5. _____
6. _____
7. _____
8. _____
9. _____
10. _____
11. _____
12. _____
13. _____
14. _____
15. _____
16. _____
17. _____
18. _____
19. _____
20. _____
21. _____
22. _____
23. _____
24. _____
25. _____
26. _____

27. _____

28. _____

29. _____

30. _____

31. _____

32. _____

33. _____

34. _____

35. _____

36. _____

37. _____

38. _____

39. _____

40. _____

41. _____

42. _____

43. _____

27. 268.19
   + 146.25

28. 860.18
   + 434.85

29. 160 − 43.19

30. 120 − 101.21

31. 15.62 × 10

32. 15.62 + 10

33. 15.62 − 10

34. 117.26 × 2.6

35. 117.26 − 2.6

36. 117.26 + 2.6

37. 0.0072 × 0.06

38. 0.0025 × 0.03

39. 0.0072 + 0.06

40. 0.03 − 0.0025

41. 0.862 × 1000

42. 2.93 × 0.01

43. Estimate the distance in miles between Garden City, Kansas, and Wichita, Kansas, by rounding each given distance to the nearest ten.

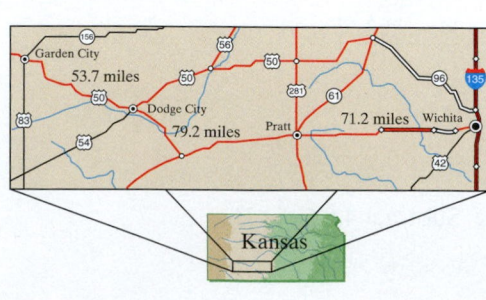

4.5 Dividing Decimals and Order of Operations

Objective A Dividing Decimals

Dividing decimal numbers is similar to dividing whole numbers. The only difference is that we place a decimal point in the quotient. If the divisor is a whole number, we place the decimal point in the quotient directly above the decimal point in the dividend, and then divide as with whole numbers. Recall that division can be checked by multiplication.

> **Dividing by a Whole Number**
>
> **Step 1:** Place the decimal point in the quotient directly above the decimal point in the dividend.
>
> **Step 2:** Divide as with whole numbers.

Example 1 Divide: $270.2 \div 7$. Check your answer.

Solution: We divide as usual. The decimal point in the quotient is directly above the decimal point in the dividend.

Write the decimal point.

$$
\begin{array}{r}
38.6 \leftarrow \text{quotient} \\
\text{divisor} \rightarrow 7\overline{)270.2} \leftarrow \text{dividend} \\
-21 \\
\hline
60 \\
-56 \\
\hline
4\,2 \\
-4\,2 \\
\hline
0
\end{array}
$$

Check:
$$
\begin{array}{r}
\overset{6\,4}{38.6} \leftarrow \text{quotient} \\
\times \quad 7 \leftarrow \text{divisor} \\
\hline
270.2 \leftarrow \text{dividend}
\end{array}
$$

The quotient is 38.6.

■ **Work Practice 1**

Example 2 Divide: $32\overline{)8.32}$

Solution: We divide as usual. The decimal point in the quotient is directly above the decimal point in the dividend.

$$
\begin{array}{r}
0.26 \leftarrow \text{quotient} \\
\text{divisor} \rightarrow 32\overline{)8.32} \leftarrow \text{dividend} \\
-64 \\
\hline
192 \\
-192 \\
\hline
0
\end{array}
$$

Check:
$$
\begin{array}{r}
0.26 \quad \text{quotient} \\
\times \ 32 \quad \text{divisor} \\
\hline
52 \\
7\,80 \\
\hline
8.32 \quad \text{dividend}
\end{array}
$$

■ **Work Practice 2**

Sometimes to continue dividing we need to insert zeros after the last digit in the dividend.

Objectives

A Divide Decimals.

B Estimate When Dividing Decimals.

C Divide Decimals by Powers of 10.

D Solve Problems by Dividing Decimals.

E Review Order of Operations by Simplifying Expressions Containing Decimals.

Practice 1

Divide: $370.4 \div 8$. Check your answer.

Practice 2

Divide: $48\overline{)34.08}$. Check your answer.

Answers
1. 46.3 **2.** 0.71

Practice 3

Divide and check.

a. $0.4 \div 8$

b. $13.62 \div 12$

Example 3   Divide and check: $0.5 \div 4$.

Solution:

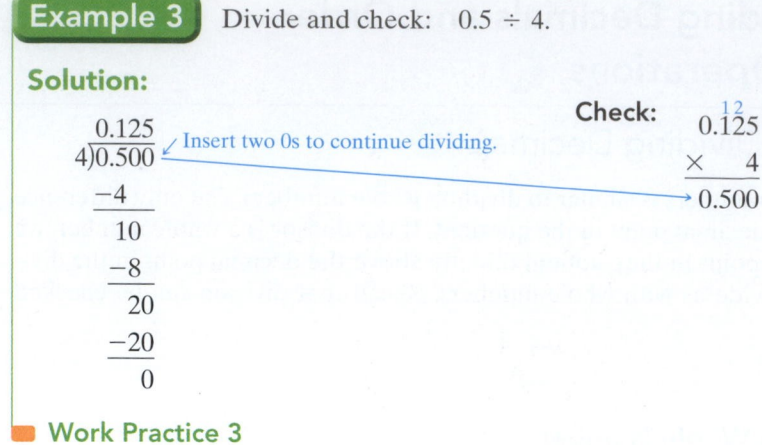

$$
\begin{array}{r}
0.125 \\
4\overline{)0.500} \\
-4 \\
\hline
10 \\
-8 \\
\hline
20 \\
-20 \\
\hline
0
\end{array}
$$

Insert two 0s to continue dividing.

Check:
$$
\begin{array}{r}
\overset{1\,2}{0.125} \\
\times \quad 4 \\
\hline
0.500
\end{array}
$$

■ **Work Practice 3**

If the divisor is not a whole number, before we divide we need to move the decimal point to the right until the divisor is a whole number.

$$1.5\overline{)64.85}$$

divisor ⌐   ⌐ dividend

To understand how this works, let's rewrite

$$1.5\overline{)64.85} \quad \text{as} \quad \frac{64.85}{1.5}$$

and then multiply by 1 in the form of $\frac{10}{10}$. We use the form $\frac{10}{10}$ so that the denominator (divisor) becomes a whole number.

$$\frac{64.85}{1.5} = \frac{64.85}{1.5} \cdot 1 = \frac{64.85}{1.5} \cdot \frac{10}{10} = \frac{64.85 \cdot 10}{1.5 \cdot 10} = \frac{648.5}{15},$$

which can be written as $15.\overline{)648.5}$. Notice that

$$1.5\overline{)64.85} \quad \text{is equivalent to} \quad 15.\overline{)648.5}$$

The decimal points in the dividend and the divisor were both moved one place to the right, and the divisor is now a whole number. This procedure is summarized next:

Dividing by a Decimal

Step 1: Move the decimal point in the divisor to the right until the divisor is a whole number.

Step 2: Move the decimal point in the dividend to the right the *same number of places* as the decimal point was moved in Step 1.

Step 3: Divide. Place the decimal point in the quotient directly over the moved decimal point in the dividend.

Answers

3. a. 0.05   **b.** 1.135

Example 4   Divide: $10.764 \div 2.3$

Solution: We move the decimal points in the divisor and the dividend one place to the right so that the divisor is a whole number.

$2.3\overline{)10.764}$ becomes

$$
\begin{array}{r}
4.68 \\
23.\overline{)107.64} \\
-92 \\
\hline
15\ 6 \\
-13\ 8 \\
\hline
1\ 84 \\
-1\ 84 \\
\hline
0
\end{array}
$$

■ **Work Practice 4**

Example 5   Divide: $5.264 \div 0.32$

Solution:

$0.32\overline{)5.264}$ becomes

$$
\begin{array}{r}
16.45 \\
32\overline{)526.40} \quad \text{Insert one 0.} \\
-32 \\
\hline
206 \\
-192 \\
\hline
14\ 4 \\
-12\ 8 \\
\hline
1\ 60 \\
-1\ 60 \\
\hline
0
\end{array}
$$

■ **Work Practice 5**

✓**Concept Check** Is it always true that the number of decimal places in a quotient equals the sum of the decimal places in the dividend and divisor?

Example 6   Divide: $17.5 \div 0.48$. Round the quotient to the nearest hundredth.

Solution: First we move the decimal points in the divisor and the dividend two places. Then we divide and round the quotient to the nearest hundredth.

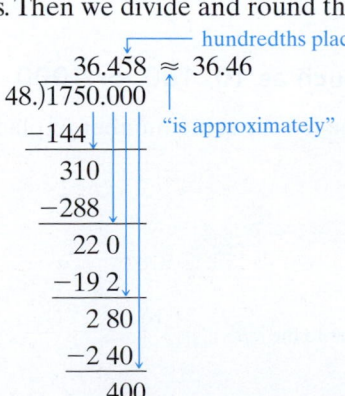

hundredths place

$36.458 \approx 36.46$

"is approximately"

When rounding to the nearest hundredth, carry the division process out to one more decimal place, the thousandths place.

■ **Work Practice 6**

Practice 4

Divide: $166.88 \div 5.6$

Practice 5

Divide: $1.976 \div 0.16$

Practice 6

Divide $23.4 \div 0.57$. Round the quotient to the nearest hundredth.

Answers
4. 29.8 **5.** 12.35 **6.** 41.05

✓**Concept Check Answer**
no

Objective B Estimating When Dividing Decimals

Just as for addition, subtraction, and multiplication of decimals, we can estimate when dividing decimals to check the reasonableness of our answer.

Example 7 Divide: 272.356 ÷ 28.4. Then estimate to see whether the proposed result is reasonable.

Solution:

Exact:	Estimate 1		Estimate 2

$$
\begin{array}{r}
9.59 \\
284.\overline{)2723.56} \\
-2556 \\
\hline
167\,5 \\
-142\,0 \\
\hline
25\,56 \\
-25\,56 \\
\hline
0
\end{array}
$$

Estimate 1:
$$
\begin{array}{r}
9 \\
30\overline{)270}
\end{array}
$$

or

Estimate 2:
$$
\begin{array}{r}
10 \\
30\overline{)300}
\end{array}
$$

The estimate is 9 or 10, so 9.59 is reasonable.

■ **Work Practice 7**

✓**Concept Check** If a quotient is to be rounded to the nearest thousandth, to what place should the division be carried out? (Assume that the division carries out to your answer.)

Objective C Dividing Decimals by Powers of 10

As with multiplication, there are patterns that occur when we divide decimals by powers of 10 such as 10, 100, 1000, and so on.

$$\frac{569.2}{10} = 56.92 \qquad \text{Move the decimal point } 1 \text{ place to the } left.$$

— 1 zero

$$\frac{569.2}{10,000} = 0.05692 \qquad \text{Move the decimal point } 4 \text{ places to the } left.$$

— 4 zeros

This pattern suggests the following rule:

Dividing Decimals by Powers of 10 such as 10, 100, or 1000

Move the decimal point of the dividend to the *left* the same number of places as there are *zeros* in the power of 10.

Practice 8–9

Divide.

8. $\dfrac{128.3}{1000}$ **9.** $\dfrac{0.56}{10}$

Answers

7. 7.8   **8.** 0.1283   **9.** 0.056

✓**Concept Check Answer**

ten-thousandths place

Examples Divide.

8. $\dfrac{786.1}{1000} = 0.7861$ Move the decimal point *3 places* to the *left.*

— 3 zeros

9. $\dfrac{0.12}{10} = 0.012$ Move the decimal point *1 place* to the *left.*

— 1 zero

■ **Work Practice 8–9**

Objective D Solving Problems by Dividing Decimals

Many real-life problems involve dividing decimals.

Example 10   Calculating Materials Needed for a Job

A gallon of paint covers a 250-square-foot area. How many gallons of paint are needed to cover a wall that measures 1450 square feet? If only gallon containers of paint are available, how many gallon containers are needed?

Solution:

1. UNDERSTAND. Read and reread the problem. We need to know how many 250s are in 1450, so we divide.

2. TRANSLATE.

In words:	number of gallons	is	square feet	divided by	square feet per gallon
	↓	↓	↓	↓	↓
Translate:	number of gallons	=	1450	÷	250

3. SOLVE. Let's see if our answer is reasonable by estimating. The dividend 1450 rounds to 1500 and divisor 250 rounds to 300. Then $1500 \div 300 = 5$.

$$
\begin{array}{r}
5.8 \\
250\overline{)1450.0} \\
-1250 \\
\hline
200\ 0 \\
-200\ 0 \\
\hline
0
\end{array}
$$

4. INTERPRET. *Check* your work. Since our estimate is close to our answer of 5, our answer is reasonable. *State* your conclusion: To paint the wall, 5.8 gallons of paint are needed. If only gallon containers of paint are available, then 6 gallon containers of paint are needed to complete the job.

■ **Work Practice 10**

Practice 10

A bag of fertilizer covers 1250 square feet of lawn. Tim Parker's lawn measures 14,800 square feet. How many bags of fertilizer does he need? If he can buy only whole bags of fertilizer, how many whole bags does he need?

Objective E Simplifying Expressions with Decimals

In the remaining examples, we will review the order of operations by simplifying expressions that contain decimals.

Order of Operations

1. Perform all operations within parentheses ( ), brackets [ ], or other grouping symbols such as square roots or fraction bars, starting with the innermost set.
2. Evaluate any expressions with exponents.
3. Multiply or divide in order from left to right.
4. Add or subtract in order from left to right.

Example 11   Simplify: $723.6 \div 1000 \times 10$

Solution: Multiply or divide in order from left to right.

$723.6 \div 1000 \times 10 = 0.7236 \times 10$ Divide.

$= 7.236$ Multiply.

■ **Work Practice 11**

Practice 11

Simplify: $897.8 \div 100 \times 10$

Answers

10. 11.84 bags; 12 bags   **11.** 89.78

Practice 12

Simplify: $8.69(3.2 - 1.8)$

Example 12   Simplify: $0.5(8.6 - 1.2)$

Solution: According to the order of operations, we simplify inside the parentheses first.

$$0.5(8.6 - 1.2) = 0.5(7.4) \quad \text{Subtract.}$$
$$= 3.7 \quad \text{Multiply.}$$

■ Work Practice 12

Practice 13

Simplify: $\dfrac{20.06 - (1.2)^2 \div 10}{0.02}$

Example 13   Simplify: $\dfrac{5.68 + (0.9)^2 \div 100}{0.2}$

Solution: First we simplify the numerator of the fraction. Then we divide.

$$\frac{5.68 + (0.9)^2 \div 100}{0.2} = \frac{5.68 + 0.81 \div 100}{0.2} \quad \text{Simplify } (0.9)^2.$$

$$= \frac{5.68 + 0.0081}{0.2} \quad \text{Divide.}$$

$$= \frac{5.6881}{0.2} \quad \text{Add.}$$

$$= 28.4405 \quad \text{Divide.}$$

Answers

12. 12.166 13. 995.8

■ Work Practice 13

📟 Calculator Explorations

Calculator errors can easily be made by pressing an incorrect key or by not pressing a correct key hard enough. Estimation is a valuable tool that can be used to check calculator results.

Example   Use estimation to determine whether the calculator result is reasonable or not. (For example, a result that is not reasonable can occur if proper keys are not pressed.)

Simplify: $82.064 \div 23$

Calculator display: $\boxed{35.68}$

Solution: Round each number to the nearest 10. Since $80 \div 20 = 4$, the calculator display 35.68 is not reasonable.

Use estimation to determine whether each result is reasonable or not.

1. 102.62×41.8 Result: 428.9516

2. $174.835 \div 47.9$ Result: 3.65

3. $1025.68 - 125.42$ Result: 900.26

4. $562.781 + 2.96$ Result: 858.781

Vocabulary, Readiness & Video Check

Use the choices below to fill in each blank. Some choices may be used more than once, and some not used at all.

dividend	divisor	quotient	true
zeros	left	right	false

1. In $6.5 \div 5 = 1.3$, the number 1.3 is called the _____, 5 is the _____, and 6.5 is the _____.

2. To check a division exercise, we can perform the following multiplication: quotient · _____ = _____.

3. To divide a decimal number by a power of 10, such as 10, 100, 1000, and so on, we move the decimal point in the number to the _____ the same number of places as there are _____ in the power of 10.

4. True or false: If $1.058 \div 0.46 = 2.3$, then $2.3 \times 0.46 = 1.058$. _____

Recall properties of division and simplify.

5. $\dfrac{5.9}{1}$

6. $\dfrac{0.7}{0.7}$

7. $\dfrac{0}{9.86}$

8. $\dfrac{2.36}{0}$

9. $\dfrac{7.261}{7.261}$

10. $\dfrac{8.25}{1}$

11. $\dfrac{11.1}{0}$

12. $\dfrac{0}{89.96}$

Martin-Gay Interactive Videos   Watch the section lecture video and answer the following questions.

See Video 4.5

Objective A 13. From the lecture before ⊞ Example 2, what must we make sure the divisor is before dividing decimals? ▶

Objective B 14. In ⊞ Example 4, what is the estimated answer and what is the exact answer? ▶

Objective C 15. In ⊞ Example 6, why don't we divide as we did in ⊞ Examples 1–3? ▶

Objective D 16. In ⊞ Example 7, why is the division carried to the hundredths place? ▶

Objective E 17. In ⊞ Example 9, besides meaning division, what other purpose does the fraction bar serve? ▶

4.5 Exercise Set MyMathLab®

Objectives A B Mixed Practice *Divide. See Examples 1 through 5 and 7. For those exercises marked, also estimate to see if the answer is reasonable.*

1. $3\overline{)13.8}$

2. $2\overline{)11.8}$

▶ **3.** $5\overline{)0.47}$

4. $6\overline{)0.51}$

5. $0.06\overline{)18}$

6. $0.04\overline{)20}$

▶ **7.** $0.82\overline{)4.756}$

8. $0.92\overline{)3.312}$

▶ **9.** $5.5\overline{)36.3}$
   Exact:
   Estimate:

10. $2.2\overline{)21.78}$
   Exact:
   Estimate:

11. $6.195 \div 15$

12. $8.823 \div 17$

13. $0.54 \div 12$

14. $1.35 \div 18$

15. Divide 4.2 by 0.6.

16. Divide 3.6 by 0.9.

17. $0.27\overline{)1.296}$

18. $0.34\overline{)2.176}$

19. $0.02\overline{)42}$

20. $0.03\overline{)24}$

21. $0.6\overline{)18}$

22. $0.4\overline{)20}$

23. $0.005\overline{)35}$

24. $0.0007\overline{)35}$

25. $7.2\overline{)70.56}$
   Exact:
   Estimate:

26. $6.3\overline{)54.18}$ **27.** $5.4\overline{)51.84}$ **28.** $7.7\overline{)33.88}$ **29.** $\dfrac{1.215}{0.027}$ **30.** $\dfrac{3.213}{0.051}$
Exact:
Estimate:

31. $0.25\overline{)13.648}$ **32.** $0.75\overline{)49.866}$ **33.** $3.78\overline{)0.02079}$ **34.** $2.96\overline{)0.01332}$

Divide. Round the quotients as indicated. See Example 6.

35. Divide 429.34 by 2.4 and round the quotient to the nearest whole number.

36. Divide 54.8 by 2.6 and round the quotient to the nearest whole number.

▶ **37.** Divide 0.549 by 0.023 and round the quotient to the nearest hundredth.

38. Divide 0.0453 by 0.98 and round the quotient to the nearest thousandth.

39. Divide 45.23 by 0.4 and round the quotient to the nearest tenth.

40. Divide 83.32 by 0.6 and round the quotient to the nearest tenth.

Objective C *Divide. See Examples 8 and 9.*

▶ **41.** $\dfrac{54.982}{100}$ **42.** $\dfrac{342.54}{100}$ **43.** $\dfrac{26.87}{10}$ **44.** $\dfrac{13.49}{10}$ ▶ **45.** $\dfrac{12.9}{1000}$ **46.** $\dfrac{0.27}{1000}$

Objectives A C Mixed Practice *Divide. See Examples 1–5, 8, and 9.*

47. $7\overline{)88.2}$ **48.** $9\overline{)130.5}$ **49.** $\dfrac{13.1}{10}$ **50.** $\dfrac{17.7}{10}$

51. $6.8\overline{)83.13}$ **52.** $4.8\overline{)123.72}$ **53.** $\dfrac{456.25}{10,000}$ **54.** $\dfrac{986.11}{10,000}$

Objective D *Solve. For Exercises 55 and 56, the solutions have been started for you. See Example 10.*

55. Josef Jones is painting the walls of a room. The walls have a total area of 546 square feet. A quart of paint covers 52 square feet. If he must buy paint in whole quarts, how many quarts does he need?

Start the solution:
1. UNDERSTAND the problem. Reread it as many times as needed.
2. TRANSLATE into an equation. (Fill in the blanks.)

number of quarts	is	square feet	divided by	square feet per quart
↓	↓	↓	↓	↓
number of quarts	=	___	÷	___

3. SOLVE. Don't forget to round up your quotient.
4. INTERPRET.

56. A shipping box can hold 36 books. If 486 books must be shipped, how many boxes are needed?

Start the solution:
1. UNDERSTAND the problem. Reread it as many times as needed.
2. TRANSLATE into an equation. (Fill in the blanks.)

number of boxes	is	number of books	divided by	books per box
↓	↓	↓	↓	↓
number of boxes	=	___	÷	___

3. SOLVE. Don't forget to round up your quotient.
4. INTERPRET.

△ **57.** A pound of fertilizer covers 39 square feet of lawn. Vivian Bulgakov's lawn measures 7883.5 square feet. How much fertilizer, to the nearest tenth of a pound, does she need to buy?

58. A page of a book contains about 1.5 kilobytes of information. If a computer disk can hold 740 kilobytes of information, how many pages of a book can be stored on one computer disk? Round to the nearest tenth of a page.

59. There are approximately 39.37 inches in 1 meter. How many meters, to the nearest tenth of a meter, are there in 200 inches?

60. There are 2.54 centimeters in 1 inch. How many inches are there in 50 centimeters? Round to the nearest tenth.

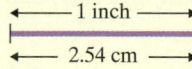

◉ **61.** In the United States, an average child will wear down 730 crayons by his or her tenth birthday. Find the number of boxes of 64 crayons this is equivalent to. Round to the nearest tenth. (*Source:* Binney & Smith Inc.)

62. During a recent year, American farmers received an average of $71.90 per hundred pounds of turkey. What was the average price per pound for turkeys? Round to the nearest cent. (*Source:* National Agricultural Statistics Service)

A child is to receive a dose of 0.5 teaspoon of cough medicine every 4 hours. If the bottle contains 4 fluid ounces, answer Exercises 63 through 66.

63. A fluid ounce equals 6 teaspoons. How many teaspoons are in 4 fluid ounces?

64. The bottle of medicine contains how many doses for the child? (*Hint:* See Exercise 63.)

65. If the child takes a dose every four hours, how many days will the medicine last?

66. If the child takes a dose every six hours, how many days will the medicine last?

Solve.

67. Americans ages 20–34 drive, on average, 15,098 miles per year. About how many miles each week is that? Round to the nearest tenth. (*Note:* There are 52 weeks in a year.) (*Source:* U.S. Office of Highway Policy Information)

68. Drake Saucier was interested in the gas mileage on his "new" used car. He filled the tank, drove 423.8 miles, and filled the tank again. When he refilled the tank, it took 19.35 gallons of gas. Calculate the miles per gallon for Drake's car. Round to the nearest tenth.

69. During the 24 Hours of the Le Mans endurance auto race in 2013, the winning team of Tom Kristensen, Allan McNish, and Loïc Duval drove a total of 2947.1 miles in 24 hours. What was their average speed in miles per hour? Round to the nearest tenth. (*Source:* Based on data from lemans-history.com)

70. During the 2012 Summer Olympics, Ethiopian runner Meseret Defar took the gold medal in the women's 5000-meter event. Her time for the event was 904.25 seconds. What was her average speed in meters per second? Round to the nearest tenth. (*Source:* International Olympic Committee)

71. Elena Delle Donne of the Chicago Sky was the WNBA's Rookie of the Year for 2013. She scored a total of 543 points in the 30 games she played in the 2013 regular season. What was the average number of points she scored per game? (*Source:* Women's National Basketball Association)

72. During the 2012 National Football League regular season, the top-scoring team was the New England Patriots with a total of 557 points throughout the season. The Patriots played 16 games. What was the average number of points the team scored per game? Round to the nearest hundredth. (*Source:* National Football League)

Objective **E** *Simplify each expression. See Examples 11 through 13.*

▶ **73.** $0.7(6 - 2.5)$

74. $1.4(2 - 1.8)$

75. $\dfrac{0.29 + 1.69}{3}$

76. $\dfrac{1.697 - 0.29}{0.7}$

77. $30.03 + 5.1 \times 9.9$

78. $60 - 6.02 \times 8.97$

79. $7.8 - 4.83 \div 2.1 + 9.2$

80. $90 - 62.1 \div 2.7 + 8.6$

81. $93.07 \div 10 \times 100$

82. $35.04 \div 100 \times 10$

▶ **83.** $\dfrac{7.8 + 1.1 \times 100 - 3.6}{0.2}$

84. $\dfrac{9.6 - 7.8 \div 10 + 1.2}{0.02}$

85. $5(20.6 - 2.06) - (0.8)^2$

86. $(10.6 - 9.8)^2 \div 0.01 + 8.6$

87. $6 \div 0.1 + 8.9 \times 10 - 4.6$

88. $8 \div 10 + 7.6 \times 0.1 - (0.1)^2$

Review

Write each decimal as a fraction. See Section 4.1.

89. 0.9

90. 0.7

91. 0.05

92. 0.08

Concept Extensions

Mixed Practice (*Sections 4.3, 4.4, 4.5*) *Perform the indicated operation.*

93. $1.278 \div 0.3$

94. 1.278×0.3

95. $1.278 + 0.3$

96. $1.278 - 0.3$

97. $\begin{array}{r} 8.6 \\ \times\ 3.1 \\ \hline \end{array}$

98. $7.2 + 0.05 + 49.1$

99. $\begin{array}{r} 1000 \\ -\ 95.71 \\ \hline \end{array}$

100. $\dfrac{87.2}{10{,}000}$

Choose the best estimate.

101. 8.62×41.7
  a. 36
  b. 32
  c. 360
  d. 3.6

102. $1.437 + 20.69$
  a. 34
  b. 22
  c. 3.4
  d. 2.2

103. $78.6 \div 97$
  a. 7.86
  b. 0.786
  c. 786
  d. 7860

104. $302.729 - 28.697$
  a. 270
  b. 20
  c. 27
  d. 300

Recall from Section 1.7 that the average of a list of numbers is their total divided by how many numbers there are in the list. Use this procedure to find the average of the test scores listed in Exercises 105 and 106. If necessary, round to the nearest tenth.

105. 86, 78, 91, 87

106. 56, 75, 80

△ **107.** The area of a rectangle is 38.7 square feet. If its width is 4.5 feet, find its length.

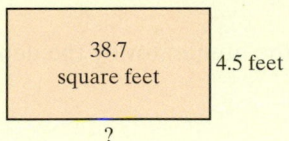

38.7 square feet | 4.5 feet

?

△ **108.** The perimeter of a square is 180.8 centimeters. Find the length of a side.

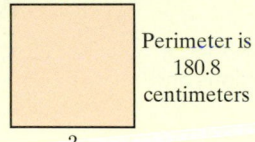

Perimeter is 180.8 centimeters

?

109. When dividing decimals, describe the process you use to place the decimal point in the quotient.

110. In your own words, describe how to quickly divide a number by a power of 10 such as 10, 100, 1000, etc.

To convert wind speeds in miles per hour to knots, divide by 1.15. Use this information and the Saffir-Simpson Hurricane Intensity chart below to answer Exercises 111 and 112. Round to the nearest tenth.

Saffir-Simpson Hurricane Intensity Scale				
Category	Wind Speed	Barometric Pressure [inches of mercury (Hg)]	Storm Surge	Damage Potential
1 (Weak)	75–95 mph	≥ 28.94 in.	4–5 ft	Minimal damage to vegetation
2 (Moderate)	96–110 mph	28.50–28.93 in.	6–8 ft	Moderate damage to houses
3 (Strong)	111–130 mph	27.91–28.49 in.	9–12 ft	Extensive damage to small buildings
4 (Very Strong)	131–155 mph	27.17–27.90 in.	13–18 ft	Extreme structural damage
5 (Devastating)	> 155 mph	< 27.17 in.	>18 ft	Catastrophic building failures possible

111. The chart gives wind speeds in miles per hour. What is the range of wind speeds for a Category 1 hurricane in knots?

112. What is the range of wind speeds for a Category 4 hurricane in knots?

113. A rancher is building a horse corral that's shaped like a rectangle with dimensions of 24.28 meters by 15.675 meters. He plans to make a four-wire fence; that is, he will string four wires around the corral. How much wire will he need?

114. A college student signed up for a new credit card that guarantees her no interest charges on transferred balances for a year. She transferred over a $2523.86 balance from her old credit card. Her minimum payment is $185.35 per month. If she only pays the minimum, will she pay off her balance before interest charges start again?

Objectives

A Write Fractions as Decimals. ▶

B Compare Fractions and Decimals. ▶

C Solve Area Problems Containing Fractions and Decimals. ▶

Objective A Writing Fractions as Decimals ▶

To write a fraction as a decimal, we interpret the fraction bar to mean division and find the quotient.

> **Writing Fractions as Decimals**
>
> To write a fraction as a decimal, divide the numerator by the denominator.

Practice 1

a. Write $\dfrac{2}{5}$ as a decimal.

b. Write $\dfrac{9}{40}$ as a decimal.

Example 1 Write $\dfrac{1}{4}$ as a decimal.

Solution: $\dfrac{1}{4} = 1 \div 4$

$$
\begin{array}{r}
0.25 \\
4\overline{)1.00} \\
-8 \\
\hline
20 \\
-20 \\
\hline
0
\end{array}
$$

Thus, $\dfrac{1}{4}$ written as a decimal is 0.25.

■ **Work Practice 1**

Practice 2

a. Write $\dfrac{5}{6}$ as a decimal.

b. Write $\dfrac{2}{9}$ as a decimal.

Example 2 Write $\dfrac{2}{3}$ as a decimal.

Solution:
$$
\begin{array}{r}
0.666\ldots \\
3\overline{)2.000} \\
-1\,8 \\
\hline
20 \\
-18 \\
\hline
20 \\
-18 \\
\hline
2
\end{array}
$$

This pattern will continue because $\dfrac{2}{3} = 0.6666\ldots$

Remainder is 2, then 0 is brought down.

Remainder is 2, then 0 is brought down.

Remainder is 2.

Notice the digit 2 keeps occurring as the remainder. This will continue so that the digit 6 will keep repeating in the quotient. We place a bar over the digit 6 to indicate that it repeats.

$$\frac{2}{3} = 0.666\ldots = 0.\overline{6}$$

We can also write a decimal approximation for $\dfrac{2}{3}$. For example, $\dfrac{2}{3}$ rounded to the nearest hundredth is 0.67. This can be written as $\dfrac{2}{3} \approx 0.67$.

■ **Work Practice 2**

Answers

1. a. 0.4 **b.** 0.225
2. a. 0.83̄ **b.** 0.2̄

Example 3 Write $\dfrac{22}{7}$ as a decimal. (The fraction $\dfrac{22}{7}$ is an approximation for π.) Round to the nearest hundredth.

Solution:

$$
\begin{array}{r}
3.142 \approx 3.14 \quad \text{\color{blue}Carry the division out to the thousandths place.} \\
7)\overline{22.000} \\
\underline{-21} \\
1\,0 \\
\underline{-\ 7} \\
30 \\
\underline{-28} \\
20 \\
\underline{-14} \\
6
\end{array}
$$

The fraction $\dfrac{22}{7}$ in decimal form is approximately 3.14. Thus, $\pi \approx \dfrac{22}{7}$ (a fraction approximation for π) and $\pi \approx 3.14$ (a decimal approximation for π).

■ **Work Practice 3**

Practice 3

Write $\dfrac{28}{13}$ as a decimal. Round to the nearest thousandth.

Example 4 Write $2\dfrac{3}{16}$ as a decimal.

Solution:

Option 1. Write the fractional part only as a decimal.

$$
\dfrac{3}{16} \longrightarrow
\begin{array}{r}
0.1875 \\
16)\overline{3.0000} \\
\underline{-1\,6} \\
1\,40 \\
\underline{-1\,28} \\
120 \\
\underline{-112} \\
80 \\
\underline{-80} \\
0
\end{array}
$$

Thus $2\dfrac{3}{16} = 2.1875$

Option 2. Write $2\dfrac{3}{16}$ as an improper fraction, and divide.

$$
2\dfrac{3}{16} = \dfrac{35}{16} \longrightarrow
\begin{array}{r}
2.1875 \\
16)\overline{35.0000} \\
\underline{-32} \\
3\,0 \\
\underline{-1\,6} \\
1\,40 \\
\underline{-1\,28} \\
120 \\
\underline{-112} \\
80 \\
\underline{-80} \\
0
\end{array}
$$

Thus $2\dfrac{3}{16} = 2.1875$

■ **Work Practice 4**

Practice 4

Write $3\dfrac{5}{16}$ as a decimal.

Some fractions may be written as decimals using our knowledge of decimals. From Section 4.1, we know that if the denominator of a fraction is 10, 100, 1000, or so on, we can immediately write the fraction as a decimal. For example,

$$
\dfrac{4}{10} = 0.4, \qquad \dfrac{12}{100} = 0.12, \text{ and so on.}
$$

Answers
3. 2.154 **4.** 3.3125

Practice 5

Write $\dfrac{3}{5}$ as a decimal.

Example 5 Write $\dfrac{4}{5}$ as a decimal.

Solution: Let's write $\dfrac{4}{5}$ as an equivalent fraction with a denominator of 10.

$$\dfrac{4}{5} = \dfrac{4}{5} \cdot \dfrac{2}{2} = \dfrac{8}{10} = 0.8$$

■ Work Practice 5

Practice 6

Write $\dfrac{3}{50}$ as a decimal.

Example 6 Write $\dfrac{1}{25}$ as a decimal.

Solution: $\dfrac{1}{25} = \dfrac{1}{25} \cdot \dfrac{4}{4} = \dfrac{4}{100} = 0.04$

■ Work Practice 6

✔ **Concept Check** Suppose you are writing the fraction $\dfrac{9}{16}$ as a decimal. How do you know you have made a mistake if your answer is 1.735?

Objective B Comparing Fractions and Decimals

Now we can compare decimals and fractions by writing fractions as equivalent decimals.

Practice 7

Insert $<$, $>$, or $=$ to form a true statement.

$$\dfrac{1}{5} \qquad 0.25$$

Example 7 Insert $<$, $>$, or $=$ to form a true statement.

$$\dfrac{1}{8} \qquad 0.12$$

Solution: First we write $\dfrac{1}{8}$ as an equivalent decimal. Then we compare decimal places.

$$\begin{array}{r} 0.125 \\ 8\overline{)1.000} \\ -8 \\ \hline 20 \\ -16 \\ \hline 40 \\ -40 \\ \hline 0 \end{array}$$

Original numbers	$\dfrac{1}{8}$	0.12
Decimals	0.125	0.120
Compare	0.125 > 0.12	

Thus, $\dfrac{1}{8} > 0.12$

■ Work Practice 7

Practice 8

Insert $<$, $>$, or $=$ to form a true statement.

a. $\dfrac{1}{2} \qquad 0.54$ b. $0.\overline{4} \qquad \dfrac{4}{9}$

c. $\dfrac{5}{7} \qquad 0.72$

Answers

5. 0.6 6. 0.06 7. $<$

8. a. $<$ b. $=$ c. $<$

✔ **Concept Check Answer**

$\dfrac{9}{}$ is less than 1 while 1.735 is greater

than 1.

Example 8 Insert $<$, $>$, or $=$ to form a true statement.

$$0.\overline{7} \qquad \dfrac{7}{9}$$

Solution: We write $\dfrac{7}{9}$ as a decimal and then compare.

$$\begin{array}{r} 0.77\ldots = 0.\overline{7} \\ 9\overline{)7.00} \\ -6\ 3 \\ \hline 70 \\ -63 \\ \hline 7 \end{array}$$

Original numbers	$0.\overline{7}$	$\dfrac{7}{9}$
Decimals	$0.\overline{7}$	$0.\overline{7}$
Compare	$0.\overline{7} = 0.\overline{7}$	

Thus, $0.\overline{7} = \dfrac{7}{9}$

■ Work Practice 8

Example 9 Write the numbers in order from smallest to largest.

$$\frac{9}{20}, \frac{4}{9}, 0.456$$

Solution:

	$\frac{9}{20}$	$\frac{4}{9}$	0.456
Original numbers			
Decimals	0.450	0.444 ...	0.456
Compare in order	2nd	1st	3rd

Written in order, we have

1st 2nd 3rd
↓ ↓ ↓
$$\frac{4}{9}, \frac{9}{20}, 0.456$$

■ **Work Practice 9**

Objective C Solving Area Problems Containing Fractions and Decimals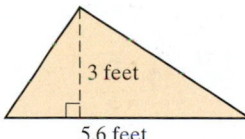

Sometimes real-life problems contain both fractions and decimals. In this section, we solve such problems concerning area. In the next example, we review the area of a triangle. This concept will be studied more in depth in a later chapter.

Example 10 The area of a triangle is Area $= \frac{1}{2} \cdot$ base $\cdot$ height. Find the area of the triangle shown.

3 feet

5.6 feet

Solution:

$$\text{Area} = \frac{1}{2} \cdot \text{base} \cdot \text{height}$$

$$= \frac{1}{2} \cdot 5.6 \cdot 3$$

$$= 0.5 \cdot 5.6 \cdot 3 \qquad \text{Write } \frac{1}{2} \text{ as the decimal } 0.5.$$

$$= 8.4$$

The area of the triangle is 8.4 square feet.

■ **Work Practice 10**

Practice 9
Write the numbers in order from smallest to largest.

a. $\frac{1}{3}, 0.302, \frac{3}{8}$ **b.** $1.26, 1\frac{1}{4}, 1\frac{2}{5}$

c. $0.4, 0.41, \frac{5}{7}$

Practice 10
Find the area of the triangle.

2.1 meters

7 meters

Vocabulary, Readiness & Video Check

Answer each exercise "true" or "false."

1. The number $0.\overline{5}$ means 0.555.

2. To write $\dfrac{9}{19}$ as a decimal, perform the division $9\overline{)19}$.

3. $(1.2)^2$ means $(1.2)(1.2)$ or 1.44.

4. The area of a figure is written in *square* units.

Martin-Gay Interactive Videos   Watch the section lecture video and answer the following questions.

See Video 4.6

Objective A 5. In Example 2, why is the bar placed over just the 6?

Objective B 6. In Example 3, why do we write the fraction as a decimal rather than the decimal as a fraction?

Objective C 7. What formula is used to solve Example 4? What is the final answer?

4.6 Exercise Set MyMathLab®

Objective A *Write each number as a decimal. See Examples 1 through 6.*

1. $\dfrac{1}{5}$

2. $\dfrac{1}{20}$

3. $\dfrac{17}{25}$

4. $\dfrac{13}{25}$

5. $\dfrac{3}{4}$

6. $\dfrac{3}{8}$

7. $\dfrac{2}{25}$

8. $\dfrac{3}{25}$

9. $\dfrac{6}{5}$

10. $\dfrac{5}{4}$

11. $\dfrac{11}{12}$

12. $\dfrac{5}{12}$

13. $\dfrac{17}{40}$

14. $\dfrac{19}{25}$

15. $\dfrac{9}{20}$

16. $\dfrac{31}{40}$

17. $\dfrac{1}{3}$

18. $\dfrac{7}{9}$

19. $\dfrac{7}{16}$

20. $\dfrac{9}{16}$

21. $\dfrac{7}{11}$

22. $\dfrac{9}{11}$

23. $5\dfrac{17}{20}$

24. $4\dfrac{7}{8}$

25. $\dfrac{78}{125}$

26. $\dfrac{159}{375}$

Round each number as indicated.

27. Round your decimal answer to Exercise 17 to the nearest hundredth.

28. Round your decimal answer to Exercise 18 to the nearest hundredth.

29. Round your decimal answer to Exercise 19 to the nearest hundredth.

30. Round your decimal answer to Exercise 20 to the nearest hundredth.

31. Round your decimal answer to Exercise 21 to the nearest tenth.

32. Round your decimal answer to Exercise 22 to the nearest tenth.

Write each fraction as a decimal. If necessary, round to the nearest hundredth. See Examples 1 through 6.

33. Of the U.S. mountains that are over 14,000 feet in elevation, $\frac{56}{91}$ are located in Colorado. (*Source:* U.S. Geological Survey)

34. About $\frac{21}{50}$ of all blood donors have type A blood. (*Source:* American Red Cross Biomedical Services)

35. As of the end of 2012, $\frac{191}{500}$ of all U.S. households were wireless-only households, meaning they no longer subscribe to landline telephone services. (*Source:* National Center for Health Statistics)

36. Porsche is the auto manufacturer with the most wins at the 24 Hours of Le Mans endurance race. By 2013, $\frac{16}{81}$ of Le Mans races had been won in Porsche vehicles. (*Source:* lemans-history.com)

37. When first launched, the Hubble Space Telescope's primary mirror was out of shape on the edges by $\frac{1}{50}$ of a human hair. This very small defect made it difficult to focus on faint objects being viewed. Because the HST was in low Earth orbit, it was serviced by a shuttle and the defect was corrected.

38. The two mirrors currently in use in the Hubble Space Telescope were ground so that they do not deviate from a perfect curve by more than $\frac{1}{800,000}$ of an inch. Do not round this number.

Objective **B** *Insert $<$, $>$, or $=$ to form a true statement. See Examples 7 and 8.*

39. 0.562 ___ 0.569

40. 0.983 ___ 0.988

41. 0.215 ___ $\frac{43}{200}$

42. $\frac{29}{40}$ ___ 0.725

43. $\frac{9}{100}$ ___ 0.0932

44. $\frac{1}{200}$ ___ 0.00563

45. $0.\overline{6}$ ___ $\frac{5}{6}$

46. $0.\overline{1}$ ___ $\frac{2}{17}$

47. $\frac{51}{91}$ ___ $0.56\overline{4}$

48. $0.58\overline{3}$ ___ $\frac{6}{11}$

49. $\frac{1}{9}$ ___ 0.1

50. 0.6 ___ $\frac{2}{3}$

▶ **51.** 1.38 ___ $\frac{18}{13}$

52. 0.372 ___ $\frac{22}{59}$

53. 7.123 ___ $\frac{456}{64}$

54. 12.713 ___ $\frac{89}{7}$

Write the numbers in order from smallest to largest. See Example 9.

55. 0.34, 0.35, 0.32

56. 0.47, 0.42, 0.40

57. 0.49, 0.491, 0.498

58. 0.72, 0.727, 0.728

59. $\frac{3}{4}$, 0.78, 0.73

60. $\frac{2}{5}$, 0.49, 0.42

61. $\frac{4}{7}$, 0.453, 0.412

62. $\frac{6}{9}$, 0.663, 0.668

63. 5.23, $\frac{42}{8}$, 5.34

64. 7.56, $\frac{67}{9}$, 7.562

65. $\frac{12}{5}$, 2.37, $\frac{17}{8}$

66. $\frac{29}{16}$, 1.75, $\frac{59}{32}$

Objective C *Find the area of each triangle or rectangle. See Example 10.*

△ **67.**

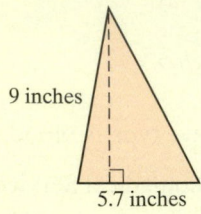

9 inches

5.7 inches

△ **68.**

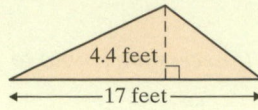

4.4 feet

17 feet

△ **69.**

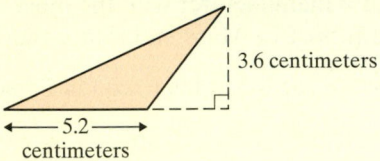

3.6 centimeters

5.2 centimeters

△ **70.**

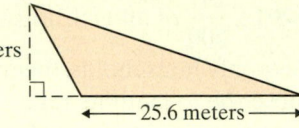

10 meters

25.6 meters

⊙ **71.**
△

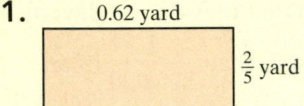

0.62 yard

$\frac{2}{5}$ yard

△ **72.**

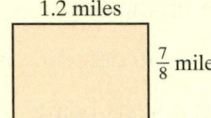

1.2 miles

$\frac{7}{8}$ mile

Review

Simplify. See Sections 1.9 and 3.5.

73. 2^3

74. 5^4

75. $6^2 \cdot 2$

76. $4 \cdot 3^4$

77. $\left(\dfrac{1}{3}\right)^4$

78. $\left(\dfrac{4}{5}\right)^3$

79. $\left(\dfrac{3}{5}\right)^2$

80. $\left(\dfrac{7}{2}\right)^2$

81. $\left(\dfrac{2}{5}\right)\left(\dfrac{5}{2}\right)^2$

82. $\left(\dfrac{2}{3}\right)^2\left(\dfrac{3}{2}\right)^3$

Concept Extensions

Without calculating, describe each number as < 1, $= 1$, or > 1. See the Concept Check in this section.

83. 1.0

84. 1.0000

85. 1.00001

86. $\dfrac{101}{99}$

87. $\dfrac{99}{100}$

88. $\dfrac{99}{99}$

In 2012, there were 15,012 broadcast radio stations in the United States. The most popular formats are shown in the graph along with their counts. Use this graph to answer Exercises 89–92. (Source: Arbitron Radio Station Information Database)

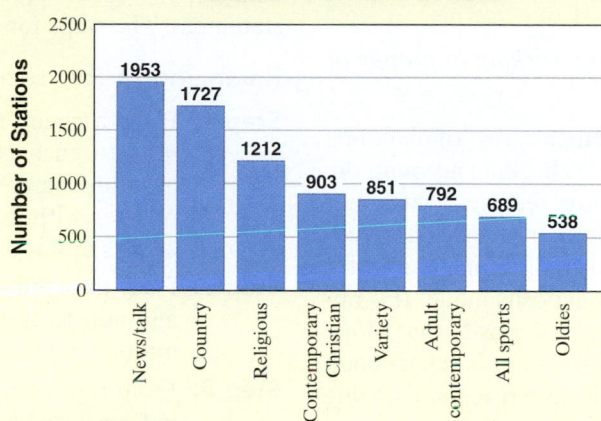

Top Broadcast Radio Station Formats in 2012

Number of Stations

News/talk: 1953
Country: 1727
Religious: 1212
Contemporary Christian: 903
Variety: 851
Adult contemporary: 792
All sports: 689
Oldies: 538

Format (Total stations: 15,012)

89. Write the fraction of radio stations with a news/talk format as a decimal. Round to the nearest hundredth.

90. Write the fraction of radio stations with a country music format as a decimal. Round to the nearest thousandth.

91. Estimate, by rounding each number in the table to the nearest hundred, the total number of stations with the top eight formats in 2012.

92. Use your estimate from Exercise 91 to write the fraction of radio stations accounted for by the top eight formats as a decimal. Round to the nearest hundredth.

93. Describe two ways to determine the larger of two fractions.

94. Describe two ways to write fractions as decimals.

95. Describe two ways to write mixed numbers as decimals.

96. Do you prefer performing operations on decimals or fractions? Why?

Find the value of each expression. Give the result as a decimal.

97. $(9.6)(5) - \dfrac{3}{4}$

98. $2(7.8) - \dfrac{1}{5}$

99. $\left(\dfrac{1}{10}\right)^2 + (1.6)(2.1)$

100. $8.25 - \left(\dfrac{1}{2}\right)^2$

101. $\dfrac{3}{8}(5.9 - 4.7)$

102. $\dfrac{1}{4}(9.6 + 5.2)$

Chapter 4 Group Activity

Maintaining a Checking Account

(Sections 4.1, 4.2, 4.3, 4.4)

This activity may be completed by working in groups or individually.

A checking account is a convenient way of handling money and paying bills. To open a checking account, the bank or savings and loan association requires a customer to make a deposit. Then the customer receives a checkbook that contains checks, deposit slips, and a register for recording checks written and deposits made. It is important to record all payments and deposits that affect the account. It is also important to keep the checkbook balance current by subtracting checks written and adding deposits made.

About once a month checking customers receive a statement from the bank listing all activity that the account has had in the last month. The statement lists a beginning balance, all checks and deposits, any service charges made against the account, and an ending balance. Because it may take several days for checks that a customer has written to clear the banking system, the check register may list checks that do not appear on the monthly bank statement. These checks are called **outstanding checks.** Deposits that are recorded in the check register but do not appear on the statement are called **deposits**

in transit. Because of these differences, it is important to balance, or reconcile, the checkbook against the monthly statement. The steps for doing so are listed below.

Balancing or Reconciling a Checkbook

Step 1: Place a check mark in the checkbook register next to each check and deposit listed on the monthly bank statement. Any entries in the register without a check mark are outstanding checks or deposits in transit.

Step 2: Find the ending checkbook register balance and add to it any outstanding checks and any interest paid on the account.

Step 3: From the total in Step 2, subtract any deposits in transit and any service charges.

Step 4: Compare the amount found in Step 3 with the ending balance listed on the bank statement. If they are the same, the checkbook balances with the bank statement. Be sure to update the check register with service charges and interest.

Step 5: If the checkbook does not balance, recheck the balancing process. Next, make sure that the running checkbook register balance was calculated correctly. Finally, compare the checkbook register with the statement to make sure that each check was recorded for the correct amount.

For the checkbook register and monthly bank statement given:

a. *update the checkbook register*   **b.** *list the outstanding checks and deposits in transit*
c. *balance the checkbook—be sure to update the register with any interest or service fees*

\	\	\	\	\	\	Balance
#	**Date**	**Description**	**Payment**	**✓**	**Deposit**	**425.86**
114	4/1	Market Basket	30.27			
115	4/3	May's Texaco	8.50			
	4/4	Cash at ATM	50.00			
116	4/6	UNO Bookstore	121.38			
	4/7	Deposit			100.00	
117	4/9	MasterCard	84.16			
118	4/10	Blockbuster	6.12			
119	4/12	Kroger	18.72			
120	4/14	Parking sticker	18.50			
	4/15	Direct deposit			294.36	
121	4/20	Rent	395.00			
122	4/25	Student fees	20.00			
	4/28	Deposit			75.00	

Table title: **Checkbook Register**

First National Bank Monthly Statement 4/30

BEGINNING BALANCE:		425.86
Date	Number	Amount
CHECKS AND ATM WITHDRAWALS		
4/3	114	30.27
4/4	ATM	50.00
4/11	117	84.16
4/13	115	8.50
4/15	119	18.72
4/22	121	395.00
DEPOSITS		
4/7		100.00
4/15	Direct deposit	294.36
SERVICE CHARGES		
Low balance fee		7.50
INTEREST		
Credited 4/30		1.15
ENDING BALANCE:		227.22

Chapter 4 Vocabulary Check

Fill in each blank with one of the choices listed below. Some choices may be used more than once or not at all.

vertically	decimal	and	right triangle	diameter
standard form	product	quotient	circumference	difference
sum	denominator	numerator		

1. Like fractional notation, _____ notation is used to denote a part of a whole.

2. To write fractions as decimals, divide the _____ by the _____.

3. To add or subtract decimals, write the decimals so that the decimal points line up _____.

4. When writing decimals in words, write "_____" for the decimal point.

5. When multiplying decimals, the decimal point in the product is placed so that the number of decimal places in the product is equal to the _____ of the number of decimal places in the factors.

6. The distance around a circle is called the _____.

7. When 2 million is written as 2,000,000, we say it is written in _____.

8. $\pi = \dfrac{\text{_____ of a circle}}{\text{_____ of the same circle}}$

9. In $3.4 - 2 = 1.4$, the number 1.4 is called the _____.

10. In $3.4 \div 2 = 1.7$, the 1.7 is called the _____.

11. In $3.4 \times 2 = 6.8$, the 6.8 is called the _____.

12. In $3.4 + 2 = 5.4$, the 5.4 is called the _____.

Helpful Hint ▶ Are you preparing for your test? Don't forget to take the Chapter 4 Test on page 312. Then check your answers at the back of the text and use the Chapter Test Prep Videos to see the fully worked-out solutions to any of the exercises you want to review.

4 Chapter Highlights

Definitions and Concepts	Examples
Section 4.1 Introduction to Decimals	

Place-Value Chart

hundreds	tens	ones	decimal point	tenths	hundredths	thousandths	ten-thousandths	hundred-thousandths
		4	.	2	6	5		
100	10	1		$\frac{1}{10}$	$\frac{1}{100}$	$\frac{1}{1000}$	$\frac{1}{10,000}$	$\frac{1}{100,000}$

4.265 means

$$4 \cdot 1 + 2 \cdot \frac{1}{10} + 6 \cdot \frac{1}{100} + 5 \cdot \frac{1}{1000}$$

or

$$4 + \frac{2}{10} + \frac{6}{100} + \frac{5}{1000}$$

(continued)

Definitions and Concepts	Examples

Section 4.1   Introduction to Decimals *(continued)*

Writing (or Reading) a Decimal in Words

Step 1: Write the whole number part in words.

Step 2: Write "and" for the decimal point.

Step 3: Write the decimal part in words as though it were a whole number, followed by the place value of the last digit.

A decimal written in words can be written in standard form by reversing the above procedure.

Write 3.08 in words.
Three and eight hundredths

Write "four and twenty-one thousandths" in standard form.
    4.021

Section 4.2   Order and Rounding

To **compare decimals,** compare digits in the same place from left to right. When two digits are not equal, the number with the larger digit is the larger decimal.

To Round Decimals to a Place Value to the Right of the Decimal Point

Step 1: Locate the digit to the right of the given place value.

Step 2: If this digit is 5 or greater, add 1 to the digit in the given place value and delete all digits to its right. If this digit is less than 5, delete all digits to the right of the given place value.

$3.0261 > 3.0186$ because

$\uparrow \qquad \uparrow$

$2 \;\; > \;\; 1$

Round 86.1256 to the nearest hundredth.

hundredths place

Step 1: 86.12⑤6

digit to the right

Step 2: Since the digit to the right is 5 or greater, we add 1 to the digit in the hundredths place and delete all digits to its right.

86.1256 rounded to the nearest hundredth is 86.13.

Section 4.3   Adding and Subtracting Decimals

To Add or Subtract Decimals

Step 1: Write the decimals so that the decimal points line up vertically.

Step 2: Add or subtract as with whole numbers.

Step 3: Place the decimal point in the sum or difference so that it lines up vertically with the decimal points in the problem.

Add: $4.6 + 0.28$

$$\begin{array}{r} 4.60 \\ +\,0.28 \\ \hline 4.88 \end{array}$$

Subtract: $2.8 - 1.04$

$$\begin{array}{r} 2.\overset{7\;10}{8\,\cancel{0}} \\ -\,1.0\,4 \\ \hline 1.7\,6 \end{array}$$

Section 4.4   Multiplying Decimals and Circumference of a Circle

To Multiply Decimals

Step 1: Multiply the decimals as though they are whole numbers.

Step 2: The decimal point in the product is placed so that the number of decimal places in the product is equal to the *sum* of the number of decimal places in the factors.

The **circumference** of a circle is the distance around the circle.

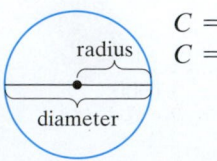

$C = 2 \cdot \pi \cdot \text{radius}$ or
$C = \pi \cdot \text{diameter},$

where $\pi \approx 3.14$ or $\dfrac{22}{7}$.

Multiply: 1.48×5.9

$$\begin{array}{r} 1.48 \quad \leftarrow 2 \text{ decimal places} \\ \times\; 5.9 \quad \leftarrow 1 \text{ decimal places} \\ \hline 1332 \quad\quad\quad \\ 7400 \quad\quad\quad \\ \hline 8.732 \quad \leftarrow 3 \text{ decimal places} \end{array}$$

Find the exact circumference of a circle with radius 5 miles and an approximation by using 3.14 for π.

$$\begin{aligned} C &= 2 \cdot \pi \cdot \text{radius} \\ &= 2 \cdot \pi \cdot 5 \\ &= 10\pi \\ &\approx 10(3.14) \\ &= 31.4 \end{aligned}$$

The circumference is exactly 10π miles and *approximately* 31.4 miles.

Definitions and Concepts	Examples

Section 4.5 Dividing Decimals and Order of Operations

To Divide Decimals

Step 1: If the divisor is not a whole number, move the decimal point in the divisor to the right until the divisor is a whole number.

Step 2: Move the decimal point in the dividend to the right the *same number of places* as the decimal point was moved in step 1.

Step 3: Divide. The decimal point in the quotient is directly over the moved decimal point in the dividend.

Divide: $1.118 \div 2.6$

$$
\begin{array}{r}
0.43 \\
2.6\overline{)1.118} \\
-1\,04 \\
\hline
78 \\
-78 \\
\hline
0
\end{array}
$$

Order of Operations

1. Perform all operations within parentheses ( ), brackets [ ], or grouping symbols such as square roots or fraction bars.

2. Evaluate any expressions with exponents.

3. Multiply or divide in order from left to right.

4. Add or subtract in order from left to right.

Simplify.

$$
\begin{aligned}
1.9(12.8 - 4.1) &= 1.9(8.7) \quad \text{Subtract.} \\
&= 16.53 \quad \text{Multiply.}
\end{aligned}
$$

Section 4.6 Fractions and Decimals

To **write fractions as decimals,** divide the numerator by the denominator.

Write $\dfrac{3}{8}$ as a decimal.

$$
\begin{array}{r}
0.375 \\
8\overline{)3.000} \\
-2\,4 \\
\hline
60 \\
-56 \\
\hline
40 \\
-40 \\
\hline
0
\end{array}
$$

Chapter 4   Review

(4.1) *Determine the place value of the digit 4 in each decimal.*

1. 23.45

2. 0.000345

Write each decimal in words.

3. 0.45

4. 0.00345

5. 109.23

6. 46.007

Write each decimal in standard form.

7. Two and fifteen hundredths

8. Five hundred three and one hundred two thousandths

Write the decimal as a fraction or a mixed number. Write your answer in simplest form.

9. 0.16   **10.** 12.023   **11.** 1.0045   **12.** 25.25

Write each fraction as a decimal.

13. $\dfrac{9}{10}$ **14.** $\dfrac{25}{100}$ **15.** $\dfrac{45}{1000}$ **16.** $\dfrac{261}{10}$

(4.2) *Insert* $<$, $>$, *or* $=$ *to make a true statement.*

17. 0.49   0.43

18. 0.973   0.9730

Write the decimals in order from smallest to largest.

19. 8.6, 8.09, 0.92

20. 0.09, 0.1, 0.091

Round each decimal to the given place value.

21. 0.623, nearest tenth

22. 0.9384, nearest hundredth

Round each money amount to the nearest cent.

23. $0.259

24. $12.461

Solve.

25. In a recent year, engaged couples in the United States spent an average of $31,304.35 on their wedding. Round this number to the nearest dollar.

26. A certain kind of chocolate candy bar contains 10.75 teaspoons of sugar. Write this number as a mixed number.

(4.3) *Add or subtract as indicated.*

27. 2.4 + 7.12   **28.** 3.9 − 1.2   **29.** 6.4 + 0.88   **30.** 19.02 + 6.98 + 0.007

31. 892.1 − 432.4

32. 100.342 − 0.064

33. Subtract 34.98 from 100.

34. Subtract 10.02 from 200.

35. Find the total distance between Grove City and Jerome.

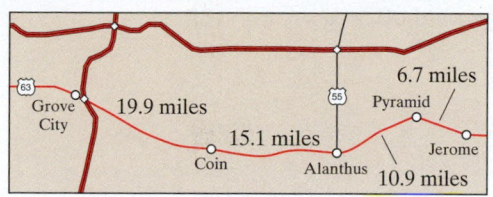

36. The price of oil was $49.02 per barrel on October 23. It was $51.46 on October 24. Find by how much the price of oil increased from the 23rd to the 24th.

△ **37.** Find the perimeter.

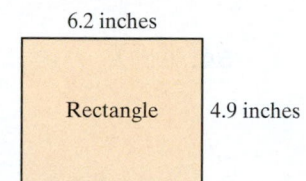

△ **38.** Find the perimeter.

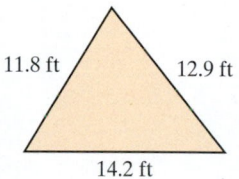

(4.4) *Multiply.*

39. 3.7
       × 5

40. 9.1
       × 6

41. 7.2 × 10

42. 9.345 × 1000

43. 4.02
       × 2.3

44.   39.02
        × 87.3

Solve.

△ **45.** Find the exact circumference of the circle. Then use the approximation 3.14 for π and approximate the circumference.

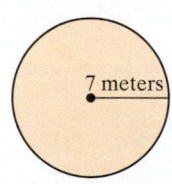

46. A kilometer is approximately 0.625 mile. It is 102 kilometers from Hays to Colby. Write 102 kilometers in miles to the nearest tenth of a mile.

Write each number in standard notation.

47. Saturn is a distance of about 887 million miles from the Sun.

48. The tail of a comet can be over 600 thousand miles long.

(4.5) *Divide. Round the quotient to the nearest thousandth if necessary.*

49. 3)0.261

50. 20)316.5

51. 21 ÷ 0.3

52. 0.0063 ÷ 0.03

53. $0.34\overline{)2.74}$

54. $19.8\overline{)601.92}$

55. $\dfrac{2.67}{100}$

56. $\dfrac{93}{10}$

57. There are approximately 3.28 feet in 1 meter. Find how many meters are in 24 feet to the nearest tenth of a meter.

←——1 meter——→

←—— ≈3.28 feet ——→

58. George Strait pays $69.71 per month to pay back a loan of $3136.95. In how many months will the loan be paid off?

Simplify each expression.

59. $7.6 \times 1.9 + 2.5$

60. $(2.3)^2 - 1.4$

61. $\dfrac{7 + 0.74}{0.06}$

62. $\dfrac{(1.5)^2 + 0.5}{0.05}$

63. $0.9(6.5 - 5.6)$

64. $0.0726 \div 10 \times 1000$

(4.6) *Write each fraction as a decimal. Round to the nearest thousandth if necessary.*

65. $\dfrac{4}{5}$

66. $\dfrac{12}{13}$

67. $2\dfrac{1}{3}$

68. $\dfrac{13}{60}$

Insert <, >, or = to make a true statement.

69. 0.392 ___ 0.3920

70. $0.\overline{4}$ ___ $\dfrac{4}{9}$

71. 0.293 ___ $\dfrac{5}{17}$

72. $\dfrac{4}{7}$ ___ 0.625

Write the numbers in order from smallest to largest.

73. $0.839, \dfrac{17}{20}, 0.837$

74. $\dfrac{18}{11}, 1.63, \dfrac{19}{12}$

Find each area.

△ **75.**

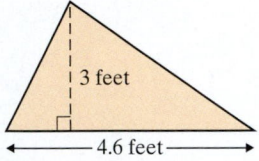

△ **76.**

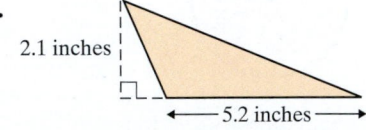

Mixed Review

77. Write 200.0032 in words.

78. Write sixteen thousand twenty-five and fourteen thousandths in standard form.

79. Write 0.00231 as a fraction or a mixed number.

80. Write the numbers $\dfrac{6}{7}, \dfrac{8}{9}, 0.75$ in order from smallest to largest.

Write each fraction as a decimal. Round to the nearest thousandth, if necessary.

81. $\dfrac{7}{100}$

82. $\dfrac{9}{80}$ (Do not round.)

83. $\dfrac{8935}{175}$

Insert <, >, or = to make a true statement.

84. 402.00032   402.000032

85. 0.230505   0.23505

86. $\dfrac{6}{11}$ 0.55

Round each decimal to the given place value.

87. 42.895, nearest hundredth

88. 16.34925, nearest thousandth

Round each money amount to the nearest dollar.

89. $123.46

90. $3645.52

Add or subtract as indicated.

91. $4.9 - 3.2$

92. $5.23 - 2.74$

93. $200.49 + 16.82 + 103.002$

94. $0.00236 + 100.45 + 48.29$

Multiply or divide as indicated. Round to the nearest thousandth, if necessary.

95. $\begin{array}{r} 2.54 \\ \times\ 3.2 \\ \hline \end{array}$

96. $\begin{array}{r} 3.45 \\ \times\ 2.1 \\ \hline \end{array}$

97. $0.005\overline{)24.5}$

98. $2.3\overline{)54.98}$

Solve.

△ **99.** Tomaso is going to fertilize his lawn, a rectangle that measures 77.3 feet by 115.9 feet. Approximate the area of the lawn by rounding each measurement to the nearest ten feet.

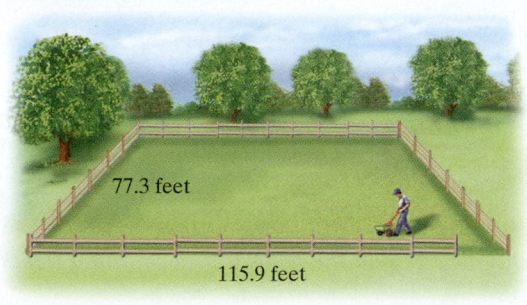

100. Estimate the cost of the items to see whether the groceries can be purchased with a $5 bill.

Simplify each expression.

101. $\dfrac{(3.2)^2}{100}$

102. $(2.6 + 1.4)(4.5 - 3.6)$

Answers

Write the decimal as indicated.

1. 45.092, in words

2. Three thousand and fifty-nine thousandths, in standard form

Round the decimal to the indicated place value.

3. 34.8923, nearest tenth

4. 0.8623, nearest thousandth

5. Insert <, >, or = to make a true statement. 25.0909 ___ 25.9090

6. Write the numbers in order from smallest to largest. $\dfrac{4}{9}$ 0.454 0.445

Write the decimal as a fraction or a mixed number in simplest form.

7. 0.345

8. 24.73

Write the fraction or mixed number as a decimal. If necessary, round to the nearest thousandth.

9. $\dfrac{13}{20}$

10. $5\dfrac{8}{9}$

11. $\dfrac{16}{17}$

Perform the indicated operations. Round the result to the nearest thousandth if necessary.

12. 2.893 + 4.2 + 10.49

13. Subtract 8.6 from 20.

14. $\begin{array}{r} 10.2 \\ \times\ 4.3 \\ \hline \end{array}$

15. $0.23\overline{)12.88}$

16. $\begin{array}{r} 0.165 \\ \times\ 0.47 \\ \hline \end{array}$

17. $7\overline{)46.71}$

1. _____

2. _____

3. _____

4. _____

5. _____

6. _____

7. _____

8. _____

9. _____

10. _____

11. _____

12. _____

13. _____

14. _____

15. _____

16. _____

17. _____

18. 126.9×100

19. $\dfrac{47.3}{10}$

20. $0.3[1.57 - (0.6)^2]$

21. $\dfrac{0.23 + 1.63}{0.3}$

22. At its farthest, Pluto is 4583 million miles from the Sun. Write this number using standard notation.

△ **23.** Find the area.

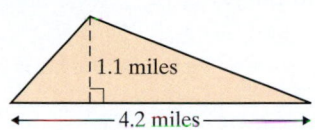

1.1 miles

4.2 miles

△ **24.** Find the exact circumference of the circle. Then use the approximation 3.14 for π and approximate the circumference.

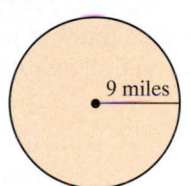

9 miles

25. Vivian Thomas is going to put insecticide on her lawn to control grubworms. The lawn is a rectangle that measures 123.8 feet by 80 feet. The amount of insecticide required is 0.02 ounce per square foot.

  a. Find the area of her lawn.

  b. Find how much insecticide Vivian needs to purchase.

26. Find the total distance from Bayette to Center City.

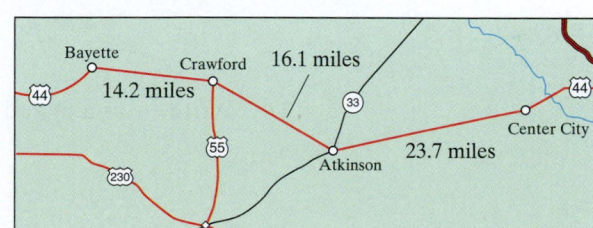

Bayette
Crawford   16.1 miles
44   14.2 miles
33
44
55
Center City
230
23.7 miles
Atkinson

18. _____

19. _____

20. _____

21. _____

22. _____

23. _____

24. _____

25. a. _____

   b. _____

26. _____

Answers

1. Write 106,052,447 in words.

2. Write two hundred seventy-six thousand, four in standard form.

3. In 2011, a total of 12,734,424 passenger vehicles were sold in the United States. In 2012, total passenger vehicle sales in the United States had increased by 1,705,636 vehicles. Find the total number of passenger vehicles sold in the United States in 2012. (*Source:* Alliance of Automobile Manufacturers)

4. There are 12 fluid ounces of soda in a can. How many fluid ounces of soda are in a case (24 cans) of soda?

5. Subtract: $900 - 174$. Then check by adding.

6. Simplify: $5^2 \cdot 2^3$

7. Round each number to the nearest hundred to find an estimated sum.

$$\begin{array}{r} 294 \\ 625 \\ 1071 \\ + \ 349 \\ \hline \end{array}$$

8. Simplify: $7 \cdot \sqrt{144}$

9. A digital video disc (DVD) can hold about 4800 megabytes (MB) of information. How many megabytes can 12 DVDs hold?

10. Find the perimeter and area of the square.

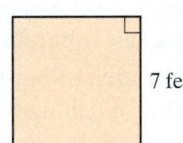

7 feet

11. Divide: $6819 \div 17$

12. Write $2\dfrac{5}{8}$ as an improper fraction.

13. Simplify: $4^3 + [3^2 - (10 \div 2)] - 7 \cdot 3$

14. Write $\dfrac{64}{5}$ as a mixed number.

15. Identify the numerator and the denominator: $\dfrac{3}{7}$

16. Simplify: $24 \div 8 \cdot 3$

17. Write $\dfrac{6}{60}$ in simplest form.

18. Simplify: $(8 - 5)^2 + (10 - 8)^3$

1. _____

2. _____

3. _____

4. _____

5. _____

6. _____

7. _____

8. _____

9. _____

10. _____

11. _____

12. _____

13. _____

14. _____

15. _____

16. _____

17. _____

18. _____

19. Multiply: $\dfrac{3}{4} \cdot 20$

20. Simplify: $1 + 2[30 \div (7 - 2)]$

21. Divide: $\dfrac{7}{8} \div \dfrac{2}{9}$

22. Find the average of 117, 125, and 142.

23. Multiply: $1\dfrac{2}{3} \cdot 2\dfrac{1}{4}$

24. A total of \$324 is paid for 36 tickets to the Audubon Zoo. How much did each ticket cost?

25. Divide: $\dfrac{3}{4} \div 5$

26. Simplify: $\left(\dfrac{3}{4} \div \dfrac{1}{2}\right) \cdot \dfrac{9}{10}$

Simplify.

27. $\dfrac{8}{9} - \dfrac{1}{9}$

28. $\dfrac{4}{15} + \dfrac{2}{15}$

29. $\dfrac{7}{8} - \dfrac{5}{8}$

30. $\dfrac{1}{20} + \dfrac{3}{20} + \dfrac{4}{20}$

Write an equivalent fraction with the indicated denominator.

31. $\dfrac{3}{4} = \dfrac{}{20}$

32. $\dfrac{7}{9} = \dfrac{}{45}$

Perform the indicated operations.

33. $\dfrac{11}{15} + \dfrac{3}{10}$

34. $\dfrac{7}{30} - \dfrac{2}{9}$

35. Two packages of ground round are purchased. One package weighs $2\dfrac{3}{8}$ pounds and the other $1\dfrac{4}{5}$ pounds. What is the combined weight of the ground round?

36. A color cartridge for a business printer weighs $2\dfrac{5}{16}$ pounds. How much do 12 cartridges weigh?

19. _____

20. _____

21. _____

22. _____

23. _____

24. _____

25. _____

26. _____

27. _____

28. _____

29. _____

30. _____

31. _____

32. _____

33. _____

34. _____

35. _____

36. _____

Evaluate each expression.

37. $\left(\dfrac{1}{4}\right)^2$ **38.** $\left(\dfrac{7}{11}\right)^2$

39. $\left(\dfrac{1}{6}\right)^2 \cdot \left(\dfrac{3}{4}\right)^3$ **40.** $\left(\dfrac{1}{2}\right)^3 \cdot \left(\dfrac{4}{9}\right)^2$

41. Write 0.43 as a fraction.   **42.** Write $\dfrac{3}{4}$ as a decimal.

43. Insert $<$, $>$, or $=$ to form a true statement.

0.378   0.368

44. Write "five and six hundredths" in standard form.

45. Subtract: $35.218 - 23.65$
Check your answer.

46. Add: $75.1 + 0.229$

Multiply.

47. 23.702×100 **48.** 1.7×0.07

49. $76{,}805 \times 0.01$ **50.** Divide: $0.1157 \div 0.013$

37. _____

38. _____

39. _____

40. _____

41. _____

42. _____

43. _____

44. _____

45. _____

46. _____

47. _____

48. _____

49. _____

50. _____

Ratio, Proportion, and Percent

5

One of many 3-D cameras

Note: 3-D viewing systems that do not require viewing glasses are currently being tested and manufactured.

How Popular Are 3-D Films Now?

A 3-D (three-dimensional) film is a film that enhances the illusion of depth perception. Believe it or not, 3-D films have existed in some form since 1890, but because of high cost and lack of a standardized format, these films are only now starting to be widely shown and produced. Although there was a decline in the total number of films released in 2013, there were 45 films released with digital 3-D versions.

In Section 5.1, Exercise 21, we calculate the ratio of 3-D to non–3-D films.

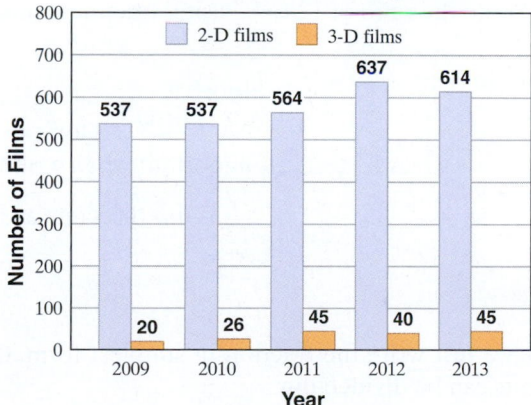

Films Released by U.S. Production Companies

Source: Motion Picture Association of America

This chapter is mainly devoted to percent, a concept used virtually every day in ordinary and business life. Understanding percent and using it efficiently depend on understanding ratios because a percent is a ratio whose denominator is 100. We present techniques to write percents as fractions and as decimals and then solve problems relating to sales tax, commission, discounts, interest, and other real-life situations that use percents.

5.1 Ratio and Proportion

Objectives

A Write Ratios as Fractions.

B Write Rates as Fractions.

C Determine Whether Proportions Are True.

D Find an Unknown Number in a Proportion.

E Solve Problems by Writing Proportions.

Objective A Writing Ratios as Fractions

A **ratio** is the quotient of two quantities. A ratio, in fact, is no different from a fraction, except that a ratio is sometimes written using notation other than fractional notation. For example, the ratio of 1 to 2 can be written as

$$1 \text{ to } 2 \quad \text{or} \quad \frac{1}{2} \quad \text{or} \quad 1:2$$

fractional notation   colon notation

These ratios are all read as, "the ratio of 1 to 2."

✓**Concept Check** How should each ratio be read aloud?

a. $\frac{8}{5}$ **b.** $\frac{5}{8}$

In this section, we write ratios using fractional notation. If the fraction happens to be an improper fraction, do not write the fraction as a mixed number. Why? The mixed number form is not a ratio or quotient of two quantities.

> **Writing a Ratio as a Fraction**
>
> The order of the quantities is important when writing ratios. To write a ratio as a fraction, write the *first number* of the ratio as the *numerator* of the fraction and the *second number* as the *denominator*.

Helpful Hint

The ratio of 6 to 11 is $\frac{6}{11}$, *not* $\frac{11}{6}$.

Practice 1

Write the ratio of 20 to 23 using fractional notation.

Example 1 Write the ratio of 12 to 17 using fractional notation.

Solution: The ratio is $\frac{12}{17}$.

Helpful Hint Don't forget that order is important when writing ratios. The ratio $\frac{17}{12}$ is *not* the same as the ratio $\frac{12}{17}$.

■ **Work Practice 1**

To simplify a ratio, we just write the fraction in simplest form. Common factors as well as common units can be divided out.

Practice 2

Write the ratio of $8 to $6 as a fraction in simplest form.

Example 2 Write the ratio of $15 to $10 as a fraction in simplest form.

Solution:

$$\frac{\$15}{\$10} = \frac{15}{10} = \frac{3 \cdot \cancel{5}}{2 \cdot \cancel{5}} = \frac{3}{2}$$

Helpful Hint In this example, although $\frac{3}{2} = 1\frac{1}{2}$, a ratio is a quotient of *two* quantities. For that reason, ratios are not written as mixed numbers.

■ **Work Practice 2**

Answers

1. $\frac{20}{23}$ **2.** $\frac{4}{3}$

✓**Concept Check Answers**

a. "the ratio of eight to five"

b. "the ratio of five to eight"

If a ratio contains decimal numbers or mixed numbers, we simplify by writing the ratio as a ratio of whole numbers.

Example 3   Write the ratio of 2.6 to 3.1 as a fraction in simplest form.

Solution: The ratio in fraction form is

$$\frac{2.6}{3.1}$$

Now let's clear the ratio of decimals.

$$\frac{2.6}{3.1} = \frac{2.6}{3.1} \cdot 1 = \frac{2.6}{3.1} \cdot \frac{10}{10} = \frac{2.6 \cdot 10}{3.1 \cdot 10} = \frac{26}{31} \quad \text{Simplest form}$$

■ **Work Practice 3**

Example 4   Writing a Ratio from a Circle Graph

The circle graph in the margin shows the part of a car's total mileage that falls into a particular category. Write the ratio of medical miles to total miles as a fraction in simplest form.

Solution: $\dfrac{\text{medical miles}}{\text{total miles}} = \dfrac{150 \text{ miles}}{15{,}000 \text{ miles}} = \dfrac{150}{15{,}000} = \dfrac{\overset{1}{\cancel{150}}}{\underset{1}{\cancel{150} \cdot 100}} = \dfrac{1}{100}$

■ **Work Practice 4**

△ **Example 5**   Given the rectangle shown:

a. Find the ratio of its width to its length.
b. Find the ratio of its length to its perimeter.

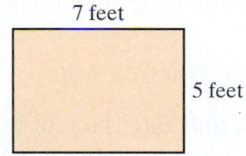
7 feet

5 feet

Solution:

a. The ratio of its width to its length is

$$\frac{\text{width}}{\text{length}} = \frac{5 \text{ feet}}{7 \text{ feet}} = \frac{5}{7}$$

b. Recall that the perimeter of the rectangle is the distance around the rectangle: $7 + 5 + 7 + 5 = 24$ feet. The ratio of its length to its perimeter is

$$\frac{\text{length}}{\text{perimeter}} = \frac{7 \text{ feet}}{24 \text{ feet}} = \frac{7}{24}$$

■ **Work Practice 5**

✓**Concept Check**   Explain why the answer $\frac{7}{5}$ would be incorrect for part (a) of Example 5.

Practice 3
Write the ratio of 1.71 to 4.56 as a fraction in simplest form.

Practice 4
Use the circle graph below to write the ratio of work miles to total miles as a fraction in simplest form.

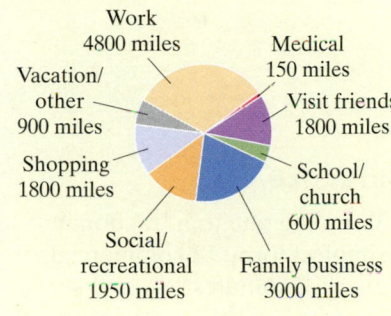

Work
4800 miles

Medical
150 miles

Vacation/
other
900 miles

Visit friends
1800 miles

Shopping
1800 miles

School/
church
600 miles

Social/
recreational
1950 miles

Family business
3000 miles

Total yearly mileage: 15,000

Sources: The American Automobile Manufacturers Association and The National Automobile Dealers Association

△ Practice 5

Given the triangle shown:

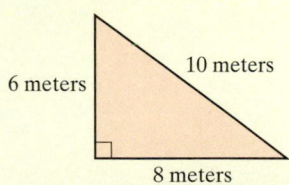
6 meters   10 meters

8 meters

a. Find the ratio of the length of the shortest side to the length of the longest side.
b. Find the ratio of the length of the longest side to the perimeter of the triangle.

Answers

3. $\frac{3}{8}$ **4.** $\frac{8}{25}$ **5. a.** $\frac{3}{5}$ **b.** $\frac{5}{12}$

✓ **Concept Check Answer**

$\frac{7}{5}$ would be the ratio of the rectangle's length to its width.

Objective B Writing Rates as Fractions

A special type of ratio is a rate. **Rates** are used to compare *different* kinds of quantities. For example, suppose that a recreational runner can run 3 miles in 33 minutes. If we write this rate as a fraction, we have

$$\frac{3 \text{ miles}}{33 \text{ minutes}} = \frac{1 \text{ mile}}{11 \text{ minutes}} \quad \text{In simplest form}$$

Helpful Hint

When comparing quantities with different units, write the units as part of the comparison. They do not divide out.

Same Units: $\dfrac{3 \cancel{\text{ inches}}}{12 \cancel{\text{ inches}}} = \dfrac{1}{4}$

Different Units: $\dfrac{2 \text{ miles}}{20 \text{ minutes}} = \dfrac{1 \text{ mile}}{10 \text{ minutes}}$ Units are still written.

Practice 6

Write the rate as a fraction in simplest form: 12 commercials every 45 minutes

Example 6   Write the rate as a fraction in simplest form: 10 nails every 6 feet

Solution: $\dfrac{10 \text{ nails}}{6 \text{ feet}} = \dfrac{5 \text{ nails}}{3 \text{ feet}}$

■ **Work Practice 6**

Practice 7–8

Write each rate as a fraction in simplest form.

7. $1680 for 8 weeks

8. 236 miles on 12 gallons of gasoline

Examples   Write each rate as a fraction in simplest form.

7. $2160 for 12 weeks is $\dfrac{2160 \text{ dollars}}{12 \text{ weeks}} = \dfrac{180 \text{ dollars}}{1 \text{ week}}$

8. 360 miles on 16 gallons of gasoline is $\dfrac{360 \text{ miles}}{16 \text{ gallons}} = \dfrac{45 \text{ miles}}{2 \text{ gallons}}$

■ **Work Practice 7–8**

Note: A **unit rate** is a rate with a denominator of 1. A familiar example of a unit rate is mph, read as **"miles per hour."** For example, 55 mph means 55 miles per 1 hour

or $\dfrac{55 \text{ miles.}}{1 \text{ hour}}$

If we write the rate in Example 8 as a unit rate, we have

Helpful Hint
In this context, the word "per" translates to division.

$$\frac{45 \text{ miles}}{2 \text{ gallons}} = \frac{22.5 \text{ miles}}{1 \text{ gallon}} \text{ or } 22.5 \text{ miles/gallon}$$

Objective C Determining Whether Proportions Are True

A **proportion** is a statement that 2 ratios or rates are equal. For example,

$$\frac{5}{6} = \frac{10}{12}$$

is a proportion. We can read this as, "5 is to 6 as 10 is to 12."

Let's write each sentence as a proportion.

"12 diamonds is to 15 rubies as 4 diamonds is to 5 rubies" translates to

$$\begin{array}{ccccc} \text{diamonds} & \rightarrow & \dfrac{12}{15} = \dfrac{4}{5} & \leftarrow & \text{diamonds} \\ \text{rubies} & \rightarrow & & \leftarrow & \text{rubies} \end{array}$$

Answers

6. $\dfrac{4 \text{ commercials}}{15 \text{ min}}$ **7.** $\dfrac{\$210}{1 \text{ wk}}$ **8.** $\dfrac{59 \text{ mi}}{3 \text{ gal}}$

"5 hits is to 9 at bats as 20 hits is to 36 at bats" translates to

$$\text{hits} \rightarrow \quad \frac{5}{9} = \frac{20}{36} \quad \leftarrow \text{hits}$$
$$\text{at bats} \rightarrow \qquad\qquad\qquad \leftarrow \text{at bats}$$

Helpful Hint

Notice in the previous proportions that the numerators contain the same units and the denominators contain the same units. In this text, proportions will be written so that this is the case.

Like other mathematical statements, a proportion may be either true or false. A proportion is true if its ratios are equal. Since ratios are fractions, one way to determine whether a proportion is true is to write both fractions in simplest form and compare them.

Another way is to compare cross products as we did in Section 2.3.

Using Cross Products to Determine Whether Proportions Are True or False

Cross products

$a \cdot d$ $b \cdot c$

$$\frac{a}{b} = \frac{c}{d}$$

If cross products are *equal*, the proportion is *true*. If $ad = bc$, then the proportion is true.

If cross products are *not equal*, the proportion is *false*. If $ad \neq bc$, then the proportion is false.

Example 9 Is $\frac{2}{3} = \frac{4}{6}$ a true proportion?

Solution:

Cross products

$2 \cdot 6$ $3 \cdot 4$

$$\frac{2}{3} = \frac{4}{6}$$

$2 \cdot 6 \stackrel{?}{=} 3 \cdot 4$ Are cross products equal?

$12 = 12$ Equal, so proportion is true.

Since the cross products are equal, the proportion is true.

■ **Work Practice 9**

Example 10 Is $\frac{4.1}{7} = \frac{2.9}{5}$ a true proportion?

Solution:

Cross products

$4.1 \cdot 5$ $7 \cdot 2.9$

$$\frac{4.1}{7} = \frac{2.9}{5}$$

$4.1 \cdot 5 \stackrel{?}{=} 7 \cdot 2.9$ Are cross products equal?

$20.5 \neq 20.3$ Not equal, so proportion is false.

Since the cross products are not equal, $\frac{4.1}{7} \neq \frac{2.9}{5}$. The proportion is false.

■ **Work Practice 10**

Practice 9

Is $\frac{3}{6} = \frac{4}{8}$ a true proportion?

Practice 10

Is $\frac{3.6}{6} = \frac{5.4}{8}$ a true proportion?

Answers

9. yes **10.** no

Objective D Finding Unknown Numbers in Proportions ▶

When one number of a proportion is unknown, we can use cross products to find the unknown number. For example, to find the unknown number n in the proportion $\dfrac{n}{30} = \dfrac{2}{3}$, we first find the cross products.

$$n \cdot 3 \qquad\qquad 30 \cdot 2 \quad \text{Find the cross products.}$$

$$\dfrac{n}{30} \diagup\!\!\!\!\diagdown \dfrac{2}{3}$$

If the proportion is true, then cross products are equal.

$n \cdot 3 = 30 \cdot 2$ Set the cross products equal to each other.

$n \cdot 3 = 60$ Write $2 \cdot 30$ as 60.

To find the unknown number n, we ask ourselves, "What number times 3 is 60?" The number is 20 and can be found by dividing 60 by 3.

$n = \dfrac{60}{3}$ Divide 60 by the number multiplied by n.

$n = 20$ Simplify.

Thus, the unknown number is 20.

To *check*, replace n with this value, 20, and verify that a true proportion results.

> **Finding an Unknown Value n in a Proportion**
>
> **Step 1:** Set the cross products equal to each other.
>
> **Step 2:** Divide the number not multiplied by n by the number multiplied by n.

Practice 11

Find the value of the unknown number n.

$$\dfrac{15}{2} = \dfrac{60}{n}$$

Example 11   Find the value of the unknown number n.

$$\dfrac{51}{34} = \dfrac{3}{n}$$

Solution:

Step 1:

$$\dfrac{51}{34} \diagup\!\!\!\!\diagdown \dfrac{3}{n}$$

$51 \cdot n = 34 \cdot 3$ Set cross products equal.

$51 \cdot n = 102$ Multiply.

Step 2:

$n = \dfrac{102}{51}$ Divide 102 by 51, the number multiplied by n.

$n = 2$ Simplify.

Check to see that 2 is the unknown number n.

▪ **Work Practice 11**

Answer

11. $n = 8$

Example 12 Find the unknown number n.

$$\frac{7}{n} = \frac{6}{5}$$

Solution:

Step 1:

$$\frac{7}{n} = \frac{6}{5}$$

$7 \cdot 5 = n \cdot 6$ Set the cross products equal to each other.

$35 = n \cdot 6$ Multiply.

Step 2:

$\frac{35}{6} = n$ Divide 35 by 6, the number multiplied by n.

$5\frac{5}{6} = n$

Check to see that $5\frac{5}{6}$ is the unknown number.

■ **Work Practice 12**

Practice 12

Find the unknown number n.

$$\frac{8}{n} = \frac{5}{9}$$

Example 13 Find the unknown number n.

$$\frac{n}{3} = \frac{0.8}{1.5}$$

Solution:

Step 1:

$$\frac{n}{3} = \frac{0.8}{1.5}$$

$n \cdot 1.5 = 3 \cdot 0.8$ Set the cross products equal to each other.

$n \cdot 1.5 = 2.4$ Multiply.

Step 2:

$n = \frac{2.4}{1.5}$ Divide 2.4 by 1.5, the number multiplied by n.

$n = 1.6$ Simplify.

Check to see that 1.6 is the unknown number.

■ **Work Practice 13**

Practice 13

Find the unknown number n.

$$\frac{n}{6} = \frac{0.7}{1.2}$$

Objective E Solving Problems by Writing Proportions ▶

Writing proportions is a powerful tool for solving problems in almost every field, including business, chemistry, biology, health sciences, and engineering, as well as in daily life. Given a specified ratio (or rate) of two quantities, a proportion can be used to determine an unknown quantity.

In this section, we use the same problem-solving steps that we have used earlier in this text.

Answers

12. $n = 14\frac{2}{5}$ **13.** $n = 3.5$

Practice 14

On an architect's blueprint, 1 inch corresponds to 4 feet. How long is a wall represented by a $4\frac{1}{4}$-inch line on the blueprint?

Example 14 Determining Distances from a Map

On a chamber of commerce map of Abita Springs, 5 miles corresponds to 2 inches. How many miles correspond to 7 inches?

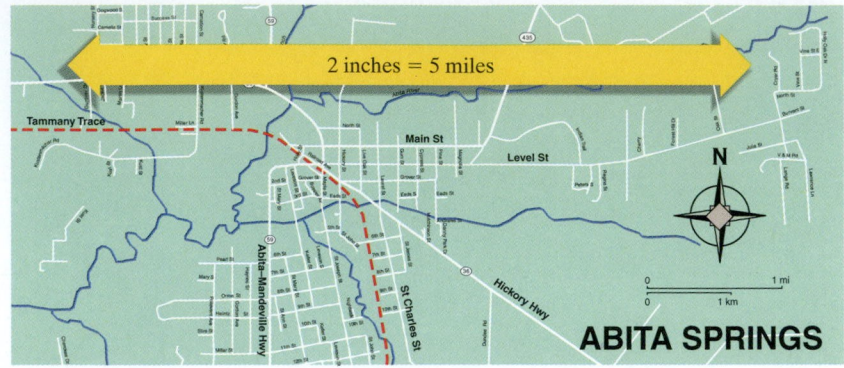

Solution:

1. UNDERSTAND. Read and reread the problem. You may want to draw a diagram.

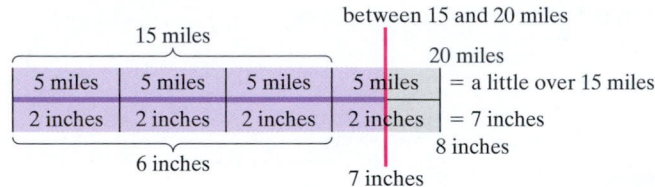

From the diagram we can see that a reasonable solution should be between 15 and 20 miles.

2. TRANSLATE. We will let n represent our unknown number. Since 5 miles corresponds to 2 inches as n miles corresponds to 7 inches, we have the proportion

$$\begin{array}{ccc}\text{miles} & \rightarrow & \dfrac{5}{2} = \dfrac{n}{7} & \leftarrow & \text{miles}\\ \text{inches} & \rightarrow & & \leftarrow & \text{inches}\end{array}$$

3. SOLVE: In earlier sections, we estimated to obtain a reasonable answer. Notice we did this in Step 1 above.

$$\frac{5}{2} = \frac{n}{7}$$

$5 \cdot 7 = 2 \cdot n$ Set the cross products equal to each other.

$35 = 2 \cdot n$ Multiply.

$\dfrac{35}{2} = n$ Divide 35 by 2, the number multiplied by n.

$n = 17\dfrac{1}{2}$ or 17.5   Simplify.

4. INTERPRET. *Check* your work. This result is reasonable since it is between 15 and 20 miles. *State* your conclusion: 7 inches corresponds to 17.5 miles.

■ **Work Practice 14**

Answer

14. 17 ft

Helpful Hint

We can also solve Example 14 by writing the proportion

$$\frac{2 \text{ inches}}{5 \text{ miles}} = \frac{7 \text{ inches}}{n \text{ miles}}$$

Although other proportions may be used to solve Example 14, we will solve by writing proportions so that the numerators have the same unit measures and the denominators have the same unit measures.

Example 15 Finding Medicine Dosage

The standard dose of an antibiotic is 4 cc (cubic centimeters) for every 25 pounds (lb) of body weight. At this rate, find the standard dose for a 140-lb woman.

Solution:

1. UNDERSTAND. Read and reread the problem. You may want to draw a diagram to estimate a reasonable solution.

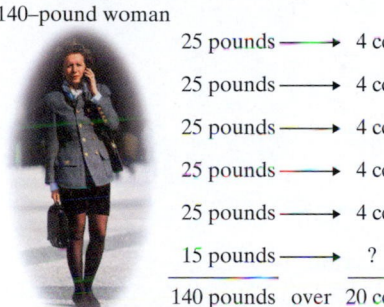

140–pound woman

25 pounds	⟶	4 cc
25 pounds	⟶	4 cc
25 pounds	⟶	4 cc
25 pounds	⟶	4 cc
25 pounds	⟶	4 cc
15 pounds	⟶	?

140 pounds over 20 cc

From the diagram, we can see that a reasonable solution is a little over 20 cc.

2. TRANSLATE. We will let n represent the unknown number. From the problem, we know that 4 cc is to 25 pounds as n cc is to 140 pounds, or

$$\begin{array}{ccc} \text{cubic centimeters} & \rightarrow & \dfrac{4}{25} = \dfrac{n}{140} \leftarrow \text{cubic centimeters} \\ \text{pounds} & \rightarrow & \phantom{\dfrac{4}{25}} \leftarrow \text{pounds} \end{array}$$

3. SOLVE:

$$\frac{4}{25} = \frac{n}{140}$$

$$4 \cdot 140 = 25 \cdot n \qquad \text{Set the cross products equal to each other.}$$

$$560 = 25 \cdot n \qquad \text{Multiply.}$$

$$\frac{560}{25} = n \qquad \text{Divide 560 by 25, the number multiplied by } n.$$

$$n = 22\frac{2}{5} \text{ or } 22.4 \qquad \text{Simplify.}$$

4. INTERPRET. *Check* your work. This result is reasonable since it is a little over 20 cc. *State* your conclusion: The standard dose for a 140-lb woman is 22.4 cc.

■ **Work Practice 15**

Practice 15

An auto mechanic recommends that 3 ounces of isopropyl alcohol be mixed with a tankful of gas (14 gallons) to increase the octane of the gasoline for better engine performance. At this rate, how many gallons of gas can be treated with a 16-ounce bottle of alcohol?

Answer

15. $74\frac{2}{3}$ or $74.\overline{6}$ gal

Vocabulary, Readiness & Video Check

Use the choices below to fill in each blank. Some choices may be used more than once, some not at all.

rate	unit	ratio	different	division	proportion
numerator	true	cross products	denominator	false	

Answer each statement true or false.

1. The quotient of two quantities is called a ratio. _____

2. The ratio $\frac{7}{5}$ means the same as the ratio $\frac{5}{7}$. _____

3. The ratio $\frac{7.2}{8.1}$ is in simplest form. _____

4. The ratio $\frac{10 \text{ feet}}{30 \text{ feet}}$ is in simplest form. _____

5. The ratio $\frac{9}{10}$ is in simplest form. _____

6. The ratio 2 to 5 equals $\frac{5}{2}$ in fractional notation. _____

7. The ratio 30 : 41 equals $\frac{30}{41}$ in fractional notation. _____

8. The ratio 15 to 45 equals $\frac{3}{1}$ in fractional notation. _____

9. A rate with a denominator of 1 is called a(n) _____ rate.

10. A(n) _____ is the quotient of two quantities.

11. The word *per* translates to "_____."

12. To write a rate as a unit rate, divide the _____ of the rate by the _____.

13. $\frac{4.2}{8.4} = \frac{1}{2}$ is called a(n) _____ while $\frac{7}{8}$ is called a(n) _____.

14. In $\frac{a}{b} = \frac{c}{d}$, $a \cdot d$ and $b \cdot c$ are called _____.

15. In a proportion, if cross products are equal, the proportion is _____.

16. In a proportion, if cross products are not equal, the proportion is _____.

Martin-Gay Interactive Videos

See Video 5.1

Watch the section lecture video and answer the following questions.

Objective A   **17.** Based on the lecture before ▦ Example 1, what three notations can we use for a ratio? For your answer, use the ratio given in the lecture. ▶

          18. Why is the ratio in ▦ Example 3 not in simplest form? ▶

Objective B   **19.** Why can't we divide out the units in ▦ Example 6? ▶

Objective C   **20.** In ▦ Example 7, what are the cross products of the proportion? Is the proportion true or false? ▶

Objective D   **21.** As briefly mentioned in ▦ Example 9, what's another word for the unknown value *n*? ▶

Objective E   **22.** In ▦ Example 13, interpret the meaning of the answer 102.9. ▶

5.1 **Exercise Set** MyMathLab®

Objective **A** *Write each ratio using fractional notation. Do not simplify. See Examples 1 through 3.*

▶ **1.** 23 to 10

2. 14 to 5

3. $3\frac{3}{4}$ to $1\frac{2}{3}$

4. $2\frac{2}{5}$ to $6\frac{1}{2}$

Write each ratio as a ratio of whole numbers using fractional notation. Write the fraction in simplest form. See Examples 1 through 3.

▶ **5.** 16 to 24

6. 25 to 150

▶ **7.** 7.7 to 10

8. 8.1 to 10

9. 10 hours to 24 hours

10. 18 quarts to 30 quarts

11. $32 to $100

12. $46 to $102

▶ **13.** 24 days to 14 days

14. 80 miles to 120 miles

15. 32,000 bytes to 46,000 bytes

16. 600 copies to 150 copies

17. 8 inches to 20 inches

18. 9 yards to 2 yards

Find the ratio described in each exercise as a fraction in simplest form. See Examples 4 and 5.

△ **19.** Find the ratio of the longest side to the perimeter of the right-triangular-shaped billboard.

△ **20.** Find the ratio of the width to the perimeter of the rectangular vegetable garden.

8 feet   15 feet

17 feet

4.5 meters   2 meters

In 2013, nearly 660 films by U.S. production companies were released. Use this information for Exercises 21 and 22.

21. In 2013, 45 digital 3-D films were released by U.S. production companies. Find the ratio of digital films to total films for 2013.

22. In 2013, 545 independent films were released by U.S. production companies. Find the ratio of independent films to total films for 2013.

23. Of the U.S. mountains that are over 14,000 feet in elevation, 57 are located in Colorado and 19 are located in Alaska. Find the ratio of the number of mountains over 14,000 feet found in Alaska to the number of mountains over 14,000 feet found in Colorado. (*Source:* U.S. Geological Survey)

24. Citizens of the United States eat an average of 25 pints of ice cream per year. Residents of the New England states eat an average of 39 pints of ice cream per year. Find the ratio of the amount of ice cream eaten by New Englanders to the amount eaten by the average U.S. citizen. (*Source:* International Dairy Foods Association)

Objective B *Write each rate as a fraction in simplest form. See Examples 6 through 8.*

25. 5 shrubs every 15 feet

26. 14 lab tables for 28 students

27. 15 returns for 100 sales

28. 150 graduate students for 8 advisors

29. 8 phone lines for 36 employees

30. 6 laser printers for 28 computers

31. 18 gallons of pesticide for 4 acres of crops

32. 4 inches of rain in 18 hours

33. 6 flight attendants for 200 passengers

34. 240 pounds of grass seed for 9 lawns

35. 355 calories in a 10-fluid-ounce chocolate milkshake (*Source: Home and Garden Bulletin No. 72,* U.S. Department of Agriculture)

36. 160 calories in an 8-fluid-ounce serving of cream of tomato soup (*Source: Home and Garden Bulletin No. 72,* U.S. Department of Agriculture)

Write each rate as a unit rate.

37. 330 calories in a 3-ounce serving

38. 275 miles in 11 hours

39. A hummingbird moves its wings at a rate of 5400 wingbeats a minute. Write this rate in wingbeats per second.

40. A bat moves its wings at a rate of 1200 wingbeats a minute. Write this rate in wingbeats per second.

Objective C *Determine whether each proportion is a true proportion. See Examples 9 and 10.*

▶ 41. $\dfrac{8}{6} = \dfrac{9}{7}$

42. $\dfrac{7}{12} = \dfrac{4}{7}$

▶ 43. $\dfrac{9}{36} = \dfrac{2}{8}$

44. $\dfrac{8}{24} = \dfrac{3}{9}$

Write each sentence as a proportion. Then determine whether the proportion is a true proportion. See Examples 9 and 10.

45. one and eight tenths is to two as four and five tenths is to five

46. fifteen hundredths is to three as thirty-five hundredths is to seven

47. two thirds is to one fifth as two fifths is to one ninth

48. ten elevenths is to three fourths as one fourth is to one half

Objective D *For each proportion, find the unknown number n. See Examples 11 through 13.*

49. $\dfrac{n}{5} = \dfrac{6}{10}$

50. $\dfrac{n}{3} = \dfrac{12}{9}$

51. $\dfrac{18}{54} = \dfrac{3}{n}$

52. $\dfrac{25}{100} = \dfrac{7}{n}$

▶ 53. $\dfrac{n}{8} = \dfrac{50}{100}$

54. $\dfrac{n}{21} = \dfrac{12}{18}$

55. $\dfrac{8}{15} = \dfrac{n}{6}$

56. $\dfrac{12}{10} = \dfrac{n}{16}$

57. $\dfrac{0.05}{12} = \dfrac{n}{0.6}$

58. $\dfrac{7.8}{13} = \dfrac{n}{2.6}$

▶ 59. $\dfrac{8}{1\frac{1}{3}} = \dfrac{24}{n}$

60. $\dfrac{12}{\frac{3}{4}} = \dfrac{48}{n}$

▶ **61.** $\dfrac{n}{1\frac{1}{5}} = \dfrac{4\frac{1}{6}}{6\frac{2}{3}}$

62. $\dfrac{n}{3\frac{1}{8}} = \dfrac{7\frac{3}{5}}{2\frac{3}{8}}$

63. $\dfrac{25}{n} = \dfrac{3}{\frac{7}{30}}$

64. $\dfrac{9}{n} = \dfrac{5}{\frac{11}{15}}$

Objective E *Solve. For Exercises 65 and 66, the solutions have been started for you. See Examples 14 and 15.*
An NBA basketball player averages 45 baskets for every 100 attempts.

65. If he attempted 800 field goals, how many baskets did he make?

Start the solution:

1. UNDERSTAND the problem. Reread it as many times as needed. Let's let
   n = how many field goals he made
2. TRANSLATE into an equation.

baskets (field goals) → $\dfrac{45}{100}$ $=$ $\dfrac{n}{800}$ ← baskets (field goals)
attempts → ← attempts

3. SOLVE the equation. Set cross products equal to each other and solve.

$$\dfrac{45}{100} \diagdown\!\!\!\!\!\diagup \dfrac{n}{800}$$

Finish by SOLVING and **4.** INTERPRET.

66. If he made 225 baskets, how many did he attempt?

Start the solution:

1. UNDERSTAND the problem. Reread it as many times as needed. Let's let
   n = how many baskets attempted
2. TRANSLATE into an equation.

baskets → $\dfrac{45}{100}$ $=$ $\dfrac{225}{n}$ ← baskets
attempts → ← attempts

3. SOLVE the equation. Set cross products equal to each other and solve.

$$\dfrac{45}{100} \diagdown\!\!\!\!\!\diagup \dfrac{225}{n}$$

Finish by SOLVING and **4.** INTERPRET.

It takes a word processor 30 minutes to word process and spell check 4 pages.

67. Find how long it takes her to word process and spell check 22 pages.

68. Find how many pages she can word process and spell check in 4.5 hours.

On an architect's blueprint, 1 inch corresponds to 8 feet.

69. Find the length of a wall represented by a line $2\frac{7}{8}$ inches long on the blueprint.

70. Find the length of a wall represented by a line $5\frac{1}{4}$ inches long on the blueprint.

A Honda Civic Hybrid car averages 627 miles on a 12.3-gallon tank of gas.

71. Manuel Lopez is planning a 1250-mile vacation trip in his Honda Civic Hybrid. Find how many gallons of gas he can expect to burn. Round to the nearest gallon.

72. Ramona Hatch has enough money to put 6.9 gallons of gas in her Honda Civic Hybrid. She is planning on driving home from college for the weekend. If her home is 290 miles away, should she make it home before she runs out of gas?

The scale on an Italian map states that 1 centimeter corresponds to 30 kilometers.

73. Find how far apart Milan and Rome are if their corresponding points on the map are 15 centimeters apart.

74. On the map, a small Italian village is located 0.4 centimeter from the Mediterranean Sea. Find the actual distance.

A bag of Scotts fertilizer covers 3000 square feet of lawn.

△ **75.** Find how many bags of fertilizer should be purchased to cover a rectangular lawn 260 feet by 180 feet.

△ **76.** Find how many bags of fertilizer should be purchased to cover a square lawn measuring 160 feet on each side.

A self-tanning lotion advertises that a 3-oz bottle will provide four applications.

77. Jen Haddad found a great deal on a 14-oz bottle of the self-tanning lotion she had been using. Based on the advertising claims, how many applications of the self-tanner should Jen expect? Round down to the smaller whole number.

78. The Community College thespians need fake tans for a play they are doing. If the play has a cast of 35, how many ounces of self-tanning lotion should the cast purchase? Round up to the next whole number of ounces.

The school's computer lab goes through 5 reams of printer paper every 3 weeks.

▶ **79.** Find out how long a case of printer paper is likely to last (a case of paper holds 8 reams of paper). Round to the nearest week.

80. How many cases of printer paper should be purchased to last the entire semester of 15 weeks? Round up to the next case.

81. In the Seattle Space Needle, the elevators whisk you to the revolving restaurant at a speed of 800 feet in 60 seconds. If the revolving restaurant is 500 feet up, how long will it take you to reach the restaurant by elevator? (*Source:* Seattle Space Needle)

82. A 16-oz grande Shaken Sweet Tea at Starbucks has 100 calories. How many calories are there in a 24-oz venti Shaken Sweet Tea? (*Source:* Starbucks Coffee Company)

83. Mosquitos are annoying insects. To eliminate mosquito larvae, a certain granular substance can be applied to standing water in a ratio of 1 tsp per 25 sq ft of standing water.

   a. At this rate, find how many teaspoons of granules must be used for 450 square feet.

   b. If 3 tsp = 1 tbsp, how many tablespoons of granules must be used?

84. Another type of mosquito control is liquid, where 3 oz of pesticide is mixed with 100 oz of water. This mixture is sprayed on roadsides to control mosquito breeding grounds hidden by tall grass.

   a. If one mixture of water with this pesticide can treat 150 feet of roadway, how many ounces of pesticide are needed to treat one mile? (*Hint:* 1 mile = 5280 feet)

   b. If 8 liquid ounces equals one cup, write your answer to part **a** in cups. Round to the nearest cup.

85. The daily supply of oxygen for one person is provided by 625 square feet of lawn. A total of 3750 square feet of lawn would provide the daily supply of oxygen for how many people? (*Source:* Professional Lawn Care Association of America)

86. In 2013, approximately 96 million of the 136 million U.S. employees worked in service industries. In a town of 6800 workers, how many would be expected to work in service-industry jobs? (*Source:* U.S. Bureau of Labor Statistics)

87. A student would like to estimate the height of the Statue of Liberty in New York City's harbor. The length of the Statue of Liberty's right arm is 42 feet. The student's right arm is 2 feet long and her height is $5\frac{1}{3}$ feet. Use this information to estimate the height of the Statue of Liberty. How close is your estimate to the statue's actual height of 111 feet, 1 inch from heel to top of head? (*Source:* National Park Service)

88. The length of the Statue of Liberty's index finger is 8 feet while the height to the top of the head is about 111 feet. Suppose your measurements are proportionally the same as this statue and your height is 5 feet.

   a. Use this information to find the proposed length of your index finger. Give an exact measurement and then a decimal rounded to the nearest hundredth.

   b. Measure your index finger and write it as a decimal in feet rounded to the nearest hundredth. How close is the length of your index finger to the answer to part **a**? Explain why.

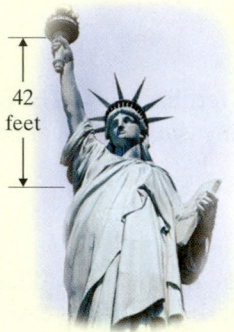
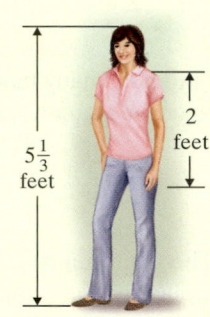

42 feet

$5\frac{1}{3}$ feet

2 feet

89. There are 72 milligrams of cholesterol in a 3.5-ounce serving of lobster. How much cholesterol is in 5 ounces of lobster? Round to the nearest tenth of a milligram. (*Source:* The National Institutes of Health)

90. There are 76 milligrams of cholesterol in a 3-ounce serving of skinless chicken. How much cholesterol is in 8 ounces of chicken? (*Source:* USDA)

91. The adult daily dosage for a certain medicine is 150 mg (milligrams) of medicine for every 20 pounds of body weight.

   a. At this rate, find the daily dose for a man who weighs 275 pounds.

   b. If the man is to receive 500 mg of this medicine every 8 hours, is he receiving the proper dosage?

92. The adult daily dosage for a certain medicine is 80 mg (milligrams) for every 25 pounds of body weight.

   a. At this rate, find the daily dose for a woman who weighs 190 pounds.

   b. If she is to receive this medicine every 6 hours, find the amount to be given every 6 hours.

93. The gas/oil ratio for a certain chainsaw is 50 to 1.

   a. How much oil (in gallons) should be mixed with 5 gallons of gasoline?

   b. If 1 gallon equals 128 fluid ounces, write the answer to part **a** in fluid ounces. Round to the nearest whole ounce.

94. The gas/oil ratio for a certain tractor mower is 20 to 1.

   a. How much oil (in gallons) should be mixed with 10 gallons of gas?

   b. If 1 gallon equals 4 quarts, write the answer to part **a** in quarts.

Review

Find the prime factorization of each number. See Section 2.2.

95. 20   **96.** 24   **97.** 200   **98.** 300   **99.** 32   **100.** 81

Concept Extensions

As we have seen earlier, proportions are often used in medicine dosage calculations. The exercises below have to do with liquid drug preparations, where the weight of the drug is contained in a volume of solution. The description of mg and ml below will help.

mg means milligrams (A paper clip weighs about a gram. A milligram is about the weight of $\frac{1}{1000}$ of a paper clip.)

ml means milliliter (A liter is about a quart. A milliliter is about the amount of liquid in $\frac{1}{1000}$ of a quart.)

One way to solve the applications below is to set up the proportion $\frac{mg}{ml} = \frac{mg}{ml}$.

A solution strength of 15 mg of medicine in 1 ml of solution is available.

101. If a patient needs 12 mg of medicine, how many ml do you administer?

102. If a patient needs 33 mg of medicine, how many ml do you administer?

A solution strength of 8 mg of medicine in 1 ml of solution is available.

103. If a patient needs 10 mg of medicine, how many ml do you administer?

104. If a patient needs 6 mg of medicine, how many ml do you administer?

105. Is the ratio $\frac{11}{15}$ the same as the ratio of $\frac{15}{11}$? Explain your answer.

106. Explain why the ratio $\frac{40}{17}$ is incorrect for Exercise 19.

107. Explain the difference between a ratio and a proportion.

108. Explain how to find the unknown number in a proportion such as $\frac{n}{18} = \frac{12}{8}$.

For each proportion, find the unknown number n.

109. $\frac{n}{1150} = \frac{588}{483}$

110. $\frac{222}{1515} = \frac{37}{n}$

Objectives

A Understand Percent.

B Write Percents as Decimals.

C Write Decimals as Percents.

5.2 Introduction to Percent

Objective A Understanding Percent

The word **percent** comes from the Latin phrase *per centum*, which means **"per 100."** For example, 53% (percent) means 53 per 100. In the square below, 53 of the 100 squares are shaded. Thus, 53% of the figure is shaded.

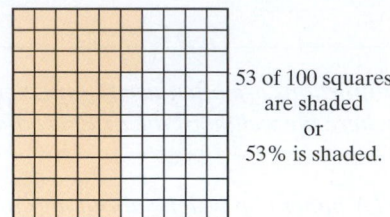

53 of 100 squares
are shaded
or
53% is shaded.

Since 53% means 53 per 100, 53% is the ratio of 53 to 100, or $\frac{53}{100}$.

$$53\% = \frac{53}{100}$$

Also,

$$7\% = \frac{7}{100} \quad \text{7 parts per 100 parts}$$

$$73\% = \frac{73}{100} \quad \text{73 parts per 100 parts}$$

$$109\% = \frac{109}{100} \quad \text{109 parts per 100 parts}$$

> **Percent**
>
> **Percent** means **per one hundred.** The "%" symbol is used to denote percent.

Percent is used in a variety of everyday situations. For example,

- 77.6% of the U.S. population uses the Internet.
- The store is having a 25%-off sale.
- 78% of us trust our local fire department.
- The enrollment in community colleges is predicted to increase 1.3% each year.
- The South is the home of 49% of all frequent paintball participants.

Practice 1

Of 100 students in a club, 23 are freshmen. What percent of the students are freshmen?

Example 1 Since 2011, white has been the world's most popular color for cars.

For 2013 model-year cars, 25 out of every 100 were painted white. What percent of model-year 2013 cars were white? (*Source:* PPG Industries)

Solution: Since 25 out of 100 cars were painted white, the fraction is $\frac{25}{100}$. Then

$$\frac{25}{100} = 25\%$$

■ Work Practice 1

Answer

1. 23%

Example 2 46 out of every 100 college students live at home. What percent of students live at home? (*Source:* Independent Insurance Agents of America)

Solution:

$$\frac{46}{100} = 46\%$$

■ **Work Practice 2**

Practice 2

29 out of 100 executives are in their forties. What percent of executives are in their forties?

Objective B Writing Percents as Decimals

Since percent means "per hundred," we have that

$$1\% = \frac{1}{100} = 0.01$$

In other words, the percent symbol means "per hundred" or, equivalently, "$\frac{1}{100}$" or "0.01." Thus

$$87\% = 87 \times \frac{1}{100} = \frac{87}{100}$$

or

$$87\% = 87 \times (0.01) = 0.87$$

> Results are the same

Of course, we know that the end results are the same, that is,

$$\frac{87}{100} = 0.87$$

The above gives us two options for converting percents. We can replace the percent symbol, %, by $\frac{1}{100}$ or 0.01 and then multiply.

> For consistency, when we
> - convert from a percent to a *decimal*, we will drop the % symbol and multiply by 0.01 (this section).
> - convert from a percent to a *fraction*, we will drop the % symbol and multiply by $\frac{1}{100}$ (next section).

Thus, to write 53% (or 53.%) as a decimal,

$$53\% = 53(0.01) = 0.53 \quad \text{Replace the percent symbol with 0.01. Then multiply.}$$

> **Writing a Percent as a Decimal**
>
> Replace the percent symbol with its decimal equivalent, 0.01; then multiply.
>
> $$43\% = 43(0.01) = 0.43$$

Helpful Hint

If it helps, think of writing a percent as a decimal by

Percent → | Remove the % symbol and move decimal point 2 places to the left | → Decimal

Answer

2. 29%

Practice 3

Write 89% as a decimal.

Example 3 Write 23% as a decimal.

Solution:

$$23\% = 23(0.01) \quad \text{Replace the percent symbol with 0.01.}$$
$$= 0.23 \quad \text{Multiply.}$$

■ **Work Practice 3**

Practice 4–7

Write each percent as a decimal.

4. 2.7%   **5.** 150%

6. 0.69%   **7.** 800%

Examples Write each percent as a decimal.

4. $4.6\% = 4.6(0.01) = 0.046$ Replace the percent symbol with 0.01. Then multiply.

5. $190\% = 190(0.01) = 1.90$ or   1.9

6. $0.74\% = 0.74(0.01) = 0.0074$

7. $100\% = 100(0.01) = 1.00$ or   1

> **Helpful Hint**
> We just learned that $100\% = 1$

■ **Work Practice 4–7**

✓**Concept Check** Why is it incorrect to write the percent 0.033% as 3.3 in decimal form?

Objective C Writing Decimals as Percents

To write a decimal as a percent, we use the result of Example 7 above. In this example, we found that $1 = 100\%$.

$$0.38 = 0.38(1) = 0.38(100\%) = 38\%$$

Notice that the result is

$$0.38 = 0.38(100\%) = 38.\%$$ Multiply by 1 in the form of 100%.

> **Writing a Decimal as a Percent**
>
> Multiply by 1 in the form of 100%.
>
> $$0.27 = 0.27(100\%) = 27.\%$$

> **Helpful Hint**
>
> If it helps, think of writing a decimal as a percent by reversing the steps in the Helpful Hint on the previous page.
>
> Percent ← | Move the decimal point 2 places to the right and attach a % symbol. | ← Decimal

Answers

3. 0.89 **4.** 0.027 **5.** 1.5
6. 0.0069 **7.** 8.00 or 8

✓**Concept Check Answer**

To write a percent as a decimal, the decimal point should be moved two places to the left, not to the right. So the correct answer is 0.00033.

Example 8 Write 0.65 as a percent.

Solution:

$$0.65\% = 0.65(100\%) = 65.\% \quad \text{Multiply by 100\%.}$$
$$= 65\%$$

■ **Work Practice 8**

Examples Write each decimal as a percent.

9. $1.25 = 1.25(100\%) = 125.\%$ or 125%

10. $0.012 = 0.012(100\%) = 001.2\%$ or 1.2%

11. $0.6 = 0.6(100\%) = 060.\%$ or 60%

Helpful Hint

A zero was inserted as a placeholder.

■ **Work Practice 9–11**

✔**Concept Check** Why is it incorrect to write the decimal 0.0345 as 34.5% in percent form?

Practice 8

Write 0.19 as a percent.

Practice 9–11

Write each decimal as a percent.

9. 1.75 **10.** 0.044 **11.** 0.7

Answers

8. 19% **9.** 175% **10.** 4.4%

11. 70%

✔**Concept Check Answer**

To change a decimal to a percent, multiply by 100%, or move the decimal point *only* two places to the right. So the correct answer is 3.45%.

Vocabulary, Readiness & Video Check

Use the choices below to fill in each blank. Some choices may be used more than once or not at all.

$$\frac{1}{100} \qquad 0.01 \qquad 100\% \qquad percent$$

1. _____ means "per hundred."

2. _____ = 1.

3. The % symbol is read as _____.

4. To write a decimal as a *percent*, multiply by 1 in the form of _____.

5. To write a percent as a *decimal*, drop the % symbol and multiply by _____.

Martin-Gay Interactive Videos

See Video 5.2 🍎

Watch the section lecture video and answer the following questions.

Objective A **6.** From the lecture before ▭ Example 1, what is the most important thing to remember about percent?

Objective B **7.** From the lecture before ▭ Example 2, what does 1% equal in fraction form? In decimal form?

Objective C **8.** Complete this statement based on ▭ Example 7: Multiplying a decimal by _____ is the same as multiplying it by 1 and does not change the _____ of the decimal.

5.2 Exercise Set MyMathLab®

Objective A *Solve. See Examples 1 and 2.*

1. In a survey of 100 college students, 96 use the Internet. What percent use the Internet?

2. A basketball player makes 81 out of 100 attempted free throws. What percent of free throws are made?

3. Michigan leads the United States in tart cherry production, producing 75 out of every 100 tart cherries each year.
   a. What percent of tart cherries are produced in Michigan?
   b. What percent of tart cherries are *not* produced in Michigan? (*Source:* Cherry Marketing Institute)

4. The United States is the world's second-largest producer of apples. Twenty-five out of every 100 apples harvested in the United States are exported (shipped to other countries). (*Source:* U.S. Apple Association)
   a. What percent of U.S.-grown apples are exported?
   b. What percent of U.S.-grown apples are *not* exported?

One hundred adults were asked to name their favorite sport, and the results are shown in the circle graph.

5. What sport was preferred by most adults? What percent preferred this sport?

6. What sport was preferred by the least number of adults? What percent preferred this sport?

7. What percent of adults preferred football or soccer?

8. What percent of adults preferred basketball or baseball?

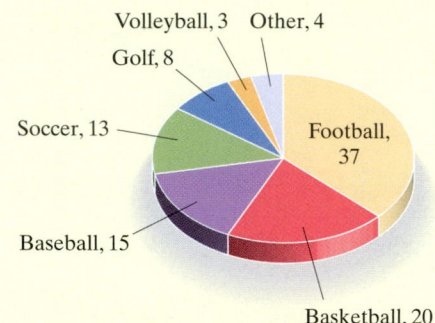

Volleyball, 3   Other, 4
Golf, 8
Soccer, 13
Football, 37
Baseball, 15
Basketball, 20

Objective B *Write each percent as a decimal. See Examples 3 through 7.*

9. 41%

10. 62%

11. 6%

12. 3%

13. 100%

14. 136%

15. 73.6%

16. 45.7%

17. 2.8%

18. 1.4%

19. 0.6%

20. 0.9%

21. 300%

22. 500%

23. 32.58%

24. 72.18%

Write each percent as a decimal. See Examples 3 through 7.

25. People take aspirin for a variety of reasons. The most common use of aspirin is to prevent heart disease, accounting for 38% of all aspirin use. (*Source:* Bayer Market Research)

26. Exports to Europe accounted for 17.7% of all of Japan's motor vehicle exports. (*Source:* Japan Automobile Manufacturers Association)

27. In 2012, 38.2% of households in the United States had no landline telephones, just cell phones. (*Source:* National Center for Health Statistics)

28. In 2012, 8.6% of households in the United States had no cell phones, just landline telephones. (*Source:* National Center for Health Statistics)

29. A $\frac{1}{2}$-cup serving of dried tart cherries delivers 45% of an adult's Daily Value of vitamin A. (*Source:* USDA Nutrient Data Laboratory)

30. In 2010, 63.5% of the paper used in the United States was recovered for recycling. (*Source:* Keep America Beautiful, Inc.)

Objective C *Write each decimal as a percent. See Examples 8 through 11.*

31. 0.98   **32.** 0.75   **33.** 3.1   **34.** 4.8   **35.** 29

36. 56   **37.** 0.003   **38.** 0.006   ▶ **39.** 0.22   **40.** 0.45

41. 5.3   **42.** 1.6   ▶ **43.** 0.056   **44.** 0.027   **45.** 0.3328

46. 0.1115   ▶ **47.** 3   **48.** 5   ▶ **49.** 0.7   **50.** 0.8

Write each decimal as a percent. See Examples 8 through 11.

51. Leisure travel accounted for 0.77 of all domestic trips in the United States. (*Source:* U.S. Travel Association)

52. According to a recent survey, 0.34 of American adults reported having used the Internet to research nutritional information about restaurant foods. (*Source:* National Restaurant Association's *2013 Restaurant Industry Forecast*)

53. In 2012, about 0.281 of full-time workers in the United States spent time volunteering in the previous year. (*Source:* Bureau of Labor Statistics)

54. In 2012, about 0.334 of part-time workers in the United States participated in volunteer activities in the previous year. (*Source:* Bureau of Labor Statistics)

55. Nearly 0.081 of the United States labor force was unemployed in 2012. (*Source:* Bureau of Labor Statistics)

56. In 2012, an estimated 0.235 of the American population was under the age of 18. (*Source:* U.S. Census Bureau)

Review

Write each fraction as a decimal. See Section 4.6.

57. $\frac{1}{4}$ **58.** $\frac{3}{5}$ **59.** $\frac{13}{20}$ **60.** $\frac{11}{40}$ **61.** $\frac{9}{10}$ **62.** $\frac{7}{10}$

Concept Extensions

Solve. See the Concept Checks in this section.

63. Which of the following are correct?
   a. 6.5% = 0.65
   b. 7.8% = 0.078
   c. 120% = 0.12
   d. 0.35% = 0.0035

64. Which of the following are correct?
   a. 0.231 = 23.1%
   b. 5.12 = 0.0512%
   c. 3.2 = 320%
   d. 0.0175 = 0.175%

Recall that 1 = 100%. This means that 1 whole is 100%. Use this for Exercises 65 and 66. (Source: Some Body by Dr. Pete Rowen)

65. The four blood types are A, B, O, and AB. (Each blood type can also be further classified as Rh-positive or Rh-negative depending upon whether your blood contains protein or not.) Given the percent blood types for the United States below, calculate the percent of U.S. population with AB blood type.

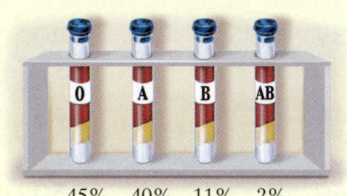

45%   40%   11%   ?%

66. The components of bone are all listed in the categories below. Find the missing percent.
   1. Minerals–45%
   2. Living tissue–30%
   3. Water–20%
   4. Other–?

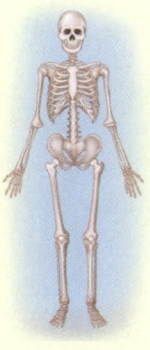

The bar graph shows the predicted fastest-growing occupations. Use the graph for Exercises 67 through 70. (Source: Bureau of Labor Statistics)

Fastest-Growing Occupations 2010–2020

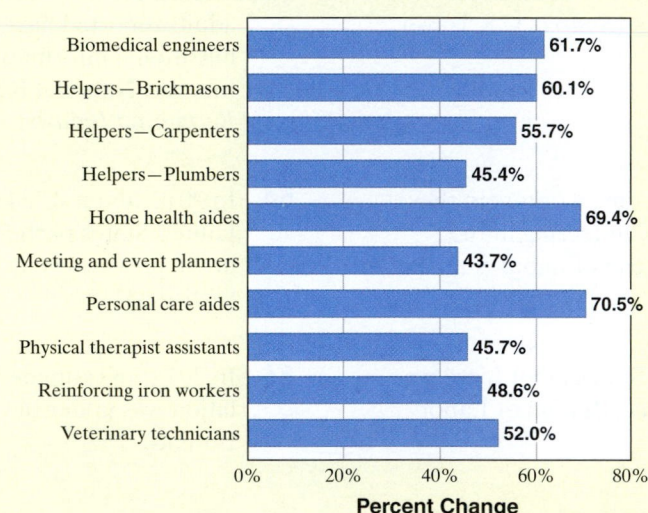

Occupation	Percent Change
Biomedical engineers	61.7%
Helpers—Brickmasons	60.1%
Helpers—Carpenters	55.7%
Helpers—Plumbers	45.4%
Home health aides	69.4%
Meeting and event planners	43.7%
Personal care aides	70.5%
Physical therapist assistants	45.7%
Reinforcing iron workers	48.6%
Veterinary technicians	52.0%

Percent Change

Source: Bureau of Labor Statistics

67. What occupation is predicted to be the fastest growing?

68. What occupation is predicted to be the second fastest growing?

69. Write the percent change for biomedical engineers as a decimal.

70. Write the percent change for meeting and event planners as a decimal.

71. In your own words, explain how to write a percent as a decimal.

72. In your own words, explain how to write a decimal as a percent.

5.3   Percents and Fractions

Objective A   Writing Percents as Fractions

Recall from Section 5.2 that percent means per hundred. Thus

$$1\% = \frac{1}{100} = 0.01$$

For example,

$$87\% = 87 \times \frac{1}{100} = \frac{87}{100} \quad \text{Writing 87\% as a fraction.}$$

or

$$87\% = 87 \times 0.01 = 0.87 \quad \text{Writing 87\% as a decimal.}$$

In this section we are writing percents as fractions, so we do the following.

> **Writing a Percent as a Fraction**
>
> Replace the percent symbol with its fraction equivalent, $\frac{1}{100}$; then multiply. Don't forget to simplify the fraction if possible.
>
> $$7\% = 7 \cdot \frac{1}{100} = \frac{7}{100}$$

Examples   Write each percent as a fraction or mixed number in simplest form.

1. $40\% = 40 \cdot \frac{1}{100} = \frac{40}{100} = \frac{2 \cdot \overset{1}{\cancel{20}}}{5 \cdot \underset{1}{\cancel{20}}} = \frac{2}{5}$

2. $1.9\% = 1.9 \cdot \frac{1}{100} = \frac{1.9}{100}$. We don't want the numerator of the fraction to contain a decimal, so we multiply by 1 in the form of $\frac{10}{10}$.

$$= \frac{1.9}{100} \cdot \frac{10}{10} = \frac{1.9 \cdot 10}{100 \cdot 10} = \frac{19}{1000}$$

3. $125\% = 125 \cdot \frac{1}{100} = \frac{125}{100} = \frac{5 \cdot \overset{1}{\cancel{25}}}{4 \cdot \underset{1}{\cancel{25}}} = \frac{5}{4} \text{ or } 1\frac{1}{4}$

4. $33\frac{1}{3}\% = 33\frac{1}{3} \cdot \frac{1}{100} = \frac{100}{3} \cdot \frac{1}{100} = \frac{\overset{1}{\cancel{100}} \cdot 1}{3 \cdot \underset{1}{\cancel{100}}} = \frac{1}{3}$

   Write as an improper fraction.

5. $100\% = 100 \cdot \frac{1}{100} = \frac{100}{100} = 1$

> **Helpful Hint**   Just as in the previous section, we confirm that $100\% = 1$

■ **Work Practice 1–5**

Objective B   Writing Fractions as Percents

Recall that to write a percent as a fraction, we replace the percent symbol by its fraction equivalent, $\frac{1}{100}$. We reverse these steps to write a fraction as a percent.

Objectives

A Write Percents as Fractions.

B Write Fractions as Percents.

C Convert Percents, Decimals, and Fractions.

Practice 1–5

Write each percent as a fraction or mixed number in simplest form.

1. 25%

2. 2.3%

3. 225%

4. $66\frac{2}{3}\%$

5. 8%

Answers

1. $\frac{1}{4}$ 2. $\frac{23}{1000}$ 3. $\frac{9}{4}$ or $2\frac{1}{4}$

4. $\frac{2}{3}$ 5. $\frac{2}{25}$

> ### Writing a Fraction as a Percent
> Multiply by 1 in the form of 100%.
> $$\frac{1}{8} = \frac{1}{8} \cdot 100\% = \frac{1}{8} \cdot \frac{100}{1}\% = \frac{100}{8}\% = 12\frac{1}{2}\% \quad \text{or} \quad 12.5\%$$

Helpful Hint

From Example 5, we know that

$$100\% = 1$$

Recall that when we multiply a number by 1, we are not changing the value of that number. This means that when we multiply a number by 100%, we are not changing its value but rather writing the number as an equivalent percent.

Practice 6–8

Write each fraction or mixed number as a percent.

6. $\frac{1}{2}$ **7.** $\frac{7}{40}$ **8.** $2\frac{1}{4}$

Examples   Write each fraction or mixed number as a percent.

6. $\frac{9}{20} = \frac{9}{20} \cdot 100\% = \frac{9}{20} \cdot \frac{100}{1}\% = \frac{900}{20}\% = 45\%$

7. $\frac{2}{3} = \frac{2}{3} \cdot 100\% = \frac{2}{3} \cdot \frac{100}{1}\% = \frac{200}{3}\% = 66\frac{2}{3}\%$

8. $1\frac{1}{2} = \frac{3}{2} \cdot 100\% = \frac{3}{2} \cdot \frac{100}{1}\% = \frac{300}{2}\% = 150\%$

Helpful Hint

$\frac{200}{3} = 66.\overline{6}.$

Thus, another way to write $\frac{200}{3}\%$ is $66.\overline{6}\%$.

■ **Work Practice 6–8**

✔**Concept Check** Which digit in the percent 76.4582% represents

a. A tenth percent?   **b.** A thousandth percent?

c. A hundredth percent?   **d.** A whole percent?

Practice 9

Write $\frac{3}{17}$ as a percent. Round to the nearest hundredth percent.

Example 9   Write $\frac{1}{12}$ as a percent. Round to the nearest hundredth percent.

Solution:

$$\frac{1}{12} = \frac{1}{12} \cdot 100\% = \frac{1}{12} \cdot \frac{100}{1}\% = \frac{100}{12}\% \approx 8.33\%$$

— "approximately"

$$\begin{array}{r} 8.333 \approx 8.33 \\ 12\overline{)100.000} \\ -96 \\ \hline 4\,0 \\ -3\,6 \\ \hline 40 \\ -36 \\ \hline 40 \\ -36 \\ \hline 4 \end{array}$$

Thus, $\frac{1}{12}$ is approximately 8.33%.

■ **Work Practice 9**

Answers

6. 50%   **7.** $17\frac{1}{2}\%$ **8.** 225%   **9.** 17.65%

✔**Concept Check Answers**

a. 4   **b.** 8   **c.** 5   **d.** 6

Objective C Converting Percents, Decimals, and Fractions

Let's summarize what we have learned so far about percents, decimals, and fractions:

> ### Summary of Converting Percents, Decimals, and Fractions
>
> - *To write a percent as a decimal*, replace the % symbol with its decimal equivalent, 0.01; then multiply.
> - *To write a percent as a fraction*, replace the % symbol with its fraction equivalent, $\frac{1}{100}$; then multiply.
> - *To write a decimal or fraction as a percent*, multiply by 100%.

If we let p represent a number, below we summarize using symbols.

Write a percent as a decimal:	Write a percent as a fraction:	Write a number as a percent:
$p\% = p(0.01)$	$p\% = p \cdot \dfrac{1}{100}$	$p = p \cdot 100\%$

Example 10 The greatest percent of automobile thefts in the continental United States, 37.8%, occur in the South. Write this percent as a decimal and as a fraction. (*Source:* Federal Bureau of Investigation's (FBI) Uniform Crime Report)

Solution:

As a decimal: $37.8\% = 37.8(0.01) = 0.378$

As a fraction: $37.8\% = 37.8 \cdot \dfrac{1}{100} = \dfrac{37.8}{100} = \dfrac{37.8}{100} \cdot \dfrac{10}{10} = \dfrac{378}{1000} = \dfrac{\overset{1}{\cancel{2}} \cdot 189}{\underset{1}{\cancel{2}} \cdot 500} = \dfrac{189}{500}.$

Thus, 37.8% written as a decimal is 0.378 and written as a fraction is $\dfrac{189}{500}$.

■ **Work Practice 10**

Practice 10

A family decides to spend no more than 22.5% of its monthly income on rent. Write 22.5% as a decimal and as a fraction.

Example 11 An advertisement for a stereo system reads "$\frac{1}{4}$ off." What percent off is this?

Solution: Write $\frac{1}{4}$ as a percent.

$\dfrac{1}{4} = \dfrac{1}{4} \cdot 100\% = \dfrac{1}{4} \cdot \dfrac{100}{1}\% = \dfrac{100}{4}\% = 25\%$

Thus, "$\frac{1}{4}$ off" is the same as "25% off."

■ **Work Practice 11**

Practice 11

Provincetown's budget for waste disposal increased by $1\frac{1}{4}$ times over the budget from last year. What percent increase is this?

Note: It is helpful to know a few basic percent conversions. Appendix B.2 contains a handy reference of percent, decimal, and fraction equivalencies.

Answers

10. $0.225, \dfrac{9}{40}$ **11.** 125%

Vocabulary, Readiness & Video Check

Use the choices below to fill in each blank. Some choices may be used more than once.

$\frac{1}{100}$ 100% percent

1. _____ means "per hundred."

2. _____ = 1.

3. To write a decimal or a fraction as a *percent*, multiply by 1 in the form of _____.

4. To write a percent as a *fraction*, drop the % symbol and multiply by _____.

Write each fraction as a percent.

5. $\frac{13}{100}$ **6.** $\frac{92}{100}$ **7.** $\frac{87}{100}$ **8.** $\frac{71}{100}$ **9.** $\frac{1}{100}$ **10.** $\frac{2}{100}$

Martin-Gay Interactive Videos Watch the section lecture video and answer the following questions.

See Video 5.3

Objective A **11.** In ⊞ Example 1, since the % symbol is replaced with $\frac{1}{100}$, why doesn't the final answer have a denominator of 100? ▶

Objective B **12.** From the lecture before ⊞ Example 4, how is writing a fraction as a percent similar to writing a decimal as a percent? ▶

Objective C **13.** From ⊞ Example 7, what is the main difference between writing a percent as an equivalent decimal and writing a percent as an equivalent fraction? ▶

5.3 Exercise Set MyMathLab®

Objective A *Write each percent as a fraction or mixed number in simplest form. See Examples 1 through 5.*

1. 12%   **2.** 24%   ▶**3.** 4%   **4.** 2%   **5.** 4.5%

6. 7.5%   ▶**7.** 175%   **8.** 250%   **9.** 73%   **10.** 86%

11. 12.5%   **12.** 62.5%   **13.** 6.25%   **14.** 3.75%   **15.** 6%

16. 16%   ▶**17.** $10\frac{1}{3}$%   **18.** $7\frac{3}{4}$%   **19.** $22\frac{3}{8}$%   **20.** $15\frac{5}{8}$%

Objective **B** *Write each fraction or mixed number as a percent. See Examples 6 through 8.*

21. $\dfrac{3}{4}$ 22. $\dfrac{1}{4}$ ▶ 23. $\dfrac{7}{10}$ 24. $\dfrac{3}{10}$ 25. $\dfrac{2}{5}$ 26. $\dfrac{4}{5}$

27. $\dfrac{59}{100}$ 28. $\dfrac{83}{100}$ 29. $\dfrac{17}{50}$ 30. $\dfrac{47}{50}$ ▶ 31. $\dfrac{3}{8}$ 32. $\dfrac{5}{8}$

33. $\dfrac{5}{16}$ 34. $\dfrac{7}{16}$ 35. $1\dfrac{3}{5}$ 36. $1\dfrac{3}{4}$ 37. $\dfrac{7}{9}$ 38. $\dfrac{1}{3}$

39. $\dfrac{13}{20}$ 40. $\dfrac{3}{20}$ 41. $2\dfrac{1}{2}$ 42. $2\dfrac{1}{5}$ 43. $1\dfrac{9}{10}$ 44. $2\dfrac{7}{10}$

Write each fraction as a percent. Round to the nearest hundredth percent. See Example 9.

45. $\dfrac{7}{11}$ 46. $\dfrac{5}{12}$ ▶ 47. $\dfrac{4}{15}$ 48. $\dfrac{10}{11}$

49. $\dfrac{1}{7}$ 50. $\dfrac{1}{9}$ 51. $\dfrac{11}{12}$ 52. $\dfrac{5}{6}$

Objective **C** *Complete each table. See Examples 10 and 11.*

53.

Percent	Decimal	Fraction
35%		
		$\dfrac{1}{5}$
	0.5	
70%		
		$\dfrac{3}{8}$

54.

Percent	Decimal	Fraction
60%		
		$\dfrac{2}{5}$
	0.25	
12.5%		
		$\dfrac{5}{8}$
		$\dfrac{7}{50}$

55.

Percent	Decimal	Fraction
40%		
	0.235	
		$\dfrac{4}{5}$
$33\dfrac{1}{3}\%$		
		$\dfrac{7}{8}$
7.5%		

56.

Percent	Decimal	Fraction
	0.525	
		$\dfrac{3}{4}$
$66\dfrac{2}{3}\%$		
		$\dfrac{5}{6}$
100%		

57.

Percent	Decimal	Fraction
200%		
	2.8	
705%		
		$4\frac{27}{50}$

58.

Percent	Decimal	Fraction
800%		
	3.2	
608%		
		$9\frac{13}{50}$

Solve. See Examples 10 and 11.

59. A 2012 poll revealed that 67% of American adults are in favor of keeping the penny in circulation. Write this percent as a decimal and a fraction. (*Source:* Americans for Common Cents)

60. China was responsible for 42% of the world's smelter production of aluminum in 2012. Write this percent as a decimal and a fraction. (*Source:* U.S. Geological Survey)

61. In 2012, 52.2% of all veterinarians in private practice were female. Write this percent as a decimal and a fraction. (*Source:* American Veterinary Medical Association)

62. The U.S. penny is 97.5% zinc. Write this percent as a decimal and a fraction. (*Source:* Americans for Common Cents)

63. According to a recent survey, $\frac{4}{5}$ of American restaurant owners started out in the restaurant industry in an entry-level position. Write this fraction as a percent. (*Source:* National Restaurant Association's *2013 Restaurant Industry Forecast*)

64. Of all U.S. veterinarians in private practice in 2012, $\frac{3}{50}$ focused exclusively on horses. Write this fraction as a percent. (*Source:* American Veterinary Medical Association)

65. The sales tax in Slidell, Louisiana, is 8.75%. Write this percent as a decimal.

66. A real estate agent receives a commission of 3% of the sale price of a house. Write this percent as a decimal.

67. The average American wastes $\frac{9}{50}$ of all grain products brought into the home. Write this fraction as a percent. (*Source:* Natural Resources Defense Council)

68. Canada produces $\frac{1}{4}$ of the uranium produced in the world. Write this fraction as a percent. (*Source:* World Nuclear Association)

Canada

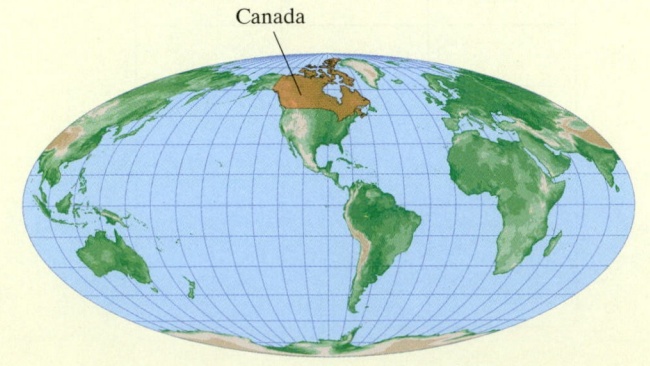

In Exercises 69 through 74, write the percent from the circle graph as a decimal and a fraction.

World Population by Continent

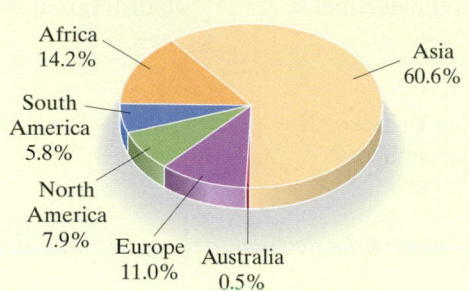

Africa
14.2%

South
America
5.8%

North
America
7.9%

Europe
11.0%

Australia
0.5%

Asia
60.6%

69. Australia: 0.5%

70. Europe: 11%

▶ 71. Africa: 14.2%

72. Asia: 60.6%

73. North America: 7.9%

74. South America: 5.8%

Review

Find the value of n. See Section 5.1.

75. $3 \cdot n = 45$

76. $7 \cdot n = 48$

77. $8 \cdot n = 80$

78. $2 \cdot n = 16$

79. $6 \cdot n = 72$

80. $5 \cdot n = 35$

Concept Extensions

Solve. See the Concept Check in this section.

81. Given the percent 52.8647%, round as indicated.
   a. Round to a tenth of a percent.
   b. Round to a hundredth of a percent.

82. Given the percent 0.5269%, round as indicated.
   a. Round to a tenth of a percent.
   b. Round to a hundredth of a percent.

83. Write 1.07835 as a percent rounded to the nearest tenth of a percent.

84. Write 1.25348 as a percent rounded to the nearest tenth of a percent.

85. Write 0.65794 as a percent rounded to the nearest hundredth of a percent.

86. Write 0.92571 as a percent rounded to the nearest hundredth of a percent.

87. Write 0.7682 as a percent rounded to the nearest percent.

88. Write 0.2371 as a percent rounded to the nearest percent.

What percent of the figure is shaded?

89.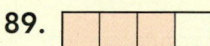

90.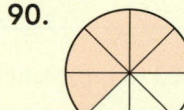

91.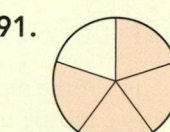

92.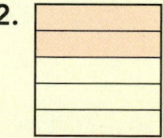

Fill in the blanks.

93. A fraction written as a percent is greater than 100% when the numerator is _____ than the denominator. (greater/less)

94. A decimal written as a percent is less than 100% when the decimal is _____ than 1. (greater/less)

95. In your own words, explain how to write a percent as a fraction.

96. In your own words, explain how to write a fraction as a decimal.

Write each fraction as a decimal and then write each decimal as a percent. Round the decimal to three decimal places (nearest thousandth) and the percent to the nearest tenth of a percent.

97. $\dfrac{21}{79}$

98. $\dfrac{56}{102}$

99. $\dfrac{850}{736}$

100. $\dfrac{506}{248}$

5.4 Solving Percent Problems Using Equations

Objectives

A Write Percent Problems as Equations.

B Solve Percent Problems.

Note: Sections 5.4 and 5.5 introduce two methods for solving percent problems. It is not necessary that you study both sections. You may want to check with your instructor for further advice.

Throughout this text, we have written mathematical statements such as $3 + 10 = 13$, or area = length · width. These statements are called "equations." An **equation** is a mathematical statement that contains an equal sign. To solve percent problems in this section, we translate the problems into such mathematical statements, or equations.

Objective A Writing Percent Problems as Equations

Recognizing key words in a percent problem is helpful in writing the problem as an equation. Three key words in the statement of a percent problem and their meanings are as follows:

of means **multiplication** (·)

is means **equals** (=)

what (or some equivalent) means **the unknown number**

In our examples, we let the letter *n* stand for the unknown number.

> **Helpful Hint**
>
> Any letter of the alphabet can be used to represent the unknown number. In this section, we mostly use the letter *n*.

Example 1   Translate to an equation.

5 is what percent of 20?

Solution: 5   is   what percent of   20?
          ↓   ↓   ↓   ↓   ↓
          5   =   n   ·   20

■ **Work Practice 1**

Helpful Hint

Remember that an equation is simply a mathematical statement that contains an equal sign ($=$).

$$5 = n \cdot 20$$
         ↑
    equal sign

Example 2   Translate to an equation.

1.2 is 30% of what number?

Solution: 1.2   is   30%   of   what number?
           ↓   ↓   ↓   ↓   ↓
          1.2   =   30%   ·   n

■ **Work Practice 2**

Practice 2

Translate: 1.8 is 20% of what number?

Example 3   Translate to an equation.

What number is 25% of 0.008?

Solution: What number   is   25%   of   0.008?
               ↓   ↓   ↓   ↓   ↓
               n   =   25%   ·   0.008

■ **Work Practice 3**

Practice 3

Translate: What number is 40% of 3.6?

Examples   Translate each of the following to an equation:

4. 38%   of   200   is   what number?
       ↓   ↓   ↓   ↓   ↓
      38%   ·   200   =   n

5. 40%   of   what number   is   80?
       ↓   ↓   ↓   ↓   ↓
      40%   ·   n   =   80

6. What percent   of   85   is   34?
            ↓   ↓   ↓   ↓   ↓
            n   ·   85   =   34

■ **Work Practice 4–6**

Practice 4–6

Translate each to an equation.
4. 42% of 50 is what number?
5. 15% of what number is 9?
6. What percent of 150 is 90?

Answers
1. $6 = n \cdot 24$ **2.** $1.8 = 20\% \cdot n$
3. $n = 40\% \cdot 3.6$ **4.** $42\% \cdot 50 = n$
5. $15\% \cdot n = 9$ **6.** $n \cdot 150 = 90$

✓**Concept Check** In the equation $2 \cdot n = 10$, what step should be taken to solve the equation for n?

Objective B Solving Percent Problems ▶

You may have noticed by now that each percent problem has contained three numbers–in our examples, two are known and one is unknown. Each of these numbers is given a special name.

15%	of	60	is	9
↓	↓	↓	↓	↓

$$\underset{\text{percent}}{15\%} \quad \cdot \quad \underset{\text{base}}{60} \quad = \quad \underset{\text{amount}}{9}$$

We call this equation the **percent equation.**

Percent Equation

percent · base = amount

Helpful Hint

Notice that the percent equation given above is a true statement. To see this, simplify the left side as shown:

$15\% \cdot 60 = 9$

$0.15 \cdot 60 = 9$ Write 15% as 0.15.

$9 = 9$ Multiply.

The statement $9 = 9$ is true.

After a percent problem has been written as a percent equation, we can use the equation to find the unknown number. This is called **solving** the equation.

Practice 7

What number is 20% of 85?

Example 7 Solving Percent Equation for the Amount

What number is 35% of 40?

Solution:

	↓	↓	↓	↓	
	n	$= 35\%$	$\cdot$	40	Translate to an equation.
	n	$= 0.35$	$\cdot$	40	Write 35% as 0.35.
	n	$= 14$			Multiply $0.35 \cdot 40 = 14$.

Thus, 14 is 35% of 40.

Is this reasonable? To see, round 35% to 40%. Then 40% of 40 or 0.40(40) is 16. Our result is reasonable since 16 is close to 14.

■ **Work Practice 7**

Answer

7. 17

✓**Concept Check Answer**

If $2 \cdot n = 10$, then $n = \dfrac{10}{2}$, or $n = 5$.

Helpful Hint

When solving a percent equation, write the percent as a decimal (or fraction).

Example 8 Solving Percent Equation for the Amount

85% of 300 is what number?

Solution: $85\% \cdot 300 = n$ Translate to an equation.

$0.85 \cdot 300 = n$ Write 85% as 0.85.

$255 = n$ Multiply $0.85 \cdot 300 = 255$.

Thus, 85% of 300 is 255.

Is this result reasonable? To see, round 85% to 90%. Then 90% of 300 or $0.90(300) = 270$, which is close to 255.

■ **Work Practice 8**

Practice 8
90% of 150 is what number?

Example 9 Solving Percent Equation for the Base

12% of what number is 0.6?

Solution: $12\% \cdot n = 0.6$ Translate to an equation.

$0.12 \cdot n = 0.6$ Write 12% as 0.12.

Recall from Section 5.3 that if "0.12 times some number is 0.6," then the number is 0.6 divided by 0.12.

$n = \dfrac{0.6}{0.12}$ Divide 0.6 by 0.12, the number multiplied by n.

$n = 5$

Thus, 12% of 5 is 0.6.

Is this reasonable? To see, round 12% to 10%. Then 10% of 5 or $0.10(5) = 0.5$, which is close to 0.6.

■ **Work Practice 9**

Practice 9
15% of what number is 1.2?

Example 10 Solving Percent Equation for the Base

13 is $6\frac{1}{2}\%$ of what number?

Solution: $13 = 6\frac{1}{2}\% \cdot n$ Translate to an equation.

$13 = 0.065 \cdot n$ $6\frac{1}{2}\% = 6.5\% = 0.065$.

$\dfrac{13}{0.065} = n$ Divide 13 by 0.065, the number multiplied by n.

$200 = n$

Thus, 13 is $6\frac{1}{2}\%$ of 200.

Check to see if this result is reasonable.

■ **Work Practice 10**

Practice 10
27 is $4\frac{1}{2}\%$ of what number?

Answers
8. 135 **9.** 8 **10.** 600

Practice 11

What percent of 80 is 8?

Example 11 Solving Percent Equation for the Percent

$$\underbrace{\text{What percent}} \quad \text{of} \quad 12 \quad \text{is} \quad 9?$$

Solution: $\quad n \quad \cdot \quad 12 = 9 \quad$ Translate to an equation.

$$n = \frac{9}{12} \quad$$ Divide 9 by 12, the number multiplied by n.

$$n = 0.75$$

Next, since we are looking for percent, we write 0.75 as a percent.

$$n = 75\%$$

So, 75% of 12 is 9. To check, see that $75\% \cdot 12 = 9$.

■ **Work Practice 11**

Helpful Hint

If your unknown in the percent equation is the percent, don't forget to convert your answer to a percent.

Practice 12

35 is what percent of 25?

Example 12 Solving Percent Equation for the Percent

$$78 \quad \text{is} \quad \underbrace{\text{what percent}} \quad \text{of} \quad 65?$$

Solution: $\quad 78 = \quad n \quad \cdot \quad 65 \quad$ Translate to an equation.

$$\frac{78}{65} = \quad n \quad$$ Divide 78 by 65, the number multiplied by n.

$$1.2 = \quad n$$

$$120\% = \quad n \quad$$ Write 1.2 as a percent.

So, 78 is 120% of 65. Check this result.

■ **Work Practice 12**

✓**Concept Check** Consider these problems.

1. 75% of 50 =
   a. 50   **b.** a number greater than 50   **c.** a number less than 50
2. 40% of a number is 10. Is the number
   a. 10?   **b.** less than 10?   **c.** greater than 10?
3. 800 is 120% of what number? Is the number
   a. 800?   **b.** less than 800?   **c.** greater than 800?

Helpful Hint

Use the following to see if your answers are reasonable.

$$(100\%) \text{ of a number} = \text{the number}$$

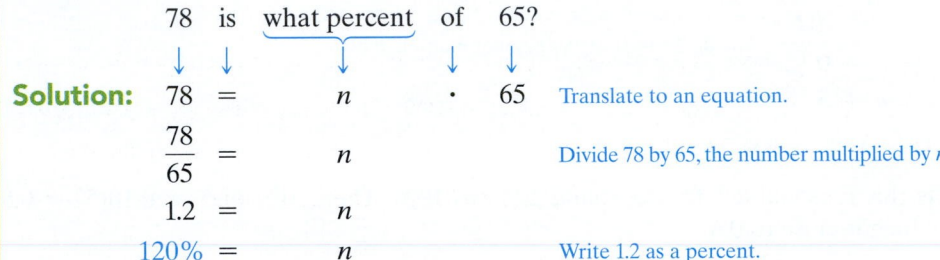

$$\begin{pmatrix} \text{a percent} \\ \text{greater than} \\ 100\% \end{pmatrix} \text{of a number} = \begin{matrix} \text{a number greater} \\ \text{than the original number} \end{matrix}$$

$$\begin{pmatrix} \text{a percent} \\ \text{less than } 100\% \end{pmatrix} \text{of a number} = \begin{matrix} \text{a number less} \\ \text{than the original number} \end{matrix}$$

Answers

11. 10%   **12.** 140%

✓**Concept Check Answers**

1. c   **2.** c   **3.** b

Vocabulary, Readiness & Video Check

Use the choices below to fill in each blank.

percent	amount	of	less
base	the number	is	greater

1. The word _____ translates to " =."

2. The word _____ usually translates to "multiplication."

3. In the statement "10% of 90 is 9," the number 9 is called the _____ , 90 is called the _____ , and 10 is called the _____ .

4. 100% of a number = _____ .

5. Any "percent greater than 100%" of "a number" = "a number _____ than the original number."

6. Any "percent less than 100%" of "a number" = "a number _____ than the original number."

Identify the percent, the base, and the amount in each equation. Recall that percent · base = amount.

7. $42\% \cdot 50 = 21$

8. $30\% \cdot 65 = 19.5$

9. $107.5 = 125\% \cdot 86$

10. $99 = 110\% \cdot 90$

Martin-Gay Interactive Videos

See Video 5.4

Watch the section lecture video and answer the following questions.

Objective A **11.** From the lecture before Example 1, what are the key words and their translations that we need to remember? ▶

Objective B **12.** What is the difference between the translated equation in  Example 5 and those in Examples 4 and 6? ▶

5.4 **Exercise Set** MyMathLab® ▶

Objective A **Translating** *Translate each to an equation. Do not solve. See Examples 1 through 6.*

▶ **1.** 18% of 81 is what number?

2. 36% of 72 is what number?

3. 20% of what number is 105?

4. 40% of what number is 6?

5. 0.6 is 40% of what number?

6. 0.7 is 20% of what number?

▶ **7.** What percent of 80 is 3.8?

8. 9.2 is what percent of 92?

9. What number is 9% of 43?

10. What number is 25% of 55?

11. What percent of 250 is 150?

12. What percent of 375 is 300?

Objective B   *Solve. See Examples 7 and 8.*

▶ 13. 10% of 35 is what number?

14. 25% of 68 is what number?

15. What number is 14% of 205?

16. What number is 18% of 425?

Solve. See Examples 9 and 10.

▶ 17. 1.2 is 12% of what number?

18. 0.22 is 44% of what number?

19. $8\frac{1}{2}\%$ of what number is 51?

20. $4\frac{1}{2}\%$ of what number is 45?

Solve. See Examples 11 and 12.

21. What percent of 80 is 88?

22. What percent of 40 is 60?

23. 17 is what percent of 50?

24. 48 is what percent of 50?

Objectives A B   Mixed Practice *Solve. See Examples 1 through 12.*

25. 0.1 is 10% of what number?

26. 0.5 is 5% of what number?

27. 150% of 430 is what number?

28. 300% of 56 is what number?

29. 82.5 is $16\frac{1}{2}\%$ of what number?

30. 7.2 is $6\frac{1}{4}\%$ of what number?

▶ 31. 2.58 is what percent of 50?

32. 2.64 is what percent of 25?

33. What number is 42% of 60?

34. What number is 36% of 80?

35. What percent of 184 is 64.4?

36. What percent of 120 is 76.8?

37. 120% of what number is 42?

38. 160% of what number is 40?

39. 2.4% of 26 is what number?

40. 4.8% of 32 is what number?

41. What percent of 600 is 3?

42. What percent of 500 is 2?

43. 6.67 is 4.6% of what number?

44. 9.75 is 7.5% of what number?

45. 1575 is what percent of 2500?

46. 2520 is what percent of 3500?

47. 2 is what percent of 50?

48. 2 is what percent of 40?

Review

Find the value of n in each proportion. See Section 5.1.

49. $\dfrac{27}{n} = \dfrac{9}{10}$

50. $\dfrac{35}{n} = \dfrac{7}{5}$

51. $\dfrac{n}{5} = \dfrac{8}{11}$

52. $\dfrac{n}{3} = \dfrac{6}{13}$

Write each phrase as a proportion.

53. 17 is to 12 as *n* is to 20   **54.** 20 is to 25 as *n* is to 10   **55.** 8 is to 9 as 14 is to *n*   **56.** 5 is to 6 as 15 is to *n*

Concept Extensions

For each equation, determine the next step taken to find the value of n. See the first Concept Check in this section.

57. $5 \cdot n = 32$

   a. $n = 5 \cdot 32$ **b.** $n = \dfrac{5}{32}$ **c.** $n = \dfrac{32}{5}$ **d.** none of these

58. $n = 0.7 \cdot 12$

   a. $n = 8.4$ **b.** $n = \dfrac{12}{0.7}$ **c.** $n = \dfrac{0.7}{12}$ **d.** none of these

59. $0.06 = n \cdot 7$

   a. $n = 0.06 \cdot 7$ **b.** $n = \dfrac{0.06}{7}$ **c.** $n = \dfrac{7}{0.06}$ **d.** none of these

60. $0.01 = n \cdot 8$

   a. $n = 0.01 \cdot 8$ **b.** $n = \dfrac{8}{0.01}$ **c.** $n = \dfrac{0.01}{8}$ **d.** none of these

61. Write a word statement for the equation $20\% \cdot n = 18.6$. Use the phrase "some number" for "*n*".

62. Write a word statement for the equation $n = 33\frac{1}{3}\% \cdot 24$. Use the phrase "some number" for "*n*".

For each exercise, determine whether the percent, n, is (a) 100%, (b) greater than 100%, or (c) less than 100%. See the last Concept Check in this section.

63. *n*% of 20 is 30   **64.** *n*% of 98 is 98   **65.** *n*% of 120 is 85   **66.** *n*% of 35 is 50

For each exercise, determine whether the number, n, is (a) equal to 45, (b) greater than 45, or (c) less than 45.

67. 55% of 45 is *n*   **68.** 230% of 45 is *n*   **69.** 100% of 45 is *n*

70. 30% of *n* is 45   **71.** 100% of *n* is 45   **72.** 180% of *n* is 45

73. In your own words, explain how to solve a percent equation.

74. Write a percent problem that uses the percent 50%.

Solve.

75. 1.5% of 45,775 is what number?   **76.** What percent of 75,528 is 27,945.36?

77. 22,113 is 180% of what number?

5.5 Solving Percent Problems Using Proportions

Objectives

A Write Percent Problems as Proportions. ▶

B Solve Percent Problems. ▶

There is more than one method that can be used to solve percent problems. (See the note at the beginning of Section 5.4.) In the last section, we used the percent equation. In this section, we will use proportions.

Objective A Writing Percent Problems as Proportions ▶

To understand the proportion method, recall that 70% means the ratio of 70 to 100, or $\frac{70}{100}$.

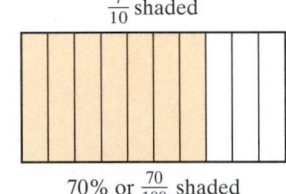

$$70\% = \frac{70}{100} = \frac{7}{10}$$

$\frac{7}{10}$ shaded

70% or $\frac{70}{100}$ shaded

Since the ratio $\frac{70}{100}$ is equal to the ratio $\frac{7}{10}$, we have the proportion

$$\frac{7}{10} = \frac{70}{100}.$$

We call this proportion the "percent proportion." In general, we can name the parts of this proportion as follows:

Percent Proportion

$$\frac{\text{amount}}{\text{base}} = \frac{\text{percent}}{100} \quad \leftarrow \text{always 100}$$

or

$$\begin{array}{c} \text{amount} \rightarrow \\ \text{base} \rightarrow \end{array} \frac{a}{b} = \frac{p}{100} \quad \leftarrow \text{percent}$$

When we translate percent problems to proportions, the **percent,** p, can be identified by looking for the symbol % or the word *percent*. The **base,** b, usually follows the word *of*. The **amount,** a, is the part compared to the whole.

Helpful Hint

Part of Proportion	How It's Identified
Percent	% or percent
Base	Appears after *of*
Amount	Part compared to whole

Example 1   Translate to a proportion.

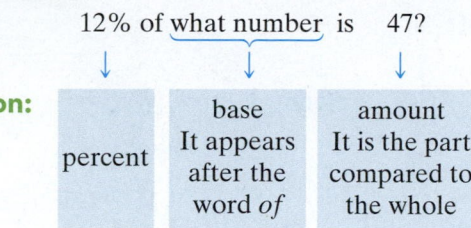

12% of <u>what number</u> is   47?

Solution:

percent	base It appears after the word *of*	amount It is the part compared to the whole

amount → $\dfrac{47}{b} = \dfrac{12}{100}$ ← percent
base →

■ **Work Practice 1**

Example 2   Translate to a proportion.

101 is <u>what percent</u> of 200?

Solution:

amount It is the part compared to the whole	percent	base It appears after the word *of*

amount → $\dfrac{101}{200} = \dfrac{p}{100}$ ← percent
base →

■ **Work Practice 2**

Example 3   Translate to a proportion.

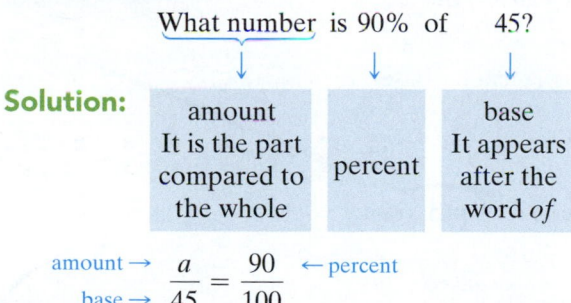

<u>What number</u> is 90% of   45?

Solution:

amount It is the part compared to the whole	percent	base It appears after the word *of*

amount → $\dfrac{a}{45} = \dfrac{90}{100}$ ← percent
base →

■ **Work Practice 3**

Example 4   Translate to a proportion.

238 is   40% of <u>what number?</u>

Solution: amount   percent   base

$\dfrac{238}{b} = \dfrac{40}{100}$

■ **Work Practice 4**

Practice 1

Translate to a proportion.
15% of what number is 55?

Practice 2

Translate to a proportion.
35 is what percent of 70?

Practice 3

Translate to a proportion.
What number is 25% of 68?

Practice 4

Translate to a proportion.
520 is 65% of what number?

Answers

1. $\dfrac{55}{b} = \dfrac{15}{100}$ **2.** $\dfrac{35}{70} = \dfrac{p}{100}$

3. $\dfrac{a}{68} = \dfrac{25}{100}$ **4.** $\dfrac{520}{b} = \dfrac{65}{100}$

Practice 5

Translate to a proportion.
What percent of 50 is 65?

Example 5 Translate to a proportion.

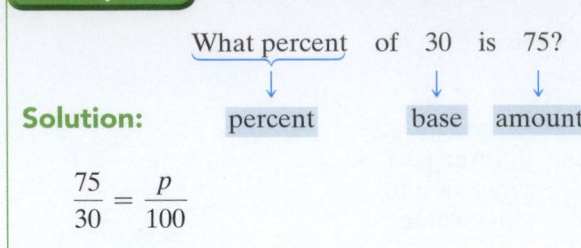

What percent of 30 is 75?

Solution: percent   base   amount

$$\frac{75}{30} = \frac{p}{100}$$

■ **Work Practice 5**

Practice 6

Translate to a proportion.
36% of 80 is what number?

Example 6 Translate to a proportion.

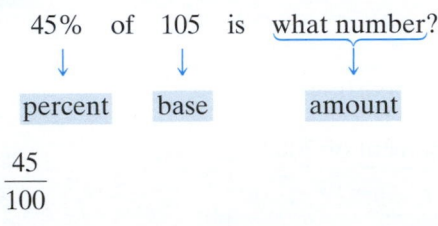

45% of 105 is what number?

Solution: percent   base   amount

$$\frac{a}{105} = \frac{45}{100}$$

■ **Work Practice 6**

Objective B Solving Percent Problems

The proportions that we have written in this section contain three values that can change: the percent, the base, and the amount. If any two of these values are known, we can find the third (the unknown value). To do this, we write a percent proportion and find the unknown value as we did in Section 5.1.

Practice 7

What number is 8% of 120?

Example 7 Solving Percent Proportions for the Amount

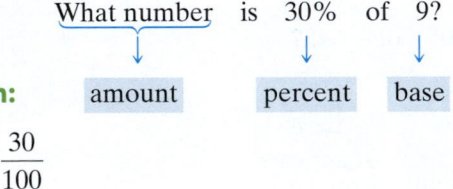

What number is 30% of 9?

Solution: amount   percent base

$$\frac{a}{9} = \frac{30}{100}$$

To solve, we set cross products equal to each other.

$$\frac{a}{9} = \frac{30}{100}$$

$a \cdot 100 = 9 \cdot 30$ Set cross products equal.
$a \cdot 100 = 270$ Multiply.

Recall from Section 5.1 that if "some number times 100 is 270," then the number is 270 divided by 100.

$$a = \frac{270}{100}$$ Divide 270 by 100, the number multiplied by a.

$a = 2.7$ Simplify.

Thus, 2.7 is 30% of 9.

■ **Work Practice 7**

> **Helpful Hint** The proportion in Example 7 contains the ratio $\frac{30}{100}$. A ratio in a proportion may be simplified before solving the proportion. The unknown number in both $\frac{a}{9} = \frac{30}{100}$ and $\frac{a}{9} = \frac{3}{10}$ is 2.7

Answers

5. $\frac{65}{50} = \frac{p}{100}$ 6. $\frac{a}{80} = \frac{36}{100}$

7. 9.6

✓**Concept Check** Consider the statement: "78 is what percent of 350?"
Which part of the percent proportion is unknown?

a. the amount   **b.** the base   **c.** the percent

Consider another statement: "14 is 10% of some number."
Which part of the percent proportion is unknown?

a. the amount   **b.** the base   **c.** the percent

Example 8 Solving Percent Proportion for the Base

$$150\% \quad \text{of} \quad \underline{\text{what number}} \quad \text{is} \quad 30?$$
$$\downarrow \qquad\qquad \downarrow \qquad\qquad \downarrow$$

Solution: percent   base   amount

$$\frac{30}{b} = \frac{150}{100} \qquad \text{Write the proportion.}$$

$$\frac{30}{b} = \frac{3}{2} \qquad \text{Write } \frac{150}{100} \text{ as } \frac{3}{2}.$$

$$30 \cdot 2 = b \cdot 3 \qquad \text{Set cross products equal.}$$

$$60 = b \cdot 3 \qquad \text{Multiply.}$$

$$\frac{60}{3} = b \qquad \text{Divide 60 by 3, the number multiplied by } b.$$

$$20 = b \qquad \text{Simplify.}$$

Thus, 150% of 20 is 30.

■ **Work Practice 8**

Practice 8

75% of what number is 60?

✓**Concept Check** When solving a percent problem by using a proportion, describe how you can check the result.

Example 9 Solving Percent Proportion for the Base

$$20.8 \quad \text{is} \quad 40\% \quad \text{of} \quad \underline{\text{what number}}?$$
$$\downarrow \qquad \downarrow \qquad\qquad \downarrow$$

Solution: amount   percent   base

$$\frac{20.8}{b} = \frac{40}{100} \quad \text{or} \quad \frac{20.8}{b} = \frac{2}{5} \qquad \text{Write the proportion and simplify } \frac{40}{100}.$$

$$20.8 \cdot 5 = b \cdot 2 \qquad \text{Set cross products equal.}$$

$$104 = b \cdot 2 \qquad \text{Multiply.}$$

$$\frac{104}{2} = b \qquad \text{Divide 104 by 2, the number multiplied by } b.$$

$$52 = b \qquad \text{Simplify.}$$

So, 20.8 is 40% of 52.

■ **Work Practice 9**

Practice 9

15.2 is 5% of what number?

Answers
8. 80 **9.** 304

✓**Concept Check Answers**
c, b; by putting the result into the proportion and checking that the proportion is true

Practice 10

What percent of 40 is 6?

Example 10 Solving Percent Proportion for the Percent

$$\underbrace{\text{What percent}}_{\downarrow} \quad \text{of} \quad \underset{\downarrow}{50} \quad \text{is} \quad \underset{\downarrow}{8?}$$

Solution: percent   base   amount

$$\frac{8}{50} = \frac{p}{100} \quad \text{or} \quad \frac{4}{25} = \frac{p}{100} \qquad \text{Write the proportion and simplify } \frac{8}{50}.$$

$$4 \cdot 100 = 25 \cdot p \qquad \text{Set cross products equal.}$$

$$400 = 25 \cdot p \qquad \text{Multiply.}$$

$$\frac{400}{25} = p \qquad \text{Divide 400 by 25, the number multiplied by } p.$$

$$16 = p \qquad \text{Simplify.}$$

So, 16% of 50 is 8.

■ **Work Practice 10**

Helpful Hint

Recall from our percent proportion that this number already is a percent. Just keep the number as is and attach a % symbol.

Practice 11

336 is what percent of 160?

Example 11 Solving Percent Proportion for the Percent

$$\underset{\downarrow}{504} \quad \text{is} \quad \underbrace{\text{what percent}}_{\downarrow} \quad \text{of} \quad \underset{\downarrow}{360?}$$

Solution: amount   percent   base

$$\frac{504}{360} = \frac{p}{100}$$

Let's choose not to simplify the ratio $\dfrac{504}{360}$.

$$504 \cdot 100 = 360 \cdot p \quad \text{Set cross products equal.}$$

$$50,400 = 360 \cdot p \quad \text{Multiply.}$$

$$\frac{50,400}{360} = p \qquad \text{Divide 50,400 by 360, the number multiplied by } p.$$

$$140 = p \qquad \text{Simplify.}$$

Notice that by choosing not to simplify $\dfrac{504}{360}$, we had larger numbers in our equation.

Either way, we find that 504 is 140% of 360.

■ **Work Practice 11**

You may have noticed the following while working examples.

Helpful Hint

Use the following to see whether your answers are reasonable.

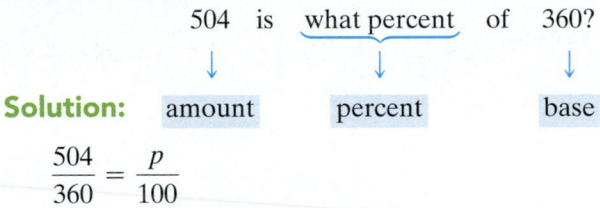

$$100\% \text{ of a number} = \text{the number}$$

$$\left(\begin{array}{c}\text{a percent} \\ \text{greater than} \\ 100\%\end{array}\right) \text{of a number} = \begin{array}{c}\text{a number larger} \\ \text{than the original number}\end{array}$$

$$\left(\begin{array}{c}\text{a percent} \\ \text{less than } 100\%\end{array}\right) \text{of a number} = \begin{array}{c}\text{a number less} \\ \text{than the original number}\end{array}$$

Answers

10. 15%   **11.** 210%

Vocabulary, Readiness & Video Check

Use the choices below to fill in each blank. These choices will be used more than once.

amount 　　　　　 base 　　　　　 percent

1. When translating the statement "20% of 15 is 3" to a proportion, the number 3 is called the _____, 15 is the _____, and 20 is the _____.

2. In the question "50% of what number is 28?", which part of the percent proportion is unknown? _____

3. In the question "What number is 25% of 200?", which part of the percent proportion is unknown? _____

4. In the question "38 is what percent of 380?", which part of the percent proportion is unknown? _____

Identify the amount, the base, and the percent in each equation. Recall that $\dfrac{amount}{base} = \dfrac{percent}{100}$.

5. $\dfrac{12.6}{42} = \dfrac{30}{100}$

6. $\dfrac{201}{300} = \dfrac{67}{100}$

7. $\dfrac{20}{100} = \dfrac{102}{510}$

8. $\dfrac{40}{100} = \dfrac{248}{620}$

Martin-Gay Interactive Videos 　 Watch the section lecture video and answer the following questions.

Objective A 　**9.** In ▦ Example 1, how did we identify what part of the percent proportion 45 is?

Objective B 　**10.** From ▦ Examples 4–6, what number is *always* part of the cross product equation of a percent proportion? ▶

See Video 5.5 🍎

5.5 Exercise Set MyMathLab® ▶

Objective A Translating *Translate each to a proportion. Do not solve. See Examples 1 through 6.*

▶**1.** 98% of 45 is what number?

2. 92% of 30 is what number?

3. What number is 4% of 150?

4. What number is 7% of 175?

5. 14.3 is 26% of what number?

6. 1.2 is 47% of what number?

7. 35% of what number is 84?

8. 85% of what number is 520?

▶**9.** What percent of 400 is 70?

10. What percent of 900 is 216?

11. 8.2 is what percent of 82?

12. 9.6 is what percent of 96?

Objective B *Solve. See Example 7.*

13. 40% of 65 is what number?

14. 25% of 84 is what number?

15. What number is 18% of 105?

16. What number is 60% of 29?

Solve. See Examples 8 and 9.

17. 15% of what number is 90?

18. 55% of what number is 55?

▶ **19.** 7.8 is 78% of what number?

20. 1.1 is 44% of what number?

Solve. See Examples 10 and 11.

21. What percent of 35 is 42?

22. What percent of 98 is 147?

23. 14 is what percent of 50?

24. 24 is what percent of 50?

Objectives A B Mixed Practice *Solve. See Examples 1 through 11.*

25. 3.7 is 10% of what number?

26. 7.4 is 5% of what number?

27. 2.4% of 70 is what number?

28. 2.5% of 90 is what number?

29. 160 is 16% of what number?

30. 30 is 6% of what number?

31. 394.8 is what percent of 188?

32. 550.4 is what percent of 172?

33. What number is 89% of 62?

34. What number is 53% of 130?

▶ **35.** What percent of 6 is 2.7?

36. What percent of 5 is 1.6?

37. 140% of what number is 105?

38. 170% of what number is 221?

39. 1.8% of 48 is what number?

40. 7.8% of 24 is what number?

41. What percent of 800 is 4?

42. What percent of 500 is 3?

43. 3.5 is 2.5% of what number?

44. 9.18 is 6.8% of what number?

▶ **45.** 20% of 48 is what number?

46. 75% of 14 is what number?

47. 2486 is what percent of 2200?

48. 9310 is what percent of 3800?

Review

Add or subtract the fractions. See Sections 3.1, 3.3, and 3.4.

49. $\dfrac{11}{16} + \dfrac{3}{16}$ **50.** $\dfrac{5}{8} - \dfrac{7}{12}$ **51.** $3\dfrac{1}{2} - \dfrac{11}{30}$ **52.** $2\dfrac{2}{3} + 4\dfrac{1}{2}$

Add or subtract the decimals. See Section 4.3.

53. $\begin{array}{r} 0.41 \\ +\,0.29 \end{array}$ **54.** $\begin{array}{r} 10.78 \\ 4.3 \\ +\ 0.21 \end{array}$ **55.** $\begin{array}{r} 2.38 \\ -\,0.19 \end{array}$ **56.** $\begin{array}{r} 16.37 \\ -\ 2.61 \end{array}$

Concept Extensions

57. Write a word statement for the proportion $\dfrac{x}{28} = \dfrac{25}{100}$. Use the phrase "what number" for "*x*."

58. Write a percent statement that translates to $\dfrac{16}{80} = \dfrac{20}{100}$.

Suppose you have finished solving four percent problems using proportions that you set up correctly. Check each answer to see if each makes the proportion a true proportion. If any proportion is not true, solve it to find the correct solution. See the Concept Checks in this section.

59. $\dfrac{a}{64} = \dfrac{25}{100}$
Is the amount equal to 17?

60. $\dfrac{520}{b} = \dfrac{65}{100}$
Is the base equal to 800?

61. $\dfrac{p}{100} = \dfrac{13}{52}$
Is the percent equal to 25 (25%)?

62. $\dfrac{36}{12} = \dfrac{p}{100}$
Is the percent equal to 50 (50%)?

63. In your own words, describe how to identify the percent, the base, and the amount in a percent problem.

64. In your own words, explain how to use a proportion to solve a percent problem.

Solve. Round to the nearest tenth, if necessary.

65. What number is 22.3% of 53,862?

66. What percent of 110,736 is 88,542?

67. 8652 is 119% of what number?

Ratio, Proportion, and Percent

Write each ratio as a ratio of whole numbers using fractional notation. Write the fraction in simplest form.

1. 1.6 to 4.6

2. $3\frac{1}{2}$ to 13

Write each rate as a unit rate.

3. 165 miles in 3 hours

4. 560 feet in 4 seconds

Write each price as a unit rate rounded to the nearest hundredth, and decide which is the better buy.

5. Dog food:
$2.16 for 8 pounds
$4.99 for 18 pounds

6. Paper plates:
$1.98 for 100
$8.99 for 500
(Round to the nearest thousandths.)

For each proportion, find the unknown number n.

7. $\dfrac{24}{n} = \dfrac{60}{96}$

8. $\dfrac{28}{49} = \dfrac{26}{n}$

Write each number as a percent.

9. 0.12

10. 0.68

11. $\dfrac{1}{8}$

12. $\dfrac{5}{2}$

13. 5.2

14. 8

15. $\dfrac{3}{50}$

16. $\dfrac{11}{25}$

17. $7\frac{1}{2}$

18. $3\frac{1}{4}$

19. 0.03

20. 0.05

Write each percent as a decimal.

21. 65%　　　**22.** 31%　　　**23.** 8%　　　**24.** 7%

25. 142%　　**26.** 400%　　**27.** 2.9%　　**28.** 6.6%

Write each percent as a decimal and as a fraction or mixed number in simplest form.
(If necessary when writing as a decimal, round to the nearest thousandth.)

29. 3%　　　**30.** 5%　　　**31.** 5.25%　　**32.** 12.75%

33. 38%　　　**34.** 45%　　　**35.** $12\frac{1}{3}$%　　**36.** $16\frac{2}{3}$%

Solve each percent problem.

37. 12% of 70 is what number?　　　**38.** 36 is 36% of what number?

39. 212.5 is 85% of what number?　　**40.** 66 is what percent of 55?

41. 23.8 is what percent of 85?　　　**42.** 38% of 200 is what number?

43. What number is 25% of 44?　　　**44.** What percent of 99 is 128.7?

45. What percent of 250 is 215?　　　**46.** What number is 45% of 84?

47. 42% of what number is 63?　　　**48.** 95% of what number is 58.9?

21. _____

22. _____

23. _____

24. _____

25. _____

26. _____

27. _____

28. _____

29. _____

30. _____

31. _____

32. _____

33. _____

34. _____

35. _____

36. _____

37. _____

38. _____

39. _____

40. _____

41. _____

42. _____

43. _____

44. _____

45. _____

46. _____

47. _____

48. _____

Objectives

A Solve Applications Involving Percent. ▶

B Find Percent of Increase and Percent of Decrease. ▶

Objective A Solving Applications Involving Percent ▶

Percent is used in a variety of everyday situations. The next examples show just a few ways that percent occurs in real-life settings. (Each of these examples shows two ways of solving these problems. If you studied Section 5.4 only, see *Method 1*. If you studied Section 5.5 only, see *Method 2*.)

The first example has to do with the Appalachian Trail, a hiking trail conceived by a forester in 1921 and diagrammed to the right.

Mount Katahdin, Maine

The Appalachian Trail

Springer Mountain, Georgia

Practice 1

If the total mileage of the Appalachian Trail is 2174, use the circle graph to determine the number of miles in the state of Virginia.

Appalachian Trail Mileage by State Percent

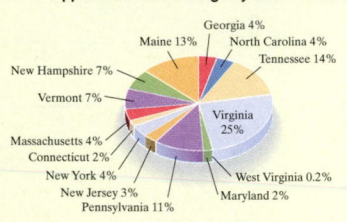

Georgia 4%
Maine 13%
North Carolina 4%
Tennessee 14%
New Hampshire 7%
Vermont 7%
Virginia 25%
Massachusetts 4%
Connecticut 2%
New York 4%
West Virginia 0.2%
New Jersey 3%
Maryland 2%
Pennsylvania 11%

Total miles: 2174
Note: The sum of the percents is 100.2% because of rounding.
Source: purebound.com

Example 1 The circle graph in the margin shows the Appalachian Trail mileage by state. If the total mileage of the trail is 2174, use the circle graph to determine the number of miles in the state of New York. Round to the nearest whole mile.

Solution: *Method 1.* First, we state the problem in words.

In words: What number is 4% of 2174?

Translate: n = 4% · 2174

To solve for n, we find 4% · 2174.

$n = 0.04 \cdot 2174$ Write 4% as a decimal.

$n = 86.96$ Multiply.

$n \approx 87$ Round to the nearest whole.

Rounded to the nearest whole mile, we have that approximately 87 miles of the Appalachian Trail are in New York state.

Method 2. State the problem in words; then translate.

In words: What number is 4% of 2174?

amount   percent   base

Translate: amount → $\dfrac{a}{2174} = \dfrac{4}{100}$ ← percent
base →

Next, we solve for a.

$a \cdot 100 = 2174 \cdot 4$ Set cross products equal.

$a \cdot 100 = 8696$ Multiply.

$\dfrac{a \cdot 100}{100} = \dfrac{8696}{100}$ Divide both sides by 100.

$a = 86.96$ Simplify.

$a \approx 87$ Round to the nearest whole.

Rounded to the nearest whole mile, we have that approximately 87 miles of the Appalachian Trail are in New York state.

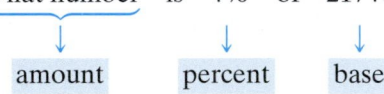
Work Practice 1

Answer

1. 543.5 mi

Example 2   Finding Percent of Nursing Job Openings
Due to Retirements

There is a worldwide shortage of nurses. It is expected that there will be 1,207,500 job openings for nurses in 2020. About 495,500 of these job openings will be to replace retiring nurses. What percent of job openings for nurses in 2020 will be due to retirements? Round to the nearest whole percent. (*Source:* Bureau of Labor Statistics)

Solution: *Method 1.*   First, we state the problem in words.

In words: 495,500   is   what percent   of   1,207,500?
                 ↓   ↓   ↓   ↓   ↓
Translate: 495,500   = n ·   1,207,500

Next, solve for n.

$$\frac{495,500}{1,207,500} = n \quad \text{Divide 495,500 by 1,207,500, the number multiplied by } n.$$

$$0.41 \approx n \quad \text{Divide and round to the nearest hundredth.}$$

$$41 \approx n \quad \text{Write as a percent.}$$

In 2020, about 41% of job openings for nurses will be due to retirements.

Method 2.

In words: 495,500   is   what percent   of   1,207,500?
                 ↓   ↓   ↓
              amount   percent   base

Translate: amount → $\dfrac{495,500}{1,207,500} = \dfrac{p}{100}$ ← percent
             base →

Next, solve for p.

$$495,500 \cdot 100 = 1,207,500 \cdot p \quad \text{Set cross products equal.}$$

$$49,550,000 = 1,207,500 \cdot p \quad \text{Multiply.}$$

$$\frac{49,550,000}{1,207,500} = p \quad \text{Divide 49,550,000 by 1,207,500, the number multiplied by } p.$$

$$41 \approx p$$

In 2020, about 41% of job openings for nurses will be due to retirements.

■ **Work Practice 2**

Example 3   Finding the Base Number of Absences

Mr. Buccaran, the principal at Slidell High School, counted 31 freshmen absent during a particular day. If this is 4% of the total number of freshmen, how many freshmen are there at Slidell High School?

Solution: *Method 1.*   First we state the problem in words; then we translate.

In words: 31   is   4%   of   what number?
             ↓   ↓   ↓   ↓   ↓
Translate: 31   =   4%   · n

(Continued on next page)

Practice 2

In Florida, about 34,000 new nurses were recently needed and hired. If there are now 130,000 nurses, what percent of new nurses were needed in Florida? Round to the nearest whole percent. (*Source*: St. Petersburg Times and The Registered Nurse Population)

Practice 3

The freshmen class of 775 students is 31% of all students at Euclid University. How many students go to Euclid University?

Answers
2. 26%   **3.** 2500

Next, we solve for n.

$$31 = 0.04 \cdot n \quad \text{Write 4\% as a decimal.}$$

$$\frac{31}{0.04} = n \quad \text{Divide 31 by 0.04, the number multiplied by } n.$$

$$775 = n \quad \text{Simplify.}$$

There are 775 freshmen at Slidell High School.

Method 2.   First we state the problem in words; then we translate.

In words: 31   is   4%   of   what number?

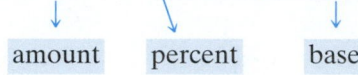

amount   percent   base

Translate: amount → $\dfrac{31}{b} = \dfrac{4}{100}$ ← percent
            base →

Next, we solve for b.

$$31 \cdot 100 = b \cdot 4 \quad \text{Set cross products equal.}$$

$$3100 = b \cdot 4 \quad \text{Multiply.}$$

$$\frac{3100}{4} = b \quad \text{Divide 3100 by 4, the number multiplied by } b.$$

$$775 = b \quad \text{Simplify.}$$

There are 775 freshmen at Slidell High School.

■ **Work Practice 3**

Practice 4

From 2010 to 2013, the number of full-time workers in the United States increased by approximately 5%. In 2010, the number of full-time workers was 110 million. (*Source:* U.S. Bureau of Labor Statistics)

a. Find the increase in the number of full-time workers from 2010 to 2013.

b. Find the total number of full-time workers in 2013.

Example 4   Finding the Base Increase in Part-Time Workers

From 2003 to 2013, the number of part-time workers in the United States increased by 12%. In 2003, there were 24.5 million part-time workers. (*Source:* U.S. Bureau of Labor Statistics)

a. Find the increase in the number of part-time workers from 2003 to 2013.

b. Find the number of part-time workers in 2013.

Solution: *Method 1.*   First we find the increase in the number of part-time workers.

In words: What number   is   12%   of   24.5?

Translate: n =   12%   ·   24.5

Answers
4. a. 5.5 million   **b.** 115.5 million

Next, we solve for *n*.

$$n = 0.12 \cdot 24.5 \quad \text{Write 12\% as a decimal.}$$
$$n = 2.94 \quad \text{Multiply.}$$

a. The increase in the number of part-time workers was 2.94 million.

b. This means that the number of part-time workers in 2013 was

$$\begin{array}{ccc} \text{Number of} & \text{Number of} & \text{Increase} \\ \text{part-time workers} = \text{part-time workers} + & \text{in number of} \\ \text{in 2013} & \text{in 2003} & \text{part-time workers} \end{array}$$
$$= 24.5 \text{ million} + 2.94 \text{ million}$$
$$= 27.44 \text{ million}$$

Method 2. First we find the increase in the number of part-time workers.

In words: What number is 12% of 24.5?

amount   percent   base

Translate:
$$\text{amount} \rightarrow \frac{a}{24.5} = \frac{12}{100} \leftarrow \text{precent}$$
$$\text{base} \rightarrow$$

Next, we solve for *a*.

$$a \cdot 100 = 24.5 \cdot 12 \quad \text{Set cross products equal.}$$
$$a \cdot 100 = 294 \quad \text{Multiply.}$$
$$\frac{a \cdot 100}{100} = \frac{294}{100} \quad \text{Divide both sides by 100.}$$
$$a = 2.94 \quad \text{Simplify.}$$

a. The increase in the number of part-time workers was 2.94 million.

b. This means that the number of part-time workers in 2013 was

$$\begin{array}{ccc} \text{Number of} & \text{Number of} & \text{Increase} \\ \text{part-time workers} = \text{part-time workers} + & \text{in number of} \\ \text{in 2013} & \text{in 2003} & \text{part-time workers} \end{array}$$
$$= 24.5 \text{ million} + 2.94 \text{ million}$$
$$= 27.44 \text{ million}$$

■ **Work Practice 4**

Objective B Finding Percent of Increase and Percent of Decrease ▶

We often use percents to show how much an amount has increased or decreased.

Suppose that the population of a town is 10,000 people and then it increases by 2000 people. The **percent of increase** is

$$\text{amount of increase} \rightarrow \frac{2000}{10,000} = 0.2 = 20\%$$
$$\text{original amount} \rightarrow$$

In general, we have the following.

Percent of Increase

$$\text{percent of increase} = \frac{\text{amount of increase}}{\text{original amount}}$$

Then write the quotient as a percent.

Practice 5

The number of people attending the local play, *Peter Pan*, increased from 285 on Friday to 333 on Saturday. Find the percent of increase in attendance. Round to the nearest tenth percent.

Helpful Hint   Make sure that this number is the original number and not the new number.

Example 5   Finding Percent of Increase

The number of applications for a mathematics scholarship at Yale increased from 34 to 45 in one year. What is the percent of increase? Round to the nearest whole percent.

Solution: First we find the amount of increase by subtracting the original number of applicants from the new number of applicants.

$$\text{amount of increase} = 45 - 34 = 11$$

The amount of increase is 11 applicants. To find the percent of increase,

$$\text{percent of increase} = \frac{\text{amount of increase}}{\text{original amount}} = \frac{11}{34} \approx 0.32 = 32\%$$

The number of applications increased by about 32%.

■ **Work Practice 5**

✓**Concept Check** A student is calculating the percent of increase in enrollment from 180 students one year to 200 students the next year. Explain what is wrong with the following calculations:

$$\frac{\text{Amount}}{\text{of increase}} = 200 - 180 = 20$$

$$\frac{\text{Percent of}}{\text{increase}} = \frac{20}{200} = 0.1 = 10\%$$

Suppose that your income was $300 a week and then it decreased by $30. The **percent of decrease** is

$$\begin{aligned}\text{amount of decrease} &\rightarrow \\ \text{original amount} &\rightarrow\end{aligned} \frac{\$30}{\$300} = 0.1 = 10\%$$

Percent of Decrease

$$\text{percent of decrease} = \frac{\text{amount of decrease}}{\text{original amount}}$$

Then write the quotient as a percent.

Practice 6

A town's population of 20,200 in 1995 decreased to 18,483 in 2005. What was the percent of decrease?

Example 6   Finding Percent of Decrease

In response to a decrease in sales, a company with 1500 employees reduces the number of employees to 1230. What is the percent of decrease?

Solution: First we find the amount of decrease by subtracting 1230 from 1500.

$$\text{amount of decrease} = 1500 - 1230 = 270$$

The amount of decrease is 270. To find the percent of decrease,

$$\frac{\text{percent of}}{\text{decrease}} = \frac{\text{amount of decrease}}{\text{original amount}} = \frac{270}{1500} = 0.18 = 18\%$$

The number of employees decreased by 18%.

■ **Work Practice 6**

Answers

5. 16.8%   **6.** 8.5%

✓**Concept Check Answers**

To find the percent of increase, you have to divide the amount of increase (20) by the original amount (180); 10% decrease.

✓**Concept Check** An ice cream stand sold 6000 ice cream cones last summer. This year the same stand sold 5400 cones. Was there a 10% increase, a 10% decrease, or neither? Explain.

Vocabulary, Readiness & Video Check

Martin-Gay Interactive Videos   Watch the section lecture video and answer the following questions.

Objective **A** **1.** How do we interpret the answer 175,000 in 📄 Example 1? ▶

Objective **B** **2.** In 📄 Example 3, what does the improper fraction tell us? ▶

See Video 5.6 🍎

5.6 Exercise Set MyMathLab® ▶

Objective **A** *Solve. For Exercises 1 and 2, the solutions have been started for you. See Examples 1 through 4. If necessary, round percents to the nearest tenth and all other answers to the nearest whole.*

1. An inspector found 24 defective bolts during an inspection. If this is 1.5% of the total number of bolts inspected, how many bolts were inspected?

Start the solution:

1. UNDERSTAND the problem. Reread it as many times as needed.

Go to *Method 1* or *Method 2*.

Method 1.

2. TRANSLATE into an equation. (Fill in the boxes.)

24   is   1.5%   of   what number?
↓   ↓   ↓   ↓   ↓
24   □   1.5%   □ n

3. SOLVE for n. (See Example 3, Method 1, for help.)

4. INTERPRET. The total number of bolts inspected was _____.

Method 2.

2. TRANSLATE into a proportion. (Fill in the blanks with "amount" or "base.")

24   is   1.5%   of   what number?
↓   ↘   ↓
_____   percent   _____

amount → $\dfrac{\_\_\_}{\_\_\_} = \dfrac{1.5}{100}$ ← percent
base →

3. SOLVE the proportion. (See Example 3, Method 2, for help.)

4. INTERPRET. The total number of bolts inspected was _____.

2. A day care worker found 28 children absent one day during an epidemic of chicken pox. If this was 35% of the total number of children attending the day care center, how many children attend this day care center?

Start the solution:

1. UNDERSTAND the problem. Reread it as many times as needed.

Go to *Method 1* or *Method 2*.

Method 1.

2. TRANSLATE into an equation. (Fill in the boxes.)

28   is   35%   of   what number?
↓   ↓   ↓   ↓   ↓
28   □   35%   □ n

3. SOLVE for n. (See Example 3, Method 1, for help.)

4. INTERPRET. The total number of children attending the day care center was _____.

Method 2.

2. TRANSLATE into a proportion. (Fill in the blanks with "amount" or "base.")

28   is   35%   of   what number?
↓   ↘   ↓
_____   percent   _____

amount → $\dfrac{\_\_\_}{\_\_\_} = \dfrac{35}{100}$ ← percent
base →

3. SOLVE the proportion. (See Example 3, Method 2, for help.)

4. INTERPRET. The total number of children attending the day care center was _____.

3. The Total Gym® provides weight resistance through adjustments of incline. The minimum weight resistance is 4% of the weight of the person using the Total Gym. Find the minimum weight resistance possible for a 220-pound man. (*Source:* Total Gym)

4. The maximum weight resistance for the Total Gym is 60% of the weight of the person using it. Find the maximum weight resistance possible for a 220-pound man. (See Exercise 3 if needed.)

5. A student's cost for last semester at her community college was $2700. She spent $378 of that on books. What percent of last semester's college costs was spent on books?

6. Pierre Sampeau belongs to his local food cooperative, where he receives a percentage of what he spends each year as a dividend. He spent $3850 last year at the food cooperative store and received a dividend of $154. What percent of his total spending at the food cooperative did he receive as a dividend?

7. The U.S. motion picture and television industry is made up of over 108,000 businesses. About 85% of these are small businesses with fewer than 10 employees. How many motion picture and television industry businesses have fewer than 10 employees? (*Source:* Motion Picture Association of America)

8. In 2012, 13% of the population of the United States and Canada were considered frequent moviegoers, purchasing movie tickets at least once per month. If the combined population of the United States and Canada was 331,700,000 in 2012, how many people were considered frequent moviegoers? (*Source:* Motion Picture Association of America)

9. The average wedding in the United States cost $28,400 in 2013. The average cost of a wedding reception venue was approximately $12,780. Determine the percent of an average wedding budget that is devoted to the reception venue. (*Source:* Wedding Channel/TheKnot.com)

10. Of the 64,500 veterinarians in private practice in the United States in 2012, approximately 33,540 were female. Determine the percent of female veterinarians in private practice in the United States. (*Source:* American Veterinary Medical Association)

11. A furniture company currently produces 6200 chairs per month. If production decreases by 8%, find the decrease and the new number of chairs produced each month.

12. The enrollment at a local college decreased by 5% over last year's enrollment of 7640. Find the decrease in enrollment and the current enrollment.

13. From 2010 to 2020, the number of people employed as physician assistants in the United States is expected to increase by 30%. The number of people employed as physician assistants in 2010 was 83,600. Find the predicted number of physician assistants in 2020. (*Source:* Bureau of Labor Statistics)

14. From 2007 to 2012, the number of U.S. veterinarians increased by 16%. The number of U.S. veterinarians in 2007 was about 83,700. Find the number of U.S. veterinarians in 2012. (*Source:* American Veterinary Medical Association)

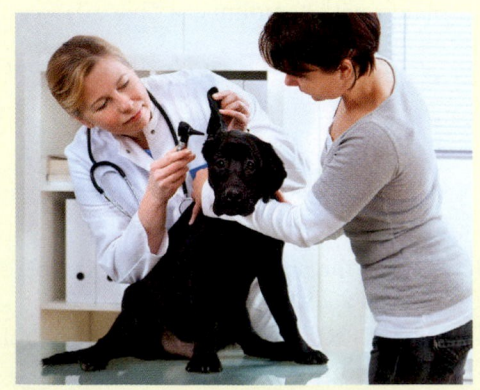

Let's look at the populations of two states, Michigan and Louisiana. Their locations are shown on the partial U.S. map below. Round each answer to the nearest thousand. (Source: U. S. Census Bureau)

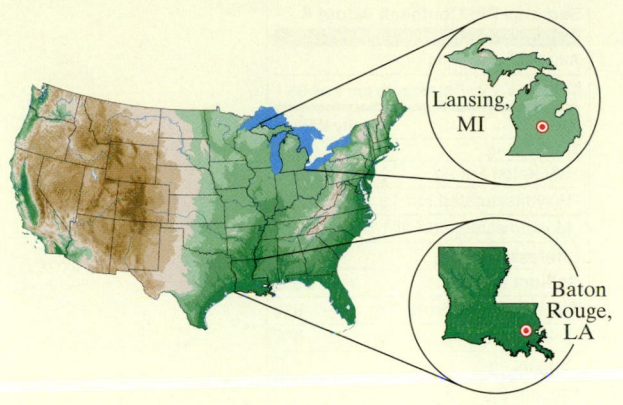

Lansing, MI

Baton Rouge, LA

15. In 2000, the population of Michigan was approximately 9940 thousand. If the population decreased by 0.6% between 2000 and 2012, find the population of Michigan in 2012.

16. In 2000, the population of Louisiana was approximately 4470 thousand. If the population increased about 3% between 2000 and 2012, find the population of Louisiana in 2012.

A popular extreme sport is snowboarding. Ski trails are marked with difficulty levels of easy ●, intermediate ■, difficult ◆, expert ◆◆, and other variations. Use this information for Exercises 17 and 18. Round each percent to the nearest whole. See Example 2.

17. At Keystone ski area in Colorado, 38 of the 131 total ski runs are rated intermediate. What percent of the runs are intermediate? (*Source:* Vail Resorts Management Company)

18. At Telluride ski area in Colorado, 29 of the 127 total ski trails are rated easy. What percent of the trails are easy? (*Source:* Telluride Ski & Golf Resort)

For each food described, find the percent of total calories from fat. If necessary, round to the nearest tenth percent. See Example 2.

19. Ranch dressing serving size of 2 tablespoons

	Calories
Total	40
From fat	20

20. Unsweetened cocoa powder serving size of 1 tablespoon

	Calories
Total	20
From fat	5

21.

Nutrition Facts

Serving Size 1 pouch (20g)
Servings Per Container 6

Amount Per Serving

Calories	80
Calories from fat	10

	% Daily Value*
Total Fat 1g	**2%**
Sodium 45mg	**2%**
Total Carbohydrate 17g	**6%**
Sugars 9g	
Protein 0g	

Vitamin C	25%

Not a significant source of saturated fat, cholesterol, dietary fiber, vitamin A, calcium and iron.

*Percent Daily Values are based on a 2,000 calorie diet.

Artificial Fruit Snacks

22.

Nutrition Facts

Serving Size $\frac{1}{4}$ cup (33g)
Servings Per Container About 9

Amount Per Serving

Calories 190 **Calories from Fat** 130

	% Daily Value
Total Fat 16g	**24%**
Saturated Fat 3g	**16%**
Cholesterol 0mg	**0%**
Sodium 135mg	**6%**
Total Carbohydrate 9g	**3%**
Dietary Fiber 1g	**5%**
Sugars 2g	
Protein 5g	

Vitamin A 0% • Vitamin C 0%
Calcium 0% • Iron 8%

Peanut Mixture

23.

Nutrition Facts

Serving Size 18 crackers (29g)
Servings Per Container About 9

Amount Per Serving

Calories 120 **Calories from Fat** 35

	% Daily Value*
Total Fat 4g	**6%**
Saturated Fat 0.5g	**3%**
Polyunsaturated Fat 0g	
Monounsaturated Fat 1.5g	
Cholesterol 0mg	**0%**
Sodium 220mg	**9%**
Total Carbohydrate 21g	**7%**
Dietary Fiber 2g	**7%**
Sugars 3g	
Protein 2g	

Vitamin A 0% • Vitamin C 0%

Calcium 2% • Iron 4%

Phosphorus 10%

Snack Crackers

24.

Nutrition Facts

Serving Size 28 crackers (31g)
Servings Per Container About 6

Amount Per Serving

Calories 130 **Calories from Fat** 35

	% Daily Value*
Total Fat 4g	**6%**
Saturated Fat 2g	**10%**
Polyunsaturated Fat 1g	
Monounsaturated Fat 1g	
Cholesterol 0mg	**0%**
Sodium 470mg	**20%**
Total Carbohydrate 23g	**8%**
Dietary Fiber 1g	**4%**
Sugars 4g	
Protein 2g	

Vitamin A 0% • Vitamin C 0%

Calcium 0% • Iron 2%

Snack Crackers

Solve. If necessary, round money amounts to the nearest cent and all other amounts to the nearest tenth. See Examples 1 through 4.

25. A family paid $26,250 as a down payment for a home. If this represents 15% of the price of the home, find the price of the home.

26. A banker learned that $842.40 is withheld from his monthly check for taxes and insurance. If this represents 18% of his total pay, find the total pay.

27. An owner of a repair service company estimates that for every 40 hours a repairperson is on the job, he can bill for only 78% of the hours. The remaining hours, the repairperson is idle or driving to or from a job. Determine the number of hours per 40-hour week the owner can bill for a repairperson.

28. A manufacturer of electronic components expects 1.04% of its products to be defective. Determine the number of defective components expected in a batch of 28,350 components. Round to the nearest whole component.

29. A car manufacturer announced that next year the price of a certain model of car will increase by 4.5%. This year the price is $19,286. Find the increase in price and the new price.

30. A union contract calls for a 6.5% salary increase for all employees. Determine the increase and the new salary that a worker currently making $58,500 under this contract can expect.

A popular extreme sport is artificial wall climbing. The photo shown is an artificial climbing wall. Exercises 31 and 32 are about the Footsloggers Climbing Tower in Boone, North Carolina.

31. A climber is resting at a height of 24 feet while on the Footsloggers Climbing Tower. If this is 60% of the tower's total height, find the height of the tower.

32. A group plans to climb the Footsloggers Climbing Tower at the group rate, once they save enough money. Thus far, $175 has been saved. If this is 70% of the total amount needed for the group, find the total price.

Solve.

33. Tuition for an Ohio resident at the Columbus campus of Ohio State University was $8406 in 2009. The tuition increased by 14.4% during the period from 2009 to 2013. Find the increase and the tuition for the 2013–2014 school year. Round the increase to the nearest whole dollar. (*Source:* The Ohio State University)

34. The population of Americans aged 65 and older was 43 million in 2012. That population is projected to increase by 69% by 2030. Find the increase and the projected 2030 population. (*Source:* Bureau of the Census)

35. From 2010–2011 to 2020–2021, the number of associate degrees awarded is projected to increase by 14.5%. If the number of associate degrees awarded in 2010–2011 was 888,000, find the increase and the projected number of associate degrees awarded in the 2020–2021 school year. (*Source:* National Center for Education Statistics)

36. From 2010–2011 to 2020–2021, the number of bachelor degrees awarded is projected to increase by 16%. If the number of bachelor degrees awarded in 2010–2011 was 1,703,000, find the increase and the projected number of bachelor degrees awarded in the 2020–2021 school year. (*Source:* National Center for Education Statistics)

Objective B *Find the amount of increase and the percent of increase. See Example 5.*

	Original Amount	New Amount	Amount of Increase	Percent of Increase
37.	50	80		
38.	8	12		
39.	65	117		
40.	68	170		

Find the amount of decrease and the percent of decrease. See Example 6.

	Original Amount	New Amount	Amount of Decrease	Percent of Decrease
41.	8	6		
42.	25	20		
43.	160	40		
44.	200	162		

Solve. Round percents to the nearest tenth, if necessary. See Examples 5 and 6.

▶ 45. There are 150 calories in a cup of whole milk and only 84 in a cup of skim milk. In switching to skim milk, find the percent of decrease in number of calories per cup.

46. In reaction to a slow economy, the number of employees at a soup company decreased from 530 to 477. What was the percent of decrease in the number of employees?

47. The number of cable TV systems recently decreased from 10,845 to 10,700. Find the percent of decrease.

48. Before taking a typing course, Geoffry Landers could type 32 words per minute. By the end of the course, he was able to type 76 words per minute. Find the percent of increase.

49. In 1940, the average size of a privately owned farm in the United States was 174 acres. In a recent year, the average size of a privately owned farm in the United States had increased to 421 acres. Find the percent of increase. (*Source:* National Agricultural Statistics Service)

50. In 2006, the average size of a privately owned farm in the United States was 443 acres. In 2012, the average size of a privately owned farm in the United States had decreased to 421 acres. Find the percent of decrease. (*Source:* National Agricultural Statistics Service)

51. Total U.S. music industry revenues fell from $7.8 billion in 2009 to $7.1 billion in 2012. Find the percent of decrease. (*Source:* Recording Industry Association of America)

52. In 2008, the number of milk cow operations in the United States was 67,000. By 2012, this number had decreased to 58,000. What was the percent of decrease? (*Source:* National Agricultural Statistics Service)

53. In 2004, 21 million people in the United States subscribed to high-speed Internet service via their cable provider. By 2012, this number had increased to 50.3 million. What was the percent of increase? (*Source:* SNL Kagan)

54. In 2010, there were 1038 thousand high school teachers employed in the United States. This number is expected to increase to 1110 thousand teachers in 2020. What is the percent of increase? (*Source:* Bureau of Labor Statistics)

55. Between 2000 and 2012, the number of indoor cinema sites in the United States decreased from 6550 to 5317. Find the percent of decrease. (*Source:* National Association of Theater Owners)

56. In 2003, the average price of a cinema ticket was $6.03. By 2012, this price had increased to $7.96. What was the percent of increase? (*Source:* National Association of Theatre Owners)

57. In 2010, there were 64,400 dieticians employed in the United States. This number is expected to increase to 77,100 dieticians in 2020. What is the percent of increase? (*Source:* Bureau of Labor Statistics)

58. In 2010, there were 242,900 coaches and scouts employed in the United States. This number is expected to increase to 314,300 in 2020. What is the percent of increase? (*Source:* Bureau of Labor Statistics)

59. The number of cell phone tower sites in the United States was 195,613 in 2006. By 2012, the number of cell sites had increased to 301,779. Find the percent of increase. (*Source:* CTIA—The Wireless Association)

60. The population of Japan is expected to decrease from 127,050 thousand in 2014 to 100,600 thousand in 2050. Find the percent of decrease. (*Source:* Department of Population Dynamics Research)

Japan

Tokyo

Review

Perform each indicated operation. See Sections 4.3 and 4.4.

61.   0.12
       $\times$ 38

62.   42
        $\times$ 0.7

63. 9.20 + 1.98

64. 46 + 7.89

65. 78 − 19.46

66. 64.80 − 10.72

Concept Extensions

67. If a number is increased by 100%, how does the increased number compare with the original number? Explain your answer.

68. In your own words, explain what is wrong with the following statement. "Last year we had 80 students attend. This year we have a 50% increase or a total of 160 students attending."

Explain what errors were made by each student when solving percent of increase or decrease problems and then correct the errors. See the Concept Checks in this section.

The population of a certain rural town was 150 in 1980, 180 in 1990, and 150 in 2000.

69. Find the percent of increase in population from 1980 to 1990.

Miranda's solution: Percent of increase = $\dfrac{30}{180}$ = $0.1\overline{6}$ ≈ 16.7%

70. Find the percent of decrease in population from 1990 to 2000.

Jeremy's solution: Percent of decrease = $\dfrac{30}{150}$ = 0.20 = 20%

71. The percent of increase from 1980 to 1990 is the same as the percent of decrease from 1990 to 2000. True or false.

Chris's answer: True because they had the same amount of increase as the amount of decrease.

72. Refer to Exercises 49 and 50. They are similar except one asks us to find the percent of increase in the size of U.S. privately owned farms and one asks us to find the percent of decrease. In your own words, explain how these can both be correct.

5.7 Percent and Problem Solving: Sales Tax, Commission, and Discount ▶

Objective A Calculating Sales Tax and Total Price ▶

Percents are frequently used in the retail trade. For example, most states charge a tax on certain items when purchased. This tax is called a **sales tax,** and retail stores collect it for the state. Sales tax is almost always stated as a percent of the purchase price.

A 9% sales tax rate on a purchase of a $10 calculator gives a sales tax of

sales tax = 9% of $10 = 0.09 · $10.00 = $0.90

Objectives

A Calculate Sales Tax and Total Price. ▶

B Calculate Commissions. ▶

C Calculate Discount and Sale Price. ▶

The total price to the customer would be

purchase price   plus   sales tax

$10.00   + $0.90 = $10.90

This example suggests the following equations:

Sales Tax and Total Price

sales tax = tax rate · purchase price

total price = purchase price + sales tax

In this section we round dollar amounts to the nearest cent.

Practice 1

If the sales tax rate is 8.5%, what is the sales tax and the total amount due on a $59.90 Goodgrip tire? (Round the sales tax to the nearest cent.)

Example 1   Finding Sales Tax and Purchase Price

Find the sales tax and the total price on the purchase of an $85.50 atlas in a city where the sales tax rate is 7.5%.

Solution: The purchase price is $85.50 and the tax rate is 7.5%.

sales tax = tax rate · purchase price

sales tax = 7.5% · $85.50

= 0.075 · $85.50   Write 7.5% as a decimal.

≈ $6.41   Round to the nearest cent

Thus, the sales tax is $6.41. Next find the total price.

total price = purchase price + sales tax

total price = $85.50 + $6.41

= $91.91

The sales tax on $85.50 is $6.41, and the total price is $91.91.

■ **Work Practice 1**

✓**Concept Check** The purchase price of a textbook is $50 and sales tax is 10%. If you are told by the cashier that the total price is $75, how can you tell that a mistake has been made?

Answer

1. tax: $5.09; total: $64.99

✓**Concept Check Answer**

Since $10\% = \dfrac{1}{10}$, the sales tax is $\dfrac{\$50}{10} = \$5$. The total price should have been $55.

Example 2 Finding a Sales Tax Rate

The sales tax on a $406 Sony flat-screen digital 27-inch television is $34.51. Find the sales tax rate.

Solution: Let r represent the unknown sales tax rate. Then

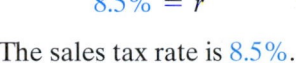

sales tax = tax rate · purchase price

$$\$34.51 = r \cdot \$406$$

$$\frac{34.51}{406} = \frac{r \cdot 406}{406} \qquad \text{Divide both sides by 406.}$$

$$0.085 = r \qquad \text{Simplify.}$$

$$8.5\% = r \qquad \text{Write 0.085 as a percent.}$$

The sales tax rate is 8.5%.

■ **Work Practice 2**

Practice 2

The sales tax on an $18,500 automobile is $1665. Find the sales tax rate.

Objective B Calculating Commissions ▶

A **wage** is payment for performing work. Hourly wage, commissions, and salary are some of the ways wages can be paid. Many people who work in sales are paid a commission. An employee who is paid a **commission** is paid a percent of his or her total sales.

Commission

commission = commission rate · sales

Example 3 Finding the Amount of Commission

Sherry Souter, a real estate broker for Wealth Investments, sold a house for $214,000 last week. If her commission rate is 1.5% of the selling price of the home, find the amount of her commission.

Solution:

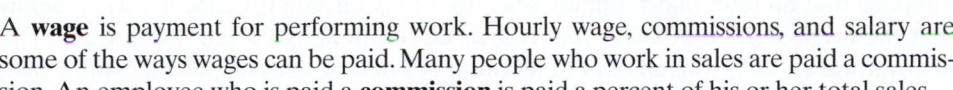

commission = commission rate · sales

commission = 1.5% · $214,000

= 0.015 · $214,000   Write 1.5% as 0.015.

= $3210   Multiply.

Her commission on the house is $3210.

■ **Work Practice 3**

Practice 3

A sales representative for Office Product Copiers sold $47,632 worth of copy equipment and supplies last month. What is his commission for the month if he is paid a commission rate of 6.6% of his total sales for the month?

Answers

2. 9% **3.** $3143.71

Practice 4

A salesperson earns $645 for selling $4300 worth of appliances. Find the commission rate.

Example 4 Finding a Commission Rate

A salesperson earned $1560 for selling $13,000 worth of electronics equipment. Find the commission rate.

Solution: Let r stand for the unknown commission rate. Then

commission = commission rate · sales

$$\$1560 \; = \; r \quad\quad\quad\quad \cdot \; \$13,000$$

$$\frac{1560}{13,000} \; = \; r \quad\quad \text{Divide 1560 by 13,000, the number multiplied by } r.$$

$$0.12 \; = \; r \quad\quad \text{Simplify.}$$

$$12\% \; = \; r \quad\quad \text{Write 0.12 as a percent.}$$

The commission rate is 12%.

■ Work Practice 4

Objective C Calculating Discount and Sale Price

Suppose that an item that normally sells for $40 is on sale for 25% off. This means that the **original price** of $40 is reduced, or **discounted,** by 25% of $40, or $10. The **discount rate** is 25%, the **amount of discount** is $10, and the **sale price** is $40 − $10, or $30. Study the diagram below to visualize these terms.

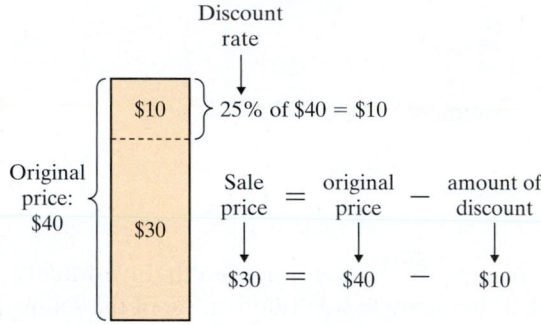

To calculate discounts and sale prices, we can use the following equations:

Discount and Sale Price

amount of discount = discount rate · original price

sale price = original price − amount of discount

Practice 5

A discontinued washer and dryer combo is advertised on sale for 35% off the regular price of $700. Find the amount of discount and the sale price.

Example 5 Finding a Discount and a Sale Price

An electric rice cooker that normally sells for $65 is on sale for 25% off. What is the amount of discount and what is the sale price?

Solution: First we find the amount of discount, or simply the discount.

amount of discount = discount rate · original price

$$\text{amount of discount} \; = \quad 25\% \quad\quad \cdot \quad\quad \$65$$

$$= \; 0.25 \cdot \$65 \quad \text{Write 25\% as 0.25.}$$

$$= \; \$16.25 \quad\quad \text{Multiply.}$$

Answers

4. 15%   5. $245; $455

The discount is $16.25. Next, find the sale price.

sale price = original price − discount

↓ ↓ ↓

sale price = $65 − $16.25

= $48.75   Subtract.

The sale price is $48.75.

■ **Work Practice 5**

Vocabulary, Readiness & Video Check

Use the choices below to fill in each blank. Some choices may be used more than once.

amount of discount   sale price   sales tax

commission   total price

1. _____ = tax rate · purchase price.

2. _____ = purchase price + sales tax.

3. _____ = commission rate · sales.

4. _____ = discount rate · original price.

5. _____ = original price − amount of discount.

6. sale price = original price − _____.

Martin-Gay Interactive Videos   *Watch the section lecture video and answer the following questions.*

Objective A 7. In ▦ Example 1, what is our first step after translating the problem into an equation?

Objective B 8. What is our final step in solving ▦ Example 2? ▶

Objective C 9. In the lecture before ▦ Example 3, since both equations shown involve the "amount of discount," how can the two equations be combined into one equation? ▶

See Video 5.7 🍎

5.7 Exercise Set MyMathLab® ▶

Objective A *Solve. See Examples 1 and 2.*

1. What is the sales tax on a jacket priced at $150 if the sales tax rate is 5%?

2. If the sales tax rate is 6%, find the sales tax on a microwave oven priced at $188.

3. The purchase price of a camcorder is $799. What is the total price if the sales tax rate is 7.5%?

4. A stereo system has a purchase price of $426. What is the total price if the sales tax rate is 8%?

5. A new large-screen television has a purchase price of $4790. If the sales tax on this purchase is $335.30, find the sales tax rate.

6. The sales tax on the purchase of a $6800 used car is $374. Find the sales tax rate.

7. The sales tax on a table saw is $10.20.

   a. What is the purchase price of the table saw (before tax) if the sales tax rate is 8.5%? (*Hint:* Use the sales tax equation and insert the replacement values.)

   b. Find the total price of the table saw.

8. The sales tax on a one-half-carat diamond ring is $76.

   a. Find the purchase price of the ring (before tax) if the sales tax rate is 9.5%. (See the hint for Exercise 7a.)

   b. Find the total price of the ring.

9. A gold and diamond bracelet sells for $1800. Find the sales tax and the total price if the sales tax rate is 6.5%.

10. The purchase price of a personal computer is $1890. If the sales tax rate is 8%, what is the sales tax and the total price?

▶ 11. The sales tax on the purchase of a futon is $24.25. If the tax rate is 5%, find the purchase price of the futon.

12. The sales tax on the purchase of a TV-DVD combination is $32.85. If the tax rate is 9%, find the purchase price of the TV-DVD.

13. The sales tax is $98.70 on a stereo sound system purchase of $1645. Find the sales tax rate.

14. The sales tax is $103.50 on a necklace purchase of $1150. Find the sales tax rate.

15. A cell phone costs $210, a battery recharger costs $15, and batteries cost $5. What is the sales tax and total price for purchasing these items if the sales tax rate is 7%?

16. Ms. Warner bought a blouse for $35, a skirt for $55, and a blazer for $95. Find the sales tax and the total price she paid, given a sales tax rate of 6.5%.

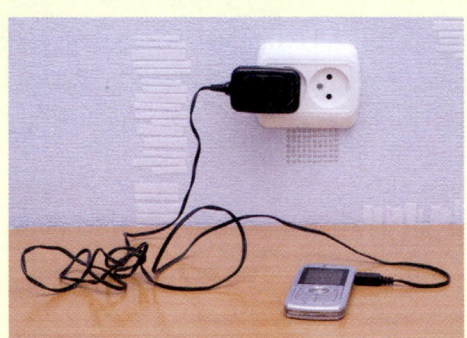

Objective B *Solve. See Examples 3 and 4.*

17. A sales representative for a large furniture warehouse is paid a commission rate of 4%. Find her commission if she sold $1,329,401 worth of furniture last year.

18. Rosie Davis-Smith is a beauty consultant for a home cosmetic business. She is paid a commission rate of 12.8%. Find her commission if she sold $1638 in cosmetics last month.

▶ 19. A salesperson earned a commission of $1380.40 for selling $9860 worth of paper products. Find the commission rate.

20. A salesperson earned a commission of $3575 for selling $32,500 worth of books to various bookstores. Find the commission rate.

21. How much commission will Jack Pruet make on the sale of a $325,900 house if he receives 1.5% of the selling price?

22. Frankie Lopez sold $9638 of jewelry this week. Find her commission for the week if she receives a commission rate of 5.6%.

23. A real estate agent earned a commission of $5565 for selling a house. If his rate is 3%, find the selling price of the house. (*Hint:* Use the commission equation and insert the replacement values.)

24. A salesperson earned $1750 for selling fertilizer. If her commission rate is 7%, find the selling price of the fertilizer. (See the hint for Exercise 23.)

Objective C *Find the amount of discount and the sale price. See Example 5.*

	Original Price	Discount Rate	Amount of Discount	Sale Price
25.	$89	10%		
26.	$74	20%		
27.	$196.50	50%		
28.	$110.60	40%		
29.	$410	35%		
30.	$370	25%		
31.	$21,700	15%		
32.	$17,800	12%		

▶ 33. A $300 fax machine is on sale for 15% off. Find the amount of discount and the sale price.

34. A $4295 designer dress is on sale for 30% off. Find the amount of discount and the sale price.

Objectives A B Mixed Practice *Complete each table.*

	Purchase Price	Tax Rate	Sales Tax	Total Price
35.	$305	9%		
36.	$243	8%		
37.	$56	5.5%		
38.	$65	8.4%		

	Sale	Commission Rate	Commission
39.	$235,800	3%	
40.	$195,450	5%	
41.	$17,900		$1432
42.	$25,600		$2304

Review

Multiply. See Sections 4.4 and 4.6.

43. $2000 \cdot \dfrac{3}{10} \cdot 2$

44. $500 \cdot \dfrac{2}{25} \cdot 3$

45. $400 \cdot \dfrac{3}{100} \cdot 11$

46. $1000 \cdot \dfrac{1}{20} \cdot 5$

47. $600 \cdot 0.04 \cdot \dfrac{2}{3}$

48. $6000 \cdot 0.06 \cdot \dfrac{3}{4}$

Concept Extensions

Solve. See the Concept Check in this section.

49. Your purchase price is $68 and the sales tax rate is 9.5%. Round each amount and use the rounded amounts to estimate the total price. Choose the best estimate.
a. $105   **b.** $58   **c.** $93   **d.** $77

50. Your purchase price is $200 and the tax rate is 10%. Choose the best estimate of the total price.
a. $190   **b.** $210   **c.** $220   **d.** $300

Tipping

One very useful application of percent is mentally calculating a tip. Recall that to find 10% of a number, simply move the decimal point one place to the left. To find 20% of a number, just double 10% of the number. To find 15% of a number, find 10% and then add to that number half of the 10% amount. Mentally fill in the chart below. To do so, start by rounding the bill amount to the nearest dollar.

	Tipping Chart		
Bill Amount	**10%**	**15%**	**20%**
51. $40.21			
52. $15.89			
53. $72.17			
54. $9.33			

55. Suppose that the original price of a shirt is $50. Which is better, a 60% discount or a discount of 30% followed by a discount of 35% of the reduced price? Explain your answer.

56. Which is better, a 30% discount followed by an additional 25% off or a 20% discount followed by an additional 40% off? To see, suppose an item costs $100 and calculate each discounted price. Explain your answer.

57. A diamond necklace sells for $24,966. If the tax rate is 7.5%, find the total price.

58. A house recently sold for $562,560. The commission rate on the sale is 5.5%. If the real estate agent is to receive 60% of the commission, find the amount received by the agent.

5.8   **Percent and Problem Solving: Interest**

Objectives

A Calculate Simple Interest.

B Calculate Compound Interest.

C Calculate Monthly Payments.

Objective A Calculating Simple Interest

Interest is money charged for using other people's money. When you borrow money, you pay interest. When you loan or invest money, you earn interest. The money borrowed, loaned, or invested is called the **principal amount,** or simply **principal.** Interest is normally stated in terms of a percent of the principal for a given period of time. The **interest rate** is the percent used in computing the interest. Unless stated otherwise, *the rate is understood to be per year.* When the interest is computed on the original principal, it is called **simple interest.** Simple interest is calculated using the following equation:

> **Simple Interest**
>
> Simple Interest = Principal · Rate · Time
> $$I = P \cdot R \cdot T$$
>
> where the rate is understood to be per year and time is in years.

Example 1 Finding Simple Interest

Find the simple interest after 2 years on $500 at an interest rate of 12%.

Solution: In this example, $P = \$500$, $R = 12\%$, and $T = 2$ years. Replace the variables with values in the formula $I = PRT$.

$$I = P \cdot R \cdot T$$
$$I = \$500 \cdot 12\% \cdot 2 \qquad \text{Let } P = \$500, R = 12\%, \text{ and } T = 2.$$
$$= \$500 \cdot (0.12) \cdot 2 \qquad \text{Write 12\% as a decimal.}$$
$$= \$120 \qquad \text{Multiply.}$$

The simple interest is $120.

■ **Work Practice 1**

If time is not given in years, we need to convert the given time to years.

Example 2 Finding Simple Interest

Ivan Borski borrowed $2400 at 10% simple interest for 8 months to buy a used Toyota Corolla. Find the simple interest he paid.

Solution: Since there are 12 months in a year, we first find what part of a year 8 months is.

$$8 \text{ months} = \frac{8}{12} \text{ year} = \frac{2}{3} \text{ year}$$

Now we find the simple interest.

simple interest	=	principal	·	rate	·	time
↓		↓		↓		↓
simple interest	=	$2400	·	10%	·	$\frac{2}{3}$
	=	$2400	·	0.10	·	$\frac{2}{3}$
	=	$160				

The interest on Ivan's loan is $160.

■ **Work Practice 2**

✓**Concept Check** Suppose in Example 2 you had obtained an answer of $16,000. How would you know that you had made a mistake in this problem?

When money is borrowed, the borrower pays the original amount borrowed, or the principal, as well as the interest. When money is invested, the investor receives the original amount invested, or the principal, as well as the interest. In either case, the **total amount** is the sum of the principal and the interest.

Finding the Total Amount of a Loan or Investment

total amount (paid or received) = principal + interest

Practice 1

Find the simple interest after 5 years on $875 at an interest rate of 7%.

Practice 2

A student borrowed $1500 for 9 months on her credit card at a simple interest rate of 20%. How much interest did she pay?

Answers
1. $306.25   **2.** $225

✓**Concept Check Answer**
$16,000 is too much interest.

Practice 3

If $2100 is borrowed at a simple interest rate of 13% for 6 months, find the total amount paid.

Example 3   Finding the Total Amount of an Investment

An accountant invested $2000 at a simple interest rate of 10% for 2 years. What total amount of money will she have from her investment in 2 years?

Solution: First we find her interest.

$$I = P \cdot R \cdot T$$
$$= \$2000 \cdot (0.10) \cdot 2 \quad \text{Let } P = \$2000, R = 10\% \text{ or } 0.10, \text{ and } T = 2.$$
$$= \$400$$

The interest is $400.

Next, we add the interest to the principal.

total amount	=	principal	+	interest
↓		↓		↓
total amount	=	$2000	+	$400
	=	$2400		

After 2 years, she will have a total amount of $2400.

■ **Work Practice 3**

✓**Concept Check** Which investment would earn more interest: an amount of money invested at 8% interest for 2 years, or the same amount of money invested at 8% for 3 years? Explain.

Objective B Calculating Compound Interest

Recall that simple interest depends on the original principal only. Another type of interest is compound interest. **Compound interest** is computed not only on the principal, but also on the interest already earned in previous compounding periods. Compound interest is used more often than simple interest.

Let's see how compound interest differs from simple interest. Suppose that $2000 is invested at 7% interest **compounded annually** for 3 years. This means that interest is added to the principal at the end of each year and that next year's interest is computed on this new amount. In this section, we round dollar amounts to the nearest cent.

	Amount at Beginning of Year	Principal	•	Rate	•	Time	= Interest	Amount at End of Year
1st year	$2000	$2000	•	0.07	•	1	= $140	$2000 + 140 = $2140
2nd year	$2140	$2140	•	0.07	•	1	= $149.80	$2140 + 149.80 = $2289.80
3rd year	$2289.80	$2289.80	•	0.07	•	1	= $160.29	$2289.80 + 160.29 = $2450.09

The compound interest earned can be found by

total amount	−	original principal	=	compound interest
↓		↓		↓
$2450.09	−	$2000	=	$450.09

The simple interest earned would have been

principal	•	rate	•	time	=	interest
↓		↓		↓		↓
$2000	•	0.07	•	3	=	$420

Answer

3. $2236.50

✓**Concept Check Answer**

8% for 3 years. Since the interest rate is the same, the longer you keep the money invested, the more interest you earn.

Since compound interest earns "interest on interest," compound interest earns more than simple interest.

Computing compound interest using the method above can be tedious. We can use a calculator and the compound interest formula below to compute compound interest more quickly.

Compound Interest Formula

The total amount A in an account is given by

$$A = P\left(1 + \frac{r}{n}\right)^{n \cdot t}$$

where P is the principal, r is the interest rate written as a decimal, t is the length of time in years, and n is the number of times compounded per year.

Example 4 $1800 is invested at 2% interest compounded annually. Find the total amount after 3 years.

Solution: "Compounded annually" means 1 time a year, so

$n = 1$. Also, $P = \$1800$, $r = 2\% = 0.02$, and $t = 3$ years.

$$A = P\left(1 + \frac{r}{n}\right)^{n \cdot t}$$

$$= 1800\left(1 + \frac{0.02}{1}\right)^{1 \cdot 3}$$

$$= 1800(1.02)^3$$

$$\approx 1910.17 \qquad \text{Round to 2 decimal places.}$$

> **Helpful Hint**
> Remember order of operations. **First** evaluate $(1.02)^3$, then multiply by 1800.

The total amount at the end of 3 years is $1910.17.

■ **Work Practice 4**

Practice 4

$3000 is invested at 4% interest compounded annually. Find the total amount after 6 years.

Example 5   Finding Total Amount Received from an Investment

$4000 is invested at 5.3% compounded quarterly for 10 years. Find the total amount at the end of 10 years.

Solution: "Compounded quarterly" means 4 times a year, so

$n = 4$. Also, $P = \$4000$, $r = 5.3\% = 0.053$, and $t = 10$ years.

$$A = P\left(1 + \frac{r}{n}\right)^{n \cdot t}$$

$$= 4000\left(1 + \frac{0.053}{4}\right)^{4 \cdot 10}$$

$$= 4000(1.01325)^{40}$$

$$\approx 6772.12$$

The total amount after 10 years is $6772.12.

■ **Work Practice 5**

Practice 5

$5500 is invested at $6\frac{1}{4}\%$ compounded *daily* for 5 years. Find the total amount at the end of 5 years. (Use 1 year = 365 days.)

Note: Part of the compound interest formula, $\left(1 + \frac{r}{n}\right)^{n \cdot t}$, is called the **compound interest factor.** Appendix B.3 contains a table of various calculated compound interest factors. Another way to calculate the total amount, A, in the compound interest

Answers
4. $3795.96   **5.** $7517.41

formula is to multiply the principal, P, by the appropriate compound interest factor found in Appendix B.3.

The Calculator Explorations box below shows how compound interest factors are calculated.

Objective C Calculating a Monthly Payment ▶

We conclude this section with a method to find the monthly payment on a loan.

> **Finding the Monthly Payment of a Loan**
>
> $$\text{monthly payment} = \frac{\text{principal} + \text{interest}}{\text{total number of payments}}$$

Practice 6

Find the monthly payment on a $3000 3-year loan if the interest on the loan is $1123.58.

Example 6   Finding a Monthly Payment

Find the monthly payment on a $2000 loan for 2 years. The interest on the 2-year loan is $435.88.

Solution: First we determine the total number of monthly payments. The loan is for 2 years. Since there are 12 months per year, the number of payments is $2 \cdot 12$, or 24. Now we calculate the monthly payment.

$$\text{monthly payment} = \frac{\text{principal} + \text{interest}}{\text{total number of payments}}$$

$$\text{monthly payment} = \frac{\$2000 + \$435.88}{24}$$

$$\approx \$101.50.$$

The monthly payment is about $101.50.

Answer

6. $114.54

■ **Work Practice 6**

🖩 Calculator Explorations Compound Interest Factor

A compound interest factor may be found by using your calculator and evaluating the formula

$$\textbf{compound interest factor} = \left(1 + \frac{r}{n}\right)^{n \cdot t}$$

where r is the interest rate, t is the time in years, and n is the number of times compounded per year. For example, the compound interest factor for 10 years at 8% compounded semiannually is about 2.19112. Let's find this factor by evaluating the compound interest factor formula when $r = 8\%$ or 0.08, $t = 10$, and $n = 2$ (compounded semiannually means 2 times per year). Thus,

$$\text{compound interest factor} = \left(1 + \frac{0.08}{2}\right)^{2 \cdot 10}$$

$$\text{or} \quad \left(1 + \frac{0.08}{2}\right)^{20}$$

To evaluate, press the keys

$(\ |\ 1\ |\ +\ |\ 0.08\ |\ \div\ |\ 2\ |\ )\ |\ y^x$ or $\land$ $|\ 20\ |$ then $|\ =\ |$ or $|\ \text{ENTER}\ |$.

The display will read $\boxed{2.1911231}$. Rounded to 5 decimal places, this is 2.19112.

Find the compound interest factors. Use the table in Appendix B.3 to check your answers. For Exercises 1–4, round to 5 decimal places. For Exercises 5 and 6, round to 2 decimal places.

1. 5 years, 9%, compounded quarterly
2. 15 years, 14%, compounded daily
3. 20 years, 11%, compounded annually
4. 1 year, 7%, compounded semiannually
5. Find the total amount after 4 years when $500 is invested at 6% compounded quarterly. (Multiply the appropriate compound interest factor by $500.)
6. Find the total amount for 19 years when $2500 is invested at 5% compounded daily.

Vocabulary, Readiness & Video Check

Use the choices below to fill in each blank. Choices may be used more than once.

    total amount   simple   principal amount   compound

1. To calculate _____ interest, use $I = P \cdot R \cdot T$.

2. To calculate _____ interest, use $A = P\left(1 + \dfrac{r}{n}\right)^{n \cdot t}$.

3. _____ interest is computed not only on the original principal, but also on interest already earned in previous compounding periods.

4. When interest is computed on the original principal only, it is called _____ interest.

5. _____ (paid or received) = principal + interest.

6. The _____ is the money borrowed, loaned, or invested.

Martin-Gay Interactive Videos   *Watch the section lecture video and answer the following questions.*

See Video 5.8

Objective A 7. Complete this statement based on the lecture before ▦ Example 1: Simple interest is charged on the _____ only.

Objective B 8. In ▦ Example 2, how often is the interest compounded and what number does this translate to in the formula? ▶

Objective C 9. In ▦ Example 3, how was the denominator of 48 determined? ▶

5.8 Exercise Set MyMathLab® ▶

Objective A *Find the simple interest. See Examples 1 and 2.*

	Principal	Rate	Time
1.	$200	8%	2 years
3.	$160	11.5%	4 years
5.	$5000	10%	$1\frac{1}{2}$ years
7.	$375	18%	6 months
9.	$2500	16%	21 months

	Principal	Rate	Time
2.	$800	9%	3 years
4.	$950	12.5%	5 years
6.	$1500	14%	$2\frac{1}{4}$ years
8.	$775	15%	8 months
10.	$1000	10%	18 months

Solve. See Examples 1 through 3.

▶**11.** A company borrows $162,500 for 5 years at a simple interest rate of 12.5%. Find the interest paid on the loan and the total amount paid back.

12. $265,000 is borrowed to buy a house. If the simple interest rate on the 30-year loan is 8.25%, find the interest paid on the loan and the total amount paid back.

13. A money market fund advertises a simple interest rate of 9%. Find the total amount received on an investment of $5000 for 15 months.

14. The Real Service Company takes out a 270-day (9-month) short-term, simple interest loan of $4500 to finance the purchase of some new equipment. If the interest rate is 14%, find the total amount that the company pays back.

arsha borrows $8500 and agrees to pay it back in years. If the simple interest rate is 17%, find the total amount she pays back.

16. An 18-year-old is given a high school graduation gift of $2000. If this money is invested at 8% simple interest for 5 years, find the total amount.

Objective B *Find the total amount in each compound interest account. See Examples 4 and 5.*

▶ **17.** $6150 is compounded semiannually at a rate of 14% for 15 years.

18. $2060 is compounded annually at a rate of 15% for 10 years.

19. $1560 is compounded daily at a rate of 8% for 5 years.

20. $1450 is compounded quarterly at a rate of 10% for 15 years.

21. $10,000 is compounded semiannually at a rate of 9% for 20 years.

22. $3500 is compounded daily at a rate of 8% for 10 years.

23. $2675 is compounded annually at a rate of 9% for 1 year.

24. $6375 is compounded semiannually at a rate of 10% for 1 year.

25. $2000 is compounded annually at a rate of 8% for 5 years.

26. $2000 is compounded semiannually at a rate of 8% for 5 years.

27. $2000 is compounded quarterly at a rate of 8% for 5 years.

28. $2000 is compounded daily at a rate of 8% for 5 years.

Objective C *Solve. See Example 6.*

29. A college student borrows $1500 for 6 months to pay for a semester of school. If the interest is $61.88, find the monthly payment.

30. Jim Tillman borrows $1800 for 9 months. If the interest is $148.90, find his monthly payment.

▶ **31.** $20,000 is borrowed for 4 years. If the interest on the loan is $10,588.70, find the monthly payment.

32. $105,000 is borrowed for 15 years. If the interest on the loan is $181,125, find the monthly payment.

Review

Find the perimeter of each figure. See Section 1.3.

△ **33.**

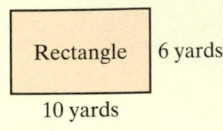

Rectangle — 6 yards — 10 yards

△ **34.**

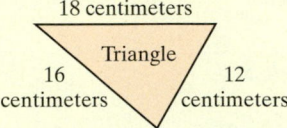

18 centimeters — Triangle — 16 centimeters — 12 centimeters

△ **35.**

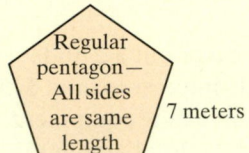

Regular pentagon — All sides are same length — 7 meters

△ **36.**

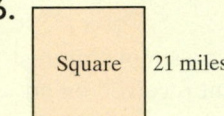

Square — 21 miles

Concept Extensions

37. Explain how to look up a compound interest factor in the compound interest table.

38. Explain how to find the amount of interest in a compounded account.

39. Compare the following accounts: Account 1: $1000 is invested for 10 years at a simple interest rate of 6%. Account 2: $1000 is compounded semiannually at a rate of 6% for 10 years. Discuss how the interest is computed for each account. Determine which account earns more interest. Why?

Chapter 5 Group Activity

Fastest-Growing Occupations

According to U.S. Bureau of Labor Statistics projections, the careers listed below are the top ten fastest-growing jobs ranked by expected percent increase through the year 2020. (*Source:* Bureau of Labor Statistics)

	Occupation	Employment in 2010	Percent Change	Expected Employment in 2020
1	Personal care aides	861,000	70.5%	
2	Home health aides	1,017,700	69.4%	
3	Biomedical engineers	15,700	61.7%	
4	Helpers—masons	29,400	60.1%	
5	Helpers—carpenters	46,500	55.7%	
6	Veterinary technicians	80,200	52.0%	
7	Reinforcing iron/rebar workers	19,100	48.6%	
8	Physical therapist assistants	67,400	45.7%	
9	Helpers—plumbers	57,900	45.4%	
10	Meeting/event planners	71,600	43.7%	

What do most of these fast-growing occupations have in common? They require knowledge of math! For some careers, such as home health aides, event planners, and biomedical engineers, the ways math is used on the job may be obvious. For other occupations, the use of math may not be quite as apparent. However, tasks common to many jobs—filling in a time sheet, writing up an expense or mileage report, planning a budget, figuring a bill, ordering supplies, and even making a work schedule—all require math.

This activity may be completed by working in groups or individually.

1. List the top five occupations by order of employment figures for 2010.

2. Using the 2010 employment figures and the percent increase from 2010 to 2020, find the expected 2020 employment figure for each occupation listed in the table. Round to the nearest thousand.

3. List the top five occupations by order of employment figures for 2020. Did the order change at all from 2010? Explain.

Chapter 5 Vocabulary Check

Fill in each blank with one of the words or phrases listed below. Some choices may be used more than once.

not equal	equal	cross products	rate	percent	sales tax
amount	ratio	unit rate	proportion	base	of
0.01	is	amount of discount	percent of decrease	total price	$\frac{1}{100}$
100%	compound interest	percent of increase	sale price	commission	

1. A(n) _____ is the quotient of two numbers. It can be written as a fraction, using a colon, or using the word *to*.

2. $\dfrac{x}{2} = \dfrac{7}{16}$ is an example of a(n) _____.

3. A(n) _____ is a rate with a denominator of 1.

4. A(n) _____ is a statement that two ratios are equal.

5. A(n) _____ is used to compare different kinds of quantities.

6. In the proportion $\dfrac{x}{2} = \dfrac{7}{16}$, $x \cdot 16$ and $2 \cdot 7$ are called _____.

7. If cross products are _____, the proportion is true.

8. If cross products are _____, the proportion is false.

9. In a mathematical statement, _____ usually means "multiplication."

10. In a mathematical statement, _____ means "equals."

11. _____ means "per hundred."

12. _____ is computed not only on the principal, but also on interest already earned in previous compounding periods.

13. In the percent proportion $\dfrac{\rule{3cm}{0.4pt}}{\rule{3cm}{0.4pt}} = \dfrac{\text{percent}}{100}$.

14. To write a decimal or fraction as a percent, multiply by _____.

15. The decimal equivalent of the % symbol is _____.

16. The fraction equivalent of the % symbol is _____.

17. The percent equation is _____ $\cdot$ percent $=$ _____.

18. _____ $= \dfrac{\text{amount of decrease}}{\text{original amount}}$.

19. _____ $= \dfrac{\text{amount of increase}}{\text{original amount}}$.

20. _____ $=$ tax rate $\cdot$ purchase price.

21. _____ $=$ purchase price $+$ sales tax.

22. _____ $=$ commission rate $\cdot$ sales.

23. _____ $=$ discount rate $\cdot$ original price.

24. _____ $=$ original price $-$ amount of discount.

Helpful Hint

▶ Are you preparing for your test? Don't forget to take the Chapter 5 Test on page 401. Then check your answers at the back of the text and use the Chapter Test Prep Videos to see the fully worked-out solutions to any of the exercises you want to review.

5 Chapter Highlights

Definitions and Concepts	Examples
Section 5.1 Ratio and Proportion	

A **ratio** is the quotient of two quantities.	The ratio of 3 to 4 can be written as $$\frac{3}{4} \quad \text{or} \quad 3:4$$ ↑ fraction notation ↑ colon notation
Rates are used to compare different kinds of quantities.	Write the rate 12 spikes every 8 inches as a fraction in simplest form. $$\frac{12 \text{ spikes}}{8 \text{ inches}} = \frac{3 \text{ spikes}}{2 \text{ inches}}$$
A **unit rate** is a rate with a denominator of 1.	Write as a unit rate: 117 miles on 5 gallons of gas $$\frac{117 \text{ miles}}{5 \text{ gallons}} = \frac{23.4 \text{ miles}}{1 \text{ gallon}} \quad \begin{array}{l} \text{or 23.4 miles per gallon} \\ \text{or 23.4 miles/gallon} \end{array}$$
A **proportion** is a statement that two ratios or rates are equal.	$\frac{1}{2} = \frac{4}{8}$ is a proportion.
Using Cross Products to Determine Whether Proportions Are True or False	Is $\frac{6}{10} = \frac{9}{15}$ a true proportion?
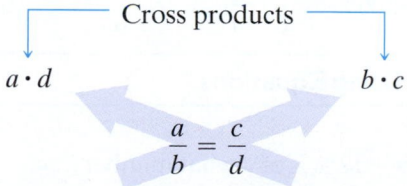	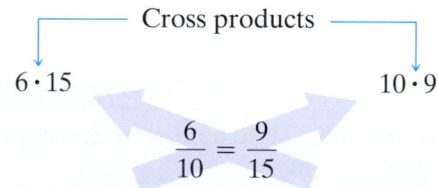
If cross products are equal, the proportion is true. If $ad = bc$, then the proportion is true. If cross products are not equal, the proportion is false. If $ad \neq bc$, then the proportion is false.	$6 \cdot 15 \overset{?}{=} 10 \cdot 9$ Are cross products equal? $90 = 90$ Since cross products are equal, the proportion is a true proportion.

(continued)

Definitions and Concepts	Examples

Section 5.1   Ratio and Proportion (continued)

Finding an Unknown Value *n* in a Proportion

Step 1: Set the cross products equal to each other.

Step 2: Divide the number not multiplied by *n* by the number multiplied by *n*.

Find *n*: $\dfrac{n}{7} = \dfrac{5}{8}$

Step 1:

$$\dfrac{n}{7} = \dfrac{5}{8}$$

$n \cdot 8 = 7 \cdot 5$ Set the cross products equal to each other.

$n \cdot 8 = 35$ Multiply.

Step 2:

$n = \dfrac{35}{8}$ Divide 35 by 8, the number multiplied by *n*.

$n = 4\dfrac{3}{8}$

Section 5.2   Introduction to Percent

Percent means "per hundred." The % symbol denotes percent.

$51\% = \dfrac{51}{100}$ 51 per 100

$7\% = \dfrac{7}{100}$ 7 per 100

To write a percent as a decimal, replace the % symbol with its decimal equivalent, 0.01, and multiply.

$32\% = 32(0.01) = 0.32$

To write a decimal as a percent, multiply by 100%.

$0.08 = 0.08(100\%) = 08.\% = 8\%$

Section 5.3   Percents and Fractions

To write a percent as a fraction, replace the % symbol with its fraction equivalent, $\dfrac{1}{100}$, and multiply.

$25\% = \dfrac{25}{100} = \dfrac{\overset{1}{\cancel{25}}}{4 \cdot \underset{1}{\cancel{25}}} = \dfrac{1}{4}$

To write a fraction as a percent, multiply by 100%.

$\dfrac{1}{6} = \dfrac{1}{6} \cdot 100\% = \dfrac{1}{6} \cdot \dfrac{100}{1}\% = \dfrac{100}{6}\% = 16\dfrac{2}{3}\%$

Section 5.4   Solving Percent Problems Using Equations

Three key words in the statement of a percent problem are

of, which means **multiplication (·)**
is, which means **equals (=)**
what (or some equivalent word or phrase), which stands for **the unknown number**

Solve:

6	is	12%	of	what number?
↓	↓	↓	↓	↓
6	=	12%	·	n
6	=	0.12	·	n Write 12% as a decimal.

$\dfrac{6}{0.12} = n$ Divide 6 by 0.12, the number multiplied by *n*.

$50 = n$

Thus, 6 is 12% of 50.

Definitions and Concepts	**Examples**

Section 5.5   Solving Percent Problems Using Proportions

Percent Proportion

$$\frac{\text{amount}}{\text{base}} = \frac{\text{percent}}{100} \leftarrow \text{always } 100$$

or

$$\text{amount} \rightarrow \frac{a}{b} = \frac{p}{100} \leftarrow \text{percent}$$

Solve:

20.4 is what percent of 85?

amount   percent   base

$$\text{amount} \rightarrow \frac{20.4}{85} = \frac{p}{100} \leftarrow \text{percent}$$

$$20.4 \cdot 100 = 85 \cdot p \quad \text{Set cross products equal.}$$
$$2040 = 85 \cdot p \quad \text{Multiply.}$$
$$\frac{2040}{85} = p \quad \text{Divide 2040 by 85, the number multiplied by } p.$$
$$24 = p \quad \text{Simplify.}$$

Thus, 20.4 is 24% of 85.

Section 5.6   Applications of Percent

Percent of Increase

$$\text{percent of increase} = \frac{\text{amount of increase}}{\text{original amount}}$$

Percent of Decrease

$$\text{percent of decrease} = \frac{\text{amount of decrease}}{\text{original amount}}$$

A town with a population of 16,480 decreased to 13,870 over a 12-year period. Find the percent of decrease. Round to the nearest whole percent.

$$\text{amount of decrease} = 16,480 - 13,870$$
$$= 2610$$

$$\text{percent of decrease} = \frac{\text{amount of decrease}}{\text{original amount}}$$
$$= \frac{2610}{16,480} \approx 0.16$$
$$= 16\%$$

The town's population decreased by 16%.

Section 5.7   Percent and Problem Solving: Sales Tax, Commission, and Discount

Sales Tax and Total Price

sales tax = sales tax rate · purchase price
total price = purchase price + sales tax

Find the sales tax and the total price of a purchase of $42 if the sales tax rate is 9%.

sales tax = sales tax rate · purchase price

sales tax = 9% · $42
= 0.09 · $42
= $3.78

The total price is

total price = purchase price + sales tax

total price = $42 + $3.78
= $45.78

(continued)

Definitions and Concepts	Examples

Section 5.7   Percent and Problem Solving: Sales Tax, Commission, and Discount (*continued*)

Commission

     commission = commission rate · total sales

A salesperson earns a commission of 3%. Find the commission from sales of $12,500 worth of appliances.

$$\text{commission} = \text{commission rate} \cdot \text{sales}$$

$$\text{commission} = 3\% \cdot \$12,500$$
$$= 0.03 \cdot \$12,500$$
$$= \$375$$

Discount and Sale Price

     amount of discount = discount rate · original price

     sale price = original price − amount of discount

A suit is priced at $320 and is on sale today for 25% off. What is the sale price?

$$\text{amount of discount} = \text{discount rate} \cdot \text{original price}$$

$$\text{amount of discount} = 25\% \cdot \$320$$
$$= 0.25 \cdot \$320$$
$$= \$80$$

$$\text{sale price} = \text{original price} - \text{amount of discount}$$

$$\text{sale price} = \$320 - \$80$$
$$= \$240$$

The sale price is $240.

Section 5.8   Percent and Problem Solving: Interest

Simple Interest

     interest = principal · rate · time

where the rate is understood to be per year.

Find the simple interest after 3 years on $800 at an interest rate of 5%.

$$\text{interest} = \text{principal} \cdot \text{rate} \cdot \text{time}$$

$$\text{interest} = \$800 \cdot 5\% \cdot 3$$
$$= \$800 \cdot 0.05 \cdot 3 \quad \text{Write 5\% as 0.05.}$$
$$= \$120 \quad \text{Multiply.}$$

The interest is $120.

Compound interest is computed not only on the principal, but also on interest already earned in previous compounding periods. (See Appendix B.3 for various compound interest factors.)

$$A = P\left(1 + \frac{r}{n}\right)^{n \cdot t}$$

where n is the number of times compounded per year.

$800 is invested at 5% compounded quarterly for 10 years. Find the total amount at the end of 10 years.

$$A = \$800\left(1 + \frac{0.05}{4}\right)^{4 \cdot 10}$$
$$= \$800\left(1.0125\right)^{40}$$
$$\approx \$1314.90$$

(5.1) *Write each ratio as a fraction in simplest form.*

1. 23 to 37

2. 6000 people to 4800 people

3. $121 to $143

4. 4.25 yards to 8.75 yards

The circle graph below shows how the top 25 movies (or films) of 2009 were rated. Use this graph to answer the questions.

Top 25 Movies of 2009

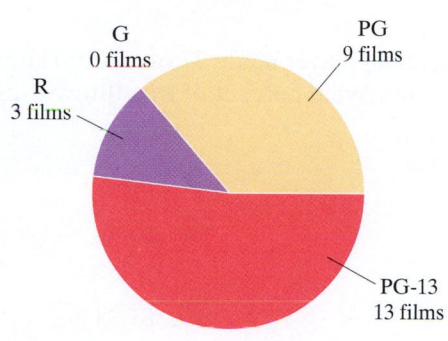

G
0 films

PG
9 films

R
3 films

PG-13
13 films

Source: MPAA

Note: There were no G-rated films in the top 25 for 2009

5. a. How many top 25 movies were rated PG?

    b. Find the ratio of top 25 PG-rated movies to total top movies for that year.

6. a. How many top 25 movies were rated R?

    b. Find the ratio of top 25 R-rated movies to total top movies for that year.

Write each rate as a fraction in simplest form.

7. 6 professors for 20 graduate research assistants

8. 15 word processing pages printed in 6 minutes

Write each rate as a unit rate.

9. 468 miles in 9 hours

10. 180 feet in 12 seconds

Determine whether each proportion is true.

11. $\dfrac{21}{8} = \dfrac{14}{6}$

12. $\dfrac{3.75}{3} = \dfrac{7.5}{6}$

Find the unknown number n in each proportion.

13. $\dfrac{n}{9} = \dfrac{5}{3}$

14. $\dfrac{4}{13} = \dfrac{10}{n}$

15. $\dfrac{27}{\frac{9}{4}} = \dfrac{n}{5}$

16. $\dfrac{0.4}{n} = \dfrac{2}{4.7}$

Solve. An owner of a Ford Escort can drive 420 miles on 11 gallons of gas.

17. If Tom Aloiso runs out of gas in an Escort and AAA comes to his rescue with $1\frac{1}{2}$ gallons of gas, determine whether Tom can then drive to a gas station 65 miles away.

18. Find how many gallons of gas Tom can expect to burn on a 3000-mile trip. Round to the nearest gallon.

Yearly homeowner property taxes are figured at a rate of $1.15 tax for every $100 of house value.

19. If a homeowner pays $627.90 in property taxes, find the value of his home.

20. Find the property taxes on a town house valued at $89,000.

(5.2) *Solve.*

21. In a survey of 100 adults, 37 preferred pepperoni on their pizzas. What percent preferred pepperoni?

22. A basketball player made 77 out of 100 attempted free throws. What percent of free throws was made?

Write each percent as a decimal.

23. 83%

24. 75%

25. 0.5%

26. 0.7%

27. 200%

28. 400%

29. 26.25%

30. 85.34%

Write each decimal as a percent.

31. 2.6

32. 0.055

33. 0.35

34. 1.02

35. 0.71

36. 0.65

37. 4

38. 9

(5.3) *Write each percent as a fraction or mixed number in simplest form.*

39. 1%

40. 10%

41. 25%

42. 8.5%

43. 10.2%

44. $16\frac{2}{3}$%

45. $33\frac{1}{3}$%

46. 110%

Write each fraction or mixed number as a percent.

47. $\frac{1}{5}$

48. $\frac{7}{10}$

49. $\frac{5}{6}$

50. $1\frac{2}{3}$

51. $1\frac{1}{4}$

52. $\frac{3}{5}$

53. $\frac{1}{16}$

54. $\frac{5}{8}$

(5.4) *Translate each to an equation and solve.*

55. 1250 is 1.25% of what number?

56. What number is $33\frac{1}{3}$% of 24,000?

57. 124.2 is what percent of 540?

58. 22.9 is 20% of what number?

59. What number is 40% of 7500?

60. 693 is what percent of 462?

(5.5) *Translate each to a proportion and solve.*

61. 104.5 is 25% of what number?

62. 16.5 is 5.5% of what number?

63. What number is 36% of 180?

64. 63 is what percent of 35?

65. 93.5 is what percent of 85?

66. What number is 33% of 500?

(5.6) *Solve.*

67. In a survey of 2000 people, it was found that 1320 have a microwave oven. Find the percent of people who own microwaves.

68. Of the 12,360 freshmen entering County College, 2000 are enrolled in basic college mathematics. Find the percent of entering freshmen who are enrolled in basic college mathematics. Round to the nearest whole percent.

69. The number of violent crimes in a city decreased from 675 to 534. Find the percent of decrease. Round to the nearest tenth of a percent.

70. The current charge for dumping waste in a local landfill is $16 per cubic foot. To cover new environmental costs, the charge will increase to $33 per cubic foot. Find the percent of increase.

71. This year the fund drive for a charity collected $215,000. Next year, a 4% decrease is expected. Find how much is expected to be collected in next year's drive.

72. A local union negotiated a new contract that increases the hourly pay 15% over last year's pay. The old hourly rate was $11.50. Find the new hourly rate rounded to the nearest cent.

(5.7) *Solve.*

73. If the sales tax rate is 5.5%, what is the total amount charged for a $250 coat?

74. Find the sales tax paid on a $25.50 purchase if the sales tax rate is 4.5%.

75. Russ James is a sales representative for a chemical company and is paid a commission rate of 5% on all sales. Find his commission if he sold $100,000 worth of chemicals last month.

76. Carol Sell is a sales clerk in a clothing store. She receives a commission of 7.5% on all sales. Find her commission for the week if her sales for the week were $4005. Round to the nearest cent.

77. A $3000 mink coat is on sale for 30% off. Find the discount and the sale price.

78. A $90 calculator is on sale for 10% off. Find the discount and the sale price.

(5.8) *Solve.*

79. Find the simple interest due on $4000 loaned for 4 months at 12% interest.

80. Find the simple interest due on $6500 loaned for 3 months at 20%.

81. Find the total amount in an account if $5500 is compounded annually at 12% for 15 years.

82. Find the total amount in an account if $6000 is compounded semiannually at 11% for 10 years.

83. Find the compound interest earned if $100 is compounded quarterly at 12% for 5 years.

84. Find the compound interest earned if $1000 is compounded quarterly at 18% for 20 years.

Mixed Review

Find the unknown number n in each proportion.

85. $\dfrac{3}{n} = \dfrac{15}{8}$

86. $\dfrac{42}{5} = \dfrac{n}{10}$

Write each percent as a decimal.

87. 3.8%

88. 24.5%

89. 0.9%

Write each decimal as a percent.

90. 0.54

91. 95.2

92. 0.3

Write each percent as a fraction or mixed number in simplest form.

93. 47%

94. $6\dfrac{2}{5}\%$

95. 5.6%

Write each fraction or mixed number as a percent.

96. $\dfrac{3}{8}$

97. $\dfrac{2}{13}$

98. $\dfrac{6}{5}$

Translate each into an equation and solve.

99. 43 is 16% of what number?

100. 27.5 is what percent of 25?

101. What number is 36% of 1968?

102. 67 is what percent of 50?

Translate each into a proportion and solve.

103. 75 is what percent of 25?

104. What number is 16% of 240?

105. 28 is 5% of what number?

106. 52 is what percent of 16?

Solve.

107. The total number of cans in a soft drink machine is 300. If 78 soft drinks have been sold, find the percent of soft drink cans that have been sold.

108. A home valued at $96,950 last year has lost 7% of its value this year. Find the loss in value.

109. A dinette set sells for $568.00. If the sales tax rate is 8.75%, find the purchase price of the dinette set.

110. The original price of a video game is $23.00. It is on sale for 15% off. What is the amount of the discount?

111. A candy salesman makes a commission of $1.60 from each case of candy he sells. If a case of candy costs $12.80, what is his rate of commission?

112. Find the total amount due on a 6-month loan of $1400 at a simple interest rate of 13%.

113. Find the total amount due on a loan of $5500 for 9 years at 12.5% simple interest.

Write each ratio or rate as a fraction in simplest form.

1. $75 to $10

2. 8.6 to 10

Find each unit rate.

3. 8 inches of rain in 12 hours

4. QRI0 (Quest for Curiosity) is the world's first bipedal robot capable of running (moving with both legs off the ground at the same time) at a rate of 108 inches each 12 seconds. (*Source: Guinness World Records*)

Find the unknown number n in each proportion.

5. $\dfrac{8}{n} = \dfrac{11}{6}$

6. $\dfrac{1.5}{5} = \dfrac{2.4}{n}$

Solve.

7. The standard dose of medicine for a dog is 10 grams for every 15 pounds of body weight. What is the standard dose for a dog that weighs 80 pounds?

8. Currently 27 out of every 50 American adults drink coffee every day. In a town with a population of 7900 adults, how many of these adults would you expect to drink coffee every day? (*Source:* National Coffee Association)

Write each percent as a decimal.

9. 85%

10. 500%

11. 0.8%

Write each decimal as a percent.

12. 0.056

13. 6.1

14. 0.39

Write each percent as a fraction or mixed number in simplest form.

15. 120%

16. 38.5%

17. 0.2%

Answers

1. _____

2. _____

3. _____

4. _____

5. _____

6. _____

7. _____

8. _____

9. _____

10. _____

11. _____

12. _____

13. _____

14. _____

15. _____

16. _____

17. _____

18. _____

19. _____

20. _____

21. _____

22. _____

23. _____

24. _____

25. _____

26. _____

27. _____

28. _____

29. _____

30. _____

31. _____

32. _____

33. _____

Write each fraction or mixed number as a percent.

18. $\frac{11}{20}$ **19.** $\frac{3}{8}$ **20.** $1\frac{5}{9}$

Solve.

21. What number is 42% of 80?

22. 0.6% of what number is 7.5?

23. 567 is what percent of 756?

Solve. Round all dollar amounts to the nearest cent.

24. An alloy is 12% copper. How much copper is contained in 320 pounds of this alloy?

25. A farmer in Nebraska estimates that 20% of his potential crop, or $11,350, has been lost to a hard freeze. Find the total value of his potential crop.

26. If the local sales tax rate is 1.25%, find the total amount charged for a stereo system priced at $354.

27. A town's population increased from 25,200 to 26,460. Find the percent of increase.

28. A $120 framed picture is on sale for 15% off. Find the discount and the sale price.

29. Randy Nguyen is paid a commission rate of 4% on all sales. Find Randy's commission if his sales were $9875.

30. A sales tax of $1.53 is added to an item's price of $152.99. Find the sales tax rate. Round to the nearest whole percent.

31. Find the simple interest earned on $2000 saved for $3\frac{1}{2}$ years at an interest rate of 9.25%.

32. $1365 is compounded annually at 8%. Find the total amount in the account after 5 years.

33. A couple borrowed $400 from a bank at 13.5% for 6 months for car repairs. Find the total amount due the bank at the end of the 6-month period.

1. How many cases can be filled with 9900 cans of jalapeños if each case holds 48 cans? How many cans will be left over? Will there be enough cases to fill an order for 200 cases?

2. Multiply: 409×76

3. Write each as a mixed number or a whole number.

   a. $\dfrac{30}{7}$ **b.** $\dfrac{16}{15}$ **c.** $\dfrac{84}{6}$

4. Write each mixed number as an improper fraction.

   a. $2\dfrac{5}{7}$ **b.** $10\dfrac{1}{10}$ **c.** $5\dfrac{3}{8}$

5. Use a factor tree to find the prime factorization of 80.

6. Find the area of the rectangle.

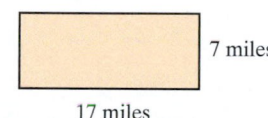

7 miles

17 miles

7. Write in simplest form: $\dfrac{10}{27}$

8. Find the average of 28, 34, and 70.

9. Multiply and simplify: $\dfrac{23}{32} \cdot \dfrac{4}{7}$

10. Round 76,498 to the nearest ten.

11. Find the reciprocal of $\dfrac{11}{8}$.

12. Write the shaded part of the figure as an improper fraction and as a mixed number.

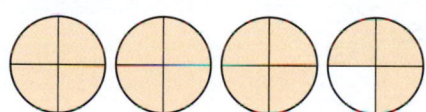

△ **13.** Find the perimeter of the rectangle.

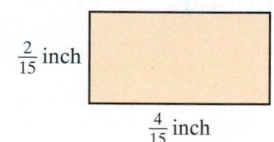

$\frac{2}{15}$ inch

$\frac{4}{15}$ inch

14. Find $2 \cdot 5^2$

15. Find the LCM of 12 and 20.

16. Subtract $\dfrac{7}{9}$ from $\dfrac{10}{9}$.

17. Add: $\dfrac{2}{5} + \dfrac{4}{15}$

18. Find $\dfrac{2}{3}$ of 510.

19. Subtract: $7\dfrac{3}{14} - 3\dfrac{6}{7}$

20. Simplify: $9 \cdot \sqrt{25} - 6 \cdot \sqrt{4}$

Perform each indicated operation.

21. $\dfrac{1}{2} \div \dfrac{8}{7}$

22. $20\dfrac{4}{5} + 12\dfrac{7}{8}$

Answers

1. _____

2. _____

3. a. _____

   b. _____

   c. _____

4. a. _____

   b. _____

   c. _____

5. _____

6. _____

7. _____

8. _____

9. _____

10. _____

11. _____

12. _____

13. _____

14. _____

15. _____

16. _____

17. _____

18. _____

19. _____

20. _____

21. _____

22. _____

403

23. _____

24. _____

25. _____

26. _____

27. _____

28. _____

29. _____

30. _____

31. _____

32. _____

33. _____

34. _____

35. _____

36. _____

37. _____

38. _____

39. _____

40. _____

41. _____

42. _____

43. _____

44. _____

23. $\dfrac{2}{9} \cdot \dfrac{3}{11}$

24. $1\dfrac{7}{8} \cdot 3\dfrac{2}{5}$

Write each fraction as a decimal.

25. $\dfrac{8}{10}$

26. $\dfrac{9}{100}$

27. $\dfrac{87}{10}$

28. $\dfrac{48}{10,000}$

29. The price of a gallon of premium gasoline in Cross City is currently $3.1779. Round this to the nearest cent.

30. Subtract: $38 - 10.06$

31. Add: $763.7651 + 22.001 + 43.89$

32. 12.483×100

33. Multiply: 23.6×0.78

34. 76.3×1000

Divide.

35. $\dfrac{786.1}{1000}$

36. $0.5\overline{)0.638}$

37. $\dfrac{0.12}{10}$

38. $0.23\overline{)11.6495}$

39. Simplify: $723.6 \div 1000 \times 10$

40. Simplify: $\dfrac{3.19 - 0.707}{13}$

41. Write $\dfrac{1}{4}$ as a decimal.

42. Write $\dfrac{5}{9}$ as a decimal. Give an exact answer and a three-decimal-place approximation.

43. Translate to an equation: What number is 25% of 0.008?

44. Write $\dfrac{3}{8}$ as a percent.

Geometry

A Swiss company created the Zaugg Pipe Monster (shown above) specifically for building superpipes.

What Is a Superpipe?

For winter sports, the term *superpipe* is used to describe a halfpipe built of snow that has walls 22 feet high from the flat bottom on both sides. The length of a superpipe ranges from 400 feet to 600 feet.

Halfpipes in snow were originally formed by hand tools or with heavy machinery. The current method of halfpipe cutting and grooming is by use of a Zaugg Pipe Monster, shown above. Because of the high expense of constructing and maintaining them, there are very few true superpipes. During the 2013–2014 northern hemisphere winter, only 14 superpipes existed globally.

Throughout this chapter, we work with lines, angles, circumferences, and other geometric concepts that give us an appreciation of the work that goes into constructing a halfpipe.

**1998 Olympics
Nagano, Japan**

11.5 feet

394 feet

49 feet

**2014 Olympics
Sochi, Russia**

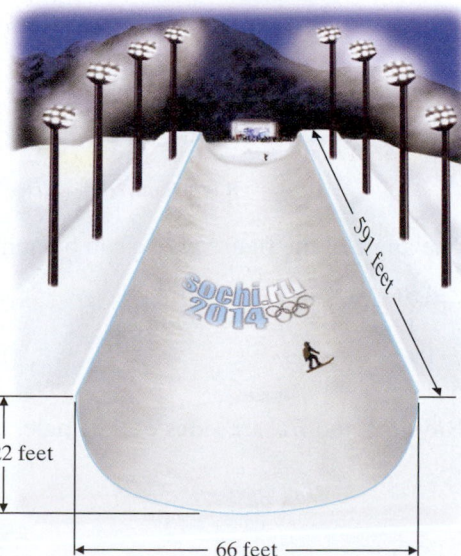

22 feet

591 feet

66 feet

Check Your Progress

The word *geometry* is formed from the Greek words *geo*, meaning Earth, and *metron*, meaning measure. Geometry literally means to measure the Earth. In this chapter we learn about various geometric figures and their properties such as perimeter, area, and volume. Knowledge of geometry can help us solve practical problems in real-life situations. For instance, knowing certain measures of a circular swimming pool allows us to calculate how much water it can hold.

Objectives

A Identify Lines, Line Segments, Rays, and Angles. ▶

B Classify Angles as Acute, Right, Obtuse, or Straight. ▶

C Identify Complementary and Supplementary Angles. ▶

D Find Measures of Angles. ▶

Objective A Identifying Lines, Line Segments, Rays, and Angles ▶

Let's begin with a review of two important concepts—space and plane.

Space extends in all directions indefinitely. Examples of objects in space are houses, grains of salt, bushes, your *Developmental Mathematics* textbook, and you.

A **plane** is a flat surface that extends indefinitely. Surfaces like a plane are a classroom floor or a blackboard or whiteboard.

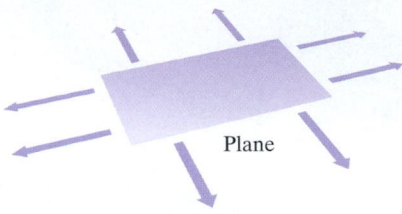

Plane

The most basic concept of geometry is the idea of a point in space. A **point** has no length, no width, and no height, but it does have location. We represent a point by a dot, and we usually label points with capital letters.

P
•
Point *P*

A **line** is a set of points extending indefinitely in two directions. A line has no width or height, but it does have length. We can name a line by any two of its points or by a single lowercase letter. A **line segment** is a piece of a line with two endpoints.

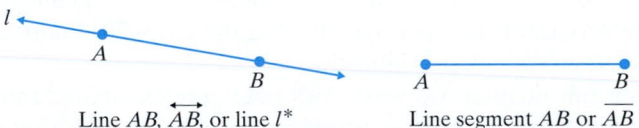

Line *AB*, $\overleftrightarrow{AB}$, or line *l**   Line segment *AB* or $\overline{AB}$

A **ray** is a part of a line with one endpoint. A ray extends indefinitely in one direction. An **angle** is made up of two rays that share the same endpoint. The common endpoint is called the **vertex.**

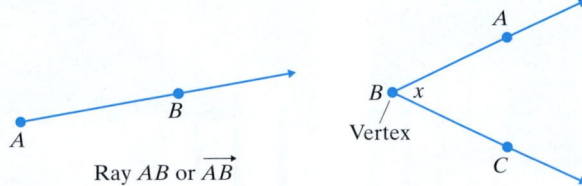

Ray *AB* or $\overrightarrow{AB}$

Vertex

The angle in the figure above can be named

∠*ABC*   ∠*CBA*   ∠*B*   or   ∠*x*

↑   ↑
The vertex is the
middle point.

Rays *BA* and *BC* are **sides** of the angle.

*Although line *l* is also line *BA* or $\overleftrightarrow{BA}$, we will use only one order of points to name a line or line segment.

Helpful Hint

Naming an Angle
When there is no confusion as to what angle is being named, you may use the vertex alone.

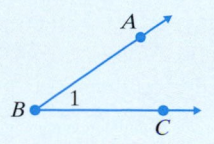

Name of ∠B is all right.
There is no confusion. ∠B means ∠1.

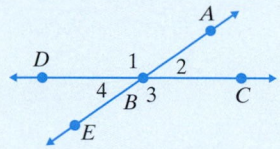

Name of ∠B is *not* all right.
There is confusion. Does ∠B mean
∠1, ∠2, ∠3, or ∠4?

Example 1 Identify each figure as a line, a ray, a line segment, or an angle. Then name the figure using the given points.

a.
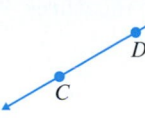

b.

c.
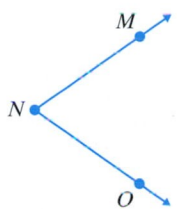

d.
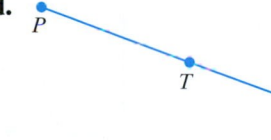

Solution:

Figure (a) extends indefinitely in two directions. It is line *CD* or $\overleftrightarrow{CD}$.
Figure (b) has two endpoints. It is line segment *EF* or $\overline{EF}$.
Figure (c) has two rays with a common endpoint. It is ∠*MNO*, ∠*ONM*, or ∠*N*.
Figure (d) is part of a line with one endpoint. It is ray *PT* or $\overrightarrow{PT}$.

■ **Work Practice 1**

Example 2 List other ways to name ∠*y*.

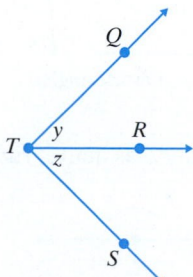

Solution: Two other ways to name ∠*y* are ∠*QTR* and ∠*RTQ*. We may *not* use the vertex alone to name this angle because three different angles have *T* as their vertex.

■ **Work Practice 2**

Practice 1

Identify each figure as a line, a ray, a line segment, or an angle. Then name the figure using the given points.

a. **b.**
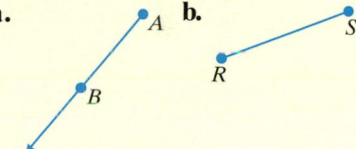

c.

d.

Practice 2

Use the figure in Example 2 to list other ways to name ∠*z*.

Answers
1. a. ray; ray *AB* or $\overrightarrow{AB}$ **b.** line segment; line segment *RS* or $\overline{RS}$ **c.** line; line *EF* or $\overleftrightarrow{EF}$ **d.** angle; ∠*TVH* or ∠*HVT* or ∠*V* **2.** ∠*RTS*, ∠*STR*

Objective B Classifying Angles as Acute, Right, Obtuse, or Straight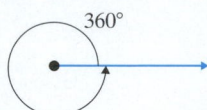

An angle can be measured in **degrees.** The symbol for degrees is a small, raised circle, °. There are 360° in a full revolution, or a full circle.

360°

$\frac{1}{2}$ of a revolution measures $\frac{1}{2}(360°) = 180°$. An angle that measures 180° is called a **straight angle.**

180°

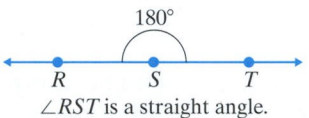

∠RST is a straight angle.

$\frac{1}{4}$ of a revolution measures $\frac{1}{4}(360°) = 90°$. An angle that measures 90° is called a **right angle.** The symbol ∟ is used to denote a right angle.

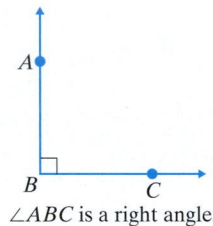

∠ABC is a right angle.

An angle whose measure is between 0° and 90° is called an **acute angle.**

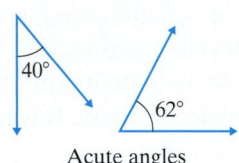

Acute angles

An angle whose measure is between 90° and 180° is called an **obtuse angle.**

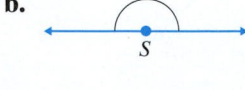

Obtuse angles

Practice 3

Classify each angle as acute, right, obtuse, or straight.

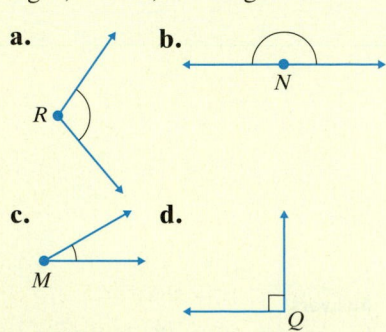

a.

b.

c.

d.

Example 3 Classify each angle as acute, right, obtuse, or straight.

a.

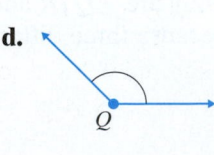

b.

c.

d.

Answers

3. **a.** obtuse **b.** straight **c.** acute

d. right

Solution:

a. $\angle R$ is a right angle, denoted by ∟. It measures 90°.

b. $\angle S$ is a straight angle. It measures 180°.

c. $\angle T$ is an acute angle. It measures between 0° and 90°.

d. $\angle Q$ is an obtuse angle. It measures between 90° and 180°.

■ **Work Practice 3**

Let's look at $\angle B$ below, whose measure is 62°.

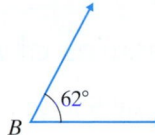

There is a shorthand notation for writing the measure of this angle. To write "The measure of $\angle B$ is 62°," we can write

$$m\angle B = 62°.$$

By the way, note that $\angle B$ is an acute angle because $m\angle B$ is between 0° and 90°.

Objective C Identifying Complementary and Supplementary Angles ▶

Two angles that have a sum of 90° are called **complementary angles.** We say that each angle is the **complement** of the other.

$\angle R$ and $\angle S$ are complementary angles because

$$m\angle R + m\angle S = 60° + 30° = 90°$$

Complementary angles
60° + 30° = 90°

Two angles that have a sum of 180° are called **supplementary angles.** We say that each angle is the **supplement** of the other.

$\angle M$ and $\angle N$ are supplementary angles because

$$m\angle M + m\angle N = 125° + 55° = 180°$$

Supplementary angles
125° + 55° = 180°

Example 4 Find the complement of a 48° angle.

Solution: Two angles that have a sum of 90° are complementary. This means that the complement of an angle that measures 48° is an angle that measures 90° − 48° = 42°.

■ **Work Practice 4**

Practice 4

Find the complement of a 29° angle.

Answer

4. 61°

Practice 5

Find the supplement of a 67° angle.

Example 5 Find the supplement of a 107° angle.

Solution: Two angles that have a sum of 180° are supplementary. This means that the supplement of an angle that measures 107° is an angle that measures $180° - 107° = 73°$.

■ **Work Practice 5**

✔ **Concept Check** True or false? The supplement of a 48° angle is 42°. Explain.

Objective D Finding Measures of Angles

Measures of angles can be added or subtracted to find measures of related angles.

Practice 6

a. Find the measure of ∠y.

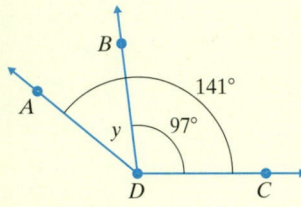

b. Find the measure of ∠x.

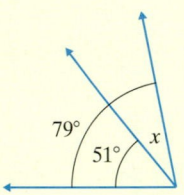

c. Classify ∠x and ∠y as acute, obtuse, or right angles.

Example 6 Find the measure of ∠x. Then classify ∠x as an acute, obtuse, or right angle.

Solution:
$$m\angle x = m\angle QTS - m\angle RTS$$
$$= 87° - 52°$$
$$= 35°$$

Thus, the measure of ∠x ($m\angle x$) is 35°.
   Since ∠x measures between 0° and 90°, it is an acute angle.

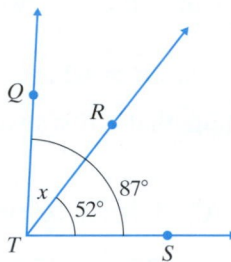

■ **Work Practice 6**

Two lines in a plane can be either parallel or intersecting. **Parallel lines** never meet. **Intersecting lines** meet at a point. The symbol ‖ is used to indicate "is parallel to." For example, in the figure, $p \parallel q$.

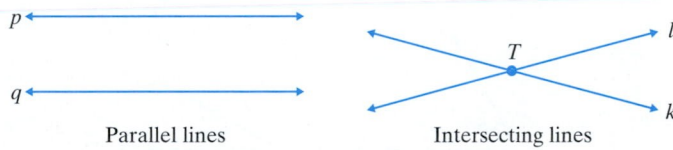

Parallel lines   Intersecting lines

Some intersecting lines are perpendicular. Two lines are **perpendicular** if they form right angles when they intersect. The symbol ⊥ is used to denote "is perpendicular to." For example, in the figure below, $m \perp n$.

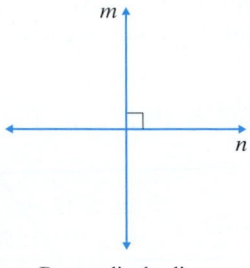

Perpendicular lines

When two lines intersect, four angles are formed. Two angles that are opposite each other are called **vertical angles.** Vertical angles have the same measure.
   Two angles that share a common side are called **adjacent angles.** Adjacent angles formed by intersecting lines are supplementary. That is, the sum of their measures is 180°.

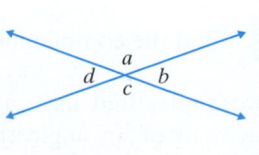

Vertical angles:
∠a and ∠c
∠d and ∠b

Adjacent angles:
∠a and ∠b
∠b and ∠c
∠c and ∠d
∠d and ∠a

Answers

5. 113° 6. a. 44° b. 28° c. both acute

✔ **Concept Check Answer**

false; the *complement* of a 48° angle is 42°; the *supplement* of a 48° angle is 132°

Here are a few real-life examples of the lines we just discussed.

Parallel lines

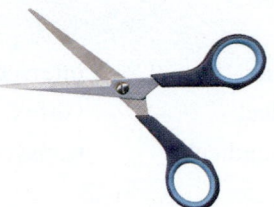

Vertical angles

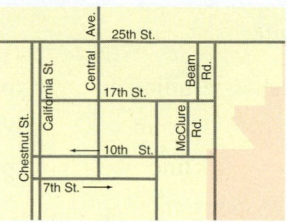

Perpendicular lines

Example 7 Find the measures of $\angle x$, $\angle y$, and $\angle z$ if the measure of $\angle t$ is $42°$.

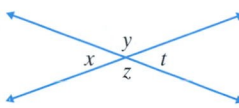

Solution: Since $\angle t$ and $\angle x$ are vertical angles, they have the same measure, so $\angle x$ measures $42°$.
Since $\angle t$ and $\angle y$ are adjacent angles, their measures have a sum of $180°$. So $\angle y$ measures $180° - 42° = 138°$.
Since $\angle y$ and $\angle z$ are vertical angles, they have the same measure. So $\angle z$ measures $138°$.

■ **Work Practice 7**

A line that intersects two or more lines at different points is called a **transversal.** Line l is a transversal that intersects lines m and n. The eight angles formed have special names. Some of these names are:

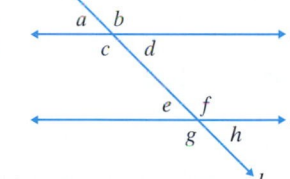

Corresponding angles: $\angle a$ and $\angle e$, $\angle c$ and $\angle g$, $\angle b$ and $\angle f$, $\angle d$ and $\angle h$

Alternate interior angles: $\angle c$ and $\angle f$, $\angle d$ and $\angle e$

When two lines cut by a transversal are *parallel*, the following statement is true:

Parallel Lines Cut by a Transversal

If two parallel lines are cut by a transversal, then the measures of **corresponding angles are equal** and the measures of the **alternate interior angles are equal.**

Example 8 Given that $m \parallel n$ and that the measure of $\angle w$ is $100°$, find the measures of $\angle x$, $\angle y$, and $\angle z$.

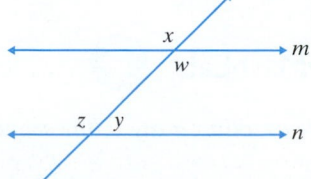

Solution:

$m\angle x = 100°$ $\angle x$ and $\angle w$ are vertical angles.

$m\angle z = 100°$ $\angle x$ and $\angle z$ are corresponding angles.

$m\angle y = 180° - 100° = 80°$ $\angle z$ and $\angle y$ are supplementary angles.

■ **Work Practice 8**

Practice 7

Find the measures of $\angle a$, $\angle b$, and $\angle c$.

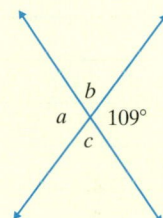

Practice 8

Given that $m \parallel n$ and that the measure of $\angle w = 45°$, find the measures of all the angles shown.

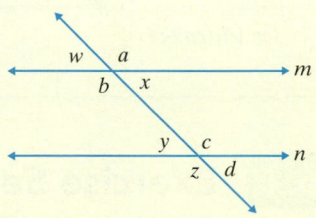

Answers

7. $m\angle a = 109°; m\angle b = 71°;$
$m\angle c = 71°$

8. $m\angle x = 45°; m\angle y = 45°;$
$m\angle z = 135°;$
$m\angle a = 135°;$
$m\angle b = 135°;$
$m\angle c = 135°;$
$m\angle d = 45°$

Vocabulary, Readiness & Video Check

Use the choices below to fill in each blank.

acute	straight	degrees	adjacent	parallel	intersecting
obtuse	space	plane	point	vertical	vertex
right	angle	ray	line	perpendicular	transversal

1. A(n) _____ is a flat surface that extends indefinitely.

2. A(n) _____ has no length, no width, and no height.

3. _____ extends in all directions indefinitely.

4. A(n) _____ is a set of points extending indefinitely in two directions.

5. A(n) _____ is part of a line with one endpoint.

6. A(n) _____ is made up of two rays that share a common endpoint. The common endpoint is called
   the _____.

7. A(n) _____ angle measures 180°.

8. A(n) _____ angle measures 90°.

9. A(n) _____ angle measures between 0° and 90°.

10. A(n) _____ angle measures between 90° and 180°.

11. _____ lines never meet and _____ lines meet at a point.

12. Two intersecting lines are _____ if they form right angles when they intersect.

13. An angle can be measured in _____.

14. A line that intersects two or more lines at different points is called a(n) _____.

15. When two lines intersect, four angles are formed. The angles that are opposite each other are called _____
    angles.

16. Two angles that share a common side are called _____ angles.

Martin-Gay Interactive Videos   Watch the section lecture video and answer the following questions.

Objective **A** 17. In the lecture after ▣ Example 2, what are the four ways we can name the angle shown? ▶

Objective **B** 18. In the lecture before ▣ Example 3, what type of angle forms a line? What is its measure? ▶

Objective **C** 19. What calculation is used to find the answer to ▣ Example 6? ▶

Objective **D** 20. In the lecture before ▣ Example 7, two lines in a plane that aren't parallel must what? ▶

See Video 6.1

6.1 Exercise Set MyMathLab® ▶

Objective **A** *Identify each figure as a line, a ray, a line segment, or an angle. Then name the figure using the given points. See Examples 1 and 2.*

▶ 1.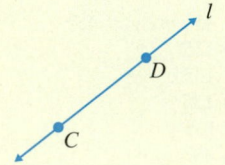

2.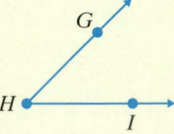

3.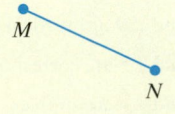

4.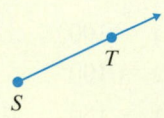

5.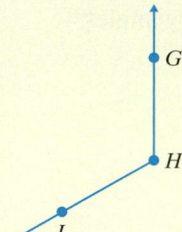

6.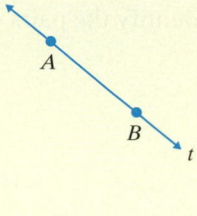

7.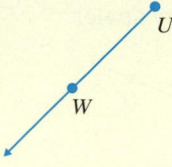

8.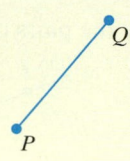

List two other ways to name each angle. See Example 2.

9. $\angle x$

10. $\angle w$

11. $\angle z$

12. $\angle y$

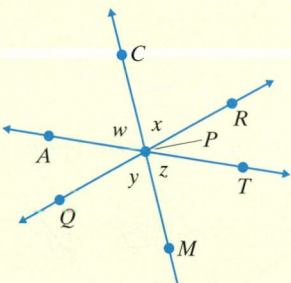

Objective B *Classify each angle as acute, right, obtuse, or straight. See Example 3.*

13.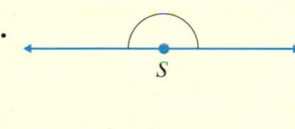

14.

15.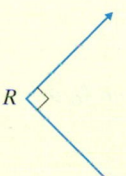

16.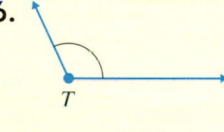

17.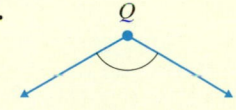

18.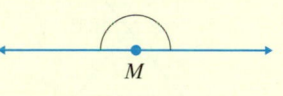

19.

20.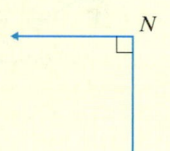

Objective C *Find each complementary or supplementary angle as indicated. See Examples 4 and 5.*

21. Find the complement of a 23° angle.

22. Find the complement of a 77° angle.

23. Find the supplement of a 17° angle.

24. Find the supplement of a 77° angle.

25. Find the complement of a 58° angle.

26. Find the complement of a 22° angle.

27. Find the supplement of a 150° angle.

28. Find the supplement of a 130° angle.

29. Identify the pairs of complementary angles.

30. Identify the pairs of complementary angles.

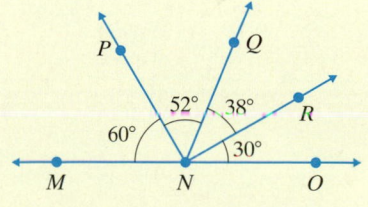

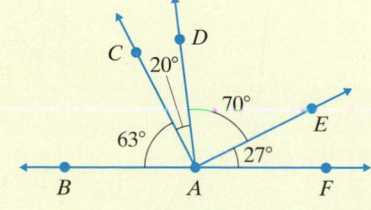

31. Identify the pairs of supplementary angles.

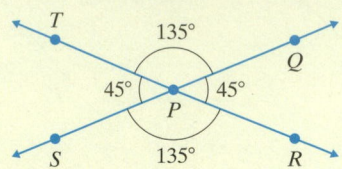

32. Identify the pairs of supplementary angles.

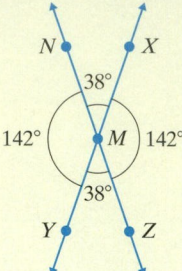

Objective D *Find the measure of ∠x in each figure. See Example 6.*

33.

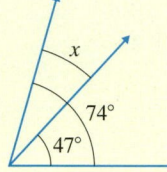

34.

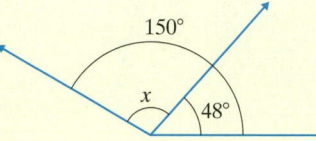

35.

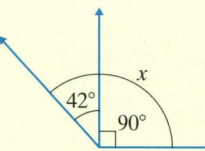

36.

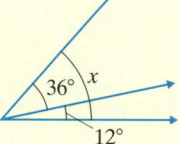

Find the measures of angles x, y, and z in each figure. See Examples 7 and 8.

37.

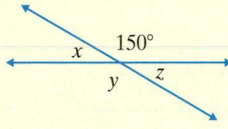

38.

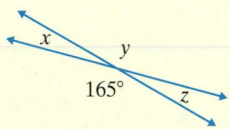

▶ 39.

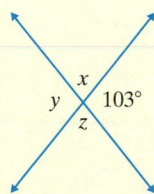

40.

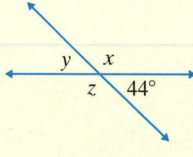

41. *m ‖ n*

42. *m ‖ n*

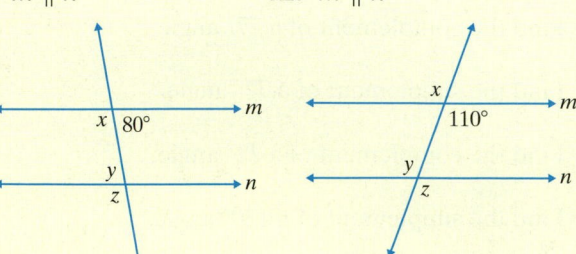

▶ 43. *m ‖ n*

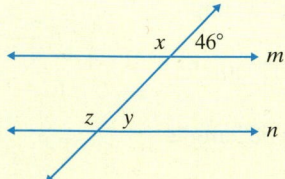

44. *m ‖ n*

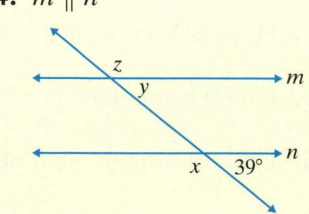

Objectives A D **Mixed Practice** *Find two other ways of naming each angle. See Example 2.*

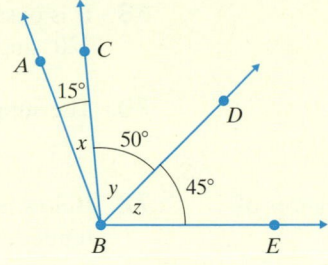

45. ∠x

46. ∠y

47. ∠z

48. ∠ABE (just name one other way)

Find the measure of each angle in the figure above. See Example 6.

▶ **49.** ∠ABC

50. ∠EBD

51. ∠CBD

52. ∠CBA

▶ **53.** ∠DBA

54. ∠EBC

55. ∠CBE

56. ∠ABE

Review

Perform each indicated operation. See Sections 2.4, 2.5, 3.3, and 3.4.

57. $\dfrac{7}{8} + \dfrac{1}{4}$

58. $\dfrac{7}{8} - \dfrac{1}{4}$

59. $\dfrac{7}{8} \cdot \dfrac{1}{4}$

60. $\dfrac{7}{8} \div \dfrac{1}{4}$

61. $3\dfrac{1}{3} - 2\dfrac{1}{2}$

62. $3\dfrac{1}{3} + 2\dfrac{1}{2}$

63. $3\dfrac{1}{3} \div 2\dfrac{1}{2}$

64. $3\dfrac{1}{3} \cdot 2\dfrac{1}{2}$

Concept Extensions

65. The angle between the two walls of the Vietnam Veterans Memorial in Washington, D.C., is 125.2°. Find the supplement of this angle. (*Source:* National Park Service)

66. The faces of Khafre's Pyramid at Giza, Egypt, are inclined at an angle of 53.13°. Find the complement of this angle. (*Source:* PBS *NOVA* Online)

Answer true or false for Exercises 67 through 70. See the Concept Check in this section. If false, explain why.

67. The complement of a 100° angle is an 80° angle.

68. It is possible to find the complement of a 120° angle.

69. It is possible to find the supplement of a 120° angle.

70. The supplement of a 5° angle is a 175° angle.

71. If lines *m* and *n* are parallel, find the measures of angles *a* through *e*.

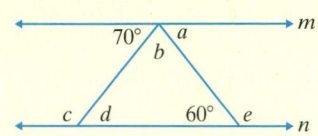

△ **72.** Below is a rectangle. List which segments, if extended, would be parallel lines.

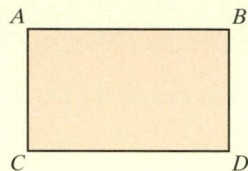

73. Can two supplementary angles both be acute? Explain why or why not.

74. In your own words, describe how to find the complement and the supplement of a given angle.

75. Find two complementary angles with the same measure.

76. Is the figure below possible? Why or why not?

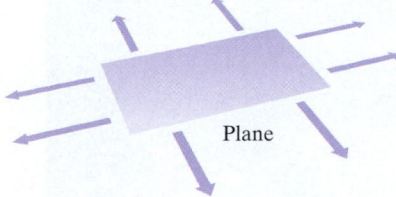

△ **6.2**   **Plane Figures and Solids**

Objectives

A Identify Plane Figures.

B Identify Solids.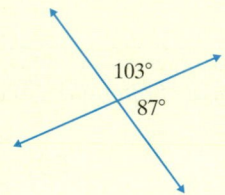

In order to prepare for the sections ahead in this chapter, we first review plane figures and solids.

Objective A Identifying Plane Figures

Recall from Section 6.1 that a **plane** is a flat surface that extends indefinitely.

Plane

A **plane figure** is a figure that lies on a plane. Plane figures, like planes, have length and width but no thickness or depth.

A **polygon** is a closed plane figure that basically consists of three or more line segments that meet at their endpoints.

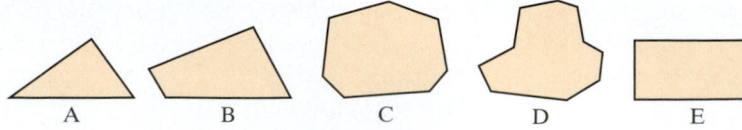

A **regular polygon** is one whose sides are all the same length and whose angles are the same measure.

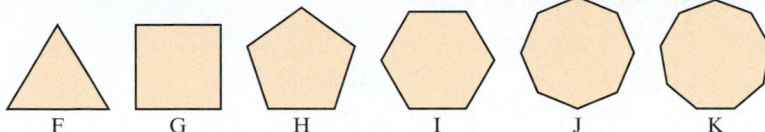

A polygon is named according to the number of its sides.

Polygons		
Number of Sides	Name	Figure Examples
3	Triangle	A, F
4	Quadrilateral	B, E, G
5	Pentagon	H
6	Hexagon	I
7	Heptagon	C
8	Octagon	J
9	Nonagon	K
10	Decagon	D

Some triangles and quadrilaterals are given special names, so let's study these polygons further. We begin with triangles.

The sum of the measures of the angles of a triangle is 180°.

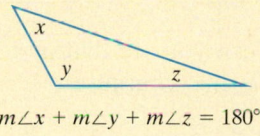

$$m\angle x + m\angle y + m\angle z = 180°$$

Example 1   Find the measure of $\angle a$.

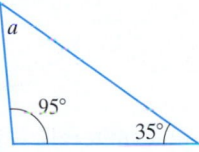

Solution: Since the sum of the measures of the three angles is 180°, we have

measure of $\angle a$, or $m\angle a = 180° - 95° - 35° = 50°$

To check, see that $95° + 35° + 50° = 180°$.

■ **Work Practice 1**

We can classify triangles according to the lengths of their sides. (We will use tick marks to denote the sides and angles of a figure that are equal.)

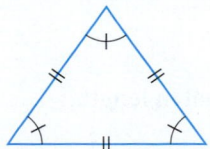

Equilateral triangle

All three sides are the same length. Also, all three angles have the same measure.

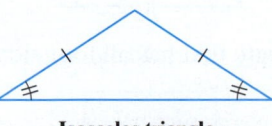

Isosceles triangle

Two sides are the same length. Also, the angles opposite the equal sides have equal measure.

Scalene triangle

No sides are the same length. No angles have the same measure.

Practice 1

Find the measure of $\angle x$.

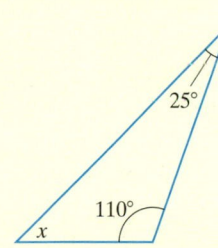

Answer

1. 45°

One other important type of triangle is a right triangle. A **right triangle** is a triangle with a right angle. The side opposite the right angle is called the **hypotenuse,** and the other two sides are called **legs.**

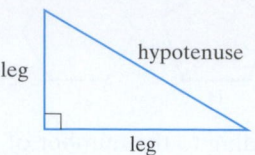

Practice 2

Find the measure of ∠*y*.

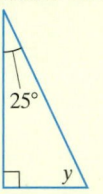

Example 2   Find the measure of ∠*b*.

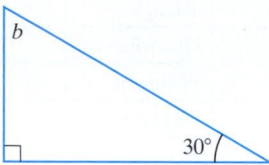

Solution: We know that the measure of the right angle, ∟, is 90°. Since the sum of the measures of the angles is 180°, we have

measure of ∠*b*, or $m\angle b = 180° - 90° - 30° = 60°$

▪ **Work Practice 2**

Helpful Hint

From the previous example, can you see that in a right triangle, the sum of the other two acute angles is 90°? This is because

$$90° \ + \ 90° \ = \ 180°$$

| right angle's measure | sum of other two angles' measures | sum of angles' measures |

Now we review some special quadrilaterals. A **parallelogram** is a special quadrilateral with opposite sides parallel and equal in length.

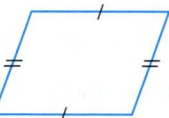

A **rectangle** is a special **parallelogram** that has four right angles.

A **square** is a special **rectangle** that has all four sides equal in length.

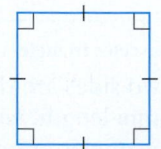

Answer

2. 65°

A **rhombus** is a special **parallelogram** that has all four sides equal in length.

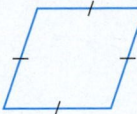

A **trapezoid** is a quadrilateral with exactly one pair of opposite sides parallel.

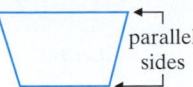

✓**Concept Check** True or false? All quadrilaterals are parallelograms. Explain.

In addition to triangles, quadrilaterals, and other polygons, circles are also plane figures. A **circle** is a plane figure that consists of all points that are the same fixed distance from a point c. The point c is called the **center** of the circle. The **radius** of a circle is the distance from the center of the circle to any point on the circle. The **diameter** of a circle is the distance across the circle passing through the center. Notice that the diameter is twice the radius, and the radius is half the diameter.

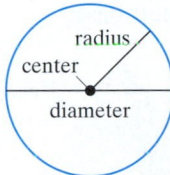

| diameter | = | 2 | · | radius | | radius | = | diameter / 2 |

$$d = 2 \cdot r \qquad\qquad r = \frac{d}{2}$$

Example 3 Find the diameter of the circle.

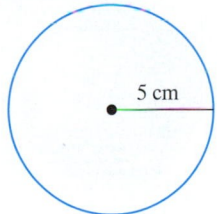

5 cm

Solution: The diameter is twice the radius.

$$d = 2 \cdot r$$
$$d = 2 \cdot 5 \text{ cm} = 10 \text{ cm}$$

The diameter is 10 centimeters.

▪ **Work Practice 3**

Objective B Identifying Solid Figures

Recall from Section 6.1 that space extends in all directions indefinitely.
A **solid** is a figure that lies in space. Solids have length, width, and height or depth.

Practice 3

Find the radius of the circle.

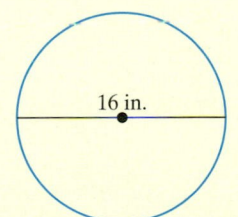

16 in.

Answer

3. 8 in.

✓**Concept Check Answer**

false

A **rectangular solid** is a solid that consists of six sides, or faces, all of which are rectangles.

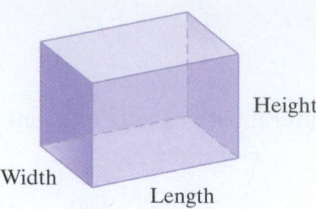

A **cube** is a rectangular solid whose six sides are squares.

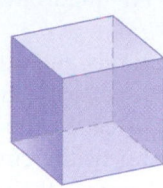

A **pyramid** is shown below. The pyramids we will study have square bases and heights that are perpendicular to their base.

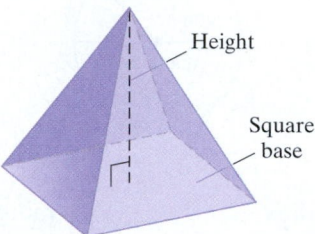

A **sphere** consists of all points in space that are the same distance from a point c. The point c is called the **center** of the sphere. The **radius** of a sphere is the distance from the center to any point on the sphere. The **diameter** of a sphere is the distance across the sphere passing through the center.

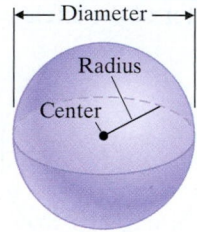

The radius and diameter of a sphere are related in the same way as the radius and diameter of a circle.

$$d = 2 \cdot r \quad \text{or} \quad r = \frac{d}{2}$$

Practice 4

Find the diameter of the sphere.

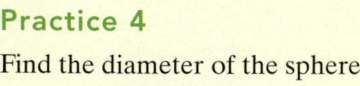

7 mi

Answer

4. 14 mi

Example 4   Find the radius of the sphere.

Solution: The radius is half the diameter.

$$r = \frac{d}{2}$$

$$r = \frac{36 \text{ feet}}{2} = 18 \text{ feet}$$

The radius is 18 feet.

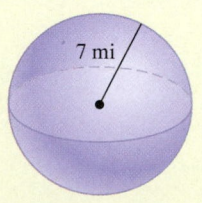

36 ft

■ **Work Practice 4**

The **cylinders** we will study have bases that are in the shape of circles and heights that are perpendicular to their base.

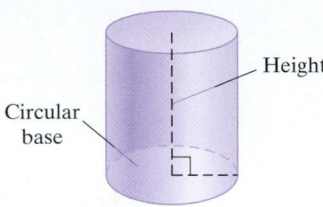

The **cones** we will study have bases that are circles and heights that are perpendicular to their base.

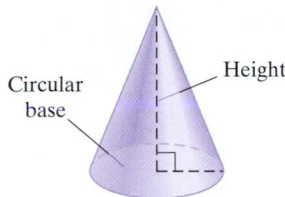

Vocabulary, Readiness & Video Check

Martin-Gay Interactive Videos   *Watch the section lecture video and answer the following questions.*

Objective A 1. From the lecture after ▣ Example 2, since all angles of an equilateral triangle have the same measure, what is the measure of each angle? ▶

Objective B 2. What solid is identified in ▣ Example 6? What two real-life examples of the solid are given? ▶

See Video 6.2 🍎

6.2 Exercise Set MyMathLab® ▶

Objective A *Identify each polygon. See the table at the beginning of this section.*

1.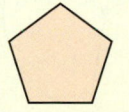

2.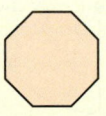

3.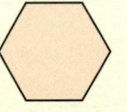

4.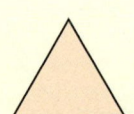

5.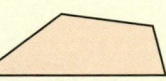

6.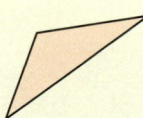

7.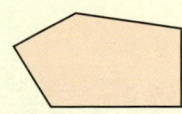

8.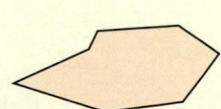

Classify each triangle as equilateral, isosceles, or scalene. Also identify any triangles that are also right triangles. See the triangle classification after Example 1.

9.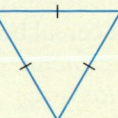

10.

11.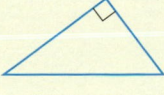

12. 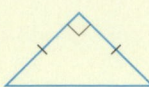   **13.** 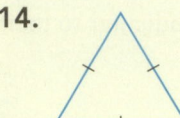   **14.**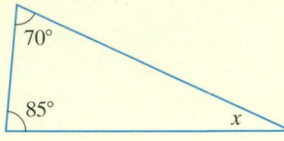

Find the measure of ∠x in each figure. See Examples 1 and 2.

15.
70°
85°
x

16.
x
112°
28°

17.
95°
72°
x

18.
x
80° 65°

19.
x
50°

20.
x
20°

Fill in each blank.

21. Twice the radius of a circle is its _____.

22. A rectangle with all four sides equal is a(n) _____.

23. A parallelogram with four right angles is a(n) _____.

24. Half the diameter of a circle is its _____.

25. A quadrilateral with opposite sides parallel is a(n) _____.

26. A quadrilateral with exactly one pair of opposite sides parallel is a(n) _____.

27. The side opposite the right angle of a right triangle is called the _____.

28. A triangle with no equal sides is a(n) _____.

Find the unknown diameter or radius in each figure. See Example 3.

29.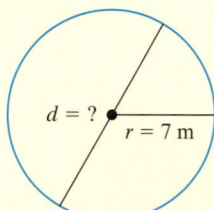
$d = ?$
$r = 7$ m

30.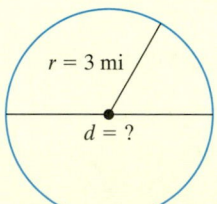
$r = 3$ mi
$d = ?$

31.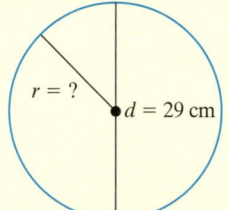
$r = ?$
$d = 29$ cm

32.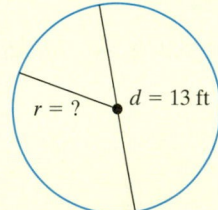
$r = ?$
$d = 13$ ft

33.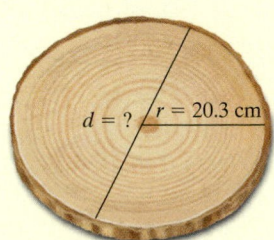
$d = ?$ $r = 20.3$ cm

34.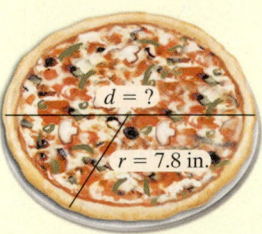
$d = ?$
$r = 7.8$ in.

35.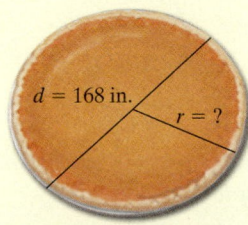
$d = 168$ in.
$r = ?$

Largest pumpkin pie (*Source:* Circleville, Ohio, Pumpkin Festival)

36.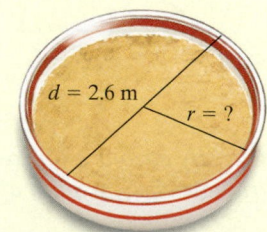
$d = 2.6$ m
$r = ?$

Largest cereal bowl (*Source: Guinness World Records*)

Objective B *Identify each solid.*

37.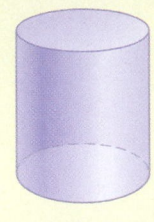

38.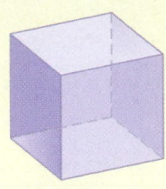

39.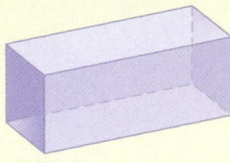

40.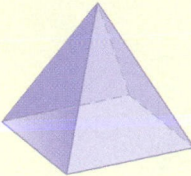

41.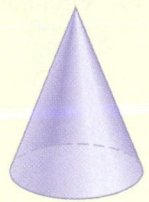

42.

Identify the basic shape of each item.

43.

44.

45.

46.

47.

48.

49.

50.

Find each unknown radius or diameter. See Example 4.

51. The radius of a sphere is 7.4 inches. Find its diameter.

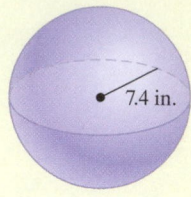

52. The radius of a sphere is 5.8 meters. Find its diameter.

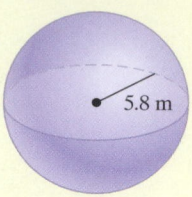

53. Find the radius of the sphere.

54. Find the radius of the sphere.

55. Saturn has a radius of approximately 36,184 miles. What is its diameter?

56. A sphere-shaped wasp nest found in Japan had a radius of approximately 15 inches. What was its diameter? (*Source: Guinness World Records*)

Review

Perform each indicated operation. See Sections 1.3, 1.6, 4.3, and 4.4.

57. $2(18) + 2(36)$

58. $4(87)$

59. $4(3.14)$

60. $2(7.8) + 2(9.6)$

Concept Extensions

Determine whether each statement is true or false. See the Concept Check in this section.

61. A square is also a rhombus.

62. A square is also a regular polygon.

63. A rectangle is also a parallelogram.

64. A trapezoid is also a parallelogram.

65. A pentagon is also a quadrilateral.

66. A rhombus is also a parallelogram.

67. Is an isosceles right triangle possible? If so, draw one.

68. In your own words, explain whether a square is also a rhombus.

69. The following demonstration is credited to the mathematician Pascal, who is said to have developed it as a young boy.

Cut a triangle from a piece of paper. The length of the sides and the size of the angles are unimportant. Tear the points off the triangle as shown in the top right figure.

Place the points of the triangle together as shown in the bottom right figure. Notice that a straight line is formed. What was Pascal trying to show?

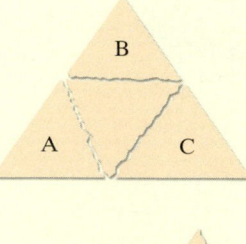

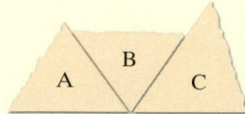

△ 6.3 Perimeter

Objective A Using Formulas to Find Perimeters

Recall from Section 1.3 that the perimeter of a polygon is the distance around the polygon. This means that the perimeter of a polygon is the sum of the lengths of its sides.

Example 1 Find the perimeter of the rectangle below.

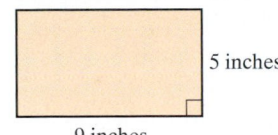

5 inches

9 inches

Solution:

$$\text{perimeter} = 9 \text{ inches} + 9 \text{ inches} + 5 \text{ inches} + 5 \text{ inches}$$
$$= 28 \text{ inches}$$

■ **Work Practice 1**

Notice that the perimeter of the rectangle in Example 1 can be written as
$2 \cdot (9 \text{ inches}) + 2 \cdot (5 \text{ inches})$.

↑   ↑

length   width

In general, we can say that the perimeter of a rectangle is always

$2 \cdot \text{length} + 2 \cdot \text{width}$

As we have just seen, the perimeters of some special figures such as rectangles form patterns. These patterns are given as **formulas.** The formula for the perimeter of a rectangle is shown next:

Perimeter of a Rectangle

$\text{perimeter} = 2 \cdot \text{length} + 2 \cdot \text{width}$

In symbols, this can be written as

$P = 2 \cdot l + 2 \cdot w$

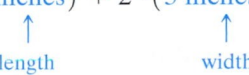

length

width   width

length

Example 2 Find the perimeter of a rectangle with a length of 11 inches and a width of 3 inches.

11 in.

3 in.

Solution: We use the formula for perimeter and replace the letters by their known lengths.

$$P = 2 \cdot l + 2 \cdot w$$
$$= 2 \cdot 11 \text{ in.} + 2 \cdot 3 \text{ in.} \quad \text{Replace } l \text{ with 11 in. and } w \text{ with 3 in.}$$
$$= 22 \text{ in.} + 6 \text{ in.}$$
$$= 28 \text{ in.}$$

The perimeter is 28 inches.

■ **Work Practice 2**

Objectives

A Use Formulas to Find Perimeters.

B Use Formulas to Find Circumferences.

Practice 1

a. Find the perimeter of the rectangle.

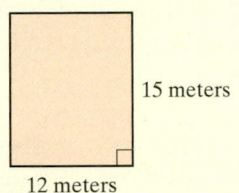

15 meters

12 meters

b. Find the perimeter of the rectangular lot shown below:

60 feet

80 feet

Practice 2

Find the perimeter of a rectangle with a length of 22 centimeters and a width of 10 centimeters.

Answers

1. a. 54 m   **b.** 280 ft

2. 64 cm

425

Recall that a square is a special rectangle with all four sides the same length. The formula for the perimeter of a square is shown next:

Perimeter of a Square

Perimeter = side + side + side + side
= 4 · side

In symbols,

$P = 4 \cdot s$

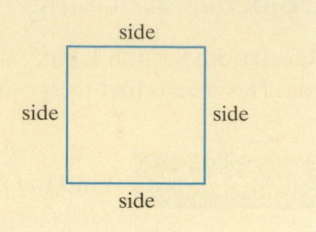

Copyright 2015 Pearson Education, Inc.

Practice 3

Find the perimeter of a square tabletop if each side is 5 feet long.

Example 3 Finding the Perimeter of a Field

How much fencing is needed to enclose a square field 50 yards on a side?

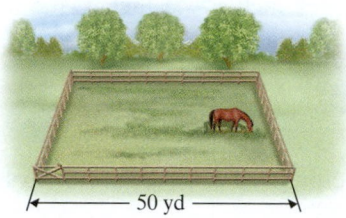

Solution: To find the amount of fencing needed, we find the distance around, or perimeter. The formula for the perimeter of a square is $P = 4 \cdot s$. We use this formula and replace s by 50 yards.

$P = 4 \cdot s$
$= 4 \cdot 50$ yd
$= 200$ yd

The amount of fencing needed is 200 yards.

■ **Work Practice 3**

The formula for the perimeter of a triangle with sides of lengths a, b, and c is given next:

Perimeter of a Triangle

Perimeter = side a + side b + side c

In symbols,

$P = a + b + c$

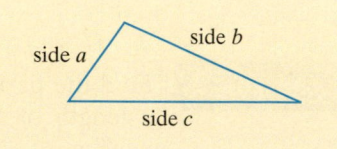

Practice 4

Find the perimeter of a triangle if the sides are 5 centimeters, 10 centimeters, and 6 centimeters in length.

Example 4 Find the perimeter of a triangle if the sides are 3 inches, 7 inches, and 6 inches.

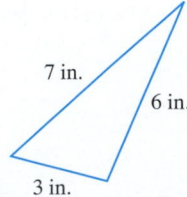

Answers

3. 20 ft **4.** 21 cm

Solution: The formula for the perimeter is $P = a + b + c$, where a, b, and c are the lengths of the sides. Thus,

$$P = a + b + c$$
$$= 3 \text{ in.} + 7 \text{ in.} + 6 \text{ in.}$$
$$= 16 \text{ in.}$$

The perimeter of the triangle is 16 inches.

■ **Work Practice 4**

Recall that to find the perimeter of other polygons, we find the sum of the lengths of their sides.

Example 5 Find the perimeter of the trapezoid shown below:

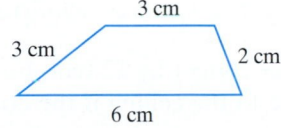

Solution: To find the perimeter, we find the sum of the lengths of its sides.

perimeter $= 3 \text{ cm} + 2 \text{ cm} + 6 \text{ cm} + 3 \text{ cm} = 14 \text{ cm}$

The perimeter is 14 centimeters.

■ **Work Practice 5**

Example 6 Finding the Perimeter of a Room

Find the perimeter of the room shown below:

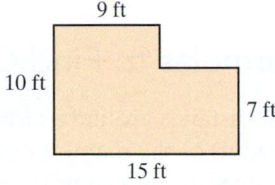

Solution: To find the perimeter of the room, we first need to find the lengths of all sides of the room.

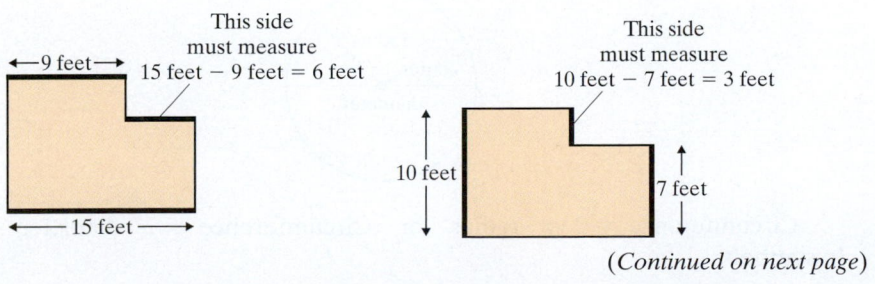

(*Continued on next page*)

Practice 5

Find the perimeter of the trapezoid shown.

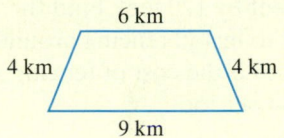

Practice 6

Find the perimeter of the room shown.

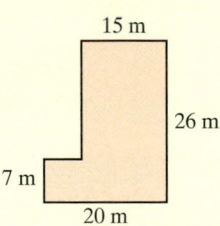

Answers

5. 23 km   **6.** 92 m

Now that we know the measures of all sides of the room, we can add the measures to find the perimeter.

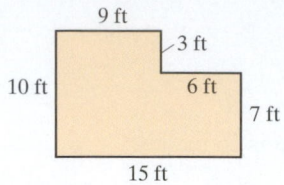

$$\text{perimeter} = 10 \text{ ft} + 9 \text{ ft} + 3 \text{ ft} + 6 \text{ ft} + 7 \text{ ft} + 15 \text{ ft}$$
$$= 50 \text{ ft}$$

The perimeter of the room is 50 feet.

■ **Work Practice 6**

Practice 7

A rectangular lot measures 60 feet by 120 feet. Find the cost to install fencing around the lot if the cost of fencing is $1.90 per foot.

Example 7   Calculating the Cost of Wallpaper Border

A rectangular room measures 10 feet by 12 feet. Find the cost to hang a wallpaper border on the walls close to the ceiling if the cost of the wallpaper border is $1.09 per foot.

Solution: First we find the perimeter of the room.

$$P = 2 \cdot l + 2 \cdot w$$
$$= 2 \cdot 12 \text{ ft} + 2 \cdot 10 \text{ ft} \quad \text{Replace } l \text{ with 12 feet and } w \text{ with 10 feet.}$$
$$= 24 \text{ ft} + 20 \text{ ft}$$
$$= 44 \text{ ft}$$

The cost of the wallpaper is

$$\text{cost} = \$1.09 \cdot 44 \text{ ft} = 47.96$$

The cost of the wallpaper is $47.96.

■ **Work Practice 7**

Objective B   Using Formulas to Find Circumferences

Recall from Section 4.4 that the distance around a circle is called the **circumference.** This distance depends on the radius or the diameter of the circle.
    The formulas for circumference are shown next:

Circumference of a Circle

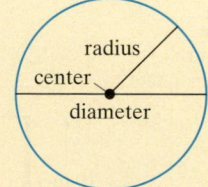

    Circumference $= 2 \cdot \pi \cdot$ radius   or   Circumference $= \pi \cdot$ diameter
In symbols,

$$C = 2 \cdot \pi \cdot r \quad \text{or} \quad C = \pi \cdot d,$$

where $\pi \approx 3.14$ or $\pi \approx \dfrac{22}{7}$.

Answer

7. $684

To better understand circumference and π (pi), try the following experiment. Take any can and measure its circumference and its diameter.

The can in the figure above has a circumference of 23.5 centimeters and a diameter of 7.5 centimeters. Now divide the circumference by the diameter.

$$\frac{\text{circumference}}{\text{diameter}} = \frac{23.5 \text{ cm}}{7.5 \text{ cm}} \approx 3.13$$

Try this with other sizes of cylinders and circles—you should always get a number close to 3.1. The exact ratio of circumference to diameter is π. (Recall that $\pi \approx 3.14$ or $\pi \approx \frac{22}{7}$.)

Example 8 Finding Circumference of Spa

Mary Catherine Dooley plans to install a border of new tiling around the circumference of her circular spa. If her spa has a diameter of 14 feet, find its exact circumference. Then use the approximation 3.14 for π to approximate the circumference.

14 feet

Solution: Because we are given the diameter, we use the formula $C = \pi \cdot d$.

$C = \pi \cdot d$
$\quad = \pi \cdot 14 \text{ ft}$ Replace d with 14 feet.
$\quad = 14\pi \text{ ft}$

The circumference of the spa is *exactly* 14π feet. By replacing π with the *approximation* 3.14, we find that the circumference is *approximately* 14 feet $\cdot$ 3.14 = 43.96 feet.

■ Work Practice 8

✔**Concept Check** The distance around which figure is greater: a square with side length 5 inches or a circle with radius 3 inches?

Practice 8

a. An irrigation device waters a circular region with a diameter of 20 yards. Find the exact circumference of the watered region, then use $\pi \approx 3.14$ to give an approximation.

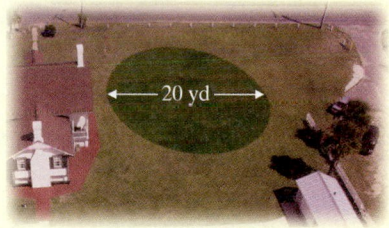
20 yd

b. A manufacturer of clocks is designing a new model. To help the designer calculate the cost of materials to make the new clock, calculate the circumference of a clock with a face diameter of 12 inches. Give an exact circumference; then use $\pi \approx 3.14$ to approximate.

Answers
8. a. exactly 20π yd ≈ 62.8 yd
b. exactly 12π in. ≈ 37.68 in.

✔**Concept Check Answer**
a square with side length 5 in.

Vocabulary, Readiness & Video Check

Use the choices below to fill in each blank.

circumference   radius π $\dfrac{22}{7}$

diameter   perimeter   3.14

1. The _____ of a polygon is the sum of the lengths of its sides.

2. The distance around a circle is called the _____.

3. The exact ratio of circumference to diameter is _____.

4. The diameter of a circle is double its _____.

5. Both _____ and _____ are approximations for π.

6. The radius of a circle is half its _____.

Martin-Gay Interactive Videos   Watch the section lecture video and answer the following questions.

Objective A   7. In ⊟ Example 1, how can the perimeter be found if we forget the formula? ▶

Objective B   8. From the lecture before ⊟ Example 6, circumference is a special name for what? ▶

See Video 6.3 🍎

6.3   Exercise Set   MyMathLab® ▶

Objective A   *Find the perimeter of each figure. See Examples 1 through 6.*

▶1.
Rectangle, 15 ft, 17 ft

2.
Rectangle, 14 m, 5 m

3.
Parallelogram, 25 cm, 35 cm

4.
Parallelogram, 3 yd, 2 yd

▶5.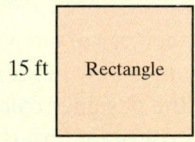
5 in., 7 in., 9 in.

6.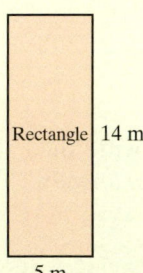
5 units, 11 units, 10 units

▶7.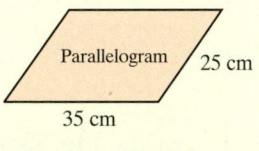
10 ft, 8 ft, 7 ft, 8 ft, 15 ft

8.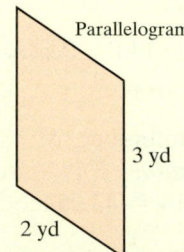
10 m, 4 m, 10 m, 13 m, 9 m, 20 m

Find the perimeter of each regular polygon. (The sides of a regular polygon have the same length.)

9. 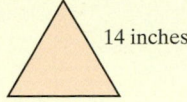 14 inches

10. 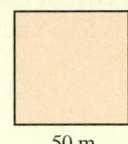 50 m

11. 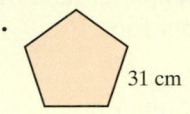 31 cm

12. 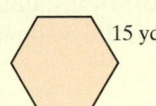 15 yd

Solve. See Examples 1 through 7.

13. A polygon has sides of length 5 feet, 3 feet, 2 feet, 7 feet, and 4 feet. Find its perimeter.

14. A triangle has sides of length 8 inches, 12 inches, and 10 inches. Find its perimeter.

15. A line-marking machine lays down lime powder to mark both foul lines on a baseball field. If each foul line for this field measures 312 feet, how many feet of lime powder will be deposited?

16. A baseball diamond has 4 sides, with each side length 90 feet. If a baseball player hits a home run, how far does the player run (home plate, around the bases, then back to home plate)?

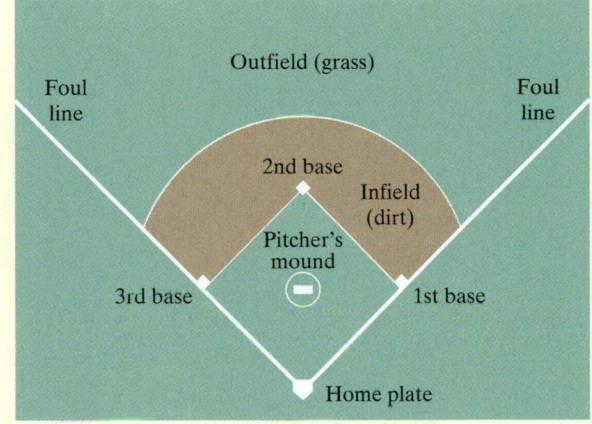

17. If a football field is 53 yards wide and 120 yards long, what is the perimeter?

18. A stop sign has eight equal sides of length 12 inches. Find its perimeter.

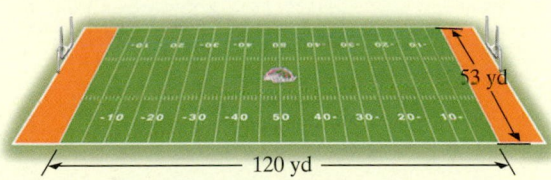

19. A metal strip is being installed around a workbench that is 8 feet long and 3 feet wide. Find how much stripping is needed for this project.

20. Find how much fencing is needed to enclose a rectangular garden 70 feet by 21 feet.

21. If the stripping in Exercise 19 costs $2.50 per foot, find the total cost of the stripping.

22. If the fencing in Exercise 20 costs $2 per foot, find the total cost of the fencing.

23. A regular octagon has a side length of 9 inches. Find its perimeter.

24. A regular pentagon has a side length of 14 meters. Find its perimeter.

▶ 25. Find the perimeter of the top of a square compact disc case if the length of one side is 7 inches.

26. Find the perimeter of a square ceramic tile with a side of length 3 inches.

27. A rectangular room measures 10 feet by 11 feet. Find the cost of installing a strip of wallpaper around the room if the wallpaper costs $0.86 per foot.

28. A rectangular house measures 85 feet by 70 feet. Find the cost of installing gutters around the house if the cost is $2.36 per foot.

Find the perimeter of each figure. See Example 6.

29.

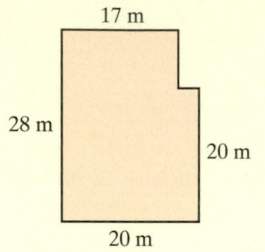

17 m
28 m
20 m
20 m

30.

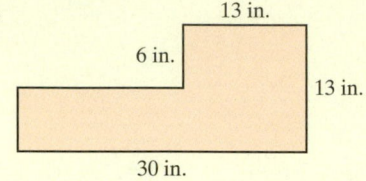

13 in.
6 in.
13 in.
30 in.

▶ **31.**

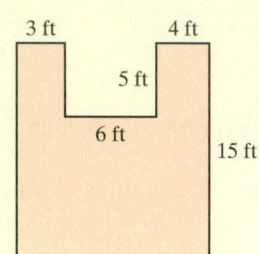

3 ft   4 ft
5 ft
6 ft
15 ft

32.

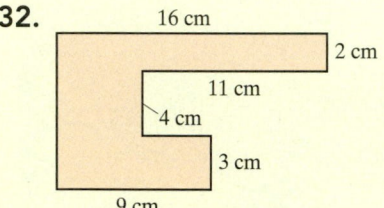

16 cm
2 cm
11 cm
4 cm
3 cm
9 cm

33.

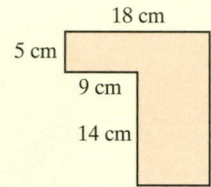

18 cm
5 cm
9 cm
14 cm

34.

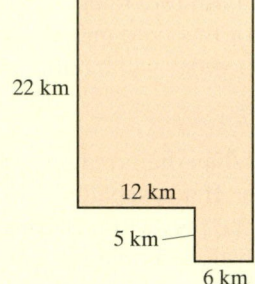

22 km
12 km
5 km
6 km

Objective B *Find the circumference of each circle. Give the exact circumference and then an approximation.*
Use $\pi \approx 3.14$. *See Example 8.*

35.

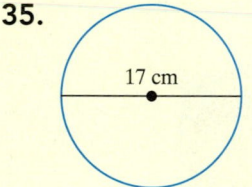

17 cm

36.

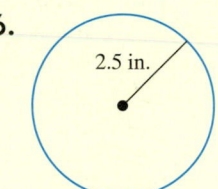

2.5 in.

37.

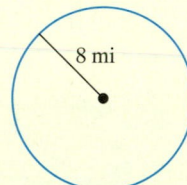

8 mi

38.

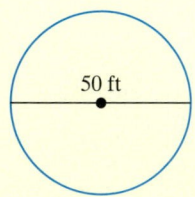

50 ft

▶ **39.**

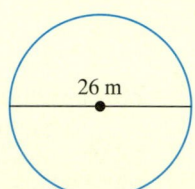

26 m

40.

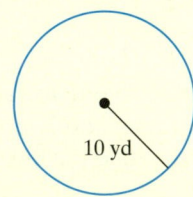

10 yd

41. Wyley Robinson just bought a trampoline for his children to use. The trampoline has a diameter of 15 feet. If Wyley wishes to buy netting to go around the outside of the trampoline, how many feet of netting does he need?

42. The largest round barn in the world is located at the Marshfield Fairgrounds in Wisconsin. The barn has a diameter of 150 ft. What is the circumference of the barn? (*Source: The Milwaukee Journal Sentinel*)

43. Meteor Crater, near Winslow, Arizona, is 4000 feet in diameter. Approximate the distance around the crater. Use 3.14 for π. (*Source: The Handy Science Answer Book*)

44. The largest pearl, the *Pearl of Lao-tze,* has a diameter of $5\frac{1}{2}$ inches. Approximate the distance around the pearl. Use $\frac{22}{7}$ for π. (*Source: The Guinness Book of World Records*)

Objectives **A** **B** **Mixed Practice** *Find the distance around each figure. For circles, give the exact circumference and then an approximation. Use $\pi \approx 3.14$. See Examples 1 through 8.*

45.
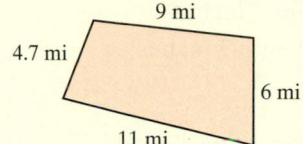

46.
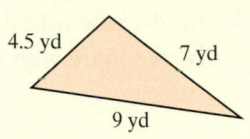

47.
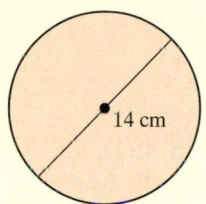

48.
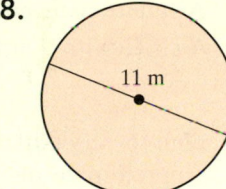

49.
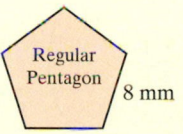

50.
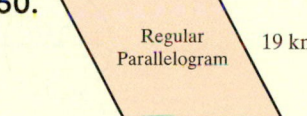

51.
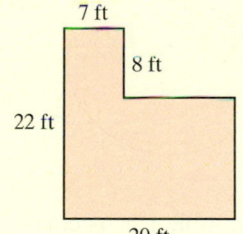

52.
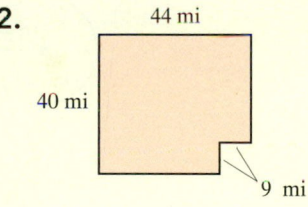

Review

Simplify. See Section 1.9.

53. $5 + 6 \cdot 3$

54. $25 - 3 \cdot 7$

55. $(20 - 16) \div 4$

56. $6 \cdot (8 + 2)$

57. $72 \div (2 \cdot 6)$

58. $(72 \div 2) \cdot 6$

59. $(18 + 8) - (12 + 4)$

60. $4^1 \cdot (2^3 - 8)$

Concept Extensions

There are a number of factors that determine the dimensions of a rectangular soccer field. Use the table below to answer Exercises 61 and 62.

Soccer Field Width and Length		
Age	Width Min–Max	Length Min–Max
Under 6/7:	15–20 yards	25–30 yards
Under 8:	20–25 yards	30–40 yards
Under 9:	30–35 yards	40–50 yards
Under 10:	40–50 yards	60–70 yards
Under 11:	40–50 yards	70–80 yards
Under 12:	40–55 yards	100–105 yards
Under 13:	50–60 yards	100–110 yards
International:	70–80 yards	110–120 yards

61. a. Find the minimum length and width of a soccer field for 8-year-old children. (Carefully consider the age.)

   b. Find the perimeter of this field.

62. a. Find the maximum length and width of a soccer field for 12-year-old children.

   b. Find the perimeter of this field.

Solve. See the Concept Check in this section. Choose the figure that has the greater distance around.

63. a. A square with side length 3 inches

   b. A circle with diameter 4 inches

64. a. A circle with diameter 7 inches

   b. A square with side length 7 inches

65. a. Find the circumference of each circle. Approximate the circumference by using 3.14 for π.

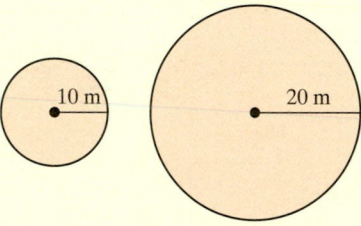

   b. If the radius of a circle is doubled, is its corresponding circumference doubled?

66. a. Find the circumference of each circle. Approximate the circumference by using 3.14 for π.

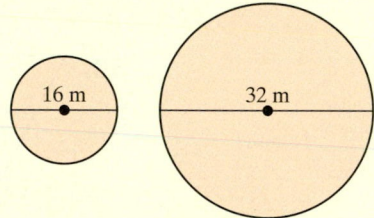

   b. If the diameter of a circle is doubled, is its corresponding circumference doubled?

67. In your own words, explain how to find the perimeter of any polygon.

68. In your own words, explain how perimeter and circumference are the same and how they are different.

Find the perimeter. Round your results to the nearest tenth.

69.

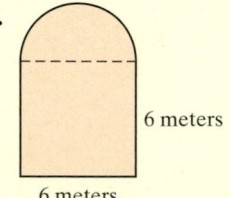

6 meters

6 meters

70.

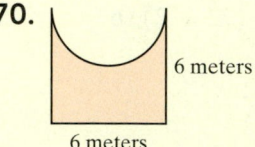

6 meters

6 meters

71.

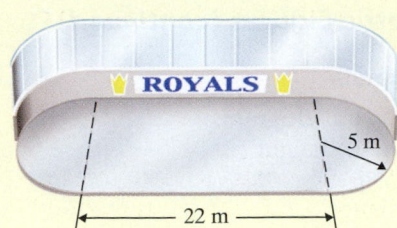

ROYALS

5 m

22 m

72.

← 5 feet →

7 feet

73. The perimeter of this rectangle is 31 feet. Find its width.

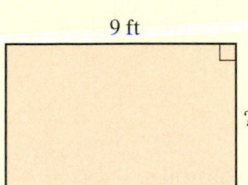

9 ft

?

74. The perimeter of this square is 18 inches. Find the length of a side.

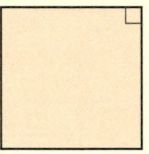

△ 6.4 Area

Objective A Finding Areas of Geometric Figures

Recall that area measures the amount of surface of a region. Thus far, we know how to find the area of a rectangle and a square. These formulas, as well as formulas for finding the areas of other common geometric figures, are given next:

Objective

A Find the Areas of Geometric Figures.

Area Formulas of Common Geometric Figures

Geometric Figure	Area Formula
RECTANGLE 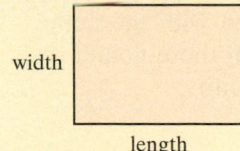 width, length	Area of a rectangle: **Area = length · width** $A = lw$
SQUARE 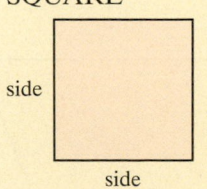 side, side	Area of a square: **Area = side · side** $A = s \cdot s = s^2$
TRIANGLE 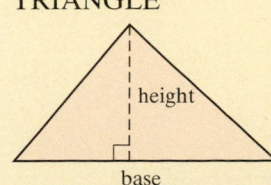 height, base	Area of a triangle: **Area** $= \dfrac{1}{2} \cdot$ **base** $\cdot$ **height** $A = \dfrac{1}{2} \cdot b \cdot h$

(Continued on next page)

Area Formulas of Common Geometric Figures (*continued*)

Geometric Figure	Area Formula
PARALLELOGRAM 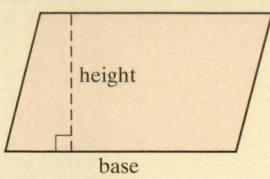	Area of a parallelogram: **Area = base · height** $A = b \cdot h$
TRAPEZOID 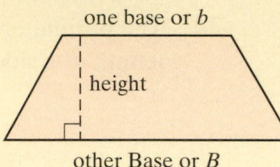	Area of a trapezoid: **Area** $= \dfrac{1}{2} \cdot ($one **base** + other **Base**$) \cdot$ **height** $A = \dfrac{1}{2} \cdot (b + B) \cdot h$

Use these formulas for the following examples.

> **Helpful Hint**
>
> Area is always measured in square units.

Practice 1

Find the area of the triangle.

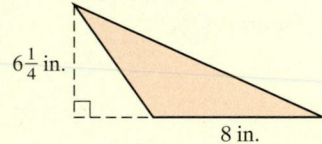

Example 1   Find the area of the triangle.

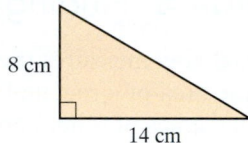

Solution:
$$A = \frac{1}{2} \cdot b \cdot h$$
$$= \frac{1}{2} \cdot 14 \text{ cm} \cdot 8 \text{ cm}$$
$$= \frac{\overset{1}{2} \cdot 7 \cdot 8}{\underset{1}{2}} \text{ sq cm}$$
$$= 56 \text{ square cm}$$

> **Helpful Hint**   You may see 56 sq cm, for example, written with the notation 56 cm². Both of these notations mean the same quantity.

The area is 56 square centimeters.

■ **Work Practice 1**

Practice 2

Find the area of the square.

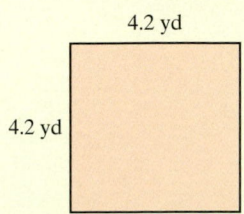

Example 2   Find the area of the parallelogram.

Solution: $A = b \cdot h$
$$= 3.4 \text{ miles} \cdot 1.5 \text{ miles}$$
$$= 5.1 \text{ square miles}$$

The area is 5.1 square miles.

■ **Work Practice 2**

Answers
1. 25 sq in.   **2.** 17.64 sq yd

When finding the area of figures, be sure all measurements are changed to the same unit before calculations are made.

Example 3 Find the area of the figure.

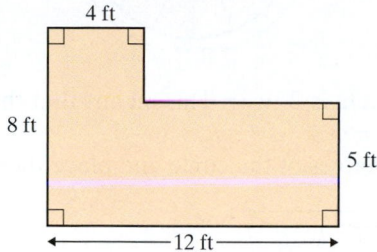

Practice 3

Find the area of the figure.

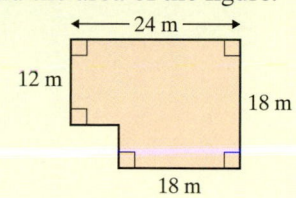

Solution: Split the figure into two rectangles. To find the area of the figure, we find the sum of the areas of the two rectangles.

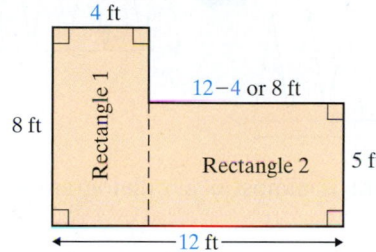

Area of Rectangle 1 $= l \cdot w$

$\qquad = 8 \text{ feet} \cdot 4 \text{ feet}$

$\qquad = 32 \text{ square feet}$

Notice that the length of Rectangle 2 is 12 feet − 4 feet, or 8 feet.

Area of Rectangle 2 $= l \cdot w$

$\qquad = 8 \text{ feet} \cdot 5 \text{ feet}$

$\qquad = 40 \text{ square feet}$

Area of the Figure $=$ Area of Rectangle 1 $+$ Area of Rectangle 2

$\qquad = 32 \text{ square feet} + 40 \text{ square feet}$

$\qquad = 72 \text{ square feet}$

■ **Work Practice 3**

The figure in Example 3 can also be split into two rectangles as shown:

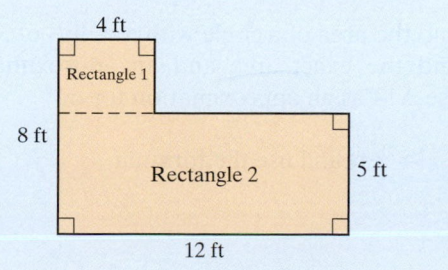

Answer

3. 396 sq m

To better understand the formula for area of a circle, try the following. Cut a circle into many pieces as shown:

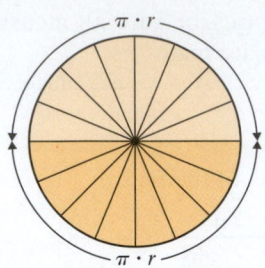

The circumference of a circle is $2 \cdot \pi \cdot r$. This means that the circumference of half a circle is half of $2 \cdot \pi \cdot r$, or $\pi \cdot r$.

Then unfold the two halves of the circle and place them together as shown:

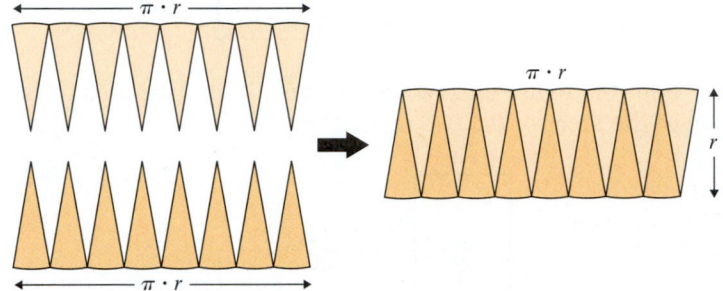

The figure on the right is almost a parallelogram with a base of $\pi \cdot r$ and a height of r. The area is

$$A = \boxed{\text{base}} \cdot \boxed{\text{height}}$$
$$= (\pi \cdot r) \cdot r$$
$$= \pi \cdot r^2$$

This is the formula for area of a circle.

Area Formula of a Circle

CIRCLE

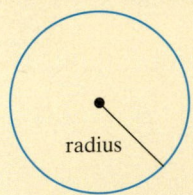

radius

Area of a circle:

Area $= \pi \cdot (\text{radius})^2$
$$A = \pi \cdot r^2$$

(A fraction approximation for π is $\dfrac{22}{7}$.)

(A decimal approximation for π is 3.14.)

Practice 4

Find the area of the given circle. Find the exact area and an approximation. Use 3.14 as an approximation for π.

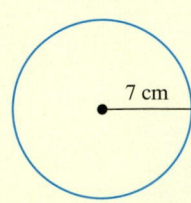

7 cm

Answer

4. 49π sq cm ≈ 153.86 sq cm

Example 4 Find the area of a circle with a radius of 3 feet. Find the exact area and an approximation. Use 3.14 as an approximation for π.

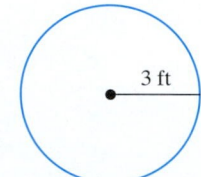

3 ft

Solution: We let $r = 3$ ft and use the formula.

$$A = \pi \cdot r^2$$
$$= \pi \cdot (3 \text{ ft})^2$$
$$= \pi \cdot 9 \text{ square ft, or } 9 \cdot \pi \text{ square ft}$$

To approximate this area, we substitute 3.14 for π.

$9 \cdot \pi$ square feet $\approx 9 \cdot 3.14$ square feet
$= 28.26$ square feet

The *exact* area of the circle is 9π square feet, which is *approximately* 28.26 square feet.

■ **Work Practice 4**

✓**Concept Check** Use diagrams to decide which figure would have a larger area: a circle of diameter 10 inches or a square 10 inches long on each side.

✓**Concept Check Answer**
a square 10 in. long on each side

Vocabulary, Readiness & Video Check

Martin-Gay Interactive Videos Watch the section lecture video and answer the following question.

Objective A 1. What formula was used to solve ⊟ Example 3 and why did we use it twice? ▶

See Video 6.4 🍎

6.4 Exercise Set MyMathLab® ▶

Objective A *Find the area of the geometric figure. If the figure is a circle, give an exact area and then use the given* **approximation** *for π to approximate the area. See Examples 1 through 4.*

▶**1.**

2 m | Rectangle
3.5 m

2.

2.75 ft | Rectangle
7 ft

▶**3.**

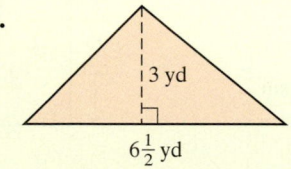
3 yd
$6\frac{1}{2}$ yd

4.

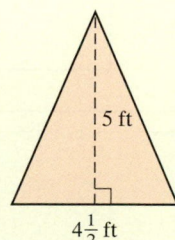
5 ft
$4\frac{1}{2}$ ft

5.

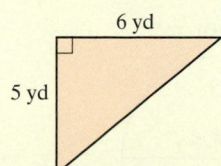
6 yd
5 yd

6.

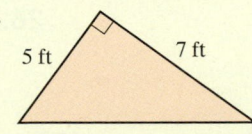
5 ft   7 ft

▶**7.** Use 3.14 for π.

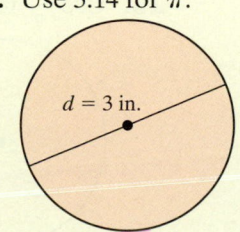
$d = 3$ in.

8. Use $\frac{22}{7}$ for π.

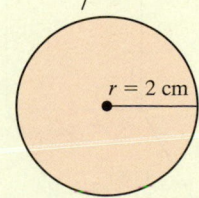
$r = 2$ cm

9.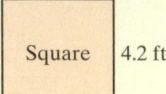
Square 4.2 ft

10.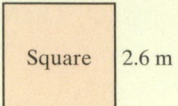
Square 2.6 m

11.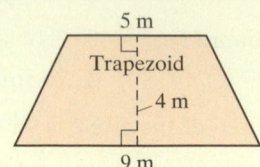
5 m
Trapezoid
4 m
9 m

12.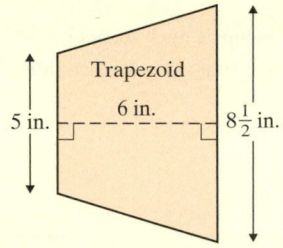
Trapezoid
6 in.
5 in. $8\frac{1}{2}$ in.

13.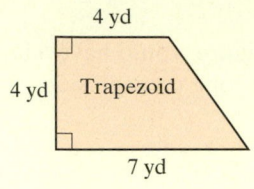
4 yd
4 yd Trapezoid
7 yd

14.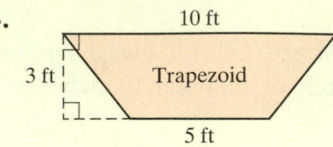
10 ft
3 ft Trapezoid
5 ft

15.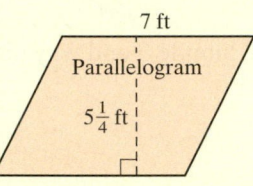
7 ft
Parallelogram
$5\frac{1}{4}$ ft

16.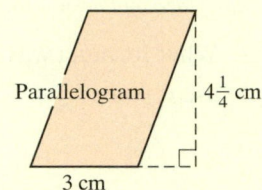
Parallelogram $4\frac{1}{4}$ cm
3 cm

17.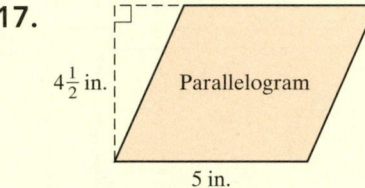
$4\frac{1}{2}$ in. Parallelogram
5 in.

18.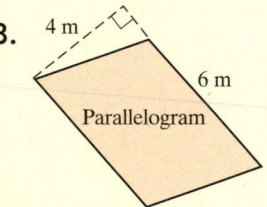
4 m
6 m
Parallelogram

19.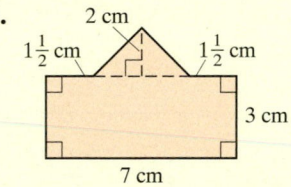
2 cm
$1\frac{1}{2}$ cm $1\frac{1}{2}$ cm
3 cm
7 cm

20.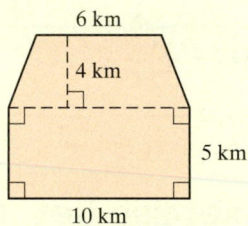
6 km
4 km
5 km
10 km

⊙ 21.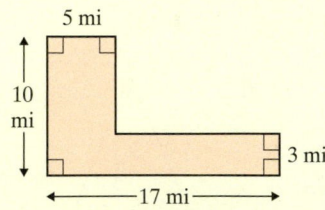
5 mi
10 mi
3 mi
17 mi

22.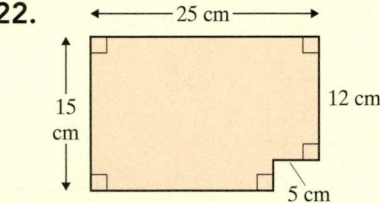
25 cm
15 cm   12 cm
5 cm

23.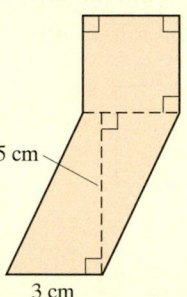
5 cm
3 cm

24.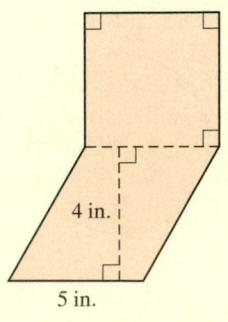
4 in.
5 in.

25. Use $\dfrac{22}{7}$ for π.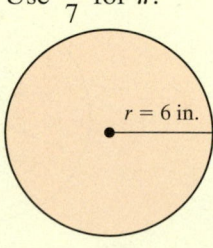
$r = 6$ in.

26. Use 3.14 for π.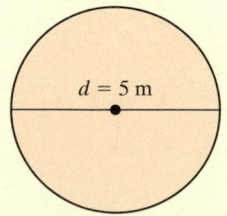
$d = 5$ m

Solve. See Examples 1 through 4.

27. A $10\frac{1}{2}$-foot by 16-foot concrete wall is to be built using concrete blocks. Find the area of the wall.

28. The floor of Terry's attic is 24 feet by 35 feet. Find how many square feet of insulation are needed to cover the attic floor.

29. The world's largest U.S. flag is the "Superflag," which measures 505 feet by 255 feet. Find its area. (*Source:* Superflag.com)

30. The world's largest illuminated indoor advertising sign is located in the Dubai International Airport in Dubai, UAE. It measures 28.0 meters in length by 6.2 meters in height. Find its area. (*Source:* World Record Academy)

31. The face of a watch has a diameter of 2 centimeters. What is its area? Give an exact answer then an approximation using 3.14 for π.

32. The world's largest commercially available pizza is sold by Big Mama's & Papa's Pizzeria in Los Angeles, CA. This huge square pizza, called "The Giant Sicilian," measures 54 inches on each side and sells for $199.99 plus tax. Find the area of the top of the pizza. (*Source: Guinness World Records*, Big Mama's & Big Papa's Pizzeria Inc.)

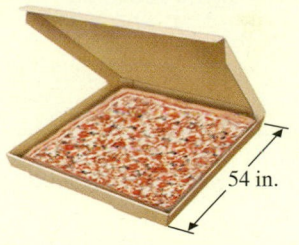

54 in.

33. One side of a concrete block measures 8 inches by 16 inches. Find the area of the side in square inches. Find the area in square feet (144 sq in. = 1 sq ft).

34. A standard *double* roll of wallpaper is $6\frac{5}{6}$ feet wide and 33 feet long. Find the area of the *double* roll.

35. A picture frame measures 20 inches by $25\frac{1}{2}$ inches. Find how many square inches of glass the frame requires.

36. A mat to go under a tablecloth is made to fit a round dining table with a 4-foot diameter. Approximate how many square feet of mat there are. Use 3.14 as an approximation for π.

▷ 37. A drapery panel measures 6 feet by 7 feet. Find how many square feet of material are needed for *four* panels.

38. A page in a book measures 27.5 centimeters by 20.5 centimeters. Find its area.

39. Find how many square feet of land are in the plot shown:

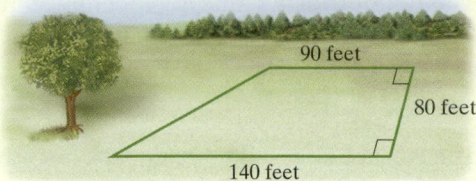

40. For Gerald Gomez to determine how much grass seed he needs to buy, he must know the size of his yard. Use the drawing to determine how many square feet are in his yard.

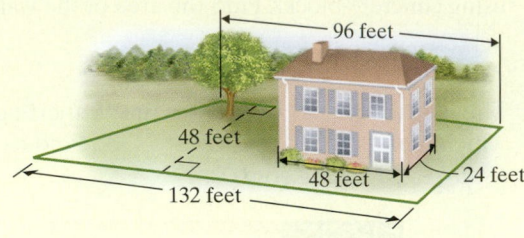

41. The outlined part of the roof shown is in the shape of a trapezoid and needs to be shingled. The number of shingles to buy depends on the area.

a. Use the dimensions given to find the area of the outlined part of the roof to the nearest whole square foot.

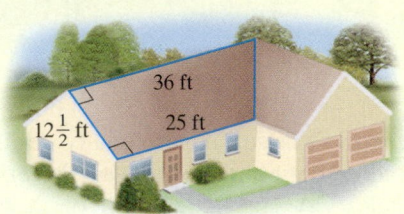

b. Shingles are packaged in a unit called a "square." If a "square" covers 100 square feet, how many whole squares need to be purchased to shingle this part of the roof?

42. The entire side of the building shaded in the drawing is to be bricked. The number of bricks to buy depends on the area.

a. Find the area.

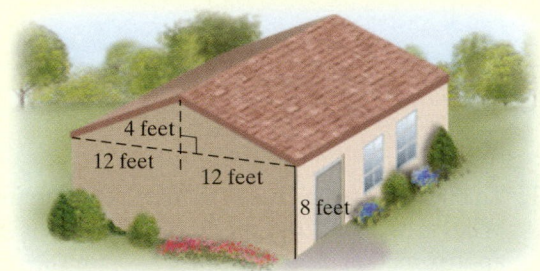

b. If the side area of each brick (including mortar room) is $\frac{1}{6}$ square foot, find the number of bricks needed to brick the end of the building.

Review

Find the perimeter or circumference of each geometric figure. See Section 6.3.

43. Give the exact circumference and an approximation. Use 3.14 for π.

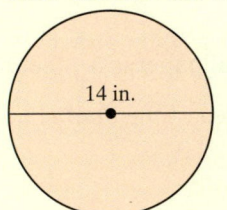

14 in.

44.

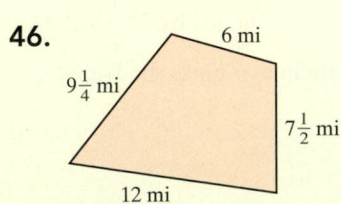

4 cm   5 cm

Rectangle

45.

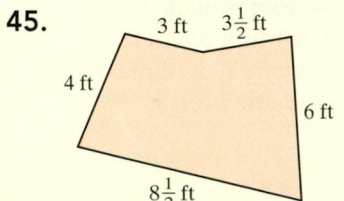

3 ft $3\frac{1}{2}$ ft

4 ft

6 ft

$8\frac{1}{2}$ ft

46.

6 mi

$9\frac{1}{4}$ mi

$7\frac{1}{2}$ mi

12 mi

47. 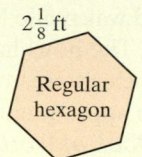 $2\frac{1}{8}$ ft

Regular hexagon

48. 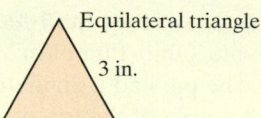 Equilateral triangle

3 in.

Concept Extensions

Given the following situations, tell whether you are more likely to be concerned with area or perimeter.

49. ordering fencing to fence a yard

50. ordering grass seed to plant in a yard

51. buying carpet to install in a room

52. buying gutters to install on a house

53. ordering paint to paint a wall

54. ordering baseboards to install in a room

55. buying a wallpaper border to go on the walls around a room

56. buying fertilizer for your yard

57. A pizza restaurant recently advertised two specials. The first special was a 12-inch diameter pizza for $10. The second special was two 8-inch diameter pizzas for $9. Determine the better buy. (*Hint:* First find and compare the areas of the pizzas in the two specials. Then find a price per square inch for the pizzas in both specials.)

58. Find the approximate area of the state of Utah.

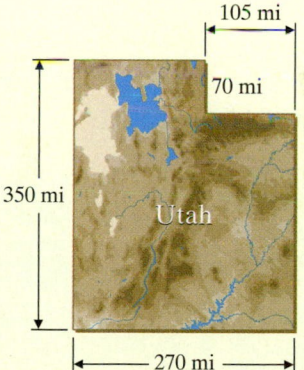

105 mi

70 mi

350 mi

270 mi

Utah

59. Find the area of a rectangle that measures 2 *feet* by 8 *inches*. Give the area in square feet and in square inches.

60. In your own words, explain why perimeter is measured in units and area is measured in square units. (*Hint:* See Section 1.6 for an introduction on the meaning of area.)

61. Find the area of the shaded region. Use the approximation 3.14 for π.

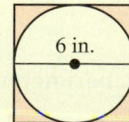

6 in.

62. Estimate the cost of a piece of carpet for a rectangular room 10 feet by 15 feet. The cost of the carpet is $6.50 per yard.

63. The largest pumpkin pie was made for the 100th anniversary of the Circleville, Ohio, Pumpkin Festival in October 2008. The pie had a diameter of 168 inches. Find the exact area of the top of the pie, and an approximation. Use $\pi \approx 3.14$. (*Source:* Circleville, Ohio, Pumpkin Festival)

64. The largest cereal bowl in the world was made by Kellogg's South Africa in July 2007. The bowl had a 2.6-meter diameter. Calculate the exact area of the circular base of the bowl and an approximation. Use $\pi \approx 3.14$. (*Source: Guinness Book of World Records*)

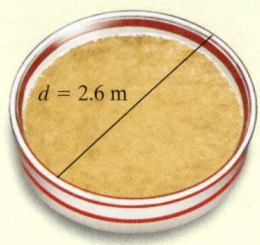

Find the area of each figure. If needed, use $\pi \approx 3.14$ and round results to the nearest tenth.

65. Find the skating area.

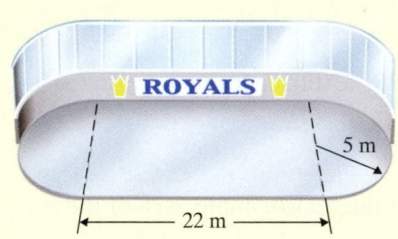

66.

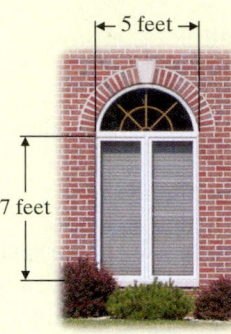

There are a number of factors that determine the dimensions of a rectangular soccer field. Use the table below to answer Exercises 67 and 68.

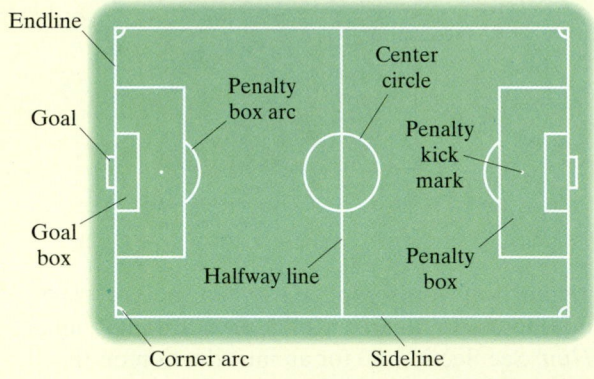

Soccer Field Width and Length		
Age	**Width Min–Max**	**Length Min–Max**
Under 6/7:	15–20 yards	25–30 yards
Under 8:	20–25 yards	30–40 yards
Under 9:	30–35 yards	40–50 yards
Under 10:	40–50 yards	60–70 yards
Under 11:	40–50 yards	70–80 yards
Under 12:	40–55 yards	100–105 yards
Under 13:	50–60 yards	100–110 yards
International:	70–80 yards	110–120 yards

67. a. Find the minimum length and width of a soccer field for 9-year-old children. (Carefully consider the age.)

  b. Find the area of this field.

68. a. Find the maximum length and width of a soccer field for 11-year-old children.

  b. Find the area of this field.

69. Do two rectangles with the same perimeter have the same area? To see, find the perimeter and the area of each rectangle.

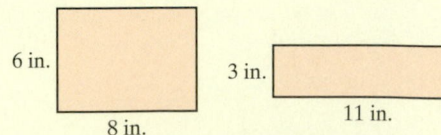

6.5 Volume

Objective A Finding Volumes of Solids

Volume is a measure of the space of a region. The volume of a box or can, for example, is the amount of space inside. Volume can be used to describe the amount of juice in a pitcher or the amount of concrete needed to pour a foundation for a house.

The volume of a solid is the number of **cubic units** in the solid. A cubic centimeter and a cubic inch are illustrated.

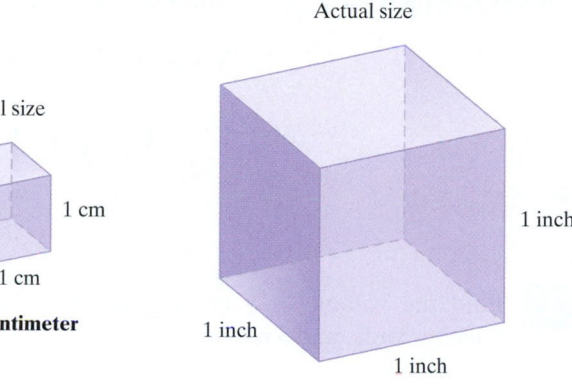

Actual size

Actual size

1 cm
1 cm
1 cm

1 cubic centimeter

1 inch

1 inch
1 inch
1 inch

1 cubic inch

Formulas for finding the volumes of some common solids are given next:

Volume Formulas of Common Solids

Solid	Volume Formulas
RECTANGULAR SOLID 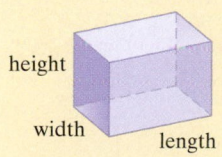 height width length	Volume of a rectangular solid: **Volume = length · width · height** $V = l \cdot w \cdot h$
CUBE 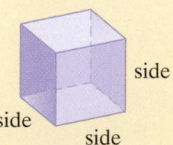 side side side	Volume of cube: **Volume = side · side · side** $V = s^3$
SPHERE 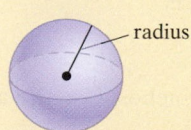 radius	Volume of sphere: **Volume** $= \dfrac{4}{3} \cdot \pi \cdot (\text{radius})^3$ $V = \dfrac{4}{3} \cdot \pi \cdot r^3$
CIRCULAR CYLINDER 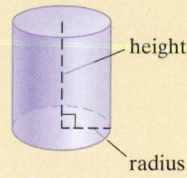 height radius	Volume of a circular cylinder: **Volume** $= \pi \cdot (\text{radius})^2 \cdot \text{height}$ $V = \pi \cdot r^2 \cdot h$

(Continued on next page)

Volume Formulas of Common Solids (*continued*)

Solid	Volume Formulas
CONE 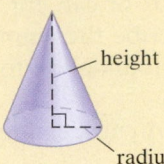 height radius	Volume of a cone: $\text{Volume} = \dfrac{1}{3} \cdot \pi \cdot (\textbf{radius})^2 \cdot \textbf{height}$ $V = \dfrac{1}{3} \cdot \pi \cdot r^2 \cdot h$
SQUARE-BASED PYRAMID height side	Volume of a square-based pyramid: $\text{Volume} = \dfrac{1}{3} \cdot (\text{side})^2 \cdot \textbf{height}$ $V = \dfrac{1}{3} \cdot s^2 \cdot h$

Helpful Hint

Volume is always measured in cubic units.

Practice 1

Find the volume of a rectangular box that is 5 feet long, 2 feet wide, and 4 feet deep.

Example 1 Find the volume of a rectangular box that is 12 inches long, 6 inches wide, and 3 inches high.

3 in.
6 in.   12 in.
FRAGILE

Solution:

$V = l \cdot w \cdot h$

$V = 12 \text{ in.} \cdot 6 \text{ in.} \cdot 3 \text{ in.} = 216 \text{ cubic in.}$

The volume of the rectangular box is 216 cubic inches.

■ **Work Practice 1**

✓**Concept Check** Juan is calculating the volume of the following rectangular solid. Find the error in his calculation.

$\text{Volume} = l + w + h$

$= 14 \text{ cm} + 8 \text{ cm} + 5 \text{ cm}$

$= 27 \text{ cu cm}$

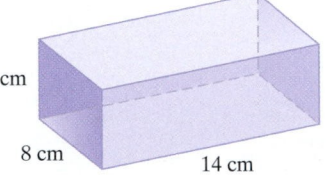

5 cm
8 cm   14 cm

Practice 2

Approximate the volume of a ball of radius $\dfrac{1}{2}$ centimeter. Use $\dfrac{22}{7}$ for π. Give an exact answer and an approximate answer.

Example 2 Approximate the volume of a ball of radius 3 inches. Use the approximation $\dfrac{22}{7}$ for π. Give an exact answer and an approximate answer.

3 in.

Answers

1. 40 cu ft   **2.** $\dfrac{1}{6} \pi$ cu cm $\approx \dfrac{11}{21}$ cu cm

✓**Concept Check Answer**

$\text{Volume} = l \cdot w \cdot h$

$= 14 \text{ cm} \cdot 8 \text{ cm} \cdot 5 \text{ cm}$

$= 560 \text{ cu cm}$

Solution:

$$V = \frac{4}{3} \cdot \pi \cdot r^3$$

$$= \frac{4}{3} \cdot \pi (3 \text{ in.})^3$$

$$= \frac{4}{3} \cdot \pi \cdot 27 \text{ cu in.}$$

$$= \frac{4 \cdot \pi \cdot \overset{1}{\cancel{3}} \cdot 9}{\underset{1}{\cancel{3}}} \text{ cu in.}$$

$$= 36\pi \text{ cu in.}$$

This is the exact volume. To approximate the volume, use the approximation $\frac{22}{7}$ for π.

$$V = 36\pi \text{ cu in.}$$

$$\approx 36 \cdot \frac{22}{7} \text{ cu in.} \quad \text{Replace } \pi \text{ with } \frac{22}{7}.$$

$$= \frac{36 \cdot 22}{7} \text{ cu in.}$$

$$= \frac{792}{7} \quad \text{or} \quad 113\frac{1}{7} \text{ cubic inches.}$$

The volume is *approximately* $113\frac{1}{7}$ cubic inches.

■ **Work Practice 2**

Example 3 Approximate the volume of a can that has a $3\frac{1}{2}$-inch radius and a height of 6 inches. Use $\frac{22}{7}$ for π. Give an exact volume and an approximate volume.

$3\frac{1}{2}$ in.

6 in.

Practice 3

Approximate the volume of a cylinder of radius 5 inches and height 7 inches. Use 3.14 for π. Give an exact answer and an approximate answer.

Solution: Using the formula for a circular cylinder, we have

$$V = \pi \cdot r^2 \cdot h$$

$$3\frac{1}{2} = \frac{7}{2}$$

$$= \pi \cdot \left(\frac{7}{2} \text{ in.}\right)^2 \cdot 6 \text{ in.}$$

$$= \pi \cdot \frac{49}{4} \text{ sq in.} \cdot 6 \text{ in.}$$

$$= \frac{\pi \cdot 49 \cdot \overset{1}{\cancel{2}} \cdot 3}{\underset{1}{\cancel{2}} \cdot 2} \text{ cu in.}$$

$$= 73\frac{1}{2} \pi \text{ cu in. or } 73.5\pi \text{ cu in.}$$

(Continued on next page)

Answer

3. 175π cu in. ≈ 549.5 cu in.

This is the exact volume. To approximate the volume, use the approximation $\frac{22}{7}$ for π.

$$V = 73\frac{1}{2}\pi \approx \frac{147}{2} \cdot \frac{22}{7} \text{ cu in.} \quad \text{\color{blue}Replace } \pi \text{ with } \frac{22}{7}.$$

$$= \frac{21 \cdot \overset{1}{\cancel{7}} \cdot \overset{1}{\cancel{2}} \cdot 11}{\underset{1}{\cancel{2}} \cdot \underset{1}{\cancel{7}}} \text{ cu in.}$$

$$= \text{\color{blue}231 cubic in.}$$

The volume is approximately **231 cubic inches**.

■ **Work Practice 3**

Practice 4

Find the volume of a square-based pyramid that has a 3-meter side and a height of 5.1 meters.

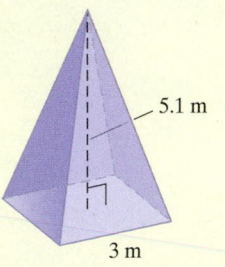

5.1 m

3 m

Answer

4. 15.3 cu m

Example 4   Approximate the volume of a cone that has a height of 14 centimeters and a radius of 3 centimeters. Use 3.14 for π. Give an exact answer and an approximate answer.

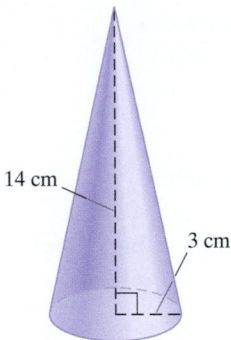

14 cm

3 cm

Solution: Using the formula for volume of a cone, we have

$$V = \frac{1}{3} \cdot \pi \cdot r^2 \cdot h$$

$$= \frac{1}{3} \cdot \pi \cdot (3 \text{ cm})^2 \cdot 14 \text{ cm} \quad \text{\color{blue}Replace } r \text{ with 3 cm and } h \text{ with 14 cm.}$$

$$= 42\pi \text{ cu cm}$$

Thus, 42π cubic centimeters is the exact volume. To approximate the volume, use the approximation 3.14 for π.

$$V \approx 42 \cdot 3.14 \text{ cu cm} \quad \text{\color{blue}Replace } \pi \text{ with 3.14.}$$

$$= 131.88 \text{ cu cm}$$

The volume is approximately **131.88 cubic centimeters**.

■ **Work Practice 4**

Vocabulary, Readiness & Video Check

Use the choices below to fill in each blank. Some exercises are from Sections 6.3 and 6.4.

cubic	perimeter	volume
units	area	square

1. The measure of the amount of space inside a solid is its _____.

2. _____ measures the amount of surface of a region.

3. Volume is measured in _____ units.

4. Area is measured in _____ units.

5. The _____ of a polygon is the sum of the lengths of its sides.

6. Perimeter is measured in _____ .

Martin-Gay Interactive Videos Watch the section lecture video and answer the following question.

Objective **A** 7. In ⊞ Examples 2 and 3, explain the difference in the two answers found for each. ▶

See Video 6.5 🍎

6.5 **Exercise Set** MyMathLab® ▶

Objective **A** *Find the volume of each solid. See Examples 1 through 4. Use $\frac{22}{7}$ for π.*

▶ **1.**

3 in.

4 in.   6 in.

2.

4 cm

4 cm   8 cm

3.

8 cm

8 cm   8 cm

4.

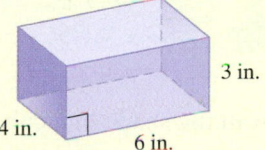

11 mi

11 mi   11 mi

5.

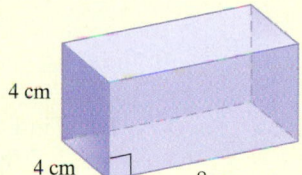

3 yd

2 yd

6.

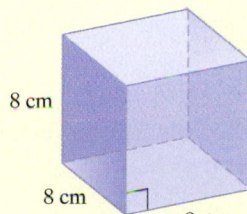

$1\frac{3}{4}$ in.

9 in.

▶ **7.**

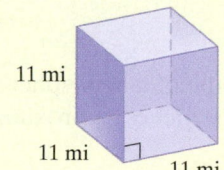

10 in.

8.

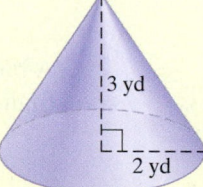

3 mi

▶ **9.**

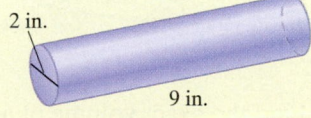

2 in.

9 in.

10.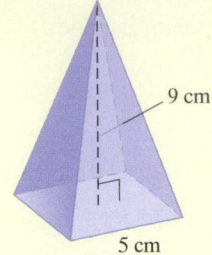
10 ft
6 ft

11.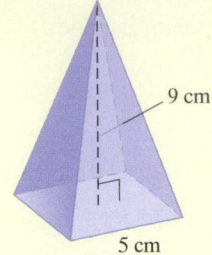
9 cm
5 cm

12.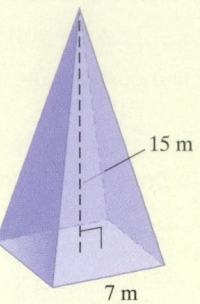
15 m
7 m

Solve.

▶ **13.** Find the volume of a cube with edges of $1\frac{1}{3}$ inches.

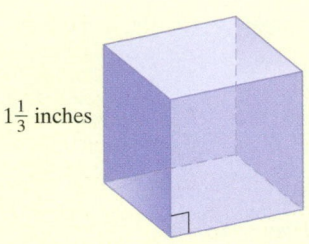

$1\frac{1}{3}$ inches

14. A water storage tank is in the shape of a cone with the pointed end down. If the radius is 14 feet and the depth of the tank is 15 feet, approximate the volume of the tank in cubic feet. Use $\frac{22}{7}$ for π.

14 ft
15 ft

15. Find the volume of a rectangular box 2 feet by 1.4 feet by 3 feet.

16. Find the volume of a box in the shape of a cube that is 5 feet on each side.

17. Find the volume of a pyramid with a square base 5 inches on a side and a height of $1\frac{3}{10}$ inches.

18. Approximate to the nearest hundredth the volume of a sphere with a radius of 2 centimeters. Use 3.14 for π.

19. A paperweight is in the shape of a square-based pyramid 20 centimeters tall. If an edge of the base is 12 centimeters, find the volume of the paperweight.

20. A birdbath is made in the shape of a hemisphere (half-sphere). If its radius is 10 inches, approximate the volume. Use $\frac{22}{7}$ for π.

10 in.

21. Find the exact volume of a sphere with a radius of 7 inches.

22. A tank is in the shape of a cylinder 8 feet tall and 3 feet in radius. Find the exact volume of the tank.

23. Find the volume of a rectangular block of ice 2 feet by $2\frac{1}{2}$ feet by $1\frac{1}{2}$ feet.

24. Find the capacity (volume in cubic feet) of a rectangular ice chest with inside measurements of 3 feet by $1\frac{1}{2}$ feet by $1\frac{3}{4}$ feet.

25. In 2013, the largest free-floating soap bubble made with a wand had a diameter between 11 and 12 feet. Calculate the exact volume of a sphere with a diameter of 12 feet. (*Source: Guinness World Records*)

26. The largest inflatable beach ball was created in Poland in 2012. It has a diameter of just under 54 feet. Calculate the exact volume of a sphere with a diameter of 54 feet. (*Source: Guinness World Records*)

27. Find the exact volume of a waffle ice cream cone with a 3-in. diameter and a height of 7 inches.

28. A snow globe has a diameter of 6 inches. Find its exact volume. Then approximate its volume using 3.14 for π.

29. The largest cereal bowl in the world was made by Kellogg's South Africa in July 2007. The bowl had a 2.6-m diameter and a height of 1.5 m. Calculate the volume of cereal you could put into this bowl. Use $\pi \approx 3.14$ and round to the nearest hundredth of a cubic meter. (*Source: Guinness Book of World Records*)

30. Mount Fuji, in Japan, is considered the most beautiful composite volcano in the world. The mountain is in the shape of a cone whose height is about 3.5 kilometers and whose base radius is about 3 kilometers. Approximate the volume of Mt. Fuji in cubic kilometers. Use $\frac{22}{7}$ for π.

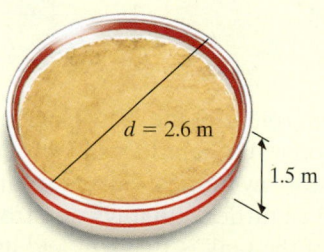

31. An ice cream cone with a 4-centimeter diameter and 3-centimeter depth is filled exactly level with the top of the cone. Approximate how much ice cream (in cubic centimeters) is in the cone. Use $\frac{22}{7}$ for π.

32. A child's toy is in the shape of a square-based pyramid 10 inches tall. If an edge of the base is 7 inches, find the volume of the toy.

The Space Cube is supposed to be the world's smallest computer, with dimensions of 2 inches by 2 inches by 2.2 inches.

33. Find the volume of the Space Cube.

34. Find the volume of an actual cube that measures 2 inches by 2 inches by 2 inches.

35. Find the volume of an actual cube that measures 2.2 inches by 2.2 inches by 2.2 inches.

36. Comment on the results of Exercises 33–35. Were you surprised when you compared volumes? Why or why not?

Review

Evaluate. See Section 1.9.

37. 5^2 **38.** 7^2 **39.** 3^2 **40.** 20^2

41. $1^2 + 2^2$ **42.** $5^2 + 3^2$ **43.** $4^2 + 2^2$ **44.** $1^2 + 6^2$

Concept Extensions

45. The Hayden Planetarium, at the Museum of Natural History in New York City, boasts a dome that has a diameter of 20 m. The dome is a hemisphere, or half a sphere. What is the volume enclosed by the dome at the Hayden Planetarium? Use 3.14 for π and round to the nearest hundredth. (*Source:* Hayden Planetarium)

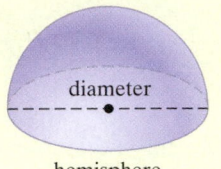

diameter

hemisphere

46. The Adler Museum in Chicago recently added a new planetarium, its StarRider Theater, which has a diameter of 55 feet. Find the volume of its hemispheric (half a sphere) dome. Use 3.14 for π and round to the nearest hundredth. (*Source:* The Adler Museum)

47. Do two rectangular solids with the same volume have the same shape? To help, find the volume of each rectangular solid.

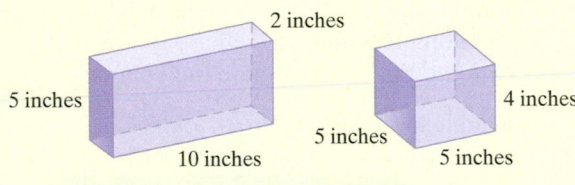

2 inches

5 inches

10 inches

4 inches

5 inches

5 inches

48. Do two rectangular solids with the same volume have the same surface area? To see, find the volume and surface area of each rectangular solid. Surface area is the area of the surface of the solid. To find the surface area of each rectangular solid, find the sum of the areas of the 6 rectangles that form each solid.

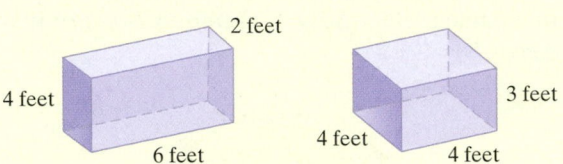

2 feet

4 feet

6 feet

3 feet

4 feet

4 feet

49. Two kennels are offered at a hotel. The kennels measure

  a. 2′1″ by 1′8″ by 1′7″ and

  b. 1′1″ by 2′ by 2′8″

What is the volume of each kennel rounded to the nearest tenth of a cubic foot? Which is larger?

50. The centerpiece of the New England Aquarium in Boston is its Giant Ocean Tank. This exhibit is a four-story cylindrical saltwater tank containing sharks, sea turtles, stingrays, and tropical fish. The radius of the tank is 16.3 feet and its height is 32 feet (assuming that a story is 8 feet). What is the volume of the Giant Ocean Tank? Use $\pi \approx 3.14$ and round to the nearest tenth of a cubic foot. (*Source:* New England Aquarium)

51. Find the volume of the figure below. Give an exact measure and then a whole number approximation.

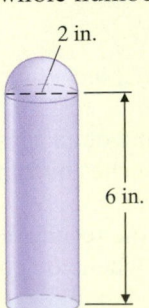

2 in.

6 in.

52. Can you compute the volume of a rectangle? Why or why not?

Geometry Concepts

1. _____

△ **1.** Find the supplement and the complement of a 27° angle.

2. _____

Find the measures of angles x, y, and z in each figure.

2.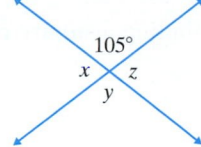

3. $m \parallel n$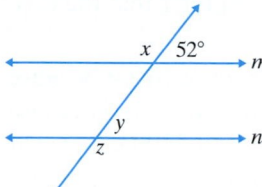

3. _____

4. _____

4. Find the measure of ∠x. 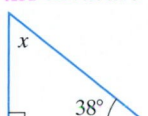　**5.** Find the diameter. 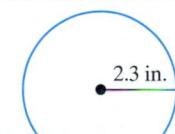　**6.** Find the radius.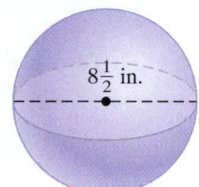

5. _____

6. _____

7. _____

For Exercises 7 through 11, find the perimeter (or circumference) and area of each figure. For the circle give an exact circumference and area. Then use $\pi \approx 3.14$ to approximate each. Don't forget to attach correct units.

8. _____

7. 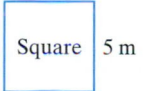　**8.** 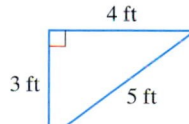　**9.** 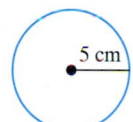　**10.**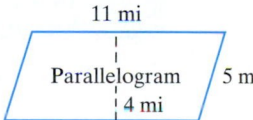

9. _____

10. _____

11.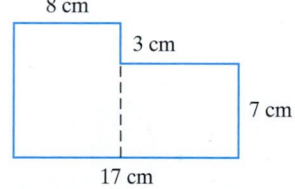

12. The smallest cathedral is in High-landville, Missouri. The rectangular floor of the cathedral measures 14 feet by 17 feet. Find its perimeter and its area. (*Source: The Guinness Book of Records*)

11. _____

12. _____

13. _____

Find the volume of each solid. Don't forget to attach correct units.

13. A cube with edges of 4 inches each.

14. A rectangular box 2 feet by 3 feet by 5.1 feet.

14. _____

15. _____

15. A pyramid with a square base 10 centimeters on a side and a height of 12 centimeters.

16. A sphere with a diameter of 3 miles. Give the exact volume and then use $\pi \approx \dfrac{22}{7}$ to approximate.

16. _____

Objectives

A Find the Square Root of a Number. ▶

B Approximate Square Roots. ▶

C Use the Pythagorean Theorem. ▶

Objective **A** Finding Square Roots ▶

The square of a number is the number times itself. For example:

The square of 5 is 25 because 5^2 or $5 \cdot 5 = 25$.
The square of 4 is 16 because 4^2 or $4 \cdot 4 = 16$.
The square of 10 is 100 because 10^2 or $10 \cdot 10 = 100$.

Recall from Chapter 1 that the reverse process of squaring is finding a **square root.** For example:

A square root of 16 is 4 because $4^2 = 16$.
A square root of 25 is 5 because $5^2 = 25$.
A square root of 100 is 10 because $10^2 = 100$.

We use the symbol $\sqrt{}$, called a **radical sign,** to name square roots. For example:

$\sqrt{16} = 4$ because $4^2 = 16$
$\sqrt{25} = 5$ because $5^2 = 25$

Square Root of a Number

A square root of a number a is a number b whose square is a. We use the radical sign $\sqrt{}$ to name square roots. In symbols,

$$\sqrt{a} = b \quad \text{if} \quad b^2 = a$$

Also,

$$\sqrt{0} = 0$$

Practice 1

Find each square root.
a. $\sqrt{100}$ **b.** $\sqrt{64}$
c. $\sqrt{121}$ **d.** $\sqrt{0}$

Example 1 Find each square root.

a. $\sqrt{49}$ **b.** $\sqrt{1}$ **c.** $\sqrt{81}$

Solution:
a. $\sqrt{49} = 7$ because $7^2 = 49$
b. $\sqrt{1} = 1$ because $1^2 = 1$
c. $\sqrt{81} = 9$ because $9^2 = 81$

■ **Work Practice 1**

Practice 2

Find: $\sqrt{\dfrac{1}{4}}$

Example 2 Find: $\sqrt{\dfrac{1}{36}}$

Solution: $\sqrt{\dfrac{1}{36}} = \dfrac{1}{6}$ because $\dfrac{1}{6} \cdot \dfrac{1}{6} = \dfrac{1}{36}$

■ **Work Practice 2**

Practice 3

Find: $\sqrt{\dfrac{9}{16}}$

Example 3 Find: $\sqrt{\dfrac{4}{25}}$

Solution: $\sqrt{\dfrac{4}{25}} = \dfrac{2}{5}$ because $\dfrac{2}{5} \cdot \dfrac{2}{5} = \dfrac{4}{25}$

■ **Work Practice 3**

Answers
1. a. 10 **b.** 8 **c.** 11 **d.** 0
2. $\dfrac{1}{2}$ **3.** $\dfrac{3}{4}$

Objective B Approximating Square Roots

Thus far, we have found square roots of perfect squares. Numbers like $\frac{1}{4}$, 36, $\frac{4}{25}$, and 1 are called **perfect squares** because their square root is a whole number or a fraction. A square root such as $\sqrt{5}$ cannot be written as a whole number or a fraction since 5 is not a perfect square.

Although $\sqrt{5}$ cannot be written as a whole number or a fraction, it can be approximated by estimating, by using a table (as in the appendix), or by using a calculator.

Example 4 Use Appendix B.1 or a calculator to approximate each square root to the nearest thousandth.

a. $\sqrt{43} \approx 6.557$ is approximately
b. $\sqrt{80} \approx 8.944$

■ **Work Practice 4**

Practice 4

Use Appendix B.1 or a calculator to approximate each square root to the nearest thousandth.
a. $\sqrt{21}$ b. $\sqrt{52}$

Helpful Hint

$\sqrt{80}$, above, is *approximately* 8.944. This means that if we multiply 8.944 by 8.944, the product is *close* to 80.

$8.944 \times 8.944 \approx 79.995$

It is possible to approximate a square root to the nearest whole number without the use of a calculator or table. To do so, study the number line below and look for patterns.

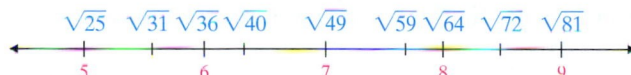

Above the number line, notice that as the numbers under the radical signs increase, their values, and thus their placement on the number line, increase also.

Example 5 Without a calculator or table:

a. Determine which two whole numbers $\sqrt{78}$ is between.
b. Use part **a** to approximate $\sqrt{78}$ to the nearest whole.

Practice 5

Without a calculator or table, approximate $\sqrt{62}$ to the nearest whole.

Solution:

a. Review perfect squares and recall that $\sqrt{64} = 8$ and $\sqrt{81} = 9$. Since 78 is between 64 and 81, $\sqrt{78}$ is between $\sqrt{64}$ (or 8) and $\sqrt{81}$ (or 9).

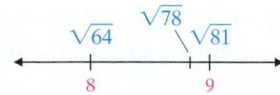

Thus, $\sqrt{78}$ is between 8 and 9.

b. Since 78 is closer to 81, then (as our number line shows) $\sqrt{78}$ is closer to $\sqrt{81}$, or 9.

■ **Work Practice 5**

Objective C Using the Pythagorean Theorem

One important application of square roots has to do with right triangles. Recall that a **right triangle** is a triangle in which one of the angles is a right angle, or measures 90° (degrees). The **hypotenuse** of a right triangle is the side opposite the right angle.

Answers

4. a. 4.583   b. 7.211
5. 8

The **legs** of a right triangle are the other two sides. These are shown in the following figure. The right angle in the triangle is indicated by the small square drawn in that angle.

The following theorem is true for all right triangles:

Pythagorean Theorem

In any right triangle,

$$(\text{leg})^2 + (\text{other leg})^2 = (\text{hypotenuse})^2$$

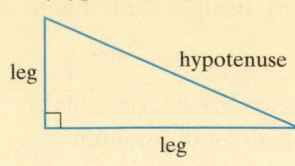

Using the Pythagorean theorem, we can use one of the following formulas to find an unknown length of a right triangle:

Finding an Unknown Length of a Right Triangle

$$\text{hypotenuse} = \sqrt{(\text{leg})^2 + (\text{other leg})^2}$$

or

$$\text{leg} = \sqrt{(\text{hypotenuse})^2 - (\text{other leg})^2}$$

Practice 6

Find the length of the hypotenuse of the given right triangle.

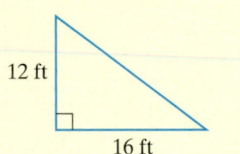

Example 6 Find the length of the hypotenuse of the given right triangle.

Solution: Since we are finding the hypotenuse, we use the formula

$$\text{hypotenuse} = \sqrt{(\text{leg})^2 + (\text{other leg})^2}$$

Putting the known values into the formula, we have

$$\text{hypotenuse} = \sqrt{(6)^2 + (8)^2} \quad \text{The legs are 6 feet and 8 feet.}$$
$$= \sqrt{36 + 64}$$
$$= \sqrt{100}$$
$$= 10$$

The hypotenuse is 10 feet long.

■ **Work Practice 6**

Practice 7

Approximate the length of the hypotenuse of the given right triangle. Round to the nearest whole unit.

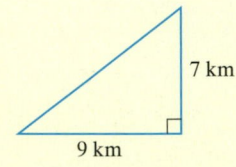

Example 7 Approximate the length of the hypotenuse of the given right triangle. Round the length to the nearest whole unit.

Solution:

$$\text{hypotenuse} = \sqrt{(\text{leg})^2 + (\text{other leg})^2}$$
$$= \sqrt{(17)^2 + (10)^2} \quad \text{The legs are 10 meters and 17 meters.}$$
$$= \sqrt{289 + 100}$$
$$= \sqrt{389}$$
$$\approx 20 \quad \text{From Appendix B.1 or a calculator}$$

The hypotenuse is exactly $\sqrt{389}$ meters, which is approximately 20 meters.

■ **Work Practice 7**

Answers

6. 20 ft   **7.** 11 km

Example 8 Find the length of the leg in the given right triangle. Give the exact length and a two-decimal-place approximation.

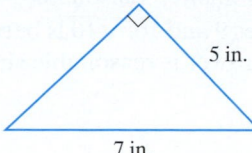

5 in.

7 in.

Solution: Notice that the hypotenuse measures 7 inches and the length of one leg measures 5 inches. Since we are looking for the length of the other leg, we use the formula

$$\text{leg} = \sqrt{(\text{hypotenuse})^2 - (\text{other leg})^2}$$

Putting the known values into the formula, we have

$$\text{leg} = \sqrt{(7)^2 - (5)^2} \quad \text{The hypotenuse is 7 inches, and the other leg is 5 inches.}$$
$$= \sqrt{49 - 25}$$
$$= \sqrt{24} \quad \text{Exact answer}$$
$$\approx 4.90 \quad \text{From Appendix B.1 or a calculator}$$

The length of the leg is exactly $\sqrt{24}$ inches, which is approximately 4.90 inches.

■ **Work Practice 8**

✓**Concept Check** The following lists are the lengths of the sides of two triangles. Which set forms a right triangle? Explain.

a. 8, 15, 17   **b.** 24, 30, 40

△ **Example 9** Finding the Dimensions of a Park

An inner-city park is in the shape of a square that measures 300 feet on a side. A sidewalk is to be constructed along the diagonal of the park. Find the length of the sidewalk rounded to the nearest whole foot.

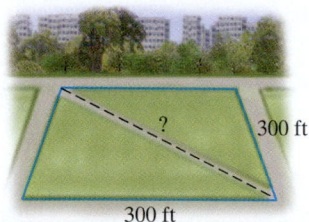

?   300 ft

300 ft

Solution: The diagonal is the hypotenuse of a right triangle, so we use the formula

$$\text{hypotenuse} = \sqrt{(\text{leg})^2 + (\text{other leg})^2}$$

Putting the known values into the formula we have

$$\text{hypotenuse} = \sqrt{(300)^2 + (300)^2} \quad \text{The legs are both 300 feet.}$$
$$= \sqrt{90,000 + 90,000}$$
$$= \sqrt{180,000}$$
$$\approx 424 \quad \text{From Appendix B.1 or a calculator}$$

The length of the sidewalk is approximately 424 feet.

■ **Work Practice 9**

Practice 8

Find the length of the leg in the given right triangle. Give the exact length and a two-decimal-place approximation.

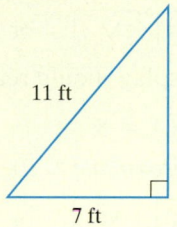

11 ft

7 ft

Practice 9

A football field is a rectangle measuring 100 yards by 53 yards. Draw a diagram and find the length of the diagonal of a football field to the nearest yard.

Answers
8. $\sqrt{72}$ ft ≈ 8.49 ft
9. 113 yd

✓**Concept Check Answer**
set (a) forms a right triangle

Calculator Explorations **Finding Square Roots**

To simplify or approximate square roots using a calculator, locate the key marked $\boxed{\sqrt{\ }}$.

To simplify $\sqrt{64}$, for example, press the keys

$\boxed{64}$ $\boxed{\sqrt{\ }}$ or $\boxed{\sqrt{\ }}$ $\boxed{64}$

The display should read $\boxed{\quad 8\ }$. Then

$\sqrt{64} = 8$

To *approximate* $\sqrt{10}$, press the keys

$\boxed{10}$ $\boxed{\sqrt{\ }}$ or $\boxed{\sqrt{\ }}$ $\boxed{10}$

The display should read $\boxed{3.16227766}$. This is an *approximation* for $\sqrt{10}$. A three-decimal-place approximation is

$\sqrt{10} \approx 3.162$

Is this answer reasonable? Since 10 is between perfect squares 9 and 16, $\sqrt{10}$ is between $\sqrt{9} = 3$ and $\sqrt{16} = 4$. Our answer is reasonable since 3.162 is between 3 and 4.

Simplify.

1. $\sqrt{1024}$
2. $\sqrt{676}$

Approximate each square root. Round each answer to the nearest thousandth.

3. $\sqrt{31}$
4. $\sqrt{19}$
5. $\sqrt{97}$
6. $\sqrt{56}$

Vocabulary, Readiness & Video Check

Use the choices below to fill in each blank. Some choices may be used more than once.

squaring   Pythagorean theorem   radical   leg
hypotenuse   perfect squares   10

1. $\sqrt{100} =$ _____ because $10 \cdot 10 = 100$.
2. The _____ sign is used to denote the square root of a number.
3. The reverse process of _____ a number is finding a square root of a number.
4. The numbers 9, 1, and $\frac{1}{25}$ are called _____.
5. Label the parts of the right triangle. ____
6. The _____ can be used for right triangles.

Martin-Gay Interactive Videos

See Video 6.6 🍎

Watch the section lecture video and answer the following questions.

Objective A 7. From the lecture before ▣ Example 1, explain why $\sqrt{49} = 7$. ▶

Objective B 8. In ▣ Example 5, how do we know $\sqrt{15}$ is closer to 4 than to 3? ▶

Objective C 9. At the beginning of ▣ Example 6, what are we reminded about regarding the Pythagorean theorem? ▶

6.6 Exercise Set MyMathLab®

Objective A *Find each square root. See Examples 1 through 3.*

▶ **1.** $\sqrt{4}$ **2.** $\sqrt{9}$ ▶ **3.** $\sqrt{121}$ **4.** $\sqrt{144}$

▶ **5.** $\sqrt{\dfrac{1}{81}}$ **6.** $\sqrt{\dfrac{1}{64}}$ **7.** $\sqrt{\dfrac{16}{64}}$ **8.** $\sqrt{\dfrac{36}{81}}$

Objective B *Use Appendix B.1 or a calculator to approximate each square root. Round the square root to the nearest thousandth. See Examples 4 and 5.*

9. $\sqrt{3}$ **10.** $\sqrt{5}$ ▶ **11.** $\sqrt{15}$ **12.** $\sqrt{17}$

13. $\sqrt{47}$ **14.** $\sqrt{85}$ **15.** $\sqrt{26}$ **16.** $\sqrt{35}$

Determine what two whole numbers each square root is between without using a calculator or table. Then use a calculator or Appendix B.1 to check. See Example 5.

▶ **17.** $\sqrt{38}$ **18.** $\sqrt{27}$ **19.** $\sqrt{101}$ **20.** $\sqrt{85}$

Objectives A B Mixed Practice *Find each square root. If necessary, round the square root to the nearest thousandth. See Examples 1 through 5.*

21. $\sqrt{256}$ **22.** $\sqrt{625}$ **23.** $\sqrt{92}$ **24.** $\sqrt{18}$

25. $\sqrt{\dfrac{49}{144}}$ **26.** $\sqrt{\dfrac{121}{169}}$ **27.** $\sqrt{71}$ **28.** $\sqrt{62}$

Objective C *Find the unknown length in each right triangle. If necessary, approximate the length to the nearest thousandth. See Examples 6 through 8.*

▶ **29.**
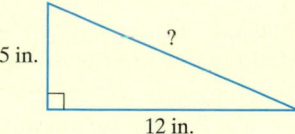
5 in., ?, 12 in.

30.
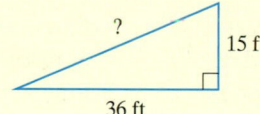
?, 15 ft, 36 ft

31.
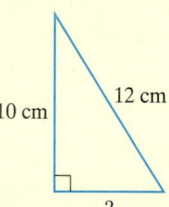
10 cm, 12 cm, ?

32.
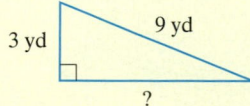
3 yd, 9 yd, ?

33.
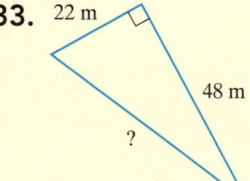
22 m, 48 m, ?

34.
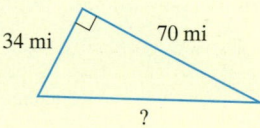
34 mi, 70 mi, ?

35.
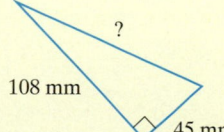
?, 108 mm, 45 mm

36.
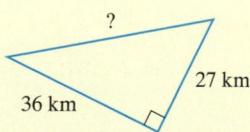
?, 27 km, 36 km

Sketch each right triangle and find the length of the side not given. If necessary, approximate the length to the nearest thousandth. (Each length is in units.) See Examples 6 through 8.

37. leg = 3, leg = 4

38. leg = 9, leg = 12

39. leg = 5, hypotenuse = 13

40. leg = 6, hypotenuse = 10

41. leg = 10, leg = 14

42. leg = 2, leg = 16

43. leg = 35, leg = 28

44. leg = 30, leg = 15

45. leg = 30, leg = 30

46. leg = 21, leg = 21

▶ **47.** hypotenuse = 2, leg = 1

48. hypotenuse = 9, leg = 8

49. leg = 7.5, leg = 4

50. leg = 12, leg = 22.5

Solve. See Example 9.

51. A standard city block is a square with each side measuring 100 yards. Find the length of the diagonal of a city block to the nearest hundredth yard.

52. A section of land is a square with each side measuring 1 mile. Find the length of the diagonal of the section of land to the nearest thousandth mile.

53. Find the height of the tree. Round the height to one decimal place.

? | 32 feet | 20 feet

54. Find the height of the antenna. Round the height to one decimal place.

168 ft | ? | 60 ft

55. The playing field for football is a rectangle that is 300 feet long by 160 feet wide. Find, to the nearest foot, the length of a straight-line run that started at one corner and went diagonally to end at the opposite corner.

? | 160 feet | 300 feet

56. A soccer field is in the shape of a rectangle and its dimensions depend on the age of the players. The dimensions of the soccer field below are the minimum dimensions for international play. Find the length of the diagonal of this rectangle. Round the answer to the nearest tenth of a yard.

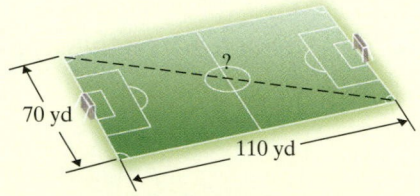

70 yd | ? | 110 yd

Review

Find the value of n in each proportion. See Section 5.1.

57. $\dfrac{n}{6} = \dfrac{2}{3}$

58. $\dfrac{8}{n} = \dfrac{4}{8}$

59. $\dfrac{9}{11} = \dfrac{n}{55}$

60. $\dfrac{5}{6} = \dfrac{35}{n}$

61. $\dfrac{3}{n} = \dfrac{7}{14}$

62. $\dfrac{n}{9} = \dfrac{4}{6}$

Concept Extensions

Use the results of Exercises 17–20 and approximate each square root to the nearest whole without using a calculator or table. Then use a calculator or Appendix B.1 to check. See Example 5.

63. $\sqrt{38}$

64. $\sqrt{27}$

65. $\sqrt{101}$

66. $\sqrt{85}$

67. Without using a calculator, explain how you know that $\sqrt{105}$ is *not* approximately 9.875.

68. Without using a calculator, explain how you know that $\sqrt{27}$ is *not* approximately 3.296.

Does the set form the lengths of the sides of a right triangle? See the Concept Check in this section.

69. 25, 60, 65

70. 20, 45, 50

△ **71.** Find the exact length of x. Then give a two-decimal-place approximation.

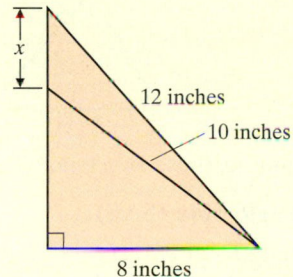

12 inches

10 inches

8 inches

6.7 Congruent and Similar Triangles

Objective A Deciding Whether Two Triangles Are Congruent.

Congruent angles are angles that have the same measure. Two triangles are **congruent** when they have the same shape and the same size. In congruent triangles, the measures of corresponding angles are equal and the lengths of corresponding sides are equal. The following triangles are congruent:

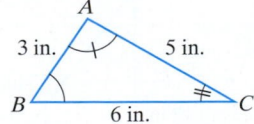
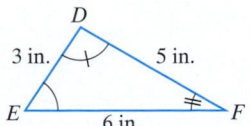

Since these triangles are congruent, the measures of corresponding angles are equal.
Angles with equal measure: $\angle A$ and $\angle D$, $\angle B$ and $\angle E$, $\angle C$ and $\angle F$. Also, the lengths of corresponding sides are equal.
Equal corresponding sides: $\overline{AB}$ and $\overline{DE}$, $\overline{BC}$ and $\overline{EF}$, $\overline{CA}$ and $\overline{FD}$

Objectives

A Deciding Whether Two Triangles Are Congruent.

B Find the Ratio of Corresponding Sides in Similar Triangles.

C Find Unknown Lengths of Sides in Similar Triangles.

Any one of the following may be used to determine whether two triangles are congruent:

Congruent Triangles

Angle-Side-Angle (ASA)

If the measures of two angles of a triangle equal the measures of two angles of another triangle, and the lengths of the sides between each pair of angles are equal, the triangles are congruent.

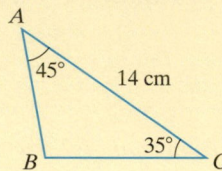 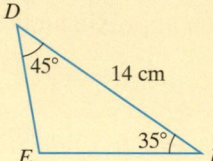

For example, these two triangles are congruent by Angle-Side-Angle.

Side-Side-Side (SSS)

If the lengths of the three sides of a triangle equal the lengths of the corresponding sides of another triangle, the triangles are congruent.

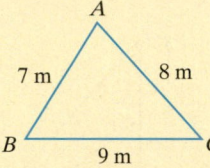 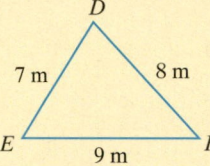

For example, these two triangles are congruent by Side-Side-Side.

Side-Angle-Side (SAS)

If the lengths of two sides of a triangle equal the lengths of corresponding sides of another triangle, and the measures of the angles between each pair of sides are equal, the triangles are congruent.

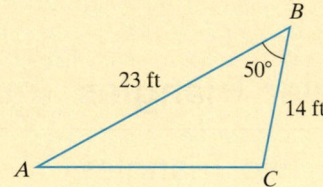 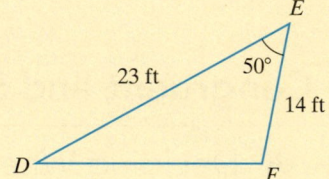

For example, these two triangles are congruent by Side-Angle-Side.

Practice 1

a. Determine whether triangle *MNO* is congruent to triangle *RQS*.

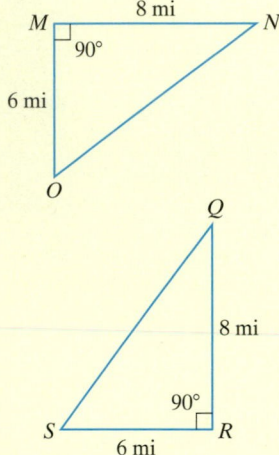

b. Determine whether triangle *GHI* is congruent to triangle *JKL*.

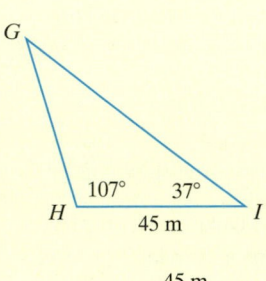

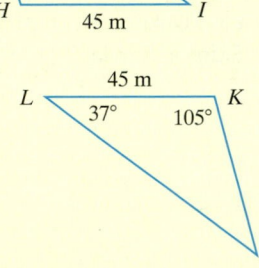

Answers

1. a. congruent   **b.** not congruent

Example 1   Determine whether triangle *ABC* is congruent to triangle *DEF*.

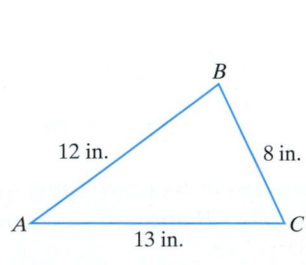 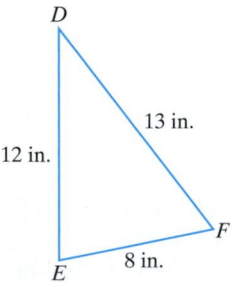

Solution: Since the lengths of all three sides of triangle *ABC* equal the lengths of all three sides of triangle *DEF*, the triangles are congruent.

■ **Work Practice 1**

In Example 1, notice that as soon as we know that the two triangles are congruent, we know that all three corresponding angles are congruent.

Objective B Finding the Ratio of Corresponding Sides in Similar Triangles ▶

Two triangles are **similar** when they have the same shape but not necessarily the same size. In similar triangles, the measures of corresponding angles are equal and corresponding sides are in proportion. The following triangles are similar:

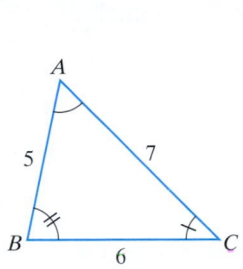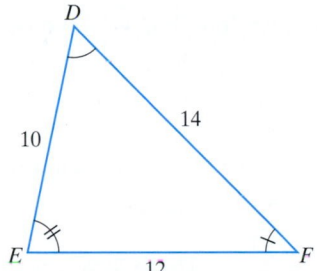

Since these triangles are similar, the measures of corresponding angles are equal.

Angles with equal measure: $\angle A$ and $\angle D$, $\angle B$ and $\angle E$, $\angle C$ and $\angle F$. Also, the lengths of corresponding sides are in proportion.

Sides in proportion: $\dfrac{AB}{DE} = \dfrac{BC}{EF} = \dfrac{CA}{FD}$ or, in this particular case,

$$\frac{AB}{DE} = \frac{5}{10} = \frac{1}{2}, \frac{BC}{EF} = \frac{6}{12} = \frac{1}{2}, \frac{CA}{FD} = \frac{7}{14} = \frac{1}{2}$$

The ratio of corresponding sides is $\dfrac{1}{2}$.

Example 2 Find the ratio of corresponding sides for the similar triangles *ABC* and *DEF*.

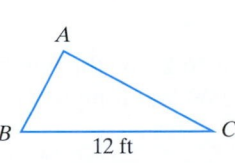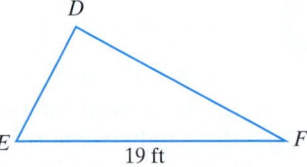

Solution: We are given the lengths of two corresponding sides. Their ratio is

$$\frac{12 \text{ feet}}{19 \text{ feet}} = \frac{12}{19}$$

■ **Work Practice 2**

Objective C Finding Unknown Lengths of Sides in Similar Triangles ▶

Because the ratios of lengths of corresponding sides are equal, we can use proportions to find unknown lengths in similar triangles.

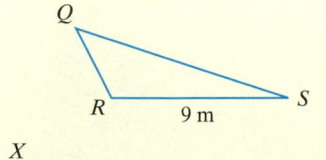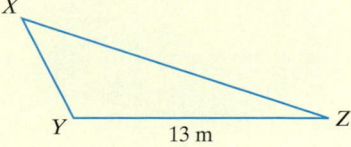

Practice 3

Given that the triangles are similar, find the missing length n.

a.

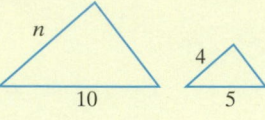

b.

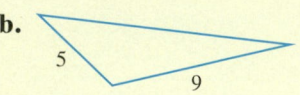

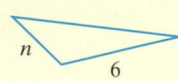

Example 3 Given that the triangles are similar, find the missing length n.

Solution: Since the triangles are similar, corresponding sides are in proportion. Thus, the ratio of 2 to 3 is the same as the ratio of 10 to n, or

$$\frac{2}{3} = \frac{10}{n}$$

To find the unknown length n, we set cross products equal.

$$\frac{2}{3} = \frac{10}{n}$$

$2 \cdot n = 3 \cdot 10$ Set cross products equal.

$2 \cdot n = 30$ Multiply.

$n = \dfrac{30}{2}$ Divide 30 by 2, the number multiplied by n.

$n = 15$

The missing length is 15 units.

■ **Work Practice 3**

✓**Concept Check** The following two triangles are similar. Which vertices of the first triangle appear to correspond to which vertices of the second triangle?

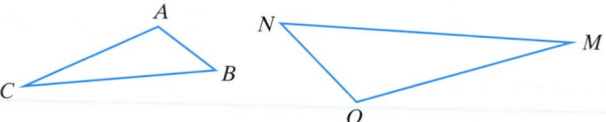

Many applications involve diagrams containing similar triangles. Surveyors, astronomers, and many other professionals continually use similar triangles in their work.

Practice 4

Tammy Shultz, a firefighter, needs to estimate the height of a burning building. She estimates the length of her shadow to be 8 feet long and the length of the building's shadow to be 60 feet long. Find the approximate height of the building if she is 5 feet tall.

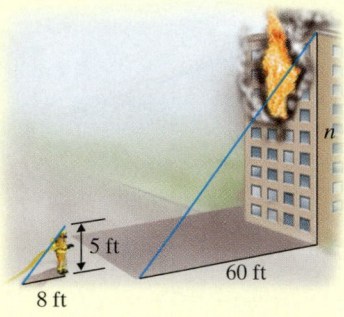

Example 4 Finding the Height of a Tree

Mel Rose is a 6-foot-tall park ranger who needs to know the height of a particular tree. He measures the shadow of the tree to be 69 feet long when his own shadow is 9 feet long. Find the height of the tree.

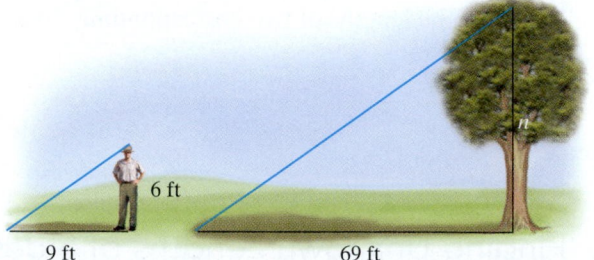

Solution:

1. UNDERSTAND. Read and reread the problem. Notice that the triangle formed by the Sun's rays, Mel, and his shadow is similar to the triangle formed by the Sun's rays, the tree, and its shadow.

Answers

3. a. $n = 8$ **b.** $n = \dfrac{10}{3}$ or $3\dfrac{1}{3}$

4. approximately 37.5 ft

✓**Concept Check Answer**

A corresponds to O; B corresponds to N; C corresponds to M

Copyright 2015 Pearson Education, Inc.

2. TRANSLATE. Write a proportion from the similar triangles formed.

$$\begin{array}{ll} \text{Mel's height} \rightarrow \\ \overline{\text{height of tree}} \rightarrow \end{array} \quad \frac{6}{n} = \frac{9}{69} \quad \begin{array}{l} \leftarrow \text{length of Mel's shadow} \\ \leftarrow \text{length of tree's shadow} \end{array}$$

$$\text{or } \frac{6}{n} = \frac{3}{23} \quad \text{Simplify } \frac{9}{69}. \text{ (ratio in lowest terms)}$$

3. SOLVE for n:

$$\frac{6}{n} = \frac{3}{23}$$

$$6 \cdot 23 = n \cdot 3 \quad \text{Set cross products equal.}$$

$$138 = n \cdot 3 \quad \text{Multiply.}$$

$$\frac{138}{3} = n \quad \text{Divide 138 by 3, the number multiplied by } n.$$

$$46 = n$$

4. INTERPRET. *Check* to see that replacing n with 46 in the proportion makes the proportion true. *State* your conclusion: The height of the tree is 46 feet.

■ **Work Practice 4**

Vocabulary, Readiness & Video Check

Answer each question true or false.

1. Two triangles that have the same shape but not necessarily the same size are congruent.
2. Two triangles are congruent if they have the same shape and size.
3. Congruent triangles are also similar.
4. Similar triangles are also congruent.
5. For the two similar triangles, the ratio of corresponding sides is $\frac{5}{6}$.

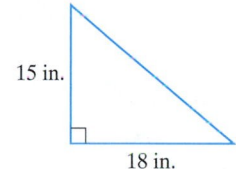
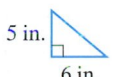

Each pair of triangles is similar. Name the congruent angles and the corresponding sides that are proportional.

6.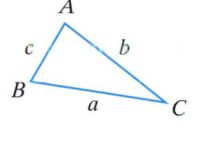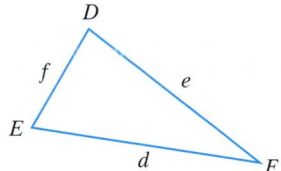

7.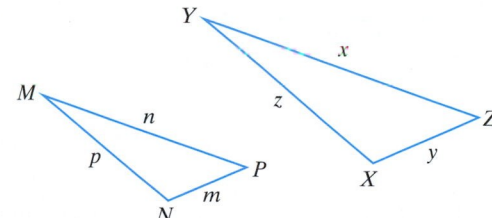

Watch the section lecture video and answer the following questions.

Objective A 8. How did we decide which congruency rule to use to determine if the two triangles in ▣ Example 1 are congruent? ▶

Objective B 9. From ▣ Example 2, what does "corresponding sides are in proportion" mean? ▶

Objective C 10. In ▣ Example 3, what is another proportion named that we could have used to solve the application? ▶

6.7 Exercise Set MyMathLab®

Objective A *Determine whether each pair of triangles is congruent. If congruent, state the reason why, such as SSS, SAS, or ASA. See Example 1.*

1.

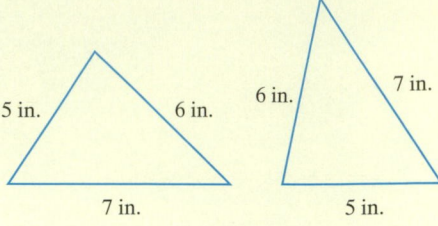

5 in.   6 in.   6 in.   7 in.
7 in.   5 in.

2.

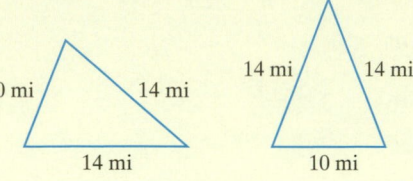

10 mi   14 mi   14 mi   14 mi
14 mi   10 mi

3.

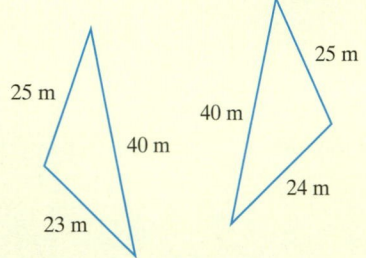

25 m   25 m
40 m   40 m
23 m   24 m

4.

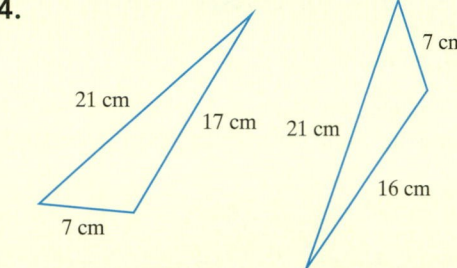

21 cm   7 cm
17 cm   21 cm
7 cm   16 cm

5.

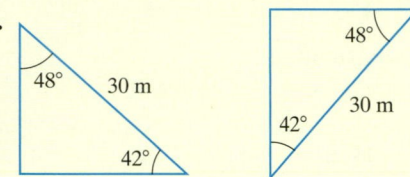

48°   30 m   48°
48°   30 m
42°   42°

6.

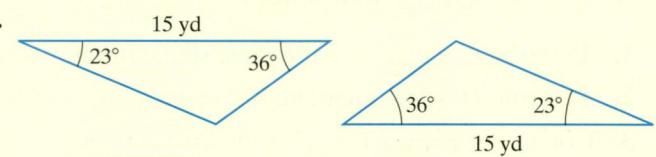

15 yd
23°   36°   36°   23°
15 yd

7.

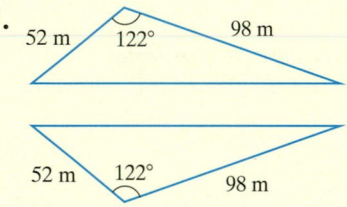

52 m   122°   98 m
52 m   122°   98 m

8.

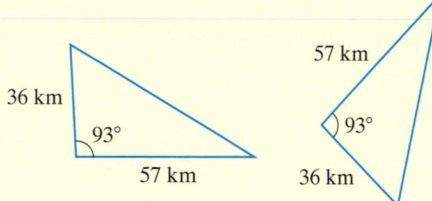

36 km   57 km
93°   93°
57 km   36 km

Objective B *Find each ratio of the corresponding sides of the given similar triangles. See Example 2.*

9.

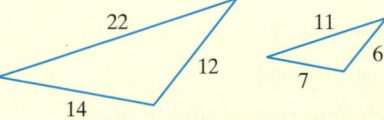

22   11
12   6
14   7

10.

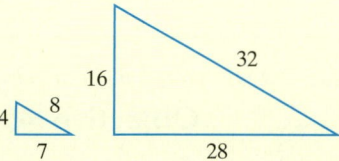

16   32
4   8
7   28

11.

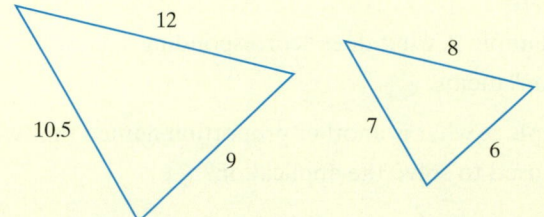

12   8
10.5   7
9   6

12.

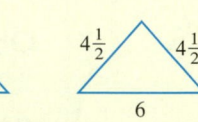

6   6 $4\frac{1}{2}$ $4\frac{1}{2}$
8   6

Objective C *Given that the pairs of triangles are similar, find the unknown length of the side labeled n.*
See Example 3.

13. 3, 6, n, 9

14. 5, 3, 60, n

▶ **15.** 12, 18, 4, n

16. 4, 7, n, 14

17. n, 12, 3.75, 9

18. 9, 15, n, 22.5

19. 40, 30, n, 18

20. 14, 8, n, 9

21. n, 17.5, 3.25, 3.25

22. 33.2, n, 8.3, 9.6

23. $18\frac{1}{3}$, $3\frac{2}{3}$, n, 2

24. n, $13\frac{1}{2}$, 6, 9

25. 32, 15, n, 60

26. 26, 13, 13, n

27. n, 100°, 15, 7, 100°, $10\frac{1}{2}$

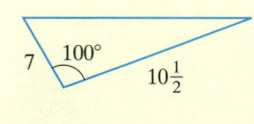

28. 20, 8, 82°, 82°, n, 37.5

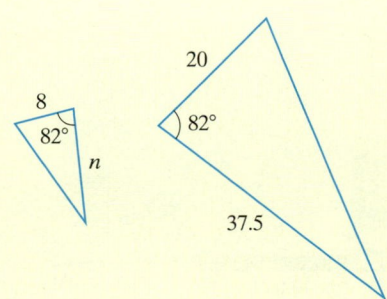

Solve. For Exercises 29 and 30, the solutions have been started for you. See Example 4.

29. Given the following diagram, approximate the height of the observation deck in the Seattle Space Needle in Seattle, Washington. (*Source:* Seattle Space Needle)

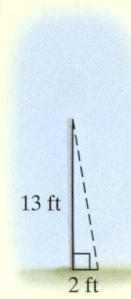

Start the solution:

1. UNDERSTAND the problem. Reread it as many times as needed.
2. TRANSLATE into a proportion using the similar triangles formed. (Fill in the blanks.)

height of
observation deck → $\dfrac{n}{13} = \dfrac{\quad}{\quad}$ ← Needle shadow
height of pole →   ← length of pole shadow

3. SOLVE by setting cross products equal.
4. INTERPRET.

30. A fountain in Fountain Hills, Arizona, sits in a 28-acre lake and shoots up a column of water every hour. Based on the diagram below, what is the height of the fountain?

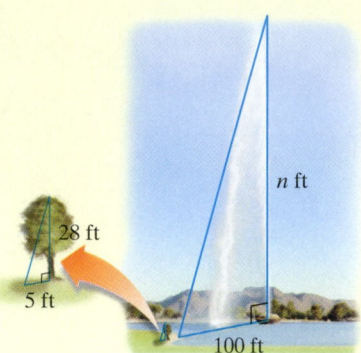

Start the solution:

1. UNDERSTAND the problem. Reread it as many times as needed.
2. TRANSLATE into a proportion using the similar triangles formed. (Fill in the blanks.)

height of tree → $\dfrac{28}{n} = \dfrac{\quad}{\quad}$ ← length of tree shadow
height of fountain →   ← length of fountain shadow

3. SOLVE by setting cross products equal.
4. INTERPRET.

31. Given the following diagram, approximate the height of the Chase Tower in Oklahoma City, Oklahoma. Here, we use *x* to represent the unknown number. (*Source:* Council on Tall Buildings and Urban Habitat)

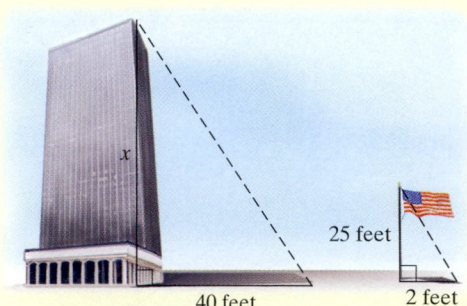

32. The tallest tree standing today is a redwood located in the Humboldt Redwoods State Park near Ukiah, California. Given the following diagram, approximate its height. Here, we use *x* to represent the unknown number. (*Source:* Guinness World Records)

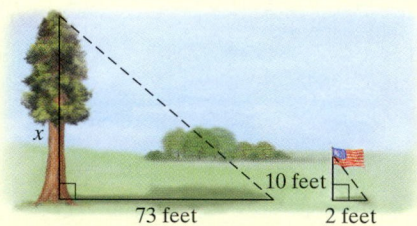

33. Samantha Black, a 5-foot-tall park ranger, needs to know the height of a tree. She notices that when the shadow of the tree is 48 feet long, her shadow is 4 feet long. Find the height of the tree.

34. Lloyd White, a firefighter, needs to estimate the height of a burning building. He estimates the length of his shadow to be 9 feet long and the length of the building's shadow to be 75 feet long. Find the approximate height of the building if he is 6 feet tall.

35. If a 30-foot tree casts an 18-foot shadow, find the length of the shadow cast by a 24-foot tree.

36. If a 24-foot flagpole casts a 32-foot shadow, find the length of the shadow cast by a 44-foot antenna. Round to the nearest tenth.

Review

Solve. See Section 5.1.

37. For the health of his fish, the owner of Pete's Sea World uses the standard that a 20-gallon tank should house only 19 neon tetras. Find the number of neon tetras that Pete would place into a 55-gallon tank.

38. A local package express deliveryman is traveling the city expressway at 45 mph when he is forced to slow down due to traffic ahead. His truck slows at the rate of 3 mph every 5 seconds. Find his speed 8 seconds after braking.

Solve. See Section 6.6.

39. Launch Umbilical Tower 1 is the name of the gantry used for the *Apollo* launch that took Neil Armstrong and Buzz Aldrin to the moon. Find the height of the gantry to the nearest whole foot.

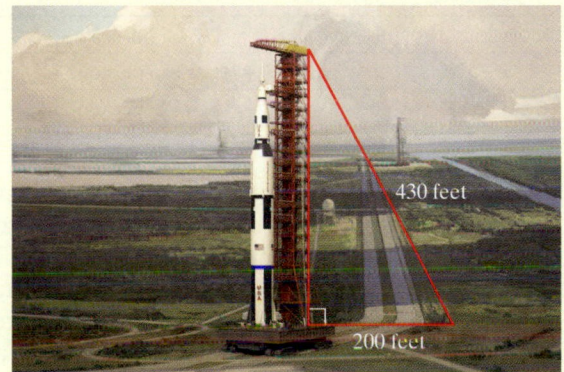

430 feet

200 feet

40. Arena polo, popular in the United States and England, is played on a field that is 100 yards long and usually 50 yards wide. Find the length, to the nearest yard, of the diagonal of this field.

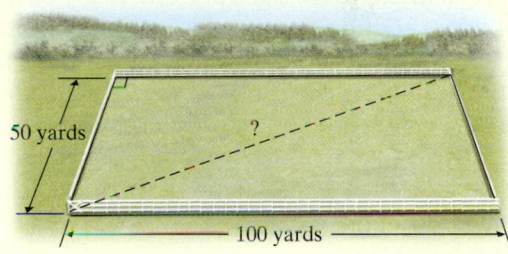

50 yards

?

100 yards

Perform the indicated operation. See Sections 4.3 through 4.5.

41. $3.6 + 0.41$

42. $3.6 - 0.41$

43. $(0.41)(3)$

44. $0.48 \div 3$

Concept Extensions

45. The print area on a particular page measures 7 inches by 9 inches. A printing shop is to copy the page and reduce the print area so that its length is 5 inches. What will its width be? Will the print now fit on a 3-by-5-inch index card?

46. The art sample for a banner measures $\frac{1}{3}$ foot in width by $1\frac{1}{2}$ feet in length. If the completed banner is to have a length of 9 feet, find its width.

Given that the pairs of triangles are similar, find the length of the side labeled n. Round your results to 1 decimal place.

47.

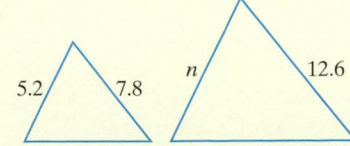

5.2   7.8   n   12.6

48.

11.6   n   20.8   58.7

49. In your own words, describe any differences in similar triangles and congruent triangles.

50. Describe a situation where similar triangles would be useful for a contractor building a house.

51. A triangular park is planned and waiting to be approved by the city zoning commission. A drawing of the park shows sides of length 5 inches, $7\frac{1}{2}$ inches, and $10\frac{5}{8}$ inches. If the scale on the drawing is $\frac{1}{4}$ in. = 10 ft, find the actual proposed dimensions of the park.

52. John and Robyn Costello draw a triangular deck on their house plans. Robyn measures sides of the deck drawing on the plans to be 3 inches, $4\frac{1}{2}$ inches, and 6 inches. If the scale on the drawing is $\frac{1}{4}$ in. = 1 foot, find the lengths of the sides of the deck they want built.

Chapter 6 Group Activity

The Cost of Road Signs

Sections 6.1, 6.2, 6.4

There are nearly 4 million miles of streets and roads in the United States. With streets, roads, and highways comes the need for traffic control, guidance, warning, and regulation. Road signs perform many of these tasks. Just in our routine travels, we see a wide variety of road signs every day. Think how many road signs must exist on the 4 million miles of roads in the United States. Have you ever wondered how much signs like these cost?

The cost of a road sign generally depends on the type of sign. Costs for several types of signs and signposts are listed in the table. Examples of various types of signs are shown below.

Road Sign Costs	
Type of Sign	**Cost**
Regulatory, warning, marker	$15–$18 per square foot
Large guide	$20–$25 per square foot
Type of Post	**Cost**
U-channel	$125–$200 each
Square tube	$10–$15 per foot
Steel breakaway posts	$15–$25 per foot

The cost of a sign is based on its area. For diamond, square, or rectangular signs, the area is found by multiplying the length (in feet) times the width (in feet). Then the area is multiplied by the cost per square foot. For signs with irregular shapes, costs are generally figured *as if* the sign were a rectangle, multiplying the height and width at the tallest and widest parts of the sign.

Regulatory   Warning   Marker   Large Guide   Posts

Group Activity

Locate four different kinds of road signs on or near your campus. Measure the dimensions of each sign, including the height of the post on which it is mounted. Using the cost data given in the table, find the minimum and maximum costs of each sign, including its post. Summarize your results in a table, and include a sketch of each sign.

Chapter 6 Vocabulary Check

Fill in each blank with one of the words or phrases listed below.

transversal	line	congruent	hypotenuse	legs	acute
right	line segment	complementary	square root	vertical	supplementary
right triangle	volume	obtuse	vertex	ray	angle
similar	perimeter	area	straight	adjacent	

1. A(n) _____ is a triangle with a right angle. The side opposite the right angle is called the _____, and the other two sides are called _____.

2. A(n) _____ is a piece of a line with two endpoints.

3. Two angles that have a sum of 90° are called _____ angles.

4. A(n) _____ is a set of points extending indefinitely in two directions.

5. The _____ of a polygon is the distance around the polygon.

6. A(n) _____ is made up of two rays that share the same endpoint. The common endpoint is called the _____.

7. _____ triangles have the same shape and the same size.

8. _____ measures the amount of surface of a region.

9. A(n) _____ is a part of a line with one endpoint. A ray extends indefinitely in one direction.

10. A(n) _____ of a number a is a number b whose square is a.

11. A line that intersects two or more lines at different points is called a(n) _____.

12. A angle that measures 180° is called a(n) _____ angle.

13. The measure of the space of a solid is called its _____.

14. When two lines intersect, four angles are formed. The angles that are opposite each other are called _____ angles.

15. When two of the four angles from Exercise 14 share a common side, they are called _____ angles.

16. An angle whose measure is between 90° and 180° is called a(n) _____ angle.

17. An angle that measures 90° is called a(n) _____ angle.

18. An angle whose measure is between 0° and 90° is called a(n) _____ angle.

19. Two angles that have a sum of 180° are called _____ angles.

20. _____ triangles have exactly the same shape but not necessarily the same size.

> **Helpful Hint**
>
> ▶ Are you preparing for your test? Don't forget to take the Chapter 6 Test on page 482. Then check your answers at the back of the text and use the Chapter Test Prep Videos to see the fully worked-out solutions to any of the exercises you want to review.

6 Chapter Highlights

Definitions and Concepts	Examples
Section 6.1 Lines and Angles	
A **line** is a set of points extending indefinitely in two directions. A line has no width or height, but it does have length. We name a line by any two of its points.	Line AB or $\overleftrightarrow{AB}$

(continued)

Definitions and Concepts	Examples

Section 6.1   Lines and Angles (*continued*)

A **line segment** is a piece of a line with two endpoints.

Line segment AB or $\overline{AB}$

A •————————• B

A **ray** is a part of a line with one endpoint. A ray extends indefinitely in one direction.

Ray AB or $\overrightarrow{AB}$

An **angle** is made up of two rays that share the same endpoint. The common endpoint is called the **vertex.**

Angle ABC, $\angle ABC$, $\angle CBA$, or $\angle B$

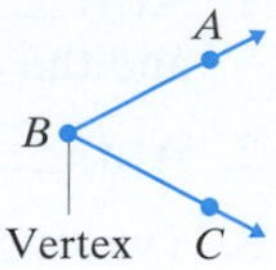

Vertex

An angle that measures 180° is called a **straight angle.**

$\angle RST$ is a straight angle.

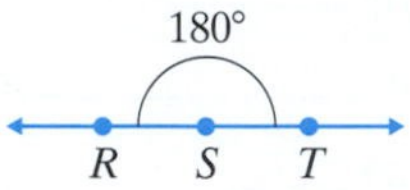

180°

R   S   T

An angle that measures 90° is called a **right angle.** The symbol ∟ is used to denote a right angle.

$\angle ABC$ is a right angle.

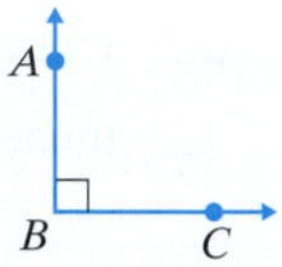

An angle whose measure is between 0° and 90° is called an **acute angle.**

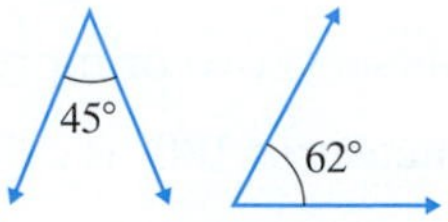

45°   62°

Acute angles

An angle whose measure is between 90° and 180° is called an **obtuse angle.**

120°   95°

Obtuse angles

Two angles that have a sum of 90° are called **complementary angles.** We say that each angle is the **complement** of the other.

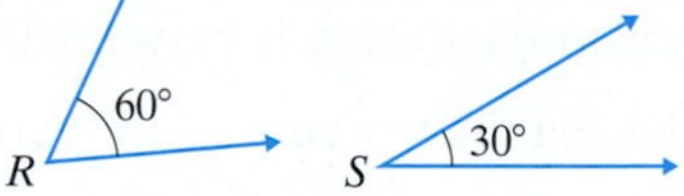

60°   30°

R   S

Complementary angles
$60° + 30° = 90°$

Two angles that have a sum of 180° are called **supplementary angles.** We say that each angle is the **supplement** of the other.

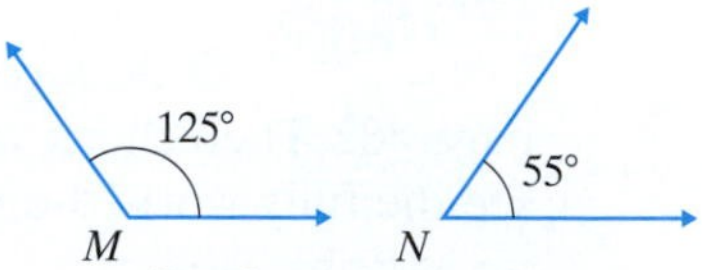

125°   55°

M   N

Supplementary angles
$125° + 55° = 180°$

Definitions and Concepts	Examples

Section 6.1   Lines and Angles (*continued*)

When two lines intersect, four angles are formed. Two of these angles that are opposite each other are called **vertical angles.** Vertical angles have the same measure.

Two of these angles that share a common side are called **adjacent angles.** Adjacent angles formed by intersecting lines are supplementary.

Vertical angles:
$\angle a$ and $\angle c$
$\angle d$ and $\angle b$

Adjacent angles:
$\angle a$ and $\angle b$
$\angle b$ and $\angle c$
$\angle c$ and $\angle d$
$\angle d$ and $\angle a$

A line that intersects two or more lines at different points is called a **transversal.** Line *l* is a transversal that intersects lines *m* and *n*. The eight angles formed have special names. Some of these names are:

Corresponding angles: $\angle a$ and $\angle e$, $\angle c$ and $\angle g$, $\angle b$ and $\angle f$, $\angle d$ and $\angle h$

Alternate interior angles: $\angle c$ and $\angle f$, $\angle d$ and $\angle e$

Parallel Lines Cut by a Transversal

If two parallel lines are cut by a transversal, then the measures of **corresponding angles are equal** and the measures of **alternate interior angles are equal.**

Section 6.2   Plane Figures and Solids

The **sum of the measures** of the angles of a triangle is 180°.

Find the measure of $\angle x$.

The measure of $\angle x = 180° - 85° - 45° = 50°$

A **right triangle** is a triangle with a right angle. The side opposite the right angle is called the **hypotenuse,** and the other two sides are called **legs.**

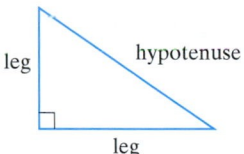

For a circle or a sphere:

diameter $= 2 \cdot$ radius
$d = 2 \cdot r$

radius $= \dfrac{\text{diameter}}{2}$

$r = \dfrac{d}{2}$

Find the diameter of the circle.

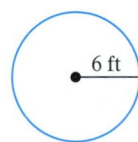

$d = 2 \cdot r$
$\quad = 2 \cdot 6 \text{ feet} = 12 \text{ feet}$

Definitions and Concepts	Examples

Section 6.3   Perimeter

Perimeter Formulas

Rectangle: $P = 2 \cdot l + 2 \cdot w$

Square: $P = 4 \cdot s$

Triangle: $P = a + b + c$

Circumference of a Circle: $C = 2 \cdot \pi \cdot r$ or $C = \pi \cdot d$,

where $\pi \approx 3.14$ or $\pi \approx \dfrac{22}{7}$

Find the perimeter of a rectangle with length 28 meters and width 15 meters.

$$P = 2 \cdot l + 2 \cdot w$$
$$= 2 \cdot 28 \text{ m} + 2 \cdot 15 \text{ m}$$
$$= 56 \text{ m} + 30 \text{ m}$$
$$= 86 \text{ m}$$

The perimeter is 86 meters.

Section 6.4   Area

Area Formulas

Rectangle: $A = l \cdot w$

Square: $A = s^2$

Triangle: $A = \dfrac{1}{2} \cdot b \cdot h$

Parallelogram: $A = b \cdot h$

Trapezoid: $A = \dfrac{1}{2} \cdot (b + B) \cdot h$

Circle: $A = \pi \cdot r^2$

Find the area of a square with side length 8 centimeters.

$$A = s^2$$
$$= (8 \text{ cm})^2$$
$$= 64 \text{ square centimeters}$$

The area of the square is 64 square centimeters.

Section 6.5   Volume

Volume Formulas

Rectangular Solid: $V = l \cdot w \cdot h$

Cube: $V = s^3$

Sphere: $V = \dfrac{4}{3} \cdot \pi \cdot r^3$

Right Circular Cylinder: $V = \pi \cdot r^2 \cdot h$

Cone: $V = \dfrac{1}{3} \cdot \pi \cdot r^2 \cdot h$

Square-Based Pyramid: $V = \dfrac{1}{3} \cdot s^2 \cdot h$

Find the volume of the sphere. Use $\dfrac{22}{7}$ for π.

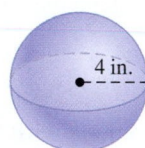
4 in.

$$V = \dfrac{4}{3} \cdot \pi \cdot r^3$$

$$\approx \dfrac{4}{3} \cdot \dfrac{22}{7} \cdot (4 \text{ inches})^3$$

$$= \dfrac{4 \cdot 22 \cdot 64}{3 \cdot 7} \text{ cubic inches}$$

$$= \dfrac{5632}{21} \quad \text{or} \quad 268\dfrac{4}{21} \text{ cubic inches}$$

Definitions and Concepts	Examples

Section 6.6   Square Roots and the Pythagorean Theorem

Square Root of a Number

A **square root** of a number a is a number b whose square is a. We use the radical sign $\sqrt{}$ to name square roots.

$\sqrt{9} = 3, \sqrt{100} = 10, \sqrt{1} = 1$

Pythagorean Theorem

$$(\text{leg})^2 + (\text{other leg})^2 = (\text{hypotenuse})^2$$

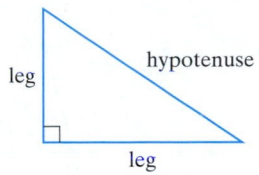

To Find an Unknown Length of a Right Triangle

$$\text{hypotenuse} = \sqrt{(\text{leg})^2 + (\text{other leg})^2}$$

$$\text{leg} = \sqrt{(\text{hypotenuse})^2 - (\text{other leg})^2}$$

Find the hypotenuse of the given triangle.

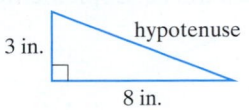

$$\begin{aligned}
\text{hypotenuse} &= \sqrt{(\text{leg})^2 + (\text{other leg})^2} \\
&= \sqrt{(3)^2 + (8)^2} \quad \text{\small The legs are 3 and 8 inches.} \\
&= \sqrt{9 + 64} \\
&= \sqrt{73} \text{ inches} \\
&\approx 8.5 \text{ inches}
\end{aligned}$$

Section 6.7   Congruent and Similar Triangles

Congruent triangles have the same shape and the same size. Corresponding angles are equal, and corresponding sides are equal.

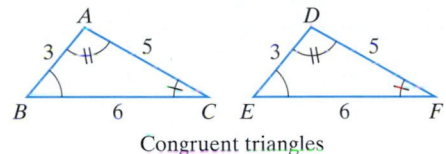

Congruent triangles

Similar triangles have exactly the same shape but not necessarily the same size. Corresponding angles are equal, and the ratios of the lengths of corresponding sides are equal.

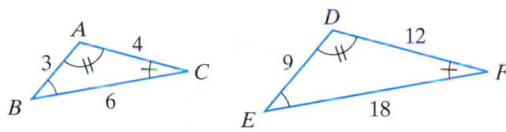

Similar triangles

$$\frac{AB}{DE} = \frac{3}{9} = \frac{1}{3}, \frac{BC}{EF} = \frac{6}{18} = \frac{1}{3},$$

$$\frac{CA}{FD} = \frac{4}{12} = \frac{1}{3}$$

Chapter 6   Review

(6.1) *Classify each angle as acute, right, obtuse, or straight.*

1.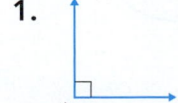

2.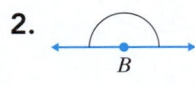

3.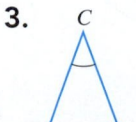

4.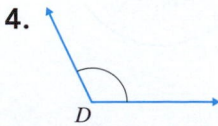

5. Find the complement of a 25° angle.

6. Find the supplement of a 105° angle.

Find the measure of angle x in each figure.

7.
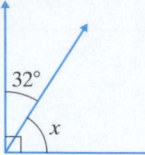
32°
x

8.
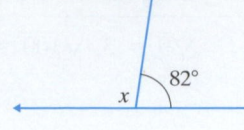
x   82°

9.
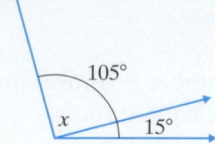
105°
x   15°

10.
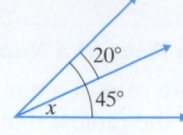
20°
x   45°

11. Identify the pairs of supplementary angles.

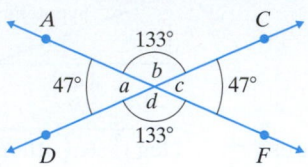

A   133°   C
47°   a b c   47°
d
133°
D   F

12. Identify the pairs of complementary angles.

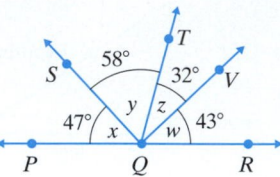

T
58°   32°
S   V
y z
47° x w 43°
P   Q   R

Find the measures of angles x, y, and z in each figure.

13.
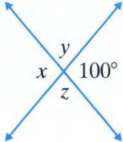
y
x   100°
z

14.
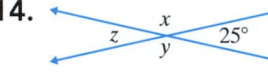
x
z   y   25°

15. Given that $m \parallel n$.

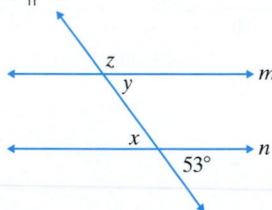

z
y   m
x   n
53°

16. Given that $m \parallel n$.

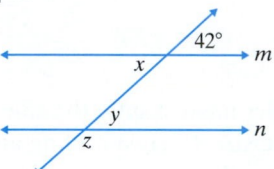

42°
x   m
y   n
z

(6.2) *Find the measure of $\angle x$ in each figure.*

17.
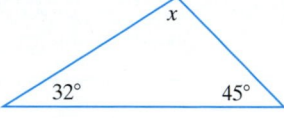
x
32°   45°

18.
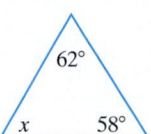
62°
x   58°

19.
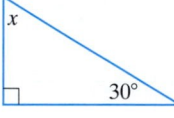
x
30°

20.
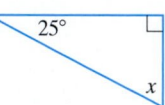
25°
x

Find the unknown diameter or radius as indicated.

21.
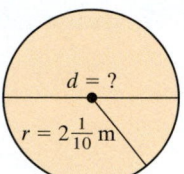
d = ?
$r = 2\frac{1}{10}$ m

22.
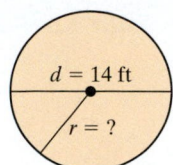
d = 14 ft
r = ?

23.
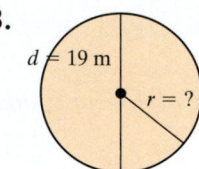
d = 19 m
r = ?

24.
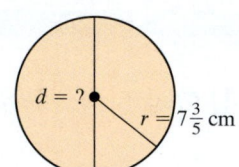
d = ?
$r = 7\frac{3}{5}$ cm

Identify each solid.

25.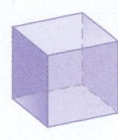

26.

27.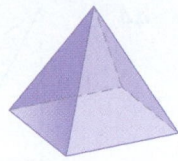

28.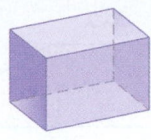

Find the unknown radius or diameter as indicated.

29. The radius of a sphere is 9 inches. Find its diameter.

30. The diameter of a sphere is 4.7 meters. Find its radius.

Identify each regular polygon.

31.

32.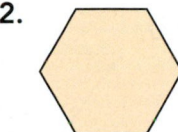

Identify each triangle as equilateral, isosceles, or scalene. Also identify any triangle that is a right triangle.

33.

34.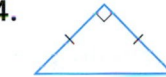

(6.3) *Find the perimeter of each figure.*

35.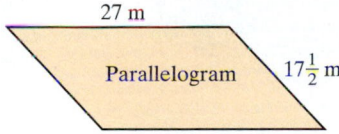

36.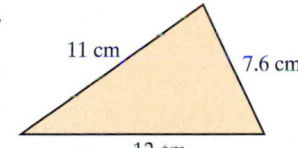

37.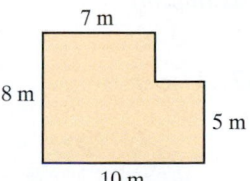

38.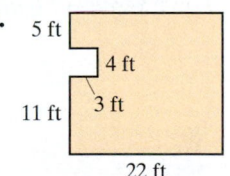

Solve.

39. Find the perimeter of a rectangular sign that measures 6 feet by 10 feet.

40. Find the perimeter of a town square that measures 110 feet on a side.

Find the circumference of each circle. Use $\pi \approx 3.14$.

41.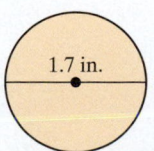

42.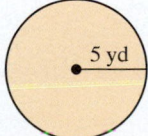

(6.4) *Find the area of each figure. For the circles, find the exact area and then use* $\pi \approx 3.14$ *to approximate the area.*

43.
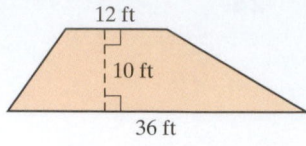
12 ft
10 ft
36 ft

44.
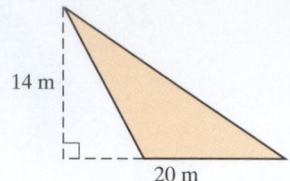
14 m
20 m

45.
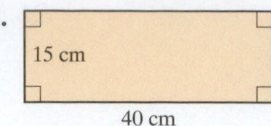
15 cm
40 cm

46.
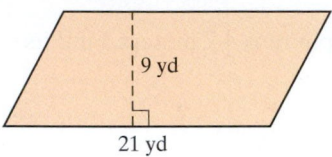
9 yd
21 yd

47.
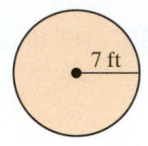
7 ft

48.
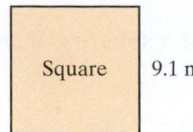
Square   9.1 m

49.
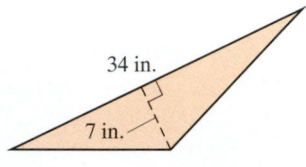
34 in.
7 in.

50.
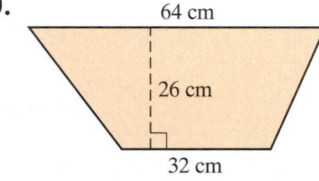
64 cm
26 cm
32 cm

51.
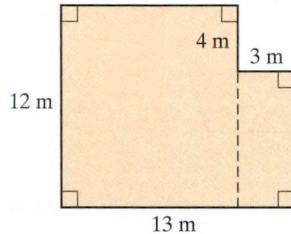
4 m
3 m
12 m
13 m

52. The amount of sealant necessary to seal a driveway depends on the area. Find the area of a rectangular driveway 36 feet by 12 feet.

53. Find how much carpet is necessary to cover the floor of the room shown.

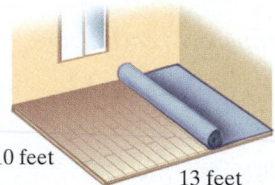
10 feet
13 feet

(6.5) *Find the volume of each solid. For Exercises 56 and 57, give an exact volume and an approximation.*

54.
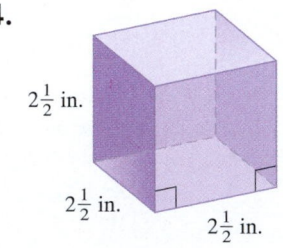
$2\frac{1}{2}$ in.
$2\frac{1}{2}$ in.
$2\frac{1}{2}$ in.

55.
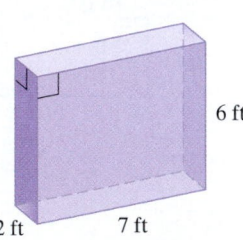
6 ft
2 ft   7 ft

56. Use $\pi \approx 3.14$.
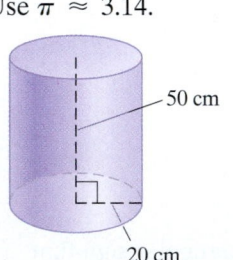
50 cm
20 cm

57. Use $\pi \approx \frac{22}{7}$.
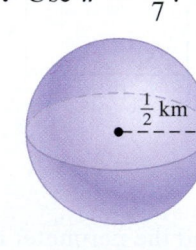
$\frac{1}{2}$ km

58. Find the volume of a pyramid with a square base 2 feet on a side and a height of 2 feet.

59. Approximate the volume of a tin can 8 inches high and 3.5 inches in radius. Use 3.14 for π.

60. A chest has 3 drawers. If each drawer has inside measurements of $2\frac{1}{2}$ feet by $1\frac{1}{2}$ feet by $\frac{2}{3}$ foot, find the total volume of the 3 drawers.

61. A cylindrical canister for a shop vacuum is 2 feet tall and 1 foot in *diameter*. Find its exact volume.

(6.6) *Simplify.*

62. $\sqrt{64}$

63. $\sqrt{144}$

64. $\sqrt{\dfrac{4}{25}}$

65. $\sqrt{\dfrac{1}{100}}$

Find the unknown length of each given right triangle. If necessary, round to the nearest tenth.

66. leg $= 12$, leg $= 5$

67. leg $= 20$, leg $= 21$

68. leg $= 9$, hypotenuse $= 14$

69. leg $= 124$, hypotenuse $= 155$

70. A baseball diamond is in the shape of a square and has sides of length 90 feet. Find the distance across the diamond from third base to first base, to the nearest tenth of a foot.

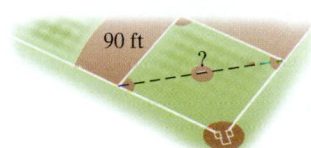

71. Find the height of the building rounded to the nearest tenth of a foot.

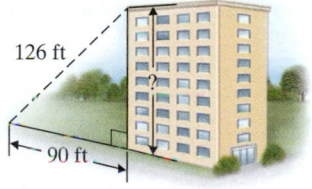

126 ft ? 90 ft

(6.7) *Given that the pairs of triangles are similar, find the unknown length n.*

72.

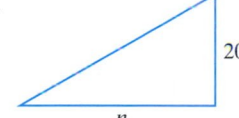

20   8
15
n

73.

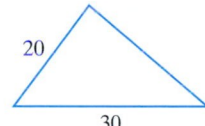

20   n
20
30

74.

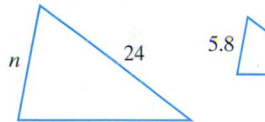

n   24   5.8   8
n

Solve.

75. A housepainter needs to estimate the height of a condominium. He estimates the length of his shadow to be 7 feet long and the length of the building's shadow to be 42 feet long. Find the approximate height of the building if the housepainter is $5\frac{1}{2}$ feet tall.

76. A toy company is making a triangular sail for a toy sailboat. The toy sail is to be the same shape as a real sailboat's sail. Use the following diagram to find the unknown lengths *x* and *y*.

26 ft   24 ft   10 ft

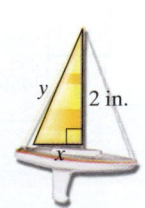

y   2 in.   x

Mixed Review

Find the following.

77. Find the supplement of a 72° angle.

78. Find the complement of a 1° angle.

Find the measure of angle x in each figure.

79.

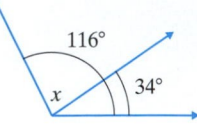

80.

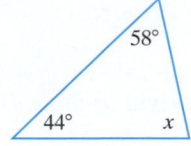

81.

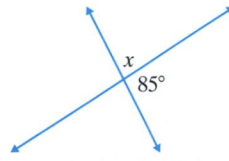

82.

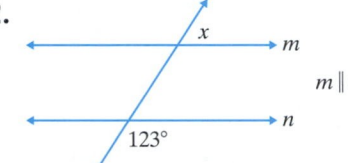

$m \parallel n$

Find the unknown diameter or radius as indicated.

83.

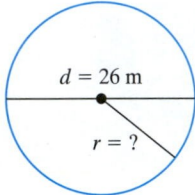

84.

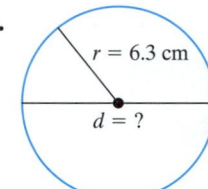

Find the perimeter of each figure.

85.

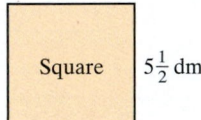

86.

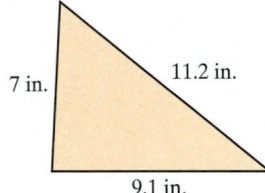

87.

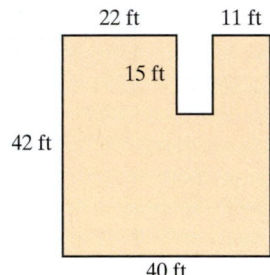

Find the area of each figure. For the circles, find the exact area and then use $\pi = 3.14$ to approximate the area.

88.

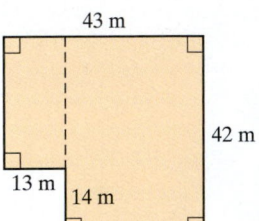

89.

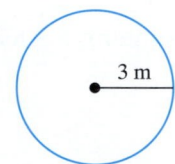

Find the volume of each solid.

90. Give an approximation using $\frac{22}{7}$ for π.

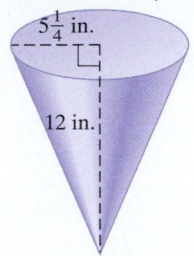

91.

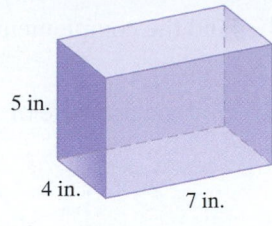

Solve.

92. Find the volume of air in a rectangular room 15 feet by 12 feet with a 7-foot ceiling.

93. A mover has two boxes left for packing. Both are cubical, one 3 feet on a side and the other 1.2 feet on a side. Find their combined volume.

Simplify.

94. $\sqrt{1}$

95. $\sqrt{36}$

96. $\sqrt{\dfrac{16}{81}}$

Find the unknown length of each given right triangle. If necessary, round to the nearest tenth.

97. leg $= 66$, leg $= 56$

98. leg $= 12$, hypotenuse $= 24$

99. leg $= 17$, hypotenuse $= 51$

100. leg $= 10$, leg $= 17$

Given that the pairs of triangles are similar, find the unknown length n.

101.

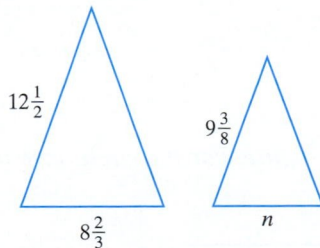

102.

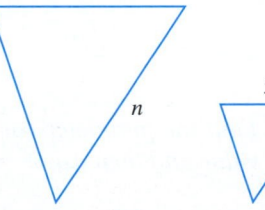

Answers

1. Find the complement of a 78° angle.

2. Find the supplement of a 124° angle.

3. Find the measure of ∠x.

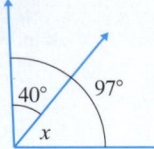

Find the measures of x, y, and z in each figure.

4.

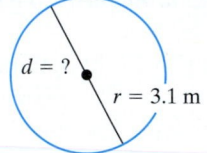

5. Given: m ‖ n.

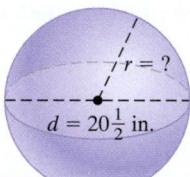

Find the unknown diameter or radius as indicated.

6.

d = ? r = 3.1 m

7.

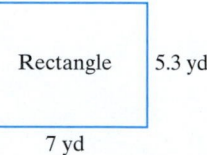

r = ? d = 20½ in.

8. Find the measure of ∠x.

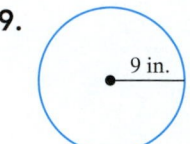

92° 62° x

Find the perimeter (or circumference) and area of each figure. For the circle, give the exact value and then use π ≈ 3.14 for an approximation.

9.

9 in.

10.

Rectangle 5.3 yd

7 yd

11.

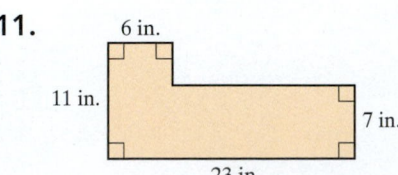

6 in. 11 in. 7 in. 23 in.

1. _____

2. _____

3. _____

4. _____

5. _____

6. _____

7. _____

8. _____

9. _____

10. _____

11. _____

482

Find the volume of each solid. For the cylinder, use $\pi \approx \dfrac{22}{7}$.

12.

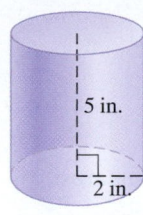

5 in.

2 in.

13.

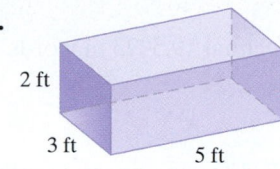

2 ft

3 ft

5 ft

Find each square root and simplify. Round the square root to the nearest thousandth if necessary.

14. $\sqrt{49}$

15. $\sqrt{79}$

16. $\sqrt{\dfrac{64}{100}}$

Solve.

17. Find the perimeter of a square photo with a side length of 4 inches.

18. How much soil is needed to fill a rectangular hole 3 feet by 3 feet by 2 feet?

19. Find how much baseboard is needed to go around a rectangular room that measures 18 feet by 13 feet. If baseboard costs $1.87 per foot, also calculate the total cost needed for materials.

20. Approximate to the nearest hundredth of a centimeter the length of the hypotenuse of a right triangle with legs of 4 centimeters each.

21. Vivian Thomas is going to put insecticide on her lawn to control grubworms. The lawn is a rectangle measuring 123.8 feet by 80 feet. The amount of insecticide required is 0.02 ounces per square foot. Find how much insecticide Vivian needs to purchase.

22. Given that the following triangles are similar, find the missing length n.

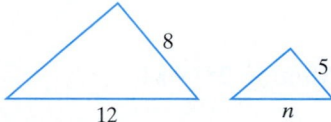

8

12

5

n

23. Tamara Watford, a surveyor, needs to estimate the height of a tower. She estimates the length of her shadow to be 4 feet long and the length of the tower's shadow to be 48 feet long. Find the approximate height of the tower if she is $5\dfrac{3}{4}$ feet tall.

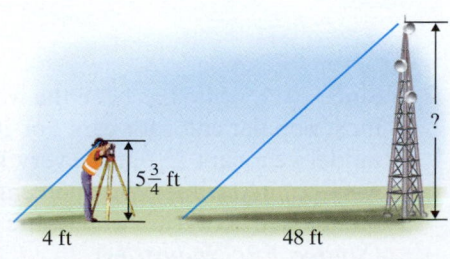

$5\dfrac{3}{4}$ ft

4 ft

48 ft

?

12. _____

13. _____

14. _____

15. _____

16. _____

17. _____

18. _____

19. _____

20. _____

21. _____

22. _____

23. _____

Answers

1. _____

2. _____

3. _____

4. _____

5. _____

6. _____

7. _____

8. _____

9. _____

10. _____

11. _____

12. _____

13. _____

14. _____

15. _____

16. _____

17. _____

18. _____

19. _____

20. _____

21. _____

22. _____

484

1. Write the decimal 19.5023 in words.

2. Add: $\dfrac{7}{11} + \dfrac{1}{6}$

3. Round 736.2359 to the nearest tenth.

4. Round 736.2359 to the nearest hundred.

5. Add: $45 + 2.06$

6. Divide: $3\dfrac{1}{3} \div 1\dfrac{5}{6}$

Multiply.

7. 7.68×10

8. $\dfrac{7}{11} \cdot \dfrac{1}{6}$

9. 76.3×1000

10. $5\dfrac{1}{2} \cdot 2\dfrac{1}{11}$

11. Divide: $270.2 \div 7$. Check your answer.

12. Divide: $\dfrac{56.7}{100}$

13. Simplify: $0.5(8.6 - 1.2)$

14. Simplify: $\dfrac{5 + 2(8 - 3)}{30 \div 6 \cdot 5}$

15. Insert $<$, $>$, or $=$ to form a true statement. $\dfrac{1}{8}$ ____ 0.12

16. Insert $<$, $>$, or $=$ to form a true statement. $\dfrac{3}{4}$ ____ $\dfrac{13}{16}$

17. Multipy: $25 \cdot 9000$

18. Find: $\dfrac{2}{9} + \dfrac{7}{15} - \dfrac{1}{3}$

19. Multipy: $20 \cdot 7000$

20. Solve for n: $\dfrac{7}{8} = \dfrac{n}{20}$

21. Since 2011, white has been the world's most popular color for cars. For 2013 model-year cars, 25 out of every 100 were painted white. What percent of model-year 2013 cars were white? (*Source: PPG Industries*)

22. In a survey of 50 people, 34 people prefer taking pictures with digital cameras. What percent is this?

Write each percent as a fraction or mixed number in simplest form.

23. 1.9%   **24.** 26%   **25.** 125%   **26.** 560%

27. 85% of 300 is what number?   **28.** What percent of 16 is 2.4?

29. 20.8 is 40% of what number?   **30.** Find: $\sqrt{\dfrac{25}{81}}$

31. Mr. Buccaran, the principal at Slidell High School, counted 31 freshmen absent during a particular day. If this is 4% of the total number of freshmen, how many freshmen are there at Slidell High School?

32. Flooring tiles cost $90 for a box with 40 tiles. Each tile is 1 square foot. Find the price in dollars per square foot.

33. Sherry Souter, a real estate broker for Wealth Investments, sold a house for $214,000 last week. If her commission is 1.5% of the selling price of the home, find the amount of her commission.

34. A student can complete 7 exercises in 6 minutes. At this rate, how many exercises can be completed in 30 minutes?

35. Simplify: $2 \cdot 4 - 3 \div 3$

36. Write seventy thousand, fifty-two in standard form.

37. Write $\dfrac{1}{12}$ as a percent. Round to the nearest hundredth percent.

38. Write $\dfrac{1}{8}$ as a percent.

39. Find the measure of $\angle a$.

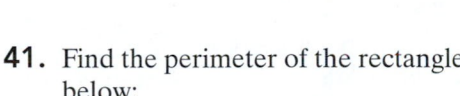

40. Find the perimeter of the triangle in Exercise 39.

41. Find the perimeter of the rectangle below:

42. Find the area of the rectangle in Exercise 41.

43. Find: $\sqrt{\dfrac{4}{25}}$

44. Find: $\sqrt{\dfrac{9}{16}}$

45. Find the ratio of corresponding sides for the similar triangles ABC and DEF.

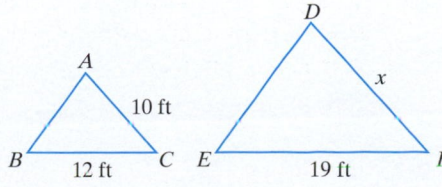

46. Use the figures in Exercise 45 and find the value of x.

23. _____

24. _____

25. _____

26. _____

27. _____

28. _____

29. _____

30. _____

31. _____

32. _____

33. _____

34. _____

35. _____

36. _____

37. _____

38. _____

39. _____

40. _____

41. _____

42. _____

43. _____

44. _____

45. _____

46. _____

Statistics and Probability

Can We Experience Weightlessness?

Today, space travel, while still not commonplace, is well within everyone's comprehension. From landing on the Moon, to probes to Mars and beyond, the Hubble Space Telescope, and an International Space Station, there are many exciting explorations in space. There are even space tourists who pay big money to travel in space.

If space travel is not for you but you'd like to experience weightlessness, there are commercial companies that offer rides on Boeing 727s with flight patterns similar to that shown below to simulate zero gravity.

In Section 7.1, Example 2, we explore lunar and planetary exploration since 1957.

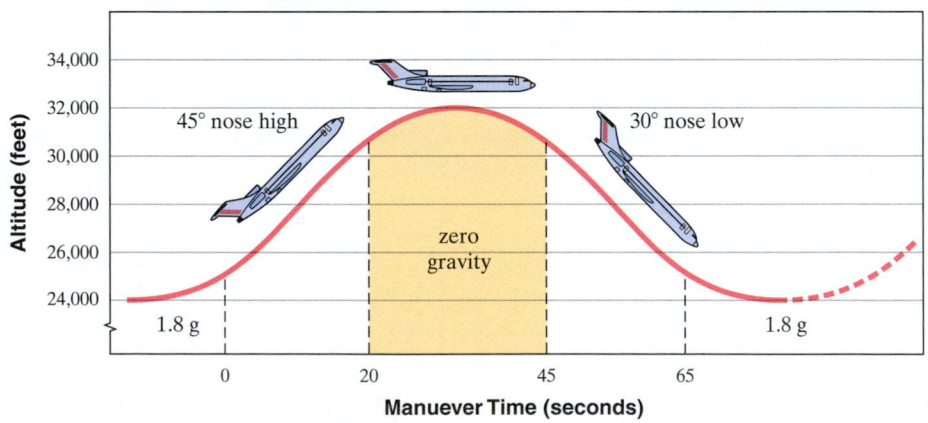

Flight Pattern to Simulate Weightlessness

Check Your Progress

We often need to make decisions based on known statistics or the probability of an event occurring. For example, we decide whether or not to bring an umbrella to work based on the probability of rain. We choose an investment based on its mean, or average, return. We can predict which football team will win based on the trend in its previous wins and losses. This chapter reviews presenting data in a usable form on a graph and the basic ideas of statistics and probability.

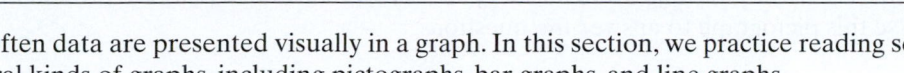
Often data are presented visually in a graph. In this section, we practice reading several kinds of graphs, including pictographs, bar graphs, and line graphs.

Objective A Reading Pictographs

A **pictograph** such as the one below is a graph in which pictures or symbols are used. This type of graph contains a key that explains the meaning of the symbol used. An advantage of using a pictograph to display information is that comparisons can easily be made. A disadvantage of using a pictograph is that it is often hard to tell what fractional part of a symbol is shown. For example, in the pictograph below, Arabic shows a part of a symbol, but it's hard to read with any accuracy what fractional part of a symbol is shown.

Objectives

A Read Pictographs.

B Read and Construct Bar Graphs.

C Read and Construct Histograms.

D Read Line Graphs.

Example 1 Calculating Languages Spoken

The following pictograph shows the top eight most-spoken (primary) languages. Use this pictograph to answer the questions.

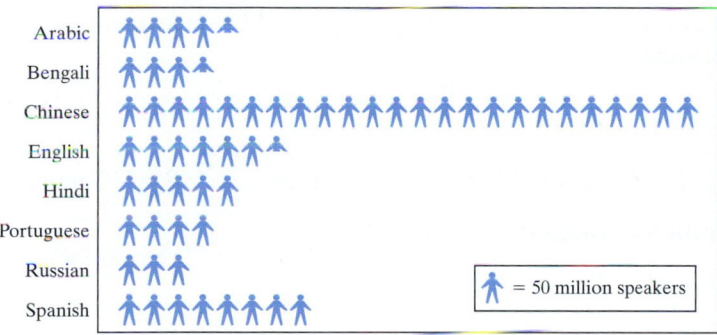

Top 8 Most-Spoken (Primary) Languages

= 50 million speakers

Source: www.ethnologue.com

a. Approximate the number of people who primarily speak Hindi.

b. Approximate how many more people primarily speak English than Hindi.

Solution:

a. Hindi corresponds to 5 symbols, and each symbol represents 50 million speakers. This means that the number of people who primarily speak Hindi is approximately 5 · (50 million) or 250 million people. This can also be written as 250,000,000 people.

b. English shows $1\frac{1}{2}$ more symbols than Hindi. This means that $1\frac{1}{2} \cdot (50 \text{ million})$ or 75 million or 75,000,000 more people primarily speak English than Hindi.

■ Work Practice 1

Practice 1

Use the pictograph shown in Example 1 to answer the following questions:

a. Approximate the number of people who primarily speak Spanish.

b. Approximate how many more people primarily speak Spanish than Arabic.

Answers

1. a. 400 million or 400,000,000 people

b. 175 million or 175,000,000 people

487

Practice 2

Use the pictograph shown in Example 2 to answer the following questions:

a. Approximate the number of solar system exploration missions undertaken by the European Space Agency.

b. Approximate the total number of solar system exploration missions undertaken by the European Space Agency and Japan.

Example 2   Calculating Solar System Exploration

The following pictograph shows the approximate number of solar system exploration missions by various countries or space consortia from 1957 to the present day. Use this pictograph to answer the questions.

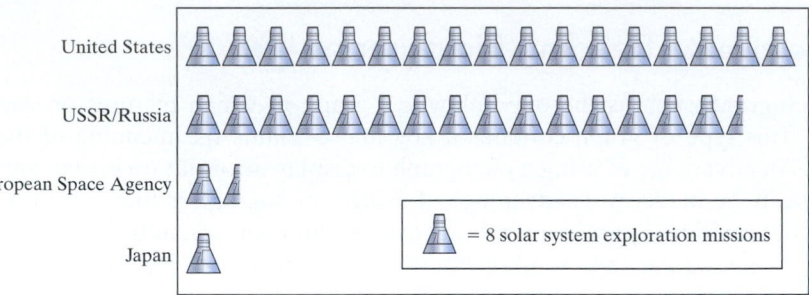

Solar System Exploration Missions

a. Approximate the number of solar system exploration missions undertaken by the United States.

b. Approximate how many more solar system exploration missions have been undertaken by the United States than by the USSR/Russia.

Solution:

a. The United States corresponds to 17 symbols, and each symbol represents 8 solar system exploration missions. This means that the United States has undertaken approximately $17 \cdot 8 = 136$ missions for solar system exploration.

b. The USSR/Russia shows $15\frac{1}{2}$ symbols, or $1\frac{1}{2}$ fewer than the United States. This means that the United States has undertaken $1\frac{1}{2} \cdot 8 = 12$ more solar system exploration missions than the USSR/Russia.

■ **Work Practice 2**

Objective B Reading and Constructing Bar Graphs

Another way to visually present data is with a **bar graph.** Bar graphs can appear with vertical bars or horizontal bars. Although we have studied bar graphs in previous sections, we now practice reading the height or length of the bars contained in a bar graph. An advantage to using bar graphs is that a scale is usually included for greater accuracy. Care must be taken when reading bar graphs, as well as other types of graphs—they may be misleading, as shown later in this section.

Answers

2. a. 12 **b.** 20

Example 3   Finding the Number of Endangered Species

The following bar graph shows the number of endangered species in the United States in 2013. Use this graph to answer the questions.

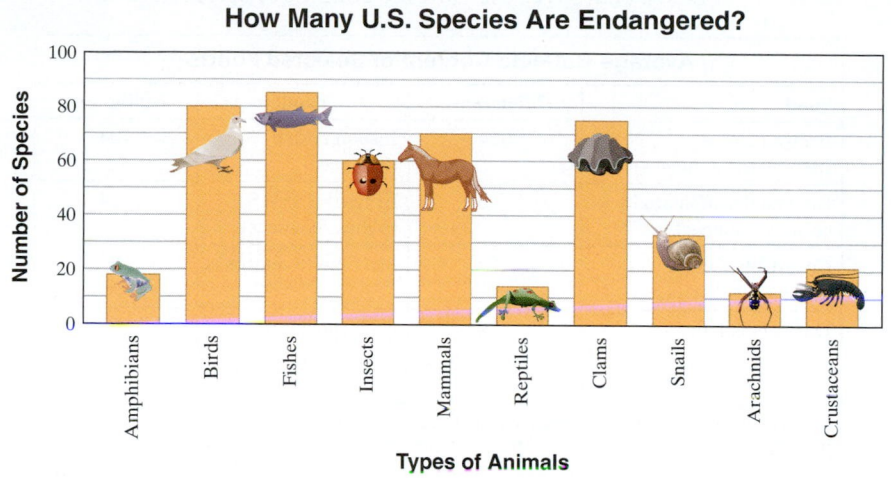

How Many U.S. Species Are Endangered?

Source: U.S. Fish and Wildlife Service

a. Approximate the number of endangered species that are clams.

b. Which category has the most endangered species?

Solution:

a. To approximate the number of endangered species that are clams, we go to the top of the bar that represents clams. From the top of this bar, we move horizontally to the left until the scale is reached. We read the height of the bar on the scale as approximately 75. There are approximately 75 clam species that are endangered, as shown.

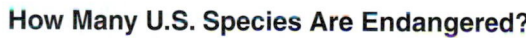

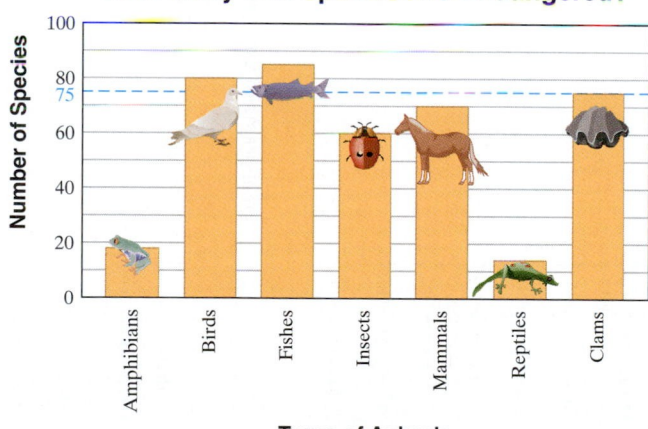

How Many U.S. Species Are Endangered?

Source: U.S. Fish and Wildlife Service

b. The most endangered species is represented by the tallest (longest) bar. The tallest bar corresponds to fishes.

■ **Work Practice 3**

Practice 3

Use the bar graph in Example 3 to answer the following questions:

a. Approximate the number of endangered species that are amphibians.

b. Which category shows the fewest endangered species?

Answers

3. a. 18   **b.** arachnids

Practice 4

Draw a vertical bar graph using the information in the table about selected states' electoral votes for President in the 2012, 2016, and 2020 presidential elections.

Total Electoral Votes by Selected States

State	Electoral Votes
Texas	38
California	55
Florida	29
Nebraska	5
Indiana	11
Georgia	16

(*Source:* U.S. Electoral College)

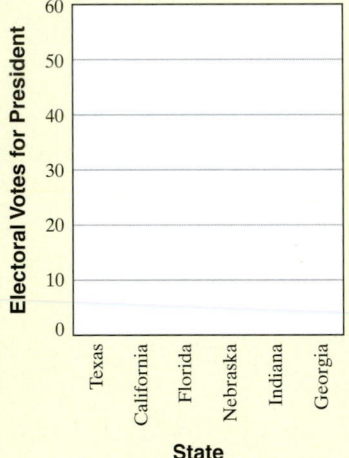

Next, we practice constructing a bar graph.

Example 4   Draw a vertical bar graph using the information in the table below that gives the caffeine content of selected foods.

Average Caffeine Content of Selected Foods

Food	Milligrams	Food	Milligrams
Brewed coffee (percolator, 8 ounces)	124	Instant coffee (8 ounces)	104
Brewed decaffeinated coffee (8 ounces)	3	Brewed tea (U.S. brands, 8 ounces)	64
Coca-Cola Classic (8 ounces)	31	Mr. Pibb (8 ounces)	27
Dark chocolate (semisweet, $1\frac{1}{2}$ ounces)	30	Milk chocolate (8 ounces)	9

(*Sources:* International Food Information Council and the Coca-Cola Company)

Solution: We draw and label a vertical line and a horizontal line as shown below on the left. These lines are also called axes. We place the different food categories along the horizontal axis. Along the vertical axis, we place a scale.

There are many choices of scales that would be appropriate. Notice that the milligrams range from a low of 3 to a high of 124. From this information, we use a scale that starts at 0 and then shows multiples of 20 so that the scale is not too cluttered. The scale stops at 140, the smallest multiple of 20 that will allow all milligrams to be graphed. It may also be helpful to draw horizontal lines along the scale markings to help draw the vertical bars at the correct height. The finished bar graph is shown below on the right.

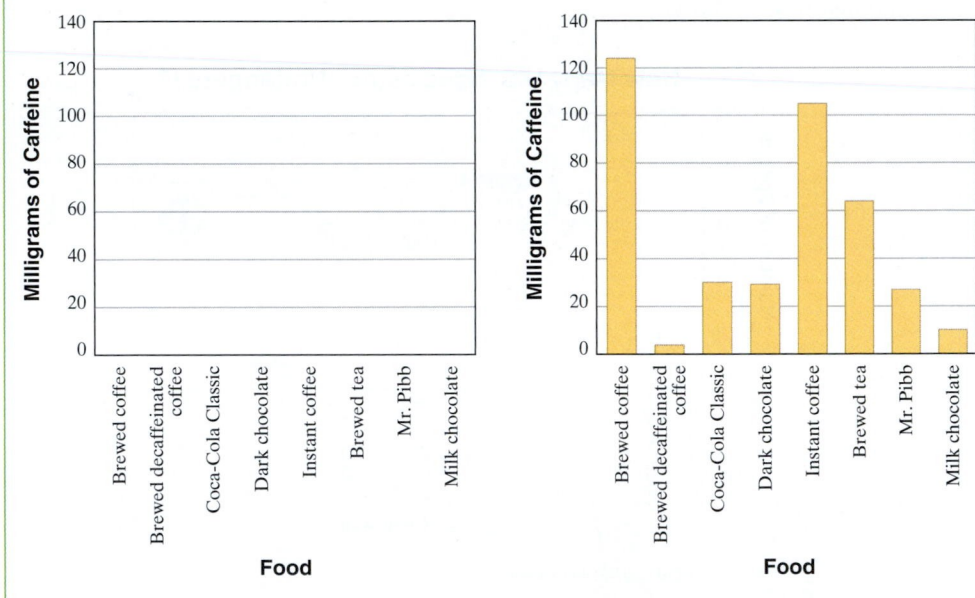

■ **Work Practice 4**

As mentioned previously, graphs can be misleading. Both graphs on the next page show the same information, but with different scales. Special care should be taken when forming conclusions from the appearance of a graph.

Answer

4.

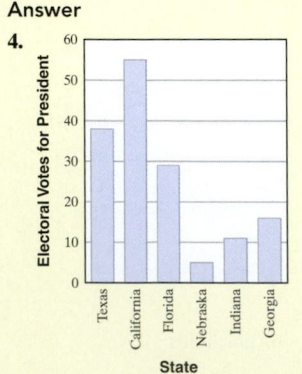

Notice the ⌇ symbol on each vertical scale on the graphs below. This symbol alerts us that numbers are missing from that scale.

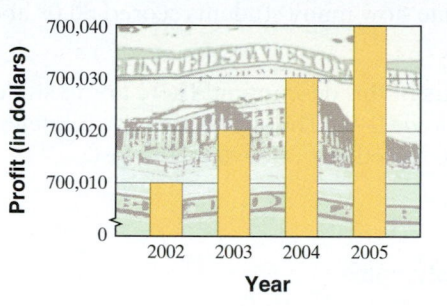

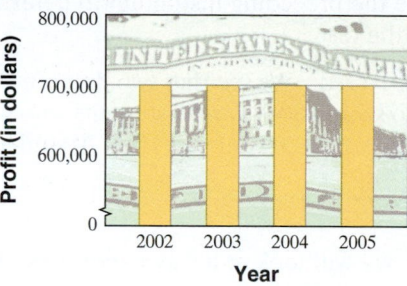

Are profits shown in the graphs above greatly increasing, or are they remaining about the same?

Objective C Reading and Constructing Histograms

Suppose that the test scores of 36 students are summarized in the table below:

Student Scores	Frequency (Number of Students)
40–49	1
50–59	3
60–69	2
70–79	10
80–89	12
90–99	8

The results in the table can be displayed in a histogram. A **histogram** is a special bar graph. The width of each bar represents a range of numbers called a **class interval.** The height of each bar corresponds to how many times a number in the class interval occurs and is called the **class frequency.** The bars in a histogram lie side by side with no space between them.

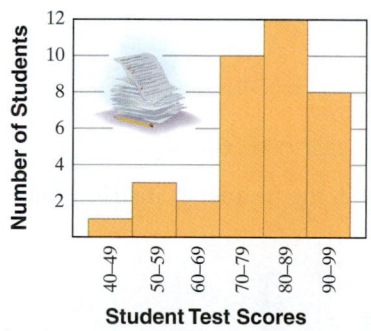

Student Test Scores

Example 5 Reading a Histogram on Student Test Scores

Use the preceding histogram to determine how many students scored 50–59 on the test.

Solution: We find the bar representing 50–59. The height of this bar is 3, which means 3 students scored 50–59 on the test.

■ **Work Practice 5**

Practice 5

Use the histogram on the left to determine how many students scored 80–89 on the test.

Answer
5. 12

Practice 6

Use the histogram above Example 5 to determine how many students scored less than 80 on the test.

Practice 7

Complete the frequency distribution table for the data below. Each number represents a credit card owner's unpaid balance for one month.

0	53	89	125
265	161	37	76
62	201	136	42

Class Intervals (Credit Card Balances)	Tally	Class Frequency (Number of Months)
$0–$49	____	____
$50–$99	____	____
$100–$149	____	____
$150–$199	____	____
$200–$249	____	____
$250–$299	____	____

Practice 8

Construct a histogram from the frequency distribution table above.

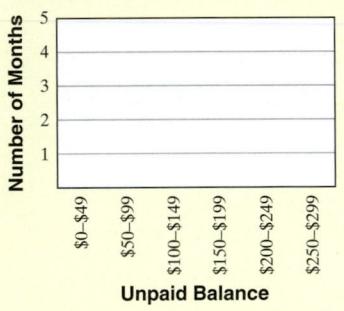

Answers

6. 16

7. table in class interval order:

Tally	Class Frequency (Number of Months)	Tally	Class Frequency (Number of Months)
				3			1
					4			1
			2			1

8.

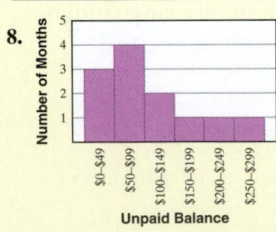

Example 6   Reading a Histogram on Student Test Scores

Use the preceding histogram to determine how many students scored 80 or above on the test.

Solution: We see that two different bars fit this description. There are 12 students who scored 80–89 and 8 students who scored 90–99. The sum of these two categories is 12 + 8 or 20 students. Thus, 20 students scored 80 or above on the test.

■ **Work Practice 6**

Now we will look at a way to construct histograms.

The daily high temperatures for 1 month in New Orleans, Louisiana, are recorded in the following list:

85°	90°	95°	89°	88°	94°
87°	90°	95°	92°	95°	94°
82°	92°	96°	91°	94°	92°
89°	89°	90°	93°	95°	91°
88°	90°	88°	86°	93°	89°

The data in this list have not been organized and can be hard to interpret. One way to organize the data is to place them in a **frequency distribution table.** We will do this in Example 7.

Example 7   Completing a Frequency Distribution on Temperature

Complete the frequency distribution table for the preceding temperature data.

Solution: Go through the data and place a tally mark in the second column of the table next to the class interval. Then count the tally marks and write each total in the third column of the table.

Class Intervals (Temperatures)	Tally	Class Frequency (Number of Days)
82°–84°			1
85°–87°	|||	3
88°–90°	ℋℋ ℋℋI	11
91°–93°	ℋℋII	7
94°–96°	ℋℋIII	8

■ **Work Practice 7**

Example 8   Constructing a Histogram

Construct a histogram from the frequency distribution table in Example 7.

Solution:

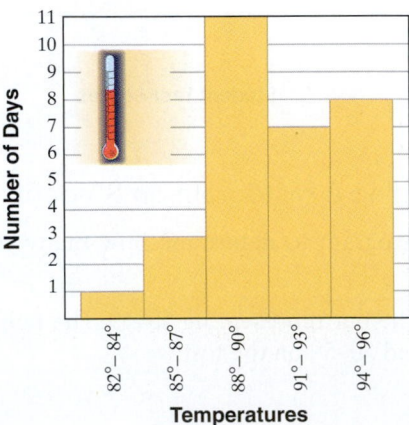

■ **Work Practice 8**

✓**Concept Check** Which of the following sets of data is better suited to representation by a histogram? Explain.

Set 1		Set 2	
Grade on Final	# of Students	Section Number	Avg. Grade on Final
51–60	12	150	78
61–70	18	151	83
71–80	29	152	87
81–90	23	153	73
91–100	25		

Objective D Reading Line Graphs

Another common way to display information with a graph is by using a **line graph.** An advantage of a line graph is that it can be used to visualize relationships between two quantities. A line graph can also be very useful in showing a change over time.

Example 9 Reading Temperatures from a Line Graph

The following line graph shows the average daily temperature for each month in Omaha, Nebraska. Use this graph to answer the questions below.

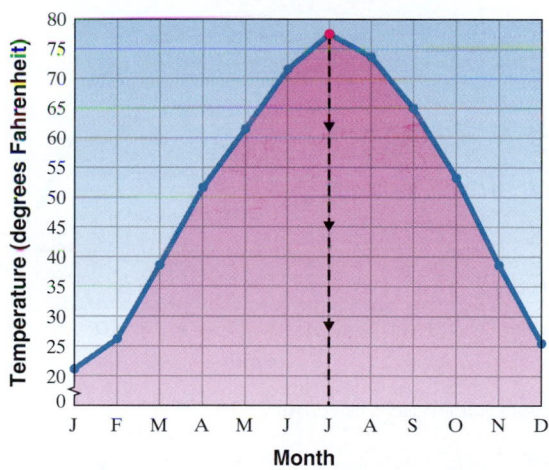

Average Daily Temperature for Omaha, Nebraska

Source: National Climatic Data Center

a. During what month is the average daily temperature the highest?

b. During what month, from July through December, is the average daily temperature 65°F?

c. During what months is the average daily temperature less than 30°F?

Solution:

a. The month with the highest temperature corresponds to the highest point. This is the red point shown on the graph above. We follow this highest point downward to the horizontal month scale and see that this point corresponds to July.

(Continued on next page)

Practice 9

Use the temperature graph in Example 9 to answer the following questions:

a. During what month is the average daily temperature the lowest?

b. During what month is the average daily temperature 25°F?

c. During what months is the average daily temperature greater than 70°F?

Answers

9. **a.** January **b.** December
c. June, July, and August

✓**Concept Check Answer**

Set 1; the grades are arranged in ranges of scores.

b. The months July through December correspond to the right side of the graph. We find the 65°F mark on the vertical temperature scale and move to the right until a point on the right side of the graph is reached. From that point, we move downward to the horizontal month scale and read the corresponding month. During the month of September, the average daily temperature is 65°F.

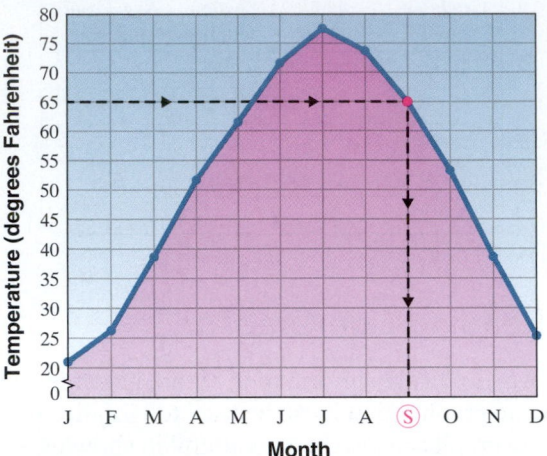

Source: National Climatic Data Center

c. To see what months the temperature is less than 30°F, we find what months correspond to points that fall below the 30°F mark on the vertical scale. These months are January, February, and December.

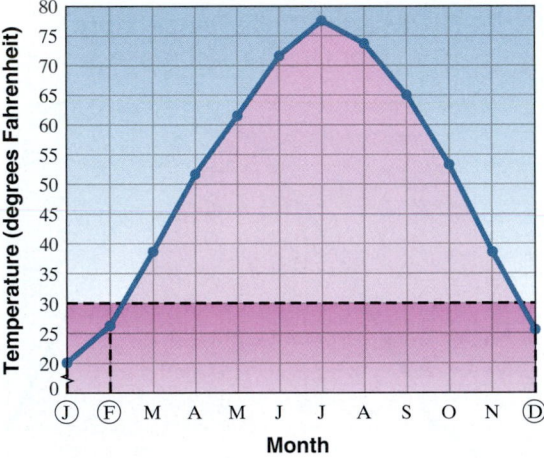

Source: National Climatic Data Center

■ **Work Practice 9**

Vocabulary, Readiness & Video Check

Fill in each blank with one of the choices below.

pictograph	bar	class frequency
histogram	line	class interval

1. A _____ graph presents data using vertical or horizontal bars.

2. A _____ is a graph in which pictures or symbols are used to visually present data.

3. A _____ graph displays information with a line that connects data points.

4. A _____ is a special bar graph in which the width of each bar represents a _____ and the height of each bar represents the _____.

Martin-Gay Interactive Videos   Watch the section lecture video and answer the following questions.

See Video 7.1

Objective **A**   **5.** From the pictograph in ▣ Example 1, how would you approximate the number of wildfires for any given year? ▶

Objective **B**   **6.** From ▣ Example 5, what is one advantage of displaying data in a bar graph? ▶

Objective **C**   **7.** Complete this statement based on the lecture before ▣ Example 6: A histogram is a special kind of _____. ▶

Objective **D**   **8.** From the line graph in ▣ Examples 10–13, during which year(s) were total points scored greater than 40? ▶

7.1 Exercise Set MyMathLab® ▶

Objective **A**   *The following pictograph shows the number of acres devoted to wheat production in selected states in 2013. Use this graph to answer Exercises 1 through 8. See Examples 1 and 2.* (*Source: National Agricultural Statistics Service*)

1. Which of the states shown planted the greatest quantity of acreage in wheat?

2. Which of the states shown planted the least amount of wheat acreage?

3. Approximate the number of acres of wheat planted in Oklahoma.

4. Approximate the number of acres of wheat planted in Kansas.

5. Which state planted about 2,500,000 acres of wheat?

6. Which states planted about 6,000,000 acres of wheat?

7. How many more acres of wheat were planted in Kansas than in Montana?

8. How many more acres of wheat were planted in Oklahoma than in Washington?

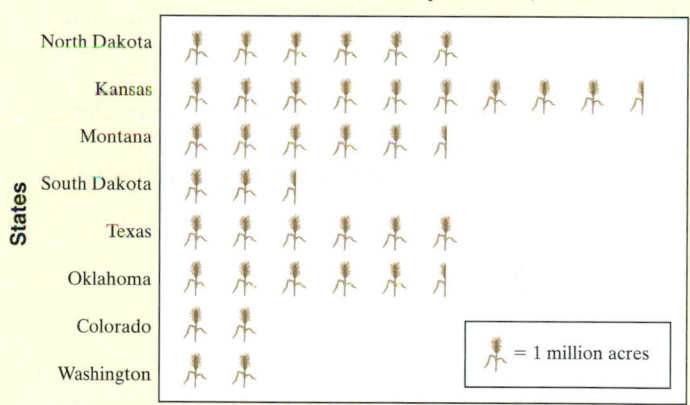

Wheat Acreage in Selected Top States, 2013

= 1 million acres

The following pictograph shows the annual number of wildfires in the United States between 2006 and 2012. Use this graph to answer Exercises 9 through 16. See Examples 1 and 2. (*Source: National Interagency Fire Center*)

▶ **9.** Approximate the number of wildfires in 2012.

10. Approximately how many wildfires were there in 2006?

▶ **11.** Which year, of the years shown, had the most wildfires?

12. In what years were the number of wildfires greater than 72,000?

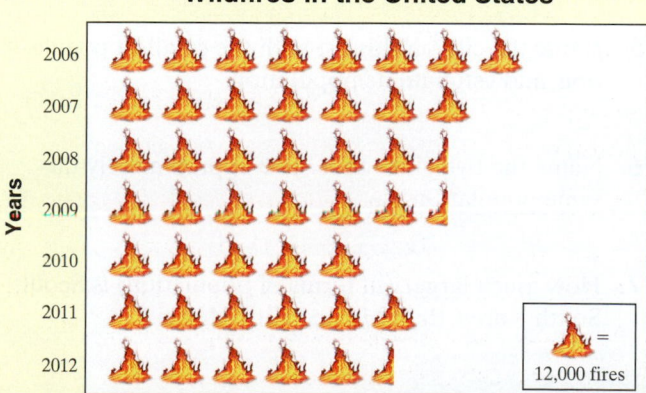

Wildfires in the United States

= 12,000 fires

13. What was the amount of increase in wildfires from 2010 to 2011?

14. What was the amount of decrease in wildfires from 2006 to 2012?

15. What was the average annual number of wildfires from 2010 to 2012? (*Hint:* How do you calculate the average?)

16. Give an explanation for the large number of wildfires in 2006.

Objective **B** *The National Weather Service has exacting definitions for hurricanes; they are tropical storms with winds in excess of 74 mph. The following bar graph shows the number of hurricanes, by month, that have made landfall on the mainland United States between 1851 and 2012. Use this graph to answer Exercises 17 through 22. See Example 3. (Source: NOAA: Hurricane Research Division)*

17. In which month did the most hurricanes make landfall in the United States?

18. In which month did the fewest hurricanes make landfall in the United States?

19. Approximate the number of hurricanes that made landfall in the United States during the month of August.

20. Approximate the number of hurricanes that made landfall in the United States in September.

21. In 2008, two hurricanes made landfall during the month of August. What fraction of all the 76 hurricanes that made landfall during August is this?

22. In 2012, only one hurricane, Hurricane Sandy, made landfall on the United States during the month of October. If there have been 51 hurricanes to make landfall in the month of October since 1851, approximately what percent of these occurred in 2012?

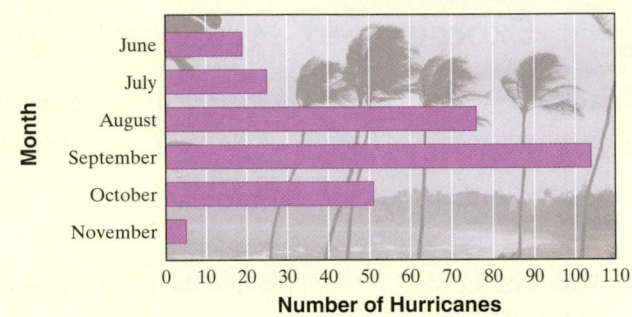

Hurricanes Making Landfall in the United States, by Month, 1851–2012

The following horizontal bar graph shows the recent population of the world's largest cities (including their suburbs). Use this graph to answer Exercises 23 through 28. See Example 3. (Source: CityPopulation)

23. Name the city with the largest population, and estimate its population.

24. Name the city whose population is between 26 million and 27 million, and estimate its population.

25. Name the city on this list with the smallest population, and estimate its population.

26. Name the two cities that have approximately the same population.

27. How much larger (in terms of population) is Seoul, South Korea, than Mexico City, Mexico?

28. How much larger (in terms of population) is Guangzhou, China, than Karachi, Pakistan?

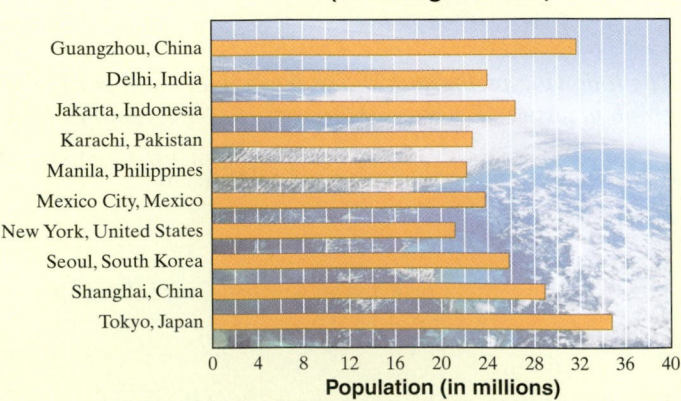

World's Largest Cities (including suburbs)

Use the information given to draw a vertical bar graph. Clearly label the bars. See Example 4.

29.

Fiber Content of Selected Foods	
Food	**Grams of Total Fiber**
Kidney beans $\left(\frac{1}{2}c\right)$	4.5
Oatmeal, cooked $\left(\frac{3}{4}c\right)$	3.0
Peanut butter, chunky (2 tbsp)	1.5
Popcorn (1 c)	1.0
Potato, baked, with skin (1 med)	4.0
Whole wheat bread (1 slice)	2.5
(*Sources:* American Dietetic Association and National Center for Nutrition and Dietetics.)	

30.

U.S. Annual Food Sales	
Year	**Sales (in billions of dollars)**
2007	1079
2008	1117
2009	1086
2010	1139
2011	1274
2012	1345
(*Source:* U.S. Department of Agriculture)	

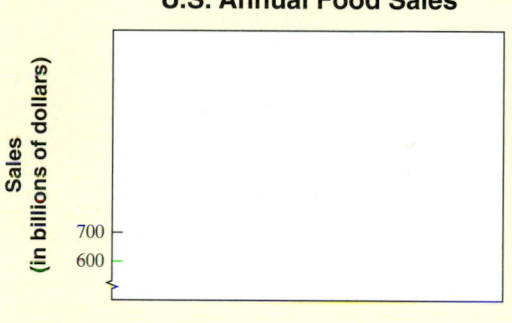

31.

Best-Selling Albums of All Time (U.S. sales)	
Album	**Estimated Units Sold (in millions)**
Led Zeppelin: *Led Zeppelin IV* (1971)	23
Eagles: *Their Greatest Hits* (1976)	29
Pink Floyd: *The Wall* (1979)	23
Michael Jackson: *Thriller* (1982)	29
Billy Joel: *Greatest Hits: Volumes I & II* (1985)	23
(*Source:* Recording Industry Association of America)	

32.

Selected Worldwide Commercial Space Launches	
Source	**Total Commercial Space Launches 1990–2012**
United States	156
Europe	153
Russia	141
China	23
Sea Launch*	39
*Sea Launch is an international venture involving 4 countries that uses its own launch facility outside national borders.	
(*Source:* Bureau of Transportation Statistics)	

Objective C *The following histogram shows the number of miles that each adult, from a survey of 100 adults, drives per week. Use this histogram to answer Exercises 33 through 42. See Examples 5 and 6.*

33. How many adults drive 100–149 miles per week?

34. How many adults drive 200–249 miles per week?

○ 35. How many adults drive fewer than 150 miles per week?

36. How many adults drive 200 miles or more per week?

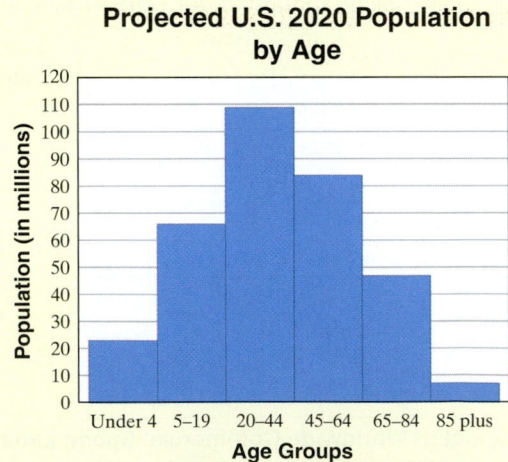

37. How many adults drive 100–199 miles per week?

38. How many adults drive 150–249 miles per week?

○ 39. How many more adults drive 250–299 miles per week than 200–249 miles per week?

40. How many more adults drive 0–49 miles per week than 50–99 miles per week?

41. What is the ratio of adults who drive 150–199 miles per week to the total number of adults surveyed?

42. What is the ratio of adults who drive 50–99 miles per week to the total number of adults surveyed?

The following histogram shows the projected population (in millions), by age groups, for the United States for the year 2020. Use this histogram to answer Exercises 43 through 50. For Exercises 45 through 48, estimate to the nearest whole million. See Examples 5 and 6.

43. What age range will be the largest population group in 2020?

44. What age range will be the smallest population group in 2020?

45. How large is the population of 20- to 44-year-olds expected to be in 2020?

46. How large is the population of 45- to 64-year-olds expected to be in 2020?

Projected U.S. 2020 Population by Age

47. How large is the population of those less than 4 years old expected to be in 2020?

48. How large is the population of 5- to 19-year-olds expected to be in 2020?

49. Which bar represents the age range you expect to be in during 2020?

50. How many more 20- to 44-year-olds are there expected to be than 45- to 64-year-olds in 2020?

The following list shows the golf scores for an amateur golfer. Use this list to complete the frequency distribution table to the right. See Example 7.

78	84	91	93	97
97	95	85	95	96
101	89	92	89	100

	Class Intervals (Scores)	Tally	Class Frequency (Number of Games)
▶ **51.**	70–79		
▶ **52.**	80–89		
▶ **53.**	90–99		
▶ **54.**	100–109		

Twenty-five people in a survey were asked to give their current checking account balances. Use the balances shown in the following list to complete the frequency distribution table to the right. See Example 7.

$53	$105	$162	$443	$109
$468	$47	$259	$316	$228
$207	$357	$15	$301	$75
$86	$77	$512	$219	$100
$192	$288	$352	$166	$292

	Class Intervals (Account Balances)	Tally	Class Frequency (Number of People)
55.	$0–$99		
56.	$100–$199		
57.	$200–$299		
58.	$300–$399		
59.	$400–$499		
60.	$500–$599		

▶ **61.** Use the frequency distribution table from Exercises 51 through 54 to construct a histogram. See Example 8.

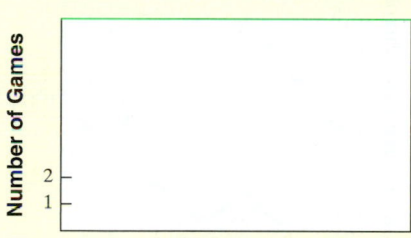

Golf Scores

62. Use the frequency distribution table from Exercises 55 through 60 to construct a histogram. See Example 8.

Account Balances

Objective D *The following line graph shows the total points scored by both teams in the NFL Super Bowl from 2006 through 2013. Use this graph to answer Exercises 63 through 70. See Example 9. (Source: superbowlhistory.net)*

▶ **63.** Find the total points scored in the Super Bowl in 2009.

64. Find the total points scored in the Super Bowl in 2013.

▶ **65.** During which year(s) shown were the total points scored in the Super Bowl greater than 50?

66. During which year(s) shown was the total score in the Super Bowl the highest?

Total Points Scored in Super Bowl

67. During which year(s) shown was the total score in the Super Bowl the lowest?

68. Between 2011 and 2012, did the total score in the Super Bowl increase or decrease?

69. During which year(s) was the total score in the Super Bowl less than 40?

70. Between 2008 and 2009, did the total score in the Super Bowl increase or decrease?

Review

Find each percent. See Sections 5.4 and 5.5.

71. 30% of 12

72. 45% of 120

73. 10% of 62

74. 95% of 50

Write each fraction as a percent. See Section 5.3.

75. $\frac{1}{4}$

76. $\frac{2}{5}$

77. $\frac{17}{50}$

78. $\frac{9}{10}$

Concept Extensions

The following double line graph shows temperature highs and lows for a week. Use this graph to answer Exercises 79 through 84.

79. What was the high temperature reading on Thursday?

80. What was the low temperature reading on Thursday?

81. What day was the temperature the lowest? What was this low temperature?

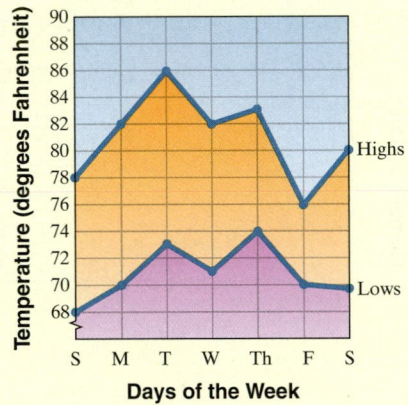

82. What day of the week was the temperature the highest? What was this high temperature?

83. On what day of the week was the difference between the high temperature and the low temperature the greatest? What was this difference in temperature?

84. On what day of the week was the difference between the high temperature and the low temperature the least? What was this difference in temperature?

85. True or false? With a bar graph, the width of the bar is just as important as the height of the bar. Explain your answer.

86. Kansas plants about 24% of the wheat acreage in the United States. About how many acres of wheat are planted in the United States, according to the pictograph for Exercises 1 through 8? Round to the nearest million acre.

7.2 Reading Circle Graphs

Objective A Reading Circle Graphs

Objectives

A Read Circle Graphs.

B Draw Circle Graphs.

In Exercise Set 5.2, the following **circle graph** was shown. This particular graph shows the favorite sport for 100 adults.

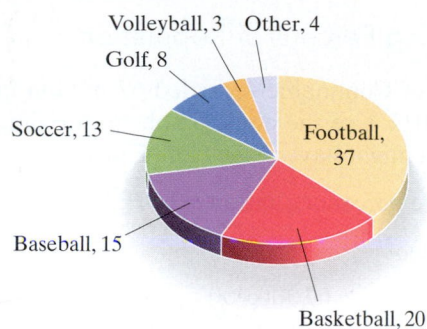

Volleyball, 3 Other, 4
Golf, 8
Soccer, 13
Football, 37
Baseball, 15
Basketball, 20

Each sector of the graph (shaped like a piece of pie) shows a category and the relative size of the category. In other words, the most popular sport is football, and it is represented by the largest sector.

Example 1 Find the ratio of adults preferring basketball to total adults. Write the ratio as a fraction in simplest form.

Solution: The ratio is

$$\frac{\text{people preferring basketball}}{\text{total adults}} = \frac{20}{100} = \frac{1}{5}$$

■ Work Practice 1

Practice 1

Find the ratio of adults preferring golf to total adults. Write the ratio as a fraction in simplest form.

A circle graph is often used to show percents in different categories, with the whole circle representing 100%.

Example 2 Using a Circle Graph

The following graph shows the percent of visitors to the United States in 2012 from various regions. Using the circle graph shown, determine the percent of visitors who came to the United States from Mexico and Canada.

Solution: To find this percent, we add the percents corresponding to Mexico and Canada. The percent of visitors to the United States that came from Mexico and Canada is

$$21.3\% + 34.1\% = 55.4\%$$

■ Work Practice 2

Visitors to U.S. by Region

Other 3.9%
Europe 18.7%
Mexico 21.3%
Asia 12.5%
South America 9.5%
Canada 34.1%

Source: Office of Travel and Tourism Industries

Practice 2

Using the circle graph shown in Example 2, determine the percent of visitors to the United States that came from Europe, Asia, and South America.

Answers

1. $\frac{2}{25}$ **2.** 40.7%

Copyright 2015 Pearson Education, Inc.

Helpful Hint

Since a circle graph represents a whole, the percents should add to 100% or 1. Notice this is true for Example 2.

Practice 3

Use the information in Example 3 and the circle graph from Example 2 to estimate the number of tourists that might have come from Mexico in 2012.

Example 3 Finding Percent of Population

The U.S. Department of Commerce recorded 67 million international visitors to the United States in 2012. Use the circle graph from Example 2 and estimate the number of tourists that might have come from Europe.

Solution: We use the percent equation.

$$\text{amount} = \boxed{\text{percent}} \cdot \boxed{\text{base}}$$

$$\text{amount} = 0.187 \cdot 67{,}000{,}000$$

$$= 0.187(67{,}000{,}000)$$

$$= 12{,}529{,}000$$

Thus, 12,529,000 tourists might have come from Europe in 2012.

■ **Work Practice 3**

✓**Concept Check** Can the following data be represented by a circle graph? Why or why not?

Responses to the Question, "In Which Activities Are You Involved?"	
Intramural sports	60%
On-campus job	42%
Fraternity/sorority	27%
Academic clubs	21%
Music programs	14%

Objective B Drawing Circle Graphs ▶

To draw a circle graph, we use the fact that a whole circle contains 360° (degrees).

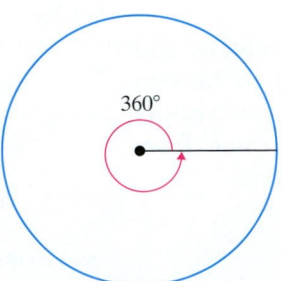

360°

Answer

3. 14,271,000 tourists from Mexico

✓**Concept Check Answer**

no; the percents add up to more than 100%

Example 4 Drawing a Circle Graph for U.S. Armed Forces Personnel

The following table shows the percent of U.S. armed forces personnel that were in each branch of service in 2011. (*Source:* U.S. Department of Defense)

Branch of Service	Percent
Army	38
Navy	23
Marine Corps	14
Air Force	22
Coast Guard	3

Draw a circle graph showing this data.

Solution: First we find the number of degrees in each sector representing each branch of service. Remember that the whole circle contains 360°. (We will round degrees to the nearest whole.)

Sector	Degrees in Each Sector
Army	38% × 360° = 0.38 × 360° = 136.8° ≈ 137°
Navy	23% × 360° = 0.23 × 360° = 82.8° ≈ 83°
Marine Corps	14% × 360° = 0.14 × 360° = 50.4° ≈ 50°
Air Force	22% × 360° = 0.22 × 360° = 79.2° ≈ 79°
Coast Guard	3% × 360° = 0.03 × 360° = 10.8° ≈ 11°

Helpful Hint

Check your calculations by finding the sum of the degrees.

$$137° + 83° + 50° + 79° + 11° = 360°$$

The sum should be 360°. (It may vary only slightly because of rounding.)

Next we draw a circle and mark its center. Then we draw a line from the center of the circle to the circle itself.

To construct the sectors, we will use a **protractor.** A protractor measures the number of degrees in an angle. We place the hole in the protractor over the center of the circle. Then we adjust the protractor so that 0° on the protractor is aligned with the line that we drew.

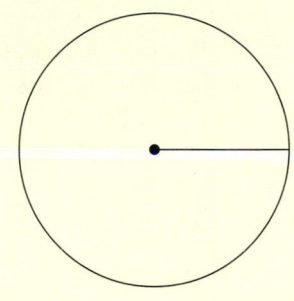

It makes no difference which sector we draw first. To construct the "Army" sector, we find 137° on the protractor and mark our circle. Then we remove the protractor and use this mark to draw a second line from the center to the circle itself.

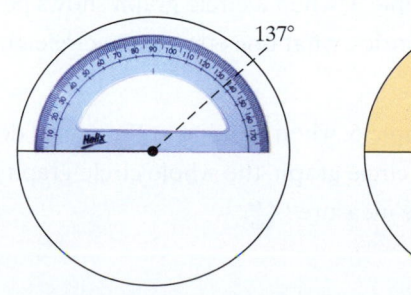

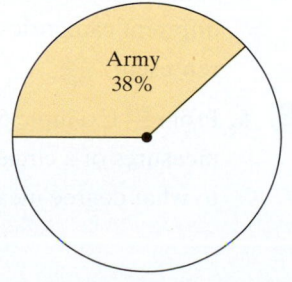

(*Continued on next page*)

Practice 4

Use the data shown to draw a circle graph.

Freshmen	30%
Sophomores	27%
Juniors	25%
Seniors	18%

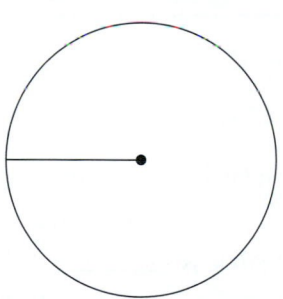

Answer

4.

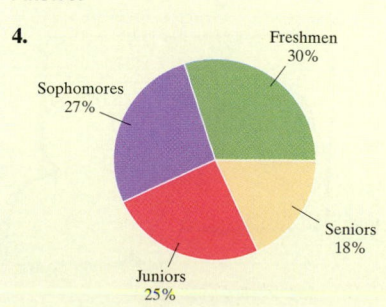

To construct the "Navy" sector, we follow the same procedure as above, except that we line up 0° with the second line we drew and mark the protractor at 83°.

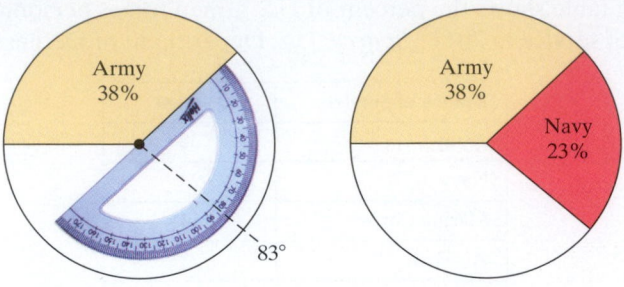

We continue in this manner until the circle graph is complete.

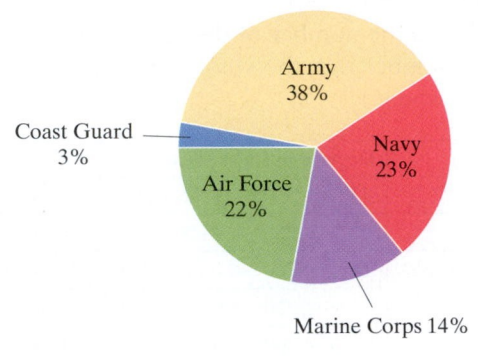

■ **Work Practice 4**

✓**Concept Check** True or false? The larger a sector in a circle graph, the larger the percent of the total it represents. Explain your answer.

Vocabulary, Readiness & Video Check

Use the choices below to fill in each blank.

   sector   circle   100   360

1. In a(n) _____ graph, each section (shaped like a piece of pie) shows a category and the relative size of the category.

2. A circle graph contains pie-shaped sections, each called a(n) _____.

3. The number of degrees in a whole circle is _____.

4. If a circle graph has percent labels, the percents should add up to _____.

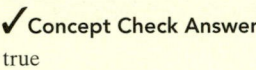 *Martin-Gay Interactive Videos*

See Video 7.2

Watch the section lecture video and answer the following questions.

Objective A **5.** From ▥ Example 3, when a circle graph shows percents in different categories, what does the whole circle graph always represent? ▶

Objective B **6.** From ▥ Example 6, when looking at the sector degree measures of a circle graph, the whole circle graph corresponds to what degree measure? ▶

7.2 Exercise Set MyMathLab®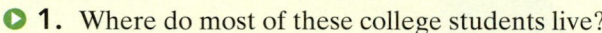

Objective A *The following circle graph is a result of surveying 700 college students. They were asked where they live while attending college. Use this graph to answer Exercises 1 through 6. Write all ratios as fractions in simplest form. See Example 1.*

▶ **1.** Where do most of these college students live?

2. Besides the category "Other arrangements," where do the fewest of these college students live?

▶ **3.** Find the ratio of students living in campus housing to total students.

4. Find the ratio of students living in off-campus rentals to total students.

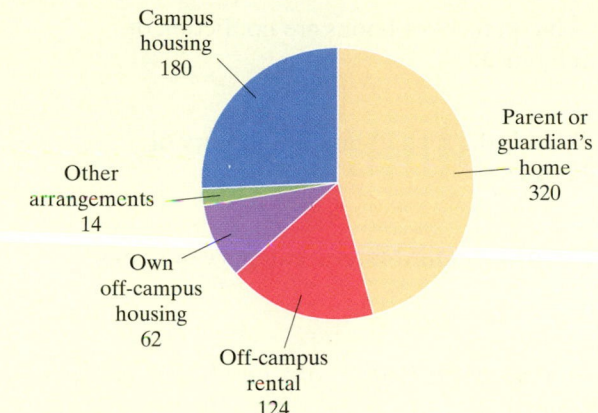

5. Find the ratio of students living in campus housing to students living in a parent or guardian's home.

6. Find the ratio of students living in off-campus rentals to students living in a parent or guardian's home.

The following circle graph shows the percent of the land area of the continents of Earth. Use this graph for Exercises 7 through 14. See Example 2.

7. Which continent is the largest?

8. Which continent is the smallest?

9. What percent of the land on Earth is accounted for by Asia and Europe together?

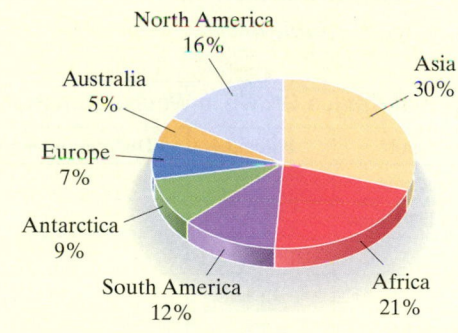

Source: National Geographic Society

10. What percent of the land on Earth is accounted for by North and South America?

The total amount of land from the continents is approximately 57,000,000 square miles. Use the graph to find the area of the continents given in Exercises 11 through 14. See Example 3.

11. Asia

12. South America

13. Australia

14. Europe

The following circle graph shows the percent of the types of books available at Midway Memorial Library. Use this graph for Exercises 15 through 24. See Example 2.

▶ **15.** What percent of books are classified as some type of fiction?

16. What percent of books are nonfiction or reference?

▶ **17.** What is the second-largest category of books?

18. What is the third-largest category of books?

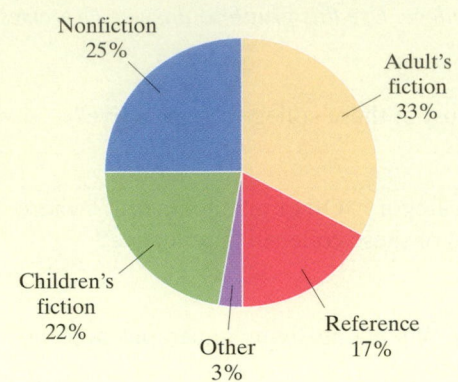

If this library has 125,600 books, find how many books are in each category given in Exercises 19 through 24. See Example 3.

▶ **19.** Nonfiction

20. Reference

21. Children's fiction

22. Adult's fiction

23. Reference or other

24. Nonfiction or other

Objective **B** *Fill in the tables. Round to the nearest degree. Then draw a circle graph to represent the information given in each table. (Remember: The total of "Degrees in Sector" column should equal 360° or very close to 360° because of rounding.) See Example 4.*

▶ **25.**

Types of Apples Grown in Washington State		
Type of Apple	**Percent**	**Degrees in Sector**
Red Delicious	37%	
Golden Delicious	13%	
Fuji	14%	
Gala	15%	
Granny Smith	12%	
Other varieties	6%	
Braeburn	3%	
(*Source:* U.S. Apple Association)		

26.

Color Distribution of M&M's Milk Chocolate		
Color	**Percent**	**Degrees in Sector**
Blue	22.1%	
Orange	16.7%	
Green	16.7%	
Red	16.7%	
Brown	16.7%	
Yellow	11.1%	
(*Source:* M&M Mars)		

27.

Distribution of Large Dams by Continent		
Continent	**Percent**	**Degrees in Sector**
Europe	19%	
North America	32%	
South America	3%	
Asia	39%	
Africa	5%	
Australia	2%	
(*Source:* International Commission on Large Dams)		

28.

Number of Times the "Are We There Yet?" Question Is Asked of Parents During Road Trips		
	Percent	**Degrees in Sector**
Never	20%	
Once	11%	
2–5 times	36%	
6–10 times	14%	
More than 10 times	19%	
(*Source:* KRC Research for Goodyear Tire & Rubber Co.)		

Review

Write the prime factorization of each number. See Section 2.2.

29. 20

30. 25

31. 40

32. 16

33. 85

34. 105

Concept Extensions

The following circle graph shows the relative sizes of the great oceans. Use this graph for Exercises 35 through 40.

35. Without calculating, determine which ocean is the largest. How can you answer this question by looking at the circle graph?

36. Without calculating, determine which ocean is the smallest. How can you answer this question by looking at the circle graph?

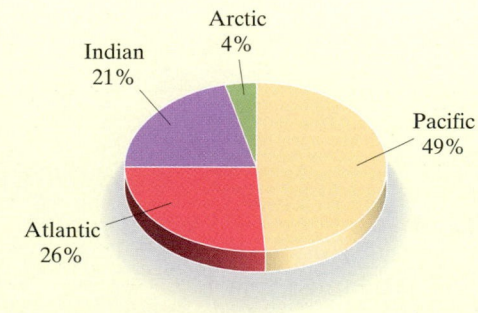

Source: Philip's World Atlas

These oceans together make up 264,489,800 square kilometers of the Earth's surface. Find the square kilometers for each ocean.

37. Pacific Ocean

38. Atlantic Ocean

39. Indian Ocean

40. Arctic Ocean

The following circle graph summarizes the results of a survey of 2800 Internet users who make purchases online. Use this graph for Exercises 41 through 46. Round to the nearest whole.

41. How many of the survey respondents said that they spend $0–$15 online each month?

42. How many of the survey repondents said that they spend $15–$175 online each month?

43. How many of the survey respondents said that they spend $0 to $175 online each month?

44. How many of the survey respondents said that they spend $15 to over $175 online each month?

Online Spending per Month

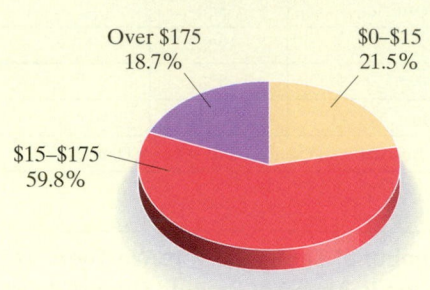

Source: UCLA Center for Communication Policy

45. Find the ratio of *number* of respondents who spend $0–$15 online to *number* of respondents who spend $15–$175 online. Write the ratio as a fraction with integers in the numerator and denominator.

46. Find the ratio of *percent* of respondents who spend $0–$15 online to *percent* of those who spend $15–$175. Write the ratio as a fraction with integers in the numerator and denominator.

See the Concept Checks in this section.

47. Can the data below be represented by a circle graph? Why or why not?

Responses to the Question, "What Classes Are You Taking?"	
Math	80%
English	72%
History	37%
Biology	21%
Chemistry	14%

48. True or false? The smaller a sector in a circle graph, the smaller the percent of the total it represents. Explain why.

Reading Graphs

The following pictograph shows the six occupations with the largest estimated numerical increase in employment in the United States between 2010 and 2020. Use this graph to answer Exercises 1 through 4.

1. _____

Jobs with Projected Highest Numerical Increase: 2010–2020

Source: Bureau of Labor Statistics

2. _____

1. Approximate the increase in the number of registered nurses from 2010 to 2020.

3. _____

2. Approximate the increase in the number of general office clerks from 2010 to 2020.

4. _____

3. Which occupations are expected to show approximately the same increase in numbers of employees between the years shown?

5. _____

4. Which of the listed occupations is expected to show the least increase in numbers of employees between the years shown?

The following bar graph shows the highest U.S. dams. Use this graph to answer Exercises 5 through 8.

6. _____

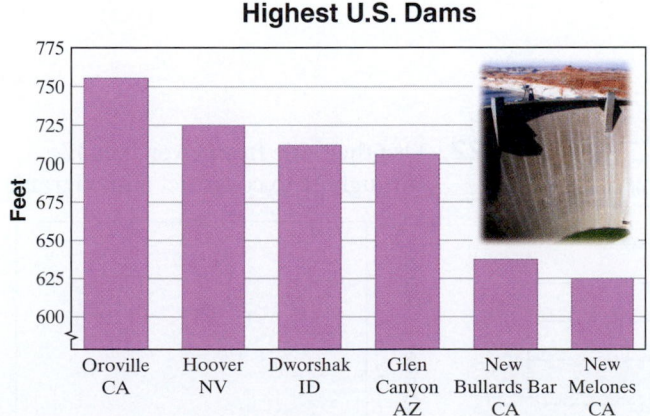

Highest U.S. Dams

Source: Committee on Register of Dams

5. Name the U.S. dam with the greatest height and estimate its height.

6. Name the U.S. dam whose height is between 625 and 650 feet and estimate its height.

7. _____

7. Estimate how much higher the Hoover Dam is than the Glen Canyon Dam.

8. _____

8. How many U.S. dams have heights over 700 feet?

9. _____

10. _____

11. _____

12. _____

13. _____

14. _____

15. _____

16. _____

17. _____

18. _____

19. _____

20. _____

21. _____

22. _____

The following line graph shows the daily high temperatures for 1 week in Annapolis, Maryland. Use this graph to answer Exercises 9 through 12.

9. Name the day(s) of the week with the highest temperature and give that high temperature.

10. Name the day(s) of the week with the lowest temperature and give that low temperature.

11. On what days of the week was the temperature less than 90° Fahrenheit?

12. On what days of the week was the temperature greater than 90° Fahrenheit?

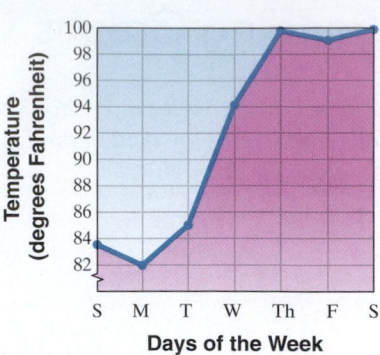

The following circle graph shows the type of beverage milk consumed in the United States. Use this graph for Exercises 13 through 16.

   If a store in Kerrville, Texas, sells 200 quart containers of milk per week, estimate how many quart containers are sold in each category below.

Types of Beverage Milk Consumed

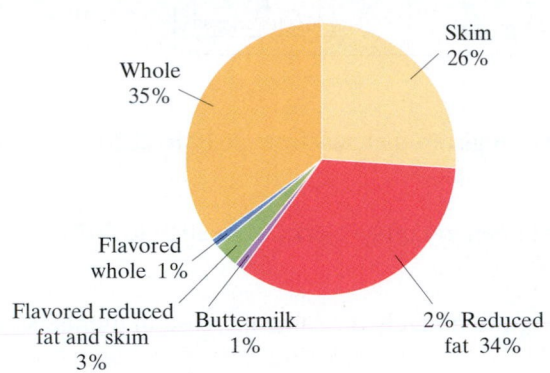

Source: U.S. Department of Agriculture

13. Whole milk

14. Skim milk

15. Buttermilk

16. Flavored reduced fat and skim milk

The following list shows weekly quiz scores for a student in basic college mathematics. Use this list to complete the frequency distribution table.

50	80	71	83	86
67	89	93	88	97
	53	90		
75	80	78	93	99

	Class Intervals (Scores)	Tally	Class Frequency (Number of Quizzes)
17.	50–59		
18.	60–69		
19.	70–79		
20.	80–89		
21.	90–99		

22. Use the table from Exercises 17 through 21 to construct a histogram.

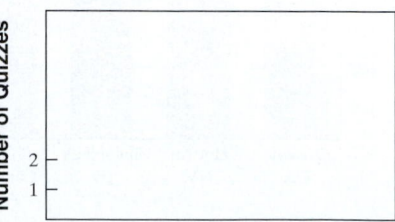

Quiz Scores

7.3 Mean, Median, and Mode

Objective A Finding the Mean

Sometimes we want to summarize data by displaying them in a graph, but sometimes it is also desirable to be able to describe a set of data, or a set of numbers, by a single "middle" number. Three such **measures of central tendency** are the **mean,** the **median,** and the **mode.**

The most common measure of central tendency is the mean (sometimes called the "arithmetic mean" or the "average"). Recall that we first introduced finding the average of a list of numbers in Section 1.7.

Objectives

A Find the Mean of a List of Numbers.

B Find the Median of a List of Numbers.

C Find the Mode of a List of Numbers.

The **mean (average)** of a set of number items is the sum of the items divided by the number of items.

$$\text{mean} = \frac{\text{sum of items}}{\text{number of items}}$$

Example 1 Finding the Mean Time in an Experiment

Seven students in a psychology class conducted an experiment on mazes. Each student was given a pencil and asked to successfully complete the same maze. The timed results are below:

Student	Ann	Thanh	Carlos	Jesse	Melinda	Ramzi	Dayni
Time (Seconds)	13.2	11.8	10.7	16.2	15.9	13.8	18.5

a. Who completed the maze in the shortest time? Who completed the maze in the longest time?
b. Find the mean time.
c. How many students took longer than the mean time? How many students took shorter than the mean time?

Solution:
a. Carlos completed the maze in 10.7 seconds, the shortest time. Dayni completed the maze in 18.5 seconds, the longest time.
b. To find the mean (or average), we find the sum of the items and divide by 7, the number of items.

$$\text{mean} = \frac{13.2 + 11.8 + 10.7 + 16.2 + 15.9 + 13.8 + 18.5}{7}$$

$$= \frac{100.1}{7} = 14.3$$

c. Three students, Jesse, Melinda, and Dayni, had times longer than the mean time. Four students, Ann, Thanh, Carlos, and Ramzi, had times shorter than the mean time.

■ **Work Practice 1**

✓**Concept Check** Estimate the mean of the following set of data:

5, 10, 10, 10, 10, 15

Often in college, the calculation of a **grade point average** (GPA) is a **weighted mean** and is calculated as shown in Example 2.

Practice 1

Find the mean of the following test scores: 87, 75, 96, 91, and 78.

511

Practice 2

Find the grade point average if the following grades were earned in one semester.

Grade	Credit Hours
A	2
B	4
C	5
D	2
A	2

Example 2   Calculating Grade Point Average (GPA)

The following grades were earned by a student during one semester. Find the student's grade point average.

Course	Grade	Credit Hours
College mathematics	A	3
Biology	B	3
English	A	3
PE	C	1
Social studies	D	2

Solution: To calculate the grade point average, we need to know the point values for the different possible grades. The point values of grades commonly used in colleges and universities are given below:

A: 4,   B: 3,   C: 2,   D: 1,   F: 0

Now, to find the grade point average, we multiply the number of credit hours for each course by the point value of each grade. The grade point average is the sum of these products divided by the sum of the credit hours.

Course	Grade	Point Value of Grade	Credit Hours	Point Value × Credit Hours
College mathematics	A	4	3	12
Biology	B	3	3	9
English	A	4	3	12
PE	C	2	1	2
Social studies	D	1	2	2
		Totals:	12	37

$$\text{grade point average} = \frac{37}{12} \approx 3.08 \text{ rounded to two decimal places}$$

The student earned a grade point average of 3.08.

■ **Work Practice 2**

Objective B Finding the Median ▶

You may have noticed that a very low number or a very high number can affect the mean of a list of numbers. Because of this, you may sometimes want to use another measure of central tendency. A second measure of central tendency is called the **median.** The median of a list of numbers is not affected by a low or high number in the list.

> The **median** of a set of numbers in numerical order is the middle number. If the number of items is odd, the median is the middle number. If the number of items is even, the median is the mean of the two middle numbers.

Practice 3

Find the median of the list of numbers: 5, 11, 14, 23, 24, 35, 38, 41, 43

Example 3   Find the median of the following list of numbers:

25, 54, 56, 57, 60, 71, 98

Solution: Because this list is in numerical order, the median is the middle number, 57.

■ **Work Practice 3**

Answers

2. 2.67   **3.** 24

The ten tallest buildings in the world, completed as of 2013, are listed in the following table. Use this table to answer Exercises 9 through 14. If necessary, round results to one decimal place. See Examples 1 and 3 through 6.

9. Find the mean height of the five tallest buildings.

Building	Height (in feet)
Burj Khalifa	2717
Makkah Royal Clock Tower Hotel	1972
Taipei 101	1667
Shanghai World Financial Center	1614
International Commerce Centre	1588
Petronas Tower 1	1483
Petronas Tower 2	1483
Zifeng Tower	1476
Willis Tower	1451
KK100	1449

(*Source:* Council on Tall Buildings and Urban Habitat)

10. Find the median height of the five tallest buildings.

11. Find the median height of the eight tallest buildings.

12. Find the mean height of the eight tallest buildings.

13. Given the building heights, explain how you know, without calculating, that the answer to Exercise 10 is greater than the answer to Exercise 11.

14. Given the building heights, explain how you know, without calculating, that the answer to Exercise 12 is less than the answer to Exercise 9.

For Exercises 15 through 18, the grades are given for a student for a particular semester. Find the grade point average. If necessary, round the grade point average to the nearest hundredth. See Example 2.

15.

Grade	Credit Hours
B	3
C	3
A	4
C	4

16.

Grade	Credit Hours
D	1
F	1
C	4
B	5

17.

Grade	Credit Hours
A	3
A	3
A	4
B	3
C	1

18.

Grade	Credit Hours
B	2
B	2
C	3
A	3
B	3

For Exercises 19 through 27, find the mean, median, and mode, as requested. See Examples 1 and 3 through 6. During an experiment, the following times (in seconds) were recorded:

7.8, 6.9, 7.5, 4.7, 6.9, 7.0.

19. Find the mean.

20. Find the median.

21. Find the mode.

In a mathematics class, the following test scores were recorded for a student: 93, 85, 89, 79, 88, 92.

22. Find the mean. Round to the nearest hundredth.

23. Find the median.

24. Find the mode.

The following pulse rates were recorded for a group of 15 students:

78, 80, 66, 68, 71, 64, 82, 71, 70, 65, 70, 75, 77, 86, 72.

25. Find the mean.

26. Find the median.

27. Find the mode.

28. How many pulse rates were higher than the mean?

29. How many pulse rates were lower than the mean?

Review

Write each fraction in simplest form. See Section 2.3.

30. $\dfrac{12}{20}$ **31.** $\dfrac{6}{18}$ **32.** $\dfrac{4}{36}$ **33.** $\dfrac{18}{30}$ **34.** $\dfrac{35}{100}$ **35.** $\dfrac{55}{75}$

Concept Extensions

Find the missing numbers in each set of numbers.

36. 16, 18, _____, _____, _____. The mode is 21. The median is 20.

37. _____, _____, _____, 40, _____. The mode is 35. The median is 37. The mean is 38.

38. Write a list of numbers for which you feel the median would be a better measure of central tendency than the mean.

39. Without making any computations, decide whether the median of the following list of numbers will be a whole number. Explain your reasoning.

36, 77, 29, 58, 43

7.4 Counting and Introduction to Probability

Objectives

A Use a Tree Diagram to Count Outcomes.

B Find the Probability of an Event.

Objective A Using a Tree Diagram

In our daily conversations, we often talk about the likelihood or **probability** of a given result occurring. For example:

The *chance* of thundershowers is 70 percent.

What are the *odds* that the New Orleans Saints will go to the Super Bowl?

What is the *probability* that you will finish cleaning your room today?

Each of these chance happenings—thundershowers, the New Orleans Saints playing in the Super Bowl, and cleaning your room today—is called an **experiment.** The possible results of an experiment are called **outcomes.** For example, flipping a coin is an experiment, and the possible outcomes are heads (H) or tails (T).

One way to picture the outcomes of an experiment is to draw a **tree diagram.** Each outcome is shown on a separate branch. For example, the outcomes of flipping a coin are

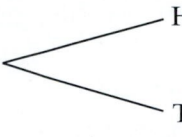

H

T

Heads

Tails

Example 1 Draw a tree diagram for tossing a coin twice. Then use the diagram to find the number of possible outcomes.

Solution:

First Coin Toss	Second Coin Toss	Outcomes
	H	H, H
H	T	H, T
	H	T, H
T	T	T, T

There are 4 possible outcomes when tossing a coin twice.

■ **Work Practice 1**

Example 2 Draw a tree diagram for an experiment consisting of rolling a die and then tossing a coin. Then use the diagram to find the number of possible outcomes.

Die

Solution: Recall that a die has six sides and that each side represents a number, 1 through 6.

Roll a Die	Toss a Coin	Outcomes
1	H	1, H
1	T	1, T
2	H	2, H
2	T	2, T
3	H	3, H
3	T	3, T
4	H	4, H
4	T	4, T
5	H	5, H
5	T	5, T
6	H	6, H
6	T	6, T

There are 12 possible outcomes for rolling a die and then tossing a coin.

■ **Work Practice 2**

Any number of outcomes considered together is called an **event.** For example, when tossing a coin twice, H, H is an event. The event is tossing heads first and tossing heads second. Another event would be tossing tails first and then heads (T, H), and so on.

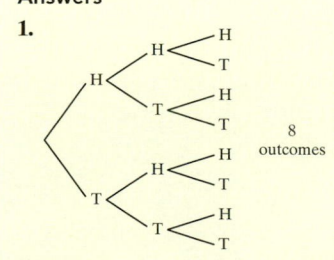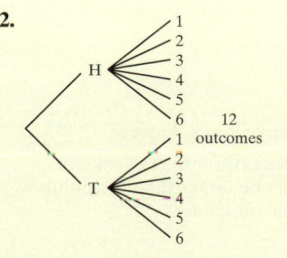

Objective B   Finding the Probability of an Event ▶

As we mentioned earlier, the **probability of an event is a measure of the chance or likelihood of it occurring.** For example, if a coin is tossed, what is the probability that heads occurs? Since one of two equally likely possible outcomes is heads, the probability is $\frac{1}{2}$.

The Probability of an Event

$$\text{probability of an event} = \frac{\text{number of ways that the event can occur}}{\text{number of possible outcomes}}$$

Note from the definition of probability that the probability of an event is always between 0 and 1, inclusive (i.e., including 0 and 1). A probability of 0 means that an event won't occur, and a probability of 1 means that an event is certain to occur.

Practice 3

If a coin is tossed three times, find the probability of tossing tails, then heads, then tails (T, H, T).

Example 3   If a coin is tossed twice, find the probability of tossing heads on the first toss and then heads again on the second toss (H, H).

Solution: 1 way the event can occur
↓

$$\underbrace{\text{H,T,} \quad \overbrace{\text{H,H,}} \quad \text{T,H,} \quad \text{T,T}}_{\text{4 possible outcomes}}$$

$$\text{probability} = \frac{1}{4} \quad \begin{array}{l}\text{Number of ways the event can occur}\\ \text{Number of possible outcomes}\end{array}$$

The probability of tossing heads and then heads is $\frac{1}{4}$.

■ **Work Practice 3**

Practice 4

If a die is rolled one time, find the probability of rolling a 2 or a 5.

Example 4   If a die is rolled one time, find the probability of rolling a 3 or a 4.

Solution: Recall that there are 6 possible outcomes when rolling a die.

2 ways that the event can occur

$$\text{possible outcomes:} \quad \underbrace{1, \quad 2, \quad 3, \quad 4, \quad 5, \quad 6}_{\text{6 possible outcomes}}$$

$$\text{probability of a 3 or a 4} = \frac{2}{6} \quad \begin{array}{l}\text{Number of ways the event can occur}\\ \text{Number of possible outcomes}\end{array}$$

$$= \frac{1}{3} \quad \text{Simplest form}$$

■ **Work Practice 4**

Answers

3. $\frac{1}{8}$ **4.** $\frac{1}{3}$

✓**Concept Check Answer**

The number of ways an event can occur can't be larger than the number of possible outcomes.

✓**Concept Check** Suppose you have calculated a probability of $\frac{11}{9}$. How do you know that you have made an error in your calculation?

Example 5　Find the probability of choosing a red marble from a box containing 1 red, 1 yellow, and 2 blue marbles.

Solution: 1 way that event can occur

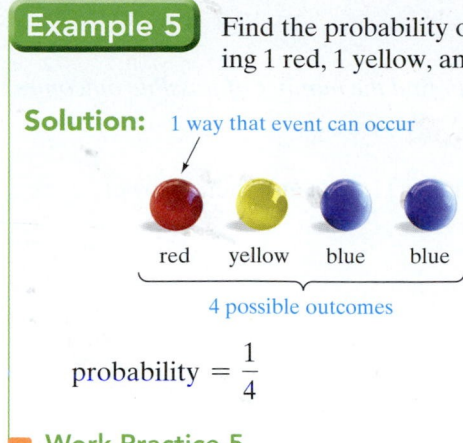

red　yellow　blue　blue

4 possible outcomes

$$\text{probability} = \frac{1}{4}$$

■ **Work Practice 5**

Practice 5

Use the diagram and information in Example 5 and find the probability of choosing a blue marble from the box.

Answer

5. $\frac{1}{2}$

Vocabulary, Readiness & Video Check

Use the choices below to fill in each blank. Choices may be used more than once.

0　　　probability　　　tree diagram

1　　　outcome

1. A possible result of an experiment is called a(n) _____.
2. A(n) _____ shows each outcome of an experiment as a separate branch.
3. The _____ of an event is a measure of the likelihood of it occurring.
4. _____ is calculated by the number of ways that the event can occur divided by the number of possible outcomes.
5. A probability of _____ means that an event won't occur.
6. A probability of _____ means that an event is certain to occur.

Martin-Gay Interactive Videos　*Watch the section lecture video and answer the following questions.*

See Video 7.4

Objective A　7. In ▤ Example 1, how was the possible number of outcomes to the experiment determined from the tree diagram? ▶

Objective B　8. In ▤ Example 2, what is the probability of getting a 7? Explain this result. ▶

7.4 Exercise Set MyMathLab®

Objective A *Draw a tree diagram for each experiment. Then use the diagram to find the number of possible outcomes. See Examples 1 and 2.*

1. Choosing a letter in the word MATH, then a number (1, 2, or 3)

2. Choosing a number (1 or 2) and then a vowel (a, e, i, o, u)

Use the following spinners to answer Exercises 3–10.

Spinner A

Spinner B

3. Spinning Spinner A once

4. Spinning Spinner B once

5. Spinning Spinner B twice

6. Spinning Spinner A twice

7. Spinning Spinner A and then Spinner B

8. Spinning Spinner B and then Spinner A

9. Tossing a coin and then spinning Spinner B

10. Spinning Spinner A and then tossing a coin

Objective B *If a single die is tossed once, find the probability of each event. See Examples 3 through 5.*

⊙ **11.** A 5

12. A 9

13. A 1 or a 6

14. A 2 or a 3

⊙ **15.** An even number

16. An odd number

17. A number greater than 2

18. A number less than 6

Suppose the spinner shown is spun once. Find the probability of each event. See Examples 3 through 5.

⊙ **19.** The result of the spin is 2.

20. The result of the spin is 3.

21. The result of the spin is 1, 2, or 3.

22. The result of the spin is not 3.

23. The result of the spin is an odd number.

24. The result of the spin is an even number.

If a single choice is made from the bag of marbles shown, find the probability of each event. See Examples 3 through 5.

25. A red marble is chosen.

26. A blue marble is chosen.

27. A yellow marble is chosen.

28. A green marble is chosen.

29. A green or red marble is chosen.

30. A blue or yellow marble is chosen.

A new drug is being tested that is supposed to lower blood pressure. This drug was given to 200 people and the results are shown below. See Examples 3 through 5.

Lower Blood Pressure	Higher Blood Pressure	Blood Pressure Not Changed
152	38	10

31. If a person is testing this drug, what is the probability that his or her blood pressure will be higher?

32. If a person is testing this drug, what is the probability that his or her blood pressure will be lower?

33. If a person is testing this drug, what is the probability that his or her blood pressure will not change?

34. What is the sum of the answers to Exercises 31, 32, and 33? In your own words, explain why.

Review

Perform each indicated operation. See Sections 2.4, 2.5, and 3.3.

35. $\dfrac{1}{2} + \dfrac{1}{3}$

36. $\dfrac{7}{10} - \dfrac{2}{5}$

37. $\dfrac{1}{2} \cdot \dfrac{1}{3}$

38. $\dfrac{7}{10} \div \dfrac{2}{5}$

39. $5 \div \dfrac{3}{4}$

40. $\dfrac{3}{5} \cdot 10$

Concept Extensions

Recall that a deck of cards contains 52 cards. These cards consist of four suits (hearts, spades, clubs, and diamonds) of each of the following: 2, 3, 4, 5, 6, 7, 8, 9, 10, jack, queen, king, and ace. If a card is chosen from a deck of cards, find the probability of each event.

41. The king of hearts

42. The 10 of spades

43. A king

44. A 10

45. A heart

46. A club

47. A card in black ink

48. A queen or ace

Two dice are tossed. Find the probability of each sum of the dice. (Hint: Draw a tree diagram of the possibilities of two tosses of a die, and then find the sum of the numbers on each branch.)

49. A sum of 6

50. A sum of 10

51. A sum of 13

52. A sum of 2

Solve. See the Concept Check in this section.

53. In your own words, explain why the probability of an event cannot be greater than 1.

54. In your own words, explain when the probability of an event is 0.

Chapter 7 Group Activity

Sections 7.1, 7.3

This activity may be completed by working in groups or individually.

How often have you read an article in a newspaper or in a magazine that included results from a survey or poll? Surveys seem to have become very popular ways of getting feedback on anything from a political candidate, to a new product, to services offered by a health club. In this activity, you will conduct a survey and analyze the results.

1. Conduct a survey of 30 students in one of your classes. Ask each student to report his or her age.

2. Classify each age according to the following categories: under 20, 20 to 24, 25 to 29, 30 to 39, 40 to 49,

and 50 or over. Tally the number of your survey respondents that fall into each category. Make a bar graph of your results. What does this graph tell you about the ages of your survey respondents?

3. Find the average age of your survey respondents.

4. Find the median age of your survey respondents.

5. Find the mode of the ages of your survey respondents.

6. Compare the mean, median, and mode of your age data. Are these measures similar? Which is largest? Which is smallest? If there is a noticeable difference between any of these measures, can you explain why?

Chapter 7 Vocabulary Check

Fill in each blank with one of the words or phrases listed below.

outcomes	bar	experiment	mean	tree diagram
pictograph	line	class interval	median	probability
histogram	circle	class frequency	mode	

1. A(n) _____ graph presents data using vertical or horizontal bars.

2. The _____ of a set of number items is $\dfrac{\text{sum of items}}{\text{number of items}}$.

3. The possible results of an experiment are the _____.

4. A(n) _____ is a graph in which pictures or symbols are used to visually present data.

5. The _____ of a set of numbers is the number that occurs most often.

6. A(n) _____ graph displays information with a line that connects data points.

7. The _____ of an ordered set of numbers is the middle number.

8. A(n) _____ is one way to picture and count outcomes.

9. A(n) _____ is an activity being considered, such as tossing a coin or rolling a die.

10. In a(n) _____ graph, each section (shaped like a piece of pie) shows a category and the relative size of the category.

11. The _____ of an event is $\dfrac{\text{number of ways that the event can occur}}{\text{number of possible outcomes}}$.

12. A(n) _____ is a special bar graph in which the width of each bar represents a(n) _____ and the height of each bar represents the _____.

Helpful Hint

○ Are you preparing for your test? Don't forget to take the Chapter 7 Test on page 531. Then check your answers at the back of your text and use the Chapter Test Prep Videos to see the fully worked-out solutions to any of the exercises you want to review.

7 Chapter Highlights

Definitions and Concepts	Examples

Section 7.1   Reading Pictographs, Bar Graphs, Histograms, and Line Graphs

A **pictograph** is a graph in which pictures or symbols are used to visually present data.

A **line graph** displays information with a line that connects data points.

A **bar graph** presents data using vertical or horizontal bars.

The bar graph on the right shows the number of acres of wheat harvested in 2013 for leading states.

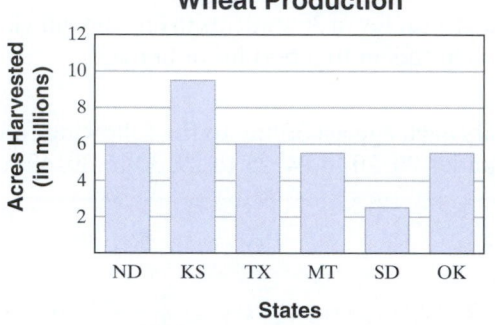

Wheat Production

Source: National Agricultural Statistics Service

1. Approximately how many acres of wheat were harvested in Kansas?

9,500,000 acres

2. About how many more acres of wheat were harvested in North Dakota than South Dakota?

$$\begin{array}{r} 6 \text{ million} \\ -2.5 \text{ million} \\ \hline 3.5 \text{ million} \end{array} \quad \text{or } 3,500,000 \text{ acres}$$

A **histogram** is a special bar graph in which the width of each bar represents a **class interval** and the height of each bar represents the **class frequency.** The histogram on the right shows student quiz scores.

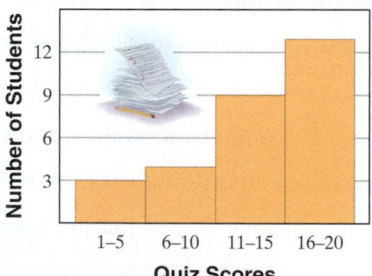

1. How many students received a score of 6–10?

4 students

2. How many students received a score of 11–20?

$9 + 13 = 22$ students

Definitions and Concepts	Examples

Section 7.2   Reading Circle Graphs

In a **circle graph,** each section (shaped like a piece of pie) shows a category and the relative size of the category.

The circle graph on the right classifies tornadoes by wind speed.

Tornado Wind Speeds

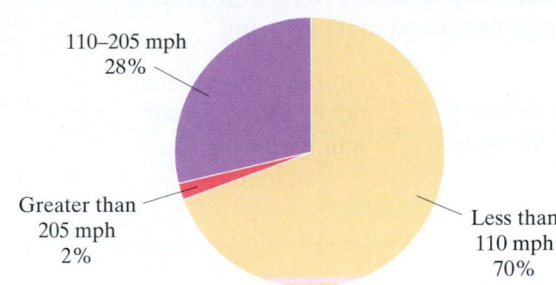

110–205 mph
28%

Greater than
205 mph
2%

Less than
110 mph
70%

Source: National Oceanic and Atmospheric Administration

1. What percent of tornadoes have wind speeds of 110 mph or greater?

   28% + 2% = 30%

2. If there were 1235 tornadoes in the United States in 1995, how many of these might we expect to have had wind speeds less than 110 mph? Find 70% of 1235.

   $70\%(1235) = 0.70(1235) = 864.5 \approx 865$

   Around 865 tornadoes would be expected to have had wind speeds of less than 110 mph.

Section 7.3   Mean, Median, and Mode

The **mean** (or **average**) of a set of number items is

$$\text{mean} = \frac{\text{sum of items}}{\text{number of items}}$$

The **median** of a set of numbers in numerical order is the middle number. If the number of items is even, the median is the mean of the two middle numbers.

The **mode** of a set of numbers is the number that occurs most often. (A set of numbers may have no mode or more than one mode.)

Find the mean, median, and mode of the following set of numbers: 33, 35, 35, 43, 68, 68

$$\text{mean} = \frac{33 + 35 + 35 + 43 + 68 + 68}{6} = 47$$

The median is the mean of the two middle numbers, 35 and 43

$$\text{median} = \frac{35 + 43}{2} = 39$$

There are two modes because there are two numbers that occur twice:

35 and 68

Section 7.4   Counting and Introduction to Probability

An **experiment** is an activity being considered, such as tossing a coin or rolling a die. The possible results of an experiment are the **outcomes.** A **tree diagram** is one way to picture and count outcomes.

Draw a tree diagram for tossing a coin and then choosing a number from 1 to 4.

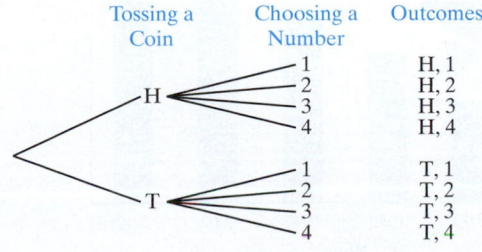

Tossing a Coin	Choosing a Number	Outcomes
H	1	H, 1
	2	H, 2
	3	H, 3
	4	H, 4
T	1	T, 1
	2	T, 2
	3	T, 3
	4	T, 4

(continued)

Definitions and Concepts	Examples
Section 7.4 Counting and Introduction to Probability (*continued*)	
Any number of outcomes considered together is called an **event**. The **probability** of an event is a measure of the chance or likelihood of it occurring. $$\text{probability of an event} = \frac{\text{number of ways that the event can occur}}{\text{number of possible outcomes}}$$	Find the probability of tossing a coin twice and tails occurring each time. 1 way the event can occur HH, HT, TH, TT 4 possible outcomes $$\text{probability} = \frac{1}{4}$$

Chapter 7   Review

(7.1) *The following pictograph shows the number of new homes constructed from August 2012 to August 2013, by region. Use this graph to answer Exercises 1 through 6.*

New Home Construction

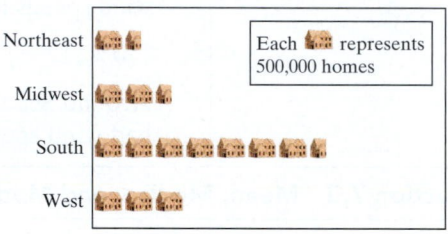

Each 🏠 represents 500,000 homes

Source: U.S. Census Bureau

1. How many new homes were constructed in the Midwest during the given year?

2. How many new homes were constructed in the West during the given year?

3. Which region had the most new homes constructed?

4. Which region had the fewest new homes constructed?

5. Which region(s) had 3,000,000 or more new homes constructed?

6. Which region(s) had fewer than 3,000,000 new homes constructed?

The following bar graph shows the percent of persons age 25 or over who completed four or more years of college. Use this graph to answer Exercises 7 through 10.

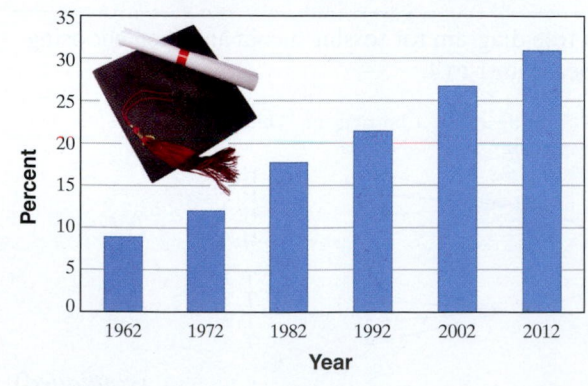

Four or More Years of College by Persons Age 25 or Over

7. Approximate the percent of persons who had completed four or more years of college in 1972.

8. What year shown had the greatest percent of persons completing four or more years of college?

9. What years shown had 20% or more of persons completing four or more years of college?

10. Describe any patterns you notice in this graph.

Source: U.S. Census Bureau

The following line graph shows the total number of Olympic medals awarded during the Summer Olympics between 1992 and 2012. Use this graph to answer Exercises 11 through 16.

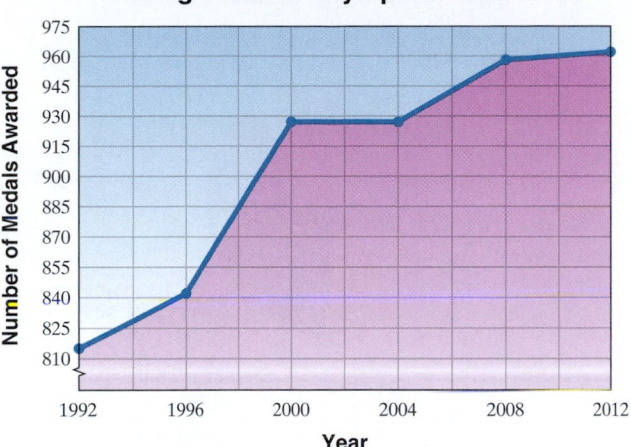

Number of Medals Awarded During Summer Olympics: 1992–2012

Source: International Olympic Committee

11. Approximate the number of medals awarded during the Summer Olympics of 2012.

12. Approximate the number of medals awarded during the Summer Olympics of 2000.

13. Between which two Summer Olympics did the number of medals awarded not change?

14. Between which two Summer Olympics did the number of medals awarded change the most?

15. How many more medals were awarded at the Summer Olympics of 1996 than at the Summer Olympics of 1992?

16. How many more medals were awarded at the Summer Olympics of 2012 than at the Summer Olympics of 1996?

The following histogram shows the hours worked per week by the employees of Southern Star Furniture. Use this histogram to answer Exercises 17 through 20.

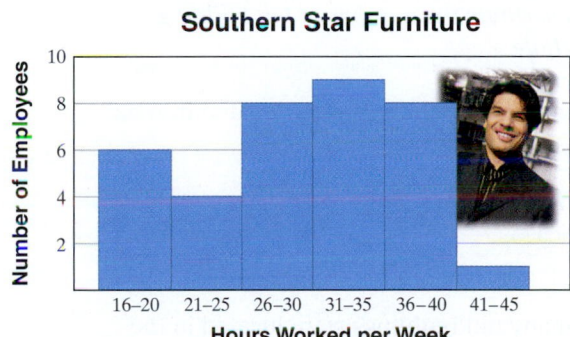

Southern Star Furniture

17. How many employees work 41–45 hours per week?

18. How many employees work 21–25 hours per week?

19. How many employees work 30 hours or less per week?

20. How many employees work 36 hours or more per week?

Following is a list of monthly record high temperatures for New Orleans, Louisiana. Use this list to complete the frequency distribution table below.

83	96	101	92
85	100	92	102
89	101	87	84

	Class Intervals (Temperatures)	Tally	Class Frequency (Number of Months)
21.	80°–89°		
22.	90°–99°		
23.	100°–109°		

24. Use the table from Exercises 21–23 to draw a histogram.

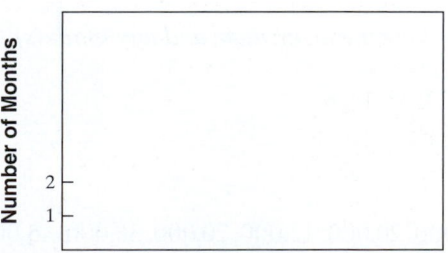

Temperatures

(7.2) *The following circle graph shows a family's $4000 monthly budget. Use this graph to answer Exercises 25 through 30. Write all ratios as fractions in simplest form.*

25. What is the largest budget item?

26. What is the smallest budget item?

27. How much money is budgeted for the mortgage payment and utilities?

28. How much money is budgeted for savings and contributions?

29. Find the ratio of the mortage payment to the total monthly budget.

30. Find the ratio of food to the total monthly budget.

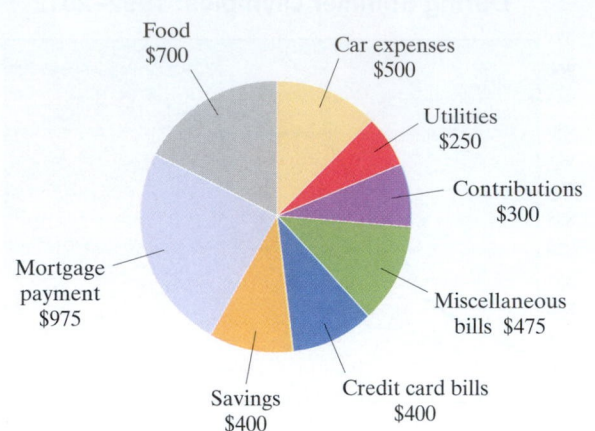

In 2013, there were 65 buildings over 1000 feet tall in the world. The following circle graph shows the percent of buildings over 1000 feet tall in the world by region in 2013. Use this graph to determine the number of tall buildings on each continent in Exercises 31 through 34. Round each answer to the nearest whole.

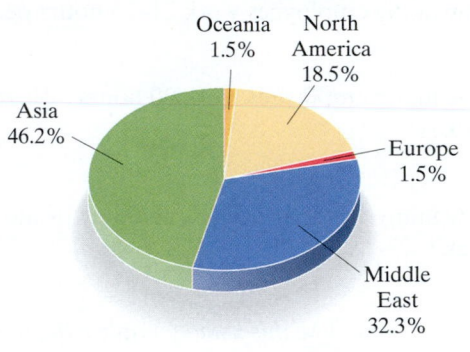

Source: Council on Tall Buildings and Urban Habitat

31. How many tall buildings were located in Asia?

32. How many tall buildings were located in North America?

33. How many tall buildings were located in the Middle East?

34. How many tall buildings were located in Europe?

(7.3) *Find the mean, median, and any mode(s) for each list of numbers. If necessary, round to the nearest tenth.*

35. 13, 23, 33, 14, 6

36. 45, 86, 21, 60, 86, 64, 45

37. 14,000, 20,000, 12,000, 20,000, 36,000, 45,000

38. 560, 620, 123, 400, 410, 300, 400, 780, 430, 450

For Exercises 39 and 40, the grades are given for a student for a particular semester. Find each grade point average. If necessary, round the grade point average to the nearest hundredth.

39.

Grade	Credit Hours
A	3
A	3
C	2
B	3
C	1

40.

Grade	Credit Hours
B	3
B	4
C	2
D	2
B	3

(7.4) *Draw a tree diagram for each experiment. Then use the diagram to determine the number of outcomes.*

Spinner 1

Spinner 2

41. Tossing a coin and then spinning Spinner 1

42. Spinning Spinner 2 and then tossing a coin

43. Spinning Spinner 1 twice

44. Spinning Spinner 2 twice

45. Spinning Spinner 1 and then Spinner 2

Find the probability of each event.

46. Rolling a 4 on a die

47. Rolling a 3 on a die

48. Spinning a 4 on the spinner

49. Spinning a 3 on the spinner

50. Spinning either a 1, 3, or 5 on the spinner

51. Spinning either a 2 or a 4 on the spinner

52. Rolling an even number on a die

53. Rolling a number greater than 3 on a die

Mixed Review

Find the mean, median, and any mode(s) for each list of numbers. If needed, round answers to two decimal places.

54. 73, 82, 95, 68, 54

55. 25, 27, 32, 98, 62

56. 750, 500, 427, 322, 500, 225

57. 952, 327, 566, 814, 327, 729

Given a bag containing 2 red marbles, 2 blue marbles, 3 yellow marbles, and 1 green marble, find the following:

58. The probability of choosing a blue marble from the bag

59. The probability of choosing a yellow marble from the bag

60. The probability of choosing a red marble from the bag

61. The probability of choosing a green marble from the bag

The following pictograph shows the money collected each week from a wrapping paper fundraiser. Use this graph to answer Exercises 1 through 3.

Answers

Weekly Wrapping Paper Sales

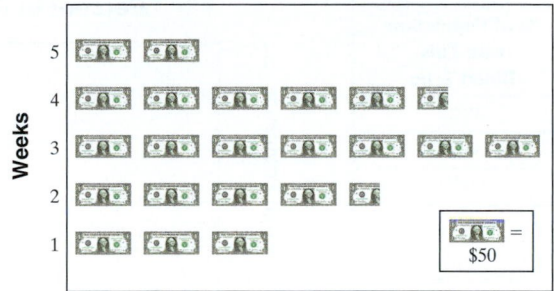

1. How much money was collected during the second week?

2. During which week was the most money collected? How much money was collected during that week?

3. What was the total money collected for the fundraiser?

The bar graph shows the normal monthly precipitation in centimeters for Chicago, Illinois. Use this graph to answer Exercises 4 through 6.

Chicago Precipitation

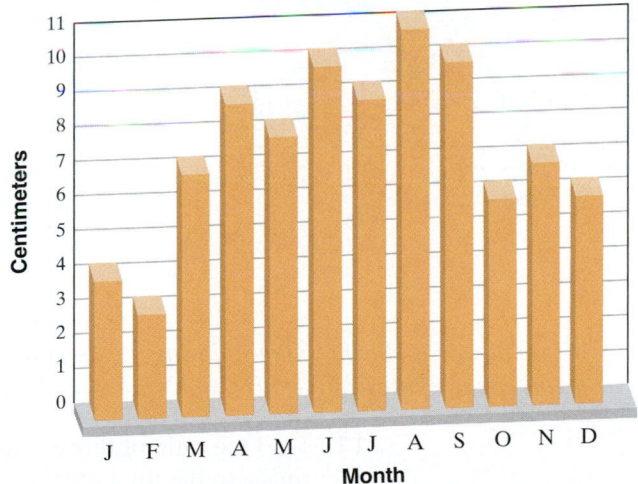

Source: U.S. National Oceanic and Atmospheric Administration, *Climatography of the United States*, No. 81

4. During which month(s) does Chicago normally have more than 9 centimeters of precipitation?

5. During which month does Chicago normally have the least amount of precipitation? How much precipitation occurs during that month?

1. _____

2. _____

3. _____

4. _____

5. _____

6. During which month(s) does 7 centimeters of precipitation normally occur?

6. _____

7. Use the information in the table to draw a bar graph. Clearly label each bar.

7. _____

Most Common Blood Types	
Blood Type	**% of Population with This Blood Type**
O+	38%
A+	34%
B+	9%
O−	7%
A−	6%
AB+	3%
B−	2%
AB−	1%

8. _____

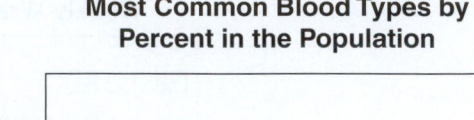

Most Common Blood Types by Percent in the Population

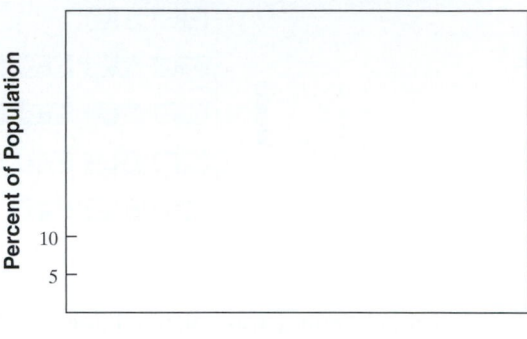

The following line graph shows the average annual inflation rate in the United States for the years 2003–2013. Use this graph to answer Exercises 8 through 10.

9. _____

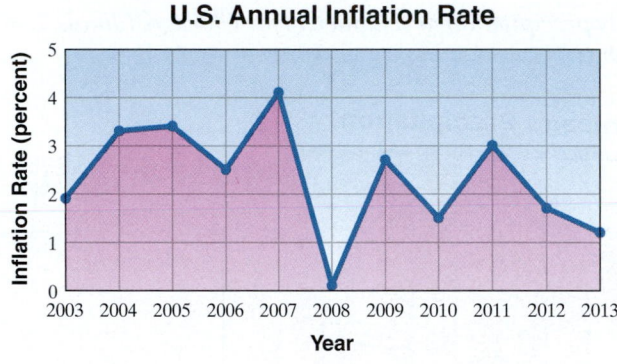

U.S. Annual Inflation Rate

Source: Bureau of Labor Statistics

8. Approximate the annual inflation rate in 2011.

9. During which of the years shown was the inflation rate greater than 3%?

10. During which sets of years was the inflation rate increasing?

10. _____

The result of a survey of 200 people is shown in the following circle graph. Each person was asked to tell his or her favorite type of music. Use this graph to answer Exercises 11 and 12.

11. _____

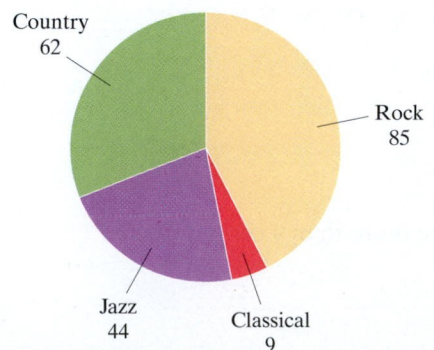

Country 62

Rock 85

Jazz 44

Classical 9

11. Find the ratio of those who prefer rock music to the total number surveyed.

12. Find the ratio of those who prefer country music to those who prefer jazz.

12. _____

The following graph (from earlier in this chapter) shows the percent of visitors to the United States in 2012 by various regions. During 2012, there were approximately 66,700,000 foreign visitors to the United States. Use the graph to find how many people visited the United States in 2012 from the world regions given in Exercises 13 and 14. (Source: Office of Travel & Tourism Industries)

Visitors to U.S. by Region

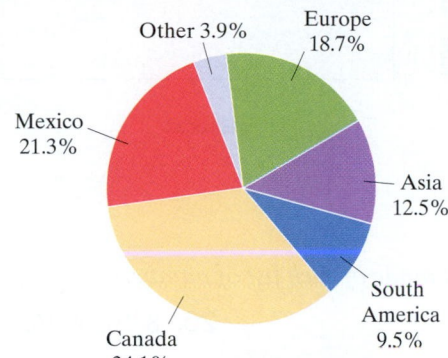

13. Canada   **14.** Asia

A professor measures the heights of the students in her class. The results are shown in the following histogram. Use this histogram to answer Exercises 15 and 16.

Student Heights

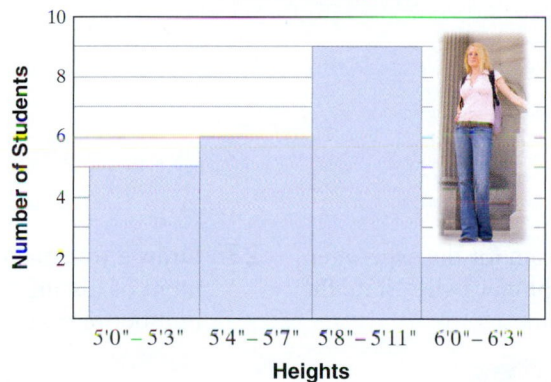

15. How many students are 5′8″–5′11″ tall?

16. How many students are 5′7″ or shorter?

17. The history test scores of 25 students are shown below. Use these scores to complete the frequency distribution table.

70	86	81	65	92
43	72	85	69	97
82	51	75	50	68
88	83	85	77	99
77	63	59	84	90

Class Intervals (Scores)	Tally	Class Frequency (Number of Students)
40–49		
50–59		
60–69		
70–79		
80–89		
90–99		

13. _____

14. _____

15. _____

16. _____

17. _____

18. _____

18. Use the results of Exercise 17 to draw a histogram.

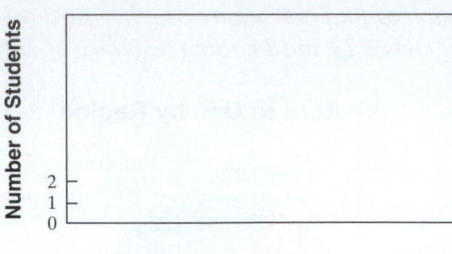

19. _____

20. _____

Find the mean, median, and mode of each list of numbers.

19. 26, 32, 42, 43, 49

20. 8, 10, 16, 16, 14, 12, 12, 13

21. _____

Find the grade point average. If necessary, round to the nearest hundredth.

21.

Grade	Credit Hours
A	3
B	3
C	3
B	4
A	1

22. _____

22. Draw a tree diagram for the experiment of spinning the spinner twice. State the number of outcomes.

23. Draw a tree diagram for the experiment of tossing a coin twice. State the number of outcomes.

23. _____

24. _____

Suppose that the numbers 1 to 10 are each written on a scrap of paper and placed in a bag. You then select one number from the bag.

25. _____

24. What is the probability of choosing a 6 from the bag?

25. What is the probability of choosing a 3 or a 4 from the bag?

1. Simplify: $(8 - 6)^2 + 2^3 \cdot 3$

2. Simplify: $48 \div 8 \cdot 2$

3. Write $\dfrac{30}{108}$ in simplest form.

4. Subtract: $\dfrac{19}{40} - \dfrac{3}{10}$

5. Add: $1\dfrac{4}{5} + 4 + 2\dfrac{1}{2}$

6. Multiply: $5\dfrac{1}{3} \cdot 2\dfrac{1}{8}$

△ **7.** The area of a triangle is
$$\text{Area} = \dfrac{1}{2} \cdot \text{base} \cdot \text{height}.$$
Find the area of a triangle with height 3 feet and base length 5.6 feet.

8. Find the perimeter of a rectangle with length $3\dfrac{1}{2}$ meters and width $1\dfrac{1}{2}$ meters.

9. Subtract. Check each answer by adding.
  a. $12 - 9$
  b. $22 - 7$
  c. $35 - 35$
  d. $70 - 0$

10. Multiply.
  a. $20 \cdot 0$
  b. $20 \cdot 1$
  c. $0 \cdot 20$
  d. $1 \cdot 20$

11. Round 248,982 to the nearest hundred.

12. Round 248,982 to the nearest thousand.

13. Multiply:
  a. $\begin{array}{r} 25 \\ \times\, 8 \\ \hline \end{array}$
  b. $\begin{array}{r} 246 \\ \times\, 5 \\ \hline \end{array}$

14. Divide: $10,468 \div 28$

Answers

1. _____

2. _____

3. _____

4. _____

5. _____

6. _____

7. _____

8. _____

9. a. _____

  b. _____

  c. _____

  d. _____

10. a. _____

  b. _____

  c. _____

  d. _____

11. _____

12. _____

13. a. _____

  b. _____

14. _____

15. _____

16. _____

17. _____

18. _____

19. _____

20. _____

21. _____

22. _____

23. _____

24. _____

25. _____

26. _____

27. _____

28. _____

29. _____

30. _____

31. _____

32. _____

15. The director of a learning lab at a local community college is working on next year's budget. Thirty-three new DVD players are needed at a cost of $187 each. What is the total cost of these DVD players?

16. A study is being conducted for erecting soundproof walls along the interstate of a metropolitan area. The following feet of walls are part of the proposal. Find their total: 4800 feet, 3270 feet, 2761 feet 5760 feet.

17. Find the prime factorization of 45.

18. Find $\sqrt{64}$.

Write each percent as a decimal.

19. 4.6%

20. 0.29%

21. 190%

22. 452%

Write each percent as a fraction in simplest form.

23. 40%

24. 27%

25. $33\frac{1}{3}\%$

26. $61\frac{1}{7}\%$

27. Translate to an equation: Five is what percent of 20?

28. Translate to a proportion: Five is what percent of 20?

29. Find the sales tax and the total price on the purchase of an $85.50 atlas in a city where the sales tax rate is 7.5%.

30. A salesperson makes a 7% commission rate on her total sales. If her total sales are $23,000, what is her commission?

31. An accountant invested $2000 at a simple interest rate of 10% for 2 years. What total amount of money will she have from her investment in 2 years?

32. Find the mean (or average) of 28, 35, 40, and 32.

33. Find the complement of a 48° angle.

34. Find the supplement of a 48° angle.

35. Find: $\sqrt{\dfrac{1}{36}}$

36. Find: $\sqrt{\dfrac{1}{25}}$

37. Find the mode of the list of numbers: 11, 14, 14, 16, 31, 56, 65, 77, 77, 78, 79

38. Find the median of the numbers in Exercise 37.

39. If a coin is tossed twice, find the probability of tossing heads on the first toss and then heads again on the second toss (H, H).

40. A bag contains 3 red marbles and 2 blue marbles. Find the probability of choosing a red marble.

33. _____

34. _____

35. _____

36. _____

37. _____

38. _____

39. _____

40. _____

8 Real Numbers and Introduction to Algebra

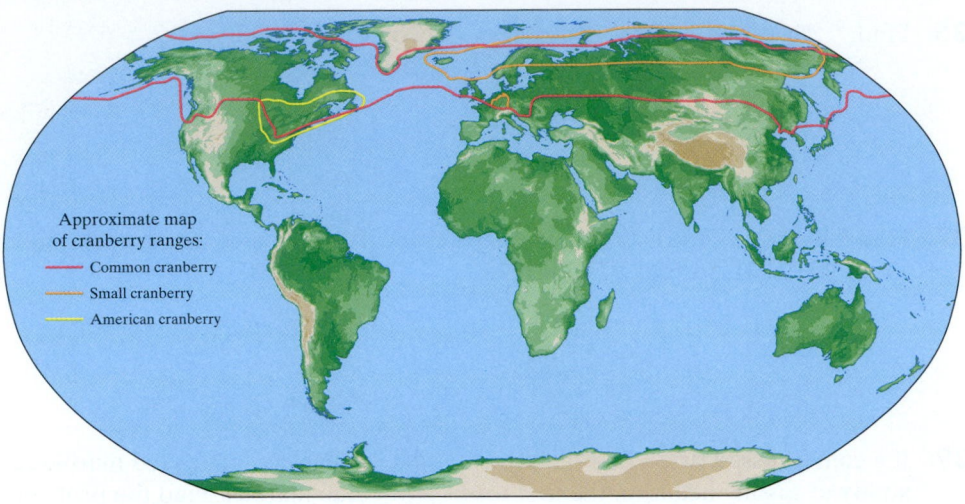

Approximate map of cranberry ranges:
— Common cranberry
— Small cranberry
— American cranberry

In this chapter, we begin with a review of the basic symbols—the language—of mathematics. We then introduce algebra by using a variable in place of a number. From there, we translate phrases to algebraic expressions and sentences to equations. This is the beginning of problem solving, which we formally study in Chapter 9.

Where Are Cranberries Produced?

The cranberry is one of only three fruits that are native to North America (the other two are the blueberry and the Concord grape). Native Americans pounded cranberries into a paste that they mixed with dried meat to create pemmican. American and Canadian sailors took cranberries on long sea voyages to combat scurvy because of their high vitamin C content.

Americans consume 400 million pounds of cranberries annually, 20% of them during Thanksgiving week. They are available in many forms: fresh, jellied, dried, and in juice. They are high in antioxidants and beneficial to the health of gums and teeth. In Exercises 81–84, Section 8.1, we will explore information about cranberry crops from the top five cranberry-producing states. (*Source:* Cape Cod Cranberry Growers Association)

Cranberry flower—named because of resemblance of Crane

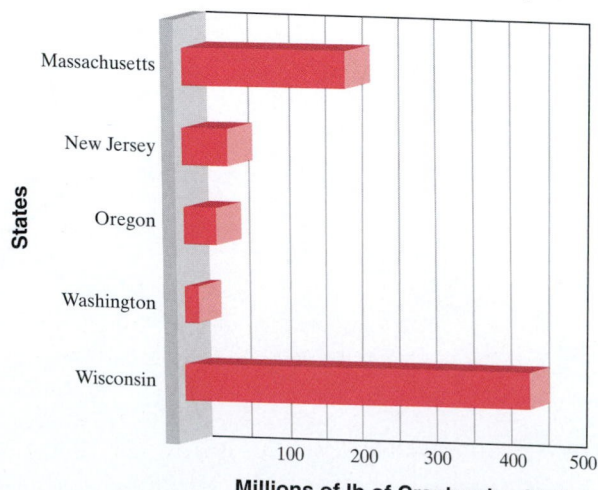

Top Cranberry-Producing States (in millions of pounds)

Source: National Agricultural Statistics Service

8.1 Symbols and Sets of Numbers ▶

We begin with a review of the set of natural numbers and the set of whole numbers and how we use symbols to compare these numbers. A **set** is a collection of objects, each of which is called a **member** or **element** of the set. A pair of brace symbols { } encloses the list of elements and is translated as "the set of" or "the set containing."

Objectives

A Define the Meaning of the Symbols =, ≠, <, >, ≤, and ≥. ▶

B Translate Sentences into Mathematical Statements. ▶

C Identify Integers, Rational Numbers, Irrational Numbers, and Real Numbers. ▶

D Find the Absolute Value of a Real Number. ▶

Natural Numbers

$\{1, 2, 3, 4, 5, 6, \ldots\}$

Whole Numbers

$\{0, 1, 2, 3, 4, 5, 6, \ldots\}$

Helpful Hint

The three dots (an ellipsis) at the end of the list of elements of a set means that the list continues in the same manner indefinitely.

Objective A Equality and Inequality Symbols ▶

Picturing natural numbers and whole numbers on a number line helps us to see the order of the numbers. Symbols can be used to describe in writing the order of two quantities. We will use equality symbols and inequality symbols to compare quantities.

Below is a review of these symbols. The letters a and b are used to represent quantities. Letters such as a and b that are used to represent numbers or quantities are called **variables.**

Equality and Inequality Symbols

		Meaning
Equality symbol:	$a = b$	a is equal to b.
Inequality symbols:	$a \neq b$	a is not equal to b.
	$a < b$	a is less than b.
	$a > b$	a is greater than b.
	$a \leq b$	a is less than or equal to b.
	$a \geq b$	a is greater than or equal to b.

These symbols may be used to form **mathematical statements** such as

$$2 = 2 \quad \text{and} \quad 2 \neq 6$$

Recall that on a number line, we see that a number **to the right of** another number is **larger.** Similarly, a number **to the left of** another number is **smaller.** For example, 3 is to the left of 5 on the number line, which means that 3 is less than 5, or $3 < 5$. Similarly, 2 is to the right of 0 on the number line, which means that 2 is greater than 0, or $2 > 0$. Since 0 is to the left of 2, we can also say that 0 is less than 2, or $0 < 2$.

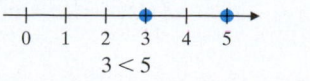
$3 < 5$

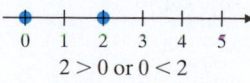
$2 > 0 \text{ or } 0 < 2$

Helpful Hint

Recall that $2 > 0$ has exactly the same meaning as $0 < 2$. Switching the order of the numbers and reversing the direction of the inequality symbol does not change the meaning of the statement.

$6 > 4$ has the same meaning as $4 < 6$.

Also notice that when the statement is true, the inequality symbol points to the smaller number.

Our discussion on the previous page can be generalized in the order property below.

Order Property for Real Numbers

For any two real numbers a and b, a is less than b if a is to the left of b on a number line.

$$a < b \text{ or also } b > a$$

Practice 1–6

Determine whether each statement is true or false.

1. $8 < 6$ **2.** $100 > 10$
3. $21 \leq 21$ **4.** $21 \geq 21$
5. $0 \geq 5$ **6.** $25 \geq 22$

Helpful Hint
If either $3 < 3$ or $3 = 3$ is true, then $3 \leq 3$ is true.

Examples   Determine whether each statement is true or false.

1. $2 < 3$ True.   Since 2 is to the left of 3 on a number line
2. $72 < 27$ False.   72 is to the right of 27 on a number line, so $72 > 27$.
3. $8 \geq 8$ True.   Since $8 = 8$ is true
4. $8 \leq 8$ True.   Since $8 = 8$ is true
5. $23 \leq 0$ False.   Since neither $23 < 0$ nor $23 = 0$ is true
6. $0 \leq 23$ True.   Since $0 < 23$ is true

■ **Work Practice 1–6**

Objective B   Translating Sentences into Mathematical Statements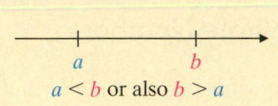

Now, let's use the symbols discussed above to translate sentences into mathematical statements.

Practice 7

Translate each sentence into a mathematical statement.

a. Fourteen is greater than or equal to fourteen.
b. Zero is less than five.
c. Nine is not equal to ten.

Example 7   Translate each sentence into a mathematical statement.

a. Nine is less than or equal to eleven.
b. Eight is greater than one.
c. Three is not equal to four.

Solution:

a.
nine	is less than or equal to	eleven
↓	↓	↓
9	$\leq$	11

b.
eight	is greater than	one
↓	↓	↓
8	$>$	1

c.
three	is not equal to	four
↓	↓	↓
3	$\neq$	4

■ **Work Practice 7**

Answers
1. false   **2.** true   **3.** true   **4.** true
5. false   **6.** true   **7. a.** $14 \geq 14$
b. $0 < 5$ **c.** $9 \neq 10$

Objective C Identifying Common Sets of Numbers

Whole numbers are not sufficient to describe many situations in the real world. For example, quantities smaller than zero must sometimes be represented, such as temperatures less than 0 degrees.

Recall that we can place numbers less than zero on a number line as follows: Numbers less than 0 are to the left of 0 and are labeled −1, −2, −3, and so on. The numbers we have labeled on the number line below are called the set of **integers.**

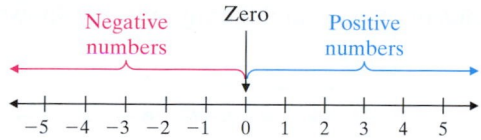

Integers to the left of 0 are called **negative integers;** integers to the right of 0 are called **positive integers.** The integer 0 is neither positive nor negative.

Integers

$$\{\ldots, -3, -2, -1, 0, 1, 2, 3, \ldots\}$$

Helpful Hint

A − sign, such as the one in −2, tells us that the number is to the left of 0 on a number line.

−2 is read "negative two."

A + sign or no sign tells us that the number lies to the right of 0 on a number line. For example, 3 and +3 both mean positive three.

Example 8 Use an integer to express the number in the following. "The lowest temperature ever recorded at South Pole Station, Antarctica, occurred during the month of June. The record-low temperature was 117 degrees below zero." (*Source:* The National Oceanic and Atmospheric Administration)

Solution: The integer −117 represents 117 degrees below zero.

■ **Work Practice 8**

Practice 8

Use an integer to express the number in the following. The average elevation of New Orleans, Louisiana, is 8 feet below sea level. (*Source: The World Almanac*)

Answer
8. −8

A problem with integers in real-life settings arises when quantities are smaller than some integer but greater than the next smallest integer. On a number line, these quantities may be visualized by points between integers. Some of these quantities between integers can be represented as a quotient of integers. For example,

The point on the number line halfway between 0 and 1 can be represented by $\frac{1}{2}$, a quotient of integers.

The point on the number line halfway between 0 and -1 can be represented by $-\frac{1}{2}$. Other quotients of integers and their graphs are shown below.

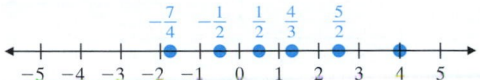

These numbers, each of which can be represented as a quotient of integers, are examples of **rational numbers.** It's not possible to list the set of rational numbers using the notation that we have been using. For this reason, we will use a different notation.

Rational Numbers

$$\left\{ \frac{a}{b} \,\middle|\, a \text{ and } b \text{ are integers and } b \neq 0 \right\}$$

We read this set as "the set of numbers $\frac{a}{b}$ such that a and b are integers and **b is not equal to 0.**"

Helpful Hint

We commonly refer to rational numbers as fractions.

Notice that every integer is also a rational number since each integer can be written as a quotient of integers. For example, the integer 5 is also a rational number since $5 = \frac{5}{1}$. For the rational number $\frac{5}{1}$, recall that the top number, 5, is called the numerator and the bottom number, 1, is called the denominator.

Let's practice **graphing** numbers on a number line.

| Example 9 | Graph the numbers on a number line.

$$-\frac{4}{3}, \quad \frac{1}{4}, \quad \frac{3}{2}, \quad -2\frac{1}{8}, \quad 3.5$$

Solution: To help graph the improper fractions in the list, we first write them as mixed numbers.

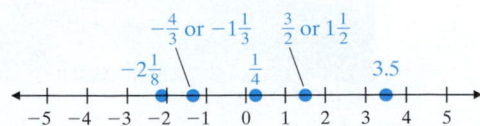

■ **Work Practice 9**

Every rational number has a point on the number line that corresponds to it. But not every point on the number line corresponds to a rational number. Those points that do not correspond to rational numbers correspond instead to **irrational numbers.**

Practice 9

Graph the numbers on the number line.

$$-2\frac{1}{2}, \quad -\frac{2}{3}, \quad \frac{1}{5}, \quad \frac{5}{4}, \quad 2.25$$

Answer

9.

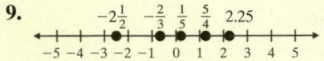

Irrational Numbers

{Nonrational numbers that correspond to points on a number line}

An irrational number that you have probably seen is π. Also, $\sqrt{2}$, the length of the diagonal of the square shown below, is an irrational number.

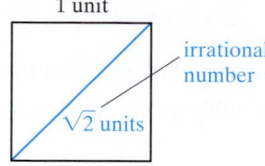

Both rational and irrational numbers can be written as decimal numbers. The decimal equivalent of a rational number will either terminate or repeat in a pattern. For example, upon dividing we find that

$$\frac{3}{4} = 0.75 \qquad \text{(Decimal number terminates or ends.)}$$

$$\frac{2}{3} = 0.66666\ldots \qquad \text{(Decimal number repeats in a pattern.)}$$

The decimal representation of an irrational number will neither terminate nor repeat. (For further review of decimals, see Chapter 4.)

The set of numbers, each of which corresponds to a point on a number line, is called the set of **real numbers.** One and only one point on a number line corresponds to each real number.

Real Numbers

{All numbers that correspond to points on a number line}

Several different sets of numbers have been discussed in this section. The following diagram shows the relationships among these sets of real numbers. Notice that, together, the rational numbers and the irrational numbers make up the real numbers.

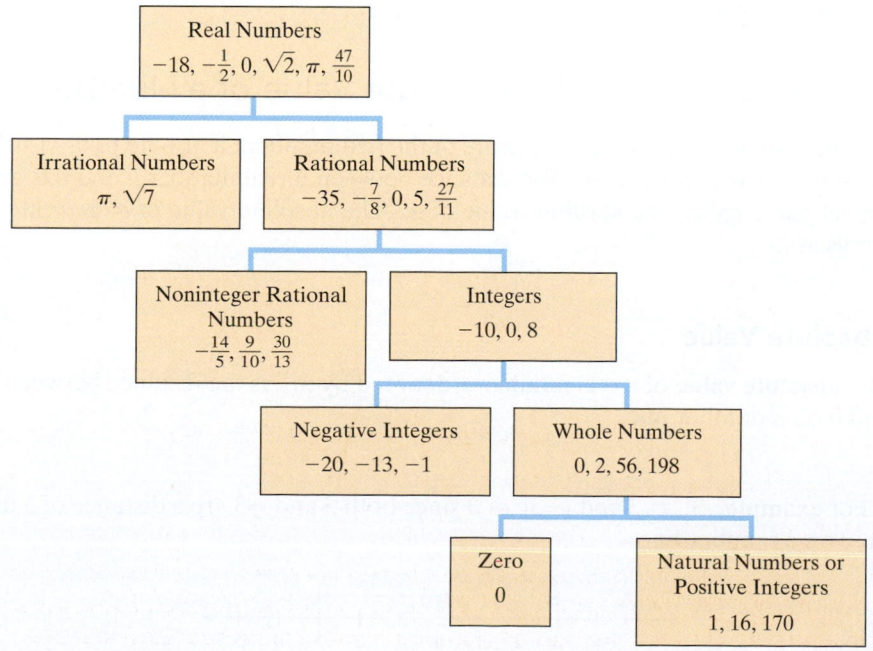

Now that other sets of numbers have been reviewed, let's continue our practice of comparing numbers.

Practice 10

Insert $<$, $>$, or $=$ between the pairs of numbers to form true statements.

a. -11 -9 **b.** 4.511 4.151

c. $\dfrac{7}{8}$ $\dfrac{2}{3}$

Example 10   Insert $<$, $>$, or $=$ between the pairs of numbers to form true statements.

a. -5 -6 **b.** 3.195 3.2 **c.** $\dfrac{1}{4}$ $\dfrac{1}{3}$

Solution:

a. $-5 > -6$ since -5 lies to the right of -6 on a number line.

b. By comparing digits in the same place values, we find that $3.195 < 3.2$, since $0.1 < 0.2$.

c. By dividing, we find that $\dfrac{1}{4} = 0.25$ and $\dfrac{1}{3} = 0.33\ldots$. Since $0.25 < 0.33\ldots$, $\dfrac{1}{4} < \dfrac{1}{3}$.

■ **Work Practice 10**

Practice 11

Given the set
$\left\{ -100, -\dfrac{2}{5}, 0, \pi, 6, 913 \right\}$, list the numbers in this set that belong to the set of:

a. Natural numbers

b. Whole numbers

c. Integers

d. Rational numbers

e. Irrational numbers

f. Real numbers

Example 11   Given the set $\left\{ -2, 0, \dfrac{1}{4}, 112, -3, 11, \sqrt{2} \right\}$, list the numbers in this set that belong to the set of:

a. Natural numbers   **b.** Whole numbers   **c.** Integers

d. Rational numbers   **e.** Irrational numbers   **f.** Real numbers

Solution:

a. The natural numbers are 11 and 112.

b. The whole numbers are $0, 11,$ and 112.

c. The integers are $-3, -2, 0, 11,$ and 112.

d. Recall that integers are also rational numbers. The rational numbers are $-3, -2, 0, \dfrac{1}{4}, 11,$ and 112.

e. The only irrational number is $\sqrt{2}$.

f. All numbers in the given set are real numbers.

■ **Work Practice 11**

Objective D   Finding the Absolute Value of a Number

A number line not only gives us a picture of the real numbers, it also helps us visualize the distance between numbers. The distance between a real number a and 0 is given a special name called the **absolute value** of a. "The absolute value of a" is written in symbols as $|a|$.

> **Absolute Value**
>
> The **absolute value** of a real number a, denoted by $|a|$, is the distance between a and 0 on a number line.

For example, $|3| = 3$ and $|-3| = 3$ since both 3 and -3 are a distance of 3 units from 0 on a number line.

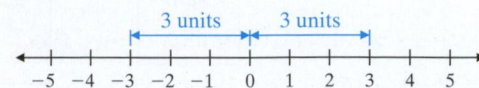

Example 12   Find the absolute value of each number.

a. $|4|$ **b.** $|-5|$ **c.** $|0|$ **d.** $\left|-\dfrac{2}{9}\right|$ **e.** $|4.93|$

Solution:

a. $|4| = 4$ since 4 is 4 units from 0 on a number line.

b. $|-5| = 5$ since -5 is 5 units from 0 on a number line.

c. $|0| = 0$ since 0 is 0 units from 0 on a number line.

d. $\left|-\dfrac{2}{9}\right| = \dfrac{2}{9}$

e. $|4.93| = 4.93$

■ **Work Practice 12**

Practice 12

Find the absolute value of each number.

a. $|7|$ **b.** $|-8|$ **c.** $\left|\dfrac{2}{3}\right|$

d. $|0|$ **e.** $|-3.06|$

Example 13   Insert $<$, $>$, or $=$ in the appropriate space to make each statement true.

a. $|0|$ 2   **b.** $|-5|$ 5   **c.** $|-3|$ $|-2|$

d. $|-9|$ $|-9.7|$ **e.** $\left|-7\dfrac{1}{6}\right|$ $|7|$

Solution:

a. $|0| < 2$ since $|0| = 0$ and $0 < 2$.

b. $|-5| = 5$.

c. $|-3| > |-2|$ since $3 > 2$.

d. $|-9| < |-9.7|$ since $9 < 9.7$.

e. $\left|-7\dfrac{1}{6}\right| > |7|$ since $7\dfrac{1}{6} > 7$.

■ **Work Practice 13**

Practice 13

Insert $<$, $>$, or $=$ in the appropriate space to make each statement true.

a. $|-4|$ 4

b. -3 $|0|$

c. $|-2.7|$ $|-2|$

d. $|-6|$ $|-16|$

e. $|10|$ $\left|-10\dfrac{1}{3}\right|$

Answers

12. a. 7   **b.** 8   **c.** $\dfrac{2}{3}$ **d.** 0   **e.** 3.06

13. a. $=$ **b.** $<$ **c.** $>$ **d.** $<$ **e.** $<$

Vocabulary, Readiness & Video Check

Use the choices below to fill in each blank. Not all choices will be used.

real	natural	absolute value	$\dfrac{1}{2}$	$\dfrac{1}{4}$	$|a|$	whole
rational	inequality	integers	0	1	$|-1|$	

1. The _____ numbers are $\{0, 1, 2, 3, 4, \ldots\}$.

2. The _____ numbers are $\{1, 2, 3, 4, 5, \ldots\}$.

3. The symbols $\neq$, $\leq$, and $>$ are called _____ symbols.

4. The _____ are $\{\ldots, -3, -2, -1, 0, 1, 2, 3, \ldots\}$.

5. The _____ numbers are $\{$ all numbers that correspond to points on a number line $\}$.

6. The _____ numbers are $\left\{\dfrac{a}{b} \,\middle|\, a \text{ and } b \text{ are integers, } b \neq 0\right\}$.

7. The integer _____ is neither positive nor negative.

8. The point on a number line halfway between 0 and $\dfrac{1}{2}$ can be represented by _____.

9. The distance between a real number a and 0 is called the _____ of a.

10. The absolute value of a is written in symbols as _____.

Martin-Gay Interactive Videos   Watch the section lecture video and answer the following questions.

Objective A 11. In ⊞ Example 2, why is the symbol < inserted between the two numbers? ▶

Objective B 12. Write the sentence given in ⊞ Example 4 and translate it to a mathematical statement, using symbols. ▶

Objective C 13. Which sets of numbers does the number in ⊞ Example 6 belong to? Why is this number not an irrational number? ▶

Objective D 14. Complete this statement based on the lecture given before ⊞ Example 8. The _____ of a real number a, denoted by $|a|$, is the distance between a and 0 on a number line. ▶

See Video 8.1 🍎

8.1 Exercise Set MyMathLab® ▶

Objectives A C Mixed Practice *Insert* <, >, *or* = *in the space between the paired numbers to make each statement true. See Examples 1 through 6 and 10.*

1. 4   10

2. 8   5

▶3. 7   3

4. 9   15

5. 6.26   6.26

6. 1.13   1.13

▶7. 0   7

8. 20   0

9. The freezing point of water is 32° Fahrenheit. The boiling point of water is 212° Fahrenheit. Write an inequality statement using < or > comparing the numbers 32 and 212.

10. The freezing point of water is 0° Celsius. The boiling point of water is 100° Celsius. Write an inequality statement using < or > comparing the numbers 0 and 100.

△11. An angle measuring 30° and an angle measuring 45° are shown. Write an inequality statement using ≤ or ≥ comparing the numbers 30 and 45.

△12. The sum of the measures of the angles of a parallelogram is 360°. The sum of the measures of the angles of a triangle is 180°. Write an inequality statement using ≤ or ≥ comparing the numbers 360 and 180.

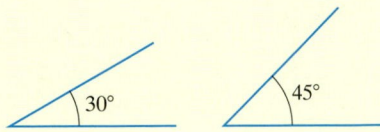

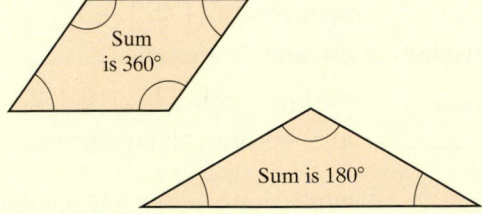

Determine whether each statement is true or false. See Examples 1 through 6 and 10.

▶13. $11 \leq 11$

14. $8 \geq 9$

15. $-11 > -10$

16. $-16 > -17$

17. $5.092 < 5.902$

18. $1.02 > 1.021$

19. $\dfrac{9}{10} \leq \dfrac{8}{9}$

20. $\dfrac{4}{5} \leq \dfrac{9}{11}$

Rewrite each inequality so that the inequality symbol points in the opposite direction and the resulting statement has the same meaning as the given one. See Examples 1 through 6 and 10.

21. $25 \geq 20$

22. $-13 \leq 13$

23. $0 < 6$

24. $5 > 3$

25. $-10 > -12$

26. $-4 < -2$

Objectives B C Mixed Practice—Translating *Write each sentence as a mathematical statement. See Example 7.*

27. Seven is less than eleven.

28. Twenty is greater than two.

▶ **29.** Five is greater than or equal to four.

30. Negative ten is less than or equal to thirty-seven.

▶ **31.** Fifteen is not equal to negative two.

32. Negative seven is not equal to seven.

Use integers to represent the value(s) in each statement. See Example 8.

33. The highest elevation in California is Mt. Whitney, with an altitude of 14,494 feet. The lowest elevation in California is Death Valley, with an altitude of 282 feet below sea level. (*Source: U.S. Geological Survey*)

34. Driskill Mountain, in Louisiana, has an altitude of 535 feet. New Orleans, Louisiana, lies 8 feet below sea level. (*Source: U.S. Geological Survey*)

35. The number of graduate students at the University of Texas at Austin was 27,724 fewer than the number of undergraduate students. (*Source: University of Texas at Austin, 2012*)

36. The number of students admitted to the class of 2011 at UCLA was 56,715 fewer students than the number that had applied. (*Source: UCLA, 2012*)

37. A community college student deposited $475 in her savings account. She later withdrew $195.

38. A deep-sea diver ascended 17 feet and later descended 15 feet.

Graph each set of numbers on the number line. See Example 9.

39. $-4, 0, 2, -2$

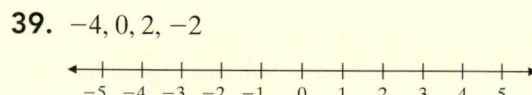

40. $-3, 0, 1, -5$

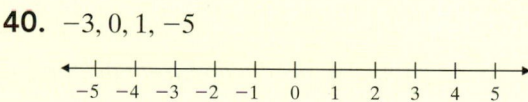

41. $-2, 4, \dfrac{1}{3}, -\dfrac{1}{4}$

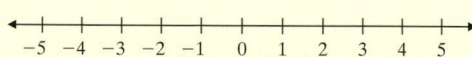

42. $-5, 3, -\dfrac{1}{3}, \dfrac{7}{8}$

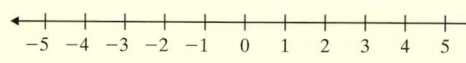

43. $-4.5, \dfrac{7}{4}, 3.25, -\dfrac{3}{2}$

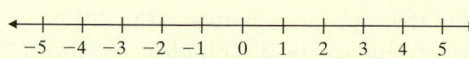

44. $4.5, -\dfrac{9}{4}, 1.75, -\dfrac{7}{2}$

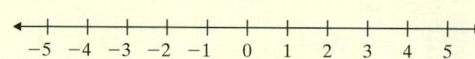

Tell which set or sets each number belongs to: natural numbers, whole numbers, integers, rational numbers, irrational numbers, or real numbers. See Example 11.

▶ **45.** 0

46. $\dfrac{1}{4}$

47. -7

48. $-\dfrac{1}{7}$

49. 265

50. 7941

▶ **51.** $\dfrac{2}{3}$

52. $\sqrt{3}$

Determine whether each statement is true or false.

53. Every rational number is also an integer.

54. Every natural number is positive.

55. 0 is a real number.

56. $\frac{1}{2}$ is an integer.

57. Every negative number is also a rational number.

58. Every rational number is also a real number.

59. Every real number is also a rational number.

60. Every whole number is an integer.

Objective **D** *Find each absolute value. See Example 12.*

61. $|8.9|$

62. $|11.2|$

63. $|-20|$

64. $|-17|$

65. $\left|\frac{9}{2}\right|$

66. $\left|\frac{10}{7}\right|$

67. $\left|-\frac{12}{13}\right|$

68. $\left|-\frac{1}{15}\right|$

Insert $<, >,$ or $=$ in the appropriate space to make each statement true. See Examples 12 and 13.

▶ 69. $|-5|$ _____ -4

70. $|-12|$ _____ $|0|$

71. $\left|-\frac{5}{8}\right|$ _____ $\left|\frac{5}{8}\right|$

72. $\left|\frac{2}{5}\right|$ _____ $\left|-\frac{2}{5}\right|$

73. $|-2|$ _____ $|-2.7|$

74. $|-5.01|$ _____ $|-5|$

▶ 75. $|0|$ _____ $|-8|$

76. $|-12|$ _____ $\frac{-24}{2}$

Review

Perform each indicated operation. See Section 1.9.

77. $90 + 12^2 - 5^3$

78. $3 \cdot (7 - 4) + 2 \cdot 5^2$

79. $12 \div 4 - 2 + 7$

80. $12 \div (4 - 2) + 7$

Concept Extensions

The bar graph shows cranberry production from the top five cranberry-producing states. (Source: National Agricultural Statistics Service)

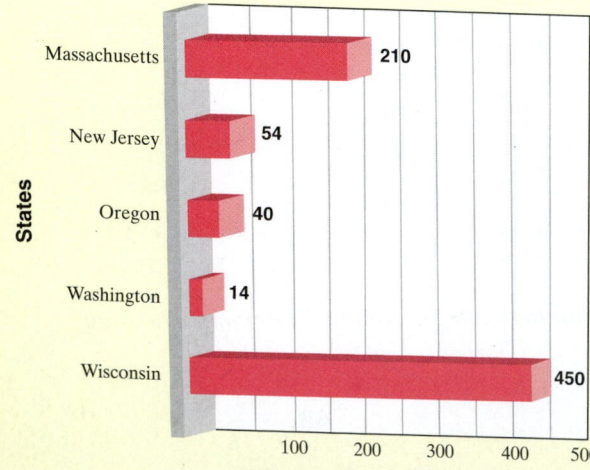

Top Cranberry-Producing States (in millions of pounds)

Massachusetts 210
New Jersey 54
Oregon 40
Washington 14
Wisconsin 450

States
100 200 300 400 500
Millions of lb of Cranberries 2012
Source: National Agricultural Statistics Service

81. Write an inequality comparing the 2012 cranberry production in Oregon with the 2012 cranberry production in Washington.

82. Write an inequality comparing the 2012 cranberry production in Massachusetts with the 2012 cranberry production in Wisconsin.

83. Determine the difference between the 2012 cranberry production in Washington and the 2012 cranberry production in New Jersey.

84. Determine the difference between the 2012 cranberry production in Massachusetts and the 2012 cranberry production in Wisconsin.

The apparent magnitude of a star is the measure of its brightness as seen by someone on Earth. The smaller the apparent magnitude, the brighter the star. Below, the apparent magnitudes of some stars are listed. Use this table to answer Exercises 85 through 90.

Star	Apparent Magnitude	Star	Apparent Magnitude
Arcturus	−0.04	Spica	0.98
Sirius	−1.46	Rigel	0.12
Vega	0.03	Regulus	1.35
Antares	0.96	Canopus	−0.72
Sun (Sol)	−26.7	Hadar	0.61

(*Source: Norton's 2000.0: Star Atlas and Reference Handbooks*, 18th ed., Longman Group, UK, 1989)

85. The apparent magnitude of the sun is −26.7. The apparent magnitude of the star Arcturus is −0.04. Write an inequality statement comparing the numbers −0.04 and −26.7.

86. The apparent magnitude of Antares is 0.96. The apparent magnitude of Spica is 0.98. Write an inequality statement comparing the numbers 0.96 and 0.98.

87. Which is brighter, the sun or Arcturus?

88. Which is dimmer, Antares or Spica?

89. Which star listed is the brightest?

90. Which star listed is the dimmest?

91. In your own words, explain how to find the absolute value of a number.

92. Give an example of a real-life situation that can be described with integers but not with whole numbers.

8.2 # Exponents, Order of Operations, and Variable Expressions

Objective A Exponents and the Order of Operations

Frequently in algebra, products occur that contain repeated multiplication of the same factor. For example, the volume of a cube whose sides each measure 2 centimeters is $(2 \cdot 2 \cdot 2)$ cubic centimeters. We may use **exponential notation** to write such products in a more compact form. For example,

$2 \cdot 2 \cdot 2$ may be written as 2^3.

Objectives

A Define and Use Exponents and the Order of Operations.

B Evaluate Algebraic Expressions, Given Replacement Values for Variables.

C Determine Whether a Number Is a Solution of a Given Equation.

D Translate Phrases into Expressions and Sentences into Equations.

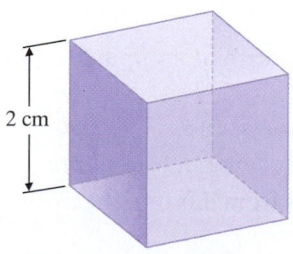

2 cm

Volume is $(2 \cdot 2 \cdot 2)$ cubic centimeters.

The 2 in 2^3 is called the **base;** it is the repeated factor. The 3 in 2^3 is called the **exponent** and is the number of times the base is used as a factor. The expression 2^3 is called an **exponential expression.**

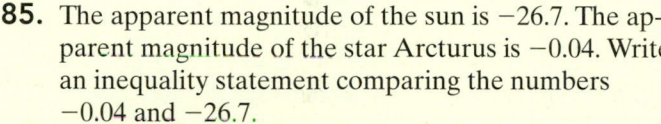

$$2^3 = 2 \cdot 2 \cdot 2 = 8$$

exponent

base

2 is a factor 3 times.

Practice 1

Evaluate each expression.

a. 4^2

b. 2^2

c. 3^4

d. 9^1

e. $\left(\dfrac{2}{5}\right)^3$

f. $(0.8)^2$

Example 1 Evaluate (find the value of) each expression.

a. 3^2 [read as "3 squared" or as "3 to the second power"]

b. 5^3 [read as "5 cubed" or as "5 to the third power"]

c. 2^4 [read as "2 to the fourth power"]

d. 7^1

e. $\left(\dfrac{3}{7}\right)^2$

f. $(0.6)^2$

Solution:

a. $3^2 = 3 \cdot 3 = 9$

b. $5^3 = 5 \cdot 5 \cdot 5 = 125$

c. $2^4 = 2 \cdot 2 \cdot 2 \cdot 2 = 16$

d. $7^1 = 7$

e. $\left(\dfrac{3}{7}\right)^2 = \left(\dfrac{3}{7}\right)\left(\dfrac{3}{7}\right) = \dfrac{3 \cdot 3}{7 \cdot 7} = \dfrac{9}{49}$

f. $(0.6)^2 = (0.6)(0.6) = 0.36$

■ **Work Practice 1**

Using symbols for mathematical operations is a great convenience. The more operation symbols presented in an expression, the more careful we must be when performing the indicated operation. For example, in the expression $2 + 3 \cdot 7$, do we add first or multiply first? To eliminate confusion, **grouping symbols** are used. Examples of grouping symbols are parentheses ( ), brackets [ ], braces { }, absolute value bars | |, and the fraction bar. If we wish $2 + 3 \cdot 7$ to be simplified by adding first, we enclose $2 + 3$ in parentheses.

$$(2 + 3) \cdot 7 = 5 \cdot 7 = 35$$

If we wish to multiply first, $3 \cdot 7$ may be enclosed in parentheses.

$$2 + (3 \cdot 7) = 2 + 21 = 23$$

To eliminate confusion when no grouping symbols are present, we use the following agreed-upon order of operations.

Order of Operations

1. Perform all operations within grouping symbols first, starting with the innermost set.
2. Evaluate exponential expressions.
3. Multiply or divide in order from left to right.
4. Add or subtract in order from left to right.

Using this order of operations, we now simplify $2 + 3 \cdot 7$. There are no grouping symbols and no exponents, so we multiply and then add.

$$2 + 3 \cdot 7 = 2 + 21 \quad \text{Multiply.}$$
$$= 23 \quad \text{Add.}$$

Answers

1. a. 16 **b.** 4 **c.** 81 **d.** 9 **e.** $\dfrac{8}{125}$

f. 0.64

Examples Simplify each expression.

2. $6 \div 3 + 5^2 = 6 \div 3 + 25$　Evaluate 5^2

$\qquad = 2 + 25$　Divide.

$\qquad = 27$　Add.

3. $\underbrace{20 \div 5} \cdot 4 = 4 \cdot 4$

$\qquad = 16$

> **Helpful Hint**
>
> Remember to multiply or divide in order from left to right.

4. $\dfrac{3}{2} \cdot \dfrac{1}{2} - \dfrac{1}{2} = \dfrac{3}{4} - \dfrac{1}{2}$　Multiply.

$\qquad = \dfrac{3}{4} - \dfrac{2}{4}$　The least common denominator is 4.

$\qquad = \dfrac{1}{4}$　Subtract.

5. $1 + 2[5(2 \cdot 3 + 1) - 10] = 1 + 2[5(7) - 10]$　Simplify the expression in the innermost set of parentheses. $2 \cdot 3 + 1 = 6 + 1 = 7$.

$\qquad = 1 + 2[35 - 10]$　Multiply 5 and 7.

$\qquad = 1 + 2[25]$　Subtract inside the brackets.

$\qquad = 1 + 50$　Multiply 2 and 25.

$\qquad = 51$　Add.

■ **Work Practice 2–5**

In the next example, the fraction bar serves as a grouping symbol and separates the numerator and denominator. Simplify each separately.

Example 6 Simplify: $\dfrac{3 + |4 - 3| + 2^2}{6 - 3}$

Solution:

$\dfrac{3 + |4 - 3| + 2^2}{6 - 3} = \dfrac{3 + |1| + 2^2}{6 - 3}$　Simplify the expression inside the absolute value bars.

$\qquad = \dfrac{3 + 1 + 2^2}{3}$　Find the absolute value and simplify the denominator.

$\qquad = \dfrac{3 + 1 + 4}{3}$　Evaluate the exponential expression.

$\qquad = \dfrac{8}{3}$　Simplify the numerator.

■ **Work Practice 6**

> **Helpful Hint**
>
> Be careful when evaluating an exponential expression.
>
> $3 \cdot 4^2 = 3 \cdot 16 = 48$　　$(3 \cdot 4)^2 = (12)^2 = 144$
>
> ↑　　　　　　　　　　↑
>
> Base is 4.　　　　　Base is $3 \cdot 4$.

Practice 2–5

Simplify each expression.

2. $3 \cdot 2 + 4^2$

3. $28 \div 7 \cdot 2$

4. $\dfrac{9}{5} \cdot \dfrac{1}{3} - \dfrac{1}{3}$

5. $5 + 3[2(3 \cdot 4 + 1) - 20]$

Practice 6

Simplify: $\dfrac{1 + |7 - 4| + 3^2}{8 - 5}$

Answers

2. 22　**3.** 8　**4.** $\dfrac{4}{15}$　**5.** 23　**6.** $\dfrac{13}{3}$

Objective B   Evaluating Algebraic Expressions

Recall that letters used to represent quantities are called **variables.** An **algebraic expression** is a collection of numbers, variables, operation symbols, and grouping symbols. For example,

$$2x, \quad -3, \quad 2x - 10, \quad 5(p^2 + 1), \quad xy, \quad \text{and} \quad \frac{3y^2 - 6y + 1}{5}$$

are algebraic expressions.

Expression	Meaning
$2x$	$2 \cdot x$
$5(p^2 + 1)$	$5 \cdot (p^2 + 1)$
$3y^2$	$3 \cdot y^2$
xy	$x \cdot y$

If we give a specific value to a variable, we can **evaluate an algebraic expression.** To evaluate an algebraic expression means to find its numerical value once we know the values of the variables.

Algebraic expressions are often used in problem solving. For example, the expression

$$16t^2$$

gives the distance in feet (neglecting air resistance) that an object will fall in t seconds.

Practice 7

Evaluate each expression when $x = 1$ and $y = 4$.

a. $3y^2$

b. $2y - x$

c. $\dfrac{11x}{3y}$

d. $\dfrac{x}{y} + \dfrac{6}{y}$

e. $y^2 - x^2$

Example 7   Evaluate each expression when $x = 3$ and $y = 2$.

a. $5x^2$ **b.** $2x - y$ **c.** $\dfrac{3x}{2y}$ **d.** $\dfrac{x}{y} + \dfrac{y}{2}$ **e.** $x^2 - y^2$

Solution:

a. Replace x with 3. Then simplify.

$$5x^2 = 5 \cdot (3)^2 = 5 \cdot 9 = 45$$

b. Replace x with 3 and y with 2. Then simplify.

$$2x - y = 2(3) - 2 \quad \text{Let } x = 3 \text{ and } y = 2.$$
$$= 6 - 2 \quad \text{Multiply.}$$
$$= 4 \quad \text{Subtract.}$$

c. Replace x with 3 and y with 2. Then simplify.

$$\frac{3x}{2y} = \frac{3 \cdot 3}{2 \cdot 2} = \frac{9}{4} \quad \text{Let } x = 3 \text{ and } y = 2.$$

d. Replace x with 3 and y with 2. Then simplify.

$$\frac{x}{y} + \frac{y}{2} = \frac{3}{2} + \frac{2}{2} = \frac{5}{2}$$

e. Replace x with 3 and y with 2. Then simplify.

$$x^2 - y^2 = 3^2 - 2^2 = 9 - 4 = 5$$

■ **Work Practice 7**

Answers

7. a. 48   **b.** 7   **c.** $\dfrac{11}{12}$ **d.** $\dfrac{7}{4}$ **e.** 15

Objective C Solutions of Equations

Many times a problem-solving situation is modeled by an equation. An **equation** is a mathematical statement that two expressions have equal value. The equal sign "=" is used to equate the two expressions. For example,

$$3 + 2 = 5, \quad 7x = 35, \quad \frac{2(x - 1)}{3} = 0, \quad \text{and} \quad I = PRT \text{ are all equations.}$$

> **Helpful Hint**
>
> An equation contains the equal sign "=". An algebraic expression does not.

✓**Concept Check** Which of the following are equations? Which are expressions?

a. $5x = 8$ **b.** $5x - 8$ **c.** $12y + 3x$ **d.** $12y = 3x$

When an equation contains a variable, deciding which value(s) of the variable make the equation a true statement is called **solving** the equation for the variable. A **solution** of an equation is a value for the variable that makes the equation a true statement. For example, 3 is a solution of the equation $x + 4 = 7$, because if x is replaced with 3 the statement is true.

$$x + 4 = 7$$
$$\downarrow$$
$$3 + 4 \overset{?}{=} 7 \quad \text{Replace } x \text{ with 3.}$$
$$7 = 7 \quad \text{True}$$

Similarly, 1 is not a solution of the equation $x + 4 = 7$, because $1 + 4 = 7$ is **not** a true statement.

Example 8 Decide whether 2 is a solution of $3x + 10 = 8x$.

Solution: Replace x with 2 and see if a true statement results.

$$3x + 10 = 8x \quad \text{Original equation}$$
$$3(2) + 10 \overset{?}{=} 8(2) \quad \text{Replace } x \text{ with 2.}$$
$$6 + 10 \overset{?}{=} 16 \quad \text{Simplify each side.}$$
$$16 = 16 \quad \text{True}$$

Since we arrived at a true statement after replacing x with 2 and simplifying both sides of the equation, 2 is a solution of the equation.

■ **Work Practice 8**

Practice 8

Decide whether 3 is a solution of $5x - 10 = x + 2$.

Objective D Translating Words to Symbols

Now that we know how to represent an unknown number by a variable, let's practice translating phrases into algebraic expressions (no "=" sign) and sentences into equations (with "=" sign). Oftentimes solving problems involves the ability to translate word phrases and sentences into symbols. A list of key words and phrases to help us translate is on the next page.

Answer

8. It is a solution.

✓**Concept Check Answers**

equations: **a, d**; expressions: **b, c**

Helpful Hint

Order matters when subtracting and also dividing, so be especially careful with these translations.

Addition (+)	Subtraction (−)	Multiplication (·)	Division (÷)	Equality (=)
Sum	Difference of	Product	Quotient	Equals
Plus	Minus	Times	Divide	Gives
Added to	Subtracted from	Multiply	Into	Is/was/should be
More than	Less than	Twice	Ratio	Yields
Increased by	Decreased by	Of	Divided by	Amounts to
Total	Less			Represents
				Is the same as

Practice 9

Write an algebraic expression that represents each phrase. Let the variable x represent the unknown number.

a. The product of 5 and a number
b. A number added to 7
c. A number divided by 11.2
d. A number subtracted from 8
e. Twice a number, plus 1

Example 9 Write an algebraic expression that represents each phrase. Let the variable x represent the unknown number.

a. The sum of a number and 3
b. The product of 3 and a number
c. The quotient of 7.3 and a number
d. 10 decreased by a number
e. 5 times a number, increased by 7

Solution:

a. $x + 3$ since "sum" means to add
b. $3 \cdot x$ and $3x$ are both ways to denote the product of 3 and x
c. $7.3 \div x$ or $\dfrac{7.3}{x}$
d. $10 - x$ because "decreased by" means to subtract
e. $\underbrace{5x}_{\substack{5 \text{ times} \\ \text{a number}}} + 7$

■ Work Practice 9

Helpful Hint

Make sure you understand the difference when translating phrases containing "decreased by," "subtracted from," and "less than."

Phrase	Translation
A number decreased by 10	$x - 10$
A number subtracted from 10	$10 - x$
10 less than a number	$x - 10$
A number less 10	$x - 10$

Notice the order.

Now let's practice translating sentences into equations.

Answers

9. a. $5 \cdot x$ or $5x$ b. $7 + x$
c. $x \div 11.2$ or $\dfrac{x}{11.2}$ d. $8 - x$
e. $2x + 1$

Example 10 Write each sentence as an equation. Let x represent the unknown number.

a. The quotient of 15 and a number is 4.

b. Three subtracted from 12 is a number.

c. 17 added to four times a number is 21.

Solution:

a. In words:

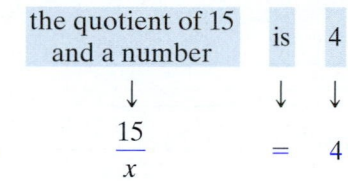

the quotient of 15 and a number   is   4

$$\downarrow \qquad\qquad\qquad \downarrow \quad \downarrow$$

Translate: $\qquad \dfrac{15}{x} \qquad\qquad = \qquad 4$

b. In words: three subtracted **from** 12   is   a number

$$\qquad\qquad\qquad\qquad \downarrow \qquad\qquad\quad \downarrow \qquad \downarrow$$

Translate: $\qquad\qquad 12 - 3 \qquad\quad = \qquad x$

Care must be taken when the operation is subtraction. The expression $3 - 12$ would be incorrect. Notice that $3 - 12 \neq 12 - 3$.

c. In words: 17   added to   four times a number   is   21

$$\quad\;\; \downarrow \qquad \downarrow \qquad\qquad \downarrow \qquad\qquad \downarrow \quad \downarrow$$

Translate: 17   + $4x$ =   21

■ **Work Practice 10**

Practice 10

Write each sentence as an equation. Let x represent the unknown number.

a. The ratio of a number and 6 is 24.

b. The difference of 10 and a number is 18.

c. One less than twice a number is 99.

Answers

10. a. $\dfrac{x}{6} = 24$,   **b.** $10 - x = 18$,

c. $2x - 1 = 99$

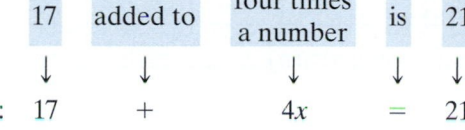

Calculator Explorations   Exponents

To evaluate exponential expressions on a calculator, find the key marked $\boxed{y^x}$ or $\boxed{\wedge}$. To evaluate, for example, 6^5, press the following keys: $\boxed{6}\ \boxed{y^x}\ \boxed{5}\ \boxed{=}$ or $\boxed{6}\ \boxed{\wedge}\ \boxed{5}\ \boxed{=}$.

$\updownarrow$ or

$\boxed{\text{ENTER}}$

The display should read $\boxed{\quad 7776 \quad}$

Order of Operations

Some calculators follow the order of operations, and others do not. To see whether or not your calculator has the order of operations built in, use your calculator to find $2 + 3 \cdot 4$. To do this, press the following sequence of keys:

$\boxed{2}\ \boxed{+}\ \boxed{3}\ \boxed{\times}\ \boxed{4}\ \boxed{=}$

$\updownarrow$ or

$\boxed{\text{ENTER}}$

    The correct answer is 14 because the order of operations is to multiply before we add. If the calculator displays $\boxed{\quad 14 \quad}$, then it has the order of operations built in.

Even if the order of operations is built in, parentheses must sometimes be inserted. For example, to simplify $\dfrac{5}{12 - 7}$, press the keys

$\boxed{5}\ \boxed{\div}\ \boxed{(}\ \boxed{1}\ \boxed{2}\ \boxed{-}\ \boxed{7}\ \boxed{)}\ \boxed{=}$.

$\updownarrow$ or

$\boxed{\text{ENTER}}$

The display should read $\boxed{\quad 1 \quad}$.

Use a calculator to evaluate each expression.

1. 5^3 **2.** 7^4

3. 9^5 **4.** 8^6

5. $2(20 - 5)$ **6.** $3(14 - 7) + 21$

7. $24(862 - 455) + 89$

8. $99 + (401 + 962)$

9. $\dfrac{4623 + 129}{36 - 34}$

10. $\dfrac{956 - 452}{89 - 86}$

Vocabulary, Readiness & Video Check

Use the choices below to fill in each blank. Some choices may be used more than once.

addition	multiplication	exponent	expression	solution	evaluating the expression
subtraction	division	base	equation	variable(s)	

1. In 2^5, the 2 is called the _____ and the 5 is called the _____.

2. True or false: 2^5 means $2 \cdot 5$. _____.

3. To simplify $8 + 2 \cdot 6$, which operation should be performed first? _____

4. To simplify $(8 + 2) \cdot 6$, which operation should be performed first? _____

5. To simplify $9(3 - 2) \div 3 + 6$, which operation should be performed first? _____

6. To simplify $8 \div 2 \cdot 6$, which operation should be performed first? _____

7. A combination of operations on letters (variables) and numbers is a(n) _____.

8. A letter that represents a number is a(n) _____.

9. $3x - 2y$ is called a(n) _____ and the letters x and y are _____.

10. Replacing a variable in an expression by a number and then finding the value of the expression is called _____.

11. A statement of the form "expression = expression" is called a(n) _____.

12. A value for the variable that makes an equation a true statement is called a(n) _____.

Martin-Gay Interactive Videos

See Video 8.2

Watch the section lecture video and answer the following questions.

Objective A   **13.** In ▣ Example 3 and the lecture before, what is the main point made about the order of operations? ▶

Objective B   **14.** What happens with the replacement value for z in ▣ Example 6 and why? ▶

Objective C   **15.** Is the value 0 a solution of the equation given in ▣ Example 9? How is this determined? ▶

Objective D   **16.** Earlier in this video the point was made that equations have =, while expressions do not. In the lecture before ▣ Example 10, translating from English to math is discussed and another difference between expressions and equations is explained. What is it? ▶

8.2   Exercise Set   MyMathLab® ▶

Objective A   *Evaluate. See Example 1.*

1. 3^5 **2.** 5^4 ▶**3.** 3^3 **4.** 4^4 **5.** 1^5 **6.** 1^8

7. 5^1 **8.** 8^1 **9.** 7^2 **10.** 9^2 ▶**11.** $\left(\dfrac{2}{3}\right)^4$ **12.** $\left(\dfrac{6}{11}\right)^2$

13. $\left(\dfrac{1}{5}\right)^3$ **14.** $\left(\dfrac{1}{2}\right)^5$ **15.** $(1.2)^2$ **16.** $(1.5)^2$ **17.** $(0.7)^3$ **18.** $(0.4)^3$

△ **19.** The area of a square whose sides each measure 5 meters is $(5 \cdot 5)$ square meters. Write this area using exponential notation.

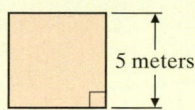

5 meters

△ **20.** The area of a circle whose radius is 9 meters is $(9 \cdot 9 \cdot \pi)$ square meters. Write this area using exponential notation.

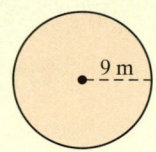

9 m

Simplify each expression. See Examples 2 through 6.

▶ **21.** $5 + 6 \cdot 2$

22. $8 + 5 \cdot 3$

23. $4 \cdot 8 - 6 \cdot 2$

24. $12 \cdot 5 - 3 \cdot 6$

25. $18 \div 3 \cdot 2$

26. $48 \div 6 \cdot 2$

27. $2 + (5 - 2) + 4^2$

28. $6 - 2 \cdot 2 + 2^5$

29. $5 \cdot 3^2$

30. $2 \cdot 5^2$

31. $\dfrac{1}{4} \cdot \dfrac{2}{3} - \dfrac{1}{6}$

32. $\dfrac{3}{4} \cdot \dfrac{1}{2} + \dfrac{2}{3}$

33. $\dfrac{6 - 4}{9 - 2}$

34. $\dfrac{8 - 5}{24 - 20}$

▶ **35.** $2[5 + 2(8 - 3)]$

36. $3[4 + 3(6 - 4)]$

37. $\dfrac{19 - 3 \cdot 5}{6 - 4}$

38. $\dfrac{14 - 2 \cdot 3}{12 - 8}$

▶ **39.** $\dfrac{|6 - 2| + 3}{8 + 2 \cdot 5}$

40. $\dfrac{15 - |3 - 1|}{12 - 3 \cdot 2}$

41. $\dfrac{3 + 3(5 + 3)}{3^2 + 1}$

42. $\dfrac{3 + 6(8 - 5)}{4^2 + 2}$

43. $\dfrac{6 + |8 - 2| + 3^2}{18 - 3}$

44. $\dfrac{16 + |13 - 5| + 4^2}{17 - 5}$

45. $2 + 3[10(4 \cdot 5 - 16) - 30]$

46. $3 + 4[8(5 \cdot 5 - 20) - 41]$

47. $\left(\dfrac{2}{3}\right)^3 + \dfrac{1}{9} + \dfrac{1}{3} \cdot \dfrac{4}{3}$

48. $\left(\dfrac{3}{8}\right)^2 + \dfrac{1}{4} + \dfrac{1}{8} \cdot \dfrac{3}{2}$

Objective **B** *Evaluate each expression when $x = 1$, $y = 3$, and $z = 5$. See Example 7.*

49. $3y$

50. $4x$

51. $\dfrac{z}{5x}$

52. $\dfrac{y}{2z}$

53. $3x - 2$

54. $6y - 8$

▶ **55.** $|2x + 3y|$

56. $|5z - 2y|$

57. $xy + z$

58. $yz - x$

59. $5y^2$

60. $2z^2$

Evaluate each expression when $x = 12$, $y = 8$, and $z = 4$. See Example 7.

61. $\dfrac{x}{z} + 3y$

62. $\dfrac{y}{z} + 8x$

63. $x^2 - 3y + x$

64. $y^2 - 3x + y$

▶ **65.** $\dfrac{x^2 + z}{y^2 + 2z}$

66. $\dfrac{y^2 + x}{x^2 + 3y}$

Objective C *Decide whether the given number is a solution of the given equation. See Example 8.*

67. $3x - 6 = 9; 5$ **68.** $2x + 7 = 3x; 6$ **69.** $2x + 6 = 5x - 1; 0$ **70.** $4x + 2 = x + 8; 2$

71. $2x - 5 = 5; 8$ ▶ **72.** $3x - 10 = 8; 6$ **73.** $x + 6 = x + 6; 2$ **74.** $x + 6 = x + 6; 10$

▶ **75.** $x = 5x + 15; 0$ **76.** $4 = 1 - x; 1$ **77.** $\frac{1}{3}x = 9; 27$ **78.** $\frac{2}{7}x = \frac{3}{14}; 6$

Objective D *Write each phrase as an algebraic expression. Let x represent the unknown number. See Example 9.*

79. Fifteen more than a number

80. A number increased by 9

81. Five subtracted from a number

82. Five decreased by a number

83. The ratio of a number and 4

84. The quotient of a number and 9

▶ **85.** Three times a number, increased by 22

86. Twice a number, decreased by 72

Write each sentence as an equation or inequality. Use x to represent any unknown number. See Example 10.

▶ **87.** One increased by two equals the quotient of nine and three.

88. Four subtracted from eight is equal to two squared.

▶ **89.** Three is not equal to four divided by two.

90. The difference of sixteen and four is greater than ten.

91. The sum of 5 and a number is 20.

92. Seven subtracted from a number is 0.

93. The product of 7.6 and a number is 17.

94. 9.1 times a number equals 4

95. Thirteen minus three times a number is 13.

96. Eight added to twice a number is 42.

Review

Add. See Section 1.3.

97. $15 + 20$ **98.** $20 + 15$ **99.** $47 + 236 + 77$ **100.** $362 + 37 + 90$

Concept Extensions

101. Are parentheses necessary in the expression $2 + (3 \cdot 5)$? Explain your answer.

102. Are parentheses necessary in the expression $(2 + 3) \cdot 5$? Explain your answer.

For Exercises 103 and 104, match each expression in the first column with its value in the second column.

103. **a.** $(6 + 2) \cdot (5 + 3)$ 19
     b. $(6 + 2) \cdot 5 + 3$ 22
     c. $6 + 2 \cdot 5 + 3$ 64
     d. $6 + 2 \cdot (5 + 3)$ 43

104. **a.** $(1 + 4) \cdot 6 - 3$ 15
     b. $1 + 4 \cdot (6 - 3)$ 13
     c. $1 + 4 \cdot 6 - 3$ 27
     d. $(1 + 4) \cdot (6 - 3)$ 22

△ *Recall that perimeter measures the distance around a plane figure and area measures the amount of surface of a plane figure. The expression $2l + 2w$ gives the perimeter of the rectangle below (measured in units), and the expression lw gives its area (measured in square units). Complete the chart below for the given lengths and widths. Be sure to include units.*

	Length: *l*	Width: *w*	Perimeter of Rectangle: $2l + 2w$	Area of Rectangle: lw
105.	4 in.	3 in.		
106.	6 in.	1 in.		
107.	5.3 in.	1.7 in.		
108.	4.6 in.	2.4 in.		

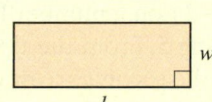

109. Study the perimeters and areas found in the chart to the left. Do you notice any trends?

110. In your own words, explain the difference between an expression and an equation.

111. Insert one set of parentheses so that the following expression simplifies to 32.

$$20 - 4 \cdot 4 \div 2$$

112. Insert parentheses so that the following expression simplifies to 28.

$$2 \cdot 5 + 3^2$$

Determine whether each is an expression or an equation. See the Concept Check in this section.

113. **a.** $5x + 6$
     b. $2a = 7$
     c. $3a + 2 = 9$
     d. $4x + 3y - 8z$
     e. $5^2 - 2(6 - 2)$

114. **a.** $3x^2 - 26$
     b. $3x^2 - 26 = 1$
     c. $2x - 5 = 7x - 5$
     d. $9y + x - 8$
     e. $3^2 - 4(5 - 3)$

115. Why is 4^3 usually read as "four cubed"?
△ (*Hint:* What is the volume of the **cube** below?)

116. Why is 8^2 usually read as "eight squared"?
△ (*Hint:* What is the area of the **square** below?)

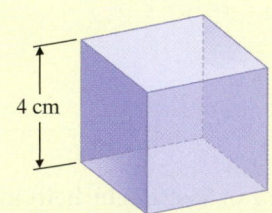

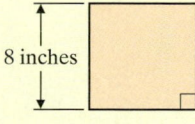

117. Write any expression, using 3 or more numbers, that simplifies to -11.

118. Write any expression, using 4 or more numbers, that simplifies to 7.

8.3 Adding Real Numbers

Objectives

A Add Real Numbers.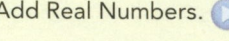

B Find the Opposite of a Number.

C Evaluate Algebraic Expressions Using Real Numbers.

D Solve Applications That Involve Addition of Real Numbers.

Real numbers can be added, subtracted, multiplied, divided, and raised to powers, just as whole numbers can.

Objective A Adding Real Numbers

Adding real numbers can be visualized by using a number line. A positive number can be represented on the number line by an arrow of appropriate length pointing to the right, and a negative number by an arrow of appropriate length pointing to the left.

Both arrows represent 2 or +2.

They both point to the right, and they are both 2 units long.

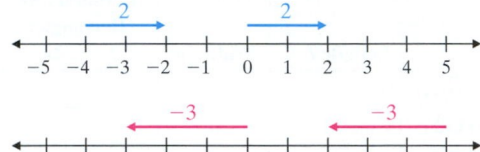

Both arrows represent −3.

They both point to the left, and they are both 3 units long.

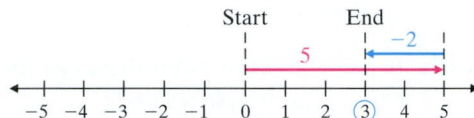

To add signed numbers such as $5 + (-2)$ on a number line, we start at 0 on the number line and draw an arrow representing 5. From the tip of this arrow, we draw another arrow representing −2. The tip of the second arrow ends at their sum, 3.

$$5 + (-2) = 3$$

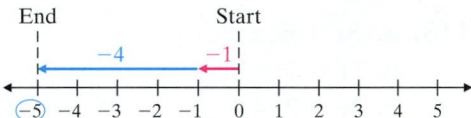

To add $-1 + (-4)$ on the number line, we start at 0 and draw an arrow representing −1. From the tip of this arrow, we draw another arrow representing −4. The tip of the second arrow ends at their sum, −5.

$$-1 + (-4) = -5$$

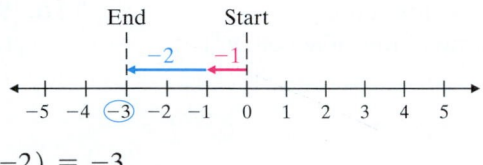

Practice 1

Add using a number line:
$-2 + (-4)$

<--+--+--+--+--+--+--+--+--+--+--+-->
−5 −4 −3 −2 −1 0 1 2 3 4 5

Example 1 Add: $-1 + (-2)$

Solution:

$$-1 + (-2) = -3$$

■ Work Practice 1

Thinking of integers as money earned or lost might help make addition more meaningful. Earnings can be thought of as positive numbers. If $1 is earned and later another $3 is earned, the total amount earned is $4. In other words, $1 + 3 = 4$.

On the other hand, losses can be thought of as negative numbers. If $1 is lost and later another $3 is lost, a total of $4 is lost. In other words, $(-1) + (-3) = -4$.

In Example 1, we added numbers with the same sign. Adding numbers whose signs are not the same can be pictured on a number line also.

Answer

1. −6

Example 2 Add: $-4 + 6$

Solution:

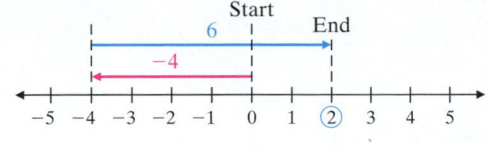

$$-4 + 6 = 2$$

■ **Work Practice 2**

Let's use temperature as an example. If the thermometer registers 4 degrees below 0 degrees and then rises 6 degrees, the new temperature is 2 degrees above 0 degrees. Thus, it is reasonable that $-4 + 6 = 2$. (See the diagram in the margin.)

Example 3 Add: $4 + (-6)$

Solution:

$$4 + (-6) = -2$$

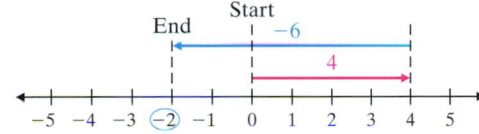

■ **Work Practice 3**

Using a number line each time we add two numbers can be time consuming. Instead, we can notice patterns in the previous examples and write rules for adding real numbers.

Adding Real Numbers

To add two real numbers

1. with the *same sign*, add their absolute values. Use their common sign as the sign of the answer.
2. with *different signs*, subtract their absolute values. Give the answer the same sign as the number with the larger absolute value.

Example 4 Add without using a number line: $(-7) + (-6)$

Solution: Here, we are adding two numbers with the same sign.

$$(-7) + (-6) = -13$$

↑ ↖ sum of absolute values ($|-7| = 7, |-6| = 6, 7 + 6 = 13$)

same sign

■ **Work Practice 4**

Example 5 Add without using a number line: $(-10) + 4$

Solution: Here, we are adding two numbers with different signs.

$$(-10) + 4 = -6$$

↑ ↖ difference of absolute values ($|-10| = 10, |4| = 4, 10 - 4 = 6$)

sign of number with larger absolute value, -10

■ **Work Practice 5**

Practice 2

Add using a number line:
$-5 + 8$

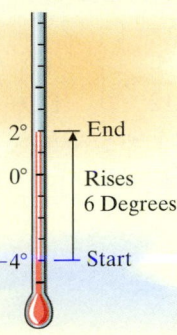

Practice 3

Add using a number line:
$5 + (-4)$

Practice 4

Add without using a number line: $(-8) + (-5)$

Practice 5

Add without using a number line: $(-14) + 6$

Answers

2. 3 **3.** 1 **4.** -13 **5.** -8

Practice 6–11

Add without using a number line.

6. $(-17) + (-10)$

7. $(-4) + 12$

8. $1.5 + (-3.2)$

9. $-\dfrac{5}{12} + \left(-\dfrac{1}{12}\right)$

10. $12.1 + (-3.6)$

11. $-\dfrac{4}{5} + \dfrac{2}{3}$

Examples   Add without using a number line.

6. $(-8) + (-11) = -19$

7. $(-2) + 10 = 8$

8. $0.2 + (-0.5) = -0.3$

9. $-\dfrac{7}{10} + \left(-\dfrac{1}{10}\right) = -\dfrac{8}{10} = -\dfrac{\overset{1}{\cancel{2}} \cdot 4}{\underset{1}{\cancel{2}} \cdot 5} = -\dfrac{4}{5}$

10. $11.4 + (-4.7) = 6.7$

11. $-\dfrac{3}{8} + \dfrac{2}{5} = -\dfrac{15}{40} + \dfrac{16}{40} = \dfrac{1}{40}$

■ Work Practice 6–11

In Example 12a, we add three numbers. Remember that by the associative and commutative properties for addition, we may add numbers in any order that we wish. For Example 12a, let's add the numbers from left to right.

Practice 12

Find each sum.

a. $16 + (-9) + (-9)$

b. $[3 + (-13)] + [-4 + (-7)]$

Example 12   Find each sum.

a. $3 + (-7) + (-8)$

b. $[7 + (-10)] + [-2 + (-4)]$

Solution:

a. Perform the additions from left to right.

$$3 + (-7) + (-8) = -4 + (-8) \quad \text{Adding numbers with different signs}$$
$$= -12 \quad \text{Adding numbers with the same sign}$$

b. Simplify inside the brackets first.

$$[7 + (-10)] + [-2 + (-4)] = [-3] + [-6]$$
$$= -9 \quad \text{Add.}$$

Helpful Hint

Don't forget that brackets are grouping symbols. We simplify within them first.

■ Work Practice 12

Objective B Finding Opposites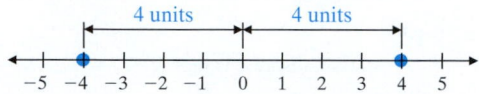

To help us subtract real numbers in the next section, we first review what we mean by opposites. The graphs of 4 and -4 are shown on the number line below.

Notice that the graphs of 4 and -4 lie on opposite sides of 0, and each is 4 units away from 0. Such numbers are known as **opposites** or **additive inverses** of each other.

Opposite or Additive Inverse

Two numbers that are the same distance from 0 but lie on opposite sides of 0 are called **opposites** or **additive inverses** of each other.

Answers

6. -27 7. 8 8. -1.7 9. $-\dfrac{1}{2}$

10. 8.5 11. $-\dfrac{2}{15}$ 12. a. -2 b. -21

Examples Find the opposite of each number.

13. 10   The opposite of 10 is -10.

14. -3 The opposite of -3 is 3.

15. $\dfrac{1}{2}$ The opposite of $\dfrac{1}{2}$ is $-\dfrac{1}{2}$.

16. -4.5 The opposite of -4.5 is 4.5.

■ **Work Practice 13–16**

We use the symbol "$-$" to represent the phrase "the opposite of" or "the additive inverse of." In general, if a is a number, we write the opposite or additive inverse of a as $-a$. We know that the opposite of -3 is 3. Notice that this translates as

the opposite of -3 is   3
    $\downarrow$ $\downarrow$ $\downarrow$ $\downarrow$
     $-$ (-3) $=$ 3

This is true in general.

> If a is a number, then $-(-a) = a$.

Example 17 Simplify each expression.

a. $-(-10)$ **b.** $-\left(-\dfrac{1}{2}\right)$ **c.** $-(-2x)$

d. $-|-6|$ **e.** $-|4.1|$

Solution:

a. $-(-10) = 10$

b. $-\left(-\dfrac{1}{2}\right) = \dfrac{1}{2}$

c. $-(-2x) = 2x$

d. $-|-6| = -6$ Since $|-6| = 6$.

e. $-|4.1| = -4.1$ Since $|4.1| = 4.1$.

■ **Work Practice 17**

Let's discover another characteristic about opposites. Notice that the sum of a number and its opposite is always 0.

$10 + (-10) = 0$ $-3 + 3 = 0$
  opposites   opposites

$\dfrac{1}{2} + \left(-\dfrac{1}{2}\right) = 0$

  opposites

In general, we can write the following:

> The sum of a number a and its opposite $-a$ is 0.
>
> $a + (-a) = 0$ Also, $-a + a = 0$.

Notice that this means that the opposite of 0 is then 0 since $0 + 0 = 0$.

Practice 13–16

Find the opposite of each number.

13. -35 **14.** 12

15. $-\dfrac{3}{11}$ **16.** 1.9

Practice 17

Simplify each expression.

a. $-(-22)$

b. $-\left(-\dfrac{2}{7}\right)$

c. $-(-x)$

d. $-|-14|$

e. $-|2.3|$

Answers

13. 35 **14.** -12 **15.** $\dfrac{3}{11}$ **16.** -1.9

17. a. 22 **b.** $\dfrac{2}{7}$ **c.** x **d.** -14 **e.** -2.3

Practice 18–19

Add.

18. $30 + (-30)$

19. $-81 + 81$

Examples Add.

18. $-56 + 56 = 0$

19. $17 + (-17) = 0$

■ **Work Practice 18–19**

✓**Concept Check** What is wrong with the following calculation?

$$5 + (-22) = 17$$

Objective C Evaluating Algebraic Expressions

We can continue our work with algebraic expressions by evaluating expressions given real-number replacement values.

Practice 20

Evaluate $x + 3y$ for $x = -6$ and $y = 2$.

Example 20 Evaluate $2x + y$ for $x = 3$ and $y = -5$.

Solution: Replace x with 3 and y with -5 in $2x + y$.

$$2x + y = 2 \cdot 3 + (-5)$$
$$= 6 + (-5)$$
$$= 1$$

■ **Work Practice 20**

Practice 21

Evaluate $x + y$ for $x = -13$ and $y = -9$.

Example 21 Evaluate $x + y$ for $x = -2$ and $y = -10$.

Solution: $x + y = (-2) + (-10)$ Replace x with -2 and y with -10.

$$= -12$$

■ **Work Practice 21**

Practice 22

If the temperature was $-7°$ Fahrenheit at 6 a.m., and it rose 4 degrees by 7 a.m. and then rose another 7 degrees in the hour from 7 a.m. to 8 a.m., what was the temperature at 8 a.m.?

Objective D Solving Applications That Involve Addition

Positive and negative numbers are used in everyday life. Stock market returns show gains and losses as positive and negative numbers. Temperatures in cold climates often dip into the negative range, commonly referred to as "below zero" temperatures. Bank statements report deposits and withdrawals as positive and negative numbers.

Example 22 Calculating Temperature

In Philadelphia, Pennsylvania, the record extreme high temperature is 104°F. Decrease this temperature by 111 degrees, and the result is the record extreme low temperature. Find this temperature. (*Source:* National Climatic Data Center)

Solution:

In words:	extreme low temperature	=	extreme high temperature	+	decrease of 111°
Translate:	extreme low temperature	=	104	+	(-111)

$$= -7$$

The record extreme low temperature in Philadelphia, Pennsylvania, is $-7°$F.

■ **Work Practice 22**

Answers

18. 0 **19.** 0 **20.** 0 **21.** -22 **22.** 4°F

✓**Concept Check Answer**

$5 + (-22) = -17$

Vocabulary, Readiness & Video Check

Use the choices below to fill in each blank. Not all choices will be used.

   $-a$ a 0 commutative   associative

1. If n is a number, then $-n + n = $ _____.
2. Since $x + n = n + x$, we say that addition is _____.
3. If a is a number, then $-(-a) = $ _____.
4. Since $n + (x + a) = (n + x) + a$, we say that addition is _____.

Martin-Gay Interactive Videos   Watch the section lecture video and answer the following questions.

See Video 8.3

Objective A 5. Complete this statement based on the lecture given before ▥ Example 1. To add two numbers with the same sign, add their _____ and use their common sign as the sign of the sum. ▶

6. What is the sign of the sum in ▥ Example 6 and why? ▶

Objective B 7. ▥ Example 11 illustrates the idea that if a is a real number, the opposite of $-a$ is a. ▥ Example 12 looks similar to ▥ Example 11, but it's actually quite different. Explain the difference. ▶

Objective C 8. Explain the difference in the algebraic expression for ▥ Example 13 and the algebraic example for ▥ Example 14. ▶

Objective D 9. What is the real-life application of negative numbers used in ▥ Example 15? The answer to ▥ Example 15 is -231. What does this number mean in the context of the problem? ▶

8.3 Exercise Set MyMathLab®

Objectives A B **Mixed Practice** *Add. See Examples 1 through 12, 18, and 19.*

1. $6 + (-3)$
2. $9 + (-12)$
▶3. $-6 + (-8)$
4. $-6 + (-14)$

5. $8 + (-7)$
6. $16 + (-4)$
7. $-14 + 2$
8. $-10 + 5$

▶9. $-2 + (-3)$
10. $-7 + (-4)$
▶11. $-9 + (-3)$
12. $-11 + (-5)$

13. $-7 + 3$
14. $-5 + 9$
15. $10 + (-3)$
16. $8 + (-6)$

▶17. $5 + (-7)$
18. $3 + (-6)$
19. $-16 + 16$
20. $23 + (-23)$

21. $27 + (-46)$
22. $53 + (-37)$
23. $-18 + 49$
24. $-26 + 14$

25. $-33 + (-14)$ **26.** $-18 + (-26)$ **27.** $6.3 + (-8.4)$ **28.** $9.2 + (-11.4)$

29. $117 + (-79)$ **30.** $144 + (-88)$ **31.** $-9.6 + (-3.5)$ **32.** $-6.7 + (-7.6)$

33. $-\dfrac{3}{8} + \dfrac{5}{8}$ **34.** $-\dfrac{5}{12} + \dfrac{7}{12}$ **35.** $-\dfrac{7}{16} + \dfrac{1}{4}$ **36.** $-\dfrac{5}{9} + \dfrac{1}{3}$

37. $-\dfrac{7}{10} + \left(-\dfrac{3}{5}\right)$ **38.** $-\dfrac{5}{6} + \left(-\dfrac{2}{3}\right)$ **39.** $|-8| + (-16)$ **40.** $|-6| + (-61)$

41. $-15 + 9 + (-2)$ **42.** $-9 + 15 + (-5)$ **43.** $-21 + (-16) + (-22)$ **44.** $-18 + (-6) + (-40)$

45. $-23 + 16 + (-2)$ **46.** $-14 + (-3) + 11$ **47.** $|5 + (-10)|$ **48.** $|7 + (-17)|$

49. $6 + (-4) + 9$ **50.** $8 + (-2) + 7$ **51.** $[-17 + (-4)] + [-12 + 15]$

52. $[-2 + (-7)] + [-11 + 22]$ **53.** $|9 + (-12)| + |-16|$ **54.** $|43 + (-73)| + |-20|$

55. $-13 + [5 + (-3) + 4]$ **56.** $-30 + [1 + (-6) + 8]$

57. Find the sum of -38 and 12.   **58.** Find the sum of -44 and 16.

Objective **B** *Find each additive inverse or opposite. See Examples 13 through 16.*

59. 6 **60.** 4 **61.** -2 **62.** -8

63. 0 **64.** $-\dfrac{1}{4}$ **65.** $|-6|$ **66.** $|-11|$

Simplify each of the following. See Example 17.

67. $-|-2|$ **68.** $-|-5|$ **69.** $-(-7)$ **70.** $-(-14)$ **71.** $-(-7.9)$

72. $-(-8.4)$ **73.** $-(-5z)$ **74.** $-(-7m)$ **75.** $\left|-\dfrac{2}{3}\right|$ **76.** $-\left|-\dfrac{2}{3}\right|$

Objective **C** *Evaluate $x + y$ for the given replacement values. See Examples 20 and 21.*

77. $x = -20$ and $y = -50$ **78.** $x = -1$ and $y = -29$

Evaluate 3x + y for the given replacement values. See Examples 20 and 21.

79. *x* = 2 and *y* = −3

80. *x* = 7 and *y* = −11

Objective D **Translating** *Translate each phrase; then simplify. See Example 22.*

81. Find the sum of −6 and 25.

82. Find the sum of −30 and 15.

83. Find the sum of −31, −9, and 30.

84. Find the sum of −49, −2, and 40.

Solve. See Example 22.

85. Suppose a deep-sea diver dives from the surface to 215 feet below the surface. He then dives down 16 more feet. Use integers to represent this situation. Then find the diver's present depth.

86. Suppose a diver dives from the surface to 248 meters below the surface and then swims up 8 meters, down 16 meters, down another 28 meters, and then up 32 meters. Use integers to represent this situation. Then find the diver's depth after these movements.

87. The lowest temperature ever recorded in Massachusetts was −35°F. The highest recorded temperature in Massachusetts was 142° higher than the record low temperature. Find Massachusetts' highest recorded temperature. (*Source:* National Climatic Data Center)

88. On January 2, 1943, the temperature was −4° at 7:30 a.m. in Spearfish, South Dakota. Incredibly, it got 49° warmer in the next 2 minutes. To what temperature did it rise by 7:32?

89. The lowest elevation on Earth is −411 meters (that is, 411 meters below sea level) at the Dead Sea. If you are standing 316 meters above the Dead Sea, what is your elevation? (*Source:* National Geographic Society)

90. The lowest elevation in Australia is −52 feet at Lake Eyre. If you are standing at a point 439 feet above Lake Eyre, what is your elevation? (*Source:* National Geographic Society)

91. During the 2014 PGA Masters Tournament, the winner, Bubba Watson, had scores of −3, +2, −4, and −3 over four rounds of golf. What was his total score for the tournament? (*Source:* Professional Golfer's Association)

92. Lizette Salas won the 2014 LPGA Kingsmill Championship Tournament with scores of −4, −3, −6, and 0 over four rounds of golf. What was her total score for the tournament? (*Source:* LPGA of America)

93. A negative net income results when a company spends more money than it brings in. Johnson Outdoors Inc. had the following quarterly net incomes during its 2013 fiscal year. (*Source:* MarketWatch, Inc.)

Quarter of Fiscal 2013	Net Income (in millions)
First	8.9
Second	13.7
Third	−3.5
Fourth	−2.2

What was the total net income for fiscal year 2013?

94. Barnes & Noble Inc. had the following quarterly net incomes during 2013. (*Source:* MarketWatch, Inc.)

Quarter of 2013	Net Income (in millions)
ended January 31	−6.8
ended April 30	−110.2
ended July 31	−87.8
ended October 31	12.6

What was the total net income for 2013?

Review

Subtract. See Sections 1.4 and 4.3.

95. $76.1 − 4.09$

96. $93.7 − 10.08$

97. $200 − 59$

98. $400 − 18$

Concept Extensions

The following bar graph shows each month's average daily low temperature in degrees Fahrenheit for Barrow, Alaska. Use this graph to answer Exercises 99 through 104.

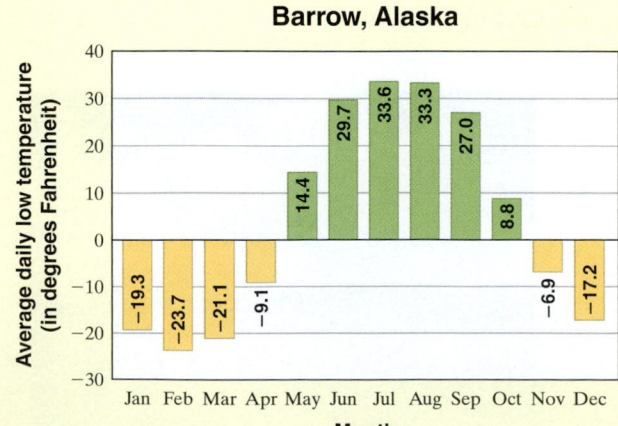

Barrow, Alaska

Average daily low temperature (in degrees Fahrenheit)

Jan −19.3, Feb −23.7, Mar −21.1, Apr −9.1, May 14.4, Jun 29.7, Jul 33.6, Aug 33.3, Sep 27.0, Oct 8.8, Nov −6.9, Dec −17.2

Month

Source: National Climatic Data Center

99. For what month is the graphed temperature the highest?

100. For what month is the graphed temperature the lowest?

101. For what month is the graphed temperature positive *and* closest to 0°?

102. For what month is the graphed temperature negative *and* closest to 0°?

103. Find the average (mean) of the temperatures shown for the months of April, May, and October.

104. Find the average (mean) of the temperatures shown for the months of January, September, and October.

105. Name 2 numbers whose sum is −17.

106. Name 2 numbers whose sum is −30.

Each calculation below is incorrect. Find the error and correct it. See the Concept Check in this section.

107. 7 + (−10) $\stackrel{?}{=}$ 17

108. −4 + 14 $\stackrel{?}{=}$ −18

109. −10 + (−12) $\stackrel{?}{=}$ −120

110. −15 + (−17) $\stackrel{?}{=}$ 32

For Exercises 111 through 114, determine whether each statement is true or false.

111. The sum of two negative numbers is always a negative number.

112. The sum of two positive numbers is always a positive number.

113. The sum of a positive number and a negative number is always a negative number.

114. The sum of zero and a negative number is always a negative number.

115. In your own words, explain how to add two negative numbers.

116. In your own words, explain how to add a positive number and a negative number.

8.4 Subtracting Real Numbers

Objective A Subtracting Real Numbers

Now that addition of real numbers has been discussed, we can explore subtraction. We know that 9 − 7 = 2. Notice that 9 + (−7) = 2, also. This means that

$$9 - 7 = 9 + (-7)$$

Notice that the *difference* of 9 and 7 is the same as the *sum* of 9 and the opposite of 7. This is how we can subtract real numbers.

> **Subtracting Real Numbers**
>
> If a and b are real numbers, then $a - b = a + (-b)$.

In other words, to find the difference of two numbers, we add the opposite of the number being subtracted.

Objectives

A Subtract Real Numbers.

B Evaluate Algebraic Expressions Using Real Numbers.

C Determine Whether a Number Is a Solution of a Given Equation.

D Solve Applications That Involve Subtraction of Real Numbers.

E Find Complementary and Supplementary Angles.

Practice 1

Subtract.
a. $-20 - 6$
b. $3 - (-5)$
c. $7 - 17$
d. $-4 - (-9)$

Example 1   Subtract.

a. $-13 - 4$ b. $5 - (-6)$ c. $3 - 6$ d. $-1 - (-7)$

Solution:

a. $-13 - 4 = -13 + (-4)$ Add -13 to the opposite of 4, which is -4.
$$= -17$$

b. $5 - (-6) = 5 + (6)$ Add 5 to the opposite of -6, which is 6.
$$= 11$$

c. $3 - 6 = 3 + (-6)$ Add 3 to the opposite of 6, which is -6.
$$= -3$$

d. $-1 - (-7) = -1 + (7) = 6$

■ **Work Practice 1**

Helpful Hint

Study the patterns indicated.

No change → Change to addition.
└ Change to opposite.
$$5 - 11 = 5 + (-11) = -6$$
$$-3 - 4 = -3 + (-4) = -7$$
$$7 - (-1) = 7 + (1) = 8$$

Practice 2–4

Subtract.
2. $9.6 - (-5.7)$

3. $-\dfrac{4}{9} - \dfrac{2}{9}$

4. $-\dfrac{1}{4} - \left(-\dfrac{2}{5}\right)$

Examples   Subtract.

2. $5.3 - (-4.6) = 5.3 + (4.6) = 9.9$

3. $-\dfrac{3}{10} - \dfrac{5}{10} = -\dfrac{3}{10} + \left(-\dfrac{5}{10}\right) = -\dfrac{8}{10} = -\dfrac{4}{5}$

4. $-\dfrac{2}{3} - \left(-\dfrac{4}{5}\right) = -\dfrac{2}{3} + \left(\dfrac{4}{5}\right) = -\dfrac{10}{15} + \dfrac{12}{15} = \dfrac{2}{15}$

■ **Work Practice 2–4**

Practice 5

Write each phrase as an expression and simplify.
a. Subtract 7 from -11.
b. Decrease 35 by -25.

Example 5   Write each phrase as an expression and simplify.

a. Subtract 8 from -4.   b. Decrease 10 by -20.

Solution: Be careful when interpreting these. The order of numbers in subtraction is important.

a. 8 is to be subtracted **from** -4.
$$-4 - 8 = -4 + (-8) = -12$$

b. To decrease 10 by -20, we find 10 **minus** -20.
$$10 - (-20) = 10 + 20 = 30$$

■ **Work Practice 5**

If an expression contains additions and subtractions, just write the subtractions as equivalent additions. Then simplify from left to right.

Answers

1. a. -26 b. 8 c. -10 d. 5
2. 15.3 3. $-\dfrac{2}{3}$ 4. $\dfrac{3}{20}$ 5. a. -18 b. 60

Example 6   Simplify each expression.

a. $-14 - 8 + 10 - (-6)$ **b.** $1.6 - (-10.3) + (-5.6)$

Solution:

a. $-14 - 8 + 10 - (-6) = -14 + (-8) + 10 + 6 = -6$

b. $1.6 - (-10.3) + (-5.6) = 1.6 + 10.3 + (-5.6) = 6.3$

■ **Work Practice 6**

When an expression contains parentheses and brackets, remember the order of operations. Start with the innermost set of parentheses or brackets and work your way outward.

Example 7   Simplify each expression.

a. $-3 + [(-2 - 5) - 2]$ **b.** $2^3 - 10 + [-6 - (-5)]$

Solution:

a. Start with the innermost set of parentheses. Rewrite $-2 - 5$ as an addition.

$$-3 + [(-2 - 5) - 2] = -3 + [(-2 + (-5)) - 2]$$
$$= -3 + [(-7) - 2] \qquad \text{Add: } -2 + (-5).$$
$$= -3 + [-7 + (-2)] \qquad \text{Write } -7 - 2 \text{ as an addition.}$$
$$= -3 + [-9] \qquad \text{Add.}$$
$$= -12 \qquad \text{Add.}$$

b. Start simplifying the expression inside the brackets by writing $-6 - (-5)$ as an addition.

$$2^3 - 10 + [-6 - (-5)] = 2^3 - 10 + [-6 + 5]$$
$$= 2^3 - 10 + [-1] \qquad \text{Add.}$$
$$= 8 - 10 + (-1) \qquad \text{Evaluate } 2^3.$$
$$= 8 + (-10) + (-1) \qquad \text{Write } 8 - 10 \text{ as an addition.}$$
$$= -2 + (-1) \qquad \text{Add.}$$
$$= -3 \qquad \text{Add.}$$

■ **Work Practice 7**

Objective B Evaluating Algebraic Expressions ▶

It is important to be able to evaluate expressions for given replacement values. This helps, for example, when checking solutions of equations.

Example 8   Find the value of each expression when $x = 2$ and $y = -5$.

a. $\dfrac{x - y}{12 + x}$ **b.** $x^2 - y$

Solution:

a. Replace x with 2 and y with -5. Be sure to put parentheses around -5 to separate signs. Then simplify the resulting expression.

$$\frac{x - y}{12 + x} = \frac{2 - (-5)}{12 + 2} = \frac{2 + 5}{14} = \frac{7}{14} = \frac{1}{2}$$

b. Replace x with 2 and y with -5 and simplify.

$$x^2 - y = 2^2 - (-5) = 4 - (-5) = 4 + 5 = 9$$

■ **Work Practice 8**

Practice 6

Simplify each expression.
a. $-20 - 5 + 12 - (-3)$
b. $5.2 - (-4.4) + (-8.8)$

Practice 7

Simplify each expression.
a. $-9 + [(-4 - 1) - 10]$
b. $5^2 - 20 + [-11 - (-3)]$

Practice 8

Find the value of each expression when $x = 1$ and $y = -4$.

a. $\dfrac{x - y}{14 + x}$ **b.** $x^2 - y$

Answers

6. **a.** -10 **b.** 0.8 7. **a.** -24
b. -3 8. **a.** $\dfrac{1}{3}$ **b.** 5

Helpful Hint

For additional help when replacing variables with replacement values, first place parentheses about any variables.

For Example 8b on the previous page, we have

$$x^2 - y = \underbrace{(x)^2 - (y)}_{\substack{\text{Place parentheses} \\ \text{about variables}}} = \underbrace{(2)^2 - (-5)}_{\substack{\text{Replace variables} \\ \text{with values}}} = 4 - (-5) = 4 + 5 = 9$$

Objective C Solutions of Equations ▶

Recall from Section 8.2 that a solution of an equation is a value for the variable that makes the equation true.

Practice 9

Determine whether -2 is a solution of $-1 + x = 1$.

Example 9 Determine whether -4 is a solution of $x - 5 = -9$.

Solution: Replace x with -4 and see if a true statement results.

$$
\begin{aligned}
x - 5 &= -9 \quad \text{Original equation} \\
-4 - 5 &\overset{?}{=} -9 \quad \text{Replace } x \text{ with } -4. \\
-4 + (-5) &\overset{?}{=} -9 \\
-9 &= -9 \quad \text{True}
\end{aligned}
$$

Thus -4 is a solution of $x - 5 = -9$.

■ **Work Practice 9**

Objective D Solving Applications That Involve Subtraction ▶

Another use of real numbers is in recording altitudes above and below sea level, as shown in the next example.

Practice 10

The highest point in Asia is the top of Mount Everest, at a height of 29,028 feet above sea level. The lowest point is the Dead Sea, which is 1312 feet below sea level. How much higher is Mount Everest than the Dead Sea? (*Source:* National Geographic Society)

Example 10 Finding a Change in Elevation

The highest point in the United States is the top of Mount McKinley, at a height of 20,320 feet above sea level. The lowest point is Death Valley, California, which is 282 feet below sea level. How much higher is Mount McKinley than Death Valley? (*Source:* U.S. Geological Survey)

Solution: To find "how much higher," we subtract. Don't forget that since Death Valley is 282 feet *below* sea level, we represent its height by -282. Draw a diagram to help visualize the problem.

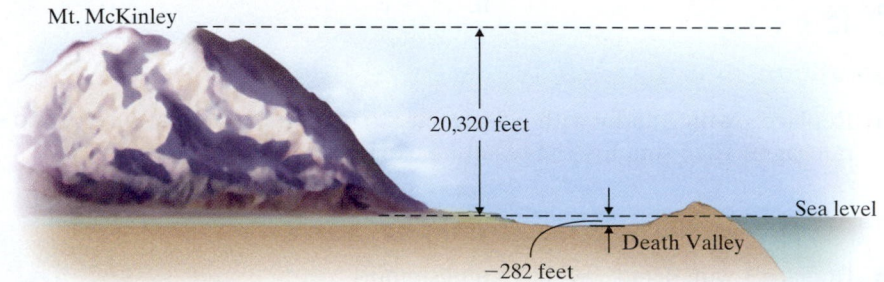

Mt. McKinley

20,320 feet

Sea level

Death Valley

-282 feet

Answers

9. -2 is not a solution.   **10.** 30,340 ft

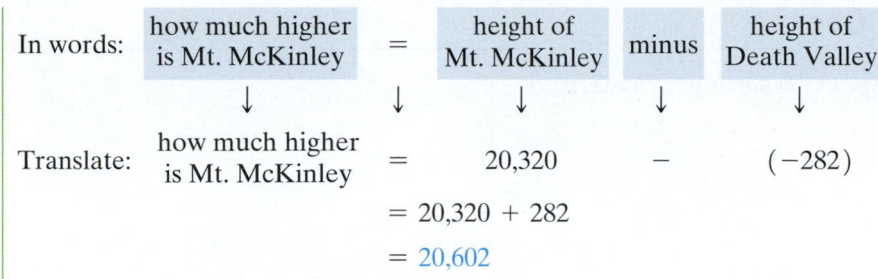

In words:	how much higher is Mt. McKinley	=	height of Mt. McKinley	minus	height of Death Valley

$$\text{Translate:} \quad \text{how much higher is Mt. McKinley} = 20{,}320 - (-282)$$

$$= 20{,}320 + 282$$

$$= 20{,}602$$

Thus, Mount McKinley is 20,602 feet higher than Death Valley.

■ Work Practice 10

Objective E Finding Complementary and Supplementary Angles

A knowledge of geometric concepts is needed by many professionals, such as doctors, carpenters, electronic technicians, gardeners, machinists, and pilots, just to name a few. With this in mind, we review the geometric concepts of **complementary** and **supplementary angles.**

Complementary and Supplementary Angles

Two angles are **complementary** if the sum of their measures is 90°.

Two angles are **supplementary** if the sum of their measures is 180°.

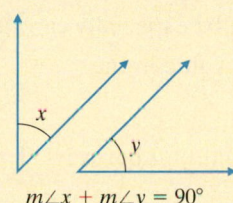

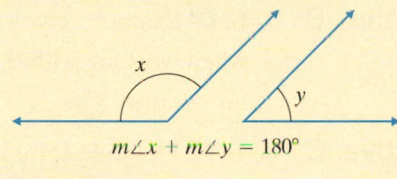

Example 11 Find the measure of each unknown complementary or supplementary angle.

a.

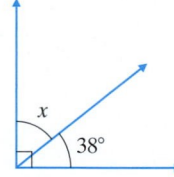

b.

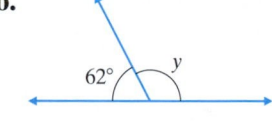

Solution:

a. These angles are complementary, so their sum is 90°. This means that the measure of angle x, $m\angle x$, is $90° - 38°$.

$$m\angle x = 90° - 38° = 52°$$

b. These angles are supplementary, so their sum is 180°. This means that $m\angle y$ is $180° - 62°$.

$$m\angle y = 180° - 62° = 118°$$

■ Work Practice 11

Practice 11

Find the measure of each unknown complementary or supplementary angle.

a.

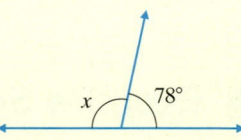

b.

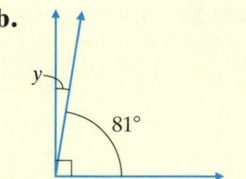

Vocabulary, Readiness & Video Check

Multiple choice: Select the correct lettered response following each exercise.

1. It is true that $a - b = $ _____.

    a. $b - a$ **b.** $a + (-b)$ **c.** $a + b$

2. The opposite of n is _____.

    a. $-n$ **b.** $-(-n)$ **c.** n

3. To evaluate $x - y$ for $x = -10$ and $y = -14$, we replace x with -10 and y with -14 and evaluate _____.

    a. $10 - 14$ **b.** $-10 - 14$ **c.** $-14 - 10$ **d.** $-10 - (-14)$

4. The expression $-5 - 10$ equals _____.

    a. $5 - 10$ **b.** $5 + 10$ **c.** $-5 + (-10)$ **d.** $10 - 5$

Martin-Gay Interactive Videos

See Video 8.4

Watch the section lecture video and answer the following questions.

Objective A   **5.** Complete this statement based on the lecture given before ▦ Example 1. To subtract two real numbers, change the operation to _____ and take the _____ of the second number.

          6. When simplifying ▦ Example 5, what is the result of the first step and why is the expression rewritten in this way? ▶

Objective B   **7.** In ▦ Example 7, why are you told to be especially careful when working with the replacement value in the numerator? ▶

Objective C   **8.** In ▦ Example 8, we learned that what number is NOT a solution of what equation? ▶

Objective D   **9.** For ▦ Example 9, why is the overall vertical change represented as a negative number? ▶

Objective E **10.** The definition of supplementary angles is given just before ▦ Example 10. Explain how this definition is used to solve ▦ Example 10. ▶

8.4 Exercise Set MyMathLab®

Objective A *Subtract. See Examples 1 through 4.*

1. $-6 - 4$ **2.** $-12 - 8$ **3.** $4 - 9$ **4.** $8 - 11$ ▶**5.** $16 - (-3)$

6. $12 - (-5)$ **7.** $7 - (-4)$ **8.** $3 - (-6)$ **9.** $-26 - (-18)$ **10.** $-60 - (-48)$

▶**11.** $-6 - 5$ **12.** $-8 - 4$ **13.** $16 - (-21)$ **14.** $15 - (-33)$ **15.** $-6 - (-11)$

16. $-4 - (-16)$ **17.** $-44 - 27$ **18.** $-36 - 51$ **19.** $-21 - (-21)$ **20.** $-17 - (-17)$

21. $-\dfrac{3}{11} - \left(-\dfrac{5}{11}\right)$ **22.** $-\dfrac{4}{7} - \left(-\dfrac{1}{7}\right)$ **23.** $9.7 - 16.1$ **24.** $8.3 - 11.2$ **25.** $-2.6 - (-6.7)$

26. $-6.1 - (-5.3)$ **27.** $\dfrac{1}{2} - \dfrac{2}{3}$ **28.** $\dfrac{3}{4} - \dfrac{7}{8}$ **29.** $-\dfrac{1}{6} - \dfrac{3}{4}$ **30.** $-\dfrac{1}{10} - \dfrac{7}{8}$

31. $8.3 - (-0.62)$ **32.** $4.3 - (-0.87)$ **33.** $0 - 8.92$ **34.** $0 - (-4.21)$

Translating *Translate each phrase to an expression and simplify. See Example 5.*

35. Subtract -5 from 8.

36. Subtract -2 from 3.

37. Find the difference between -6 and -1.

38. Find the difference between -17 and -1.

39. Subtract 8 from 7.

40. Subtract 9 from -4.

41. Decrease -8 by 15.

42. Decrease 11 by -14.

Mixed Practice (*Sections 8.2, 8.3, 8.4*) *Simplify each expression. (Remember the order of operations.) See Examples 6 and 7.*

43. $-10 - (-8) + (-4) - 20$

44. $-16 - (-3) + (-11) - 14$

45. $5 - 9 + (-4) - 8 - 8$

46. $7 - 12 + (-5) - 2 + (-2)$

47. $-6 - (2 - 11)$

48. $-9 - (3 - 8)$

49. $3^3 - 8 \cdot 9$

50. $2^3 - 6 \cdot 3$

51. $2 - 3(8 - 6)$

52. $4 - 6(7 - 3)$

53. $(3 - 6) + 4^2$

54. $(2 - 3) + 5^2$

55. $-2 + [(8 - 11) - (-2 - 9)]$

56. $-5 + [(4 - 15) - (-6) - 8]$

57. $|-3| + 2^2 + [-4 - (-6)]$

58. $|-2| + 6^2 + (-3 - 8)$

Objective B *Evaluate each expression when $x = -5$, $y = 4$, and $t = 10$. See Example 8.*

59. $x - y$

60. $y - x$

▶ 61. $\dfrac{9 - x}{y + 6}$

62. $\dfrac{15 - x}{y + 2}$

63. $|x| + 2t - 8y$

64. $|y| + 3x - 2t$

▶ 65. $y^2 - x$

66. $t^2 - x$

67. $\dfrac{|x - (-10)|}{2t}$

68. $\dfrac{|5y - x|}{6t}$

Objective C *Decide whether the given number is a solution of the given equation. See Example 9.*

▶ 69. $x - 9 = 5$; -4

70. $x - 10 = -7$; 3

71. $-x + 6 = -x - 1$; -2

72. $-x - 6 = -x - 1$; -10

73. $-x - 13 = -15$; 2

74. $4 = 1 - x$; 5

Objectives D E **Mixed Practice** *Solve. See Examples 10 and 11.*

75. The coldest temperature ever recorded on Earth was −129°F in Antarctica. The warmest temperature ever recorded was 134°F in Death Valley, California. How many degrees warmer is 134°F than −129°F? (*Source: The World Almanac*, 2013)

76. The coldest temperature ever recorded in the United States was −80°F in Alaska. The warmest temperature ever recorded was 134°F in California. How many degrees warmer is 134°F than −80°F? (*Source: The World Almanac*, 2013)

77. Mauna Kea in Hawaii has an elevation of 13,796 feet above sea level. The Mid-America Trench in the Pacific Ocean has an elevation of 21,857 feet below sea level. Find the difference in elevation between those two points. (*Source:* National Geographic Society and Defense Mapping Agency)

78. A woman received a statement of her charge account at Old Navy. She spent $93 on purchases last month. She returned an $18 top because she didn't like the color. She also returned a $26 nightshirt because it was damaged. What does she actually owe on her account?

▶ 79. Find x if the angles below are complementary angles.

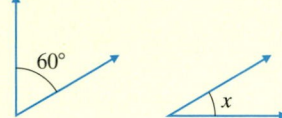

△ 80. Find y if the angles below are supplementary angles.

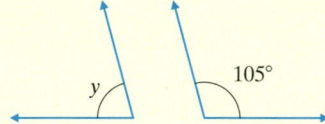

▶ 81. A commercial jetliner hits an air pocket and drops 250 feet. After climbing 120 feet, it drops another 178 feet. What is its overall vertical change?

82. In some card games, it is possible to have a negative score. Lavonne Schultz currently has a score of 15 points. She then loses 24 points. What is her new score?

83. The highest point in Africa is Mt. Kilimanjaro, Tanzania, at an elevation of 19,340 feet. The lowest point is Lake Assal, Djibouti, at 512 feet below sea level. How much higher is Mt. Kilimanjaro than Lake Assal? (*Source:* National Geographic Society)

84. The airport in Bishop, California, is at an elevation of 4101 feet above sea level. The nearby Furnace Creek Airport in Death Valley, California, is at an elevation of 226 feet below sea level. How much higher in elevation is the Bishop Airport than the Furnace Creek Airport? (*Source:* National Climatic Data Center)

Find each unknown complementary or supplementary angle.

85.

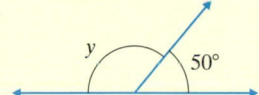

86.

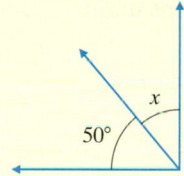

Mixed Practice—Translating (*Sections 8.3, 8.4*) *Translate each phrase to an algebraic expression. Use "x" to represent "a number."*

87. The sum of -5 and a number.

88. The difference of -3 and a number.

89. Subtract a number from -20.

90. Add a number and -36.

Review

Multiply or divide as indicated. See Sections 2.4 and 2.5.

91. $\dfrac{5}{8} \cdot 0$

92. $\dfrac{2}{3} \div \dfrac{3}{2}$

93. $1\dfrac{2}{3} \div 2\dfrac{1}{6}$

94. $3\dfrac{1}{2} \cdot \dfrac{11}{14}$

Concept Extensions

Recall the bar graph from Section 8.3. It shows each month's average daily low temperature in degrees Fahrenheit for Barrow, Alaska. Use this graph to answer Exercises 95 through 98.

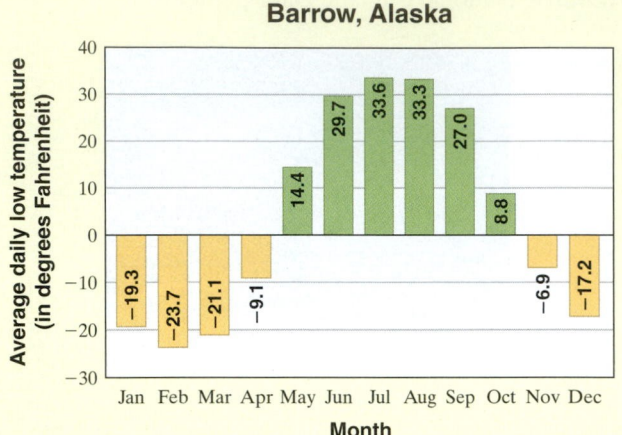

Barrow, Alaska

Source: National Climatic Data Center

95. Record the monthly increases and decreases in the low temperature from the previous month.

Month	Monthly Increase or Decrease (from the previous month)
February	
March	
April	
May	
June	

96. Record the monthly increases and decreases in the low temperature from the previous month.

Month	Monthly Increase or Decrease (from the previous month)
July	
August	
September	
October	
November	
December	

97. Which month had the greatest increase in temperature?

98. Which month had the greatest decrease in temperature?

99. Find two numbers whose difference is −5.

100. Find two numbers whose difference is −9.

*Each calculation below is **incorrect.** Find the error and correct it.*

101. $9 - (-7) \overset{?}{=} 2$

102. $-4 - 8 \overset{?}{=} 4$

103. $10 - 30 \overset{?}{=} 20$

104. $-3 - (-10) \overset{?}{=} -13$

If p is a positive number and n is a negative number, determine whether each statement is true or false. Explain your answer.

105. $p - n$ is always a positive number.

106. $n - p$ is always a negative number.

107. $|n| - |p|$ is always a positive number.

108. $|n - p|$ is always a positive number.

Without calculating, determine whether each answer is positive or negative. Then use a calculator to find the exact difference.

109. $56,875 - 87,262$

110. $4.362 - 7.0086$

Integrated Review

Operations on Real Numbers

Answer the following with positive, negative, or 0.

1. The opposite of a positive number is a _____ number.

2. The sum of two negative numbers is a _____ number.

3. The absolute value of a negative number is a _____ number.

4. The absolute value of zero is _____.

5. The sum of two positive numbers is a _____ number.

6. The sum of a number and its opposite is _____.

7. The absolute value of a positive number is a _____ number.

8. The opposite of a negative number is a _____ number.

Fill in the chart.

	Number	Opposite	Absolute Value
9.	$\dfrac{1}{7}$		
10.	$-\dfrac{12}{5}$		
11.		-3	
12.		$\dfrac{9}{11}$	

Perform each indicated operation and simplify. Don't forget to use order of operations if needed.

13. $-19 + (-23)$ **14.** $7 - (-3)$ **15.** $-15 + 17$ **16.** $-8 - 10$

17. $18 + (-25)$ **18.** $-2 + (-37)$ **19.** $-14 - (-12)$ **20.** $5 - 14$

21. $4.5 - 7.9$ **22.** $-8.6 - 1.2$ **23.** $-\dfrac{3}{4} - \dfrac{1}{7}$ **24.** $\dfrac{2}{3} - \dfrac{7}{8}$

Answers

1. _____
2. _____
3. _____
4. _____
5. _____
6. _____
7. _____
8. _____
9. _____
10. _____
11. _____
12. _____
13. _____
14. _____
15. _____
16. _____
17. _____
18. _____
19. _____
20. _____
21. _____
22. _____
23. _____
24. _____

25. _____

26. _____

27. _____

28. _____

29. _____

30. _____

31. _____

32. _____

33. _____

34. _____

35. _____

36. _____

37. _____

38. _____

39. _____

40. _____

41. _____

42. _____

43. _____

44. _____

45. _____

46. _____

25. $-9 - (-7) + 4 - 6$ **26.** $11 - 20 + (-3) - 12$ **27.** $24 - 6(14 - 11)$

28. $30 - 5(10 - 8)$ **29.** $(7 - 17) + 4^2$ **30.** $9^2 + (10 - 30)$

31. $|-9| + 3^2 + (-4 - 20)$ **32.** $|-4 - 5| + 5^2 + (-50)$

33. $-7 + [(1 - 2) + (-2 - 9)]$ **34.** $-6 + [(-3 + 7) + (4 - 15)]$

35. Subtract 5 from 1.   **36.** Subtract -2 from -3.

37. Subtract $-\dfrac{2}{5}$ from $\dfrac{1}{4}$.   **38.** Subtract $\dfrac{1}{10}$ from $-\dfrac{5}{8}$.

39. $2(19 - 17)^3 - 3(-7 + 9)^2$ **40.** $3(10 - 9)^2 + 6(20 - 19)^3$

Evaluate each expression when $x = -2$, $y = -1$, and $z = 9$.

41. $x - y$ **42.** $x + y$

43. $y + z$ **44.** $z - y$

45. $\dfrac{|5z - x|}{y - x}$ **46.** $\dfrac{|-x - y + z|}{2z}$

8.5 Multiplying and Dividing Real Numbers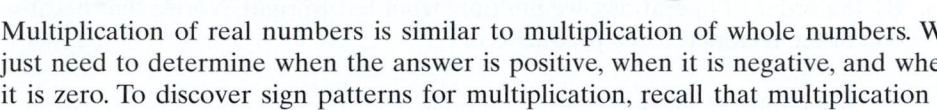

Objective A Multiplying Real Numbers

Multiplication of real numbers is similar to multiplication of whole numbers. We just need to determine when the answer is positive, when it is negative, and when it is zero. To discover sign patterns for multiplication, recall that multiplication is repeated addition. For example, $3(2)$ means that 2 is added to itself three times, or

$$3(2) = 2 + 2 + 2 = 6$$

Also,

$$3(-2) = (-2) + (-2) + (-2) = -6$$

Since $3(-2) = -6$, this suggests that the product of a positive number and a negative number is a negative number.

What about the product of two negative numbers? To find out, consider the following pattern.

Factor decreases by 1 each time.

$-3 \cdot 2 = -6$
$-3 \cdot 1 = -3$ Product increases by 3 each time.
$-3 \cdot 0 = 0$
$-3 \cdot -1 = 3$
$-3 \cdot -2 = 6$

This suggests that the product of two negative numbers is a positive number. Our results are given below.

Multiplying Real Numbers

1. The product of two numbers with the *same* sign is a positive number.
2. The product of two numbers with *different* signs is a negative number.

Examples Multiply.

1. $-7(6) = -42$ Different signs, so the product is negative.
2. $2(-10) = -20$
3. $-2(-14) = 28$ Same sign, so the product is positive.
4. $-\dfrac{2}{3} \cdot \dfrac{4}{7} = -\dfrac{2 \cdot 4}{3 \cdot 7} = -\dfrac{8}{21}$
5. $5(-1.7) = -8.5$
6. $-18(-3) = 54$

■ **Work Practice 1–6**

We already know that the product of 0 and any whole number is 0. This is true of all real numbers.

Products Involving Zero

If b is a real number, then $b \cdot 0 = 0$. Also $0 \cdot b = 0$.

Objectives

A Multiply Real Numbers.

B Find the Reciprocal of a Real Number.

C Divide Real Numbers.

D Evaluate Expressions Using Real Numbers.

E Determine Whether a Number Is a Solution of a Given Equation.

F Solve Applications That Involve Multiplication or Division of Real Numbers.

Practice 1–6

Multiply.

1. $-8(3)$ 2. $5(-30)$ 3. $-4(-12)$
4. $-\dfrac{5}{6} \cdot \dfrac{1}{4}$ 5. $6(-2.3)$ 6. $-15(-2)$

Answers

1. -24 2. -150 3. 48 4. $-\dfrac{5}{24}$
5. -13.8 6. 30

Practice 7

Multiply.
a. $5(0)(-3)$
b. $(-1)(-6)(-7)$
c. $(-2)(4)(-8)(-1)$

Example 7   Multiply.

a. $7(0)(-6)$ b. $(-2)(-3)(-4)$ c. $(-1)(-5)(-9)(-2)$

Solution:

a. By the order of operations, we multiply from left to right. Notice that because one of the factors is 0, the product is 0.

$$7(0)(-6) = 0(-6) = 0$$

b. Multiply two factors at a time, from left to right.

$$(-2)(-3)(-4) = (6)(-4) \quad \text{Multiply } (-2)(-3).$$
$$= -24$$

c. Multiply from left to right.

$$(-1)(-5)(-9)(-2) = (5)(-9)(-2) \quad \text{Multiply } (-1)(-5).$$
$$= -45(-2) \quad \text{Multiply } 5(-9).$$
$$= 90$$

■ Work Practice 7

✔**Concept Check**   What is the sign of the product of five negative numbers? Explain.

> **Helpful Hint**
>
> Have you noticed a pattern when multiplying signed numbers?
> If we let $(-)$ represent a negative number and $(+)$ represent a positive number, then
>
> $$(-)(-) = (+)$$
> The product of an even $$(-)(-)(-) = (-)$$ The product of an odd
> number of negative $$(-)(-)(-)(-) = (+)$$ number of negative numbers
> numbers is a positive   is a negative result.
> result. $$(-)(-)(-)(-)(-) = (-)$$

Now that we know how to multiply positive and negative numbers, let's see how we find the values of $(-5)^2$ and -5^2, for example. Although these two expressions look similar, the difference between the two is the parentheses. In $(-5)^2$, the parentheses tell us that the base, or repeated factor, is -5. In -5^2, only 5 is the base. Thus,

$$(-5)^2 = (-5)(-5) = 25 \quad \text{The base is } -5.$$
$$-5^2 = -(5 \cdot 5) = -25 \quad \text{The base is } 5.$$

Practice 8

Evaluate.
▶ a. $(-2)^4$ ▶ b. -2^4
c. $(-1)^5$ d. -1^5
e. $\left(-\dfrac{7}{9}\right)^2$

Example 8   Evaluate.

a. $(-2)^3$ b. -2^3 c. $(-3)^2$ d. -3^2 e. $\left(-\dfrac{2}{3}\right)^2$

Solution:

a. $(-2)^3 = (-2)(-2)(-2) = -8$ The base is -2.
b. $-2^3 = -(2 \cdot 2 \cdot 2) = -8$ The base is 2.
c. $(-3)^2 = (-3)(-3) = 9$ The base is -3.
d. $-3^2 = -(3 \cdot 3) = -9$ The base is 3.
e. $\left(-\dfrac{2}{3}\right)^2 = \left(-\dfrac{2}{3}\right)\left(-\dfrac{2}{3}\right) = \dfrac{4}{9}$ The base is $-\dfrac{2}{3}$.

■ Work Practice 8

Answers

7. a. 0 b. -42 c. -64 8. a. 16
b. -16 c. -1 d. -1 e. $\dfrac{49}{81}$

✔**Concept Check Answer**

negative

Helpful Hint

Be careful when identifying the base of an exponential expression.

$$(-3)^2 \qquad\qquad\qquad -3^2$$

$$\text{Base is } -3 \qquad\qquad\qquad \text{Base is } 3$$

$$(-3)^2 = (-3)(-3) = 9 \qquad\qquad -3^2 = -(3 \cdot 3) = -9$$

Objective B Finding Reciprocals ▶

Addition and subtraction are related. Every difference of two numbers $a - b$ can be written as the sum $a + (-b)$. Multiplication and division are related also. For example, the quotient $6 \div 3$ can be written as the product $6 \cdot \dfrac{1}{3}$. Recall that the pair of numbers 3 and $\dfrac{1}{3}$ has a special relationship. Their product is 1 and they are called **reciprocals** or **multiplicative inverses** of each other.

Reciprocal or Multiplicative Inverse

Two numbers whose product is 1 are called **reciprocals** or **multiplicative inverses** of each other.

Example 9 Find the reciprocal of each number.

a. 22   Reciprocal is $\dfrac{1}{22}$ since $22 \cdot \dfrac{1}{22} = 1$.

b. $\dfrac{3}{16}$ Reciprocal is $\dfrac{16}{3}$ since $\dfrac{3}{16} \cdot \dfrac{16}{3} = 1$.

c. -10 Reciprocal is $-\dfrac{1}{10}$ since $-10 \cdot -\dfrac{1}{10} = 1$.

d. $-\dfrac{9}{13}$ Reciprocal is $-\dfrac{13}{9}$ since $-\dfrac{9}{13} \cdot -\dfrac{13}{9} = 1$.

e. 1.7   Reciprocal is $\dfrac{1}{1.7}$ since $1.7 \cdot \dfrac{1}{1.7} = 1$.

■ **Work Practice 9**

Helpful Hint

The fraction $\dfrac{1}{1.7}$ is not simplified since the denominator is a decimal number. For the purpose of finding a reciprocal, we will leave the fraction as is.

Does the number 0 have a reciprocal? If it does, it is a number n such that $0 \cdot n = 1$. Notice that this can never be true since $0 \cdot n = 0$. This means that 0 has no reciprocal.

Quotients Involving Zero

The number 0 does not have a reciprocal.

Practice 9

Find the reciprocal of each number.

a. 13   **b.** $\dfrac{7}{15}$ **c.** -5

d. $-\dfrac{8}{11}$ **e.** 7.9

Answers

9. a. $\dfrac{1}{13}$ **b.** $\dfrac{15}{7}$ **c.** $-\dfrac{1}{5}$

d. $-\dfrac{11}{8}$ **e.** $\dfrac{1}{7.9}$

Objective C Dividing Real Numbers ▶

We may now write a quotient as an equivalent product.

> ### Quotient of Two Real Numbers
>
> If a and b are real numbers and b is not 0, then
>
> $$a \div b = \frac{a}{b} = a \cdot \frac{1}{b}$$

In other words, the quotient of two real numbers is the product of the first number and the multiplicative inverse or reciprocal of the second number.

Practice 10

Use the definition of the quotient of two numbers to find each quotient.

a. $-12 \div 4$ **b.** $\dfrac{-20}{-10}$

c. $\dfrac{36}{-4}$

Example 10 Use the definition of the quotient of two numbers to find each quotient. $\left(a \div b = a \cdot \dfrac{1}{b} \right)$

a. $-18 \div 3$ **b.** $\dfrac{-14}{-2}$ **c.** $\dfrac{20}{-4}$

Solution:

a. $-18 \div 3 = -18 \cdot \dfrac{1}{3} = -6$

b. $\dfrac{-14}{-2} = -14 \cdot -\dfrac{1}{2} = 7$

c. $\dfrac{20}{-4} = 20 \cdot -\dfrac{1}{4} = -5$

■ Work Practice 10

Since the quotient $a \div b$ can be written as the product $a \cdot \dfrac{1}{b}$, it follows that sign patterns for dividing two real numbers are the same as sign patterns for multiplying two real numbers.

> ### Dividing Real Numbers
>
> 1. The quotient of two numbers with the *same* sign is a positive number.
> 2. The quotient of two numbers with *different* signs is a negative number.

Practice 11

Divide.

a. $\dfrac{-25}{5}$ **b.** $\dfrac{-48}{-6}$

c. $\dfrac{50}{-2}$ **d.** $\dfrac{-72}{0.2}$

Example 11 Divide.

a. $\dfrac{-30}{-10} = 3$ Same sign, so the quotient is positive.

b. $\dfrac{-100}{5} = -20$

c. $\dfrac{20}{-2} = -10$ Different signs, so the quotient is negative.

d. $\dfrac{42}{-0.6} = -70$ $0.6\overline{)42.0}$ → $70.$

■ Work Practice 11

Answers

10. a. -3 **b.** 2 **c.** -9 **11. a.** -5
b. 8 **c.** -25 **d.** -360

✓**Concept Check** What is wrong with the following calculation?

$$\frac{-36}{-9} = -4$$

In the examples on the previous page, we divided mentally or by long division. When we divide by a fraction, it is usually easier to multiply by its reciprocal.

Examples Divide.

12. $\frac{2}{3} \div \left(-\frac{5}{4}\right) = \frac{2}{3} \cdot \left(-\frac{4}{5}\right) = -\frac{8}{15}$

13. $-\frac{1}{6} \div \left(-\frac{2}{3}\right) = -\frac{1}{6} \cdot \left(-\frac{3}{2}\right) = \frac{3}{12} = \frac{\overset{1}{\cancel{3}}}{\cancel{3} \cdot 4} = \frac{1}{4}$

■ **Work Practice 12–13**

Our definition of the quotient of two real numbers does not allow for division by 0 because 0 does not have a reciprocal. How then do we interpret $\frac{3}{0}$? We say that an expression such as this one is **undefined.** Can we divide 0 by a number other than 0? Yes; for example,

$$\frac{0}{3} = 0 \cdot \frac{1}{3} = 0$$

Division Involving Zero

If a is a nonzero number, then $\frac{0}{a} = 0$ and $\frac{a}{0}$ is undefined.

Example 14 Divide, if possible.

a. $\frac{1}{0}$ is undefined. **b.** $\frac{0}{-3} = 0$

■ **Work Practice 14**

Notice that $\frac{12}{-2} = -6, -\frac{12}{2} = -6,$ and $\frac{-12}{2} = -6.$ This means that

$$\frac{12}{-2} = -\frac{12}{2} = \frac{-12}{2}$$

In other words, a single negative sign in a fraction can be written in the denominator, in the numerator, or in front of the fraction without changing the value of the fraction.

If a and b are real numbers, and $b \neq 0$, then $\frac{a}{-b} = \frac{-a}{b} = -\frac{a}{b}$.

Objective D Evaluating Expressions ▶

Examples combining basic arithmetic operations along with the principles of the order of operations help us to review these concepts of multiplying and dividing real numbers.

Practice 12–13
Divide.
12. $-\frac{5}{9} \div \frac{2}{3}$ **13.** $-\frac{2}{7} \div \left(-\frac{1}{5}\right)$

Practice 14
Divide if possible.
a. $\frac{-7}{0}$ **b.** $\frac{0}{-2}$

Answers
12. $-\frac{5}{6}$ **13.** $\frac{10}{7}$ **14. a.** undefined **b.** 0

✓**Concept Check Answer**
$\frac{-36}{-9} = 4$

Practice 15

Use order of operations to evaluate each expression.

a. $\dfrac{0(-5)}{3}$

b. $-3(-9) - 4(-4)$

c. $(-3)^2 + 2[(5 - 15) - |-4 - 1|]$

d. $\dfrac{-7(-4) + 2}{-10 - (-5)}$

e. $\dfrac{5(-2)^3 + 52}{-4 + 1}$

Example 15 Use the order of operations to evaluate each expression.

a. $\dfrac{0(-8)}{2}$

b. $-4(-11) - 5(-2)$

c. $(-2)^2 + 3[(-3 - 2) - |4 - 6|]$

d. $\dfrac{(-12)(-3) + 4}{-7 - (-2)}$

e. $\dfrac{2(-3)^2 - 20}{|-5| + 4}$

Solution:

a. $\dfrac{0(-8)}{2} = \dfrac{0}{2} = 0$

b. $(-4)(-11) - 5(-2) = 44 - (-10)$ Find the products.

$= 44 + 10$ Add 44 to the opposite of -10.

$= 54$ Add.

c. $(-2)^2 + 3[(-3 - 2) - |4 - 6|] = (-2)^2 + 3[(-5) - |-2|]$ Simplify within innermost sets of grouping symbols.

$= (-2)^2 + 3[-5 - 2]$ Write $|-2|$ as 2.

$= (-2)^2 + 3(-7)$ Combine.

$= 4 + (-21)$ Evaluate $(-2)^2$ and multiply $3(-7)$.

$= -17$ Add.

For parts d and e, first simplify the numerator and denominator separately; then divide.

d. $\dfrac{(-12)(-3) + 4}{-7 - (-2)} = \dfrac{36 + 4}{-7 + 2}$

$= \dfrac{40}{-5}$

$= -8$ Divide.

e. $\dfrac{2(-3)^2 - 20}{|-5| + 4} = \dfrac{2 \cdot 9 - 20}{5 + 4} = \dfrac{18 - 20}{9} = \dfrac{-2}{9} = -\dfrac{2}{9}$

■ **Work Practice 15**

Using what we have learned about multiplying and dividing real numbers, we continue to practice evaluating algebraic expressions.

Practice 16

Evaluate each expression when $x = -1$ and $y = -5$.

a. $\dfrac{3y}{45x}$

b. $x^2 - y^3$

c. $\dfrac{x + y}{3x}$

Example 16 Evaluate each expression when $x = -2$ and $y = -4$.

a. $\dfrac{3x}{2y}$

b. $x^3 - y^2$

c. $\dfrac{x - y}{-x}$

Solution: Replace x with -2 and y with -4 and simplify.

a. $\dfrac{3x}{2y} = \dfrac{3(-2)}{2(-4)} = \dfrac{-6}{-8} = \dfrac{6}{8} = \dfrac{\overset{1}{\cancel{2}} \cdot 3}{\underset{1}{\cancel{2}} \cdot 4} = \dfrac{3}{4}$

Answers

15. a. 0   **b.** 43   **c.** -21   **d.** -6

e. -4   **16. a.** $\dfrac{1}{3}$ **b.** 126   **c.** 2

b. $x^3 - y^2 = (-2)^3 - (-4)^2$ Substitute the given values for the variables.

$\qquad\qquad = -8 - (16)$ Evaluate $(-2)^3$ and $(-4)^2$.

$\qquad\qquad = -8 + (-16)$ Write as a sum.

$\qquad\qquad = -24$ Add.

c. $\dfrac{x - y}{-x} = \dfrac{-2 - (-4)}{-(-2)} = \dfrac{-2 + 4}{2} = \dfrac{2}{2} = 1$

■ **Work Practice 16**

Helpful Hint

Remember: For additional help when replacing variables with replacement values, first place parentheses about any variables.

Evaluate $3x - y^2$ when $x = 5$ and $y = -4$.

$3x - y^2 = 3(x) - (y)^2$ Place parentheses about variables only.

$\qquad\quad = 3(5) - (-4)^2$ Replace variables with values.

$\qquad\quad = 15 - 16$ Simplify.

$\qquad\quad = -1$

Objective E Solutions of Equations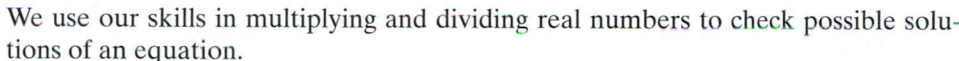

We use our skills in multiplying and dividing real numbers to check possible solutions of an equation.

Example 17   Determine whether -10 is a solution of $\dfrac{-20}{x} + 15 = 2x$.

Solution: $\dfrac{-20}{x} + 15 = 2x$ Original equation

$\dfrac{-20}{-10} + 15 \overset{?}{=} 2(-10)$ Replace x with -10.

$2 + 15 \overset{?}{=} -20$ Divide and multiply.

$17 = -20$ False

Since we have a false statement, -10 is *not* a solution of the equation.

■ **Work Practice 17**

Practice 17

Determine whether -8 is a solution of $\dfrac{x}{4} - 3 = x + 3$.

Objective F Solving Applications That Involve Multiplying or Dividing Numbers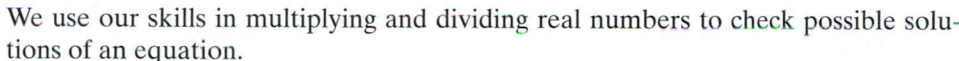

Many real-life problems involve multiplication and division of numbers.

Answer

17. -8 is a solution

Practice 18

A card player had a score of −13 for each of four games. Find the total score.

Answer

18. −52

Example 18 Calculating a Total Golf Score

A professional golfer finished seven strokes under par (−7) for each of three days of a tournament. What was her total score for the tournament?

Solution: Although the key word is "total," since this is repeated addition of the same number, we multiply.

In words:

golfer's total score	=	number of days	·	score each day
↓	↓	↓	↓	↓

Translate:

$$\text{golfer's total} = 3 \cdot (-7)$$
$$= -21$$

Thus, the golfer's total score was −21, or 21 strokes under par.

■ **Work Practice 18**

🖩 Calculator Explorations

Entering Negative Numbers on a Scientific Calculator

To enter a negative number on a scientific calculator, find a key marked $\boxed{+/-}$. (On some calculators, this key is marked $\boxed{\text{CHS}}$ for "change sign.") To enter −8, for example, press the keys $\boxed{8}\ \boxed{+/-}$. The display will read $\boxed{-8}$.

Entering Negative Numbers on a Graphing Calculator

To enter a negative number on a graphing calculator, find a key marked $\boxed{(-)}$. Do not confuse this key with the key $\boxed{-}$, which is used for subtraction. To enter −8, for example, press the keys $\boxed{(-)}\boxed{8}$. The display will read $\boxed{-8}$.

Operations with Real Numbers

To evaluate −2(7 − 9) − 20 on a calculator, press the keys

$\boxed{2}\ \boxed{+/-}\ \boxed{\times}\ \boxed{(}\boxed{7}\boxed{-}\boxed{9}\boxed{)}\boxed{-}\boxed{2}\boxed{0}\ \boxed{=}$,

or

$\boxed{(-)}\boxed{2}\boxed{(}\boxed{7}\boxed{-}\boxed{9}\boxed{)}\boxed{-}\boxed{2}\boxed{0}\ \boxed{\text{ENTER}}$.

The display will read $\boxed{-16}$ or

$-2(7 - 9) - 20$
-16

Use a calculator to simplify each expression.

1. $-38(26 - 27)$
2. $-59(-8) + 1726$
3. $134 + 25(68 - 91)$
4. $45(32) - 8(218)$
5. $\dfrac{-50(294)}{175 - 205}$
6. $\dfrac{-444 - 444.8}{-181 - (-181)}$
7. $9^5 - 4550$
8. $5^8 - 6259$
9. $(-125)^2$ (Be careful.)
10. -125^2 (Be careful.)

Vocabulary, Readiness & Video Check

Use the choices below to fill in each blank. Each choice may be used more than once.

negative   0

positive   undefined

1. The product of a negative number and a positive number is a(n) _____ number.
2. The product of two negative numbers is a(n) _____ number.
3. The quotient of two negative numbers is a(n) _____ number.
4. The quotient of a negative number and a positive number is a(n) _____ number.
5. The product of a negative number and zero is _____.
6. The reciprocal of a negative number is a _____ number.
7. The quotient of 0 and a negative number is _____.
8. The quotient of a negative number and 0 is _____.

Martin-Gay Interactive Videos   Watch the section lecture video and answer the following questions.

See Video 8.5 🍎

Objective A 9. Explain the significance of the use of parentheses when comparing ▥ Examples 6 and 7. ▶

Objective B 10. In ▥ Example 9, why is the reciprocal equal to $\frac{3}{2}$ and not $-\frac{3}{2}$? ▶

Objective C 11. Before ▥ Example 11, the sign rules for division of real numbers are discussed. Are the sign rules for division the same as for multiplication? Why or why not? ▶

Objective D 12. In ▥ Example 17, the importance of placing the replacement values in parentheses when evaluating is emphasized. Why? ▶

Objective E 13. In ▥ Example 18, is 5 a solution of $-3x - 5 = -20$? Why or why not? ▶

Objective F 14. In ▥ Example 19, explain why each loss of 4 yards is represented by -4 and not 4. ▶

8.5 Exercise Set MyMathLab®

Objective A *Multiply. See Examples 1 through 7.*

▶**1.** $-6(4)$

2. $-8(5)$

▶**3.** $2(-1)$

4. $7(-4)$

▶**5.** $-5(-10)$

6. $-6(-11)$

7. $-3 \cdot 15$

8. $-2 \cdot 37$

9. $-\frac{1}{2}\left(-\frac{3}{5}\right)$

10. $-\frac{1}{8}\left(-\frac{1}{3}\right)$

11. $5(-1.4)$

12. $6(-2.5)$

13. $(-1)(-3)(-5)$

14. $(-2)(-3)(-4)$

15. $(2)(-1)(-3)(0)$

16. $(3)(-5)(-2)(0)$

Evaluate. See Example 8.

17. $(-4)^2$

18. $(-3)^3$

19. -4^2

20. -6^2

21. $\left(-\dfrac{3}{4}\right)^2$

22. $\left(-\dfrac{2}{7}\right)^2$

23. -0.7^2

24. -0.8^2

Objective **B** *Find each reciprocal. See Example 9.*

▶**25.** $\dfrac{2}{3}$

26. $\dfrac{1}{7}$

▶**27.** -14

28. -8

29. $-\dfrac{3}{11}$

30. $-\dfrac{6}{13}$

31. 0.2

32. 1.5

Objective **C** *Divide. See Examples 10 through 14.*

▶**33.** $\dfrac{18}{-2}$

34. $\dfrac{36}{-9}$

35. $-48 \div 12$

36. $-60 \div 5$

▶**37.** $\dfrac{0}{-4}$

38. $\dfrac{0}{-9}$

▶**39.** $\dfrac{5}{0}$

40. $\dfrac{8}{0}$

41. $\dfrac{6}{7} \div \left(-\dfrac{1}{3}\right)$

42. $\dfrac{4}{5} \div \left(-\dfrac{1}{2}\right)$

43. $-3.2 \div -0.02$

44. $-4.9 \div -0.07$

Objectives **A** **C** **Mixed Practice** *Perform the indicated operation. See Examples 1 through 14.*

45. $(-8)(-8)$

46. $(-7)(-7)$

▶**47.** $\dfrac{2}{3}\left(-\dfrac{4}{9}\right)$

48. $\dfrac{2}{7}\left(-\dfrac{2}{11}\right)$

▶**49.** $\dfrac{-12}{-4}$

50. $\dfrac{-45}{-9}$

51. $\dfrac{30}{-2}$

52. $\dfrac{14}{-2}$

53. $(-5)^3$

54. $(-2)^5$

55. $(-0.2)^3$

56. $(-0.3)^3$

57. $-\dfrac{3}{4}\left(-\dfrac{8}{9}\right)$

58. $-\dfrac{5}{6}\left(-\dfrac{3}{10}\right)$

▶**59.** $-\dfrac{5}{9} \div \left(-\dfrac{3}{4}\right)$

60. $-\dfrac{1}{10} \div \left(-\dfrac{8}{11}\right)$

61. $-2.1(-0.4)$

62. $-1.3(-0.6)$

63. $\dfrac{-48}{1.2}$

64. $\dfrac{-86}{2.5}$

65. $(-3)^4$

66. -3^4

67. -1^7

68. $(-1)^7$

69. Multiply -11 by 11.

70. Multiply -12 by 12.

71. Find the quotient of $-\dfrac{4}{9}$ and $\dfrac{4}{9}$.

72. Find the quotient of $-\dfrac{5}{12}$ and $\dfrac{5}{12}$.

Mixed Practice (Sections 8.3, 8.4, 8.5) *Perform the indicated operation.*

73. $-9 - 10$

74. $-8 - 11$

75. $-9(-10)$

76. $-8(-11)$

77. $7(-12)$

78. $6(-15)$

79. $7 + (-12)$

80. $6 + (-15)$

Objective D *Evaluate each expression. See Example 15.*

81. $\dfrac{-9(-3)}{-6}$

82. $\dfrac{-6(-3)}{-4}$

83. $-3(2 - 8)$

84. $-4(3 - 9)$

85. $-7(-2) - 3(-1)$

86. $-8(-3) - 4(-1)$

87. $2^2 - 3[(2 - 8) - (-6 - 8)]$

88. $3^2 - 2[(3 - 5) - (2 - 9)]$

89. $\dfrac{-6^2 + 4}{-2}$

90. $\dfrac{3^2 + 4}{5}$

91. $\dfrac{-3 - 5^2}{2(-7)}$

92. $\dfrac{-2 - 4^2}{3(-6)}$

93. $\dfrac{22 + (3)(-2)^2}{-5 - 2}$

94. $\dfrac{-20 + (-4)^2(3)}{1 - 5}$

95. $\dfrac{(-4)^2 - 16}{4 - 12}$

96. $\dfrac{(-2)^2 - 4}{4 - 9}$

97. $\dfrac{6 - 2(-3)}{4 - 3(-2)}$

98. $\dfrac{8 - 3(-2)}{2 - 5(-4)}$

99. $\dfrac{|5 - 9| + |10 - 15|}{|2(-3)|}$

100. $\dfrac{|-3 + 6| + |-2 + 7|}{|-2 \cdot 2|}$

101. $\dfrac{-7(-1) + (-3)4}{(-2)(5) + (-6)(-8)}$

102. $\dfrac{8(-7) + (-2)(-6)}{(-9)(3) + (-10)(-11)}$

Evaluate each expression when $x = -5$ and $y = -3$. See Example 16.

103. $\dfrac{2x - 5}{y - 2}$

104. $\dfrac{2y - 12}{x - 4}$

105. $\dfrac{6 - y}{x - 4}$

106. $\dfrac{10 - y}{x - 8}$

107. $\dfrac{4 - 2x}{y + 3}$

108. $\dfrac{2y + 3}{-5 - x}$

109. $\dfrac{x^2 + y}{3y}$

110. $\dfrac{y^2 - x}{2x}$

Objective E *Decide whether the given number is a solution of the given equation. See Example 17.*

111. $-3x - 5 = -20; \quad 5$

112. $17 - 4x = x + 27; \quad -2$

113. $\dfrac{x}{5} + 2 = -1; \quad 15$

114. $\dfrac{x}{6} - 3 = 5; \quad 48$

115. $\dfrac{x - 3}{7} = -2; \quad -11$

116. $\dfrac{x + 4}{5} = -6; \quad -30$

Objective **F** **Translating** *Translate each phrase to an expression. Use x to represent "a number." See Example 18.*

117. The product of −71 and a number

118. The quotient of −8 and a number

119. Subtract a number from −16.

120. The sum of a number and −12

121. −29 increased by a number

122. The difference of a number and −10

123. Divide a number by −33.

124. Multiply a number by −17.

Solve. See Example 18.

▶ **125.** A football team lost four yards on each of three consecutive plays. Represent the total loss as a product of signed numbers and find the total loss.

126. A stock market broker lost $400 on each of seven consecutive days in the stock market. Represent his total loss as a product of signed numbers and find his total loss.

127. A deep-sea diver must move up or down in the water in short steps in order to keep from getting a physical condition called the "bends." Suppose a diver moves down from the surface in five steps of 20 feet each. Represent his total movement as a product of signed numbers and find the depth.

128. A weather forecaster predicts that the temperature will drop five degrees each hour for the next six hours. Represent this drop as a product of signed numbers and find the total drop in temperature.

Review

Find the perimeter of each figure. See Section 6.3.

△ **129.** Square
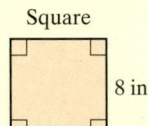
8 in.

△ **130.** Parallelogram
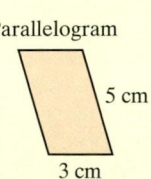
5 cm

3 cm

△ **131.** Rectangle
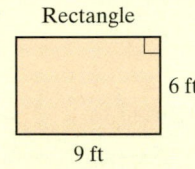
6 ft

9 ft

△ **132.** Triangle
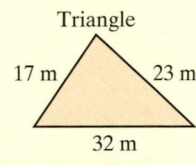
17 m   23 m

32 m

Concept Extensions

State whether each statement is true or false.

133. The product of three negative integers is negative.

134. The product of three positive integers is positive.

135. The product of four negative integers is negative.

136. The product of four positive integers is positive.

Study the bar graph below showing the average surface temperatures of planets. Use Exercises 137 and 138 to complete the planet temperatures on the graph. (Pluto is now classified as a dwarf planet.)

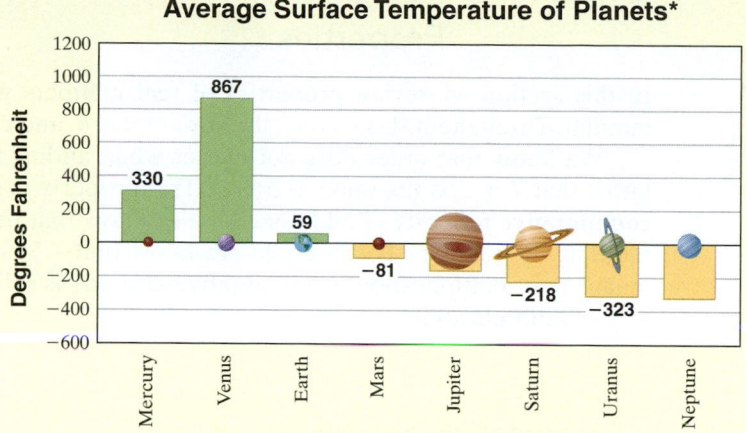

Average Surface Temperature of Planets*

*(For some planets, the temperature given is the temperature where the atmospheric pressure equals 1 Earth atmosphere; Source: The World Almanac)

137. The surface temperature of Jupiter is twice the temperature of Mars. Find this temperature.

138. The surface temperature of Neptune is equal to the temperature of Mercury divided by -1. Find this temperature.

139. For the first quarter of 2013, Wal-Mart, Inc. posted a loss of $33 million in membership and other income. If this trend was consistent for each month of the quarter, how much would you expect this loss to have been for each month? (*Source:* Wal-Mart Stores, Inc.)

140. For the first quarter of 2013, Chrysler Group LLC, maker of Jeep vehicles, posted a loss of about 30,000 Jeep Liberty shipments because it had stopped producing the vehicle in 2012. If this trend was consistent for each month of the quarter, how much would you expect this loss to have been for each month? (*Source:* Chrysler Group, LLC)

141. Explain why the product of an even number of negative numbers is a positive number.

142. If a and b are any real numbers, is the statement $a \cdot b = b \cdot a$ always true? Why or why not?

143. Find two real numbers that are their own reciprocal. Explain why there are only two.

144. Explain why 0 has no reciprocal.

Mixed Practice (8.3, 8.4, 8.5) *Write each as an algebraic expression. Then simplify the expression.*

145. 7 subtracted from the quotient of 0 and 5

146. Twice the sum of -3 and -4

147. -1 added to the product of -8 and -5

148. The difference of -9 and the product of -4 and -6

Objectives

A Use the Commutative and Associative Properties.

B Use the Distributive Property.

C Use the Identity and Inverse Properties.

Objective A Using the Commutative and Associative Properties

In this section we review properties of real numbers with which we are already familiar. Throughout this section, the variables a, b, and c represent real numbers.

We know that order does not matter when adding numbers. For example, we know that $7 + 5$ is the same as $5 + 7$. This property is given a special name—the **commutative property of addition.** We also know that order does not matter when multiplying numbers. For example, we know that $-5(6) = 6(-5)$. This property means that multiplication is commutative also and is called the **commutative property of multiplication.**

Commutative Properties

Addition:	$a + b = b + a$
Multiplication:	$a \cdot b = b \cdot a$

These properties state that the *order* in which any two real numbers are added or multiplied does not change their sum or product. For example, if we let $a = 3$ and $b = 5$, then the commutative properties guarantee that

$$3 + 5 = 5 + 3 \quad \text{and} \quad 3 \cdot 5 = 5 \cdot 3$$

Helpful Hint

Is subtraction also commutative? Try an example. Is $3 - 2 = 2 - 3$? **No!** The left side of this statement equals 1; the right side equals -1. There is no commutative property of subtraction. Similarly, there is no commutative property of division. For example, $10 \div 2$ does not equal $2 \div 10$.

Practice 1

Use a commutative property to complete each statement.

a. $7 \cdot y =$ _____

b. $4 + x =$ _____

Example 1   Use a commutative property to complete each statement.

a. $x + 5 =$ _____   **b.** $3 \cdot x =$ _____

Solution:

a. $x + 5 = 5 + x$ By the commutative property of addition

b. $3 \cdot x = x \cdot 3$ By the commutative property of multiplication

■ **Work Practice 1**

✓**Concept Check** Which of the following pairs of actions are commutative?

a. "raking the leaves" and "bagging the leaves"

b. "putting on your left glove" and "putting on your right glove"

c. "putting on your coat" and "putting on your shirt"

d. "reading a novel" and "reading a newspaper"

Answers

1. a. $y \cdot 7$ **b.** $x + 4$

✓**Concept Check Answers**

b, d

594

Let's now discuss grouping numbers. When we add three numbers, the way in which they are grouped or associated does not change their sum. For example, we know that $2 + (3 + 4) = 2 + 7 = 9$. This result is the same if we group the numbers differently. In other words, $(2 + 3) + 4 = 5 + 4 = 9$, also. Thus, $2 + (3 + 4) = (2 + 3) + 4$. This property is called the **associative property of addition.**

In the same way, changing the grouping of numbers when multiplying does not change their product. For example, $2 \cdot (3 \cdot 4) = (2 \cdot 3) \cdot 4$ (check it). This is the **associative property of multiplication.**

Associative Properties

Addition:	$(a + b) + c = a + (b + c)$
Multiplication:	$(a \cdot b) \cdot c = a \cdot (b \cdot c)$

These properties state that the way in which three numbers are *grouped* does not change their sum or their product.

Example 2   Use an associative property to complete each statement.

a. $5 + (4 + 6) =$ _____

b. $(-1 \cdot 2) \cdot 5 =$ _____

c. $(m + n) + 9 =$ _____

d. $(xy) \cdot 12 =$ _____

Solution:

a. $5 + (4 + 6) = (5 + 4) + 6$ By the associative property of addition

b. $(-1 \cdot 2) \cdot 5 = -1 \cdot (2 \cdot 5)$ By the associative property of multiplication

c. $(m + n) + 9 = m + (n + 9)$ By the associative property of addition

d. $(xy) \cdot 12 = x \cdot (y \cdot 12)$ Recall that xy means $x \cdot y$.

■ **Work Practice 2**

Practice 2

Use an associative property to complete each statement.

a. $5 \cdot (-3 \cdot 6) =$ _____

b. $(-2 + 7) + 3 =$ _____

c. $(q + r) + 17 =$ _____

d. $(ab) \cdot 21 =$ _____

Helpful Hint

Remember the difference between the commutative properties and the associative properties. The commutative properties have to do with the *order* of numbers and the associative properties have to do with the *grouping* of numbers.

Examples   Determine whether each statement is true by an associative property or a commutative property.

3. $(7 + 10) + 4 = (10 + 7) + 4$ Since the order of two numbers was changed and their grouping was not, this is true by the commutative property of addition.

4. $2 \cdot (3 \cdot 1) = (2 \cdot 3) \cdot 1$ Since the grouping of the numbers was changed and their order was not, this is true by the associative property of multiplication.

■ **Work Practice 3–4**

Let's now illustrate how these properties can help us simplify expressions.

Practice 3–4

Determine whether each statement is true by an associative property or a commutative property.

3. $5 \cdot (4 \cdot 7) = 5 \cdot (7 \cdot 4)$

4. $-2 + (4 + 9)$
$= (-2 + 4) + 9$

Answers
2. a. $(5 \cdot -3) \cdot 6$ **b.** $-2 + (7 + 3)$
c. $q + (r + 17)$ **d.** $a \cdot (b \cdot 21)$
3. commutative   **4.** associative

Practice 5–6

Simplify each expression.

5. $(-3 + x) + 17$

6. $4(5x)$

Examples   Simplify each expression.

5.
$$10 + (x + 12) = 10 + (12 + x) \quad \text{By the commutative property of addition}$$
$$= (10 + 12) + x \quad \text{By the associative property of addition}$$
$$= 22 + x \quad \text{Add.}$$

6.
$$-3(7x) = (-3 \cdot 7)x \quad \text{By the associative property of multiplication}$$
$$= -21x \quad \text{Multiply.}$$

■ **Work Practice 5–6**

Objective B Using the Distributive Property ▶

The **distributive property of multiplication over addition** is used repeatedly throughout algebra. It is useful because it allows us to write a product as a sum or a sum as a product.

We know that $7(2 + 4) = 7(6) = 42$. Compare that with

$$7(2) + 7(4) = 14 + 28 = 42$$

Since both original expressions equal 42, they must equal each other, or

$$7(2 + 4) = 7(2) + 7(4)$$

This is an example of the distributive property. The product on the left side of the equal sign is equal to the sum on the right side. We can think of the 7 as being distributed to each number inside the parentheses.

Distributive Property of Multiplication Over Addition

$$a(b + c) = ab + ac$$

Since multiplication is commutative, this property can also be written as

$$(b + c)a = ba + ca$$

The distributive property can also be extended to more than two numbers inside the parentheses. For example,

$$3(x + y + z) = 3(x) + 3(y) + 3(z)$$
$$= 3x + 3y + 3z$$

Since we define subtraction in terms of addition, the distributive property is also true for subtraction. For example,

$$2(x - y) = 2(x) - 2(y)$$
$$= 2x - 2y$$

Practice 7–12

Use the distributive property to write each expression without parentheses. Then simplify the result.

7. $5(x + y)$

8. $-3(2 + 7x)$

9. $4(x + 6y - 2z)$

10. $-1(3 - a)$

11. $-(8 + a - b)$

12. $\dfrac{1}{2}(2x + 4) + 9$

Answers

5. $14 + x$ **6.** $20x$ **7.** $5x + 5y$

8. $-6 - 21x$ **9.** $4x + 24y - 8z$

10. $-3 + a$ **11.** $-8 - a + b$

12. $x + 11$

Examples   Use the distributive property to write each expression without parentheses. Then simplify the result.

7.
$$2(x + y) = 2(x) + 2(y)$$
$$= 2x + 2y$$

8.
$$-5(-3 + 2z) = -5(-3) + (-5)(2z)$$
$$= 15 - 10z$$

9.
$$5(x + 3y - z) = 5(x) + 5(3y) - 5(z)$$
$$= 5x + 15y - 5z$$

10. $-1(2 - y) = (-1)(2) - (-1)(y)$
$= -2 + y$

11. $-(3 + x - w) = -1(3 + x - w)$
$= (-1)(3) + (-1)(x) - (-1)(w)$
$= -3 - x + w$

> **Helpful Hint**
> Notice in Example 11 that $-(3 + x - w)$ can be rewritten as $-1(3 + x - w)$.

12. $\dfrac{1}{2}(6x + 14) + 10 = \dfrac{1}{2}(6x) + \dfrac{1}{2}(14) + 10$ Apply the distributive property.

$= 3x + 7 + 10$ Multiply.

$= 3x + 17$ Add.

■ **Work Practice 7–12**

The distributive property can also be used to write a sum as a product.

> **Examples** Use the distributive property to write each sum as a product.
>
> **13.** $8 \cdot 2 + 8 \cdot x = 8(2 + x)$
> **14.** $7s + 7t = 7(s + t)$

■ **Work Practice 13–14**

Practice 13–14

Use the distributive property to write each sum as a product.

13. $9 \cdot 3 + 9 \cdot y$

14. $4x + 4y$

Objective C Using the Identity and Inverse Properties ▶

Next, we look at the **identity properties**.

The number 0 is called the identity for addition because when 0 is added to any real number, the result is the same real number. In other words, the *identity* of the real number is not changed.

The number 1 is called the identity for multiplication because when a real number is multiplied by 1, the result is the same real number. In other words, the *identity* of the real number is not changed.

> **Identities for Addition and Multiplication**
>
> 0 is the identity element for addition.
>
> $a + 0 = a$ and $0 + a = a$
>
> 1 is the identity element for multiplication.
>
> $a \cdot 1 = a$ and $1 \cdot a = a$

Notice that 0 is the *only* number that can be added to any real number with the result that the sum is the same real number. Also, 1 is the *only* number that can be multiplied by any real number with the result that the product is the same real number.

Additive inverses or **opposites** were introduced in Section 8.3. Two numbers are called additive inverses or opposites if their sum is 0. The additive inverse or opposite of 6 is -6 because $6 + (-6) = 0$. The additive inverse or opposite of -5 is 5 because $-5 + 5 = 0$.

Reciprocals or **multiplicative inverses** were introduced in Section 8.5. Two nonzero numbers are called reciprocals or multiplicative inverses if their product is 1. The reciprocal or multiplicative inverse of $\dfrac{2}{3}$ is $\dfrac{3}{2}$ because $\dfrac{2}{3} \cdot \dfrac{3}{2} = 1$. Likewise, the reciprocal of -5 is $-\dfrac{1}{5}$ because $-5\left(-\dfrac{1}{5}\right) = 1$.

Answers

13. $9(3 + y)$ **14.** $4(x + y)$

Additive or Multiplicative Inverses

The numbers a and $-a$ are additive inverses or opposites of each other because their sum is 0; that is,

$$a + (-a) = 0$$

The numbers b and $\dfrac{1}{b}$ (for $b \neq 0$) are reciprocals or multiplicative inverses of each other because their product is 1; that is,

$$b \cdot \dfrac{1}{b} = 1$$

Practice 15–21

Name the property illustrated by each true statement.

15. $7(a + b) = 7 \cdot a + 7 \cdot b$

16. $12 + y = y + 12$

17. $-4 \cdot (6 \cdot x) = (-4 \cdot 6) \cdot x$

18. $6 + (z + 2) = 6 + (2 + z)$

19. $3\left(\dfrac{1}{3}\right) = 1$

20. $(x + 0) + 23 = x + 23$

21. $(7 \cdot y) \cdot 10 = y \cdot (7 \cdot 10)$

Answers

15. distributive property
16. commutative property of addition
17. associative property of multiplication
18. commutative property of addition
19. multiplicative inverse property
20. identity element for addition
21. commutative and associative properties of multiplication

✓**Concept Check Answers**

a. $\dfrac{3}{10}$ **b.** $-\dfrac{10}{3}$

✓**Concept Check** Which of the following is

a. the opposite of $-\dfrac{3}{10}$, and

b. the reciprocal of $-\dfrac{3}{10}$?

$$1, -\dfrac{10}{3}, \dfrac{3}{10}, 0, \dfrac{10}{3}, -\dfrac{3}{10}$$

Examples   Name the property illustrated by each true statement.

15. $3(x + y) = 3 \cdot x + 3 \cdot y$ Distributive property

16. $(x + 7) + 9 = x + (7 + 9)$ Associative property of addition (grouping changed)

17. $(b + 0) + 3 = b + 3$ Identity element for addition

18. $2 \cdot (z \cdot 5) = 2 \cdot (5 \cdot z)$ Commutative property of multiplication (order changed)

19. $-2 \cdot \left(-\dfrac{1}{2}\right) = 1$ Multiplicative inverse property

20. $-2 + 2 = 0$ Additive inverse property

21. $-6 \cdot (y \cdot 2) = (-6 \cdot 2) \cdot y$ Commutative and associative properties of multiplication (order and grouping changed)

■ **Work Practice 15–21**

Vocabulary, Readiness & Video Check

Use the choices below to fill in each blank.

distributive property   associative property of multiplication   commutative property of addition
opposites or additive inverses   associative property of addition
reciprocals or multiplicative inverses   commutative property of multiplication

1. $x + 5 = 5 + x$ is a true statement by the _____.

2. $x \cdot 5 = 5 \cdot x$ is a true statement by the _____.

3. $3(y + 6) = 3 \cdot y + 3 \cdot 6$ is true by the _____.

4. $2 \cdot (x \cdot y) = (2 \cdot x) \cdot y$ is a true statement by the _____.

5. $x + (7 + y) = (x + 7) + y$ is a true statement by the _____.

6. The numbers $-\dfrac{2}{3}$ and $-\dfrac{3}{2}$ are called _____.

7. The numbers $-\dfrac{2}{3}$ and $\dfrac{2}{3}$ are called _____.

Martin-Gay Interactive Videos   Watch the section lecture video and answer the following questions.

See Video 8.6

Objective A   **8.** The commutative properties are discussed in ⊞ Examples 1 and 2, and the associative properties are discussed in ⊞ Examples 3–7. What's the one word used again and again to describe the commutative property? The associative property? ▶

Objective B   **9.** In ⊞ Example 10, what point is made about the term 2? ▶

Objective C   **10.** Complete these statements based on the lecture given before ⊞ Example 12. ▶

- The identity element for addition is _____ because if we add _____ to any real number, the result is that real number.
- The identity element for multiplication is _____ because any real number times _____ gives a result of that original real number.

8.6   Exercise Set   MyMathLab®

Objective A   *Use a commutative property to complete each statement. See Examples 1 and 3.*

▶**1.** $x + 16 =$ _____
2. $8 + y =$ _____
3. $-4 \cdot y =$ _____
4. $-2 \cdot x =$ _____

▶**5.** $xy =$ _____
6. $ab =$ _____
7. $2x + 13 =$ _____
8. $19 + 3y =$ _____

Use an associative property to complete each statement. See Examples 2 and 4.

▶**9.** $(xy) \cdot z =$ _____
10. $3 \cdot (x \cdot y) =$ _____
11. $2 + (a + b) =$ _____

12. $(y + 4) + z =$ _____
13. $4 \cdot (ab) =$ _____
14. $(-3y) \cdot z =$ _____

▶**15.** $(a + b) + c =$ _____
16. $6 + (r + s) =$ _____

Use the commutative and associative properties to simplify each expression. See Examples 5 and 6.

▶**17.** $8 + (9 + b)$
18. $(r + 3) + 11$
▶**19.** $4(6y)$
20. $2(42x)$
21. $\dfrac{1}{5}(5y)$

22. $\dfrac{1}{8}(8z)$
23. $(13 + a) + 13$
24. $7 + (x + 4)$
25. $-9(8x)$
26. $-3(12y)$

27. $\dfrac{3}{4}\left(\dfrac{4}{3}s\right)$
28. $\dfrac{2}{7}\left(\dfrac{7}{2}r\right)$
29. $-\dfrac{1}{2}(5x)$
30. $-\dfrac{1}{3}(7x)$

Objective B *Use the distributive property to write each expression without parentheses. Then simplify the result, if possible. See Examples 7 through 12.*

31. $4(x + y)$ **32.** $7(a + b)$ **33.** $9(x - 6)$ **34.** $11(y - 4)$

35. $2(3x + 5)$ **36.** $5(7 + 8y)$ **37.** $7(4x - 3)$ **38.** $3(8x - 1)$

▶ **39.** $3(6 + x)$ **40.** $2(x + 5)$ **41.** $-2(y - z)$ **42.** $-3(z - y)$

43. $-\dfrac{1}{3}(3y + 5)$ **44.** $-\dfrac{1}{2}(2r + 11)$ **45.** $5(x + 4m + 2)$ **46.** $8(3y + z - 6)$

47. $-4(1 - 2m + n) + 4$ **48.** $-4(4 + 2p + 5r) + 16$ **49.** $-(5x + 2)$ **50.** $-(9r + 5)$

▶ **51.** $-(r - 3 - 7p)$ **52.** $-(q - 2 + 6r)$ **53.** $\dfrac{1}{2}(6x + 7) + \dfrac{1}{2}$ **54.** $\dfrac{1}{4}(4x - 2) - \dfrac{7}{2}$

55. $-\dfrac{1}{3}(3x - 9y)$ **56.** $-\dfrac{1}{5}(10a - 25b)$ **57.** $3(2r + 5) - 7$ **58.** $10(4s + 6) - 40$

▶ **59.** $-9(4x + 8) + 2$ **60.** $-11(5x + 3) + 10$ **61.** $-0.4(4x + 5) - 0.5$ **62.** $-0.6(2x + 1) - 0.1$

Use the distributive property to write each sum as a product. See Examples 13 and 14.

63. $4 \cdot 1 + 4 \cdot y$ **64.** $14 \cdot z + 14 \cdot 5$ ▶ **65.** $11x + 11y$ **66.** $9a + 9b$

67. $(-1) \cdot 5 + (-1) \cdot x$ **68.** $(-3)a + (-3)y$ **69.** $30a + 30b$ **70.** $25x + 25y$

Objectives A C Mixed Practice *Name the property illustrated by each true statement. See Examples 15 through 21.*

71. $3 \cdot 5 = 5 \cdot 3$ **72.** $4(3 + 8) = 4 \cdot 3 + 4 \cdot 8$

73. $2 + (x + 5) = (2 + x) + 5$ **74.** $9 \cdot (x \cdot 7) = (9 \cdot x) \cdot 7$

75. $(x + 9) + 3 = (9 + x) + 3$ ▶ **76.** $1 \cdot 9 = 9$

77. $(4 \cdot y) \cdot 9 = 4 \cdot (y \cdot 9)$ **78.** $-4 \cdot (8 \cdot 3) = (8 \cdot 3) \cdot (-4)$

▶ **79.** $0 + 6 = 6$ **80.** $(a + 9) + 6 = a + (9 + 6)$

81. $-4(y + 7) = -4 \cdot y + (-4) \cdot 7$ ▶ **82.** $(11 + r) + 8 = (r + 11) + 8$

83. $6 \cdot \dfrac{1}{6} = 1$

84. $r + 0 = r$

85. $-6 \cdot 1 = -6$

86. $-\dfrac{3}{4}\left(-\dfrac{4}{3}\right) = 1$

Review

Perform each indicated operation. See Sections 1.3, 1.4, 1.6, and 1.7.

87. $45 \cdot 90$

88. $90 \div 45$

89. $90 - 45$

90. $45 + 90$

Concept Extensions

Fill in the table with the opposite (additive inverse), the reciprocal (multiplicative inverse), or the expression. Assume that the value of each expression is not 0.

	91.	92.	93.	94.	95.	96.
Expression	8	$-\dfrac{2}{3}$	x	$4y$		
Opposite						$7x$
Reciprocal					$\dfrac{1}{2x}$	

Decide whether each statement is true or false. See the second Concept Check in this section.

97. The opposite of $-\dfrac{a}{2}$ is $-\dfrac{2}{a}$.

98. The reciprocal of $-\dfrac{a}{2}$ is $\dfrac{a}{2}$.

Determine which pairs of actions are commutative. See the first Concept Check in this section.

99. "taking a test" and "studying for the test"

100. "putting on your shoes" and "putting on your socks"

101. "putting on your left shoe" and "putting on your right shoe"

102. "reading the sports section" and "reading the comics section"

103. "mowing the lawn" and "trimming the hedges"

104. "baking a cake" and "eating the cake"

105. "feeding the dog" and "feeding the cat"

106. "dialing a number" and "turning on the cell phone"

Name the property illustrated by each step.

107. **a.** $\triangle + (\square + \bigcirc) = (\square + \bigcirc) + \triangle$

b. $\qquad = (\bigcirc + \square) + \triangle$

c. $\qquad = \bigcirc + (\square + \triangle)$

108. **a.** $(x + y) + z = x + (y + z)$

b. $\qquad = (y + z) + x$

c. $\qquad = (z + y) + x$

109. Explain why 0 is called the identity element for addition.

110. Explain why 1 is called the identity element for multiplication.

111. Write an example that shows that division is not commutative.

112. Write an example that shows that subtraction is not commutative.

8.7 Simplifying Expressions

Objectives

A Identify Terms, Like Terms, and Unlike Terms. ▶

B Combine Like Terms. ▶

C Simplify Expressions Containing Parentheses. ▶

D Write Word Phrases as Algebraic Expressions. ▶

As we explore in this section, we will see that an expression such as $3x + 2x$ is not written as simply as possible. This is because—even without replacing x by a value—we can perform the indicated addition.

Objective A Identifying Terms, Like Terms, and Unlike Terms ▶

Before we practice simplifying expressions, we must learn some new language. A **term** is a number or the product of a number and variables raised to powers.

Terms

$$-y, \quad 2x^3, \quad -5, \quad 3xz^2, \quad \frac{2}{y}, \quad 0.8z$$

The **numerical coefficient** of a term is the numerical factor. The numerical coefficient of $3x$ is 3. Recall that $3x$ means $3 \cdot x$.

Term	Numerical Coefficient
$3x$	3
$\dfrac{y^3}{5}$	$\dfrac{1}{5}$ since $\dfrac{y^3}{5}$ means $\dfrac{1}{5} \cdot y^3$
$-0.7ab^3c^5$	-0.7
z	1
$-y$	-1
-5	-5

Helpful Hint

The term z means $1z$ and thus has a numerical coefficient of 1.
The term $-y$ means $-1y$ and thus has a numerical coefficient of -1.

Practice 1

Identify the numerical coefficient of each term.

a. $-4x$ **b.** $15y^3$ **c.** x

d. $-y$ **e.** $\dfrac{z}{4}$

Example 1 Identify the numerical coefficient of each term.

a. $-3y$ **b.** $22z^4$ **c.** y **d.** $-x$ **e.** $\dfrac{x}{7}$

Solution:

a. The numerical coefficient of $-3y$ is -3.

b. The numerical coefficient of $22z^4$ is 22.

c. The numerical coefficient of y is 1, since y is $1y$.

d. The numerical coefficient of $-x$ is -1, since $-x$ is $-1x$.

e. The numerical coefficient of $\dfrac{x}{7}$ is $\dfrac{1}{7}$, since $\dfrac{x}{7}$ is $\dfrac{1}{7} \cdot x$.

■ **Work Practice 1**

Answers

1. a. -4 **b.** 15 **c.** 1 **d.** -1 **e.** $\dfrac{1}{4}$

Terms with the same variables raised to exactly the same powers are called **like terms**. Terms that aren't like terms are called **unlike terms**.

Like Terms	Unlike Terms	Reason Why
$3x$, $2x$	$5x$, $5x^2$	Why? Same variable x, but different powers of x and x^2
$-6x^2y$, $2x^2y$, $4x^2y$	$7y$, $3z$, $8x^2$	Why? Different variables
$2ab^2c^3$, ac^3b^2	$6abc^3$, $6ab^2$	Why? Different variables and different powers

Helpful Hint

In like terms, each variable and its exponent must match exactly, but these factors don't need to be in the same order.

$2x^2y$ and $3yx^2$ are like terms.

Example 2 Determine whether the terms are like or unlike.

a. $2x$, $3x^2$ b. $4x^2y$, x^2y, $-2x^2y$ c. $-2yz$, $-3zy$

d. $-x^4$, x^4 e. $-8a^5$, $8a^5$

Solution:

a. Unlike terms, since the exponents on x are not the same.

b. Like terms, since each variable and its exponent match.

c. Like terms, since $zy = yz$ by the commutative property of multiplication.

d. Like terms. The variable and its exponent match.

e. Like terms. The variable and its exponent match.

■ **Work Practice 2**

Practice 2

Determine whether the terms are like or unlike.

a. $7x^2$, $-6x^3$

b. $3x^2y^2$, $-x^2y^2$, $4x^2y^2$

c. $-5ab$, $3ba$

d. $2x^3$, $4y^3$

e. $-7m^4$, $7m^4$

Objective B Combining Like Terms

An algebraic expression containing the sum or difference of like terms can be simplified by applying the distributive property. For example, by the distributive property, we rewrite the sum of the like terms $6x + 2x$ as

$$6x + 2x = (6 + 2)x = 8x$$

Also,

$$-y^2 + 5y^2 = (-1 + 5)y^2 = 4y^2$$

Simplifying the sum or difference of like terms is called **combining like terms**.

Example 3 Simplify each expression by combining like terms.

a. $7x - 3x$ b. $10y^2 + y^2$

c. $8x^2 + 2x - 3x$ d. $9n^2 - 5n^2 + n^2$

Solution:

a. $7x - 3x = (7 - 3)x = 4x$

b. $10y^2 + y^2 = (10 + 1)y^2 = 11y^2$

c. $8x^2 + 2x - 3x = 8x^2 + (2 - 3)x = 8x^2 - 1x$ or $8x^2 - x$

d. $9n^2 - 5n^2 + n^2 = (9 - 5 + 1)n^2 = 5n^2$

■ **Work Practice 3**

Practice 3

Simplify each expression by combining like terms.

a. $9y - 4y$

b. $11x^2 + x^2$

c. $5y - 3x + 4x$

d. $14m^2 - m^2 + 3m^2$

Answers

2. **a.** unlike **b.** like **c.** like
d. unlike **e.** like **3. a.** $5y$ **b.** $12x^2$
c. $5y + x$ **d.** $16m^2$

The preceding examples suggest the following.

> **Combining Like Terms**
>
> To **combine like terms,** combine the numerical coefficients and multiply the result by the common variable factors.

Practice 4–7

Simplify each expression by combining like terms.

4. $7y + 2y + 6 + 10$

5. $-2x + 4 + x - 11$

6. $3z - 3z^2$

7. $8.9y + 4.2y - 3$

Examples   Simplify each expression by combining like terms.

4. $2x + 3x + 5 + 2 = (2 + 3)x + (5 + 2)$
$$= 5x + 7$$

5. $-5a - 3 + a + 2 = -5a + 1a + (-3 + 2)$
$$= (-5 + 1)a + (-3 + 2)$$
$$= -4a - 1$$

6. $4y - 3y^2$

These two terms cannot be combined because they are unlike terms.

7. $2.3x + 5x - 6 = (2.3 + 5)x - 6$
$$= 7.3x - 6$$

■ **Work Practice 4–7**

Objective C Simplifying Expressions Containing Parentheses ▶

In simplifying expressions we make frequent use of the distributive property to remove parentheses.

It may be helpful to study the examples below.

$$+(3a + 2) = +1(3a + 2) = +1(3a) + (+1)(2) = 3a + 2$$
means

$$-(3a + 2) = -1(3a + 2) = -1(3a) + (-1)(2) = -3a - 2$$
means

Practice 8–10

Find each product by using the distributive property to remove parentheses.

8. $3(11y + 6)$

9. $-4(x + 0.2y - 3)$

10. $-(3x + 2y + z - 1)$

Examples   Find each product by using the distributive property to remove parentheses.

8. $5(3x + 2) = 5(3x) + 5(2)$ Apply the distributive property.
$$= 15x + 10$$ Multiply.

9. $-2(y + 0.3z - 1) = -2(y) + (-2)(0.3z) - (-2)(1)$ Apply the distributive property.
$$= -2y - 0.6z + 2$$ Multiply.

10. $-(9x + y - 2z + 6) = -1(9x + y - 2z + 6)$ Distribute -1 over each term.
$$= -1(9x) + (-1)(y) - (-1)(2z) + (-1)(6)$$
$$= -9x - y + 2z - 6$$

■ **Work Practice 8–10**

Answers

4. $9y + 16$ **5.** $-x - 7$
6. $3z - 3z^2$ **7.** $13.1y - 3$
8. $33y + 18$ **9.** $-4x - 0.8y + 12$
10. $-3x - 2y - z + 1$

Helpful Hint

If a "−" sign precedes parentheses, the sign of each term inside the parentheses is changed when the distributive property is applied to remove the parentheses.

Examples:

$$-(2x + 1) = -2x - 1$$
$$-(x - 2y) = -x + 2y$$
$$-(-5x + y - z) = 5x - y + z$$
$$-(-3x - 4y - 1) = 3x + 4y + 1$$

When simplifying an expression containing parentheses, we often use the distributive property first to remove parentheses and then again to combine any like terms.

Examples   Simplify each expression.

11. $3(2x - 5) + 1 = 6x - 15 + 1$ Apply the distributive property.
$$= 6x - 14$$ Combine like terms.

12. $8 - (7x + 2) + 3x = 8 - 7x - 2 + 3x$ Apply the distributive property.
$$= -7x + 3x + 8 - 2$$
$$= -4x + 6$$ Combine like terms.

13. $-2(4x + 7) - (3x - 1) = -8x - 14 - 3x + 1$ Apply the distributive property.
$$= -11x - 13$$ Combine like terms.

14. $9 + 3(4x - 10) = 9 + 12x - 30$ Apply the distributive property.
$$= -21 + 12x$$ Combine like terms.
$$\text{or } 12x - 21$$

◼ **Work Practice 11–14**

Practice 11–14

Simplify each expression.
11. $4(4x - 6) + 20$
12. $5 - (3x + 9) + 6x$
13. $-3(7x + 1) - (4x - 2)$
14. $8 + 11(2y - 9)$

Helpful Hint   Don't forget to use the distributive property and multiply before adding or subtracting like terms.

Example 15   Subtract $4x - 2$ from $2x - 3$.

Solution: We first note that "subtract $4x - 2$ **from** $2x - 3$" translates to $(2x - 3) - (4x - 2)$. Notice that parentheses were placed around each given expression. This is to ensure that the entire expression after the subtraction sign is subtracted. Next, we simplify the algebraic expression.

$$(2x - 3) - (4x - 2) = 2x - 3 - 4x + 2$$ Apply the distributive property.
$$= -2x - 1$$ Combine like terms.

◼ **Work Practice 15**

Practice 15

Subtract $9x - 10$ from $4x - 3$.

Answers
11. $16x - 4$ **12.** $3x - 4$ **13.** $-25x - 1$
14. $-91 + 22y$ **15.** $-5x + 7$

Practice 16–19

Write each phrase as an algebraic expression and simplify if possible. Let x represent the unknown number.

16. Three times a number, subtracted from 10

17. The sum of a number and 2, divided by 5

18. Three times the sum of a number and 6

19. Seven times the difference of a number and 4.

Objective D Writing Algebraic Expressions

To prepare for problem solving, we next practice writing word phrases as algebraic expressions.

Examples Write each phrase as an algebraic expression and simplify if possible. Let x represent the unknown number.

16. Twice a number, plus 6

This expression cannot be simplified.

17. The difference of a number and 4, divided by 7

$$(x - 4) \div 7 \text{ or } \frac{x - 4}{7}$$

This expression cannot be simplified.

18. Five plus the sum of a number and 1

$$5 + (x + 1)$$

We can simplify this expression.

$$5 + (x + 1) = 5 + x + 1$$
$$= 6 + x$$

19. Four times the sum of a number and 3

$$4 \cdot (x + 3)$$

Use the distributive property to simplify the expression.

$$4 \cdot (x + 3) = 4(x + 3)$$
$$= 4 \cdot x + 4 \cdot 3$$
$$= 4x + 12$$

■ Work Practice 16–19

Answers

16. $10 - 3x$ 17. $(x + 2) \div 5$ or $\dfrac{x + 2}{5}$

18. $3x + 18$ 19. $7x - 28$

Vocabulary, Readiness & Video Check

Use the choices below to fill in each blank. Some choices may be used more than once.

numerical coefficient	expression	unlike	distributive
combine like terms	like	term	

1. $14y^2 + 2x - 23$ is called a(n) _____ while $14y^2$, $2x$, and -23 are each called a(n) _____.

2. To multiply $3(-7x + 1)$, we use the _____ property.

3. To simplify an expression like $y + 7y$, we _____.

4. The term z has an understood _____ of 1.

5. The terms $-x$ and $5x$ are _____ terms and the terms $5x$ and $5y$ are _____ terms.

6. For the term $-3x^2y$, -3 is called the _____.

Martin-Gay Interactive Videos   Watch the section lecture video and answer the following questions.

See Video 8.7

Objective A   7. Example 7 shows two terms with exactly the same variables. Why are these terms not considered like terms?

Objective B   8. Example 8 shows us that when combining like terms, we are actually applying what property?

Objective C   9. The expression in Example 11 shows a minus sign before parentheses. When using the distributive property to multiply and remove parentheses, what number are we actually distributing to each term within the parentheses?

Objective D   10. Write the phrase given in Example 14, translate it to an algebraic expression, then simplify it. Why are we able to simplify it?

8.7 Exercise Set MyMathLab®

Objective A *Identify the numerical coefficient of each term. See Example 1.*

1. $-7y$ 2. $3x$ 3. x 4. $-y$ 5. $17x^2y$ 6. $1.2xyz$

Indicate whether the terms in each list are like or unlike. See Example 2.

7. $5y, -y$

8. $-2x^2y, 6xy$

9. $2z, 3z^2$

10. $ab^2, -7ab^2$

11. $8wz, \frac{1}{7}zw$

12. $7.4p^3q^2, 6.2p^3q^2r$

Objective B *Simplify each expression by combining any like terms. See Examples 3 through 7.*

13. $7y + 8y$

14. $3x + 2x$

15. $8w - w + 6w$

16. $c - 7c + 2c$

17. $3b - 5 - 10b - 4$

18. $6g + 5 - 3g - 7$

19. $m - 4m + 2m - 6$

20. $a + 3a - 2 - 7a$

21. $5g - 3 - 5 - 5g$

22. $8p + 4 - 8p - 15$

23. $6.2x - 4 + x - 1.2$

24. $7.9y - 0.7 - y + 0.2$

25. $2k - k - 6$

26. $7c - 8 - c$

27. $-9x + 4x + 18 - 10x$

28. $5y - 14 + 7y - 20y$

29. $6x - 5x + x - 3 + 2x$

30. $8h + 13h - 6 + 7h - h$

31. $7x^2 + 8x^2 - 10x^2$

32. $8x^3 + x^3 - 11x^3$

33. $3.4m - 4 - 3.4m - 7$

34. $2.8w - 0.9 - 0.5 - 2.8w$

35. $6x + 0.5 - 4.3x - 0.4x + 3$

36. $0.4y - 6.7 + y - 0.3 - 2.6y$

Objective C *Simplify each expression. Use the distributive property to remove any parentheses. See Examples 8 through 10.*

37. $5(y + 4)$

38. $7(r + 3)$

39. $-2(x + 2)$

40. $-4(y + 6)$

41. $-5(2x - 3y + 6)$

42. $-2(4x - 3z - 1)$

43. $-(3x - 2y + 1)$

44. $-(y + 5z - 7)$

Objectives B C **Mixed Practice** *Remove parentheses and simplify each expression. See Examples 8 through 14.*

45. $7(d - 3) + 10$

46. $9(z + 7) - 15$

47. $-4(3y - 4) + 12y$

48. $-3(2x + 5) + 6x$

49. $3(2x - 5) - 5(x - 4)$

50. $2(6x - 1) - (x - 7)$

51. $-2(3x - 4) + 7x - 6$

52. $8y - 2 - 3(y + 4)$

53. $5k - (3k - 10)$

54. $-11c - (4 - 2c)$

55. $(3x + 4) - (6x - 1)$

56. $(8 - 5y) - (4 + 3y)$

▶ **57.** $5(x + 2) - (3x - 4)$

58. $4(2x - 3) - (x + 1)$

59. $\frac{1}{3}(7y - 1) + \frac{1}{6}(4y + 7)$

60. $\frac{1}{5}(9y + 2) + \frac{1}{10}(2y - 1)$

61. $2 + 4(6x - 6)$

62. $8 + 4(3x - 4)$

63. $0.5(m + 2) + 0.4m$

64. $0.2(k + 8) - 0.1k$

65. $10 - 3(2x + 3y)$

66. $14 - 11(5m + 3n)$

67. $6(3x - 6) - 2(x + 1) - 17x$

68. $7(2x + 5) - 4(x + 2) - 20x$

69. $\frac{1}{2}(12x - 4) - (x + 5)$

70. $\frac{1}{3}(9x - 6) - (x - 2)$

Perform each indicated operation. Don't forget to simplify if possible. See Example 15.

71. Add $6x + 7$ to $4x - 10$.

72. Add $3y - 5$ to $y + 16$.

73. Subtract $7x + 1$ from $3x - 8$.

74. Subtract $4x - 7$ from $12 + x$.

▶ **75.** Subtract $5m - 6$ from $m - 9$.

76. Subtract $m - 3$ from $2m - 6$.

Objective D *Write each phrase as an algebraic expression and simplify if possible. Let x represent the unknown number. See Examples 16 through 19.*

▶ **77.** Twice a number, decreased by four

78. The difference of a number and two, divided by five

79. Three-fourths of a number, increased by twelve

80. Eight more than triple a number

81. The sum of 5 times a number and −2, added to 7 times the number

82. The sum of 3 times a number and 10, **subtracted from** 9 times the number

83. Eight times the sum of a number and six

84. Six times the difference of a number and five

85. Double a number minus the sum of the number and ten

86. Half a number minus the product of the number and eight

Review

Evaluate each expression for the given values. See Section 8.2 and 8.5.

87. If $x = -1$ and $y = 3$, find $y - x^2$

88. If $g = 0$ and $h = -4$, find $gh - h^2$

89. If $a = 2$ and $b = -5$, find $a - b^2$

90. If $x = -3$, find $x^3 - x^2 + 4$

91. If $y = -5$ and $z = 0$, find $yz - y^2$

92. If $x = -2$, find $x^3 - x^2 - x$

Concept Extensions

Given the following information, determine whether each scale is balanced or not.

1 cone balances 1 cube

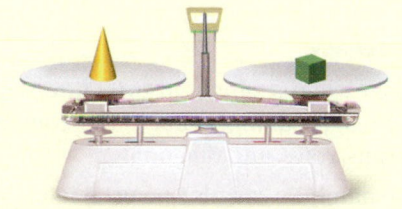

1 cylinder balances 2 cubes

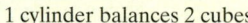
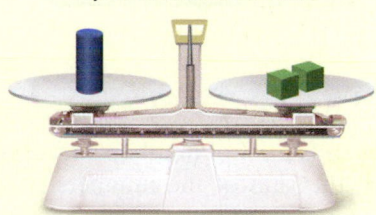

93.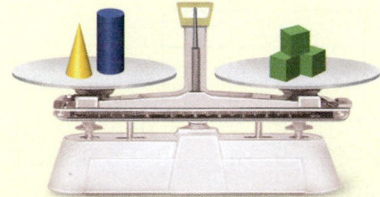

94.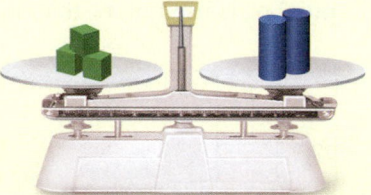

95.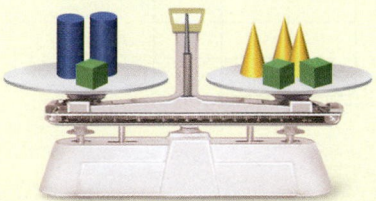

96.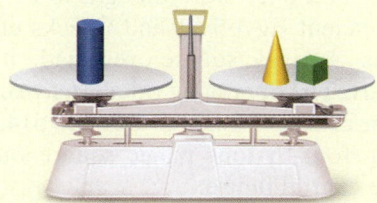

Write each algebraic expression described.

97. Write an expression with 4 terms that simplifies to $3x - 4$.

98. Write an expression of the form
_____(_____ +_____) whose product is $6x + 24$.

△ **99.** Given the following rectangle, express the perimeter as an algebraic expression containing the variable x.

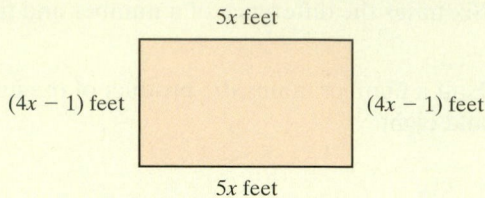

5x feet

(4x − 1) feet   (4x − 1) feet

5x feet

△ **100.** Given the following triangle, express its perimeter as an algebraic expression containing the variable x.

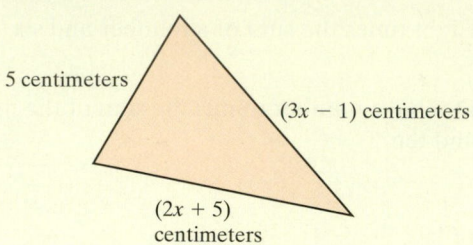

5 centimeters

(3x − 1) centimeters

(2x + 5) centimeters

△ **101.** To convert from feet to inches, we multiply by 12. For example, the number of inches in 2 feet is $12 \cdot 2$ inches. If one board has a length of $(x + 2)$ *feet* and a second board has a length of $(3x - 1)$ *inches*, express their total length in inches as an algebraic expression.

102. The value of 7 nickels is $5 \cdot 7$ cents. Likewise, the value of x nickels is $5x$ cents. If the money box in a drink machine contains x *nickels,* $3x$ *dimes,* and $(30x - 1)$ *quarters,* express their total value in cents as an algebraic expression.

✎ **103.** In your own words, explain how to combine like terms.

✎ **104.** Do like terms always contain the same numerical coefficients? Explain your answer.

Chapter 8 Group Activity

Magic Squares

Sections 8.2, 8.3, 8.4

A magic square is a set of numbers arranged in a square table so that the sum of the numbers in each column, row, and diagonal is the same. For instance, in the magic square below, the sum of each column, row, and diagonal is 15. Notice that no number is used more than once in the magic square.

2	9	4
7	5	3
6	1	8

The properties of magic squares have been known for a very long time and once were thought to be good luck charms. The ancient Egyptians and Greeks understood their patterns. A magic square even made it into a famous work of art. The engraving titled *Melencolia I,* created by German artist Albrecht Dürer in 1514, features the following four-by-four magic square on the building behind the central figure.

16	3	2	13
5	10	11	8
9	6	7	12
4	15	14	1

Group Exercises

1. Verify that what is shown in the Dürer engraving is, in fact, a magic square. What is the common sum of the columns, rows, and diagonals?

2. Negative numbers can also be used in magic squares. Complete the following magic square:

		−2
	−1	
0		−4

3. Use the numbers $-12, -9, -6, -3, 0, 3, 6, 9,$ and 12 to form a magic square.

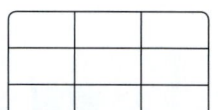

Chapter 8 Vocabulary Check

Fill in each blank with one of the words or phrases listed below.

inequality symbols	exponent	term	numerical coefficient
grouping symbols	solution	like terms	unlike terms
equation	absolute value	numerator	denominator
opposites	base	reciprocals	variable

1. The symbols $\neq$, $<$, and $>$ are called _____.

2. A mathematical statement that two expressions are equal is called a(n) _____.

3. The _____ of a number is the distance between that number and 0 on a number line.

4. A symbol used to represent a number is called a(n) _____.

5. Two numbers that are the same distance from 0 but lie on opposite sides of 0 are called _____.

6. The number in a fraction above the fraction bar is called the _____.

7. A(n) _____ of an equation is a value for the variable that makes the equation a true statement.

8. Two numbers whose product is 1 are called _____.

9. In 2^3, the 2 is called the _____ and the 3 is called the _____.

10. The _____ of a term is its numerical factor.

11. The number in a fraction below the fraction bar is called the _____.

12. Parentheses and brackets are examples of _____.

13. A(n) _____ is a number or the product of a number and variables raised to powers.

14. Terms with the same variables raised to the same powers are called _____.

15. If terms are not like terms, then they are _____.

Helpful Hint

▶ Are you preparing for your test? Don't forget to take the Chapter 8 Test on page 619. Then check your answers at the back of the text and use the Chapter Test Prep Videos to see the fully worked-out solutions to any of the exercises you want to review.

8 Chapter Highlights

Definitions and Concepts	Examples
Section 8.1 Symbols and Sets of Numbers	
A **set** is a collection of objects, called **elements,** enclosed in braces.	$\{a, c, e\}$ Given the set $\left\{-3.4, \sqrt{3}, 0, \frac{2}{3}, 5, -4\right\}$ list the numbers that belong to the set of
Natural numbers: $\{1, 2, 3, 4, \dots\}$	Natural numbers: 5
Whole numbers: $\{0, 1, 2, 3, 4, \dots\}$	Whole numbers: 0, 5
Integers: $\{\dots, -3, -2, -1, 0, 1, 2, 3, \dots\}$	Integers: $-4, 0, 5$
Rational numbers: {real numbers that can be expressed as quotients of integers}	Rational numbers: $-3.4, 0, \frac{2}{3}, 5, -4$
Irrational numbers: {real numbers that cannot be expressed as quotients of integers}	Irrational numbers: $\sqrt{3}$
Real numbers: {all numbers that correspond to points on the number line}	Real numbers: $-3.4, \sqrt{3}, 0, \frac{2}{3}, 5, -4$

(continued)

Definitions and Concepts	Examples

Section 8.1   Symbols and Sets of Numbers (*continued*)

A line used to picture numbers is called a **number line.**

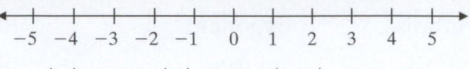

The **absolute value** of a real number a denoted by $|a|$ is the distance between a and 0 on a number line.

$|5| = 5$ $|0| = 0$ $|-2| = 2$

Symbols: $=$ is equal to
$\quad\quad\;\; \neq$ is not equal to
$\quad\quad\;\; >$ is greater than
$\quad\quad\;\; <$ is less than
$\quad\quad\;\; \leq$ is less than or equal to
$\quad\quad\;\; \geq$ is greater than or equal to

$-7 = -7$
$3 \neq -3$
$4 > 1$
$1 < 4$
$6 \leq 6$

$18 \geq -\dfrac{1}{3}$

Order Property for Real Numbers

For any two real numbers a and b, a is less than b if a is to the left of b on a number line.

$$-3 < 0 \quad\quad \begin{matrix} 0 > -3 \\ 0 < 2.5 \end{matrix} \quad 2.5 > 0$$

Section 8.2   Exponents, Order of Operations, and Variable Expressions

The expression a^n is an **exponential expression.** The number a is called the **base;** it is the repeated factor. The number n is called the **exponent;** it is the number of times that the base is a factor.

$4^3 = 4 \cdot 4 \cdot 4 = 64$
$7^2 = 7 \cdot 7 = 49$

Order of Operations

1. Perform all operations within grouping symbols first, starting with the innermost set.
2. Evaluate exponential expressions.
3. Multiply or divide in order from left to right.
4. Add or subtract in order from left to right.

$$\frac{8^2 + 5(7 - 3)}{3 \cdot 7} = \frac{8^2 + 5(4)}{21}$$
$$= \frac{64 + 5(4)}{21}$$
$$= \frac{64 + 20}{21}$$
$$= \frac{84}{21}$$
$$= 4$$

A symbol used to represent a number is called a **variable.**

Examples of variables are
$\quad q, x, z$

An **algebraic expression** is a collection of numbers, variables, operation symbols, and grouping symbols.

Examples of algebraic expressions are
$$5x, \quad 2(y - 6), \quad \frac{q^2 - 3q + 1}{6}$$

To **evaluate an algebraic expression** containing a variable, substitute a given number for the variable and simplify.

Evaluate $x^2 - y^2$ when $x = 5$ and $y = 3$.
$$x^2 - y^2 = (5)^2 - 3^2$$
$$= 25 - 9$$
$$= 16$$

A mathematical statement that two expressions are equal is called an **equation.**

Equations:
$\quad 3x - 9 = 20$
$\quad A = \pi r^2$

A **solution** of an equation is a value for the variable that makes the equation a true statement.

Determine whether 4 is a solution of $5x + 7 = 27$.
$$5x + 7 = 27$$
$$5(4) + 7 \stackrel{?}{=} 27$$
$$20 + 7 \stackrel{?}{=} 27$$
$$27 = 27 \quad \text{True}$$
4 is a solution.

Definitions and Concepts	**Examples**

Section 8.3   Adding Real Numbers

To Add Two Numbers with the Same Sign

1. Add their absolute values.
2. Use their common sign as the sign of the sum.

To Add Two Numbers with Different Signs

1. Subtract their absolute values.
2. Use the sign of the number whose absolute value is larger as the sign of the sum.

Two numbers that are the same distance from 0 but lie on opposite sides of 0 are called **opposites** or **additive inverses.** The opposite of a number a is denoted by $-a$.

Add.

$$10 + 7 = 17$$
$$-3 + (-8) = -11$$

$$-25 + 5 = -20$$
$$14 + (-9) = 5$$

The opposite of -7 is 7.
The opposite of 123 is -123.

Section 8.4   Subtracting Real Numbers

To subtract two numbers a and b, add the first number a to the opposite of the second number, b.

$$a - b = a + (-b)$$

Subtract.

$$3 - (-44) = 3 + 44 = 47$$
$$-5 - 22 = -5 + (-22) = -27$$
$$-30 - (-30) = -30 + 30 = 0$$

Section 8.5   Multiplying and Dividing Real Numbers

Multiplying Real Numbers

The product of two numbers with the same sign is a positive number. The product of two numbers with different signs is a negative number.

Products Involving Zero

The product of 0 and any number is 0.

$$b \cdot 0 = 0 \quad \text{and} \quad 0 \cdot b = 0$$

Quotient of Two Real Numbers

$$\frac{a}{b} = a \cdot \frac{1}{b}, b \neq 0$$

Dividing Real Numbers

The quotient of two numbers with the same sign is a positive number. The quotient of two numbers with different signs is a negative number.

Quotients Involving Zero

Let a be a nonzero number. $\dfrac{0}{a} = 0$ and $\dfrac{a}{0}$ is undefined.

Multiply.

$$7 \cdot 8 = 56 \qquad -7 \cdot (-8) = 56$$
$$-2 \cdot 4 = -8 \qquad 2 \cdot (-4) = -8$$

$$-4 \cdot 0 = 0 \qquad 0 \cdot \left(-\frac{3}{4}\right) = 0$$

Divide.

$$\frac{42}{2} = 42 \cdot \frac{1}{2} = 21$$

$$\frac{90}{10} = 9 \qquad \frac{-90}{-10} = 9$$

$$\frac{42}{-6} = -7 \qquad \frac{-42}{6} = -7$$

$$\frac{0}{18} = 0 \qquad \frac{0}{-47} = 0 \qquad \frac{-85}{0} \text{ is undefined.}$$

Definitions and Concepts	Examples

Section 8.6   Properties of Real Numbers

Commutative Properties

Addition: $a + b = b + a$

Multiplication: $a \cdot b = b \cdot a$

$3 + (-7) = -7 + 3$

$-8 \cdot 5 = 5 \cdot (-8)$

Associative Properties

Addition: $(a + b) + c = a + (b + c)$

Multiplication: $(a \cdot b) \cdot c = a \cdot (b \cdot c)$

$(5 + 10) + 20 = 5 + (10 + 20)$

$(-3 \cdot 2) \cdot 11 = -3 \cdot (2 \cdot 11)$

Two numbers whose product is 1 are called **multiplicative inverses** or **reciprocals.** The reciprocal of a nonzero number a is $\dfrac{1}{a}$ because $a \cdot \dfrac{1}{a} = 1.$

The reciprocal of 3 is $\dfrac{1}{3}$.

The reciprocal of $-\dfrac{2}{5}$ is $-\dfrac{5}{2}$.

Distributive Property

$a(b + c) = a \cdot b + a \cdot c$

$5(6 + 10) = 5 \cdot 6 + 5 \cdot 10$

$-2(3 + x) = -2 \cdot 3 + (-2)(x)$

Identities

$a + 0 = a \qquad 0 + a = a$

$a \cdot 1 = a \qquad 1 \cdot a = a$

$5 + 0 = 5 \qquad 0 + (-2) = -2$

$-14 \cdot 1 = -14 \qquad 1 \cdot 27 = 27$

Inverses

Additive or opposite: $a + (-a) = 0$

Multiplicative or reciprocal: $b \cdot \dfrac{1}{b} = 1, \qquad b \neq 0$

$7 + (-7) = 0$

$3 \cdot \dfrac{1}{3} = 1$

Section 8.7   Simplifying Expressions

The **numerical coefficient** of a **term** is its numerical factor.

Term	Numerical Coefficient
$-7y$	-7
x	1
$\dfrac{1}{5}a^2b$	$\dfrac{1}{5}$

Terms with the same variables raised to exactly the same powers are **like terms.**

Like Terms	Unlike Terms
$12x, -x$	$3y, 3y^2$
$-2xy, 5yx$	$7a^2b, -2ab^2$

To combine like terms, add the numerical coefficients and multiply the result by the common variable factor.

$9y + 3y = 12y$

$-4z^2 + 5z^2 - 6z^2 = -5z^2$

To remove parentheses, apply the distributive property.

$-4(x + 7) + 10(3x - 1)$

$= -4x - 28 + 30x - 10$

$= 26x - 38$

(8.1) *Insert* $<$, $>$, *or* $=$ *in the appropriate space to make each statement true.*

1. 8   10

2. 7   2

3. -4 -5

4. $\dfrac{12}{2}$ -8

5. $|-7|$ $|-8|$

6. $|-9|$ -9

7. $-|-1|$ -1

8. $|-14|$ $-(-14)$

9. 1.2   1.02

10. $-\dfrac{3}{2}$ $-\dfrac{3}{4}$

Translate each statement into symbols.

11. Four is greater than or equal to negative three.

12. Six is not equal to five.

13. 0.03 is less than 0.3.

14. The United States is home to 1729 two-year colleges and 2870 four-year colleges. Write an inequality comparing the numbers 1729 and 2870. (*Source:* National Center for Education Statistics)

Given the sets of numbers below, list the numbers in each set that also belong to the set of:

a. Natural numbers
b. Whole numbers
c. Integers
d. Rational numbers
e. Irrational numbers
f. Real numbers

15. $\left\{-6, 0, 1, 1\dfrac{1}{2}, 3, \pi, 9.62\right\}$

16. $\left\{-3, -1.6, 2, 5, \dfrac{11}{2}, 15.1, \sqrt{5}, 2\pi\right\}$

The following chart shows the gains and losses in dollars of Density Oil and Gas stock for a particular week. Use this chart to answer Exercises 17 and 18.

Day	Gain or Loss (in dollars)
Monday	+1
Tuesday	−2
Wednesday	+5
Thursday	+1
Friday	−4

17. Which day showed the greatest loss?

18. Which day showed the greatest gain?

(8.2) *Choose the correct answer for each statement.*

19. The expression $6 \cdot 3^2 + 2 \cdot 8$ simplifies to
   a. -52 **b.** 448   **c.** 70   **d.** 64

20. The expression $68 - 5 \cdot 2^3$ simplifies to
   a. -232 **b.** 28   **c.** 38   **d.** 504

Simplify each expression.

21. $3(1 + 2 \cdot 5) + 4$ **22.** $8 + 3(2 \cdot 6 - 1)$ **23.** $\dfrac{4 + |6 - 2| + 8^2}{4 + 6 \cdot 4}$ **24.** $5[3(2 + 5) - 5]$

Translate each word statement to symbols.

25. The difference of twenty and twelve is equal to the product of two and four.

26. The quotient of nine and two is greater than negative five.

Evaluate each expression when $x = 6$, $y = 2$, and $z = 8$.

27. $2x + 3y$ **28.** $x(y + 2z)$ **29.** $\dfrac{x}{y} + \dfrac{z}{2y}$ **30.** $x^2 - 3y^2$

△ **31.** The expression $180 - a - b$ represents the measure of the unknown angle of the given triangle. Replace a with 37 and b with 80 to find the measure of the unknown angle.

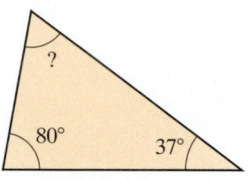

△ **32.** The expression $360 - a - b - c$ represents the measure of the unknown angle of the given quadrilateral. Replace a with 93, b with 80, and c with 82 to find the measure of the unknown angle.

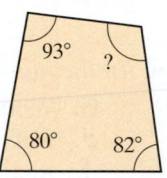

Decide whether the given number is a solution to the given equation.

33. $7x - 3 = 18$; 3   **34.** $3x^2 + 4 = x - 1$; 1

(8.3) *Find the additive inverse or opposite of each number.*

35. -9 **36.** $\dfrac{2}{3}$ **37.** $|-2|$ **38.** $-|-7|$

Add.

39. $-15 + 4$ **40.** $-6 + (-11)$ **41.** $\dfrac{1}{16} + \left(-\dfrac{1}{4}\right)$

42. $-8 + |-3|$ **43.** $-4.6 + (-9.3)$ **44.** $-2.8 + 6.7$

(8.4) *Perform each indicated operation.*

45. $6 - 20$

46. $-3.1 - 8.4$

47. $-6 - (-11)$

48. $4 - 15$

49. $-21 - 16 + 3(8 - 2)$

50. $\dfrac{11 - (-9) + 6(8 - 2)}{2 + 3 \cdot 4}$

Evaluate each expression for $x = 3$, $y = -6$, and $z = -9$. Then choose the correct evaluation.

51. $2x^2 - y + z$
 a. 15 **b.** 3 **c.** 27 **d.** -3

52. $\dfrac{|y - 4x|}{2x}$
 a. 3 **b.** 1 **c.** -1 **d.** -3

53. At the beginning of the week the price of Density Oil and Gas stock from Exercises 17 and 18 is $50 per share. Find the price of a share of stock at the end of the week.

54. Find the price of a share of stock by the end of the day on Wednesday.

(8.5) *Find each multiplicative inverse or reciprocal.*

55. -6

56. $\dfrac{3}{5}$

Simplify each expression.

57. $6(-8)$

58. $(-2)(-14)$

59. $\dfrac{-18}{-6}$

60. $\dfrac{42}{-3}$

61. $-3(-6)(-2)$

62. $(-4)(-3)(0)(-6)$

63. $\dfrac{4(-3) + (-8)}{2 + (-2)}$

64. $\dfrac{3(-2)^2 - 5}{-14}$

(8.6) *Name the property illustrated in each equation.*

65. $-6 + 5 = 5 + (-6)$

66. $6 \cdot 1 = 6$

67. $3(8 - 5) = 3 \cdot 8 + 3 \cdot (-5)$

68. $4 + (-4) = 0$

69. $2 + (3 + 9) = (2 + 3) + 9$

70. $2 \cdot 8 = 8 \cdot 2$

71. $6(8 + 5) = 6 \cdot 8 + 6 \cdot 5$

72. $(3 \cdot 8) \cdot 4 = 3 \cdot (8 \cdot 4)$

73. $4 \cdot \frac{1}{4} = 1$

74. $8 + 0 = 8$

75. $4(8 + 3) = 4(3 + 8)$

76. $5(2 + 1) = 5 \cdot 2 + 5 \cdot 1$

(8.7) *Simplify each expression.*

77. $5x - x + 2x$

78. $0.2z - 4.6z - 7.4z$

79. $\frac{1}{2}x + 3 + \frac{7}{2}x - 5$

80. $\frac{4}{5}y + 1 + \frac{6}{5}y + 2$

81. $2(n - 4) + n - 10$

82. $3(w + 2) - (12 - w)$

83. Subtract $7x - 2$ from $x + 5$.

84. Subtract $1.4y - 3$ from $y - 0.7$.

Write each phrase as an algebraic expression. Simplify if possible.

85. Three times a number decreased by 7

86. Twice the sum of a number and 2.8, added to 3 times the number

Mixed Review

Insert $<, >,$ or $=$ in the space between each pair of numbers.

87. $-|-11| \quad |11.4|$

88. $-1\frac{1}{2} \quad -2\frac{1}{2}$

Perform the indicated operations.

89. $-7.2 + (-8.1)$

90. $14 - 20$

91. $4(-20)$

92. $\frac{-20}{4}$

93. $-\frac{4}{5}\left(\frac{5}{16}\right)$

94. $-0.5(-0.3)$

95. $8 \div 2 \cdot 4$

96. $(-2)^4$

97. $\frac{-3 - 2(-9)}{-15 - 3(-4)}$

98. $5 + 2[(7 - 5)^2 + (1 - 3)]$

99. $-\frac{5}{8} \div \frac{3}{4}$

100. $\frac{-15 + (-4)^2 + |-9|}{10 - 2 \cdot 5}$

Remove parentheses and simplify each expression.

101. $7(3x - 3) - 5(x + 4)$

102. $8 + 2(9x - 10)$

Translate each statement into symbols.

1. The absolute value of negative seven is greater than five.

2. The sum of nine and five is greater than or equal to four.

Simplify each expression.

3. $-13 + 8$

4. $-13 - (-2)$

5. $6 \cdot 3 - 8 \cdot 4$

6. $13(-3)$

7. $(-6)(-2)$

8. $\dfrac{|-16|}{-8}$

9. $\dfrac{-8}{0}$

10. $\dfrac{|-6| + 2}{5 - 6}$

11. $\dfrac{1}{2} - \dfrac{5}{6}$

12. $-1\dfrac{1}{8} + 5\dfrac{3}{4}$

13. $-\dfrac{3}{5} + \dfrac{15}{8}$

14. $3(-4)^2 - 80$

15. $6[5 + 2(3 - 8) - 3]$

16. $\dfrac{-12 + 3 \cdot 8}{4}$

17. $\dfrac{(-2)(0)(-3)}{-6}$

Insert $<, >,$ *or* $=$ *in the appropriate space to make each statement true.*

18. $-3 \quad -7$

19. $4 \quad -8$

20. $|-3| \quad 2$

21. $|-2| \quad -1 - (-3)$

Answers

1. _____

2. _____

3. _____

4. _____

5. _____

6. _____

7. _____

8. _____

9. _____

10. _____

11. _____

12. _____

13. _____

14. _____

15. _____

16. _____

17. _____

18. _____

19. _____

20. _____

21. _____

22. a. _____

b. _____

c. _____

d. _____

e. _____

f. _____

23. _____

24. _____

25. _____

26. _____

27. _____

28. _____

29. _____

30. _____

31. _____

32. _____

33. _____

34. _____

35. _____

36. _____

37. _____

38. _____

39. _____

40. _____

22. Given $\left\{-5, -1, \frac{1}{4}, 0, 1, 7, 11.6, \sqrt{7}, 3\pi\right\}$, list the numbers in this set that also belong to the set of:

   a. Natural numbers   **b.** Whole numbers

   c. Integers   **d.** Rational numbers

   e. Irrational numbers   **f.** Real numbers

Evaluate each expression when $x = 6$, $y = -2$, and $z = -3$.

23. $x^2 + y^2$ **24.** $x + yz$ **25.** $2 + 3x - y$ **26.** $\dfrac{y + z - 1}{x}$

Identify the property illustrated by each equation.

27. $8 + (9 + 3) = (8 + 9) + 3$ **28.** $6 \cdot 8 = 8 \cdot 6$

29. $-6(2 + 4) = -6 \cdot 2 + (-6) \cdot 4$ **30.** $\dfrac{1}{6}(6) = 1$

31. Find the opposite of -9.   **32.** Find the reciprocal of $-\dfrac{1}{3}$.

The New Orleans Saints were 22 yards from the goal when the series of gains and losses shown in the chart occurred. Use this chart to answer Exercises 33 and 34.

	Gains and Losses (in yards)
First down	5
Second down	−10
Third down	−2
Fourth down	29

33. During which down did the greatest loss of yardage occur?

34. Was a touchdown scored?

35. The temperature at the Winter Olympics was a frigid 14° below zero in the morning, but by noon it had risen 31°. What was the temperature at noon?

36. A stockbroker decided to sell 280 shares of stock, which decreased in value by $1.50 per share yesterday. How much money did she lose?

Simplify each expression.

37. $2y - 6 - y - 4$ **38.** $2.7x + 6.1 + 3.2x - 4.9$

39. $4(x - 2) - 3(2x - 6)$ **40.** $-5(y + 1) + 2(3 - 5y)$

1. Add: $1647 + 246 + 32 + 85$

2. Subtract: $2000 - 469$

3. Find the prime factorization of 945.

4. Find the area of the rectangle.

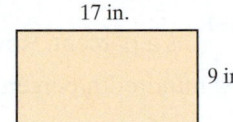

17 in.

9 in.

5. Find the LCM of 11 and 33.

6. Subtract: $\dfrac{8}{21} - \dfrac{2}{9}$

7. Add: $3\dfrac{4}{5} + 1\dfrac{4}{15}$

8. Multiply: $2\dfrac{1}{2} \cdot 4\dfrac{2}{15}$

Write each decimal as a fraction or mixed number in simplest form.

9. 0.125

10. 1.2

11. 105.083

12. Evaluate: $\left(\dfrac{2}{3}\right)^3$

13. Insert $<$, $>$, or $=$ to form a true statement.
0.052   0.236

14. Evaluate: $30 \div 6 \cdot 5$

15. Subtract $85 - 17.31$. Check your answer.

16. Add: $27.9 + 8.07 + 103.261$

Multiply.

17. 42.1×0.1

18. 186.04×1000

19. 9.2×0.001

20. Find the average of 6.8, 9.7, and 0.9.

21. Divide: $32\overline{)8.32}$

22. Add: $\dfrac{3}{10} + \dfrac{3}{4}$

23. Write $2\dfrac{3}{16}$ as a decimal.

24. Round 7.2846 to the nearest tenth.

25. Write $\dfrac{2}{3}$ as a decimal.

26. Simplify: $\dfrac{0.12 + 0.96}{0.5}$

Answers

1. _____
2. _____
3. _____
4. _____
5. _____
6. _____
7. _____
8. _____
9. _____
10. _____
11. _____
12. _____
13. _____
14. _____
15. _____
16. _____
17. _____
18. _____
19. _____
20. _____
21. _____
22. _____
23. _____
24. _____
25. _____
26. _____

27. _____

28. _____

29. _____

30. _____

31. _____

32. _____

33. _____

34. _____

35. _____

36. _____

37. _____

38. _____

39. _____

40. _____

27. Write 23% as a decimal.

28. Write $\dfrac{7}{8}$ as a percent.

29. Write $\dfrac{1}{12}$ as a percent. Round to the nearest hundredth percent.

30. 108 is what percent of 450?

31. What number is 35% of 40?

32. Write 23% as a fraction.

33. Translate to a proportion. What percent of 30 is 75?

34. Add: $-1.8 + (-2.7)$

35. In response to a decrease in sales, a company with 1500 employees reduces the number of employees to 1230. What is the percent of decrease?

36. Subtract: $1.8 - 2.7$

37. An electric rice cooker that normally sells for $65 is on sale for 25% off. What is the discount and what is the sale price?

38. Find 47% of 200.

39. Find the simple interest after 2 years on $500 at an interest rate of 12%.

40. The number of faculty at a local community college was recently increased from 240 to 276. What is the percent increase?

Equations, Inequalities, and Problem Solving

Top 5 Countries by Number of Internet-Crime Complaints

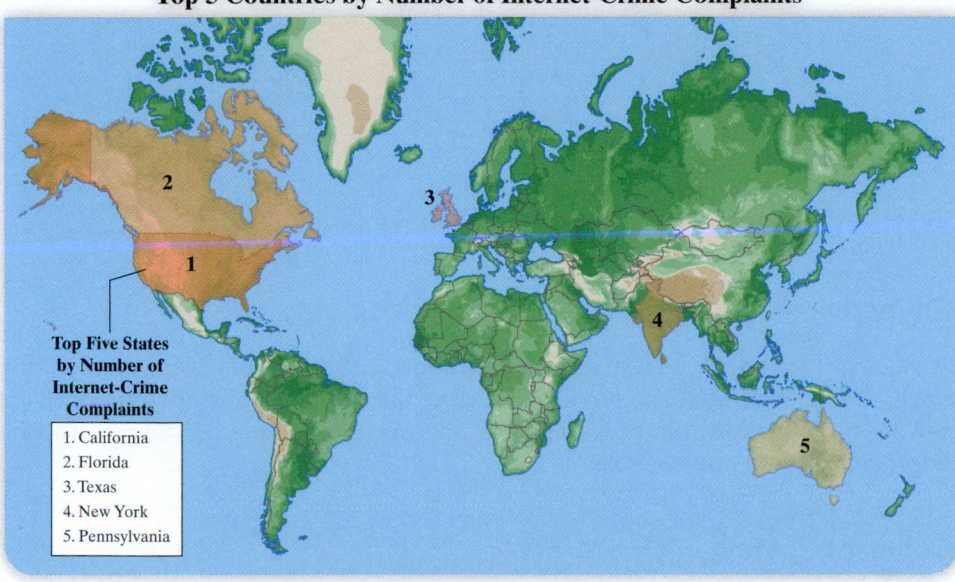

Top Five States by Number of Internet-Crime Complaints

1. California
2. Florida
3. Texas
4. New York
5. Pennsylvania

Internet Crime

The Internet Crime Complaint Center (IC3) is a joint operation between the FBI and the National White-Collar Crime Center. The IC3 receives and refers criminal complaints occurring on the Internet. Of course, nondelivery of merchandise or payment are the highest reported offenses.

In Section 9.6, Exercises 15 and 16, we analyze a bar graph on the yearly number of complaints received by the IC3.

Ages of Persons Filing Complaints

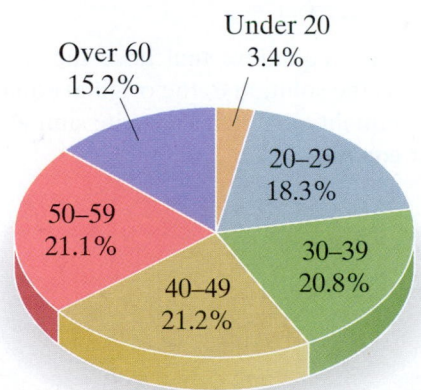

Under 20
3.4%

Over 60
15.2%

20–29
18.3%

50–59
21.1%

30–39
20.8%

40–49
21.2%

In this chapter, we solve equations and inequalities. Once we know how to solve equations and inequalities, we may solve word problems. Of course, problem solving is an integral topic in algebra and its discussion is continued throughout this text.

623

Objectives

A Use the Addition Property of Equality to Solve Linear Equations.

B Simplify an Equation and Then Use the Addition Property of Equality.

C Write Word Phrases as Algebraic Expressions.

> **Helpful Hint** Simply stated, an equation contains "=" while an expression does not. Also, we *simplify* expressions and *solve* equations.

Let's recall from Section 8.2 the difference between an equation and an expression. A combination of operations on variables and numbers is an expression, and an equation is of the form "expression = expression."

Equations	Expressions
$3x - 1 = -17$	$3x - 1$
area = length · width	$5(20 - 3) + 10$
$8 + 16 = 16 + 8$	y^3
$-9a + 11b = 14b + 3$	$-x^2 + y - 2$

Now, let's concentrate on equations.

Objective A Using the Addition Property

A value of the variable that makes an equation a true statement is called a **solution** or **root** of the equation. The process of finding the solution of an equation is called **solving** the equation for the variable. In this section, we concentrate on solving *linear equations* in one variable.

> **Linear Equation in One Variable**
>
> A **linear equation in one variable** can be written in the form
>
> $$Ax + B = C$$
>
> where A, B, and C are real numbers and $A \neq 0$.

Evaluating each side of a linear equation for a given value of the variable, as we did in Section 8.2, can tell us whether that value is a solution. But we can't rely on this as our method of solving it—with what value would we start?

Instead, to solve a linear equation in x, we write a series of simpler equations, all *equivalent* to the original equation, so that the final equation has the form

$$x = \text{number} \qquad \text{or} \qquad \text{number} = x$$

Equivalent equations are equations that have the same solution. This means that the "number" above is the solution to the original equation.

The first property of equality that helps us write simpler equivalent equations is the **addition property of equality.**

> **Addition Property of Equality**
>
> Let a, b, and c represent numbers. Then
>
> $$a = b \qquad\qquad \text{Also,} \quad a = b$$
> $$\text{and } a + c = b + c \qquad\qquad \text{and } a - c = b - c$$
>
> are equivalent equations.   are equivalent equations.

In other words, **the same number may be added to or subtracted from both sides** of an equation without changing the solution of the equation. (We may subtract the same number from both sides since subtraction is defined in terms of addition.)

Let's visualize how we use the addition property of equality to solve an equation. Picture the equation $x - 2 = 1$ as a balanced scale. The left side of the equation has the same value (weight) as the right side.

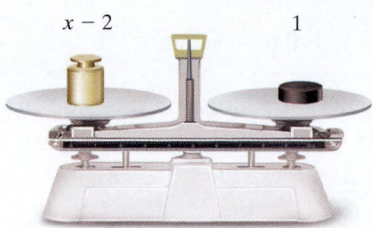

If the same weight is added to each side of a scale, the scale remains balanced. Likewise, if the same number is added to each side of an equation, the left side continues to have the same value as the right side.

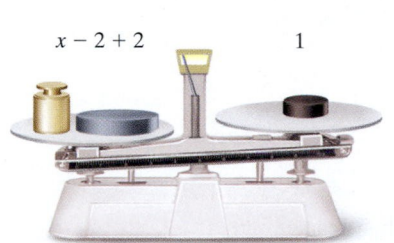

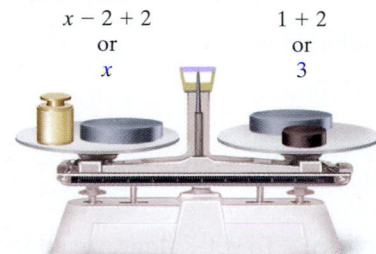

We use the addition property of equality to write equivalent equations until the variable is alone (by itself on one side of the equation) and the equation looks like "x = number" or "number = x."

✔**Concept Check** Use the addition property to fill in the blanks so that the middle equation simplifies to the last equation.

$$x - 5 = 3$$
$$x - 5 + \underline{} = 3 + \underline{}$$
$$x = 8$$

Example 1 Solve $x - 7 = 10$ for x.

Solution: To solve for x, we first get x alone on one side of the equation. To do this, we add 7 to both sides of the equation.

$$x - 7 = 10$$
$$x - 7 + 7 = 10 + 7 \quad \text{Add 7 to both sides.}$$
$$x = 17 \qquad\qquad \text{Simplify.}$$

The solution of the equation $x = 17$ is obviously 17.
Since we are writing equivalent equations, the solution of the equation $x - 7 = 10$ is also 17.

Check: To check, replace x with 17 in the original equation.

$$x - 7 = 10 \quad \text{Original equation.}$$
$$17 - 7 \stackrel{?}{=} 10 \quad \text{Replace } x \text{ with 17.}$$
$$10 = 10 \quad \text{True}$$

Since the statement is true, 17 is the solution.

■ **Work Practice 1**

Practice 1

Solve: $x - 5 = 8$ for x.

Answer
1. $x = 13$

✔**Concept Check Answer**
5

Practice 2

Solve: $y + 1.7 = 0.3$

Example 2   Solve: $y + 0.6 = -1.0$

Solution: To solve for y (get y alone on one side of the equation), we subtract 0.6 from both sides of the equation.

$$y + 0.6 = -1.0$$
$$y + 0.6 - 0.6 = -1.0 - 0.6 \quad \text{Subtract 0.6 from both sides.}$$
$$y = -1.6 \quad \text{Combine like terms.}$$

Check: $y + 0.6 = -1.0$ Original equation.

$$-1.6 + 0.6 \overset{?}{=} -1.0 \quad \text{Replace } y \text{ with } -1.6.$$
$$-1.0 = -1.0 \quad \text{True}$$

The solution is -1.6.

■ **Work Practice 2**

Practice 3

Solve: $\dfrac{7}{8} = y - \dfrac{1}{3}$

Example 3   Solve: $\dfrac{1}{2} = x - \dfrac{3}{4}$

Solution: To get x alone, we add $\dfrac{3}{4}$ to both sides.

$$\frac{1}{2} = x - \frac{3}{4}$$

$$\frac{1}{2} + \frac{3}{4} = x - \frac{3}{4} + \frac{3}{4} \quad \text{Add } \frac{3}{4} \text{ to both sides.}$$

$$\frac{1}{2} \cdot \frac{2}{2} + \frac{3}{4} = x \quad \text{The LCD is 4.}$$

$$\frac{2}{4} + \frac{3}{4} = x \quad \text{Add the fractions.}$$

$$\frac{5}{4} = x$$

Check: $\dfrac{1}{2} = x - \dfrac{3}{4}$ Original equation.

$$\frac{1}{2} \overset{?}{=} \frac{5}{4} - \frac{3}{4} \quad \text{Replace } x \text{ with } \frac{5}{4}.$$

$$\frac{1}{2} \overset{?}{=} \frac{2}{4} \quad \text{Subtract.}$$

$$\frac{1}{2} = \frac{1}{2} \quad \text{True}$$

The solution is $\dfrac{5}{4}$.

■ **Work Practice 3**

> **Helpful Hint**   We may solve an equation so that the variable is alone on *either* side of the equation. For example, $\dfrac{5}{4} = x$ is equivalent to $x = \dfrac{5}{4}$.

Practice 4

Solve: $3x + 10 = 4x$

Example 4   Solve: $5t - 5 = 6t$

Solution: To solve for t, we first want all terms containing t on one side of the equation and numbers on the other side. Notice that if we subtract $5t$ from both sides of the equation, then variable terms will be on one side of the equation and the number -5 will be alone on the other side.

$$5t - 5 = 6t$$
$$5t - 5 - 5t = 6t - 5t \quad \text{Subtract } 5t \text{ from both sides.}$$
$$-5 = t \quad \text{Combine like terms.}$$

Answers

2. $y = -1.4$ **3.** $y = \dfrac{29}{24}$ **4.** $x = 10$

Check:

$$5t - 5 = 6t \quad \text{Original equation.}$$

$$5(-5) - 5 \stackrel{?}{=} 6(-5) \quad \text{Replace } t \text{ with } -5.$$

$$-25 - 5 \stackrel{?}{=} -30$$

$$-30 = -30 \quad \text{True}$$

The solution is -5.

■ **Work Practice 4**

Helpful Hint   For Example 4, why not subtract $6t$ from both sides? The addition property allows us to do this, and we would have $-t - 5 = 0$. We are just no closer to our goal of having variable terms on one side of the equation and numbers on the other.

Objective B Simplifying Equations ▶

Many times, it is best to simplify one or both sides of an equation before applying the addition property of equality.

Example 5 Solve: $2x + 3x - 5 + 7 = 10x + 3 - 6x - 4$

Solution: First we simplify both sides of the equation.

$$2x + 3x - 5 + 7 = 10x + 3 - 6x - 4$$

$$5x + 2 = 4x - 1 \quad \text{Combine like terms on each side of the equation.}$$

Next, we want all terms with a variable on one side of the equation and all numbers on the other side.

$$5x + 2 - 4x = 4x - 1 - 4x \quad \text{Subtract } 4x \text{ from both sides.}$$

$$x + 2 = -1 \quad \text{Combine like terms.}$$

$$x + 2 - 2 = -1 - 2 \quad \text{Subtract 2 from both sides to get } x \text{ alone.}$$

$$x = -3 \quad \text{Combine like terms.}$$

Check:

$$2x + 3x - 5 + 7 = 10x + 3 - 6x - 4 \quad \text{Original equation.}$$

$$2(-3) + 3(-3) - 5 + 7 \stackrel{?}{=} 10(-3) + 3 - 6(-3) - 4 \quad \text{Replace } x \text{ with } -3.$$

$$-6 - 9 - 5 + 7 \stackrel{?}{=} -30 + 3 + 18 - 4 \quad \text{Multiply.}$$

$$-13 = -13 \quad \text{True}$$

The solution is -3.

■ **Work Practice 5**

Practice 5

Solve:
$$10w + 3 - 4w + 4 = -2w + 3 + 7w$$

If an equation contains parentheses, we use the distributive property to remove them, as before. Then we combine any like terms.

Example 6 Solve: $6(2a - 1) - (11a + 6) = 7$

Solution:

$$6(2a - 1) - 1(11a + 6) = 7$$

$$6(2a) + 6(-1) - 1(11a) - 1(6) = 7 \quad \text{Apply the distributive property.}$$

$$12a - 6 - 11a - 6 = 7 \quad \text{Multiply.}$$

$$a - 12 = 7 \quad \text{Combine like terms.}$$

$$a - 12 + 12 = 7 + 12 \quad \text{Add 12 to both sides.}$$

$$a = 19 \quad \text{Simplify.}$$

Check: Check by replacing a with 19 in the original equation.

■ **Work Practice 6**

Practice 6

Solve:
$$3(2w - 5) - (5w + 1) = -3$$

Answers

5. $w = -4$ **6.** $w = 13$

Practice 7

Solve: $12 - y = 9$

Example 7 Solve: $3 - x = 7$

Solution: First we subtract 3 from both sides.

$$3 - x = 7$$
$$3 - x - 3 = 7 - 3 \quad \text{Subtract 3 from both sides.}$$
$$-x = 4 \quad \text{Simplify.}$$

We have not yet solved for x since x is not alone. However, this equation does say that the opposite of x is 4. If the opposite of x is 4, then x is the opposite of 4, or $x = -4$.

$$\text{If} \quad -x = 4,$$
$$\text{then} \quad x = -4.$$

Check:
$$3 - x = 7 \quad \text{Original equation.}$$
$$3 - (-4) \stackrel{?}{=} 7 \quad \text{Replace } x \text{ with } -4.$$
$$3 + 4 \stackrel{?}{=} 7 \quad \text{Add.}$$
$$7 = 7 \quad \text{True}$$

The solution is -4.

■ **Work Practice 7**

Objective C Writing Algebraic Expressions

In this section, we continue to practice writing algebraic expressions.

Practice 8

a. The sum of two numbers is 11. If one number is 4, find the other number.

b. The sum of two numbers is 11. If one number is x, write an expression representing the other number.

c. The sum of two numbers is 56. If one number is a, write an expression representing the other number.

Example 8

a. The sum of two numbers is 8. If one number is 3, find the other number.

b. The sum of two numbers is 8. If one number is x, write an expression representing the other number.

Solution:

a. If the sum of two numbers is 8 and one number is 3, we find the other number by subtracting 3 from 8. The other number is $8 - 3$, or 5.

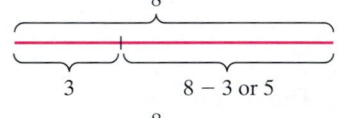

b. If the sum of two numbers is 8 and one number is x, we find the other number by subtracting x from 8. The other number is represented by $8 - x$.

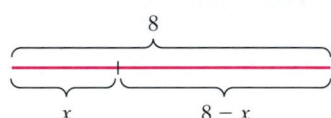

■ **Work Practice 8**

Practice 9

In a recent House of Representatives race in California, Mike Thompson received 100,445 more votes than Zane Starkewolf. If Zane received n votes, how many did Mike receive? (*Source:* Voter News Service)

Example 9 The Verrazano-Narrows Bridge in New York City is the longest suspension bridge in North America. The Golden Gate Bridge in San Francisco is 60 feet shorter than the Verrazano-Narrows Bridge. If the length of the Verrazano-Narrows Bridge is m feet, express the length of the Golden Gate Bridge as an algebraic expression in m. (*Source:* Survey of State Highway Engineers)

Answers

7. $y = 3$ **8. a.** $11 - 4$ or 7 **b.** $11 - x$
c. $56 - a$ **9.** $(n + 100,445)$ votes

Solution: Since the Golden Gate Bridge is 60 feet shorter than the Verrazano-Narrows Bridge, we have that its length is

In words:	Length of Verrazano-Narrows Bridge	minus	60
Translate:	m	$-$	60

The Golden Gate Bridge is $(m - 60)$ feet long.

■ **Work Practice 9**

Vocabulary, Readiness & Video Check

Use the choices below to fill in each blank. Some choices may be used more than once or not at all.

equation	multiplication	addition
expression	solution	equivalent

1. A combination of operations on variables and numbers is called a(n) _____.

2. A statement of the form "expression = expression" is called a(n) _____.

3. A(n) _____ contains an equal sign ($=$).

4. A(n) _____ does not contain an equal sign ($=$).

5. A(n) _____ may be simplified and evaluated while a(n) _____ may be solved.

6. A(n) _____ of an equation is a number that when substituted for a variable makes the equation a true statement.

7. _____ equations have the same solution.

8. By the _____ property of equality, the same number may be added to or subtracted from both sides of an equation without changing the solution of the equation.

Solve each equation mentally. See Examples 1 and 2.

9. $x + 4 = 6$

10. $x + 7 = 17$

11. $n + 18 = 30$

12. $z + 22 = 40$

13. $b - 11 = 6$

14. $d - 16 = 5$

Martin-Gay Interactive Videos

See Video 9.1

Watch the section lecture video and answer the following questions.

Objective A 15. Complete this statement based on the lecture given before ▣ Example 1. The addition property of equality means that if we have an equation, we can add the same real number to _____ of an equation and have an equivalent equation. ▶

Objective B 16. After both sides of ▣ Example 5 are simplified, write down the simplified equation. ▶

Objective C 17. Suppose we were to solve ▣ Example 8 again, this time letting the area of the Sahara Desert be x square miles. Use this to express the area of the Gobi Desert as an algebraic expression in x. ▶

9.1 Exercise Set MyMathLab®

Objective **A** *Solve each equation. Check each solution. See Examples 1 through 4.*

1. $x + 7 = 10$

2. $x + 14 = 25$

3. $x - 2 = -4$

4. $y - 9 = 1$

5. $-11 = 3 + x$

6. $-8 = 8 + z$

7. $r - 8.6 = -8.1$

8. $t - 9.2 = -6.8$

9. $x - \dfrac{2}{5} = -\dfrac{3}{20}$

10. $y - \dfrac{4}{7} = -\dfrac{3}{14}$

11. $\dfrac{1}{3} + f = \dfrac{3}{4}$

12. $c + \dfrac{1}{6} = \dfrac{3}{8}$

Objective **B** *Solve each equation. Don't forget to first simplify each side of the equation, if possible. Check each solution. See Examples 5 through 7.*

13. $7x + 2x = 8x - 3$

14. $3n + 2n = 7 + 4n$

15. $\dfrac{5}{6}x + \dfrac{1}{6}x = -9$

16. $\dfrac{13}{11}y - \dfrac{2}{11}y = -3$

17. $2y + 10 = 5y - 4y$

18. $4x - 4 = 10x - 7x$

19. $-5(n - 2) = 8 - 4n$

20. $-4(z - 3) = 2 - 3z$

21. $\dfrac{3}{7}x + 2 = -\dfrac{4}{7}x - 5$

22. $\dfrac{1}{5}x - 1 = -\dfrac{4}{5}x - 13$

23. $5x - 6 = 6x - 5$

24. $2x + 7 = x - 10$

25. $8y + 2 - 6y = 3 + y - 10$

26. $4p - 11 - p = 2 + 2p - 20$

27. $-3(x - 4) = -4x$

28. $-2(x - 1) = -3x$

29. $\dfrac{3}{8}x - \dfrac{1}{6} = -\dfrac{5}{8}x - \dfrac{2}{3}$

30. $\dfrac{2}{5}x - \dfrac{1}{12} = -\dfrac{3}{5}x - \dfrac{3}{4}$

31. $2(x - 4) = x + 3$

32. $3(y + 7) = 2y - 5$

33. $3(n - 5) - (6 - 2n) = 4n$

34. $5(3 + z) - (8z + 9) = -4z$

35. $-2(x + 6) + 3(2x - 5) = 3(x - 4) + 10$

36. $-5(x + 1) + 4(2x - 3) = 2(x + 2) - 8$

Objectives **A** **B** **Mixed Practice** *Solve. See Examples 1 through 7.*

37. $13x - 3 = 14x$

38. $18x - 9 = 19x$

39. $5b - 0.7 = 6b$

40. $9x + 5.5 = 10x$

41. $3x - 6 = 2x + 5$

42. $7y + 2 = 6y + 2$

43. $13x - 9 + 2x - 5 = 12x - 1 + 2x$

44. $15x + 20 - 10x - 9 = 25x + 8 - 21x - 7$

45. $7(6 + w) = 6(2 + w)$

46. $6(5 + c) = 5(c - 4)$

47. $n + 4 = 3.6$

48. $m + 2 = 7.1$

49. $10 - (2x - 4) = 7 - 3x$

50. $15 - (6 - 7k) = 2 + 6k$

51. $\dfrac{1}{3} = x + \dfrac{2}{3}$

52. $\dfrac{1}{11} = y + \dfrac{10}{11}$

53. $-6.5 - 4x - 1.6 - 3x = -6x + 9.8$

54. $-1.4 - 7x - 3.6 - 2x = -8x + 4.4$

Objective C *Write each algebraic expression described. See Examples 8 and 9.*

55. Two numbers have a sum of 20. If one number is p, express the other number in terms of p.

56. Two numbers have a sum of 13. If one number is y, express the other number in terms of y.

57. A 10-foot board is cut into two pieces. If one piece is x feet long, express the other length in terms of x.

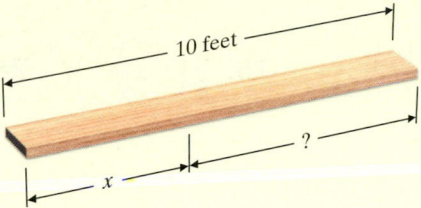

58. A 5-foot piece of string is cut into two pieces. If one piece is x feet long, express the other length in terms of x.

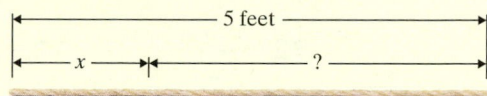

△ **59.** Recall that two angles are *supplementary* if their sum is 180°. If one angle measures $x°$, express the measure of its supplement in terms of x.

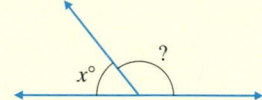

△ **60.** Recall that two angles are *complementary* if their sum is 90°. If one angle measures $x°$, express the measure of its complement in terms of x.

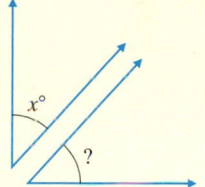

61. In 2013, the number of graduate students at the University of Texas at Austin was approximately 29,000 fewer than the number of undergraduate students. If the number of undergraduate students was n, how many graduate students attended UT Austin? (*Source:* University of Texas at Austin)

62. The longest interstate highway in the U.S. is I-90, which connects Seattle, Washington, and Boston, Massachusetts. The second longest interstate highway, I-80 (connecting San Francisco, California, and Teaneck, New Jersey), is 121 miles shorter than I-90. If the length of I-80 is m miles, express the length of I-90 as an algebraic expression in m. (*Source:* U.S. Department of Transportation—Federal Highway Administration)

63. The area of the Sahara Desert in Africa is 7 times the area of the Gobi Desert in Asia. If the area of the Gobi Desert is x square miles, express the area of the Sahara Desert as an algebraic expression in x.

64. The largest meteorite in the world is the Hoba West located in Namibia. Its weight is 3 times the weight of the Armanty meteorite located in Outer Mongolia. If the weight of the Armanty meteorite is y kilograms, express the weight of the Hoba West meteorite as an algebraic expression in y.

Review

Find each multiplicative inverse or reciprocal. See Section 8.6.

65. $\dfrac{5}{8}$ **66.** $\dfrac{7}{6}$ **67.** 2 **68.** 5 **69.** $-\dfrac{1}{9}$ **70.** $-\dfrac{3}{5}$

Perform each indicated operation and simplify. See Sections 8.5 and 8.6.

71. $\dfrac{3x}{3}$ **72.** $\dfrac{-2y}{-2}$ **73.** $-5\left(-\dfrac{1}{5}y\right)$ **74.** $7\left(\dfrac{1}{7}r\right)$ **75.** $\dfrac{3}{5}\left(\dfrac{5}{3}x\right)$ **76.** $\dfrac{9}{2}\left(\dfrac{2}{9}x\right)$

Concept Extensions

77. Write two terms whose sum is $-3x$.

78. Write four terms whose sum is $2y - 6$.

Use the addition property to fill in the blank so that the middle equation simplifies to the last equation. See the Concept Check in this section.

79.
$$x - 4 = -9$$
$$x - 4 + (\quad) = -9 + (\quad)$$
$$x = -5$$

80.
$$a + 9 = 15$$
$$a + 9 + (\quad) = 15 + (\quad)$$
$$a = 6$$

Fill in the blanks with numbers of your choice so that each equation has the given solution. Note: Each blank will be replaced with a different number.

81. _____ $+ x =$ _____ ; Solution: -3

82. $x -$ _____ $=$ _____ ; Solution: -10

Solve.

△ **83.** The sum of the angles of a triangle is 180°. If one angle of a triangle measures $x°$ and a second angle measures $(2x + 7)°$, express the measure of the third angle in terms of x. Simplify the expression.

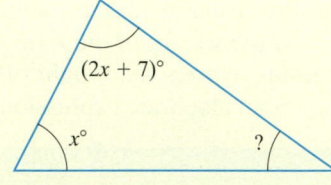

△ **84.** A quadrilateral is a four-sided figure (like the one shown in the figure) whose angle sum is 360°. If one angle measures $x°$, a second angle measures $3x°$, and a third angle measures $5x°$, express the measure of the fourth angle in terms of x. Simplify the expression.

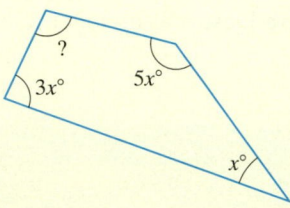

85. In your own words, explain what is meant by the solution of an equation.

86. In your own words, explain how to check a solution of an equation.

Use a calculator to determine the solution of each equation.

87. $36.766 + x = -108.712$

88. $-85.325 = x - 97.985$

Objective A Using the Multiplication Property

As useful as the addition property of equality is, it cannot help us solve every type of linear equation in one variable. For example, adding or subtracting a value on both sides of the equation does not help solve

$$\frac{5}{2}x = 15$$

because the variable x is being multiplied by a number (other than 1). Instead, we apply another important property of equality, the **multiplication property of equality.**

Multiplication Property of Equality

Let a, b, and c represent numbers and let $c \neq 0$. Then

$a = b$	Also, $a = b$
and $a \cdot c = b \cdot c$	and $\dfrac{a}{c} = \dfrac{b}{c}$
are equivalent equations.	are equivalent equations.

In other words, **both sides** of an equation **may be multiplied or divided by the same nonzero number** without changing the solution of the equation. (We may divide both sides by the same nonzero number since division is defined in terms of multiplication.)

Picturing again our balanced scale, if we multiply or divide the weight on each side by the same nonzero number, the scale (or equation) remains balanced.

$2x$ 6

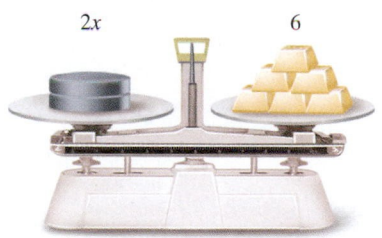

$\frac{2x}{2}$ or x $\frac{6}{2}$ or 3

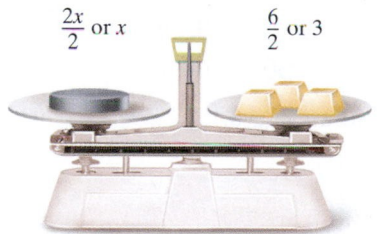

Example 1   Solve: $\dfrac{5}{2}x = 15$

Solution: To get x alone, we multiply both sides of the equation by the reciprocal (or multiplicative inverse) of $\dfrac{5}{2}$, which is $\dfrac{2}{5}$.

$$\frac{5}{2}x = 15$$

$$\frac{2}{5} \cdot \left(\frac{5}{2}x\right) = \frac{2}{5} \cdot 15 \quad \text{Multiply both sides by } \frac{2}{5}.$$

$$\left(\frac{2}{5} \cdot \frac{5}{2}\right)x = \frac{2}{5} \cdot 15 \quad \text{Apply the associative property.}$$

$$1x = 6 \qquad \text{Simplify.}$$

or

$$x = 6$$

(Continued on next page)

Objectives

A Use the Multiplication Property of Equality to Solve Linear Equations.

B Use Both the Addition and Multiplication Properties of Equality to Solve Linear Equations.

C Write Word Phrases as Algebraic Expressions.

Practice 1

Solve: $\dfrac{3}{7}x = 9$

Answer
1. $x = 21$

Check: Replace x with 6 in the original equation.

$$\frac{5}{2}x = 15 \quad \text{Original equation.}$$

$$\frac{5}{2}(6) \stackrel{?}{=} 15 \quad \text{Replace } x \text{ with 6.}$$

$$15 = 15 \quad \text{True}$$

The solution is 6.

■ **Work Practice 1**

In the equation $\frac{5}{2}x = 15$, $\frac{5}{2}$ is the coefficient of x. When the coefficient of x is a *fraction,* we will get x alone by multiplying by the reciprocal. When the coefficient of x is an integer or a decimal, it is usually more convenient to divide both sides by the coefficient. (Dividing by a number is, of course, the same as multiplying by the reciprocal of the number.)

Practice 2

Solve: $7x = 42$

Example 2 Solve: $5x = 30$

Solution: To get x alone, we divide both sides of the equation by 5, the coefficient of x.

$$5x = 30$$

$$\frac{5x}{5} = \frac{30}{5} \quad \text{Divide both sides by 5.}$$

$$1 \cdot x = 6 \quad \text{Simplify.}$$

$$x = 6$$

Check: $5x = 30$ Original equation.

$$5 \cdot 6 \stackrel{?}{=} 30 \quad \text{Replace } x \text{ with 6.}$$

$$30 = 30 \quad \text{True}$$

The solution is 6.

■ **Work Practice 2**

Practice 3

Solve: $-4x = 52$

Example 3 Solve: $-3x = 33$

Solution: Recall that $-3x$ means $-3 \cdot x$. To get x alone, we divide both sides by the coefficient of x, that is, -3.

$$-3x = 33$$

$$\frac{-3x}{-3} = \frac{33}{-3} \quad \text{Divide both sides by } -3.$$

$$1x = -11 \quad \text{Simplify.}$$

$$x = -11$$

Check: $-3x = 33$ Original equation.

$$-3(-11) \stackrel{?}{=} 33 \quad \text{Replace } x \text{ with } -11.$$

$$33 = 33 \quad \text{True}$$

The solution is -11.

■ **Work Practice 3**

Answers

2. $x = 6$ **3.** $x = -13$

Example 4 Solve: $\dfrac{y}{7} = 20$

Solution: Recall that $\dfrac{y}{7} = \dfrac{1}{7}y$. To get y alone, we multiply both sides of the equation by 7, the reciprocal of $\dfrac{1}{7}$.

$$\dfrac{y}{7} = 20$$

$$\dfrac{1}{7}y = 20$$

$$7 \cdot \dfrac{1}{7}y = 7 \cdot 20 \quad \text{Multiply both sides by 7.}$$

$$1y = 140 \quad \text{Simplify.}$$

$$y = 140$$

Check: $\dfrac{y}{7} = 20 \quad$ Original equation.

$$\dfrac{140}{7} \stackrel{?}{=} 20 \quad \text{Replace } y \text{ with 140.}$$

$$20 = 20 \quad \text{True}$$

The solution is 140.

■ **Work Practice 4**

Example 5 Solve: $3.1x = 4.96$

Solution:
$$3.1x = 4.96$$

$$\dfrac{3.1x}{3.1} = \dfrac{4.96}{3.1} \quad \text{Divide both sides by 3.1.}$$

$$1x = 1.6 \quad \text{Simplify.}$$

$$x = 1.6$$

Check: Check by replacing x with 1.6 in the original equation. The solution is 1.6.

■ **Work Practice 5**

Example 6 Solve: $-\dfrac{2}{3}x = -\dfrac{5}{2}$

Solution: To get x alone, we multiply both sides of the equation by $-\dfrac{3}{2}$, the reciprocal of the coefficient of x.

$$-\dfrac{2}{3}x = -\dfrac{5}{2}$$

$$-\dfrac{3}{2} \cdot -\dfrac{2}{3}x = -\dfrac{3}{2} \cdot -\dfrac{5}{2} \quad \text{Multiply both sides by } -\dfrac{3}{2}, \text{ the reciprocal of } -\dfrac{2}{3}.$$

$$x = \dfrac{15}{4} \quad \text{Simplify.}$$

Check: Check by replacing x with $\dfrac{15}{4}$ in the original equation. The solution is $\dfrac{15}{4}$.

■ **Work Practice 6**

Objective B Using Both the Addition and Multiplication Properties ▶

We are now ready to combine the skills learned in the last section with the skills learned in this section to solve equations by applying more than one property.

Practice 7

Solve: $-x + 7 = -12$

Example 7 Solve: $-z - 4 = 6$

Solution: First, let's get $-z$, the term containing the variable, alone. To do so, we add 4 to both sides of the equation.

$$-z - 4 + 4 = 6 + 4 \quad \text{Add 4 to both sides.}$$
$$-z = 10 \quad \text{Simplify.}$$

Next, recall that $-z$ means $-1 \cdot z$. Thus to get z alone, we either multiply or divide both sides of the equation by -1. In this example, we divide.

$$-z = 10$$
$$\frac{-z}{-1} = \frac{10}{-1} \quad \text{Divide both sides by the coefficient } -1.$$
$$1z = -10 \quad \text{Simplify.}$$
$$z = -10$$

Check: $-z - 4 = 6 \quad$ Original equation.
$$-(-10) - 4 \overset{?}{=} 6 \quad \text{Replace } z \text{ with } -10.$$
$$10 - 4 \overset{?}{=} 6$$
$$6 = 6 \quad \text{True}$$

The solution is -10.

■ **Work Practice 7**

Don't forget to first simplify one or both sides of an equation, if possible.

Practice 8

Solve:
$-7x + 2x + 3 - 20 = -2$

Example 8 Solve: $a + a - 10 + 7 = -13$

Solution: First, we simplify the left side of the equation by combining like terms.

$$a + a - 10 + 7 = -13$$
$$2a - 3 = -13 \quad \text{Combine like terms.}$$
$$2a - 3 + 3 = -13 + 3 \quad \text{Add 3 to both sides.}$$
$$2a = -10 \quad \text{Simplify.}$$
$$\frac{2a}{2} = \frac{-10}{2} \quad \text{Divide both sides by 2.}$$
$$a = -5 \quad \text{Simplify.}$$

Check: To check, replace a with -5 in the original equation. The solution is -5.

■ **Work Practice 8**

Practice 9

Solve: $10x - 4 = 7x + 14$

Example 9 Solve: $7x - 3 = 5x + 9$

Solution: To get x alone, let's first use the addition property to get variable terms on one side of the equation and numbers on the other side. One way to get variable terms on one side is to subtract $5x$ from both sides.

$$7x - 3 = 5x + 9$$
$$7x - 3 - 5x = 5x + 9 - 5x \quad \text{Subtract } 5x \text{ from both sides.}$$
$$2x - 3 = 9 \quad \text{Simplify.}$$

Answers

7. $x = 19$ **8.** $x = -3$ **9.** $x = 6$

Now, to get numbers on the other side, let's add 3 to both sides.

$2x - 3 + 3 = 9 + 3$ Add 3 to both sides.

$2x = 12$ Simplify.

Use the multiplication property to get x alone.

$$\frac{2x}{2} = \frac{12}{2}$$ Divide both sides by 2.

$x = 6$ Simplify.

Check: To check, replace x with 6 in the original equation to see that a true statement results. The solution is 6.

■ Work Practice 9

If an equation has parentheses, don't forget to use the distributive property to remove them. Then combine any like terms.

Example 10   Solve: $5(2x + 3) = -1 + 7$

Solution:

$$5(\overset{\frown}{2x + 3}) = -1 + 7$$

$5(2x) + 5(3) = -1 + 7$ Apply the distributive property.

$10x + 15 = 6$ Multiply and write $-1 + 7$ as 6.

$10x + 15 - 15 = 6 - 15$ Subtract 15 from both sides.

$10x = -9$ Simplify.

$$\frac{10x}{10} = -\frac{9}{10}$$ Divide both sides by 10.

$$x = -\frac{9}{10}$$ Simplify.

Check: To check, replace x with $-\dfrac{9}{10}$ in the original equation to see that a true

statement results. The solution is $-\dfrac{9}{10}$.

■ Work Practice 10

Objective C Writing Algebraic Expressions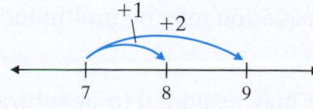

We continue to sharpen our problem-solving skills by writing algebraic expressions.

Example 11   Writing an Expression for Consecutive Integers

If x is the first of three consecutive integers, express the sum of the three integers in terms of x. Simplify if possible.

Solution: An example of three consecutive integers is 7, 8, and 9.

```
        +1
          +2
   ←——┬———┬———┬——→
      7   8   9
```

(Continued on next page)

Practice 10

Solve: $4(3x - 2) = -1 + 4$

Practice 11

a. If x is the first of two consecutive integers, express the sum of the two integers in terms of x. Simplify if possible.

b. If x is the first of two consecutive odd integers (see next page), express the sum of the two integers in terms of x. Simplify if possible.

Answers

10. $x = \dfrac{11}{12}$ **11. a.** $2x + 1$ **b.** $2x + 2$

The second consecutive integer is always 1 more than the first, and the third consecutive integer is 2 more than the first. If x is the first of three consecutive integers, the three consecutive integers are $x, x + 1$, and $x + 2$.

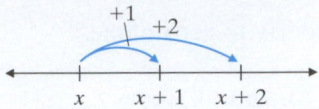

Their sum is shown below.

In words:

first integer	+	second integer	+	third integer

Translate: x + $(x + 1)$ + $(x + 2)$

This simplifies to $3x + 3$.

■ **Work Practice 11**

Study these examples of consecutive even and consecutive odd integers.

Consecutive even integers:

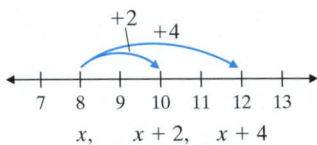

Consecutive odd integers:

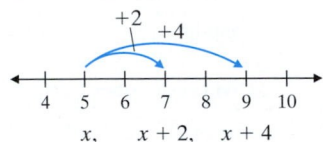

Helpful Hint

If x is an odd integer, then $x + 2$ is the next odd integer. This 2 simply means that odd integers are always 2 units from each other.

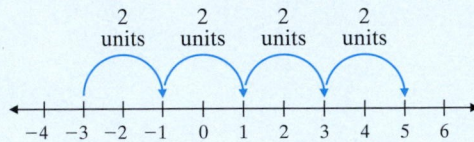

Vocabulary, Readiness & Video Check

Use the choices below to fill in each blank. Some choices may be used more than once. Many of these exercises contain an important review of Section 9.1 also.

equation	multiplication	addition
expression	solution	equivalent

1. By the _____ property of equality, both sides of an equation may be multiplied or divided by the same nonzero number without changing the solution of the equation.

2. By the _____ property of equality, the same number may be added to or subtracted from both sides of an equation without changing the solution of the equation.

3. A(n) _____ may be solved while a(n) _____ may be simplified and evaluated.

4. A(n) _____ contains an equal sign (=) while a(n) _____ does not.

5. _____ equations have the same solution.

6. A(n) _____ of an equation is a number that when substituted for the variable makes the equation a true statement.

Solve each equation mentally. See Examples 2 and 3.

7. $3a = 27$ **8.** $9c = 54$ **9.** $5b = 10$ **10.** $7t = 14$ **11.** $6x = -30$ **12.** $8r = -64$

Martin-Gay Interactive Videos   *Watch the section lecture video and answer the following questions.*

Objective A **13.** Complete this statement based on the lecture given before ⊞ Example 1. We can multiply both sides of an equation by the _____ nonzero number and have an equivalent equation. ▶

Objective B **14.** Both the addition and multiplication properties of equality are used to solve ⊞ Examples 4–6. In each of these exercises, what property is applied first? What property is applied last? What conclusion, if any, can you make? ▶

See Video 9.2

Objective C **15.** Let x be the first of four consecutive integers, as in ⊞ Example 8. Now express the sum of the second integer and the fourth integer as an algebraic expression containing x. ▶

9.2 **Exercise Set** MyMathLab®

Objective A *Solve each equation. Check each solution. See Examples 1 through 6.*

▶1. $-5x = -20$ **2.** $-7x = -49$ **3.** $3x = 0$ **4.** $2x = 0$

5. $-x = -12$ **6.** $-y = 8$ **▶7.** $\dfrac{2}{3}x = -8$ **8.** $\dfrac{3}{4}n = -15$

9. $\dfrac{1}{6}d = \dfrac{1}{2}$ **10.** $\dfrac{1}{8}v = \dfrac{1}{4}$ **11.** $\dfrac{a}{2} = 1$ **▶12.** $\dfrac{d}{15} = 2$

13. $\dfrac{k}{-7} = 0$ **14.** $\dfrac{f}{-5} = 0$ **15.** $1.7x = 10.71$ **16.** $8.5y = 19.55$

Objective B *Solve each equation. Check each solution. See Examples 7 and 8.*

17. $2x - 4 = 16$ **18.** $3x - 1 = 26$ **19.** $-x + 2 = 22$ **20.** $-x + 4 = -24$

▶21. $6a + 3 = 3$ **22.** $8t + 5 = 5$ **23.** $\dfrac{x}{3} - 2 = -5$ **24.** $\dfrac{b}{4} - 1 = -7$

25. $6z - 8 - z + 3 = 0$ **26.** $4a + 1 + a - 11 = 0$ **27.** $1 = 0.4x - 0.6x - 5$ **28.** $19 = 0.4x - 0.9x - 6$

29. $\frac{2}{3}y - 11 = -9$ **30.** $\frac{3}{5}x - 14 = -8$ **31.** $\frac{3}{4}t - \frac{1}{2} = \frac{1}{3}$ **32.** $\frac{2}{7}z - \frac{1}{5} = \frac{1}{2}$

Solve each equation. See Examples 9 and 10.

▶ **33.** $8x + 20 = 6x + 18$ **34.** $11x + 13 = 9x + 9$ **35.** $3(2x + 5) = -18 + 9$ **36.** $2(4x + 1) = -12 + 6$

37. $2x - 5 = 20x + 4$ **38.** $6x - 4 = -2x - 10$ **39.** $2 + 14 = -4(3x - 4)$ **40.** $8 + 4 = -6(5x - 2)$

41. $-6y - 3 = -5y - 7$ **42.** $-17z - 4 = -16z - 20$ **43.** $\frac{1}{2}(2x - 1) = -\frac{1}{7} - \frac{3}{7}$

44. $\frac{1}{3}(3x - 1) = -\frac{1}{10} - \frac{2}{10}$ ▶ **45.** $-10z - 0.5 = -20z + 1.6$ **46.** $-14y - 1.8 = -24y + 3.9$

47. $-4x + 20 = 4x - 20$ **48.** $-3x + 15 = 3x - 15$

Objectives A B Mixed Practice *See Examples 1 through 10.*

49. $42 = 7x$ **50.** $81 = 3x$ **51.** $4.4 = -0.8x$

52. $6.3 = -0.6x$ **53.** $6x + 10 = -20$ **54.** $10y + 15 = -5$

55. $5 - 0.3k = 5$ **56.** $2 - 0.4p = 2$ **57.** $13x - 5 = 11x - 11$

58. $20x - 20 = 16x - 40$ ▶ **59.** $9(3x + 1) = 4x - 5x$ **60.** $7(2x + 1) = 18x - 19x$

61. $-\frac{3}{7}p = -2$ **62.** $-\frac{4}{5}r = -5$ **63.** $-\frac{4}{3}x = 12$

64. $-\frac{10}{3}x = 30$ **65.** $-2x - \frac{1}{2} = \frac{7}{2}$ **66.** $-3n - \frac{1}{3} = \frac{8}{3}$

67. $10 = 2x - 1$ **68.** $12 = 3j - 4$ **69.** $10 - 3x - 6 - 9x = 7$

70. $12x + 30 + 8x - 6 = 10$ **71.** $z - 5z = 7z - 9 - z$ **72.** $t - 6t = -13 + t - 3t$

73. $-x - \frac{4}{5} = x + \frac{1}{2} + \frac{2}{5}$ **74.** $x + \frac{3}{7} = -x + \frac{1}{3} + \frac{4}{7}$

75. $-15 + 37 = -2(x + 5)$ **76.** $-19 + 74 = -5(x + 3)$

Objective C   *Write each algebraic expression described. Simplify if possible. See Example 11.*

77. If x represents the first of two consecutive odd integers, express the sum of the two integers in terms of x.

78. If x is the first of three consecutive even integers, write their sum as an algebraic expression in x.

79. If x is the first of four consecutive integers, express the sum of the first integer and the third integer as an algebraic expression containing the variable x.

80. If x is the first of two consecutive integers, express the sum of 20 and the second consecutive integer as an algebraic expression containing the variable x.

81. Classrooms on one side of the science building are all numbered with consecutive even integers. If the first room on this side of the building is numbered x, write an expression in x for the sum of five classroom numbers in a row. Then simplify this expression.

82. Two sides of a quadrilateral have the same length, x, while the other two sides have the same length, both being the next consecutive odd integer. Write the sum of these lengths. Then simplify this expression.

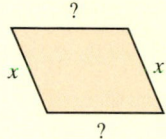

Review

Simplify each expression. See Section 8.7.

83. $5x + 2(x - 6)$

84. $-7y + 2y - 3(y + 1)$

85. $6(2z + 4) + 20$

86. $-(3a - 3) + 2a - 6$

87. $-(x - 1) + x$

88. $8(z - 6) + 7z - 1$

Concept Extensions

For Exercises 89 and 90, fill in the blank with a number of your choice so that each equation has the given solution.

89. $6x = $ _____ ; solution: -8

90. _____ $x = 10$; solution: $\dfrac{1}{2}$

91. The equation $3x + 6 = 2x + 10 + x - 4$ is true for all real numbers. Substitute a few real numbers for x to see that this is so and then try solving the equation. Describe what happens.

92. The equation $6x + 2 - 2x = 4x + 1$ has no solution. Try solving this equation for x and describe what happens.

93. From the results of Exercises 91 and 92, when do you think an equation has all real numbers as its solutions?

94. From the results of Exercises 91 and 92, when do you think an equation has no solution?

Solve.

95. $0.07x - 5.06 = -4.92$

96. $0.06y + 2.63 = 2.5562$

Objectives

A Apply the General Strategy for Solving a Linear Equation.

B Solve Equations Containing Fractions or Decimals.

C Recognize Identities and Equations with No Solution.

Objective A Solving Linear Equations

Let's begin by restating the formal definition of a linear equation in one variable.

A **linear equation in one variable** can be written in the form

$$Ax + B = C$$

where A, B, and C are real numbers and $A \neq 0$.

We now combine our knowledge from the previous sections into a general strategy for solving linear equations.

To Solve Linear Equations in One Variable

Step 1: If an equation contains fractions, multiply both sides by the LCD to clear the equation of fractions.

Step 2: Use the distributive property to remove parentheses if they are present.

Step 3: Simplify each side of the equation by combining like terms.

Step 4: Get all variable terms on one side and all numbers on the other side by using the addition property of equality.

Step 5: Get the variable alone by using the multiplication property of equality.

Step 6: Check the solution by substituting it into the original equation.

We will use these steps as we solve the equations in Examples 1–5.

Practice 1

Solve:
$$5(3x - 1) + 2 = 12x + 6$$

Example 1 Solve: $4(2x - 3) + 7 = 3x + 5$

Solution: There are no fractions, so we begin with Step 2.

$$4(2x - 3) + 7 = 3x + 5$$

Step 2: $\quad 8x - 12 + 7 = 3x + 5 \quad$ Use the distributive property.

Step 3: $\qquad 8x - 5 = 3x + 5 \quad$ Combine like terms.

Step 4: Get all variable terms on one side of the equation and all numbers on the other side. One way to do this is by subtracting $3x$ from both sides and then adding 5 to both sides.

$$8x - 5 - 3x = 3x + 5 - 3x \quad \text{Subtract } 3x \text{ from both sides.}$$
$$5x - 5 = 5 \qquad\qquad\quad \text{Simplify.}$$
$$5x - 5 + 5 = 5 + 5 \qquad\quad \text{Add 5 to both sides.}$$
$$5x = 10 \qquad\qquad\quad \text{Simplify.}$$

Step 5: Use the multiplication property of equality to get x alone.

$$\frac{5x}{5} = \frac{10}{5} \quad \text{Divide both sides by 5.}$$
$$x = 2 \quad \text{Simplify.}$$

Step 6: Check.

$$4(2x - 3) + 7 = 3x + 5 \qquad \text{Original equation}$$
$$4[2(2) - 3] + 7 \stackrel{?}{=} 3(2) + 5 \quad \text{Replace } x \text{ with 2.}$$
$$4(4 - 3) + 7 \stackrel{?}{=} 6 + 5$$
$$4(1) + 7 \stackrel{?}{=} 11$$
$$4 + 7 \stackrel{?}{=} 11$$
$$11 = 11 \qquad\qquad \text{True}$$

The solution is 2.

■ **Work Practice 1**

Answer

1. $x = 3$

Example 2 Solve: $8(2 - t) = -5t$

Solution: First, we apply the distributive property.

$$\overset{\frown}{8(2 - t)} = -5t$$

Step 2: $\quad 16 - 8t = -5t$ Use the distributive property.

Step 4: $16 - 8t + 8t = -5t + 8t$ Add $8t$ to both sides.

$$16 = 3t \qquad \text{Combine like terms.}$$

Step 5: $\quad \dfrac{16}{3} = \dfrac{3t}{3}$ Divide both sides by 3.

$$\dfrac{16}{3} = t \qquad \text{Simplify.}$$

Step 6: Check.

$$8(2 - t) = -5t \qquad \text{Original equation}$$

$$8\left(2 - \dfrac{16}{3}\right) \overset{?}{=} -5\left(\dfrac{16}{3}\right) \qquad \text{Replace } t \text{ with } \dfrac{16}{3}.$$

$$8\left(\dfrac{6}{3} - \dfrac{16}{3}\right) \overset{?}{=} -\dfrac{80}{3} \qquad \text{The LCD is 3.}$$

$$8\left(-\dfrac{10}{3}\right) \overset{?}{=} -\dfrac{80}{3} \qquad \text{Subtract fractions.}$$

$$-\dfrac{80}{3} = -\dfrac{80}{3} \qquad \text{True}$$

The solution is $\dfrac{16}{3}$.

■ **Work Practice 2**

Objective B Solving Equations Containing Fractions or Decimals ▶

If an equation contains fractions, we can clear the equation of fractions by multiplying both sides by the LCD of all denominators. By doing this, we avoid working with time-consuming fractions.

Example 3 Solve: $\dfrac{x}{2} - 1 = \dfrac{2}{3}x - 3$

Solution: We begin by clearing fractions. To do this, we multiply both sides of the equation by the LCD, which is 6.

$$\dfrac{x}{2} - 1 = \dfrac{2}{3}x - 3$$

Step 1: $\quad 6\left(\dfrac{x}{2} - 1\right) = 6\left(\dfrac{2}{3}x - 3\right)$ Multiply both sides by the LCD, 6.

Step 2: $6\left(\dfrac{x}{2}\right) - 6(1) = 6\left(\dfrac{2}{3}x\right) - 6(3)$ Use the distributive property.

$$3x - 6 = 4x - 18 \qquad \text{Simplify.}$$

There are no longer grouping symbols and no like terms on either side of the equation, so we continue with Step 4.

(Continued on next page)

Practice 2

Solve: $9(5 - x) = -3x$

> **Helpful Hint** When checking solutions, use the original equation.

Practice 3

Solve: $\dfrac{5}{2}x - 1 = \dfrac{3}{2}x - 4$

> **Helpful Hint** Don't forget to multiply *each* term by the LCD.

Answers

2. $x = \dfrac{15}{2}$ **3.** $x = -3$

$$3x - 6 = 4x - 18$$

Step 4: $3x - 6 - 3x = 4x - 18 - 3x$ Subtract $3x$ from both sides.

$$-6 = x - 18$$ Simplify.

$$-6 + 18 = x - 18 + 18$$ Add 18 to both sides.

$$12 = x$$ Simplify.

Step 5: The variable is now alone, so there is no need to apply the multiplication property of equality.

Step 6: Check.

$$\frac{x}{2} - 1 = \frac{2}{3}x - 3$$ Original equation

$$\frac{12}{2} - 1 \stackrel{?}{=} \frac{2}{3} \cdot 12 - 3$$ Replace x with 12.

$$6 - 1 \stackrel{?}{=} 8 - 3$$ Simplify.

$$5 = 5$$ True

The solution is 12.

■ **Work Practice 3**

Practice 4

Solve: $\dfrac{3(x - 2)}{5} = 3x + 6$

Example 4   Solve: $\dfrac{2(a + 3)}{3} = 6a + 2$

Solution: We clear the equation of fractions first.

$$\frac{2(a + 3)}{3} = 6a + 2$$

Step 1: $3 \cdot \dfrac{2(a + 3)}{3} = 3(6a + 2)$ Clear the fraction by multiplying both sides by the LCD, 3.

$$2(a + 3) = 3(6a + 2)$$ Simplify.

Step 2: Next, we use the distributive property to remove parentheses.

$$2a + 6 = 18a + 6$$ Use the distributive property.

Step 4: $2a + 6 - 18a = 18a + 6 - 18a$ Subtract $18a$ from both sides.

$$-16a + 6 = 6$$ Simplify.

$$-16a + 6 - 6 = 6 - 6$$ Subtract 6 from both sides.

$$-16a = 0$$

Step 5: $\dfrac{-16a}{-16} = \dfrac{0}{-16}$ Divide both sides by -16.

$$a = 0$$ Simplify.

Step 6: To check, replace a with 0 in the original equation. The solution is 0.

■ **Work Practice 4**

Helpful Hint

Remember: When solving an equation, it makes no difference on which side of the equation variable terms lie. Just make sure that constant terms lie on the other side.

When solving a problem about money, you may need to solve an equation containing decimals. If you choose, you may multiply to clear the equation of decimals.

Answer

4. $x = -3$

Example 5 Solve: $0.25x + 0.10(x - 3) = 1.1$

Solution: First we clear this equation of decimals by multiplying both sides of the equation by 100. Recall that multiplying a decimal number by 100 has the effect of moving the decimal point 2 places to the right.

$$0.25x + 0.10(x - 3) = 1.1$$

Step 1: $0.25x + 0.10(x - 3) = 1.10$ Multiply both sides by 100.

$$25x + 10(x - 3) = 110$$

Step 2: $25x + 10x - 30 = 110$ Apply the distributive property.

Step 3: $35x - 30 = 110$ Combine like terms.

Step 4: $35x - 30 + 30 = 110 + 30$ Add 30 to both sides.

$$35x = 140$$ Combine like terms.

Step 5: $\dfrac{35x}{35} = \dfrac{140}{35}$ Divide both sides by 35.

$$x = 4$$

Step 6: To check, replace x with 4 in the original equation. The solution is 4.

■ **Work Practice 5**

Objective C Recognizing Identities and Equations with No Solution ▶

So far, each equation that we have solved has had a single solution. However, not every equation in one variable has a single solution. Some equations have no solution, while others have an infinite number of solutions. For example,

$$x + 5 = x + 7$$

has **no solution** since no matter which real number we replace x with, the equation is false.

real number $+ 5 =$ same real number $+ 7$ FALSE

On the other hand,

$$x + 6 = x + 6$$

has infinitely many solutions since x can be replaced by any real number and the equation will always be true.

real number $+ 6 =$ same real number $+ 6$ TRUE

The equation $x + 6 = x + 6$ is called an **identity.** The next two examples illustrate special equations like these.

Example 6 Solve: $-2(x - 5) + 10 = -3(x + 2) + x$

Solution:

$$-2(x - 5) + 10 = -3(x + 2) + x$$
$$-2x + 10 + 10 = -3x - 6 + x$$ Apply the distributive property on both sides.
$$-2x + 20 = -2x - 6$$ Combine like terms.
$$-2x + 20 + 2x = -2x - 6 + 2x$$ Add $2x$ to both sides.
$$20 = -6$$ Combine like terms.

The final equation contains no variable terms, and the result is the false statement $20 = -6$. This means that there is no value for x that makes $20 = -6$ a true equation. Thus, we conclude that there is **no solution** to this equation.

■ **Work Practice 6**

Practice 5

Solve:
$0.06x - 0.10(x - 2) = -0.16$

Helpful Hint   If you have trouble with this step, try removing parentheses first.

$$0.25x + 0.10(x - 3) = 1.1$$
$$0.25x + 0.10x - 0.3 = 1.1$$
$$0.25x + 0.10x - 0.30 = 1.10$$
$$25x + 10x - 30 = 110$$

Then continue.

Practice 6

Solve:
$5(2 - x) + 8x = 3(x - 6)$

Practice 7

Solve: $-6(2x + 1) - 14$
$\qquad = -10(x + 2) - 2x$

Example 7 Solve: $3(x - 4) = 3x - 12$

Solution: $3(x - 4) = 3x - 12$

$\qquad 3x - 12 = 3x - 12$ Apply the distributive property.

The left side of the equation is now identical to the right side. Every real number may be substituted for x and a true statement will result. We arrive at the same conclusion if we continue.

$$3x - 12 = 3x - 12$$
$$3x - 12 - 3x = 3x - 12 - 3x \quad \text{Subtract } 3x \text{ from both sides.}$$
$$-12 = -12 \qquad\qquad \text{Combine like terms.}$$

Again, the final equation contains no variables, but this time the result is the true statement $-12 = -12$. This means that one side of the equation is identical to the other side. Thus, $3(x - 4) = 3x - 12$ is an **identity** and **every real number** is a solution.

■ **Work Practice 7**

Answer

7. Every real number is a solution.

✓ Concept Check Answers

a. Every real number is a solution.

b. The solution is 0.

c. There is no solution.

✓ Concept Check Suppose you have simplified several equations and obtained the following results. What can you conclude about the solutions to the original equation?

a. $7 = 7$ \qquad b. $x = 0$ \qquad c. $7 = -4$

🖩 Calculator Explorations Checking Equations

We can use a calculator to check possible solutions of equations. To do this, replace the variable by the possible solution and evaluate each side of the equation separately.

$\qquad$ Equation: $\quad 3x - 4 = 2(x + 6)$ $\quad$ Solution: $\quad x = 16$

$$3x - 4 = 2(x + 6) \qquad \text{Original equation}$$
$$3(16) - 4 \stackrel{?}{=} 2(16 + 6) \qquad \text{Replace } x \text{ with 16.}$$

Now evaluate each side with your calculator.

Evaluate left side: $\boxed{3}\ \boxed{\times}\ \boxed{16}\ \boxed{-}\ \boxed{4}\ \boxed{=}$

$\qquad\qquad\qquad$ or

$\qquad\qquad\qquad$ $\boxed{\text{ENTER}}$

Display: $\boxed{\qquad 44}$

Evaluate right side: $\boxed{2}\ \boxed{(}\ \boxed{16}\ \boxed{+}\ \boxed{6}\ \boxed{)}\ \boxed{=}$

$\qquad\qquad\qquad$ or

$\qquad\qquad\qquad$ $\boxed{\text{ENTER}}$

Display: $\boxed{\qquad 44}$

Since the left side equals the right side, the equation checks.

Use a calculator to check the possible solutions to each equation.

1. $2x = 48 + 6x$; $\quad x = -12$

2. $-3x - 7 = 3x - 1$; $\quad x = -1$

3. $5x - 2.6 = 2(x + 0.8)$; $\quad x = 4.4$

4. $-1.6x - 3.9 = -6.9x - 25.6$; $\quad x = 5$

5. $\dfrac{564x}{4} = 200x - 11(649)$; $\quad x = 121$

6. $20(x - 39) = 5x - 432$; $\quad x = 23.2$

Vocabulary, Readiness & Video Check

Throughout algebra, it is important to be able to identify equations and expressions.

Remember,

- an equation contains an equal sign and
- an expression does not.

Among other things,

- we solve equations and
- we simplify or perform operations on expressions.

Identify each as an equation or an expression.

1. $x = -7$ _____

2. $x - 7$ _____

3. $4y - 6 + 9y + 1$ _____

4. $4y - 6 = 9y + 1$ _____

5. $\dfrac{1}{x} - \dfrac{x-1}{8}$ _____

6. $\dfrac{1}{x} - \dfrac{x-1}{8} = 6$ _____

7. $0.1x + 9 = 0.2x$ _____

8. $0.1x^2 + 9y - 0.2x^2$ _____

Martin-Gay Interactive Videos

See Video 9.3 🍎

Watch the section lecture video and answer the following questions.

Objective A **9.** The general strategy for solving linear equations in one variable is discussed after  Example 1. How many properties are mentioned in this strategy and what are they? ▶

Objective B **10.** In the first step for solving  Example 2, both sides of the equation are being multiplied by the LCD. Why is the distributive property mentioned? ▶

11. In Example 3, why is the number of decimal places in each term of the equation important? ▶

Objective C **12.** Complete each statement based on Examples 4 and 5.

When solving an equation and all variable terms subtract out:

a. If you have a true statement, then the equation has _____ solution(s).

b. If you have a false statement, then the equation has _____ solution(s). ▶

9.3 **Exercise Set** MyMathLab® ▶

Objective A *Solve each equation. See Examples 1 and 2.*

1. $-4y + 10 = -2(3y + 1)$

2. $-3x + 1 = -2(4x + 2)$

3. $15x - 8 = 10 + 9x$

4. $15x - 5 = 7 + 12x$

5. $-2(3x - 4) = 2x$

6. $-(5x - 10) = 5x$

▶**7.** $5(2x - 1) - 2(3x) = 1$

8. $3(2 - 5x) + 4(6x) = 12$

9. $-6(x - 3) - 26 = -8$

10. $-4(n - 4) - 23 = -7$

11. $8 - 2(a + 1) = 9 + a$

12. $5 - 6(2 + b) = b - 14$

13. $4x + 3 = -3 + 2x + 14$

14. $6y - 8 = -6 + 3y + 13$

15. $-2y - 10 = 5y + 18$

16. $-7n + 5 = 8n - 10$

Objective **B** *Solve each equation. See Examples 3 through 5.*

17. $\dfrac{2}{3}x + \dfrac{4}{3} = -\dfrac{2}{3}$

18. $\dfrac{4}{5}x - \dfrac{8}{5} = -\dfrac{16}{5}$

19. $\dfrac{3}{4}x - \dfrac{1}{2} = 1$

20. $\dfrac{2}{9}x - \dfrac{1}{3} = 1$

▶ **21.** $0.50x + 0.15(70) = 35.5$

22. $0.40x + 0.06(30) = 9.8$

23. $\dfrac{2(x + 1)}{4} = 3x - 2$

24. $\dfrac{3(y + 3)}{5} = 2y + 6$

25. $x + \dfrac{7}{6} = 2x - \dfrac{7}{6}$

26. $\dfrac{5}{2}x - 1 = x + \dfrac{1}{4}$

27. $0.12(y - 6) + 0.06y = 0.08y - 0.70$

28. $0.60(z - 300) + 0.05z = 0.70z - 205$

Objective **C** *Solve each equation. See Examples 6 and 7.*

29. $4(3x + 2) = 12x + 8$

30. $14x + 7 = 7(2x + 1)$

31. $\dfrac{x}{4} + 1 = \dfrac{x}{4}$

32. $\dfrac{x}{3} - 2 = \dfrac{x}{3}$

33. $3x - 7 = 3(x + 1)$

34. $2(x - 5) = 2x + 10$

35. $-2(6x - 5) + 4 = -12x + 14$

36. $-5(4y - 3) + 2 = -20y + 17$

Objectives **A** **B** **C** **Mixed Practice** *Solve. See Examples 1 through 7.*

37. $\dfrac{6(3 - z)}{5} = -z$

38. $\dfrac{4(5 - w)}{3} = -w$

39. $-3(2t - 5) + 2t = 5t - 4$

40. $-(4a - 7) - 5a = 10 + a$

41. $5y + 2(y - 6) = 4(y + 1) - 2$

42. $9x + 3(x - 4) = 10(x - 5) + 7$

43. $\dfrac{3(x - 5)}{2} = \dfrac{2(x + 5)}{3}$

44. $\dfrac{5(x - 1)}{4} = \dfrac{3(x + 1)}{2}$

45. $0.7x - 2.3 = 0.5$

46. $0.9x - 4.1 = 0.4$

▶ **47.** $5x - 5 = 2(x + 1) + 3x - 7$

48. $3(2x - 1) + 5 = 6x + 2$

49. $4(2n + 1) = 3(6n + 3) + 1$

50. $4(4y + 2) = 2(1 + 6y) + 8$

51. $x + \dfrac{5}{4} = \dfrac{3}{4}x$

52. $\dfrac{7}{8}x + \dfrac{1}{4} = \dfrac{3}{4}x$

▶ **53.** $\dfrac{x}{2} - 1 = \dfrac{x}{5} + 2$

54. $\dfrac{x}{5} - 7 = \dfrac{x}{3} - 5$

55. $2(x + 3) - 5 = 5x - 3(1 + x)$

56. $4(2 + x) + 1 = 7x - 3(x - 2)$

57. $0.06 - 0.01(x + 1) = -0.02(2 - x)$

58. $-0.01(5x + 4) = 0.04 - 0.01(x + 4)$

59. $\dfrac{9}{2} + \dfrac{5}{2}y = 2y - 4$

60. $3 - \dfrac{1}{2}x = 5x - 8$

61. $\dfrac{3}{4}x - 1 + \dfrac{1}{2}x = \dfrac{5}{12}x + \dfrac{1}{6}$

62. $\dfrac{5}{9}x + 2 - \dfrac{1}{6}x = \dfrac{11}{18}x + \dfrac{1}{3}$

63. $3x + \dfrac{5}{16} = \dfrac{3}{4} - \dfrac{1}{8}x - \dfrac{1}{2}$

64. $2x - \dfrac{1}{10} = \dfrac{2}{5} - \dfrac{1}{4}x - \dfrac{17}{20}$

Review

Translating *Write each algebraic expression described. See Section 8.7. Recall that the perimeter of a figure is the total distance around the figure.*

65. A plot of land is in the shape of a triangle. If one side is x meters, a second side is $(2x - 3)$ meters, and a third side is $(3x - 5)$ meters, express the perimeter of the lot as a simplified expression in x.

66. A portion of a board has length x feet. The other part has length $(7x - 9)$ feet. Express the total length of the board as a simplified expression in x.

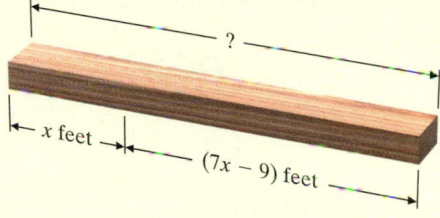

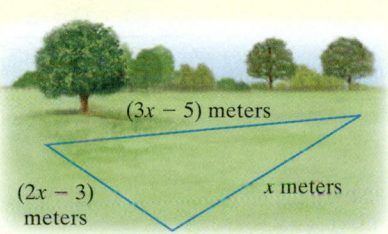

Translating *Write each phrase as an algebraic expression. Use x for the unknown number. See Section 8.7.*

67. A number subtracted from -8

68. Three times a number

69. The sum of -3 and twice a number

70. The difference of 8 and twice a number

71. The product of 9 and the sum of a number and 20

72. The quotient of -12 and the difference of a number and 3

Concept Extensions

See the Concept Check in this section.

73. a. Solve: $x + 3 = x + 3$

  b. If you simplify an equation (such as the one in part a) and get a true statement such as $3 = 3$ or $0 = 0$, what can you conclude about the solution(s) of the original equation?

  c. On your own, construct an equation for which every real number is a solution.

74. a. Solve: $x + 3 = x + 5$

  b. If you simplify an equation (such as the one in part a) and get a false statement such as $3 = 5$ or $10 = 17$, what can you conclude about the solution(s) of the original equation?

  c. On your own, construct an equation that has no solution.

Match each equation in the first column with its solution in the second column. Items in the second column may be used more than once.

75. $5x + 1 = 5x + 1$

76. $3x + 1 = 3x + 2$

77. $2x - 6x - 10 = -4x + 3 - 10$

78. $x - 11x - 3 = -10x - 1 - 2$

79. $9x - 20 = 8x - 20$

80. $-x + 15 = x + 15$

a. all real numbers

b. no solution

c. 0

81. Explain the difference between simplifying an expression and solving an equation.

82. On your own, write an expression and then an equation. Label each.

*For Exercises 83 and 84, **a.** Write an equation for perimeter. (Recall that the perimeter of a geometric figure is the sum of the lengths of its sides.) **b.** Solve the equation in part (a). **c.** Find the length of each side.*

△ **83.** The perimeter of the following pentagon (five-sided figure) is 28 centimeters.

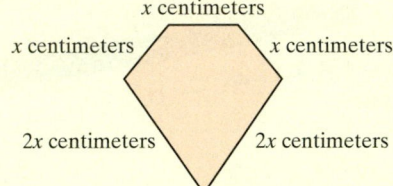

x centimeters

x centimeters x centimeters

$2x$ centimeters $2x$ centimeters

△ **84.** The perimeter of the following triangle is 35 meters.

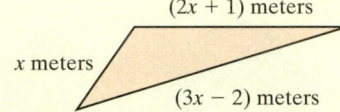

$(2x + 1)$ meters

x meters

$(3x - 2)$ meters

Fill in the blanks with numbers of your choice so that each equation has the given solution. Note: Each blank will be replaced by a different number.

85. $x +$ ____ $= 2x -$ ____ ; solution: 9

86. $-5x -$ ____ $=$ ____ ; solution: 2

Solve.

87. $1000(7x - 10) = 50(412 + 100x)$

88. $1000(x + 40) = 100(16 + 7x)$

89. $0.035x + 5.112 = 0.010x + 5.107$

90. $0.127x - 2.685 = 0.027x - 2.38$

Solving Linear Equations

Solve. Feel free to use the steps given in Section 9.3.

1. $x - 10 = -4$

2. $y + 14 = -3$

3. $9y = 108$

4. $-3x = 78$

5. $-6x + 7 = 25$

6. $5y - 42 = -47$

7. $\dfrac{2}{3}x = 9$

8. $\dfrac{4}{5}z = 10$

9. $\dfrac{r}{-4} = -2$

10. $\dfrac{y}{-8} = 8$

11. $6 - 2x + 8 = 10$

12. $-5 - 6y + 6 = 19$

13. $2x - 7 = 6x - 27$

14. $3 + 8y = 3y - 2$

15. $9(3x - 1) = -4 + 49$

16. $12(2x + 1) = -6 + 66$

17. $-3a + 6 + 5a = 7a - 8a$

18. $4b - 8 - b = 10b - 3b$

19. $-\dfrac{2}{3}x = \dfrac{5}{9}$

20. $-\dfrac{3}{8}y = -\dfrac{1}{16}$

21. $10 = -6n + 16$

22. $-5 = -2m + 7$

1. _____

2. _____

3. _____

4. _____

5. _____

6. _____

7. _____

8. _____

9. _____

10. _____

11. _____

12. _____

13. _____

14. _____

15. _____

16. _____

17. _____

18. _____

19. _____

20. _____

21. _____

22. _____

23. _____

24. _____

25. _____

26. _____

27. _____

28. _____

29. _____

30. _____

31. _____

32. _____

33. _____

34. _____

35. _____

36. _____

37. _____

38. _____

23. $3(5c - 1) - 2 = 13c + 3$

24. $4(3t + 4) - 20 = 3 + 5t$

25. $\dfrac{2(z + 3)}{3} = 5 - z$

26. $\dfrac{3(w + 2)}{4} = 2w + 3$

27. $-2(2x - 5) = -3x + 7 - x + 3$

28. $-4(5x - 2) = -12x + 4 - 8x + 4$

29. $0.02(6t - 3) = 0.04(t - 2) + 0.02$

30. $0.03(m + 7) = 0.02(5 - m) + 0.03$

31. $-3y = \dfrac{4(y - 1)}{5}$

32. $-4x = \dfrac{5(1 - x)}{6}$

33. $\dfrac{5}{3}x - \dfrac{7}{3} = x$

34. $\dfrac{7}{5}n + \dfrac{3}{5} = -n$

35. $\dfrac{1}{10}(3x - 7) = \dfrac{3}{10}x + 5$

36. $\dfrac{1}{7}(2x - 5) = \dfrac{2}{7}x + 1$

37. $5 + 2(3x - 6) = -4(6x - 7)$

38. $3 + 5(2x - 4) = -7(5x + 2)$

Further Problem Solving

First, let's review a list of key words and phrases from Section 8.2 to help us translate.

Helpful Hint

Order matters when subtracting and also dividing, so be especially careful with these translations.

Addition (+)	Subtraction (−)	Multiplication (·)	Division (÷)	Equality (=)
Sum	Difference of	Product	Quotient	Equals
Plus	Minus	Times	Divide	Gives
Added to	Subtracted from	Multiply	Into	Is/was/should be
More than	Less than	Twice	Ratio	Yields
Increased by	Decreased by	Of	Divided by	Amounts to
Total	Less			Represents
				Is the same as

We are now ready to put all our translating skills to practical use. To begin, we present a general strategy for problem solving.

General Strategy for Problem Solving

1. UNDERSTAND the problem. During this step, become comfortable with the problem. Some ways of doing this are:

   Read and reread the problem.

   Choose a variable to represent the unknown.

   Construct a drawing.

   Propose a solution and check. Pay careful attention to how you check your proposed solution. This will help when writing an equation to model the problem.

2. TRANSLATE the problem into an equation.
3. SOLVE the equation.
4. INTERPRET the results: *Check* the proposed solution in the stated problem and *state* your conclusion.

Objective A Solving Direct Translation Problems

Much of problem solving involves a direct translation from a sentence to an equation.

Example 1 Finding an Unknown Number

Twice a number, added to seven, is the same as three subtracted from the number. Find the number.

Solution: Translate the sentence into an equation and solve.

In words:	twice a number	added to	seven	is the same as	three subtracted from the number
	↓	↓	↓	↓	↓
Translate:	$2x$	$+$	7	$=$	$x - 3$

(Continued on next page)

Objectives

A Solve Problems Involving Direct Translations.

B Solve Problems Involving Relationships Among Unknown Quantities.

C Solve Problems Involving Consecutive Integers.

Practice 1

Three times a number, minus 6, is the same as two times the number, plus 3. Find the number.

Answer
1. The number is 9.

To solve, begin by subtracting x from both sides to isolate the variable term.

$$2x + 7 = x - 3$$

$2x + 7 - x = x - 3 - x$	Subtract x from both sides.
$x + 7 = -3$	Combine like terms.
$x + 7 - 7 = -3 - 7$	Subtract 7 from both sides.
$x = -10$	Combine like terms.

Check the solution in the problem as it was originally stated. To do so, replace "number" in the sentence with -10. Twice "-10" added to 7 is the same as 3 subtracted from "-10."

$$2(-10) + 7 = -10 - 3$$
$$-13 = -13$$

The unknown number is -10.

■ **Work Practice 1**

Helpful Hint

When checking solutions, go back to the original stated problem rather than to your equation in case errors have been made in translating to an equation.

Practice 2

Three times the difference of a number and 5 is the same as twice the number decreased by 3. Find the number.

Example 2   Finding an Unknown Number

Twice the sum of a number and 4 is the same as four times the number decreased by 12. Find the number.

Solution:

1. **UNDERSTAND.** Read and reread the problem. If we let x = the unknown number, then

   "the sum of a number and 4" translates to "$x + 4$" and
   "four times the number" translates to "$4x$"

2. **TRANSLATE.**

twice	sum of a number and 4	is the same as	four times the number	decreased by	12
↓	↓	↓	↓	↓	↓
2	$(x + 4)$	$=$	$4x$	$-$	12

3. **SOLVE**

$2(x + 4) = 4x - 12$	
$2x + 8 = 4x - 12$	Apply the distributive property.
$2x + 8 - 4x = 4x - 12 - 4x$	Subtract $4x$ from both sides.
$-2x + 8 = -12$	
$-2x + 8 - 8 = -12 - 8$	Subtract 8 from both sides.
$-2x = -20$	
$\dfrac{-2x}{-2} = \dfrac{-20}{-2}$	Divide both sides by -2.
$x = 10$	

4. **INTERPRET.**

Check: Check this solution in the problem as it was originally stated. To do so, replace "number" with 10. Twice the sum of "10" and 4 is 28, which is the same as 4 times "10" decreased by 12.

State: The number is 10.

■ **Work Practice 2**

Answer

2. The number is 12.

Objective B Solving Problems Involving Relationships Among Unknown Quantities

Example 3 Finding the Length of a Board

A 10-foot board is to be cut into two pieces so that the length of the longer piece is 4 times the length of the shorter. Find the length of each piece.

Solution:

1. UNDERSTAND the problem. To do so, read and reread the problem. You may also want to propose a solution. For example, if 3 feet represents the length of the shorter piece, then $4(3) = 12$ feet is the length of the longer piece, since it is 4 times the length of the shorter piece. This guess gives a total board length of 3 feet + 12 feet = 15 feet, which is too long. However, the purpose of proposing a solution is not to guess correctly, but to help better understand the problem and how to model it.

In general, if we let

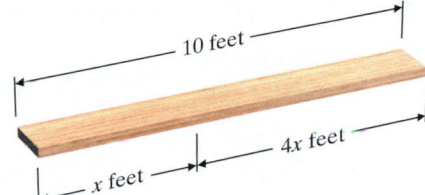

x = length of shorter piece, then
$4x$ = length of longer piece

2. TRANSLATE the problem. First, we write the equation in words.

length of shorter piece	added to	length of longer piece	equals	total length of board
↓	↓	↓	↓	↓
x	$+$	$4x$	$=$	10

3. SOLVE.

$$x + 4x = 10$$
$$5x = 10 \quad \text{Combine like terms.}$$
$$\frac{5x}{5} = \frac{10}{5} \quad \text{Divide both sides by 5.}$$
$$x = 2$$

4. INTERPRET.

Check: Check the solution in the stated problem. If the length of the shorter piece of board is 2 feet, the length of the longer piece is $4 \cdot (2 \text{ feet}) = 8$ feet and the sum of the lengths of the two pieces is 2 feet + 8 feet = 10 feet.

State: The shorter piece of board is 2 feet and the longer piece of board is 8 feet.

▪ **Work Practice 3**

> **Helpful Hint**
>
> Make sure that units are included in your answer, if appropriate.

Example 4 Finding the Number of Republican and Democratic Senators

As of May 2014, the total number of Democrats and Republicans in the U.S. House of Representatives was 432. There were 34 more Republican representatives than Democratic. Find the number of representatives from each party. (*Source:* Office of the Clerk, U.S. Capitol)

(Continued on next page)

Practice 3

An 18-foot wire is to be cut so that the length of the longer piece is 5 times the length of the shorter piece. Find the length of each piece.

Practice 4

Through the year 2020, the state of California will have 17 more electoral votes for president than the state of Texas. If the total electoral votes for these two states is 93, find the number of electoral votes for each state. (*Source*: U.S. Census Bureau)

Answers

3. shorter piece: 3 feet; longer piece: 15 feet   **4.** Texas: 38 electoral votes; California: 55 electoral votes

Solution:

1. UNDERSTAND the problem. Read and reread the problem. Let's suppose that there are 100 Democratic representatives. Since there are 34 more Republicans than Democrats, there must be $100 + 34 = 134$ Republicans. The total number of Republicans and Democrats is then $134 + 100 = 234$. This is incorrect since the total should be 432, but we now have a better understanding of the problem.

In general, if we let

$$x = \text{number of Democrats, then}$$
$$x + 34 = \text{number of Republicans}$$

2. TRANSLATE the problem. First, we write the equation in words.

number of Democrats	added to	number of Republicans	equals	432
↓	↓	↓	↓	↓
x	$+$	$(x + 34)$	$=$	432

3. SOLVE.

$$x + (x + 34) = 432$$
$$2x + 34 = 432 \qquad \text{Combine like terms.}$$
$$2x + 34 - 34 = 432 - 34 \qquad \text{Subtract 34 from both sides.}$$
$$2x = 398$$
$$\frac{2x}{2} = \frac{398}{2} \qquad \text{Divide both sides by 2.}$$
$$x = 199$$

4. INTERPRET.

Check: If there were 199 Democratic representatives, then there were $199 + 34 = 233$ Republican representatives. The total number of representatives is then $199 + 233 = 432$. The results check.

State: There were 199 Democratic and 233 Republican representatives in the U.S. House of Representatives.

■ Work Practice 4

Practice 5

A car rental agency charges $28 a day and $0.15 a mile. If you rent a car for a day and your bill (before taxes) is $52, how many miles did you drive?

Example 5 Calculating Hours on the Job

A computer science major at a local university has a part-time job working on computers for his clients. He charges $20 to come to your home or office and then $25 per hour. During one month he visited 10 homes or offices and his total income was $575. How many hours did he spend working on computers?

Solution:

1. UNDERSTAND. Read and reread the problem. Let's propose that the student spent 20 hours working on computers. Pay careful attention as to how his income is calculated. For 20 hours and 10 visits, his income is $20(\$25) + 10(\$20) = \$700$, which is more than $575. We now have a better understanding of the problem and know that the time working on computers is less than 20 hours.

Let's let

$$x = \text{hours working on computers. Then}$$
$$25x = \text{amount of money made while working on computers}$$

Answer

5. 160 miles

2. TRANSLATE.

money made while working on computers	plus	money made for visits	is equal to	575
↓	↓	↓	↓	↓
$25x$	$+$	$10(20)$	$=$	575

3. SOLVE.

$$25x + 200 = 575$$

$$25x + 200 - 200 = 575 - 200 \quad \text{Subtract 200 from both sides.}$$

$$25x = 375 \quad \text{Simplify.}$$

$$\frac{25x}{25} = \frac{375}{25} \quad \text{Divide both sides by 25.}$$

$$x = 15 \quad \text{Simplify.}$$

4. INTERPRET.

Check: If the student works 15 hours and makes 10 visits, his income is $15(\$25) + 10(\$20) = \$575$.

State: The student spent 15 hours working on computers.

◼ **Work Practice 5**

Example 6   Finding Angle Measures

If the two walls of the Vietnam Veterans Memorial in Washington, D.C., were connected, an isosceles triangle would be formed. The measure of the third angle is 97.5° more than the measure of either of the two equal angles. Find the measure of the third angle. (*Source:* National Park Service)

Solution:

1. UNDERSTAND. Read and reread the problem. We then draw a diagram (recall that an isosceles triangle has two angles with the same measure) and let

$$x = \text{degree measure of one angle}$$
$$x = \text{degree measure of the second equal angle}$$
$$x + 97.5 = \text{degree measure of the third angle}$$

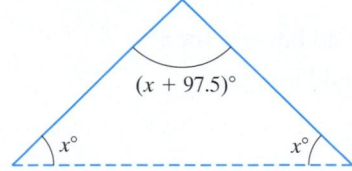

2. TRANSLATE. Recall that the sum of the measures of the angles of a triangle equals 180.

measure of first angle	+	measure of second angle	+	measure of third angle	equal	180
↓		↓		↓	↓	↓
x	$+$	x	$+$	$(x + 97.5)$	$=$	180

(Continued on next page)

Practice 6

The measure of the second angle of a triangle is twice the measure of the smallest angle. The measure of the third angle of the triangle is three times the measure of the smallest angle. Find the measures of the angles.

Answer

6. smallest: 30°; second: 60°; third: 90°

3. SOLVE.

$$x + x + (x + 97.5) = 180$$

$$3x + 97.5 = 180 \quad \text{Combine like terms.}$$

$$3x + 97.5 - 97.5 = 180 - 97.5 \quad \text{Subtract 97.5 from both sides.}$$

$$3x = 82.5$$

$$\frac{3x}{3} = \frac{82.5}{3} \quad \text{Divide both sides by 3.}$$

$$x = 27.5$$

4. INTERPRET.

Check: If $x = 27.5$, then the measure of the third angle is $x + 97.5 = 125$. The sum of the angles is then $27.5 + 27.5 + 125 = 180$, the correct sum.

State: The third angle measures $125°$.*

■ **Work Practice 6**

Objective C Solving Consecutive Integer Problems

The next example has to do with consecutive integers. Recall what we have learned thus far about these integers.

	Example	General Representation	
Consecutive Integers	11, 12, 13 $\underset{+1\quad +1}{\frown\frown}$	Let x be an integer.	$x,\ x + 1,\ x + 2$ $\underset{+1\quad +1}{\frown\frown}$
Consecutive Even Integers	38, 40, 42 $\underset{+2\quad +2}{\frown\frown}$	Let x be an even integer.	$x,\ x + 2,\ x + 4$ $\underset{+2\quad +2}{\frown\frown}$
Consecutive Odd Integers	57, 59, 61 $\underset{+2\quad +2}{\frown\frown}$	Let x be an odd integer.	$x,\ x + 2,\ x + 4$ $\underset{+2\quad +2}{\frown\frown}$

The next example has to do with consecutive integers.

Practice 7

The sum of three consecutive even integers is 144. Find the integers.

Helpful Hint Remember, the 2 here means that odd integers are 2 units apart, for example, the odd integers 13 and $13 + 2 = 15$.

Example 7 Some states have a single area code for the entire state. Two such states have area codes that are consecutive odd integers. If the sum of these integers is 1208, find the two area codes. (*Source: World Almanac*)

Solution:

1. UNDERSTAND. Read and reread the problem. If we let

$$x = \text{the first odd integer, then}$$

$$x + 2 = \text{the next odd integer}$$

2. TRANSLATE.

first odd integer	added to	next odd integer	is	1208
↓	↓	↓		
x	$+$	$(x + 2)$	$=$	1208

Answer

7. 46, 48, 50

*The two walls actually meet at an angle of 125 degrees 12 minutes. The measurement of $97.5°$ given in the problem is an approximation.

3. SOLVE.

$$x + x + 2 = 1208$$
$$2x + 2 = 1208$$
$$2x + 2 - 2 = 1208 - 2$$
$$2x = 1206$$
$$\frac{2x}{2} = \frac{1206}{2}$$
$$x = 603$$

4. INTERPRET.

Check: If $x = 603$, then the next odd integer $x + 2 = 603 + 2 = 605$. Notice their sum, $603 + 605 = 1208$, as needed.

State: The area codes are 603 and 605.

Note: New Hampshire's area code is 603 and South Dakota's area code is 605.

■ **Work Practice 7**

Vocabulary, Readiness & Video Check

Fill in the table.

1.	A number: x	→ Double the number:	→ Double the number, decreased by 31:
2.	A number: x	→ Three times the number:	→ Three times the number, increased by 17:
3.	A number: x	→ The sum of the number and 5:	→ Twice the sum of the number and 5:
4.	A number: x	→ The difference of the number and 11:	→ Seven times the difference of a number and 11:
5.	A number: y	→ The difference of 20 and the number:	→ The difference of 20 and the number, divided by 3:
6.	A number: y	→ The sum of -10 and the number:	→ The sum of -10 and the number, divided by 9:

Martin-Gay Interactive Videos   *Watch the section lecture video and answer the following questions.*

Objective A **7.** At the end of ▥ Example 1, where are you told is the best place to check the solution of an application problem? ⊙

Objective B **8.** The solution of the equation for ▥ Example 3 is $x = 43$. Why is this not the solution to the application? ⊙

Objective C **9.** What are two things that should be checked to make sure the solution of ▥ Example 4 is correct? ⊙

See Video 9.4 🍎

9.4 Exercise Set MyMathLab®

Objective A *Solve. For Exercises 1 through 4, write each of the following as equations. Then solve. See Examples 1 and 2.*

1. The sum of twice a number and 7 is equal to the sum of the number and 6. Find the number.

2. The difference of three times a number and 1 is the same as twice the number. Find the number.

3. Three times a number, minus 6, is equal to two times the number, plus 8. Find the number.

4. The sum of 4 times a number and −2 is equal to the sum of 5 times the number and −2. Find the number.

5. Twice the difference of a number and 8 is equal to three times the sum of the number and 3. Find the number.

6. Five times the sum of a number and −1 is the same as 6 times the number. Find the number.

7. The product of twice a number and three is the same as the difference of five times the number and $\frac{3}{4}$. Find the number.

8. If the difference of a number and four is doubled, the result is $\frac{1}{4}$ less than the number. Find the number.

Objective B *Solve. For Exercises 9 and 10, the solutions have been started for you. See Examples 3 and 4.*

9. A 25-inch piece of steel is cut into three pieces so that the second piece is twice as long as the first piece, and the third piece is one inch more than five times the length of the first piece. Find the lengths of the pieces.

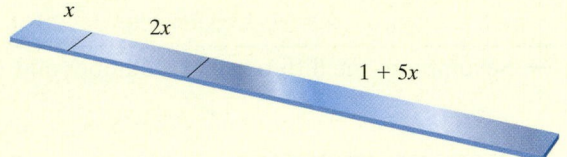

Start the solution:

1. UNDERSTAND the problem. Reread it as many times as needed.

2. TRANSLATE into an equation. (Fill in the blanks below.)

total length of steel	equals	length of first piece	plus	length of second piece	plus	length of third piece
↓	↓	↓	↓	↓	↓	↓
25	=	___	+	___	+	___

Finish with:

3. SOLVE and **4.** INTERPRET

10. A 46-foot piece of rope is cut into three pieces so that the second piece is three times as long as the first piece, and the third piece is two feet more than seven times the length of the first piece. Find the lengths of the pieces.

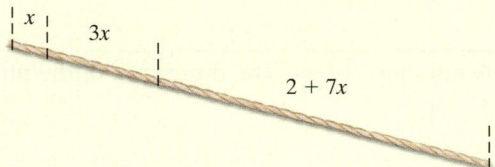

Start the solution:

1. UNDERSTAND the problem. Reread it as many times as needed.

2. TRANSLATE into an equation. (Fill in the blanks below.)

total length of rope	equals	length of first piece	plus	length of second piece	plus	length of third piece
↓	↓	↓	↓	↓	↓	↓
46	=	___	+	___	+	___

Finish with:

3. SOLVE and **4.** INTERPRET

11. A 40-inch board is to be cut into three pieces so that the second piece is twice as long as the first piece and the third piece is 5 times as long as the first piece. If x represents the length of the first piece, find the lengths of all three pieces.

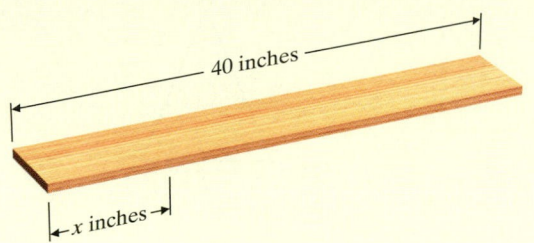

40 inches

x inches

12. A 21-foot beam is to be divided so that the longer piece is 1 foot more than 3 times the length of the shorter piece. If x represents the length of the shorter piece, find the lengths of both pieces.

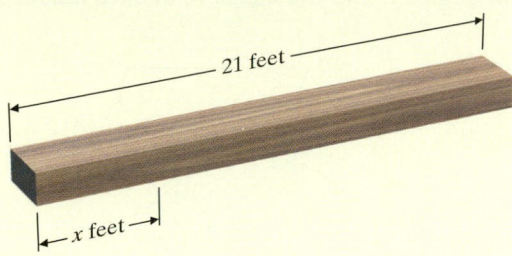

21 feet

x feet

13. In 2012, New York produced 226 million pounds more apples than Pennsylvania. Together, the two states produced 1214 million pounds of apples. Find the amount of apples grown in New York and Pennsylvania in 2012. (*Source:* National Agriculture Statistics Service)

14. In the 2012 Summer Olympics in London, the U.S. team won 8 more gold medals than the Chinese team. If the total number of gold medals won by both teams was 84, find the number of gold medals won by each team. (*Source:* International Olympic Committee)

Solve. See Example 5.

15. A car rental agency advertised renting a Buick Century for $24.95 per day and $0.29 per mile. If you rent this car for 2 days, how many whole miles can you drive on a $100 budget?

16. A plumber gave an estimate for the renovation of a kitchen. Her hourly pay is $27 per hour and the plumbing parts will cost $80. If her total estimate is $404, how many hours does she expect this job to take?

17. In one U.S. city, the taxi cost is $3 plus $0.80 per mile. If you are traveling from the airport, there is an additional charge of $4.50 for tolls. How far can you travel from the airport by taxi for $27.50?

18. A professional carpet cleaning service charges $30 plus $25.50 per hour to come to your home. If your total bill from this company is $119.25 before taxes, for how many hours were you charged?

Solve. See Example 6.

△ **19.** The flag of Equatorial Guinea contains an isosceles triangle. (Recall that an isosceles triangle contains two angles with the same measure.) If the measure of the third angle of the triangle is 30° more than twice the measure of either of the other two angles, find the measure of each angle of the triangle. (*Hint:* Recall that the sum of the measures of the angles of a triangle is 180°.)

△ **20.** The flag of Brazil contains a parallelogram. One angle of the parallelogram is 15° less than twice the measure of the angle next to it. Find the measure of each angle of the parallelogram. (*Hint:* Recall that opposite angles of a parallelogram have the same measure and that the sum of the measures of the angles is 360°.)

21. The sum of the measures of the angles of a parallelogram is 360°. In the parallelogram below, angles *A* and *D* have the same measure as well as angles *C* and *B*. If the measure of angle *C* is twice the measure of angle *A*, find the measure of each angle.

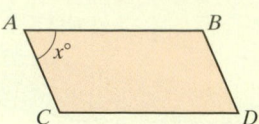

22. Recall that the sum of the measures of the angles of a triangle is 180°. In the triangle below, angle *C* has the same measure as angle *B*, and angle *A* measures 42° less than angle *B*. Find the measure of each angle.

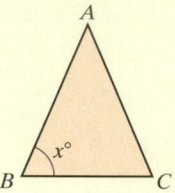

Objective C   *Solve. See Example 7. Fill in the table. Most of the first row has been completed for you.*

23. Three consecutive integers:

24. Three consecutive integers:

25. Three consecutive even integers:

26. Three consecutive odd integers:

27. Four consecutive integers:

28. Four consecutive integers:

29. Three consecutive odd integers:

30. Three consecutive even integers:

First Integer →	Next Integers	→	Indicated Sum
Integer: x	$x + 1$	$x + 2$	Sum of the three consecutive integers, simplified:
Integer: x			Sum of the second and third consecutive integers, simplified:
Even integer: x			Sum of the first and third even consecutive integers, simplified:
Odd integer: x			Sum of the three consecutive odd integers, simplified:
Integer: x			Sum of the four consecutive integers, simplified:
Integer: x			Sum of the first and fourth consecutive integers, simplified:
Odd integer: x			Sum of the second and third consecutive odd integers, simplified:
Even integer: x			Sum of the three consecutive even integers, simplified:

Solve. See Example 7.

31. The left and right page numbers of an open book are two consecutive integers whose sum is 469. Find these page numbers.

32. The room numbers of two adjacent classrooms are two consecutive even numbers. If their sum is 654, find the classroom numbers.

33. To make an international telephone call, you need the code for the country you are calling. The codes for Belgium, France, and Spain are three consecutive integers whose sum is 99. Find the code for each country. (*Source: The World Almanac and Book of Facts*)

34. The code to unlock a student's combination lock happens to be three consecutive odd integers whose sum is 51. Find the integers.

Objectives A B C Mixed Practice *Solve. See Examples 1 through 7.*

35. A 17-foot piece of string is cut into two pieces so that the longer piece is 2 feet longer than twice the length of the shorter piece. Find the lengths of both pieces.

36. A 25-foot wire is to be cut so that the longer piece is one foot longer than 5 times the length of the shorter piece. Find the length of each piece.

37. Currently, the two fastest trains are the Japanese Maglev and the French TGV. The sum of their fastest speeds is 718.2 miles per hour. If the speed of the Maglev is 3.8 mph faster than the speed of the TGV, find the speeds of each.

38. The Pentagon is the world's largest office building in terms of floor space. It has three times the amount of floor space as the Empire State Building. If the total floor space for these two buildings is approximately 8700 thousand square feet, find the floor space of each building.

39. Two angles are supplementary if their sum is 180°. The larger angle below measures eight degrees more than three times the measure of the smaller angle. If x represents the measure of the smaller angle and these two angles are supplementary, find the measure of each angle.

40. Two angles are complementary if their sum is 90°. Given the measures of the complementary angles shown, find the measure of each angle.

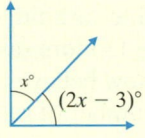

41. The measures of the angles of a triangle are 3 consecutive even integers. Find the measure of each angle.

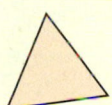

42. A quadrilateral is a polygon with 4 sides. The sum of the measures of the 4 angles in a quadrilateral is 360°. If the measures of the angles of a quadrilateral are consecutive odd integers, find the measures.

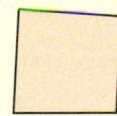

43. The sum of $\frac{1}{5}$ and twice a number is equal to $\frac{4}{5}$ subtracted from three times the number. Find the number.

44. The sum of $\frac{2}{3}$ and four times a number is equal to $\frac{5}{6}$ subtracted from five times the number. Find the number.

45. Hertz Car Rental charges a daily rate of $39 plus $0.20 per mile for a certain car. Suppose that you rent that car for a day and your bill (before taxes) is $95. How many miles did you drive?

46. A woman's $15,000 estate is to be divided so that her husband receives twice as much as her son. Find the amount of money that her husband receives and the amount of money that her son receives.

47. During the 2013 Rose Bowl, Stanford University beat University of Wisconsin by 6 points. If their combined scores totaled 34, find the individual team scores. (*Source:* Tournament of Roses Association)

48. In June 2013, there were 10 more Republican governors than Democratic governors in the United States. How many Democrats and how many Republicans held governors' offices at that point in time?

49. The number of counties in California and the number of counties in Montana are consecutive even integers whose sum is 114. If California has more counties than Montana, how many counties does each state have? (*Source: The World Almanac and Book of Facts*)

50. A student is building a bookcase with stepped shelves for her dorm room. She buys a 48-inch board and wants to cut the board into three pieces with lengths equal to three consecutive even integers. Find the three board lengths.

51. Scientists are continually updating information about the planets in our solar system, including the number of satellites orbiting each. Uranus is now believed to have 13 more satellites than Neptune. Also, Saturn is now believed to have 6 more than four times the number of satellites of Neptune. If the total number of satellites for these planets is 103, find the number of satellites for each planet. (*Source: National Space Science Data Center*)

52. Apple's iPad Mini tablet computer was launched in 2012. The height of each iPad Mini is 70 millimeters less than twice the width. If the sum of the height and width of an iPad Mini is 335 millimeters, find each dimension. (*Source:* Apple Inc.)

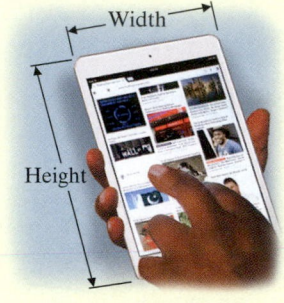

53. If the sum of a number and five is tripled, the result is one less than twice the number. Find the number.

54. Twice the sum of a number and six equals three times the sum of the number and four. Find the number.

55. The area of the Sahara Desert is 7 times the area of the Gobi Desert. If the sum of their areas is 4,000,000 square miles, find the area of each desert.

56. The largest meteorite in the world is the Hoba West, located in Namibia. Its weight is 3 times the weight of the Armanty meteorite, located in Outer Mongolia. If the sum of their weights is 88 tons, find the weight of each.

57. In the 2012 Summer Olympics in London, New Zealand won more gold medals than Cuba, which won more gold medals than Jamaica. If the number of gold medals won by these three countries is three consecutive integers whose sum is 15, find the number of gold medals won by each. (*Source: International Olympic Committee*)

58. To make an international telephone call, you need the code for the country you are calling. The codes for Mali Republic, Côte d'Ivoire, and Niger are three consecutive odd integers whose sum is 675. Find the code for each country.

59. In the fall of 2012, there were 1580 more female students enrolled at Rutgers University than male students. If the total student enrollment at Rutgers was 58,788 that fall, find the numbers of female students and male students who were enrolled. (*Source:* Rutgers, The State University of New Jersey)

60. The middle-sized car category saw the greatest percentage increase in sales of any vehicle category in the United States in 2012 over the previous year. Approximately 1.1 million fewer pickups were sold than middle-sized cars in 2012. If the total number of pickups and middle-sized cars sold was 4.9 million, find the number of vehicles sold in each category. (*Source:* Alliance of Automobile Manufacturers)

61. A geodesic dome, based on the design by Buckminster Fuller, is composed of two different types of triangular panels. One of these is an isosceles triangle. In one geodesic dome, the measure of the third angle is 76.5° more than the measure of either of the two equal angles. Find the measure of the three angles. (*Source:* Buckminster Fuller Institute)

62. The measures of the angles of a particular triangle are such that the second and third angles are each four times the measure of the smallest angle. Find the measures of the angles of this triangle.

The graph below shows the states with the highest tourist budgets for the 2012–2013 fiscal year. Use this graph for Exercises 63 through 68.

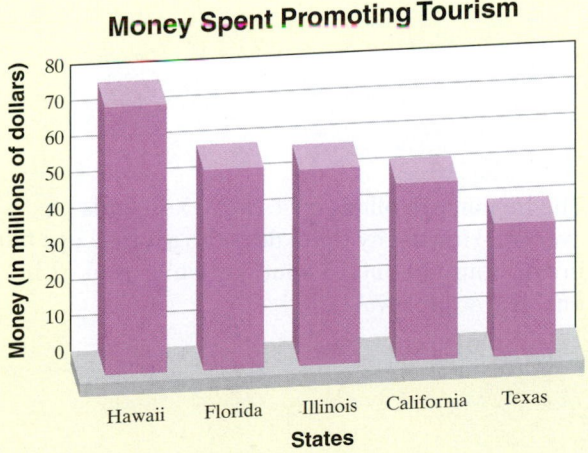

Money Spent Promoting Tourism

Money (in millions of dollars) / States

Source: U.S. Travel Association

63. Which state spent the most money on tourism?

64. Which state(s) spent between $30 million and $40 million on tourism?

65. The states of Florida and California spent a total of $106 million on tourism. The state of Florida spent $6 million more than the state of California. Find the amount that each state spent on tourism.

66. The states of Illinois and Texas spent a total of $92 million on tourism. The state of Illinois spent $19 million less than twice the amount of money that the state of Texas spent. Find the amount that each state spent on tourism.

Compare the heights of the bars in the graph on the previous page with your results for the exercises below. Are your answers reasonable?

67. Exercise 65

68. Exercise 66

Review

Evaluate each expression for the given values. See Section 8.7.

69. $2W + 2L$; $W = 7$ and $L = 10$

70. $\frac{1}{2}Bh$; $B = 14$ and $h = 22$

71. πr^2; $r = 15$

72. $r \cdot t$; $r = 15$ and $t = 2$

Concept Extensions

△ **73.** A golden rectangle is a rectangle whose length is approximately 1.6 times its width. The early Greeks thought that a rectangle with these dimensions was the most pleasing to the eye and examples of the golden rectangle are found in many early works of art. For example, the Parthenon in Athens contains many examples of golden rectangles.

Mike Hallahan would like to plant a rectangular garden in the shape of a golden rectangle. If he has 78 feet of fencing available, find the dimensions of the garden.

△ **74.** Dr. Dorothy Smith gave the students in her geometry class at the University of New Orleans the following question. Is it possible to construct a triangle such that the second angle of the triangle has a measure that is twice the measure of the first angle and the measure of the third angle is 5 times the measure of the first? If so, find the measure of each angle. (*Hint:* Recall that the sum of the measures of the angles of a triangle is 180°.)

75. Only male crickets chirp. They chirp at different rates depending on their species and the temperature of their environment. Suppose a certain species is currently chirping at a rate of 90 chirps per minute. At this rate, how many chirps occur in one hour? In one 24-hour day? In one year?

76. The human eye blinks once every 5 seconds on average. How many times does the average eye blink in one hour? In one 16-hour day while awake? In one year while awake?

77. In your own words, explain why a solution of a word problem should be checked using the original wording of the problem and not the equation written from the wording.

78. Give an example of how you recently solved a problem using mathematics.

Recall from Exercise 73 that a golden rectangle is a rectangle whose length is approximately 1.6 times its width.

 79. It is thought that for about 75% of adults, a rectangle in the shape of the golden rectangle is the most pleasing to the eye. Draw three rectangles, one in the shape of the golden rectangle, and poll your class. Do the results agree with the percentage given above?

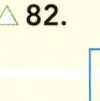 **80.** Examples of golden rectangles can be found today in architecture and manufacturing packaging. Find an example of a golden rectangle in your home. A few suggestions: the front face of a book, the floor of a room, the front of a box of food.

For Exercises 81 and 82, measure the dimensions of each rectangle and decide which one best approximates the shape of a golden rectangle.

81.

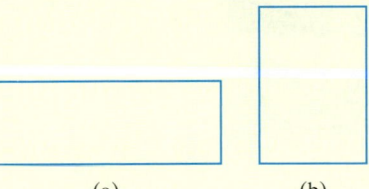

   (a)   (b)   (c)

 82.

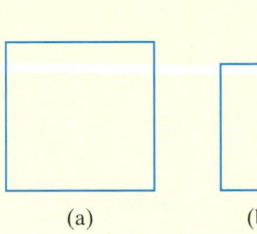

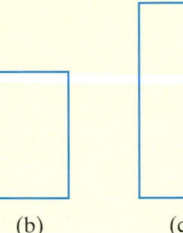

        (a)   (b)   (c)

9.5 Formulas and Problem Solving

Objective A Using Formulas to Solve Problems

A **formula** describes a known relationship among quantities. Many formulas are given as equations. For example, the formula

$$d = r \cdot t$$

stands for the relationship

$$\text{distance} = \text{rate} \cdot \text{time}$$

Let's look at one way that we can use this formula.

   If we know we traveled a distance of 100 miles at a rate of 40 miles per hour, we can replace the variables d and r in the formula $d = rt$ and find our travel time, t.

$d = rt$ Formula

$100 = 40t$ Replace d with 100 and r with 40.

To solve for t, we divide both sides of the equation by 40.

$\dfrac{100}{40} = \dfrac{40t}{40}$ Divide both sides by 40.

$\dfrac{5}{2} = t$ Simplify.

The travel times was $\dfrac{5}{2}$ hours, or $2\dfrac{1}{2}$ hours, or 2.5 hours.

   In this section, we solve problems that can be modeled by known formulas. We use the same problem-solving strategy that was introduced in the previous section.

Objectives

A Use Formulas to Solve Problems.

B Solve a Formula or Equation for One of Its Variables.

Practice 1

A family is planning their vacation to visit relatives. They will drive from Cincinnati, Ohio, to Rapid City, South Dakota, a distance of 1180 miles. They plan to average a rate of 50 miles per hour. How much time will they spend driving?

Example 1   Finding Time Given Rate and Distance

A glacier is a giant mass of rocks and ice that flows downhill like a river. Portage Glacier in Alaska is about 6 miles, or 31,680 *feet*, long and moves 400 *feet* per year. Icebergs are created when the front end of the glacier flows into Portage Lake. How long does it take for ice at the head (beginning) of the glacier to reach the lake?

Solution:

1. **UNDERSTAND.** Read and reread the problem. The appropriate formula needed to solve this problem is the distance formula, $d = rt$. To become familiar with this formula, let's find the distance that ice traveling at a rate of 400 feet per year travels in 100 years. To do so, we let time t be 100 years and rate r be the given 400 feet per year, and substitute these values into the formula $d = rt$. We then have that distance $d = 400(100) = 40,000$ feet. Since we are interested in finding how long it takes ice to travel 31,680 feet, we now know that it is less than 100 years.

   Since we are using the formula $d = rt$, we let

   t = the time in years for ice to reach the lake

   r = rate or speed of ice

   d = distance from beginning of glacier to lake

2. **TRANSLATE.** To translate to an equation, we use the formula $d = rt$ and let distance $d = 31,680$ feet and rate $r = 400$ feet per year.

   $$d = r \cdot t$$
   $$31{,}680 = 400 \cdot t \quad \text{Let } d = 31{,}680 \text{ and } r = 400.$$

3. **SOLVE.** Solve the equation for t. To solve for t, we divide both sides by 400.

   $$\frac{31{,}680}{400} = \frac{400 \cdot t}{400} \quad \text{Divide both sides by 400.}$$
   $$79.2 = t \quad \text{Simplify.}$$

4. **INTERPRET.**

   Check: To check, substitute 79.2 for t and 400 for r in the distance formula and check to see that the distance is 31,680 feet.

   State: It takes 79.2 years for the ice at the head of Portage Glacier to reach the lake.

◼ Work Practice 1

> **Helpful Hint**
> Don't forget to include units, if appropriate.

Answer

1. 23.6 hours

△ **Example 2** Calculating the Length of a Garden

Charles Pecot can afford enough fencing to enclose a rectangular garden with a perimeter of 140 feet. If the width of his garden is to be 30 feet, find the length.

$w = 30$ feet

l

△ **Practice 2**

A wood deck is being built behind a house. The width of the deck must be 18 feet because of the shape of the house. If there is 450 square feet of decking material, find the length of the deck.

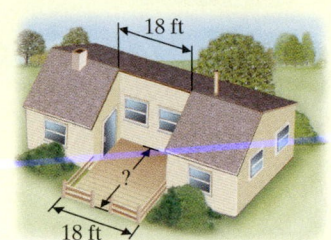

18 ft

18 ft

Solution:

1. **UNDERSTAND.** Read and reread the problem. The formula needed to solve this problem is the formula for the perimeter of a rectangle, $P = 2l + 2w$. Before continuing, let's become familar with this formula.

   $l = $ the length of the rectangular garden

   $w = $ the width of the rectangular garden

   $P = $ perimeter of the garden

2. **TRANSLATE.** To translate to an equation, we use the formula $P = 2l + 2w$ and let perimeter $P = 140$ feet and width $w = 30$ feet.

   $$P = 2l + 2w \qquad \text{Let } P = 140 \text{ and } w = 30.$$
   $$140 = 2l + 2(30)$$

3. **SOLVE.**

   $$140 = 2l + 2(30)$$
   $$140 = 2l + 60 \qquad \text{Multiply } 2(30).$$
   $$140 - 60 = 2l + 60 - 60 \qquad \text{Subtract 60 from both sides.}$$
   $$80 = 2l \qquad \text{Combine like terms.}$$
   $$40 = l \qquad \text{Divide both sides by 2.}$$

4. **INTERPRET.**

Check: Substitute 40 for l and 30 for w in the perimeter formula and check to see that the perimeter is 140 feet.

State: The length of the rectangular garden is 40 feet.

■ **Work Practice 2**

Example 3 Finding an Equivalent Temperature

The average maximum temperature for the month of January in Algiers, Algeria, is 59° Fahrenheit. Find the equivalent temperature in degrees Celsius.

Solution:

1. **UNDERSTAND.** Read and reread the problem. A formula that can be used to solve this problem is the formula for converting degrees Celsius to degrees Fahrenheit, $F = \dfrac{9}{5}C + 32$. Before continuing, become familiar with this formula. Using this formula, we let

   $C = $ temperature in degrees Celsius, and

   $F = $ temperature in degrees Fahrenheit.

(Continued on next page)

Practice 3

Convert the temperature 5°C to Fahrenheit.

2. TRANSLATE. To translate to an equation, we use the formula $F = \frac{9}{5}C + 32$ and let degrees Fahrenheit $F = 59$.

Formula: $F = \frac{9}{5}C + 32$

Substitute: $59 = \frac{9}{5}C + 32$ Let $F = 59$.

3. SOLVE.

$$59 = \frac{9}{5}C + 32$$

$$59 - 32 = \frac{9}{5}C + 32 - 32 \quad \text{Subtract 32 from both sides.}$$

$$27 = \frac{9}{5}C \quad \text{Combine like terms.}$$

$$\frac{5}{9} \cdot 27 = \frac{5}{9} \cdot \frac{9}{5}C \quad \text{Multiply both sides by } \frac{5}{9}.$$

$$15 = C \quad \text{Simplify.}$$

4. INTERPRET.

Check: To check, replace C with 15 and F with 59 in the formula and see that a true statement results.

State: Thus, $59°$ Fahrenheit is equivalent to $15°$ Celsius.

■ **Work Practice 3**

In the next example, we again use the formula for perimeter of a rectangle as in Example 2. In Example 2, we knew the width of the rectangle. In this example, both the length and width are unknown.

Practice 4

The length of a rectangle is one meter more than 4 times its width. Find the dimensions if the perimeter is 52 meters.

△ **Example 4**   Finding Road Sign Dimensions

The length of a rectangular road sign is 2 feet less than three times its width. Find the dimensions if the perimeter is 28 feet.

Solution:

1. UNDERSTAND. Read and reread the problem. Recall that the formula for the perimeter of a rectangle is $P = 2l + 2w$. Draw a rectangle and guess the solution. If the width of the rectangular sign is 5 feet, its length is 2 feet less than 3 times the width, or $3(5 \text{ feet}) - 2 \text{ feet} = 13 \text{ feet}$. The perimeter P of the rectangle is then $2(13 \text{ feet}) + 2(5 \text{ feet}) = 36 \text{ feet}$, too much. We now know that the width is less than 5 feet.

Proposed rectangle:

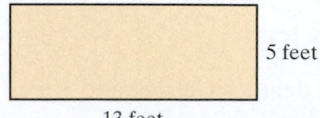

5 feet

13 feet

Answer

4. length: 21 m; width: 5 m

Let

   w = the width of the rectangular sign; then

 $3w - 2$ = the length of the sign.

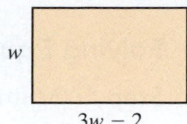

$3w - 2$

Draw a rectangle and label it with the assigned variables.

2. **TRANSLATE.**

  Formula: $P = 2l + 2w$

  Substitute: $28 = 2(3w - 2) + 2w$

3. **SOLVE.**

$$28 = 2(3w - 2) + 2w$$
$$28 = 6w - 4 + 2w \qquad \text{Apply the distributive property.}$$
$$28 = 8w - 4$$
$$28 + 4 = 8w - 4 + 4 \qquad \text{Add 4 to both sides.}$$
$$32 = 8w$$
$$\frac{32}{8} = \frac{8w}{8} \qquad \text{Divide both sides by 8.}$$
$$4 = w$$

4. **INTERPRET.**

Check: If the width of the sign is 4 feet, the length of the sign is $3(4 \text{ feet}) - 2 \text{ feet} = 10$ feet. This gives the rectangular sign a perimeter of $P = 2(4 \text{ feet}) + 2(10 \text{ feet}) = 28$ feet, the correct perimeter.

State: The width of the sign is 4 feet and the length of the sign is 10 feet.

■ **Work Practice 4**

Objective B Solving a Formula for a Variable

We say that the formula

 $d = rt$

is solved for d because d is alone on one side of the equation and the other side contains no d's. Suppose that we have a large number of problems to solve where we are given distance d and rate r and asked to find time t. In this case, it may be easier to first solve the formula $d = rt$ for t. To solve for t, we divide both sides of the equation by r.

 $d = rt$

 $\dfrac{d}{r} = \dfrac{rt}{r}$ Divide both sides by r.

 $\dfrac{d}{r} = t$ Simplify.

To solve a formula or an equation for a specified variable, we use the same steps as for solving a linear equation except that we treat the specified variable as the only variable in the equation. These steps are listed next.

> **Solving Equations for a Specified Variable**
>
> **Step 1:** Multiply on both sides to clear the equation of fractions if they appear.
>
> **Step 2:** Use the distributive property to remove parentheses if they appear.
>
> **Step 3:** Simplify each side of the equation by combining like terms.
>
> **Step 4:** Get all terms containing the specified variable on one side and all other terms on the other side by using the addition property of equality.
>
> **Step 5:** Get the specified variable alone by using the multiplication property of equality.

Practice 5

Solve $C = 2\pi r$ for r.
(This formula is used to find the circumference, C, of a circle given its radius, r.)

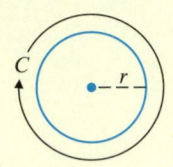

Example 5 Solve $V = lwh$ for l.

Solution: This formula is used to find the volume of a box. To solve for l, divide both sides by wh.

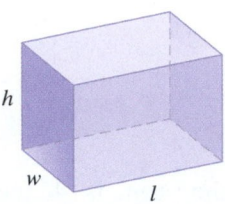

$$V = lwh$$

$$\frac{V}{wh} = \frac{lwh}{wh} \quad \text{Divide both sides by } wh.$$

$$\frac{V}{wh} = l \quad \text{Simplify.}$$

Since we have l alone on one side of the equation, we have solved for l in terms of V, w, and h. Remember that it does not matter on which side of the equation we get the variable alone.

■ **Work Practice 5**

Practice 6

Solve $P = 2l + 2w$ for l.

Example 6 Solve $y = mx + b$ for x.

Solution: First we get mx alone by subtracting b from both sides.

$$y = mx + b$$

$$y - b = mx + b - b \quad \text{Subtract } b \text{ from both sides.}$$

$$y - b = mx \quad \text{Combine like terms.}$$

Next we solve for x by dividing both sides by m.

$$\frac{y - b}{m} = \frac{mx}{m}$$

$$\frac{y - b}{m} = x \quad \text{Simplify.}$$

■ **Work Practice 6**

Answers

5. $r = \dfrac{C}{2\pi}$ 6. $l = \dfrac{P - 2w}{2}$

✔ **Concept Check Answer**

a.  b.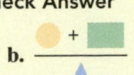

✔ **Concept Check** Solve:

a.  for ▨ b.   for ▨

Example 7 Solve $P = 2l + 2w$ for w.

Solution: This formula relates the perimeter of a rectangle to its length and width. Find the term containing the variable w. To get this term, $2w$, alone, subtract $2l$ from both sides.

$$P = 2l + 2w$$
$$P - 2l = 2l + 2w - 2l \quad \text{Subtract } 2l \text{ from both sides.}$$
$$P - 2l = 2w \quad \text{Combine like terms.}$$
$$\frac{P - 2l}{2} = \frac{2w}{2} \quad \text{Divide both sides by 2.}$$
$$\frac{P - 2l}{2} = w \quad \text{Simplify.}$$

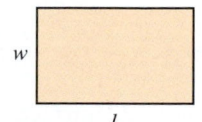

■ **Work Practice 7**

Practice 7

Solve $P = 2a + b - c$ for a.

Helpful Hint The 2s may *not* be divided out here. Although 2 is a factor of the denominator, 2 is *not* a factor of the numerator since it is not a factor of both terms in the numerator.

The next example has an equation containing a fraction. We will first clear the equation of fractions and then solve for the specified variable.

Example 8 Solve $F = \frac{9}{5}C + 32$ for C.

Solution:
$$F = \frac{9}{5}C + 32$$
$$5(F) = 5\left(\frac{9}{5}C + 32\right) \quad \text{Clear the fraction by multiplying both sides by the LCD.}$$
$$5F = 9C + 160 \quad \text{Distribute the 5.}$$
$$5F - 160 = 9C + 160 - 160 \quad \text{To get the term containing the variable } C \text{ alone, subtract 160 from both sides.}$$
$$5F - 160 = 9C \quad \text{Combine like terms.}$$
$$\frac{5F - 160}{9} = \frac{9C}{9} \quad \text{Divide both sides by 9.}$$
$$\frac{5F - 160}{9} = C \quad \text{Simplify.}$$

■ **Work Practice 8**

Practice 8

Solve $A = \frac{a + b}{2}$ for b.

Answers

7. $a = \dfrac{P - b + c}{2}$ **8.** $b = 2A - a$

Vocabulary, Readiness & Video Check

Martin-Gay Interactive Videos

See Video 9.5

Watch the section lecture video and answer the following questions.

Objective A **1.** Complete this statement based on the lecture given before ⊞ Example 1. A formula is an equation that describes known _____ among quantities. ▶

2. In ⊞ Example 2, how are the units for the solution determined? ▶

Objective B **3.** During ⊞ Example 4, the equation $5x = 30$ is shown to demonstrate what process? ▶

9.5 Exercise Set MyMathLab®

Objective A *Substitute the given values into each given formula and solve for the unknown variable. See Examples 1 through 4.*

△ **1.** $A = bh$; $A = 45, b = 15$ (Area of a parallelogram)

2. $d = rt$; $d = 195, t = 3$ (Distance formula)

△ **3.** $S = 4lw + 2wh$; $S = 102, l = 7, w = 3$ (Surface area of a special rectangular box)

△ **4.** $V = lwh$; $l = 14, w = 8, h = 3$ (Volume of a rectangular box)

△ **5.** $A = \frac{1}{2}h(B + b)$; $A = 180, B = 11, b = 7$ (Area of a trapezoid)

△ **6.** $A = \frac{1}{2}h(B + b)$; $A = 60, B = 7, b = 3$ (Area of a trapezoid)

△ **7.** $P = a + b + c$; $P = 30, a = 8, b = 10$ (Perimeter of a triangle)

△ **8.** $V = \frac{1}{3}Ah$; $V = 45, h = 5$ (Volume of a pyramid)

△ **9.** $C = 2\pi r$; $C = 15.7$ (Circumference of a circle) (Use the approximation 3.14 for π.)

△ **10.** $A = \pi r^2$; $r = 4$ (Area of a circle) (Use the approximation 3.14 for π.)

Objective B *Solve each formula for the specified variable. See Examples 5 through 8.*

11. $f = 5gh$ for h

△ **12.** $x = 4\pi y$ for y

△ **13.** $V = lwh$ for w

14. $T = mnr$ for n

15. $3x + y = 7$ for y

16. $-x + y = 13$ for y

17. $A = P + PRT$ for R

18. $A = P + PRT$ for T

△ **19.** $V = \frac{1}{3}Ah$ for A

20. $D = \frac{1}{4}fk$ for k

△ **21.** $P = a + b + c$ for a

22. $PR = x + y + z + w$ for z

△ **23.** $S = 2\pi rh + 2\pi r^2$ for h

△ **24.** $S = 4lw + 2wh$ for h

Objective A *Solve. For Exercises 25 and 26, the solutions have been started for you. See Examples 1 through 4.*

△ **25.** The iconic NASDAQ sign in New York's Times Square has a width of 84 feet and an area of 10,080 square feet. Find the height (or length) of the sign. (*Source:* livedesignonline.com)

Start the solution:

1. UNDERSTAND the problem. Reread it as many times as needed.
2. TRANSLATE into an equation. (Fill in the blanks below.)

Area	=	length	times	width
↓	↓	↓	↓	↓
___	=	x	·	___

Finish with:

3. SOLVE and **4.** INTERPRET

△ **26.** The world's largest sign for Coca-Cola is located in Arica, Chile. The rectangular sign has a length of 400 feet and an area of 52,400 square feet. Find the width of the sign. (*Source:* Fabulous Facts about Coca-Cola, Atlanta, GA)

Start the solution:

1. UNDERSTAND the problem. Reread it as many times as needed.
2. TRANSLATE into an equation. (Fill in the blanks below.)

Area	=	length	times	width
↓	↓	↓	↓	↓
___	=	___	·	x

Finish with:

3. SOLVE and **4.** INTERPRET

△ **27.** A frame shop charges according to both the amount of framing needed to surround the picture and the amount of glass needed to cover the picture.

  a. Find the area and perimeter of the picture below.
  b. Identify whether the frame has to do with perimeter or area and the same with the glass.

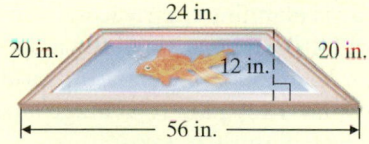

△ **28.** A decorator is painting and placing a border completely around the parallelogram-shaped wall.

  a. Find the area and perimeter of the wall below. ($A = bh$)
  b. Identify whether the border has to do with perimeter or area and the same with paint.

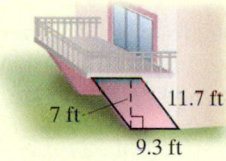

△ **29.** For the purpose of purchasing new baseboard and carpet,

  a. Find the area and perimeter of the room below (neglecting doors).
  b. Identify whether baseboard has to do with area or perimeter and the same with carpet.

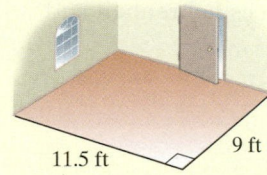

△ **30.** For the purpose of purchasing lumber for a new fence and seed to plant grass,

  a. Find the area and perimeter of the yard below.

$$\left(A = \frac{1}{2}bh\right)$$

  b. Identify whether a fence has to do with area or perimeter and the same with grass seed.

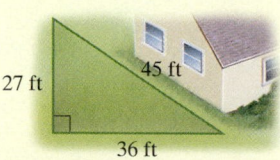

▶ **31.** Convert Nome, Alaska's 14°F high temperature to Celsius.

32. Convert Paris, France's low temperature of −5°C to Fahrenheit.

33. The X-30 is a "space plane" that skims the edge of space at 4000 miles per hour. Neglecting altitude, if the circumference of Earth is approximately 25,000 miles, how long will it take for the X-30 to travel around Earth?

34. In the United States, a notable hang glider flight was a 303-mile, $8\frac{1}{2}$-hour flight from New Mexico to Kansas. What was the average rate during this flight?

35. An architect designs a rectangular flower garden such that the width is exactly two-thirds of the length. If 260 feet of antique picket fencing are to be used to enclose the garden, find the dimensions of the garden.

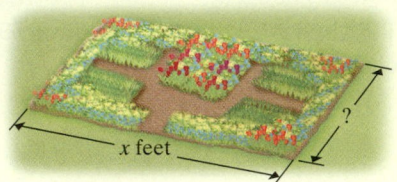

x feet

36. If the length of a rectangular parking lot is 10 meters less than twice its width, and the perimeter is 400 meters, find the length of the parking lot.

x meters

37. A flower bed is in the shape of a triangle with one side twice the length of the shortest side, and the third side is 30 feet more than the length of the shortest side. Find the dimensions if the perimeter is 102 feet.

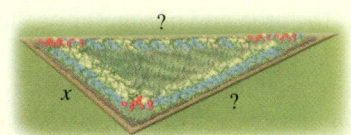

?

x

?

38. The perimeter of a yield sign in the shape of an isosceles triangle is 22 feet. If the shortest side is 2 feet less than the other two sides, find the length of the shortest side. (*Hint:* An isosceles triangle has two sides the same length.)

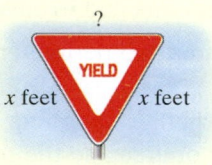

?

x feet   *x* feet

39. The Cat is a high-speed catamaran auto ferry that operates between Bar Harbor, Maine, and Yarmouth, Nova Scotia. The Cat can make the trip in about $2\frac{1}{2}$ hours at a speed of 55 mph. About how far apart are Bar Harbor and Yarmouth? (*Source:* Bay Ferries)

40. A family is planning their vacation to Disney World. They will drive from a small town outside New Orleans, Louisiana, to Orlando, Florida, a distance of 700 miles. They plan to average a rate of 55 mph. How long will this trip take?

Dolbear's Law states the relationship between the rate at which Snowy Tree Crickets chirp and the air temperature of their environment. The formula is

$$T = 50 + \frac{N - 40}{4}, \text{ where}$$

$T =$ temperature in degrees Fahrenheit and
$N =$ number of chirps per minute

41. If $N = 86$, find the temperature in degrees Fahrenheit, T.

42. If $N = 94$, find the temperature in degrees Fahrenheit, T.

43. If $T = 55°F$, find the number of chirps per minute.

44. If $T = 65°F$, find the number of chirps per minute.

Use the results of Exercises 41–44 to complete each sentence with "increases" or "decreases."

45. As the number of cricket chirps per minute increases, the air temperature of their environment _____.

46. As the air temperature of their environment decreases, the number of cricket chirps per minute _____.

Solve. See Examples 1 through 4.

△ **47.** Piranha fish require 1.5 cubic feet of water per fish to maintain a healthy environment. Find the maximum number of piranha you could put in a tank measuring 8 feet by 3 feet by 6 feet.

6 feet

3 feet   8 feet

△ **48.** Find the maximum number of goldfish you can put in a cylindrical tank whose diameter is 8 meters and whose height is 3 meters, if each goldfish needs 2 cubic meters of water. ($V = \pi r^2 h$)

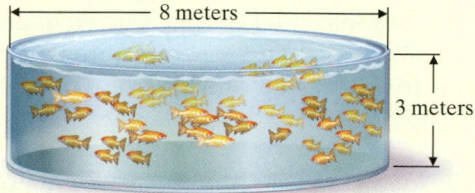

8 meters

3 meters

△ **49.** A lawn is in the shape of a trapezoid with a height of 60 feet and bases of 70 feet and 130 feet. How many bags of fertilizer must be purchased to cover the lawn if each bag covers 4000 square feet?
$$\left(A = \frac{1}{2}h(B + b) \right)$$

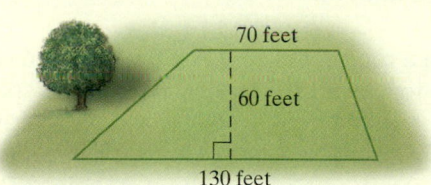

70 feet

60 feet

130 feet

△ **50.** If the area of a right-triangularly shaped sail is 20 square feet and its base is 5 feet, find the height of the sail. $\left(A = \frac{1}{2}bh \right)$

?

5 feet

△ **51.** Maria's Pizza sells one 16-inch cheese pizza or two 10-inch cheese pizzas for $9.99. Determine which size gives more pizza. ($A = \pi r^2$)

16 inches   10 inches   10 inches

△ **52.** Find how much rope is needed to wrap around Earth at the equator, if the radius of Earth is 4000 miles. (*Hint:* Use 3.14 for π and the formula for circumference.)

53. A Japanese "bullet" train set a new world record for train speed at 552 kilometers per hour during a manned test run on the Yamanashi Maglev Test Line in April 1999. The Yamanashi Maglev Test Line is 42.8 kilometers long. How many *minutes* would a test run on the Yamanashi Line last at this record-setting speed? Round to the nearest hundredth of a minute. (*Source:* Japan Railways Central Co.)

54. In 1983, the Hawaiian volcano Kilauea began erupting in a series of episodes still occurring at the time of this writing. At times, the lava flows advanced at speeds of up to 0.5 kilometer per hour. In 1983 and 1984 lava flows destroyed 16 homes in the Royal Gardens subdivision, about 6 km away from the eruption site. Roughly how long did it take the lava to reach Royal Gardens? (*Source:* U.S. Geological Survey Hawaiian Volcano Observatory)

△ **55.** The perimeter of an equilateral triangle is 7 inches more than the perimeter of a square, and the side of the triangle is 5 inches longer than the side of the square. Find the side of the triangle. (*Hint:* An equilateral triangle has three sides the same length.)

△ **56.** A square animal pen and a pen shaped like an equilateral triangle have equal perimeters. Find the length of the sides of each pen if the sides of the triangular pen are fifteen less than twice a side of the square pen. (*Hint:* An equilateral triangle has three sides the same length.)

57. Find how long it takes Tran Nguyen to drive 135 miles on I-10 if he merges onto I-10 at 10 a.m. and drives nonstop with his cruise control set on 60 mph.

58. Beaumont, Texas, is about 150 miles from Toledo Bend. If Leo Miller leaves Beaumont at 4 a.m. and averages 45 mph, when should he arrive at Toledo Bend?

△ **59.** The longest runway at Los Angeles International Airport has the shape of a rectangle and an area of 1,813,500 square feet. This runway is 150 feet wide. How long is the runway? (*Source:* Los Angeles World Airports)

60. The return stroke of a bolt of lightning can travel at a speed of 87,000 miles per second (almost half the speed of light). At this speed, how many times can an object travel around the world in one second? (See Exercise 52.) Round to the nearest tenth. (*Source: The Handy Science Answer Book*)

61. The highest temperature ever recorded in Europe was 122°F in Seville, Spain, in August of 1881. Convert this record high temperature to Celsius. (*Source:* National Climatic Data Center)

62. The lowest temperature ever recorded in Oceania was −10°C at the Haleakala Summit in Maui, Hawaii, in January 1961. Convert this record low temperature to Fahrenheit. (*Source:* National Climatic Data Center)

△ **63.** The CART FedEx Championship Series is an open-wheeled race car competition based in the United States. A CART car has a maximum length of 199 inches, a maximum width of 78.5 inches, and a maximum height of 33 inches. When the CART series travels to another country for a grand prix, teams must ship their cars. Find the volume of the smallest shipping crate needed to ship a CART car of maximum dimensions. (*Source:* Championship Auto Racing Teams, Inc.)

64. On a road course, a CART car's speed can average up to around 105 mph. Based on this speed, how long would it take a CART driver to travel from Los Angeles to New York City, a distance of about 2810 miles by road, without stopping? Round to the nearest tenth of an hour.

CART Racing Car

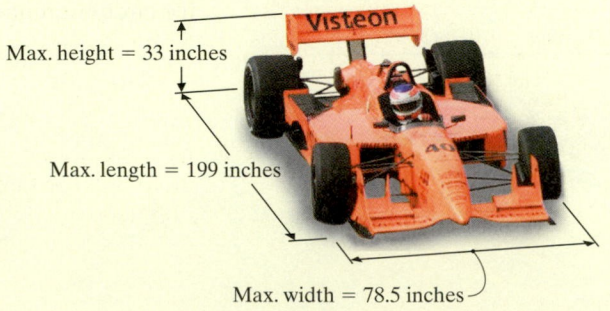

Max. height = 33 inches

Max. length = 199 inches

Max. width = 78.5 inches

△ **65.** The Hoberman Sphere is a toy ball that expands and contracts. When it is completely closed, it has a diameter of 9.5 inches. Find the volume of the Hoberman Sphere when it is completely closed. Use 3.14 for π. Round to the nearest whole cubic inch. (*Hint:* volume of a sphere $= \frac{4}{3}\pi r^3$. *Source:* Hoberman Designs, Inc.)

△ **66.** When the Hoberman Sphere (see Exercise 65) is completely expanded, its diameter is 30 inches. Find the volume of the Hoberman Sphere when it is completely expanded. Use 3.14 for π. (*Source:* Hoberman Designs, Inc.)

67. The average temperature on the planet Mercury is 167°C. Convert this temperature to degrees Fahrenheit. Round to the nearest degree. (*Source:* National Space Science Data Center)

68. The average temperature on the planet Jupiter is −227°F. Convert this temperature to degrees Celsius. Round to the nearest degree. (*Source:* National Space Science Data Center)

Review

Write each percent as a decimal. See Section 5.2.

69. 32%   **70.** 8%   **71.** 200%   **72.** 0.5%

Write each decimal as a percent. See Section 5.2.

73. 0.17   **74.** 0.03   **75.** 7.2   **76.** 5

Concept Extensions

Solve.

77. $N = R + \dfrac{V}{G}$ for V (Urban forestry: tree plantings per year)

78. $B = \dfrac{F}{P - V}$ for V (Business: break-even point)

△ **79.** The formula $V = lwh$ is used to find the volume of a box. If the length of a box is doubled, the width is doubled, and the height is doubled, how does this affect the volume? Explain your answer.

△ **80.** The formula $A = bh$ is used to find the area of a parallelogram. If the base of a parallelogram is doubled and its height is doubled, how does this affect the area? Explain your answer.

81. Use the Dolbear's Law formula for Exercises 41–46 and calculate when the number of cricket chirps per minute is the same as the temperature in degrees Fahrenheit. (*Hint:* Replace T with N and solve for N or replace N with T and solve for T.)

82. Find the temperature at which the Celsius measurement and the Fahrenheit measurement are the same number.

Solve. See the Concept Check in this section.

83. △ − ● · ▮ = ▮ for ●

84. ⬠ · ▮ + △ = ● for ▮

85. Flying fish do not *actually* fly, but glide. They have been known to travel a distance of 1300 feet at a rate of 20 miles per hour. How many seconds would it take to travel this distance? (*Hint:* First convert miles per hour to feet per second. Recall that 1 mile = 5280 feet.) Round to the nearest tenth of a second.

86. A glacier is a giant mass of rocks and ice that flows downhill like a river. Exit Glacier, near Seward, Alaska, moves at a rate of 20 inches a day. Find the distance in feet the glacier moves in a year. (Assume 365 days a year.) Round to two decimal places.

Substitute the given values into each given formula and solve for the unknown variable. If necessary, round to one decimal place.

87. $I = PRT$; $I = 1,056,000, R = 0.055, T = 6$
(Simple interest formula)

88. $I = PRT$; $I = 3750, P = 25,000, R = 0.05$
(Simple interest formula)

89. $V = \frac{4}{3}\pi r^3$; $r = 3$ (Volume of a sphere)
(Use a calculator approximation for π.)

90. $V = \frac{1}{3}\pi r^2 h$; $V = 565.2, r = 6$ (Volume of a cone)
(Use a calculator approximation for π.)

9.6   Percent and Mixture Problem Solving

Objectives

A Solve Percent Equations.

B Solve Discount and Mark-Up Problems.

C Solve Percent of Increase and Percent of Decrease Problems.

D Solve Mixture Problems.

This section is devoted to solving problems in the categories listed. The same problem-solving steps used in previous sections are also followed in this section. They are listed below for review.

> ### General Strategy for Problem Solving
>
> 1. UNDERSTAND the problem. During this step, become comfortable with the problem. Some ways of doing this are as follows:
>   Read and reread the problem.
>   Choose a variable to represent the unknown.
>   Construct a drawing, whenever possible.
>   Propose a solution and check. Pay careful attention to how you check your proposed solution. This will help writing an equation to model the problem.
> 2. TRANSLATE the problem into an equation.
> 3. SOLVE the equation.
> 4. INTERPRET the results: *Check* the proposed solution in the stated problem and *state* your conclusion.

Objective A   Solving Percent Equations

Many of today's statistics are given in terms of percent: a basketball player's free throw percent, current interest rates, stock market trends, and nutrition labeling, just to name a few. In this section, we first explore percent, percent equations, and applications involving percents. See Chapter 5 if a further review of percents is needed.

2, 3. TRANSLATE and SOLVE.

$$\text{discount} = \text{percent} \cdot \text{original price}$$
$$= 20\% \cdot \$140$$
$$= 0.20 \cdot \$140$$
$$= \$28$$

Thus, the discount in price is $28.

$$\text{new price} = \text{original price} - \text{discount}$$
$$= \$140 - \$28$$
$$= \$112$$

4. INTERPRET.

Check: Check your calculations in the formulas, and also see if our results are reasonable. They are.

State: The discount in price is $28 and the new price is $112.

■ **Work Practice 4**

A concept similar to discount is mark-up. What is the difference between the two? A discount is subtracted from the original price while a mark-up is added to the original price. For mark-ups,

$$\text{mark-up} = \text{percent} \cdot \text{original price}$$

$$\text{new price} = \text{original price} + \text{mark-up}$$

Mark-up exercises can be found in Exercise Set 9.6.

Objective C Solving Percent of Increase and Percent of Decrease Problems ▶

Percent of increase or percent of decrease is a common way to describe how some measurement has increased or decreased. For example, crime increased by 8%, teachers received a 5.5% increase in salary, or a company decreased its employees by 10%. The next example is a review of percent of increase.

Example 5   Calculating the Percent of Increase of Attending College

The average cost of tuition and fees for attending a four-year public college rose from $4650 during the 1998–1999 academic year to $8890 during the 2013–2014 year. Find the percent of increase. (*Source:* The College Board)

Solution:

1. **UNDERSTAND. Read and reread the problem.** Notice that the new tuition, $8890, is almost double the old tuition of $4650. Because of that, we know that the percent of increase is close to 100%. To see this, let's guess that the percent of increase is 100%. To check, we find 100% of $4650 to find the *increase* in cost. Then we add this increase to $4650 to find the *new cost*. In other words, $100\%(\$4650) = 1.00(\$4650) = \$4650$, the *increase* in cost. The *new cost* would be old cost + increase = $4650 + $4650 = $9300, close to the actual new cost of $8890. We now know that the increase is close to, but less than, 100% and we know how to check our proposed solution.

Let x = the percent of increase.

(*Continued on next page*)

Practice 5

If a number increases from 120 to 200, find the percent of increase. Round to the nearest tenth of a percent.

Answer
5. 66.7%

2. **TRANSLATE.** First, find the **increase**, and then the **percent of increase.** The increase in cost is found by:

In words:	increase	=	new cost	−	old cost	or
Translate:	increase	=	$8890	−	$4650	
		=	$4240			

Next, find the percent of increase. The percent of increase or percent of decrease is always a percent of the original number or, in this case, the old cost.

In words:	increase	is	what percent	of	old cost
Translate:	$4240	=	x	·	$4650

3. **SOLVE.**

$$4240 = 4650x$$
$$0.912 \approx x \qquad \text{Divide both sides by 4650 and round to 3 decimal places.}$$
$$91.2\% \approx x \qquad \text{Write as a percent.}$$

4. **INTERPRET.**

Check: Check the proposed solution

State: The percent of increase in cost is approximately 91.2%.

■ **Work Practice 5**

Percent of decrease is found using a similar method. First find the decrease, then determine what percent of the original or first amount is that decrease.

Read the next example carefully. For Example 5, we were asked to find percent of increase. In Example 6, we are given the percent of increase and asked to find the number before the increase.

Practice 6

Find the original price of a suit if the sale price is $46 after a 20% discount.

Answer

6. $57.50

Example 6 Growth in digital 3-D theater screens is fastest in the Asia/Pacific entertainment market. Find the number of digital 3-D screens in Asia/Pacific in 2011 if, after a 106% increase, the number in 2013 was 17,726. Round to the nearest whole. (*Source:* MPAA)

Solution:

1. **UNDERSTAND.** Read and reread the problem. Let's guess a solution and see how we would check our guess. If the number of digital 3-D screens in 2011 was 8000, we would see if 8000 plus the increase is 17,726; that is,

$$8000 + 106\%(8000) = 8000 + 1.06(8000) = 8000 + 8480 = 16,480$$

Since 16,480 is too small, we know that our guess of 8000 is too small. We also have a better understanding of the problem. Let

x = number of digital 3-D screens in 2011

2. **TRANSLATE.** To translate an equation, we remember that

In words:	number of digital 3-D screens in 2011	plus	increase	equals	number of digital 3-D screens in 2013
Translate:	x	+	$1.06x$	=	17,726

3. **SOLVE.**

$$2.06x = 17,726$$
$$x = \frac{17,726}{2.06}$$
$$x \approx 8605$$

4. INTERPRET.

Check: Recall that x represents the number of digital 3-D screens in 2011. If this number is approximately 8605, let's see if 8605 plus the increase is close to 17,726. (We use the word "close" since 8605 is rounded.)

$$8605 + 106\%(8605) = 8605 + 1.06(8605) = 8605 + 9121.3 = 17{,}726.3$$

which is close to 17,726.

State: There were approximately 8605 digital 3-D screens in the Asia/Pacific region in 2011.

■ **Work Practice 6**

Objective D Solving Mixture Problems ▶

Mixture problems involve two or more different quantities being combined to form a new mixture. These applications range from Dow Chemical's need to form a chemical mixture of a required strength to Planter's Peanut Company's need to find the correct mixture of peanuts and cashews, given taste and price constraints.

Example 7   Calculating Percent for a Lab Experiment

A chemist working on his doctoral degree at Massachusetts Institute of Technology needs 12 liters of a 50% acid solution for a lab experiment. The stockroom has only 40% and 70% solutions. How much of each solution should be mixed together to form 12 liters of a 50% solution?

Solution:

1. **UNDERSTAND.** First, read and reread the problem a few times. Next, guess a solution. Suppose that we need 7 liters of the 40% solution. Then we need $12 - 7 = 5$ liters of the 70% solution. To see if this is indeed the solution, find the amount of pure acid in 7 liters of the 40% solution, in 5 liters of the 70% solution, and in 12 liters of a 50% solution, the required amount and strength.

number of liters	×	acid strength	=	amount of pure acid
7 liters	×	40%	=	$7(0.40)$ or 2.8 liters
5 liters	×	70%	=	$5(0.70)$ or 3.5 liters
12 liters	×	50%	=	$12(0.50)$ or 6 liters

Since 2.8 liters + 3.5 liters = 6.3 liters and not 6, our guess is incorrect, but we have gained some valuable insight into how to model and check this problem.
   Let

$$x = \text{number of liters of 40\% solution; then}$$
$$12 - x = \text{number of liters of 70\% solution.}$$

2. **TRANSLATE.** To help us translate to an equation, the following table summarizes the information given. Recall that the amount of acid in each solution is found by multiplying the acid strength of each solution by the number of liters.

	No. of Liters	·	Acid Strength	=	Amount of Acid
40% Solution	x		40%		$0.40x$
70% Solution	$12 - x$		70%		$0.70(12 - x)$
50% Solution Needed	12		50%		$0.50(12)$

The amount of acid in the final solution is the sum of the amounts of acid in the two beginning solutions.

In words: acid in 40% solution + acid in 70% solution = acid in 50% mixture

Translate: $0.40x$ + $0.70(12 - x)$ = $0.50(12)$

(Continued on next page)

Practice 7

How much 20% dye solution and 50% dye solution should be mixed to obtain 6 liters of a 40% solution?

x liters $(12 - x)$ liters

$(12 - x)$ liters + x liters

12 liters

40% solution   70% solution   50% solution

Answer

7. 2 liters of the 20% solution; 4 liters of the 50% solution

3. SOLVE.

$$0.40x + 0.70(12 - x) = 0.50(12)$$

$$0.4x + 8.4 - 0.7x = 6 \qquad \text{Apply the distributive property.}$$

$$-0.3x + 8.4 = 6 \qquad \text{Combine like terms.}$$

$$-0.3x = -2.4 \qquad \text{Subtract 8.4 from both sides.}$$

$$x = 8 \qquad \text{Divide both sides by } -0.3.$$

4. INTERPRET.

Check: To check, recall how we checked our guess.

State: If 8 liters of the 40% solution are mixed with 12 − 8 or 4 liters of the 70% solution, the result is 12 liters of a 50% solution.

■ **Work Practice 7**

Vocabulary, Readiness & Video Check

Tell whether the percent labels in the circle graphs are correct.

1.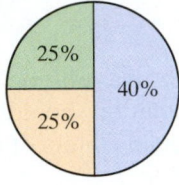

2.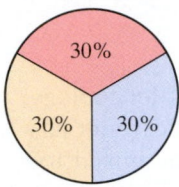

3.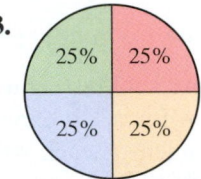

4.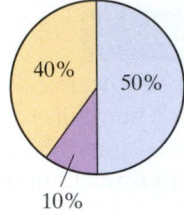

Martin-Gay Interactive Videos

See Video 9.6 🍎

Watch the section lecture video and answer the following questions.

Objective A **5.** Answer these questions based on how the ▣ Example 2 was translated to an equation.

    a. What does "is" translate to?

    b. What does "of" translate to?

    c. How do you write a percent as an equivalent decimal?

Objective B **6.** At the end of ▣ Example 3 you are told that the process for finding discount is *almost* the same as finding mark-up. ▶

    a. How is discount similar?

    b. How does discount differ?

Objective C **7.** According to ▣ Example 4, what amount must you find before you can find a percent of increase in price? How do you find this amount? ▶

Objective D **8.** The following problem is worded like ▣ Example 6 in the video, but using different quantities.

How much of an alloy that is 10% copper should be mixed with 400 ounces of an alloy that is 30% copper in order to get an alloy that is 20% copper? Fill in the table and set up an equation that could be used to solve for the unknowns (do not actually solve). Use ▣ Example 6 in the video as a model for your work. ▶

Alloy	Ounces	Copper Percent	Amount of Copper

9.6 Exercise Set MyMathLab®

Objective A *Find each number described. For Exercises 1 and 2, the solutions have been started for you. See Examples 1 and 2.*

1. What number is 16% of 70?

   Start the solution:

   1. UNDERSTAND the problem. Reread it as many times as needed.
   2. TRANSLATE into an equation. (Fill in the blanks below.)

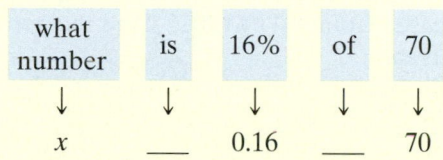

what number	is	16%	of	70
↓	↓	↓	↓	↓
x	___	0.16	___	70

   Finish with:

   3. SOLVE and
   4. INTERPRET

2. What number is 88% of 1000?

   Start the solution:

   1. UNDERSTAND the problem. Reread it as many times as needed.
   2. TRANSLATE into an equation. (Fill in the blanks below.)

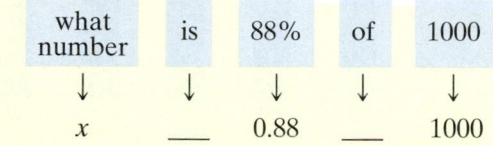

what number	is	88%	of	1000
↓	↓	↓	↓	↓
x	___	0.88	___	1000

   Finish with:

   3. SOLVE and
   4. INTERPRET

3. The number 28.6 is what percent of 52?

4. The number 87.2 is what percent of 436?

5. The number 45 is 25% of what number?

6. The number 126 is 35% of what number?

The circle graph below shows the types of accommodations that overnight visitors to national parks used in 2012. Use this graph for Exercises 7 through 10. See Example 3.

Overnight Stays at National Parks, 2012

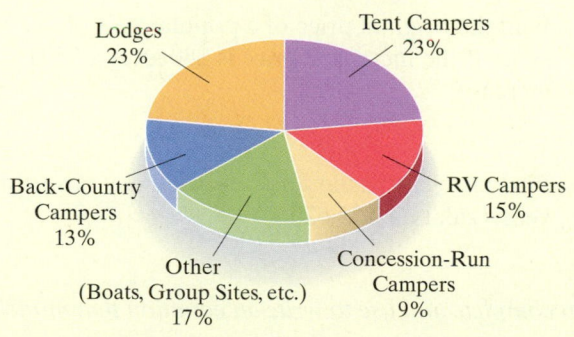

Lodges 23%
Tent Campers 23%
Back-Country Campers 13%
Other (Boats, Group Sites, etc.) 17%
Concession-Run Campers 9%
RV Campers 15%

Source: National Park Service

7. What percent of overnight stays were in RVs?

8. What percent of overnight stays involved tent camping?

9. In 2012, Yellowstone National Park reported approximately 1,350,000 overnight stays. How many of these stays might you expect were in lodges?

10. In 2012, Yosemite National Park reported approximately 1,732,000 overnight stays. How many of these stays might you expect involved back-country camping?

Objective B *Solve. If needed, round answers to the nearest cent. See Example 4.*

11. A used automobile dealership recently reduced the price of a used sports car by 8%. If the price of the car before discount was $18,500, find the discount and the new price.

12. A music store is advertising a 25%-off sale on all new releases. Find the discount and the sale price of a newly released CD that regularly sells for $12.50.

13. A birthday celebration meal is $40.50 including tax. Find the total cost if a 15% tip is added to the cost.

14. A retirement dinner for two is $65.40 including tax. Find the total cost if a 20% tip is added to the cost.

Objective C *Solve. Round percents to the nearest tenth. See Example 5.*

Use the graph below for Exercises 15 and 16.

Yearly Complaints Received by the Internet Crime Complaint Center

Source: Data from Internet Crime Complaint Center (www.ic3.gov)

15. The number of Internet-crime complaints decreased from 2012 to 2013. Find the percent of decrease.

16. The number of Internet-crime complaints decreased from 2011 to 2012. Find the percent of decrease.

17. By decreasing each dimension by 1 unit, the area of a rectangle decreased from 40 square feet (on the left) to 28 square feet (on the right). Find the percent of decrease in area.

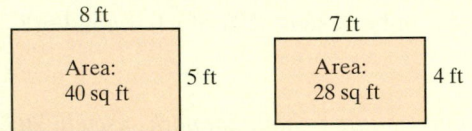

18. By decreasing the length of the side by one unit, the area of a square decreased from 100 square meters to 81 square meters. Find the percent of decrease in area.

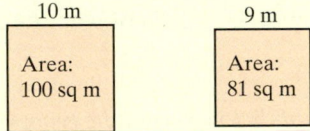

Solve. See Example 6.

19. Find the original price of a pair of shoes if the sale price is $78 after a 25% discount.

20. Find the original price of a popular pair of shoes if the increased price is $80 after a 25% increase.

▶**21.** Find last year's salary if after a 4% pay raise, this year's salary is $44,200.

22. Find last year's salary if after a 3% pay raise, this year's salary is $55,620.

Objective D *Solve. For each exercise, a table is given for you to complete and use to write an equation that models the situation. See Example 7.*

23. How much pure acid should be mixed with 2 gallons of a 40% acid solution in order to get a 70% acid solution?

	Number of Gallons	·	Acid Strength	=	Amount of Acid
Pure Acid			100%		
40% Acid Solution					
70% Acid Solution Needed					

24. How many cubic centimeters (cc) of a 25% antibiotic solution should be added to 10 cubic centimeters of a 60% antibiotic solution in order to get a 30% antibiotic solution?

	Number of Cubic cm	·	Antibiotic Strength	=	Amount of Antibiotic
25% Antibiotic Solution					
60% Antibiotic Solution					
30% Antibiotic Solution Needed					

25. Community Coffee Company wants a new flavor of Cajun coffee. How many pounds of coffee worth $7 a pound should be added to 14 pounds of coffee worth $4 a pound to get a mixture worth $5 a pound?

	Number of Pounds	·	Cost per Pound	=	Value
$7 per lb Coffee					
$4 per lb Coffee					
$5 per lb Coffee Wanted					

26. Planter's Peanut Company wants to mix 20 pounds of peanuts worth $3 a pound with cashews worth $5 a pound in order to make an experimental mix worth $3.50 a pound. How many pounds of cashews should be added to the peanuts?

	Number of Pounds	·	Cost per Pound	=	Value
$3 per lb Peanuts					
$5 per lb Cashews					
$3.50 per lb Mixture Wanted					

Objectives A B C D Mixed Practice *Solve. If needed, round money amounts to two decimal places and all other amounts to one decimal place. See Examples 1 through 7.*

27. Find 23% of 20.

28. Find 140% of 86.

29. The number 40 is 80% of what number?

30. The number 56.25 is 45% of what number?

31. The number 144 is what percent of 480?

32. The number 42 is what percent of 35?

The graph shows the communities in the United States that have the highest percentages of citizens that shop by catalog. Use the graph to answer Exercises 33 through 36.

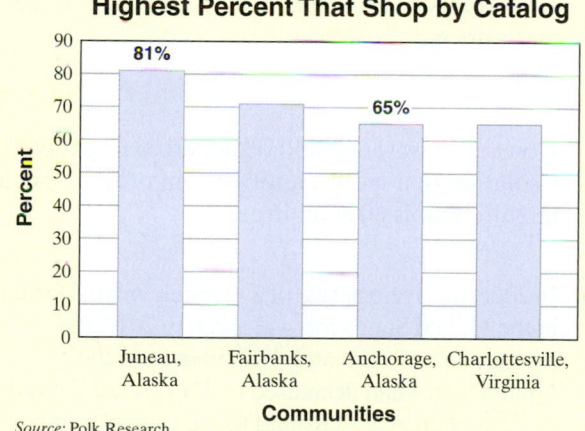

Highest Percent That Shop by Catalog

Source: Polk Research

33. Estimate the percent of the population in Fairbanks, Alaska, who shop by catalog.

34. Estimate the percent of the population in Charlottesville, Virginia, who shop by catalog.

35. According to the U.S. Census Bureau, in 2014, Anchorage had a population of about 291,800. How many catalog shoppers might we predict lived in Anchorage?

36. According to the U.S. Census Bureau, in 2014, Juneau had a population of 31,275. How many catalog shoppers might we predict lived in Juneau? Round to the nearest whole number.

For Exercises 37 and 38, fill in the percent column in each table. Each table contains a worked-out example.

37.

Top Cranberry-Producing States in 2012 (in millions of pounds)

	Millions of Pounds	Percent of Total (rounded to nearest percent)
Wisconsin	450	
Oregon	40	
Massachusetts	210	
Washington	14	
New Jersey	54	Example: $\frac{54}{768} \approx 7\%$
Total	768	

Source: National Agricultural Statistics Service

38.

The Gap, Inc. Brands North American Stores in 2014

Store Brand	Number of Stores	Percent of Total (rounded to nearest percent)
Gap	968	
Athleta	65	
Banana Republic	596	Example: $\frac{596}{2670} \approx 22\%$
Intermix	37	
Old Navy	1004	
Total	2670	

39. Iceberg lettuce is grown and shipped to stores for about 40 cents a head, and consumers purchase it for about 70 cents a head. Find the percent of increase.

40. U.S. macadamia nut production in 2011 was 24,440 tons, and in 2012 the production dropped to 22,000 tons. Find the percent of decrease. (*Source:* Agricultural Marketing Resource Center)

41. A student at the University of New Orleans makes money by buying and selling used cars. Charles bought a used car and later sold it for a 20% profit. If he sold it for $4680, how much did Charles pay for the car?

42. From 2010 to 2020, the number of people employed as physician assistants in the United States is expected to increase by 30%. The number of people employed as physician assistants in 2010 was 83,600. Find the predicted number of physician assistants in 2020. (*Source:* Bureau of Labor Statistics)

43. By doubling each dimension, the area of a parallelogram increased from 36 square centimeters to 144 square centimeters. Find the percent of increase in area.

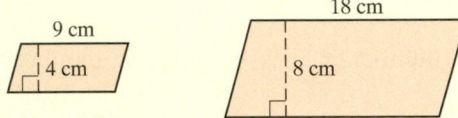

44. By doubling each dimension, the area of a triangle increased from 6 square miles to 24 square miles. Find the percent of increase in area.

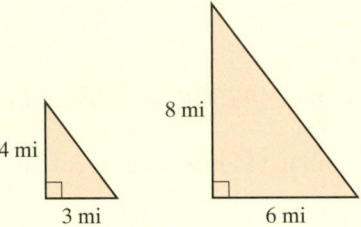

45. A gasoline station recently increased the price of one grade of gasoline by 5%. If this gasoline originally cost $2.20 per gallon, find the mark-up and the new price.

46. The price of a biology book recently increased by 10%. If this book originally cost $89.90, find the mark-up and the new price.

47. How much of an alloy that is 20% copper should be mixed with 200 ounces of an alloy that is 50% copper in order to get an alloy that is 30% copper?

48. How much water should be added to 30 gallons of a solution that is 70% antifreeze in order to get a mixture that is 60% antifreeze?

49. In 2008, the number of milk cow operations in the United States was 67,000. By 2012, this number had decreased to 58,000. What was the percent of decrease? Round to the nearest tenth of a percent. (*Source:* National Agricultural Statistics Service)

50. In 2006, the average size of a privately owned farm in the United States was 443 acres. In 2012, the average size of a privately owned farm in the United States had decreased to 421 acres. What is this percent of decrease? Round to the nearest tenth of a percent. (*Source:* National Agricultural Statistics Service)

51. A company recently downsized its number of employees by 35%. If there are still 78 employees, how many employees were there prior to the layoffs?

52. The average number of children born to each U.S. woman has decreased by 44% since 1920. If this average is now 1.9, find the average in 1920. Round to the nearest tenth.

53. Nordstrom advertised a 25%-off sale. If a London Fog coat originally sold for $256, find the decrease in price and the sale price.

54. A gasoline station decreased the price of a $0.95 cola by 15%. Find the decrease in price and the new price.

55. Scoville units are used to measure the hotness of a pepper. Measuring 577 thousand Scoville units, the "Red Savina" habañero pepper was known as the hottest chili pepper. That has recently changed with the discovery of Naga Jolokia pepper from India. It measures 48% hotter than the habañero. Find the measure of the Naga Jolokia pepper. Round to the nearest thousand units.

56. The number of cell phone tower sites in the United States was 195,613 in 2006. By 2012, the number of cell sites had increased by 54.3%. Find the number of cell towers in 2012. Round to the nearest whole number. (*Source:* CTIA—The Wireless Association)

57. In 2013, a survey found that about 55% of all adults in the United States owned a smartphone. There were roughly 240 million U.S. adults at that time. How many U.S. adults owned a smartphone in 2013? (*Source:* Pew Research Center, U.S. Census Bureau)

58. In 2012, there were approximately 225 million moviegoers in the United States and Canada. Of these, about 52% were female. Find the approximate number of females who attended the movies in that year. (*Source:* Motion Picture Association of America)

59. A new self-tanning lotion for everyday use is to be sold. First, an experimental lotion mixture is made by mixing 800 ounces of everyday moisturizing lotion worth $0.30 an ounce with self-tanning lotion worth $3 per ounce. If the experimental lotion is to cost $1.20 per ounce, how many ounces of the self-tanning lotion should be in the mixture?

60. The owner of a local chocolate shop wants to develop a new trail mix. How many pounds of chocolate-covered peanuts worth $5 a pound should be mixed with 10 pounds of granola bites worth $2 a pound to get a mixture worth $3 per pound?

Review

Place $<$, $>$, *or* $=$ *in the appropriate space to make each a true statement. See Sections 8.1, 8.2, and 8.5.*

61. -5 ___ -7

62. $\dfrac{12}{3}$ ___ 2^2

63. $|-5|$ ___ $-(-5)$

64. -3^3 ___ $(-3)^3$

65. $(-3)^2$ ___ -3^2

66. $|-2|$ ___ $-|-2|$

Concept Extensions

67. Is it possible to mix a 10% acid solution and a 40% acid solution to obtain a 60% acid solution? Why or why not?

68. Must the percents in a circle graph have a sum of 100%? Why or why not?

Standardized nutrition labels like the one below have been displayed on food items since 1994. The percent column on the right shows the percent of daily values (based on a 2000-calorie diet) shown at the bottom of the label. For example, a serving of this food contains 4 grams of total fat, where the recommended daily fat based on a 2000-calorie diet is less than 65 grams of fat. This means that $\frac{4}{65}$ or approximately 6% (as shown) of your daily recommended fat is taken in by eating a serving of this food. Use this nutrition label to answer Exercises 69 through 71.

Nutrition Facts

Serving Size 18 Crackers (31g)
Servings Per Container About 9

Amount Per Serving

Calories 130   Calories from Fat 35

% Daily Value*

Total Fat 4g	6%
Saturated Fat 0.5g	3%
Polyunsaturated Fat 0g	
Monounsaturated Fat 1.5g	
Cholesterol 0mg	0%
Sodium 230mg	*x*
Total Carbohydrate 23g	*y*
Dietary Fiber 2g	8%
Sugars 3g	
Protein 2g	

Vitamin A 0%   •   Vitamin C 0%
Calcium 2%   •   Iron 6%

* Percent Daily Values are based on a 2,000 calorie diet. Your daily values may be higher or lower depending on your calorie needs.

	Calories	2,000	2,500
Total Fat	Less than	65g	80g
Sat. Fat	Less than	20g	25g
Cholesterol	Less than	300mg	300mg
Sodium	Less than	2400mg	2400mg
Total Carbohydrate		300g	375g
Dietary Fiber		25g	30g

69. Based on a 2000-calorie diet, what percent of daily value of sodium is contained in a serving of this food? In other words, find *x* in the label. (Round to the nearest tenth of a percent.)

70. Based on a 2000-calorie diet, what percent of daily value of total carbohydrate is contained in a serving of this food? In other words, find *y* in the label. (Round to the nearest tenth of a percent.)

71. Notice on the nutrition label that one serving of this food contains 130 calories and 35 of these calories are from fat. Find the percent of calories from fat. (Round to the nearest tenth of a percent.) It is recommended that no more than 30% of calorie intake come from fat. Does this food satisfy this recommendation?

Use the nutrition label below to answer Exercises 72 through 74.

NUTRITIONAL INFORMATION PER SERVING

Serving Size: 9.8 oz.   Servings Per Container: 1

Calories 280	Polyunsaturated Fat 1g
Protein 12g	Saturated Fat 3g
Carbohydrate 45g	Cholesterol 20mg
Fat .6g	Sodium 520mg
Percent of Calories from Fat....?	Potassium 220mg

72. If fat contains approximately 9 calories per gram, find the percent of calories from fat in one serving of this food. (Round to the nearest tenth of a percent.)

73. If protein contains approximately 4 calories per gram, find the percent of calories from protein from one serving of this food. (Round to the nearest tenth of a percent.)

74. Find a food that contains more than 30% of its calories per serving from fat. Analyze the nutrition label and verify that the percents shown are correct.

Linear Inequalities and Problem Solving

In Chapter 8, we reviewed these inequality symbols and their meanings:

$<$ means "is less than"　　　$\leq$ means "is less than or equal to"

$>$ means "is greater than"　　$\geq$ means "is greater than or equal to"

An **inequality** is a statement that contains one of the symbols above.

Equations	Inequalities
$x = 3$	$x \leq 3$
$5n - 6 = 14$	$5n - 6 > 14$
$12 = 7 - 3y$	$12 \leq 7 - 3y$
$\dfrac{x}{4} - 6 = 1$	$\dfrac{x}{4} - 6 > 1$

Objective A Graphing Inequalities on a Number Line

Recall that the single solution to the equation $x = 3$ is 3. The solutions of the inequality $x \leq 3$ include 3 and *all real numbers less than 3* (for example, $-10, \dfrac{1}{2}, 2,$ and 2.9).

Because we can't list all numbers less than 3, we show instead a picture of the solutions by graphing them on a number line.

To graph the solutions of $x \leq 3$, we shade the numbers to the left of 3 since they are less than 3. Then we place a closed circle on the point representing 3. The closed circle indicates that 3 *is* a solution: 3 *is* less than or equal to 3.

To graph the solutions of $x < 3$, we shade the numbers to the left of 3. Then we place an open circle on the point representing 3. The open circle indicates that 3 *is not* a solution: 3 *is not* less than 3.

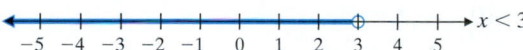

Example 1 Graph: $x \geq -1$

Solution: To graph the solutions of $x \geq -1$, we place a closed circle at -1 since the inequality symbol is $\geq$ and -1 is greater than or equal to -1. Then we shade to the right of -1.

■ Work Practice 1

Example 2 Graph: $-1 > x$

Solution: Recall from Section 8.1 that $-1 > x$ means the same as $x < -1$. The graph of the solutions of $x < -1$ is shown below.

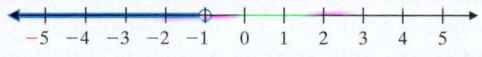

■ Work Practice 2

Objectives

A Graph Inequalities on a Number Line.

B Use the Addition Property of Inequality to Solve Inequalities.

C Use the Multiplication Property of Inequality to Solve Inequalities.

D Use Both Properties to Solve Inequalities.

E Solve Problems Modeled by Inequalities.

Practice 1

Graph: $x \geq -2$

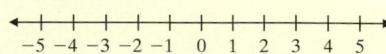

Practice 2

Graph: $5 > x$

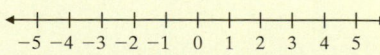

Answers

1.
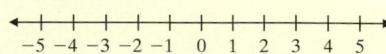

2.

693

Practice 3

Graph: $-3 \le x < 1$

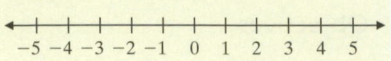

Example 3   Graph: $-4 < x \le 2$

Solution: We read as $-4 < x \le 2$ as "-4 is less than x and x is less than or equal to 2," or as "x is greater than -4 and x is less than or equal to 2." To graph the solutions of this inequality, we place an open circle at -4 (-4 is not part of the graph), a closed circle at 2 (2 is part of the graph), and we shade all numbers between -4 and 2. Why? All numbers between -4 and 2 are greater than -4 *and* also less than 2.

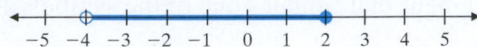

■ Work Practice 3

Objective B Using the Addition Property

When solutions of a linear inequality are not immediately obvious, they are found through a process similar to the one used to solve a linear equation. Our goal is to get the variable alone on one side of the inequality. We use properties of inequality similar to properties of equality.

> **Addition Property of Inequality**
>
> If a, b, and c are real numbers, then
>
> $$a < b \quad \text{and} \quad a + c < b + c$$
>
> are equivalent inequalities.

   This property also holds true for subtracting values, since subtraction is defined in terms of addition. In other words, adding or subtracting the same quantity from both sides of an inequality does not change the solutions of the inequality.

Practice 4

Solve $x - 6 \ge -11$. Graph the solutions.

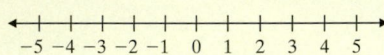

Example 4   Solve $x + 4 \le -6$. Graph the solutions.

Solution: To solve for x, subtract 4 from both sides of the inequality.

$$x + 4 \le -6 \qquad \text{Original inequality}$$
$$x + 4 - 4 \le -6 - 4 \qquad \text{Subtract 4 from both sides.}$$
$$x \le -10 \qquad \text{Simplify.}$$

The graph of the solutions is shown below.

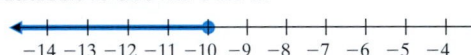

■ Work Practice 4

> **Helpful Hint**
>
> Notice that any number less than or equal to -10 is a solution to $x \le -10$. For example, solutions include
>
> $$-10, \quad -200, \quad -11\frac{1}{2}, \quad -\sqrt{130}, \quad \text{and} \quad -50.3$$

Objective C Using the Multiplication Property

Answers

3.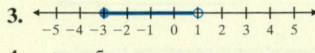

4. $x \ge -5$

An important difference between solving linear equations and solving linear inequalities is shown when we multiply or divide both sides of an inequality by a nonzero real number. For example, start with the true statement $6 < 8$ and multiply both sides by 2. As we see on the next page, the resulting inequality is also true.

$$6 < 8 \quad \text{True}$$

$$2(6) < 2(8) \quad \text{Multiply both sides by 2.}$$

$$12 < 16 \quad \text{True}$$

But if we start with the same true statement $6 < 8$ and multiply both sides by -2, the resulting inequality is not a true statement.

$$6 < 8 \quad \text{True}$$

$$-2(6) < -2(8) \quad \text{Multiply both sides by } -2.$$

$$-12 < -16 \quad \text{False}$$

Notice, however, that if we reverse the direction of the inequality symbol, the resulting inequality is true.

$$-12 < -16 \quad \text{False}$$

$$-12 > -16 \quad \text{True}$$

This demonstrates the multiplication property of inequality.

Multiplication Property of Inequality

1. If a, b, and c are real numbers, and c is **positive,** then

$$a < b \quad \text{and} \quad ac < bc$$

are equivalent inequalities.

2. If a, b, and c are real numbers, and c is **negative,** then

$$a < b \quad \text{and} \quad ac > bc$$

are equivalent inequalities.

Because division is defined in terms of multiplication, this property also holds true when dividing both sides of an inequality by a nonzero number: If we multiply or divide both sides of an inequality by a negative number, **the direction of the inequality sign must be reversed for the inequalities to remain equivalent.**

✓**Concept Check** Fill in the box with $<, >, \leq,$ or $\geq$.

a. Since $-8 < -4$, then $3(-8) \,\square\, 3(-4)$.

b. Since $5 \geq -2$, then $\dfrac{5}{-7} \,\square\, \dfrac{-2}{-7}$.

c. If $a < b$, then $2a \,\square\, 2b$.

d. If $a \geq b$, then $\dfrac{a}{-3} \,\square\, \dfrac{b}{-3}$.

Example 5 Solve $-2x \leq -4$. Graph the solutions.

Solution: Remember to reverse the direction of the inequality symbol when dividing by a negative number.

$$-2x \leq -4$$

$$\frac{-2x}{-2} \geq \frac{-4}{-2} \quad \text{Divide both sides by } -2 \text{ and reverse the inequality sign.}$$

$$x \geq 2 \quad \text{Simplify.}$$

The graph of the solutions is shown.

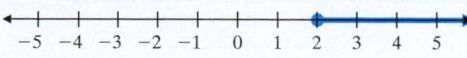

■ **Work Practice 5**

Practice 5

Solve $-3x \leq 12$. Graph the solutions.

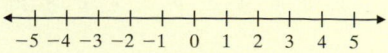

Answer

5. $x \geq -4$

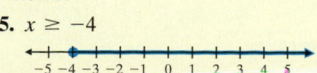

✓**Concept Check Answer**

a. $<$ **b.** $\leq$ **c.** $<$ **d.** $\leq$

Practice 6

Solve $5x > -20$. Graph the solutions.

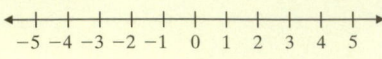

Example 6 Solve $2x < -4$. Graph the solutions.

Solution: $2x < -4$

$$\frac{2x}{2} < \frac{-4}{2} \qquad \text{Divide both sides by 2. Do not reverse the inequality sign.}$$

$$x < -2 \qquad \text{Simplify.}$$

The graph of the solutions is shown.

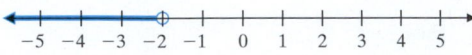

■ **Work Practice 6**

Since we cannot list all solutions to an inequality such as $x < -2$, we will use the set notation $\{x | x < -2\}$. Recall from Section 8.1 that this is read "the set of all x such that x is less than -2." We will use this notation when solving inequalities.

Objective D Using Both Properties of Inequality

The following steps may be helpful when solving inequalities in one variable. Notice that these steps are similar to the ones given in Section 9.3 for solving equations.

> ### To Solve Linear Inequalities in One Variable
>
> **Step 1:** If an inequality contains fractions, multiply both sides by the LCD to clear the inequality of fractions.
>
> **Step 2:** Use the distributive property to remove parentheses if they appear.
>
> **Step 3:** Simplify each side of the inequality by combining like terms.
>
> **Step 4:** Get all variable terms on one side and all numbers on the other side by using the addition property of inequality.
>
> **Step 5:** Get the variable alone by using the multiplication property of inequality.

Helpful Hint

Don't forget that if both sides of an inequality are multiplied or divided by a negative number, the direction of the inequality sign must be reversed.

Practice 7

Solve $-3x + 11 \le -13$. Graph the solution set.

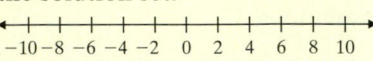

Example 7 Solve $-4x + 7 \ge -9$. Graph the solution set.

Solution:

$$-4x + 7 \ge -9$$

$$-4x + 7 - 7 \ge -9 - 7 \qquad \text{Subtract 7 from both sides.}$$

$$-4x \ge -16 \qquad \text{Simplify.}$$

$$\frac{-4x}{-4} \le \frac{-16}{-4} \qquad \text{Divide both sides by } -4 \text{ and reverse the direction of the inequality sign.}$$

$$x \le 4 \qquad \text{Simplify.}$$

The graph of the solution set $\{x | x \le 4\}$ is shown.

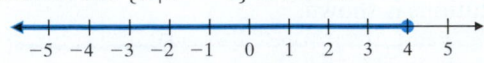

■ **Work Practice 7**

Answers

6. $x > -4$

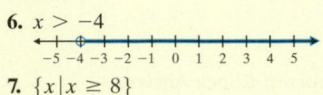

7. $\{x | x \ge 8\}$

Example 8 Solve $-5x + 7 < 2(x - 3)$. Graph the solution set.

Solution: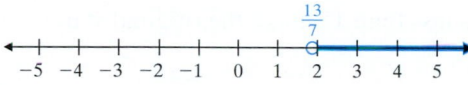
$$-5x + 7 < 2(x - 3)$$

$-5x + 7 < 2x - 6$	Apply the distributive property.
$-5x + 7 - 2x < 2x - 6 - 2x$	Subtract $2x$ from both sides.
$-7x + 7 < -6$	Combine like terms.
$-7x + 7 - 7 < -6 - 7$	Subtract 7 from both sides.
$-7x < -13$	Combine like terms.
$\dfrac{-7x}{-7} > \dfrac{-13}{-7}$	Divide both sides by -7 and reverse the direction of the inequality sign.
$x > \dfrac{13}{7}$	Simplify.

The graph of the solution set $\left\{ x \middle| x > \dfrac{13}{7} \right\}$ is shown.

$$\overset{\frac{13}{7}}{\underset{-5 \ -4 \ -3 \ -2 \ -1 \ \ 0 \ \ 1 \ \ 2 \ \ 3 \ \ 4 \ \ 5}{\longleftarrow | \ | \ | \ | \ | \ | \ | \ \circ\!\!\!-\!\!-\!\!-\!\!\longrightarrow}}$$

■ **Work Practice 8**

Example 9 Solve: $2(x - 3) - 5 \leq 3(x + 2) - 18$

Solution: $2(x - 3) - 5 \leq 3(x + 2) - 18$

$2x - 6 - 5 \leq 3x + 6 - 18$	Apply the distributive property.
$2x - 11 \leq 3x - 12$	Combine like terms.
$-x - 11 \leq -12$	Subtract $3x$ from both sides.
$-x \leq -1$	Add 11 to both sides.
$\dfrac{-x}{-1} \geq \dfrac{-1}{-1}$	Divide both sides by -1 and reverse the direction of the inequality sign.
$x \geq 1$	Simplify.

The solution set is $\{ x \mid x \geq 1 \}$.

■ **Work Practice 9**

Objective E Solving Problems Modeled by Inequalities

Problems containing words such as "at least," "at most," "between," "no more than," and "no less than" usually indicate that an inequality should be solved instead of an equation. In solving applications involving linear inequalities, we use the same procedure we used to solve applications involving linear equations.

Some Inequality Translations			
$\geq$	$\leq$	$<$	$>$
at least	at most	is less than	is greater than
no less than	no more than		

Example 10 12 subtracted from 3 times a number is less than 21. Find all numbers that make this statement true.

Solution:

1. **UNDERSTAND.** Read and reread the problem. This is a direct translation problem, and let's let

$x =$ the unknown number

(Continued on next page)

Practice 8

Solve $2x - 3 > 4(x - 1)$. Graph the solution set.

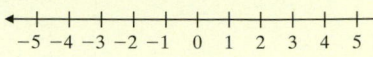

Practice 9

Solve:
$3(x + 5) - 1 \geq 5(x - 1) + 7$

Practice 10

Twice a number, subtracted from 35, is greater than 15. Find all numbers that make this true.

Answers

8. $\left\{ x \middle| x < \dfrac{1}{2} \right\}$

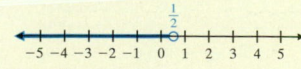

9. $\{ x \mid x \leq 6 \}$

10. all numbers less than 10

2. TRANSLATE.

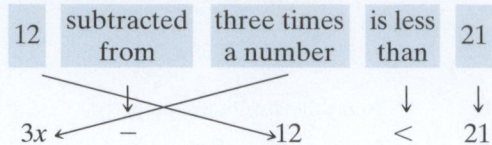

| 12 | subtracted from | three times a number | is less than | 21 |

$$3x \quad - \quad 12 \quad < \quad 21$$

3. SOLVE. $3x - 12 < 21$

$$3x < 33 \qquad \text{Add 12 to both sides.}$$

$$\frac{3x}{3} < \frac{33}{3} \qquad \text{Divide both sides by 3 and do not reverse the direction of the inequality sign.}$$

$$x < 11 \qquad \text{Simplify.}$$

4. INTERPRET.

Check: Check the translation; then let's choose a number less than 11 to see if it checks. For example, let's check 10. 12 subtracted from 3 times 10 is 12 subtracted from 30, or 18. Since 18 is less than 21, the number 10 checks.

State: All numbers less than 11 make the original statement true.

■ **Work Practice 10**

Practice 11

Alex earns $600 per month plus 4% of all his sales. Find the minimum sales that will allow Alex to earn at least $3000 per month.

Example 11   Budgeting for a Wedding

Marie Chase and Jonathan Edwards are having their wedding reception at the Gallery reception hall. They may spend at most $1000 for the reception. If the reception hall charges a $100 cleanup fee plus $14 per person, find the greatest number of people that they can invite and still stay within their budget.

Solution:

1. UNDERSTAND. Read and reread the problem. Suppose that 50 people attend the reception. The cost is then $100 + \$14(50) = \$100 + \$700 = \$800$.

Let $x =$ the number of people who attend the reception.

2. TRANSLATE.

cleanup fee	+	cost per person	times	number of people	must be less than or equal to	$1000
100	+	14	·	x	≤	1000

3. SOLVE.

$$100 + 14x \leq 1000$$

$$14x \leq 900 \qquad \text{Subtract 100 from both sides.}$$

$$x \leq 64\frac{2}{7} \qquad \text{Divide both sides by 14.}$$

4. INTERPRET.

Check: Since x represents the number of people, we round down to the nearest whole, or 64. Notice that if 64 people attend, the cost is $100 + \$14(64) = \$996$. If 65 people attend, the cost is $100 + \$14(65) = \$1010$, which is more than the given $1000.

State: Marie Chase and Jonathan Edwards can invite at most 64 people to the reception.

■ **Work Practice 11**

Answer

11. $60,000

Vocabulary, Readiness & Video Check

Identify each as an equation, expression, or inequality.

1. $6x - 7(x + 9)$ _____

2. $6x = 7(x + 9)$ _____

3. $6x < 7(x + 9)$ _____

4. $5y - 2 \geq -38$ _____

5. $\dfrac{9}{7} = \dfrac{x + 2}{14}$ _____

6. $\dfrac{9}{7} - \dfrac{x + 2}{14}$ _____

Decide which number listed is not a solution to each given inequality.

7. $x \geq -3;\quad -3, 0, -5, \pi$ _____

8. $x < 6;\quad -6, |-6|, 0, -3.2$ _____

9. $x < 4.01;\quad 4, -4.01, 4.1, -4.1$ _____

10. $x \geq -3;\quad -4, -3, -2, -(-2)$ _____

Martin-Gay Interactive Videos   Watch the section lecture video and answer the following questions.

See Video 9.7

Objective **A** **11.** From Example 1, when graphing an inequality, what inequality symbol(s) does an open circle indicate? What inequality symbol(s) does a closed circle indicate? ▶

Objective **B** **12.** From the lecture before ⊞ Example 2, which property is the addition property of inequality very similar to? ▶

Objective **C** **13.** What is the answer to ⊞ Example 3, written in solution set notation? ▶

Objective **D** **14.** When solving ⊞ Example 4, why is special attention given to the coefficient of x in the last step? ▶

Objective **E** **15.** What is the phrase in ⊞ Example 5 that tells us to translate to an inequality? What does this phrase translate to? ▶

9.7 Exercise Set MyMathLab® ▶

Objective **A** *Graph each inequality on the number line. See Examples 1 and 2.*

▶ **1.** $x \leq -1$

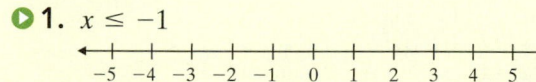

2. $y < 0$

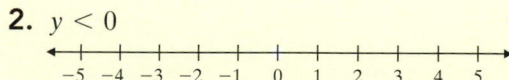

3. $x > \dfrac{1}{2}$

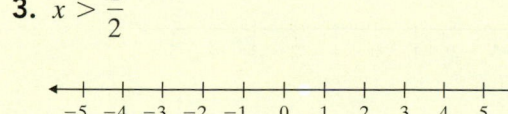

4. $z \geq -\dfrac{2}{3}$

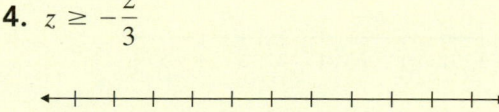

5. $y < 4$

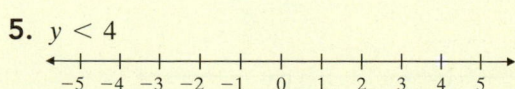

6. $x > 3$

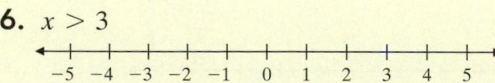

7. $-2 \leq m$

8. $-5 \geq x$

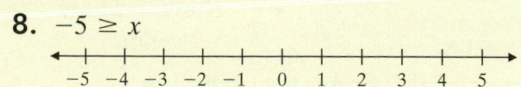

Graph each inequality on the number line. See Example 3.

9. $-1 < x < 3$

10. $-2 \le x \le 3$

11. $0 \le y < 2$

12. $-4 < x \le 0$

Objective B *Solve each inequality. Graph the solution set. Write each answer using solution set notation. See Example 4.*

13. $x - 2 \ge -7$

14. $x + 4 \le 1$
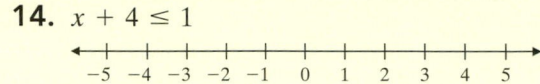

15. $-9 + y < 0$

16. $-3 + m > 5$
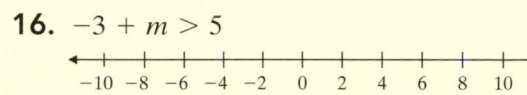

17. $3x - 5 > 2x - 8$

18. $3 - 7x \ge 10 - 8x$

19. $4x - 1 \le 5x - 2x$

20. $7x + 3 < 9x - 3x$
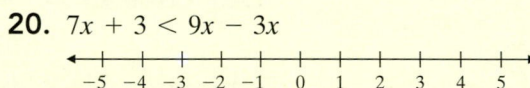

Objective C *Solve each inequality. Graph the solution set. See Examples 5 and 6.*

21. $2x < -6$

22. $3x > -9$

23. $-8x \le 16$
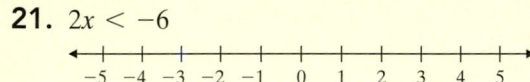

24. $-5x < 20$
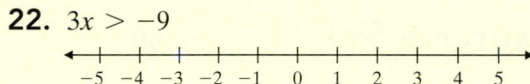

25. $-x > 0$
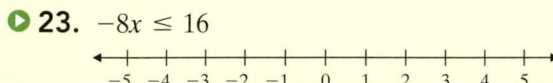

26. $-y \ge 0$
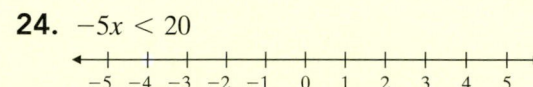

27. $\dfrac{3}{4}y \ge -2$

28. $\dfrac{5}{6}x \le -8$
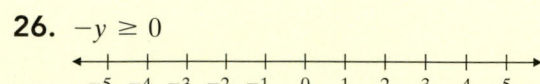

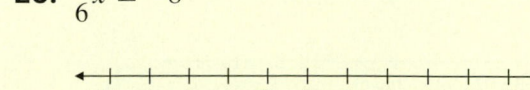

29. $-0.6y < -1.8$

30. $-0.3x > -2.4$
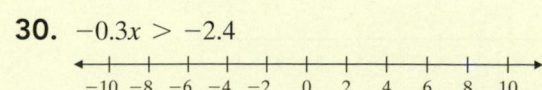

Objectives **B** **C** **D** **Mixed Practice** *Solve each inequality. Write each answer using solution set notation.*
See Examples 4 through 9.

31. $-8 < x + 7$

32. $-11 > x + 4$

33. $7(x + 1) - 6x \geq -4$

34. $10(x + 2) - 9x \leq -1$

35. $4x > 1$

36. $6x < 5$

37. $-\dfrac{2}{3}y \leq 8$

38. $-\dfrac{3}{4}y \geq 9$

39. $4(2z + 1) < 4$

40. $6(2 - z) \geq 12$

41. $3x - 7 < 6x + 2$

42. $2x - 1 \geq 4x - 5$

43. $5x - 7x \leq x + 2$

44. $4 - x < 8x + 2x$

45. $-6x + 2 \geq 2(5 - x)$

46. $-7x + 4 > 3(4 - x)$

47. $3(x - 5) < 2(2x - 1)$

48. $5(x - 2) \leq 3(2x - 1)$

49. $4(3x - 1) \leq 5(2x - 4)$

50. $3(5x - 4) \leq 4(3x - 2)$

51. $3(x + 2) - 6 > -2(x - 3) + 14$

52. $7(x - 2) + x \leq -4(5 - x) - 12$

53. $-5(1 - x) + x \leq -(6 - 2x) + 6$

54. $-2(x - 4) - 3x < -(4x + 1) + 2x$

55. $\dfrac{1}{4}(x + 4) < \dfrac{1}{5}(2x + 3)$

56. $\dfrac{1}{2}(x - 5) < \dfrac{1}{3}(2x - 1)$

57. $-5x + 4 \leq -4(x - 1)$

58. $-6x + 2 < -3(x + 4)$

Objective **E** *Solve the following. For Exercises 61 and 62, the solutions have been started for you. See Examples 10 and 11.*

59. Six more than twice a number is greater than negative fourteen. Find all numbers that make this statement true.

60. One more than five times a number is less than or equal to ten. Find all such numbers.

△ **61.** The perimeter of a rectangle is to be no greater than 100 centimeters and the width must be 15 centimeters. Find the maximum length of the rectangle.

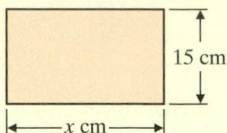

15 cm

x cm

Start the solution:

1. UNDERSTAND the problem. Reread it as many times as needed.

2. TRANSLATE into an equation. (Fill in the blanks below.)

the perimeter of the rectangle	is less than or equal to	100
↓	↓	↓
$x + 15 + x + 15$	_____	100

Finish with:

3. SOLVE and 4. INTERPRET

△ **62.** One side of a triangle is three times as long as another side, and the third side is 12 inches long. If the perimeter can be no longer than 32 inches, find the maximum lengths of the other two sides.

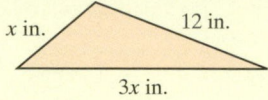

x in.   12 in.

3x in.

Start the solution:

1. UNDERSTAND the problem. Reread it as many times as needed.

2. TRANSLATE into an equation. (Fill in the blanks below.)

the perimeter of the triangle	is less than or equal to	32
↓	↓	↓
$12 + 3x + x$	_____	32

Finish with:

3. SOLVE and 4. INTERPRET

63. Ben Holladay bowled 146 and 201 in his first two games. What must he bowl in his third game to have an average of at least 180? (*Hint:* The average of a list of numbers is their sum divided by the number of numbers in the list.)

64. On an NBA team the two forwards measure 6′ 8″ and 6′ 6″ tall and the two guards measure 6′ 0″ and 5′ 9″ tall. How tall should the center be if they wish to have a starting team average height of at least 6′ 5″?

65. Dennis and Nancy Wood are celebrating their 30th wedding anniversary by having a reception at Tiffany Oaks reception hall. They have budgeted $3000 for their reception. If the reception hall charges a $50 cleanup fee plus $34 per person, find the greatest number of people that they may invite and still stay within their budget.

66. A surprise retirement party is being planned for Pratap Puri. A total of $860 has been collected for the event, which is to be held at a local reception hall. This reception hall charges a cleanup fee of $40 and $15 per person for drinks and light snacks. Find the greatest number of people that may be invited and still stay within the $860 budget.

67. A 150-pound person uses 5.8 calories per minute when walking at a speed of 4 mph. How long must a person walk at this speed to use at least 200 calories? Round up to the nearest minute. (*Source:* Home & Garden Bulletin No. 72)

68. A 170-pound person uses 5.3 calories per minute when bicycling at a speed of 5.5 mph. How long must a person ride a bike at this speed in order to use at least 200 calories? Round up to the nearest minute. (*Source:* Same as Exercise 67)

Review

Evaluate each expression. See Section 8.2.

69. 3^4　　　　**70.** 4^3　　　　**71.** 1^8　　　　**72.** 0^7　　　　**73.** $\left(\dfrac{7}{8}\right)^2$　　　　**74.** $\left(\dfrac{2}{3}\right)^3$

The graph shows the number of U.S. Starbucks locations from 2007 to 2013. The height of the graph for each year shown corresponds to the number of Starbucks locations in the United States. Use this graph to answer Exercises 75 through 80. See Section 7.1.

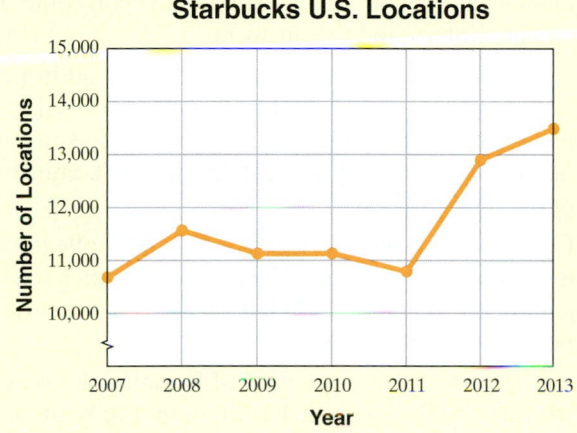

Starbucks U.S. Locations

75. Approximate the number of Starbucks locations in 2013.

76. Approximate the number of Starbucks locations in 2011.

77. Between which two years did the greatest increase in the number of Starbucks locations occur?

78. Between which two years did the number of Starbucks locations appear to remain the same?

79. During which year did the number of Starbucks locations rise above 12,000?

80. During which year did the number of Starbucks locations rise above 13,000?

Concept Extensions

Fill in the box with $<$, $>$, $\leq$, or $\geq$. See the Concept Check in this section.

81. Since $3 < 5$, then $3(-4) \,\square\, 5(-4)$.

82. If $m \leq n$, then $2m \,\square\, 2n$.

83. If $m \leq n$, then $-2m \,\square\, -2n$.

84. If $-x < y$, then $x \,\square\, -y$.

85. When solving an inequality, when must you reverse the direction of the inequality symbol?

86. If both sides of the inequality $-3x < -30$ are divided by 3, do you reverse the direction of the inequality symbol? Why or why not?

Solve.

87. Eric Daly has scores of 75, 83, and 85 on his history tests. Use an inequality to find the scores he can make on his final exam to receive a B in the class. The final exam counts as **two** tests, and a B is received if the final course average is greater than or equal to 80.

88. Maria Lipco has scores of 85, 95, and 92 on her algebra tests. Use an inequality to find the scores she can make on her final exam to receive an A in the course. The final exam counts as **three** tests, and an A is received if the final course average is greater than or equal to 90. Round to one decimal place.

Chapter 9 Group Activity

Investigating Averages

Sections 9.1–9.6

Materials:

- small rubber ball or crumpled paper ball
- bucket or waste can

This activity may be completed by working in groups or individually.

1. Try shooting the ball into the bucket or waste can 5 times. Record your results below.

   Shots Made   **Shots Missed**

2. Find your shooting percent for the 5 shots (that is, the percent of the shots you actually made out of the number you tried).

3. Suppose you are going to try an additional 5 shots. How many of the next 5 shots will you have to make to have a 50% shooting percent for all 10 shots? An 80% shooting percent?

4. Did you solve an equation in Question 3? If so, explain what you did. If not, explain how you could use an equation to find the answers.

5. Now suppose you are going to try an additional 22 shots. How many of the next 22 shots will you have to make to have at least a 50% shooting percent for all 27 shots? At least a 70% shooting percent?

6. Choose one of the sports played at your college that is currently in season. How many regular-season games are scheduled? What is the team's current percent of games won?

7. Suppose the team has a goal of finishing the season with a winning percent better than 110% of their current wins. At least how many of the remaining games must they win to achieve their goal?

Chapter 9 Vocabulary Check

Fill in each blank with one of the words or phrases listed below.

no solution	all real numbers	linear equation in one variable
equivalent equations	formula	reversed
linear inequality in one variable	the same	

1. A(n) _____ can be written in the form $Ax + B = C$.

2. Equations that have the same solution are called _____.

3. An equation that describes a known relationship among quantities is called a(n) _____.

4. A(n) _____ can be written in the form $ax + b < c$, (or $>, \leq, \geq$).

5. The solution(s) to the equation $x + 5 = x + 5$ is/are _____.

6. The solution(s) to the equation $x + 5 = x + 4$ is/are _____.

7. If both sides of an inequality are multiplied or divided by the same positive number, the direction of the inequality symbol is _____.

8. If both sides of an inequality are multiplied or divided by the same negative number, the direction of the inequality symbol is _____.

Helpful Hint

⊳ Are you preparing for your test? Don't forget to take the Chapter 9 Test on page 713. Then check your answers at the back of the text and use the Chapter Test Prep Videos to see the fully worked-out solutions to any of the exercises you want to review.

9 Chapter Highlights

Definitions and Concepts	Examples
Section 9.1 The Addition Property of Equality	

A **linear equation in one variable** can be written in the form $Ax + B = C$ where $A, B,$ and C are real numbers and $A \neq 0$.	$-3x + 7 = 2$ $3(x - 1) = -8(x + 5) + 4$
Equivalent equations are equations that have the same solution.	$x - 7 = 10$ and $x = 17$ are equivalent equations.
Addition Property of Equality Adding the same number to or subtracting the same number from both sides of an equation does not change its solution.	$y + 9 = 3$ $y + 9 - 9 = 3 - 9$ $y = -6$

Section 9.2 The Multiplication Property of Equality	

Multiplication Property of Equality Multiplying both sides or dividing both sides of an equation by the same nonzero number does not change its solution.	$\dfrac{2}{3}a = 18$ $\dfrac{3}{2}\left(\dfrac{2}{3}a\right) = \dfrac{3}{2}(18)$ $a = 27$

Section 9.3 Further Solving Linear Equations	

To Solve Linear Equations	*Solve:* $\dfrac{5(-2x + 9)}{6} + 3 = \dfrac{1}{2}$
1. Clear the equation of fractions.	1. $6 \cdot \dfrac{5(-2x + 9)}{6} + 6 \cdot 3 = 6 \cdot \dfrac{1}{2}$
2. Remove any grouping symbols such as parentheses.	2. $5(-2x + 9) + 18 = 3$ *Apply the distributive property.* $-10x + 45 + 18 = 3$
3. Simplify each side by combining like terms.	3. $-10x + 63 = 3$ *Combine like terms.*
4. Get all variable terms on one side and all numbers on the other side by using the addition property of equality.	4. $-10x + 63 - 63 = 3 - 63$ *Subtract 63.* $-10x = -60$
5. Get the variable alone by using the multiplication property of equality.	5. $\dfrac{-10x}{-10} = \dfrac{-60}{-10}$ *Divide by -10.* $x = 6$
6. Check the solution by substituting it into the original equation.	

Definitions and Concepts	Examples

Section 9.4   Further Problem Solving

Problem-Solving Steps

1. UNDERSTAND the problem.

The height of the Hudson volcano in Chile is twice the height of the Kiska volcano in the Aleutian Islands. If the sum of their heights is 12,870 feet, find the height of each.

1. Read and reread the problem. Guess a solution and check your guess.
Let x be the height of the Kiska volcano. Then $2x$ is the height of the Hudson volcano.

 x  $2x$

2. TRANSLATE the problem.

2.

height of Kiska	added to	height of Hudson	is	12,870
↓	↓	↓	↓	↓
x	$+$	$2x$	$=$	12,870

3. SOLVE the equation.

3. $x + 2x = 12{,}870$
$3x = 12{,}870$
$x = 4290$

4. INTERPRET the results.

4. Check: If x is 4290, then $2x$ is $2(4290)$ or 8580. Their sum is $4290 + 8580$ or 12,870, the required amount.

State: The Kiska volcano is 4290 feet tall, and the Hudson volcano is 8580 feet tall.

Section 9.5   Formulas and Problem Solving

An equation that describes a known relationship among quantities is called a **formula.**

To solve a formula for a specified variable, use the same steps as for solving a linear equation. Treat the specified variable as the only variable of the equation.

$A = lw$ (area of a rectangle)
$I = PRT$ (simple interest)

Solve: $P = 2l + 2w$ for l.
$P = 2l + 2w$
$P - 2w = 2l + 2w - 2w$ Subtract $2w$.
$P - 2w = 2l$
$\dfrac{P - 2w}{2} = \dfrac{2l}{2}$ Divide by 2.
$\dfrac{P - 2w}{2} = l$

Section 9.6   Percent and Mixture Problem Solving

Use the same problem-solving steps to solve a problem containing percents.

1. UNDERSTAND.

32% of what number is 36.8?

1. Read and reread. Propose a solution and check.
Let x = the unknown number.

2. TRANSLATE.

2.

32%	of	what number	is	36.8
↓	↓	↓	↓	↓
32%	$\cdot$	x	$=$	36.8

Definitions and Concepts	Examples
Section 9.6 Percent and Mixture Problem Solving (*continued*)	

3. SOLVE.

3. *Solve:* $32\% \cdot x = 36.8$

$$0.32x = 36.8$$

$$\frac{0.32x}{0.32} = \frac{36.8}{0.32} \quad \text{Divide by 0.32.}$$

$$x = 115 \quad \text{Simplify.}$$

4. INTERPRET.

4. *Check, then state:* 32% of 115 is 36.8.

How many liters of a 20% acid solution must be mixed with a 50% acid solution in order to obtain 12 liters of a 30% solution?

1. UNDERSTAND.

1. Read and reread. Guess a solution and check.
Let x = number of liters of 20% solution.
Then $12 - x$ = number of liters of 50% solution.

2. TRANSLATE.

2.

	No. of Liters · Acid Strength = Amount of Acid		
20% Solution	x	20%	$0.20x$
50% Solution	$12 - x$	50%	$0.50(12 - x)$
30% Solution Needed	12	30%	$0.30(12)$

In words: acid in 20% solution $+$ acid in 50% solution $=$ acid in 30% solution

Translate: $0.20x \quad + 0.50(12 - x) = \quad 0.30(12)$

3. SOLVE.

3. *Solve:* $0.20x + 0.50(12 - x) = 0.30(12)$

$$0.20x + 6 - 0.50x = 3.6 \quad \text{Apply the distributive property.}$$

$$-0.30x + 6 = 3.6 \quad \text{Combine like terms.}$$

$$-0.30x = -2.4 \quad \text{Subtract 6.}$$

$$x = 8 \quad \text{Divide by } -0.30.$$

4. INTERPRET.

4. *Check, then state:*

If 8 liters of a 20% acid solution are mixed with $12 - 8$ or 4 liters of a 50% acid solution, the result is 12 liters of a 30% solution.

Definitions and Concepts	Examples

Section 9.7   Linear Inequalities and Problem Solving

Properties of inequalities are similar to properties of equations. However, if you multiply or divide both sides of an inequality by the same *negative* number, you must reverse the direction of the inequality symbol.	$-2x \leq 4$ $\dfrac{-2x}{-2} \geq \dfrac{4}{-2}$ Divide by -2; reverse the inequality symbol. $x \geq -2$ 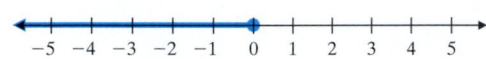

To Solve Linear Inequalities

1. Clear the inequality of fractions.

2. Remove grouping symbols.

3. Simplify each side by combining like terms.

4. Write all variable terms on one side and all numbers on the other side using the addition property of inequality.

5. Get the variable alone by using the multiplication property of inequality.

Solve: $3(x + 2) \leq -2 + 8$

1. $3(x + 2) \leq -2 + 8$ No fractions to clear.

2. $3x + 6 \leq -2 + 8$ Apply the distributive property.

3. $3x + 6 \leq 6$ Combine like terms.

4. $3x + 6 - 6 \leq 6 - 6$ Subtract 6.

      $3x \leq 0$

5. $\dfrac{3x}{3} \leq \dfrac{0}{3}$ Divide by 3.

      $x \leq 0$

The solution set is $\{x \mid x \leq 0\}$.

Chapter 9   Review

(9.1) *Solve each equation.*

1. $8x + 4 = 9x$

2. $5y - 3 = 6y$

3. $\dfrac{2}{7}x + \dfrac{5}{7}x = 6$

4. $3x - 5 = 4x + 1$

5. $2x - 6 = x - 6$

6. $4(x + 3) = 3(1 + x)$

7. $6(3 + n) = 5(n - 1)$

8. $5(2 + x) - 3(3x + 2) = -5(x - 6) + 2$

Choose the correct algebraic expression.

9. The sum of two numbers is 10. If one number is x, express the other number in terms of x.
 a. $x - 10$
 b. $10 - x$
 c. $10 + x$
 d. $10x$

10. Mandy is 5 inches taller than Melissa. If x inches represents the height of Mandy, express Melissa's height in terms of x.
 a. $x - 5$
 b. $5 - x$
 c. $5 + x$
 d. $5x$

△ **11.** If one angle measures $x°$, express the measure of its complement in terms of x.
    a. $(180 - x)°$
    b. $(90 - x)°$
    c. $(x - 180)°$
    d. $(x - 90)°$

△ **12.** If one angle measures $(x + 5)°$, express the measure of its supplement in terms of x.
    a. $(185 + x)°$
    b. $(95 + x)°$
    c. $(175 - x)°$
    d. $(x - 170)°$

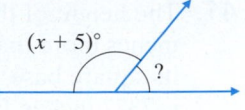

(9.2) *Solve each equation.*

13. $\dfrac{3}{4}x = -9$

14. $\dfrac{x}{6} = \dfrac{2}{3}$

15. $-5x = 0$

16. $-y = 7$

17. $0.2x = 0.15$

18. $\dfrac{-x}{3} = 1$

19. $-3x + 1 = 19$

20. $5x + 25 = 20$

21. $7(x - 1) + 9 = 5x$

22. $7x - 6 = 5x - 3$

23. $-5x + \dfrac{3}{7} = \dfrac{10}{7}$

24. $5x + x = 9 + 4x - 1 + 6$

25. Write the sum of three consecutive integers as an expression in x. Let x be the first integer.

26. Write the sum of the first and fourth of four consecutive even integers. Let x be the first even integer.

(9.3) *Solve each equation.*

27. $\dfrac{5}{3}x + 4 = \dfrac{2}{3}x$

28. $\dfrac{7}{8}x + 1 = \dfrac{5}{8}x$

29. $-(5x + 1) = -7x + 3$

30. $-4(2x + 1) = -5x + 5$

31. $-6(2x - 5) = -3(9 + 4x)$

32. $3(8y - 1) = 6(5 + 4y)$

33. $\dfrac{3(2 - z)}{5} = z$

34. $\dfrac{4(n + 2)}{5} = -n$

35. $0.5(2n - 3) - 0.1 = 0.4(6 + 2n)$

36. $-9 - 5a = 3(6a - 1)$

37. $\dfrac{5(c + 1)}{6} = 2c - 3$

38. $\dfrac{2(8 - a)}{3} = 4 - 4a$

39. $200(70x - 3560) = -179(150x - 19,300)$

40. $1.72y - 0.04y = 0.42$

(9.4) *Solve each of the following.*

41. The height of the Washington Monument is 50.5 inches more than 10 times the length of a side of its square base. If the sum of these two dimensions is 7327 inches, find the height of the Washington Monument. (*Source:* National Park Service)

42. A 12-foot board is to be divided into two pieces so that one piece is twice as long as the other. If *x* represents the length of the shorter piece, find the length of each piece.

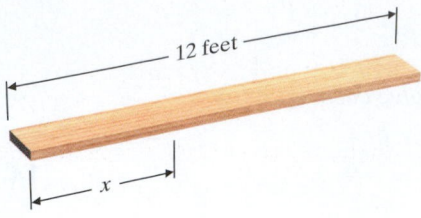

43. The national park system in the United States includes a variety of park unit types. In 2013, there were a total of 40 national battlefields and national memorials. The number of national memorials was four less than three times the number of national battlefields. How many of each park unit were there? (*Source:* National Park System)

44. Find three consecutive integers whose sum is −114.

45. The quotient of a number and 3 is the same as the difference of the number and two. Find the number.

46. Double the sum of a number and 6 is the opposite of the number. Find the number.

(9.5) *Substitute the given values into the given formulas and solve for the unknown variable.*

47. $P = 2l + 2w$; $P = 46, l = 14$

48. $V = lwh$; $V = 192, l = 8, w = 6$

Solve each equation for the indicated variable or constant.

49. $y = mx + b$ for m

50. $r = vst - 5$ for s

51. $2y - 5x = 7$ for x

52. $3x - 6y = -2$ for y

△**53.** $C = \pi D$ for π

△**54.** $C = 2\pi r$ for π

△**55.** A swimming pool holds 900 cubic meters of water. If its length is 20 meters and its height is 3 meters, find its width.

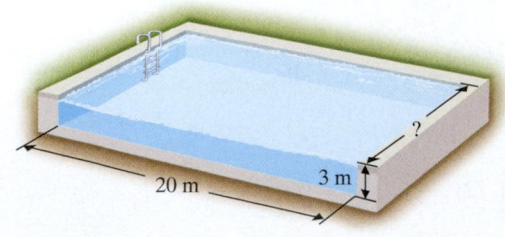

56. The perimeter of a rectangular billboard is 60 feet and the billboard has a length 6 feet longer than its width. Find the dimensions of the billboard.

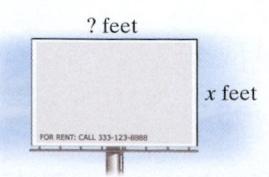

57. A charity 10K race is given annually to benefit a local hospice organization. How long will it take to run/walk a 10K race (10 kilometers or 10,000 meters) if your average pace is 125 **meters** per minute? Give your time in hours and minutes.

58. On September 14, 2013, the highest temperature recorded in the United States was 113°F, which occurred in Death Valley, California. Convert this temperature to degrees Celsius. (*Source:* National Weather Service)

(9.6) *Find each of the following.*

59. The number 9 is what percent of 45?

60. The number 59.5 is what percent of 85?

61. The number 137.5 is 125% of what number?

62. The number 768 is 60% of what number?

63. The price of a small diamond ring was recently increased by 11%. If the ring originally cost $1900, find the mark-up and the new price of the ring.

64. The U.S. motion picture and television industry is made up of over 108,000 businesses. About 85% of these are small businesses with fewer than 10 employees. How many motion picture and television industry businesses have fewer than 10 employees? (*Source:* Motion Picture Association of America)

65. Thirty gallons of a 20% acid solution are needed for an experiment. Only 40% and 10% acid solutions are available. How much of each should be mixed to form the needed solution?

66. In 2003, the average price of a cinema ticket was $6.03. By 2012, this price had increased to $7.96. What was the percent of increase? (*Source:* National Association of Theatre Owners)

The graph below shows the percent(s) of cell phone users who have engaged in various behaviors while driving and talking on their cell phones. Use this graph to answer Exercises 67 through 70.

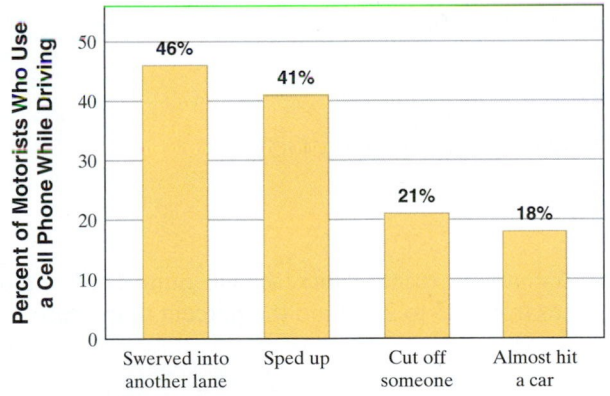

Effects of Cell Phone Use on Driving

Source: Progressive Insurance

67. What percent of motorists who use a cell phone while driving have almost hit another car?

68. What is the most common effect of cell phone use on driving?

69. If a cell phone service has an estimated 4600 customers who use their cell phones while driving, how many of these customers would you expect to have cut someone off while driving and talking on their cell phones?

70. Do the percents in the graph to the left have a sum of 100%? Why or why not?

(9.7) *Graph on a number line.*

71. $x \leq -2$

     -5 -4 -3 -2 -1 0 1 2 3 4 5

72. $0 < x \leq 5$

     -5 -4 -3 -2 -1 0 1 2 3 4 5

Solve each inequality.

73. $x - 5 \leq -4$

74. $x + 7 > 2$

75. $-2x \geq -20$

76. $-3x > 12$

77. $5x - 7 > 8x + 5$

78. $x + 4 \geq 6x - 16$

79. $\frac{2}{3}y > 6$

80. $-0.5y \leq 7.5$

81. $-2(x - 5) > 2(3x - 2)$

82. $4(2x - 5) \leq 5x - 1$

83. Carol Abolafia earns \$175 per week plus a 5% commission on all her sales. Find the minimum amount of sales she must make to ensure that she earns at least \$300 per week.

84. Joseph Barrow shot rounds of 76, 82, and 79 golfing. What must he shoot on his next round so that his average will be below 80?

Mixed Review

Solve each equation.

85. $6x + 2x - 1 = 5x + 11$

86. $2(3y - 4) = 6 + 7y$

87. $4(3 - a) - (6a + 9) = -12a$

88. $\frac{x}{3} - 2 = 5$

89. $2(y + 5) = 2y + 10$

90. $7x - 3x + 2 = 2(2x - 1)$

Solve.

91. The sum of six and twice a number is equal to seven less than the number. Find the number.

92. A 23-inch piece of string is to be cut into two pieces so that the length of the longer piece is three more than four times the shorter piece. If x represents the length of the shorter piece, find the lengths of both pieces.

93. Solve $V = \frac{1}{3}Ah$ for h.

94. What number is 26% of 85?

95. The number 72 is 45% of what number?

96. A company recently increased its number of employees from 235 to 282. Find the percent of increase.

Solve each inequality. Graph the solution set.

97. $4x - 7 > 3x + 2$

$$\begin{array}{c} \longleftrightarrow \\ -10\ -8\ -6\ -4\ -2\ \ 0\ \ 2\ \ 4\ \ 6\ \ 8\ \ 10 \end{array}$$

98. $-5x < 20$

$$\begin{array}{c} \longleftrightarrow \\ -5\ -4\ -3\ -2\ -1\ \ 0\ \ 1\ \ 2\ \ 3\ \ 4\ \ 5 \end{array}$$

99. $-3(1 + 2x) + x \geq -(3 - x)$

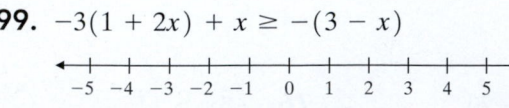

Solve each equation.

Answers

1. $-\dfrac{4}{5}x = 4$

2. $4(n - 5) = -(4 - 2n)$

3. $5y - 7 + y = -(y + 3y)$

4. $4z + 1 - z = 1 + z$

5. $\dfrac{2(x + 6)}{3} = x - 5$

6. $\dfrac{4(y - 1)}{5} = 2y + 3$

7. $\dfrac{1}{2} - x + \dfrac{3}{2} = x - 4$

8. $\dfrac{1}{3}(y + 3) = 4y$

9. $-0.3(x - 4) + x = 0.5(3 - x)$

10. $-4(a + 1) - 3a = -7(2a - 3)$

11. $-2(x - 3) = x + 5 - 3x$

Solve each application.

12. A number increased by two-thirds of the number is 35. Find the number.

△ **13.** A gallon of water seal covers 200 square feet. How many gallons are needed to paint two coats of water seal on a deck that measures 20 feet by 35 feet?

20 feet   35 feet

14. Find the value of x if $y = -14$, $m = -2$, and $b = -2$ in the formula $y = mx + b$.

Solve each equation for the indicated variable.

15. $V = \pi r^2 h$ for h

16. $3x - 4y = 10$ for y

1. _____

2. _____

3. _____

4. _____

5. _____

6. _____

7. _____

8. _____

9. _____

10. _____

11. _____

12. _____

13. _____

14. _____

15. _____

16. _____

713

17. _____

Solve each inequality. Graph the solution set.

17. $3x - 5 \geq 7x + 3$

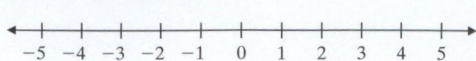

18. $x + 6 > 4x - 6$

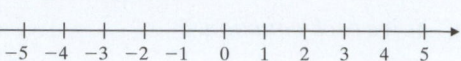

18. _____

Solve each inequality.

19. $-0.3x \geq 2.4$

20. $-5(x - 1) + 6 \leq -3(x + 4) + 1$

19. _____

21. $\dfrac{2(5x + 1)}{3} > 2$

20. _____

The following graph shows the breakdown of tornadoes occurring in the United States by strength. The corresponding Fujita Tornado Scale categories are shown in parentheses. Use this graph to answer Exercise 22.

21. _____

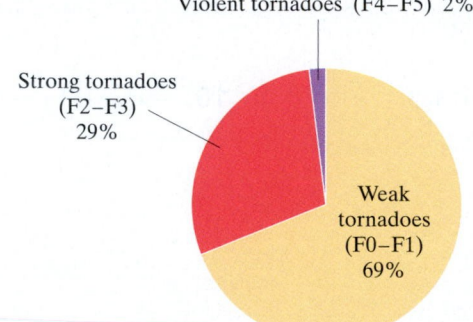

Violent tornadoes (F4–F5) 2%

Strong tornadoes (F2–F3) 29%

Weak tornadoes (F0–F1) 69%

Source: National Climatic Data Center

22. _____

23. _____

22. According to the National Climatic Data Center, in an average year, about 800 tornadoes are reported in the United States. How many of these would you expect to be classified as "weak" tornadoes?

23. The number 72 is what percent of 180?

24. _____

24. Some states have a single area code for the entire state. Two such states have area codes where one is double the other. If the sum of these integers is 1203, find the two area codes.

25. California has more public libraries than any other state. It has 387 more public libraries than Ohio. If the total number of public libraries for these states is 1827, find the number of public libraries in California and the number in Ohio. (*Source:* Institute of Museum and Library Services)

25. _____

Determine whether each statement is true or false.

1. $8 \geq 8$

2. $-4 < -6$

3. $8 \leq 8$

4. $3 > -3$

5. $23 \leq 0$

6. $-8 \geq -8$

7. $0 \leq 23$

8. $-8 \leq -8$

9. Add: $2\frac{1}{3} + 5\frac{3}{8}$

10. Perform the indicated operation.

  a. $\dfrac{2}{5} + \dfrac{3}{10}$

  b. $\dfrac{7}{8} - \dfrac{1}{3}$

Simplify.

11. $\dfrac{3 + |4 - 3| + 2^2}{6 - 3}$

12. $1 + 2(9 - 7)^3 + 4^2$

Add without using number lines.

13. $(-8) + (-11)$

14. $-2 + (-8)$

15. $(-2) + 10$

16. $-10 + 20$

17. $0.2 + (-0.5)$

18. $1.2 + (-1.2)$

19. Simplify each expression.
  a. $-3 + [(-2 - 5) - 2]$
  b. $2^3 - 10 + [-6 - (-5)]$

20. Simplify each expression.
  a. $-(-5)$ **b.** $-\left(-\dfrac{2}{3}\right)$
  c. $-(-a)$ **d.** $-|-3|$

Answers

1. _____

2. _____

3. _____

4. _____

5. _____

6. _____

7. _____

8. _____

9. _____

10. a. _____

   b. _____

11. _____

12. _____

13. _____

14. _____

15. _____

16. _____

17. _____

18. _____

19. a. _____

   b. _____

20. a. _____

   b. _____

   c. _____

   d. _____

21. a. _____

b. _____

c. _____

22. a. _____

b. _____

c. _____

23. a. _____

b. _____

c. _____

24. a. _____

b. _____

25. _____

26. _____

27. _____

28. _____

29. a. _____

b. _____

c. _____

d. _____

e. _____

30. a. _____

b. _____

c. _____

31. _____

32. _____

33. _____

21. Multiply.
   a. $7(0)(-6)$
   b. $(-2)(-3)(-4)$
   c. $(-1)(-5)(-9)(-2)$

22. Subtract.
   a. $-2.7 - 8.4$
   b. $-\dfrac{4}{5} - \left(-\dfrac{3}{5}\right)$
   c. $\dfrac{1}{4} - \left(-\dfrac{1}{2}\right)$

23. Use the definition of the quotient of two numbers to find each quotient.
   a. $-18 \div 3$
   b. $\dfrac{-14}{-2}$
   c. $\dfrac{20}{-4}$

24. Find each product.
   a. $(4.5)(-0.08)$
   b. $-\dfrac{3}{4} \cdot -\dfrac{8}{17}$

Use the distributive property to write each expression without parentheses. Then simplify the result.

25. $-5(-3 + 2z)$

26. $2(y - 3x + 4)$

27. $\dfrac{1}{2}(6x + 14) + 10$

28. $-(x + 4) + 3(x + 4)$

29. Determine whether the terms are like or unlike.
   a. $2x, 3x^2$
   b. $4x^2y, x^2y, -2x^2y$
   c. $-2yz, -3zy$
   d. $-x^4, x^4$
   e. $-8a^5, 8a^5$

30. Find each quotient.
   a. $\dfrac{-32}{8}$ **b.** $\dfrac{-108}{-12}$
   c. $-\dfrac{5}{7} \div \left(-\dfrac{9}{2}\right)$

31. Subtract $4x - 2$ from $2x - 3$.

32. Subtract $10x + 3$ from $-5x + 1$.

33. Solve $x - 7 = 10$ for x.

Solve.

34. $\dfrac{5}{6} + x = \dfrac{2}{3}$

35. $-z - 4 = 6$

36. $-3x + 1 - (-4x - 6) = 10$

37. $\dfrac{2(a + 3)}{3} = 6a + 2$

38. $\dfrac{x}{4} = 18$

39. As of May 2014, the total number of Democrats and Republicans in the U.S. House of Representatives was 432. There were 34 more Republican representatives than Democratic. Find the number of representatives from each party. (*Source:* Office of the Clerk, U.S. Capitol)

40. $6x + 5 = 4(x + 4) - 1$

41. A glacier is a giant mass of rocks and ice that flows downhill like a river. Portage Glacier in Alaska is about 6 miles, or 31,680 feet, long and moves 400 feet per year. Icebergs are created when the front end of the glacier flows into Portage Lake. How long does it take for ice at the head (beginning) of the glacier to reach the lake?

42. A number increased by 4 is the same as 3 times the number decreased by 8. Find the number.

43. The number 63 is what percent of 72?

44. Solve $C = 2\pi r$ for r.

45. Solve: $5(2x + 3) = -1 + 7$

46. Solve: $x - 3 > 2$

47. Graph $-1 > x$.

48. Solve: $3x - 4 \le 2x - 14$

49. Solve: $2(x - 3) - 5 \le 3(x + 2) - 18$

50. Solve: $-3x \ge 9$

34. _____

35. _____

36. _____

37. _____

38. _____

39. _____

40. _____

41. _____

42. _____

43. _____

44. _____

45. _____

46. _____

47. _____

48. _____

49. _____

50. _____

In Chapter 9 we learned to solve and graph the solutions of linear equations and inequalities in one variable on number lines. Now we define and present techniques for solving and graphing linear equations and inequalities in two variables on grids. Two-variable equations lead directly to the concept of *function*, perhaps the most important concept in all of mathematics. Functions are introduced in Section 10.6.

What Is *Tourism Towards 2030*?

Tourism 2020 Vision was the World Tourism Organization's long-term forecast and study of the growth and economic impact of world tourism through the year 2020. As we approach the year 2020, the World Tourism Organization's new long-term study and forecast is entitled *Tourism Towards 2030*. The broken line graph below shows some actual trends and forecasts from 2000 to 2030. In Section 10.1, Exercises 45–50, we read a bar graph showing the top tourist destinations by country.

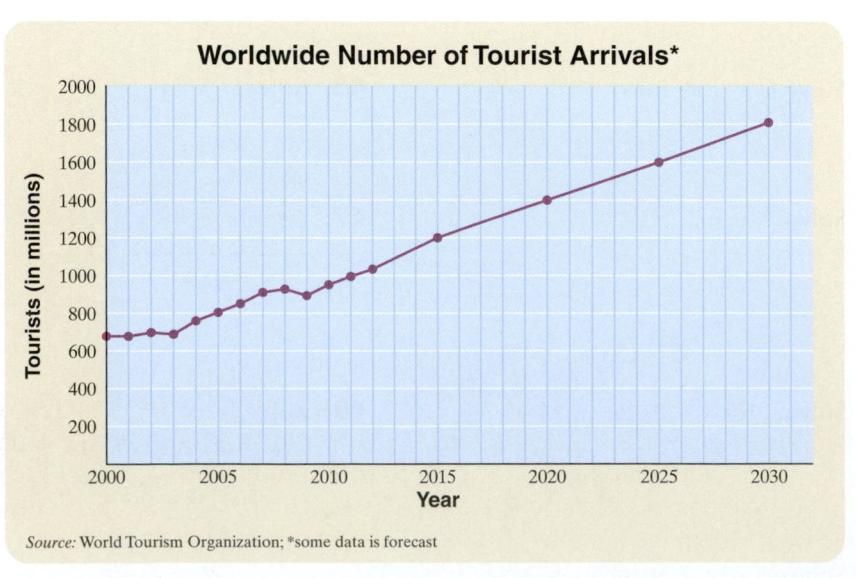

Source: World Tourism Organization; *some data is forecast

In Sections 7.1 and 7.2, we learned how to read graphs. The broken line graph below shows the relationship between the time before and after smoking a cigarette and pulse rate. The horizontal line or axis shows time in minutes and the vertical line or axis shows the pulse rate in heartbeats per minute. Notice that there are two numbers associated with each point of the graph. For example, the graph shows that 15 minutes after "lighting up," the pulse rate is 80 beats per minute. If we agree to write the time first and the pulse rate second, we can say there is a point on the graph corresponding to the **ordered pair** of numbers (15, 80). A few more ordered pairs are shown alongside their corresponding points.

Objectives

A Plot Ordered Pairs of Numbers on the Rectangular Coordinate System.

B Graph Paired Data to Create a Scatter Diagram.

C Find the Missing Coordinate of an Ordered Pair Solution, Given One Coordinate of the Pair.

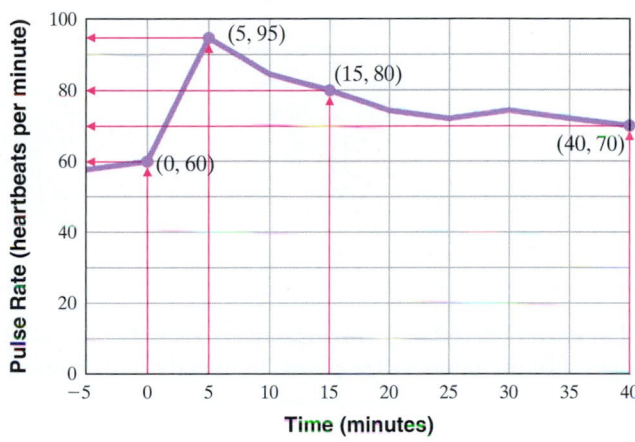

Objective A Plotting Ordered Pairs of Numbers

In general, we use the idea of ordered pairs to describe the location of a point in a plane (such as a piece of paper). We start with a horizontal and a vertical axis. Each axis is a number line, and for the sake of consistency we construct our axes to intersect at the 0 coordinate of both. This point of intersection is called the **origin**. Notice that these two number lines or axes divide the plane into four regions called **quadrants.** The quadrants are usually numbered with Roman numerals as shown. The axes are not considered to be in any quadrant.

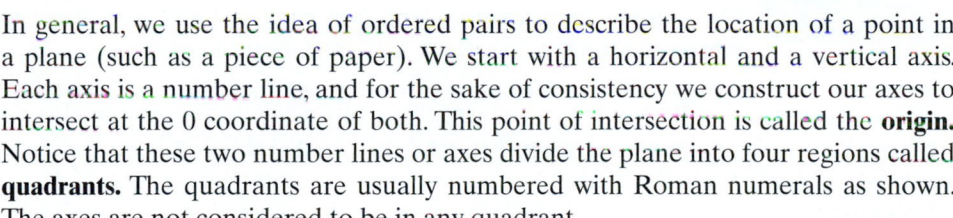

It is helpful to label axes, so we label the horizontal axis the *x*-axis and the vertical axis the *y*-axis. We call the system described above the **rectangular coordinate system,** or the **coordinate plane.** Just as with other graphs shown, we can then describe the locations of points by ordered pairs of numbers. We list the horizontal, *x*-axis measurement first and the vertical, *y*-axis measurement second.

Practice 1

On a single coordinate system, plot each ordered pair. State in which quadrant, or on which axis, each point lies.

a. $(4, 2)$ **b.** $(-1, -3)$

c. $(2, -2)$ **d.** $(-5, 1)$

e. $(0, 3)$ **f.** $(3, 0)$

g. $(0, -4)$ **h.** $\left(-2\frac{1}{2}, 0\right)$

i. $\left(1, -3\frac{3}{4}\right)$

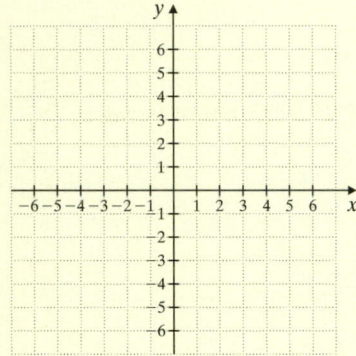

To plot or graph the point corresponding to the ordered pair (a, b) we start at the origin. We then move a units left or right (right if a is positive, left if a is negative). From there, we move b units up or down (up if b is positive, down if b is negative). For example, to plot the point corresponding to the ordered pair $(3, 2)$, we start at the origin, move 3 units right, and from there move 2 units up. (See the figure below.) The x-value, 3, is also called the **x-coordinate** and the y-value, 2, is also called the **y-coordinate.** From now on, we will call the point with coordinates $(3, 2)$ simply the point $(3, 2)$. The point $(-2, 5)$ is also graphed below.

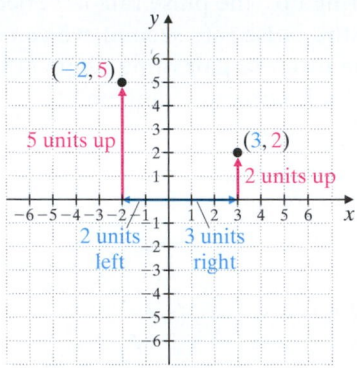

Helpful Hint

Don't forget that **each ordered pair corresponds to exactly one point in the plane and that each point in the plane corresponds to exactly one ordered pair.**

✓**Concept Check** Is the graph of the point $(-5, 1)$ in the same location as the graph of the point $(1, -5)$? Explain.

Example 1

On a single coordinate system, plot each ordered pair. State in which quadrant, or on which axis, each point lies.

a. $(5, 3)$ **b.** $(-2, -4)$ **c.** $(1, -2)$ **d.** $(-5, 3)$ **e.** $(0, 0)$

f. $(0, 2)$ **g.** $(-5, 0)$ **h.** $\left(0, -5\frac{1}{2}\right)$ **i.** $\left(4\frac{2}{3}, -3\right)$

Solution:

a. Point $(5, 3)$ lies in quadrant I.

b. Point $(-2, -4)$ lies in quadrant III.

c. Point $(1, -2)$ lies in quadrant IV.

d. Point $(-5, 3)$ lies in quadrant II.

e–h. Points $(0, 0)$, $(0, 2)$, and $\left(0, -5\frac{1}{2}\right)$ lie on the y-axis. Points $(0, 0)$ and $(-5, 0)$ lie on the x-axis.

i. Point $\left(4\frac{2}{3}, -3\right)$ lies in quadrant IV.

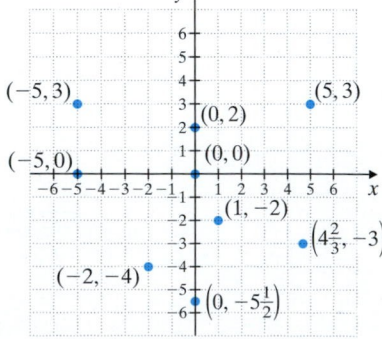

■ **Work Practice 1**

Helpful Hint

In Example 1, notice that the point $(0, 0)$ lies on both the x-axis and the y-axis. It is the only point in the entire rectangular coordinate system that has this feature. Why? It is the only point of intersection of the x-axis and the y-axis.

Answers

1.

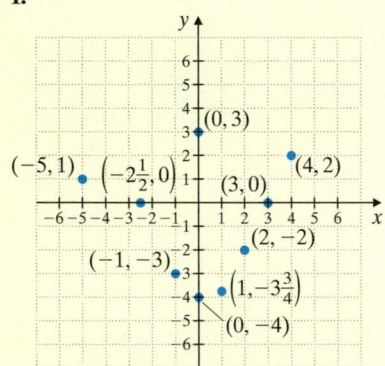

a. Point $(4, 2)$ lies in quadrant I.

b. Point $(-1, -3)$ lies in quadrant III.

c. Point $(2, -2)$ lies in quadrant IV.

d. Point $(-5, 1)$ lies in quadrant II.

e–h. Points $(3, 0)$ and $\left(-2\frac{1}{2}, 0\right)$ lie on the x-axis. Points $(0, 3)$ and $(0, -4)$ lie on the y-axis.

i. Point $\left(1, -3\frac{3}{4}\right)$ lies in quadrant IV.

✓**Concept Check Answer**

The graph of point $(-5, 1)$ lies in quadrant II and the graph of point $(1, -5)$ lies in quadrant IV. They are *not* in the same location.

✔**Concept Check** For each description of a point in the rectangular coordinate system, write an ordered pair that represents it.

a. Point *A* is located three units to the left of the *y*-axis and five units above the *x*-axis.

b. Point *B* is located six units below the origin.

From Example 1, notice that the *y*-coordinate of any point on the *x*-axis is 0. For example, the point $(-5, 0)$ lies on the *x*-axis. Also, the *x*-coordinate of any point on the *y*-axis is 0. For example, the point $(0, 2)$ lies on the *y*-axis.

Objective B Creating Scatter Diagrams

Data that can be represented as ordered pairs are called **paired data.** Many types of data collected from the real world are paired data. For instance, the annual measurements of a child's height can be written as ordered pairs of the form (year, height in inches) and are paired data. The graph of paired data as points in the rectangular coordinate system is called a **scatter diagram.** Scatter diagrams can be used to look for patterns and trends in paired data.

> **Example 2** The table gives the annual net sales for PetSmart for the years shown. (*Source:* PetSmart)
>
> **a.** Write this paired data as a set of ordered pairs of the form (year, net sales in billions of dollars).
>
> **b.** Create a scatter diagram of the paired data.
>
> **c.** What trend in the paired data does the scatter diagram show?

Year	PetSmart Net Sales (in billions of dollars)
2007	4.7
2008	5.1
2009	5.3
2010	5.7
2011	6.1
2012	6.8

Solution:

a. The ordered pairs are $(2007, 4.7)$, $(2008, 5.1)$, $(2009, 5.3)$, $(2010, 5.7)$, $(2011, 6.1)$, and $(2012, 6.8)$.

b. We begin by plotting the ordered pairs. Because the *x*-coordinate in each ordered pair is a year, we label the *x*-axis "Year" and mark the horizontal axis with the years given. Then we label the *y*-axis or vertical axis "Net Sales (in billions of dollars)." In this case, it is convenient to mark the vertical axis in increments of 1, starting with 0.

PetSmart Net Sales

(scatter diagram: Net Sales (in billions of dollars) vs. Year, points plotted at (2007, 4.7), (2008, 5.1), (2009, 5.3), (2010, 5.7), (2011, 6.1), (2012, 6.8))

c. The scatter diagram shows that PetSmart net sales steadily increased over the years 2007–2012.

■ **Work Practice 2**

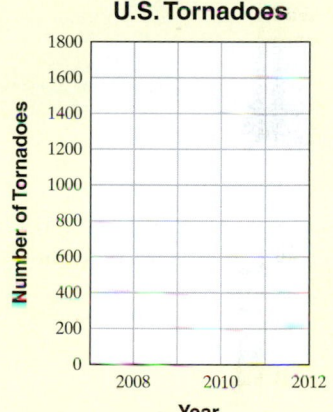
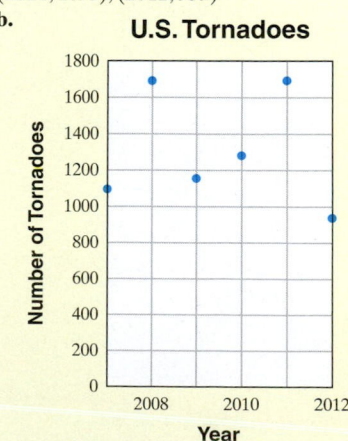

Objective C Completing Ordered Pair Solutions

Let's see how we can use ordered pairs to record solutions of equations containing two variables. An equation in one variable such as $x + 1 = 5$ has one solution, 4: The number 4 is the value of the variable x that makes the equation true.

An equation in two variables, such as $2x + y = 8$, has solutions consisting of two values, one for x and one for y. For example, $x = 3$ and $y = 2$ is a solution of $2x + y = 8$ because, if x is replaced with 3 and y with 2, we get a true statement.

$$2x + y = 8$$
$$2(3) + 2 \stackrel{?}{=} 8 \quad \text{Replace } x \text{ with 3 and } y \text{ with 2.}$$
$$8 = 8 \quad \text{True}$$

The solution $x = 3$ and $y = 2$ can be written as $(3, 2)$, an ordered pair of numbers.

> In general, an ordered pair is a **solution** of an equation in two variables if replacing the variables by the values of the ordered pair results in a *true statement*.

For example, another ordered pair solution of $2x + y = 8$ is $(5, -2)$. Replacing x with 5 and y with -2 results in a true statement.

$$2x + y = 8$$
$$2(5) + (-2) \stackrel{?}{=} 8 \quad \text{Replace } x \text{ with 5 and } y \text{ with } -2.$$
$$10 - 2 \stackrel{?}{=} 8$$
$$8 = 8 \quad \text{True}$$

Practice 3

Complete each ordered pair so that it is a solution to the equation $x + 2y = 8$.

a. $(0, \ )$
b. $(\ , 3)$
c. $(-4, \ )$

Example 3 Complete each ordered pair so that it is a solution to the equation $3x + y = 12$.

a. $(0, \ )$ **b.** $(\ , 6)$ **c.** $(-1, \ )$

Solution:

a. In the ordered pair $(0, \ )$, the x-value is 0. We let $x = 0$ in the equation and solve for y.

$$3x + y = 12$$
$$3(0) + y = 12 \quad \text{Replace } x \text{ with 0.}$$
$$0 + y = 12$$
$$y = 12$$

The completed ordered pair is $(0, 12)$.

b. In the ordered pair $(\ , 6)$, the y-value is 6. We let $y = 6$ in the equation and solve for x.

$$3x + y = 12$$
$$3x + 6 = 12 \quad \text{Replace } y \text{ with 6.}$$
$$3x = 6 \quad \text{Subtract 6 from both sides.}$$
$$x = 2 \quad \text{Divide both sides by 3.}$$

The ordered pair is $(2, 6)$.

c. In the ordered pair $(-1, \ )$, the x-value is -1. We let $x = -1$ in the equation and solve for y.

$$3x + y = 12$$
$$3(-1) + y = 12 \quad \text{Replace } x \text{ with } -1.$$
$$-3 + y = 12$$
$$y = 15 \quad \text{Add 3 to both sides.}$$

The ordered pair is $(-1, 15)$.

■ **Work Practice 3**

Answers

3. a. $(0, 4)$ **b.** $(2, 3)$ **c.** $(-4, 6)$

Solutions of equations in two variables can also be recorded in a **table of paired values,** as shown in the next example.

Example 4 Complete the table for the equation $y = 3x$.

x	y
a. -1	
b.	0
c.	-9

Solution:

a. We replace x with -1 in the equation and solve for y.

$$y = 3x$$
$$y = 3(-1) \quad \text{Let } x = -1.$$
$$y = -3$$

The ordered pair is $(-1, -3)$.

b. We replace y with 0 in the equation and solve for x.

$$y = 3x$$
$$0 = 3x \quad \text{Let } y = 0.$$
$$0 = x \quad \text{Divide both sides by 3.}$$

The ordered pair is $(0, 0)$.

c. We replace y with -9 in the equation and solve for x.

$$y = 3x$$
$$-9 = 3x \quad \text{Let } y = -9.$$
$$-3 = x \quad \text{Divide both sides by 3.}$$

The ordered pair is $(-3, -9)$. The completed table is shown to the right.

x	y
-1	-3
0	0
-3	-9

■ **Work Practice 4**

Example 5 Complete the table for the equation

$$y = \frac{1}{2}x - 5.$$

x	y
a. -2	
b. 0	
c.	0

Solution:

a. Let $x = -2$.

$$y = \frac{1}{2}x - 5$$
$$y = \frac{1}{2}(-2) - 5$$
$$y = -1 - 5$$
$$y = -6$$

b. Let $x = 0$.

$$y = \frac{1}{2}x - 5$$
$$y = \frac{1}{2}(0) - 5$$
$$y = 0 - 5$$
$$y = -5$$

c. Let $y = 0$.

$$y = \frac{1}{2}x - 5$$
$$0 = \frac{1}{2}x - 5 \quad \text{Now, solve for } x.$$
$$5 = \frac{1}{2}x \quad \text{Add 5.}$$
$$10 = x \quad \text{Multiply by 2.}$$

Ordered pairs: $(-2, -6)$ $(0, -5)$ $(10, 0)$

The completed table is

x	-2	0	10
y	-6	-5	0

■ **Work Practice 5**

Practice 4

Complete the table for the equation $y = -2x$.

x	y
a. -3	
b.	0
c.	10

Practice 5

Complete the table for the equation $y = \frac{1}{3}x - 1$.

x	y
a. -3	
b. 0	
c.	0

Answers

4.

x	y
a. -3	6
b. 0	0
c. -5	10

5.

x	y
a. -3	-2
b. 0	-1
c. 3	0

By now, you have noticed that equations in two variables often have more than one solution. We discuss this more in the next section.

A table showing ordered pair solutions may be written vertically or horizontally, as shown in the next example.

Practice 6

A company purchased a fax machine for $400. The business manager of the company predicts that the fax machine will be used for 7 years and the value in dollars y of the machine in x years is $y = -50x + 400$. Complete the table.

x	1	2	3	4	5	6	7
y							

Example 6   A small business purchased a computer for $2000. The business predicts that the computer will be used for 5 years and the value in dollars y of the computer in x years is $y = -300x + 2000$. Complete the table.

x	0	1	2	3	4	5
y						

Solution:

To find the value of y when x is 0, we replace x with 0 in the equation. We use this same procedure to find y when x is 1 and when x is 2.

When $x = 0$,	When $x = 1$,	When $x = 2$,
$y = -300x + 2000$	$y = -300x + 2000$	$y = -300x + 2000$
$y = -300 \cdot 0 + 2000$	$y = -300 \cdot 1 + 2000$	$y = -300 \cdot 2 + 2000$
$y = 0 + 2000$	$y = -300 + 2000$	$y = -600 + 2000$
$y = 2000$	$y = 1700$	$y = 1400$

We have the ordered pairs (0, 2000), (1, 1700), and (2, 1400). This means that in 0 years the value of the computer is $2000, in 1 year the value of the computer is $1700, and in 2 years the value is $1400. To complete the table of values, we continue the procedure for $x = 3$, $x = 4$, and $x = 5$.

When $x = 3$,	When $x = 4$,	When $x = 5$,
$y = -300x + 2000$	$y = -300x + 2000$	$y = -300x + 2000$
$y = -300 \cdot 3 + 2000$	$y = -300 \cdot 4 + 2000$	$y = -300 \cdot 5 + 2000$
$y = -900 + 2000$	$y = -1200 + 2000$	$y = -1500 + 2000$
$y = 1100$	$y = 800$	$y = 500$

The completed table is shown below.

x	0	1	2	3	4	5
y	2000	1700	1400	1100	800	500

■ **Work Practice 6**

The ordered pair solutions recorded in the completed table for Example 6 are another set of paired data. They are graphed next. Notice that this scatter diagram gives a visual picture of the decrease in value of the computer.

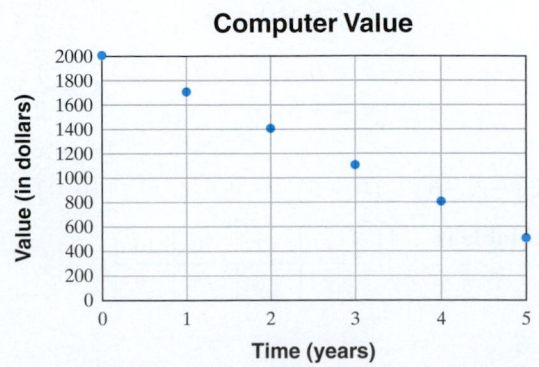

Answer

6.

x	1	2	3	4	5	6	7
y	350	300	250	200	150	100	50

Vocabulary, Readiness & Video Check

Use the choices below to fill in each blank. The exercises below all have to do with the rectangular coordinate system.

origin	*x*-coordinate	*x*-axis	scatter diagram	four
quadrants	*y*-coordinate	*y*-axis	solution	one

1. The horizontal axis is called the _____.

2. The vertical axis is called the _____.

3. The intersection of the horizontal axis and the vertical axis is a point called the _____.

4. The axes divide the plane into regions, called _____. There are _____ of these regions.

5. In the ordered pair of numbers $(-2, 5)$, the number -2 is called the _____ and the number 5 is called the _____.

6. Each ordered pair of numbers corresponds to _____ point in the plane.

7. An ordered pair is a(n) _____ of an equation in two variables if replacing the variables by the coordinates of the ordered pair results in a true statement.

8. The graph of paired data as points in a rectangular coordinate system is called a(n) _____.

Martin-Gay Interactive Videos   *Watch the section lecture video and answer the following questions.*

See Video 10.1

Objective A 9. Several points are plotted in ▯ Examples 1–6. Where do we always start when plotting a point? How does the first coordinate tell us to move? How does the second coordinate tell us to move?

Objective B 10. From ▯ Example 7, what kind of data can be graphed in a scatter diagram?

Objective C 11. In ▯ Example 8, when finding the missing value in an ordered pair solution of a linear equation in two variables, how can we check our solution?

10.1 Exercise Set MyMathLab®

Objective A *Plot each ordered pair. State in which quadrant or on which axis each point lies. See Example 1.*

1. a. $(1, 5)$ b. $(-5, -2)$ c. $(-3, 0)$ d. $(0, -1)$
   e. $(2, -4)$ f. $\left(-1, 4\frac{1}{2}\right)$ g. $(3.7, 2.2)$ h. $\left(\frac{1}{2}, -3\right)$

2. a. $(2, 4)$ b. $(0, 2)$ c. $(-2, 1)$ d. $(-3, -3)$
   e. $\left(3\frac{3}{4}, 0\right)$ f. $(5, -4)$ g. $(-3.4, 4.8)$ h. $\left(\frac{1}{3}, -5\right)$

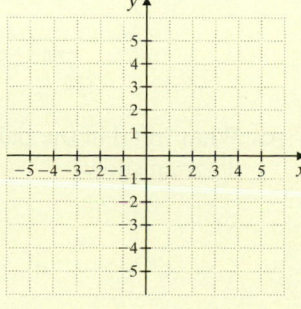

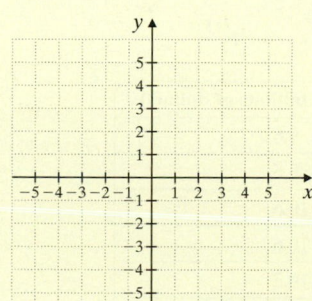

Find the x- and y-coordinates of each labeled point. See Example 1.

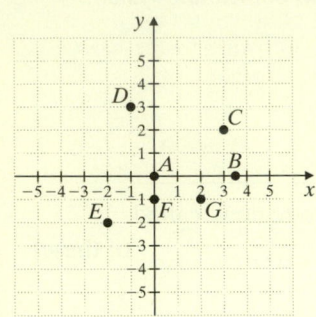

3. *A*

4. *B*

5. *C*

6. *D*

7. *E*

8. *F*

9. *G*

10. *A*

11. *B*

12. *C*

13. *D*

14. *E*

15. *F*

16. *G*

Objective B *Solve. See Example 2.*

17. The table shows the domestic box office (in billions of dollars) for the U.S. movie industry during the years shown. (*Source:* Motion Picture Association of America)

Year	Box Office (in billions of dollars)
2007	9.6
2008	9.6
2009	10.6
2010	10.6
2011	10.2
2012	10.8

a. Write this paired data as a set of ordered pairs of the form (year, box office).

b. In your own words, write the meaning of the ordered pair (2010, 10.6).

c. Create a scatter diagram of the paired data. Be sure to label the axes appropriately.

Domestic Box Office

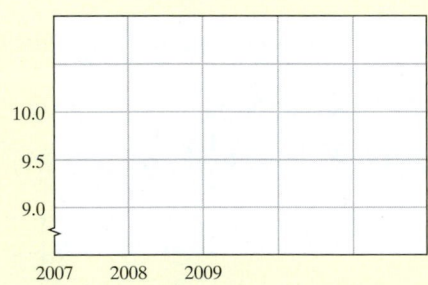

d. What trend in the paired data does the scatter diagram show?

18. The table shows the amount of money (in billions of dollars) that Americans spent on their pets for the years shown. (*Source:* American Pet Products Association, Inc.)

Year	Pet-Related Expenditures (in billions of dollars)
2007	41
2008	43
2009	46
2010	48
2011	51
2012	53

a. Write this paired data as a set of ordered pairs of the form (year, pet-related expenditures).

b. In your own words, write the meaning of the ordered pair (2012, 53).

c. Create a scatter diagram of the paired data. Be sure to label the axes appropriately.

Pet-Related Expenditures

d. What trend in the paired data does the scatter diagram show?

19. Minh, a psychology student, kept a record of how much time she spent studying for each of her 20-point psychology quizzes and her score on each quiz.

Hours Spent Studying	0.50	0.75	1.00	1.25	1.50	1.50	1.75	2.00
Quiz Score	10	12	15	16	18	19	19	20

a. Write the data as ordered pairs of the form (hours spent studying, quiz score).

b. In your own words, write the meaning of the ordered pair (1.25, 16).

c. Create a scatter diagram of the paired data. Be sure to label the axes appropriately.

d. What might Minh conclude from the scatter diagram?

Minh's Chart for Psychology

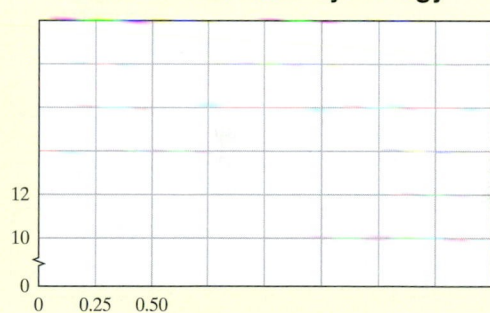

20. A local lumberyard uses quantity pricing. The table shows the price per board for different amounts of lumber purchased.

Price per Board (in dollars)	Number of Boards Purchased
8.00	1
7.50	10
6.50	25
5.00	50
2.00	100

a. Write the data as ordered pairs of the form (price per board, number of boards purchased).

b. In your own words, write the meaning of the ordered pair (2.00, 100).

c. Create a scatter diagram of the paired data. Be sure to label the axes appropriately.

Lumberyard Board Pricing

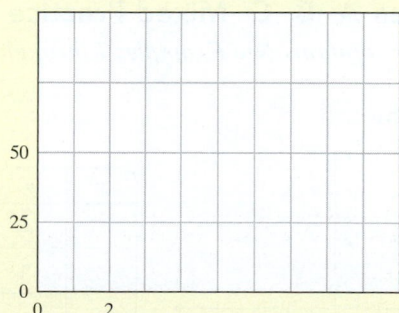

d. What trend in the paired data does the scatter diagram show?

Objective C *Complete each ordered pair so that it is a solution to the given linear equation. See Example 3.*

21. $x - 4y = 4$; $(\ \ , -2), (4, \ \ )$

22. $x - 5y = -1$; $(\ \ , -2), (4, \ \ )$

23. $y = \dfrac{1}{4}x - 3$; $(-8, \ \ ), (\ \ , 1)$

24. $y = \dfrac{1}{5}x - 2$; $(-10, \ \ ), (\ \ , 1)$

Complete the table of ordered pairs for each linear equation. See Examples 4 and 5.

25. $y = -7x$

x	y
0	
−1	
	2

26. $y = -9x$

x	y
	0
−3	
	2

27. $x = -y + 2$

x	y
0	
	0
−3	

28. $x = -y + 4$

x	y
	0
0	
	−3

29. $y = \dfrac{1}{2}x$

x	y
0	
−6	
	1

30. $y = \dfrac{1}{3}x$

x	y
0	
−6	
	1

31. $x + 3y = 6$

x	y
0	
	0
	1

32. $2x + y = 4$

x	y
0	
	0
	2

33. $y = 2x - 12$

x	y
0	
	−2
3	

34. $y = 5x + 10$

x	y
	0
	5
0	

35. $2x + 7y = 5$

x	y
0	
	0
	1

36. $x - 6y = 3$

x	y
0	
1	
	−1

Objectives A B C Mixed Practice *Complete the table of ordered pairs for each equation. Then plot the ordered pair solutions. See Examples 1 through 5.*

37. $x = -5y$

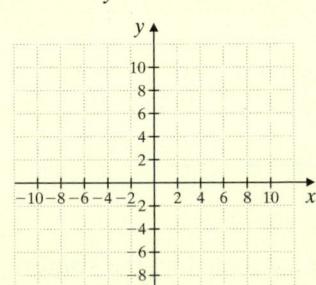

x	y
	0
	1
10	

38. $y = -3x$

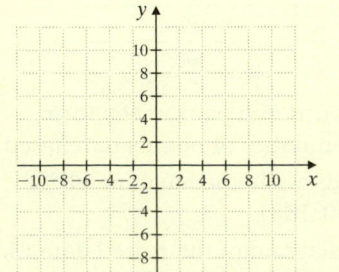

x	y
0	
−2	
	9

39. $y = \dfrac{1}{3}x + 2$

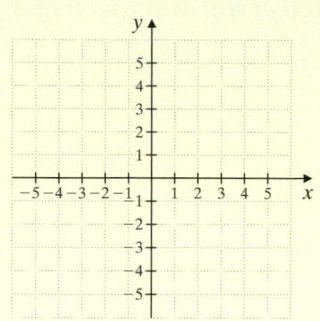

x	y
0	
−3	
	0

40. $y = \dfrac{1}{2}x + 3$

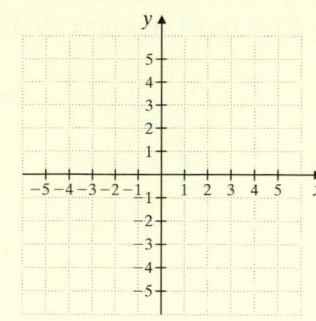

x	y
0	
−4	
	0

Solve. See Example 6.

41. The cost in dollars y of producing x computer desks is given by $y = 80x + 5000$.

   a. Complete the table.

x	100	200	300
y			

   b. Find the number of computer desks that can be produced for $8600. (*Hint:* Find x when $y = 8600$.)

42. The hourly wage y of an employee at a certain production company is given by $y = 0.25x + 9$, where x is the number of units produced by the employee in an hour.

   a. Complete the table.

x	0	1	5	10
y				

   b. Find the number of units that the employee must produce each hour to earn an hourly wage of $12.25. (*Hint:* Find x when $y = 12.25$.)

43. The average annual cinema admission price y (in dollars) from 2003 through 2012 is given by $y = 0.24x + 5.96$. In this equation, x represents the number of years after 2003. (*Source:* Motion Picture Association of America)

   a. Complete the table.

x	2	5	8
y			

   b. Find the year in which the average cinema admission price was approximately $7.40. (*Hint:* Find x when $y = 7.40$ and round to the nearest whole number.)

   c. Use the given equation to predict when the cinema admission price might be $9.00. (Use the hint for part **b.**)

44. The number of farms y in the United States from 2009 through 2012 is given by $y = -10,000x + 2,201,000$. In the equation, x represents the number of years after 2009. (*Source:* Based on data from the National Agricultural Statistics Service)

   a. Complete the table.

x	0	2	4
y			

   b. Find the year in which there were approximately 2,170,000 farms. (*Hint:* Find x when $y = 2,170,000$ and round to the nearest whole number.)

   c. Use the given equation to predict when the number of farms might be 2,100,000. (Use the hint for part **b.**)

Review

The following bar graph shows the top 10 tourist destinations and the number of tourists that visit each country per year. Use this graph to answer Exercises 45 through 50. See Section 7.1.

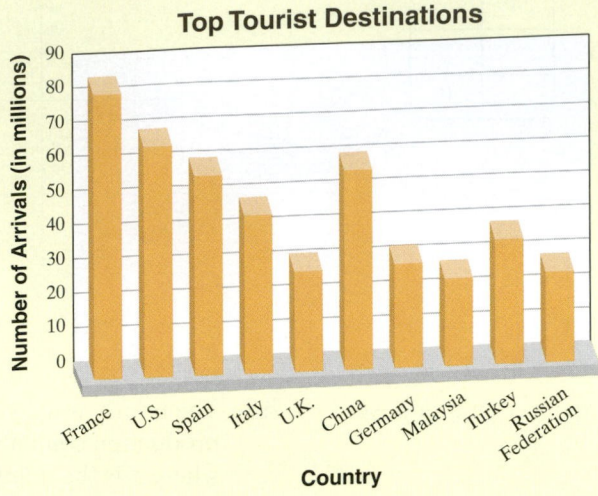

Top Tourist Destinations

Source: World Tourism Organization

45. Which country shown has the most tourist arrivals?

46. Which country shown has the least tourist arrivals?

47. Which countries shown have more than 50 million tourists per year?

48. Which countries shown have fewer than 30 million tourists per year?

49. Estimate the number of tourists per year whose destination is Germany.

50. Estimate the number of tourists per year whose destination is Malaysia.

Solve each equation for y. See Section 9.5.

51. $x + y = 5$

52. $x - y = 3$

53. $2x + 4y = 5$

54. $5x + 2y = 7$

55. $10x = -5y$

56. $4y = -8x$

Concept Extensions

Answer each exercise with true or false.

57. Point $(-1, 5)$ lies in quadrant IV.

58. Point $(3, 0)$ lies on the y-axis.

59. For the point $\left(-\dfrac{1}{2}, 1.5\right)$, the first value, $-\dfrac{1}{2}$, is the x-coordinate and the second value, 1.5, is the y-coordinate.

60. The ordered pair $\left(2, \dfrac{2}{3}\right)$ is a solution of $2x - 3y = 6$.

For Exercises 61 through 65, fill in each blank with "0," "positive," or "negative." For Exercises 66 and 67, fill in each blank with "x" or "y."

	Point	Location
61.	(_____ , _____)	quadrant III
62.	(_____ , _____)	quadrant I
63.	(_____ , _____)	quadrant IV
64.	(_____ , _____)	quadrant II
65.	(_____ , _____)	origin
66.	(number, 0)	__ -axis
67.	(0, number)	__ -axis

68. Give an example of an ordered pair whose location is in (or on)
   a. quadrant I   **b.** quadrant II   **c.** quadrant III
   d. quadrant IV   **e.** *x*-axis   **f.** *y*-axis

Solve. See the Concept Checks in this section.

69. Is the graph of (3, 0) in the same location as the graph of (0, 3)? Explain why or why not.

70. Give the coordinates of a point such that if the coordinates are reversed, their location is the same.

71. In general, what points can have coordinates reversed and still have the same location?

72. In your own words, describe how to plot or graph an ordered pair of numbers.

73. Discuss any similarities in the graphs of the ordered pair solutions for Exercises **37–40**.

74. Discuss any differences in the graphs of the ordered pair solutions for Exercises **37–40**.

Write an ordered pair for each point described.

75. Point *C* is four units to the right of the *y*-axis and seven units below the *x*-axis.

76. Point *D* is three units to the left of the origin.

Solve.

77. Find the perimeter of the rectangle whose vertices are the points with coordinates (−1, 5), (3, 5), (3, −4), and (−1, −4).

78. Find the area of the rectangle whose vertices are the points with coordinates (5, 2), (5, −6), (0, −6), and (0, 2).

The scatter diagram below shows the annual number of people enrolled as Gold Star Members at Costco Wholesale. The horizontal axis represents the number of years after 2008.

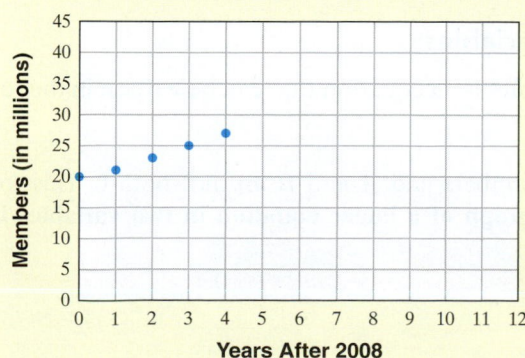

Costco's Annual Gold Star Membership

Members (in millions) — Years After 2008

Source: Costco Wholesale Corporation

79. Estimate the annual Gold Star Membership for years 1, 2, 3, and 4.

80. Use a straightedge or ruler and this scatter diagram to predict Costco's Gold Star Membership in the year 2018.

Objective

A Graph a Linear Equation by Finding and Plotting Ordered Pair Solutions.

In the previous section, we found that equations in two variables may have more than one solution. For example, both $(2, 2)$ and $(0, 4)$ are solutions of the equation $x + y = 4$. In fact, this equation has an infinite number of solutions. Other solutions include $(-2, 6)$, $(4, 0)$, and $(6, -2)$. Notice the pattern that appears in the graph of these solutions.

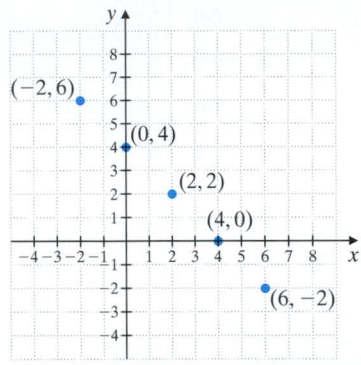

These solutions all appear to lie on the same line, as seen in the second graph. It can be shown that every ordered pair solution of the equation corresponds to a point on this line and that every point on this line corresponds to an ordered pair solution. Thus, we say that this line is the **graph of the equation** $x + y = 4$. Notice that we can show only a part of a line on a graph. The arrowheads on each end of the line below remind us that the line actually extends indefinitely in both directions.

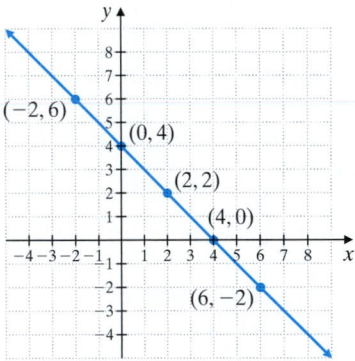

The equation $x + y = 4$ is called a *linear equation in two variables* and *the graph of every linear equation in two variables is a straight line.*

Linear Equation in Two Variables

A **linear equation in two variables** is an equation that can be written in the form

$$Ax + By = C$$

where A, B, and C are real numbers and A and B are not both 0. This form is called **standard form. The graph of a linear equation in two variables is a straight line.**

A linear equation in two variables may be written in many forms. Standard form, $Ax + By = C$, is just one of these many forms.

Following are examples of linear equations in two variables.

$$2x + y = 8 \qquad -2x = 7y \qquad y = \frac{1}{3}x + 2 \qquad y = 7$$

(Standard form)

Objective A Graphing Linear Equations

From geometry, we know that a straight line is determined by just two points. Thus, to graph a linear equation in two variables, we need to find just two of its infinitely many solutions. Once we do so, we plot the solution points and draw the line connecting the points. Usually, we find a third solution as well, as a check.

Example 1 Graph the linear equation $2x + y = 5$.

Solution: To graph this equation, we find three ordered pair solutions of $2x + y = 5$. To do this, we choose a value for one variable, x or y, and solve for the other variable. For example, if we let $x = 1$, then $2x + y = 5$ becomes

$$2x + y = 5$$
$$2(1) + y = 5 \qquad \text{Replace } x \text{ with 1.}$$
$$2 + y = 5 \qquad \text{Multiply.}$$
$$y = 3 \qquad \text{Subtract 2 from both sides.}$$

Since $y = 3$ when $x = 1$, the ordered pair $(1, 3)$ is a solution of $2x + y = 5$. Next, we let $x = 0$.

$$2x + y = 5$$
$$2(0) + y = 5 \qquad \text{Replace } x \text{ with 0.}$$
$$0 + y = 5$$
$$y = 5$$

The ordered pair $(0, 5)$ is a second solution.

The two solutions found so far allow us to draw the straight line that is the graph of all solutions of $2x + y = 5$. However, we will find a third ordered pair as a check. Let $y = -1$.

$$2x + y = 5$$
$$2x + (-1) = 5 \qquad \text{Replace } y \text{ with } -1.$$
$$2x - 1 = 5$$
$$2x = 6 \qquad \text{Add 1 to both sides.}$$
$$x = 3 \qquad \text{Divide both sides by 2.}$$

The third solution is $(3, -1)$. These three ordered pair solutions are listed in the table and plotted on the coordinate plane. The graph of $2x + y = 5$ is the line through the three points.

x	y
1	3
0	5
3	-1

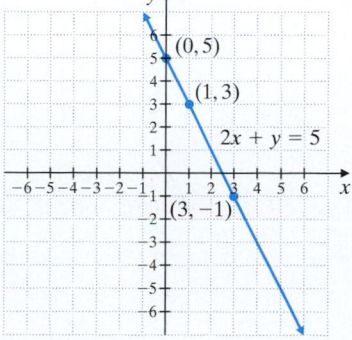

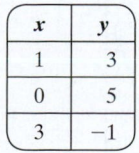 **Work Practice 1**

Practice 1

Graph the linear equation $x + 3y = 6$.

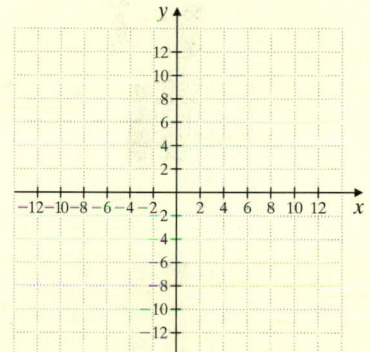

Helpful Hint All three points should fall on the same straight line. If not, check your ordered pair solutions for a mistake.

Answer

1.

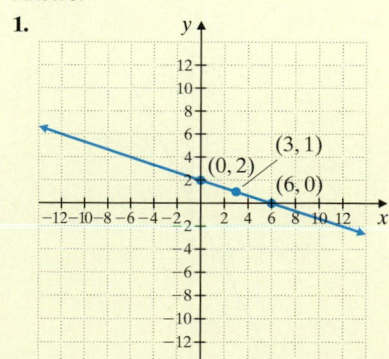

Practice 2

Graph the linear equation
$-2x + 4y = 8$.

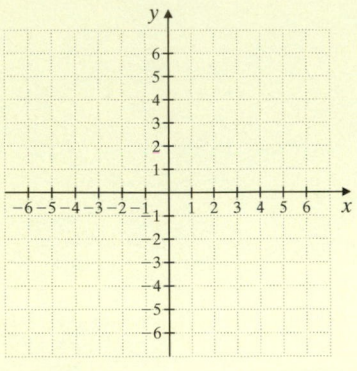

Practice 3

Graph the linear equation
$y = 2x$.

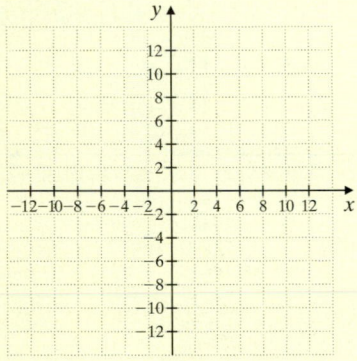

Answers

2.

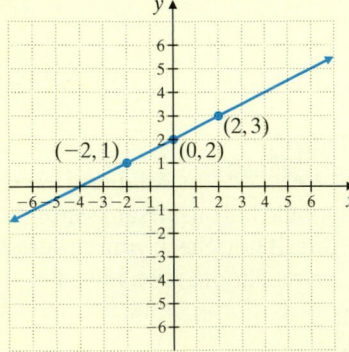

Example 2   Graph the linear equation $-5x + 3y = 15$.

Solution: We find three ordered pair solutions of $-5x + 3y = 15$.

Let $x = 0$.	**Let $y = 0$.**	**Let $x = -2$.**
$-5x + 3y = 15$	$-5x + 3y = 15$	$-5x + 3y = 15$
$-5 \cdot 0 + 3y = 15$	$-5x + 3 \cdot 0 = 15$	$-5 \cdot -2 + 3y = 15$
$0 + 3y = 15$	$-5x + 0 = 15$	$10 + 3y = 15$
$3y = 15$	$-5x = 15$	$3y = 5$
$y = 5$	$x = -3$	$y = \dfrac{5}{3}$ or $1\dfrac{2}{3}$

The ordered pairs are $(0, 5)$, $(-3, 0)$, and $\left(-2, 1\dfrac{2}{3}\right)$. The graph of $-5x + 3y = 15$ is the line through the three points.

x	y
0	5
-3	0
-2	$1\dfrac{2}{3}$

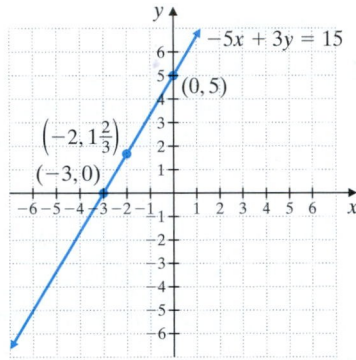

■ **Work Practice 2**

Example 3   Graph the linear equation $y = 3x$.

Solution: We find three ordered pair solutions. Since this equation is solved for y, we'll choose three x-values.

If $x = 2$, $y = 3 \cdot 2 = 6$.

If $x = 0$, $y = 3 \cdot 0 = 0$.

If $x = -1$, $y = 3 \cdot -1 = -3$.

Next, we plot the ordered pair solutions and draw a line through the plotted points. The line is the graph of $y = 3x$.

Think about the following for a moment: A line is made up of an infinite number of points. Every point on the line defined by $y = 3x$ represents an ordered pair solution of the equation, and every ordered pair solution is a point on this line.

x	y
2	6
0	0
-1	-3

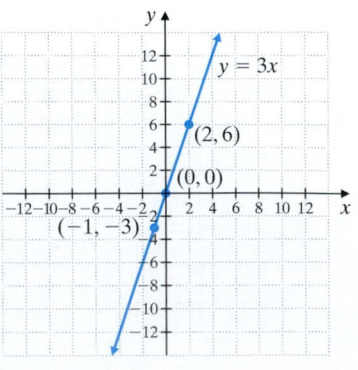

■ **Work Practice 3**

Helpful Hint

When graphing a linear equation in two variables, if it is

- solved for y, it may be easier to find ordered pair solutions by choosing x-values. If it is
- solved for x, it may be easier to find ordered pair solutions by choosing y-values.

Example 4 Graph the linear equation $y = -\dfrac{1}{3}x + 2$.

Solution: We find three ordered pair solutions, plot the solutions, and draw a line through the plotted solutions. To avoid fractions, we'll choose x-values that are multiples of 3 to substitute into the equation.

If $x = 6$, then $y = -\dfrac{1}{3} \cdot 6 + 2 = -2 + 2 = 0$.

If $x = 0$, then $y = -\dfrac{1}{3} \cdot 0 + 2 = 0 + 2 = 2$.

If $x = -3$, then $y = -\dfrac{1}{3} \cdot -3 + 2 = 1 + 2 = 3$.

x	y
6	0
0	2
-3	3

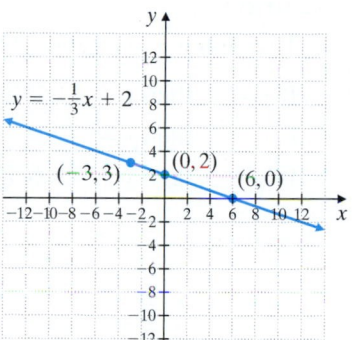

■ **Work Practice 4**

Let's take a moment and compare the graphs in Examples 3 and 4. The graph of $y = 3x$ tilts upward (as we follow the line from left to right) and the graph of $y = -\dfrac{1}{3}x + 2$ tilts downward (as we follow the line from left to right). We will learn more about the tilt, or slope, of a line in Section 10.4.

Example 5 Graph the linear equation $y = -2$.

Solution: The equation $y = -2$ can be written in standard form as $0x + y = -2$. No matter what value we replace x with, y is always -2.

x	y
0	-2
3	-2
-2	-2

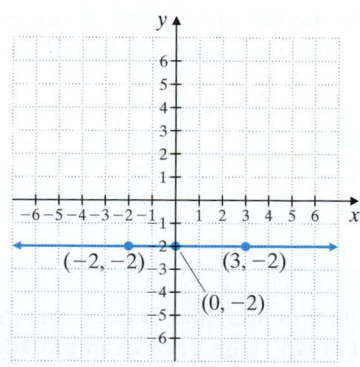

Notice that the graph of $y = -2$ is a horizontal line.

■ **Work Practice 5**

Linear equations are often used to model real data, as seen in the next example.

Practice 4

Graph the linear equation $y = -\dfrac{1}{2}x + 4$.

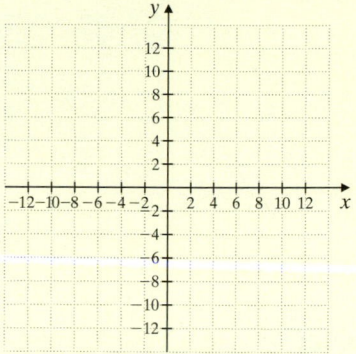

Practice 5

Graph the linear equation $x = 3$.

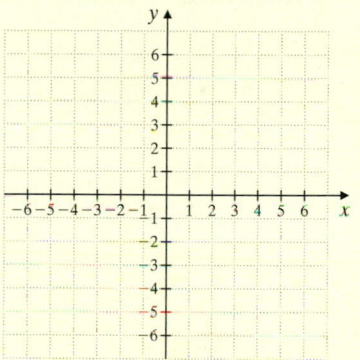

Answers

4.

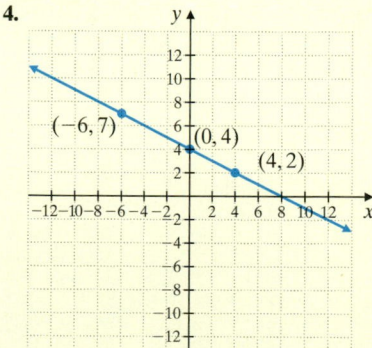

5.

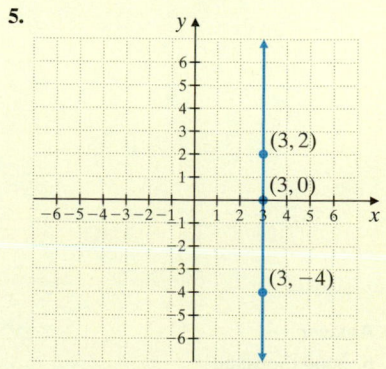

Practice 6
Use the graph in Example 6 to predict the number of registered nurses in 2022.

Helpful Hint   From Example 5, we learned that equations such as $y = -2$ are linear equations since $y = -2$ can be written as $0x + y = -2$.

Example 6   Estimating the Number of Registered Nurses

One of the occupations expected to have the most growth in the next few years is registered nurse. The number of people y (in thousands) employed as registered nurses in the United States can be estimated by the linear equation $y = 71.2x + 2734.4$, where x is the number of years after the year 2010. (*Source:* Based on data from the Bureau of Labor Statistics)

a. Graph the equation.

b. Use the graph to predict the number of registered nurses in the year 2021.

Solution:

a. To graph $y = 71.2x + 2734.4$, choose x-values and substitute into the equation.

If $x = 0$, then $y = 71.2(0) + 2734.4 = 2734.4$.

If $x = 2$, then $y = 71.2(2) + 2734.4 = 2876.8$.

If $x = 5$, then $y = 71.2(5) + 2734.4 = 3090.4$.

x	y
0	2734.4
2	2876.8
5	3090.4

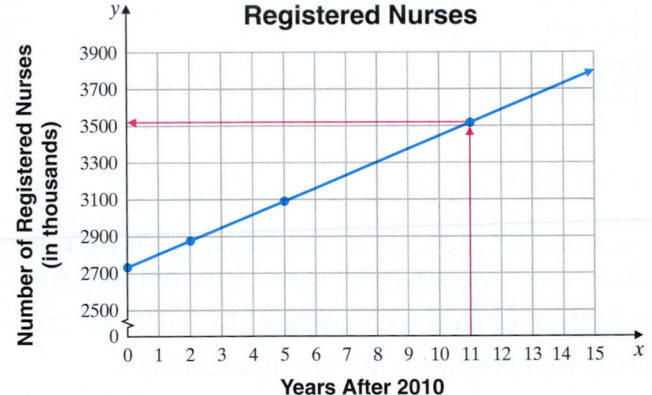

b. To use the graph to *predict* the number of registered nurses in the year 2021, we need to find the y-coordinate that corresponds to $x = 11$. (11 years after 2010 is the year 2021.) To do so, find 11 on the x-axis. Move vertically upward to the graphed line and then horizontally to the left. We approximate the number on the y-axis to be 3500. Thus, in the year 2021, we predict that there will be 3500 thousand registered nurses. (The actual value, using 11 for x, is 3517.6.)

■ **Work Practice 6**

Helpful Hint

Make sure you understand that models are mathematical approximations of the data for the known years. (For example, see the model in Example 6.) Any number of unknown factors can affect future years, so be cautious when using models to make predictions.

Answer

6. 3600 thousand

Calculator Explorations Graphing

In this section, we begin an optional study of graphing calculators and graphing software packages for computers. These graphers use the same point plotting technique that was introduced in this section. The advantage of this graphing technology is, of course, that graphing calculators and computers can find and plot ordered pair solutions much faster than we can. Note, however, that the features described in these boxes may not be available on all graphing calculators.

The rectangular screen where a portion of the rectangular coordinate system is displayed is called a **window.** We call it a **standard window** for graphing when both the x- and y-axes show coordinates between -10 and 10. This information is often displayed in the window menu on a graphing calculator as follows.

Xmin = -10
Xmax = 10
  Xscl = 1 The scale on the x-axis is one unit per tick mark.
Ymin = -10
Ymax = 10
  Yscl = 1 The scale on the y-axis is one unit per tick mark.

To use a graphing calculator to graph the equation $y = 2x + 3$, press the $\boxed{Y =}$ key and enter the keystrokes $\boxed{2}\ \boxed{x}\ \boxed{+}\ \boxed{3}$. The top row should now read $Y_1 = 2x + 3$. Next press the $\boxed{\text{GRAPH}}$ key, and the display should look like this:

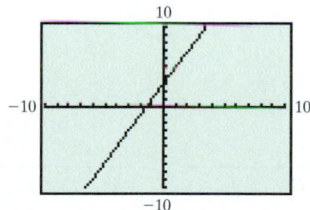

Graph the following linear equations. (Unless otherwise stated, use a standard window when graphing.)

1. $y = -3x + 7$

2. $y = -x + 5$

3. $y = 2.5x - 7.9$

4. $y = -1.3x + 5.2$

5. $y = -\dfrac{3}{10}x + \dfrac{32}{5}$

6. $y = \dfrac{2}{9}x - \dfrac{22}{3}$

Vocabulary, Readiness & Video Check

Martin-Gay Interactive Videos

See Video 10.2 🍎

Watch the section lecture video and answer the following questions.

Objective A **1.** In the lecture before ⊞ Example 1, it's mentioned that we need only two points to determine a line. Why, then, are three ordered pair solutions found in ⊞ Examples 1–3? ▶

2. What does a graphed line represent, as discussed at the end of ⊞ Examples 1 and 3? ▶

10.2 Exercise Set MyMathLab®

Objective A *For each equation, find three ordered pair solutions by completing the table. Then use the ordered pairs to graph the equation. See Examples 1 through 5.*

1. $x - y = 6$

x	y
	0
4	
	-1

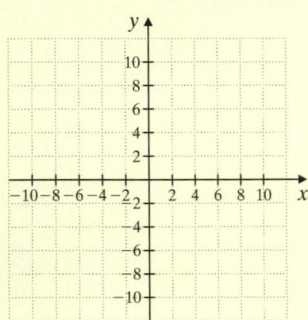

2. $x - y = 4$

x	y
0	
	2
-1	

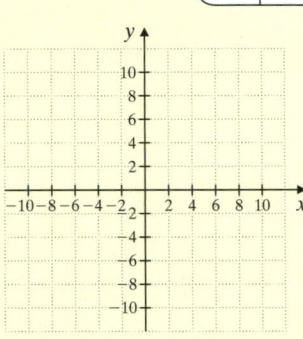

3. $y = -4x$

x	y
1	
0	
-1	

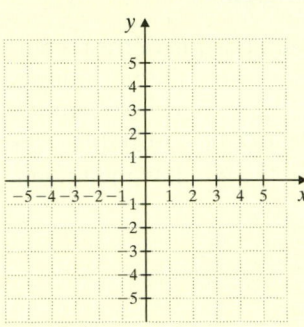

4. $y = -5x$

x	y
1	
0	
-1	

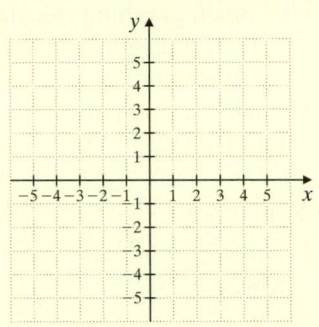

5. $y = \dfrac{1}{3}x$

x	y
0	
6	
-3	

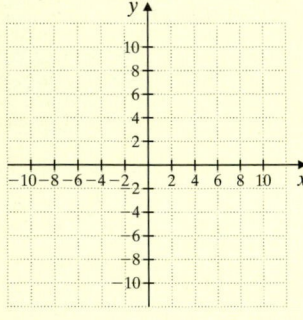

6. $y = \dfrac{1}{2}x$

x	y
0	
-4	
2	

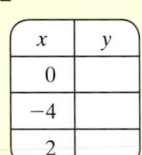

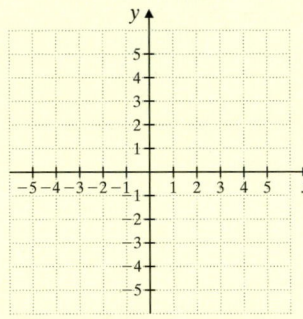

7. $y = -4x + 3$

x	y
0	
1	
2	

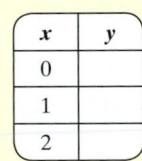

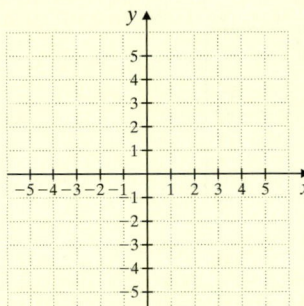

8. $y = -5x + 2$

x	y
0	
1	
2	

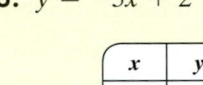

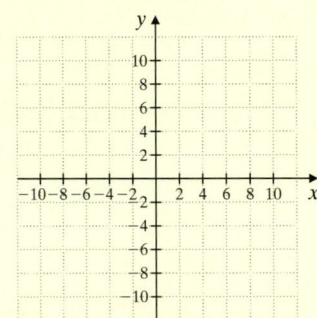

Graph each linear equation. See Examples 1 through 5.

9. $x + y = 1$

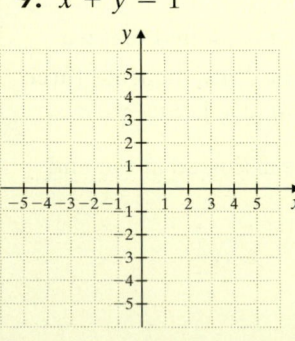

10. $x + y = 7$

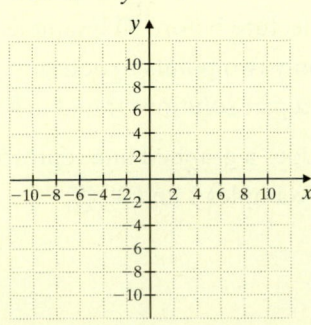

11. $x - y = -2$

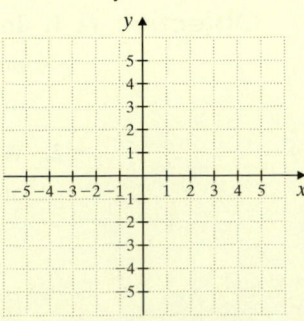

12. $-x + y = 6$

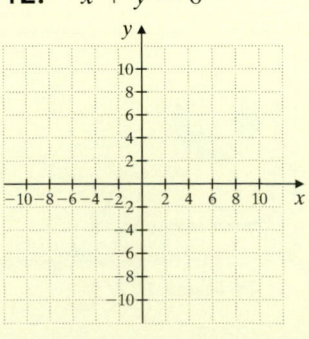

13. $x - 2y = 6$

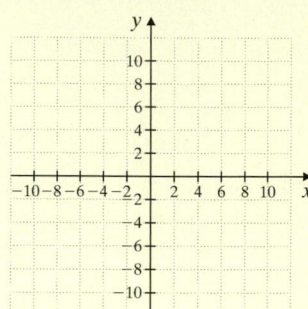

14. $-x + 5y = 5$

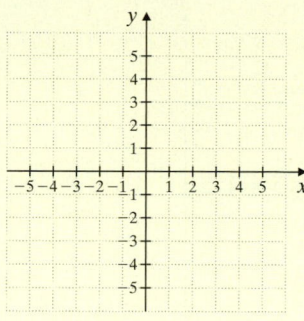

15. $y = 6x + 3$

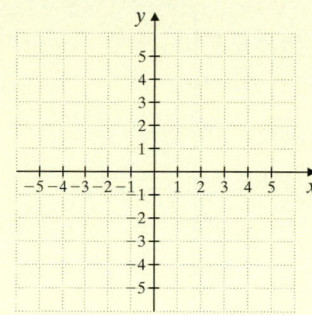

16. $y = -2x + 7$

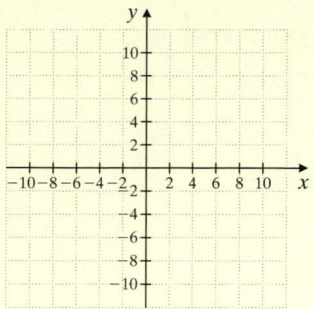

17. $x = -4$

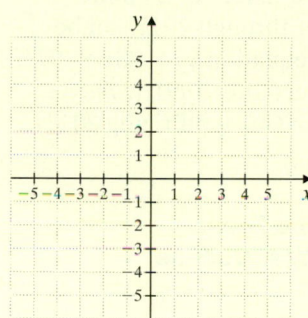

18. $y = 5$

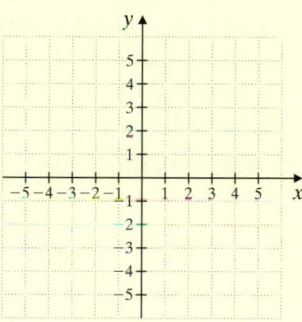

19. $y = 3$

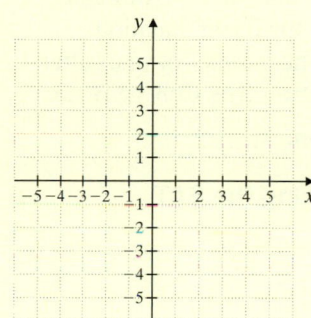

20. $x = -1$

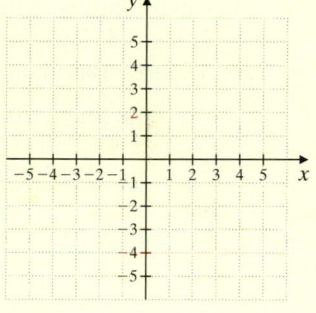

21. $y = x$

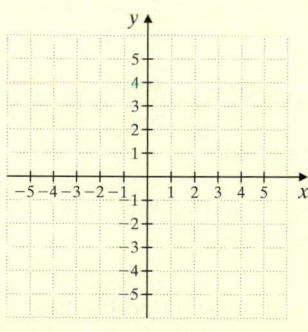

22. $y = -x$

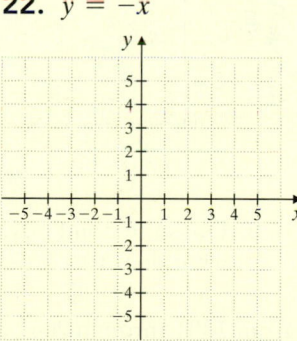

23. $x = -3y$

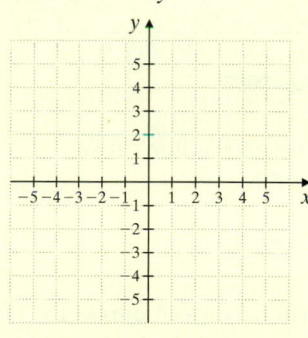

24. $x = 4y$

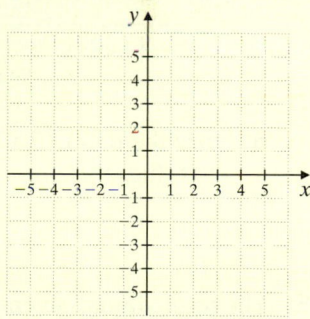

25. $x + 3y = 9$

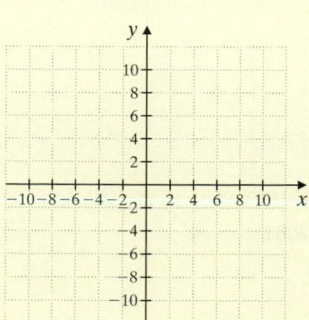

26. $2x + y = 2$

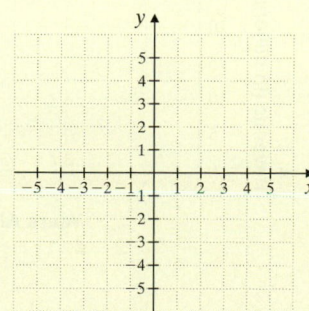

27. $y = \frac{1}{2}x + 2$

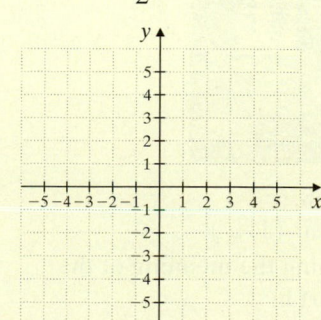

28. $y = \frac{1}{4}x + 3$

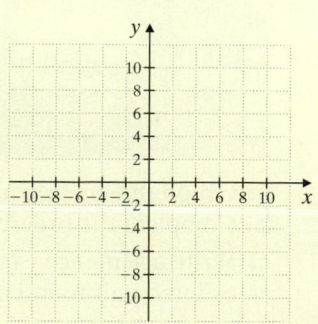

29. $3x - 2y = 12$

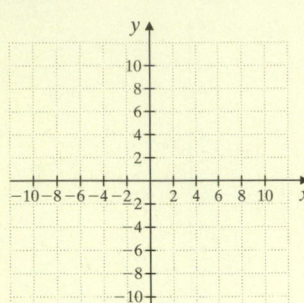

30. $2x - 7y = 14$

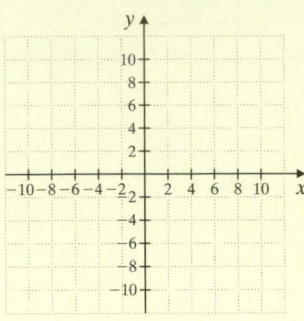

31. $y = -3.5x + 4$

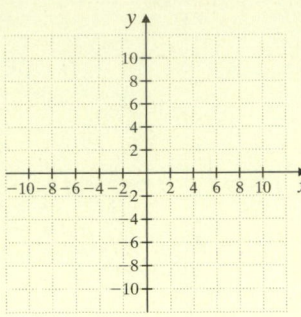

32. $y = -1.5x - 3$

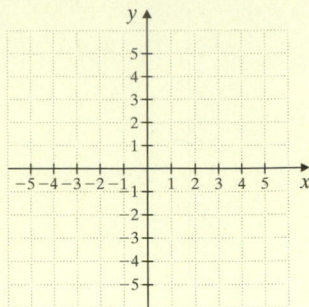

Solve. See Example 6.

33. The number of students y (in thousands) taking the SAT college entrance exam each year from 2008 through 2012 can be approximated by the linear equation $y = 28x + 1552$, where x is the number of years after 2008. (*Source:* Based on data from The College Board)

  a. Graph the linear equation.

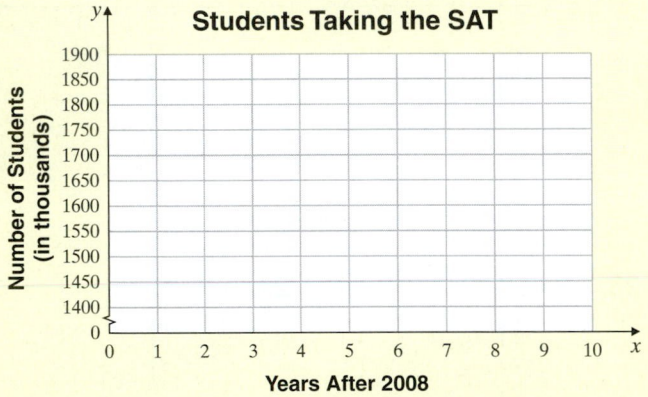

  b. Does the point $(7, 1748)$ lie on the line? If so, what does this ordered pair mean?

34. College is getting more expensive every year. The average cost for tuition and fees at a public two-year college y from 1991 through 2012 can be approximated by the linear equation $y = 88x + 973$, where x is the number of years after 1991. (*Source:* The College Board: Trends in College Pricing 2012)

  a. Graph the linear equation.

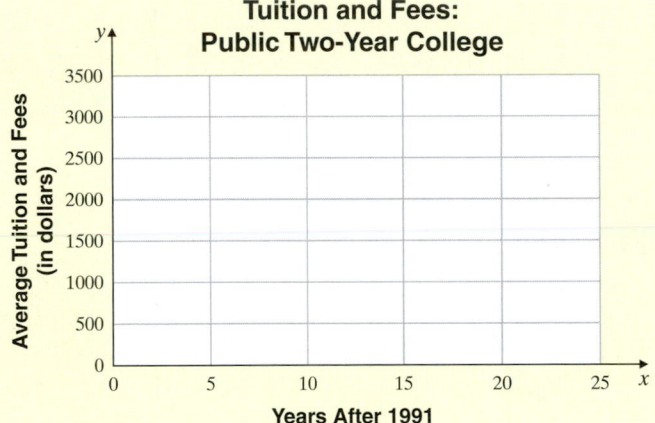

  b. Does the point $(21, 2821)$ lie on the line? If so, what does this ordered pair mean?

35. The total annual revenue y (in billions of euros) for IKEA from 2001 through 2012 can be approximated by the equation $y = 1.6x + 9.1$, where x is the number of years after 2001. (*Source:* Based on data from IKEA Group)

  a. Graph the linear equation.
  b. Complete the ordered pair $(11, \ )$.
  c. Write a sentence explaining the meaning of the ordered pair found in part **b.**

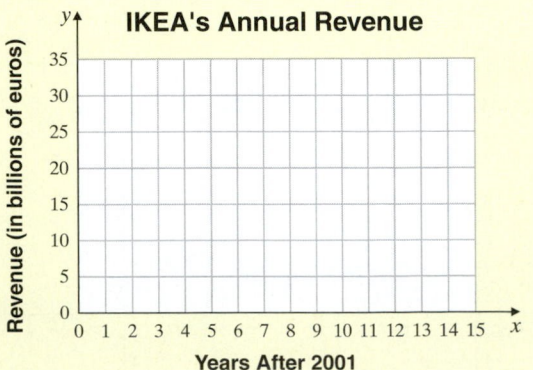

36. For the period 1970 through 2013, the annual level of sales for restaurants in the United States can be estimated by $y = 14.3x - 5.3$, where x is the number of years after 1970 and y is the sales in billions of dollars. (*Source:* Based on data from the National Restaurant Association)

a. Graph the linear equation.

b. Complete the ordered pair (43,　).

c. Write a sentence explaining the meaning of the ordered pair found in part **b**.

Review

△ **37.** The coordinates of three vertices of a rectangle are $(-2, 5)$, $(4, 5)$, and $(-2, -1)$. Find the coordinates of the fourth vertex. See Section 10.1.

△ **38.** The coordinates of two vertices of a square are $(-3, -1)$ and $(2, -1)$. Find the coordinates of two pairs of possible points for the third and fourth vertices. See Section 10.1.

Complete each table. See Section 10.1.

39. $x - y = -3$

x	y
0	
	0

40. $y - x = 5$

x	y
0	
	0

41. $y = 2x$

x	y
0	
	0

42. $x = -3y$

x	y
0	
	0

Concept Extensions

Graph each pair of linear equations on the same set of axes. Discuss how the graphs are similar and how they are different.

43. $y = 5x$
$y = 5x + 4$

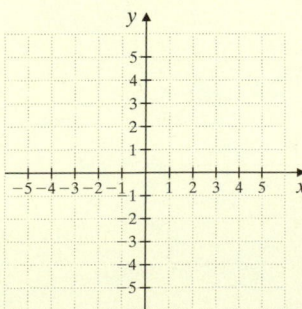

44. $y = 2x$
$y = 2x + 5$

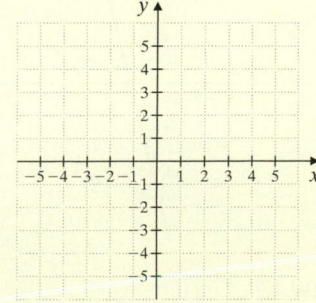

45. $y = -2x$
$y = -2x - 3$

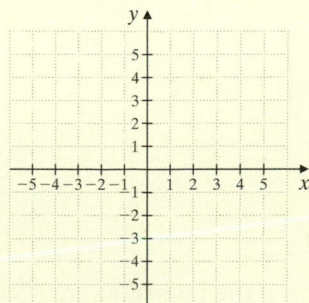

46. $y = x$
$y = x - 7$

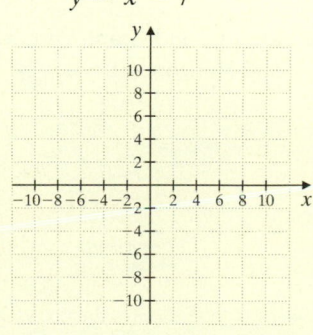

47. Graph the nonlinear equation $y = x^2$ by completing the table shown. Plot the ordered pairs and connect them with a smooth curve.

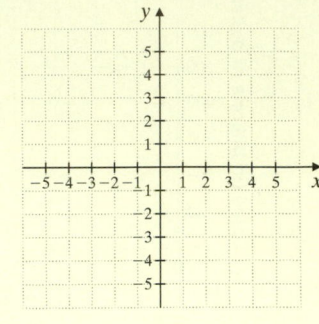

x	y
0	
1	
−1	
2	
−2	

48. Graph the nonlinear equation $y = |x|$ by completing the table shown. Plot the ordered pairs and connect them. This curve is "V" shaped.

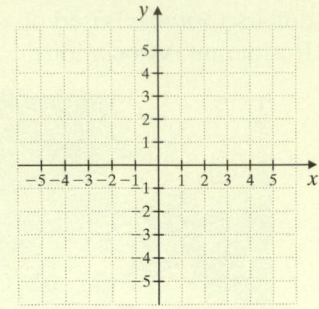

x	y
0	
1	
−1	
2	
−2	

△ **49.** The perimeter of the trapezoid below is 22 centimeters. Write a linear equation in two variables for the perimeter. Find y if x is 3 centimeters.

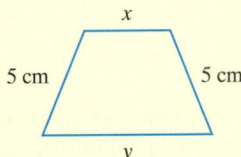

5 cm   5 cm

△ **50.** The perimeter of the rectangle below is 50 miles. Write a linear equation in two variables for the perimeter. Use this equation to find x when y is 20.

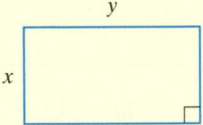

✎ **51.** If (a, b) is an ordered pair solution of $x + y = 5$, is (b, a) also a solution? Explain why or why not.

✎ **52.** If (a, b) is an ordered pair solution of $x - y = 5$, is (b, a) also a solution? Explain why or why not.

10.3 Intercepts

Objectives

A Identify Intercepts of a Graph.

B Graph a Linear Equation by Finding and Plotting Intercept Points. ▶

C Identify and Graph Vertical and Horizontal Lines. ▶

Objective **A** Identifying Intercepts ▶

The graph of $y = 4x - 8$ is shown below. Notice that this graph crosses the y-axis at the point $(0, -8)$. This point is called the **y-intercept.** Likewise the graph crosses the x-axis at $(2, 0)$. This point is called the **x-intercept.**

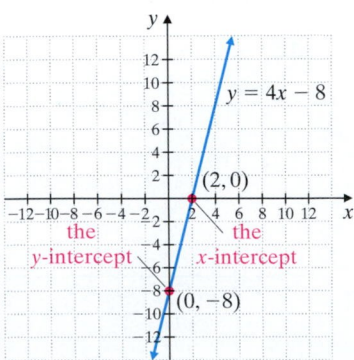

The intercepts are $(2, 0)$ and $(0, -8)$.

Helpful Hint

If a graph crosses the *x*-axis at $(2, 0)$ and the *y*-axis at $(0, -8)$, then

$\underbrace{(2, 0)}_{}$ $\underbrace{(0, -8)}_{}$

↑   ↑

x-intercept   *y*-intercept

Notice that for the *x*-intercept, the *y*-value is 0 and that for the *y*-intercept, the *x*-value is 0.

Note: Sometimes in mathematics, you may see just the number -8 stated as the *y*-intercept, and 2 stated as the *x*-intercept.

Examples   Identify the *x*- and *y*-intercepts.

1.

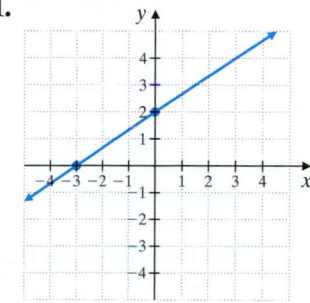

Solution:

x-intercept: $(-3, 0)$

y-intercept: $(0, 2)$

2.

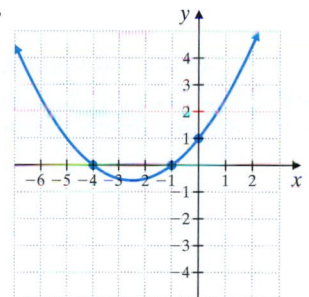

Solution:

x-intercepts: $(-4, 0)$, $(-1, 0)$

y-intercept: $(0, 1)$

3.

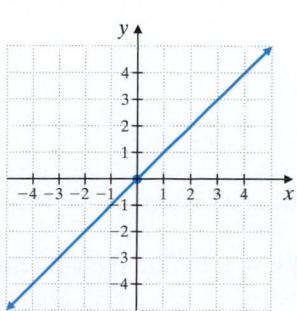

Helpful Hint

Notice that any time $(0, 0)$ is a point of a graph, then it is an *x*-intercept and a *y*-intercept. Why? It is the *only* point that lies on both axes.

Practice 1–3

Identify the *x*- and *y*-intercepts.

1.

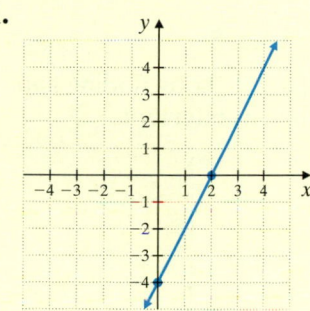

2.

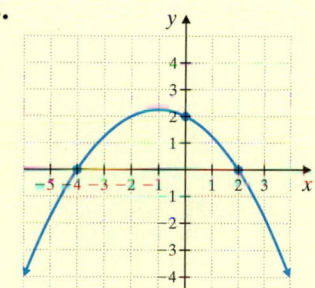

3.

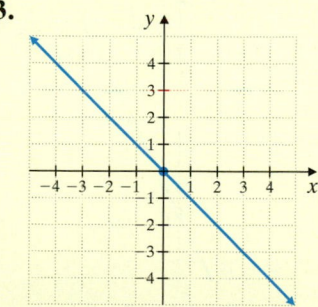

Answers

1. *x*-intercept: $(2, 0)$; *y*-intercept: $(0, -4)$

2. *x*-intercepts: $(-4, 0)$, $(2, 0)$; *y*-intercept: $(0, 2)$

3. *x*-intercept and *y*-intercept: $(0, 0)$

(*Continued on next page*)

Practice 4

Graph $2x - y = 4$ by finding and plotting its intercepts.

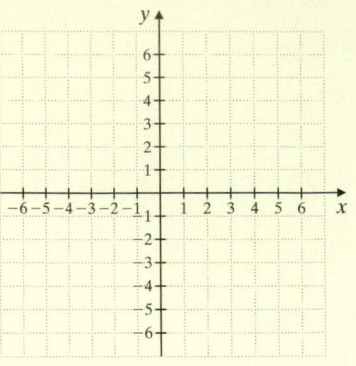

Practice 5

Graph $y = 3x$ by finding and plotting its intercepts.

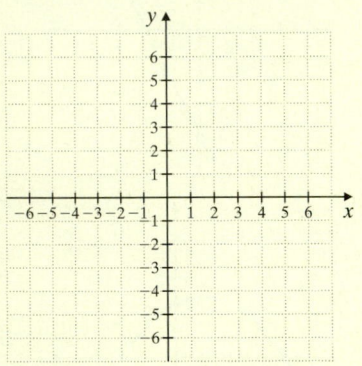

Answers

4.

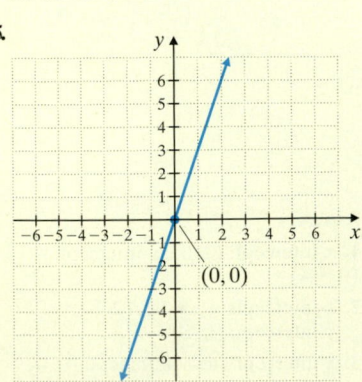

5.

Solution:

x-intercept: $(0, 0)$

y-intercept: $(0, 0)$

Here, the x- and y-intercepts happen to be the same point.

■ **Work Practice 1–3**

Objective B Finding and Plotting Intercepts ▶

Given an equation of a line, we can usually find intercepts easily since one coordinate is 0.

To find the x-intercept of a line from its equation, let $y = 0$, since a point on the x-axis has a y-coordinate of 0. To find the y-intercept of a line from its equation, let $x = 0$, since a point on the y-axis has an x-coordinate of 0.

> **Finding x- and y-Intercepts**
>
> To find the x-intercept, let $y = 0$ and solve for x.
> To find the y-intercept, let $x = 0$ and solve for y.

Example 4   Graph $x - 3y = 6$ by finding and plotting its intercepts.

Solution: We let $y = 0$ to find the x-intercept and $x = 0$ to find the y-intercept.

Let $y = 0$.	Let $x = 0$.
$x - 3y = 6$	$x - 3y = 6$
$x - 3(0) = 6$	$0 - 3y = 6$
$x - 0 = 6$	$-3y = 6$
$x = 6$	$y = -2$

The x-intercept is $(6, 0)$ and the y-intercept is $(0, -2)$. We find a third ordered pair solution to check our work. If we let $y = -1$, then $x = 3$. We plot the points $(6, 0)$, $(0, -2)$, and $(3, -1)$. The graph of $x - 3y = 6$ is the line drawn through these points, as shown.

x	y
6	0
0	-2
3	-1

■ **Work Practice 4**

Example 5   Graph $x = -2y$ by finding and plotting its intercepts.

Solution: We let $y = 0$ to find the x-intercept and $x = 0$ to find the y-intercept.

Let $y = 0$.	Let $x = 0$.
$x = -2y$	$x = -2y$
$x = -2(0)$	$0 = -2y$
$x = 0$	$0 = y$

Both the *x*-intercept and *y*-intercept are $(0, 0)$. In other words, when $x = 0$, then $y = 0$, which gives the ordered pair $(0, 0)$. Also, when $y = 0$, then $x = 0$, which gives the same ordered pair, $(0, 0)$. This happens when the graph passes through the origin. Since two points are needed to determine a line, we must find at least one more ordered pair that satisfies $x = -2y$. Since the equation is solved for *x*, we choose *y*-values so that there is no need to solve to find the corresponding *x*-value. We let $y = -1$ to find a second ordered pair solution and let $y = 1$ as a check point.

Let $y = -1$.

$x = -2(-1)$

$x = 2$ Multiply.

Let $y = 1$.

$x = -2(1)$

$x = -2$ Multiply.

The ordered pairs are $(0, 0)$, $(2, -1)$, and $(-2, 1)$. We plot these points to graph $x = -2y$.

x	*y*
0	0
2	-1
-2	1

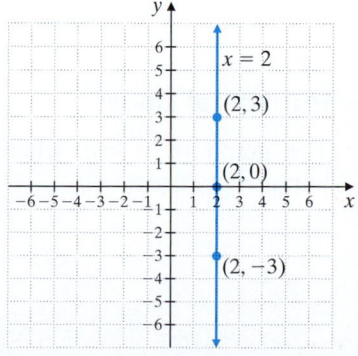

■ **Work Practice 5**

Objective C Graphing Vertical and Horizontal Lines

From Section 10.2, recall that the equation $x = 2$, for example, is a linear equation in two variables because it can be written in the form $x + 0y = 2$. The graph of this equation is a vertical line, as reviewed in the next example.

Example 6   Graph: $x = 2$

Solution: The equation $x = 2$ can be written as $x + 0y = 2$. For any *y*-value chosen, notice that *x* is 2. No other value for *x* satisfies $x + 0y = 2$. Any ordered pair whose *x*-coordinate is 2 is a solution of $x + 0y = 2$. We will use the ordered pair solutions $(2, 3)$, $(2, 0)$, and $(2, -3)$ to graph $x = 2$.

x	*y*
2	3
2	0
2	-3

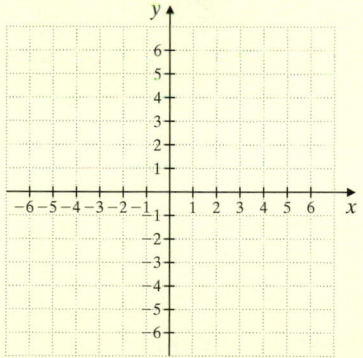

The graph is a vertical line with *x*-intercept 2. Note that this graph has no *y*-intercept because *x* is never 0.

■ **Work Practice 6**

Practice 6

Graph: $x = -3$

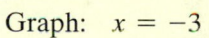

Answer

6.

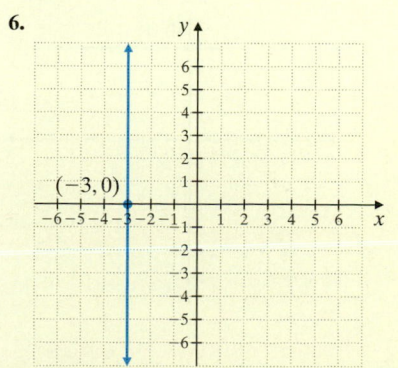

In general, we have the following.

Vertical Lines

The graph of $x = c$, where c is a real number, is a **vertical line** with x-intercept $(c, 0)$.

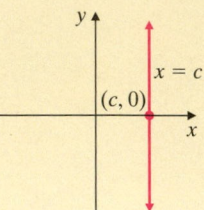

Practice 7

Graph: $y = 4$

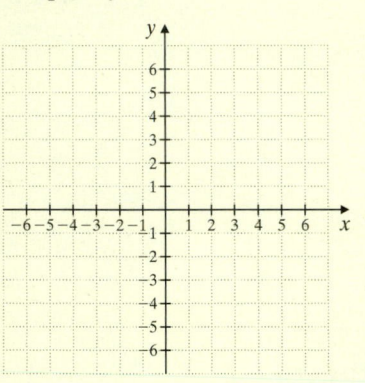

Example 7   Graph: $y = -3$

Solution: The equation $y = -3$ can be written as $0x + y = -3$. For any x-value chosen, y is -3. If we choose 4, 1, and -2 as x-values, the ordered pair solutions are $(4, -3)$, $(1, -3)$, and $(-2, -3)$. We use these ordered pairs to graph $y = -3$. The graph is a horizontal line with y-intercept -3 and no x-intercept.

x	y
4	-3
1	-3
-2	-3

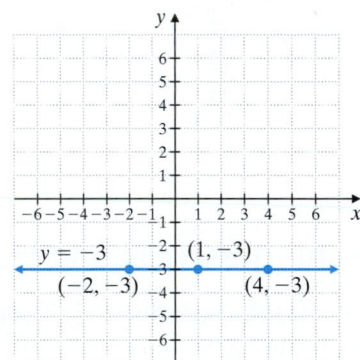

■ **Work Practice 7**

In general, we have the following.

Answer

7.

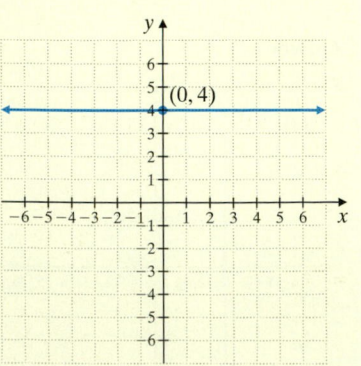

Horizontal Lines

The graph of $y = c$, where c is a real number, is a **horizontal line** with y-intercept $(0, c)$.

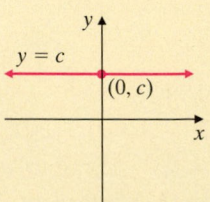

Calculator Explorations Graphing

You may have noticed that to use the $\boxed{Y =}$ key on a graphing calculator to graph an equation, the equation must be solved for y. For example, to graph $2x + 3y = 7$, we solve the equation for y.

$$2x + 3y = 7$$
$$3y = -2x + 7 \quad \text{Subtract } 2x \text{ from both sides.}$$
$$\frac{3y}{3} = -\frac{2x}{3} + \frac{7}{3} \quad \text{Divide both sides by 3.}$$
$$y = -\frac{2}{3}x + \frac{7}{3} \quad \text{Simplify.}$$

To graph $2x + 3y = 7$ or $y = -\frac{2}{3}x + \frac{7}{3}$, press the $\boxed{Y =}$ key and enter

$$Y_1 = -\frac{2}{3}x + \frac{7}{3}$$

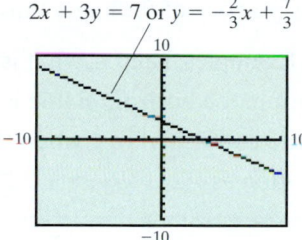

Graph each linear equation.

1. $x = 3.78y$

2. $-2.61y = x$

3. $3x + 7y = 21$

4. $-4x + 6y = 12$

5. $-2.2x + 6.8y = 15.5$

6. $5.9x - 0.8y = -10.4$

Vocabulary, Readiness & Video Check

Use the choices below to fill in each blank. Some choices may be used more than once. Exercises 1 and 2 come from Section 10.2.

x	vertical	x-intercept	linear
y	horizontal	y-intercept	standard

1. An equation that can be written in the form $Ax + By = C$ is called a(n) _____ equation in two variables.

2. The form $Ax + By = C$ is called _____ form.

3. The graph of the equation $y = -1$ is a(n) _____ line.

4. The graph of the equation $x = 5$ is a(n) _____ line.

5. A point where a graph crosses the y-axis is called a(n) _____.

6. A point where a graph crosses the x-axis is called a(n) _____.

7. Given an equation of a line, to find the x-intercept (if there is one), let _____ $= 0$ and solve for _____.

8. Given an equation of a line, to find the y-intercept (if there is one), let _____ $= 0$ and solve for _____.

Answer the following true or false.

9. All lines have an *x*-intercept *and* a *y*-intercept. _____

10. The graph of $y = 4x$ contains the point $(0, 0)$. _____

11. The graph of $x + y = 5$ has an *x*-intercept of $(5, 0)$ and a *y*-intercept of $(0, 5)$. _____

12. The graph of $y = 5x$ contains the point $(5, 1)$. _____

Martin-Gay Interactive Videos   *Watch the section lecture video and answer the following questions.*

See Video 10.3

Objective A 13. At the end of ⊞ Example 2, patterns are discussed. What reason is given for why *x*-intercepts have *y*-values of 0? For why *y*-intercepts have *x*-values of 0? ▶

Objective B 14. In ⊞ Example 3, the goal is to use the *x*- and *y*-intercepts to graph a line. Yet once the two intercepts are found, a third point is also found before the line is graphed. Why do you think this practice of finding a third point is continued? ▶

Objective C 15. From ⊞ Examples 5 and 6, what is the coefficient of *x* when the equation of a horizontal line is written as $Ax + By = C$? What is the coefficient of *y* when the equation of a vertical line is written as $Ax + By = C$? ▶

10.3 **Exercise Set** MyMathLab®

Objective A *Identify the intercepts. See Examples 1 through 3.*

▶ 1.
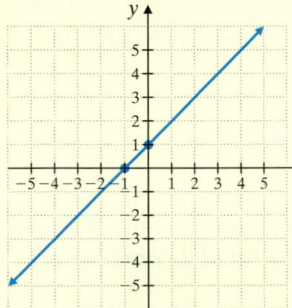

2.
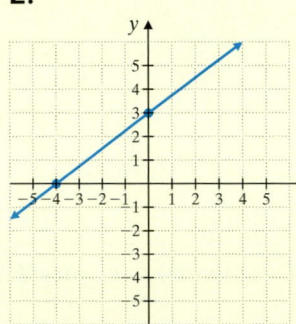

▶ 3.
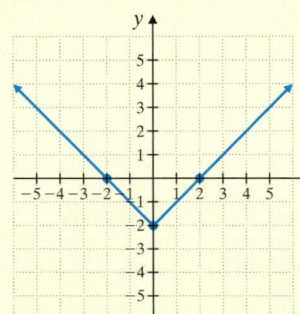

4.
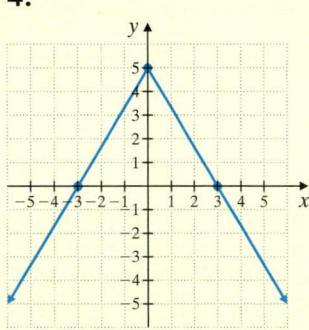

5.
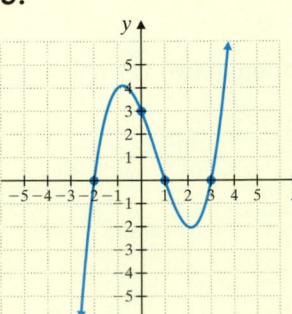

6.
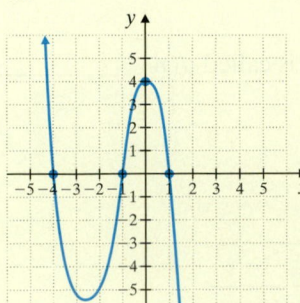

7.
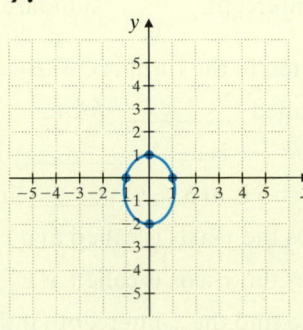

8.
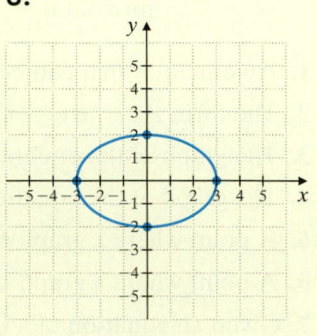

Objective B *Graph each linear equation by finding and plotting its intercepts. See Examples 4 and 5.*

9. $x - y = 3$

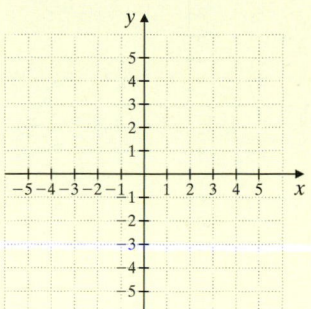

10. $x - y = -4$

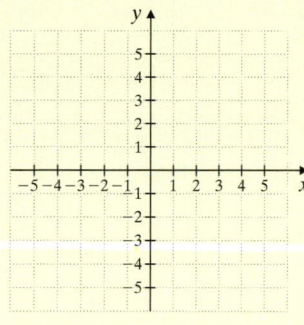

11. $x = 5y$

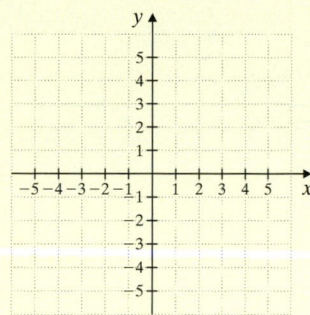

12. $x = 2y$

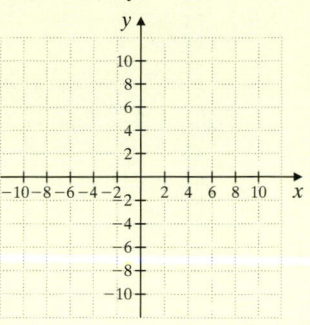

13. $-x + 2y = 6$

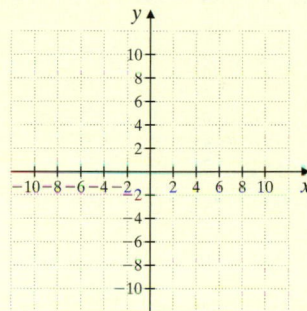

14. $x - 2y = -8$

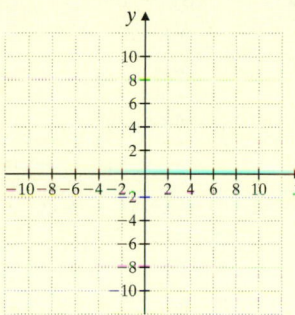

15. $2x - 4y = 8$

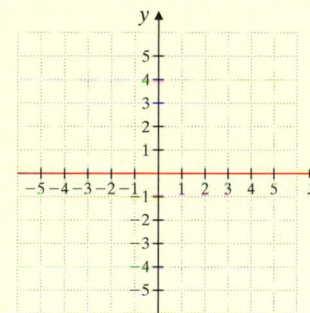

16. $2x + 3y = 6$

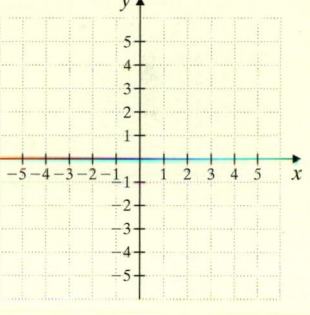

17. $y = 2x$

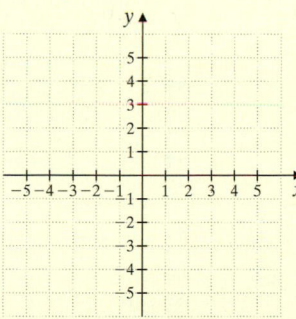

18. $y = -2x$

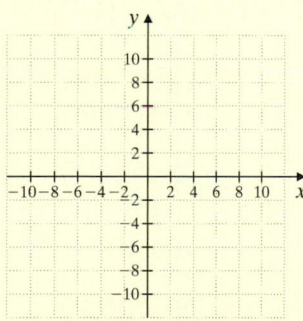

19. $y = 3x + 6$

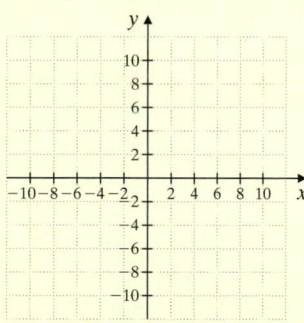

20. $y = 2x + 10$

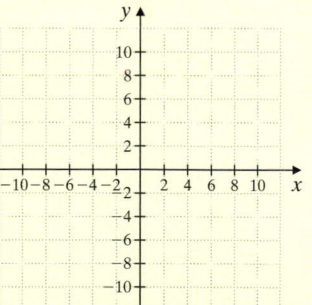

Objective C *Graph each linear equation. See Examples 6 and 7.*

21. $x = -1$

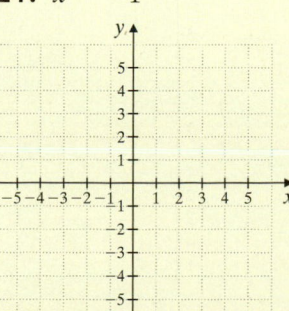

22. $y = 5$

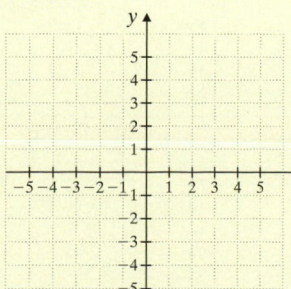

23. $y = 0$

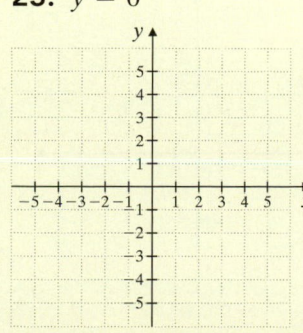

24. $x = 0$

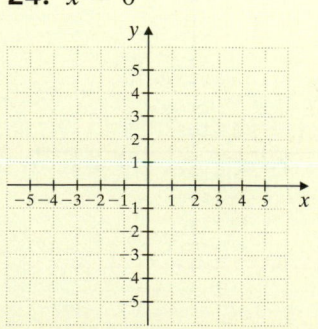

25. $y + 7 = 0$

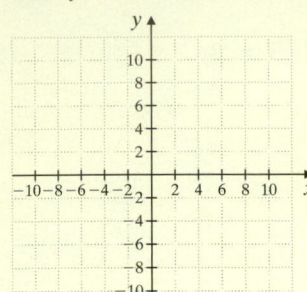

26. $x - 2 = 0$

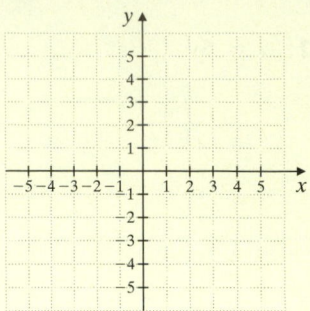

○ **27.** $x + 3 = 0$

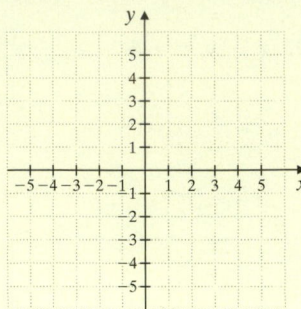

28. $y - 6 = 0$

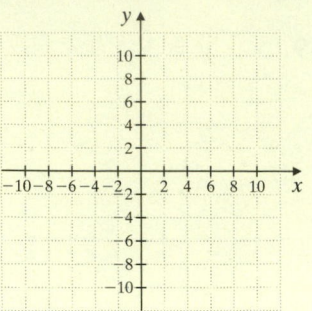

Objectives B C Mixed Practice *Graph each linear equation. See Examples 4 through 7.*

29. $x = y$

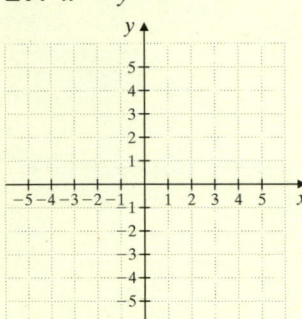

30. $x = -y$

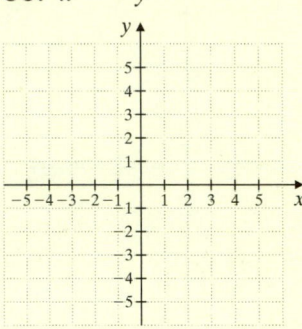

31. $x + 8y = 8$

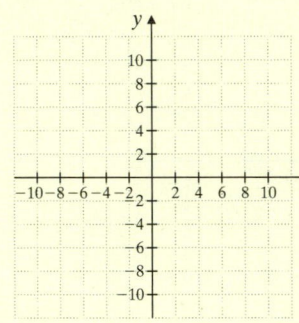

32. $x + 3y = 9$

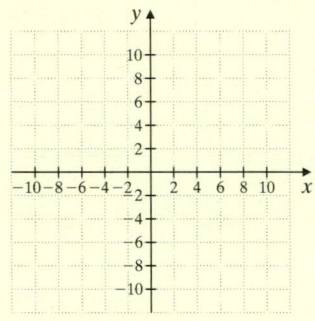

33. $5 = 6x - y$

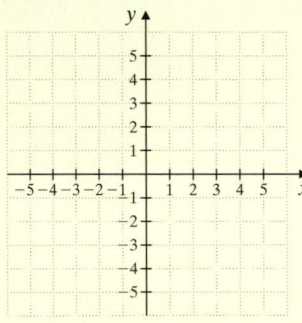

34. $4 = x - 3y$

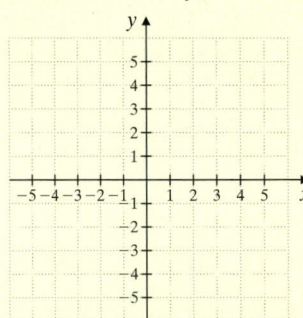

35. $-x + 10y = 11$

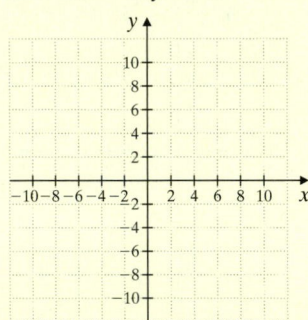

36. $-x + 9y = 10$

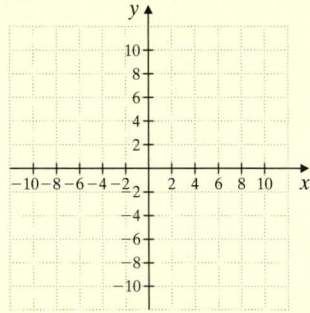

37. $x = -4\frac{1}{2}$

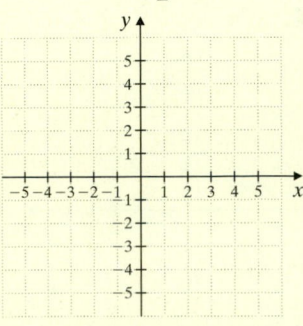

38. $x = -1\frac{3}{4}$

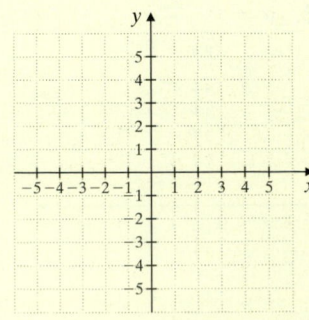

39. $y = 3\frac{1}{4}$

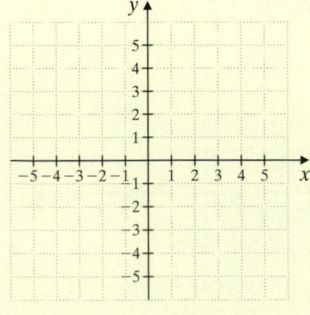

40. $y = 2\frac{1}{2}$

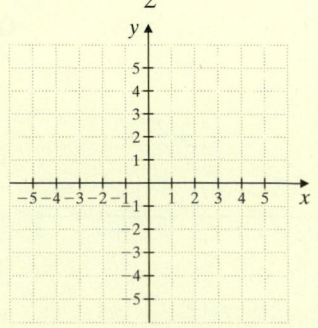

41. $y = -\dfrac{2}{3}x + 1$

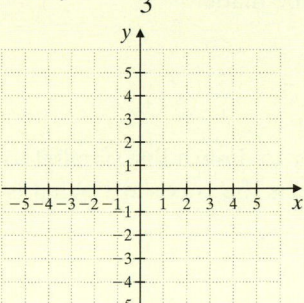

42. $y = -\dfrac{3}{5}x + 3$

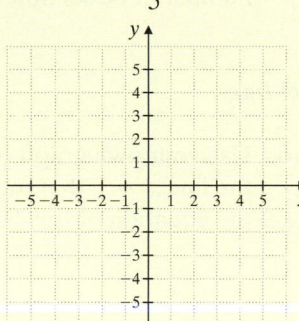

43. $4x - 6y + 2 = 0$

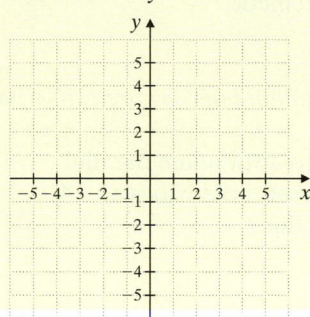

44. $9x - 6y + 3 = 0$

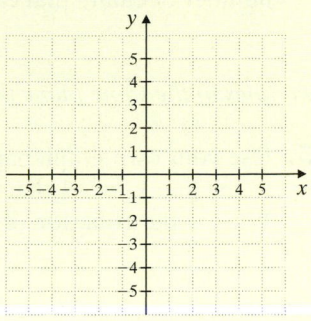

Review

Simplify. See Sections 8.2, 8.4, and 8.5.

45. $\dfrac{-6 - 3}{2 - 8}$

46. $\dfrac{4 - 5}{-1 - 0}$

47. $\dfrac{-8 - (-2)}{-3 - (-2)}$

48. $\dfrac{12 - 3}{10 - 9}$

49. $\dfrac{0 - 6}{5 - 0}$

50. $\dfrac{2 - 2}{3 - 5}$

Concept Extensions

Match each equation with its graph.

51. $y = 3$
a.

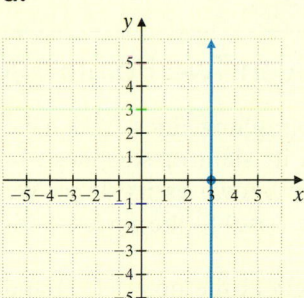

52. $y = 2x + 2$
b.

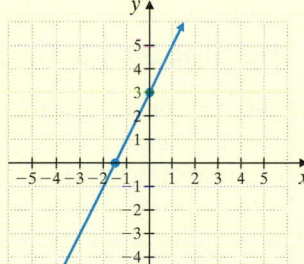

53. $x = 3$
c.

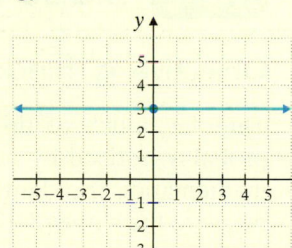

54. $y = 2x + 3$
d.

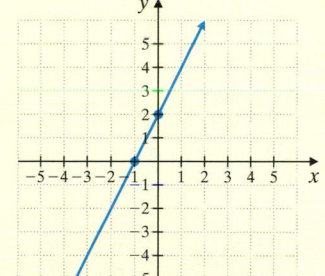

55. What is the greatest number of x- and y-intercepts that a line can have?

56. What is the smallest number of x- and y-intercepts that a line can have?

57. What is the smallest number of x- and y-intercepts that a circle can have?

58. What is the greatest number of x- and y-intercepts that a circle can have?

59. Discuss whether a vertical line ever has a y-intercept.

60. Discuss whether a horizontal line ever has an x-intercept.

The production supervisor at Alexandra's Office Products finds that it takes 3 hours to manufacture a particular office chair and 6 hours to manufacture an office desk. A total of 1200 hours is available to produce office chairs and desks of this style. The linear equation that models this situation is $3x + 6y = 1200$, where x represents the number of chairs produced and y represents the number of desks manufactured.

61. Complete the ordered pair solution $(0, \ )$ of this equation. Describe the manufacturing situation that corresponds to this solution.

62. Complete the ordered pair solution $(\ , 0)$ of this equation. Describe the manufacturing situation that corresponds to this solution.

63. If 50 desks are manufactured, find the greatest number of chairs that can be made.

64. If 50 chairs are manufactured, find the greatest number of desks that can be made.

*Two lines in the same plane that do not intersect are called **parallel lines.***

65. Use your own graph paper to draw a line parallel to the line $y = -1$ that intersects the y-axis at -4. What is the equation of this line?

66. Use your own graph paper to draw a line parallel to the line $x = 5$ that intersects the x-axis at 1. What is the equation of this line?

Solve.

67. It has been said that newspapers are disappearing, replaced by various electronic media. The average circulation of newspapers in the United States y, in millions, from 2006 to 2011 can be modeled by the equation $y = -1.7x + 52$, where x represents the number of years after 2006. (*Source:* Newspaper Association of America)

  a. Find the x-intercept of this equation (round to the nearest tenth).

  b. What does this x-intercept mean?

68. The number y of Barnes & Noble retail stores in operation for the years 2008–2012 can be modeled by the equation $y = -8.6x + 730$, where x represents the number of years after 2008. (*Source:* Based on data from Barnes & Noble, Inc.)

  a. Find the y-intercept of this equation.

  b. What does this y-intercept mean?

10.4   Slope and Rate of Change

Objectives

A Find the Slope of a Line Given Two Points of the Line.

B Find the Slope of a Line Given Its Equation.

C Find the Slopes of Horizontal and Vertical Lines.

D Compare the Slopes of Parallel and Perpendicular Lines.

E Slope as a Rate of Change.

Objective A   Finding the Slope of a Line Given Two Points

Thus far, much of this chapter has been devoted to graphing lines. You have probably noticed by now that a key feature of a line is its slant or steepness. In mathematics, the slant or steepness of a line is formally known as its **slope.** We measure the slope of a line by the ratio of vertical change (rise) to the corresponding horizontal change (run) as we move along the line.

On the line below, for example, suppose that we begin at the point $(1, 2)$ and move to the point $(4, 6)$. The vertical change is the change in y-coordinates: $6 - 2$ or 4 units. The corresponding horizontal change is the change in x-coordinates: $4 - 1 = 3$ units. The ratio of these changes is

$$\text{slope} = \frac{\text{change in } y \text{ (vertical change or rise)}}{\text{change in } x \text{ (horizontal change or run)}} = \frac{4}{3}$$

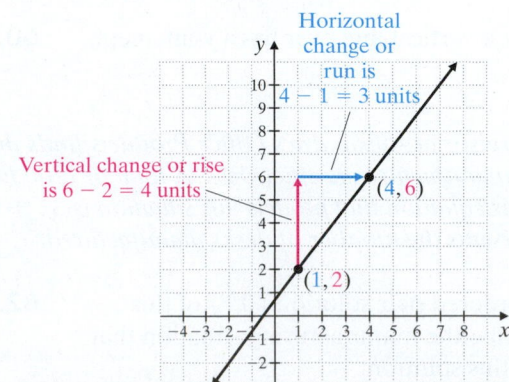

The slope of this line, then, is $\frac{4}{3}$. This means that for every 4 units of change in y-coordinates, there is a corresponding change of 3 units in x-coordinates.

Helpful Hint

It makes no difference what two points of a line are chosen to find its slope. The slope of a line is the same everywhere on the line.

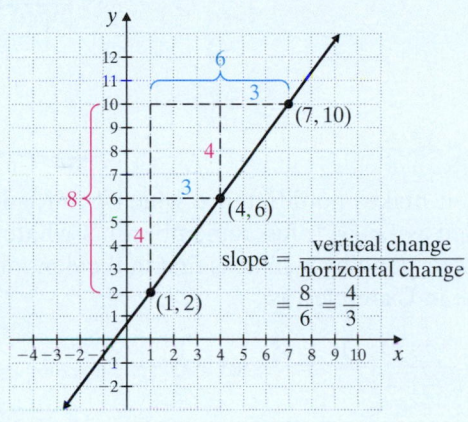

To find the slope of a line, then, choose two points of the line. Label the two x-coordinates of the two points x_1 and x_2 (read "x sub one" and "x sub two"), and label the corresponding y-coordinates y_1 and y_2.

The vertical change or **rise** between these points is the difference in the y-coordinates: $y_2 - y_1$. The horizontal change or **run** between the points is the difference of the x-coordinates: $x_2 - x_1$. The slope of the line is the ratio of $y_2 - y_1$ to $x_2 - x_1$, and we traditionally use the letter m to denote slope: $m = \dfrac{y_2 - y_1}{x_2 - x_1}$.

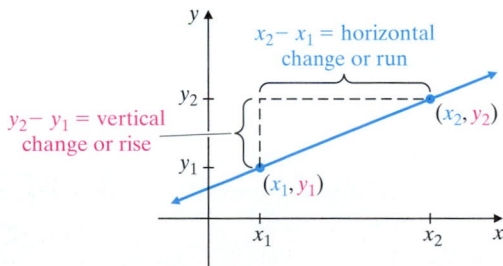

Slope of a Line

The slope m of the line containing the points (x_1, y_1) and (x_2, y_2) is given by

$$m = \frac{\text{rise}}{\text{run}} = \frac{\text{change in } y}{\text{change in } x} = \frac{y_2 - y_1}{x_2 - x_1}, \qquad \text{as long as } x_2 \neq x_1$$

Example 1 Find the slope of the line through $(-1, 5)$ and $(2, -3)$. Graph the line.

Solution: Let (x_1, y_1) be $(-1, 5)$ and (x_2, y_2) be $(2, -3)$. Then, by the definition of slope, we have the following.

(Continued on next page)

Practice 1

Find the slope of the line through $(-2, 3)$ and $(4, -1)$. Graph the line.

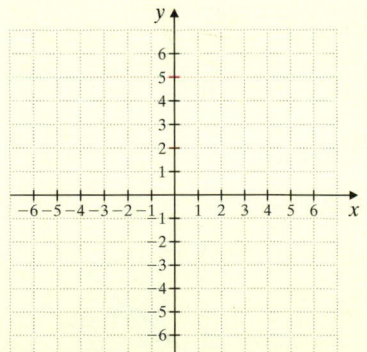

Answer

1. $m = -\dfrac{2}{3}$

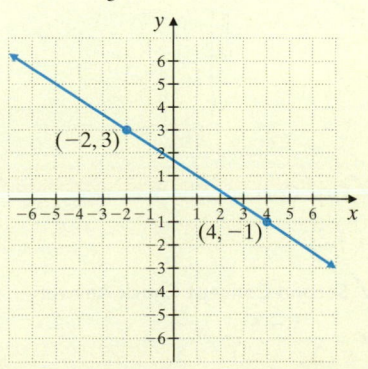

$$m = \frac{y_2 - y_1}{x_2 - x_1}$$

$$= \frac{-3 - 5}{2 - (-1)}$$

$$= \frac{-8}{3} = -\frac{8}{3}$$

The slope of the line is $-\frac{8}{3}$.

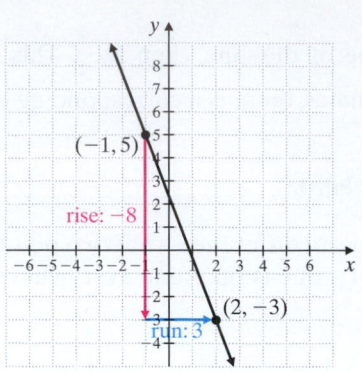

🟧 **Work Practice 1**

Helpful Hint

When finding slope, it makes no difference which point is identified as (x_1, y_1) and which is identified as (x_2, y_2). Just remember that whatever y-value is first in the numerator, its corresponding x-value is first in the denominator. Another way to calculate the slope in Example 1 is

$$m = \frac{y_2 - y_1}{x_2 - x_1} = \frac{5 - (-3)}{-1 - 2} = \frac{8}{-3} \text{ or } -\frac{8}{3} \leftarrow \text{Same slope as found in Example 1}$$

Practice 2

Find the slope of the line through $(-2, 1)$ and $(3, 5)$. Graph the line.

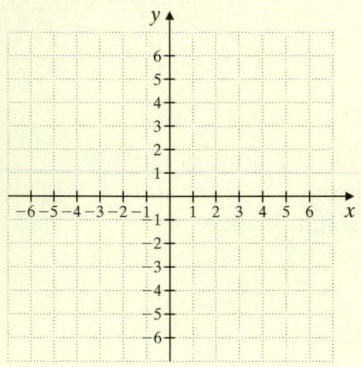

✔ **Concept Check** The points $(-2, -5)$, $(0, -2)$, $(4, 4)$, and $(10, 13)$ all lie on the same line. Work with a partner and verify that the slope is the same no matter which points are used to find slope.

Example 2   Find the slope of the line through $(-1, -2)$ and $(2, 4)$. Graph the line.

Solution: Let (x_1, y_1) be $(2, 4)$ and (x_2, y_2) be $(-1, -2)$.

$$m = \frac{y_2 - y_1}{x_2 - x_1}$$

$$= \frac{-2 - 4}{-1 - 2} \quad \begin{array}{l} y\text{-value} \\ \text{corresponding } x\text{-value} \end{array}$$

$$= \frac{-6}{-3} = 2$$

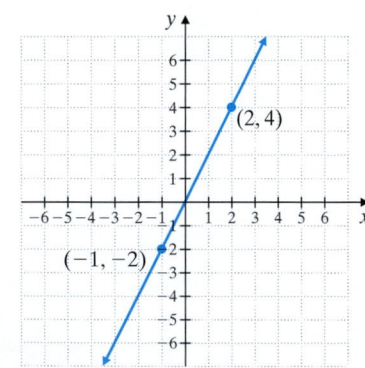

The slope is 2.

🟧 **Work Practice 2**

Answer

2. $m = \frac{4}{5}$

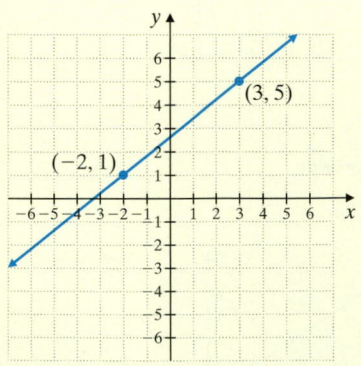

✔ **Concept Check** What is wrong with the following slope calculation for the points $(3, 5)$ and $(-2, 6)$?

$$m = \frac{5 - 6}{-2 - 3} = \frac{-1}{-5} = \frac{1}{5}$$

Notice that the slope of the line in Example 1 is negative and that the slope of the line in Example 2 is positive. Let your eye follow the line with negative slope from left to right and notice that the line "goes down." If you follow the line with positive slope from left to right, you will notice that the line "goes up." This is true in general.

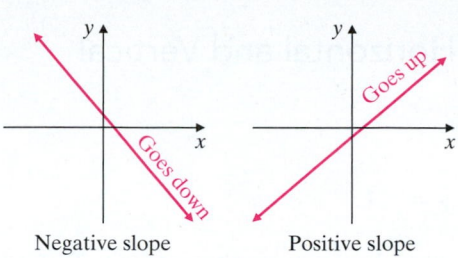

Negative slope   Positive slope

Helpful Hint To decide whether a line "goes up" or "goes down," always follow the line from left to right.

Objective B Finding the Slope of a Line Given Its Equation

As we have seen, the slope of a line is defined by two points on the line. Thus, if we know the equation of a line, we can find its slope by finding two of its points. For example, let's find the slope of the line

$$y = 3x - 2$$

To find two points, we can choose two values for x and substitute to find corresponding y-values. If $x = 0$, for example, $y = 3 \cdot 0 - 2$ or $y = -2$. If $x = 1$, $y = 3 \cdot 1 - 2$ or $y = 1$. This gives the ordered pairs $(0, -2)$ and $(1, 1)$. Using the definition for slope, we have

$$m = \frac{1 - (-2)}{1 - 0} = \frac{3}{1} = 3 \quad \text{The slope is 3.}$$

Notice that the slope, 3, is the same as the coefficient of x in the equation $y = 3x - 2$. This is true in general.

If a linear equation is solved for y, the coefficient of x is the line's slope. In other words, the slope of the line given by $y = mx + b$ is m, the coefficient of x.

$$y = \underset{\underset{\text{slope}}{\uparrow}}{m}x + b$$

Example 3 Find the slope of the line $-2x + 3y = 11$.

Solution: When we solve for y, the coefficient of x is the slope.

$$-2x + 3y = 11$$
$$3y = 2x + 11 \quad \text{Add } 2x \text{ to both sides.}$$
$$y = \frac{2}{3}x + \frac{11}{3} \quad \text{Divide both sides by 3.}$$

The slope is $\frac{2}{3}$.

■ **Work Practice 3**

Example 4 Find the slope of the line $-y = 5x - 2$.

Solution: Remember, the equation must be solved for y (not $-y$) in order for the coefficient of x to be the slope.

To solve for y, let's divide both sides of the equation by -1.

$$-y = 5x - 2$$
$$\frac{-y}{-1} = \frac{5x}{-1} - \frac{2}{-1} \quad \text{Divide both sides by } -1.$$
$$y = -5x + 2 \quad \text{Simplify.}$$

The slope is -5.

■ **Work Practice 4**

Practice 3

Find the slope of the line $5x + 4y = 10$.

Practice 4

Find the slope of the line $-y = -2x + 7$.

Answers

3. $m = -\frac{5}{4}$ **4.** $m = 2$

Objective C Finding Slopes of Horizontal and Vertical Lines ▶

Practice 5

Find the slope of $y = 3$.

Example 5 Find the slope of the line $y = -1$.

Solution: Recall that $y = -1$ is a horizontal line with y-intercept -1. To find the slope, we find two ordered pair solutions of $y = -1$, knowing that solutions of $y = -1$ must have a y-value of -1. We will use $(2, -1)$ and $(-3, -1)$. We let (x_1, y_1) be $(2, -1)$ and (x_2, y_2) be $(-3, -1)$.

$$m = \frac{y_2 - y_1}{x_2 - x_1} = \frac{-1 - (-1)}{-3 - 2} = \frac{0}{-5} = 0$$

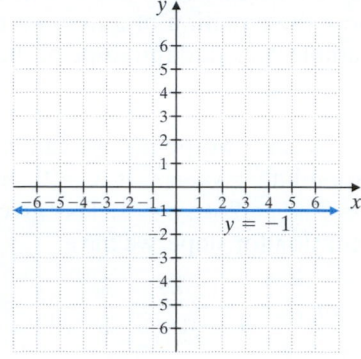

The slope of the line $y = -1$ is 0. Since the y-values will have a difference of 0 for every horizontal line, we can say that all **horizontal lines have a slope of 0.**

■ **Work Practice 5**

Practice 6

Find the slope of the line $x = -2$.

Example 6 Find the slope of the line $x = 5$.

Solution: Recall that the graph of $x = 5$ is a vertical line with x-intercept 5. To find the slope, we find two ordered pair solutions of $x = 5$. Ordered pair solutions of $x = 5$ must have an x-value of 5. We will use $(5, 0)$ and $(5, 4)$. We let $(x_1, y_1) = (5, 0)$ and $(x_2, y_2) = (5, 4)$.

$$m = \frac{y_2 - y_1}{x_2 - x_1} = \frac{4 - 0}{5 - 5} = \frac{4}{0}$$

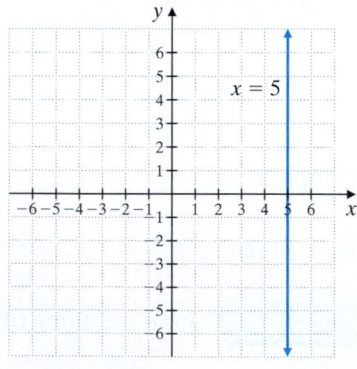

Helpful Hint Slope of 0 and undefined slope are not the same. Vertical lines have undefined slope, while horizontal lines have a slope of 0.

Since $\frac{4}{0}$ is undefined, we say that the slope of the vertical line $x = 5$ is undefined.

Since the x-values will have a difference of 0 for every vertical line, we can say that all **vertical lines have undefined slope.**

■ **Work Practice 6**

Answers

5. $m = 0$ **6.** undefined slope

Here is a general review of slope.

Summary of Slope

Slope m of the line through (x_1, y_1) and (x_2, y_2) is given by the equation

$$m = \frac{y_2 - y_1}{x_2 - x_1}.$$

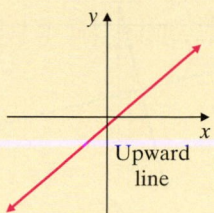

Upward
line

Positive slope: $m > 0$

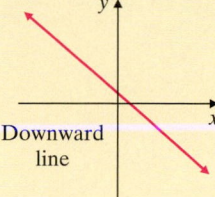

Downward
line

Negative slope: $m < 0$

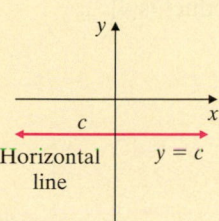

Horizontal
line $\quad y = c$

Zero slope: $m = 0$

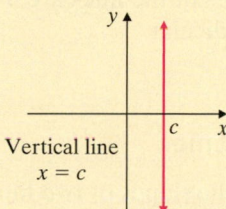

Vertical line
$x = c$

No slope or undefined slope

Objective D Slopes of Parallel and Perpendicular Lines

Two lines in the same plane are **parallel** if they do not intersect. Slopes of lines can help us determine whether lines are parallel. Since parallel lines have the same steepness, it follows that they have the same slope.

For example, the graphs of

$$y = -2x + 4$$

and

$$y = -2x - 3$$

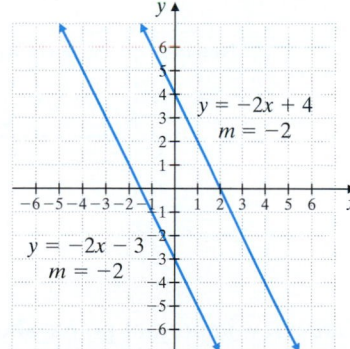

are shown. These lines have the same slope, -2. They also have different y-intercepts, so the lines are parallel. (If the y-intercepts were the same also, the lines would be the same.)

Parallel Lines

Nonvertical parallel lines have the same slope and different y-intercepts.

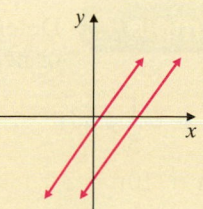

Two lines are **perpendicular** if they lie in the same plane and meet at a 90° (right) angle. How do the slopes of perpendicular lines compare? The product of the slopes of two perpendicular lines is −1.

For example, the graphs of

$$y = 4x + 1$$

and

$$y = -\frac{1}{4}x - 3$$

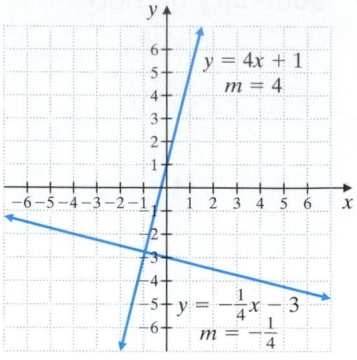

are shown. The slopes of the lines are 4 and $-\frac{1}{4}$. Their product is $4\left(-\frac{1}{4}\right) = -1$, so the lines are perpendicular.

Perpendicular Lines

If the product of the slopes of two lines is −1, then the lines are perpendicular.

(Two nonvertical lines are perpendicular if the slope of one is the negative reciprocal of the slope of the other.)

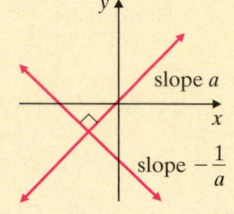

Helpful Hint

Here are examples of numbers that are negative (opposite) reciprocals.

Number	Negative Reciprocal	Their product is −1.
$\dfrac{2}{3}$	$-\dfrac{3}{2}$	$\dfrac{2}{3} \cdot -\dfrac{3}{2} = -\dfrac{6}{6} = -1$
-5 or $-\dfrac{5}{1}$	$\dfrac{1}{5}$	$-5 \cdot \dfrac{1}{5} = -\dfrac{5}{5} = -1$

Practice 7

Determine whether each pair of lines is parallel, perpendicular, or neither.

a. $x + y = 5$
  $2x + y = 5$

b. $5y = 2x - 3$
  $5x + 2y = 1$

c. $y = 2x + 1$
  $4x - 2y = 8$

Answers

7. **a.** neither **b.** perpendicular
c. parallel

Here are a few important points about vertical and horizontal lines.

- Two distinct vertical lines are parallel.
- Two distinct horizontal lines are parallel.
- A horizontal line and a vertical line are always perpendicular.

△ **Example 7** Determine whether each pair of lines is parallel, perpendicular, or neither.

a. $y = -\dfrac{1}{5}x + 1$ **b.** $x + y = 3$ **c.** $3x + y = 5$
    $2x + 10y = 3$ $-x + y = 4$ $2x + 3y = 6$

Solution:

a. The slope of the line $y = -\frac{1}{5}x + 1$ is $-\frac{1}{5}$. We find the slope of the second line by solving its equation for y.

$$2x + 10y = 3$$
$$10y = -2x + 3 \qquad \text{Subtract } 2x \text{ from both sides.}$$
$$y = \frac{-2}{10}x + \frac{3}{10} \qquad \text{Divide both sides by 10.}$$
$$y = -\frac{1}{5}x + \frac{3}{10} \qquad \text{Simplify.}$$

The slope of this line is $-\frac{1}{5}$ also. Since the lines have the same slope and different y-intercepts, they are parallel, as shown below on the left.

b. To find each slope, we solve each equation for y.

$$x + y = 3 \qquad\qquad\qquad -x + y = 4$$
$$y = -x + 3 \qquad\qquad\qquad y = x + 4$$

The slope is -1.   The slope is 1.

The slopes are not the same, so the lines are not parallel. Next we check the product of the slopes: $(-1)(1) = -1$. Since the product is -1, the lines are perpendicular, as shown below on the right.

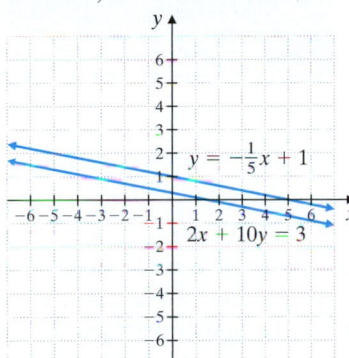
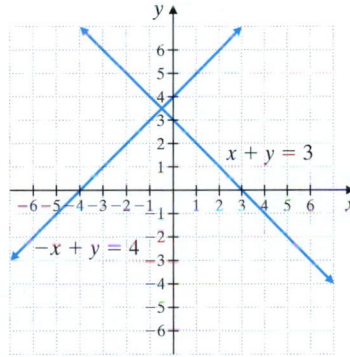

c. We solve each equation for y to find each slope. The slopes are -3 and $-\frac{2}{3}$. The slopes are not the same and their product is not -1. Thus, the lines are neither parallel nor perpendicular.

■ **Work Practice 7**

✓**Concept Check** Consider the line $-6x + 2y = 1$.

a. Write the equations of two lines parallel to this line.

b. Write the equations of two lines perpendicular to this line.

Objective E Slope as a Rate of Change

Slope can also be interpreted as a rate of change. In other words, slope tells us how fast y is changing with respect to x. To see this, let's look at a few of the many real-world applications of slope. For example, the pitch of a roof, used by builders and architects, is its slope. The pitch of the roof on the right is $\frac{7}{10}\left(\frac{\text{rise}}{\text{run}}\right)$. This means that the roof rises vertically 7 feet for every horizontal 10 feet. The rate of change for the roof is 7 vertical feet (y) per 10 horizontal feet (x).

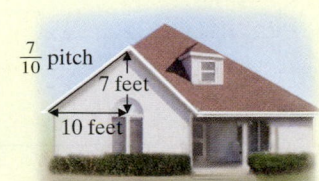

✓**Concept Check Answers**

Answers may vary; for example,

a. $y = 3x - 3$, $y = 3x - 1$

b. $y = -\frac{1}{3}x$, $y = -\frac{1}{3}x + 1$

The grade of a road is its slope written as a percent. A 7% grade, as shown below, means that the road rises (or falls) 7 feet for every horizontal 100 feet. $\left(\text{Recall that } 7\% = \dfrac{7}{100}.\right)$ Here, the slope of $\dfrac{7}{100}$ gives us the rate of change. The road rises (in our diagram) 7 vertical feet (y) for every 100 horizontal feet (x).

$\frac{7}{100} = 7\%$ grade
7 feet
100 feet

Practice 8

Find the grade of the road shown.

3 feet
20 feet

Example 8   Finding the Grade of a Road

At one part of the road to the summit of Pike's Peak, the road rises 15 feet for a horizontal distance of 250 feet. Find the grade of the road.

Solution: Recall that the grade of a road is its slope written as a percent.

$$\text{grade} = \frac{\text{rise}}{\text{run}} = \frac{15}{250} = 0.06 = 6\%$$

15 feet
250 feet

The grade is 6%.

■ **Work Practice 8**

Practice 9

Find the slope of the line and write the slope as a rate of change. This graph represents annual restaurant-industry employment y (in billions of workers) for year x. Write a sentence explaining the meaning of slope in this application.

U.S. Restaurant-Industry Employment

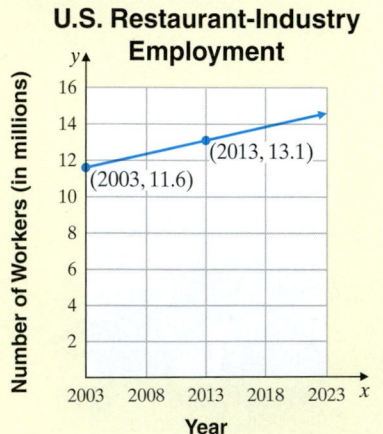

Number of Workers (in millions)

(2003, 11.6)
(2013, 13.1)

Year

Source: National Restaurant Association

Example 9   Finding the Slope of a Line

The following graph shows the cost y (in cents) of a nationwide long-distance telephone call from Texas with a certain telephone-calling plan, where x is the length of the call in minutes. Find the slope of the line and attach the proper units for the rate of change. Then write a sentence explaining the meaning of slope in this application.

Solution: Use $(2, 34)$ and $(6, 62)$ to calculate slope.

Cost of Long-Distance Telephone Call

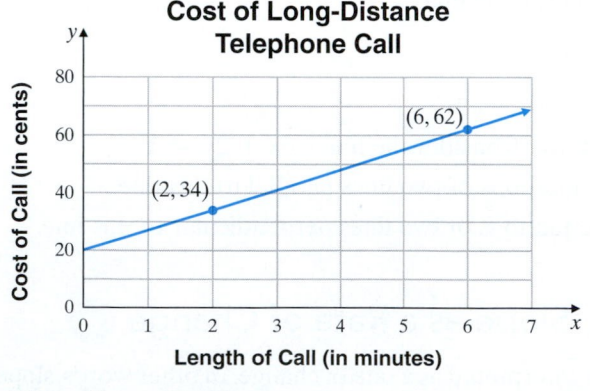

Cost of Call (in cents)

(2, 34)
(6, 62)

Length of Call (in minutes)

$$m = \frac{62 - 34}{6 - 2} = \frac{28}{4} = \frac{7 \text{ cents}}{1 \text{ minute}}$$

This means that the rate of change of the phone call is 7 cents per 1 minute, or the cost of the phone call is 7 cents per minute.

■ **Work Practice 9**

Answers

8. 15%   **9.** $m = 0.15$; Each year the number of workers employed in the U.S. restaurant industry increases by 0.15 million, or 150,000, workers per year.

Calculator Explorations Graphing

It is possible to use a graphing calculator to sketch the graph of more than one equation on the same set of axes. This feature can be used to see parallel lines with the same slope. For example, graph the equations $y = \frac{2}{5}x$, $y = \frac{2}{5}x + 7$, and $y = \frac{2}{5}x - 4$ on the same set of axes. To do so, press the $\boxed{Y =}$ key and enter the equations on the first three lines.

$$Y_1 = \frac{2}{5}x$$

$$Y_2 = \frac{2}{5}x + 7$$

$$Y_3 = \frac{2}{5}x - 4$$

The displayed equations should look like this:

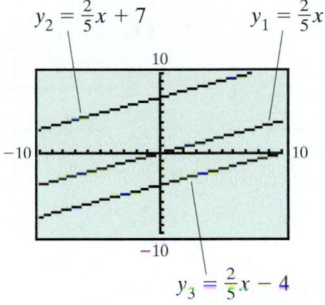

These lines are parallel, as expected, since they all have a slope of $\frac{2}{5}$. The graph of $y = \frac{2}{5}x + 7$ is the graph of $y = \frac{2}{5}x$ moved 7 units upward with a y-intercept of 7. Also, the graph of $y = \frac{2}{5}x - 4$ is the graph of $y = \frac{2}{5}x$ moved 4 units downward with a y-intercept of -4.

Graph the parallel lines on the same set of axes. Describe the similarities and differences in their graphs.

1. $y = 3.8x$, $y = 3.8x - 3$, $y = 3.8x + 9$

2. $y = -4.9x$, $y = -4.9x + 1$, $y = -4.9x + 8$

3. $y = \frac{1}{4}x$, $y = \frac{1}{4}x + 5$, $y = \frac{1}{4}x - 8$

4. $y = -\frac{3}{4}x$, $y = -\frac{3}{4}x - 5$, $y = -\frac{3}{4}x + 6$

Vocabulary, Readiness & Video Check

Use the choices below to fill in each blank. Not all choices will be used.

m	*x*	0	positive	undefined
b	*y*	slope	negative	

1. The measure of the steepness or tilt of a line is called _____.

2. If an equation is written in the form $y = mx + b$, the value of the letter _____ is the value of the slope of the graph.

3. The slope of a horizontal line is _____.

4. The slope of a vertical line is _____.

5. If the graph of a line moves upward from left to right, the line has _____ slope.

6. If the graph of a line moves downward from left to right, the line has _____ slope.

7. Given two points of a line, slope $= \dfrac{\text{change in } \underline{\hspace{2cm}}}{\text{change in } \underline{\hspace{2cm}}}$.

State whether the slope of the line is positive, negative, 0, or undefined.

8. 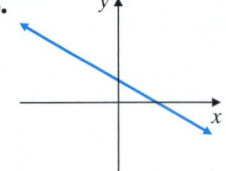   9. 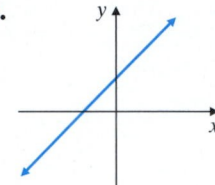   10. 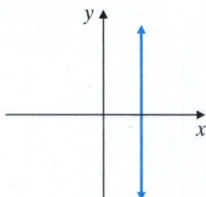   11.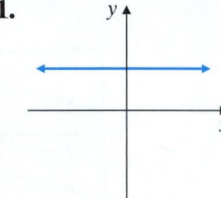

Decide whether a line with the given slope slants upward or downward or is horizontal or vertical.

12. $m = \dfrac{7}{6}$ _____   13. $m = -3$ _____   14. $m = 0$ _____   15. *m* is undefined. _____

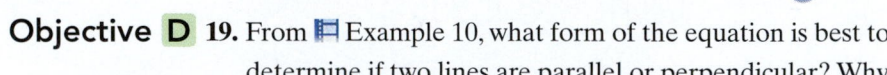

Martin-Gay Interactive Videos

Watch the section lecture video and answer the following questions.

See Video 10.4

Objective A 16. What important point is made during ▣ Example 1 having to do with the order of the points in the slope formula? ▶

Objective B 17. From ▣ Example 5, how do we write an equation in "slope-intercept form"? Once the equation is in slope-intercept form, how do we identify the slope? ▶

Objective C 18. In the lecture after ▣ Example 8, different slopes are summarized. What is the difference between zero slope and undefined slope? What does "no slope" mean? ▶

Objective D 19. From ▣ Example 10, what form of the equation is best to determine if two lines are parallel or perpendicular? Why? ▶

Objective E 20. Writing the slope as a rate of change in ▣ Example 11 gave real-life meaning to the slope. What step in the general strategy for problem solving does this correspond to? ▶

10.4 **Exercise Set** MyMathLab®

Objective A *Find the slope of the line that passes through the given points. See Examples 1 and 2.*

▶ **1.** $(-1, 5)$ and $(6, -2)$ **2.** $(-1, 16)$ and $(3, 4)$ **3.** $(1, 4)$ and $(5, 3)$ **4.** $(3, 1)$ and $(2, 6)$

▶ **5.** $(5, 1)$ and $(-2, 1)$ **6.** $(-8, 3)$ and $(-2, 3)$ ▶ **7.** $(-4, 3)$ and $(-4, 5)$ **8.** $(-2, -3)$ and $(-2, 5)$

Use the points shown on each graph to find the slope of each line. See Examples 1 and 2.

9. 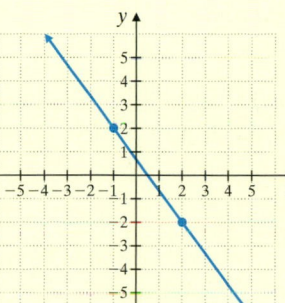   ▶ **10.** 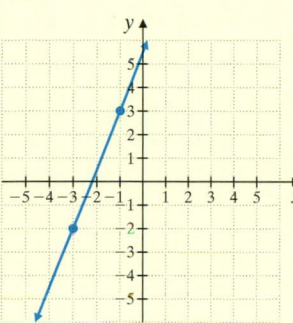   **11.** 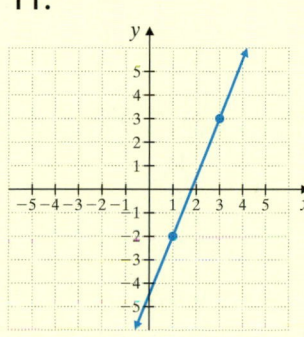   **12.**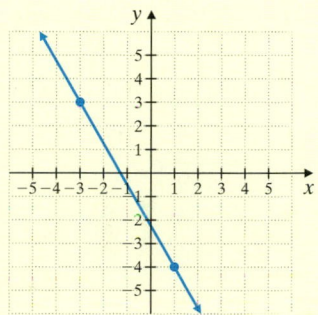

For each graph, determine which line has the greater slope.

13. 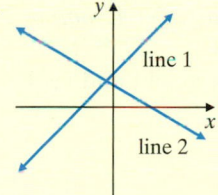   **14.** 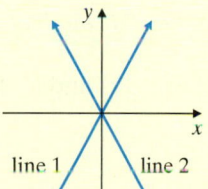   **15.** 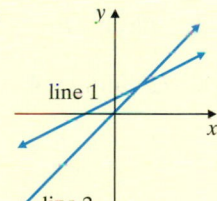   **16.**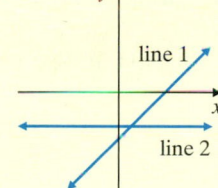

Objectives B C Mixed Practice *Find the slope of each line. See Examples 3 through 6.*

17. $y = 5x - 2$ **18.** $y = -2x + 6$ **19.** $y = -0.3x + 2.5$

20. $y = -7.6x - 0.1$ ▶ **21.** $2x + y = 7$ **22.** $-5x + y = 10$

23. 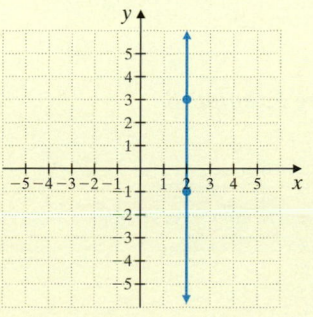   **24.**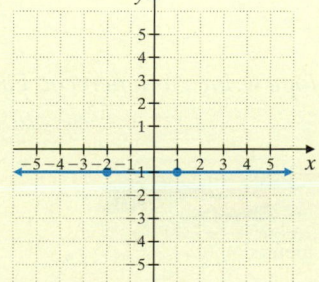

▶ 25. $2x - 3y = 10$ **26.** $3x - 5y = 1$ **▶ 27.** $x = 1$ **28.** $y = -2$

29. $x = 2y$ **30.** $x = -4y$ **▶ 31.** $y = -3$ **32.** $x = 5$

33. $-3x - 4y = 6$ **34.** $-4x - 7y = 9$ **35.** $20x - 5y = 1.2$ **36.** $24x - 3y = 5.7$

△ **Objective D** *Determine whether each pair of lines is parallel, perpendicular, or neither. See Example 7.*

▶ 37. $y = \dfrac{2}{9}x + 3$

$y = -\dfrac{2}{9}x$

38. $y = \dfrac{1}{5}x + 20$

$y = -\dfrac{1}{5}x$

39. $x - 3y = -6$

$y = 3x - 9$

40. $y = 4x - 2$

$4x + y = 5$

41. $6x = 5y + 1$

$-12x + 10y = 1$

42. $-x + 2y = -2$

$2x = 4y + 3$

43. $6 + 4x = 3y$

$3x + 4y = 8$

▶ 44. $10 + 3x = 5y$

$5x + 3y = 1$

△ *Find the slope of a line that is (a) parallel and (b) perpendicular to the line through each pair of points. See Example 7.*

45. $(-3, -3)$ and $(0, 0)$ **46.** $(6, -2)$ and $(1, 4)$ **47.** $(-8, -4)$ and $(3, 5)$ **48.** $(6, -1)$ and $(-4, -10)$

Objective E *The pitch of a roof is its slope. Find the pitch of each roof shown. See Example 8.*

49.

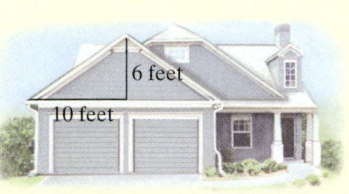

50.

The grade of a road is its slope written as a percent. Find the grade of each road shown. See Example 8.

51.

52.

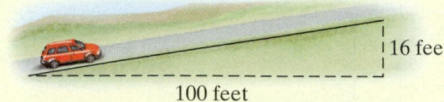

53. One of Japan's superconducting "bullet" trains is researched and tested at the Yamanashi Maglev Test Line near Otsuki City. The steepest section of the track has a rise of 2580 meters for a horizontal distance of 6450 meters. What is the grade (slope written as a percent) of this section of track? (*Source:* Japan Railways Central Co.)

2580 meters

6450 meters

54. Professional plumbers suggest that a sewer pipe rise 0.25 inch for every horizontal foot. Find the recommended slope for a sewer pipe and write the slope as a grade, or percent. Round to the nearest percent.

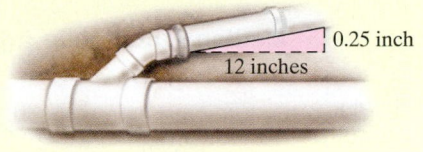

0.25 inch

12 inches

55. There has been controversy over the past few years about the world's steepest street. *The Guinness Book of Records* listed Baldwin Street, in Dunedin, New Zealand, as the world's steepest street, but Canton Avenue in the Pittsburgh neighborhood of Beechview may actually be steeper. Calculate each grade to the nearest percent.

Canton Avenue	For every 30 meters of horizontal distance, the vertical change is 11 meters.	
Baldwin Street	For every 2.86 meters of horizontal distance, the vertical change is 1 meter.	

56. According to federal regulations, a wheelchair ramp should rise no more than 1 foot for a horizontal distance of 12 feet. Write the slope as a grade. Round to the nearest tenth of a percent.

Find the slope of each line and write a sentence using the slope as a rate of change. Don't forget to attach the proper units. See Example 9.

57. This graph approximates the number of U.S. households that have televisions *y* (in millions) for year *x*.

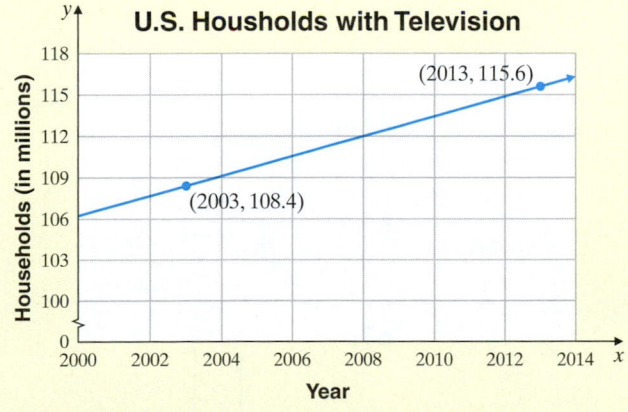

U.S. Houatholds with Television

Households (in millions)

(2013, 115.6)

(2003, 108.4)

Year

Source: The Nielson Company

58. The graph approximates the amount of money *y* (in billions of dollars) spent worldwide on tourism for year *x*.

Money Spent on World Tourism

Dollars (in billions)

(2012, 1075)

(2000, 475)

Year

Source: World Tourism Organization

59. Americans are keeping their cars longer. The graph below shows the median age y (in years) of automobiles in the United States for the years shown.

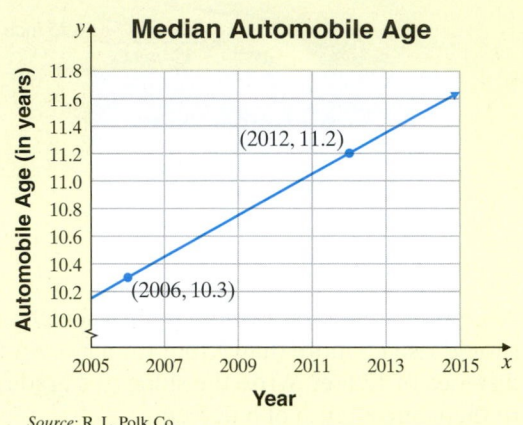

Median Automobile Age

(2012, 11.2)

(2006, 10.3)

Source: R. L. Polk Co.

60. The graph below shows the total cost y (in dollars) of owning and operating a large sedan in the United States in 2013, where x is the annual number of miles driven.

Owning and Operating a Large Sedan

(20,000, 12,700)

(10,000, 9750)

Source: AAA

Review

Solve each equation for y. See Section 9.5.

61. $y - (-6) = 2(x - 4)$

62. $y - 7 = -9(x - 6)$

63. $y - 1 = -6(x - (-2))$

64. $y - (-3) = 4(x - (-5))$

Concept Extensions

Match each line with its slope.

a. $m = 0$

b. undefined slope

c. $m = 3$

d. $m = 1$

e. $m = -\dfrac{1}{2}$

f. $m = -\dfrac{3}{4}$

65.

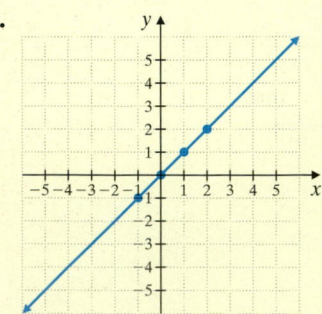

66.

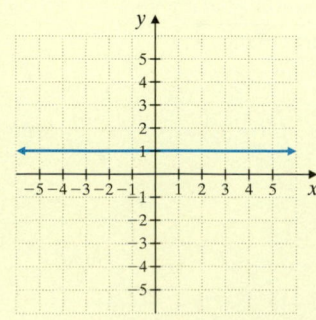

67.

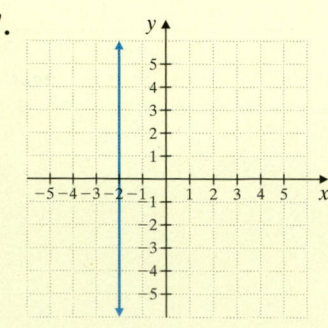

68.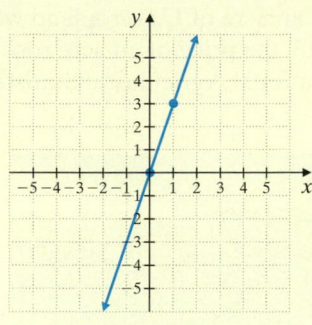

69.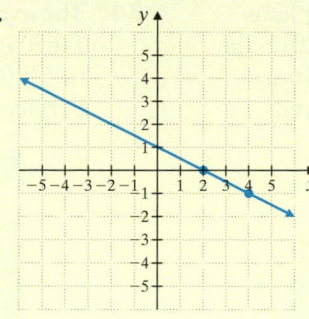

70.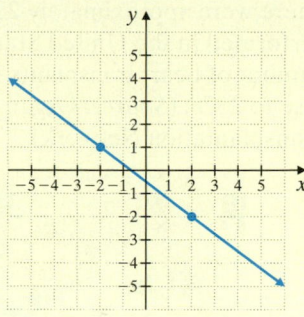

Solve. See a Concept Check in this section.

71. Verify that the points $(2, 1), (0, 0), (-2, -1)$, and $(-4, -2)$ are all on the same line by computing the slope between each pair of points. (See the first Concept Check.)

72. Given the points $(2, 3)$ and $(-5, 1)$, can the slope of the line through these points be calculated by $\dfrac{1 - 3}{2 - (-5)}$? Why or why not? (See the second Concept Check.)

73. Write the equations of three lines parallel to $10x - 5y = -7$. (See the third Concept Check.)

74. Write the equations of two lines perpendicular to $10x - 5y = -7$. (See the third Concept Check.)

The following line graph shows the average fuel economy (in miles per gallon) of passenger automobiles produced during each of the model years shown. Use this graph to answer Exercises 75 through 80.

75. What was the average fuel economy (in miles per gallon) for automobiles produced during 2008?

76. Find the decrease in average fuel economy for automobiles between the years 2010 and 2011.

77. During which of the model year(s) shown was average fuel economy the lowest?
What was the average fuel economy for that year/those years?

78. During which of the model years shown was average fuel economy the highest?
What was the average fuel economy for that year?

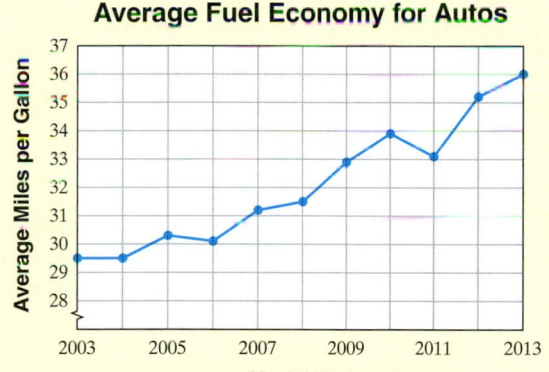

Average Fuel Economy for Autos

Source: U.S. Department of Transportation

79. Of the following line segments, which has the greatest slope: from 2007 to 2008, from 2009 to 2010, or from 2011 to 2012?

80. What line segment has a slope of 0?

81. Find x so that the pitch of the roof is $\dfrac{2}{5}$.

82. Find x so that the pitch of the roof is $\dfrac{1}{3}$.

83. There were approximately 2209 heart transplants performed in the United States in 2007. In 2012, the number of heart transplants in the United States rose to 2378. (*Source:* Organ Procurement and Transplantation Network)

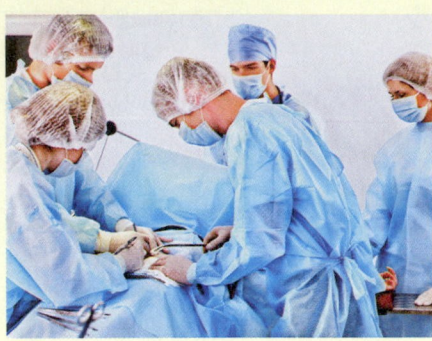

    a. Write two ordered pairs of the form (year, number of heart transplants).

    b. Find the slope of the line between the two points.

    c. Write a sentence explaining the meaning of the slope as a rate of change.

84. The average price of an acre of U.S. cropland was $2670 in 2009. In 2013, the price of an acre rose to $4000. (*Source:* National Agricultural Statistics Service)

    a. Write two ordered pairs of the form (year, price of acre).

    b. Find the slope of the line through the two points.

    c. Write a sentence explaining the meaning of the slope as a rate of change.

85. Show that the quadrilateral with vertices $(1, 3)$, $(2, 1)$, $(-4, 0)$, and $(-3, -2)$ is a parallelogram.

86. Show that a triangle with vertices at the points $(1, 1)$, $(-4, 4)$, and $(-3, 0)$ is a right triangle.

Find the slope of the line through the given points.

87. $(-3.8, 1.2)$ and $(-2.2, 4.5)$

88. $(2.1, 6.7)$ and $(-8.3, 9.3)$

89. $(14.3, -10.1)$ and $(9.8, -2.9)$

90. $(2.3, 0.2)$ and $(7.9, 5.1)$

91. The graph of $y = \frac{1}{2}x$ has a slope of $\frac{1}{2}$. The graph of $y = 3x$ has a slope of 3. The graph of $y = 5x$ has a slope of 5. Graph all three equations on a single coordinate system. As the slope becomes larger, how does the steepness of the line change?

92. The graph of $y = -\frac{1}{3}x + 2$ has a slope of $-\frac{1}{3}$. The graph of $y = -2x + 2$ has a slope of -2. The graph of $y = -4x + 2$ has a slope of -4. Graph all three equations on a single coordinate system. As the absolute value of the slope becomes larger, how does the steepness of the line change?

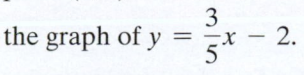
We know that when a linear equation is solved for y, the coefficient of x is the slope of the line. For example, the slope of the line whose equation is $y = 3x + 1$ is 3. In the equation $y = 3x + 1$, what does 1 represent? To find out, let $x = 0$ and watch what happens.

$$y = 3x + 1$$
$$y = 3 \cdot 0 + 1 \quad \text{Let } x = 0.$$
$$y = 1$$

We now have the ordered pair $(0, 1)$, which means that 1 is the y-intercept.
   This is true in general. To see this, let $x = 0$ and solve for y in $y = mx + b$.

$$y = m \cdot 0 + b \quad \text{Let } x = 0.$$
$$y = b$$

We obtain the ordered pair $(0, b)$, which means that point is the y-intercept.
   The form $y = mx + b$ is appropriately called the *slope-intercept form* of a linear equation.
$$\underset{\text{slope}}{\uparrow} \qquad \underset{y\text{-intercept is } (0, b).}{\uparrow}$$

Objectives

A Use the Slope-Intercept Form to Graph a Linear Equation. ▶

B Use the Slope-Intercept Form to Write an Equation of a Line. ▶

C Use the Point-Slope Form to Find an Equation of a Line Given Its Slope and a Point of the Line. ▶

D Use the Point-Slope Form to Find an Equation of a Line Given Two Points of the Line. ▶

E Use the Point-Slope Form to Solve Problems. ▶

Slope-Intercept Form

When a linear equation in two variables is written in **slope-intercept form,**
$$y = \underset{\uparrow}{m}x + \underset{\uparrow}{b}$$
$$\quad \text{slope} \quad (0, b), y\text{-intercept}$$
then m is the slope of the line and $(0, b)$ is the y-intercept of the line.

Objective **A** Using the Slope-Intercept Form to Graph an Equation ▶

We can use the slope-intercept form of the equation of a line to graph a linear equation.

Example 1 Use the slope-intercept form to graph the equation $y = \dfrac{3}{5}x - 2$.

Solution: Since the equation $y = \dfrac{3}{5}x - 2$ is written in slope-intercept form $y = mx + b$, the slope of its graph is $\dfrac{3}{5}$ and the y-intercept is $(0, -2)$. To graph this equation, we begin by plotting the point $(0, -2)$. From this point, we can find another point of the graph by using the slope $\dfrac{3}{5}$ and recalling that slope is $\dfrac{\text{rise}}{\text{run}}$. We start at the y-intercept and move 3 units up since the numerator of the slope is 3; then we move 5 units to the right since the denominator of the slope is 5. We stop at the point $(5, 1)$. The line through $(0, -2)$ and $(5, 1)$ is the graph of $y = \dfrac{3}{5}x - 2$.

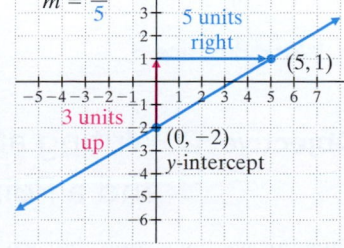

Work Practice 1

Practice 1

Use the slope-intercept form to graph the equation $y = \dfrac{2}{3}x - 4$.

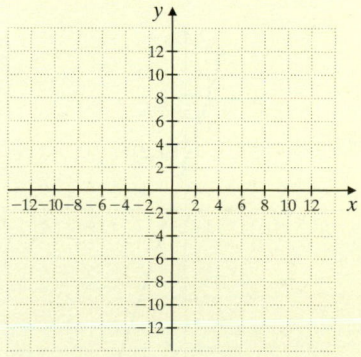

Answer
1. See page 770.

769

Practice 2

Use the slope-intercept form to graph $3x + y = 2$.

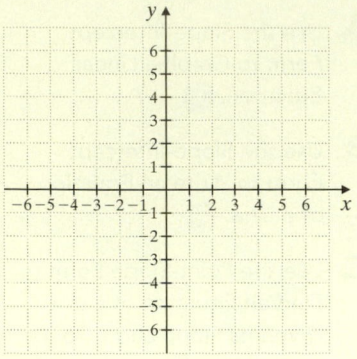

Example 2

Use the slope-intercept form to graph the equation $4x + y = 1$.

Solution: First we write the given equation in slope-intercept form.

$$4x + y = 1$$
$$y = -4x + 1$$

The graph of this equation will have slope -4 and y-intercept $(0, 1)$. To graph this line, we first plot the point $(0, 1)$. To find another point of the graph, we use the slope -4, which can be written as $\dfrac{-4}{1}\left(\dfrac{4}{-1} \text{ could also be used}\right)$. We start at the point $(0, 1)$ and move 4 units down (since the numerator of the slope is -4), and then 1 unit to the right (since the denominator of the slope is 1).

We arrive at the point $(1, -3)$. The line through $(0, 1)$ and $(1, -3)$ is the graph of $4x + y = 1$.

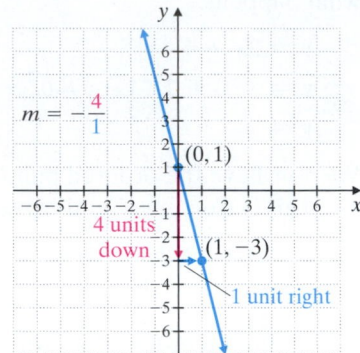

■ **Work Practice 2**

Practice 3

Find an equation of the line with y-intercept $(0, -2)$ and slope of $\dfrac{3}{5}$.

Helpful Hint

In Example 2, if we interpret the slope of -4 as $\dfrac{4}{-1}$, we arrive at $(-1, 5)$ for a second point. Notice that this point is also on the line.

Objective B Using the Slope-Intercept Form to Write an Equation ▶

The slope-intercept form can also be used to write the equation of a line when we know its slope and y-intercept.

Example 3

Find an equation of the line with y-intercept $(0, -3)$ and slope of $\dfrac{1}{4}$.

Solution: We are given the slope and the y-intercept. We let $m = \dfrac{1}{4}$ and $b = -3$ and write the equation in slope-intercept form, $y = mx + b$.

$$y = mx + b$$
$$y = \dfrac{1}{4}x + (-3) \quad \text{Let } m = \dfrac{1}{4} \text{ and } b = -3.$$
$$y = \dfrac{1}{4}x - 3 \quad \text{Simplify.}$$

■ **Work Practice 3**

Answers

1.

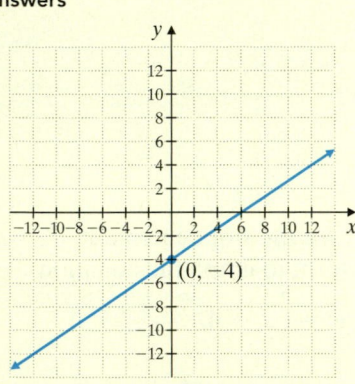

2.

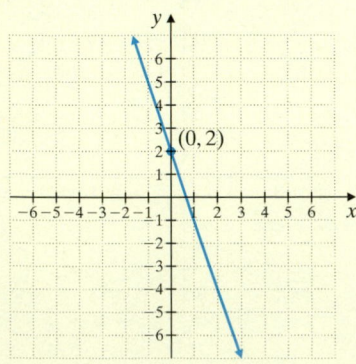

3. $y = \dfrac{3}{5}x - 2$

Objective C Writing an Equation Given Its Slope and a Point ▶

Thus far, we have seen that we can write an equation of a line if we know its slope and y-intercept. We can also write an equation of a line if we know its slope and any point on the line. To see how we do this, let m represent slope and (x_1, y_1) represent a point on the line. Then if (x, y) is any other point on the line, we have that

$$\frac{y - y_1}{x - x_1} = m$$

$$y - y_1 = m(x - x_1) \qquad \text{Multiply both sides by } (x - x_1).$$

     ↑
    slope

This is the *point-slope form* of the equation of a line.

> ### Point-Slope Form of the Equation of a Line
>
> The **point-slope form** of the equation of a line is $y - y_1 = m(x - x_1)$, where m is the slope of the line and (x_1, y_1) is a point on the line.

Example 4   Find an equation of the line with slope -2 that passes through $(-1, 5)$. Write the equation in slope-intercept form, $y = mx + b$, and in standard form, $Ax + By = C$.

Solution: Since the slope and a point on the line are given, we use point-slope form, $y - y_1 = m(x - x_1)$, to write the equation. Let $m = -2$ and $(-1, 5) = (x_1, y_1)$.

$$y - y_1 = m(x - x_1)$$
$$y - 5 = -2[x - (-1)] \qquad \text{Let } m = -2 \text{ and } (x_1, y_1) = (-1, 5).$$
$$y - 5 = -2(x + 1) \qquad \text{Simplify.}$$
$$y - 5 = -2x - 2 \qquad \text{Use the distributive property.}$$

To write the equation in slope-intercept form, $y = mx + b$, we simply solve the equation for y. To do this, we add 5 to both sides.

$$y - 5 = -2x - 2$$
$$y = -2x + 3 \qquad \text{Slope-intercept form}$$
$$2x + y = 3 \qquad \text{Add } 2x \text{ to both sides and we have standard form.}$$

■ **Work Practice 4**

Objective D   Writing an Equation Given Two Points

We can also find the equation of a line when we are given any two points of the line.

Example 5   Find an equation of the line through $(2, 5)$ and $(-3, 4)$. Write the equation in the form $Ax + By = C$.

Solution: First, use the two given points to find the slope of the line.

$$m = \frac{4 - 5}{-3 - 2} = \frac{-1}{-5} = \frac{1}{5}$$

Next we use the slope $\frac{1}{5}$ and either one of the given points to write the equation in point-slope form. We use $(2, 5)$. Let $x_1 = 2$, $y_1 = 5$, and $m = \frac{1}{5}$.

$$y - y_1 = m(x - x_1) \qquad \text{Use point-slope form.}$$
$$y - 5 = \frac{1}{5}(x - 2) \qquad \text{Let } x_1 = 2, y_1 = 5, \text{ and } m = \frac{1}{5}.$$
$$5(y - 5) = 5 \cdot \frac{1}{5}(x - 2) \qquad \text{Multiply both sides by 5 to clear fractions.}$$
$$5y - 25 = x - 2 \qquad \text{Use the distributive property and simplify.}$$
$$-x + 5y - 25 = -2 \qquad \text{Subtract } x \text{ from both sides.}$$
$$-x + 5y = 23 \qquad \text{Add 25 to both sides.}$$

■ **Work Practice 5**

Practice 4

Find an equation of the line with slope -3 that passes through $(2, -4)$. Write the equation in slope-intercept form, $y = mx + b$.

Practice 5

Find an equation of the line through $(1, 3)$ and $(5, -2)$. Write the equation in the form $Ax + By = C$.

Answers

4. $y = -3x + 2$ **5.** $5x + 4y = 17$

Helpful Hint

When you multiply both sides of the equation from Example 5, $-x + 5y = 23$, by -1, it becomes $x - 5y = -23$.

Both $-x + 5y = 23$ and $x - 5y = -23$ are in the form $Ax + By = C$ and both are equations of the same line.

Objective E Using the Point-Slope Form to Solve Problems

Problems occurring in many fields can be modeled by linear equations in two variables. The next example is from the field of marketing and shows how consumer demand of a product depends on the price of the product.

Example 6 The Whammo Company has learned that by pricing a newly released Frisbee at $6, sales will reach 2000 Frisbees per day. Raising the price to $8 will cause the sales to fall to 1500 Frisbees per day.

a. Assume that the relationship between sales price and number of Frisbees sold is linear, and write an equation describing this relationship. Write the equation in slope-intercept form. Use ordered pairs of the form (sales price, number sold).

b. Predict the daily sales of Frisbees if the price is $7.50.

Solution:

a. We use the given information and write two ordered pairs. Our ordered pairs are $(6, 2000)$ and $(8, 1500)$. To use the point-slope form to write an equation, we find the slope of the line that contains these points.

$$m = \frac{2000 - 1500}{6 - 8} = \frac{500}{-2} = -250$$

Next we use the slope and either one of the points to write the equation in point-slope form. We use $(6, 2000)$.

$$y - y_1 = m(x - x_1) \quad \text{Use point-slope form.}$$
$$y - 2000 = -250(x - 6) \quad \text{Let } x_1 = 6, y_1 = 2000, \text{ and } m = -250.$$
$$y - 2000 = -250x + 1500 \quad \text{Use the distributive property.}$$
$$y = -250x + 3500 \quad \text{Write in slope-intercept form.}$$

b. To predict the sales if the price is $7.50, we find y when $x = 7.50$.

$$y = -250x + 3500$$
$$y = -250(7.50) + 3500 \quad \text{Let } x = 7.50.$$
$$y = -1875 + 3500$$
$$y = 1625$$

If the price is $7.50, sales will reach 1625 Frisbees per day.

■ **Work Practice 6**

Practice 6

The Pool Entertainment Company learned that by pricing a new pool toy at $10, local sales will reach 200 a week. Lowering the price to $9 will cause sales to rise to 250 a week.

a. Assume that the relationship between sales price and number of toys sold is linear, and write an equation describing this relationship. Write the equation in slope-intercept form. Use ordered pairs of the form (sales price, number sold).

b. Predict the weekly sales of the toy if the price is $7.50.

Answers

6. a. $y = -50x + 700$ **b.** 325

We also could have solved Example 6 by using ordered pairs of the form (number sold, sales price).

Here is a summary of our discussion on linear equations thus far.

Forms of Linear Equations

$Ax + By = C$	**Standard form** of a linear equation. A and B are not both 0.
$y = mx + b$	**Slope-intercept form** of a linear equation. The slope is m and the y-intercept is $(0, b)$.
$y - y_1 = m(x - x_1)$	**Point-slope form** of a linear equation. The slope is m and (x_1, y_1) is a point on the line.
$y = c$	**Horizontal line** The slope is 0 and the y-intercept is $(0, c)$.
$x = c$	**Vertical line** The slope is undefined and the x-intercept is $(c, 0)$.

Parallel and Perpendicular Lines

Nonvertical parallel lines have the same slope.
The product of the slopes of two nonvertical perpendicular lines is -1.

 ## Calculator Explorations   Graphing

A graphing calculator is a very useful tool for discovering patterns. To discover the change in the graph of a linear equation caused by a change in slope, try the following: Use a standard window and graph a linear equation in the form $y = mx + b$. Recall that the graph of such an equation will have slope m and y-intercept $(0, b)$.

First graph $y = x + 3$. To do so, press the $\boxed{Y =}$ key and enter $Y_1 = x + 3$. Notice that this graph has slope 1 and that the y-intercept is 3. Next, on the same set of axes, graph $y = 2x + 3$ and $y = 3x + 3$ by pressing $\boxed{Y =}$ and entering $Y_2 = 2x + 3$ and $Y_3 = 3x + 3$.

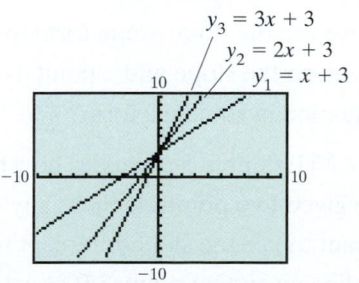

Notice the difference in the graph of each equation as the slope changes from 1 to 2 to 3. How would the graph of $y = 5x + 3$ appear? To see the change in the graph caused by a change to negative slope, try graphing $y = -x + 3$, $y = -2x + 3$, and $y = -3x + 3$ on the same set of axes.

Use a graphing calculator to graph the following equations. For each exercise, graph the first equation and use its graph

to predict the appearance of the other equations. Then graph the other equations on the same set of axes and check your prediction.

1. $y = x; y = 6x, y = -6x$

2. $y = -x; y = -5x, y = -10x$

3. $y = \dfrac{1}{2}x + 2; y = \dfrac{3}{4}x + 2, y = x + 2$

4. $y = x + 1; y = \dfrac{5}{4}x + 1, y = \dfrac{5}{2}x + 1$

Vocabulary, Readiness & Video Check

Use the choices below to fill in each blank. Some choices may be used more than once and some not at all.

b (y_1, x_1) point-slope   vertical   standard

m (x_1, y_1) slope-intercept   horizontal

1. The form $y = mx + b$ is called _____
   form. When a linear equation in two variables is
   written in this form, _____ is the slope of its
   graph and $(0,$ _____ $)$ is its y-intercept.

2. The form $y - y_1 = m(x - x_1)$ is called
   _____ form. When a linear equation in two
   variables is written in this form, _____ is the
   slope of its graph and _____ is a point on
   the graph.

For Exercises 3 through 6, identify the form that the linear equation in two variables is written in. For Exercises 7 and 8, identify the appearance of the graph of the equation.

3. $y - 7 = 4(x + 3)$; _____ form

4. $5x - 9y = 11$; _____ form

5. $y = \dfrac{3}{4}x - \dfrac{1}{3}$; _____ form

6. $y + 2 = \dfrac{-1}{3}(x - 2)$ _____ form

7. $y = \dfrac{1}{2}$; _____ line

8. $x = -17$; _____ line

Martin-Gay Interactive Videos

See Video 10.5

Watch the section lecture video and answer the following questions.

Objective A 9. We can use the slope-intercept form to graph a line.
Complete these statements based on ⊞ Example 1:
Start by graphing the _____.
From this point, find another point by applying the slope–
if necessary, rewrite the slope as a(n) _____ ▶

Objective B 10. In ⊞ Example 3, what is the y-intercept? ▶

Objective C 11. In ⊞ Example 4, we use the point-slope form to find the
equation of a line given the slope and a point. How do we
then write this equation in standard form? ▶

Objective D 12. The lecture before ⊞ Example 5 discusses how to find the
equation of a line given two points. Is there any circumstance
when we might want to use the slope-intercept form to find
the equation of a line given two points? If so, when? ▶

Objective E 13. In ⊞ Example 8, we are told to use ordered pairs of the form
(time, speed). Explain why it is important to keep track of
how we define our ordered pairs and/or our variables. ▶

10.5 Exercise Set MyMathLab®

Objective A *Use the slope-intercept form to graph each equation. See Examples 1 and 2.*

1. $y = 2x + 1$

2. $y = -4x - 1$

3. $y = \frac{2}{3}x + 5$

4. $y = \frac{1}{4}x - 3$

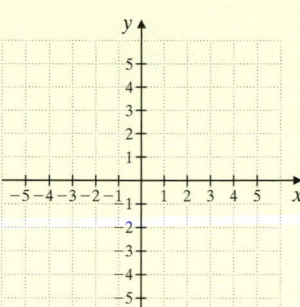 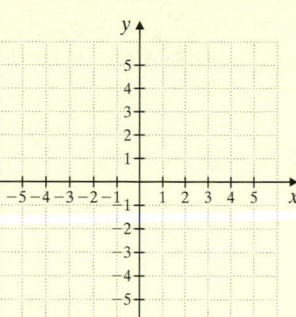 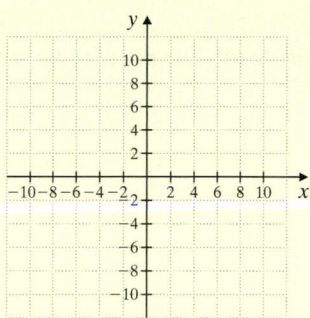 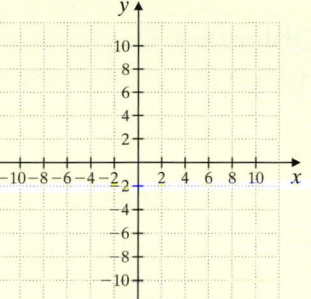

5. $y = -5x$

6. $y = -6x$

7. $4x + y = 6$

8. $-3x + y = 2$

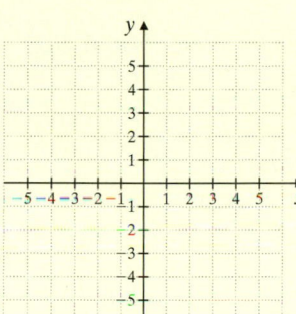 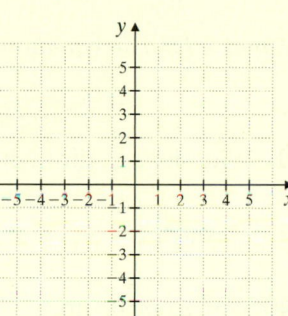 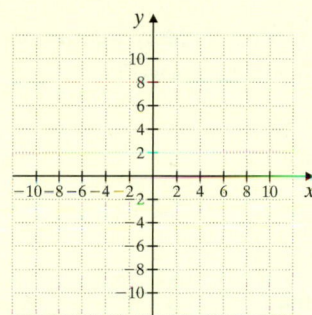

9. $4x - 7y = -14$

10. $3x - 4y = 4$

11. $x = \frac{5}{4}y$

12. $x = \frac{3}{2}y$

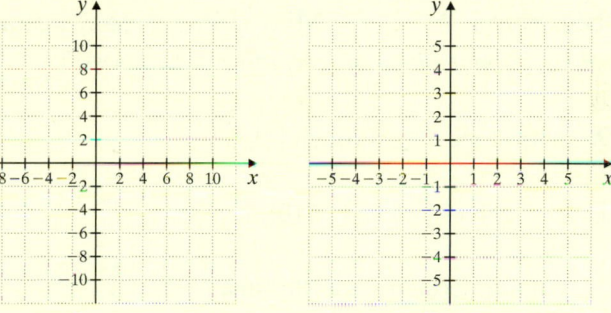 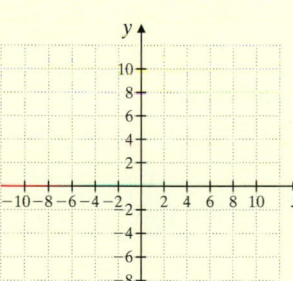 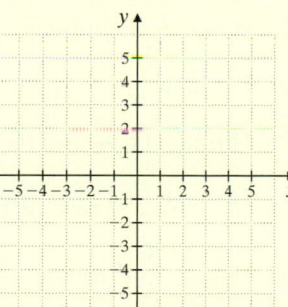 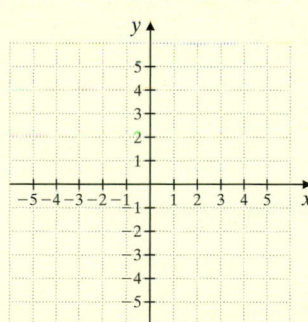 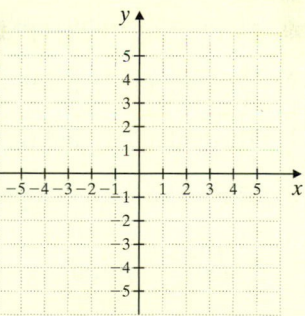

Objective B *Write an equation of the line with each given slope, m, and y-intercept, (0, b). See Example 3.*

13. $m = 5, b = 3$

14. $m = -3, b = -3$

15. $m = -4, b = -\frac{1}{6}$

16. $m = 2, b = \frac{3}{4}$

17. $m = \frac{2}{3}, b = 0$

18. $m = -\frac{4}{5}, b = 0$

19. $m = 0, b = -8$

20. $m = 0, b = -2$

21. $m = -\frac{1}{5}, b = \frac{1}{9}$

22. $m = \frac{1}{2}, b = -\frac{1}{3}$

Objective C *Find an equation of each line with the given slope that passes through the given point. Write the equation in the form Ax + By = C. See Example 4.*

23. $m = 6;$ $(2, 2)$

24. $m = 4;$ $(1, 3)$

▶ **25.** $m = -8$; $(-1, -5)$

26. $m = -2$; $(-11, -12)$

27. $m = \dfrac{3}{2}$; $(5, -6)$

28. $m = \dfrac{2}{3}$; $(-8, 9)$

29. $m = -\dfrac{1}{2}$; $(-3, 0)$

30. $m = -\dfrac{1}{5}$; $(4, 0)$

Objective D   *Find an equation of the line passing through each pair of points. Write the equation in the form* $Ax + By = C$. *See Example 5.*

31. $(3, 2)$ and $(5, 6)$

32. $(6, 2)$ and $(8, 8)$

33. $(-1, 3)$ and $(-2, -5)$

34. $(-4, 0)$ and $(6, -1)$

▶ **35.** $(2, 3)$ and $(-1, -1)$

36. $(7, 10)$ and $(-1, -1)$

37. $(0, 0)$ and $\left(-\dfrac{1}{8}, \dfrac{1}{13}\right)$

38. $(0, 0)$ and $\left(-\dfrac{1}{2}, \dfrac{1}{3}\right)$

Objectives A C D   **Mixed Practice** *See Examples 1, 4, and 5. Find an equation of each line described. Write each equation in slope-intercept form when possible.*

39. With slope $-\dfrac{1}{2}$, through $\left(0, \dfrac{5}{3}\right)$

40. With slope $\dfrac{5}{7}$, through $(0, -3)$

41. Through $(10, 7)$ and $(7, 10)$

42. Through $(5, -6)$ and $(-6, 5)$

▶ **43.** With undefined slope, through $\left(-\dfrac{3}{4}, 1\right)$

44. With slope 0, through $(6.7, 12.1)$

45. Slope 1, through $(-7, 9)$

46. Slope 5, through $(6, -8)$

47. Slope -5, y-intercept $(0, 7)$

48. Slope -2, y-intercept $(0, -4)$

▶ **49.** Through $(1, 2)$, parallel to $y = 5$

50. Through $(1, -5)$, parallel to the y-axis

51. Through $(2, 3)$ and $(0, 0)$

52. Through $(4, 7)$ and $(0, 0)$

53. Through $(-2, -3)$, perpendicular to the y-axis

54. Through $(0, 12)$, perpendicular to the x-axis

55. Slope $-\dfrac{4}{7}$, through $(-1, -2)$

56. Slope $-\dfrac{3}{5}$, through $(4, 4)$

Objective E   *Solve. Assume each exercise describes a linear relationship. Write the equations in slope-intercept form. See Example 6.*

57. In 2007, a total of 370 million magazines were sold in the United States. By 2012, this number was 312 million. (*Source:* MPA—The Association of Magazine Media)
   a. Write two ordered pairs of the form (years after 2007, millions of magazines sold) for this situation.
   b. Assume the relationship between years after 2007 and millions of magazines sold is linear over this period. Use the ordered pairs from part **a** to write an equation for the line relating year after 2007 to millions of magazines sold.
   c. Use the linear equation in part **b** to estimate the millions of magazines sold in 2010.

58. In 2008, crude oil field production in the United States was 1830 thousand barrels. In 2012, U.S. crude oil field production increased to 2374 thousand barrels. (*Source:* Energy Information Administration)
   a. Write two ordered pairs of the form (years after 2008, crude oil production).
   b. Assume the relationship between years after 2008 and crude oil production is linear over this period. Use the ordered pairs from part **a** to write an equation of the line relating years after 2008 to crude oil production.
   c. Use the linear equation from part **b** to estimate crude oil production in the United States in 2015, if this trend were to continue.

59. A rock is dropped from the top of a 400-foot cliff. After 1 second, the rock is traveling 32 feet per second. After 3 seconds, the rock is traveling 96 feet per second.

400 feet

a. Assume that the relationship between time and speed is linear and write an equation describing this relationship. Use ordered pairs of the form (time, speed).

b. Use this equation to determine the speed of the rock 4 seconds after it is dropped.

60. A Hawaiian fruit company is studying the sales of a pineapple sauce to see if this product is to be continued. At the end of its first year, profits on this product amounted to $30,000. At the end of the fourth year, profits were $66,000.

a. Assume that the relationship between years on the market and profit is linear and write an equation describing this relationship. Use ordered pairs of the form (years on the market, profit).

b. Use this equation to predict the profit at the end of 7 years.

61. In 2008 there were approximately 314,000 hybrid vehicles sold in the United States. In 2012, there were approximately 434,000 such vehicles sold. (*Source:* HybridCars.com)

a. Write an equation describing the relationship between time and the number of hybrid vehicles sold. Use ordered pairs of the form (years past 2008, number of vehicles sold).

b. Use this equation to estimate the number of hybrid sales in 2014.

62. In 2008, there were approximately 945 thousand restaurants in the United States. In 2012, there were 980 thousand restaurants. (*Source:* National Restaurant Association)

a. Write an equation describing the relationship between time and the number of restaurants. Use ordered pairs of the form (years past 2008, number of restaurants in thousands).

b. Use this equation to predict the number of eating establishments in 2016.

63. In 2007 there were 5545 indoor cinema sites in the United States. In 2012, there were approximately 5320 indoor cinema sites. (*Source:* National Association of Theater Owners)

a. Write an equation describing this relationship. Use ordered pairs of the form (years past 2007, number of indoor cinema sites).

b. Use this equation to predict the number of indoor cinema sites in 2015.

64. In 2010, the U.S. population per square mile of land area was approximately 87.4. In 2000, this person-per-square-mile population was 79.7. (*Source:* U.S. Census Bureau)

a. Write an equation describing the relationship between year and person per square mile. Use ordered pairs of the form (years past 2000, person per square mile).

b. Use this equation to predict the person-per-square-mile population in 2016.

65. The Pool Fun Company has learned that, by pricing a newly released Fun Noodle at $3, sales will reach 10,000 Fun Noodles per day during the summer. Raising the price to $5 will cause the sales to fall to 8000 Fun Noodles per day.

  a. Assume that the relationship between sales price and number of Fun Noodles sold is linear and write an equation describing this relationship. Use ordered pairs of the form (sales price, number sold).

  b. Predict the daily sales of Fun Noodles if the price is $3.50.

66. The value of a building bought in 1995 may be depreciated (or decreased) as time passes for income tax purposes. Seven years after the building was bought, this value was $225,000 and 12 years after it was bought, this value was $195,000.

  a. If the relationship between number of years past 1995 and the depreciated value of the building is linear, write an equation describing this relationship. Use ordered pairs of the form (years past 1995, value of building).

  b. Use this equation to estimate the depreciated value of the building in 2013.

Review

Find the value of $x^2 - 3x + 1$ for each given value of x. See Section 8.2.

67. 2   **68.** 5   **69.** -1 **70.** -3

Concept Extensions

Match each linear equation with its graph.

71. $y = 2x + 1$ **72.** $y = -x + 1$ **73.** $y = -3x - 2$ **74.** $y = \dfrac{5}{3}x - 2$

a.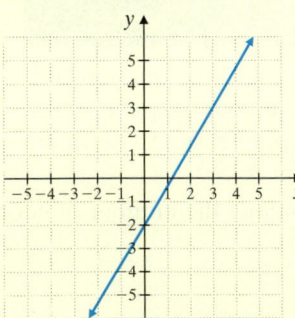

b.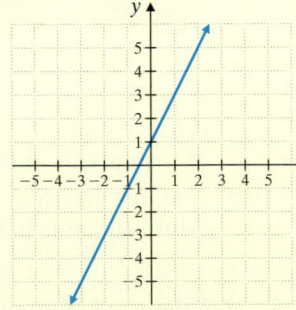

c.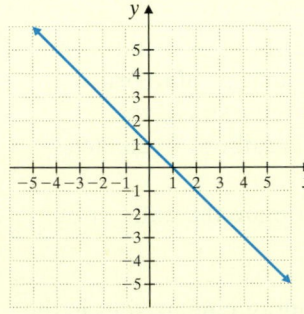

d.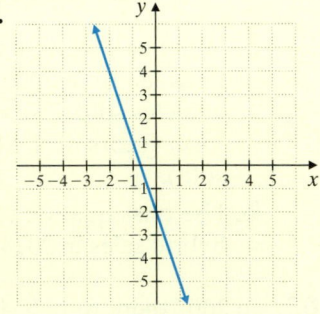

75. Write an equation in standard form of the line that contains the point $(-2, 4)$ and has the same slope as the line $y = 2x + 5$.

76. Write an equation in standard form of the line that contains the point $(3, 0)$ and has the same slope as the line $y = -3x - 1$.

△**77.** Write an equation in standard form of the line that contains the point $(-1, 2)$ and is
  a. parallel to the line $y = 3x - 1$.
  b. perpendicular to the line $y = 3x - 1$.

△**78.** Write an equation in standard form of the line that contains the point $(4, 0)$ and is
  a. parallel to the line $y = -2x + 3$.
  b. perpendicular to the line $y = -2x + 3$.

Integrated Review

Summary on Linear Equations

Find the slope of each line.

1.

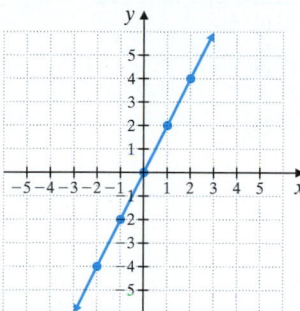

2.

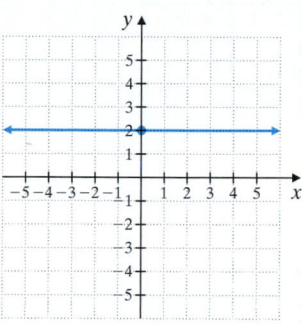

1. _____

2. _____

3.

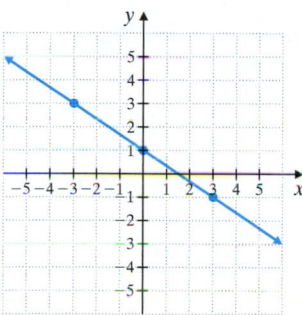

4.

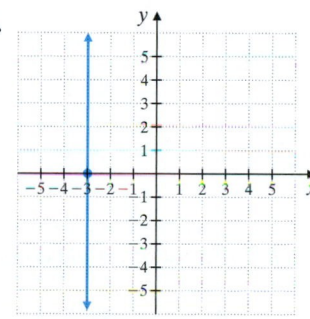

3. _____

4. _____

Graph each linear equation. For Exercises 11 and 12, label the intercepts.

5. $y = -2x$

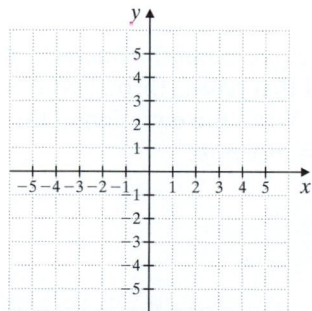

6. $x + y = 3$

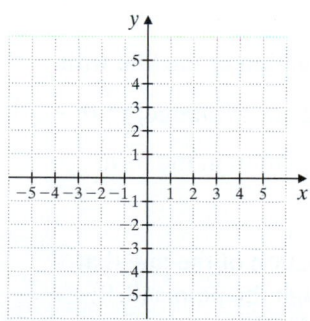

5. _____

6. _____

7. $x = -1$

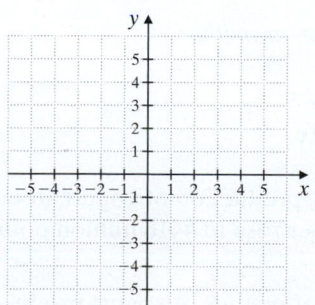

8. $y = 4$

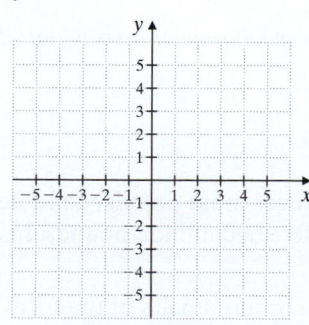

7. _____

8. _____

9. _____

10. _____

11. _____

12. _____

13. _____

14. _____

15. _____

16. _____

17. _____

18. _____

19. _____

20. _____

21. _____

22. _____

23. _____

24. a. _____

b. _____

c. _____

9. $x - 2y = 6$

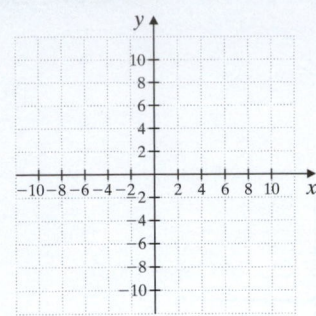

10. $y = 3x + 2$

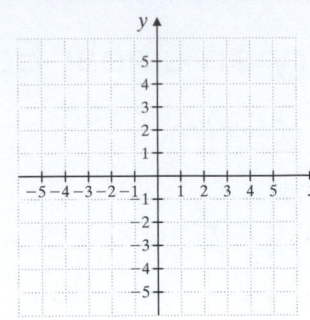

11. $y = -\dfrac{3}{4}x + 3$

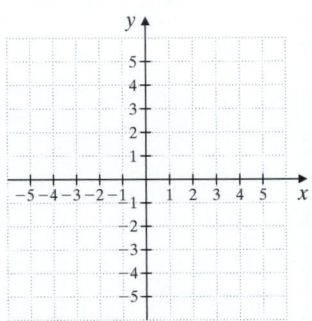

12. $5x - 2y = 8$

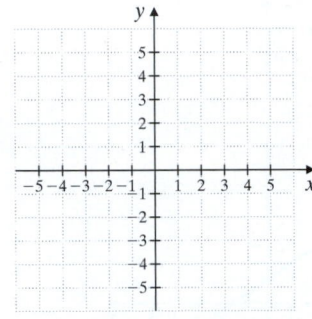

Find the slope of each line by writing the equation in slope-intercept form.

13. $y = 3x - 1$ **14.** $y = -6x + 2$ **15.** $7x + 2y = 11$ **16.** $2x - y = 0$

Find the slope of each line.

17. $x = 2$

18. $y = -4$

19. Write an equation of the line with slope $m = 2$ and y-intercept $\left(0, -\dfrac{1}{3}\right)$. Write the equation in the form $y = mx + b$.

20. Find an equation of the line with slope $m = -4$ that passes through the point $(-1, 3)$. Write the equation in the form $y = mx + b$.

21. Find an equation of the line that passes through the points $(2, 0)$ and $(-1, -3)$. Write the equation in the form $Ax + By = C$.

Determine whether each pair of lines is parallel, perpendicular, or neither.

22. $6x - y = 7$
$2x + 3y = 4$

23. $3x - 6y = 4$
$y = -2x$

24. Yogurt is an ever more popular food item. In 2008, U.S. production of yogurt stood at approximately 3600 million pounds. In 2012, this number rose to 4416 million pounds of yogurt.

   a. Write two ordered pairs of the form (year, millions of pounds of yogurt produced).
   b. Find the slope of the line between these two points.
   c. Write a sentence explaining the meaning of the slope as a rate of change.

10.6 Introduction to Functions

Objective A Identifying Relations, Domains, and Ranges

In previous sections, we have discussed the relationships between two quantities. For example, the relationship between the length of the side of a square x and its area y is described by the equation $y = x^2$. Ordered pairs can be used to write down solutions of this equation. For example, $(2, 4)$ is a solution of $y = x^2$, and this notation tells us that the x-value 2 is related to the y-value 4 for this equation. In other words, when the length of the side of a square is 2 units, its area is 4 square units.

Examples of Relationships Between Two Quantities		
Area of Square: $y = x^2$	**Equation of Line: $y = x + 2$**	**Online Advertising Revenue**

Some Ordered Pairs		**Some Ordered Pairs**		**Ordered Pairs**	
x	**y**	**x**	**y**	**Year**	**Billions of Dollars**
2	4	-3	-1	2006	16.7
5	25	0	2	2007	20.3
7	49	2	4	2008	23.5
12	144	9	11	2009	26.6
				2010	29.4

A set of ordered pairs is called a **relation.** The set of all x-coordinates is called the **domain** of a relation, and the set of all y-coordinates is called the **range** of a relation. Equations such as $y = x^2$ are also called relations since equations in two variables define a set of ordered pair solutions.

Example 1 Find the domain and the range of the relation
$\{(0, 2), (3, 3), (-1, 0), (3, -2)\}$.

Solution: The domain is the set of all x-coordinates, or $\{-1, 0, 3\}$, and the range is the set of all y-coordinates, or $\{-2, 0, 2, 3\}$.

■ **Work Practice 1**

Objective B Identifying Functions

Paired data occur often in real-life applications. Some special sets of paired data, or ordered pairs, are called *functions*.

Objectives

A Identify Relations, Domains, and Ranges.

B Identify Functions.

C Use the Vertical Line Test.

D Use Function Notation.

Practice 1

Find the domain and range of the relation $\{(-3, 5), (-3, 1), (4, 6), (7, 0)\}$.

Answer
1. domain: $\{-3, 4, 7\}$; range: $\{0, 1, 5, 6\}$

Function

A **function** is a set of ordered pairs in which each x-coordinate has exactly one y-coordinate.

In other words, a function cannot have two ordered pairs with the same x-coordinate but different y-coordinates.

Example 2   Which of the following relations is also a function?

a. $\{(-1, 1), (2, 3), (7, 3), (8, 6)\}$

b. $\{(0, -2), (1, 5), (0, 3), (7, 7)\}$

Solution:

a. Although the ordered pairs $(2, 3)$ and $(7, 3)$ have the same y-value, each x-value is assigned to only one y-value, so this set of ordered pairs is a function.

b. The x-value 0 is paired with two y-values, -2 and 3, so this set of ordered pairs is not a function.

■ **Work Practice 2**

Practice 2

Are the following relations also functions?

a. $\{(2, 5), (-3, 7), (4, 5), (0, -1)\}$

b. $\{(1, 4), (6, 6), (1, -3), (7, 5)\}$

Relations and functions can be described by graphs of their ordered pairs.

Example 3   Which graph is the graph of a function?

a.
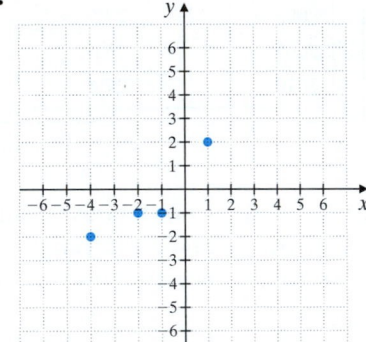

b.
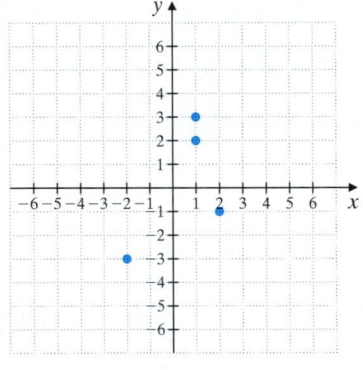

Solution:

a. This is the graph of the relation $\{(-4, -2), (-2, -1), (-1, -1), (1, 2)\}$. Each x-coordinate has exactly one y-coordinate, so this is the graph of a function.

b. This is the graph of the relation $\{(-2, -3), (1, 2), (1, 3), (2, -1)\}$. The x-coordinate 1 is paired with two y-coordinates, 2 and 3, so this is not the graph of a function.

■ **Work Practice 3**

Practice 3

Is each graph the graph of a function?

a.
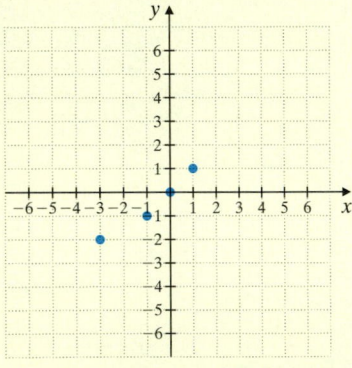

b.
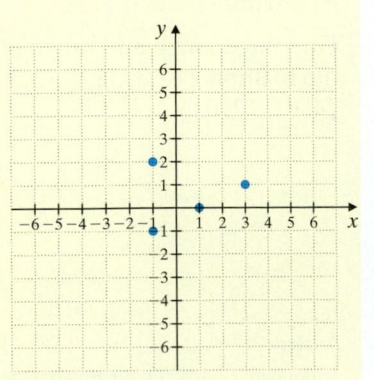

Objective C Using the Vertical Line Test ▶

The graph in Example 3(b) was not the graph of a function because the x-coordinate 1 was paired with two y-coordinates, 2 and 3. Notice that when an x-coordinate is paired with more than one y-coordinate, a vertical line can be drawn that will

---

Answers

2. a. a function   **b.** not a function

3. a. a function   **b.** not a function

intersect the graph at more than one point. We can use this fact to determine whether a relation is also a function. We call this the vertical line test.

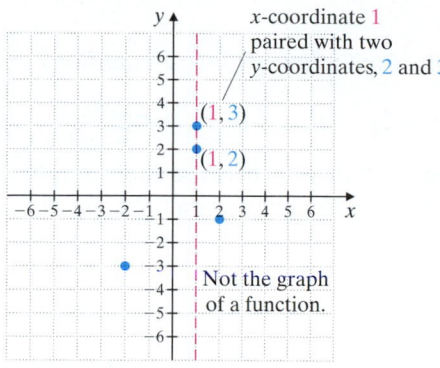

Vertical Line Test

If a vertical line can be drawn so that it intersects a graph more than once, the graph is not the graph of a function. (If no such vertical line can be drawn, the graph is that of a function.)

This vertical line test works for all types of graphs on the rectangular coordinate system.

▶ **Example 4**   Use the vertical line test to determine whether each graph is the graph of a function.

a.   **b.**   **c.**   **d.**

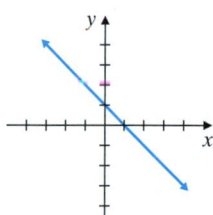   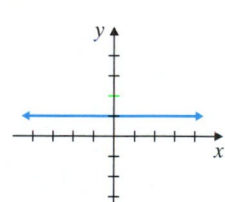   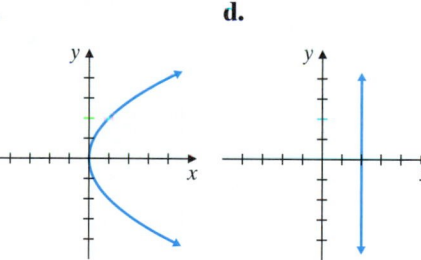

Solution:

a. This graph is the graph of a function since no vertical line will intersect this graph more than once.

b. This graph is also the graph of a function; no vertical line will intersect it more than once.

c. This graph is not the graph of a function. Vertical lines can be drawn that intersect the graph in two points. An example of one such line is shown.

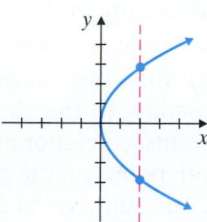

d. This graph is not the graph of a function. A vertical line can be drawn that intersects this line at every point.

🟧 **Work Practice 4**

Practice 4

Determine whether each graph is the graph of a function.

a.

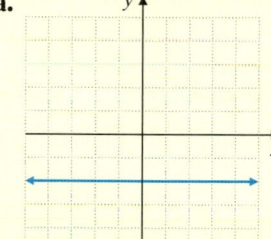

b.

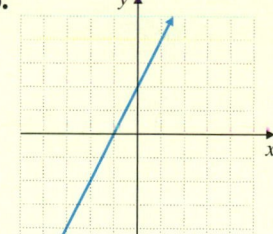

c.

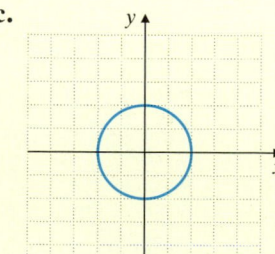

d.

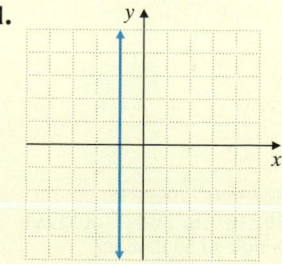

Answers

4. a. a function   **b.** a function
c. not a function   **d.** not a function

Recall that the graph of a linear equation is a line, and a line that is not vertical will pass the vertical line test. **Thus, all linear equations are functions except those of the form $x = c$, which are vertical lines.**

Practice 5

Which of the following linear equations are functions?

a. $y = 2x$ **b.** $y = -3x - 1$

c. $y = 8$ **d.** $x = 2$

Example 5 Which of the following linear equations are functions?

a. $y = x$ **b.** $y = 2x + 1$ **c.** $y = 5$ **d.** $x = -1$

Solution: **a, b,** and **c** are functions because their graphs are nonvertical lines. **d** is not a function because its graph is a vertical line.

■ Work Practice 5

Examples of functions can often be found in magazines, newspapers, books, and other printed material in the form of tables or graphs such as that in Example 6.

Practice 6

Use the graph in Example 6 to answer the questions.

a. Approximate the time of sunrise on March 1.

b. Approximate the date(s) when the sun rises at 6 a.m.

Example 6 The graph shows the sunrise time for Indianapolis, Indiana, for the first of each month for one year. Use this graph to answer the questions.

a. Approximate the time of sunrise on February 1.

b. Approximate the date(s) when the sun rises at 5 a.m.

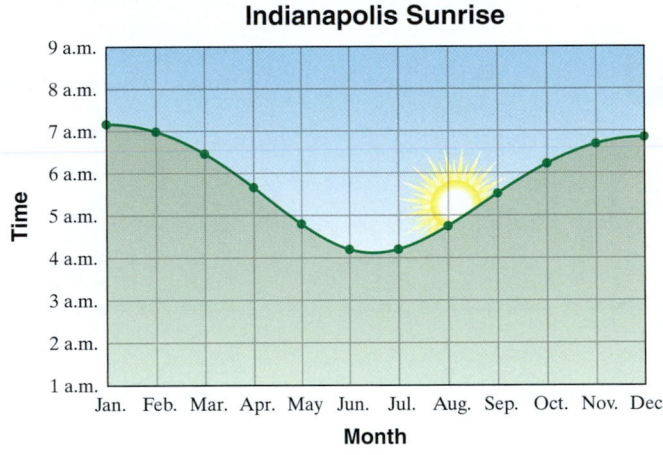

Indianapolis Sunrise

Source: Wolff World Atlas

c. Is this the graph of a function?

Solution:

a. To approximate the time of sunrise on February 1, we find the mark on the horizontal axis that corresponds to February 1. From this mark, we move vertically upward (shown in blue) until the graph is reached. From that point on the graph, we move horizontally to the left until the vertical axis is reached. The vertical axis there reads 7 a.m., as shown on the next page.

b. To approximate the date(s) when the sun rises at 5 a.m., we find 5 a.m. on the time axis and move horizontally to the right (shown in red). Notice that we will hit the graph at two points, corresponding to two dates for which the sun rises at 5 a.m. We follow both points on the graph vertically downward until the horizontal axis is reached. The sun rises at 5 a.m. at approximately the end of the month of April and the middle of the month of August.

Answers

5. a, b, and **c** are functions.

6. a. 6:30 a.m. **b.** middle of March and middle of September

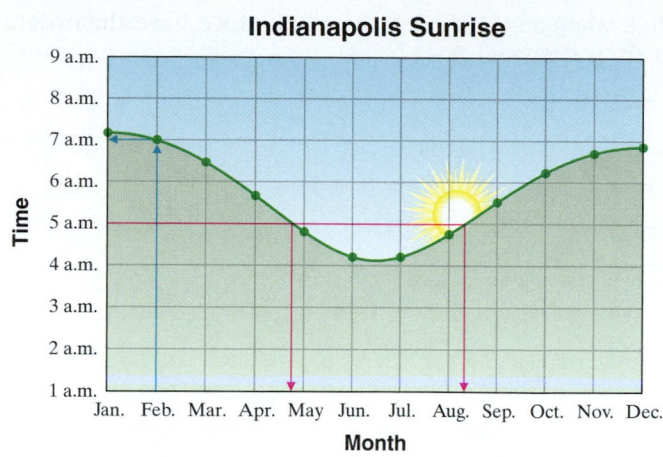

Indianapolis Sunrise

Source: Wolff World Atlas

c. The graph is the graph of a function since it passes the vertical line test. In other words, for every day of the year in Indianapolis, there is exactly one sunrise time.

■ **Work Practice 6**

Objective D Using Function Notation ▶

The graph of the linear equation $y = 2x + 1$ passes the vertical line test, so we say that $y = 2x + 1$ is a function. In other words, $y = 2x + 1$ gives us a rule for writing ordered pairs where every x-coordinate is paired with at most one y-coordinate.

The variable y is a function of the variable x. For each value of x, there is only one value of y. Thus, we say the variable x is the **independent variable** because any value in the domain can be assigned to x. The variable y is the **dependent variable** because its value depends on x.

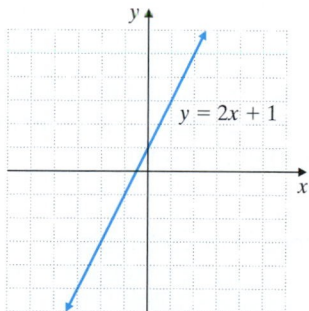

We often use letters such as f, g, and h to name functions. For example, the symbol $f(x)$ means *function of x* and is read "f of x." This notation is called **function notation.** The equation $y = 2x + 1$ can be written as $f(x) = 2x + 1$ using function notation, and these equations mean the same thing. In other words, $y = f(x)$.

The notation $f(1)$ means to replace x with 1 and find the resulting y or function value. Since

$$f(x) = 2x + 1$$

then

$$f(1) = 2(1) + 1 = 3$$

Helpful Hint

Note that, for example, if $f(2) = 5$, the corresponding ordered pair is $(2, 5)$.

This means that, when $x = 1$, y or $f(x) = 3$, and we have the ordered pair $(1, 3)$. Now let's find $f(2)$, $f(0)$, and $f(-1)$.

$$f(x) = 2x + 1 \qquad\qquad f(x) = 2x + 1 \qquad\qquad f(x) = 2x + 1$$
$$f(2) = 2(2) + 1 \qquad f(0) = 2(0) + 1 \qquad f(-1) = 2(-1) + 1$$
$$= 4 + 1 \qquad\qquad\quad = 0 + 1 \qquad\qquad\quad\; = -2 + 1$$
$$= 5 \qquad\qquad\qquad = 1 \qquad\qquad\qquad = -1$$

Ordered Pair: $\qquad (2, 5) \qquad\qquad\qquad (0, 1) \qquad\qquad\qquad (-1, -1)$

Helpful Hint

Note that $f(x)$ is a special symbol in mathematics used to denote a function. The symbol $f(x)$ is read "f of x." It does **not** mean $f \cdot x$ (f times x).

Practice 7

Given $f(x) = x^2 + 1$, find the following and list the corresponding ordered pairs.

a. $f(1)$ **b.** $f(-3)$ **c.** $f(0)$

Example 7   Given $g(x) = x^2 - 3$, find the following and list the corresponding ordered pairs generated.

a. $g(2)$ **b.** $g(-2)$ **c.** $g(0)$

Solution:

a. $g(x) = x^2 - 3$ **b.** $g(x) = x^2 - 3$ **c.** $g(x) = x^2 - 3$

$\; g(2) = 2^2 - 3 \qquad\quad g(-2) = (-2)^2 - 3 \qquad g(0) = 0^2 - 3$

$= 4 - 3 \qquad\qquad\quad\; = 4 - 3 \qquad\qquad\quad\; = 0 - 3$

$= 1 \qquad\qquad\qquad\quad = 1 \qquad\qquad\qquad\;\; = -3$

Ordered Pairs:	$g(2) = 1$ gives $(2, 1)$	$g(-2) = 1$ gives $(-2, 1)$	$g(0) = -3$ gives $(0, -3)$

■ **Work Practice 7**

We now practice finding the domain and the range of a function. The domain of our functions will be the set of all possible real numbers that x can be replaced by. The range is the set of corresponding y-values.

Practice 8

Find the domain of each function.

a. $h(x) = 6x + 3$

b. $f(x) = \dfrac{1}{x^2}$

Example 8   Find the domain of each function.

a. $g(x) = \dfrac{1}{x}$ **b.** $f(x) = 2x + 1$

Solution:

a. Recall that we cannot divide by 0, so the domain of $g(x)$ is the set of all real numbers except 0.

b. In this function, x can be any real number. The domain of $f(x)$ is the set of all real numbers.

■ **Work Practice 8**

Answers

7. a. $2; (1, 2)$ **b.** $10; (-3, 10)$
c. $1; (0, 1)$
8. a. Domain: all real numbers
   b. Domain: all real numbers except 0

✓ **Concept Check Answer**

$f(2) = -7$

✓ **Concept Check**   Suppose that the value of f is -7 when the function is evaluated at 2. Write this situation in function notation.

Example 9 Find the domain and the range of each function graphed.

a.

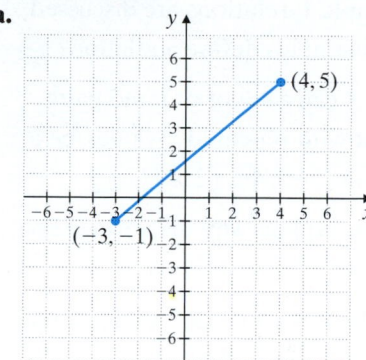

b.

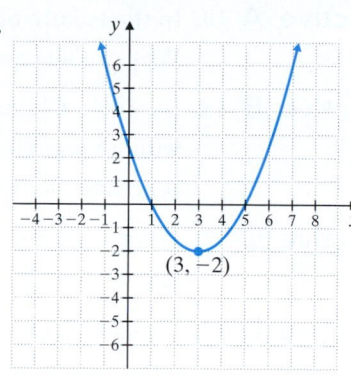

Solution:

a.

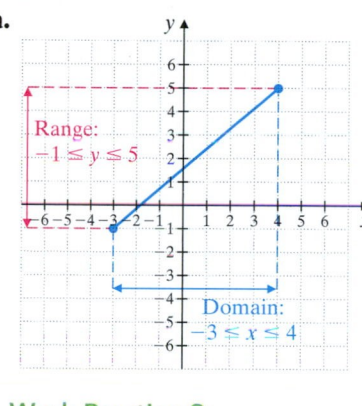

Range: $-1 \leq y \leq 5$

Domain: $-3 \leq x \leq 4$

b.

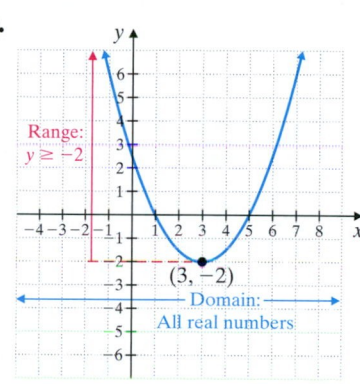

Range: $y \geq -2$

Domain: All real numbers

■ **Work Practice 9**

Practice 9

Find the domain and the range of each function graphed.

a.

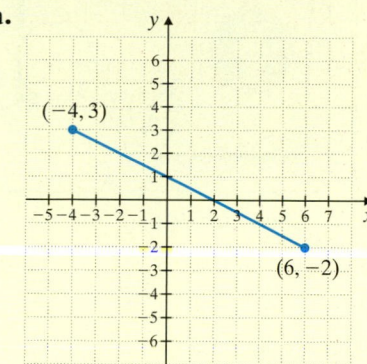

b.

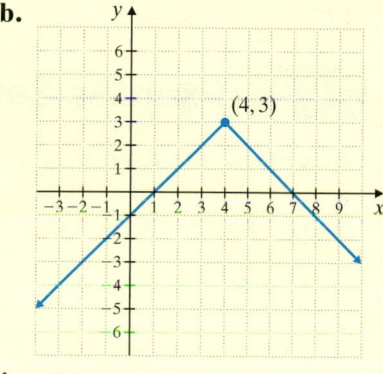

Answers

9. a. Domain: $-4 \leq x \leq 6$
Range: $-2 \leq y \leq 3$

  b. Domain: all real numbers
Range: $y \leq 3$

Vocabulary, Readiness & Video Check

Use the choices below to fill in each blank. Some choices may not be used.

$x = c$	horizontal	domain	relation	$(7, 3)$	x	$\{x \mid x \leq 5\}$
$y = c$	vertical	range	function	$(3, 7)$	y	all real numbers

1. A set of ordered pairs is called a(n) _____.

2. A set of ordered pairs that assigns to each x-value exactly one y-value is called a(n) _____.

3. The set of all y-coordinates of a relation is called the _____.

4. The set of all x-coordinates of a relation is called the _____.

5. All linear equations are functions except those whose graphs are _____ lines.

6. All linear equations are functions except those whose equations are of the form _____.

7. If $f(3) = 7$, the corresponding ordered pair is _____.

8. The domain of $f(x) = x + 5$ is _____.

9. For the function $y = mx + b$, the dependent variable is _____ and the independent variable is

_____.

Martin-Gay Interactive Videos   *Watch the section lecture video and answer the following questions.*

Objective A **10.** In the lecture before ▤ Example 1, relations are discussed. Why can an equation in two variables define a relation? ▶

Objective B **11.** Based on ▤ Examples 2 and 3, can a set of ordered pairs with no repeated *x*-values, but with repeated *y*-values, be a function? For example: $\{(0,4), (-3,4), (2,4)\}$. ▶

Objective C **12.** After reviewing ▤ Example 8, explain why the vertical line test works. ▶

Objective D **13.** Using ▤ Example 10, write the three function values found and their corresponding ordered pairs. One example is: $f(0) = 2$ corresponds to $(0,2)$. ▶

See Video 10.6 🍎

10.6 Exercise Set MyMathLab® ▶

Objective A *Find the domain and the range of each relation. See Example 1.*

1. $\{(2,4), (0,0), (-7,10), (10,-7)\}$

2. $\{(3,-6), (1,4), (-2,-2)\}$

▶ **3.** $\{(0,-2), (1,-2), (5,-2)\}$

4. $\{(5,0), (5,-3), (5,4), (5,3)\}$

Objective B *Determine whether each relation is also a function. See Example 2.*

▶ **5.** $\{(1,1), (2,2), (-3,-3), (0,0)\}$

6. $\{(11,6), (-1,-2), (0,0), (3,-2)\}$

▶ **7.** $\{(-1,0), (-1,6), (-1,8)\}$

8. $\{(1,2), (3,2), (1,4)\}$

Objectives B C Mixed Practice *Determine whether each graph is the graph of a function. For Exercises 9 through 12, either write down the ordered pairs or use the vertical line test. See Examples 3 and 4.*

▶ **9.**

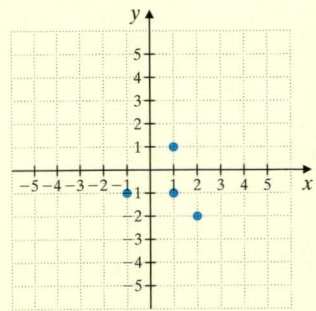

10.

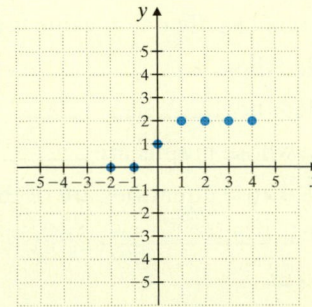

11.

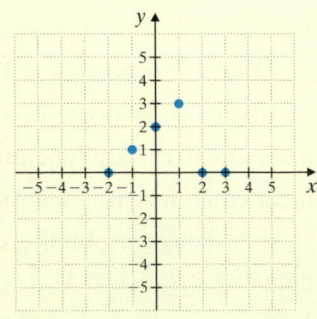

12.
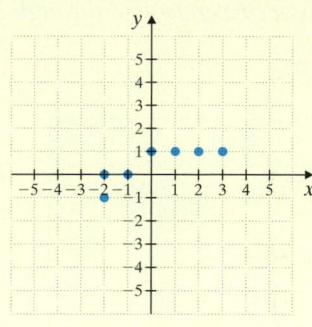

13.
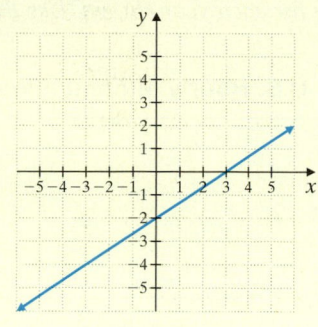

14.
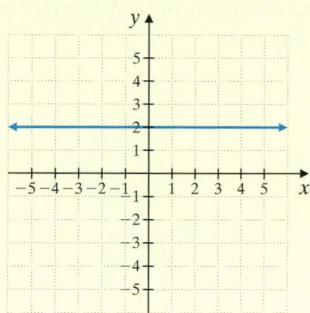

15.
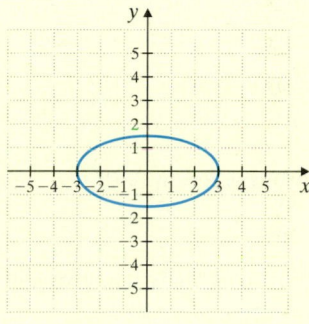

16.
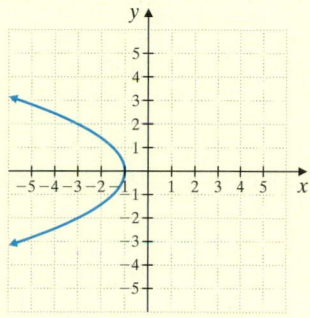

For each exercise, choose the value of x so that the relation is NOT also a function.

17. $\{(2,3), (-1,7), (x,9)\}$

   a. −1   **b.** 1   **c.** 9   **d.** 7

18. $\{(-8,0), (x,1), (5,-3)\}$

   a. 0   **b.** −3   **c.** −5   **d.** 5

Decide whether the equation describes a function. See Example 5.

19. $y - x = 7$

20. $2x - 3y = 9$

21. $y = 6$

22. $x = 3$

23. $x = -2$

24. $y = -9$

25. $x = y^2$

26. $y = x^2 - 3$

(*Hint:* For Exercises **25** and **26**, check to see whether each x-value pairs with exactly one y-value.)

The graph shows the sunset times for Seward, Alaska, for the first of each month for one year. Use this graph to answer Exercises 27 through 32. See Example 6.

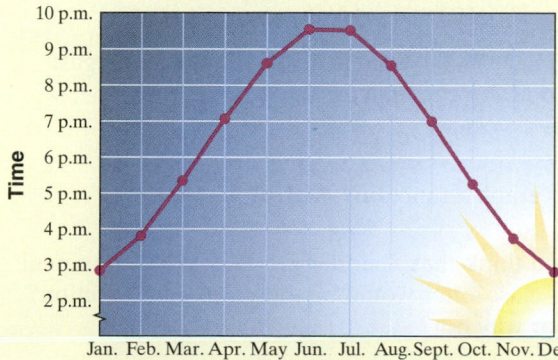

Seward, Alaska, Sunsets

Jan. Feb. Mar. Apr. May Jun. Jul. Aug. Sept. Oct. Nov. Dec.

Month

27. Approximate the time of sunset on June 1.

28. Approximate the time of sunset on November 1.

29. Approximate the date(s) when the sunset is at 3 p.m.

30. Approximate the date(s) when the sunset is at 9 p.m.

31. Is this graph the graph of a function? Why or why not?

32. Do you think a graph of sunset times for any location will always be a function? Why or why not?

This graph shows the U.S. hourly minimum wage for each year shown. Use this graph to answer Exercises 33 through 38. See Example 6.

U.S. Hourly Minimum Wage

Source: U.S. Department of Labor

33. Approximate the minimum wage before October 1996.

34. Approximate the minimum wage in 2006.

35. Find the year when the minimum wage increased to over $7.00 per hour.

36. According to the graph, what hourly wage was in effect for the greatest number of years?

37. Is this graph the graph of a function? Why or why not?

38. Do you think that a similar graph of your hourly wage on January 1 of every year (whether you are working or not) would be the graph of a function? Why or why not?

This graph shows the cost of mailing a large envelope through the U.S. Postal Service by weight. Use this graph to answer Exercises 39 through 44. See Example 6.

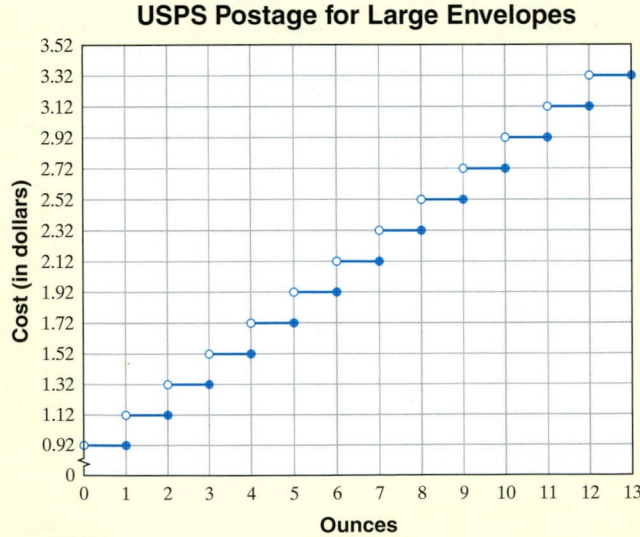

USPS Postage for Large Envelopes

Source: United States Postal Service

39. Approximate the postage to mail a large envelope weighing more than 4 ounces but not more than 5 ounces.

40. Approximate the postage to mail a large envelope weighing more than 7 ounces but not more than 8 ounces.

41. Give the weight of a large envelope that costs $1.12 to mail.

42. If you have $3.00, what is the weight of the largest envelope you can mail for that amount of money?

43. Is this graph a function? Why or why not?

44. Do you think that a similar graph of postage to mail a first-class letter would be the graph of a function? Why or why not?

Objective D *Find* $f(-2), f(0),$ *and* $f(3)$ *for each function. See Example 7.*

45. $f(x) = 2x - 5$ **46.** $f(x) = 3 - 7x$ ▶ **47.** $f(x) = x^2 + 2$ **48.** $f(x) = x^2 - 4$

49. $f(x) = 3x$ **50.** $f(x) = -3x$ **51.** $f(x) = |x|$ **52.** $f(x) = |2 - x|$

Find $h(-1)$, $h(0)$, and $h(4)$ for each function. See Example 7.

53. $h(x) = -5x$ **54.** $h(x) = -3x$ **55.** $h(x) = 2x^2 + 3$ **56.** $h(x) = 3x^2$

For each given function value, write a corresponding ordered pair.

57. $f(3) = 6$ **58.** $f(7) = -2$ **59.** $g(0) = -\dfrac{1}{2}$

60. $g(0) = -\dfrac{7}{8}$ **61.** $h(-2) = 9$ **62.** $h(-10) = 1$

Objectives **A** **D** **Mixed Practice** *Find the domain of each function. See Example 8.*

63. $f(x) = 3x - 7$ **64.** $g(x) = 5 - 2x$ **65.** $h(x) = \dfrac{1}{x + 5}$ **66.** $f(x) = \dfrac{1}{x - 6}$

Objectives **A** **D** **Mixed Practice** *Find the domain and the range of each relation graphed. See Example 9.*

67.

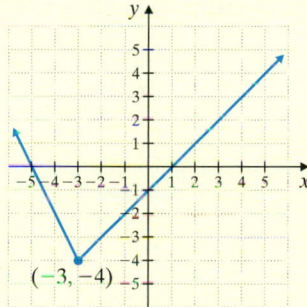

68.

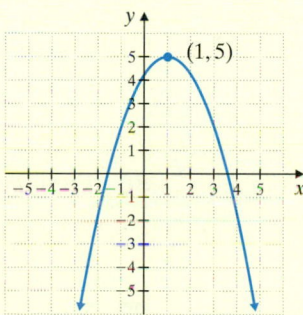

69.

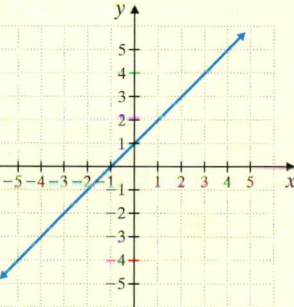

70.

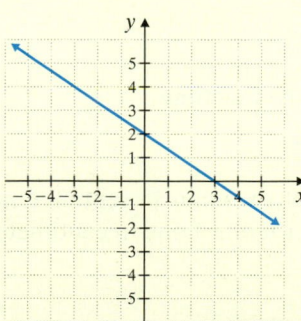

71.

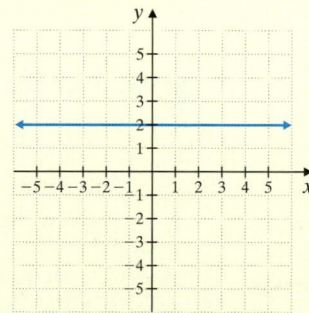

72.

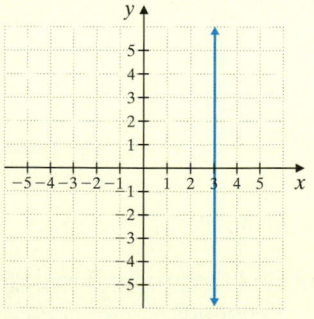

Use the graph of f below to answer Exercises 73 through 78.

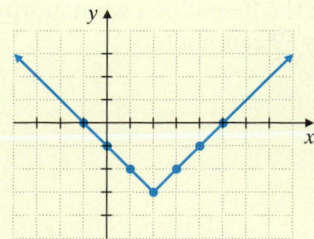

73. Complete the ordered pair solution for f. $(0, \ )$

74. Complete the ordered pair solution for f. $(3, \ )$

75. $f(0) = $ _____?

76. $f(3) = $ _____?

77. If $f(x) = 0$, find the value(s) of x.

78. If $f(x) = -1$, find the value(s) of x.

Review

Solve each inequality. See Section 9.7.

79. $2x + 5 < 7$ **80.** $3x - 1 \geq 11$ **81.** $-x + 6 \leq 9$ **82.** $-2x + 3 > 3$

Find the perimeter of each figure. See Section 6.3.

△**83.**

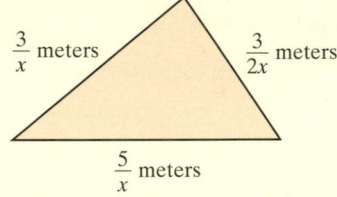

$\frac{3}{x}$ meters $\frac{3}{2x}$ meters

$\frac{5}{x}$ meters

△**84.**

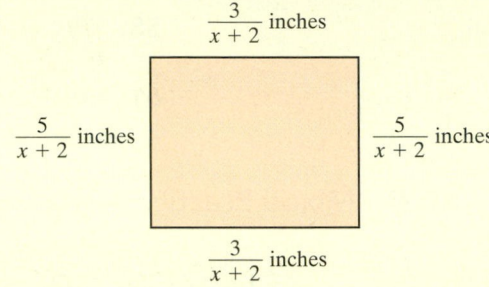

$\frac{3}{x + 2}$ inches

$\frac{5}{x + 2}$ inches $\frac{5}{x + 2}$ inches

$\frac{3}{x + 2}$ inches

Concept Extensions

Solve. See the Concept Check in this section.

85. If a function f is evaluated at -5, the value of the function is 12. Write this situation using function notation.

86. Suppose $(9, 20)$ is an ordered pair solution for the function g. Write this situation using function notation.

Solve.

87. In your own words define (a) function; (b) domain; (c) range.

88. Explain the vertical line test and how it is used.

89. Since $y = x + 7$ is a function, rewrite the equation using function notation.

90. Since $y = 3$ is a function, rewrite the equation using function notation.

91. The dosage in milligrams of Ivermectin, a heartworm preventive for a dog who weighs x pounds, is given by the function

$$f(x) = \frac{136}{25}x$$

    a. Find the proper dosage for a dog who weighs 35 pounds.

    b. Find the proper dosage for a dog who weighs 70 pounds.

92. Forensic scientists use the function

$$f(x) = 2.59x + 47.24$$

to estimate the height of a woman, in centimeters, given the length x of her femur bone in centimeters.

    a. Estimate the height of a woman whose femur measures 46 centimeters.

    b. Estimate the height of a woman whose femur measures 39 centimeters.

10.7 Graphing Linear Inequalities in Two Variables

Recall that a linear equation in two variables is an equation that can be written in the form $Ax + By = C$, where A, B, and C are real numbers and A and B are not both 0. A **linear inequality in two variables** is an inequality that can be written in one of the forms

$$Ax + By < C \qquad Ax + By \leq C$$
$$Ax + By > C \qquad Ax + By \geq C$$

where A, B, and C are real numbers and A and B are not both 0.

Objectives

A Determine Whether an Ordered Pair Is a Solution of a Linear Inequality in Two Variables. ▶

B Graph a Linear Inequality in Two Variables. ▶

Objective A Determining Solutions of Linear Inequalities in Two Variables ▶

Just as for linear equations in x and y, an ordered pair is a **solution** of an inequality in x and y if replacing the variables with the coordinates of the ordered pair results in a true statement.

Example 1 Determine whether each ordered pair is a solution of the inequality $2x - y < 6$.

a. $(5, -1)$ **b.** $(2, 7)$

Solution:

a. We replace x with 5 and y with -1 and see if a true statement results.

$$2x - y < 6$$
$$2(5) - (-1) < 6 \quad \text{Replace } x \text{ with 5 and } y \text{ with } -1.$$
$$10 + 1 < 6$$
$$11 < 6 \quad \text{False}$$

The ordered pair $(5, -1)$ is not a solution since $11 < 6$ is a false statement.

b. We replace x with 2 and y with 7 and see if a true statement results.

$$2x - y < 6$$
$$2(2) - (7) < 6 \quad \text{Replace } x \text{ with 2 and } y \text{ with 7.}$$
$$4 - 7 < 6$$
$$-3 < 6 \quad \text{True}$$

The ordered pair $(2, 7)$ is a solution since $-3 < 6$ is a true statement.

■ **Work Practice 1**

Objective B Graphing Linear Inequalities in Two Variables ▶

The linear equation $x - y = 1$ is graphed next. Recall that all points on the line correspond to ordered pairs that satisfy the equation $x - y = 1$.

Notice that the line defined by $x - y = 1$ divides the rectangular coordinate system plane into 2 sides. All points on one side of the line satisfy the inequality $x - y < 1$, and all points on the other side satisfy the inequality $x - y > 1$. The graph on the next page shows a few examples of this.

Practice 1

Determine whether each ordered pair is a solution of $x - 4y > 8$.

a. $(-3, 2)$ **b.** $(9, 0)$

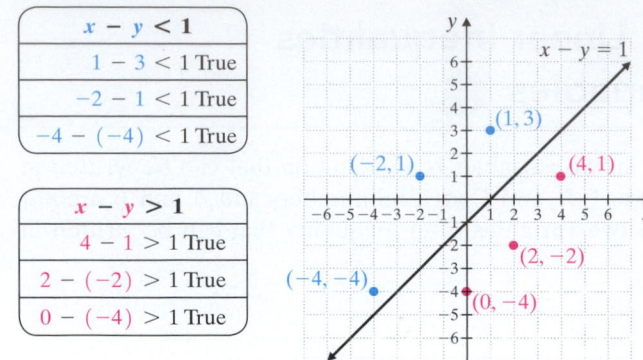

The graph of $x - y < 1$ is the region shaded blue and the graph of $x - y > 1$ is the region shaded red below.

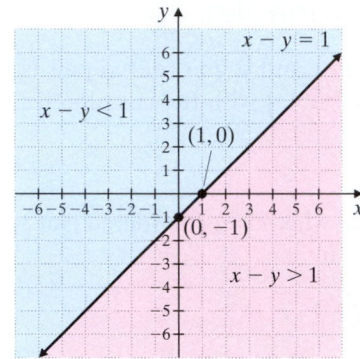

The region to the left of the line and the region to the right of the line are called **half-planes.** Every line divides the plane (similar to a sheet of paper extending indefinitely in all directions) into two half-planes; the line is called the **boundary.**

Recall that the inequality $x - y \leq 1$ means

$$x - y = 1 \quad \text{or} \quad x - y < 1$$

Thus, the graph of $x - y \leq 1$ is the half-plane $x - y < 1$ along with the boundary line $x - y = 1$.

To Graph a Linear Inequality in Two Variables

Step 1: Graph the boundary line found by replacing the inequality sign with an equal sign. If the inequality sign is $>$ or $<$, graph a dashed boundary line (indicating that the points on the line are not solutions of the inequality). If the inequality sign is $\geq$ or $\leq$, graph a solid boundary line (indicating that the points on the line are solutions of the inequality).

Step 2: Choose a point *not* on the boundary line as a test point. Substitute the coordinates of this test point into the *original* inequality.

Step 3: If a true statement is obtained in Step 2, shade the half-plane that contains the test point. If a false statement is obtained, shade the half-plane that does not contain the test point.

Example 2   Graph: $x + y < 7$

Solution:

Step 1: First we graph the boundary line by graphing the equation $x + y = 7$. We graph this boundary as a *dashed line* because the inequality sign is $<$, and thus the points on the line are not solutions of the inequality $x + y < 7$.

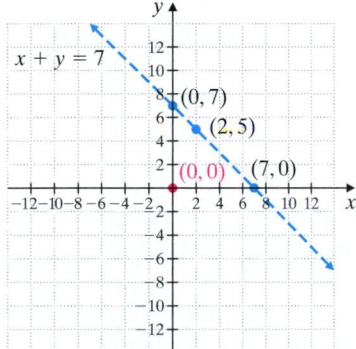

Step 2: Next we choose a test point, being careful *not* to choose a point on the boundary line. We choose $(0, 0)$, and substitute the coordinates of $(0, 0)$ into $x + y < 7$.

$x + y < 7$ Original inequality

$0 + 0 < 7$ Replace x with 0 and y with 0.

$0 < 7$ True

Step 3: Since the result is a true statement, $(0, 0)$ is a solution of $x + y < 7$, and every point in the same half-plane as $(0, 0)$ is also a solution. To indicate this, we shade the entire half-plane containing $(0, 0)$, as shown.

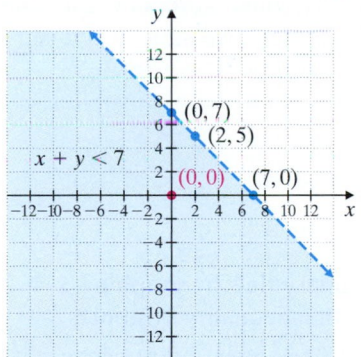

■ **Work Practice 2**

✓**Concept Check** Determine whether $(0, 0)$ is included in the graph of

a. $y \geq 2x + 3$

b. $x < 7$

c. $2x - 3y < 6$

Practice 2

Graph: $x - y > 3$

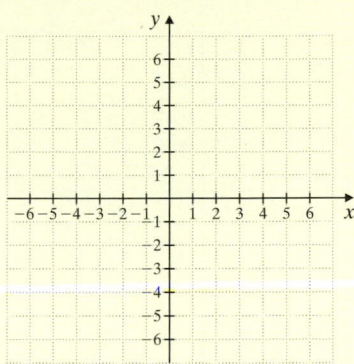

Answer

2.

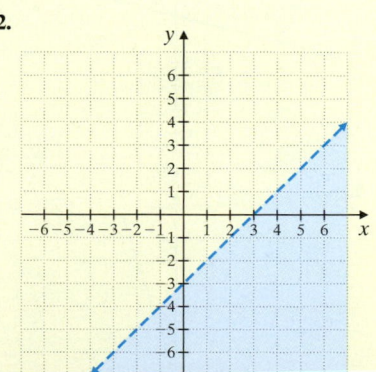

✓**Concept Check Answers**

a. no   **b.** yes   **c.** yes

Practice 3

Graph: $x - 4y \leq 4$

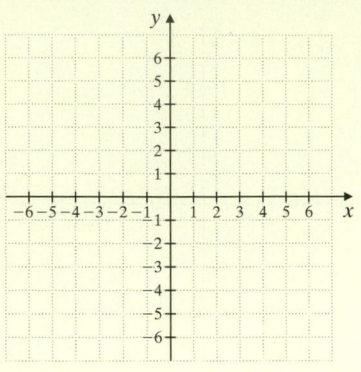

Practice 4

Graph: $y < 3x$

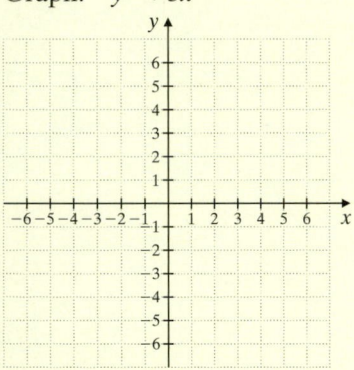

Answers

3.

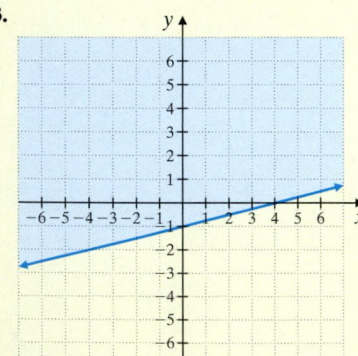

4.

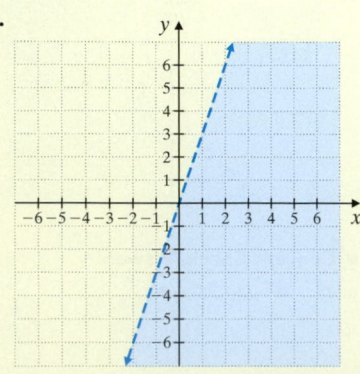

Example 3   Graph: $2x - y \geq 3$

Solution:

Step 1: We graph the boundary line by graphing $2x - y = 3$. We draw this line as a solid line because the inequality sign is $\geq$, and thus the points on the line are solutions of $2x - y \geq 3$.

Step 2: Once again, $(0, 0)$ is a convenient test point since it is not on the boundary line.

We substitute 0 for x and 0 for y into the original inequality.

$$2x - y \geq 3$$
$$2(0) - 0 \geq 3 \quad \text{Let } x = 0 \text{ and } y = 0.$$
$$0 \geq 3 \quad \text{False}$$

Step 3: Since the statement is false, no point in the half-plane containing $(0, 0)$ is a solution. Therefore, we shade the half-plane that does not contain $(0, 0)$. Every point in the shaded half-plane and every point on the boundary line is a solution of $2x - y \geq 3$.

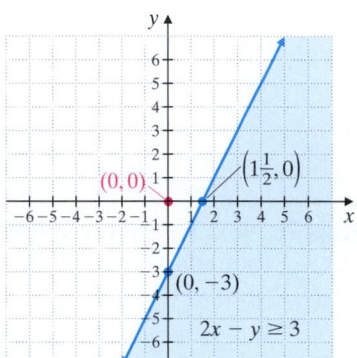

■ **Work Practice 3**

Helpful Hint

When graphing an inequality, make sure the test point is substituted into the **original inequality.** For Example 3, we substituted the test point $(0, 0)$ into the **original inequality** $2x - y \geq 3$, *not* $2x - y = 3$.

Example 4   Graph: $x > 2y$

Solution:

Step 1: We find the boundary line by graphing $x = 2y$. The boundary line is a dashed line since the inequality symbol is $>$.

Step 2: We cannot use $(0, 0)$ as a test point because it is a point on the boundary line. We choose instead $(0, 2)$.

$$x > 2y$$
$$0 > 2(2) \quad \text{Let } x = 0 \text{ and } y = 2.$$
$$0 > 4 \quad \text{False}$$

Step 3: Since the statement is false, we shade the half-plane that does not contain the test point $(0, 2)$, as shown.

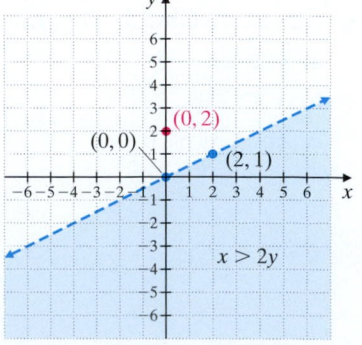

■ **Work Practice 4**

Example 5   Graph: $5x + 4y \leq 20$

Solution: We graph the solid boundary line $5x + 4y = 20$ and choose $(0, 0)$ as the test point.

$$5x + 4y \leq 20$$
$$5(0) + 4(0) \leq 20 \quad \text{Let } x = 0 \text{ and } y = 0.$$
$$0 \leq 20 \quad \text{True}$$

We shade the half-plane that contains $(0, 0)$, as shown.

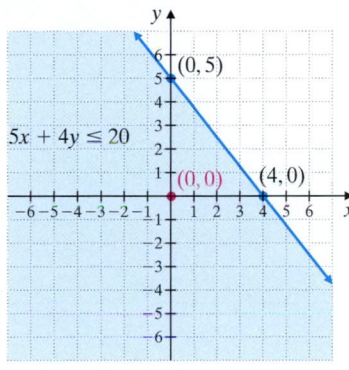

■ **Work Practice 5**

Example 6   Graph: $y > 3$

Solution: We graph the dashed boundary line $y = 3$ and choose $(0, 0)$ as the test point. (Recall that the graph of $y = 3$ is a horizontal line with y-intercept 3.)

$$y > 3$$
$$0 > 3 \quad \text{Let } y = 0.$$
$$0 > 3 \quad \text{False}$$

We shade the half-plane that does not contain $(0, 0)$, as shown.

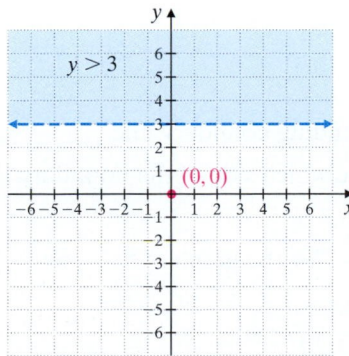

■ **Work Practice 6**

Practice 5

Graph: $3x + 2y \geq 12$

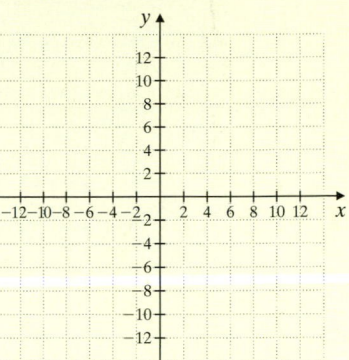

Practice 6

Graph: $x < 2$

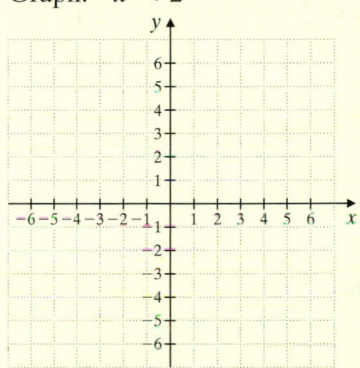

Answers

5.

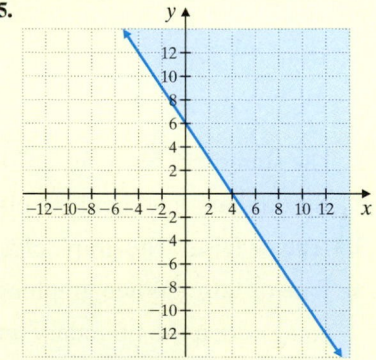

6.

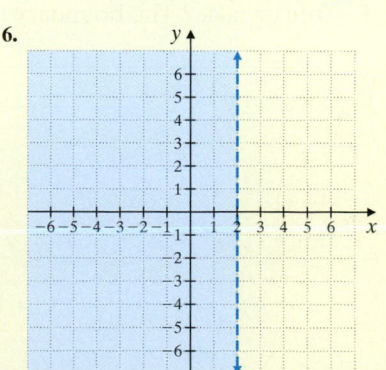

Practice 7

Graph: $y \geq \dfrac{1}{4}x + 3$

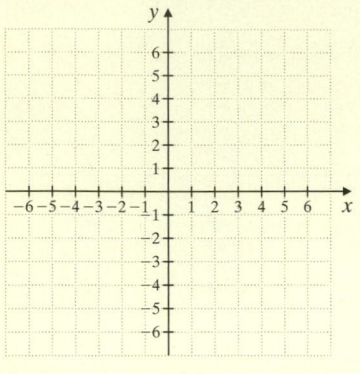

Answer

7.

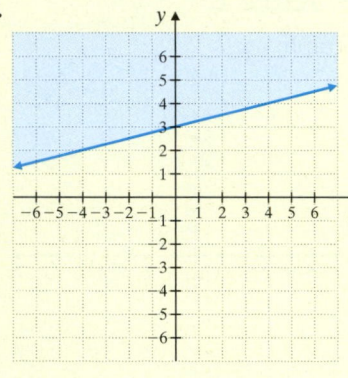

Example 7   Graph: $y \leq \dfrac{2}{3}x - 4$

Solution: Graph the solid boundary line $y = \dfrac{2}{3}x - 4$. This equation is in slope-intercept form, with slope $\dfrac{2}{3}$ and *y*-intercept -4 .

We use this information to graph the line. Then we choose $(0, 0)$ as our test point.

$$y \leq \frac{2}{3}x - 4$$

$$0 \leq \frac{2}{3} \cdot 0 - 4$$

$$0 \leq -4 \quad \text{False}$$

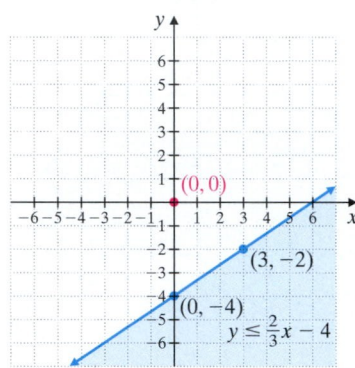

We shade the half-plane that does not contain $(0, 0)$, as shown.

■ **Work Practice 7**

Vocabulary, Readiness & Video Check

Use the choices below to fill in each blank. Some choices may be used more than once, and some not at all.

true	$x < 2$	$y < 2$	half-planes
false	$x \leq 2$	$y \leq 2$	linear inequality in two variables

1. The statement $5x - 6y < 7$ is an example of a _____.

2. A boundary line divides a plane into two regions called _____.

3. True or false? The graph of $5x - 6y < 7$ includes its corresponding boundary line. _____

4. True or false? When graphing a linear inequality, to determine which side of the boundary line to shade, choose a point *not* on the boundary line. _____

5. True or false? The boundary line for the inequality $5x - 6y < 7$ is the graph of $5x - 6y = 7$.

   _____

6. The graph of _____ is

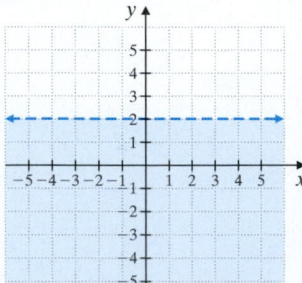

Martin-Gay Interactive Videos   *Watch the section lecture video and answer the following questions.*

Objective A 7. From Example 1, how do we determine whether an ordered pair in x and y is a solution of an inequality in x and y? ▶

Objective B 8. From Example 3, how do we find the equation of the boundary line? How do we determine if the points on the boundary line are solutions of the inequality? ▶

See Video 10.7 🍎

10.7 Exercise Set MyMathLab® ▶

Objective A *Determine whether the ordered pairs given are solutions of the linear inequality in two variables. See Example 1.*

1. $x - y > 3$; $(0, 3)$, $(2, -1)$

2. $y - x < -2$; $(2, 1)$, $(5, -1)$

3. $3x - 5y \leq -4$; $(2, 3)$, $(-1, -1)$

4. $2x + y \geq 10$; $(0, 11)$, $(5, 0)$

▶ **5.** $x < -y$; $(0, 2)$, $(-5, 1)$

6. $y > 3x$; $(0, 0)$, $(1, 4)$

Objective B *Graph each inequality. See Examples 2 through 7.*

7. $x + y \leq 1$

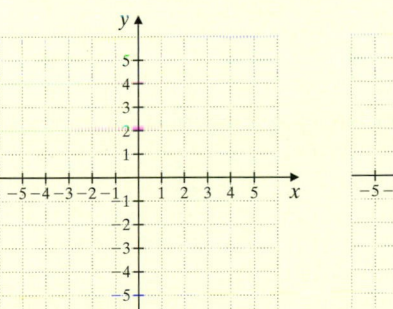

8. $x + y \geq -2$

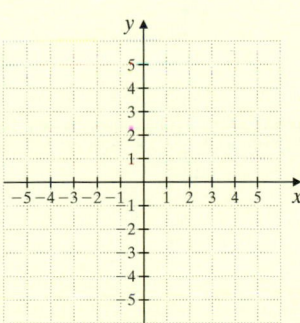

9. $2x - y > -4$

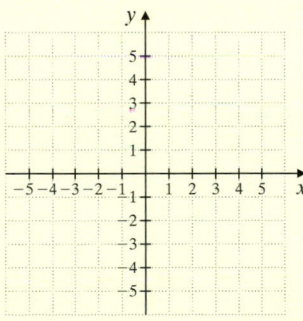

10. $x - 3y < 3$

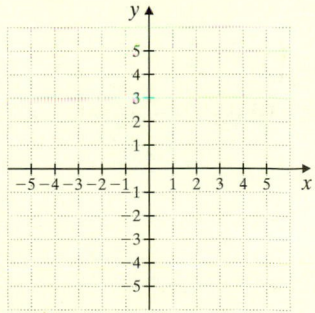

▶ **11.** $y \geq 2x$

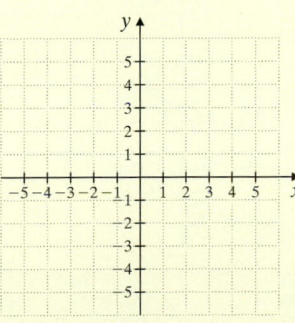

12. $y \leq 3x$

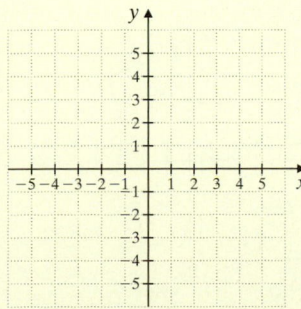

13. $x < -3y$

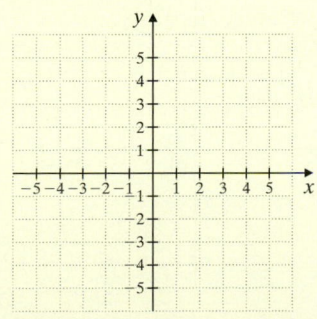

14. $x > -2y$

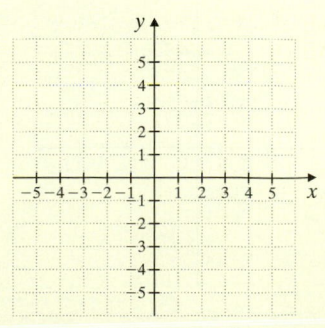

15. $y \geq x + 5$

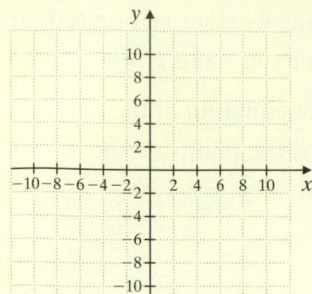

16. $y \leq x + 1$

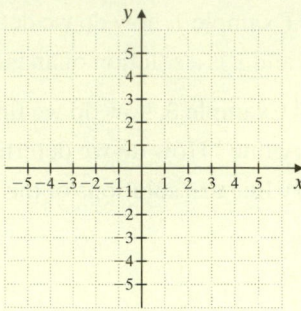

17. $y < 4$

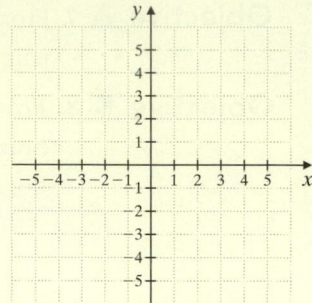

18. $y > 2$

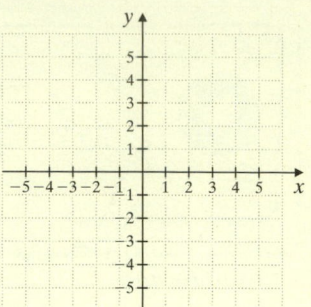

19. $x \geq -3$

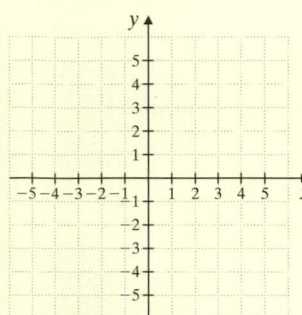

20. $x \leq -1$

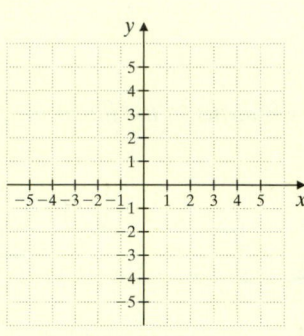

21. $5x + 2y \leq 10$

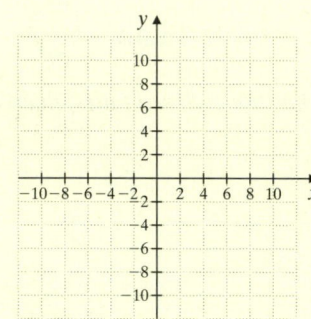

22. $4x + 3y \geq 12$

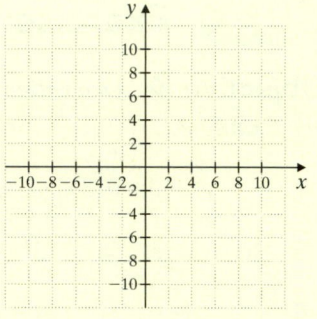

23. $x > y$

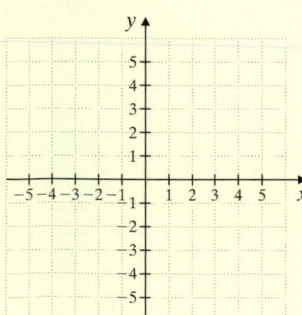

24. $x \leq -y$

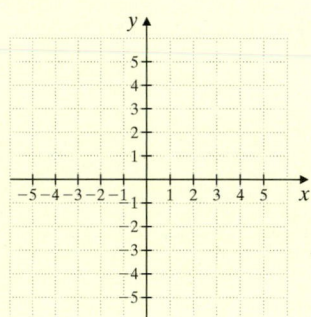

25. $x - y \leq 6$

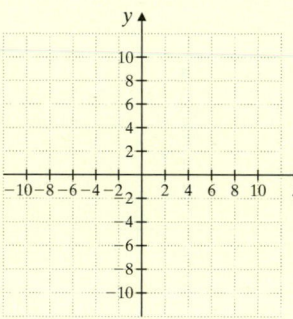

26. $x - y > 10$

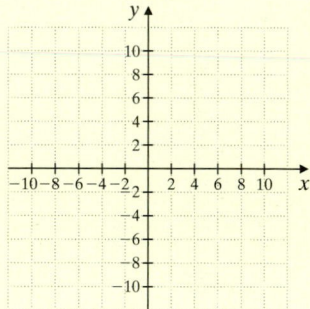

27. $x \geq 0$

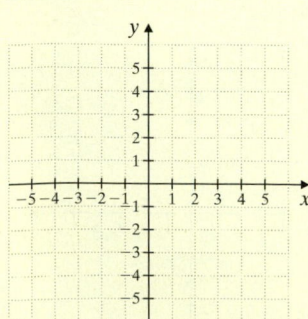

28. $y \leq 0$

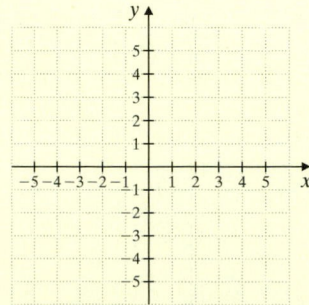

29. $2x + 7y > 5$

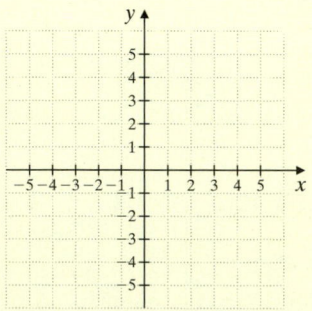

30. $3x + 5y \leq -2$

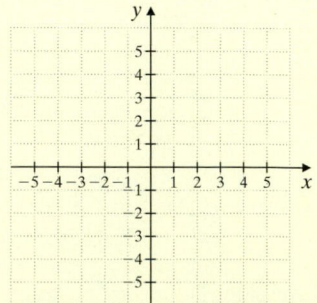

31. $y \geq \dfrac{1}{2}x - 4$

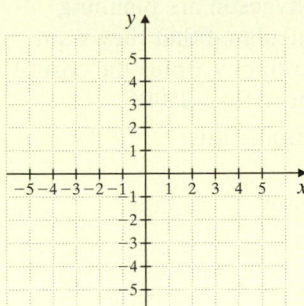

32. $y < \dfrac{2}{5}x - 3$

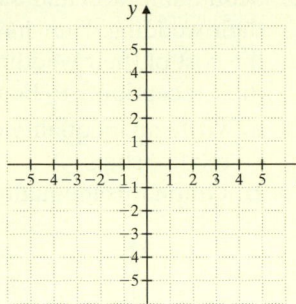

33. $y < -\dfrac{3}{4}x + 2$

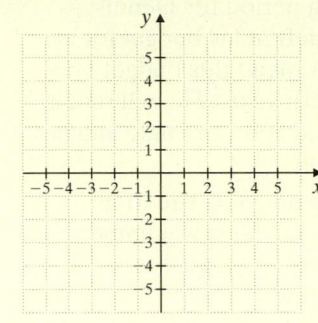

34. $y > -\dfrac{1}{3}x + 4$

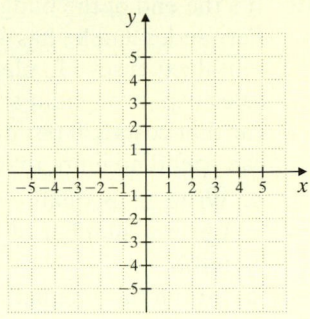

Review

Approximate the coordinates of each point of intersection. See Section 10.1.

35.

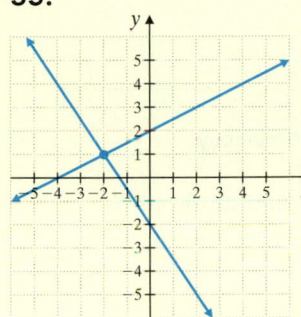

36.

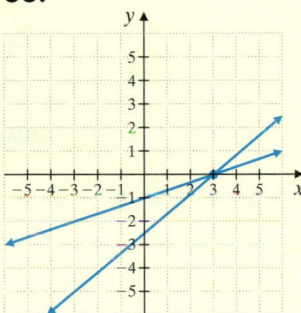

37.

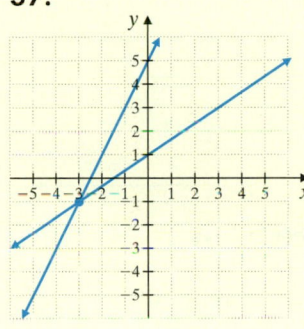

38.

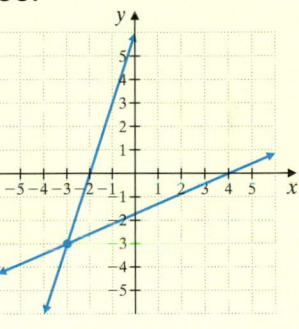

Concept Extensions

Match each inequality with its graph.

a. $x > 2$

b. $y < 2$

c. $y \leq 2x$

d. $y \leq -3x$

39.

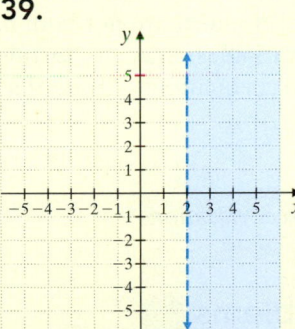

40.

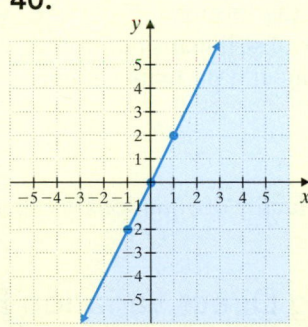

41.

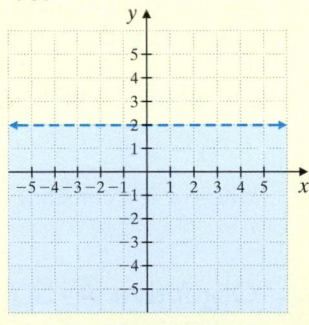

42.

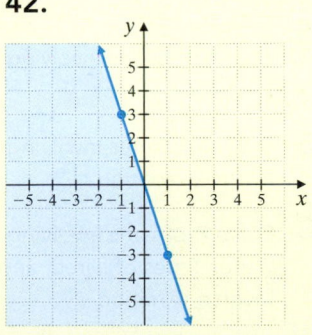

43. Explain why a point on the boundary line should not be chosen as the test point.

44. Write an inequality whose solutions are all points with coordinates whose sum is at least 13.

Determine whether $(1, 1)$ *is included in each graph. See the Concept Check in this section.*

45. $3x + 4y < 8$

46. $y > 5x$

47. $y \geq -\dfrac{1}{2}x$

48. $x > 3$

Solve.

49. It's the end of the budgeting period for Dennis Fernandes and he has $500 left in his budget for car rental expenses. He plans to spend this budget on a sales trip throughout southern Texas. He will rent a car that costs $30 per day and $0.15 per mile and he can spend no more than $500.

   a. Write an inequality describing this situation. Let x = number of days and let y = number of miles.

   b. Graph this inequality below.

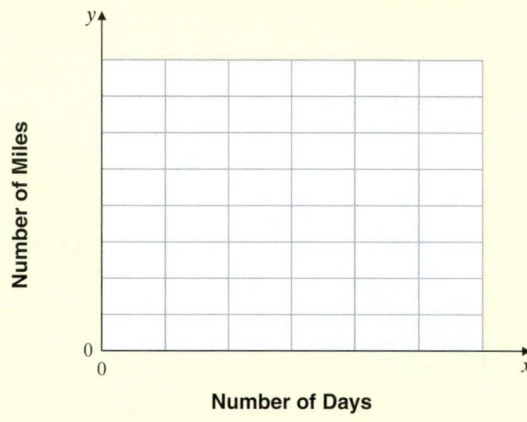

c. Why is the grid showing quadrant I only?

50. Scott Sambracci and Sara Thygeson are planning their wedding. They have calculated that they want the cost of their wedding ceremony x plus the cost of their reception y to be no more than $5000.

   a. Write an inequality describing this relationship.

   b. Graph this inequality below.

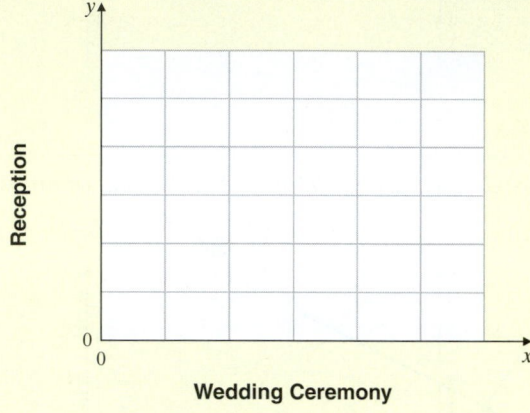

c. Why is the grid showing quadrant I only?

10.8   Direct and Inverse Variation ⊙

Objectives

A Solve Problems Involving Direct Variation. ⊙

B Solve Problems Involving Inverse Variation. ⊙

C Solve Problems Involving Other Types of Direct and Inverse Variation. ⊙

D Solve Applications of Variation. ⊙

Thus far, we have studied linear equations in two variables. Recall that such an equation can be written in the form $Ax + By = C$, where A and B are not both 0. Also recall that the graph of a linear equation in two variables is a line. In this section, we begin by looking at a particular family of linear equations—those that can be written in the form

$$y = kx$$

where k is a constant. This family of equations is called *direct variation*.

Objective A   Solving Direct Variation Problems ⊙

Let's suppose that you are earning $7.25 per hour at a part-time job. The amount of money you earn depends on the number of hours you work. This is illustrated by the following table:

Hours Worked	0	1	2	3	4	
Money Earned (before deductions)	0	7.25	14.50	21.75	29.00	and so on

In general, to calculate your earnings (before deductions), multiply the constant $7.25 by the number of hours you work. If we let y represent the amount of money earned and x represent the number of hours worked, we get the direct variation equation

$$y = 7.25 \cdot x$$

earnings $= \$7.25 \cdot$ hours worked

Notice that in this direct variation equation, as the number of hours increases, the pay increases as well.

Direct Variation

y varies directly as x, or **y is directly proportional to x,** if there is a nonzero constant k such that

$$y = kx$$

The number k is called the **constant of variation** or the **constant of proportionality.**

In our direct variation example, $y = 7.25x$, the constant of variation is 7.25.

Let's use the previous table to graph $y = 7.25x$. We begin our graph at the ordered pair solution $(0, 0)$. Why? We assume that the least amount of hours worked is 0. If 0 hours are worked, then the pay is $0.

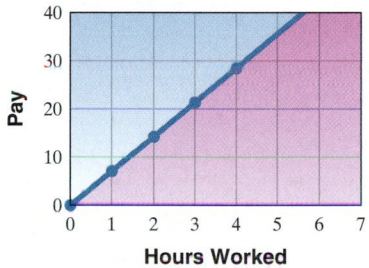

As illustrated in this graph, a direct variation equation $y = kx$ is linear. Also notice that $y = 7.25x$ is a function since its graph passes the vertical line test.

Example 1 Write a direct variation equation of the form $y = kx$ that satisfies the ordered pairs in the table below.

x	2	9	1.5	-1
y	6	27	4.5	-3

Solution: We are given that there is a direct variation relationship between x and y. This means that

$$y = kx$$

By studying the given values, you may be able to mentally calculate k. If not, to find k, we simply substitute one given ordered pair into this equation and solve for k. We'll use the given pair $(2, 6)$.

$$y = kx$$
$$6 = k \cdot 2$$
$$\frac{6}{2} = \frac{k \cdot 2}{2}$$
$$3 = k \qquad \text{Solve for } k.$$

Since $k = 3$, we have the equation $y = 3x$.
To check, see that each given y is 3 times the given x.

■ **Work Practice 1**

Practice 1

Write a direct variation equation that satisfies:

x	4	$\frac{1}{2}$	1.5	6
y	8	1	3	12

Answer

1. $y = 2x$

Let's try another type of direct variation example.

Practice 2

Suppose that y varies directly as x. If y is 15 when x is 45, find the constant of variation and the direct variation equation. Then find y when x is 3.

Example 2 Suppose that y varies directly as x. If y is 17 when x is 34, find the constant of variation and the direct variation equation. Then find y when x is 12.

Solution: Let's use the same method as in Example 1 to find k. Since we are told that y varies directly as x, we know the relationship is of the form

$$y = kx$$

Let $y = 17$ and $x = 34$ and solve for k.

$$17 = k \cdot 34$$

$$\frac{17}{34} = \frac{k \cdot 34}{34}$$

$$\frac{1}{2} = k \qquad \text{Solve for } k.$$

Thus, the constant of variation is $\frac{1}{2}$ and the equation is $y = \frac{1}{2}x$.

To find y when $x = 12$, use $y = \frac{1}{2}x$ and replace x with 12.

$$y = \frac{1}{2}x$$

$$y = \frac{1}{2} \cdot 12 \qquad \text{Replace } x \text{ with 12.}$$

$$y = 6$$

Thus, when x is 12, y is 6.

■ **Work Practice 2**

Let's review a few facts about linear equations of the form $y = kx$.

Direct Variation: $y = kx$

- There is a direct variation relationship between x and y.
- The graph is a line.
- The line will always go through the origin $(0, 0)$. Why?

  Let $x = 0$. Then $y = k \cdot 0$ or $y = 0$.

- The slope of the graph of $y = kx$ is k, the constant of variation. Why? Remember that the slope of an equation of the form $y = mx + b$ is m, the coefficient of x.
- The equation $y = kx$ describes a function. Each x has a unique y and its graph passes the vertical line test.

Practice 3

Find the constant of variation and the direct variation equation for the line below.

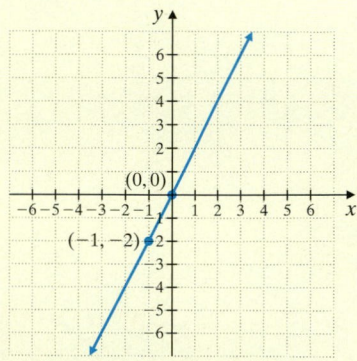

Example 3 The line is the graph of a direct variation equation. Find the constant of variation and the direct variation equation.

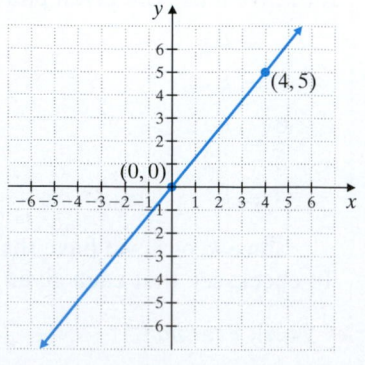

Answers

2. $k = \frac{1}{3}$; $y = \frac{1}{3}x$; $y = 1$

3. $k = 2$; $y = 2x$

Solution: Recall that k, the constant of variation, is the same as the slope of the line. Thus, to find k, we use the slope formula and find slope.

Using the given points $(0, 0)$ and $(4, 5)$, we have

$$\text{slope} = \frac{5 - 0}{4 - 0} = \frac{5}{4}$$

Thus, $k = \frac{5}{4}$ and the variation equation is $y = \frac{5}{4}x$.

■ **Work Practice 3**

Objective B Solving Inverse Variation Problems ▶

In this section, we introduce another type of variation called inverse variation.

Let's suppose you need to drive a distance of 40 miles. You know that the faster you drive the distance, the sooner you arrive at your destination. Recall that there is a mathematical relationship between distance, rate, and time. It is $d = r \cdot t$. In our example, distance is a constant 40 miles, so we have $40 = r \cdot t$ or $t = \frac{40}{r}$.

For example, if you drive 10 mph, the time to drive the 40 miles is

$$t = \frac{40}{r} = \frac{40}{10} = 4 \text{ hours}$$

If you drive 20 mph, the time is

$$t = \frac{40}{r} = \frac{40}{20} = 2 \text{ hours}$$

Again, notice that as speed increases, time decreases. Below are some ordered pair solutions of $t = \frac{40}{r}$ and its graph.

Rate (mph)	r	5	10	20	40	60	80
Time (hr)	t	8	4	2	1	$\frac{2}{3}$	$\frac{1}{2}$

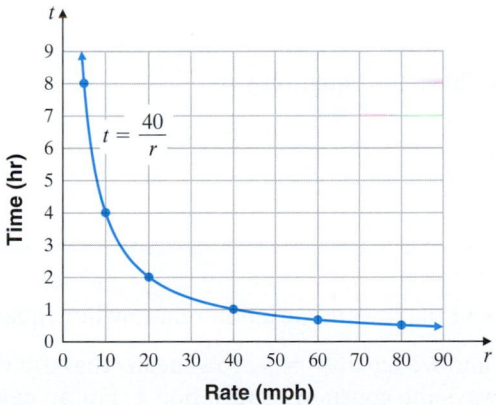

Notice that the graph of this variation is not a line, but it passes the vertical line test so $t = \frac{40}{r}$ does describe a function. This is an example of inverse variation.

Inverse Variation

y varies inversely as x, or y is inversely proportional to x, if there is a nonzero constant k such that

$$y = \frac{k}{x}$$

The number k is called the **constant of variation** or the **constant of proportionality.**

In our inverse variation example, $t = \dfrac{40}{r}$ or $y = \dfrac{40}{x}$, the constant of variation is 40.

We can immediately see differences and similarities in direct variation and inverse variation.

Direct Variation	$y = kx$	linear equation	both functions
Inverse Variation	$y = \frac{k}{x}$	rational equation	

In Chapter 14 we will see that $y = \dfrac{k}{x}$ is a rational equation and not a linear equation. Also notice that because x is in the denominator, x can be any value except 0.

We can still derive an inverse variation equation from a table of values.

Practice 4

Write an inverse variation equation of the form $y = \dfrac{k}{x}$ that satisfies:

x	4	10	40	−2
y	5	2	$\frac{1}{2}$	−10

Example 4 Write an inverse variation equation of the form $y = \dfrac{k}{x}$ that satisfies the ordered pairs in the table below.

x	2	4	$\frac{1}{2}$
y	6	3	24

Solution: Since there is an inverse variation relationship between x and y, we know that $y = \dfrac{k}{x}$.

To find k, choose one given ordered pair and substitute the values into the equation. We'll use $(2, 6)$.

$$y = \frac{k}{x}$$

$$6 = \frac{k}{2}$$

$$2 \cdot 6 = 2 \cdot \frac{k}{2} \quad \text{Multiply both sides by 2.}$$

$$12 = k \quad \text{Solve.}$$

Since $k = 12$, we have the equation $y = \dfrac{12}{x}$.

■ **Work Practice 4**

Helpful Hint

Multiply both sides of the inverse variation relationship equation $y = \dfrac{k}{x}$ by x (as long as x is not 0), and we have $xy = k$. This means that if y varies inversely as x, their product is always the constant of variation k. For an example of this, check the table from Example 4:

x	2	4	$\frac{1}{2}$
y	6	3	24

$$2 \cdot 6 = 12 \qquad 4 \cdot 3 = 12 \qquad \tfrac{1}{2} \cdot 24 = 12$$

Answer

4. $y = \dfrac{20}{x}$

Example 5   Suppose that y varies inversely as x. If $y = 0.02$ when $x = 75$, find the constant of variation and the inverse variation equation. Then find y when x is 30.

Solution: Since y varies inversely as x, the constant of variation may be found by simply finding the product of the given x and y.

$$k = xy = 75(0.02) = 1.5$$

To check, we will use the inverse variation equation

$$y = \frac{k}{x}$$

Let $y = 0.02$ and $x = 75$ and solve for k.

$$0.02 = \frac{k}{75}$$

$$75(0.02) = 75 \cdot \frac{k}{75} \qquad \text{Multiply both sides by 75.}$$

$$1.5 = k \qquad \text{Solve for } k.$$

Thus, the constant of variation is 1.5 and the equation is $y = \dfrac{1.5}{x}$. To find y when $x = 30$, use $y = \dfrac{1.5}{x}$ and replace x with 30.

$$y = \frac{1.5}{x}$$

$$y = \frac{1.5}{30} \qquad \text{Replace } x \text{ with 30.}$$

$$y = 0.05$$

Thus, when x is 30, y is 0.05.

■ **Work Practice 5**

Objective C Solving Other Types of Direct and Inverse Variation Problems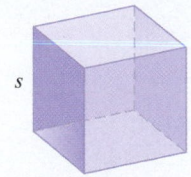

It is possible for y to vary directly or inversely as powers of x.

Direct and Inverse Variation as *n*th Powers of *x*

y varies directly as a power of x if there is a nonzero constant k and a natural number n such that

$$y = kx^n$$

y varies inversely as a power of x if there is a nonzero constant k and a natural number n such that

$$y = \frac{k}{x^n}$$

Example 6   The surface area of a cube A varies directly as the square of a length of its sides. If A is 54 when s is 3, find A when $s = 4.2$.

s

(Continued on next page)

Solution: Since the surface area A varies directly as the square of side s, we have

$$A = ks^2$$

To find k, let $A = 54$ and $s = 3$.

$$A = k \cdot s^2$$

$54 = k \cdot 3^2$	Let $A = 54$ and $s = 3$.
$54 = 9k$	$3^2 = 9$.
$6 = k$	Divide by 9.

The formula for surface area of a cube is then

$A = 6s^2$, where s is the length of a side.

To find the surface area when $s = 4.2$, substitute.

$$A = 6s^2$$
$$A = 6 \cdot (4.2)^2$$
$$A = 105.84$$

The surface area of a cube whose side measures 4.2 units is 105.84 sq units.

■ **Work Practice 6**

Objective D Solving Applications of Variation

There are many real-life applications of direct and inverse variation.

Practice 7

The distance d that an object falls is directly proportional to the square of the time of the fall, t. If an object falls 144 feet in 3 seconds, find how far the object falls in 5 seconds.

Example 7   The weight of a body w varies inversely with the square of its distance from the center of Earth, d. If a person weighs 160 pounds on the surface of Earth, what is the person's weight 200 miles above the surface? (Assume that the radius of Earth is 4000 miles.)

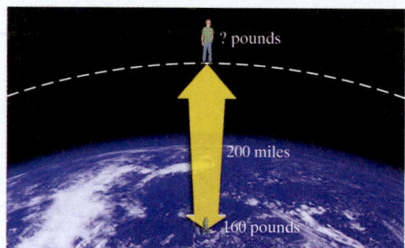
? pounds
200 miles
160 pounds

Solution:

1. UNDERSTAND. Make sure you read and reread the problem.
2. TRANSLATE. Since we are told that weight w varies inversely with the square of its distance from the center of Earth, d, we have

$$w = \frac{k}{d^2}$$

3. SOLVE. To solve the problem, we first find k. To do so, we use the fact that the person weighs 160 pounds on Earth's surface, which is a distance of 4000 miles from the Earth's center.

$$w = \frac{k}{d^2}$$

$$160 = \frac{k}{(4000)^2}$$

$$2,560,000,000 = k$$

Thus, we have $w = \dfrac{2,560,000,000}{d^2}$.

Answer

7. 400 feet

Since we want to know the person's weight 200 miles above the Earth's surface, we let $d = 4200$ and find w.

$$w = \frac{2,560,000,000}{d^2}$$

$$w = \frac{2,560,000,000}{(4200)^2}$$ A person 200 miles above the Earth's surface is 4200 miles from the Earth's center.

$$w \approx 145$$ Simplify.

4. INTERPRET. *Check:* Your answer is reasonable since the farther a person is from Earth, the less the person weighs. *State:* Thus, 200 miles above the surface of the Earth, a 160-pound person weighs approximately 145 pounds.

■ Work Practice 7

Vocabulary, Readiness & Video Check

State whether each equation represents direct or inverse variation.

1. $y = \dfrac{k}{x}$, where k is a constant. _____

2. $y = kx$, where k is a constant. _____

3. $y = 5x$ _____

4. $y = \dfrac{5}{x}$ _____

5. $y = \dfrac{7}{x^2}$ _____

6. $y = 6.5x^4$ _____

7. $y = \dfrac{11}{x}$ _____

8. $y = 18x$ _____

9. $y = 12x^2$ _____

10. $y = \dfrac{20}{x^3}$ _____

Martin-Gay Interactive Videos

See Video 10.8

Watch the section lecture video and answer the following questions.

Objective A 11. Based on the lecture before ▥ Example 1, what kind of equation is a direct variation equation? What does k represent in this equation? ▶

Objective B 12. In ▥ Example 4, why is it not necessary to place the given values of x and y into the inverse variation equation in order to find k? ▶

Objective C 13. From ▥ Examples 5–7, does a power on x change the basic direct and inverse variation formula relationships? ▶

Objective D 14. In ▥ Example 8, why is it reasonable to expect our answer to be a greater distance than the original distance given in the problem? ▶

10.8 Exercise Set MyMathLab®

Objective A *Write a direct variation equation, $y = kx$, that satisfies the ordered pairs in each table. See Example 1.*

1.

x	0	6	10
y	0	3	5

2.

x	0	2	−1	3
y	0	14	−7	21

3.

x	−2	2	4	5
y	−12	12	24	30

4.

x	3	9	−2	12
y	1	3	$-\frac{2}{3}$	4

Write a direct variation equation, $y = kx$, that describes each graph. See Example 3.

5.

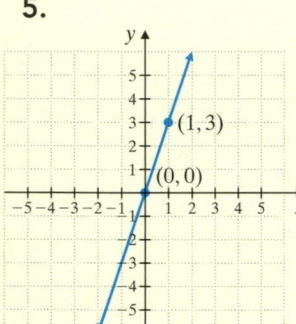

6.

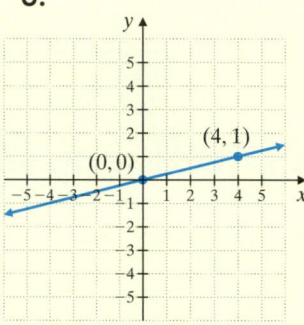

7.

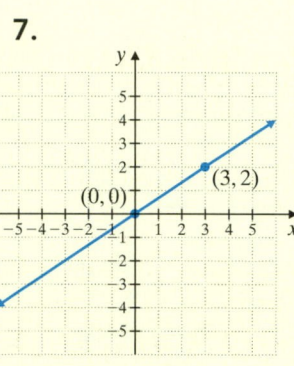

8.

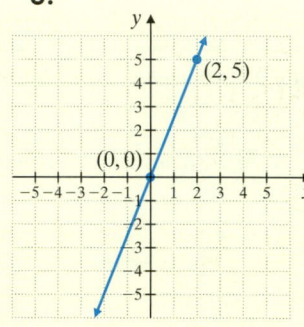

Objective B *Write an inverse variation equation, $y = \dfrac{k}{x}$, that satisfies the ordered pairs in each table. See Example 4.*

9.

x	1	−7	3.5	−2
y	7	−1	2	−3.5

10.

x	2	−11	4	−4
y	11	−2	5.5	−5.5

11.

x	10	$\frac{1}{2}$	$-\frac{1}{4}$
y	0.05	1	−2

12.

x	4	$\frac{1}{5}$	−8
y	0.1	2	−0.05

Objectives A B C **Translating** *Write an equation to describe each variation. Use k for the constant of proportionality. See Examples 1 through 6.*

13. y varies directly as x.　**14.** a varies directly as b.　**15.** h varies inversely as t.　**16.** s varies inversely as t.

17. z varies directly as x^2.　**18.** p varies inversely as x^2.　**19.** y varies inversely as z^3.　**20.** x varies directly as y^4.

21. x varies inversely as $\sqrt{y}$.　　　　　　　**22.** y varies directly as d^2.

Objectives A B C *Solve. See Examples 2, 5, and 6.*

23. y varies directly as x. If $y = 20$ when $x = 5$, find y when x is 10.

24. y varies directly as x. If $y = 27$ when $x = 3$, find y when x is 2.

25. y varies inversely as x. If $y = 5$ when $x = 60$, find y when x is 100.

26. y varies inversely as x. If $y = 200$ when $x = 5$, find y when x is 4.

27. z varies directly as x^2. If $z = 96$ when $x = 4$, find z when $x = 3$.

28. s varies directly as t^3. If $s = 270$ when $t = 3$, find s when $t = 1$.

▶ 29. a varies inversely as b^3. If $a = \dfrac{3}{2}$ when $b = 2$, find a when b is 3.

30. p varies inversely as q^2. If $p = \dfrac{5}{16}$ when $q = 8$, find p when $q = \dfrac{1}{2}$.

Objectives A B C D *Solve. See Examples 1 through 7.*

31. Your paycheck (before deductions) varies directly as the number of hours you work. If your paycheck is $166.50 for 18 hours, find your pay for 10 hours.

32. If your paycheck (before deductions) is $304.50 for 30 hours, find your pay for 34 hours. (See Exercise **31**.)

33. The cost of manufacturing a certain type of headphone varies inversely as the number of headphones increases. If 5000 headphones can be manufactured for $9.00 each, find the cost per headphone to manufacture 7500 headphones.

34. The cost of manufacturing a certain composition notebook varies inversely as the number of notebooks increases. If 10,000 notebooks can be manufactured for $0.50 each, find the cost per notebook to manufacture 18,000 notebooks. Round your answer to the nearest cent.

▶ 35. The distance a spring stretches varies directly with the weight attached to the spring. If a 60-pound weight stretches the spring 4 inches, find the distance that an 80-pound weight stretches the spring.

36. If a 30-pound weight stretches a spring 10 inches, find the distance a 20-pound weight stretches the spring. (See Exercise **35**.)

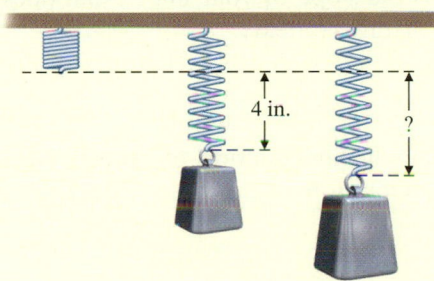

4 in.

?

37. The weight of an object varies inversely as the square of its distance from the center of Earth. If a person weighs 180 pounds on Earth's surface, what is his weight 10 miles above the surface of Earth? (Assume that Earth's radius is 4000 miles and round your answer to one decimal place.)

38. For a constant distance, the rate of travel varies inversely as the time traveled. If a family travels 55 mph and arrives at a destination in 4 hours, how long will the return trip take traveling at 60 mph?

39. The distance d that an object falls is directly proportional to the square of the time of the fall, t. A person who is parachuting for the first time is told to wait 10 seconds before opening the parachute. If the person falls 64 feet in 2 seconds, find how far he falls in 10 seconds.

40. The distance needed for a car to stop, d, is directly proportional to the square of its rate of travel, r. Under certain driving conditions, a car traveling 60 mph needs 300 feet to stop. With these same driving conditions, how long does it take a car to stop if the car is traveling 30 mph when the brakes are applied?

Review

Add the equations. See Section 8.7.

41. $-3x + 4y = 7$
$\underline{3x - 2y = 9}$

42. $x - y = -9$
$\underline{-x - y = -14}$

43. $5x - 0.4y = 0.7$
$\underline{-9x + 0.4y = -0.2}$

44. $1.9x - 2y = 5.7$
$\underline{-1.9x - 0.1y = 2.3}$

Concept Extensions

45. Suppose that y varies directly as x. If x is tripled, what is the effect on y?

46. Suppose that y varies directly as x^2. If x is tripled, what is the effect on y?

47. The period of a pendulum p (the time of one complete back-and-forth swing) varies directly with the square root of its length, ℓ. If the length of the pendulum is quadrupled, what is the effect on the period, p?

48. For a constant distance, the rate of travel r varies inversely with the time traveled, t. If a car traveling 100 mph completes a test track in 6 minutes, find the rate needed to complete the same test track in 4 minutes. (*Hint:* Convert minutes to hours.)

Chapter 10 Group Activity

Finding a Linear Model

This activity may be completed by working in groups or individually.

The following table shows the actual number of international tourist arrivals to the United States for the years 2009 through 2012.

Year	International Tourist Arrivals to the United States (in millions)
2009	55
2010	60
2011	63
2012	67

Source: World Tourism Organization

1. Make a scatter diagram of the paired data in the table.

2. Use what you have learned in this chapter to write an equation of the line representing the paired data in the table. Explain how you found the equation, and what each variable represents.

3. What is the slope of your line? What does the slope mean in this context?

4. Use your linear equation to predict the number of international tourist arrivals to the United States in 2018.

5. Compare your linear equation to that found by other students or groups. Is it the same, similar, or different? How?

6. Compare your prediction from question **4** to that of other students or groups. Describe what you find.

7. Suppose that the number of international tourist arrivals to the United States for 2013 was estimated to be 65 million. If this data point is added to the chart, how does it affect your results?

Chapter 10 Vocabulary Check

Fill in each blank with one of the words listed below.

y-axis	x-axis	solution	linear	standard	point-slope
x-intercept	y-intercept	y	x	slope	relation
domain	range	direct	inverse	slope-intercept	function

1. An ordered pair is a(n) _____ of an equation in two variables if replacing the variables by the coordinates of the ordered pair results in a true statement.

2. The vertical number line in the rectangular coordinate system is called the _____.

3. A(n) _____ equation can be written in the form $Ax + By = C$.

4. A(n) _____ is a point of the graph where the graph crosses the x-axis.

5. The form $Ax + By = C$ is called _____ form.

6. A(n) _____ is a point of the graph where the graph crosses the y-axis.

7. A set of ordered pairs that assigns to each x-value exactly one y-value is called a(n) _____.

8. The equation $y = 7x - 5$ is written in _____ form.

9. The set of all x-coordinates of a relation is called the _____ of the relation.

10. The set of all y-coordinates of a relation is called the _____ of the relation.

11. A set of ordered pairs is called a(n) _____.

12. The equation $y + 1 = 7(x - 2)$ is written in _____ form.

13. To find an x-intercept of a graph, let _____ = 0.

14. The horizontal number line in the rectangular coordinate system is called the _____.

15. To find a y-intercept of a graph, let _____ = 0.

16. The _____ of a line measures the steepness or tilt of the line.

17. The equation $y = kx$ is an example of _____ variation.

18. The equation $y = \dfrac{k}{x}$ is an example of _____ variation.

> **Helpful Hint**
> ▶ Are you preparing for your test? Don't forget to take the Chapter 10 Test on page 824. Then check your answers at the back of the text and use the Chapter Test Prep Videos to see the fully worked-out solutions to any of the exercises you want to review.

10 Chapter Highlights

Definitions and Concepts	Examples
Section 10.1　The Rectangular Coordinate System	

The **rectangular coordinate system** consists of a plane and a vertical and a horizontal number line intersecting at their 0 coordinates. The vertical number line is called the **y-axis** and the horizontal number line is called the **x-axis.** The point of intersection of the axes is called the **origin.**

To **plot** or **graph** an ordered pair means to find its corresponding point on a rectangular coordinate system.

To plot or graph an ordered pair such as $(3, -2)$, start at the origin. Move 3 units to the right and from there, 2 units down.

To plot or graph $(-3, 4)$, start at the origin. Move 3 units to the left and from there, 4 units up.

An ordered pair is a **solution** of an equation in two variables if replacing the variables with the coordinates of the ordered pair results in a true statement.

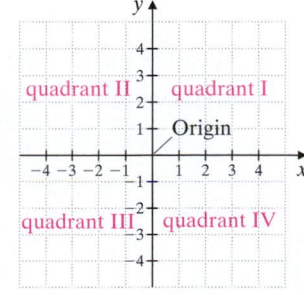

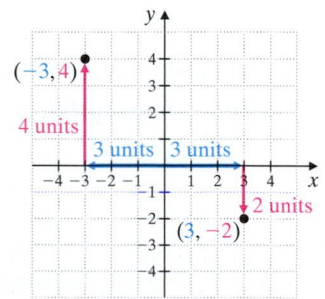

(continued)

Definitions and Concepts	Examples

Section 10.1   The Rectangular Coordinate System (*continued*)

If one coordinate of an ordered pair solution of an equation is known, the other value can be determined by substitution.

Complete the ordered pair $(0, \ )$ for the equation $x - 6y = 12$.

$$x - 6y = 12$$
$$0 - 6y = 12 \quad \text{Let } x = 0.$$
$$\frac{-6y}{-6} = \frac{12}{-6} \quad \text{Divide by } -6.$$
$$y = -2$$

The ordered pair solution is $(0, -2)$.

Section 10.2   Graphing Linear Equations

A **linear equation in two variables** is an equation that can be written in the form $Ax + By = C$, where A and B are not both 0. The form $Ax + By = C$ is called **standard form.**

To graph a linear equation in two variables, find three ordered pair solutions. Plot the solution points and draw the line connecting the points.

$$3x + 2y = -6 \qquad x = -5$$
$$y = 3 \qquad y = -x + 10$$

$x + y = 10$ is in standard form.

Graph: $x - 2y = 5$

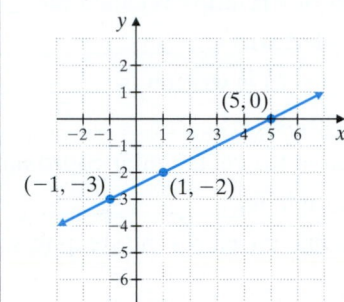

x	y
5	0
1	-2
-1	-3

Section 10.3   Intercepts

An **intercept** of a graph is a point where the graph intersects an axis. If a graph intersects the x-axis at a, then $(a, 0)$ is an **x-intercept.** If a graph intersects the y-axis at b, then $(0, b)$ is a **y-intercept.**

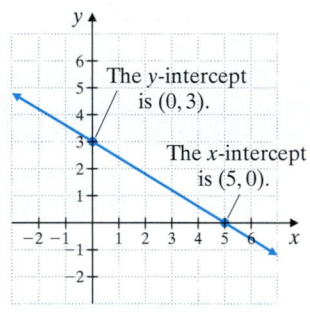

The y-intercept is $(0, 3)$.

The x-intercept is $(5, 0)$.

To find the x-intercept(s), let $y = 0$ and solve for x.

To find the y-intercept(s), let $x = 0$ and solve for y.

Find the intercepts for $2x - 5y = -10$.

If $y = 0$, then
$$2x - 5 \cdot 0 = -10$$
$$2x = -10$$
$$\frac{2x}{2} = \frac{-10}{2}$$
$$x = -5$$

If $x = 0$, then
$$2 \cdot 0 - 5y = -10$$
$$-5y = -10$$
$$\frac{-5y}{-5} = \frac{-10}{-5}$$
$$y = 2$$

Definitions and Concepts	Examples

Section 10.3 Intercepts (*continued*)

The *x*-intercept is $(-5, 0)$. The *y*-intercept is $(0, 2)$.

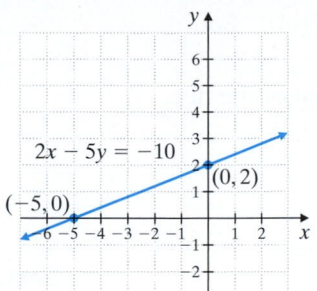

The graph of $x = c$ is a vertical line with *x*-intercept $(c, 0)$.

The graph of $y = c$ is a horizontal line with *y*-intercept $(0, c)$.

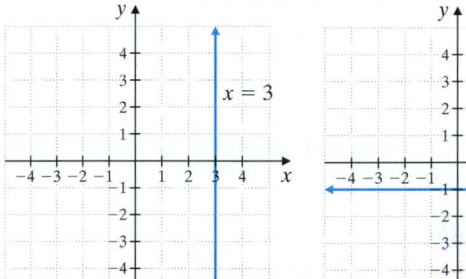

Section 10.4 Slope and Rate of Change

The **slope** *m* of the line through points (x_1, y_1) and (x_2, y_2) is given by

$$m = \frac{y_2 - y_1}{x_2 - x_1} \quad \text{as long as } x_2 \neq x_1$$

A horizontal line has slope 0.
The slope of a vertical line is undefined.
Nonvertical parallel lines have the same slope.
Two nonvertical lines are perpendicular if the slope of one is the negative reciprocal of the slope of the other.

The slope of the line through points $(-1, 6)$ and $(-5, 8)$ is

$$m = \frac{y_2 - y_1}{x_2 - x_1} = \frac{8 - 6}{-5 - (-1)} = \frac{2}{-4} = -\frac{1}{2}$$

The slope of the line $y = -5$ is 0.
The line $x = 3$ has undefined slope.

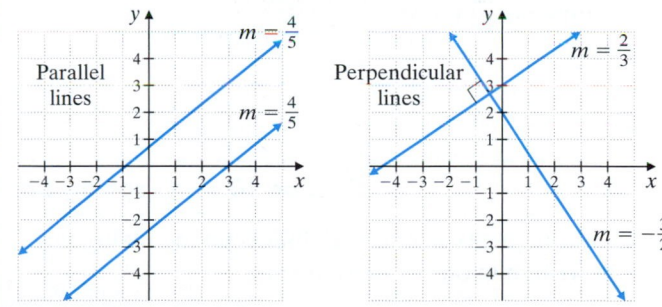

Section 10.5 Equations of Lines

Slope-Intercept Form

$$y = mx + b$$

m is the slope of the line.
$(0, b)$ is the *y*-intercept.

Find the slope and the *y*-intercept of the line $2x + 3y = 6$.
Solve for *y*:

$$2x + 3y = 6$$
$$3y = -2x + 6 \quad \text{Subtract } 2x.$$
$$y = -\frac{2}{3}x + 2 \quad \text{Divide by 3.}$$

The slope of the line is $-\dfrac{2}{3}$ and the *y*-intercept is $(0, 2)$.

(continued)

Definitions and Concepts	Examples

Section 10.5 Equations of Lines (*continued*)

Point-Slope Form

$$y - y_1 = m(x - x_1)$$

m is the slope.
(x_1, y_1) is a point of the line.

Find an equation of the line with slope $\frac{3}{4}$ that contains the point $(-1, 5)$.

$$y - 5 = \frac{3}{4}[x - (-1)]$$

$4(y - 5) = 3(x + 1)$	Multiply by 4.
$4y - 20 = 3x + 3$	Distribute.
$-3x + 4y = 23$	Subtract $3x$ and add 20.

Section 10.6 Introduction to Functions

A set of ordered pairs is a **relation.** The set of all x-coordinates is called the **domain** of the relation and the set of all y-coordinates is called the **range** of the relation.

A **function** is a set of ordered pairs that assigns to each x-value exactly one y-value.

Vertical Line Test

If a vertical line can be drawn so that it intersects a graph more than once, the graph is not the graph of a function. (If no such line can be drawn, the graph is that of a function.)

The domain of the relation

$$\{(0, 5), (2, 5), (4, 5), (5, -2)\}$$

is $\{0, 2, 4, 5\}$. The range is $\{-2, 5\}$.

Which are graphs of functions?

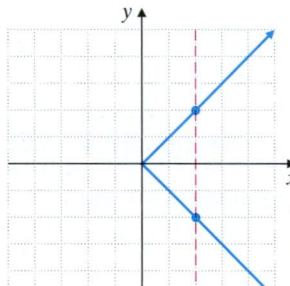 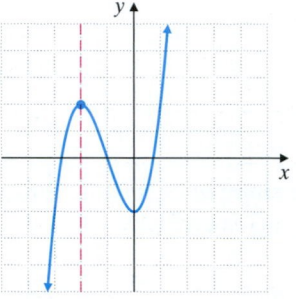

This graph is not the graph of a function. This graph is the graph of a function.

The symbol $f(x)$ means **function of x.** This notation is called **function notation.**

If $f(x) = 3x - 7$, then

$$\begin{aligned} f(-1) &= 3(-1) - 7 \\ &= -3 - 7 \\ &= -10 \end{aligned}$$

Section 10.7 Graphing Linear Inequalities in Two Variables

A **linear inequality in two variables** is an inequality that can be written in one of these forms:

$$\begin{array}{ll} Ax + By < C & Ax + By \leq C \\ Ax + By > C & Ax + By \geq C \end{array}$$

where A and B are not both 0.

$$\begin{array}{ll} 2x - 5y < 6 & x \geq -5 \\ y > -8x & y \leq 2 \end{array}$$

Definitions and Concepts	Examples

Section 10.7   Graphing Linear Inequalities in Two Variables (*continued*)

To Graph a Linear Inequality

1. Graph the boundary line by graphing the related equation. Draw the line solid if the inequality symbol is $\leq$ or $\geq$. Draw the line dashed if the inequality symbol is $<$ or $>$.

2. Choose a test point not on the line. Substitute its coordinates into the original inequality.

3. If the resulting inequality is true, shade the half-plane that contains the test point. If the inequality is not true, shade the half-plane that does not contain the test point.

Graph: $2x - y \leq 4$

1. Graph $2x - y = 4$. Draw a solid line because the inequality symbol is $\leq$.

2. Check the test point $(0, 0)$ in the original inequality, $2x - y \leq 4$.

$$2 \cdot 0 - 0 \leq 4 \quad \text{Let } x = 0 \text{ and } y = 0.$$
$$0 \leq 4 \quad \text{True}$$

3. The inequality is true, so shade the half-plane containing $(0, 0)$, as shown.

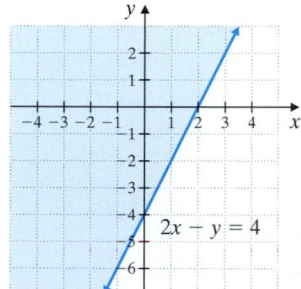

Section 10.8   Direct and Inverse Variation

y **varies directly as** *x*, or *y* is **directly proportional to** *x*, if there is a nonzero constant *k* such that

$$y = kx$$

y **varies inversely as** *x*, or *y* is **inversely proportional to** *x*, if there is a nonzero constant *k* such that

$$y = \frac{k}{x}$$

The circumference of a circle C varies directly as its radius r.

$$C = \underset{k}{\underbrace{2\pi}} r$$

Pressure P varies inversely with volume V.

$$P = \frac{k}{V}$$

Chapter 10   Review

(10.1) *Plot each ordered pair on the same rectangular coordinate system.*

1. $(-7, 0)$

2. $\left(0, 4\dfrac{4}{5}\right)$

3. $(-2, -5)$

4. $(1, -3)$

5. $(0.7, 0.7)$

6. $(-6, 4)$

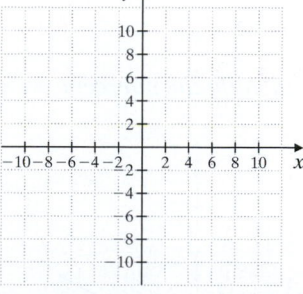

Complete each ordered pair so that it is a solution of the given equation.

7. $-2 + y = 6x;\ (7, \ \ )$

8. $y = 3x + 5;\ (\ \ , -8)$

Complete the table of values for each given equation.

9. $9 = -3x + 4y$

x	y
	0
	3
9	

10. $y = 5$

x	y
7	
-7	
0	

11. $x = 2y$

x	y
	0
	5
	-5

12. The cost in dollars of producing x compact disc holders is given by $y = 5x + 2000$.

    a. Complete the table.

x	1	100	1000
y			

b. Find the number of compact disc holders that can be produced for $6430.

(10.2) *Graph each linear equation.*

13. $x - y = 1$

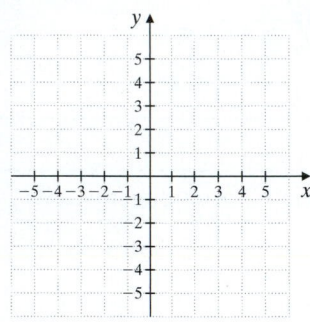

14. $x + y = 6$

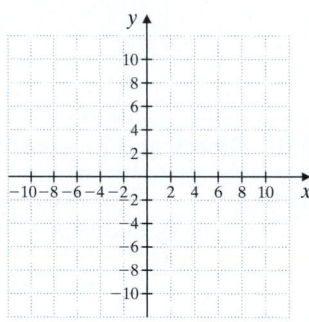

15. $x - 3y = 12$

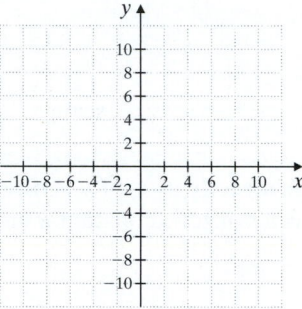

16. $5x - y = -8$

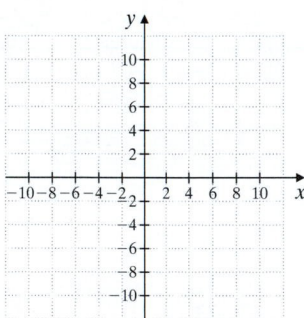

17. $x = 3y$

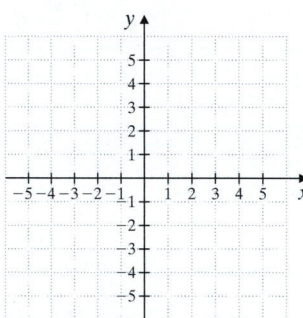

18. $y = -2x$

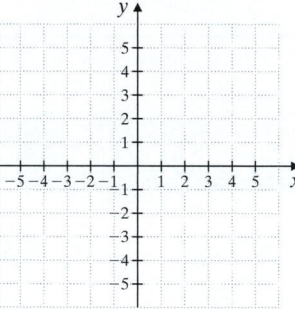

(10.3) *Identify the intercepts in each graph.*

19.

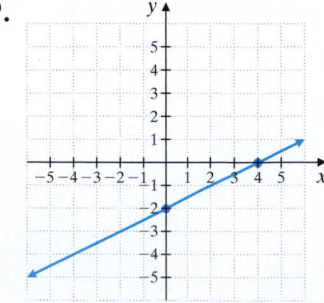

20.

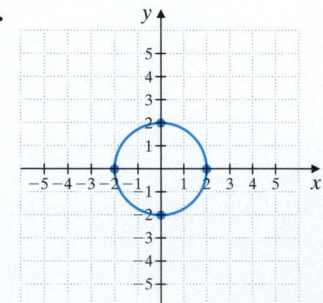

Graph each linear equation.

21. $y = -3$

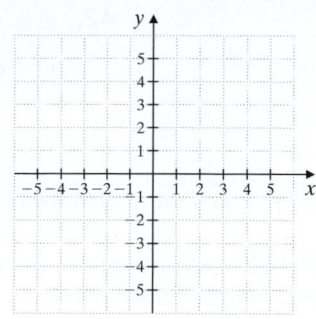

22. $x = 5$

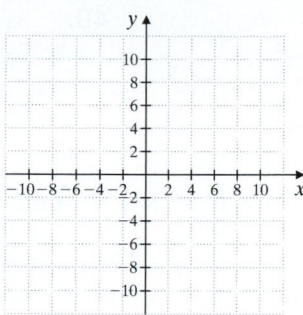

Find the intercepts of each equation.

23. $x - 3y = 12$

24. $-4x + y = 8$

(10.4) *Find the slope of each line.*

25.

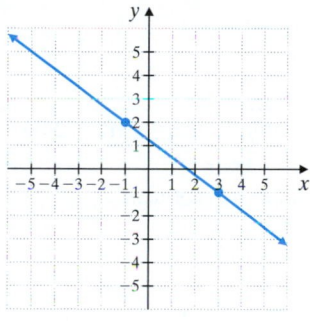

26.

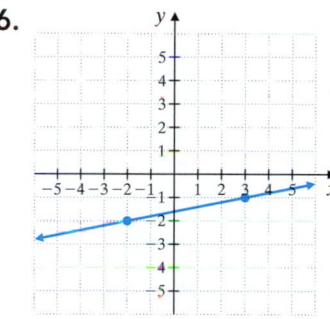

Match each line with its slope.

a.

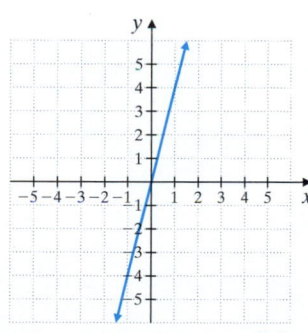

b.

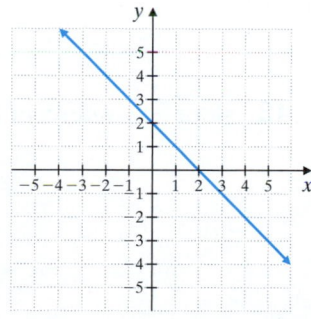

c.

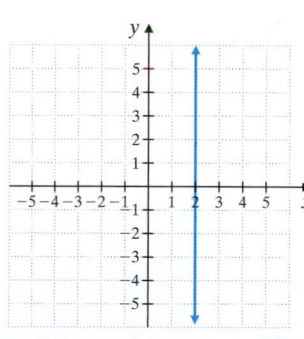

d.

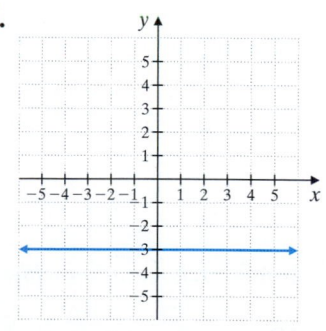

27. $m = 0$

28. $m = -1$

29. undefined slope

30. $m = 4$

Find the slope of the line that passes through each pair of points.

31. $(2, 5)$ and $(6, 8)$

32. $(4, 7)$ and $(1, 2)$

33. $(1, 3)$ and $(-2, -9)$

34. $(-4, 1)$ and $(3, -6)$

Find the slope of each line.

35. $y = 3x + 7$

36. $x - 2y = 4$

37. $y = -2$

38. $x = 0$

△ *Determine whether each pair of lines is parallel, perpendicular, or neither.*

39. $x - y = -6$
$x + y = 3$

40. $3x + y = 7$
$-3x - y = 10$

41. $y = 4x + \dfrac{1}{2}$
$4x + 2y = 1$

42. $y = 6x - \dfrac{1}{3}$
$x + 6y = 6$

Find the slope of each line and write the slope as a rate of change. Don't forget to attach the proper units.

43. The graph below approximates the total number of U.S. magazines in print for each year *x*.

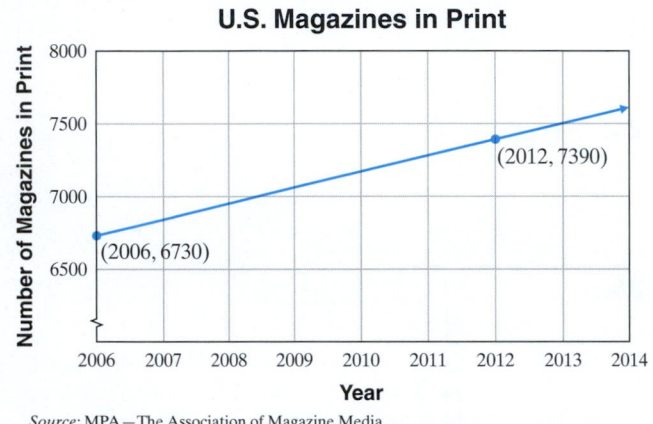

U.S. Magazines in Print

Number of Magazines in Print

(2012, 7390)

(2006, 6730)

Year

Source: MPA—The Association of Magazine Media

44. The graph below approximates the number of lung transplants *y* in the United States for year *x*.

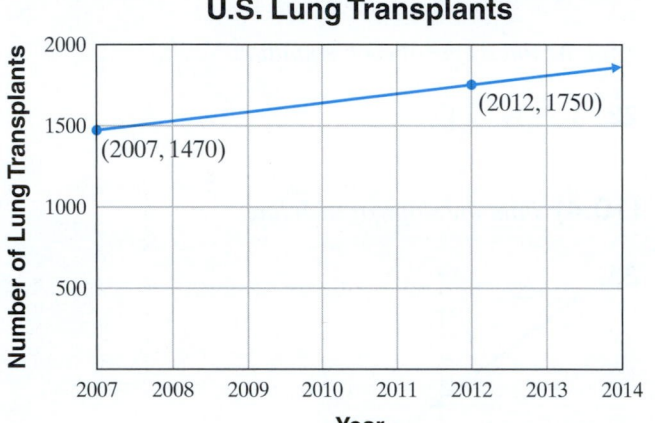

U.S. Lung Transplants

Number of Lung Transplants

(2012, 1750)

(2007, 1470)

Year

Source: Organ Procurement and Transplantation Network

(10.5) *Determine the slope and the y-intercept of the graph of each equation.*

45. $x - 6y = -1$

46. $3x + y = 7$

Write an equation of each line.

47. slope -5; y-intercept $\left(0, \dfrac{1}{2}\right)$

48. slope $\dfrac{2}{3}$; y-intercept $(0, 6)$

Match each equation with its graph.

49. $y = 2x + 1$

50. $y = -4x$

51. $y = 2x$

52. $y = 2x - 1$

a.
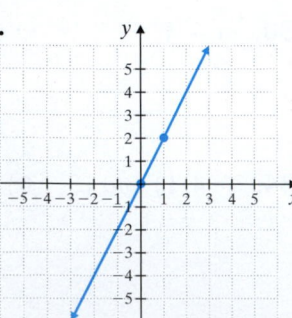

b.
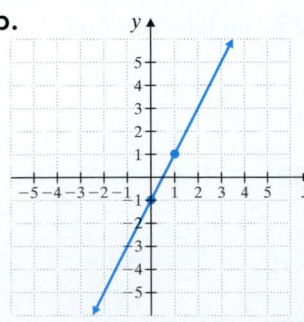

c.
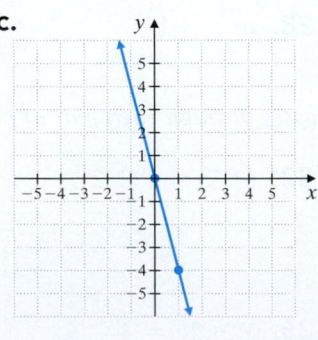

d.
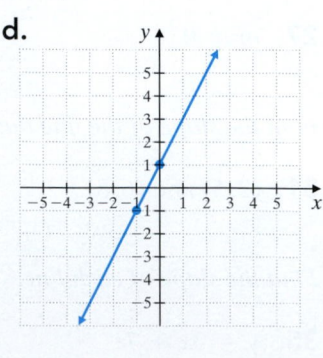

Write an equation of the line with the given slope that passes through the given point. Write the equation in the form
$Ax + By = C.$

53. $m = 4; (2, 0)$ **54.** $m = -3; (0, -5)$ **55.** $m = \dfrac{3}{5}; (1, 4)$ **56.** $m = -\dfrac{1}{3}; (-3, 3)$

Write an equation of the line passing through each pair of points. Write the equation in the form $y = mx + b$.

57. $(1, 7)$ and $(2, -7)$ **58.** $(-2, 5)$ and $(-4, 6)$

(10.6) *Determine whether each relation or graph is a function.*

59. $\{(7, 1), (7, 5), (2, 6)\}$ **60.** $\{(0, -1), (5, -1), (2, 2)\}$

61. 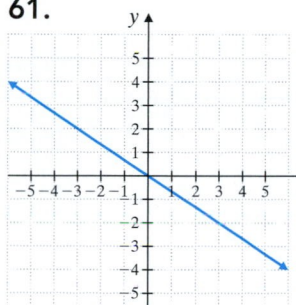 **62.** 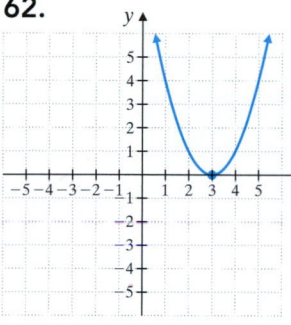 **63.** 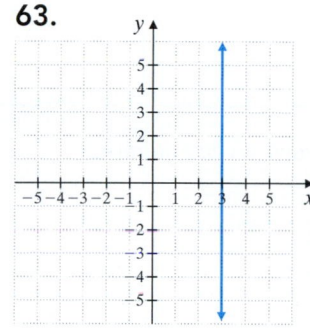 **64.**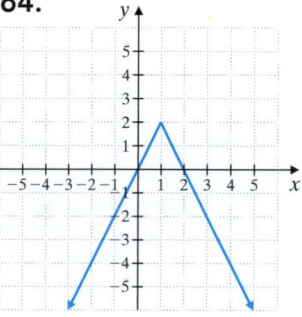

Find each indicated function value for the function $f(x) = -2x + 6$.

65. $f(0)$ **66.** $f(-2)$ **67.** $f\left(\dfrac{1}{2}\right)$ **68.** $f\left(-\dfrac{1}{2}\right)$

(10.7) *Graph each inequality.*

69. $x + 6y < 6$ **70.** $x + y > -2$ **71.** $y \geq -7$

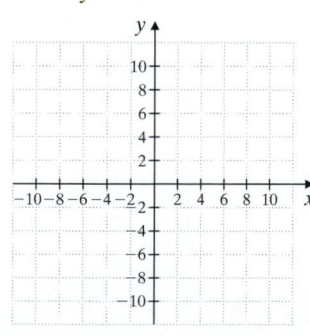 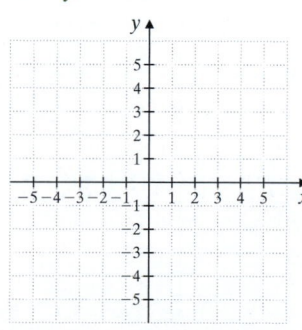 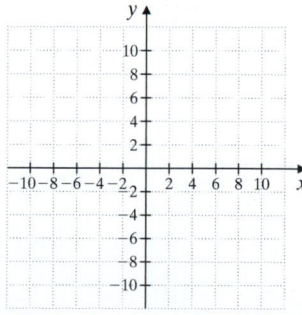

72. $y \leq -4$ **73.** $-x \leq y$ **74.** $x \geq -y$

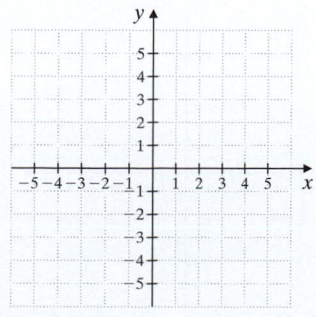 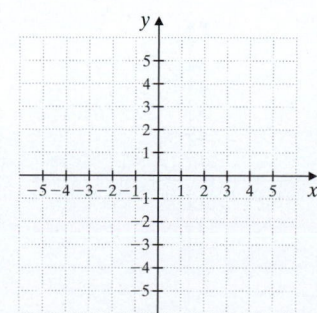 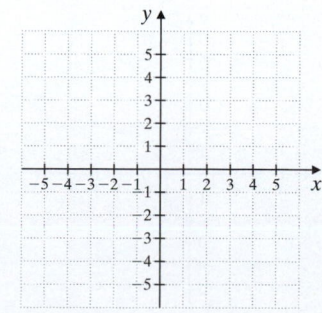

(10.8) *Solve.*

75. y varies directly as x. If $y = 40$ when $x = 4$, find y when x is 11.

76. y varies inversely as x. If $y = 4$ when $x = 6$, find y when x is 48.

77. y varies inversely as x^3. If $y = 12.5$ when $x = 2$, find y when x is 3.

78. y varies directly as x^2. If $y = 175$ when $x = 5$, find y when $x = 10$.

79. The cost of manufacturing a certain medicine varies inversely as the amount of medicine manufactured increases. If 3000 milliliters can be manufactured for $6600, find the cost to manufacture 5000 milliliters.

80. The distance a spring stretches varies directly with the weight attached to the spring. If a 150-pound weight stretches the spring 8 inches, find the distance that a 90-pound weight stretches the spring.

Mixed Review

Complete the table of values for each given equation.

81. $2x - 5y = 9$

x	y
	1
2	
	-3

82. $x = -3y$

x	y
0	
	1
6	

Find the intercepts for each equation.

83. $2x - 3y = 6$

84. $-5x + y = 10$

Graph each linear equation.

85. $x - 5y = 10$

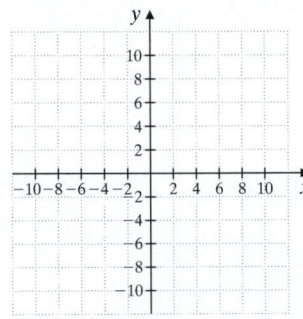

86. $x + y = 4$

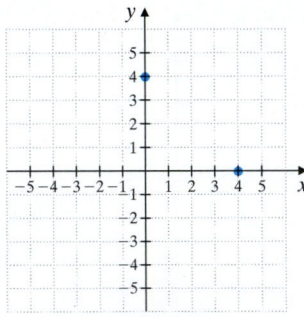

87. $y = -4x$

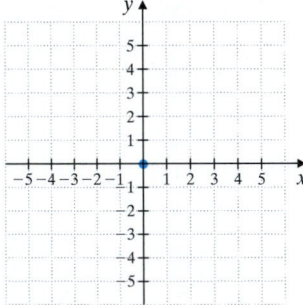

88. $2x + 3y = -6$

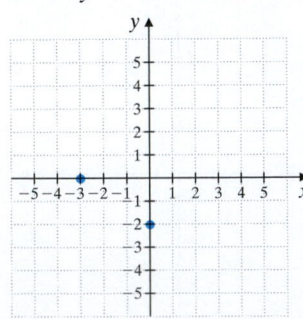

89. $x = 3$

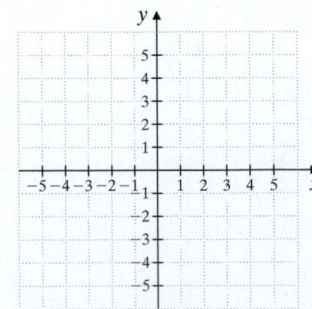

90. $y = -2$

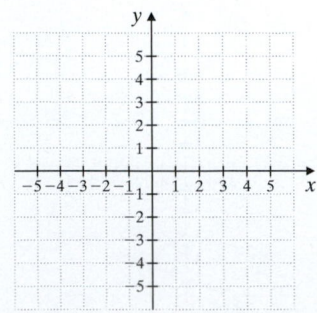

Find the slope of the line that passes through each pair of points.

91. $(3, -5)$ and $(-4, 2)$

92. $(1, 3)$ and $(-6, -8)$

Find the slope of each line.

93.

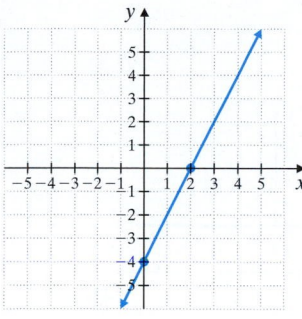

94.

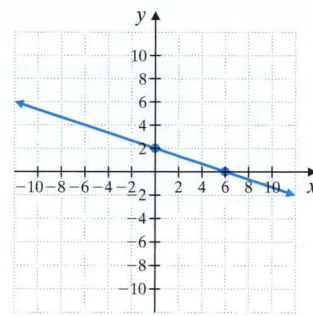

Determine the slope and y-intercept of the graph of each equation.

95. $-2x + 3y = -15$

96. $6x + y - 2 = 0$

Write an equation of the line with the given slope that passes through the given point. Write the equation in the form $Ax + By = C$.

97. $m = -5; (3, -7)$

98. $m = 3; (0, 6)$

Write an equation of the line passing through each pair of points. Write the equation in the form $Ax + By = C$.

99. $(-3, 9)$ and $(-2, 5)$

100. $(3, 1)$ and $(5, -9)$

Chapter 10

Test   Step-by-step test solutions are found on the Chapter Test Prep Videos. Where available: **MyMathLab®** or **YouTube™**

Answers

Complete each ordered pair so that it is a solution of the given equation.

1. $12y - 7x = 5; (1, \quad)$

2. $y = 17; (-4, \quad)$

Find the slope of each line.

3.

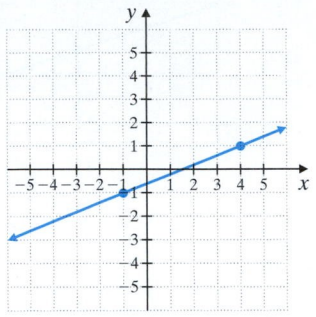

4.

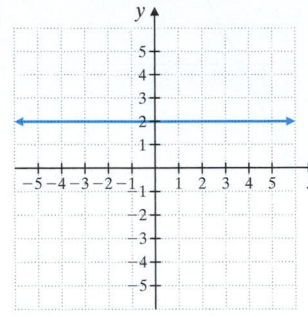

5. Passes through $(6, -5)$ and $(-1, 2)$

6. Passes through $(0, -8)$ and $(-1, -1)$

7. $-3x + y = 5$

8. $x = 6$

Graph.

9. $2x + y = 8$

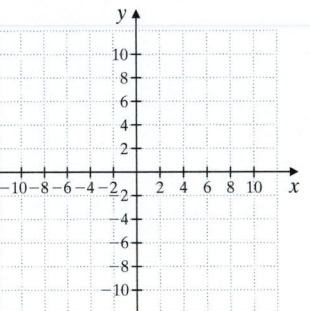

10. $-x + 4y = 5$

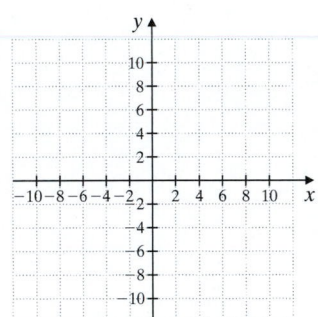

11. $x - y \geq -2$

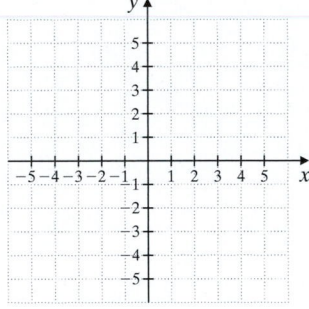

12. $y \geq -4x$

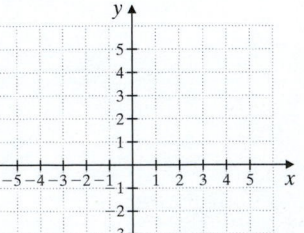

13. $5x - 7y = 10$

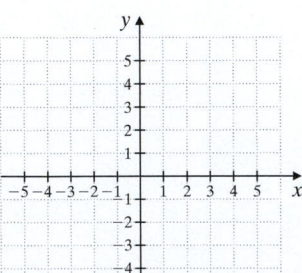

14. $2x - 3y > -6$

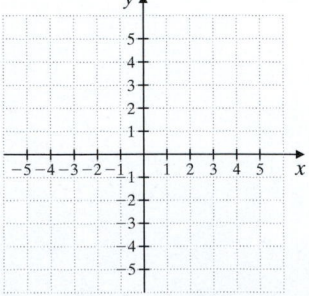

1. _____

2. _____

3. _____

4. _____

5. _____

6. _____

7. _____

8. _____

9. _____

10. _____

11. _____

12. _____

13. _____

14. _____

15. $6x + y > -1$

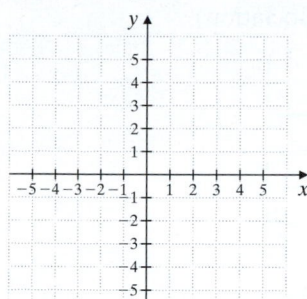

16. $y = -1$

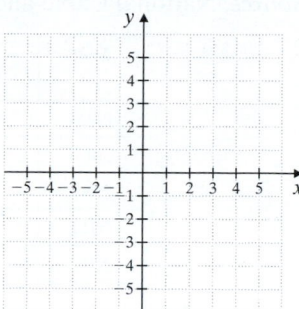

17. Determine whether the graphs of $y = 2x - 6$ and $-4x = 2y$ are parallel lines, perpendicular lines, or neither.

Find the equation of each line. Write the equation in the form $Ax + By = C$.

18. Slope $-\dfrac{1}{4}$, passes through $(2, 2)$

19. Passes through the origin and $(6, -7)$

20. Passes through $(2, -5)$ and $(1, 3)$

21. Slope $\dfrac{1}{8}$; y-intercept $(0, 12)$

Determine whether each relation is a function.

22. $\{(-1, 2), (-2, 4), (-3, 6), (-4, 8)\}$

23. $\{(-3, -3), (0, 5), (-3, 2), (0, 0)\}$

24. The graph shown in Exercise **3**

25. The graph shown in Exercise **4**

Find the indicated function values for each function.

26. $f(x) = 2x - 4$
  a. $f(-2)$
  b. $f(0.2)$
  c. $f(0)$

27. $f(x) = x^3 - x$
  a. $f(-1)$
  b. $f(0)$
  c. $f(4)$

△**28.** The perimeter of the parallelogram below is 42 meters. Write a linear equation in two variables for the perimeter. Use this equation to find x when y is 8 meters.

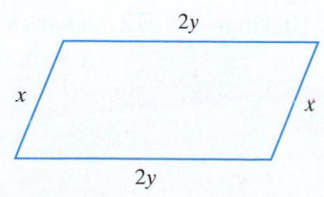

15. _____

16. _____

17. _____

18. _____

19. _____

20. _____

21. _____

22. _____

23. _____

24. _____

25. _____

26. a. _____

b. _____

c. _____

27. a. _____

b. _____

c. _____

28. _____

29. The table gives the number of cable phone customers (in millions) for the years shown. (*Source:* National Cable and Telecommunications Association)

Year	Cable Phone Customers (in millions)
2008	19.6
2009	22.2
2010	23.9
2011	25.3
2012	26.7

a. Write this data as a set of ordered pairs of the form (year, number of cable phone customers in millions).

b. Create a scatter diagram of the data. Be sure to label the axes properly.

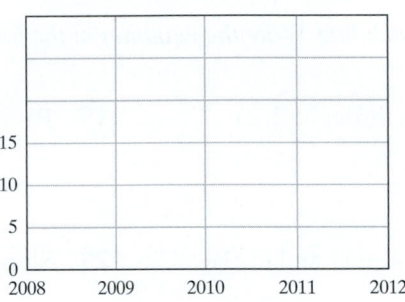

Cable Phone Customers

30. This graph approximates the movie ticket sales *y* (in millions) for the year *x*. Find the slope of the line and write the slope as a rate of change. Don't forget to attach the proper units.

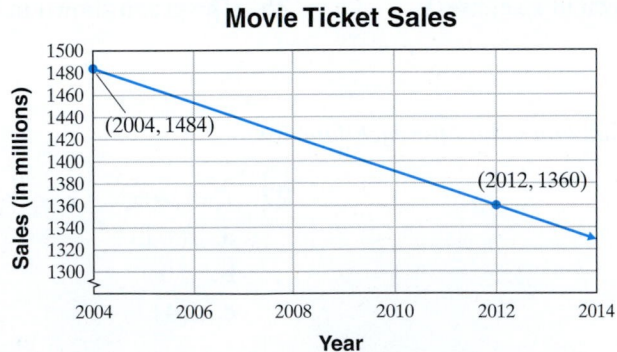

Movie Ticket Sales

Source: National Association of Theater Owners; Annual U.S./Canada Admissions

29. a. _____

b. _____

30. _____

31. _____

32. _____

31. *y* varies directly as *x*. If $y = 10$ when $x = 15$, find *y* when *x* is 42.

32. *y* varies inversely as x^2. If $y = 8$ when $x = 5$, find *y* when *x* is 15.

1. Multiply: 631×125

2. Multiply: $\dfrac{5}{8} \cdot \dfrac{10}{11}$

3. Divide: $\dfrac{2}{5} \div \dfrac{1}{2}$

4. Divide: $2124 \div 9$

5. Add: $\dfrac{2}{3} + \dfrac{1}{7}$

6. Subtract: $9\dfrac{2}{7} - 7\dfrac{1}{2}$

For Exercises 7 through 9, write each decimal in standard form.

7. Forty-eight and twenty-six hundredths

8. Eight hundredths

9. Six and ninety-five thousandths

10. Multiply: 563.21×100

11. Subtract: $3.5 - 0.068$

12. Divide: $0.27 \div 0.02$

13. Simplify: $\dfrac{5.68 + (0.9)^2 \div 100}{0.2}$

14. Simplify: $50 \div 5 \cdot 2$

15. 46 out of every 100 college students live at home. What percent of students live at home? (*Source:* Independent Insurance Agents of America)

16. A basketball player made 4 out of 5 free throws. What percent of free throws were made?

Simplify each expression.

17. $6 \div 3 + 5^2$

18. $\dfrac{10}{3} + \dfrac{5}{21}$

19. $1 + 2[5(2 \cdot 3 + 1) - 10]$

20. $16 - 3 \cdot 3 + 2^4$

Answers

1. _____

2. _____

3. _____

4. _____

5. _____

6. _____

7. _____

8. _____

9. _____

10. _____

11. _____

12. _____

13. _____

14. _____

15. _____

16. _____

17. _____

18. _____

19. _____

20. _____

21. _____

22. _____

23. _____

24. _____

25. _____

26. _____

27. _____

28. _____

29. _____

30. _____

31. _____

32. _____

33. a. _____

    b. _____

    c. _____

34. _____

21. The highest point in the United States is the top of Mount McKinley, at a height of 20,320 feet above sea level. The lowest point is Death Valley, California, which is 282 feet below sea level. How much higher is Mount McKinley than Death Valley? (*Source:* U.S. Geological Survey)

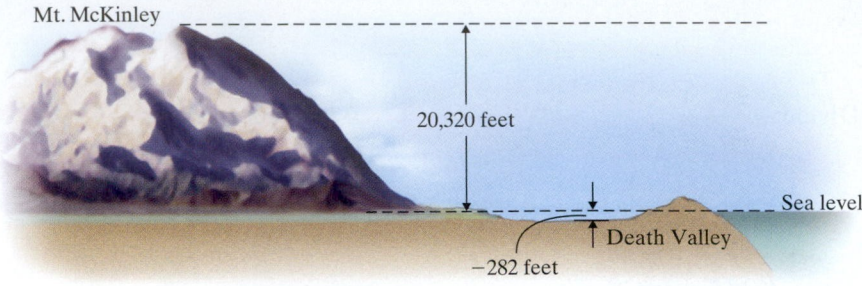

22. Simplify: $1.7x - 11 - 0.9x - 25$

Write each phrase as an algebraic expression and simplify if possible. Let x represent the unknown number.

23. Twice a number, plus 6.

24. The product of -15 and the sum of a number and $\frac{2}{3}$.

25. The difference of a number and 4, divided by 7.

26. The quotient of -9 and twice a number.

27. Five plus the sum of a number and 1.

28. A number subtracted from -86.

29. Solve: $\frac{5}{2}x = 15$

30. Solve: $\frac{x}{4} - 1 = -7$

31. Solve $2x < -4$. Graph the solutions.

$$\begin{array}{ccccccccccc} & & & & & & & & & & \\ \hline -5 & -4 & -3 & -2 & -1 & 0 & 1 & 2 & 3 & 4 & 5 \end{array}$$

32. Solve: $5(x + 4) \geq 4(2x + 3)$

33. Complete each ordered pair so that it is a solution to the equation $3x + y = 12$.

    a. $(0, \quad)$

    b. $(\quad, 6)$

    c. $(-1, \quad)$

34. Complete the table for $y = -5x$.

x	y
	0
-1	
	10

35. Graph the linear equation $2x + y = 5$.

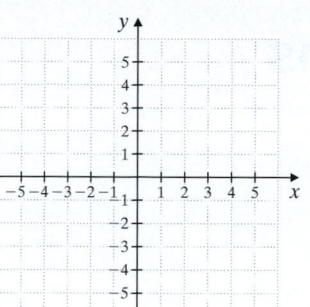

36. Find the slope of the line through $(0, 5)$ and $(-5, 4)$.

37. Find the slope of the line $-2x + 3y = 11$.

38. Find the slope of the line $x = -10$.

39. Find an equation of the line with slope -2 that passes through $(-1, 5)$. Write the equation in slope-intercept form, $y = mx + b$, and in standard form, $Ax + Bx = C$

40. Find the slope and y-intercept of the line whose equation is $2x - 5y = 10$.

41. Given $g(x) = x^2 - 3$, find each function value and list the corresponding ordered pairs generated.

   a. $g(2)$

   b. $g(-2)$

   c. $g(0)$

42. Write an equation of the line through $(2, 3)$ and $(0, 0)$. Write the equation in standard form.

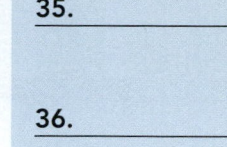

35. _____

36. _____

37. _____

38. _____

39. _____

40. _____

41. a. _____

   b. _____

   c. _____

42. _____

Systems of Equations

In Chapter 10, we graphed equations containing two variables. As we have seen, equations like these are often needed to represent relationships between two different quantities. There are also many opportunities to compare and contrast two such equations, called a *system of equations*. This chapter presents *linear systems* and ways we solve these systems and apply them to real-life situations.

Delivery Methods for Watching Movies Outside the Theater

Movies are not watched just in movie theaters anymore. Consumers can purchase movies for home viewing in such varied formats as DVD, Blu-ray, and digital downloads to television, computer, or other electronic device. Technology is changing so fast that there may well be new ways of obtaining videos that are in the development stream now. The graph below represents consumer spending statistics for two different home entertainment formats, DVD and Blu-ray. In Section 11.2, Exercise 57, we will explore the relationship between consumer spending on these two home entertainment formats.

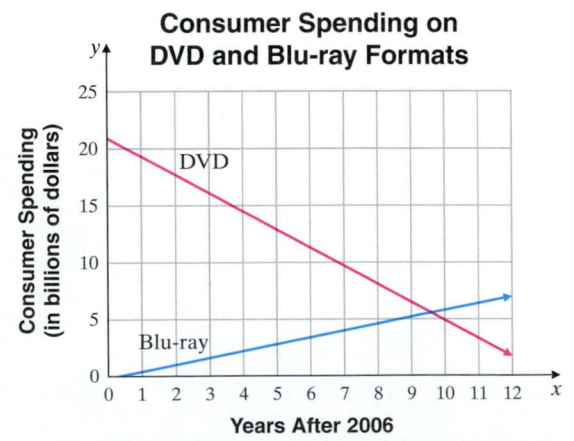

Source: Based on data from DEG: The Digital Entertainment Group

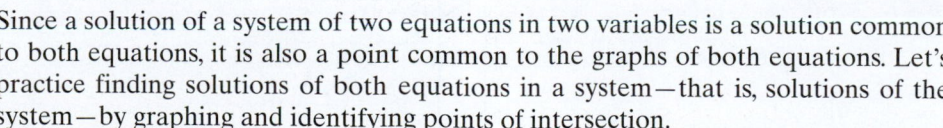
A **system of linear equations** consists of two or more linear equations. In this section, we focus on solving systems of linear equations containing two equations in two variables. Examples of such linear systems are

$$\begin{cases} 3x - 3y = 0 \\ x = 2y \end{cases} \qquad \begin{cases} x - y = 0 \\ 2x + y = 10 \end{cases} \qquad \begin{cases} y = 7x - 1 \\ y = 4 \end{cases}$$

Objective A Deciding Whether an Ordered Pair Is a Solution

A **solution** of a system of two equations in two variables is an ordered pair of numbers that is a solution of both equations in the system.

Example 1 Determine whether $(12, 6)$ is a solution of the system

$$\begin{cases} 2x - 3y = 6 \\ x = 2y \end{cases}$$

Solution: To determine whether $(12, 6)$ is a solution of the system, we replace x with 12 and y with 6 in both equations.

$2x - 3y = 6$	First equation	$x = 2y$	Second equation
$2(12) - 3(6) \stackrel{?}{=} 6$	Let $x = 12$ and $y = 6$.	$12 \stackrel{?}{=} 2(6)$	Let $x = 12$ and $y = 6$.
$24 - 18 \stackrel{?}{=} 6$	Simplify.	$12 = 12$	True
$6 = 6$	True		

Since $(12, 6)$ is a solution of both equations, it is a solution of the system.

■ **Work Practice 1**

Example 2 Determine whether $(-1, 2)$ is a solution of the system

$$\begin{cases} x + 2y = 3 \\ 4x - y = 6 \end{cases}$$

Solution: We replace x with -1 and y with 2 in both equations.

$x + 2y = 3$	First equation	$4x - y = 6$	Second equation
$-1 + 2(2) \stackrel{?}{=} 3$	Let $x = -1$ and $y = 2$.	$4(-1) - 2 \stackrel{?}{=} 6$	Let $x = -1$ and $y = 2$.
$-1 + 4 \stackrel{?}{=} 3$	Simplify.	$-4 - 2 \stackrel{?}{=} 6$	Simplify.
$3 = 3$	True	$-6 = 6$	False

$(-1, 2)$ is not a solution of the second equation, $4x - y = 6$, so it is not a solution of the system.

■ **Work Practice 2**

Objective B Solving Systems of Equations by Graphing

Since a solution of a system of two equations in two variables is a solution common to both equations, it is also a point common to the graphs of both equations. Let's practice finding solutions of both equations in a system—that is, solutions of the system—by graphing and identifying points of intersection.

Objectives

A Decide Whether an Ordered Pair Is a Solution of a System of Linear Equations.

B Solve a System of Linear Equations by Graphing.

C Identify Special Systems of Linear Equations.

D Without Graphing, Determine the Number of Solutions of a System.

Practice 1

Determine whether $(3, 9)$ is a solution of the system
$$\begin{cases} 5x - 2y = -3 \\ y = 3x \end{cases}$$

Practice 2

Determine whether $(3, -2)$ is a solution of the system
$$\begin{cases} 2x - y = 8 \\ x + 3y = 4 \end{cases}$$

Answers

1. $(3, 9)$ is a solution of the system.
2. $(3, -2)$ is not a solution of the system.

Practice 3

Solve the system of equations by graphing.

$$\begin{cases} -3x + y = -10 \\ x - y = 6 \end{cases}$$

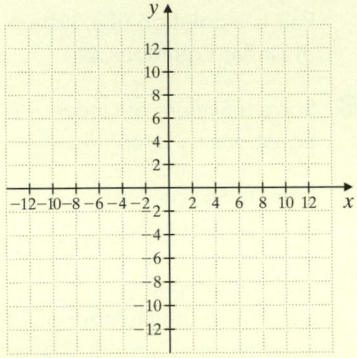

Practice 4

Solve the system of equations by graphing.

$$\begin{cases} x + 3y = -1 \\ y = 1 \end{cases}$$

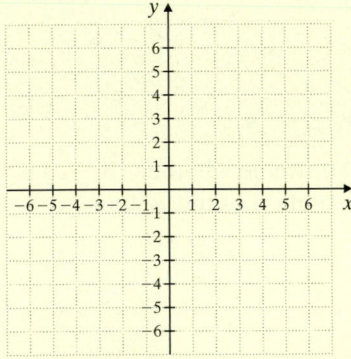

Answers

3. $(2, -4)$;

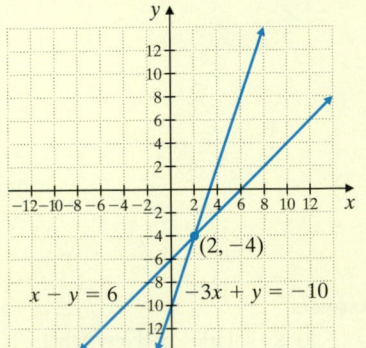

4. See page 834.

Example 3 Solve the system of equations by graphing.

$$\begin{cases} -x + 3y = 10 \\ x + y = 2 \end{cases}$$

Solution: On a single set of axes, graph each linear equation.

$-x + 3y = 10$

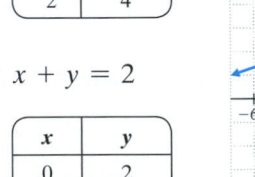

x	y
0	$\dfrac{10}{3}$
-4	2
2	4

$x + y = 2$

x	y
0	2
2	0
1	1

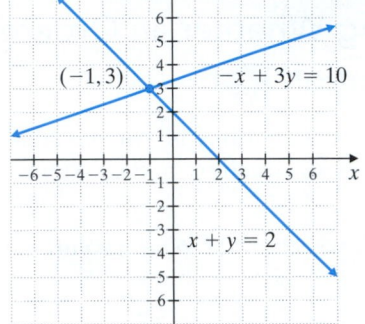

> **Helpful Hint** The point of intersection gives the solution of the system.

The two lines appear to intersect at the point $(-1, 3)$. To check, we replace x with -1 and y with 3 in both equations.

$-x + 3y = 10$	First equation	$x + y = 2$	Second equation
$-(-1) + 3(3) \stackrel{?}{=} 10$	Let $x = -1$ and $y = 3$.	$-1 + 3 \stackrel{?}{=} 2$	Let $x = -1$ and $y = 3$.
$1 + 9 \stackrel{?}{=} 10$	Simplify.	$2 = 2$	True
$10 = 10$	True		

$(-1, 3)$ checks, so it is the solution of the system.

■ **Work Practice 3**

> **Helpful Hint**
>
> Neatly drawn graphs can help when "guessing" the solution of a system of linear equations by graphing.

Example 4 Solve the system of equations by graphing.

$$\begin{cases} 2x + 3y = -2 \\ x = 2 \end{cases}$$

Solution: We graph each linear equation on a single set of axes.

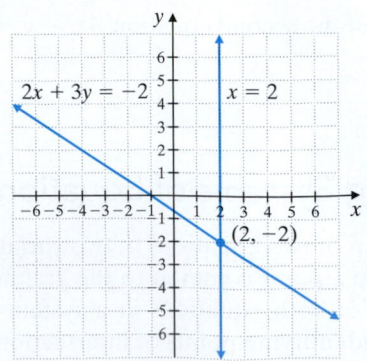

The two lines appear to intersect at the point $(2, -2)$. To determine whether $(2, -2)$ is the solution, we replace x with 2 and y with -2 in both equations.

$$2x + 3y = -2 \quad \text{First equation} \qquad\qquad x = 2 \quad \text{Second equation}$$
$$2(2) + 3(-2) \overset{?}{=} -2 \quad \text{Let } x = 2 \text{ and } y = -2. \qquad 2 \overset{?}{=} 2 \quad \text{Let } x = 2.$$
$$4 + (-6) \overset{?}{=} -2 \quad \text{Simplify.} \qquad\qquad 2 = 2 \quad \text{True}$$
$$-2 = -2 \quad \text{True}$$

Since a true statement results in both equations, $(2, -2)$ is the solution of the system.

■ **Work Practice 4**

Objective C Identifying Special Systems of Linear Equations ▶

Not all systems of linear equations have a single solution. Some systems have no solution and some have an infinite number of solutions.

Example 5 Solve the system of equations by graphing.

$$\begin{cases} 2x + y = 7 \\ 2y = -4x \end{cases}$$

Solution: We graph the two equations in the system. The equations in slope-intercept form are $y = -2x + 7$ and $y = -2x$. Notice from the equations that the lines have the same slope, -2, and different y-intercepts. This means that the lines are parallel.

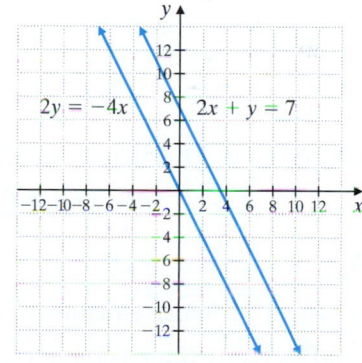

Since the lines are parallel, they do not intersect. This means that the system has *no solution*.

■ **Work Practice 5**

Example 6 Solve the system of equations by graphing.

$$\begin{cases} x - y = 3 \\ -x + y = -3 \end{cases}$$

Solution: We graph each equation. The graphs of the equations are the same line. To see this, notice that if both sides of the first equation in the system are multiplied by -1, the result is the second equation.

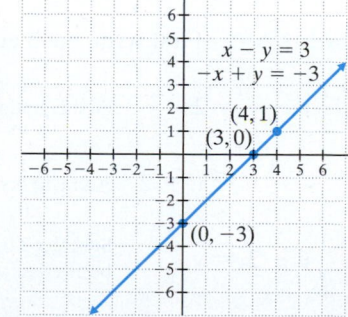

$$x - y = 3 \qquad \text{First equation}$$
$$-1(x - y) = -1(3) \qquad \text{Multiply both sides by } -1.$$
$$-x + y = -3 \qquad \text{Simplify. This is the second equation.}$$

Any ordered pair that is a solution of one equation is a solution of the other equation and is then a solution of the system. This means that the system has an infinite number of solutions.

■ **Work Practice 6**

Practice 5

Solve the system of equations by graphing.

$$\begin{cases} 3x - y = 6 \\ 6x = 2y \end{cases}$$

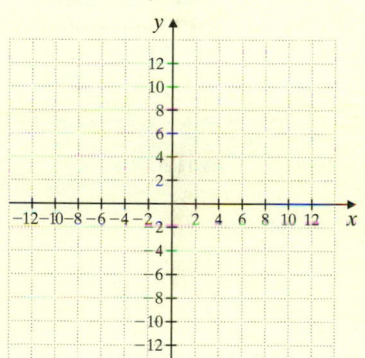

Practice 6

Solve the system of equations by graphing.

$$\begin{cases} x + y = -4 \\ -2x - 2y = 8 \end{cases}$$

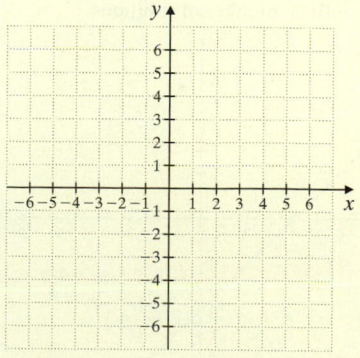

Answers

5. See page 834.   **6.** See page 834.

Examples 5 and 6 are special cases of systems of linear equations. A system that has no solution is said to be an **inconsistent system.** If the graphs of the two equations of a system are identical, we call the equations **dependent equations.** Thus, the system in Example 5 is an inconsistent system and the equations in the system in Example 6 are dependent equations.

As we have seen, three different situations can occur when graphing the two lines associated with the equations in a linear system. These situations are shown in the figures.

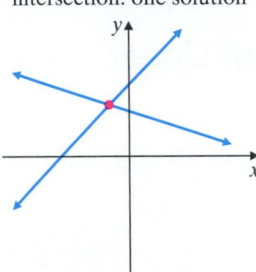

One point of intersection: one solution

Consistent system
(at least one solution)
Independent equations
(graphs of equations differ)

Parallel lines: no solution

Inconsistent system
(no solution)
Independent equations
(graphs of equations differ)

Same line: infinite number of solutions

Consistent system
(at least one solution)
Dependent equations
(graphs of equations identical)

Objective D Finding the Number of Solutions of a System Without Graphing ▶

You may have suspected by now that graphing alone is not an accurate way to solve a system of linear equations. For example, a solution of $\left(\dfrac{1}{2}, \dfrac{2}{9}\right)$ is unlikely to be read correctly from a graph. The next two sections present two accurate methods of solving these systems. In the meantime, we can decide how many solutions a system has by writing each equation in slope-intercept form.

Example 7   Without graphing, determine the number of solutions of the system.

$$\begin{cases} \dfrac{1}{2}x - y = 2 \\ x = 2y + 5 \end{cases}$$

Solution: First write each equation in slope-intercept form.

$\dfrac{1}{2}x - y = 2$ First equation

$\dfrac{1}{2}x = y + 2$ Add y to both sides.

$\dfrac{1}{2}x - 2 = y$ Subtract 2 from both sides.

$x = 2y + 5$ Second equation

$x - 5 = 2y$ Subtract 5 from both sides.

$\dfrac{x}{2} - \dfrac{5}{2} = \dfrac{2y}{2}$ Divide both sides by 2.

$\dfrac{1}{2}x - \dfrac{5}{2} = y$ Simplify.

The slope of each line is $\dfrac{1}{2}$, but they have different y-intercepts. This tells us that the lines representing these equations are parallel. Since the lines are parallel, the system has no solution and is inconsistent.

■ **Work Practice 7**

Practice 7

Without graphing, determine the number of solutions of the system.

$$\begin{cases} 5x + 4y = 6 \\ x - y = 3 \end{cases}$$

Answers

4. $(-4, 1)$;

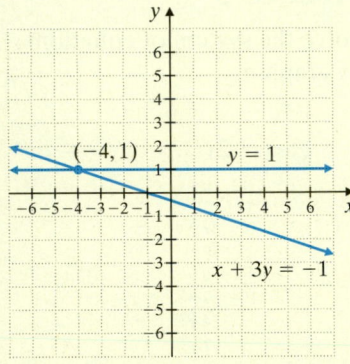

5. no solution;

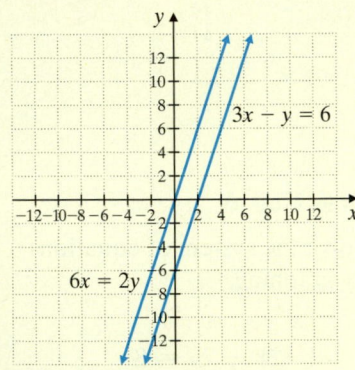

6. infinite number of solutions;

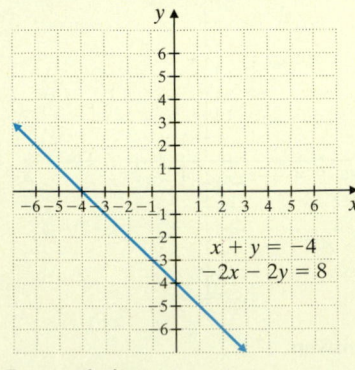

7. one solution

Example 8 Without graphing, determine the number of solutions of the system.

$$\begin{cases} 3x - y = 4 \\ x + 2y = 8 \end{cases}$$

Solution: Once again, the slope-intercept form helps determine how many solutions this system has.

$3x - y = 4$	First equation	$x + 2y = 8$	Second equation
$3x = y + 4$	Add y to both sides.	$x = -2y + 8$	Subtract $2y$ from both sides.
$3x - 4 = y$	Subtract 4 from both sides.	$x - 8 = -2y$	Subtract 8 from both sides.
		$\dfrac{x}{-2} - \dfrac{8}{-2} = \dfrac{-2y}{-2}$	Divide both sides by -2.
		$-\dfrac{1}{2}x + 4 = y$	Simplify.

The slope of the second line is $-\dfrac{1}{2}$, whereas the slope of the first line is 3. Since the slopes are not equal, the two lines are neither parallel nor identical and must intersect. Therefore, this system has one solution and is consistent.

■ **Work Practice 8**

Practice 8

Without graphing, determine the number of solutions of the system.

$$\begin{cases} -\dfrac{2}{3}x + y = 6 \\ 3y = 2x + 5 \end{cases}$$

Answer

8. no solution

📟 **Calculator Explorations** **Graphing**

A graphing calculator may be used to approximate solutions of systems of equations. For example, to approximate the solution of the system

$$\begin{cases} y = -3.14x - 1.35 \\ y = 4.88x + 5.25, \end{cases}$$

first graph each equation on the same set of axes. Then use the Intersect feature of your calculator to approximate the point of intersection.

The approximate point of intersection is $(-0.82, 1.23)$.

Solve each system of equations. Approximate the solutions to two decimal places.

1. $\begin{cases} y = -2.68x + 1.21 \\ y = 5.22x - 1.68 \end{cases}$

2. $\begin{cases} y = 4.25x + 3.89 \\ y = -1.88x + 3.21 \end{cases}$

3. $\begin{cases} 4.3x - 2.9y = 5.6 \\ 8.1x + 7.6y = -14.1 \end{cases}$

4. $\begin{cases} -3.6x - 8.6y = 10 \\ -4.5x + 9.6y = -7.7 \end{cases}$

Vocabulary, Readiness & Video Check

Fill in each blank with one of the words or phrases listed below.

system of linear equations	solution	consistent
dependent	inconsistent	independent

1. In a system of linear equations in two variables, if the graphs of the equations are the same, the equations are _____ equations.

2. Two or more linear equations are called a(n) _____.

3. A system of equations that has at least one solution is called a(n) _____ system.

4. A(n) _____ of a system of two equations in two variables is an ordered pair of numbers that is a solution of both equations in the system.

5. A system of equations that has no solution is called a(n) _____ system.

6. In a system of linear equations in two variables, if the graphs of the equations are different, the equations are _____ equations.

Each rectangular coordinate system shows the graph of the equations in a system of equations. Use each graph to determine the number of solutions for each associated system. If the system has only one solution, give its coordinates.

7.

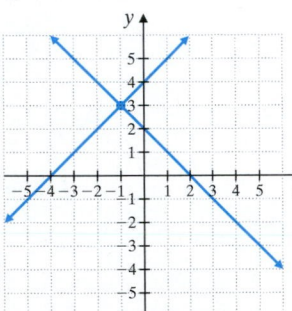

8.

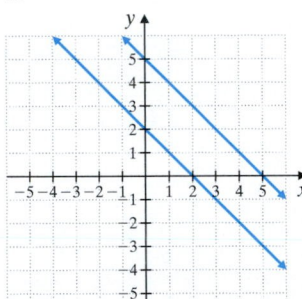

9.

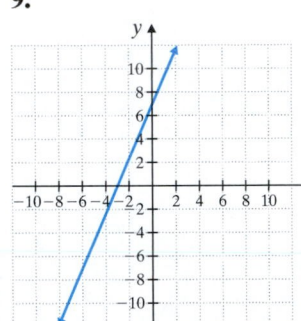

10.

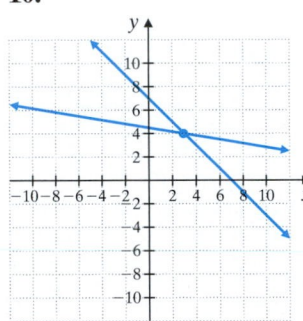

Martin-Gay Interactive Videos   *Watch the section lecture video and answer the following questions.*

Objective A 11. In Example 1, the first ordered pair is a solution of the first equation of the system. Why is this not enough to determine whether this ordered pair is a solution of the system? ▶

**Objectives B
C 12.** From ⊞ Examples 2 and 3, why is finding the solution of a system of equations from a graph considered "guessing" and this proposed solution checked algebraically? ▶

See Video 11.1

Objective D 13. From ⊞ Examples 5–7, explain how the slope-intercept form tells us how many solutions a system of equations has. ▶

11.1 Exercise Set MyMathLab®

Objective A *Determine whether each ordered pair is a solution of the system of linear equations. See Examples 1 and 2.*

1. $\begin{cases} x + y = 8 \\ 3x + 2y = 21 \end{cases}$
  a. $(2, 4)$
  b. $(5, 3)$

2. $\begin{cases} 2x + y = 5 \\ x + 3y = 5 \end{cases}$
  a. $(5, 0)$
  b. $(2, 1)$

3. $\begin{cases} 3x - y = 5 \\ x + 2y = 11 \end{cases}$
  a. $(3, 4)$
  b. $(0, -5)$

4. $\begin{cases} 2x - 3y = 8 \\ x - 2y = 6 \end{cases}$
  a. $(-2, -4)$
  b. $(7, 2)$

5. $\begin{cases} 2y = 4x + 6 \\ 2x - y = -3 \end{cases}$
  a. $(-3, -3)$
  b. $(0, 3)$

6. $\begin{cases} x + 5y = -4 \\ -2x = 10y + 8 \end{cases}$
  a. $(-4, 0)$
  b. $(6, -2)$

7. $\begin{cases} -2 = x - 7y \\ 6x - y = 13 \end{cases}$
  a. $(-2, 0)$
  b. $\left(\dfrac{1}{2}, \dfrac{5}{14}\right)$

8. $\begin{cases} 4x = 1 - y \\ x - 3y = -8 \end{cases}$
  a. $(0, 1)$
  b. $\left(\dfrac{1}{6}, \dfrac{1}{3}\right)$

Objectives B C *Solve each system of linear equations by graphing. See Examples 3 through 6.*

9. $\begin{cases} x + y = 4 \\ x - y = 2 \end{cases}$

10. $\begin{cases} x + y = 3 \\ x - y = 5 \end{cases}$

11. $\begin{cases} x + y = 6 \\ -x + y = -6 \end{cases}$

12. $\begin{cases} x + y = 1 \\ -x + y = -3 \end{cases}$

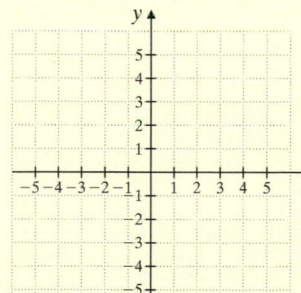
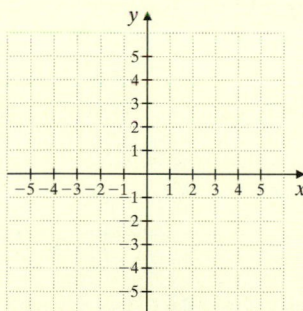
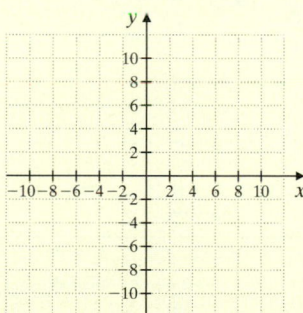
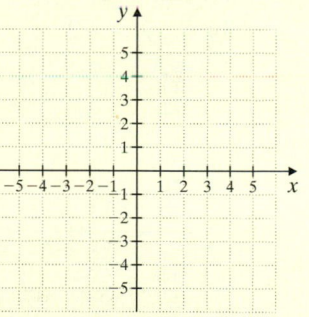

13. $\begin{cases} y = 2x \\ 3x - y = -2 \end{cases}$

14. $\begin{cases} y = -3x \\ 2x - y = -5 \end{cases}$

15. $\begin{cases} y = x + 1 \\ y = 2x - 1 \end{cases}$

16. $\begin{cases} y = 3x - 4 \\ y = x + 2 \end{cases}$

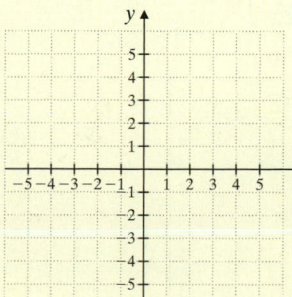
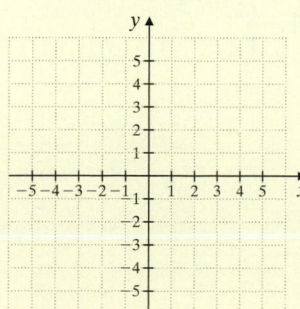
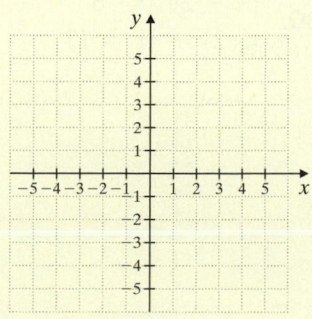
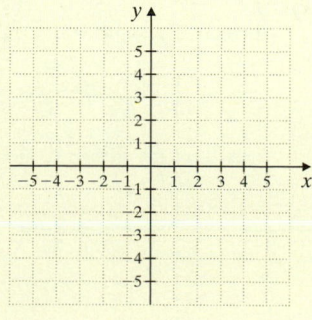

17. $\begin{cases} 2x + y = 0 \\ 3x + y = 1 \end{cases}$

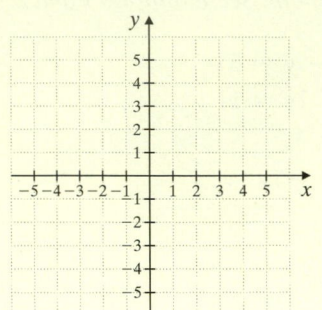

18. $\begin{cases} 2x + y = 1 \\ 3x + y = 0 \end{cases}$

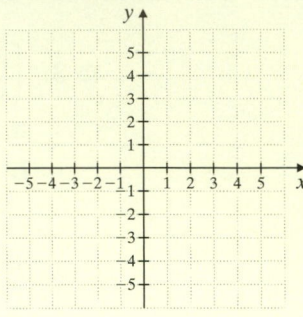

19. $\begin{cases} y = -x - 1 \\ y = 2x + 5 \end{cases}$

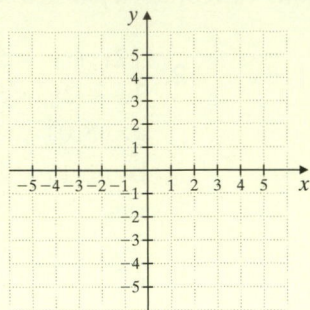

20. $\begin{cases} y = x - 1 \\ y = -3x - 5 \end{cases}$

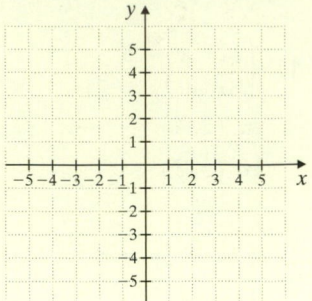

21. $\begin{cases} x + y = 5 \\ x + y = 6 \end{cases}$

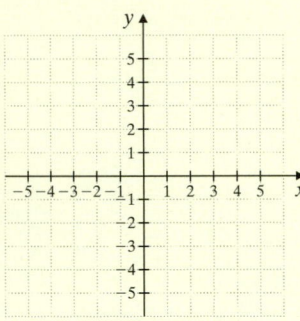

22. $\begin{cases} x - y = 4 \\ x - y = 1 \end{cases}$

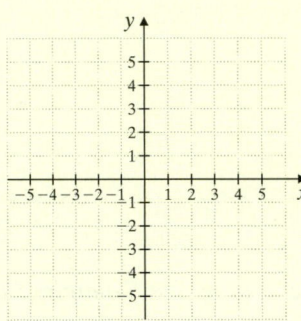

23. $\begin{cases} 2x - y = 6 \\ y = 2 \end{cases}$

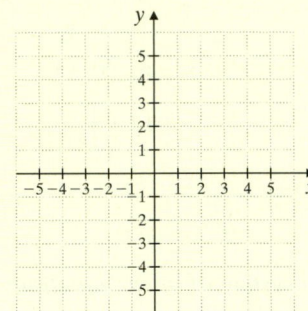

24. $\begin{cases} x + y = 5 \\ x = 4 \end{cases}$

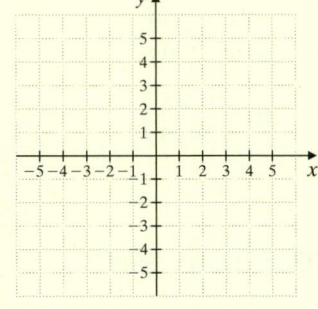

25. $\begin{cases} x - 2y = 2 \\ 3x + 2y = -2 \end{cases}$

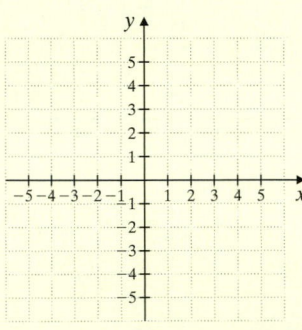

26. $\begin{cases} x + 3y = 7 \\ 2x - 3y = -4 \end{cases}$

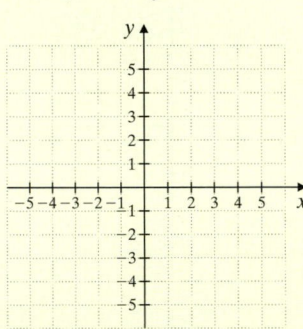

27. $\begin{cases} 2x + y = 4 \\ 6x = -3y + 6 \end{cases}$

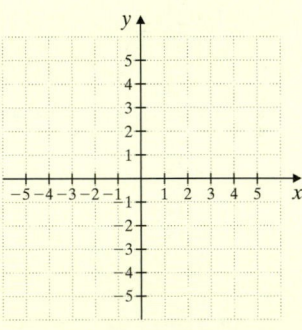

28. $\begin{cases} y + 2x = 3 \\ 4x = 2 - 2y \end{cases}$

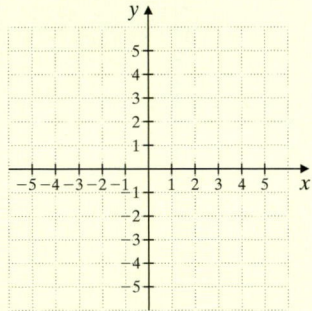

29. $\begin{cases} y - 3x = -2 \\ 6x - 2y = 4 \end{cases}$

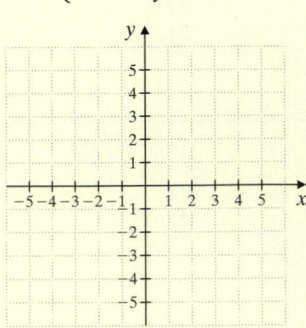

30. $\begin{cases} x - 2y = -6 \\ -2x + 4y = 12 \end{cases}$

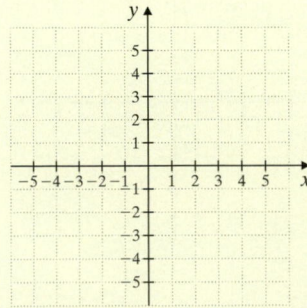

31. $\begin{cases} x = 3 \\ y = -1 \end{cases}$

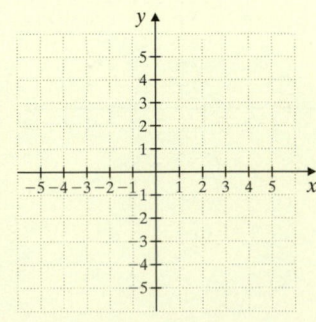

32. $\begin{cases} x = -5 \\ y = 3 \end{cases}$

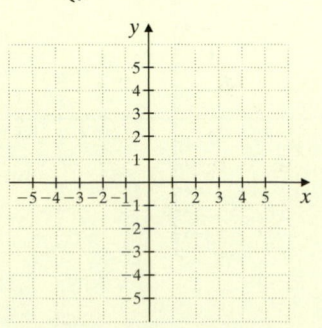

33. $\begin{cases} y = x - 2 \\ y = 2x + 3 \end{cases}$

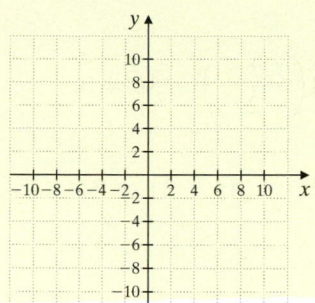

34. $\begin{cases} y = x + 5 \\ y = -2x - 4 \end{cases}$

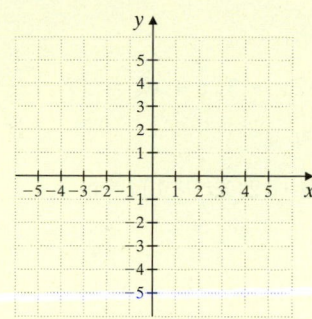

35. $\begin{cases} 2x - 3y = -2 \\ -3x + 5y = 5 \end{cases}$

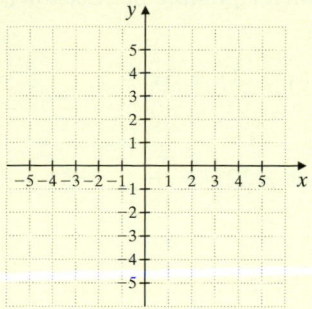

36. $\begin{cases} 4x - y = 7 \\ 2x - 3y = -9 \end{cases}$

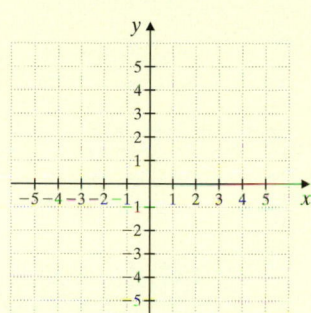

37. $\begin{cases} 6x - y = 4 \\ \dfrac{1}{2}y = -2 + 3x \end{cases}$

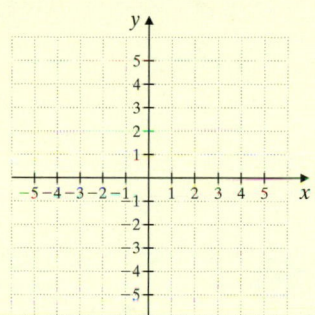

38. $\begin{cases} 3x - y = 6 \\ \dfrac{1}{3}y = -2 + x \end{cases}$

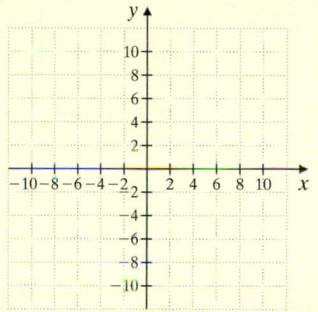

Objective D *Without graphing, decide:*

a. *Are the graphs of the equations identical lines, parallel lines, or lines intersecting at a single point?*

b. *How many solutions does the system have? See Examples 7 and 8.*

39. $\begin{cases} 4x + y = 24 \\ x + 2y = 2 \end{cases}$

40. $\begin{cases} 3x + y = 1 \\ 3x + 2y = 6 \end{cases}$

41. $\begin{cases} 2x + y = 0 \\ 2y = 6 - 4x \end{cases}$

42. $\begin{cases} 3x + y = 0 \\ 2y = -6x \end{cases}$

43. $\begin{cases} 6x - y = 4 \\ \dfrac{1}{2}y = -2 + 3x \end{cases}$

44. $\begin{cases} 3x - y = 2 \\ \dfrac{1}{3}y = -2 + 3x \end{cases}$

45. $\begin{cases} x = 5 \\ y = -2 \end{cases}$

46. $\begin{cases} y = 3 \\ x = -4 \end{cases}$

47. $\begin{cases} 3y - 2x = 3 \\ x + 2y = 9 \end{cases}$

48. $\begin{cases} 2y = x + 2 \\ y + 2x = 3 \end{cases}$

49. $\begin{cases} 6y + 4x = 6 \\ 3y - 3 = -2x \end{cases}$

50. $\begin{cases} 8y + 6x = 4 \\ 4y - 2 = 3x \end{cases}$

51. $\begin{cases} x + y = 4 \\ x + y = 3 \end{cases}$

52. $\begin{cases} 2x + y = 0 \\ y = -2x + 1 \end{cases}$

Review

Solve each equation. See Section 9.3.

53. $5(x - 3) + 3x = 1$

54. $-2x + 3(x + 6) = 17$

55. $4\left(\dfrac{y + 1}{2}\right) + 3y = 0$

56. $-y + 12\left(\dfrac{y - 1}{4}\right) = 3$

57. $8a - 2(3a - 1) = 6$

58. $3z - (4z - 2) = 9$

Concept Extensions

59. Draw a graph of two linear equations whose associated system has the solution $(-1, 4)$.

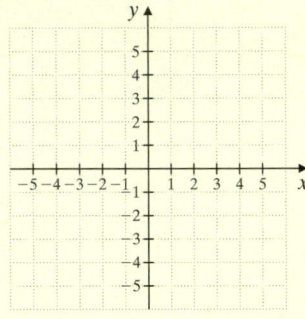

60. Draw a graph of two linear equations whose associated system has the solution $(3, -2)$.

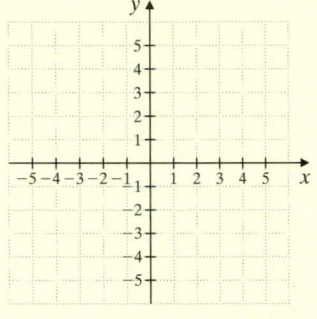

61. Draw a graph of two linear equations whose associated system has no solution.

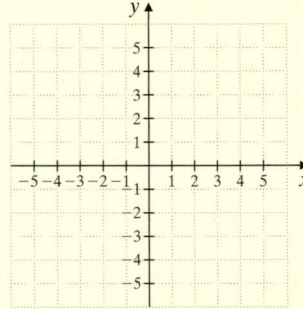

62. Draw a graph of two linear equations whose associated system has an infinite number of solutions.

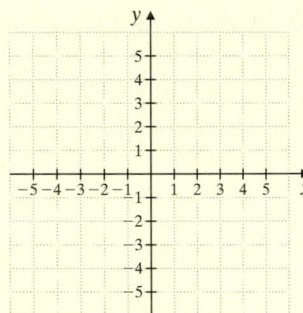

The double line graph below shows the number of digital and analog movie screens in U.S. cinemas for the years shown. Use this graph to answer Exercises 63 and 64. (Source: Motion Picture Association of America, Inc.)

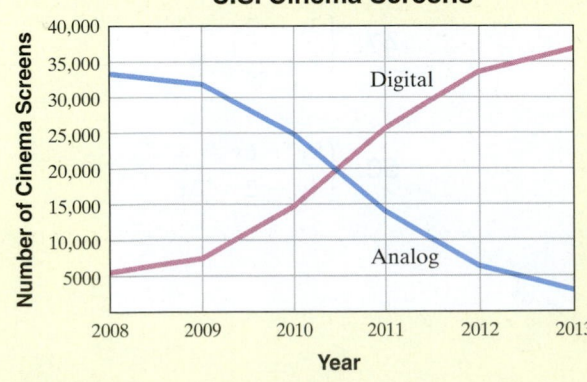

U.S. Cinema Screens

Source: Motion Picture Association of America, Inc.

63. Between what pairs of years did the number of digital cinema screens equal the number of analog cinema screens?

64. For what year(s) was the number of digital cinema screens less than the number of analog cinema screens?

The double line graph below shows the average attendance per game for the Cleveland Indians and the Pittsburgh Pirates baseball teams for the years shown. Use this for Exercises 65 and 66. (Source: Baseball Almanac)

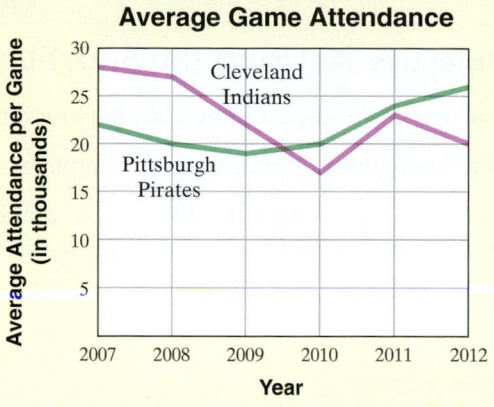

Average Game Attendance

Source: Baseball Almanac

65. In what year(s) was the average attendance per game for the Pittsburgh Pirates greater than the average attendance per game for the Cleveland Indians?

66. In what year was the average attendance per game for the Cleveland Indians closest to the average attendance per game for the Pittsburgh Pirates?

67. Construct a system of two linear equations that has $(2, 5)$ as a solution.

68. Construct a system of two linear equations that has $(0, 1)$ as a solution.

69. The ordered pair $(-2, 3)$ is a solution of the three linear equations below:

$$x + y = 1$$
$$2x - y = -7$$
$$x + 3y = 7$$

If each equation has a distinct graph, describe the graph of all three equations on the same axes.

70. Explain how to use a graph to determine the number of solutions of a system.

71. Below are tables of values for two linear equations.
  a. Find a solution of the corresponding system.
  b. Graph several ordered pairs from each table and sketch the two lines.

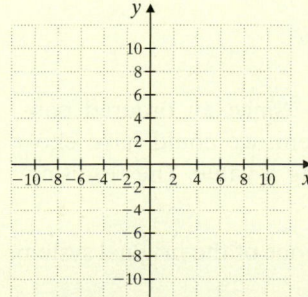

x	y
1	3
2	5
3	7
4	9
5	11

x	y
1	6
2	7
3	8
4	9
5	10

72. Below are tables of values for two linear equations.
  a. Find a solution of the corresponding system.
  b. Graph several ordered pairs from each table and sketch the two lines.

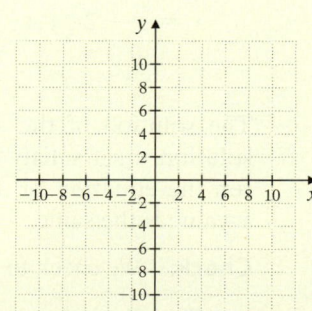

x	y
-3	5
-1	1
0	-1
1	-3
2	-5

x	y
-3	7
-1	1
0	-2
1	-5
2	-8

c. Does your graph confirm the solution from part **a**?

c. Does your graph confirm the solution from part **a**?

Objective

A Use the Substitution Method to Solve a System of Linear Equations.

Objective A Using the Substitution Method ▶

You may have suspected by now that graphing alone is not an accurate way to solve a system of linear equations. For example, a solution of $\left(\frac{1}{2}, \frac{2}{9}\right)$ is unlikely to be read correctly from a graph. In this section, we discuss a second, more accurate method for solving systems of equations. This method is called the **substitution method** and is introduced in the next example.

Practice 1

Use the substitution method to solve the system:
$$\begin{cases} 2x + 3y = 13 \\ x = y + 4 \end{cases}$$

Example 1 Solve the system:

$$\begin{cases} 2x + y = 10 & \text{First equation} \\ x = y + 2 & \text{Second equation} \end{cases}$$

Solution: The second equation in this system is $x = y + 2$. This tells us that x and $y + 2$ have the same value. This means that we may substitute $y + 2$ for x in the first equation.

$$2x + y = 10 \quad \text{First equation}$$

$$2(y + 2) + y = 10 \quad \text{Substitute } y + 2 \text{ for } x \text{ since } x = y + 2.$$

Notice that this equation now has one variable, y. Let's now solve this equation for y.

Helpful Hint Don't forget the distributive property.

$$\begin{aligned}
2(y + 2) + y &= 10 \\
2y + 4 + y &= 10 \quad \text{Apply the distributive property.} \\
3y + 4 &= 10 \quad \text{Combine like terms.} \\
3y &= 6 \quad \text{Subtract 4 from both sides.} \\
y &= 2 \quad \text{Divide both sides by 3.}
\end{aligned}$$

Now we know that the y-value of the ordered pair solution of the system is 2. To find the corresponding x-value, we replace y with 2 in the second equation, $x = y + 2$, and solve for x.

$$\begin{aligned}
x &= y + 2 \quad \text{Second equation} \\
x &= 2 + 2 \quad \text{Let } y = 2. \\
x &= 4
\end{aligned}$$

The solution of the system is the ordered pair $(4, 2)$. Since an ordered pair solution must satisfy both linear equations in the system, we could have chosen the equation $2x + y = 10$ to find the corresponding x-value. The resulting x-value is the same.

Check: We check to see that $(4, 2)$ satisfies both equations of the original system.

First Equation	Second Equation
$2x + y = 10$	$x = y + 2$
$2(4) + 2 \stackrel{?}{=} 10$	$4 \stackrel{?}{=} 2 + 2$ Let $x = 4$ and $y = 2$.
$10 = 10$ True	$4 = 4$ True

Answer

1. $(5, 1)$

The solution of the system is $(4, 2)$.

A graph of the two equations shows the two lines intersecting at the point $(4, 2)$.

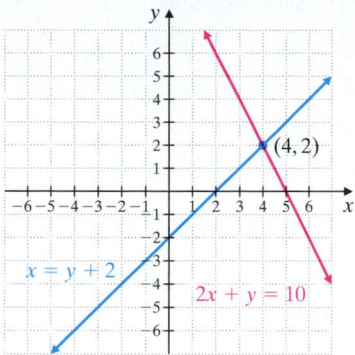

■ **Work Practice 1**

Example 2 Solve the system:

$$\begin{cases} 5x - y = -2 \\ y = 3x \end{cases}$$

Solution: The second equation is solved for y in terms of x. We substitute $3x$ for y in the first equation.

$$5x - y = -2 \quad \text{First equation}$$
$$5x - (3x) = -2 \quad \text{Substitute } 3x \text{ for } y.$$

Now we solve for x.

$$5x - 3x = -2$$
$$2x = -2 \quad \text{Combine like terms.}$$
$$x = -1 \quad \text{Divide both sides by 2.}$$

The x-value of the ordered pair solution is -1. To find the corresponding y-value, we replace x with -1 in the second equation, $y = 3x$.

$$y = 3x \quad \text{Second equation}$$
$$y = 3(-1) \quad \text{Let } x = -1.$$
$$y = -3$$

Check to see that the solution of the system is $(-1, -3)$.

■ **Work Practice 2**

   To solve a system of equations by substitution, we first need an equation solved for one of its variables, as in Examples 1 and 2. If neither equation in a system is solved for x or y, this will be our first step.

Example 3 Solve the system:

$$\begin{cases} x + 2y = 7 \\ 2x + 2y = 13 \end{cases}$$

Solution: Notice that neither equation is solved for x or y. Thus, we choose one of the equations and solve for x or y. We will solve the first equation for x so that we will not introduce tedious fractions when solving. To solve the first equation for x, we subtract $2y$ from both sides.

$$x + 2y = 7 \quad \text{First equation}$$
$$x = 7 - 2y \quad \text{Subtract } 2y \text{ from both sides.}$$

(Continued on next page)

Practice 2

Use the substitution method to solve the system:

$$\begin{cases} 4x - y = 2 \\ y = 5x \end{cases}$$

Practice 3

Solve the system:

$$\begin{cases} 3x + y = 5 \\ 3x - 2y = -7 \end{cases}$$

Answers

2. $(-2, -10)$ **3.** $\left(\dfrac{1}{3}, 4\right)$

Don't forget to insert parentheses when substituting $7 - 2y$ for x.

Since $x = 7 - 2y$, we now substitute $7 - 2y$ for x in the second equation and solve for y.

$$2x + 2y = 13 \qquad \text{Second equation}$$
$$2(7 - 2y) + 2y = 13 \qquad \text{Let } x = 7 - 2y.$$
$$14 - 4y + 2y = 13 \qquad \text{Apply the distributive property.}$$
$$14 - 2y = 13 \qquad \text{Simplify.}$$
$$-2y = -1 \qquad \text{Subtract 14 from both sides.}$$
$$y = \frac{1}{2} \qquad \text{Divide both sides by } -2.$$

To find x, any equation in two variables equivalent to the original equations of the system may be used. We used this equation since it is solved for x.

To find x, we let $y = \dfrac{1}{2}$ in the equation $x = 7 - 2y$.

$$x = 7 - 2y$$
$$x = 7 - 2\left(\frac{1}{2}\right) \qquad \text{Let } y = \frac{1}{2}.$$
$$x = 7 - 1$$
$$x = 6$$

Check the solution in both equations of the original system. The solution is $\left(6, \dfrac{1}{2}\right)$.

◼ **Work Practice 3**

The following steps summarize how to solve a system of equations by the substitution method.

> **To Solve a System of Two Linear Equations by the Substitution Method**
>
> **Step 1:** Solve one of the equations for one of its variables.
>
> **Step 2:** Substitute the expression for the variable found in Step 1 into the other equation.
>
> **Step 3:** Solve the equation from Step 2 to find the value of one variable.
>
> **Step 4:** Substitute the value found in Step 3 into any equation containing both variables to find the value of the other variable.
>
> **Step 5:** Check the proposed solution in the original system.

✓**Concept Check** As you solve the system

$$\begin{cases} 2x + y = -5 \\ x - y = 5 \end{cases}$$

you find that $y = -5$. Is this the solution of the system?

Practice 4

Solve the system:
$$\begin{cases} 5x - 2y = 6 \\ -3x + y = -3 \end{cases}$$

Answer

4. $(0, -3)$

✓**Concept Check Answer**

no, the solution will be an ordered pair

Example 4 Solve the system:

$$\begin{cases} 7x - 3y = -14 \\ -3x + y = 6 \end{cases}$$

Solution: Since the coefficient of y is 1 in the second equation, we will solve the second equation for y. This way, we avoid introducing tedious fractions.

$$-3x + y = 6 \qquad \text{Second equation}$$
$$y = 3x + 6$$

Next, we substitute $3x + 6$ for y in the first equation.

$$7x - 3y = -14 \quad \text{First equation}$$
$$7x - 3(3x + 6) = -14 \quad \text{Let } y = 3x + 6.$$
$$7x - 9x - 18 = -14 \quad \text{Use the distributive property.}$$
$$-2x - 18 = -14 \quad \text{Simplify.}$$
$$-2x = 4 \quad \text{Add 18 to both sides.}$$
$$x = -2 \quad \text{Divide both sides by } -2.$$

To find the corresponding y-value, we substitute -2 for x in the equation $y = 3x + 6$. Then $y = 3(-2) + 6$ or $y = 0$. The solution of the system is $(-2, 0)$. Check this solution in both equations of the system.

■ **Work Practice 4**

✓**Concept Check** To avoid fractions, which of the equations below would you use to solve for x?

a. $3x - 4y = 15$ **b.** $14 - 3y = 8x$ **c.** $7y + x = 12$

Helpful Hint

When solving a system of equations by the substitution method, begin by solving an equation for one of its variables. If possible, solve for a variable that has a coefficient of 1 or -1 to avoid working with time-consuming fractions.

Example 5 Solve the system:
$$\begin{cases} \dfrac{1}{2}x - y = 3 \\ x = 6 + 2y \end{cases}$$

Solution: The second equation is already solved for x in terms of y. Thus we substitute $6 + 2y$ for x in the first equation and solve for y.

$$\frac{1}{2}x - y = 3 \quad \text{First equation}$$
$$\frac{1}{2}(6 + 2y) - y = 3 \quad \text{Let } x = 6 + 2y.$$
$$3 + y - y = 3 \quad \text{Apply the distributive property.}$$
$$3 = 3 \quad \text{Simplify.}$$

Arriving at a true statement such as $3 = 3$ indicates that the two linear equations in the original system are equivalent. This means that their graphs are identical, as shown in the figure. There is an infinite number of solutions to the system, and any solution of one equation is also a solution of the other.

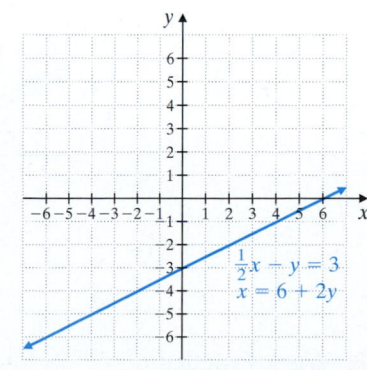

$\frac{1}{2}x - y = 3$
$x = 6 + 2y$

■ **Work Practice 5**

Practice 5
Solve the system:
$$\begin{cases} -x + 3y = 6 \\ y = \dfrac{1}{3}x + 2 \end{cases}$$

Answer
5. infinite number of solutions

✓**Concept Check Answer**
c

Practice 6

Solve the system:

$$\begin{cases} 2x - 3y = 6 \\ -4x + 6y = 12 \end{cases}$$

Example 6   Solve the system:

$$\begin{cases} 6x + 12y = 5 \\ -4x - 8y = 0 \end{cases}$$

Solution: We choose the second equation and solve for y. (*Note:* Although you might not see this beforehand, if you solve the second equation for x, the result is $x = -2y$ and no fractions are introduced. Either way will lead to the correct solution.)

$$-4x - 8y = 0 \qquad \text{Second equation}$$
$$-8y = 4x \qquad \text{Add } 4x \text{ to both sides.}$$
$$\frac{-8y}{-8} = \frac{4x}{-8} \qquad \text{Divide both sides by } -8.$$
$$y = -\frac{1}{2}x \qquad \text{Simplify.}$$

Now we replace y with $-\dfrac{1}{2}x$ in the first equation.

$$6x + 12y = 5 \qquad \text{First equation}$$
$$6x + 12\left(-\frac{1}{2}x\right) = 5 \qquad \text{Let } y = -\frac{1}{2}x.$$
$$6x + (-6x) = 5 \qquad \text{Simplify.}$$
$$0 = 5 \qquad \text{Combine like terms.}$$

The false statement $0 = 5$ indicates that this system has no solution. The graph of the linear equations in the system is a pair of parallel lines, as shown in the figure.

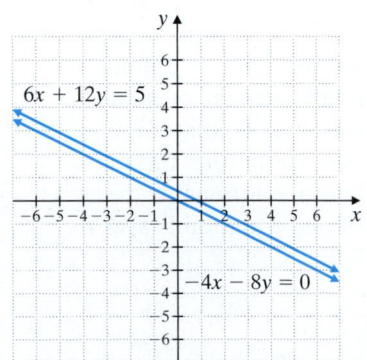

Work Practice 6

✔**Concept Check**   Describe how the graphs of the equations in a system appear if the system has

    a. no solution   **b.** one solution   **c.** an infinite number of solutions

Vocabulary, Readiness & Video Check

Give the solution of each system. If the system has no solution or an infinite number of solutions, say so. If the system has one solution, find it.

1. $\begin{cases} y = 4x \\ -3x + y = 1 \end{cases}$

When solving, you obtain $x = 1$.

2. $\begin{cases} 4x - y = 17 \\ -8x + 2y = 0 \end{cases}$

When solving, you obtain $0 = 34$.

3. $\begin{cases} 4x - y = 17 \\ -8x + 2y = -34 \end{cases}$

When solving, you obtain $0 = 0$.

4. $\begin{cases} 5x + 2y = 25 \\ x = y + 5 \end{cases}$

When solving, you obtain $y = 0$.

5. $\begin{cases} x + y = 0 \\ 7x - 7y = 0 \end{cases}$

When solving, you obtain $x = 0$.

6. $\begin{cases} y = -2x + 5 \\ 4x + 2y = 10 \end{cases}$

When solving, you obtain $0 = 0$.

Martin-Gay Interactive Videos Watch the section lecture video and answer the following question.

See Video 11.2

Objective A 7. The systems in ▢ Examples 2–4 all need one of their equations solved for a variable as a first step. What important part of the substitution method is emphasized in each example? ◉

11.2 Exercise Set MyMathLab®

Objective A *Solve each system of equations by the substitution method. See Examples 1 and 2.*

1. $\begin{cases} x + y = 3 \\ x = 2y \end{cases}$

2. $\begin{cases} x + y = 20 \\ x = 3y \end{cases}$

▷ 3. $\begin{cases} x + y = 6 \\ y = -3x \end{cases}$

4. $\begin{cases} x + y = 6 \\ y = -4x \end{cases}$

5. $\begin{cases} y = 3x + 1 \\ 4y - 8x = 12 \end{cases}$

6. $\begin{cases} y = 2x + 3 \\ 5y - 7x = 18 \end{cases}$

7. $\begin{cases} y = 2x + 9 \\ y = 7x + 10 \end{cases}$

8. $\begin{cases} y = 5x - 3 \\ y = 8x + 4 \end{cases}$

Solve each system of equations by the substitution method. See Examples 1 through 6.

9. $\begin{cases} 3x - 4y = 10 \\ y = x - 3 \end{cases}$

10. $\begin{cases} 4x - 3y = 10 \\ y = x - 5 \end{cases}$

11. $\begin{cases} x + 2y = 6 \\ 2x + 3y = 8 \end{cases}$

12. $\begin{cases} x + 3y = -5 \\ 2x + 2y = 6 \end{cases}$

13. $\begin{cases} 3x + 2y = 16 \\ x = 3y - 2 \end{cases}$

14. $\begin{cases} 2x + 3y = 18 \\ x = 2y - 5 \end{cases}$

15. $\begin{cases} 2x - 5y = 1 \\ 3x + y = -7 \end{cases}$

16. $\begin{cases} 3y - x = 6 \\ 4x + 12y = 0 \end{cases}$

17. $\begin{cases} 4x + 2y = 5 \\ -2x = y + 4 \end{cases}$

18. $\begin{cases} 2y = x + 2 \\ 6x - 12y = 0 \end{cases}$

19. $\begin{cases} 4x + y = 11 \\ 2x + 5y = 1 \end{cases}$

20. $\begin{cases} 3x + y = -14 \\ 4x + 3y = -22 \end{cases}$

21. $\begin{cases} x + 2y + 5 = -4 + 5y - x \\ 2x + x = y + 4 \end{cases}$

(*Hint:* First simplify each equation.)

22. $\begin{cases} 5x + 4y - 2 = -6 + 7y - 3x \\ 3x + 4x = y + 3 \end{cases}$

(*Hint:* See Exercise **21**.)

23. $\begin{cases} 6x - 3y = 5 \\ x + 2y = 0 \end{cases}$

24. $\begin{cases} 10x - 5y = -21 \\ x + 3y = 0 \end{cases}$

▶ 25. $\begin{cases} 3x - y = 1 \\ 2x - 3y = 10 \end{cases}$

26. $\begin{cases} 2x - y = -7 \\ 4x - 3y = -11 \end{cases}$

27. $\begin{cases} -x + 2y = 10 \\ -2x + 3y = 18 \end{cases}$

28. $\begin{cases} -x + 3y = 18 \\ -3x + 2y = 19 \end{cases}$

29. $\begin{cases} 5x + 10y = 20 \\ 2x + 6y = 10 \end{cases}$

30. $\begin{cases} 6x + 3y = 12 \\ 9x + 6y = 15 \end{cases}$

▶ 31. $\begin{cases} 3x + 6y = 9 \\ 4x + 8y = 16 \end{cases}$

32. $\begin{cases} 2x + 4y = 6 \\ 5x + 10y = 16 \end{cases}$

▶ 33. $\begin{cases} \dfrac{1}{3}x - y = 2 \\ x - 3y = 6 \end{cases}$

34. $\begin{cases} \dfrac{1}{4}x - 2y = 1 \\ x - 8y = 4 \end{cases}$

35. $\begin{cases} x = \dfrac{3}{4}y - 1 \\ 8x - 5y = -6 \end{cases}$

36. $\begin{cases} x = \dfrac{5}{6}y - 2 \\ 12x - 5y = -9 \end{cases}$

Review

Write equivalent equations by multiplying both sides of each given equation by the given nonzero number. See Section 9.2.

37. $3x + 2y = 6$ by -2

38. $-x + y = 10$ by 5

39. $-4x + y = 3$ by 3

40. $5a - 7b = -4$ by -4

Add the binomials. See Section 8.7.

41. $\begin{array}{r} 3n + 6m \\ 2n - 6m \\ \hline \end{array}$

42. $\begin{array}{r} -2x + 5y \\ 2x + 11y \\ \hline \end{array}$

43. $\begin{array}{r} -5a - 7b \\ 5a - 8b \\ \hline \end{array}$

44. $\begin{array}{r} 9q + p \\ -9q - p \\ \hline \end{array}$

Concept Extensions

Solve each system by the substitution method. First simplify each equation by combining like terms.

45. $\begin{cases} -5y + 6y = 3x + 2(x - 5) - 3x + 5 \\ 4(x + y) - x + y = -12 \end{cases}$

46. $\begin{cases} 5x + 2y - 4x - 2y = 2(2y + 6) - 7 \\ 3(2x - y) - 4x = 1 + 9 \end{cases}$

47. Explain how to identify a system with no solution when using the substitution method.

48. Occasionally, when using the substitution method, we obtain the equation $0 = 0$. Explain how this result indicates that the graphs of the equations in the system are identical.

Solve. See a Concept Check in this section.

49. As you solve the system $\begin{cases} 3x - y = -6 \\ -3x + 2y = 7 \end{cases}$, you find that $y = 1$. Is this the solution of the system?

50. As you solve the system $\begin{cases} x = 5y \\ y = 2x \end{cases}$, you find that $x = 0$ and $y = 0$. What is the solution of this system?

51. To avoid fractions, which of the equations below would you use if solving for y? Explain why.

a. $\frac{1}{2}x - 4y = \frac{3}{4}$ b. $8x - 5y = 13$ c. $7x - y = 19$

52. Give the number of solutions for a system if the graphs of the equations in the system are

a. lines intersecting in one point
b. parallel lines
c. same line

Use a graphing calculator to solve each system.

53. $\begin{cases} y = 5.1x + 14.56 \\ y = -2x - 3.9 \end{cases}$

54. $\begin{cases} y = 3.1x - 16.35 \\ y = -9.7x + 28.45 \end{cases}$

55. $\begin{cases} 3x + 2y = 14.04 \\ 5x + y = 18.5 \end{cases}$

56. $\begin{cases} x + y = -15.2 \\ -2x + 5y = -19.3 \end{cases}$

Solve.

57. U.S. consumer spending y (in billions of dollars) on DVD-format home entertainment from 2006 to 2010 is given by $y = -1.6x + 20.9$, where x is the number of years after 2006. U.S. consumer spending y (in billions of dollars) on Blu-ray-format home entertainment from 2006 to 2010 is given by $y = 0.6x - 0.2$, where x is the number of years after 2006. (*Source:* Based on data from DEG: The Digital Entertainment Group)

a. Use the substitution method to solve this system of equations.

$$\begin{cases} y = -1.6x + 20.9 \\ y = 0.6x - 0.2 \end{cases}$$

Round x to the nearest tenth and y to the nearest whole.

b. Explain the meaning of your answer to part **a**.

c. Sketch a graph of the system of equations. Write a sentence describing the trends in the popularity of these two types of home entertainment formats.

58. Promoting tourism in a state can significantly boost jobs and tax revenue for that state. For that reason, most states operate a tourism office to attract travelers and their spending. For fiscal years 2008 through 2012, Maryland's state tourism office budget y (in millions of dollars) is given by the equation $y = 0.61x + 7.62$, where x is the number of years after fiscal year 2008. For the same period, Georgia's state tourism office budget y (in millions of dollars) is given by the equation $y = -0.74x + 9.32$, where x is the number of years after fiscal year 2008. (*Source:* Based on data from the U.S. Travel Association)

a. Use the substitution method to solve this system of equations.

$$\begin{cases} y = 0.61x + 7.62 \\ y = -0.74x + 9.32 \end{cases}$$

Round x and y to the nearest tenth.

b. Explain the meaning of your answer to part **a**.

c. Sketch a graph of the system of equations. Write a sentence describing the trends in Maryland's and Georgia's state tourism office budgets between fiscal years 2008 and 2012.

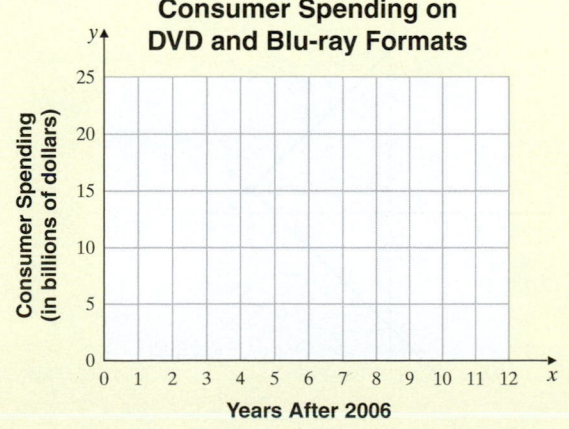

Consumer Spending on DVD and Blu-ray Formats

Consumer Spending (in billions of dollars) vs. Years After 2006

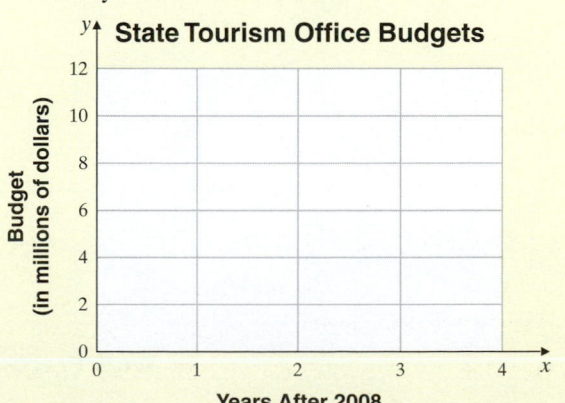

State Tourism Office Budgets

Budget (in millions of dollars) vs. Years After 2008

Solving Systems of Linear Equations by Addition

A Use the Addition Method to Solve a System of Linear Equations.

Objective A Using the Addition Method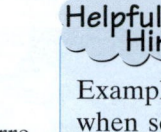

We have seen that substitution is an accurate method for solving a system of linear equations. Another accurate method is the **addition** or **elimination method.** The addition method is based on the addition property of equality: Adding equal quantities to both sides of an equation does not change the solution of the equation. In symbols,

if $A = B$ and $C = D$, then $A + C = B + D$

To see how we use this to solve a system of equations, study Example 1.

Practice 1

Use the addition method to solve the system:

$$\begin{cases} x + y = 13 \\ x - y = 5 \end{cases}$$

Example 1 Solve the system: $\begin{cases} x + y = 7 \\ x - y = 5 \end{cases}$

Solution: Since the left side of each equation is equal to its right side, we are adding equal quantities when we add the left sides of the equations together and add the right sides of the equations together. This adding eliminates the variable y and gives us an equation in one variable, x. We can then solve for x.

$$\begin{aligned} x + y &= 7 \quad \text{First equation} \\ \underline{x - y} &= \underline{5} \quad \text{Second equation} \\ 2x &= 12 \quad \text{Add the equations to eliminate } y. \\ x &= 6 \quad \text{Divide both sides by 2.} \end{aligned}$$

The x-value of the solution is 6. To find the corresponding y-value, we let $x = 6$ in either equation of the system. We will use the first equation.

$$\begin{aligned} x + y &= 7 \quad \text{First equation} \\ 6 + y &= 7 \quad \text{Let } x = 6. \\ y &= 1 \quad \text{Solve for } y. \end{aligned}$$

Helpful Hint Notice in Example 1 that our goal when solving a system of equations by the addition method is to eliminate a variable when adding the equations.

The solution is $(6, 1)$.

Check: Check the solution in both equations of the original system.

First Equation	**Second Equation**
$x + y = 7$	$x - y = 5$
$6 + 1 \stackrel{?}{=} 7$ Let $x = 6$ and $y = 1$.	$6 - 1 \stackrel{?}{=} 5$ Let $x = 6$ and $y = 1$.
$7 = 7$ True	$5 = 5$ True

Thus, the solution of the system is $(6, 1)$.

If we graph the two equations in the system, we have two lines that intersect at the point $(6, 1)$, as shown.

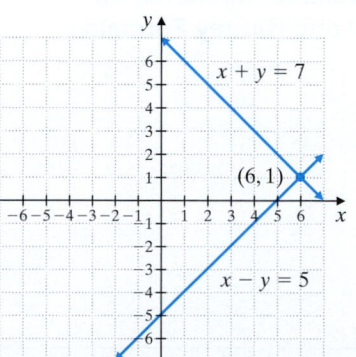

■ **Work Practice 1**

Example 2 Solve the system: $\begin{cases} -2x + y = 2 \\ -x + 3y = -4 \end{cases}$

Solution: If we simply add these two equations, the result is still an equation in two variables. However, from Example 1, remember that our goal is to eliminate one of the variables so that we have an equation in the other variable. To do this, notice what happens if we multiply *both sides* of the first equation by -3. We are allowed to do this by the multiplication property of equality. Then the system

$$\begin{cases} -3(-2x + y) = -3(2) \\ -x + 3y = -4 \end{cases} \quad \text{simplifies to} \quad \begin{cases} 6x - 3y = -6 \\ -x + 3y = -4 \end{cases}$$

When we add the resulting equations, the y-variable is eliminated.

$$\begin{array}{rl} 6x - 3y &= -6 \\ \underline{-x + 3y} &= \underline{-4} \\ 5x &= -10 \quad \text{Add.} \\ x &= -2 \quad \text{Divide both sides by 5.} \end{array}$$

To find the corresponding y-value, we let $x = -2$ in either of the original equations. We use the first equation of the original system.

$$\begin{array}{rl} -2x + y = 2 & \quad \text{First equation} \\ -2(-2) + y = 2 & \quad \text{Let } x = -2. \\ 4 + y = 2 & \\ y = -2 & \end{array}$$

Check the ordered pair $(-2, -2)$ in both equations of the *original* system. The solution is $(-2, -2)$.

■ **Work Practice 2**

Helpful Hint

When finding the second value of an ordered pair solution, any equation equivalent to one of the original equations in the system may be used.

In Example 2, the decision to multiply the first equation by -3 was no accident. **To eliminate a variable** when adding two equations, **the coefficient of the variable in one equation must be the opposite of its coefficient in the other equation.**

Helpful Hint

Be sure to multiply *both sides* of an equation by the chosen number when solving by the addition method. A common mistake is to multiply only the side containing the variables.

Example 3 Solve the system: $\begin{cases} 2x - y = 7 \\ 8x - 4y = 1 \end{cases}$

Solution: When we multiply both sides of the first equation by -4, the resulting coefficient of x is -8. This is the opposite of 8, the coefficient of x in the second equation. Then the system

$$\begin{cases} -4(2x - y) = -4(7) \\ 8x - 4y = 1 \end{cases} \quad \text{simplifies to}$$

Helpful Hint Don't forget to multiply both sides by -4.

$$\begin{array}{rl} \begin{cases} -8x + 4y = -28 \\ 8x - 4y = 1 \end{cases} \\ 0 = -27 \quad \text{Add the equations.} \end{array}$$

(Continued on next page)

Practice 2

Solve the system:
$\begin{cases} 2x - y = -6 \\ -x + 4y = 17 \end{cases}$

Practice 3

Solve the system:
$\begin{cases} x - 3y = -2 \\ -3x + 9y = 5 \end{cases}$

Answers

2. $(-1, 4)$ **3.** no solution

When we add the equations, both variables are eliminated and we have $0 = -27$, a false statement. This means that the system has no solution. The equations, if graphed, would represent parallel lines.

■ Work Practice 3

Example 4 Solve the system: $\begin{cases} 3x - 2y = 2 \\ -9x + 6y = -6 \end{cases}$

Solution: First we multiply both sides of the first equation by 3 and then we add the resulting equations.

$$\begin{cases} 3(3x - 2y) = 3(2) \\ -9x + 6y = -6 \end{cases} \text{ simplifies to } \begin{cases} 9x - 6y = 6 \\ -9x + 6y = -6 \quad \text{Add the equations.} \\ \hline 0 = 0 \end{cases}$$

Both variables are eliminated and we have $0 = 0$, a true statement. This means that the system has an infinite number of solutions. The equations, if graphed, would be the same line.

■ Work Practice 4

✔**Concept Check** Suppose you are solving the system

$$\begin{cases} 3x + 8y = -5 \\ 2x - 4y = 3 \end{cases}$$

You decide to use the addition method by multiplying both sides of the second equation by 2. In which of the following was the multiplication performed correctly? Explain.

a. $4x - 8y = 3$ **b.** $4x - 8y = 6$

In the next example, we multiply both equations by numbers so that coefficients of a variable are opposites.

Example 5 Solve the system: $\begin{cases} 3x + 4y = 13 \\ 5x - 9y = 6 \end{cases}$

Solution: We can eliminate the variable y by multiplying the first equation by 9 and the second equation by 4. Then we add the resulting equations.

$$\begin{cases} 9(3x + 4y) = 9(13) \\ 4(5x - 9y) = 4(6) \end{cases} \text{ simplifies to } \begin{cases} 27x + 36y = 117 \\ 20x - 36y = 24 \\ \hline 47x = 141 \quad \text{Add the equations.} \\ x = 3 \quad \text{Solve for } x. \end{cases}$$

To find the corresponding y-value, we let $x = 3$ in one of the original equations of the system. Doing so in either of these equations will give $y = 1$. Check to see that $(3, 1)$ satisfies each equation in the original system. The solution is $(3, 1)$.

■ Work Practice 5

If we had decided to eliminate x instead of y in Example 5, the first equation could have been multiplied by 5 and the second by -3. Try solving the original system this way to check that the solution is $(3, 1)$.

The following steps summarize how to solve a system of linear equations by the addition method.

Practice 4

Solve the system:

$$\begin{cases} 2x + 5y = 1 \\ -4x - 10y = -2 \end{cases}$$

Practice 5

Solve the system:

$$\begin{cases} 4x + 5y = 14 \\ 3x - 2y = -1 \end{cases}$$

Answers

4. infinite number of solutions

5. $(1, 2)$

✔**Concept Check Answer**

b; answers may vary

> ## To Solve a System of Two Linear Equations by the Addition Method
>
> **Step 1:** Rewrite each equation in standard form, $Ax + By = C$.
>
> **Step 2:** If necessary, multiply one or both equations by a nonzero number so that the coefficients of the chosen variable in the system are opposites.
>
> **Step 3:** Add the equations.
>
> **Step 4:** Find the value of one variable by solving the resulting equation from Step 3.
>
> **Step 5:** Find the value of the second variable by substituting the value found in Step 4 into either of the original equations.
>
> **Step 6:** Check the proposed solution in the original system.

✔**Concept Check** Suppose you are solving the system

$$\begin{cases} -4x + 7y = 6 \\ x + 2y = 5 \end{cases}$$

by the addition method.

a. What step(s) should you take if you wish to eliminate x when adding the equations?

b. What step(s) should you take if you wish to eliminate y when adding the equations?

Example 6 Solve the system: $\begin{cases} -x - \dfrac{y}{2} = \dfrac{5}{2} \\ \dfrac{x}{6} - \dfrac{y}{2} = 0 \end{cases}$

Solution: We begin by clearing each equation of fractions. To do so, we multiply both sides of the first equation by the LCD, 2, and both sides of the second equation by the LCD, 6. Then the system

$$\begin{cases} 2\left(-x - \dfrac{y}{2}\right) = 2\left(\dfrac{5}{2}\right) \\ 6\left(\dfrac{x}{6} - \dfrac{y}{2}\right) = 6(0) \end{cases} \quad \text{simplifies to} \quad \begin{cases} -2x - y = 5 \\ x - 3y = 0 \end{cases}$$

We can now eliminate the variable x by multiplying the second equation by 2.

$$\begin{cases} -2x - y = 5 \\ 2(x - 3y) = 2(0) \end{cases} \quad \text{simplifies to} \quad \begin{cases} -2x - y = 5 \\ 2x - 6y = 0 \end{cases}$$
$$\overline{{-7y} = 5} \quad \text{Add the equations.}$$
$$y = -\frac{5}{7} \quad \text{Solve for } y.$$

To find x, we could replace y with $-\dfrac{5}{7}$ in one of the equations with two variables.

Instead, let's go back to the simplified system and multiply by appropriate factors to eliminate the variable y and solve for x. To do this, we multiply the first equation by -3. Then the system

$$\begin{cases} -3(-2x - y) = -3(5) \\ x - 3y = 0 \end{cases} \quad \text{simplifies to} \quad \begin{cases} 6x + 3y = -15 \\ x - 3y = 0 \end{cases}$$
$$\overline{7x = -15} \quad \text{Add the equations.}$$
$$x = -\frac{15}{7} \quad \text{Solve for } x.$$

Check the ordered pair $\left(-\dfrac{15}{7}, -\dfrac{5}{7}\right)$ in both equations of the original system. The solution is $\left(-\dfrac{15}{7}, -\dfrac{5}{7}\right)$.

■ **Work Practice 6**

Practice 6

Solve the system:

$$\begin{cases} -\dfrac{x}{3} + y = \dfrac{4}{3} \\ \dfrac{x}{2} - \dfrac{5}{2}y = -\dfrac{1}{2} \end{cases}$$

Answer

6. $\left(-\dfrac{17}{2}, -\dfrac{3}{2}\right)$

✔**Concept Check Answers**

a. multiply the second equation by 4

b. possible answer: multiply the first equation by -2 and the second equation by 7

Vocabulary, Readiness & Video Check

Given the system $\begin{cases} 3x - 2y = -9 \\ x + 5y = 14 \end{cases}$ read each row (Step 1, Step 2, and Result). Then answer whether the result is true or false.

	Step 1	Step 2	Result	True or False?
1.	Multiply 2nd equation through by −3.	Add the resulting equation to the 1st equation.	The y's are eliminated.	
2.	Multiply 2nd equation through by −3.	Add the resulting equation to the 1st equation.	The x's are eliminated.	
3.	Multiply 1st equation by 5 and 2nd equation by 2.	Add the two new equations.	The y's are eliminated.	
4.	Multiply 1st equation by 5 and 2nd equation by −2.	Add the two new equations.	The y's are eliminated.	

Martin-Gay Interactive Videos

See Video 11.3 🍎

Watch the section lecture video and answer the following question.

Objective A 5. For the addition/elimination method, sometimes we need to multiply an equation through by a nonzero number so that the coefficients of a variable are opposites, as is shown in Example 2. What property allows us to do this? What important reminder is made at this step? ▶

11.3 Exercise Set MyMathLab®

Objective A *Solve each system of equations by the addition method. See Example 1.*

1. $\begin{cases} 3x + y = 5 \\ 6x - y = 4 \end{cases}$

2. $\begin{cases} 4x + y = 13 \\ 2x - y = 5 \end{cases}$

▶ **3.** $\begin{cases} x - 2y = 8 \\ -x + 5y = -17 \end{cases}$

4. $\begin{cases} x - 2y = -11 \\ -x + 5y = 23 \end{cases}$

Solve each system of equations by the addition method. If a system contains fractions or decimals, you may want to first clear each equation of the fractions or decimals. See Examples 2 through 6.

5. $\begin{cases} 3x + y = -11 \\ 6x - 2y = -2 \end{cases}$

6. $\begin{cases} 4x + y = -13 \\ 6x - 3y = -15 \end{cases}$

7. $\begin{cases} 3x + 2y = 11 \\ 5x - 2y = 29 \end{cases}$

8. $\begin{cases} 4x + 2y = 2 \\ 3x - 2y = 12 \end{cases}$

9. $\begin{cases} x + 5y = 18 \\ 3x + 2y = -11 \end{cases}$

10. $\begin{cases} x + 4y = 14 \\ 5x + 3y = 2 \end{cases}$

11. $\begin{cases} x + y = 6 \\ x - y = 6 \end{cases}$

12. $\begin{cases} x - y = 1 \\ -x + 2y = 0 \end{cases}$

13. $\begin{cases} 2x + 3y = 0 \\ 4x + 6y = 3 \end{cases}$

14. $\begin{cases} 3x + y = 4 \\ 9x + 3y = 6 \end{cases}$

15. $\begin{cases} -x + 5y = -1 \\ 3x - 15y = 3 \end{cases}$

16. $\begin{cases} 2x + y = 6 \\ 4x + 2y = 12 \end{cases}$

▶ **17.** $\begin{cases} 3x - 2y = 7 \\ 5x + 4y = 8 \end{cases}$

18. $\begin{cases} 6x - 5y = 25 \\ 4x + 15y = 13 \end{cases}$

19. $\begin{cases} 8x = -11y - 16 \\ 2x + 3y = -4 \end{cases}$

20. $\begin{cases} 10x + 3y = -12 \\ 5x = -4y - 16 \end{cases}$

21. $\begin{cases} 4x - 3y = 7 \\ 7x + 5y = 2 \end{cases}$

22. $\begin{cases} -2x + 3y = 10 \\ 3x + 4y = 2 \end{cases}$

23. $\begin{cases} 4x - 6y = 8 \\ 6x - 9y = 12 \end{cases}$

24. $\begin{cases} 9x - 3y = 12 \\ 12x - 4y = 18 \end{cases}$

25. $\begin{cases} 2x - 5y = 4 \\ 3x - 2y = 4 \end{cases}$

26. $\begin{cases} 6x - 5y = 7 \\ 4x - 6y = 7 \end{cases}$

27. $\begin{cases} \dfrac{x}{3} + \dfrac{y}{6} = 1 \\ \dfrac{x}{2} - \dfrac{y}{4} = 0 \end{cases}$

28. $\begin{cases} \dfrac{x}{2} + \dfrac{y}{8} = 3 \\ x - \dfrac{y}{4} = 0 \end{cases}$

29. $\begin{cases} \dfrac{10}{3}x + 4y = -4 \\ 5x + 6y = -6 \end{cases}$

30. $\begin{cases} \dfrac{3}{2}x + 4y = 1 \\ 9x + 24y = 5 \end{cases}$

31. $\begin{cases} x - \dfrac{y}{3} = -1 \\ -\dfrac{x}{2} + \dfrac{y}{8} = \dfrac{1}{4} \end{cases}$

32. $\begin{cases} 2x - \dfrac{3y}{4} = -3 \\ x + \dfrac{y}{9} = \dfrac{13}{3} \end{cases}$

33. $\begin{aligned} -4(x + 2) &= 3y \\ 2x - 2y &= 3 \end{aligned}$

34. $\begin{aligned} -9(x + 3) &= 8y \\ 3x - 3y &= 8 \end{aligned}$

⊙ 35. $\begin{cases} \dfrac{x}{3} - y = 2 \\ -\dfrac{x}{2} + \dfrac{3y}{2} = -3 \end{cases}$

36. $\begin{cases} \dfrac{x}{2} + \dfrac{y}{4} = 1 \\ -\dfrac{x}{4} - \dfrac{y}{8} = 1 \end{cases}$

37. $\begin{cases} \dfrac{3}{5}x - y = -\dfrac{4}{5} \\ 3x + \dfrac{y}{2} = -\dfrac{9}{5} \end{cases}$

38. $\begin{cases} 3x + \dfrac{7}{2}y = \dfrac{3}{4} \\ -\dfrac{x}{2} + \dfrac{5}{3}y = -\dfrac{5}{4} \end{cases}$

⊙ 39. $\begin{cases} 3.5x + 2.5y = 17 \\ -1.5x - 7.5y = -33 \end{cases}$

40. $\begin{cases} -2.5x - 6.5y = 47 \\ 0.5x - 4.5y = 37 \end{cases}$

41. $\begin{cases} 0.02x + 0.04y = 0.09 \\ -0.1x + 0.3y = 0.8 \end{cases}$

42. $\begin{cases} 0.04x - 0.05y = 0.105 \\ 0.2x - 0.6y = 1.05 \end{cases}$

Review

Translating *Rewrite each sentence using mathematical symbols. Do not solve the equations. See Sections 8.2, 9.4, and 9.5.*

43. Twice a number, added to 6, is 3 less than the number.

44. The sum of three consecutive integers is 66.

45. Three times a number, subtracted from 20, is 2.

46. Twice the sum of 8 and a number is the difference of the number and 20.

47. The product of 4 and the sum of a number and 6 is twice the number.

48. If the quotient of twice a number and 7 is subtracted from the reciprocal of the number, the result is 2.

Concept Extensions

Solve. See a Concept Check in this section.

49. To solve this system by the addition method and eliminate the variable y,

$$\begin{cases} 4x + 2y = -7 \\ 3x - y = -12 \end{cases}$$

by what value would you multiply the second equation? What do you get when you complete the multiplication?

Given the system of linear equations $\begin{cases} 3x - y = -8 \\ 5x + 3y = 2 \end{cases}$:

50. Use the addition method and
   a. Solve the system by eliminating x.
   b. Solve the system by eliminating y.

Solve.

51. Suppose you are solving the system

$$\begin{cases} 3x + 8y = -5 \\ 2x - 4y = 3 \end{cases}$$

You decide to use the addition method by multiplying both sides of the second equation by 2. In which of the following was the multiplication performed correctly? Explain.
   a. $4x - 8y = 3$
   b. $4x - 8y = 6$

52. Suppose you are solving the system

$$\begin{cases} -2x - y = 0 \\ -2x + 3y = 6 \end{cases}$$

You decide to use the addition method by multiplying both sides of the first equation by 3, then adding the resulting equation to the second equation. Which of the following is the correct sum? Explain.
   a. $-8x = 6$
   b. $-8x = 9$

53. When solving a system of equations by the addition method, how do we know when the system has no solution?

54. Explain why the addition method might be preferred over the substitution method for solving the system $\begin{cases} 2x - 3y = 5 \\ 5x + 2y = 6. \end{cases}$

55. Use the system of linear equations below to answer the questions.

$$\begin{cases} x + y = 5 \\ 3x + 3y = b \end{cases}$$

   a. Find the value of b so that the system has an infinite number of solutions.
   b. Find a value of b so that there are no solutions to the system.

56. Use the system of linear equations below to answer the questions.

$$\begin{cases} x + y = 4 \\ 2x + by = 8 \end{cases}$$

   a. Find the value of b so that the system has an infinite number of solutions.
   b. Find a value of b so that the system has a single solution.

Solve each system by the addition method.

57. $\begin{cases} 2x + 3y = 14 \\ 3x - 4y = -69.1 \end{cases}$

58. $\begin{cases} 5x - 2y = -19.8 \\ -3x + 5y = -3.7 \end{cases}$

59. As the economy and job marketplace change, demand for certain types of workers changes. The number of jobs for postal service mail carriers that is predicted for 2010 through 2020 can be approximated by $38x + 10y = 3167$. The number of jobs for market research analysts that is predicted for the same period can be approximated by $117x - 10y = -2827$. For both equations, x is the number of years since 2010, and y is the number of jobs in the thousands. (*Source:* Based on data from the U.S. Bureau of Labor Statistics)

a. Use the addition method to solve this system of equations.

$$\begin{cases} 38x + 10y = 3167 \\ 117x - 10y = -2827 \end{cases}$$

(Eliminate y first and solve for x. Round this result to the nearest whole number.)

b. Interpret your solution from part **a**.

c. Using the year in your answer to part **b**, estimate the number of mail carrier jobs and market research analyst jobs in that year.

60. In recent years, the number of newspapers printed as morning editions has been increasing and the number of newspapers printed as evening editions has been decreasing. The number y of daily morning newspapers in existence from 1995 through 2011 is approximated by the equation $153x - 10y = -6720$, where x is the number of years since 1995. The number y of daily evening newspapers in existence from 1995 through 2011 is approximated by $125x + 5y = 4350$, where x is the number of years since 1995. (*Source:* Based on data from Newspaper Association of America)

a. Use the addition method to solve this system of equations.

$$\begin{cases} 153x - 10y = -6720 \\ 125x + 5y = 4350 \end{cases}$$

(Round to the nearest whole number. Because of rounding, the y-value of your ordered pair solution may vary.)

b. Interpret your solution from part **a**.

c. Use the year in your answer to part **b** to find how many of each type of newspaper were in existence that year.

Summary on Solving Systems of Equations

Answers

Solve each system by either the addition method or the substitution method.

1. _____

2. _____

3. _____

4. _____

5. _____

6. _____

7. _____

8. _____

9. _____

10. _____

11. _____

12. _____

13. _____

14. _____

15. _____

16. _____

17. _____

18. _____

19. _____

20. _____

21. _____

22. _____

1. $\begin{cases} 2x - 3y = -11 \\ y = 4x - 3 \end{cases}$

2. $\begin{cases} 4x - 5y = 6 \\ y = 3x - 10 \end{cases}$

3. $\begin{cases} x + y = 3 \\ x - y = 7 \end{cases}$

4. $\begin{cases} x - y = 20 \\ x + y = -8 \end{cases}$

5. $\begin{cases} x + 2y = 1 \\ 3x + 4y = -1 \end{cases}$

6. $\begin{cases} x + 3y = 5 \\ 5x + 6y = -2 \end{cases}$

7. $\begin{cases} y = x + 3 \\ 3x = 2y - 6 \end{cases}$

8. $\begin{cases} y = -2x \\ 2x - 3y = -16 \end{cases}$

9. $\begin{cases} y = 2x - 3 \\ y = 5x - 18 \end{cases}$

10. $\begin{cases} y = 6x - 5 \\ y = 4x - 11 \end{cases}$

11. $\begin{cases} x + \dfrac{1}{6}y = \dfrac{1}{2} \\ 3x + 2y = 3 \end{cases}$

12. $\begin{cases} x + \dfrac{1}{3}y = \dfrac{5}{12} \\ 8x + 3y = 4 \end{cases}$

13. $\begin{cases} x - 5y = 1 \\ -2x + 10y = 3 \end{cases}$

14. $\begin{cases} -x + 2y = 3 \\ 3x - 6y = -9 \end{cases}$

15. $\begin{cases} 0.2x - 0.3y = -0.95 \\ 0.4x + 0.1y = 0.55 \end{cases}$

16. $\begin{cases} 0.08x - 0.04y = -0.11 \\ 0.02x - 0.06y = -0.09 \end{cases}$

17. $\begin{cases} x = 3y - 7 \\ 2x - 6y = -14 \end{cases}$

18. $\begin{cases} y = \dfrac{x}{2} - 3 \\ 2x - 4y = 0 \end{cases}$

19. $\begin{cases} 2x + 5y = -1 \\ 3x - 4y = 33 \end{cases}$

20. $\begin{cases} 7x - 3y = 2 \\ 6x + 5y = -21 \end{cases}$

21. Which method, substitution or addition, would you prefer to use to solve the system below? Explain your reasoning.
$$\begin{cases} 3x + 2y = -2 \\ y = -2x \end{cases}$$

22. Which method, substitution or addition, would you prefer to use to solve the system below? Explain your reasoning.
$$\begin{cases} 3x - 2y = -3 \\ 6x + 2y = 12 \end{cases}$$

11.4 Systems of Linear Equations and Problem Solving ▶

Objective A Using a System of Equations for Problem Solving ▶

Many of the word problems solved earlier with one-variable equations can also be solved with two equations in two variables. We use the same problem-solving steps that we have used throughout this text. The only difference is that two variables are assigned to represent the two unknown quantities and that the problem is translated into two equations.

Objective

A Use a System of Equations to Solve Problems.

Problem-Solving Steps

1. UNDERSTAND the problem. During this step, become comfortable with the problem. Some ways of doing this are to

   Read and reread the problem.

   Choose two variables to represent the two unknowns.

   Construct a drawing.

   Propose a solution and check. Pay careful attention to how you check your proposed solution. This will help when writing equations to model the problem.

2. TRANSLATE the problem into two equations.

3. SOLVE the system of equations.

4. INTERPRET the results: *Check* the proposed solution in the stated problem and *state* your conclusion.

Example 1 Finding Unknown Numbers

Find two numbers whose sum is 37 and whose difference is 21.

Solution:

1. UNDERSTAND. Read and reread the problem. Suppose that one number is 20. If their sum is 37, the other number is 17 because $20 + 17 = 37$. Is their difference 21? No; $20 - 17 = 3$. Our proposed solution is incorrect, but we now have a better understanding of the problem.

   Since we are looking for two numbers, we let

   x = first number and

   y = second number

2. TRANSLATE. Since we have assigned two variables to this problem, we translate our problem into two equations.

   In words: | two number whose sum | is | 37 |

   Translate: $x + y = 37$

   In words: | two number whose difference | is | 21 |

   Translate: $x - y = 21$

(*Continued on next page*)

Practice 1

Find two numbers whose sum is 50 and whose difference is 22.

Answer

1. 36 and 14

859

3. SOLVE. Now we solve the system.

$$\begin{cases} x + y = 37 \\ x - y = 21 \end{cases}$$

Notice that the coefficients of the variable y are opposites. Let's then solve by the addition method and begin by adding the equations.

$$\begin{array}{rl} x + y = 37 & \\ \underline{x - y = 21} & \text{Add the equations.} \\ 2x = 58 & \\ x = 29 & \text{Divide both sides by 2.} \end{array}$$

Now we let $x = 29$ in the first equation to find y.

$$\begin{array}{rl} x + y = 37 & \text{First equation} \\ 29 + y = 37 & \\ y = 8 & \text{Subtract 29 from both sides.} \end{array}$$

4. INTERPRET. The solution of the system is $(29, 8)$.

Check: Notice that the sum of 29 and 8 is $29 + 8 = 37$, the required sum. Their difference is $29 - 8 = 21$, the required difference.

State: The numbers are 29 and 8.

▪ **Work Practice 1**

Practice 2

Admission prices at a local weekend fair were $5 for children and $7 for adults. The total money collected was $3379, and 587 people attended the fair. How many children and how many adults attended the fair?

Example 2   Solving a Problem About Prices

The Cirque du Soleil show Varekai is performing locally. Matinee admission for 4 adults and 2 children is $374, while admission for 2 adults and 3 children is $285.

a. What is the price of an adult's ticket?

b. What is the price of a child's ticket?

c. Suppose that a special rate of $1000 is offered for groups of 20 persons. Should a group of 4 adults and 16 children use the group rate? Why or why not?

Solution:

1. UNDERSTAND. Read and reread the problem and guess a solution. Let's suppose that the price of an adult's ticket is $50 and the price of a child's ticket is $40. To check our proposed solution, let's see if admission for 4 adults and 2 children is $374. Admission for 4 adults is 4($50) or $200 and admission for 2 children is 2($40) or $80. This gives a total admission of $200 + $80 = $280, not the required $374. Again, though, we have accomplished the purpose of this process: We have a better understanding of the problem. To continue, we let

$A =$ the price of an adult's ticket and

$C =$ the price of a child's ticket

Answer

2. 365 children and 222 adults

2. TRANSLATE. We translate the problem into two equations using both variables.

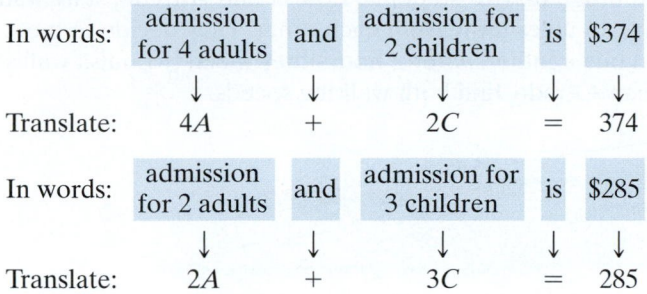

In words: admission for 4 adults and admission for 2 children is $374

Translate: $4A$ + $2C$ = 374

In words: admission for 2 adults and admission for 3 children is $285

Translate: $2A$ + $3C$ = 285

3. SOLVE. We solve the system.

$$\begin{cases} 4A + 2C = 374 \\ 2A + 3C = 285 \end{cases}$$

Since both equations are written in standard form, we solve by the addition method. First we multiply the second equation by -2 so that when we add the equations, we eliminate the variable A. Then the system

$$\begin{cases} 4A + 2C = 374 \\ -2(2A + 3C) = -2(285) \end{cases}$$

simplifies to

$$\begin{cases} 4A + 2C = 374 \\ -4A - 6C = -570 \end{cases}$$

Add the equations.

$$\begin{array}{r} -4C = -196 \\ C = 49 \text{ or } \$49, \text{ the child's ticket price} \end{array}$$

To find A, we replace C with 49 in the first equation.

$4A + 2C = 374$ First equation

$4A + 2(49) = 374$ Let $C = 49$.

$4A + 98 = 374$

$4A = 276$

$A = 69$ or $\$69$, the adult's ticket price

4. INTERPRET.

Check: Notice that 4 adults and 2 children will pay $4(\$69) + 2(\$49) = \$276 + \$98 = \$374$, the required amount. Also, the price for 2 adults and 3 children is $2(\$69) + 3(\$49) = \$138 + \$147 = \$285$, the required amount.

State: Answer the three original questions.

a.   Since $A = 69$, the price of an adult's ticket is $69.

b.   Since $C = 49$, the price of a child's ticket is $49.

c.   The regular admission price for 4 adults and 16 children is

$4(\$69) + 16(\$49) = \$276 + \$784$

$= \$1060$

This is $60 more than the special group rate of $1000, so they should request the group rate.

■ **Work Practice 2**

Practice 3

Two cars are 440 miles apart and traveling toward each other. They meet in 3 hours. If one car's speed is 10 miles per hour faster than the other car's speed, find the speed of each car.

	r · t = d	
Faster Car		
Slower Car		

Answer

3. One car's speed is $68\frac{1}{3}$ mph and the other car's speed is $78\frac{1}{3}$ mph.

Example 3   Finding Rates

As part of an exercise program, two students, Louisa and Alfredo, start walking each morning. They live 15 miles away from each other. They decide to meet one day by walking toward one another. After 2 hours they meet. If Louisa walks one mile per hour faster than Alfredo, find both walking speeds.

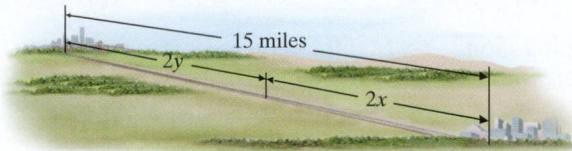

Solution:

1. **UNDERSTAND.** Read and reread the problem. Let's propose a solution and use the formula $d = r \cdot t$ to check. Suppose that Louisa's rate is 4 miles per hour. Since Louisa's rate is 1 mile per hour faster, Alfredo's rate is 3 miles per hour. To check, see if they can walk a total of 15 miles in 2 hours. Louisa's distance is rate · time = $4(2)$ = 8 miles and Alfredo's distance is rate · time = $3(2)$ = 6 miles. Their total distance is 8 miles + 6 miles = 14 miles, not the required 15 miles. Now that we have a better understanding of the problem, let's model it with a system of equations.

First, we let

x = Alfredo's rate in miles per hour and

y = Louisa's rate in miles per hour

Now we use the facts stated in the problem and the formula $d = rt$ to fill in the following chart.

	r · t = d		
Alfredo	x	2	$2x$
Louisa	y	2	$2y$

2. **TRANSLATE.** We translate the problem into two equations using both variables.

In words: Alfredo's distance + Louisa's distance = 15 miles

Translate: $2x$ + $2y$ = 15

In words: Louisa's rate   is   1 mile per hour faster than Alfredo's

Translate: y = $x + 1$

3. **SOLVE.** The system of equations we are solving is

$$\begin{cases} 2x + 2y = 15 \\ y = x + 1 \end{cases}$$

Let's use substitution to solve the system since the second equation is solved for y.

$$2x + 2y = 15 \qquad \text{First equation}$$
$$2x + 2(x + 1) = 15 \qquad \text{Replace } y \text{ with } x + 1.$$
$$2x + 2x + 2 = 15$$
$$4x = 13$$
$$x = \frac{13}{4} = 3\frac{1}{4} \text{ or } 3.25$$
$$y = x + 1 = 3\frac{1}{4} + 1 = 4\frac{1}{4} \text{ or } 4.25$$

4. **INTERPRET.** Alfredo's proposed rate is $3\frac{1}{4}$ miles per hour and Louisa's proposed rate is $4\frac{1}{4}$ miles per hour.

Check: Use the formula $d = rt$ and find that in 2 hours, Alfredo's distance is $(3.25)(2)$ miles or 6.5 miles. In 2 hours, Louisa's distance is $(4.25)(2)$ miles or 8.5 miles. The total distance walked is 6.5 miles + 8.5 miles or 15 miles, the given distance.

State: Alfredo walks at a rate of 3.25 miles per hour and Louisa walks at a rate of 4.25 miles per hour.

■ **Work Practice 3**

Example 4   Finding Amounts of Solutions

A chemistry teaching assistant needs 10 liters of a 20% saline solution (salt water) for his 2 p.m. laboratory class. Unfortunately, the only mixtures on hand are a 5% saline solution and a 25% saline solution. How much of each solution should he mix to produce the 20% solution?

Solution:

1. **UNDERSTAND.** Read and reread the problem. Suppose that we need 4 liters of the 5% solution. Then we need $10 - 4 = 6$ liters of the 25% solution. To see if this gives us 10 liters of a 20% saline solution, let's find the amount of pure salt in each solution.

	concentration rate	×	amount of solution	=	amount of pure salt
	↓		↓		↓
5% solution:	0.05	×	4 liters	=	0.2 liter
25% solution:	0.25	×	6 liters	=	1.5 liters
20% solution:	0.20	×	10 liters	=	2 liters

Since 0.2 liter + 1.5 liters = 1.7 liters, not 2 liters, our proposed solution is incorrect. But we have gained some insight into how to model and check this problem.

We let

x = number of liters of 5% solution and
y = number of liters of 25% solution

x liters

5% saline   25% saline   20% saline
solution   solution   solution

$+$ y liters $=$ $x + y$ or 10 liters

(Continued on next page)

Practice 4

A pharmacist needs 50 liters of a 60% alcohol solution. She currently has available a 20% solution and a 70% solution. How many liters of each must she use to make the needed 50 liters of 60% alcohol solution?

Answer
4. 10 liters of the 20% alcohol solution and 40 liters of the 70% alcohol solution

Now we use a table to organize the given data.

	Concentration Rate	Liters of Solution	Liters of Pure Salt
First Solution	5%	x	$0.05x$
Second Solution	25%	y	$0.25y$
Mixture Needed	20%	10	$(0.20)(10)$

2. TRANSLATE. We translate into two equations using both variables.

In words: liters of 5% solution $+$ liters of 25% solution $=$ 10 liters

Translate: x $+$ y $=$ 10

In words: salt in 5% solution $+$ salt in 25% solution $=$ salt in mixture

Translate: $0.05x$ $+$ $0.25y$ $=$ $(0.20)(10)$

3. SOLVE. Here we solve the system

$$\begin{cases} x + y = 10 \\ 0.05x + 0.25y = 2 \end{cases}$$

To solve by the addition method, we first multiply the first equation by -25 and the second equation by 100. Then the system

$$\begin{cases} -25(x + y) = -25(10) \\ 100(0.05x + 0.25y) = 100(2) \end{cases}$$ simplifies to $\begin{cases} -25x - 25y = -250 \\ \underline{5x + 25y = 200} \\ -20x = -50 \quad \text{Add.} \\ x = 2.5 \end{cases}$

To find y, we let $x = 2.5$ in the first equation of the original system.

$$x + y = 10$$
$$2.5 + y = 10 \quad \text{Let } x = 2.5.$$
$$y = 7.5$$

4. INTERPRET. Thus, we propose that he needs to mix 2.5 liters of 5% saline solution with 7.5 liters of 25% saline solution.

Check: Notice that $2.5 + 7.5 = 10$, the required number of liters. Also, the sum of the liters of salt in the two solutions equals the liters of salt in the required mixture:

$$0.05(2.5) + 0.25(7.5) = 0.20(10)$$
$$0.125 + 1.875 = 2$$

State: He needs 2.5 liters of the 5% saline solution and 7.5 liters of the 25% saline solution.

◼ **Work Practice 4**

✓**Concept Check** Suppose you mix an amount of a 30% acid solution with an amount of a 50% acid solution. Which of the following acid strengths would be possible for the resulting acid mixture?

a. 22%   **b.** 44%   **c.** 63%

✓Concept Check Answer
b

Vocabulary, Readiness & Video Check

Martin-Gay Interactive Videos *Watch the section lecture video and answer the following question.*

Objective **A** **1.** In the lecture before 🏳 Example 1, the problem-solving steps for solving applications involving systems are discussed. How do these steps differ from the general problem-solving strategy steps?

See Video 11.4 🍎

11.4 Exercise Set MyMathLab® ▶

Without actually solving each problem, choose the correct solution by deciding which choice satisfies the given conditions.

△ **1.** The length of a rectangle is 3 feet longer than the width. The perimeter is 30 feet. Find the dimensions of the rectangle.

   a. length = 8 feet; width = 5 feet

   b. length = 8 feet; width = 7 feet

   c. length = 9 feet; width = 6 feet

△ **2.** An isosceles triangle, a triangle with two sides of equal length, has a perimeter of 20 inches. Each of the equal sides is one inch longer than the third side. Find the lengths of the three sides.

   a. 6 inches, 6 inches, and 7 inches

   b. 7 inches, 7 inches, and 6 inches

   c. 6 inches, 7 inches, and 8 inches

3. Two computer disks and three notebooks cost $17. However, five computer disks and four notebooks cost $32. Find the price of each.

   a. notebook = $4; computer disk = $3

   b. notebook = $3; computer disk = $4

   c. notebook = $5; computer disk = $2

4. Two music CDs and four DVDs cost a total of $40. However, three music CDs and five DVDs cost $55. Find the price of each.

   a. CD = $12; DVD = $4

   b. CD = $15; DVD = $2

   c. CD = $10; DVD = $5

5. Kesha has a total of 100 coins, all of which are either dimes or quarters. The total value of the coins is $13.00. Find the number of each type of coin.

   a. 80 dimes; 20 quarters

   b. 20 dimes; 44 quarters

   c. 60 dimes; 40 quarters

6. Samuel has 28 gallons of saline solution available in two large containers at his pharmacy. One container holds three times as much as the other container. Find the capacity of each container.

   a. 15 gallons; 5 gallons

   b. 20 gallons; 8 gallons

   c. 21 gallons; 7 gallons

Objective **A** *Write a system of equations describing each situation. Do not solve the system. See Example 1.*

7. Two numbers add up to 15 and have a difference of 7.

8. The total of two numbers is 16. The first number plus 2 more than 3 times the second equals 18.

9. Keiko has a total of $6500, which she has invested in two accounts. The larger account is $800 greater than the smaller account.

10. Dominique has four times as much money in his savings account as in his checking account. The total amount is $2300.

Solve. See Examples 1 through 4.

11. Two numbers total 83 and have a difference of 17. Find the two numbers.

12. The sum of two numbers is 76 and their difference is 52. Find the two numbers.

13. A first number plus twice a second number is 8. Twice the first number plus the second totals 25. Find the numbers.

14. One number is 4 more than twice a second number. Their total is 25. Find the numbers.

15. Miguel Cabrera of the Detroit Tigers led Major League Baseball in runs batted in for the 2012 regular season. Josh Hamilton of the Texas Rangers, who came in second to Cabrera, had 11 fewer runs batted in for the 2012 regular season. Together, these two players brought home 267 runs during the 2012 regular season. How many runs batted in each did Cabrera and Hamilton account for? (*Source: Baseball Almanac*)

16. The highest scorer during the WNBA 2013 regular season was Angel McCoughtry of the Atlanta Dream. Over the season, McCoughtry scored 60 more points than the second-highest scorer, Diana Taurasi, of the Phoenix Mercury. Together, McCoughtry and Taurasi scored 1362 points during the 2013 regular season. How many points did each player score over the course of the season? (*Source:* Women's National Basketball Association)

17. Ann Marie Jones has been pricing Amtrak train fares for a group trip to New York. Three adults and four children must pay $159. Two adults and three children must pay $112. Find the price of an adult's ticket, and find the price of a child's ticket.

18. Last month, Jerry Papa purchased five DVDs and two CDs at Wall-to-Wall Sound for $65. This month he bought three DVDs and four CDs for $81. Find the price of each DVD, and find the price of each CD.

19. Johnston and Betsy Waring have a jar containing 80 coins, all of which are either quarters or nickels. The total value of the coins is $14.60. How many of each type of coin do they have?

20. Sarah and Keith Robinson purchased 40 stamps, a mixture of 44¢ and 28¢ stamps. Find the number of each type of stamp if they spent $16.80.

21. Norman and Suzanne Scarpulla own 30 shares of McDonald's Corp. stock and 68 shares of Ford Motor Co. stock. As the New York Stock Exchange opened on a day in 2013, their stock portfolio consisting of these two stocks was worth $4107. The McDonald's stock was $80.55 more per share than the Ford stock. What was the price of each stock on that day? (*Source:* SIX Financial Information)

22. Saralee Rose has investments in Google and Facebook stock. As the NASDAQ exchange opened on a day in 2013, Google stock was at $886.50 per share, and Facebook stock was at $50 per share. Saralee's portfolio made up of these two stocks was worth $32,964 at that time. If Saralee owns 15 more shares of Google stock than she owns of Facebook stock, how many shares of each type of stock does she own? (*Source:* SIX Financial Information)

23. Twice last month, Judy Carter rented a car from Enterprise in Fresno, California, and traveled around the Southwest on business. Enterprise rents this car for a daily fee, plus an additional charge per mile driven. Judy recalls that her first trip lasted 4 days, she drove 450 miles, and the rental cost her $240.50. On her second business trip she drove the same level of car 200 miles in 3 days, and paid $146.00 for the rental. Find the daily fee and the mileage charge.

24. Joan Gundersen rented the same car model twice from Hertz, which rents this car model for a daily fee plus an additional charge per mile driven. Joan recalls that the car rented for 5 days and driven for 300 miles cost her $178, while the same model car rented for 4 days and driven for 500 miles cost $197. Find the daily fee, and find the mileage charge.

25. Pratap Puri rowed 18 miles down the Delaware River in 2 hours, but the return trip took him $4\frac{1}{2}$ hours. Find the rate Pratap can row in still water, and find the rate of the current.

Let x = rate Pratap can row in still water and
$\quad y$ = rate of the current

	$d =$	r	$\cdot$	t
Downstream		$x + y$		
Upstream		$x - y$		

26. The Jonathan Schultz family took a canoe 10 miles down the Allegheny River in $1\frac{1}{4}$ hours. After lunch it took them 4 hours to return. Find the rate of the current.

Let x = rate the family can row in still water and
$\quad y$ = rate of the current

	$d =$	r	$\cdot$	t
Downstream		$x + y$		
Upstream		$x - y$		

27. Dave and Sandy Hartranft are frequent flyers with Delta Airlines. They often fly from Philadelphia to Chicago, a distance of 780 miles. On one particular trip they fly into the wind, and the flight takes 2 hours. The return trip, with the wind behind them, takes only $1\frac{1}{2}$ hours. Find the speed of the wind and find the speed of the plane in still air.

28. With a strong wind behind it, a United Airlines jet flies 2400 miles from Los Angeles to Orlando in $4\frac{3}{4}$ hours. The return trip takes 6 hours, as the plane flies into the wind. Find the speed of the plane in still air, and find the wind speed to the nearest tenth of a mile per hour.

29. Kevin Briley began a 186-mile bicycle trip to build up stamina for a triathlon competition. Unfortunately, his bicycle chain broke, so he finished the trip walking. The whole trip took 6 hours. If Kevin walks at a rate of 4 miles per hour and rides at 40 miles per hour, find the amount of time he spent on the bicycle.

30. In Canada, eastbound and westbound trains travel along the same track, with sidings to pull onto to avoid accidents. Two trains are now 150 miles apart, with the westbound train traveling twice as fast as the eastbound train. A warning must be issued to pull one train onto a siding, or else the trains will crash in $1\frac{1}{4}$ hours. Find the speed of the eastbound train and the speed of the westbound train.

31. Dorren Schmidt is a chemist with Gemco Pharmaceutical. She needs to prepare 12 ounces of a 9% hydrochloric acid solution. Find the amount of a 4% solution and the amount of a 12% solution she should mix to get this solution.

Concentration Rate	Liters of Solution	Liters of Pure Acid
0.04	x	0.04x
0.12	y	?
0.09	12	?

32. Elise Everly is preparing 15 liters of a 25% saline solution. Elise has two other saline solutions with strengths of 40% and 10%. Find the amount of 40% solution and the amount of 10% solution she should mix to get 15 liters of a 25% solution.

Concentration Rate	Liters of Solution	Liters of Pure Salt
0.40	x	0.40x
0.10	y	?
0.25	15	?

33. Wayne Osby blends coffee for a local coffee café. He needs to prepare 200 pounds of blended coffee beans selling for $3.95 per pound. He intends to do this by blending together a high-quality bean costing $4.95 per pound and a cheaper bean costing $2.65 per pound. To the nearest pound, find how much of the high-quality coffee beans and how much of the cheaper coffee beans he should blend.

34. Macadamia nuts cost an astounding $16.50 per pound, but research by an independent firm says that mixed nuts sell better if macadamias are included. The standard mix costs $9.25 per pound. Find how many pounds of macadamias and how many pounds of the standard mix should be combined to produce 40 pounds that will cost $10 per pound. Find the amounts to the nearest tenth of a pound.

35. Recall that two angles are complementary if the sum of their measures is 90°. Find the measures of two complementary angles if one angle is twice the other.

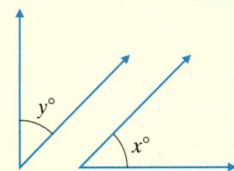

36. Recall that two angles are supplementary if the sum of their measures is 180°. Find the measures of two supplementary angles if one angle is 20° more than four times the other.

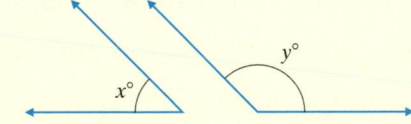

37. Find the measures of two complementary angles if one angle is 10° more than three times the other.

38. Find the measures of two supplementary angles if one angle is 18° more than twice the other.

39. Kathi and Robert Hawn had a pottery stand at the annual Skippack Craft Fair. They sold some of their pottery at the original price of $9.50 each, but later decreased the price of each by $2. If they sold all 90 pieces and took in $721, find how many they sold at the original price and how many they sold at the reduced price.

40. A charity fundraiser consisted of a spaghetti supper where a total of 387 people were fed. They charged $6.80 for adults and half price for children. If they took in $2444.60, find how many adults and how many children attended the supper.

41. The Santa Fe National Historic Trail is approximately 1200 miles between Old Franklin, Missouri, and Santa Fe, New Mexico. Suppose that a group of hikers start from each town and walk the trail toward each other. They meet after a total hiking time of 240 hours. If one group travels $\frac{1}{2}$ mile per hour slower than the other group, find the rate of each group. (*Source:* National Park Service)

42. California 1 South is a historic highway that stretches 123 miles along the coast from Monterey to Morro Bay. Suppose that two cars start driving this highway, one from each town. They meet after 3 hours. Find the rate of each car if one car travels 1 mile per hour faster than the other car. (*Source: National Geographic*)

43. A 30% solution of fertilizer is to be mixed with a 60% solution of fertilizer in order to get 150 gallons of a 50% solution. How many gallons of the 30% solution and 60% solution should be mixed?

44. A 10% acid solution is to be mixed with a 50% acid solution in order to get 120 ounces of a 20% acid solution. How many ounces of the 10% solution and 50% solution should be mixed?

45. Traffic signs are regulated by the *Manual on Uniform Traffic Control Devices* (MUTCD). According to this manual, if the sign below is placed on a freeway, its perimeter must be 144 inches. Also, its length must be 12 inches longer than its width. Find the dimensions of this sign.

46. According to the MUTCD (see Exercise **45**), this sign must have a perimeter of 60 inches. Also, its length must be 6 inches longer than its width. Find the perimeter of this sign.

Review

Find the square of each expression. For example, the square of 7 is 7^2 or 49. The square of 5x is $(5x)^2$ or $25x^2$. See Section 8.2.

47. 4   **48.** 3   **49.** $6x$ **50.** $11y$ **51.** $10y^3$ **52.** $8x^5$

Concept Extensions

Solve. See the Concept Check in this section.

53. Suppose you mix an amount of candy costing $0.49 a pound with candy costing $0.65 a pound. Which of the following costs per pound could result?
a. $0.58   **b.** $0.72   **c.** $0.29

54. Suppose you mix a 50% acid solution with pure acid (100%). Which of the following acid strengths are possible for the resulting acid mixture?
a. 25%   **b.** 150%   **c.** 62%   **d.** 90%

Solve.

△ **55.** Dale and Sharon Mahnke have decided to fence off a garden plot behind their house, using their house as the "fence" along one side of the garden. The length (which runs parallel to the house) is 3 feet less than twice the width. Find the dimensions if 33 feet of fencing is used along the three sides requiring it.

△ **56.** Judy McElroy plans to erect 152 feet of fencing to make a rectangular horse pasture. A river bank serves as one side length of the rectangle. If each width is 4 feet longer than half the length, find the dimensions.

Chapter 11 Group Activity

Break-Even Point

Sections 11.1, 11.2, 11.3, 11.4

When a business sells a new product, it generally does not start making a profit right away. There are usually many expenses associated with creating a new product. These expenses might include an advertising blitz to introduce the product to the public. These start-up expenses might also include the cost of market research and product development or any brand-new equipment needed to manufacture the product. Start-up costs like these are generally called *fixed costs* because they don't depend on the number of items manufactured. Expenses that do depend on the number of items manufactured, such as the cost of materials and shipping, are called *variable costs*. The total cost of manufacturing the new product is given by the cost equation Total cost = Fixed costs + Variable costs.

For instance, suppose a greeting card company is launching a new line of greeting cards. The company spent $7000 doing product research and development for the new line and spent $15,000 advertising the new line. The company does not need to buy any new equipment to manufacture the cards, but the paper and ink needed to make each card will cost $0.20 per card. The total cost y in dollars for manufacturing x cards is $y = 22,000 + 0.20x$.

Once a business sets a price for a new product, the company can find the product's expected *revenue*. Revenue is the amount of money the company takes in from the sales of its product. The revenue from selling a product is given by the revenue equation Revenue = Price per item × Number of items sold.

For instance, suppose that the card company plans to sell its new cards for $1.50 each. The revenue y, in dollars, that the company can expect to receive from the sales of x cards is $y = 1.50x$.

If the total cost and revenue equations are graphed on the same coordinate system, the graphs should intersect. The point of intersection is where total cost equals revenue and is called the *break-even point*. The break-even point gives the number of items x that must be manufactured and sold for the company to recover its expenses. If fewer than this number of items are produced and sold, the company loses money. If more than this number of items are produced and sold, the company makes a profit. In the case of the greeting card company, approximately 16,923 cards must be manufactured and sold for the company to break-even on this new card line. The total cost and revenue of producing and selling 16,923 cards is the same. It is approximately $25,385.

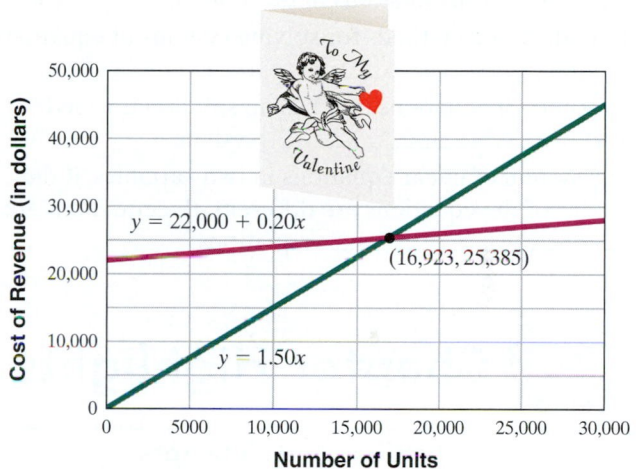

Group Activity

Suppose your group is starting a small business near your campus.

a. Choose a business and decide what campus-related product or service you will provide.

b. Research the fixed costs of starting up such a business.

c. Research the variable costs of producing such a product or providing such a service.

d. Decide how much you will charge per unit of your product or service.

e. Find a system of equations for the total cost and revenue of your product or service.

f. How many units of your product or service must be sold before your business will break even?

Chapter 11 Vocabulary Check

Fill in each blank with one of the words or phrases listed below.

system of linear equations	solution	consistent	independent
dependent	inconsistent	substitution	addition

1. In a system of linear equations in two variables, if the graphs of the equations are the same, the equations are _____ equations.

2. Two or more linear equations are called a(n) _____.

3. A system of equations that has at least one solution is called a(n) _____ system.

4. A(n) _____ of a system of two equations in two variables is an ordered pair of numbers that is a solution of both equations in the system.

5. Two algebraic methods for solving systems of equations are _____ and _____.

6. A system of equations that has no solution is called a(n) _____ system.

7. In a system of linear equations in two variables, if the graphs of the equations are different, the equations are _____ equations.

Helpful Hint

▶ Are you preparing for your test? Don't forget to take the Chapter 11 Test on page 878. Then check your answers at the back of the text and use the Chapter Test Prep Videos to see the fully worked-out solutions to any of the exercises you want to review.

11 Chapter Highlights

Definitions and Concepts	Examples
Section 11.1 Solving Systems of Linear Equations by Graphing	

A **system of linear equations** consists of two or more linear equations.

A **solution** of a system of two equations in two variables is an ordered pair of numbers that is a solution of both equations in the system.

$$\begin{cases} 2x + y = 6 \\ x = -3y \end{cases} \quad \begin{cases} -3x + 5y = 10 \\ x - 4y = -2 \end{cases}$$

Determine whether $(-1, 3)$ is a solution of the system.

$$\begin{cases} 2x - y = -5 \\ x = 3y - 10 \end{cases}$$

Replace x with -1 and y with 3 in both equations.

$$2x - y = -5$$
$$2(-1) - 3 \overset{?}{=} -5$$
$$-5 = -5 \qquad \text{True}$$

$$x = 3y - 10$$
$$-1 \overset{?}{=} 3(3) - 10$$
$$-1 = -1 \qquad \text{True}$$

$(-1, 3)$ is a solution of the system.

Graphically, a solution of a system is a point common to the graphs of both equations.

Solve by graphing: $\begin{cases} 3x - 2y = -3 \\ x + y = 4 \end{cases}$

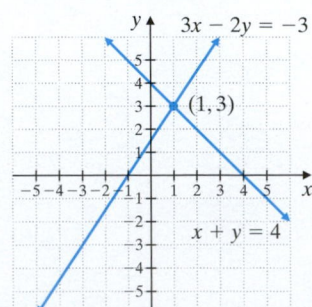

Definitions and Concepts	Examples

Section 11.1   Solving Systems of Linear Equations by Graphing (*continued*)

Three different situations can occur when graphing the two lines associated with the equations in a linear system.

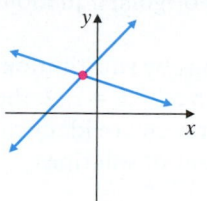

One point of inter-section; one solution

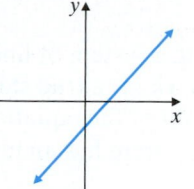

Same line; infinite number of solutions

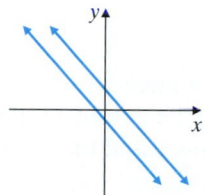

Parallel lines; no solution

Section 11.2   Solving Systems of Linear Equations by Substitution

To Solve a System of Linear Equations by the Substitution Method

Step 1: Solve one equation for a variable.

Step 2: Substitute the expression for the variable into the other equation.

Step 3: Solve the equation from Step 2 to find the value of one variable.

Step 4: Substitute the value from Step 3 in either original equation to find the value of the other variable.

Step 5: Check the solution in both original equations.

Solve by substitution.

$$\begin{cases} 3x + 2y = 1 \\ x = y - 3 \end{cases}$$

Substitute $y - 3$ for x in the first equation.

$$3x + 2y = 1$$
$$3(y - 3) + 2y = 1$$
$$3y - 9 + 2y = 1$$
$$5y = 10$$
$$y = 2 \quad \text{Divide by 5.}$$

To find x, substitute 2 for y in $x = y - 3$ so that $x = 2 - 3$ or -1. The solution $(-1, 2)$ checks.

Section 11.3   Solving Systems of Linear Equations by Addition

To Solve a System of Linear Equations by the Addition Method

Step 1: Rewrite each equation in standard form, $Ax + By = C$.

Step 2: Multiply one or both equations by a nonzero number so that the coefficients of a variable are opposites.

Step 3: Add the equations.

Step 4: Find the value of one variable by solving the resulting equation.

Step 5: Substitute the value from Step 4 into either original equation to find the value of the other variable.

Solve by addition.

$$\begin{cases} x - 2y = 8 \\ 3x + y = -4 \end{cases}$$

Multiply both sides of the first equation by -3.

$$\begin{cases} -3x + 6y = -24 \\ \underline{3x + y = -4} \end{cases}$$
$$7y = -28 \quad \text{Add.}$$
$$y = -4 \quad \text{Divide by 7.}$$

To find x, let $y = -4$ in an original equation.

$$x - 2(-4) = 8 \quad \text{First equation}$$
$$x + 8 = 8$$
$$x = 0$$

(*continued*)

Definitions and Concepts	**Examples**

Section 11.3 Solving Systems of Linear Equations by Addition (*continued*)

Step 6: Check the solution in both original equations.

If solving a system of linear equations by substitution or addition yields a true statement such as $-2 = -2$, then the graphs of the equations in the system are identical and the system has an infinite number of solutions.

The solution $(0, -4)$ checks.

Solve: $\begin{cases} 2x - 6y = -2 \\ x = 3y - 1 \end{cases}$

Substitute $3y - 1$ for x in the first equation.

$$2(3y - 1) - 6y = -2$$
$$6y - 2 - 6y = -2$$
$$-2 = -2 \quad \text{True}$$

The system has an infinite number of solutions.

If solving a system of linear equations yields a false statement such as $0 = 3$, the graphs of the equations in the system are parallel lines and the system has no solution.

Solve: $\begin{cases} 5x - 2y = 6 \\ -5x + 2y = -3 \end{cases}$

$$0 = 3 \quad \text{False}$$

The system has no solution.

Section 11.4 Systems of Linear Equations and Problem Solving

Problem-Solving Steps

1. UNDERSTAND. Read and reread the problem.

Two angles are supplementary if the sum of their measures is $180°$. The larger of two supplementary angles is three times the smaller, decreased by twelve. Find the measure of each angle. Let

$x =$ measure of smaller angle and

$y =$ measure of larger angle

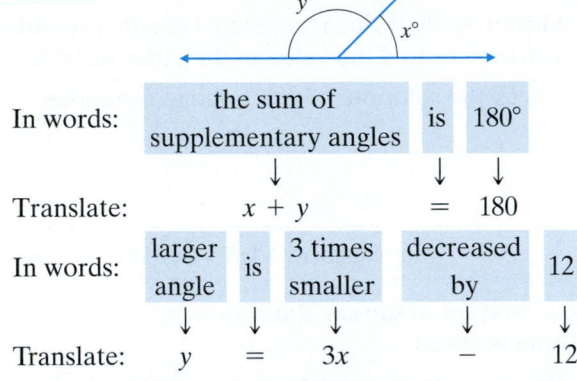

2. TRANSLATE.

In words: the sum of supplementary angles   is $180°$

Translate: $x + y$ $=$ 180

In words: larger angle   is   3 times smaller   decreased by   12

Translate: y $=$ $3x$ $-$ 12

3. SOLVE.

Solve the system. $\begin{cases} x + y = 180 \\ y = 3x - 12 \end{cases}$

Use the substitution method and replace y with $3x - 12$ in the first equation.

$$x + y = 180$$
$$x + (3x - 12) = 180$$
$$4x = 192$$
$$x = 48$$

4. INTERPRET.

Since $y = 3x - 12$, then $y = 3 \cdot 48 - 12$ or 132.

The solution checks. The smaller angle measures $48°$ and the larger angle measures $132°$.

(11.1) *Determine whether each ordered pair is a solution of the system of linear equations.*

1. $\begin{cases} 2x - 3y = 12 \\ 3x + 4y = 1 \end{cases}$

   a. $(12, 4)$

   b. $(3, -2)$

2. $\begin{cases} 2x + 3y = 1 \\ 3y - x = 4 \end{cases}$

   a. $(2, 2)$

   b. $(-1, 1)$

3. $\begin{cases} 5x - 6y = 18 \\ 2y - x = -4 \end{cases}$

   a. $(-6, -8)$

   b. $\left(3, \dfrac{5}{2}\right)$

4. $\begin{cases} 4x + y = 0 \\ -8x - 5y = 9 \end{cases}$

   a. $\left(\dfrac{3}{4}, -3\right)$

   b. $(-2, 8)$

Solve each system of equations by graphing.

5. $\begin{cases} x + y = 5 \\ x - y = 1 \end{cases}$

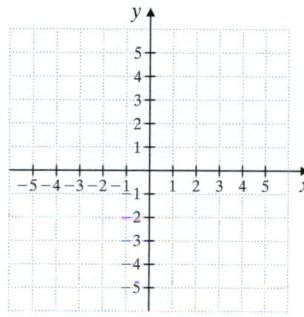

6. $\begin{cases} x + y = 3 \\ x - y = -1 \end{cases}$

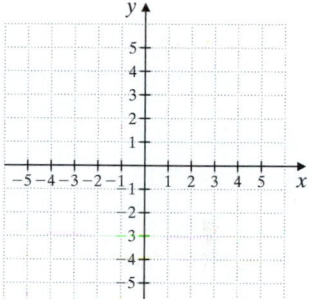

7. $\begin{cases} x = 5 \\ y = -1 \end{cases}$

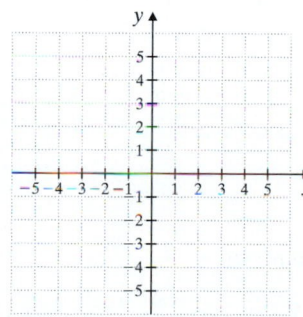

8. $\begin{cases} x = -3 \\ y = 2 \end{cases}$

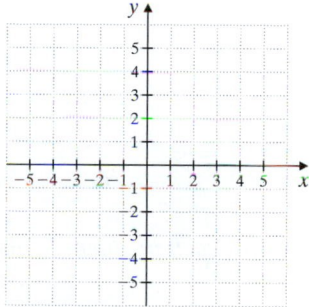

9. $\begin{cases} 2x + y = 5 \\ x = -3y \end{cases}$

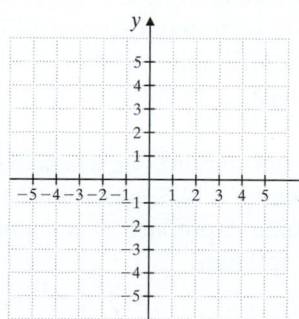

10. $\begin{cases} 3x + y = -2 \\ y = -5x \end{cases}$

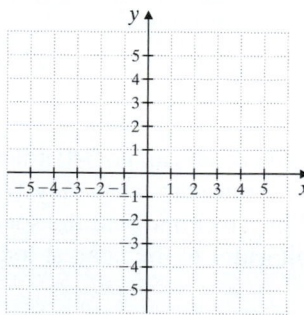

11. $\begin{cases} y = 3x \\ -6x + 2y = 6 \end{cases}$

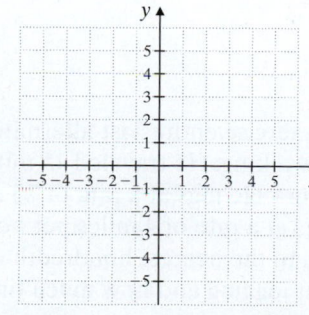

12. $\begin{cases} x - 2y = 2 \\ -2x + 4y = -4 \end{cases}$

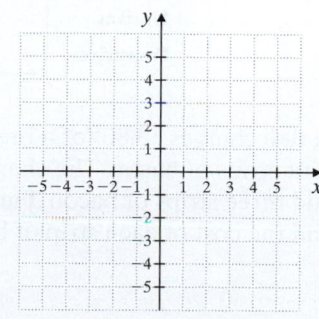

(11.2) *Solve each system of equations by the substitution method.*

13. $\begin{cases} y = 2x + 6 \\ 3x - 2y = -11 \end{cases}$

14. $\begin{cases} y = 3x - 7 \\ 2x - 3y = 7 \end{cases}$

15. $\begin{cases} x + 3y = -3 \\ 2x + y = 4 \end{cases}$

16. $\begin{cases} 3x + y = 11 \\ x + 2y = 12 \end{cases}$

17. $\begin{cases} 4y = 2x + 6 \\ x - 2y = -3 \end{cases}$

18. $\begin{cases} 9x = 6y + 3 \\ 6x - 4y = 2 \end{cases}$

19. $\begin{cases} x + y = 6 \\ y = -x - 4 \end{cases}$

20. $\begin{cases} -3x + y = 6 \\ y = 3x + 2 \end{cases}$

(11.3) *Solve each system of equations by the addition method.*

21. $\begin{cases} 2x + 3y = -6 \\ x - 3y = -12 \end{cases}$

22. $\begin{cases} 4x + y = 15 \\ -4x + 3y = -19 \end{cases}$

23. $\begin{cases} 2x - 3y = -15 \\ x + 4y = 31 \end{cases}$

24. $\begin{cases} x - 5y = -22 \\ 4x + 3y = 4 \end{cases}$

25. $\begin{cases} 2x - 6y = -1 \\ -x + 3y = \dfrac{1}{2} \end{cases}$

26. $\begin{cases} 0.6x - 0.3y = -1.5 \\ 0.04x - 0.02y = -0.1 \end{cases}$

27. $\begin{cases} \dfrac{3}{4}x + \dfrac{2}{3}y = 2 \\ x + \dfrac{y}{3} = 6 \end{cases}$

28. $\begin{cases} 10x + 2y = 0 \\ 3x + 5y = 33 \end{cases}$

(11.4) *Solve each problem by writing and solving a system of linear equations.*

29. The sum of two numbers is 16. Three times the larger number decreased by the smaller number is 72. Find the two numbers.

30. The Forrest Theater can seat a total of 360 people. They take in $15,150 when every seat is sold. If orchestra section tickets cost $45 and balcony tickets cost $35, find the number of seats in the orchestra section and the number of seats in the balcony.

31. A riverboat can go 340 miles upriver in 19 hours, but the return trip takes only 14 hours. Find the current of the river and find the speed of the riverboat in still water to the nearest tenth of a mile.

32. Find the amount of a 6% acid solution and the amount of a 14% acid solution Pat Mayfield should combine to prepare 50 cc (cubic centimeters) of a 12% solution.

$d =$	r	$\cdot$	t
Upriver	$x - y$		
Downriver	$x + y$		

33. A deli charges $3.80 for a breakfast of three eggs and four strips of bacon. The charge is $2.75 for two eggs and three strips of bacon. Find the cost of each egg and the cost of each strip of bacon.

34. An exercise enthusiast alternates between jogging and walking. He traveled 15 miles during the past 3 hours. He jogs at a rate of 7.5 miles per hour and walks at a rate of 4 miles per hour. Find how much time, to the nearest hundredth of an hour, he actually spent jogging and how much time he spent walking.

Mixed Review

Solve each system of equations by graphing.

35. $\begin{cases} x - 2y = 1 \\ 2x + 3y = -12 \end{cases}$

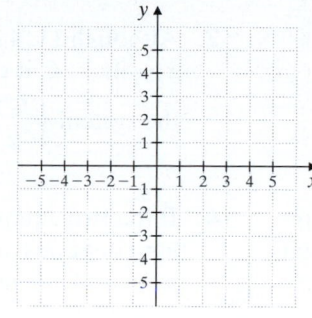

36. $\begin{cases} 3x - y = -4 \\ 6x - 2y = -8 \end{cases}$

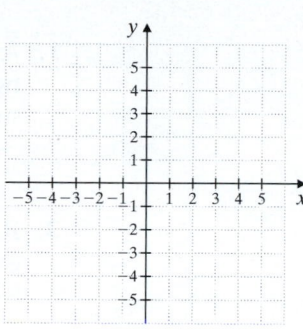

Solve each system of equations.

37. $\begin{cases} x + 4y = 11 \\ 5x - 9y = -3 \end{cases}$

38. $\begin{cases} x + 9y = 16 \\ 3x - 8y = 13 \end{cases}$

39. $\begin{cases} y = -2x \\ 4x + 7y = -15 \end{cases}$

40. $\begin{cases} 3y = 2x + 15 \\ -2x + 3y = 21 \end{cases}$

41. $\begin{cases} 3x - y = 4 \\ 4y = 12x - 16 \end{cases}$

42. $\begin{cases} x + y = 19 \\ x - y = -3 \end{cases}$

43. $\begin{cases} x - 3y = -11 \\ 4x + 5y = -10 \end{cases}$

44. $\begin{cases} -x - 15y = 44 \\ 2x + 3y = 20 \end{cases}$

45. $\begin{cases} 2x + y = 3 \\ 6x + 3y = 9 \end{cases}$

46. $\begin{cases} -3x + y = 5 \\ -3x + y = -2 \end{cases}$

Solve each problem by writing and solving a system of linear equations.

47. The sum of two numbers is 12. Three times the smaller number increased by the larger number is 20. Find the numbers.

48. The difference of two numbers is −18. Twice the smaller decreased by the larger is −23. Find the two numbers.

49. Emma Hodges has a jar containing 65 coins, all of which are either nickels or dimes. The total value of the coins is $5.30. How many of each type does she have?

50. Sarah and Owen Hebert purchased 26 stamps, a mixture of 13¢ and 22¢ stamps. Find the number of each type of stamp if they spent $4.19.

Answers

Answer each question true or false.

1. A system of two linear equations in two variables can have exactly two solutions.

2. Although $(1, 4)$ is not a solution of $x + 2y = 6$, it can still be a solution of the system $\begin{cases} x + 2y = 6 \\ x + y = 5 \end{cases}$.

3. If the two equations in a system of linear equations are added and the result is $3 = 0$, the system has no solution.

4. If the two equations in a system of linear equations are added and the result is $3x = 0$, the system has no solution.

1. _____

2. _____

3. _____

Is the ordered pair a solution of the given linear system?

5. $\begin{cases} 2x - 3y = 5 \\ 6x + y = 1 \end{cases}; (1, -1)$

6. $\begin{cases} 4x - 3y = 24 \\ 4x + 5y = -8 \end{cases}; (3, -4)$

4. _____

Solve each system by graphing.

7. $\begin{cases} x - y = 2 \\ 3x - y = -2 \end{cases}$

8. $\begin{cases} y = -3x \\ 3x + y = 6 \end{cases}$

5. _____

6. _____

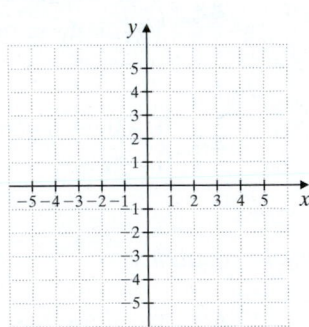

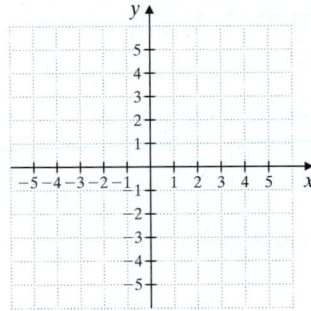

7. _____

8. _____

9. _____

Solve each system by the substitution method.

9. $\begin{cases} 3x - 2y = -14 \\ y = x + 5 \end{cases}$

10. $\begin{cases} \dfrac{1}{2}x + 2y = -\dfrac{15}{4} \\ 4x = -y \end{cases}$

10. _____

Solve each system by the addition method.

11. _____

11. $\begin{cases} x + y = 28 \\ x - y = 12 \end{cases}$

12. $\begin{cases} 4x - 6y = 7 \\ -2x + 3y = 0 \end{cases}$

12. _____

13. _____

Solve each system using the substitution method or the addition method.

13. $\begin{cases} 3x + y = 7 \\ 4x + 3y = 1 \end{cases}$

14. $\begin{cases} 3(2x + y) = 4x + 20 \\ x - 2y = 3 \end{cases}$

14. _____

Objective A Evaluating Exponential Expressions

In this section, we continue our work with integer exponents. Recall from Section 1.9 that repeated multiplication of the same factor can be written using exponents. For example,

$$2 \cdot 2 \cdot 2 \cdot 2 \cdot 2 = 2^5$$

The exponent 5 tells us how many times 2 is a factor. The expression 2^5 is called an **exponential expression.** It is also called the fifth **power** of 2, or we can say that 2 is **raised** to the fifth power.

$$5^6 = \underbrace{5 \cdot 5 \cdot 5 \cdot 5 \cdot 5 \cdot 5}_{\text{6 factors; each factor is 5}} \quad \text{and} \quad (-3)^4 = \underbrace{(-3) \cdot (-3) \cdot (-3) \cdot (-3)}_{\text{4 factors; each factor is }-3}$$

The base of an exponential expression is the repeated factor. The exponent is the number of times that the base is used as a factor.

$$\overset{\text{exponent or power}}{a^n} = \underbrace{a \cdot a \cdot a \cdots a}_{n \text{ factors; each factor is } a}$$
base

Examples Evaluate each expression.

1. $2^3 = 2 \cdot 2 \cdot 2 = 8$

2. $3^1 = 3$. To raise 3 to the first power means to use 3 as a factor only once. When no exponent is shown, the exponent is assumed to be 1.

3. $(-4)^2 = (-4)(-4) = 16$

4. $-4^2 = -(4 \cdot 4) = -16$

5. $\left(\dfrac{1}{2}\right)^4 = \dfrac{1}{2} \cdot \dfrac{1}{2} \cdot \dfrac{1}{2} \cdot \dfrac{1}{2} = \dfrac{1}{16}$

6. $4 \cdot 3^2 = 4 \cdot 9 = 36$

■ **Work Practice 1–6**

Notice how similar -4^2 is to $(-4)^2$ in the examples above. The difference between the two is the parentheses. In $(-4)^2$, the parentheses tell us that the base, or the repeated factor, is -4. In -4^2, only 4 is the base.

Helpful Hint

Be careful when identifying the base of an exponential expression. Pay close attention to the use of parentheses.

$$(-3)^2 \qquad\qquad -3^2 \qquad\qquad 2 \cdot 3^2$$

The base is -3. The base is 3. The base is 3.

$(-3)^2 = (-3)(-3) = 9 \quad -3^2 = -(3 \cdot 3) = -9 \quad 2 \cdot 3^2 = 2 \cdot 3 \cdot 3 = 18$

An exponent has the same meaning whether the base is a number or a variable. If x is a real number and n is a positive integer, then x^n is the product of n factors, each of which is x.

$$x^n = \underbrace{x \cdot x \cdot x \cdot x \cdot x \cdots x}_{n \text{ factors; each factor is } x}$$

Objectives

A Evaluate Exponential Expressions.

B Use the Product Rule for Exponents.

C Use the Power Rule for Exponents.

D Use the Power Rules for Products and Quotients.

E Use the Quotient Rule for Exponents, and Define a Number Raised to the 0 Power.

F Decide Which Rule(s) to Use to Simplify an Expression.

Practice 1–6

Evaluate each expression.

1. 3^4 **2.** 7^1 **3.** $(-2)^3$

4. -2^3 **5.** $\left(\dfrac{2}{3}\right)^2$ **6.** $5 \cdot 6^2$

Answers

1. 81   **2.** 7   **3.** -8 **4.** -8 **5.** $\dfrac{4}{9}$
6. 180

Practice 7

Evaluate each expression for the given value of x.

a. $3x^2$ when x is 4

b. $\dfrac{x^4}{-8}$ when x is -2

Example 7   Evaluate each expression for the given value of x.

a. $2x^3$ when x is 5   **b.** $\dfrac{9}{x^2}$ when x is -3

Solution:

a. When x is 5, $2x^3 = 2 \cdot 5^3$

$$= 2 \cdot (5 \cdot 5 \cdot 5)$$
$$= 2 \cdot 125$$
$$= 250$$

b. When x is -3, $\dfrac{9}{x^2} = \dfrac{9}{(-3)^2}$

$$= \dfrac{9}{(-3)(-3)}$$
$$= \dfrac{9}{9} = 1$$

■ **Work Practice 7**

Objective B Using the Product Rule ▶

Exponential expressions can be multiplied, divided, added, subtracted, and themselves raised to powers. Let's see if we can discover a shortcut method for multiplying exponential expressions with the same base. By our definition of an exponent,

$$5^4 \cdot 5^3 = \underbrace{(5 \cdot 5 \cdot 5 \cdot 5)}_{4 \text{ factors of } 5} \cdot \underbrace{(5 \cdot 5 \cdot 5)}_{3 \text{ factors of } 5}$$
$$= \underbrace{5 \cdot 5 \cdot 5 \cdot 5 \cdot 5 \cdot 5 \cdot 5}_{7 \text{ factors of } 5}$$
$$= 5^7$$

Also,

▶ $x^2 \cdot x^3 = (x \cdot x) \cdot (x \cdot x \cdot x)$
$$= x \cdot x \cdot x \cdot x \cdot x$$
$$= x^5$$

In both cases, notice that the result is exactly the same if the exponents are added.

$$5^4 \cdot 5^3 = 5^{4+3} = 5^7 \quad \text{and} \quad x^2 \cdot x^3 = x^{2+3} = x^5$$

This suggests the following rule.

Product Rule for Exponents

If m and n are positive integers and a is a real number, then

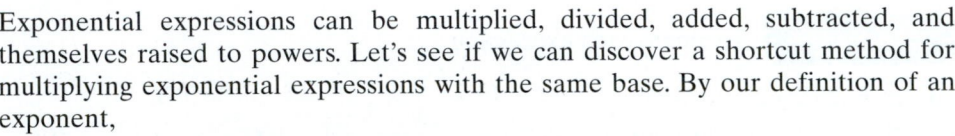

$a^m \cdot a^n = a^{m+n}$ ← Add exponents.
            ↑
            └——— Keep common base.

For example,

$3^5 \cdot 3^7 = 3^{5+7} = 3^{12}$ ← Add exponents.
          ↑
          └——— Keep common base.

Answers

7. a. 48   **b.** -2

Helpful Hint

Don't forget that

$$3^5 \cdot 3^7 \neq 9^{12} \leftarrow \text{Add exponents.}$$

$\qquad$ Common base *not* kept.

$$3^5 \cdot 3^7 = \underbrace{3 \cdot 3 \cdot 3 \cdot 3 \cdot 3}_{5 \text{ factors of } 3} \cdot \underbrace{3 \cdot 3 \cdot 3 \cdot 3 \cdot 3 \cdot 3 \cdot 3}_{7 \text{ factors of } 3}$$

$$= 3^{12} \quad 12 \text{ factors of } 3, \text{ not } 9$$

In other words, to multiply two exponential expressions with the **same base,** we keep the base and add the exponents. We call this **simplifying** the exponential expression.

Examples Use the product rule to simplify each expression.

8. $4^2 \cdot 4^5 = 4^{2+5} = 4^7 \leftarrow$ Add exponents.

$\qquad$ Keep common base.

9. $x^2 \cdot x^5 = x^{2+5} = x^7$

10. $y^3 \cdot y = y^3 \cdot y^1$

$\quad = y^{3+1}$

$\quad = y^4$

Helpful Hint Don't forget that if no exponent is written, it is assumed to be 1.

11. $y^3 \cdot y^2 \cdot y^7 = y^{3+2+7} = y^{12}$

12. $(-5)^7 \cdot (-5)^8 = (-5)^{7+8} = (-5)^{15}$

■ **Work Practice 8–12**

Practice 8–12

Use the product rule to simplify each expression.
8. $7^3 \cdot 7^2$ **9.** $x^4 \cdot x^9$
10. $r^5 \cdot r$ **11.** $s^6 \cdot s^2 \cdot s^3$
12. $(-3)^9 \cdot (-3)$

✔**Concept Check** Where possible, use the product rule to simplify the expression.

a. $z^2 \cdot z^{14}$ **b.** $x^2 \cdot z^{14}$ **c.** $9^8 \cdot 9^3$ **d.** $9^8 \cdot 2^7$

Example 13 Use the product rule to simplify $(2x^2)(-3x^5)$.

Solution: Recall that $2x^2$ means $2 \cdot x^2$ and $-3x^5$ means $-3 \cdot x^5$.

$$(2x^2)(-3x^5) = (2 \cdot x^2) \cdot (-3 \cdot x^5)$$

$$= (2 \cdot -3) \cdot (x^2 \cdot x^5) \quad \text{Group factors with common bases (using commutative and associative properties).}$$

$$= -6x^7 \quad \text{Simplify.}$$

■ **Work Practice 13**

Practice 13

Use the product rule to simplify $(6x^3)(-2x^9)$.

Examples Simplify.

14. $(x^2y)(x^3y^2) = (x^2 \cdot x^3) \cdot (y^1 \cdot y^2)$ Group like bases and write y as y^1.

$\qquad = x^5 \cdot y^3$ or x^5y^3 Multiply.

15. $(-a^7b^4)(3ab^9) = (-1 \cdot 3) \cdot (a^7 \cdot a^1) \cdot (b^4 \cdot b^9)$

$\qquad = -3a^8b^{13}$

■ **Work Practice 14–15**

Practice 14–15

Simplify.
14. $(m^5n^{10})(mn^8)$
15. $(-x^9y)(4x^2y^{11})$

Answers
8. 7^5 **9.** x^{13} **10.** r^6 **11.** s^{11}
12. $(-3)^{10}$ **13.** $-12x^{12}$ **14.** m^6n^{18}
15. $-4x^{11}y^{12}$

✔**Concept Check Answers**

a. z^{16} **b.** cannot be simplified

c. 9^{11} **d.** cannot be simplified

Objective C Using the Power Rule

Exponential expressions can themselves be raised to powers. Let's try to discover a rule that simplifies an expression like $(x^2)^3$. By the definition of a^n,

$$(x^2)^3 = (x^2)(x^2)(x^2)$$ $(x^2)^3$ means 3 factors of (x^2).

which can be simplified by the product rule for exponents.

$$(x^2)^3 = (x^2)(x^2)(x^2) = x^{2+2+2} = x^6$$

Notice that the result is exactly the same if we multiply the exponents.

▶ $$(x^2)^3 = x^{2 \cdot 3} = x^6$$

The following rule states this result.

Power Rule for Exponents

If m and n are positive integers and a is a real number, then

$$(a^m)^n = a^{mn} \leftarrow \text{Multiply exponents.}$$
$$\uparrow\!\!\underline{\qquad} \text{Keep the base.}$$

For example,

$$(7^2)^5 = 7^{2 \cdot 5} = 7^{10} \leftarrow \text{Multiply exponents.}$$
$$\uparrow\!\!\underline{\qquad} \text{Keep the base.}$$

$$[(-5)^3]^7 = (-5)^{3 \cdot 7} = (\underline{-5})^{21} \leftarrow \text{Multiply exponents.}$$
$$\uparrow\!\!\underline{\qquad} \text{Keep the base.}$$

In other words, to raise an exponential expression to a power, we keep the base and multiply the exponents.

Practice 16–17

Use the power rule to simplify each expression.

16. $(9^4)^{10}$ **17.** $(z^6)^3$

> **Examples**   Use the power rule to simplify each expression.
>
> **16.** $(5^3)^6 = 5^{3 \cdot 6} = 5^{18}$ **17.** $(y^8)^2 = y^{8 \cdot 2} = y^{16}$
>
> ■ **Work Practice 16–17**

Answers

16. 9^{40} **17.** z^{18}

Objective D Using the Power Rules for Products and Quotients ▶

When the base of an exponential expression is a product, the definition of a^n still applies. For example, simplify $(xy)^3$ as follows.

$$(xy)^3 = (xy)(xy)(xy) \qquad (xy)^3 \text{ means 3 factors of } (xy).$$
$$= x \cdot x \cdot x \cdot y \cdot y \cdot y \quad \text{Group factors with common bases.}$$
$$= x^3y^3 \qquad \text{Simplify.}$$

Notice that to simplify the expression $(xy)^3$, we raise each factor within the parentheses to a power of 3.

$$(xy)^3 = x^3y^3$$

In general, we have the following rule.

Power of a Product Rule

If n is a positive integer and a and b are real numbers, then

$$(ab)^n = a^nb^n$$

For example,

$$(3x)^5 = 3^5x^5$$

In other words, to raise a product to a power, we raise each factor to the power.

Examples Simplify each expression.

18. $(st)^4 = s^4 \cdot t^4 = s^4t^4$ 　　　Use the power of a product rule.

19. $(2a)^3 = 2^3 \cdot a^3 = 8a^3$ 　　　Use the power of a product rule.

20. $(-5x^2y^3z)^2 = (-5)^2 \cdot (x^2)^2 \cdot (y^3)^2 \cdot (z^1)^2$ 　Use the power of a product rule.
$$= 25x^4y^6z^2$$

21. $(-xy^3)^5 = (-1xy^3)^5 = (-1)^5 \cdot x^5 \cdot (y^3)^5$ 　Use the power of a product rule.
$$= -1x^5y^{15} \quad \text{or} \quad -x^5y^{15}$$

■ **Work Practice 18–21**

Let's see what happens when we raise a quotient to a power. For example, we simplify $\left(\dfrac{x}{y}\right)^3$ as follows.

$$\left(\frac{x}{y}\right)^3 = \left(\frac{x}{y}\right)\left(\frac{x}{y}\right)\left(\frac{x}{y}\right) \quad \left(\frac{x}{y}\right)^3 \text{ means 3 factors of } \left(\frac{x}{y}\right).$$
$$= \frac{x \cdot x \cdot x}{y \cdot y \cdot y} \qquad \text{Multiply fractions.}$$
$$= \frac{x^3}{y^3} \qquad \text{Simplify.}$$

Notice that to simplify the expression $\left(\dfrac{x}{y}\right)^3$, we raise both the numerator and the denominator to a power of 3.

$$\left(\frac{x}{y}\right)^3 = \frac{x^3}{y^3}$$

Practice 18–21

Simplify each expression.

18. $(xy)^7$ 　　　**19.** $(3y)^4$

20. $(-2p^4q^2r)^3$ 　**21.** $(-a^4b)^7$

In general, we have the following rule.

Power of a Quotient Rule

If n is a positive integer and a and c are real numbers, then

$$\left(\frac{a}{c}\right)^n = \frac{a^n}{c^n}, \quad c \neq 0$$

For example,

$$\left(\frac{y}{7}\right)^3 = \frac{y^3}{7^3}$$

In other words, to raise a quotient to a power, we raise both the numerator and the denominator to the power.

Practice 22–23

Simplify each expression.

22. $\left(\dfrac{r}{s}\right)^6$ 23. $\left(\dfrac{5x^6}{9y^3}\right)^2$

Examples   Simplify each expression.

22. $\left(\dfrac{m}{n}\right)^7 = \dfrac{m^7}{n^7}, \quad n \neq 0$ Use the power of a quotient rule.

23. $\left(\dfrac{2x^4}{3y^5}\right)^4 = \dfrac{2^4 \cdot \left(x^4\right)^4}{3^4 \cdot \left(y^5\right)^4}$ Use the power of a quotient rule.

$\qquad\qquad = \dfrac{16x^{16}}{81y^{20}}, \quad y \neq 0$ Use the power rule for exponents.

■ **Work Practice 22–23**

Objective E Using the Quotient Rule and Defining the Zero Exponent ▶

Another pattern for simplifying exponential expressions involves quotients.

$$\frac{x^5}{x^3} = \frac{x \cdot x \cdot x \cdot x \cdot x}{x \cdot x \cdot x}$$

$$= \frac{x \cdot x \cdot x \cdot x \cdot x}{x \cdot x \cdot x}$$

$$= 1 \cdot 1 \cdot 1 \cdot x \cdot x$$

$$= x \cdot x$$

$$= x^2$$

Notice that the result is exactly the same if we subtract exponents of the common bases.

$$\frac{x^5}{x^3} = x^{5-3} = x^2$$

The following rule states this result in a general way.

Quotient Rule for Exponents

If m and n are positive integers and a is a real number, then

$$\frac{a^m}{a^n} = a^{m-n}, \quad a \neq 0$$

For example,

$$\frac{x^6}{x^2} = x^{6-2} = x^4, \quad x \neq 0$$

Answers

22. $\dfrac{r^6}{s^6}, \ s \neq 0$ 23. $\dfrac{25x^{12}}{81y^6}, \ y \neq 0$

In other words, to divide one exponential expression by another with a common base, we keep the base and subtract the exponents.

> **Examples** Simplify each quotient.

▶ **24.** $\dfrac{x^5}{x^2} = x^{5-2} = x^3$ Use the quotient rule.

25. $\dfrac{4^7}{4^3} = 4^{7-3} = 4^4 = 256$ Use the quotient rule.

26. $\dfrac{(-3)^5}{(-3)^2} = (-3)^3 = -27$ Use the quotient rule.

27. $\dfrac{2x^5y^2}{xy} = 2 \cdot \dfrac{x^5}{x^1} \cdot \dfrac{y^2}{y^1}$

$\qquad\quad = 2 \cdot \left(x^{5-1}\right) \cdot \left(y^{2-1}\right)$ Use the quotient rule.

$\qquad\quad = 2x^4y^1 \quad \text{or} \quad 2x^4y$

■ **Work Practice 24–27**

Practice 24–27

Simplify each quotient.

24. $\dfrac{y^7}{y^3}$ **25.** $\dfrac{5^9}{5^6}$

26. $\dfrac{(-2)^{14}}{(-2)^{10}}$ **27.** $\dfrac{7a^4b^{11}}{ab}$

Let's now give meaning to an expression such as x^0. To do so, we will simplify $\dfrac{x^3}{x^3}$ in two ways and compare the results.

▶ $\dfrac{x^3}{x^3} = x^{3-3} = x^0$ Apply the quotient rule.

$\dfrac{x^3}{x^3} = \dfrac{x \cdot x \cdot x}{x \cdot x \cdot x} = 1$ Apply the fundamental principle for fractions.

Since $\dfrac{x^3}{x^3} = x^0$ and $\dfrac{x^3}{x^3} = 1$, we define that $x^0 = 1$ as long as x is not 0.

Zero Exponent

$a^0 = 1$, as long as a is not 0.

For example, $5^0 = 1$.

In other words, a base raised to the 0 power is 1, as long as the base is not 0.

> **Examples** Simplify each expression.

28. $3^0 = 1$

29. $\left(5x^3y^2\right)^0 = 1$

30. $(-4)^0 = 1$

31. $-4^0 = -1 \cdot 4^0 = -1 \cdot 1 = -1$

32. $5x^0 = 5 \cdot x^0 = 5 \cdot 1 = 5$

■ **Work Practice 28–32**

Practice 28–32

Simplify each expression.

28. 8^0 **29.** $(2r^2s)^0$

30. $(-7)^0$ **31.** -7^0

32. $7y^0$

Answers

24. y^4 **25.** 125 **26.** 16 **27.** $7a^3b^{10}$

28. 1 **29.** 1 **30.** 1 **31.** -1 **32.** 7

✓**Concept Check** Suppose you are simplifying each expression. Tell whether you would *add* the exponents, *subtract* the exponents, *multiply* the exponents, *divide* the exponents, or *none of these*.

a. $\left(x^{63}\right)^{21}$ **b.** $\dfrac{y^{15}}{y^3}$ **c.** $z^{16} + z^8$ **d.** $w^{45} \cdot w^9$

Objective F Deciding Which Rule to Use ▶

Let's practice deciding which rule to use to simplify an expression. We will continue this discussion with more examples in the next section.

Practice 33

Simplify each expression.

a. $\dfrac{x^7}{x^4}$ **b.** $\left(3y^4\right)^4$ **c.** $\left(\dfrac{x}{4}\right)^3$

Example 33   Simplify each expression.

a. $x^7 \cdot x^4$ **b.** $\left(\dfrac{t}{2}\right)^4$ **c.** $\left(9y^5\right)^2$

Solution:

a. Here, we have a product, so we use the product rule to simplify.

$$x^7 \cdot x^4 = x^{7+4} = x^{11}$$

b. This is a quotient raised to a power, so we use the power of a quotient rule.

$$\left(\frac{t}{2}\right)^4 = \frac{t^4}{2^4} = \frac{t^4}{16}$$

c. This is a product raised to a power, so we use the power of a product rule.

$$\left(9y^5\right)^2 = 9^2\left(y^5\right)^2 = 81y^{10}$$

Answers

33. a. x^3 **b.** $81y^{16}$ **c.** $\dfrac{x^3}{64}$

✓**Concept Check Answers**

a. multiply   **b.** subtract
c. none of these   **d.** add

■ **Work Practice 33**

Vocabulary, Readiness & Video Check

Use the choices below to fill in each blank. Some choices may be used more than once.

0	base	add
1	exponent	multiply

1. Repeated multiplication of the same factor can be written using a(n) _____.

2. In 5^2, the 2 is called the _____ and the 5 is called the _____.

3. To simplify $x^2 \cdot x^7$, keep the base and _____ the exponents.

4. To simplify $\left(x^3\right)^6$, keep the base and _____ the exponents.

5. The understood exponent on the term y is _____.

6. If $x^{\square} = 1$, the exponent is _____.

State the bases and the exponents for each of the following expressions.

7. 3^2 _____

8. $(-3)^6$ _____

9. -4^2 _____

10. $5 \cdot 3^4$ _____

11. $5x^2$ _____

12. $(5x)^2$ _____

Martin-Gay Interactive Videos   Watch the section lecture video and answer the following questions.

See Video 12.1 🍎

Objective A **13.** ⊞ Examples 3 and 4 illustrate how to find the base of an exponential expression both with and without parentheses. Explain how identifying the base of ⊞ Example 7 is similar to identifying the base of ⊞ Example 4. ▶

Objective B **14.** Why were the commutative and associative properties applied in ⊞ Example 12? ▶

Objective C **15.** What point is made at the end of ⊞ Example 15? ▶

Objective D **16.** Although it's not especially emphasized in ⊞ Example 20, what is helpful to remind ourselves about the −2 in the problem? ▶

Objective E **17.** In ⊞ Example 24, which exponent rule is used to show that any nonzero base raised to the power of zero is 1? ▶

Objective F **18.** When simplifying an exponential expression that's a fraction, will we always use the quotient rule? Refer to ⊞ Example 30 to support your answer. ▶

12.1 Exercise Set MyMathLab®

Objective A *Evaluate each expression. See Examples 1 through 6.*

1. 7^2

2. -3^2

3. $(-5)^1$

4. $(-3)^2$

5. -2^4

6. -4^3

7. $(-2)^4$

8. $(-4)^3$

9. $\left(\dfrac{1}{3}\right)^3$

10. $\left(-\dfrac{1}{9}\right)^2$

11. $7 \cdot 2^4$

12. $9 \cdot 2^2$

Evaluate each expression with the given replacement values. See Example 7.

13. x^2 when $x = -2$

14. x^3 when $x = -2$

15. $5x^3$ when $x = 3$

16. $4x^2$ when $x = 5$

17. $2xy^2$ when $x = 3$ and $y = -5$

18. $-4x^2y^3$ when $x = 2$ and $y = -1$

▶ **19.** $\dfrac{2z^4}{5}$ when $z = -2$

20. $\dfrac{10}{3y^3}$ when $y = -3$

Objective B *Use the product rule to simplify each expression. Write the results using exponents. See Examples 8 through 15.*

21. $x^2 \cdot x^5$

22. $y^2 \cdot y$

23. $(-3)^3 \cdot (-3)^9$

24. $(-5)^7 \cdot (-5)^6$

▶ **25.** $(5y^4)(3y)$

26. $(-2z^3)(-2z^2)$

▶ **27.** $(x^9y)(x^{10}y^5)$

28. $(a^2b)(a^{13}b^{17})$

29. $(-8mn^6)(9m^2n^2)$

30. $(-7a^3b^3)(7a^{19}b)$

31. $(4z^{10})(-6z^7)(z^3)$

32. $(12x^5)(-x^6)(x^4)$

33. The rectangle below has width $4x^2$ feet and length $5x^3$ feet. Find its area as an expression in x.

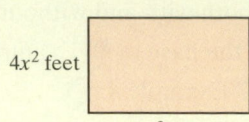

$4x^2$ feet

$5x^3$ feet

△ **34.** The parallelogram below has base length $9y^7$ meters and height $2y^{10}$ meters. Find its area as an expression in y.

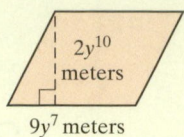

$2y^{10}$ meters

$9y^7$ meters

Objectives C D **Mixed Practice** *Use the power rule and the power of a product or quotient rule to simplify each expression. See Examples 16 through 23.*

▶ **35.** $(x^9)^4$

36. $(y^7)^5$

▶ **37.** $(pq)^8$

38. $(ab)^6$

39. $(2a^5)^3$

40. $(4x^6)^2$

▶ **41.** $(x^2y^3)^5$

42. $(a^4b)^7$

43. $(-7a^2b^5c)^2$

44. $(-3x^7yz^2)^3$

▶ **45.** $\left(\dfrac{r}{s}\right)^9$

46. $\left(\dfrac{q}{t}\right)^{11}$

47. $\left(\dfrac{mp}{n}\right)^9$

48. $\left(\dfrac{xy}{7}\right)^2$

▶ **49.** $\left(\dfrac{-2xz}{y^5}\right)^2$

50. $\left(\dfrac{xy^4}{-3z^3}\right)^3$

△ **51.** The square shown has sides of length $8z^5$ decimeters. Find its area.

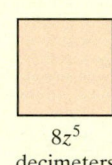

$8z^5$ decimeters

△ **52.** Given the circle below with radius $5y$ centimeters, find its area. Do not approximate π.

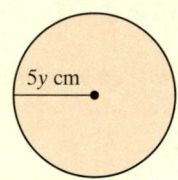

$5y$ cm

53. The vault below is in the shape of a cube. If each side is $3y^4$ feet, find its volume.

$3y^4$ feet $3y^4$ feet

$3y^4$ feet

54. The silo shown is in the shape of a cylinder. If its radius is $4x$ meters and its height is $5x^3$ meters, find its volume. Do not approximate π.

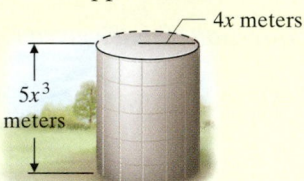

$4x$ meters

$5x^3$ meters

Objective E *Use the quotient rule and simplify each expression. See Examples 24 through 27.*

55. $\dfrac{x^3}{x}$

56. $\dfrac{y^{10}}{y^9}$

▶ **57.** $\dfrac{(-4)^6}{(-4)^3}$

58. $\dfrac{(-6)^{13}}{(-6)^{11}}$

59. $\dfrac{p^7q^{20}}{pq^{15}}$

60. $\dfrac{x^8y^6}{xy^5}$

61. $\dfrac{7x^2y^6}{14x^2y^3}$

▶ **62.** $\dfrac{9a^4b^7}{27ab^2}$

Simplify each expression. See Examples 28 through 32.

63. 7^0

64. 23^0

▶ **65.** $(2x)^0$

66. $(4y)^0$

▶ **67.** $-7x^0$

68. $-2x^0$

▶ **69.** $5^0 + y^0$

70. $-3^0 + 4^0$

Objectives A B C D E F Mixed Practice *Simplify each expression. See Examples 1 through 6, and 8 through 33.*

71. -9^2

72. $(-9)^2$

73. $\left(\dfrac{1}{4}\right)^3$

74. $\left(\dfrac{2}{3}\right)^3$

75. $b^4 b^2$

76. $y^4 y$

77. $a^2 a^3 a^4$

78. $x^2 x^{15} x^9$

▶ **79.** $(2x^3)(-8x^4)$

80. $(3y^4)(-5y)$

81. $(a^7 b^{12})(a^4 b^8)$

82. $(y^2 z^2)(y^{15} z^{13})$

83. $(-2mn^6)(-13m^8 n)$

84. $(-3s^5 t)(-7st^{10})$

85. $(z^4)^{10}$

86. $(t^5)^{11}$

87. $(4ab)^3$

88. $(2ab)^4$

89. $(-6xyz^3)^2$

90. $(-3xy^2 a^3)^3$

91. $\dfrac{z^{12}}{z^4}$

92. $\dfrac{b^6}{b^3}$

▶ **93.** $\dfrac{3x^5}{x}$

94. $\dfrac{5x^9}{x}$

95. $(6b)^0$

96. $(5ab)^0$

97. $(9xy)^2$

98. $(2ab)^5$

99. $2^3 + 2^5$

100. $7^2 - 7^0$

▶ **101.** $\left(\dfrac{3y^5}{6x^4}\right)^3$

102. $\left(\dfrac{2ab}{6yz}\right)^4$

103. $\dfrac{2x^3 y^2 z}{xyz}$

104. $\dfrac{5x^{12} y^{13}}{x^5 y^7}$

Review

Subtract. See Section 8.4.

105. $5 - 7$

106. $9 - 12$

107. $3 - (-2)$

108. $5 - (-10)$

109. $-11 - (-4)$

110. $-15 - (-21)$

Concept Extensions

Solve. See the Concept Checks in this section. For Exercises 111 through 114, match the expression with the operation needed to simplify each. A letter may be used more than once and a letter may not be used at all.

111. $(x^{14})^{23}$

112. $x^{14} \cdot x^{23}$

113. $x^{14} + x^{23}$

114. $\dfrac{x^{35}}{x^{17}}$

a. Add the exponents.
b. Subtract the exponents.
c. Multiply the exponents.
d. Divide the exponents.
e. None of these

Fill in the boxes so that each statement is true. (More than one answer is possible for each exercise.)

115. $x^\square \cdot x^\square = x^{12}$

116. $\left(x^\square\right)^\square = x^{20}$

117. $\dfrac{y^\square}{y^\square} = y^7$

118. $\left(y^\square\right)^\square \cdot \left(y^\square\right)^\square = y^{30}$

△ **119.** The formula $V = x^3$ can be used to find the volume V of a cube with side length x. Find the volume of a cube with side length 7 meters. (Volume is the capacity of a solid such as a cube and is measured in cubic units.)

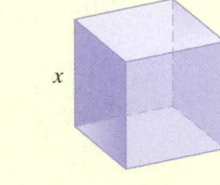

△ **120.** The formula $S = 6x^2$ can be used to find the surface area S of a cube with side length x. Find the surface area of a cube with side length 5 meters. (Surface area is the area of the surface of the cube and is measured in square units.)

△ **121.** To find the amount of water that a swimming pool in the shape of a cube can hold, do we use the formula for volume of the cube or surface area of the cube? (See Exercises **119** and **120**.)

△ **122.** To find the amount of material needed to cover an ottoman in the shape of a cube, do we use the formula for volume of the cube or surface area of the cube? (See Exercises **119** and **120**.)

123. Explain why $(-5)^4 = 625$, while $-5^4 = -625$.

124. Explain why $5 \cdot 4^2 = 80$, while $(5 \cdot 4)^2 = 400$.

125. In your own words, explain why $5^0 = 1$.

126. In your own words, explain when $(-3)^n$ is positive and when it is negative.

Simplify each expression. Assume that variables represent positive integers.

127. $x^{5a}x^{4a}$

128. $b^{9a}b^{4a}$

129. $\left(a^b\right)^5$

130. $\left(2a^{4b}\right)^4$

131. $\dfrac{x^{9a}}{x^{4a}}$

132. $\dfrac{y^{15b}}{y^{6b}}$

12.2 Negative Exponents and Scientific Notation ▶

Objective A Simplifying Expressions Containing Negative Exponents ▶

Our work with exponential expressions so far has been limited to exponents that are positive integers or 0. Here we will also give meaning to an expression like x^{-3}.

Suppose that we wish to simplify the expression $\dfrac{x^2}{x^5}$. If we use the quotient rule for exponents, we subtract exponents:

$$\frac{x^2}{x^5} = x^{2-5} = x^{-3}, \quad x \neq 0$$

But what does x^{-3} mean? Let's simplify $\dfrac{x^2}{x^5}$ using the definition of a^n.

$$\frac{x^2}{x^5} = \frac{x \cdot x}{x \cdot x \cdot x \cdot x \cdot x}$$

$$= \frac{x \cdot x}{x \cdot x \cdot x \cdot x \cdot x} \qquad \text{Divide numerator and denominator by common factors by applying the fundamental principle for fractions.}$$

$$= \frac{1}{x^3}$$

If the quotient rule is to hold true for negative exponents, then x^{-3} must equal $\dfrac{1}{x^3}$. From this example, we state the definition for negative exponents.

Negative Exponents

If a is a real number other than 0 and n is an integer, then

$$a^{-n} = \frac{1}{a^n}$$

For example,

$$x^{-3} = \frac{1}{x^3}$$

In other words, another way to write a^{-n} is to take its reciprocal and change the sign of its exponent.

▶ **Examples** Simplify by writing each expression with positive exponents only.

1. $3^{-2} = \dfrac{1}{3^2} = \dfrac{1}{9}$ Use the definition of negative exponents.

2. $2x^{-3} = 2^1 \cdot \dfrac{1}{x^3} = \dfrac{2^1}{x^3}$ or $\dfrac{2}{x^3}$ Use the definition of negative exponents.

3. $2^{-1} + 4^{-1} = \dfrac{1}{2} + \dfrac{1}{4} = \dfrac{2}{4} + \dfrac{1}{4} = \dfrac{3}{4}$

4. $(-2)^{-4} = \dfrac{1}{(-2)^4} = \dfrac{1}{(-2)(-2)(-2)(-2)} = \dfrac{1}{16}$

Helpful Hint Don't forget that since there are no parentheses, only x is the base for the exponent -3.

■ **Work Practice 1–4**

Objectives

A Simplify Expressions Containing Negative Exponents. ▶

B Use the Rules and Definitions for Exponents to Simplify Exponential Expressions. ▶

C Write Numbers in Scientific Notation. ▶

D Convert Numbers in Scientific Notation to Standard Form. ▶

E Perform Operations on Numbers Written in Scientific Notation. ▶

Practice 1–4

Simplify by writing each expression with positive exponents only.

1. 5^{-3} **2.** $7x^{-4}$

3. $5^{-1} + 3^{-1}$ **4.** $(-3)^{-4}$

Answers

1. $\dfrac{1}{125}$ **2.** $\dfrac{7}{x^4}$ **3.** $\dfrac{8}{15}$ **4.** $\dfrac{1}{81}$

▶ **Helpful Hint**

A negative exponent *does not affect* the sign of its base.
Remember: Another way to write a^{-n} is to take its reciprocal and change the sign
of its exponent: $a^{-n} = \dfrac{1}{a^n}$. For example,

$$x^{-2} = \dfrac{1}{x^2}, \qquad 2^{-3} = \dfrac{1}{2^3} \quad \text{or} \quad \dfrac{1}{8}$$

▶ $\dfrac{1}{y^{-4}} = \dfrac{1}{\frac{1}{y^4}} = y^4, \qquad \dfrac{1}{5^{-2}} = 5^2 \quad \text{or} \quad 25$

From the preceding Helpful Hint, we know that $x^{-2} = \dfrac{1}{x^2}$ and $\dfrac{1}{y^{-4}} = y^4$. We can
use this to include another statement in our definition of negative exponents.

Negative Exponents

If a is a real number other than 0 and n is an integer, then

$$a^{-n} = \dfrac{1}{a^n} \quad \text{and} \quad \dfrac{1}{a^{-n}} = a^n$$

Practice 5–8

Simplify each expression. Write
each result using positive
exponents only.

5. $\left(\dfrac{6}{7}\right)^{-2}$ **6.** $\dfrac{x}{x^{-4}}$

7. $\dfrac{y^{-9}}{z^{-5}}$ **8.** $\dfrac{y^{-4}}{y^6}$

Examples Simplify each expression. Write each result using positive exponents
only.

5. $\left(\dfrac{2}{x}\right)^{-3} = \dfrac{2^{-3}}{x^{-3}} = \dfrac{2^{-3}}{1} \cdot \dfrac{1}{x^{-3}} = \dfrac{1}{2^3} \cdot \dfrac{x^3}{1} = \dfrac{x^3}{2^3} = \dfrac{x^3}{8}$ Use the negative exponents rule.

6. $\dfrac{y}{y^{-2}} = \dfrac{y^1}{y^{-2}} = y^{1-(-2)} = y^3$ Use the quotient rule.

7. $\dfrac{p^{-4}}{q^{-9}} = p^{-4} \cdot \dfrac{1}{q^{-9}} = \dfrac{1}{p^4} \cdot q^9 = \dfrac{q^9}{p^4}$ Use the negative exponents rule.

8. $\dfrac{x^{-5}}{x^7} = x^{-5-7} = x^{-12} = \dfrac{1}{x^{12}}$

■ **Work Practice 5–8**

Objective B Simplifying Exponential Expressions ▶

All the previously stated rules for exponents apply for negative exponents also.
Here is a summary of the rules and definitions for exponents.

Summary of Exponent Rules

If m and n are integers and a, b, and c are real numbers, then

Product rule for exponents:	$a^m \cdot a^n = a^{m+n}$
Power rule for exponents:	$(a^m)^n = a^{m \cdot n}$
Power of a product:	$(ab)^n = a^n b^n$
Power of a quotient:	$\left(\dfrac{a}{c}\right)^n = \dfrac{a^n}{c^n}, \quad c \neq 0$
Quotient rule for exponents:	$\dfrac{a^m}{a^n} = a^{m-n}, \quad a \neq 0$
Zero exponent:	$a^0 = 1, \quad a \neq 0$
Negative exponent:	$a^{-n} = \dfrac{1}{a^n}, \quad a \neq 0$

Answers

5. $\dfrac{49}{36}$ **6.** x^5 **7.** $\dfrac{z^5}{y^9}$ **8.** $\dfrac{1}{y^{10}}$

Examples | Simplify each expression. Write each result using positive exponents only.

9. $\dfrac{(x^3)^4 x}{x^7} = \dfrac{x^{12} \cdot x}{x^7} = \dfrac{x^{12+1}}{x^7} = \dfrac{x^{13}}{x^7} = x^{13-7} = x^6$ Use the power rule.

10. $\left(\dfrac{3a^2}{b}\right)^{-3} = \dfrac{3^{-3}(a^2)^{-3}}{b^{-3}}$ Raise each factor in the numerator and the denominator to the -3 power.

$= \dfrac{3^{-3}a^{-6}}{b^{-3}}$ Use the power rule.

$= \dfrac{b^3}{3^3 a^6}$ Use the negative exponent rule.

$= \dfrac{b^3}{27a^6}$ Write 3^3 as 27.

11. $(y^{-3}z^6)^{-6} = (y^{-3})^{-6}(z^6)^{-6}$ Raise each factor to the -6 power.

$= y^{18}z^{-36} = \dfrac{y^{18}}{z^{36}}$

12. $\dfrac{(2x)^5}{x^3} = \dfrac{2^5 \cdot x^5}{x^3} = 2^5 \cdot x^{5-3} = 32x^2$ Raise each factor in the numerator to the fifth power.

13. $\dfrac{x^{-7}}{(x^4)^3} = \dfrac{x^{-7}}{x^{12}} = x^{-7-12} = x^{-19} = \dfrac{1}{x^{19}}$

14. $(5y^3)^{-2} = 5^{-2}(y^3)^{-2} = 5^{-2}y^{-6} = \dfrac{1}{5^2 y^6} = \dfrac{1}{25y^6}$

15. $-\dfrac{22a^7 b^{-5}}{11a^{-2}b^3} = -\dfrac{22}{11} \cdot a^{7-(-2)}b^{-5-3} = -2a^9 b^{-8} = -\dfrac{2a^9}{b^8}$

16. $\dfrac{(2xy)^{-3}}{(x^2 y^3)^2} = \dfrac{2^{-3}x^{-3}y^{-3}}{(x^2)^2(y^3)^2} = \dfrac{2^{-3}x^{-3}y^{-3}}{x^4 y^6} = 2^{-3}x^{-3-4}y^{-3-6}$

$= 2^{-3}x^{-7}y^{-9} = \dfrac{1}{2^3 x^7 y^9}$ or $\dfrac{1}{8x^7 y^9}$

■ **Work Practice 9–16**

Objective C Writing Numbers in Scientific Notation ▶

Both very large and very small numbers frequently occur in many fields of science. For example, the distance between the sun and the dwarf planet Pluto is approximately 5,906,000,000 kilometers, and the mass of a proton is approximately 0.0000000000000000000000165 gram. It can be tedious to write these numbers in this standard decimal notation, so **scientific notation** is used as a convenient shorthand for expressing very large and very small numbers.

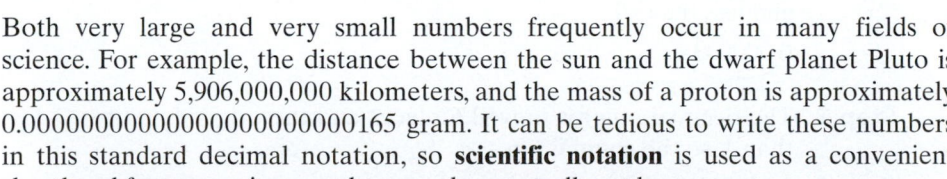

5,906,000,000 kilometers | Pluto

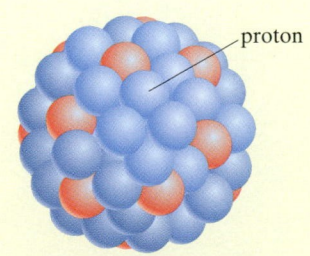

Scientific Notation

A positive number is written in scientific notation if it is written as the product of a number a, where $1 \leq a < 10$, and an integer power r of 10: $a \times 10^r$.

The following numbers are written in scientific notation. The $\times$ sign for multiplication is used as part of the notation.

$$2.03 \times 10^2 \quad 7.362 \times 10^7 \quad 5.906 \times 10^9 \quad \text{(Distance between the sun and Pluto)}$$
$$1 \times 10^{-3} \quad 8.1 \times 10^{-5} \quad 1.65 \times 10^{-24} \quad \text{(Mass of a proton)}$$

The following steps are useful when writing numbers in scientific notation.

To Write a Number in Scientific Notation

Step 1: Move the decimal point in the original number so that the new number has a value between 1 and 10.

Step 2: Count the number of decimal places the decimal point is moved in Step 1. If the original number is 10 or greater, the count is positive. If the original number is less than 1, the count is negative.

Step 3: Multiply the new number in Step 1 by 10 raised to an exponent equal to the count found in Step 2.

Practice 17

Write each number in scientific notation.

a. 420,000   **b.** 0.00017
c. 9,060,000,000   **d.** 0.000007

Example 17 Write each number in scientific notation.

a. 367,000,000   **b.** 0.000003

c. 20,520,000,000   **d.** 0.00085

Solution:

a. Step 1: Move the decimal point until the number is greater than or equal to 1 and less than 10.

$$367,000,000.$$
8 places

Step 2: The decimal point is moved 8 places and the original number is 10 or greater, so the count is positive 8.

Step 3: $367,000,000 = 3.67 \times 10^8$

b. Step 1: Move the decimal point until the number is greater than or equal to 1 and less than 10.

$$0.000003$$
6 places

Step 2: The decimal point is moved 6 places and the original number is less than 1, so the count is -6.

Step 3: $0.000003 = 3.0 \times 10^{-6}$

c. $20,520,000,000 = 2.052 \times 10^{10}$

d. $0.00085 = 8.5 \times 10^{-4}$

■ Work Practice 17

Objective D Converting Numbers to Standard Form

A number written in scientific notation can be rewritten in standard form. For example, to write 8.63×10^3 in standard form, recall that $10^3 = 1000$.

$$8.63 \times 10^3 = 8.63(1000) = 8630$$

Answers

17. a. 4.2×10^5 **b.** 1.7×10^{-4}
c. 9.06×10^9 **d.** 7×10^{-6}

Notice that the exponent on the 10 is positive 3, and we moved the decimal point 3 places to the right.

To write 7.29×10^{-3} in standard form, recall that $10^{-3} = \dfrac{1}{10^3} = \dfrac{1}{1000}$.

$$7.29 \times 10^{-3} = 7.29\left(\dfrac{1}{1000}\right) = \dfrac{7.29}{1000} = 0.00729$$

The exponent on the 10 is negative 3, and we moved the decimal to the left 3 places.

In general, **to write a scientific notation number in standard form,** move the decimal point the same number of places as the exponent on 10. If the exponent is positive, move the decimal point to the right; if the exponent is negative, move the decimal point to the left.

Example 18 Write each number in standard form, without exponents.

a. 1.02×10^5 **b.** 7.358×10^{-3}

c. 8.4×10^7 **d.** 3.007×10^{-5}

Solution:

a. Move the decimal point 5 places to the right.
$1.02 \times 10^5 = 102{,}000.$

b. Move the decimal point 3 places to the left.
$7.358 \times 10^{-3} = 0.007358$

c. $8.4 \times 10^7 = 84{,}000{,}000.$ 7 places to the right

d. $3.007 \times 10^{-5} = 0.00003007$ 5 places to the left

■ **Work Practice 18**

✓**Concept Check** Which number in each pair is larger?
a. 7.8×10^3 or 2.1×10^5
b. 9.2×10^{-2} or 2.7×10^4
c. 5.6×10^{-4} or 6.3×10^{-5}

Objective E Performing Operations with Scientific Notation ▶

Performing operations on numbers written in scientific notation makes use of the rules and definitions for exponents.

Example 19 Perform each indicated operation. Write each result in standard decimal form.

a. $\left(8 \times 10^{-6}\right)\left(7 \times 10^3\right)$

b. $\dfrac{12 \times 10^2}{6 \times 10^{-3}}$

Solution:

a. $\left(8 \times 10^{-6}\right)\left(7 \times 10^3\right) = 8 \cdot 7 \cdot 10^{-6} \cdot 10^3$
$= 56 \times 10^{-3}$
$= 0.056$

b. $\dfrac{12 \times 10^2}{6 \times 10^{-3}} = \dfrac{12}{6} \times 10^{2-(-3)} = 2 \times 10^5 = 200{,}000$

■ **Work Practice 19**

Practice 18

Write the numbers in standard form, without exponents.
a. 3.062×10^{-4}
b. 5.21×10^4
c. 9.6×10^{-5}
d. 6.002×10^6

Practice 19

Perform each indicated operation. Write each result in standard decimal form.
a. $\left(9 \times 10^7\right)\left(4 \times 10^{-9}\right)$
b. $\dfrac{8 \times 10^4}{2 \times 10^{-3}}$

Answers
18. a. 0.0003062   **b.** 52,100
c. 0.000096   **d.** 6,002,000
19. a. 0.36   **b.** 40,000,000

✓**Concept Check Answers**
a. 2.1×10^5 **b.** 2.7×10^4
c. 5.6×10^{-4}

Calculator Explorations Scientific Notation

To enter a number written in scientific notation on a scientific calculator, locate the scientific notation key, which may be marked $\boxed{EE}$ or $\boxed{EXP}$. To enter 3.1×10^7, press $\boxed{3.1}\boxed{EE}\boxed{7}$. The display should read $\boxed{3.1 \quad 07}$.

Enter each number written in scientific notation on your calculator.

1. 5.31×10^3
2. -4.8×10^{14}
3. 6.6×10^{-9}
4. -9.9811×10^{-2}

Multiply each of the following on your calculator. Notice the form of the result.

5. $3,000,000 \times 5,000,000$
6. $230,000 \times 1,000$

Multiply each of the following on your calculator. Write the product in scientific notation.

7. $(3.26 \times 10^6)(2.5 \times 10^{13})$
8. $(8.76 \times 10^{-4})(1.237 \times 10^9)$

Vocabulary, Readiness & Video Check

Fill in each blank with the correct choice.

1. The expression x^{-3} equals _____.

   a. $-x^3$ **b.** $\dfrac{1}{x^3}$ **c.** $\dfrac{-1}{x^3}$ **d.** $\dfrac{1}{x^{-3}}$

2. The expression 5^{-4} equals _____.

   a. -20 **b.** -625 **c.** $\dfrac{1}{20}$ **d.** $\dfrac{1}{625}$

3. The number 3.021×10^{-3} is written in _____.

   a. standard form   **b.** expanded form

   c. scientific notation

4. The number 0.0261 is written in _____.

   a. standard form   **b.** expanded form

   c. scientific notation

Write each expression using positive exponents only.

5. $5x^{-2}$ 6. $3x^{-3}$ 7. $\dfrac{1}{y^{-6}}$ 8. $\dfrac{1}{x^{-3}}$ 9. $\dfrac{4}{y^{-3}}$ 10. $\dfrac{16}{y^{-7}}$

Martin-Gay Interactive Videos

See Video 12.2 🍎

Watch the section lecture video and answer the following questions.

Objective A 11. What important reminder is given at the end of ⊞ Example 1? ▶

Objective B 12. Name all the rules and definitions used to simplify ⊞ Example 8. ▶

Objective C 13. From ⊞ Examples 9 and 10, explain how the movement of the decimal point in Step 1 suggests the sign of the exponent on the number 10. ▶

Objective D 14. From ⊞ Example 11, what part of a number written in scientific notation is key in telling us how to write the number in standard form? ▶

Objective E 15. In ⊞ Example 13, what exponent rules were needed to evaluate the expression? ▶

12.2 Exercise Set MyMathLab®

Objective A *Simplify each expression. Write each result using positive exponents only. See Examples 1 through 8.*

1. 4^{-3}

2. 6^{-2}

3. $7x^{-3}$

4. $(7x)^{-3}$

5. $\left(-\dfrac{1}{4}\right)^{-3}$

6. $\left(-\dfrac{1}{8}\right)^{-2}$

7. $3^{-1} + 2^{-1}$

8. $4^{-1} + 4^{-2}$

9. $\dfrac{1}{p^{-3}}$

10. $\dfrac{1}{q^{-5}}$

11. $\dfrac{p^{-5}}{q^{-4}}$

12. $\dfrac{r^{-5}}{s^{-2}}$

13. $\dfrac{x^{-2}}{x}$

14. $\dfrac{y}{y^{-3}}$

15. $\dfrac{z^{-4}}{z^{-7}}$

16. $\dfrac{x^{-4}}{x^{-1}}$

17. $3^{-2} + 3^{-1}$

18. $4^{-2} - 4^{-3}$

19. $(-3)^{-2}$

20. $(-2)^{-6}$

21. $\dfrac{-1}{p^{-4}}$

22. $\dfrac{-1}{y^{-6}}$

23. $-2^0 - 3^0$

24. $5^0 + (-5)^0$

Objective B *Simplify each expression. Write each result using positive exponents only. See Examples 9 through 16.*

25. $\dfrac{x^2 x^5}{x^3}$

26. $\dfrac{y^4 y^5}{y^6}$

27. $\dfrac{p^2 p}{p^{-1}}$

28. $\dfrac{y^3 y}{y^{-2}}$

29. $\dfrac{(m^5)^4 m}{m^{10}}$

30. $\dfrac{(x^2)^8 x}{x^9}$

31. $\dfrac{r}{r^{-3} r^{-2}}$

32. $\dfrac{p}{p^{-3} q^{-5}}$

33. $(x^5 y^3)^{-3}$

34. $(z^5 x^5)^{-3}$

35. $\dfrac{(x^2)^3}{x^{10}}$

36. $\dfrac{(y^4)^2}{y^{12}}$

37. $\dfrac{(a^5)^2}{(a^3)^4}$

38. $\dfrac{(x^2)^5}{(x^4)^3}$

39. $\dfrac{8k^4}{2k}$

40. $\dfrac{27r^6}{3r^4}$

41. $\dfrac{-6m^4}{-2m^3}$

42. $\dfrac{15a^4}{-15a^5}$

43. $\dfrac{-24a^6 b}{6ab^2}$

44. $\dfrac{-5x^4 y^5}{15x^4 y^2}$

45. $\dfrac{6x^2 y^3}{-7x^2 y^5}$

46. $\dfrac{-8xa^2 b}{-5xa^5 b}$

47. $(3a^2 b^{-4})^3$

48. $(5x^3 y^{-2})^2$

49. $(a^{-5} b^2)^{-6}$

50. $(4^{-1} x^5)^{-2}$

51. $\left(\dfrac{x^{-2} y^4}{x^3 y^7}\right)^{-2}$

52. $\left(\dfrac{a^5 b}{a^7 b^{-2}}\right)^{-3}$

53. $\dfrac{4^2 z^{-3}}{4^3 z^{-5}}$

54. $\dfrac{5^{-1} z^7}{5^{-2} z^9}$

55. $\dfrac{3^{-1} x^4}{3^3 x^{-7}}$

56. $\dfrac{2^{-3} x^{-4}}{2^2 x}$

57. $\dfrac{7ab^{-4}}{7^{-1} a^{-3} b^2}$

58. $\dfrac{6^{-5} x^{-1} y^2}{6^{-2} x^{-4} y^4}$

59. $\dfrac{-12m^5 n^{-7}}{4m^{-2} n^{-3}}$

60. $\dfrac{-15r^{-6} s}{5r^{-4} s^{-3}}$

61. $\left(\dfrac{a^{-5} b}{ab^3}\right)^{-4}$

62. $\left(\dfrac{r^{-2} s^{-3}}{r^{-4} s^{-3}}\right)^{-3}$

63. $(5^2)(8)(2^0)$

64. $(3^4)(7^0)(2)$

65. $\dfrac{(xy^3)^5}{(xy)^{-4}}$

66. $\dfrac{(rs)^{-3}}{(r^2 s^3)^2}$

67. $\dfrac{(-2xy^{-3})^{-3}}{(xy^{-1})^{-1}}$

68. $\dfrac{(-3x^2 y^2)^{-2}}{(xyz)^{-2}}$

69. $\dfrac{(a^4 b^{-7})^{-5}}{(5a^2 b^{-1})^{-2}}$

70. $\dfrac{(a^6 b^{-2})^4}{(4a^{-3} b^{-3})^3}$

71. Find the volume of the cube.

$\frac{3x^{-2}}{z}$ inches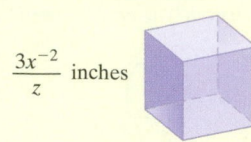

72. Find the area of the triangle.

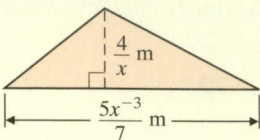

Objective C *Write each number in scientific notation. See Example 17.*

▶ **73.** 78,000

74. 9,300,000,000

▶ **75.** 0.00000167

76. 0.00000017

77. 0.00635

78. 0.00194

79. 1,160,000

80. 700,000

81. When it is completed in 2022, the Thirty Meter Telescope is expected to be the world's largest optical telescope. Located in an observatory complex at the summit of Mauna Kea in Hawaii, the elevation of the Thirty Meter Telescope will be roughly 4200 meters above sea level. Write 4200 in scientific notation.

82. The Thirty Meter Telescope (see Exercise **81**) will have the ability to view objects 13,000,000,000 light-years away. Write 13,000,000,000 in scientific notation.

Objective D *Write each number in standard form. See Example 18.*

83. 8.673×10^{-10}

84. 9.056×10^{-4}

▶ **85.** 3.3×10^{-2}

86. 4.8×10^{-6}

▶ **87.** 2.032×10^{4}

88. 9.07×10^{10}

89. Each second, the Sun converts 7.0×10^{8} tons of hydrogen into helium and energy in the form of gamma rays. Write this number in standard form. (*Source:* Students for the Exploration and Development of Space)

90. In chemistry, Avogadro's number is the number of atoms in one mole of an element. Avogadro's number is $6.02214199 \times 10^{23}$. Write this number in standard form. (*Source:* National Institute of Standards and Technology)

Objectives C D Mixed Practice *See Examples 17 and 18. If a number is written in standard form, write it in scientific notation. If a number is written in scientific notation, write it in standard form. (Source: CIA World Factbook) The bar graph below shows estimates of the top six national debts as of December 31, 2012.*

Top Six National Debts

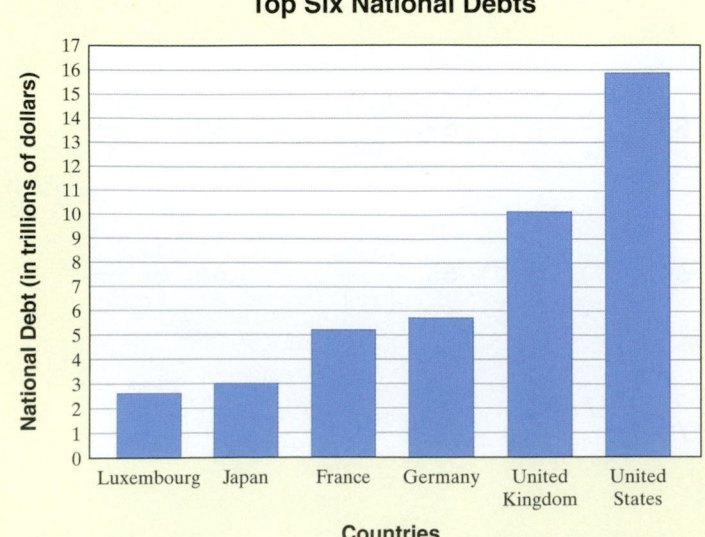

Source: CIA World Factbook

91. Germany's national debt as of the end of 2012 was $5,700,000,000,000.

92. Luxembourg's national debt as of the end of 2012 was $2,600,000,000,000.

93. The United Kingdom's national debt as of the end of 2012 was 1.01×10^{13}.

94. France's national debt as of the end of 2012 was 5.2×10^{12}.

95. Use the bar graph to estimate the national debt of Japan and then express it in both standard and scientific notation.

96. Use the bar graph to estimate the national debt of the United States and then express it in both standard and scientific notation.

Objective **E** *Evaluate each expression using exponential rules. Write each result in standard form. See Example 19.*

97. $(1.2 \times 10^{-3})(3 \times 10^{-2})$

98. $(2.5 \times 10^{6})(2 \times 10^{-6})$

99. $(4 \times 10^{-10})(7 \times 10^{-9})$

100. $(5 \times 10^{6})(4 \times 10^{-8})$

101. $\dfrac{8 \times 10^{-1}}{16 \times 10^{5}}$

102. $\dfrac{25 \times 10^{-4}}{5 \times 10^{-9}}$

▶ **103.** $\dfrac{1.4 \times 10^{-2}}{7 \times 10^{-8}}$

104. $\dfrac{0.4 \times 10^{5}}{0.2 \times 10^{11}}$

105. Although the actual amount varies by season and time of day, the average volume of water that flows over Niagara Falls (the American and Canadian falls combined) each second is 7.5×10^{5} gallons. How much water flows over Niagara Falls in an hour? Write the result in scientific notation. (*Hint:* 1 hour equals 3600 seconds.) (*Source:* http://niagarafallslive.com)

106. A beam of light travels 9.460×10^{12} kilometers per year. How far does light travel in 10,000 years? Write the result in scientific notation.

Review

Simplify each expression by combining any like terms. See Section 8.7.

107. $3x - 5x + 7$

108. $7w + w - 2w$

109. $y - 10 + y$

110. $-6z + 20 - 3z$

111. $7x + 2 - 8x - 6$

112. $10y - 14 - y - 14$

Concept Extensions

For Exercises 113–120, write each number in standard form. Then write the number in scientific notation.

113. In September 2013, Google received an estimated 900 million unique monthly visitors. (*Source:* eBizMBA Inc.)

114. In September 2013, YouTube received an estimated 0.45 billion unique monthly visitors. (*Source:* eBizMBA Inc.)

115. The surface of the Arctic Ocean encompasses 14.056 million square kilometers of water. (*Source:* CIA World Factbook)

116. The surface of the Pacific Ocean encompasses 155.557 million square kilometers of water. (*Source:* CIA World Factbook)

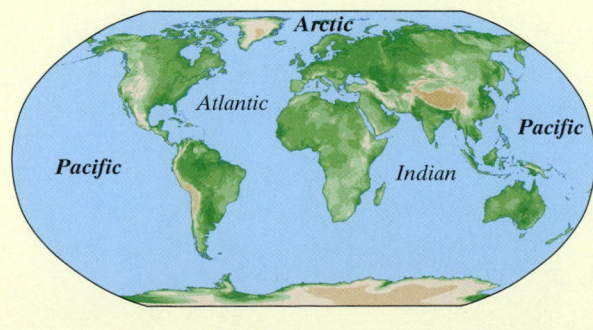

Solve.

117. A nanometer is one-billionth, or 10^{-9}, of a meter. A strand of DNA is about 2.5 nanometers in diameter. Use scientific notation to write the diameter of a DNA strand in terms of meters. (*Source:* United States National Nanotechnology Initiative)

118. A micrometer (sometimes referred to as a micron) is one-millionth, or 10^{-6}, of a meter. A single red blood cell is about 7 micrometers in diameter. Use scientific notation to write the diameter of a red blood cell in terms of meters. (*Source:* National Institute of Standards and Technology)

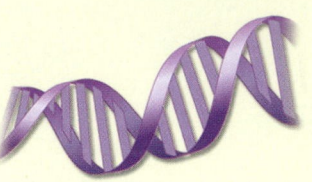

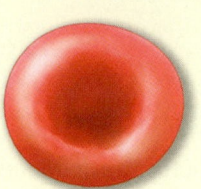

119. The Thirty Meter Telescope, described in Exercises **81–82**, will be capable of observing ultraviolet wavelengths measuring 310 nanometers. Express this wavelength in terms of meters using both standard form and scientific notation. (See Exercise **117** for a definition of nanometer.) (*Source:* TMT Observatory Corporation)

120. The Thirty Meter Telescope, described in Exercises **81–82**, will be capable of observing infrared wavelengths measuring 28 micrometers. Express this wavelength in terms of meters using both standard form and scientific notation. (See Exercise **118** for a definition of micrometer.) (*Source:* TMT Observatory Corporation)

Simplify.

121. $\left(2a^3\right)^3 a^4 + a^5 a^8$

122. $\left(2a^3\right)^3 a^{-3} + a^{11} a^{-5}$

Fill in the boxes so that each statement is true. (More than one answer is possible for these exercises.)

123. $x^{\square} = \dfrac{1}{x^5}$

124. $7^{\square} = \dfrac{1}{49}$

125. $z^{\square} \cdot z^{\square} = z^{-10}$

126. $\left(x^{\square}\right)^{\square} = x^{-15}$

127. Which is larger? See the Concept Check in this section.
  a. 9.7×10^{-2} or 1.3×10^1
  b. 8.6×10^5 or 4.4×10^7
  c. 6.1×10^{-2} or 5.6×10^{-4}

128. Determine whether each statement is true or false.
  a. $5^{-1} < 5^{-2}$
  b. $\left(\dfrac{1}{5}\right)^{-1} < \left(\dfrac{1}{5}\right)^{-2}$
  c. $a^{-1} < a^{-2}$ for all nonzero numbers.

129. It was stated earlier that for an integer n,

$$x^{-n} = \dfrac{1}{x^n}, \quad x \neq 0.$$

Explain why x may not equal 0.

130. The quotient rule states that

$$\dfrac{a^m}{a^n} = a^{m-n}, \; a \neq 0.$$

Explain why a may not equal 0.

Simplify each expression. Assume that variables represent positive integers.

131. $\left(x^{-3s}\right)^3$

132. $a^{-4m} \cdot a^{5m}$

133. $a^{4m+1} \cdot a^4$

134. $\left(3y^{2z}\right)^3$

Objective A Defining Term and Coefficient

In this section, we introduce a special algebraic expression called a polynomial. Let's first review some definitions presented in Section 8.7.

Recall that a term is a number or the product of a number and variables raised to powers. The terms of an expression are separated by plus signs. The terms of the expression $4x^2 + 3x$ are $4x^2$ and $3x$. The terms of the expression $9x^4 - 7x - 1$, or $9x^4 + (-7x) + (-1)$, are $9x^4$, $-7x$, and -1.

Expression	Terms
$4x^2 + 3x$	$4x^2, 3x$
$9x^4 - 7x - 1$	$9x^4, -7x, -1$
$7y^3$	$7y^3$

The **numerical coefficient** of a term, or simply the **coefficient,** is the numerical factor of each term. If no numerical factor appears in the term, then the coefficient is understood to be 1. If the term is a number only, it is called a **constant term** or simply a **constant.**

Term	Coefficient
x^5	1
$3x^2$	3
$-4x$	-4
$-x^2y$	-1
3 (constant)	3

Example 1 Complete the table for the expression $7x^5 - 8x^4 + x^2 - 3x + 5$.

Term	Coefficient
x^2	
	-8
$-3x$	
	7
5	

Solution: The completed table is shown below.

Term	Coefficient
x^2	1
$-8x^4$	-8
$-3x$	-3
$7x^5$	7
5	5

■ Work Practice 1

Objectives

A Define Term and Coefficient of a Term.

B Define Polynomial, Monomial, Binomial, Trinomial, and Degree.

C Evaluate a Polynomial for Given Replacement Values.

D Simplify a Polynomial by Combining Like Terms.

E Simplify a Polynomial in Several Variables.

F Write a Polynomial in Descending Powers of the Variable and with No Missing Powers of the Variable.

Practice 1

Complete the table for the expression
$-6x^6 + 4x^5 + 7x^3 - 9x^2 - 1$.

Term	Coefficient
$7x^3$	
	-9
$-6x^6$	
	4
-1	

Answer
1. term: $-9x^2, 4x^5$; coefficient: 7, -6, -1

Objective B Defining Polynomial, Monomial, Binomial, Trinomial, and Degree ▶

Now we are ready to define what we mean by a polynomial.

Polynomial

A **polynomial in x** is a finite sum of terms of the form ax^n, where a is a real number and n is a whole number.

For example,

$$x^5 - 3x^3 + 2x^2 - 5x + 1$$

is a polynomial in x. Notice that this polynomial is written in **descending powers** of x, because the powers of x decrease from left to right. (Recall that the term 1 can be thought of as $1x^0$.)

On the other hand,

$$x^{-5} + 2x - 3$$

is **not** a polynomial because one of its terms contains a variable with an exponent, -5, that is not a whole number.

Types of Polynomials

A **monomial** is a polynomial with exactly one term.
A **binomial** is a polynomial with exactly two terms.
A **trinomial** is a polynomial with exactly three terms.

The following are examples of monomials, binomials, and trinomials. Each of these examples is also a polynomial.

Polynomials			
Monomials	**Binomials**	**Trinomials**	**More than Three Terms**
ax^2	$x + y$	$x^2 + 4xy + y^2$	$5x^3 - 6x^2 + 3x - 6$
$-3z$	$3p + 2$	$x^5 + 7x^2 - x$	$-y^5 + y^4 - 3y^3 - y^2 + y$
4	$4x^2 - 7$	$-q^4 + q^3 - 2q$	$x^6 + x^4 - x^3 + 1$

Each term of a polynomial has a degree. The **degree of a term in one variable** is the exponent on the variable.

Practice 2

Identify the degree of each term of the trinomial $-15x^3 + 2x^2 - 5$.

Example 2 Identify the degree of each term of the trinomial $12x^4 - 7x + 3$.

Solution: The term $12x^4$ has degree 4.
The term $-7x$ has degree 1 since $-7x$ is $-7x^1$.
The term 3 has degree 0 since 3 is $3x^0$.

■ Work Practice 2

Each polynomial also has a degree.

Degree of a Polynomial

The **degree of a polynomial** is the greatest degree of any term of the polynomial.

Answer
2. 3; 2; 0

Example 3 Find the degree of each polynomial and tell whether the polynomial is a monomial, binomial, trinomial, or none of these.

a. $-2t^2 + 3t + 6$ **b.** $15x - 10$ **c.** $7x + 3x^3 + 2x^2 - 1$

Solution:

a. The degree of the trinomial $-2t^2 + 3t + 6$ is 2, the greatest degree of any of its terms.

b. The degree of the binomial $15x - 10$ or $15x^1 - 10$ is 1.

c. The degree of the polynomial $7x + 3x^3 + 2x^2 - 1$ is 3. The polynomial is neither a monomial, binomial, nor trinomial.

■ **Work Practice 3**

Objective C Evaluating Polynomials ▶

Polynomials have different values depending on the replacement values for the variables. When we find the value of a polynomial for a given replacement value, we are evaluating the polynomial for that value.

Example 4 Evaluate each polynomial when $x = -2$.

a. $-5x + 6$ **b.** $3x^2 - 2x + 1$

Solution:

a. $-5x + 6 = -5(-2) + 6$ Replace x with -2.

$$= 10 + 6$$
$$= 16$$

b. $3x^2 - 2x + 1 = 3(-2)^2 - 2(-2) + 1$ Replace x with -2.

$$= 3(4) - 2(-2) + 1$$
$$= 12 + 4 + 1$$
$$= 17$$

■ **Work Practice 4**

Many physical phenomena can be modeled by polynomials.

Example 5 Finding Free-Fall Time

The Swiss Re Building in London is a unique building. Londoners often refer to it as the "pickle building." The building is 592.1 feet tall. An object is dropped from the highest point of this building. Neglecting air resistance, the height in feet of the object above ground at time t seconds is given by the polynomial $-16t^2 + 592.1$. Find the height of the object when $t = 1$ second and when $t = 6$ seconds.

Solution: To find each height, we evaluate the polynomial when $t = 1$ and when $t = 6$.

$$-16t^2 + 592.1 = -16(1)^2 + 592.1 \quad \text{Replace } t \text{ with 1.}$$
$$= -16(1) + 592.1$$
$$= -16 + 592.1$$
$$= 576.1$$

The height of the object at 1 second is 576.1 feet.

$$-16t^2 + 592.1 = -16(6)^2 + 592.1 \quad \text{Replace } t \text{ with 6.}$$
$$= -16(36) + 592.1$$
$$= -576 + 592.1 = 16.1$$

(Continued on next page)

Practice 3

Find the degree of each polynomial and tell whether the polynomial is a monomial, binomial, trinomial, or none of these.

a. $-6x + 14$

b. $9x - 3x^6 + 5x^4 + 2$

c. $10x^2 - 6x - 6$

Practice 4

Evaluate each polynomial when $x = -1$.

a. $-2x + 10$

b. $6x^2 + 11x - 20$

Practice 5

Find the height of the object in Example 5 when $t = 2$ seconds and $t = 4$ seconds.

Answers

3. a. binomial, 1   **b.** none of these, 6
c. trinomial, 2   **4. a.** 12   **b.** -25
5. 528.1 feet, 336.1 feet

The height of the object at 6 seconds is 16.1 feet.

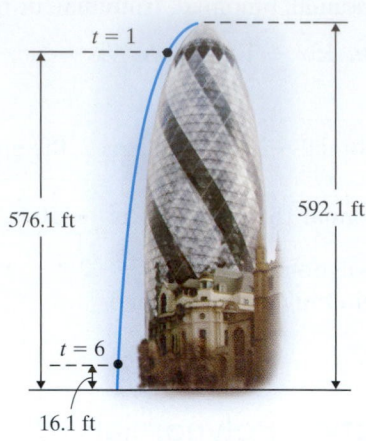

■ **Work Practice 5**

Objective D Simplifying Polynomials by Combining Like Terms ▶

We can simplify polynomials with like terms by combining the like terms. Recall from Section 8.7 that like terms are terms that contain exactly the same variables raised to exactly the same powers.

Like Terms	Unlike Terms
$5x^2, -7x^2$	$3x, 3y$
$y, 2y$	$-2x^2, -5x$
$\frac{1}{2}a^2 b, -a^2 b$	$6st^2, 4s^2 t$

Only like terms can be combined. We combine like terms by applying the distributive property.

Practice 6–10

Simplify each polynomial by combining any like terms.

6. $-6y + 8y$
7. $14y^2 + 3 - 10y^2 - 9$
8. $7x^3 + x^3$
9. $23x^2 - 6x - x - 15$
10. $\frac{2}{7}x^3 - \frac{1}{4}x + 2 - \frac{1}{2}x^3 + \frac{3}{8}x$

Examples Simplify each polynomial by combining any like terms.

6. $-3x + 7x = (-3 + 7)x = 4x$

7. $11x^2 + 5 + 2x^2 - 7 = 11x^2 + 2x^2 + 5 - 7$
$$= 13x^2 - 2$$

8. $9x^3 + x^3 = 9x^3 + 1x^3$ Write x^3 as $1x^3$.
$$= 10x^3$$

9. $5x^2 + 6x - 9x - 3 = 5x^2 - 3x - 3$ Combine like terms $6x$ and $-9x$.

10. $\frac{2}{5}x^4 + \frac{2}{3}x^3 - x^2 + \frac{1}{10}x^4 - \frac{1}{6}x^3$

$$= \left(\frac{2}{5} + \frac{1}{10}\right)x^4 + \left(\frac{2}{3} - \frac{1}{6}\right)x^3 - x^2$$

$$= \left(\frac{4}{10} + \frac{1}{10}\right)x^4 + \left(\frac{4}{6} - \frac{1}{6}\right)x^3 - x^2$$

$$= \frac{5}{10}x^4 + \frac{3}{6}x^3 - x^2$$

$$= \frac{1}{2}x^4 + \frac{1}{2}x^3 - x^2$$

■ **Work Practice 6–10**

Answers

6. $2y$ 7. $4y^2 - 6$ 8. $8x^3$

9. $23x^2 - 7x - 15$

10. $-\frac{3}{14}x^3 + \frac{1}{8}x + 2$

Example 11   Write a polynomial that describes the total area of the squares and rectangles shown below. Then simplify the polynomial.

Solution: Recall that the area of a rectangle is length times width.

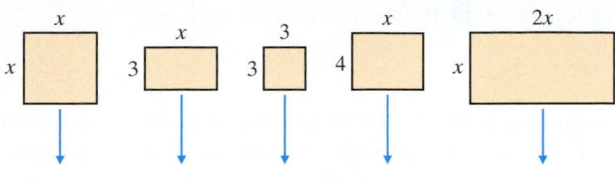

Area: $x \cdot x \; + \; 3 \cdot x \; + \; 3 \cdot 3 \; + \; 4 \cdot x \; + \; x \cdot 2x$

$= x^2 + 3x + 9 + 4x + 2x^2$

$= 3x^2 + 7x + 9$ Combine like terms.

■ **Work Practice 11**

Objective E Simplifying Polynomials Containing Several Variables

A polynomial may contain more than one variable. One example is

$$5x + 3xy^2 - 6x^2y^2 + x^2y - 2y + 1$$

We call this expression a polynomial in several variables.

The **degree of a term** with more than one variable is the sum of the exponents on the variables. The **degree of a polynomial** in several variables is still the greatest degree of the terms of the polynomial.

Example 12   Identify the degrees of the terms and the degree of the polynomial $5x + 3xy^2 - 6x^2y^2 + x^2y - 2y + 1$.

Solution: To organize our work, we use a table.

Terms of Polynomial	Degree of Term	Degree of Polynomial
$5x$	1	
$3xy^2$	1 + 2, or 3	
$-6x^2y^2$	2 + 2, or 4	4 (greatest degree)
x^2y	2 + 1, or 3	
$-2y$	1	
1	0	

■ **Work Practice 12**

To simplify a polynomial containing several variables, we combine any like terms.

Examples   Simplify each polynomial by combining any like terms.

13. $3xy - 5y^2 + 7yx - 9x^2 = (3 + 7)xy - 5y^2 - 9x^2$

$= 10xy - 5y^2 - 9x^2$

14. $9a^2b - 6a^2 + 5b^2 + a^2b - 11a^2 + 2b^2$

$= 10a^2b - 17a^2 + 7b^2$

Helpful Hint   This term can be written as $7yx$ or $7xy$.

■ **Work Practice 13–14**

Practice 11

Write a polynomial that describes the total area of the squares and rectangles shown below. Then simplify the polynomial.

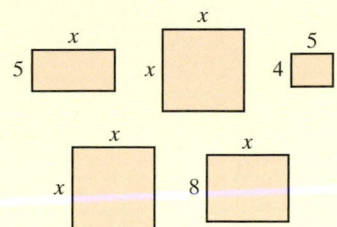

Practice 12

Identify the degrees of the terms and the degree of the polynomial $-2x^3y^2 + 4 - 8xy + 3x^3y + 5xy^2$.

Practice 13–14

Simplify each polynomial by combining any like terms.
13. $11ab - 6a^2 - ba + 8b^2$
14. $7x^2y^2 + 2y^2 - 4y^2x^2 + x^2 - y^2 + 5x^2$

Answers

11. $5x + x^2 + 20 + x^2 + 8x$;
$2x^2 + 13x + 20$ **12.** $5, 0, 2, 4, 3; 5$
13. $10ab - 6a^2 + 8b^2$
14. $3x^2y^2 + y^2 + 6x^2$

Objective F Inserting "Missing" Terms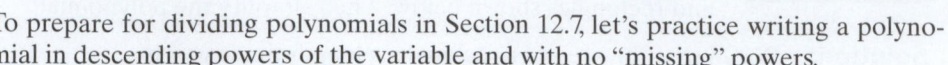

To prepare for dividing polynomials in Section 12.7, let's practice writing a polynomial in descending powers of the variable and with no "missing" powers.

Recall from Objective **B** that a polynomial such as

$$x^5 - 3x^3 + 2x^2 - 5x + 1$$

is written in descending powers of x because the powers of x decrease from left to right. Study the decreasing powers of x and notice that there is a "missing" power of x. This missing power is x^4. Writing a polynomial in decreasing powers of the variable helps you immediately determine important features of the polynomial, such as its degree. It is also sometimes helpful to write a polynomial so that there are no "missing" powers of x. For our polynomial above, if we simply insert a term of $0x^4$, which equals 0, we have an equivalent polynomial with no missing powers of x.

$$x^5 - 3x^3 + 2x^2 - 5x + 1 = x^5 + 0x^4 - 3x^3 + 2x^2 - 5x + 1$$

Practice 15

Write each polynomial in descending powers of the variable with no missing powers.

a. $x^2 + 9$

b. $9m^3 + m^2 - 5$

c. $-3a^3 + a^4$

Example 15 Write each polynomial in descending powers of the variable with no missing powers.

a. $x^2 - 4$ **b.** $3m^3 - m + 1$ **c.** $2x + x^4$

Solution:

a. $x^2 - 4 = x^2 + 0x^1 - 4$ or $x^2 + 0x - 4$ Insert a missing term of $0x^1$ or $0x$.

b. $3m^3 - m + 1 = 3m^3 + 0m^2 - m + 1$ Insert a missing term of $0m^2$.

c. $2x + x^4 = x^4 + 2x$ Write in descending powers of variable.

$\qquad = x^4 + 0x^3 + 0x^2 + 2x + 0x^0$ Insert missing terms of $0x^3, 0x^2$, and $0x^0$ (or 0).

■ Work Practice 15

Helpful Hint

Since there is no constant as a last term, we insert a $0x^0$. This $0x^0$ (or 0) is the final power of x in our polynomial.

Answers

15. a. $x^2 + 0x + 9$

b. $9m^3 + m^2 + 0m - 5$

c. $a^4 - 3a^3 + 0a^2 + 0a + 0a^0$

Vocabulary, Readiness & Video Check

Use the choices below to fill in each blank. Not all choices will be used.

least	monomial	trinomial	coefficient
greatest	binomial	constant	

1. A _____ is a polynomial with exactly two terms.

2. A _____ is a polynomial with exactly one term.

3. A _____ is a polynomial with exactly three terms.

4. The numerical factor of a term is called the _____.

5. A number term is also called a _____.

6. The degree of a polynomial is the _____ degree of any term of the polynomial.

Martin-Gay Interactive Videos   Watch the section lecture video and answer the following questions.

See Video 12.3 🍎

Objective A 7. How many terms does the polynomial in ▥ Example 1 have? What are they? ▶

Objective B 8. For ▥ Example 2, why is the degree of each **term** found when the example asks for the degree of the **polynomial** only? ▶

Objective C 9. From ▥ Example 3, what does the value of a polynomial depend on? ▶

Objective D 10. When combining any like terms in a polynomial, as in ▥ Example 5, what are we doing to the polynomial? ▶

Objective E 11. In ▥ Example 6, after combining like terms what is the degree of the binomial? Which term determines this? ▶

Objective F 12. In ▥ Example 7, what power is "missing" from the original polynomial? What term is inserted to replace this missing power? ▶

12.3 **Exercise Set** MyMathLab® ▶

Objective A *Complete each table for each polynomial. See Example 1.*

▶ **1.** $x^2 - 3x + 5$

Term	Coefficient
x^2	
	-3
5	

2. $2x^3 - x + 4$

Term	Coefficient
	2
$-x$	
4	

3. $-5x^4 + 3.2x^2 + x - 5$

Term	Coefficient
$-5x^4$	
$3.2x^2$	
x	
-5	

4. $9.7x^7 - 3x^5 + x^3 - \frac{1}{4}x^2$

Term	Coefficient
$9.7x^7$	
$-3x^5$	
x^3	
$-\frac{1}{4}x^2$	

Objective B *Find the degree of each polynomial and determine whether it is a monomial, binomial, trinomial, or none of these. See Examples 2 and 3.*

5. $x + 2$

6. $-6y + 4$

▶ **7.** $9m^3 - 5m^2 + 4m - 8$

8. $a + 5a^2 + 3a^3 - 4a^4$

9. $12x^4 - x^6 - 12x^2$

10. $7r^2 + 2r - 3r^5$

11. $3z - 5z^4$

12. $5y^6 + 2$

Objective C *Evaluate each polynomial when* **a.** $x = 0$ *and* **b.** $x = -1$. *See Examples 4 and 5.*

13. $5x - 6$

14. $2x - 10$

▶ **15.** $x^2 - 5x - 2$

16. $x^2 + 3x - 4$

17. $-x^3 + 4x^2 - 15$

18. $-2x^3 + 3x^2 - 6$

A rocket is fired upward from the ground with an initial velocity of 200 feet per second. Neglecting air resistance, the height of the rocket at any time t can be described in feet by the polynomial $-16t^2 + 200t$. Find the height of the rocket at the time given in Exercises 19 through 22. See Example 5.

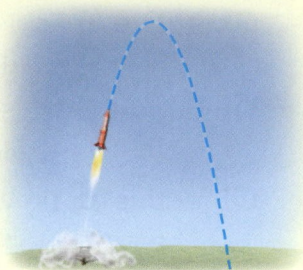

	Time, t (in seconds)	Height $-16t^2 + 200t$
19.	1	
20.	5	
21.	7.6	
22.	10.3	

Pictured Rocks National Lakeshore

Scotts Bluff National Monument

23. The polynomial $-15x^2 + 77x + 499$ models the yearly number of visitors (in thousands) x years after 2010 at Pictured Rocks National Lakeshore. Use this polynomial to estimate the number of visitors to the park in 2012 ($x = 2$).

24. The polynomial $-0.5x^2 - 4.5x + 134$ models the yearly number of visitors (in thousands) x years after 2010 at Scotts Bluff National Monument. Use this polynomial to estimate the number of visitors to the monument in 2011 ($x = 1$).

25. The number of cell sites (in thousands) in the United States x years after 2008 is given by the polynomial $3.7x^2 + 0.7x + 241.6$ for 2008 through 2012. Use this model to predict the number of cell sites in the United States in 2015 ($x = 7$). (*Source:* Based on data from CTIA—The Wireless Association)

26. The number of active wireless devices (in millions) in the United States x years after 2004 is given by the polynomial $-x^2 + 26x + 184$ for 2004 through 2012. Use this model to predict the number of active wireless devices in the United States in 2016 ($x = 12$). (*Source:* Based on data from CTIA—The Wireless Association)

Objective D *Simplify each expression by combining like terms. See Examples 6 through 10.*

27. $9x - 20x$

28. $14y - 30y$

29. $14x^3 + 9x^3$

30. $18x^3 + 4x^3$

31. $7x^2 + 3 + 9x^2 - 10$

32. $8x^2 + 4 + 11x^2 - 20$

33. $15x^2 - 3x^2 - 13$

34. $12k^3 - 9k^3 + 11$

35. $8s - 5s + 4s$

36. $5y + 7y - 6y$

37. $0.1y^2 - 1.2y^2 + 6.7 - 1.9$

38. $7.6y + 3.2y^2 - 8y - 2.5y^2$

39. $\frac{2}{3}x^4 + 12x^3 + \frac{1}{6}x^4 - 19x^3 - 19$

40. $\frac{2}{5}x^4 - 23x^2 + \frac{1}{15}x^4 + 5x^2 - 5$

41. $\frac{3}{20}x^3 + \frac{1}{10} - \frac{3}{10}x - \frac{1}{5} - \frac{7}{20}x + 6x^2$

42. $\frac{5}{16}x^3 - \frac{1}{8} + \frac{3}{8}x + \frac{1}{4} - \frac{9}{16}x - 14x^2$

Write a polynomial that describes the total area of each set of rectangles and squares shown in Exercises 43 and 44. Then simplify the polynomial. See Example 11.

△ **43.**

△ **44.**

Recall that the perimeter of a figure such as the ones shown in Exercises 45 and 46 is the sum of the lengths of its sides. Write each perimeter as a polynomial. Then simplify the polynomial.

△ **45.**

△ **46.**

Objective **E** *Identify the degrees of the terms and the degree of the polynomial. See Example 12.*

47. $9ab - 6a + 5b - 3$

48. $y^4 - 6y^3x + 2x^2y^2 - 5y^2 + 3$

49. $x^3y - 6 + 2x^2y^2 + 5y^3$

50. $2a^2b + 10a^4b - 9ab + 6$

Simplify each polynomial by combining any like terms. See Examples 13 and 14.

▶ **51.** $3ab - 4a + 6ab - 7a$

52. $-9xy + 7y - xy - 6y$

53. $4x^2 - 6xy + 3y^2 - xy$

54. $3a^2 - 9ab + 4b^2 - 7ab$

55. $5x^2y + 6xy^2 - 5yx^2 + 4 - 9y^2x$

56. $17a^2b - 16ab^2 + 3a^3 + 4ba^3 - b^2a$

57. $14y^3 - 9 + 3a^2b^2 - 10 - 19b^2a^2$

58. $18x^4 + 2x^3y^3 - 1 - 2y^3x^3 - 17x^4$

Objective **F** *Write each polynomial in descending powers of the variable and with no missing powers. See Example 15.*

59. $7x^2 + 3$

60. $5x^2 - 2$

61. $x^3 - 64$

62. $x^3 - 8$

63. $5y^3 + 2y - 10$

64. $6m^3 - 3m + 4$

65. $8y + 2y^4$

66. $11z + 4z^4$

67. $6x^5 + x^3 - 3x + 15$

68. $9y^5 - y^2 + 2y - 11$

Review

Simplify each expression. See Section 8.7.

69. $4 + 5(2x + 3)$ **70.** $9 - 6(5x + 1)$ **71.** $2(x - 5) + 3(5 - x)$ **72.** $-3(w + 7) + 5(w + 1)$

Concept Extensions

73. Describe how to find the degree of a term.

74. Describe how to find the degree of a polynomial.

75. Explain why xyz is a monomial while $x + y + z$ is a trinomial.

76. Explain why the degree of the term $5y^3$ is 3 and the degree of the polynomial $2y + y + 2y$ is 1.

Simplify, if possible.

77. $x^4 \cdot x^9$

78. $x^4 + x^9$

79. $a \cdot b^3 \cdot a^2 \cdot b^7$

80. $a + b^3 + a^2 + b^7$

81. $\left(y^5\right)^4 + \left(y^2\right)^{10}$

82. $x^5y^2 + y^2x^5$

Fill in the boxes so that the terms in each expression can be combined. Then simplify. Each exercise has more than one solution.

83. $7x^\square + 2x^\square$

84. $\left(3y^2\right)^\square + \left(4y^3\right)^\square$

85. Explain why the height of the rocket in Exercises **19** through **22** increases and then decreases as time passes.

86. Approximate (to the nearest tenth of a second) how long before the rocket in Exercises **19** through **22** hits the ground.

Simplify each polynomial by combining like terms.

87. $1.85x^2 - 3.76x + 9.25x^2 + 10.76 - 4.21x$

88. $7.75x + 9.16x^2 - 1.27 - 14.58x^2 - 18.34$

12.4 Adding and Subtracting Polynomials

Objective A Adding Polynomials

To add polynomials, we use commutative and associative properties and then combine like terms. To see if you are ready to add polynomials, try the Concept Check.

✓**Concept Check** When combining like terms in the expression $5x - 8x^2 - 8x$, which of the following is the proper result?

a. $-11x^2$ **b.** $-3x - 8x^2$ **c.** $-11x$ **d.** $-11x^4$

To Add Polynomials

To add polynomials, combine all like terms.

Objectives

A Add Polynomials.

B Subtract Polynomials.

C Add or Subtract Polynomials in One Variable.

D Add or Subtract Polynomials in Several Variables.

Examples Add.

1. $\left(4x^3 - 6x^2 + 2x + 7\right) + \left(5x^2 - 2x\right)$

$= 4x^3 - 6x^2 + 2x + 7 + 5x^2 - 2x$ Remove parentheses.

$= 4x^3 + \left(-6x^2 + 5x^2\right) + (2x - 2x) + 7$ Group like terms.

$= 4x^3 - x^2 + 7$ Simplify.

2. $\left(-2x^2 + 5x - 1\right) + \left(-2x^2 + x + 3\right)$

$= -2x^2 + 5x - 1 - 2x^2 + x + 3$ Remove parentheses.

$= \left(-2x^2 - 2x^2\right) + (5x + 1x) + (-1 + 3)$ Group like terms.

$= -4x^2 + 6x + 2$ Simplify.

■ **Work Practice 1–2**

Practice 1–2

Add.

1. $\left(3x^5 - 7x^3 + 2x - 1\right)$ $+ \left(3x^3 - 2x\right)$

2. $\left(5x^2 - 2x + 1\right)$ $+ \left(-6x^2 + x - 1\right)$

Just as we can add numbers vertically, polynomials can be added vertically if we line up like terms underneath one another.

Example 3 Add $\left(7y^3 - 2y^2 + 7\right)$ and $\left(6y^2 + 1\right)$ using a vertical format.

Solution: Vertically line up like terms and add.

$$
\begin{array}{r}
7y^3 - 2y^2 + 7 \\
6y^2 + 1 \\
\hline
7y^3 + 4y^2 + 8
\end{array}
$$

■ **Work Practice 3**

Practice 3

Add $\left(9y^2 - 6y + 5\right)$ and $(4y + 3)$ using a vertical format.

Objective B Subtracting Polynomials

To subtract one polynomial from another, recall the definition of subtraction. To subtract a number, we add its opposite: $a - b = a + (-b)$. To subtract a polynomial, we also add its opposite. Just as $-b$ is the opposite of b, $-\left(x^2 + 5\right)$ is the opposite of $\left(x^2 + 5\right)$.

To Subtract Polynomials

To subtract two polynomials, change the signs of the terms of the polynomial being subtracted and then add.

Answers

1. $3x^5 - 4x^3 - 1$ **2.** $-x^2 - x$

3. $9y^2 - 2y + 8$

✓**Concept Check Answer**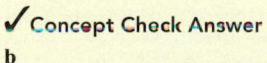

b

Practice 4

Subtract:

$(9x + 5) - (4x - 3)$

Example 4   Subtract: $(5x - 3) - (2x - 11)$

Solution: From the definition of subtraction, we have

$$(5x - 3) - (2x - 11) = (5x - 3) + [-(2x - 11)] \quad \text{Add the opposite.}$$
$$= (5x - 3) + (-2x + 11) \quad \text{Apply the distributive property.}$$
$$= 5x - 3 - 2x + 11 \quad \text{Remove parentheses.}$$
$$= 3x + 8 \quad \text{Combine like terms.}$$

■ **Work Practice 4**

Practice 5

Subtract:

$(4x^3 - 10x^2 + 1)$
$- (-4x^3 + x^2 - 11)$

Example 5   Subtract: $(2x^3 + 8x^2 - 6x) - (2x^3 - x^2 + 1)$

Solution: First, we change the sign of each term of the second polynomial; then we add.

$$(2x^3 + 8x^2 - 6x) - (2x^3 - x^2 + 1)$$
$$= (2x^3 + 8x^2 - 6x) + (-2x^3 + x^2 - 1)$$
$$= 2x^3 + 8x^2 - 6x - 2x^3 + x^2 - 1$$
$$= 2x^3 - 2x^3 + 8x^2 + x^2 - 6x - 1$$
$$= 9x^2 - 6x - 1 \quad \text{Combine like terms.}$$

■ **Work Practice 5**

Just as polynomials can be added vertically, so can they be subtracted vertically.

Practice 6

Subtract $(6y^2 - 3y + 2)$ from $(2y^2 - 2y + 7)$ using a vertical format.

Example 6   Subtract $(5y^2 + 2y - 6)$ from $(-3y^2 - 2y + 11)$ using a vertical format.

Solution: Arrange the polynomials in a vertical format, lining up like terms.

$$\begin{array}{r} -3y^2 - 2y + 11 \\ -(5y^2 + 2y - 6) \\ \hline \end{array} \qquad \begin{array}{r} -3y^2 - 2y + 11 \\ -5y^2 - 2y + 6 \\ \hline -8y^2 - 4y + 17 \end{array}$$

■ **Work Practice 6**

> **Helpful Hint**
>
> Don't forget to change the sign of each term in the polynomial being subtracted.

Objective C Adding and Subtracting Polynomials in One Variable ▶

Let's practice adding and subtracting polynomials in one variable.

Practice 7

Subtract $(3x + 1)$ from the sum of $(4x - 3)$ and $(12x - 5)$.

Example 7   Subtract $(5z - 7)$ from the sum of $(8z + 11)$ and $(9z - 2)$.

Solution: Notice that $(5z - 7)$ is to be subtracted **from** a sum. The translation is

$$[(8z + 11) + (9z - 2)] - (5z - 7)$$
$$= 8z + 11 + 9z - 2 - 5z + 7 \quad \text{Remove grouping symbols.}$$
$$= 8z + 9z - 5z + 11 - 2 + 7 \quad \text{Group like terms.}$$
$$= 12z + 16 \quad \text{Combine like terms.}$$

■ **Work Practice 7**

Answers

4. $5x + 8$ **5.** $8x^3 - 11x^2 + 12$
6. $-4y^2 + y + 5$ **7.** $13x - 9$

Objective D Adding and Subtracting Polynomials in Several Variables

Now that we know how to add or subtract polynomials in one variable, we can also add and subtract polynomials in several variables.

Examples   Add or subtract as indicated.

8. $\left(3x^2 - 6xy + 5y^2\right) + \left(-2x^2 + 8xy - y^2\right)$
$= 3x^2 - 6xy + 5y^2 - 2x^2 + 8xy - y^2$
$= x^2 + 2xy + 4y^2$ Combine like terms.

9. $\left(9a^2b^2 + 6ab - 3ab^2\right) - \left(5b^2a + 2ab - 3 - 9b^2\right)$
$= 9a^2b^2 + 6ab - 3ab^2 - 5b^2a - 2ab + 3 + 9b^2$
$= 9a^2b^2 + 4ab - 8ab^2 + 9b^2 + 3$ Combine like terms.

■ **Work Practice 8–9**

Practice 8–9

Add or subtract as indicated.

8. $\left(2a^2 - ab + 6b^2\right)$
$+ \left(-3a^2 + ab - 7b^2\right)$

9. $\left(5x^2y^2 + 3 - 9x^2y + y^2\right)$
$- \left(-x^2y^2 + 7 - 8xy^2 + 2y^2\right)$

✓**Concept Check** If possible, simplify each expression by performing the indicated operation.

a. $2y + y$

b. $2y \cdot y$

c. $-2y - y$

d. $(-2y)(-y)$

e. $2x + y$

Answers

8. $-a^2 - b^2$

9. $6x^2y^2 - 4 - 9x^2y + 8xy^2 - y^2$

✓**Concept Check Answers**

a. $3y$ **b.** $2y^2$ **c.** $-3y$ **d.** $2y^2$

e. cannot be simplified

Vocabulary, Readiness & Video Check

Simplify by combining like terms if possible.

1. $-9y - 5y$

2. $6m^5 + 7m^5$

3. $x + 6x$

4. $7z - z$

5. $5m^2 + 2m$

6. $8p^3 + 3p^2$

Martin-Gay Interactive Videos

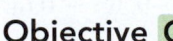

Watch the section lecture video and answer the following questions.

Objective A **7.** In ▣ Example 1, like terms are combined when adding the polynomials. What are the two sets of like terms? ▶

Objective B **8.** In ▣ Example 2, why can't parentheses just be removed as they were in ▣ Example 1? ▶

Objective C **9.** In ▣ Example 3, why are we told to be careful when translating to an expression? ▶

Objective D **10.** In ▣ Example 5, why aren't any signs changed when parentheses are removed? ▶

See Video 12.4 🍎

12.4 Exercise Set MyMathLab®

Objective A *Add. See Examples 1 and 2.*

1. $(3x + 7) + (9x + 5)$

2. $(-y - 2) + (3y + 5)$

3. $(-7x + 5) + (-3x^2 + 7x + 5)$

4. $(3x - 8) + (4x^2 - 3x + 3)$

5. $(-5x^2 + 3) + (2x^2 + 1)$

6. $(3x^2 + 7) + (3x^2 + 9)$

▶ 7. $(-3y^2 - 4y) + (2y^2 + y - 1)$

8. $(7x^2 + 2x - 9) + (-3x^2 + 5)$

9. $(1.2x^3 - 3.4x + 7.9) + (6.7x^3 + 4.4x^2 - 10.9)$

10. $(9.6y^3 + 2.7y^2 - 8.6) + (1.1y^3 - 8.8y + 11.6)$

11. $\left(\frac{3}{4}m^2 - \frac{2}{5}m + \frac{1}{8}\right) + \left(-\frac{1}{4}m^2 - \frac{3}{10}m + \frac{11}{16}\right)$

12. $\left(-\frac{4}{7}n^2 + \frac{5}{6}m - \frac{1}{20}\right) + \left(\frac{3}{7}n^2 - \frac{5}{12}m - \frac{3}{10}\right)$

Add using a vertical format. See Example 3.

13. $\begin{array}{r} 3t^2 + 4 \\ 5t^2 - 8 \\ \hline \end{array}$

14. $\begin{array}{r} 7x^3 + 3 \\ 2x^3 - 7 \\ \hline \end{array}$

15. $\begin{array}{r} 10a^3 - 8a^2 + 4a + 9 \\ 15a^3 + 9a^2 - 7a + 7 \\ \hline \end{array}$

16. $\begin{array}{r} 2x^3 - 3x^2 + x - 4 \\ 5x^3 + 2x^2 - 3x + 2 \\ \hline \end{array}$

Objective B *Subtract. See Examples 4 and 5.*

17. $(2x + 5) - (3x - 9)$

18. $(4 + 5a) - (-a - 5)$

19. $(5x^2 + 4) - (-2x^2 + 4)$

20. $(-7y^2 + 5) - (-8y^2 + 12)$

21. $3x - (5x - 9)$

22. $4 - (-y - 4)$

23. $(2x^2 + 3x - 9) - (-4x + 7)$

24. $(-7x^2 + 4x + 7) - (-8x + 2)$

▶ 25. $(5x + 8) - (-2x^2 - 6x + 8)$

26. $(-6y^2 + 3y - 4) - (9y^2 - 3y)$

27. $(0.7x^2 + 0.2x - 0.8) - (0.9x^2 + 1.4)$

28. $(-0.3y^2 + 0.6y - 0.3) - (0.5y^2 + 0.3)$

29. $\left(\frac{1}{4}z^2 - \frac{1}{5}z\right) - \left(-\frac{3}{20}z^2 + \frac{1}{10}z - \frac{7}{20}\right)$

30. $\left(\frac{1}{3}x^2 - \frac{2}{7}x\right) - \left(\frac{4}{21}x^2 + \frac{1}{21}x - \frac{2}{3}\right)$

Subtract using a vertical format. See Example 6.

31.
$$\begin{array}{r} 4z^2 - 8z + 3 \\ -(6z^2 + 8z - 3) \\ \hline \end{array}$$

32.
$$\begin{array}{r} 7a^2 - 9a + 6 \\ -(11a^2 - 4a + 2) \\ \hline \end{array}$$

33.
$$\begin{array}{r} 5u^5 - 4u^2 + 3u - 7 \\ -(3u^5 + 6u^2 - 8u + 2) \\ \hline \end{array}$$

34.
$$\begin{array}{r} 5x^3 - 4x^2 + 6x - 2 \\ -(3x^3 - 2x^2 - x - 4) \\ \hline \end{array}$$

Objectives A B C Mixed Practice *Add or subtract as indicated. See Examples 1 through 7.*

35. $(3x + 5) + (2x - 14)$

36. $(2y + 20) + (5y - 30)$

37. $(9x - 1) - (5x + 2)$

38. $(7y + 7) - (y - 6)$

39. $(14y + 12) + (-3y - 5)$

40. $(26y + 17) + (-20y - 10)$

41. $(x^2 + 2x + 1) - (3x^2 - 6x + 2)$

42. $(5y^2 - 3y - 1) - (2y^2 + y + 1)$

43. $(3x^2 + 5x - 8) + (5x^2 + 9x + 12) - (8x^2 - 14)$

44. $(2x^2 + 7x - 9) + (x^2 - x + 10) - (3x^2 - 30)$

45. $(-a^2 + 1) - (a^2 - 3) + (5a^2 - 6a + 7)$

46. $(-m^2 + 3) - (m^2 - 13) + (6m^2 - m + 1)$

Translating *Perform each indicated operation. See Examples 3, 6, and 7.*

47. Subtract $4x$ from $(7x - 3)$.

48. Subtract y from $(y^2 - 4y + 1)$.

49. Add $(4x^2 - 6x + 1)$ and $(3x^2 + 2x + 1)$.

50. Add $(-3x^2 - 5x + 2)$ and $(x^2 - 6x + 9)$.

▶ **51.** Subtract $(5x + 7)$ from $(7x^2 + 3x + 9)$.

52. Subtract $(5y^2 + 8y + 2)$ from $(7y^2 + 9y - 8)$.

53. Subtract $(4y^2 - 6y - 3)$ from the sum of $(8y^2 + 7)$ and $(6y + 9)$.

54. Subtract $(4x^2 - 2x + 2)$ from the sum of $(x^2 + 7x + 1)$ and $(7x + 5)$.

55. Subtract $(3x^2 - 4)$ from the sum of $(x^2 - 9x + 2)$ and $(2x^2 - 6x + 1)$.

56. Subtract $(y^2 - 9)$ from the sum of $(3y^2 + y + 4)$ and $(2y^2 - 6y - 10)$.

Objective D *Add or subtract as indicated. See Examples 8 and 9.*

57. $(9a + 6b - 5) + (-11a - 7b + 6)$

58. $(3x - 2 + 6y) + (7x - 2 - y)$

59. $(4x^2 + y^2 + 3) - (x^2 + y^2 - 2)$

60. $(7a^2 - 3b^2 + 10) - (-2a^2 + b^2 - 12)$

61. $\left(x^2 + 2xy - y^2\right) + \left(5x^2 - 4xy + 20y^2\right)$

62. $\left(a^2 - ab + 4b^2\right) + \left(6a^2 + 8ab - b^2\right)$

63. $\left(11r^2s + 16rs - 3 - 2r^2s^2\right) - \left(3sr^2 + 5 - 9r^2s^2\right)$

64. $\left(3x^2y - 6xy + x^2y^2 - 5\right) - \left(11x^2y^2 - 1 + 5yx^2\right)$

Objectives **A** **B** **C** **Mixed Practice** *For Exercises 65 through 68, find the perimeter of each figure.*

65.

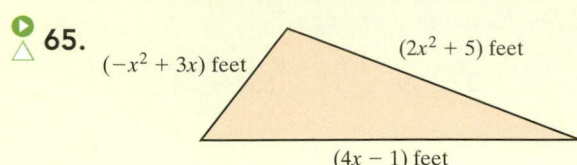

$(-x^2 + 3x)$ feet $(2x^2 + 5)$ feet $(4x - 1)$ feet

66.

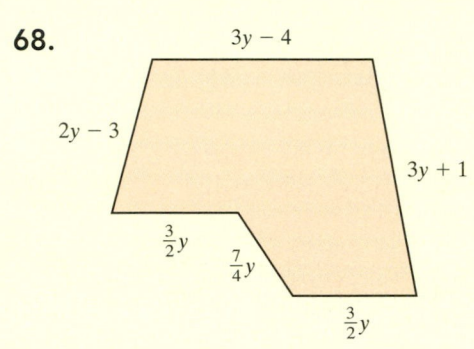

$(-x + 4)$ centimeters $5x$ centimeters x^2 centimeters $(x^2 - 6x - 2)$ centimeters

67.

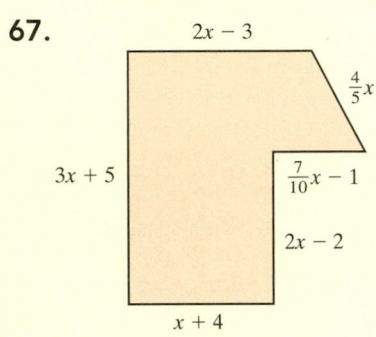

$2x - 3$ $\frac{4}{5}x$ $3x + 5$ $\frac{7}{10}x - 1$ $2x - 2$ $x + 4$

68.

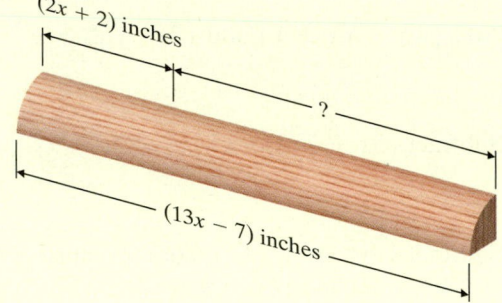

$3y - 4$ $2y - 3$ $3y + 1$ $\frac{3}{2}y$ $\frac{7}{4}y$ $\frac{3}{2}y$

69. A wooden beam is $(4y^2 + 4y + 1)$ meters long. If a piece $(y^2 - 10)$ meters is cut off, express the length of the remaining piece of beam as a polynomial in y.

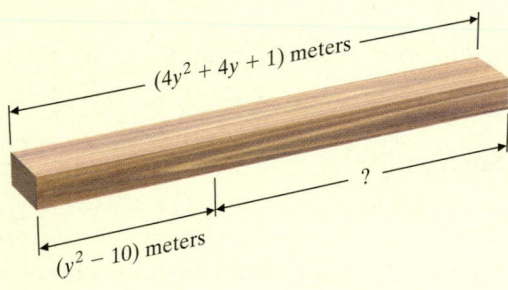

$(4y^2 + 4y + 1)$ meters $(y^2 - 10)$ meters   ?

70. A piece of quarter-round molding is $(13x - 7)$ inches long. If a piece $(2x + 2)$ inches long is removed, express the length of the remaining piece of molding as a polynomial in x.

$(2x + 2)$ inches   ? $(13x - 7)$ inches

Perform each indicated operation.

71. $\left[\left(1.2x^2 - 3x + 9.1\right) - \left(7.8x^2 - 3.1 + 8\right)\right] + (1.2x - 6)$

72. $\left[\left(7.9y^4 - 6.8y^3 + 3.3y\right) + \left(6.1y^3 - 5\right)\right] - \left(4.2y^4 + 1.1y - 1\right)$

Review

Multiply. See Section 12.1.

73. $3x(2x)$ **74.** $-7x(x)$ **75.** $(12x^3)(-x^5)$ **76.** $6r^3(7r^{10})$ **77.** $10x^2(20xy^2)$ **78.** $-z^2y(11zy)$

Concept Extensions

Fill in the squares so that each is a true statement.

79. $3x^\square + 4x^2 = 7x^\square$

80. $9y^7 + 3y^\square = 12y^7$

81. $2x^\square + 3x^\square - 5x^\square + 4x^\square = 6x^4 - 2x^3$

82. $3y^\square + 7y^\square - 2y^\square - y^\square = 10y^5 - 3y^2$

Match each expression on the left with its simplification on the right. Not all letters on the right must be used and a letter may be used more than once.

83. $10y - 6y^2 - y$

84. $5x + 5x$

85. $(5x - 3) + (5x - 3)$

86. $(15x - 3) - (5x - 3)$

a. $3y$
b. $9y - 6y^2$
c. $10x$
d. $25x^2$
e. $10x - 6$
f. none of these

Simplify each expression by performing the indicated operation. Explain how you arrived at each answer. See the second Concept Check in this section.

87. a. $z + 3z$
 b. $z \cdot 3z$
 c. $-z - 3z$
 d. $(-z)(-3z)$

88. a. $2y + y$
 b. $2y \cdot y$
 c. $-2y - y$
 d. $(-2y)(-y)$

89. a. $m \cdot m \cdot m$
 b. $m + m + m$
 c. $(-m)(-m)(-m)$
 d. $-m - m - m$

90. a. $x + x$
 b. $x \cdot x$
 c. $-x - x$
 d. $(-x)(-x)$

91. The polynomial $377x^2 - 720x + 1003$ represents the electricity generated (in thousand megawatthours) by solar sources in the United States during 2008–2012. The polynomial $538x^2 + 19{,}421x + 54{,}762$ represents the electricity generated (in thousand megawatthours) by wind power in the United States during 2008–2012. In both polynomials, x represents the number of years after 2008. Find a polynomial for the total electricity generated by both solar and wind power during 2008–2012. (*Source:* Based on information from the Energy Information Administration)

92. The polynomial $-0.4x^2 + 4.8x + 48.5$ represents the number of Americans (in millions) under age 65 covered by government health insurance during 2007–2012. The polynomial $0.8x^2 - 5.8x + 183.3$ represents the number of Americans (in millions) under age 65 covered by private health insurance during 2007–2012. In both polynomials, x represents the number of years since 2007. Find a polynomial for the total number of Americans (in millions) under age 65 with some form of health insurance during this period. (*Source:* Based on data from U.S. Census Bureau)

12.5 Multiplying Polynomials

Objectives

A Multiply Monomials.

B Multiply a Monomial by a Polynomial.

C Multiply Two Polynomials.

D Multiply Polynomials Vertically.

Objective A Multiplying Monomials

Recall from Section 12.1 that to multiply two monomials such as $\left(-5x^3\right)$ and $\left(-2x^4\right)$, we use the associative and commutative properties and regroup. Remember also that to multiply exponential expressions with a common base, we use the product rule for exponents and add exponents.

$$\left(-5x^3\right)\left(-2x^4\right) = (-5)(-2)\left(x^3 \cdot x^4\right) \quad \text{Use the commutative and associative properties.}$$
$$= 10x^7 \quad \text{Multiply.}$$

Practice 1–3

Multiply.

1. $10x \cdot 9x$
2. $8x^3\left(-11x^7\right)$
3. $\left(-5x^4\right)(-x)$

Examples Multiply.

1. $6x \cdot 4x = (6 \cdot 4)(x \cdot x)$ Use the commutative and associative properties.
         $= 24x^2$ Multiply.

2. $-7x^2 \cdot 2x^5 = (-7 \cdot 2)\left(x^2 \cdot x^5\right)$
           $= -14x^7$

3. $\left(-12x^5\right)(-x) = \left(-12x^5\right)(-1x)$
                $= (-12)(-1)\left(x^5 \cdot x\right)$
                $= 12x^6$

■ **Work Practice 1–3**

✓**Concept Check** Simplify.
   a. $3x \cdot 2x$ **b.** $3x + 2x$

Objective B Multiplying Monomials by Polynomials

To multiply a monomial such as $7x$ by a trinomial such as $x^2 + 2x + 5$, we use the distributive property.

Practice 4–6

Multiply.

4. $4x\left(x^2 + 4x + 3\right)$
5. $8x\left(7x^4 + 1\right)$
6. $-2x^3\left(3x^2 - x + 2\right)$

Examples Multiply.

4. $7x\left(x^2 + 2x + 5\right) = 7x\left(x^2\right) + 7x(2x) + 7x(5)$ Apply the distributive property.
                  $= 7x^3 + 14x^2 + 35x$ Multiply.

5. $5x\left(2x^3 + 6\right) = 5x\left(2x^3\right) + 5x(6)$ Apply the distributive property.
            $= 10x^4 + 30x$ Multiply.

6. $-3x^2\left(5x^2 + 6x - 1\right)$
    $= \left(-3x^2\right)\left(5x^2\right) + \left(-3x^2\right)(6x) + \left(-3x^2\right)(-1)$ Apply the distributive property.
    $= -15x^4 - 18x^3 + 3x^2$ Multiply.

■ **Work Practice 4–6**

Answers

1. $90x^2$ **2.** $-88x^{10}$ **3.** $5x^5$
4. $4x^3 + 16x^2 + 12x$ **5.** $56x^5 + 8x$
6. $-6x^5 + 2x^4 - 4x^3$

✓**Concept Check Answers**

a. $6x^2$ **b.** $5x$

Objective C Multiplying Two Polynomials

We also use the distributive property to multiply two binomials.

Example 7 Multiply.

a. $(m + 4)(m + 6)$ $\qquad$ **b.** $(3x + 2)(2x - 5)$

Solution:

a. $(m + 4)(m + 6) = m(m + 6) + 4(m + 6)$ $\qquad$ Use the distributive property.

$\qquad\qquad\qquad = m \cdot m + m \cdot 6 + 4 \cdot m + 4 \cdot 6$ $\qquad$ Use the distributive property.

$\qquad\qquad\qquad = m^2 + 6m + 4m + 24$ $\qquad$ Multiply.

$\qquad\qquad\qquad = m^2 + 10m + 24$ $\qquad$ Combine like terms.

b. $(3x + 2)(2x - 5) = 3x(2x - 5) + 2(2x - 5)$ $\qquad$ Use the distributive property.

$\qquad\qquad\qquad = 3x(2x) + 3x(-5) + 2(2x) + 2(-5)$

$\qquad\qquad\qquad = 6x^2 - 15x + 4x - 10$ $\qquad$ Multiply.

$\qquad\qquad\qquad = 6x^2 - 11x - 10$ $\qquad$ Combine like terms.

■ **Work Practice 7**

Practice 7

Multiply:

a. $(x + 5)(x + 10)$

b. $(4x + 5)(3x - 4)$

This idea can be expanded so that we can multiply any two polynomials.

To Multiply Two Polynomials

Multiply each term of the first polynomial by each term of the second polynomial, and then combine like terms.

Examples Multiply.

8. $(2x - y)^2$

$= (2x - y)(2x - y)$ $\qquad$ Using the meaning of an exponent, we have 2 factors of $(2x - y)$.

$= 2x(2x) + 2x(-y) + (-y)(2x) + (-y)(-y)$

$= 4x^2 - 2xy - 2xy + y^2$ $\qquad$ Multiply.

$= 4x^2 - 4xy + y^2$ $\qquad$ Combine like terms.

9. $(t + 2)(3t^2 - 4t + 2)$

$= t(3t^2) + t(-4t) + t(2) + 2(3t^2) + 2(-4t) + 2(2)$

$= 3t^3 - 4t^2 + 2t + 6t^2 - 8t + 4$

$= 3t^3 + 2t^2 - 6t + 4$ $\qquad$ Combine like terms.

■ **Work Practice 8–9**

Practice 8–9

Multiply.

8. $(3x - 2y)^2$

9. $(x + 3)(2x^2 - 5x + 4)$

✓**Concept Check** Square where indicated. Simplify if possible.

a. $(4a)^2 + (3b)^2$ $\qquad$ **b.** $(4a + 3b)^2$

Objective D Multiplying Polynomials Vertically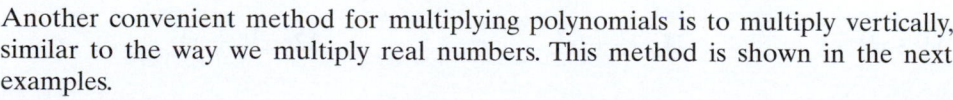

Another convenient method for multiplying polynomials is to multiply vertically, similar to the way we multiply real numbers. This method is shown in the next examples.

Answers

7. a. $x^2 + 15x + 50$

b. $12x^2 - x - 20$

8. $9x^2 - 12xy + 4y^2$

9. $2x^3 + x^2 - 11x + 12$

✓**Concept Check Answers**

a. $16a^2 + 9b^2$ $\quad$ **b.** $16a^2 + 24ab + 9b^2$

Practice 10

Multiply vertically:

$(3y^2 + 1)(y^2 - 4y + 5)$

Example 10 Multiply vertically: $(2y^2 + 5)(y^2 - 3y + 4)$

Solution:

$$
\begin{array}{r}
y^2 - 3y + 4 \\
2y^2 + 5 \\
\hline
5y^2 - 15y + 20 \\
2y^4 - 6y^3 + 8y^2 \\
\hline
2y^4 - 6y^3 + 13y^2 - 15y + 20
\end{array}
$$

Multiply $y^2 - 3y + 4$ by 5.

Multiply $y^2 - 3y + 4$ by $2y^2$.

Combine like terms.

■ **Work Practice 10**

Practice 11

Find the product of
$(4x^2 - x - 1)$ and
$(3x^2 + 6x - 2)$ using a
vertical format.

Example 11 Find the product of $(2x^2 - 3x + 4)$ and $(x^2 + 5x - 2)$ using a vertical format.

Solution: First, we arrange the polynomials in a vertical format. Then we multiply each term of the second polynomial by each term of the first polynomial.

$$
\begin{array}{r}
2x^2 - 3x + 4 \\
x^2 + 5x - 2 \\
\hline
-4x^2 + 6x - 8 \\
10x^3 - 15x^2 + 20x \\
2x^4 - 3x^3 + 4x^2 \\
\hline
2x^4 + 7x^3 - 15x^2 + 26x - 8
\end{array}
$$

Multiply $2x^2 - 3x + 4$ by -2.

Multiply $2x^2 - 3x + 4$ by $5x$.

Multiply $2x^2 - 3x + 4$ by x^2.

Combine like terms.

■ **Work Practice 11**

Answers

10. $3y^4 - 12y^3 + 16y^2 - 4y + 5$

11. $12x^4 + 21x^3 - 17x^2 - 4x + 2$

Vocabulary, Readiness & Video Check

Fill in each blank with the correct choice.

1. The expression $5x(3x + 2)$ equals $5x \cdot 3x + 5x \cdot 2$ by the _____ property.

   a. commutative   **b.** associative   **c.** distributive

2. The expression $(x + 4)(7x - 1)$ equals $x(7x - 1) + 4(7x - 1)$ by the _____ property.

   a. commutative   **b.** associative   **c.** distributive

3. The expression $(5y - 1)^2$ equals _____.

   a. $2(5y - 1)$ **b.** $(5y - 1)(5y + 1)$ **c.** $(5y - 1)(5y - 1)$

4. The expression $9x \cdot 3x$ equals _____.

   a. $27x$ **b.** $27x^2$ **c.** $12x$ **d.** $12x^2$

Perform the indicated operation, if possible.

5. $x^3 \cdot x^5$

6. $x^2 \cdot x^6$

7. $x^3 + x^5$

8. $x^2 + x^6$

9. $x^7 \cdot x^7$

10. $x^{11} \cdot x^{11}$

11. $x^7 + x^7$

12. $x^{11} + x^{11}$

Martin-Gay Interactive Videos   Watch the section lecture video and answer the following questions.

Objective A **13.** For Example 1, we use the product property to multiply the monomials. Is it possible to add the same two monomials? Why or why not? ▶

Objective B **14.** What property and what exponent rule are used in ▣ Examples 3 and 4? ▶

Objective C **15.** In ▣ Example 5, how many times is the distributive property actually applied? Explain. ▶

Objective D **16.** Would you say the vertical format used in ▣ Example 8 also applies the distributive property? Explain. ▶

See Video 12.5 🍎

12.5 **Exercise Set** MyMathLab® ▶

Objective A *Multiply. See Examples 1 through 3.*

1. $8x^2 \cdot 3x$

▶ **2.** $6x \cdot 3x^2$

3. $(-x^3)(-x)$

4. $(-x^6)(-x)$

5. $-4n^3 \cdot 7n^7$

6. $9t^6(-3t^5)$

7. $(-3.1x^3)(4x^9)$

8. $(-5.2x^4)(3x^4)$

▶ **9.** $\left(-\frac{1}{3}y^2\right)\left(\frac{2}{5}y\right)$

10. $\left(-\frac{3}{4}y^7\right)\left(\frac{1}{7}y^4\right)$

11. $(2x)(-3x^2)(4x^5)$

12. $(x)(5x^4)(-6x^7)$

Objective B *Multiply. See Examples 4 through 6.*

▶ **13.** $3x(2x + 5)$

14. $2x(6x + 3)$

15. $7x(x^2 + 2x - 1)$

16. $5y(y^2 + y - 10)$

17. $-2a(a + 4)$

18. $-3a(2a + 7)$

19. $3x(2x^2 - 3x + 4)$

20. $4x(5x^2 - 6x - 10)$

21. $3a^2(4a^3 + 15)$

22. $9x^3(5x^2 + 12)$

23. $-2a^2(3a^2 - 2a + 3)$

24. $-4b^2(3b^3 - 12b^2 - 6)$

25. $3x^2y(2x^3 - x^2y^2 + 8y^3)$

26. $4xy^2(7x^3 + 3x^2y^2 - 9y^3)$

▶ **27.** $-y(4x^3 - 7x^2y + xy^2 + 3y^3)$

28. $-x(6y^3 - 5xy^2 + x^2y - 5x^3)$

29. $\frac{1}{2}x^2(8x^2 - 6x + 1)$

30. $\frac{1}{3}y^2(9y^2 - 6y + 1)$

Objective C *Multiply. See Examples 7 through 9.*

31. $(x + 4)(x + 3)$ **32.** $(x + 2)(x + 9)$ ▶ **33.** $(a + 7)(a - 2)$ **34.** $(y - 10)(y + 11)$

35. $\left(x + \dfrac{2}{3}\right)\left(x - \dfrac{1}{3}\right)$ **36.** $\left(x + \dfrac{3}{5}\right)\left(x - \dfrac{2}{5}\right)$ **37.** $(3x^2 + 1)(4x^2 + 7)$ **38.** $(5x^2 + 2)(6x^2 + 2)$

39. $(4x - 3)(3x - 5)$ **40.** $(8x - 3)(2x - 4)$ **41.** $(1 - 3a)(1 - 4a)$ **42.** $(3 - 2a)(2 - a)$

43. $(2y - 4)^2$ **44.** $(6x - 7)^2$ **45.** $(x - 2)(x^2 - 3x + 7)$ **46.** $(x + 3)(x^2 + 5x - 8)$

47. $(x + 5)(x^3 - 3x + 4)$ **48.** $(a + 2)(a^3 - 3a^2 + 7)$ ▶ **49.** $(2a - 3)(5a^2 - 6a + 4)$

50. $(3 + b)(2 - 5b - 3b^2)$ ▶ **51.** $(7xy - y)^2$ **52.** $(x^2 - 4)^2$

Objective D *Multiply vertically. See Examples 10 and 11.*

53. $(2x - 11)(6x + 1)$ **54.** $(4x - 7)(5x + 1)$ ▶ **55.** $(x + 3)(2x^2 + 4x - 1)$

56. $(4x - 5)(8x^2 + 2x - 4)$ **57.** $(x^2 + 5x - 7)(2x^2 - 7x - 9)$ **58.** $(3x^2 - x + 2)(x^2 + 2x + 1)$

Objectives A B C D **Mixed Practice** *Multiply. See Examples 1 through 11.*

59. $-1.2y(-7y^6)$ **60.** $-4.2x(-2x^5)$ **61.** $-3x(x^2 + 2x - 8)$ **62.** $-5x(x^2 - 3x + 10)$

63. $(x + 19)(2x + 1)$ **64.** $(3y + 4)(y + 11)$ **65.** $\left(x + \dfrac{1}{7}\right)\left(x - \dfrac{3}{7}\right)$ **66.** $\left(m + \dfrac{2}{9}\right)\left(m - \dfrac{1}{9}\right)$

67. $(3y + 5)^2$ **68.** $(7y + 2)^2$ **69.** $(a + 4)(a^2 - 6a + 6)$ **70.** $(t + 3)(t^2 - 5t + 5)$

Express as the product of polynomials. Then multiply.

71. Find the area of the rectangle.

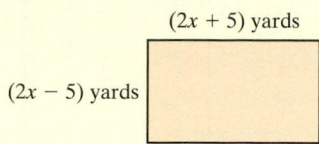

(2x + 5) yards

(2x − 5) yards

72. Find the area of the square field.

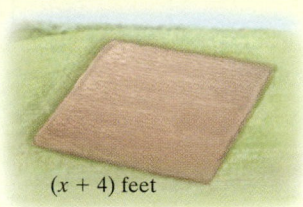

(x + 4) feet

73. Find the area of the triangle.

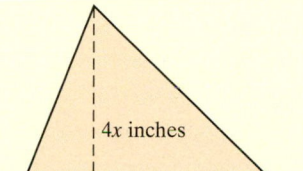

4x inches

(3x − 2) inches

△ **74.** Find the volume of the cube-shaped glass block.

(y − 1) meters

Review

Perform each indicated operation. See Section 12.1.

75. $(5x)^2$

76. $(4p)^2$

77. $(-3y^3)^2$

78. $(-7m^2)^2$

Concept Extensions

79. Perform each indicated operation. Explain the difference between the two expressions.
  a. $(3x + 5) + (3x + 7)$
  b. $(3x + 5)(3x + 7)$

80. Perform each indicated operation. Explain the difference between the two expressions.
  a. $(8x − 3) − (5x − 2)$
  b. $(8x − 3)(5x − 2)$

Mixed Practice *Perform the indicated operations. See Sections 12.4 and 12.5.*

81. $(3x − 1) + (10x − 6)$

82. $(2x − 1) + (10x − 7)$

83. $(3x − 1)(10x − 6)$

84. $(2x − 1)(10x − 7)$

85. $(3x − 1) − (10x − 6)$

86. $(2x − 1) − (10x − 7)$

△ **87.** The area of the largest rectangle below is $x(x + 3)$. Find another expression for this area by finding the sum of the areas of the smaller rectangles.

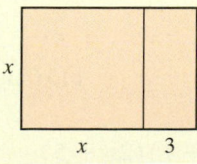

x

x   3

△ **88.** The area of the figure below is $(x + 2)(x + 3)$. Find another expression for this area by finding the sum of the areas of the smaller rectangles.

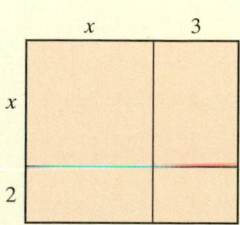

x   3

x

2

89. Write an expression for the area of the largest rectangle below in two different ways.

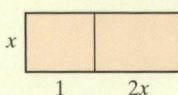

90. Write an expression for the area of the figure below in two different ways.

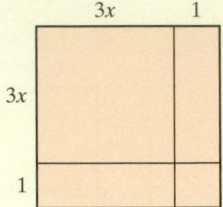

Simplify. See the Concept Checks in this section.

91. $5a + 6a$

92. $5a \cdot 6a$

Square where indicated. Simplify if possible.

93. $(5x)^2 + (2y)^2$

94. $(5x + 2y)^2$

95. Multiply each of the following polynomials.
  a. $(a + b)(a - b)$
  b. $(2x + 3y)(2x - 3y)$
  c. $(4x + 7)(4x - 7)$
  d. Can you make a general statement about all products of the form $(x + y)(x - y)$?

96. Evaluate each of the following.
  a. $(2 + 3)^2; 2^2 + 3^2$
  b. $(8 + 10)^2; 8^2 + 10^2$
  c. Does $(a + b)^2 = a^2 + b^2$ no matter what the values of a and b are? Why or why not?

12.6 Special Products ▶

Objectives

A Multiply Two Binomials Using the FOIL Method.

B Square a Binomial.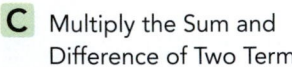

C Multiply the Sum and Difference of Two Terms.

D Use Special Products to Multiply Binomials. ▶

Objective **A** Using the FOIL Method ▶

In this section, we multiply binomials using special products. First, we introduce a special order for multiplying binomials called the FOIL order or method. This order, or pattern, is a result of the distributive property. We demonstrate by multiplying $(3x + 1)$ by $(2x + 5)$.

The FOIL Method

F stands for the product of the **First** terms.

$$(3x + 1)(2x + 5)$$
$$(3x)(2x) = 6x^2 \qquad \text{F}$$

O stands for the product of the **Outer** terms.

$$(3x + 1)(2x + 5)$$
$$(3x)(5) = 15x \qquad \text{O}$$

I stands for the product of the **Inner** terms.

$$(3x + 1)(2x + 5)$$
$$(1)(2x) = 2x \qquad \text{I}$$

L stands for the product of the **Last** terms.

$$(3x + 1)(2x + 5)$$
$$(1)(5) = 5 \qquad \text{L}$$

$$\overset{F\qquad O\qquad I\qquad L}{(3x + 1)(2x + 5) = 6x^2 + 15x + 2x + 5}$$
$$= 6x^2 + 17x + 5 \qquad \text{Combine like terms.}$$

Let's practice multiplying binomials using the FOIL method.

Example 1 Multiply: $(x - 3)(x + 4)$

Solution:

$$\overset{F\qquad\qquad O\qquad\qquad I\qquad\qquad L}{(x - 3)(x + 4) = (x)(x) + (x)(4) + (-3)(x) + (-3)(4)}$$
$$= x^2 + 4x - 3x - 12$$
$$= x^2 + x - 12 \qquad \text{Combine like terms.}$$

■ Work Practice 1

Practice 1

Multiply: $(x + 7)(x - 5)$

Example 2 Multiply: $(5x - 7)(x - 2)$

Solution:

$$\overset{F\qquad\quad O\qquad\quad I\qquad\quad L}{(5x - 7)(x - 2) = 5x(x) + 5x(-2) + (-7)(x) + (-7)(-2)}$$
$$= 5x^2 - 10x - 7x + 14$$
$$= 5x^2 - 17x + 14 \qquad \text{Combine like terms.}$$

■ Work Practice 2

Practice 2

Multiply: $(6x - 1)(x - 4)$

Helpful Hint Remember that the FOIL order for multiplying can be used only for the product of 2 binomials.

Example 3 Multiply: $(y^2 + 6)(2y - 1)$

Solution:
$$\overset{F\qquad O\qquad I\qquad L}{(y^2 + 6)(2y - 1) = 2y^3 - 1y^2 + 12y - 6}$$

Notice in this example that there are no like terms that can be combined, so the product is $2y^3 - y^2 + 12y - 6$.

■ Work Practice 3

Practice 3

Multiply: $(2y^2 + 3)(y - 4)$

Objective B Squaring Binomials ▶

An expression such as $(3y + 1)^2$ is called the square of a binomial. Since $(3y + 1)^2 = (3y + 1)(3y + 1)$, we can use the FOIL method to find this product.

Example 4 Multiply: $(3y + 1)^2$

Solution: $(3y + 1)^2 = (3y + 1)(3y + 1)$

$$\overset{F\qquad\quad O\qquad\quad I\qquad\quad L}{= (3y)(3y) + (3y)(1) + 1(3y) + 1(1)}$$
$$= 9y^2 + 3y + 3y + 1$$
$$= 9y^2 + 6y + 1$$

■ Work Practice 4

Practice 4

Multiply: $(2x + 9)^2$

Answers
1. $x^2 + 2x - 35$ **2.** $6x^2 - 25x + 4$
3. $2y^3 - 8y^2 + 3y - 12$
4. $4x^2 + 36x + 81$

Notice the pattern that appears in Example 4.

$$(3y + 1)^2 = 9y^2 + 6y + 1$$

$9y^2$ is the first term of the binomial squared: $(3y)^2 = 9y^2$.

$6y$ is 2 times the product of both terms of the binomial: $(2)(3y)(1) = 6y$.

1 is the second term of the binomial squared: $(1)^2 = 1$.

This pattern leads to the formulas below, which can be used when squaring a binomial. We call these **special products.**

Squaring a Binomial

A binomial squared is equal to the square of the first term plus or minus twice the product of both terms plus the square of the second term.

$$(a + b)^2 = a^2 + 2ab + b^2$$
$$(a - b)^2 = a^2 - 2ab + b^2$$

This product can be visualized geometrically.

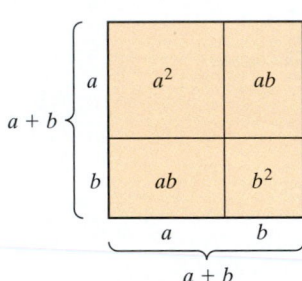

The area of the large square is side $\cdot$ side.
Area $= (a + b)(a + b) = (a + b)^2$
The area of the large square is also the sum of the areas of the smaller rectangles.
Area $= a^2 + ab + ab + b^2 = a^2 + 2ab + b^2$
Thus, $(a + b)^2 = a^2 + 2ab + b^2$.

Practice 5–8

Use a special product to square each binomial.

5. $(y + 3)^2$

6. $(r - s)^2$

7. $(6x + 5)^2$

8. $(x^2 - 3y)^2$

Examples   Use a special product to square each binomial.

first term squared	plus or minus	twice the product of the terms	plus	second term squared

5. $\quad (t + 2)^2 = \quad t^2 + \quad 2(t)(2) \quad + \quad 2^2 = t^2 + 4t + 4$

6. $\quad (p - q)^2 = \quad p^2 - \quad 2(p)(q) \quad + \quad q^2 = p^2 - 2pq + q^2$

7. $\quad (2x + 5)^2 = (2x)^2 + \quad 2(2x)(5) \quad + \quad 5^2 = 4x^2 + 20x + 25$

8. $\quad (x^2 - 7y)^2 = (x^2)^2 - 2(x^2)(7y) \quad + \quad (7y)^2 = x^4 - 14x^2y + 49y^2$

■ **Work Practice 5–8**

Answers

5. $y^2 + 6y + 9$ **6.** $r^2 - 2rs + s^2$
7. $36x^2 + 60x + 25$ **8.** $x^4 - 6x^2y + 9y^2$

Helpful Hint

Notice that

$(a + b)^2 \neq a^2 + b^2$ The middle term, $2ab$, is missing.

$(a + b)^2 = (a + b)(a + b) = a^2 + 2ab + b^2$

Likewise,

$(a - b)^2 \neq a^2 - b^2$

$(a - b)^2 = (a - b)(a - b) = a^2 - 2ab + b^2$

Objective C Multiplying the Sum and Difference of Two Terms

Another special product is the product of the sum and difference of the same two terms, such as $(x + y)(x - y)$. Finding this product by the FOIL method, we see a pattern emerge.

$$(x + y)(x - y) = x^2 - xy + xy - y^2$$
$$= x^2 - y^2$$

Notice that the two middle terms subtract out. This is because the **Outer** product is the opposite of the **Inner** product. Only the **difference of squares** remains.

Multiplying the Sum and Difference of Two Terms

The product of the sum and difference of two terms is the square of the first term minus the square of the second term.

$$(a + b)(a - b) = a^2 - b^2$$

Examples Use a special product to multiply.

first term squared   minus   second term squared

9. $(x + 4)(x - 4) = x^2 \quad - \quad 4^2 = x^2 - 16$

10. $(6t + 7)(6t - 7) = (6t)^2 \quad - \quad 7^2 = 36t^2 - 49$

11. $\left(x - \dfrac{1}{4}\right)\left(x + \dfrac{1}{4}\right) = x^2 \quad - \quad \left(\dfrac{1}{4}\right)^2 = x^2 - \dfrac{1}{16}$

12. $(2p - q)(2p + q) = (2p)^2 - q^2 = 4p^2 - q^2$

13. $\left(3x^2 - 5y\right)\left(3x^2 + 5y\right) = \left(3x^2\right)^2 - (5y)^2 = 9x^4 - 25y^2$

■ **Work Practice 9–13**

Practice 9–13

Use a special product to multiply.

9. $(x + 9)(x - 9)$

10. $(5 + 4y)(5 - 4y)$

11. $\left(x - \dfrac{1}{3}\right)\left(x + \dfrac{1}{3}\right)$

12. $(3a - b)(3a + b)$

13. $\left(2x^2 - 6y\right)\left(2x^2 + 6y\right)$

Answers

9. $x^2 - 81$ **10.** $25 - 16y^2$

11. $x^2 - \dfrac{1}{9}$ **12.** $9a^2 - b^2$

13. $4x^4 - 36y^2$

✓**Concept Check** Match each expression on the left to the equivalent expression or expressions in the list on the right.

$(a + b)^2$

$(a + b)(a - b)$

a. $(a + b)(a + b)$

b. $a^2 - b^2$

c. $a^2 + b^2$

d. $a^2 - 2ab + b^2$

e. $a^2 + 2ab + b^2$

Objective D Using Special Products

Let's now practice using our special products on a variety of multiplication problems. This practice will help us recognize when to apply what special product formula.

Practice 14–17

Use a special product to multiply, if possible.

14. $(7x - 1)^2$

15. $(5y + 3)(2y - 5)$

16. $(2a - 1)(2a + 1)$

17. $\left(5y - \dfrac{1}{9}\right)^2$

Examples   Use a special product to multiply, if possible.

14. $(4x - 9)(4x + 9)$ This is the sum and difference of the same two terms.

$= (4x)^2 - 9^2 = 16x^2 - 81$

15. $(3y + 2)^2$ This is a binomial squared.

$= (3y)^2 + 2(3y)(2) + 2^2$

$= 9y^2 + 12y + 4$

16. $(6a + 1)(a - 7)$ No special product applies.

    F   **O**   **I**   **L**   Use the FOIL method.

$= 6a \cdot a + 6a(-7) + 1 \cdot a + 1(-7)$

$= 6a^2 - 42a + a - 7$

$= 6a^2 - 41a - 7$

17. $\left(4x - \dfrac{1}{11}\right)^2$ This is a binomial squared.

$= (4x)^2 - 2(4x)\left(\dfrac{1}{11}\right) + \left(\dfrac{1}{11}\right)^2$

$= 16x^2 - \dfrac{8}{11}x + \dfrac{1}{121}$

■ **Work Practice 14–17**

Helpful Hint

- When multiplying two binomials, you may always use the FOIL order or method.
- When multiplying any two polynomials, you may always use the distributive property to find the product.

Answers

14. $49x^2 - 14x + 1$

15. $10y^2 - 19y - 15$ **16.** $4a^2 - 1$

17. $25y^2 - \dfrac{10}{9}y + \dfrac{1}{81}$

✓**Concept Check Answer**

a and **e**, **b**

Vocabulary, Readiness & Video Check

Answer each exercise true or false.

1. $(x + 4)^2 = x^2 + 16$ _____
2. For $(x + 6)(2x - 1)$, the product of the first terms is $2x^2$. _____
3. $(x + 4)(x - 4) = x^2 + 16$ _____
4. The product $(x - 1)(x^3 + 3x - 1)$ is a polynomial of degree 5. _____

Martin-Gay Interactive Videos

See Video 12.6

Watch the section lecture video and answer the following questions.

Objective A 5. From ▦ Examples 1–3, for what type of multiplication problem is the FOIL order of multiplication used? ▶

Objective B 6. Name at least one other method we can use to multiply ▦ Example 4. ▶

Objective C 7. From ▦ Example 5, why does multiplying the sum and difference of the same two terms always give us a binomial answer? ▶

Objective D 8. At the end of ▦ Example 8, what three special products for multiplying binomials are summarized? ▶

12.6 Exercise Set MyMathLab®

Objective A *Multiply using the FOIL method. See Examples 1 through 3.*

1. $(x + 3)(x + 4)$

2. $(x + 5)(x + 1)$

3. $(x - 5)(x + 10)$

4. $(y - 12)(y + 4)$

5. $(5x - 6)(x + 2)$

6. $(3y - 5)(2y + 7)$

7. $(y - 6)(4y - 1)$

8. $(2x - 9)(x - 11)$

9. $(2x + 5)(3x - 1)$

10. $(6x + 2)(x - 2)$

11. $(y^2 + 7)(6y + 4)$

12. $(y^2 + 3)(5y + 6)$

13. $\left(x - \dfrac{1}{3}\right)\left(x + \dfrac{2}{3}\right)$

14. $\left(x - \dfrac{2}{5}\right)\left(x + \dfrac{1}{5}\right)$

15. $(0.4 - 3a)(0.2 - 5a)$

16. $(0.3 - 2a)(0.6 - 5a)$

17. $(x + 5y)(2x - y)$

18. $(x + 4y)(3x - y)$

Objective B *Multiply. See Examples 4 through 8.*

19. $(x + 2)^2$

20. $(x + 7)^2$

21. $(2a - 3)^2$

22. $(7x - 3)^2$

23. $(3a - 5)^2$

24. $(5a - 2)^2$

25. $(x^2 + 0.5)^2$

26. $(x^2 + 0.3)^2$

27. $\left(y - \dfrac{2}{7}\right)^2$

28. $\left(y - \dfrac{3}{4}\right)^2$

⊙ **29.** $(2x - 1)^2$

30. $(5b - 4)^2$

⊙ **31.** $(5x + 9)^2$

32. $(6s + 2)^2$

33. $(3x - 7y)^2$

34. $(4s - 2y)^2$

35. $(4m + 5n)^2$

36. $(3n + 5m)^2$

37. $(5x^4 - 3)^2$

38. $(7x^3 - 6)^2$

Objective C *Multiply. See Examples 9 through 13.*

⊙ **39.** $(a - 7)(a + 7)$

40. $(b + 3)(b - 3)$

41. $(x + 6)(x - 6)$

42. $(x - 8)(x + 8)$

43. $(3x - 1)(3x + 1)$

44. $(7x - 5)(7x + 5)$

45. $(x^2 + 5)(x^2 - 5)$

46. $(a^2 + 6)(a^2 - 6)$

47. $(2y^2 - 1)(2y^2 + 1)$

48. $(3x^2 + 1)(3x^2 - 1)$

49. $(4 - 7x)(4 + 7x)$

50. $(8 - 7x)(8 + 7x)$

51. $\left(3x - \dfrac{1}{2}\right)\left(3x + \dfrac{1}{2}\right)$

52. $\left(10x + \dfrac{2}{7}\right)\left(10x - \dfrac{2}{7}\right)$

⊙ **53.** $(9x + y)(9x - y)$

54. $(2x - y)(2x + y)$

55. $(2m + 5n)(2m - 5n)$

56. $(5m + 4n)(5m - 4n)$

Objective D **Mixed Practice** *Multiply. See Examples 14 through 17.*

57. $(a + 5)(a + 4)$

58. $(a + 5)(a + 7)$

59. $(a - 7)^2$

60. $(b - 2)^2$

61. $(4a + 1)(3a - 1)$

62. $(6a + 7)(6a + 5)$

63. $(x + 2)(x - 2)$

64. $(x - 10)(x + 10)$

65. $(3a + 1)^2$

66. $(4a + 2)^2$

67. $(x + y)(4x - y)$

68. $(3x + 2)(4x - 2)$

69. $\left(\dfrac{1}{3}a^2 - 7\right)\left(\dfrac{1}{3}a^2 + 7\right)$ **70.** $\left(\dfrac{a}{2} + 4y\right)\left(\dfrac{a}{2} - 4y\right)$ **71.** $(3b + 7)(2b - 5)$ **72.** $(3y - 13)(y - 3)$

73. $(x^2 + 10)(x^2 - 10)$ **74.** $(x^2 + 8)(x^2 - 8)$ **75.** $(4x + 5)(4x - 5)$ **76.** $(3x + 5)(3x - 5)$

77. $(5x - 6y)^2$ **78.** $(4x - 9y)^2$ **79.** $(2r - 3s)(2r + 3s)$ **80.** $(6r - 2x)(6r + 2x)$

Express each as a product of polynomials in x. Then multiply and simplify.

81. Find the area of the square rug if its side is $(2x + 1)$ feet.

$(2x + 1)$ feet

$(2x + 1)$ feet

82. Find the area of the rectangular canvas if its length is $(3x - 2)$ inches and its width is $(x - 4)$ inches.

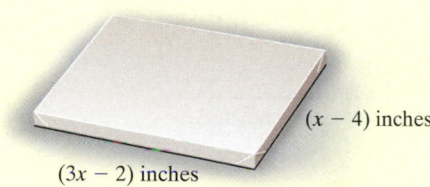

$(x - 4)$ inches

$(3x - 2)$ inches

Review

Simplify each expression. See Sections 12.1 and 12.2.

83. $\dfrac{50b^{10}}{70b^5}$ **84.** $\dfrac{60y^6}{80y^2}$ **85.** $\dfrac{8a^{17}b^5}{-4a^7b^{10}}$ **86.** $\dfrac{-6a^8y}{3a^4y}$ **87.** $\dfrac{2x^4y^{12}}{3x^4y^4}$ **88.** $\dfrac{-48ab^6}{32ab^3}$

Concept Extensions

Match each expression on the left to the equivalent expression on the right. See the Concept Check in this section. (Not all choices will be used.)

89. $(a - b)^2$

90. $(a - b)(a + b)$

91. $(a + b)^2$

92. $(a + b)^2(a - b)^2$

a. $a^2 - b^2$

b. $a^2 + b^2$

c. $a^2 - 2ab + b^2$

d. $a^2 + 2ab + b^2$

e. none of these

Fill in the squares so that a true statement forms.

93. $(x^{\square} + 7)(x^{\square} + 3) = x^4 + 10x^2 + 21$

94. $(5x^{\square} - 2)^2 = 25x^6 - 20x^3 + 4$

Find the area of the shaded figure. To do so, subtract the area of the smaller square(s) from the area of the larger geometric figure.

△ 95.

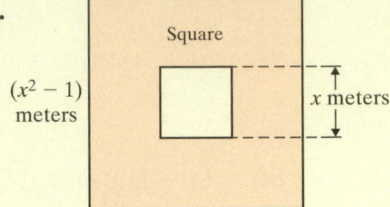

△ 96.

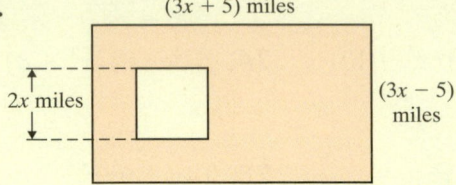

△ 97.

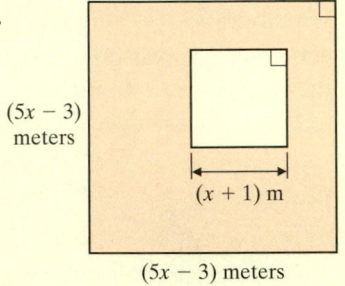

△ 98.

99. In your own words, describe the different methods that can be used to find the product: $(2x - 5)(3x + 1)$.

100. In your own words, describe the different methods that can be used to find the product: $(5x + 1)^2$.

101. Suppose that a classmate asked you why $(2x + 1)^2$ is **not** $(4x^2 + 1)$. Write down your response to this classmate.

102. Suppose that a classmate asked you why $(2x + 1)^2$ **is** $(4x^2 + 4x + 1)$. Write down your response to this classmate.

Exponents and Operations on Polynomials

Answers

Perform operations and simplify.

1. $(5x^2)(7x^3)$

2. $(4y^2)(-8y^7)$

3. -4^2

4. $(-4)^2$

5. $(x-5)(2x+1)$

6. $(3x-2)(x+5)$

7. $(x-5)+(2x+1)$

8. $(3x-2)+(x+5)$

9. $\dfrac{7x^9y^{12}}{x^3y^{10}}$

10. $\dfrac{20a^2b^8}{14a^2b^2}$

11. $(12m^7n^6)^2$

12. $(4y^9z^{10})^3$

13. $(4y-3)(4y+3)$

14. $(7x-1)(7x+1)$

15. $(x^{-7}y^5)^9$

16. 8^{-2}

17. $(3^{-1}x^9)^3$

18. $\dfrac{(r^7s^{-5})^6}{(2r^{-4}s^{-4})^4}$

19. $(7x^2-2x+3)-(5x^2+9)$

20. $(10x^2+7x-9)-(4x^2-6x+2)$

1. _____

2. _____

3. _____

4. _____

5. _____

6. _____

7. _____

8. _____

9. _____

10. _____

11. _____

12. _____

13. _____

14. _____

15. _____

16. _____

17. _____

18. _____

19. _____

20. _____

21. _____

22. _____

23. _____

24. _____

25. _____

26. _____

27. _____

28. _____

29. _____

30. _____

31. _____

32. _____

33. _____

34. _____

35. _____

36. _____

37. _____

38. _____

21. $0.7y^2 - 1.2 + 1.8y^2 - 6y + 1$

22. $7.8x^2 - 6.8x - 3.3 + 0.6x^2 - 0.9$

23. Subtract $(y^2 + 2)$ from $(3y^2 - 6y + 1)$.

24. $(z^2 + 5) - (3z^2 - 1) + \left(8z^2 + 2z - \dfrac{1}{2}\right)$

25. $(x + 4)^2$

26. $(y - 9)^2$

27. $(x + 4) + (x + 4)$

28. $(y - 9) + (y - 9)$

29. $7x^2 - 6xy + 4(y^2 - xy)$

30. $5a^2 - 3ab + 6(b^2 - a^2)$

31. $(x - 3)(x^2 + 5x - 1)$

32. $(x + 1)(x^2 - 3x - 2)$

33. $(2x - 7)(3x + 10)$

34. $(5x - 1)(4x + 5)$

35. $(2x - 7)(x^2 - 6x + 1)$

36. $(5x - 1)(x^2 + 2x - 3)$

37. $\left(2x + \dfrac{5}{9}\right)\left(2x - \dfrac{5}{9}\right)$

38. $\left(12y + \dfrac{3}{7}\right)\left(12y - \dfrac{3}{7}\right)$

Dividing Polynomials

Objective A Dividing by a Monomial

To divide a polynomial by a monomial, recall addition of fractions. Fractions that have a common denominator are added by adding the numerators:

$$\frac{a}{c} + \frac{b}{c} = \frac{a + b}{c}$$

If we read this equation from right to left and let a, b, and c be monomials, $c \neq 0$, we have the following.

> **To Divide a Polynomial by a Monomial**
>
> Divide each term of the polynomial by the monomial.
>
> $$\frac{a + b}{c} = \frac{a}{c} + \frac{b}{c}, \quad c \neq 0$$

Throughout this section, we assume that denominators are not 0.

Example 1 Divide: $(6m^2 + 2m) \div 2m$

Solution: We begin by writing the quotient in fraction form. Then we divide each term of the polynomial $6m^2 + 2m$ by the monomial $2m$ and use the quotient rule for exponents to simplify.

$$\frac{6m^2 + 2m}{2m} = \frac{6m^2}{2m} + \frac{2m}{2m}$$

$$= 3m + 1 \qquad \text{Simplify.}$$

Check: To check, we multiply.

$$2m(3m + 1) = 2m(3m) + 2m(1) = 6m^2 + 2m$$

The quotient $3m + 1$ checks.

■ **Work Practice 1**

✓**Concept Check** In which of the following is $\dfrac{x + 5}{5}$ simplified correctly?

a. $\dfrac{x}{5} + 1$ **b.** x **c.** $x + 1$

Example 2 Divide: $\dfrac{9x^5 - 12x^2 + 3x}{3x^2}$

Solution: $\dfrac{9x^5 - 12x^2 + 3x}{3x^2} = \dfrac{9x^5}{3x^2} - \dfrac{12x^2}{3x^2} + \dfrac{3x}{3x^2}$ Divide each term by $3x^2$.

$$= 3x^3 - 4 + \frac{1}{x} \qquad \text{Simplify.}$$

Notice that the quotient is not a polynomial because of the term $\dfrac{1}{x}$. This expression is called a rational expression—we will study rational expressions in Chapter 14. Although the quotient of two polynomials is not always a polynomial, we may still check by multiplying.

(Continued on next page)

Objectives

A Divide a Polynomial by a Monomial.

B Use Long Division to Divide a Polynomial by a Polynomial Other than a Monomial.

Practice 1

Divide: $(25x^3 + 5x^2) \div 5x^2$

Practice 2

Divide: $\dfrac{24x^7 + 12x^2 - 4x}{4x^2}$

Answers

1. $5x + 1$ **2.** $6x^5 + 3 - \dfrac{1}{x}$

✓**Concept Check Answer**

a

$$Check: 3x^2\left(3x^3 - 4 + \frac{1}{x}\right) = 3x^2(3x^3) - 3x^2(4) + 3x^2\left(\frac{1}{x}\right)$$
$$= 9x^5 - 12x^2 + 3x$$

■ **Work Practice 2**

Practice 3

Divide: $\dfrac{12x^3y^3 - 18xy + 6y}{3xy}$

Example 3 Divide: $\dfrac{8x^2y^2 - 16xy + 2x}{4xy}$

Solution: $\dfrac{8x^2y^2 - 16xy + 2x}{4xy} = \dfrac{8x^2y^2}{4xy} - \dfrac{16xy}{4xy} + \dfrac{2x}{4xy}$ Divide each term by $4xy$.

$$= 2xy - 4 + \frac{1}{2y}$$ Simplify.

$$Check: 4xy\left(2xy - 4 + \frac{1}{2y}\right) = 4xy(2xy) - 4xy(4) + 4xy\left(\frac{1}{2y}\right)$$
$$= 8x^2y^2 - 16xy + 2x$$

■ **Work Practice 3**

Objective B Dividing by a Polynomial Other than a Monomial ▶

To divide a polynomial by a polynomial other than a monomial, we use a process known as long division. Polynomial long division is similar to number long division, so we review long division by dividing 13 into 3660.

$$\begin{array}{r} 281 \\ 13\overline{)3660} \\ -26 \\ \hline 106 \\ -104 \\ \hline 20 \\ -13 \\ \hline 7 \end{array}$$

> **Helpful Hint**
> Recall that 3660 is called the dividend.

$2 \cdot 13 = 26$

Subtract and bring down the next digit in the dividend.

$8 \cdot 13 = 104$

Subtract and bring down the next digit in the dividend.

$1 \cdot 13 = 13$

Subtract. There are no more digits to bring down, so the remainder is 7.

The quotient is 281 R 7, which can be written as $281\dfrac{7}{13}$. $\leftarrow$ remainder $\leftarrow$ divisor

Recall that division can be checked by multiplication. To check this division problem, we see that

$$13 \cdot 281 + 7 = 3660, \text{ the dividend.}$$

Now we demonstrate long division of polynomials.

Practice 4

Divide $x^2 + 12x + 35$ by $x + 5$ using long division.

Example 4 Divide $x^2 + 7x + 12$ by $x + 3$ using long division.

Solution:

How many times does x divide x^2?

$\dfrac{x^2}{x} = x$.

$$\begin{array}{r} x \\ x + 3\overline{)x^2 + 7x + 12} \\ \underline{x^2 + 3x} \\ 4x + 12 \end{array}$$

To subtract, change the signs of these terms and add.

Multiply: $x(x + 3)$

Subtract and bring down the next term.

Answers

3. $4x^2y^2 - 6 + \dfrac{2}{x}$ 4. $x + 7$

Now we repeat this process.

$$x + 4$$

$$x + 3\overline{)x^2 + 7x + 12}$$

How many times does x divide $4x$? $\dfrac{4x}{x} = 4$.

$$\underline{x^2 + 3x}$$

To subtract, change the signs of these terms and add.

$$4x + 12 \quad \text{Multiply: } 4(x + 3)$$

$$\underline{4x + 12} \quad \text{Subtract. The remainder is 0.}$$

$$0$$

The quotient is $x + 4$.

Check: We check by multiplying.

$$\text{divisor} \cdot \text{quotient} + \text{remainder} = \text{dividend}$$

or

$$(x + 3) \cdot (x + 4) + 0 = x^2 + 7x + 12$$

The quotient checks.

■ **Work Practice 4**

Example 5 Divide $6x^2 + 10x - 5$ by $3x - 1$ using long division.

Solution:

$$2x + 4$$

$$3x - 1\overline{)6x^2 + 10x - 5}$$

$$\dfrac{6x^2}{3x} = 2x, \text{ so } 2x \text{ is a term of the quotient.}$$

$$\underline{6x^2 - 2x} \quad \text{Multiply: } 2x(3x - 1)$$

$$12x - 5 \quad \text{Subtract and bring down the next term.}$$

$$\underline{12x - 4} \quad \dfrac{12x}{3x} = 4. \text{ Multiply: } 4(3x - 1)$$

$$-1 \quad \text{Subtract. The remainder is } -1.$$

Thus $\left(6x^2 + 10x - 5\right)$ divided by $(3x - 1)$ is $(2x + 4)$ with a remainder of -1. This can be written as follows.

$$\dfrac{6x^2 + 10x - 5}{3x - 1} = 2x + 4 + \dfrac{-1}{3x - 1} \quad \leftarrow \text{remainder} \atop \leftarrow \text{divisor}$$

$$\text{or } 2x + 4 - \dfrac{1}{3x - 1}$$

Check: To check, we multiply $(3x - 1)(2x + 4)$. Then we add the remainder, -1, to this product.

$$(3x - 1)(2x + 4) + (-1) = \left(6x^2 + 12x - 2x - 4\right) - 1$$
$$= 6x^2 + 10x - 5$$

The quotient checks.

■ **Work Practice 5**

Notice that the division process is continued until the degree of the remainder polynomial is less than the degree of the divisor polynomial.

Recall that in Section 12.3 we practiced writing polynomials in descending order of powers and with no missing terms. For example, $2 - 4x^2$ written in this form is $-4x^2 + 0x + 2$. Writing the dividend and divisor in this form is helpful when dividing polynomials.

Practice 5

Divide: $8x^2 + 2x - 7$ by $2x - 1$

Answer

5. $4x + 3 + \dfrac{-4}{2x - 1}$ or $4x + 3 - \dfrac{4}{2x - 1}$

Practice 6

Divide: $(15 - 2x^2) \div (x - 3)$

Example 6 Divide: $(2 - 4x^2) \div (x + 1)$

Solution: We use the rewritten form of $2 - 4x^2$ from the previous page.

$$
\begin{array}{r}
-4x + 4 \\
x + 1 \overline{)-4x^2 + 0x + 2} \\
-4x^2 - 4x \\
\hline
4x + 2 \\
-4x - 4 \\
\hline
-2
\end{array}
$$

$\dfrac{-4x^2}{x} = -4x$, so $-4x$ is a term of the quotient.

Multiply: $-4x(x + 1)$

Subtract and bring down the next term.

$\dfrac{4x}{x} = 4$. Multiply: $4(x + 1)$

Remainder

Thus, $\dfrac{-4x^2 + 0x + 2}{x + 1}$ or $\dfrac{2 - 4x^2}{x + 1} = -4x + 4 + \dfrac{-2}{x + 1}$ or $-4x + 4 - \dfrac{2}{x + 1}$.

Check: To check, see that $(x + 1)(-4x + 4) + (-2) = 2 - 4x^2$.

■ **Work Practice 6**

Practice 7

Divide: $\dfrac{5 - x + 9x^3}{3x + 2}$

Example 7 Divide: $\dfrac{4x^2 + 7 + 8x^3}{2x + 3}$

Solution: Before we begin the division process, we rewrite $4x^2 + 7 + 8x^3$ as $8x^3 + 4x^2 + 0x + 7$. Notice that we have written the polynomial in descending order and have represented the missing x-term by $0x$.

$$
\begin{array}{r}
4x^2 - 4x + 6 \\
2x + 3 \overline{)8x^3 + 14x^2 + 0x + 7} \\
-8x^3 + 12x^2 \\
\hline
-8x^2 + 0x \\
+8x^2 + 12x \\
\hline
12x + 7 \\
-12x + 18 \\
\hline
-11
\end{array}
$$

Remainder

Thus, $\dfrac{4x^2 + 7 + 8x^3}{2x + 3} = 4x^2 - 4x + 6 + \dfrac{-11}{2x + 3}$ or $4x^2 - 4x + 6 - \dfrac{11}{2x + 3}$.

■ **Work Practice 7**

Practice 8

Divide: $x^3 - 1$ by $x - 1$

Example 8 Divide $x^3 - 8$ by $x - 2$.

Solution: Notice that the polynomial $x^3 - 8$ is missing an x^2-term and an x-term. We'll represent these terms by inserting $0x^2$ and $0x$.

$$
\begin{array}{r}
x^2 + 2x + 4 \\
x - 2 \overline{)x^3 + 0x^2 + 0x - 8} \\
-x^3 + 2x^2 \\
\hline
2x^2 + 0x \\
-2x^2 + 4x \\
\hline
4x - 8 \\
-4x + 8 \\
\hline
0
\end{array}
$$

Answers

6. $-2x - 6 + \dfrac{-3}{x - 3}$

or $-2x - 6 - \dfrac{3}{x - 3}$

7. $3x^2 - 2x + 1 + \dfrac{3}{3x + 2}$

8. $x^2 + x + 1$

Thus, $\dfrac{x^3 - 8}{x - 2} = x^2 + 2x + 4$.

Check: To check, see that $(x^2 + 2x + 4)(x - 2) = x^3 - 8$.

■ **Work Practice 8**

Vocabulary, Readiness & Video Check

Use the choices below to fill in each blank. Choices may be used more than once.

    dividend   divisor   quotient

1. In $6\overline{)18}^{\,3}$, the 18 is the _____, the 3 is the _____, and the 6 is the _____.

2. In $x+1\overline{)x^2+3x+2}^{\,x+2}$, the $x+1$ is the _____, the x^2+3x+2 is the _____, and the $x+2$ is the _____.

Simplify each expression mentally.

3. $\dfrac{a^6}{a^4}$
        4. $\dfrac{p^8}{p^3}$
        5. $\dfrac{y^2}{y}$
        6. $\dfrac{a^3}{a}$

Martin-Gay Interactive Videos   Watch the section lecture video and answer the following questions.

Objective A **7.** The lecture before ▥ Example 1 begins with adding two fractions with the same denominator. From there, the lecture continues to a method for dividing a polynomial by a monomial. What role does the monomial play in the fraction example? ▶

Objective B **8.** In ▥ Example 5, we're told that although we don't have to fill in missing powers in the divisor and the dividend, it really is a good idea to do so. Why? ▶

See Video 12.7 🍎

12.7   Exercise Set   MyMathLab® ▶

Objective A *Perform each division. See Examples 1 through 3.*

▶ **1.** $\dfrac{12x^4+3x^2}{x}$
    2. $\dfrac{15x^2-9x^5}{x}$
    3. $\dfrac{20x^3-30x^2+5x+5}{5}$
    4. $\dfrac{8x^3-4x^2+6x+2}{2}$

5. $\dfrac{15p^3+18p^2}{3p}$
    6. $\dfrac{6x^5+3x^4}{3x^4}$
    7. $\dfrac{-9x^4+18x^5}{6x^5}$
    8. $\dfrac{14m^2-27m^3}{7m}$

▶ **9.** $\dfrac{-9x^5+3x^4-12}{3x^3}$
    10. $\dfrac{6a^2-4a+12}{-2a^2}$
    11. $\dfrac{4x^4-6x^3+7}{-4x^4}$
    12. $\dfrac{-12a^3+36a-15}{3a}$

Objective B *Find each quotient using long division. See Examples 4 and 5.*

▶ 13. $\dfrac{x^2 + 4x + 3}{x + 3}$

14. $\dfrac{x^2 + 7x + 10}{x + 5}$

15. $\dfrac{2x^2 + 13x + 15}{x + 5}$

16. $\dfrac{3x^2 + 8x + 4}{x + 2}$

17. $\dfrac{2x^2 - 7x + 3}{x - 4}$

18. $\dfrac{3x^2 - x - 4}{x - 1}$

19. $\dfrac{9a^3 - 3a^2 - 3a + 4}{3a + 2}$

20. $\dfrac{4x^3 + 12x^2 + x - 14}{2x + 3}$

21. $\dfrac{8x^2 + 10x + 1}{2x + 1}$

22. $\dfrac{3x^2 + 17x + 7}{3x + 2}$

23. $\dfrac{2x^3 + 2x^2 - 17x + 8}{x - 2}$

24. $\dfrac{4x^3 + 11x^2 - 8x - 10}{x + 3}$

Find each quotient using long division. Don't forget to write the polynomials in descending order and fill in any missing terms. See Examples 6 through 8.

25. $\dfrac{x^2 - 36}{x - 6}$

26. $\dfrac{a^2 - 49}{a - 7}$

▶ 27. $\dfrac{x^3 - 27}{x - 3}$

28. $\dfrac{x^3 + 64}{x + 4}$

29. $\dfrac{1 - 3x^2}{x + 2}$

30. $\dfrac{7 - 5x^2}{x + 3}$

31. $\dfrac{-4b + 4b^2 - 5}{2b - 1}$

32. $\dfrac{-3y + 2y^2 - 15}{2y + 5}$

Objectives A B Mixed Practice *Divide. If the divisor contains 2 or more terms, use long division. See Examples 1 through 8.*

33. $\dfrac{a^2b^2 - ab^3}{ab}$

34. $\dfrac{m^3n^2 - mn^4}{mn}$

35. $\dfrac{8x^2 + 6x - 27}{2x - 3}$

36. $\dfrac{18w^2 + 18w - 8}{3w + 4}$

37. $\dfrac{2x^2y + 8x^2y^2 - xy^2}{2xy}$

38. $\dfrac{11x^3y^3 - 33xy + x^2y^2}{11xy}$

▶ 39. $\dfrac{2b^3 + 9b^2 + 6b - 4}{b + 4}$

40. $\dfrac{2x^3 + 3x^2 - 3x + 4}{x + 2}$

41. $\dfrac{y^3 + 3y^2 + 4}{y - 2}$

42. $\dfrac{3x^3 + 11x + 12}{x + 4}$

43. $\dfrac{5 - 6x^2}{x - 2}$

44. $\dfrac{3 - 7x^2}{x - 3}$

45. $\dfrac{x^5 + x^2}{x^2 + x}$

46. $\dfrac{x^6 - x^3}{x^3 - x^2}$

Review

Fill in each blank. See Section 12.1.

47. $12 = 4 \cdot$ ___

48. $12 = 2 \cdot$ ___

49. $20 = -5 \cdot$ ___

50. $20 = -4 \cdot$ ___

51. $9x^2 = 3x \cdot$ ___

52. $9x^2 = 9x \cdot$ ___

53. $36x^2 = 4x \cdot$ ___

54. $36x^2 = 2x \cdot$ ___

Concept Extensions

Solve.

55. The perimeter of a square is $\left(12x^3 + 4x - 16\right)$ feet. Find the length of its side.

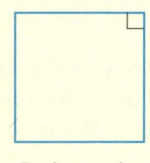

Perimeter is
$(12x^3 + 4x - 16)$ feet

△ **56.** The volume of the swimming pool shown is $\left(36x^5 - 12x^3 + 6x^2\right)$ cubic feet. If its height is $2x$ feet and its width is $3x$ feet, find its length.

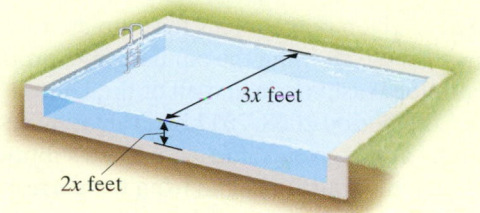

3x feet

2x feet

57. The area of the parallelogram shown is $\left(10x^2 + 31x + 15\right)$ square meters. If its base is $(5x + 3)$ meters, find its height.

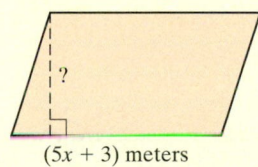

?

$(5x + 3)$ meters

58. The area of the top of the Ping-Pong table shown is $\left(49x^2 + 70x - 200\right)$ square inches. If its length is $(7x + 20)$ inches, find its width.

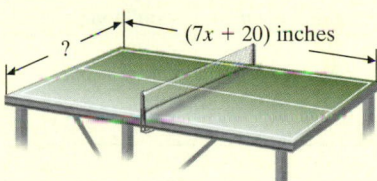

$(7x + 20)$ inches

?

59. Explain how to check a polynomial long division result when the remainder is 0.

60. Explain how to check a polynomial long division result when the remainder is not 0.

61. In which of the following is $\dfrac{a + 7}{7}$ simplified correctly? See the Concept Check in this section.

a. $a + 1$

b. a

c. $\dfrac{a}{7} + 1$

62. In which of the following is $\dfrac{5x + 15}{5}$ simplified correctly? See the Concept Check in this section.

a. $x + 15$

b. $x + 3$

c. $x + 1$

Chapter 12 Group Activity

Modeling with Polynomials

Materials:

- calculator

This activity may be completed by working in groups or individually.

Washington state is the leading producer of apples in the United States. The polynomial model $184x^2 - 545x + 5649$ gives Washington's annual apple production (in million pounds) for the period 2008–2012. The polynomial model $-23x^2 - 50x + 9662$ gives the total U.S. annual apple production (in million pounds) for the same period. In both models, x is the number of years after 2008. (*Source*: Based on data from the National Agricultural Statistics Service)

1. Use the given polynomials to complete the following table showing the annual apple production (both for Washington and all of the United States) over the period 2008–2012 by evaluating each polynomial at the given values of x. Then subtract each value in the fourth column from the corresponding value in the third column. Record the result in the last column, labeled "Difference." What do you think these values represent?

Year	x	Total U.S. Annual Apple Production (million pounds)	Washington's Annual Apple Production (million pounds)	Difference
2008	0			
2009	1			
2010	2			
2011	3			
2012	4			

2. Use the polynomial models to find a new polynomial model representing the annual apple production of *all other* U.S. states, excluding Washington. Then evaluate your new polynomial model to complete the accompanying table.

Year	x	Other States' Annual Apple Production (million pounds)
2008	0	
2009	1	
2010	2	
2011	3	
2012	4	

3. Compare the values in the last column of the table in Question **1** to the values in the last column of the table in Question **2**. What do you notice? What can you conclude?

4. Make a bar graph of the data in the table in Question **2**. Describe what you see.

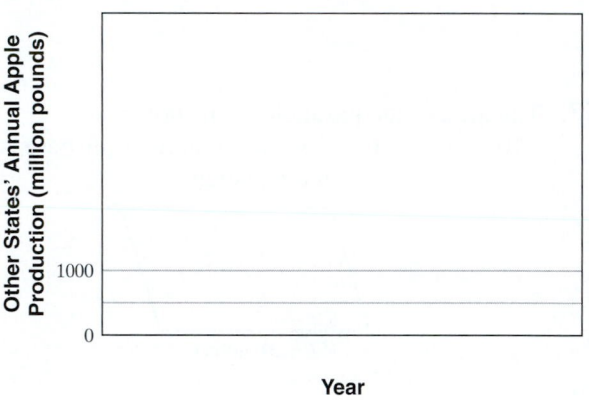

Chapter 12 Vocabulary Check

Fill in each blank with one of the words or phrases listed below.

term	coefficient	monomial	binomial	trinomial
polynomials	degree of a term	degree of a polynomial	distributive	FOIL

1. A _____ is a number or the product of a number and variables raised to powers.

2. The _____ method may be used when multiplying two binomials.

3. A polynomial with exactly 3 terms is called a _____.

4. The _____ is the greatest degree of any term of the polynomial.

5. A polynomial with exactly 2 terms is called a _____.

6. The _____ of a term is its numerical factor.

7. The _____ is the sum of the exponents on the variables in the term.

8. A polynomial with exactly 1 term is called a _____.

9. Monomials, binomials, and trinomials are all examples of _____.

10. The _____ property is used to multiply $2x(x - 4)$.

> **Helpful Hint**
> ▶ Are you preparing for your test? Don't forget to take the Chapter 12 Test on page 955. Then check your answers at the back of the text and use the Chapter Test Prep Videos to see the fully worked-out solutions to any of the exercises you want to review.

12 Chapter Highlights

Definitions and Concepts	Examples
Section 12.1 Exponents	

<table>
<tr><td>a^n means the product of n factors, each of which is a.</td><td>$3^2 = 3 \cdot 3 = 9$

$(-5)^3 = (-5)(-5)(-5) = -125$

$\left(\dfrac{1}{2}\right)^4 = \dfrac{1}{2} \cdot \dfrac{1}{2} \cdot \dfrac{1}{2} \cdot \dfrac{1}{2} = \dfrac{1}{16}$</td></tr>
<tr><td>Let m and n be integers and no denominators be 0.

Product Rule: $a^m \cdot a^n = a^{m+n}$

Power Rule: $(a^m)^n = a^{mn}$

Power of a Product Rule: $(ab)^n = a^n b^n$

Power of a Quotient Rule: $\left(\dfrac{a}{b}\right)^n = \dfrac{a^n}{b^n}$

Quotient Rule: $\dfrac{a^m}{a^n} = a^{m-n}$

Zero Exponent: $a^0 = 1, a \neq 0$</td><td>$x^2 \cdot x^7 = x^{2+7} = x^9$

$\left(5^3\right)^8 = 5^{3 \cdot 8} = 5^{24}$

$(7y)^4 = 7^4 y^4$

$\left(\dfrac{x}{8}\right)^3 = \dfrac{x^3}{8^3}$

$\dfrac{x^9}{x^4} = x^{9-4} = x^5$

$5^0 = 1; x^0 = 1, x \neq 0$</td></tr>
</table>

Section 12.2 Negative Exponents and Scientific Notation	

<table>
<tr><td>If $a \neq 0$ and n is an integer,

$a^{-n} = \dfrac{1}{a^n}$</td><td>$3^{-2} = \dfrac{1}{3^2} = \dfrac{1}{9}; 5x^{-2} = \dfrac{5}{x^2}$

Simplify: $\left(\dfrac{x^{-2}y}{x^5}\right)^{-2} = \dfrac{x^4 y^{-2}}{x^{-10}}$

$= x^{4-(-10)}y^{-2}$

$= \dfrac{x^{14}}{y^2}$</td></tr>
<tr><td>A positive number is written in scientific notation if it is written as the product of a number a, where $1 \leq a < 10$, and an integer power r of 10.

$a \times 10^r$</td><td>$1200 = 1.2 \times 10^3$

$0.000000568 = 5.68 \times 10^{-7}$</td></tr>
</table>

Definitions and Concepts	Examples

Section 12.3   Introduction to Polynomials

A **term** is a number or the product of a number and variables raised to powers.

$-5x, 7a^2b, \dfrac{1}{4}y^4, 0.2$

The **numerical coefficient,** or **coefficient,** of a term is its numerical factor.

Term	Coefficient
$7x^2$	7
y	1
$-a^2b$	-1

A **polynomial** is a finite sum of terms of the form ax^n where a is a real number and n is a whole number.

$5x^3 - 6x^2 + 3x - 6$ (Polynomial)

A **monomial** is a polynomial with exactly 1 term.

$\dfrac{5}{6}y^3$ (Monomial)

A **binomial** is a polynomial with exactly 2 terms.

$-0.2a^2b - 5b^2$ (Binomial)

A **trinomial** is a polynomial with exactly 3 terms.

$3x^2 - 2x + 1$ (Trinomial)

The **degree of a polynomial** is the greatest degree of any term of the polynomial.

Polynomial	Degree
$5x^2 - 3x + 2$	2
$7y + 8y^2z^3 - 12$	$2 + 3 = 5$

Section 12.4   Adding and Subtracting Polynomials

To add polynomials, combine like terms.

Add.

$$(7x^2 - 3x + 2) + (-5x - 6)$$
$$= 7x^2 - 3x + 2 - 5x - 6$$
$$= 7x^2 - 8x - 4$$

To subtract two polynomials, change the signs of the terms of the second polynomial, and then add.

Subtract.

$$(17y^2 - 2y + 1) - (-3y^3 + 5y - 6)$$
$$= (17y^2 - 2y + 1) + (3y^3 - 5y + 6)$$
$$= 17y^2 - 2y + 1 + 3y^3 - 5y + 6$$
$$= 3y^3 + 17y^2 - 7y + 7$$

Section 12.5   Multiplying Polynomials

To multiply two polynomials, multiply each term of one polynomial by each term of the other polynomial, and then combine like terms.

Multiply.

$$(2x + 1)(5x^2 - 6x + 2)$$
$$= 2x(5x^2 - 6x + 2) + 1(5x^2 - 6x + 2)$$
$$= 10x^3 - 12x^2 + 4x + 5x^2 - 6x + 2$$
$$= 10x^3 - 7x^2 - 2x + 2$$

Definitions and Concepts	Examples

Section 12.6   Special Products

The **FOIL method** may be used when multiplying two binomials.

Multiply: $(5x - 3)(2x + 3)$

$$
\begin{array}{c}
\text{First} \quad \text{Last} \\
(5x - 3)(2x + 3) \\
\text{Inner} \\
\text{Outer}
\end{array}
$$

$$
\begin{aligned}
& \quad F \qquad\quad O \qquad\quad I \qquad\quad L \\
&= (5x)(2x) + (5x)(3) + (-3)(2x) + (-3)(3) \\
&= 10x^2 + 15x - 6x - 9 \\
&= 10x^2 + 9x - 9
\end{aligned}
$$

Squaring a Binomial

$$(a + b)^2 = a^2 + 2ab + b^2$$

$$(a - b)^2 = a^2 - 2ab + b^2$$

Square each binomial.

$$
\begin{aligned}
(x + 5)^2 &= x^2 + 2(x)(5) + 5^2 \\
&= x^2 + 10x + 25
\end{aligned}
$$

$$
\begin{aligned}
(3x - 2y)^2 &= (3x)^2 - 2(3x)(2y) + (2y)^2 \\
&= 9x^2 - 12xy + 4y^2
\end{aligned}
$$

Multiplying the Sum and Difference of Two Terms

$$(a + b)(a - b) = a^2 - b^2$$

Multiply.

$$
\begin{aligned}
(6y + 5)(6y - 5) &= (6y)^2 - 5^2 \\
&= 36y^2 - 25
\end{aligned}
$$

Section 12.7   Dividing Polynomials

To divide a polynomial by a monomial,

$$\frac{a + b}{c} = \frac{a}{c} + \frac{b}{c}, c \neq 0$$

Divide.

$$\frac{15x^5 - 10x^3 + 5x^2 - 2x}{5x^2}$$

$$
= \frac{15x^5}{5x^2} - \frac{10x^3}{5x^2} + \frac{5x^2}{5x^2} - \frac{2x}{5x^2}
$$

$$
= 3x^3 - 2x + 1 - \frac{2}{5x}
$$

To divide a polynomial by a polynomial other than a monomial, use long division.

$$
\begin{array}{r}
5x - 1 + \dfrac{-4}{2x + 3} \\
2x + 3 \overline{) 10x^2 + 13x - 7} \\
\underline{10x^2 + 15x} \\
-2x - 7 \\
\underline{-2x - 3} \\
-4
\end{array}
$$

$$\text{or } 5x - 1 - \frac{4}{2x + 3}$$

(12.1) *State the base and the exponent for each expression.*

1. 3^2

2. $(-5)^4$

3. -5^4

4. x^6

Evaluate each expression.

5. 8^3

6. $(-6)^2$

7. -6^2

8. $-4^3 - 4^0$

9. $(3b)^0$

10. $\dfrac{8b}{8b}$

Simplify each expression.

11. $y^2 \cdot y^7$

12. $x^9 \cdot x^5$

13. $(2x^5)(-3x^6)$

14. $(-5y^3)(4y^4)$

15. $(x^4)^2$

16. $(y^3)^5$

17. $(3y^6)^4$

18. $(2x^3)^3$

19. $\dfrac{x^9}{x^4}$

20. $\dfrac{z^{12}}{z^5}$

21. $\dfrac{3x^4y^{10}}{12xy^6}$

22. $\dfrac{2x^7y^8}{8xy^2}$

23. $5a^7(2a^4)^3$

24. $(2x)^2(9x)$

25. $\dfrac{(4a^5b)^2}{-16ab^2}$

26. $\dfrac{(2x^3y)^4}{-16x^5y^4}$

27. $(-5a)^0 + 7^0 + 8^0$

28. $8x^0 + 9^0$

Simplify the given expression and choose the correct result.

29. $\left(\dfrac{3x^4}{4y}\right)^3$

a. $\dfrac{27x^{64}}{64y^3}$

b. $\dfrac{27x^{12}}{64y^3}$

c. $\dfrac{9x^{12}}{12y^3}$

d. $\dfrac{3x^{12}}{4y^3}$

30. $\left(\dfrac{5a^6}{b^3}\right)^2$

a. $\dfrac{10a^{12}}{b^6}$

b. $\dfrac{25a^{36}}{b^9}$

c. $\dfrac{25a^{12}}{b^6}$

d. $25a^{12}b^6$

(12.2) *Simplify each expression.*

31. 7^{-2}

32. -7^{-2}

33. $2x^{-4}$

34. $(2x)^{-4}$

35. $\left(\dfrac{1}{5}\right)^{-3}$

36. $\left(\dfrac{-2}{3}\right)^{-2}$

37. $2^0 + 2^{-4}$

38. $6^{-1} - 7^{-1}$

Simplify each expression. Write each answer using positive exponents only.

39. $\dfrac{r^{-3}}{r^{-4}}$

40. $\dfrac{y^{-2}}{y^{-5}}$

41. $\left(\dfrac{bc^{-2}}{bc^{-3}}\right)^4$

42. $\left(\dfrac{x^{-3}y^{-4}}{x^{-2}y^{-5}}\right)^{-3}$

43. $\dfrac{10a^3b^4c^0}{50ab^{11}c^3}$

44. $\dfrac{8a^0b^4c^5}{40a^6bc^{12}}$

45. $\dfrac{9x^{-4}y^{-6}}{x^2y^7}$

46. $\dfrac{3a^5b^{-5}}{a^{-5}b^5}$

Write each number in scientific notation.

47. 0.00027

48. 0.8868

49. 80,800,000

50. 868,000

51. In November 2012, approximately 130,300,000 Americans voted in the U.S. presidential election. Write this number in scientific notation. (*Source:* Nonprofit VOTE)

52. The approximate diameter of the Milky Way galaxy is 150,000 light years. Write this number in scientific notation. (*Source:* NASA IMAGE/POETRY Education and Public Outreach Program)

150,000 light-years

Write each number in standard form.

53. 8.67×10^5

54. 3.86×10^{-3}

55. 8.6×10^{-4}

56. 8.936×10^5

57. The volume of the planet Jupiter is 1.43128×10^{15} cubic kilometers. Write this number in standard form. (*Source:* National Space Science Data Center)

58. An angstrom is a unit of measure, equal to 1×10^{-10} meter, used for measuring wavelengths or the diameters of atoms. Write this number in standard form. (*Source:* National Institute of Standards and Technology)

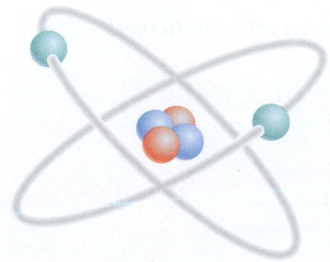

Simplify. Express each result in standard form.

59. $(8 \times 10^4)(2 \times 10^{-7})$

60. $\dfrac{8 \times 10^4}{2 \times 10^{-7}}$

(12.3) *Find the degree of each polynomial.*

61. $y^5 + 7x - 8x^4$

62. $9y^2 + 30y + 25$

63. $-14x^2y - 28x^2y^3 - 42x^2y^2$

64. $6x^2y^2z^2 + 5x^2y^3 - 12xyz$

65. The Glass Bridge Skywalk is suspended 4000 feet over the Colorado River at the very edge of the Grand Canyon. Neglecting air resistance, the height of an object dropped from the Skywalk at time t seconds is given by the polynomial $-16t^2 + 4000$. Find the height of the object at the given times below.

t	0 seconds	1 second	3 seconds	5 seconds
$-16t^2 + 4000$				

△ **66.** The surface area of a box with a square base and a height of 5 units is given by the polynomial $2x^2 + 20x$. Fill in the table below by evaluating $2x^2 + 20x$ for the given values of x.

x	1	3	5.1	10
$2x^2 + 20x$				

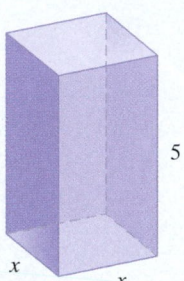

Combine like terms in each expression.

67. $7a^2 - 4a^2 - a^2$

68. $9y + y - 14y$

69. $6a^2 + 4a + 9a^2$

70. $21x^2 + 3x + x^2 + 6$

71. $4a^2b - 3b^2 - 8q^2 - 10a^2b + 7q^2$

72. $2s^{14} + 3s^{13} + 12s^{12} - s^{10}$

(12.4) *Add or subtract as indicated.*

73. $(3x^2 + 2x + 6) + (5x^2 + x)$

74. $(2x^5 + 3x^4 + 4x^3 + 5x^2) + (4x^2 + 7x + 6)$

75. $(-5y^2 + 3) - (2y^2 + 4)$

76. $(2m^7 + 3x^4 + 7m^6) - (8m^7 + 4m^2 + 6x^4)$

77. $(3x^2 - 7xy + 7y^2) - (4x^2 - xy + 9y^2)$

78. $(8x^6 - 5xy - 10y^2) - (7x^6 - 9xy - 12y^2)$

Translating *Perform the indicated operations.*

79. Add $(-9x^2 + 6x + 2)$ and $(4x^2 - x - 1)$.

80. Subtract $(4x^2 + 8x - 7)$ from the sum of $(x^2 + 7x + 9)$ and $(x^2 + 4)$.

(12.5) *Multiply each expression.*

81. $6(x + 5)$　　**82.** $9(x - 7)$　　**83.** $4(2a + 7)$　　**84.** $9(6a - 3)$

85. $-7x(x^2 + 5)$　　**86.** $-8y(4y^2 - 6)$　　**87.** $-2(x^3 - 9x^2 + x)$　　**88.** $-3a(a^2b + ab + b^2)$

89. $(-2a)(3a^3 - 4a + 1)$　　**90.** $(7b)(6b^3 - 4b + 2)$　　**91.** $(2x + 2)(x - 7)$

92. $(2x - 5)(3x + 2)$　　**93.** $(4a - 1)(a + 7)$　　**94.** $(6a - 1)(7a + 3)$

95. $(x + 7)(x^3 + 4x - 5)$　　**96.** $(x + 2)(x^5 + x + 1)$　　**97.** $(x^2 + 2x + 4)(x^2 + 2x - 4)$

98. $(x^3 + 4x + 4)(x^3 + 4x - 4)$　　**99.** $(x + 7)^3$　　**100.** $(2x - 5)^3$

(12.6) *Use special products to multiply each of the following.*

101. $(x + 7)^2$　　**102.** $(x - 5)^2$　　**103.** $(3x - 7)^2$　　**104.** $(4x + 2)^2$

105. $(5x - 9)^2$　　**106.** $(5x + 1)(5x - 1)$　　**107.** $(7x + 4)(7x - 4)$　　**108.** $(a + 2b)(a - 2b)$

109. $(2x - 6)(2x + 6)$　　**110.** $(4a^2 - 2b)(4a^2 + 2b)$

Express each as a product of polynomials in x. Then multiply and simplify.

111. Find the area of the square if its side is $(3x - 1)$ meters.

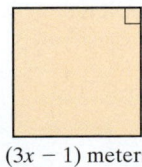

$(3x - 1)$ meters

112. Find the area of the rectangle.

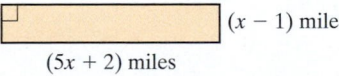

$(x - 1)$ miles

$(5x + 2)$ miles

(12.7) *Divide.*

113. $\dfrac{x^2 + 21x + 49}{7x^2}$

114. $\dfrac{5a^3b - 15ab^2 + 20ab}{-5ab}$

115. $(a^2 - a + 4) \div (a - 2)$

116. $(4x^2 + 20x + 7) \div (x + 5)$

117. $\dfrac{a^3 + a^2 + 2a + 6}{a - 2}$

118. $\dfrac{9b^3 - 18b^2 + 8b - 1}{3b - 2}$

119. $\dfrac{4x^4 - 4x^3 + x^2 + 4x - 3}{2x - 1}$

120. $\dfrac{-10x^2 - x^3 - 21x + 18}{x - 6}$

△ **121.** The area of the rectangle below is $\left(15x^3 - 3x^2 + 60\right)$ square feet. If its length is $3x^2$ feet, find its width.

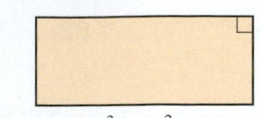

Area is $(15x^3 - 3x^2 + 60)$ sq feet

122. The perimeter of the equilateral triangle below is $\left(21a^3b^6 + 3a - 3\right)$ units. Find the length of a side.

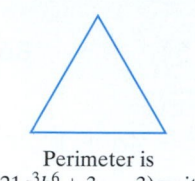

Perimeter is
$(21a^3b^6 + 3a - 3)$ units

Mixed Review

Evaluate.

123. 3^3

124. $\left(-\dfrac{1}{2}\right)^3$

Simplify each expression. Write each answer using positive exponents only.

125. $\left(4xy^2\right)\left(x^3y^5\right)$

126. $\dfrac{18x^9}{27x^3}$

127. $\left(\dfrac{3a^4}{b^2}\right)^3$

128. $\left(2x^{-4}y^3\right)^{-4}$

129. $\dfrac{a^{-3}b^6}{9^{-1}a^{-5}b^{-2}}$

Perform the indicated operations and simplify.

130. $\left(-y^2 - 4\right) + \left(3y^2 - 6\right)$

131. $\left(6x + 2\right) + \left(5x - 7\right)$

132. $\left(5x^2 + 2x - 6\right) - \left(-x - 4\right)$

133. $\left(8y^2 - 3y + 1\right) - \left(3y^2 + 2\right)$

134. $\left(2x + 5\right)\left(3x - 2\right)$

135. $4x\left(7x^2 + 3\right)$

136. $\left(7x - 2\right)\left(4x - 9\right)$

137. $\left(x - 3\right)\left(x^2 + 4x - 6\right)$

Use special products to multiply.

138. $\left(5x + 4\right)^2$

139. $\left(6x + 3\right)\left(6x - 3\right)$

Divide.

140. $\dfrac{8a^4 - 2a^3 + 4a - 5}{2a^3}$

141. $\dfrac{x^2 + 2x + 10}{x + 5}$

142. $\dfrac{4x^3 + 8x^2 - 11x + 4}{2x - 3}$

Answers

Evaluate each expression.

1. 2^5 **2.** $(-3)^4$ **3.** -3^4 **4.** 4^{-3}

Simplify each expression. Write the result using only positive exponents.

5. $(3x^2)(-5x^9)$ **6.** $\dfrac{y^7}{y^2}$ **7.** $\dfrac{r^{-8}}{r^{-3}}$

8. $\left(\dfrac{4x^2y^3}{x^3y^{-4}}\right)^2$ **9.** $\dfrac{6^2x^{-4}y^{-1}}{6^3x^{-3}y^7}$

Express each number in scientific notation.

10. 563,000   **11.** 0.0000863

Write each number in standard form.

12. 1.5×10^{-3} **13.** 6.23×10^4

14. Simplify. Write the answer in standard form.

$$(1.2 \times 10^5)(3 \times 10^{-7})$$

15. a. Complete the table for the polynomial $4xy^2 + 7xyz + x^3y - 2$.

Term	Numerical Coefficient	Degree of Term
$4xy^2$		
$7xyz$		
x^3y		
-2		

    b. What is the degree of the polynomial?

16. Simplify by combining like terms.
$5x^2 + 4x - 7x^2 + 11 + 8x$

Perform each indicated operation.

17. $(8x^3 + 7x^2 + 4x - 7) + (8x^3 - 7x - 6)$ **18.** $\begin{array}{r} 5x^3 + x^2 + 5x - 2 \\ -(8x^3 - 4x^2 + 5x - 7) \\ \hline \end{array}$

19. Subtract $(4x + 2)$ from the sum of $(8x^2 + 7x + 5)$ and $(x^3 - 8)$.

1. _____

2. _____

3. _____

4. _____

5. _____

6. _____

7. _____

8. _____

9. _____

10. _____

11. _____

12. _____

13. _____

14. _____

15. a. _____

   b. _____

16. _____

17. _____

18. _____

19. _____

Multiply in Exercises 20 through 26.

20. $(3x + 7)(x^2 + 5x + 2)$

21. $3x^2(2x^2 - 3x + 7)$

22. $(x + 7)(3x - 5)$

23. $\left(3x - \dfrac{1}{5}\right)\left(3x + \dfrac{1}{5}\right)$

24. $(4x - 2)^2$

25. $(8x + 3)^2$

26. $(x^2 - 9b)(x^2 + 9b)$

27. The height of the Bank of China in Hong Kong is 1001 feet. Neglecting air resistance, the height of an object dropped from this building at time t seconds is given by the polynomial $-16t^2 + 1001$. Find the height of the object at the given times below.

t	**0** seconds	**1** second	**3** seconds	**5** seconds
$-16t^2 + 1001$				

△ **28.** Find the area of the top of the table. Express the area as a product, then multiply and simplify.

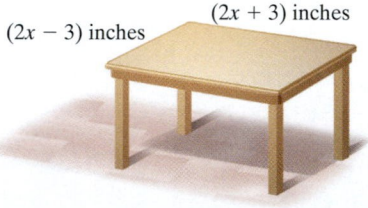

$(2x - 3)$ inches $(2x + 3)$ inches

Divide.

29. $\dfrac{4x^2 + 2xy - 7x}{8xy}$

30. $(x^2 + 7x + 10) \div (x + 5)$

31. $\dfrac{27x^3 - 8}{3x + 2}$

20. _____

21. _____

22. _____

23. _____

24. _____

25. _____

26. _____

27. _____

28. _____

29. _____

30. _____

31. _____

1. Multiply: 0.0531×16

2. Multiply: 0.0531×1000

3. Given the rectangle shown:

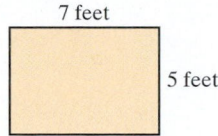

7 feet

5 feet

   a. Find the ratio of its width to its length.

   b. Find the ratio of its length to its perimeter.

4. Add: $\dfrac{5}{12} + \dfrac{2}{9}$

5. 12% of what number is 0.6?

6. Multiply: $\dfrac{7}{8} \cdot \dfrac{2}{3}$

7. What percent of 12 is 9?

8. Divide: $1\dfrac{4}{5} \div 2\dfrac{3}{10}$

9. Identify each figure as a line, a ray, a line segment, or an angle. Then name the figure using the given points.

   a.

   b.

   c.

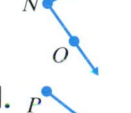

   d.

10. Find the supplement of a 12° angle.

△**11.** Find the diameter of the circle.

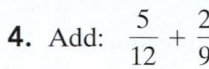

5 cm

12. Find the measure of the unknown angle.

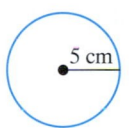

92°

54°

?

△**13.** Find the perimeter of the room shown below.

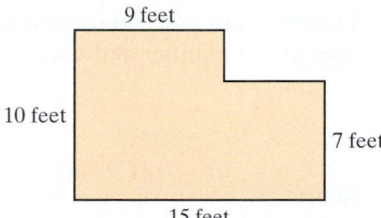

9 feet

10 feet

7 feet

15 feet

△**14.** Find the area of the room in Exercise 13.

15. Given the set $\left\{ -2, 0, \dfrac{1}{4}, 112, -3, 11, \sqrt{2} \right\}$, list the numbers in this set that belong to the set of:

   a. Natural numbers
   b. Whole numbers
   c. Integers
   d. Rational numbers
   e. Irrational numbers
   f. Real numbers

16. Find the absolute value of each number.

   a. $|-7.2|$
   b. $|0|$
   c. $\left| -\dfrac{1}{2} \right|$

Answers

1. _____
2. _____
3. a. _____
   b. _____
4. _____
5. _____
6. _____
7. _____
8. _____
9. a. _____
   b. _____
   c. _____
   d. _____
10. _____
11. _____
12. _____
13. _____
14. _____
15. a. _____
   b. _____
   c. _____
   d. _____
   e. _____
   f. _____
16. a. _____
   b. _____
   c. _____

17. _____

18. _____

19. a. _____

 b. _____

 c. _____

 d. _____

 e. _____

20. _____

21. _____

22. _____

23. _____

24. _____

25. _____

26. _____

27. _____

28. _____

29. _____

30. _____

31. _____

32. _____

33. _____

34. _____

17. Simplify: $\dfrac{3}{2} \cdot \dfrac{1}{2} - \dfrac{1}{2}$

18. Evaluate $\dfrac{2x - 7y}{x^2}$ for $x = 5$ and $y = 1$.

19. Write an algebraic expression that represents each phrase. Let the variable x represent the unknown number.
   a. The sum of a number and 3
   b. The product of 3 and a number
   c. The quotient of 7.3 and a number
   d. 10 decreased by a number
   e. 5 times a number, increased by 7

20. Simplify: $8 + 3(2 \cdot 6 - 1)$

Find each product by using the distributive property to remove parentheses.

21. $-(9x + y - 2z + 6)$

22. $-(-4xy + 6y - 2)$

23. Solve: $6(2a - 1) - (11a + 6) = 7$

24. Solve: $2x + \dfrac{1}{8} = x - \dfrac{3}{8}$

25. Solve: $\dfrac{y}{7} = 20$

26. Solve: $10 = 5j - 2$

27. Solve: $0.25x + 0.10(x - 3) = 1.1$

28. Solve: $\dfrac{7x + 5}{3} = x + 3$

29. Twice the sum of a number and 4 is the same as four times the number decreased by 12. Find the number.

30. Write the phrase as an algebraic expression and simplify if possible. Double a number, subtracted from the sum of the number and seven.

△ **31.** Charles Pecot can afford enough fencing to enclose a rectangular garden with a perimeter of 140 feet. If the width of his garden is to be 30 feet, find the length.

32. Simplify: $\dfrac{4(-3) + (-8)}{5 + (-5)}$

33. The number 120 is 15% of what number?

34. Graph $x < 5$.

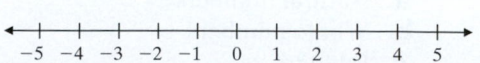

Simplify the following expressions. Write each result using positive exponents only.

35. $\left(\dfrac{3a^2}{b}\right)^{-3}$

36. $(5x^7)(-3x^9)$

37. $(5y^3)^{-2}$

38. $(-3)^{-2}$

Simplify each polynomial by combining any like terms.

39. $9x^3 + x^3$

40. $(5y^2 - 6) - (y^2 + 2)$

41. Multiply: $7x(x^2 + 2x + 5)$

42. Multiply: $(10x^2 + 3)^2$

35. _____

36. _____

37. _____

38. _____

39. _____

40. _____

41. _____

42. _____

13

Factoring Polynomials

In Chapter 12, we learned how to multiply polynomials. Now we will deal with an operation that is the reverse process of multiplying—factoring. Factoring is an important algebraic skill because it allows us to write a sum as a product. As we will see in Sections 13.6 and 13.7, factoring can be used to solve equations other than linear equations. In Chapter 14, we will also use factoring to simplify and perform arithmetic operations on rational expressions.

500,000 Units

1,000,000 Units

2,000,000 Units

Downloaded Music Now Calculated in Gold, Platinum, or Multi-Platinum Program

In 2012, digital album sales set a new high. Also, digital track sales were up 5%, and for the first time, two digital songs each had more than 6,000,000 downloads for the calendar year. In response to the continued digital market, the Recording Industry Association of America (RIAA) made some changes. In 2013, the RIAA announced the integration of downloaded music into the calculation of its Gold, Platinum, or Multi-Platinum Program for artistic achievement. Simply put, units are defined as:

- Each permanent digital download counts as 1 unit for certification purposes.
- 100 on-demand audio and/or video streams will count as 1 unit for certification purposes.

In Section 13.1, Exercises 99 and 100, we continue to study the increase in popularity of digital music.

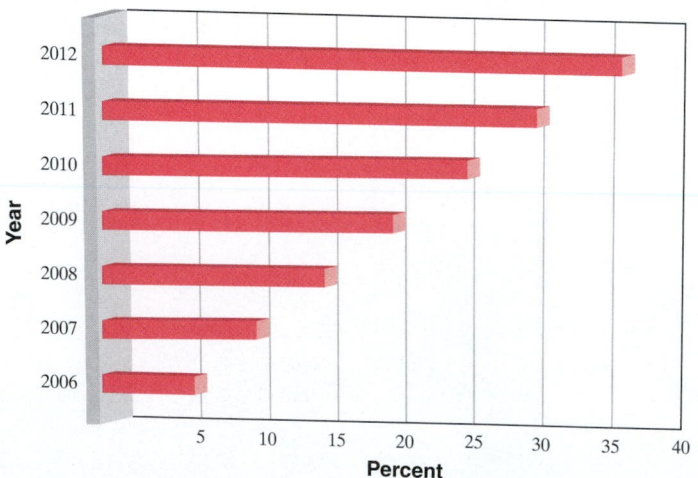
Percent of Digital Album Sales Compared to All Album Purchases

Source: The Nielsen Company & Billboard

13.1 The Greatest Common Factor and Factoring by Grouping

In the product $2 \cdot 3 = 6$, the numbers 2 and 3 are called **factors** of 6 and $2 \cdot 3$ is a **factored form** of 6. This is true of polynomials also. Since $(x + 2)(x + 3) = x^2 + 5x + 6$, then $(x + 2)$ and $(x + 3)$ are factors of $x^2 + 5x + 6$, and $(x + 2)(x + 3)$ is a factored form of the polynomial.

a factored form of 6

$$\underset{\text{factor}}{2} \cdot \underset{\text{factor}}{3} = \underset{\text{product}}{6}$$

a factored form of x^5

$$\underset{\text{factor}}{x^2} \cdot \underset{\text{factor}}{x^3} = \underset{\text{product}}{x^5}$$

a factored form of $x^2 + 5x + 6$

$$\underset{\text{factor}}{(x + 2)} \underset{\text{factor}}{(x + 3)} = \underset{\text{product}}{x^2 + 5x + 6}$$

> The process of writing a polynomial as a product is called **factoring** the polynomial.

Study the examples below and look for a pattern.

Multiplying: $5(x^2 + 3) = 5x^2 + 15$ $2x(x - 7) = 2x^2 - 14x$

Factoring: $5x^2 + 15 = 5(x^2 + 3)$ $2x^2 - 14x = 2x(x - 7)$

Do you see that factoring is the reverse process of multiplying?

$$x^2 + 5x + 6 = (x + 2)(x + 3)$$
factoring
multiplying

✓Concept Check Multiply: $2(x - 4)$
What do you think the result of factoring $2x - 8$ would be? Why?

Objective A Finding the Greatest Common Factor of a List of Numbers

The first step in factoring a polynomial is to see whether the terms of the polynomial have a common factor. If there is one, we can write the polynomial as a product by **factoring out** the common factor. We will usually factor out the *greatest* common factor (GCF).

The GCF of a list of integers is the largest integer that is a factor of all the integers in the list. For example, the GCF of 12 and 20 is 4 because 4 is the largest integer that is a factor of both 12 and 20. With large integers, the GCF may not be easily found by inspection. When this happens, use the following steps.

Objectives

A Find the Greatest Common Factor of a List of Numbers. ▶

B Find the Greatest Common Factor of a List of Terms. ▶

C Factor Out the Greatest Common Factor from the Terms of a Polynomial. ▶

D Factor a Polynomial by Grouping. ▶

✓Concept Check Answer
$2x - 8$; the result would be $2(x - 4)$ because factoring is the reverse process of multiplying.

> **Finding the GCF of a List of Integers**
>
> **Step 1:** Write each number as a product of prime numbers.
>
> **Step 2:** Identify the common prime factors.
>
> **Step 3:** The product of all common prime factors found in Step 2 is the greatest common factor. If there are no common prime factors, the greatest common factor is 1.

Recall from Section 2.2 that a prime number is a whole number other than 1 whose only factors are 1 and itself.

Practice 1

Find the GCF of each list of numbers.

a. 45 and 75

b. 32 and 33

c. 14, 24, and 60

Example 1   Find the GCF of each list of numbers.

a. 28 and 40   **b.** 55 and 21   **c.** 15, 18, and 66

Solution:

a. Write each number as a product of primes.

$$28 = 2 \cdot 2 \cdot 7 = 2^2 \cdot 7$$
$$40 = 2 \cdot 2 \cdot 2 \cdot 5 = 2^3 \cdot 5$$

There are two common factors, each of which is 2, so the GCF is

$$\text{GCF} = 2 \cdot 2 = 4$$

b. $55 = 5 \cdot 11$
$21 = 3 \cdot 7$

There are no common prime factors; thus, the GCF is 1.

c. $15 = 3 \cdot 5$
$18 = 2 \cdot 3 \cdot 3 = 2 \cdot 3^2$
$66 = 2 \cdot 3 \cdot 11$

The only prime factor common to all three numbers is 3, so the GCF is

$$\text{GCF} = 3$$

■ **Work Practice 1**

Objective B Finding the Greatest Common Factor of a List of Terms ▶

The greatest common factor of a list of variables raised to powers is found in a similar way. For example, the GCF of x^2, x^3, and x^5 is x^2 because each term contains a factor of x^2 and no higher power of x is a factor of each term.

$$x^2 = x \cdot x$$
$$x^3 = x \cdot x \cdot x$$
$$x^5 = x \cdot x \cdot x \cdot x \cdot x$$

Practice 2

Find the GCF of each list of terms.

a. y^4, y^5, and y^8

b. x and x^{10}

There are two common factors, each of which is x, so the GCF $= x \cdot x$ or x^2. From this example, we see that **the GCF of a list of common variables raised to powers is the variable raised to the smallest exponent in the list.**

Example 2   Find the GCF of each list of terms.

a. x^3, x^7, and x^5

b. y, y^4, and y^7

Answers

1. a. 15   **b.** 1   **c.** 2

2. a. y^4 **b.** x

Solution:

a. The GCF is x^3, since 3 is the smallest exponent to which x is raised.

b. The GCF is y^1 or y, since 1 is the smallest exponent on y.

■ Work Practice 2

The **greatest common factor (GCF) of a list of terms** is the product of the GCF of the numerical coefficients and the GCF of the variable factors.

$$20x^2y^2 = 2 \cdot 2 \cdot 5 \cdot x \cdot x \cdot y \cdot y$$
$$6xy^3 = 2 \cdot 3 \cdot x \cdot y \cdot y \cdot y$$
$$\text{GCF} = 2 \cdot x \cdot y \cdot y = 2xy^2$$

Helpful Hint

Remember that the GCF of a list of terms contains the smallest exponent on each common variable.

The GCF of x^5y^6, x^2y^7, and x^3y^4 is x^2y^4.

———— Smallest exponent on x
———— Smallest exponent on y

Example 3 Find the greatest common factor of each list of terms.

a. $6x^2$, $10x^3$, and $-8x$

b. $-18y^2$, $-63y^3$, and $27y^4$

c. a^3b^2, a^5b, and a^6b^2

Solution:

a. $6x^2 = 2 \cdot 3 \cdot x^2$

$10x^3 = 2 \cdot 5 \cdot x^3$ $\quad\longrightarrow$ The GCF of x^2, x^3, and x^1 is x^1 or x.

$-8x = -1 \cdot 2 \cdot 2 \cdot 2 \cdot x^1$

$\text{GCF} = 2 \cdot x^1 \quad$ or $\quad 2x$

b. $-18y^2 = -1 \cdot 2 \cdot 3 \cdot 3 \cdot y^2$

$-63y^3 = -1 \cdot 3 \cdot 3 \cdot 7 \cdot y^3$ $\quad\longrightarrow$ The GCF of y^2, y^3, and y^4 is y^2.

$27y^4 = 3 \cdot 3 \cdot 3 \cdot y^4$

$\text{GCF} = 3 \cdot 3 \cdot y^2 \quad$ or $\quad 9y^2$

c. The GCF of a^3, a^5, and a^6 is a^3.

The GCF of b^2, b, and b^2 is b. Thus,

the GCF of a^3b^2, a^5b, and a^6b^2 is a^3b.

■ Work Practice 3

Objective C Factoring Out the Greatest Common Factor ▶

To factor a polynomial such as $8x + 14$, we first see whether the terms have a greatest common factor other than 1. In this case, they do: The GCF of $8x$ and 14 is 2.

We factor out 2 from each term by writing each term as the product of 2 and the term's remaining factors.

$$8x + 14 = 2 \cdot 4x + 2 \cdot 7$$

Using the distributive property, we can write

$$8x + 14 = 2 \cdot 4x + 2 \cdot 7$$
$$= 2(4x + 7)$$

Practice 3

Find the greatest common factor of each list of terms.

a. $6x^2$, $9x^4$, and $-12x^5$

b. $-16y$, $-20y^6$, and $40y^4$

c. a^5b^4, ab^3, and a^3b^2

Answer

3. **a.** $3x^2$ **b.** $4y$ **c.** ab^2

Thus, a factored form of $8x + 14$ is $2(4x + 7)$. We can check by multiplying:

$$2(4x + 7) = 2 \cdot 4x + 2 \cdot 7 = 8x + 14$$

Helpful Hint

A factored form of $8x + 14$ is *not*

$$2 \cdot 4x + 2 \cdot 7$$

Although the *terms* have been factored (written as products), the *polynomial* $8x + 14$ has not been factored. A factored form of $8x + 14$ is the *product* $2(4x + 7)$.

✓**Concept Check** Which of the following is/are factored form(s) of $6t + 18$?

a. 6 **b.** $6 \cdot t + 6 \cdot 3$ **c.** $6(t + 3)$ **d.** $3(t + 6)$

Practice 4

Factor each polynomial by factoring out the greatest common factor (GCF).

a. $10y + 25$

b. $x^4 - x^9$

Example 4 Factor each polynomial by factoring out the greatest common factor (GCF).

a. $5ab + 10a$ **b.** $y^5 - y^{12}$

Solution:

a. The GCF of terms $5ab$ and $10a$ is $5a$. Thus,

$$5ab + 10a = 5a \cdot b + 5a \cdot 2$$
$$= 5a(b + 2) \qquad \text{Apply the distributive property.}$$

We can check our work by multiplying $5a$ and $(b + 2)$.
$5a(b + 2) = 5a \cdot b + 5a \cdot 2 = 5ab + 10a$, the original polynomial.

b. The GCF of y^5 and y^{12} is y^5. Thus,

$$y^5 - y^{12} = y^5(1) - y^5(y^7)$$
$$= y^5(1 - y^7)$$

Helpful Hint

Don't forget the 1.

■ **Work Practice 4**

Practice 5

Factor: $-10x^3 + 8x^2 - 2x$

Example 5 Factor: $-9a^5 + 18a^2 - 3a$

Solution:

$$-9a^5 + 18a^2 - 3a = 3a(-3a^4) + 3a(6a) + 3a(-1)$$
$$= 3a(-3a^4 + 6a - 1)$$

■ **Work Practice 5**

Helpful Hint

Don't forget the -1.

In Example 5, we could have chosen to factor out $-3a$ instead of $3a$. If we factor out $-3a$, we have

$$-9a^5 + 18a^2 - 3a = (-3a)(3a^4) + (-3a)(-6a) + (-3a)(1)$$
$$= -3a(3a^4 - 6a + 1)$$

Helpful Hint

Notice the changes in signs when factoring out $-3a$.

Answers

4. a. $5(2y + 5)$ **b.** $x^4(1 - x^5)$

5. $2x(-5x^2 + 4x - 1)$

✓**Concept Check Answer**

c

Examples   Factor.

6. $6a^4 - 12a = 6a(a^3 - 2)$

7. $\dfrac{3}{7}x^4 + \dfrac{1}{7}x^3 - \dfrac{5}{7}x^2 = \dfrac{1}{7}x^2(3x^2 + x - 5)$

8. $15p^2q^4 + 20p^3q^5 + 5p^3q^3 = 5p^2q^3(3q + 4pq^2 + p)$

◼ **Work Practice 6–8**

Factor.

6. $4x^3 + 12x$

7. $\dfrac{2}{5}a^5 - \dfrac{4}{5}a^3 + \dfrac{1}{5}a^2$

8. $6a^3b + 3a^3b^2 + 9a^2b^4$

Example 9   Factor: $5(x + 3) + y(x + 3)$

Solution: The binomial $(x + 3)$ is present in both terms and is the greatest common factor. We use the distributive property to factor out $(x + 3)$.

$$5(x + 3) + y(x + 3) = (x + 3)(5 + y)$$

◼ **Work Practice 9**

Practice 9

Factor: $7(p + 2) + q(p + 2)$

Example 10   Factor: $3m^2n(a + b) - (a + b)$

Solution: The greatest common factor is $(a + b)$.

$$3m^2n(a + b) - 1(a + b) = (a + b)(3m^2n - 1)$$

◼ **Work Practice 10**

Practice 10

Factor $7xy^3(p + q) - (p + q)$

Objective D   Factoring by Grouping ▶

Once the GCF is factored out, we can often continue to factor the polynomial using a variety of techniques. We discuss here a technique called **factoring by grouping.** This technique can be used to factor some polynomials with four terms.

Example 11   Factor $xy + 2x + 3y + 6$ by grouping.

Solution: Notice that the first two terms of this polynomial have a common factor of x and that the second two terms have a common factor of 3. Because of this, group the first two terms, then the last two terms, and then factor out these common factors.

$$xy + 2x + 3y + 6 = (xy + 2x) + (3y + 6) \quad \text{Group terms.}$$
$$= x(y + 2) + 3(y + 2) \quad \text{Factor out GCF from each grouping.}$$

Next we factor out the common binomial factor, $(y + 2)$.

$$x(y + 2) + 3(y + 2) = (y + 2)(x + 3)$$

Now the result is a factored form because it is a product. We were able to write the polynomial as a product because of the common binomial factor, $(y + 2)$, that appeared. If this does not happen, try rearranging the terms of the original polynomial.

Check: Multiply $(y + 2)$ by $(x + 3)$.

$$(y + 2)(x + 3) = xy + 2x + 3y + 6,$$

the original polynomial.

Thus, a factored form of $xy + 2x + 3y + 6$ is the product $(y + 2)(x + 3)$.

◼ **Work Practice 11**

Practice 11

Factor $ab + 7a + 2b + 14$ by grouping.

Helpful Hint   Notice that this form, $x(y + 2) + 3(y + 2)$, is *not* a factored form of the original polynomial. It is a sum, not a product.

Answers

6. $4x(x^2 + 3)$ **7.** $\dfrac{1}{5}a^2(2a^3 - 4a + 1)$

8. $3a^2b(2a + ab + 3b^3)$

9. $(p + 2)(7 + q)$

10. $(p + q)(7xy^3 - 1)$

11. $(b + 7)(a + 2)$

You may want to try these steps when factoring by grouping.

To Factor a Four-Term Polynomial by Grouping

Step 1: Group the terms in two groups of two terms so that each group has a common factor.

Step 2: Factor out the GCF from each group.

Step 3: If there is a common binomial factor, factor it out.

Step 4: If not, rearrange the terms and try these steps again.

Practice 12–14

Factor by grouping.

12. $28x^3 - 7x^2 + 12x - 3$

13. $2xy + 5y^2 - 4x - 10y$

14. $3x^2 + 4xy + 3x + 4y$

Examples Factor by grouping.

12. $15x^3 - 10x^2 + 6x - 4$
$= (15x^3 - 10x^2) + (6x - 4)$ Group the terms.
$= 5x^2(3x - 2) + 2(3x - 2)$ Factor each group.
$= (3x - 2)(5x^2 + 2)$ Factor out the common factor, $(3x - 2)$.

13. $3x^2 + 4xy - 3x - 4y$
$= (3x^2 + 4xy) + (-3x - 4y)$
$= x(3x + 4y) - 1(3x + 4y)$ Factor each group. A -1 is factored from the second pair of terms so that there is a common factor, $(3x + 4y)$.
$= (3x + 4y)(x - 1)$ Factor out the common factor, $(3x + 4y)$.

14. $2a^2 + 5ab + 2a + 5b$
$= (2a^2 + 5ab) + (2a + 5b)$ Factor each group. An understood 1 is written before $(2a + 5b)$ to help remember that $(2a + 5b)$ is $1(2a + 5b)$.
$= a(2a + 5b) + 1(2a + 5b)$ Factor out the common factor, $(2a + 5b)$.
$= (2a + 5b)(a + 1)$

Helpful Hint Notice that the factor of 1 is written when $(2a + 5b)$ is factored out.

Work Practice 12–14

Practice 15–17

Factor by grouping.

15. $4x^3 + x - 20x^2 - 5$

16. $3xy - 4 + x - 12y$

17. $2x - 2 + x^3 - 3x^2$

Examples Factor by grouping.

15. $3x^3 - 2x - 9x^2 + 6$
$= x(3x^2 - 2) - 3(3x^2 - 2)$ Factor each group. A -3 is factored from the second pair of terms so that there is a common factor, $(3x^2 - 2)$.
$= (3x^2 - 2)(x - 3)$ Factor out the common factor, $(3x^2 - 2)$.

16. $3xy + 2 - 3x - 2y$

Notice that the first two terms have no common factor other than 1. However, if we rearrange these terms, a grouping emerges that does lead to a common factor.

$3xy + 2 - 3x - 2y$
$= (3xy - 3x) + (-2y + 2)$
$= 3x(y - 1) - 2(y - 1)$ Factor -2 from the second group.
$= (y - 1)(3x - 2)$ Factor out the common factor, $(y - 1)$.

17. $5x - 10 + x^3 - x^2 = 5(x - 2) + x^2(x - 1)$

There is no common binomial factor that can now be factored out. No matter how we rearrange the terms, no grouping will lead to a common factor. Thus, this polynomial is not factorable by grouping.

Work Practice 15–17

Helpful Hint

Throughout this chapter, we will be factoring polynomials. Even when the instructions do not so state, it is always a good idea to check your answers by multiplying.

Answers

12. $(4x - 1)(7x^2 + 3)$
13. $(2x + 5y)(y - 2)$
14. $(3x + 4y)(x + 1)$
15. $(4x^2 + 1)(x - 5)$
16. $(3y + 1)(x - 4)$
17. cannot be factored by grouping

Vocabulary, Readiness & Video Check

Use the choices below to fill in each blank. Some choices may be used more than once and some may not be used at all.

greatest common factor   factors   factoring   true   false   least   greatest

1. Since $5 \cdot 4 = 20$, the numbers 5 and 4 are called _____ of 20.
2. The _____ of a list of integers is the largest integer that is a factor of all the integers in the list.
3. The greatest common factor of a list of common variables raised to powers is the variable raised to the _____ exponent in the list.
4. The process of writing a polynomial as a product is called _____.
5. True or false? A factored form of $7x + 21 + xy + 3y$ is $7(x + 3) + y(x + 3)$. _____
6. True or false? A factored form of $3x^3 + 6x + x^2 + 2$ is $3x(x^2 + 2)$. _____

Write the prime factorization of the following integers.

7. 14   **8.** 15

Write the GCF of the following pairs of integers.

9. 18, 3   **10.** 7, 35   **11.** 20, 15   **12.** 6, 15

Martin-Gay Interactive Videos

See Video 13.1

Watch the section lecture video and answer the following questions.

Objective A **13.** Based on ⊞ Example 1, give a general definition for the greatest common factor (GCF) of a list of numbers. ▶

Objective B **14.** When finding the GCF of the terms in ⊞ Example 3, why are the numerical parts of the terms factored out, but not the variable parts? ▶

Objective C **15.** From ⊞ Example 5, once we factor out the GCF, how can the number of terms in the other factor help us determine if our factorization is correct? ▶

Objective D **16.** In ⊞ Examples 7 and 8, what are we reminded to always do first when factoring a polynomial? Also, a polynomial with how many terms suggests it might be factored by grouping? ▶

13.1 Exercise Set MyMathLab® ▶

Objectives A B Mixed Practice *Find the GCF for each list. See Examples 1 through 3.*

1. 32, 36   ▶ **2.** 36, 90   **3.** 18, 42, 84   **4.** 30, 75, 135

5. 24, 14, 21   **6.** 15, 25, 27   **7.** y^2, y^4, y^7 ▶ **8.** x^3, x^2, x^5

9. z^7, z^9, z^{11} **10.** y^8, y^{10}, y^{12} **11.** $x^{10}y^2, xy^2, x^3y^3$ **12.** p^7q, p^8q^2, p^9q^3

13. $14x, 21$ **14.** $20y, 15$ ▶ **15.** $12y^4, 20y^3$ **16.** $32x^5, 18x^2$

17. $-10x^2, 15x^3$ **18.** $-21x^3, 14x$ **19.** $12x^3, -6x^4, 3x^5$ **20.** $15y^2, 5y^7, -20y^3$

21. $-18x^2y, 9x^3y^3, 36x^3y$ **22.** $7x^3y^3, -21x^2y^2, 14xy^4$ **23.** $20a^6b^2c^8, 50a^7b$ **24.** $40x^7y^2z, 64x^9y$

Objective **C** *Factor out the GCF from each polynomial. See Examples 4 through 10.*

25. $3a + 6$ **26.** $18a + 12$ ▶ **27.** $30x - 15$ **28.** $42x - 7$ **29.** $x^3 + 5x^2$

30. $y^5 + 6y^4$ **31.** $6y^4 + 2y^3$ **32.** $5x^2 + 10x^6$ **33.** $32xy - 18x^2$ **34.** $10xy - 15x^2$

35. $4x - 8y + 4$ **36.** $7x + 21y - 7$ **37.** $6x^3 - 9x^2 + 12x$ **38.** $12x^3 + 16x^2 - 8x$

39. $a^7b^6 - a^3b^2 + a^2b^5 - a^2b^2$ **40.** $x^9y^6 + x^3y^5 - x^4y^3 + x^3y^3$ **41.** $5x^3y - 15x^2y + 10xy$

▶ **42.** $14x^3y + 7x^2y - 7xy$ **43.** $8x^5 + 16x^4 - 20x^3 + 12$ **44.** $9y^6 - 27y^4 + 18y^2 + 6$

45. $\dfrac{1}{3}x^4 + \dfrac{2}{3}x^3 - \dfrac{4}{3}x^5 + \dfrac{1}{3}x$ **46.** $\dfrac{2}{5}y^7 - \dfrac{4}{5}y^5 + \dfrac{3}{5}y^2 - \dfrac{2}{5}y$ ▶ **47.** $y(x^2 + 2) + 3(x^2 + 2)$

48. $x(y^2 + 1) - 3(y^2 + 1)$ **49.** $z(y + 4) + 3(y + 4)$ **50.** $8(x + 2) - y(x + 2)$

51. $r(z^2 - 6) + (z^2 - 6)$ **52.** $q(b^3 - 5) + (b^3 - 5)$

Factor a negative number or a GCF with a negative coefficient from each polynomial. See Example 5.

53. $-2x - 14$ **54.** $-7y - 21$ **55.** $-2x^5 + x^7$

56. $-5y^3 + y^6$ **57.** $-6a^4 + 9a^3 - 3a^2$ **58.** $-5m^6 + 10m^5 - 5m^3$

Objective **D** *Factor each four-term polynomial by grouping. If this is not possible, write "not factorable by grouping."*
See Examples 11 through 17.

59. $x^3 + 2x^2 + 5x + 10$ **60.** $x^3 + 4x^2 + 3x + 12$ **61.** $5x + 15 + xy + 3y$

62. $xy + y + 2x + 2$ **63.** $6x^3 - 4x^2 + 15x - 10$ **64.** $16x^3 - 28x^2 + 12x - 21$

65. $5m^3 + 6mn + 5m^2 + 6n$ **66.** $8w^2 + 7wv + 8w + 7v$ **67.** $2y - 8 + xy - 4x$

68. $6x - 42 + xy - 7y$ **69.** $2x^3 + x^2 + 8x + 4$ **70.** $2x^3 - x^2 - 10x + 5$

71. $3x - 3 + x^3 - 4x^2$ **72.** $7x - 21 + x^3 - 2x^2$ **73.** $4x^2 - 8xy - 3x + 6y$

▶ **74.** $5xy - 15x - 6y + 18$ **75.** $5q^2 - 4pq - 5q + 4p$ **76.** $6m^2 - 5mn - 6m + 5n$

Objectives C D Mixed Practice *Factor out the GCF from each polynomial. Then factor by grouping.*

77. $12x^2y - 42x^2 - 4y + 14$

78. $90 + 15y^2 - 18x - 3xy^2$

▶ **79.** $6a^2 + 9ab^2 + 6ab + 9b^3$

80. $16x^2 + 4xy^2 + 8xy + 2y^3$

Review

Multiply. See Section 12.5 or 12.6.

81. $(x + 2)(x + 5)$ **82.** $(y + 3)(y + 6)$ **83.** $(b + 1)(b - 4)$ **84.** $(x - 5)(x + 10)$

Fill in the chart by finding two numbers that have the given product and sum. The first column is filled in for you.

		85.	**86.**	**87.**	**88.**	**89.**	**90.**	**91.**	**92.**
Two Numbers	4, 7								
Their Product	28	12	20	8	16	−10	−9	−24	−36
Their Sum	11	8	9	−9	−10	3	0	−5	−5

Concept Extensions

See the Concept Checks in this section.

93. Which of the following is/are factored form(s) of $-2x + 14$?
  a. $-2(x + 7)$ **b.** $-2 \cdot x + 14$
  c. $-2(x - 14)$ **d.** $-2(x - 7)$

94. Which of the following is/are factored form(s) of $8a - 24$?
  a. $8 \cdot a - 24$ **b.** $8(a - 3)$
  c. $4(2a - 12)$ **d.** $8 \cdot a - 2 \cdot 12$

Which of the following expressions are factored?

95. $(a + 6)(a + 2)$

96. $(x + 5)(x + y)$

97. $5(2y + z) - b(2y + z)$

98. $3x(a + 2b) + 2(a + 2b)$

99. The annual digital music track units sold (in millions) in the United States for the period 2009–2012 can be approximated by the polynomial $15x^2 - 250x + 2200$ where x is the number of years after 2000. (*Source:* The Nielsen Company & Billboard)
  a. Find the approximate U.S. annual digital music track units sold in 2011. To do so, let $x = 11$ and evaluate $15x^2 - 250x + 2200$.
  b. Find the approximate U.S. annual digital music track units sold in 2012.
  c. Suppose the annual digital music track units sold continues to be approximated by the polynomial $15x^2 - 250x + 2200$. Use this polynomial to predict the track music units sold in 2018.
  d. Factor out the GCF from the polynomial $15x^2 - 250x + 2200$.

100. The annual percent of digital album sales when compared to all album purchases in the United States for the period 2006–2012 can be approximated by the polynomial $\frac{1}{10}x^2 + \frac{32}{10}x - \frac{180}{10}$ where x is the number of years after 2000. (*Source:* The Nielsen Company & Billboard)
  a. Find the approximate U.S. percent of digital album sales in 2010. To do so, let $x = 10$ and evaluate $\frac{1}{10}x^2 + \frac{32}{10}x - \frac{180}{10}$.
  b. Find the approximate percent of U.S. digital album sales in 2012.
  c. Suppose the percent of digital album sales continues to be approximated by the polynomial $\frac{1}{10}x^2 + \frac{32}{10}x - \frac{180}{10}$. Use this polynomial to predict the percent of digital album sales in 2020.
  d. Factor out a factor of $\frac{1}{10}$ from the polynomial $\frac{1}{10}x^2 + \frac{32}{10}x - \frac{180}{10}$.

101. The annual orange production (in thousand tons) in the United States for the period 2010–2013 can be approximated by the polynomial $-322x^2 + 966x + 8372$, where x is the number of years after 2010. (*Source:* Based on data from the National Agricultural Statistics Service)

   a. Find the approximate U.S. orange production in 2011. To do so, let $x = 1$ and evaluate $-322x^2 + 966x + 8372$.

   b. Find the approximate U.S. orange production in 2013.

   c. Factor out the GCF from the polynomial $-322x^2 + 966x + 8372$.

102. The polynomial $4x^2 - 28x + 580$ represents the approximate number of visitors (in thousands) per year to Lowell National Historical Park in Lowell, Massachusetts, during 2008–2012. In this polynomial, x represents the years since 2008. (*Source:* Based on data from the National Park Service)

   a. Find the approximate number of visitors to Lowell National Historical Park in 2010. To do so, let $x = 2$ and evaluate $4x^2 - 28x + 580$.

   b. Find the approximate number of visitors to Lowell National Historical Park in 2012.

   c. Factor out the GCF from the polynomial $4x^2 - 28x + 580$.

Write an expression for the area of each shaded region. Then write the expression as a factored polynomial.

△**103.**

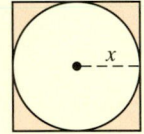

△**104.**

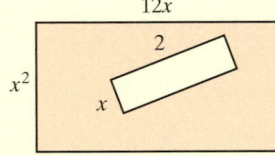

Write an expression for the length of each rectangle. (Hint: Factor the area binomial and recall that Area = width · length.)

△**105.**

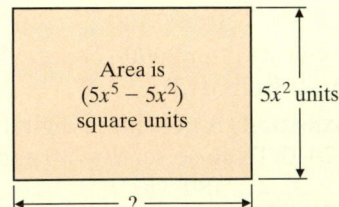

△**106.**

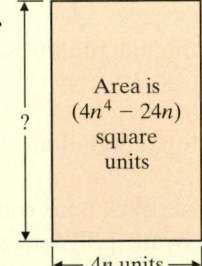

107. Construct a binomial whose greatest common factor is $5a^3$. (*Hint:* Multiply $5a^3$ by a binomial whose terms contain no common factor other than 1: $5a^3(\square + \square)$.)

108. Construct a trinomial whose greatest common factor is $2x^2$. See the hint for Exercise **105**.

109. Explain how you can tell whether a polynomial is written in factored form.

110. Construct a four-term polynomial that can be factored by grouping. Explain how you constructed the polynomial.

13.2 Factoring Trinomials of the Form $x^2 + bx + c$ ▶

Objective A Factoring Trinomials of the Form $x^2 + bx + c$ ▶

In this section, we factor trinomials of the form $x^2 + bx + c$, such as

$$x^2 + 7x + 12, \quad x^2 - 12x + 35, \quad x^2 + 4x - 12, \quad \text{and} \quad r^2 - r - 42$$

Notice that for these trinomials, the coefficient of the squared variable is 1.

Recall that factoring means to write as a product and that factoring and multiplying are reverse processes. Using the FOIL method of multiplying binomials, we have the following.

$$\begin{array}{cccc} F & O & I & L \end{array}$$
$$(x + 3)(x + 1) = x^2 + 1x + 3x + 3$$
$$= x^2 + 4x + 3$$

Thus, a factored form of $x^2 + 4x + 3$ is $(x + 3)(x + 1)$.

Notice that the product of the first terms of the binomials is $x \cdot x = x^2$, the first term of the trinomial. Also, the product of the last two terms of the binomials is $3 \cdot 1 = 3$, the third term of the trinomial. The sum of these same terms is $3 + 1 = 4$, the coefficient of the middle, x, term of the trinomial.

The product of these numbers is 3.

$$x^2 + 4x + 3 = (x + 3)(x + 1)$$

The sum of these numbers is 4.

Many trinomials, such as the one above, factor into two binomials. To factor $x^2 + 7x + 10$, let's assume that it factors into two binomials and begin by writing two pairs of parentheses. The first term of the trinomial is x^2, so we use x and x as the first terms of the binomial factors.

$$x^2 + 7x + 10 = (x + \square)(x + \square)$$

To determine the last term of each binomial factor, we look for two integers whose product is 10 and whose sum is 7. The integers are 2 and 5. Thus,

$$x^2 + 7x + 10 = (x + 2)(x + 5)$$

Check: To see if we have factored correctly, we multiply.

$$(x + 2)(x + 5) = x^2 + 5x + 2x + 10$$
$$= x^2 + 7x + 10 \qquad \text{Combine like terms.}$$

Helpful Hint

Since multiplication is commutative, the factored form of $x^2 + 7x + 10$ can be written as either $(x + 2)(x + 5)$ or $(x + 5)(x + 2)$.

To Factor a Trinomial of the Form $x^2 + bx + c$

The product of these numbers is c.

$$x^2 + bx + c = (x + \square)(x + \square)$$

The sum of these numbers is b.

Practice 1

Factor: $x^2 + 12x + 20$

Example 1   Factor: $x^2 + 7x + 12$

Solution: We begin by writing the first terms of the binomial factors.

$$(x + \Box)(x + \Box)$$

Next we look for two numbers whose product is 12 and whose sum is 7. Since our numbers must have a positive product and a positive sum, we look at pairs of positive factors of 12 only.

Factors of 12	Sum of Factors
1, 12	13
2, 6	8
3, 4	7

Correct sum, so the numbers are 3 and 4.

Thus, $x^2 + 7x + 12 = (x + 3)(x + 4)$

Check: $(x + 3)(x + 4) = x^2 + 4x + 3x + 12 = x^2 + 7x + 12$

■ **Work Practice 1**

Practice 2

Factor each trinomial.
a. $x^2 - 23x + 22$
b. $x^2 - 27x + 50$

Example 2   Factor: $x^2 - 12x + 35$

Solution: Again, we begin by writing the first terms of the binomials.

$$(x + \Box)(x + \Box)$$

Now we look for two numbers whose product is 35 and whose sum is -12. Since our numbers must have a positive product and a negative sum, we look at pairs of negative factors of 35 only.

Factors of 35	Sum of Factors
$-1, -35$	-36
$-5, -7$	-12

Correct sum, so the numbers are -5 and -7.

$$x^2 - 12x + 35 = (x - 5)(x - 7)$$

Check: To check, multiply $(x - 5)(x - 7)$.

■ **Work Practice 2**

Practice 3

Factor: $x^2 + 5x - 36$

Example 3   Factor: $x^2 + 4x - 12$

Solution: $x^2 + 4x - 12 = (x + \Box)(x + \Box)$
We look for two numbers whose product is -12 and whose sum is 4. Since our numbers must have a negative product, we look at pairs of factors with opposite signs.

Factors of -12	Sum of Factors
$-1, 12$	11
$1, -12$	-11
$-2, 6$	4
$2, -6$	-4
$-3, 4$	1
$3, -4$	-1

Correct sum, so the numbers are -2 and 6.

$$x^2 + 4x - 12 = (x - 2)(x + 6)$$

■ **Work Practice 3**

Answers

1. $(x + 10)(x + 2)$
2. a. $(x - 1)(x - 22)$
b. $(x - 2)(x - 25)$
3. $(x + 9)(x - 4)$

Example 4   Factor: $r^2 - r - 42$

Solution: Because the variable in this trinomial is r, the first term of each binomial factor is r.

$$r^2 - r - 42 = (r + \square)(r + \square)$$

Now we look for two numbers whose product is -42 and whose sum is -1, the numerical coefficient of r. The numbers are 6 and -7. Therefore,

$$r^2 - r - 42 = (r + 6)(r - 7)$$

■ **Work Practice 4**

Example 5   Factor: $a^2 + 2a + 10$

Solution: Look for two numbers whose product is 10 and whose sum is 2. Neither 1 and 10 nor 2 and 5 give the required sum, 2. We conclude that $a^2 + 2a + 10$ is not factorable with integers. A polynomial such as $a^2 + 2a + 10$ is called a **prime polynomial.**

■ **Work Practice 5**

Example 6   Factor: $x^2 + 5xy + 6y^2$

Solution: $x^2 + 5xy + 6y^2 = (x + \square)(x + \square)$

Recall that the middle term, $5xy$, is the same as $5yx$. Thus, we can see that $5y$ is the "coefficient" of x. We then look for two terms whose product is $6y^2$ and whose sum is $5y$. The terms are $2y$ and $3y$ because $2y \cdot 3y = 6y^2$ and $2y + 3y = 5y$. Therefore,

$$x^2 + 5xy + 6y^2 = (x + 2y)(x + 3y)$$

■ **Work Practice 6**

Example 7   Factor: $x^4 + 5x^2 + 6$

Solution: As usual, we begin by writing the first terms of the binomials. Since the greatest power of x in this polynomial is x^4, we write

$$(x^2 + \square)(x^2 + \square) \quad \text{Since } x^2 \cdot x^2 = x^4$$

Now we look for two factors of 6 whose sum is 5. The numbers are 2 and 3. Thus,

$$x^4 + 5x^2 + 6 = (x^2 + 2)(x^2 + 3)$$

■ **Work Practice 7**

If the terms of a polynomial are not written in descending powers of the variable, you may want to rearrange the terms before factoring.

Example 8   Factor: $40 - 13t + t^2$

Solution: First, we rearrange terms so that the trinomial is written in descending powers of t.

$$40 - 13t + t^2 = t^2 - 13t + 40$$

Next, try to factor.

$$t^2 - 13t + 40 = (t + \square)(t + \square)$$

Now we look for two factors of 40 whose sum is -13. The numbers are -8 and -5. Thus,

$$t^2 - 13t + 40 = (t - 8)(t - 5)$$

■ **Work Practice 8**

Practice 4

Factor each trinomial.
a. $q^2 - 3q - 40$
b. $y^2 + 2y - 48$

Practice 5

Factor: $x^2 + 6x + 15$

Practice 6

Factor each trinomial.
a. $x^2 + 9xy + 14y^2$
b. $a^2 - 13ab + 30b^2$

Practice 7

Factor: $x^4 + 8x^2 + 12$

Practice 8

Factor: $48 - 14x + x^2$

Answers
4. a. $(q - 8)(q + 5)$
b. $(y + 8)(y - 6)$
5. prime polynomial
6. a. $(x + 2y)(x + 7y)$
b. $(a - 3b)(a - 10b)$
7. $(x^2 + 6)(x^2 + 2)$
8. $(x - 6)(x - 8)$

The following sign patterns may be useful when factoring trinomials.

Helpful Hint

A positive constant in a trinomial tells us to look for two numbers with the same sign. The sign of the coefficient of the middle term tells us whether the signs are both positive or both negative.

both positive   same sign

$$x^2 + 10x + 16 = (x + 2)(x + 8)$$

both negative   same sign

$$x^2 - 10x + 16 = (x - 2)(x - 8)$$

A negative constant in a trinomial tells us to look for two numbers with opposite signs.

opposite signs

$$x^2 + 6x - 16 = (x + 8)(x - 2)$$

opposite signs

$$x^2 - 6x - 16 = (x - 8)(x + 2)$$

Objective B Factoring Out the Greatest Common Factor ▶

Remember that the first step in factoring any polynomial is to factor out the greatest common factor (if there is one other than 1 or −1).

Practice 9

Factor each trinomial.
a. $4x^2 - 24x + 36$
b. $x^3 + 3x^2 - 4x$

Example 9   Factor: $3m^2 - 24m - 60$

Solution: First we factor out the greatest common factor, 3, from each term.

$$3m^2 - 24m - 60 = 3(m^2 - 8m - 20)$$

Now we factor $m^2 - 8m - 20$ by looking for two factors of −20 whose sum is −8. The factors are −10 and 2. Therefore, the complete factored form is

$$3m^2 - 24m - 60 = 3(m + 2)(m - 10)$$

■ **Work Practice 9**

Helpful Hint

Remember to write the common factor, 3, as part of the factored form.

Practice 10

Factor: $5x^5 - 25x^4 - 30x^3$

Example 10   Factor: $2x^4 - 26x^3 + 84x^2$

Solution:

$$2x^4 - 26x^3 + 84x^2 = 2x^2(x^2 - 13x + 42) \quad \text{Factor out common factor, } 2x^2.$$
$$= 2x^2(x - 6)(x - 7) \quad \text{Factor } x^2 - 13x + 42.$$

■ **Work Practice 10**

Answers

9. a. $4(x - 3)(x - 3)$
b. $x(x + 4)(x - 1)$
10. $5x^3(x + 1)(x - 6)$

Vocabulary, Readiness & Video Check

Fill in each blank with "true" or "false."

1. To factor $x^2 + 7x + 6$, we look for two numbers whose product is 6 and whose sum is 7. _____

2. We can write the factorization $(y + 2)(y + 4)$ also as $(y + 4)(y + 2)$. _____

3. The factorization $(4x - 12)(x - 5)$ is completely factored. _____

4. The factorization $(x + 2y)(x + y)$ may also be written as $(x + 2y)^2$. _____

Complete each factored form.

5. $x^2 + 9x + 20 = (x + 4)(x \qquad)$

6. $x^2 + 12x + 35 = (x + 5)(x \qquad)$

7. $x^2 - 7x + 12 = (x - 4)(x \qquad)$

8. $x^2 - 13x + 22 = (x - 2)(x \qquad)$

9. $x^2 + 4x + 4 = (x + 2)(x \qquad)$

10. $x^2 + 10x + 24 = (x + 6)(x \qquad)$

Martin-Gay Interactive Videos   Watch the section lecture video and answer the following questions.

Objective A 11. In Example 2, why are only negative factors of 15 considered?

Objective B 12. In Example 5, we know we need a positive and a negative factor of -10. How do we determine which factor is negative?

See Video 13.2

13.2 Exercise Set MyMathLab®

Objective A *Factor each trinomial completely. If a polynomial can't be factored, write "prime." See Examples 1 through 8.*

1. $x^2 + 7x + 6$

2. $x^2 + 6x + 8$

3. $y^2 - 10y + 9$

4. $y^2 - 12y + 11$

5. $x^2 - 6x + 9$

6. $x^2 - 10x + 25$

7. $x^2 - 3x - 18$

8. $x^2 - x - 30$

9. $x^2 + 3x - 70$

10. $x^2 + 4x - 32$

11. $x^2 + 5x + 2$

12. $x^2 - 7x + 5$

13. $x^2 + 8xy + 15y^2$

14. $x^2 + 6xy + 8y^2$

15. $a^4 - 2a^2 - 15$

16. $y^4 - 3y^2 - 70$

17. $13 + 14m + m^2$

18. $17 + 18n + n^2$

19. $10t - 24 + t^2$

20. $6q - 27 + q^2$

21. $a^2 - 10ab + 16b^2$

22. $a^2 - 9ab + 18b^2$

Objectives **A** **B** **Mixed Practice** *Factor each trinomial completely. Some of these trinomials contain a greatest common factor (other than 1). Don't forget to factor out the GCF first. See Examples 1 through 10.*

23. $2z^2 + 20z + 32$

24. $3x^2 + 30x + 63$

25. $2x^3 - 18x^2 + 40x$

26. $3x^3 - 12x^2 - 36x$

▶ **27.** $x^2 - 3xy - 4y^2$

28. $x^2 - 4xy - 77y^2$

29. $x^2 + 15x + 36$

30. $x^2 + 19x + 60$

31. $x^4 - x^2 - 2$

32. $x^4 - 5x^2 - 14$

33. $r^2 - 16r + 48$

34. $r^2 - 10r + 21$

35. $x^2 + xy - 2y^2$

36. $x^2 - xy - 6y^2$

▶ **37.** $3x^2 + 9x - 30$

38. $4x^2 - 4x - 48$

39. $3x^4 - 60x^2 + 108$

40. $2x^4 - 24x^2 + 70$

41. $x^2 - 18x - 144$

42. $x^2 + x - 42$

43. $r^2 - 3r + 6$

44. $x^2 + 4x - 10$

▶ **45.** $x^2 - 8x + 15$

46. $x^2 - 9x + 14$

47. $6x^3 + 54x^2 + 120x$

48. $3x^3 + 3x^2 - 126x$

49. $4x^2y + 4xy - 12y$

50. $3x^2y - 9xy + 45y$

51. $x^2 - 4x - 21$

52. $x^2 - 4x - 32$

53. $x^2 + 7xy + 10y^2$

54. $x^2 - 3xy - 4y^2$

55. $64 + 24t + 2t^2$

56. $50 + 20t + 2t^2$

57. $x^3 - 2x^2 - 24x$

58. $x^3 - 3x^2 - 28x$

59. $2t^5 - 14t^4 + 24t^3$

60. $3x^6 + 30x^5 + 72x^4$

▶ **61.** $5x^3y - 25x^2y^2 - 120xy^3$

62. $7a^3b - 35a^2b^2 + 42ab^3$

63. $162 - 45m + 3m^2$

64. $48 - 20n + 2n^2$

65. $-x^2 + 12x - 11$
(Factor out -1 first.)

66. $-x^2 + 8x - 7$
(Factor out -1 first.)

67. $\frac{1}{2}y^2 - \frac{9}{2}y - 11$
(Factor out $\frac{1}{2}$ first.)

68. $\frac{1}{3}y^2 - \frac{5}{3}y - 8$
(Factor out $\frac{1}{3}$ first.)

69. $x^3y^2 + x^2y - 20x$

70. $a^2b^3 + ab^2 - 30b$

Review

Multiply. See Section 12.5 or 12.6.

71. $(2x + 1)(x + 5)$

72. $(3x + 2)(x + 4)$

73. $(5y - 4)(3y - 1)$

74. $(4z - 7)(7z - 1)$

75. $(a + 3b)(9a - 4b)$

76. $(y - 5x)(6y + 5x)$

Concept Extensions

77. Write a polynomial that factors as $(x - 3)(x + 8)$.

78. To factor $x^2 + 13x + 42$, think of two numbers whose _____ is 42 and whose _____ is 13.

Complete each sentence in your own words.

79. If $x^2 + bx + c$ is factorable and c is negative, then the signs of the last-term factors of the binomials are opposite because …

80. If $x^2 + bx + c$ is factorable and c is positive, then the signs of the last-term factors of the binomials are the same because …

Remember that perimeter means distance around. Write the perimeter of each rectangle as a simplified polynomial. Then factor the polynomial completely.

△**81.**

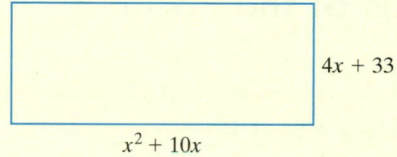

$4x + 33$

$x^2 + 10x$

△**82.**

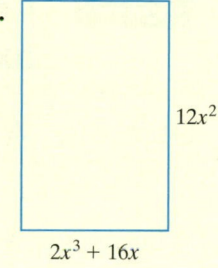

$12x^2$

$2x^3 + 16x$

83. An object is thrown upward from the top of an 80-foot building with an initial velocity of 64 feet per second. Neglecting air resistance, the height of the object after t seconds is given by $-16t^2 + 64t + 80$. Factor this polynomial.

84. An object is thrown upward from the top of a 112-foot building with an initial velocity of 96 feet per second. Neglecting air resistance, the height of the object after t seconds is given by $-16t^2 + 96t + 112$. Factor this polynomial.

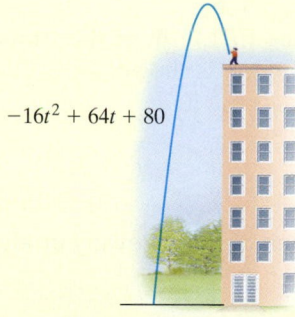

$-16t^2 + 64t + 80$

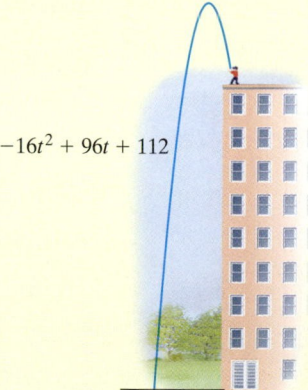

$-16t^2 + 96t + 112$

Factor each trinomial completely.

85. $x^2 + \dfrac{1}{2}x + \dfrac{1}{16}$

86. $x^2 + x + \dfrac{1}{4}$

87. $z^2(x + 1) - 3z(x + 1) - 70(x + 1)$

88. $y^2(x + 1) - 2y(x + 1) - 15(x + 1)$

Find a positive value of c so that each trinomial is factorable.

89. $n^2 - 16n + c$

90. $y^2 - 4y + c$

Find a positive value of b so that each trinomial is factorable.

91. $y^2 + by + 20$

92. $x^2 + bx + 15$

Factor each trinomial. (Hint: Notice that $x^{2n} + 4x^n + 3$ factors as $(x^n + 1)(x^n + 3)$. Remember: $x^n \cdot x^n = x^{n+n}$ or x^{2n}.)

93. $x^{2n} + 8x^n - 20$

94. $x^{2n} + 5x^n + 6$

13.3 Factoring Trinomials of the Form $ax^2 + bx + c$ ▶

Objectives

A Factor Trinomials of the Form $ax^2 + bx + c$, where $a \neq 1$. ▶

B Factor Out the GCF Before Factoring a Trinomial of the Form $ax^2 + bx + c$. ▶

Objective **A** Factoring Trinomials of the Form $ax^2 + bx + c$ ▶

In this section, we factor trinomials of the form $ax^2 + bx + c$, such as

$$3x^2 + 11x + 6, \quad 8x^2 - 22x + 5, \quad \text{and} \quad 2x^2 + 13x - 7$$

Notice that the coefficient of the squared variable in these trinomials is a number other than 1. We will factor these trinomials using a trial-and-check method based on our work in the last section.

To begin, let's review the relationship between the numerical coefficients of the trinomial and the numerical coefficients of its factored form. For example, since

$$(2x + 1)(x + 6) = 2x^2 + 13x + 6,$$

a factored form of $2x^2 + 13x + 6$ is $(2x + 1)(x + 6)$.

Notice that $2x$ and x are factors of $2x^2$, the first term of the trinomial. Also, 6 and 1 are factors of 6, the last term of the trinomial, as shown:

$$2x^2 + 13x + 6 = (2x + 1)(x + 6)$$

with $2x \cdot x$ (the first terms) and $1 \cdot 6$ (the last terms).

Also notice that $13x$, the middle term, is the sum of the following products:

$$2x^2 + 13x + 6 = (2x + 1)(x + 6)$$

$$\begin{array}{c} 1x \\ + 12x \\ \hline 13x \quad \text{Middle term} \end{array}$$

Let's use this pattern to factor $5x^2 + 7x + 2$. First, we find factors of $5x^2$. Since all numerical coefficients in this trinomial are positive, we will use factors with positive numerical coefficients only. Thus, the factors of $5x^2$ are $5x$ and x. Let's try these factors as first terms of the binomials. Thus far, we have

$$5x^2 + 7x + 2 = (5x + \square)(x + \square)$$

Next, we need to find positive factors of 2. Positive factors of 2 are 1 and 2. Now we try possible combinations of these factors as second terms of the binomials until we obtain a middle term of $7x$.

$$(5x + 1)(x + 2) = 5x^2 + 11x + 2$$

$$\begin{array}{r} 1x \\ + 10x \\ \hline 11x \end{array} \longrightarrow \textbf{Incorrect} \text{ middle term}$$

Let's try switching factors 2 and 1.

$$(5x + 2)(x + 1) = 5x^2 + 7x + 2$$

$$\begin{array}{r} 2x \\ + 5x \\ \hline 7x \end{array} \longrightarrow \textbf{Correct} \text{ middle term}$$

Thus a factored form of $5x^2 + 7x + 2$ is $(5x + 2)(x + 1)$. To check, we multiply $(5x + 2)$ and $(x + 1)$. The product is $5x^2 + 7x + 2$.

Example 1 Factor: $3x^2 + 11x + 6$

Solution: Since all numerical coefficients are positive, we use factors with positive numerical coefficients. We first find factors of $3x^2$.

Factors of $3x^2$: $3x^2 = 3x \cdot x$

If factorable, the trinomial will be of the form

$$3x^2 + 11x + 6 = (3x + \square)(x + \square)$$

Next we factor 6.

Factors of 6: $6 = 1 \cdot 6,$ $6 = 2 \cdot 3$

Now we try combinations of factors of 6 until a middle term of $11x$ is obtained. Let's try 1 and 6 first.

$$(3x + 1)(x + 6) = 3x^2 + 19x + 6$$

$$\begin{array}{r} 1x \\ + 18x \\ \hline 19x \end{array} \longrightarrow \textbf{Incorrect} \text{ middle term}$$

Now let's next try 6 and 1.

$$(3x + 6)(x + 1)$$

Before multiplying, notice that the terms of the factor $3x + 6$ have a common factor of 3. The terms of the original trinomial $3x^2 + 11x + 6$ have no common factor other than 1, so the terms of its factors will also contain no common factor other than 1. This means that $(3x + 6)(x + 1)$ is not a factored form.
Next let's try 2 and 3 as last terms.

$$(3x + 2)(x + 3) = 3x^2 + 11x + 6$$

$$\begin{array}{r} 2x \\ + 9x \\ \hline 11x \end{array} \longrightarrow \textbf{Correct} \text{ middle term}$$

Thus a factored form of $3x^2 + 11x + 6$ is $(3x + 2)(x + 3)$.

■ **Work Practice 1**

Practice 1

Factor each trinomial.
a. $5x^2 + 27x + 10$
b. $4x^2 + 12x + 5$

Helpful Hint   This is true in general: If the terms of a trinomial have no common factor (other than 1), then the terms of each of its binomial factors will contain no common factor (other than 1).

Answer
1. a. $(5x + 2)(x + 5)$
b. $(2x + 5)(2x + 1)$

✓**Concept Check** Do the terms of $3x^2 + 29x + 18$ have a common factor? Without multiplying, decide which of the following factored forms could not be a factored form of $3x^2 + 29x + 18$.

a. $(3x + 18)(x + 1)$ **b.** $(3x + 2)(x + 9)$

c. $(3x + 6)(x + 3)$ **d.** $(3x + 9)(x + 2)$

Practice 2

Factor each trinomial.
a. $2x^2 - 11x + 12$
b. $6x^2 - 5x + 1$

Example 2   Factor: $8x^2 - 22x + 5$

Solution: Factors of $8x^2$: $8x^2 = 8x \cdot x$, $8x^2 = 4x \cdot 2x$

We'll try $8x$ and x.

$$8x^2 - 22x + 5 = (8x + \square)(x + \square)$$

Since the middle term, $-22x$, has a negative numerical coefficient, we factor 5 into negative factors.

Factors of 5: $5 = -1 \cdot -5$

Let's try -1 and -5.

$$(8x - 1)(x - 5) = 8x^2 - 41x + 5$$

$$\begin{array}{r} -1x \\ +(-40x) \\ \hline -41x \end{array} \longrightarrow \textbf{Incorrect} \text{ middle term}$$

Now let's try -5 and -1.

$$(8x - 5)(x - 1) = 8x^2 - 13x + 5$$

$$\begin{array}{r} -5x \\ +(-8x) \\ \hline -13x \end{array} \longrightarrow \textbf{Incorrect} \text{ middle term}$$

Don't give up yet! We can still try other factors of $8x^2$. Let's try $4x$ and $2x$ with -1 and -5.

$$(4x - 1)(2x - 5) = 8x^2 - 22x + 5$$

$$\begin{array}{r} -2x \\ +(-20x) \\ \hline -22x \end{array} \longrightarrow \textbf{Correct} \text{ middle term}$$

A factored form of $8x^2 - 22x + 5$ is $(4x - 1)(2x - 5)$.

■ **Work Practice 2**

Practice 3

Factor each trinomial.
a. $3x^2 + 14x - 5$
b. $35x^2 + 4x - 4$

Answers
2. a. $(2x - 3)(x - 4)$
b. $(3x - 1)(2x - 1)$
3. a. $(3x - 1)(x + 5)$
b. $(5x + 2)(7x - 2)$

✓**Concept Check Answer**
no; **a, c, d**

Example 3   Factor: $2x^2 + 13x - 7$

Solution: Factors of $2x^2$: $2x^2 = 2x \cdot x$

Factors of -7: $-7 = -1 \cdot 7$, $-7 = 1 \cdot -7$

We try possible combinations of these factors:

$$(2x + 1)(x - 7) = 2x^2 - 13x - 7 \quad \textbf{Incorrect} \text{ middle term}$$
$$(2x - 1)(x + 7) = 2x^2 + 13x - 7 \quad \textbf{Correct} \text{ middle term}$$

A factored form of $2x^2 + 13x - 7$ is $(2x - 1)(x + 7)$.

■ **Work Practice 3**

Example 4   Factor: $10x^2 - 13xy - 3y^2$

Solution: Factors of $10x^2$: $10x^2 = 10x \cdot x$, $10x^2 = 2x \cdot 5x$

Factors of $-3y^2$: $-3y^2 = -3y \cdot y$, $-3y^2 = 3y \cdot -y$

We try some combinations of these factors:

              Correct   Correct
                ↓   ↓

$(10x - 3y)(x + y) = 10x^2 + 7xy - 3y^2$
$(x + 3y)(10x - y) = 10x^2 + 29xy - 3y^2$
$(5x + 3y)(2x - y) = 10x^2 + xy - 3y^2$
$(2x - 3y)(5x + y) = 10x^2 - 13xy - 3y^2$ **Correct** middle term

A factored form of $10x^2 - 13xy - 3y^2$ is $(2x - 3y)(5x + y)$.

■ **Work Practice 4**

Example 5   Factor: $3x^4 - 5x^2 - 8$

Solution: Factors of $3x^4$: $3x^4 = 3x^2 \cdot x^2$

Factors of -8: $-8 = -2 \cdot 4, 2 \cdot -4, -1 \cdot 8, 1 \cdot -8$

Try combinations of these factors:

              Correct   Correct
                ↓   ↓

$(3x^2 - 2)(x^2 + 4) = 3x^4 + 10x^2 - 8$
$(3x^2 + 4)(x^2 - 2) = 3x^4 - 2x^2 - 8$
$(3x^2 + 8)(x^2 - 1) = 3x^4 + 5x^2 - 8$ **Incorrect sign** on middle term, so switch signs in binomial factors.

$(3x^2 - 8)(x^2 + 1) = 3x^4 - 5x^2 - 8$ **Correct** middle term

■ **Work Practice 5**

Helpful Hint

Study the last two lines of Example 5. If a factoring attempt gives you a middle term whose numerical coefficient is the opposite of the desired numerical coefficient, try switching the signs of the last terms in the binomials.

Switched signs $(3x^2 + 8)(x^2 - 1) = 3x^4 + 5x^2 - 8$ Middle term: $+5x$
               $(3x^2 - 8)(x^2 + 1) = 3x^4 - 5x^2 - 8$ Middle term: $-5x$

Objective B   Factoring Out the Greatest Common Factor ▶

Don't forget that the first step in factoring any polynomial is to look for a common factor to factor out.

Example 6   Factor: $24x^4 + 40x^3 + 6x^2$

Solution: Notice that all three terms have a common factor of $2x^2$. Thus we factor out $2x^2$ first.

$24x^4 + 40x^3 + 6x^2 = 2x^2(12x^2 + 20x + 3)$

(Continued on next page)

Practice 4

Factor each trinomial.
a. $14x^2 - 3xy - 2y^2$
b. $12a^2 - 16ab - 3b^2$

Practice 5

Factor: $2x^4 - 5x^2 - 7$

Practice 6

Factor each trinomial.
a. $3x^3 + 17x^2 + 10x$
b. $6xy^2 + 33xy - 18x$

Answers
4. a. $(7x + 2y)(2x - y)$
b. $(6a + b)(2a - 3b)$
5. $(2x^2 - 7)(x^2 + 1)$
6. a. $x(3x + 2)(x + 5)$
b. $3x(2y - 1)(y + 6)$

Next we factor $12x^2 + 20x + 3$.

Factors of $12x^2$: $12x^2 = 4x \cdot 3x,$ $12x^2 = 12x \cdot x,$ $12x^2 = 6x \cdot 2x$

Since all terms in the trinomial have positive numerical coefficients, we factor 3 using positive factors only.

Factors of 3: $3 = 1 \cdot 3$

We try some combinations of the factors.

$$2x^2(4x + 3)(3x + 1) = 2x^2(12x^2 + 13x + 3)$$
$$2x^2(12x + 1)(x + 3) = 2x^2(12x^2 + 37x + 3)$$
$$2x^2(2x + 3)(6x + 1) = 2x^2(12x^2 + 20x + 3) \quad \textbf{Correct} \text{ middle term}$$

A factored form of $24x^4 + 40x^3 + 6x^2$ is $2x^2(2x + 3)(6x + 1)$.

> **Helpful Hint**
> Don't forget to include the common factor in the factored form.

■ Work Practice 6

When the term containing the squared variable has a negative coefficient, you may want to first factor out a common factor of -1.

Practice 7

Factor: $-5x^2 - 19x + 4$

Example 7 Factor: $-6x^2 - 13x + 5$

Solution: We begin by factoring out a common factor of -1.

$$-6x^2 - 13x + 5 = -1(6x^2 + 13x - 5) \quad \text{Factor out } -1.$$
$$= -1(3x - 1)(2x + 5) \quad \text{Factor } 6x^2 + 13x - 5.$$

Answer
7. $-1(x + 4)(5x - 1)$

■ Work Practice 7

Vocabulary, Readiness & Video Check

Complete each factorization.

1. $2x^2 + 5x + 3$ factors as $(2x + 3)(\underline{\hspace{1cm}})$.
 a. $(x + 3)$ **b.** $(2x + 1)$ **c.** $(3x + 4)$ **d.** $(x + 1)$

2. $7x^2 + 9x + 2$ factors as $(7x + 2)(\underline{\hspace{1cm}})$.
 a. $(3x + 1)$ **b.** $(x + 1)$ **c.** $(x + 2)$ **d.** $(7x + 1)$

3. $3x^2 + 31x + 10$ factors as $\underline{\hspace{1cm}}$.
 a. $(3x + 2)(x + 5)$ **b.** $(3x + 5)(x + 2)$ **c.** $(3x + 1)(x + 10)$

4. $5x^2 + 61x + 12$ factors as $\underline{\hspace{1cm}}$.
 a. $(5x + 1)(x + 12)$ **b.** $(5x + 3)(x + 4)$ **c.** $(5x + 2)(x + 6)$

Martin-Gay Interactive Videos   Watch the section lecture video and answer the following questions.

Objective A **5.** From ⊞ Example 1, explain in general terms how we would go about factoring a trinomial with a first-term coefficient $\neq 1$. ▶

Objective B **6.** From ⊞ Examples 3 and 5, how can factoring the GCF from a trinomial help us save time when trying to factor the remaining trinomial? ▶

See Video 13.3 🍎

13.3 **Exercise Set** MyMathLab®

Objective **A** *Complete each factored form. See Examples 1 through 5.*

1. $5x^2 + 22x + 8 = (5x + 2)(\quad)$

2. $2y^2 + 15y + 25 = (2y + 5)(\quad)$

3. $50x^2 + 15x - 2 = (5x + 2)(\quad)$

4. $6y^2 + 11y - 10 = (2y + 5)(\quad)$

5. $20x^2 - 7x - 6 = (5x + 2)(\quad)$

6. $8y^2 - 2y - 55 = (2y + 5)(\quad)$

Factor each trinomial completely. See Examples 1 through 5.

7. $2x^2 + 13x + 15$

8. $3x^2 + 8x + 4$

9. $8y^2 - 17y + 9$

10. $21x^2 - 41x + 10$

11. $2x^2 - 9x - 5$

12. $36r^2 - 5r - 24$

13. $20r^2 + 27r - 8$

14. $3x^2 + 20x - 63$

15. $10x^2 + 31x + 3$

16. $12x^2 + 17x + 5$

17. $x + 3x^2 - 2$

18. $y + 8y^2 - 9$

19. $6x^2 - 13xy + 5y^2$

20. $8x^2 - 14xy + 3y^2$

21. $15m^2 - 16m - 15$

22. $25n^2 - 5n - 6$

23. $-9x + 20 + x^2$

24. $-7x + 12 + x^2$

25. $2x^2 - 7x - 99$

26. $2x^2 + 7x - 72$

27. $-27t + 7t^2 - 4$

28. $-3t + 4t^2 - 7$

29. $3a^2 + 10ab + 3b^2$

30. $2a^2 + 11ab + 5b^2$

31. $49p^2 - 7p - 2$

32. $3r^2 + 10r - 8$

33. $18x^2 - 9x - 14$

34. $42a^2 - 43a + 6$

35. $2m^2 + 17m + 10$

36. $3n^2 + 20n + 5$

37. $24x^2 + 41x + 12$

38. $24x^2 - 49x + 15$

Objectives **A** **B** **Mixed Practice** *Factor each trinomial completely. See Examples 1 through 7.*

39. $12x^3 + 11x^2 + 2x$

40. $8a^3 + 14a^2 + 3a$

41. $21b^2 - 48b - 45$

42. $12x^2 - 14x - 10$

43. $7z + 12z^2 - 12$

44. $16t + 15t^2 - 15$

45. $6x^2y^2 - 2xy^2 - 60y^2$

46. $8x^2y + 34xy - 84y$

47. $4x^2 - 8x - 21$

48. $6x^2 - 11x - 10$

49. $3x^2 - 42x + 63$

50. $5x^2 - 75x + 60$

51. $8x^2 + 6xy - 27y^2$

52. $54a^2 + 39ab - 8b^2$

53. $-x^2 + 2x + 24$

54. $-x^2 + 4x + 21$

55. $4x^3 - 9x^2 - 9x$

56. $6x^3 - 31x^2 + 5x$

57. $24x^2 - 58x + 9$

58. $36x^2 + 55x - 14$

59. $40a^2b + 9ab - 9b$ **60.** $24y^2x + 7yx - 5x$ ▶ **61.** $30x^3 + 38x^2 + 12x$ **62.** $6x^3 - 28x^2 + 16x$

63. $6y^3 - 8y^2 - 30y$ **64.** $12x^3 - 34x^2 + 24x$ **65.** $10x^4 + 25x^3y - 15x^2y^2$ **66.** $42x^4 - 99x^3y - 15x^2y^2$

▶ **67.** $-14x^2 + 39x - 10$ **68.** $-15x^2 + 26x - 8$ **69.** $16p^4 - 40p^3 + 25p^2$ **70.** $9q^4 - 42q^3 + 49q^2$

71. $-2x^2 + 9x + 5$ **72.** $-3x^2 + 8x + 16$ **73.** $-4 + 52x - 48x^2$ **74.** $-5 + 55x - 50x^2$

75. $2t^4 + 3t^2 - 27$ **76.** $4r^4 - 17r^2 - 15$ **77.** $5x^2y^2 + 20xy + 1$ **78.** $3a^2b^2 + 12ab + 1$

79. $6a^5 + 37a^3b^2 + 6ab^4$ **80.** $5m^5 + 26m^3h^2 + 5mh^4$

Review

Multiply. See Section 12.6.

81. $(x - 4)(x + 4)$ **82.** $(2x - 9)(2x + 9)$ **83.** $(x + 2)^2$

84. $(x + 3)^2$ **85.** $(2x - 1)^2$ **86.** $(3x - 5)^2$

The following graph shows the percent of text message users in each age group. See Section 7.1.

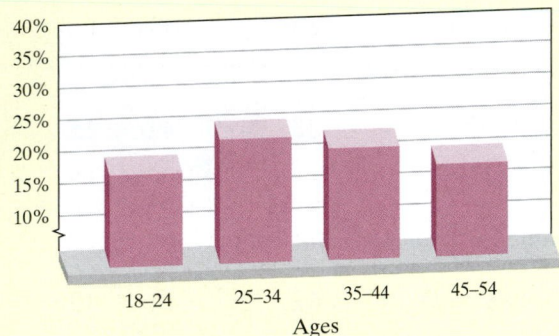

Source: Data from CellSigns, Inc.

87. What range of ages shown has the highest percent of text message users?

88. What range of ages shown has the lowest percent of text message users?

89. Describe any trend you see.

90. Why don't the percents shown in the graph add to 100%?

Concept Extensions

See the Concept Check in this section.

91. Do the terms of $4x^2 + 19x + 12$ have a common factor (other than 1)?

92. Without multiplying, decide which of the following factored forms is not a factored form of $4x^2 + 19x + 12$.
   a. $(2x + 4)(2x + 3)$
   b. $(4x + 4)(x + 3)$
   c. $(4x + 3)(x + 4)$
   d. $(2x + 2)(2x + 6)$

Write the perimeter of each figure as a simplified polynomial. Then factor the polynomial completely.

93.

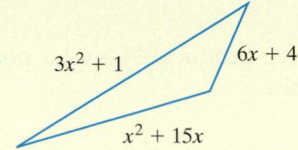

$3x^2 + 1$ $6x + 4$

$x^2 + 15x$

94.

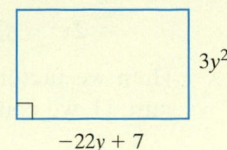

$3y^2$

$-22y + 7$

Factor each trinomial completely.

95. $4x^2 + 2x + \dfrac{1}{4}$

96. $27x^2 + 2x - \dfrac{1}{9}$

97. $4x^2(y - 1)^2 + 25x(y - 1)^2 + 25(y - 1)^2$

98. $3x^2(a + 3)^3 - 28x(a + 3)^3 + 25(a + 3)^3$

Find a positive value of b so that each trinomial is factorable.

99. $3x^2 + bx - 5$

100. $2z^2 + bz - 7$

Find a positive value of c so that each trinomial is factorable.

101. $5x^2 + 7x + c$

102. $3x^2 - 8x + c$

103. In your own words, describe the steps you use to factor a trinomial.

104. A student in your class factored $6x^2 + 7x + 1$ as $(3x + 1)(2x + 1)$. Write down how you would explain the student's error.

13.4 Factoring Trinomials of the Form $ax^2 + bx + c$ by Grouping

Objective A Using the Grouping Method

There is an alternative method that can be used to factor trinomials of the form $ax^2 + bx + c, a \neq 1$. This method is called the **grouping method** because it uses factoring by grouping as we learned in Section 13.1.

To see how this method works, recall from Section 13.1 that to factor a trinomial such as $x^2 + 11x + 30$, we find two numbers such that

Product is 30.
↓
$x^2 + 11x + 30$
↓
Sum is 11.

To factor a trinomial such as $2x^2 + 11x + 12$ by grouping, we use an extension of the method in Section 13.1. Here we look for two numbers such that

Product is $2 \cdot 12 = 24$.
↓   ↓
$2x^2 + 11x + 12$
↓
Sum is 11.

Objective

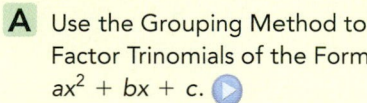

A Use the Grouping Method to Factor Trinomials of the Form $ax^2 + bx + c$.

This time, we use the two numbers to write

$2x^2 + 11x + 12$ as

$= 2x^2 + \square x + \square x + 12$

Then we factor by grouping. Since we want a positive product, 24, and a positive sum, 11, we consider pairs of positive factors of 24 only.

Factors of 24	Sum of Factors	
1, 24	25	
2, 12	14	
3, 8	11	**Correct** sum

The factors are 3 and 8. Now we use these factors to write the middle term, $11x$, as $3x + 8x$ (or $8x + 3x$). We replace $11x$ with $3x + 8x$ in the original trinomial and then we can factor by grouping.

$$2x^2 + 11x + 12 = 2x^2 + 3x + 8x + 12$$
$$= (2x^2 + 3x) + (8x + 12) \quad \text{Group the terms.}$$
$$= x(2x + 3) + 4(2x + 3) \quad \text{Factor each group.}$$
$$= (2x + 3)(x + 4) \quad \text{Factor out } (2x + 3).$$

In general, we have the following procedure.

To Factor Trinomials by Grouping

Step 1: Factor out a greatest common factor, if there is one other than 1.

Step 2: For the resulting trinomial $ax^2 + bx + c$, find two numbers whose product is $a \cdot c$ and whose sum is b.

Step 3: Write the middle term, bx, using the factors found in Step 2.

Step 4: Factor by grouping.

Practice 1

Factor each trinomial by grouping.

a. $3x^2 + 14x + 8$

b. $12x^2 + 19x + 5$

Practice 2

Factor each trinomial by grouping.

a. $30x^2 - 26x + 4$

b. $6x^2y - 7xy - 5y$

Answers

1. a. $(x + 4)(3x + 2)$

b. $(4x + 5)(3x + 1)$

2. a. $2(5x - 1)(3x - 2)$

b. $y(2x + 1)(3x - 5)$

Example 1 Factor $8x^2 - 14x + 5$ by grouping.

Solution:

Step 1: The terms of this trinomial contain no greatest common factor other than 1.

Step 2: This trinomial is of the form $ax^2 + bx + c$, with $a = 8$, $b = -14$, and $c = 5$. Find two numbers whose product is $a \cdot c$ or $8 \cdot 5 = 40$ and whose sum is b or -14.

The numbers are -4 and -10.

Factors of 40	Sum of Factors
$-40, -1$	-41
$-20, -2$	-22
$-10, -4$	-14

Correct sum

Step 3: Write $-14x$ as $-4x - 10x$ so that

$8x^2 - 14x + 5 = 8x^2 - 4x - 10x + 5$

Step 4: Factor by grouping.

$$8x^2 - 4x - 10x + 5 = 4x(2x - 1) - 5(2x - 1)$$
$$= (2x - 1)(4x - 5)$$

■ **Work Practice 1**

Example 2 Factor $6x^2 - 2x - 20$ by grouping.

Solution:

Step 1: First factor out the greatest common factor, 2.

$$6x^2 - 2x - 20 = 2(3x^2 - x - 10)$$

Note that the binomial $x^2 + 4$ is the *sum* of two squares since we can write $x^2 + 4$ as $x^2 + 2^2$. We might try to factor using $(x + 2)(x + 2)$ or $(x - 2)(x - 2)$. But when we multiply to check, we find that neither factoring is correct.

$$(x + 2)(x + 2) = x^2 + 4x + 4$$
$$(x - 2)(x - 2) = x^2 - 4x + 4$$

In both cases, the product is a trinomial, not the required binomial. In fact, $x^2 + 4$ is a prime polynomial.

■ Work Practice 9–12

Examples   Factor each difference of two squares.

13. $4x^2 - 1 = (2x)^2 - 1^2 = (2x + 1)(2x - 1)$
14. $25a^2 - 9b^2 = (5a)^2 - (3b)^2 = (5a + 3b)(5a - 3b)$
15. $y^4 - 16 = \left(y^2\right)^2 - 4^2$
   $= \left(y^2 + 4\right)\left(y^2 - 4\right)$ *Factor the difference of two squares.*
   $= \left(y^2 + 4\right)(y + 2)(y - 2)$ *Factor the difference of two squares.*

■ Work Practice 13–15

Helpful Hint

1. Don't forget to first see whether there's a greatest common factor (other than 1) that can be factored out.
2. Factor completely. In other words, check to see whether any factors can be factored further (as in Example 15).

Examples   Factor each difference of two squares.

16. $4x^3 - 49x = x\left(4x^2 - 49\right)$ *Factor out the common factor, x.*
   $= x\left[(2x)^2 - 7^2\right]$
   $= x(2x + 7)(2x - 7)$ *Factor the difference of two squares.*
17. $162x^4 - 2 = 2\left(81x^4 - 1\right)$ *Factor out the common factor, 2.*
   $= 2\left(9x^2 + 1\right)\left(9x^2 - 1\right)$ *Factor the difference of two squares.*
   $= 2\left(9x^2 + 1\right)(3x + 1)(3x - 1)$ *Factor the difference of two squares.*
18. $-49x^2 + 16 = -1\left(49x^2 - 16\right)$ *Factor out -1.*
   $= -1(7x + 4)(7x - 4)$ *Factor the difference of two squares.*

■ Work Practice 16–18

Example 19   Factor: $36 - x^2$

Solution: This is the difference of two squares. Factor as is or, if you like, first write the binomial with the variable term first.

Factor as is: $36 - x^2 = 6^2 - x^2 = (6 + x)(6 - x)$
Rewrite binomial: $36 - x^2 = -x^2 + 36 = -1\left(x^2 - 36\right)$
                 $= -1(x + 6)(x - 6)$

Both factorizations are correct and are equal. To see this, factor -1 from $(6 - x)$ in the first factorization.

■ Work Practice 19

Helpful Hint

When rearranging terms, keep in mind that the sign of a term is in front of the term.

Practice 13–15

Factor each difference of two squares.

13. $9s^2 - 1$
14. $16x^2 - 49y^2$
15. $p^4 - 81$

Practice 16–18

Factor each difference of two squares.

16. $9x^3 - 25x$ **17.** $48x^4 - 3$
18. $-9x^2 + 100$

Practice 19

Factor: $121 - m^2$

Answers
13. $(3s - 1)(3s + 1)$
14. $(4x - 7y)(4x + 7y)$
15. $(p^2 + 9)(p + 3)(p - 3)$
16. $x(3x - 5)(3x + 5)$
17. $3(4x^2 + 1)(2x + 1)(2x - 1)$
18. $-1(3x - 10)(3x + 10)$
19. $(11 + m)(11 - m)$ or $-1(m + 11)(m - 11)$

Calculator Explorations **Graphing**

A graphing calculator is a convenient tool for evaluating an expression at a given replacement value. For example, let's evaluate $x^2 - 6x$ when $x = 2$. To do so, store the value 2 in the variable x and then enter and evaluate the algebraic expression.

```
2→X
            2
X²-6X
           -8
```

The value of $x^2 - 6x$ when $x = 2$ is -8. You may want to use this method for evaluating expressions as you explore the following.

We can use a graphing calculator to explore factoring patterns numerically. Use your calculator to evaluate

$x^2 - 2x + 1$, $x^2 - 2x - 1$, and $(x - 1)^2$ for each value of x given in the table. What do you observe?

	$x^2 - 2x + 1$	$x^2 - 2x - 1$	$(x - 1)^2$
$x = 5$			
$x = -3$			
$x = 2.7$			
$x = -12.1$			
$x = 0$			

Notice in each case that $x^2 - 2x - 1 \neq (x - 1)^2$. Because for each x in the table the value of $x^2 - 2x + 1$ and the value of $(x - 1)^2$ are the same, we might guess that $x^2 - 2x + 1 = (x - 1)^2$. We can verify our guess algebraically with multiplication:

$$(x - 1)(x - 1) = x^2 - x - x + 1 = x^2 - 2x + 1$$

Vocabulary, Readiness & Video Check

Use the choices below to fill in each blank. Some choices may be used more than once and some choices may not be used at all.

perfect square trinomial   true $(5y)^2$ $(x + 5y)^2$

difference of two squares   false $(x - 5y)^2$ $5y^2$

1. A _____ is a trinomial that is the square of a binomial.

2. The term $25y^2$ written as a square is _____.

3. The expression $x^2 + 10xy + 25y^2$ is called a _____.

4. The expression $x^2 - 49$ is called a _____.

5. The factorization $(x + 5y)(x + 5y)$ may also be written as _____.

6. True or false? The factorization $(x - 5y)(x + 5y)$ may also be written as $(x - 5y)^2$. _____

7. The trinomial $x^2 - 6x - 9$ is a perfect square trinomial. _____

8. The binomial $y^2 + 9$ factors as $(y + 3)^2$. _____

Write each number or term as a square. For example, 16 written as a square is 4^2.

9. 64   **10.** 9   **11.** $121a^2$

12. $81b^2$ **13.** $36p^4$ **14.** $4q^4$

Martin-Gay Interactive Videos   Watch the section lecture video and answer the following questions.

See Video 13.5

Objective A **15.** The polynomial in ▣ Example 2 is shown to *not* be a perfect square trinomial. Does this necessarily mean it can't be factored? ▶

Objective B **16.** Describe in words the special patterns that the trinomials in ▣ Examples 3 and 4 have that identify them as perfect square trinomials. ▶

Objective C **17.** ▣ In Examples 5 and 6, what are two reasons the original binomial is rewritten so that each term is a square? ▶

18. For ▣ Example 7, what is a prime polynomial? ▶

13.5 Exercise Set MyMathLab® ▶

Objective A *Determine whether each trinomial is a perfect square trinomial. See Examples 1 through 3.*

▶ **1.** $x^2 + 16x + 64$

2. $x^2 + 22x + 121$

▶ **3.** $y^2 + 5y + 25$

4. $y^2 + 4y + 16$

5. $m^2 - 2m + 1$

6. $p^2 - 4p + 4$

7. $a^2 - 16a + 49$

8. $n^2 - 20n + 144$

9. $4x^2 + 12xy + 8y^2$

10. $25x^2 + 20xy + 2y^2$

11. $25a^2 - 40ab + 16b^2$

12. $36a^2 - 12ab + b^2$

Objective B *Factor each trinomial completely. See Examples 4 through 8.*

▶ **13.** $x^2 + 22x + 121$

14. $x^2 + 18x + 81$

15. $x^2 - 16x + 64$

16. $x^2 - 12x + 36$

17. $16a^2 - 24a + 9$

18. $25x^2 - 20x + 4$

19. $x^4 + 4x^2 + 4$

20. $m^4 + 10m^2 + 25$

21. $2n^2 - 28n + 98$

22. $3y^2 - 6y + 3$

23. $16y^2 + 40y + 25$

24. $9y^2 + 48y + 64$

25. $x^2y^2 - 10xy + 25$

26. $4x^2y^2 - 28xy + 49$

27. $m^3 + 18m^2 + 81m$

28. $y^3 + 12y^2 + 36y$

29. $1 + 6x^2 + x^4$

30. $1 + 16x^2 + x^4$

▶ **31.** $9x^2 - 24xy + 16y^2$

32. $25x^2 - 60xy + 36y^2$

Objective C   *Factor each binomial completely. See Examples 9 through 19.*

▶ **33.** $x^2 - 4$ **34.** $x^2 - 36$ **35.** $81 - p^2$ **36.** $100 - t^2$

37. $-4r^2 + 1$ **38.** $-9t^2 + 1$ **39.** $9x^2 - 16$ **40.** $36y^2 - 25$

▶ **41.** $16r^2 + 1$ **42.** $49y^2 + 1$ **43.** $-36 + x^2$ **44.** $-1 + y^2$

45. $m^4 - 1$ **46.** $n^4 - 16$ **47.** $x^2 - 169y^2$ **48.** $x^2 - 225y^2$

49. $18r^2 - 8$ **50.** $32t^2 - 50$ **51.** $9xy^2 - 4x$ **52.** $36x^2y - 25y$

53. $16x^4 - 64x^2$ **54.** $25y^4 - 100y^2$ ▶ **55.** $xy^3 - 9xyz^2$ **56.** $x^3y - 4xy^3$

57. $36x^2 - 64y^2$ **58.** $225a^2 - 81b^2$ **59.** $144 - 81x^2$ **60.** $12x^2 - 27$

61. $25y^2 - 9$ **62.** $49a^2 - 16$ ▶ **63.** $121m^2 - 100n^2$ **64.** $169a^2 - 49b^2$

65. $x^2y^2 - 1$ **66.** $a^2b^2 - 16$ **67.** $x^2 - \dfrac{1}{4}$

68. $y^2 - \dfrac{1}{16}$ ▶ **69.** $49 - \dfrac{9}{25}m^2$ **70.** $100 - \dfrac{4}{81}n^2$

Objectives B C   Mixed Practice   *Factor each binomial or trinomial completely. See Examples 4 through 19.*

71. $81a^2 - 25b^2$ **72.** $49y^2 - 100z^2$ **73.** $x^2 + 14xy + 49y^2$ **74.** $x^2 + 10xy + 25y^2$

75. $32n^4 - 112n^2 + 98$ **76.** $162a^4 - 72a^2 + 8$ **77.** $x^6 - 81x^2$

78. $n^9 - n^5$ **79.** $64p^3q - 81pq^3$ **80.** $100x^3y - 49xy^3$

Review

Solve each equation. See Section 9.3.

81. $x - 6 = 0$

82. $y + 5 = 0$

83. $2m + 4 = 0$

84. $3x - 9 = 0$

85. $5z - 1 = 0$

86. $4a + 2 = 0$

Concept Extensions

Factor each expression completely.

87. $x^2 - \dfrac{2}{3}x + \dfrac{1}{9}$

88. $x^2 - \dfrac{1}{25}$

89. $(x + 2)^2 - y^2$

90. $(y - 6)^2 - z^2$

91. $a^2(b - 4) - 16(b - 4)$

92. $m^2(n + 8) - 9(n + 8)$

93. $(x^2 + 6x + 9) - 4y^2$ (*Hint:* Factor the trinomial in parentheses first.)

94. $(x^2 + 2x + 1) - 36y^2$ (See the hint for Exercise **93**.)

95. $x^{2n} - 100$

96. $x^{2n} - 81$

97. Fill in the blank so that $x^2 + $ _____ $x + 16$ is a perfect square trinomial.

98. Fill in the blank so that $9x^2 + $ _____ $x + 25$ is a perfect square trinomial.

99. Describe a perfect square trinomial.

100. Write a perfect square trinomial that factors as $(x + 3y)^2$.

101. What binomial multiplied by $(x - 6)$ gives the difference of two squares?

102. What binomial multiplied by $(5 + y)$ gives the difference of two squares?

The area of the largest square in the figure is $(a + b)^2$. Use this figure to answer Exercises 103 and 104.

103. Write the area of the largest square as the sum of the areas of the smaller squares and rectangles.

104. What factoring formula from this section is visually represented by this square?

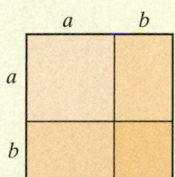

105. The Toroweap Overlook, on the North Rim of the Grand Canyon, lies 3000 vertical feet above the Colorado River. The view is spectacular, and the sheer drop is dramatic. A film crew creating a documentary about the Grand Canyon has suspended a camera platform 296 feet below the Overlook. A camera filter comes loose and falls to the river below. The height of the filter above the river, after t seconds, is given by the expression $2704 - 16t^2$.

   a. Find the height of the filter above the river after 3 seconds.

   b. Find the height of the filter above the river after 7 seconds.

   c. To the nearest whole second, estimate when the filter lands in the river.

   d. Factor $2704 - 16t^2$.

106. An object is dropped from the top of Pittsburgh's U.S. Steel Tower, which is 841 feet tall. (*Source: World Almanac* research) The height of the object after t seconds is given by the expression $841 - 16t^2$.

   a. Find the height of the object after 2 seconds.

   b. Find the height of the object after 5 seconds.

   c. To the nearest whole second, estimate when the object hits the ground.

   d. Factor $841 - 16t^2$.

841 feet

107. The world's tallest building is the Burj Khalifa in Dubai, United Arab Emirates, at a height of 2723 feet. (*Source:* Council on Tall Buildings and Urban Habitat) Suppose a worker is suspended 419 feet below the tip of the building, at a height of 2304 feet above the ground. If the worker accidentally drops a bolt, the height of the bolt after t seconds is given by the expression $2304 - 16t^2$.

   a. Find the height of the bolt after 3 seconds.

   b. Find the height of the bolt after 7 seconds.

   c. To the nearest whole second, estimate when the bolt hits the ground.

   d. Factor $2304 - 16t^2$.

108. A performer with the Moscow Circus is planning a stunt involving a free fall from the top of the MV Lomonosov State University building, which is 784 feet tall. (*Source:* Council on Tall Buildings and Urban Habitat) Neglecting air resistance, the performer's height above gigantic cushions positioned at ground level after t seconds is given by the expression $784 - 16t^2$.

   a. Find the performer's height after 2 seconds.

   b. Find the performer's height after 5 seconds.

   c. To the nearest whole second, estimate when the performer reaches the cushions positioned at ground level.

   d. Factor $784 - 16t^2$.

Integrated Review

Choosing a Factoring Strategy

The following steps may be helpful when factoring polynomials.

To Factor a Polynomial

Step 1: Are there any common factors? If so, factor out the GCF.

Step 2: How many terms are in the polynomial?

    a. Two terms: Is it the difference of two squares? $a^2 - b^2 = (a - b)(a + b)$

    b. Three terms: Try one of the following.

        i Perfect square trinomial: $a^2 + 2ab + b^2 = (a + b)^2$
                                  $a^2 - 2ab + b^2 = (a - b)^2$

        ii If not a perfect square trinomial, factor using the methods presented in Sections 13.2 through 13.4.

    c. Four terms: Try factoring by grouping.

Step 3: See if any factors in the factored polynomial can be factored further.

Step 4: Check by multiplying.

Factor each polynomial completely.

1. $x^2 + x - 12$

2. $x^2 - 10x + 16$

3. $x^2 + 2x + 1$

4. $x^2 - 6x + 9$

5. $x^2 - x - 6$

6. $x^2 + x - 2$

7. $x^2 + x - 6$

8. $x^2 + 7x + 12$

9. $x^2 - 7x + 10$

10. $x^2 - x - 30$

11. $2x^2 - 98$

12. $3x^2 - 75$

13. $x^2 + 3x + 5x + 15$

14. $3y - 21 + xy - 7x$

15. $x^2 + 6x - 16$

16. $x^2 - 3x - 28$

▶ **17.** $4x^3 + 20x^2 - 56x$

18. $6x^3 - 6x^2 - 120x$

19. $12x^2 + 34x + 24$

20. $24a^2 + 18ab - 15b^2$

21. $4a^2 - b^2$

22. $x^2 - 25y^2$

23. $28 - 13x - 6x^2$

24. $20 - 3x - 2x^2$

25. $4 - 2x + x^2$

26. $a + a^2 - 3$

27. $6y^2 + y - 15$

28. $4x^2 - x - 5$

29. $18x^3 - 63x^2 + 9x$

30. $12a^3 - 24a^2 + 4a$

31. $16a^2 - 56a + 49$

32. $25p^2 - 70p + 49$

33. $14 + 5x - x^2$

34. $3 - 2x - x^2$

35. $3x^4y + 6x^3y - 72x^2y$

36. $2x^3y + 8x^2y^2 - 10xy^3$

Answers

1. _____
2. _____
3. _____
4. _____
5. _____
6. _____
7. _____
8. _____
9. _____
10. _____
11. _____
12. _____
13. _____
14. _____
15. _____
16. _____
17. _____
18. _____
19. _____
20. _____
21. _____
22. _____
23. _____
24. _____
25. _____
26. _____
27. _____
28. _____
29. _____
30. _____
31. _____
32. _____
33. _____
34. _____
35. _____
36. _____

37. _____

38. _____

39. _____

40. _____

41. _____

42. _____

43. _____

44. _____

45. _____

46. _____

47. _____

48. _____

49. _____

50. _____

51. _____

52. _____

53. _____

54. _____

55. _____

56. _____

57. _____

58. _____

59. _____

60. _____

61. _____

62. _____

63. _____

64. _____

65. _____

66. _____

67. _____

68. _____

69. _____

70. _____

71. _____

72. _____

73. _____

74. _____

75. _____

76. _____

37. $12x^3y + 243xy$ **38.** $6x^3y^2 + 8xy^2$ ▶ **39.** $2xy - 72x^3y$

40. $2x^3 - 18x$ **41.** $x^3 + 6x^2 - 4x - 24$ **42.** $x^3 - 2x^2 - 36x + 72$

43. $6a^3 + 10a^2$ **44.** $4n^2 - 6n$ **45.** $3x^3 - x^2 + 12x - 4$

46. $x^3 - 2x^2 + 3x - 6$ **47.** $6x^2 + 18xy + 12y^2$ **48.** $12x^2 + 46xy - 8y^2$

49. $5(x + y) + x(x + y)$ **50.** $7(x - y) + y(x - y)$ **51.** $14t^2 - 9t + 1$

52. $3t^2 - 5t + 1$ **53.** $-3x^2 - 2x + 5$ **54.** $-7x^2 - 19x + 6$

55. $1 - 8a - 20a^2$ **56.** $1 - 7a - 60a^2$ **57.** $x^4 - 10x^2 + 9$

58. $x^4 - 13x^2 + 36$ **59.** $x^2 - 23x + 120$ **60.** $y^2 + 22y + 96$

61. $25p^2 - 70pq + 49q^2$ **62.** $16a^2 - 56ab + 49b^2$ **63.** $x^2 - 14x - 48$

64. $7x^2 + 24xy + 9y^2$ **65.** $-x^2 - x + 30$ **66.** $-x^2 + 6x - 8$

67. $3rs - s + 12r - 4$ **68.** $x^3 - 2x^2 + x - 2$ ▶ **69.** $4x^2 - 8xy - 3x + 6y$

70. $4x^2 - 2xy - 7yz + 14xz$ **71.** $x^2 + 9xy - 36y^2$ **72.** $3x^2 + 10xy - 8y^2$

73. $x^4 - 14x^2 - 32$ **74.** $x^4 - 22x^2 - 75$

✏ **75.** Explain why it makes good sense to factor out the GCF first, before using other methods of factoring.

✏ **76.** The sum of two squares usually does not factor. Is the sum of two squares $9x^2 + 81y^2$ factorable?

13.6 Solving Quadratic Equations by Factoring ▶

In this section, we introduce a new type of equation—the **quadratic equation.**

> ### Quadratic Equation
>
> A quadratic equation is one that can be written in the form
>
> $$ax^2 + bx + c = 0$$
>
> where a, b, and c are real numbers and $a \neq 0$.

Objectives

A Solve Quadratic Equations by Factoring. ▶

B Solve Equations with Degree Greater than Two by Factoring. ▶

Some examples of quadratic equations are shown below.

$$x^2 - 9x - 22 = 0 \qquad 4x^2 - 28 = -49 \qquad x(2x - 7) = 4$$

The form $ax^2 + bx + c = 0$ is called the **standard form** of a quadratic equation. The quadratic equation $x^2 - 9x - 22 = 0$ is the only equation above that is in standard form.

Quadratic equations model many real-life situations. For example, let's suppose we want to know how long before a person diving from a 144-foot cliff reaches the ocean. The answer to this question is found by solving the quadratic equation $-16t^2 + 144 = 0$. (See Example 1 in Section 13.7.)

144 feet

Objective A Solving Quadratic Equations by Factoring ▶

Some quadratic equations can be solved by making use of factoring and the **zero-factor property.**

> ### Zero-Factor Property
>
> If a and b are real numbers and if $ab = 0$, then $a = 0$ or $b = 0$.

In other words, if the product of two numbers is 0, then at least one of the numbers must be 0.

Example 1 Solve: $(x - 3)(x + 1) = 0$

Solution: If this equation is to be a true statement, then either the factor $x - 3$ must be 0 or the factor $x + 1$ must be 0. In other words, either

$$x - 3 = 0 \qquad \text{or} \qquad x + 1 = 0$$

If we solve these two linear equations, we have

$$x = 3 \qquad \text{or} \qquad x = -1$$

(Continued on next page)

Practice 1

Solve: $(x - 7)(x + 2) = 0$

Answer

1. 7 and −2

Thus, 3 and -1 are both solutions of the equation $(x - 3)(x + 1) = 0$. To check, we replace x with 3 in the original equation. Then we replace x with -1 in the original equation.

Check:

$$(x - 3)(x + 1) = 0 \qquad\qquad (x - 3)(x + 1) = 0$$
$$(3 - 3)(3 + 1) \stackrel{?}{=} 0 \quad \text{Replace } x \text{ with 3.} \qquad (-1 - 3)(-1 + 1) \stackrel{?}{=} 0 \quad \text{Replace } x \text{ with } -1.$$
$$0(4) = 0 \quad \text{True} \qquad\qquad (-4)(0) = 0 \quad \text{True}$$

The solutions are 3 and -1.

■ **Work Practice 1**

Helpful Hint

The zero-factor property says that *if a product is 0, then a factor is 0.*

If $a \cdot b = 0$, then $a = 0$ or $b = 0$.

If $x(x + 5) = 0$, then $x = 0$ or $x + 5 = 0$.

If $(x + 7)(2x - 3) = 0$, then $x + 7 = 0$ or $2x - 3 = 0$.

Use this property only when the product is 0. For example, if $a \cdot b = 8$, we do not know the value of a or b. The values may be $a = 2, b = 4$ or $a = 8, b = 1$, or any other two numbers whose product is 8.

Practice 2

Solve: $(x - 10)(3x + 1) = 0$

Example 2 Solve: $(x - 5)(2x + 7) = 0$

Solution: The product is 0. By the zero-factor property, this is true only when a factor is 0. To solve, we set each factor equal to 0 and solve the resulting linear equations.

$$(x - 5)(2x + 7) = 0$$
$$x - 5 = 0 \quad \text{or} \quad 2x + 7 = 0$$
$$x = 5 \qquad\qquad 2x = -7$$
$$x = -\frac{7}{2}$$

Check: Let $x = 5$.

$$(x - 5)(2x + 7) = 0$$
$$(5 - 5)(2 \cdot 5 + 7) \stackrel{?}{=} 0 \quad \text{Replace } x \text{ with 5.}$$
$$0 \cdot 17 \stackrel{?}{=} 0$$
$$0 = 0 \quad \text{True}$$

Let $x = -\frac{7}{2}$.

$$(x - 5)(2x + 7) = 0$$
$$\left(-\frac{7}{2} - 5\right)\left(2\left(-\frac{7}{2}\right) + 7\right) \stackrel{?}{=} 0 \quad \text{Replace } x \text{ with } -\frac{7}{2}.$$
$$\left(-\frac{17}{2}\right)(-7 + 7) \stackrel{?}{=} 0$$
$$\left(-\frac{17}{2}\right) \cdot 0 \stackrel{?}{=} 0$$
$$0 = 0 \quad \text{True}$$

The solutions are 5 and $-\frac{7}{2}$.

■ **Work Practice 2**

Answer

2. 10 and $-\frac{1}{3}$

Example 3 Solve: $x(5x - 2) = 0$

Solution:

$$x(5x - 2) = 0$$
$$x = 0 \quad \text{or} \quad 5x - 2 = 0 \quad \text{Use the zero-factor property.}$$
$$5x = 2$$
$$x = \frac{2}{5}$$

Check these solutions in the original equation. The solutions are 0 and $\frac{2}{5}$.

■ **Work Practice 3**

Practice 3

Solve each equation.
a. $y(y + 3) = 0$
b. $x(4x - 3) = 0$

Example 4 Solve: $x^2 - 9x - 22 = 0$

Solution: One side of the equation is 0. However, to use the zero-factor property, one side of the equation must be 0 *and* the other side must be written as a product (must be factored). Thus, we must first factor this polynomial.

$$x^2 - 9x - 22 = 0$$
$$(x - 11)(x + 2) = 0 \quad \text{Factor.}$$

Now we can apply the zero-factor property.

$$x - 11 = 0 \quad \text{or} \quad x + 2 = 0$$
$$x = 11 \qquad\qquad x = -2$$

Check: Let $x = 11$.   Let $x = -2$.

$$x^2 - 9x - 22 = 0 \qquad\qquad x^2 - 9x - 22 = 0$$
$$11^2 - 9 \cdot 11 - 22 \stackrel{?}{=} 0 \qquad (-2)^2 - 9(-2) - 22 \stackrel{?}{=} 0$$
$$121 - 99 - 22 \stackrel{?}{=} 0 \qquad\qquad 4 + 18 - 22 \stackrel{?}{=} 0$$
$$22 - 22 \stackrel{?}{=} 0 \qquad\qquad 22 - 22 \stackrel{?}{=} 0$$
$$0 = 0 \quad \text{True} \qquad\qquad 0 = 0 \quad \text{True}$$

The solutions are 11 and −2.

■ **Work Practice 4**

Practice 4

Solve: $x^2 - 3x - 18 = 0$

Example 5 Solve: $4x^2 - 28x = -49$

Solution: First we rewrite the equation in standard form so that one side is 0. Then we factor the polynomial.

$$4x^2 - 28x = -49$$
$$4x^2 - 28x + 49 = 0 \quad \text{Write in standard form by adding 49 to both sides.}$$
$$(2x - 7)(2x - 7) = 0 \quad \text{Factor.}$$

Next we use the zero-factor property and set each factor equal to 0. Since the factors are the same, the related equations will give the same solution.

$$2x - 7 = 0 \quad \text{or} \quad 2x - 7 = 0 \quad \text{Set each factor equal to 0.}$$
$$2x = 7 \qquad\qquad 2x = 7 \quad \text{Solve.}$$
$$x = \frac{7}{2} \qquad\qquad x = \frac{7}{2}$$

Check this solution in the original equation. The solution is $\frac{7}{2}$.

■ **Work Practice 5**

Practice 5

Solve: $9x^2 - 24x = -16$

Answers
3. a. 0 and −3 **b.** 0 and $\frac{3}{4}$
4. 6 and −3 **5.** $\frac{4}{3}$

The following steps may be used to solve a quadratic equation by factoring.

> ### To Solve Quadratic Equations by Factoring
>
> **Step 1:** Write the equation in standard form so that one side of the equation is 0.
>
> **Step 2:** Factor the quadratic equation completely.
>
> **Step 3:** Set each factor containing a variable equal to 0.
>
> **Step 4:** Solve the resulting equations.
>
> **Step 5:** Check each solution in the original equation.

Since it is not always possible to factor a quadratic polynomial, not all quadratic equations can be solved by factoring. Other methods of solving quadratic equations are presented in Chapter 16.

Practice 6

Solve each equation.

a. $x(x - 4) = 5$

b. $x(3x + 7) = 6$

Example 6   Solve: $x(2x - 7) = 4$

Solution: First we write the equation in standard form; then we factor.

$$x(2x - 7) = 4$$
$$2x^2 - 7x = 4 \qquad \text{Multiply.}$$
$$2x^2 - 7x - 4 = 0 \qquad \text{Write in standard form.}$$
$$(2x + 1)(x - 4) = 0 \qquad \text{Factor.}$$
$$2x + 1 = 0 \quad \text{or} \quad x - 4 = 0 \qquad \text{Set each factor equal to zero.}$$
$$2x = -1 \qquad\qquad x = 4 \qquad \text{Solve.}$$
$$x = -\frac{1}{2}$$

Check the solutions in the original equation. The solutions are $-\dfrac{1}{2}$ and 4.

■ **Work Practice 6**

Helpful Hint

To solve the equation $x(2x - 7) = 4$, do **not** set each factor equal to 4. Remember that to apply the zero-factor property, one side of the equation must be 0 and the other side of the equation must be in factored form.

✔**Concept Check** Explain the error and solve the equation correctly.

$$(x - 3)(x + 1) = 5$$
$$x - 3 = 0 \quad \text{or} \quad x + 1 = 0$$
$$x = 3 \quad \text{or} \qquad x = -1$$

Answer

6. a. 5 and -1 **b.** $\dfrac{2}{3}$ and -3

✔**Concept Check Answer**

To use the zero-factor property, one side of the equation must be 0, not 5. Correctly, $(x - 3)(x + 1) = 5$, $x^2 - 2x - 3 = 5$, $x^2 - 2x - 8 = 0$, $(x - 4)(x + 2) = 0$, $x - 4 = 0$ or $x + 2 = 0$, $x = 4$ or $x = -2$.

Objective B Solving Equations with Degree Greater than Two by Factoring ▶

Some equations with degree greater than 2 can be solved by factoring and then using the zero-factor property.

Example 7 Solve: $3x^3 - 12x = 0$

Solution: To factor the left side of the equation, we begin by factoring out the greatest common factor, $3x$.

$$3x^3 - 12x = 0$$
$$3x(x^2 - 4) = 0 \quad \text{Factor out the GCF, } 3x.$$
$$3x(x + 2)(x - 2) = 0 \quad \text{Factor } x^2 - 4, \text{ a difference of two squares.}$$
$$3x = 0 \quad \text{or} \quad x + 2 = 0 \quad \text{or} \quad x - 2 = 0 \quad \text{Set each factor equal to 0.}$$
$$x = 0 \qquad\qquad x = -2 \qquad\qquad x = 2 \quad \text{Solve.}$$

Thus, the equation $3x^3 - 12x = 0$ has three solutions: $0, -2,$ and 2.

Check: Replace x with each solution in the original equation.
Let $x = 0$.

$$3(0)^3 - 12(0) \stackrel{?}{=} 0$$
$$0 = 0 \quad \text{True}$$

Let $x = -2$.

$$3(-2)^3 - 12(-2) \stackrel{?}{=} 0$$
$$3(-8) + 24 \stackrel{?}{=} 0$$
$$0 = 0 \quad \text{True}$$

Let $x = 2$.

$$3(2)^3 - 12(2) \stackrel{?}{=} 0$$
$$3(8) - 24 \stackrel{?}{=} 0$$
$$0 = 0 \quad \text{True}$$

The solutions are $0, -2,$ and 2.

■ **Work Practice 7**

Practice 7

Solve: $2x^3 - 18x = 0$

Example 8 Solve: $(5x - 1)(2x^2 + 15x + 18) = 0$

Solution:

$$(5x - 1)(2x^2 + 15x + 18) = 0$$
$$(5x - 1)(2x + 3)(x + 6) = 0 \qquad \text{Factor the trinomial.}$$
$$5x - 1 = 0 \quad \text{or} \quad 2x + 3 = 0 \quad \text{or} \quad x + 6 = 0 \qquad \text{Set each factor equal to 0.}$$
$$5x = 1 \qquad\qquad 2x = -3 \qquad\qquad x = -6 \quad \text{Solve.}$$
$$x = \frac{1}{5} \qquad\qquad x = -\frac{3}{2}$$

Check each solution in the original equation. The solutions are $\frac{1}{5}, -\frac{3}{2},$ and -6.

■ **Work Practice 8**

Practice 8

Solve:
$(x + 3)(3x^2 - 20x - 7) = 0$

Answers

7. $0, 3,$ and -3 **8.** $-3, -\dfrac{1}{3},$ and 7

Vocabulary, Readiness & Video Check

Use the choices below to fill in each blank. Not all choices will be used.

$-3, 5$	$a = 0$ or $b = 0$	0	linear
$3, -5$	quadratic	1	

1. An equation that can be written in the form $ax^2 + bx + c = 0$ (with $a \neq 0$) is called a _____ equation.

2. If the product of two numbers is 0, then at least one of the numbers must be _____.

3. The solutions to $(x - 3)(x + 5) = 0$ are _____.

4. If $a \cdot b = 0$, then _____.

Martin-Gay Interactive Videos   Watch the section lecture video and answer the following questions.

Objective A 5. As shown in Examples 1–3, what two things have to be true in order to use the zero-factor theorem? ▶

Objective B 6. Example 4 implies that the zero-factor theorem can be used with any number of factors on one side of the equation as long as the other side of the equation is zero. Why do you think this is true? ▶

See Video 13.6

13.6 Exercise Set MyMathLab®

Objective A *Solve each equation. See Examples 1 through 3.*

1. $(x - 2)(x + 1) = 0$

2. $(x + 3)(x + 2) = 0$

3. $(x - 6)(x - 7) = 0$

4. $(x + 4)(x - 10) = 0$

5. $(x + 9)(x + 17) = 0$

6. $(x - 11)(x - 1) = 0$

7. $x(x + 6) = 0$

8. $x(x - 7) = 0$

9. $3x(x - 8) = 0$

10. $2x(x + 12) = 0$

▶ 11. $(2x + 3)(4x - 5) = 0$

12. $(3x - 2)(5x + 1) = 0$

13. $(2x - 7)(7x + 2) = 0$

14. $(9x + 1)(4x - 3) = 0$

15. $\left(x - \frac{1}{2}\right)\left(x + \frac{1}{3}\right) = 0$

16. $\left(x + \frac{2}{9}\right)\left(x - \frac{1}{4}\right) = 0$

17. $(x + 0.2)(x + 1.5) = 0$

18. $(x + 1.7)(x + 2.3) = 0$

Solve. See Examples 4 through 6.

19. $x^2 - 13x + 36 = 0$

20. $x^2 + 2x - 63 = 0$

▶ 21. $x^2 + 2x - 8 = 0$

22. $x^2 - 5x + 6 = 0$

23. $x^2 - 7x = 0$

24. $x^2 - 3x = 0$

25. $x^2 + 20x = 0$

26. $x^2 + 15x = 0$

27. $x^2 = 16$

28. $x^2 = 9$

29. $x^2 - 4x = 32$

30. $x^2 - 5x = 24$

31. $(x + 4)(x - 9) = 4x$

32. $(x + 3)(x + 8) = x$

▶ 33. $x(3x - 1) = 14$

34. $x(4x - 11) = 3$

35. $3x^2 + 19x - 72 = 0$

36. $36x^2 + x - 21 = 0$

Objectives A B and Section 9.3 Mixed Practice *Solve each equation. See Examples 1 through 8. (A few exercises are linear equations.)*

37. $4x^3 - x = 0$

38. $4y^3 - 36y = 0$

39. $4(x - 7) = 6$

40. $5(3 - 4x) = 9$

41. $(4x - 3)(16x^2 - 24x + 9) = 0$

42. $(2x + 5)(4x^2 + 20x + 25) = 0$

43. $4y^2 - 1 = 0$

44. $4y^2 - 81 = 0$

▶ **45.** $(2x + 3)(2x^2 - 5x - 3) = 0$

46. $(2x - 9)(x^2 + 5x - 36) = 0$

47. $x^2 - 15 = -2x$

48. $x^2 - 26 = -11x$

49. $30x^2 - 11x = 30$

50. $9x^2 + 7x = 2$

51. $5x^2 - 6x - 8 = 0$

52. $12x^2 + 7x - 12 = 0$

53. $6y^2 - 22y - 40 = 0$

54. $3x^2 - 6x - 9 = 0$

55. $(y - 2)(y + 3) = 6$

56. $(y - 5)(y - 2) = 28$

57. $x^3 - 12x^2 + 32x = 0$

58. $x^3 - 14x^2 + 49x = 0$

59. $x^2 + 14x + 49 = 0$

60. $x^2 + 22x + 121 = 0$

61. $12y = 8y^2$

62. $9y = 6y^2$

63. $7x^3 - 7x = 0$

64. $3x^3 - 27x = 0$

65. $3x^2 + 8x - 11 = 13 - 6x$

66. $2x^2 + 12x - 1 = 4 + 3x$

67. $3x^2 - 20x = -4x^2 - 7x - 6$

68. $4x^2 - 20x = -5x^2 - 6x - 5$

Review

Perform each indicated operation. Write all results in lowest terms. See Sections 2.4 and 3.3.

69. $\dfrac{3}{5} + \dfrac{4}{9}$

70. $\dfrac{2}{3} + \dfrac{3}{7}$

71. $\dfrac{7}{10} - \dfrac{5}{12}$

72. $\dfrac{5}{9} - \dfrac{5}{12}$

73. $\dfrac{4}{5} \cdot \dfrac{7}{8}$

74. $\dfrac{3}{7} \cdot \dfrac{12}{17}$

Concept Extensions

For Exercises 75 and 76, see the Concept Check in this section.

75. Explain the error and solve correctly:

$$x(x - 2) = 8$$
$$x = 8 \quad \text{or} \quad x - 2 = 8$$
$$x = 10$$

76. Explain the error and solve correctly:

$$(x - 4)(x + 2) = 0$$
$$x = -4 \quad \text{or} \quad x = 2$$

77. Write a quadratic equation that has two solutions, 6 and -1. Leave the polynomial in the equation in factored form.

78. Write a quadratic equation that has two solutions, 0 and -2. Leave the polynomial in the equation in factored form.

79. Write a quadratic equation in standard form that has two solutions, 5 and 7.

80. Write an equation that has three solutions, 0, 1, and 2.

81. A compass is accidentally thrown upward and out of a hot-air balloon at a height of 300 feet. The height, y, of the compass at time x is given by the equation $y = -16x^2 + 20x + 300$.

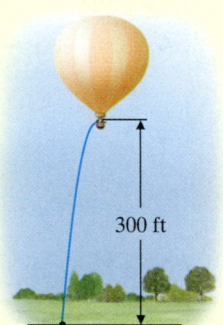

300 ft

a. Find the height of the compass at the given times by filling in the table below.

Time, x (in seconds)	0	1	2	3	4	5	6
Height, y (in feet)							

b. Use the table to determine when the compass strikes the ground.
c. Use the table to approximate the maximum height of the compass.

82. A rocket is fired upward from the ground with an initial velocity of 100 feet per second. The height, y, of the rocket at any time x is given by the equation $y = -16x^2 + 100x$.

y

a. Find the height of the rocket at the given times by filling in the table below.

Time, x (in seconds)	0	1	2	3	4	5	6	7
Height, y (in feet)								

b. Use the table to determine between what two whole numbered seconds the rocket strikes the ground.
c. Use the table to approximate the maximum height of the rocket.

Solve each equation.

83. $(x - 3)(3x + 4) = (x + 2)(x - 6)$

84. $(2x - 3)(x + 6) = (x - 9)(x + 2)$

85. $(2x - 3)(x + 8) = (x - 6)(x + 4)$

86. $(x + 6)(x - 6) = (2x - 9)(x + 4)$

13.7 Quadratic Equations and Problem Solving ▶

Objective A Solving Problems Modeled by Quadratic Equations ▶

Objective

A Solve Problems That Can Be Modeled by Quadratic Equations. ▶

Some problems may be modeled by quadratic equations. To solve these problems, we use the same problem-solving steps that were introduced in Section 1.8. When solving these problems, keep in mind that a solution of an equation that models a problem may not be a solution to the problem. For example, a person's age or the length of a rectangle is always a positive number. Thus we discard solutions that do not make sense as solutions of the problem.

Example 1 Finding Free-Fall Time

Since the 1940s, one of the top tourist attractions in Acapulco, Mexico, is watching the cliff divers off La Quebrada. The divers' platform is about 144 feet above the sea. These divers must time their descent just right, since they land in the crashing Pacific, in an inlet that is at most $9\frac{1}{2}$ feet deep. Neglecting air resistance, the height h in feet of a cliff diver above the ocean after t seconds is given by the quadratic equation $h = -16t^2 + 144$.

Find out how long it takes the diver to reach the ocean.

Practice 1

Cliff divers also frequent the falls at Waimea Falls Park in Oahu, Hawaii. Here, a diver can jump from a ledge 64 feet up the waterfall into a rocky pool below. Neglecting air resistance, the height of a diver above the pool after t seconds is $h = -16t^2 + 64$. Find how long it takes the diver to reach the pool.

Solution:

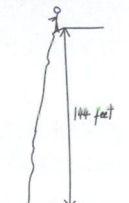

1. **UNDERSTAND.** Read and reread the problem. Then draw a picture of the problem.

   The equation $h = -16t^2 + 144$ models the height of the falling diver at time t. Familiarize yourself with this equation by finding the height of the diver at time $t = 1$ second and $t = 2$ seconds.

   When $t = 1$ second, the height of the diver is $h = -16(1)^2 + 144 = 128$ feet.
   When $t = 2$ seconds, the height of the diver is $h = -16(2)^2 + 144 = 80$ feet.

2. **TRANSLATE.** To find out how long it takes the diver to reach the ocean, we want to know the value of t for which $h = 0$.

3. **SOLVE.** Solve the equation.

$$0 = -16t^2 + 144$$
$$0 = -16(t^2 - 9) \qquad \text{Factor out } -16.$$
$$0 = -16(t - 3)(t + 3) \qquad \text{Factor completely.}$$
$$t - 3 = 0 \quad \text{or} \quad t + 3 = 0 \qquad \text{Set each factor containing a variable equal to 0.}$$
$$t = 3 \quad \text{or} \qquad t = -3 \qquad \text{Solve.}$$

4. **INTERPRET.** Since the time t cannot be negative, the proposed solution is 3 seconds.

   Check: Verify that the height of the diver when t is 3 seconds is 0.

   When $t = 3$ seconds, $h = -16(3)^2 + 144 = -144 + 144 = 0$.

■ **Work Practice 1**

Answer
1. 2 sec

The square of a number minus twice the number is 63. Find the number.

Example 2   Finding a Number

The square of a number plus three times the number is 70. Find the number.

Solution:

1. UNDERSTAND. Read and reread the problem. Suppose that the number is 5. The square of 5 is 5^2 or 25. Three times 5 is 15. Then $25 + 15 = 40$, not 70, so the number must be greater than 5. Remember, the purpose of proposing a number, such as 5, is to better understand the problem. Now that we do, we will let x = the number.

2. TRANSLATE.

the square of a number	plus	three times the number	is	70
↓	↓	↓	↓	↓
x^2	$+$	$3x$	$=$	70

3. SOLVE.

$$x^2 + 3x = 70$$
$$x^2 + 3x - 70 = 0 \qquad \text{Subtract 70 from both sides.}$$
$$(x + 10)(x - 7) = 0 \qquad \text{Factor.}$$
$$x + 10 = 0 \quad \text{or} \quad x - 7 = 0 \quad \text{Set each factor equal to 0.}$$
$$x = -10 \qquad\qquad x = 7 \quad \text{Solve.}$$

4. INTERPRET.

Check: The square of -10 is $(-10)^2$, or 100. Three times -10 is $3(-10)$ or -30. Then $100 + (-30) = 70$, the correct sum, so -10 checks.
   The square of 7 is 7^2 or 49. Three times 7 is $3(7)$, or 21. Then $49 + 21 = 70$, the correct sum, so 7 checks.

State: There are two numbers. They are -10 and 7.

■ Work Practice 2

The length of a rectangular garden is 5 feet more than its width. The area of the garden is 176 square feet. Find the length and the width of the garden.

Height = $2x - 2$

Base = x

△ Example 3   Finding the Dimensions of a Sail

The height of a triangular sail is 2 meters less than twice the length of the base. If the sail has an area of 30 square meters, find the length of its base and the height.

Solution:

1. UNDERSTAND. Read and reread the problem. Since we are finding the length of the base and the height, we let

x = the length of the base

Since the height is 2 meters less than twice the length of the base,

$2x - 2$ = the height

An illustration is shown in the margin.

2. TRANSLATE. We are given that the area of the triangle is 30 square meters, so we use the formula for area of a triangle.

area of triangle	=	$\frac{1}{2}$	·	base	·	height
↓		↓	↓	↓		↓
30	$=$	$\frac{1}{2}$	·	x	·	$(2x - 2)$

Answers

2. 9 and -7

3. length: 16 ft; width: 11 ft

3. SOLVE. Now we solve the quadratic equation.

$$30 = \frac{1}{2}x(2x - 2)$$

$30 = x^2 - x$ Multiply.

$0 = x^2 - x - 30$ Write in standard form.

$0 = (x - 6)(x + 5)$ Factor.

$x - 6 = 0$ or $x + 5 = 0$ Set each factor equal to 0.

$x = 6$ $x = -5$

4. INTERPRET. Since x represents the length of the base, we discard the solution -5. The base of a triangle cannot be negative. The base is then 6 meters and the height is $2(6) - 2 = 10$ meters.

Check: To check this problem, we recall that

$$\text{area} = \frac{1}{2}\,\text{base} \cdot \text{height or}$$

$$30 \stackrel{?}{=} \frac{1}{2}(6)(10)$$

$30 = 30$ True

State: The base of the triangular sail is 6 meters and the height is 10 meters.

■ **Work Practice 3**

The next example has to do with consecutive integers. Study the following diagrams for a review of consecutive integers.

Examples

If x is the first integer, then consecutive integers are
$x, x + 1, x + 2, \ldots$.

If x is the first even integer, then consecutive even integers are
$x, x + 2, x + 4, \ldots$.

If x is the first odd integer, then consecutive odd integers are
$x, x + 2, x + 4, \ldots$.

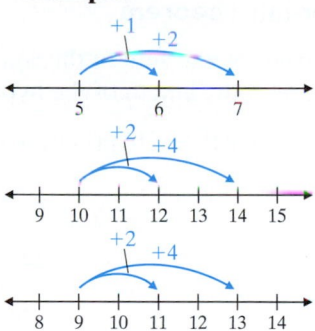

Example 4   Finding Consecutive Even Integers

Find two consecutive even integers whose product is 34 more than their sum.

Solution:

1. UNDERSTAND. Read and reread the problem. Let's just choose two consecutive even integers to help us better understand the problem. Let's choose 10 and 12. Their product is $10(12) = 120$ and their sum is $10 + 12 = 22$. The product is $120 - 22$, or 98 greater than the sum. Thus our guess is incorrect, but we have a better understanding of this example.

     Let's let x and $x + 2$ be the consecutive even integers.

2. TRANSLATE.

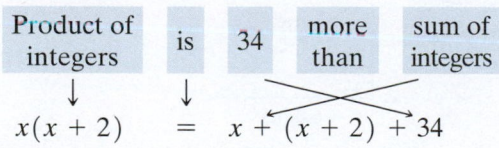

$$x(x + 2) = x + (x + 2) + 34$$

(Continued on next page)

Practice 4

Find two consecutive odd integers whose product is 23 more than their sum.

Answer

4. 5 and 7 or -5 and -3

3. SOLVE. Now we solve the equation.

$$x(x + 2) = x + (x + 2) + 34$$
$$x^2 + 2x = x + x + 2 + 34 \qquad \text{Multiply.}$$
$$x^2 + 2x = 2x + 36 \qquad \text{Combine like terms.}$$
$$x^2 - 36 = 0 \qquad \text{Write in standard form.}$$
$$(x + 6)(x - 6) = 0 \qquad \text{Factor.}$$
$$x + 6 = 0 \quad \text{or} \quad x - 6 = 0 \qquad \text{Set each factor equal to 0.}$$
$$x = -6 \qquad\qquad x = 6 \qquad \text{Solve.}$$

4. INTERPRET. If $x = -6$, then $x + 2 = -6 + 2$, or -4.

If $x = 6$, then $x + 2 = 6 + 2$, or 8.

Check: $-6, -4$ $\qquad\qquad\qquad\qquad\qquad$ $6, 8$

$$-6(-4) \stackrel{?}{=} -6 + (-4) + 34 \qquad\qquad 6(8) \stackrel{?}{=} 6 + 8 + 34$$
$$24 \stackrel{?}{=} -10 + 34 \qquad\qquad\qquad 48 \stackrel{?}{=} 14 + 34$$
$$24 = 24 \qquad \text{True} \qquad\qquad 48 = 48 \qquad \text{True}$$

State: The two consecutive even integers are -6 and -4 or 6 and 8.

■ Work Practice 4

The next example makes use of the **Pythagorean theorem.** Before we review this theorem, recall that a **right triangle** is a triangle that contains a 90° or right angle. The **hypotenuse** of a right triangle is the side opposite the right angle and is the longest side of the triangle. The **legs** of a right triangle are the other sides of the triangle.

Pythagorean Theorem

In a right triangle, the sum of the squares of the lengths of the two legs is equal to the square of the length of the hypotenuse.

$$(\text{leg})^2 + (\text{leg})^2 = (\text{hypotenuse})^2 \quad \text{or} \quad a^2 + b^2 = c^2$$

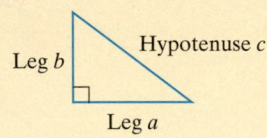

Leg b $\qquad$ Hypotenuse c

Leg a

Practice 5

The length of one leg of a right triangle is 7 meters less than the length of the other leg. The length of the hypotenuse is 13 meters. Find the lengths of the legs.

△ **Example 5**   Finding the Dimensions of a Triangle

Find the lengths of the sides of a right triangle if the lengths can be expressed as three consecutive even integers.

Solution:

1. UNDERSTAND. Read and reread the problem. Let's suppose that the length of one leg of the right triangle is 4 units. Then the other leg is the next even integer, or 6 units, and the hypotenuse of the triangle is the next even integer, or 8 units. Remember that the hypotenuse is the longest side. Let's see if a triangle with sides of these lengths forms a right triangle. To do this, we check to see whether the Pythagorean theorem holds true.

$$4^2 + 6^2 \stackrel{?}{=} 8^2$$
$$16 + 36 \stackrel{?}{=} 64$$
$$52 = 64 \quad \text{False}$$

4 units $\qquad$ 8 units

6 units

Our proposed numbers do not check, but we now have a better understanding of the problem.

Answer

5. 5 meters, 12 meters

We let x, $x + 2$, and $x + 4$ be three consecutive even integers. Since these integers represent lengths of the sides of a right triangle, we have the following.

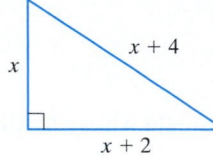

$x = $ one leg

$x + 2 = $ other leg

$x + 4 = $ hypotenuse (longest side)

2. **TRANSLATE.** By the Pythagorean theorem, we have that

$$(\text{leg})^2 + (\text{leg})^2 = (\text{hypotenuse})^2$$
$$(x)^2 + (x + 2)^2 = (x + 4)^2$$

3. **SOLVE.** Now we solve the equation.

$$x^2 + (x + 2)^2 = (x + 4)^2$$

$x^2 + x^2 + 4x + 4 = x^2 + 8x + 16$ Multiply.

$2x^2 + 4x + 4 = x^2 + 8x + 16$ Combine like terms.

$x^2 - 4x - 12 = 0$ Write in standard form.

$(x - 6)(x + 2) = 0$ Factor.

$x - 6 = 0 \quad \text{or} \quad x + 2 = 0$ Set each factor equal to 0.

$x = 6 \qquad\qquad x = -2$

4. **INTERPRET.** We discard $x = -2$ since length cannot be negative. If $x = 6$, then $x + 2 = 8$ and $x + 4 = 10$.

Check: Verify that

$$(\text{leg})^2 + (\text{leg})^2 = (\text{hypotenuse})^2$$

$$6^2 + 8^2 \stackrel{?}{=} 10^2$$

$$36 + 64 \stackrel{?}{=} 100$$

$$100 = 100 \qquad \text{True}$$

State: The sides of the right triangle have lengths 6 units, 8 units, and 10 units.

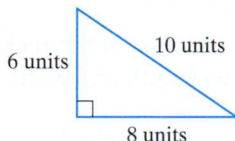

■ **Work Practice 5**

Vocabulary, Readiness & Video Check

Martin-Gay Interactive Videos   Watch the section lecture video and answer the following question.

Objective A 1. In each of ▤ Examples 1–3, why aren't both solutions of the translated equation accepted as solutions to the application? ▶

See Video 13.7 🍎

13.7 Exercise Set MyMathLab®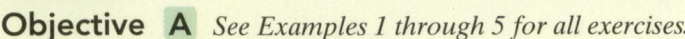

Objective A *See Examples 1 through 5 for all exercises.*

Translating *For Exercises 1 through 6, represent each given condition using a single variable, x.*

△ **1.** The length and width of a rectangle whose length is 4 centimeters more than its width

△ **2.** The length and width of a rectangle whose length is twice its width

3. Two consecutive odd integers

4. Two consecutive even integers

△ **5.** The base and height of a triangle whose height is one more than four times its base

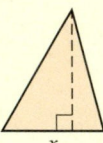

△ **6.** The base and height of a trapezoid whose base is three less than five times its height

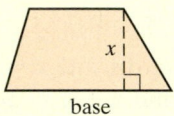

Use the information given to find the dimensions of each figure.

△ **7.**

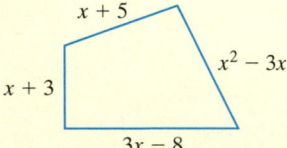

The *area* of the square is 121 square units. Find the length of its sides.

△ **8.**

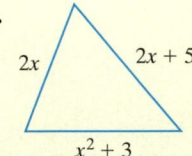

The *area* of the rectangle is 84 square inches. Find its length and width.

△ **9.**

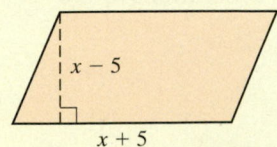

The *perimeter* of the quadrilateral is 120 centimeters. Find the lengths of its sides.

▶ **10.**

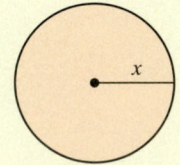

The *perimeter* of the triangle is 85 feet. Find the lengths of its sides.

△ **11.**

The *area* of the parallelogram is 96 square miles. Find its base and height.

△ **12.**

The *area* of the circle is 25π square kilometers. Find its radius.

Solve.

▶ **13.** An object is thrown upward from the top of an 80-foot building with an initial velocity of 64 feet per second. The height h of the object after t seconds is given by the quadratic equation $h = -16t^2 + 64t + 80$. When will the object hit the ground?

14. A hang glider accidentally drops her compass from the top of a 400-foot cliff. The height h of the compass after t seconds is given by the quadratic equation $h = -16t^2 + 400$. When will the compass hit the ground?

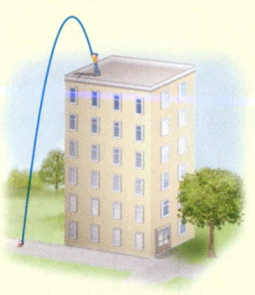

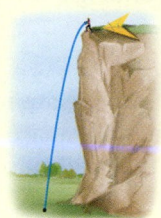

15. The width of a rectangle is 7 centimeters less than twice its length. Its area is 30 square centimeters. Find the dimensions of the rectangle.

16. The length of a rectangle is 9 inches more than its width. Its area is 112 square inches. Find the dimensions of the rectangle.

△ *The equation $D = \dfrac{1}{2}n(n - 3)$ gives the number of diagonals D for a polygon with n sides. For example, a polygon with 6 sides has $D = \dfrac{1}{2} \cdot 6(6 - 3)$ or $D = 9$ diagonals. (See if you can count all 9 diagonals. Some are shown in the figure.)*

Use this equation, $D = \dfrac{1}{2}n(n - 3)$, for Exercises 17 through 20.

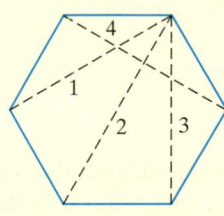

17. Find the number of diagonals for a polygon that has 12 sides.

18. Find the number of diagonals for a polygon that has 15 sides.

19. Find the number of sides n for a polygon that has 35 diagonals.

20. Find the number of sides n for a polygon that has 14 diagonals.

21. The sum of a number and its square is 132. Find the number.

▶ **22.** The sum of a number and its square is 182. Find the number.

23. The product of two consecutive room numbers is 210. Find the room numbers.

▶ **24.** The product of two consecutive page numbers is 420. Find the page numbers.

25. A ladder is leaning against a building so that the distance from the ground to the top of the ladder is one foot less than the length of the ladder. Find the length of the ladder if the distance from the bottom of the ladder to the building is 5 feet.

26. Use the given figure to find the length of the guy wire.

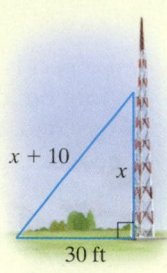

△ **27.** If the sides of a square are increased by 3 inches, the area becomes 64 square inches. Find the length of the sides of the original square.

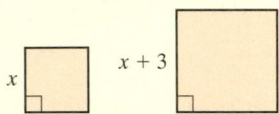

△ **28.** If the sides of a square are increased by 5 meters, the area becomes 100 square meters. Find the length of the sides of the original square.

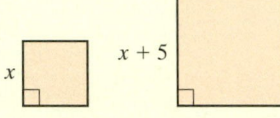

△ **29.** One leg of a right triangle is 4 millimeters longer than the shorter leg and the hypotenuse is 8 millimeters longer than the shorter leg. Find the lengths of the sides of the triangle.

△ **30.** One leg of a right triangle is 9 centimeters longer than the other leg and the hypotenuse is 45 centimeters. Find the lengths of the legs of the triangle.

△ **31.** The length of the base of a triangle is twice its height. If the area of the triangle is 100 square kilometers, find the height.

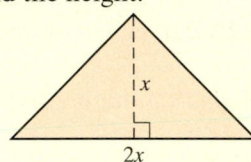

△ **32.** The height of a triangle is 2 millimeters less than the base. If the area is 60 square millimeters, find the base.

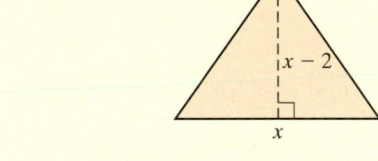

△ **33.** Find the length of the shorter leg of a right triangle if the longer leg is 12 feet more than the shorter leg and the hypotenuse is 12 feet less than twice the shorter leg.

△ **34.** Find the length of the shorter leg of a right triangle if the longer leg is 10 miles more than the shorter leg and the hypotenuse is 10 miles less than twice the shorter leg.

35. An object is dropped from 39 feet below the tip of the pinnacle atop one of the 1483-foot-tall Petronas Twin Towers in Kuala Lumpur, Malaysia. (*Source:* Council on Tall Buildings and Urban Habitat) The height h of the object after t seconds is given by the equation $h = -16t^2 + 1444$. Find how many seconds pass before the object reaches the ground.

36. An object is dropped from the top of 311 South Wacker Drive, a 961-foot-tall office building in Chicago. (*Source:* Council on Tall Buildings and Urban Habitat) The height h of the object after t seconds is given by the equation $h = -16t^2 + 961$. Find how many seconds pass before the object reaches the ground.

37. At the end of 2 years, P dollars invested at an interest rate r compounded annually increases to an amount, A dollars, given by

$$A = P(1 + r)^2$$

Find the interest rate if $100 increased to $144 in 2 years. Write your answer as a percent.

38. At the end of 2 years, P dollars invested at an interest rate r compounded annually increases to an amount, A dollars, given by

$$A = P(1 + r)^2$$

Find the interest rate if $2000 increased to $2420 in 2 years. Write your answer as a percent.

△ **39.** Find the dimensions of a rectangle whose width is 7 miles less than its length and whose area is 120 square miles.

△ **40.** Find the dimensions of a rectangle whose width is 2 inches less than half its length and whose area is 160 square inches.

41. If the cost, C, for manufacturing x units of a certain product is given by $C = x^2 - 15x + 50$, find the number of units manufactured at a cost of $9500.

42. If a switchboard handles n telephones, the number C of telephone connections it can make simultaneously is given by the equation $C = \dfrac{n(n-1)}{2}$. Find how many telephones are handled by a switchboard making 120 telephone connections simultaneously.

Review

The following double line graph shows a comparison of the number of annual visitors (in millions) to Acadia National Park and Cuyahoga Valley National Park for the years shown. Use this graph to answer Exercises 43 through 50. See Section 11.1.

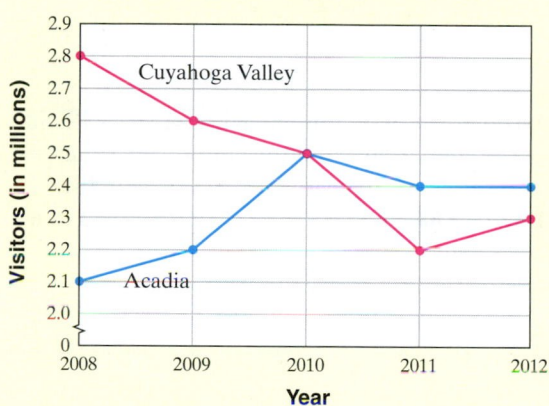

Annual Visitors to Acadia and Cuyahoga Valley National Parks

Source: National Park Service

43. Find the number of visitors to Acadia National Park in 2009.

44. Find the number of visitors to Cuyahoga Valley National Park in 2009.

45. Find the number of visitors to Acadia National Park in 2012.

46. Find the number of visitors to Cuyahoga Valley National Park in 2012.

47. Determine the year that the colored lines in this graph intersect.

48. For what year(s) on the graph is the number of visitors to Cuyahoga Valley Park greater than the number of visitors to Acadia Park?

49. In your own words, explain the meaning of the point of intersection in the graph.

50. Describe the trends shown in this graph and speculate as to why these trends have occurred.

Write each fraction in simplest form. See Section 2.3.

51. $\dfrac{20}{35}$ **52.** $\dfrac{24}{32}$ **53.** $\dfrac{27}{18}$ **54.** $\dfrac{15}{27}$ **55.** $\dfrac{14}{42}$ **56.** $\dfrac{45}{50}$

Concept Extensions

△ **57.** The side of a square equals the width of a rectangle. The length of the rectangle is 6 meters longer than its width. The sum of the areas of the square and the rectangle is 176 square meters. Find the side of the square.

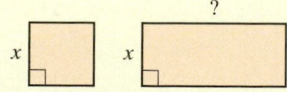

△ **58.** Two boats travel at right angles to each other after leaving the same dock at the same time. One hour later the boats are 17 miles apart. If one boat travels 7 miles per hour faster than the other boat, find the rate of each boat.

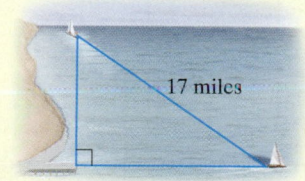

17 miles

59. The sum of two numbers is 25, and the sum of their squares is 325. Find the numbers.

60. The sum of two numbers is 20, and the sum of their squares is 218. Find the numbers.

△ **61.** A rectangular pool is surrounded by a walk 4 meters wide. The pool is 6 meters longer than its width. If the total area of the pool and walk is 576 square meters more than the area of the pool, find the dimensions of the pool.

△ **62.** A rectangular garden is surrounded by a walk of uniform width. The area of the garden is 180 square yards. If the dimensions of the garden plus the walk are 16 yards by 24 yards, find the width of the walk.

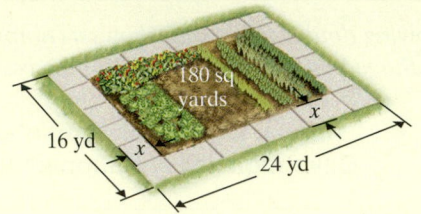

63. Write down two numbers whose sum is 10. Square each number and find the sum of the squares. Use this work to write a word problem like Exercise **59.** Then give the word problem to a classmate to solve.

64. Write down two numbers whose sum is 12. Square each number and find the sum of the squares. Use this work to write a word problem like Exercise **60.** Then give the word problem to a classmate to solve.

Chapter 13 Group Activity

Factoring polynomials can be visualized using areas of rectangles. To see this, let's first find the areas of the following squares and rectangles. (Recall that Area = Length · Width.)

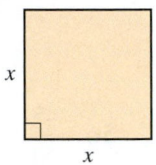 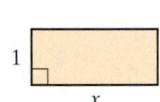

To use these areas to visualize factoring the polynomial $x^2 + 3x + 2$, for example, use the shapes below to form a rectangle. The factored form is found by reading the length and the width of the rectangle, as shown below.

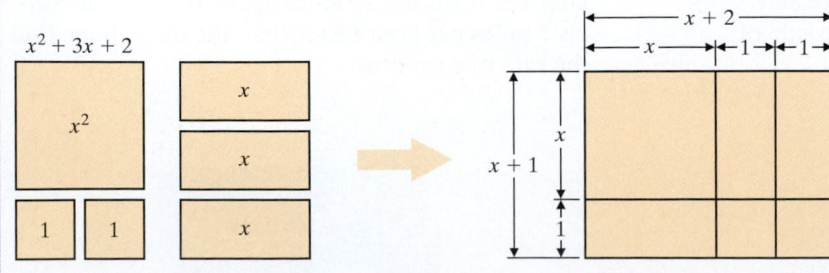

Thus, $x^2 + 3x + 2 = (x + 2)(x + 1)$.

Try using this method to visualize the factored form of each polynomial below.

Work in a group and use tiles to find a factored form for the polynomials below. (Tiles can be handmade from index cards.)

1. $x^2 + 6x + 5$

2. $x^2 + 5x + 6$

3. $x^2 + 5x + 4$

4. $x^2 + 4x + 3$

5. $x^2 + 6x + 9$

6. $x^2 + 4x + 4$

Chapter 13 Vocabulary Check

Fill in each blank with one of the words or phrases listed below. Some words or phrases may be used more than once.

factoring	hypotenuse	quadratic equation
greatest common factor	leg	perfect square trinomial

1. An equation that can be written in the form $ax^2 + bx + c = 0$ (with a not 0) is called a _____.

2. _____ is the process of writing an expression as a product.

3. The _____ of a list of terms is the product of all common factors.

4. A trinomial that is the square of some binomial is called a _____.

5. In a right triangle, the side opposite the right angle is called the _____.

6. In a right triangle, each side adjacent to the right angle is called a _____.

7. The Pythagorean theorem states that $(\text{leg})^2 + (\text{leg})^2 = ($ _____ $)^2$.

> **Helpful Hint**
>
> ▶ Are you preparing for your test? Don't forget to take the Chapter 13 Test on page 1026. Then check your answers at the back of the text and use the Chapter Test Prep Videos to see the fully worked-out solutions to any of the exercises you want to review.

13 Chapter Highlights

Definitions and Concepts	Examples
Section 13.1 The Greatest Common Factor and Factoring by Grouping	

Definitions and Concepts	Examples
Factoring is the process of writing an expression as a product.	Factor: $6 = 2 \cdot 3$ Factor: $x^2 + 5x + 6 = (x + 2)(x + 3)$
The GCF of a list of variable terms contains the smallest exponent on each common variable.	The GCF of z^5, z^3, and z^{10} is z^3.
The GCF of a list of terms is the product of all common factors.	Find the GCF of $8x^2y$, $10x^3y^2$, and $50x^2y^3$. $8x^2y = 2 \cdot 2 \cdot 2 \cdot x^2 \cdot y$ $10x^3y^2 = 2 \cdot 5 \cdot x^3 \cdot y^2$ $50x^2y^3 = 2 \cdot 5 \cdot 5 \cdot x^2 \cdot y^3$ $\text{GCF} = 2 \cdot x^2 \cdot y \quad \text{or} \quad 2x^2y$
To Factor by Grouping **Step 1:** Group the terms in two groups so that each group has a common factor. **Step 2:** Factor out the GCF from each group. **Step 3:** If there is a common binomial factor, factor it out. **Step 4:** If not, rearrange the terms and try these steps again.	Factor: $10ax + 15a - 6xy - 9y$ **Step 1:** $(10ax + 15a) + (-6xy - 9y)$ **Step 2:** $5a(2x + 3) - 3y(2x + 3)$ **Step 3:** $(2x + 3)(5a - 3y)$

Definitions and Concepts	Examples

Section 13.2   Factoring Trinomials of the Form $x^2 + bx + c$

The product of these numbers is c.

$$x^2 + bx + c = (x + \Box)(x + \Box)$$

The sum of these numbers is b.

Factor: $x^2 + 7x + 12$

$3 + 4 = 7$ $3 \cdot 4 = 12$

$x^2 + 7x + 12 = (x + 3)(x + 4)$

Section 13.3   Factoring Trinomials of the Form $ax^2 + bx + c$

To factor $ax^2 + bx + c$, try various combinations of factors of ax^2 and c until a middle term of bx is obtained when checking.

Factor: $3x^2 + 14x - 5$

Factors of $3x^2$: $3x, x$

Factors of -5: $-1, 5$ and $1, -5$

$(3x - 1)(x + 5)$

$-1x$

$+15x$

$14x$ Correct middle term

Section 13.4   Factoring Trinomials of the Form $ax^2 + bx + c$ by Grouping

To Factor $ax^2 + bx + c$ by Grouping

Step 1: Find two numbers whose product is $a \cdot c$ and whose sum is b.

Step 2: Rewrite bx, using the factors found in Step 1.

Step 3: Factor by grouping.

Factor: $3x^2 + 14x - 5$

Step 1: Find two numbers whose product is $3 \cdot (-5)$ or -15 and whose sum is 14. They are 15 and -1.

Step 2: $3x^2 + 14x - 5$
$= 3x^2 + 15x - 1x - 5$

Step 3: $= 3x(x + 5) - 1(x + 5)$
$= (x + 5)(3x - 1)$

Section 13.5   Factoring Perfect Square Trinomials and the Difference of Two Squares

A **perfect square trinomial** is a trinomial that is the square of some binomial.

Perfect Square Trinomial = Square of Binomial

$$x^2 + 4x + 4 = (x + 2)^2$$
$$25x^2 - 10x + 1 = (5x - 1)^2$$

Factoring Perfect Square Trinomials

$$a^2 + 2ab + b^2 = (a + b)^2$$
$$a^2 - 2ab + b^2 = (a - b)^2$$

Factor.

$$x^2 + 6x + 9 = x^2 + 2 \cdot x \cdot 3 + 3^2 = (x + 3)^2$$
$$4x^2 - 12x + 9 = (2x)^2 - 2 \cdot 2x \cdot 3 + 3^2$$
$$= (2x - 3)^2$$

Difference of Two Squares

$$a^2 - b^2 = (a + b)(a - b)$$

Factor.

$$x^2 - 9 = x^2 - 3^2 = (x + 3)(x - 3)$$

Definitions and Concepts	**Examples**

Section 13.6   Solving Quadratic Equations by Factoring

A **quadratic equation** is an equation that can be written in the form $ax^2 + bx + c = 0$ with a not 0.	**Quadratic Equation** **Standard Form**
The form $ax^2 + bx + c = 0$ is called the **standard form** of a quadratic equation.	$x^2 = 16$ $x^2 - 16 = 0$
	$y = -2y^2 + 5$ $2y^2 + y - 5 = 0$

Zero-Factor Property
If a and b are real numbers and if $ab = 0$, then $a = 0$ or $b = 0$.

If $(x + 3)(x - 1) = 0$, then $x + 3 = 0$ or $x - 1 = 0$.

To Solve Quadratic Equations by Factoring

Step 1: Write the equation in standard form so that one side of the equation is 0.

Step 2: Factor completely.

Step 3: Set each factor containing a variable equal to 0.

Step 4: Solve the resulting equations.

Step 5: Check solutions in the original equation.

Solve: $3x^2 = 13x - 4$

Step 1: $3x^2 - 13x + 4 = 0$

Step 2: $(3x - 1)(x - 4) = 0$

Step 3: $3x - 1 = 0$ or $x - 4 = 0$

Step 4: $3x = 1$ $x = 4$

$$x = \frac{1}{3}$$

Step 5: Check both $\frac{1}{3}$ and 4 in the original equation.

Section 13.7   Quadratic Equations and Problem Solving

Problem-Solving Steps

A garden is in the shape of a rectangle whose length is two feet more than its width. If the area of the garden is 35 square feet, find its dimensions.

1. UNDERSTAND the problem.

1. Read and reread the problem. Guess a solution and check your guess. Draw a diagram.
   Let x be the width of the rectangular garden. Then $x + 2$ is the length.

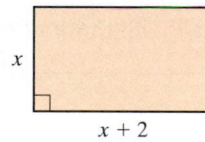

2. TRANSLATE.

2. length   ·   width   =   area
   ↓   ↓   ↓
   $(x + 2)$ · x =   35

3. SOLVE.

3.
$$(x + 2)x = 35$$
$$x^2 + 2x - 35 = 0$$
$$(x - 5)(x + 7) = 0$$
$$x - 5 = 0 \quad \text{or} \quad x + 7 = 0$$
$$x = 5 \qquad\qquad x = -7$$

4. INTERPRET.

4. Discard the solution $x = -7$ since x represents width.

   Check: If x is 5 feet, then $x + 2 = 5 + 2 = 7$ feet. The area of a rectangle whose width is 5 feet and whose length is 7 feet is (5 feet)(7 feet) or 35 square feet.

   State: The garden is 5 feet by 7 feet.

(13.1) *Complete each factoring.*

1. $6x^2 - 15x = 3x($ $)$

2. $4x^5 + 2x - 10x^4 = 2x($ $)$

Factor out the GCF from each polynomial.

3. $5m + 30$

4. $20x^3 + 12x^2 + 24x$

5. $3x(2x + 3) - 5(2x + 3)$

6. $5x(x + 1) - (x + 1)$

Factor each polynomial by grouping.

7. $3x^2 - 3x + 2x - 2$

8. $3a^2 + 9ab + 3b^2 + ab$

9. $10a^2 + 5ab + 7b^2 + 14ab$

10. $6x^2 + 10x - 3x - 5$

(13.2) *Factor each trinomial.*

11. $x^2 + 6x + 8$
12. $x^2 - 11x + 24$
13. $x^2 + x + 2$
14. $x^2 - 5x - 6$

15. $x^2 + 2x - 8$
16. $x^2 + 4xy - 12y^2$
17. $x^2 + 8xy + 15y^2$
18. $72 - 18x - 2x^2$

19. $32 + 12x - 4x^2$
20. $5y^3 - 50y^2 + 120y$

21. To factor $x^2 + 2x - 48$, think of two numbers whose product is _____ and whose sum is _____.

22. What is the first step in factoring $3x^2 + 15x + 30$?

(13.3) **or** **(13.4)** *Factor each trinomial.*

23. $2x^2 + 13x + 6$
24. $4x^2 + 4x - 3$
25. $6x^2 + 5xy - 4y^2$
26. $x^2 - x + 2$

27. $2x^2 - 23x - 39$
28. $18x^2 - 9xy - 20y^2$
29. $10y^3 + 25y^2 - 60y$
30. $60y^3 - 39y^2 + 6y$

Write the perimeter of each figure as a simplified polynomial. Then factor each polynomial completely.

△ **31.**

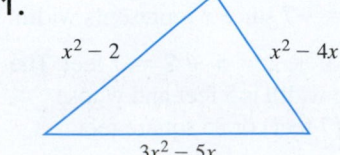

△ **32.**

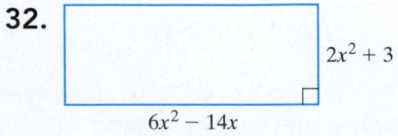

(13.5) *Determine whether each polynomial is a perfect square trinomial.*

33. $x^2 + 6x + 9$ **34.** $x^2 + 8x + 64$ **35.** $9m^2 - 12m + 16$ **36.** $4y^2 - 28y + 49$

Determine whether each binomial is a difference of two squares.

37. $x^2 - 9$ **38.** $x^2 + 16$ **39.** $4x^2 - 25y^2$ **40.** $9a^3 - 1$

Factor each polynomial completely.

41. $x^2 - 81$ **42.** $x^2 + 12x + 36$ **43.** $4x^2 - 9$ **44.** $9t^2 - 25s^2$

45. $16x^2 + y^2$ **46.** $n^2 - 18n + 81$ **47.** $3r^2 + 36r + 108$ **48.** $9y^2 - 42y + 49$

49. $5m^8 - 5m^6$ **50.** $4x^2 - 28xy + 49y^2$ **51.** $3x^2y + 6xy^2 + 3y^3$ **52.** $16x^4 - 1$

(13.6) *Solve each equation.*

53. $(x + 6)(x - 2) = 0$ **54.** $(x - 7)(x + 11) = 0$ **55.** $3x(x + 1)(7x - 2) = 0$

56. $4(5x + 1)(x + 3) = 0$ **57.** $x^2 + 8x + 7 = 0$ **58.** $x^2 - 2x - 24 = 0$ **59.** $x^2 + 10x = -25$

60. $x(x - 10) = -16$ **61.** $(3x - 1)(9x^2 + 3x + 1) = 0$ **62.** $56x^2 - 5x - 6 = 0$

63. $m^2 = 6m$ **64.** $r^2 = 25$ **65.** Write a quadratic equation that has the two solutions 4 and 5.   **66.** Write a quadratic equation that has two solutions, both -1.

(13.7) *Use the given information to choose the correct dimensions.*

△ **67.** The perimeter of a rectangle is 24 inches. The length is twice the width. Find the dimensions of the rectangle.

    a. 5 inches by 7 inches   **b.** 5 inches by 10 inches
    c. 4 inches by 8 inches   **d.** 2 inches by 10 inches

△ **68.** The area of a rectangle is 80 meters. The length is one more than three times the width. Find the dimensions of the rectangle.

    a. 8 meters by 10 meters   **b.** 4 meters by 13 meters
    c. 4 meters by 20 meters   **d.** 5 meters by 16 meters

Use the given information to find the dimensions of each figure.

△ **69.** The *area* of the square is 81 square units. Find the length of a side.

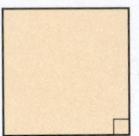

x

△ **70.** The *perimeter* of the quadrilateral is 47 units. Find the lengths of the sides.

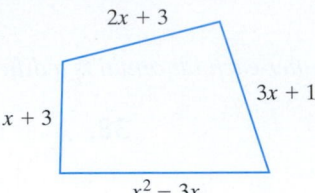

$2x + 3$

$3x + 1$

$x + 3$

$x^2 - 3x$

Solve.

△ **71.** A flag for a local organization is in the shape of a rectangle whose length is 15 inches less than twice its width. If the area of the flag is 500 square inches, find its dimensions.

x

△ **72.** The base of a triangular sail is four times its height. If the area of the triangle is 162 square yards, find the base.

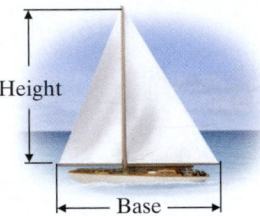

Height

Base

73. Find two consecutive positive integers whose product is 380.

74. Find two consecutive positive even integers whose product is 440.

75. A rocket is fired from the ground with an initial velocity of 440 feet per second. Its height h after t seconds is given by the equation $h = -16t^2 + 440t$.

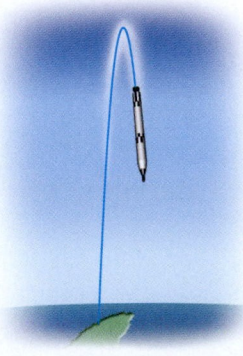

     a. Find how many seconds pass before the rocket reaches a height of 2800 feet. Explain why two answers are obtained.

     b. Find how many seconds pass before the rocket reaches the ground again.

△ **76.** An architect's squaring instrument is in the shape of a right triangle. Find the length of the longer leg of the right triangle if the hypotenuse is 8 centimeters longer than the longer leg and the shorter leg is 8 centimeters shorter than the longer leg.

Mixed Review

Factor completely.

77. $6x + 24$

78. $7x - 63$

79. $11x(4x - 3) - 6(4x - 3)$

80. $2x(x - 5) - (x - 5)$

81. $3x^3 - 4x^2 + 6x - 8$

82. $xy + 2x - y - 2$

83. $2x^2 + 2x - 24$

84. $3x^3 - 30x^2 + 27x$

85. $4x^2 - 81$

86. $2x^2 - 18$

87. $16x^2 - 24x + 9$

88. $5x^2 + 20x + 20$

Solve.

89. $2x^2 - x - 28 = 0$

90. $x^2 - 2x = 15$

91. $2x(x + 7)(x + 4) = 0$

92. $x(x - 5) = -6$

93. $x^2 = 16x$

94. The perimeter of the following triangle is 48 inches. Find the lengths of its sides.

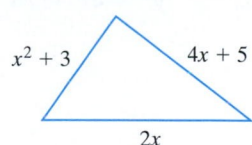

$x^2 + 3$ $4x + 5$

$2x$

95. The width of a rectangle is 4 inches less than its length. Its area is 12 square inches. Find the dimensions of the rectangle.

Answers

Factor each polynomial completely. If a polynomial cannot be factored, write "prime."

1. $9x^2 - 3x$

2. $x^2 + 11x + 28$

3. $49 - m^2$

4. $y^2 + 22y + 121$

5. $x^4 - 16$

6. $4(a + 3) - y(a + 3)$

7. $x^2 + 4$

8. $y^2 - 8y - 48$

9. $3a^2 + 3ab - 7a - 7b$

10. $3x^2 - 5x + 2$

11. $180 - 5x^2$

12. $3x^3 - 21x^2 + 30x$

13. $6t^2 - t - 5$

14. $xy^2 - 7y^2 - 4x + 28$

15. $x - x^5$

16. $x^2 + 14xy + 24y^2$

Solve each equation.

17. $(x - 3)(x + 9) = 0$

18. $x^2 + 5x = 14$

19. $x(x + 6) = 7$

20. $3x(2x - 3)(3x + 4) = 0$

21. $5t^3 - 45t = 0$

22. $t^2 - 2t - 15 = 0$

23. $6x^2 = 15x$

Answers

1. _____
2. _____
3. _____
4. _____
5. _____
6. _____
7. _____
8. _____
9. _____
10. _____
11. _____
12. _____
13. _____
14. _____
15. _____
16. _____
17. _____
18. _____
19. _____
20. _____
21. _____
22. _____
23. _____

24. _____

Solve.

△ **24.** A deck for a home is in the shape of a triangle. The length of the base of the triangle is 9 feet longer than its height. If the area of the triangle is 68 square feet, find the length of the base.

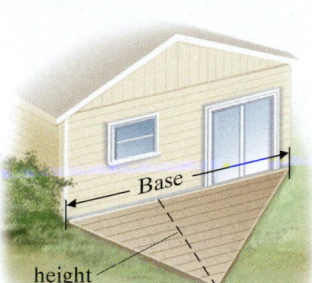

△ **25.** The *area* of the rectangle is 54 square units. Find the dimensions of the rectangle.

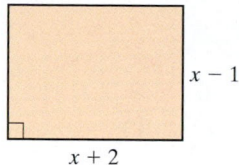

25. _____

26. An object is dropped from the top of the Woolworth Building on Broadway in New York City. The height h of the object after t seconds is given by the equation

$$h = -16t^2 + 784$$

Find how many seconds pass before the object reaches the ground.

△ **27.** Find the lengths of the sides of a right triangle if the hypotenuse is 10 centimeters longer than the shorter leg and 5 centimeters longer than the longer leg.

26. _____

27. _____

28. A window washer is suspended 38 feet below the roof of the 1127-foot-tall John Hancock Center in Chicago. (*Source:* Council on Tall Buildings and Urban Habitat) If the window washer drops an object from this height, the object's height h after t seconds is given by the equation $h = -16t^2 + 1089$. Find how many seconds pass before the object reaches the ground.

28. _____

Answers

Perform the indicated operation and simplify.

1. _____

2. _____

3. _____

4. _____

5. _____

6. _____

7. _____

8. _____

9. _____

10. _____

11. _____

12. _____

13. _____

14. _____

15. _____

16. _____

1. $\dfrac{2}{7} + \dfrac{3}{7}$

2. $\dfrac{26}{30} - \dfrac{7}{30}$

3. $\dfrac{7}{13} + \dfrac{6}{13} + \dfrac{3}{13}$

4. $\dfrac{7}{10} - \dfrac{3}{10} + \dfrac{4}{10}$

5. Find the LCM of 9 and 12.

6. Add: $\dfrac{17}{25} + \dfrac{3}{10}$

7. Write an equivalent fraction with the indicated denominator.

$$\dfrac{1}{2} = \dfrac{}{24}$$

8. Determine whether these fractions are equivalent.

$$\dfrac{10}{55}, \quad \dfrac{6}{33}$$

9. Subtract: $\dfrac{10}{11} - \dfrac{2}{3}$

10. Subtract: $17\dfrac{5}{24} - 9\dfrac{5}{9}$

△ 11. Find the area of the triangle.

8 centimeters

14 centimeters

12. Simplify: $\dfrac{(4 + \sqrt{4})^2}{\sqrt{100} - \sqrt{64}}$

13. Use Appendix B.1 or a calculator to approximate $\sqrt{43}$ to the nearest thousandth.

14. Divide: $0.1156 \div 0.02$

15. Mel Rose is a 6-foot-tall park ranger who needs to know the height of a particular tree. He measures the shadow of the tree to be 69 feet long when his own shadow is 9 feet long. Find the height of the tree.

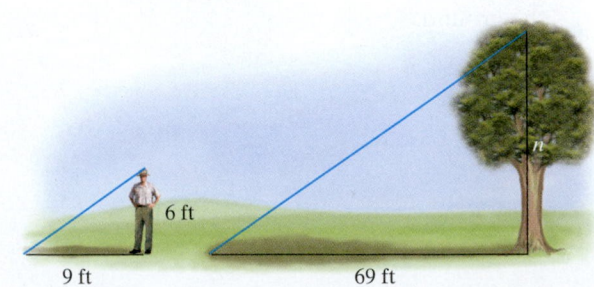

6 ft

9 ft

69 ft

16. What percent of 120 is 28.8?

17. The following bar graph shows the number of endangered species in 2013. Use this graph to answer the questions.

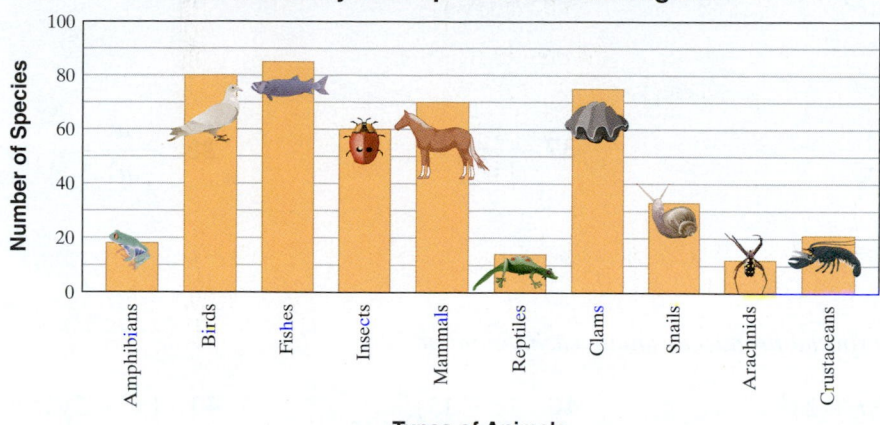

How Many U.S. Species Are Endangered?

Source: U.S. Fish and Wildlife Service

  a. Approximate the number of endangered species that are clams.
  b. Which category has the most endangered species?

18. Find the mean, median, and mode of 1, 7, 8, 10, 11, 11.

19. Translate each sentence into a mathematical statement.
  a. Nine is less than or equal to eleven.
  b. Eight is greater than one.
  c. Three is not equal to four.

20. Insert $<$ or $>$ in the space to make each statement true.
  a. $|-5|$ $|-3|$
  b. $|0|$ $|-2|$

21. Decide whether 2 is a solution of $3x + 10 = 8x$.

22. Evaluate $\dfrac{x}{y} + 5x$ if $x = 20$ and $y = 10$.

23. Write as an expression and simplify: subtract 8 from -4.

24. Evaluate $\dfrac{x}{y} + 5x$ if $x = -20$ and $y = 10$.

25. Evaluate each expression when $x = -2$ and $y = -4$.
  a. $\dfrac{3x}{2y}$
  b. $x^3 - y^2$
  c. $\dfrac{x - y}{-x}$

26. Evaluate $\dfrac{x}{y} + 5x$ if $x = -20$ and $y = -10$.

Solve each equation.

27. $-3x = 33$

28. $\dfrac{x}{-7} = -4$

29. $3(x - 4) = 3x - 12$

30. $-\dfrac{2}{3}x = -22$

31. Solve for *l*: $V = lwh$

32. Solve for *y*: $3x + 2y = -7$

17. a. _____

b. _____

18. _____

19. a. _____

b. _____

c. _____

20. a. _____

b. _____

21. _____

22. _____

23. _____

24. _____

25. a. _____

b. _____

c. _____

26. _____

27. _____

28. _____

29. _____

30. _____

31. _____

32. _____

33. _____

34. _____

35. _____

36. _____

37. _____

38. _____

39. _____

40. _____

41. _____

42. _____

43. _____

44. _____

45. _____

46. _____

47. _____

48. _____

49. _____

50. _____

Simplify the following expressions. Write each result using positive exponents only.

33. $\dfrac{(x^3)^4 x}{x^7}$

34. 3^{-2}

35. $(y^{-3}z^6)^{-6}$

36. $\dfrac{x^{-3}}{x^{-7}}$

37. $\dfrac{x^{-7}}{(x^4)^3}$

38. $\dfrac{(5a^7)^2}{a^5}$

Use a special product to square each binomial.

39. $(t + 2)^2$

40. $(x - 13)^2$

41. $(x^2 - 7y)^2$

42. $(7x + y)^2$

43. Divide: $\dfrac{8x^2y^2 - 16xy + 2x}{4xy}$

Factor each polynomial.

44. $z^3 + 7z + z^2 + 7$

45. $5(x + 3) + y(x + 3)$

46. $2x^3 + 2x^2 - 84x$

47. $x^4 + 5x^2 + 6$

48. $9xy^2 - 16x$

49. The platform for the cliff divers in Acapulco, Mexico, is about 144 feet above the sea. Neglecting air resistance, the height h in feet of a cliff diver above the ocean after t seconds is given by the quadratic equation $h = -16t^2 + 144$. Find how long it takes the diver to reach the ocean.

50. Solve $x^2 - 13x = -36$.

Rational Expressions

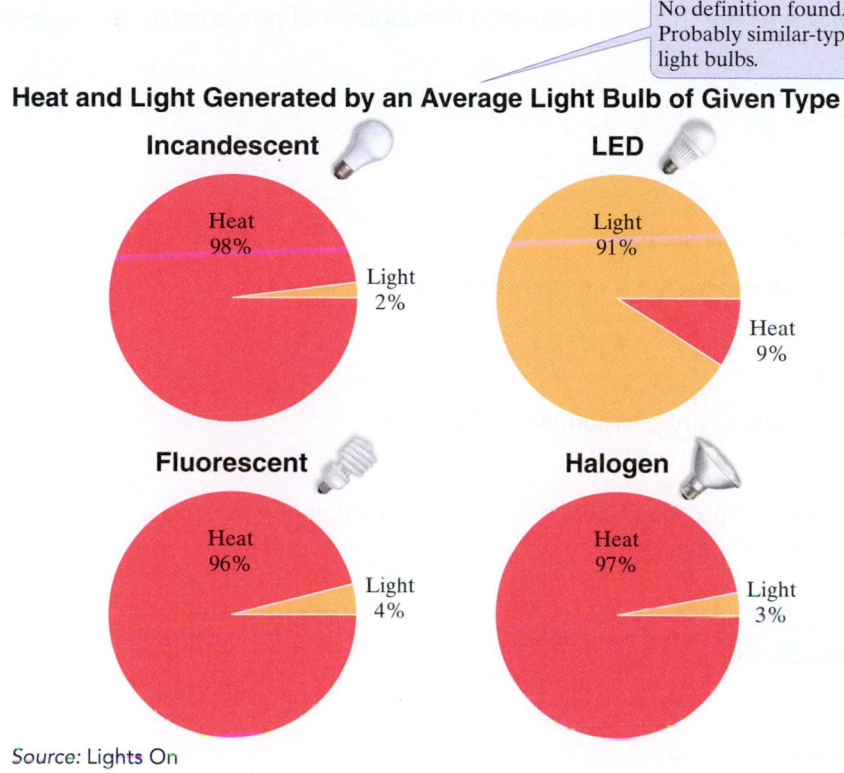

Heat and Light Generated by an Average Light Bulb of Given Type

No definition found. Probably similar-type light bulbs.

Incandescent — Heat 98%, Light 2%

LED — Light 91%, Heat 9%

Fluorescent — Heat 96%, Light 4%

Halogen — Heat 97%, Light 3%

Source: Lights On

Check Your Progress

Vocabulary Check

Chapter Highlights

Chapter Review

Chapter Test

Cumulative Review

In this chapter, we expand our knowledge of algebraic expressions to include algebraic fractions, called *rational expressions*. We explore the operations of addition, subtraction, multiplication, and division using principles similar to the principles for numerical fractions.

Is An Average Always An Average?

In mathematics, the average or mean of a set of data values is defined by the sum of the data values divided by the number of data values. For example, the average of two numbers, a and b, is $\frac{a+b}{2}$. This is an excellent example of a rational expression and sometimes a complex rational expression as we see in Section 14.7. Unfortunately, the word *average* is often used and defined in ways that we might not expect, as noted above and below. Sometimes the average is defined and sometimes not. In Section 14.7, Exercises 47–50, we find mathematical averages.

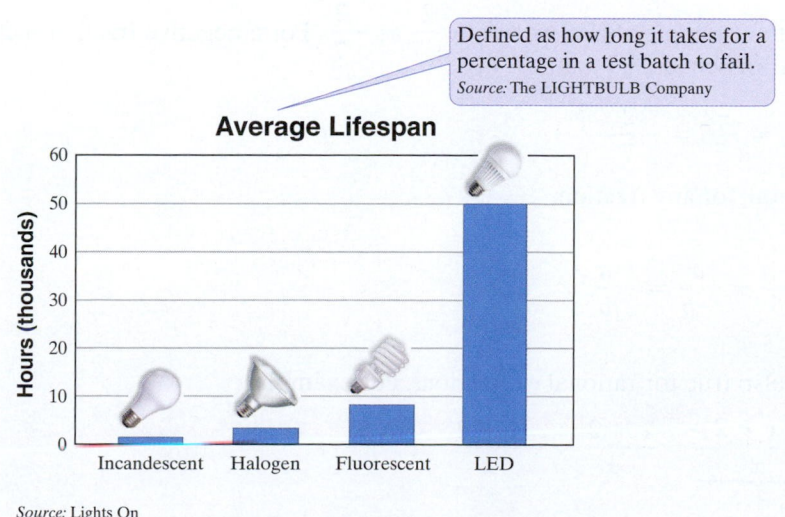

Defined as how long it takes for a percentage in a test batch to fail.
Source: The LIGHTBULB Company

Average Lifespan

Hours (thousands) — Incandescent, Halogen, Fluorescent, LED

Source: Lights On

Objectives

A Find the Value of a Rational Expression Given a Replacement Number.

B Identify Values for Which a Rational Expression Is Undefined.

C Simplify, or Write Rational Expressions in Lowest Terms.

D Write Equivalent Rational Expressions of the Forms $-\dfrac{a}{b} = \dfrac{-a}{b} = \dfrac{a}{-b}$.

Objective A Evaluating Rational Expressions

A rational number is a number that can be written as a quotient of integers. A *rational expression* is also a quotient; it is a quotient of polynomials. Examples are

$$\frac{2}{3}, \quad \frac{3y^3}{8}, \quad \frac{-4p}{p^3 + 2p + 1}, \quad \text{and} \quad \frac{5x^2 - 3x + 2}{3x + 7}$$

Rational Expression

A **rational expression** is an expression that can be written in the form

$$\frac{P}{Q}$$

where P and Q are polynomials and $Q \neq 0$.

Rational expressions have different numerical values depending on what values replace the variables.

Example 1 Find the numerical value of $\dfrac{x + 4}{2x - 3}$ for each replacement value.

a. $x = 5$ **b.** $x = -2$

Solution:

a. We replace each x in the expression with 5 and then simplify.

$$\frac{x + 4}{2x - 3} = \frac{5 + 4}{2(5) - 3} = \frac{9}{10 - 3} = \frac{9}{7}$$

b. We replace each x in the expression with -2 and then simplify.

$$\frac{x + 4}{2x - 3} = \frac{-2 + 4}{2(-2) - 3} = \frac{2}{-7} \quad \text{or} \quad -\frac{2}{7}$$

■ **Work Practice 1**

In the example above, we wrote $\dfrac{2}{-7}$ as $-\dfrac{2}{7}$. For a negative fraction such as $\dfrac{2}{-7}$, recall from Section 8.5 that

$$\frac{2}{-7} = \frac{-2}{7} = -\frac{2}{7}$$

In general, for any fraction,

$$\frac{-a}{b} = \frac{a}{-b} = -\frac{a}{b}, \qquad b \neq 0$$

This is also true for rational expressions. For example,

$$\underbrace{\frac{-(x + 2)}{x}}_{} = \frac{x + 2}{-x} = -\frac{x + 2}{x}$$

↑
Notice the parentheses.

Practice 1

Find the value of $\dfrac{x - 3}{5x + 1}$ for each replacement value.

a. $x = 4$

b. $x = -3$

Answers

1. a. $\dfrac{1}{21}$ **b.** $\dfrac{3}{7}$

Objective B Identifying When a Rational Expression Is Undefined ▶

In the definition of rational expression (first "box" in this section), notice that we wrote $Q \neq 0$ for the denominator Q. The denominator of a rational expression must not equal 0 since division by 0 is not defined. (See the Helpful Hint to the right.) This means we must be careful when replacing the variable in a rational expression by a number. For example, suppose we replace x with 5 in the rational expression $\frac{3 + x}{x - 5}$. The expression becomes

$$\frac{3 + x}{x - 5} = \frac{3 + 5}{5 - 5} = \frac{8}{0}$$

But division by 0 is undefined. Therefore, in this expression we can allow x to be any real number *except* 5. **A rational expression is undefined for values that make the denominator 0.** Thus,

> To find values for which a rational expression is undefined, find values for which the denominator is 0.

Example 2 Are there any values for x for which each expression is undefined?

a. $\dfrac{x}{x - 3}$ **b.** $\dfrac{x^2 + 2}{x^2 - 3x + 2}$ **c.** $\dfrac{x^3 - 6x^2 - 10x}{3}$

Solution: To find values for which a rational expression is undefined, we find values that make the denominator 0.

a. The denominator of $\dfrac{x}{x - 3}$ is 0 when $x - 3 = 0$ or when $x = 3$. Thus, when $x = 3$, the expression $\dfrac{x}{x - 3}$ is undefined.

b. We set the denominator equal to 0.

$$x^2 - 3x + 2 = 0$$
$$(x - 2)(x - 1) = 0 \qquad \text{Factor.}$$
$$x - 2 = 0 \quad \text{or} \quad x - 1 = 0 \qquad \text{Set each factor equal to 0.}$$
$$x = 2 \qquad\qquad x = 1 \qquad \text{Solve.}$$

Thus, when $x = 2$ or $x = 1$, the denominator $x^2 - 3x + 2$ is 0. So the rational expression $\dfrac{x^2 + 2}{x^2 - 3x + 2}$ is undefined when $x = 2$ or when $x = 1$.

c. The denominator of $\dfrac{x^3 - 6x^2 - 10x}{3}$ is never 0, so there are no values of x for which this expression is undefined.

■ **Work Practice 2**

Note: Unless otherwise stated, we will now assume that variables in rational expressions are replaced only by values for which the expressions are defined.

Practice 2

Are there any values for x for which each rational expression is undefined?

a. $\dfrac{x}{x + 8}$

b. $\dfrac{x - 3}{x^2 + 5x + 4}$

c. $\dfrac{x^2 - 3x + 2}{5}$

Answers

2. a. $x = -8$ **b.** $x = -4, x = -1$
c. no

Objective C Simplifying Rational Expressions

A fraction is said to be written in lowest terms or simplest form when the numerator and denominator have no common factors other than 1 (or −1). For example, the fraction $\dfrac{7}{10}$ is written in lowest terms since the numerator and denominator have no common factors other than 1 (or −1).

The process of writing a rational expression in lowest terms or simplest form is called **simplifying** the rational expression.

Simplifying a rational expression is similar to simplifying a fraction. Recall from Section 2.3 that to simplify a fraction, we essentially "remove factors of 1." Our ability to do this comes from these facts:

- Any nonzero number over itself simplifies to 1 $\left(\dfrac{5}{5} = 1, \dfrac{-7.26}{-7.26} = 1, \text{ and } \dfrac{c}{c} = 1 \right.$ as long as c is not $0 \Big)$, and

- The product of any number and 1 is that number $\left(19 \cdot 1 = 19, -8.9 \cdot 1 = -8.9, \right.$ $\dfrac{a}{b} \cdot 1 = \dfrac{a}{b} \Big).$

In other words, we have the following: ————

$$\dfrac{a \cdot c}{b \cdot c} = \dfrac{a}{b} \cdot \dfrac{c}{c} = \dfrac{a}{b}$$

Since $\dfrac{a}{b} \cdot 1 = \dfrac{a}{b}$

Simplify: $\dfrac{15}{20}$

$$\dfrac{15}{20} = \dfrac{3 \cdot 5}{2 \cdot 2 \cdot 5} \qquad \text{Factor the numerator and the denominator.}$$

$$= \dfrac{3 \cdot 5}{2 \cdot 2 \cdot 5} \qquad \text{Look for common factors.}$$

$$= \dfrac{3}{2 \cdot 2} \cdot \dfrac{5}{5} \qquad \text{Common factors in the numerator and denominator form factors of 1.}$$

$$= \dfrac{3}{2 \cdot 2} \cdot 1 \qquad \text{Write } \dfrac{5}{5} \text{ as 1.}$$

$$= \dfrac{3}{2 \cdot 2} = \dfrac{3}{4} \qquad \text{Multiply to remove a factor of 1.}$$

Before we use the same technique to simplify a rational expression, remember that as long as the denominator is not 0, $\dfrac{a^3 b}{a^3 b} = 1, \dfrac{x+3}{x+3} = 1,$ and $\dfrac{7x^2 + 5x - 100}{7x^2 + 5x - 100} = 1.$

Simplify: $\dfrac{x^2 - 9}{x^2 + x - 6}$

$$\dfrac{x^2 - 9}{x^2 + x - 6} = \dfrac{(x-3)(x+3)}{(x-2)(x+3)} \qquad \text{Factor the numerator and the denominator.}$$

$$= \dfrac{(x-3)(x+3)}{(x-2)(x+3)} \qquad \text{Look for common factors.}$$

$$= \dfrac{x-3}{x-2} \cdot \dfrac{x+3}{x+3} $$

$$= \dfrac{x-3}{x-2} \cdot 1 \qquad \text{Write } \dfrac{x+3}{x+3} \text{ as 1.}$$

$$= \dfrac{x-3}{x-2} \qquad \text{Multiply to remove a factor of 1.}$$

Just as for numerical fractions, we can use a shortcut notation. Remember that as long as exact factors in both the numerator and denominator are divided out, we are "removing a factor of 1." We will use the following notation to show this:

$$\frac{x^2 - 9}{x^2 + x - 6} = \frac{(x - 3)(x + 3)}{(x - 2)(x + 3)}$$ A factor of 1 is identified by the shading.

$$= \frac{x - 3}{x - 2}$$ Remove a factor of 1.

Thus, the rational expression $\dfrac{x^2 - 9}{x^2 + x - 6}$ has the same value as the rational expression $\dfrac{x - 3}{x - 2}$ for all values of x except 2 and -3. (Remember that when x is 2, the denominator of both rational expressions is 0 and that when x is -3, the original rational expression has a denominator of 0.)

As we simplify rational expressions, we will assume that the simplified rational expression is equal to the original rational expression for all real numbers except those for which either denominator is 0. The following steps may be used to simplify rational expressions.

> **To Simplify a Rational Expression**
>
> **Step 1:** Completely factor the numerator and denominator.
>
> **Step 2:** Divide out factors common to the numerator and denominator. (This is the same as "removing a factor of 1.")

Example 3 Simplify: $\dfrac{5x - 5}{x^3 - x^2}$

Solution: To begin, we factor the numerator and denominator if possible. Then we look for common factors.

$$\frac{5x - 5}{x^3 - x^2} = \frac{5(x - 1)}{x^2(x - 1)} = \frac{5}{x^2}$$

■ **Work Practice 3**

Practice 3

Simplify: $\dfrac{x^4 + x^3}{5x + 5}$

Example 4 Simplify: $\dfrac{x^2 + 8x + 7}{x^2 - 4x - 5}$

Solution: We factor the numerator and denominator and then look for common factors.

$$\frac{x^2 + 8x + 7}{x^2 - 4x - 5} = \frac{(x + 7)(x + 1)}{(x - 5)(x + 1)} = \frac{x + 7}{x - 5}$$

■ **Work Practice 4**

Practice 4

Simplify: $\dfrac{x^2 + 11x + 18}{x^2 + x - 2}$

Example 5 Simplify: $\dfrac{x^2 + 4x + 4}{x^2 + 2x}$

Solution: We factor the numerator and denominator and then look for common factors.

$$\frac{x^2 + 4x + 4}{x^2 + 2x} = \frac{(x + 2)(x + 2)}{x(x + 2)} = \frac{x + 2}{x}$$

■ **Work Practice 5**

Practice 5

Simplify: $\dfrac{x^2 + 10x + 25}{x^2 + 5x}$

Answers

3. $\dfrac{x^3}{5}$ **4.** $\dfrac{x + 9}{x - 1}$ **5.** $\dfrac{x + 5}{x}$

Copyright 2015 Pearson Education, Inc.

Helpful Hint

When simplifying a rational expression, we look for **common** *factors,* **not** common *terms.*

$$\frac{x \cdot (x + 2)}{x \cdot x} = \frac{x + 2}{x} \qquad \frac{x + 2}{x}$$

Common factors. These can be divided out.   Common terms. There is no factor of 1 that can be generated.

✓**Concept Check** Recall that we can remove only *factors* of 1. Which of the following are *not* true? Explain why.

a. $\frac{3 - 1}{3 + 5}$ simplifies to $-\frac{1}{5}$.   **b.** $\frac{2x + 10}{2}$ simplifies to $x + 5$.

c. $\frac{37}{72}$ simplifies to $\frac{3}{2}$.   **d.** $\frac{2x + 3}{2}$ simplifies to $x + 3$.

Practice 6

Simplify: $\frac{x + 5}{x^2 - 25}$

Example 6 Simplify: $\frac{x + 9}{x^2 - 81}$

Solution: We factor and then apply the fundamental principle. Remember that this principle allows us to divide the numerator and denominator by all common factors.

$$\frac{x + 9}{x^2 - 81} = \frac{x + 9}{(x + 9)(x - 9)} = \frac{1}{x - 9}$$

■ **Work Practice 6**

Practice 7

Simplify each rational expression.

a. $\frac{x + 4}{4 + x}$

b. $\frac{x - 4}{4 - x}$

Example 7 Simplify each rational expression.

a. $\frac{x + y}{y + x}$ **b.** $\frac{x - y}{y - x}$

Solution:

a. The expression $\frac{x + y}{y + x}$ can be simplified by using the commutative property of addition to rewrite the denominator $y + x$ as $x + y$.

$$\frac{x + y}{y + x} = \frac{x + y}{x + y} = 1$$

b. The expression $\frac{x - y}{y - x}$ can be simplified by recognizing that $y - x$ and $x - y$ are opposites. In other words, $y - x = -1(x - y)$. We proceed as follows:

$$\frac{x - y}{y - x} = \frac{1 \cdot (x - y)}{(-1)(x - y)} = \frac{1}{-1} = -1$$

■ **Work Practice 7**

Answers

6. $\frac{1}{x - 5}$ 7. **a.** 1 **b.** −1

✓**Concept Check Answer**

a, c, d

Example 8 Simplify: $\dfrac{4 - x^2}{3x^2 - 5x - 2}$

Solution:

$$\dfrac{4 - x^2}{3x^2 - 5x - 2} = \dfrac{(2 - x)(2 + x)}{(x - 2)(3x + 1)} \qquad \text{Factor.}$$

$$= \dfrac{(-1)(x - 2)(2 + x)}{(x - 2)(3x + 1)} \qquad \text{Write } 2 - x \text{ as } -1(x - 2).$$

$$= \dfrac{(-1)(2 + x)}{3x + 1} \quad \text{or} \quad \dfrac{-2 - x}{3x + 1} \qquad \text{Simplify.}$$

■ **Work Practice 8**

Practice 8

Simplify: $\dfrac{2x^2 - 5x - 12}{16 - x^2}$

Objective D Writing Equivalent Forms of Rational Expressions ▶

From Example 7(a), we have $y + x = x + y$. $\qquad y + x$ and $x + y$ are equivalent.
From Example 7(b), we have $y - x = -1(x - y)$. $\quad y - x$ and $x - y$ are opposites.

Thus, $\dfrac{x + y}{y + x} = \dfrac{x + y}{x + y} = 1 \quad$ and $\quad \dfrac{x - y}{y - x} = \dfrac{x - y}{-1(x - y)} = \dfrac{1}{-1} = -1.$

When performing operations on rational expressions, equivalent forms of answers often result. For this reason, it is very important to be able to recognize equivalent answers.

Example 9 List some equivalent forms of $-\dfrac{5x - 1}{x + 9}$.

Solution: To do so, recall that $-\dfrac{a}{b} = \dfrac{-a}{b} = \dfrac{a}{-b}$. Thus,

$$-\dfrac{5x - 1}{x + 9} = \dfrac{-(5x - 1)}{x + 9} = \dfrac{-5x + 1}{x + 9} \quad \text{or} \quad \dfrac{1 - 5x}{x + 9}$$

Also,

$$-\dfrac{5x - 1}{x + 9} = \dfrac{5x - 1}{-(x + 9)} = \dfrac{5x - 1}{-x - 9} \quad \text{or} \quad \dfrac{5x - 1}{-9 - x}$$

Thus $-\dfrac{5x - 1}{x + 9} = \dfrac{-(5x - 1)}{x + 9} = \dfrac{-5x + 1}{x + 9} = \dfrac{5x - 1}{-(x + 9)} = \dfrac{5x - 1}{-x - 9}$

■ **Work Practice 9**

Practice 9

List 4 equivalent forms of $-\dfrac{3x + 7}{x - 6}$.

> **Helpful Hint** Remember, a negative sign in front of a fraction or rational expression may be moved to the numerator or the denominator, but *not* both.

Keep in mind that many rational expressions may look different but in fact are equivalent.

Answers

8. $-\dfrac{2x + 3}{x + 4}$ or $\dfrac{-2x - 3}{x + 4}$

9. $\dfrac{-(3x + 7)}{x - 6}; \dfrac{-3x - 7}{x - 6}; \dfrac{3x + 7}{-(x - 6)};$

$\dfrac{3x + 7}{-x + 6}$

Vocabulary, Readiness & Video Check

Use the choices below to fill in each blank. Not all choices will be used.

-1	0	simplifying	$\dfrac{-a}{-b}$	$\dfrac{-a}{b}$	$\dfrac{a}{-b}$
1	2	rational expression			

1. A _____ is an expression that can be written in the form $\dfrac{P}{Q}$, where P and Q are polynomials and $Q \neq 0$.

2. The expression $\dfrac{x+3}{3+x}$ simplifies to _____.

3. The expression $\dfrac{x-3}{3-x}$ simplifies to _____.

4. A rational expression is undefined for values that make the denominator _____.

5. The expression $\dfrac{7x}{x-2}$ is undefined for $x =$ _____.

6. The process of writing a rational expression in lowest terms is called _____.

7. For a rational expression, $-\dfrac{a}{b} =$ _____ $=$ _____.

Decide which rational expression(s) can be simplified. (Do not actually simplify.)

8. $\dfrac{x}{x+7}$

9. $\dfrac{3+x}{x+3}$

10. $\dfrac{5-x}{x-5}$

11. $\dfrac{x+2}{x+8}$

 *Martin-Gay Interactive Videos*

See Video 14.1 🍎

Watch the section lecture video and answer the following questions.

Objective A 12. From the lecture before ⊞ Example 1, what do the different values of a rational expression depend on? How are these different values found? ▶

Objective B 13. Why can't the denominators of rational expressions be zero? How can we find the numbers for which a rational expression is undefined? ▶

Objective C 14. In ⊞ Example 7, why isn't a factor of x divided out of the expression at the end? ▶

Objective D 15. From ⊞ Example 9, if we move a negative sign from in front of a rational expression to either the numerator or the denominator, when would we need to use parentheses and why? ▶

14.1 Exercise Set MyMathLab®

Objective A *Find the value of the following expressions when $x = 2$, $y = -2$, and $z = -5$. See Example 1.*

1. $\dfrac{x+5}{x+2}$

2. $\dfrac{x+8}{x+1}$

3. $\dfrac{y^3}{y^2-1}$

▶ 4. $\dfrac{z}{z^2-5}$

5. $\dfrac{x^2+8x+2}{x^2-x-6}$

6. $\dfrac{x+5}{x^2+4x-8}$

7. The average cost per DVD, in dollars, for a company to produce x DVDs on exercising is given by the formula $A = \dfrac{3x + 400}{x}$, where A is the average cost per DVD and x is the number of DVDs produced.

   a. Find the cost for producing 1 DVD.

   b. Find the average cost for producing 100 DVDs.

   c. Does the cost per DVD decrease or increase when more DVDs are produced? Explain your answer.

8. For a certain model of fax machine, the manufacturing cost C per machine is given by the equation

$$C = \frac{250x + 10{,}000}{x}$$

where x is the number of fax machines manufactured and cost C is in dollars per machine.

   a. Find the cost per fax machine when manufacturing 100 fax machines.

   b. Find the cost per fax machine when manufacturing 1000 fax machines.

   c. Does the cost per machine decrease or increase when more machines are manufactured? Explain why this is so.

Objective B *Find any numbers for which each rational expression is undefined. See Example 2.*

9. $\dfrac{7}{2x}$

10. $\dfrac{3}{5x}$

11. $\dfrac{x + 3}{x + 2}$

12. $\dfrac{5x + 1}{x - 9}$

13. $\dfrac{x - 4}{2x - 5}$

14. $\dfrac{x + 1}{5x - 2}$

15. $\dfrac{9x^3 + 4}{15x^2 + 30x}$

16. $\dfrac{19x^3 + 2}{x^2 - x}$

17. $\dfrac{x^2 - 5x - 2}{4}$

18. $\dfrac{9y^5 + y^3}{9}$

19. $\dfrac{3x^2 + 9}{x^2 - 5x - 6}$

20. $\dfrac{11x^2 + 1}{x^2 - 5x - 14}$

21. $\dfrac{x}{3x^2 + 13x + 14}$

22. $\dfrac{x}{2x^2 + 15x + 27}$

Objective C *Simplify each expression. See Examples 3 through 8.*

23. $\dfrac{x + 7}{7 + x}$

24. $\dfrac{y + 9}{9 + y}$

25. $\dfrac{x - 7}{7 - x}$

26. $\dfrac{y - 9}{9 - y}$

27. $\dfrac{2}{8x + 16}$

28. $\dfrac{3}{9x + 6}$

29. $\dfrac{x - 2}{x^2 - 4}$

30. $\dfrac{x + 5}{x^2 - 25}$

31. $\dfrac{2x - 10}{3x - 30}$

32. $\dfrac{3x - 9}{4x - 16}$

33. $\dfrac{-5a - 5b}{a + b}$

34. $\dfrac{-4x - 4y}{x + y}$

35. $\dfrac{7x + 35}{x^2 + 5x}$

36. $\dfrac{9x + 99}{x^2 + 11x}$

37. $\dfrac{x + 5}{x^2 - 4x - 45}$

38. $\dfrac{x - 3}{x^2 - 6x + 9}$

39. $\dfrac{5x^2 + 11x + 2}{x + 2}$

40. $\dfrac{12x^2 + 4x - 1}{2x + 1}$

▶ **41.** $\dfrac{x^3 + 7x^2}{x^2 + 5x - 14}$

42. $\dfrac{x^4 - 10x^3}{x^2 - 17x + 70}$

43. $\dfrac{14x^2 - 21x}{2x - 3}$

44. $\dfrac{4x^2 + 24x}{x + 6}$

45. $\dfrac{x^2 + 7x + 10}{x^2 - 3x - 10}$

46. $\dfrac{2x^2 + 7x - 4}{x^2 + 3x - 4}$

47. $\dfrac{3x^2 + 7x + 2}{3x^2 + 13x + 4}$

48. $\dfrac{4x^2 - 4x + 1}{2x^2 + 9x - 5}$

49. $\dfrac{2x^2 - 8}{4x - 8}$

50. $\dfrac{5x^2 - 500}{35x + 350}$

▶ **51.** $\dfrac{4 - x^2}{x - 2}$

52. $\dfrac{49 - y^2}{y - 7}$

53. $\dfrac{x^2 - 1}{x^2 - 2x + 1}$

54. $\dfrac{x^2 - 16}{x^2 - 8x + 16}$

Simplify each expression. Each exercise contains a four-term polynomial that should be factored by grouping. See Examples 3 through 8.

55. $\dfrac{x^2 + xy + 2x + 2y}{x + 2}$

56. $\dfrac{ab + ac + b^2 + bc}{b + c}$

57. $\dfrac{5x + 15 - xy - 3y}{2x + 6}$

58. $\dfrac{xy - 6x + 2y - 12}{y^2 - 6y}$

59. $\dfrac{2xy + 5x - 2y - 5}{3xy + 4x - 3y - 4}$

60. $\dfrac{2xy + 2x - 3y - 3}{2xy + 4x - 3y - 6}$

Objective **D** *Study Example 9. Then list four equivalent forms for each rational expression.*

61. $-\dfrac{x - 10}{x + 8}$

▶ **62.** $-\dfrac{x + 11}{x - 4}$

63. $-\dfrac{5y - 3}{y - 12}$

64. $-\dfrac{8y - 1}{y - 15}$

Objectives **C** **D** **Mixed Practice** *Simplify each expression. Then determine whether the given answer is correct. See Examples 3 through 9.*

65. $\dfrac{9 - x^2}{x - 3}$; Answer: $-3 - x$?

66. $\dfrac{100 - x^2}{x - 10}$; Answer: $-10 - x$?

67. $\dfrac{7 - 34x - 5x^2}{25x^2 - 1}$; Answer: $\dfrac{x + 7}{-5x - 1}$?

68. $\dfrac{2 - 15x - 8x^2}{64x^2 - 1}$; Answer: $\dfrac{x + 2}{-8x - 1}$?

Review

Perform each indicated operation. See Sections 2.4 and 2.5.

69. $\dfrac{1}{3} \cdot \dfrac{9}{11}$

70. $\dfrac{5}{27} \cdot \dfrac{2}{5}$

71. $\dfrac{1}{3} \div \dfrac{1}{4}$

72. $\dfrac{7}{8} \div \dfrac{1}{2}$

73. $\dfrac{13}{20} \div \dfrac{2}{9}$

74. $\dfrac{8}{15} \div \dfrac{5}{8}$

Concept Extensions

Which of the following are incorrect and why? See the Concept Check in this section.

75. $\dfrac{5a - 15}{5}$ simplifies to $a - 3$?

76. $\dfrac{7m - 9}{7}$ simplifies to $m - 9$?

77. $\dfrac{1 + 2}{1 + 3}$ simplifies to $\dfrac{2}{3}$?

78. $\dfrac{46}{54}$ simplifies to $\dfrac{6}{5}$?

79. Explain how to write a fraction in lowest terms.

80. Explain how to write a rational expression in lowest terms.

81. Explain why the denominator of a fraction or a rational expression must not equal 0.

82. Does $\dfrac{(x - 3)(x + 3)}{x - 3}$ have the same value as $x + 3$ for all real numbers? Explain why or why not.

83. The dose of medicine prescribed for a child depends on the child's age A in years and the adult dose D for the medication. Young's Rule is a formula used by pediatricians that gives a child's dose C as

$$C = \frac{DA}{A + 12}$$

Suppose that an 8-year-old child needs medication, and the normal adult dose is 1000 mg. What size dose should the child receive?

84. Calculating body-mass index is a way to gauge whether a person should lose weight. Doctors recommend that body-mass index values fall between 19 and 25. The formula for body-mass index B is

$$B = \frac{705w}{h^2}$$

where w is weight in pounds and h is height in inches. Should a 148-pound person who is 5 feet 6 inches tall lose weight?

85. Anthropologists and forensic scientists use a measure called the cephalic index to help classify skulls. The cephalic index of a skull with width W and length L from front to back is given by the formula

$$C = \frac{100W}{L}$$

A long skull has an index value less than 75, a medium skull has an index value between 75 and 85, and a broad skull has an index value over 85. Find the cephalic index of a skull that is 5 inches wide and 6.4 inches long. Classify the skull.

86. A company's gross profit margin P can be computed with the formula $P = \dfrac{R - C}{R}$, where

$R = $ the company's revenue and $C = $ cost of goods sold. For the fiscal year 2012, Ford Motor Company had revenues of \$134.25 billion and cost of goods sold \$115.1 billion. (*Source:* Ford Motor Company) What was Ford's gross profit margin in 2012? Express the answer as a percent, rounded to the nearest tenth of a percent.

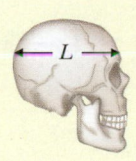
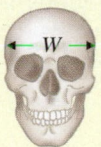

87. A baseball player's slugging percentage S can be calculated with the following formula:

$$S = \frac{h + d + 2t + 3r}{b},$$ where h = number of hits,

d = number of doubles, t = number of triples, r = number of home runs, and b = number of at bats. In 2012, Giancarlo Stanton of the Miami Marlins led Major League Baseball in slugging percentage. During the 2012 season, Stanton had 449 at bats, 130 hits, 30 doubles, 1 triple, and 37 home runs. (*Source:* Major League Baseball) Calculate Stanton's 2012 slugging percentage. Round to the nearest tenth of a percent.

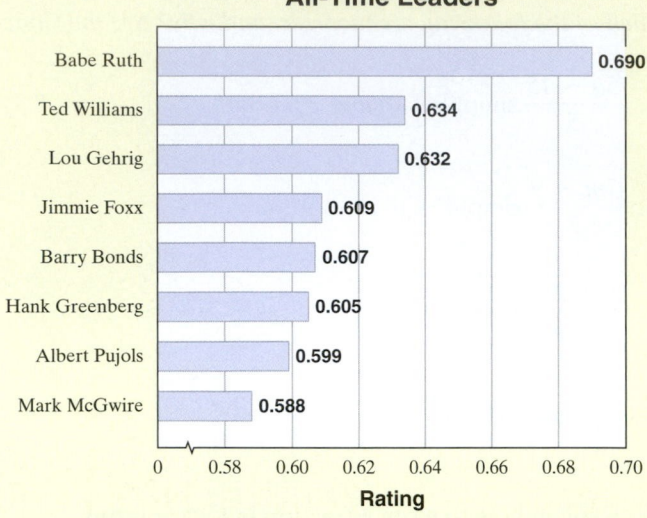

Baseball Slugging Percentage— All-Time Leaders

Player	Rating
Babe Ruth	0.690
Ted Williams	0.634
Lou Gehrig	0.632
Jimmie Foxx	0.609
Barry Bonds	0.607
Hank Greenberg	0.605
Albert Pujols	0.599
Mark McGwire	0.588

Source: Baseball Almanac

88. To calculate a quarterback's rating in NCAA football, you may use the formula

$$\frac{100C + 330T - 200I + 8.4Y}{A},$$ where C = the

number of completed passes, A = the number of attempted passes, T = the number of touchdown passes, Y = the number of yards in the completed passes, and I = the number of interceptions.

Jameis Winston of Florida State University was selected as the 2013 winner of the Heisman Memorial Trophy as the Most Outstanding Football Player. Winston, a freshman quarterback with the Seminoles, ended the season with 384 attempted passes, 257 completed passes, 4057 yards, 40 touchdowns, and only 10 interceptions. Calculate Winston's quarterback rating for the 2013 season. (*Source*: NCAA) Round the answer to the nearest tenth.

14.2 Multiplying and Dividing Rational Expressions

Objectives

A Multiply Rational Expressions.

B Divide Rational Expressions.

C Multiply and Divide Rational Expressions.

D Convert Between Units of Measure.

Objective A Multiplying Rational Expressions

Just as simplifying rational expressions is similar to simplifying number fractions, multiplying and dividing rational expressions is similar to multiplying and dividing number fractions.

Fractions	Rational Expressions
Multiply: $\dfrac{3}{5} \cdot \dfrac{10}{11}$	Multiply: $\dfrac{x-3}{x+5} \cdot \dfrac{2x+10}{x^2-9}$

Multiply numerators and then multiply denominators.

$$\frac{3}{5} \cdot \frac{10}{11} = \frac{3 \cdot 10}{5 \cdot 11} \qquad \frac{x-3}{x+5} \cdot \frac{2x+10}{x^2-9} = \frac{(x-3) \cdot (2x+10)}{(x+5) \cdot (x^2-9)}$$

Simplify by factoring numerators and denominators.

$$= \frac{3 \cdot 2 \cdot 5}{5 \cdot 11} \qquad = \frac{(x-3) \cdot 2\,(x+5)}{(x+5)\,(x+3)\,(x-3)}$$

Apply the fundamental principle.

$$= \frac{3 \cdot 2}{11} \text{ or } \frac{6}{11} \qquad = \frac{2}{x+3}$$

> ### Multiplying Rational Expressions
>
> If $\dfrac{P}{Q}$ and $\dfrac{R}{S}$ are rational expressions, then
>
> $$\frac{P}{Q} \cdot \frac{R}{S} = \frac{PR}{QS}$$
>
> To multiply rational expressions, multiply the numerators and then multiply the denominators.

Note: Recall that for Sections 14.1 through 14.4, we assume variables in rational expressions have only those replacement values for which the expressions are defined.

Example 1 Multiply.

a. $\dfrac{25x}{2} \cdot \dfrac{1}{y^3}$ 　　　　**b.** $\dfrac{-7x^2}{5y} \cdot \dfrac{3y^5}{14x^2}$

Solution: To multiply rational expressions, we first multiply the numerators and then multiply the denominators of both expressions. Then we write the product in lowest terms.

a. $\dfrac{25x}{2} \cdot \dfrac{1}{y^3} = \dfrac{25x \cdot 1}{2 \cdot y^3} = \dfrac{25x}{2y^3}$

The expression $\dfrac{25x}{2y^3}$ is in lowest terms.

b. $\dfrac{-7x^2}{5y} \cdot \dfrac{3y^5}{14x^2} = \dfrac{-7x^2 \cdot 3y^5}{5y \cdot 14x^2}$ 　　Multiply.

The expression $\dfrac{-7x^2 \cdot 3y^5}{5y \cdot 14x^2}$ is not in lowest terms, so we factor the numerator and the denominator and apply the fundamental principle to "remove factors of 1."

$$= \frac{-1 \cdot 7 \cdot 3 \cdot x^2 \cdot y \cdot y^4}{5 \cdot 2 \cdot 7 \cdot x^2 \cdot y}$$ 　　Common factors in the numerator and denominator form factors of 1.

$$= -\frac{3y^4}{10}$$ 　　Divide out common factors. (This is the same as "removing a factor of 1.")

■ **Work Practice 1**

When multiplying rational expressions, it is usually best to factor each numerator and denominator first. This will help us when we apply the fundamental principle to write the product in lowest terms.

Example 2 Multiply: $\dfrac{x^2 + x}{3x} \cdot \dfrac{6}{5x + 5}$

Solution:

$$\frac{x^2 + x}{3x} \cdot \frac{6}{5x + 5} = \frac{x(x + 1)}{3x} \cdot \frac{2 \cdot 3}{5(x + 1)}$$ 　　Factor numerators and denominators.

$$= \frac{x(x + 1) \cdot 2 \cdot 3}{3x \cdot 5(x + 1)}$$ 　　Multiply.

$$= \frac{2}{5}$$ 　　Divide out common factors.

■ **Work Practice 2**

The following steps may be used to multiply rational expressions.

To Multiply Rational Expressions

Step 1: Completely factor numerators and denominators.

Step 2: Multiply numerators and multiply denominators.

Step 3: Simplify or write the product in lowest terms by dividing out common factors.

✓ **Concept Check** Which of the following is a true statement?

a. $\dfrac{1}{3} \cdot \dfrac{1}{2} = \dfrac{1}{5}$ **b.** $\dfrac{2}{x} \cdot \dfrac{5}{x} = \dfrac{10}{x}$ **c.** $\dfrac{3}{x} \cdot \dfrac{1}{2} = \dfrac{3}{2x}$ **d.** $\dfrac{x}{7} \cdot \dfrac{x+5}{4} = \dfrac{2x+5}{28}$

Practice 3

Multiply:

$$\dfrac{4x+8}{7x^2-14x} \cdot \dfrac{3x^2-5x-2}{9x^2-1}$$

Example 3 Multiply: $\dfrac{3x+3}{5x^2-5x} \cdot \dfrac{2x^2+x-3}{4x^2-9}$

Solution:

$$\dfrac{3x+3}{5x^2-5x} \cdot \dfrac{2x^2+x-3}{4x^2-9} = \dfrac{3(x+1)}{5x(x-1)} \cdot \dfrac{(2x+3)(x-1)}{(2x-3)(2x+3)} \qquad \text{Factor.}$$

$$= \dfrac{3(x+1)\ (2x+3)(x-1)}{5x\ (x-1)\ (2x-3)\ (2x+3)} \qquad \text{Multiply.}$$

$$= \dfrac{3(x+1)}{5x(2x-3)} \qquad \text{Simplify.}$$

■ **Work Practice 3**

Objective B Dividing Rational Expressions ▶

We can divide by a rational expression in the same way we divide by a number fraction. Recall that to divide by a fraction, we multiply by its reciprocal.

For example, to divide $\dfrac{3}{2}$ by $\dfrac{7}{8}$, we multiply $\dfrac{3}{2}$ by $\dfrac{8}{7}$.

$$\dfrac{3}{2} \div \dfrac{7}{8} = \dfrac{3}{2} \cdot \dfrac{8}{7} = \dfrac{3 \cdot 4 \cdot 2}{2 \cdot 7} = \dfrac{12}{7}$$

Helpful Hint

Don't forget how to find reciprocals. The reciprocal of $\dfrac{a}{b}$ is $\dfrac{b}{a}$, $a \neq 0, b \neq 0$.

Dividing Rational Expressions

If $\dfrac{P}{Q}$ and $\dfrac{R}{S}$ are rational expressions and $\dfrac{R}{S}$ is not 0, then

$$\dfrac{P}{Q} \div \dfrac{R}{S} = \dfrac{P}{Q} \cdot \dfrac{S}{R} = \dfrac{PS}{QR}$$

To divide two rational expressions, multiply the first rational expression by the reciprocal of the second rational expression.

Answer

3. $\dfrac{4(x+2)}{7x(3x-1)}$

✓ **Concept Check Answer**

c

Example 4 　Divide: $\dfrac{3x^3}{40} \div \dfrac{4x^3}{y^2}$

Solution:

$$\dfrac{3x^3}{40} \div \dfrac{4x^3}{y^2} = \dfrac{3x^3}{40} \cdot \dfrac{y^2}{4x^3} \quad \text{Multiply by the reciprocal of } \dfrac{4x^3}{y^2}.$$

$$= \dfrac{3 \; x^3 \; \cdot y^2}{160 \; x^3}$$

$$= \dfrac{3y^2}{160} \qquad \text{Simplify.}$$

■ **Work Practice 4**

Practice 4

Divide: $\dfrac{7x^2}{6} \div \dfrac{x}{2y}$

Example 5 　Divide: $\dfrac{(x+2)^2}{10} \div \dfrac{2x+4}{5}$

Solution:

$$\dfrac{(x+2)^2}{10} \div \dfrac{2x+4}{5} = \dfrac{(x+2)^2}{10} \cdot \dfrac{5}{2x+4} \quad \text{Multiply by the reciprocal of } \dfrac{2x+4}{5}.$$

$$= \dfrac{(x+2)\,(x+2) \cdot 5}{5 \cdot 2 \cdot 2 \cdot (x+2)} \qquad \text{Factor and multiply.}$$

$$= \dfrac{x+2}{4} \qquad \text{Simplify.}$$

■ **Work Practice 5**

Practice 5

Divide: $\dfrac{(x-4)^2}{6} \div \dfrac{3x-12}{2}$

> **Helpful Hint** 　Remember, **to Divide by a Rational Expression,** multiply by its reciprocal.

Example 6 　Divide: $\dfrac{6x+2}{x^2-1} \div \dfrac{3x^2+x}{x-1}$

Solution:

$$\dfrac{6x+2}{x^2-1} \div \dfrac{3x^2+x}{x-1} = \dfrac{6x+2}{x^2-1} \cdot \dfrac{x-1}{3x^2+x} \qquad \text{Multiply by the reciprocal.}$$

$$= \dfrac{2\,(3x+1)(x-1)}{(x+1)\,(x-1) \cdot x\,(3x+1)} \qquad \text{Factor and multiply.}$$

$$= \dfrac{2}{x(x+1)} \qquad \text{Simplify.}$$

■ **Work Practice 6**

Practice 6

Divide: $\dfrac{10x+4}{x^2-4} \div \dfrac{5x^3+2x^2}{x+2}$

Example 7 　Divide: $\dfrac{2x^2-11x+5}{5x-25} \div \dfrac{4x-2}{10}$

Solution:

$$\dfrac{2x^2-11x+5}{5x-25} \div \dfrac{4x-2}{10} = \dfrac{2x^2-11x+5}{5x-25} \cdot \dfrac{10}{4x-2} \qquad \text{Multiply by the reciprocal.}$$

$$= \dfrac{(2x-1)(x-5) \cdot 2 \cdot 5}{5(x-5) \cdot 2(2x-1)} \qquad \text{Factor and multiply.}$$

$$= \dfrac{1}{1} \quad \text{or} \quad 1 \qquad \text{Simplify.}$$

■ **Work Practice 7**

Practice 7

Divide:

$\dfrac{3x^2-10x+8}{7x-14} \div \dfrac{9x-12}{21}$

Answers

4. $\dfrac{7xy}{3}$ 　**5.** $\dfrac{x-4}{9}$

6. $\dfrac{2}{x^2(x-2)}$ 　**7.** 1

Objective C Multiplying and Dividing Rational Expressions

Let's make sure that we understand the difference between multiplying and dividing rational expressions.

Rational Expressions	
Multiplication	Multiply the numerators and multiply the denominators.
Division	Multiply by the reciprocal of the divisor.

Practice 8

Multiply or divide as indicated.

a. $\dfrac{x + 3}{x} \cdot \dfrac{7}{x + 3}$

b. $\dfrac{x + 3}{x} \div \dfrac{7}{x + 3}$

c. $\dfrac{3 - x}{x^2 + 6x + 5} \cdot \dfrac{2x + 10}{x^2 - 7x + 12}$

Example 8 Multiply or divide as indicated.

a. $\dfrac{x - 4}{5} \cdot \dfrac{x}{x - 4}$

b. $\dfrac{x - 4}{5} \div \dfrac{x}{x - 4}$

c. $\dfrac{x^2 - 4}{2x + 6} \cdot \dfrac{x^2 + 4x + 3}{2 - x}$

Solution:

a. $\dfrac{x - 4}{5} \cdot \dfrac{x}{x - 4} = \dfrac{(x - 4) \cdot x}{5 \cdot (x - 4)} = \dfrac{x}{5}$

b. $\dfrac{x - 4}{5} \div \dfrac{x}{x - 4} = \dfrac{x - 4}{5} \cdot \dfrac{x - 4}{x} = \dfrac{(x - 4)^2}{5x}$

c. $\dfrac{x^2 - 4}{2x + 6} \cdot \dfrac{x^2 + 4x + 3}{2 - x} = \dfrac{(x - 2)(x + 2) \cdot (x + 1)(x + 3)}{2(x + 3) \cdot (2 - x)}$ Factor and multiply.

Recall from Section 14.1 that $x - 2$ and $2 - x$ are opposites. This means that $\dfrac{x - 2}{2 - x} = -1$. Thus,

$$\dfrac{(x - 2)(x + 2) \cdot (x + 1)(x + 3)}{2(x + 3) \cdot (2 - x)} = \dfrac{-1(x + 2)(x + 1)}{2}$$

$$= -\dfrac{(x + 2)(x + 1)}{2}$$

■ **Work Practice 8**

Objective D Converting Between Units of Measure

How many square inches are in 1 square foot?
How many cubic feet are in a cubic yard?

If you have trouble answering these questions, this section will be helpful to you.

Now that we know how to multiply fractions and rational expressions, we can use this knowledge to help us convert between units of measure. To do so, we will use **unit fractions.** A unit fraction is a fraction that equals 1. For example, since 12 in. = 1 ft, we have the unit fractions

$$\dfrac{12 \text{ in.}}{1 \text{ ft}} = 1 \quad \text{and} \quad \dfrac{1 \text{ ft}}{12 \text{ in.}} = 1$$

Answers

8. a. $\dfrac{7}{x}$ **b.** $\dfrac{(x + 3)^2}{7x}$

c. $-\dfrac{2}{(x + 1)(x - 4)}$

Example 9   18 square feet = _____ square yards

Solution: Let's multiply 18 square feet by a unit fraction that has square feet in the denominator and square yards in the numerator. From the diagram, you can see that

     1 square yard = 9 square feet

Thus,

$$18 \text{ sq ft} = \frac{18 \text{ sq ft}}{1} \cdot 1 = \frac{\overset{2}{\cancel{18} \text{ sq ft}}}{1} \cdot \frac{1 \text{ sq yd}}{\underset{1}{\cancel{9} \text{ sq ft}}}$$

$$= \frac{2 \cdot 1}{1 \cdot 1} \text{ sq yd} = 2 \text{ sq yd}$$

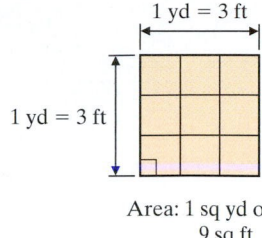

1 yd = 3 ft

1 yd = 3 ft

Area: 1 sq yd or 9 sq ft

Thus, 18 sq ft = 2 sq yd.

     Draw a diagram of 18 sq ft to help you see that this is reasonable.

■ **Work Practice 9**

Practice 9

288 square inches = _____ square feet

Example 10   5.2 square yards = _____ square feet

Solution:

$$5.2 \text{ sq yd} = \frac{5.2 \text{ sq yd}}{1} \cdot 1 = \frac{5.2 \text{ sq yd}}{1} \cdot \frac{9 \text{ sq ft}}{1 \text{ sq yd}} \quad \begin{array}{l} \leftarrow \text{Units converting to} \\ \leftarrow \text{Units given} \end{array}$$

$$= \frac{5.2 \cdot 9}{1 \cdot 1} \text{ sq ft}$$

$$= 46.8 \text{ sq ft}$$

Thus, 5.2 sq yd = 46.8 sq ft.

     Draw a diagram to see that this is reasonable.

■ **Work Practice 10**

Practice 10

3.5 square feet = _____ square inches

Example 11   Converting from Cubic Feet to Cubic Yards

The largest building in the world by volume is The Boeing Company's Everett, Washington, factory complex, where Boeing's wide-body jetliners, the 747, 767, and 777, are built. The volume of this factory complex is 472,370,319 cubic feet. Find the volume of this Boeing facility in cubic yards. (*Source:* The Boeing Company)

(*Continued on next page*)

Practice 11

The largest casino in the world is the Venetian, in Macau, on the southern tip of China. The gaming area for this casino is approximately 61,000 *square yards*. Find the size of the gaming area in *square feet*. (*Source: USA Today*)

Answers
9. 2 sq ft   **10.** 504 sq in.
11. 549,000 sq ft

Solution: There are 27 cubic feet in 1 cubic yard. (See the diagram.)

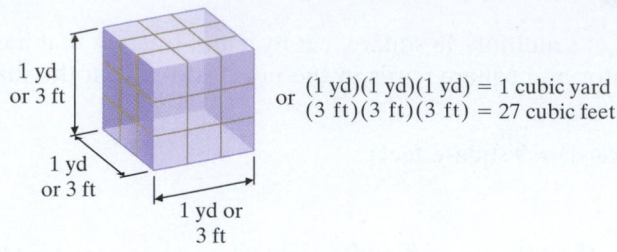

1 yd
or 3 ft

1 yd
or 3 ft

1 yd or
3 ft

or (1 yd)(1 yd)(1 yd) = 1 cubic yard
(3 ft)(3 ft)(3 ft) = 27 cubic feet

$$472{,}370{,}319 \text{ cu ft} = 472{,}370{,}319 \text{ cu ft} \cdot \frac{1 \text{ cu yd}}{27 \text{ cu ft}}$$

$$= \frac{472{,}370{,}319}{27} \text{ cu yd}$$

$$= 17{,}495{,}197 \text{ cu yd}$$

■ **Work Practice 11**

Helpful Hint

When converting between units of measurement, if possible, write the unit fraction so that **the numerator contains the units you are converting to** and **the denominator contains the original units.**

Unit fraction

$$48 \text{ in.} = \frac{48 \text{ in.}}{1} \cdot \frac{1 \text{ ft}}{12 \text{ in.}} \quad \begin{array}{l} \leftarrow \text{Units converting to} \\ \leftarrow \text{Original units} \end{array}$$

$$= \frac{48}{12} \text{ ft} = 4 \text{ ft}$$

Practice 12

The cheetah is the fastest land animal, being clocked at about 102.7 feet per second. Convert this to miles per hour. Round to the nearest tenth. (*Source: World Almanac and Book of Facts*)

Example 12

At the Summer Olympics, Jamaican athlete Usain Bolt won the gold medal in the men's 100-meter track event. He ran the distance at an average speed of 34.1 feet per second. Convert this speed to miles per hour. (*Source: International Olympic Committee*)

Solution: Recall that 1 mile = 5280 feet and 1 hour = 3600 seconds (60 · 60).

Unit fractions

$$34.1 \text{ feet/second} = \frac{34.1 \text{ feet}}{1 \text{ second}} \cdot \frac{3600 \text{ seconds}}{1 \text{ hour}} \cdot \frac{1 \text{ mile}}{5280 \text{ feet}}$$

$$= \frac{34.1 \cdot 3600}{5280} \text{ miles/hour}$$

$$= 23.25 \text{ miles/hour}$$

■ **Work Practice 12**

Answer

12. 70.0 miles per hour

Vocabulary, Readiness & Video Check

Use the choices below to fill in each blank. Not all choices will be used.

opposites $\dfrac{a \cdot d}{b \cdot c}$ $\dfrac{a \cdot c}{b \cdot d}$ $\dfrac{x}{42}$ $\dfrac{x^2}{42}$ $\dfrac{2x}{42}$ $\dfrac{6}{7}$ $\dfrac{7}{6}$

reciprocals

1. The expressions $\dfrac{x}{2y}$ and $\dfrac{2y}{x}$ are called _____.

2. $\dfrac{a}{b} \cdot \dfrac{c}{d} =$ _____

3. $\dfrac{a}{b} \div \dfrac{c}{d} =$ _____

4. $\dfrac{x}{7} \cdot \dfrac{x}{6} =$ _____

5. $\dfrac{x}{7} \div \dfrac{x}{6} =$ _____

Martin-Gay Interactive Videos

See Video 14.2

Watch the section lecture video and answer the following questions.

Objective A 6. Would you say a person needs to be quite comfortable with factoring polynomials in order to be successful with multiplying rational expressions? Explain, referencing ⊞ Example 2 in your answer. ▶

Objective B 7. Based on the lecture before ⊞ Example 3, complete the following statements: Dividing rational expressions is exactly like dividing _____. Therefore, to divide by a rational expression, multiply by its _____. ▶

Objective C 8. In ⊞ Examples 4 and 5, determining the operation is the first step in deciding how to perform the operation. Why is this so? ▶

Objective D 9. In ⊞ Example 6, why is the unit fraction $\dfrac{27 \text{ cu ft}}{1 \text{ cu yd}}$ used? ▶

14.2 Exercise Set MyMathLab® ▶

Objective A *Find each product and simplify if possible. See Examples 1 through 3.*

1. $\dfrac{3x}{y^2} \cdot \dfrac{7y}{4x}$

2. $\dfrac{9x^2}{y} \cdot \dfrac{4y}{3x^3}$

▶ 3. $\dfrac{8x}{2} \cdot \dfrac{x^5}{4x^2}$

4. $\dfrac{6x^2}{10x^3} \cdot \dfrac{5x}{12}$

5. $-\dfrac{5a^2 b}{30a^2 b^2} \cdot b^3$

6. $-\dfrac{9x^3 y^2}{18xy^5} \cdot y^3$

7. $\dfrac{x}{2x - 14} \cdot \dfrac{x^2 - 7x}{5}$

8. $\dfrac{4x - 24}{20x} \cdot \dfrac{5}{x - 6}$

9. $\dfrac{6x + 6}{5} \cdot \dfrac{10}{36x + 36}$

10. $\dfrac{x^2 + x}{8} \cdot \dfrac{16}{x + 1}$

11. $\dfrac{(m + n)^2}{m - n} \cdot \dfrac{m}{m^2 + mn}$

12. $\dfrac{(m - n)^2}{m + n} \cdot \dfrac{m}{m^2 - mn}$

13. $\dfrac{x^2 - 25}{x^2 - 3x - 10} \cdot \dfrac{x + 2}{x}$

14. $\dfrac{a^2 - 4a + 4}{a^2 - 4} \cdot \dfrac{a + 3}{a - 2}$

15. $\dfrac{x^2 + 6x + 8}{x^2 + x - 20} \cdot \dfrac{x^2 + 2x - 15}{x^2 + 8x + 16}$

16. $\dfrac{x^2 + 9x + 20}{x^2 - 15x + 44} \cdot \dfrac{x^2 - 11x + 28}{x^2 + 12x + 35}$

Objective B   *Find each quotient and simplify. See Examples 4 through 7.*

17. $\dfrac{5x^7}{2x^5} \div \dfrac{15x}{4x^3}$

18. $\dfrac{9y^4}{6y} \div \dfrac{y^2}{3}$

19. $\dfrac{8x^2}{y^3} \div \dfrac{4x^2y^3}{6}$

20. $\dfrac{7a^2b}{3ab^2} \div \dfrac{21a^2b^2}{14ab}$

21. $\dfrac{(x - 6)(x + 4)}{4x} \div \dfrac{2x - 12}{8x^2}$

22. $\dfrac{(x + 3)^2}{5} \div \dfrac{5x + 15}{25}$

23. $\dfrac{3x^2}{x^2 - 1} \div \dfrac{x^5}{(x + 1)^2}$

24. $\dfrac{9x^5}{a^2 - b^2} \div \dfrac{27x^2}{3b - 3a}$

25. $\dfrac{m^2 - n^2}{m + n} \div \dfrac{m}{m^2 + nm}$

26. $\dfrac{(m - n)^2}{m + n} \div \dfrac{m^2 - mn}{m}$

🔘 **27.** $\dfrac{x + 2}{7 - x} \div \dfrac{x^2 - 5x + 6}{x^2 - 9x + 14}$

28. $\dfrac{x - 3}{2 - x} \div \dfrac{x^2 + 3x - 18}{x^2 + 2x - 8}$

29. $\dfrac{x^2 + 7x + 10}{x - 1} \div \dfrac{x^2 + 2x - 15}{x - 1}$

30. $\dfrac{x + 1}{2x^2 + 5x + 3} \div \dfrac{20x + 100}{2x + 3}$

Objective C   **Mixed Practice** *Multiply or divide as indicated. See Example 8.*

🔘 **31.** $\dfrac{5x - 10}{12} \div \dfrac{4x - 8}{8}$

32. $\dfrac{6x + 6}{5} \div \dfrac{9x + 9}{10}$

33. $\dfrac{x^2 + 5x}{8} \cdot \dfrac{9}{3x + 15}$

34. $\dfrac{3x^2 + 12x}{6} \cdot \dfrac{9}{2x + 8}$

35. $\dfrac{7}{6p^2 + q} \div \dfrac{14}{18p^2 + 3q}$

36. $\dfrac{3x + 6}{20} \div \dfrac{4x + 8}{8}$

37. $\dfrac{3x + 4y}{x^2 + 4xy + 4y^2} \cdot \dfrac{x + 2y}{2}$

38. $\dfrac{x^2 - y^2}{3x^2 + 3xy} \cdot \dfrac{3x^2 + 6x}{3x^2 - 2xy - y^2}$

39. $\dfrac{(x + 2)^2}{x - 2} \div \dfrac{x^2 - 4}{2x - 4}$

40. $\dfrac{x + 3}{x^2 - 9} \div \dfrac{5x + 15}{(x - 3)^2}$

41. $\dfrac{x^2 - 4}{24x} \div \dfrac{2 - x}{6xy}$

42. $\dfrac{3y}{3 - x} \div \dfrac{12xy}{x^2 - 9}$

43. $\dfrac{a^2 + 7a + 12}{a^2 + 5a + 6} \cdot \dfrac{a^2 + 8a + 15}{a^2 + 5a + 4}$

44. $\dfrac{b^2 + 2b - 3}{b^2 + b - 2} \cdot \dfrac{b^2 - 4}{b^2 + 6b + 8}$

45. $\dfrac{5x - 20}{3x^2 + x} \cdot \dfrac{3x^2 + 13x + 4}{x^2 - 16}$

46. $\dfrac{9x + 18}{4x^2 - 3x} \cdot \dfrac{4x^2 - 11x + 6}{x^2 - 4}$

47. $\dfrac{8n^2 - 18}{2n^2 - 5n + 3} \div \dfrac{6n^2 + 7n - 3}{n^2 - 9n + 8}$

48. $\dfrac{36n^2 - 64}{3n^2 - 10n + 8} \div \dfrac{3n^2 - 5n - 12}{n^2 - 9n + 14}$

Objective D *Convert as indicated. See Examples 9 through 12.*

49. 10 square feet = _____ square inches.

50. 1008 square inches = _____ square feet.

51. 45 square feet = _____ square yards.

52. 2 square yards = _____ square inches.

53. 3 cubic yards = _____ cubic feet.

54. 2 cubic yards = _____ cubic inches.

55. 50 miles per hour = _____ feet per second (round to the nearest whole).

56. 10 feet per second = _____ miles per hour (round to the nearest tenth).

57. 6.3 square yards = _____ square feet.

58. 3.6 square yards = _____ square feet.

59. In January 2010, the Burj Khalifa Tower officially became the tallest building in the world. This tower has a curtain wall (the exterior skin of the building) that is approximately 133,500 square yards. Convert this to square feet. (*Source:* Burj Khalifa)

60. The Pentagon, headquarters for the Department of Defense, contains 3,705,793 square feet of office and storage space. Convert this to square yards. Round to the nearest square yard. (*Source:* U.S. Department of Defense)

61. On January 7, 2011, Australian driver Barton Mawer set a new solar-powered-car land speed record of 80.9 feet per second in the Sunswift IV, built by a student team at the University of New South Wales. Convert this speed to miles per hour. Round to the nearest tenth. (*Source:* University of New South Wales)

62. Peregrine falcons are among the fastest birds in the world. When engaged in a high-speed dive for prey, a peregrine falcon can reach speeds over 200 miles per hour. Find this speed in feet per second. Round to the nearest tenth. (*Source:* Ohio Department of Natural Resources)

Review

Perform each indicated operation. See Section 3.1.

63. $\dfrac{1}{5} + \dfrac{4}{5}$

64. $\dfrac{3}{15} + \dfrac{6}{15}$

65. $\dfrac{9}{9} - \dfrac{19}{9}$

66. $\dfrac{4}{3} - \dfrac{8}{3}$

67. $\dfrac{6}{5} + \left(\dfrac{1}{5} - \dfrac{8}{5}\right)$

68. $-\dfrac{3}{2} + \left(\dfrac{1}{2} - \dfrac{3}{2}\right)$

Concept Extensions

Identify each statement as true or false. If false, correct the multiplication. See the Concept Check in this section.

69. $\dfrac{4}{a} \cdot \dfrac{1}{b} = \dfrac{4}{ab}$

70. $\dfrac{2}{3} \cdot \dfrac{2}{4} = \dfrac{2}{7}$

71. $\dfrac{x}{5} \cdot \dfrac{x+3}{4} = \dfrac{2x+3}{20}$

72. $\dfrac{7}{a} \cdot \dfrac{3}{a} = \dfrac{21}{a}$

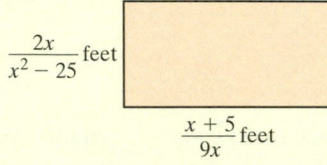 **73.** Find the area of the rectangle.

$\dfrac{2x}{x^2 - 25}$ feet

$\dfrac{x+5}{9x}$ feet

△ **74.** Find the area of the square.

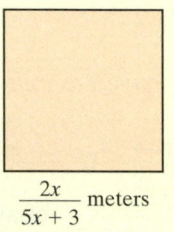

$\dfrac{2x}{5x+3}$ meters

Multiply or divide as indicated.

75. $\left(\dfrac{x^2 - y^2}{x^2 + y^2} \div \dfrac{x^2 - y^2}{3x}\right) \cdot \dfrac{x^2 + y^2}{6}$

76. $\left(\dfrac{x^2 - 9}{x^2 - 1} \cdot \dfrac{x^2 + 2x + 1}{2x^2 + 9x + 9}\right) \div \dfrac{2x + 3}{1 - x}$

77. $\left(\dfrac{2a + b}{b^2} \cdot \dfrac{3a^2 - 2ab}{ab + 2b^2}\right) \div \dfrac{a^2 - 3ab + 2b^2}{5ab - 10b^2}$

78. $\left(\dfrac{x^2 y^2 - xy}{4x - 4y} \div \dfrac{3y - 3x}{8x - 8y}\right) \cdot \dfrac{y - x}{8}$

79. In your own words, explain how you multiply rational expressions.

80. Explain how dividing rational expressions is similar to dividing rational numbers.

81. On a day in August 2014, 1 euro was equivalent to 1.3387 American dollars. If you had wanted to exchange $2000 U.S. for euros on that day for a European vacation, how many would you have received? Round to the nearest hundredth. (*Source: Barclay's Bank*)

82. An environmental technician finds that warm water from an industrial process is being discharged into a nearby pond at a rate of 30 gallons per minute. Plant regulations state that the flow rate should be no more than 0.1 cubic foot per second. Is the flow rate of 30 gallons per minute in violation of the plant regulations? (*Hint:* 1 cubic foot is equivalent to 7.48 gallons.)

14.3 Adding and Subtracting Rational Expressions with the Same Denominator and Least Common Denominator

Objective A Adding and Subtracting Rational Expressions with the Same Denominator

Like multiplication and division, addition and subtraction of rational expressions are similar to addition and subtraction of rational numbers. In this section, we add and subtract rational expressions with a common denominator.

Add: $\dfrac{6}{5} + \dfrac{2}{5}$ | Add: $\dfrac{9}{x+2} + \dfrac{3}{x+2}$

Add the numerators and place the sum over the common denominator.

$$\frac{6}{5} + \frac{2}{5} = \frac{6+2}{5}$$

$$= \frac{8}{5} \quad \text{Simplify.}$$

$$\frac{9}{x+2} + \frac{3}{x+2} = \frac{9+3}{x+2}$$

$$= \frac{12}{x+2} \quad \text{Simplify.}$$

> **Adding and Subtracting Rational Expressions with Common Denominators**
>
> If $\dfrac{P}{R}$ and $\dfrac{Q}{R}$ are rational expressions, then
>
> $$\frac{P}{R} + \frac{Q}{R} = \frac{P+Q}{R} \quad \text{and} \quad \frac{P}{R} - \frac{Q}{R} = \frac{P-Q}{R}$$
>
> To add or subtract rational expressions, add or subtract numerators and place the sum or difference over the common denominator.

Example 1 Add: $\dfrac{5m}{2n} + \dfrac{m}{2n}$

Solution:

$$\frac{5m}{2n} + \frac{m}{2n} = \frac{5m + m}{2n} \quad \text{Add the numerators.}$$

$$= \frac{6m}{2n} \quad \text{Simplify the numerator by combining like terms.}$$

$$= \frac{3m}{n} \quad \text{Simplify by applying the fundamental principle.}$$

■ Work Practice 1

Example 2 Subtract: $\dfrac{2y}{2y-7} - \dfrac{7}{2y-7}$

Solution:

$$\frac{2y}{2y-7} - \frac{7}{2y-7} = \frac{2y-7}{2y-7} \quad \text{Subtract the numerators.}$$

$$= \frac{1}{1} \text{ or } 1 \quad \text{Simplify.}$$

■ Work Practice 2

Objectives

A Add and Subtract Rational Expressions with Common Denominators.

B Find the Least Common Denominator of a List of Rational Expressions.

C Write a Rational Expression as an Equivalent Expression Whose Denominator Is Given.

Practice 1

Add: $\dfrac{8x}{3y} + \dfrac{x}{3y}$

Practice 2

Subtract: $\dfrac{3x}{3x-7} - \dfrac{7}{3x-7}$

Answers

1. $\dfrac{3x}{y}$ **2.** 1

1053

Practice 3

Subtract: $\dfrac{2x^2 + 5x}{x + 2} - \dfrac{4x + 6}{x + 2}$

Example 3   Subtract: $\dfrac{3x^2 + 2x}{x - 1} - \dfrac{10x - 5}{x - 1}$

Solution:

$$\dfrac{3x^2 + 2x}{x - 1} - \dfrac{10x - 5}{x - 1} = \dfrac{3x^2 + 2x - (10x - 5)}{x - 1} \qquad \text{Subtract the numerators.}$$
$$\text{Notice the parentheses.}$$

$$= \dfrac{3x^2 + 2x - 10x + 5}{x - 1} \qquad \text{Use the distributive property.}$$

$$= \dfrac{3x^2 - 8x + 5}{x - 1} \qquad \text{Combine like terms.}$$

$$= \dfrac{(x - 1)(3x - 5)}{x - 1} \qquad \text{Factor.}$$

$$= 3x - 5 \qquad \text{Simplify.}$$

■ **Work Practice 3**

Helpful Hint

Notice how the numerator $10x - 5$ was subtracted in Example 3.

This − sign applies to the entire numerator $10x - 5$.

So parentheses are inserted here to indicate this.

$$\dfrac{3x^2 + 2x}{x - 1} - \dfrac{10x - 5}{x - 1} = \dfrac{3x^2 + 2x - (10x - 5)}{x - 1}$$

Objective B   Finding the Least Common Denominator

Recall from Chapter 3 that to add and subtract fractions with different denominators, we first find the least common denominator (LCD). Then we write all fractions as equivalent fractions with the LCD.

For example, suppose we want to add $\dfrac{3}{8}$ and $\dfrac{1}{6}$. To find the LCD of the denominators, factor 8 and 6. Remember, the LCD is the same as the least common multiple, LCM. It is the smallest number that is a multiple of 6 and also 8.

$$8 = 2 \cdot 2 \cdot 2$$
$$6 = 2 \cdot 3$$

The LCM is a multiple of 6.

$$\text{LCM} = 2 \cdot 2 \cdot \overset{\frown}{2 \cdot 3} = 24$$

The LCM is a multiple of 8.

In the next section, we will find the sum $\dfrac{3}{8} + \dfrac{1}{6}$, but for now, let's concentrate on the LCD.

To add or subtract rational expressions with different denominators, we also first find the LCD and then write all rational expressions as equivalent expressions with the LCD. The **least common denominator (LCD) of a list of rational expressions** is a polynomial of least degree whose factors include all the factors of the denominators in the list.

To Find the Least Common Denominator (LCD)

Step 1: Factor each denominator completely.

Step 2: The least common denominator (LCD) is the product of all unique factors found in Step 1, each raised to a power equal to the greatest number of times that the factor appears in any one factored denominator.

Answer

3. $2x - 3$

Example 4 Find the LCD for each pair.

a. $\dfrac{1}{8}, \dfrac{3}{22}$ **b.** $\dfrac{7}{5x}, \dfrac{6}{15x^2}$

Solution:

a. We start by finding the prime factorization of each denominator.

$$8 = 2^3 \quad \text{and}$$
$$22 = 2 \cdot 11$$

Next we write the product of all the unique factors, each raised to a power equal to the greatest number of times that the factor appears.

The greatest number of times that the factor 2 appears is 3.
The greatest number of times that the factor 11 appears is 1.

$$\text{LCD} = 2^3 \cdot 11^1 = 8 \cdot 11 = 88$$

b. We factor each denominator.

$$5x = 5 \cdot x \quad \text{and}$$
$$15x^2 = 3 \cdot 5 \cdot x^2$$

The greatest number of times that the factor 5 appears is 1.
The greatest number of times that the factor 3 appears is 1.
The greatest number of times that the factor x appears is 2.

$$\text{LCD} = 3^1 \cdot 5^1 \cdot x^2 = 15x^2$$

■ **Work Practice 4**

Example 5 Find the LCD of $\dfrac{7x}{x+2}$ and $\dfrac{5x^2}{x-2}$.

Solution: The denominators $x + 2$ and $x - 2$ are completely factored already. The factor $x + 2$ appears once and the factor $x - 2$ appears once.

$$\text{LCD} = (x+2)(x-2)$$

■ **Work Practice 5**

Example 6 Find the LCD of $\dfrac{6m^2}{3m+15}$ and $\dfrac{2}{(m+5)^2}$.

Solution: We factor each denominator.

$$3m + 15 = 3(m+5)$$
$$(m+5)^2 = (m+5)^2 \quad \text{This denominator is already factored.}$$

The greatest number of times that the factor 3 appears is 1.
The greatest number of times that the factor $m + 5$ appears *in any one denominator* is 2.

$$\text{LCD} = 3(m+5)^2$$

■ **Work Practice 6**

✔**Concept Check** Choose the correct LCD of $\dfrac{x}{(x+1)^2}$ and $\dfrac{5}{x+1}$.

  a. $x + 1$ **b.** $(x+1)^2$ **c.** $(x+1)^3$ **d.** $5x(x+1)^2$

Practice 4

Find the LCD for each pair.

a. $\dfrac{2}{9}, \dfrac{7}{15}$

b. $\dfrac{5}{6x^3}, \dfrac{11}{8x^5}$

Practice 5

Find the LCD of $\dfrac{3a}{a+5}$ and $\dfrac{7a}{a-5}$.

Practice 6

Find the LCD of $\dfrac{7x^2}{(x-4)^2}$ and $\dfrac{5x}{3x-12}$.

Answers

4. a. 45   **b.** $24x^5$
5. $(a+5)(a-5)$ **6.** $3(x-4)^2$

✔**Concept Check Answer**

b

Practice 7

Find the LCD of $\dfrac{y + 5}{y^2 + 2y - 3}$

and $\dfrac{y + 4}{y^2 - 3y + 2}$.

Example 7 Find the LCD of $\dfrac{t - 10}{2t^2 + t - 6}$ and $\dfrac{t + 5}{t^2 + 3t + 2}$.

Solution:

$$2t^2 + t - 6 = (2t - 3)(t + 2)$$
$$t^2 + 3t + 2 = (t + 1)(t + 2)$$
$$\text{LCD} = (2t - 3)(t + 2)(t + 1)$$

■ **Work Practice 7**

Practice 8

Find the LCD of $\dfrac{6}{x - 4}$ and $\dfrac{9}{4 - x}$.

Example 8 Find the LCD of $\dfrac{2}{x - 2}$ and $\dfrac{10}{2 - x}$.

Solution: The denominators $x - 2$ and $2 - x$ are opposites. That is, $2 - x = -1(x - 2)$. We can use either $x - 2$ or $2 - x$ as the LCD.

$$\text{LCD} = x - 2 \quad \text{or} \quad \text{LCD} = 2 - x$$

■ **Work Practice 8**

Objective C Writing Equivalent Rational Expressions

Next we practice writing a rational expression as an equivalent rational expression with a given denominator. To do this, we multiply by a form of 1. Recall that multiplying an expression by 1 produces an equivalent expression. In other words,

$$\frac{P}{Q} = \frac{P}{Q} \cdot 1 = \frac{P}{Q} \cdot \frac{R}{R} = \frac{PR}{QR}$$

Practice 9

Write the rational expression as an equivalent rational expression with the given denominator.

$$\frac{2x}{5y} = \frac{}{20x^2y^2}$$

Example 9 Write each rational expression as an equivalent rational expression with the given denominator.

a. $\dfrac{4b}{9a} = \dfrac{}{27a^2b}$ **b.** $\dfrac{7x}{2x + 5} = \dfrac{}{6x + 15}$

Solution:

a. We can ask ourselves: "What do we multiply $9a$ by to get $27a^2b$?" The answer is $3ab$, since $9a(3ab) = 27a^2b$. So we multiply by 1 in the form of $\dfrac{3ab}{3ab}$.

$$\frac{4b}{9a} = \frac{4b}{9a} \cdot 1 = \frac{4b}{9a} \cdot \frac{3ab}{3ab}$$
$$= \frac{4b(3ab)}{9a(3ab)} = \frac{12ab^2}{27a^2b}$$

b. First, factor the denominator on the right.

$$\frac{7x}{2x + 5} = \frac{}{3(2x + 5)}$$

To obtain the denominator on the right from the denominator on the left, we multiply by 1 in the form of $\dfrac{3}{3}$.

$$\frac{7x}{2x + 5} = \frac{7x}{2x + 5} \cdot \frac{3}{3} = \frac{7x \cdot 3}{(2x + 5) \cdot 3} = \frac{21x}{3(2x + 5)}$$

■ **Work Practice 9**

Answers

7. $(y + 3)(y - 1)(y - 2)$

8. $x - 4$ or $4 - x$ 9. $\dfrac{8x^3y}{20x^2y^2}$

Example 10 Write the rational expression as an equivalent rational expression with the given denominator.

$$\frac{5}{x^2 - 4} = \frac{}{(x - 2)(x + 2)(x - 4)}$$

Solution: First we factor the denominator $x^2 - 4$ as $(x - 2)(x + 2)$. If we multiply the original denominator $(x - 2)(x + 2)$ by $x - 4$, the result is the new denominator $(x - 2)(x + 2)(x - 4)$. Thus, we multiply by 1 in the form of $\frac{x - 4}{x - 4}$:

$$\frac{5}{x^2 - 4} = \frac{5}{(x - 2)(x + 2)} = \frac{5}{(x - 2)(x + 2)} \cdot \frac{x - 4}{x - 4}$$

$$= \frac{5(x - 4)}{(x - 2)(x + 2)(x - 4)}$$

$$= \frac{5x - 20}{(x - 2)(x + 2)(x - 4)}$$

■ **Work Practice 10**

Vocabulary, Readiness & Video Check

Use the choices below to fill in each blank. Not all choices will be used.

$$\frac{9}{22} \qquad \frac{5}{22} \qquad \frac{9}{11} \qquad \frac{5}{11} \qquad \frac{ac}{b} \qquad \frac{a - c}{b} \qquad \frac{a + c}{b} \qquad \frac{5 - 6 + x}{x} \qquad \frac{5 - (6 + x)}{x}$$

1. $\dfrac{7}{11} + \dfrac{2}{11} = $ _____

2. $\dfrac{7}{11} - \dfrac{2}{11} = $ _____

3. $\dfrac{a}{b} + \dfrac{c}{b} = $ _____

4. $\dfrac{a}{b} - \dfrac{c}{b} = $ _____

5. $\dfrac{5}{x} - \dfrac{6 + x}{x} = $ _____

Martin-Gay Interactive Videos

See Video 14.3 🍎

Watch the section lecture video and answer the following questions.

Objective A **6.** In ▦ Example 3, why is it important to place parentheses around the second numerator when writing as one expression? ▶

Objective B **7.** In ▦ Examples 4 and 5, we factor the denominators completely. How does this help determine the LCD? ▶

Objective C **8.** Based on ▦ Example 6, complete the following statements: To write an equivalent rational expression, we multiply the _____ of a rational expression by the same expression as the denominator. This means we're multiplying the original rational expression by a factor of _____ and therefore not changing the _____ of the original expression. ▶

14.3 Exercise Set MyMathLab®

Objective A *Add or subtract as indicated. Simplify the result if possible. See Examples 1 through 3.*

1. $\dfrac{a}{13} + \dfrac{9}{13}$

2. $\dfrac{x+1}{7} + \dfrac{6}{7}$

3. $\dfrac{4m}{3n} + \dfrac{5m}{3n}$

4. $\dfrac{3p}{2q} + \dfrac{11p}{2q}$

5. $\dfrac{4m}{m-6} - \dfrac{24}{m-6}$

6. $\dfrac{8y}{y-2} - \dfrac{16}{y-2}$

▶ 7. $\dfrac{9}{3+y} + \dfrac{y+1}{3+y}$

8. $\dfrac{9}{y+9} + \dfrac{y-5}{y+9}$

9. $\dfrac{5x^2+4x}{x-1} - \dfrac{6x+3}{x-1}$

10. $\dfrac{x^2+9x}{x+7} - \dfrac{4x+14}{x+7}$

11. $\dfrac{4a}{a^2+2a-15} - \dfrac{12}{a^2+2a-15}$

12. $\dfrac{3y}{y^2+3y-10} - \dfrac{6}{y^2+3y-10}$

▶ 13. $\dfrac{2x+3}{x^2-x-30} - \dfrac{x-2}{x^2-x-30}$

14. $\dfrac{3x-1}{x^2+5x-6} - \dfrac{2x-7}{x^2+5x-6}$

15. $\dfrac{2x+1}{x-3} + \dfrac{3x+6}{x-3}$

16. $\dfrac{4p-3}{2p+7} + \dfrac{3p+8}{2p+7}$

17. $\dfrac{2x^2}{x-5} - \dfrac{25+x^2}{x-5}$

18. $\dfrac{6x^2}{2x-5} - \dfrac{25+2x^2}{2x-5}$

19. $\dfrac{5x+4}{x-1} - \dfrac{2x+7}{x-1}$

20. $\dfrac{7x+1}{x-4} - \dfrac{2x+21}{x-4}$

Objective B *Find the LCD for each list of rational expressions. See Examples 4 through 8.*

21. $\dfrac{19}{2x}, \dfrac{5}{4x^3}$

22. $\dfrac{17x}{4y^5}, \dfrac{2}{8y}$

▶ 23. $\dfrac{9}{8x}, \dfrac{3}{2x+4}$

24. $\dfrac{1}{6y}, \dfrac{3x}{4y+12}$

25. $\dfrac{2}{x+3}, \dfrac{5}{x-2}$

26. $\dfrac{-6}{x-1}, \dfrac{4}{x+5}$

27. $\dfrac{x}{x+6}, \dfrac{10}{3x+18}$

28. $\dfrac{12}{x+5}, \dfrac{x}{4x+20}$

29. $\dfrac{8x^2}{(x-6)^2}, \dfrac{13x}{5x-30}$

30. $\dfrac{9x^2}{7x-14}, \dfrac{6x}{(x-2)^2}$

▶ 31. $\dfrac{1}{3x+3}, \dfrac{8}{2x^2+4x+2}$

32. $\dfrac{19x+5}{4x-12}, \dfrac{3}{2x^2-12x+18}$

33. $\dfrac{5}{x-8}, \dfrac{3}{8-x}$

34. $\dfrac{2x+5}{3x-7}, \dfrac{5}{7-3x}$

35. $\dfrac{5x+1}{x^2+3x-4}, \dfrac{3x}{x^2+2x-3}$

36. $\dfrac{4}{x^2+4x+3}, \dfrac{4x-2}{x^2+10x+21}$

37. $\dfrac{2x}{3x^2 + 4x + 1}$, $\dfrac{7}{2x^2 - x - 1}$

38. $\dfrac{3x}{4x^2 + 5x + 1}$, $\dfrac{5}{3x^2 - 2x - 1}$

39. $\dfrac{1}{x^2 - 16}$, $\dfrac{x + 6}{2x^3 - 8x^2}$

40. $\dfrac{5}{x^2 - 25}$, $\dfrac{x + 9}{3x^3 - 15x^2}$

Objective **C** *Rewrite each rational expression as an equivalent rational expression with the given denominator.*
See Examples 9 and 10.

41. $\dfrac{3}{2x} = \dfrac{}{4x^2}$

42. $\dfrac{3}{9y^5} = \dfrac{}{72y^9}$

▶ **43.** $\dfrac{6}{3a} = \dfrac{}{12ab^2}$

44. $\dfrac{5}{4y^2x} = \dfrac{}{32y^3x^2}$

45. $\dfrac{9}{2x + 6} = \dfrac{}{2y(x + 3)}$

46. $\dfrac{4x + 1}{3x + 6} = \dfrac{}{3y(x + 2)}$

▶ **47.** $\dfrac{9a + 2}{5a + 10} = \dfrac{}{5b(a + 2)}$

48. $\dfrac{5 + y}{2x^2 + 10} = \dfrac{}{4(x^2 + 5)}$

49. $\dfrac{x}{x^3 + 6x^2 + 8x} = \dfrac{}{x(x + 4)(x + 2)(x + 1)}$

50. $\dfrac{5x}{x^3 + 2x^2 - 3x} = \dfrac{}{x(x - 1)(x - 5)(x + 3)}$

51. $\dfrac{9y - 1}{15x^2 - 30} = \dfrac{}{30x^2 - 60}$

52. $\dfrac{6m - 5}{3x^2 - 9} = \dfrac{}{12x^2 - 36}$

Mixed Practice (*Sections 14.2, 14.3*) *Perform the indicated operations.*

53. $\dfrac{5x}{7} + \dfrac{9x}{7}$

54. $\dfrac{5x}{7} \cdot \dfrac{9x}{7}$

55. $\dfrac{x + 3}{4} \div \dfrac{2x - 1}{4}$

56. $\dfrac{x + 3}{4} - \dfrac{2x - 1}{4}$

57. $\dfrac{x^2}{x - 6} - \dfrac{5x + 6}{x - 6}$

58. $\dfrac{-2x}{x^3 - 8x} + \dfrac{3x}{x^3 - 8x}$

59. $\dfrac{x^2 + 5x}{x^2 - 25} \cdot \dfrac{3x - 15}{x^2}$

60. $\dfrac{-2x}{x^3 - 8x} \div \dfrac{3x}{x^3 - 8x}$

61. $\dfrac{x^3 + 7x^2}{3x^3 - x^2} \div \dfrac{5x^2 + 36x + 7}{9x^2 - 1}$

62. $\dfrac{12x - 6}{x^2 + 3x} \cdot \dfrac{4x^2 + 13x + 3}{4x^2 - 1}$

Review

Perform each indicated operation. See Section 3.3.

63. $\dfrac{2}{3} + \dfrac{5}{7}$

64. $\dfrac{9}{10} - \dfrac{3}{5}$

65. $\dfrac{2}{6} - \dfrac{3}{4}$

66. $\dfrac{11}{15} + \dfrac{5}{9}$

67. $\dfrac{1}{12} + \dfrac{3}{20}$

68. $\dfrac{7}{30} + \dfrac{3}{18}$

Concept Extensions

For Exercises 69 and 70, see the Concept Check in this section.

69. Choose the correct LCD of $\dfrac{11a^3}{4a - 20}$ and $\dfrac{15a^3}{(a - 5)^2}$.

    a. $4a(a - 5)(a + 5)$ **b.** $a - 5$

    c. $(a - 5)^2$ **d.** $4(a - 5)^2$

    e. $(4a - 20)(a - 5)^2$

70. Choose the correct LCD of $\dfrac{5}{14x^2}$ and $\dfrac{y}{6x^3}$.

    a. $84x^5$ **b.** $84x^3$

    c. $42x^3$ **d.** $42x^5$

For Exercises 71 and 72, an algebra student approaches you with each incorrect solution. Find the error and correct the work shown below.

71.
$$\frac{2x - 6}{x - 5} - \frac{x + 4}{x - 5}$$
$$= \frac{2x - 6 - x + 4}{x - 5}$$
$$= \frac{x - 2}{x - 5}$$

72.
$$\frac{x}{x + 3} + \frac{2}{x + 3}$$
$$= \frac{x + 2}{x + 3}$$
$$= \frac{2}{3}$$

△ **73.** A square has a side of length $\dfrac{5}{x - 2}$ meters. Express its perimeter as a rational expression.

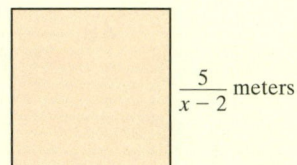

$\dfrac{5}{x - 2}$ meters

△ **74.** A trapezoid has sides of the indicated lengths. Find its perimeter.

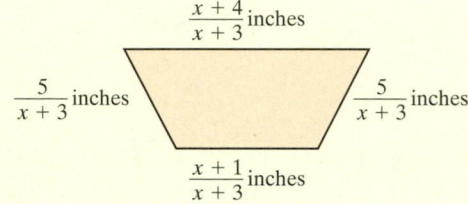

$\dfrac{x + 4}{x + 3}$ inches

$\dfrac{5}{x + 3}$ inches $\dfrac{5}{x + 3}$ inches

$\dfrac{x + 1}{x + 3}$ inches

75. Write two rational expressions with the same denominator whose sum is $\dfrac{5}{3x - 1}$.

76. Write two rational expressions with the same denominator whose difference is $\dfrac{x - 7}{x^2 + 1}$.

77. The planet Mercury revolves around the Sun in 88 Earth days. It takes Jupiter 4332 Earth days to make one revolution around the Sun. (*Source:* National Space Science Data Center) If the two planets are aligned as shown in the figure, how long will it take for them to align again?

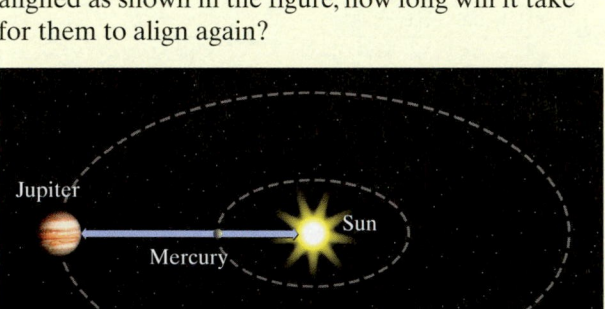

78. You are throwing a barbecue and you want to make sure that you purchase the same number of hot dogs as hot dog buns. Hot dogs come 8 to a package and hot dog buns come 12 to a package. What is the least number of each type of package you should buy?

79. Write some instructions to help a friend who is having difficulty finding the LCD of two rational expressions.

80. In your own words, describe how to add or subtract two rational expressions with the same denominator.

81. Explain why the LCD of the rational expressions $\dfrac{7}{x + 1}$ and $\dfrac{9x}{(x + 1)^2}$ is $(x + 1)^2$ and not $(x + 1)^3$.

82. Explain the similarities between subtracting $\dfrac{3}{8}$ from $\dfrac{7}{8}$ and subtracting $\dfrac{6}{x + 3}$ from $\dfrac{9}{x + 3}$.

14.4 Adding and Subtracting Rational Expressions with Different Denominators

Objective A Adding and Subtracting Rational Expressions with Different Denominators

Let's add $\frac{3}{8}$ and $\frac{1}{6}$. In the previous section, we found the LCD of 8 and 6 to be 24. Now let's write equivalent fractions with denominator 24 by multiplying by different forms of 1.

$$\frac{3}{8} = \frac{3}{8} \cdot 1 = \frac{3}{8} \cdot \frac{3}{3} = \frac{3 \cdot 3}{8 \cdot 3} = \frac{9}{24}$$

$$\frac{1}{6} = \frac{1}{6} \cdot 1 = \frac{1}{6} \cdot \frac{4}{4} = \frac{1 \cdot 4}{6 \cdot 4} = \frac{4}{24}$$

Now that the denominators are the same, we may add.

$$\frac{3}{8} + \frac{1}{6} = \frac{9}{24} + \frac{4}{24} = \frac{9 + 4}{24} = \frac{13}{24}$$

We add or subtract rational expressions the same way. You may want to use the steps below.

To Add or Subtract Rational Expressions with Different Denominators

Step 1: Find the LCD of the rational expressions.

Step 2: Rewrite each rational expression as an equivalent expression whose denominator is the LCD found in Step 1.

Step 3: Add or subtract numerators and write the sum or difference over the common denominator.

Step 4: Simplify or write the rational expression in lowest terms.

Example 1 Perform each indicated operation.

a. $\dfrac{a}{4} - \dfrac{2a}{8}$

b. $\dfrac{3}{10x^2} + \dfrac{7}{25x}$

Solution:

a. First, we must find the LCD. Since $4 = 2^2$ and $8 = 2^3$, the LCD $= 2^3 = 8$. Next we write each fraction as an equivalent fraction with the denominator 8, and then we subtract.

$$\frac{a}{4} = \frac{a}{4} \cdot 1 = \frac{a}{4} \cdot \frac{2}{2} = \frac{a \cdot 2}{4 \cdot 2} = \frac{2a}{8}$$

$$\frac{a}{4} - \frac{2a}{8} = \frac{2a}{8} - \frac{2a}{8} = \frac{2a - 2a}{8} = \frac{0}{8} = 0$$

Notice that we wrote $\dfrac{a}{4}$ as the equivalent expression $\dfrac{2a}{8}$. Multiplying by a form of 1 means we multiply the numerator and the denominator by the same number. Since this is so, we will start using the shorthand notation on the next page.

(Continued on next page)

(Continued on next page)

Objective

A Add and Subtract Rational Expressions with Different Denominators.

Practice 1

Perform each indicated operation.

a. $\dfrac{y}{5} - \dfrac{3y}{15}$

b. $\dfrac{5}{8x} + \dfrac{11}{10x^2}$

Answer

1. a. 0 **b.** $\dfrac{25x + 44}{40x^2}$

$$\frac{a}{4} = \frac{a(2)}{4(2)} = \frac{2a}{8}$$

Multiplying the numerator and denominator by 2 is the same as multiplying by $\frac{2}{2}$ or 1.

b. Since $10x^2 = 2 \cdot 5 \cdot x \cdot x$ and $25x = 5 \cdot 5 \cdot x$, the LCD $= 2 \cdot 5^2 \cdot x^2 = 50x^2$. We write each fraction as an equivalent fraction with a denominator of $50x^2$.

$$\frac{3}{10x^2} + \frac{7}{25x} = \frac{3(5)}{10x^2(5)} + \frac{7(2x)}{25x(2x)}$$

$$= \frac{15}{50x^2} + \frac{14x}{50x^2}$$

$$= \frac{15 + 14x}{50x^2} \qquad \text{Add numerators. Write the sum over the common denominator.}$$

■ **Work Practice 1**

Practice 2

Subtract: $\dfrac{10x}{x^2 - 9} - \dfrac{5}{x + 3}$

Example 2 Subtract: $\dfrac{6x}{x^2 - 4} - \dfrac{3}{x + 2}$

Solution: Since $x^2 - 4 = (x + 2)(x - 2)$, the LCD $= (x + 2)(x - 2)$. We write equivalent expressions with the LCD as denominators.

$$\frac{6x}{x^2 - 4} - \frac{3}{x + 2} = \frac{6x}{(x + 2)(x - 2)} - \frac{3(x - 2)}{(x + 2)(x - 2)}$$

$$= \frac{6x - 3(x - 2)}{(x + 2)(x - 2)} \qquad \text{Subtract numerators. Write the difference over the common denominator.}$$

$$= \frac{6x - 3x + 6}{(x + 2)(x - 2)} \qquad \text{Apply the distributive property in the numerator.}$$

$$= \frac{3x + 6}{(x + 2)(x - 2)} \qquad \text{Combine like terms in the numerator.}$$

Next we factor the numerator to see if this rational expression can be simplified.

$$\frac{3x + 6}{(x + 2)(x - 2)} = \frac{3\,(x + 2)}{(x + 2)\,(x - 2)} \qquad \text{Factor.}$$

$$= \frac{3}{x - 2} \qquad \text{Apply the fundamental principle to simplify.}$$

■ **Work Practice 2**

Practice 3

Add: $\dfrac{5}{7x} + \dfrac{2}{x + 1}$

Example 3 Add: $\dfrac{2}{3t} + \dfrac{5}{t + 1}$

Solution: The LCD is $3t(t + 1)$. We write each rational expression as an equivalent rational expression with a denominator of $3t(t + 1)$.

$$\frac{2}{3t} + \frac{5}{t + 1} = \frac{2(t + 1)}{3t(t + 1)} + \frac{5(3t)}{(t + 1)(3t)}$$

$$= \frac{2(t + 1) + 5(3t)}{3t(t + 1)} \qquad \text{Add numerators. Write the sum over the common denominator.}$$

$$= \frac{2t + 2 + 15t}{3t(t + 1)} \qquad \text{Apply the distributive property in the numerator.}$$

$$= \frac{17t + 2}{3t(t + 1)} \qquad \text{Combine like terms in the numerator.}$$

■ **Work Practice 3**

Answers

2. $\dfrac{5}{x - 3}$ **3.** $\dfrac{19x + 5}{7x(x + 1)}$

Example 4 Subtract: $\dfrac{7}{x-3} - \dfrac{9}{3-x}$

Solution: To find a common denominator, we notice that $x-3$ and $3-x$ are opposites. That is, $3-x = -(x-3)$. We write the denominator $3-x$ as $-(x-3)$ and simplify.

$$\dfrac{7}{x-3} - \dfrac{9}{3-x} = \dfrac{7}{x-3} - \dfrac{9}{-(x-3)}$$

$$= \dfrac{7}{x-3} - \dfrac{-9}{x-3} \qquad \text{Apply } \dfrac{a}{-b} = \dfrac{-a}{b}.$$

$$= \dfrac{7-(-9)}{x-3} \qquad \text{Subtract numerators. Write the difference over the common denominator.}$$

$$= \dfrac{16}{x-3}$$

■ **Work Practice 4**

Practice 4

Subtract: $\dfrac{10}{x-6} - \dfrac{15}{6-x}$

Example 5 Add: $1 + \dfrac{m}{m+1}$

Solution: Recall that 1 is the same as $\dfrac{1}{1}$. The LCD of $\dfrac{1}{1}$ and $\dfrac{m}{m+1}$ is $m+1$.

$$1 + \dfrac{m}{m+1} = \dfrac{1}{1} + \dfrac{m}{m+1} \qquad \text{Write 1 as } \dfrac{1}{1}.$$

$$= \dfrac{1(m+1)}{1(m+1)} + \dfrac{m}{m+1} \qquad \text{Multiply both the numerator and the denominator of } \dfrac{1}{1} \text{ by } m+1.$$

$$= \dfrac{m+1+m}{m+1} \qquad \text{Add numerators. Write the sum over the common denominator.}$$

$$= \dfrac{2m+1}{m+1} \qquad \text{Combine like terms in the numerator.}$$

■ **Work Practice 5**

Practice 5

Add: $2 + \dfrac{x}{x+5}$

Example 6 Subtract: $\dfrac{3}{2x^2+x} - \dfrac{2x}{6x+3}$

Solution: First, we factor the denominators.

$$\dfrac{3}{2x^2+x} - \dfrac{2x}{6x+3} = \dfrac{3}{x(2x+1)} - \dfrac{2x}{3(2x+1)}$$

The LCD is $3x(2x+1)$. We write equivalent expressions with denominator $3x(2x+1)$.

$$\dfrac{3}{x(2x+1)} - \dfrac{2x}{3(2x+1)} = \dfrac{3(3)}{x(2x+1)(3)} - \dfrac{2x(x)}{3(2x+1)(x)}$$

$$= \dfrac{9-2x^2}{3x(2x+1)} \qquad \text{Subtract numerators. Write the difference over the common denominator.}$$

■ **Work Practice 6**

Practice 6

Subtract: $\dfrac{4}{3x^2+2x} - \dfrac{3x}{12x+8}$

Answers

4. $\dfrac{25}{x-6}$ **5.** $\dfrac{3x+10}{x+5}$ **6.** $\dfrac{16-3x^2}{4x(3x+2)}$

Practice 7

Add: $\dfrac{6x}{x^2 + 4x + 4} + \dfrac{x}{x^2 - 4}$

Answer

7. $\dfrac{x(7x - 10)}{(x + 2)^2(x - 2)}$

Example 7 Add: $\dfrac{2x}{x^2 + 2x + 1} + \dfrac{x}{x^2 - 1}$

Solution: First we factor the denominators.

$$\dfrac{2x}{x^2 + 2x + 1} + \dfrac{x}{x^2 - 1}$$

$$= \dfrac{2x}{(x + 1)(x + 1)} + \dfrac{x}{(x + 1)(x - 1)} \qquad \text{Rewrite each expression with LCD } (x + 1)(x + 1)(x - 1).$$

$$= \dfrac{2x(x - 1)}{(x + 1)(x + 1)(x - 1)} + \dfrac{x(x + 1)}{(x + 1)(x - 1)(x + 1)}$$

$$= \dfrac{2x(x - 1) + x(x + 1)}{(x + 1)^2(x - 1)} \qquad \text{Add numerators. Write the sum over the common denominator.}$$

$$= \dfrac{2x^2 - 2x + x^2 + x}{(x + 1)^2(x - 1)} \qquad \text{Apply the distributive property in the numerator.}$$

$$= \dfrac{3x^2 - x}{(x + 1)^2(x - 1)} \quad \text{or} \quad \dfrac{x(3x - 1)}{(x + 1)^2(x - 1)}$$

The numerator was factored as a last step to see if the rational expression could be simplified further. Since there are no factors common to the numerator and the denominator, we can't simplify further.

■ **Work Practice 7**

Vocabulary, Readiness & Video Check

Multiple choice. Choose the correct response.

1. $\dfrac{3}{7x} + \dfrac{5}{7} =$

    a. $\dfrac{3}{7x} + \dfrac{5}{7x} = \dfrac{8}{7x}$
    b. $\dfrac{3}{7x} + \dfrac{5}{7} \cdot \dfrac{x}{x} = \dfrac{3 + 5x}{7x}$
    c. $\dfrac{3}{7x} + \dfrac{5}{7} \cdot \dfrac{x}{x} = \dfrac{8x}{7x}$ or $\dfrac{8}{7}$

2. $\dfrac{1}{x} + \dfrac{2}{x^2} =$

    a. $\dfrac{1}{x} \cdot \dfrac{x}{x} + \dfrac{2}{x^2} = \dfrac{x + 2}{x^2}$
    b. $\dfrac{3}{x^3}$
    c. $\dfrac{1}{x} \cdot \dfrac{x}{x} + \dfrac{2}{x^2} = \dfrac{3x}{x^2}$ or $\dfrac{3}{x}$

Martin-Gay Interactive Videos Watch the section lecture video and answer the following question.

Objective A 3. What special case is shown in ▯ Example 2, and what's the purpose of presenting it? ▶

See Video 14.4 🍎

14.4 Exercise Set MyMathLab®

Objective A *Perform each indicated operation. Simplify if possible. See Examples 1 through 7.*

1. $\dfrac{4}{2x} + \dfrac{9}{3x}$

2. $\dfrac{15}{7a} + \dfrac{8}{6a}$

3. $\dfrac{15a}{b} + \dfrac{6b}{5}$

4. $\dfrac{4c}{d} - \dfrac{8d}{5}$

5. $\dfrac{3}{x} + \dfrac{5}{2x^2}$

6. $\dfrac{14}{3x^2} + \dfrac{6}{x}$

7. $\dfrac{6}{x+1} + \dfrac{10}{2x+2}$

8. $\dfrac{8}{x+4} - \dfrac{3}{3x+12}$

9. $\dfrac{3}{x+2} - \dfrac{2x}{x^2-4}$

10. $\dfrac{5}{x-4} + \dfrac{4x}{x^2-16}$

11. $\dfrac{3}{4x} + \dfrac{8}{x-2}$

12. $\dfrac{5}{y^2} - \dfrac{y}{2y+1}$

13. $\dfrac{6}{x-3} + \dfrac{8}{3-x}$

14. $\dfrac{15}{y-4} + \dfrac{20}{4-y}$

15. $\dfrac{9}{x-3} + \dfrac{9}{3-x}$

16. $\dfrac{5}{a-7} + \dfrac{5}{7-a}$

17. $\dfrac{-8}{x^2-1} - \dfrac{7}{1-x^2}$

18. $\dfrac{-9}{25x^2-1} + \dfrac{7}{1-25x^2}$

19. $\dfrac{5}{x} + 2$

20. $\dfrac{7}{x^2} - 5x$

21. $\dfrac{5}{x-2} + 6$

22. $\dfrac{6y}{y+5} + 1$

23. $\dfrac{y+2}{y+3} - 2$

24. $\dfrac{7}{2x-3} - 3$

25. $\dfrac{-x+2}{x} - \dfrac{x-6}{4x}$

26. $\dfrac{-y+1}{y} - \dfrac{2y-5}{3y}$

27. $\dfrac{5x}{x+2} - \dfrac{3x-4}{x+2}$

28. $\dfrac{7x}{x-3} - \dfrac{4x+9}{x-3}$

29. $\dfrac{3x^4}{7} - \dfrac{4x^2}{21}$

30. $\dfrac{5x}{6} + \dfrac{11x^2}{2}$

31. $\dfrac{1}{x+3} - \dfrac{1}{(x+3)^2}$

32. $\dfrac{5x}{(x-2)^2} - \dfrac{3}{x-2}$

33. $\dfrac{4}{5b} + \dfrac{1}{b-1}$

34. $\dfrac{1}{y+5} + \dfrac{2}{3y}$

35. $\dfrac{2}{m} + 1$

36. $\dfrac{6}{x} - 1$

37. $\dfrac{2x}{x-7} - \dfrac{x}{x-2}$

38. $\dfrac{9x}{x-10} - \dfrac{x}{x-3}$

39. $\dfrac{6}{1-2x} - \dfrac{4}{2x-1}$

40. $\dfrac{10}{3n-4} - \dfrac{5}{4-3n}$

41. $\dfrac{7}{(x+1)(x-1)} + \dfrac{8}{(x+1)^2}$

42. $\dfrac{5}{(x+1)(x+5)} - \dfrac{2}{(x+5)^2}$

43. $\dfrac{x}{x^2-1} - \dfrac{2}{x^2-2x+1}$

44. $\dfrac{x}{x^2-4} - \dfrac{5}{x^2-4x+4}$

▶ **45.** $\dfrac{3a}{2a+6} - \dfrac{a-1}{a+3}$

46. $\dfrac{1}{2x+2y} - \dfrac{y}{x+y}$

47. $\dfrac{y-1}{2y+3} + \dfrac{3}{(2y+3)^2}$

48. $\dfrac{x-6}{5x+1} + \dfrac{6}{(5x+1)^2}$

49. $\dfrac{5}{2-x} + \dfrac{x}{2x-4}$

50. $\dfrac{-1}{a-2} + \dfrac{4}{4-2a}$

51. $\dfrac{15}{x^2+6x+9} + \dfrac{2}{x+3}$

52. $\dfrac{2}{x^2+4x+4} + \dfrac{1}{x+2}$

53. $\dfrac{13}{x^2-5x+6} - \dfrac{5}{x-3}$

54. $\dfrac{-7}{y^2-3y+2} - \dfrac{2}{y-1}$

55. $\dfrac{70}{m^2-100} + \dfrac{7}{2(m+10)}$

56. $\dfrac{27}{y^2-81} + \dfrac{3}{2(y+9)}$

▶ **57.** $\dfrac{x+8}{x^2-5x-6} + \dfrac{x+1}{x^2-4x-5}$

58. $\dfrac{x+4}{x^2+12x+20} + \dfrac{x+1}{x^2+8x-20}$

59. $\dfrac{5}{4n^2-12n+8} - \dfrac{3}{3n^2-6n}$

60. $\dfrac{6}{5y^2-25y+30} - \dfrac{2}{4y^2-8y}$

Mixed Practice (*Sections 14.2, 14.3, 14.4*) *Perform the indicated operations. Addition, subtraction, multiplication, and division of rational expressions are included here.*

61. $\dfrac{15x}{x+8} \cdot \dfrac{2x+16}{3x}$

62. $\dfrac{9z+5}{15} \cdot \dfrac{5z}{81z^2-25}$

63. $\dfrac{8x+7}{3x+5} - \dfrac{2x-3}{3x+5}$

64. $\dfrac{2z^2}{4z-1} - \dfrac{z-2z^2}{4z-1}$

65. $\dfrac{5a+10}{18} \div \dfrac{a^2-4}{10a}$

66. $\dfrac{9}{x^2-1} \div \dfrac{12}{3x+3}$

67. $\dfrac{5}{x^2-3x+2} + \dfrac{1}{x-2}$

68. $\dfrac{4}{2x^2+5x-3} + \dfrac{2}{x+3}$

Review

Solve each linear or quadratic equation. See Sections 9.3 and 13.6.

69. $3x + 5 = 7$

70. $5x - 1 = 8$

71. $2x^2 - x - 1 = 0$

72. $4x^2 - 9 = 0$

73. $4(x + 6) + 3 = -3$

74. $2(3x + 1) + 15 = -7$

Concept Extensions

Perform each indicated operation.

75. $\dfrac{3}{x} - \dfrac{2x}{x^2 - 1} + \dfrac{5}{x + 1}$

76. $\dfrac{5}{x - 2} + \dfrac{7x}{x^2 - 4} - \dfrac{11}{x}$

77. $\dfrac{5}{x^2 - 4} + \dfrac{2}{x^2 - 4x + 4} - \dfrac{3}{x^2 - x - 6}$

78. $\dfrac{8}{x^2 + 6x + 5} - \dfrac{3x}{x^2 + 4x - 5} + \dfrac{2}{x^2 - 1}$

79. $\dfrac{9}{x^2 + 9x + 14} - \dfrac{3x}{x^2 + 10x + 21} + \dfrac{x + 4}{x^2 + 5x + 6}$

80. $\dfrac{x + 10}{x^2 - 3x - 4} - \dfrac{8}{x^2 + 6x + 5} - \dfrac{9}{x^2 + x - 20}$

81. A board of length $\dfrac{3}{x + 4}$ inches was cut into two pieces. If one piece is $\dfrac{1}{x - 4}$ inches, express the length of the other piece as a rational expression.

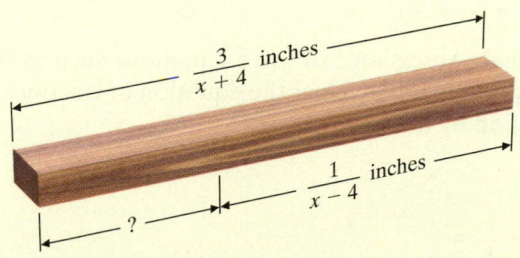

△ 82. The length of a rectangle is $\dfrac{3}{y - 5}$ feet, while its width is $\dfrac{2}{y}$ feet. Find its perimeter and then find its area.

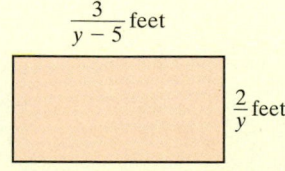

83. In ice hockey, penalty killing percentage is a statistic calculated as $1 - \dfrac{G}{P}$, where $G =$ opponent's power play goals and $P =$ opponent's power play opportunities. Simplify this expression.

84. The dose of medicine prescribed for a child depends on the child's age A in years and the adult dose D for the medication. Two expressions that give a child's dose are Young's Rule, $\dfrac{DA}{A + 12}$, and Cowling's Rule, $\dfrac{D(A + 1)}{24}$. Find an expression for the difference in the doses given by these expressions.

85. Explain when the LCD of the rational expressions in a sum is the product of the denominators.

86. Explain when the LCD is the same as one of the denominators of a rational expression to be added or subtracted.

△ **87.** Two angles are said to be complementary if the sum of their measures is 90°. If one angle measures $\frac{40}{x}$ degrees, find the measure of its complement.

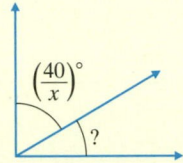

△ **88.** Two angles are said to be supplementary if the sum of their measures is 180°. If one angle measures $\frac{x + 2}{x}$ degrees, find the measure of its supplement.

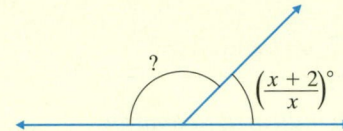

89. In your own words, explain how to add two rational expressions with different denominators.

90. In your own words, explain how to subtract two rational expressions with different denominators.

14.5 Solving Equations Containing Rational Expressions

Objectives

A Solve Equations Containing Rational Expressions.

B Solve Equations Containing Rational Expressions for a Specified Variable.

Objective A Solving Equations Containing Rational Expressions

In Chapter 9, we solved equations containing fractions. In this section, we continue the work we began in those chapters by solving equations containing rational expressions. For example,

$$\frac{x}{2} + \frac{8}{3} = \frac{1}{6} \quad \text{and} \quad \frac{4x}{x^2 + x - 30} + \frac{2}{x - 5} = \frac{1}{x + 6}$$

are equations containing rational expressions. To solve equations such as these, we use the multiplication property of equality to clear the equation of fractions by multiplying both sides of the equation by the LCD.

Practice 1

Solve: $\frac{x}{4} + \frac{4}{5} = \frac{1}{20}$

Example 1 Solve: $\frac{x}{2} + \frac{8}{3} = \frac{1}{6}$

Solution: The LCD of denominators 2, 3, and 6 is 6, so we multiply both sides of the equation by 6.

$$6\left(\frac{x}{2} + \frac{8}{3}\right) = 6\left(\frac{1}{6}\right)$$

Helpful Hint
Make sure that *each* term is multiplied by the LCD.

$$6\left(\frac{x}{2}\right) + 6\left(\frac{8}{3}\right) = 6\left(\frac{1}{6}\right) \quad \text{Apply the distributive property.}$$

$$3 \cdot x + 16 = 1 \quad \text{Multiply and simplify.}$$

$$3x = -15 \quad \text{Subtract 16 from both sides.}$$

$$x = -5 \quad \text{Divide both sides by 3.}$$

Answer

1. $x = -3$

Check: To check, we replace x with -5 in the original equation.

$$\frac{-5}{2} + \frac{8}{3} \stackrel{?}{=} \frac{1}{6} \qquad \text{Replace } x \text{ with } -5.$$

$$\frac{1}{6} = \frac{1}{6} \qquad \text{True}$$

This number checks, so the solution is -5.

■ **Work Practice 1**

Example 2 Solve: $\dfrac{t-4}{2} - \dfrac{t-3}{9} = \dfrac{5}{18}$

Solution: The LCD of denominators 2, 9, and 18 is 18, so we multiply both sides of the equation by 18.

$$18\left(\frac{t-4}{2} - \frac{t-3}{9}\right) = 18\left(\frac{5}{18}\right)$$

$$18\left(\frac{t-4}{2}\right) - 18\left(\frac{t-3}{9}\right) = 18\left(\frac{5}{18}\right) \qquad \text{Apply the distributive property.}$$

$$9(t-4) - 2(t-3) = 5 \qquad \text{Simplify.}$$

$$9t - 36 - 2t + 6 = 5 \qquad \text{Use the distributive property.}$$

$$7t - 30 = 5 \qquad \text{Combine like terms.}$$

$$7t = 35$$

$$t = 5 \qquad \text{Solve for } t.$$

Check:

$$\frac{t-4}{2} - \frac{t-3}{9} = \frac{5}{18}$$

$$\frac{5-4}{2} - \frac{5-3}{9} \stackrel{?}{=} \frac{5}{18} \qquad \text{Replace } t \text{ with 5.}$$

$$\frac{1}{2} - \frac{2}{9} \stackrel{?}{=} \frac{5}{18} \qquad \text{Simplify.}$$

$$\frac{5}{18} = \frac{5}{18} \qquad \text{True}$$

The solution is 5.

■ **Work Practice 2**

Practice 2

Solve: $\dfrac{x+2}{3} - \dfrac{x-1}{5} = \dfrac{1}{15}$

> **Helpful Hint**
> Multiply *each* term by 18.

Recall from Section 14.1 that a rational expression is defined for all real numbers except those that make the denominator of the expression 0. This means that if an equation contains *rational expressions with variables in the denominator,* we must be certain that the proposed solution does not make the denominator 0. If replacing the variable with the proposed solution makes the denominator 0, the rational expression is undefined and this proposed solution must be rejected.

Answer

2. $x = -6$

Practice 3

Solve: $2 + \dfrac{6}{x} = x + 7$

Example 3   Solve: $3 - \dfrac{6}{x} = x + 8$

Helpful Hint   Notice that Example 3 contains our first equation with a variable in the denominator.

Solution: In this equation, 0 cannot be a solution because if x is 0, the rational expression $\dfrac{6}{x}$ is undefined. The LCD is x, so we multiply both sides of the equation by x.

Helpful Hint   Multiply *each* term by x.

$$x\left(3 - \dfrac{6}{x}\right) = x(x + 8)$$

$$x(3) - x\left(\dfrac{6}{x}\right) = x \cdot x + x \cdot 8 \quad \text{Apply the distributive property.}$$

$$3x - 6 = x^2 + 8x \quad \text{Simplify.}$$

Now we write the quadratic equation in standard form and solve for x.

$$0 = x^2 + 5x + 6$$

$$0 = (x + 3)(x + 2) \quad \text{Factor.}$$

$$x + 3 = 0 \quad \text{or} \quad x + 2 = 0 \quad \text{Set each factor equal to 0 and solve.}$$

$$x = -3 \qquad\qquad x = -2$$

Notice that neither -3 nor -2 makes the denominator in the original equation equal to 0.

Check: To check these solutions, we replace x in the original equation by -3, and then by -2.

If $x = -3$:

$$3 - \dfrac{6}{x} = x + 8$$

$$3 - \dfrac{6}{-3} \stackrel{?}{=} -3 + 8$$

$$3 - (-2) \stackrel{?}{=} 5$$

$$5 = 5 \quad \text{True}$$

If $x = -2$:

$$3 - \dfrac{6}{x} = x + 8$$

$$3 - \dfrac{6}{-2} \stackrel{?}{=} -2 + 8$$

$$3 - (-3) \stackrel{?}{=} 6$$

$$6 = 6 \quad \text{True}$$

Both -3 and -2 are solutions.

■ **Work Practice 3**

The following steps may be used to solve an equation containing rational expressions.

To Solve an Equation Containing Rational Expressions

Step 1: Multiply both sides of the equation by the LCD of all rational expressions in the equation.

Step 2: Remove any grouping symbols and solve the resulting equation.

Step 3: Check the solution in the original equation.

Answer

3. $x = -6, x = 1$

Vocabulary, Readiness & Video Check

Multiple choice. Choose the correct response.

1. Multiply both sides of the equation $\dfrac{3x}{2} + 5 = \dfrac{1}{4}$ by 4. The result is:

   a. $3x + 5 = 1$ **b.** $6x + 5 = 1$ **c.** $6x + 20 = 1$ **d.** $6x + 9 = 1$

2. Multiply both sides of the equation $\dfrac{1}{x} - \dfrac{3}{5x} = 2$ by $5x$. The result is:

   a. $1 - 3 = 10x$ **b.** $5 - 3 = 10x$ **c.** $5x - 3 = 10x$ **d.** $5 - 3 = 7x$

Choose the correct LCD for the fractions in each equation.

3. Equation: $\dfrac{9}{x} + \dfrac{3}{4} = \dfrac{1}{12}$; LCD: _____

   a. $4x$ **b.** $12x$ **c.** $48x$ **d.** x

4. Equation: $\dfrac{8}{3x} - \dfrac{1}{x} = \dfrac{7}{9}$; LCD: _____

   a. x **b.** $3x$ **c.** $27x$ **d.** $9x$

5. Equation: $\dfrac{9}{x - 1} = \dfrac{7}{(x - 1)^2}$; LCD: _____

   a. $(x - 1)^2$ **b.** $(x - 1)$ **c.** $(x - 1)^3$ **d.** 63

6. Equation: $\dfrac{1}{x - 2} - \dfrac{3}{x^2 - 4} = 8$; LCD: _____

   a. $(x - 2)$ **b.** $(x + 2)$ **c.** $(x^2 - 4)$ **d.** $(x - 2)(x^2 - 4)$

Martin-Gay Interactive Videos

See Video 14.5 🍎

Watch the section lecture video and answer the following questions.

Objective A   7. After multiplying through by the LCD and then simplifying, why is it important to take a moment and determine whether we have a linear or a quadratic equation before we finish solving the problem? ▶

     8. From ⊞ Examples 2–5, what extra step is needed when checking solutions to an equation containing rational expressions? ▶

Objective B   9. The steps for solving ⊞ Example 6 for a specified variable are the same as what other steps? How do we treat this specified variable? ▶

14.5 Exercise Set MyMathLab®

Objective A *Solve each equation and check each solution. See Examples 1 through 3.*

1. $\dfrac{x}{5} + 3 = 9$

2. $\dfrac{x}{5} - 2 = 9$

3. $\dfrac{x}{2} + \dfrac{5x}{4} = \dfrac{x}{12}$

4. $\dfrac{x}{6} + \dfrac{4x}{3} = \dfrac{x}{18}$

5. $2 - \dfrac{8}{x} = 6$

6. $5 + \dfrac{4}{x} = 1$

7. $2 + \dfrac{10}{x} = x + 5$

8. $6 + \dfrac{5}{y} = y - \dfrac{2}{y}$

9. $\dfrac{a}{5} = \dfrac{a-3}{2}$

10. $\dfrac{b}{5} = \dfrac{b+2}{6}$

◐ 11. $\dfrac{x-3}{5} + \dfrac{x-2}{2} = \dfrac{1}{2}$

12. $\dfrac{a+5}{4} + \dfrac{a+5}{2} = \dfrac{a}{8}$

Solve each equation and check each proposed solution. See Examples 4 through 6.

13. $\dfrac{3}{2a-5} = -1$

14. $\dfrac{6}{4-3x} = -3$

15. $\dfrac{4y}{y-4} + 5 = \dfrac{5y}{y-4}$

16. $\dfrac{2a}{a+2} - 5 = \dfrac{7a}{a+2}$

◐ 17. $2 + \dfrac{3}{a-3} = \dfrac{a}{a-3}$

18. $\dfrac{2y}{y-2} - \dfrac{4}{y-2} = 4$

19. $\dfrac{1}{x+3} + \dfrac{6}{x^2-9} = 1$

20. $\dfrac{1}{x+2} + \dfrac{4}{x^2-4} = 1$

21. $\dfrac{2y}{y+4} + \dfrac{4}{y+4} = 3$

22. $\dfrac{5y}{y+1} - \dfrac{3}{y+1} = 4$

23. $\dfrac{2x}{x+2} - 2 = \dfrac{x-8}{x-2}$

24. $\dfrac{4y}{y-3} - 3 = \dfrac{3y-1}{y+3}$

Solve each equation. See Examples 1 through 6.

◐ 25. $\dfrac{2}{y} + \dfrac{1}{2} = \dfrac{5}{2y}$

26. $\dfrac{6}{3y} + \dfrac{3}{y} = 1$

27. $\dfrac{a}{a-6} = \dfrac{-2}{a-1}$

28. $\dfrac{5}{x-6} = \dfrac{x}{x-2}$

29. $\dfrac{11}{2x} + \dfrac{2}{3} = \dfrac{7}{2x}$

30. $\dfrac{5}{3} - \dfrac{3}{2x} = \dfrac{3}{2}$

31. $\dfrac{2}{x-2} + 1 = \dfrac{x}{x+2}$

32. $1 + \dfrac{3}{x+1} = \dfrac{x}{x-1}$

33. $\dfrac{x+1}{3} - \dfrac{x-1}{6} = \dfrac{1}{6}$

34. $\dfrac{3x}{5} - \dfrac{x-6}{3} = -\dfrac{2}{5}$

◐ 35. $\dfrac{t}{t-4} = \dfrac{t+4}{6}$

36. $\dfrac{15}{x+4} = \dfrac{x-4}{x}$

37. $\dfrac{y}{2y+2} + \dfrac{2y-16}{4y+4} = \dfrac{2y-3}{y+1}$

38. $\dfrac{1}{x+2} = \dfrac{4}{x^2-4} - \dfrac{1}{x-2}$

39. $\dfrac{4r - 4}{r^2 + 5r - 14} + \dfrac{2}{r + 7} = \dfrac{1}{r - 2}$

40. $\dfrac{3}{x + 3} = \dfrac{12x + 19}{x^2 + 7x + 12} - \dfrac{5}{x + 4}$

41. $\dfrac{x + 1}{x + 3} = \dfrac{x^2 - 11x}{x^2 + x - 6} - \dfrac{x - 3}{x - 2}$

42. $\dfrac{2t + 3}{t - 1} - \dfrac{2}{t + 3} = \dfrac{5 - 6t}{t^2 + 2t - 3}$

Objective B *Solve each equation for the indicated variable. See Example 7.*

43. $R = \dfrac{E}{I}$ for I (Electronics: resistance of a circuit)

44. $T = \dfrac{V}{Q}$ for Q (Water purification: settling time)

45. $T = \dfrac{2U}{B + E}$ for B (Merchandising: stock turnover rate)

46. $i = \dfrac{A}{t + B}$ for t (Hydrology: rainfall intensity)

47. $B = \dfrac{705w}{h^2}$ for w (Health: body-mass index)

△ **48.** $\dfrac{A}{W} = L$ for W (Geometry: area of a rectangle)

49. $N = R + \dfrac{V}{G}$ for G (Urban forestry: tree plantings per year)

50. $C = \dfrac{D(A + 1)}{24}$ for A (Medicine: Cowling's Rule for child's dose)

△ **51.** $\dfrac{C}{\pi r} = 2$ for r (Geometry: circumference of a circle)

52. $W = \dfrac{CE^2}{2}$ for C (Electronics: energy stored in a capacitor)

53. $\dfrac{1}{y} + \dfrac{1}{3} = \dfrac{1}{x}$ for x

54. $\dfrac{1}{5} + \dfrac{2}{y} = \dfrac{1}{x}$ for x

Review

Translating *Write each phrase as an expression. See Section 8.2.*

55. The reciprocal of x

56. The reciprocal of $x + 1$

57. The reciprocal of x, added to the reciprocal of 2

58. The reciprocal of x, subtracted from the reciprocal of 5

Answer each question.

59. If a tank is filled in 3 hours, what part of the tank is filled in 1 hour?

60. If a strip of beach is cleaned in 4 hours, what part of the beach is cleaned in 1 hour?

Concept Extensions

61. Explain the difference between solving an equation such as $\dfrac{x}{2} + \dfrac{3}{4} = \dfrac{x}{4}$ for x and performing an operation such as adding $\dfrac{x}{2} + \dfrac{3}{4}$.

62. When solving an equation such as $\dfrac{y}{4} = \dfrac{y}{2} - \dfrac{1}{4}$, we may multiply all terms by 4. When subtracting two rational expressions such as $\dfrac{y}{2} - \dfrac{1}{4}$, we may not. Explain why.

Determine whether each of the following is an equation or an expression. If it is an equation, then solve it for its variable. If it is an expression, perform the indicated operation.

△ **63.** $\dfrac{1}{x} + \dfrac{5}{9}$

64. $\dfrac{1}{x} + \dfrac{5}{9} = \dfrac{2}{3}$

65. $\dfrac{5}{x-1} - \dfrac{2}{x} = \dfrac{5}{x(x-1)}$

66. $\dfrac{5}{x-1} - \dfrac{2}{x}$

Recall that two angles are supplementary if the sum of their measures is 180°. Find the measures of the supplementary angles.

△ **67.**

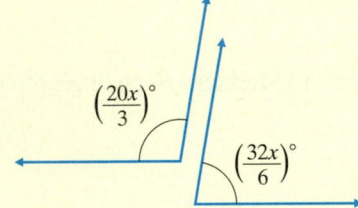

△ **68.**

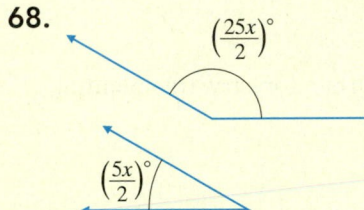

Recall that two angles are complementary if the sum of their measures is 90°. Find the measures of the complementary angles.

△ **69.**

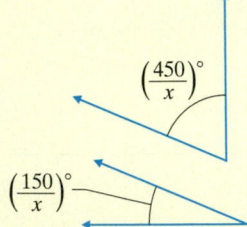

△ **70.**

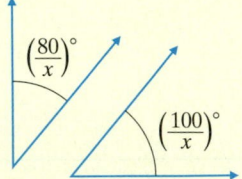

Solve each equation.

71. $\dfrac{4}{a^2 + 4a + 3} + \dfrac{2}{a^2 + a - 6} - \dfrac{3}{a^2 - a - 2} = 0$

72. $\dfrac{-4}{a^2 + 2a - 8} + \dfrac{1}{a^2 + 9a + 20} = \dfrac{-4}{a^2 + 3a - 10}$

Summary on Rational Expressions

It is important to know the difference between performing operations with rational expressions and solving an equation containing rational expressions. Study the examples below.

Performing Operations with Rational Expressions

Adding: $\dfrac{1}{x} + \dfrac{1}{x+5} = \dfrac{1 \cdot (x+5)}{x(x+5)} + \dfrac{1 \cdot x}{x(x+5)} = \dfrac{x+5+x}{x(x+5)} = \dfrac{2x+5}{x(x+5)}$

Subtracting: $\dfrac{3}{x} - \dfrac{5}{x^2 y} = \dfrac{3 \cdot xy}{x \cdot xy} - \dfrac{5}{x^2 y} = \dfrac{3xy - 5}{x^2 y}$

Multiplying: $\dfrac{2}{x} \cdot \dfrac{5}{x-1} = \dfrac{2 \cdot 5}{x(x-1)} = \dfrac{10}{x(x-1)}$

Dividing: $\dfrac{4}{2x+1} \div \dfrac{x-3}{x} = \dfrac{4}{2x+1} \cdot \dfrac{x}{x-3} = \dfrac{4x}{(2x+1)(x-3)}$

Solving an Equation Containing Rational Expressions

To solve an equation containing rational expressions, we clear the equation of fractions by multiplying both sides by the LCD.

$$\frac{3}{x} - \frac{5}{x-1} = \frac{1}{x(x-1)} \qquad \text{Note that } x \text{ can't be 0 or 1.}$$

$$x(x-1)\left(\frac{3}{x}\right) - x(x-1)\left(\frac{5}{x-1}\right) = x(x-1) \cdot \frac{1}{x(x-1)} \qquad \text{Multiply both sides by the LCD.}$$

$$3(x-1) - 5x = 1 \qquad \text{Simplify.}$$

$$3x - 3 - 5x = 1 \qquad \text{Use the distributive property.}$$

$$-2x - 3 = 1 \qquad \text{Combine like terms.}$$

$$-2x = 4 \qquad \text{Add 3 to both sides.}$$

$$x = -2 \qquad \text{Divide both sides by } -2.$$

Don't forget to check to make sure our proposed solution of -2 does not make any denominators 0. If it does, this proposed solution is *not* a solution of the equation. -2 checks and is the solution.

Determine whether each of the following is an equation or an expression. If it is an equation, solve it for its variable. If it is an expression, perform the indicated operation.

1. $\dfrac{1}{x} + \dfrac{2}{3}$

2. $\dfrac{3}{a} + \dfrac{5}{6}$

3. $\dfrac{1}{x} + \dfrac{2}{3} = \dfrac{3}{x}$

4. $\dfrac{3}{a} + \dfrac{5}{6} = 1$

5. $\dfrac{2}{x+1} - \dfrac{1}{x}$

6. $\dfrac{4}{x-3} - \dfrac{1}{x}$

7. $\dfrac{2}{x+1} - \dfrac{1}{x} = 1$

8. $\dfrac{4}{x-3} - \dfrac{1}{x} = \dfrac{6}{x(x-3)}$

9. $\dfrac{15x}{x+8} \cdot \dfrac{2x+16}{3x}$

10. $\dfrac{9z+5}{15} \cdot \dfrac{5z}{81z^2 - 25}$

Answers

1. _____

2. _____

3. _____

4. _____

5. _____

6. _____

7. _____

8. _____

9. _____

10. _____

11. _____

12. _____

13. _____

14. _____

15. _____

16. _____

17. _____

18. _____

19. _____

20. _____

21. _____

22. _____

23. _____

24. _____

11. $\dfrac{2x + 1}{x - 3} + \dfrac{3x + 6}{x - 3}$

12. $\dfrac{4p - 3}{2p + 7} + \dfrac{3p + 8}{2p + 7}$

13. $\dfrac{x + 5}{7} = \dfrac{8}{2}$

14. $\dfrac{1}{2} = \dfrac{x + 1}{8}$

15. $\dfrac{5a + 10}{18} \div \dfrac{a^2 - 4}{10a}$

16. $\dfrac{9}{x^2 - 1} \div \dfrac{12}{3x + 3}$

17. $\dfrac{x + 2}{3x - 1} + \dfrac{5}{(3x - 1)^2}$

18. $\dfrac{4}{(2x - 5)^2} + \dfrac{x + 1}{2x - 5}$

19. $\dfrac{x - 7}{x} - \dfrac{x + 2}{5x}$

20. $\dfrac{10x - 9}{x} - \dfrac{x - 4}{3x}$

21. $\dfrac{3}{x + 3} = \dfrac{5}{x^2 - 9} - \dfrac{2}{x - 3}$

22. $\dfrac{9}{x^2 - 4} + \dfrac{2}{x + 2} = \dfrac{-1}{x - 2}$

23. Explain the difference between solving an equation such as $\dfrac{x}{3} + \dfrac{1}{6} = \dfrac{x}{6}$ for x and performing an operation such as adding $\dfrac{x}{3} + \dfrac{1}{6}$.

24. When solving an equation such as $\dfrac{y}{6} = \dfrac{y}{3} - \dfrac{1}{6}$, we may multiply all terms by 6. When subtracting two rational expressions such as $\dfrac{y}{3} - \dfrac{1}{6}$, we may not. Explain why.

Objective A Solving Problems About Numbers

In this section, we solve problems that can be modeled by equations containing rational expressions. To solve these problems, we use the same problem-solving steps that were first introduced in Section 1.8. In our first example, our goal is to find an unknown number.

Example 1 Finding an Unknown Number

The quotient of a number and 6, minus $\frac{5}{3}$, is the quotient of the number and 2. Find the number.

Solution:

1. UNDERSTAND. Read and reread the problem. Suppose that the unknown number is 2; then we see if the quotient of 2 and 6, or $\frac{2}{6}$, minus $\frac{5}{3}$ is equal to the quotient of 2 and 2, or $\frac{2}{2}$.

$$\frac{2}{6} - \frac{5}{3} = \frac{1}{3} - \frac{5}{3} = -\frac{4}{3}, \text{ not } \frac{2}{2}$$

Don't forget that the purpose of a proposed solution is to better understand the problem.

Let x = the unknown number.

2. TRANSLATE.

In words: | the quotient of x and 6 | minus | $\frac{5}{3}$ | is | the quotient of x and 2 |

Translate: $\frac{x}{6}$ $-$ $\frac{5}{3}$ $=$ $\frac{x}{2}$

3. SOLVE. Here, we solve the equation $\frac{x}{6} - \frac{5}{3} = \frac{x}{2}$. We begin by multiplying both sides of the equation by the LCD, 6.

$$6\left(\frac{x}{6} - \frac{5}{3}\right) = 6\left(\frac{x}{2}\right)$$

$$6\left(\frac{x}{6}\right) - 6\left(\frac{5}{3}\right) = 6\left(\frac{x}{2}\right) \quad \text{Apply the distributive property.}$$

$$x - 10 = 3x \quad \text{Simplify.}$$

$$-10 = 2x \quad \text{Subtract } x \text{ from both sides.}$$

$$\frac{-10}{2} = \frac{2x}{2} \quad \text{Divide both sides by 2.}$$

$$-5 = x \quad \text{Simplify.}$$

4. INTERPRET.

Check: To check, we verify that "the quotient of -5 and 6 minus $\frac{5}{3}$ is the quotient of -5 and 2," or $-\frac{5}{6} - \frac{5}{3} = -\frac{5}{2}$.

State: The unknown number is -5.

■ **Work Practice 1**

Objectives

A Solve Problems About Numbers.

B Solve Problems About Work.

C Solve Problems About Distance.

Practice 1

The quotient of a number and 2, minus $\frac{1}{3}$, is the quotient of the number and 6. Find the number.

Answer
1. 1

Objective B Solving Problems About Work ▶

The next example is often called a work problem. Work problems usually involve people or machines doing a certain task.

Example 2 Finding Work Rates

Sam Waterton and Frank Schaffer work in a plant that manufactures automobiles. Sam can complete a quality control tour of the plant in 3 hours while his assistant, Frank, needs 7 hours to complete the same job. The regional manager is coming to inspect the plant facilities, so both Sam and Frank are directed to complete a quality control tour together. How long will this take?

Solution:

1. UNDERSTAND. Read and reread the problem. The key idea here is the relationship between the **time** (hours) it takes to complete the job and the **part of the job** completed in 1 unit of time (hour). For example, if the **time** it takes Sam to complete the job is 3 hours, the **part of the job** he can complete in 1 hour is $\frac{1}{3}$. Similarly, Frank can complete $\frac{1}{7}$ of the job in 1 hour.

Let x = the **time** in hours it takes Sam and Frank to complete the job together.

Then $\frac{1}{x}$ = the **part of the job** they complete in 1 hour.

	Hours to Complete Total Job	Part of Job Completed in 1 Hour
Sam	3	$\frac{1}{3}$
Frank	7	$\frac{1}{7}$
Together	x	$\frac{1}{x}$

2. TRANSLATE.

In words:	part of job job Sam completes in 1 hour	added to	part of job Frank completes in 1 hour	is equal to	part of job they complete together in in 1 hour
	↓	↓	↓	↓	↓
Translate:	$\frac{1}{3}$	$+$	$\frac{1}{7}$	$=$	$\frac{1}{x}$

3. SOLVE. Here, we solve the equation $\frac{1}{3} + \frac{1}{7} = \frac{1}{x}$. We begin by multiplying both sides of the equation by the LCD, $21x$.

$$21x\left(\frac{1}{3}\right) + 21x\left(\frac{1}{7}\right) = 21x\left(\frac{1}{x}\right)$$

$$7x + 3x = 21 \qquad \text{Simplify.}$$

$$10x = 21$$

$$x = \frac{21}{10} \quad \text{or} \quad 2\frac{1}{10} \text{ hours}$$

4. INTERPRET.

Check: Our proposed solution is $2\frac{1}{10}$ hours. This proposed solution is reasonable since $2\frac{1}{10}$ hours is more than half of Sam's time and less than half of Frank's time. Check this solution in the originally *stated* problem.

State: Sam and Frank can complete the quality control tour in $2\frac{1}{10}$ hours.

■ **Work Practice 2**

✔**Concept Check**   Solve $E = mc^2$
   a. for m **b.** for c^2

Objective C Solving Problems About Distance

Next we look at a problem solved by the distance formula,

   $d = r \cdot t$

| **Example 3** | Finding Speeds of Vehicles |

A car travels 180 miles in the same time that a truck travels 120 miles. If the car's speed is 20 miles per hour faster than the truck's, find the car's speed and the truck's speed.

Solution:

1. UNDERSTAND. Read and reread the problem. Suppose that the truck's speed is 45 miles per hour. Then the car's speed is 20 miles per hour faster, or 65 miles per hour.

We are given that the car travels 180 miles in the same time that the truck travels 120 miles. To find the time it takes the car to travel 180 miles, remember that since $d = rt$, we know that $\frac{d}{r} = t$.

Car's Time	*Truck's Time*
$t = \dfrac{d}{r} = \dfrac{180}{65} = 2\dfrac{50}{65} = 2\dfrac{10}{13}$ hours	$t = \dfrac{d}{r} = \dfrac{120}{45} = 2\dfrac{30}{45} = 2\dfrac{2}{3}$ hours

Since the times are not the same, our proposed solution is not correct. But we have a better understanding of the problem.

Let $x =$ the speed of the truck.

Since the car's speed is 20 miles per hour faster than the truck's, then

   $x + 20 =$ the speed of the car

Use the formula $d = r \cdot t$ or **distance = rate · time.** Prepare a chart to organize the information in the problem.

	Distance	**=**	**Rate**	**·**	**Time**
Truck	120		x		$\dfrac{120}{x}$ ← distance / rate
Car	180		$x + 20$		$\dfrac{180}{x + 20}$ ← distance / rate

(Continued on next page)

(Continued on next page)

Practice 3

A car travels 600 miles in the same time that a motorcycle travels 450 miles. If the car's speed is 15 miles per hour faster than the motorcycle's, find the speed of the car and the speed of the motorcycle.

Helpful Hint   If $d = r \cdot t$,
then $t = \dfrac{d}{r}$
or *time* $= \dfrac{distance}{rate}$.

Answer
3. car: 60 mph; motorcycle: 45 mph

✔**Concept Check Answers**
a. $m = \dfrac{E}{c^2}$ **b.** $c^2 = \dfrac{E}{m}$

2. TRANSLATE. Since the car and the truck traveled the same amount of time, we have that

In words: car's time   =   truck's time

Translate: $\dfrac{180}{x + 20}$ = $\dfrac{120}{x}$

3. SOLVE. We begin by multiplying both sides of the equation by the LCD, $x(x + 20)$, or cross multiplying.

$$\frac{180}{x + 20} = \frac{120}{x}$$

$$180x = 120(x + 20)$$

$$180x = 120x + 2400 \quad \text{Use the distributive property.}$$

$$60x = 2400 \quad \text{Subtract } 120x \text{ from both sides.}$$

$$x = 40 \quad \text{Divide both sides by 60.}$$

4. INTERPRET. The speed of the truck is 40 miles per hour. The speed of the car must then be $x + 20$ or 60 miles per hour.

Check: Find the time it takes the car to travel 180 miles and the time it takes the truck to travel 120 miles.

Car's Time

$$t = \frac{d}{r} = \frac{180}{60} = 3 \text{ hours}$$

Truck's Time

$$t = \frac{d}{r} = \frac{120}{40} = 3 \text{ hours}$$

Since both travel the same amount of time, the proposed solution is correct.

State: The car's speed is 60 miles per hour and the truck's speed is 40 miles per hour.

■ **Work Practice 3**

Vocabulary, Readiness & Video Check

Without solving algebraically, select the best choice for each exercise.

1. One person can complete a job in 7 hours. A second person can complete the same job in 5 hours. How long will it take them to complete the job if they work together?

  a. more than 7 hours
  b. between 5 and 7 hours
  c. less than 5 hours

2. One inlet pipe can fill a pond in 30 hours. A second inlet pipe can fill the same pond in 25 hours. How long before the pond is filled if both inlet pipes are on?

  a. less than 25 hours
  b. between 25 and 30 hours
  c. more than 30 hours

Fill in a Table *Given the variable in the first column, use the phrase in the second column to translate to an expression, and then continue to the phrase in the third column to translate to another expression.*

3.	A number: x	The reciprocal of the number:	The reciprocal of the number, decreased by 3:
4.	A number: y	The reciprocal of the number:	The reciprocal of the number, increased by 2:
5.	A number: z	The sum of the number and 5:	The reciprocal of the sum of the number and 5:
6.	A number: x	The difference of the number and 1:	The reciprocal of the difference of the number and 1:
7.	A number: y	Twice the number:	Eleven divided by twice the number:
8.	A number: z	Triple the number:	Negative ten divided by triple the number:

Martin-Gay Interactive Videos *Watch the section lecture video and answer the following questions.*

See Video 14.6 🍎

Objective A **9.** What words or phrases in ⊟ Example 1 told you to translate to an equation containing rational expressions? ▶

Objective B **10.** From ⊟ Example 2, how can you determine a somewhat reasonable answer to a work problem before you begin to solve it? ▶

Objective C **11.** The following problem is worded like ⊟ Example 3 in the video, but uses different quantities.

A car travels 325 miles in the same time that a motorcycle travels 290 miles. If the car's speed is 7 miles per hour faster than the motorcycle's, find the speed of the car and the speed of the motorcycle. Fill in the table and set up an equation based on this problem (do not solve). Use ⊟ Example 3 in the video as a model for your work.

	d	$=$	r	$\cdot$	t
Car					
Motorcycle					

14.6 **Exercise Set** MyMathLab® ▶

Objective A *Solve the following. See Example 1.*

1. Three times the reciprocal of a number equals 9 times the reciprocal of 6. Find the number.

▶ **2.** Twelve divided by the sum of x and 2 equals the quotient of 4 and the difference of x and 2. Find x.

3. If twice a number added to 3 is divided by the number plus 1, the result is three halves. Find the number.

4. A number added to the product of 6 and the reciprocal of the number equals -5. Find the number.

Objective B *See Example 2.*

5. Smith Engineering found that an experienced surveyor surveys a roadbed in 4 hours. An apprentice surveyor needs 5 hours to survey the same stretch of road. If the two work together, find how long it takes them to complete the job.

6. An experienced bricklayer constructs a small wall in 3 hours. The apprentice completes the job in 6 hours. Find how long it takes if they work together.

▶ **7.** In 2 minutes, a conveyor belt moves 300 pounds of recyclable aluminum from the delivery truck to a storage area. A smaller belt moves the same amount of cans the same distance in 6 minutes. If both belts are used, find how long it takes to move the cans to the storage area.

8. Find how long it takes the conveyor belts described in Exercise **7** to move 1200 pounds of cans. (*Hint:* Think of 1200 pounds as four 300-pound jobs.)

Objective C *See Example 3.*

9. A jogger begins her workout by jogging to the park, a distance of 12 miles. She rests, then jogs home at the same speed but along a different route. This return trip is 18 miles and her time is one hour longer. Find her jogging speed. Complete the accompanying chart and use it to find her jogging speed.

	Distance	=	Rate	·	Time
Trip to Park	12				
Return Trip	18				

10. A boat can travel 9 miles upstream in the same amount of time it takes to travel 11 miles downstream. If the current of the river is 3 miles per hour, complete the chart below and use it to find the speed of the boat in still water.

	Distance	=	Rate	·	Time
Upstream	9		$r - 3$		
Downstream	11		$r + 3$		

11. A cyclist rode the first 20-mile portion of his workout at a constant speed. For the 16-mile cooldown portion of his workout, he reduced his speed by 2 miles per hour. Each portion of the workout took the same time. Find the cyclist's speed during the first portion and find his speed during the cooldown portion.

12. A semi truck travels 300 miles through the flatland in the same amount of time that it travels 180 miles through mountains. The rate of the truck is 20 miles per hour slower in the mountains than in the flatland. Find both the flatland rate and the mountain rate.

Objectives A B C Mixed Practice *Solve the following. See Examples 1 through 3. (Note: Some exercises can be modeled by equations without rational expressions.)*

13. One-fourth equals the quotient of a number and 8. Find the number.

14. Four times a number added to 5 is divided by 6. The result is $\frac{7}{2}$. Find the number.

15. Marcus and Tony work for Lombardo's Pipe and Concrete. Mr. Lombardo is preparing an estimate for a customer. He knows that Marcus lays a slab of concrete in 6 hours. Tony lays the same-size slab in 4 hours. If both work on the job and the cost of labor is $45.00 per hour, decide what the labor estimate should be.

16. Mr. Dodson can paint his house by himself in 4 days. His son needs an additional day to complete the job if he works by himself. If they work together, find how long it takes to paint the house.

17. A pilot can travel 400 miles with the wind in the same amount of time as 336 miles against the wind. Find the speed of the wind if the pilot's speed in still air is 230 miles per hour.

18. A fisherman on Pearl River rows 9 miles downstream in the same amount of time he rows 3 miles upstream. If the current is 6 miles per hour, find how long it takes him to cover the 12 miles.

19. Two divided by the difference of a number and 3, minus 4 divided by the sum of the number and 3, equals 8 times the reciprocal of the difference of the number squared and 9. What is the number?

20. If 15 times the reciprocal of a number is added to the ratio of 9 times the number minus 7 and the number plus 2, the result is 9. What is the number?

21. A pilot flies 630 miles with a tail wind of 35 miles per hour. Against the wind, he flies only 455 miles in the same amount of time. Find the rate of the plane in still air.

22. A marketing manager travels 1080 miles in a corporate jet and then an additional 240 miles by car. If the car ride takes one hour longer than the jet ride takes, and if the rate of the jet is 6 times the rate of the car, find the time the manager travels by jet and find the time the manager travels by car.

23. The quotient of a number and 3, minus 1, equals $\frac{5}{3}$. Find the number.

24. The quotient of a number and 5, minus 1, equals $\frac{7}{5}$. Find the number.

25. Two hikers are 11 miles apart and walking toward each other. They meet in 2 hours. Find the rate of each hiker if one hiker walks 1.1 mph faster than the other.

26. On a 255-mile trip, Gary Alessandrini traveled at an average speed of 70 mph, got a speeding ticket, and then traveled at 60 mph for the remainder of the trip. If the entire trip took 4.5 hours and the speeding ticket stop took 30 minutes, how long did Gary speed before getting stopped?

27. One custodian cleans a suite of offices in 3 hours. When a second worker is asked to join the regular custodian, the job takes only $1\frac{1}{2}$ hours. How long does it take the second worker to do the same job alone?

28. One person proofreads copy for a small newspaper in 4 hours. If a second proofreader is also employed, the job can be done in $2\frac{1}{2}$ hours. How long does it take the second proofreader to do the same job alone?

29. A boater travels 16 miles per hour on the water on a still day. During one particularly windy day, he finds that he travels 48 miles with the wind behind him in the same amount of time that he travels 16 miles into the wind. Find the rate of the wind.

Let x be the rate of the wind.

	r	$\cdot$	t	$=$	d
With wind	$16 + x$				48
Into wind	$16 - x$				16

30. The current on a portion of the Mississippi River is 3 miles per hour. A barge can go 6 miles upstream in the same amount of time it takes to go 10 miles downstream. Find the speed of the boat in still water.

Let x be the speed of the boat in still water.

	r	$\cdot$	t	$=$	d
Upstream	$x - 3$				6
Downstream	$x + 3$				10

31. Currently, the Ford Focus is the best-selling car in the world. A driver of this car took a day trip around the California coastline driving at two different speeds. He drove 70 miles at a slower speed and 300 miles at a speed 40 miles per hour faster. If the time spent driving at the faster speed was twice that spent driving at the slower speed, find the two speeds during the trip. (*Source: Top Ten of Everything*)

32. The second best-selling car in the world is the Toyota Corolla. Suppose that during a test drive of two Corollas, one car travels 224 miles in the same time that the second car travels 175 miles. If the speed of the first car is 14 miles per hour faster than the speed of the second car, find the speed of both cars. (*Source: R. L. Polk*)

33. A pilot can fly an MD-11 2160 miles with the wind in the same time she can fly 1920 miles against the wind. If the speed of the wind is 30 mph, find the speed of the plane in still air. (*Source: Air Transport Association of America*)

34. A pilot can fly a DC-10 1365 miles against the wind in the same time he can fly 1575 miles with the wind. If the speed of the plane in still air is 490 miles per hour, find the speed of the wind. (*Source: Air Transport Association of America*)

35. A jet plane traveling at 500 mph overtakes a propeller plane traveling at 200 mph that had a 2-hour head start. How far from the starting point are the planes?

36. How long will it take a bus traveling at 60 miles per hour to overtake a car traveling at 40 miles per hour if the car had a 1.5-hour head start?

37. One pipe fills a storage pool in 20 hours. A second pipe fills the same pool in 15 hours. When a third pipe is added and all three are used to fill the pool, it takes only 6 hours. Find how long it takes the third pipe to do the job.

38. One pump fills a tank 2 times as fast as another pump. If the pumps work together, they fill the tank in 18 minutes. How long does it take each pump to fill the tank?

39. A car travels 280 miles in the same time that a motorcycle travels 240 miles. If the car's speed is 10 miles per hour faster than the motorcycle's, find the speed of the car and the speed of the motorcycle.

40. A bus traveled on a level road for 3 hours at an average speed 20 miles per hour faster than it traveled on a winding road. The time spent on the winding road was 4 hours. Find the average speed on the level road if the entire trip was 305 miles.

41. In 6 hours, an experienced cook prepares enough pies to supply a local restaurant's daily order. Another cook prepares the same number of pies in 7 hours. Together with a third cook, they prepare the pies in 2 hours. Find how long it takes the third cook to prepare the pies alone.

42. Mrs. Smith balances the company books in 8 hours. It takes her assistant 12 hours to do the same job. If they work together, find how long it takes them to balance the books.

43. Suppose two trains leave Holbrook, Arizona, at the same time, traveling in opposite directions. One train travels 10 mph faster than the other. In 3.5 hours, the trains are 322 miles apart. Find the speed of each train.

44. Suppose two cars leave Brinkley, Arkansas, at the same time, traveling in opposite directions. One car travels 8 mph faster than the other car. In 2.5 hours, the cars are 280 miles apart. Find the speed of each car.

Review

Simplify. Follow the circled steps in the order shown. See Sections 2.5 and 3.1.

45. $\dfrac{\frac{3}{4} + \frac{1}{4} \left.\right\} \leftarrow ① \text{ Add.}}{\frac{3}{8} + \frac{13}{8} \left.\right\} \leftarrow ② \text{ Add.}}$

46. $\dfrac{\frac{9}{5} + \frac{6}{5} \left.\right\} \leftarrow ① \text{ Add.}}{\frac{17}{6} + \frac{7}{6} \left.\right\} \leftarrow ② \text{ Add.}}$

47. $\dfrac{\frac{2}{5} + \frac{1}{5} \left.\right\} \quad ① \text{ Add.}}{\frac{7}{10} + \frac{7}{10} \left.\right\} \quad ② \text{ Add.}} \leftarrow ③ \text{ Divide.}$

48. $\dfrac{\frac{1}{4} + \frac{5}{4} \left.\right\} \quad ① \text{ Add.}}{\frac{3}{8} + \frac{7}{8} \left.\right\} \quad ② \text{ Add.}} \leftarrow ③ \text{ Divide.}$

Concept Extensions

49. One pump fills a tank 3 times as fast as another pump. If the pumps work together, they fill the tank in 21 minutes. How long does it take each pump to fill the tank?

50. It takes 9 hours for pump A to fill a tank alone. Pump B takes 15 hours to fill the same tank alone. If pumps A, B, and C are used, the tank fills in 5 hours. How long does it take pump C to fill the tank alone?

51. Person A can complete a job in 5 hours, and person B can complete the same job in 3 hours. Without solving algebraically, discuss reasonable and unreasonable answers for how long it would take them to complete the job together.

52. For which of the following equations can we immediately use cross products to solve for x?

a. $\dfrac{2 - x}{5} = \dfrac{1 + x}{3}$ **b.** $\dfrac{2}{5} - x = \dfrac{1 + x}{3}$

Solve. See the Concept Check in this section.

Solve D = RT

53. for *R*

54. for *T*

55. A hyena spots a giraffe 0.5 mile away and begins running toward it. The giraffe starts running away from the hyena just as the hyena begins running toward it. A hyena can run at a speed of 40 mph and a giraffe can run at 32 mph. How long will it take the hyena to overtake the giraffe? (*Source: The World Almanac and Book of Facts*)

56. During the 2013 Formula 1 Grand Prix of Spain, Esteban Gutierrez posted the fastest lap of the race while Romain Grosjean's best race lap was the slowest fast lap in the field. The Spanish Grand Prix circuit is 4,655 kilometers long. When traveling at their fastest lap speeds, Grosjean drove 4,404 kilometers in the same time that Gutierrez completed an entire 4,655-kilometer lap. Gutierrez's fastest lap speed was 10.491 km/hr faster than Grosjean's fastest lap speed. Find each driver's fastest lap speed. Round each speed to the nearest tenth. (*Source: Formula One World Championship Limited*)

14.7 Simplifying Complex Fractions

A rational expression whose numerator or denominator or both numerator and denominator contain fractions is called a **complex rational expression** or a **complex fraction.** Some examples are

$$
\frac{4}{2 - \dfrac{1}{2}} \qquad \frac{\dfrac{3}{2}}{\dfrac{4}{7} - x} \qquad \frac{\dfrac{1}{x + 2}}{x + 2 - \dfrac{1}{x}}
$$

← Numerator of complex fraction
← Main fraction bar
← Denominator of complex fraction

Objectives

A Simplify Complex Fractions Using Method 1.

B Simplify Complex Fractions Using Method 2.

Our goal in this section is to write complex fractions in simplest form. A complex fraction is in simplest form when it is in the form $\dfrac{P}{Q}$, where *P* and *Q* are polynomials that have no common factors.

Objective **A** Simplifying Complex Fractions—Method 1

In this section, two methods of simplifying complex fractions are presented. The first method presented uses the fact that the main fraction bar indicates division.

> **Method 1: To Simplify a Complex Fraction**
>
> **Step 1:** Add or subtract fractions in the numerator or denominator so that the numerator is a single fraction and the denominator is a single fraction.
>
> **Step 2:** Perform the indicated division by multiplying the numerator of the complex fraction by the reciprocal of the denominator of the complex fraction.
>
> **Step 3:** Write the rational expression in lowest terms.

Practice 1

Simplify the complex fraction $\dfrac{\frac{3}{7}}{\frac{5}{9}}$.

Example 1 Simplify the complex fraction $\dfrac{\frac{5}{8}}{\frac{2}{3}}$.

Solution: Since the numerator and denominator of the complex fraction are already single fractions, we proceed to Step 2: Perform the indicated division by multiplying the numerator $\dfrac{5}{8}$ by the reciprocal of the denominator $\dfrac{2}{3}$.

$$\frac{\frac{5}{8}}{\frac{2}{3}} = \frac{5}{8} \div \frac{2}{3} = \frac{5}{8} \cdot \frac{3}{2} = \frac{15}{16}$$

The reciprocal of $\dfrac{2}{3}$ is $\dfrac{3}{2}$.

■ **Work Practice 1**

Practice 2

Simplify: $\dfrac{\frac{3}{4} - \frac{2}{3}}{\frac{1}{2} + \frac{3}{8}}$

Example 2 Simplify: $\dfrac{\frac{2}{3} + \frac{1}{5}}{\frac{2}{3} - \frac{2}{9}}$

Solution: We simplify the numerator and denominator of the complex fraction separately. First we add $\dfrac{2}{3}$ and $\dfrac{1}{5}$ to obtain a single fraction in the numerator. Then we subtract $\dfrac{2}{9}$ from $\dfrac{2}{3}$ to obtain a single fraction in the denominator.

$$\frac{\frac{2}{3} + \frac{1}{5}}{\frac{2}{3} - \frac{2}{9}} = \frac{\frac{2(5)}{3(5)} + \frac{1(3)}{5(3)}}{\frac{2(3)}{3(3)} - \frac{2}{9}} \quad \text{The LCD of the numerator's fractions is 15.}$$

The LCD of the denominator's fractions is 9.

$$= \frac{\frac{10}{15} + \frac{3}{15}}{\frac{6}{9} - \frac{2}{9}} \quad \text{Simplify.}$$

$$= \frac{\frac{13}{15}}{\frac{4}{9}} \quad \begin{array}{l}\text{Add the numerator's fractions.}\\[4pt]\text{Subtract the denominator's fractions.}\end{array}$$

Answers

1. $\dfrac{27}{35}$ **2.** $\dfrac{2}{21}$

Next we perform the indicated division by multiplying the numerator of the complex fraction by the reciprocal of the denominator of the complex fraction.

$$\dfrac{\dfrac{13}{15}}{\dfrac{4}{9}} = \dfrac{13}{15} \cdot \dfrac{9}{4} \qquad \text{The reciprocal of } \dfrac{4}{9} \text{ is } \dfrac{9}{4}.$$

$$= \dfrac{13 \cdot 3 \cdot 3}{3 \cdot 5 \cdot 4} = \dfrac{39}{20}$$

■ **Work Practice 2**

Example 3 Simplify: $\dfrac{\dfrac{1}{z} - \dfrac{1}{2}}{\dfrac{1}{3} - \dfrac{z}{6}}$

Solution: Subtract to get a single fraction in the numerator and a single fraction in the denominator of the complex fraction.

$$\dfrac{\dfrac{1}{z} - \dfrac{1}{2}}{\dfrac{1}{3} - \dfrac{z}{6}} = \dfrac{\dfrac{2}{2z} - \dfrac{z}{2z}}{\dfrac{2}{6} - \dfrac{z}{6}} \qquad \begin{array}{l}\text{The LCD of the numerator's fractions is } 2z. \\ \\ \text{The LCD of the denominator's fractions is } 6.\end{array}$$

$$= \dfrac{\dfrac{2 - z}{2z}}{\dfrac{2 - z}{6}}$$

$$= \dfrac{2 - z}{2z} \cdot \dfrac{6}{2 - z} \qquad \text{Multiply by the reciprocal of } \dfrac{2 - z}{6}.$$

$$= \dfrac{2 \cdot 3 \cdot (2 - z)}{2 \cdot z \cdot (2 - z)} \qquad \text{Factor.}$$

$$= \dfrac{3}{z} \qquad \text{Write in lowest terms.}$$

■ **Work Practice 3**

Practice 3

Simplify: $\dfrac{\dfrac{2}{5} - \dfrac{1}{x}}{\dfrac{2x}{15} - \dfrac{1}{3}}$

Objective B Simplifying Complex Fractions— Method 2 ▶

Next we study a second method for simplifying complex fractions. In this method, we multiply the numerator and the denominator of the complex fraction by the LCD of all fractions in the complex fraction.

Method 2: To Simplify a Complex Fraction

Step 1: Find the LCD of all the fractions in the complex fraction.

Step 2: Multiply both the numerator and the denominator of the complex fraction by the LCD from Step 1.

Step 3: Perform the indicated operations and write the result in lowest terms.

We use Method 2 to rework Example 2.

Answer

3. $\dfrac{3}{x}$

Practice 4

Use Method 2 to simplify the complex fraction in Practice 2:

$$\dfrac{\dfrac{3}{4} - \dfrac{2}{3}}{\dfrac{1}{2} + \dfrac{3}{8}}$$

Example 4 Simplify: $\dfrac{\dfrac{2}{3} + \dfrac{1}{5}}{\dfrac{2}{3} - \dfrac{2}{9}}$

Solution: The LCD of $\dfrac{2}{3}, \dfrac{1}{5}, \dfrac{2}{3}$, and $\dfrac{2}{9}$ is 45, so we multiply the numerator and the denominator of the complex fraction by 45. Then we perform the indicated operations, and write in lowest terms.

$$\frac{\dfrac{2}{3} + \dfrac{1}{5}}{\dfrac{2}{3} - \dfrac{2}{9}} = \frac{45\left(\dfrac{2}{3} + \dfrac{1}{5}\right)}{45\left(\dfrac{2}{3} - \dfrac{2}{9}\right)}$$

$$= \frac{45\left(\dfrac{2}{3}\right) + 45\left(\dfrac{1}{5}\right)}{45\left(\dfrac{2}{3}\right) - 45\left(\dfrac{2}{9}\right)} \quad \text{Apply the distributive property.}$$

$$= \frac{30 + 9}{30 - 10} = \frac{39}{20} \quad \text{Simplify.}$$

■ **Work Practice 4**

Helpful Hint

The same complex fraction was simplified using two different methods in Examples 2 and 4. Notice that the simplified results are the same.

Practice 5

Simplify: $\dfrac{1 + \dfrac{x}{y}}{\dfrac{2x + 1}{y}}$

Example 5 Simplify: $\dfrac{\dfrac{x + 1}{y}}{\dfrac{x}{y} + 2}$

Solution: The LCD of $\dfrac{x + 1}{y}, \dfrac{x}{y}$ and $\dfrac{2}{1}$ is y, so we multiply the numerator and the denominator of the complex fraction by y.

$$\frac{\dfrac{x + 1}{y}}{\dfrac{x}{y} + 2} = \frac{y\left(\dfrac{x + 1}{y}\right)}{y\left(\dfrac{x}{y} + 2\right)}$$

$$= \frac{y\left(\dfrac{x + 1}{y}\right)}{y\left(\dfrac{x}{y}\right) + y \cdot 2} \quad \text{Apply the distributive property in the denominator.}$$

$$= \frac{x + 1}{x + 2y} \quad \text{Simplify.}$$

■ **Work Practice 5**

Answers

4. $\dfrac{2}{21}$ **5.** $\dfrac{y + x}{2x + 1}$

Example 6   Simplify: $\dfrac{\dfrac{x}{y} + \dfrac{3}{2x}}{\dfrac{x}{2} + y}$

Solution: The LCD of $\dfrac{x}{y}, \dfrac{3}{2x}, \dfrac{x}{2},$ and $\dfrac{y}{1}$ is $2xy$, so we multiply both the numerator and the denominator of the complex fraction by $2xy$.

$$\frac{\dfrac{x}{y} + \dfrac{3}{2x}}{\dfrac{x}{2} + y} = \frac{2xy\left(\dfrac{x}{y} + \dfrac{3}{2x}\right)}{2xy\left(\dfrac{x}{2} + y\right)}$$

$$= \frac{2xy\left(\dfrac{x}{y}\right) + 2xy\left(\dfrac{3}{2x}\right)}{2xy\left(\dfrac{x}{2}\right) + 2xy(y)} \quad \text{Apply the distributive property.}$$

$$= \frac{2x^2 + 3y}{x^2y + 2xy^2}$$

$$\text{or} \quad \frac{2x^2 + 3y}{xy(x + 2y)}$$

■ **Work Practice 6**

Practice 6

Simplify: $\dfrac{\dfrac{5}{6y} + \dfrac{y}{x}}{\dfrac{y}{3} - x}$

Answer

6. $\dfrac{5x + 6y^2}{2xy^2 - 6x^2y}$ or $\dfrac{5x + 6y^2}{2xy(y - 3x)}$

Vocabulary, Readiness & Video Check

One method for simplifying a complex fraction is to multiply the fraction's numerator and denominator by the LCD of all fractions in the complex fraction. For each complex fraction, choose the LCD of its fractions.

1. $\dfrac{\dfrac{1}{4} + \dfrac{1}{2}}{\dfrac{1}{3} + \dfrac{1}{2}}$ The LCD for $\dfrac{1}{4}, \dfrac{1}{2}, \dfrac{1}{3},$ and $\dfrac{1}{2}$ is
  a. 4   **b.** 2   **c.** 12   **d.** 6

2. $\dfrac{\dfrac{3}{5} + \dfrac{2}{3}}{\dfrac{1}{10} + \dfrac{1}{6}}$ The LCD for $\dfrac{3}{5}, \dfrac{2}{3}, \dfrac{1}{10},$ and $\dfrac{1}{6}$ is
  a. 15   **b.** 30   **c.** 60   **d.** 180

3. $\dfrac{\dfrac{5}{2x^2} + \dfrac{3}{16x}}{\dfrac{x}{8} + \dfrac{3}{4x}}$ The LCD for $\dfrac{5}{2x^2}, \dfrac{3}{16x}, \dfrac{x}{8},$ and $\dfrac{3}{4x}$ is
  a. $16x^2$ **b.** $32x^3$ **c.** $16x$ **d.** $16x^3$

4. $\dfrac{\dfrac{11}{6} + \dfrac{10}{x^2}}{\dfrac{7}{9} + \dfrac{5}{x}}$ The LCD for $\dfrac{11}{6}, \dfrac{10}{x^2}, \dfrac{7}{9},$ and $\dfrac{5}{x}$ is
  a. 18   **b.** x^2 **c.** $18x^2$ **d.** $54x^3$

Martin-Gay Interactive Videos   *Watch the section lecture video and answer the following questions.*

Objective A   **5.** From ▤ Example 2, before we can rewrite the complex fraction as a division problem, what must we make sure we have? ▶

Objective B   **6.** How does finding an LCD in Method 2, as in ▤ Examples 4 and 5, differ from finding an LCD in Method 1? Mention the purpose of the LCD in each method. ▶

See Video 14.7 🍎

14.7 Exercise Set MyMathLab®

Objectives **A** **B** **Mixed Practice** *Simplify each complex fraction. See Examples 1 through 6.*

1. $\dfrac{\dfrac{1}{2}}{\dfrac{3}{4}}$

2. $\dfrac{\dfrac{1}{8}}{-\dfrac{5}{12}}$

3. $\dfrac{\dfrac{4x}{9}}{\dfrac{2x}{3}}$

4. $\dfrac{-\dfrac{6y}{11}}{\dfrac{4y}{9}}$

5. $\dfrac{\dfrac{1+x}{6}}{\dfrac{1+x}{3}}$

6. $\dfrac{\dfrac{6x-3}{5x^2}}{\dfrac{2x-1}{10x}}$

7. $\dfrac{\dfrac{1}{2}+\dfrac{2}{3}}{\dfrac{5}{9}-\dfrac{5}{6}}$

8. $\dfrac{\dfrac{3}{4}-\dfrac{1}{2}}{\dfrac{3}{8}+\dfrac{1}{6}}$

9. $\dfrac{2+\dfrac{7}{10}}{1+\dfrac{3}{5}}$

10. $\dfrac{4-\dfrac{11}{12}}{5+\dfrac{1}{4}}$

11. $\dfrac{\dfrac{1}{3}}{\dfrac{1}{2}-\dfrac{1}{4}}$

12. $\dfrac{\dfrac{7}{10}-\dfrac{3}{5}}{\dfrac{1}{2}}$

13. $\dfrac{-\dfrac{2}{9}}{-\dfrac{14}{3}}$

14. $\dfrac{\dfrac{3}{8}}{\dfrac{4}{15}}$

15. $\dfrac{\dfrac{5}{12x^2}}{\dfrac{25}{16x^3}}$

16. $\dfrac{-\dfrac{7}{8y}}{\dfrac{21}{4y}}$

17. $\dfrac{\dfrac{m}{n}-1}{\dfrac{m}{n}+1}$

18. $\dfrac{\dfrac{x}{2}+2}{\dfrac{x}{2}-2}$

19. $\dfrac{\dfrac{1}{5}-\dfrac{1}{x}}{\dfrac{7}{10}+\dfrac{1}{x^2}}$

20. $\dfrac{\dfrac{1}{y^2}+\dfrac{2}{3}}{\dfrac{1}{y}-\dfrac{5}{6}}$

21. $\dfrac{1+\dfrac{1}{y-2}}{y+\dfrac{1}{y-2}}$

22. $\dfrac{x-\dfrac{1}{2x+1}}{1-\dfrac{x}{2x+1}}$

23. $\dfrac{\dfrac{4y-8}{16}}{\dfrac{6y-12}{4}}$

24. $\dfrac{\dfrac{7y+21}{3}}{\dfrac{3y+9}{8}}$

25. $\dfrac{\dfrac{x}{y}+1}{\dfrac{x}{y}-1}$

26. $\dfrac{\dfrac{3}{5y}+8}{\dfrac{3}{5y}-8}$

27. $\dfrac{1}{2+\dfrac{1}{3}}$

28. $\dfrac{3}{1-\dfrac{4}{3}}$

29. $\dfrac{\dfrac{ax+ab}{x^2-b^2}}{\dfrac{x+b}{x-b}}$

30. $\dfrac{\dfrac{m+2}{m-2}}{\dfrac{2m+4}{m^2-4}}$

31. $\dfrac{\dfrac{-3+y}{4}}{\dfrac{8+y}{28}}$

32. $\dfrac{\dfrac{-x+2}{18}}{\dfrac{8}{9}}$

33. $\dfrac{3+\dfrac{12}{x}}{1-\dfrac{16}{x^2}}$

34. $\dfrac{2+\dfrac{6}{x}}{1-\dfrac{9}{x^2}}$

35. $\dfrac{\dfrac{8}{x+4}+2}{\dfrac{12}{x+4}-2}$

36. $\dfrac{\dfrac{25}{x+5}+5}{\dfrac{3}{x+5}-5}$

37. $\dfrac{\dfrac{s}{r} + \dfrac{r}{s}}{\dfrac{s}{r} - \dfrac{r}{s}}$

38. $\dfrac{\dfrac{2}{x} + \dfrac{x}{2}}{\dfrac{2}{x} - \dfrac{x}{2}}$

39. $\dfrac{\dfrac{6}{x-5} + \dfrac{x}{x-2}}{\dfrac{3}{x-6} - \dfrac{2}{x-5}}$

40. $\dfrac{\dfrac{4}{x} + \dfrac{x}{x+1}}{\dfrac{1}{2x} + \dfrac{1}{x+6}}$

Review

Use the bar graph below to answer Exercises 41 through 44. See Section 7.1. Note: Some of these players are still competing; thus, their total prize money may increase.

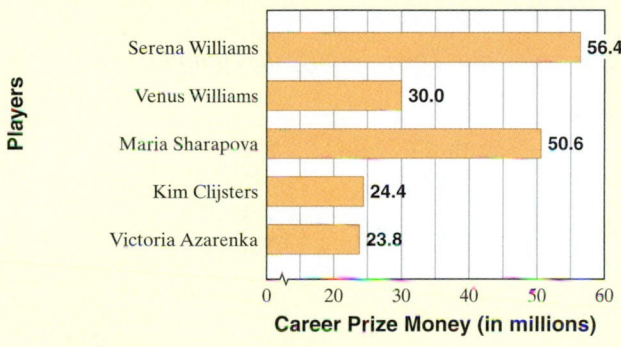

Women's Tennis Career Prize Money Leaders

Source: WTA Media Information System, as of August 2014

41. Which women's tennis player has earned the most prize money in her career?

42. How much more prize money has Maria Sharapova earned in her career than Victoria Azarkenka?

43. What is the difference in lifetime prize money between Kim Clijsters and Venus Williams?

44. To date in her career, Serena Williams has won 82 doubles and singles tournament titles. Assuming her prize money is earned only for tournament titles, how much prize money has she earned, on average, per tournament title?

Concept Extensions

45. Explain how to simplify a complex fraction using Method 1.

46. Explain how to simplify a complex fraction using Method 2.

To find the average of two numbers, we find their sum and divide by 2. For example, the average of 65 and 81 is found by simplifying $\dfrac{65 + 81}{2}$. This simplifies to $\dfrac{146}{2} = 73$. Use this for Exercises 47–50.

47. Find the average of $\dfrac{1}{3}$ and $\dfrac{3}{4}$.

48. Write the average of $\dfrac{3}{n}$ and $\dfrac{5}{n^2}$ as a simplified rational expression.

49. A carpenter needs to drill a hole halfway between the two marked points. An intersecting board keeps him from measuring between the marked points, but he does have earlier measurements as shown. How far from the left side of the marked board should he drill?

50. Use the same diagram as for Exercise **49**. Suppose the measurements are 7.2 inches and 10.3 inches. How far from the left side of the marked board should he drill?

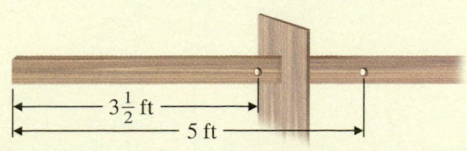

$3\frac{1}{2}$ ft

5 ft

Solve.

51. In electronics, when two resistors R_1 (read R sub 1) and R_2 (read R sub 2) are connected in parallel, the total resistance is given by the complex fraction

$$\frac{1}{\dfrac{1}{R_1} + \dfrac{1}{R_2}}.$$

Simplify this expression.

Resistance R_1 R_2

52. Astronomers occasionally need to know the day of the week a particular date fell on. The complex fraction

$$\frac{J + \dfrac{3}{2}}{7}$$

where J is the *Julian day number,* is used to make this calculation. Simplify this expression.

Simplify each of the following. First, write each expression with positive exponents. Then simplify the complex fraction. The first step has been completed for Exercise 53.

53. 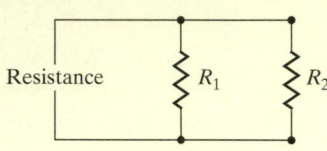 $\dfrac{x^{-1} + 2^{-1}}{x^{-2} - 4^{-1}} = \dfrac{\dfrac{1}{x} + \dfrac{1}{2}}{\dfrac{1}{x^2} - \dfrac{1}{4}}$

54. $\dfrac{3^{-1} - x^{-1}}{9^{-1} - x^{-2}}$

55. $\dfrac{y^{-2}}{1 - y^{-2}}$

56. $\dfrac{4 + x^{-1}}{3 + x^{-1}}$

57. If the distance formula $d = r \cdot t$ is solved for t, then $t = \dfrac{d}{r}$. Use this formula to find t if distance d is $\dfrac{20x}{3}$ miles and rate r is $\dfrac{5x}{9}$ miles per hour. Write t in simplified form.

△ **58.** If the formula for the area of a rectangle, $A = l \cdot w$, is solved for w, then $w = \dfrac{A}{l}$. Use this formula to find w if area A is $\dfrac{4x - 2}{3}$ square meters and length l is $\dfrac{6x - 3}{5}$ meters. Write w in simplified form.

Chapter 14 Group Activity

Fast-Growing Careers

According to U.S. Bureau of Labor Statistics projections, the careers listed below will have the largest job growth in the years shown.

Occupation	Employment (number in thousands)		
	2010	2020	Change
1. Registered nurses	2737.4	3449.3	+711.9
2. Retail salespersons	4261.6	4968.4	+706.8
3. Home health aides	1017.7	1723.9	+706.3
4. Personal care aides	861.0	1468.0	+607.0
5. Office clerks, general	2950.7	3440.2	+489.5
6. Combined food preparation and serving workers, including fast food	2682.1	3080.1	+398.0
7. Customer service representatives	2187.3	2525.6	+338.3
8. Heavy and tractor-trailer truck drivers	1604.8	1934.9	+330.1
9. Laborers and freight, stock, and material movers, hand	2068.2	2387.3	+319.1
10. Postsecondary teachers	1756.0	2061.7	+305.7

What do all of these in-demand occupations have in common? They all require a knowledge of math! For some careers, like nurses, postsecondary teachers, and salespersons, the ways math is used on the job may be obvious. For other occupations, the use of math may not be quite as obvious. However, tasks common to many jobs, such as filling in a time sheet or a medication log, writing up an expense report, planning a budget, figuring a bill, ordering supplies, and even making a work schedule, all require math.

Activity

Suppose that your college placement office is planning to publish an occupational handbook on math in popular occupations. Choose one of the occupations from the given list that interests you. Research the occupation. Then write a brief entry for the occupational handbook that describes how a person in that career would use math in his or her job. Include an example if possible.

Chapter 14 Vocabulary Check

Fill in each blank with one of the words or phrases listed below. Not all choices will be used.

least common denominator   simplifying   reciprocals   numerator $\dfrac{-a}{b}$

rational expression   unit   complex fraction   denominator $\dfrac{-a}{-b}$ $\dfrac{a}{-b}$

1. A _____, is an expression that can be written in the form $\dfrac{P}{Q}$, where P and Q are polynomials and Q is not 0.

2. In a _____ the numerator or denominator or both may contain fractions.

3. For a rational expression, $-\dfrac{a}{b} = $ _____ $ = $ _____ .

4. A rational expression is undefined when the _____ is 0.

5. The process of writing a rational expression in lowest terms is called _____ .

6. The expressions $\dfrac{2x}{7}$ and $\dfrac{7}{2x}$ are called _____ .

7. The _____ of a list of rational expressions is a polynomial of least degree whose factors include all factors of the denominators in the list.

8. A _____ fraction is a fraction that equals 1.

Helpful Hint ▶ Are you preparing for your test? Don't forget to take the Chapter 14 Test on page 1104. Then check your answers at the back of the text and use the Chapter Test Prep Videos to see the fully worked-out solutions to any of the exercises you want to review.

14 Chapter Highlights

Definitions and Concepts	Examples

Section 14.1 Simplifying Rational Expressions

A **rational expression** is an expression that can be written in the form $\dfrac{P}{Q}$, where P and Q are polynomials and Q does not equal 0.

$$\frac{7y^3}{4}, \quad \frac{x^2 + 6x + 1}{x - 3}, \quad \frac{-5}{s^3 + 8}$$

To find values for which a rational expression is undefined, find values for which the denominator is 0.

Find any values for which the expression $\dfrac{5y}{y^2 - 4y + 3}$ is undefined.

$$
\begin{aligned}
y^2 - 4y + 3 &= 0 &&\text{Set the denominator equal to 0.}\\
(y - 3)(y - 1) &= 0 &&\text{Factor.}\\
y - 3 = 0 \quad &\text{or} \quad y - 1 = 0 &&\text{Set each factor equal to 0.}\\
y = 3 \qquad & \qquad\quad y = 1 &&\text{Solve.}
\end{aligned}
$$

The expression is undefined when y is 3 and when y is 1.

To Simplify a Rational Expression

Step 1: Factor the numerator and denominator.

Step 2: Divide out factors common to the numerator and denominator. (This is the same as removing a factor of 1.)

Simplify: $\dfrac{4x + 20}{x^2 - 25}$

$$\frac{4x + 20}{x^2 - 25} = \frac{4\,(x + 5)}{(x + 5)\,(x - 5)} = \frac{4}{x - 5}$$

Section 14.2 Multiplying and Dividing Rational Expressions

To Multiply Rational Expressions

Step 1: Factor numerators and denominators.

Step 2: Multiply numerators and multiply denominators.

Step 3: Write the product in lowest terms.

$$\frac{P}{Q} \cdot \frac{R}{S} = \frac{PR}{QS}$$

Multiply: $\dfrac{4x + 4}{2x - 3} \cdot \dfrac{2x^2 + x - 6}{x^2 - 1}$

$$
\begin{aligned}
&\frac{4x + 4}{2x - 3} \cdot \frac{2x^2 + x - 6}{x^2 - 1}\\[4pt]
&= \frac{4(x + 1)}{2x - 3} \cdot \frac{(2x - 3)(x + 2)}{(x + 1)(x - 1)}\\[4pt]
&= \frac{4\,(x + 1)(2x - 3)\,(x + 2)}{(2x - 3)(x + 1)\,(x - 1)}\\[4pt]
&= \frac{4(x + 2)}{x - 1}
\end{aligned}
$$

Definitions and Concepts	Examples

Section 14.2 Multiplying and Dividing Rational Expressions (*continued*)

To divide by a rational expression, multiply by the reciprocal.

$$\frac{P}{Q} \div \frac{R}{S} = \frac{P}{Q} \cdot \frac{S}{R} = \frac{PS}{QR}$$

Divide: $\dfrac{15x + 5}{3x^2 - 14x - 5} \div \dfrac{15}{3x - 12}$

$$\frac{15x + 5}{3x^2 - 14x - 5} \div \frac{15}{3x - 12}$$

$$= \frac{5(3x + 1)}{(3x + 1)\,(x - 5)} \cdot \frac{3\,(x - 4)}{3 \cdot 5}$$

$$= \frac{x - 4}{x - 5}$$

Section 14.3 Adding and Subtracting Rational Expressions with the Same Denominator and Least Common Denominator

To add or subtract rational expressions with the same denominator, add or subtract numerators, and place the sum or difference over the common denominator.

$$\frac{P}{R} + \frac{Q}{R} = \frac{P + Q}{R}$$

$$\frac{P}{R} - \frac{Q}{R} = \frac{P - Q}{R}$$

Perform each indicated operation.

$$\frac{5}{x + 1} + \frac{x}{x + 1} = \frac{5 + x}{x + 1}$$

$$\frac{2y + 7}{y^2 - 9} - \frac{y + 4}{y^2 - 9}$$

$$= \frac{2y + 7 - (y + 4)}{y^2 - 9}$$

$$= \frac{2y + 7 - y - 4}{y^2 - 9}$$

$$= \frac{y + 3}{(y + 3)\,(y - 3)}$$

$$= \frac{1}{y - 3}$$

To Find the Least Common Denominator (LCD)

Step 1: Factor the denominators.

Step 2: The LCD is the product of all unique factors, each raised to a power equal to the greatest number of times that it appears in any one factored denominator.

Find the LCD for

$$\frac{7x}{x^2 + 10x + 25} \quad \text{and} \quad \frac{11}{3x^2 + 15x}$$

$$x^2 + 10x + 25 = (x + 5)(x + 5)$$

$$3x^2 + 15x = 3x(x + 5)$$

$$\text{LCD} = 3x(x + 5)(x + 5) \text{ or}$$

$$3x(x + 5)^2$$

Definitions and Concepts	Examples

Section 14.4   Adding and Subtracting Rational Expressions with Different Denominators

To Add or Subtract Rational Expressions with Different Denominators

Step 1: Find the LCD.

Step 2: Rewrite each rational expression as an equivalent expression whose denominator is the LCD.

Step 3: Add or subtract numerators and place the sum or difference over the common denominator.

Step 4: Write the result in lowest terms.

Perform the indicated operation.

$$\frac{9x + 3}{x^2 - 9} - \frac{5}{x - 3}$$

$$= \frac{9x + 3}{(x + 3)(x - 3)} - \frac{5}{x - 3}$$

LCD is $(x + 3)(x - 3)$.

$$= \frac{9x + 3}{(x + 3)(x - 3)} - \frac{5(x + 3)}{(x - 3)(x + 3)}$$

$$= \frac{9x + 3 - 5(x + 3)}{(x + 3)(x - 3)}$$

$$= \frac{9x + 3 - 5x - 15}{(x + 3)(x - 3)}$$

$$= \frac{4x - 12}{(x + 3)(x - 3)}$$

$$= \frac{4(x - 3)}{(x + 3)(x - 3)} = \frac{4}{x + 3}$$

Section 14.5   Solving Equations Containing Rational Expressions

To Solve an Equation Containing Rational Expressions

Step 1: Multiply both sides of the equation by the LCD of all rational expressions in the equation.

Step 2: Remove any grouping symbols and solve the resulting equation.

Step 3: Check the solution in the original equation.

Solve: $\dfrac{5x}{x + 2} + 3 = \dfrac{4x - 6}{x + 2}$ The LCD is $x + 2$.

$$(x + 2)\left(\frac{5x}{x + 2} + 3\right) = (x + 2)\left(\frac{4x - 6}{x + 2}\right)$$

$$(x + 2)\left(\frac{5x}{x + 2}\right) + (x + 2)(3) = (x + 2)\left(\frac{4x - 6}{x + 2}\right)$$

$$5x + 3x + 6 = 4x - 6$$

$$4x = -12$$

$$x = -3$$

The solution checks; the solution is -3.

Section 14.6   Rational Equations and Problem Solving

Problem-Solving Steps

1. UNDERSTAND. Read and reread the problem.

A small plane and a car leave Kansas City, Missouri, and head for Minneapolis, Minnesota, a distance of 450 miles. The speed of the plane is 3 times the speed of the car, and the plane arrives 6 hours ahead of the car. Find the speed of the car.

Let $x =$ the speed of the car.
Then $3x =$ the speed of the plane.

	Distance =	Rate	·	Time
Car	450	x		$\dfrac{450}{x}\left(\dfrac{\text{distance}}{\text{rate}}\right)$
Plane	450	$3x$		$\dfrac{450}{3x}\left(\dfrac{\text{distance}}{\text{rate}}\right)$

Definitions and Concepts	Examples

Section 14.6 Rational Equations and Problem Solving (continued)

2. TRANSLATE.

In words:
plane's time	+	6 hours	=	car's time

3. SOLVE.

Translate:
$$\frac{450}{3x} + 6 = \frac{450}{x}$$

$$\frac{450}{3x} + 6 = \frac{450}{x}$$

$$3x\left(\frac{450}{3x}\right) + 3x(6) = 3x\left(\frac{450}{x}\right)$$

$$450 + 18x = 1350$$

$$18x = 900$$

$$x = 50$$

4. INTERPRET.

Check this solution in the originally stated problem.
State the conclusion: The speed of the car is 50 miles per hour.

Section 14.7 Simplifying Complex Fractions

Method 1: To Simplify a Complex Fraction

Step 1: Add or subtract fractions in the numerator and the denominator of the complex fraction.

Step 2: Perform the indicated division.

Step 3: Write the result in lowest terms.

Simplify:

$$\frac{\frac{1}{x} + 2}{\frac{1}{x} - \frac{1}{y}} = \frac{\frac{1}{x} + \frac{2x}{x}}{\frac{y}{xy} - \frac{x}{xy}}$$

$$= \frac{\frac{1 + 2x}{x}}{\frac{y - x}{xy}}$$

$$= \frac{1 + 2x}{x} \cdot \frac{x\, y}{y - x}$$

$$= \frac{y(1 + 2x)}{y - x}$$

Method 2: To Simplify a Complex Fraction

Step 1: Find the LCD of all fractions in the complex fraction.

Step 2: Multiply the numerator and the denominator of the complex fraction by the LCD.

Step 3: Perform the indicated operations and write the result in lowest terms.

$$\frac{\frac{1}{x} + 2}{\frac{1}{x} - \frac{1}{y}} = \frac{xy\left(\frac{1}{x} + 2\right)}{xy\left(\frac{1}{x} - \frac{1}{y}\right)}$$

$$= \frac{xy\left(\frac{1}{x}\right) + xy(2)}{xy\left(\frac{1}{x}\right) - xy\left(\frac{1}{y}\right)}$$

$$= \frac{y + 2xy}{y - x} \quad \text{or} \quad \frac{y(1 + 2x)}{y - x}$$

(14.1) *Find any real number for which each rational expression is undefined.*

1. $\dfrac{x+5}{x^2-4}$

2. $\dfrac{5x+9}{4x^2-4x-15}$

Find the value of each rational expression when $x=5$, $y=7$, and $z=-2$.

3. $\dfrac{2-z}{z+5}$

4. $\dfrac{x^2+xy-y^2}{x+y}$

Simplify each rational expression.

5. $\dfrac{2x+6}{x^2+3x}$

6. $\dfrac{3x-12}{x^2-4x}$

7. $\dfrac{x+2}{x^2-3x-10}$

8. $\dfrac{x+4}{x^2+5x+4}$

9. $\dfrac{x^3-4x}{x^2+3x+2}$

10. $\dfrac{5x^2-125}{x^2+2x-15}$

11. $\dfrac{x^2-x-6}{x^2-3x-10}$

12. $\dfrac{x^2-2x}{x^2+2x-8}$

Simplify each expression. First, factor the four-term polynomials by grouping.

13. $\dfrac{x^2+xa+xb+ab}{x^2-xc+bx-bc}$

14. $\dfrac{x^2+5x-2x-10}{x^2-3x-2x+6}$

(14.2) *Perform each indicated operation and simplify.*

15. $\dfrac{15x^3y^2}{z}\cdot\dfrac{z}{5xy^3}$

16. $\dfrac{-y^3}{8}\cdot\dfrac{9x^2}{y^3}$

17. $\dfrac{x^2-9}{x^2-4}\cdot\dfrac{x-2}{x+3}$

18. $\dfrac{2x+5}{x-6}\cdot\dfrac{2x}{-x+6}$

19. $\dfrac{x^2-5x-24}{x^2-x-12}\div\dfrac{x^2-10x+16}{x^2+x-6}$

20. $\dfrac{4x+4y}{xy^2}\div\dfrac{3x+3y}{x^2y}$

21. $\dfrac{x^2+x-42}{x-3}\cdot\dfrac{(x-3)^2}{x+7}$

22. $\dfrac{2a+2b}{3}\cdot\dfrac{a-b}{a^2-b^2}$

23. $\dfrac{2x^2-9x+9}{8x-12}\div\dfrac{x^2-3x}{2x}$

24. $\dfrac{x^2-y^2}{x^2+xy}\div\dfrac{3x^2-2xy-y^2}{3x^2+6x}$

(14.3) *Perform each indicated operation and simplify.*

25. $\dfrac{x}{x^2 + 9x + 14} + \dfrac{7}{x^2 + 9x + 14}$

26. $\dfrac{x}{x^2 + 2x - 15} + \dfrac{5}{x^2 + 2x - 15}$

27. $\dfrac{4x - 5}{3x^2} - \dfrac{2x + 5}{3x^2}$

28. $\dfrac{9x + 7}{6x^2} - \dfrac{3x + 4}{6x^2}$

Find the LCD of each pair of rational expressions.

29. $\dfrac{x + 4}{2x}, \dfrac{3}{7x}$

30. $\dfrac{x - 2}{x^2 - 5x - 24}, \dfrac{3}{x^2 + 11x + 24}$

Rewrite each rational expression as an equivalent expression whose denominator is the given polynomial.

31. $\dfrac{5}{7x} = \dfrac{}{14x^3y}$

32. $\dfrac{9}{4y} = \dfrac{}{16y^3x}$

33. $\dfrac{x + 2}{x^2 + 11x + 18} = \dfrac{}{(x + 2)(x - 5)(x + 9)}$

34. $\dfrac{3x - 5}{x^2 + 4x + 4} = \dfrac{}{(x + 2)^2(x + 3)}$

(14.4) *Perform each indicated operation and simplify.*

35. $\dfrac{4}{5x^2} + \dfrac{6}{y}$

36. $\dfrac{2}{x - 3} - \dfrac{4}{x - 1}$

37. $\dfrac{4}{x + 3} - 2$

38. $\dfrac{3}{x^2 + 2x - 8} + \dfrac{2}{x^2 - 3x + 2}$

39. $\dfrac{2x - 5}{6x + 9} - \dfrac{4}{2x^2 + 3x}$

40. $\dfrac{x - 1}{x^2 - 2x + 1} - \dfrac{x + 1}{x - 1}$

(14.5) *Solve each equation.*

41. $\dfrac{n}{10} = 9 - \dfrac{n}{5}$

42. $\dfrac{2}{x + 1} - \dfrac{1}{x - 2} = -\dfrac{1}{2}$

43. $\dfrac{y}{2y + 2} + \dfrac{2y - 16}{4y + 4} = \dfrac{y - 3}{y + 1}$

44. $\dfrac{2}{x - 3} - \dfrac{4}{x + 3} = \dfrac{8}{x^2 - 9}$

45. $\dfrac{x - 3}{x + 1} - \dfrac{x - 6}{x + 5} = 0$

46. $x + 5 = \dfrac{6}{x}$

(14.6) *Solve.*

47. Five times the reciprocal of a number equals the sum of $\frac{3}{2}$ the reciprocal of the number and $\frac{7}{6}$. What is the number?

48. The reciprocal of a number equals the reciprocal of the difference of 4 and the number. Find the number.

49. A car travels 90 miles in the same time that a car traveling 10 miles per hour slower travels 60 miles. Find the speed of each car.

50. The current in a bayou near Lafayette, Louisiana, is 4 miles per hour. A paddleboat travels 48 miles upstream in the same amount of time it takes to travel 72 miles downstream. Find the speed of the boat in still water.

51. When Mark and Maria manicure Mr. Stergeon's lawn, it takes them 5 hours. If Mark works alone, it takes 7 hours. Find how long it takes Maria alone.

52. It takes pipe A 20 days to fill a fish pond. Pipe B takes 15 days. Find how long it takes both pipes together to fill the pond.

(14.7) *Simplify each complex fraction.*

53. $\dfrac{\dfrac{5x}{27}}{-\dfrac{10xy}{21}}$

54. $\dfrac{\dfrac{3}{5}+\dfrac{2}{7}}{\dfrac{1}{5}+\dfrac{5}{6}}$

55. $\dfrac{3-\dfrac{1}{y}}{2-\dfrac{1}{y}}$

56. $\dfrac{\dfrac{6}{x+2}+4}{\dfrac{8}{x+2}-4}$

Mixed Review

Simplify each rational expression.

57. $\dfrac{4x+12}{8x^2+24x}$

58. $\dfrac{x^3-6x^2+9x}{x^2+4x-21}$

Perform the indicated operations and simplify.

59. $\dfrac{x^2+9x+20}{x^2-25}\cdot\dfrac{x^2-9x+20}{x^2+8x+16}$

60. $\dfrac{x^2-x-72}{x^2-x-30}\div\dfrac{x^2+6x-27}{x^2-9x+18}$

61. $\dfrac{x}{x^2-36}+\dfrac{6}{x^2-36}$

62. $\dfrac{5x-1}{4x}-\dfrac{3x-2}{4x}$

63. $\dfrac{3x}{x^2+9x+14}-\dfrac{6x}{x^2+4x-21}$

64. $\dfrac{4}{3x^2+8x-3}+\dfrac{2}{3x^2-7x+2}$

Solve.

65. $\dfrac{4}{a-1} + 2 = \dfrac{3}{a-1}$

66. $\dfrac{x}{x+3} + 4 = \dfrac{x}{x+3}$

Solve.

67. The quotient of twice a number and three, minus one-sixth, is the quotient of the number and two. Find the number.

68. Mr. Crocker can paint his shed by himself in three days. His son will need an additional day to complete the job if he works alone. If they work together, find how long it takes to paint the shed.

Simplify each complex fraction.

69. $\dfrac{\dfrac{1}{4}}{\dfrac{1}{3} + \dfrac{1}{2}}$

70. $\dfrac{4 + \dfrac{2}{x}}{6 + \dfrac{3}{x}}$

Chapter 14   Test

Step-by-step test solutions are found on the Chapter Test Prep Videos. Where available: MyMathLab® or You Tube™

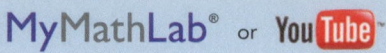

Answers

1. Find any real numbers for which the following expression is undefined.

$$\frac{x + 5}{x^2 + 4x + 3}$$

2. For a certain computer desk, the average manufacturing cost C per desk (in dollars) is

$$C = \frac{100x + 3000}{x}$$

where x is the number of desks manufactured.

a. Find the average cost per desk when manufacturing 200 computer desks.
b. Find the average cost per desk when manufacturing 1000 computer desks.

1. _____

2. a. _____

   b. _____

Simplify each rational expression.

3. _____

3. $\dfrac{3x - 6}{5x - 10}$ **4.** $\dfrac{x + 6}{x^2 + 12x + 36}$ **5.** $\dfrac{7 - x}{x - 7}$

4. _____

5. _____

6. $\dfrac{y - x}{x^2 - y^2}$ **7.** $\dfrac{2m^3 - 2m^2 - 12m}{m^2 - 5m + 6}$ **8.** $\dfrac{ay + 3a + 2y + 6}{ay + 3a + 5y + 15}$

6. _____

7. _____

Perform each indicated operation and simplify if possible.

8. _____

9. $\dfrac{x^2 - 13x + 42}{x^2 + 10x + 21} \div \dfrac{x^2 - 4}{x^2 + x - 6}$ **10.** $\dfrac{3}{x - 1} \cdot (5x - 5)$

9. _____

10. _____

11. $\dfrac{y^2 - 5y + 6}{2y + 4} \cdot \dfrac{y + 2}{2y - 6}$ **12.** $\dfrac{5}{2x + 5} - \dfrac{6}{2x + 5}$

11. _____

12. _____

13. $\dfrac{5a}{a^2 - a - 6} - \dfrac{2}{a - 3}$ **14.** $\dfrac{6}{x^2 - 1} + \dfrac{3}{x + 1}$

13. _____

14. _____

15. $\dfrac{x^2 - 9}{x^2 - 3x} \div \dfrac{x^2 + 4x + 1}{2x + 10}$ **16.** $\dfrac{x + 2}{x^2 + 11x + 18} + \dfrac{5}{x^2 - 3x - 10}$

15. _____

16. _____

17. $\dfrac{4y}{y^2 + 6y + 5} - \dfrac{3}{y^2 + 5y + 4}$

17. _____

Solve each equation.

18. $\dfrac{4}{y} - \dfrac{5}{3} = -\dfrac{1}{5}$

19. $\dfrac{5}{y+1} = \dfrac{4}{y+2}$

20. $\dfrac{a}{a-3} = \dfrac{3}{a-3} - \dfrac{3}{2}$

21. $\dfrac{10}{x^2 - 25} = \dfrac{3}{x+5} + \dfrac{1}{x-5}$

22. $x - \dfrac{14}{x-1} = 4 - \dfrac{2x}{x-1}$

Simplify each complex fraction.

23. $\dfrac{\dfrac{5x^2}{yz^2}}{\dfrac{10x}{z^3}}$

24. $\dfrac{\dfrac{b}{a} - \dfrac{a}{b}}{\dfrac{1}{b} + \dfrac{1}{a}}$

25. $\dfrac{5 - \dfrac{1}{y^2}}{\dfrac{1}{y} + \dfrac{2}{y^2}}$

26. One number plus five times its reciprocal is equal to six. Find the number.

27. A pleasure boat traveling down the Red River takes the same time to go 14 miles upstream as it takes to go 16 miles downstream. If the current of the river is 2 miles per hour, find the speed of the boat in still water.

28. An inlet pipe can fill a tank in 12 hours. A second pipe can fill the tank in 15 hours. If both pipes are used, find how long it takes to fill the tank.

18. _____

19. _____

20. _____

21. _____

22. _____

23. _____

24. _____

25. _____

26. _____

27. _____

28. _____

Answers

1. _____

2. _____

3. _____

4. _____

5. _____

6. _____

7. _____

8. _____

9. _____

10. _____

11. _____

12. _____

13. _____

14. _____

15. _____

16. _____

△ **1.** Find the area of the parallelogram:

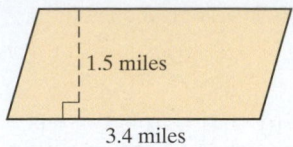

1.5 miles

3.4 miles

△ **2.** Find the area of the circle. Give an exact area, then use 3.14 for π to approximate the area.

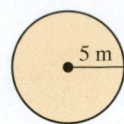

5 m

△ **3.** Approximate the volume of a ball of radius 3 inches. Use the approximation $\dfrac{22}{7}$ for π. Give an exact answer and an approximate answer.

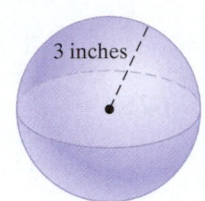

3 inches

△ **4.** Find the volume of the box.

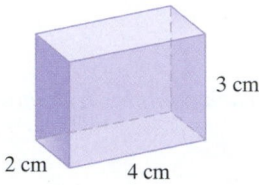

3 cm

2 cm   4 cm

5. Find the median of the list of numbers: 25, 54, 56, 57, 60, 71, 98

6. Find the mean or average of 36, 25, 18, and 19.

7. If a die is rolled one time, find the probability of rolling a 3 or a 4.

8. Subtract: $-9 - (-4.1)$

Simplify each expression.

9. $\left(\dfrac{m}{n}\right)^7$

10. $\dfrac{a^7 b^{10}}{ab^{15}}$

11. Subtract: $(2x^3 + 8x^2 - 6x) - (2x^3 - x^2 + 1)$

12. Add: $\left(5x^2 + 6x + \dfrac{1}{2}\right) + \left(x^2 - \dfrac{4}{3}x - \dfrac{10}{21}\right)$

13. Solve: $x(2x - 7) = 4$

14. Solve: $x(2x - 7) = 0$

15. Subtract: $\dfrac{2y}{2y - 7} - \dfrac{7}{2y - 7}$

16. Add: $\dfrac{2}{x - 6} + \dfrac{3}{x + 1}$

17. Write each sentence as an equation. Let x represent the unknown number.

   a. The quotient of 15 and a number is 4.

   b. Three subtracted from 12 is a number.

   c. 17 added to four times a number is 21.

18. Write each sentence as an equation. Let x represent the unknown number.

   a. The difference of 12 and a number is -45.

   b. The product of 12 and a number is -45.

   c. A number less 10 is twice the number.

19. Find each sum.

   a. $3 + (-7) + (-8)$

   b. $[7 + (-10)] + [-2 + (-4)]$

20. Find each difference.

   a. $28 - 6 - 30$

   b. $7 - 2 - 22$

For Exercises 21 through 24, name the property illustrated by each true statement.

21. $3(x + y) = 3 \cdot x + 3 \cdot y$

22. $3 + y = y + 3$

23. $(x + 7) + 9 = x + (7 + 9)$

24. $(x \cdot 7) \cdot 9 = x \cdot (7 \cdot 9)$

25. Solve: $3 - x = 7$

26. Solve: $7x - 6 = 6x - 6$

27. A 10-foot board is to be cut into two pieces so that the length of the longer piece is 4 times the length of the shorter. Find the length of each piece.

28. Find two consecutive even integers whose sum is 382.

29. Solve $y = mx + b$ for x.

30. Solve $3x - 2y = 6$ for x.

31. Factor: $25x^2 + 20xy + 4y^2$

32. Factor: $x^2 - 4$

17. a. _____

   b. _____

   c. _____

18. a. _____

   b. _____

   c. _____

19. a. _____

   b. _____

20. a. _____

   b. _____

21. _____

22. _____

23. _____

24. _____

25. _____

26. _____

27. _____

28. _____

29. _____

30. _____

31. _____

32. _____

33. _____

34. _____

35. _____

36. _____

37. _____

38. _____

39. _____

40. _____

41. _____

42. _____

43. _____

44. _____

33. Solve: $x^2 - 9x - 22 = 0$

34. Solve: $3x^2 + 5x = 2$

35. Multiply: $\dfrac{x^2 + x}{3x} \cdot \dfrac{6}{5x + 5}$

36. Simplify: $\dfrac{2x^2 - 50}{4x^4 - 20x^3}$

37. Subtract: $\dfrac{3x^2 + 2x}{x - 1} - \dfrac{10x - 5}{x - 1}$

38. Factor: $7x^6 - 7x^5 + 7x^4$

39. Subtract: $\dfrac{6x}{x^2 - 4} - \dfrac{3}{x + 2}$

40. Factor: $4x^2 + 12x + 9$

41. Solve: $\dfrac{t - 4}{2} - \dfrac{t - 3}{9} = \dfrac{5}{18}$

42. Multiply: $\dfrac{6x^2 - 18x}{3x^2 - 2x} \cdot \dfrac{15x - 10}{x^2 - 9}$

43. Sam Waterton and Frank Schaffer work in a plant that manufactures automobiles. Sam can complete a quality control tour of the plant in 3 hours while his assistant, Frank, needs 7 hours to complete the same job. The regional manager is coming to inspect the plant facilities, so both Sam and Frank are directed to complete a quality control tour together. How long will this take?

44. Simplify: $\dfrac{\dfrac{m}{3} + \dfrac{n}{6}}{\dfrac{m + n}{12}}$

Roots and Radicals

Pendulums Can Demonstrate Earth's Rotation on Its Axis

In 1851, French physicist Léon Foucault developed a special pendulum in an experiment to demonstrate that the earth rotates on its axis. He connected his tall pendulum, capable of running for many hours, to the roof of the Paris Observatory. The pendulum's bob was able to swing back and forth in one plane but not to twist in other directions. So, when the pendulum bob appeared to move in a circle over time, he demonstrated that it was not the pendulum but the building that moved. And since the building was firmly attached to the earth, it must be the earth rotating that created the apparent circular motion of the bob. In Section 15.1, Exercise 93, roots are used to explore the time it takes Foucault's pendulum to complete one swing of its bob.

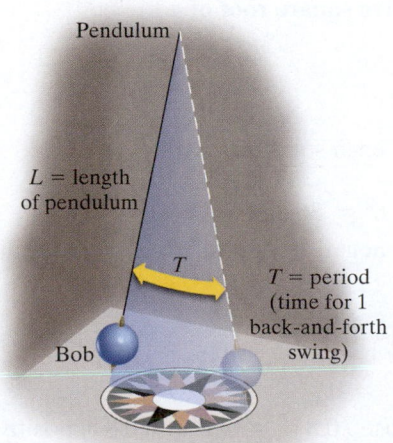

Pendulum

L = length of pendulum

T

Bob

T = period (time for 1 back-and-forth swing)

Having spent the last chapter studying rational expressions, we now turn to another kind of algebraic expression. We expand on our skills of operating on expressions—adding, subtracting, multiplying, dividing, and raising to powers—to include finding roots. Just as subtraction is defined by addition and division by multiplication, finding roots is defined by raising to powers. As we master finding roots, we will work with equations that contain roots and solve problems that can be modeled by such equations.

1109

Objectives

A Find Square Roots.

B Find Cube Roots.

C Find nth Roots.

D Approximate Square Roots.

E Simplify Radicals Containing Variables.

Objective A Finding Square Roots

In this section, we define finding the **root** of a number by its reverse operation, raising a number to a power. We begin with squares and square roots.

The *square* of 5 is $5^2 = 25$.

The *square* of -5 is $(-5)^2 = 25$.

The *square* of $\frac{1}{2}$ is $\left(\frac{1}{2}\right)^2 = \frac{1}{4}$.

The reverse operation of squaring a number is finding a **square root** of a number. For example,

A *square root* of 25 is 5, because $5^2 = 25$.

A *square root* of 25 is also -5, because $(-5)^2 = 25$.

A *square root* of $\frac{1}{4}$ is $\frac{1}{2}$, because $\left(\frac{1}{2}\right)^2 = \frac{1}{4}$.

> In general, the number b is a square root of a number a if $b^2 = a$.

The symbol $\sqrt{}$ is used to denote the **positive** or **principal square root** of a number. For example,

$\sqrt{25} = 5$ only, since $5^2 = 25$ and 5 is positive.

The symbol $-\sqrt{}$ is used to denote the **negative square root.** For example,

$-\sqrt{25} = -5$

The symbol $\sqrt{}$ is called a **radical** or **radical sign.** The expression within or under a radical sign is called the **radicand.** An expression containing a radical is called a **radical expression.**

$$\sqrt{a}$$

radical sign

radicand

> **Square Root**
>
> If a is a positive number, then
>
> $\sqrt{a}$ is the **positive square root** of a and
>
> $-\sqrt{a}$ is the **negative square root** of a.
>
> Also, $\sqrt{0} = 0$.

Practice 1–5

Find each square root.

1. $\sqrt{100}$ **2.** $-\sqrt{81}$

3. $\sqrt{\dfrac{25}{81}}$ **4.** $\sqrt{1}$

5. $\sqrt{0.81}$

Examples   Find each square root.

1. $\sqrt{36} = 6$, because $6^2 = 36$ and 6 is positive.

2. $-\sqrt{16} = -4$. The negative sign in front of the radical indicates the negative square root of 16.

3. $\sqrt{\dfrac{9}{100}} = \dfrac{3}{10}$ because $\left(\dfrac{3}{10}\right)^2 = \dfrac{9}{100}$ and $\dfrac{3}{10}$ is positive.

4. $\sqrt{0} = 0$ because $0^2 = 0$.

5. $\sqrt{0.64} = 0.8$ because $(0.8)^2 = 0.64$ and 0.8 is positive.

■ **Work Practice 1–5**

Answers

1. 10 **2.** -9 **3.** $\dfrac{5}{9}$ **4.** 1 **5.** 0.9

Is the square root of a negative number a real number? For example, is $\sqrt{-4}$ a real number? To answer this question, we ask ourselves, is there a real number whose square is -4? Since there is no real number whose square is -4, we say that $\sqrt{-4}$ is not a real number. In general,

> A square root of a negative number is not a real number.

Study the following table to make sure you understand the differences discussed earlier.

Number	Square Roots of Number	$\sqrt{\text{number}}$	$-\sqrt{\text{number}}$
25	$-5, 5$	$\sqrt{25} = 5$ only	$-\sqrt{25} = -5$
$\dfrac{1}{4}$	$-\dfrac{1}{2}, \dfrac{1}{2}$	$\sqrt{\dfrac{1}{4}} = \dfrac{1}{2}$ only	$-\sqrt{\dfrac{1}{4}} = -\dfrac{1}{2}$
-9	No real square roots.	$\sqrt{-9}$ is not a real number.	

Objective B Finding Cube Roots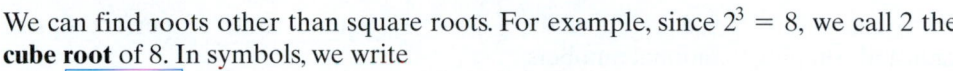

We can find roots other than square roots. For example, since $2^3 = 8$, we call 2 the **cube root** of 8. In symbols, we write

▶ $\sqrt[3]{8} = 2$ The number 3 is called the **index.**

Also,

$\sqrt[3]{-64} = -4$ Since $(-4)^3 = -64$

Notice that unlike the square root of a negative number, the cube root of a negative number is a real number. This is so because while we cannot find a real number whose *square* is negative, we *can* find a real number whose *cube* is negative. In fact, the cube of a negative number is a negative number. Therefore, the cube root of a negative number is a negative number.

Examples   Find each cube root.

6. $\sqrt[3]{1} = 1$ because $1^3 = 1$.
7. $\sqrt[3]{-27} = -3$ because $(-3)^3 = -27$.
8. $\sqrt[3]{\dfrac{1}{125}} = \dfrac{1}{5}$ because $\left(\dfrac{1}{5}\right)^3 = \dfrac{1}{125}$.

■ **Work Practice 6–8**

Objective C Finding nth Roots

Just as we can raise a real number to powers other than 2 or 3, we can find roots other than square roots and cube roots. In fact, we can take the nth root of a number where n is any natural number. An **nth root** of a number a is a number whose nth power is a.

In symbols, the nth root of a is written as $\sqrt[n]{a}$. Recall that n is called the **index.** The index 2 is usually omitted for square roots.

Helpful Hint

If the index is even, as it is in $\sqrt{}$, $\sqrt[4]{}$, $\sqrt[6]{}$, and so on, the radicand must be nonnegative for the root to be a real number. For example,

$\sqrt[4]{81} = 3$ but $\sqrt[4]{-81}$ is not a real number.
$\sqrt[6]{64} = 2$ but $\sqrt[6]{-64}$ is not a real number.

Practice 6–8

Find each cube root.
6. $\sqrt[3]{27}$
7. $\sqrt[3]{-8}$
8. $\sqrt[3]{\dfrac{1}{64}}$

Answers

6. 3 **7.** -2 **8.** $\dfrac{1}{4}$

Copyright 2015 Pearson Education, Inc.

✓**Concept Check** Which of the following is a real number?

    a. $\sqrt{-64}$ **b.** $\sqrt[4]{-64}$ **c.** $\sqrt[5]{-64}$ **d.** $\sqrt[6]{-64}$

> **Examples** Find each root.

9. $\sqrt[4]{16} = 2$ because $2^4 = 16$ and 2 is positive.

▶ **10.** $\sqrt[5]{-32} = -2$ because $(-2)^5 = -32$.

11. $-\sqrt[6]{1} = -1$ because $\sqrt[6]{1} = 1$.

▶ **12.** $\sqrt[4]{-81}$ is not a real number since the index, 4, is even and the radicand, -81, is negative. In other words, there is no real number that when raised to the 4th power gives -81.

■ **Work Practice 9–12**

Objective D Approximating Square Roots

Recall that numbers such as 1, 4, 9, 25, and $\dfrac{4}{25}$ are called **perfect squares,** since $1^2 = 1, 2^2 = 4, 3^2 = 9, 5^2 = 25,$ and $\left(\dfrac{2}{5}\right)^2 = \dfrac{4}{25}$. Square roots of perfect square radicands simplify to rational numbers.

What happens when we try to simplify a root such as $\sqrt{3}$? Since 3 is not a perfect square, $\sqrt{3}$ is not a rational number. It cannot be written as a quotient of integers. It is called an **irrational number** and we can find a decimal **approximation** of it. To find decimal approximations, use a calculator or Appendix B.1. (For calculator help, see the next example or the box at the end of this section.)

> **Example 13** Use a calculator or Appendix B.1 to approximate $\sqrt{3}$ to three decimal places.

Solution: We may use Appendix B.1 or a calculator to approximate $\sqrt{3}$. To use a calculator, find the square root key $\boxed{\sqrt{\ }}$.

    $\sqrt{3} \approx 1.732050808$

To three decimal places, $\sqrt{3} \approx 1.732$.

■ **Work Practice 13**

From Example 13, we found that

    $\sqrt{3} \approx 1.732$

To see if the approximation is reasonable, notice that since

    $1 < 3 < 4$, then
    $\sqrt{1} < \sqrt{3} < \sqrt{4}$, or
    $1 < \sqrt{3} < 2$.

Since $\sqrt{3}$ is a number between 1 and 2, our result of $\sqrt{3} \approx 1.732$ is reasonable.

Objective E Simplifying Radicals Containing Variables

Radicals can also contain variables. To simplify radicals containing variables, special care must be taken. To see how we simplify $\sqrt{x^2}$, let's look at a few examples in this form.

    If $x = 3$, we have $\sqrt{3^2} = \sqrt{9} = 3$, or x.
    If x is 5, we have $\sqrt{5^2} = \sqrt{25} = 5$, or x.

From these two examples, you may think that $\sqrt{x^2}$ simplifies to x. Let's now look at an example where x is a negative number. If $x = -3$, we have $\sqrt{(-3)^2} = \sqrt{9} = 3$, not -3, our original x. To make sure that $\sqrt{x^2}$ simplifies to a nonnegative number, we have the following.

Practice 9–12

Find each root.

  9. $\sqrt[4]{-16}$

10. $\sqrt[5]{-1}$

11. $\sqrt[4]{256}$

12. $\sqrt[6]{-1}$

Practice 13

Use a calculator or Appendix B.1 to approximate $\sqrt{22}$ to three decimal places.

Answers

9. not a real number   **10.** -1 **11.** 4

12. not a real number   **13.** 4.690

✓**Concept Check Answer**

c

For any real number a,
$$\sqrt{a^2} = |a|$$

Thus,

▸ $\sqrt{x^2} = |x|$,
$\sqrt{(-8)^2} = |-8| = 8$
$\sqrt{(7y)^2} = |7y|$,   and so on.

To avoid this confusion, for the rest of the chapter we assume that **if a variable appears in the radicand of a radical expression, it represents positive numbers only.** Then

$\sqrt{x^2} = |x| = x$ since x is a positive number.

$\sqrt{y^2} = y$ Because $(y)^2 = y^2$

$\sqrt{x^8} = x^4$ Because $(x^4)^2 = x^8$

$\sqrt{9x^2} = 3x$ Because $(3x)^2 = 9x^2$

Examples   Simplify each expression. Assume that all variables represent positive numbers.

14. $\sqrt{z^2} = z$ because $(z)^2 = z^2$.

15. $\sqrt{x^6} = x^3$ because $(x^3)^2 = x^6$.

16. $\sqrt[3]{27y^6} = 3y^2$ because $(3y^2)^3 = 27y^6$.

17. $\sqrt{16x^{16}} = 4x^8$ because $(4x^8)^2 = 16x^{16}$.

18. $\sqrt{\dfrac{x^4}{25}} = \dfrac{x^2}{5}$ because $\left(\dfrac{x^2}{5}\right)^2 = \dfrac{x^4}{25}$.

19. $\sqrt[3]{-125a^{12}b^{15}} = -5a^4b^5$ because $(-5a^4b^5)^3 = -125a^{12}b^{15}$.

■ **Work Practice 14–19**

Practice 14–19

Simplify each expression. Assume that all variables represent positive numbers.

14. $\sqrt{z^8}$ **15.** $\sqrt{x^{20}}$

16. $\sqrt{4x^6}$ **17.** $\sqrt[3]{8y^{12}}$

18. $\sqrt{\dfrac{z^8}{81}}$ **19.** $\sqrt[3]{-64x^9y^{24}}$

Answers

14. z^4 **15.** x^{10} **16.** $2x^3$ **17.** $2y^4$

18. $\dfrac{z^4}{9}$ **19.** $-4x^3y^8$

Calculator Explorations **Simplifying Square Roots**

To simplify or approximate square roots using a calculator, locate the key marked $\boxed{\sqrt{}}$. To simplify $\sqrt{25}$ using a scientific calculator, press $\boxed{25}$ $\boxed{\sqrt{}}$. The display should read $\boxed{5}$. To simplify $\sqrt{25}$ using a graphing calculator, press $\boxed{\sqrt{}}$ $\boxed{25}$ $\boxed{\text{ENTER}}$.

To approximate $\sqrt{30}$, press $\boxed{30}$ $\boxed{\sqrt{}}$ (or $\boxed{\sqrt{}}$ $\boxed{30}$). The display should read $\boxed{5.477225575}$. This is an approximation for $\sqrt{30}$. A three-decimal-place approximation is

$$\sqrt{30} \approx 5.477$$

Is this answer reasonable? Since 30 is between perfect squares 25 and 36, $\sqrt{30}$ is between $\sqrt{25} = 5$ and $\sqrt{36} = 6$. The calculator result is then reasonable since 5.477225575 is between 5 and 6.

Use a calculator to approximate each expression to three decimal places. Decide whether each result is reasonable.

1. $\sqrt{6}$ **2.** $\sqrt{14}$

3. $\sqrt{11}$ **4.** $\sqrt{200}$

5. $\sqrt{82}$ **6.** $\sqrt{46}$

Many scientific calculators have a key, such as $\boxed{\sqrt[x]{y}}$, that can be used to approximate roots other than square roots. To approximate these roots using a graphing calculator, look under the $\boxed{\text{MATH}}$ menu or consult your manual. To use a $\boxed{\sqrt[x]{y}}$ key to find $\sqrt[3]{8}$, press $\boxed{3}$ $\boxed{\sqrt[x]{y}}$ $\boxed{8}$ (press $\boxed{\text{ENTER}}$ if needed). The display should read $\boxed{2}$.

Use a calculator to approximate each expression to three decimal places. Decide whether each result is reasonable.

7. $\sqrt[3]{40}$ **8.** $\sqrt[3]{71}$

9. $\sqrt[4]{20}$ **10.** $\sqrt[4]{15}$

11. $\sqrt[5]{18}$ **12.** $\sqrt[6]{2}$

Vocabulary, Readiness & Video Check

Use the choices below to fill in each blank. All choices are used.

positive	index	radical sign	power
negative	principal	square root	radicand

1. The symbol $\sqrt{}$ is used to denote the positive, or _____, square root.
2. In the expression $\sqrt[4]{16}$, the number 4 is called the _____, the number 16 is called the _____, and $\sqrt{}$ is called the _____.
3. The reverse operation of squaring a number is finding a(n) _____ of a number.
4. For a positive number a,

   $-\sqrt{a}$ is the _____ square root of a and

   $\sqrt{a}$ is the _____ square root of a.
5. An nth root of a number a is a number whose nth _____ is a.

Answer each true or false.

6. $\sqrt{4} = -2$ _____
7. $\sqrt{-9} = -3$ _____
8. $\sqrt{1000} = 100$ _____
9. $\sqrt{1} = 1$ and $\sqrt{0} = 0$ _____
10. $\sqrt{64} = 8$ and $\sqrt[3]{64} = 4$ _____

Martin-Gay Interactive Videos

See Video 15.1

Watch the section lecture video and answer the following questions.

Objective A 11. Explain the differences between ▥ Examples 1 and 2, including how we know which one simplifies to a positive number and which one simplifies to a negative number. ▶

Objective B 12. From ▥ Example 11, what is an important difference between the square root and the cube root of a negative number? ▶

Objective C 13. From ▥ Examples 12–15, given a negative radicand, what kind of index must we have to be a real number? ▶

Objective D 14. From ▥ Example 16, how do we determine if an approximate answer is reasonable? ▶

Objective E 15. As explained in ▥ Example 19, when simplifying radicals containing variables, what is a shortcut you can use when dealing with exponents? ▶

15.1 Exercise Set MyMathLab®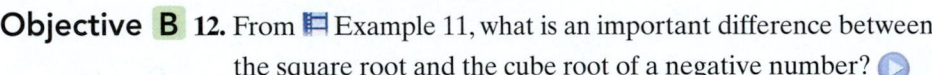

Objective A *Find each square root. See Examples 1 through 5.*

1. $\sqrt{16}$
2. $\sqrt{64}$
3. $\sqrt{\dfrac{1}{25}}$
4. $\sqrt{\dfrac{1}{64}}$
5. $-\sqrt{100}$

6. $-\sqrt{36}$
▶ 7. $\sqrt{-4}$
8. $\sqrt{-25}$
9. $-\sqrt{121}$
▶ 10. $-\sqrt{49}$

11. $\sqrt{\dfrac{9}{25}}$ **12.** $\sqrt{\dfrac{4}{81}}$ **13.** $\sqrt{900}$ **14.** $\sqrt{400}$ **15.** $\sqrt{144}$

16. $\sqrt{169}$ **17.** $\sqrt{\dfrac{1}{100}}$ **18.** $\sqrt{\dfrac{1}{121}}$ **19.** $\sqrt{0.25}$ **20.** $\sqrt{0.49}$

Objective B   *Find each cube root. See Examples 6 through 8.*

▶ 21. $\sqrt[3]{125}$ **22.** $\sqrt[3]{64}$ **23.** $\sqrt[3]{-64}$ **▶ 24.** $\sqrt[3]{-27}$ **25.** $-\sqrt[3]{8}$

26. $-\sqrt[3]{27}$ **27.** $\sqrt[3]{\dfrac{1}{8}}$ **28.** $\sqrt[3]{\dfrac{1}{64}}$ **29.** $\sqrt[3]{-125}$ **30.** $\sqrt[3]{-1}$

Objectives A B C   Mixed Practice   *Find each root. See Examples 1 through 12.*

31. $\sqrt[5]{32}$ **▶ 32.** $\sqrt[4]{81}$ **33.** $\sqrt{81}$ **▶ 34.** $\sqrt{49}$

35. $\sqrt[4]{-16}$ **36.** $\sqrt{-9}$ **37.** $\sqrt[3]{-\dfrac{27}{64}}$ **38.** $\sqrt[3]{-\dfrac{8}{27}}$

39. $-\sqrt[4]{625}$ **▶ 40.** $-\sqrt[5]{32}$ **41.** $\sqrt[6]{1}$ **42.** $\sqrt[5]{1}$

Objective D   *Approximate each square root to three decimal places. See Example 13.*

43. $\sqrt{7}$ **44.** $\sqrt{10}$ **45.** $\sqrt{37}$ **46.** $\sqrt{27}$ **▶ 47.** $\sqrt{136}$ **48.** $\sqrt{8}$

49. A standard baseball diamond is a square with 90-foot sides connecting the bases. The distance from home plate to second base is $90 \cdot \sqrt{2}$ feet. Approximate $\sqrt{2}$ to two decimal places and use your result to approximate the distance $90 \cdot \sqrt{2}$ feet.

50. The roof of the warehouse shown needs to be shingled. The total area of the roof is exactly $480 \cdot \sqrt{29}$ square feet. Approximate $\sqrt{29}$ to two decimal places and use your result to approximate the area $480 \cdot \sqrt{29}$ square feet. Approximate this area to the nearest whole number.

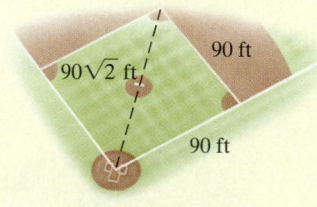

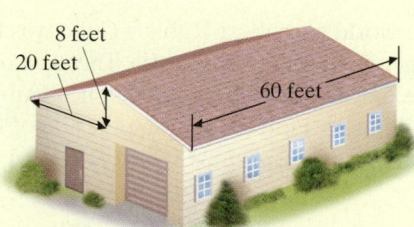

Objective E   *Find each root. Assume that all variables represent positive numbers. See Examples 14 through 19.*

51. $\sqrt{m^2}$ **52.** $\sqrt{y^{10}}$ **▶ 53.** $\sqrt{x^4}$ **54.** $\sqrt{x^6}$

55. $\sqrt{9x^8}$ **▶ 56.** $\sqrt{36x^{12}}$ **57.** $\sqrt{81x^2}$ **58.** $\sqrt{100z^4}$

59. $\sqrt{a^2 b^4}$ **60.** $\sqrt{x^{12} y^{20}}$ **61.** $\sqrt{16 a^6 b^4}$ **62.** $\sqrt{4 m^{14} n^2}$

⊙ **63.** $\sqrt[3]{a^6 b^{18}}$ **64.** $\sqrt[3]{x^{12} y^{18}}$ **65.** $\sqrt[3]{-8 x^3 y^{27}}$ **66.** $\sqrt[3]{-27 a^6 b^{30}}$

⊙ **67.** $\sqrt{\dfrac{x^6}{36}}$ **68.** $\sqrt{\dfrac{y^8}{49}}$ **69.** $\sqrt{\dfrac{25 y^2}{9}}$ **70.** $\sqrt{\dfrac{4 x^2}{81}}$

Review

Write each integer as a product of two integers such that one of the factors is a perfect square. For example, we can write
18 = 9 · 2, where 9 is a perfect square. See Section 2.2.

71. 50 **72.** 8 **73.** 32 **74.** 75

75. 28 **76.** 44 **77.** 27 **78.** 90

Concept Extensions

Solve. See the Concept Check in this section.

79. Which of the following is a real number?
    a. $\sqrt[7]{-1}$ **b.** $\sqrt[3]{-125}$
    c. $\sqrt[6]{-128}$ **d.** $\sqrt[8]{-1}$

80. Which of the following is a real number?
    a. $\sqrt{-1}$ **b.** $\sqrt[3]{-1}$
    c. $\sqrt[4]{-1}$ **d.** $\sqrt[5]{-1}$

The length of a side of a square is given by the expression $\sqrt{A}$, where A is the square's area. Use this expression for Exercises 81 through 84. Be sure to attach the appropriate units.

△ **81.** The area of a square is 49 square miles. Find the length of a side of the square.

△ **82.** The area of a square is $\dfrac{1}{81}$ square meters. Find the length of a side of the square.

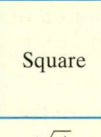

Square

$\sqrt{A}$

△ **83.** The world's smallest Rubik's Cube was built by Evgeniy Grigoriev of Russia. The area of one square face of this fully functional Rubik's Cube is 100 square millimeters. Find the length of a side of the square face. (*Source: Guinness World Records*)

△ **84.** A parking lot is in the shape of a square with area 2500 square yards. Find the length of a side.

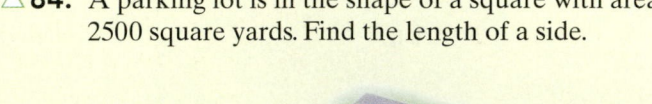

85. Simplify $\sqrt{\sqrt{81}}$.

86. Simplify $\sqrt[3]{\sqrt[3]{1}}$.

87. Simplify $\sqrt{\sqrt{10,000}}$.

88. Simplify $\sqrt{\sqrt{1,600,000,000}}$.

For each square root below, give two whole numbers that the square root lies between. For example, since 11 is between 9 and 16, then $\sqrt{11}$ is between 3 and 4.

89. $\sqrt{18}$

90. $\sqrt{28}$

91. $\sqrt{80}$

92. $\sqrt{98}$

93. The formula for calculating the period (one back-and-forth swing) of a pendulum is $T = 2\pi\sqrt{\dfrac{L}{g}}$, where T is time of the period of the swing, L is the length of the pendulum, and g is the acceleration of gravity. At the California Academy of Sciences, one can see a Foucault's pendulum with length = 30 ft and $g = 32$ ft/sec^2. Using $\pi \approx 3.14$, find the period of this pendulum. (Round to the nearest tenth of a second.)

94. If the amount of gold discovered by humankind could be assembled in one place, it would be a cube with a volume of 195,112 cubic feet. Each side of the cube would be $\sqrt[3]{195{,}112}$ feet long. How long would one side of the cube be? (*Source: Reader's Digest*)

95. Explain why the square root of a negative number is not a real number.

96. Explain why the cube root of a negative number is a real number.

97. Graph $y = \sqrt{x}$. (Complete the table below, plot the ordered pair solutions, and draw a smooth curve through the points. Remember that since the radicand cannot be negative, this particular graph begins at the point with coordinates $(0, 0)$.)

x	y
0	0
1	
3	(approximate)
4	
9	

98. Graph $y = \sqrt[3]{x}$. (Complete the table below, plot the ordered pair solutions, and draw a smooth curve through the points.)

x	y
-8	
-2	(approximate)
-1	
0	
1	
2	(approximate)
8	

Recall from this section that $\sqrt{a^2} = |a|$ for any real number a. Simplify the following given that x represents any real number.

99. $\sqrt{x^2}$

100. $\sqrt{4x^2}$

101. $\sqrt{(x + 2)^2}$

102. $\sqrt{x^2 + 6x + 9}$
(*Hint:* First factor $x^2 + 6x + 9$.)

Use a graphing calculator and graph each function. Observe the graph from left to right and give the ordered pair that corresponds to the "beginning" of the graph. Then tell why the graph starts at that point.

103. $y = \sqrt{x - 2}$

104. $y = \sqrt{x + 3}$

105. $y = \sqrt{x + 4}$

106. $y = \sqrt{x - 5}$

15.2 Simplifying Radicals

Objectives

A Use the Product Rule to Simplify Radicals. ▶

B Use the Quotient Rule to Simplify Radicals. ▶

C Use Both Rules to Simplify Radicals Containing Variables. ▶

D Simplify Cube Roots. ▶

Objective A Simplifying Radicals Using the Product Rule ▶

A square root is simplified when the radicand contains no perfect square factors (other than 1). For example, $\sqrt{20}$ is not simplified because $\sqrt{20} = \sqrt{4 \cdot 5}$ and 4 is a perfect square.

To begin simplifying square roots, we notice the following pattern.

$$\sqrt{9 \cdot 16} = \sqrt{144} = 12$$
$$\sqrt{9} \cdot \sqrt{16} = 3 \cdot 4 = 12$$

Since both expressions simplify to 12, we can write

$$\sqrt{9 \cdot 16} = \sqrt{9} \cdot \sqrt{16}$$

This suggests the following product rule for square roots.

> **Product Rule for Square Roots**
>
> If $\sqrt{a}$ and $\sqrt{b}$ are real numbers, then
> $$\sqrt{a \cdot b} = \sqrt{a} \cdot \sqrt{b}$$

In other words, the square root of a product is equal to the product of the square roots.

To simplify $\sqrt{45}$, for example, we factor 45 so that one of its factors is a perfect square factor.

$$\sqrt{45} = \sqrt{9 \cdot 5} \qquad \text{Factor 45.}$$
$$= \sqrt{9} \cdot \sqrt{5} \qquad \text{Use the product rule.}$$
$$= 3\sqrt{5} \qquad \text{Write } \sqrt{9} \text{ as 3.}$$

The notation $3\sqrt{5}$ means $3 \cdot \sqrt{5}$. Since the radicand 5 has no perfect square factor other than 1, the expression $3\sqrt{5}$ is in simplest form.

Helpful Hint

A radical expression in simplest form *does not mean* a decimal approximation. The simplest form of a radical expression is an exact form and may still contain a radical.

$$\underset{\text{exact}}{\sqrt{45} = 3\sqrt{5}} \qquad \underset{\text{decimal approximation}}{\sqrt{45} \approx 6.71}$$

Practice 1–4

Simplify.
1. $\sqrt{40}$
2. $\sqrt{18}$
3. $\sqrt{500}$
4. $\sqrt{15}$

Answers
1. $2\sqrt{10}$
2. $3\sqrt{2}$
3. $10\sqrt{5}$
4. $\sqrt{15}$

Examples Simplify.

1. $\sqrt{54} = \sqrt{9 \cdot 6}$ Factor 54 so that one factor is a perfect square. 9 is a perfect square.

$\qquad = \sqrt{9} \cdot \sqrt{6}$ Use the product rule.

$\qquad = 3\sqrt{6}$ Write $\sqrt{9}$ as 3.

2. $\sqrt{12} = \sqrt{4 \cdot 3}$ Factor 12 so that one factor is a perfect square. 4 is a perfect square.

$\qquad = \sqrt{4} \cdot \sqrt{3}$ Use the product rule.

$\qquad = 2\sqrt{3}$ Write $\sqrt{4}$ as 2.

3. $\sqrt{200} = \sqrt{100 \cdot 2}$ $\qquad$ Factor 200 so that one factor is a perfect square. 100 is a perfect square.

$\qquad\quad = \sqrt{100} \cdot \sqrt{2}$ $\qquad$ Use the product rule.

$\qquad\quad = 10\sqrt{2}$ $\qquad$ Write $\sqrt{100}$ as 10.

4. $\sqrt{35}$ The radicand 35 contains no perfect square factors other than 1. Thus $\sqrt{35}$ is in simplest form.

■ **Work Practice 1–4**

In Example 3, 100 is the largest perfect square factor of 200. What happens if we don't use the largest perfect square factor? Although using the largest perfect square factor saves time, the result is the same no matter what perfect square factor is used. For example, it is also true that $200 = 4 \cdot 50$. Then

$$\sqrt{200} = \sqrt{4} \cdot \sqrt{50}$$
$$= 2 \cdot \sqrt{50}$$

Since $\sqrt{50}$ is not in simplest form, we continue.

$$\sqrt{200} = 2 \cdot \sqrt{50}$$
$$= 2 \cdot \sqrt{25 \cdot 2}$$
$$= 2 \cdot \sqrt{25} \cdot \sqrt{2}$$
$$= 2 \cdot 5 \cdot \sqrt{2}$$
$$= 10\sqrt{2}$$

Example 5 $\quad$ Simplify $3\sqrt{8}$.

Solution: $\quad$ Remember that $3\sqrt{8}$ means $3 \cdot \sqrt{8}$.

$\quad 3 \cdot \sqrt{8} = 3 \cdot \sqrt{4 \cdot 2}$ $\qquad$ Factor 8 so that one factor is a perfect square.

$\qquad\quad = 3 \cdot \sqrt{4} \cdot \sqrt{2}$ $\qquad$ Use the product rule.

$\qquad\quad = 3 \cdot 2 \cdot \sqrt{2}$ $\qquad$ Write $\sqrt{4}$ as 2.

$\qquad\quad = 6 \cdot \sqrt{2}$ or $6\sqrt{2}$ $\quad$ Write $3 \cdot 2$ as 6.

■ **Work Practice 5**

Objective B Simplifying Radicals Using the Quotient Rule ▶

Next, let's examine the square root of a quotient.

$$\sqrt{\frac{16}{4}} = \sqrt{4} = 2$$

Also,

$$\frac{\sqrt{16}}{\sqrt{4}} = \frac{4}{2} = 2$$

Since both expressions equal 2, we can write

$$\sqrt{\frac{16}{4}} = \frac{\sqrt{16}}{\sqrt{4}}$$

This suggests the following quotient rule.

Practice 5
Simplify $7\sqrt{75}$.

Answer
5. $35\sqrt{3}$

Quotient Rule for Square Roots

If $\sqrt{a}$ and $\sqrt{b}$ are real numbers and $b \neq 0$, then

$$\sqrt{\frac{a}{b}} = \frac{\sqrt{a}}{\sqrt{b}}$$

In other words, the square root of a quotient is equal to the quotient of the square roots.

Practice 6–8

Use the quotient rule to simplify.

6. $\sqrt{\dfrac{16}{81}}$

7. $\sqrt{\dfrac{2}{25}}$

8. $\sqrt{\dfrac{45}{49}}$

Examples   Use the quotient rule to simplify.

6. $\sqrt{\dfrac{25}{36}} = \dfrac{\sqrt{25}}{\sqrt{36}} = \dfrac{5}{6}$

7. $\sqrt{\dfrac{3}{64}} = \dfrac{\sqrt{3}}{\sqrt{64}} = \dfrac{\sqrt{3}}{8}$

8. $\sqrt{\dfrac{40}{81}} = \dfrac{\sqrt{40}}{\sqrt{81}}$ Use the quotient rule.

$\qquad = \dfrac{\sqrt{4} \cdot \sqrt{10}}{9}$ Use the product rule and write $\sqrt{81}$ as 9.

$\qquad = \dfrac{2\sqrt{10}}{9}$ Write $\sqrt{4}$ as 2.

■ **Work Practice 6–8**

Objective C Simplifying Radicals Containing Variables

Recall that $\sqrt{x^6} = x^3$ because $(x^3)^2 = x^6$. If a variable radicand in a square root has an odd exponent, we write the exponential expression so that one factor is the greatest even power contained in the expression. Then we use the product rule to simplify.

Practice 9–12

Simplify each radical. Assume that all variables represent positive numbers.

9. $\sqrt{x^{11}}$ **10.** $\sqrt{18x^4}$

11. $\sqrt{\dfrac{27}{x^8}}$ **12.** $\sqrt{\dfrac{7y^7}{25}}$

Answers

6. $\dfrac{4}{9}$ **7.** $\dfrac{\sqrt{2}}{5}$ **8.** $\dfrac{3\sqrt{5}}{7}$ **9.** $x^5\sqrt{x}$

10. $3x^2\sqrt{2}$ **11.** $\dfrac{3\sqrt{3}}{x^4}$ **12.** $\dfrac{y^3\sqrt{7y}}{5}$

Examples   Simplify each radical. Assume that all variables represent positive numbers.

9. $\sqrt{x^5} = \sqrt{x^4 \cdot x} = \sqrt{x^4} \cdot \sqrt{x} = x^2\sqrt{x}$

10. $\sqrt{8y^2} = \sqrt{4 \cdot 2 \cdot y^2} = \sqrt{4y^2 \cdot 2} = \sqrt{4y^2} \cdot \sqrt{2} = 2y\sqrt{2}$ 4 and y^2 are both perfect square factors so we grouped them under one radical.

11. $\sqrt{\dfrac{45}{x^6}} = \dfrac{\sqrt{45}}{\sqrt{x^6}} = \dfrac{\sqrt{9 \cdot 5}}{x^3} = \dfrac{\sqrt{9} \cdot \sqrt{5}}{x^3} = \dfrac{3\sqrt{5}}{x^3}$

12. $\sqrt{\dfrac{5p^3}{9}} = \dfrac{\sqrt{5p^3}}{\sqrt{9}} = \dfrac{\sqrt{p^2 \cdot 5p}}{3} = \dfrac{\sqrt{p^2} \cdot \sqrt{5p}}{3} = \dfrac{p\sqrt{5p}}{3}$

■ **Work Practice 9–12**

Objective D Simplifying Cube Roots

The product and quotient rules also apply to roots other than square roots. For example, to simplify cube roots, we look for perfect cube factors of the radicand. Recall that 8 is a perfect cube since $2^3 = 8$. Therefore, to simplify $\sqrt[3]{48}$, we factor 48 as $8 \cdot 6$.

$$\sqrt[3]{48} = \sqrt[3]{8 \cdot 6} \qquad \text{Factor 48.}$$
$$= \sqrt[3]{8} \cdot \sqrt[3]{6} \qquad \text{Use the product rule.}$$
$$= 2\sqrt[3]{6} \qquad \text{Write } \sqrt[3]{8} \text{ as 2.}$$

$2\sqrt[3]{6}$ is in simplest form since the radicand, 6, contains no perfect cube factors other than 1.

Examples   Simplify each radical.

13. $\sqrt[3]{54} = \sqrt[3]{27 \cdot 2} = \sqrt[3]{27} \cdot \sqrt[3]{2} = 3\sqrt[3]{2}$

14. $\sqrt[3]{18}$ The number 18 contains no perfect cube factors, so $\sqrt[3]{18}$ cannot be simplified further.

15. $\sqrt[3]{\dfrac{7}{8}} = \dfrac{\sqrt[3]{7}}{\sqrt[3]{8}} = \dfrac{\sqrt[3]{7}}{2}$

16. $\sqrt[3]{\dfrac{40}{27}} = \dfrac{\sqrt[3]{40}}{\sqrt[3]{27}} = \dfrac{\sqrt[3]{8 \cdot 5}}{3} = \dfrac{\sqrt[3]{8} \cdot \sqrt[3]{5}}{3} = \dfrac{2\sqrt[3]{5}}{3}$

■ **Work Practice 13–16**

Practice 13–16

Simplify each radical.

13. $\sqrt[3]{88}$ **14.** $\sqrt[3]{50}$

15. $\sqrt[3]{\dfrac{10}{27}}$ **16.** $\sqrt[3]{\dfrac{81}{8}}$

Answers

13. $2\sqrt[3]{11}$ **14.** $\sqrt[3]{50}$ **15.** $\dfrac{\sqrt[3]{10}}{3}$

16. $\dfrac{3\sqrt[3]{3}}{2}$

Vocabulary, Readiness & Video Check

Use the choices below to fill in each blank. Not all choices will be used.

$$a \cdot b \qquad \dfrac{a}{b} \qquad \dfrac{\sqrt{a}}{\sqrt{b}} \qquad \sqrt{a} \cdot \sqrt{b}$$

1. If $\sqrt{a}$ and $\sqrt{b}$ are real numbers, then $\sqrt{a \cdot b} = $ _____.

2. If $\sqrt{a}$ and $\sqrt{b}$ are real numbers, then $\sqrt{\dfrac{a}{b}} = $ _____.

For Exercises 3 and 4, fill in the blanks using the example: $\sqrt{4 \cdot 9} = \underline{\sqrt{4}} \cdot \underline{\sqrt{9}} = \underline{2} \cdot \underline{3} = \underline{6}.$

3. $\sqrt{16 \cdot 25} = \sqrt{\underline{}} \cdot \sqrt{\underline{}} = \underline{} \cdot \underline{} = \underline{}.$

4. $\sqrt{36 \cdot 3} = \sqrt{\underline{}} \cdot \sqrt{\underline{}} = \underline{} \cdot \sqrt{\underline{}} = \underline{}.$

True or False? Decide whether each radical is completely simplified.

5. $\sqrt{48} = 2\sqrt{12}$ Completely simplified: _____

6. $\sqrt[3]{48} = 2\sqrt[3]{6}$ Completely simplified: _____

Martin-Gay Interactive Videos   Watch the section lecture video and answer the following questions.

Objective A 7. From ▥ Example 3, if we have trouble finding a perfect square factor in the radicand, what is recommended? ▶

Objective B 8. Based on the lecture before ▥ Example 5, complete the following statement: In words, the quotient rule for square roots says that the square root of a quotient is equal to the square root of the _____ over the square root of the _____. ▶

Objective C 9. From ▥ Examples 6–8, we know that even powers of a variable are perfect square factors of the variable. Therefore, what must be true about the power of any variable left in the radicand of a simplified square root? Explain. ▶

Objective D 10. From ▥ Example 9, how does factoring the radicand as a product of primes help simplify higher roots also? ▶

See Video 15.2

15.2 Exercise Set MyMathLab® ▶

Objective A *Use the product rule to simplify each radical. See Examples 1 through 4.*

▶ **1.** $\sqrt{20}$ **2.** $\sqrt{44}$ **3.** $\sqrt{50}$ **4.** $\sqrt{28}$ ▶ **5.** $\sqrt{33}$

6. $\sqrt{21}$ **7.** $\sqrt{98}$ **8.** $\sqrt{125}$ **9.** $\sqrt{60}$ **10.** $\sqrt{90}$

▶ **11.** $\sqrt{180}$ **12.** $\sqrt{150}$ **13.** $\sqrt{52}$ **14.** $\sqrt{75}$

Use the product rule to simplify each radical. See Example 5.

15. $3\sqrt{25}$ **16.** $9\sqrt{36}$ **17.** $7\sqrt{63}$

18. $11\sqrt{99}$ ▶ **19.** $-5\sqrt{27}$ **20.** $-6\sqrt{75}$

Objective B *Use the quotient rule and the product rule to simplify each radical. See Examples 6 through 8.*

21. $\sqrt{\dfrac{8}{25}}$ **22.** $\sqrt{\dfrac{63}{16}}$ ▶ **23.** $\sqrt{\dfrac{27}{121}}$ **24.** $\sqrt{\dfrac{24}{169}}$

25. $\sqrt{\dfrac{9}{4}}$ **26.** $\sqrt{\dfrac{100}{49}}$ **27.** $\sqrt{\dfrac{125}{9}}$ **28.** $\sqrt{\dfrac{27}{100}}$

29. $\sqrt{\dfrac{11}{36}}$ **30.** $\sqrt{\dfrac{30}{49}}$ **31.** $-\sqrt{\dfrac{27}{144}}$ **32.** $-\sqrt{\dfrac{84}{121}}$

Objective C *Simplify each radical. Assume that all variables represent positive numbers. See Examples 9 through 12.*

33. $\sqrt{x^7}$ 　　　　**34.** $\sqrt{y^3}$ 　　　　▶ **35.** $\sqrt{x^{13}}$ 　　　　**36.** $\sqrt{y^{17}}$

37. $\sqrt{36a^3}$ 　　　　▶ **38.** $\sqrt{81b^5}$ 　　　　**39.** $\sqrt{96x^4}$ 　　　　**40.** $\sqrt{40y^{10}}$

▶ **41.** $\sqrt{\dfrac{12}{m^2}}$ 　　　**42.** $\sqrt{\dfrac{63}{p^2}}$ 　　　**43.** $\sqrt{\dfrac{9x}{y^{10}}}$ 　　　**44.** $\sqrt{\dfrac{6y^2}{z^{16}}}$

45. $\sqrt{\dfrac{88}{x^{12}}}$ 　　　**46.** $\sqrt{\dfrac{500}{y^{22}}}$

Objectives A B C **Mixed Practice** *Simplify each radical. See Examples 1 through 12.*

47. $8\sqrt{4}$ 　　　**48.** $6\sqrt{49}$ 　　　**49.** $\sqrt{\dfrac{36}{121}}$ 　　　**50.** $\sqrt{\dfrac{25}{144}}$

51. $\sqrt{175}$ 　　　**52.** $\sqrt{700}$ 　　　**53.** $\sqrt{\dfrac{20}{9}}$ 　　　**54.** $\sqrt{\dfrac{45}{64}}$

55. $\sqrt{24m^7}$ 　　　**56.** $\sqrt{50n^{13}}$ 　　　**57.** $\sqrt{\dfrac{23y^3}{4x^6}}$ 　　　**58.** $\sqrt{\dfrac{41x^5}{9y^8}}$

Objective D *Simplify each radical. See Examples 13 through 16.*

59. $\sqrt[3]{24}$ 　　　**60.** $\sqrt[3]{81}$ 　　　▶ **61.** $\sqrt[3]{250}$ 　　　**62.** $\sqrt[3]{56}$

▶ **63.** $\sqrt[3]{\dfrac{5}{64}}$ 　　　**64.** $\sqrt[3]{\dfrac{32}{125}}$ 　　　**65.** $\sqrt[3]{\dfrac{23}{8}}$ 　　　**66.** $\sqrt[3]{\dfrac{37}{27}}$

67. $\sqrt[3]{\dfrac{15}{64}}$ 　　　**68.** $\sqrt[3]{\dfrac{4}{27}}$ 　　　**69.** $\sqrt[3]{80}$ 　　　**70.** $\sqrt[3]{108}$

Review

Perform each indicated operation. See Sections 12.4 and 12.5.

71. $6x + 8x$ 　　　**72.** $(6x)(8x)$ 　　　**73.** $(2x + 3)(x - 5)$

74. $(2x + 3) + (x - 5)$ 　　　**75.** $9y^2 - 9y^2$ 　　　**76.** $(9y^2)(-8y^2)$

Concept Extensions

Simplify each radical. Assume that all variables represent positive numbers.

77. $\sqrt{x^6y^3}$ 　　　**78.** $\sqrt{a^{13}b^{14}}$ 　　　**79.** $\sqrt{98x^5y^4}$

80. $\sqrt{27x^8y^{11}}$ 　　　**81.** $\sqrt[3]{-8x^6}$ 　　　**82.** $\sqrt[3]{27x^{12}}$

Solve.

83. If a cube is to have a volume of 80 cubic inches, then each side must be $\sqrt[3]{80}$ inches long. Simplify the radical representing the side length.

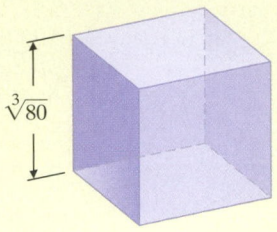

84. Jeannie Boswell is swimming across a 40-foot-wide river, trying to head straight across to the opposite shore. However, the current is strong enough to move her downstream 100 feet by the time she reaches land. (See the figure.) Because of the current, the actual distance she swims is $\sqrt{11,600}$ feet. Simplify this radical.

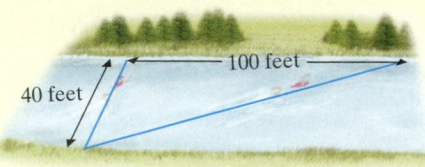

100 feet

40 feet

85. By using replacement values for a and b, show that $\sqrt{a^2 + b^2}$ does not equal $a + b$.

86. By using replacement values for a and b, show that $\sqrt{a + b}$ does not equal $\sqrt{a} + \sqrt{b}$.

87. The "Water Cube" was the swimming and diving venue for the 2008 Beijing Summer Olympics. It is not actually a cube, because it is only 31 meters tall, which is not the same as its width and length. However, the roof of it is a square. If the area of the roof of the Water Cube is 31,329 square meters, find the dimensions of the roof of the Water Cube.

△ **88.** The competition diving pool in the Water Cube at the Beijing Summer Olympics is not a cube either. It has a square footprint, but is only 5 meters deep. If the volume of the diving pool is 3125 cubic meters, find the length and width of the competition diving pool.

The length of a side of a cube is given by the expression $\sqrt{6A}/6$, where A is the cube's surface area. Use this expression for Exercises 89 through 92. Be sure to attach the appropriate units.

△ **89.** The surface area of a cube is 120 square inches. Find the exact length of a side of the cube.

△ **90.** The surface area of a cube is 594 square feet. Find the exact length of a side of the cube.

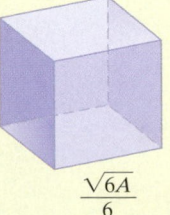

$\dfrac{\sqrt{6A}}{6}$

91. Rubik's Cube, named after its inventor, Erno Rubik, was first imagined by him in 1974, and by 1980 was a worldwide phenomenon. A standard Rubik's Cube has a surface area of 30.375 square inches. Find the length of one side of a Rubik's Cube. (A few world records are listed below. *Source: Guinness World Records*)

Fastest time to solve 1 Rubik's Cube: 5.5 sec by Mats Valk (Netherlands) in 2013.

Most Rubik's Cubes solved in 1 hour: 260 by Max Iovane (Italy) in 2010.

92. In 2011, Apple renovated its flagship Apple Store on Fifth Avenue in New York City, taking advantage of advances in glass manufacturing to simplify the giant glass cube that serves as the store's entrance. A cube of this size has a surface area of 6144 square feet. Find the length of one side of the Apple Store glass cube. (*Source:* Based on data from AppleInsider.com)

The cost C in dollars per day to operate a small delivery service is given by $C = 100\sqrt[3]{n} + 700$, where n is the number of deliveries per day.

93. Find the cost if the number of deliveries is 1000.

94. Approximate the cost if the number of deliveries is 500.

The Mosteller formula for calculating body surface area is $B = \sqrt{\dfrac{hw}{3600}}$, where B is an individual's body surface area in square meters, h is the individual's height in centimeters, and w is the individual's weight in kilograms. Use this formula in Exercises 95 and 96. Round answers to the nearest tenth.

95. Find the body surface area of a person who is 169 cm tall and weighs 64 kilograms.

96. Approximate the body surface area of a person who is 183 cm tall and weighs 85 kilograms.

15.3 Adding and Subtracting Radicals

Objectives

A Add or Subtract Like Radicals.

B Simplify Square Root Radical Expressions, and Then Add or Subtract Any Like Radicals.

C Simplify Cube Root Radical Expressions, and Then Add or Subtract Any Like Radicals.

Practice 1–4

Add or subtract as indicated.

1. $6\sqrt{11} + 9\sqrt{11}$
2. $\sqrt{7} - 3\sqrt{7}$
3. $\sqrt{2} + \sqrt{2} - \sqrt{15}$
4. $3\sqrt{3} - 3\sqrt{2}$

Practice 5–8

Simplify each radical expression.

5. $\sqrt{27} + \sqrt{75}$
6. $3\sqrt{20} - 7\sqrt{45}$
7. $\sqrt{36} - \sqrt{48} - 4\sqrt{3} - \sqrt{9}$
8. $\sqrt{9x^4} - \sqrt{36x^3} + \sqrt{x^3}$

Answers

1. $15\sqrt{11}$ 2. $-2\sqrt{7}$ 3. $2\sqrt{2} - \sqrt{15}$
4. $3\sqrt{3} - 3\sqrt{2}$ 5. $8\sqrt{3}$ 6. $-15\sqrt{5}$
7. $3 - 8\sqrt{3}$ 8. $3x^2 - 5x\sqrt{x}$

✓ Concept Check Answer

e

1126

Objective A Adding and Subtracting Radicals

Recall that to combine like terms, we use the distributive property.

$$5x + 3x = (5 + 3)x = 8x$$

The distributive property can also be applied to expressions containing the same radicals. For example,

$$5\sqrt{2} + 3\sqrt{2} = (5 + 3)\sqrt{2} = 8\sqrt{2}$$

Also,

$$9\sqrt{5} - 6\sqrt{5} = (9 - 6)\sqrt{5} = 3\sqrt{5}$$

Radical terms such as $5\sqrt{2}$ and $3\sqrt{2}$ are **like radicals,** as are $9\sqrt{5}$ and $6\sqrt{5}$. Like radicals have the same index and the same radicand.

Examples Add or subtract as indicated.

1. $4\sqrt{5} + 3\sqrt{5} = (4 + 3)\sqrt{5} = 7\sqrt{5}$
2. $\sqrt{10} - 6\sqrt{10} = 1\sqrt{10} - 6\sqrt{10} = (1 - 6)\sqrt{10} = -5\sqrt{10}$
3. $2\sqrt{6} + 2\sqrt{5}$ cannot be simplified further since the radicands are not the same.
4. $\sqrt{15} + \sqrt{15} - \sqrt{2} = 1\sqrt{15} + 1\sqrt{15} - \sqrt{2}$
   $$= (1 + 1)\sqrt{15} - \sqrt{2}$$
   $$= 2\sqrt{15} - \sqrt{2}$$

This expression cannot be simplified further since the radicands are not the same.

■ Work Practice 1–4

✓ **Concept Check** Which is true?

  a. $2 + 3\sqrt{5} = 5\sqrt{5}$

  b. $2\sqrt{3} + 2\sqrt{7} = 2\sqrt{10}$

  c. $\sqrt{3} + \sqrt{5} = \sqrt{8}$

  d. $\sqrt{3} + \sqrt{3} = 3$

  e. None of the above is true. In each case, the left-hand side cannot be simplified further.

Objective B Simplifying Square Root Radicals Before Adding or Subtracting

At first glance, it appears that the expression $\sqrt{50} + \sqrt{8}$ cannot be simplified further because the radicands are different. However, the product rule can be used to simplify each radical, and then further simplification might be possible.

Examples Simplify each radical expression.

5. $\sqrt{50} + \sqrt{8} = \sqrt{25 \cdot 2} + \sqrt{4 \cdot 2}$ Factor radicands.
   $$= \sqrt{25} \cdot \sqrt{2} + \sqrt{4} \cdot \sqrt{2}$$ Use the product rule.
   $$= 5\sqrt{2} + 2\sqrt{2}$$ Simplify $\sqrt{25}$ and $\sqrt{4}$.
   $$= 7\sqrt{2}$$ Add like radicals.

6. $7\sqrt{12} - 2\sqrt{75} = 7\sqrt{4 \cdot 3} - 2\sqrt{25 \cdot 3}$ Factor radicands.

$\qquad\qquad\quad = 7\sqrt{4} \cdot \sqrt{3} - 2\sqrt{25} \cdot \sqrt{3}$ Use the product rule.

$\qquad\qquad\quad = 7 \cdot 2\sqrt{3} - 2 \cdot 5\sqrt{3}$ Simplify $\sqrt{4}$ and $\sqrt{25}$.

$\qquad\qquad\quad = 14\sqrt{3} - 10\sqrt{3}$ Multiply.

$\qquad\qquad\quad = 4\sqrt{3}$ Subtract like radicals.

7. $\sqrt{25} - \sqrt{27} - 2\sqrt{18} - \sqrt{16}$

$\quad = 5 - \sqrt{9 \cdot 3} - 2\sqrt{9 \cdot 2} - 4$ Factor radicands and simplify $\sqrt{25}$ and $\sqrt{16}$.

$\quad = 5 - \sqrt{9} \cdot \sqrt{3} - 2\sqrt{9} \cdot \sqrt{2} - 4$ Use the product rule.

$\quad = 5 - 3\sqrt{3} - 2 \cdot 3\sqrt{2} - 4$ Simplify $\sqrt{9}$.

$\quad = 1 - 3\sqrt{3} - 6\sqrt{2}$ Write $5 - 4$ as 1 and $2 \cdot 3$ as 6.

8. $2\sqrt{x^2} - \sqrt{25x^5} + \sqrt{x^5}$

$\quad = 2x - \sqrt{25x^4 \cdot x} + \sqrt{x^4 \cdot x}$ Factor radicands so that one factor is a perfect square. Simplify $\sqrt{x^2}$.

$\quad = 2x - \sqrt{25x^4} \cdot \sqrt{x} + \sqrt{x^4} \cdot \sqrt{x}$ Use the product rule.

$\quad = 2x - 5x^2\sqrt{x} + x^2\sqrt{x}$ Write $\sqrt{25x^4}$ as $5x^2$ and $\sqrt{x^4}$ as x^2.

$\quad = 2x - 4x^2\sqrt{x}$ Add like radicals.

■ **Work Practice 5–8**

Objective C   Simplifying Cube Root Radicals Before Adding or Subtracting ▶

Example 9   Simplify the radical expression.

$5\sqrt[3]{16x^3} - \sqrt[3]{54x^3}$

$\quad = 5\sqrt[3]{8x^3 \cdot 2} - \sqrt[3]{27x^3 \cdot 2}$ Factor radicands so that one factor is a perfect cube.

$\quad = 5 \cdot \sqrt[3]{8x^3} \cdot \sqrt[3]{2} - \sqrt[3]{27x^3} \cdot \sqrt[3]{2}$ Use the product rule.

$\quad = 5 \cdot 2x\sqrt[3]{2} - 3x\sqrt[3]{2}$ Write $\sqrt[3]{8x^3}$ as $2x$ and $\sqrt[3]{27x^3}$ as $3x$.

$\quad = 10x\sqrt[3]{2} - 3x\sqrt[3]{2}$ Write $5 \cdot 2x$ as $10x$.

$\quad = 7x\sqrt[3]{2}$ Subtract like radicands.

■ **Work Practice 9**

Practice 9

Simplify the radical expression.
$10\sqrt[3]{81p^6} - \sqrt[3]{24p^6}$

Answer
9. $28p^2\sqrt[3]{3}$

Vocabulary, Readiness & Video Check

Fill in each blank.

1. Radicals that have the same index and same radicand are called _____ .

2. The expressions $7\sqrt[3]{2x}$ and $-\sqrt[3]{2x}$ are called _____ .

3. $11\sqrt{2} + 6\sqrt{2} =$ _____ .
   a. $66\sqrt{2}$ **b.** $17\sqrt{2}$ **c.** $17\sqrt{4}$

4. $\sqrt{5}$ is the same as _____ .
   a. $0\sqrt{5}$ **b.** $1\sqrt{5}$ **c.** $5\sqrt{5}$

5. $\sqrt{5} + \sqrt{5} =$ _____
   a. $\sqrt{10}$ **b.** 5 **c.** $2\sqrt{5}$

6. $9\sqrt{7} - \sqrt{7} =$ _____
   a. $8\sqrt{7}$ **b.** 9 **c.** 0

Martin-Gay Interactive Videos   *Watch the section lecture video and answer the following questions.*

Objective A 7. From ▦ Examples 1–4, how is combining like radicals similar to combining like terms?

Objective B 8. From ▦ Example 5, why should we always check to see if all radical terms in our expression are simplified before attempting to add or subtract the radicals? ▶

Objective C 9. In ▦ Example 8, what property is used during the simplification of the expression? ▶

See Video 15.3 🍎

15.3 Exercise Set MyMathLab® ▶

Objective A *Add or subtract as indicated. See Examples 1 through 4.*

▶ **1.** $4\sqrt{3} - 8\sqrt{3}$

2. $\sqrt{5} - 9\sqrt{5}$

▶ **3.** $3\sqrt{6} + 8\sqrt{6} - 2\sqrt{6} - 5$

4. $12\sqrt{2} - 3\sqrt{2} + 8\sqrt{2} + 10$

5. $6\sqrt{5} - 5\sqrt{5} + \sqrt{2}$

6. $4\sqrt{3} + \sqrt{5} - 3\sqrt{3}$

7. $2\sqrt{3} + 5\sqrt{3} - \sqrt{2}$

8. $8\sqrt{14} + 2\sqrt{14} + \sqrt{5}$

9. $2\sqrt{2} - 7\sqrt{2} - 6$

10. $5\sqrt{7} + 2 - 11\sqrt{7}$

Objective B *Add or subtract by first simplifying each radical and then combining any like radicals. Assume that all variables represent positive numbers. See Examples 5 through 8.*

▶ **11.** $\sqrt{12} + \sqrt{27}$

12. $\sqrt{50} + \sqrt{18}$

13. $\sqrt{45} + 3\sqrt{20}$

14. $5\sqrt{32} - \sqrt{72}$

15. $2\sqrt{54} - \sqrt{20} + \sqrt{45} - \sqrt{24}$

16. $2\sqrt{8} - \sqrt{128} + \sqrt{48} + \sqrt{18}$

17. $4x - 3\sqrt{x^2} + \sqrt{x}$

18. $x - 6\sqrt{x^2} + 2\sqrt{x}$

19. $\sqrt{25x} + \sqrt{36x} - 11\sqrt{x}$

20. $\sqrt{9x} - \sqrt{16x} + 2\sqrt{x}$

21. $\sqrt{\dfrac{5}{9}} + \sqrt{\dfrac{5}{81}}$

▶ **22.** $\sqrt{\dfrac{3}{64}} + \sqrt{\dfrac{3}{16}}$

23. $\sqrt{\dfrac{3}{4}} - \sqrt{\dfrac{3}{64}}$

24. $\sqrt{\dfrac{2}{25}} + \sqrt{\dfrac{2}{9}}$

Objectives A B Mixed Practice *See Examples 1 through 8.*

25. $12\sqrt{5} - \sqrt{5} - 4\sqrt{5}$

26. $\sqrt{6} + 3\sqrt{6} + \sqrt{6}$

27. $\sqrt{75} + \sqrt{48}$

28. $2\sqrt{80} - \sqrt{45}$

29. $\sqrt{5} + \sqrt{15}$

30. $\sqrt{5} + \sqrt{5}$

31. $3\sqrt{x^3} - x\sqrt{4x}$

32. $x\sqrt{16x} - \sqrt{x^3}$

33. $\sqrt{8} + \sqrt{9} + \sqrt{18} + \sqrt{81}$

34. $\sqrt{6} + \sqrt{16} + \sqrt{24} + \sqrt{25}$

35. $4 + 8\sqrt{2} - 9$

36. $11 - 5\sqrt{7} - 8$

37. $2\sqrt{45} - 2\sqrt{20}$ 　　　**38.** $5\sqrt{18} + 2\sqrt{32}$ 　　　**39.** $\sqrt{35} - \sqrt{140}$ 　　　**40.** $\sqrt{15} - \sqrt{135}$

41. $6 - 2\sqrt{3} - \sqrt{3}$ 　　　**42.** $8 - \sqrt{2} - 5\sqrt{2}$ 　　　**43.** $3\sqrt{9x} + 2\sqrt{x}$ 　　　▶ **44.** $5\sqrt{2x} + \sqrt{98x}$

45. $\sqrt{9x^2} + \sqrt{81x^2} - 11\sqrt{x}$ 　　　**46.** $\sqrt{100x^2} + 3\sqrt{x} - \sqrt{36x^2}$ 　　　**47.** $\sqrt{3x^3} + 3x\sqrt{x}$

48. $x\sqrt{4x} + \sqrt{9x^3}$ 　　　**49.** $\sqrt{32x^2} + \sqrt{32x^2} + \sqrt{4x^2}$ 　　　**50.** $\sqrt{18x^2} + \sqrt{24x^3} + \sqrt{2x^2}$

51. $\sqrt{40x} + \sqrt{40x^4} - 2\sqrt{10x} - \sqrt{5x^4}$ 　　　　　**52.** $\sqrt{72x^2} + \sqrt{54x} - x\sqrt{50} - 3\sqrt{2x}$

Objective C *Simplify each radical expression. See Example 9.*

53. $2\sqrt[3]{9} + 5\sqrt[3]{9} - \sqrt[3]{25}$ 　　**54.** $8\sqrt[3]{4} + 2\sqrt[3]{4} - \sqrt[3]{49}$ 　▶ **55.** $2\sqrt[3]{2} - 7\sqrt[3]{2} - 6$ 　　**56.** $5\sqrt[3]{9} + 2 - 11\sqrt[3]{9}$

57. $\sqrt[3]{81} + \sqrt[3]{24}$ 　　**58.** $\sqrt[3]{32} + \sqrt[3]{4}$ 　▶ **59.** $\sqrt[3]{8} + \sqrt[3]{54} - 5$ 　　**60.** $\sqrt[3]{64} + \sqrt[3]{14} - 9$

61. $2\sqrt[3]{8x^3} + 2\sqrt[3]{16x^3}$ 　　**62.** $3\sqrt[3]{27z^3} + 3\sqrt[3]{81z^3}$ 　　**63.** $12\sqrt[3]{y^7} - y^2\sqrt[3]{8y}$ 　　**64.** $19\sqrt[3]{z^{11}} - z^3\sqrt[3]{125z^2}$

65. $\sqrt{40x} + x\sqrt[3]{40} - 2\sqrt{10x} - x\sqrt[3]{5}$ 　　　　**66.** $\sqrt{72x^2} + \sqrt[3]{54} - x\sqrt{50} - 3\sqrt[3]{2}$

Review

Square each binomial. See Section 12.6.

67. $(x + 6)^2$ 　　　**68.** $(3x + 2)^2$ 　　　**69.** $(2x - 1)^2$ 　　　**70.** $(x - 5)^2$

Concept Extensions

71. In your own words, describe like radicals. 　　　**72.** In the expression $\sqrt{5} + 2 - 3\sqrt{5}$, explain why 2 and -3 cannot be combined.

△ **73.** Find the perimeter of the rectangular picture frame. 　　　△ **74.** Find the perimeter of the plot of land.

$\sqrt{5}$ inches

$3\sqrt{5}$ inches

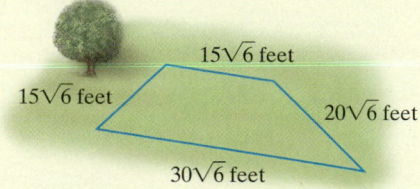

$15\sqrt{6}$ feet

$15\sqrt{6}$ feet

$20\sqrt{6}$ feet

$30\sqrt{6}$ feet

△ **75.** A water trough is to be made of wood. Each of the two triangular end pieces has an area of $\dfrac{3\sqrt{27}}{4}$ square feet. The two side panels are both rectangular. In simplest radical form, find the total area of the wood needed.

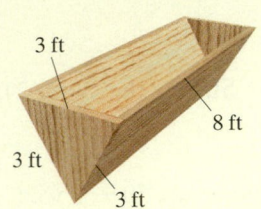

76. Four wooden braces are to be attached along diagonals of the vertical sides of a storage bin. Each of two of these diagonals has a length of $\sqrt{52}$ feet, while each of the other two has a length of $\sqrt{80}$ feet. In simplest radical form, find the total length of the wood needed for these braces.

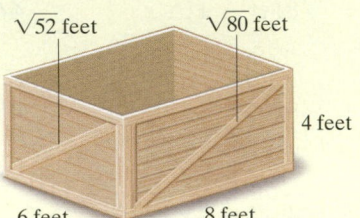

Determine whether each expression can be simplified. If yes, then simplify. See the Concept Check in this section.

77. $4\sqrt{2} + 3\sqrt{2}$

78. $3\sqrt{7} + 3\sqrt{6}$

79. $6 + 7\sqrt{6}$

80. $5x\sqrt{2} + 8x\sqrt{2}$

81. $\sqrt{7} + \sqrt{7} + \sqrt{7}$

82. $6\sqrt{5} - \sqrt{5}$

Simplify.

83. $\sqrt{\dfrac{x^3}{16}} - x\sqrt{\dfrac{9x}{25}} + \dfrac{\sqrt{81x^3}}{2}$

84. $7\sqrt{x^{11}y^7} - x^2y\sqrt{25x^7y^5} + \sqrt{8x^8y^2}$

15.4 Multiplying and Dividing Radicals

Objectives

A Multiply Radicals. ▶

B Divide Radicals. ▶

C Rationalize Denominators. ▶

D Rationalize Denominators Using Conjugates. ▶

Objective A Multiplying Radicals ▶

In Section 15.2, we used the product and quotient rules for radicals to help us simplify radicals. In this section, we use these rules to simplify products and quotients of radicals.

> **Product Rule for Radicals**
>
> If $\sqrt{a}$ and $\sqrt{b}$ are real numbers, then
> $$\sqrt{a} \cdot \sqrt{b} = \sqrt{a \cdot b}$$

In other words, the product of the square roots of two numbers is the square root of the product of the two numbers. For example,
$$\sqrt{3} \cdot \sqrt{2} = \sqrt{3 \cdot 2} = \sqrt{6}$$

Examples Multiply. Then simplify each product if possible.

1. $\sqrt{7} \cdot \sqrt{3} = \sqrt{7 \cdot 3}$
　　　　　　$= \sqrt{21}$

2. $\sqrt{3} \cdot \sqrt{3} = \sqrt{3 \cdot 3} = \sqrt{9} = 3$

3. $\sqrt{3} \cdot \sqrt{15} = \sqrt{45}$　　　　　Use the product rule.
　　　　　　$= \sqrt{9 \cdot 5}$　　　　Factor the radicand.
　　　　　　$= \sqrt{9} \cdot \sqrt{5}$　　　Use the product rule.
　　　　　　$= 3\sqrt{5}$　　　　　Simplify $\sqrt{9}$.

4. $\sqrt{2x^3} \cdot \sqrt{6x} = \sqrt{2x^3 \cdot 6x}$　　Use the product rule.
　　　　　　$= \sqrt{12x^4}$　　　Multiply.
　　　　　　$= \sqrt{4x^4 \cdot 3}$　　Write $12x^4$ so that one factor is a perfect square.
　　　　　　$= \sqrt{4x^4} \cdot \sqrt{3}$　Use the product rule.
　　　　　　$= 2x^2\sqrt{3}$　　　Simplify.

■ **Work Practice 1–4**

From Example 2, we found that

$$\sqrt{3} \cdot \sqrt{3} = 3 \quad \text{or} \quad (\sqrt{3})^2 = 3$$

This is true in general.

> If a is a positive number,
> $$\sqrt{a} \cdot \sqrt{a} = a \quad \text{or} \quad (\sqrt{a})^2 = a$$

✓**Concept Check** Identify the true statement(s).

a. $\sqrt{7} \cdot \sqrt{7} = 7$　　　　　**b.** $\sqrt{2} \cdot \sqrt{3} = 6$

c. $(\sqrt{131})^2 = 131$　　　**d.** $\sqrt{5x} \cdot \sqrt{5x} = 5x$ (Here x is a positive number.)

When multiplying radical expressions containing more than one term, we use the same techniques we use to multiply other algebraic expressions with more than one term.

Example 5 Multiply.

a. $\sqrt{5}(\sqrt{5} - \sqrt{2})$　　　　　**b.** $\sqrt{3x}(\sqrt{x} - 5\sqrt{3})$

c. $(\sqrt{x} + \sqrt{2})(\sqrt{x} - \sqrt{7})$

Solution:

a. Using the distributive property, we have

$$\sqrt{5}(\sqrt{5} - \sqrt{2}) = \sqrt{5} \cdot \sqrt{5} - \sqrt{5} \cdot \sqrt{2}$$
$$= 5 - \sqrt{10} \quad \text{Since } \sqrt{5} \cdot \sqrt{5} = 5 \text{ and } \sqrt{5} \cdot \sqrt{2} = \sqrt{10}$$

b. $\sqrt{3x}(\sqrt{x} - 5\sqrt{3}) = \sqrt{3x} \cdot \sqrt{x} - \sqrt{3x} \cdot 5\sqrt{3}$　Use the distributive property.
　　　　　　$= \sqrt{3x \cdot x} - 5\sqrt{3x \cdot 3}$　Use the product rule.
　　　　　　$= \sqrt{3 \cdot x^2} - 5\sqrt{9 \cdot x}$　Factor each radicand so that one factor is a perfect square.
　　　　　　$= \sqrt{3} \cdot \sqrt{x^2} - 5 \cdot \sqrt{9} \cdot \sqrt{x}$　Use the product rule.
　　　　　　$= x\sqrt{3} - 5 \cdot 3 \cdot \sqrt{x}$　Simplify.
　　　　　　$= x\sqrt{3} - 15\sqrt{x}$　Simplify.

(Continued on next page)

(Continued on next page)

Practice 1–4

Multiply. Then simplify each product if possible.

1. $\sqrt{5} \cdot \sqrt{2}$

2. $\sqrt{7} \cdot \sqrt{7}$

3. $\sqrt{6} \cdot \sqrt{3}$

4. $\sqrt{10x} \cdot \sqrt{2x}$

Practice 5

Multiply.

a. $\sqrt{7}(\sqrt{7} - \sqrt{3})$

b. $\sqrt{5x}(\sqrt{x} - 3\sqrt{5})$

c. $(\sqrt{x} + \sqrt{5})(\sqrt{x} - \sqrt{3})$

Answers

1. $\sqrt{10}$　**2.** 7　**3.** $3\sqrt{2}$　**4.** $2x\sqrt{5}$

5. a. $7 - \sqrt{21}$　**b.** $x\sqrt{5} - 15\sqrt{x}$

c. $x - \sqrt{3x} + \sqrt{5x} - \sqrt{15}$

✓**Concept Check Answer**

a, c, d

c. Using the FOIL method of multiplication, we have

$$\left(\sqrt{x} + \sqrt{2}\right)\left(\sqrt{x} - \sqrt{7}\right)$$

$$\quad\quad\quad\quad F \quad\quad O \quad\quad I \quad\quad L$$

$$= \sqrt{x} \cdot \sqrt{x} - \sqrt{x} \cdot \sqrt{7} + \sqrt{2} \cdot \sqrt{x} - \sqrt{2} \cdot \sqrt{7}$$

$$= x - \sqrt{7x} + \sqrt{2x} - \sqrt{14} \quad\quad\quad\quad \text{Use the product rule.}$$

■ **Work Practice 5**

The special product formulas also can be used to multiply expressions containing radicals.

Practice 6

Multiply.
a. $\left(\sqrt{3} + 8\right)\left(\sqrt{3} - 8\right)$
b. $\left(\sqrt{5x} + 4\right)^2$

Example 6   Multiply.

a. $\left(\sqrt{5} - 7\right)\left(\sqrt{5} + 7\right)$ **b.** $\left(\sqrt{7x} + 2\right)^2$

Solution:

a. $\left(\sqrt{5} - 7\right)\left(\sqrt{5} + 7\right) = \left(\sqrt{5}\right)^2 - 7^2$ Recall that $(a - b)(a + b) = a^2 - b^2$.

$$\quad\quad\quad\quad\quad\quad\quad\quad\quad = 5 - 49$$

$$\quad\quad\quad\quad\quad\quad\quad\quad\quad = -44$$

b. $\left(\sqrt{7x} + 2\right)^2$

$$= \left(\sqrt{7x}\right)^2 + 2\left(\sqrt{7x}\right)(2) + (2)^2 \quad\quad \text{Recall that } (a + b)^2 = a^2 + 2ab + b^2.$$

$$= 7x + 4\sqrt{7x} + 4$$

■ **Work Practice 6**

Objective B Dividing Radicals

To simplify quotients of rational expressions, we use the quotient rule.

> ### Quotient Rule for Radicals
>
> If $\sqrt{a}$ and $\sqrt{b}$ are real numbers and $b \neq 0$, then
>
> $$\frac{\sqrt{a}}{\sqrt{b}} = \sqrt{\frac{a}{b}}$$

Practice 7–9

Divide. Then simplify the quotient if possible.

7. $\dfrac{\sqrt{15}}{\sqrt{3}}$

8. $\dfrac{\sqrt{90}}{\sqrt{2}}$

9. $\dfrac{\sqrt{125x^3}}{\sqrt{5x}}$

Examples   Divide. Then simplify the quotient if possible.

7. $\dfrac{\sqrt{14}}{\sqrt{2}} = \sqrt{\dfrac{14}{2}} = \sqrt{7}$

8. $\dfrac{\sqrt{100}}{\sqrt{5}} = \sqrt{\dfrac{100}{5}} = \sqrt{20} = \sqrt{4 \cdot 5} = \sqrt{4} \cdot \sqrt{5} = 2\sqrt{5}$

9. $\dfrac{\sqrt{12x^3}}{\sqrt{3x}} = \sqrt{\dfrac{12x^3}{3x}} = \sqrt{4x^2} = 2x$

■ **Work Practice 7–9**

Objective C Rationalizing Denominators ▶

It is sometimes easier to work with radical expressions if the denominator does not contain a radical. To rewrite an expression so that the denominator does not contain a radical expression, we use the fact that we can multiply the numerator and the denominator of a fraction by the same nonzero number without changing the value of the

Answers

6. a. -61 **b.** $5x + 8\sqrt{5x} + 16$

7. $\sqrt{5}$ **8.** $3\sqrt{5}$ **9.** $5x$

expression. This is the same as multiplying the fraction by 1. For example, to get rid of the radical in the denominator of $\dfrac{\sqrt{5}}{\sqrt{2}}$, we multiply by 1 in the form of $\dfrac{\sqrt{2}}{\sqrt{2}}$. Then

$$\frac{\sqrt{5}}{\sqrt{2}} = \frac{\sqrt{5}}{\sqrt{2}} \cdot 1 = \frac{\sqrt{5}}{\sqrt{2}} \cdot \frac{\sqrt{2}}{\sqrt{2}} = \frac{\sqrt{5} \cdot \sqrt{2}}{\sqrt{2} \cdot \sqrt{2}} = \frac{\sqrt{10}}{2}$$

This process is called **rationalizing** the denominator.

Example 10 Rationalize the denominator of $\dfrac{2}{\sqrt{7}}$.

Solution: To rewrite $\dfrac{2}{\sqrt{7}}$ so that there is no radical in the denominator, we multiply by 1 in the form of $\dfrac{\sqrt{7}}{\sqrt{7}}$.

$$\frac{2}{\sqrt{7}} = \frac{2}{\sqrt{7}} \cdot \frac{\sqrt{7}}{\sqrt{7}} = \frac{2 \cdot \sqrt{7}}{\sqrt{7} \cdot \sqrt{7}} = \frac{2\sqrt{7}}{7}$$

■ **Work Practice 10**

Example 11 Rationalize the denominator of $\dfrac{\sqrt{5}}{\sqrt{12}}$.

Solution: We can multiply by $\dfrac{\sqrt{12}}{\sqrt{12}}$, but see what happens if we simplify first.

$$\frac{\sqrt{5}}{\sqrt{12}} = \frac{\sqrt{5}}{\sqrt{4 \cdot 3}} = \frac{\sqrt{5}}{2\sqrt{3}}$$

To rationalize the denominator now, we multiply by $\dfrac{\sqrt{3}}{\sqrt{3}}$.

$$\frac{\sqrt{5}}{2\sqrt{3}} = \frac{\sqrt{5}}{2\sqrt{3}} \cdot \frac{\sqrt{3}}{\sqrt{3}} = \frac{\sqrt{5} \cdot \sqrt{3}}{2\sqrt{3} \cdot \sqrt{3}} = \frac{\sqrt{15}}{2 \cdot 3} = \frac{\sqrt{15}}{6}$$

■ **Work Practice 11**

Example 12 Rationalize the denominator of $\sqrt{\dfrac{1}{18x}}$.

Solution: First we simplify.

$$\sqrt{\frac{1}{18x}} = \frac{\sqrt{1}}{\sqrt{18x}} = \frac{1}{\sqrt{9} \cdot \sqrt{2x}} = \frac{1}{3\sqrt{2x}}$$

Now to rationalize the denominator, we multiply by $\dfrac{\sqrt{2x}}{\sqrt{2x}}$.

$$\frac{1}{3\sqrt{2x}} = \frac{1}{3\sqrt{2x}} \cdot \frac{\sqrt{2x}}{\sqrt{2x}} = \frac{1 \cdot \sqrt{2x}}{3\sqrt{2x} \cdot \sqrt{2x}} = \frac{\sqrt{2x}}{3 \cdot 2x} = \frac{\sqrt{2x}}{6x}$$

■ **Work Practice 12**

Objective D Rationalizing Denominators Using Conjugates ▶

To rationalize a denominator that is a sum or a difference, such as the denominator in

$$\frac{2}{4 + \sqrt{3}}$$

we multiply the numerator and the denominator by $4 - \sqrt{3}$. The expressions $4 + \sqrt{3}$ and $4 - \sqrt{3}$ are called conjugates of each other. When a radical expression

Practice 10

Rationalize the denominator of $\dfrac{5}{\sqrt{3}}$.

Practice 11

Rationalize the denominator of $\dfrac{\sqrt{7}}{\sqrt{20}}$.

Practice 12

Rationalize the denominator of $\sqrt{\dfrac{2}{45x}}$.

such as $4 + \sqrt{3}$ is multiplied by its conjugate, $4 - \sqrt{3}$, the product simplifies to an expression that contains no radicals.

In general, the expressions $a + b$ and $a - b$ are **conjugates** of each other.

$$(a + b)(a - b) = a^2 - b^2$$
$$(4 + \sqrt{3})(4 - \sqrt{3}) = 4^2 - (\sqrt{3})^2 = 16 - 3 = 13$$

Then

$$\frac{2}{4 + \sqrt{3}} = \frac{2(4 - \sqrt{3})}{(4 + \sqrt{3})(4 - \sqrt{3})} = \frac{2(4 - \sqrt{3})}{13}$$

Example 13   Rationalize the denominator of $\dfrac{2}{1 + \sqrt{3}}$.

Solution: We multiply the numerator and the denominator of this fraction by the conjugate of $1 + \sqrt{3}$, that is, by $1 - \sqrt{3}$.

$$\frac{2}{1 + \sqrt{3}} = \frac{2(1 - \sqrt{3})}{(1 + \sqrt{3})(1 - \sqrt{3})}$$

$$= \frac{2(1 - \sqrt{3})}{1^2 - (\sqrt{3})^2}$$

$$= \frac{2(1 - \sqrt{3})}{1 - 3}$$

> **Helpful Hint**
> Don't forget that $(\sqrt{3})^2 = 3$.

$$= \frac{2(1 - \sqrt{3})}{-2}$$

$$= -\frac{2(1 - \sqrt{3})}{2} \qquad \frac{a}{-b} = -\frac{a}{b}$$

$$= -1(1 - \sqrt{3}) \qquad \text{Simplify.}$$

$$= -1 + \sqrt{3} \qquad \text{Multiply.}$$

■ **Work Practice 13**

Practice 13

Rationalize the denominator of $\dfrac{3}{2 + \sqrt{7}}$.

Example 14   Rationalize the denominator of $\dfrac{\sqrt{5} + 4}{\sqrt{5} - 1}$.

Solution:

$$\frac{\sqrt{5} + 4}{\sqrt{5} - 1} = \frac{(\sqrt{5} + 4)(\sqrt{5} + 1)}{(\sqrt{5} - 1)(\sqrt{5} + 1)} \quad \text{Multiply the numerator and denominator by } \sqrt{5} + 1, \text{ the conjugate of } \sqrt{5} - 1.$$

$$= \frac{5 + \sqrt{5} + 4\sqrt{5} + 4}{5 - 1} \quad \text{Multiply.}$$

$$= \frac{9 + 5\sqrt{5}}{4} \quad \text{Simplify.}$$

■ **Work Practice 14**

Practice 14

Rationalize the denominator of $\dfrac{\sqrt{2} + 5}{\sqrt{2} - 1}$.

Example 15   Rationalize the denominator of $\dfrac{3}{1 + \sqrt{x}}$.

Solution:

$$\frac{3}{1 + \sqrt{x}} = \frac{3(1 - \sqrt{x})}{(1 + \sqrt{x})(1 - \sqrt{x})} \quad \text{Multiply the numerator and denominator by } 1 - \sqrt{x}, \text{ the conjugate of } 1 + \sqrt{x}.$$

$$= \frac{3(1 - \sqrt{x})}{1 - x}$$

■ **Work Practice 15**

Practice 15

Rationalize the denominator of $\dfrac{7}{2 - \sqrt{x}}$.

Answers

13. $-2 + \sqrt{7}$ 14. $7 + 6\sqrt{2}$

15. $\dfrac{7(2 + \sqrt{x})}{4 - x}$

Vocabulary, Readiness & Video Check

Fill in each blank.

1. $\sqrt{7} \cdot \sqrt{3} =$ _____

2. $\sqrt{10} \cdot \sqrt{10} =$ _____

3. $\dfrac{\sqrt{15}}{\sqrt{3}} =$ _____

4. The process of eliminating the radical in the denominator of a radical expression is called _____.

5. The conjugate of $2 + \sqrt{3}$ is _____.

 *Martin-Gay Interactive Videos*

See Video 15.4 🍎

Watch the section lecture video and answer the following questions.

Objective A 6. In ▣ Examples 1 and 3, the product rule for radicals is applied twice, but in different ways. Explain. ▶

7. Starting with ▣ Example 2, what important reminder is made repeatedly about the square root of a positive number that is squared? ▶

Objective B 8. From ▣ Examples 5 and 6, when we're looking at a quotient of two radicals, what would make us think to apply the quotient rule in order to simplify? ▶

Objective C 9. From the lecture before ▣ Example 7, what is the goal of rationalizing a denominator? ▶

Objective D 10. From ▣ Example 9, why will multiplying a denominator by its conjugate rationalize the denominator? ▶

15.4 Exercise Set MyMathLab®

Objective A *Multiply and simplify. Assume that all variables represent positive real numbers. See Examples 1 through 6.*

1. $\sqrt{8} \cdot \sqrt{2}$

2. $\sqrt{3} \cdot \sqrt{12}$

▶ 3. $\sqrt{10} \cdot \sqrt{5}$

4. $\sqrt{2} \cdot \sqrt{14}$

5. $\left(\sqrt{6}\right)^2$

6. $\left(\sqrt{10}\right)^2$

7. $\sqrt{2x} \cdot \sqrt{2x}$

8. $\sqrt{5y} \cdot \sqrt{5y}$

9. $\left(2\sqrt{5}\right)^2$

10. $\left(3\sqrt{10}\right)^2$

▶ 11. $\left(6\sqrt{x}\right)^2$

12. $\left(8\sqrt{y}\right)^2$

13. $\sqrt{3x^5} \cdot \sqrt{6x}$

14. $\sqrt{21y^7} \cdot \sqrt{3y}$

15. $\sqrt{2xy^2} \cdot \sqrt{8xy}$

16. $\sqrt{18x^2y^2} \cdot \sqrt{2x^2y}$

▶ 17. $\sqrt{6}\left(\sqrt{5} + \sqrt{7}\right)$

18. $\sqrt{10}\left(\sqrt{3} - \sqrt{7}\right)$

19. $\sqrt{10}\left(\sqrt{2} + \sqrt{5}\right)$

20. $\sqrt{6}\left(\sqrt{3} + \sqrt{2}\right)$

21. $\sqrt{7y}\left(\sqrt{y} - 2\sqrt{7}\right)$

22. $\sqrt{5b}\left(2\sqrt{b} + \sqrt{5}\right)$

23. $\left(\sqrt{3} + 6\right)\left(\sqrt{3} - 6\right)$

24. $\left(\sqrt{5} + 2\right)\left(\sqrt{5} - 2\right)$

25. $\left(\sqrt{3} + \sqrt{5}\right)\left(\sqrt{2} - \sqrt{5}\right)$

26. $\left(\sqrt{7} + \sqrt{5}\right)\left(\sqrt{2} - \sqrt{5}\right)$

27. $\left(2\sqrt{11} + 1\right)\left(\sqrt{11} - 6\right)$

28. $\left(5\sqrt{3} + 2\right)\left(\sqrt{3} - 1\right)$

29. $\left(\sqrt{x}+6\right)\left(\sqrt{x}-6\right)$ **30.** $\left(\sqrt{y}+5\right)\left(\sqrt{y}-5\right)$ **31.** $\left(\sqrt{x}-7\right)^2$

32. $\left(\sqrt{x}+4\right)^2$ **33.** $\left(\sqrt{6y}+1\right)^2$ **34.** $\left(\sqrt{3y}-2\right)^2$

Objective B   *Divide and simplify. Assume that all variables represent positive real numbers. See Examples 7 through 9.*

35. $\dfrac{\sqrt{32}}{\sqrt{2}}$ **36.** $\dfrac{\sqrt{40}}{\sqrt{10}}$ **37.** $\dfrac{\sqrt{21}}{\sqrt{3}}$ **38.** $\dfrac{\sqrt{55}}{\sqrt{5}}$ **39.** $\dfrac{\sqrt{90}}{\sqrt{5}}$

40. $\dfrac{\sqrt{96}}{\sqrt{8}}$ **41.** $\dfrac{\sqrt{75y^5}}{\sqrt{3y}}$ **42.** $\dfrac{\sqrt{24x^7}}{\sqrt{6x}}$ **43.** $\dfrac{\sqrt{150}}{\sqrt{2}}$ **44.** $\dfrac{\sqrt{120}}{\sqrt{3}}$

45. $\dfrac{\sqrt{72y^5}}{\sqrt{3y^3}}$ **46.** $\dfrac{\sqrt{54x^3}}{\sqrt{2x}}$ **47.** $\dfrac{\sqrt{24x^3y^4}}{\sqrt{2xy}}$ **48.** $\dfrac{\sqrt{96x^5y^3}}{\sqrt{3x^2y}}$

Objective C   *Rationalize each denominator and simplify. Assume that all variables represent positive real numbers. See Examples 10 through 12.*

49. $\dfrac{\sqrt{3}}{\sqrt{5}}$ **50.** $\dfrac{\sqrt{2}}{\sqrt{3}}$ **51.** $\dfrac{7}{\sqrt{2}}$ **52.** $\dfrac{8}{\sqrt{11}}$

53. $\dfrac{1}{\sqrt{6y}}$ **54.** $\dfrac{1}{\sqrt{10z}}$ **55.** $\sqrt{\dfrac{5}{18}}$ **56.** $\sqrt{\dfrac{7}{12}}$

57. $\sqrt{\dfrac{3}{x}}$ **58.** $\sqrt{\dfrac{5}{x}}$ **59.** $\sqrt{\dfrac{1}{8}}$ **60.** $\sqrt{\dfrac{1}{27}}$

61. $\sqrt{\dfrac{2}{15}}$ **62.** $\sqrt{\dfrac{11}{14}}$ **63.** $\sqrt{\dfrac{3}{20}}$ **64.** $\sqrt{\dfrac{3}{50}}$

65. $\dfrac{3x}{\sqrt{2x}}$ **66.** $\dfrac{5y}{\sqrt{3y}}$ **67.** $\dfrac{8y}{\sqrt{5}}$ **68.** $\dfrac{7x}{\sqrt{2}}$

69. $\sqrt{\dfrac{x}{36y}}$ **70.** $\sqrt{\dfrac{z}{49y}}$ **71.** $\sqrt{\dfrac{y}{12x}}$ **72.** $\sqrt{\dfrac{x}{20y}}$

Objective D   *Rationalize each denominator and simplify. Assume that all variables represent positive real numbers. See Examples 13 through 15.*

73. $\dfrac{3}{\sqrt{2}+1}$ **74.** $\dfrac{6}{\sqrt{5}+2}$

75. $\dfrac{4}{2-\sqrt{5}}$ **76.** $\dfrac{2}{\sqrt{10}-3}$

77. $\dfrac{\sqrt{5}+1}{\sqrt{6}-\sqrt{5}}$ **78.** $\dfrac{\sqrt{3}+1}{\sqrt{3}-\sqrt{2}}$

79. $\dfrac{\sqrt{3}+1}{\sqrt{2}-1}$ **80.** $\dfrac{\sqrt{2}-2}{2-\sqrt{3}}$

81. $\dfrac{5}{2 + \sqrt{x}}$

82. $\dfrac{9}{3 + \sqrt{x}}$

83. $\dfrac{3}{\sqrt{x} - 4}$

84. $\dfrac{4}{\sqrt{x} - 1}$

Review

Solve each equation. See Sections 9.3 and 13.6.

85. $x + 5 = 7^2$

86. $2y - 1 = 3^2$

87. $4z^2 + 6z - 12 = (2z)^2$

88. $16x^2 + x + 9 = (4x)^2$

89. $9x^2 + 5x + 4 = (3x + 1)^2$

90. $x^2 + 3x + 4 = (x + 2)^2$

Concept Extensions

△ **91.** Find the area of a rectangular room whose length is $13\sqrt{2}$ meters and width is $5\sqrt{6}$ meters.

13√2 meters   5√6 meters

△ **92.** Find the volume of a microwave oven whose length is $\sqrt{3}$ feet, width is $\sqrt{2}$ feet, and height is $\sqrt{2}$ feet.

√3 feet

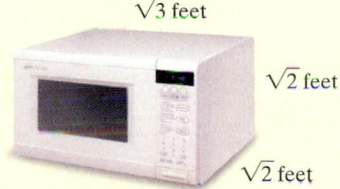

√2 feet

√2 feet

△ **93.** If a circle has area A, then the formula for the radius r of the circle is

$$r = \sqrt{\dfrac{A}{\pi}}$$

Rationalize the denominator of this expression.

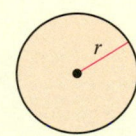

△ **94.** If a round ball has volume V, then the formula for the radius r of the ball is

$$r = \sqrt[3]{\dfrac{3V}{4\pi}}$$

Simplify this expression by rationalizing the denominator.

Identify each statement as true or false. See the Concept Check in this section.

95. $\sqrt{5} \cdot \sqrt{5} = 5$

96. $\sqrt{5} \cdot \sqrt{3} = 15$

97. $\sqrt{3x} \cdot \sqrt{3x} = 2\sqrt{3x}$

98. $\sqrt{3x} + \sqrt{3x} = 2\sqrt{3x}$

99. $\sqrt{11} + \sqrt{2} = \sqrt{13}$

100. $\sqrt{11} \cdot \sqrt{2} = \sqrt{22}$

Solve.

101. When rationalizing the denominator of $\dfrac{\sqrt{2}}{\sqrt{3}}$, explain why both the numerator and the denominator must be multiplied by $\sqrt{3}$.

102. In your own words, explain why $\sqrt{6} + \sqrt{2}$ cannot be simplified further, but $\sqrt{6} \cdot \sqrt{2}$ can be.

103. When rationalizing the denominator of $\dfrac{\sqrt[3]{2}}{\sqrt[3]{3}}$, explain why both the numerator and the denominator must be multiplied by $\sqrt[3]{9}$.

104. When rationalizing the denominator of $\dfrac{5}{1 + \sqrt{2}}$, explain why multiplying by $\dfrac{\sqrt{2}}{\sqrt{2}}$ will not accomplish this, but multiplying by $\dfrac{1 - \sqrt{2}}{1 - \sqrt{2}}$ will.

It is often more convenient to work with a radical expression whose numerator is rationalized. Rationalize the numerator of each expression by multiplying the numerator and denominator by the conjugate of the numerator.

105. $\dfrac{\sqrt{3} + 1}{\sqrt{2} - 1}$

106. $\dfrac{\sqrt{2} - 2}{2 - \sqrt{3}}$

Integrated Review

Simplifying Radicals

Simplify. Assume that all variables represent positive numbers.

1. $\sqrt{36}$

2. $\sqrt{48}$

3. $\sqrt{x^4}$

4. $\sqrt{y^7}$

5. $\sqrt{16x^2}$

6. $\sqrt{18x^{11}}$

7. $\sqrt[3]{8}$

8. $\sqrt[4]{81}$

9. $\sqrt[3]{-27}$

10. $\sqrt{-4}$

11. $\sqrt{\dfrac{11}{9}}$

12. $\sqrt[3]{\dfrac{7}{64}}$

13. $-\sqrt{16}$

14. $-\sqrt{25}$

15. $\sqrt{\dfrac{9}{49}}$

16. $\sqrt{\dfrac{1}{64}}$

17. $\sqrt{a^8 b^2}$

18. $\sqrt{x^{10} y^{20}}$

19. $\sqrt{25m^6}$

20. $\sqrt{9n^{16}}$

Add or subtract as indicated.

21. $5\sqrt{7} + \sqrt{7}$

22. $\sqrt{50} - \sqrt{8}$

1.

2.

3.

4.

5.

6.

7.

8.

9.

10.

11.

12.

13.

14.

15.

16.

17.

18.

19.

20.

21.

22.

23. _____	**23.** $5\sqrt{2} - 5\sqrt{3}$
24. _____	**24.** $2\sqrt{x} + \sqrt{25x} - \sqrt{36x} + 3x$
25. _____	

Multiply and simplify if possible.

26. _____	**25.** $\sqrt{2} \cdot \sqrt{15}$	**26.** $\sqrt{3} \cdot \sqrt{3}$
27. _____		
28. _____	**27.** $\left(2\sqrt{7}\right)^2$	**28.** $\left(3\sqrt{5}\right)^2$
29. _____		
30. _____		
31. _____	**29.** $\sqrt{3}\left(\sqrt{11} + 1\right)$	**30.** $\sqrt{6}\left(\sqrt{3} - 2\right)$
32. _____		
33. _____	**31.** $\sqrt{8y} \cdot \sqrt{2y}$	**32.** $\sqrt{15x^2} \cdot \sqrt{3x^2}$
34. _____		
35. _____		
36. _____	**33.** $\left(\sqrt{x} - 5\right)\left(\sqrt{x} + 2\right)$	**34.** $\left(3 + \sqrt{2}\right)^2$
37. _____		

Divide and simplify if possible.

35. $\dfrac{\sqrt{8}}{\sqrt{2}}$	**36.** $\dfrac{\sqrt{45}}{\sqrt{15}}$	**37.** $\dfrac{\sqrt{24x^5}}{\sqrt{2x}}$	**38.** $\dfrac{\sqrt{75a^4b^5}}{\sqrt{5ab}}$

38. _____
39. _____
40. _____
41. _____
42. _____

Rationalize each denominator.

39. $\sqrt{\dfrac{1}{6}}$	**40.** $\dfrac{x}{\sqrt{20}}$	**41.** $\dfrac{4}{\sqrt{6} + 1}$	**42.** $\dfrac{\sqrt{2} + 1}{\sqrt{x} - 5}$

Solving Equations Containing Radicals

Objective A Using the Squaring Property of Equality Once

In this section, we solve **radical equations** such as

$$\sqrt{x + 3} = 5 \quad \text{and} \quad \sqrt{2x + 1} = \sqrt{3x}$$

Radical equations contain variables in the radicand. To solve these equations, we rely on the following squaring property.

> ### The Squaring Property of Equality
>
> If $a = b$, then $a^2 = b^2$.

Unfortunately, this squaring property does not guarantee that all solutions of the new equation are solutions of the original equation. For example, if we square both sides of the equation

$$x = 2$$

we have

$$x^2 = 4$$

This new equation has two solutions, 2 and -2, while the original equation, $x = 2$, has only one solution. For this reason, we must **always check proposed solutions of radical equations in the original equation.**

Example 1 Solve: $\sqrt{x + 3} = 5$

Solution: To solve this radical equation, we use the squaring property of equality and square both sides of the equation.

$$\sqrt{x + 3} = 5$$
$$(\sqrt{x + 3})^2 = 5^2 \quad \text{Square both sides.}$$
$$x + 3 = 25 \quad \text{Simplify.}$$
$$x = 22 \quad \text{Subtract 3 from both sides.}$$

Check: We replace x with 22 in the original equation.

$$\sqrt{x + 3} = 5 \quad \text{Original equation}$$
$$\sqrt{22 + 3} \overset{?}{=} 5 \quad \text{Let } x = 22.$$
$$\sqrt{25} \overset{?}{=} 5$$
$$5 = 5 \quad \text{True}$$

Since a true statement results, 22 is the solution.

■ **Work Practice 1**

When solving radical equations, if possible, move radicals so that at least one radical is by itself on one side of the equation.

Example 2 Solve: $\sqrt{x} = \sqrt{5x - 2}$

Solution: Each radical is by itself on one side of the equation. Let's begin solving by squaring both sides.

$$\sqrt{x} = \sqrt{5x - 2} \quad \text{Original equation}$$
$$(\sqrt{x})^2 = (\sqrt{5x - 2})^2 \quad \text{Square both sides.}$$
$$x = 5x - 2 \quad \text{Simplify.}$$
$$-4x = -2 \quad \text{Subtract } 5x \text{ from both sides.}$$
$$x = \frac{-2}{-4} = \frac{1}{2} \quad \text{Divide both sides by } -4 \text{ and simplify.}$$

(Continued on next page)

Objectives

A Solve Radical Equations by Using the Squaring Property of Equality Once.

B Solve Radical Equations by Using the Squaring Property of Equality Twice.

Practice 1

Solve: $\sqrt{x - 2} = 7$

> **Helpful Hint**
> Don't forget to check the proposed solutions of radical equations in the original equation.

Practice 2

Solve: $\sqrt{6x - 1} = \sqrt{x}$

Answers

1. $x = 51$ **2.** $x = \frac{1}{5}$

1141

Check: We replace x with $\frac{1}{2}$ in the original equation.

$$\sqrt{x} = \sqrt{5x - 2} \qquad \text{Original equation}$$

$$\sqrt{\frac{1}{2}} \overset{?}{=} \sqrt{5 \cdot \frac{1}{2} - 2} \qquad \text{Let } x = \frac{1}{2}.$$

$$\sqrt{\frac{1}{2}} \overset{?}{=} \sqrt{\frac{5}{2} - 2} \qquad \text{Multiply.}$$

$$\sqrt{\frac{1}{2}} \overset{?}{=} \sqrt{\frac{5}{2} - \frac{4}{2}} \qquad \text{Write 2 as } \frac{4}{2}.$$

$$\sqrt{\frac{1}{2}} = \sqrt{\frac{1}{2}} \qquad \text{True}$$

This statement is true, so the solution is $\frac{1}{2}$.

■ **Work Practice 2**

Practice 3
Solve: $\sqrt{x} + 9 = 2$

Example 3   Solve: $\sqrt{x} + 6 = 4$

Solution: First we write the equation so that the radical is by itself on one side of the equation.

$$\sqrt{x} + 6 = 4$$

$$\sqrt{x} = -2 \qquad \text{Subtract 6 from both sides to get the radical by itself.}$$

Normally we would now square both sides. Recall, however, that $\sqrt{x}$ is the principal or nonnegative square root of x, so $\sqrt{x}$ cannot equal -2 and thus this equation has no solution. We arrive at the same conclusion if we continue by applying the squaring property.

$$\sqrt{x} = -2$$

$$(\sqrt{x})^2 = (-2)^2 \qquad \text{Square both sides.}$$

$$x = 4 \qquad \text{Simplify.}$$

Check: We replace x with 4 in the original equation.

$$\sqrt{x} + 6 = 4 \qquad \text{Original equation}$$

$$\sqrt{4} + 6 \overset{?}{=} 4 \qquad \text{Let } x = 4.$$

$$2 + 6 = 4 \qquad \text{False}$$

Since 4 *does not* satisfy the original equation, this equation has no solution.

■ **Work Practice 3**

Example 3 makes it very clear that we *must* check proposed solutions in the original equation to determine if they are truly solutions. If a proposed solution does not work, we say that the value is an **extraneous solution.**

The following steps can be used to solve radical equations containing square roots.

> ### To Solve a Radical Equation Containing Square Roots
>
> **Step 1:** Arrange terms so that one radical is by itself on one side of the equation. That is, isolate a radical.
>
> **Step 2:** Square both sides of the equation.
>
> **Step 3:** Simplify both sides of the equation.
>
> **Step 4:** If the equation still contains a radical term, repeat Steps 1 through 3.
>
> **Step 5:** Solve the equation.
>
> **Step 6:** Check all solutions in the original equation for extraneous solutions.

Answer

3. no solution

Example 4 Solve: $\sqrt{4y^2 + 5y - 15} = 2y$

Solution: The radical is already isolated, so we start by squaring both sides.

$$\sqrt{4y^2 + 5y - 15} = 2y$$

$$\left(\sqrt{4y^2 + 5y - 15}\right)^2 = (2y)^2 \qquad \text{Square both sides.}$$

$$4y^2 + 5y - 15 = 4y^2 \qquad \text{Simplify.}$$

$$5y - 15 = 0 \qquad \text{Subtract } 4y^2 \text{ from both sides.}$$

$$5y = 15 \qquad \text{Add 15 to both sides.}$$

$$y = 3 \qquad \text{Divide both sides by 5.}$$

Check: We replace y with 3 in the original equation.

$$\sqrt{4y^2 + 5y - 15} = 2y \qquad \text{Original equation}$$

$$\sqrt{4 \cdot 3^2 + 5 \cdot 3 - 15} \stackrel{?}{=} 2 \cdot 3 \qquad \text{Let } y = 3.$$

$$\sqrt{4 \cdot 9 + 15 - 15} \stackrel{?}{=} 6 \qquad \text{Simplify.}$$

$$\sqrt{36} \stackrel{?}{=} 6$$

$$6 = 6 \qquad \text{True}$$

This statement is true, so the solution is 3.

◼ **Work Practice 4**

Practice 4
Solve: $\sqrt{9y^2 + 2y - 10} = 3y$

Example 5 Solve: $\sqrt{x + 3} - x = -3$

Solution: First we isolate the radical by adding x to both sides. Then we square both sides.

$$\sqrt{x + 3} - x = -3$$

$$\sqrt{x + 3} = x - 3 \qquad \text{Add } x \text{ to both sides.}$$

$$\left(\sqrt{x + 3}\right)^2 = (x - 3)^2 \qquad \text{Square both sides.}$$

$$x + 3 = x^2 - 6x + 9 \qquad \text{Simplify.}$$

To solve the resulting quadratic equation, we write the equation in standard form by subtracting x and 3 from both sides.

$$x + 3 = x^2 - 6x + 9$$

$$3 = x^2 - 7x + 9 \qquad \text{Subtract } x \text{ from both sides.}$$

$$0 = x^2 - 7x + 6 \qquad \text{Subtract 3 from both sides.}$$

$$0 = (x - 6)(x - 1) \qquad \text{Factor.}$$

$$0 = x - 6 \quad \text{or} \quad 0 = x - 1 \qquad \text{Set each factor equal to zero.}$$

$$6 = x \qquad\qquad 1 = x \qquad \text{Solve for } x.$$

Check: We replace x with 6 and then x with 1 in the original equation.

Let $x = 6$.	Let $x = 1$.
$\sqrt{x + 3} - x = -3$	$\sqrt{x + 3} - x = -3$
$\sqrt{6 + 3} - 6 \stackrel{?}{=} -3$	$\sqrt{1 + 3} - 1 \stackrel{?}{=} -3$
$\sqrt{9} - 6 \stackrel{?}{=} -3$	$\sqrt{4} - 1 \stackrel{?}{=} -3$
$3 - 6 \stackrel{?}{=} -3$	$2 - 1 \stackrel{?}{=} -3$
$-3 = -3$ True	$1 = -3$ False

Since replacing x with 1 resulted in a false statement, 1 is an extraneous solution. The only solution is 6.

◼ **Work Practice 5**

Practice 5
Solve: $\sqrt{x + 1} - x = -5$

Helpful Hint
Don't forget that $(x - 3)^2 = (x - 3)(x - 3) = x^2 - 6x + 9$.

Answers
4. $y = 5$ **5.** $x = 8$

Objective B Using the Squaring Property of Equality Twice ▶

If a radical equation contains two radicals, we may need to use the squaring property twice.

Practice 6

Solve: $\sqrt{x} + 3 = \sqrt{x + 15}$

> **Example 6** Solve: $\sqrt{x - 4} = \sqrt{x} - 2$
>
> **Solution:**
>
> $$\sqrt{x - 4} = \sqrt{x} - 2$$
> $$(\sqrt{x - 4})^2 = (\sqrt{x} - 2)^2 \qquad \text{Square both sides.}$$
> $$x - 4 = \underbrace{x - 4\sqrt{x} + 4}$$
>
> $$-8 = -4\sqrt{x} \qquad \text{To get the radical term alone, subtract } x \text{ and } 4 \text{ from both sides.}$$
> $$2 = \sqrt{x} \qquad \text{Divide both sides by } -4.$$
> $$4 = x \qquad \text{Square both sides again.}$$
>
> Check the proposed solution in the original equation. The solution is 4.
>
> ■ **Work Practice 6**

Answer

6. $x = 1$

> **Helpful Hint**
>
> Don't forget:
> $$(\sqrt{x} - 2)^2 = (\sqrt{x} - 2)(\sqrt{x} - 2)$$
> $$= \sqrt{x} \cdot \sqrt{x} - 2\sqrt{x} - 2\sqrt{x} + 4$$
> $$= x - 4\sqrt{x} + 4$$

Vocabulary, Readiness & Video Check

Martin-Gay Interactive Videos

See Video 15.5

Watch the section lecture video and answer the following questions.

Objective A **1.** From ▥ Examples 1 and 2, why must we be sure to check our proposed solution(s) in the original equation? ▶

Objective B **2.** Solving ▥ Example 5 requires using the squaring property twice. Is anything else done differently to solve these equations as compared to equations where the property is used only once? ▶

15.5 Exercise Set MyMathLab®

Objective A *Solve each equation. See Examples 1 through 3.*

1. $\sqrt{x} = 9$

2. $\sqrt{x} = 4$

▶ **3.** $\sqrt{x + 5} = 2$

4. $\sqrt{x + 12} = 3$

5. $\sqrt{x} - 2 = 5$

6. $4\sqrt{x} - 7 = 5$

▶ **7.** $3\sqrt{x} + 5 = 2$

8. $3\sqrt{x} + 8 = 5$

9. $\sqrt{x} = \sqrt{3x - 8}$

10. $\sqrt{x} = \sqrt{4x - 3}$

11. $\sqrt{4x - 3} = \sqrt{x + 3}$

12. $\sqrt{5x - 4} = \sqrt{x + 8}$

Solve each equation. See Examples 4 and 5.

13. $\sqrt{9x^2 + 2x - 4} = 3x$

14. $\sqrt{4x^2 + 3x - 9} = 2x$

15. $\sqrt{x} = x - 6$

16. $\sqrt{x} = x - 2$

17. $\sqrt{x + 7} = x + 5$

18. $\sqrt{x + 5} = x - 1$

19. $\sqrt{3x + 7} - x = 3$

20. $x = \sqrt{4x - 7} + 1$

21. $\sqrt{16x^2 + 2x + 2} = 4x$

22. $\sqrt{4x^2 + 3x + 2} = 2x$

23. $\sqrt{2x^2 + 6x + 9} = 3$

24. $\sqrt{3x^2 + 6x + 4} = 2$

Objective B *Solve each equation. See Example 6.*

▶ **25.** $\sqrt{x - 7} = \sqrt{x} - 1$

26. $\sqrt{x - 8} = \sqrt{x} - 2$

27. $\sqrt{x} + 2 = \sqrt{x + 24}$

28. $\sqrt{x} + 5 = \sqrt{x + 55}$

29. $\sqrt{x + 8} = \sqrt{x} + 2$

30. $\sqrt{x} + 1 = \sqrt{x + 15}$

Objectives A B Mixed Practice *Solve each equation. See Examples 1 through 6.*

31. $\sqrt{2x + 6} = 4$

32. $\sqrt{3x + 7} = 5$

▶ **33.** $\sqrt{x + 6} + 1 = 3$

34. $\sqrt{x + 5} + 2 = 5$

35. $\sqrt{x + 6} + 5 = 3$

36. $\sqrt{2x - 1} + 7 = 1$

37. $\sqrt{16x^2 - 3x + 6} = 4x$

38. $\sqrt{9x^2 - 2x + 8} = 3x$

39. $-\sqrt{x} = -6$

40. $-\sqrt{y} = -8$

41. $\sqrt{x + 9} = \sqrt{x} - 3$

42. $\sqrt{x} - 6 = \sqrt{x + 36}$

43. $\sqrt{2x + 1} + 3 = 5$

44. $\sqrt{3x - 1} + 1 = 4$

45. $\sqrt{x + 3} = 7$

46. $\sqrt{x + 5} = 10$

47. $\sqrt{4x} = \sqrt{2x + 6}$

48. $\sqrt{5x + 6} = \sqrt{8x}$

49. $\sqrt{2x + 1} = x - 7$

50. $\sqrt{2x + 5} = x - 5$

51. $x = \sqrt{2x - 2} + 1$

52. $\sqrt{2x - 4} + 2 = x$

▶ **53.** $\sqrt{1 - 8x} - x = 4$

54. $\sqrt{2x + 5} - 1 = x$

Review

Translating   *Translate each sentence into an equation and then solve. See Section 9.4.*

55. If 8 is subtracted from the product of 3 and x, the result is 19. Find x.

56. If 3 more than x is subtracted from twice x, the result is 11. Find x.

57. The length of a rectangle is twice the width. The perimeter is 24 inches. Find the length.

58. The length of a rectangle is 2 inches longer than the width. The perimeter is 24 inches. Find the length.

Concept Extensions

Solve each equation.

59. $\sqrt{x - 3} + 3 = \sqrt{3x + 4}$

60. $\sqrt{2x + 3} = \sqrt{x - 2} + 2$

Solve.

61. Explain why proposed solutions of radical equations must be checked in the original equation.

62. Is 8 a solution of the equation $\sqrt{x - 4} - 5 = \sqrt{x + 1}$? Explain why or why not.

63. The formula $b = \sqrt{\dfrac{V}{2}}$ can be used to determine the length b of a side of the base of a square-based pyramid with height 6 units and volume V cubic units.

  a. Find the length of the side of the base that produces a pyramid with each volume. (Round to the nearest tenth of a unit.)

V	20	200	2000
b			

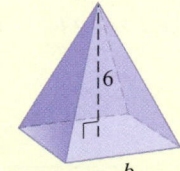

  b. Notice in the table that volume V has been increased by a factor of 10 each time. Does the corresponding length b of a side increase by a factor of 10 each time also?

64. The formula $r = \sqrt{\dfrac{V}{2\pi}}$ can be used to determine the radius r of a cylinder with height 2 units and volume V cubic units.

  a. Find the radius needed to manufacture a cylinder with each volume. (Round to the nearest tenth of a unit.)

V	10	100	1000
r			

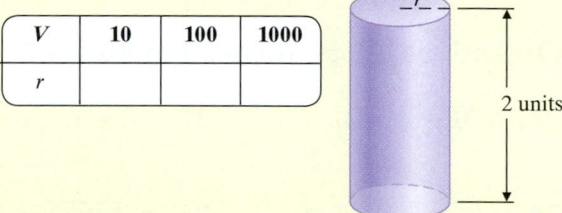

  b. Notice in the table that volume V has been increased by a factor of 10 each time. Does the corresponding radius increase by a factor of 10 each time also?

Graphing calculators can be used to solve equations. To solve $\sqrt{x - 2} = x - 5$, for example, graph $y_1 = \sqrt{x - 2}$ and $y_2 = x - 5$ on the same set of axes. Use the Trace and Zoom features or an Intersect feature to find the point of intersection of the graphs. The x-value of the point is the solution of the equation. Use a graphing calculator to solve the equations below. Approximate solutions to the nearest hundredth.

65. $\sqrt{x - 2} = x - 5$

66. $\sqrt{x + 1} = 2x - 3$

67. $-\sqrt{x + 4} = 5x - 6$

68. $-\sqrt{x + 5} = -7x + 1$

15.6 Radical Equations and Problem Solving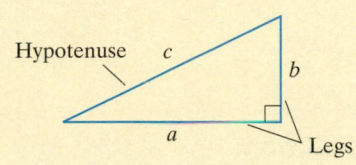

Objective A Using the Pythagorean Theorem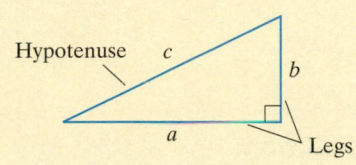

Applications of radicals can be found in geometry, finance, science, and other areas of technology. Our first application involves the Pythagorean theorem, which gives a formula that relates the lengths of the three sides of a right triangle. We studied the Pythagorean theorem in Chapters 6 and 13, and we review it here.

Objectives

A Use the Pythagorean Theorem to Solve Problems.

B Solve Problems Using Formulas Containing Radicals.

The Pythagorean Theorem

If a and b are lengths of the legs of a right triangle and c is the length of the hypotenuse, then $a^2 + b^2 = c^2$.

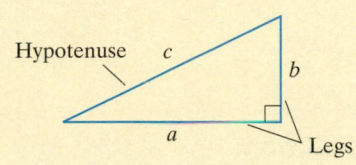

 Example 1 Find the length of the hypotenuse of a right triangle whose legs are 6 inches and 8 inches long.

Solution: Because this is a right triangle, we use the Pythagorean theorem. We let $a = 6$ inches and $b = 8$ inches. Length c must be the length of the hypotenuse.

$a^2 + b^2 = c^2$ Use the Pythagorean theorem.
$6^2 + 8^2 = c^2$ Substitute the lengths of the legs.
$36 + 64 = c^2$ Simplify.
$100 = c^2$

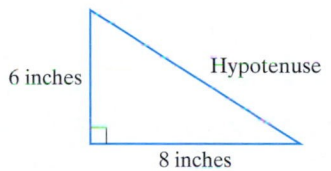

Since c represents a length, we know that c is positive and is the principal square root of 100.

$100 = c^2$
$\sqrt{100} = c$ Use the definition of principal square root.
$10 = c$ Simplify.

The hypotenuse has a length of 10 inches.

■ **Work Practice 1**

Practice 1

Find the length of the hypotenuse of the right triangle shown.

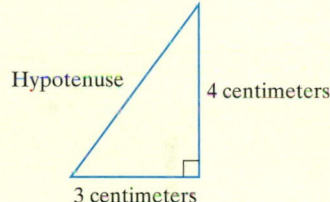

 Example 2 Find the length of the leg of the right triangle shown. Give the exact length and a two-decimal-place approximation.

Solution: We let $a = 2$ meters and b be the unknown length of the other leg. The hypotenuse is $c = 5$ meters.

$a^2 + b^2 = c^2$ Use the Pythagorean theorem.
$2^2 + b^2 = 5^2$ Let $a = 2$ and $c = 5$.
$4 + b^2 = 25$
$b^2 = 21$
$b = \sqrt{21} \approx 4.58$ meters

The length of the leg is exactly $\sqrt{21}$ meters or approximately 4.58 meters.

■ **Work Practice 2**

Practice 2

Find the length of the leg of the right triangle shown. Give the exact length and a two-decimal-place approximation.

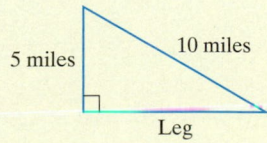

Answers

1. 5 cm   **2.** $5\sqrt{3}$ mi; 8.66 mi

Practice 3

Evan Saacks wants to determine the distance at certain points across a pond on his property. He is able to measure the distances shown on the following diagram. Find how wide the pond is to the nearest tenth of a foot.

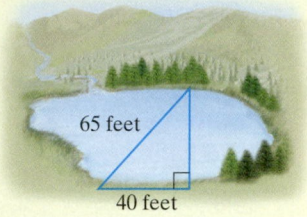

65 feet

40 feet

△ **Example 3**   Finding a Distance

A surveyor must determine the distance across a lake at points P and Q, as shown in the figure. To do this, she finds a third point, R, perpendicular to line PQ. If the length of $\overline{PR}$ is 320 feet and the length of $\overline{QR}$ is 240 feet, what is the distance across the lake? Approximate this distance to the nearest whole foot.

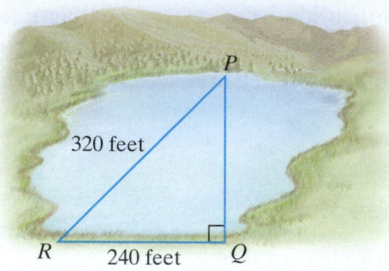

P

320 feet

R 240 feet Q

Solution:

1. **UNDERSTAND.** Read and reread the problem. We will set up the problem using the Pythagorean theorem. By creating a line perpendicular to line PQ, the surveyor deliberately constructed a right triangle. The hypotenuse, $\overline{PR}$, has a length of 320 feet, so we let $c = 320$ in the Pythagorean theorem. The side $\overline{QR}$ is one of the legs, so we let $a = 240$ and $b =$ the unknown length.

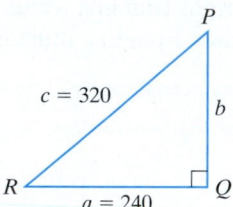

P

$c = 320$

b

R $a = 240$ Q

2. **TRANSLATE.**

$$a^2 + b^2 = c^2 \qquad \text{Use the Pythagorean theorem.}$$
$$240^2 + b^2 = 320^2 \qquad \text{Let } a = 240 \text{ and } c = 320.$$

3. **SOLVE.**

$$57{,}600 + b^2 = 102{,}400$$
$$b^2 = 44{,}800 \qquad \text{Subtract 57,600 from both sides.}$$
$$b = \sqrt{44{,}800} \qquad \text{Use the definition of principal square root.}$$
$$= 80\sqrt{7} \qquad \text{Simplify.}$$

4. **INTERPRET.**

Check: See that $240^2 + (\sqrt{44{,}800})^2 = 320^2$.

State: The distance across the lake is *exactly* $\sqrt{44{,}800}$ or $80\sqrt{7}$ feet. The surveyor can now use a calculator to find that $80\sqrt{7}$ feet is *approximately* 211.6601 feet, so the distance across the lake is roughly 212 feet.

■ **Work Practice 3**

Objective B Using Formulas Containing Radicals ▶

The Pythagorean theorem is an extremely important formula in mathematics and should be memorized. But there are other applications involving formulas containing radicals that are not quite as well known, such as the velocity formula used in the next example.

Answer

3. 51.2 feet

Example 4 Finding the Velocity of an Object

A formula used to determine the velocity v, in feet per second, of an object after it has fallen a certain height (neglecting air resistance) is $v = \sqrt{2gh}$, where g is the acceleration due to gravity and h is the height the object has fallen. On Earth, the acceleration g due to gravity is approximately 32 feet per second per second. Find the velocity of a person after falling 5 feet. Round to the nearest tenth.

Solution: We are told that $g = 32$ feet per second per second. To find the velocity v when $h = 5$ feet, we use the velocity formula.

$$v = \sqrt{2gh} \qquad \text{Use the velocity formula.}$$
$$ = \sqrt{2 \cdot 32 \cdot 5} \qquad \text{Substitute known values.}$$
$$ = \sqrt{320}$$
$$ = 8\sqrt{5} \qquad \text{Simplify the radicand.}$$

The velocity of the person after falling 5 feet is *exactly* $8\sqrt{5}$ feet per second, or *approximately* 17.9 feet per second.

■ Work Practice 4

Practice 4

Use the formula from Example 4 to find the velocity of an object after it has fallen 20 feet. Round to the nearest tenth.

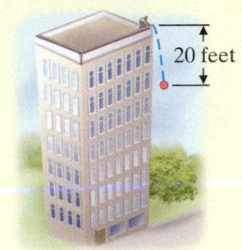

20 feet

Answer

4. $16\sqrt{5}$ ft per sec ≈ 35.8 ft per sec

Vocabulary, Readiness & Video Check

Martin-Gay Interactive Videos

See Video 15.6 🍎

Watch the section lecture video and answer the following questions.

Objective A **1.** From ▢ Examples 1 and 2, when solving exercises using the Pythagorean theorem, what two things must we keep in mind? ▶

2. What very important point is made as the final answer to ▢ Example 1 is being found? ▶

Objective B **3.** In ▢ Examples 3 and 4, how do we know to give an estimated answer instead of an exact answer? In what form would the exact answer be given? ▶

15.6 Exercise Set   MyMathLab®

Objective A *Use the Pythagorean theorem to find the length of the unknown side of each right triangle. Give an exact answer and a two-decimal-place approximation. See Examples 1 and 2.*

1.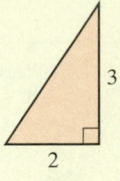

2.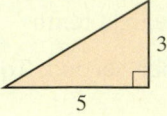

3.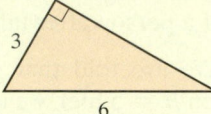

4.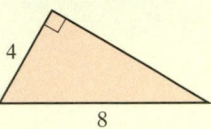

5.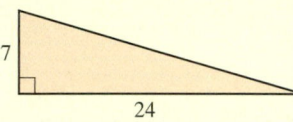

6.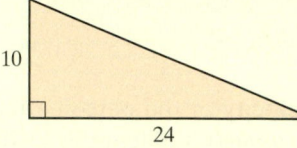

7.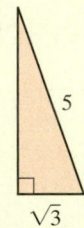

8.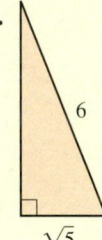

9.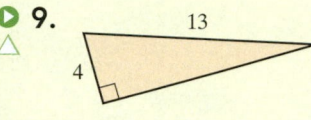

10.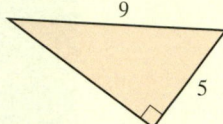

Find the length of the unknown side of each right triangle with sides a, b, and c, where c is the hypotenuse. See Examples 1 and 2. Give an exact answer and a two-decimal-place approximation.

11. $a = 4, b = 5$

12. $a = 2, b = 7$

13. $b = 2, c = 6$

14. $b = 1, c = 5$

15. $a = \sqrt{10}, c = 10$

16. $a = \sqrt{7}, c = \sqrt{35}$

Solve each problem. See Example 3.

17. A wire is used to anchor a 20-foot-tall pole. One end of the wire is attached to the top of the pole. The other end is fastened to a stake five feet away from the bottom of the pole. Find the length of the wire rounded to the nearest tenth of a foot.

18. Jim Spivey needs to connect two underground pipelines, which are offset by 3 feet, as pictured in the diagram. Neglecting the joints needed to join the pipes, find the length of the shortest possible connecting pipe rounded to the nearest hundredth of a foot.

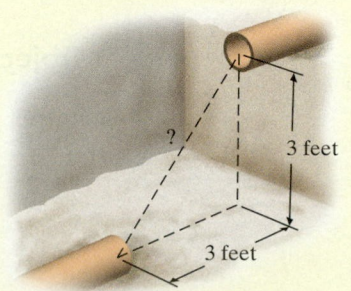

△**19.** Robert Weisman needs to attach a diagonal brace to a rectangular frame in order to make it structurally sound. If the framework is 6 feet by 10 feet, find how long the brace needs to be to the nearest tenth of a foot.

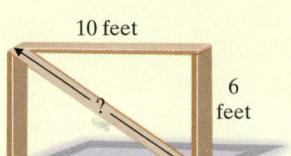

10 feet

? 6 feet

△**20.** Elizabeth Kaster is flying a kite. She let out 80 feet of string and attached the string to a stake in the ground. The kite is now directly above her brother Mike, who is 32 feet away from the stake. Find the height of the kite to the nearest foot.

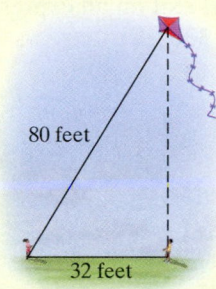

80 feet

32 feet

Objective B *Solve each problem. See Example 4.*

△**21.** For a square-based pyramid, the formula $b = \sqrt{\dfrac{3V}{h}}$ describes the relationship among the length b of one side of the base, the volume V, and the height h. Find the volume if each side of the base is 6 feet long, and the pyramid is 2 feet high.

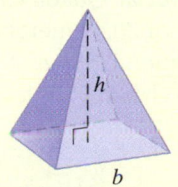

h

b

22. The formula $t = \dfrac{\sqrt{d}}{4}$ relates the distance d, in feet, that an object falls in t seconds, assuming that air resistance does not slow down the object. Find how long, to the nearest hundredth of a second, it takes an object to reach the ground from the top of the Willis Tower in Chicago, a distance of 1730 feet. (*Source:* Council on Tall Buildings and Urban Habitat)

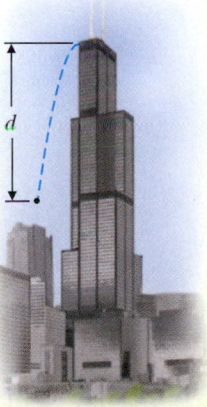

d

23. Police use the formula $s = \sqrt{30fd}$ to estimate the speed s of a car just before it skidded. In this formula, the speed s is measured in miles per hour, d represents the distance the car skidded in feet, and f represents the coefficient of friction. The value of f depends on the type of road surface, and for wet concrete f is 0.35. Find how fast a car was moving if it skidded 280 feet on wet concrete. Round your result to the nearest mile per hour.

d

24. The coefficient of friction of a certain dry road is 0.95. Use the formula in Exercise **23** to find how far a car will skid on this dry road if it is traveling at a rate of 60 mph. Round the length to the nearest foot.

25. The formula $v = \sqrt{2.5r}$ can be used to estimate the maximum safe velocity, v, in miles per hour, at which a car can travel if it is driven along a curved road with a **radius of curvature** r in feet. Find the maximum safe speed to the nearest whole number if a cloverleaf exit on an expressway has a radius of curvature of 300 feet.

26. Use the formula from Exercise **25** to find the radius of curvature if the safe velocity is 30 mph.

The maximum distance d in kilometers that you can see from a height of h meters is given by $d = 3.5\sqrt{h}$. Use this equation for Exercises 27 through 30.

27. Find how far you can see from the top of Trump International Hotel & Tower in Toronto, Ontario, a height of 276.9 meters. Round to the nearest tenth of a kilometer. (*Source:* Council on Tall Buildings and Urban Habitat)

28. Find how far you can see from the top of Great American Tower at Queen City Square in Cincinnati, Ohio, a height of 202.7 meters. Round to the nearest tenth of a kilometer. (*Source:* Council on Tall Buildings and Urban Habitat)

29. The newly built One World Trade Center, in New York City, is the tallest building in the Western Hemisphere. Its height, including the spire at the top of the building, is 541.3 meters. Find how far you could see from the top of One World Trade Center's spire. Round to the nearest tenth of a kilometer. (*Source:* Council on Tall Buildings and Urban Habitat)

30. Guests can take in the views from One World Trade Center by visiting the building's One World Observatory, located at a height of 386.1 meters. Find how far a visitor to One World Observatory could see. Round to the nearest tenth of a kilometer. (*Source:* Council on Tall Buildings and Urban Habitat)

Review

Find two numbers whose square is the given number. See Section 15.1.

31. 9

32. 25

33. 100

34. 49

35. 64

36. 121

Concept Extensions

For each triangle, find the length of y, then x.

△ **37.**

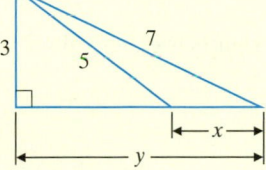

△ **38.**

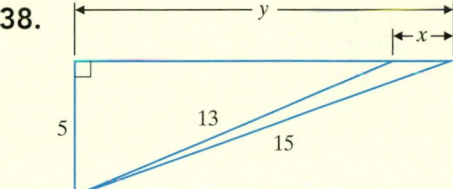

Solve.

△ **39.** Mike and Sandra Hallahan leave the seashore at the same time. Mike drives northward at a rate of 30 miles per hour, while Sandra drives west at 60 mph. Find how far apart they are after 3 hours to the nearest mile.

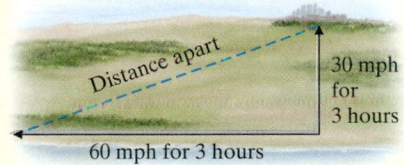

△ **40.** Railroad tracks are invariably made up of relatively short sections of rail connected by expansion joints. To see why this construction is necessary, consider a single rail 100 feet long (or 1200 inches). On an extremely hot day, suppose it expands 1 inch in the hot sun to a new length of 1201 inches. Theoretically, the track would bow upward as pictured.

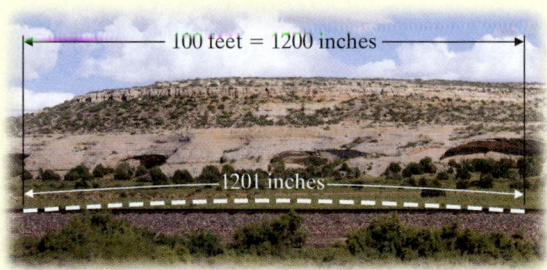

Let us approximate the bulge in the railroad this way.

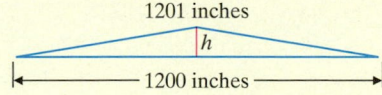

Calculate the height *h* of the bulge to the nearest tenth of an inch.

41. Based on the results of Exercise **40**, explain why railroads use short sections of rail connected by expansion joints.

Chapter 15 Group Activity

Graphing and the Distance Formula

One application of radicals is finding the distance between two points in the coordinate plane. This can be very useful in graphing.

The distance d between two points with coordinates (x_1, y_1) and (x_2, y_2) is given by the **distance formula** $d = \sqrt{(x_2 - x_1)^2 + (y_2 - y_1)^2}$.

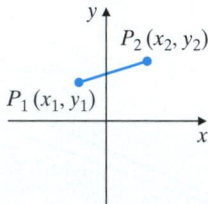

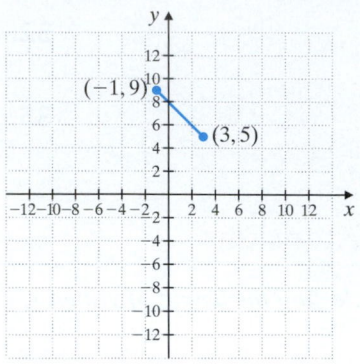

The distance between the two points is exactly $4\sqrt{2}$ units or approximately 5.66 units.

Group Activity

Brainstorm to come up with several disciplines or activities in which the distance formula might be useful. Make up an example that shows how the distance formula would be used in one of the activities on your list. Then present your example to the rest of the class.

Suppose we want to find the distance between the two points $(-1, 9)$ and $(3, 5)$. We can use the distance formula with $(x_1, y_1) = (-1, 9)$ and $(x_2, y_2) = (3, 5)$. Then we have

$$
\begin{aligned}
d &= \sqrt{(x_2 - x_1)^2 + (y_2 - y_1)^2} \\
&= \sqrt{[3 - (-1)]^2 + (5 - 9)^2} \\
&= \sqrt{(4)^2 + (-4)^2} \\
&= \sqrt{16 + 16} \\
&= \sqrt{32} = 4\sqrt{2}
\end{aligned}
$$

Chapter 15 Vocabulary Check

Fill in each blank with one of the words or phrases listed below. Not all choices will be used.

index	radicand	like radicals
rationalizing the denominator	conjugate	leg
principal square root	radical	hypotenuse

1. The expressions $5\sqrt{x}$ and $7\sqrt{x}$ are examples of _____.
2. In the expression $\sqrt[3]{45}$, the number 3 is the _____, the number 45 is the _____, and $\sqrt{}$ is called the _____ sign.
3. The _____ of $a + b$ is $a - b$.
4. The _____ of 25 is 5.
5. The process of eliminating the radical in the denominator of a radical expression is called _____.
6. The Pythagorean theorem states that for a right triangle, $(\text{leg})^2 + (\text{leg})^2 = (\underline{\hspace{2cm}})^2$.

Helpful Hint ▶ Are you preparing for your test? Don't forget to take the Chapter 15 Test on page 1161. Then check your answers at the back of the text and use the Chapter Test Prep Videos to see the fully worked-out solutions to any of the exercises you want to review.

15 Chapter Highlights

Definitions and Concepts	Examples
Section 15.1 Introduction to Radicals	
The **positive** or **principal square root** of a positive number a is written as $\sqrt{a}$. The **negative square root** of a is written as $-\sqrt{a}$. $\sqrt{a} = b$ only if $b^2 = a$ and $b > 0$.	$\sqrt{25} = 5 \qquad\qquad \sqrt{100} = 10$ $-\sqrt{9} = -3 \qquad \sqrt{\dfrac{4}{49}} = \dfrac{2}{7}$
A square root of a negative number is not a real number.	$\sqrt{-4}$ is not a real number.
The **cube root** of a real number a is written as $\sqrt[3]{a}$. $\sqrt[3]{a} = b$ only if $b^3 = a$. The ***n*th root** of a number a is written as $\sqrt[n]{a}$. $\sqrt[n]{a} = b$ only if $b^n = a$. In $\sqrt[n]{a}$, the natural number n is called the **index,** the symbol $\sqrt{}$ is called a **radical,** and the expression within the radical is called the **radicand.** (*Note:* If the index is even, the radicand must be nonnegative for the root to be a real number.)	$\sqrt[3]{64} = 4 \qquad\qquad \sqrt[3]{-8} = -2$ $\sqrt[4]{81} = 3$ $\sqrt[5]{-32} = -2$ $\overset{\text{index}}{\downarrow}$ $\sqrt[n]{a}$ $\underset{\text{radicand}}{\uparrow}$
Section 15.2 Simplifying Radicals	
Product Rule for Radicals If $\sqrt{a}$ and $\sqrt{b}$ are real numbers, then $\sqrt{a \cdot b} = \sqrt{a} \cdot \sqrt{b}$ A square root is in **simplified form** if the radicand contains no perfect square factors other than 1. To simplify a square root, factor the radicand so that one of its factors is a perfect square factor.	$\begin{aligned} \sqrt{45} &= \sqrt{9 \cdot 5} \\ &= \sqrt{9} \cdot \sqrt{5} \\ &= 3\sqrt{5} \end{aligned}$

(continued)

Definitions and Concepts	Examples

Section 15.2   Simplifying Radicals (*continued*)

Quotient Rule for Radicals

If $\sqrt{a}$ and $\sqrt{b}$ are real numbers and $b \neq 0$, then

$$\sqrt{\frac{a}{b}} = \frac{\sqrt{a}}{\sqrt{b}}$$

$$\sqrt{\frac{18}{x^6}} = \frac{\sqrt{9 \cdot 2}}{\sqrt{x^6}} = \frac{\sqrt{9} \cdot \sqrt{2}}{x^3} = \frac{3\sqrt{2}}{x^3}$$

Section 15.3   Adding and Subtracting Radicals

Like radicals are radical expressions that have the same index and the same radicand.

To **combine like radicals** use the distributive property.

$$5\sqrt{2}, \; -7\sqrt{2}, \; \sqrt{2}$$

$$2\sqrt{7} - 13\sqrt{7} = (2 - 13)\sqrt{7} = -11\sqrt{7}$$
$$\sqrt{8} + \sqrt{50} = 2\sqrt{2} + 5\sqrt{2} = 7\sqrt{2}$$

Section 15.4   Multiplying and Dividing Radicals

The product and quotient rules for radicals may be used to simplify products and quotients of radicals.

Perform each indicated operation and simplify.
Multiply.

$$\sqrt{2} \cdot \sqrt{8} = \sqrt{16} = 4$$
$$(\sqrt{3x} + 1)(\sqrt{5} - \sqrt{3})$$
$$= \sqrt{15x} - \sqrt{9x} + \sqrt{5} - \sqrt{3}$$
$$= \sqrt{15x} - 3\sqrt{x} + \sqrt{5} - \sqrt{3}$$

Divide.

$$\frac{\sqrt{20}}{\sqrt{2}} = \sqrt{\frac{20}{2}} = \sqrt{10}$$

The process of eliminating the radical in the denominator of a radical expression is called **rationalizing the denominator.**

Rationalize the denominator.

$$\frac{5}{\sqrt{11}} = \frac{5 \cdot \sqrt{11}}{\sqrt{11} \cdot \sqrt{11}} = \frac{5\sqrt{11}}{11}$$

The **conjugate** of $a + b$ is $a - b$.

To rationalize a denominator that is a sum or difference of radicals, multiply the numerator and the denominator by the conjugate of the denominator.

The conjugate of $2 + \sqrt{3}$ is $2 - \sqrt{3}$.

Rationalize the denominator.

$$\frac{5}{6 - \sqrt{5}} = \frac{5(6 + \sqrt{5})}{(6 - \sqrt{5})(6 + \sqrt{5})}$$

$$= \frac{5(6 + \sqrt{5})}{36 - 5}$$

$$= \frac{5(6 + \sqrt{5})}{31}$$

Definitions and Concepts	**Examples**

Section 15.5   Solving Equations Containing Radicals

To Solve a Radical Equation Containing Square Roots

Step 1: Get one radical by itself on one side of the equation.

Step 2: Square both sides of the equation.

Step 3: Simplify both sides of the equation.

Step 4: If the equation still contains a radical term, repeat Steps 1 through 3.

Step 5: Solve the equation.

Step 6: Check solutions in the original equation.

Solve:

$$\sqrt{2x - 1} - x = -2$$
$$\sqrt{2x - 1} = x - 2$$
$$(\sqrt{2x - 1})^2 = (x - 2)^2 \qquad \text{Square both sides.}$$
$$2x - 1 = x^2 - 4x + 4$$
$$0 = x^2 - 6x + 5$$
$$0 = (x - 1)(x - 5) \qquad \text{Factor.}$$
$$x - 1 = 0 \quad \text{or} \quad x - 5 = 0$$
$$x = 1 \qquad\qquad x = 5 \quad \text{Solve.}$$

Check both proposed solutions in the original equation. Here, 5 checks but 1 does not. The only solution is 5.

Section 15.6   Radical Equations and Problem Solving

Problem-Solving Steps

1. UNDERSTAND. Read and reread the problem.

A rain gutter is to be mounted on the eaves of a house 15 feet above the ground. A garden is adjacent to the house, so the closest a ladder can be placed to the house is 6 feet. How long a ladder is needed for installing the gutter?

Let $x =$ the length of the ladder.

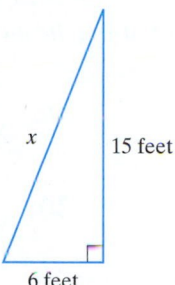

15 feet

x

6 feet

Here, we use the Pythagorean theorem. The unknown length x is the hypotenuse.

In words:

$$(\text{leg})^2 \; + \; (\text{leg})^2 \; = \; (\text{hypotenuse})^2$$

2. TRANSLATE.

Translate:

3. SOLVE.

$$6^2 + 15^2 = x^2$$
$$36 + 225 = x^2$$
$$261 = x^2$$
$$\sqrt{261} = x \quad \text{or} \quad x = 3\sqrt{29}$$

4. INTERPRET.

Check and state. The ladder needs to be $3\sqrt{29}$ feet or approximately 16.2 feet long.

(15.1) *Find each root.*

1. $\sqrt{81}$ **2.** $-\sqrt{49}$ **3.** $\sqrt[3]{27}$ **4.** $\sqrt[4]{81}$

5. $-\sqrt{\dfrac{9}{64}}$ **6.** $\sqrt{\dfrac{36}{81}}$ **7.** $\sqrt[4]{16}$ **8.** $\sqrt[3]{-8}$

9. Which radical(s) is not a real number?

   a. $\sqrt{4}$ **b.** $-\sqrt{4}$ **c.** $\sqrt{-4}$ **d.** $\sqrt[3]{-4}$

10. Which radical(s) is not a real number?

   a. $\sqrt{-5}$ **b.** $\sqrt[3]{-5}$ **c.** $\sqrt[4]{-5}$ **d.** $\sqrt[5]{-5}$

Find each root. Assume that all variables represent positive numbers.

11. $\sqrt{x^{12}}$ **12.** $\sqrt{x^8}$ **13.** $\sqrt{9y^2}$ **14.** $\sqrt{25x^4}$

(15.2) *Simplify each expression using the product rule. Assume that all variables represent positive numbers.*

15. $\sqrt{40}$ **16.** $\sqrt{24}$ **17.** $\sqrt{54}$ **18.** $\sqrt{88}$ **19.** $\sqrt{x^5}$

20. $\sqrt{y^7}$ **21.** $\sqrt{20x^2}$ **22.** $\sqrt{50y^4}$ **23.** $\sqrt[3]{54}$ **24.** $\sqrt[3]{88}$

Simplify each expression using the quotient rule. Assume that all variables represent positive numbers.

25. $\sqrt{\dfrac{18}{25}}$ **26.** $\sqrt{\dfrac{75}{64}}$ **27.** $-\sqrt{\dfrac{50}{9}}$ **28.** $-\sqrt{\dfrac{12}{49}}$

29. $\sqrt{\dfrac{11}{x^2}}$ **30.** $\sqrt{\dfrac{7}{y^4}}$ **31.** $\sqrt{\dfrac{y^5}{100}}$ **32.** $\sqrt{\dfrac{x^3}{81}}$

(15.3) *Add or subtract by combining like radicals.*

33. $5\sqrt{2} - 8\sqrt{2}$ **34.** $\sqrt{3} - 6\sqrt{3}$

35. $6\sqrt{5} + 3\sqrt{6} - 2\sqrt{5} + \sqrt{6}$ **36.** $-\sqrt{7} + 8\sqrt{2} - \sqrt{7} - 6\sqrt{2}$

Add or subtract by simplifying each radical and then combining like terms. Assume that all variables represent positive numbers.

37. $\sqrt{28} + \sqrt{63} + \sqrt{56}$ **38.** $\sqrt{75} + \sqrt{48} - \sqrt{16}$

39. $\sqrt{\dfrac{5}{9}} - \sqrt{\dfrac{5}{36}}$ **40.** $\sqrt{\dfrac{11}{25}} + \sqrt{\dfrac{11}{16}}$

41. $\sqrt{45x^2} + 3\sqrt{5x^2} - 7x\sqrt{5} + 10$ **42.** $\sqrt{50x} - 9\sqrt{2x} + \sqrt{72x} - \sqrt{3x}$

(15.4) *Multiply and simplify if possible. Assume that all variables represent positive numbers.*

43. $\sqrt{3} \cdot \sqrt{6}$ **44.** $\sqrt{5} \cdot \sqrt{15}$

45. $\sqrt{2}(\sqrt{5} - \sqrt{7})$

46. $\sqrt{5}(\sqrt{11} + \sqrt{3})$

47. $(\sqrt{3} + 2)(\sqrt{6} - 5)$

48. $(\sqrt{5} + 1)(\sqrt{5} - 3)$

49. $(\sqrt{x} - 2)^2$

50. $(\sqrt{y} + 4)^2$

Divide and simplify if possible. Assume that all variables represent positive numbers.

51. $\dfrac{\sqrt{27}}{\sqrt{3}}$

52. $\dfrac{\sqrt{20}}{\sqrt{5}}$

53. $\dfrac{\sqrt{160}}{\sqrt{8}}$

54. $\dfrac{\sqrt{96}}{\sqrt{3}}$

55. $\dfrac{\sqrt{30x^6}}{\sqrt{2x^3}}$

56. $\dfrac{\sqrt{54x^5y^2}}{\sqrt{3xy^2}}$

Rationalize each denominator and simplify.

57. $\dfrac{\sqrt{2}}{\sqrt{11}}$

58. $\dfrac{\sqrt{3}}{\sqrt{13}}$

59. $\sqrt{\dfrac{5}{6}}$

60. $\sqrt{\dfrac{7}{10}}$

61. $\dfrac{1}{\sqrt{5x}}$

62. $\dfrac{5}{\sqrt{3y}}$

63. $\sqrt{\dfrac{3}{x}}$

64. $\sqrt{\dfrac{6}{y}}$

65. $\dfrac{3}{\sqrt{5} - 2}$

66. $\dfrac{8}{\sqrt{10} - 3}$

67. $\dfrac{\sqrt{2} + 1}{\sqrt{3} - 1}$

68. $\dfrac{\sqrt{3} - 2}{\sqrt{5} + 2}$

69. $\dfrac{10}{\sqrt{x} + 5}$

70. $\dfrac{8}{\sqrt{x} - 1}$

(15.5) *Solve each radical equation.*

71. $\sqrt{2x} = 6$

72. $\sqrt{x + 3} = 4$

73. $\sqrt{x} + 3 = 8$

74. $\sqrt{x} + 8 = 3$

75. $\sqrt{2x + 1} = x - 7$

76. $\sqrt{3x + 1} = x - 1$

77. $\sqrt{x + 3} = \sqrt{x + 15}$

78. $\sqrt{x - 5} = \sqrt{x} - 1$

(15.6) *Use the Pythagorean theorem to find the length of each unknown side. Give an exact answer and a two-decimal-place approximation.*

△ **79.**
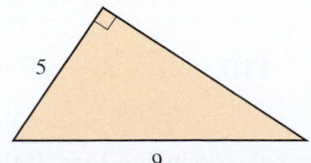
5
9

△ **80.**
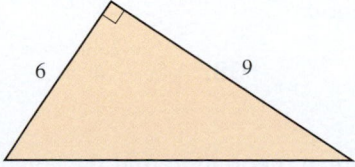
6
9

△ **81.** Romeo is standing 20 feet away from the wall below Juliet's balcony during a school play. Juliet is on the balcony, 12 feet above the ground. Find how far apart Romeo and Juliet are.

△ **82.** The diagonal of a rectangle is 10 inches long. If the width of the rectangle is 5 inches, find the length of the rectangle.

Use the formula $r = \sqrt{\dfrac{S}{4\pi}}$, where r = the radius of a sphere and S = the surface area of the sphere, for Exercises 83 and 84.

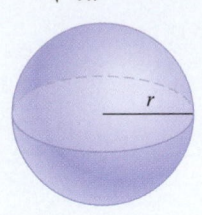

△ **83.** Find the radius of a sphere to the nearest tenth of an inch if the surface area is 72 square inches.

△ **84.** Find the exact surface area of a sphere if its radius is 6 inches. (Do not approximate π.)

Mixed Review

Find each root. Assume all variables represent positive numbers.

85. $\sqrt{144}$

86. $-\sqrt[3]{64}$

87. $\sqrt{16x^{16}}$

88. $\sqrt{4x^{24}}$

Simplify each expression. Assume all variables represent positive numbers.

89. $\sqrt{18x^7}$

90. $\sqrt{48y^6}$

91. $\sqrt{\dfrac{y^4}{81}}$

92. $\sqrt{\dfrac{x^9}{9}}$

Add or subtract by simplifying and then combining like terms. Assume all variables represent positive numbers.

93. $\sqrt{12} + \sqrt{75}$

94. $\sqrt{63} + \sqrt{28} - \sqrt{9}$

95. $\sqrt{\dfrac{3}{16}} - \sqrt{\dfrac{3}{4}}$

96. $\sqrt{45x^3} + x\sqrt{20x} - \sqrt{5x^3}$

Multiply and simplify if possible. Assume all variables represent positive numbers.

97. $\sqrt{7} \cdot \sqrt{14}$

98. $\sqrt{3}(\sqrt{9} - \sqrt{2})$

99. $(\sqrt{2} + 4)(\sqrt{5} - 1)$

100. $(\sqrt{x} + 3)^2$

Divide and simplify if possible. Assume all variables represent positive numbers.

101. $\dfrac{\sqrt{120}}{\sqrt{5}}$

102. $\dfrac{\sqrt{60x^9}}{\sqrt{15x^7}}$

Rationalize each denominator and simplify.

103. $\sqrt{\dfrac{2}{7}}$

104. $\dfrac{3}{\sqrt{2x}}$

105. $\dfrac{3}{\sqrt{x} - 6}$

106. $\dfrac{\sqrt{7} - 5}{\sqrt{5} + 3}$

Solve each radical equation.

107. $\sqrt{4x} = 2$

108. $\sqrt{x - 4} = 3$

109. $\sqrt{4x + 8} + 6 = x$

110. $\sqrt{x - 8} = \sqrt{x} - 2$

111. Use the Pythagorean theorem to find the length of the unknown side. Give an exact answer and a two-decimal-place approximation.

112. The diagonal of a rectangle is 6 inches long. If the width of the rectangle is 2 inches, find the length of the rectangle. Give an exact answer and a two-decimal approximation.

3

7

Simplify each radical. Indicate if the radical is not a real number. Assume that x represents a positive number.

1. $\sqrt{16}$

2. $\sqrt[3]{125}$

3. $\sqrt[4]{81}$

4. $\sqrt{\dfrac{9}{16}}$

5. $\sqrt[4]{-81}$

6. $\sqrt{x^{10}}$

Simplify each radical. Assume that all variables represent positive numbers.

7. $\sqrt{54}$

8. $\sqrt{92}$

9. $\sqrt{y^7}$

10. $\sqrt{24x^8}$

11. $\sqrt[3]{27}$

12. $\sqrt[3]{16}$

13. $\sqrt{\dfrac{5}{16}}$

14. $\sqrt{\dfrac{y^3}{25}}$

Perform each indicated operation. Assume that all variables represent positive numbers.

15. $\sqrt{13} + \sqrt{13} - 4\sqrt{13}$

16. $\sqrt{18} - \sqrt{75} + 7\sqrt{3} - \sqrt{8}$

17. $\sqrt{\dfrac{3}{4}} + \sqrt{\dfrac{3}{25}}$

18. $\sqrt{7} \cdot \sqrt{14}$

19. $\sqrt{2}\left(\sqrt{6} - \sqrt{5}\right)$

Answers

1. _____

2. _____

3. _____

4. _____

5. _____

6. _____

7. _____

8. _____

9. _____

10. _____

11. _____

12. _____

13. _____

14. _____

15. _____

16. _____

17. _____

18. _____

19. _____

20. _____

21. _____

22. _____

23. _____

24. _____

25. _____

26. _____

27. _____

28. _____

29. _____

30. _____

31. _____

20. $(\sqrt{x} + 2)(\sqrt{x} - 3)$ **21.** $\dfrac{\sqrt{50}}{\sqrt{10}}$ **22.** $\dfrac{\sqrt{40x^4}}{\sqrt{2x}}$

Rationalize each denominator. Assume that all variables represent positive numbers.

23. $\sqrt{\dfrac{2}{3}}$ **24.** $\dfrac{8}{\sqrt{5y}}$ **25.** $\dfrac{8}{\sqrt{6} + 2}$ **26.** $\dfrac{1}{3 - \sqrt{x}}$

Solve each radical equation.

27. $\sqrt{x} + 8 = 11$ **28.** $\sqrt{3x - 6} = \sqrt{x + 4}$ **29.** $\sqrt{2x - 2} = x - 5$

△ **30.** Find the length of the unknown leg of the right triangle shown. Give an exact answer.

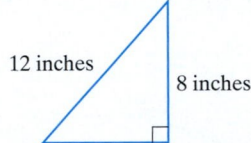

12 inches   8 inches

△ **31.** The formula $r = \sqrt{\dfrac{A}{\pi}}$ can be used to find the radius r of a circle given its area A. Use this formula to approximate the radius of the given circle. Round to two decimal places.

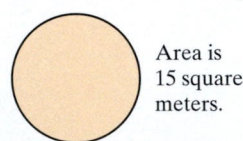

Area is 15 square meters.

1. A flight from Tucson to Phoenix, Arizona, requires $\frac{5}{12}$ of an hour. If the plane has been flying $\frac{1}{4}$ of an hour, find how much time remains before landing.

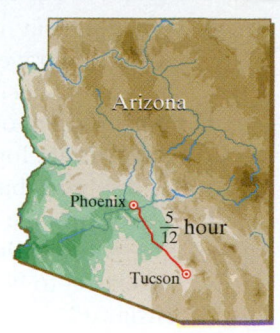

2. Simplify: $80 \div 8 \cdot 2 + 7$

3. Add: $2\frac{1}{3} + 5\frac{3}{8}$

4. Find the average of $\frac{3}{5}, \frac{4}{9}$, and $\frac{11}{15}$.

5. Insert $<$ or $>$ to form a true statement.

   $\frac{3}{10}$ $\frac{2}{7}$

6. Multiply: $28{,}000 \times 500$

7. Write the decimal 1.3 in words.

8. Write "seventy-five thousandths" in standard form.

9. Round 736.2359 to the nearest tenth.

10. Round 736.2359 to the nearest hundredth.

Multiply.

11. $-2(-14)$

12. $9(-5.2)$

13. $-\frac{2}{3} \cdot \frac{4}{7}$

14. $-3\frac{3}{8} \cdot 5\frac{1}{3}$

15. Solve: $4(2x - 3) + 7 = 3x + 5$

16. Solve: $6y - 11 + 4 + 2y = 8 + 15y - 8y$

Answers

1. _____

2. _____

3. _____

4. _____

5. _____

6. _____

7. _____

8. _____

9. _____

10. _____

11. _____

12. _____

13. _____

14. _____

15. _____

16. _____

17. a. _____

b. _____

c. _____

18. a. _____

b. _____

19. a. _____

b. _____

c. _____

d. _____

20. a. _____

b. _____

21. _____

22. _____

23. _____

24. _____

25. _____

26. _____

27. a. _____

b. _____

c. _____

28. _____

29. _____

30. _____

17. The circle graph below shows the purpose of trips made by American travelers. Use this graph to answer the questions below.

    a. What percent of trips made by American travelers are solely for the purpose of business?

    b. What percent of trips made by American travelers are for the purpose of business or combined business/pleasure?

    c. On an airplane flight of 253 Americans, how many of these people might we expect to be traveling solely for business?

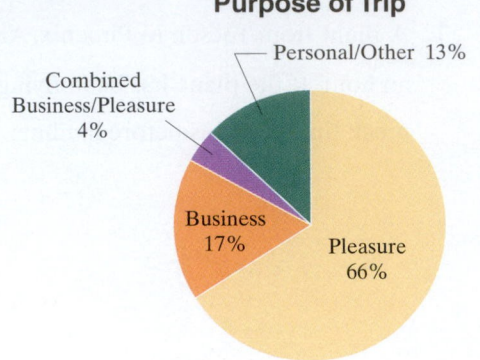

Purpose of Trip

Source: Travel Industry Association of America

18. Simplify each expression.

    a. $\dfrac{4(-3) - (-6)}{-8 + 4}$

    b. $\dfrac{3 + (-3)(-2)^3}{-1 - (-4)}$

19. Write each number in standard form, without exponents.

    a. 1.02×10^5

    b. 7.358×10^{-3}

    c. 8.4×10^7

    d. 3.007×10^{-5}

20. Write the following in scientific notation:

    a. 7,200,000

    b. 0.000308

21. Multiply: $(3x + 2)(2x - 5)$

22. Multiply: $(7x + 1)^2$

23. Factor $xy + 2x + 3y + 6$ by grouping.

24. Factor $xy^2 + 5x - y^2 - 5$ by grouping.

25. Factor: $3x^2 + 11x + 6$

26. Factor: $3x^2 + 15x + 18$

27. Are there any values for x for which each expression is undefined?

    a. $\dfrac{x}{x - 3}$

    b. $\dfrac{x^2 + 2}{x^2 - 3x + 2}$

    c. $\dfrac{x^3 - 6x^2 - 10x}{3}$

28. Simplify: $\dfrac{2x^2 + 7x + 3}{x^2 - 9}$

29. Solve: $\dfrac{4x}{x^2 + x - 30} + \dfrac{2}{x - 5} = \dfrac{1}{x + 6}$

30. Find an equation of the line with y-intercept $(0, 4)$ and slope of -2.

31. Graph $y = -3$.

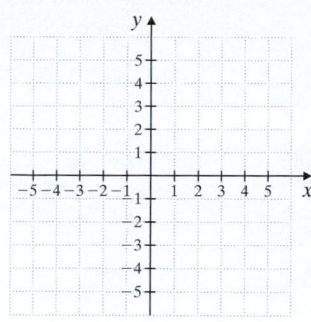

32. Complete the table for the equation $2x + y = 6$.

x	y
0	
	-2
3	

33. Find an equation of the line with y-intercept $(0, -3)$ and slope of $\frac{1}{4}$.

34. Find an equation of the line perpendicular to $y = 2x + 4$ and passing through $(1, 5)$.

35. Solve the system:
$$\begin{cases} 3x + 4y = 13 \\ 5x - 9y = 6 \end{cases}$$

36. Solve the system:
$$\begin{cases} \dfrac{x}{2} + y = \dfrac{5}{6} \\ 2x - y = \dfrac{5}{6} \end{cases}$$

37. As part of an exercise program, two students, Louisa and Alfredo, start walking each morning. They live 15 miles away from each other. They decide to meet one day by walking toward one another. After 2 hours they meet. If Louisa walks one mile per hour faster than Alfredo, find both walking speeds.

38. Two streetcars are 11 miles apart and traveling toward each other on parallel tracks. They meet in 12 minutes. Find the speed of each streetcar if one travels 15 miles per hour faster than the other.

Simplify.

39. $\sqrt{54}$

40. $\sqrt{63}$

41. $\sqrt{200}$

42. $\sqrt{500}$

Perform indicated operations. If possible, first simplify each radical.

43. $7\sqrt{12} - 2\sqrt{75}$

44. $(\sqrt{x} + 5)(\sqrt{x} - 5)$

45. $2\sqrt{x^2} - \sqrt{25x^5} + \sqrt{x^5}$

46. $(\sqrt{6} + 2)^2$

47. Rationalize the denominator of $\dfrac{2}{\sqrt{7}}$.

48. Simplify: $\dfrac{x + 3}{\dfrac{1}{x} + \dfrac{1}{3}}$

49. Solve: $\sqrt{x} = \sqrt{5x - 2}$

50. Solve: $\sqrt{x + 4} = \sqrt{3x - 1}$

31. _____

32. _____

33. _____

34. _____

35. _____

36. _____

37. _____

38. _____

39. _____

40. _____

41. _____

42. _____

43. _____

44. _____

45. _____

46. _____

47. _____

48. _____

49. _____

50. _____

16

Quadratic Equations and Nonlinear Graphs

An important part of the study of algebra is learning to use methods for solving equations. Starting in Chapter 9, we presented techniques for solving linear equations in one variable. In Chapter 13, we solved quadratic equations in one variable by factoring the quadratic expressions. We now present other methods for solving quadratic equations in one variable.

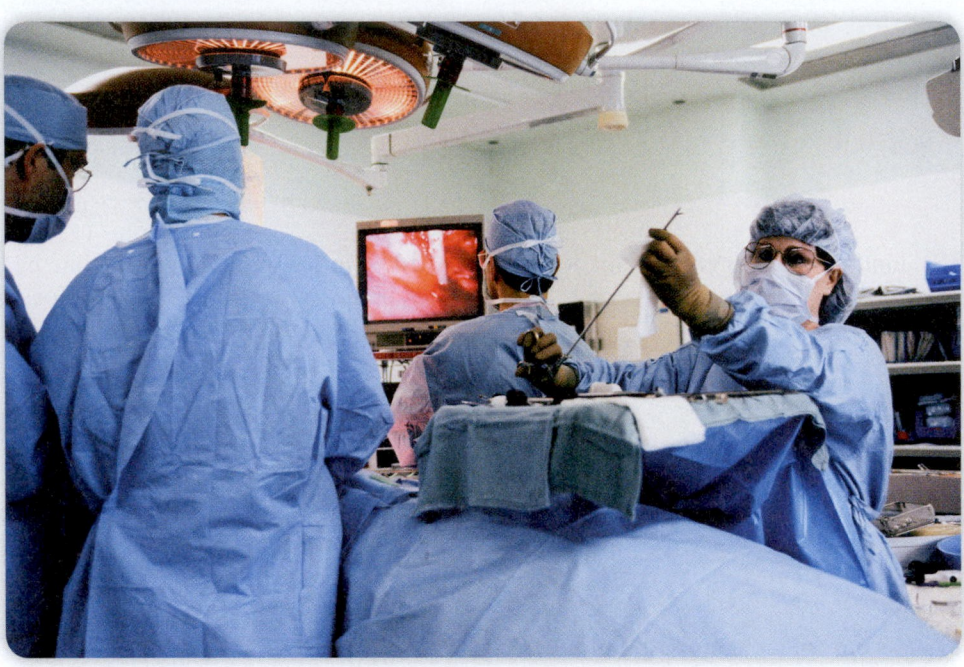

Transplants Performed Versus Waiting List in the United States

There are currently more than 123,000 people living in the United States waiting for an organ transplant, and that number is increasing every year. The bars on the graph below show the yearly number of transplants performed, and the broken line shows the number of patients waiting for a transplant. Physicians use these data to help plan for the future. In Section 16.2, Exercise 47, we see that the revenues from a company that manufactures an artificial heart are modeled by a quadratic equation.

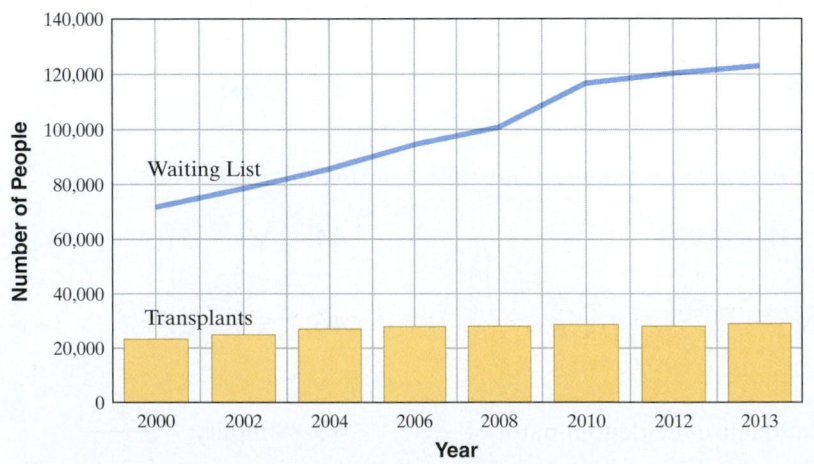

Date from optn.transplant.hrsa.gov and OPTN/SRTR Annual Report

16.1 Solving Quadratic Equations by the Square Root Property

Recall that a quadratic equation is an equation that can be written in the form

$$ax^2 + bx + c = 0$$

where a, b, and c are real numbers and $a \neq 0$.

Solving Quadratic Equations by Factoring

Recall from Section 13.6 that to solve quadratic equations by factoring, we use the **zero-factor property:** If the product of two numbers is zero, then at least one of the two numbers is zero. Examples 1 and 2 review the process of solving quadratic equations by factoring.

Example 1 Solve: $x^2 - 4 = 0$

Solution:

$$x^2 - 4 = 0$$
$$(x + 2)(x - 2) = 0 \quad \text{Factor.}$$
$$x + 2 = 0 \quad \text{or} \quad x - 2 = 0 \quad \text{Use the zero-factor property.}$$
$$x = -2 \qquad\qquad x = 2 \quad \text{Solve each equation.}$$

The solutions are -2 and 2.

■ **Work Practice 1**

Example 2 Solve: $3y^2 + 13y = 10$

Solution: Recall that to use the zero-factor property, one side of the equation must be 0 and the other side must be factored.

$$3y^2 + 13y = 10$$
$$3y^2 + 13y - 10 = 0 \quad \text{Subtract 10 from both sides.}$$
$$(3y - 2)(y + 5) = 0 \quad \text{Factor.}$$
$$3y - 2 = 0 \quad \text{or} \quad y + 5 = 0 \quad \text{Use the zero-factor property.}$$
$$3y = 2 \qquad\qquad y = -5 \quad \text{Solve each equation.}$$
$$y = \frac{2}{3}$$

The solutions are $\frac{2}{3}$ and -5.

■ **Work Practice 2**

Objective A Using the Square Root Property ▶

Consider solving Example 1, $x^2 - 4 = 0$, another way. First, add 4 to both sides of the equation.

$$x^2 - 4 = 0$$
$$x^2 = 4 \quad \text{Add 4 to both sides.}$$

Now we see that the value for x must be a number whose square is 4. Therefore $x = \sqrt{4} = 2$ or $x = -\sqrt{4} = -2$. This reasoning is an example of the square root property.

Objectives

A Use the Square Root Property to Solve Quadratic Equations. ▶

B Use the Square Root Property to Solve Applications. ▶

Practice 1
Solve: $x^2 - 25 = 0$

Practice 2
Solve: $2x^2 - 3x = 9$

Answers
1. 5 and -5 **2.** $-\frac{3}{2}$ and 3

Square Root Property

If $x^2 = a$ for $a \geq 0$, then

$$x = \sqrt{a} \quad \text{or} \quad x = -\sqrt{a}$$

Practice 3

Use the square root property to solve $x^2 - 16 = 0$.

Example 3   Use the square root property to solve $x^2 - 9 = 0$.

Solution: First we solve for x^2 by adding 9 to both sides.

$$x^2 - 9 = 0$$
$$x^2 = 9 \quad \text{Add 9 to both sides.}$$

Next we use the square root property.

$$x = \sqrt{9} \quad \text{or} \quad x = -\sqrt{9}$$
$$x = 3 \qquad\qquad x = -3$$

Check:

$$x^2 - 9 = 0 \quad \text{Original equation} \qquad\qquad x^2 - 9 = 0 \quad \text{Original equation}$$
$$3^2 - 9 \stackrel{?}{=} 0 \quad \text{Let } x = 3. \qquad\qquad (-3)^2 - 9 \stackrel{?}{=} 0 \quad \text{Let } x = -3.$$
$$0 = 0 \quad \text{True} \qquad\qquad\qquad\qquad 0 = 0 \quad \text{True}$$

The solutions are 3 and -3.

■ **Work Practice 3**

Practice 4

Use the square root property to solve $3x^2 = 11$.

Example 4   Use the square root property to solve $2x^2 = 7$.

Solution: First we solve for x^2 by dividing both sides by 2. Then we use the square root property.

$$2x^2 = 7$$
$$x^2 = \frac{7}{2} \qquad\qquad \text{Divide both sides by 2.}$$
$$x = \sqrt{\frac{7}{2}} \quad \text{or} \quad x = -\sqrt{\frac{7}{2}} \qquad \text{Use the square root property.}$$
$$x = \frac{\sqrt{7} \cdot \sqrt{2}}{\sqrt{2} \cdot \sqrt{2}} \qquad x = -\frac{\sqrt{7} \cdot \sqrt{2}}{\sqrt{2} \cdot \sqrt{2}} \qquad \text{Rationalize the denominator.}$$
$$x = \frac{\sqrt{14}}{2} \qquad\qquad x = -\frac{\sqrt{14}}{2} \qquad \text{Simplify.}$$

Remember to check both solutions in the original equation. The solutions are $\frac{\sqrt{14}}{2}$ and $-\frac{\sqrt{14}}{2}$.

■ **Work Practice 4**

Practice 5

Use the square root property to solve $(x - 4)^2 = 49$.

Example 5   Use the square root property to solve $(x - 3)^2 = 16$.

Solution: Instead of x^2, here we have $(x - 3)^2$. But the square root property can still be used.

$$(x - 3)^2 = 16$$
$$x - 3 = \sqrt{16} \quad \text{or} \quad x - 3 = -\sqrt{16} \qquad \text{Use the square root property.}$$
$$x - 3 = 4 \qquad\qquad x - 3 = -4 \qquad \text{Write } \sqrt{16} \text{ as 4 and } -\sqrt{16} \text{ as } -4.$$
$$x = 7 \qquad\qquad\quad x = -1 \qquad \text{Solve.}$$

Answers

3. 4 and -4 **4.** $\frac{\sqrt{33}}{3}$ and $-\frac{\sqrt{33}}{3}$

5. 11 and -3

Check:

$(x - 3)^2 = 16$ Original equation $(x - 3)^2 = 16$ Original equation

$(7 - 3)^2 \stackrel{?}{=} 16$ Let $x = 7$. $(-1 - 3)^2 \stackrel{?}{=} 16$ Let $x = -1$.

$4^2 \stackrel{?}{=} 16$ Simplify. $(-4)^2 \stackrel{?}{=} 16$ Simplify.

$16 = 16$ True $16 = 16$ True

Both 7 and −1 are solutions.

■ **Work Practice 5**

Example 6 Use the square root property to solve $(x + 1)^2 = 8$.

Solution: $(x + 1)^2 = 8$

$x + 1 = \sqrt{8}$ or $x + 1 = -\sqrt{8}$ Use the square root property.

$x + 1 = 2\sqrt{2}$ $x + 1 = -2\sqrt{2}$ Simplify the radical.

$x = -1 + 2\sqrt{2}$ $x = -1 - 2\sqrt{2}$ Solve for x.

Check both solutions in the original equation. The solutions are $-1 + 2\sqrt{2}$ and $-1 - 2\sqrt{2}$. This can be written compactly as $-1 \pm 2\sqrt{2}$. The notation $\pm$ is read as "plus or minus."

■ **Work Practice 6**

Example 7 Use the square root property to solve $(x - 1)^2 = -2$.

Solution: This equation has no real solution because the square root of −2 is not a real number.

■ **Work Practice 7**

Example 8 Use the square root property to solve $(5x - 2)^2 = 10$.

Solution: $(5x - 2)^2 = 10$

$5x - 2 = \sqrt{10}$ or $5x - 2 = -\sqrt{10}$ Use the square root property.

$5x = 2 + \sqrt{10}$ $5x = 2 - \sqrt{10}$ Add 2 to both sides.

$x = \dfrac{2 + \sqrt{10}}{5}$ $x = \dfrac{2 - \sqrt{10}}{5}$ Divide both sides by 5.

Check both solutions in the original equation. The solutions are $\dfrac{2 + \sqrt{10}}{5}$ and $\dfrac{2 - \sqrt{10}}{5}$, which can be written as $\dfrac{2 \pm \sqrt{10}}{5}$.

■ **Work Practice 8**

Helpful Hint

For some applications and graphing purposes, decimal approximations of exact solutions to quadratic equations may be desired.

Exact Solutions from Example 8		Decimal Approximations
$\dfrac{2 + \sqrt{10}}{5}$	$\approx$	1.032
$\dfrac{2 - \sqrt{10}}{5}$	$\approx$	−0.232

Practice 6

Use the square root property to solve $(x - 5)^2 = 18$.

Helpful Hint

read "plus or minus"
↓

The notation $-1 \pm \sqrt{5}$, for example, is just a shorthand notation for both $-1 + \sqrt{5}$ and $-1 - \sqrt{5}$.

Practice 7

Use the square root property to solve $(x + 3)^2 = -5$.

Practice 8

Use the square root property to solve $(4x + 1)^2 = 15$.

Answers

6. $5 \pm 3\sqrt{2}$ **7.** no real solution

8. $\dfrac{-1 \pm \sqrt{15}}{4}$

Objective B Using the Square Root Property to Solve Applications

Many real-world applications are modeled by quadratic equations. In the next example, we use the quadratic formula $h = 16t^2$. This formula gives the distance h traveled by a free-falling object in time t. One important note is that this formula does not take into account any air resistance.

Practice 9

Use the formula $h = 16t^2$ (see Example 9) to find how long, to the nearest tenth of a second, it takes a free-falling body to fall 650 feet.

Example 9   Finding the Length of Time of a Dive

The record for the highest dive into a lake was made by Harry Froboess of Switzerland. In 1936 he dove 394 feet from the airship Hindenburg into Lake Constance. To the nearest tenth of a second, how long did his dive take? (*Source: Guinness World Records*)

Solution:

1. UNDERSTAND. To approximate the time of the dive, we use the formula*
   $h = 16t^2$, where t is time in seconds and h is the distance in feet traveled by a free-falling body or object. For example, to find the distance traveled in 1 second, or 3 seconds, we let $t = 1$ and then $t = 3$.

   If $t = 1, h = 16(1)^2 = 16 \cdot 1 = 16$ feet.

   If $t = 3, h = 16(3)^2 = 16 \cdot 9 = 144$ feet.

   Since a body travels 144 feet in 3 seconds, we now know the dive of 394 feet lasted longer than 3 seconds.

2. TRANSLATE. Use the formula $h = 16t^2$, let the distance $h = 394$, and we have the equation $394 = 16t^2$.

3. SOLVE. To solve $394 = 16t^2$ for t, we will use the square root property.

$$394 = 16t^2$$

$$\frac{394}{16} = t^2 \qquad \text{Divide both sides by 16.}$$

$$24.625 = t^2 \qquad \text{Simplify.}$$

$$\sqrt{24.625} = t \quad \text{ or } \quad -\sqrt{24.625} = t \quad \text{Use the square root property.}$$

$$5.0 \approx t \quad \text{ or } \quad -5.0 \approx t \quad \text{Approximate.}$$

4. INTERPRET.

   Check: We reject the solution -5.0 since the length of the dive is not a negative number.

   State: The dive lasted approximately 5 seconds.

■ Work Practice 9

Answer

9. 6.4 sec

*The formula $h = 16t^2$ does not take into account air resistance.

Vocabulary, Readiness & Video Check

Martin-Gay Interactive Videos   *Watch the section lecture video and answer the following questions.*

Objective **A** **1.** As explained in ▯ Example 2, why is $a \geq 0$ in the statement of the square root property? ▶

Objective **B** **2.** In ▯ Example 6, how can we tell by looking at the translated equation that the square root property can be used to solve it? Why is the negative square root not considered? ▶

See Video 16.1 🍎

16.1 Exercise Set MyMathLab® ▶

Solve each equation by factoring. See Examples 1 and 2.

1. $k^2 - 49 = 0$ $\qquad$ **2.** $k^2 - 9 = 0$ $\qquad$ **3.** $m^2 + 2m = 15$ $\qquad$ **4.** $m^2 + 6m = 7$ $\qquad$ **5.** $2x^2 - 32 = 0$

6. $2x^2 - 98 = 0$ $\qquad$ **7.** $4a^2 - 36 = 0$ $\qquad$ **8.** $7a^2 - 175 = 0$ $\qquad$ **9.** $x^2 + 7x = -10$ $\qquad$ **10.** $x^2 + 10x = -24$

Objective **A** *Use the square root property to solve each quadratic equation. See Examples 3 and 4.*

▶ **11.** $x^2 = 64$ $\qquad$ **12.** $x^2 = 121$ $\qquad$ **13.** $x^2 = 21$ $\qquad$ **14.** $x^2 = 22$ $\qquad$ **15.** $x^2 = \dfrac{1}{25}$

16. $x^2 = \dfrac{1}{16}$ $\qquad$ ▶ **17.** $x^2 = -4$ $\qquad$ **18.** $x^2 = -25$ $\qquad$ **19.** $3x^2 = 13$

20. $5x^2 = 2$ $\qquad$ **21.** $7x^2 = 4$ $\qquad$ **22.** $2x^2 = 9$ $\qquad$ ▶ **23.** $2x^2 - 10 = 0$ $\qquad$ **24.** $3x^2 - 45 = 0$

Use the square root property to solve each quadratic equation. See Examples 5 through 8.

25. $(x - 5)^2 = 49$ $\qquad$ **26.** $(x + 2)^2 = 25$ $\qquad$ **27.** $(x + 2)^2 = 7$ $\qquad$ **28.** $(x - 7)^2 = 2$

29. $\left(m - \dfrac{1}{2}\right)^2 = \dfrac{1}{4}$ $\qquad$ **30.** $\left(m + \dfrac{1}{3}\right)^2 = \dfrac{1}{9}$ $\qquad$ ▶ **31.** $(p + 2)^2 = 10$ $\qquad$ **32.** $(p - 7)^2 = 13$

33. $(3y + 2)^2 = 100$ $\qquad$ **34.** $(4y - 3)^2 = 81$ $\qquad$ **35.** $(z - 4)^2 = -9$ $\qquad$ **36.** $(z + 7)^2 = -20$

37. $(2x - 11)^2 = 50$ $\qquad$ **38.** $(3x - 17)^2 = 28$ $\qquad$ ▶ **39.** $(3x - 7)^2 = 32$ $\qquad$ **40.** $(5x - 11)^2 = 54$

Use the square root property to solve. See Examples 3 through 8.

41. $x^2 - 29 = 0$

42. $x^2 - 35 = 0$

43. $(x + 6)^2 = 24$

44. $(x + 5)^2 = 20$

45. $\frac{1}{2}n^2 = 5$

46. $\frac{1}{5}y^2 = 2$

47. $(4x - 1)^2 = 5$

48. $(7x - 2)^2 = 11$

49. $3z^2 = 36$

50. $3z^2 = 24$

51. $(8 - 3x)^2 - 45 = 0$

52. $(10 - 9x)^2 - 75 = 0$

Objective B   *Solve. For Exercises 53 through 56, use the formula $h = 16t^2$. See Example 9. Round each answer to the nearest tenth of a second.*

53. The highest regularly performed dives are made by professional divers from La Quebrada. If this cliff in Acapulco has a height of 87.6 feet, determine the time of a dive. (*Source: Guinness World Records*)

54. In 1988, Eddie Turner saved Frank Fanan, who became unconscious after an injury while jumping out of an airplane. Fanan fell 11,136 feet before Turner pulled his ripcord. Determine the time of Fanan's unconscious free fall.

55. In 2007, the Hualapai Indian Tribe allowed the Grand Canyon Skywalk to be built over the rim of the Grand Canyon on its tribal land. The skywalk extends 70 feet beyond the canyon's edge and is 4000 feet above the canyon floor. Determine the time, to the nearest tenth of a second, it would take an object, dropped off the skywalk, to land at the bottom of the Grand Canyon. (*Source: Boston Globe*)

56. If a sandblaster drops his goggles from a bridge 400 feet from the water below, find how long it takes for the goggles to hit the water.

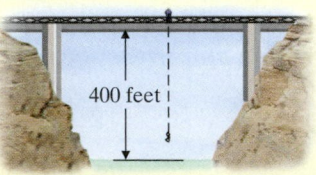

400 feet

57. The area of a circle is found by the equation $A = \pi r^2$. If the area A of a certain circle is 36π square inches, find its radius r.

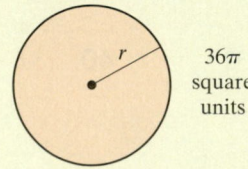

36π
square
units

r

58. If the area of the circle below is 10π square units, find its exact radius. (See Exercise **57**.)

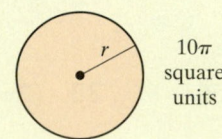

r

10π
square
units

The formula for the area of a square is $A = s^2$, where s is the length of a side. Use this formula for Exercises 59 through 62. For each exercise, give an exact answer and a two-decimal-place approximation.

△ **59.** If the area of a square is 20 square inches, find the length of a side.

△ **60.** If the area of a square is 32 square meters, find the length of a side.

△ **61.** The Washington Monument has a square base whose area is approximately 3039 square feet. Find the length of a side. (*Source: The World Almanac*)

△ **62.** The largest Buddhist temple is Borobudur, near Yogyakarta, central Java, Indonesia. Its square base has an area of 162,409 square feet. Find the dimensions of the base of the temple.

Review

Factor each perfect square trinomial. See Section 13.5.

63. $x^2 + 6x + 9$

64. $y^2 + 10y + 25$

65. $x^2 - 4x + 4$

66. $x^2 - 20x + 100$

Concept Extensions

67. Explain why the equation $x^2 = -9$ has no real solution.

68. Explain why the equation $x^2 = 9$ has two solutions.

Solve each quadratic equation by first factoring the perfect square trinomial on the left side. Then apply the square root property.

69. $x^2 + 4x + 4 = 16$

70. $y^2 - 10y + 25 = 11$

Solve each quadratic equation by using the square root property. Use a calculator and round each solution to the nearest hundredth.

71. $x^2 = 1.78$

72. $(x - 1.37)^2 = 5.71$

Solve.

73. The number of U.S. highway bridges for the years 2006 to 2012 can be modeled by the equation $y = -84(x - 12)^2 + 609,652$, where $x = 0$ represents the year 2006. Assume that this trend continued and find the first year in which there were 608,900 highway bridges in the United States. (*Hint:* Replace y with 608,900 in the equation and solve for x. Round to the nearest year.) (*Source:* Based on data from the U.S. Department of Transportation, Federal Highway Administration)

74. The number y (in millions) of active wireless devices in the United States from 2002 through 2012 can be estimated by $y = -0.75(x - 17.8)^2 + 374$, where $x = 0$ represents the year 2002. Assume that this trend continues, and determine the first year in which the number of active wireless devices will surpass 368 million. (*Hint:* Replace y with 368 in the equation and solve for x. Round to the nearest year.) (*Source:* Based on data from CTIA—The Wireless Association)

Objectives

A Solve Quadratic Equations of the Form $x^2 + bx + c = 0$ by Completing the Square.

B Solve Quadratic Equations of the Form $ax^2 + bx + c = 0$ by Completing the Square.

16.2 Solving Quadratic Equations by Completing the Square

Objective A Completing the Square to Solve $x^2 + bx + c = 0$

In the last section, we used the square root property to solve equations such as

$$(x + 1)^2 = 8 \quad \text{and} \quad (5x - 2)^2 = 3$$

Notice that one side of each equation is a quantity squared and that the other side is a constant. To solve

$$x^2 + 2x = 4$$

notice that if we add 1 to both sides of the equation, the left side is a perfect square trinomial that can be factored.

$$x^2 + 2x + 1 = 4 + 1 \quad \text{Add 1 to both sides.}$$
$$(x + 1)^2 = 5 \quad \text{Factor.}$$

Now we can solve this equation as we did in the previous section, by using the square root property.

$$x + 1 = \sqrt{5} \quad \text{or} \quad x + 1 = -\sqrt{5} \quad \text{Use the square root property.}$$
$$x = -1 + \sqrt{5} \qquad x = -1 - \sqrt{5} \quad \text{Solve.}$$

The solutions are $-1 \pm \sqrt{5}$.

Adding a number to $x^2 + 2x$ to form a perfect square trinomial is called **completing the square** on $x^2 + 2x$.

In general, we have the following:

Completing the Square

To complete the square on $x^2 + bx$, add $\left(\dfrac{b}{2}\right)^2$. To find $\left(\dfrac{b}{2}\right)^2$, **find half the coefficient of x, and then square the result.**

Practice 1

Solve $x^2 + 8x + 1 = 0$ by completing the square.

Example 1 Solve $x^2 + 6x + 3 = 0$ by completing the square.

Solution: First we get the variable terms alone by subtracting 3 from both sides of the equation.

$$x^2 + 6x + 3 = 0$$
$$x^2 + 6x = -3 \quad \text{Subtract 3 from both sides.}$$

Next we find half the coefficient of the x-term, and then square it. We add this result to *both sides* of the equation. This will make the left side a perfect square trinomial. The coefficient of x is 6, and half of 6 is 3. So we add 3^2 or 9 to both sides.

$$x^2 + 6x + 9 = -3 + 9 \quad \text{Complete the square.}$$
$$(x + 3)^2 = 6 \quad \text{Factor the trinomial } x^2 + 6x + 9.$$
$$x + 3 = \sqrt{6} \quad \text{or} \quad x + 3 = -\sqrt{6} \quad \text{Use the square root property.}$$
$$x = -3 + \sqrt{6} \qquad x = -3 - \sqrt{6} \quad \text{Subtract 3 from both sides.}$$

Check by substituting $-3 + \sqrt{6}$ and $-3 - \sqrt{6}$ in the original equation. The solutions are $-3 \pm \sqrt{6}$.

■ Work Practice 1

Answer
1. $-4 \pm \sqrt{15}$

1174

Copyright 2015 Pearson Education, Inc.

Example 2 Solve $x^2 - 10x = -14$ by completing the square.

Solution: The variable terms are already alone on one side of the equation. The coefficient of x is -10. Half of -10 is -5, and $(-5)^2 = 25$. So we add 25 to both sides.

$$x^2 - 10x = -14$$
$$x^2 - 10x + 25 = -14 + 25$$

Helpful Hint Add 25 to *both* sides of the equation.

$$(x - 5)^2 = 11 \qquad \text{Factor the trinomial and simplify } -14 + 25.$$

$$x - 5 = \sqrt{11} \quad \text{or} \quad x - 5 = -\sqrt{11} \qquad \text{Use the square root property.}$$

$$x = 5 + \sqrt{11} \qquad x = 5 - \sqrt{11} \qquad \text{Add 5 to both sides.}$$

The solutions are $5 \pm \sqrt{11}$.

■ **Work Practice 2**

Objective B Completing the Square to Solve $ax^2 + bx + c = 0$ ▶

The method of completing the square can be used to solve *any* quadratic equation whether the coefficient of the squared variable is 1 or not. When the coefficient of the squared variable is not 1, we first divide both sides of the equation by the coefficient of the squared variable so that the new coefficient is 1. Then we complete the square.

Example 3 Solve $4x^2 - 8x - 5 = 0$ by completing the square.

Solution: Since the coefficient of x^2 is 4, not 1, we first divide both sides of the equation by 4 so that the coefficient of x^2 is 1.

$$4x^2 - 8x - 5 = 0$$

$$x^2 - 2x - \frac{5}{4} = 0 \qquad \text{Divide both sides by 4.}$$

$$x^2 - 2x = \frac{5}{4} \qquad \text{Get the variable terms alone on one side of the equation.}$$

The coefficient of x is -2. Half of -2 is -1, and $(-1)^2 = 1$. So we add 1 to both sides.

$$x^2 - 2x + 1 = \frac{5}{4} + 1$$

$$(x - 1)^2 = \frac{9}{4} \qquad \text{Factor } x^2 - 2x + 1 \text{ and simplify } \frac{5}{4} + 1.$$

$$x - 1 = \sqrt{\frac{9}{4}} \quad \text{or} \quad x - 1 = -\sqrt{\frac{9}{4}} \qquad \text{Use the square root property.}$$

$$x = 1 + \frac{3}{2} \qquad x = 1 - \frac{3}{2} \qquad \text{Add 1 to both sides and simplify the radical.}$$

$$x = \frac{5}{2} \qquad x = -\frac{1}{2} \qquad \text{Simplify.}$$

Both $\frac{5}{2}$ and $-\frac{1}{2}$ are solutions.

■ **Work Practice 3**

Practice 2

Solve $x^2 - 14x = -32$ by completing the square.

Practice 3

Solve $4x^2 - 16x - 9 = 0$ by completing the square.

Answers

2. $7 \pm \sqrt{17}$ **3.** $\frac{9}{2}$ and $-\frac{1}{2}$

The following steps may be used to solve a quadratic equation in x by completing the square.

> ### To Solve a Quadratic Equation in x by Completing the Square
>
> **Step 1:** If the coefficient of x^2 is 1, go to Step 2. If not, divide both sides of the equation by the coefficient of x^2.
>
> **Step 2:** Get all terms with variables on one side of the equation and constants on the other side.
>
> **Step 3:** Find half the coefficient of x and then square the result. Add this number to both sides of the equation.
>
> **Step 4:** Factor the resulting perfect square trinomial.
>
> **Step 5:** Use the square root property to solve the equation.

Practice 4

Solve $2x^2 + 10x = -13$ by completing the square.

Example 4   Solve $2x^2 + 6x = -7$ by completing the square.

Solution: The coefficient of x^2 is not 1. We divide both sides by 2, the coefficient of x^2.

$$2x^2 + 6x = -7$$

$$x^2 + 3x = -\frac{7}{2} \qquad \text{Divide both sides by 2.}$$

$$x^2 + 3x + \frac{9}{4} = -\frac{7}{2} + \frac{9}{4} \qquad \text{Add } \left(\frac{3}{2}\right)^2 \text{ or } \frac{9}{4} \text{ to both sides.}$$

$$\left(x + \frac{3}{2}\right)^2 = -\frac{5}{4} \qquad \text{Factor the left side and simplify the right.}$$

There is no real solution to this equation since the square root of a negative number is not a real number.

■ **Work Practice 4**

Practice 5

Solve $2x^2 = -6x + 5$ by completing the square.

Example 5   Solve $2x^2 = 10x + 1$ by completing the square.

Solution: First we divide both sides of the equation by 2, the coefficient of x^2.

$$2x^2 = 10x + 1$$

$$x^2 = 5x + \frac{1}{2} \qquad \text{Divide both sides by 2.}$$

Next we get the variable terms alone by subtracting $5x$ from both sides.

$$x^2 - 5x = \frac{1}{2}$$

$$x^2 - 5x + \frac{25}{4} = \frac{1}{2} + \frac{25}{4} \qquad \text{Add } \left(-\frac{5}{2}\right)^2 \text{ or } \frac{25}{4} \text{ to both sides.}$$

$$\left(x - \frac{5}{2}\right)^2 = \frac{27}{4} \qquad \text{Factor the left side and simplify the right side.}$$

$$x - \frac{5}{2} = \sqrt{\frac{27}{4}} \quad \text{or} \quad x - \frac{5}{2} = -\sqrt{\frac{27}{4}} \qquad \text{Use the square root property.}$$

$$x - \frac{5}{2} = \frac{3\sqrt{3}}{2} \qquad x - \frac{5}{2} = -\frac{3\sqrt{3}}{2} \qquad \text{Simplify.}$$

$$x = \frac{5}{2} + \frac{3\sqrt{3}}{2} \qquad x = \frac{5}{2} - \frac{3\sqrt{3}}{2}$$

The solutions are $\dfrac{5 \pm 3\sqrt{3}}{2}$.

■ **Work Practice 5**

Answers

4. no real solution **5.** $\dfrac{-3 \pm \sqrt{19}}{2}$

Vocabulary, Readiness & Video Check

Use the choices below to fill in each blank. Not all choices will be used, and these exercises come from Sections 16.1 and 16.2.

$\sqrt{a}$	linear equation	zero	$\left(\dfrac{b}{2}\right)^2$	$\dfrac{b}{2}$	6
$\pm\sqrt{a}$	quadratic equation	one	completing the square	9	3

1. By the zero-factor property, if the product of two numbers is zero, then at least one of these two numbers must be _____.

2. If a is a positive number, and if $x^2 = a$, then $x =$ _____.

3. An equation that can be written in the form $ax^2 + bx + c = 0$ where a, b, and c are real numbers and a is not zero is called a(n) _____.

4. The process of solving a quadratic equation by writing it in the form $(x + a)^2 = c$ is called _____.

5. To complete the square on $x^2 + 6x$, add _____.

6. To complete the square on $x^2 + bx$, add _____.

Fill in the blank with the number needed to make each expression a perfect square trinomial. See Example 1.

7. $p^2 + 8p +$ _____

8. $p^2 + 6p +$ _____

9. $x^2 + 20x +$ _____

10. $x^2 + 18x +$ _____

11. $y^2 + 14y +$ _____

12. $y^2 + 2y +$ _____

 *Martin-Gay Interactive Videos*    Watch the section lecture video and answer the following questions.

Objective A **13.** In ▣ Examples 3 and 4, explain why the constant that completes the square is added to both sides of the equation.

Objective B **14.** In ▣ Example 5, why is the equation first divided through by 2? ▶

See Video 16.2 🍎

16.2 Exercise Set MyMathLab® ▶

Objective A *Solve each quadratic equation by completing the square. See Examples 1 and 2.*

▶ **1.** $x^2 + 8x = -12$

2. $x^2 - 10x = -24$

3. $x^2 + 2x - 7 = 0$

4. $z^2 + 6z - 9 = 0$

5. $x^2 - 6x = 0$

6. $y^2 + 4y = 0$

7. $y^2 + 5y + 4 = 0$

8. $y^2 - 5y + 6 = 0$

▶ **9.** $x^2 - 2x - 1 = 0$

10. $x^2 - 4x + 2 = 0$

11. $z^2 + 5z = 7$

12. $x^2 - 7x = 5$

Objective B *Solve each quadratic equation by completing the square. See Examples 3 through 5.*

13. $3x^2 - 6x = 24$

14. $2x^2 + 18x = -40$

15. $5x^2 + 10x + 6 = 0$

16. $3x^2 - 12x + 14 = 0$

17. $2x^2 = 6x + 5$ **18.** $4x^2 = -20x + 3$ ▶ **19.** $2y^2 + 8y + 5 = 0$ **20.** $4z^2 - 8z + 1 = 0$

Objectives A B **Mixed Practice** *Solve each quadratic equation by completing the square. See Examples 1 through 5.*

21. $x^2 + 6x - 25 = 0$ **22.** $x^2 - 6x + 7 = 0$ **23.** $x^2 - 3x - 3 = 0$

24. $x^2 - 9x + 3 = 0$ **25.** $2y^2 - 3y + 1 = 0$ **26.** $2y^2 - y - 1 = 0$

27. $x(x + 3) = 18$ (*Hint:* First use the distributive property and multiply.)   **28.** $x(x - 3) = 18$ (See hint for #27.)   **29.** $3z^2 + 6z + 4 = 0$

30. $2y^2 + 8y + 9 = 0$ **31.** $4x^2 + 16x = 48$ **32.** $6x^2 - 30x = -36$

Review

Simplify each expression. See Section 15.2.

33. $\dfrac{3}{4} - \sqrt{\dfrac{25}{16}}$ **34.** $\dfrac{3}{5} + \sqrt{\dfrac{16}{25}}$ **35.** $\dfrac{1}{2} + \sqrt{\dfrac{9}{4}}$ **36.** $\dfrac{9}{10} - \sqrt{\dfrac{49}{100}}$

Simplify each expression. See Section 15.4.

37. $\dfrac{6 + 4\sqrt{5}}{2}$ **38.** $\dfrac{10 + 20\sqrt{3}}{2}$ **39.** $\dfrac{3 - 9\sqrt{2}}{6}$ **40.** $\dfrac{12 - 8\sqrt{7}}{16}$

Concept Extensions

Solve.

41. In your own words, describe a perfect square trinomial.

42. Describe how to find the number to add to $x^2 - 7x$ to make a perfect square trinomial.

43. Write your own quadratic equation to be solved by completing the square. Write it in the form

     perfect square trinomial = a number that is not a perfect square
         $x^2 + 6x + 9 = 11$ (An example)

  a. Solve the example $x^2 + 6x + 9 = 11$.
  b. Solve your quadratic equation by completing the square.

44. Follow the directions of Exercise **43**, except write your equation in the form

     perfect square trinomial = negative number

  a. For example, solve $x^2 - 2x + 1 = -7$.
  b. Solve your quadratic equation by completing the square.

45. Find a value of k that will make $x^2 + kx + 16$ a perfect square trinomial.

46. Find a value of k that will make $x^2 + kx + 25$ a perfect square trinomial.

47. The revenues from product sales y (in millions of dollars) of Abiomed, Inc., maker of the AbioCor artificial heart, during fiscal years 2009 through 2013 can be modeled by the equation $y = 3.5x^2 + 7x + 74$, where $x = 0$ represents 2009. Assume that this trend continues and predict the year after 2009 in which Abiomed's revenues from product sales will be $354 million. (*Source:* Based on data from Abiomed, Inc.)

48. The average price y of gold (in dollars per ounce) from 2008 through 2012 is given by the equation $y = 10x^2 + 180x + 841$, where x is the number of years after 2008. Assume that this trend continued and find the year after 2008 in which the price of gold was $2280 per ounce. (Round to the nearest whole number.) (*Source:* Based on data from the U.S. Geological Survey, Minerals Information)

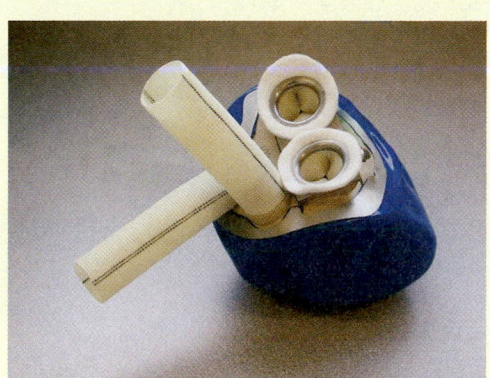

Recall that a graphing calculator may be used to solve an equation. For example, to solve $x^2 + 8x = -12$ (Exercise 1), graph

$$y_1 = x^2 + 8x$$
$$y_2 = -12$$

The x-coordinate of the point of intersection of the graphs is the solution. Use a graphing calculator to solve each equation. Round solutions to the nearest hundredth.

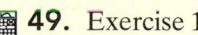 **49.** Exercise 1 **50.** Exercise 2 **51.** Exercise 17 **52.** Exercise 12

16.3 Solving Quadratic Equations by the Quadratic Formula

Objective A Using the Quadratic Formula

We can use the technique of completing the square to develop a formula to find solutions of any quadratic equation. We develop and use the **quadratic formula** in this section.

Recall that a quadratic equation in **standard form** is

$$ax^2 + bx + c = 0, \quad a \neq 0$$

To develop the quadratic formula, let's complete the square for this quadratic equation in standard form.

First we divide both sides of the equation by the coefficient of x^2 and then get the variable terms alone on one side of the equation.

$$x^2 + \frac{b}{a}x + \frac{c}{a} = 0 \qquad \text{Divide by } a; \text{ recall that } a \text{ cannot be 0.}$$

$$x^2 + \frac{b}{a}x = -\frac{c}{a} \qquad \text{Get the variable terms alone on one side of the equation.}$$

Objectives

A Use the Quadratic Formula to Solve Quadratic Equations.

B Approximate Solutions to Quadratic Equations.

The coefficient of x is $\dfrac{b}{a}$. Half of $\dfrac{b}{a}$ is $\dfrac{b}{2a}$ and $\left(\dfrac{b}{2a}\right)^2 = \dfrac{b^2}{4a^2}$. So we add $\dfrac{b^2}{4a^2}$ to both sides of the equation.

$$x^2 + \frac{b}{a}x + \frac{b^2}{4a^2} = -\frac{c}{a} + \frac{b^2}{4a^2} \qquad \text{Add } \frac{b^2}{4a^2} \text{ to both sides.}$$

$$\left(x + \frac{b}{2a}\right)^2 = -\frac{c}{a} + \frac{b^2}{4a^2} \qquad \text{Factor the left side.}$$

$$\left(x + \frac{b}{2a}\right)^2 = -\frac{4ac}{4a^2} + \frac{b^2}{4a^2} \qquad \text{Multiply } -\frac{c}{a} \text{ by } \frac{4a}{4a} \text{ so that the terms on the right side have a common denominator.}$$

$$\left(x + \frac{b}{2a}\right)^2 = \frac{b^2 - 4ac}{4a^2} \qquad \text{Simplify the right side.}$$

Now we use the square root property.

$$x + \frac{b}{2a} = \sqrt{\frac{b^2 - 4ac}{4a^2}} \quad \text{or} \quad x + \frac{b}{2a} = -\sqrt{\frac{b^2 - 4ac}{4a^2}} \qquad \begin{array}{l}\text{Use the square root} \\ \text{property.}\end{array}$$

$$x + \frac{b}{2a} = \frac{\sqrt{b^2 - 4ac}}{2a} \qquad x + \frac{b}{2a} = -\frac{\sqrt{b^2 - 4ac}}{2a} \qquad \text{Simplify the radical.}$$

$$x = -\frac{b}{2a} + \frac{\sqrt{b^2 - 4ac}}{2a} \qquad x = -\frac{b}{2a} - \frac{\sqrt{b^2 - 4ac}}{2a} \qquad \begin{array}{l}\text{Subtract } \frac{b}{2a} \text{ from} \\ \text{both sides.}\end{array}$$

$$x = \frac{-b + \sqrt{b^2 - 4ac}}{2a} \qquad x = \frac{-b - \sqrt{b^2 - 4ac}}{2a} \qquad \text{Simplify.}$$

The solutions are $\dfrac{-b \pm \sqrt{b^2 - 4ac}}{2a}$. This final equation is called the **quadratic formula** and gives the solutions of any quadratic equation.

Quadratic Formula

If a, b, and c are real numbers and $a \neq 0$, a quadratic equation written in the standard form $ax^2 + bx + c = 0$ has solutions

$$x = \frac{-b \pm \sqrt{b^2 - 4ac}}{2a}$$

Helpful Hint

Don't forget that to correctly identify a, b, and c in the quadratic formula, you should write the equation in standard form.

Quadratic Equations in Standard Form

$$5x^2 - 6x + 2 = 0 \qquad a = 5, b = -6, c = 2$$

$$4y^2 - 9 = 0 \qquad a = 4, b = 0, c = -9$$

$$x^2 + x = 0 \qquad a = 1, b = 1, c = 0$$

$$\sqrt{2}x^2 + \sqrt{5}x + \sqrt{3} = 0 \qquad a = \sqrt{2}, b = \sqrt{5}, c = \sqrt{3}$$

Example 1 Solve $3x^2 + x - 3 = 0$ using the quadratic formula.

Solution: This equation is in standard form with $a = 3$, $b = 1$, and $c = -3$. By the quadratic formula, we have

$$x = \frac{-b \pm \sqrt{b^2 - 4ac}}{2a}$$

$$x = \frac{-1 \pm \sqrt{1^2 - 4 \cdot 3 \cdot (-3)}}{2 \cdot 3} \quad \text{Let } a = 3, b = 1, \text{ and } c = -3.$$

$$= \frac{-1 \pm \sqrt{1 + 36}}{6} \quad \text{Simplify.}$$

$$= \frac{-1 \pm \sqrt{37}}{6}$$

Check both solutions in the original equation. The solutions are $\frac{-1 + \sqrt{37}}{6}$ and $\frac{-1 - \sqrt{37}}{6}$.

■ **Work Practice 1**

Example 2 Solve $2x^2 - 9x = 5$ using the quadratic formula.

Solution: First we write the equation in standard form by subtracting 5 from both sides.

$$2x^2 - 9x = 5$$
$$2x^2 - 9x - 5 = 0$$

Next we note that $a = 2$, $b = -9$, and $c = -5$. We substitute these values into the quadratic formula.

$$x = \frac{-b \pm \sqrt{b^2 - 4ac}}{2a}$$

$$x = \frac{-(-9) \pm \sqrt{(-9)^2 - 4 \cdot 2 \cdot (-5)}}{2 \cdot 2} \quad \text{Substitute into the formula.}$$

$$= \frac{9 \pm \sqrt{81 + 40}}{4} \quad \text{Simplify.}$$

$$= \frac{9 \pm \sqrt{121}}{4} = \frac{9 \pm 11}{4}$$

Then,

$$x = \frac{9 - 11}{4} = -\frac{1}{2} \quad \text{or} \quad x = \frac{9 + 11}{4} = 5$$

Check $-\frac{1}{2}$ and 5 in the original equation. Both $-\frac{1}{2}$ and 5 are solutions.

■ **Work Practice 2**

The following steps may be useful when solving a quadratic equation by the quadratic formula.

Practice 1

Solve $2x^2 - x - 5 = 0$ using the quadratic formula.

Practice 2

Solve $3x^2 + 8x = 3$ using the quadratic formula.

Helpful Hint Notice that the fraction bar is under the entire numerator $-b \pm \sqrt{b^2 - 4ac}$.

Answers

1. $\frac{1 + \sqrt{41}}{4}$ and $\frac{1 - \sqrt{41}}{4}$

2. $\frac{1}{3}$ and -3

> **To Solve a Quadratic Equation by the Quadratic Formula**
>
> **Step 1:** Write the quadratic equation in standard form: $ax^2 + bx + c = 0$.
>
> **Step 2:** If necessary, clear the equation of fractions to simplify calculations.
>
> **Step 3:** Identify a, b, and c.
>
> **Step 4:** Replace a, b, and c in the quadratic formula with the identified values, and simplify.

✔**Concept Check** For the quadratic equation $2x^2 - 5 = 7x$, if $a = 2$ and $c = -5$ in the quadratic formula, the value of b is which of the following?

a. $\dfrac{7}{2}$ **b.** 7 **c.** -5 **d.** -7

Practice 3

Solve $5x^2 = 2$ using the quadratic formula.

Example 3   Solve $7x^2 = 1$ using the quadratic formula.

Solution: First we write the equation in standard form by subtracting 1 from both sides.

$$7x^2 = 1$$
$$7x^2 - 1 = 0$$

> **Helpful Hint**
>
> $7x^2 - 1 = 0$ can be written as $7x^2 + 0x - 1 = 0$. This form helps you see that $b = 0$.

Next we replace a, b, and c with the identified values: $a = 7$, $b = 0$, $c = -1$.

$$x = \frac{0 \pm \sqrt{0^2 - 4 \cdot 7 \cdot (-1)}}{2 \cdot 7} \qquad \text{Substitute into the formula.}$$

$$= \frac{\pm \sqrt{28}}{14} \qquad \text{Simplify.}$$

$$= \frac{\pm 2\sqrt{7}}{14}$$

$$= \pm \frac{2 \cdot \sqrt{7}}{2 \cdot 7}$$

$$= \pm \frac{\sqrt{7}}{7}$$

The solutions are $\dfrac{\sqrt{7}}{7}$ and $-\dfrac{\sqrt{7}}{7}$.

■ **Work Practice 3**

Notice that we could have solved the equation $7x^2 = 1$ in Example 3 by dividing both sides by 7 and then using the square root property. We solved the equation by the quadratic formula to show that this formula can be used to solve any quadratic equation.

Practice 4

Solve $x^2 = -2x - 3$ using the quadratic formula.

Example 4   Solve $x^2 = -x - 1$ using the quadratic formula.

Solution: First we write the equation in standard form.

$$x^2 + x + 1 = 0$$

Next we replace a, b, and c in the quadratic formula with $a = 1$, $b = 1$, and $c = 1$.

Answers

3. $\dfrac{\sqrt{10}}{5}$ and $-\dfrac{\sqrt{10}}{5}$

4. no real solution

✔**Concept Check Answer**

d

$$x = \frac{-1 \pm \sqrt{1^2 - 4 \cdot 1 \cdot 1}}{2 \cdot 1} \qquad \text{Substitute into the formula.}$$

$$= \frac{-1 \pm \sqrt{-3}}{2} \qquad \text{Simplify.}$$

There is no real number solution because $\sqrt{-3}$ is not a real number.

■ **Work Practice 4**

Example 5   Solve $\frac{1}{2}x^2 - x = 2$ using the quadratic formula.

Solution: We write the equation in standard form and then clear the equation of fractions by multiplying both sides by the LCD, 2.

$$\frac{1}{2}x^2 - x = 2$$

$$\frac{1}{2}x^2 - x - 2 = 0 \qquad \text{Write in standard form.}$$

$$x^2 - 2x - 4 = 0 \qquad \text{Multiply both sides by 2.}$$

Here, $a = 1$, $b = -2$, and $c = -4$, so we substitute these values into the quadratic formula.

$$x = \frac{-(-2) \pm \sqrt{(-2)^2 - 4 \cdot 1 \cdot (-4)}}{2 \cdot 1}$$

$$= \frac{2 \pm \sqrt{20}}{2} = \frac{2 \pm 2\sqrt{5}}{2} \qquad \text{Simplify.}$$

$$= \frac{2\left(1 \pm \sqrt{5}\right)}{2} = 1 \pm \sqrt{5} \qquad \text{Factor and simplify.}$$

The solutions are $1 - \sqrt{5}$ and $1 + \sqrt{5}$.

■ **Work Practice 5**

Notice that in Example 5, although we cleared the equation of fractions, the coefficients $a = \frac{1}{2}$, $b = -1$, and $c = -2$ will give the same results.

Helpful Hint

When simplifying an expression such as

$$\frac{3 \pm 6\sqrt{2}}{6}$$

first factor out a common factor from the terms of the numerator and then simplify.

$$\frac{3 \pm 6\sqrt{2}}{6} = \frac{3\left(1 \pm 2\sqrt{2}\right)}{2 \cdot 3} = \frac{1 \pm 2\sqrt{2}}{2}$$

Practice 5

Solve $\frac{1}{3}x^2 - x = 1$ using the quadratic formula.

Answer

5. $\dfrac{3 + \sqrt{21}}{2}$ and $\dfrac{3 - \sqrt{21}}{2}$

Objective B Approximating Solutions to Quadratic Equations ▶

Sometimes approximate solutions for quadratic equations are appropriate.

Practice 6

Approximate the exact solutions of the quadratic equation in Practice 1. Round the approximations to the nearest tenth.

Example 6 Approximate the exact solutions of the quadratic equation in Example 1. Round the approximations to the nearest tenth.

Solution: From Example 1, we have exact solutions $\dfrac{-1 \pm \sqrt{37}}{6}$. Thus,

$$\frac{-1 + \sqrt{37}}{6} \approx 0.847127088 \approx 0.8 \text{ to the nearest tenth.}$$

$$\frac{-1 - \sqrt{37}}{6} \approx -1.180460422 \approx -1.2 \text{ to the nearest tenth.}$$

Thus approximate solutions to the quadratic equation in Example 1 are 0.8 and −1.2.

Answer

6. $\dfrac{1 + \sqrt{41}}{4} \approx 1.9, \dfrac{1 - \sqrt{41}}{4} \approx -1.4$

■ **Work Practice 6**

Vocabulary, Readiness & Video Check

Fill in each blank.

1. The quadratic formula is _____.

Identify the values of a, b, and c in each quadratic equation.

2. $5x^2 - 7x + 1 = 0$; $a =$ _____, $b =$ _____, $c =$ _____

3. $x^2 + 3x - 7 = 0$; $a =$ _____, $b =$ _____, $c =$ _____

4. $x^2 - 6 = 0$; $a =$ _____, $b =$ _____, $c =$ _____

5. $x^2 + x - 1 = 0$; $a =$ _____, $b =$ _____, $c =$ _____

6. $9x^2 - 4 = 0$; $a =$ _____, $b =$ _____, $c =$ _____

Simplify the following.

7. $\dfrac{-1 \pm \sqrt{1^2 - 4(1)(-2)}}{2(1)}$

8. $\dfrac{-(-5) \pm \sqrt{(-5)^2 - 4(2)(3)}}{2(2)}$

9. $\dfrac{-5 \pm \sqrt{5^2 - 4(1)(2)}}{2(1)}$

10. $\dfrac{-7 \pm \sqrt{7^2 - 4(2)(1)}}{2(2)}$

Martin-Gay Interactive Videos   Watch the section lecture video and answer the following questions.

See Video 16.3

Objective A 11. Based on the lectures and ⊟ Examples 1–3, answer the following.

    a. Must a quadratic equation be written in standard form in order to use the quadratic formula? Why or why not?

    b. Must fractions be cleared from the equation before using the quadratic formula? Why or why not? ▶

Objective B 12. From ⊟ Example 4, how are approximate solutions found? ▶

16.3 Exercise Set MyMathLab®

Objective **A** *Use the quadratic formula to solve each quadratic equation. See Examples 1 through 4.*

1. $x^2 - 3x + 2 = 0$

2. $x^2 - 5x - 6 = 0$

3. $3k^2 + 7k + 1 = 0$

4. $7k^2 + 3k - 1 = 0$

5. $4x^2 - 3 = 0$

6. $25x^2 - 15 = 0$

7. $5z^2 - 4z + 3 = 0$

8. $3x^2 + 2x + 1 = 0$

9. $y^2 = 7y + 30$

10. $y^2 = 5y + 36$

11. $2x^2 = 10$

12. $5x^2 = 15$

13. $m^2 - 12 = m$

14. $m^2 - 14 = 5m$

15. $3 - x^2 = 4x$

16. $10 - x^2 = 2x$

17. $6x^2 + 9x = 2$

18. $3x^2 - 9x = 8$

19. $7p^2 + 2 = 8p$

20. $11p^2 + 2 = 10p$

21. $x^2 - 6x + 2 = 0$

22. $x^2 - 10x + 19 = 0$

23. $2x^2 - 6x + 3 = 0$

24. $5x^2 - 8x + 2 = 0$

25. $3x^2 = 1 - 2x$

26. $5y^2 = 4 - y$

27. $4y^2 = 6y + 1$

28. $6z^2 = 2 - 3z$

29. $x^2 + x + 2 = 0$

30. $k^2 + 2k + 5 = 0$

31. $20y^2 = 3 - 11y$

32. $2z^2 = z + 3$

33. $x^2 - 5x - 2 = 0$

34. $x^2 - 2x - 5 = 0$

35. $3x^2 - x - 14 = 0$

36. $5x^2 - 13x - 6 = 0$

Use the quadratic formula to solve each quadratic equation. See Example 5.

37. $\dfrac{m^2}{2} = m + \dfrac{1}{2}$

38. $\dfrac{m^2}{2} = 3m - 1$

39. $3p^2 - \dfrac{2}{3}p + 1 = 0$

40. $\dfrac{5}{2}p^2 - p + \dfrac{1}{2} = 0$

41. $4p^2 + \dfrac{3}{2} = -5p$

42. $4p^2 + \dfrac{3}{2} = 5p$

43. $5x^2 = \dfrac{7}{2}x + 1$

44. $2x^2 = \dfrac{5}{2}x + \dfrac{7}{2}$

45. $x^2 - \dfrac{11}{2}x - \dfrac{1}{2} = 0$

46. $\dfrac{2}{3}x^2 - 2x - \dfrac{2}{3} = 0$

47. $5z^2 - 2z = \dfrac{1}{5}$

48. $9z^2 + 12z = -1$

Objectives **A** **B** **Mixed Practice** *Use the quadratic formula to solve each quadratic equation. Find the exact solutions; then approximate these solutions to the nearest tenth. See Examples 1 through 6.*

49. $3x^2 = 21$

50. $2x^2 = 26$

51. $x^2 + 6x + 1 = 0$

52. $x^2 + 4x + 2 = 0$

53. $x^2 = 9x + 4$

54. $x^2 = 7x + 5$

55. $3x^2 - 2x - 2 = 0$

56. $5x^2 - 3x - 1 = 0$

Review

Graph the following linear equations in two variables. See Sections 10.2 and 10.3.

57. $y = -3$

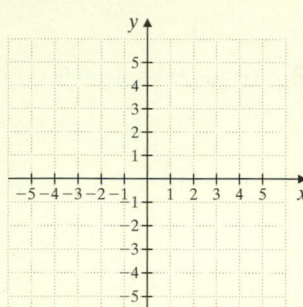

58. $x = 4$

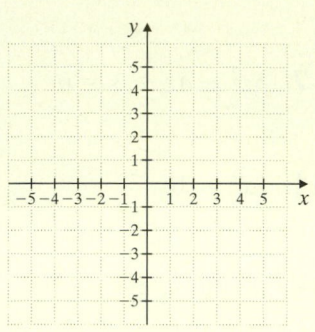

59. $y = 3x - 2$

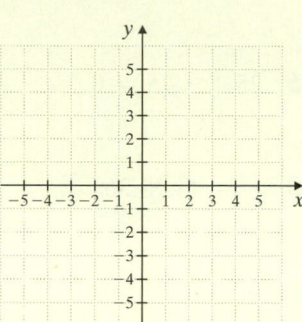

60. $y = 2x + 3$

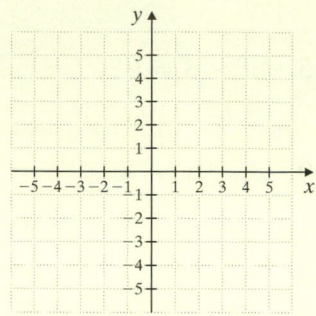

Concept Extensions

Solve. See the Concept Check in this section. For the quadratic equation $5x^2 + 2 = x$, if $a = 5$,

61. What is the value of b?

    a. $\dfrac{1}{5}$ **b.** 0 **c.** -1 **d.** 1

62. What is the value of c?

    a. 5 **b.** x **c.** -2 **d.** 2

For the quadratic equation $7y^2 = 3y$, if $b = 3$,

63. What is the value of a?

    a. 7 **b.** -7 **c.** 0 **d.** 1

64. What is the value of c?

    a. 7 **b.** 3 **c.** 0 **d.** 1

Solve.

△**65.** In a recent year, Nestle created a chocolate bar that the company claimed weighed more than 2 tons. The rectangular bar had a base area of approximately 34.65 square feet, and its length was 0.6 foot shorter than three times its width. Find the length and width of the bar. (*Source:* Nestle)

△**66.** The area of a rectangular conference room table is 95 square feet. If its length is six feet longer than its width, find the dimensions of the table. Round each dimension to the nearest tenth.

Solve each equation using the quadratic formula.

67. $x^2 + 3\sqrt{2}x - 5 = 0$

68. $y^2 - 2\sqrt{5}y - 1 = 0$

69. Explain how to identify a, b, and c correctly when solving a quadratic equation by the quadratic formula.

70. Explain how the quadratic formula is developed and why it is useful.

Use the quadratic formula and a calculator to solve each equation. Round solutions to the nearest tenth.

71. $7.3z^2 + 5.4z - 1.1 = 0$

72. $1.2x^2 - 5.2x - 3.9 = 0$

A rocket is launched from the top of an 80-foot cliff with an initial velocity of 120 feet per second. The height, h, of the rocket after t seconds is given by the equation $h = -16t^2 + 120t + 80$.

73. How long after the rocket is launched will it be 30 feet from the ground? Round to the nearest tenth of a second.

74. How long after the rocket is launched will it strike the ground? Round to the nearest tenth of a second. (*Hint:* The rocket will strike the ground when its height $h = 0$.)

80 feet

Solve.

75. Total revenue y (in millions of dollars) for Target Corporation for the years 2008 through 2012 is approximated by the equation $y = 450x^2 + 330x + 64{,}820$, where $x = 0$ represents the year 2008. Assume that this trend continues and predict the year after 2008 in which Target's total revenue will be approximately $96,000 million. (Round to the nearest whole number.) (*Source:* Based on data from Target Corporation)

76. Retail sales y (in billions of dollars) from online shopping and mail order in the United States from 2007 through 2011 can be approximated by the equation $y = 3.2x^2 + 1.4x + 105.8$, where x is the number of years since 2007. Assume this trend continues and predict the year after 2007 in which the retail sales for online and mail order shopping will be $440 billion. (Round to the nearest whole number.) (*Source:* Based on data from the U.S. Bureau of the Census, Annual Retail Trade Survey)

Summary on Solving Quadratic Equations

An important skill in mathematics is learning when to use one technique in favor of another. We now practice this by deciding which method to use when solving quadratic equations. Although both the quadratic formula and completing the square can be used to solve any quadratic equation, the quadratic formula is usually less tedious and thus preferred. The following steps may be used to solve a quadratic equation.

To Solve a Quadratic Equation

Step 1: If the equation is in the form $ax^2 = c$ or $(ax + b)^2 = c$, use the square root property and solve. If not, go to Step 2.

Step 2: Write the equation in standard form: $ax^2 + bx + c = 0$.

Step 3: Try to solve the equation by factoring. If not possible, go to Step 4.

Step 4: Solve the equation by the quadratic formula.

Study the examples below to help you review these steps.

Practice 1

Solve $y^2 - 4y - 6 = 0$.

Example 1 Solve $m^2 - 2m - 7 = 0$.

Solution: The equation is in standard form, but the quadratic expression $m^2 - 2m - 7$ is not factorable, so use the quadratic formula with $a = 1, b = -2$, and $c = -7$.

$$m^2 - 2m - 7 = 0$$

$$m = \frac{-(-2) \pm \sqrt{(-2)^2 - 4 \cdot 1 \cdot (-7)}}{2 \cdot 1} = \frac{2 \pm \sqrt{32}}{2}$$

$$m = \frac{2 \pm 4\sqrt{2}}{2} = \frac{2(1 \pm 2\sqrt{2})}{2} = 1 \pm 2\sqrt{2}$$

The solutions are $1 - 2\sqrt{2}$ and $1 + 2\sqrt{2}$.

■ **Work Practice 1**

Practice 2

Solve $(2x + 5)^2 = 45$.

Example 2 Solve $(3x + 1)^2 = 20$.

Solution: This equation is in a form that makes the square root property easy to apply.

$$(3x + 1)^2 = 20$$

$$3x + 1 = \pm\sqrt{20} \quad \text{Apply the square root property.}$$

$$3x + 1 = \pm 2\sqrt{5} \quad \text{Simplify } \sqrt{20}.$$

$$3x = -1 \pm 2\sqrt{5}$$

$$x = \frac{-1 \pm 2\sqrt{5}}{3}$$

The solutions are $\frac{-1 - 2\sqrt{5}}{3}$ and $\frac{-1 + 2\sqrt{5}}{3}$.

Answers

1. $-1, 4$ **2.** $2 \pm \sqrt{10}$

■ **Work Practice 2**

Example 3 Solve $x^2 - \dfrac{11}{2}x = -\dfrac{5}{2}$.

Solution: The fractions make factoring more difficult and complicate the calculations for using the quadratic formula. Clear the equation of fractions by multiplying both sides of the equation by the LCD 2.

$$x^2 - \frac{11}{2}x = -\frac{5}{2}$$

$$x^2 - \frac{11}{2}x + \frac{5}{2} = 0 \qquad \text{Write in standard form.}$$

$$2x^2 - 11x + 5 = 0 \qquad \text{Multiply both sides by 2.}$$

$$(2x - 1)(x - 5) = 0 \qquad \text{Factor.}$$

$$2x - 1 = 0 \quad \text{or} \quad x - 5 = 0 \qquad \text{Apply the zero factor property.}$$

$$2x = 1 \qquad\qquad x = 5$$

$$x = \frac{1}{2} \qquad\qquad x = 5$$

The solutions are $\dfrac{1}{2}$ and 5.

■ **Work Practice 3**

Choose and use a method to solve each equation.

1. $5x^2 - 11x + 2 = 0$

2. $5x^2 + 13x - 6 = 0$

3. $x^2 - 1 = 2x$

4. $x^2 + 7 = 6x$

5. $a^2 = 20$

6. $a^2 = 72$

7. $x^2 - x + 4 = 0$

8. $x^2 - 2x + 7 = 0$

9. $3x^2 - 12x + 12 = 0$

10. $5x^2 - 30x + 45 = 0$

11. $9 - 6p + p^2 = 0$

12. $49 - 28p + 4p^2 = 0$

13. $4y^2 - 16 = 0$

14. $3y^2 - 27 = 0$

15. $x^2 - 3x + 2 = 0$

16. $x^2 + 7x + 12 = 0$

Practice 3

Solve $x^2 - \dfrac{5}{2}x = -\dfrac{3}{2}$.

Answer

3. $\dfrac{3}{2}, 1$

Answers

1. _____

2. _____

3. _____

4. _____

5. _____

6. _____

7. _____

8. _____

9. _____

10. _____

11. _____

12. _____

13. _____

14. _____

15. _____

16. _____

17. _____

18. _____

19. _____

20. _____

21. _____

22. _____

23. _____

24. _____

25. _____

26. _____

27. _____

28. _____

29. _____

30. _____

31. _____

32. _____

33. _____

34. _____

35. _____

36. _____

37. _____

38. _____

39. _____

40. _____

41. _____

▶ **17.** $(2z + 5)^2 = 25$

18. $(3z - 4)^2 = 16$

19. $30x = 25x^2 + 2$

20. $12x = 4x^2 + 4$

21. $\frac{2}{3}m^2 - \frac{1}{3}m - 1 = 0$

22. $\frac{5}{8}m^2 + m - \frac{1}{2} = 0$

▶ **23.** $x^2 - \frac{1}{2}x - \frac{1}{5} = 0$

24. $x^2 + \frac{1}{2}x - \frac{1}{8} = 0$

25. $4x^2 - 27x + 35 = 0$

26. $9x^2 - 16x + 7 = 0$

27. $(7 - 5x)^2 = 18$

28. $(5 - 4x)^2 = 75$

29. $3z^2 - 7z = 12$

30. $6z^2 + 7z = 6$

31. $x = x^2 - 110$

32. $x = 56 - x^2$

33. $\frac{3}{4}x^2 - \frac{5}{2}x - 2 = 0$

34. $x^2 - \frac{6}{5}x - \frac{8}{5} = 0$

35. $x^2 - 0.6x + 0.05 = 0$

36. $x^2 - 0.1x - 0.06 = 0$

37. $10x^2 - 11x + 2 = 0$

38. $20x^2 - 11x + 1 = 0$

39. $\frac{1}{2}z^2 - 2z + \frac{3}{4} = 0$

40. $\frac{1}{5}z^2 - \frac{1}{2}z - 2 = 0$

41. Explain how you will decide what method to use when solving quadratic equations.

Graphing Quadratic Equations in Two Variables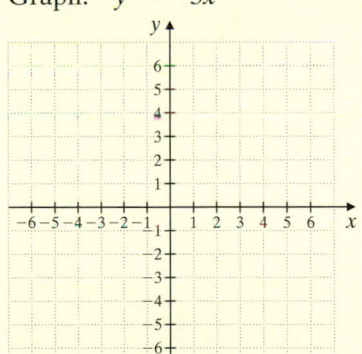

Recall from Section 10.2 that the graph of a linear equation in two variables, $Ax + By = C$, is a straight line. In this section, we will find that the graph of a quadratic equation in the form $y = ax^2 + bx + c$ is a parabola.

Objectives

A Graph Quadratic Equations of the Form $y = ax^2$.

B Graph Quadratic Equations of the Form $y = ax^2 + bx + c$.

C Use the Vertex Formula to Determine the Vertex of a Parabola.

Objective **A** Graphing $y = ax^2$

We begin our work by graphing $y = x^2$. To do so, we will find and plot ordered pair solutions of this equation. Let's select a few values for x, find the corresponding y-values, and record them in a table of values to keep track. Then we can plot the points corresponding to these solutions on a coordinate plane.

If $x = -3$, then $y = (-3)^2$, or 9.

If $x = -2$, then $y = (-2)^2$, or 4.

If $x = -1$, then $y = (-1)^2$, or 1.

If $x = 0$, then $y = 0^2$, or 0.

If $x = 1$, then $y = 1^2$, or 1.

If $x = 2$, then $y = 2^2$, or 4.

If $x = 3$, then $y = 3^2$, or 9.

x	y
-3	9
-2	4
-1	1
0	0
1	1
2	4
3	9

The graph of $y = x^2$ is a smooth curve through the plotted points. This curve is called a **parabola.** The lowest point on a parabola opening upward is called the **vertex.** The vertex is $(0, 0)$ for the parabola $y = x^2$. If we fold the graph paper along the y-axis, the two pieces of the parabola match perfectly. For this reason, we say the graph is **symmetric about the y-axis,** and we call the y-axis the **line of symmetry.**

Notice that the parabola that corresponds to the equation $y = x^2$ opens upward. This happens when the coefficient of x^2 is positive. In the equation $y = x^2$, the coefficient of x^2 is 1. Example 1 shows the graph of a quadratic equation where the coefficient of x^2 is negative.

Example 1 Graph: $y = -2x^2$

Solution: We begin by selecting x-values and calculating the corresponding y-values. Then we plot the ordered pairs found and draw a smooth curve through those points. Notice that when the coefficient of x^2 is negative, the corresponding

(Continued on next page)

Practice 1

Graph: $y = -3x^2$

Answer

1.

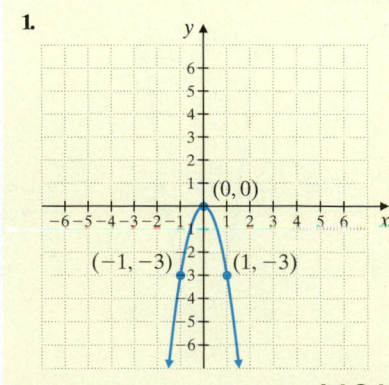

1191

parabola opens downward. When a parabola opens downward, the vertex is the highest point of the parabola. The vertex of this parabola is $(0, 0)$.

$y = -2x^2$

x	y
0	0
1	-2
2	-8
3	-18
-1	-2
-2	-8
-3	-18

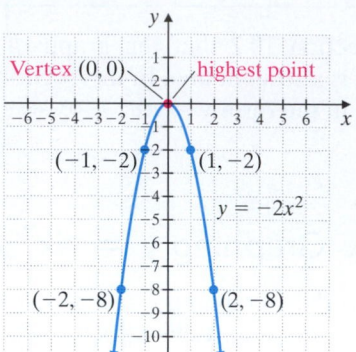

■ **Work Practice 1**

Objective B Graphing $y = ax^2 + bx + c$ ▶

Just as for linear equations, we can use x- and y-intercepts to help graph quadratic equations. Recall from Chapter 10 that an x-intercept is the point where the graph crosses the x-axis. A y-intercept is the point where the graph crosses the y-axis. We find intercepts just as we did in Chapter 10.

> **Helpful Hint**
>
> Recall that:
>
>   To find x-intercepts, let $y = 0$ and solve for x.
>   To find y-intercepts, let $x = 0$ and solve for y.

Practice 2

Graph: $y = x^2 - 9$

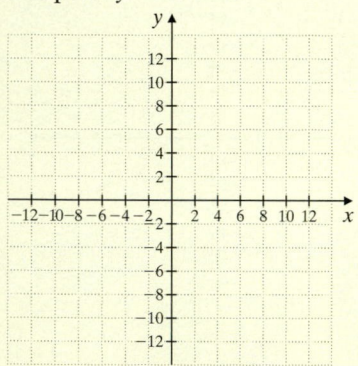

Answer

2.

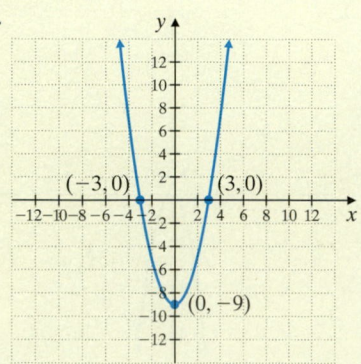

Example 2 Graph: $y = x^2 - 4$

Solution: If we write this equation as $y = x^2 + 0x + (-4)$, we can see that it is in the form $y = ax^2 + bx + c$. To graph it, we first find the intercepts. To find the y-intercept, we let $x = 0$. Then

$$y = 0^2 - 4 = -4$$

To find x-intercepts, we let $y = 0$.

$$0 = x^2 - 4$$
$$0 = (x - 2)(x + 2)$$
$$x - 2 = 0 \quad \text{or} \quad x + 2 = 0$$
$$x = 2 \qquad\qquad x = -2$$

Thus far, we have the y-intercept $(0, -4)$ and the x-intercepts $(2, 0)$ and $(-2, 0)$. Now we can select additional x-values, find the corresponding y-values, plot the points, and draw a smooth curve through the points.

$y = x^2 - 4$

x	y
0	−4
1	−3
2	0
3	5
−1	−3
−2	0
−3	5

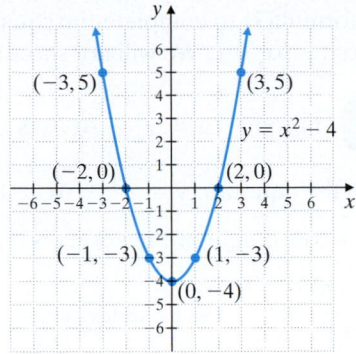

Notice that the vertex of this parabola is $(0, -4)$.

■ **Work Practice 2**

✓**Concept Check** Tell whether the graph of each equation opens upward or downward.

a. $y = 2x^2$ **b.** $y = 3x^2 + 4x - 5$ **c.** $y = -5x^2 + 2$

Helpful Hint

For the graph of $y = ax^2 + bx + c$,

If a is positive, the parabola opens upward.

If a is negative, the parabola opens downward.

✓**Concept Check** For which of the following graphs of $y = ax^2 + bx + c$ would the value of a be negative?

a.
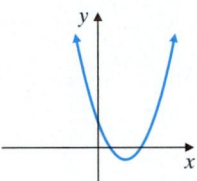

b.
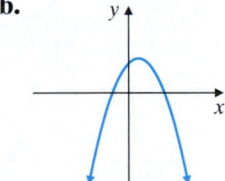

Objective C Using the Vertex Formula

Thus far, we have accidentally stumbled upon the vertex of each parabola that we have graphed. However, our choice of values for x may not yield an ordered pair for the vertex of the parabola. It would be helpful if we could first find the vertex of a parabola. Next we would determine whether the parabola opens upward or downward. Finally we would calculate additional points such as x- and y-intercepts as needed. In fact, there is a formula that may be used to find the vertex of a parabola.

Vertex Formula

The vertex of the parabola $y = ax^2 + bx + c$ has x-coordinate

$$\frac{-b}{2a}$$

The corresponding y-coordinate of the vertex is obtained by substituting the x-coordinate into the equation and finding y.

✓**First Concept Check Answer**

a. upward **b.** upward **c.** downward

✓**Second Concept Check Answer**

b

One way to develop this formula is to notice that the *x*-value of the vertex of the parabolas that we are considering lies halfway between its *x*-intercepts. Another way to develop this formula is to complete the square on the general form of a quadratic equation: $y = ax^2 + bx + c$. We will not show the development of this formula here.

Practice 3

Graph: $y = x^2 - 2x - 3$

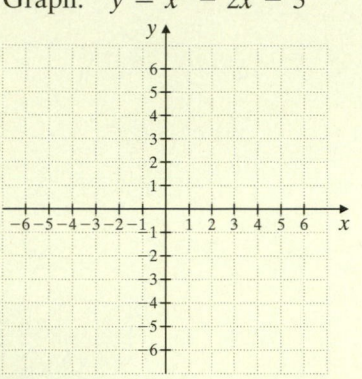

> **Example 3** Graph: $y = x^2 - 6x + 8$

Solution: In the equation $y = x^2 - 6x + 8$, $a = 1$ and $b = -6$.

Vertex: The *x*-coordinate of the vertex is

$$\frac{-b}{2a} = \frac{-(-6)}{2 \cdot 1} = 3 \quad \text{Use the vertex formula, } \frac{-b}{2a}.$$

To find the corresponding *y*-coordinate, we let $x = 3$ in the original equation.

$$y = x^2 - 6x + 8 = 3^2 - 6 \cdot 3 + 8 = -1$$

The vertex is $(3, -1)$ and the parabola opens upward since *a* is positive. We now find and plot the intercepts.

Intercepts: To find the *x*-intercepts, we let $y = 0$.

$$0 = x^2 - 6x + 8$$

We factor the expression $x^2 - 6x + 8$ to find $(x - 4)(x - 2) = 0$. The *x*-intercepts are $(4, 0)$ and $(2, 0)$.

If we let $x = 0$ in the original equation, then $y = 8$ gives us the *y*-intercept $(0, 8)$. Now we plot the vertex $(3, -1)$ and the intercepts $(4, 0)$, $(2, 0)$, and $(0, 8)$. Then we can sketch the parabola.

These and two additional points are shown in the table.

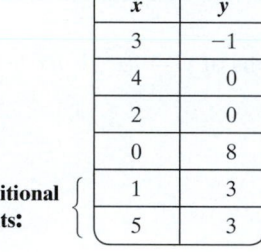

x	*y*
3	-1
4	0
2	0
0	8
1	3
5	3

Additional points: the last two rows (1, 3) and (5, 3)

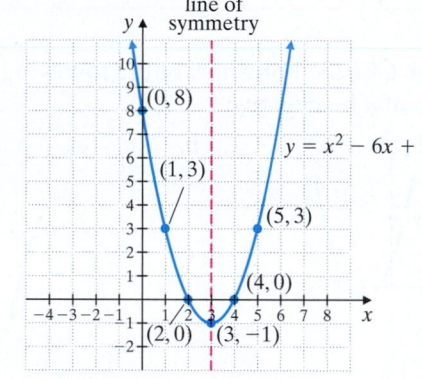

Work Practice 3

Study Example 3 and let's use it to write down a general procedure for graphing quadratic equations.

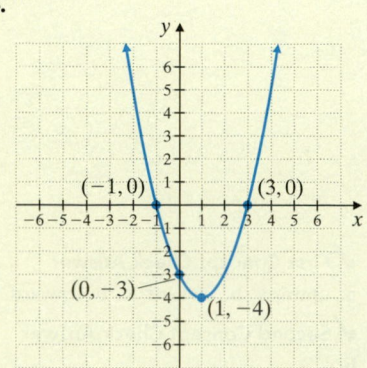

Graphing Parabolas Defined by $y = ax^2 + bx + c$

1. **Find the vertex by using the formula $x = \dfrac{-b}{2a}$.** Don't forget to find the *y*-value of the vertex.

2. **Find the intercepts.**

   - Let $x = 0$ and solve for *y* to find the *y*-intercept. There will be only one.
   - Let $y = 0$ and solve for *x* to find any *x*-intercepts. There may be 0, 1, or 2.

3. **Plot the vertex and the intercepts.**

4. **Find and plot additional points on the graph.** Then draw a smooth curve through the plotted points. Keep in mind that if $a > 0$, the parabola opens upward and that if $a < 0$, the parabola opens downward.

Example 4   Graph: $y = x^2 + 2x - 5$

Solution: In the equation $y = x^2 + 2x - 5$, $a = 1$ and $b = 2$. Using the vertex formula, we find that the x-coordinate of the vertex is

$$x = \frac{-b}{2a} = \frac{-2}{2 \cdot 1} = -1$$

The y-coordinate is

$$y = (-1)^2 + 2(-1) - 5 = -6$$

Thus the vertex is $(-1, -6)$.
To find the x-intercepts, we let $y = 0$.

$$0 = x^2 + 2x - 5$$

This cannot be solved by factoring, so we use the quadratic formula.

$$x = \frac{-2 \pm \sqrt{2^2 - 4(1)(-5)}}{2 \cdot 1} \qquad \text{Let } a = 1, b = 2, \text{ and } c = -5.$$

$$x = \frac{-2 \pm \sqrt{24}}{2}$$

$$x = \frac{-2 \pm 2\sqrt{6}}{2} \qquad \text{Simplify the radical.}$$

$$x = \frac{2(-1 \pm \sqrt{6})}{2} = -1 \pm \sqrt{6}$$

The x-intercepts are $(-1 + \sqrt{6}, 0)$ and $(-1 - \sqrt{6}, 0)$. We use a calculator to approximate these so that we can easily graph these intercepts.

$$-1 + \sqrt{6} \approx 1.4 \qquad \text{and} \qquad -1 - \sqrt{6} \approx -3.4$$

To find the y-intercept, we let $x = 0$ in the original equation and find that $y = -5$. Thus the y-intercept is $(0, -5)$. You will find, because of symmetry, that $(-2, -5)$ is also an ordered pair solution.

x	y
-1	-6
$-1 + \sqrt{6} \approx 1.4$	0
$-1 - \sqrt{6} \approx -3.4$	0
0	-5
-2	-5

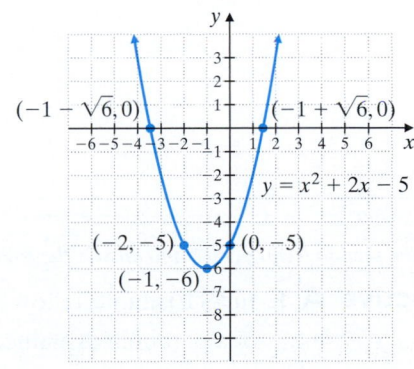

■ **Work Practice 4**

Practice 4
Graph: $y = x^2 - 4x + 1$

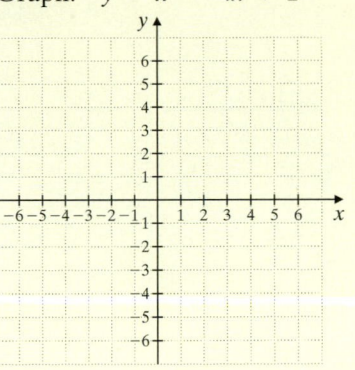

Answer

4.

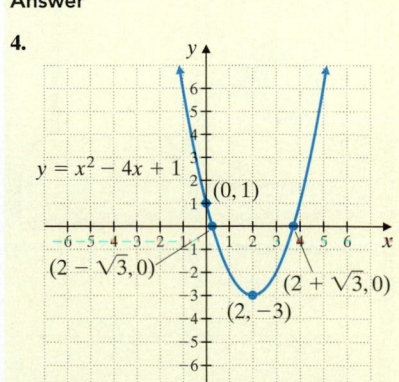

Helpful Hint

Notice that the number of x-intercepts of the graph of the parabola $y = ax^2 + bx + c$ is the same as the number of real solutions of $0 = ax^2 + bx + c$.

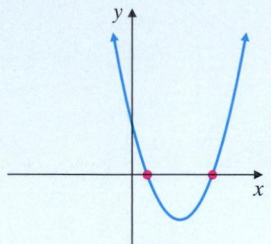

$y = ax^2 + bx + c$
$a > 0$

Two x-intercepts
Two real solutions of
$0 = ax^2 + bx + c$

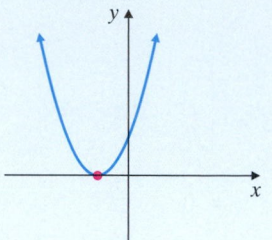

$y = ax^2 + bx + c$
$a > 0$

One x-intercept
One real solution of
$0 = ax^2 + bx + c$

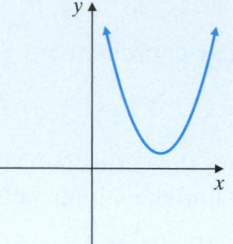

$y = ax^2 + bx + c$
$a > 0$

No x-intercepts
No real solutions of
$0 = ax^2 + bx + c$

Calculator Explorations   Graphing

Recall that a graphing calculator may be used to solve quadratic equations. The x-intercepts of the graph of $y = ax^2 + bx + c$ are solutions of $0 = ax^2 + bx + c$. To solve $x^2 - 7x - 3 = 0$, for example, graph $y_1 = x^2 - 7x - 3$. The x-intercepts of the graph are the solutions of the equation.

Use a graphing calculator to solve each quadratic equation. Round solutions to two decimal places.

1. $x^2 - 7x - 3 = 0$

2. $2x^2 - 11x - 1 = 0$

3. $-1.7x^2 + 5.6x - 3.7 = 0$

4. $-5.8x^2 + 2.3x - 3.9 = 0$

5. $5.8x^2 - 2.6x - 1.9 = 0$

6. $7.5x^2 - 3.7x - 1.1 = 0$

Vocabulary, Readiness & Video Check

Martin-Gay Interactive Videos

See Video 16.4

Watch the section lecture video and answer the following questions.

Objective A 1. In ⊞ Example 1, how are the vertex and line of symmetry of a parabola explained? ▶

Objective B 2. In ⊞ Example 2, what important point was accidentally found? Why would it be useful to have an algebraic way to find this point? ▶

Objective C 3. From ⊞ Example 3, how can finding the vertex and noting whether the parabola opens up or down possibly help save us time and work? Explain using an example. ▶

16.4 Exercise Set MyMathLab®

Objective A *Graph each quadratic equation by finding and plotting ordered pair solutions. See Example 1.*

1. $y = 2x^2$

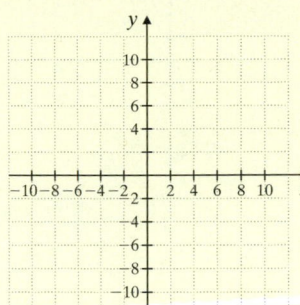

2. $y = 3x^2$

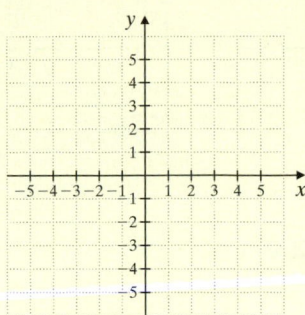

3. $y = -x^2$

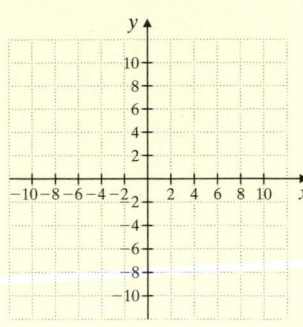

4. $y = -4x^2$

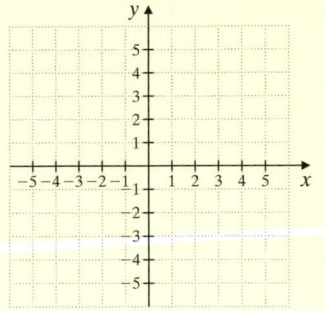

Objective B *Sketch the graph of each equation. Label the vertex and the intercepts. See Example 2.*

5. $y = x^2 - 1$

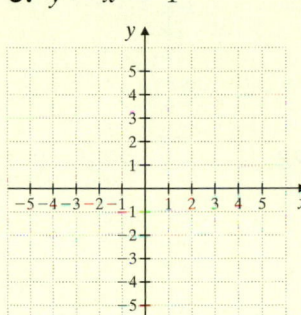

6. $y = x^2 - 16$

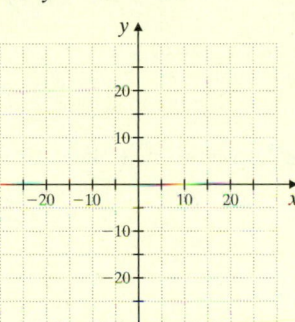

7. $y = x^2 + 4$

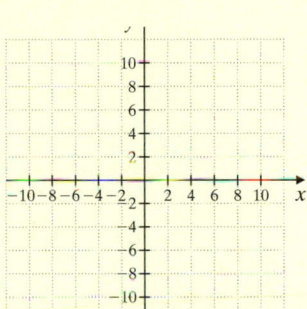

8. $y = x^2 + 9$

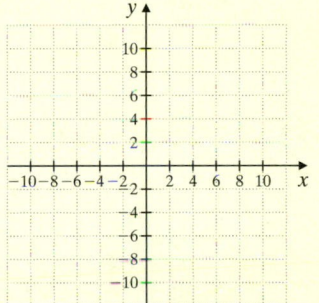

Objectives A B C *Sketch the graph of each equation. Label the vertex and the intercepts. See Examples 1 through 4.*

9. $y = -x^2 + 4x - 4$

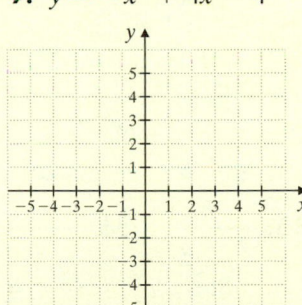

10. $y = -x^2 - 2x - 1$

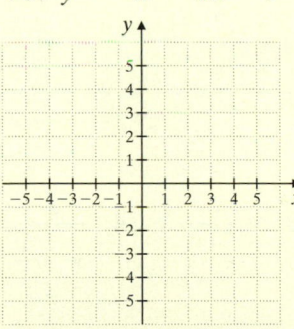

11. $y = x^2 + 5x + 4$

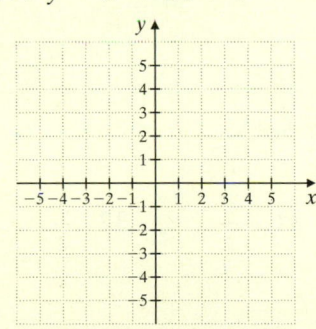

12. $y = x^2 + 7x + 10$

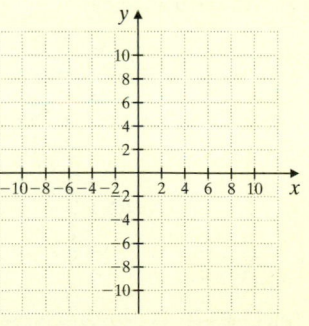

13. $y = x^2 - 4x + 5$

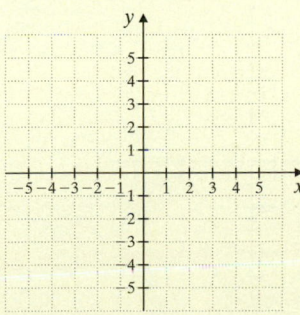

14. $y = x^2 - 6x + 10$

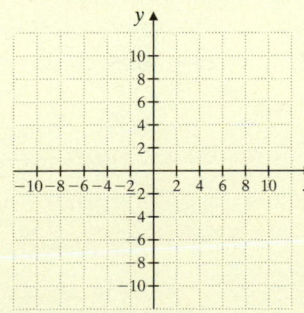

15. $y = 2 - x^2$

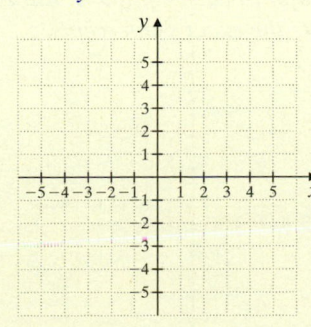

16. $y = 3 - x^2$

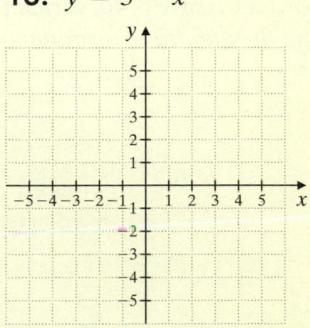

17. $y = \frac{1}{3}x^2$

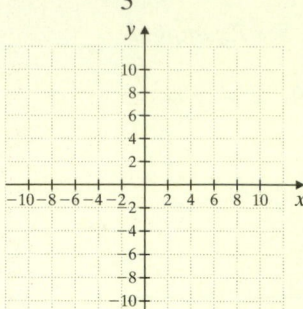

18. $y = \frac{1}{2}x^2$

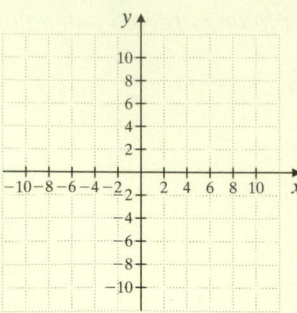

19. $y = x^2 + 6x$

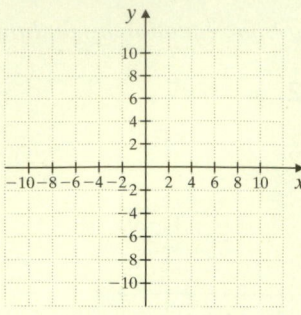

20. $y = x^2 - 4x$

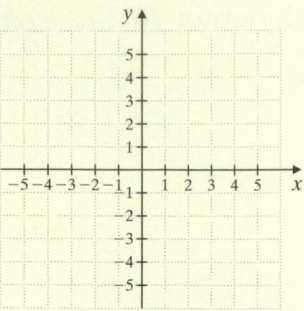

21. $y = x^2 + 2x - 8$

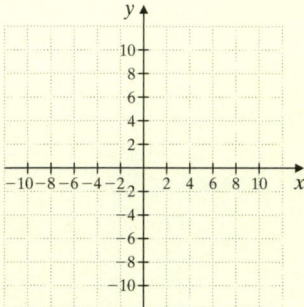

22. $y = x^2 - 2x - 3$

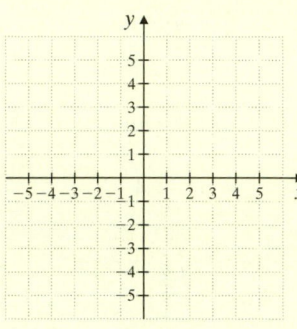

23. $y = -\frac{1}{2}x^2$

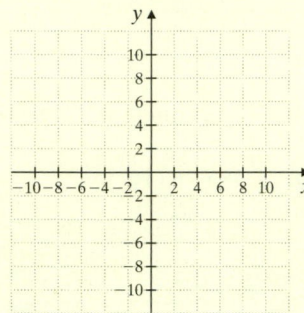

24. $y = -\frac{1}{3}x^2$

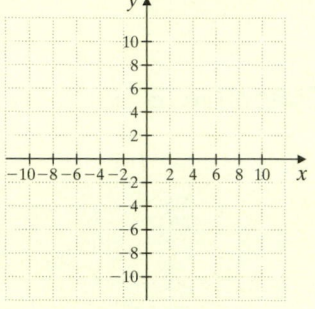

○ 25. $y = 2x^2 - 11x + 5$

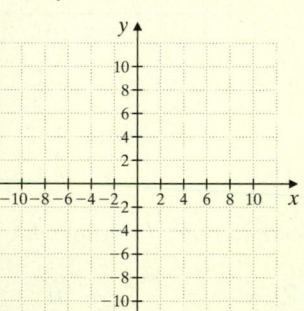

26. $y = 2x^2 + x - 3$

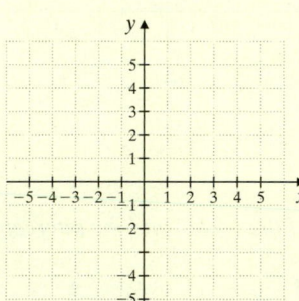

○ 27. $y = -x^2 + 4x - 3$

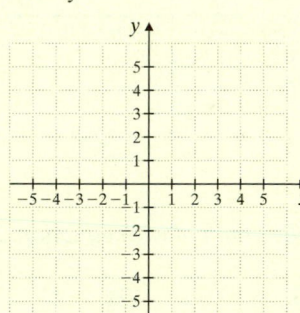

28. $y = -x^2 + 6x - 8$

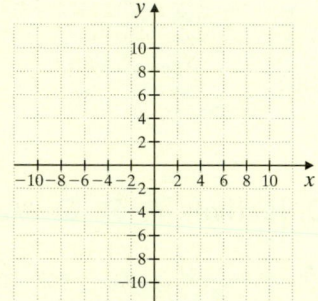

Review

Simplify each complex fraction. See Section 14.7.

29. $\dfrac{\frac{1}{7}}{\frac{2}{5}}$

30. $\dfrac{\frac{3}{8}}{\frac{1}{7}}$

31. $\dfrac{\frac{1}{x}}{\frac{2}{x^2}}$

32. $\dfrac{\frac{x}{5}}{\frac{2}{x}}$

33. $\dfrac{2x}{1 - \frac{1}{x}}$

34. $\dfrac{x}{x - \frac{1}{x}}$

35. $\dfrac{\frac{a - b}{2b}}{\frac{b - a}{8b^2}}$

36. $\dfrac{\frac{2a^2}{a - 3}}{\frac{a}{3 - a}}$

Concept Extensions

For Exercises 37–40, sketch the graph of each equation. Label the vertex and intercepts. Use the quadratic formula to locate the exact x-intercepts.

37. $y = x^2 + 2x - 2$

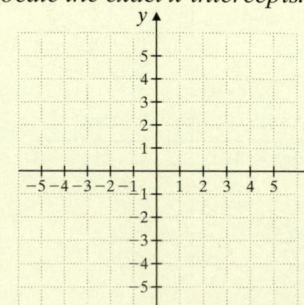

38. $y = x^2 - 4x - 3$

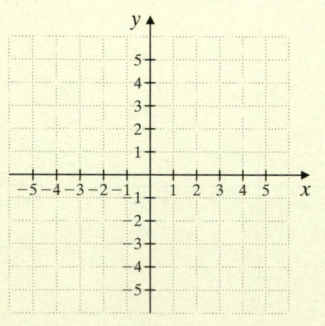

39. $y = x^2 - 3x + 1$

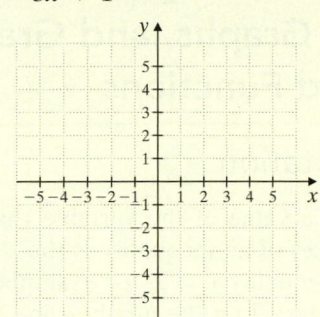

40. $y = x^2 - 2x - 5$

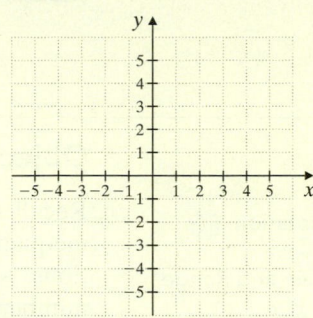

41. The height h of a fireball launched from a Roman candle with an initial velocity of 128 feet per second is given by the equation $h = -16t^2 + 128t$, where t is time in seconds after launch.

Use the graph of this equation to answer each question.

a. Estimate the maximum height of the fire ball.

b. Estimate the time when the fireball is at its maximum height.

c. Estimate the time when the fireball would return to the ground.

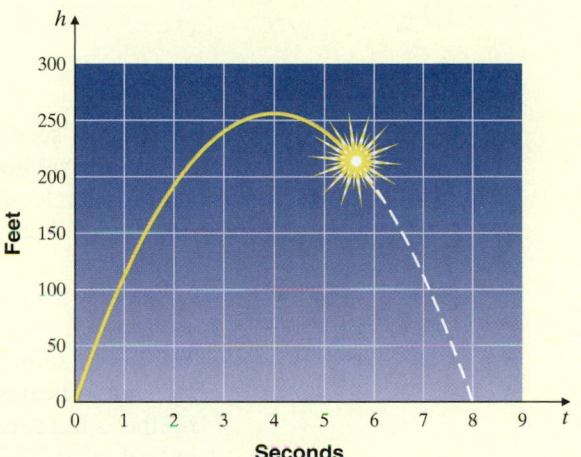

42. Determine the maximum number and the minimum number of x-intercepts for a parabola. Explain your answers.

Match the graph of each equation of the form $y = ax^2 + bx + c$ with the given description.

43. $a > 0$, two x-intercepts

44. $a < 0$, one x-intercept

45. $a < 0$, no x-intercept

46. $a > 0$, no x-intercept

47. $a > 0$, one x-intercept

48. $a < 0$, two x-intercepts

A

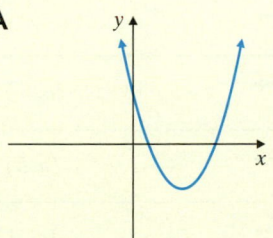

B

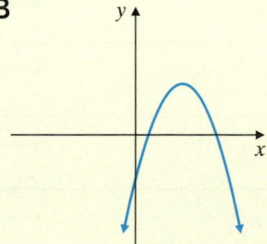

C

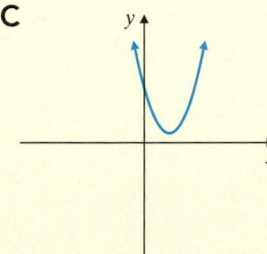

D

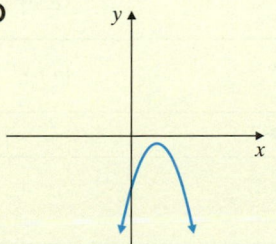

E

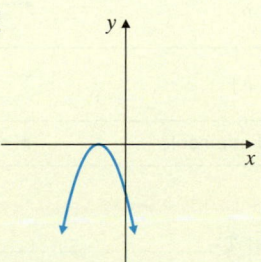

F

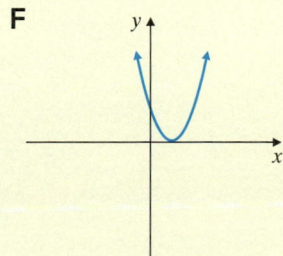

16.5 Interval Notation, Finding Domains and Ranges from Graphs, and Graphing Piecewise-Defined Functions

Objectives

A Use Interval Notation.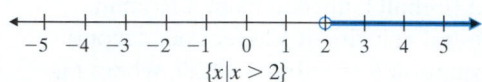

B Find the Domain and Range from a Graph.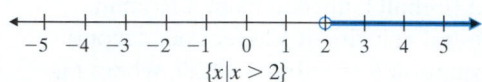

C Graph Piecewise-Defined Functions.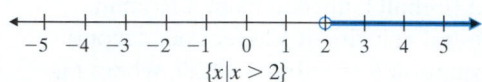

Objective A Using Interval Notation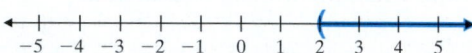

Recall that a **solution** of an inequality is a value of the variable that makes the inequality a true statement. The **solution set** of an inequality is the set of all solutions. Notice that the solution set of the inequality $x > 2$, for example, contains all numbers greater than 2. Its graph is an interval on the number line since an infinite number of values satisfy the variable. If we use open/closed-circle notation, the graph of $\{x \mid x > 2\}$ looks like:

$$\{x \mid x > 2\}$$

In this section, a different graphing notation will be used to help us understand **interval notation.** Instead of an open circle, we use a parenthesis; instead of a closed circle, we use a bracket. With this new notation, the graph of $\{x \mid x > 2\}$ now looks like:

and can be represented in interval notation as $(2, \infty)$. The symbol ∞ is read "infinity" and indicates that the interval includes *all* numbers greater than 2. The left parenthesis indicates that 2 *is not* included in the interval. Using a left bracket, [, would indicate that 2 *is* included in the interval. The following table shows three equivalent ways to describe an interval: in set notation, as a graph, and in interval notation.

Set Notation	Graph	Interval Notation
$\{x \mid x < a\}$	*a*	$(-\infty, a)$
$\{x \mid x > a\}$	*a*	(a, ∞)
$\{x \mid x \le a\}$	*a*	$(-\infty, a]$
$\{x \mid x \ge a\}$	*a*	$[a, \infty)$
$\{x \mid a < x < b\}$	*a* *b*	(a, b)
$\{x \mid a \le x \le b\}$	*a* *b*	$[a, b]$
$\{x \mid a < x \le b\}$	*a* *b*	$(a, b]$
$\{x \mid a \le x < b\}$	*a* *b*	$[a, b)$
$\{x \mid x \text{ is a real number}\}$		$(-\infty, \infty)$

Helpful Hint

Notice that a parenthesis is always used to enclose ∞ and $-\infty$.

Examples   Graph each set on a number line and then write it in interval notation.

1. $\{x \mid x \geq 2\}$ 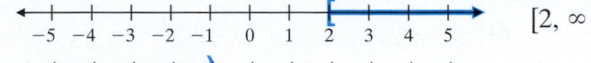 $[2, \infty)$

2. $\{x \mid x < -1\}$ $(-\infty, -1)$

3. $\{x \mid 0.5 < x \leq 3\}$ $(0.5, 3]$

■ **Work Practice 1–3**

✓**Concept Check**   Explain what is wrong with writing the interval $(5, \infty]$.

Objective B   Finding the Domain and Range from a Graph ▶

Recall from Section 10.6 that the

> **domain** of a relation is the set of all first components of the ordered pairs of the relation and the
> **range** of a relation is the set of all second components of the ordered pairs of the relation.

In this section we use the graph of a relation to find its domain and range. Let's use interval notation to write these domains and ranges. Remember, we use a parenthesis to indicate that a number is not part of the domain and we use a bracket to indicate that a number is part of the domain. Of course, as usual, parentheses are placed about infinity symbols indicating that we approach but never reach infinity.

To find the domain of a function (or relation) from its graph, recall that on the rectangular coordinate system, "domain" is the set of first components of the ordered pairs, so this means the x-values that are graphed. Similarly, "range" is the set of second components of the ordered pairs, so this means the y-values that are graphed.

Examples   Find the domain and range of each relation.

4.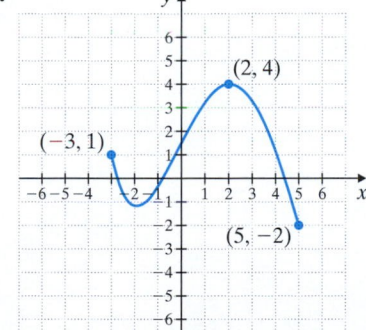

5.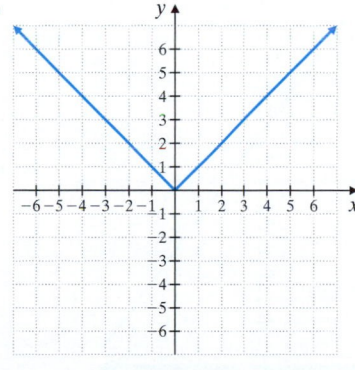

6.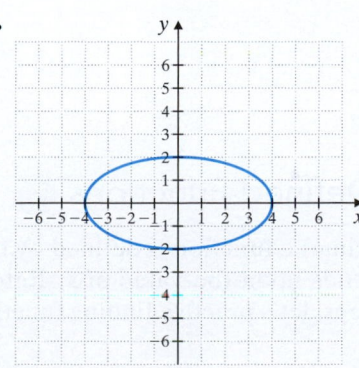

7.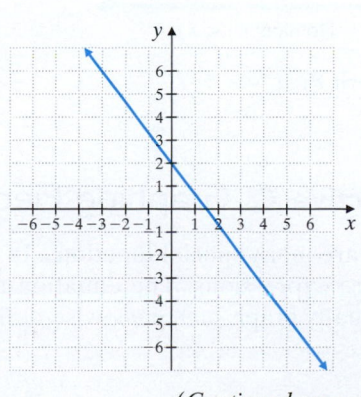

(Continued on next page)

Practice 1–3

Graph each set on a number line and then write it in interval notation.

1. $\{x \mid x > -3\}$

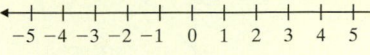

2. $\{x \mid x \leq 0\}$

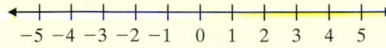

3. $\{x \mid -0.5 \leq x < 2\}$

Practice 4–7

Find the domain and range of each relation.

4.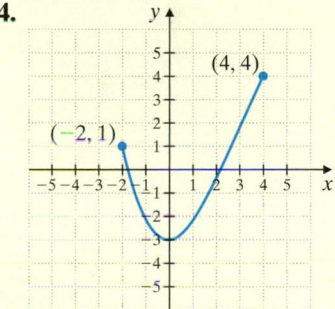

5.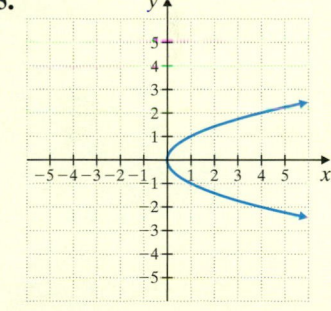

Answers

1. $(-3, \infty)$,

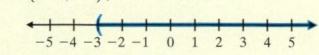

2. $(-\infty, 0]$,

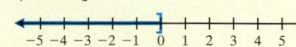

3. $[-0.5, 2)$,

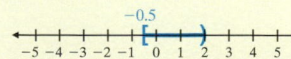

4. domain: $[-2, 4]$; range: $[-3, 4]$

5. domain: $[0, \infty)$; range: $(-\infty, \infty)$

✓**Concept Check Answer**

should be $(5, \infty)$ since a parenthesis is always used to enclose ∞

6.

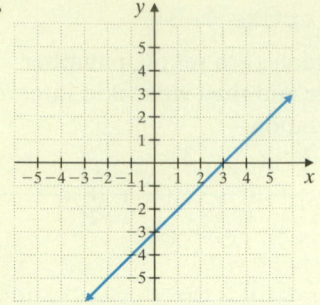

7.

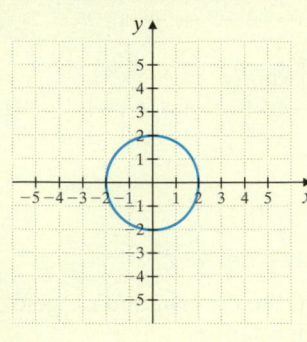

Solution: Notice that the graphs for Examples 4, 5, and 7 are graphs of functions because each passes the vertical line test.

4.

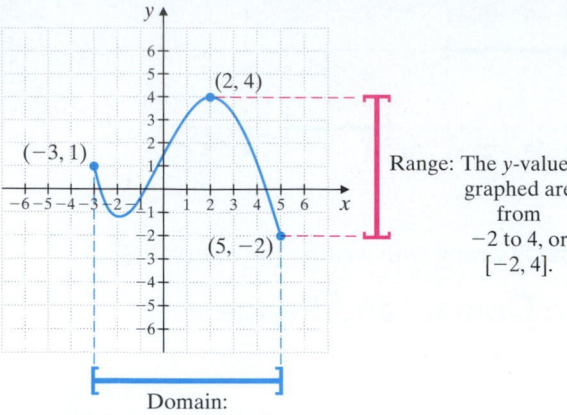

Range: The y-values graphed are from −2 to 4, or [−2, 4].

Domain: The x-values graphed are from −3 to 5, or [−3, 5].

5.

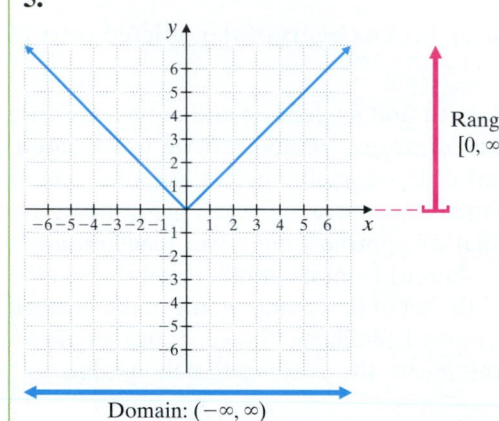

Range: [0, ∞)

Domain: (−∞, ∞)

6.

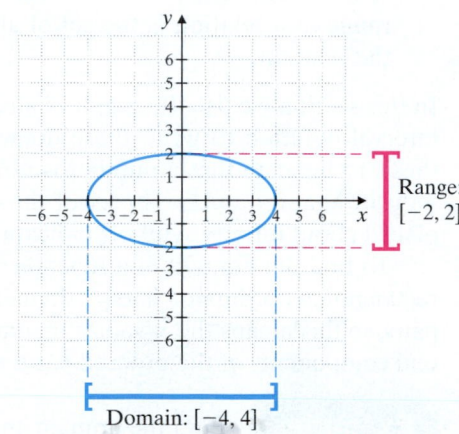

Range: [−2, 2]

Domain: [−4, 4]

7.

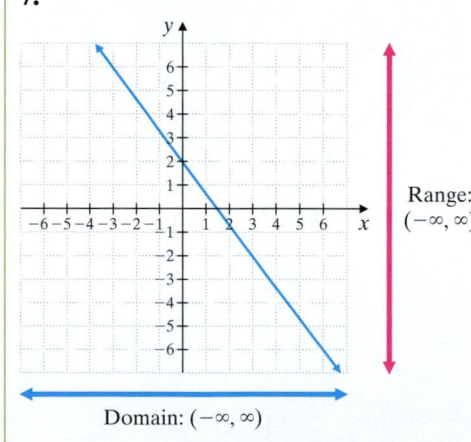

Range: (−∞, ∞)

Domain: (−∞, ∞)

■ **Work Practice 4–7**

Objective C Graphing Piecewise-Defined Functions ▶

There are many special functions. In fact, sometimes a function is defined by two or more expressions. The equation to use depends upon the value of x. Before we actually graph such piecewise-defined functions, let's practice finding function values.

Answers

6. domain: (−∞, ∞); range: (−∞, ∞)

7. domain: [−2, 2]; range: [−2, 2]

Example 8   Evaluate $f(2)$, $f(-6)$, and $f(0)$ for the function

$$f(x) = \begin{cases} 2x + 3 & \text{if } x \le 0 \\ -x - 1 & \text{if } x > 0 \end{cases}$$

Then write your results in ordered-pair form.

Solution: Take a moment and study this function. It is a single function defined by two expressions depending on the value of x. From above, if $x \le 0$, use $f(x) = 2x + 3$. If $x > 0$, use $f(x) = -x - 1$. Thus

$$f(2) = -(2) - 1 \qquad f(-6) = 2(-6) + 3 \qquad f(0) = 2(0) + 3$$
$$= -3 \text{ since } 2 > 0 \qquad = -9 \text{ since } -6 \le 0 \qquad = 3 \text{ since } 0 \le 0$$
$$f(2) = -3 \qquad\qquad f(-6) = -9 \qquad\qquad f(0) = 3$$

Ordered pairs: $(2, -3)$ $\qquad\qquad$ $(-6, -9)$ $\qquad\qquad$ $(0, 3)$

■ **Work Practice 8**

Now, let's graph a piecewise-defined function.

Example 9   Graph $f(x) = \begin{cases} 2x + 3 & \text{if } x \le 0 \\ -x - 1 & \text{if } x > 0 \end{cases}$

Solution: Let's graph each piece.

If $x \le 0$, $\qquad\qquad\qquad\qquad\qquad\qquad$ If $x > 0$,
$f(x) = 2x + 3$ $\qquad\qquad\qquad\qquad\qquad$ $f(x) = -x - 1$

Values ≤ 0
x	$f(x) = 2x + 3$
0	3 Closed circle
−1	1
−2	−1

Values > 0
x	$f(x) = -x - 1$
1	−2
2	−3
3	−4

The graph of the first part of $f(x)$ listed will look like a ray with a closed-circle endpoint at $(0, 3)$. The graph of the second part of $f(x)$ listed will look like a ray with an open-circle endpoint. To find the exact location of the open-circle endpoint, use $f(x) = -x - 1$ and find $f(0)$. Since $f(0) = -0 - 1 = -1$, we graph the values from the second table and place an open circle at $(0, -1)$.

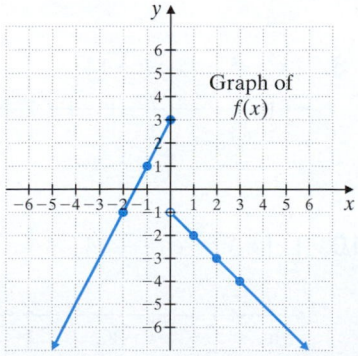

Graph of $f(x)$

Notice that this graph is the graph of a function because it passes the vertical line test. The domain of this function is $(-\infty, \infty)$ and the range is $(-\infty, 3]$.

■ **Work Practice 9**

Practice 8

Evaluate $f(-4)$, $f(3)$, and $f(0)$ for the function

$$f(x) = \begin{cases} 3x + 4 & \text{if } x < 0 \\ -x + 2 & \text{if } x \ge 0 \end{cases}$$

Then write your results in ordered-pair form.

Practice 9

Graph

$$f(x) = \begin{cases} 3x + 4 & \text{if } x < 0 \\ -x + 2 & \text{if } x \ge 0 \end{cases}$$

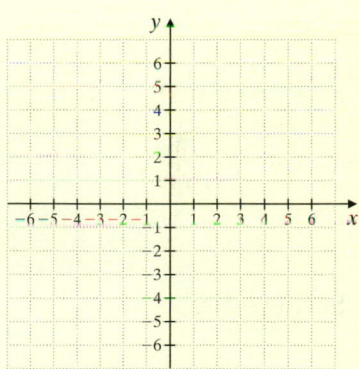

Answers

8. $f(-4) = -8; f(3) = -1; f(0) = 2;$
$(-4, -8); (3, -1); (0, 2),$

9.

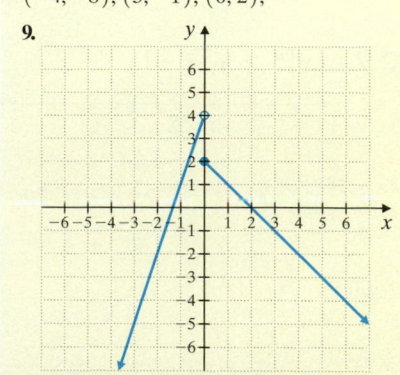

Vocabulary, Readiness & Video Check

Martin-Gay Interactive Videos

See Video 16.5 🍎

Watch the section lecture video and answer the following questions.

Objective A   **1.** Using ⊞ Example 1 as a reference, explain how the graph of the solution set of an inequality can help you write the solution set in interval notation. ▶

Objective B   **2.** In ⊞ Example 4, why is the range not $[3, \infty)$? What is the range? ▶

Objective C   **3.** In ⊞ Example 7, only one piece of the function is defined for the value $x = -1$. Why do we find $f(-1)$ for $f(x) = x + 3$? ▶

16.5   Exercise Set   MyMathLab® ▶

Objective A   *Graph the solution set of each inequality on a number line and then write it in interval notation. See Examples 1 through 3.*

▶ **1.** $\{x \mid x < -3\}$

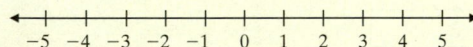

2. $\{x \mid x > 5\}$

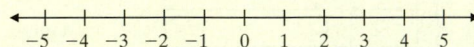

3. $\{x \mid x \geq 0.3\}$

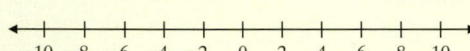

4. $\{x \mid x < -0.2\}$

5. $\{x \mid -7 \leq x\}$

6. $\{x \mid -7 \geq x\}$

7. $\{x \mid -2 < x < 5\}$

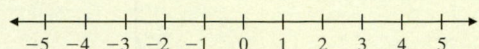

8. $\{x \mid -5 \leq x \leq -1\}$

▶ **9.** $\{x \mid 5 \geq x > -1\}$

10. $\{x \mid -3 > x \geq -7\}$

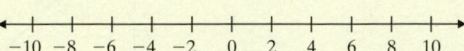

Objective B *Find the domain and the range of each relation. See Examples 4 through 7.*

11.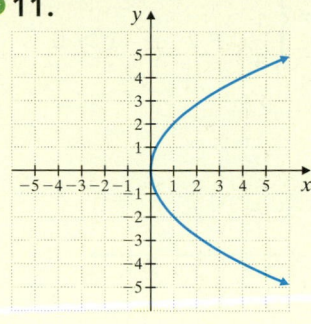

12.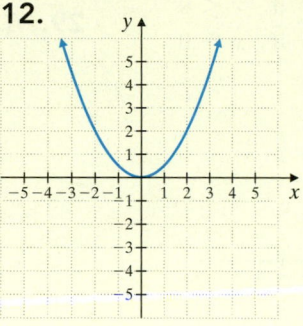

13.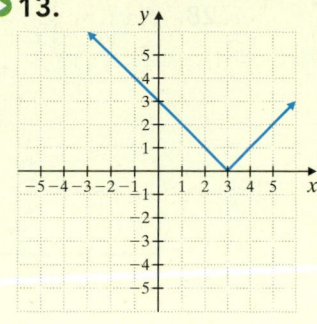

14.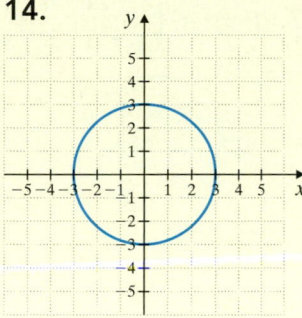

15.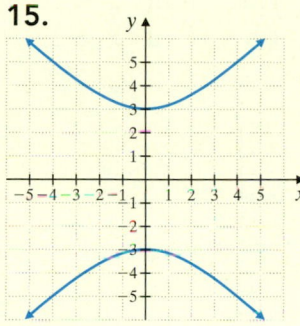

16.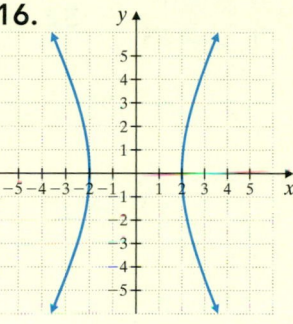

17.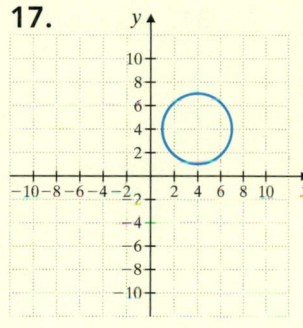

18.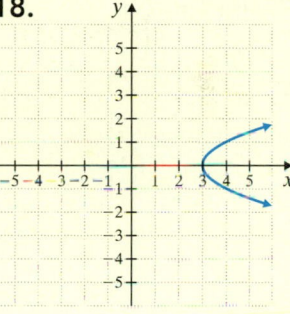

19.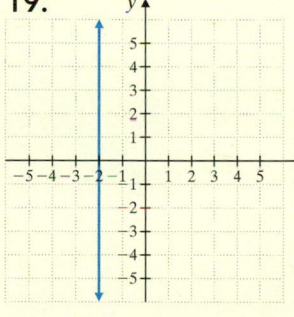

20.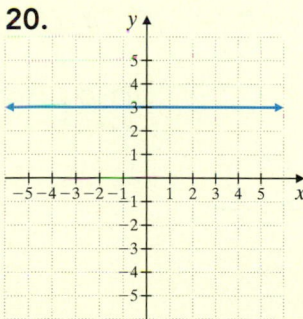

21.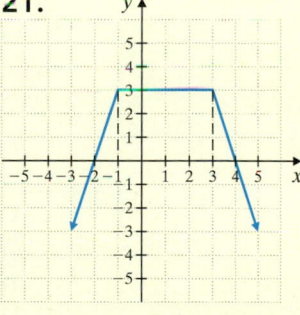

22.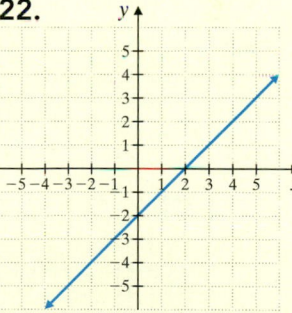

23.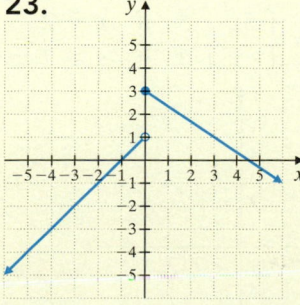

24.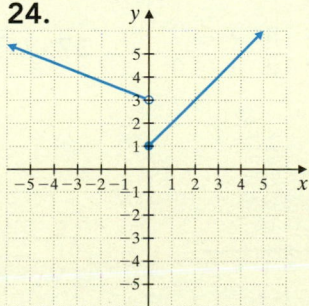

25.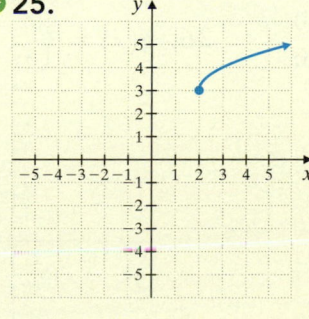

26.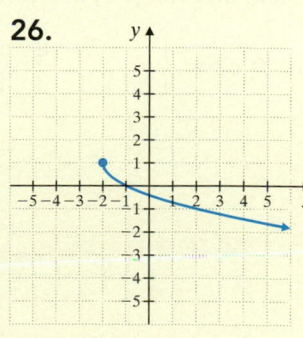

Objective C *Graph each piecewise-defined function. See Examples 8 and 9.*

27. $f(x) = \begin{cases} 2x & \text{if } x < 0 \\ x + 1 & \text{if } x \geq 0 \end{cases}$

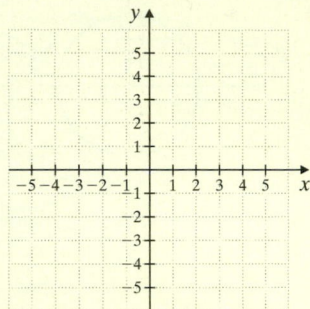

28. $f(x) = \begin{cases} 3x & \text{if } x < 0 \\ x + 2 & \text{if } x \geq 0 \end{cases}$

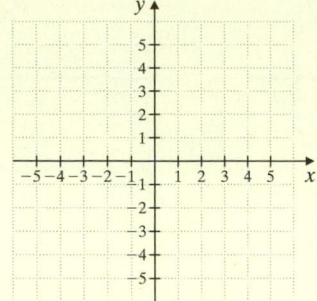

29. $f(x) = \begin{cases} 4x + 5 & \text{if } x \leq 0 \\ \dfrac{1}{4}x + 2 & \text{if } x > 0 \end{cases}$

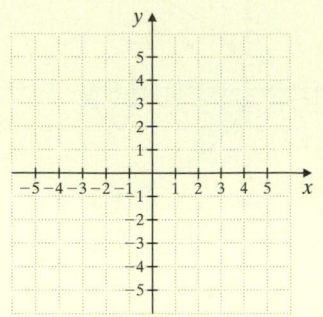

30. $f(x) = \begin{cases} 5x + 4 & \text{if } x \leq 0 \\ \dfrac{1}{3}x - 1 & \text{if } x > 0 \end{cases}$

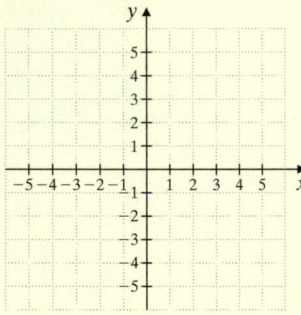

31. $g(x) = \begin{cases} -x & \text{if } x \leq 1 \\ 2x + 1 & \text{if } x > 1 \end{cases}$

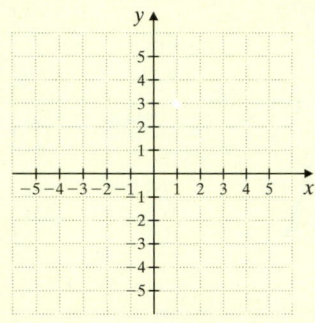

32. $g(x) = \begin{cases} 3x - 1 & \text{if } x \leq 2 \\ -x & \text{if } x > 2 \end{cases}$

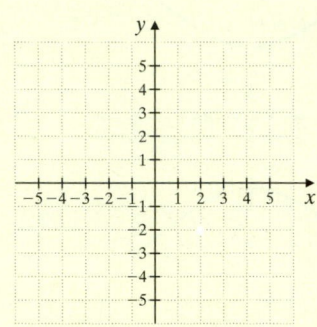

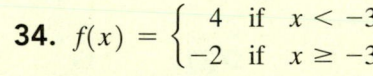

33. $f(x) = \begin{cases} 5 & \text{if } x < -2 \\ 3 & \text{if } x \geq -2 \end{cases}$

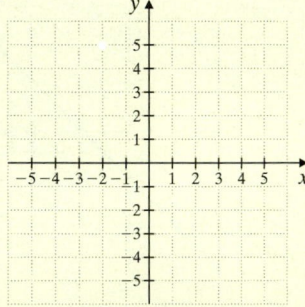

34. $f(x) = \begin{cases} 4 & \text{if } x < -3 \\ -2 & \text{if } x \geq -3 \end{cases}$

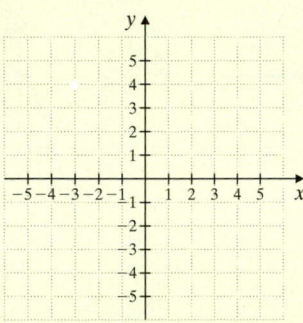

Objectives A B C **Mixed Practice** *Graph each piecewise-defined function. Use the graph to determine the domain and range of the function. See Examples 1 through 9.*

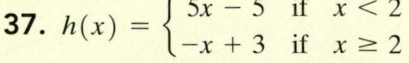

35. $f(x) = \begin{cases} -2x & \text{if } x \leq 0 \\ 2x + 1 & \text{if } x > 0 \end{cases}$

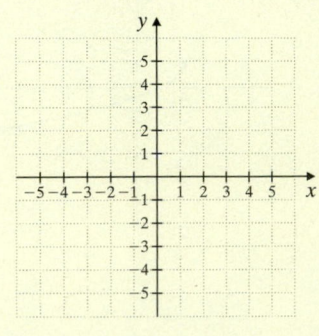

36. $g(x) = \begin{cases} -3x & \text{if } x \leq 0 \\ 3x + 2 & \text{if } x > 0 \end{cases}$

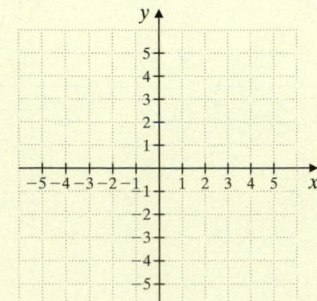

37. $h(x) = \begin{cases} 5x - 5 & \text{if } x < 2 \\ -x + 3 & \text{if } x \geq 2 \end{cases}$

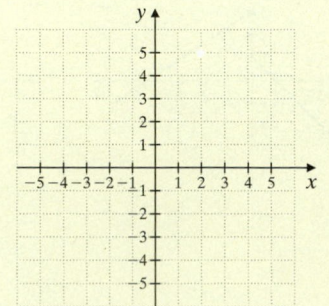

38. $f(x) = \begin{cases} 4x - 4 & \text{if } x < 2 \\ -x + 1 & \text{if } x \geq 2 \end{cases}$

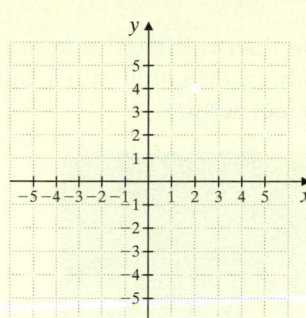

▶39. $f(x) = \begin{cases} x + 3 & \text{if } x < -1 \\ -2x + 4 & \text{if } x \geq -1 \end{cases}$

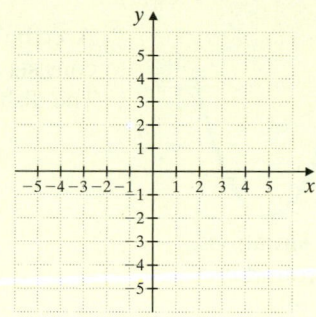

40. $h(x) = \begin{cases} x + 2 & \text{if } x < 1 \\ 2x + 1 & \text{if } x \geq 1 \end{cases}$

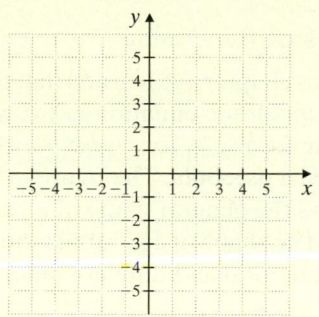

41. $g(x) = \begin{cases} -2 & \text{if } x \leq 0 \\ -4 & \text{if } x \geq 1 \end{cases}$

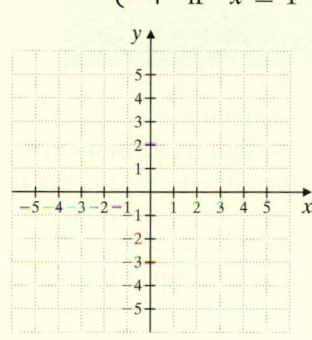

42. $f(x) = \begin{cases} -1 & \text{if } x \leq 0 \\ -3 & \text{if } x \geq 2 \end{cases}$

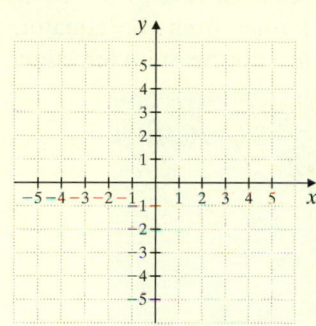

Review

Match each equation with its graph. See Section 10.5.

43. $y = -1$
A
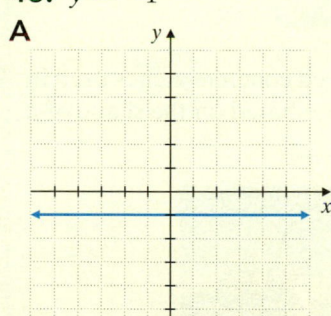

44. $x = -1$
B
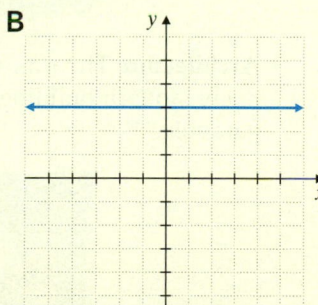

45. $x = 3$
C
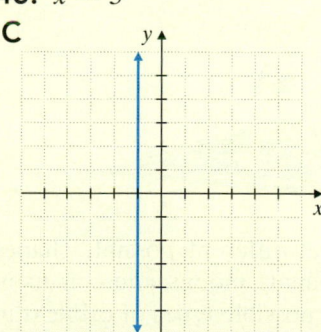

46. $y = 3$
D
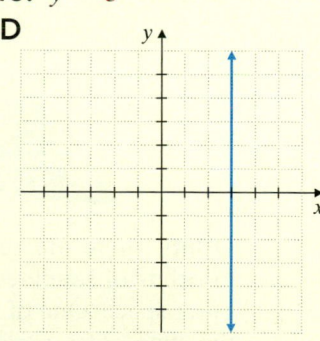

Concept Extensions

47. Draw a graph whose domain is $(-\infty, 5]$ and whose range is $[2, \infty)$.

48. In your own words, describe how to graph a piecewise-defined function.

49. Graph: $f(x) = \begin{cases} -\dfrac{1}{2}x & \text{if } x \leq 0 \\ x + 1 & \text{if } 0 < x \leq 2 \\ 2x - 1 & \text{if } x > 2 \end{cases}$

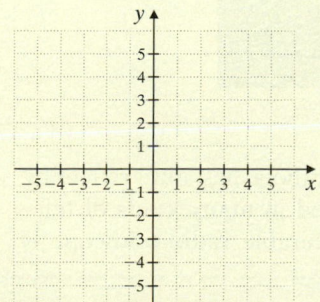

Uses of Parabolas

In this chapter, we learned that the graph of a quadratic equation in two variables of the form $y = ax^2 + bx + c$ is a shape called a **parabola.** The figure to the right shows the general shape of a parabola.

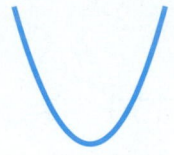

The shape of a parabola shows up in many situations, both natural and human-made, in the world around us.

Natural Situations

- **Hurricanes** The paths of many hurricanes are roughly shaped like a parabola. In the northern hemisphere, hurricanes generally begin moving to the northwest. Then, as they move farther from the equator, they swing around to head in a northeastern direction.

- **Projectiles** The force of the Earth's gravity acts on a projectile launched into the air. The resulting path of the projectile, anything from a bullet to a football, is generally shaped like a parabola.

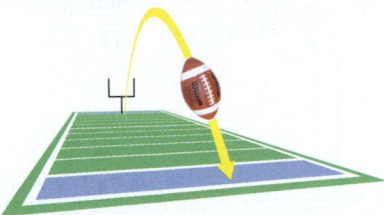

- **Orbits** There are several different possible shapes of orbits of satellites, planets, moons, and comets in outer space. One of the possible types of orbits is in the shape of a parabola. A parabolic orbit is most often seen with comets.

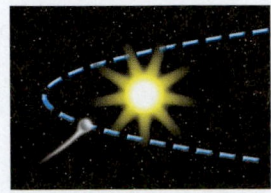

Human-Made Situations

- **Telescopes** Because a parabola has nice reflecting properties, its shape is used in many kinds of telescopes. The largest nonsteerable radio telescope is

Arecibo Observatory in Puerto Rico. This telescope consists of a huge parabolic dish built into a valley. The dish is about 1000 feet across.

- **Training Astronauts** Astronauts must be able to work in zero-gravity conditions on missions in space. However, it's nearly impossible to escape the force of gravity on Earth. To help astronauts train to work in weightlessness, a specially modified jet can be flown in a parabolic path. At the top of the parabola, weightlessness can be simulated for up to 30 seconds at a time.

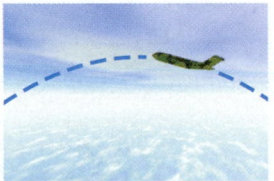

- **Architecture** The reinforced concrete arches used in many modern buildings are based on the shape of a parabola.

- **Music** The design of the modern flute incorporates a parabolic head joint.

Group Activity

There are many other physical applications of parabolas. For example, satellite dishes often have parabolic shapes. Choose a physical example of a parabola given here or use one of your own and write a report (with diagrams).

Chapter 16 Vocabulary Check

Fill in each blank with one of the words or phrases listed below. Some choices may be used more than once and some may not be used at all.

square root	vertex	one	parabola
completing the square	quadratic	zero	

1. If $x^2 = a$, then $x = \sqrt{a}$ or $x = -\sqrt{a}$. This property is called the _____ property.

2. The graph of $y = x^2$ is called a(n) _____.

3. The formula $x = \dfrac{-b}{2a}$ where $y = ax^2 + bx + c$ is called the _____ formula.

4. The process of solving a quadratic equation by writing it in the form $(x + a)^2 = c$ is called _____.

5. The formula $x = \dfrac{-b \pm \sqrt{b^2 - 4ac}}{2a}$ is called the _____ formula.

6. The lowest point on a parabola that opens upward is called the _____.

7. The zero-factor property states that if the product of two numbers is zero, then at least one of the two numbers is _____.

> **Helpful Hint**
> ▶ Are you preparing for your test? Don't forget to take the Chapter 16 Test on page 1216. Then check your answers at the back of the text and use the Chapter Test Prep Videos to see the fully worked-out solutions to any of the exercises you want to review.

16 Chapter Highlights

Definitions and Concepts	Examples
Section 16.1 Solving Quadratic Equations by the Square Root Property	

Square Root Property

If $x^2 = a$ for $a \geq 0$, then $x = \sqrt{a}$ or $x = -\sqrt{a}$.

Solve the equation.

$$(x - 1)^2 = 15$$
$$x - 1 = \sqrt{15} \qquad \text{or} \quad x - 1 = -\sqrt{15}$$
$$x = 1 + \sqrt{15} \qquad\qquad x = 1 - \sqrt{15}$$

| **Section 16.2   Solving Quadratic Equations by Completing the Square** | |

To Solve a Quadratic Equation by Completing the Square

Step 1: If the coefficient of x^2 is not 1, divide both sides of the equation by the coefficient.

Step 2: Get all terms with variables alone on one side.

Step 3: Complete the square by adding the square of half of the coefficient of x to both sides.

Step 4: Factor the perfect square trinomial.

Step 5: Use the square root property to solve.

Solve $2x^2 + 12x - 10 = 0$ by completing the square.

$$\frac{2x^2}{2} + \frac{12x}{2} - \frac{10}{2} = \frac{0}{2} \quad \text{Divide by 2.}$$
$$x^2 + 6x - 5 = 0 \quad \text{Simplify.}$$
$$x^2 + 6x = 5 \quad \text{Add 5.}$$

The coefficient of x is 6. Half of 6 is 3 and $3^2 = 9$. Add 9 to both sides.

$$x^2 + 6x + 9 = 5 + 9$$
$$(x + 3)^2 = 14 \quad \text{Factor.}$$
$$x + 3 = \sqrt{14} \qquad \text{or} \quad x + 3 = -\sqrt{14}$$
$$x = -3 + \sqrt{14} \qquad\qquad x = -3 - \sqrt{14}$$

Definitions and Concepts	Examples

Section 16.3   Solving Quadratic Equations by the Quadratic Formula

Quadratic Formula

If a, b, and c are real numbers and $a \neq 0$, the quadratic equation $ax^2 + bx + c = 0$ has solutions

$$x = \frac{-b \pm \sqrt{b^2 - 4ac}}{2a}$$

To Solve a Quadratic Equation by the Quadratic Formula

Step 1: Write the equation in standard form: $ax^2 + bx + c = 0$.

Step 2: If necessary, clear the equation of fractions.

Step 3: Identify a, b, and c.

Step 4: Replace a, b, and c in the quadratic formula with the identified values, and simplify.

Identify a, b, and c in the quadratic equation

$$4x^2 - 6x = 5$$

First, subtract 5 from both sides.

$$4x^2 - 6x - 5 = 0$$

$a = 4$, $b = -6$, and $c = -5$

Solve $3x^2 - 2x - 2 = 0$.

In this equation, $a = 3$, $b = -2$, and $c = -2$.

$$x = \frac{-(-2) \pm \sqrt{(-2)^2 - 4(3)(-2)}}{2 \cdot 3}$$

$$= \frac{2 \pm \sqrt{4 - (-24)}}{6}$$

$$= \frac{2 \pm \sqrt{28}}{6} = \frac{2 \pm \sqrt{4 \cdot 7}}{6} = \frac{2 \pm 2\sqrt{7}}{6}$$

$$= \frac{2(1 \pm \sqrt{7})}{2 \cdot 3} = \frac{1 \pm \sqrt{7}}{3}$$

Section 16.4   Graphing Quadratic Equations in Two Variables

The graph of a quadratic equation $y = ax^2 + bx + c$, $a \neq 0$, is called a **parabola**. The lowest point on a parabola opening upward or the highest point on a parabola opening downward is called the **vertex**. The vertical line through the vertex is the **line of symmetry**.

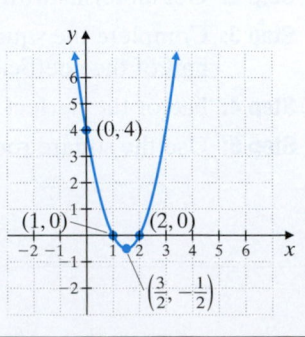

Vertex Formula

The vertex of the parabola $y = ax^2 + bx + c$ has x-coordinate $\dfrac{-b}{2a}$. To find the corresponding y-coordinate, substitute the x-coordinate into the original equation and solve for y.

Graph: $y = 2x^2 - 6x + 4$

The x-coordinate of the vertex is

$$x = \frac{-b}{2a} = \frac{-(-6)}{2(2)} = \frac{6}{4} = \frac{3}{2}$$

The y-coordinate is

$$y = 2\left(\frac{3}{2}\right)^2 - 6\left(\frac{3}{2}\right) + 4 = 2\left(\frac{9}{4}\right) - 9 + 4 = -\frac{1}{2}$$

The vertex is $\left(\dfrac{3}{2}, -\dfrac{1}{2}\right)$.

The y-intercept is

$$y = 2 \cdot 0^2 - 6 \cdot 0 + 4 = 4$$

The x-intercepts are

$$0 = 2x^2 - 6x + 4$$

$$0 = 2(x - 2)(x - 1)$$

$$x = 2 \quad \text{or} \quad x = 1$$

Definitions and Concepts	Examples

Section 16.5 Interval Notation, Finding Domains and Ranges from Graphs, and Graphing Piecewise-Defined Functions

To find the domain of a function (or relation) from its graph, recall that on the rectangular coordinate system, "domain" means the x-values that are graphed. Similarly, "range" means the y-values that are graphed.	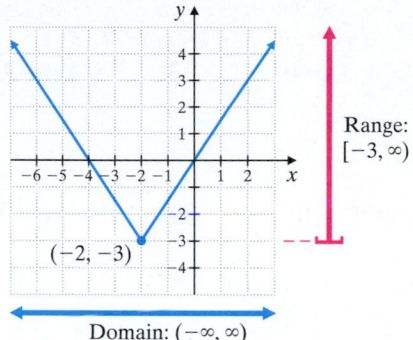

Chapter 16 Review

(16.1) *Solve each quadradic equation by factoring.*

1. $x^2 - 121 = 0$

2. $y^2 - 100 = 0$

3. $3m^2 - 5m = 2$

4. $7m^2 + 2m = 5$

Use the square root property to solve each quadratic equation.

5. $x^2 = 36$

6. $x^2 = 81$

7. $k^2 = 50$

8. $k^2 = 45$

9. $(x - 11)^2 = 49$

10. $(x + 3)^2 = 100$

11. $(4p + 5)^2 = 41$

12. $(3p + 7)^2 = 37$

Solve. For Exercises 13 and 14, use the formula $h = 16t^2$, where h is the height in feet at time t seconds.

13. If Kara Washington dives from a height of 100 feet, how long before she hits the water?

14. How long does a 5-mile free fall take? Round your result to the nearest tenth of a second. (*Hint:* 1 mi = 5280 ft)

(16.2) *Solve each quadratic equation by completing the square.*

15. $x^2 - 9x = -8$

16. $x^2 + 8x = 20$

17. $x^2 + 4x = 1$

18. $x^2 - 8x = 3$

19. $x^2 - 6x + 7 = 0$

20. $x^2 + 6x + 7 = 0$

21. $2y^2 + y - 1 = 0$

22. $4y^2 + 3y - 1 = 0$

(16.3) *Use the quadratic formula to solve each quadratic equation.*

23. $9x^2 + 30x + 25 = 0$ **24.** $16x^2 - 72x + 81 = 0$ **25.** $7x^2 = 35$ **26.** $11x^2 = 33$

27. $x^2 - 10x + 7 = 0$ **28.** $x^2 + 4x - 7 = 0$ **29.** $3x^2 + x - 1 = 0$ **30.** $x^2 + 3x - 1 = 0$

31. $2x^2 + x + 5 = 0$ **32.** $7x^2 - 3x + 1 = 0$

For the exercise numbers given, approximate the exact solutions to the nearest tenth.

33. Exercise 29   **34.** Exercise 30

35. The annual number of visitors y (in thousands) to Yosemite National Park in California can be modeled by the equation $y = -66x^2 + 368x + 3432$. In this equation, x is the number of years since 2008. Assume that this trend continues and find the year after 2008 in which 3264 thousand people visit Yosemite National Park. (*Source:* Based on data from the National Park Service)

36. The amount y of electricity generated by wind power (in thousand megawatt hours) in the United States from 2008 through 2012 can be modeled by the equation $y = 538x^2 + 19,421x + 54,762$, where x represents the number of years after 2008. Assume that this trend continues and find the year after 2008 in which the amount of electricity generated by wind power is 302,772 thousand megawatt hours. (*Source:* Based on information from the Energy Information Administration)

(16.4) *Graph each quadratic equation and find and plot any intercepts.*

37. $y = 5x^2$

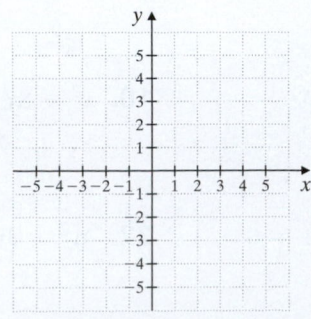

38. $y = -\dfrac{1}{2}x^2$

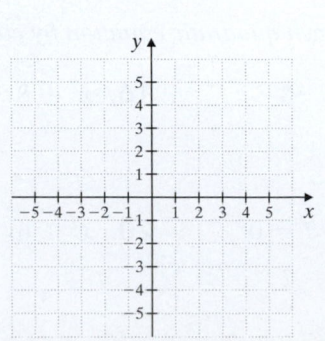

Graph each quadratic equation. Label the vertex and the intercepts with their coordinates.

39. $y = x^2 - 25$

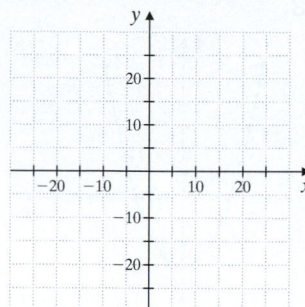

40. $y = x^2 - 36$

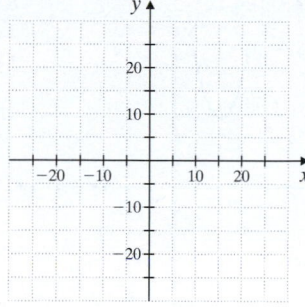

41. $y = x^2 + 3$

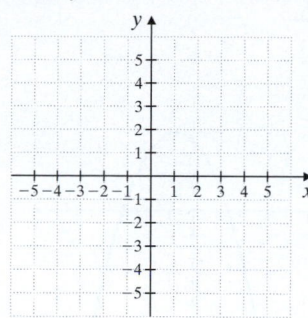

42. $y = x^2 + 8$

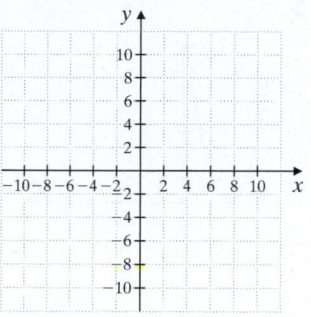

43. $y = -4x^2 + 8$

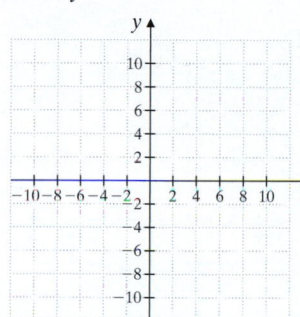

44. $y = -3x^2 + 9$

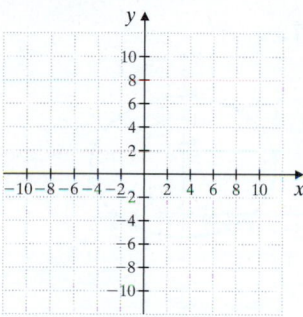

45. $y = x^2 + 3x - 10$

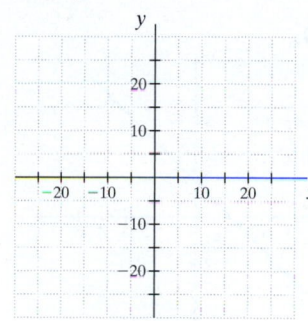

46. $y = x^2 + 3x - 4$

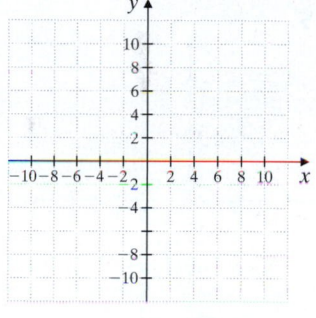

47. $y = -x^2 - 5x - 6$

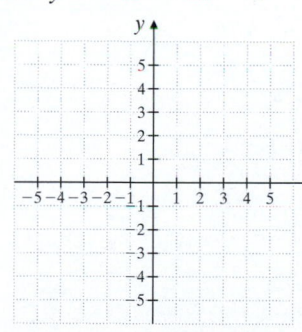

48. $y = 3x^2 - x - 2$

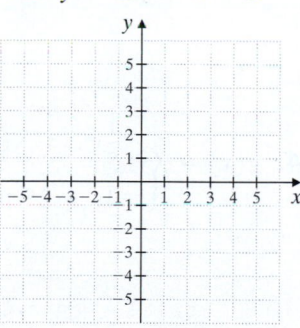

49. $y = 2x^2 - 11x - 6$

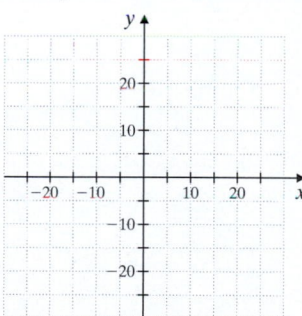

50. $y = -x^2 + 4x + 8$

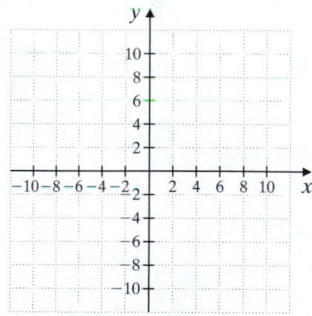

Match each quadratic equation with its graph.

51. $y = 2x^2$

A

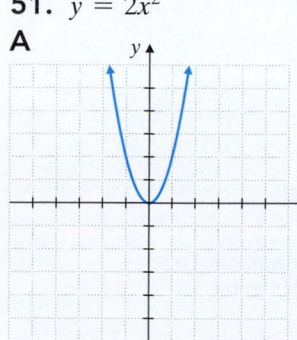

52. $y = -x^2$

B

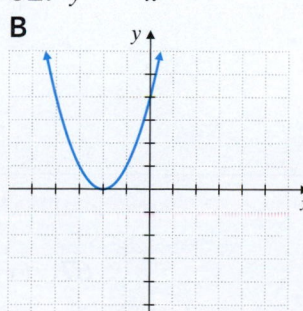

53. $y = x^2 + 4x + 4$

C

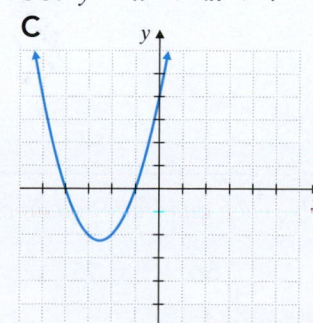

54. $y = x^2 + 5x + 4$

D

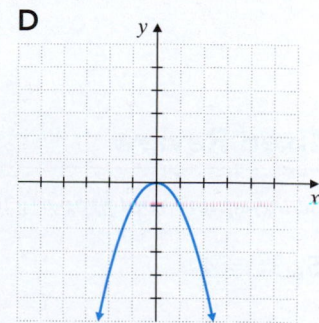

Quadratic equations in the form $y = ax^2 + bx + c$ are graphed below. Determine the number of real solutions for the related equation $0 = ax^2 + bx + c$ from each graph.

55.

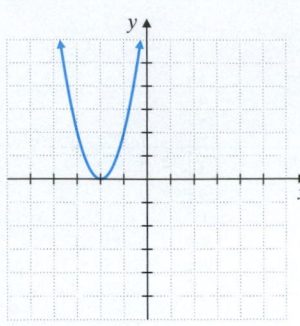

56.

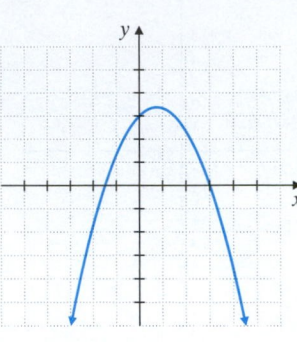

57.

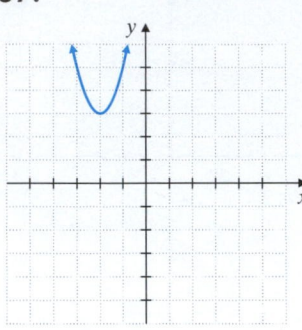

58.

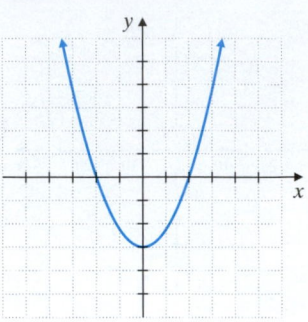

(16.5) *Find the domain and range of each relation. Use interval notation to write your answers.*

59.

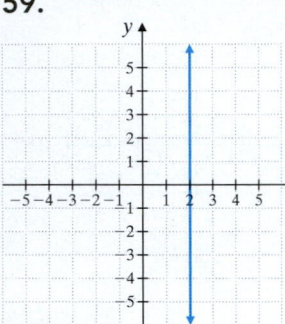

60.

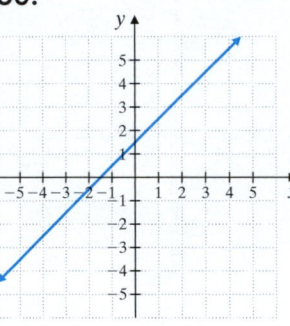

61.

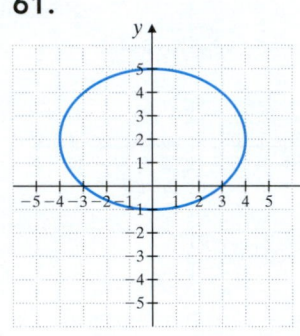

62.

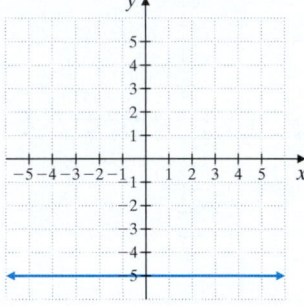

Graph each function.

63. $f(x) = \begin{cases} -3x & \text{if } x < 0 \\ x - 3 & \text{if } x \geq 0 \end{cases}$

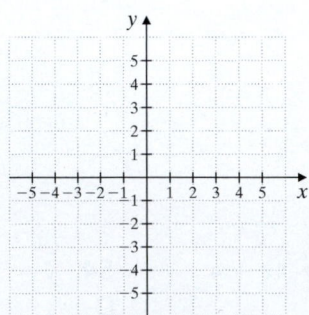

64. $g(x) = \begin{cases} -\dfrac{1}{5}x & \text{if } x \leq -1 \\ -4x + 2 & \text{if } x > -1 \end{cases}$

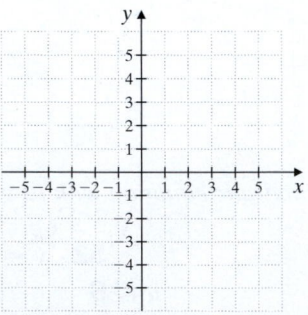

Mixed Review

Use the square root property to solve each quadratic equation.

65. $x^2 = 49$

66. $y^2 = 75$

67. $(x - 7)^2 = 64$

Solve each quadratic equation by completing the square.

68. $x^2 + 4x = 6$

69. $3x^2 + x = 2$

70. $4x^2 - x - 2 = 0$

Use the quadratic formula to solve each quadratic equation.

71. $4x^2 - 3x - 2 = 0$

72. $5x^2 + x - 2 = 0$

73. $4x^2 + 12x + 9 = 0$

74. $2x^2 + x + 4 = 0$

Graph each quadratic equation. Label the vertex and the intercepts with their coordinates.

75. $y = 4 - x^2$

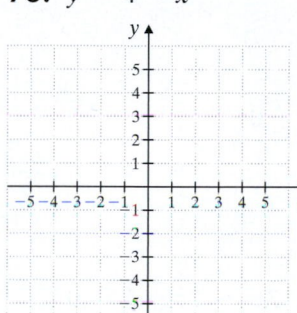

76. $y = x^2 + 4$

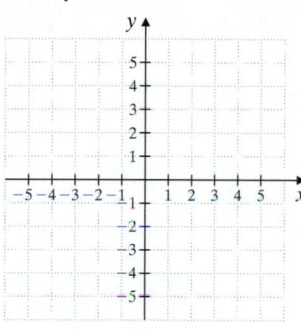

77. $y = x^2 + 6x + 8$

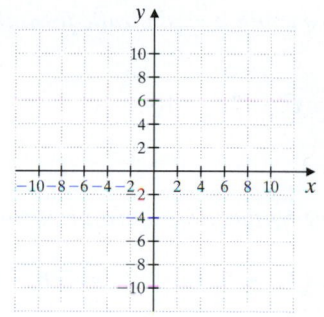

78. $y = x^2 - 2x - 4$

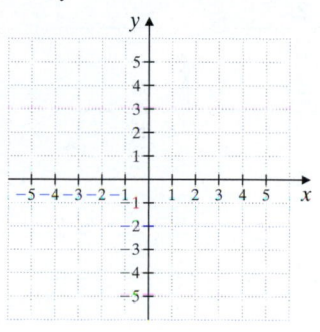

Chapter 16   Test

Step-by-step test solutions are found on the Chapter Test Prep Videos. Where available: MyMathLab® or YouTube™

Answers

Solve by factoring.

1. $x^2 - 400 = 0$

2. $2x^2 - 11x = 21$

Solve using the square root property.

3. $5k^2 = 80$

4. $(3m - 5)^2 = 8$

Solve by completing the square.

5. $x^2 - 26x + 160 = 0$

6. $3x^2 + 12x - 4 = 0$

Solve using the quadratic formula.

7. $x^2 - 3x - 10 = 0$

8. $p^2 - \dfrac{5}{3}p - \dfrac{1}{3} = 0$

Solve by the most appropriate method.

9. $(3x - 5)(x + 2) = -6$ **10.** $(3x - 1)^2 = 16$ **11.** $3x^2 - 7x - 2 = 0$

12. $x^2 - 4x - 5 = 0$ **13.** $3x^2 - 7x + 2 = 0$ **14.** $2x^2 - 6x + 1 = 0$

△ **15.** The height of a triangle is 4 times the length of the base. The area of the triangle is 18 square feet. Find the height and base of the triangle.

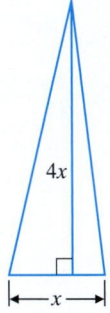

Graph each quadratic equation. Label the vertex and the intercept points with their coordinates.

16. $y = -5x^2$

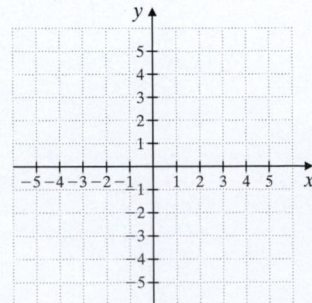

17. $y = x^2 - 4$

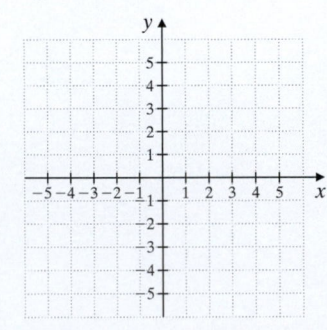

1. _____

2. _____

3. _____

4. _____

5. _____

6. _____

7. _____

8. _____

9. _____

10. _____

11. _____

12. _____

13. _____

14. _____

15. _____

16. _____

17. _____

18. $y = x^2 - 7x + 10$

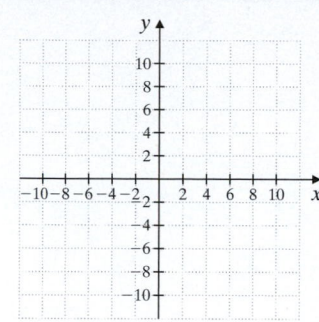

19. $y = 2x^2 + 4x - 1$

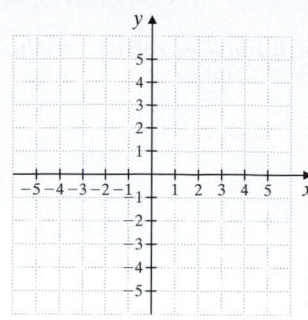

18. _____

19. _____

△ **20.** The number of diagonals d that a polygon with n sides has is given by the formula

$$d = \frac{n^2 - 3n}{2}$$

Find the number of sides of a polygon if it has 9 diagonals.

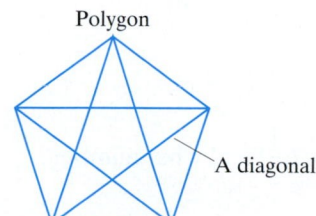

Polygon

A diagonal

20. _____

Solve.

21. The highest dive from a diving board by a woman was made by Lucy Wardle of the United States. She dove from a height of 120.75 feet at Ocean Park, Hong Kong, in 1985. To the nearest tenth of a second, how long did the dive take? Use the formula $h = 16t^2$.

21. _____

Find the domain and range of each relation. Also determine whether the relation is a function.

22.

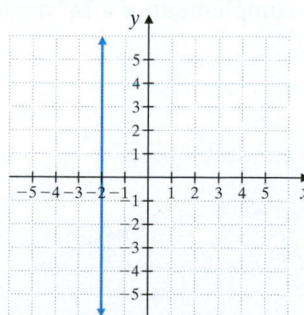

23.

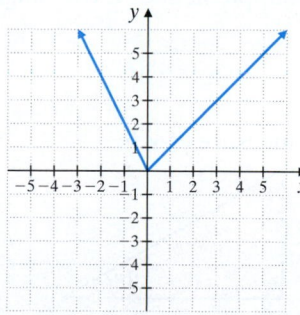

22. _____

23. _____

Graph the function. State the domain and the range of the function.

24. $f(x) = \begin{cases} -\dfrac{1}{2}x & \text{if } x \le 0 \\ 2x - 3 & \text{if } x > 0 \end{cases}$

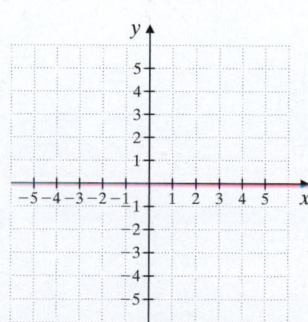

24. _____

Answers

Write each fraction, mixed, or whole number as a percent.

1. $\dfrac{9}{20}$

2. $\dfrac{53}{50}$

3. $1\dfrac{1}{2}$

4. 5

Solve.

5. 13 is $6\dfrac{1}{2}$% of what number?

6. What is 110% of 220?

7. Translate to a proportion. 101 is what percent of 200?

8. Translate to an equation. 101 is what percent of 200?

9. Ivan Borski borrowed $2400 at 10% simple interest for 8 months to buy a used Toyota Corolla. Find the simple interest he paid.

10. C. J. Dufour wants to buy a digital camera. She has $762 in her savings account. If the camera costs $237, how much money will she have in her account after buying the camera?

11. Find the supplement of a 107° angle.

12. Find the complement of a 34° angle.

△13. Find the measure of ∠b.

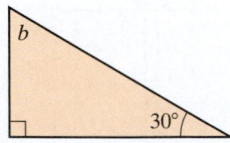

△14. Find the measure of ∠a.

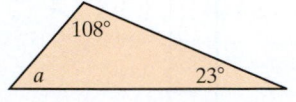

Solve each equation.

15. $y + 0.6 = -1.0$

16. $8x - 14 = 6x - 20$

17. $8(2 - t) = -5t$

18. $2(x + 7) = 5(2x - 3)$

---

1. _____
2. _____
3. _____
4. _____
5. _____
6. _____
7. _____
8. _____
9. _____
10. _____
11. _____
12. _____
13. _____
14. _____
15. _____
16. _____
17. _____
18. _____

19. As of May 2014, the total number of Democrats and Republicans in the U.S. House of Representatives was 432. There were 34 more Republican representatives than Democratic. Find the number of representatives from each party. (*Source:* Congressional Research Service)

20. The sum of three consecutive integers is 438. Find the integers.

Simplify the following expressions.

21. 3^0

22. $\left(\dfrac{-6x}{y^3}\right)^3$

23. $(5x^3y^2)^0$

24. $\dfrac{a^2b^7}{(2b^2)^5}$

25. -4^0

26. $\dfrac{(3y)^2}{y^2}$

27. Multiply: $(3y + 2)^2$

28. Multiply: $(x^2 + 5)(y - 1)$

29. Divide $x^2 + 7x + 12$ by $x + 3$ using long division.

30. Simplify by combining like terms: $2 + 8.1a + a - 6$

31. Factor: $r^2 - r - 42$

32. Find the value of each expression when $x = -4$ and $y = 7$.
   a. $\dfrac{x - y}{7 - x}$ b. $x^2 + 2y$

33. Factor: $10x^2 - 13xy - 3y^2$

34. Add: $\dfrac{1}{x + 2} + \dfrac{7}{x - 1}$

35. Factor $8x^2 - 14x + 5$ by grouping.

36. Multiply: $\dfrac{x^2 + 7x}{5x} \cdot \dfrac{10x + 25}{x^2 - 49}$

37. Factor each difference of two squares.
   a. $4x^3 - 49x$ b. $162x^4 - 2$

19. _____

20. _____

21. _____

22. _____

23. _____

24. _____

25. _____

26. _____

27. _____

28. _____

29. _____

30. _____

31. _____

32. a. _____

    b. _____

33. _____

34. _____

35. _____

36. _____

37. a. _____

    b. _____

38. Solve $\dfrac{2x + 7}{3} = \dfrac{x - 6}{2}$.

39. Solve: $(5x - 1)(2x^2 + 15x + 18) = 0$

40. Simplify each expression by combining like terms.
   a. $4x - 3 + 7 - 5x$
   b. $-6y + 3y - 8 + 8y$
   c. $7 + 10.1a - a - 11$
   d. $2x^2 - 2x$

41. Simplify: $\dfrac{x^2 + 8x + 7}{x^2 - 4x - 5}$

42. Solve $2x^2 + 5x = 7$.

43. The quotient of a number and 6, minus $\dfrac{5}{3}$, is the quotient of the number and 2. Find the number.

44. Find the distance between $(-2, 8)$ and $(4, 6)$. (See the Chapter 15 Group Activity.)

45. Complete the table for the equation $y = 3x$.

	x	y
a.	-1	
b.		0
c.		-9

46. Identify the x- and y-intercepts.

   a.

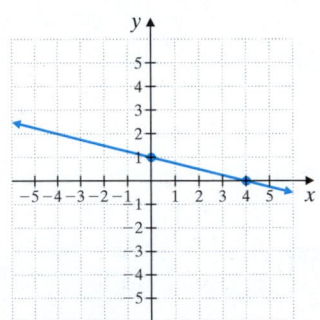

   b.

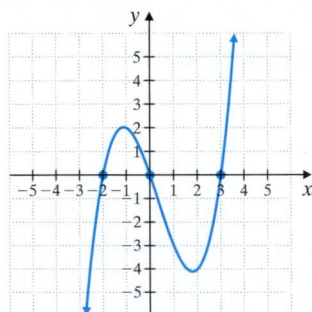

47. Determine whether each pair of lines is parallel, perpendicular, or neither.

   a. $y = -\dfrac{1}{5}x + 1$
   $2x + 10y = 3$

   b. $x + y = 3$
   $-x + y = 4$

   c. $3x + y = 5$
   $2x + 3y = 6$

48. Determine whether the graphs of $y = 3x + 7$ and $x + 3y = -15$ are parallel lines, perpendicular lines, or neither.

49. Which of the following relations is also a function?
   a. $\{(-1, 1), (2, 3), (7, 3), (8, 6)\}$
   b. $\{(0, -2), (1, 5), (0, 3), (7, 7)\}$

38. _____

39. _____

40. a. _____

   b. _____

   c. _____

   d. _____

41. _____

42. _____

43. _____

44. _____

45. _____

46. a. _____

   b. _____

47. a. _____

   b. _____

   c. _____

48. _____

49. a. _____

   b. _____

50. Add or subtract by first simplifying each radical.

    a. $\sqrt{80} + \sqrt{20}$

    b. $2\sqrt{98} - 2\sqrt{18}$

    c. $\sqrt{32} + \sqrt{121} - \sqrt{12}$

51. Solve the system: $\begin{cases} 2x + y = 10 \\ x = y + 2 \end{cases}$

52. Solve the system: $\begin{cases} 5x + y = 3 \\ y = -5x \end{cases}$

53. Solve the system: $\begin{cases} 2x - y = 7 \\ 8x - 4y = 1 \end{cases}$

54. Solve the system: $\begin{cases} -2x + y = 7 \\ 6x - 3y = -21 \end{cases}$

Find each square root.

55. $\sqrt{36}$

56. $\sqrt{\dfrac{4}{25}}$

57. $\sqrt{\dfrac{9}{100}}$

58. $\sqrt{\dfrac{16}{121}}$

59. Rationalize the denominator of $\dfrac{2}{1 + \sqrt{3}}$.

60. Rationalize the denominator of $\dfrac{5}{\sqrt{8}}$

61. Use the square root property to solve $(x - 3)^2 = 16$.

62. Use the square root property to solve $3(x - 4)^2 = 9$.

63. Solve $\dfrac{1}{2}x^2 - x = 2$ using the quadratic formula.

64. Solve $x^2 + 4x = 8$ using the quadratic formula.

50. a. _____

   b. _____

   c. _____

51. _____

52. _____

53. _____

54. _____

55. _____

56. _____

57. _____

58. _____

59. _____

60. _____

61. _____

62. _____

63. _____

64. _____

Further Algebraic Topics

A.1 Factoring Sums and Differences of Cubes

Objective

A Factor Sums and Differences of Cubes.

Objective A Factoring Sums and Differences of Cubes

Although the sum of two squares usually does not factor, the sum or difference of two cubes can be factored and reveal factoring patterns. The pattern for the sum of cubes can be checked by multiplying the binomial $x + y$ and the trinomial $x^2 - xy + y^2$. The pattern for the difference of two cubes can be checked by multiplying the binomial $x - y$ by the trinomial $x^2 + xy + y^2$.

> **Sum or Difference of Two Cubes**
>
> $$a^3 + b^3 = (a + b)(a^2 - ab + b^2)$$
> $$a^3 - b^3 = (a - b)(a^2 + ab + b^2)$$

Practice 1

Factor $x^3 + 27$

Example 1 Factor $x^3 + 8$.

Solution: First, write the binomial in the form $a^3 + b^3$.

$$x^3 + 8 = x^3 + 2^3 \quad \text{Write in the form } a^3 + b^3.$$

If we replace a with x and b with 2 in the formula above, we have

$$x^3 + 2^3 = (x + 2)[x^2 - (x)(2) + 2^2]$$
$$= (x + 2)(x^2 - 2x + 4)$$

■ **Work Practice 1**

Helpful Hint

When factoring sums or differences of cubes, notice the sign patterns.

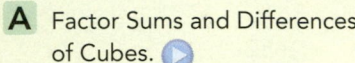

Answer

1. $(x + 3)(x^2 - 3x + 9)$

Example 2 Factor $y^3 - 27$.

Solution:

$$y^3 - 27 = y^3 - 3^3 \qquad \text{Write in the form } a^3 - b^3.$$
$$= (y - 3)[y^2 + (y)(3) + 3^2]$$
$$= (y - 3)(y^2 + 3y + 9)$$

■ **Work Practice 2**

Example 3 Factor $64x^3 + 1$.

Solution:

$$64x^3 + 1 = (4x)^3 + 1^3$$
$$= (4x + 1)[(4x)^2 - (4x)(1) + 1^2]$$
$$= (4x + 1)(16x^2 - 4x + 1)$$

■ **Work Practice 3**

Example 4 Factor $54a^3 - 16b^3$.

Solution: Remember to factor out common factors first before using other factoring methods.

$$54a^3 - 16b^3 = 2(27a^3 - 8b^3) \qquad \text{Factor out the GCF, 2.}$$
$$= 2[(3a)^3 - (2b)^3] \qquad \text{Difference of two cubes}$$
$$= 2(3a - 2b)[(3a)^2 + (3a)(2b) + (2b)^2]$$
$$= 2(3a - 2b)(9a^2 + 6ab + 4b^2)$$

■ **Work Practice 4**

Practice 2

Factor $y^3 - 8$

Practice 3

Factor $125z^3 + 1$

Practice 4

Factor $16a^3 - 250b^3$

Answers

2. $(y - 2)(y^2 + 2y + 4)$
3. $(5z + 1)(25z^2 - 5z + 1)$
4. $2(2a - 5b)(4a^2 + 10ab + 25b^2)$

A.1 **Exercise Set** MyMathLab®

Factor the binomials completely. See Examples 1 through 4.

1. $a^3 + 27$

2. $b^3 - 8$

3. $8a^3 + 1$

4. $64x^3 - 1$

5. $5k^3 + 40$

6. $6r^3 - 162$

7. $x^3y^3 - 64$

8. $8x^3 - y^3$

9. $x^3 + 125$

10. $a^3 - 216$

11. $24x^4 - 81xy^3$

12. $375y^6 - 24y^3$

13. $27 - t^3$

14. $125 + r^3$

15. $8r^3 - 64$

16. $54r^3 + 2$

17. $t^3 - 343$

18. $s^3 + 216$

19. $s^3 - 64t^3$

20. $8t^3 + s^3$

Objectives

A Find the Intersection of Two Sets.

B Solve Compound Inequalities Containing "and."

C Find the Union of Two Sets.

D Solve Compound Inequalities Containing "or."

Two inequalities joined by the words **and** or **or** are called **compound inequalities.**

Compound Inequalities

$x + 3 < 8$ and $x > 2$

$\dfrac{2x}{3} \geq 5$ or $-x + 10 < 7$

Objective A Finding the Intersection of Two Sets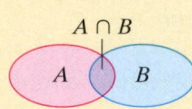

The solution set of a compound inequality formed by the word **and** is the *intersection* of the solution sets of the two inequalities. We use the symbol $\cap$ to denote "intersection."

> ### Intersection of Two Sets
>
> The **intersection** of two sets, A and B, is the set of all elements common to both sets. A intersect B is denoted by
>
> $A \cap B$
>
> A B

Practice 1

Find the intersection:

$\{1, 2, 3, 4, 5\} \cap \{3, 4, 5, 6\}$

Example 1   Find the intersection: $\{2, 4, 6, 8\} \cap \{3, 4, 5, 6\}$

Solution: The numbers 4 and 6 are in both sets. The intersection is $\{4, 6\}$.

■ **Work Practice 1**

Objective B Solving Compound Inequalities Containing "and"

A value is a **solution** of a compound inequality formed by the word **and** if it is a solution of *both* inequalities. For example, the solution set of the compound inequality $x \leq 5$ **and** $x \geq 3$ contains all values of x that make the inequality $x \leq 5$ a true statement **and** the inequality $x \geq 3$ a true statement. The first graph shown below is the graph of $x \leq 5$, the second graph is the graph of $x \geq 3$, and the third graph shows the intersection of the two graphs. The third graph is the graph of $x \leq 5$ **and** $x \geq 3$.

$\{x \mid x \leq 5\}$ 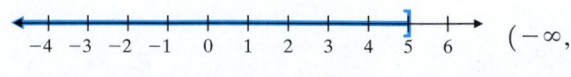 $(-\infty, 5]$

$\{x \mid x \geq 3\}$ 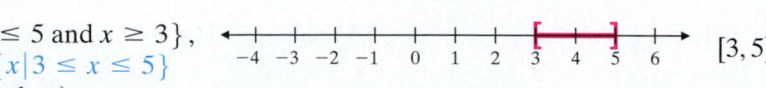 $[3, \infty)$

$\{x \mid x \leq 5 \text{ and } x \geq 3\}$, also $\{x \mid 3 \leq x \leq 5\}$ (see below) $[3, 5]$

Since $x \geq 3$ is the same as $3 \leq x$, the compound inequality $3 \leq x$ and $x \leq 5$ can be written in a more compact form as $3 \leq x \leq 5$. The solution set $\{x \mid 3 \leq x \leq 5\}$ includes all numbers that are greater than or equal to 3 and at the same time less than or equal to 5.

In interval notation, the set $\{x \mid x \leq 5 \text{ and } x \geq 3\}$ or $\{x \mid 3 \leq x \leq 5\}$ is written as $[3, 5]$.

Answer

1. $\{3, 4, 5\}$

Example 2   Solve: $x - 7 < 2$ and $2x + 1 < 9$

Solution: First we solve each inequality separately.

$$x - 7 < 2 \quad \text{and} \quad 2x + 1 < 9$$
$$x < 9 \quad \text{and} \quad 2x < 8$$
$$x < 9 \quad \text{and} \quad x < 4$$

Now we can graph the two intervals on two number lines and find their intersection.

$\{x | x < 9\}$ $(-\infty, 9)$

$\{x | x < 4\}$ $(-\infty, 4)$

$\{x | x < 9 \text{ and } x < 4\}$
$= \{x | x < 4\}$ $(-\infty, 4)$

The solution set is $(-\infty, 4)$.

■ **Work Practice 2**

Example 3   Solve: $2x \geq 0$ and $4x - 1 \leq -9$

Solution: First we solve each inequality separately.

$$2x \geq 0 \quad \text{and} \quad 4x - 1 \leq -9$$
$$x \geq 0 \quad \text{and} \quad 4x \leq -8$$
$$x \geq 0 \quad \text{and} \quad x \leq -2$$

Now we can graph the two intervals and find their intersection.

$\{x | x \geq 0\}$ $[0, \infty)$

$\{x | x \leq -2\}$ $(-\infty, -2]$

$\{x | x \geq 0 \text{ and } x \leq -2\} = \varnothing$

There is no number that is greater than or equal to 0 **and** less than or equal to -2. The solution set is $\varnothing$.

■ **Work Practice 3**

Practice 2

Solve: $x + 5 < 9$ and $3x - 1 < 2$

Practice 3

Solve: $4x \geq 0$ and $2x + 4 \leq 2$

Answers

2. $(-\infty, 1)$ **3.** $\varnothing$

To solve a compound inequality like $2 < 4 - x < 7$, we get x alone in the middle. Since a compound inequality is really two inequalities in one statement, we must perform the same operation to all three parts of the inequality.

Practice 4

Solve: $5 < 1 - x < 9$

Example 4   Solve: $2 < 4 - x < 7$

Solution: To get x alone, we first subtract 4 from all three parts.

$$2 < 4 - x < 7$$
$$2 - 4 < 4 - x - 4 < 7 - 4 \quad \text{Subtract 4 from all three parts.}$$
$$-2 < -x < 3 \quad \text{Simplify.}$$
$$\frac{-2}{-1} > \frac{-x}{-1} > \frac{3}{-1} \quad \text{Divide all three parts by } -1 \text{ and reverse the inequality symbols.}$$
$$2 > x > -3$$

> **Helpful Hint**
> Don't forget to reverse both inequality symbols.

This is equivalent to $-3 < x < 2$, and its graph is shown.

The solution set in interval notation is $(-3, 2)$.

■ **Work Practice 4**

Practice 5

Solve: $-2 < \frac{3}{4}x + 2 \le 5$

Example 5   Solve: $-1 \le \frac{2}{3}x + 5 < 2$

Solution: First we clear the inequality of fractions by multiplying all three parts by the LCD, 3.

$$-1 \le \frac{2}{3}x + 5 < 2$$
$$3(-1) \le 3\left(\frac{2}{3}x + 5\right) < 3(2) \quad \text{Multiply all three parts by the LCD, 3.}$$
$$-3 \le 2x + 15 < 6 \quad \text{Use the distributive property and multiply.}$$
$$-3 - 15 \le 2x + 15 - 15 < 6 - 15 \quad \text{Subtract 15 from all three parts.}$$
$$-18 \le 2x < -9 \quad \text{Simplify.}$$
$$\frac{-18}{2} \le \frac{2x}{2} < \frac{-9}{2} \quad \text{Divide all three parts by 2.}$$
$$-9 \le x < -\frac{9}{2} \quad \text{Simplify.}$$

The graph of the solution is shown.

The solution set in interval notation is $\left[-9, -\frac{9}{2}\right)$.

■ **Work Practice 5**

Objective C Finding the Union of Two Sets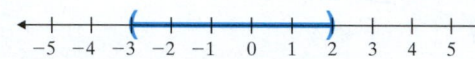

The solution set of a compound inequality formed by the word **or** is the **union** of the solution sets of the two inequalities. We use the symbol ∪ to denote "union."

Answers

4. $(-8, -4)$ **5.** $\left(-\frac{16}{3}, 4\right]$

Union of Two Sets

The **union** of two sets, A and B, is the set of elements that belong to *either* of the sets. A union B is denoted by

A   B

$A \cup B$

Helpful Hint   The word "either" in this definition means "one or the other or both."

Example 6   Find the union: $\{2, 4, 6, 8\} \cup \{3, 4, 5, 6\}$

Solution: The numbers in either set are $\{2, 3, 4, 5, 6, 8\}$. This set is the union.

■ **Work Practice 6**

Practice 6

Find the union:
$\{1, 2, 3, 4, 5\} \cup \{3, 4, 5, 6\}$

Objective D Solving Compound Inequalities Containing "or"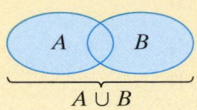

A value of x is a solution of a compound inequality formed by the word **or** if it is a solution of **either** inequality. For example, the solution set of the compound inequality $x \le 1$ **or** $x \ge 3$ contains all numbers that make the inequality $x \le 1$ a true statement **or** the inequality $x \ge 3$ a true statement. In other words, the solution of such an inequality is the *union* of the solutions of the individual inequalities.

$\{x \mid x \le 1\}$ $(-\infty, 1]$

$\{x \mid x \ge 3\}$ $[3, \infty)$

$\{x \mid x \le 1$ or $x \ge 3\}$ $(-\infty, 1] \cup [3, \infty)$

In interval notation, the set $\{x \mid x \le 1 \text{ or } x \ge 3\}$ is written as $(-\infty, 1] \cup [3, \infty)$.

Example 7   Solve: $5x - 3 \le 10 \text{ or } x + 1 \ge 5$

Solution: First we solve each inequality separately.

$$5x - 3 \le 10 \quad \text{or} \quad x + 1 \ge 5$$
$$5x \le 13 \quad \text{or} \quad x \ge 4$$
$$x \le \frac{13}{5} \quad \text{or} \quad x \ge 4$$

Now we can graph each interval and find their union.

$\left\{ x \mid x \le \dfrac{13}{5} \right\}$ $\left(-\infty, \dfrac{13}{5} \right]$

$\{x \mid x \ge 4\}$ $[4, \infty)$

$\left\{ x \mid x \le \dfrac{13}{5} \text{ or } x \ge 4 \right\}$

     $\left(-\infty, \dfrac{13}{5} \right] \cup [4, \infty)$

The solution set is $\left(-\infty, \dfrac{13}{5} \right] \cup [4, \infty)$.

■ **Work Practice 7**

Practice 7

Solve:
$3x - 2 \ge 10 \text{ or } x - 6 \le -4$

Answers

6. $\{1, 2, 3, 4, 5, 6\}$

7. $(-\infty, 2] \cup [4, \infty)$

Practice 8

Solve:

$x - 7 \le -1$ or $2x - 6 \ge 2$

Example 8 Solve: $-2x - 5 < -3$ or $6x < 0$

Solution: First we solve each inequality separately.

$$-2x - 5 < -3 \quad \text{or} \quad 6x < 0$$
$$-2x < 2 \quad \text{or} \quad x < 0$$
$$x > -1 \quad \text{or} \quad x < 0$$

Now we can graph each interval and find their union.

$\{x \mid x > -1\}$  $(-1, \infty)$

$\{x \mid x < 0\}$ 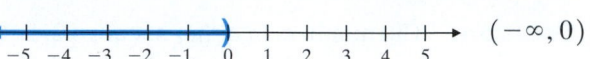 $(-\infty, 0)$

$\{x \mid x > -1 \text{ or } x < 0\}$ $(-\infty, \infty)$
= all real numbers

The solution set is $(-\infty, \infty)$.

■ **Work Practice 8**

Concept Check Which of the following is *not* a correct way to represent the set of all numbers between -3 and 5?

a. $\{x \mid -3 < x < 5\}$

b. $-3 < x$ or $x < 5$

c. $(-3, 5)$

d. $x > -3$ and $x < 5$

Answer

8. $(-\infty, \infty)$

✓ **Concept Check Answer**

b is not correct

A.2 Exercise Set MyMathLab®

Objectives A C Mixed Practice *If $A = \{x \mid x \text{ is an even integer}\}$, $B = \{x \mid x \text{ is an odd integer}\}$, $C = \{2, 3, 4, 5\}$, and $D = \{4, 5, 6, 7\}$, list the elements of each set. See Examples 1 and 6.*

1. $C \cup D$

2. $C \cap D$

3. $A \cap D$

4. $A \cup D$

5. $A \cup B$

6. $A \cap B$

7. $B \cap D$

8. $B \cup D$

9. $B \cup C$

10. $B \cap C$

11. $A \cap C$

12. $A \cup C$

Objective B *Solve each compound inequality. Graph the two inequalities on the first two number lines and the solution set on the third number line. See Examples 2 and 3.*

13. $x < 1$ and $x > -3$

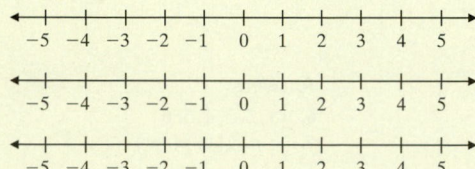

14. $x \le 0$ and $x \ge -2$

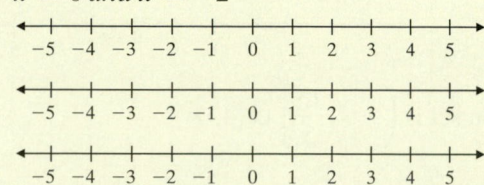

15. $x \le -3$ *and* $x \ge -2$

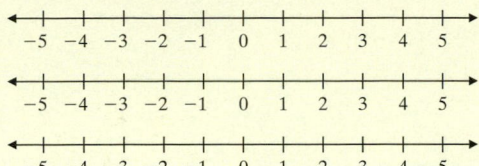

16. $x < 2$ *and* $x > 4$

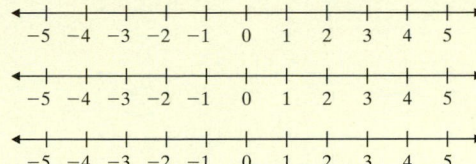

17. $x < -1$ *and* $x < 1$

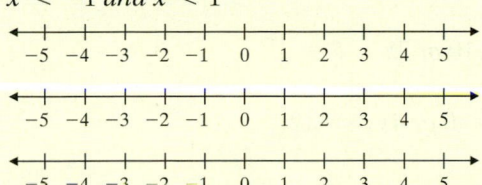

18. $x \ge -4$ *and* $x > 1$

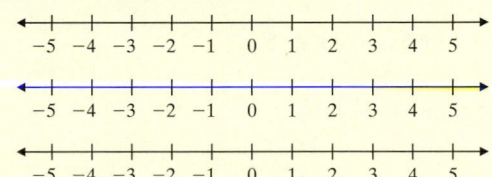

Solve each compound inequality. Write solutions in interval notation. See Examples 2 and 3.

19. $x + 1 \ge 7$ *and* $3x - 1 \ge 5$

20. $x + 2 \ge 3$ *and* $5x - 1 \ge 9$

21. $4x + 2 \le -10$ *and* $2x \le 0$

22. $2x + 4 > 0$ *and* $4x > 0$

23. $-2x < -8$ *and* $x - 5 < 5$

24. $-7x \le -21$ *and* $x - 20 \le -15$

Solve each compound inequality. See Examples 4 and 5.

25. $5 < x - 6 < 11$

26. $-2 \le x + 3 \le 0$

27. $-2 \le 3x - 5 \le 7$

28. $1 < 4 + 2x < 7$

29. $1 \le \dfrac{2}{3}x + 3 \le 4$

30. $-2 < \dfrac{1}{2}x - 5 < 1$

31. $-5 \le \dfrac{-3x + 1}{4} \le 2$

32. $-4 \le \dfrac{-2x + 5}{3} \le 1$

Objective D *Solve each compound inequality. Graph the two given inequalities on the first two number lines and the solution set on the third number line. See Examples 7 and 8.*

33. $x < 4$ *or* $x < 5$

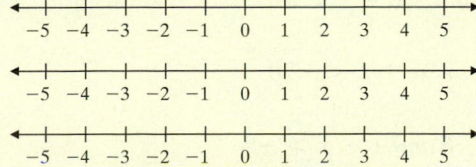

34. $x < 0$ *or* $x < 1$

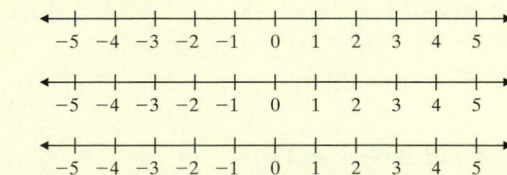

35. $x \le -4$ *or* $x \ge 1$

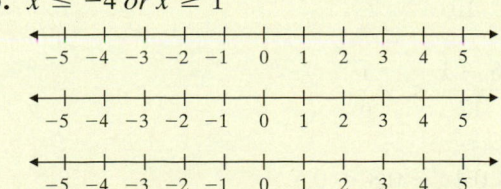

36. $x \ge -3$ *or* $x \le -4$

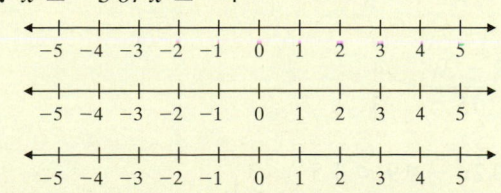

37. $x > 0 \text{ or } x < 3$

38. $x \geq -2 \text{ or } x \leq 2$

Solve each compound inequality. Write answers in interval notation. See Examples 7 and 8.

39. $-2x \leq -4 \text{ or } 5x - 20 \geq 5$

40. $-5x \leq 10 \text{ or } 3x - 5 \geq 1$

41. $x + 4 < 0 \text{ or } 6x > -12$

42. $x + 9 < 0 \text{ or } 4x > -12$

43. $3(x - 1) < 12 \text{ or } x + 7 > 10$

44. $5(x - 1) \leq -5 \text{ or } 5 - x \leq 11$

Objectives **B** **D** **Mixed Practice** *Solve each compound inequality. Write solutions in interval notation. See Examples 2 through 5, 7, and 8.*

45. $x < \dfrac{2}{3} \text{ and } x > -\dfrac{1}{2}$

46. $x < \dfrac{5}{7} \text{ and } x < 1$

47. $x < \dfrac{2}{3} \text{ or } x > -\dfrac{1}{2}$

48. $x < \dfrac{5}{7} \text{ or } x < 1$

49. $0 \leq 2x - 3 \leq 9$

50. $3 < 5x + 1 < 11$

51. $\dfrac{1}{2} < x - \dfrac{3}{4} < 2$

52. $\dfrac{2}{3} < x + \dfrac{1}{2} < 4$

53. $x + 3 \geq 3 \text{ and } x + 3 \leq 2$

54. $2x - 1 \geq 3 \text{ and } -x > 2$

55. $3x \geq 5 \text{ or } -\dfrac{5}{8}x - 6 > 1$

56. $\dfrac{3}{8}x + 1 \leq 0 \text{ or } -2x < -4$

57. $0 < \dfrac{5 - 2x}{3} < 5$

58. $-2 < \dfrac{-2x - 1}{3} < 2$

59. $-6 < 3(x - 2) \leq 8$

60. $-5 < 2(x + 4) < 8$

61. $-x + 5 > 6 \text{ and } 1 + 2x \leq -5$

62. $5x \leq 0 \text{ and } -x + 5 < 8$

▶63. $3x + 2 \leq 5 \text{ or } 7x > 29$

64. $-x < 7 \text{ or } 3x + 1 < -20$

65. $5 - x > 7 \text{ and } 2x + 3 \geq 13$

66. $-2x < -6 \text{ and } 1 - x > -2$

67. $-\dfrac{1}{2} \leq \dfrac{4x - 1}{6} < \dfrac{5}{6}$

68. $-\dfrac{1}{2} \leq \dfrac{3x - 1}{10} < \dfrac{1}{2}$

69. $\dfrac{1}{15} < \dfrac{8 - 3x}{15} < \dfrac{4}{5}$

70. $-\dfrac{1}{4} < \dfrac{6 - x}{12} < -\dfrac{1}{6}$

71. $0.3 < 0.2x - 0.9 < 1.5$

72. $-0.7 \leq 0.4x + 0.8 < 0.5$

Use the graph to answer Exercises 73 and 74.

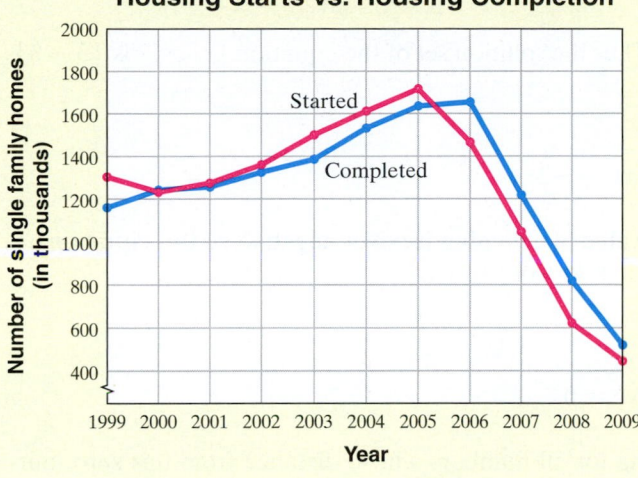

United States – Single Family Homes Housing Starts vs. Housing Completion

Source: U.S. Census Bureau

73. For which years were the number of single family housing starts greater than 1500 and the number of single-family home completions greater than 1500?

74. For which years were the number of single family housing starts less than 1000 or the number of single family housing completions greater than 1500?

75. In your own words, describe how to find the union of two sets.

76. In your own words, describe how to find the intersection of two sets.

A.3 Absolute Value Equations and Inequalities

In Chapter 8, we defined the absolute value of a number as its distance from 0 on a number line.

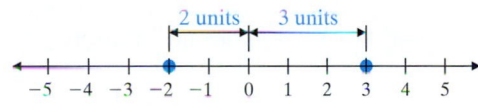

$$|-2| = 2 \quad \text{and} \quad |3| = 3$$

In this section, we concentrate on solving equations and inequalities containing the absolute value of a variable or a variable expression. Examples of absolute value equations and inequalities are

$$|x| = 3 \quad -5 \geq |2y + 7| \quad |z - 6.7| = |3z + 1.2| \quad |x - 3| > 7$$

Absolute value equations and inequalities are extremely useful in data analysis, especially for calculating acceptable measurement error and errors that result from the way numbers are sometimes represented in computers.

Objective A Solving Absolute Value Equations

To begin, let's solve a few absolute value equations by inspection.

Example 1 Solve: $|x| = 3$

Solution: The solution set of this equation will contain all numbers whose distance from 0 is 3 units. Two numbers are 3 units away from 0 on the number line; 3 and −3.

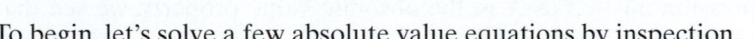

(Continued on next page)

Objectives

A Solve Absolute Value Equations.

B Solve Absolute Value Inequalities.

Practice 1

Solve: $|y| = 5$

Answer

1. $\{-5, 5\}$

Check: To check, let $x = 3$ and $x = -3$ in the original equation.

$$|x| = 3 \qquad\qquad\qquad |x| = 3$$
$$|3| \stackrel{?}{=} 3 \quad \text{Let } x = 3. \qquad |-3| \stackrel{?}{=} 3 \quad \text{Let } x = -3.$$
$$3 = 3 \quad \text{True} \qquad\qquad 3 = 3 \quad \text{True}$$

Both solutions check. Thus the solution set of the equation $|x| = 3$ is $\{3, -3\}$.

■ Work Practice 1

Practice 2

Solve: $|p| = -4$

Example 2 Solve: $|x| = -2$

Solution: The absolute value of a number is never negative, so this equation has no solution. The solution set is $\{\ \}$ or $\varnothing$.

■ Work Practice 2

Practice 3

Solve: $|x| = 0$

Example 3 Solve: $|y| = 0$

Solution: We are looking for all numbers whose distance from 0 is zero units. The only number is 0. The solution set is $\{0\}$.

■ Work Practice 3

From the above examples, we have the following.

> **Absolute Value Property**
>
> To solve $|X| = a$,
>
> If a is positive, then solve $X = a$ or $X = -a$.
>
> If a is 0, then $X = 0$.
>
> If a is negative, the equation $|X| = a$ has no solution.

> **Helpful Hint**
>
> For the equation $|X| = a$ in the box above, X can be a single variable or a variable expression.

When we are solving absolute value equations, if $|X|$ is not alone on one side of the equation we first use properties of equality to get $|X|$ alone.

Practice 4

Solve: $|4x + 2| = 6$.

> **Helpful Hint** If the equation has a single absolute value expression containing variables, get the absolute value expression alone. Then use the absolute value property.

Example 4 Solve: $|5w + 3| = 7$

Solution: Here the expression inside the absolute value bars is $5w + 3$. If we think of the expression $5w + 3$ as X in the absolute value property, we see that $|X| = 7$ is equivalent to

$$X = 7 \quad \text{or} \quad X = -7$$

Then substitute $5w + 3$ for X, and we have

$$5w + 3 = 7 \quad \text{or} \quad 5w + 3 = -7$$

Solve these two equations for w.

$$5w + 3 = 7 \quad \text{or} \quad 5w + 3 = -7$$
$$5w = 4 \quad \text{or} \quad 5w = -10$$
$$w = \frac{4}{5} \quad \text{or} \quad w = -2$$

Answers

2. $\varnothing$ **3.** $\{0\}$ **4.** $\{1, -2\}$

Check: To check, let $w = -2$ and then $w = \dfrac{4}{5}$ in the original equation.

Let $w = -2$

$$|5(-2) + 3| = 7$$

$$|-10 + 3| = 7$$

$$|-7| = 7$$

$$7 = 7 \qquad \text{True}$$

Let $w = \dfrac{4}{5}$

$$\left|5\left(\dfrac{4}{5}\right) + 3\right| = 7$$

$$|4 + 3| = 7$$

$$|7| = 7$$

$$7 = 7 \qquad \text{True}$$

Both solutions check, and the solution set is $\left\{-2, \dfrac{4}{5}\right\}$.

■ **Work Practice 4**

Example 5 Solve: $\left|\dfrac{x}{2} - 1\right| = 11$

Solution: $\left|\dfrac{x}{2} - 1\right| = 11$ is equivalent to

$$\dfrac{x}{2} - 1 = 11 \qquad \text{or} \qquad \dfrac{x}{2} - 1 = -11$$

$$2\left(\dfrac{x}{2} - 1\right) = 2(11) \quad \text{or} \quad 2\left(\dfrac{x}{2} - 1\right) = 2(-11) \qquad \text{Clear fractions.}$$

$$x - 2 = 22 \qquad \text{or} \qquad x - 2 = -22 \qquad \text{Apply the distributive property.}$$

$$x = 24 \qquad \text{or} \qquad x = -20$$

The solution set is $\{-20, 24\}$.

■ **Work Practice 5**

Don't forget that to use the absolute value property you must first make sure that the absolute value expression is alone on one side of the equation.

Example 6 Solve: $|2x - 1| + 5 = 6$

Solution: We want the absolute value expression alone on one side of the equation, so we begin by subtracting 5 from both sides. Then we use the absolute value property.

$$|2x - 1| + 5 = 6$$

$$|2x - 1| = 1 \qquad\qquad \text{Subtract 5 from both sides.}$$

$$2x - 1 = 1 \qquad \text{or} \qquad 2x - 1 = -1 \qquad \text{Use the absolute value property.}$$

$$2x = 2 \qquad \text{or} \qquad 2x = 0$$

$$x = 1 \qquad \text{or} \qquad x = 0 \qquad \text{Solve.}$$

The solution set is $\{0, 1\}$.

■ **Work Practice 6**

Given two absolute value expressions, we might ask, when are the absolute values of two expressions equal? To see the answer, notice that

$$|2| = |2| \qquad |-2| = |-2| \qquad |-2| = |2| \qquad |2| = |-2|$$

same   same   opposites   opposites

Practice 5

Solve: $\left|\dfrac{x}{3} + 4\right| = 1$

Practice 6

Solve: $|4x + 2| + 1 = 7$

Answers

5. $\{-9, -15\}$ **6.** $\{1, -2\}$

Two absolute value expressions are equal when the expressions inside the absolute value bars are equal to or are opposites of each other. In other words,

> To solve $|X| = |Y|$, solve $X = Y$ or $X = -Y$.

Practice 7

Solve: $|4x - 5| = |3x + 5|$

Example 7 Solve: $|3x + 2| = |5x - 8|$

Solution: This equation is true if the expressions inside the absolute value bars are equal to or are opposites of each other.

$$3x + 2 = 5x - 8 \quad \text{or} \quad 3x + 2 = -5x + 8$$

Next we solve each equation.

$$3x + 2 = 5x - 8 \quad \text{or} \quad 3x + 2 = -5x + 8$$
$$-2x + 2 = -8 \quad \text{or} \quad 8x + 2 = 8$$
$$-2x = -10 \quad \text{or} \quad 8x = 6$$
$$x = 5 \quad \text{or} \quad x = \frac{3}{4}$$

Check to see that replacing x with 5 or with $\frac{3}{4}$ results in a true statement. The solution set is $\left\{ \frac{3}{4}, 5 \right\}$.

■ **Work Practice 7**

Practice 8

Solve: $|x + 2| = |4 - x|$

Example 8 Solve: $|x - 3| = |5 - x|$

Solution:

$$x - 3 = 5 - x \quad \text{or} \quad x - 3 = -(5 - x)$$
$$2x - 3 = 5 \quad \text{or} \quad x - 3 = -5 + x$$
$$2x = 8 \quad \text{or} \quad x - 3 - x = -5 + x - x$$
$$x = 4 \quad \text{or} \quad -3 = -5 \qquad \text{False}$$

Recall from Section 9.3 that when an equation simplifies to a false statement, the equation has no solution. Thus the only solution for the original absolute value equation is 4, and the solution set is $\{4\}$.

■ **Work Practice 8**

✔**Concept Check** True or false? Absolute value equations always have two solutions. Explain your answer.

Objective B Solving Absolute Value Inequalities ▶

To begin, let's solve a few absolute value inequalities by inspection.

Answers

7. $\{0, 10\}$ **8.** $\{1\}$

✔**Concept Check Answer**
false; answers may vary

Example 9 Solve $|x| < 2$ using a number line.

Solution: The solution set contains all numbers whose distance from 0 is less than 2 units on the number line.

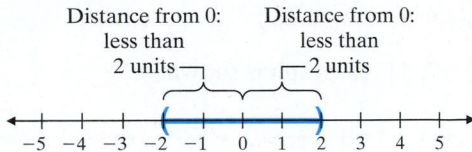

The solution set is $\{x | -2 < x < 2\}$, or $(-2, 2)$ in interval notation.

◼ **Work Practice 9**

Example 10 Solve $|x| \geq 3$ using a number line.

Solution: The solution set contains all numbers whose distance from 0 is 3 or more units. Thus the graph of the solution set contains 3 and all points to the right of 3 on the number line or -3 and all points to the left of -3 on the number line.

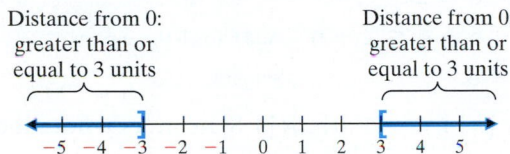

This solution set is $\{x | x \leq -3 \text{ or } x \geq 3\}$. In interval notation, the solution set is $(-\infty, -3] \cup [3, \infty)$, since **or** means union.

◼ **Work Practice 10**

The following box summarizes solving absolute value equations and inequalities.

Solving Absolute Value Equations and Inequalities

If a is a positive number,

To solve $|X| = a$, solve $X = a$ or $X = -a$.

To solve $|X| = |Y|$, solve $X = Y$ or $X = -Y$.

To solve $|X| < a$, solve $-a < X < a$.

To solve $|X| > a$, solve $X < -a$ or $X > a$.

Example 11 Solve: $|x - 3| > 7$

Solution: Since 7 is positive, to solve $|x - 3| > 7$, we solve the compound inequality $x - 3 < -7$ or $x - 3 > 7$.

$$x - 3 < -7 \quad \text{or} \quad x - 3 > 7$$
$$x < -4 \quad \text{or} \quad x > 10 \quad \text{Add 3 to both sides.}$$

The solution set is $\{x | x < -4 \text{ or } x > 10\}$ or $(-\infty, -4) \cup (10, \infty)$ in interval notation. Its graph is shown.

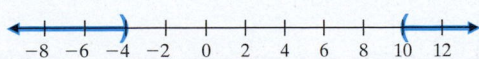

◼ **Work Practice 11**

Let's remember the differences between solving absolute value equations and inequalities by solving an absolute value equation.

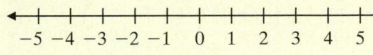
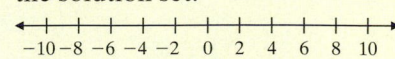
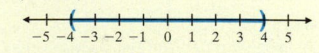
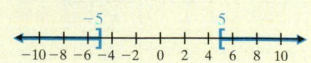

Practice 12

Solve: $|x - 3| = 5$. Graph the solution set.

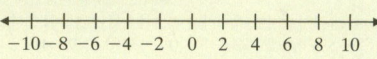

Example 12 Solve: $|x + 1| = 6$

Solution: This is an equation, so we solve

$$x + 1 = 6 \quad \text{or} \quad x + 1 = -6$$
$$x = 5 \quad \text{or} \quad x = -7$$

The solution set is $\{-7, 5\}$. Its graph is shown.

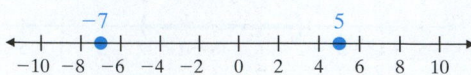

■ **Work Practice 12**

Notice that the next example is an absolute value inequality.

Practice 13

Solve: $|x - 2| \leq 1$. Graph the solution set.

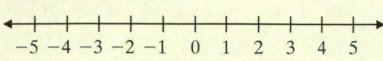

Example 13 Solve: $|x - 6| \leq 2$

Solution: To solve $|x - 6| \leq 2$, we solve

$$-2 \leq x - 6 \leq 2$$
$$-2 + 6 \leq x - 6 + 6 \leq 2 + 6 \quad \text{Add 6 to all three parts.}$$
$$4 \leq x \leq 8 \quad \text{Simplify.}$$

The solution set is $\{x \,|\, 4 \leq x \leq 8\}$, or $[4, 8]$ in interval notation. Its graph is shown.

■ **Work Practice 13**

Helpful Hint

As with absolute value equations, before using an absolute value inequality property, get an absolute value expression alone on one side of the inequality.

Practice 14

Solve: $|2x - 5| + 2 \leq 9$

Example 14 Solve: $|5x + 1| + 1 \leq 10$

Solution: First we get the absolute value expression alone by subtracting 1 from both sides.

$$|5x + 1| + 1 \leq 10$$
$$|5x + 1| \leq 10 - 1 \quad \text{Subtract 1 from both sides.}$$
$$|5x + 1| \leq 9 \quad \text{Simplify.}$$

Since 9 is positive, to solve $|5x + 1| \leq 9$, we solve

$$-9 \leq 5x + 1 \leq 9$$
$$-9 - 1 \leq 5x + 1 - 1 \leq 9 - 1 \quad \text{Subtract 1 from all three parts.}$$
$$-10 \leq 5x \leq 8 \quad \text{Simplify.}$$
$$-2 \leq x \leq \frac{8}{5} \quad \text{Divide all three parts by 5.}$$

The solution set is $\left[-2, \dfrac{8}{5}\right]$.

■ **Work Practice 14**

The next few examples are special cases of absolute value inequalities.

Answers

12. $\{-2, 8\}$

13. $[1, 3]$

14. $[-1, 6]$

Example 15 Solve: $|x| \le -3$

Solution: The absolute value of a number is never negative. Thus it will then never be less than or equal to -3. The solution set is $\{\ \}$ or $\varnothing$.

■ **Work Practice 15**

Example 16 Solve: $|x - 1| > -2$

Solution: The absolute value of a number is always nonnegative. Thus it will always be greater than -2. The solution set contains all real numbers, or $(-\infty, \infty)$.

■ **Work Practice 16**

✓**Concept Check** Without taking any solution steps, how do you know that the absolute value inequality $|3x - 2| > -9$ has a solution? What is its solution?

Practice 15

Solve: $|x| < -1$

Practice 16

Solve: $|x + 1| \ge -3$

Answers

15. $\varnothing$ **16.** $(-\infty, \infty)$

✓**Concept Check Answer**
$(-\infty, \infty)$ since an absolute value is always nonnegative

Vocabulary and Readiness Check

Match each absolute value equation or inequality with an equivalent statement.

1. $|2x + 1| = 3$

2. $|2x + 1| \le 3$

3. $|2x + 1| < 3$

4. $|2x + 1| \ge 3$

5. $|2x + 1| > 3$

a. $2x + 1 > 3$ or $2x + 1 < -3$

b. $2x + 1 \ge 3$ or $2x + 1 \le -3$

c. $-3 < 2x + 1 < 3$

d. $2x + 1 = 3$ or $2x + 1 = -3$

e. $-3 \le 2x + 1 \le 3$

A.3 **Exercise Set** MyMathLab®

Objective A *Solve. See Examples 1 through 6.*

▶ 1. $|x| = 7$

2. $|y| = 15$

▶ 3. $|x| = -4$

4. $|x| = -20$

5. $|3x| = 12.6$

6. $|6n| = 12.6$

7. $3|x| - 5 = 7$

8. $5|x| - 12 = 8$

9. $-6|x| + 44 = -10$

10. $-4|x| + 18 = -22$

11. $|x - 9| = 14$

12. $|x + 2| = 8$

13. $|2x - 5| = 9$

14. $|6 + 2n| = 4$

▶ 15. $\left|\dfrac{x}{2} - 3\right| = 1$

16. $\left|\dfrac{n}{3} + 2\right| = 4$

17. $|z| + 4 = 9$

18. $|x| + 1 = 3$

19. $|3x| + 5 = 14$

20. $|2x| - 6 = 4$

21. $\left|\dfrac{4x - 6}{3}\right| = 6$

22. $\left|\dfrac{2x + 1}{5}\right| = 7$

23. $|2x| = 0$

24. $|7z| = 0$

25. $|4n + 1| + 10 = 4$

26. $|3z - 2| + 8 = 1$

27. $3|x - 1| + 19 = 23$

28. $5|x + 1| - 1 = 3$

Solve. See Examples 7 and 8.

29. $|5x - 7| = |3x + 11|$

30. $|9y + 1| = |6y + 4|$

31. $|z + 8| = |z - 3|$

32. $|2x - 5| = |2x + 5|$

▶ **33.** $|2y - 3| = |9 - 4y|$

34. $|5z - 1| = |7 - z|$

35. $\left|\dfrac{3}{4}x - 2\right| = \left|\dfrac{1}{4}x + 6\right|$

36. $\left|\dfrac{2}{3}x - 5\right| = \left|\dfrac{1}{3}x + 4\right|$

37. $|2x - 6| = |10 - 2x|$

38. $|4n + 5| = |4n + 3|$

39. $|x + 4| = |7 - x|$

40. $|8 - y| = |y + 2|$

41. $\left|\dfrac{2x + 1}{5}\right| = \left|\dfrac{3x - 7}{3}\right|$

42. $\left|\dfrac{5x - 1}{2}\right| = \left|\dfrac{4x + 5}{6}\right|$

43. $|5x + 1| = |4x - 7|$

44. $|3 + 6n| = |4n + 11|$

Objective B   *Solve. Graph the solution set. See Examples 9 through 16.*

▶ **45.** $|x| \leq 4$

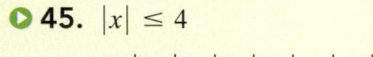

46. $|x| < 6$

▶ **47.** $|x| > 3$

48. $|y| \geq 4$

49. $|x + 3| < 2$

50. $|x + 4| < 6$

51. $|y - 6| \geq 7$

52. $|x - 3| \geq 10$

53. $\left|\dfrac{x + 2}{3}\right| < 1$

54. $\left|\dfrac{x - 6}{4}\right| < 1$

55. $|x| + 7 \leq 12$

56. $|x| + 6 \leq 7$

57. $|2x + 3| \leq 0$

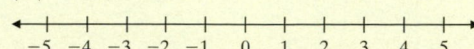

58. $|7x + 1| \leq 0$

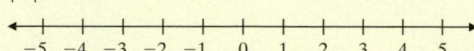

59. $|x| + 2 > 6$

60. $|x| - 1 > 3$

61. $|2x + 7| \leq 13$

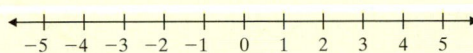

62. $|5x - 3| \leq 18$

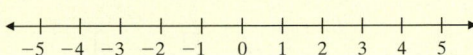

63. $|8 - 3x| < 5$

64. $|7 - 4x| < 5$

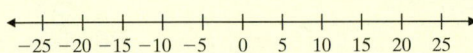

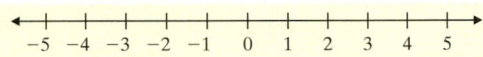

▶ **65.** $|x + 10| \geq 14$

66. $|x - 9| \geq 2$

67. $|2x - 7| \leq 11$

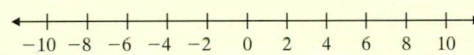

68. $|5x + 2| < 8$

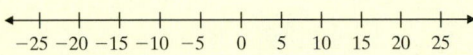

69. $|x| > -4$

70. $|4 + 9x| \geq -6$

▶ **71.** $6 + |4x - 1| \leq 9$

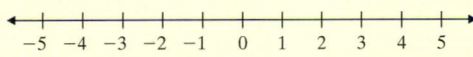

72. $-3 + |5x - 2| \leq 4$

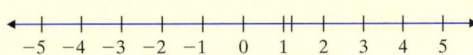

73. $|6x - 8| - 7 > -3$

74. $|10 + 3x| - 2 > -1$

75. $|5x + 3| < -6$

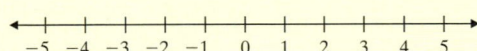

76. $|x| \leq -7$

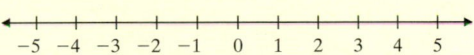

77. $\left|\dfrac{x + 6}{3}\right| > 2$

78. $\left|\dfrac{7 + x}{2}\right| \geq 4$

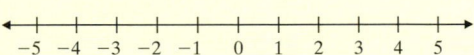

Objectives **A** **B** **Mixed Practice** *Solve each equation or inequality for x. See Examples 1 through 16.*

79. $|x| = 13$

80. $|x| > 13$

81. $|x| < 13$

82. $|3x| = 12$

83. $|x| + 12 = 9$

84. $|x| - 4 = -9$

85. $2|x| - 9 \leq 11$

86. $4|x| - 2 \geq 6$

87. $|2x - 3| = 7$ **88.** $|5 - 6x| = 29$ **89.** $|x - 5| \geq 12$ **90.** $|x + 4| \geq 20$

91. $|9 + 4x| = 0$ **92.** $|9 + 4x| \geq 0$ **93.** $|2x + 1| - 7 < -4$ **94.** $-11 + |5x - 3| \geq -8$

95. $\left|\dfrac{1}{3}x + 1\right| > 5$ **96.** $\left|\dfrac{1}{4}x - 2\right| < 1$ **97.** $|3x - 5| + 4 = 5$ **98.** $|5x - 1| + 7 = 11$

99. $|6x + 11| = -1$ **100.** $|4x - 4| = -3$

101. $\left|\dfrac{1 - 2x}{3}\right| = 6$ **102.** $\left|\dfrac{6 - 3x}{4}\right| = 5$

103. $\left|\dfrac{3x - 5}{6}\right| > 5$ **104.** $\left|\dfrac{4x - 7}{5}\right| < 2$

105. $|6x - 3| = |4x + 5|$ **106.** $|3x + 1| = |4x + 10|$

107. $\left|\dfrac{1 + 3x}{4}\right| = |-4|$ **108.** $\left|\dfrac{5x + 2}{2}\right| = |-6|$

Without going through a solution procedure, determine the solution of each absolute value equation or inequality.

109. $|x - 7| = -4$ **110.** $|x - 7| < -4$

111. $|x - 7| > -4$ **112.** $\left|\dfrac{3x - 2}{7}\right| \geq -7$

113. Write an absolute value equation representing all numbers x whose distance from 0 is 5 units.

114. Write an absolute value equation representing all numbers x whose distance from 0 is 2 units.

115. Write an absolute value inequality representing all numbers x whose distance from 0 is less than 7 units.

116. Write an absolute value inequality representing all numbers x whose distance from 0 is greater than 4 units.

117. Write $-5 \leq x \leq 5$ as an equivalent inequality containing an absolute value. Explain your answer.

118. Write $x > 1$ or $x < -1$ as an equivalent inequality containing an absolute value. Explain your answer.

The Distance and Midpoint Formulas

Objective A Using the Distance and Midpoint Formulas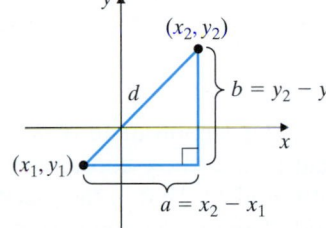

If we know how to simplify radicals, we can derive and use the distance formula. The midpoint formula is often confused with the distance formula, so to clarify both, we will also review the midpoint formula.

The Cartesian coordinate system helps us visualize the distance between points. To find the distance between two points, we use the distance formula, which is derived from the Pythagorean theorem.

To find the distance d between two points (x_1, y_1) and (x_2, y_2), draw vertical and horizontal lines so that a right triangle is formed, as shown. Notice that the length of leg a is $x_2 - x_1$ and that the length of leg b is $y_2 - y_1$. Thus, the Pythagorean theorem tells us that

$$d^2 = a^2 + b^2$$

or

$$d^2 = (x_2 - x_1)^2 + (y_2 - y_1)^2$$

or

$$d = \sqrt{(x_2 - x_1)^2 + (y_2 - y_1)^2}$$

This formula gives us the distance between any two points on the real plane.

Distance Formula

The distance d between two points (x_1, y_1) and (x_2, y_2) is given by

$$d = \sqrt{(x_2 - x_1)^2 + (y_2 - y_1)^2}$$

Example 1 Find the distance between $(2, -5)$ and $(1, -4)$. Give the exact distance and a three-decimal-place approximation.

Solution: To use the distance formula, it makes no difference which point we call (x_1, y_1) and which point we call (x_2, y_2). We will let $(x_1, y_1) = (2, -5)$ and $(x_2, y_2) = (1, -4)$.

$$d = \sqrt{(x_2 - x_1)^2 + (y_2 - y_1)^2}$$
$$= \sqrt{(1 - 2)^2 + [-4 - (-5)]^2}$$
$$= \sqrt{(-1)^2 + (1)^2}$$
$$= \sqrt{1 + 1}$$
$$= \sqrt{2} \approx 1.414$$

The distance between the two points is exactly $\sqrt{2}$ units, or approximately 1.414 units.

Work Practice 1

Practice 1

Find the distance between $(-1, 3)$ and $(-2, 6)$. Give the exact distance and a three-decimal-place approximation.

Answer
1. $\sqrt{10} \approx 3.162$

The **midpoint** of a line segment is the **point** located exactly halfway between the two endpoints of the line segment. On the following graph, the point M is the midpoint of line segment PQ. Thus, the distance between M and P equals the distance between M and Q. *Note:* We usually need no knowledge of roots to calculate the midpoint of a line segment. We review midpoint here only because it is often confused with the distance between two points.

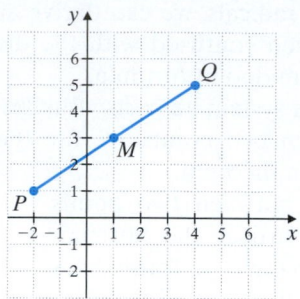

The x-coordinate of M is at half the distance between the x-coordinates of P and Q, and the y-coordinate of M is at half the distance between the y-coordinates of P and Q. That is, the x-coordinate of M is the average of the x-coordinates of P and Q; the y-coordinate of M is the average of the y-coordinates of P and Q.

Midpoint Formula

The midpoint of the line segment whose endpoints are (x_1, y_1) and (x_2, y_2) is the point with coordinates

$$\left(\frac{x_1 + x_2}{2}, \frac{y_1 + y_2}{2} \right)$$

Practice 2

Find the midpoint of the line segment that joins points $P(-2, 5)$ and $Q(4, -6)$.

Example 2   Find the midpoint of the line segment that joins points $P(-3, 3)$ and $Q(1, 0)$.

Solution: To use the midpoint formula, it makes no difference which point we call (x_1, y_1) and which point we call (x_2, y_2). We will let $(x_1, y_1) = (-3, 3)$ and $(x_2, y_2) = (1, 0)$.

$$\text{midpoint} = \left(\frac{x_1 + x_2}{2}, \frac{y_1 + y_2}{2} \right)$$
$$= \left(\frac{-3 + 1}{2}, \frac{3 + 0}{2} \right)$$
$$= \left(\frac{-2}{2}, \frac{3}{2} \right)$$
$$= \left(-1, \frac{3}{2} \right)$$

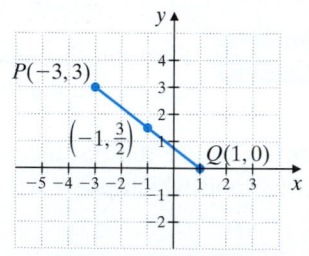

The midpoint of the segment is $\left(-1, \frac{3}{2} \right)$.

■ **Work Practice 2**

Helpful Hint

The distance between two points is a distance. The midpoint of a line segment is the point halfway between the endpoints of the segment.

distance—measured in units

midpoint—it is a point

Answer

2. $\left(1, -\frac{1}{2} \right)$

A.4 Exercise Set MyMathLab®

Find the distance between each pair of points. Give the exact distance and a three-decimal-place approximation. See Example 1.

1. $(5, 1)$ and $(8, 5)$

2. $(2, 3)$ and $(14, 8)$

▶**3.** $(-3, 2)$ and $(1, -3)$

4. $(3, -2)$ and $(-4, 1)$

5. $(0, -\sqrt{2})$ and $(\sqrt{3}, 0)$

6. $(-\sqrt{5}, 0)$ and $(0, \sqrt{7})$

7. $(1.7, -3.6)$ and $(-8.6, 5.7)$

8. $(9.6, 2.5)$ and $(-1.9, -3.7)$

Find the midpoint of each line segment whose endpoints are given. See Example 2.

9. $(6, -8)$; $(2, 4)$

10. $(3, 9)$; $(7, 11)$

▶**11.** $(-2, -1)$; $(-8, 6)$

12. $(-3, -4)$; $(6, -8)$

13. $\left(\frac{1}{2}, \frac{3}{8}\right)$; $\left(-\frac{3}{2}, \frac{5}{8}\right)$

14. $\left(-\frac{2}{5}, \frac{7}{15}\right)$; $\left(-\frac{2}{5}, -\frac{4}{15}\right)$

15. $(\sqrt{2}, 3\sqrt{5})$; $(\sqrt{2}, -2\sqrt{5})$

16. $(\sqrt{8}, -\sqrt{12})$; $(3\sqrt{2}, 7\sqrt{3})$

A.5 Writing Equations of Parallel and Perpendicular Lines

Objective A Writing Equations of Parallel and Perpendicular Lines

Objective

A Write Equations of Parallel and Perpendicular Lines.

In this appendix, we practice writing equations of parallel and perpendicular lines.

Example 1 Write an equation of the line containing the point $(4, 4)$ and parallel to the line $2x + y = -6$. Write the equation in slope-intercept form, $y = mx + b$.

Practice 1

Write an equation of the line containing the point $(-1, 2)$ and parallel to the line $3x + y = 5$. Write the equation in slope-intercept form $y = mx + b$.

Solution: Because the line we want to find is *parallel* to the line $2x + y = -6$, the two lines must have equal slopes. So we first find the slope of $2x + y = -6$ by solving the equation for y to write it in the form $y = mx + b$. Here $y = -2x - 6$, so the slope is -2.

Now we use the point-slope form to write an equation of the line through $(4, 4)$ with slope -2.

$$y - y_1 = m(x - x_1)$$
$$y - 4 = -2(x - 4) \quad \text{Let } m = -2, x_1 = 4, \text{ and } y_1 = 4.$$
$$y - 4 = -2x + 8 \quad \text{Use the distributive property.}$$
$$y = -2x + 12$$

(Continued on next page)

Answer

1. $y = -3x - 1$

The equation $y = -2x - 6$ and the new equation $y = -2x + 12$ have the same slope but different y-intercepts, so their graphs are parallel. Also, the graph of $y = -2x + 12$ contains the point $(4, 4)$, as desired.

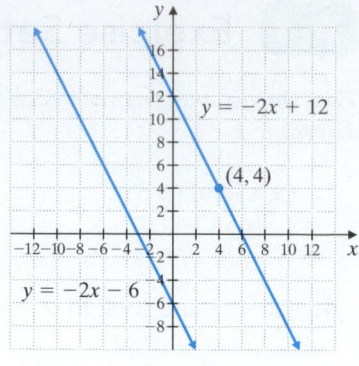

■ **Work Practice 1**

Practice 2

Write an equation of the line containing the point $(3, 4)$ and perpendicular to the line $2x + 4y = 5$. Write the equation in slope-intercept form, $y = mx + b$.

Example 2   Write an equation of the line containing the point $(-2, 1)$ and perpendicular to the line $3x + 5y = 4$. Write the equation in slope-intercept form, $y = mx + b$.

Solution: First we find the slope of $3x + 5y = 4$ by solving the equation for y.

$$5y = -3x + 4$$

$$y = -\frac{3}{5}x + \frac{4}{5}$$

The slope of the given line is $-\frac{3}{5}$. A line perpendicular to this line will have a slope that is the negative reciprocal of $-\frac{3}{5}$, or $\frac{5}{3}$. We use the point-slope form to write an equation of the new line through $(-2, 1)$ with slope $\frac{5}{3}$.

$$y - 1 = \frac{5}{3}[x - (-2)]$$

$$y - 1 = \frac{5}{3}(x + 2) \qquad \text{Simplify.}$$

$$y - 1 = \frac{5}{3}x + \frac{10}{3} \qquad \text{Use the distributive property.}$$

$$y = \frac{5}{3}x + \frac{13}{3} \qquad \text{Add 1 to both sides.}$$

The equation $y = -\frac{3}{5}x + \frac{4}{5}$ and the new equation $y = \frac{5}{3}x + \frac{13}{3}$ have negative reciprocal slopes, so their graphs are perpendicular. Also, the graph of $y = \frac{5}{3}x + \frac{13}{3}$ contains the point $(-2, 1)$.

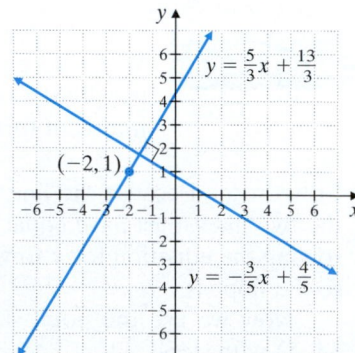

■ **Work Practice 2**

A.5 Exercise Set MyMathLab®

Write an equation of each line. Write the equation in the form $x = a$ (vertical line), $y = b$ (horizontal line), or $y = mx + b$. See Examples 1 and 2.

1. Through $(3, 8)$; parallel to $y = 4x - 2$

2. Through $(1, 5)$; parallel to $y = 3x - 4$

▶ **3.** Through $(2, -5)$; perpendicular to $3y = x - 6$

4. Through $(-4, 8)$; perpendicular to $2x - 3y = 1$

5. Through $(1, 4)$; parallel to $y = 7$

6. Through $(-2, 6)$; perpendicular to $y = 7$

7. Through $(-2, -3)$; parallel to $3x + 2y = 5$

8. Through $(-2, -3)$; perpendicular to $3x + 2y = 5$

9. Through $(-1, -5)$; perpendicular to $x = 3$

10. Through $(4, -6)$; parallel to $x = -2$

11. Through $(-1, 5)$; perpendicular to $x - 4y = 4$

12. Through $(2, -3)$; perpendicular to $x - 5y = 10$

13. Through $(6, -2)$; parallel to the line $2x + 4y = 9$

14. Through $(8, -3)$; parallel to the line $6x + 2y = 5$

15. Through $(6, 1)$; parallel to the line $8x - y = 9$

16. Through $(3, 5)$; perpendicular to the line $2x - y = 8$

17. Through $(5, -6)$; perpendicular to $y = 9$

18. Through $(-3, -5)$; parallel to $y = 9$

A.6 Nonlinear Inequalities in One Variable

Objective A Solving Polynomial Inequalities ▶

Just as we can solve linear inequalities in one variable, we can also solve quadratic and higher-degree inequalities in one variable. Let's begin with quadratic inequalities. A **quadratic inequality** is an inequality that can be written so that one side is a quadratic expression and the other side is 0. Here are examples of quadratic inequalities in one variable. Each is written in **standard form.**

$$x^2 - 10x + 7 \leq 0 \qquad 3x^2 + 2x - 6 > 0$$
$$2x^2 + 9x - 2 < 0 \qquad x^2 - 3x + 11 \geq 0$$

A solution of a quadratic inequality in one variable is a value of the variable that makes the inequality a true statement.

The value of an expression such as $x^2 - 3x - 10$ will sometimes be positive, sometimes negative, and sometimes 0, depending on the value substituted for x. To solve the inequality $x^2 - 3x - 10 < 0$, we look for all values of x that make the

Objectives

A Solve Polynomial Inequalities of Degree 2 or Greater. ▶

B Solve Inequalities That Contain Rational Expressions with Variables in the Denominator. ▶

expression $x^2 - 3x - 10$ **less than 0,** or **negative.** To understand how we find these values, we'll study the graph of the quadratic function $y = x^2 - 3x - 10$.

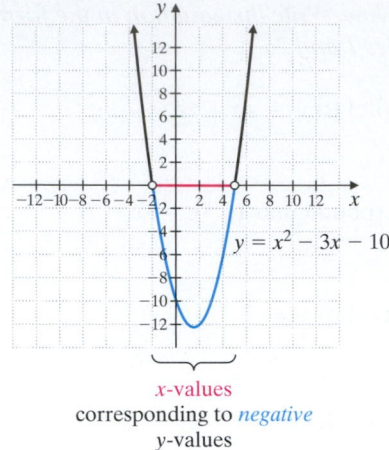

x-values
corresponding to *negative*
y-values

Notice that the *x*-values for which *y* or $x^2 - 3x - 10$ is positive are separated from the *x*-values for which *y* or $x^2 - 3x - 10$ is negative by the values for which *y* or $x^2 - 3x - 10$ is 0, the *x*-intercepts. Thus, the solution set of $x^2 - 3x - 10 < 0$ consists of all real numbers from -2 to 5 or, in interval notation, $(-2, 5)$.

It is not necessary to graph $y = x^2 - 3x - 10$ to solve the related inequality $x^2 - 3x - 10 < 0$. Instead, we can draw a number line representing the *x*-axis and keep the following in mind: *A region on the number line for which the value of* $x^2 - 3x - 10$ *is positive is separated from a region on the number line for which the value of* $x^2 - 3x - 10$ *is negative by a value for which the expression is 0.*

Let's find these values for which the expression is 0 by solving the related equation, $x^2 - 3x - 10 = 0$.

$$x^2 - 3x - 10 = 0$$
$$(x - 5)(x + 2) = 0 \qquad \text{Factor.}$$
$$x - 5 = 0 \quad \text{or} \quad x + 2 = 0 \qquad \text{Set each factor equal to 0.}$$
$$x = 5 \qquad\qquad x = -2 \qquad \text{Solve.}$$

These two numbers -2 and 5 divide the number line into three regions. We will call the regions *A*, *B*, and *C*. These regions are important because if the value of $x^2 - 3x - 10$ is negative when a number from a region is substituted for *x*, then $x^2 - 3x - 10$ is negative when any number in that region is substituted for *x*. Similarly, if the value of $x^2 - 3x - 10$ is positive when a number from a region is substituted for *x*, then $x^2 - 3x - 10$ is positive when any number in that region is substituted for *x*.

To see whether the inequality $x^2 - 3x - 10 < 0$ is true or false in each region, we choose a test point from each region and substitute its value for *x* in the inequality $x^2 - 3x - 10 < 0$. If the resulting inequality is true, the region containing the test point is a solution region.

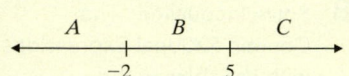

Region	Test Point Value	$(x - 5)(x + 2) < 0$	Result
A	-3	$(-8)(-1) < 0$	False
B	0	$(-5)(2) < 0$	True
C	6	$(1)(8) < 0$	False

The values in region *B* satisfy the inequality. The numbers -2 and 5 are not included in the solution set since the inequality symbol is $<$. The solution set is $(-2, 5)$, and its graph is shown.

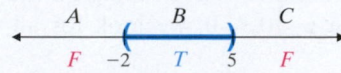

Example 1   Solve: $(x + 3)(x - 3) > 0$

Solution: First we solve the related equation, $(x + 3)(x - 3) = 0$.

$$(x + 3)(x - 3) = 0$$
$$x + 3 = 0 \quad \text{or} \quad x - 3 = 0$$
$$x = -3 \qquad\qquad x = 3$$

The two numbers -3 and 3 separate the number line into three regions, A, B, and C.

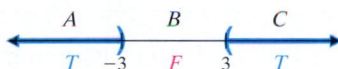

Now we substitute the value of a test point from each region. If the test value satisfies the inequality, every value in the region containing the test value is a solution.

Region	Test Point Value	$(x + 3)(x - 3) > 0$	Result
A	-4	$(-1)(-7) > 0$	True
B	0	$(3)(-3) > 0$	False
C	4	$(7)(1) > 0$	True

The points in regions A and C satisfy the inequality. The numbers -3 and 3 are not included in the solution since the inequality symbol is $>$. The solution set is $(-\infty, -3) \cup (3, \infty)$, and its graph is shown.

Work Practice 1

The steps below may be used to solve a polynomial inequality of degree 2 or greater.

> ### Solving a Polynomial Inequality of Degree 2 or Greater
>
> **Step 1:** Write the inequality in standard form and then solve the related equation.
>
> **Step 2:** Separate the number line into regions with the solutions from Step 1.
>
> **Step 3:** For each region, choose a test point and determine whether its value satisfies the *original inequality*.
>
> **Step 4:** The solution set includes the regions whose test point value is a solution. If the inequality symbol is $\leq$ or $\geq$, the values from Step 1 are solutions; if $<$ or $>$, they are not.

✔**Concept Check**   When choosing a test point in Step 4, why would the solutions from Step 2 not make good choices for test points?

Example 2   Solve: $x^2 - 4x \leq 0$

Solution: First we solve the related equation, $x^2 - 4x = 0$.

$$x^2 - 4x = 0$$
$$x(x - 4) = 0$$
$$x = 0 \quad \text{or} \quad x = 4$$

(Continued on next page)

The numbers 0 and 4 separate the number line into three regions, A, B, and C.

We check a test value in each region in the original inequality. Values in region B satisfy the inequality. The numbers 0 and 4 are included in the solution since the inequality symbol is $\leq$. The solution set is $[0, 4]$, and its graph is shown.

■ **Work Practice 2**

Practice 3

Solve:

$(x - 2)(x + 1)(x + 5) \leq 0$

Example 3 Solve: $(x + 2)(x - 1)(x - 5) \leq 0$

Solution: First we solve $(x + 2)(x - 1)(x - 5) = 0$. By inspection, we see that the solutions are $-2, 1$ and 5. They separate the number line into four regions, A, B, C, and D. Next we check test points from each region.

Region	Test Point Value	$(x + 2)(x - 1)(x - 5) \leq 0$	Result
A	-3	$(-1)(-4)(-8) \leq 0$	True
B	0	$(2)(-1)(-5) \leq 0$	False
C	2	$(4)(1)(-3) \leq 0$	True
D	6	$(8)(5)(1) \leq 0$	False

The solution set is $(-\infty, -2] \cup [1, 5]$, and its graph is shown. We include the numbers $-2, 1$, and 5 because the inequality symbol is $\leq$.

■ **Work Practice 3**

Objective B Solving Rational Inequalities ▶

Inequalities containing rational expressions with variables in the denominator are solved by using a similar procedure. Notice as we solve an example that unlike quadratic inequalities, we must also consider values for which the rational inequality is undefined. Why? As usual, these values may not be solution values for the inequality.

Practice 4

Solve: $\dfrac{x - 3}{x + 5} \leq 0$

Example 4 Solve: $\dfrac{x + 2}{x - 3} \leq 0$

Solution: First we find all values that make the denominator equal to 0. To do this, we solve $x - 3 = 0$ or $x = 3$.

Next, we solve the related equation, $\dfrac{x + 2}{x - 3} = 0$.

$$\frac{x + 2}{x - 3} = 0 \qquad \text{Multiply both sides by the LCD, } x - 3.$$

$$x + 2 = 0$$

$$x = -2$$

Now we place these numbers on a number line and proceed as before, checking test point values in the original inequality.

Answers

3. $(-\infty, -5] \cup [-1, 2]$ **4.** $(-5, 3]$

Choose −3 from region A.

$$\frac{x + 2}{x - 3} \le 0$$

$$\frac{-3 + 2}{-3 - 3} \le 0$$

$$\frac{-1}{-6} \le 0$$

$$\frac{1}{6} \le 0 \quad \text{False}$$

Choose 0 from region B.

$$\frac{x + 2}{x - 3} \le 0$$

$$\frac{0 + 2}{0 - 3} \le 0$$

$$-\frac{2}{3} \le 0 \quad \text{True}$$

Choose 4 from region C.

$$\frac{x + 2}{x - 3} \le 0$$

$$\frac{4 + 2}{4 - 3} \le 0$$

$$6 \le 0 \quad \text{False}$$

The solution set is $[-2, 3)$. This interval includes -2 because -2 satisfies the original inequality. This interval does not include 3 because 3 would make the denominator 0.

■ **Work Practice 4**

The steps shown below may be used to solve a rational inequality with variables in the denominator.

Solving a Rational Inequality

Step 1: Find values that make any denominators 0.

Step 2: Solve the related equation.

Step 3: Separate the number line into regions with the solutions from Steps 1 and 2.

Step 4: For each region, choose a test point and determine whether its value satisfies the *original inequality*.

Step 5: The solution set includes the regions whose test point value is a solution. Check whether to include values from Step 2. Be sure *not* to include values that make any denominator 0.

Example 5   Solve: $\dfrac{5}{x + 1} < -2$

Solution: First we find values for x that make the denominator equal to 0.

$$x + 1 = 0$$

$$x = -1$$

Next we solve $\dfrac{5}{x + 1} = -2$.

$$(x + 1) \cdot \frac{5}{x + 1} = (x + 1) \cdot -2 \quad \text{Multiply both sides by the LCD, } x + 1.$$

$$5 = -2x - 2 \quad \text{Simplify.}$$

$$7 = -2x$$

$$-\frac{7}{2} = x$$

We use these two solutions to divide a number line into three regions and choose test points. Only a test point value from region B satisfies the *original inequality*.

The solution set is $\left(-\dfrac{7}{2}, -1\right)$, and its graph is shown to the right.

■ **Work Practice 5**

Practice 5

Solve: $\dfrac{3}{x - 2} < 2$

Answer

5. $(-\infty, 2) \cup \left(\dfrac{7}{2}, \infty\right)$

A.6 Exercise Set MyMathLab®

Objective A *Solve. See Examples 1 through 3.*

1. $(x + 1)(x + 5) > 0$

2. $(x + 1)(x + 5) \le 0$

3. $(x - 3)(x + 4) \le 0$

4. $(x + 4)(x - 1) > 0$

5. $x^2 + 8x + 15 \ge 0$

6. $x^2 - 7x + 10 \le 0$

7. $3x^2 + 16x < -5$

8. $2x^2 - 5x < 7$

9. $(x - 6)(x - 4)(x - 2) > 0$

10. $(x - 6)(x - 4)(x - 2) \le 0$

11. $x(x - 1)(x + 4) \le 0$

12. $x(x - 6)(x + 2) > 0$

13. $(x^2 - 9)(x^2 - 4) > 0$

14. $(x^2 - 16)(x^2 - 1) \le 0$

Objective B *Solve. See Examples 4 and 5.*

15. $\dfrac{x + 7}{x - 2} < 0$

16. $\dfrac{x - 5}{x - 6} > 0$

17. $\dfrac{5}{x + 1} > 0$

18. $\dfrac{3}{y - 5} < 0$

19. $\dfrac{x + 1}{x - 4} \ge 0$

20. $\dfrac{x + 1}{x - 4} \le 0$

21. $\dfrac{3}{x - 2} < 4$

22. $\dfrac{-2}{y + 3} > 2$

23. $\dfrac{x^2 + 6}{5x} \ge 1$

24. $\dfrac{y^2 + 15}{8y} \le 1$

25. $\dfrac{x + 2}{x - 3} < 1$

26. $\dfrac{x - 1}{x + 4} > 2$

Objectives A B Mixed Practice *Solve each inequality. Write the solution set in interval notation. See Examples 1 through 5.*

27. $(2x - 3)(4x + 5) \le 0$

28. $(6x + 7)(7x - 12) > 0$

29. $x^2 > x$

30. $x^2 < 25$

31. $\dfrac{x}{x - 10} < 0$

32. $\dfrac{x + 10}{x - 10} > 0$

33. $(2x - 8)(x + 4)(x - 6) \le 0$

34. $(3x - 12)(x + 5)(2x - 3) \ge 0$

35. $6x^2 - 5x \ge 6$

36. $12x^2 + 11x \le 15$

37. $\dfrac{x - 5}{x + 4} \ge 0$

38. $\dfrac{x - 3}{x + 2} \le 0$

39. $\dfrac{-1}{x-1} > -1$

40. $\dfrac{4}{y+2} < -2$

41. $4x^3 + 16x^2 - 9x - 36 > 0$

42. $x^3 + 2x^2 - 4x - 8 < 0$

43. $x^4 - 26x^2 + 25 \geq 0$

44. $16x^4 - 40x^2 + 9 \leq 0$

45. $\dfrac{x(x+6)}{(x-7)(x+1)} \geq 0$

46. $\dfrac{(x-2)(x+2)}{(x+1)(x-4)} \leq 0$

47. $\dfrac{x}{x+4} \leq 2$

48. $\dfrac{4x}{x-3} \geq 5$

49. $(2x-7)(3x+5) > 0$

50. $(4x-9)(2x+5) < 0$

51. $\dfrac{z}{z-5} \geq 2z$

52. $\dfrac{p}{p+4} \leq 3p$

53. $\dfrac{(x+1)^2}{5x} > 0$

54. $\dfrac{(2x-3)^2}{x} < 0$

55. Explain why $\dfrac{x+2}{x-3} > 0$ and $(x+2)(x-3) > 0$ have the same solution sets.

56. Explain why $\dfrac{x+2}{x-3} \geq 0$ and $(x+2)(x-3) \geq 0$ do not have the same solution sets.

A.7 Rational Exponents

Objective A Understanding $a^{1/n}$

So far in this text, we have not defined expressions with rational exponents such as $3^{1/2}$, $x^{2/3}$, and $-9^{-1/4}$. We will define these expressions so that the rules for exponents apply to these rational exponents as well.

Suppose that $x = 5^{1/3}$. Then

$$x^3 = (5^{1/3})^3 = 5^{1/3 \cdot 3} = 5^1 \text{ or } 5$$

$\qquad\qquad\uparrow\!\!\!\!\text{--- using rules ↑}$
$\qquad\qquad\quad\text{for exponents}$

Since $x^3 = 5$, then x is the number whose cube is 5, or $x = \sqrt[3]{5}$. Notice that we also know that $x = 5^{1/3}$. This means that

$$5^{1/3} = \sqrt[3]{5}$$

> **Definition of $a^{1/n}$**
>
> If n is a positive integer greater than 1 and $\sqrt[n]{a}$ is a real number, then
>
> $$a^{1/n} = \sqrt[n]{a}$$

Notice that the denominator of the rational exponent corresponds to the index of the radical.

Objectives

A Understand the Meaning of $a^{1/n}$.

B Understand the Meaning of $a^{m/n}$.

C Understand the Meaning of $a^{-m/n}$.

D Use Rules for Exponents to Simplify Expressions That Contain Rational Exponents.

E Use Rational Exponents to Simplify Radical Expressions.

Practice 1–6

Use radical notation to rewrite each expression. Simplify if possible.

1. $25^{1/2}$ **2.** $125^{1/3}$

3. $x^{1/5}$ **4.** $-25^{1/2}$

5. $(-27y^6)^{1/3}$ **6.** $7x^{1/5}$

Examples Use radical notation to rewrite each expression. Simplify if possible.

1. $4^{1/2} = \sqrt{4} = 2$

2. $64^{1/3} = \sqrt[3]{64} = 4$

3. $x^{1/4} = \sqrt[4]{x}$

4. $-9^{1/2} = -\sqrt{9} = -3$

5. $(81x^8)^{1/4} = \sqrt[4]{81x^8} = 3x^2$

6. $5y^{1/3} = 5\sqrt[3]{y}$

■ **Work Practice 1–6**

Objective B Understanding $a^{m/n}$

As we expand our use of exponents to include $\dfrac{m}{n}$, we define their meaning so that rules for exponents still hold true. For example, by properties of exponents,

$$8^{2/3} = (8^{1/3})^2 = (\sqrt[3]{8})^2 \quad \text{or} \quad 8^{2/3} = (8^2)^{1/3} = \sqrt[3]{8^2}$$

Definition of $a^{m/n}$

If m and n are positive integers greater than 1 with $\dfrac{m}{n}$ in simplest form, then

$$a^{m/n} = \sqrt[n]{a^m} = (\sqrt[n]{a})^m$$

as long as $\sqrt[n]{a}$ is a real number.

Helpful Hint Most of the time, $(\sqrt[n]{a})^m$ will be easier to calculate than $\sqrt[n]{a^m}$.

Notice that the denominator n of the rational exponent corresponds to the index of the radical. The numerator m of the rational exponent indicates that the base is to be raised to the mth power. This means that

$$8^{2/3} = \sqrt[3]{8^2} = \sqrt[3]{64} = 4 \quad \text{or} \quad 8^{2/3} = (\sqrt[3]{8})^2 = 2^2 = 4$$

Practice 7–11

Use radical notation to rewrite each expression. Simplify if possible.

7. $9^{3/2}$ **8.** $-256^{3/4}$

9. $(-32)^{2/5}$ **10.** $\left(\dfrac{1}{4}\right)^{3/2}$

11. $(2x + 1)^{2/7}$

Examples Use radical notation to rewrite each expression. Simplify if possible.

7. $4^{3/2} = (\sqrt{4})^3 = 2^3 = 8$

8. $-16^{3/4} = -(\sqrt[4]{16})^3 = -(2)^3 = -8$

9. $(-27)^{2/3} = (\sqrt[3]{-27})^2 = (-3)^2 = 9$

10. $\left(\dfrac{1}{9}\right)^{3/2} = \left(\sqrt{\dfrac{1}{9}}\right)^3 = \left(\dfrac{1}{3}\right)^3 = \dfrac{1}{27}$

11. $(4x - 1)^{3/5} = \sqrt[5]{(4x - 1)^3}$

■ **Work Practice 7–11**

Helpful Hint The *denominator* of a rational exponent is the index of the corresponding radical. For example, $x^{1/5} = \sqrt[5]{x}$, and $z^{2/3} = \sqrt[3]{z^2}$ or $z^{2/3} = (\sqrt[3]{z})^2$.

Objective C Understanding $a^{-m/n}$

The rational exponents we have given meaning to exclude negative rational numbers. To complete the set of definitions, we define $a^{-m/n}$.

Answers

1. 5 **2.** 5 **3.** $\sqrt[5]{x}$ **4.** -5 **5.** $-3y^2$

6. $7\sqrt[5]{x}$ **7.** 27 **8.** -64 **9.** 4 **10.** $\dfrac{1}{8}$

11. $\sqrt[7]{(2x + 1)^2}$

Definition of $a^{-m/n}$

$$a^{-m/n} = \frac{1}{a^{m/n}}$$

as long as $a^{m/n}$ is a nonzero real number.

Examples Write each expression with a positive exponent. Then simplify.

12. $16^{-3/4} = \dfrac{1}{16^{3/4}} = \dfrac{1}{(\sqrt[4]{16})^3} = \dfrac{1}{2^3} = \dfrac{1}{8}$

13. $(-27)^{-2/3} = \dfrac{1}{(-27)^{2/3}} = \dfrac{1}{(\sqrt[3]{-27})^2} = \dfrac{1}{(-3)^2} = \dfrac{1}{9}$

■ **Work Practice 12–13**

Helpful Hint

If an expression contains a negative rational exponent, you may want to first write the expression with a positive exponent, then interpret the rational exponent. Notice that the sign of the base is not affected by the sign of its exponent. For example,

$$9^{-3/2} = \frac{1}{9^{3/2}} = \frac{1}{(\sqrt{9})^3} = \frac{1}{27}$$

Also,

$$(-27)^{-1/3} = \frac{1}{(-27)^{1/3}} = -\frac{1}{3}$$

✓ Concept Check Which one is correct?

a. $-8^{2/3} = \dfrac{1}{4}$ **b.** $8^{-2/3} = -\dfrac{1}{4}$ **c.** $8^{-2/3} = -4$ **d.** $-8^{-2/3} = -\dfrac{1}{4}$

Objective D Using Rules for Exponents ▶

It can be shown that the properties of integer exponents hold for rational exponents. By using these properties and definitions, we can now simplify expressions that contain rational exponents. These rules are repeated here for review.

Summary of Exponent Rules

If m and n are rational numbers, and a, b, and c are numbers for which the expressions below exist, then

Product rule for exponents: $a^m \cdot a^n = a^{m+n}$

Power rule for exponents: $(a^m)^n = a^{m \cdot n}$

Power rules for products and quotients: $(ab)^n = a^n b^n$ and

$$\left(\frac{a}{c}\right)^n = \frac{a^n}{c^n}, \quad c \neq 0$$

Quotient rule for exponents: $\dfrac{a^m}{a^n} = a^{m-n}, \quad a \neq 0$

Zero exponent: $a^0 = 1, \quad a \neq 0$

Negative exponent: $a^{-n} = \dfrac{1}{a^n}, \quad a \neq 0$

Practice 12–13

Write each expression with a positive exponent. Then simplify.

12. $27^{-2/3}$ **13.** $-256^{-3/4}$

Answers

12. $\dfrac{1}{9}$ **13.** $-\dfrac{1}{64}$

✓ Concept Check Answer

d

Practice 14–18

Use the properties of exponents to simplify.

14. $x^{1/3}x^{1/4}$ **15.** $\dfrac{9^{2/5}}{9^{12/5}}$

16. $y^{-3/10} \cdot y^{6/10}$ **17.** $(11^{2/9})^3$

18. $\dfrac{(3x^{2/3})^3}{x^2}$

Examples   Use the properties of exponents to simplify.

14. $x^{1/2}x^{1/3} = x^{1/2+1/3} = x^{3/6+2/6} = x^{5/6}$ Use the product rule.

15. $\dfrac{7^{1/3}}{7^{4/3}} = 7^{1/3-4/3} = 7^{-3/3} = 7^{-1} = \dfrac{1}{7}$ Use the quotient rule.

16. $y^{-4/7} \cdot y^{6/7} = y^{-4/7+6/7} = y^{2/7}$ Use the product rule.

17. $(5^{3/8})^4 = 5^{3/8 \cdot 4} = 5^{12/8} = 5^{3/2}$ Use the power rule.

18. $\dfrac{(2x^{2/5})^5}{x^2} = \dfrac{2^5(x^{2/5})^5}{x^2}$ Use the power rule.

$\qquad = \dfrac{32x^2}{x^2}$ Simplify.

$\qquad = 32x^{2-2}$ Use the quotient rule.

$\qquad = 32x^0$ Simplify.

$\qquad = 32 \cdot 1 \quad \text{or} \quad 32$ Substitute 1 for x^0.

■ Work Practice 14–18

Objective E Using Rational Exponents to Simplify Radical Expressions ▶

We can simplify some radical expressions by first writing the expression with rational exponents. Use the properties of exponents to simplify, and then convert back to radical notation.

Practice 19–21

Use rational exponents to simplify. Assume that all variables represent positive real numbers.

19. $\sqrt[10]{y^5}$ **20.** $\sqrt[4]{9}$

21. $\sqrt[9]{a^6b^3}$

Examples   Use rational exponents to simplify. Assume that all variables represent positive real numbers.

19. $\sqrt[8]{x^4} = x^{4/8}$ Write with rational exponents.

$\qquad = x^{1/2}$ Simplify the exponent.

$\qquad = \sqrt{x}$ Write with radical notation.

20. $\sqrt[6]{25} = 25^{1/6}$ Write with rational exponents.

$\qquad = (5^2)^{1/6}$ Write 25 as 5^2.

$\qquad = 5^{2/6}$ Use the power rule.

$\qquad = 5^{1/3}$ Simplify the exponent.

$\qquad = \sqrt[3]{5}$ Write with radical notation.

21. $\sqrt[6]{r^2s^4} = (r^2s^4)^{1/6}$ Write with rational exponents.

$\qquad = r^{2/6}s^{4/6}$ Use the power rule.

$\qquad = r^{1/3}s^{2/3}$ Simplify the exponents.

$\qquad = (rs^2)^{1/3}$ Use $a^nb^n = (ab)^n$.

$\qquad = \sqrt[3]{rs^2}$ Write with radical notation.

■ Work Practice 19–21

Practice 22–24

Use rational exponents to write as a single radical.

22. $\sqrt{y} \cdot \sqrt[3]{y}$ **23.** $\dfrac{\sqrt[3]{x}}{\sqrt[4]{x}}$

24. $\sqrt{5} \cdot \sqrt[3]{2}$

Examples   Use rational exponents to write as a single radical.

22. $\sqrt{x} \cdot \sqrt[4]{x} = x^{1/2} \cdot x^{1/4} = x^{1/2+1/4}$

$\qquad = x^{3/4} = \sqrt[4]{x^3}$

23. $\dfrac{\sqrt{x}}{\sqrt[3]{x}} = \dfrac{x^{1/2}}{x^{1/3}} = x^{1/2-1/3} = x^{3/6-2/6}$

$\qquad = x^{1/6} = \sqrt[6]{x}$

Answers

14. $x^{7/12}$ **15.** $\dfrac{1}{81}$ **16.** $y^{3/10}$ **17.** $11^{2/3}$

18. 27   **19.** $\sqrt{y}$ **20.** $\sqrt{3}$ **21.** $\sqrt[3]{a^2b}$

22. $\sqrt[6]{y^5}$ **23.** $\sqrt[12]{x}$ **24.** $\sqrt[6]{500}$

24. $\sqrt[3]{3} \cdot \sqrt{2} = 3^{1/3} \cdot 2^{1/2}$ Write with rational exponents.

$= 3^{2/6} \cdot 2^{3/6}$ Write the exponents so that they have the same denominator.

$= (3^2 \cdot 2^3)^{1/6}$ Use $a^n b^n = (ab)^n$.

$= \sqrt[6]{3^2 \cdot 2^3}$ Write with radical notation.

$= \sqrt[6]{72}$ Multiply $3^2 \cdot 2^3$.

■ **Work Practice 22–24**

A.7 Exercise Set MyMathLab®

Objective A *Use radical notation to rewrite each expression. Simplify if possible. See Examples 1 through 6.*

1. $49^{1/2}$ 　　 **2.** $64^{1/3}$ 　　 **3.** $27^{1/3}$ 　　 **4.** $8^{1/3}$ 　　 **5.** $\left(\dfrac{1}{16}\right)^{1/4}$ 　　 **6.** $\left(\dfrac{1}{64}\right)^{1/2}$

7. $169^{1/2}$ 　　 **8.** $81^{1/4}$ 　　 **9.** $2m^{1/3}$ 　　 **10.** $(2m)^{1/3}$ 　　 **11.** $(9x^4)^{1/2}$ 　　 **12.** $(16x^8)^{1/2}$

13. $(-27)^{1/3}$ 　　 **14.** $-64^{1/2}$ 　　 **15.** $-16^{1/4}$ 　　 **16.** $(-32)^{1/5}$

Objective B *Use radical notation to rewrite each expression. Simplify if possible. See Examples 7 through 11.*

17. $16^{3/4}$ 　　 **18.** $4^{5/2}$ 　　 **19.** $(-64)^{2/3}$ 　　 **20.** $(-8)^{4/3}$ 　　 **21.** $(-16)^{3/4}$ 　　 **22.** $(-9)^{3/2}$

23. $(2x)^{3/5}$ 　　 **24.** $2x^{3/5}$ 　　 **25.** $(7x+2)^{2/3}$ 　　 **26.** $(x-4)^{3/4}$ 　　 **27.** $\left(\dfrac{16}{9}\right)^{3/2}$ 　　 **28.** $\left(\dfrac{49}{25}\right)^{3/2}$

Objective C *Write with positive exponents. Simplify if possible. See Examples 12 and 13.*

29. $8^{-4/3}$ 　　 **30.** $64^{-2/3}$ 　　 **31.** $(-64)^{-2/3}$ 　　 **32.** $(-8)^{-4/3}$ 　　 **33.** $(-4)^{-3/2}$ 　　 **34.** $(-16)^{-5/4}$

35. $x^{-1/4}$ 　　 **36.** $y^{-1/6}$ 　　 **37.** $\dfrac{1}{a^{-2/3}}$ 　　 **38.** $\dfrac{1}{n^{-8/9}}$ 　　 **39.** $\dfrac{5}{7x^{-3/4}}$ 　　 **40.** $\dfrac{2}{3y^{-5/7}}$

Objective D *Use the properties of exponents to simplify each expression. Write with positive exponents. See Examples 14 through 18.*

41. $a^{2/3}a^{5/3}$ 　　 **42.** $b^{9/5}b^{8/5}$ 　　 **43.** $x^{-2/5} \cdot x^{7/5}$ 　　 **44.** $y^{4/3} \cdot y^{-1/3}$ 　　 **45.** $3^{1/4} \cdot 3^{3/8}$

46. $5^{1/2} \cdot 5^{1/6}$ 　　 **47.** $\dfrac{y^{1/3}}{y^{1/6}}$ 　　 **48.** $\dfrac{x^{3/4}}{x^{1/8}}$ 　　 **49.** $(4u^2)^{3/2}$ 　　 **50.** $(32^{1/5}x^{2/3})^3$

51. $\dfrac{b^{1/2}b^{3/4}}{-b^{1/4}}$ **52.** $\dfrac{a^{1/4}a^{-1/2}}{a^{2/3}}$ **53.** $\dfrac{(x^3)^{1/2}}{x^{7/2}}$ **54.** $\dfrac{y^{11/3}}{(y^5)^{1/3}}$ **55.** $\dfrac{(3x^{1/4})^3}{x^{1/12}}$

56. $\dfrac{(2x^{1/5})^4}{x^{3/10}}$ **57.** $\dfrac{(y^3z)^{1/6}}{y^{-1/2}z^{1/3}}$ **58.** $\dfrac{(m^2n)^{1/4}}{m^{-1/2}n^{5/8}}$ **59.** $\dfrac{(x^3y^2)^{1/4}}{(x^{-5}y^{-1})^{-1/2}}$ **60.** $\dfrac{(a^{-2}b^3)^{1/8}}{(a^{-3}b)^{-1/4}}$

Objective E   *Use rational exponents to simplify each radical. Assume that all variables represent positive real numbers. See Examples 19 through 21.*

◉ 61. $\sqrt[6]{x^3}$ **62.** $\sqrt[9]{a^3}$ **63.** $\sqrt[6]{4}$ **64.** $\sqrt[4]{36}$ **◉ 65.** $\sqrt[4]{16x^2}$ **66.** $\sqrt[8]{4y^2}$

67. $\sqrt[4]{(x+3)^2}$ **68.** $\sqrt[8]{(y+1)^4}$ **69.** $\sqrt[8]{x^4y^4}$ **70.** $\sqrt[9]{y^6z^3}$ **71.** $\sqrt[12]{a^8b^4}$ **72.** $\sqrt[10]{a^5b^5}$

Use rational expressions to write as a single radical expression. See Examples 22 through 24.

73. $\sqrt[3]{y}\cdot\sqrt[5]{y^2}$ **74.** $\sqrt[3]{y^2}\cdot\sqrt[6]{y}$ **75.** $\dfrac{\sqrt[3]{b^2}}{\sqrt[4]{b}}$ **76.** $\dfrac{\sqrt[4]{a}}{\sqrt[5]{a}}$ **77.** $\sqrt[3]{x}\cdot\sqrt[4]{x}\cdot\sqrt[8]{x^3}$

78. $\sqrt[6]{y}\cdot\sqrt[3]{y}\cdot\sqrt[5]{y^2}$ **79.** $\dfrac{\sqrt[3]{a^2}}{\sqrt[6]{a}}$ **80.** $\dfrac{\sqrt[5]{b^2}}{\sqrt[10]{b^3}}$ **81.** $\sqrt{3}\cdot\sqrt[3]{4}$ **82.** $\sqrt[3]{5}\cdot\sqrt{2}$

83. $\sqrt[5]{7}\cdot\sqrt[3]{y}$ **84.** $\sqrt[4]{5}\cdot\sqrt[3]{x}$ **85.** $\sqrt{5r}\cdot\sqrt[3]{s}$ **86.** $\sqrt[3]{b}\cdot\sqrt[5]{4a}$

Basal metabolic rate (BMR) is the number of calories per day a person needs to maintain life. A person's basal metabolic rate $B(w)$ in calories per day can be estimated with the function $B(w) = 70w^{3/4}$, where w is the person's weight in kilograms. Use this information to answer Exercises 87 and 88.

87. Estimate the BMR for a person who weighs 60 kilograms. Round to the nearest calorie. (*Note:* 60 kilograms is approximately 132 pounds.)

88. Estimate the BMR for a person who weighs 90 kilograms. Round to the nearest calorie. (*Note:* 90 kilograms is approximately 198 pounds.)

✎ 89. Explain how writing x^{-7} with positive exponents is similar to writing $x^{-1/4}$ with positive exponents.

✎ 90. Explain how writing $2x^{-5}$ with positive exponents is similar to writing $2x^{-3/4}$ with positive exponents.

Fill in each box with the correct expression.

91. $\square\cdot a^{2/3} = a^{3/3}$, or a **92.** $\square\cdot x^{1/8} = x^{4/8}$, or $x^{1/2}$ **93.** $\dfrac{\square}{x^{-2/5}} = x^{3/5}$ **94.** $\dfrac{\square}{y^{-3/4}} = y^{4/4}$, or y

Systems of Linear Inequalities

Objective A Graphing Systems of Linear Inequalities

A **solution of a system of linear inequalities** is an ordered pair that satisfies each inequality in the system. The set of all such ordered pairs is the solution set of the system. Graphing this set gives us a picture of the solution set. We can graph a system of inequalities by graphing each inequality in the system and identifying the region of overlap.

Objective

A Graph a System of Linear Inequalities.

Graphing the Solutions of a System of Linear Inequalities

Step 1: Graph each inequality in the system on the same set of axes.

Step 2: The solutions of the system are the points common to the graphs of all the inequalities in the system.

Example 1 Graph the solutions of the system: $\begin{cases} 3x \geq y \\ x + 2y \leq 8 \end{cases}$

Solution: We begin by graphing each inequality on the *same* set of axes. The graph of the solutions of the system is the region contained in the graphs of both inequalities. In other words, it is their intersection.

First let's graph $3x \geq y$. The boundary line is the graph of $3x = y$. We sketch a solid boundary line since the inequality $3x \geq y$ means $3x > y$ or $3x = y$. The test point $(1, 0)$ satisfies the inequality, so we shade the half-plane that includes $(1, 0)$.

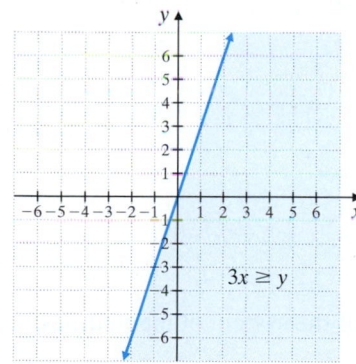

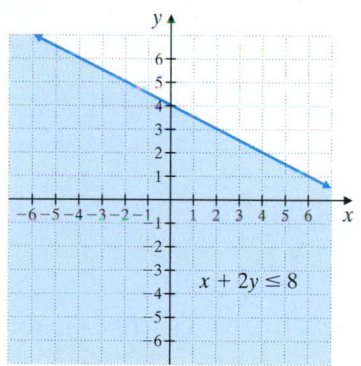

Next we sketch a solid boundary line $x + 2y = 8$ on the same set of axes. The test point $(0, 0)$ satisfies the inequality $x + 2y \leq 8$, so we shade the half-plane that includes $(0, 0)$. (For clarity, the graph of $x + 2y \leq 8$ is shown here on a separate set of axes.) An ordered pair solution of the system must satisfy both inequalities. These solutions are points that lie in both shaded regions. The solution of the system is the darkest shaded region. This solution includes parts of both boundary lines.

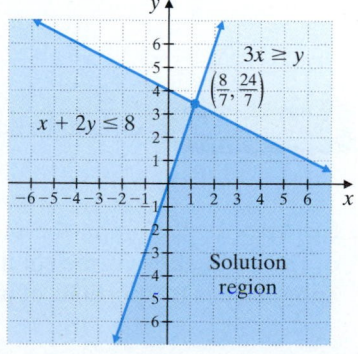

■ **Work Practice 1**

Practice 1

Graph the solutions of the system:

$$\begin{cases} 2x \leq y \\ x + 4y \geq 4 \end{cases}$$

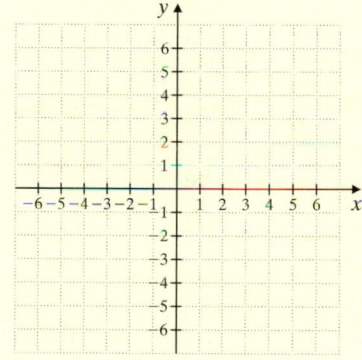

Answer

1.

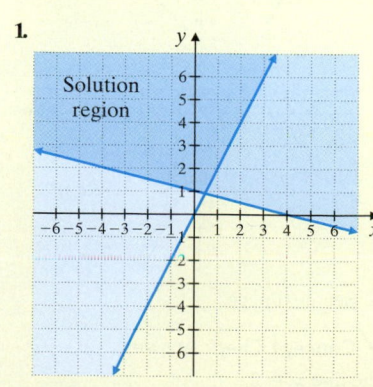

1258

Practice 2

Graph the solutions of the system:

$$\begin{cases} -x + y < 3 \\ \qquad y < 1 \\ 2x + y > -2 \end{cases}$$

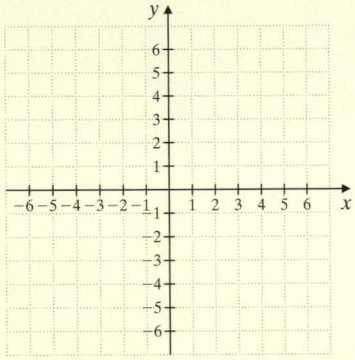

Practice 3

Graph the solutions of the system:

$$\begin{cases} 2x - 3y \le 6 \\ \qquad y \ge 0 \\ \qquad y \le 4 \\ \qquad x \ge 0 \end{cases}$$

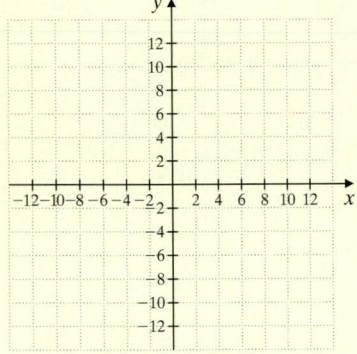

Answers

2

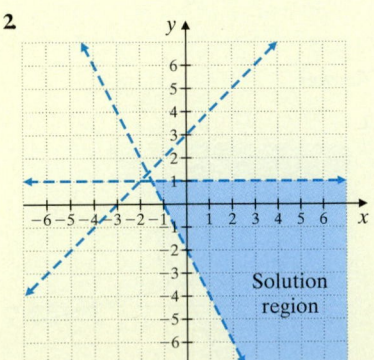

3. See next page.

✓ **Concept Check Answer**

the line $x = 2$

Appendix A | Further Algebraic Topics

In linear programming, it is sometimes necessary to find the coordinates of the **corner point:** the point at which the two boundary lines intersect. To find the corner point for the system of Example 1, we solve the related linear system

$$\begin{cases} 3x = y \\ x + 2y = 8 \end{cases}$$

using either the substitution or the elimination method. The lines intersect at $\left(\dfrac{8}{7}, \dfrac{24}{7} \right)$, the corner point of the graph.

Example 2 Graph the solutions of the system:

$$\begin{cases} x - y < 2 \\ x + 2y > -1 \\ \qquad y < 2 \end{cases}$$

Solution: First we graph all three inequalities on the same set of axes. All boundary lines in the graph below are dashed lines since the inequality symbols are $<$ and $>$. The solution of the system is the region shown by the shading. In this example, the boundary lines are *not* a part of the solution.

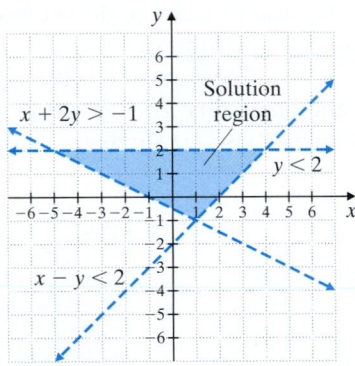

▪ **Work Practice 2**

✓ **Concept Check** Describe the solution of the system of inequalities:

$$\begin{cases} x \le 2 \\ x \ge 2 \end{cases}$$

Example 3 Graph the solutions of the system: $\begin{cases} -3x + 4y \le 12 \\ \qquad x \le 3 \\ \qquad x \ge 0 \\ \qquad y \ge 0 \end{cases}$

Solution: We graph the inequalities on the same set of axes. The intersection of the inequalities is the solution region. It is the only region shaded in this graph and includes the portions of all four boundary lines that border the shaded region.

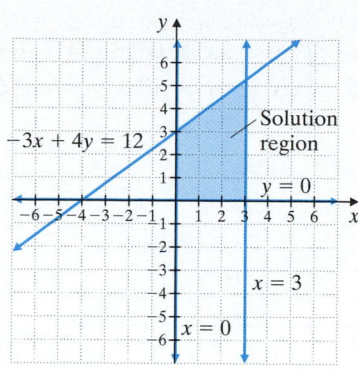

$-3x + 4y = 12$

Solution region

$y = 0$

$x = 3$

$x = 0$

Answer

3.

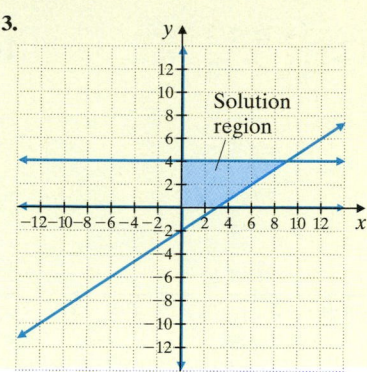

Solution region

Work Practice 3

A.8 Exercise Set MyMathLab® ▶

Objective A *Graph the solutions of each system of linear inequalities. See Examples 1 through 3.*

1. $\begin{cases} y \ge x + 1 \\ y \ge 3 - x \end{cases}$

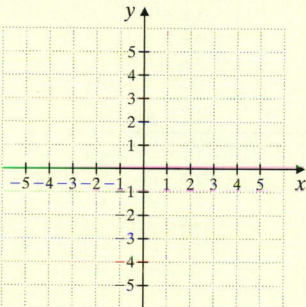

2. $\begin{cases} y \ge x - 3 \\ y \ge -1 - x \end{cases}$

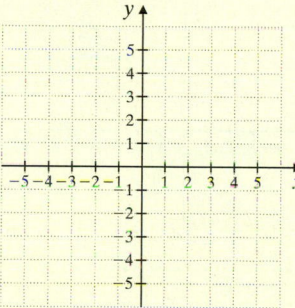

3. $\begin{cases} y < 3x - 4 \\ y \le x + 2 \end{cases}$

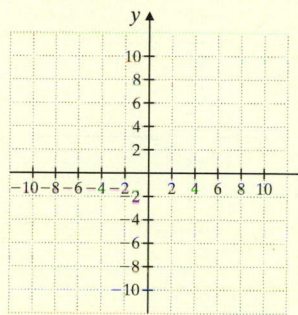

4. $\begin{cases} y \le 2x + 1 \\ y > x + 2 \end{cases}$

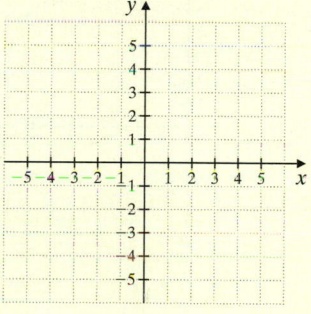

5. $\begin{cases} y < -2x - 2 \\ y > x + 4 \end{cases}$

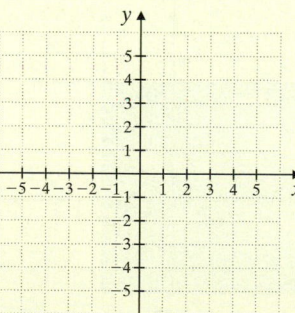

6. $\begin{cases} y \le 2x + 4 \\ y \ge -x - 5 \end{cases}$

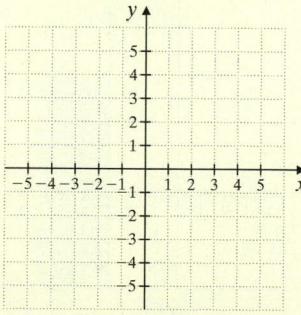

7. $\begin{cases} y \ge -x + 2 \\ y \le 2x + 5 \end{cases}$

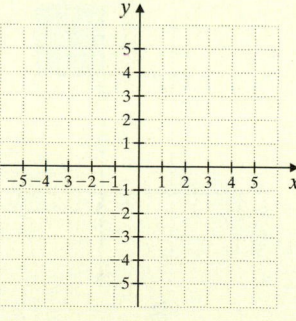

8. $\begin{cases} y \ge x - 5 \\ y \le -3x + 3 \end{cases}$

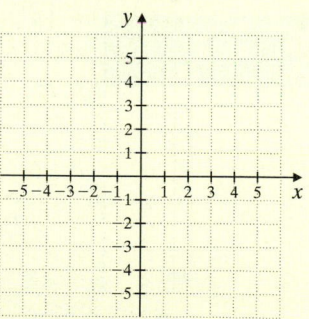

▶ **9.** $\begin{cases} x \ge 3y \\ x + 3y \le 6 \end{cases}$

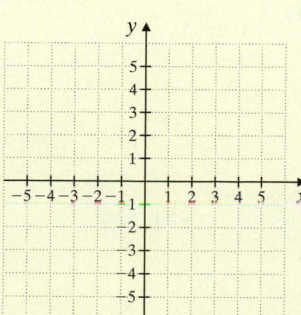

10. $\begin{cases} -2x < y \\ x + 2y < 3 \end{cases}$

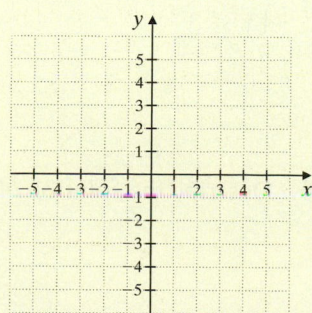

11. $\begin{cases} x \le 2 \\ y \ge -3 \end{cases}$

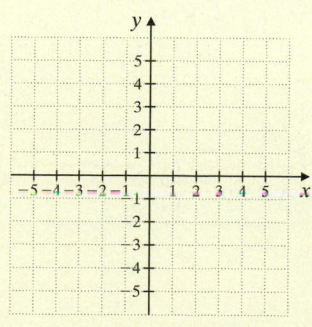

12. $\begin{cases} x \ge -3 \\ y \ge -2 \end{cases}$

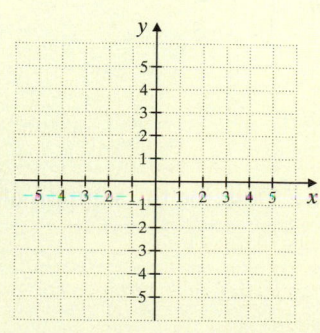

13. $\begin{cases} y \geq 1 \\ x < -3 \end{cases}$

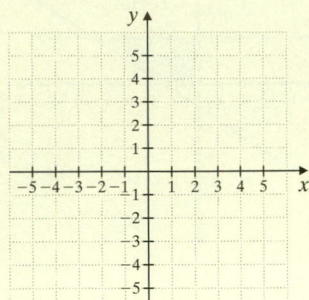

14. $\begin{cases} y > 2 \\ x \geq -1 \end{cases}$

15. $\begin{cases} y + 2x \geq 0 \\ 5x - 3y \leq 12 \\ y \leq 2 \end{cases}$

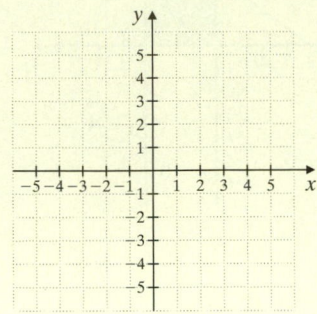

16. $\begin{cases} y + 2x \leq 0 \\ 5x + 3y \geq -2 \\ y \leq 4 \end{cases}$

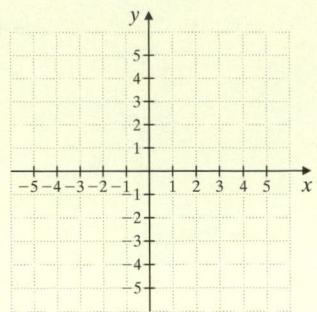

17. $\begin{cases} 3x - 4y \geq -6 \\ 2x + y \leq 7 \\ y \geq -3 \end{cases}$

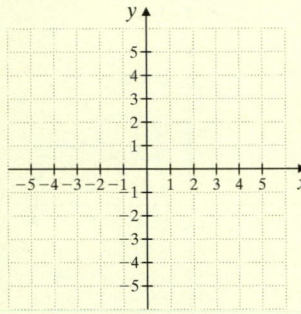

18. $\begin{cases} 4x - y \geq -2 \\ 2x + 3y \leq -8 \\ y \geq -5 \end{cases}$

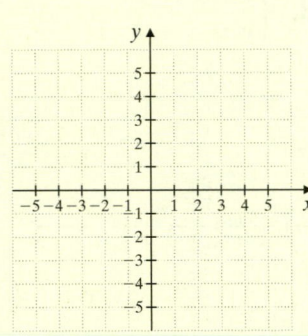

19. $\begin{cases} 2x + y \leq 5 \\ x \leq 3 \\ x \geq 0 \\ y \geq 0 \end{cases}$

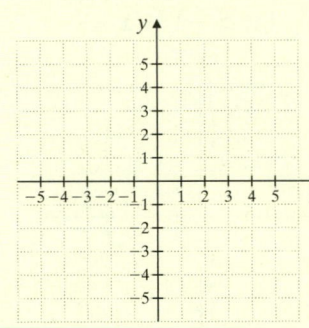

20. $\begin{cases} 3x + y \leq 4 \\ x \leq 4 \\ x \geq 0 \\ y \geq 0 \end{cases}$

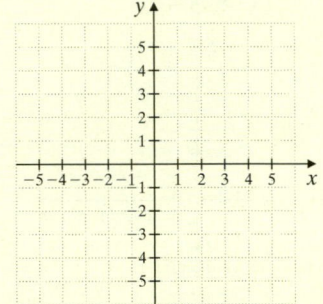

Match each system of inequalities to the corresponding graph.

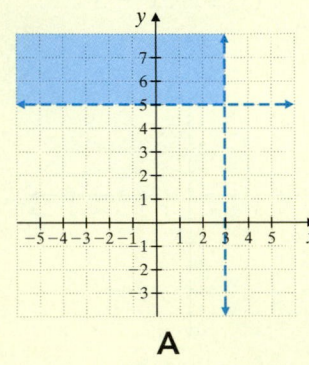
A

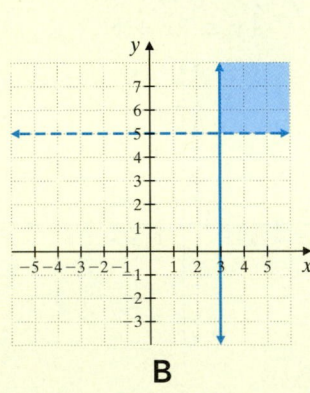
B

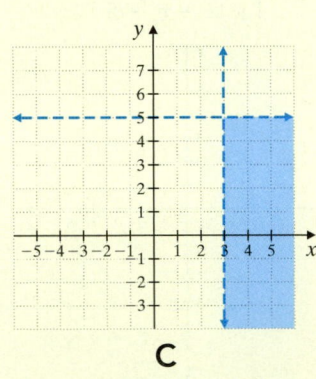
C

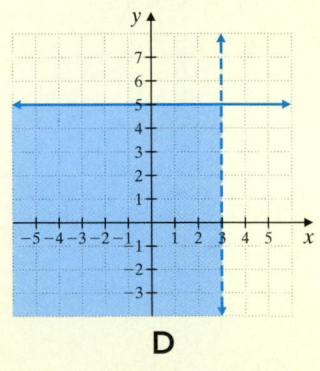
D

21. $\begin{cases} y < 5 \\ x > 3 \end{cases}$

22. $\begin{cases} y > 5 \\ x < 3 \end{cases}$

23. $\begin{cases} y \leq 5 \\ x < 3 \end{cases}$

24. $\begin{cases} y > 5 \\ x \geq 3 \end{cases}$

Solve.

25. Describe the solution of the system: $\begin{cases} y \leq 3 \\ y \geq 3 \end{cases}$

26. Describe the solution of the system: $\begin{cases} x \leq 5 \\ x \leq 3 \end{cases}$

Tables

B.1 Table of Squares and Square Roots ▶

n	n^2	$\sqrt{n}$	n	n^2	$\sqrt{n}$
1	1	1.000	51	2601	7.141
2	4	1.414	52	2704	7.211
3	9	1.732	53	2809	7.280
4	16	2.000	54	2916	7.348
5	25	2.236	55	3025	7.416
6	36	2.449	56	3136	7.483
7	49	2.646	57	3249	7.550
8	64	2.828	58	3364	7.616
9	81	3.000	59	3481	7.681
10	100	3.162	60	3600	7.746
11	121	3.317	61	3721	7.810
12	144	3.464	62	3844	7.874
13	169	3.606	63	3969	7.937
14	196	3.742	64	4096	8.000
15	225	3.873	65	4225	8.062
16	256	4.000	66	4356	8.124
17	289	4.123	67	4489	8.185
18	324	4.243	68	4624	8.246
19	361	4.359	69	4761	8.307
20	400	4.472	70	4900	8.367
21	441	4.583	71	5041	8.426
22	484	4.690	72	5184	8.485
23	529	4.796	73	5329	8.544
24	576	4.899	74	5476	8.602
25	625	5.000	75	5625	8.660
26	676	5.099	76	5776	8.718
27	729	5.196	77	5929	8.775
28	784	5.292	78	6084	8.832
29	841	5.385	79	6241	8.888
30	900	5.477	80	6400	8.944
31	961	5.568	81	6561	9.000
32	1024	5.657	82	6724	9.055
33	1089	5.745	83	6889	9.110
34	1156	5.831	84	7056	9.165
35	1225	5.916	85	7225	9.220
36	1296	6.000	86	7396	9.274
37	1369	6.083	87	7569	9.327
38	1444	6.164	88	7744	9.381
39	1521	6.245	89	7921	9.434
40	1600	6.325	90	8100	9.487
41	1681	6.403	91	8281	9.539
42	1764	6.481	92	8464	9.592
43	1849	6.557	93	8649	9.644
44	1936	6.633	94	8836	9.695
45	2025	6.708	95	9025	9.747
46	2116	6.782	96	9216	9.798
47	2209	6.856	97	9409	9.849
48	2304	6.928	98	9604	9.899
49	2401	7.000	99	9801	9.950
50	2500	7.071	100	10,000	10.000

B.2 Table of Percents, Decimals, and Fraction Equivalents

Percent	Decimal	Fraction
1%	0.01	$\frac{1}{100}$
5%	0.05	$\frac{1}{20}$
10%	0.1	$\frac{1}{10}$
12.5% or $12\frac{1}{2}$%	0.125	$\frac{1}{8}$
$16.\overline{6}$% or $16\frac{2}{3}$%	$0.1\overline{6}$	$\frac{1}{6}$
20%	0.2	$\frac{1}{5}$
25%	0.25	$\frac{1}{4}$
30%	0.3	$\frac{3}{10}$
$33.\overline{3}$% or $33\frac{1}{3}$%	$0.\overline{3}$	$\frac{1}{3}$
37.5% or $37\frac{1}{2}$%	0.375	$\frac{3}{8}$
40%	0.4	$\frac{2}{5}$
50%	0.5	$\frac{1}{2}$
60%	0.6	$\frac{3}{5}$
62.5% or $62\frac{1}{2}$%	0.625	$\frac{5}{8}$
$66.\overline{6}$% or $66\frac{2}{3}$%	$0.\overline{6}$	$\frac{2}{3}$
70%	0.7	$\frac{7}{10}$
75%	0.75	$\frac{3}{4}$
80%	0.8	$\frac{4}{5}$
$83.\overline{3}$% or $83\frac{1}{3}$%	$0.8\overline{3}$	$\frac{5}{6}$
87.5% or $87\frac{1}{2}$%	0.875	$\frac{7}{8}$
90%	0.9	$\frac{9}{10}$
100%	1.0	1
110%	1.1	$1\frac{1}{10}$
125%	1.25	$1\frac{1}{4}$
$133.\overline{3}$% or $133\frac{1}{3}$%	$1.\overline{3}$	$1\frac{1}{3}$
150%	1.5	$1\frac{1}{2}$
$166.\overline{6}$% or $166\frac{2}{3}$%	$1.\overline{6}$	$1\frac{2}{3}$
175%	1.75	$1\frac{3}{4}$
200%	2.0	2

 **Compound Interest**

Compounded Annually

	5%	6%	7%	8%	9%	10%	11%	12%	13%	14%	15%	16%	17%	18%
1 year	1.05000	1.06000	1.07000	1.08000	1.09000	1.10000	1.11000	1.12000	1.13000	1.14000	1.15000	1.16000	1.17000	1.18000
5 years	1.27628	1.33823	1.40255	1.46933	1.53862	1.61051	1.68506	1.76234	1.84244	1.92541	2.01136	2.10034	2.19245	2.28776
10 years	1.62889	1.79085	1.96715	2.15892	2.36736	2.59374	2.83942	3.10585	3.39457	3.70722	4.04556	4.41144	4.80683	5.23384
15 years	2.07893	2.39656	2.75903	3.17217	3.64248	4.17725	4.78459	5.47357	6.25427	7.13794	8.13706	9.26552	10.53872	11.97375
20 years	2.65330	3.20714	3.86968	4.66096	5.60441	6.72750	8.06231	9.64629	11.52309	13.74349	16.36654	19.46076	23.10560	27.39303

Compounded Semiannually

	5%	6%	7%	8%	9%	10%	11%	12%	13%	14%	15%	16%	17%	18%
1 year	1.05063	1.06090	1.07123	1.08160	1.09203	1.10250	1.11303	1.12360	1.13423	1.14490	1.15563	1.16640	1.17723	1.18810
5 years	1.28008	1.34392	1.41060	1.48024	1.55297	1.62889	1.70814	1.79085	1.87714	1.96715	2.06103	2.15892	2.26098	2.36736
10 years	1.63862	1.80611	1.98979	2.19112	2.41171	2.65330	2.91776	3.20714	3.52365	3.86968	4.24785	4.66096	5.11205	5.60441
15 years	2.09757	2.42726	2.80679	3.24340	3.74532	4.32194	4.98395	5.74349	6.61437	7.61226	8.75496	10.06266	11.55825	13.26768
20 years	2.68506	3.26204	3.95926	4.80102	5.81636	7.03999	8.51331	10.28572	12.41607	14.97446	18.04424	21.72452	26.13302	31.40942

Compounded Quarterly

	5%	6%	7%	8%	9%	10%	11%	12%	13%	14%	15%	16%	17%	18%
1 year	1.05095	1.06136	1.07186	1.08243	1.09308	1.10381	1.11462	1.12551	1.13648	1.14752	1.15865	1.16986	1.18115	1.19252
5 years	1.28204	1.34686	1.41478	1.48595	1.56051	1.63862	1.72043	1.80611	1.89584	1.98979	2.08815	2.19112	2.29891	2.41171
10 years	1.64362	1.81402	2.00160	2.20804	2.43519	2.68506	2.95987	3.26204	3.59420	3.95926	4.36038	4.80102	5.28497	5.81636
15 years	2.10718	2.44322	2.83182	3.28103	3.80013	4.39979	5.09225	5.89160	6.81402	7.87809	9.10513	10.51963	12.14965	14.02741
20 years	2.70148	3.29066	4.00639	4.87544	5.93015	7.20957	8.76085	10.64089	12.91828	15.67574	19.01290	23.04980	27.93091	33.83010

Compounded Daily

	5%	6%	7%	8%	9%	10%	11%	12%	13%	14%	15%	16%	17%	18%
1 year	1.05127	1.06183	1.07250	1.08328	1.09416	1.10516	1.11626	1.12747	1.13880	1.15024	1.16180	1.17347	1.18526	1.19716
5 years	1.28400	1.34983	1.41902	1.49176	1.56823	1.64861	1.73311	1.82194	1.91532	2.01348	2.11667	2.22515	2.33918	2.45906
10 years	1.64866	1.82203	2.01362	2.22535	2.45933	2.71791	3.00367	3.31946	3.66845	4.05411	4.48031	4.95130	5.47178	6.04696
15 years	2.11689	2.45942	2.85736	3.31968	3.85678	4.48077	5.20569	6.04786	7.02625	8.16288	9.48335	11.01738	12.79950	14.86983
20 years	2.71810	3.31979	4.05466	4.95216	6.04831	7.38703	9.02202	11.01883	13.45751	16.43582	20.07316	24.51533	29.94039	36.56577

Contents of Student Resources

Student Resources

Study Skills Builders

Attitude and Study Tips

Study Skills Builder 1

Have You Decided to Complete This Course Successfully?

Ask yourself if one of your current goals is to complete this course successfully.

If it is not a goal of yours, ask yourself why. One common reason is fear of failure. Amazingly enough, fear of failure alone can be strong enough to keep many of us from doing our best in any endeavor.

Another common reason is that you simply haven't taken the time to think about or write down your goals for this course. To help accomplish this, answer the questions below.

Exercises

1. Write down your goal(s) for this course.

2. Now list steps you will take to make sure your goal(s) in Exercise 1 are accomplished.

3. Rate your commitment to this course with a number between 1 and 5. Use the diagram below to help.

High Commitment		Average Commitment		Not Committed at All
5	4	3	2	1

4. If you have rated your personal commitment level (from the exercise above) as a 1, 2, or 3, list the reasons why this is so. Then determine whether it is possible to increase your commitment level to a 4 or 5.

Good luck, and don't forget that a positive attitude will make a big difference.

Study Skills Builder 2

Tips for Studying for an Exam

To prepare for an exam, try the following study techniques:

- Start the study process days before your exam.
- Make sure that you are up to date on your assignments.
- If there is a topic that you are unsure of, use one of the many resources that are available to you. For example,

  See your instructor.
  View a lecture video on the topic.
  Visit a learning resource center on campus.
  Read the textbook material and examples on the topic.

- Reread your notes and carefully review the Chapter Highlights at the end of any chapter.
- Work the review exercises at the end of the chapter.
- Find a quiet place to take the Chapter Test found at the end of the chapter. Do not use any resources when taking this sample test. This way, you will have a clear indication of how prepared you are for your exam. Check your answers and use the Chapter Test Prep Videos to make sure that you correct any missed exercises.

Good luck, and keep a positive attitude.

Exercises

Let's see how you did on your last exam.

1. How many days before your last exam did you start studying for that exam?

2. Were you up to date on your assignments at that time or did you need to catch up on assignments?

3. List the most helpful text supplement (if you used one).

4. List the most helpful campus supplement (if you used one).

5. List your process for preparing for a mathematics test.

6. Was this process helpful? In other words, were you satisfied with your performance on your exam?

7. If not, what changes can you make in your process that will make it more helpful to you?

Study Skills Builder 3

What to Do the Day of an Exam

Your first exam may be soon. On the day of an exam, don't forget to try the following:

- Allow yourself plenty of time to arrive.
- Read the directions on the test carefully.
- Read each problem carefully as you take your test. Make sure that you answer the question asked.
- Watch your time and pace yourself so that you may attempt each problem on your test.
- Check your work and answers.
- ***Do not turn your test in early.*** If you have extra time, spend it double-checking your work.

Good luck!

Exercises

Answer the following questions based on your most recent mathematics exam, whenever that was.

1. How soon before class did you arrive?

2. Did you read the directions on the test carefully?

3. Did you make sure you answered the question asked for each problem on the exam?

4. Were you able to attempt each problem on your exam?

5. If your answer to Exercise 4 is no, list reasons why.

6. Did you have extra time on your exam?

7. If your answer to Exercise 6 is yes, describe how you spent that extra time.

Study Skills Builder 4

Are You Satisfied with Your Performance on a Particular Quiz or Exam?

If not, don't forget to analyze your quiz or exam and look for common errors. Were most of your errors a result of:

- *Carelessness?* Did you turn in your quiz or exam before the allotted time expired? If so, resolve to use any extra time to check your work.

- *Running out of time?* Answer the questions you are sure of first. Then attempt the questions you are unsure of, and delay checking your work until all questions have been answered.

- *Not understanding a concept?* If so, review that concept and correct your work so that you make sure you understand it before the next quiz or the final exam.

- *Test conditions?* When studying for a quiz or exam, make sure you place yourself in conditions similar to test conditions. For example, before your next quiz or exam, take a sample test without the aid of your notes or text. (For a sample test, see your instructor or use the Chapter Test at the end of each chapter.)

Exercises

1. Have you corrected all your previous quizzes and exams?

2. List any errors you have found common to two or more of your graded papers.

3. Is one of your common errors not understanding a concept? If so, are you making sure you understand all the concepts for the next quiz or exam?

4. Is one of your common errors making careless mistakes? If so, are you now taking all the time allotted to check over your work so that you can minimize the number of careless mistakes?

5. Are you satisfied with your grades thus far on quizzes and tests?

6. If your answer to Exercise 5 is no, are there any more suggestions you can make to your instructor or yourself to help? If so, list them here and share these with your instructor.

Study Skills Builder 5

How Are You Doing?

If you haven't done so yet, take a few moments and think about how you are doing in this course. Are you working toward your goal of successfully completing this course? Is your performance on homework, quizzes, and tests satisfactory? If not, you might want to see your instructor to see if he/she has any suggestions on how you can improve your performance. Reread Section 1.1 for ideas on places to get help with your mathematics course.

Exercises

Answer the following.

1. List any textbook supplements you are using to help you through this course.

2. List any campus resources you are using to help you through this course.

3. Write a short paragraph describing how you are doing in your mathematics course.

4. If improvement is needed, list ways that you can work toward improving your situation as described in Exercise 3.

Study Skills Builder 6

Are You Preparing for Your Final Exam?

To prepare for your final exam, try the following study techniques:

- Review the material that you will be responsible for on your exam. This includes material from your textbook, your notebook, and any handouts from your instructor.
- Review any formulas that you may need to memorize.
- Check to see if your instructor or mathematics department will be conducting a final exam review.
- Check with your instructor to see whether final exams from previous semesters/quarters are available to students for review.

- Use your previously taken exams as a practice final exam. To do so, rewrite the test questions in mixed order on blank sheets of paper. This will help you prepare for exam conditions.
- If you are unsure of a few concepts, see your instructor or visit a learning lab for assistance. Also, view the video segment of any troublesome sections.
- If you need further exercises to work, try the Cumulative Reviews at the end of the chapters.

Once again, good luck! I hope you are enjoying this textbook and your mathematics course.

Organizing Your Work

Study Skills Builder 7

Learning New Terms

Many of the terms used in this text may be new to you. It will be helpful to make a list of new mathematical terms and symbols as you encounter them and to review them frequently. Placing these new terms (including page references) on 3×5 index cards might help you later when you're preparing for a quiz.

Exercises

1. Name one way you might place a word and its definition on a 3×5 card.

2. How do new terms stand out in this text so that they can be found?

Study Skills Builder 8

Are You Organized?

Have you ever had trouble finding a completed assignment? When it's time to study for a test, are your notes neat and organized? Have you ever had trouble reading your own mathematics handwriting? (Be honest—I have.)

When any of these things happens, it's time to get organized. Here are a few suggestions:

- Write your notes and complete your homework assignments in a notebook with pockets (spiral or ring binder).

- Take class notes in this notebook, and then follow the notes with your completed homework assignment.

- When you receive graded papers or handouts, place them in the notebook pocket so that you will not lose them.

- Mark (possibly with an exclamation point) any note(s) that seem extra important to you.

- Mark (possibly with a question mark) any notes or homework that you are having trouble with.

- See your instructor or a math tutor to help you with the concepts or exercises that you are having trouble understanding.

- If you are having trouble reading your own handwriting, *slow down* and write your mathematics work clearly!

Exercises

1. Have you been completing your assignments on time?

2. Have you been correcting any exercises you may be having difficulty with?

3. If you are having trouble with a mathematical concept or correcting any homework exercises, have you visited your instructor, a tutor, or your campus math lab?

4. Are you taking lecture notes in your mathematics course? (By the way, these notes should include worked-out examples solved by your instructor.)

5. Is your mathematics course material (handouts, graded papers, lecture notes) organized?

6. If your answer to Exercise 5 is no, take a moment and review your course material. List at least two ways that you might better organize it.

Study Skills Builder 9

Organizing a Notebook

It's never too late to get organized. If you need ideas about organizing a notebook for your mathematics course, try some of these:

- Use a spiral or ring binder notebook with pockets and use it for mathematics only.
- Start each page by writing the book's section number you are working on at the top.
- When your instructor is lecturing, take notes. *Always* include any examples your instructor works for you.
- Place your worked-out homework exercises in your notebook immediately after the lecture notes from that section. This way, a section's worth of material is together.
- Homework exercises: Attempt and check all assigned homework.
- Place graded quizzes in the pockets of your notebook or a special section of your binder.

Exercises

Check your notebook organization by answering the following questions.

1. Do you have a spiral or ring binder notebook for your mathematics course only?

2. Have you ever had to flip through several sheets of notes and work in your mathematics notebook to determine what section's work you are in?

3. Are you now writing the textbook's section number at the top of each notebook page?

4. Have you ever lost or had trouble finding a graded quiz or test?

5. Are you now placing all your graded work in a dedicated place in your notebook?

6. Are you attempting all of your homework and placing all of your work in your notebook?

7. Are you checking and correcting your homework in your notebook? If not, why not?

8. Are you writing in your notebook the examples your instructor works for you in class?

Study Skills Builder 10

How Are Your Homework Assignments Going?

It is very important in mathematics to keep up with homework. Why? Many concepts build on each other. Often your understanding of a day's concepts depends on an understanding of the previous day's material.

Remember that completing your homework assignment involves a lot more than attempting a few of the problems assigned.

To complete a homework assignment, remember these four things:

- Attempt all of it.
- Check it.
- Correct it.
- If needed, ask questions about it.

Exercises

Take a moment and review your completed homework assignments. Answer the questions below based on this review.

1. Approximate the fraction of your homework you have attempted.

2. Approximate the fraction of your homework you have checked (if possible).

3. If you are able to check your homework, have you corrected it when errors have been found?

4. When working homework, if you do not understand a concept, what do you do?

MyMathLab and MathXL

Study Skills Builder 11

Tips for Turning In Your Homework on Time

It is very important to keep up with your mathematics homework assignments. Why? Many concepts in mathematics build upon each other.

Remember these 4 tips to help ensure your work is completed on time:

- Know the assignments and due dates set by your instructor.
- Do not wait until the last minute to submit your homework.
- Set a goal to submit your homework 6–8 hours before the scheduled due date in case you have unexpected technology trouble.
- Schedule enough time to complete each assignment.

Following the tips above will also help you avoid potentially losing points for late or missed assignments.

Exercises

Take a moment to consider your work on your homework assignments to date and answer the following questions:

1. What percentage of your assignments have you turned in on time?

2. Why might it be a good idea to submit your homework 6–8 hours before the scheduled deadline?

3. If you have missed submitting any homework by the due date, list some of the reasons why this occurred.

4. What steps do you plan to take in the future to ensure your homework is submitted on time?

Study Skills Builder 12

Tips for Doing Your Homework Online

Practice is one of the main keys to success in any mathematics course. Did you know that MyMathLab/MathXL provides you with **immediate feedback** for each exercise? If you are incorrect, you are given hints to work the exercise correctly. You have **unlimited practice opportunities** and can rework any exercises you have trouble with until you master them, and submit homework assignments unlimited times before the deadline.

Remember these success tips when doing your homework online:

- Attempt all assigned exercises.
- Write down (neatly) your step-by-step work for each exercise before entering your answer.
- Use the immediate feedback provided by the program to help you check and correct your work for each exercise.
- Rework any exercises you have trouble with until you master them.
- Work through your homework assignment as many times as necessary until you are satisfied.

Exercises

Take a moment to think about your homework assignments to date and answer the following:

1. Have you attempted all assigned exercises?

2. Of the exercises attempted, have you also written out your work before entering your answer—so that you can check it?

3. Are you familiar with how to enter answers using the MathXL player so that you avoid answer entry type errors?

4. List some ways the immediate feedback and practice supports have helped you with your homework. If you have not used these supports, how do you plan to use them with the success tips above on your next assignment?

Study Skills Builder 13

Organizing Your Work

Have you ever used any readily available paper (such as the back of a flyer, another course assignment, post-its, etc.) to work out homework exercises before entering the answer in MathXL? To save time, have you ever entered answers directly into MathXL without working the exercises on paper? When it's time to study, have you ever been unable to find your completed work or read and follow your own mathematics handwriting?

When any of these things happen, it's time to get organized. Here are some suggestions:

- Write your step-by-step work for each homework exercise, (neatly) on lined, loose-leaf paper and keep this in a 3-ring binder.

- Refer to your step-by-step work when you receive feedback that your answer is incorrect in MathXL. Double-check against the steps and hints provided by the program and correct your work accordingly.

- Keep your written homework with your class notes for that section.

- Identify any exercises you are having trouble with and ask questions about them.

- Keep all graded quizzes and tests in this binder as well, to study later.

If you follow the suggestions above, you and your instructor or tutor will be able to follow your steps and correct any mistakes. You will have a written copy of your work to refer to later to ask questions and study for tests.

Exercises

1. Why is it important that you write out your step-by-step work on homework exercises and keep a hard copy of all work submitted online?

2. If you have gotten an incorrect answer, are you able to follow your steps and find your error?

3. If you were asked today to review your previous homework assignments and first test, could you find them? If not, list some ways you might better organize your work.

Study Skills Builder 14

Getting Help with Your Homework Assignments

There are many helpful resources available to you through MathXL to help you work through any homework exercises you may have trouble with. It is important that you know what these resources are and know when and how to use them.

Let's review these features found in the homework exercises:

- **Help Me Solve This**—provides step-by-step help for the exercise you are working. You must work an additional exercise of the same type (without this help) before you can get credit for having worked it correctly.

- **View an Example**—allows you to view a correctly worked exercise similar to the one you are having trouble with. You can go back to your original exercise and work it on your own.

- **E-Book**—allows you to read examples from your text and find similar exercises.

- **Video**—your text author, Elayn Martin-Gay, works an exercise similar to the one you need help with. **Not all exercises have an accompanying video clip.

- **Ask My Instructor**—allows you to email your instructor for help with an exercise.

Exercises

1. How does the "Help Me Solve This" feature work?

2. If the "View an Example" feature is used, is it necessary to work an additional problem before continuing the assignment?

3. When might be a good time to use the "Video" feature? Do all exercises have an accompanying video clip?

4. Which of the features above have you used? List those you found the most helpful to you.

5. If you haven't used the features discussed, list those you plan to try on your next homework assignment.

Study Skills Builder 15

Tips for Preparing for an Exam

Did you know that you can rework your previous homework assignments in MyMathLab and MathXL? This is a great way to prepare for tests. To do this, open a previous homework assignment and click "similar exercise." This will generate new exercises similar to the homework you have submitted. You can then rework the exercises and assignments until you feel confident that you understand them.

    To prepare for an exam, follow these tips:

- Review your written work for your previous homework assignments along with your class notes.
- Identify any exercises or topics that you have questions on or have difficulty understanding.
- Rework your previous assignments in MyMathLab and MathXL until you fully understand them and can do them without help.
- Get help for any topics you feel unsure of or for which you have questions.

Exercises

1. Are your current homework assignments up to date and is your written work for them organized in a binder or notebook? If the answer is no, it's time to get organized. For tips on this, see Study Skills Builder 13—Organizing Your Work.

2. How many days in advance of an exam do you usually start studying?

3. List some ways you think that practicing previous homework assignments can help you prepare for your test.

4. List two or three resources you can use to get help for any topics you are unsure of or have questions on.

Good luck!

Study Skills Builder 16

How Well Do You Know the Resources Available to You in MyMathLab?

There are many helpful resources available to you in MyMathLab. Let's take a moment to locate and explore a few of them now. Go into your MyMathLab course, and visit the multimedia library, tools for success, and E-book.

    Let's see what you found.

Exercises

1. List the resources available to you in the Multimedia Library.

2. List the resources available to you in the Tools for Success folder.

3. Where did you find the English/Spanish Audio Glossary?

4. Can you view videos from the E-book?

5. Did you find any resources you did not know about? If so, which ones?

6. Which resources have you used most often or found most helpful?

Additional Help Inside and Outside Your Textbook

Study Skills Builder 17

How Well Do You Know Your Textbook?

The questions below will help determine whether you are familiar with your textbook. For additional information, see Section 1.1 in this text.

Exercises

1. What does the ▶ icon mean?

2. What does the ✎ icon mean?

3. What does the △ icon mean?

4. Where can you find a review for each chapter? What answers to this review can be found in the back of your text?

5. Each chapter contains an overview of the chapter along with examples. What is this feature called?

6. Each chapter contains a review of vocabulary. What is this feature called?

7. There are practice exercises that are contained in this text. What are they and how can they be used?

8. This text contains a student section in the back entitled Student Resources. List the contents of this section and how they might be helpful.

9. What exercise answers are available in this text? Where are they located?

Study Skills Builder 18

Are You Familiar with Your Textbook Supplements?

Below is a review of some of the student supplements available for additional study. Check to see if you are using the ones most helpful to you.

- Chapter Test Prep Videos. These videos provide video clip solutions to the Chapter Test exercises in this text. You will find this extremely useful when studying for tests or exams.

- Interactive DVD Lecture Series. These are keyed to each section of the text. The material is presented by me, Elayn Martin-Gay, and I have placed a ▶ by the exercises in the text that I have worked on the video.

- The *Student Solutions Manual*. This contains worked-out solutions to odd-numbered exercises as well as every exercise in the Practice Exercises, Integrated Reviews, Chapter Reviews, Chapter Tests, and Cumulative Reviews.

- Pearson Tutor Center. Mathematics questions may be phoned, faxed, or e-mailed to this center.

- MyMathLab is a text-specific online course. MathXL is an online homework, tutorial, and assessment system.

Take a moment and determine whether these are available to you.

As usual, your instructor is your best source of information.

Exercises

Let's see how you are doing with textbook supplements.

1. Name one way the Lecture Videos can be helpful to you.

2. Name one way the Chapter Test Prep Video can help you prepare for a chapter test.

3. List any textbook supplements that you have found useful.

4. Have you located and visited a learning resource lab located on your campus?

5. List the textbook supplements that are currently housed in your campus's learning resource lab.

Study Skills Builder 19

Are You Getting All the Mathematics Help That You Need?

Remember that, in addition to your instructor, there are many places to get help with your mathematics course. For example:

- This text has an accompanying video lesson for every section and the CD in this text contains worked-out solutions to every Chapter Test exercise.

- The back of the book contains answers to odd-numbered exercises.

- A *Student Solutions Manual* is available that contains worked-out solutions to odd-numbered exercises as well as solutions to every exercise in the Practice Exercises, Integrated Reviews, Chapter Reviews, Chapter Tests, and Cumulative Reviews.

- Don't forget to check with your instructor for other local resources available to you, such as a tutoring center.

Exercises

1. List items you find helpful in the text and all student supplements to this text.

2. List all the campus help that is available to you for this course.

3. List any help (besides the textbook) from Exercises 1 and 2 above that you are using.

4. List any help (besides the textbook) that you feel you should try.

5. Write a goal for yourself that includes trying everything you listed in Exercise 4 during the next week.

The Bigger Picture– Study Guide Outline

OUTLINE: PART 1

Operations on Sets of Numbers and Solving Equations

I. Some Operations on Sets of Numbers

  A. Whole Numbers

    1. Add or Subtract:
(Sec. 1.3 and 1.4)

$$\begin{array}{r} 14 \\ +39 \\ \hline 53 \end{array} \qquad \begin{array}{r} 300 \\ -27 \\ \hline 273 \end{array}$$

    2. Multiply or Divide:
(Sec. 1.6 and 1.7)

$$\begin{array}{r} 238 \\ \times\ 47 \\ \hline 1666 \\ 9520 \\ \hline 11{,}186 \end{array} \qquad \begin{array}{r} 127\ \text{R2} \\ 7\overline{)891} \\ \underline{-7} \\ 19 \\ \underline{-14} \\ 51 \\ \underline{-49} \\ 2 \end{array}$$

    3. Exponent: (Sec. 1.9)   4 factors of 3

$$3^4 = \overbrace{3 \cdot 3 \cdot 3 \cdot 3} = 81$$

    4. Square Root: (Sec. 1.9)

$$\sqrt{25} = 5 \ \textit{because} \ 5 \cdot 5 = 25 \text{ and 5 is a positive number}$$

    5. Order of Operations: (Sec. 1.9)

$$
\begin{aligned}
24 \div 3 \cdot 2 - (2 + 8) &= 24 \div 3 \cdot 2 - (10) && \text{Parentheses.}\\
&= 8 \cdot 2 - 10 && \text{Multiply or divide from left to right.}\\
&= 16 - 10 && \text{Multiply or divide from left to right.}\\
&= 6 && \text{Add or subtract from left to right.}
\end{aligned}
$$

  B. Fractions

    1. Simplify: (Sec. 2.3) Factor the numerator and denominator. Then remove factors of 1 by dividing out common factors in the numerator and denominator.

    Simplify: $\dfrac{20}{28} = \dfrac{4 \cdot 5}{4 \cdot 7} = \dfrac{5}{7}$

    2. Multiply: (Sec. 2.4) Numerator times numerator over denominator times denominator.

    $\dfrac{5}{9} \cdot \dfrac{2}{7} = \dfrac{10}{63}$

    3. Divide: (Sec. 2.5) First fraction times the reciprocal of the second fraction.

    $\dfrac{2}{11} \div \dfrac{3}{4} = \dfrac{2}{11} \cdot \dfrac{4}{3} = \dfrac{8}{33}$

4. **Add or Subtract:** (Sec. 3.3) Must have same denominator. If not, find the LCD, and write each fraction as an equivalent fraction with the LCD as denominator.

$$\frac{2}{5} + \frac{1}{15} = \frac{2}{5} \cdot \frac{3}{3} + \frac{1}{15} = \frac{6}{15} + \frac{1}{15} = \frac{7}{15}$$

C. Decimals

1. **Add or Subtract:** (Sec. 4.3) Line up decimal points.

$$\begin{array}{r} 1.27 \\ + \ 0.6 \\ \hline 1.87 \end{array}$$

2. **Multiply:** (Sec. 4.4)

$$\begin{array}{r} 2.56 \\ \times \ 3.2 \\ \hline 512 \\ 7\,680 \\ \hline 8.192 \end{array}$$

2 decimal places
1 decimal place
$2 + 1 = 3$
3 decimal places

3. **Divide:** (Sec. 4.5)

$$8\overline{)5.6}^{\,0.7} \qquad 0.6\overline{)0.786}^{\,1.31}$$

II. Solving Equations

A. Proportions: (Sec. 5.1) Set cross products equal to each other. Then solve.

$$\frac{14}{3} = \frac{2}{n} \text{ or } 14 \cdot n = 3 \cdot 2 \text{ or } 14 \cdot n = 6 \text{ or } n = \frac{6}{14} = \frac{3}{7}$$

B. Percent Problems

1. **Solved by Equations:** (Sec. 5.4) Remember that "of" means multiplication and "is" means equals.

   "12% of some number is 6" translates to

   $$12\% \cdot n = 6 \text{ or } 0.12 \cdot n = 6 \text{ or } n = \frac{6}{0.12} \text{ or } n = 50$$

2. **Solved by Proportions:** (Sec. 5.5) Remember that percent, p, is identified by % or percent, base, b, usually appears after "of" and amount, a, is the part compared to the whole.

   "12% of some number is 6" translates to

   $$\frac{6}{b} = \frac{12}{100} \text{ or } 6 \cdot 100 = b \cdot 12 \text{ or } \frac{600}{12} = b \text{ or } 50 = b$$

OUTLINE: PART 2

Simplifying Expressions and Solving Equations and Inequalities

I. Simplifying Expressions

A. Real Numbers

1. **Add:** (Sec. 8.3)

   $$-1.7 + (-0.21) = -1.91 \qquad \text{Adding like signs.}$$
   Add absolute value. Attach common sign.

   $$-7 + 3 = -4 \qquad \text{Adding different signs.}$$
   Subtract absolute values. Attach the sign of the number with the larger absolute value.

2. **Subtract:** Add the first number to the opposite of the second number. (Sec. 8.4)

   $$17 - 25 = 17 + (-25) = -8$$

3. Multiply or Divide: Multiply or divide the two numbers as usual. If the signs are the same, the answer is positive. If the signs are different, the answer is negative. (Sec. 8.5)

$$-10 \cdot 3 = -30, \qquad -81 \div (-3) = 27$$

B. Exponents (Sec. 12.2)

$$x^7 \cdot x^5 = x^{12}; \; (x^7)^5 = x^{35}; \frac{x^7}{x^5} = x^2; x^0 = 1; 8^{-2} = \frac{1}{8^2} = \frac{1}{64}$$

C. Polynomials

1. Add: Combine like terms. (Sec. 12.4)

$$(3y^2 + 6y + 7) + (9y^2 - 11y - 15) = 3y^2 + 6y + 7 + 9y^2 - 11y - 15$$
$$= 12y^2 - 5y - 8$$

2. Subtract: Change the sign of the terms of the polynomial being subtracted, then add. (Sec. 12.4)

$$(3y^2 + 6y + 7) - (9y^2 - 11y - 15) = 3y^2 + 6y + 7 - 9y^2 + 11y + 15$$
$$= -6y^2 + 17y + 22$$

3. Multiply: Multiply each term of one polynomial by each term of the other polynomial. (Sec. 12.5)

$$(x + 5)(2x^2 - 3x + 4) = x(2x^2 - 3x + 4) + 5(2x^2 - 3x + 4)$$
$$= 2x^3 - 3x^2 + 4x + 10x^2 - 15x + 20$$
$$= 2x^3 + 7x^2 - 11x + 20$$

4. Divide: (Sec. 12.7)

a. To divide by a monomial, divide each term of the polynomial by the monomial.

$$\frac{8x^2 + 2x - 6}{2x} = \frac{8x^2}{2x} + \frac{2x}{2x} - \frac{6}{2x} = 4x + 1 - \frac{3}{x}$$

b. To divide by a polynomial other than a monomial, use long division.

$$
\begin{array}{r}
x - 6 + \dfrac{40}{2x + 5} \\[2pt]
2x + 5\overline{)2x^2 - 7x + 10} \\
\underline{2x^2 + 5x} \\
-12x + 10 \\
\underline{12x - 30} \\
40
\end{array}
$$

D. Factoring Polynomials

See the Chapter 13 Integrated Review for steps.

$$3x^4 - 78x^2 + 75 = 3(x^4 - 26x^2 + 25) \qquad \text{Factor out GCF—always first step.}$$
$$= 3(x^2 - 25)(x^2 - 1) \qquad \text{Factor trinomial.}$$
$$= 3(x + 5)(x - 5)(x + 1)(x - 1) \qquad \text{Factor further—each difference of squares.}$$

E. Rational Expressions

1. Simplify: Factor the numerator and denominator. Then remove factors of 1 by dividing out common factors in the numerator and denominator. (Sec. 14.1)

$$\frac{x^2 - 9}{7x^2 - 21x} = \frac{(x + 3)(x - 3)}{7x(x - 3)} = \frac{x + 3}{7x}$$

2. **Multiply:** Multiply numerators, then multiply denominators. (Sec. 14.2)

$$\frac{5z}{2z^2 - 9z - 18} \cdot \frac{22z + 33}{10z} = \frac{5 \cdot z}{(2z + 3)(z - 6)} \cdot \frac{11(2z + 3)}{2 \cdot 5 \cdot z} = \frac{11}{2(z - 6)}$$

3. **Divide:** First fraction times the reciprocal of the second fraction. (Sec. 14.2)

$$\frac{14}{x + 5} \div \frac{x + 1}{2} = \frac{14}{x + 5} \cdot \frac{2}{x + 1} = \frac{28}{(x + 5)(x + 1)}$$

4. **Add or Subtract:** Must have same denominator. If not, find the LCD and write each fraction as an equivalent fraction with the LCD as denominator. (Sec. 14.4)

$$\frac{9}{10} - \frac{x + 1}{x + 5} = \frac{9(x + 5)}{10(x + 5)} - \frac{10(x + 1)}{10(x + 5)}$$

$$= \frac{9x + 45 - 10x - 10}{10(x + 5)} = \frac{-x + 35}{10(x + 5)}$$

F. Radicals

1. **Simplify square roots:** If possible, factor the radicand so that one factor is a perfect square. Then use the product rule and simplify. (Sec. 15.2)

$$\sqrt{75} = \sqrt{25 \cdot 3} = \sqrt{25} \cdot \sqrt{3} = 5\sqrt{3}$$

2. **Add or Subtract:** Only like radicals (same index and radicand) can be added or subtracted. (Sec. 15.3)

$$8\sqrt{10} - \sqrt{40} + \sqrt{5} = 8\sqrt{10} - 2\sqrt{10} + \sqrt{5} = 6\sqrt{10} + \sqrt{5}$$

3. **Multiply or divide:** $\sqrt{a} \cdot \sqrt{b} = \sqrt{ab}; \frac{\sqrt{a}}{\sqrt{b}} = \sqrt{\frac{a}{b}}$. (Sec. 15.4)

$$\sqrt{11} \cdot \sqrt{3} = \sqrt{33}; \frac{\sqrt{140}}{\sqrt{7}} = \sqrt{\frac{140}{7}} = \sqrt{20} = \sqrt{4 \cdot 5} = 2\sqrt{5}$$

4. **Rationalizing the denominator:** (Sec. 15.4)

   a. If denominator is one term,

   $$\frac{5}{\sqrt{11}} = \frac{5 \cdot \sqrt{11}}{\sqrt{11} \cdot \sqrt{11}} = \frac{5\sqrt{11}}{11}$$

   b. If denominator is two terms, multiply by 1 in the form of $\frac{\text{conjugate of denominator}}{\text{conjugate of denominator}}$.

   $$\frac{13}{3 + \sqrt{2}} = \frac{13}{3 + \sqrt{2}} \cdot \frac{3 - \sqrt{2}}{3 - \sqrt{2}} = \frac{13(3 - \sqrt{2})}{9 - 2} = \frac{39 - 13\sqrt{2}}{7}$$

II. Solving Equations and Inequalities

A. Linear Equations: Power on variable is 1 and there are no variables in denominator. (Sec. 9.3)

$7(x - 3) = 4x + 6$ Linear equation. (If fractions, multiply by LCD.)

$7x - 21 = 4x + 6$ Use the distributive property.

$7x = 4x + 27$ Add 21 to both sides.

$3x = 27$ Subtract $4x$ from both sides.

$x = 9$ Divide both sides by 3.

B. Linear Inequalities: Same as linear equation except if you multiply or divide by a negative number, then reverse direction of inequality. (Sec. 9.7)

$$-4x + 11 \leq -1 \qquad \text{Linear inequality.}$$

$$-4x \leq -12 \qquad \text{Subtract 11 from both sides.}$$

$$\frac{-4x}{-4} \geq \frac{-12}{-4} \qquad \begin{array}{l}\text{Divide both sides by 4 and reverse}\\\text{the direction of the inequality symbol.}\end{array}$$

$$x \geq 3 \qquad \text{Simplify.}$$

C. Quadratic and Higher Degree Equations: First write the equation in standard form (one side is 0).

1. If the polynomial on one side factors, solve by factoring. (Sec. 16.1)

2. If the polynomial does not factor, solve by the quadratic formula. (Sec. 16.3)

By factoring:	**By quadratic formula:**
$x^2 + x = 6$	$x^2 + x = 5$
$x^2 + x - 6 = 0$	$x^2 + x - 5 = 0$
$(x - 2)(x + 3) = 0$	$a = 1, b = 1, c = -5$
$x - 2 = 0 \text{ or } x + 3 = 0$	$x = \dfrac{-1 \pm \sqrt{1^2 - 4(1)(-5)}}{2 \cdot 1}$
$x = 2 \text{ or } \qquad x = -3$	$= \dfrac{-1 \pm \sqrt{21}}{2}$

D. Equations with Rational Expressions: Make sure the proposed solution does not make any denominator 0. (Sec. 14.5)

$$\frac{3}{x} - \frac{1}{x - 1} = \frac{4}{x - 1} \qquad \text{Equation with rational expressions}$$

$$x(x - 1) \cdot \frac{3}{x} - x(x - 1) \cdot \frac{1}{x - 1} = x(x - 1) \cdot \frac{4}{x - 1} \qquad \begin{array}{l}\text{Multiply through by}\\x(x - 1).\end{array}$$

$$3(x - 1) - x \cdot 1 = x \cdot 4 \qquad \text{Simplify.}$$

$$3x - 3 - x = 4x \qquad \text{Use the distributive property.}$$

$$-3 = 2x \qquad \begin{array}{l}\text{Simplify and move variable terms}\\\text{to right side.}\end{array}$$

$$-\frac{3}{2} = x \qquad \text{Divide both sides by 2.}$$

E. Proportions: An equation with two ratios equal. Set cross products equal, then solve. Make sure the proposed solution does not make any denominator 0. (Sec. 14.6)

$$\frac{5}{x} \underset{\nearrow}{\overset{\searrow}{\bowtie}} \frac{9}{2x - 3}$$

$$5(2x - 3) = 9 \cdot x \qquad \text{Set cross products equal.}$$

$$10x - 15 = 9x \qquad \text{Multiply.}$$

$$x = 15 \qquad \begin{array}{l}\text{Write equation with variable terms on}\\\text{one side and constants on the other.}\end{array}$$

F. Equations with Radicals: To solve, isolate a radical, then square both sides. You may have to repeat this. Check possible solution in the original equation. (Sec. 15.5)

$$\sqrt{x + 49} + 7 = x$$

$$\sqrt{x + 49} = x - 7 \qquad \text{Subtract 7 from both sides.}$$

$$x + 49 = x^2 - 14x + 49 \qquad \text{Square both sides.}$$

$$0 = x^2 - 15x \qquad \text{Set terms equal to 0.}$$

$$0 = x(x - 15) \qquad \text{Factor.}$$

$$\cancel{x = 0} \text{ or } x = 15 \qquad \text{Set each factor equal to 0 and solve.}$$

Practice Final Exam

1. _____

2. _____

3. _____

4. _____

5. _____

6. _____

7. _____

8. _____

9. _____

10. _____

11. _____

12. _____

13. _____

14. _____

15. _____

16. _____

17. _____

18. _____

19. _____

20. _____

21. _____

22. _____

Note: Exercises 1–36 review operations with nonnegative numbers. Simplify by performing the indicated operations.

Chapters 1–8

1. $600 - 487$

2. $(2^4 - 5) \cdot 3$

3. $\dfrac{16}{3} \div \dfrac{3}{12}$

4. $\dfrac{11}{12} + \dfrac{3}{8} + \dfrac{5}{24}$

5. $\dfrac{64 \div 8 \cdot 2}{(\sqrt{9} - \sqrt{4})^2 + 1}$

6. $\begin{array}{r} 10.2 \\ \times\ 4.3 \\ \hline \end{array}$

7. $\dfrac{0.23 + 1.63}{0.3}$

8. $\begin{array}{r} 5\frac{1}{6} \\ -\ 3\frac{7}{8} \\ \hline \end{array}$

9. $3\dfrac{1}{3} \cdot 6\dfrac{3}{4}$

10. 126.9×100

11. $\left(\dfrac{3}{4}\right)^2 \div \left(\dfrac{2}{3} + \dfrac{5}{6}\right)$

12. Round 0.8623 to the nearest thousandth.

13. Round 34.8923 to the nearest tenth.

14. Write $\dfrac{16}{17}$ as a decimal. Round to the nearest thousandth.

15. Write 85% as a decimal.

16. Write 6.1 as a percent.

17. Write $\dfrac{3}{8}$ as a percent.

18. Write 0.2% as a fraction in simplest form.

19. Find the perimeter and the area of the rectangle below.

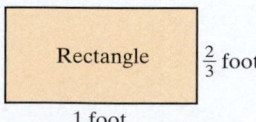

Rectangle — $\frac{2}{3}$ foot — 1 foot

20. Write the ratio as a fraction in simplest form: $75 to $10

21. Find the unit rate: 8 inches of rain in 12 hours

22. Find the unknown number, n, in the proportion: $\dfrac{8}{n} = \dfrac{11}{6}$

Solve.

23. Subtract 8.6 from 20.

24. A small airplane used $58\frac{3}{4}$ gallons of fuel on a $7\frac{1}{2}$-hour trip. How many gallons of fuel were used for each hour?

25. The standard dose of medicine for a dog is 10 grams for every 15 pounds of body weight. What is the standard dose for a dog that weighs 80 pounds?

26. Twenty-nine cans of Sherwin-Williams paint costs $493. How much was each can?

27. 0.6% of what number is 7.5?

28. 567 is what percent of 756?

29. An alloy is 12% copper. How much copper is contained in 320 pounds of this alloy?

30. A $120 framed picture is on sale for 15% off. Find the discount and the sale price.

31. Find the complement of a 78° angle.

32. Find the measures of angles x, y, and z.

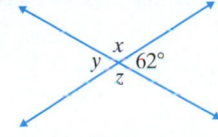

33. Find the measure of $\angle x$.

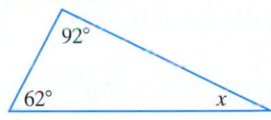

34. Given that the following triangles are similar, find the missing length n.

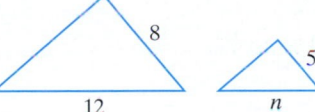

Find the mean, median, and mode of the list of numbers.

35. 26, 32, 42, 43, 49

A professor measures the heights of the students in her class. The results are shown in the following histogram. Use this histogram to answer Exercise 36.

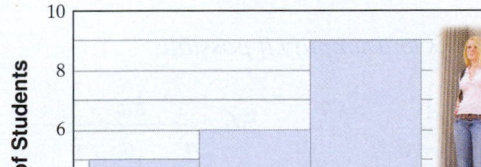

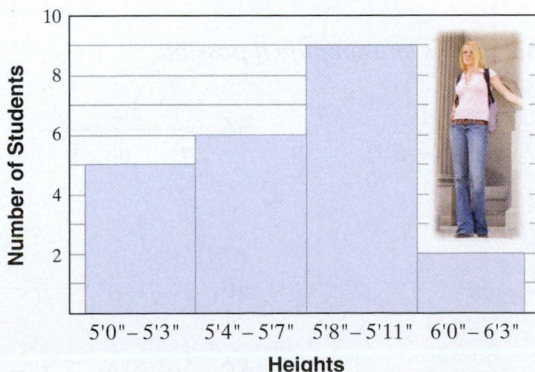

Student Heights

36. How many students are 5′7″ or shorter?

23. _____

24. _____

25. _____

26. _____

27. _____

28. _____

29. _____

30. _____

31. _____

32. _____

33. _____

34. _____

35. _____

36. _____

37. _____

38. _____

39. _____

40. _____

41. _____

42. _____

43. _____

44. _____

45. _____

46. _____

47. _____

48. _____

49. _____

50. _____

51. _____

52. _____

53. _____

54. _____

55. _____

56. _____

57. _____

58. _____

59. _____

60. _____

61. _____

62. _____

Note: Exercises 37–40 contain signed numbers. Simplify by performing the indicated operations.

37. $-13 - (-2)$

38. $3(-4)^2 - 80$

39. $\dfrac{-12 + 3 \cdot 8}{4}$

40. $6[5 + 2(3 - 8) - 3]$

41. Evaluate $\dfrac{y + z - 1}{x}$ when $x = 6$, $y = -2$, and $z = -3$.

42. Multiply, then simplify:
$-5(y + 1) + 2(3 - 5y)$

Chapters 9–16

Evaluate.

43. -3^4

44. 4^{-3}

Perform the indicated operations and simplify if possible.

45. $(5x^3 + x^2 + 5x - 2) - (8x^3 - 4x^2 + x - 7)$

46. $(4x - 2)^2$

47. $(3x + 7)(x^2 + 5x + 2)$

Factor.

48. $6t^2 - t - 5$

49. $3x^3 - 21x^2 + 30x$

50. $180 - 5x^2$

51. $3a^2 + 3ab - 7a - 7b$

52. $x - x^5$

Simplify. Write answers with positive exponents only.

53. $\left(\dfrac{4x^2y^3}{x^3y^{-4}}\right)^2$

54. $\dfrac{5 - \dfrac{1}{y^2}}{\dfrac{1}{y} + \dfrac{2}{y^2}}$

Perform the indicated operations and simplify if possible.

55. $\dfrac{x^2 - 13x + 42}{x^2 + 10x + 21} \div \dfrac{x^2 - 4}{x^2 + x - 6}$

56. $\dfrac{5a}{a^2 - a - 6} - \dfrac{2}{a - 3}$

Solve each equation or inequality.

57. $4(n - 5) = -(4 - 2n)$

58. $x(x + 6) = 7$

59. $3x - 5 \geq 7x + 3$

60. $2x^2 - 6x + 1 = 0$

61. $\dfrac{4}{y} - \dfrac{5}{3} = -\dfrac{1}{5}$

62. $\dfrac{5}{y + 1} = \dfrac{4}{y + 2}$

63. $\dfrac{a}{a-3} = \dfrac{3}{a-3} - \dfrac{3}{2}$

64. $\sqrt{2x-2} = x - 5$

Graph the following.

65. $5x - 7y = 10$

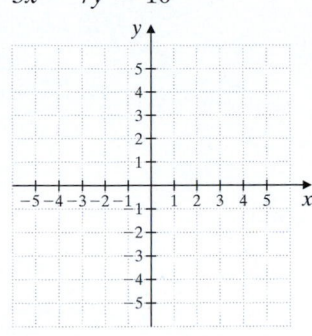

66. $y = -1$

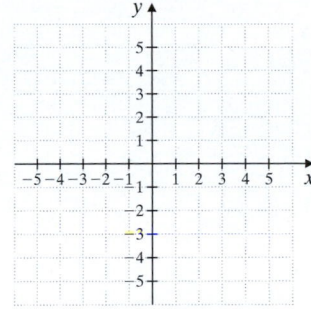

67. $y \geq -4x$

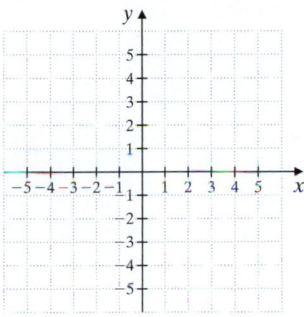

Find the slope of each line.

68. Passes through $(6, -5)$ and $(-1, 2)$

69. $-3x + y = 5$

Write equations of the following lines. Write each equation in standard form.

70. Passes through $(2, -5)$ and $(1, 3)$

71. Slope $\dfrac{1}{8}$; y-intercept $(0, 12)$

Solve each system of equations.

72. $\begin{cases} 3x - 2y = -14 \\ y = x + 5 \end{cases}$

73. $\begin{cases} 4x - 6y = 7 \\ -2x + 3y = 0 \end{cases}$

Answer the questions about functions.

74. If $f(x) = x^3 - x$, find
   a. $f(-1)$ **b.** $f(0)$ **c.** $f(4)$

75. Determine whether the relation is also a function. If a function, find its domain and range.

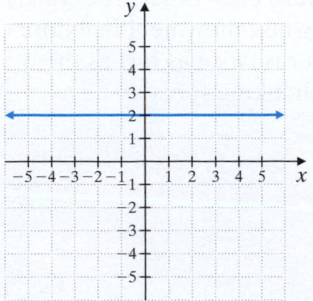

63. _____

64. _____

65. _____

66. _____

67. _____

68. _____

69. _____

70. _____

71. _____

72. _____

73. _____

74. a. _____

    b. _____

    c. _____

75. _____

Evaluate.

76. $\sqrt{16}$

77. $\sqrt[3]{125}$

78. $\sqrt{\dfrac{9}{16}}$

Simplify.

79. $\sqrt{54}$

80. $\sqrt{24x^8}$

Perform the indicated operations and simplify if possible.

81. $\sqrt{18} - \sqrt{75} + 7\sqrt{3} - \sqrt{8}$

82. $\dfrac{\sqrt{40x^4}}{\sqrt{2x}}$

83. $\sqrt{2}(\sqrt{6} - \sqrt{5})$

Rationalize each denominator.

84. $\dfrac{8}{\sqrt{5y}}$

85. $\dfrac{8}{\sqrt{6} + 2}$

Solve each application.

86. One number plus five times its reciprocal is equal to six. Find the number.

87. Some states have a single area code for the entire state. Two such states have area codes where one is double the other. If the sum of these integers is 1203, find the two area codes.

88. Two hikers start at opposite ends of the St. Tammany Trails and walk toward each other. The trail is 36 miles long and they meet in 4 hours. If one hiker is twice as fast as the other, find both hiking speeds.

89. Find the amount of a 12% saline solution a lab assistant should add to 80 cc (cubic centimeters) of a 22% saline solution in order to have a 16% solution.

76. _____

77. _____

78. _____

79. _____

80. _____

81. _____

82. _____

83. _____

84. _____

85. _____

86. _____

87. _____

88. _____

89. _____

Answers to Selected Exercises

Chapter 1 The Whole Numbers

Section 1.2

Vocabulary, Readiness & Video Check   1. whole   **3.** words   **5.** period   **7.** hundreds   **9.** 80,000

Exercise Set 1.2   1. tens   **3.** thousands   **5.** hundred-thousands   **7.** millions   **9.** three hundred fifty-four   **11.** eight thousand, two hundred seventy-nine   **13.** twenty-six thousand, nine hundred ninety   **15.** two million, three hundred eighty-eight thousand **17.** twenty-four million, three hundred fifty thousand, one hundred eighty-five   **19.** three hundred fifteen thousand, two hundred eighty-one   **21.** two thousand, seven hundred seventeen   **23.** thirty-two million, one hundred thousand   **25.** fourteen thousand, four hundred thirty-three   **27.** seventeen million   **29.** 6587   **31.** 59,800   **33.** 13,601,011   **35.** 7,000,017   **37.** 260,997   **39.** 418 **41.** 16,732   **43.** $91,071,000   **45.** 101,000   **47.** 400 + 6   **49.** 3000 + 400 + 70   **51.** 80,000 + 700 + 70 + 4 **53.** 60,000 + 6000 + 40 + 9   **55.** 30,000,000 + 9,000,000 + 600,000 + 80,000   **57.** 5532; five thousand, five hundred thirty-two **59.** 5000 + 400 + 90 + 2   **61.** Mt. Washington   **63.** National Gallery   **65.** five million, sixty-five thousand   **67.** 4 **69.** 9861   **71.** no; one hundred five   **73.** answers may vary   **75.** 34,000,000,000,000,000   **77.** Canton

Section 1.3

Calculator Explorations   1. 134   **3.** 340   **5.** 2834

Vocabulary, Readiness & Video Check   1. number   **3.** sum; addend   **5.** grouping; associative   **7.** place; right; left **9.** increased by

Exercise Set 1.3   1. 36   **3.** 292   **5.** 49   **7.** 5399   **9.** 117   **11.** 512   **13.** 209,078   **15.** 25   **17.** 62   **19.** 212   **21.** 94   **23.** 910 **25.** 8273   **27.** 11,926   **29.** 1884   **31.** 16,717   **33.** 1110   **35.** 8999   **37.** 35,901   **39.** 632,389   **41.** 42 in.   **43.** 25 ft   **45.** 24 in. **47.** 8 yd   **49.** 29 in.   **51.** 44 m   **53.** 2093   **55.** 266   **57.** 544   **59.** 3452   **61.** 21,141 thousand   **63.** 6684 ft   **65.** 340 ft   **67.** 2425 ft **69.** 249,849   **71.** 1,474,711 vehicles   **73.** 124 ft   **75.** 3095   **77.** California   **79.** 529 stores   **81.** New York and Pennsylvania **83.** 6358 mi   **85.** answers may vary   **87.** answers may vary   **89.** 1,044,473,765   **91.** correct   **93.** incorrect: 530

Section 1.4

Calculator Explorations   1. 770   **3.** 109   **5.** 8978

Vocabulary, Readiness & Video Check   1. 0   **3.** minuend; subtrahend   **5.** 0   **7.** 600   **9.** We cannot take 7 from 2 in the ones place, so we borrow one ten from the tens place and move it over to the ones place to give us 10 + 2 or 12.

Exercise Set 1.4   1. 44   **3.** 265   **5.** 135   **7.** 2254   **9.** 5545   **11.** 600   **13.** 25   **15.** 45   **17.** 146   **19.** 288   **21.** 168   **23.** 106 **25.** 447   **27.** 5723   **29.** 504   **31.** 89   **33.** 79   **35.** 39,914   **37.** 32,711   **39.** 5041   **41.** 31,213   **43.** 4   **45.** 20   **47.** 7   **49.** 72 **51.** 88   **53.** 264 pages   **55.** 7 million sq km   **57.** 264,000 sq mi   **59.** 283,000 sq mi   **61.** 6065 ft   **63.** 28 ft   **65.** 358 mi   **67.** $619 **69.** 2630 thousand   **71.** 100 dB   **73.** 58 dB   **75.** 346   **77.** 5920 sq ft   **79.** Hartsfield-Jackson Atlanta International   **81.** 14 million **83.** Jo; by 271 votes   **85.** 1034   **87.** 9   **89.** 8518   **91.** 22,876   **93.** minuend: 48; subtrahend: 1   **95.** minuend: 70; subtrahend: 7 **97.** incorrect: 685   **99.** correct   **101.** 5269 − 2385 = 2884   **103.** no; answers may vary   **105.** no: 1089 more pages

Section 1.5

Vocabulary, Readiness & Video Check   1. graph   **3.** 70; 60   **5.** 3 is in the place value we're rounding to (tens), and the digit to the right of this place value is 5 or greater, so we need to add 1 to the 3.   **7.** Each circled digit is to the right of the place value being rounded to and is used to determine whether or not we add 1 to the digit in the place value being rounded to.

Exercise Set 1.5   1. 420   **3.** 640   **5.** 2800   **7.** 500   **9.** 21,000   **11.** 34,000   **13.** 328,500   **15.** 36,000   **17.** 39,990   **19.** 30,000,000 **21.** 5280; 5300; 5000   **23.** 9440; 9400; 9000   **25.** 14,880; 14,900; 15,000   **27.** 28,000   **29.** 38,000 points   **31.** $98,000,000,000 **33.** $5,022,000   **35.** $129,000,000,000   **37.** 130   **39.** 80   **41.** 5700   **43.** 300   **45.** 11,400   **47.** incorrect   **49.** correct   **51.** correct **53.** $3400   **55.** 900 mi   **57.** 6000 ft   **59.** 8,000,000,000   **61.** 3000 children   **63.** $3,067,000,000; $3,070,000,000; $3,100,000,000 **65.** $2,910,000,000; $2,910,000,000; $2,900,000,000   **67.** 5723, for example   **69. a.** 8550   **b.** 8649   **71.** answers may vary   **73.** 140 m

Section 1.6

Calculator Explorations   1. 3456   **3.** 15,322   **5.** 272,291

Vocabulary, Readiness & Video Check   1. 0   **3.** product; factor   **5.** grouping; associative   **7.** length   **9.** distributive property   **11.** Think of the problem as 50 times 9 and then attach the two zeros from 900, or think of the problem as 5 times 9 and then attach the three zeros at the end of 50 and 900. Both approaches give us 45,000.   **13.** Multiplication is also an application of addition since it is addition of the same addend.

Exercise Set 1.6   1. 24   **3.** 0   **5.** 0   **7.** 87   **9.** $6 \cdot 3 + 6 \cdot 8$ **11.** $4 \cdot 3 + 4 \cdot 9$ **13.** $20 \cdot 14 + 20 \cdot 6$ **15.** 512   **17.** 3678   **19.** 1662
21. 6444   **23.** 1157   **25.** 24,418   **27.** 24,786   **29.** 15,600   **31.** 0   **33.** 6400   **35.** 48,126   **37.** 142,506   **39.** 2,369,826   **41.** 64,790
43. 3,949,935   **45.** 800   **47.** 11,000   **49.** 74,060   **51.** 24,000   **53.** 45,000   **55.** 3,280,000   **57.** area: 63 sq m; perimeter: 32 m
59. area: 680 sq ft; perimeter: 114 ft   **61.** 240,000   **63.** 300,000   **65.** c   **67.** c   **69.** 880   **71.** 4200   **73.** 4480   **75.** 375 cal   **77.** $3290
79. a. 20 boxes   **b.** 100 boxes   **c.** 2000 lb   **81.** 8800 sq ft   **83.** 56,000 sq ft   **85.** 5828 pixels   **87.** 2100 characters   **89.** 1360 cal
91. $10, $60; $10, $200; $12, $36, $12, $36; total cost: $372   **93.** 1,440,000 tea bags   **95.** 135   **97.** 2144   **99.** 23   **101.** 15   **103.** $4 \cdot 7$ or $7 \cdot 4$
105. a. $5 + 5 + 5$ or $3 + 3 + 3 + 3 + 3$ **b.** answers may vary   **107.** $\begin{array}{r} 203 \\ \times\ 14 \\ \hline 812 \\ 2030 \\ \hline 2842 \end{array}$ **109.** $\underline{42} \\ \times\ 93$ **111.** answers may vary
113. 506 windows

Section 1.7

Calculator Explorations   1. 53   **3.** 62   **5.** 261   **7.** 0

Vocabulary, Readiness & Video Check   1. quotient; dividend; divisor   **3.** 1   **5.** undefined   **7.** 0   **9.** $202 \cdot 102 + 15 = 20,619$
11. addition and division

Exercise Set 1.7   1. 6   **3.** 12   **5.** 0   **7.** 31   **9.** 1   **11.** 8   **13.** undefined   **15.** 1   **17.** 0   **19.** 9   **21.** 29   **23.** 74   **25.** 338
27. undefined   **29.** 9   **31.** 25   **33.** 68 R 3   **35.** 236 R 5   **37.** 38 R 1   **39.** 326 R 4   **41.** 13   **43.** 49   **45.** 97 R 8   **47.** 209 R 11
49. 506   **51.** 202 R 7   **53.** 54   **55.** 99 R 100   **57.** 202 R 15   **59.** 579 R 72   **61.** 17   **63.** 511 R 3   **65.** 2132 R 32   **67.** 6080
69. 23 R 2   **71.** 5 R 25   **73.** 20 R 2   **75.** 33 students   **77.** 165 lb   **79.** 310 yd   **81.** 89 bridges   **83.** 11 light poles   **85.** 5 mi
87. 1760 yd   **89.** 20   **91.** 387   **93.** 79   **95.** 74°   **97.** 9278   **99.** 15,288   **101.** 679   **103.** undefined   **105.** 9 R 12   **107.** c
109. b   **111.** 84   **113.** increase; answers may vary   **115.** no; answers may vary   **117.** 12 ft   **119.** answers may vary   **121.** 5 R 1

Integrated Review   1. 148   **2.** 6555   **3.** 1620   **4.** 562   **5.** 79   **6.** undefined   **7.** 9   **8.** 1   **9.** 0   **10.** 0   **11.** 0   **12.** 3   **13.** 2433
14. 9826   **15.** 213 R 3   **16.** 79,317   **17.** 27   **18.** 9   **19.** 138   **20.** 276   **21.** 1099 R 2   **22.** 111 R 1   **23.** 663 R 6   **24.** 1076 R 60
25. 1024   **26.** 9899   **27.** 30,603   **28.** 47,500   **29.** 65   **30.** 456   **31.** 6 R 8   **32.** 53   **33.** 183   **34.** 231   **35.** 9740; 9700; 10,000
36. 1430; 1400; 1000   **37.** 20,800; 20,800; 21,000   **38.** 432,200; 432,200; 432,000   **39.** perimeter: 24 ft; area: 36 sq ft
40. perimeter: 42 in.; area: 98 sq in.   **41.** 28 mi   **42.** 26 m   **43.** 24   **44.** 124   **45.** Lake Pontchartrain Bridge; 2175 ft   **46.** 730 qt

Section 1.8

Vocabulary, Readiness & Video Check   1. The George Washington Bridge has a length of 3500 feet.

Exercise Set 1.8   1. 49   **3.** 237   **5.** 42   **7.** 600   **9. a.** 400 ft   **b.** 9600 sq ft   **11.** $15,500   **13.** 168 hr   **15.** 3500 ft   **17.** 141 yr
19. 372 billion bricks   **21.** 719 towns   **23.** $21   **25.** 55 cal   **27.** 22 hot dogs   **29.** $32,842,230   **31.** 1,215,051 people   **33.** 3987 mi
35. 13 paychecks   **37.** $239   **39.** $1045   **41.** b will be cheaper by $3   **43.** Asia   **45.** 1053 million   **47.** 19 million   **49.** 798 million
51. $14,754   **53.** 16,800 mg   **55. a.** 3750 sq ft   **b.** 375 sq ft   **c.** 3375 sq ft   **57.** $2   **59.** answers may vary

Section 1.9

Calculator Explorations   1. 4096   **3.** 3125   **5.** 2048   **7.** 2526   **9.** 4295   **11.** 8

Vocabulary, Readiness & Video Check   1. base; exponent   **3.** addition   **5.** division   **7.** exponent; base   **9.** Because $8 \cdot 8 = 64$.
11. The area of a rectangle is length $\cdot$ width. A square is a special rectangle where length = width. Thus, the area of a square is
side $\cdot$ side or $(\text{side})^2$.

Exercise Set 1.9   1. 4^3 **3.** 7^6 **5.** 12^3 **7.** $6^2 \cdot 5^3$ **9.** $9 \cdot 8^2$ **11.** $3 \cdot 2^4$ **13.** $3 \cdot 2^4 \cdot 5^5$ **15.** 64   **17.** 125   **19.** 32   **21.** 1
23. 7   **25.** 128   **27.** 256   **29.** 256   **31.** 729   **33.** 144   **35.** 100   **37.** 20   **39.** 729   **41.** 192   **43.** 162   **45.** 3   **47.** 8   **49.** 12
51. 4   **53.** 21   **55.** 7   **57.** 5   **59.** 16   **61.** 46   **63.** 8   **65.** 64   **67.** 83   **69.** 2   **71.** 48   **73.** 4   **75.** undefined   **77.** 59   **79.** 52
81. 44   **83.** 12   **85.** 21   **87.** 24   **89.** 28   **91.** 3   **93.** 25   **95.** 23   **97.** 13   **99.** area: 49 sq m; perimeter: 28 m   **101.** area: 529 sq mi;
perimeter: 92 mi   **103.** true   **105.** false   **107.** $(2 + 3) \cdot 6 - 2$ **109.** $24 \div (3 \cdot 2) + 2 \cdot 5$ **111.** 1260 ft   **113.** 6,384,814
115. answers may vary; $(20 - 10) \cdot 5 \div 25 + 3$

Chapter 1 Vocabulary Check   1. whole numbers   **2.** perimeter   **3.** place value   **4.** exponent   **5.** area   **6.** square root
7. digits   **8.** average   **9.** divisor   **10.** dividend   **11.** quotient   **12.** factor   **13.** product   **14.** minuend   **15.** subtrahend
16. difference   **17.** addend   **18.** sum

Chapter 1 Review   1. tens   **2.** ten-millions   **3.** seven thousand, six hundred forty   **4.** forty-six million, two hundred thousand,
one hundred twenty   **5.** $3000 + 100 + 50 + 8$ **6.** $400,000,000 + 3,000,000 + 200,000 + 20,000 + 5000$ **7.** 81,900
8. 6,304,000,000   **9.** 518,512,109   **10.** 184,177,220   **11.** Asia   **12.** Oceania/Australia   **13.** 63   **14.** 67   **15.** 48   **16.** 77   **17.** 956
18. 840   **19.** 7950   **20.** 7250   **21.** 4211   **22.** 1967   **23.** 1326   **24.** 886   **25.** 27,346   **26.** 39,300   **27.** 8032 mi   **28.** $197,699
29. 276 ft   **30.** 66 km   **31.** 14   **32.** 34   **33.** 65   **34.** 304   **35.** 3914   **36.** 7908   **37.** 17,897   **38.** 34,658   **39.** 238,305   **40.** 249,795
41. 397 pages   **42.** $25,626   **43.** May   **44.** August   **45.** $110   **46.** $240   **47.** 90   **48.** 50   **49.** 470   **50.** 500   **51.** 4800   **52.** 58,000

53. 50,000,000 **54.** 800,000 **55.** 126,000,000 **56.** 99,000 **57.** 7400 **58.** 4100 **59.** 2500 mi **60.** 900,000 **61.** 1911 **62.** 1396
63. 1410 **64.** 2898 **65.** 800 **66.** 900 **67.** 3696 **68.** 1694 **69.** 0 **70.** 0 **71.** 16,994 **72.** 8954 **73.** 113,634 **74.** 44,763
75. 411,426 **76.** 636,314 **77.** 375,000 **78.** 108,000 **79.** 12,000 **80.** 35,000 **81.** 5,100,000 **82.** 7,600,000 **83.** 1150 **84.** 4920
85. 108 **86.** 112 **87.** 24 g **88.** $152,340 **89.** 60 sq mi **90.** 500 sq cm **91.** 3 **92.** 4 **93.** 6 **94.** 7 **95.** 5 R 2 **96.** 4 R 2
97. undefined **98.** 0 **99.** 1 **100.** 10 **101.** 0 **102.** undefined **103.** 33 R 2 **104.** 19 R 7 **105.** 24 R 2 **106.** 35 R 15 **107.** 506 R 10
108. 907 R 40 **109.** 2793 R 140 **110.** 2012 R 60 **111.** 18 R 2 **112.** 21 R 2 **113.** 458 ft **114.** 13 mi **115.** 51 **116.** 59
117. 27 boxes **118.** $192 **119.** $5 billion **120.** 75¢ **121.** $898 **122.** 23,150 sq ft **123.** 49 **124.** 125 **125.** 45 **126.** 400 **127.** 13
128. 10 **129.** 15 **130.** 7 **131.** 12 **132.** 9 **133.** 42 **134.** 33 **135.** 9 **136.** 2 **137.** 1 **138.** 0 **139.** 6 **140.** 29 **141.** 40
142. 72 **143.** 5 **144.** 7 **145.** 49 sq m **146.** 9 sq in. **147.** 307 **148.** 682 **149.** 2169 **150.** 2516 **151.** 901 **152.** 1411 **153.** 458 R 8
154. 237 R 1 **155.** 70,848 **156.** 95,832 **157.** 1644 **158.** 8481 **159.** 740 **160.** 258,000 **161.** 2000 **162.** 40,000 **163.** thirty-six
thousand, nine hundred eleven **164.** one hundred fifty-four thousand, eight hundred sixty-three **165.** 70,943 **166.** 43,401 **167.** 64
168. 125 **169.** 12 **170.** 10 **171.** 12 **172.** 1 **173.** 2 **174.** 6 **175.** 4 **176.** 24 **177.** 24 **178.** 14 **179.** $4,205,000 **180.** $2,129,000
181. 53 full boxes with 18 left over **182.** $86

Chapter 1 Test 1. eighty-two thousand, four hundred twenty-six **2.** 402,550 **3.** 141 **4.** 113 **5.** 14,880 **6.** 766 R 42 **7.** 200
8. 10 **9.** 0 **10.** undefined **11.** 33 **12.** 21 **13.** 8 **14.** 36 **15.** 5,698,000 **16.** 11,200,000 **17.** 52,000 **18.** 13,700 **19.** 1600
20. 92 **21.** 122 **22.** 1605 **23.** 7 R 2 **24.** $17 **25.** $126 **26.** 360 cal **27.** $7905 **28.** 20 cm; 25 sq cm **29.** 60 yd; 200 sq yd

Chapter 2 Multiplying and Dividing Fractions

Section 2.1

Vocabulary, Readiness & Video Check 1. fraction; denominator; numerator **3.** improper; proper; mixed number **5.** The fraction is equal to 1. **7.** Each shape is divided into 3 equal parts. **9.** division

Exercise Set 2.1 1. numerator: 1; denominator: 2; proper **3.** numerator: 10; denominator: 3; improper **5.** numerator: 15; denominator: 15; improper **7.** 1 **9.** undefined **11.** 13 **13.** 0 **15.** undefined **17.** 16 **19.** $\frac{5}{6}$ **21.** $\frac{7}{12}$ **23.** $\frac{3}{7}$ **25.** $\frac{4}{9}$ **27.** $\frac{1}{6}$ **29.** $\frac{5}{8}$

31. **33.** **35.** **37.**

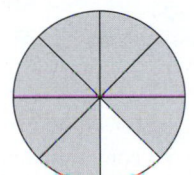

39. $\frac{42}{131}$ **41. a.** 89 **b.** $\frac{89}{131}$ **43.** $\frac{7}{44}$ **45.** $\frac{10}{19}$ **47.** $\frac{11}{31}$ **49.** $\frac{10}{31}$ **51. a.** $\frac{33}{50}$ **b.** 17 **c.** $\frac{17}{50}$ **53. a.** $\frac{21}{50}$ **b.** 29 **c.** $\frac{29}{50}$ **55. a.** $\frac{11}{4}$
b. $2\frac{3}{4}$ **57. a.** $\frac{23}{6}$ **b.** $3\frac{5}{6}$ **59. a.** $\frac{4}{3}$ **b.** $1\frac{1}{3}$ **61. a.** $\frac{11}{2}$ **b.** $5\frac{1}{2}$ **63.** $\frac{7}{3}$ **65.** $\frac{18}{5}$ **67.** $\frac{53}{8}$ **69.** $\frac{41}{15}$ **71.** $\frac{83}{7}$ **73.** $\frac{84}{13}$ **75.** $\frac{109}{24}$ **77.** $\frac{211}{12}$
79. $\frac{187}{20}$ **81.** $\frac{265}{107}$ **83.** $\frac{500}{3}$ **85.** $3\frac{2}{5}$ **87.** $4\frac{5}{8}$ **89.** $3\frac{2}{15}$ **91.** $2\frac{4}{21}$ **93.** 33 **95.** 15 **97.** $66\frac{2}{3}$ **99.** $10\frac{17}{23}$ **101.** $17\frac{13}{18}$ **103.** $1\frac{7}{175}$
105. $6\frac{65}{112}$ **107.** 9 **109.** 125 **111.** 7^5 **113.** $2^3 \cdot 3$ **115.** answers may vary **117.** $\frac{2}{3}$ **119.**

121. $\frac{18,000}{139,000}$ of employees **123.** $\frac{6}{36}$ of the countries

Section 2.2

Vocabulary, Readiness & Video Check 1. prime factorization **3.** prime **5.** factors **7.** Because order doesn't matter when we multiply, so switching the order doesn't give us any new factors of 12. **9.** order; one

Exercise Set 2.2 1. 1, 2, 4, 8 **3.** 1, 5, 25 **5.** 1, 2, 4 **7.** 1, 2, 3, 6, 9, 18 **9.** 1, 29 **11.** 1, 2, 4, 5, 8, 10, 16, 20, 40, 80 **13.** 1, 2, 3, 4, 6, 12
15. 1, 2, 17, 34 **17.** prime **19.** composite **21.** prime **23.** composite **25.** prime **27.** composite **29.** prime **31.** composite
33. composite **35.** 2^5 **37.** $3 \cdot 5$ **39.** $2^3 \cdot 5$ **41.** $2^2 \cdot 3^2$ **43.** $3 \cdot 13$ **45.** $2^2 \cdot 3 \cdot 5$ **47.** $2 \cdot 5 \cdot 11$ **49.** $5 \cdot 17$ **51.** 2^7 **53.** $2 \cdot 7 \cdot 11$
55. $2^2 \cdot 3 \cdot 5^2$ **57.** $2^4 \cdot 3 \cdot 5$ **59.** $2^2 \cdot 3^2 \cdot 23$ **61.** $2 \cdot 3^2 \cdot 7^2$ **63.** $7^2 \cdot 13$ **65.** $3 \cdot 11$ **67.** $2 \cdot 7^2$ **69.** prime **71.** $3^3 \cdot 17$ **73.** prime

75. $2^2 \cdot 5^2 \cdot 7$ **77.** 4300 **79.** 7,660,000 **81.** 20,000 **83.** 2375 **85.** $\frac{1136}{2375}$ **87.** $2^2 \cdot 3^5 \cdot 5 \cdot 7$ **89.** answers may vary

91. answers may vary

Section 2.3

Calculator Explorations   **1.** $\frac{4}{7}$ **3.** $\frac{20}{27}$ **5.** $\frac{15}{8}$ **7.** $\frac{9}{2}$

Vocabulary, Readiness & Video Check   **1.** simplest form   **3.** cross products   **5.** 0   **7.** equivalent   **9.** $\frac{10}{24}$ is not in simplest form; $\frac{5}{12}$

Exercise Set 2.3   **1.** $\frac{1}{4}$ **3.** $\frac{2}{21}$ **5.** $\frac{7}{8}$ **7.** $\frac{2}{3}$ **9.** $\frac{7}{10}$ **11.** $\frac{7}{9}$ **13.** $\frac{3}{5}$ **15.** $\frac{27}{64}$ **17.** $\frac{5}{8}$ **19.** $\frac{5}{8}$ **21.** $\frac{14}{17}$ **23.** $\frac{3}{2}$ or $1\frac{1}{2}$ **25.** $\frac{3}{4}$ **27.** $\frac{5}{14}$
29. $\frac{3}{14}$ **31.** $\frac{11}{17}$ **33.** $\frac{3}{14}$ **35.** $\frac{7}{8}$ **37.** $\frac{3}{5}$ **39.** 14   **41.** equivalent   **43.** not equivalent   **45.** equivalent   **47.** equivalent
49. not equivalent   **51.** not equivalent   **53.** $\frac{1}{4}$ of a shift   **55.** $\frac{1}{2}$ mi   **57. a.** $\frac{8}{25}$ **b.** 34 states   **c.** $\frac{17}{25}$ **59.** $\frac{5}{12}$ of the wall   **61. a.** 30
b. $\frac{3}{5}$ **63.** $\frac{2}{5}$ of top-selling games   **65.** 364   **67.** 2322   **69.** 2520   **71.** answers may vary   **73.** $\frac{3}{5}$ **75.** $\frac{9}{25}$ **77.** $\frac{1}{25}$ **79.** $\frac{1}{10}$
81. answers may vary   **83.** $\frac{7}{100}$ **85.** answers may vary   **87.** 786, 222, 900, 1470   **89.** 6; answers may vary

Integrated Review   **1.** $\frac{3}{6}$ (or $\frac{1}{2}$ simplified)   **2.** $\frac{7}{4}$ or $1\frac{3}{4}$ **3.** $\frac{73}{85}$ **4.**

5. 1   **6.** 17   **7.** 0   **8.** undefined   **9.** $\frac{25}{8}$ **10.** $\frac{28}{5}$ **11.** $\frac{69}{7}$ **12.** $\frac{141}{7}$ **13.** $2\frac{6}{7}$ **14.** 5   **15.** $4\frac{7}{8}$ **16.** $8\frac{10}{11}$ **17.** 1, 5, 7, 35   **18.** 1, 2, 4, 5,
8, 10, 20, 40   **19.** composite   **20.** prime   **21.** $5 \cdot 13$ **22.** $2 \cdot 5 \cdot 7$ **23.** $2^5 \cdot 3$ **24.** $2^2 \cdot 3 \cdot 11$ **25.** $2^2 \cdot 3^2 \cdot 7$ **26.** prime
27. $3^2 \cdot 5 \cdot 7$ **28.** $3^2 \cdot 7^2$ **29.** $2 \cdot 11 \cdot 13$ **30.** prime   **31.** $\frac{1}{7}$ **32.** $\frac{6}{5}$ or $1\frac{1}{5}$ **33.** $\frac{9}{19}$ **34.** $\frac{21}{55}$ **35.** $\frac{14}{15}$ **36.** $\frac{9}{10}$ **37.** $\frac{2}{5}$ **38.** $\frac{3}{8}$ **39.** $\frac{11}{14}$
40. $\frac{7}{11}$ **41.** not equivalent   **42.** equivalent   **43. a.** $\frac{1}{25}$ **b.** 48   **c.** $\frac{24}{25}$ **44. a.** $\frac{29}{108}$ **b.** 395   **c.** $\frac{79}{108}$

Section 2.4

Vocabulary, Readiness & Video Check   **1.** $\frac{a \cdot c}{b \cdot d}$ **3.** multiplication   **5.** There's a common factor of 2 in the numerator and

denominator that can be divided out first.   **7.** radius is $\frac{1}{2}$ of diameter

Exercise Set 2.4   **1.** $\frac{2}{15}$ **3.** $\frac{6}{35}$ **5.** $\frac{9}{80}$ **7.** $\frac{5}{28}$ **9.** $\frac{12}{5}$ or $2\frac{2}{5}$ **11.** $\frac{1}{70}$ **13.** 0   **15.** $\frac{1}{110}$ **17.** $\frac{18}{55}$ **19.** $\frac{27}{80}$ **21.** $\frac{1}{56}$ **23.** $\frac{2}{105}$ **25.** 0
27. $\frac{1}{90}$ **29.** 8   **31.** 6   **33.** 20   **35.** 3   **37.** $\frac{5}{2}$ or $2\frac{1}{2}$ **39.** $\frac{1}{5}$ **41.** $\frac{5}{3}$ or $1\frac{2}{3}$ **43.** $\frac{2}{3}$ **45.** Exact: $\frac{77}{10}$ or $7\frac{7}{10}$; Estimate: 8   **47.** Exact:
$\frac{836}{35}$ or $23\frac{31}{35}$; Estimate: 24   **49.** $\frac{25}{2}$ or $21\frac{1}{2}$ **51.** 15   **53.** 6   **55.** $\frac{45}{4}$ or $11\frac{1}{4}$ **57.** $\frac{49}{3}$ or $16\frac{1}{3}$ **59.** $\frac{1}{30}$ **61.** 0   **63.** $\frac{16}{5}$ or $3\frac{1}{5}$ **65.** $\frac{7}{2}$ or
$3\frac{1}{2}$ **67.** $\frac{1}{8}$ **69.** $\frac{1}{56}$ **71.** $\frac{55}{3}$ or $18\frac{1}{3}$ **73.** 0   **75.** $\frac{208}{7}$ or $29\frac{5}{7}$ **77.** 50   **79.** 20   **81.** 128 freshmen   **83.** 117 million   **85.** 868 mi
87. $\frac{3}{16}$ in.   **89.** 30 gal   **91.** $\frac{17}{2}$ or $8\frac{1}{2}$ in.   **93.** $\frac{39}{2}$ or $19\frac{1}{2}$ in.   **95.** $\frac{2242}{625}$ or $3\frac{367}{625}$ sq in.   **97.** $\frac{1}{14}$ sq ft   **99.** $\frac{7}{2}$ or $3\frac{1}{2}$ sq yd   **101.** 3840 mi
103. 2400 mi   **105.** 206   **107.** 56 R 12   **109.** answers may vary   **111.** $3\frac{2}{3} \cdot 1\frac{1}{7} = \frac{11}{3} \cdot \frac{8}{7} = \frac{11 \cdot 8}{3 \cdot 7} = \frac{88}{21}$ or $4\frac{4}{21}$ **113.** b   **115.** a
117. 37 students   **119.** 664,950 Māori

Section 2.5

Vocabulary, Readiness & Video Check   **1.** reciprocals   **3.** $\frac{a \cdot d}{b \cdot c}$ **5.** $\frac{1}{n}$ **7.** Because we still have a division problem and we can't divide out common factors until we rewrite the division as a multiplication.

Exercise Set 2.5   **1.** $\frac{7}{4}$ **3.** 11   **5.** $\frac{1}{15}$ **7.** $\frac{7}{12}$ **9.** $\frac{4}{5}$ **11.** $\frac{16}{9}$ or $1\frac{7}{9}$ **13.** $\frac{18}{35}$ **15.** $\frac{3}{4}$ **17.** $\frac{1}{100}$ **19.** $\frac{1}{3}$ **21.** $\frac{5}{3}$ or $1\frac{2}{3}$ **23.** $\frac{35}{36}$ **25.** $\frac{14}{37}$
27. $\frac{8}{45}$ **29.** 1   **31.** undefined   **33.** 0   **35.** $\frac{7}{10}$ **37.** $\frac{1}{6}$ **39.** $\frac{40}{3}$ or $13\frac{1}{3}$ **41.** 5   **43.** $\frac{5}{28}$ **45.** $\frac{36}{35}$ or $1\frac{1}{35}$ **47.** $\frac{26}{51}$ **49.** 0

51. $\frac{17}{13}$ or $1\frac{4}{13}$ **53.** $\frac{35}{18}$ or $1\frac{17}{18}$ **55.** $\frac{19}{30}$ **57.** $\frac{1}{6}$ **59.** $\frac{121}{60}$ or $2\frac{1}{60}$ **61.** 96 **63.** $\frac{3}{4}$ **65.** undefined **67.** $\frac{11}{119}$ **69.** $\frac{35}{11}$ or $3\frac{2}{11}$ **71.** $\frac{9}{5}$ or $1\frac{4}{5}$

73. $3\frac{3}{16}$ miles **75.** $\frac{5}{6}$ Tbsp **77.** $\frac{19}{30}$ in. **79.** 20 lb **81.** $4\frac{2}{3}$ m **83.** $\frac{8}{35}$ **85.** $\frac{128}{51}$ or $2\frac{26}{51}$ **87.** $\frac{16}{15}$ or $1\frac{1}{15}$ **89.** $\frac{121}{400}$ **91.** 201 **93.** 196

95. 1569 **97.** $20\frac{2}{3} \div 10\frac{1}{2} = \frac{62}{3} \div \frac{21}{2} = \frac{62}{3} \cdot \frac{2}{21} = \frac{124}{63}$ or $1\frac{61}{63}$ **99.** c **101.** d **103.** 5 **105.** 634 aircraft **107.** answers may vary

Chapter 2 Vocabulary Check 1. reciprocals **2.** composite number **3.** equivalent **4.** improper fraction **5.** prime number
6. simplest form **7.** proper fraction **8.** mixed number **9.** numerator; denominator **10.** prime factorization **11.** undefined
12. 0 **13.** cross products

Chapter 2 Review 1. proper **2.** improper **3.** proper **4.** mixed number **5.** $\frac{2}{6}$ **6.** $\frac{4}{7}$ **7.** $\frac{7}{3}$ **8.** $\frac{13}{4}$ **9.** $\frac{11}{12}$ **10. a.** 108 **b.** $\frac{108}{131}$

11. $3\frac{3}{4}$ **12.** $45\frac{5}{6}$ **13.** 3 **14.** 5 **15.** $\frac{6}{5}$ **16.** $\frac{22}{21}$ **17.** $\frac{26}{9}$ **18.** $\frac{47}{12}$ **19.** composite **20.** prime **21.** 1, 2, 3, 6, 7, 14, 21, 42

22. 1, 2, 4, 5, 10, 20 **23.** $2^2 \cdot 17$ **24.** $2 \cdot 3^2 \cdot 5$ **25.** $5 \cdot 157$ **26.** $3 \cdot 5 \cdot 17$ **27.** $\frac{3}{7}$ **28.** $\frac{5}{9}$ **29.** $\frac{1}{3}$ **30.** $\frac{1}{2}$ **31.** $\frac{29}{32}$ **32.** $\frac{18}{23}$ **33.** 8

34. 6 **35.** $\frac{2}{3}$ of a foot **36.** $\frac{3}{5}$ of the cars **37.** no **38.** yes **39.** $\frac{3}{10}$ **40.** $\frac{5}{14}$ **41.** 9 **42.** $\frac{1}{2}$ **43.** $\frac{35}{8}$ or $4\frac{3}{8}$ **44.** $\frac{5}{2}$ or $2\frac{1}{2}$ **45.** $\frac{5}{3}$ or $1\frac{2}{3}$

46. $\frac{49}{3}$ or $16\frac{1}{3}$ **47.** Exact: $\frac{26}{5}$ or $5\frac{1}{5}$; Estimate: 6 **48.** Exact: $\frac{60}{11}$ or $5\frac{5}{11}$; Estimate: 8 **49.** $\frac{99}{4}$ or $24\frac{3}{4}$ **50.** $\frac{1}{6}$ **51.** $\frac{110}{3}$ or $36\frac{2}{3}$ g

52. $\frac{135}{4}$ or $33\frac{3}{4}$ in. **53.** $\frac{119}{80}$ or $1\frac{39}{80}$ sq in. **54.** $\frac{275}{8}$ or $34\frac{3}{8}$ sq m **55.** $\frac{1}{7}$ **56.** 8 **57.** $\frac{23}{14}$ **58.** $\frac{5}{17}$ **59.** 2 **60.** $\frac{15}{4}$ or $3\frac{3}{4}$ **61.** $\frac{5}{6}$

62. $\frac{8}{3}$ or $2\frac{2}{3}$ **63.** $\frac{21}{4}$ or $5\frac{1}{4}$ **64.** $\frac{121}{46}$ or $2\frac{29}{46}$ **65.** 22 mi **66.** $\frac{21}{20}$ or $1\frac{1}{20}$ mi **67.** proper **68.** improper **69.** mixed number

70. improper **71.** $31\frac{1}{4}$ **72.** 6 **73.** $\frac{95}{17}$ **74.** $\frac{47}{6}$ **75.** composite **76.** prime **77.** $2^2 \cdot 3^2 \cdot 5$ **78.** $2 \cdot 7^2$ **79.** $\frac{9}{10}$ **80.** $\frac{5}{7}$ **81.** $\frac{14}{15}$

82. $\frac{3}{5}$ **83.** $\frac{7}{12}$ **84.** $\frac{1}{4}$ **85.** 9 **86.** $\frac{27}{2}$ or $13\frac{1}{2}$ **87.** Exact: 10; Estimate: 8 **88.** Exact: $\frac{51}{4}$ or $12\frac{3}{4}$; Estimate: 12 **89.** $\frac{7}{3}$ or $2\frac{1}{3}$

90. $\frac{32}{5}$ or $6\frac{2}{5}$ **91.** $\frac{81}{2}$ or $40\frac{1}{2}$ sq ft **92.** $\frac{47}{61}$ in.

Chapter 2 Test 1. $\frac{7}{16}$ **2.** $\frac{13}{5}$ **3.** $\frac{23}{3}$ **4.** $\frac{39}{11}$ **5.** $4\frac{3}{5}$ **6.** $18\frac{3}{4}$ **7.** $\frac{4}{35}$ **8.** $\frac{3}{5}$ **9.** not equivalent **10.** equivalent **11.** $2^2 \cdot 3 \cdot 7$

12. $3^2 \cdot 5 \cdot 11$ **13.** $\frac{4}{3}$ or $1\frac{1}{3}$ **14.** $\frac{4}{3}$ or $1\frac{1}{3}$ **15.** $\frac{1}{4}$ **16.** $\frac{16}{45}$ **17.** 16 **18.** $\frac{9}{2}$ or $4\frac{1}{2}$ **19.** $\frac{4}{11}$ **20.** 9 **21.** $\frac{64}{3}$ or $21\frac{1}{3}$ **22.** $\frac{45}{2}$ or $22\frac{1}{2}$

23. $\frac{18}{5}$ or $3\frac{3}{5}$ **24.** $\frac{20}{3}$ or $6\frac{2}{3}$ **25.** $\frac{34}{27}$ or $1\frac{7}{27}$ sq mi **26.** 24 mi **27.** $\frac{16,000}{3}$ or $5333\frac{1}{3}$ sq yd **28.** $90 per share

Cumulative Review 1. hundred-thousands (Sec. 1.2, Ex. 1) **2.** two thousand, thirty-six **3.** 805 (Sec. 1.2, Ex. 9) **4.** 31
5. 184,046 (Sec. 1.3, Ex. 2) **6.** 39 **7.** 13 in. (Sec. 1.3, Ex. 5) **8.** 17 **9.** 14,440,060 (Sec. 1.3, Ex. 7) **10.** 5 **11.** 7321 (Sec. 1.4, Ex. 2)
12. 64 **13. a.** Indonesia **b.** 331 (Sec. 1.3, Ex. 8) **14.** 25 R 5 **15.** 570 (Sec. 1.5, Ex. 1) **16.** 2400 **17.** 1800 (Sec. 1.5, Ex. 5) **18.** 300
19. a. 6 **b.** 0 **c.** 45 **d.** 0 (Sec. 1.6, Ex. 1) **20.** 20 **21. a.** $3 \cdot 4 + 3 \cdot 5$ **b.** $10 \cdot 6 + 10 \cdot 8$ **c.** $2 \cdot 7 + 2 \cdot 3$ (Sec. 1.6, Ex. 2) **22.** 180
23. a. 0 **b.** 0 **c.** 0 **d.** undefined (Sec. 1.7, Ex. 3) **24.** 154 sq mi **25.** 208 (Sec. 1.7, Ex. 5) **26.** 4014 **27.** 12 cards; 10 cards left
over (Sec. 1.7, Ex. 11) **28.** 63 **29.** 40 ft (Sec. 1.8, Ex. 5) **30.** 16 **31.** 7^3 (Sec. 1.9, Ex. 1) **32.** 7^4 **33.** $3^4 \cdot 17^3$ (Sec. 1.9, Ex. 4)
34. $2^2 \cdot 3^4$ **35.** 7 (Sec. 1.9, Ex. 12) **36.** 0 **37.** $\frac{2}{5}$ (Sec. 2.1, Ex. 7) **38.** $2^2 \cdot 3 \cdot 13$ **39. a.** $\frac{38}{9}$ **b.** $\frac{19}{11}$ (Sec. 2.1, Ex. 17) **40.** $\frac{39}{5}$

41. 1, 2, 4, 5, 10, 20 (Sec. 2.2, Ex. 1) **42.** equivalent **43.** $\frac{7}{11}$ (Sec. 2.3, Ex. 2) **44.** $\frac{2}{3}$ **45.** $\frac{35}{12}$ or $2\frac{11}{12}$ (Sec. 2.4, Ex. 8)

46. $\frac{8}{3}$ or $2\frac{2}{3}$ **47.** $\frac{3}{1}$ or 3 (Sec. 2.5, Ex. 3) **48.** $\frac{1}{9}$ **49.** $\frac{5}{12}$ (Sec. 2.5, Ex. 6) **50.** $\frac{11}{56}$

Chapter 3 Adding and Subtracting Fractions

Section 3.1

Vocabulary, Readiness & Video Check 1. like; unlike **3.** $\frac{a-c}{b}$ **5.** unlike **7.** like **9.** like **11.** unlike **13.** We can simplify
by dividing out a common factor of 3 from the numerator and denominator to get $\frac{2}{3}$. **15.** $P = \frac{4}{20} + \frac{9}{20} + \frac{7}{20}$; 1 in.

Exercise Set 3.1 **1.** $\frac{3}{7}$ **3.** $\frac{1}{5}$ **5.** $\frac{2}{3}$ **7.** $\frac{7}{20}$ **9.** $\frac{1}{2}$ **11.** $\frac{13}{11}$ or $1\frac{2}{11}$ **13.** $\frac{7}{13}$ **15.** $\frac{2}{3}$ **17.** $\frac{6}{11}$ **19.** $\frac{3}{5}$ **21.** 1 **23.** $\frac{3}{4}$ **25.** $\frac{5}{6}$ **27.** $\frac{4}{5}$

29. $\frac{1}{90}$ **31.** $\frac{19}{33}$ **33.** $\frac{13}{21}$ **35.** $\frac{9}{10}$ **37.** 0 **39.** $\frac{3}{4}$ **41.** 1 in. **43.** 2 m **45.** $\frac{7}{10}$ mi **47.** $\frac{3}{2}$ or $1\frac{1}{2}$ hr **49.** $\frac{43}{100}$ **51.** $\frac{1}{10}$ **53.** $\frac{9}{20}$ **55.** $\frac{13}{50}$

57. $\frac{7}{25}$ **59.** $\frac{1}{50}$ **61.** $\frac{21}{25}$ **63.** $2 \cdot 5$ **65.** 2^3 **67.** $5 \cdot 11$ **69.** $\frac{5}{8}$ **71.** $\frac{8}{11}$ **73.** $\frac{2}{7} + \frac{9}{7} = \frac{11}{7}$ or $1\frac{4}{7}$ **75.** answers may vary

77. 1; answers may vary **79.** $\frac{1}{4}$ mi

Section 3.2

Vocabulary, Readiness & Video Check **1.** equivalent **3.** multiple **5.** Because 24 is a multiple of 8.

Exercise Set 3.2 **1.** 12 **3.** 45 **5.** 36 **7.** 72 **9.** 126 **11.** 75 **13.** 24 **15.** 42 **17.** 216 **19.** 150 **21.** 68 **23.** 588

25. 900 **27.** 1800 **29.** 363 **31.** 60 **33.** $\frac{20}{35}$ **35.** $\frac{14}{21}$ **37.** $\frac{15}{3}$ **39.** $\frac{15}{30}$ **41.** $\frac{30}{21}$ **43.** $\frac{21}{28}$ **45.** $\frac{30}{45}$ **47.** $\frac{36}{81}$ **49.** $\frac{90}{78}$ **51.** $\frac{56}{68}$

53. $\frac{86}{100}, \frac{80}{100}, \frac{58}{100}, \frac{68}{100}, \frac{12}{100}, \frac{84}{100}, \frac{68}{100}, \frac{81}{100}, \frac{90}{100}, \frac{79}{100}, \frac{74}{100}, \frac{78}{100}$ **55.** drugs, health and beauty aids **57.** $\frac{1}{2}$ **59.** $\frac{2}{5}$ **61.** $\frac{4}{9}$ **63.** 1

65. $\frac{814}{3630}$ **67.** answers may vary **69.** a, b, and d

Section 3.3

Calculator Explorations **1.** $\frac{37}{80}$ **3.** $\frac{95}{72}$ **5.** $\frac{394}{323}$

Vocabulary, Readiness & Video Check **1.** equivalent; least common denominator **3.** $\frac{15}{24}, \frac{4}{24}, \frac{19}{24}$ **5.** Multiplying by $\frac{3}{3}$ is

multiplying by a form of 1. Thus, the result is an equivalent fraction. **7.** $\frac{34}{15}$ cm and $2\frac{4}{15}$ cm

Exercise Set 3.3 **1.** $\frac{5}{6}$ **3.** $\frac{5}{6}$ **5.** $\frac{8}{33}$ **7.** $\frac{9}{14}$ **9.** $\frac{3}{5}$ **11.** $\frac{13}{25}$ **13.** $\frac{53}{60}$ **15.** $\frac{1}{6}$ **17.** $\frac{67}{99}$ **19.** $\frac{98}{143}$ **21.** $\frac{13}{27}$ **23.** $\frac{75}{56}$ or $1\frac{19}{56}$

25. $\frac{19}{18}$ or $1\frac{1}{18}$ **27.** $\frac{19}{12}$ or $1\frac{7}{12}$ **29.** $\frac{11}{16}$ **31.** $\frac{17}{42}$ **33.** $\frac{33}{56}$ **35.** $\frac{37}{99}$ **37.** $\frac{1}{35}$ **39.** $\frac{11}{36}$ **41.** $\frac{1}{20}$ **43.** $\frac{1}{84}$ **45.** $\frac{9}{1000}$ **47.** $\frac{17}{99}$

49. $\frac{19}{36}$ **51.** $\frac{1}{5}$ **53.** $\frac{69}{280}$ **55.** $\frac{14}{9}$ or $1\frac{5}{9}$ **57.** $\frac{34}{15}$ or $2\frac{4}{15}$ cm **59.** $\frac{17}{10}$ or $1\frac{7}{10}$ m **61.** $\frac{7}{100}$ mph **63.** $\frac{5}{8}$ in. **65.** $\frac{31}{32}$ in.

67. $\frac{19}{100}$ of Girl Scout cookies **69.** $\frac{19}{25}$ **71.** $\frac{17}{50}$ **73.** $\frac{81}{100}$ **75.** 5 **77.** $\frac{16}{29}$ **79.** $\frac{19}{3}$ or $6\frac{1}{3}$ **81.** $\frac{3}{5} + \frac{4}{5} = \frac{7}{5}$ or $1\frac{2}{5}$ **83.** $\frac{223}{540}$

85. $\frac{49}{44}$ or $1\frac{5}{44}$ **87.** answers may vary

Integrated Review **1.** 30 **2.** 21 **3.** 14 **4.** 25 **5.** 100 **6.** 90 **7.** $\frac{9}{24}$ **8.** $\frac{28}{36}$ **9.** $\frac{10}{40}$ **10.** $\frac{12}{30}$ **11.** $\frac{55}{75}$ **12.** $\frac{40}{48}$ **13.** $\frac{1}{2}$ **14.** $\frac{2}{5}$

15. $\frac{7}{12}$ **16.** $\frac{13}{15}$ **17.** $\frac{3}{4}$ **18.** $\frac{2}{15}$ **19.** $\frac{17}{45}$ **20.** $\frac{19}{50}$ **21.** $\frac{37}{40}$ **22.** $\frac{11}{36}$ **23.** 0 **24.** $\frac{1}{17}$ **25.** $\frac{5}{33}$ **26.** $\frac{1}{42}$ **27.** $\frac{5}{18}$ **28.** $\frac{5}{13}$ **29.** $\frac{11}{18}$

30. $\frac{37}{50}$ **31.** $\frac{47}{30}$ or $1\frac{17}{30}$ **32.** $\frac{7}{30}$ **33.** $\frac{3}{5}$ **34.** $\frac{27}{20}$ or $1\frac{7}{20}$ **35.** $\frac{279}{350}$ **36.** $\frac{309}{350}$ **37.** $\frac{98}{5}$ or $19\frac{3}{5}$ **38.** $\frac{9}{250}$ **39.** $\frac{31}{3}$ or $10\frac{1}{3}$

40. $\frac{93}{64}$ or $1\frac{29}{64}$ **41.** $\frac{49}{54}$ **42.** $\frac{83}{48}$ or $1\frac{35}{48}$ **43.** $\frac{390}{101}$ or $3\frac{87}{101}$ **44.** $\frac{116}{5}$ or $23\frac{1}{5}$ **45.** $\frac{106}{135}$ **46.** $\frac{67}{224}$

Section 3.4

Vocabulary, Readiness & Video Check **1.** mixed number **3.** round **5.** a **7.** c **9.** The fractional part of a mixed number should not be an improper fraction. **11.** Because we need to borrow first.

Exercise Set 3.4 **1.** Exact: $6\frac{4}{5}$; Estimate: 7 **3.** Exact: $13\frac{11}{14}$; Estimate: 14 **5.** $17\frac{7}{25}$ **7.** $7\frac{5}{8}$ **9.** $7\frac{5}{24}$ **11.** $20\frac{1}{15}$ **13.** 19 **15.** $56\frac{53}{270}$

17. $13\frac{13}{24}$ **19.** $47\frac{53}{84}$ **21.** Exact: $2\frac{3}{5}$; Estimate: 3 **23.** Exact: $7\frac{5}{14}$; Estimate: 7 **25.** $\frac{24}{25}$ **27.** $2\frac{7}{15}$ **29.** $5\frac{11}{14}$ **31.** $23\frac{31}{72}$ **33.** $1\frac{4}{5}$

35. $1\frac{13}{15}$ **37.** $3\frac{5}{9}$ **39.** $15\frac{3}{4}$ **41.** $28\frac{7}{12}$ **43.** $15\frac{7}{8}$ **45.** 8   **47.** $17\frac{11}{12}$ **49.** $\frac{1}{16}$ in.   **51.** no; she will be $\frac{1}{12}$ of a foot short   **53.** $7\frac{13}{20}$ in.

55. $10\frac{1}{4}$ hr   **57.** $2\frac{3}{8}$ hr   **59.** $92\frac{99}{100}$ m   **61.** $319\frac{1}{3}$ yd   **63.** $9\frac{13}{30}$ min   **65.** $1\frac{4}{5}$ min   **67.** 7 mi   **69.** $21\frac{5}{24}$ m   **71.** 8   **73.** 25   **75.** 4

77. 167   **79.** 4   **81.** $9\frac{5}{8}$ **83.** a, b, c   **85.** answers may vary   **87.** Supreme is heavier by $\frac{1}{8}$ lb

Section 3.5

Vocabulary, Readiness & Video Check   1. multiplication   **3.** subtraction   **5.** denominators; numerators   **7.** We need to make sure we have the same denominator when adding and subtracting fractions; this is not necessary when multiplying and dividing fractions.

Exercise Set 3.5   1. >   **3.** <   **5.** <   **7.** >   **9.** >   **11.** <   **13.** >   **15.** <   **17.** $\frac{1}{16}$ **19.** $\frac{8}{125}$ **21.** $\frac{64}{343}$ **23.** $\frac{4}{81}$

25. $\frac{1}{6}$ **27.** $\frac{18}{125}$ **29.** $\frac{11}{15}$ **31.** $\frac{3}{35}$ **33.** $\frac{5}{9}$ **35.** $10\frac{4}{99}$ **37.** $\frac{1}{12}$ **39.** $\frac{9}{11}$ **41.** 0   **43.** 0   **45.** $\frac{2}{5}$ **47.** $\frac{2}{77}$ **49.** $\frac{17}{60}$ **51.** $\frac{5}{8}$ **53.** $\frac{1}{2}$

55. $\frac{29}{10}$ or $2\frac{9}{10}$ **57.** $\frac{27}{32}$ **59.** $\frac{1}{81}$ **61.** $\frac{5}{6}$ **63.** $\frac{3}{5}$ **65.** $\frac{1}{2}$ **67.** $\frac{19}{7}$ or $2\frac{5}{7}$ **69.** $\frac{9}{64}$ **71.** $\frac{3}{4}$ **73.** $\frac{13}{60}$ **75.** $\frac{13}{25}$ **77.** A   **79.** M   **81.** S

83. D   **85.** M   **87.** A   **89.** no; answers may vary   **91.** subtraction, multiplication, addition, division   **93.** division, multiplication, subtraction, addition   **95.** standard mail   **97.** discretionary purchases

Section 3.6

Vocabulary, Readiness & Video Check   1. To make sure we answer the question asked in the original problem.

Exercise Set 3.6   1. $\frac{1}{2}+\frac{1}{3}$ **3.** $20\div6\frac{2}{5}$ **5.** $\frac{15}{16}-\frac{5}{8}$ **7.** $\frac{21}{68}+\frac{7}{34}$ **9.** $8\frac{1}{3}\cdot\frac{7}{9}$ **11.** $3\frac{1}{3}$ c   **13.** $12\frac{1}{2}$ in.   **15.** $21\frac{1}{2}$ mi per gal

17. $1\frac{1}{2}$ yr   **19.** $9\frac{2}{5}$ in.   **21.** no; $\frac{1}{4}$ yd   **23.** 5 pieces   **25.** $\frac{9}{8}$ or $1\frac{1}{8}$ in.   **27.** $3\frac{3}{4}$ c   **29.** $11\frac{1}{4}$ sq in.   **31.** $1\frac{29}{60}$ min   **33.** $1\frac{13}{50}$ cu in.

35. 67 sheets   **37. a.** yes   **b.** 1 ft left over   **39.** $2\frac{15}{16}$ lb   **41.** area: $\frac{9}{128}$ sq in.; perimeter: $1\frac{1}{8}$ in.   **43.** area: $\frac{25}{81}$ sq m; perimeter: $2\frac{2}{9}$ m

45. $4\frac{3}{4}$ ft   **47.** $\frac{5}{26}$ ft   **49.** 3   **51.** 81   **53.** 4   **55.** 30   **57.** 35   **59.** no; no; answers may vary   **61.** $36\frac{44}{81}$ sq ft   **63.** 68 customers

65. 22 hr

Chapter 3 Vocabulary Check   1. like   **2.** least common multiple   **3.** equivalent   **4.** mixed number   **5.** >   **6.** <
7. least common denominator   **8.** unlike   **9.** exponent

Chapter 3 Review   1. $\frac{10}{11}$ **2.** $\frac{3}{25}$ **3.** $\frac{2}{3}$ **4.** $\frac{1}{7}$ **5.** $\frac{3}{5}$ **6.** $\frac{3}{5}$ **7.** 1   **8.** 1   **9.** $\frac{19}{25}$ **10.** $\frac{16}{21}$ **11.** $\frac{3}{4}$ of his homework   **12.** $\frac{3}{2}$ or $1\frac{1}{2}$ mi

13. 55   **14.** 60   **15.** 120   **16.** 80   **17.** 252   **18.** 72   **19.** $\frac{56}{64}$ **20.** $\frac{20}{30}$ **21.** $\frac{21}{33}$ **22.** $\frac{20}{26}$ **23.** $\frac{16}{60}$ **24.** $\frac{25}{60}$ **25.** $\frac{11}{18}$ **26.** $\frac{7}{15}$ **27.** $\frac{7}{26}$

28. $\frac{17}{36}$ **29.** $\frac{41}{42}$ **30.** $\frac{43}{72}$ **31.** $\frac{13}{45}$ **32.** $\frac{39}{70}$ **33.** $\frac{19}{9}$ or $2\frac{1}{9}$ m   **34.** $\frac{3}{2}$ or $1\frac{1}{2}$ ft   **35.** $\frac{1}{4}$ of a yd   **36.** $\frac{7}{10}$ has been cleaned   **37.** $45\frac{16}{21}$

38. 60   **39.** $32\frac{13}{22}$ **40.** $3\frac{19}{60}$ **41.** $111\frac{5}{18}$ **42.** $20\frac{7}{24}$ **43.** $5\frac{16}{35}$ **44.** $3\frac{4}{55}$ **45.** $7\frac{4}{5}$ in.   **46.** $\frac{1}{40}$ oz   **47.** 5 ft   **48.** $11\frac{1}{6}$ ft   **49.** <

50. >   **51.** <   **52.** >   **53.** >   **54.** >   **55.** $\frac{9}{49}$ **56.** $\frac{64}{125}$ **57.** $\frac{9}{400}$ **58.** $\frac{9}{100}$ **59.** $\frac{8}{13}$ **60.** 2   **61.** $\frac{81}{196}$ **62.** $\frac{1}{27}$ **63.** $\frac{13}{18}$

64. $\frac{11}{15}$ **65.** $\frac{1}{7}$ **66.** $\frac{18}{5}$ or $3\frac{3}{5}$ **67.** $\frac{45}{28}$ or $1\frac{17}{28}$ **68.** $\frac{5}{6}$ **69.** $\frac{99}{56}$ or $1\frac{43}{56}$ **70.** $\frac{29}{110}$ **71.** $\frac{29}{54}$ **72.** $\frac{37}{60}$ **73.** 50 moons   **74.** $15\frac{5}{8}$ acres

75. each measurement is $4\frac{1}{4}$ in.   **76.** $\frac{7}{10}$ yd   **77.** perimeter: $\frac{17}{11}$ or $1\frac{6}{11}$ mi; area: $\frac{3}{22}$ sq mi   **78.** perimeter: $\frac{7}{3}$ or $2\frac{1}{3}$ m; area: $\frac{5}{16}$ sq m

79. 90   **80.** 60   **81.** $\frac{40}{48}$ **82.** $\frac{63}{72}$ **83.** $\frac{1}{6}$ **84.** $\frac{1}{5}$ **85.** $\frac{11}{12}$ **86.** $\frac{27}{55}$ **87.** $13\frac{5}{12}$ **88.** $12\frac{3}{8}$ **89.** $3\frac{16}{35}$ **90.** $8\frac{1}{21}$ **91.** $\frac{11}{25}$ **92.** $\frac{1}{8}$

93. $\frac{1}{144}$ **94.** $\frac{64}{27}$ or $2\frac{10}{27}$ **95.** $\frac{5}{17}$ **96.** $\frac{1}{12}$ **97.** <   **98.** >   **99.** $\frac{1}{2}$ hr   **100.** $6\frac{7}{20}$ lb   **101.** $44\frac{1}{2}$ yd   **102.** $2\frac{2}{15}$ ft

103. $7\frac{1}{2}$ tablespoons   **104.** $\frac{3}{8}$ gal

Chapter 3 Test **1.** 60 **2.** 72 **3.** < **4.** < **5.** $\frac{8}{9}$ **6.** $\frac{2}{5}$ **7.** $\frac{13}{10}$ or $1\frac{3}{10}$ **8.** $\frac{8}{21}$ **9.** $\frac{13}{24}$ **10.** $\frac{2}{3}$ **11.** $\frac{67}{60}$ or $1\frac{7}{60}$ **12.** $\frac{7}{50}$

13. $\frac{3}{2}$ or $1\frac{1}{2}$ **14.** $14\frac{1}{40}$ **15.** $30\frac{13}{45}$ **16.** $1\frac{7}{24}$ **17.** $16\frac{8}{11}$ **18.** $\frac{5}{3}$ or $1\frac{2}{3}$ **19.** $\frac{16}{81}$ **20.** $\frac{9}{16}$ **21.** $\frac{153}{200}$ **22.** $\frac{3}{8}$ **23.** $\frac{11}{12}$ **24.** $3\frac{3}{4}$ ft

25. $7\frac{5}{6}$ gal **26.** $\frac{23}{50}$ **27.** $\frac{13}{50}$ **28.** \$2820 **29.** perimeter: $\frac{10}{3}$ or $3\frac{1}{3}$ ft; area: $\frac{2}{3}$ sq ft **30.** $\frac{5}{3}$ or $1\frac{2}{3}$ in.

Cumulative Review **1.** eighty-five (Sec. 1.2, Ex. 4) **2.** one hundred seven **3.** one hundred twenty-six (Sec. 1.2, Ex. 5)
4. five thousand, twenty-six **5.** 159 (Sec. 1.3, Ex. 1) **6.** 19 in. **7.** 514 (Sec. 1.4, Ex. 3) **8.** 121 R 1 **9.** 278,000 (Sec. 1.5, Ex. 2)
10. 1, 2, 3, 5, 6, 10, 15, 30 **11.** 20,296 (Sec. 1.6, Ex. 4) **12.** 0 **13. a.** 7 **b.** 12 **c.** 1 **d.** 1 **e.** 20 **f.** 1 (Sec. 1.7, Ex. 2) **14.** 25
15. 1038 mi (Sec. 1.8, Ex. 1) **16.** 11 **17.** 81 (Sec. 1.9, Ex. 5) **18.** 125 **19.** 81 (Sec. 1.9, Ex. 7) **20.** 1000 **21.** $\frac{4}{3}$ or $1\frac{1}{3}$ (Sec. 2.1, Ex. 15)
22. $\frac{11}{4}$ or $2\frac{3}{4}$ **23.** $\frac{5}{2}$ or $2\frac{1}{2}$ (Sec. 2.1, Ex. 16) **24.** $\frac{14}{3}$ or $4\frac{2}{3}$ **25.** 3, 11, 17 are prime; 9, 26 are composite; (Sec. 2.2, Ex. 2)
26. 5 **27.** $2^2 \cdot 3^2 \cdot 5$ (Sec. 2.2, Ex. 4) **28.** 62 **29.** $\frac{36}{13}$ or $2\frac{10}{13}$ (Sec. 2.3, Ex. 5) **30.** $\frac{79}{8}$ **31.** equivalent (Sec. 2.3, Ex. 8) **32.** >
33. $\frac{10}{33}$ (Sec. 2.4, Ex. 1) **34.** $\frac{3}{2}$ or $1\frac{1}{2}$ **35.** $\frac{1}{8}$ (Sec. 2.4, Ex. 2) **36.** 37 **37.** $\frac{11}{51}$ (Sec. 2.5, Ex. 9) **38.** $\frac{25}{19}$ or $1\frac{6}{19}$
39. $\frac{51}{23}$ or $2\frac{5}{23}$ (Sec. 2.5, Ex. 10) **40.** 16 **41.** $\frac{5}{8}$ (Sec. 3.1, Ex. 2) **42.** $\frac{1}{5}$ **43.** 24 (Sec. 3.2, Ex. 1) **44.** 35 **45.** 2 (Sec. 3.3, Ex. 4)
46. $\frac{25}{81}$ **47.** $4\frac{1}{3}$ (Sec. 3.4, Ex. 4) **48.** $\frac{11}{100}$ **49.** $\frac{6}{13}$ (Sec. 3.5, Ex. 11) **50.** $\frac{8}{175}$

Chapter 4 Decimals

Section 4.1

Vocabulary, Readiness & Video Check **1.** words; standard form **3.** and **5.** tens **7.** tenths **9.** as "and" **11.** Reading a decimal correctly gives you the correct place value, which tells you the denominator of your equivalent fraction.

Exercise Set 4.1 **1.** six and fifty-two hundredths **3.** sixteen and twenty-three hundredths **5.** two hundred five thousandths
7. one hundred sixty-seven and nine thousandths **9.** two hundred and five thousandths **11.** one hundred five and six
tenths **13.** two and forty-three hundredths **15.** eighty-seven and ninety-seven hundredths **17.** one hundred fifteen and six
tenths **19.** R. W. Financial; 321.42; Three hundred twenty-one and 42/100 **21.** Bell South; 59.68; Fifty-nine and 68/100
23. 6.5 **25.** 9.08 **27.** 705.625 **29.** 0.0046 **31.** 32.52 **33.** 1.3 **35.** $\frac{3}{10}$ **37.** $\frac{27}{100}$ **39.** $\frac{4}{5}$ **41.** $\frac{3}{20}$ **43.** $5\frac{47}{100}$ **45.** $\frac{6}{125}$ **47.** $7\frac{1}{125}$
49. $15\frac{401}{500}$ **51.** $\frac{601}{2000}$ **53.** $487\frac{8}{25}$ **55.** 0.6 **57.** 0.45 **59.** 3.7 **61.** 0.268 **63.** 0.09 **65.** 4.026 **67.** 0.028 **69.** 56.3 **71.** 0.43;
forty-three hundredths **73.** 0.8; $\frac{8}{10}$ or $\frac{4}{5}$ **75.** seventy-seven thousandths; $\frac{77}{1000}$ **77.** 47,260 **79.** 47,000 **81.** answers may vary
83. twenty-six million, eight hundred forty-nine thousand, five hundred seventy-six hundred-billionths **85.** 17.268

Section 4.2

Vocabulary, Readiness & Video Check **1.** circumference **3.** after **5.** left to right

Exercise Set 4.2 **1.** < **3.** > **5.** < **7.** = **9.** < **11.** > **13.** 0.006, 0.0061, 0.06 **15.** 0.03, 0.042, 0.36 **17.** 1.01, 1.09, 1.1, 1.16
19. 20.905, 21.001, 21.03, 21.12 **21.** 0.6 **23.** 0.23 **25.** 0.594 **27.** 98,210 **29.** 12.3 **31.** 17.67 **33.** 0.5 **35.** 0.130 **37.** 3830
39. \$0.07 **41.** \$42,650 **43.** \$27 **45.** \$0.20 **47.** 0.3 cm **49.** 1.43 hr **51.** \$48 **53.** 106.5 people per sq mi **55.** 24.623 hr **57.** 0.5 min
59. 5766 **61.** 71 **63.** 243 **65.** b **67.** a **69.** 225.228; $225\frac{57}{250}$; Audi Sport North America **71.** 225.228; 214.500; 201.265; 197.400
73. answers may vary **75.** answers may vary **77.** 0.26499, 0.25786 **79.** 0.10299, 0.1037, 0.1038, 0.9 **81.** \$3600 million

Section 4.3

Calculator Explorations **1.** 328.742 **3.** 5.2414 **5.** 865.392

Vocabulary, Readiness & Video Check **1.** 37.0 **3.** difference; minuend; subtrahend **5.** false **7.** Lining up the decimal points also lines up place values; to make sure we only add digits in the same place. **9.** Check subtraction by addition.

Exercise Set 4.3 **1.** 3.5 **3.** 6.83 **5.** 0.094 **7.** 622.012 **9.** 583.09 **11.** Exact: 465.56; Estimate: $\begin{array}{r} 230 \\ +\ 230 \\ \hline 460 \end{array}$

13. Exact: 115.123; Estimate: 100   **15.** 27.0578   **17.** 56.432   **19.** 6.5   **21.** 15.3   **23.** 598.23   **25.** Exact: 1.83; Estimate: $6 - 4 = 2$

$$\begin{array}{r} 6 \\ +\ 9 \\ \hline 115 \end{array}$$

27. 861.6   **29.** 376.89   **31.** Exact: 876.6; Estimate: 1000   **33.** 194.4   **35.** 2.9988   **37.** 16.3   **39.** 88.028   **41.** 84.072   **43.** 243.17

$$\begin{array}{r} -\ 100 \\ \hline 900 \end{array}$$

45. 56.83   **47.** 3.16   **49.** $7.52   **51.** $454.71   **53.** $0.14   **55.** 28.56 m   **57.** 9.14 in.   **59.** 196.3 mph   **61.** 64.8 degrees Fahrenheit
63. 763.035 mph   **65.** 14.46 in.   **67.** $0.03   **69.** 21.8 billion (21,800,000,000)   **71.** 326.0 in.   **73.** 67.44 ft   **75.** $1.294   **77.** 715.05 hr
79. Switzerland   **81.** 7.94 lb   **83.**

Country	Pounds of Chocolate per Person
Switzerland	26.24
Ireland	21.83
UK	20.94
Austria	19.40
Belgium	18.30

85. 46   **87.** 3870   **89.** $\dfrac{4}{9}$ **91.** incorrect;

$$\begin{array}{r} 9.200 \\ 8.630 \\ +\ 4.005 \\ \hline 21.835 \end{array}$$

93. 6.08 in.   **95.** $1.20   **97.** 1 dime, 1 nickel, and 2 pennies; 3 nickels and 2 pennies; 1 dime and 7 pennies; 2 nickels and 7 pennies
99. answers may vary   **101.** answers may vary

Section 4.4

Vocabulary, Readiness & Video Check   **1.** sum   **3.** right; zeros   **5.** circumference   **7.** 3   **9.** 4   **11.** 8   **13.** Whether we placed the decimal point correctly in our product.   **15.** We used an approximation for π. The exact answer is 8π meters.

Exercise Set 4.4   **1.** 0.12   **3.** 0.6   **5.** 1.3   **7.** Exact: 22.26; Estimate: $5 \times 4 = 20$ **9.** 0.4032   **11.** Exact: 8.23854; Estimate:

$$\begin{array}{r} 1 \\ \times\ 8 \\ \hline 8 \end{array}$$

13. 11.2746   **15.** 84.97593   **17.** 65   **19.** 0.65   **21.** 0.072   **23.** 709.3   **25.** 6046   **27.** 0.03762   **29.** 0.0492   **31.** 12.3   **33.** 1.29
35. 0.096   **37.** 0.5623   **39.** 43.274   **41.** 5,500,000,000   **43.** 97,800,000   **45.** 292,000   **47.** 8π m ≈ 25.12 m   **49.** 10π cm ≈ 31.4 cm
51. 18.2π yd ≈ 57.148 yd   **53.** $715.20   **55.** $9150   **57.** 24.8 g   **59.** 11.417 sq in.   **61.** 250π ft ≈ 785 ft   **63.** 135π m ≈ 423.9 m
65. 64.9605 in.   **67. a.** 62.8 m and 125.6 m   **b.** yes   **69.** 4.70 sq in.   **71.** 26   **73.** 36   **75.** 8   **77.** 9   **79.** 3.64   **81.** 3.56   **83.** 0.1105
85. 3,831,600 mi   **87.** answers may vary   **89.** answers may vary

Integrated Review   **1.** 2.57   **2.** 4.05   **3.** 8.9   **4.** 3.5   **5.** 0.16   **6.** 0.24   **7.** 11.06   **8.** 9.72   **9.** 4.8   **10.** 6.09   **11.** 75.56   **12.** 289.12
13. 25.026   **14.** 44.125   **15.** 82.7   **16.** 273.9   **17.** 280   **18.** 1600   **19.** 224.938   **20.** 145.079   **21.** 6   **22.** 6.2   **23.** 27.6092
24. 145.6312   **25.** 5.4   **26.** 17.74   **27.** 414.44   **28.** 1295.03   **29.** 116.81   **30.** 18.79   **31.** 156.2   **32.** 25.62   **33.** 5.62   **34.** 304.876
35. 114.66   **36.** 119.86   **37.** 0.000432   **38.** 0.000075   **39.** 0.0672   **40.** 0.0275   **41.** 862   **42.** 0.0293   **43.** 200 mi

Section 4.5

Calculator Explorations   **1.** not reasonable   **3.** reasonable

Vocabulary, Readiness & Video Check   **1.** quotient; divisor; dividend   **3.** left; zeros   **5.** 5.9   **7.** 0   **9.** 1   **11.** undefined
13. a whole number   **15.** We just need to know how to move the decimal point. 1000 has three zeros, so we move the decimal point in the decimal number three places to the left.   **17.** The fraction bar serves as a grouping symbol.

Exercise Set 4.5   **1.** 4.6   **3.** 0.094   **5.** 300   **7.** 5.8   **9.** Exact: 6.6; Estimate: $6\overline{)36}$ **11.** 0.413   **13.** 0.045   **15.** 7   **17.** 4.8   **19.** 2100

21. 30   **23.** 7000   **25.** Exact: 9.8; Estimate: $7\overline{)70}$ **27.** 9.6   **29.** 45   **31.** 54.592   **33.** 0.0055   **35.** 179   **37.** 23.87   **39.** 113.1
41. 0.54982   **43.** 2.687   **45.** 0.0129   **47.** 12.6   **49.** 1.31   **51.** 12.225   **53.** 0.045625   **55.** 11 qt   **57.** 202.1 lb   **59.** 5.1 m
61. 11.4 boxes   **63.** 24 tsp   **65.** 8 days   **67.** 290.3 mi   **69.** 122.8 mph   **71.** 18.1 points per game   **73.** 2.45   **75.** 0.66   **77.** 80.52
79. 14.7   **81.** 930.7   **83.** 571   **85.** 92.06   **87.** 144.4   **89.** $\dfrac{9}{10}$ **91.** $\dfrac{1}{20}$ **93.** 4.26   **95.** 1.578   **97.** 26.66   **99.** 904.29   **101.** c
103. b   **105.** 85.5   **107.** 8.6 ft   **109.** answers may vary   **111.** 65.2–82.6 knots   **113.** 319.64 m

Section 4.6

Vocabulary, Readiness & Video Check   **1.** false   **3.** true   **5.** We place a bar over just the repeating digits and only 6 repeats in our decimal answer.   **7.** $A = l \cdot w$; 0.248 sq yd

Exercise Set 4.6   **1.** 0.2   **3.** 0.68   **5.** 0.75   **7.** 0.08   **9.** 1.2   **11.** $0.91\overline{6}$ **13.** 0.425   **15.** 0.45   **17.** $0.\overline{3}$ **19.** 0.4375   **21.** $0.\overline{63}$
23. 5.85   **25.** 0.624   **27.** 0.33   **29.** 0.44   **31.** 0.6   **33.** 0.62   **35.** 0.38   **37.** 0.02   **39.** <   **41.** =   **43.** <   **45.** <   **47.** <   **49.** >

51. < **53.** < **55.** 0.32, 0.34, 0.35 **57.** 0.49, 0.491, 0.498 **59.** 0.73, $\frac{3}{4}$, 0.78 **61.** 0.412, 0.453, $\frac{4}{7}$ **63.** 5.23, $\frac{42}{8}$, 5.34 **65.** $\frac{17}{8}$, 2.37, $\frac{12}{5}$

67. 25.65 sq in. **69.** 9.36 sq cm **71.** 0.248 sq yd **73.** 8 **75.** 72 **77.** $\frac{1}{81}$ **79.** $\frac{9}{25}$ **81.** $\frac{5}{2}$ **83.** $= 1$ **85.** > 1 **87.** < 1 **89.** 0.13

91. 8700 stations **93.** answers may vary **95.** answers may vary **97.** 47.25 **99.** 3.37 **101.** 0.45

Chapter 4 Vocabulary Check **1.** decimal **2.** numerator; denominator **3.** vertically **4.** and **5.** sum **6.** circumference
7. standard form **8.** circumference; diameter **9.** difference **10.** quotient **11.** product **12.** sum

Chapter 4 Review **1.** tenths **2.** hundred-thousandths **3.** forty-five hundredths **4.** three hundred forty-five hundred-
thousandths **5.** one hundred nine and twenty-three hundredths **6.** forty-six and seven thousandths **7.** 2.15 **8.** 503.102
9. $\frac{4}{25}$ **10.** $12\frac{23}{1000}$ **11.** $1\frac{9}{2000}$ **12.** $25\frac{1}{4}$ **13.** 0.9 **14.** 0.25 **15.** 0.045 **16.** 26.1 **17.** > **18.** = **19.** 0.92, 8.09, 8.6 **20.** 0.09,
0.091, 0.1 **21.** 0.6 **22.** 0.94 **23.** $0.26 **24.** $12.46 **25.** $31,304 **26.** $10\frac{3}{4}$ **27.** 9.52 **28.** 2.7 **29.** 7.28 **30.** 26.007 **31.** 459.7
32. 100.278 **33.** 65.02 **34.** 189.98 **35.** 52.6 mi **36.** $2.44 **37.** 22.2 in. **38.** 38.9 ft **39.** 18.5 **40.** 54.6 **41.** 72 **42.** 9345
43. 9.246 **44.** 3406.446 **45.** 14π m; 43.96 m **46.** 63.8 mi **47.** 887,000,000 **48.** 600,000 **49.** 0.087 **50.** 15.825 **51.** 70
52. 0.21 **53.** 8.059 **54.** 30.4 **55.** 0.0267 **56.** 9.3 **57.** 7.3 m **58.** 45 mo **59.** 16.94 **60.** 3.89 **61.** 129 **62.** 55 **63.** 0.81
64. 7.26 **65.** 0.8 **66.** 0.923 **67.** $2.\overline{3}$ or 2.333 **68.** $0.21\overline{6}$ or 0.217 **69.** = **70.** = **71.** < **72.** < **73.** 0.837, 0.839, $\frac{17}{20}$
74. $\frac{19}{12}$, 1.63, $\frac{18}{11}$ **75.** 6.9 sq ft **76.** 5.46 sq in. **77.** two hundred and thirty-two ten-thousandths **78.** 16,025.014 **79.** $\frac{231}{100,000}$
80. 0.75, $\frac{6}{7}$, $\frac{8}{9}$ **81.** 0.07 **82.** 0.1125 **83.** 51.057 **84.** > **85.** < **86.** < **87.** 42.90 **88.** 16.349 **89.** $123 **90.** $3646 **91.** 1.7
92. 2.49 **93.** 320.312 **94.** 148.74236 **95.** 8.128 **96.** 7.245 **97.** 4900 **98.** 23.904 **99.** 9600 sq ft **100.** yes **101.** 0.1024 **102.** 3.6

Chapter 4 Test **1.** forty-five and ninety-two thousandths **2.** 3000.059 **3.** 34.9 **4.** 0.862 **5.** < **6.** $\frac{4}{9}$, 0.445, 0.454 **7.** $\frac{69}{200}$
8. $24\frac{73}{100}$ **9.** 0.65 **10.** $5.\overline{8}$ or 5.889 **11.** 0.941 **12.** 17.583 **13.** 11.4 **14.** 43.86 **15.** 56 **16.** 0.07755 **17.** 6.673 **18.** 12,690
19. 4.73 **20.** 0.363 **21.** 6.2 **22.** 4,583,000,000 **23.** 2.31 sq mi **24.** 18π mi, 56.52 mi **25. a.** 9904 sq ft **b.** 198.08 oz **26.** 54 mi

Cumulative Review **1.** one hundred six million, fifty-two thousand, four hundred forty-seven (Sec. 1.2, Ex. 7) **2.** 276,004
3. 14,440,060 (Sec 1.3, Ex. 7) **4.** 288 **5.** 726 (Sec. 1.4, Ex. 4) **6.** 200 **7.** 2300 (Sec. 1.5, Ex. 4) **8.** 84 **9.** 57,600 megabytes
(Sec. 1.6, Ex. 11) **10.** perimeter: 28 ft; area: 49 sq ft **11.** 401 R 2 (Sec. 1.7, Ex. 8) **12.** $\frac{21}{8}$ **13.** 47 (Sec. 1.9, Ex. 15) **14.** $12\frac{4}{5}$
15. numerator: 3; denominator: 7 (Sec. 2.1, Ex. 1) **16.** 9 **17.** $\frac{1}{10}$ (Sec. 2.3, Ex. 6) **18.** 17 **19.** $\frac{15}{1}$ or 15 (Sec. 2.4, Ex. 9) **20.** 13
21. $\frac{63}{16}$ (Sec. 2.5, Ex. 5) **22.** 128 **23.** $\frac{15}{4}$ or $3\frac{3}{4}$ (Sec. 2.4, Ex. 10) **24.** $9 **25.** $\frac{3}{20}$ (Sec. 2.5, Ex. 8) **26.** $\frac{27}{20}$ or $1\frac{7}{20}$ **27.** $\frac{7}{9}$ (Sec. 3.1, Ex. 4)
28. $\frac{2}{5}$ **29.** $\frac{1}{4}$ (Sec. 3.1, Ex. 5) **30.** $\frac{2}{5}$ **31.** $\frac{15}{20}$ (Sec. 3.2, Ex. 8) **32.** $\frac{35}{45}$ **33.** $\frac{31}{30}$ or $1\frac{1}{30}$ (Sec. 3.3, Ex. 2) **34.** $\frac{1}{90}$ **35.** $4\frac{7}{40}$ lb (Sec. 3.4, Ex. 7)
36. $27\frac{3}{4}$ lb **37.** $\frac{1}{16}$ (Sec. 3.5, Ex. 3) **38.** $\frac{49}{121}$ **39.** $\frac{3}{256}$ (Sec. 3.5, Ex. 5) **40.** $\frac{2}{81}$ **41.** $\frac{43}{100}$ (Sec. 4.1, Ex. 8) **42.** 0.75
43. > (Sec. 4.2, Ex. 1) **44.** 5.06 **45.** 11.568 (Sec. 4.3, Ex. 4) **46.** 75.329 **47.** 2370.2 (Sec. 4.4, Ex. 6) **48.** 0.119
49. 768.05 (Sec. 4.4, Ex. 9) **50.** 8.9

Chapter 5 Ratio, Proportion, and Percent

Section 5.1

Vocabulary, Readiness & Video Check **1.** true **3.** false **5.** true **7.** true **9.** unit **11.** division **13.** proportion; ratio **15.** true
17. We can use "to" as in 1 to 2, a colon as in 1 : 2, or a fraction as in $\frac{1}{2}$. **19.** We can't divide out the units because they are different
(shrubs and feet). **21.** a variable

Exercise Set 5.1 **1.** $\frac{23}{10}$ **3.** $\frac{3\frac{3}{4}}{1\frac{2}{3}}$ **5.** $\frac{2}{3}$ **7.** $\frac{77}{100}$ **9.** $\frac{5}{12}$ **11.** $\frac{8}{25}$ **13.** $\frac{12}{7}$ **15.** $\frac{16}{23}$ **17.** $\frac{2}{5}$ **19.** $\frac{17}{40}$ **21.** $\frac{3}{44}$ **23.** $\frac{1}{3}$ **25.** $\frac{1\,\text{shrub}}{3\,\text{ft}}$

27. $\frac{3\,\text{returns}}{20\,\text{sales}}$ **29.** $\frac{2\,\text{phone lines}}{9\,\text{employees}}$ **31.** $\frac{9\,\text{gal}}{2\,\text{acres}}$ **33.** $\frac{3\,\text{flight attendants}}{100\,\text{passengers}}$ **35.** $\frac{71\,\text{cal}}{2\,\text{fl oz}}$ **37.** 110 cal/oz **39.** 90 wingbeats/sec **41.** false

43. true   **45.** $\dfrac{1.8}{2} = \dfrac{4.5}{5}$; true   **47.** $\dfrac{\frac{2}{3}}{\frac{1}{5}} = \dfrac{\frac{2}{5}}{\frac{1}{9}}$; false   **49.** 3   **51.** 9   **53.** 4   **55.** 3.2   **57.** 0.0025   **59.** 1   **61.** $\dfrac{3}{4}$ **63.** $\dfrac{35}{18}$ **65.** 360 baskets

67. 165 min   **69.** 23 ft   **71.** 25 gal   **73.** 450 km   **75.** 16 bags   **77.** 18 applications   **79.** 5 weeks   **81.** 37.5 sec   **83. a.** 18 tsp
b. 6 tbsp   **85.** 6 people   **87.** 112 ft; 11-in. difference   **89.** 102.9 mg   **91. a.** 2062.5 mg   **b.** no   **93. a.** 0.1 gal   **b.** 13 fl oz
95. $2^2 \cdot 5$ **97.** $2^3 \cdot 5^2$ **99.** 2^5 **101.** 0.8 ml   **103.** 1.25 ml   **105.** no; answers may vary   **107.** answers may vary   **109.** 1400

Section 5.2

Vocabulary, Readiness & Video Check   **1.** Percent   **3.** percent   **5.** 0.01   **7.** $\dfrac{1}{100}$; 0.01

Exercise Set 5.2   **1.** 96%   **3. a.** 75%   **b.** 25%   **5.** football; 37%   **7.** 50%   **9.** 0.41   **11.** 0.06   **13.** 1.00 or 1   **15.** 0.736
17. 0.028   **19.** 0.006   **21.** 3.00 or 3   **23.** 0.3258   **25.** 0.38   **27.** 0.382   **29.** 0.45   **31.** 98%   **33.** 310%   **35.** 2900%   **37.** 0.3%
39. 22%   **41.** 530%   **43.** 5.6%   **45.** 33.28%   **47.** 300%   **49.** 70%   **51.** 77%   **53.** 28.1%   **55.** 8.1%   **57.** 0.25   **59.** 0.65
61. 0.9   **63.** b, d   **65.** 4%   **67.** personal care aides   **69.** 0.617   **71.** answers may vary

Section 5.3

Vocabulary, Readiness & Video Check   **1.** Percent   **3.** 100%   **5.** 13%   **7.** 87%   **9.** 1%   **11.** The fraction of $\dfrac{4}{100}$ can be
simplified to $\dfrac{1}{25}$.   **13.** The difference is in how the percent symbol is replaced. For a decimal, replace % with the equivalent decimal
form 0.01 and for a fraction, replace % with the equivalent fraction form $\dfrac{1}{100}$.

Exercise Set 5.3   **1.** $\dfrac{3}{25}$ **3.** $\dfrac{1}{25}$ **5.** $\dfrac{9}{200}$ **7.** $\dfrac{7}{4}$ or $1\dfrac{3}{4}$ **9.** $\dfrac{73}{100}$ **11.** $\dfrac{1}{8}$ **13.** $\dfrac{1}{16}$ **15.** $\dfrac{3}{50}$ **17.** $\dfrac{31}{300}$ **19.** $\dfrac{179}{800}$ **21.** 75%   **23.** 70%

25. 40%   **27.** 59%   **29.** 34%   **31.** $37\dfrac{1}{2}$%   **33.** $31\dfrac{1}{4}$%   **35.** 160%   **37.** $77\dfrac{7}{9}$%   **39.** 65%   **41.** 250%   **43.** 190%   **45.** 63.64%

47. 26.67%   **49.** 14.29%   **51.** 91.67%   **53.** 0.35, $\dfrac{7}{20}$; 20%, 0.2; 50%, $\dfrac{1}{2}$; 0.7, $\dfrac{7}{10}$; 37.5%, 0.375   **55.** 0.4, $\dfrac{2}{5}$; $23\dfrac{1}{2}$%, $\dfrac{47}{200}$; 80%, 0.8; $0.33\overline{33}$,

$\dfrac{1}{3}$; 87.5%, 0.875; 0.075, $\dfrac{3}{40}$ **57.** 2, 2; 280%, $2\dfrac{4}{5}$; 7.05, $7\dfrac{1}{20}$; 454%, 4.54   **59.** 0.67; $\dfrac{67}{100}$ **61.** 0.522; $\dfrac{261}{500}$ **63.** 80%   **65.** 0.0875   **67.** 18%

69. 0.005; $\dfrac{1}{200}$ **71.** 0.142; $\dfrac{71}{500}$ **73.** 0.079; $\dfrac{79}{1000}$ **75.** $n = 15$ **77.** $n = 10$ **79.** $n = 12$ **81. a.** 52.9%   **b.** 52.86%   **83.** 107.8%
85. 65.79%   **87.** 77%   **89.** 75%   **91.** 80%   **93.** greater   **95.** answers may vary   **97.** 0.266; 26.6%   **99.** 1.155; 115.5%

Section 5.4

Vocabulary, Readiness & Video Check   **1.** is   **3.** amount; base; percent   **5.** greater   **7.** percent: 42%; base: 50; amount: 21
9. percent: 125%; base: 86; amount: 107.5   **11.** "of" means multiplication; "is" means equal; "what" (or some equivalent) means the
unknown number

Exercise Set 5.4   **1.** $18\% \cdot 81 = n$ **3.** $20\% \cdot n = 105$ **5.** $0.6 = 40\% \cdot n$ **7.** $n \cdot 80 = 3.8$ **9.** $n = 9\% \cdot 43$ **11.** $n \cdot 250 = 150$
13. 3.5   **15.** 28.7   **17.** 10   **19.** 600   **21.** 110%   **23.** 34%   **25.** 1   **27.** 645   **29.** 500   **31.** 5.16%   **33.** 25.2   **35.** 35%   **37.** 35
39. 0.624   **41.** 0.5%   **43.** 145   **45.** 63%   **47.** 4%   **49.** $n = 30$ **51.** $n = 3\dfrac{7}{11}$ **53.** $\dfrac{17}{12} = \dfrac{n}{20}$ **55.** $\dfrac{8}{9} = \dfrac{14}{n}$ **57.** c   **59.** b

61. Twenty percent of some number is eighteen and six tenths.   **63.** b   **65.** c   **67.** c   **69.** a   **71.** a   **73.** answers may vary
75. 686.625   **77.** 12,285

Section 5.5

Vocabulary, Readiness & Video Check   **1.** amount; base; percent   **3.** amount   **5.** amount: 12.6; base: 42; percent: 30   **7.** amount:
102; base: 510; percent: 20   **9.** 45 follows the word "of," so it is the base

Exercise Set 5.5   **1.** $\dfrac{a}{45} = \dfrac{98}{100}$ **3.** $\dfrac{a}{150} = \dfrac{4}{100}$ **5.** $\dfrac{14.3}{b} = \dfrac{26}{100}$ **7.** $\dfrac{84}{b} = \dfrac{35}{100}$ **9.** $\dfrac{70}{400} = \dfrac{p}{100}$ **11.** $\dfrac{8.2}{82} = \dfrac{p}{100}$ **13.** 26   **15.** 18.9
17. 600   **19.** 10   **21.** 120%   **23.** 28%   **25.** 37   **27.** 1.68   **29.** 1000   **31.** 210%   **33.** 55.18   **33.** 45%   **37.** 75   **39.** 0.864
41. 0.5%   **43.** 140   **45.** 9.6   **47.** 113%   **49.** $\dfrac{7}{8}$ **51.** $3\dfrac{2}{15}$ **53.** 0.7   **55.** 2.19   **57.** answers may vary   **59.** no; $a = 16$ **61.** yes
63. answers may vary   **65.** 12,011.2   **67.** 7270.6

Integrated Review 1. $\frac{8}{23}$ **2.** $\frac{7}{26}$ **3.** 55 mi/hr **4.** 140 ft/sec **5.** 8 lb: $0.27 per lb; 18 lb: $0.28 per lb; 8 lb **6.** 100: $0.020 per plate; 500: $0.018 per plate; 500 paper plates **7.** 38.4 **8.** 45.5 **9.** 12% **10.** 68% **11.** 12.5% **12.** 250% **13.** 520% **14.** 800% **15.** 6% **16.** 44% **17.** 750% **18.** 325% **19.** 3% **20.** 5% **21.** 0.65 **22.** 0.31 **23.** 0.08 **24.** 0.07 **25.** 1.42 **26.** 4 **27.** 0.029 **28.** 0.066 **29.** 0.03; $\frac{3}{100}$ **30.** 0.05; $\frac{1}{20}$ **31.** 0.0525; $\frac{21}{400}$ **32.** 0.1275; $\frac{51}{400}$ **33.** 0.38; $\frac{19}{50}$ **34.** 0.45; $\frac{9}{20}$ **35.** 0.123; $\frac{37}{300}$ **36.** 0.167; $\frac{1}{6}$ **37.** 8.4 **38.** 100 **39.** 250 **40.** 120% **41.** 28% **42.** 76 **43.** 11 **44.** 130% **45.** 86% **46.** 37.8 **47.** 150 **48.** 62

Section 5.6

Vocabulary, Readiness & Video Check 1. The price of the home is $175,000.

Exercise Set 5.6 1. 1600 bolts **3.** 8.8 lb **5.** 14% **7.** 91,800 businesses **9.** 45% **11.** 496 chairs; 5704 chairs **13.** 108,680 physician assistants **15.** 9880.36 thousand or approximately 9880 thousand **17.** 29% **19.** 50% **21.** 12.5% **23.** 29.2% **25.** $175.000 **27.** 31.2 hr **29.** $867.87; $20,153.87 **31.** 40 ft **33.** increase: $1210; tuition in 2013–2014: $9616 **35.** increase: 128,760; 2020–2021: 1,016,760 associate degrees **37.** 30; 60% **39.** 52; 80% **41.** 2; 25% **43.** 120; 75% **45.** 44% **47.** 1.3% **49.** 142.0% **51.** 9.0% **53.** 139.5% **55.** 18.8% **57.** 19.7% **59.** 54.3% **61.** 4.56 **63.** 11.18 **65.** 58.54 **67.** The increased number is double the original number. **69.** percent increase $= \frac{30}{150} = 20\%$ **71.** False; the percents are different.

Section 5.7

Vocabulary, Readiness & Video Check 1. sales tax **3.** commission **5.** sale price **7.** We rewrite the percent as an equivalent decimal. **9.** Replace "amount of discount" in the second equation with "discount rate · original price": sale price = original price − (discount rate · original price).

Exercise Set 5.7 1. $7.50 **3.** $858.93 **5.** 7% **7. a.** $120 **b.** $130.20 **9.** $117; $1917 **11.** $485 **13.** 6% **15.** $16.10; $246.10 **17.** $53,176.04 **19.** 14% **21.** $4888.50 **23.** $185,500 **25.** $8.90; $80.10 **27.** $98.25; $98.25 **29.** $143.50; $266.50 **31.** $3255; $18,445 **33.** $45; $255 **35.** $27.45; $332.45 **37.** $3.08; $59.08 **39.** $7074 **41.** 8% **43.** 1200 **45.** 132 **47.** 16 **49.** d **51.** $4.00; $6.00; $8.00 **53.** $7.20; $10.80; $14.40 **55.** a discount of 60% is better; answers may vary **57.** $26,838.45

Section 5.8

Calculator Explorations 1. 1.56051 **3.** 8.06231 **5.** $634.49

Vocabulary, Readiness & Video Check 1. simple **3.** Compound **5.** Total amount **7.** principal **9.** The denominator is the total number of payments. We are asked to find the monthly payment for a 4-year loan, and since there are 48 months in 4 years, there are 48 total payments.

Exercise Set 5.8 1. $32 **3.** $73.60 **5.** $750 **7.** $33.75 **9.** $700 **11.** $101,562.50; $264,062.50 **13.** $5562.50 **15.** $14,280 **17.** $46,815.37 **19.** $2327.14 **21.** $58,163.65 **23.** 2915.75 **25.** $2938.66 **27.** $2971.89 **29.** $260.31 **31.** $637.26 **33.** 32 yd **35.** 35 m **37.** answers may vary **39.** answers may vary

Chapter 5 Vocabulary Check 1. ratio **2.** proportion **3.** unit rate **4.** proportion **5.** rate **6.** cross products **7.** equal **8.** not equal **9.** of **10.** is **11.** Percent **12.** Compound interest **13.** $\frac{\text{amount}}{\text{base}}$ **14.** 100% **15.** 0.01 **16.** $\frac{1}{100}$ **17.** base; amount **18.** Percent of decrease **19.** Percent of increase **20.** Sales tax **21.** Total price **22.** Commission **23.** Amount of discount **24.** Sale price

Chapter 5 Review 1. $\frac{23}{37}$ **2.** $\frac{5}{4}$ **3.** $\frac{11}{13}$ **4.** $\frac{17}{35}$ **5. a.** 9 **b.** $\frac{9}{25}$ **6. a.** 3 **b.** $\frac{3}{25}$ **7.** $\frac{3 \text{ professors}}{10 \text{ assistants}}$ **8.** $\frac{5 \text{ pages}}{2 \text{ min}}$ **9.** 52 mi/hr **10.** 15 ft/sec **11.** no **12.** yes **13.** 15 **14.** 32.5 **15.** 60 **16.** 0.94 **17.** no **18.** 79 gal **19.** $54,600 **20.** $1023.50 **21.** 37% **22.** 77% **23.** 0.83 **24.** 0.75 **25.** 0.005 **26.** 0.007 **27.** 2.00 or 2 **28.** 4.00 or 4 **29.** 0.2625 **30.** 0.8534 **31.** 260% **32.** 5.5% **33.** 35% **34.** 102% **35.** 71% **36.** 65% **37.** 400% **38.** 900% **39.** $\frac{1}{100}$ **40.** $\frac{1}{10}$ **41.** $\frac{1}{4}$ **42.** $\frac{17}{200}$ **43.** $\frac{51}{500}$ **44.** $\frac{1}{6}$ **45.** $\frac{1}{3}$ **46.** $1\frac{1}{10}$ **47.** 20% **48.** 70% **49.** $83\frac{1}{3}\%$ **50.** $166\frac{2}{3}\%$ **51.** 125% **52.** 60% **53.** 6.25% **54.** 62.5% **55.** 100,000 **56.** 8000 **57.** 23% **58.** 114.5 **59.** 3000 **60.** 150% **61.** 418 **62.** 300 **63.** 64.8 **64.** 180% **65.** 110% **66.** 165 **67.** 66% **68.** 16% **69.** 20.9% **70.** 106.25% **71.** $206,400 **72.** $13.23 **73.** $263.75 **74.** $1.15 **75.** $5000 **76.** $300.38 **77.** discount: $900; sale price: $2100 **78.** discount: $9; sale price: $81 **79.** $160 **80.** $325 **81.** $30,104.61 **82.** $17,506.54 **83.** $80.61 **84.** $32,830.10

85. 1.6 **86.** 84 **87.** 0.038 **88.** 0.245 **89.** 0.009 **90.** 54% **91.** 9520% **92.** 30% **93.** $\frac{47}{100}$ **94.** $\frac{8}{125}$ **95.** $\frac{7}{125}$ **96.** $37\frac{1}{2}$%
97. $15\frac{5}{13}$% **98.** 120% **99.** 268.75 **100.** 110% **101.** 708.48 **102.** 134% **103.** 300% **104.** 38.4 **105.** 560 **106.** 325%
107. 26% **108.** $6786.50 **109.** $617.70 **110.** $3.45 **111.** 12.5% **112.** $1491 **113.** $11,687.50

Chapter 5 Test **1.** $\frac{15}{2}$ **2.** $\frac{43}{50}$ **3.** $\frac{2}{3}$ in./hr **4.** 9 in./sec **5.** $4\frac{4}{11}$ **6.** 8 **7.** $53\frac{1}{3}$ g **8.** 4266 adults **9.** 0.85 **10.** 5 **11.** 0.008
12. 5.6% **13.** 610% **14.** 39% **15.** $\frac{6}{5}$ or $1\frac{1}{5}$ **16.** $\frac{77}{200}$ **17.** $\frac{1}{500}$ **18.** 55% **19.** 37.5% **20.** $155\frac{5}{9}$% **21.** 33.6 **22.** 1250 **23.** 75%
24. 38.4 lb **25.** $56,750 **26.** $358.43 **27.** 5% **28.** discount: $18; sale price: $102 **29.** $395 **30.** 1% **31.** $647.50 **32.** $2005.63
33. $427

Cumulative Review **1.** 206 cases; 12 cans; yes (Sec. 1.8, Ex. 2) **2.** 31,084 **3. a.** $4\frac{2}{7}$ **b.** $1\frac{1}{15}$ **c.** 14 (Sec. 2.1, Ex. 18) **4. a.** $\frac{19}{7}$
b. $\frac{101}{10}$ **c.** $\frac{43}{8}$ **5.** $2 \cdot 2 \cdot 2 \cdot 2 \cdot 5$ or $2^4 \cdot 5$ (Sec. 2.2, Ex. 7) **6.** 119 sq mi **7.** $\frac{10}{27}$ (Sec. 2.3, Ex. 3) **8.** 44 **9.** $\frac{23}{56}$ (Sec. 2.4, Ex. 4)
10. 76,500 **11.** $\frac{8}{11}$ (Sec. 2.5, Ex. 2) **12.** $\frac{15}{4}$ or $3\frac{3}{4}$ **13.** $\frac{4}{5}$ in. (Sec. 3.1, Ex. 6) **14.** 50 **15.** 60 (Sec. 3.2, Ex. 4) **16.** $\frac{1}{3}$
17. $\frac{2}{3}$ (Sec. 3.3, Ex. 1) **18.** 340 **19.** $3\frac{5}{14}$ (Sec. 3.4, Ex. 5) **20.** 33 **21.** $\frac{7}{16}$ (Sec. 3.5, Ex. 6) **22.** $33\frac{27}{40}$ **23.** $\frac{2}{33}$ (Sec. 3.5, Ex. 8)
24. $6\frac{3}{8}$ **25.** 0.8 (Sec. 4.1, Ex. 13) **26.** 0.09 **27.** 8.7 (Sec. 4.1, Ex. 14) **28.** 0.0048 **29.** $3.18 (Sec. 4.2, Ex. 7) **30.** 27.94
31. 829.6561 (Sec. 4.3, Ex. 2) **32.** 1248.3 **33.** 18.408 (Sec. 4.4, Ex. 1) **34.** 76,300 **35.** 0.7861 (Sec. 4.5, Ex. 8) **36.** 1.276 **37.** 0.012
(Sec. 4.5, Ex. 9) **38.** 50.65 **39.** 7.236 (Sec. 4.5, Ex. 11) **40.** 0.191 **41.** 0.25 (Sec. 4.6, Ex. 1) **42.** $0.\overline{5} \approx 0.556$ **43.** $n = 25\% \cdot 0.008$
(Sec. 5.4, Ex. 3) **44.** 37.5% or $37\frac{1}{2}$%

Chapter 6 Geometry
Section 6.1

Vocabulary, Readiness & Video Check **1.** plane **3.** Space **5.** ray **7.** straight **9.** acute **11.** Parallel; intersecting **13.** degrees
15. vertical **17.** $\angle WUV$, $\angle VUW$, $\angle U$, $\angle x$ **19.** $180° - 17° = 163°$

Exercise Set 6.1 **1.** line; line CD or line l or $\overleftrightarrow{CD}$ **3.** line segment; line segment MN or $\overline{MN}$ **5.** angle; $\angle GHI$ or $\angle IHG$ or $\angle H$
7. ray; ray UW or $\overrightarrow{UW}$ **9.** $\angle CPR$, $\angle RPC$ **11.** $\angle TPM$, $\angle MPT$ **13.** straight **15.** right **17.** obtuse **19.** acute **21.** 67° **23.** 163°
25. 32° **27.** 30° **29.** $\angle MNP$ and $\angle RNO$; $\angle PNQ$ and $\angle QNR$ **31.** $\angle SPT$ and $\angle TPQ$; $\angle SPR$ and $\angle RPQ$; $\angle SPT$ and $\angle SPR$; $\angle TPQ$
and $\angle QPR$ **33.** 27° **35.** 132° **37.** $m\angle x = 30°$; $m\angle y = 150°$; $m\angle z = 30°$ **39.** $m\angle x = 77°$; $m\angle y = 103°$; $m\angle z = 77°$
41. $m\angle x = 100°$; $m\angle y = 80°$; $m\angle z = 100°$ **43.** $m\angle x = 134°$; $m\angle y = 46°$; $m\angle z = 134°$ **45.** $\angle ABC$ or $\angle CBA$ **47.** $\angle DBE$ or
$\angle EBD$ **49.** 15° **51.** 50° **53.** 65° **55.** 95° **57.** $\frac{9}{8}$ or $1\frac{1}{8}$ **59.** $\frac{7}{32}$ **61.** $\frac{5}{6}$ **63.** $\frac{4}{3}$ or $1\frac{1}{3}$ **65.** 54.8° **67.** false; answers may vary
69. true **71.** $m\angle a = 60°$; $m\angle b = 50°$; $m\angle c = 110°$; $m\angle d = 70°$; $m\angle e = 120°$ **73.** no; answers may vary **75.** 45°; 45°

Section 6.2

Vocabulary, Readiness & Video Check **1.** Because the sum of the measures of the angles of a triangle equals 180°, each angle in
an equilateral triangle must measure 60°.

Exercise Set 6.2 **1.** pentagon **3.** hexagon **5.** quadrilateral **7.** pentagon **9.** equilateral **11.** scalene; right **13.** isosceles
15. 25° **17.** 13° **19.** 40° **21.** diameter **23.** rectangle **25.** parallelogram **27.** hypotenuse **29.** 14 m **31.** 14.5 cm **33.** 40.6 cm
35. 84 in. **37.** cylinder **39.** rectangular solid **41.** cone **43.** cube **45.** rectangular solid **47.** sphere **49.** pyramid **51.** 14.8 in.
53. 13 mi **55.** 72,368 mi **57.** 108 **59.** 12.56 **61.** true **63.** true **65.** false **67.** yes; answers may vary **69.** answers may vary

Section 6.3

Vocabulary, Readiness & Video Check **1.** perimeter **3.** π **5.** $\frac{22}{7}$ (or 3.14); 3.14 $\left(\text{or } \frac{22}{7}\right)$ **7.** Opposite sides of a rectangle have
the same measure, so we can just find the sum of the measures of all four sides.

Exercise Set 6.3 **1.** 64 ft **3.** 120 cm **5.** 21 in. **7.** 48 ft **9.** 42 in. **11.** 155 cm **13.** 21 ft **15.** 624 ft **17.** 346 yd **19.** 22 ft
21. $55 **23.** 72 in. **25.** 28 in. **27.** $36.12 **29.** 96 m **31.** 66 ft **33.** 74 cm **35.** 17π cm; 53.38 cm **37.** 16π mi; 50.24 mi
39. 26π m; 81.64 m **41.** 15π ft; 47.1 ft **43.** 12,560 ft **45.** 30.7 mi **47.** 14π cm $\approx$ 43.96 cm **49.** 40 mm **51.** 84 ft **53.** 23
55. 1 **57.** 6 **59.** 10 **61. a.** width: 30 yd; length: 40 yd **b.** 140 yd **63.** b **65. a.** 62.8 m; 125.6 m **b.** yes **67.** answers may vary
69. 27.4 m **71.** 75.4 m **73.** 6.5 ft

Section 6.4

Vocabulary, Readiness & Video Check   **1.** The formula for the area of a rectangle; we split the L-shaped figure into two rectangles, used the area formula twice to find the area of each, and then added these two areas.

Exercise Set 6.4   **1.** 7 sq m   **3.** $9\frac{3}{4}$ sq yd   **5.** 15 sq yd   **7.** 2.25π sq in. $\approx$ 7.065 sq in.   **9.** 17.64 sq ft   **11.** 28 sq m   **13.** 22 sq yd   **15.** $36\frac{3}{4}$ sq ft   **17.** $22\frac{1}{2}$ sq in.   **19.** 25 sq cm   **21.** 86 sq mi   **23.** 24 sq cm   **25.** 36π sq in. $\approx 113\frac{1}{7}$ sq in.   **27.** 168 sqft   **29.** 128,775 sq ft   **31.** 1π sq cm $\approx$ 3.14 sq cm   **33.** 128 sq in.; $\frac{8}{9}$ sq ft   **35.** 510 sq in.   **37.** 168 sq ft   **39.** 9200 sq ft   **41. a.** 381 sq ft   **b.** 4 squares   **43.** 14π in. $\approx$ 43.96 in.   **45.** 25 ft   **47.** $12\frac{3}{4}$ ft   **49.** perimeter   **51.** area   **53.** area   **55.** perimeter   **57.** 12-in. pizza   **59.** $1\frac{1}{3}$ sq ft; 192 sq in.   **61.** 7.74 sq in.   **63.** 7056π sq in. $\approx$ 22,155.84 sq in.   **65.** 298.5 sq m   **67. a.** width: 40 yd; length: 60 yd   **b.** 2400 sq yd   **69.** no; answers may vary

Section 6.5

Vocabulary, Readiness & Video Check   **1.** volume   **3.** cubic   **5.** perimeter   **7.** Exact answers are in terms of π, and approximate answers use an approximation for π.

Exercise Set 6.5   **1.** 72 cu in.   **3.** 512 cu cm   **5.** $12\frac{4}{7}$ cu yd   **7.** $523\frac{17}{21}$ cu in.   **9.** $28\frac{2}{7}$ cu in.   **11.** 75 cu cm   **13.** $2\frac{10}{27}$ cu in.   **15.** 8.4 cu ft   **17.** $10\frac{5}{6}$ cu in.   **19.** 960 cu cm   **21.** $\frac{1372}{3}\pi$ cu in. or $457\frac{1}{3}\pi$ cu in.   **23.** $7\frac{1}{2}$ cu ft   **25.** 288π cu ft   **27.** 5.25π cu in.   **29.** 7.96 cu m   **31.** $12\frac{4}{7}$ cu cm   **33.** 8.8 cu in.   **35.** 10.648 cu in.   **37.** 25   **39.** 9   **41.** 5   **43.** 20   **45.** 2093.33 cu m   **47.** no; answers may vary   **49.** 5.5 cu ft; 5.8 cu ft; (b) is larger   **51.** $6\frac{2}{3}\pi$ cu in. $\approx$ 21 cu in.

Integrated Review   **1.** 153°; 63°   **2.** $m\angle x = 75°$; $m\angle y = 105°$; $m\angle z = 75°$ **3.** $m\angle x = 128°$; $m\angle y = 52°$; $m\angle z = 128°$ **4.** $m\angle x = 52°$ **5.** 4.6 in.   **6.** $4\frac{1}{4}$ in.   **7.** 20 m; 25 sq m   **8.** 12 ft; 6 sq ft   **9.** 10π cm $\approx$ 31.4 cm; 25π sq cm $\approx$ 78.5 sq cm   **10.** 32 mi; 44 sq mi   **11.** 54 cm; 143 sq cm   **12.** 62 ft; 238 sq ft   **13.** 64 cu in.   **14.** 30.6 cu ft   **15.** 400 cu cm   **16.** $4\frac{1}{2}\pi$ cu mi $\approx 14\frac{1}{7}$ cu mi

Section 6.6

Calculator Explorations   **1.** 32   **3.** 5.568   **5.** 9.849

Vocabulary, Readiness & Video Check   **1.** 10   **3.** squaring   **5.**

7. $\sqrt{49} = 7$ because 7^2 or $7 \cdot 7 = 49$.   **9.** The Pythagorean theorem works only for right triangles.

Exercise Set 6.6   **1.** 2   **3.** 11   **5.** $\frac{1}{9}$ **7.** $\frac{4}{8} = \frac{1}{2}$ **9.** 1.732   **11.** 3.873   **13.** 6.856   **15.** 5.099   **17.** 6, 7   **19.** 10, 11   **21.** 16   **23.** 9.592   **25.** $\frac{7}{12}$ **27.** 8.426   **29.** 13 in.   **31.** 6.633 cm   **33.** 52.802 m   **35.** 117 mm   **37.** 5   **39.** 12   **41.** 17.205   **43.** 44.822   **45.** 42.426   **47.** 1.732   **49.** 8.5   **51.** 141.42 yd   **53.** 25.0 ft   **55.** 340 ft   **57.** $n = 4$ **59.** $n = 45$ **61.** $n = 6$ **63.** 6   **65.** 10   **67.** answers may vary   **69.** yes   **71.** $\sqrt{80} - 6 \approx$ 2.94 in.

Section 6.7

Vocabulary, Readiness & Video Check   **1.** false   **3.** true   **5.** false   **7.** $\angle M$ and $\angle Y$, $\angle N$ and $\angle X$, $\angle P$ and $\angle Z$, $\frac{p}{z} = \frac{m}{y} = \frac{n}{x}$ **9.** The ratios of corresponding sides are the same.

Exercise Set 6.7   **1.** congruent; SSS   **3.** not congruent   **5.** congruent; ASA   **7.** congruent; SAS   **9.** $\frac{2}{1}$ **11.** $\frac{3}{2}$ **13.** 4.5   **15.** 6   **17.** 5   **19.** 13.5   **21.** 17.5   **23.** 10   **25.** 28.125   **27.** 10   **29.** 520 ft   **31.** 500 ft   **33.** 60 ft   **35.** 14.4 ft   **37.** 52 neon tetras   **39.** 381 ft   **41.** 4.01   **43.** 1.23   **45.** $3\frac{8}{9}$ in.; no   **47.** 8.4   **49.** answers may vary   **51.** 200 ft, 300 ft, 425 ft

Chapter 6 Vocabulary Check   1. right triangle; hypotenuse; legs   **2.** line segment   **3.** complementary   **4.** line   **5.** perimeter   **6.** angle; vertex   **7.** Congruent   **8.** Area   **9.** ray   **10.** square root   **11.** transversal   **12.** straight   **13.** volume   **14.** vertical   **15.** adjacent   **16.** obtuse   **17.** right   **18.** acute   **19.** supplementary   **20.** Similar

Chapter 6 Review   1. right   **2.** straight   **3.** acute   **4.** obtuse   **5.** 65°   **6.** 75°   **7.** 58°   **8.** 98°   **9.** 90°   **10.** 25°   **11.** $\angle a$ and $\angle b$; $\angle b$ and $\angle c$; $\angle c$ and $\angle d$; $\angle d$ and $\angle a$ **12.** $\angle x$ and $\angle w$; $\angle y$ and $\angle z$ **13.** $m\angle x = 100°$; $m\angle y = 80°$; $m\angle z = 80°$ **14.** $m\angle x = 155°$; $m\angle y = 155°$; $m\angle z = 25°$ **15.** $m\angle x = 53°$; $m\angle y = 53°$; $m\angle z = 127°$ **16.** $m\angle x = 42°$; $m\angle y = 42°$; $m\angle z = 138°$ **17.** 103°   **18.** 60°   **19.** 60°   **20.** 65°   **21.** $4\frac{1}{5}$ m   **22.** 7 ft   **23.** 9.5 m   **24.** $15\frac{1}{5}$ cm   **25.** cube   **26.** cylinder   **27.** pyramid   **28.** rectangular solid   **29.** 18 in.   **30.** 2.35 m   **31.** pentagon   **32.** hexagon   **33.** equilateral   **34.** isosceles, right   **35.** 89 m   **36.** 30.6 cm   **37.** 36 m   **38.** 90 ft   **39.** 32 ft   **40.** 440 ft   **41.** 5.338 in.   **42.** 31.4 yd   **43.** 240 sq ft   **44.** 140 sq m   **45.** 600 sq cm   **46.** 189 sq yd   **47.** 49π sq ft $\approx$ 153.86 sq ft   **48.** 82.81 sq m   **49.** 119 sq in.   **50.** 1248 sq cm   **51.** 144 sq m   **52.** 432 sq ft   **53.** 130 sq ft   **54.** $15\frac{5}{8}$ cu in.   **55.** 84 cu ft   **56.** $20{,}000\pi$ cu cm $\approx$ 62,800 cu cm   **57.** $\frac{1}{6}\pi$ cu km $\approx \frac{11}{21}$ cu km   **58.** $2\frac{2}{3}$ cu ft   **59.** 307.72 cu in.   **60.** $7\frac{1}{2}$ cu ft   **61.** 0.5π cu ft or $\frac{1}{2}\pi$ cu ft   **62.** 8   **63.** 12   **64.** $\frac{2}{5}$ **65.** $\frac{1}{10}$ **66.** 13   **67.** 29   **68.** 10.7   **69.** 93   **70.** 127.3 ft   **71.** 88.2 ft   **72.** $37\frac{1}{2}$ **73.** $13\frac{1}{3}$ **74.** 17.4   **75.** 33 ft   **76.** $x = \frac{5}{6}$ in.; $y = 2\frac{1}{6}$ in.   **77.** 108°   **78.** 89°   **79.** 82°   **80.** 78°   **81.** 95°   **82.** 57°   **83.** 13 m   **84.** 12.6 cm   **85.** 22 dm   **86.** 27.3 in.   **87.** 194 ft   **88.** 1624 sq m   **89.** 9π sq m $\approx$ 28.26 sq m   **90.** $346\frac{1}{2}$ cu in.   **91.** 140 cu in.   **92.** 1260 cu ft   **93.** 28.728 cu ft   **94.** 1   **95.** 6   **96.** $\frac{4}{9}$ **97.** 86.6   **98.** 20.8   **99.** 48.1   **100.** 19.7   **101.** $6\frac{1}{2}$ **102.** 12

Chapter 6 Test   1. 12°   **2.** 56°   **3.** 57°   **4.** $m\angle x = 118°$; $m\angle y = 62°$; $m\angle z = 118°$ **5.** $m\angle x = 73°$; $m\angle y = 73°$; $m\angle z = 73°$ **6.** 6.2 m   **7.** $10\frac{1}{4}$ in.   **8.** 26°   **9.** circumference $= 18\pi$ in. $\approx$ 56.52 in.; area $= 81\pi$ sq in. $\approx$ 254.34 sq in.   **10.** perimeter $=$ 24.6 yd; area $=$ 37.1 sq yd   **11.** perimeter $=$ 68 in.; area $=$ 185 sq in.   **12.** $62\frac{6}{7}$ cu in.   **13.** 30 cu ft   **14.** 7   **15.** 8.888   **16.** $\frac{8}{10} = \frac{4}{5}$ **17.** 16 in.   **18.** 18 cu ft   **19.** 62 ft; $115.94   **20.** 5.66 cm   **21.** 198.08 oz   **22.** 7.5   **23.** approximately 69 ft

Cumulative Review   1. nineteen and five thousand twenty-three ten-thousandths (Sec. 4.1, Ex. 3)   **2.** $\frac{53}{66}$ **3.** 736.2 (Sec. 4.2, Ex. 5)   **4.** 700   **5.** 47.06 (Sec. 4.3, Ex. 3)   **6.** $\frac{20}{11}$ or $1\frac{9}{11}$ **7.** 76.8 (Sec. 4.4, Ex. 5)   **8.** $\frac{7}{66}$ **9.** 76,300 (Sec. 4.4, Ex. 7)   **10.** $\frac{23}{2}$ or $11\frac{1}{2}$ **11.** 38.6 (Sec. 4.5, Ex. 1)   **12.** 0.567   **13.** 3.7 (Sec. 4.5, Ex. 12)   **14.** $\frac{3}{5}$ **15.** $>$ (Sec. 4.6, Ex. 7)   **16.** $<$ **17.** 225,000 (Sec. 1.6, Ex. 8)   **18.** $\frac{16}{45}$ **19.** 140,000 (Sec. 1.6, Ex. 9)   **20.** $\frac{35}{2}$ or $17\frac{1}{2}$ **21.** 25% (Sec. 5.2, Ex. 1)   **22.** 68%   **23.** $\frac{19}{1000}$ (Sec. 5.3, Ex. 2)   **24.** $\frac{13}{50}$ **25.** $\frac{5}{4}$ or $1\frac{1}{4}$ (Sec. 5.3, Ex. 3)   **26.** $\frac{28}{5}$ or $5\frac{3}{5}$ **27.** 255 (Sec. 5.4, Ex. 8)   **28.** 15%   **29.** 52 (Sec. 5.5, Ex. 9)   **30.** $\frac{5}{9}$ **31.** 775 freshmen (Sec. 5.6, Ex. 3)   **32.** $2.25/sq ft   **33.** $3210 (Sec. 5.7, Ex. 3)   **34.** 35 exercises   **35.** 7 (Sec. 1.9, Ex. 12)   **36.** 70,052   **37.** 8.33% (Sec. 5.3, Ex. 9)   **38.** 12.5%   **39.** 50° (Sec. 6.2, Ex. 1)   **40.** 33 m   **41.** 28 in. (Sec. 6.3, Ex. 1)   **42.** 45 sq in.   **43.** $\frac{2}{5}$ (Sec. 6.6, Ex. 3)   **44.** $\frac{3}{4}$ **45.** $\frac{12}{19}$ (Sec. 6.7, Ex. 2)   **46.** $15\frac{5}{6}$

Chapter 7 Statistics and Probability

Section 7.1

Vocabulary, Readiness & Video Check   1. bar   **3.** line   **5.** Count the number of symbols and multiply this number by how much each symbol stands for (from the key).   **7.** bar graph

Exercise Set 7.1   1. Kansas   **3.** 5.5 million or 5,500,000 acres   **5.** South Dakota   **7.** 4.0 million or 4,000,000 acres   **9.** 66,000   **11.** 2006   **13.** 12,000   **15.** 66,000 wildfires/year   **17.** September   **19.** 76   **21.** $\frac{1}{38}$ **23.** Tokyo, Japan; about 34.8 million or 34,800,000   **25.** New York, U.S.; 21.6 million or 21,600,000   **27.** approximately 2 million or 2,000,000

29.

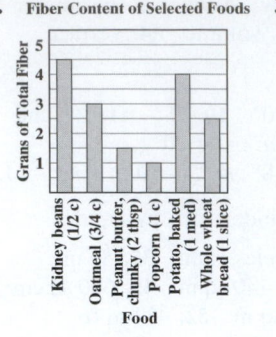

Fiber Content of Selected Foods

31.

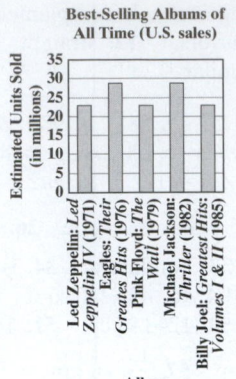

Best-Selling Albums of All Time (U.S. sales)

33. 15 adults   **35.** 61 adults   **37.** 24 adults   **39.** 12 adults

41. $\frac{9}{100}$ **43.** 20 to 44   **45.** 109 million   **47.** 23 million   **49.** answers may vary

51. |; 1   **53.** ||||| |||; 8   **55.** ||||| |; 6   **57.** ||||| |; 6   **59.** ||; 2

61.

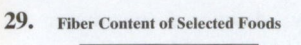

63. 50   **65.** 2011, 2013   **67.** 2006, 2008
69. 2006, 2008, 2012   **71.** 3.6   **73.** 6.2   **75.** 25%
77. 34%   **79.** 83°F   **81.** Sunday; 68°F
83. Tuesday; 13°F   **85.** answers may vary

Section 7.2

Vocabulary, Readiness & Video Check   **1.** circle   **3.** 360   **5.** 100%

Exercise Set 7.2   **1.** parent or guardian's home   **3.** $\frac{9}{35}$ **5.** $\frac{9}{16}$ **7.** Asia   **9.** 37%   **11.** 17,100,000 sq mi   **13.** 2,850,000 sq mi

15. 55%   **17.** nonfiction   **19.** 31,400 books   **21.** 27,632 books   **23.** 25,120 books

25.

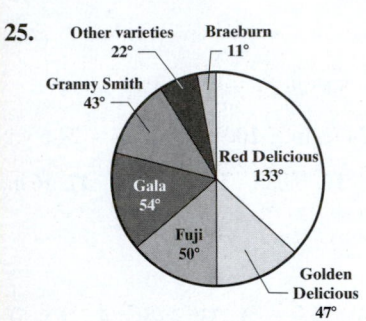

27.

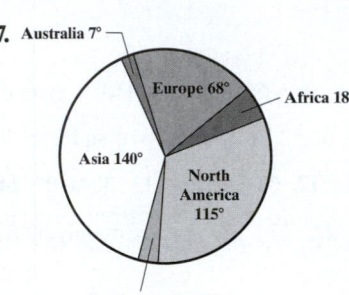

29. $2^2 \times 5$ **31.** $2^3 \times 5$ **33.** 5×17 **35.** Pacific; answers may vary   **37.** 129,600,002 sq km
39. 55,542,858 sq km   **41.** 602 respondents
43. 2276 respondents   **45.** $\frac{301}{837}$
47. no; answers may vary

Integrated Review   **1.** 700,000   **2.** 500,000   **3.** registered nurses, retail salespersons, home health aides   **4.** food workers
5. Oroville Dam; 755 ft   **6.** New Bullards Bar Dam; 635 ft   **7.** 15 ft   **8.** 4 dams   **9.** Thursday and Saturday; 100°F   **10.** Monday; 82°F
11. Sunday, Monday, and Tuesday   **12.** Wednesday, Thursday, Friday, and Saturday   **13.** 70 qt containers   **14.** 52 qt containers
15. 2 qt containers   **16.** 6 qt containers   **17.** ||; 2   **18.** |; 1   **19.** |||; 3   **20.** ||||| |; 6   **21.** |||||; 5
22.

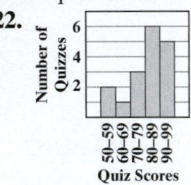

Section 7.3

Vocabulary, Readiness & Video Check   **1.** average   **3.** mean (or average)   **5.** grade point average   **7.** Place the data numbers in numerical order (or verify that they already are).

Exercise Set 7.3   **1.** mean: 21; median: 23; no mode   **3.** mean: 8.1; median: 8.2; mode: 8.2   **5.** mean: 0.5; median: 0.5; mode: 0.2 and 0.5
7. mean: 370.9; median: 313.5; no mode   **9.** 1911.6 ft   **11.** 1601 ft   **13.** answers may vary   **15.** 2.79   **17.** 3.64   **19.** 6.8
21. 6.9   **23.** 88.5   **25.** 73   **27.** 70 and 71   **29.** 9 rates   **31.** $\frac{1}{3}$ **33.** $\frac{3}{5}$ **35.** $\frac{11}{15}$ **37.** 35, 35, 37, 43   **39.** yes; answers may vary

Section 7.4

Vocabulary, Readiness & Video Check   1. outcome   **3.** probability   **5.** 0   **7.** The number of outcomes equals the ending number of branches drawn.

Exercise Set 7.4   1.

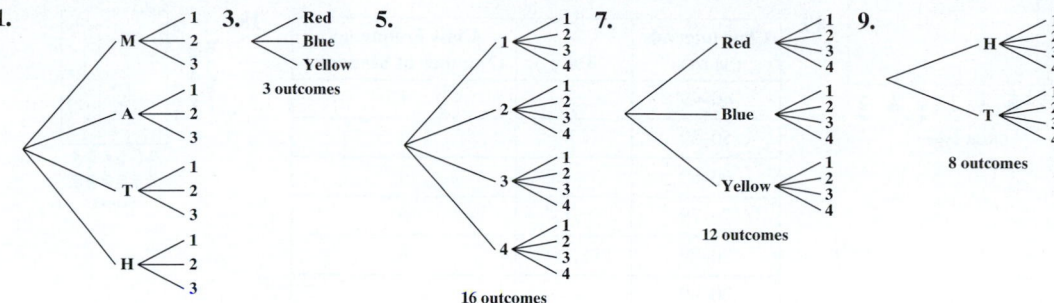

11. $\frac{1}{6}$ **13.** $\frac{1}{3}$ **15.** $\frac{1}{2}$ **17.** $\frac{2}{3}$ **19.** $\frac{1}{3}$ **21.** 1   **23.** $\frac{2}{3}$ **25.** $\frac{1}{7}$ **27.** $\frac{2}{7}$ **29.** $\frac{4}{7}$ **31.** $\frac{19}{100}$ **33.** $\frac{1}{20}$ **35.** $\frac{5}{6}$ **37.** $\frac{1}{6}$ **39.** $\frac{20}{3}$ or $6\frac{2}{3}$
41. $\frac{1}{52}$ **43.** $\frac{1}{13}$ **45.** $\frac{1}{4}$ **47.** $\frac{1}{2}$ **49.** $\frac{5}{36}$ **51.** 0   **53.** answers may vary

Chapter 7 Vocabulary Check   1. bar   **2.** mean   **3.** outcomes   **4.** pictograph   **5.** mode   **6.** line   **7.** median   **8.** tree diagram
9. experiment   **10.** circle   **11.** probability   **12.** histogram; class interval; class frequency

Chapter 7 Review   1. 1,250,000   **2.** 1,500,000   **3.** South   **4.** Northeast   **5.** South   **6.** Northeast, Midwest, West   **7.** 12%
8. 2012   **9.** 1992, 2002, 2012   **10.** answers may vary   **11.** 962   **12.** 927   **13.** 2000 and 2004   **14.** 1996 and 2000   **15.** 27
16. 120   **17.** 1 employee   **18.** 4 employees   **19.** 18 employees   **20.** 9 employees   **21.** ⊦⊦⊦⊦; 5   **22.** ⏐⏐⏐; 3   **23.** ⏐⏐⏐⏐; 4

24.

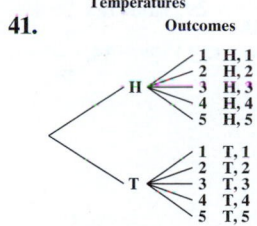

25. mortgage payment   **26.** utilities   **27.** $1225   **28.** $700   **29.** $\frac{39}{160}$ **30.** $\frac{7}{40}$ **31.** 30   **32.** 12   **33.** 21
34. 1   **35.** mean: 17.8; median: 14; no mode   **36.** mean: 58.1; median: 60; mode: 45 and 86   **37.** mean: 24,500; median: 20,000; mode: 20,000   **38.** mean: 447.3; median: 420; mode: 400   **39.** 3.25   **40.** 2.57

41.

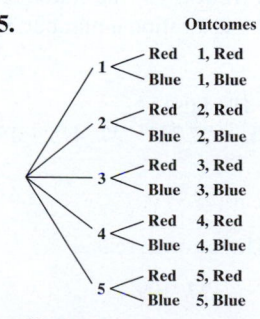

42.

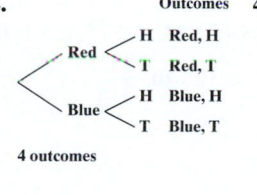

43.

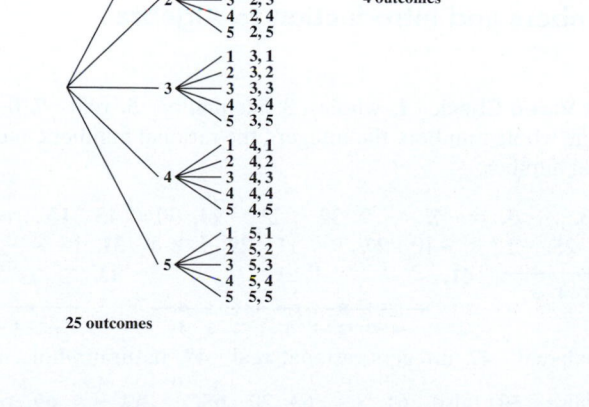

44.

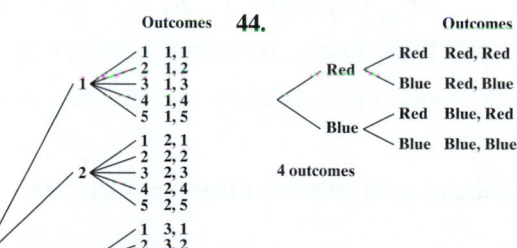

45.

46. $\frac{1}{6}$ **47.** $\frac{1}{6}$ **48.** $\frac{1}{5}$ **49.** $\frac{1}{5}$ **50.** $\frac{3}{5}$ **51.** $\frac{2}{5}$ **52.** $\frac{1}{2}$ **53.** $\frac{1}{2}$ **54.** mean: 74.4; median: 73; mode: none   **55.** mean: 48.8; median: 32;
mode: none   **56.** mean: 454; median: 463.5; mode: 500   **57.** mean: 619.17; median: 647.5; mode: 327   **58.** $\frac{1}{4}$ **59.** $\frac{3}{8}$ **60.** $\frac{1}{4}$ **61.** $\frac{1}{8}$

Chapter 7 Test   **1.** $225   **2.** 3rd week; $350   **3.** $1100   **4.** June, August, September   **5.** February; 3 cm   **6.** March and November

7.

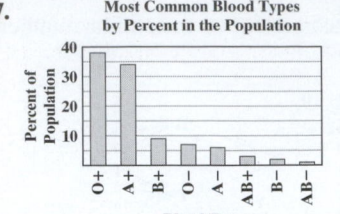

Most Common Blood Types by Percent in the Population

8. 3.0%   **9.** 2004, 2005, 2007   **10.** 2003–2004, 2004–2005, 2006–2007, 2008–2009, 2010–2011

11. $\dfrac{17}{40}$ **12.** $\dfrac{31}{22}$ **13.** 22,744,700 people   **14.** 8,337,500 people   **15.** 9 students   **16.** 11 students

17.

Class Intervals (Scores)	Tally	Class Frequency (Number of Students)
40–49	|	1
50–59	|||	3
60–69	||||	4
70–79	⊤⊦⊦	5
80–89	⊤⊦⊦ |||	8
90–99	||||	4

18.

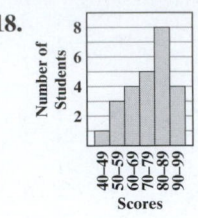

19. mean: 38.4; median: 42; no mode   **20.** mean: 12.625; median: 12.5; mode: 12 and 16   **21.** 3.07

22.

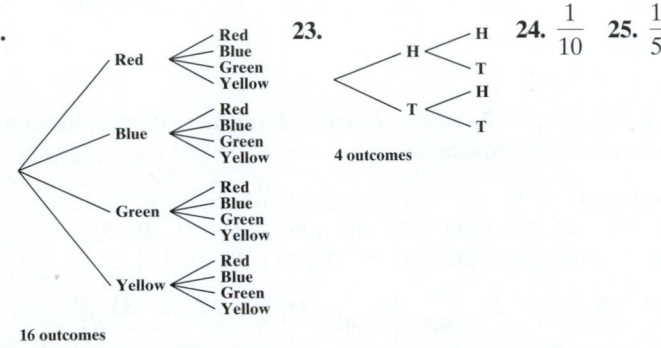

16 outcomes

23.

H ⟨ H / T ; T ⟨ H / T

4 outcomes

24. $\dfrac{1}{10}$ **25.** $\dfrac{1}{5}$

Cumulative Review   **1.** 28 (Sec. 1.9, Ex. 14)   **2.** 12   **3.** $\dfrac{5}{18}$ (Sec. 2.3, Ex. 4)   **4.** $\dfrac{7}{40}$ **5.** $8\dfrac{3}{10}$ (Sec. 3.4, Ex. 3)   **6.** $11\dfrac{1}{3}$

7. 8.4 sq ft (Sec. 4.6, Ex. 10)   **8.** 10 m   **9. a.** 3   **b.** 15   **c.** 0   **d.** 70 (Sec. 1.4, Ex. 1)   **10. a.** 0   **b.** 20   **c.** 0   **d.** 20   **11.** 249,000 (Sec. 1.5, Ex. 3)   **12.** 249,000   **13. a.** 200   **b.** 1230 (Sec. 1.6, Ex. 3)   **14.** 373 R 24   **15.** $6171 (Sec. 1.8, Ex. 3)   **16.** 16,591 ft   **17.** $3 \cdot 3 \cdot 5$ or $3^2 \cdot 5$ (Sec. 2.2, Ex. 3)   **18.** 8   **19.** 0.046 (Sec. 5.2, Ex. 4)   **20.** 0.0029   **21.** 1.9 (Sec. 5.2, Ex. 5)   **22.** 4.52

23. $\dfrac{2}{5}$ (Sec. 5.3, Ex. 1)   **24.** $\dfrac{27}{100}$ **25.** $\dfrac{1}{3}$ (Sec. 5.3, Ex. 4)   **26.** $\dfrac{107}{175}$ **27.** $5 = n \cdot 20$; (Sec. 5.4, Ex. 1)   **28.** $\dfrac{5}{20} = \dfrac{p}{100}$ **29.** sales tax: $6.41; total price: $91.91 (Sec. 5.7, Ex. 1)   **30.** $1610   **31.** $2400 (Sec. 5.8, Ex. 3)   **32.** 33.75   **33.** 42° (Sec. 6.1, Ex. 4)   **34.** 132°   **35.** $\dfrac{1}{6}$ (Sec. 6.6, Ex. 2)   **36.** $\dfrac{1}{5}$ **37.** 14 and 77 (Sec. 7.3, Ex. 5)   **38.** 56   **39.** $\dfrac{1}{4}$ (Sec. 7.4, Ex. 3)   **40.** $\dfrac{3}{5}$

Chapter 8   Real Numbers and Introduction to Algebra

Section 8.1

Vocabulary, Readiness & Video Check   **1.** whole   **3.** inequality   **5.** real   **7.** 0   **9.** absolute value   **11.** To form a true statement: $0 < 7$ **13.** 0 belongs to the whole numbers, the integers, the rational numbers, and the real numbers; since 0 is a rational number, it cannot also be an irrational number.

Exercise Set 8.1   **1.** <   **3.** >   **5.** =   **7.** <   **9.** $32 < 212$ **11.** $30 \le 45$ **13.** true   **15.** false   **17.** true   **19.** false   **21.** $20 \le 25$ **23.** $6 > 0$ **25.** $-12 < -10$ **27.** $7 < 11$ **29.** $5 \ge 4$ **31.** $15 \ne -2$ **33.** 14,494; -282 **35.** $-27,724$ **37.** 475; -195

39.

‹–•–•–•–•–|–•–•–•–•–›
$-4\ -3\ -2\ -1\ \ 0\ \ 1\ \ 2\ \ 3\ \ 4\ \ 5$

41.

$-1\frac{1}{4}\ \frac{1}{3}$

$-4\ -3\ -2\ -1\ \ 0\ \ 1\ \ 2\ \ 3\ \ 4$

43.

$-4.5\quad -\frac{3}{2}\quad \frac{7}{4}\ 3.25$

$-5\ -4\ -3\ -2\ -1\ \ 0\ \ 1\ \ 2\ \ 3\ \ 4$

45. whole, integers, rational, real   **47.** integers, rational, real   **49.** natural, whole, integers, rational, real   **51.** rational, real

53. false   **55.** true   **57.** false   **59.** false   **61.** 8.9   **63.** 20   **65.** $\dfrac{9}{2}$ **67.** $\dfrac{12}{13}$ **69.** >   **71.** =   **73.** <   **75.** <   **77.** 109

79. 8   **81.** 40 million $>$ 14 million or $40,000,000 > 14,000,000$ **83.** 40 million pounds less, or -40 million   **85.** $-0.04 > -26.7$ **87.** sun   **89.** sun   **91.** answers may vary

Section 8.2

Calculator Explorations   **1.** 125   **3.** 59,049   **5.** 30   **7.** 9857   **9.** 2376

Vocabulary, Readiness & Video Check   1. base; exponent   **3.** multiplication   **5.** subtraction   **7.** expression   **9.** expression; variables   **11.** equation   **13.** The order in which we perform operations does matter! We came up with an order of operations to avoid getting more than one answer when evaluating an expression.   **15.** No; the variable was replaced with 0 in the equation to see if a true statement occurred, and it did not.

Exercise Set 8.2   1. 243   **3.** 27   **5.** 1   **7.** 5   **9.** 49   **11.** $\frac{16}{81}$ **13.** $\frac{1}{125}$ **15.** 1.44   **17.** 0.343   **19.** 5^2 sq m   **21.** 17   **23.** 20   **25.** 12

27. 21   **29.** 45   **31.** 0   **33.** $\frac{2}{7}$ **35.** 30   **37.** 2   **39.** $\frac{7}{18}$ **41.** $\frac{27}{10}$ **43.** $\frac{7}{5}$ **45.** 32   **47.** $\frac{23}{27}$ **49.** 9   **51.** 1   **53.** 1   **55.** 11   **57.** 8

59. 45   **61.** 27   **63.** 132   **65.** $\frac{37}{18}$ **67.** solution   **69.** not a solution   **71.** not a solution   **73.** solution   **75.** not a solution

77. solution   **79.** $x + 15$ **81.** $x - 5$ **83.** $\frac{x}{4}$ **85.** $3x + 22$ **87.** $1 + 2 = 9 \div 3$ **89.** $3 \neq 4 \div 2$ **91.** $5 + x = 20$ **93.** $7.6x = 17$

95. $13 - 3x = 13$ **97.** 35   **99.** 360   **101.** no; answers may vary   **103. a.** 64   **b.** 43   **c.** 19   **d.** 22   **105.** 14 in.; 12 sq in.
107. 14 in.; 9.01 sq in.   **109.** Rectangles with the same perimeter can have different areas.   **111.** $(20 - 4) \cdot 4 \div 2$
113. a. expression   **b.** equation   **c.** equation   **d.** expression   **e.** expression   **115.** answers may vary   **117.** answers may vary, for example, $2(6) - 1$.

Section 8.3

Vocabulary, Readiness & Video Check   1. 0   **3.** a **5.** absolute values   **7.** Example 12 is an example of the opposite of the *absolute value* of $-a$, not the opposite of $-a$. The absolute value of $-a$ is positive, so its opposite is negative. Therefore, the answers to Examples 11 and 12 have different signs.   **9.** Depths below the surface; the diver's position is 231 feet below the surface.

Exercise Set 8.3   1. 3   **3.** -14 **5.** 1   **7.** -12 **9.** -5 **11.** -12 **13.** -4 **15.** 7   **17.** -2 **19.** 0   **21.** -19 **23.** 31   **25.** -47

27. -2.1 **29.** 38   **31.** -13.1 **33.** $\frac{1}{4}$ **35.** $-\frac{3}{16}$ **37.** $-\frac{13}{10}$ **39.** -8 **41.** -8 **43.** -59 **45.** -9 **47.** 5   **49.** 11   **51.** -18

53. 19   **55.** -7 **57.** -26 **59.** -6 **61.** 2   **63.** 0   **65.** -6 **67.** -2 **69.** 7   **71.** 7.9   **73.** $5z$ **75.** $\frac{2}{3}$ **77.** -70 **79.** 3   **81.** 19

83. -10 **85.** $0 + (-215) + (-16) = -231$; 231 ft below the surface   **87.** $107°F$ **89.** -95 m   **91.** -8 **93.** \$16.9 million
95. 72.01   **97.** 141   **99.** July   **101.** October   **103.** $4.7°F$ **105.** answers may vary   **107.** -3 **109.** -22 **111.** true
113. false   **115.** answers may vary

Section 8.4

Vocabulary, Readiness & Video Check   1. $a + (-b)$; b   **3.** $-10 - (-14)$; d   **5.** addition, opposite   **7.** There's a minus sign in the numerator and the replacement value is negative (notice parentheses are used around the replacement value), and it's always good to be careful when working with negative signs.   **9.** This means that the overall vertical altitude change of the jet is actually a decrease in altitude from when the example started.

Exercise Set 8.4   1. -10 **3.** -5 **5.** 19   **7.** 11   **9.** -8 **11.** -11 **13.** 37   **15.** 5   **17.** -71 **19.** 0   **21.** $\frac{2}{11}$ **23.** -6.4 **25.** 4.1

27. $-\frac{1}{6}$ **29.** $-\frac{11}{12}$ **31.** 8.92   **33.** -8.92 **35.** 13   **37.** -5 **39.** -1 **41.** -23 **43.** -26 **45.** -24 **47.** 3   **49.** -45 **51.** -4

53. 13   **55.** 6   **57.** 9   **59.** -9 **61.** $\frac{7}{5}$ **63.** -7 **65.** 21   **67.** $\frac{1}{4}$ **69.** not a solution   **71.** not a solution   **73.** solution

75. $263°F$ **77.** 35,653 ft   **79.** $30°$ **81.** -308 ft   **83.** 19,852 ft   **85.** $130°$ **87.** $-5 + x$ **89.** $-20 - x$ **91.** 0   **93.** $\frac{10}{13}$

95. $-4.4°, 2.6°, 12°, 23.5°, 15.3°$ **97.** May   **99.** answers may vary   **101.** 16   **103.** -20 **105.** true; answers may vary
107. false; answers may vary   **109.** negative, $-30,387$

Integrated Review   1. negative   **2.** negative   **3.** positive   **4.** 0   **5.** positive   **6.** 0   **7.** positive   **8.** positive   **9.** $-\frac{1}{7}; \frac{1}{7}$

10. $\frac{12}{5}; \frac{12}{5}$ **11.** $3; 3$ **12.** $-\frac{9}{11}; \frac{9}{11}$ **13.** -42 **14.** 10   **15.** 2   **16.** -18 **17.** -7 **18.** -39 **19.** -2 **20.** -9 **21.** -3.4 **22.** -9.8

23. $-\frac{25}{28}$ **24.** $-\frac{5}{24}$ **25.** -4 **26.** -24 **27.** 6   **28.** 20   **29.** 6   **30.** 61   **31.** -6 **32.** -16 **33.** -19 **34.** -13 **35.** -4 **36.** -1

37. $\frac{13}{20}$ **38.** $-\frac{29}{40}$ **39.** 4   **40.** 9   **41.** -1 **42.** -3 **43.** 8   **44.** 10   **45.** 47   **46.** $\frac{2}{3}$

Section 8.5

Calculator Explorations   1. 38   **3.** -441 **5.** 490   **7.** 54,499   **9.** 15,625

Vocabulary, Readiness & Video Check   1. negative   **3.** positive   **5.** 0   **7.** 0   **9.** The parentheses, or lack of them, determine the base of the expression. In Example 6, $(-2)^4$, the base is -2 and all of -2 is raised to the 4th power. In Example 7, -2^4, the base is 2 and only 2 is raised to the 4th power.   **11.** Yes; because division of real numbers is defined in terms of multiplication.   **13.** Yes; a true statement results when x is replaced with 5.

Exercise Set 8.5 **1.** -24 **3.** -2 **5.** 50 **7.** -45 **9.** $\dfrac{3}{10}$ **11.** -7 **13.** -15 **15.** 0 **17.** 16 **19.** -16 **21.** $\dfrac{9}{16}$ **23.** -0.49 **25.** $\dfrac{3}{2}$

27. $-\dfrac{1}{14}$ **29.** $-\dfrac{11}{3}$ **31.** $\dfrac{1}{0.2}$ **33.** -9 **35.** -4 **37.** 0 **39.** undefined **41.** $-\dfrac{18}{7}$ **43.** 160 **45.** 64 **47.** $-\dfrac{8}{27}$ **49.** 3 **51.** -15

53. -125 **55.** -0.008 **57.** $\dfrac{2}{3}$ **59.** $\dfrac{20}{27}$ **61.** 0.84 **63.** -40 **65.** 81 **67.** -1 **69.** -121 **71.** -1 **73.** -19 **75.** 90 **77.** -84

79. -5 **81.** $-\dfrac{9}{2}$ **83.** 18 **85.** 17 **87.** -20 **89.** 16 **91.** 2 **93.** $-\dfrac{34}{7}$ **95.** 0 **97.** $\dfrac{6}{5}$ **99.** $\dfrac{3}{2}$ **101.** $-\dfrac{5}{38}$ **103.** 3 **105.** -1

107. undefined **109.** $-\dfrac{22}{9}$ **111.** solution **113.** not a solution **115.** solution **117.** $-71\cdot x$ or $-71x$ **119.** $-16-x$

121. $-29+x$ **123.** $\dfrac{x}{-33}$ or $x\div(-33)$ **125.** $3\cdot(-4)=-12$; a loss of 12 yd **127.** $5(-20)=-100$; a depth of 100 ft

129. 32 in. **131.** 30 ft **133.** true **135.** false **137.** $-162°F$ **139.** $-\$11$ million per month **141.** answers may vary

143. $1,-1$; answers may vary **145.** $\dfrac{0}{5}-7=-7$ **147.** $-8(-5)+(-1)=39$

Section 8.6

Vocabulary, Readiness & Video Check **1.** commutative property of addition **3.** distributive property **5.** associative property of addition **7.** opposites or additive inverses **9.** 2 is outside the parentheses, so the point is made that you should only distribute the -9 to the terms within the parentheses and not also to the 2.

Exercise Set 8.6 **1.** $16+x$ **3.** $y\cdot(-4)$ **5.** yx **7.** $13+2x$ **9.** $x\cdot(yz)$ **11.** $(2+a)+b$ **13.** $(4a)\cdot b$ **15.** $a+(b+c)$

17. $17+b$ **19.** $24y$ **21.** y **23.** $26+a$ **25.** $-72x$ **27.** s **29.** $-\dfrac{5}{2}x$ **31.** $4x+4y$ **33.** $9x-54$ **35.** $6x+10$ **37.** $28x-21$

39. $18+3x$ **41.** $-2y+2z$ **43.** $-y-\dfrac{5}{3}$ **45.** $5x+20m+10$ **47.** $8m-4n$ **49.** $-5x-2$ **51.** $-r+3+7p$ **53.** $3x+4$

55. $-x+3y$ **57.** $6r+8$ **59.** $-36x-70$ **61.** $-1.6x-2.5$ **63.** $4(1+y)$ **65.** $11(x+y)$ **67.** $-1(5+x)$ **69.** $30(a+b)$

71. commutative property of multiplication **73.** associative property of addition **75.** commutative property of addition **77.** associative property of multiplication **79.** identity element for addition **81.** distributive property **83.** multiplicative inverse property **85.** identity element for multiplication **87.** 4050 **89.** 45 **91.** $-8;\dfrac{1}{8}$ **93.** $-x;\dfrac{1}{x}$ **95.** $2x;-2x$ **97.** false **99.** no

101. yes **103.** yes **105.** yes **107.** **a.** commutative property of addition **b.** commutative property of addition **c.** associative property of addition **109.** answers may vary **111.** answers may vary

Section 8.7

Vocabulary, Readiness & Video Check **1.** expression **3.** combine like term **5.** like; unlike **7.** Although these terms have exactly the same variables, the exponents on each are not exactly the same—the exponents on x differ in each term. **9.** -1

Exercise Set 8.7 **1.** -7 **3.** 1 **5.** 17 **7.** like **9.** unlike **11.** like **13.** $15y$ **15.** $13w$ **17.** $-7b-9$ **19.** $-m-6$ **21.** -8
23. $7.2x-5.2$ **25.** $k-6$ **27.** $-15x+18$ **29.** $4x-3$ **31.** $5x^2$ **33.** -11 **35.** $1.3x+3.5$ **37.** $5y+20$ **39.** $-2x-4$
41. $-10x+15y-30$ **43.** $-3x+2y-1$ **45.** $7d-11$ **47.** 16 **49.** $x+5$ **51.** $x+2$ **53.** $2k+10$ **55.** $-3x+5$ **57.** $2x+14$

59. $3y+\dfrac{5}{6}$ **61.** $-22+24x$ **63.** $0.9m+1$ **65.** $10-6x-9y$ **67.** $-x-38$ **69.** $5x-7$ **71.** $10x-3$ **73.** $-4x-9$

75. $-4m-3$ **77.** $2x-4$ **79.** $\dfrac{3}{4}x+12$ **81.** $12x-2$ **83.** $8x+48$ **85.** $x-10$ **87.** 2 **89.** -23 **91.** -25 **93.** balanced

95. balanced **97.** answers may vary **99.** $(18x-2)$ ft **101.** $(15x+23)$ in. **103.** answers may vary

Chapter 8 Vocabulary Check **1.** inequality symbols **2.** equation **3.** absolute value **4.** variable **5.** opposites **6.** numerator **7.** solution **8.** reciprocals **9.** base; exponent **10.** numerical coefficient **11.** denominator **12.** grouping symbols **13.** term **14.** like terms **15.** unlike terms

Chapter 8 Review **1.** $<$ **2.** $>$ **3.** $>$ **4.** $>$ **5.** $<$ **6.** $>$ **7.** $=$ **8.** $=$ **9.** $>$ **10.** $<$ **11.** $4\ge-3$ **12.** $6\ne5$ **13.** $0.03<0.3$
14. $1729<2870$ **15.** **a.** $1,3$ **b.** $0,1,3$ **c.** $-6,0,1,3$ **d.** $-6,0,1,1\dfrac{1}{2},3,9.62$ **e.** π **f.** all numbers in set **16.** **a.** $2,5$

b. $2,5$ **c.** $-3,2,5$ **d.** $-3,-1.6,2,5,\dfrac{11}{2},15.1$ **e.** $\sqrt{5},2\pi$ **f.** all numbers in set **17.** Friday **18.** Wednesday **19.** c **20.** b

21. 37 **22.** 41 **23.** $\dfrac{18}{7}$ **24.** 80 **25.** $20-12=2\cdot4$ **26.** $\dfrac{9}{2}>-5$ **27.** 18 **28.** 108 **29.** 5 **30.** 24 **31.** $63°$ **32.** $105°$

33. solution **34.** not a solution **35.** 9 **36.** $-\dfrac{2}{3}$ **37.** -2 **38.** 7 **39.** -11 **40.** -17 **41.** $-\dfrac{3}{16}$ **42.** -5 **43.** -13.9 **44.** 3.9

45. -14 **46.** -11.5 **47.** 5 **48.** -11 **49.** -19 **50.** 4 **51.** a **52.** a **53.** $\$51$ **54.** $\$54$ **55.** $-\dfrac{1}{6}$ **56.** $\dfrac{5}{3}$ **57.** -48 **58.** 28

59. 3 **60.** -14 **61.** -36 **62.** 0 **63.** undefined **64.** $-\dfrac{1}{2}$ **65.** commutative property of addition **66.** identity element for multiplication **67.** distributive property **68.** additive inverse property **69.** associative property of addition **70.** commutative property of multiplication **71.** distributive property **72.** associative property of multiplication **73.** multiplicative inverse property **74.** identity element for addition **75.** commutative property of addition **76.** distributive property **77.** $6x$ **78.** $-11.8z$ **79.** $4x - 2$ **80.** $2y + 3$ **81.** $3n - 18$ **82.** $4w - 6$ **83.** $-6x + 7$ **84.** $-0.4y + 2.3$ **85.** $3x - 7$ **86.** $5x + 5.6$ **87.** $<$ **88.** $>$ **89.** -15.3 **90.** -6 **91.** -80 **92.** -5 **93.** $-\dfrac{1}{4}$ **94.** 0.15 **95.** 16 **96.** 16 **97.** -5 **98.** 9 **99.** $-\dfrac{5}{6}$ **100.** undefined **101.** $16x - 41$ **102.** $18x - 12$

Chapter 8 Test **1.** $|-7| > 5$ **2.** $9 + 5 \geq 4$ **3.** -5 **4.** -11 **5.** -14 **6.** -39 **7.** 12 **8.** -2 **9.** undefined **10.** -8 **11.** $-\dfrac{1}{3}$ **12.** $4\dfrac{5}{8}$ **13.** $\dfrac{51}{40}$ or $1\dfrac{11}{40}$ **14.** -32 **15.** -48 **16.** 3 **17.** 0 **18.** $>$ **19.** $>$ **20.** $>$ **21.** $=$ **22. a.** $1, 7$ **b.** $0, 1, 7$ **c.** $-5, -1, 0, 1, 7$ **d.** $-5, -1, \dfrac{1}{4}, 0, 1, 7, 11.6$ **e.** $\sqrt{7}, 3\pi$ **f.** $-5, -1, \dfrac{1}{4}, 0, 1, 7, 11.6, \sqrt{7}, 3\pi$ **23.** 40 **24.** 12 **25.** 22 **26.** -1 **27.** associative property of addition **28.** commutative property of multiplication **29.** distributive property **30.** multiplicative inverse **31.** 9 **32.** -3 **33.** second down **34.** yes **35.** $17°$ **36.** \$420 **37.** $y - 10$ **38.** $5.9x + 1.2$ **39.** $-2x + 10$ **40.** $-15y + 1$

Cumulative Review **1.** 2010 (Sec. 1.3, Ex. 4) **2.** 1531 **3.** $3 \cdot 3 \cdot 3 \cdot 5 \cdot 7$ or $3^3 \cdot 5 \cdot 7$ (Sec. 2.2, Ex. 5) **4.** 153 sq in. **5.** 33 (Sec. 3.2, Ex. 7) **6.** $\dfrac{10}{63}$ **7.** $5\dfrac{1}{15}$ (Sec. 3.4, Ex. 2) **8.** $\dfrac{31}{3}$ or $10\dfrac{1}{3}$ **9.** $\dfrac{1}{8}$ (Sec. 4.1, Ex. 10) **10.** $1\dfrac{1}{5}$ **11.** $105\dfrac{83}{1000}$ (Sec. 4.1, Ex. 12) **12.** $\dfrac{8}{27}$ **13.** $<$ (Sec. 4.2, Ex. 2) **14.** 25 **15.** 67.69 (Sec. 4.3, Ex. 6) **16.** 139.231 **17.** 4.21 (Sec. 4.4, Ex. 8) **18.** 186,040 **19.** 0.0092 (Sec. 4.4, Ex. 10) **20.** 5.8 **21.** 0.26 (Sec. 4.5, Ex. 2) **22.** $\dfrac{21}{20}$ or $1\dfrac{1}{20}$ **23.** 2.1875 (Sec. 4.6, Ex. 4) **24.** 7.3 **25.** $0.\overline{6}$ (Sec. 4.6, Ex. 2) **26.** 2.16 **27.** 0.23 (Sec. 5.2, Ex. 3) **28.** 87.5% or $87\dfrac{1}{2}\%$ **29.** 8.33% (Sec. 5.3, Ex. 9) **30.** 24% **31.** 14 (Sec. 5.4, Ex. 7) **32.** $\dfrac{23}{100}$ **33.** $\dfrac{75}{30} = \dfrac{p}{100}$ (Sec. 5.5, Ex. 5) **34.** -4.5 **35.** 18% (Sec. 5.6, Ex. 6) **36.** -0.9 **37.** discount: \$16.25; sale price: \$48.75 (Sec. 5.7, Ex. 5) **38.** 94 **39.** \$120 (Sec. 5.8, Ex. 1) **40.** 15%

Chapter 9 Equations, Inequalities, and Problem Solving

Section 9.1

Vocabulary, Readiness & Video Check **1.** expression **3.** equation **5.** expression; equation **7.** Equivalent **9.** 2 **11.** 12 **13.** 17 **15.** both sides **17.** $\dfrac{1}{7}x$

Exercise Set 9.1 **1.** 3 **3.** -2 **5.** -14 **7.** 0.5 **9.** $\dfrac{1}{4}$ **11.** $\dfrac{5}{12}$ **13.** -3 **15.** -9 **17.** -10 **19.** 2 **21.** -7 **23.** -1 **25.** -9 **27.** -12 **29.** $-\dfrac{1}{2}$ **31.** 11 **33.** 21 **35.** 25 **37.** -3 **39.** -0.7 **41.** 11 **43.** 13 **45.** -30 **47.** -0.4 **49.** -7 **51.** $-\dfrac{1}{3}$ **53.** -17.9 **55.** $20 - p$ **57.** $(10 - x)$ ft **59** $(180 - x)°$ **61.** $n - 29{,}000$ **63.** $7x$ sq mi **65.** $\dfrac{8}{5}$ **67.** $\dfrac{1}{2}$ **69.** -9 **71.** x **73.** y **75.** x **77.** answers may vary **79.** 4 **81.** answers may vary **83.** $(173 - 3x)°$ **85.** answers may vary **87.** -145.478

Section 9.2

Vocabulary, Readiness & Video Check **1.** multiplication **3.** equation; expression **5.** Equivalent **7.** 9 **9.** 2 **11.** -5 **13.** same **15.** $(x + 1) + (x + 3) = 2x + 4$

Exercise Set 9.2 **1.** 4 **3.** 0 **5.** 12 **7.** -12 **9.** 3 **11.** 2 **13.** 0 **15.** 6.3 **17.** 10 **19.** -20 **21.** 0 **23.** -9 **25.** 1 **27.** -30 **29.** 3 **31.** $\dfrac{10}{9}$ **33.** -1 **35.** -4 **37.** $-\dfrac{1}{2}$ **39.** 0 **41.** 4 **43.** $-\dfrac{1}{14}$ **45.** 0.21 **47.** 5 **49.** 6 **51.** -5.5 **53.** -5 **55.** 0 **57.** -3 **59.** $-\dfrac{9}{28}$ **61.** $\dfrac{14}{3}$ **63.** -9 **65.** -2 **67.** $\dfrac{11}{2}$ **69.** $-\dfrac{1}{4}$ **71.** $\dfrac{9}{10}$ **73.** $-\dfrac{17}{20}$ **75.** -16 **77.** $2x + 2$ **79.** $2x + 2$ **81.** $5x + 20$ **83.** $7x - 12$ **85.** $12z + 44$ **87.** 1 **89.** -48 **91.** answers may vary **93.** answers may vary **95.** 2

Section 9.3

Calculator Explorations **1.** solution **3.** not a solution **5.** solution

Vocabulary, Readiness & Video Check **1.** equation **3.** expression **5.** expression **7.** equation **9.** 3; distributive property, addition property of equality, multiplication property of equality **11.** The number of decimal places in each number helps you determine what power of 10 you can multiply through by so you are no longer dealing with decimals.

Exercise Set 9.3   **1.** −6   **3.** 3   **5.** 1   **7.** $\dfrac{3}{2}$ **9.** 0   **11.** −1   **13.** 4   **15.** −4   **17.** −3   **19.** 2   **21.** 50   **23.** 1   **25.** $\dfrac{7}{3}$ **27.** 0.2

29. all real numbers   **31.** no solution   **33.** no solution   **35.** all real numbers   **37.** 18   **39.** $\dfrac{19}{9}$ **41.** $\dfrac{14}{3}$ **43.** 13   **45.** 4

47. all real numbers   **49.** $-\dfrac{3}{5}$ **51.** −5   **53.** 10   **55.** no solution   **57.** 3   **59.** −17   **61.** $\dfrac{7}{5}$ **63.** $-\dfrac{1}{50}$ **65.** $(6x - 8)$ m   **67.** $-8 - x$

69. $-3 + 2x$ **71.** $9(x + 20)$ **73. a.** all real numbers   **b.** answers may vary   **c.** answers may vary   **75.** a   **77.** b   **79.** c

81. answers may vary   **83. a.** $x + x + x + 2x + 2x = 28$ **b.** $x = 4$ **c.** $x = 4$ cm; $2x$ cm $= 8$ cm   **85.** answers may vary

87. 15.3   **89.** −0.2

Integrated Review   **1.** 6   **2.** −17   **3.** 12   **4.** −26   **5.** −3   **6.** −1   **7.** $\dfrac{27}{2}$ **8.** $\dfrac{25}{2}$ **9.** 8   **10.** −64   **11.** 2   **12.** −3   **13.** 5   **14.** −1

15. 2   **16.** 2   **17.** −2   **18.** −2   **19.** $-\dfrac{5}{6}$ **20.** $\dfrac{1}{6}$ **21.** 1   **22.** 6   **23.** 4   **24.** 1   **25.** $\dfrac{9}{5}$ **26.** $-\dfrac{6}{5}$ **27.** all real numbers   **28.** all real numbers

29. 0   **30.** −1.6   **31.** $\dfrac{4}{19}$ **32.** $-\dfrac{5}{19}$ **33.** $\dfrac{7}{2}$ **34.** $-\dfrac{1}{4}$ **35.** no solution   **36.** no solution   **37.** $\dfrac{7}{6}$ **38.** $\dfrac{1}{15}$

Section 9.4

Vocabulary, Readiness & Video Check   **1.** $2x$; $2x - 31$ **3.** $x + 5$; $2(x + 5)$ **5.** $20 - y$; $\dfrac{20 - y}{3}$ or $(20 - y) \div 3$ **7.** in the statement of the application   **9.** That the 3 angle measures are consecutive even integers and that they sum to 180°

Exercise Set 9.4   **1.** $2x + 7 = x + 6$; −1   **3.** $3x - 6 = 2x + 8$; 14   **5.** −25   **7.** $-\dfrac{3}{4}$ **9.** 3 in.; 6 in.; 16 in.   **11.** 1st piece: 5 in.;

2nd piece: 10 in.; 3rd piece: 25 in.   **13.** Pennsylvania: 494 million lb; New York: 720 million lb   **15.** 172 mi   **17.** 25 mi

19. 1st angle: 37.5°; 2nd angle: 37.5°; 3rd angle: 105°   **21.** A: 60°; B: 120°; C: 120°; D: 60°   **23.** $3x + 3$ **25.** $x + 2$; $x + 4$; $2x + 4$

27. $x + 1$; $x + 2$; $x + 3$; $4x + 6$ **29.** $x + 2$; $x + 4$; $2x + 6$ **31.** 234, 235   **33.** Belgium: 32; France: 33; Spain: 34   **35.** 5 ft, 12 ft

37. Maglev: 361 mph; TGV: 357.2 mph   **39.** 43°, 137°   **41.** 58°, 60°, 62°   **43.** 1   **45.** 280 mi   **47.** Stanford: 20; Wisconsin: 14

49. Montana: 56 counties; California: 58 counties   **51.** Neptune: 14 satellites; Uranus: 27 satellites; Saturn: 62 satellites   **53.** −16

55. Sahara: 3,500,000 sq mi; Gobi: 500,000 sq mi   **57.** Jamaica: 4; Cuba: 5; New Zealand: 6   **59.** females: 30,184; males: 28,604

61. 34.5°; 34.5°; 111°   **63.** Hawaii   **65.** Florida: $56 million; California: $50 million   **67.** answers may vary   **69.** 34   **71.** 225π

73. 15 ft by 24 ft   **75.** 5400 chirps per hour; 129,600 chirps per day; 47,304,000 chirps per year   **77.** answers may vary   **79.** answers may vary   **81.** c

Section 9.5

Vocabulary, Readiness & Video Check   **1.** relationships   **3.** To show that the process of solving this equation for x—dividing both sides by 5, the coefficient of x—is the same process used to solve a formula for a specific variable. Treat whatever is multiplied by that specific variable as the coefficient—the coefficient is all the factors except that specific variable.

Exercise Set 9.5   **1.** $h = 3$ **3.** $h = 3$ **5.** $h = 20$ **7.** $c = 12$ **9.** $r = 2.5$ **11.** $h = \dfrac{f}{5g}$ **13.** $w = \dfrac{V}{lh}$ **15.** $y = 7 - 3x$

17. $R = \dfrac{A - P}{PT}$ **19.** $A = \dfrac{3V}{h}$ **21.** $a = P - b - c$ **23.** $h = \dfrac{S - 2\pi r^2}{2\pi r}$ **25.** 120 ft   **27. a.** area: 480 sq in.; perimeter: 120 in.

b. frame: perimeter; glass: area   **29. a.** area: 103.5 sq ft; perimeter: 41 ft   **b.** baseboard: perimeter; carpet: area   **31.** −10°C

33. 6.25 hr   **35.** length: 78 ft; width: 52 ft   **37.** 18 ft, 36 ft, 48 ft   **39.** 137.5 mi   **41.** 61.5°F   **43.** 60 chirps per minute   **45.** increases

47. 96 piranha   **49.** 2 bags   **51.** one 16-in. pizza   **53.** 4.65 min   **55.** 13 in.   **57.** 2.25 hr   **59.** 12,090 ft   **61.** 50°C   **63.** 515,509.5 cu in.

65. 449 cu in.   **67.** 333°F   **69.** 0.32   **71.** 2.00 or 2   **73.** 17%   **75.** 720%   **77.** $V = G(N - R)$ **79.** multiplies the volume by 8; answers may vary   **81.** $53\dfrac{1}{3}$ **83.** $\bigcirc = \dfrac{\triangle - \square}{\blacksquare}$ **85.** 44.3 sec   **87.** $P = 3,200,000$ **89.** $V = 113.1$

Section 9.6

Vocabulary, Readiness & Video Check   **1.** no   **3.** yes   **5. a.** equals; =   **b.** multiplication; ·   **c.** Drop the percent symbol and move the decimal point two places to the left.   **7.** You must first find the actual amount of increase in price by subtracting the original price from the new price.

Exercise Set 9.6   **1.** 11.2   **3.** 55%   **5.** 180   **7.** 15%   **9.** 310,500   **11.** discount: $1480; new price: $17,020   **13.** $46.58   **15.** 9.3%

17. 30%   **19.** $104   **21.** $42,500   **23.** 2 gal   **25.** 7 lb   **27.** 4.6   **29.** 50   **31.** 30%   **33.** 71%   **35.** 189,670   **37.** 59%, 5%, 27%, 2%   **39.** 75%   **41.** $3900   **43.** 300%   **45.** mark-up: $0.11; new price: $2.31   **47.** 400 oz   **49.** 13.4%   **51.** 120 employees

53. decrease: $64; sale price: $192   **55.** 854 thousand Scoville units   **57.** 132 million adults   **59.** 400 oz   **61.** >   **63.** =   **65.** >

67. no; answers may vary   **69.** 9.6%   **71.** 26.9%; yes   **73.** 17.1%

Section 9.7

Vocabulary, Readiness & Video Check **1.** expression **3.** inequality **5.** equation **7.** -5 **9.** 4.1 **11.** An open circle indicates $>$ or $<$; a closed circle indicates $\geq$ or $\leq$. **13.** $\{x \mid x \geq -2\}$ **15.** is greater than; $>$

Exercise Set 9.7 **1.** **3.** **5.** **7.**
9. **11.** **13.** $\{x \mid x \geq -5\}$
15. $\{y \mid y < 9\}$ **17.** $\{x \mid x > -3\}$ **19.** $\{x \mid x \leq 1\}$
21. $\{x \mid x < -3\}$ **23.** $\{x \mid x \geq -2\}$ **25.** $\{x \mid x < 0\}$
27. $\left\{y \,\middle|\, y \geq -\dfrac{8}{3}\right\}$ **29.** $\{y \mid y > 3\}$ **31.** $\{x \mid x > -15\}$ **33.** $\{x \mid x \geq -11\}$
35. $\left\{x \,\middle|\, x > \dfrac{1}{4}\right\}$ **37.** $\{y \mid y \geq -12\}$ **39.** $\{z \mid z < 0\}$ **41.** $\{x \mid x > -3\}$ **43.** $\left\{x \,\middle|\, x \geq -\dfrac{2}{3}\right\}$ **45.** $\{x \mid x \leq -2\}$
47. $\{x \mid x > -13\}$ **49.** $\{x \mid x \leq -8\}$ **51.** $\{x \mid x > 4\}$ **53.** $\left\{x \,\middle|\, x \leq \dfrac{5}{4}\right\}$ **55.** $\left\{x \,\middle|\, x > \dfrac{8}{3}\right\}$ **57.** $\{x \mid x \geq 0\}$ **59.** all numbers
greater than -10 **61.** 35 cm **63.** at least 193 **65.** 86 people **67.** at least 35 min **69.** 81 **71.** 1 **73.** $\dfrac{49}{64}$ **75.** about 13,500
77. 2011 and 2012 **79.** 2012 **81.** $>$ **83.** $\geq$ **85.** when multiplying or dividing by a negative number **87.** final exam score ≥ 78.5

Chapter 9 Vocabulary Check **1.** linear equation in one variable **2.** equivalent equations **3.** formula **4.** linear inequality in
one variable **5.** all real numbers **6.** no solution **7.** the same **8.** reversed

Chapter 9 Review **1.** 4 **2.** -3 **3.** 6 **4.** -6 **5.** 0 **6.** -9 **7.** -23 **8.** 28 **9.** b **10.** a **11.** b **12.** c **13.** -12 **14.** 4 **15.** 0
16. -7 **17.** 0.75 **18.** -3 **19.** -6 **20.** -1 **21.** -1 **22.** $\dfrac{3}{2}$ **23.** $-\dfrac{1}{5}$ **24.** 7 **25.** $3x + 3$ **26.** $2x + 6$ **27.** -4 **28.** -4 **29.** 2
30. -3 **31.** no solution **32.** no solution **33.** $\dfrac{3}{4}$ **34.** $-\dfrac{8}{9}$ **35.** 20 **36.** $-\dfrac{6}{23}$ **37.** $\dfrac{23}{7}$ **38.** $-\dfrac{2}{5}$ **39.** 102 **40.** 0.25 **41.** 6665.5 in.
42. short piece: 4 ft; long piece: 8 ft **43.** national battlefields: 11; national memorials: 29 **44.** $-39, -38, -37$ **45.** 3 **46.** -4
47. $w = 9$ **48.** $h = 4$ **49.** $m = \dfrac{y - b}{x}$ **50.** $s = \dfrac{r + 5}{vt}$ **51.** $x = \dfrac{2y - 7}{5}$ **52.** $y = \dfrac{2 + 3x}{6}$ **53.** $\pi = \dfrac{C}{D}$ **54.** $\pi = \dfrac{C}{2r}$ **55.** 15 m
56. 18 ft by 12 ft **57.** 1 hr 20 min **58.** 45°C **59.** 20% **60.** 70% **61.** 110 **62.** 1280 **63.** mark-up: $209; new price: $2109
64. 91,800 businesses **65.** 40% solution: 10 gal; 10% solution: 20 gal **66.** 32.0% increase **67.** 18% **68.** swerving into another lane
69. 966 customers **70.** no; answers may vary **71.** **72.** **73.** $\{x \mid x \leq 1\}$
74. $\{x \mid x > -5\}$ **75.** $\{x \mid x \leq 10\}$ **76.** $\{x \mid x < -4\}$ **77.** $\{x \mid x < -4\}$ **78.** $\{x \mid x \leq 4\}$ **79.** $\{y \mid y > 9\}$
80. $\{y \mid y \geq -15\}$ **81.** $\left\{x \,\middle|\, x < \dfrac{7}{4}\right\}$ **82.** $\left\{x \,\middle|\, x \leq \dfrac{19}{3}\right\}$ **83.** $2500 **84.** score must be less than 83 **85.** 4 **86.** -14 **87.** $-\dfrac{3}{2}$
88. 21 **89.** all real numbers **90.** no solution **91.** -13 **92.** shorter piece: 4 in.; longer piece: 19 in. **93.** $h = \dfrac{3V}{A}$ **94.** 22.1
95. 160 **96.** 20% **97.** $\{x \mid x > 9\}$ **98.** $\{x \mid x > -4\}$
99. $\{x \mid x \leq 0\}$

Chapter 9 Test **1.** -5 **2.** 8 **3.** $\dfrac{7}{10}$ **4.** 0 **5.** 27 **6.** $-\dfrac{19}{6}$ **7.** 3 **8.** $\dfrac{3}{11}$ **9.** 0.25 **10.** $\dfrac{25}{7}$ **11.** no solution **12.** 21 **13.** 7 gal
14. $x = 6$ **15.** $h = \dfrac{V}{\pi r^2}$ **16.** $y = \dfrac{3x - 10}{4}$ **17.** $\{x \mid x \leq -2\}$ **18.** $\{x \mid x < 4\}$
19. $\{x \mid x \leq -8\}$ **20.** $\{x \mid x \geq 11\}$ **21.** $\left\{x \,\middle|\, x > \dfrac{2}{5}\right\}$ **22.** 552 **23.** 40% **24.** 401, 802 **25.** California: 1107; Ohio: 720

Cumulative Review **1.** True (Sec. 8.1, Ex. 3) **2.** False **3.** True (Sec. 8.1, Ex. 4) **4.** True **5.** False (Sec. 8.1, Ex. 5) **6.** True
7. True (Sec. 8.1, Ex. 6) **8.** True **9.** $7\dfrac{17}{24}$ (Sec. 3.4, Ex. 1) **10. a.** $\dfrac{7}{10}$ **b.** $\dfrac{13}{24}$ **11.** $\dfrac{8}{3}$ (Sec. 8.2, Ex. 6) **12.** 33 **13.** -19 (Sec. 8.3, Ex. 6)
14. -10 **15.** 8 (Sec. 8.3, Ex. 7) **16.** 10 **17.** -0.3 (Sec. 8.3, Ex. 8) **18.** 0 **19. a.** -12 **b.** -3 (Sec. 8.4, Ex. 7) **20. a.** 5 **b.** $\dfrac{2}{3}$
c. a **d.** -3 **21. a.** 0 **b.** -24 **c.** 90 (Sec. 8.5, Ex. 7) **22. a.** -11.1 **b.** $-\dfrac{1}{5}$ **c.** $\dfrac{3}{4}$ **23. a.** -6 **b.** 7 **c.** -5 (Sec. 8.5, Ex. 10)
24. a. -0.36 **b.** $\dfrac{6}{17}$ **25.** $15 - 10z$ (Sec. 8.6, Ex. 8) **26.** $2y - 6x + 8$ **27.** $3x + 17$ (Sec. 8.6, Ex. 12) **28.** $2x + 8$ **29. a.** unlike
b. like **c.** like **d.** like **e.** like (Sec. 8.7, Ex. 2) **30. a.** -4 **b.** 9 **c.** $\dfrac{10}{63}$ **31.** $-2x - 1$ (Sec. 8.7, Ex. 15) **32.** $-15x - 2$
33. 17 (Sec. 9.1, Ex. 1) **34.** $-\dfrac{1}{6}$ **35.** -10 (Sec. 9.2, Ex. 7) **36.** 3 **37.** 0 (Sec. 9.3, Ex. 4) **38.** 72 **39.** Republicans: 233;

Democrats: 199 (Sec. 9.4, Ex. 4)   **40.** 5   **41.** 79.2 yr (Sec. 9.5, Ex. 1)   **42.** 6   **43.** 87.5% (Sec. 9.6, Ex. 1)   **44.** $\dfrac{C}{2\pi} = r$

45. $-\dfrac{9}{10}$ (Sec. 9.2, Ex. 10)   **46.** $\{x \mid x > 5\}$ **47.** ←————○————→ (Sec. 9.7, Ex. 2)   **48.** $\{x \mid x \le -10\}$
$\qquad\qquad\qquad\qquad\qquad\qquad\qquad\qquad\qquad\;\; -1$

49. $\{x \mid x \ge 1\}$ (Sec. 9.7, Ex. 9)   **50.** $\{x \mid x \le -3\}$

Chapter 10 Graphing Equations and Inequalities

Section 10.1

Vocabulary, Readiness & Video Check   1. x-axis   **3.** origin   **5.** x-coordinate; y-coordinate   **7.** solution   **9.** origin; left or right; up or down   **11.** Replace both values of the ordered pair in the linear equation and see if a true statement results.

Exercise Set 10.1   1. 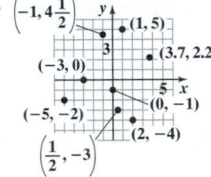 $(1, 5)$ and $(3.7, 2.2)$ are in quadrant I, $\left(-1, 4\dfrac{1}{2}\right)$ is in quadrant II, $(-5, -2)$ is in quadrant III, $(2, -4)$ and $\left(\dfrac{1}{2}, -3\right)$ are in quadrant IV, $(-3, 0)$ lies on the x-axis, $(0, -1)$ lies on the y-axis   **3.** $(0, 0)$ **5.** $(3, 2)$ **7.** $(-2, -2)$ **9.** $(2, -1)$ **11.** $(0, -3)$ **13.** $(1, 3)$ **15.** $(-3, -1)$ **17. a.** $(2007, 9.6), (2008, 9.6), (2009, 10.6), (2010, 10.6), (2011, 10.2), (2012, 10.8)$ **b.** In the year 2010, the domestic box office was $10.6 billion.

c. 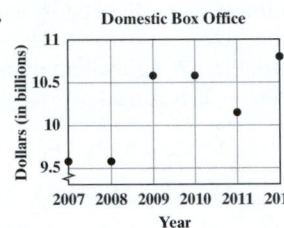   **d.** answers may vary

19. a. $(0.50, 10), (0.75, 12), (1.00, 15), (1.25, 16), (1.50, 18), (1.50, 19), (1.75, 19), (2.00, 20)$ **b.** When Minh studied 1.25 hours, her quiz score was 16.   **c.** 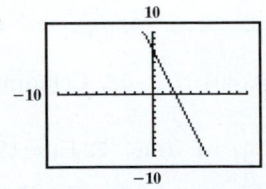   **d.** answers may vary   **21.** $(-4, -2), (4, 0)$ **23.** $(-8, -5), (16, 1)$ **25.** $0; 7; -\dfrac{2}{7}$ **27.** $2; 2; 5$ **29.** $0; -3; 2$ **31.** $2; 6; 3$ **33.** $-12; 5; -6$ **35.** $\dfrac{5}{7}; \dfrac{5}{2}; -1$

37. $0; -5; -2$ 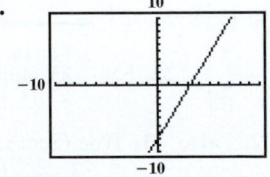   **39.** $2; 1; -6$ 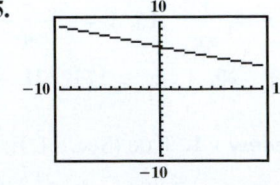   **41. a.** $13{,}000; 21{,}000; 29{,}000$ **b.** 45 desks   **43. a.** $6.44; 7.16; 7.88$ **b.** 2009   **c.** 2016

45. France   **47.** France, U.S., Spain, China   **49.** 30 million   **51.** $y = 5 - x$ **53.** $y = \dfrac{5 - 2x}{4}$ **55.** $y = -2x$ **57.** false   **59.** true   **61.** negative; negative   **63.** positive; negative   **65.** $0; 0$ **67.** y **69.** no; answers may vary   **71.** answers may vary   **73.** answers may vary   **75.** $(4, -7)$ **77.** 26 units   **79.** 21 million; 23 million; 25 million; 27 million

Section 10.2

Calculator Explorations   1. 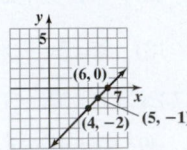   **3.** 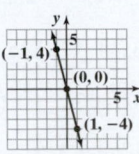   **5.**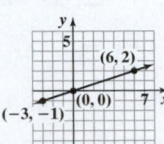

Vocabulary, Readiness & Video Check   1. It is always good practice to use a third point as a check to see that your points lie along a straight line.

Exercise Set 10.2   1. $6; -2; 5$ **3.** $-4; 0; 4$ **5.** $0; 2; -1$ **7.** $3; -1; -5$ **9.**

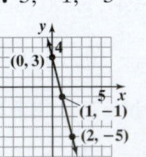 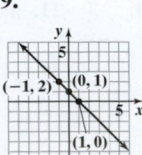

11.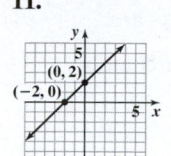

13.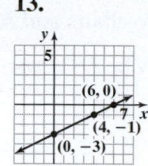

15.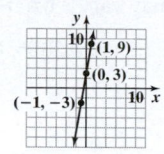

17.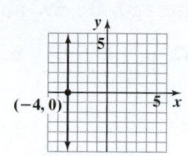

19.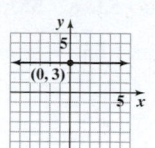

21.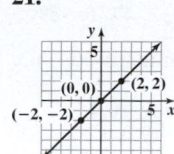

23.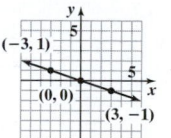

25.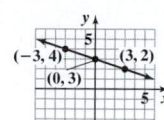

27.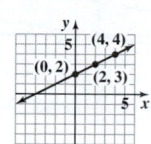

29.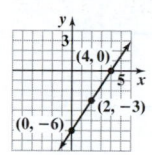

31.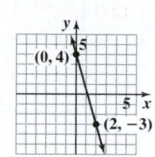

33. a. 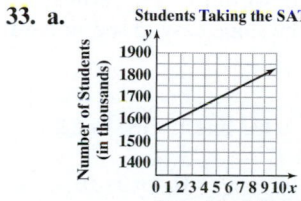   **b.** yes; answers may vary   **35. a.** 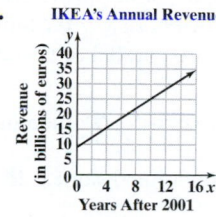 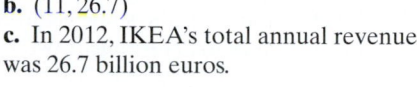   **b.** $(11, 26.7)$
c. In 2012, IKEA's total annual revenue was 26.7 billion euros.

37. $(4, -1)$ **39.** $3; -3$ **41.** $0; 0$ **43.** 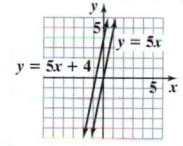   **45.** 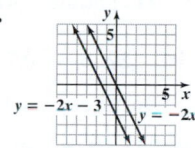   **47.** $0; 1; 1; 4; 4$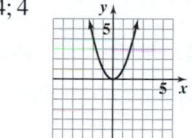

49. $x + y = 12; 9$ cm   **51.** yes; answers may vary

Section 10.3

Calculator Explorations 1. 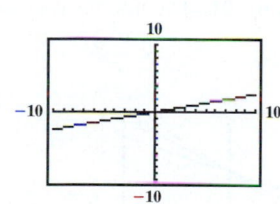   **3.** 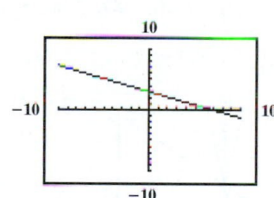   **5.**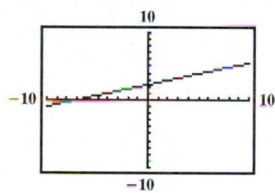

Vocabulary, Readiness & Video Check 1. linear   **3.** horizontal   **5.** y-intercept   **7.** $y; x$ **9.** false   **11.** true   **13.** because x-intercepts lie on the x-axis; because y-intercepts lie on the y-axis.   **15.** For a horizontal line, the coefficient of x will be 0; for a vertical line, the coefficient of y will be 0.

Exercise Set 10.3 1. $(-1, 0); (0, 1)$ **3.** $(-2, 0); (2, 0); (0, -2)$ **5.** $(-2, 0); (1, 0); (3, 0); (0, 3)$ **7.** $(-1, 0); (1, 0); (0, 1); (0, -2)$

9.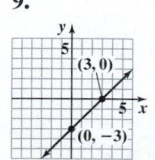

11.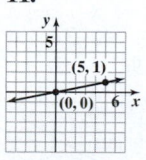

13.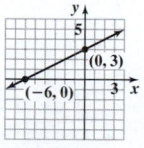

15.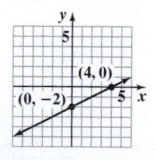

17.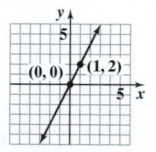

19.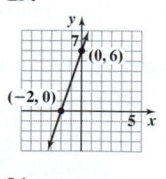

21.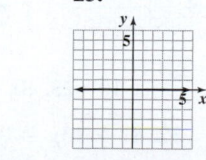

23.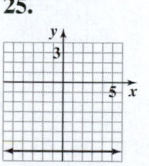

25.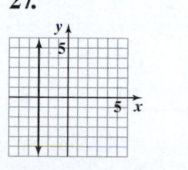

27.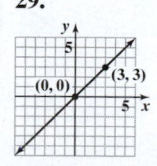

29.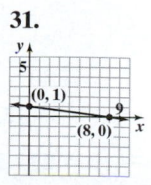

31.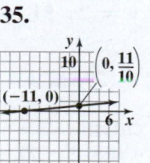

33.

35.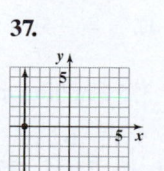

37.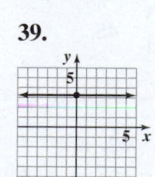

39.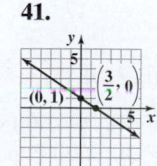

41.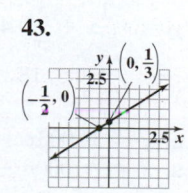

43.

45. $\dfrac{3}{2}$ **47.** 6   **49.** $-\dfrac{6}{5}$ **51.** c   **53.** a   **55.** infinite   **57.** 0   **59.** answers may vary   **61.** $(0, 200)$; no chairs and 200 desks are manufactured.   **63.** 300 chairs   **65.** $y = -4$ **67. a.** $(30.6, 0)$ **b.** 30.6 years after 2006, there may be no newspaper circulation.

Section 10.4

Calculator Explorations   **1.**

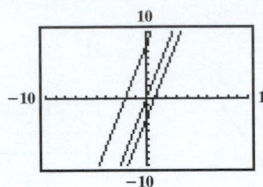

3.

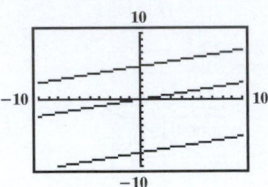

Vocabulary, Readiness & Video Check   **1.** slope   **3.** 0   **5.** positive   **7.** $y; x$ **9.** positive   **11.** 0   **13.** downward   **15.** vertical
17. Solve the equation for y; the slope is the coefficient of x.   **19.** Slope-intercept form; this form makes the slope easy to see, and we need to compare slopes to determine if two lines are parallel or perpendicular.

Exercise Set 10.4   **1.** $m = -1$ **3.** $m = -\dfrac{1}{4}$ **5.** $m = 0$ **7.** undefined slope   **9.** $m = -\dfrac{4}{3}$ **11.** $m = \dfrac{5}{2}$ **13.** line 1   **15.** line 2
17. $m = 5$ **19.** $m = -0.3$ **21.** $m = -2$ **23.** undefined slope   **25.** $m = \dfrac{2}{3}$ **27.** undefined slope   **29.** $m = \dfrac{1}{2}$ **31.** $m = 0$
33. $m = -\dfrac{3}{4}$ **35.** $m = 4$ **37.** neither   **39.** neither   **41.** parallel   **43.** perpendicular   **45. a.** 1   **b.** -1 **47. a.** $\dfrac{9}{11}$
b. $-\dfrac{11}{9}$ **49.** $\dfrac{3}{5}$ **51.** 12.5%   **53.** 40%   **55.** 37%; 35%   **57.** $m = 0.72$; Every year, the number of U.S. households with televisions increases by 0.72 million households.   **59.** $m = 0.15$; Every year, the median age of automobiles in the United States increases by 0.15 year.   **61.** $y = 2x - 14$ **63.** $y = -6x - 11$ **65.** d   **67.** b   **69.** e   **71.** $m = \dfrac{1}{2}$ **73.** answers may vary   **75.** 31.5 mi per gal
77. 2003 and 2004; 29.5 mi per gallon   **79.** from 2011 to 2012   **81.** $x = 20$ **83. a.** $(2007, 2209), (2012, 2378)$ **b.** 33.8   **c.** For the years 2007 through 2012, the number of heart transplants increased at a rate of 33.8 per year.   **85.** Opposite sides are parallel since their slopes are equal, so the figure is a parallelogram.   **87.** 2.0625   **89.** -1.6 **91.** The line becomes steeper.

Section 10.5

Calculator Explorations   **1.**

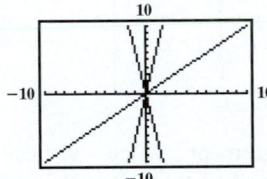

3.

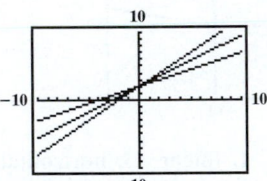

Vocabulary, Readiness & Video Check   **1.** slope-intercept; $m; b$ **3.** point-slope   **5.** slope-intercept   **7.** horizontal
9. y-intercept, fraction   **11.** Write the equation with x- and y-terms on one side of the equal sign and a constant on the other side.
13. We need to know what our variables stand for in order to solve part (b) of the example, and that depends on how we set up our ordered pairs in part (a).

Exercise Set 10.5   **1.**

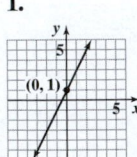

3.

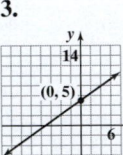

5.

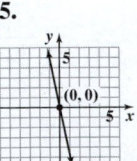

7.

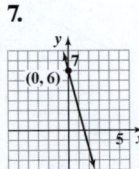

9.

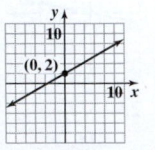

11.

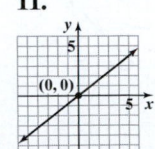

13. $y = 5x + 3$ **15.** $y = -4x - \dfrac{1}{6}$ **17.** $y = \dfrac{2}{3}x$ **19.** $y = -8$ **21.** $y = -\dfrac{1}{5}x + \dfrac{1}{9}$ **23.** $-6x + y = -10$ **25.** $8x + y = -13$
27. $3x - 2y = 27$ **29.** $x + 2y = -3$ **31.** $2x - y = 4$ **33.** $8x - y = -11$ **35.** $4x - 3y = -1$ **37.** $8x + 13y = 0$
39. $y = -\dfrac{1}{2}x + \dfrac{5}{3}$ **41.** $y = -x + 17$ **43.** $x = -\dfrac{3}{4}$ **45.** $y = x + 16$ **47.** $y = -5x + 7$ **49.** $y = 2$ **51.** $y = \dfrac{3}{2}x$ **53.** $y = -3$
55. $y = -\dfrac{4}{7}x - \dfrac{18}{7}$ **57. a.** $(0, 370), (5, 312)$ **b.** $y = -11.6x + 370$ **c.** 335.2 million magazines   **59. a.** $s = 32t$ **b.** 128 ft/sec
61. a. $y = 30,000x + 314,000$ **b.** 494,000 vehicles   **63. a.** $y = -45x + 5545$ **b.** 5185 indoor cinema sites
65. a. $S = -1000p + 13,000$ **b.** 9500 Fun Noodles   **67.** -1 **69.** 5   **71.** b   **73.** d   **75.** $3x - y = -5$ **77. a.** $3x - y = -5$
b. $x + 3y = 5$

Integrated Review   **1.** $m = 2$ **2.** $m = 0$ **3.** $m = -\dfrac{2}{3}$ **4.** slope is undefined

5. 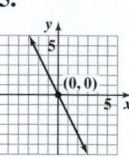   **6.** 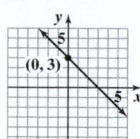   **7.** 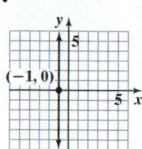   **8.**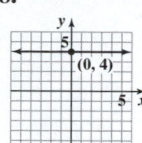

9. 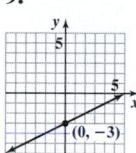   **10.** 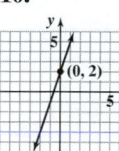   **11.** 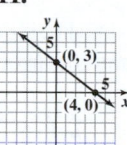   **12.**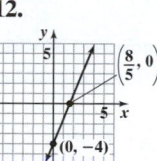

13. $m = 3$ **14.** $m = -6$ **15.** $m = -\dfrac{7}{2}$ **16.** $m = 2$ **17.** undefined slope   **18.** $m = 0$ **19.** $y = 2x - \dfrac{1}{3}$ **20.** $y = -4x - 1$
21. $-x + y = -2$ **22.** neither   **23.** perpendicular   **24. a.** $(2008, 3600); (2012, 4416)$ **b.** 204 **c.** For the years 2008 through 2012, the amount of yogurt produced increased at a rate of 204 million pounds per year.

Section 10.6

Vocabulary, Readiness & Video Check   **1.** relation   **3.** range   **5.** vertical   **7.** $(3, 7)$ **9.** $y; x$ **11.** Yes, this is a function. The definition restricts x-values to be assigned to exactly one y-value, but it makes no such restriction on the y-values.
13. $f(-2) = 6$ corresponds to $(-2, 6)$ and $f(3) = 11$ corresponds to $(3, 11)$.

Exercise Set 10.6   **1.** domain: $\{-7, 0, 2, 10\}$; range: $\{-7, 0, 4, 10\}$ **3.** domain: $\{0, 1, 5\}$; range: $\{-2\}$ **5.** yes   **7.** no   **9.** no
11. yes   **13.** yes   **15.** no   **17.** a   **19.** yes   **21.** yes   **23.** no   **25.** no   **27.** 9:30 p.m.   **29.** January 1 and December 1   **31.** yes; it
passes the vertical line test   **33.** \$4.25 per hour   **35.** 2009   **37.** yes; answers may vary   **39.** \$1.72   **41.** more than 1 ounce and less
than or equal to 2 ounces   **43.** yes; answers may vary   **45.** $-9, -5, 1$ **47.** $6, 2, 11$ **49.** $-6, 0, 9$ **51.** $2, 0, 3$ **53.** $5, 0, -20$ **55.** $5, 3, 35$
57. $(3, 6)$ **59.** $\left(0, -\dfrac{1}{2}\right)$ **61.** $(-2, 9)$ **63.** all real numbers   **65.** all real number except -5 **67.** domain: all real numbers; range: $y \geq -4$

69. domain: all real numbers; range; all real numbers   **71.** domain: all real numbers; range: $\{2\}$ **73.** -1 **75.** -1 **77.** $-1, 5$

79. $x < 1$ **81.** $x \geq -3$ **83.** $\dfrac{19}{2x}$ m   **85.** $f(-5) = 12$ **87.** answers may vary   **89.** $f(x) = x + 7$ **91. a.** 190.4 mg   **b.** 380.8 mg

Section 10.7

Vocabulary, Readiness & Video Check   **1.** linear inequality in two variables   **3.** false   **5.** true   **7.** An ordered pair is a solution of an inequality if replacing the variables with the coordinates of the ordered pair results in a true statement.

Exercise Set 10.7   **1.** no; no   **3.** yes; no   **5.** no; yes   **7.** 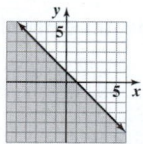   **9.** 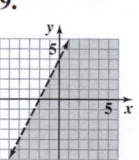   **11.** 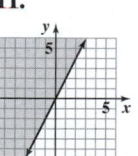   **13.**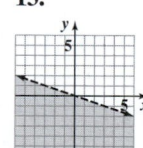

15. 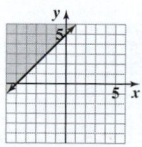   **17.** 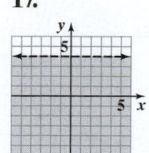   **19.** 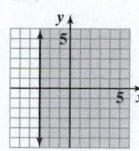   **21.** 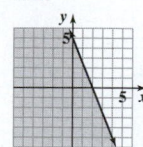   **23.**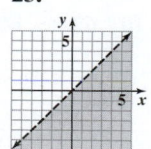

25. 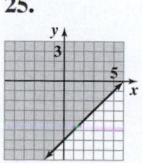   **27.** 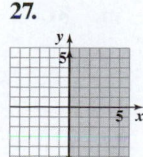   **29.** 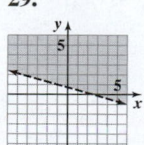   **31.** 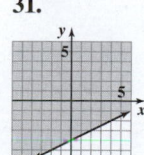   **33.**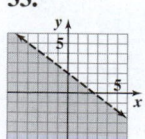

35. $(-2, 1)$ **37.** $(-3, -1)$ **39.** a   **41.** b   **43.** answers may vary   **45.** yes   **47.** yes

49. a. $30x + 0.15y \le 500$ **b.**   **c.** answers may vary

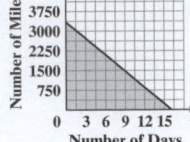

Section 10.8

Vocabulary, Readiness & Video Check   **1.** inverse   **3.** direct   **5.** inverse   **7.** inverse   **9.** direct   **11.** linear; slope   **13.** No. The direct relationship is the power of x times a constant, and the inverse relationship is the reciprocal of the power of x times a constant.

Exercise Set 10.8   **1.** $y = \frac{1}{2}x$ **3.** $y = 6x$ **5.** $y = 3x$ **7.** $y = \frac{2}{3}x$ **9.** $y = \frac{7}{x}$ **11.** $y = \frac{0.5}{x}$ **13.** $y = kx$ **15.** $h = \frac{k}{t}$ **17.** $z = kx^2$

19. $y = \frac{k}{z^3}$ **21.** $x = \frac{k}{\sqrt{y}}$ **23.** $y = 40$ **25.** $y = 3$ **27.** $z = 54$ **29.** $a = \frac{4}{9}$ **31.** \$92.50   **33.** \$6   **35.** $5\frac{1}{3}$ in.   **37.** 179.1 lb

39. 1600 feet   **41.** $2y = 16$ **43.** $-4x = 0.5$ **45.** multiplied by 3   **47.** It is doubled.

Chapter 10 Vocabulary Check   **1.** solution   **2.** y-axis   **3.** linear   **4.** x-intercept   **5.** standard   **6.** y-intercept   **7.** function
8. slope-intercept   **9.** domain   **10.** range   **11.** relation   **12.** point-slope   **13.** y **14.** x-axis   **15.** x **16.** slope   **17.** direct
18. inverse

Chapter 10 Review   **1–6.**   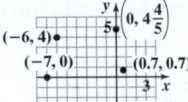   **7.** $(7, 44)$ **8.** $\left(-\frac{13}{3}, -8\right)$ **9.** $-3; 1; 9$ **10.** $5; 5; 5$ **11.** $0; 10; -10$
12. a. $2005; 2500; 7000$ **b.** 886 compact disc holders

13. 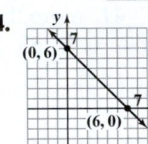   **14.** 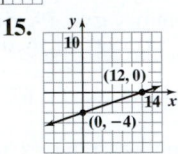   **15.** 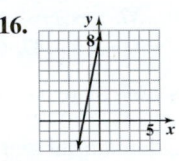   **16.** 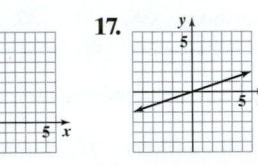   **17.** 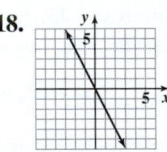   **18.**

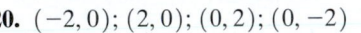

19. $(4, 0); (0, -2)$ **20.** $(-2, 0); (2, 0); (0, 2); (0, -2)$ **21.** 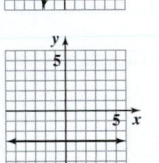   **22.** 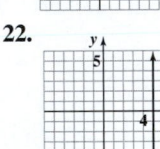   **23.** $(12, 0), (0, -4)$

24. $(-2, 0), (0, 8)$ **25.** $m = -\frac{3}{4}$
26. $m = \frac{1}{5}$ **27.** d   **28.** b   **29.** c

30. a   **31.** $m = \frac{3}{4}$ **32.** $m = \frac{5}{3}$ **33.** $m = 4$ **34.** $m = -1$ **35.** $m = 3$ **36.** $m = \frac{1}{2}$ **37.** $m = 0$ **38.** undefined slope

39. perpendicular   **40.** parallel   **41.** neither   **42.** perpendicular   **43.** $m = 110$; The total number of U.S. magazines in print increases by 110 magazines per year.   **44.** $m = 56$; Every 1 year, 56 more people get a lung transplant.   **45.** $m = \frac{1}{6}; \left(0, \frac{1}{6}\right)$

46. $m = -3; (0, 7)$ **47.** $y = -5x + \frac{1}{2}$ **48.** $y = \frac{2}{3}x + 6$ **49.** d   **50.** c   **51.** a   **52.** b   **53.** $-4x + y = -8$ **54.** $3x + y = -5$

55. $-3x + 5y = 17$ **56.** $x + 3y = 6$ **57.** $y = -14x + 21$ **58.** $y = -\frac{1}{2}x + 4$ **59.** no   **60.** yes   **61.** yes   **62.** yes   **63.** no   **64.** yes

65. 6   **66.** 10   **67.** 5   **68.** 7   **69.** 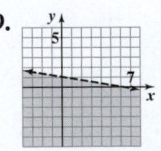   **70.** 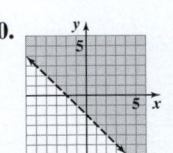   **71.** 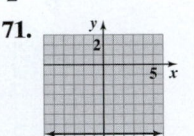   **72.** 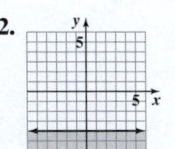   **73.**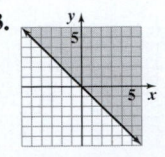

74. 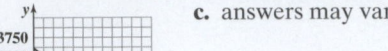   **75.** $y = 110$ **76.** $y = \frac{1}{2}$ **77.** $y = \frac{100}{27}$ **78.** $y = 700$ **79.** \$3960   **80.** $4\frac{4}{5}$ in.   **81.** $7; -1; -3$ **82.** $0; -3; -2$
83. $(3, 0); (0, -2)$ **84.** $(-2, 0); (0, 10)$

85. 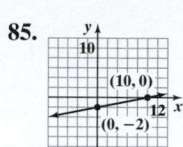 **86.** 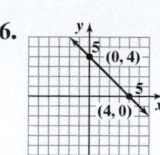 **87.** 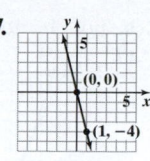 **88.** 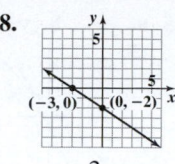 **89.** 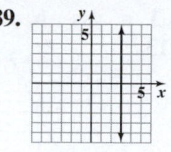 **90.**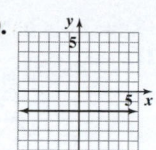

91. $m = -1$ **92.** $m = \dfrac{11}{7}$ **93.** $m = 2$ **94.** $m = -\dfrac{1}{3}$ **95.** $m = \dfrac{2}{3}; (0, -5)$ **96.** $m = -6; (0, 2)$ **97.** $5x + y = 8$

98. $3x - y = -6$ **99.** $4x + y = -3$ **100.** $5x + y = 16$

Chapter 10 Test **1.** $(1, 1)$ **2.** $(-4, 17)$ **3.** $m = \dfrac{2}{5}$ **4.** $m = 0$ **5.** $m = -1$ **6.** $m = -7$ **7.** $m = 3$ **8.** undefined slope

9. 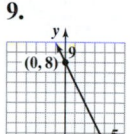 **10.** 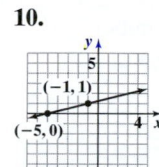 **11.** 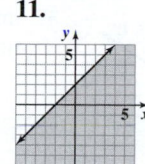 **12.** 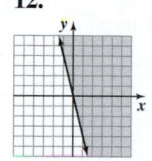 **13.** 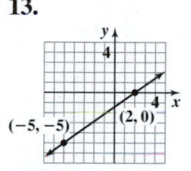 **14.**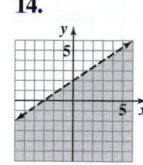

15. 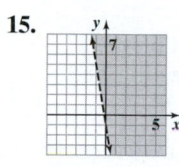 **16.** 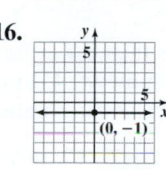 **17.** neither **18.** $x + 4y = 10$ **19.** $7x + 6y = 0$ **20.** $8x + y = 11$
21. $x - 8y = -96$ **22.** yes **23.** no **24.** yes **25.** yes
26. a. -8 **b.** -3.6 **c.** -4 **27. a.** 0 **b.** 0 **c.** 60

28. $x + 2y = 21; x = 5$ m **29. a.** $(2008, 19.6); (2009, 22.2); (2010, 23.9); (2011, 25.3); (2012, 26.7)$
b. 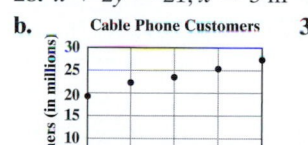 **30.** $m = -15.5;$ For every 1 year, 15.5 million fewer movie tickets are sold. **31.** $y = 28$ **32.** $y = \dfrac{8}{9}$

Cumulative Review **1.** $78{,}875$ (Sec. 1.6, Ex. 5) **2.** $\dfrac{25}{44}$ **3.** $\dfrac{4}{5}$ (Sec. 2.5, Ex. 7) **4.** 236 **5.** $\dfrac{17}{21}$ (Sec. 3.3, Ex. 3) **6.** $1\dfrac{11}{14}$

7. 48.26 (Sec. 4.1, Ex. 6) **8.** 0.08 **9.** 6.095 (Sec. 4.1, Ex. 7) **10.** $56{,}321$ **11.** 3.432 (Sec. 4.3, Ex. 5) **12.** 13.5 **13.** 28.4405 (Sec. 4.5, Ex. 13)

14. 20 **15.** 46% (Sec. 5.2, Ex. 2) **16.** 80% **17.** 27 (Sec. 8.2, Ex. 2) **18.** $\dfrac{25}{7}$ **19.** 51 (Sec. 8.2, Ex. 5) **20.** 23

21. $20{,}602$ feet (Sec. 8.4, Ex. 10) **22.** $0.8x - 36$ **23.** $2x + 6$ (Sec. 8.7, Ex. 16) **24.** $-15\left(x + \dfrac{2}{3}\right) = -15x - 10$ **25.** $(x - 4) \div 7$ or

$\dfrac{x - 4}{7}$ (Sec. 8.7, Ex. 17) **26.** $\dfrac{-9}{2x}$ **27.** $5 + (x + 1) = 6 + x$ (Sec. 8.7, Ex. 18) **28.** $-86 - x$ **29.** 6 (Sec. 9.2, Ex. 1) **30.** -24

31. 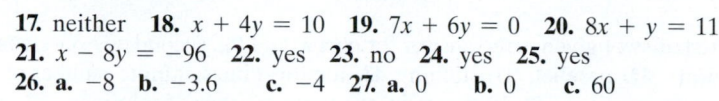 $x < -2$ (Sec. 9.7, Ex. 6) **32.** $\left\{ x \,\middle|\, x \le \dfrac{8}{3} \right\}$ **33. a.** $(0, 12)$ **b.** $(2, 6)$ **c.** $(-1, 15)$ (Sec. 10.1, Ex. 3)

34. $0; 5; -2$ **35.** 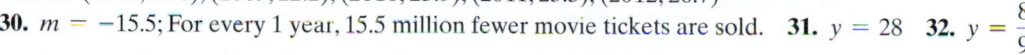 (Sec. 10.2, Ex. 1) **36.** $\dfrac{1}{5}$ **37.** $\dfrac{2}{3}$ (Sec. 10.4, Ex. 3) **38.** undefined slope **39.** $y = -2x + 3;$
$2x + y = 3$ (Sec. 10.5, Ex. 4) **40.** $m = \dfrac{2}{5},$ y-intercept: $(0, -2)$ **41. a.** $1; (2, 1)$
b. $1; (-2, 1)$ **c.** $-3; (0, -3)$ (Sec. 10.6, Ex. 7) **42.** $3x - 2y = 0$

Chapter 11 Systems of Equations

Section 11.1

Calculator Explorations **1.** $(0.37, 0.23)$ **3.** $(0.03, -1.89)$

Vocabulary, Readiness & Video Check **1.** dependent **3.** consistent **5.** inconsistent **7.** 1 solution, $(-1, 3)$ **9.** infinite number of solutions **11.** The ordered pair must satisfy all equations of the system in order to be a solution of the system, so we must check that the ordered pair is a solution of both equations. **13.** Writing the equations of a system in slope-intercept form lets us see and compare their slopes and y-intercepts. Different slopes mean one solution; same slopes with different y-intercepts mean no solution; same slopes with same y-intercepts mean an infinite number of solutions.

Exercise Set 11.1   **1. a.** no   **b.** yes   **3. a.** yes   **b.** no   **5. a.** yes   **b.** yes   **7. a.** no   **b.** no

9. 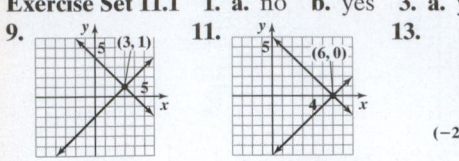   **11.**   **13.** 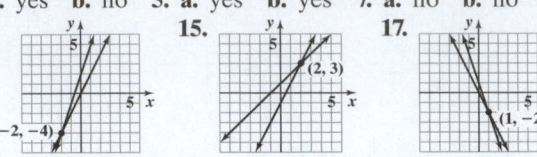   **15.**   **17.**

19. 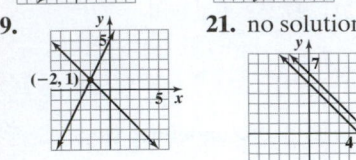   **21.** no solution   **23.** 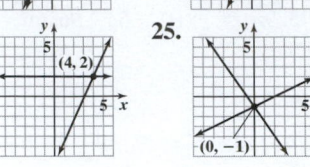   **25.** 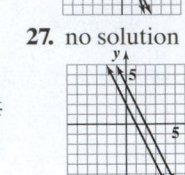   **27.** no solution

29. infinite number of solutions   **31.** 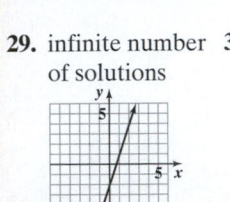   **33.** 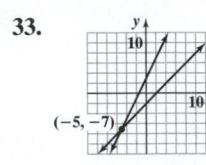   **35.** 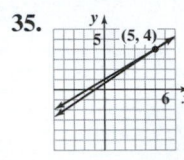   **37.** infinite number of solutions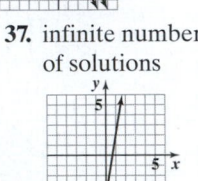

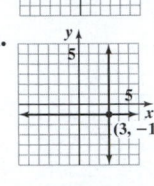

For Exercises 39–51, the first answer given is the answer for part **a**, and the second answer given is the answer for part **b**.
39. intersecting; one solution   **41.** parallel; no solution   **43.** identical lines; infinite number of solutions   **45.** intersecting; one solution   **47.** intersecting; one solution   **49.** identical lines; infinite number of solutions   **51.** parallel; no solution   **53.** 2

55. $-\dfrac{2}{5}$ **57.** 2   **59.** answers may vary   **61.** answers may vary   **63.** 2010–2011   **65.** 2010, 2011, 2012   **67.** answers may vary

69. answers may vary   **71. a.** $(4, 9)$ **b.** 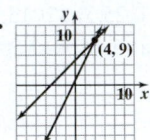   **c.** yes

Section 11.2

Vocabulary, Readiness & Video Check   **1.** $(1, 4)$ **3.** infinite number of solutions   **5.** $(0, 0)$ **7.** We solved one equation for a variable. Next, be sure to substitute this expression for the variable into the *other* equation.

Exercise Set 11.2   **1.** $(2, 1)$ **3.** $(-3, 9)$ **5.** $(2, 7)$ **7.** $\left(-\dfrac{1}{5}, \dfrac{43}{5}\right)$ **9.** $(2, -1)$ **11.** $(-2, 4)$ **13.** $(4, 2)$ **15.** $(-2, -1)$

17. no solution   **19.** $(3, -1)$ **21.** $(3, 5)$ **23.** $\left(\dfrac{2}{3}, -\dfrac{1}{3}\right)$ **25.** $(-1, -4)$ **27.** $(-6, 2)$ **29.** $(2, 1)$ **31.** no solution

33. infinite number of solutions   **35.** $\left(\dfrac{1}{2}, 2\right)$ **37.** $-6x - 4y = -12$ **39.** $-12x + 3y = 9$ **41.** $5n$ **43.** $-15b$ **45.** $(1, -3)$

47. answers may vary   **49.** no   **51.** c; answers may vary   **53.** $(-2.6, 1.3)$ **55.** $(3.28, 2.1)$ **57. a.** $(9.6, 6)$
b. In about 9.6 years after 2006, U.S. consumer spending on DVD- and Blu-ray-format home entertainment will be approximately \$6 billion for each.   **c.** answers may vary;

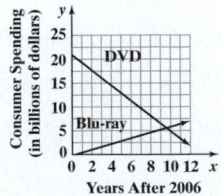

Section 11.3

Vocabulary, Readiness & Video Check   **1.** false   **3.** true   **5.** The multiplication property of equality; be sure to multiply *both* sides of the equation by the nonzero number chosen.

Exercise Set 11.3   **1.** $(1, 2)$ **3.** $(2, -3)$ **5.** $(-2, -5)$ **7.** $(5, -2)$ **9.** $(-7, 5)$ **11.** $(6, 0)$ **13.** no solution   **15.** infinite number of solutions   **17.** $\left(2, -\dfrac{1}{2}\right)$ **19.** $(-2, 0)$ **21.** $(1, -1)$ **23.** infinite number of solutions   **25.** $\left(\dfrac{12}{11}, -\dfrac{4}{11}\right)$ **27.** $\left(\dfrac{3}{2}, 3\right)$

29. infinite number of solutions   **31.** $(1, 6)$ **33.** $\left(-\dfrac{1}{2}, -2\right)$ **35.** infinite number of solutions   **37.** $\left(-\dfrac{2}{3}, \dfrac{2}{5}\right)$ **39.** $(2, 4)$

41. $(-0.5, 2.5)$ **43.** $2x + 6 = x - 3$ **45.** $20 - 3x = 2$ **47.** $4(x + 6) = 2x$ **49.** 2; $6x - 2y = -24$ **51.** b; answers may vary

53. answers may vary **55. a.** $b = 15$ **b.** any real number except 15 **57.** $(-8.9, 10.6)$ **59. a.** $(2, 309)$ or $(2, 306)$ **b.** In about 2012 $(2010 + 2)$, the number of mail carrier jobs was approximately equal to the number of market research analyst jobs. **c.** 309 thousand or 306 thousand

Integrated Review 1. $(2, 5)$ **2.** $(4, 2)$ **3.** $(5, -2)$ **4.** $(6, -14)$ **5.** $(-3, 2)$ **6.** $(-4, 3)$ **7.** $(0, 3)$ **8.** $(-2, 4)$ **9.** $(5, 7)$
10. $(-3, -23)$ **11.** $\left(\frac{1}{3}, 1\right)$ **12.** $\left(-\frac{1}{4}, 2\right)$ **13.** no solution **14.** infinite number of solutions **15.** $(0.5, 3.5)$ **16.** $(-0.75, 1.25)$
17. infinite number of solutions **18.** no solution **19.** $(7, -3)$ **20.** $(-1, -3)$ **21.** answers may vary **22.** answers may vary

Section 11.4

Vocabulary, Readiness & Video Check 1. Up to now we've been working with one variable/unknown and one equation. Because systems involve two equations with two unknowns, for these applications we need to choose two variables to represent two unknowns and translate the problem into two equations.

Exercise Set 11.4 1. c **3.** b **5.** a **7.** $\begin{cases} x + y = 15 \\ x - y = 7 \end{cases}$ **9.** $\begin{cases} x + y = 6500 \\ x = y + 800 \end{cases}$ **11.** 33 and 50 **13.** 14 and -3 **15.** Cabrera: 139;
Hamilton: 128 **17.** child's ticket: \$18; adult's ticket: \$29 **19.** quarters: 53; nickels: 27 **21.** McDonald's: \$97.80; Ford: \$17.25
23. daily fee: \$32; mileage charge: \$0.25 per mi **25.** distance downstream = distance upstream = 18 mi; time downstream: 2 hr;
time upstream: $4\frac{1}{2}$ hr; still water: 6.5 mph; current: 2.5 mph **27.** still air: 455 mph; wind: 65 mph **29.** $4\frac{1}{2}$ hr **31.** 12% solution:
$7\frac{1}{2}$ oz; 4% solution: $4\frac{1}{2}$ oz **33.** \$4.95 beans: 113 lb; \$2.65 beans: 87 lb **35.** $60°, 30°$ **37.** $20°, 70°$ **39.** number sold at \$9.50: 23;
number sold at \$7.50: 67 **41.** $2\frac{1}{4}$ mph and $2\frac{3}{4}$ mph **43.** 30%: 50 gal; 60%: 100 gal **45.** length: 42 in.; width: 30 in. **47.** 16 **49.** $36x^2$
51. $100y^6$ **53.** a **55.** width: 9 ft; length: 15 ft

Chapter 11 Vocabulary Check 1. dependent **2.** system of linear equations **3.** consistent **4.** solution **5.** addition; substitution
6. inconsistent **7.** independent

Chapter 11 Review 1. a. no **b.** yes **2. a.** no **b.** yes **3. a.** no **b.** no **4. a.** yes **b.** no
5. 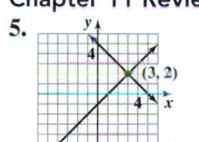 **6.** 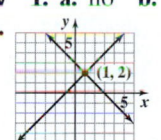 **7.** 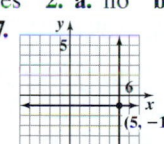 **8.** 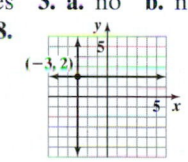 **9.**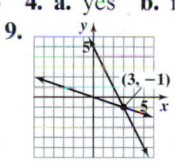

10. 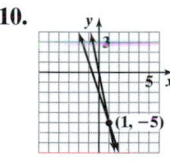 **11.** no solution **12.** infinite number of solutions **13.** $(-1, 4)$ **14.** $(2, -1)$ **15.** $(3, -2)$
16. $(2, 5)$ **17.** infinite number of solutions
18. infinite number of solutions **19.** no solution
20. no solution **21.** $(-6, 2)$ **22.** $(4, -1)$ **23.** $(3, 7)$
24. $(-2, 4)$ **25.** infinite number of solutions
26. infinite number of solutions **27.** $(8, -6)$

28. $\left(-\frac{3}{2}, \frac{15}{2}\right)$ **29.** -6 and 22 **30.** orchestra: 255 seats; balcony: 105 seats **31.** distance upriver = distance downriver = 340 mi;
time upriver: 19 hr; time downriver: 14 hr; current of river: 3.2 mph; speed in still water: 21.1 mph
32. 6% solution: $12\frac{1}{2}$ cc; 14% solution: $37\frac{1}{2}$ cc **33.** egg: \$0.40; strip of bacon: \$0.65 **34.** jogging: 0.86 hr; walking: 2.14 hr
35. 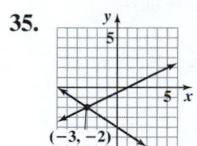 **36.** infinite number of solutions **37.** $(3, 2)$ **38.** $(7, 1)$ **39.** $\left(\frac{3}{2}, -3\right)$ **40.** no solution
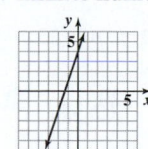
41. infinite number of solutions **42.** $(8, 11)$ **43.** $(-5, 2)$ **44.** $(16, -4)$
45. infinite number of solutions **46.** no solution **47.** 4 and 8
48. -5 and 13 **49.** 24 nickels and 41 dimes **50.** 13¢ stamps: 17; 22¢ stamps: 9

Chapter 11 Test 1. false **2.** false **3.** true **4.** false **5.** no **6.** yes
7. 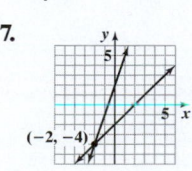 **8.** no solution **9.** $(-4, 1)$ **10.** $\left(\frac{1}{2}, -2\right)$ **11.** $(20, 8)$ **12.** no solution **13.** $(4, -5)$ **14.** $(7, 2)$
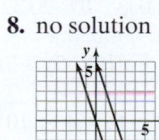
15. $(5, -2)$ **16.** infinite number of solutions **17.** $(-5, 3)$ **18.** $\left(\frac{47}{5}, \frac{48}{5}\right)$ **19.** 78, 46
20. 120 cc **21.** Texas: 245 thousand; Missouri: 106 thousand **22.** 3 mph; 6 mph
23. 2008–2009, 2011–2012 **24.** 2008, 2012

Cumulative Review **1.** $\frac{6}{5}$ (Sec. 2.4, Ex. 5) **2.** $\frac{123}{8}$ or $15\frac{3}{8}$ **3.** $\frac{2}{5}$ (Sec. 2.4, Ex. 6) **4.** $\frac{5}{54}$ **5.** 25.454 (Sec. 4.3, Ex. 1) **6.** 681.24

7. 0.0849 (Sec. 4.4, Ex. 2) **8.** 0.375 **9.** 0.125 (Sec. 4.5, Ex. 3) **10.** $\frac{79}{10}$ **11.** 3.7 (Sec. 4.5, Ex. 12) **12.** 3 **13.** $\frac{4}{9}, \frac{9}{20}$, 0.456 (Sec. 4.6, Ex. 9)

14. 140 m/sec **15. a.** -6 **b.** 6.3 (Sec. 8.4, Ex. 6) **16. a.** 25 **b.** 32 **17.** $\frac{1}{22}$ (Sec. 8.5, Ex. 9a) **18.** -22 **19.** $\frac{16}{3}$ (Sec. 8.5, Ex. 9b)

20. $-\frac{3}{16}$ **21.** $-\frac{1}{10}$ (Sec. 8.5, Ex. 9c) **22.** 10 **23.** $-\frac{13}{9}$ (Sec. 8.5, Ex. 9d) **24.** $\frac{9}{13}$ **25.** $\frac{1}{1.7}$ (Sec. 8.5, Ex. 9e) **26.** -1.7 **27. a.** 5

b. $8 - x$ (Sec. 9.1, Ex. 8) **28.** -5 **29.** no solution (Sec. 9.3, Ex. 6) **30.** no solution **31.** 12 (Sec. 9.3, Ex. 3) **32.** 40

33. $\left\{ x \,\middle|\, x > \frac{13}{7} \right\}$ ⟵———○———→ (Sec. 9.7, Ex. 8) **34.** $b = P - a - c$ **35.** $m = 0$ (Sec. 10.4, Ex. 5) **36.** undefined slope
　　　　　　　　　　　　　　$\frac{13}{7}$

37. $-x + 5y = 23$ (Sec. 10.5, Ex. 5) **38.** $y = -5x - 7$ **39.** domain: $\{-1, 0, 3\}$; range: $\{-2, 0, 2, 3\}$ (Sec. 10.6, Ex. 1)

40. $-6; 14$ **41.** It is a solution (Sec. 11.1, Ex. 1) **42. a.** yes **b.** no **c.** no **43.** $\left(6, \frac{1}{2} \right)$ (Sec. 11.2, Ex. 3) **44.** $(-2, -4)$

45. $\left(-\frac{15}{7}, -\frac{5}{7} \right)$ (Sec. 11.3, Ex. 6) **46.** $\left(-\frac{44}{3}, -\frac{7}{3} \right)$ **47.** 29 and 8 (Sec. 11.4, Ex. 1) **48. a.** no **b.** yes **c.** no

Chapter 12 Exponents and Polynomials

Section 12.1

Vocabulary, Readiness & Video Check **1.** exponent **3.** add **5.** 1 **7.** base: 3; exponent: 2 **9.** base: 4; exponent: 2 **11.** base: 5; exponent: 1; base: x; exponent: 2 **13.** Example 4 can be written as $-4^2 = -1 \cdot 4^2$, which is similar to Example 7, $4 \cdot 3^2$, and shows why the negative sign should not be considered part of the base when there are no parentheses. **15.** Be careful not to confuse the power rule with the product rule. The power rule involves a power raised to a power (exponents are multiplied), and the product rule involves a product (exponents are added). **17.** the quotient rule

Exercise Set 12.1 **1.** 49 **3.** -5 **5.** -16 **7.** 16 **9.** $\frac{1}{27}$ **11.** 112 **13.** 4 **15.** 135 **17.** 150 **19.** $\frac{32}{5}$ **21.** x^7 **23.** $(-3)^{12}$

25. $15y^5$ **27.** $x^{19}y^6$ **29.** $-72m^3n^8$ **31.** $-24z^{20}$ **33.** $20x^5$ sq ft **35.** x^{36} **37.** p^8q^8 **39.** $8a^{15}$ **41.** $x^{10}y^{15}$ **43.** $49a^4b^{10}c^2$

45. $\frac{r^9}{s^9}$ **47.** $\frac{m^9p^9}{n^9}$ **49.** $\frac{4x^2z^2}{y^{10}}$ **51.** $64z^{10}$ sq dm **53.** $27y^{12}$ cu ft **55.** x^2 **57.** -64 **59.** p^6q^5 **61.** $\frac{y^3}{2}$ **63.** 1 **65.** 1 **67.** -7

69. 2 **71.** -81 **73.** $\frac{1}{64}$ **75.** b^6 **77.** a^9 **79.** $-16x^7$ **81.** $a^{11}b^{20}$ **83.** $26m^9n^7$ **85.** z^{40} **87.** $64a^3b^3$ **89.** $36x^2y^2z^6$ **91.** z^8

93. $3x^4$ **95.** 1 **97.** $81x^2y^2$ **99.** 40 **101.** $\frac{y^{15}}{8x^{12}}$ **103.** $2x^2y$ **105.** -2 **107.** 5 **109.** -7 **111.** c **113.** e **115.** answers may vary

117. answers may vary **119.** 343 cu m **121.** volume **123.** answers may vary **125.** answers may vary **127.** x^{9a} **129.** a^{5b} **131.** x^{5a}

Section 12.2

Calculator Explorations **1.** 5.31 EE 3 **3.** 6.6 EE -9 **5.** 1.5×10^{13} **7.** 8.15×10^{19}

Vocabulary, Readiness & Video Check **1.** $\frac{1}{x^3}$ **3.** scientific notation **5.** $\frac{5}{x^2}$ **7.** y^6 **9.** $4y^3$ **11.** A negative exponent has nothing to do with the sign of the simplified result. **13.** When you move the decimal point to the left, the sign of the exponent will be positive; when you move the decimal point to the right, the sign of the exponent will be negative. **15.** the quotient rule

Exercise Set 12.2 **1.** $\frac{1}{64}$ **3.** $\frac{7}{x^3}$ **5.** -64 **7.** $\frac{5}{6}$ **9.** p^3 **11.** $\frac{q^4}{p^5}$ **13.** $\frac{1}{x^3}$ **15.** z^3 **17.** $\frac{4}{9}$ **19.** $\frac{1}{9}$ **21.** $-p^4$ **23.** -2 **25.** x^4 **27.** p^4

29. m^{11} **31.** r^6 **33.** $\frac{1}{x^{15}y^9}$ **35.** $\frac{1}{x^4}$ **37.** $\frac{1}{a^2}$ **39.** $4k^3$ **41.** $3m$ **43.** $-\frac{4a^5}{b}$ **45.** $-\frac{6}{7y^2}$ **47.** $\frac{27a^6}{b^{12}}$ **49.** $\frac{a^{30}}{b^{12}}$ **51.** $\frac{1}{x^{10}y^6}$ **53.** $\frac{z^2}{4}$

55. $\frac{x^{11}}{81}$ **57.** $\frac{49a^4}{b^6}$ **59.** $-\frac{3m^7}{n^4}$ **61.** $a^{24}b^8$ **63.** 200 **65.** x^9y^{19} **67.** $-\frac{y^8}{8x^2}$ **69.** $\frac{25b^{33}}{a^{16}}$ **71.** $\frac{27}{z^3x^6}$ cu in. **73.** 7.8×10^4

75. 1.67×10^{-6} **77.** 6.35×10^{-3} **79.** 1.16×10^6 **81.** 4.2×10^3 **83.** 0.0000000008673 **85.** 0.033 **87.** 20,320
89. 700,000,000 **91.** 5.7×10^{12} **93.** 10,100,000,000,000 **95.** 3,000,000,000,000; 3×10^{12} **97.** 0.000036
99. 0.0000000000000000028 **101.** 0.0000005 **103.** 200,000 **105.** 2.7×10^9 gal **107.** $-2x + 7$ **109.** $2y - 10$
111. $-x - 4$ **113.** 900,000,000; 9×10^8 **115.** 14,056,000; 1.4056×10^7 **117.** 2.5×10^{-9} m
119. 0.00000031 m; 3.1×10^{-7} m **121.** $9a^{13}$ **123.** -5 **125.** answers may vary **127. a.** 1.3×10^1 **b.** 4.4×10^7

c. 6.1×10^{-2} **129.** answers may vary **131.** $\frac{1}{x^{9s}}$ **133.** a^{4m+5}

Section 12.3

Vocabulary, Readiness & Video Check   1. binomial   **3.** trinomial   **5.** constant   **7.** $3; x^2, -3x, 5$ **9.** the replacement value for the variable   **11.** $2; 9ab$

Exercise Set 12.3   1. $1; -3x; 5$ **3.** $-5; 3.2; 1; -5$ **5.** 1; binomial   **7.** 3; none of these   **9.** 6; trinomial   **11.** 4; binomial
13. a. -6 **b.** -11 **15. a.** -2 **b.** 4   **17. a.** -15 **b.** -10 **19.** 184 ft   **21.** 595.84 ft   **23.** 593 thousand   **25.** 427.8 thousand

27. $-11x$ **29.** $23x^3$ **31.** $16x^2 - 7$ **33.** $12x^2 - 13$ **35.** $7s$ **37.** $-1.1y^2 + 4.8$ **39.** $\frac{5}{6}x^4 - 7x^3 - 19$

41. $\frac{3}{20}x^3 + 6x^2 - \frac{13}{20}x - \frac{1}{10}$ **43.** $4x^2 + 7x + x^2 + 5x; 5x^2 + 12x$ **45.** $5x + 3 + 4x + 3 + 2x + 6 + 3x + 7x; 21x + 12$

47. $2, 1, 1, 0; 2$ **49.** $4, 0, 4, 3; 4$ **51.** $9ab - 11a$ **53.** $4x^2 - 7xy + 3y^2$ **55.** $-3xy^2 + 4$ **57.** $14y^3 - 19 - 16a^2b^2$
59. $7x^2 + 0x + 3$ **61.** $x^3 + 0x^2 + 0x - 64$ **63.** $5y^3 + 0y^2 + 2y - 10$ **65.** $2y^4 + 0y^3 + 0y^2 + 8y + 0y^0$ or
$2y^4 + 0y^3 + 0y^2 + 8y + 0$ **67.** $6x^5 + 0x^4 + x^3 + 0x^2 - 3x + 15$ **69.** $10x + 19$ **71.** $-x + 5$ **73.** answers may vary
75. answers may vary   **77.** x^{13} **79.** a^3b^{10} **81.** $2y^{20}$ **83.** answers may vary   **85.** answers may vary   **87.** $11.1x^2 - 7.97x + 10.76$

Section 12.4

Vocabulary, Readiness & Video Check   1. $-14y$ **3.** $7x$ **5.** $5m^2 + 2m$ **7.** $-3y^2$ and $2y^2; -4y$ and y **9.** We're translating a subtraction problem. Order matters when subtracting, so we need to be careful that the order of the expressions is correct.

Exercise Set 12.4   1. $12x + 12$ **3.** $-3x^2 + 10$ **5.** $-3x^2 + 4$ **7.** $-y^2 - 3y - 1$ **9.** $7.9x^3 + 4.4x^2 - 3.4x - 3$

11. $\frac{1}{2}m^2 - \frac{7}{10}m + \frac{13}{16}$ **13.** $8t^2 - 4$ **15.** $15a^3 + a^2 - 3a + 16$ **17.** $-x + 14$ **19.** $7x^2$ **21.** $-2x + 9$ **23.** $2x^2 + 7x - 16$

25. $2x^2 + 11x$ **27.** $-0.2x^2 + 0.2x - 2.2$ **29.** $\frac{2}{5}z^2 - \frac{3}{10}z + \frac{7}{20}$ **31.** $-2z^2 - 16z + 6$ **33.** $2u^5 - 10u^2 + 11u - 9$

35. $5x - 9$ **37.** $4x - 3$ **39.** $11y + 7$ **41.** $-2x^2 + 8x - 1$ **43.** $14x + 18$ **45.** $3a^2 - 6a + 11$ **47.** $3x - 3$
49. $7x^2 - 4x + 2$ **51.** $7x^2 - 2x + 2$ **53.** $4y^2 + 12y + 19$ **55.** $-15x + 7$ **57.** $-2a - b + 1$ **59.** $3x^2 + 5$

61. $6x^2 - 2xy + 19y^2$ **63.** $8r^2s + 16rs - 8 + 7r^2s^2$ **65.** $(x^2 + 7x + 4)$ ft   **67.** $\left(\frac{19}{2}x + 3\right)$ units   **69.** $(3y^2 + 4y + 11)$ m

71. $-6.6x^2 - 1.8x - 1.8$ **73.** $6x^2$ **75.** $-12x^8$ **77.** $200x^3y^2$ **79.** $2; 2$ **81.** $4; 3; 3; 4$ **83.** b   **85.** e   **87. a.** $4z$ **b.** $3z^2$ **c.** $-4z$
d. $3z^2$; answers may vary   **89. a.** m^3 **b.** $3m$ **c.** $-m^3$ **d.** $-3m$; answers may vary   **91.** $915x^2 + 18,701x + 55,765$

Section 12.5

Vocabulary, Readiness & Video Check   1. distributive   **3.** $(5y - 1)(5y - 1)$ **5.** x^8 **7.** cannot simplify   **9.** x^{14} **11.** $2x^7$
13. No. The monomials are unlike terms.   **15.** Three times: First $(a - 2)$ is distributed to a and 7, and then a is distributed to $(a - 2)$ and 7 is distributed to $(a - 2)$.

Exercise Set 12.5   1. $24x^3$ **3.** x^4 **5.** $-28n^{10}$ **7.** $-12.4x^{12}$ **9.** $-\frac{2}{15}y^3$ **11.** $-24x^8$ **13.** $6x^2 + 15x$ **15.** $7x^3 + 14x^2 - 7x$

17. $-2a^2 - 8a$ **19.** $6x^3 - 9x^2 + 12x$ **21.** $12a^5 + 45a^2$ **23.** $-6a^4 + 4a^3 - 6a^2$ **25.** $6x^5y - 3x^4y^3 + 24x^2y^4$

27. $-4x^3y + 7x^2y^2 - xy^3 - 3y^4$ **29.** $4x^4 - 3x^3 + \frac{1}{2}x^2$ **31.** $x^2 + 7x + 12$ **33.** $a^2 + 5a - 14$ **35.** $x^2 + \frac{1}{3}x - \frac{2}{9}$

37. $12x^4 + 25x^2 + 7$ **39.** $12x^2 - 29x + 15$ **41.** $1 - 7a + 12a^2$ **43.** $4y^2 - 16y + 16$ **45.** $x^3 - 5x^2 + 13x - 14$
47. $x^4 + 5x^3 - 3x^2 - 11x + 20$ **49.** $10a^3 - 27a^2 + 26a - 12$ **51.** $49x^2y^2 - 14xy^2 + y^2$ **53.** $12x^2 - 64x - 11$
55. $2x^3 + 10x^2 + 11x - 3$ **57.** $2x^4 + 3x^3 - 58x^2 + 4x + 63$ **59.** $8.4y^7$ **61.** $-3x^3 - 6x^2 + 24x$ **63.** $2x^2 + 39x + 19$

65. $x^2 - \frac{2}{7}x - \frac{3}{49}$ **67.** $9y^2 + 30y + 25$ **69.** $a^3 - 2a^2 - 18a + 24$ **71.** $(4x^2 - 25)$ sq yd   **73.** $(6x^2 - 4x)$ sq in.   **75.** $25x^2$

77. $9y^6$ **79. a.** $6x + 12$ **b.** $9x^2 + 36x + 35$; answers may vary   **81.** $13x - 7$ **83.** $30x^2 - 28x + 6$ **85.** $-7x + 5$ **87.** $x^2 + 3x$
89. $x + 2x^2; x(1 + 2x)$ **91.** $11a$ **93.** $25x^2 + 4y^2$ **95. a.** $a^2 - b^2$ **b.** $4x^2 - 9y^2$ **c.** $16x^2 - 49$ **d.** answers may vary

Section 12.6

Vocabulary, Readiness & Video Check   1. false   **3.** false   **5.** a binomial times a binomial   **7.** Multiplying gives you four terms, and the two like terms will always subtract out.

Exercise Set 12.6   1. $x^2 + 7x + 12$ **3.** $x^2 + 5x - 50$ **5.** $5x^2 + 4x - 12$ **7.** $4y^2 - 25y + 6$ **9.** $6x^2 + 13x - 5$

11. $6y^3 + 4y^2 + 42y + 28$ **13.** $x^2 + \frac{1}{3}x - \frac{2}{9}$ **15.** $0.08 - 2.6a + 15a^2$ **17.** $2x^2 + 9xy - 5y^2$ **19.** $x^2 + 4x + 4$

21. $4a^2 - 12a + 9$ **23.** $9a^2 - 30a + 25$ **25.** $x^4 + x^2 + 0.25$ **27.** $y^2 - \frac{4}{7}y + \frac{4}{49}$ **29.** $4x^2 - 4x + 1$ **31.** $25x^2 + 90x + 81$

33. $9x^2 - 42xy + 49y^2$ **35.** $16m^2 + 40mn + 25n^2$ **37.** $25x^8 - 30x^4 + 9$ **39.** $a^2 - 49$ **41.** $x^2 - 36$ **43.** $9x^2 - 1$

45. $x^4 - 25$ **47.** $4y^4 - 1$ **49.** $16 - 49x^2$ **51.** $9x^2 - \frac{1}{4}$ **53.** $81x^2 - y^2$ **55.** $4m^2 - 25n^2$ **57.** $a^2 + 9a + 20$

59. $a^2 - 14a + 49$ **61.** $12a^2 - a - 1$ **63.** $x^2 - 4$ **65.** $9a^2 + 6a + 1$ **67.** $4x^2 + 3xy - y^2$ **69.** $\frac{1}{9}a^4 - 49$

71. $6b^2 - b - 35$ **73.** $x^4 - 100$ **75.** $16x^2 - 25$ **77.** $25x^2 - 60xy + 36y^2$ **79.** $4r^2 - 9s^2$ **81.** $(4x^2 + 4x + 1)$ sq ft

83. $\frac{5b^5}{7}$ **85.** $-\frac{2a^{10}}{b^5}$ **87.** $\frac{2y^8}{3}$ **89.** c   **91.** d   **93.** 2   **95.** $(x^4 - 3x^2 + 1)$ sq m   **97.** $(24x^2 - 32x + 8)$ sq m

99. answers may vary   **101.** answers may vary

Integrated Review   **1.** $35x^5$ **2.** $-32y^9$ **3.** -16 **4.** 16   **5.** $2x^2 - 9x - 5$ **6.** $3x^2 + 13x - 10$ **7.** $3x - 4$ **8.** $4x + 3$

9. $7x^6y^2$ **10.** $\frac{10b^6}{7}$ **11.** $144m^{14}n^{12}$ **12.** $64y^{27}z^{30}$ **13.** $16y^2 - 9$ **14.** $49x^2 - 1$ **15.** $\frac{y^{45}}{x^{63}}$ **16.** $\frac{1}{64}$ **17.** $\frac{x^{27}}{27}$ **18.** $\frac{r^{58}}{16s^{14}}$

19. $2x^2 - 2x - 6$ **20.** $6x^2 + 13x - 11$ **21.** $2.5y^2 - 6y - 0.2$ **22.** $8.4x^2 - 6.8x - 4.2$ **23.** $2y^2 - 6y - 1$

24. $6z^2 + 2z + \frac{11}{2}$ **25.** $x^2 + 8x + 16$ **26.** $y^2 - 18y + 81$ **27.** $2x + 8$ **28.** $2y - 18$ **29.** $7x^2 - 10xy + 4y^2$

30. $-a^2 - 3ab + 6b^2$ **31.** $x^3 + 2x^2 - 16x + 3$ **32.** $x^3 - 2x^2 - 5x - 2$ **33.** $6x^2 - x - 70$ **34.** $20x^2 + 21x - 5$

35. $2x^3 - 19x^2 + 44x - 7$ **36.** $5x^3 + 9x^2 - 17x + 3$ **37.** $4x^2 - \frac{25}{81}$ **38.** $144y^2 - \frac{9}{49}$

Section 12.7

Vocabulary, Readiness & Video Check   **1.** dividend; quotient; divisor   **3.** a^2 **5.** y **7.** the common denominator

Exercise Set 12.7   **1.** $12x^3 + 3x$ **3.** $4x^3 - 6x^2 + x + 1$ **5.** $5p^2 + 6p$ **7.** $-\frac{3}{2x} + 3$ **9.** $-3x^2 + x - \frac{4}{x^3}$ **11.** $-1 + \frac{3}{2x} - \frac{7}{4x^4}$

13. $x + 1$ **15.** $2x + 3$ **17.** $2x + 1 + \frac{7}{x-4}$ **19.** $3a^2 - 3a + 1 + \frac{2}{3a+2}$ **21.** $4x + 3 - \frac{2}{2x+1}$ **23.** $2x^2 + 6x - 5 - \frac{2}{x-2}$

25. $x + 6$ **27.** $x^2 + 3x + 9$ **29.** $-3x + 6 - \frac{11}{x+2}$ **31.** $2b - 1 - \frac{6}{2b-1}$ **33.** $ab - b^2$ **35.** $4x + 9$ **37.** $x + 4xy - \frac{y}{2}$

39. $2b^2 + b + 2 - \frac{12}{b+4}$ **41.** $y^2 + 5y + 10 + \frac{24}{y-2}$ **43.** $-6x - 12 - \frac{19}{x-2}$ **45.** $x^3 - x^2 + x$ **47.** 3   **49.** -4 **51.** $3x$

53. $9x$ **55.** $(3x^3 + x - 4)$ ft   **57.** $(2x + 5)$ m   **59.** answers may vary   **61.** c

Chapter 12 Vocabulary Check   **1.** term   **2.** FOIL   **3.** trinomial   **4.** degree of a polynomial   **5.** binomial   **6.** coefficient
7. degree of a term   **8.** monomial   **9.** polynomials   **10.** distributive

Chapter 12 Review   **1.** base: 3; exponent: 2   **2.** base: -5; exponent: 4   **3.** base: 5; exponent: 4   **4.** base: x; exponent: 6
5. 512   **6.** 36   **7.** -36 **8.** -65 **9.** 1   **10.** 1   **11.** y^9 **12.** x^{14} **13.** $-6x^{11}$ **14.** $-20y^7$ **15.** x^8 **16.** y^{15} **17.** $81y^{24}$ **18.** $8x^9$

19. x^5 **20.** z^7 **21.** $\frac{x^3y^4}{4}$ **22.** $\frac{x^6y^6}{4}$ **23.** $40a^{19}$ **24.** $36x^3$ **25.** $-a^9$ **26.** $-x^7$ **27.** 3   **28.** 9   **29.** b   **30.** c   **31.** $\frac{1}{49}$ **32.** $-\frac{1}{49}$

33. $\frac{2}{x^4}$ **34.** $\frac{1}{16x^4}$ **35.** 125   **36.** $\frac{9}{4}$ **37.** $\frac{17}{16}$ **38.** $\frac{1}{42}$ **39.** r **40.** y^3 **41.** c^4 **42.** $\frac{x^3}{y^3}$ **43.** $\frac{a^2}{5b^7c^3}$ **44.** $\frac{b^3}{5a^6c^7}$ **45.** $\frac{9}{x^6y^{13}}$

46. $\frac{3a^{10}}{b^{10}}$ **47.** 2.7×10^{-4} **48.** 8.868×10^{-1} **49.** 8.08×10^7 **50.** 8.68×10^5 **51.** 1.303×10^8 **52.** 1.5×10^5

53. 867,000   **54.** 0.00386   **55.** 0.00086   **56.** 893,600   **57.** 1,431,280,000,000,000   **58.** 0.0000000001   **59.** 0.016   **60.** 400,000,000,000
61. 5   **62.** 2   **63.** 5   **64.** 6   **65.** 4000 ft; 3984 ft; 3856 ft; 3600 ft   **66.** 22; 78; 154.02; 400   **67.** $2a^2$ **68.** $-4y$ **69.** $15a^2 + 4a$
70. $22x^2 + 3x + 6$ **71.** $-6a^2b - 3b^2 - q^2$ **72.** cannot be combined   **73.** $8x^2 + 3x + 6$ **74.** $2x^5 + 3x^4 + 4x^3 + 9x^2 + 7x + 6$
75. $-7y^2 - 1$ **76.** $-6m^7 - 3x^4 + 7m^6 - 4m^2$ **77.** $-x^2 - 6xy - 2y^2$ **78.** $x^6 + 4xy + 2y^2$ **79.** $-5x^2 + 5x + 1$
80. $-2x^2 - x + 20$ **81.** $6x + 30$ **82.** $9x - 63$ **83.** $8a + 28$ **84.** $54a - 27$ **85.** $-7x^3 - 35x$ **86.** $-32y^3 + 48y$
87. $-2x^3 + 18x^2 - 2x$ **88.** $-3a^3b - 3a^2b - 3ab^2$ **89.** $-6a^4 + 8a^2 - 2a$ **90.** $42b^4 - 28b^2 + 14b$ **91.** $2x^2 - 12x - 14$
92. $6x^2 - 11x - 10$ **93.** $4a^2 + 27a - 7$ **94.** $42a^2 + 11a - 3$ **95.** $x^4 + 7x^3 + 4x^2 + 23x - 35$ **96.** $x^6 + 2x^5 + x^2 + 3x + 2$
97. $x^4 + 4x^3 + 4x^2 - 16$ **98.** $x^6 + 8x^4 + 16x^2 - 16$ **99.** $x^3 + 21x^2 + 147x + 343$ **100.** $8x^3 - 60x^2 + 150x - 125$
101. $x^2 + 14x + 49$ **102.** $x^2 - 10x + 25$ **103.** $9x^2 - 42x + 49$ **104.** $16x^2 + 16x + 4$ **105.** $25x^2 - 90x + 81$
106. $25x^2 - 1$ **107.** $49x^2 - 16$ **108.** $a^2 - 4b^2$ **109.** $4x^2 - 36$ **110.** $16a^4 - 4b^2$ **111.** $(9x^2 - 6x + 1)$ sq m

112. $(5x^2 - 3x - 2)$ sq mi   **113.** $\frac{1}{7} + \frac{3}{x} + \frac{7}{x^2}$ **114.** $-a^2 + 3b - 4$ **115.** $a + 1 + \frac{6}{a-2}$ **116.** $4x + \frac{7}{x+5}$

117. $a^2 + 3a + 8 + \frac{22}{a-2}$ **118.** $3b^2 - 4b - \frac{1}{3b-2}$ **119.** $2x^3 - x^2 + 2 - \frac{1}{2x-1}$ **120.** $-x^2 - 16x - 117 - \frac{684}{x-6}$

121. $\left(5x - 1 + \frac{20}{x^2}\right)$ ft   **122.** $(7a^3b^6 + a - 1)$ units   **123.** 27   **124.** $-\frac{1}{8}$ **125.** $4x^4y^7$ **126.** $\frac{2x^6}{3}$ **127.** $\frac{27a^{12}}{b^6}$ **128.** $\frac{x^{16}}{16y^{12}}$

129. $9a^2b^8$ **130.** $2y^2 - 10$ **131.** $11x - 5$ **132.** $5x^2 + 3x - 2$ **133.** $5y^2 - 3y - 1$ **134.** $6x^2 + 11x - 10$
135. $28x^3 + 12x$ **136.** $28x^2 - 71x + 18$ **137.** $x^3 + x^2 - 18x + 18$ **138.** $25x^2 + 40x + 16$ **139.** $36x^2 - 9$

140. $4a - 1 + \frac{2}{a^2} - \frac{5}{2a^3}$ **141.** $x - 3 + \frac{25}{x+5}$ **142.** $2x^2 + 7x + 5 + \frac{19}{2x-3}$

Chapter 12 Test 1. 32 **2.** 81 **3.** -81 **4.** $\dfrac{1}{64}$ **5.** $-15x^{11}$ **6.** y^5 **7.** $\dfrac{1}{r^5}$ **8.** $\dfrac{16y^{14}}{x^2}$ **9.** $\dfrac{1}{6xy^8}$ **10.** 5.63×10^5 **11.** 8.63×10^{-5}
12. 0.0015 **13.** 62,300 **14.** 0.036 **15. a.** 4, 3; 7, 3; 1, 4; -2, 0 **b.** 4 **16.** $-2x^2 + 12x + 11$ **17.** $16x^3 + 7x^2 - 3x - 13$
18. $-3x^3 + 5x^2 + 4x + 5$ **19.** $x^3 + 8x^2 + 3x - 5$ **20.** $3x^3 + 22x^2 + 41x + 14$ **21.** $6x^4 - 9x^3 + 21x^2$ **22.** $3x^2 + 16x - 35$
23. $9x^2 - \dfrac{1}{25}$ **24.** $16x^2 - 16x + 4$ **25.** $64x^2 + 48x + 9$ **26.** $x^4 - 81b^2$ **27.** 1001 ft; 985 ft; 857 ft; 601 ft **28.** $(4x^2 - 9)$ sq in.
29. $\dfrac{x}{2y} + \dfrac{1}{4} - \dfrac{7}{8y}$ **30.** $x + 2$ **31.** $9x^2 - 6x + 4 - \dfrac{16}{3x + 2}$

Cumulative Review 1. 0.8496 (Sec. 4.4, Ex. 3) **2.** 53.1 **3. a.** $\dfrac{5}{7}$ **b.** $\dfrac{7}{24}$ (Sec. 5.1, Ex. 5) **4.** $\dfrac{23}{36}$ **5.** 5 (Sec. 5.4, Ex. 9) **6.** $\dfrac{7}{12}$

7. 75% (Sec. 5.4, Ex. 11) **8.** $\dfrac{18}{23}$ **9. a.** line CD or $\overleftrightarrow{CD}$ **b.** line segment EF or $\overline{EF}$ **c.** $\angle MNO, \angle ONM,$ or $\angle N$

d. ray PT or $\overrightarrow{PT}$ (Sec. 6.1, Ex. 1) **10.** 168° **11.** 10 cm (Sec. 6.2 Ex. 3) **12.** 34° **13.** 50 ft (Sec. 6.3, Ex. 6) **14.** 132 sq ft

15. a. 11, 112 **b.** 0, 11, 112 **c.** $-3, -2, 0, 11, 112$ **d.** $-3, -2, 0, \dfrac{1}{4}, 11, 112$ **e.** $\sqrt{2}$ **f.** $-2, 0, \dfrac{1}{4}, 112, -3, 11, \sqrt{2}$ (Sec. 8.1, Ex. 11)

16. a. 7.2 **b.** 0 **c.** $\dfrac{1}{2}$ **17.** $\dfrac{1}{4}$ (Sec. 8.2, Ex. 4) **18.** $\dfrac{3}{25}$ **19. a.** $x + 3$ **b.** $3x$ **c.** $7.3 \div x$ or $\dfrac{7.3}{x}$ **d.** $10 - x$
e. $5x + 7$ (Sec. 8.2, Ex. 9) **20.** 41 **21.** $-9x - y + 2z - 6$ (Sec. 8.7, Ex. 10) **22.** $4xy - 6y + 2$ **23.** $a = 19$ (Sec. 9.1, Ex. 6)
24. $x = -\dfrac{1}{2}$ **25.** $y = 140$ (Sec. 9.2, Ex. 4) **26.** $x = \dfrac{12}{5}$ **27.** $x = 4$ (Sec. 9.3, Ex. 5) **28.** $x = 1$ **29.** 10 (Sec. 9.4, Ex. 2)
30. $(x + 7) - 2x$ or $-x + 7$ **31.** 40 feet (Sec. 9.5, Ex. 2) **32.** undefined **33.** 800 (Sec. 9.6, Ex. 2) **34.**
35. $\dfrac{b^3}{27a^6}$ (Sec. 12.2, Ex. 10) **36.** $-15x^{16}$ **37.** $\dfrac{1}{25y^6}$ (Sec. 12.2, Ex. 14) **38.** $\dfrac{1}{9}$ **39.** $10x^3$ (Sec. 12.3, Ex. 8) **40.** $4y^2 - 8$
41. $7x^3 + 14x^2 + 35x$ (Sec. 12.5, Ex. 4) **42.** $100x^4 + 60x^2 + 9$

Chapter 13 Factoring Polynomials
Section 13.1

Vocabulary, Readiness & Video Check 1. factors **3.** least **5.** false **7.** $2 \cdot 7$ **9.** 3 **11.** 5 **13.** The GCF of a list of numbers is the largest number that is a factor of all numbers in the list. **15.** When factoring out a GCF, the number of terms in the other factor should have the same number of terms as your original polynomial.

Exercise Set 13.1 1. 4 **3.** 6 **5.** 1 **7.** y^2 **9.** z^7 **11.** xy^2 **13.** 7 **15.** $4y^3$ **17.** $5x^2$ **19.** $3x^3$ **21.** $9x^2y$ **23.** $10a^6b$ **25.** $3(a + 2)$
27. $15(2x - 1)$ **29.** $x^2(x + 5)$ **31.** $2y^3(3y + 1)$ **33.** $2x(16y - 9x)$ **35.** $4(x - 2y + 1)$ **37.** $3x(2x^2 - 3x + 4)$

39. $a^2b^2(a^5b^4 - a + b^3 - 1)$ **41.** $5xy(x^2 - 3x + 2)$ **43.** $4(2x^5 + 4x^4 - 5x^3 + 3)$ **45.** $\dfrac{1}{3}x(x^3 + 2x^2 - 4x^4 + 1)$

47. $(x^2 + 2)(y + 3)$ **49.** $(y + 4)(z + 3)$ **51.** $(z^2 - 6)(r + 1)$ **53.** $-2(x + 7)$ **55.** $-x^5(2 - x^2)$ **57.** $-3a^2(2a^2 - 3a + 1)$
59. $(x + 2)(x^2 + 5)$ **61.** $(x + 3)(5 + y)$ **63.** $(3x - 2)(2x^2 + 5)$ **65.** $(5m^2 + 6n)(m + 1)$ **67.** $(y - 4)(2 + x)$
69. $(2x + 1)(x^2 + 4)$ **71.** not factorable by grouping **73.** $(x - 2y)(4x - 3)$ **75.** $(5q - 4p)(q - 1)$ **77.** $2(2y - 7)(3x^2 - 1)$
79. $3(2a + 3b^2)(a + b)$ **81.** $x^2 + 7x + 10$ **83.** $b^2 - 3b - 4$ **85.** 2, 6 **87.** $-1, -8$ **89.** $-2, 5$ **91.** $-8, 3$ **93.** d
95. factored **97.** not factored **99. a.** 1265 million units **b.** 1360 million units **c.** 2560 million units **d.** $5(3x^2 - 50x + 440)$
101. a. 9016 thousand tons **b.** 8372 thousand tons **c.** $-322(x^2 - 3x - 26)$ or $322(-x^2 + 3x + 26)$
103. $4x^2 - \pi x^2; x^2(4 - \pi)$ **105.** $(x^3 - 1)$ units **107.** answers may vary **109.** answers may vary

Section 13.2

Vocabulary, Readiness & Video Check 1. true **3.** false **5.** $+5$ **7.** -3 **9.** $+2$ **11.** 15 is positive, so its factors would have to be either both positive or both negative. Since the factors need to sum to -8, both factors must be negative.

Exercise Set 13.2 1. $(x + 6)(x + 1)$ **3.** $(y - 9)(y - 1)$ **5.** $(x - 3)(x - 3)$ or $(x - 3)^2$ **7.** $(x - 6)(x + 3)$
9. $(x + 10)(x - 7)$ **11.** prime **13.** $(x + 5y)(x + 3y)$ **15.** $(a^2 - 5)(a^2 + 3)$ **17.** $(m + 13)(m + 1)$
19. $(t - 2)(t + 12)$ **21.** $(a - 2b)(a - 8b)$ **23.** $2(z + 8)(z + 2)$ **25.** $2x(x - 5)(x - 4)$ **27.** $(x - 4y)(x + y)$
29. $(x + 12)(x + 3)$ **31.** $(x^2 - 2)(x^2 + 1)$ **33.** $(r - 12)(r - 4)$ **35.** $(x + 2y)(x - y)$ **37.** $3(x + 5)(x - 2)$
39. $3(x^2 - 18)(x^2 - 2)$ **41.** $(x - 24)(x + 6)$ **43.** prime **45.** $(x - 5)(x - 3)$ **47.** $6x(x + 4)(x + 5)$
49. $4y(x^2 + x - 3)$ **51.** $(x - 7)(x + 3)$ **53.** $(x + 5y)(x + 2y)$ **55.** $2(t + 8)(t + 4)$ **57.** $x(x - 6)(x + 4)$
59. $2t^3(t - 4)(t - 3)$ **61.** $5xy(x - 8y)(x + 3y)$ **63.** $3(m - 9)(m - 6)$ **65.** $-1(x - 11)(x - 1)$ **67.** $\dfrac{1}{2}(y - 11)(y + 2)$

69. $x(xy - 4)(xy + 5)$ **71.** $2x^2 + 11x + 5$ **73.** $15y^2 - 17y + 4$ **75.** $9a^2 + 23ab - 12b^2$ **77.** $x^2 + 5x - 24$
79. answers may vary **81.** $2x^2 + 28x + 66$; $2(x + 3)(x + 11)$ **83.** $-16(t - 5)(t + 1)$ **85.** $\left(x + \dfrac{1}{4}\right)\left(x + \dfrac{1}{4}\right)$ or $\left(x + \dfrac{1}{4}\right)^2$
87. $(x + 1)(z - 10)(z + 7)$ **89.** $15; 28; 39; 48; 55; 60; 63; 64$ **91.** $9; 12; 21$ **93.** $(x^n + 10)(x^n - 2)$

Section 13.3

Vocabulary, Readiness & Video Check 1. d **3.** c **5.** Consider the factors of the first and last terms and the signs of the trinomial. Continue to check by multiplying until you get the middle term of the trinomial.

Exercise Set 13.3 1. $x + 4$ **3.** $10x - 1$ **5.** $4x - 3$ **7.** $(2x + 3)(x + 5)$ **9.** $(y - 1)(8y - 9)$ **11.** $(2x + 1)(x - 5)$
13. $(4r - 1)(5r + 8)$ **15.** $(10x + 1)(x + 3)$ **17.** $(3x - 2)(x + 1)$ **19.** $(3x - 5y)(2x - y)$ **21.** $(3m - 5)(5m + 3)$
23. $(x - 4)(x - 5)$ **25.** $(2x + 11)(x - 9)$ **27.** $(7t + 1)(t - 4)$ **29.** $(3a + b)(a + 3b)$ **31.** $(7p + 1)(7p - 2)$
33. $(6x - 7)(3x + 2)$ **35.** prime **37.** $(8x + 3)(3x + 4)$ **39.** $x(3x + 2)(4x + 1)$ **41.** $3(7b + 5)(b - 3)$
43. $(3z + 4)(4z - 3)$ **45.** $2y^2(3x - 10)(x + 3)$ **47.** $(2x - 7)(2x + 3)$ **49.** $3(x^2 - 14x + 21)$ **51.** $(4x + 9y)(2x - 3y)$
53. $-1(x - 6)(x + 4)$ **55.** $x(4x + 3)(x - 3)$ **57.** $(4x - 9)(6x - 1)$ **59.** $b(8a - 3)(5a + 3)$ **61.** $2x(3x + 2)(5x + 3)$
63. $2y(3y + 5)(y - 3)$ **65.** $5x^2(2x - y)(x + 3y)$ **67.** $-1(2x - 5)(7x - 2)$ **69.** $p^2(4p - 5)(4p - 5)$ or $p^2(4p - 5)^2$
71. $-1(2x + 1)(x - 5)$ **73.** $-4(12x - 1)(x - 1)$ **75.** $(2t^2 + 9)(t^2 - 3)$ **77.** prime **79.** $a(6a^2 + b^2)(a^2 + 6b^2)$
81. $x^2 - 16$ **83.** $x^2 + 4x + 4$ **85.** $4x^2 - 4x + 1$ **87.** 25–34 **89.** answers may vary **91.** no **93.** $4x^2 + 21x + 5$; $(4x + 1)(x + 5)$
95. $\left(2x + \dfrac{1}{2}\right)\left(2x + \dfrac{1}{2}\right)$ or $\left(2x + \dfrac{1}{2}\right)^2$ **97.** $(y - 1)^2(4x + 5)(x + 5)$ **99.** $2; 14$ **101.** 2 **103.** answers may vary

Section 13.4

Vocabulary, Readiness & Video Check 1. a **3.** b **5.** This gives us a four-term polynomial, which may be factored by grouping.

Exercise Set 13.4 1. $(x + 3)(x + 2)$ **3.** $(y + 8)(y - 2)$ **5.** $(8x - 5)(x - 3)$ **7.** $(5x^2 - 3)(x^2 + 5)$ **9. a.** $9, 2$
b. $9x + 2x$ **c.** $(2x + 3)(3x + 1)$ **11. a.** $-20, -3$ **b.** $-20x - 3x$ **c.** $(3x - 4)(5x - 1)$ **13.** $(3y + 2)(7y + 1)$
15. $(7x - 11)(x + 1)$ **17.** $(5x - 2)(2x - 1)$ **19.** $(2x - 5)(x - 1)$ **21.** $(2x + 3)(2x + 3)$ or $(2x + 3)^2$
23. $(2x + 3)(2x - 7)$ **25.** $(5x - 4)(2x - 3)$ **27.** $x(2x + 3)(x + 5)$ **29.** $2(8y - 9)(y - 1)$ **31.** $(2x - 3)(3x - 2)$
33. $3(3a + 2)(6a - 5)$ **35.** $a(4a + 1)(5a + 8)$ **37.** $3x(4x + 3)(x - 3)$ **39.** $y(3x + y)(x + y)$ **41.** prime
43. $6(a + b)(4a - 5b)$ **45.** $p^2(15p + q)(p + 2q)$ **47.** $(7 + x)(5 + x)$ or $(x + 7)(x + 5)$
49. $(6 - 5x)(1 - x)$ or $(5x - 6)(x - 1)$ **51.** $x^2 - 4$ **53.** $y^2 + 8y + 16$ **55.** $81z^2 - 25$ **57.** $16x^2 - 24x + 9$
59. $10x^2 + 45x + 45$; $5(2x + 3)(x + 3)$ **61.** $(x^n + 2)(x^n + 3)$ **63.** $(3x^n - 5)(x^n + 7)$ **65.** answers may vary

Section 13.5

Calculator Explorations

	$x^2 - 2x + 1$	$x^2 - 2x - 1$	$(x - 1)^2$
$x = 5$	16	14	16
$x = -3$	16	14	16
$x = 2.7$	2.89	0.89	2.89
$x = -12.1$	171.61	169.61	171.61
$x = 0$	1	-1	1

Vocabulary, Readiness & Video Check 1. perfect square trinomial **3.** perfect square trinomial **5.** $(x + 5y)^2$ **7.** false
9. 8^2 **11.** $(11a)^2$ **13.** $(6p^2)^2$ **15.** No, it just means it won't factor into a binomial squared. It may or may not be factorable.
17. In order to recognize the binomial as a difference of squares and also to identify the terms to use in the special factoring formula.

Exercise Set 13.5 1. yes **3.** no **5.** yes **7.** no **9.** no **11.** yes **13.** $(x + 11)^2$ **15.** $(x - 8)^2$ **17.** $(4a - 3)^2$
19. $(x^2 + 2)^2$ **21.** $2(n - 7)^2$ **23.** $(4y + 5)^2$ **25.** $(xy - 5)^2$ **27.** $m(m + 9)^2$ **29.** prime **31.** $(3x - 4y)^2$
33. $(x + 2)(x - 2)$ **35.** $(9 + p)(9 - p)$ or $-1(p + 9)(p - 9)$ **37.** $-1(2r + 1)(2r - 1)$ **39.** $(3x + 4)(3x - 4)$
41. prime **43.** $-1(6 + x)(6 - x)$ or $(x + 6)(x - 6)$ **45.** $(m^2 + 1)(m + 1)(m - 1)$ **47.** $(x + 13y)(x - 13y)$
49. $2(3r + 2)(3r - 2)$ **51.** $x(3y + 2)(3y - 2)$ **53.** $16x^2(x + 2)(x - 2)$ **55.** $xy(y - 3z)(y + 3z)$
57. $4(3x - 4y)(3x + 4y)$ **59.** $9(4 - 3x)(4 + 3x)$ **61.** $(5y - 3)(5y + 3)$ **63.** $(11m + 10n)(11m - 10n)$
65. $(xy - 1)(xy + 1)$ **67.** $\left(x - \dfrac{1}{2}\right)\left(x + \dfrac{1}{2}\right)$ **69.** $\left(7 - \dfrac{3}{5}m\right)\left(7 + \dfrac{3}{5}m\right)$ **71.** $(9a + 5b)(9a - 5b)$ **73.** $(x + 7y)^2$
75. $2(4n^2 - 7)^2$ **77.** $x^2(x^2 + 9)(x + 3)(x - 3)$ **79.** $pq(8p + 9q)(8p - 9q)$ **81.** 6 **83.** -2 **85.** $\dfrac{1}{5}$ **87.** $\left(x - \dfrac{1}{3}\right)^2$
89. $(x + 2 + y)(x + 2 - y)$ **91.** $(b - 4)(a + 4)(a - 4)$ **93.** $(x + 3 + 2y)(x + 3 - 2y)$ **95.** $(x^n + 10)(x^n - 10)$
97. 8 **99.** answers may vary **101.** $(x + 6)$ **103.** $a^2 + 2ab + b^2$ **105. a.** 2560 ft **b.** 1920 ft **c.** 13 sec **d.** $16(13 - t)(13 + t)$
107. a. 2160 feet **b.** 1520 feet **c.** 12 seconds **d.** $16(12 + t)(12 - t)$

Integrated Review **1.** $(x-3)(x+4)$ **2.** $(x-8)(x-2)$ **3.** $(x+1)^2$ **4.** $(x-3)^2$ **5.** $(x+2)(x-3)$
6. $(x+2)(x-1)$ **7.** $(x+3)(x-2)$ **8.** $(x+3)(x+4)$ **9.** $(x-5)(x-2)$ **10.** $(x-6)(x+5)$
11. $2(x-7)(x+7)$ **12.** $3(x-5)(x+5)$ **13.** $(x+3)(x+5)$ **14.** $(y-7)(3+x)$ **15.** $(x+8)(x-2)$
16. $(x-7)(x+4)$ **17.** $4x(x+7)(x-2)$ **18.** $6x(x-5)(x+4)$ **19.** $2(3x+4)(2x+3)$ **20.** $3(2a-b)(4a+5b)$
21. $(2a+b)(2a-b)$ **22.** $(x+5y)(x-5y)$ **23.** $(4-3x)(7+2x)$ **24.** $(5-2x)(4+x)$ **25.** prime **26.** prime
27. $(3y+5)(2y-3)$ **28.** $(4x-5)(x+1)$ **29.** $9x(2x^2-7x+1)$ **30.** $4a(3a^2-6a+1)$ **31.** $(4a-7)^2$
32. $(5p-7)^2$ **33.** $(7-x)(2+x)$ **34.** $(3+x)(1-x)$ **35.** $3x^2y(x+6)(x-4)$ **36.** $2xy(x+5y)(x-y)$
37. $3xy(4x^2+81)$ **38.** $2xy^2(3x^2+4)$ **39.** $2xy(1+6x)(1-6x)$ **40.** $2x(x-3)(x+3)$ **41.** $(x+6)(x+2)(x-2)$
42. $(x-2)(x+6)(x-6)$ **43.** $2a^2(3a+5)$ **44.** $2n(2n-3)$ **45.** $(3x-1)(x^2+4)$ **46.** $(x-2)(x^2+3)$
47. $6(x+2y)(x+y)$ **48.** $2(x+4y)(6x-y)$ **49.** $(x+y)(5+x)$ **50.** $(x-y)(7+y)$ **51.** $(7t-1)(2t-1)$
52. prime **53.** $-1(3x+5)(x-1)$ **54.** $-1(7x-2)(x+3)$ **55.** $(1-10a)(1+2a)$ **56.** $(1+5a)(1-12a)$
57. $(x+3)(x-3)(x-1)(x+1)$ **58.** $(x+3)(x-3)(x+2)(x-2)$ **59.** $(x-15)(x-8)$ **60.** $(y+16)(y+6)$
61. $(5p-7q)^2$ **62.** $(4a-7b)^2$ **63.** prime **64.** $(7x+3y)(x+3y)$ **65.** $-1(x-5)(x+6)$ **66.** $-1(x-2)(x-4)$
67. $(3r-1)(s+4)$ **68.** $(x-2)(x^2+1)$ **69.** $(x-2y)(4x-3)$ **70.** $(2x-y)(2x+7z)$ **71.** $(x+12y)(x-3y)$
72. $(3x-2y)(x+4y)$ **73.** $(x^2+2)(x+4)(x-4)$ **74.** $(x^2+3)(x+5)(x-5)$ **75.** answers may vary **76.** yes; $9(x^2+9y^2)$

Section 13.6

Vocabulary, Readiness & Video Check **1.** quadratic **3.** $3, -5$ **5.** One side of the equation must be a factored polynomial and the other side must be zero.

Exercise Set 13.6 **1.** $2, -1$ **3.** $6, 7$ **5.** $-9, -17$ **7.** $0, -6$ **9.** $0, 8$ **11.** $-\dfrac{3}{2}, \dfrac{5}{4}$ **13.** $\dfrac{7}{2}, -\dfrac{2}{7}$ **15.** $\dfrac{1}{2}, -\dfrac{1}{3}$ **17.** $-0.2, -1.5$

19. $9, 4$ **21.** $-4, 2$ **23.** $0, 7$ **25.** $0, -20$ **27.** $4, -4$ **29.** $8, -4$ **31.** $-3, 12$ **33.** $\dfrac{7}{3}, -2$ **35.** $\dfrac{8}{3}, -9$ **37.** $0, -\dfrac{1}{2}, \dfrac{1}{2}$ **39.** $\dfrac{17}{2}$ **41.** $\dfrac{3}{4}$

43. $-\dfrac{1}{2}, \dfrac{1}{2}$ **45.** $-\dfrac{3}{2}, -\dfrac{1}{2}, 3$ **47.** $-5, 3$ **49.** $-\dfrac{5}{6}, \dfrac{6}{5}$ **51.** $2, -\dfrac{4}{5}$ **53.** $-\dfrac{4}{3}, 5$ **55.** $-4, 3$ **57.** $0, 8, 4$ **59.** -7 **61.** $0, \dfrac{3}{2}$

63. $0, 1, -1$ **65.** $-6, \dfrac{4}{3}$ **67.** $\dfrac{6}{7}, 1$ **69.** $\dfrac{47}{45}$ **71.** $\dfrac{17}{60}$ **73.** $\dfrac{7}{10}$ **75.** didn't write equation in standard form; should be $x=4$ or $x=-2$

77. answers may vary, for example, $(x-6)(x+1)=0$ **79.** answers may vary, for example, $x^2-12x+35=0$

81. a. $300; 304; 276; 216; 124; 0; -156$ **b.** 5 sec **c.** 304 ft **83.** $0, \dfrac{1}{2}$ **85.** $0, -15$

Section 13.7

Vocabulary, Readiness & Video Check **1.** In applications, the context of the problem needs to be considered. Each exercise resulted in both a positive and a negative solution, and a negative solution is not appropriate for any of the problems.

Exercise Set 13.7 **1.** width: x; length: $x+4$ **3.** x and $x+2$ if x is an odd integer **5.** base: x; height: $4x+1$ **7.** 11 units
9. 15 cm, 13 cm, 22 cm, 70 cm **11.** base: 16 mi; height: 6 mi **13.** 5 sec **15.** width: 5 cm; length: 6 cm **17.** 54 diagonals **19.** 10 sides
21. -12 or 11 **23.** 14, 15 **25.** 13 feet **27.** 5 in. **29.** 12 mm, 16 mm, 20 mm **31.** 10 km **33.** 36 ft **35.** 9.5 sec **37.** 20%
39. length: 15 mi; width: 8 mi **41.** 105 units **43.** 2.2 million or 2,200,000 **45.** 2.4 million or 2,400,000 **47.** 2010 **49.** answers
may vary **51.** $\dfrac{4}{7}$ **53.** $\dfrac{3}{2}$ **55.** $\dfrac{1}{3}$ **57.** 8 m **59.** 10 and 15 **61.** width of pool: 29 m; length of pool: 35 m **63.** answers may vary

Chapter 13 Vocabulary Check **1.** quadratic equation **2.** Factoring **3.** greatest common factor **4.** perfect square trinomial
5. hypotenuse **6.** leg **7.** hypotenuse

Chapter 13 Review **1.** $2x-5$ **2.** $2x^4+1-5x^3$ **3.** $5(m+6)$ **4.** $4x(5x^2+3x+6)$ **5.** $(2x+3)(3x-5)$
6. $(x+1)(5x-1)$ **7.** $(x-1)(3x+2)$ **8.** $(a+3b)(3a+b)$ **9.** $(2a+b)(5a+7b)$ **10.** $(3x+5)(2x-1)$
11. $(x+4)(x+2)$ **12.** $(x-8)(x-3)$ **13.** prime **14.** $(x-6)(x+1)$ **15.** $(x+4)(x-2)$ **16.** $(x+6y)(x-2y)$
17. $(x+5y)(x+3y)$ **18.** $2(3-x)(12+x)$ **19.** $4(8+3x-x^2)$ **20.** $5y(y-6)(y-4)$ **21.** $-48, 2$
22. factor out the GCF, 3 **23.** $(2x+1)(x+6)$ **24.** $(2x+3)(2x-1)$ **25.** $(3x+4y)(2x-y)$ **26.** prime
27. $(2x+3)(x-13)$ **28.** $(6x+5y)(3x-4y)$ **29.** $5y(2y-3)(y+4)$ **30.** $3y(4y-1)(5y-2)$
31. $5x^2-9x-2; (5x+1)(x-2)$ **32.** $16x^2-28x+6; 2(4x-1)(2x-3)$ **33.** yes **34.** no **35.** no **36.** yes
37. yes **38.** no **39.** yes **40.** no **41.** $(x+9)(x-9)$ **42.** $(x+6)^2$ **43.** $(2x+3)(2x-3)$ **44.** $(3t+5s)(3t-5s)$
45. prime **46.** $(n-9)^2$ **47.** $3(r+6)^2$ **48.** $(3y-7)^2$ **49.** $5m^6(m+1)(m-1)$ **50.** $(2x-7y)^2$ **51.** $3y(x+y)^2$
52. $(4x^2+1)(2x+1)(2x-1)$ **53.** $-6, 2$ **54.** $-11, 7$ **55.** $0, -1, \dfrac{2}{7}$ **56.** $-\dfrac{1}{5}, -3$ **57.** $-7, -1$ **58.** $-4, 6$

59. -5 **60.** $2, 8$ **61.** $\dfrac{1}{3}$ **62.** $-\dfrac{2}{7}, \dfrac{3}{8}$ **63.** $0, 6$ **64.** $5, -5$ **65.** $x^2-9x+20=0$ **66.** $x^2+2x+1=0$ **67.** c **68.** d

69. 9 units **70.** 8 units, 13 units, 16 units, 10 units **71.** width: 20 in.; length: 25 in. **72.** 36 yd **73.** 19 and 20 **74.** 20 and 22
75. a. 17.5 sec and 10 sec; answers may vary **b.** 27.5 sec **76.** 32 cm **77.** $6(x+4)$ **78.** $7(x-9)$ **79.** $(4x-3)(11x-6)$
80. $(x-5)(2x-1)$ **81.** $(3x-4)(x^2+2)$ **82.** $(y+2)(x-1)$ **83.** $2(x+4)(x-3)$ **84.** $3x(x-9)(x-1)$

85. $(2x + 9)(2x - 9)$ **86.** $2(x + 3)(x - 3)$ **87.** $(4x - 3)^2$ **88.** $5(x + 2)^2$ **89.** $-\dfrac{7}{2}, 4$ **90.** $-3, 5$ **91.** $0, -7, -4$

92. $3, 2$ **93.** $0, 16$ **94.** 19 in.; 8 in.; 21 in. **95.** length: 6 in.; width: 2 in.

Chapter 13 Test 1. $3x(3x - 1)$ **2.** $(x + 7)(x + 4)$ **3.** $(7 + m)(7 - m)$ **4.** $(y + 11)^2$ **5.** $(x^2 + 4)(x + 2)(x - 2)$ **6.** $(a + 3)(4 - y)$ **7.** prime **8.** $(y - 12)(y + 4)$ **9.** $(a + b)(3a - 7)$ **10.** $(3x - 2)(x - 1)$ **11.** $5(6 + x)(6 - x)$ **12.** $3x(x - 5)(x - 2)$ **13.** $(6t + 5)(t - 1)$ **14.** $(x - 7)(y - 2)(y + 2)$ **15.** $x(1 + x^2)(1 + x)(1 - x)$

16. $(x + 12y)(x + 2y)$ **17.** $3, -9$ **18.** $-7, 2$ **19.** $-7, 1$ **20.** $0, \dfrac{3}{2}, -\dfrac{4}{3}$ **21.** $0, 3, -3$ **22.** $-3, 5$ **23.** $0, \dfrac{5}{2}$ **24.** 17 ft

25. width: 6 units; length: 9 units **26.** 7 sec **27.** hypotenuse: 25 cm; legs: 15 cm, 20 cm **28.** 8.25 sec

Cumulative Review 1. $\dfrac{5}{7}$ (Sec. 3.1, Ex. 1) **2.** $\dfrac{19}{30}$ **3.** $\dfrac{16}{13}$ or $1\dfrac{3}{13}$ (Sec. 3.1, Ex. 3) **4.** $\dfrac{4}{5}$ **5.** 36 (Sec. 3.2, Ex. 2) **6.** $\dfrac{49}{50}$

7. $\dfrac{12}{24}$ (Sec. 3.2, Ex. 9) **8.** yes **9.** $\dfrac{8}{33}$ (Sec. 3.3, Ex. 6) **10.** $7\dfrac{47}{72}$ **11.** 56 sq cm (Sec. 6.4, Ex. 1) **12.** 18 **13.** 6.557 (Sec. 6.6, Ex. 4a)

14. 5.78 **15.** 46 ft (Sec. 6.7, Ex. 4) **16.** 24% **17. a.** 75 clam species **b.** fishes (Sec. 7.1, Ex. 3) **18.** mean: 8; median: 9; mode: 11 **19. a.** $9 \le 11$ **b.** $8 > 1$ **c.** $3 \ne 4$ (Sec. 8.1, Ex. 7) **20. a.** $>$ **b.** $<$ **21.** solution (Sec. 8.2, Ex. 8)

22. 102 **23.** -12 (Sec. 8.4, Ex. 5a) **24.** -102 **25. a.** $\dfrac{3}{4}$ **b.** -24 **c.** 1 (Sec. 8.5, Ex. 16) **26.** -98 **27.** -11 (Sec. 9.3, Ex. 3)

28. 28 **29.** every real number (Sec. 9.3, Ex. 7) **30.** 33 **31.** $l = \dfrac{V}{wh}$ (Sec. 9.5, Ex. 5) **32.** $y = \dfrac{-3x - 7}{2}$ or $y = -\dfrac{3}{2}x - \dfrac{7}{2}$

33. x^6 (Sec. 12.2, Ex. 9) **34.** $\dfrac{1}{9}$ **35.** $\dfrac{y^{18}}{z^{36}}$ (Sec. 12.2, Ex. 11) **36.** x^4 **37.** $\dfrac{1}{x^{19}}$ (Sec. 12.2, Ex. 13) **38.** $25a^9$

39. $t^2 + 4t + 4$ (Sec. 12.6, Ex. 5) **40.** $x^2 - 26x + 169$ **41.** $x^4 - 14x^2y + 49y^2$ (Sec. 12.6, Ex. 8) **42.** $49x^2 + 14xy + y^2$

43. $2xy - 4 + \dfrac{1}{2y}$ (Sec. 12.7, Ex. 3) **44.** $(z^2 + 7)(z + 1)$ **45.** $(x + 3)(5 + y)$ (Sec. 13.1, Ex. 9) **46.** $2x(x + 7)(x - 6)$

47. $(x^2 + 2)(x^2 + 3)$ (Sec. 13.2, Ex. 7) **48.** $x(3y + 4)(3y - 4)$ **49.** 3 sec (Sec. 13.7, Ex. 1) **50.** $9, 4$

Chapter 14 Rational Expressions

Section 14.1

Vocabulary, Readiness & Video Check 1. rational expression **3.** -1 **5.** 2 **7.** $\dfrac{-a}{b}, \dfrac{a}{-b}$ **9.** yes **11.** no **13.** Rational expressions are fractions and are therefore undefined if the denominator is zero; if a denominator contains variables, set it equal to zero and solve. **15.** We would need to write parentheses around the numerator or denominator if it had more than one term because the negative sign needs to apply to the entire numerator or denominator.

Exercise Set 14.1 1. $\dfrac{7}{4}$ **3.** $-\dfrac{8}{3}$ **5.** $-\dfrac{11}{2}$ **7. a.** \$403 **b.** \$7 **c.** decrease; answers may vary **9.** $x = 0$ **11.** $x = -2$ **13.** $x = \dfrac{5}{2}$

15. $x = 0, x = -2$ **17.** none **19.** $x = 6, x = -1$ **21.** $x = -2, x = -\dfrac{7}{3}$ **23.** 1 **25.** -1 **27.** $\dfrac{1}{4(x + 2)}$ **29.** $\dfrac{1}{x + 2}$

31. can't simplify **33.** -5 **35.** $\dfrac{7}{x}$ **37.** $\dfrac{1}{x - 9}$ **39.** $5x + 1$ **41.** $\dfrac{x^2}{x - 2}$ **43.** $7x$ **45.** $\dfrac{x + 5}{x - 5}$ **47.** $\dfrac{x + 2}{x + 4}$ **49.** $\dfrac{x + 2}{2}$

51. $-(x + 2)$ **53.** $\dfrac{x + 1}{x - 1}$ **55.** $x + y$ **57.** $\dfrac{5 - y}{2}$ **59.** $\dfrac{2y + 5}{3y + 4}$ **61.** $\dfrac{-(x - 10)}{x + 8}; \dfrac{-x + 10}{x + 8}; \dfrac{x - 10}{-(x + 8)}; \dfrac{x - 10}{-x - 8}$

63. $\dfrac{-(5y - 3)}{y - 12}; \dfrac{-5y + 3}{y - 12}; \dfrac{5y - 3}{-(y - 12)}; \dfrac{5y - 3}{-y + 12}$ **65.** correct **67.** correct **69.** $\dfrac{3}{11}$ **71.** $\dfrac{4}{3}$ **73.** $\dfrac{117}{40}$ **75.** correct

77. incorrect; $\dfrac{1 + 2}{1 + 3} = \dfrac{3}{4}$ **79.** answers may vary **81.** answers may vary **83.** 400 mg **85.** $C = 78.125$; medium **87.** 60.8%

Section 14.2

Vocabulary, Readiness & Video Check 1. reciprocals **3.** $\dfrac{a \cdot d}{b \cdot c}$ **5.** $\dfrac{6}{7}$ **7.** fractions; reciprocal **9.** We're converting to cubic feet so we want cubic feet in the numerator. We want cubic yards to divide out, so cubic yards is in the denominator.

Exercise Set 14.2 1. $\dfrac{21}{4y}$ **3.** x^4 **5.** $-\dfrac{b^2}{6}$ **7.** $\dfrac{x^2}{10}$ **9.** $\dfrac{1}{3}$ **11.** $\dfrac{m + n}{m - n}$ **13.** $\dfrac{x + 5}{x}$ **15.** $\dfrac{(x + 2)(x - 3)}{(x - 4)(x + 4)}$ **17.** $\dfrac{2x^4}{3}$ **19.** $\dfrac{12}{y^6}$

21. $x(x + 4)$ **23.** $\dfrac{3(x + 1)}{x^3(x - 1)}$ **25.** $m^2 - n^2$ **27.** $-\dfrac{x + 2}{x - 3}$ **29.** $\dfrac{x + 2}{x - 3}$ **31.** $\dfrac{5}{6}$ **33.** $\dfrac{3x}{8}$ **35.** $\dfrac{3}{2}$ **37.** $\dfrac{3x + 4y}{2(x + 2y)}$ **39.** $\dfrac{2(x + 2)}{x - 2}$

41. $-\dfrac{y(x + 2)}{4}$ **43.** $\dfrac{(a + 5)(a + 3)}{(a + 2)(a + 1)}$ **45.** $\dfrac{5}{x}$ **47.** $\dfrac{2(n - 8)}{3n - 1}$ **49.** 1440 **51.** 5 **53.** 81 **55.** 73 **57.** 56.7 **59.** 1,201,500 sq ft

61. 55.2 miles/hour **63.** 1 **65.** $-\dfrac{10}{9}$ **67.** $-\dfrac{1}{5}$ **69.** true **71.** false; $\dfrac{x^2 + 3x}{20}$ **73.** $\dfrac{2}{9(x - 5)}$ sq ft **75.** $\dfrac{x}{2}$ **77.** $\dfrac{5a(2a + b)(3a - 2b)}{b^2(a - b)(a + 2b)}$

79. answers may vary **81.** 1493.99 euros

Section 14.3

Vocabulary, Readiness & Video Check **1.** $\dfrac{9}{11}$ **3.** $\dfrac{a + c}{b}$ **5.** $\dfrac{5 - (6 + x)}{x}$ **7.** We factor denominators into the smallest

factors—including coefficients—so we can determine the greatest number of times each unique factor occurs in any one denominator for the LCD.

Exercise Set 14.3 **1.** $\dfrac{a + 9}{13}$ **3.** $\dfrac{3m}{n}$ **5.** 4 **7.** $\dfrac{y + 10}{3 + y}$ **9.** $5x + 3$ **11.** $\dfrac{4}{a + 5}$ **13.** $\dfrac{1}{x - 6}$ **15.** $\dfrac{5x + 7}{x - 3}$ **17.** $x + 5$ **19.** 3

21. $4x^3$ **23.** $8x(x + 2)$ **25.** $(x + 3)(x - 2)$ **27.** $3(x + 6)$ **29.** $5(x - 6)^2$ **31.** $6(x + 1)^2$ **33.** $x - 8$ or $8 - x$

35. $(x - 1)(x + 4)(x + 3)$ **37.** $(3x + 1)(x + 1)(x - 1)(2x + 1)$ **39.** $2x^2(x + 4)(x - 4)$ **41.** $\dfrac{6x}{4x^2}$ **43.** $\dfrac{24b^2}{12ab^2}$

45. $\dfrac{9y}{2y(x + 3)}$ **47.** $\dfrac{9ab + 2b}{5b(a + 2)}$ **49.** $\dfrac{x^2 + x}{x(x + 4)(x + 2)(x + 1)}$ **51.** $\dfrac{18y - 2}{30x^2 - 60}$ **53.** $2x$ **55.** $\dfrac{x + 3}{2x - 1}$ **57.** $x + 1$ **59.** $\dfrac{3}{x}$

61. $\dfrac{3x + 1}{5x + 1}$ **63.** $\dfrac{29}{21}$ **65.** $-\dfrac{5}{12}$ **67.** $\dfrac{7}{30}$ **69.** d **71.** answers may vary **73.** $\dfrac{20}{x - 2}$ m **75.** answers may vary **77.** 95,304 Earth days

79. answers may vary **81.** answers may vary

Section 14.4

Vocabulary, Readiness & Video Check **1.** b **3.** The exercise is adding two rational expressions with denominators that are opposites of each other. Recognizing this special case can save us time and effort. If we recognize that one denominator is -1 times the other denominator, we may save many steps.

Exercise Set 14.4 **1.** $\dfrac{5}{x}$ **3.** $\dfrac{75a + 6b^2}{5b}$ **5.** $\dfrac{6x + 5}{2x^2}$ **7.** $\dfrac{11}{x + 1}$ **9.** $\dfrac{x - 6}{(x - 2)(x + 2)}$ **11.** $\dfrac{35x - 6}{4x(x - 2)}$ **13.** $-\dfrac{2}{x - 3}$ **15.** 0

17. $-\dfrac{1}{x^2 - 1}$ **19.** $\dfrac{5 + 2x}{x}$ **21.** $\dfrac{6x - 7}{x - 2}$ **23.** $-\dfrac{y + 4}{y + 3}$ **25.** $\dfrac{-5x + 14}{4x}$ or $-\dfrac{5x - 14}{4x}$ **27.** 2 **29.** $\dfrac{9x^4 - 4x^2}{21}$ **31.** $\dfrac{x + 2}{(x + 3)^2}$

33. $\dfrac{9b - 4}{5b(b - 1)}$ **35.** $\dfrac{2 + m}{m}$ **37.** $\dfrac{x(x + 3)}{(x - 7)(x - 2)}$ **39.** $\dfrac{10}{1 - 2x}$ **41.** $\dfrac{15x - 1}{(x + 1)^2(x - 1)}$ **43.** $\dfrac{x^2 - 3x - 2}{(x - 1)^2(x + 1)}$ **45.** $\dfrac{a + 2}{2(a + 3)}$

47. $\dfrac{y(2y + 1)}{(2y + 3)^2}$ **49.** $\dfrac{x - 10}{2(x - 2)}$ **51.** $\dfrac{2x + 21}{(x + 3)^2}$ **53.** $\dfrac{-5x + 23}{(x - 2)(x - 3)}$ **55.** $\dfrac{7}{2(m - 10)}$ **57.** $\dfrac{2(x^2 - x - 23)}{(x + 1)(x - 6)(x - 5)}$

59. $\dfrac{n + 4}{4n(n - 1)(n - 2)}$ **61.** 10 **63.** 2 **65.** $\dfrac{25a}{9(a - 2)}$ **67.** $\dfrac{x + 4}{(x - 2)(x - 1)}$ **69.** $\dfrac{2}{3}$ **71.** $-\dfrac{1}{2}, 1$ **73.** $-\dfrac{15}{2}$

75. $\dfrac{6x^2 - 5x - 3}{x(x + 1)(x - 1)}$ **77.** $\dfrac{4x^2 - 15x + 6}{(x - 2)^2(x + 2)(x - 3)}$ **79.** $\dfrac{-2x^2 + 14x + 55}{(x + 2)(x + 7)(x + 3)}$ **81.** $\dfrac{2(x - 8)}{(x + 4)(x - 4)}$ in. **83.** $\dfrac{P - G}{P}$

85. answers may vary **87.** $\left(\dfrac{90x - 40}{x}\right)^\circ$ **89.** answers may vary

Section 14.5

Vocabulary, Readiness & Video Check **1.** c **3.** b **5.** a **7.** These equations are solved in very different ways, so we need to determine the next correct move to make. For a linear equation, we first "move" variable terms to one side and numbers to the other; for a quadratic equation, we first set the equation equal to 0. **9.** the steps for solving an equation containing rational expressions; as if it's the only variable in the equation

Exercise Set 14.5 **1.** 30 **3.** 0 **5.** -2 **7.** $-5, 2$ **9.** 5 **11.** 3 **13.** 1 **15.** 5 **17.** no solution **19.** 4 **21.** -8 **23.** $6, -4$ **25.** 1

27. $3, -4$ **29.** -3 **31.** 0 **33.** -2 **35.** $8, -2$ **37.** no solution **39.** 3 **41.** $-11, 1$ **43.** $I = \dfrac{E}{R}$ **45.** $B = \dfrac{2U - TE}{T}$

47. $w = \dfrac{Bh^2}{705}$ **49.** $G = \dfrac{V}{N - R}$ **51.** $r = \dfrac{C}{2\pi}$ **53.** $x = \dfrac{3y}{3 + y}$ **55.** $\dfrac{1}{x}$ **57.** $\dfrac{1}{x} + \dfrac{1}{2}$ **59.** $\dfrac{1}{3}$ **61.** answers may vary **63.** $\dfrac{5x + 9}{9x}$

65. no solution **67.** $100^\circ, 80^\circ$ **69.** $22.5^\circ, 67.5^\circ$ **71.** 5

Integrated Review **1.** expression; $\dfrac{3 + 2x}{3x}$ **2.** expression; $\dfrac{18 + 5a}{6a}$ **3.** equation; 3 **4.** equation; 18 **5.** expression; $\dfrac{x - 1}{x(x + 1)}$

6. expression; $\dfrac{3(x + 1)}{x(x - 3)}$ **7.** equation; no solution **8.** equation; 1 **9.** expression; 10 **10.** expression; $\dfrac{z}{3(9z - 5)}$

11. expression; $\dfrac{5x + 7}{x - 3}$ **12.** expression; $\dfrac{7p + 5}{2p + 7}$ **13.** equation; 23 **14.** equation; 3 **15.** expression; $\dfrac{25a}{9(a - 2)}$

16. expression; $\dfrac{9}{4(x-1)}$ **17.** expression; $\dfrac{3x^2+5x+3}{(3x-1)^2}$ **18.** expression; $\dfrac{2x^2-3x-1}{(2x-5)^2}$ **19.** expression; $\dfrac{4x-37}{5x}$

20. expression; $\dfrac{29x-23}{3x}$ **21.** equation; $\dfrac{8}{5}$ **22.** equation; $-\dfrac{7}{3}$ **23.** answers may vary   **24.** answers may vary

Section 14.6

Vocabulary, Readiness & Video Check   **1.** c   **3.** $\dfrac{1}{x}; \dfrac{1}{x}-3$ **5.** $z+5; \dfrac{1}{z+5}$ **7.** $2y; \dfrac{11}{2y}$ **9.** divided by, quotient

11.

	d	$=$	r	$\cdot$	t
Car	325		$x+7$		$\dfrac{325}{x+7}$
Motorcycle	290		x		$\dfrac{290}{x}$

$\dfrac{325}{x+7}=\dfrac{290}{x}$

Exercise Set 14.6   **1.** 2   **3.** -3 **5.** $2\dfrac{2}{9}$ hr   **7.** $1\dfrac{1}{2}$ min   **9.** trip to park rate: r; to park time: $\dfrac{12}{r}$; return trip rate: r; return time:

$\dfrac{18}{r}=\dfrac{12}{r}+1; r=6$ mph   **11.** 1st portion: 10 mph; cooldown: 8 mph   **13.** 2   **15.** $108.00   **17.** 20 mph   **19.** 5   **21.** 217 mph

23. 8   **25.** 2.2 mph; 3.3 mph   **27.** 3 hr   **29.** 8 mph   **31.** 35 mph; 75 mph   **33.** 510 mph   **35.** $666\dfrac{2}{3}$ mi   **37.** 20 hr   **39.** car: 70 mph;

motorcycle: 60 mph   **41.** $5\dfrac{1}{4}$ hr   **43.** 41 mph; 51 mph   **45.** $\dfrac{1}{2}$ **47.** $\dfrac{3}{7}$ **49.** faster pump: 28 min; slower pump: 84 min   **51.** answers

may vary   **53.** $R=\dfrac{D}{T}$ **55.** 3.75 min

Section 14.7

Vocabulary, Readiness & Video Check   **1.** c   **3.** a   **5.** a single fraction in the numerator and in the denominator

Exercise Set 14.7   **1.** $\dfrac{2}{3}$ **3.** $\dfrac{2}{3}$ **5.** $\dfrac{1}{2}$ **7.** $-\dfrac{21}{5}$ **9.** $\dfrac{27}{16}$ **11.** $\dfrac{4}{3}$ **13.** $\dfrac{1}{21}$ **15.** $-\dfrac{4x}{15}$ **17.** $\dfrac{m-n}{m+n}$ **19.** $\dfrac{2x(x-5)}{7x^2+10}$ **21.** $\dfrac{1}{y-1}$ **23.** $\dfrac{1}{6}$

25. $\dfrac{x+y}{x-y}$ **27.** $\dfrac{3}{7}$ **29.** $\dfrac{a}{x+b}$ **31.** $\dfrac{7(y-3)}{8+y}$ **33.** $\dfrac{3x}{x-4}$ **35.** $-\dfrac{x+8}{x-2}$ **37.** $\dfrac{s^2+r^2}{s^2-r^2}$ **39.** $\dfrac{(x-6)(x+4)}{x-2}$ **41.** Serena Williams

43. $5.6 million   **45.** answers may vary   **47.** $\dfrac{13}{24}$ **49.** $4\dfrac{1}{4}$ ft or 4.25 ft   **51.** $\dfrac{R_1 R_2}{R_2+R_1}$ **53.** $\dfrac{2x}{2-x}$ **55.** $\dfrac{1}{y^2-1}$ **57.** 12 hr

Chapter 14 Vocabulary Check   **1.** rational expression   **2.** complex fraction   **3.** $\dfrac{-a}{b}; \dfrac{a}{-b}$ **4.** denominator   **5.** simplifying

6. reciprocals   **7.** least common denominator   **8.** unit

Chapter 14 Review   **1.** $x=2, x=-2$ **2.** $x=\dfrac{5}{2}, x=-\dfrac{3}{2}$ **3.** $\dfrac{4}{3}$ **4.** $\dfrac{11}{12}$ **5.** $\dfrac{2}{x}$ **6.** $\dfrac{3}{x}$ **7.** $\dfrac{1}{x-5}$ **8.** $\dfrac{1}{x+1}$ **9.** $\dfrac{x(x-2)}{x+1}$

10. $\dfrac{5(x-5)}{x-3}$ **11.** $\dfrac{x-3}{x-5}$ **12.** $\dfrac{x}{x+4}$ **13.** $\dfrac{x+a}{x-c}$ **14.** $\dfrac{x+5}{x-3}$ **15.** $\dfrac{3x^2}{y}$ **16.** $-\dfrac{9x^2}{8}$ **17.** $\dfrac{x-3}{x+2}$ **18.** $-\dfrac{2x(2x+5)}{(x-6)^2}$ **19.** $\dfrac{x+3}{x-4}$

20. $\dfrac{4x}{3y}$ **21.** $(x-6)(x-3)$ **22.** $\dfrac{2}{3}$ **23.** $\dfrac{1}{2}$ **24.** $\dfrac{3(x+2)}{3x+y}$ **25.** $\dfrac{1}{x+2}$ **26.** $\dfrac{1}{x-3}$ **27.** $\dfrac{2(x-5)}{3x^2}$ **28.** $\dfrac{2x+1}{2x^2}$ **29.** $14x$

30. $(x-8)(x+8)(x+3)$ **31.** $\dfrac{10x^2 y}{14x^3 y}$ **32.** $\dfrac{36y^2 x}{16y^3 x}$ **33.** $\dfrac{x^2-3x-10}{(x+2)(x-5)(x+9)}$ **34.** $\dfrac{3x^2+4x-15}{(x+2)^2(x+3)}$ **35.** $\dfrac{4y+30x^2}{5x^2 y}$

36. $\dfrac{-2(x-5)}{(x-3)(x-1)}$ **37.** $\dfrac{-2(x+1)}{x+3}$ **38.** $\dfrac{5(x+1)}{(x+4)(x-2)(x-1)}$ **39.** $\dfrac{x-4}{3x}$ **40.** $-\dfrac{x}{x-1}$ **41.** 30   **42.** 3, -4

43. no solution   **44.** 5   **45.** $\dfrac{9}{7}$ **46.** $-6, 1$ **47.** 3   **48.** 2   **49.** fast car speed: 30 mph; slow car speed: 20 mph   **50.** 20 mph

51. $17\dfrac{1}{2}$ hr   **52.** $8\dfrac{4}{7}$ days   **53.** $-\dfrac{7}{18y}$ **54.** $\dfrac{6}{7}$ **55.** $\dfrac{3y-1}{2y-1}$ **56.** $-\dfrac{7+2x}{2x}$ **57.** $\dfrac{1}{2x}$ **58.** $\dfrac{x(x-3)}{x+7}$ **59.** $\dfrac{x-4}{x+4}$

60. $\dfrac{(x-9)(x+8)}{(x+5)(x+9)}$ **61.** $\dfrac{1}{x-6}$ **62.** $\dfrac{2x+1}{4x}$ **63.** $-\dfrac{3x}{(x+2)(x-3)}$ **64.** $\dfrac{2}{(x+3)(x-2)}$ **65.** $\dfrac{1}{2}$ **66.** no solution

67. 1   **68.** $1\dfrac{5}{7}$ days   **69.** $\dfrac{3}{10}$ **70.** $\dfrac{2}{3}$

Chapter 14 Test **1.** $x = -1, x = -3$ **2. a.** $115 **b.** $103 **3.** $\dfrac{3}{5}$ **4.** $\dfrac{1}{x+6}$ **5.** -1 **6.** $-\dfrac{1}{x+y}$ **7.** $\dfrac{2m(m+2)}{m-2}$ **8.** $\dfrac{a+2}{a+5}$

9. $\dfrac{(x-6)(x-7)}{(x+7)(x+2)}$ **10.** 15 **11.** $\dfrac{y-2}{4}$ **12.** $-\dfrac{1}{2x+5}$ **13.** $\dfrac{3a-4}{(a-3)(a+2)}$ **14.** $\dfrac{3}{x-1}$ **15.** $\dfrac{2(x+3)(x+5)}{x(x^2+4x+1)}$

16. $\dfrac{x^2+2x+35}{(x+9)(x+2)(x-5)}$ **17.** $\dfrac{4y^2+13y-15}{(y+5)(y+1)(y+4)}$ **18.** $\dfrac{30}{11}$ **19.** -6 **20.** no solution **21.** no solution **22.** $-2, 5$

23. $\dfrac{xz}{2y}$ **24.** $b-a$ **25.** $\dfrac{5y^2-1}{y+2}$ **26.** 1 or 5 **27.** 30 mph **28.** $6\dfrac{2}{3}$ hr

Cumulative Review **1.** 5.1 sq mi (Sec. 6.4, Ex. 2) **2.** 25π sq m ≈ 78.5 sq m **3.** 36π cu in. $\approx 113\dfrac{1}{7}$ cu in. (Sec. 6.5, Ex. 2)

4. 24 cu cm **5.** 57 (Sec. 7.3, Ex. 3) **6.** 24.5 **7.** $\dfrac{1}{3}$ (Sec. 7.4, Ex. 4) **8.** -4.9 **9.** $\dfrac{m^7}{n^7}, n \neq 0$ (Sec. 12.1, Ex. 22) **10.** $\dfrac{a^6}{b^5}, a \neq 0, b \neq 0$

11. $9x^2 - 6x - 1$ (Sec. 12.4, Ex. 5) **12.** $6x^2 + \dfrac{14}{3}x + \dfrac{1}{42}$ **13.** $-\dfrac{1}{2}, 4$ (Sec. 13.6, Ex. 6) **14.** $0, \dfrac{7}{2}$ **15.** 1 (Sec. 14.3, Ex. 2)

16. $\dfrac{5x-16}{(x-6)(x+1)}$ **17. a.** $\dfrac{15}{x} = 4$ **b.** $12 - 3 = x$ **c.** $4x + 17 = 21$ (Sec. 8.2, Ex. 10) **18. a.** $12 - x = -45$ **b.** $12x = -45$

c. $x - 10 = 2x$ **19. a.** -12 **b.** -9 (Sec. 8.3, Ex. 12) **20. a.** -8 **b.** -17 **21.** distributive property (Sec. 8.6, Ex. 15)

22. commutative property of addition **23.** associative property of addition (Sec. 8.6, Ex. 16) **24.** associative property of multiplication **25.** -4 (Sec. 9.1, Ex. 7) **26.** 0 **27.** shorter piece, 2 ft; longer piece, 8 ft (Sec. 9.4, Ex. 3) **28.** 190, 192

29. $\dfrac{y-b}{m} = x$ (Sec. 9.5, Ex. 6) **30.** $x = \dfrac{2y+6}{3}$ **31.** $(5x+2y)^2$ (Sec. 13.5, Ex. 5) **32.** $(x+2)(x-2)$ **33.** $11, -2$ (Sec. 13.6, Ex. 4)

34. $-2, \dfrac{1}{3}$ **35.** $\dfrac{2}{5}$ (Sec. 14.2, Ex. 2) **36.** $\dfrac{x+5}{2x^3}$ **37.** $3x - 5$ (Sec. 14.3, Ex. 3) **38.** $7x^4(x^2 - x + 1)$ **39.** $\dfrac{3}{x-2}$ (Sec. 14.4, Ex. 2)

40. $(2x+3)^2$ **41.** 5 (Sec. 14.5, Ex. 2) **42.** $\dfrac{30}{x+3}$ **43.** $2\dfrac{1}{10}$ hr (Sec. 14.6, Ex. 2) **44.** $\dfrac{4m+2n}{m+n}$ or $\dfrac{2(2m+n)}{m+n}$

Chapter 15 Roots and Radicals

Section 15.1

Calculator Explorations **1.** 2.449 **3.** 3.317 **5.** 9.055 **7.** 3.420 **9.** 2.115 **11.** 1.783

Vocabulary, Readiness & Video Check **1.** principal **3.** square root **5.** power **7.** false **9.** true **11.** The radical sign, $\sqrt{}$, indicates a positive square root only. A negative sign before the radical sign, $-\sqrt{}$, indicates a negative square root. **13.** an odd-numbered index **15.** Divide the index into each exponent in the radicand—but still check by squaring your answer.

Exercise Set 15.1 **1.** 4 **3.** $\dfrac{1}{5}$ **5.** -10 **7.** not a real number **9.** -11 **11.** $\dfrac{3}{5}$ **13.** 30 **15.** 12 **17.** $\dfrac{1}{10}$ **19.** 0.5 **21.** 5 **23.** -4

25. -2 **27.** $\dfrac{1}{2}$ **29.** -5 **31.** 2 **33.** 9 **35.** not a real number **37.** $-\dfrac{3}{4}$ **39.** -5 **41.** 1 **43.** 2.646 **45.** 6.083 **47.** 11.662

49. $\sqrt{2} \approx 1.41$; 126.90 ft **51.** m **53.** x^2 **55.** $3x^4$ **57.** $9x$ **59.** ab^2 **61.** $4a^3b^2$ **63.** a^2b^6 **65.** $-2xy^9$ **67.** $\dfrac{x^3}{6}$ **69.** $\dfrac{5y}{3}$ **71.** $25 \cdot 2$

73. $16 \cdot 2$ or $4 \cdot 8$ **75.** $4 \cdot 7$ **77.** $9 \cdot 3$ **79.** a, b **81.** 7 mi **83.** 10 mm **85.** 3 **87.** 10 **89.** 4, 5 **91.** 8, 9 **93.** $T \approx 6.1$ seconds

95. answers may vary **97.** 1; 1.7; 2; 3 **99.** $|x|$ **101.** $|x+2|$ **103.** $(2, 0)$ **105.** $(-4, 0)$

Section 15.2

Vocabulary, Readiness & Video Check **1.** $\sqrt{a} \cdot \sqrt{b}$ **3.** $16; 25; 4; 5; 20$ **5.** false **7.** Factor until we have a product of primes. A repeated prime factor means a perfect square—if more than one factor is repeated, we can multiply all the repeated factors together to get one larger perfect square factor. **9.** The power must be 1. Any even power is a perfect square and can be simplified; any higher odd power is the product of an even power times the variable with a power of 1.

Exercise Set 15.2 **1.** $2\sqrt{5}$ **3.** $5\sqrt{2}$ **5.** $\sqrt{33}$ **7.** $7\sqrt{2}$ **9.** $2\sqrt{15}$ **11.** $6\sqrt{5}$ **13.** $2\sqrt{13}$ **15.** 15 **17.** $21\sqrt{7}$ **19.** $-15\sqrt{3}$

21. $\dfrac{2\sqrt{2}}{5}$ **23.** $\dfrac{3\sqrt{3}}{11}$ **25.** $\dfrac{3}{2}$ **27.** $\dfrac{5\sqrt{5}}{3}$ **29.** $\dfrac{\sqrt{11}}{6}$ **31.** $-\dfrac{\sqrt{3}}{4}$ **33.** $x^3\sqrt{x}$ **35.** $x^6\sqrt{x}$ **37.** $6a\sqrt{a}$ **39.** $4x^2\sqrt{6}$ **41.** $\dfrac{2\sqrt{3}}{m}$

43. $\dfrac{3\sqrt{x}}{y^5}$ **45.** $\dfrac{2\sqrt{22}}{x^6}$ **47.** 16 **49.** $\dfrac{6}{11}$ **51.** $5\sqrt{7}$ **53.** $\dfrac{2\sqrt{5}}{3}$ **55.** $2m^3\sqrt{6m}$ **57.** $\dfrac{y\sqrt{23y}}{2x^3}$ **59.** $2\sqrt[3]{3}$ **61.** $5\sqrt[3]{2}$ **63.** $\dfrac{\sqrt[3]{5}}{4}$

65. $\dfrac{\sqrt[3]{23}}{2}$ **67.** $\dfrac{\sqrt[3]{15}}{4}$ **69.** $2\sqrt[3]{10}$ **71.** $14x$ **73.** $2x^2 - 7x - 15$ **75.** 0 **77.** $x^3y\sqrt{y}$ **79.** $7x^2y^2\sqrt{2x}$ **81.** $-2x^2$ **83.** $2\sqrt[3]{10}$ in.

85. answers may vary **87.** 177 m by 177 m **89.** $2\sqrt{5}$ in. **91.** 2.25 in. **93.** $1700 **95.** 1.7 sq m

Section 15.3

Vocabulary, Readiness & Video Check **1.** like radicals **3.** $17\sqrt{2}$ **5.** $2\sqrt{5}$ **7.** Both like terms and like radicals are combined using the distributive property; also, only like (vs. unlike) terms can be combined, as with like radicals (same index and same radicand). **9.** the product rule for radicals

Exercise Set 15.3 **1.** $-4\sqrt{3}$ **3.** $9\sqrt{6} - 5$ **5.** $\sqrt{5} + \sqrt{2}$ **7.** $7\sqrt{3} - \sqrt{2}$ **9.** $-5\sqrt{2} - 6$ **11.** $5\sqrt{3}$ **13.** $9\sqrt{5}$

15. $4\sqrt{6} + \sqrt{5}$ **17.** $x + \sqrt{x}$ **19.** 0 **21.** $\dfrac{4\sqrt{5}}{9}$ **23.** $\dfrac{3\sqrt{3}}{8}$ **25.** $7\sqrt{5}$ **27.** $9\sqrt{3}$ **29.** $\sqrt{5} + \sqrt{15}$ **31.** $x\sqrt{x}$ **33.** $5\sqrt{2} + 12$

35. $8\sqrt{2} - 5$ **37.** $2\sqrt{5}$ **39.** $-\sqrt{35}$ **41.** $6 - 3\sqrt{3}$ **43.** $11\sqrt{x}$ **45.** $12x - 11\sqrt{x}$ **47.** $x\sqrt{3x} + 3x\sqrt{x}$ **49.** $8x\sqrt{2} + 2x$

51. $2x^2\sqrt{10} - x^2\sqrt{5}$ **53.** $7\sqrt[3]{9} - \sqrt[3]{25}$ **55.** $-5\sqrt[3]{2} - 6$ **57.** $5\sqrt[3]{3}$ **59.** $-3 + 3\sqrt[3]{2}$ **61.** $4x + 4x\sqrt[3]{2}$ **63.** $10y^2\sqrt[3]{y}$ **65.** $x\sqrt[3]{5}$

67. $x^2 + 12x + 36$ **69.** $4x^2 - 4x + 1$ **71.** answers may vary **73.** $8\sqrt{5}$ in. **75.** $\left(48 + \dfrac{9\sqrt{3}}{2}\right)$ sq ft **77.** yes; $7\sqrt{2}$ **79.** no

81. yes; $3\sqrt{7}$ **83.** $\dfrac{83x\sqrt{x}}{20}$

Section 15.4

Vocabulary, Readiness & Video Check **1.** $\sqrt{21}$ **3.** $\sqrt{\dfrac{15}{3}}$ or $\sqrt{5}$ **5.** $2 - \sqrt{3}$ **7.** The square root of a positive number times the square root of the same positive number (or the square root of a positive number squared) is that positive number. **9.** To write an equivalent expression without a radical in the denominator.

Exercise Set 15.4 **1.** 4 **3.** $5\sqrt{2}$ **5.** 6 **7.** $2x$ **9.** 20 **11.** $36x$ **13.** $3x^3\sqrt{2}$ **15.** $4xy\sqrt{y}$ **17.** $\sqrt{30} + \sqrt{42}$ **19.** $2\sqrt{5} + 5\sqrt{2}$

21. $y\sqrt{7} - 14\sqrt{y}$ **23.** -33 **25.** $\sqrt{6} - \sqrt{15} + \sqrt{10} - 5$ **27.** $16 - 11\sqrt{11}$ **29.** $x - 36$ **31.** $x - 14\sqrt{x} + 49$

33. $6y + 2\sqrt{6y} + 1$ **35.** 4 **37.** $\sqrt{7}$ **39.** $3\sqrt{2}$ **41.** $5y^2$ **43.** $5\sqrt{3}$ **45.** $2y\sqrt{6}$ **47.** $2xy\sqrt{3y}$ **49.** $\dfrac{\sqrt{15}}{5}$ **51.** $\dfrac{7\sqrt{2}}{2}$ **53.** $\dfrac{\sqrt{6y}}{6y}$

55. $\dfrac{\sqrt{10}}{6}$ **57.** $\dfrac{\sqrt{3x}}{x}$ **59.** $\dfrac{\sqrt{2}}{4}$ **61.** $\dfrac{\sqrt{30}}{15}$ **63.** $\dfrac{\sqrt{15}}{10}$ **65.** $\dfrac{3\sqrt{2x}}{2}$ **67.** $\dfrac{8y\sqrt{5}}{5}$ **69.** $\dfrac{\sqrt{xy}}{6y}$ **71.** $\dfrac{\sqrt{3xy}}{6x}$ **73.** $3\sqrt{2} - 3$

75. $-8 - 4\sqrt{5}$ **77.** $\sqrt{30} + 5 + \sqrt{6} + \sqrt{5}$ **79.** $\sqrt{6} + \sqrt{3} + \sqrt{2} + 1$ **81.** $\dfrac{10 - 5\sqrt{x}}{4 - x}$ **83.** $\dfrac{3\sqrt{x} + 12}{x - 16}$ **85.** 44 **87.** 2

89. 3 **91.** $130\sqrt{3}$ sq m **93.** $\dfrac{\sqrt{A\pi}}{\pi}$ **95.** true **97.** false **99.** false **101.** answers may vary **103.** answers may vary

105. $\dfrac{2}{\sqrt{6} - \sqrt{2} - \sqrt{3} + 1}$

Integrated Review **1.** 6 **2.** $4\sqrt{3}$ **3.** x^2 **4.** $y^3\sqrt{y}$ **5.** $4x$ **6.** $3x^5\sqrt{2x}$ **7.** 2 **8.** 3 **9.** -3 **10.** not a real number **11.** $\dfrac{\sqrt{11}}{3}$

12. $\dfrac{\sqrt[3]{7}}{4}$ **13.** -4 **14.** -5 **15.** $\dfrac{3}{7}$ **16.** $\dfrac{1}{8}$ **17.** a^4b **18.** x^5y^{10} **19.** $5m^3$ **20.** $3n^8$ **21.** $6\sqrt{7}$ **22.** $3\sqrt{2}$ **23.** cannot be simplified

24. $\sqrt{x} + 3x$ **25.** $\sqrt{30}$ **26.** 3 **27.** 28 **28.** 45 **29.** $\sqrt{33} + \sqrt{3}$ **30.** $3\sqrt{2} - 2\sqrt{6}$ **31.** $4y$ **32.** $3x^2\sqrt{5}$ **33.** $x - 3\sqrt{x} - 10$

34. $11 + 6\sqrt{2}$ **35.** 2 **36.** $\sqrt{3}$ **37.** $2x^2\sqrt{3}$ **38.** $ab^2\sqrt{15a}$ **39.** $\dfrac{\sqrt{6}}{6}$ **40.** $\dfrac{x\sqrt{5}}{10}$ **41.** $\dfrac{4\sqrt{6} - 4}{5}$ **42.** $\dfrac{\sqrt{2x} + 5\sqrt{2} + \sqrt{x} + 5}{x - 25}$

Section 15.5

Vocabulary, Readiness & Video Check **1.** The squaring property can result in extraneous solutions, so we need to check our solutions in the original equation—before the squaring property was applied—to make sure they are actual solutions.

Exercise Set 15.5 **1.** 81 **3.** -1 **5.** 49 **7.** no solution **9.** 4 **11.** 2 **13.** 2 **15.** 9 **17.** -3 **19.** $-1, -2$ **21.** no solution

23. $0, -3$ **25.** 16 **27.** 25 **29.** 1 **31.** 5 **33.** -2 **35.** no solution **37.** 2 **39.** 36 **41.** no solution **43.** $\dfrac{3}{2}$ **45.** 16 **47.** 3

49. 12 **51.** 3, 1 **53.** -1 **55.** $3x - 8 = 19; x = 9$ **57.** $2(2x) + 2x = 24$; length $= 8$ in. **59.** 4, 7 **61.** answers may vary

63. a. 3.2, 10, 31.6 **b.** no **65.** 7.30 **67.** 0.76

Section 15.6

Vocabulary, Readiness & Video Check **1.** The Pythagorean theorem applies to right triangles only, and in the formula $a^2 + b^2 = c^2$, c is the length of the hypotenuse. **3.** Both examples ask for an answer rounded to a given place, meaning an estimated answer is expected rather than an exact answer. An exact answer would be given in radical form.

Exercise Set 15.6 **1.** $\sqrt{13}$; 3.61 **3.** $3\sqrt{3}$; 5.20 **5.** 25 **7.** $\sqrt{22}$; 4.69 **9.** $3\sqrt{17}$; 12.37 **11.** $\sqrt{41}$; 6.40 **13.** $4\sqrt{2}$; 5.66 **15.** $3\sqrt{10}$; 9.49 **17.** 20.6 ft **19.** 11.7 ft **21.** 24 cu ft **23.** 54 mph **25.** 27 mph **27.** 58.2 km **29.** 81.4 km **31.** 3, -3 **33.** 10, -10 **35.** 8, -8 **37.** $y = 2\sqrt{10}$; $x = 2\sqrt{10} - 4$ **39.** 201 miles **41.** answers may vary

Chapter 15 Vocabulary Check **1.** like radicals **2.** index; radicand; radical **3.** conjugate **4.** principal square root **5.** rationalizing the denominator **6.** hypotenuse

Chapter 15 Review **1.** 9 **2.** -7 **3.** 3 **4.** 3 **5.** $-\dfrac{3}{8}$ **6.** $\dfrac{2}{3}$ **7.** 2 **8.** -2 **9.** c **10.** a, c **11.** x^6 **12.** x^4 **13.** $3y$ **14.** $5x^2$ **15.** $2\sqrt{10}$ **16.** $2\sqrt{6}$ **17.** $3\sqrt{6}$ **18.** $2\sqrt{22}$ **19.** $x^2\sqrt{x}$ **20.** $y^3\sqrt{y}$ **21.** $2x\sqrt{5}$ **22.** $5y^2\sqrt{2}$ **23.** $3\sqrt[3]{2}$ **24.** $2\sqrt[3]{11}$ **25.** $\dfrac{3\sqrt{2}}{5}$ **26.** $\dfrac{5\sqrt{3}}{8}$ **27.** $-\dfrac{5\sqrt{2}}{3}$ **28.** $-\dfrac{2\sqrt{3}}{7}$ **29.** $\dfrac{\sqrt{11}}{x}$ **30.** $\dfrac{\sqrt{7}}{y^2}$ **31.** $\dfrac{y^2\sqrt{y}}{10}$ **32.** $\dfrac{x\sqrt{x}}{9}$ **33.** $-3\sqrt{2}$ **34.** $-5\sqrt{3}$ **35.** $4\sqrt{5} + 4\sqrt{6}$ **36.** $-2\sqrt{7} + 2\sqrt{2}$ **37.** $5\sqrt{7} + 2\sqrt{14}$ **38.** $9\sqrt{3} - 4$ **39.** $\dfrac{\sqrt{5}}{6}$ **40.** $\dfrac{9\sqrt{11}}{20}$ **41.** $10 - x\sqrt{5}$ **42.** $2\sqrt{2x} - \sqrt{3x}$ **43.** $3\sqrt{2}$ **44.** $5\sqrt{3}$ **45.** $\sqrt{10} - \sqrt{14}$ **46.** $\sqrt{55} + \sqrt{15}$ **47.** $3\sqrt{2} - 5\sqrt{3} + 2\sqrt{6} - 10$ **48.** $2 - 2\sqrt{5}$ **49.** $x - 4\sqrt{x} + 4$ **50.** $y + 8\sqrt{y} + 16$ **51.** 3 **52.** 2 **53.** $2\sqrt{5}$ **54.** $4\sqrt{2}$ **55.** $x\sqrt{15x}$ **56.** $3x^2\sqrt{2}$ **57.** $\dfrac{\sqrt{22}}{11}$ **58.** $\dfrac{\sqrt{39}}{13}$ **59.** $\dfrac{\sqrt{30}}{6}$ **60.** $\dfrac{\sqrt{70}}{10}$ **61.** $\dfrac{\sqrt{5x}}{5x}$ **62.** $\dfrac{5\sqrt{3y}}{3y}$ **63.** $\dfrac{\sqrt{3x}}{x}$ **64.** $\dfrac{\sqrt{6y}}{y}$ **65.** $3\sqrt{5} + 6$ **66.** $8\sqrt{10} + 24$ **67.** $\dfrac{\sqrt{6} + \sqrt{2} + \sqrt{3} + 1}{2}$ **68.** $\sqrt{15} - 2\sqrt{3} - 2\sqrt{5} + 4$ **69.** $\dfrac{10\sqrt{x} - 50}{x - 25}$ **70.** $\dfrac{8\sqrt{x} + 8}{x - 1}$ **71.** 18 **72.** 13 **73.** 25 **74.** no solution **75.** 12 **76.** 5 **77.** 1 **78.** 9 **79.** $2\sqrt{14}$; 7.48 **80.** $3\sqrt{13}$; 10.82 **81.** $4\sqrt{34}$ ft; 23.32 ft **82.** $5\sqrt{3}$ in.; 8.66 in. **83.** 2.4 in. **84.** 144π sq in. **85.** 12 **86.** -4 **87.** $4x^8$ **88.** $2x^{12}$ **89.** $3x^3\sqrt{2x}$ **90.** $4y^3\sqrt{3}$ **91.** $\dfrac{y^2}{9}$ **92.** $\dfrac{x^4\sqrt{x}}{3}$ **93.** $7\sqrt{3}$ **94.** $5\sqrt{7} - 3$ **95.** $-\dfrac{\sqrt{3}}{4}$ **96.** $4x\sqrt{5x}$ **97.** $7\sqrt{2}$ **98.** $3\sqrt{3} - \sqrt{6}$ **99.** $\sqrt{10} - \sqrt{2} + 4\sqrt{5} - 4$ **100.** $x + 6\sqrt{x} + 9$ **101.** $2\sqrt{6}$ **102.** $2x$ **103.** $\dfrac{\sqrt{14}}{7}$ **104.** $\dfrac{3\sqrt{2x}}{2x}$ **105.** $\dfrac{3\sqrt{x} + 18}{x - 36}$ **106.** $\dfrac{\sqrt{35} - 3\sqrt{7} - 5\sqrt{5} + 15}{-4}$ **107.** 1 **108.** 13 **109.** 14 **110.** 9 **111.** $\sqrt{58}$; 7.62 **112.** $4\sqrt{2}$ in.; 5.66 in.

Chapter 15 Test **1.** 4 **2.** 5 **3.** 3 **4.** $\dfrac{3}{4}$ **5.** not a real number **6.** x^5 **7.** $3\sqrt{6}$ **8.** $2\sqrt{23}$ **9.** $y^3\sqrt{y}$ **10.** $2x^4\sqrt{6}$ **11.** 3 **12.** $2\sqrt[3]{2}$ **13.** $\dfrac{\sqrt{5}}{4}$ **14.** $\dfrac{y\sqrt{y}}{5}$ **15.** $-2\sqrt{13}$ **16.** $\sqrt{2} + 2\sqrt{3}$ **17.** $\dfrac{7\sqrt{3}}{10}$ **18.** $7\sqrt{2}$ **19.** $2\sqrt{3} - \sqrt{10}$ **20.** $x - \sqrt{x} - 6$ **21.** $\sqrt{5}$ **22.** $2x\sqrt{5x}$ **23.** $\dfrac{\sqrt{6}}{3}$ **24.** $\dfrac{8\sqrt{5y}}{5y}$ **25.** $4\sqrt{6} - 8$ **26.** $\dfrac{3 + \sqrt{x}}{9 - x}$ **27.** 9 **28.** 5 **29.** 9 **30.** $4\sqrt{5}$ in. **31.** 2.19 m

Cumulative Review **1.** $\dfrac{1}{6}$ of an hour (Sec. 3.3, Ex. 9) **2.** 27 **3.** $7\dfrac{17}{24}$ (Sec. 3.4, Ex. 1) **4.** $\dfrac{16}{27}$ **5.** > (Sec. 3.5, Ex. 1) **6.** 14,000,000 **7.** one and three tenths (Sec. 4.1, Ex. 1) **8.** 0.075 **9.** 736.2 (Sec. 4.2, Ex. 5) **10.** 736.24 **11.** 28 (Sec. 8.5, Ex. 3) **12.** -46.8 **13.** $-\dfrac{8}{21}$ (Sec. 8.5, Ex. 4) **14.** -18 **15.** 2 (Sec. 9.3, Ex. 1) **16.** 15 **17. a.** 17% **b.** 21% **c.** 43 American travelers (Sec. 9.6, Ex. 3) **18. a.** $\dfrac{3}{2}$ **b.** 9 **19. a.** 102,000 **b.** 0.007358 **c.** 84,000,000 **d.** 0.00003007 (Sec. 12.2, Ex. 18) **20. a.** 7.2×10^6 **b.** 3.08×10^{-4} **21.** $6x^2 - 11x - 10$ (Sec. 12.5, Ex. 7b) **22.** $49x^2 + 14x + 1$ **23.** $(y + 2)(x + 3)$ (Sec. 13.1, Ex. 1) **24.** $(y^2 + 5)(x - 1)$ **25.** $(3x + 2)(x + 3)$ (Sec. 13.3, Ex. 1) **26.** $3(x + 2)(x + 3)$ **27. a.** $x = 3$ **b.** $x = 2, x = 1$ **c.** none (Sec. 14.1, Ex. 2) **28.** $\dfrac{2x + 1}{x - 3}$ **29.** $-\dfrac{17}{5}$ (Sec. 14.5, Ex. 4) **30.** $y = -2x + 4$ **31.** (Sec. 10.3, Ex. 7) **32.** 6; 4; 0

33. $y = \frac{1}{4}x - 3$ (Sec. 10.5, Ex. 1)   **34.** $y = -\frac{1}{2}x + \frac{11}{2}$ **35.** $(3, 1)$ (Sec. 11.3, Ex. 5)   **36.** $\left(\frac{2}{3}, \frac{1}{2}\right)$ **37.** Alfredo: 3.25 mph; Louisa: 4.25 mph (Sec. 11.4, Ex. 3)   **38.** 20 mph, 35 mph   **39.** $3\sqrt{6}$ (Sec. 15.2, Ex. 1)   **40.** $3\sqrt{7}$ **41.** $10\sqrt{2}$ (Sec. 15.2, Ex. 3)   **42.** $10\sqrt{5}$

43. $4\sqrt{3}$ (Sec. 15.3, Ex. 6)   **44.** $x - 25$ **45.** $2x - 4x^2\sqrt{x}$ (Sec. 15.3, Ex. 8)   **46.** $10 + 4\sqrt{6}$ **47.** $\frac{2\sqrt{7}}{7}$ (Sec. 15.4, Ex. 10)   **48.** $3x$

49. $\frac{1}{2}$ (Sec. 15.5, Ex. 2)   **50.** $\frac{5}{2}$

Chapter 16 Quadratic Equations and Nonlinear Graphs

Section 16.1

Vocabulary, Readiness & Video Check   **1.** To solve, a becomes the radicand and the square root of a negative number is not a real number.

Exercise Set 16.1   **1.** ± 7 **3.** $-5, 3$ **5.** ± 4 **7.** ± 3 **9.** $-5, -2$ **11.** ± 8 **13.** $\pm\sqrt{21}$ **15.** $\pm\frac{1}{5}$ **17.** no real solution   **19.** $\pm\frac{\sqrt{39}}{3}$

21. $\pm\frac{2\sqrt{7}}{7}$ **23.** $\pm\sqrt{2}$ **25.** $12, -2$ **27.** $-2 \pm \sqrt{7}$ **29.** $1, 0$ **31.** $-2 \pm \sqrt{10}$ **33.** $\frac{8}{3}, -4$ **35.** no real solution   **37.** $\frac{11 \pm 5\sqrt{2}}{2}$

39. $\frac{7 \pm 4\sqrt{2}}{3}$ **41.** $\pm\sqrt{29}$ **43.** $-6 \pm 2\sqrt{6}$ **45.** $\pm\sqrt{10}$ **47.** $\frac{1 \pm \sqrt{5}}{4}$ **49.** $\pm 2\sqrt{3}$ **51.** $\frac{-8 \pm 3\sqrt{5}}{-3}$ or $\frac{8 \pm 3\sqrt{5}}{3}$ **53.** 2.3 sec

55. 15.8 sec   **57.** 6 in.   **59.** $2\sqrt{5}$ in. ≈ 4.47 in.   **61.** $\sqrt{3039}$ ft ≈ 55.13 ft   **63.** $(x + 3)^2$ **65.** $(x - 2)^2$ **67.** answers may vary

69. $2, -6$ **71.** ± 1.33 **73.** $x = 9$, which is 2015

Section 16.2

Vocabulary, Readiness & Video Check   **1.** zero   **3.** quadratic equation   **5.** 9   **7.** 16   **9.** 100   **11.** 49   **13.** When working with equations, whatever is added to one side must also be added to the other side to keep equality.

Exercise Set 16.2   **1.** $-6, -2$ **3.** $-1 \pm 2\sqrt{2}$ **5.** $0, 6$ **7.** $-1, -4$ **9.** $1 \pm \sqrt{2}$ **11.** $\frac{-5 \pm \sqrt{53}}{2}$ **13.** $-2, 4$ **15.** no real solution

17. $\frac{3 \pm \sqrt{19}}{2}$ **19.** $-2 \pm \frac{\sqrt{6}}{2}$ **21.** $-3 \pm \sqrt{34}$ **23.** $\frac{3 \pm \sqrt{21}}{2}$ **25.** $\frac{1}{2}, 1$ **27.** $-6, 3$ **29.** no real solution   **31.** $2, -6$ **33.** $-\frac{1}{2}$

35. 2   **37.** $3 + 2\sqrt{5}$ **39.** $\frac{1 - 3\sqrt{2}}{2}$ **41.** answers may vary   **43. a.** $-3 \pm \sqrt{11}$ **b.** answers may vary   **45.** $k = 8$ or $k = -8$

47. $x = 8, 2017$ **49.** $-6, -2$ **51.** $\approx -0.68, 3.68$

Section 16.3

Vocabulary, Readiness & Video Check   **1.** $x = \frac{-b \pm \sqrt{b^2 - 4ac}}{2a}$ **3.** $1; 3; -7$ **5.** $1; 1; -1$ **7.** $-2, 1$ **9.** $\frac{-5 \pm \sqrt{17}}{2}$

11. a. Yes, in order to make sure you have correct values for a, b, and c.   **b.** No; it simplifies calculations, but you would still get a correct answer using fraction values in the formula.

Exercise Set 16.3   **1.** $2, 1$ **3.** $\frac{-7 \pm \sqrt{37}}{6}$ **5.** $\pm\frac{\sqrt{3}}{2}$ **7.** no real solution   **9.** $10, -3$ **11.** $\pm\sqrt{5}$ **13.** $-3, 4$ **15.** $-2 \pm \sqrt{7}$

17. $\frac{-9 \pm \sqrt{129}}{12}$ **19.** $\frac{4 \pm \sqrt{2}}{7}$ **21.** $3 \pm \sqrt{7}$ **23.** $\frac{3 \pm \sqrt{3}}{2}$ **25.** $\frac{1}{3}, -1$ **27.** $\frac{3 \pm \sqrt{13}}{4}$ **29.** no real solution   **31.** $\frac{1}{5}, -\frac{3}{4}$

33. $\frac{5 \pm \sqrt{33}}{2}$ **35.** $-2, \frac{7}{3}$ **37.** $1 \pm \sqrt{2}$ **39.** no real solution   **41.** $-\frac{1}{2}, -\frac{3}{4}$ **43.** $\frac{7 \pm \sqrt{129}}{20}$ **45.** $\frac{11 \pm \sqrt{129}}{4}$ **47.** $\frac{1 \pm \sqrt{2}}{5}$

49. $\pm\sqrt{7}; -2.6, 2.6$ **51.** $-3 \pm 2\sqrt{2}; -5.8, -0.2$ **53.** $\frac{9 \pm \sqrt{97}}{2}; 9.4, -0.4$ **55.** $\frac{1 \pm \sqrt{7}}{3}; 1.2, -0.5$

57. 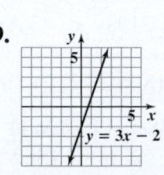   **59.**   **61.** c   **63.** b   **65.** width: 3.5 ft; length: 9.9 ft   **67.** $\frac{-3\sqrt{2} \pm \sqrt{38}}{2}$

69. answers may vary   **71.** $-0.9, 0.2$ **73.** 7.9 sec   **75.** $x = 8$, or 2016

Integrated Review   **1.** $2, \frac{1}{5}$ **2.** $\frac{2}{5}, -3$ **3.** $1 \pm \sqrt{2}$ **4.** $3 \pm \sqrt{2}$ **5.** $\pm 2\sqrt{5}$ **6.** $\pm 6\sqrt{2}$ **7.** no real solution

8. no real solution   **9.** 2   **10.** 3   **11.** 3   **12.** $\frac{7}{2}$ **13.** ± 2 **14.** ± 3 **15.** $1, 2$ **16.** $-3, -4$ **17.** $0, -5$ **18.** $\frac{8}{3}, 0$ **19.** $\frac{3 \pm \sqrt{7}}{5}$

20. $\frac{3 \pm \sqrt{5}}{2}$ **21.** $\frac{3}{2}, -1$ **22.** $\frac{2}{5}, -2$ **23.** $\frac{5 \pm \sqrt{105}}{20}$ **24.** $\frac{-1 \pm \sqrt{3}}{4}$ **25.** $5, \frac{7}{4}$ **26.** $1, \frac{7}{9}$ **27.** $\frac{-7 \pm 3\sqrt{2}}{-5}$ or $\frac{7 \pm 3\sqrt{2}}{5}$

28. $\dfrac{-5 \pm 5\sqrt{3}}{-4}$ or $\dfrac{5 \pm 5\sqrt{3}}{4}$ **29.** $\dfrac{7 \pm \sqrt{193}}{6}$ **30.** $\dfrac{-7 \pm \sqrt{193}}{12}$ **31.** $11, -10$ **32.** $7, -8$ **33.** $4, -\dfrac{2}{3}$ **34.** $2, -\dfrac{4}{5}$ **35.** $0.5, 0.1$

36. $0.3, -0.2$ **37.** $\dfrac{11 \pm \sqrt{41}}{20}$ **38.** $\dfrac{11 \pm \sqrt{41}}{40}$ **39.** $\dfrac{4 \pm \sqrt{10}}{2}$ **40.** $\dfrac{5 \pm \sqrt{185}}{4}$ **41.** answers may vary

Section 16.4

Calculator Explorations **1.** $-0.41, 7.41$ **3.** $0.91, 2.38$ **5.** $-0.39, 0.84$

Vocabulary, Readiness & Video Check **1.** If a parabola opens upward, the lowest point is called the vertex; if a parabola opens downward, the highest point is called the vertex. If a graph can be folded along a line such that the two sides coincide or form mirror images of each other, we say the graph is symmetric about that line and that line is the line of symmetry. **3.** For example, if the vertex is in quadrant III or IV and the parabola opens downward, then there won't be any x-intercepts and there's no need to let $y = 0$ and solve the equation for x.

Exercise Set 16.4 **1.**

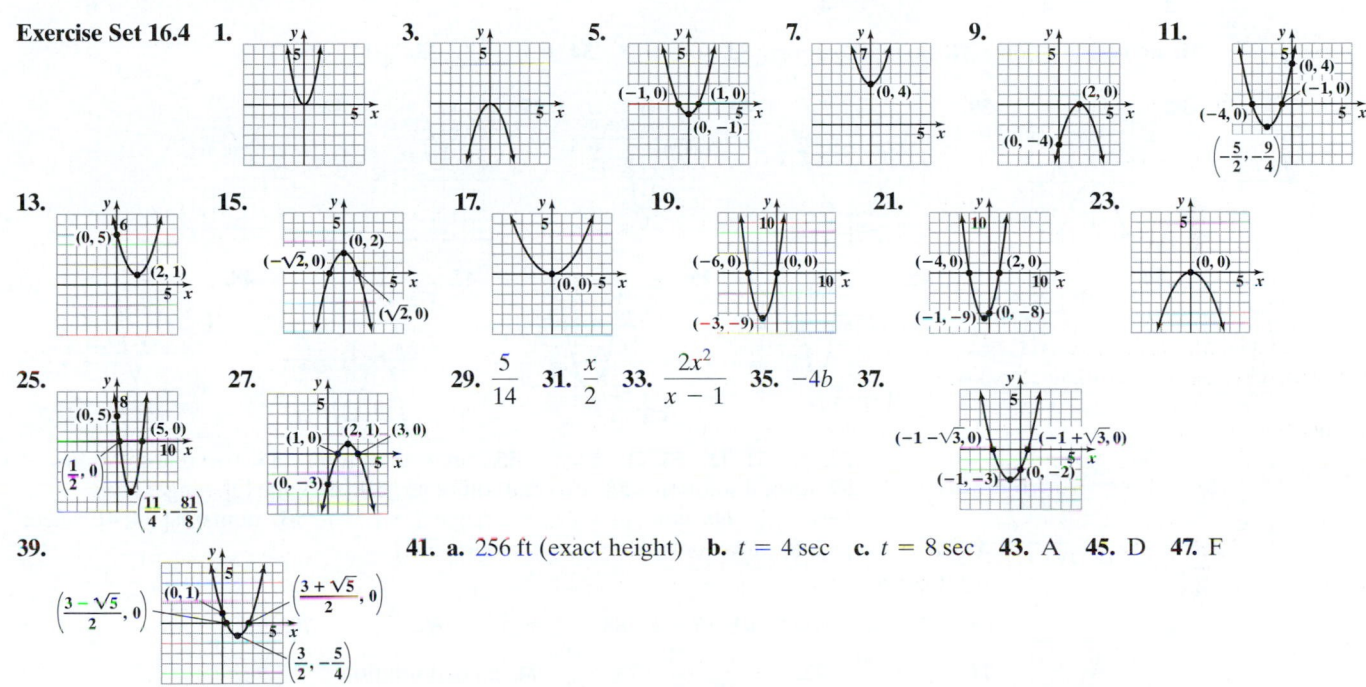

29. $\dfrac{5}{14}$ **31.** $\dfrac{x}{2}$ **33.** $\dfrac{2x^2}{x-1}$ **35.** $-4b$ **37.**

39. **41. a.** 256 ft (exact height) **b.** $t = 4$ sec **c.** $t = 8$ sec **43.** A **45.** D **47.** F

Section 16.5

Vocabulary, Readiness & Video Check **1.** The graph of Example 1 is shaded from $-\infty$ to, but not including, -3, as indicated by a parenthesis. To write interval notation, write down what is shaded for the inequality from left to right. A parenthesis is always used with $-\infty$, so from the graph, the interval notation is $(-\infty, -3)$. **3.** Although $f(x) = x + 3$ isn't defined for $x = -1$, we need to clearly indicate the point where this piece of the graph ends. Therefore, we find this point and graph it as an open circle.

Exercise Set 16.5 **1.** $(-\infty, -3)$ **3.** $[0.3, \infty)$ **5.** $[-7, \infty)$

7. $(-2, 5)$ **9.** $(-1, 5]$ **11.** domain: $[0, \infty)$; range: $(-\infty, \infty)$ **13.** domain: $(-\infty, \infty)$; range: $[0, \infty)$ **15.** domain: $(-\infty, \infty)$; range: $(-\infty, -3] \cup [3, \infty)$ **17.** domain: $[1, 7]$; range: $[1, 7]$ **19.** domain: $\{-2\}$; range: $(-\infty, \infty)$ **21.** domain: $(-\infty, \infty)$; range: $(-\infty, 3]$ **23.** domain: $(-\infty, \infty)$; range: $(-\infty, 3]$ **25.** domain: $[2, \infty)$; range: $[3, \infty)$

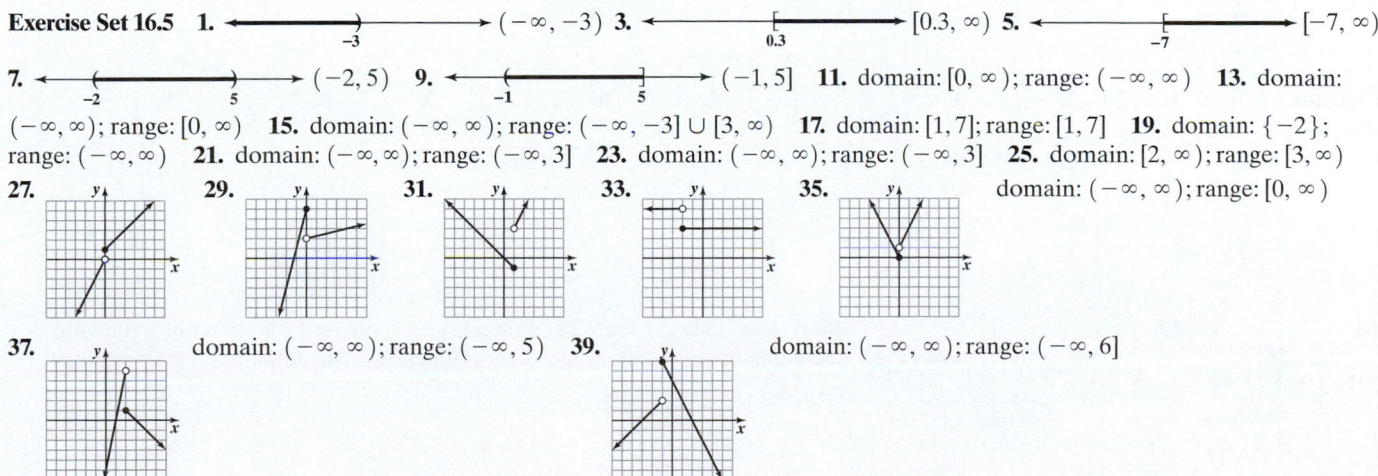

27. **29.** **31.** **33.** **35.** domain: $(-\infty, \infty)$; range: $[0, \infty)$

37. domain: $(-\infty, \infty)$; range: $(-\infty, 5)$ **39.** domain: $(-\infty, \infty)$; range: $(-\infty, 6]$

41. 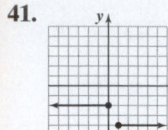   domain: $(-\infty, 0] \cup [1, \infty)$; range: $\{-4, -2\}$ **43.** A   **45.** D   **47.** answers may vary   **49.**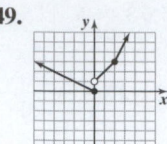

Chapter 16 Vocabulary Check   **1.** square root   **2.** parabola   **3.** vertex   **4.** completing the square   **5.** quadratic   **6.** vertex   **7.** zero

Chapter 16 Review   **1.** ± 11 **2.** ± 10 **3.** $-\dfrac{1}{3}, 2$ **4.** $\dfrac{5}{7}, -1$ **5.** ± 6 **6.** ± 9 **7.** $\pm 5\sqrt{2}$ **8.** $\pm 3\sqrt{5}$ **9.** $4, 18$ **10.** $7, -13$

11. $\dfrac{-5 \pm \sqrt{41}}{4}$ **12.** $\dfrac{-7 \pm \sqrt{37}}{3}$ **13.** 2.5 sec   **14.** 40.6 sec   **15.** $1, 8$ **16.** $-10, 2$ **17.** $-2 \pm \sqrt{5}$ **18.** $4 \pm \sqrt{19}$ **19.** $3 \pm \sqrt{2}$

20. $-3 \pm \sqrt{2}$ **21.** $\dfrac{1}{2}, -1$ **22.** $\dfrac{1}{4}, -1$ **23.** $-\dfrac{5}{3}$ **24.** $\dfrac{9}{4}$ **25.** $\pm\sqrt{5}$ **26.** $\pm\sqrt{3}$ **27.** $5 \pm 3\sqrt{2}$ **28.** $-2 \pm \sqrt{11}$ **29.** $\dfrac{-1 \pm \sqrt{13}}{6}$

30. $\dfrac{-3 \pm \sqrt{13}}{2}$ **31.** no real solution   **32.** no real solution   **33.** $0.4, -0.8$ **34.** $0.3, -3.3$ **35.** 2014   **36.** 2011

37. 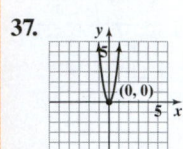   **38.** 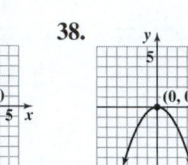   **39.** 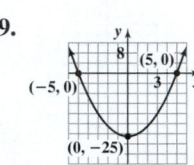   **40.** 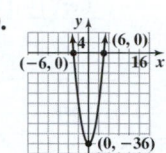   **41.** 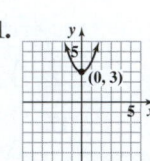   **42.**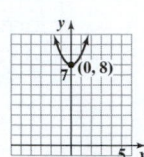

43. 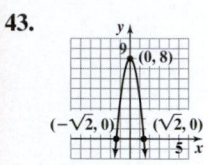   **44.** 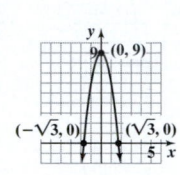   **45.** 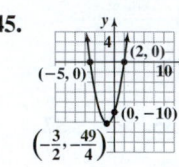   **46.** 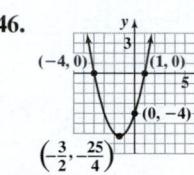   **47.** $\left(-\dfrac{5}{2}, \dfrac{1}{4}\right)$ 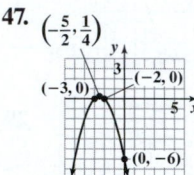   **48.**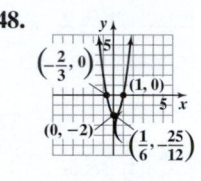

49. $\left(-\dfrac{1}{2}, 0\right)$ 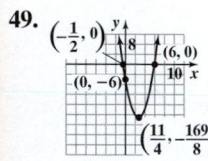   **50.** 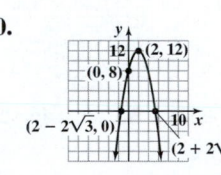   **51.** A   **52.** D   **53.** B   **54.** C   **55.** one real solution   **56.** two real solutions   **57.** no real solution   **58.** two real solutions   **59.** domain: $\{2\}$; range: $(-\infty, \infty)$ **60.** domain: $(-\infty, \infty)$; range: $(-\infty, \infty)$ **61.** domain: $[-4, 4]$; range: $[-1, 5]$ **62.** domain: $(-\infty, \infty)$; range: $\{-5\}$

63. 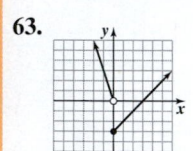   **64.** 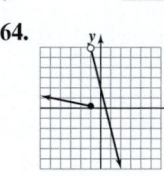   **65.** ± 7 **66.** $\pm 5\sqrt{3}$ **67.** $15, -1$ **68.** $-2 \pm \sqrt{10}$ **69.** $\dfrac{2}{3}, -1$ **70.** $\dfrac{1 \pm \sqrt{33}}{8}$

71. $\dfrac{3 \pm \sqrt{41}}{8}$ **72.** $\dfrac{-1 \pm \sqrt{41}}{10}$ **73.** $-\dfrac{3}{2}$ **74.** no real solution

75. 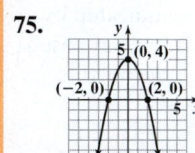   **76.** 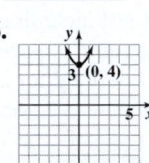   **77.** 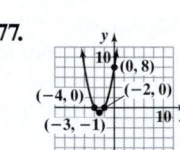   **78.**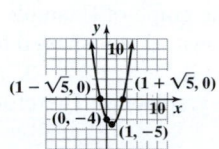

Chapter 16 Test   **1.** ± 20 **2.** $-\dfrac{3}{2}, 7$ **3.** ± 4 **4.** $\dfrac{5 \pm 2\sqrt{2}}{3}$ **5.** $10, 16$ **6.** $-2 \pm \dfrac{4\sqrt{3}}{3}$ **7.** $-2, 5$ **8.** $\dfrac{5 \pm \sqrt{37}}{6}$ **9.** $1, -\dfrac{4}{3}$

10. $-1, \dfrac{5}{3}$ **11.** $\dfrac{7 \pm \sqrt{73}}{6}$ **12.** $-1, 5$ **13.** $2, \dfrac{1}{3}$ **14.** $\dfrac{3 \pm \sqrt{7}}{2}$ **15.** base: 3 ft; height: 12 ft   **16.** 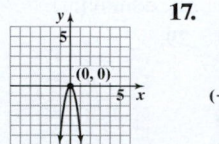   **17.**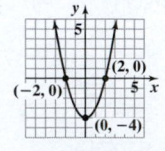

18. 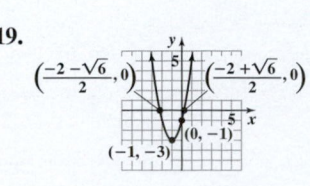   **19.** 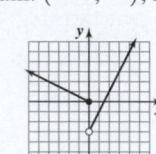   **20.** 6 sides   **21.** 2.7 sec   **22.** domain: $\{-2\}$; range: $(-\infty, \infty)$; not a function   **23.** domain: $(-\infty, \infty)$; range: $[0, \infty)$; function   **24.** domain: $(-\infty, \infty)$; range: $(-3, \infty)$

Cumulative Review **1.** 45% (Sec. 5.3, Ex. 6) **2.** 106% **3.** 150% (Sec. 5.3, Ex. 8) **4.** 500% **5.** 200 (Sec. 5.4, Ex. 10) **6.** 242

7. $\dfrac{101}{200} = \dfrac{p}{100}$ (Sec. 5.5, Ex. 2) **8.** $101 = n \cdot 200$ **9.** \$160 (Sec. 5.8, Ex. 2) **10.** \$525 **11.** 73° (Sec. 6.1, Ex. 5) **12.** 56°

13. 60° (Sec. 6.2, Ex. 2) **14.** 49° **15.** $y = -1.6$ (Sec. 9.1, Ex. 2) **16.** $x = -3$ **17.** $t = \dfrac{16}{3}$ (Sec. 9.3, Ex. 2) **18.** $x = \dfrac{29}{8}$

19. Democratic: 199; Republican: 233 (Sec. 9.4, Ex. 4) **20.** 145, 146, 147 **21.** 1 (Sec. 12.1, Ex. 28) **22.** $-\dfrac{216x^3}{y^9}$ **23.** 1 (Sec. 12.1,

Ex. 29) **24.** $\dfrac{a^2}{32b^3}$ **25.** -1 (Sec. 12.1, Ex. 31) **26.** 9 **27.** $9y^2 + 12y + 4$ (Sec. 12.6, Ex. 15) **28.** $x^2y - x^2 + 5y - 5$ **29.** $x + 4$

(Sec. 12.7, Ex. 4) **30.** $9.1a - 4$ **31.** $(r + 6)(r - 7)$ (Sec. 13.2, Ex. 4) **32. a.** -1 **b.** 30 **33.** $(2x - 3y)(5x + y)$

(Sec. 13.3, Ex. 4) **34.** $\dfrac{8x + 13}{(x + 2)(x - 1)}$ **35.** $(2x - 1)(4x - 5)$ (Sec. 13.4, Ex. 1) **36.** $\dfrac{2x + 5}{x - 7}$ **37. a.** $x(2x + 7)(2x - 7)$

(Sec. 13.5, Ex. 16) **b.** $2(9x^2 + 1)(3x + 1)(3x - 1)$ (Sec. 13.5, Ex. 17) **38.** -32 **39.** $\dfrac{1}{5}, -\dfrac{3}{2}, -6$ (Sec. 13.6, Ex. 8) **40. a.** $-x + 4$

b. $5y - 8$ **c.** $9.1a - 4$ **d.** $2x^2 - 2x$ **41.** $\dfrac{x + 7}{x - 5}$ (Sec. 14.1, Ex. 4) **42.** $x = -\dfrac{7}{2}, 1$ **43.** -5 (Sec. 14.6, Ex. 1) **44.** $2\sqrt{10}$ units

45. a. -3 **b.** 0 **c.** -3 (Sec. 10.1, Ex. 4) **46. a.** x-int: $(4, 0)$; y-int: $(0, 1)$ **b.** x-int: $(-2, 0), (0, 0), (3, 0)$; y-int: $(0, 0)$ **47. a.** parallel

b. perpendicular **c.** neither (Sec. 10.4, Ex. 7) **48.** perpendicular **49. a.** function **b.** not a function (Sec. 10.6, Ex. 2) **50. a.** $6\sqrt{5}$

b. $8\sqrt{2}$ **c.** $4\sqrt{2} + 11 - 2\sqrt{3}$ **51.** $(4, 2)$ (Sec. 11.2, Ex. 1) **52.** no solution **53.** no solution (Sec. 11.3, Ex. 3) **54.** infinite number

of solutions **55.** 6 (Sec. 15.1, Ex. 1) **56.** $\dfrac{2}{5}$ **57.** $\dfrac{3}{10}$ (Sec. 15.1, Ex. 3) **58.** $\dfrac{4}{11}$ **59.** $-1 + \sqrt{3}$ (Sec. 15.4, Ex. 13) **60.** $\dfrac{5\sqrt{2}}{4}$

61. $x = 7, -1$ (Sec. 16.1, Ex. 5) **62.** $x = 4 \pm \sqrt{3}$ **63.** $x = 1 \pm \sqrt{5}$ (Sec. 16.3, Ex. 5) **64.** $x = -2 \pm 2\sqrt{3}$

Appendices

Exercise Set Appendix A.1 **1.** $(a + 3)(a^2 - 3a + 9)$ **3.** $(2a + 1)(4a^2 - 2a + 1)$ **5.** $5(k + 2)(k^2 - 2k + 4)$
7. $(xy - 4)(x^2y^2 + 4xy + 16)$ **9.** $(x + 5)(x^2 - 5x + 25)$ **11.** $3x(2x - 3y)(4x^2 + 6xy + 9y^2)$ **13.** $(3 - t)(9 + 3t + t^2)$
15. $8(r - 2)(r^2 + 2r + 4)$ **17.** $(t - 7)(t^2 + 7t + 49)$ **19.** $(s - 4t)(s^2 + 4st + 16t^2)$

Exercise Set Appendix A.2 **1.** $\{2, 3, 4, 5, 6, 7\}$ **3.** $\{4, 6\}$ **5.** $\{\ldots, -2, -1, 0, 1, \ldots\}$ **7.** $\{5, 7\}$ **9.** $\{x \mid x$ is an odd integer or
$x = 2$ or $x = 4\}$ **11.** $\{2, 4\}$ **13.** $(-3, 1)$ ⟵———( )———⟶ **15.** ∅ ⟵————————⟶
        −3 1

17. $(-\infty, -1)$ ⟵———————) ⟶ **19.** $[6, \infty)$ **21.** $(-\infty, -3]$ **23.** $(4, 10)$ **25.** $(11, 17)$ **27.** $[1, 4]$ **29.** $\left[-3, \dfrac{3}{2}\right]$
        −1

31. $\left[-\dfrac{7}{3}, 7\right]$ **33.** $(-\infty, 5)$ ⟵——————) ⟶ **35.** $(-\infty, 4] \cup [1, \infty)$ ⟵——] [——⟶
        5   −4 1

37. $(-\infty, \infty)$ ⟵——————————⟶ **39.** $[2, \infty)$ **41.** $(-\infty, -4) \cup (-2, \infty)$ **43.** $(-\infty, \infty)$ **45.** $\left(-\dfrac{1}{2}, \dfrac{2}{3}\right)$ **47.** $(-\infty, \infty)$

49. $\left[\dfrac{3}{2}, 6\right]$ **51.** $\left(\dfrac{5}{4}, \dfrac{11}{4}\right)$ **53.** ∅ **55.** $\left(-\infty, -\dfrac{56}{5}\right) \cup \left(\dfrac{5}{3}, \infty\right)$ **57.** $\left(-5, \dfrac{5}{2}\right)$ **59.** $\left(0, \dfrac{14}{3}\right]$ **61.** $(-\infty, -3]$ **63.** $(-\infty, 1] \cup \left(\dfrac{29}{7}, \infty\right)$

65. ∅ **67.** $\left[-\dfrac{1}{2}, \dfrac{3}{2}\right)$ **69.** $\left(-\dfrac{4}{3}, \dfrac{7}{3}\right)$ **71.** $(6, 12)$ **73.** 2004, 2005 **75.** answers may vary

Appendix A.3 Vocabulary and Readiness Check **1.** d **3.** c **5.** a

Exercise Set Appendix A.3 **1.** $\{7, -7\}$ **3.** ∅ **5.** $\{4.2, -4.2\}$ **7.** $\{-4, 4\}$ **9.** $\{-9, 9\}$ **11.** $\{-5, 23\}$ **13.** $\{7, -2\}$

15. $\{8, 4\}$ **17.** $\{5, -5\}$ **19.** $\{3, -3\}$ **21.** $\{-3, 6\}$ **23.** $\{0\}$ **25.** ∅ **27.** $\left\{-\dfrac{1}{3}, \dfrac{7}{3}\right\}$ **29.** $\left\{-\dfrac{1}{2}, 9\right\}$ **31.** $\left\{-\dfrac{5}{2}\right\}$ **33.** $\{3, 2\}$

35. $\{-4, 16\}$ **37.** $\{4\}$ **39.** $\left\{\dfrac{3}{2}\right\}$ **41.** $\left\{\dfrac{32}{21}, \dfrac{38}{9}\right\}$ **43.** $\left\{-8, \dfrac{2}{3}\right\}$ **45.** ⟵——[ ]——⟶ $[-4, 4]$
                                                                                                          −4 4

47. ⟵—) (——⟶ $(-\infty, -3) \cup (3, \infty)$ **49.** ⟵——( )——⟶ $(-5, -1)$ **51.** ⟵——] [——⟶
        −3 3   −5 −1   −1 13
$(-\infty, -1] \cup [13, \infty)$ **53.** ⟵——( )——⟶ $(-5, 1)$ **55.** ⟵——[ ]——⟶ $[-5, 5]$
                          −5 1   −5 5

57. ⟵——●——⟶ $\left\{-\dfrac{3}{2}\right\}$ **59.** ⟵—) (——⟶ $(-\infty, -4) \cup (4, \infty)$ **61.** ⟵—[ ]——⟶ $[-10, 3]$
        $-\frac{3}{2}$ −4 4   −10 3

63. ⟵——( )——⟶ $\left(1, \dfrac{13}{3}\right)$ **65.** ⟵—] [——⟶ $(-\infty, -24] \cup [4, \infty)$ **67.** ⟵—[ ]——⟶ $[-2, 9]$
        1   $\frac{13}{3}$ −24 4   −2 9

69. ⟵———————————⟶ $(-\infty, \infty)$ **71.** ⟵—[————]—⟶ $\left[-\frac{1}{2}, 1\right]$ **73.** ⟵——)————(—⟶ $\left(-\infty, \frac{2}{3}\right) \cup (2, \infty)$
                                                          $-\frac{1}{2}$ 1 $\frac{2}{3}$ 2

75. ⟵———————————⟶ $\varnothing$ **77.** ⟵——————)———⟶ $(-\infty, -12) \cup (0, \infty)$ **79.** $\{-13, 13\}$ **81.** $(-13, 13)$ **83.** $\varnothing$
                                                          -12 0

85. $[-10, 10]$ **87.** $\{5, -2\}$ **89.** $(-\infty, -7] \cup [17, \infty)$ **91.** $\left\{-\frac{9}{4}\right\}$ **93.** $(-2, 1)$ **95.** $(-\infty, -18) \cup (12, \infty)$ **97.** $\left\{2, \frac{4}{3}\right\}$

99. $\varnothing$ **101.** $\left\{-\frac{17}{2}, \frac{19}{2}\right\}$ **103.** $\left(-\infty, -\frac{25}{3}\right) \cup \left(\frac{35}{3}, \infty\right)$ **105.** $\left\{4, -\frac{1}{5}\right\}$ **107.** $\left\{-\frac{17}{3}, 5\right\}$ **109.** $\varnothing$ **111.** $(-\infty, \infty)$ **113.** $|x| = 5$

115. $|x| < 7$ **117.** $|x| \le 5$, answers may vary

Exercise Set Appendix A.4   **1.** 5 units   **3.** $\sqrt{41}$ units $\approx$ 6.403 units   **5.** $\sqrt{5}$ units $\approx$ 2.236 units   **7.** $\sqrt{192.58}$ units $\approx$ 13.877 units

9. $(4, -2)$ **11.** $\left(-5, \frac{5}{2}\right)$ **13.** $\left(-\frac{1}{2}, \frac{1}{2}\right)$ **15.** $\left(\sqrt{2}, \frac{\sqrt{5}}{2}\right)$

Exercise Set Appendix A.5   **1.** $y = 4x - 4$ **3.** $y = -3x + 1$ **5.** $y = 4$ **7.** $y = -\frac{3}{2}x - 6$ **9.** $y = -5$ **11.** $y = -4x + 1$

13. $y = -\frac{1}{2}x + 1$ **15.** $y = 8x - 47$ **17.** $x = 5$

Exercise Set Appendix A.6   **1.** $(-\infty, -5) \cup (-1, \infty)$ **3.** $[-4, 3]$ **5.** $(-\infty, -5] \cup [-3, \infty)$ **7.** $\left(-5, -\frac{1}{3}\right)$ **9.** $(2, 4) \cup (6, \infty)$

11. $(-\infty, -4] \cup [0, 1]$ **13.** $(-\infty, -3) \cup (-2, 2) \cup (3, \infty)$ **15.** $(-7, 2)$ **17.** $(-1, \infty)$ **19.** $(-\infty, -1] \cup (4, \infty)$

21. $(-\infty, 2) \cup \left(\frac{11}{4}, \infty\right)$ **23.** $(0, 2] \cup [3, \infty)$ **25.** $(-\infty, 3)$ **27.** $\left[-\frac{5}{4}, \frac{3}{2}\right]$ **29.** $(-\infty, 0) \cup (1, \infty)$ **31.** $(0, 10)$

33. $(-\infty, -4] \cup [4, 6]$ **35.** $\left(-\infty, -\frac{2}{3}\right] \cup \left[\frac{3}{2}, \infty\right)$ **37.** $(-\infty, -4) \cup [5, \infty)$ **39.** $(-\infty, 1) \cup (2, \infty)$ **41.** $\left(-4, -\frac{3}{2}\right) \cup \left(\frac{3}{2}, \infty\right)$

43. $(-\infty, -5] \cup [-1, 1] \cup [5, \infty)$ **45.** $(-\infty, -6] \cup (-1, 0] \cup (7, \infty)$ **47.** $(-\infty, -8] \cup (-4, \infty)$ **49.** $\left(-\infty, -\frac{5}{3}\right) \cup \left(\frac{7}{2}, \infty\right)$

51. $(-\infty, 0] \cup \left(5, \frac{11}{2}\right]$ **53.** $(0, \infty)$ **55.** answers may vary

Exercise Set Appendix A.7   **1.** 7   **3.** 3   **5.** $\frac{1}{2}$ **7.** 13   **9.** $2\sqrt[3]{m}$ **11.** $3x^2$ **13.** -3 **15.** -2 **17.** 8   **19.** 16   **21.** not a real number

23. $\sqrt[5]{(2x)^3}$ **25.** $\sqrt[3]{(7x + 2)^2}$ **27.** $\frac{64}{27}$ **29.** $\frac{1}{16}$ **31.** $\frac{1}{16}$ **33.** not a real number   **35.** $\frac{1}{x^{1/4}}$ **37.** $a^{2/3}$ **39.** $\frac{5x^{3/4}}{7}$ **41.** $a^{7/3}$

43. x **45.** $3^{5/8}$ **47.** $y^{1/6}$ **49.** $8u^3$ **51.** $-b$ **53.** $\frac{1}{x^2}$ **55.** $27x^{2/3}$ **57.** $\frac{y}{z^{1/6}}$ **59.** $\frac{1}{x^{7/4}}$ **61.** $\sqrt{x}$ **63.** $\sqrt[3]{2}$ **65.** $2\sqrt{x}$ **67.** $\sqrt{x + 3}$

69. $\sqrt{xy}$ **71.** $\sqrt[3]{a^2b}$ **73.** $\sqrt[15]{y^{11}}$ **75.** $\sqrt[12]{b^5}$ **77.** $\sqrt[24]{x^{23}}$ **79.** $\sqrt{a}$ **81.** $\sqrt[6]{432}$ **83.** $\sqrt[15]{343y^5}$ **85.** $\sqrt[6]{125r^3s^2}$

87. 1509 calories   **89.** answers may vary   **91.** $a^{1/3}$ **93.** $x^{1/5}$

Exercise Set Appendix A.8

1. 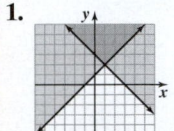   **3.** 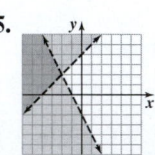   **5.** 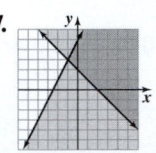   **7.** 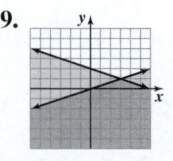   **9.** 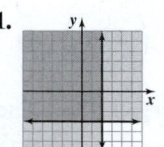   **11.** 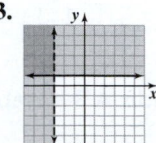   **13.**

15. 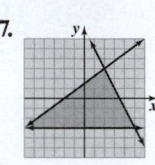   **17.** 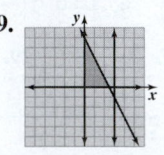   **19.**   **21.** C   **23.** D   **25.** the line $y = 3$

Practice Final Exam   **1.** 113   **2.** 33   **3.** $\frac{64}{3}$ or $21\frac{1}{3}$ **4.** $\frac{3}{2}$ or $1\frac{1}{2}$ **5.** 8   **6.** 43.86   **7.** 6.2   **8.** $1\frac{7}{24}$ **9.** $\frac{45}{2}$ or $22\frac{1}{2}$ **10.** 12,690

11. $\frac{3}{8}$ **12.** 0.862   **13.** 34.9   **14.** 0.941   **15.** 0.85   **16.** 610%   **17.** 37.5%   **18.** $\frac{1}{500}$ **19.** perimeter: $3\frac{1}{3}$ ft; area: $\frac{2}{3}$ sq ft   **20.** $\frac{15}{2}$

21. $\frac{2}{3}$ in./hr   **22.** $\frac{48}{11}$ or $4\frac{4}{11}$ **23.** 11.4   **24.** $7\frac{5}{6}$ gal   **25.** $53\frac{1}{3}$ g   **26.** $17   **27.** 1250   **28.** 75%   **29.** 38.4 lb   **30.** discount: $18;

sale price: $102   **31.** 12°   **32.** $m\angle x = 118°; m\angle y = 62°; m\angle z = 118°;$ **33.** 26°   **34.** 7.5 or $7\frac{1}{2}$ **35.** mean: 38.4; median: 42; no mode

Photo Credits

Chapter 1 **p. 1** Rachel Youdelman/Pearson Education, Inc. **p. 6** Maciej Noskowski/Vetta/Getty Images **p. 14** NASA **p. 17** Igor Terekhov/Dreamstime **p. 20** Mediagram/Shutterstock **p. 26** Cachou34/Fotolia **p. 37** (l) JoseIgnacioSoto/iStockphoto; (r) James Steidl/Dreamstime **p. 45** Auttapon Moonsawad/Fotolia **p. 55** Greg Henry/Shutterstock **p. 83** (tr) Icholakov/Fotolia; (bl) Krzysztof Wiktor/Shutterstock; (br) Ryan McVay/Photodisc/Thinkstock **p. 105** Photogolfer/Shutterstock

Chapter 2 **p. 110** Inc/Shutterstock **p. 129** Dotshock/Shutterstock **p. 138** Keith Brofsky/Photodisc/Getty Images **p. 146** Pefkos/Fotolia **p. 151** Vicnt/Thinkstock

Chapter 3 **p. 173** Lucky Business/Shutterstock **p. 189** Andres Rodriguez/Fotolia **p. 211** (l) Marcio Jose Bastos Silva/Shutterstock; (r) Chase B/Shutterstock **p. 212** Thomas Nord/Shutterstock **p. 224** Chad McDermott/Shutterstock

Chapter 4 **p. 247** Altiso/Dreamstime **p. 251** (l) SeanPavonePhoto/Fotolia; (r) Bill Ross/Corbis **p. 259** (bl) Dziewul/Shutterstock; (br) Baloncici/Shutterstock **p. 270** Apple, Inc. **p. 279** (l) Digidreamgrafix/Fotolia; (r) Carabay/Fotolia **p. 281** Debbie Rowe/Pearson Education, Inc. **p. 293** Iofoto/Shutterstock

Chapter 5 **p. 317** Dny3d/Fotolia **p. 328** (tl) Andrew Buckin/Shutterstock; (bl) John S. Sfondilias/Shutterstock; (br) Jupiterimages/Stockbyte/Thinkstock **p. 334** Peepo/iStockphoto **p. 339** (l) Bergamont/Fotolia; (r) Gmg9130/Fotolia **p. 346** (t) Pavel Losevsky/Fotolia; (b) Elena Schweitzer/Shutterstock **p. 368** CandyBox Images/Shutterstock **p. 372** (l) Tyler Olson/Shutterstock; (r) Alexander Raths/Shutterstock **p. 374** Kubais/Fotolia **p. 376** (tl) Alina Isakovich/Fotolia; (tr) Dennis Donohue/Shutterstock; (b) Rafa Irusta/Fotolia **p. 378** Stephen VanHorn/Shutterstock **p. 382** (tr) Evemilla/iStockphoto; (bl) studio online/Shutterstock; (br) LuckyPhoto/Shutterstock **p. 401** Nyul/Fotolia

Chapter 6 **p. 415** (l) ES James/Shutterstock; (r) Sculpies/iStockphoto **p. 433** StephanHoerold/iStockphoto **p. 451** Craig Hanson/Shutterstock

Chapter 7 **p. 486** ZUMA Press, Inc/Alamy

Chapter 8 **p. 541** (l) Tim Davis/Corbis; (r) Calee Allen/Fotolia **p. 552** ElementalImaging/iStockphoto **p. 564** Elen_studio/Shutterstock **p. 567** (l) Smellme/Dreamstime; (r) Midkhat Izmaylov/Shutterstock **p. 568** (l) Photogolfer/Shutterstock; (r) Sgcallaway1994/Dreamstime **p. 577** (l) Oversnap/iStockphoto; (r) Paylessimages/Fotolia **p. 588** Stockbyte/Thinkstock

Chapter 9 **p. 628** (l) Dreamstime; (r) Brandon Holmes/Dreamstime **p. 631** (l) Simone Matteo Giuseppe Manzoni/Dreamstime; (r) Dmitry Pichugin/Dreamstime **p. 656** Andre Nantel/Dreamstime **p. 657** (t) Andrzej Thiel/Dreamstime; (b) Zack Frank/Shutterstock **p. 659** Felix Mizioznikov/Fotolia **p. 664** Ewg3D/iStockphoto **p. 666** Jenifoto/Fotolia **p. 676** Steve Geer/Getty Images **p. 679** Thomson Reuters (Markets) LLC **p. 683** longimanus/Shutterstock **p. 684** Asiseeit/iStockphoto **p. 690** (l) Jelle-vd-Wolf/Shutterstock; (r) Phillip Minnis/Shutterstock **p. 710** W.Scott/Fotolia

Chapter 10 **p. 718** Murat Taner/Photographer's Choice RF/Getty Images **p. 726** Anton Prado PHOTO/Shutterstock **p. 727** WilleeCole Photography/Shutterstock **p. 736** Sunabesyou/Fotolia **p. 740** Chris Howes/Wild Places Photography/Alamy **p. 741** Demetrio Carrasco/Dorling Kindersley Limited **p. 760** Fotolia **p. 768** (l) Gennadiy Poznyakov/Fotolia; (r) Elena Volkova/Fotolia **p. 772** Paul Sutherland/Photodisc/Getty Images **p. 777** (l) deusexlupus/Fotolia; (r) Igor Mojzes/Fotolia **p. 778** Dmitry Naumov/Fotolia

Chapter 11 **p. 830** Piotr Adamowicz/Fotolia **p. 849** (l) Piotr Adamowicz/Fotolia; (r) Aigarsr/Fotolia **p. 857** (l) Petert2/Fotolia; (r) Goss Vitalij/Fotolia **p. 860** Nathan King/Alamy **p. 866** (l) Rose Palmisano/The Orange County Register/ZUMA Press/Alamy; (r) Dmitry Argunov/Alamy **p. 867** Zjk/Fotolia **p. 879** Hramovnick/Fotolia

Chapter 12 **p. 882** Arska N/Fotolia **p. 921** (l) Wajan/Fotolia; (r) Monkey Business/Fotolia **p. 951** DPA Picture Alliance Archive/Alamy **p. 952** Forcdan/Fotolia

Chapter 13 **p. 970** (l) Edu1971/Fotolia; (r) Jeffrey M. Frank/Shutterstock **p. 998** (t) Katrina Brown/Fotolia; (bl) Fotolia; (br) Vvoe/Fotolia **p. 1009** Jeremy Woodhouse/Photodisc/Getty Images

Chapter 14 **p. 1047** (l) Matthew Grant/Alamy; (r) Mediacolor's/Alamy **p. 1060** Ronnie Kaufman/Larry Hirshowitz/Blend Images/Corbis **p. 1080** Loraks/Fotolia **p. 1087** Speedpix/Alamy

Chapter 15 **p. 1109** Jose Ignacio Soto/Fotolia **p. 1117** Richard Wong/Alamy **p. 1124** Henry Westheim Photography/Alamy **p. 1125** (tl) Pearson Education; (tr) Age fotostock Spain, S.L./Alamy; (b) Kurhan/Fotolia **p. 1149** Vetal1983/Fotolia

Chapter 16 **p. 1166** Keith Brofsky/Photodisc/Getty Images **p. 1170** PF-(sdasm1)/Alamy **p. 1179** MCT/Newscom **p. 1208** (t) National Astronomy and Ionosphere Center; (b) Rudi1976/Fotolia **p. 1212** Wajan/Fotolia